Standard Handbook for
Civil Engineers

Standard Handbook for Civil Engineers

Editors:

Jonathan T. Ricketts

M. Kent Loftin

Frederick S. Merritt

Fifth Edition

McGraw-Hill

New York   Chicago   San Francisco   Lisbon   London   Madrid
Mexico City   Milan   New Delhi   San Juan   Seoul
Singapore   Sydney   Toronto

The **McGraw-Hill** Companies

Library of Congress Cataloging in Publication Data

Standard handbook for civil engineers / editors, Jonathan T. Ricketts, M. Kent Loftin,
   Frederick S. Merritt — 5th ed.
      p.   cm.
   Includes bibliographical references and index.
   ISBN 0-07-136473-0
   1. Civil engineering—Handbooks, manuals, etc. I. Ricketts, Jonathan T. II. Loftin, M.
Kent. III. Merritt, Frederick S.

  TA151.S8   2003
  624—dc22
                                          2003061440

4 5 6 7 8 9 0 DOC/DOC 0 1 0 9 8

ISBN 0-07-136473-0

*The sponsoring editor for this book was Larry S. Hager and the production supervisor
was Pamela A. Pelton. It was set in Palatino by Techset Composition Ltd.*

Printed and bound by RR Donnelley.

This book is printed on acid-free paper.

McGraw-Hill books are available at special quantity discounts to use as premiums and sales
promotions, or for use in corporate training programs. For more information, please write to
the Director of Special Sales, McGraw-Hill Professional, Two Penn Plaza, New York, NY
10121-2298. Or contact your local bookstore.

Contents

Section 4.   Construction Management   *by Jonathan T. Ricketts*   **4.1**

Section 5.   Construction Materials   *by Ruth T. Brantley and L. Reed Brantley*   **5.1**

Cementitious Materials

Metallic Materials

Masonry Units and Tiles

Organic Materials

Joint Seals

Paints and Other Coatings

Composite Materials

Environmental Influences

Section 6. Structural Theory   *by Akbar Tamboli, Mohsin Ahmed, and Michael Xing*   6.1

Equilibrium

Stress and Strain

Stresses at a Point

Straight Beams

Structural Concrete

Design of Concrete Flexural Members

Concrete Compression Members

Prestressed Concrete

Retaining Walls

Footings

Frames and Shells

Section 9.   Structural Steel Design and Construction
by Roger L. Brockenbrough   **9.1**

Section 10.   Cold-Formed-Steel Design and Construction   *by Don S. Wolford and Wei-Wen Yu*   **10.1**

Steel Roof and Floor Deck

Open-Web Steel Joists

Section 14. Community and Regional Planning *by William N. Lane*   **14.1**

Basic Approach and Organization of Planning

Resources and Environmental Quality

Land-Use Planning

Utility and Transportation Systems

Implementation Tools and Techniques

Section 15. Building Engineering *by M. Myint Lwin*   **15.1**

Section 16.   Highway Engineering   *by Richard Garrabrant*   16.1

Elements of Highway Transverse Cross Sections

Highway Alignments

Section 18.   Airport Engineering   *by Richard Harding*   18.1

Section 19.   Rail-Transportation Engineering   *by Roger S. Boraas*   19.1

Section 20.   Tunnel Engineering   *by Lars Christian F. Ingerslev and Arthur G. Bendelius*

20.1

Hydrology

Water Supply

Water Treatment

Water Collection, Storage, and Distribution

Hydroelectric Power and Dams

Section 22.   Environmental Engineering   *by G. Raymond Schulte and Thomas E. Wilson*   **22.1**

Section 23.   Coastal and Port Engineering   *by Scott L. Douglass, Robert A. Nathan, and Jeffrey D. Malyszek*   **23.1**

Coastal Hydraulics and Sediments

Harbor and Marina Engineering

Contributors

Mohsin Ahmed   *Thornton-Tomasetti Engineers, Newark, N.J.* (SEC. 6 STRUCTURAL THEORY)

Arthur G. Bendelius   *Parsons Brinckerhoff, New York, N.Y.* (SEC. 20 TUNNEL ENGINEERING)

Roger S. Boraas   *Consulting Engineer, HNTB Corporation, Denver, Colorado* (SEC. 19 RAIL-TRANSPORTATION ENGINEERING)

L. Reed Brantley   *Emeritus Professor, University of Hawaii, Honolulu, Hawaii* (SEC. 5 CONSTRUCTION MATERIALS)

Ruth T. Brantley   *Senior Lecturer (Retired), University of Hawaii, Honolulu, Hawaii* (SEC. 5 CONSTRUCTION MATERIALS)

Roger L. Brockenbrough   *President, R. L. Brockenbrough & Associates, Inc., Pittsburgh, PA* (SEC. 9 STRUCTURAL STEEL DESIGN AND CONSTRUCTION)

Jerry A. DiMaggio   *Senior Geotechnical Engineer, Federal Highway Administration, Washington, DC* (SEC. 7 GEOTECHNICAL ENGINEERING)

Scott L. Douglass   *Professor, Department of Civil Engineering, University of South Alabama, Mobile, Ala.* (SEC. 23 COASTAL AND PORT ENGINEERING)

Anne M. Ellis   *Earth Tech., Inc., Alexandria, VA* (SEC. 8 CONCRETE DESIGN AND CONSTRUCTION)

David A. Fanella   *Dir. of Engineering, S. K. Ghosh Associates Inc., Northbrook, IL* (SEC. 8 CONCRETE DESIGN AND CONSTRUCTION)

Jay Gewirtzman   *Vice President, URS Corp., New York, N.Y.* (SEC. 2 DESIGN MANAGEMENT)

Richard Garrabrandt   *Consulting Engineer, Buffalo, N.Y.* (SEC. 16 HIGHWAY ENGINEERING)

S. K. Ghosh   *President, S. K. Ghosh Associates Inc., Northbrook, IL* (SEC. 8 CONCRETE DESIGN AND CONSTRUCTION)

Richard Harding   *Air Transportation & Facilities Consultant, Harrisburg, PA* (SEC. 18 AIRPORT ENGINEERING)

Mohamed H. Hussein   *Vice President, GRL Engineers Inc., Orlando, Fla.* (SEC. 7 GEOTECHNICAL ENGINEERING)

Lars Christian F. Ingerslev   *Parsons Brinckerhoff, New York, N.Y.* (SEC. 20 TUNNEL ENGINEERING)

William N. Lane   *Director, Environmental Resource Planning, Dane County Regional Planning Commission, Madison, Wis.* (SEC. 14 COMMUNITY AND REGIONAL PLANNING)

M. Kent Loftin   *President, Syint, Inc., Hobe Sound, Fla.* (SEC. 21 WATER RESOURCES ENGINEERING)

M. Myint Lwin   *Structural Engineer* (SEC. 15 BUILDING ENGINEERING)

Jeffrey D. Malyszek   *Moffat and Nichol Engineers, Tampa, Fla.* (SEC. 23 COASTAL AND PORT ENGINEERING)

Steven L. Mellon   *Bridge Design Engineer, Quincy Engineering, Inc., Sacramento, Calif.* (SEC. 17 BRIDGE ENGINEERING)

Frank Muller   *President, Metro Mediation Services, Ltd., New York, N.Y.* (SEC. 2 DESIGN MANAGEMENT)

Robert A. Nathan   *Moffat & Nichol Engineers, Tampa, Fla.* (SEC. 23 COASTAL AND PORT ENGINEERING)

Thomas A. Ostrom   *Chief, Office of Structures-North, Calif. Department of Transportation, Sacramento, California* (SEC. 17 BRIDGE ENGINEERING)

G. William Quinby   *Consulting Engineer, Golden, Colarado* (SEC. 13 EARTHWORK)

Maurice J. Rhude   *President, Sentinel Structures, Inc., Peshtigo, Wis.* (SEC. 11 WOOD DESIGN AND CONSTRUCTION)

Jonathan T. Ricketts   *Consulting Engineer, Palm Beach Gardens, Fla.* (SEC. 1 SYSTEMS DESIGN; SEC. 4 CONSTRUCTION MANAGEMENT; SEC. 12 SURVEYING)

Ted E. Robbins   *Technical Services Administrator, Martin County Utilities and Solid Waste Dept., Stuart, Fla.* (SEC. 3 SPECIFICATIONS)

Charles H. Sain   *Consulting Engineer, Sain Associations, Inc., Birmingham, Ala.* (SEC. 13 EARTHWORK)

G. Raymond Schulte,   *Consulting Engineer, Johnson, Mirmiran and Thompson, Inc, Baltimore, MD* (SEC. 22 ENVIRONMENTAL ENGINEERING)

Akbar Tamboli   *Thornton-Tomasetti Engineers, Newark, N.J.* (SEC. 6 STRUCTURAL THEORY)

Don S. Wolford   *Consulting Engineer, Middletown, Ohio* (SEC. 10 COLD-FORMED-STEEL DESIGN AND CONSTRUCTION)

Thomas E. Wilson   *Consulting Engineer, Barrington, IL* (SEC. 22 ENVIRONMENTAL ENGINEERING)

Michael Xing   *Thornton-Tomasetti Engineers, Newark, N.J.* (SEC. 6 STRUCTURAL THEORY)

Wei-Wen Yu   *University of Missouri-Rolla, Rolla, Missouri* (SEC. 10 COLD-FORMED-STEEL DESIGN AND CONSTRUCTION)

Preface

For the fifth edition of the *Standard Handbook for Civil Engineers*, the editors and contributors have continued the objectives of the previous editions of these well-received handbooks. To bring together in a single volume the best ideas for current practice in the professional field of civil engineering – planning, design, and construction of buildings, bridges, tunnels, transportation facilities, and other structures required for the health, welfare, safety, employment, and pleasure of all people and for environmental control and use of natural resources. It provides information that would be of the greatest usefulness to those who make decisions affecting selection of engineering approaches and methods and of construction materials and methods.

Emphasis is placed on fundamental principles and the application of them, with special attention to simplified procedures. Frequent reference is made to other sources where additional authoritative information may be obtained. In keeping with current practice, numerous listings of websites containing additional information for the use by the professional have been included. The handbook provides an extensive index and a detailed table of contents to assist the reader in locating topics. These, combined with separate glossaries in many sections, make the handbook useful as a dictionary of civil engineering as well.

As a result of the latest edition, more than half of the contributors are new, reflecting the editors desire to bring the reader information from many of the leading practitioners in each of the many fields covered by the handbook. This edition's contributors have provided extensive new and updated information that was not a part of the previous edition including three dimensional modeling of pile groups, cracking in high strength steel beams, intelligent transportation and system maintenance, a various types of open wharf designs.

Many of the revisions were due to changes in the specifications for design and construction with standard construction materials. In addition, there have been substantial new developments and code revisions since the previous edition. These include ACI 318 pertaining to tension, and compression development lengths in reinforcing steel, development lengths in hooked bars for torsion design, deflections in concrete members, allowable stresses in prestressed concrete, extensive revisions from the American Iron and Steel Institute and updates from the latest Federal Aviation Administration's advisories on airport design.

As in the preceding editions, more than half of this edition is devoted to the specialty fields: such as building engineering (Section 15), highway engineering (16), bridge (17), and environment engineering (22). The rest of the book deals with engineering common to those fields, such as systems design and value engineering (Section 1), design management (2), specifications (3), construction management (4), properties of construction materials (5), structural theory (6), and geotechnical engineering (7).

The gathering of this much knowledge came from many sources and have been credited or been given as references throughout the book. Space limitations, however, preclude listing all of the sources here, but the editors and contributors wish to acknowledge their indebtedness to these sources and to express their gratitude.

The contributor's to the handbook have provided a wealth of knowledge that makes the fifth edition of the handbook so valuable. This contribution of knowledge is greatly appreciated by the editors, not only for the value of the material contained, but also for the sacrifice of the time and effort made to put forth for each section. We all hope that you will find this updated edition just as valuable as the previous ones and that the application of its information will provide a catalyst for the production of cost-effective, energy-efficient, and environmentally sound engineering works.

Jonathan T. Ricketts
M. Kent Loftin
Frederick S. Merritt

Editors

About the Editors

Jonathan T. Ricketts is a Consulting Engineer in Palm Beach Gardens, Florida, a registered engineer in several states, and the editor of McGraw-Hill's *Building Design and Construction Handbook.*

M. Kent Loftin is a Consulting Engineer in Hobe Sound, Florida.

Frederick S. Merritt (deceased) was a Consulting Engineer in West Palm Beach, Florida, and a former editor with *Engineering News-Record.*

1

Jonathan T. Ricketts*
Consulting Engineer
Palm Beach Gardens, Florida

SYSTEMS DESIGN

Civil engineering is that field of engineering concerned with planning, design and construction of natural resource development, regional and local water supply and storm water facilities, waste management facilities, transportation facilities, tunnels, buildings, bridges, and other structures for the needs of people. Persons who are qualified by education and experience and who meet state requirements for practicing the profession of civil engineering are called civil engineers.

1.1 Performance Criteria for Civil Engineers

As professionals, civil engineers should conform to the following canons as they perform their duties:

1. Hold paramount the safety, health, and welfare of the public. (Welfare of the public implies a commitment to sustainable development which is meeting the current needs and goals of the project while protecting the natural resource base for future generations.)

2. Act for every employer or client as faithful agents or trustees and avoid conflict of interest.

3. Apply to the fullest extent their knowledge and skill to every client's project.

4. Maintain life-long learning, always willing to participate in the professional exchange of ideas and technical information.

5. Perform services only in areas of competence; in other areas, engineers may engage or collaborate with qualified associates, consultants, or employees for performing assignments.

Accordingly, civil engineering projects should be planned, designed, and constructed to satisfy the following criteria:

1. They should serve the purposes specified by the owner or client.

2. They should be constructable by known techniques and with available labor and equipment within a time acceptable to the owner or client.

3. They should be capable of withstanding the elements and normal usage for a reasonable period of time.

4. Projects when completed should be optimum— lowest cost for the purposes intended or the best for the money spent—as required by the owner or client. Construction cost should not exceed the client's construction budget, and operation, maintenance, and repair, when properly executed, should not be excessively costly.

5. Projects should be designed and constructed to meet pertinent legal requirements, conform with generally accepted engineering standards, and avoid endangering the health and safety of construction workers, operators of the projects, and the general public.

6. Projects should be designed to meet the goals of sustainable development which are meeting project needs while conserving and protecting environmental quality and the natural resource base for future generations.

*Revised and updated from "System Design" by Frederick S. Merritt.

7. Projects, when properly operated, should be energy efficient.

8. To the extent possible, projects should display aesthetic qualities.

The ultimate objective of design is to provide, in precise, concise, easy-to-comprehend form, all the information necessary for construction of the project. Traditionally, designers provide this information in drawings or plans that show what is to be constructed and in specifications that describe materials and equipment to be incorporated into the project. Designers usually also prepare, with legal assistance, a construction contract between the client and a general contractor or two or more prime contractors. In addition, designers generally observe or inspect construction of the project. This should be done not only to help the client ensure that the project is constructed in accordance with plans and specifications but to obtain information that will be useful for designing future projects.

1.2 Systems

Systems design of a project comprises a rational, orderly series of steps that leads to the best decision for a given set of conditions (Art. 1.9). The procedure requires:

Analysis of a project as a system

Synthesis, or selection of components to form a system that meets specific objectives

Appraisal of system performance, including comparisons with alternative systems

Feedback to analysis and synthesis of information obtained in system evaluation, to improve the design

The prime advantage of the procedure is that, through comparisons of alternatives and data feedback to the design process, system design converges on an optimum, or best, system for the given conditions. Another advantage is that the procedure enables designers to clarify the requirements for the project being designed. Still another advantage is that the procedure provides a common basis of understanding and promotes cooperation between the specialists in various aspects of project design.

For a project to be treated as a system, as required in systems design, it is necessary to know what a system is and what its basic characteristics are:

A system is an assemblage formed to satisfy specific objectives and subject to constraints or restrictions and consisting of two or more components that are interrelated and compatible, each component being essential to the required performance of the system.

Because the components are required to be interrelated, the operation, or even the mere existence, of one component affects in some way the performance of other components. Also, the required performance of the system as a whole and the constraints on the system impose restrictions on each component.

Examples of civil engineering systems include buildings, highways, bridges, airports, railroads, tunnels, water supply to meet human needs, and wastewater collection, treatment, and disposal.

A building is a system because it is an assemblage constructed to serve specific purposes, such as shelter for human activities or enclosure of stored materials. It is subject to such restrictions as building code limitations on height and floor area. Constraints include ability to withstand loads from human activities and from natural forces like wind and earthquakes. The assemblage generally consists of a roof, floors, walls, doors, windows, structural framing for supporting the other components, and means for heating, ventilating, and cooling the interior.

A highway or a railroad is a system constructed for the specific purpose of providing a suitable surface, or road, for movement of vehicles. The restrictions are imposed by the terrain to be traversed by the highway or railroad, vehicle characteristics, and volume of traffic. A highway is used primarily by rubber-tired vehicles whose velocity and direction of travel are controlled by human drivers. A railroad is used by vehicles equipped with steel wheels designed to ride on rails that control direction of travel, while velocity is controlled directly by a human driver or indirectly by remote controls. Both highway and railroad assemblages consist of a right-of-way and road between points to be served, entrances and exits for vehicles, traffic-control devices, safety devices, bridges, tunnels, stations for refueling and servicing vehicles, stations for embarking or disembarking passengers or loading or unloading freight, and convenience stations for drivers and passengers.

A tunnel is an underground system and a bridge is an aboveground system constructed for the specific purpose of providing passage for pedestrians, vehicles, pipes, cables, or conveyors past obstructions. A tunnel is subject to such restrictions as exclusion of earth, rock, and unwanted water from the passageway, whereas a bridge must carry the passageway at required distances above obstructions. A tunnel assemblage consists primarily of the passageway and supports or lining for housing the passageway. The assemblage may also include drainage, ventilation, and lighting provisions. A bridge assemblage consists primarily of the passageway, structural framing for supporting it, and piers and abutments for holding the other components at suitable heights above the obstructions.

Water supply is a system with the specific purpose of providing water to meet human needs. The restrictions on the system are generally criteria for quantity and quality of water. The assemblage usually consists of a water source; means for extracting water in desired quantities from the source and conveying it to points where it is needed; a plant for treating the water to meet quality criteria; pipes with diameters adequate for passing the desired quantities without excessive loss of pressure; valves; reservoirs; dams; and fixtures and other devices for flow control at points of use.

Sewage collection, treatment, and disposal is a system with the specific purpose of removing wastewater from points where it is created and discharging the wastes in such condition and in such locations that human health and welfare are not endangered and there is little or no adverse effect on the environment. The restrictions on the system generally are quantity and characteristics of the wastes, quantity of water needed for conveyance of the wastes, and criteria for the products to be discharged from the system. The assemblage consists of fixtures or other means for collecting wastes at the source and removing them with water; means for conveying the wastewater to a treatment plant and then transporting the treated products to points of disposal or reuse; the treatment plant where the wastes are removed or rendered innocuous; means for safe disposal or reuse of the treated wastes and water; pipes; valves; and various devices for flow control.

Note that in all the preceding examples the system consists of two or more interrelated, compatible components. Every component is essential to the required performance of the system. Also, every component affects the performance of at least one other component, and the required performance of the whole system imposes restrictions on every component.

Subsystems ■ A group of components of a system may also be a system called a subsystem. It too may be designed as a system, but its goal must be to assist the system of which it is a component to meet the system objectives. Similarly, a group of components of a subsystem may also be a system called a subsubsystem.

For brevity, a project's major subsystems often are referred to as systems. For example, in a building, such major subsystems as structural framing, walls, or plumbing are called systems. Their components that meet the definition of a system are referred to as subsystems. For instance, plumbing consists of water-supply, wastewater, and gas-supply subsystems. The wastewater subsystem in turn includes various fixtures for collecting and discharging wastewater; soil and waste pipes; pipe supports; traps; drains; sewers; and vents. In a complex system, such as a building, subsystems and other components may be combined in various ways to form different systems.

1.3 Systems Analysis

In systems analysis, a system is resolved into its basic components. Subsystems are determined, and then the system is investigated to determine the nature, interaction, and performance of the system as a whole. The investigation should answer such questions as:

What does each component (or subsystem) do?

What does the component do it to?

How does the component serve its function?

What else does the component do?

Why does the component do the things it does?

What must the component really do?

Can the component be eliminated because it is not essential or because another component can assume its tasks?

1.4 Goals, Objectives, and Criteria

Before design of a system can commence, the designer should establish the owner's goals for the system. These goals state what the system is to accomplish, how it will affect the environment and other systems, and how other systems and the environment will affect the project. Goals should be generalized but brief statements, encompassing all the design objectives. They should be sufficiently specific, however, to guide generation of initial and alternative designs and control selection of the best alternative.

A simple example of a goal is: Design a branch post-office building with 100 employees that is to be constructed on a site owned by the client. The building should harmonize with neighboring structures. Design must be completed within 120 days and construction within 1 year. Construction cost is not to exceed $1,250,000.

The goals for a systems design applied to a subsystem serve the same purpose as for a system. They indicate the required function of the subsystem and how it affects and is affected by other systems.

Objectives ▪ With the goals known, the designers can define the system objectives. These objectives are similar to goals but supply in detail the requirements that the system must satisfy to attain the goals.

When listing objectives, the designers may start with broad generalizations that they will later develop at more detailed levels to guide design of the system. Certain objectives, such as minimization of initial costs, life-cycle costs, or construction time, should be listed. Other objectives that apply to the design of almost every similar project, such as the health, safety, and welfare objectives of building codes, zoning, and Occupational Safety and Health Administration regulations, are too numerous to list and may be adopted by reference. Objectives that are listed should be sufficiently specific to guide planning of the project and selection of components with specific characteristics. Also, some objectives should specify the degree of control needed for operation of systems provided to meet the other objectives.

Criteria ▪ At least one criterion must be associated with each objective. The criterion is a range of values within which the performance of the system must lie for the objective to be met. The criterion should be capable of serving as a guide in evaluation of alternative systems. For example, for fire resistance of a building wall, the criterion might be 2-h fire rating.

Weights ▪ In addition to establishing criteria, the designers should weight the objectives in accordance with the relative importance of the objectives to the client (see also Art. 1.10). These weights also should serve as guides in comparisons of alternatives.

1.5 Constraints and Standards

Besides establishing goals and objectives for a system at the start of design, the designers should also define constraints on the system. Constraints are restrictions on the values of design variables that represent properties of the system and that are controllable by the designers.

Designers are seldom completely free to choose any values desired for properties of a system component. One reason is that a component with a desired property may not be readily available, for instance, a 9-in-long brick. Another reason is that there usually are various restrictions, which may be legal, such as building or zoning code requirements, or economic, physical, chemical, temporal, psychological, sociological, or esthetic. Such restrictions may fix the values of the component properties or establish a range in which they must lie.

Standards ▪ At least one standard must be associated with each constraint. A standard is a value or range of values governing a property of the system. The standard specifying a fixed value may be a minimum or maximum value.

For example, a designer may be seeking to determine the thickness of a load-bearing concrete masonry wall. The governing building code may state that the wall, based on wind load requirements and the height of the wall, shall be no less than 8 in thick. This requirement is a minimum standard. The designer may then select a wall thickness of 8 in or more. The requirements of other adjoining systems, however, indicate that for the wall to be compatible, wall thickness may not exceed 16 in. This is a maximum standard. Bricks, however, may be available only in nominal widths of 4 in. Hence, the constraints limit the values of the controllable variable, in this case wall thickness, to 8, 12, or 16 in.

1.6 Construction Costs

Construction cost of a project usually is a dominant design concern. One reason is that if construction cost exceeds the owner's or client's construction budget, the project may be canceled. Another reason is that some costs, such as interest on the investment, which occur after completion of the project often are proportional to the initial cost. Hence, owners usually try to keep that cost low. Designing a project to minimize construction cost, however, may not be in the owner's best interests. There are many other costs the owner incurs during the anticipated life of the project that should be taken into account.

For example, after a project has been completed, the owner incurs operation and maintenance costs. Such costs are a consequence of decisions made during project design. Often, postconstruction costs are permitted to be high so that initial costs can be kept within the owner's construction budget; otherwise, the project will not be built.

Life-cycle cost is the sum of initial, operating, and maintenance costs. Ideally, life-cycle cost should be minimized, rather than initial or construction cost, because this enables the owner to receive the greatest return on the investment in the project.

Nevertheless, a client usually establishes a construction budget independent of life-cycle cost. This often is necessary because the client does not have adequate capital for an optimum project and places too low a limit on construction cost. The client hopes to have sufficient capital later to pay for the higher operating and maintenance costs or for replacement of undesirable, inefficient components. Sometimes, the client establishes a low construction budget because the goal is a quick profit on early sale of the project, in which case the client has little or no concern with the project's future high operating and maintenance costs. For these reasons, construction cost frequently is a dominant concern in design.

1.7 Models

For convenience in evaluating the performance of a system and for comparison with alternative designs, designers may represent the system by a model that enables them to analyze the system and evaluate its performance. The model should be simple, consistent with the role for which it is selected, for practical reasons. The cost of formulating and using the model should be negligible compared with the cost of assembling and testing the actual system.

For every input to a system, there must be a known, corresponding input to the model such that the model's responses (output) to that input are determinable and correspond to the system's responses to its input. The correlation may be approximate but nevertheless should be close enough to serve the purposes for which the model is to be used. For example, for cost estimates during the conceptual phase of design, a cost model may be used that yields only reasonable guesses of construction costs. The cost model used in the contract documents phase, however, should be accurate.

Models may be classified as iconic, symbolic, or analog. The iconic type may be the actual system or a part of it or merely bear a physical resemblance to the actual system. The iconic model is often used for physical tests of a system's performance, such as load or wind-tunnel tests or adjustment of controls for air or water flow in the actual system.

Symbolic models represent by symbols a system's input and output and are usually amenable to mathematical analysis of a system. They enable relationships to be generally, yet compactly, expressed, are less costly to develop and use than other types of models, and are easy to manipulate.

Analog models are real systems but with physical properties different from those of the actual system. Examples include dial watches for measuring time, thermometers for measuring temperature (heat changes), dial gauges for measuring small movements, flow of electric current for measuring heat flow through a metal plate, and soap membranes for measuring torsion in an elastic shaft.

Variables representing a system's input and properties may be considered independent variables, of two types:

1. Variables that the designers can control: $x_1, x_2, x_3, \ldots$

2. Variables that are uncontrollable: $y_1, y_2, y_3, \ldots$

Variables representing system output, or performance, may be considered dependent variables: $z_1, z_2, z_3, \ldots$ These variables are functions of the independent variables. The functions also contain parameters, whose values can be adjusted to calibrate the model to the behavior of the actual system.

Cost Models ▪ As an example of the use of models in systems design, consider the following cost models:

$$C = Ap \qquad (1.1)$$

where C = construction cost of project

A = convenient parameter for a project, such as floor area (square feet) in a building, length (miles) of a highway, population (persons) served by a water-supply or sewage system

p = unit construction cost, dollars per unit (square feet, miles, persons)

This is a symbolic model applicable only in the early stages of design when systems and subsystems are specified only in general form. Both A and p are estimated, usually on the basis of past experience with similar systems.

$$C = \sum A_i p_i \qquad (1.2)$$

where A_i = convenient unit of measurement for ith system

p_i = cost per unit for ith system

This symbolic model is suitable for estimating project construction cost in preliminary design stages after types of major systems have been selected. Equation (1.2) gives the cost as the sum of the cost of the major systems, to which should be added the estimated costs of other systems and contractor's overhead and profit.

$$C = \sum A_j p_j \qquad (1.3)$$

where A_j = convenient unit of measurement for jth subsystem

p_j = cost per unit for jth subsystem

This symbolic model may be used in the design development phase and later after components of the major systems have been selected and greater accuracy of the cost estimate is feasible. Equation (1.3) gives the construction cost as the sum of the

costs of all the subsystems, to which should be added contractor's overhead and profit.

For more information on cost estimating, see Art. 4.7.

1.8 Optimization

The objective of systems design is to select the best system for a given set of conditions; this process is known as optimization. When more than one property of the system is to be optimized or when there is a single characteristic to be optimized but it is nonquantifiable, an optimum solution may or may not exist. If it does exist, it may have to be found by trial and error with a model or by methods such as those described in Art. 1.10.

When one characteristic, such as construction cost, of a system is to be optimized, the criterion may be expressed as

$$\text{Optimize } z_r = f_r(x_1, x_2, x_3, \ldots, y_1, y_2, y_3, \ldots)$$
$$(1.4)$$

where z_r = dependent variable to be maximized or minimized

x = controllable variable, identified by subscript

y = uncontrollable variable, identified by subscript

f_r = objective function

Generally, however, there are restrictions on the values of the independent variables. These restrictions may be expressed as

$$f_1(x_1, x_2, x_3, \ldots, y_1, y_2, y_3, \ldots) \geq 0$$
$$f_2(x_1, x_2, x_3, \ldots, y_1, y_2, y_3, \ldots) \geq 0 \qquad (1.5)$$
$$f_n(x_1, x_2, x_3, \ldots, y_1, y_2, y_3, \ldots) \geq 0$$

Simultaneous solution of Eqs. (1.4) and (1.5) yields the optimum values of the variables. The solution may be obtained by use of such techniques as calculus, linear programming, or dynamic programming, depending on the nature of the variables and the characteristics of the equations.

Direct application of Eqs. (1.4) and (1.5) to a whole civil engineering project, its systems and its larger subsystems, usually is impractical because of the large number of variables and the complexity of their relationships. Hence, optimization generally has to be attained differently, usually by such methods as suboptimization or simulation.

Simulation ▪ Systems with large numbers of variables may sometimes be optimized by a process called simulation, which involves trial and error with the actual system or a model. In simulation, the properties of the system or model are adjusted with a specific input or range of inputs to the system, and outputs or performance are measured until an optimum result is obtained.

When the variables are quantifiable and models are used, the solution usually can be expedited by use of computers. The actual system may be used when it is available and accessible, and changes in it will have little or no effect on construction costs. For example, after installation of air ducts in a building, an air conditioning system may be operated for a variety of conditions to determine the optimum damper position for control of air flow for each condition.

Suboptimization ▪ This is a trial-and-error process in which designers try to optimize a system by first optimizing its subsystems. Suboptimization is suitable when components influence each other in series.

Consider, for example, a structural system for a building consisting only of roof, columns, and footings. The roof has a known load (input), exclusive of its own weight. Design of the roof affects the columns and footings because its output equals the loads on the columns. Design of the columns affects only the footings because the column output equals the loads on the footings. Design of the footings, however, has no effect on any of the other structural components. Therefore, the structural components are in series, and they may be designed by suboptimization to obtain the minimum construction cost or least weight of the system.

Suboptimization of the system may be achieved by first optimizing the footings, for example, designing the lowest-cost footings. Next, the design of both the columns and the footings should be optimized. (Optimization of the columns alone will not yield an optimum structural system because of the effect of the column weight on the footings.) Finally, roof, columns, and footings together should be optimized. (Optimization of the roof alone will not yield an optimum structural

system because of the effect of its weight on columns and footings. A low-cost roof may be very heavy, requiring costly columns and footings. Cost of a lightweight roof, however, may be so high as to offset any savings from less expensive columns and footings. An alternative roof may provide optimum results.)

1.9 Systems Design Procedure

Article 1.2 defines systems and explains that systems design comprises a rational, orderly series of steps which leads to the best decision for a given set of conditions. Article 1.2 also lists the basic components of the procedure as analysis, synthesis, appraisal, and feedback. Following is a more formal definition:

Systems design is the application of the scientific method to selection and assembly of components to form the optimum system to attain specified goals and objectives while subject to given constraints or restrictions.

The **scientific method**, which is incorporated into the definitions of value engineering and systems design, consists of the following steps:

1. Collecting data and observations of natural phenomena

2. Formulating a hypothesis capable of predicting future observations

3. Testing the hypothesis to verify the accuracy of its predictions and abandoning or improving the hypothesis if it is inaccurate

Systems design should provide answers to the following questions:

1. What does the client or owner actually want the project to accomplish (goals, objectives, and associated criteria)?

2. What conditions exist, or will exist after construction, that are beyond the designers' control?

3. What requirements for the project or conditions affecting system performance does design control (constraints and associated standards)?

4. What performance requirements and time and cost criteria can the client and designers use to appraise system performance?

Collection of information necessary for design of the project starts at the inception of design and may continue through the contract documents phase. Data collection is an essential part of systems design, but because it is continuous throughout design, it is not listed in the following as one of the basic steps.

To illustrate, the systems design procedure is resolved into nine basic steps in Fig. 1.1. Because value analysis is applied in steps 5 and 6, steps 4 through 8 covering synthesis, analysis, and appraisal may be repeated several times. Each iteration should bring the design closer to the optimum.

To prepare for step 1, the designers should draw up a project program, or list of the client's requirements, and information on existing conditions that will affect project design. In steps 1 and 2, the designers use the available information to define goals, objectives, and constraints to be satisfied by the system (see Arts. 1.4 and 1.5).

Synthesis ▪ In step 3, the designers must conceive at least one system that satisfies the objectives and constraints. To do so, they rely on their past experience, knowledge, imagination, and creative skills and advice from consultants, including value engineers, construction experts, and experienced operators of the type of facilities to be designed.

In addition, the designers should develop alternative systems that may be more cost-effective and can be built quicker. To save design time in obtaining an optimum system, the designers should investigate alternative systems in a logical sequence for potential for achieving optimum results. As an example, the following is a possible sequence for a building:

1. Selection of a pre-engineered building, a system that is prefabricated in a factory. Such a system is likely to be low cost because of the use of mass-production techniques and factory wages, which usually are lower than those for field personnel. Also, the quality of materials and construction may be better than for custom-built structures because of assembly under controlled conditions and close supervision.

2. Design of a pre-engineered building (if the client needs several of the same type of structure).

3. Assembling a building with prefabricated components or systems. This type of construction is similar to that used for pre-engineered buildings except that the components preassembled are much smaller parts of the building system.

4. Specification of as many prefabricated and standard components as feasible. Standard components are off-the-shelf items, readily available from building supply companies.

5. Repetition of the same component as many times as possible. This may permit mass production of some nonstandard components. Also, repetition may speed construction because field personnel will work faster as they become familiar with components.

6. Design of components for erection so that building trades will be employed continuously on the site. Work that compels one trade to wait for completion of work by another trade delays construction and is costly.

Modeling ▪ In step 4, the designers should represent the system by a simple model of acceptable accuracy. In this step, the designers should determine or estimate the values of the independent variables representing properties of the system and its components. The model should then be applied to determine optimum system performance (dependent variables) and corresponding values of controllable variables (see Arts. 1.7 and 1.8). For example, if desired system performance is minimum construction cost, the model should be used to estimate this cost and to select components and construction methods for the system that will yield this optimum result.

Appraisal ▪ In step 5 of systems design, the designers should evaluate the results obtained in step 4. The designers should verify that construction and life-cycle costs will be acceptable to the client and that the proposed system satisfies all objectives and constraints.

Value Analysis and Decision ▪ During the preceding steps, value analysis may have been applied to parts of the project (see Art. 1.10). In step 6, however, value analysis should be applied to the

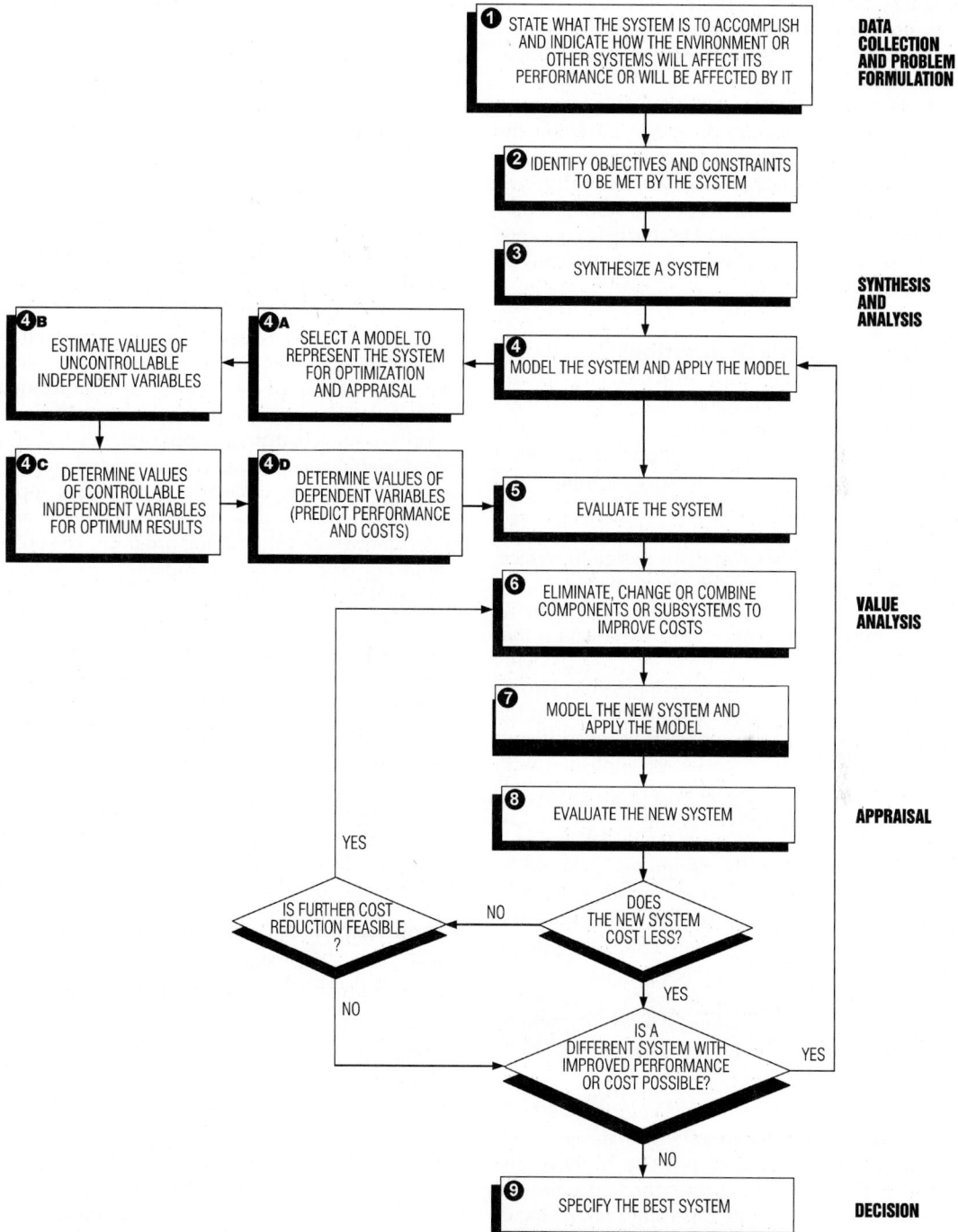

Fig. 1.1 Basic steps in systems design in addition to collection of necessary information.

whole system. This process may result in changes only to parts of the system, producing a new system, or several alternatives to the original design may be proposed.

In steps 7 and 8, therefore, the new systems, or at least those with good prospects for being the optimum, should be modeled and evaluated. During and after this process, completely different alternatives may be conceived. As a result, steps 4 through 8 should be repeated for the new concepts.

Finally, in step 9, the best of the systems studied should be selected.

Design by Team (Partnering) ▪ For efficient execution of systems design of a civil engineering project, a design organization superior to that used for traditional design is highly desirable. For systems design, the various specialists required should form a design team, to contribute their knowledge and skills in concert.

One reason why the specialists should work closely together is that in systems design the effects of each component on the performance of the whole project and the interaction of components must be taken into account. Another reason is that for cost-effectiveness, unnecessary components should be eliminated and, where possible, two or more components should be combined. When the components are the responsibility of different specialists, these tasks can be accomplished with ease only when the specialists are in direct and immediate communication.

In addition to the design consultants required for traditional design, the design team should be staffed with value engineers, cost estimators, construction experts, and building operators and users experienced in operation of the type of project to be constructed. Because of the diversity of skills present on such a team, it is highly probable that all ramifications of a decision will be considered and chances for mistakes and omissions will be small.

Project Peer Review ▪ The design team should make it standard practice to have the output of the various disciplines checked at the end of each design step and especially before incorporation in the contract documents. Checking of the work of each discipline should be performed by a competent practitioner of that discipline other than the original designer and reviewed by principals and other senior professionals. Checkers should seek to ensure that calculations, drawings, and specifications are free of errors, omissions, and conflicts between building components.

For projects that are complicated, unique, or likely to have serious effects if failure should occur, the client or the design team may find it advisable to request a peer review of critical elements of the project or of the whole project. In such cases, the review should be conducted by professionals with expertise equal to or greater than that of the original designers; that is, by peers, and they should be independent of the design team, whether part of the same firm or an outside organization. The review should be paid for by the organization that requests it. The scope may include investigation of site conditions, applicable codes and governmental regulations, environmental impact, design assumptions, calculations, drawings, specifications, alternative designs, constructability, and conformance with the building program. The peers should not be considered competitors or replacements of the original designers and there should be a high level of respect and communication between both groups. A report of the results of the review should be submitted to the authorizing agency and the leader of the design team.

(For additional information on peer review contact the American Consulting Engineering Council, 1015 15th Street, N.W., Washington, DC 20005, website www.acec.org or the American Society of Civil Engineers, 1801 Alexander Bell Drive, Reston Virginia 20191-4400, www.asce.org).

Application of Systems Design ▪ Systems design may be used profitably in all phases of project design, but it is most advantageous in the early design stages. One system may be substituted for another, and components may be eliminated or combined in those stages with little or no cost.

In the contract documents phase, systems design preferably should be applied only to the details being worked out then. Major changes are likely to be costly. Value analysis, though, should be applied to the specifications and construction contract because such studies may achieve significant cost savings.

Systems design should be applied in the construction stage only when design is required because of changes necessary in plans and specifications at that time. The amount of time available during that stage, however, may not be sufficient for thorough studies. Nevertheless, value analysis should be applied to the extent feasible.

1.10 Value Engineering

In systems design, the designers' goal is to select an optimum, or best system that meets the needs of the owner or client. Before the designers start designing a system, however, they should question whether the requirements represent the client's actual needs. Can the criteria and standards affecting the design be made less stringent? This is the first step in applying value engineering to a project.

After the criteria and standards have been reconsidered and approved or revised, the designers design one or more systems to satisfy the requirements and then select a system for value analysis. Next, the designers should question whether the system chosen provides the best value at the lowest cost. Value engineering is a useful procedure for answering this question and selecting a better alternative if the answer indicates this is desirable.

Value engineering is the application of the scientific method to the study of values of systems. (The scientific method is described in Art. 1.9.)

The major objective of value engineering as applied to civil engineering projects is reduction of initial and life-cycle costs (Art. 1.6). Thus, value engineering has one of the objectives of systems design, which has the overall goal of production of an optimum, or best, project (not necessarily the lowest cost), and should be incorporated into the systems design procedure, as indicated in Art. 1.9.

Those who conduct or administer value studies are often called value engineers or value analysts. They generally are organized into an interdisciplinary team, headed by a team coordinator, for value studies for a specific project. Sometimes, however, an individual, such as an experienced contractor, performs value engineering services for the client for a fee or a percentage of savings achieved by the services.

Value Analysis ▪ Value is a measure of benefits anticipated from a system or from the contribution of a component to system performance. This measure must be capable of serving as a guide when choosing among alternatives in evaluations of system performance. Because in comparisons of systems generally only relative values need be considered, value takes into account both advantages and disadvantages, the former being considered positive and the latter negative. It is therefore possible in comparisons of systems that the value of a component of a system will be negative and subtract from the system's overall performance.

System evaluations would be relatively easy if a monetary value could always be placed on performance; then benefits and costs could be compared directly. Value, however, often must be based on a subjective decision of the client. For example, how much extra is the client willing to pay for beauty, prestige, or better labor or community relations? Consequently, other nonmonetary values must be considered in value analysis. Such considerations require determination of the relative importance of the client's requirements and weighting values accordingly.

Value analysis is the part of the value engineering procedure devoted to investigation of the relationship between costs and values of components and systems and alternatives to these. The objective is to provide a rational guide for selection of the lowest-cost system that meets the client's actual needs.

Measurement Scales ▪ For the purpose of value analysis, it is essential that characteristics of a component or system on which a value is to be placed be distinguishable. An analyst should be able to assign different numbers, not necessarily monetary, to values that are different. These numbers may be ordinates of any one of the following four measurement scales: ratio, interval, ordinal, nominal.

Ratio Scale ▪ This scale has the property that, if any characteristic of a system is assigned a value number k, any characteristic that is n times as large must be assigned a value number nk. Absence of the characteristic is assigned the value zero. This type of scale is commonly used in engineering, especially in cost comparisons. For example, if a value of $10,000 is assigned to system A and $5000 to system B, then A is said to cost twice as much as B.

Interval Scale ▪ This scale has the property that equal intervals between assigned values represent equal differences in the characteristic being measured. The scale zero is assigned arbitrarily. The Celsius scale of temperature measurement is a good example of an interval scale. Zero is arbitrarily established at the temperature at which water freezes and does not indicate absence of heat. The boiling point of water is arbitrarily assigned the value of 100. The scale between 0 and 100 is then divided into 100 equal intervals called degrees ($°C$). Despite the arbitrariness of the selection of the zero point, the scale is useful in heat measurement. For example, changing the temperature of an object from 40 to 60 $°C$, an increase of 20 $°C$, requires twice as much heat as changing the temperature from 45 to 55 $°C$, an increase of 10 $°C$.

Ordinal Scale ▪ This scale has the property that the magnitude of a value number assigned to a characteristic indicates whether a system has more or less of the characteristic than another system has or is the same with respect to that characteristic. For example, in a comparison of the privacy afforded by different types of partitions in a building, each partition may be assigned a number that ranks it according to the degree of privacy it provides. Partitions with better privacy are given larger numbers. Ordinal scales are commonly used when values must be based an subjective judgments of nonquantifiable differences between systems.

Nominal Scale ▪ This scale has the property that the value numbers assigned to a characteristic of systems being compared merely indicate whether the systems differ in this characteristic. But no value can be assigned to the difference. This type of scale is often used to indicate the presence or absence of a characteristic or component. For example, the absence of means of access to maintenance equipment may be represented by zero or a blank space, whereas the presence of such access may be denoted by 1 or $\times$.

Weighting ▪ In practice, construction cost is only one factor, perhaps the only one with a monetary value, of several factors that must be evaluated in a comparison of systems. In some cases, some of the system's other characteristics may be more important to the owner than cost.

Under such circumstances, the comparison may be made by use of an ordinal scale for ranking each characteristic and then weighting the rankings according to the importance of the characteristic to the client.

As an example of the use of this procedure, calculations for comparison of two partitions for a building are shown in Table 1.1. Alternative 1 is an all-metal partition; alternative 2 is made of glass and metal.

In Table 1.1 the first column lists characteristics of concern in the comparison. The numbers in the second column indicate the relative importance to the client of each characteristic: 1 denotes lowest priority and 10 highest priority. These are weights. In addition, each partition is ranked on an ordinal scale, with 10 as the highest value, in accordance with the degree to which it possesses each characteristic. These rankings are listed as relative values in Table 1.1. For construction cost, for instance, the metal partition is assigned a relative value of 10 and the glass-metal partition a value of 8 because the metal partition costs a little less than the other one. In contrast, the glass-metal partition is given a relative value of 8 for visibility because the upper portion is transparent, whereas the metal partition has a value of 0 because it is opaque.

To complete the comparison, the weight of each characteristic is multiplied by the relative value of the characteristic for each partition and entered in Table 1.1 as weighted value. For construction cost, for example, the weighted values are $8 \times 10 = 80$ for the metal partition and $8 \times 8 = 64$ for the glass-metal partition. The weighted values for each partition are then added, yielding 360 for alternative 1 and 397 for alternative 2. Although this indicates that the glass-metal partition is better, it may not be the best for the money. To determine whether it is, the weighted value of each partition is divided by its cost. This yields 0.0300 for the metal partition and 0.0265 for the other. Thus, the metal partition appears to offer more value for the money and would be recommended.

The preceding calculation makes an important point: In a choice between alternative systems, only the differences between system values are significant and need be compared.

Suppose, for example, the economic effect of adding thermal insulation to a building is to be investigated. In a comparison, it is not necessary to compute the total cost of the building with and without the insulation. Generally, the value analyst

Table 1.1   Comparison of Alternative Partitions*

Characteristics	Relative Importance	Alternatives			
		1 All Metal		2 Glass and Metal	
		Relative Value	Weighted Value	Relative Value	Weighted Value
Construction cost	8	10	80	8	64
Appearance	9	7	63	9	81
Sound transmission	5	5	25	4	20
Privacy	3	10	30	2	6
Visibility	10	0	0	8	80
Movability	2	8	16	8	16
Power outlets	4	0	0	0	0
Durability	10	9	90	9	90
Low Maintenance	8	7	56	5	40
Total Weighted values			360		397
Cost			$12,000		$15,000
Ratio of values to cost			0.0300		0.0265

* Reprinted with permission from F. S. Merritt, "Building Engineering and Systems Design," Van Nostrand Reinhold Company, New York, N.Y.

need only subtract the added cost of insulation from the decrease in heating and cooling costs resulting from addition of insulation. A net saving would encourage addition of insulation. Thus, a decision can be reached without the complex computation of total building cost.

Value Analysis Procedure ■ For value analysis of a civil engineering project or one of its subsystems, it is advisable that the client or a client's representative appoint an interdisciplinary team and a team coordinator with the assignment of either recommending the project or proposing a more economical alternative. The team coordinator sets the study's goals and priorities and may appoint task groups to study parts of the system in accordance with the priorities. The value analysts should follow a systematic, scientific procedure for accomplishing all the necessary tasks that comprise a value analysis. The procedure should provide:

An expedient format for recording the study as it progresses

An assurance that consideration has been given to all information, some of which may have

been overlooked in development of the proposed system

A logical resolution of the analysis into components that can be planned, scheduled, budgeted, and appraised

The greatest cost reduction can be achieved by analysis of every component of the proposed project. This, however, is not generally practical because of the short time usually available for the study and the cost of the study increases with time. Hence, the study should concentrate on those project subsystems whose cost is a relatively high percentage of the total cost because those components have good possibilities for substantial cost reduction.

During the initial phase of value analysis, the analysts should obtain a complete understanding of the project and its major systems by rigorously reviewing the program, or list of requirements, the proposed design, and all other pertinent information. They should also define the functions, or purposes, of each component to be studied and estimate the cost of accomplishing the functions. Thus, the analysts should perform a systems analysis, as indicated in Art. 1.3, answer the questions listed in Art. 1.3 for the items to be studied, and estimate the items' initial and life-cycle costs.

In the second phase of value analysis, the analysts should question the cost-effectiveness of each component to be studied (see Art. 1.11). Also, by using imagination and creative techniques, they should generate several alternatives for accomplishing the required functions of the component. Then, in addition to answers to the questions in Art. 1.3, the analysts should obtain answers to the following questions:

Do the original design and each alternative meet performance requirements?

What does each cost installed and over the life cycle?

Will it be available when needed? Will skilled labor be available?

Can any component be eliminated?

What other components will be affected by adoption of an alternative? What will the resulting changes in the other components cost? Will there be a net saving in cost?

When investigating the elimination of a component, the analysts also should see if any part of it can be eliminated, if two or more parts can be combined into one, and if the number of different sizes and types of an element can be reduced. If costs might be increased by use of a nonstandard or unavailable item, the analysts should consider substituting a more appropriate alternative. In addition, the simplification of construction or installation of components and ease of maintenance and repair should be considered.

In the following phase of value analysis, the analysts should critically evaluate the original design and alternatives. The ultimate goal should be recommendation of the original design or an alternative, whichever offers the greatest value and cost-savings potential. The analysts should also submit estimated costs for the original design and the alternatives.

In the final phase, the analysts should prepare and submit to the client or to the client's representative who appointed them a written report on the study and resulting recommendations and a workbook containing detailed backup information.

1.11 Economic Comparisons of Alternative Systems

When evaluating systems, designers or value engineers should take into account not only initial and life-cycle costs but the return the client wishes to make on the investment in the project. Primarily, a client would like to maximize profit—the benefits or revenues accruing from use of the project less total costs. Also, the client usually would like to ensure that the rate of return, the ratio of profit to investment, is larger than all the following:

Rate of return expected from other available investment opportunities

Interest rate for borrowed money

Rate for government bonds or notes

Rate for highly rated corporate bonds

The client is concerned with interest rates because all costs represent money that either must be borrowed or could otherwise be invested at a current interest rate. The client also has to be concerned with time, measured from the date on which an investment is made, because interest cost increases with time. Therefore, economic comparisons of systems must take into account both interest rates and time. (Effects of monetary inflation can be taken into account in much the same way as interest.)

An economic comparison of alternatives usually requires evaluation of initial capital investments, salvage values after several years, annual disbursements, and annual revenues. Because each element in such a comparison may have associated with it an expected useful life different from that of the other elements, the different types of costs and revenues, or benefits, must be made commensurable by reduction to a common basis. This is done by either:

1. Converting all costs to equivalent uniform annual costs and income

2. Converting all costs and revenues to present worth at time zero

Present worth is the money that, invested at time zero, would yield at later times required costs and revenues at a specified interest rate. (In economic comparisons, the conversions should be based on a rate of return on investment that is attractive to the client. It should not be less than the interest rate the client would have to pay if the amount of the investment had to be borrowed. For this reason, the desired rate of return is called interest rate in conversions.) Calculations also should be based on actual or reasonable estimates of useful life. Salvage values should be taken as the expected return on sale or trade-in of an item after a specific number of years of service. Interest may be considered compounded annually.

Future Value ▪ Based on the preceding assumptions, a sum invested at time zero increases in time to

$$S = P(1 + i)^n \qquad (1.6)$$

where S = future amount of money, equivalent to P, at end of n periods of time with interest rate i

i = interest rate

n = number of interest periods (years)

P = sum of money invested at time zero

= present worth of S

Present Worth ▪ Solution of Eq. (1.6) for P yields the present worth of a sum of money S at a future date:

$$P = S(1 + i)^{-n} \qquad (1.7)$$

The present worth of payment R made annually for n years is

$$P = R\frac{1 - (1 + i)^{-n}}{i} \qquad (1.8)$$

The present worth of the payments R continued indefinitely can be obtained from Eq. (1.8) by making n infinitely large:

$$P = \frac{R}{i} \qquad (1.9)$$

Capital Recovery ▪ A capital investment P at time zero can be recovered in n years by making annual payments of

$$R = P\frac{i}{1 - (1 + i)^{-n}} = P\left[\frac{i}{(1 + i)^n - 1} + i\right] \qquad (1.10)$$

When an item has salvage value V after n years, capital recovery R can be computed from Eq. (1.10)

by subtracting the present worth of the salvage value from the capital investment P:

$$R = [P - V(1 + i)^{-n}]\left[\frac{i}{(1 + i)^n - 1} + i\right] \qquad (1.11)$$

Example: To illustrate the use of the preceding formulas, following is an economic comparison for two pumps. Costs are estimated as follows:

	Pump 1	Pump 2
Initial cost	$30,000	$50,000
Life, years	10	20
Salvage value	$5,000	$10,000
Annual costs	$3,000	$2,000

Cost of operation, maintenance, repairs, property taxes, and insurance are included in the annual costs. The present-worth method is used for the comparison, with interest rate $i = 8\%$.

Conversion of all costs and revenues to present worth must be based on a common service life, although the two pumps have different service lives, 10 and 20 years, respectively. For the purpose of the conversion, it may be assumed that replacement pumps will repeat the investment and annual costs predicted for the initial pumps. (Future values, however, should be corrected for monetary inflation.) In some cases, it is convenient to select for the common service life the least common multiple of the lives of the units being compared. In other cases, it may be more convenient to assume that investment and annual costs continue indefinitely. The present worth of such annual costs is called **capitalized cost**.

For this example, a common service life of 20 years, the least common multiple of 10 and 20 is selected. Hence, it is assumed that pump 1 will be replaced at the end of the tenth period at a cost of $30,000 less the salvage value. Similarly, the replacement unit will be assumed to have the same salvage value after 20 years.

The calculations in Table 1.2 indicate that the present worth of the net cost of pump 2 is less than that for pump 1. If cost were the sole consideration, purchase of pump 2 would be recommended.

1.12 Risk Management

Throughout all stages of design and construction, but especially during conceptual design of a

Table 1.2 Example Cost Comparison of Two Pumps

	Pump 1	Pump 2
Initial investment	$30,000	$50,000
Present worth of replacement cost in 10 years $P - V$ at 8% interest [Eq. (1.7)]	11,580	
Present worth of annual costs for 20 years at 8% interest [Eq. (1.8)]	29,454	19,636
Present worth of all costs	71,034	69,636
Revenue:		
Present worth of salvage value after 20 years at 8% interest [Eq. (1.8)]	1,073	2,145
Net cost:		
Present worth of net cost in 20 years at 8% interest	$69,961	$67,491

project, the possibility should be considered that the project at any stage, from excavation and grading to long after completion, may endanger public health or safety or cause economic loss to neighbors or the community. Not only the effects of identifiable hazards should be taken into account but also the consequences of unforeseen events, such as component failure, accidental explosions or fire, mechanical breakdowns, and terrorist attacks during occupancy of the project.

A **hazard** poses the threat that an unwanted event, possibly catastrophic, may occur. **Risk** is the probability that the event will occur. The responsibility of estimating both the probability of hazards occurring and the magnitudes of the consequences should the events be realized lies principally with project owners, designers, and contractors. They also are responsible for risk management. This requires establishment of an acceptable level for each risk, generally with input from government agencies and the public, and selection of cost-effective ways of avoiding the hazards, if possible, or protecting against them so as to reduce the risks of hazards occurring to within the acceptable levels.

Studies of construction failures provide information that designers should use to prevent similar catastrophes. Many of the lessons learned from failures have led to establishment of safety rules in standard design specifications and regulations of various government agencies. These rules, however, generally are minimum requirements and apply to ordinary structures. Designers, therefore, should use judgment in applying such requirements and should adopt more stringent design criteria where conditions dictate.

Designers also should use judgment in determining the degree of protection to be provided against specific hazards. Protection costs should be commensurate with probable losses from an unwanted event. In many cases, for example, it is uneconomical to construct a project that will be immune to extreme earthquakes, tornadoes, arson, bombs, burst dams, or very unusual floods. Full protection, however, should always be provided against hazards with a high probability of occurrence accompanied by personal injuries or high property losses. Such hazards include hurricanes and gales, fire, vandals, and overloading.

Design Life of Projects ■ Design criteria for natural phenomena may be based on the probability of occurrence of extreme conditions, as determined from statistical studies of events in specific localities. These probabilities are often expressed as mean recurrence intervals.

Mean recurrence interval of an extreme condition is the average time, in years, between occurrences of a condition equal to or worse than the specified extreme condition. For example, the mean occurrence interval of a wind of 60 mi/h or more may be reported for a locality as 50 years. Thus, after a structure has been constructed in that locality, chances are that in the next 50 years it will be subjected only once to a wind of 60 mi/h or more. Consequently, if the structure was assumed to have a 50-year life, designers might design it basically for a 60-mi/h wind, with a safety factor included in the design to protect against low-probability faster winds. Mean recurrence intervals are the basis for many minimum design loads in standard design specifications.

Safety Factors ▪ Design of projects for both normal and emergency conditions should always incorporate a safety factor against failure or component damage. The magnitude of the safety factor should be selected in accordance with the importance of the structure, the consequences of personal injury or property loss that might result from a failure or breakdown, and the degree of uncertainty as to the magnitude or nature of loads and the properties and behavior of project components or construction equipment.

As usually incorporated in design codes, a safety factor for quantifiable system variables is a number greater than unity. The factor may be applied in either of two ways.

One way is to relate the maximum permissible load, or demand, on a system under service conditions to design capacity. This system property is calculated by dividing by the safety factor the ultimate capacity, or capacity at failure, for sustaining that load. For example, suppose a structural member assigned a safety factor of 2 can carry 1000 lb before failure occurs. The design capacity then is $1000/2 = 500$ lb.

The second way in which codes apply safety factors is to relate the ultimate capacity of a system to a design load. This load is calculated by multiplying the maximum load under service conditions by a safety factor, often referred to as a load factor. For example, suppose a structural member assigned a load factor of 1.4 for dead loads and 1.7 for live loads is required to carry a dead load of 200 lb and a live load of 300 lb. Then, the member should have a capacity of $1.4 \times 200 + 1.7 \times 300 = 790$ lb, without failing.

While both methods achieve the objective of providing reserve capacity against unforeseen conditions, use of load factors offers the advantage of greater flexibility in design of a system for a combination of different loadings, because a different load factor can be assigned to each type of loading. The factors can be selected in accordance with the probability of occurrence of overloads and effects of other uncertainties.

Frank Muller
*President, Metro
Mediation Services, Ltd.
New York, New York*

Jay Gewirtzman
*Vice President
URS Corp., New York
New York*

2

DESIGN MANAGEMENT

Design management is concerned with an engineer's sphere of activity. It is therefore important to consider the variety and types of design activities to which professionals devote their efforts.

The engineer's basic role is to harness scientific principles and other knowledge to practical applications that benefit humanity. In fulfillment of this role, design management is concerned with proper utilization of human labor, energy, and technical skills to serve present and future needs of the economy.

The design manager's goal is to complete a project on schedule and within budget while meeting standards of quality in order to meet the client's needs.

2.1 Where Engineers Are Employed

Principal fields of employment for engineers include:

Academic ▪ For many engineers, the teaching profession is both the first and final career. Many others, however, devote to teaching a few years of their careers or sometimes part of their time, for example, teaching evening courses.

Many educators also serve as advisers to industry and consulting firms. Thus, they move into the designer's sphere of activity. Furthermore, many university departments are retained by government and industry for research projects. As a consequence, the departments, in essence, act as private firms performing professional services. The university administrators have to work within budgets and have contracts to negotiate, reimbursable expenses to determine, and schedules to meet. They also have to contend with other administrative matters that are part of design management.

Industry ▪ Industrial firms that handle any substantial volume of business have engineers on their staff. The role of such engineers, however, varies. A firm with productive capacity and thus plant facilities must have a plant engineer and staff to ensure proper maintenance and operation of the plant. In many industries, the plant engineers also serve their employers in the design field. For instance, if new equipment is to be installed in an existing plant, not only must space be provided but engineering questions must be addressed. Typical questions include: Are the foundations adequate to carry the added loads? Are new utility services required? Is the present power supply adequate? Furthermore, building may have to be constructed to house the new equipment. Thus, a plant engineer's normal activities and responsibilities often lead to the design field.

Because of their size, growth, and specialized needs, many industries have their own engineering and design departments. Such a department fulfills the same professional function as a private engineering firm, with one basic difference: The industry engineer serves one client, whereas the design firm serves many. Concerned with many of the same administrative matters as a design firm, an engineering department can be organized like a design firm. The engineering department will be organized to operate efficiently in meeting the specialized needs of only its industrial employer.

Government ▪ Like engineers in industry, government engineers serve only one client, their employer. The federal government is the largest single employer of architects and engineers. In addition, most states, counties, cities, towns, and public bodies have engineers and architects on their staffs or in their employ. These professionals perform a variety of functions encompassing both design and administrative activity.

The agencies or authorities maintain engineering and architectural departments that perform basic design work and thus act as in-house professional service firms. Such organizations do not need to retain outside private consultants, except for specialized tasks or when the volume of design work to be performed exceeds their in-house capabilities. In addition, these agencies, whether or not they have in-house design capability, employ professionals who work on a variety of different administrative levels, including administration and supervision of projects as well as review of basic design and construction activities. Administration of the engineering projects requires the services of professionals on all levels, starting with junior staff members and extending up to top-level administrators and officials charged with responsibility for implementation of the public projects.

In public service, the engineer may be either the designer or the client.

Engineer-Contractor ▪ The term as used here refers to the construction firm that identifies itself as both an engineer designer and contractor. Although many use the title engineer-contractor and perform only the actual construction, we are concerned here with the firm that truly undertakes either design-build or most frequently "turnkey" projects—both design and construction under a single contract.

Process and utility industries generally use the turnkey contract. These industries are primarily interested in the final product, such as number of barrels of oil refined or number of kilowatthours produced. The engineering staff of the company building a plant establishes design criteria that the engineer-contractor has to meet. Because of the specialized nature of these industries, the engineer-contractor employs designers with knowledge of particular processes to develop the most economical and efficient design. Engineer-contractors normally bid on performance specifications and prepare the detail designs necessary for construction. Other turnkey operations include those that combine land acquisition, design, and construction for commercial and industrial buildings and can even include financing.

The design is accomplished by the same organization, or division within the organization, that constructs the building or facility. Depending on a variety of factors, there are advantages and disadvantages of this combined service as compared with the division of responsibility between a design firm and a construction company.

Contractor ▪ A traditional construction project team consists of three parties: owner, or client; designer; and general contractor (GC). After being awarded a general construction contract by the client, the GC hires the subcontractors and the trades. Some forms of contracting, however, require several "prime" contractors instead of one GC. In such instances, the owner usually contracts directly with the major trades, such as heating, ventilating, and air conditioning (HVAC); electrical; plumbing; and vertical transportation installations. Also, in some situations, such as for a project administered by a construction manager (CM), the owner may engage several prime contractors, whose separate contracts will be coordinated and managed by the CM. Most contractors operate in a regional or limited geographic area.

Whether performing construction as a GC, prime contractor, or subcontractor, these companies employ engineers from a wide variety of disciplines. Engineers may serve as project managers who have responsibility for bringing the project to a successful completion while meeting the time, cost and quality goals; Project Engineers who schedule and coordinate construction; and Superintendents who plan and supervise the work in the field. Hence, there are many employment opportunities for engineers with contractors. Furthermore, the nature of construction contracting is such that it provides many opportunities for engineers to assume proprietorship roles.

Consulting Engineer ▪ A consulting engineer has been defined as a "professional experienced in the application of scientific principles to engineering problems." As professionals, consulting engineers owe a duty to the public as well as to their clients. In addition to rendering a professional service, the consulting engineer also operates a business. Consulting engineering is practiced by sole practitioners, partnerships, and corporations, many with large staffs of professionals, CAD operators, and other supporting personnel. Regardless of the form of the engineer's organization, the final product a client receives retains the same professional characteristics and meets the same professional standards. Consulting engineers usually have several clients, and they must select methods of operation to suit their own and their clients' needs best.

Consulting engineers are paid a fee by clients to provide professional design services on diverse projects, types including but not limited to transportation, industrial, education, institution and environmental facilities.

Construction Manager (CM) ■ Managing and coordinating construction projects as an agent of owners, i.e., acting as the CM, is the prime specialty or discipline of many firms. Although engineers and architects are the traditional professionals operating or employed by such firms, construction management is a separately defined technical field. The tasks and functions of construction managers, whether part of a professional service agreement or a guaranteed-maximum-price (GMP) contract, are well-established areas of practice.

While the primary goal of construction managers is to construct a project with the time, cost and quality goals established, they are increasingly being hired during the design phase to ensure that the design is constructible and cost effective.

Others ■ There are numerous specialty firms that practice in private industry but limit their activity to specific or specialized fields. These firms or individual practitioners may be appropriately classified under any of the broad definitions above but, as engineers, limit their professional activities. For instance, some specialty firms perform only cost-estimating services (consulting engineers or construction managers); act as construction consultants, serving as troubleshooters; or specialize in one technical area for the sole purpose of serving as expert witnesses in construction litigation.

2.2 Forms of Consulting Engineering Organizations

Consulting engineers may practice as individuals, partnerships, or corporations.

Individual Proprietorship ■ This form of organization is the simplest, has the fewest legal complications, and enables the proprietor to exercise direct control over the operation. As a one-person operation, however, this type of practice has distinct limitations because its activity essentially can be restricted to the efforts of the individual.

Although conducting a business as a sole proprietor, a consulting engineer may have several employees. Thus, as an employer, the consulting engineer is operating a business and has to handle the problems associated with a business enterprise. Also, because consulting engineers represent the legal entities conducting their businesses, they are responsible for all obligations of a business and all contracts are entered into in their names. Consulting engineers are personally responsible for all debts and can be liable for these to the extent of all their assets, business or personal. All profits, however, are earned by the consulting engineers, and they are not required to distribute earnings, as in a partnership, or be concerned with the declaration of dividends, as with a corporation.

Partnership ■ Another form for a consulting engineering organization is a *partnership*, that is, an association of two or more professionals who combine forces and talents to serve their clients on a more comprehensive scale and, by offering more services, to serve a wider clientele. Typically, each partner is responsible for a specific area. The management of the business, depending on its complexity, is assigned to one partner, the *managing partner*.

A partnership retains the identity of the individual professional, and basically its legal structure is similar to that of the individual proprietorship. Instead of one individual assuming all contractual obligations, liabilities, and earnings, all profits are shared by the partners. The partners, however, may not necessarily share equally in the business. Interest can be worked out among the partners as desired. For instance, one partner may own more than 50% and thus have a position comparable to that of the majority stockholder of a corporation.

Partnerships, although once predominant in the engineering profession as in other fields, such as architecture, accounting, and law, are rarely used. Most large engineering organizations that operated as partnerships have reorganized into corporations. From the business point of view, partnerships have several disadvantages that cause many firms to incorporate in states where such corporate practice is not restricted.

One disadvantage of partnerships is that each partner is legally liable to the extent of total personal assets for the wrongful act of any partner in the ordinary course of business. Another disadvantage is that a partnership terminates on death or retirement of one partner unless other provisions are made in the partnership agreement. Furthermore, a partnership does not have the flexibility of a corporation for comprehensive employee-benefit programs and provision for key employee participation.

Although a partnership as an entity does not pay taxes, the partners as individuals pay taxes on the profits. This is not necessarily a disadvantage, but it can be a prime consideration in the choice of an operating organization.

Also, although a professional cannot limit personal liability for professional errors or omissions in a corporate structure, the proliferation of litigation in the industry has made it more advantageous for engineers to operate as corporations or as Limited Liability Partnerships (LLPs) or Limited Liability Corporations (LLCs) rather than individual proprietorships or partnerships.

Corporations ▪ Most firms with several employees practice either as general or professional corporations (PC), depending on the laws of the state in which they practice. Practitioners who perform engineering in more than one state must take into account the variation in states' requirements, to ensure compliance not only with professional requirements (licensing) but with business practices (registering to do business, certification, and tax filings).

Most states permit the formation of professional engineering corporations. But usually a corporation can be formed for the purpose of practicing engineering only under certain conditions: ownership and management of the company must be totally vested in professionals or, at least, majority interests be held by professionals. Many states have passed legislation permitting the formation of such corporations to give professionals, not only in engineering but in other professions, the benefits and protection of conducting business as a corporation. Although permitting such corporate practice, the legislation includes requirements so structured that the public is protected from unqualified persons conducting a professional practice under a corporate guise.

With such protective requirements, professional identity can be maintained in corporate practice. Therefore, if conditions warrant and state law permits, engineering organizations should consider the corporate form of practice. The advantages that are attained, however, are mainly business ones. The management structure of the organization is clarified. Responsibility is defined. The area of employee fringe benefits becomes more diversified. Opportunities exist for profit sharing, for realistic retirement plans, and for employees to buy into the firm. Also, the principal's personal liability is limited to the assets of the corporations although the principals continue to be responsible for their own professional acts and cannot use the corporate structure as a shield from liability for professional errors and omissions.

Each form of practice has to be evaluated on its own merits. A corporate structure for an individual practitioner with a small practice may not be warranted, but one with a large volume of business that can be assigned to subordinates may find a corporation advantageous. For some firms, the tax advantages of a corporation may be more beneficial than operating as a partnership. (For federal income tax purposes, a small business corporation, meeting certain requirements, can elect to be taxed as a partnership, a practice advantageous for a small corporation.)

Limited Liability Companies and Partnerships ▪ The majority of states provide for the formation of limited liability companies (LLCs) and partnerships (LLPs). Statutes provide for the formation of LLCs for most business purpose except special areas, such as banking and insurance for which there are other controlling statutes. Professional limited liability companies (PLLCs) can also be formed. Members of a PLLC, however, must be registered professional engineers.

As the name implies the objective of doing business as an LLC and LLP is to "limit liability". Members, managers and agents of such entities are not personally liable for debts, obligations, and liabilities of the LLC or of each other. However, members, managers and employees of an LLP are personally liable for negligent acts (professional errors and omissions), as in any professional business entity. There is no business shield for any professional for professional misconduct or negligence.

The advantages of limited liability entities as compared with other business forms of organizations are the unique combination of limited liability and pass through taxation. Namely, taxes are only incurred at the ownership level not both for the business entity and for the distribution of profits (dividends).

Organizations conducting business as general or professional partnerships, or S corporations may find LLCs, PLLCs and LLPs to be advantageous business structures. Consideration, however, must be given not only to the laws of the State where the business entity is organized but also to other states where the business is to be conducted.

2.3 Clients for Engineering Services

Each client and each project has particular needs. Clients include:

Federal Government ▪ As the largest single employer of engineers and largest contractor for services and products, the federal government is a potential client for most design firms. To qualify for consideration by any government branch, a firm must file periodically with agencies from which work may be obtained a questionnaire detailing the firm's organization, key personnel (education and experience), special areas of competence, and experience (including completed projects). Preparation of such data is time-consuming, but most agencies have standardized their requirements so that the same form can be used for many filings.

Within the federal government, a standard questionnaire for architects and engineers is used by most of the agencies retaining professional services. This form, identified as standard form S.F. 254, presents, in summary fashion, data describing both the experience and qualifications of individual professionals and the firm, together with project descriptions and areas of expertise. In addition, many agencies have established computerized data banks utilizing the information contained in these standard qualification forms, to simplify both their records and the search for qualified professional firms to service specific needs.

In addition to S.F. 254, these agencies utilize S.F. 255, which is a subsequent submission of qualifications for specific assignments. This form requires identification of key personnel who would be utilized on a specific project and also requires evidence of specific experience related to the planned program or project.

When an agency needs outside design services, it is able to search its qualification file to identify firms that have the particular capabilities and professional expertise necessary for a particular project. All new projects are advertised in the *Commerce Business Daily* (CBD), to give all interested parties the opportunity to submit qualification data for consideration. After review of the qualification data, an agency may request more detailed qualification material from a select list of firms and then follow up with individual interviews prior to consultant selection.

Other-than-Federal Public Work ▪ Public work other than that performed for the federal government is in the province of states, counties, cities, and municipalities. The contracting party varies, depending on the nature of the work and its scope. Usually, engineering work is under the jurisdiction of an agency's engineering department. Sometimes, however, states or cities establish authorities to administer, construct, operate, and maintain projects. Many states, for example, have separate authorities for construction and operation of limited-access toll roads, for ports, for bridges and tunnels, and for public buildings such as schools and colleges. These authorities, as well as the public bodies, have different methods of operation. Some perform all or nearly all design in-house; they engage outside consultants infrequently. Others retain consulting engineers for most design.

Considerable areas of engineering activity lie within public authorities or regional public agencies, such as transportation, sewer, or water authorities, established within regions for the implementation of specific tasks. Such agencies either retain consultants to perform the necessary engineering for implementation of their public projects or establish in-house capability to perform the same functions.

Industry, Commerce, Residential, and Institutional ▪ Construction for these purposes varies with economic conditions and other factors, and opportunities for employment in these areas vary accordingly. Residential construction constitutes a substantial portion of the U.S. gross domestic product. It provides many employment opportunities for engineers and construction workers. Although single-family houses, which comprise a major segment of the residential market, are constructed by individual builders and small businesses, engineers play a role in this field, either as builders or in related work such as survey, utility, and support services.

Architects and other Professionals ▪ Many consulting engineers have architects and other professionals as their primary clients. Architects are employed to design a wide variety of facilities that include but are not limited to buildings, parks and waterfront facilities. Buildings may range from one-story residences to commercial high rise towers. They design the shell of the building including the exterior walls, windows, doors and roofs and the interior including wall, ceiling and floor layouts and finishes. While some architectural firms have in-house engineering capabilities, many have made a business decision not to provide that service and have decided to subcontract it to engineering firms. The consulting engineers who enter contracts with architects typically specialize in providing one facet of engineering although some of the large firms may provide services in more than one specialty. This may include the design of foundations, civil engineering site features such as roads and drainage, structural engineering including building framing systems, and mechanical, electrical and plumbing engineering to provide heating, cooling, fire alarm and sanitary system design.

In addition, professionals serve each other within their own fields of competence. Engineers may retain other engineers as consultants to supplement their own capabilities, either to take advantage of specialized knowledge or experience or for independent checks on their firm's analysis and calculations. Engineers team together to win large and complex projects. Although a consulting engineering firm may have capability to provide foundation engineering, a project may have difficult and unique subsurface conditions, which warrant the use of a firm specializing in geotechnical and foundation engineering that has provided designs for similar types of conditions.

Contractors ▪ Contractors provide a large market for consulting engineering firms. They have utilized the services of consulting engineers to provide the means and methods of construction that may be specifically excluded from the design engineers' scope of services. Means and methods may include the design of temporary support structures such as sheeting to protect excavations or scaffolding to support decks while concrete is being placed. Contractors may also employ engineers to perform re-engineering to provide an alternate system that is more economical to construct than the one included in the design documents, which the contractor bid and is using to build.

Design-Build is a project delivery system that is being used more extensively than in the past. Under this system, contractors provide a proposal to both design and build a project. Since most contractors do not have an in-house engineering staff to provide major design services, they use the services of consulting engineers.

Other Clients ▪ Sometimes, an owner may engage an engineer for projects that may require a few hours' attention or for the design of an entire facility. Professionals such as lawyers consult engineers much as engineers seek professional counsel from lawyers. Also, engineers are often called on as expert witnesses to give testimony on technical matters.

2.4 Scope of Engineering Services

The range of activity of engineers in design covers a broad spectrum from brief advice to inspection of construction and includes preparation of plans and specifications. Many firms, although qualified to render a variety of services, may limit the scope of services offered as well as specialize in a particular field. For example, some engineers offer only structural design services or foundation consulting.

Following is a brief summary of services rendered by engineering firms:

Advice and Consultation ▪ This phase may comprise no more than an expression of the consultant's opinion based on experience and technical knowledge. Normally, detailed engineering design is not an element in this phase, but the engineer may advise a client on the merits of undertaking a new project and its related technical consideration; or this phase may just be the rendering of an opinion on the advisability of undertaking further studies to determine the need for repairs or rehabilitation of an existing structure.

Technical Investigation and Analysis ▪ After consultation, the engineer may undertake detailed studies, such as physical exploration, including soil borings, topographic surveys, and hydrographic studies. Possible methods of construction may be considered. Preparation of a feasibility report may follow. This report usually considers economic as well as engineering aspects; both aspects have to be explored to enable an owner to decide whether to undertake a project.

Environmental Analysis ▪ The National Environmental Protection Act of 1969 caused a dramatic change in engineering practice. As a result of this legislation, an Environmental Impact Statement (EIS) must be filed before design implementation. Preparation of an EIS requires detailed studies and analyses in which the *impact* of the proposed improvement is determined and evaluated. Both short- and long-term impacts have to be considered, in addition to evaluation of a no-build alternative. Preparation and development of an EIS may require the effort of numerous specialists, such as archaeologists, biologists, hydrologists, and economists, for development of all the necessary plans and studies. Conversely, some improvements proceed quickly to the design phase with the filing of a negative impact statement. Such a statement is based on a determination

that there is no impact as a result of the proposed improvement.

Federal and state legislation and regulations as well as court rulings impact development of most sites and new designs. Federal legislation includes the Comprehensive Environmental Response, Compensation, and Liability Act (CERCLA), commonly known as the **Superfund** law; the Water Pollution Control Act, known as the Clean Water Act; and the Resource Conservation and Recovery Act. These laws and subsequent regulations not only affect design development for new projects but also may require modifications and alterations of existing facilities, as was the case with the removal of asbestos that had been installed in buildings.

Planning ▪ If, on the basis of a feasibility report or other information, the owner decides to proceed with the construction project, the planning phase is started. Planning must be considered separately from design. If, for instance, a plant or complex of structures is being developed, planning includes rough preliminary sketches and a master plan of the proposed project. With master plans, owners can develop a project in stages and schedule construction according to available funds.

Design ▪ The scope of engineering services varies depending upon the project delivery system used by the owner. In a typical design-bid-build process, the engineer is charged with preparing a design before the contract is awarded for construction. Under this system, the design is subdivided into schematic, preliminary and final phases. There can be a review with the owner at the end of each phase, or the review can be continuous to enable the owner to visualize the implementation of requirements and allow additions and changes to be made as the need arises. The completed design documents consist of detailed plans and specifications and contracts for construction (Arts. 3.2 and 3.4). The designer's role, however, does not end at completion of final design. Normally, the designer acts as the owner's representative in taking construction bids, awarding contracts, and administering construction contracts.

Fast track design and construction may be used under the construction management delivery system. An engineer may be required to phase the design into bid packages before the entire project is completed. An example is that the design of the foundations of a structure may have to be completed and bid for construction while the rest of the building is still being designed. The engineer will have to make accurate assumptions regarding the completion of the design to avoid changes to earlier design packages under this delivery system.

Design-Build ▪ This method brings separate challenges. During the competition for these contracts, the engineer provides a preliminary design that the contractor uses as the basis of the construction bid. If successful and the team is awarded the contract, the consulting engineer then completes the design of the project, which is built by the contractor. The design engineer may have additional challenges under this delivery method including fast tracking the design. In addition, there are added pressures on the engineer to provide an extremely economical design to allow the contractor to complete the project within the construction bid price. The engineer is required to provide support services during the construction phase to assure that the project is being built in accordance with the design documents.

Construction Administration and Inspection ▪ After the bid and contract award phase, the engineer's role involves general administration of the construction contract and acting as the owner's representative. The scope of services is defined in the design contract general conditions. This phase of the work usually includes limited on-site representation or provision of a full-time resident project engineer, inspector, or representatives. Periodic inspection and consultation during construction are normally part of the designer's obligation under the design contract. These tasks include periodic visits to the site; issuance of clarifying drawings, if required; and checks of equipment catalogs and contractor's shop drawings for compliance with contract documents. Full-time on-site representation, contracted separately as an addition to the design services, consists of an engineer and a staff, the size of which depends on the nature, magnitude, and complexity of the project. The prime function of the site engineer and the staff is to inspect the work for conformance with the contract requirements, documents, and design concept.

Construction Management ▪ Due to the growth, complexity, and inflationary costs' spiral of construction, construction management services have evolved both as traditional consulting or contracting services and as management of construction projects. A construction manager, often retained at about the same time as the project designer, may commence tasks at the beginning of design. The services of a construction manager may include basic program review and analysis, design review and evaluation, scheduling (CPM and PERT), cost estimating, value engineering, bid analysis, contractor selection, detailed construction inspection, coordination of trades and separate construction contractors, cost control, and program management. Acting as an owner's agent, the construction manager can perform all or some of these tasks to assure the owner of project and budgetary controls.

Other Services ▪ Among the other services rendered by engineering firms are preparation of technical reports; investigation surveys, such as land and property surveys to establish title to property; evaluation and rate studies; appraisal of property and building values; expert testimony in court; and services to industry, financial institutions, and public bodies in the economic field.

2.5 Selection of Consultants

A consultant prefers not to submit bids for services. The logic for this is self-evident. Because consultants render professional services, it is impossible to set a comparative basis for evaluating competitive bids. Furthermore, if consultants were selected on a price basis, the owner, by retaining the lowest bidder regardless of professional qualifications, would risk purchasing an incomplete or incompetent service. Because the fee paid a consultant is a small percentage of the total cost of a project, an owner should pay properly for such services and obtain the best professional services available. For many years, professional organizations published standards and yardsticks for fee schedules. Also, certain municipalities such as New York City continue to maintain fee curves and schedules, which are utilized to establish maximum fees paid to designers and consultants for various types of work.

Fee negotiation and competitive pricing have been studied by various government agencies and challenged in the courts as a result of antitrust administrative rulings issued by the Justice Department. One consequence has been that the American Society of Civil Engineers removed from its Code of Ethics a provision making bidding for the supply of professional services unethical. The following sequence of steps in selection of a professional consultant by an owner, however is preferred:

1. Review the capabilities of several firms and evaluate their qualifications with respect to requirements for the project. Many owners maintain lists of pre-qualified engineering firms, which resulted from invitations or advertisements to provide qualifications to perform the required engineering services. Public agencies use these lists to begin the solicitation process while large private organizations procuring engineering services may use alliances with engineering firms they have had success with in the past. An owner may have knowledge from past experiences of such firms; if not, the owner may contact professional organizations, such as the American Consulting Engineers Council or American Society of Civil Engineers, for a recommended list of firms. Owners without past experience in selecting consultants should confer with associates in their own industries for a list of recommended firms.

2. Select up to six (normally three) firms with the experience and knowledge for undertaking the assignment.

3. Request from the selected firms an indication of interest and detailed data pertaining to their qualifications and ability to undertake the project. With this submission, the firms are also asked to submit information concerning size of staff, availability of personnel to be assigned to the particular project, their understanding, approach and unique insights into the project, and their experience in similar lines of work. The firms are also interviewed.

4. Select the firm most qualified to undertake the project. In addition, the owner should list one or two additional firms, in order of their desirability, in case a contract cannot be negotiated with the first choice.

5. Notify the firm chosen of its selection, negotiate a fee, and execute an agreement for professional services to be rendered. If a mutually agreeable fee cannot be arrived at, negotiation with this firm terminates and negotiations then begin with the no. 2 selection. (For ethical reasons, to avoid conflict of interest, a consultant will not negotiate with a prospective owner if negotiations are still pending with another firm. As a consequence, the negotiations with the first firm must be terminated.)

In many cases, especially in the public sector, the owner may require that cost of services be established before selection of a consultant. Owners typically describe the firm to be selected as the one that provides the best combination of cost and quality. This allows the owner the leeway to select a firm that has provided the most technically outstanding proposal but not necessarily the lowest in price. In such instances, there are many ways in which cost may be included as part of the evaluation process. One approach is to include estimated cost as one of several weighted evaluation factors with other technical and professional qualifications. Another approach is to utilize a two-envelope system. This requires submission to the owner of the cost of services in one envelope and technical qualification data in another envelope. The owner opens the envelope with the technical qualification information first and rates the submission. Then, the owner opens the envelope with cost data and takes cost into account in the total selection process.

When determining the firm most qualified to undertake a project, an owner should consider technical qualifications, ability to absorb the additional workload in relation to the firm's capability and existing workload, experience, reputation, financial standing, and past accomplishments in related fields.

Because the cost of any service is important to an owner, an equitable fee for the services to be rendered has to be established. A caveat for owners is: "You receive only the professional services you pay for." If the fee is cut, services rendered are reduced. In the development of a project, it is important for an owner to receive complete and competent professional advice. If this is done, owners can be assured that their projects will be designed economically and efficiently. The fee paid for proper professional services will be a wise investment.

2.6 Contracts and Fees for Design Services

The interests of owner, or client, and design professional are reflected in the design contract, or agreement, which should be in writing. It should define the duties and responsibilities of each party to the agreement. It should also describe the overall project requirements.

Several standard agreements are available for contracting for design services—for example, those developed by the American Institute of Architects and those developed under the auspices of several engineering organizations. The latter standard agreements include documents issued by the Engineers Joint Contract Document Committee (EJCDC), formed by the National Society of Professional Engineers (NSPE), American Consulting Engineer's Council (ACEC), and the American Society of Civil Engineers. Representatives of the Construction Specifications Institute (CSI) also participate in the development of these documents and the CSI endorses them.

The basic methods for determining fees for design services are lump sum, cost plus fixed fee, and percentage of construction. The last method is the least often used.

Lump Sum Fee ■ A fixed fee is arrived at by estimating the man-hours and expenses anticipated for rendering the service. When the scope of a project is specifically outlined, the consultant can evaluate anticipated costs for services by analyzing the demands of the project and drawing on experience and knowledge of the firm's capabilities. The consultant can translate the project into man-hours required and compute the cost. To the cost of labor must be added overhead, any expenses beyond those normally included in the overhead factor, any unusual elements that might add to costs, and anticipated profit. Although the fixed fee may be established by using accepted industry percentages as a yardstick, the contract is negotiated for a lump sum regardless of the project's eventual construction cost. Only if there is a change in the scope of services initially agreed on will there be a possible change in the fee.

A variation of this form of payment is the lump-sum fee plus expenses, which is used if there are extraordinary expenses, for instance, a more-than-normal amount of travel to a distant site, or if subsoil investigation and surveys are included in the consultant's scope of work.

Cost Plus Fee ■ The cost-plus type of contract is normally used when the scope of work cannot be readily defined. Then, the owner agrees to reimburse the consultant for costs plus a fee. The reimbursable costs consist of technical payroll and actual expenditures, such as travel, subsistence while away from home, long-distance telephone calls, and other costs incurred directly for the project. Normally, the fee is determined by a factor applied against payroll cost. The factor compensates the consultant for management, overhead, indirect costs, and fee. Principals, partners, or officers, if engaged in actual production work (technical, as differentiated from administrative), are reimbursed for their services in the same manner as employees on the payroll.

A variation of this payment method uses a time factor (hourly or daily) with wage rates to reimburse a consultant for costs, overhead, and fee. For example, owner and consultant may agree on a rate of pay for a category of employee and multiply this rate by an overhead-and-fee factor. If a designer's average rate were set at $15 per hour and the overhead factor at 150%, the payment provision in the contract would state that reimbursement to the consultant for the designer's time would be at $37.50 per hour ($15 + 1.5 × 15). Rates also would be set for other categories of personnel to be employed on the project.

Additional cost-plus arrangements most commonly used by federal and other public agencies establish both a basis for identifying all allowable costs and for setting a fixed fee at the time of contract negotiation. Although calculated as a percentage (frequently 10%) of estimated costs, this fee remains fixed (a lump sum) for the contract unless there is a change in the scope of work. The fixed fee covers profit and nonallowable costs. Allowable costs are reimbursed as incurred for the prosecution of the work. Such costs include direct labor, direct project costs, and overhead and indirect costs attributed to the labor base. Federal Procurement Regulations spell out in great detail categories of costs, both allowable and nonallowable. All such costs are subject to audit and verification by government audit agencies. Contractors or consultants who contract with the federal government conduct yearly audits in which they verify and agree on the cost basis to be utilized.

Such cost bases are traditionally labor costs (actual payroll costs) plus indirect costs (allowable overhead) and are translated into a percentage of the technical labor cost base. This percentage is reevaluated and recalculated periodically, normally consistent with the time of an audit or fiscal year.

Percentage of Construction Value ■ This percentage may be used as a guide by parties in determining a fee. If a percentage fee is negotiated between the parties, it is of great importance to define what amount will be used for the construction value. Will it be the estimated value or the actual construction value based on the contractor's low bid? If the fee is to be based on the estimated value, will the preliminary or detailed estimate govern? If the fee is to be based on the low bid, the design contract must state that the contractor's bid be bona fide since contractors sometimes make mistakes and submit improper bids. Furthermore, the design contract should provide for a payment method if, for some reason, construction does not proceed and no bids are available to establish a construction value for fee-payment purposes.

The percentage fee is now rarely used to establish the basis of a designer's compensation. Percentage values remain a viable yardstick for establishing or evaluating design costs. But, thereafter, it is more advantageous to a designer and owner to translate the percentage value to a lump sum fee for contract purposes.

Other Types of Fees ■ Some owners engage consultants on a retainer. However, this reimbursement method is not a substitute for payment of fees as previously described. An owner who has a continuing need for engineering advice and consultation may retain a professional engineer for a period of time, normally annually. The owner is free to call on the consultant for professional assistance on a continuing basis, such as attending periodic planning and development meetings. If, however, the service required becomes more than consultation and design of a project is called for, the retainer would not be sufficient compensation; a separate fee would be negotiated.

2.7 Managing Project Design

Managing the design of a project is similar to managing a business, only on a smaller scale. The design project manager must be able to control the cost to perform the work, the time it takes to complete and the quality of the finished design.

Critical to completion of the work in a professional manner is the development of a project management plan that defines the project and how it will be managed. The project management plan should include the scope of work, the hours by discipline that it will take to perform, the schedule and any milestone deliverables that are required, the budget allocated to perform each phase of the work, and a system for providing quality control. In order to properly prepare the plan, the project manager should review the contract that has been signed with the owner to see how the project has been defined and what services the owner required to be provided in order to complete the project. Budgets should be allocated to perform all facets of the work as defined by the owner in the contract. In addition to preparing the scope of work, areas of concern should be noted (such as unique conditions that exist on the project that must get special design attention) and an approach to managing them.

It is also necessary to identify the project organization and relationships within the organization. If subconsultants are to be used, their scope and costs should be included in the plan. The simplest way to do this is to draw a project organization chart, which defines responsibilities and reporting relationships.

The organization for a new design project generally is drawn from existing staff. Operating procedures depend on the size of the project and management's philosophy.

A professional staff, to function effectively and efficiently, should be able to draw on standardized procedures and up-to-date reference materials. The latter include design codes, standards, and design manuals.

A critical element in maintenance of design standards and design quality is the use of computers. Use of computers is changing the way projects are designed, how information is shared and ultimately how it is transmitted for quality assurance reviews and bidding. Computer programs have for years been used to simplify the drudgery of massive calculations used for design purposes. Programs exist to perform a wide diversity of calculations and designs from structural all the way to lighting. The CADD technician has replaced the manual draftsperson. Alternatively, engineers prepare design documents on the computer themselves. Recently, the internet has become a great source of information allowing engineers access to the latest codes and standards. To allow engineers working at diverse locations to update drawings, companies have begun using web sites, which simplifies and expedites the coordination process. Owners have begun to require the development of web sites so that they can view progress on the design of their projects. Finally, some owners are now requiring digital copies of their projects on compact discs rather than blueprints. These same owners are providing contractors with a digital copy of their project which allows them to make their own prints.

Value engineering (VE), or value analysis, may be incorporated as part of the design process. Value engineering is a formalized and organized procedure in which a separate design team reviews the design at various stages to assess proposed designs. The team makes recommendations, as appropriate, for revisions that will both improve the design, increase value or affect cost savings. Value engineering often is utilized by some owners before start of construction to identify possibilities for reducing costs.

The earlier value engineering occurs, the more effective it is in reducing costs. If value engineering is performed during the schematic phase, it is relatively simple for the engineer to make changes to the design documents. As the design is developed and coordinated, it becomes more and more difficult to implement value engineering changes as it may affect many different parts or systems already incorporated into the design.

2.8 Project Methods and Standards

For efficient operation, a firm should establish standard methods and systems. This does not mean that once a procedure is established it is inviolate; it is subject to improvement and refinement. But within reason, the standard procedures should be adhered to on all projects. Without standardization, the result would be more than

wasting time: The firm would be unable to operate efficiently within available budgets.

A code number should be assigned to identify each project. A commonly used system identifies the project by a series of numbers, including the year (calendar or fiscal) in which a project is started. This number should be used on all work, whether a final drawing, rough calculation, or correspondence. All costs and charges pertaining to the project should also be identified by this number.

A standard procedure for the performance of all work should be established. This includes a procedure for checking calculations and a system for preparation and approval of drawings, from drafter's work to the final authorized signature. Regardless of what internal procedure is established, the ultimate objective is the same: to operate economically and efficiently. After a design problem has been evaluated and analyzed and a method of solution established, a typical design procedure would be as indicated in Fig. 2.1.

Because many specifications are similar to each other in outline and technical provisions, standardization of specifications can be most useful. This does not necessarily mean that the firm should

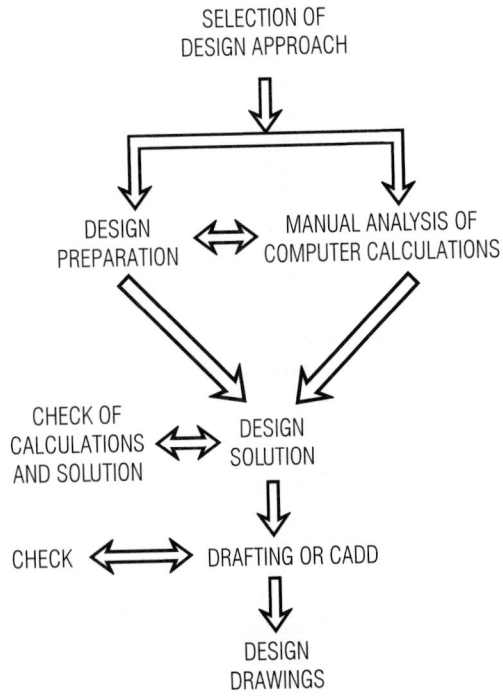

Fig. 2.1 Typical design procedure.

prepare "canned" specifications for use interchangeably on all projects. Each project has different requirements, but the various sections of the specifications should be prepared in a consistent manner on all projects. For instance, in a concrete specification, a typical section might contain the following major paragraphs: scope of work, related work (cross-referenced to other specification sections), general, material (cement, sand, aggregates, and so on), reinforcing steel, formwork, concrete strength and mixing, and concrete placement. Each paragraph has to be tailored to fit the requirements of a project—pier, bridge, or building. Many of the provisions, however, may be essentially the same in many instances, for example, the provisions for the quality of material in one geographic area.

For simplification, the firm may adopt standard specifications prepared by technical societies for an item, such as structural concrete. These specifications require the designer to insert requirements for a specific project but eliminate the necessity of writing anew for each project sections that are substantially the same for all projects.

2.9 Project Quality Control

The quality of a firm's product should be of continuing concern to all members of the firm. Achievement of quality requires sound engineering practices, especially compliance with codes, standards, and legal regulations.

Quality control (QC) is a continuing process that can be part of a quality-assurance (QA) program. Whether or not formal programs are instituted for the purpose, good engineering practice requires procedures to be established to check product quality. These should comprise reviews at various stages of design development to evaluate the quality of the work.

Interim reviews often are required as part of a designer's scope of services. Designers generally submit the work formally to the owner at various stages of completion, such as at completion of preliminary plans (30%), design plans and details (75%), and final bid plans (100%). A firm may utilize separate review teams to check work performed by others before issuance and use of design drawings and specifications for construction.

Designers should assure that products comply with applicable codes and standards. This requires familiarity with the latest statutory requirements and awareness of the latest regulations issued by the various agencies that have jurisdiction. This is especially significant for any work that has potential environmental impact, even though environmental impact statements may have been completed under prior contracts.

To assist in maintenance of quality in construction, engineering societies have promulgated programs such as total-quality management (TQM), which addresses and reviews a firm's practices. The objective of TQM is to promote quality within a design organization and of its products. TQM is implemented internally through ongoing training of all members of the organization to continuously seek quality in the firm's work practices and product and thus to achieve desired quality of results.

Many engineering firms and owners are becoming ISO 9000 certified. ISO 9000 is a quality assurance system developed a little more than a decade ago that establishes international standards for producing a high quality project. This is formal recognition that a firm has developed and maintained a Total Quality Management system to perform its design work. In order to maintain ISO 9000 certification, a firm must be regularly audited to demonstrate that it has complied with the standards. Major elements of the system include the preparation and use of a quality assurance manual, active senior management involvement, and implementation of a non-conformance documentation system.

Peer Review ▪ This is a procedure employed by a firm for a specific project wherein the firm contracts with an outside group, the "peer", to review policies and practice for the purpose of achieving the highest level of quality in design of the project.

A peer review is conducted by designers with the same expertise as those who prepared the design and who have no relationship with the designers and are totally independent. Peers can be individuals from other departments of the firm or other organizations. The designer of record, however, is not replaced by the peers. The review should result in a report of the findings of the peers. It should not be considered a criticism of the designers or their work. A peer review, unlike other design reviews, does not have a specific objective other than quality, such as cutting construction or life-cycle costs, value engineering, or a constructability review performed as part of construction management.

2.10 Scheduling Design

Without proper scheduling, a firm may find that its operation is as inefficient as if no standard procedures were used. To accomplish a design, the firm is essentially scheduling workforce needs. This task becomes more important with the number of projects to be handled at the same time. A properly run firm should be able to schedule its work so as not to take on more than it can adequately handle with a stable size of staff.

For scheduling total workload, individual project scheduling is essential. The simplest and most common device for this purpose is the **bar chart**, a graphic representation of workforce (represented by bars) plotted against time. By studying such a chart, one can quickly determine job start and completion dates and when and for what workforce needs will be greatest.

Scheduling devices such as the critical-path method (CPM) and program evaluation and review technique (PERT) have a definite place in programming design-workforce requirements. Although the design project for which a complete CPM or PERT study would be employed is unusual, modification or limited use of these programming devices is warranted in many cases. A complete computer CPM program, including scheduling costs as well as time and evaluating the economics of "crash" programs, would be used only on the most complex projects. Because more thorough planning is required, use of the basic CPM and PERT activity diagram can often result in a better scheduled project than if a bar chart were used. With the use of a bar chart, the start or completion of activities represented by a bar can be extended a week or more without affecting the basic schedule. A CPM or PERT diagram does not permit this since the diagramming of the activities interrelates them all and the change in time in one activity can affect all.

2.11 Production Control

Once a project is undertaken, the work involved has to be completed regardless of time or cost. Still,

the firm must operate within a budget so design can be performed efficiently. A designer does not deal with a tangible product for which the firm can establish a cost per unit and operate on a production line basis. Nor should the firm go to the extreme of establishing a control in such a manner that the cost becomes more important than the product.

Cost control in its simplest form is a matter of bookkeeping. The firm should keep records of all costs relating to each project. At the end of a project, therefore, the firm should know the costs and income received and whether the work was done at a profit or loss. When a firm undertakes a new project similar in nature and size to one completed previously, a record is available to guide the new activities. Such cost accounting can be refined to varying degrees.

Also, it is well to know one's financial standing as to work on hand before completion of the work since it may be years before some projects are completed. During the course of a project, the firm should project the costs and income based on percent completion at a particular time to determine whether they are in line. Such projections should be made periodically to gain a picture of the financial condition of the firm's operation at a particular time.

Many engineering firms use computerized financial control systems to monitor costs and help to manage projects. These systems allow a project manager to monitor personnel and expense costs charged to their specific project on a weekly or monthly basis. This allows the project manager to compare the actual cost expended to date versus the budgeted cost. If the actual costs are less than the budgeted costs, the project manager can be assured that the project is in good financial health. However, if the actual costs exceed the budgeted costs then the project manager must determine what elements are over budget and develop a corrective action plan to bring the project back to within budget.

Cost accounting serves an additional purpose: it establishes controls during the programming of the work. These controls enable the firm to determine where productivity and efficiency need improvement before the end of the project when it is too late.

A professional firm, like any business, is concerned with making a profit. Maintaining a proper profit margin is essential to survival and growth. Such a profit margin varies with the size of a firm and number of principals; whether principals are on a salary, as in a corporation, or not, as in a partnership. Cost control is an important tool for helping managers ensure the required profit margin to keep the firm operating efficiently.

(T. G. Hicks, "Standard Handbook of Consulting Engineering Practice" McGraw-Hill Book Company, New York.)

2.12 Internal Organization of a Design Firm

Basically, an engineering firm consists of technical departments and administrative and support staff. Figures 2.2 to 2.4 illustrate typical consulting firm organizations.

Technical Departments ▪ Depending on the size of the firm, the technical department can be divided into divisions, such as structural, civil, mechanical, and electrical engineering and architectural. These divisions can be subdivided and overlapped under the direction of a job captain, project manager, or project partner for particular projects. (In very small firms, many functions are performed by one individual, including the proprietor.)

There are numerous ways of organizing a technical department (see, for example, Figs. 2.2 to 2.4). The most important consideration in any organization is communication. Whenever a firm is formed or expanded or new departments are established, communication should be considered of prime importance. The flow of information between line levels should be well-defined. Furthermore, there should always be one individual who acts as project manager or captain in a position to coordinate all activities whether they are only those of departments within the organization or those of outside contractors or consultants involved in the project.

Many firms also have a separate construction or construction-management department, which consists of the project and construction managers, resident engineers and inspectors required on a project site, and project engineers rendering field consultation services and coordinating the efforts

SMALL FIRM OR (INDIVIDUAL PROPRIETORSHIP)

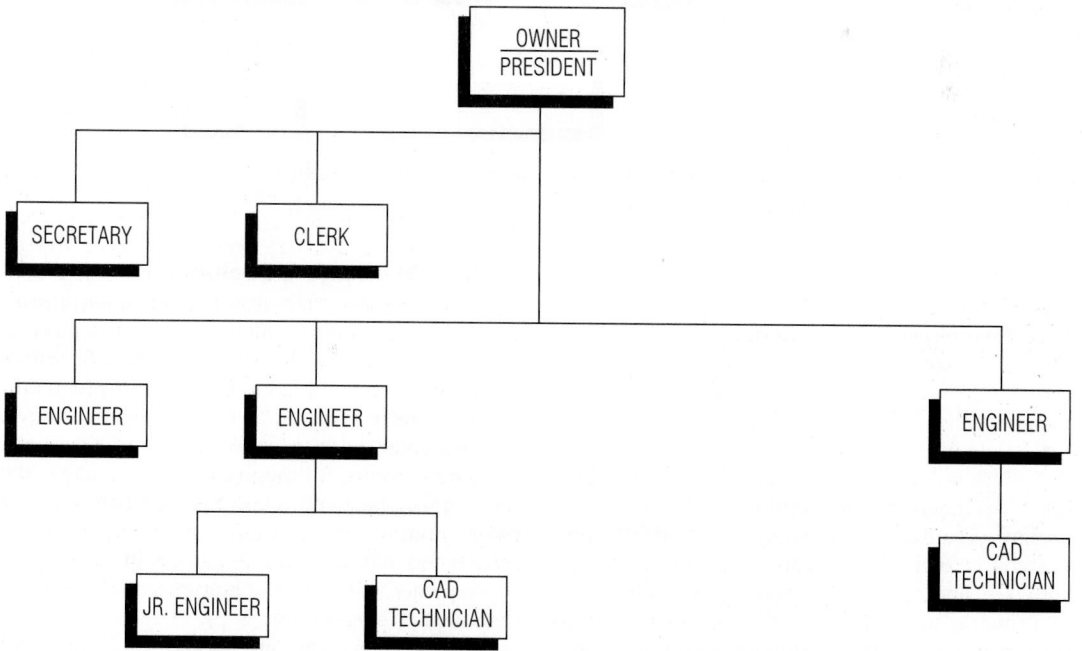

Fig. 2.2 Typical organization of a sole proprietorship.

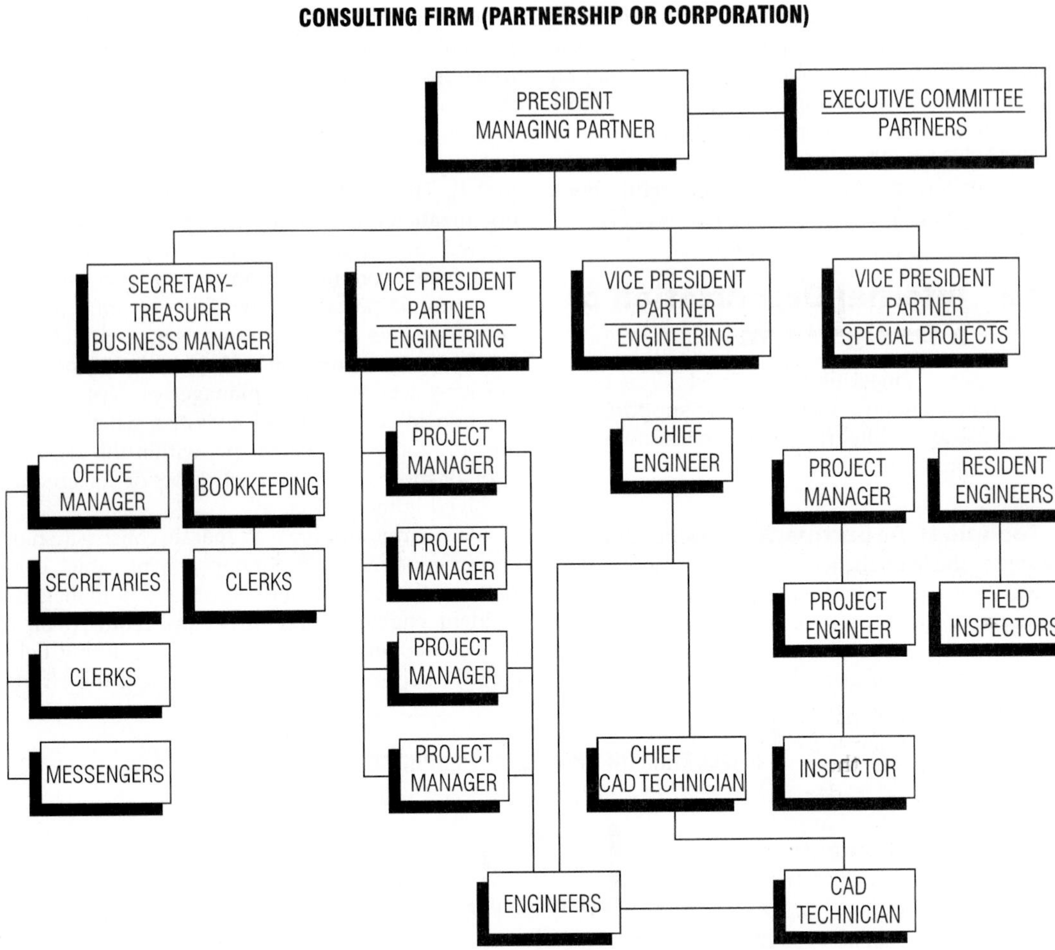

CONSULTING FIRM (PARTNERSHIP OR CORPORATION)

Fig. 2.3 Typical organization for a consulting firm.

of field personnel. Instead of establishing a separate department for this function, some firms have the project engineers for design in the various design divisions continue in the same capacity through the construction phase; they draw on a nucleus of field personnel for backup, as necessary, for on-site inspection.

Computer-aided design and drafting (CADD) offers designers multiple options and flexibility in design organization. Designers can draft their designs at their desks, using appropriate design software, and need not rely on drafting support.

Primary support functions are new business development, human resources, accounting, and office support services.

New Business Development ▪ Professionals do not sell services directly; they must apprise the market of their availability. The firm has to prepare qualification data (Arts. 2.3 and 2.5), which can range from completion of standard prequalification forms to preparation of elaborate brochures, supplemented with extensive project descriptions and photographs. Although a new client may make the initial contact and retain a design firm without prior communication, a design organization cannot rely on this manner of receiving new business. As a consequence, client contact is an essential part of an organization's operation.

Client contact can be limited to impersonal contact by mail or range to active sales efforts,

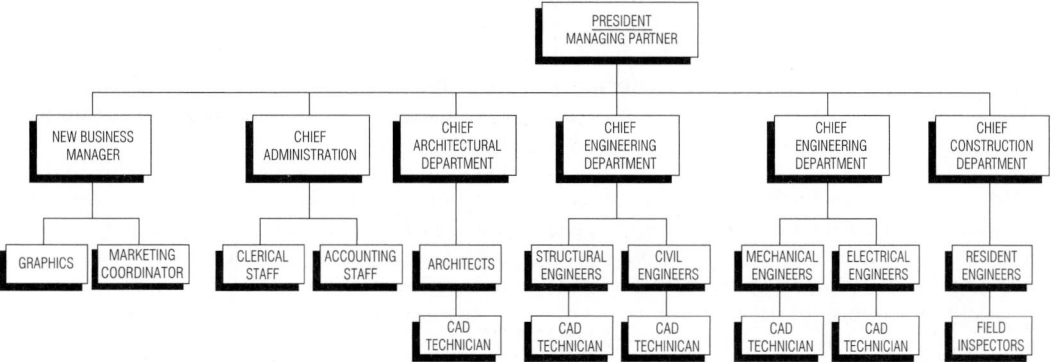

Fig. 2.4 Alternative organization for a consulting firm.

where an employee or principal (or even a staff if the size of the firm warrants it) makes personal calls on potential clients. The name of a firm has to be promoted continuously, which requires good public relations. Sales efforts, however, should not be a substitute for quality of service.

In the face of intense competition and the need for growth and diversity of a firm, the search for new markets and development of new business are vital functions.

Employee Compensation ▪ Employers have specific legal obligations. They must pay payroll taxes, such as Social Security and state unemployment and disability, and they must withhold taxes from employees' earnings. These requirements result in administrative burdens involving the filing of forms and reports. There are insurance obligations and statutory requirements, such as workers' compensation. Also, an employer has obligations mandated by federal and state laws, including labor laws affecting minimum wages and overtime and regulations governing working conditions, equal employment, and safety.

Employers may wish to give employees the opportunity to subscribe to medical and other forms of insurance as a group, and they may pay all or part of the costs for other benefits including 401K and pension plans. In the competitive market for skilled personnel, such fringe benefits must be added to the basic wage.

Employers should have firm wage and salary policies. Besides paying a competitive wage, they must establish policies for salary reviews and

increases, salary ranges for various types of positions, bonuses, and whether to include a profit-sharing plan. Primarily, however, employers should give employees opportunity for advancement. Also, they should give recognition for efforts on behalf of the firm. If employers can instill the pride of accomplishment and profession, they will have efficient and happy workforces.

Accounting ▪ To operate efficiently, a firm must be able at all times to evaluate and analyze its financial position. For this, the firm has to maintain proper accounts. The compiling and recording of all transactions relating to the financial aspects of a business are the basic responsibility of accounting. The recording of financial transactions has to be orderly for proper interpretation, to make possible preparation of financial statements and to provide information on the economic health of the business. (See also Art. 2.11.)

The method or extent of bookkeeping varies with each firm's size and needs. Normally, the double-entry system (classification of accounts into assets, liabilities, and net worth) is used. Each firm maintains journals and ledgers. The journal is a daily record of all transactions, debits, and credits; the ledgers carry the journal entries in specific accounts. Again, the number and extent of ledgers required vary with the firm.

A consulting firm has to decide how it is going to maintain its books for tax purposes, whether on a cash or accrual basis. On a cash basis, income is recorded when cash is received and expenditures are recorded when they are made. On an accrual basis, income is reported when earned and expenditures (or debits) when incurred, regardless of the time the cash transaction takes place. When tax considerations are significant in the business operation, the choice of accounting system is of primary importance; as is evident, a firm's cash and accrued statement at the same time could be quite different.

Although it is poor business practice to take a particular action solely because of the tax consequences, tax considerations are important in a consulting firm's business practice. The initial decision of which form of organization to operate under should take into account the different tax consequences on individuals, partnerships, and corporations. Depending on income, a corporation may pay a large federal income tax; in addition, its dividends are taxed. A partnership pays no tax on its income, but the partners, who receive no salaries, are taxed as individuals on their share of the firm's earnings. State and local taxes should also be considered when establishing and operating a design practice.

Payroll is a consulting firm's largest expenditure. Payroll costs should be identified as direct (technical) and indirect (administrative). Records of direct costs, preferably by department, should be maintained for each project. Also, identifiable direct expenses, such as travel, subsistence, and other allowances, long-distance telephone and telegraph, and reproduction costs, should be accounted for and identified as a job expense. Major indirect or overhead expenses should also be identifiable to enable management to analyze indirect costs and their relation to fees earned during a specified period.

In addition to internal accounting, it is customary and advisable to have an audited financial report prepared by a certified public accounting firm at the end of each fiscal year. For firms of any size and especially those not closely held (public), such certified audits are essential. Also, firms, regardless of size or type of ownership, that work in the public sector must undergo independent and certified audits.

From its inception on, an engineering firm is concerned with finances. For one thing, a consultant is not reimbursed for a firm's services the day after they are rendered. Terms of payment depend on contract conditions. Payments may be monthly, or the first payment may not be due until 25% (or another percentage) of the work has been completed. Also, the final payment may not be received for a long time after all expenditures have been made. This sums up to one basic need: capital.

Consulting engineers must have capital to start and operate their organizations. The source of capital may be a loan or earnings. But regardless of the source, there must be proper financing to meet financial obligations that cannot be deferred until accounts are paid. In particular, when interest rates are high, financial management becomes a critical aspect of all businesses, including that of a design firm.

Insurance ▪ A firm's insurance portfolio normally includes coverage for general liability, property damage, automobile accidents, and professional (errors and omissions) liability. For design firms, all insurance requirements are dominated by the professional coverage. This insurance, which is written by a few insurance carriers, protects a designer from liability resulting from a design error or omission. Because of the extensive litigation prevalent in the construction industry, with designers being named as defendants for alleged design error or as third-party defendants, the cost of this insurance is high. (This is also true in other professions, such as medicine.) This has resulted in a need for many practitioners to reevaluate the extent of their activity, to increase their fees to cover such costs, and, in limited instances, to forego this liability coverage.

Office Support Services ▪ The administrative staff's primary function is internal operation of the firm. Personnel on the administrative staff may include an office manager, secretaries, word processors, receptionist, file clerks, and office employees. The number of employees and degrees of responsibility vary with the size of firm. However small the firm may be, the basic administrative duties have to be fulfilled: Letters have to be word processed; so do reports. Files have to be maintained, telephones answered and messages taken, and plans reproduced. Although all the elements that constitute office management are secondary to design, the primary function of the firm, they should not be neglected. Even with electronic communication, an unattractive letter can make a poor first impression on its recipient, who may be a potential client. A first impression of a firm can also be made by the manner in which the telephone is answered. So although the administrative duties are routine in most offices, they should be handled as competently as the technical work. The administrative positions should be filled by competent, properly trained personnel.

In an engineering firm, there is a substantial amount of reproduction of plans and specifications and duplication of reports. The mechanics of providing the necessary reproductions is best handled by a separate department within the firm. Whether the work is done on office-owned equipment or sent out to a printing company is a matter of economics determined by the firm's volume. In addition, office services must encompass selection of the most economical and efficient office systems for the firm. For efficient, economical operation, a design office should be equipped at least with computers (personal or servers and workstations), plotters, high speed Internet connections, modems, fax (telefax) machines, and copiers, in addition to the usual desks, record-storage facilities, telephones, and good illumination. Office managers should be familiar with current electronic systems, innovations, and be able to judge their applicability to the firm's needs.

2.13 Professional Societies

The role of professional societies, such as the American Society of Civil Engineers and various associations of consulting engineers, initially was determined by their existence as organizations of individuals rather than firms. At first these societies were concerned mainly with technical matters and very little with business affairs. Although the medical, legal, and accounting professions each have one major society that speaks for them, this is not the case for civil engineers, who are generally represented by the American Society of Civil Engineers, American Consulting Engineers Council, or National Society of Professional Engineers. These societies, however, collaborate with each other on matters of common interest.

In a complex and progressive economic society, few firms other than industry giants have the resources to stay abreast of all the latest developments; keep informed of all current legislation, state and federal; and be aware of all administrative rules, regulations, and factors influencing their day-to-day activities. An association can fill these needs, and by serving these needs, professional associations are playing a more important role than previously.

In prior years also, a design firm was "on its own." It had little knowledge, if any, of the activities of its competitors or even of its closest associates. Today a firm still is on its own in the competitive marketplace, but it can pool its resources in associations that represent the profession and industry. United action and sharing of information advance the interests of individual firms.

Activities of professional groups now include:

Legislation ▪ Maintaining database on current legislation; representing and filing position papers with Congress and state legislatures on pending bills in which association members have a vital interest.

Government Relations ▪ Liaison with various administrative agencies, federal, state, and municipal. This area could include assistance to member firms interested in capitalizing on opportunities abroad.

Liaison with Industry ▪ Maintaining contact with other organizations and establishing joint committees to study and evaluate areas of common interest.

Publications ▪ Initiating and distributing to members documents reporting current activities and areas of importance and concern.

Insurance ▪ Establishing group insurance policies (life, accident, health, and so on) to give smaller members advantages of larger group plans; advising member firms in fields of common concern, such as professional-liability insurance, an area of increasing concern because of numerous third-party suits against consulting engineers.

Engineering Practice ▪ Acting as a pool and distribution center for information on the latest technical developments and areas of interest to the profession; sponsoring continuing education programs.

3

Ted E. Robbins
Technical Services Administrator
Martin County Utilities and Solid Waste Dept.
Stuart, Florida

SPECIFICATIONS

Specifications are an important tool for communicating with sufficient detail how, where, and when a particular item or project is to be manufactured or constructed to meet an owner's needs. On civil engineering projects, the specifications are part of the contract documents and usually are supplemental to a set of drawings. If the assemblage of contract documents were to be considered a body, then the drawings should be viewed as the skeleton and the specifications as such body parts as muscle, sinew, and skin, which together add up to the whole.

The term *specifications* is often used to describe a portion of the contract documents that include the bid documents, agreement between owner and contractor, general provisions, special provisions, and technical specifications. The complete document that includes all these subjects is sometimes called the **project manual**. Throughout this section, the term *specifications* is used interchangeably with project manual.

3.1 Composition of Specifications

Specifications describe the particular requirements that are to be used to bid, contract for, build, test, start up, and guarantee an engineering project. Typically, specifications include:

1. Sections that describe how a prospective bidder must prepare the bid.
2. A copy of the agreement (contract) between the owner and contractor to be executed.

3. A division called **general conditions**. This division describes procedures generally required to be followed during construction of all projects, including procedures to be followed by all parties; that is, the owner, engineer or architect, and contractor. A well crafted general conditions is likely to be reused many years for similar projects.

4. A division called **supplemental conditions**, which modifies the general conditions to the specific or special requirements of the project. Using this method to modify the general conditions ensures the integrity of the general conditions and encourages familiarity with the general conditions. Contractors can focus their attention on the supplemental conditions with confidence when they are fully aware of the standard general conditions that were used to administer their past projects.

5. A division called technical specifications. This division is organized into logically arranged sections that describe comprehensively the material, equipment, or performance of items that must be incorporated into the completed work.

This combination of requirements, together with the contract drawings and bidding documents, comprises the **contract documents**. When faced with the task of preparing specifications for an engineered project, the engineer must consider many factors, among which the most important are:

Nature of the owner's business—private industry or public body.

Magnitude of the project.

Estimated duration of construction period.

Does the owner require the engineer to adhere to a set of standard specifications, or will the engineer have a free hand in preparing the type of specifications?

Does the owner have an attorney who will review the legal aspects of the specifications?

Does the owner have an insurance advisor who will review the insurance requirements included in the specifications?

Does the owner have an engineering staff, such as that for a state department of transportation, which will review the specifications?

Also, the engineer should realize that courts of law recognize the status of contractual relations between owner and contractor as that between free and independent individuals, not as that between a principal and agent. The specifications must support this relationship by refraining from prescribing construction methods and exercising control over the contractor's work.

After the basic conditions for a project have been established, the engineer is obligated to prepare complete contract documents for the project. The principal parts of these documents usually consist of the following:

Advertisement for bids (notice to contractors, or invitation to bid)

Information to bidders

Proposal form

Contract-agreement form

Bond forms

General provisions, or general conditions

Construction drawings

Special provisions, or special conditions

Technical specifications

For general guidance, forms for all but the last three are available from sponsoring agencies, such as the Engineers Joint Contract Documents Committee, American Consulting Engineers Council, American Institute of Architects, American Society of Civil Engineers, National Society of Professional Engineers, Associated General Contractors of America, Construction Specifications Institute, and General Services Administration. Article 3.11 is an example of a technical specification prepared for a public agency having standard documents. (For a discussion of general provisions, see Art. 3.6.)

3.2 Contract Documents and Contracting Procedures

The implementation of contracts between owners and contractors for construction work requires that the parties observe certain legal formalities. Such steps are evidenced by executed written documents that, together with the plans and specifications, constitute the contract documents. The nature and content of the contract documents vary with the owner agency that sponsors the improvement and the procedure employed for the receipt of bids.

It is standard practice for government and other public agencies at all levels to provide for public letting of contracts for public works. In such cases, sealed bids are invited by advertising in various news media for stated periods. After bids are opened, publicly read aloud, tabulated, and evaluated, the low bidder is determined.

It is customary to issue the plans and specifications to prospective bidders who apply and pay stated charges. In most cases, proposals must be accompanied by a proposal guaranty in the form of either a certified check or a surety bond, to ensure that the successful bidder will enter into a contract. If an award is made, the proposal guaranty is returned. If the low bidder fails to execute the contract, the amount of the certified check will be forfeited as liquidated damages or obligations of the surety under the bond will be enforced as compensation to the owner for the cost of awarding the contract to the next lowest bidder or for the added cost of readvertising. As a general rule, proposals are acceptable from competent bidders (evidenced by statements of experience and financial responsibility submitted to the owner). Forms for these usually are included in the project manual.

Under the foregoing procedure, the contract documents generally comprise the advertisement (instruction to bidders may be either included or separately provided); proposal, properly executed; contractor's progress schedule; resolution of award of contract; executed form of contract; contract payment and performance bonds, plans and specifications; supplementary agreements; change orders; letters or other information, including addenda (Art. 3.2.3); and all provisions required by law to be inserted into the contract, whether actually inserted or not. All the documents constitute one legal instrument.

3.2.1 Adoption of Standards by Reference

Sometimes standard specifications, such as a state department of transportation specification, are made part of the contract by reference to their title only. By this reference, the standard specifications effectively become a part of the contract documents as if a copy of them were included with the contract documents. Language stipulating this should be included in the general or supplemental conditions. (See Art. 3.9.3.)

3.2.2 Noncollusion Affidavits

When required by law, a noncollusion affidavit must accompany the submission of the proposal. This affidavit certifies that the bid has been submitted without collusion or fraud and that no member of the government agency or officer or employee of the owner is directly or indirectly interested in the bid.

3.2.3 Contract Revisions

For various reasons, revisions of the contract documents become necessary between issuance of the invitation or advertisement for proposals and the termination of the contract. Such revisions may be classified as addenda, stipulations, change orders, or supplementary agreements.

Addenda are revisions of the contract documents made during the bidding period. They mainly are concerned with changes in the contract drawings and specifications due to errors or omissions, with the necessity for clarification of parts of these documents, as revealed by questions raised by prospective bidders, or with changes required by the owner. An addendum is also issued to notify bidders when a bid-opening date has been postponed.

Addenda should be delivered sufficiently in advance of the bid-opening date to permit all persons to whom contract documents have been issued to make the necessary adjustments in their proposals. Bidders must acknowledge receipt of all addenda; otherwise, their bids should never be accepted.

Stipulation is a written instrument in which the successful bidder agrees, at the time of execution of the contract, to a modification of the contract terms proposed by the owner.

Change order is a written order to the contractor, approved by the owner and signed by the contractor and the engineer, for a change in the work from that originally shown by the drawings and specifications. Usually, under a change order, the work is considered as being within the general scope of the contract. The owner, represented by the engineer, may issue the order to the contractor unilaterally, with payment provided for by contract unit prices, negotiated price, or force account.

A change order may apply to changes affecting lump-sum work or to increases and decreases in quantities of work to be performed under the various items in a unit-price contract. The changes in quantity will be evaluated at the contract unit prices and the contract total amount adjusted accordingly. But if the total cost change amounts to more than a specified percentage, say 25%, of the total contract price, a supplementary agreement acceptable to both parties to the contract should be executed before the contractor proceeds with the affected work.

Supplementary agreement is a written agreement used for modifying work considered outside the general scope and terms of the contract or for changes in work within the scope of the contract but exceeding a stipulated percentage of the original amount of the contract. The agreement must be signed by both parties to the contract and the written consent of the company that issued the payment and performance bonds for the project should be obtained.

3.3 Types of Contracts

Construction contracts for public works are almost always let on a competitive-bid basis. Usually, such contracts are of either of two types—unit price or lump sum—depending on the method of paying the contractor. Contracts for construction for private owners may be either competitive-bid or negotiated, but in either case, they generally are of the same two types (see also Art. 4.4).

3.3.1 Unit-Price Contract

When it is not possible to delineate on the drawings the exact limits for the various items of work in the contract, the work is broken down for payment purposes into major elements with respect to the kind of work and trades involved. Each element designated as a payment item, with its number of estimated units, called estimated quantity, is listed, in the proposal, and the bidders are required to write in a bid price for each unit. An example is the number of cubic yards of concrete to be bid at a unit price per cubic yard.

The **total bid** is obtained by summing the amounts, in dollars, for all items listed in the proposal, arrived at by multiplying the estimated number of units for each item by the corresponding unit-price bid. The total bid becomes the basis for comparison of all bids received to establish the low bid upon which the award of contract will be made. Payment to the contractor will be made on the basis of the measured actual quantity of each item incorporated into the work, at the contract unit price (see also Art. 4.7.6).

3.3.2 Lump-Sum Contract

When it is possible to delineate accurately on the drawings the limits of work comprised in the contract, whereby the bidder can make a precise quantity survey as the basis for the bid, a lump-sum contract is employed. For such a contract, it is imperative that the drawings and specifications be comprehensive and show in complete detail all features and requirements of the work. Compensation to the contractor is made on the basis of the lump-sum bid to cover all work and services required by the drawings and specifications (see also Art. 4.4).

3.3.3 Contract with Lump Sum and Unit Prices

It is not unusual to combine unit and lump-sum prices in the same contract; for example, an entire structure completely detailed on the drawings will be listed in the proposal as a lump-sum item, whereas unit prices may be required for features of variable quantities, such as excavations, or lengths of bearing piles.

3.3.4 Negotiated Contract

On occasion, public-works contracts and, more often, private-works contracts, are negotiated. These contracts may be prepared on the basis of one or more of several different payment methods. Some of the more widely used are:

Lump sum or unit price or combination

Cost reimbursable with a ceiling price and fixed fee

Cost reimbursable plus a fixed fee

Cost reimbursable plus a percentage of cost

Construction-management contract

In addition, incentives may be added.

For a negotiated contract, the owner chooses a contractor recognized for dependability, experience, and skill, and in direct negotiation establishes the terms of the agreement between them and the amount of the fee to be paid. For public agencies, factors contributing to the selection of a contractor are ordinarily determined by the prequalification or qualification procedures, using questionnaires and investigation. Such questionnaires are readily adaptable for use on contracts to be negotiated by private owners.

A negotiated lump-sum or unit-price agreement is negotiated around the engineer's estimate. A fixed percentage for overhead and profit is determined and agreed to, and the labor and material prices of the contractor and those of the engineer's estimate are adjusted by mutual agreement.

In a **cost-reimbursable agreement with a ceiling price**, the contractor receives reimbursement for all costs as prescribed in the agreement up to a maximum cost. The contractor receives a fixed fee, which will not vary with the cost of the work and otherwise is negotiated similarly to the cost-plus-fixed-fee type of agreement.

The determination of the fee to be paid the contractor under **a cost-plus-fixed-fee agreement**, which will be fair and reasonable to both parties to the contract, requires definitive plans, an estimate of the construction cost, and a knowledge of the magnitude and complexities of the work, the estimated time of completion, and the amount of work to be done under subcontracts. The terms of the contract must therefore set forth the methods for control and approval of expenditures and determination of the actual cost.

Under **a cost-plus-percentage-of-cost contract**, the contractor's profit is based on a fixed percentage of the actual cost of the work. This form is less desirable than the fixed fee since the contractor's compensation increases with increase in construction cost. This creates a situation where there would be no incentive for the contractor to effect any economies during construction.

A **construction-management agreement** requires the contractor to divide the work into segments, usually by trade. The contractor takes bids for the work from a group of subcontractors and awards the work to them. The prime contractor usually performs a certain prescribed segment of the work and coordinates the work of others. The owner reimburses the prime contractor for all the subcontractors' work and for the contractor's work plus a small profit and pays a negotiated fee for management of the subcontracts.

In some states, public-agency projects of larger size are required to be bid by separate trades, such as general civil; mechanical; heating, ventilating, and air conditioning (HVAC); and electrical. To accommodate this and to ensure proper contract management, some specifications have been written to require the general civil contractor to include an item for construction contract administration of the other trades. Bids for all major trades are taken by the owner with direct assignment of the mechanical, HVAC, and electrical subcontractors to the general civil contractor. In effect, the general civil contractor signs a construction management agreement along with an agreement for completion of the general civil work. The specifications require the bid of the civil contractor to include costs to account for coordination and control of the subcontractors to the same degree as if the civil contractor had taken direct bids and signed agreements with the various trade subcontractors.

Incentive-type contracts vary. The basic premise is that the owner will pay bonuses for economic construction and earlier completion and that the contractor may have to suffer for inefficiency and late completion.

3.3.5 Specialty Contracts

Special situations sometimes dictate a departure from the ordinary contract-letting procedure (Art. 3.2). Examples are contracts for the procurement and installation of highly specialized equipment and machinery, such as toll-collection facilities and communication systems.

For projects in the private sector, instead of advertising publicly for bids, the owner in such cases usually invites proposals from a selected group of contractors especially qualified and generally recognized as specialists in the manufacture and installation of such facilities. When competition is possible, it is so arranged. The contract documents prepared by the owner's engineer in such instances are as described in Art. 3.2, with certain exceptions. Since advertisement is not used, this and related items of the documents are not included, but the contracting procedure is substantially that followed for contracts publicly bid. Public agencies can use a modified procedure that involves preparation of and public bid on a prequalification bid package, prepared by their engineers.

See also Art. 3.8.

3.4 Standard Specifications

Government agencies and many other public bodies sponsoring public works publish "standard specifications," which establish a uniformity of administrative procedure and quality of constructed facilities, as evidenced by specific requirements of materials and workmanship. A sponsor's standard specifications usually contain information for prospective bidders, general requirements governing contractual procedures and performance of work by a contractor, and technical specifications covering construction of the particular work that lies within their jurisdiction. Highways, bridges, buildings, and water and sanitary works are examples of the types of improvements for which agencies may have standard specifications. Standard specifications, published periodically, may be updated in the interim by issuance of amendments, revisions, or, supplements.

So that the specifications for a particular contract are completely adapted to the work of that contract, the standard specifications almost always require modifications and additions. The assembled modifications and additions are known as supplementary specifications, special provisions, or special conditions. In conjunction with the standard specifications, they comprise the specifications for the work (see also Art. 3.11).

3.5 Master Specifications

Whereas published standard specifications are commonplace with government and other agencies (Art. 3.4), master specifications are useful tools for design organizations that serve private clients. A master specification covers a particular item of construction, such as excavation and embankment, concrete structures, or structural steel. It contains requirements for most possible conditions and construction that can be anticipated for that particular item. Master specifications are prepared in-house. (Engineers who work primarily for agencies that impose their own standards as the basic text for project specifications will find only limited uses for master specifications.)

When applying a master specification to a specific project, the specifications engineer deletes those requirements that do not apply to the project. Thus, use of a master specification not only effects a reduction in the time required to produce a contract specification but serves as a checklist and minimizes errors and omissions. Another important advantage of a master specification is that the edited text can be used directly for review without waiting for typing to be completed. When editing a master specification, however, failure to delete non-applicable provisions results in both encumbering and increasing the length of the project specifications. In addition, non-applicable provisions are confusing to contractors and others using the final documents.

To remain effective, a master specification must be periodically updated to incorporate current practices or new developments. Out-of-date information can never be considered acceptable in project specifications.

3.6 General Provisions of Specifications

The general provisions set forth the rights and responsibilities of the parties to the construction contract (owner and contractor) and the surety, the requirements governing their business and legal relationships, and the authority and responsibilities of the engineer. These articles are often mistakenly called "the legals" or the "boilerplate."

When a contracting agency maintains published standard specifications, the specifications for a project comprise these standards and, in addition, the modifications and additions necessary for the particular requirements of the project, generally called the **special provisions**.

On privately owned work, where generally there are no owner-published standard specifications, the specifications are especially tailored to fit the requirements of the project. A substantial part of standard general provisions is pertinent to such contracts. Requirements peculiar to the nature of the work are added, as necessary. Parts of the general provisions that pertain to legal requirements inherent in a public agency's corporate existence naturally are not included in a contract for privately owned construction. For example, most public-agency charters require protection with performance and payment bonds, while private owners can contract for work without any bonds. This saves cost for the private owner but puts that owner at greater risk in the event the contractor fails to perform or pay suppliers, workers, or subcontractors.

The general provisions may be set forth as detailed under the following subsections:

Definitions and Abbreviations ▪ This section covers abbreviations and definitions of terms used in the specifications.

Bidding Requirements ▪ This section deals with preparation and submission of bids and other pertinent information for bidders (Art. 3.8.1 & 3.8.2).

Contract and Subcontract Procedure ▪ This section includes award and execution of the contract, requirements for contract bonds, submission of progress schedule, recourse for failure to execute the contract, and provisions for subletting and assigning contracts.

Scope of the Work ▪ This section presents a statement describing the work to be performed; requirements for maintenance and protection of highway and railroad traffic, where involved; cleaning up before final acceptance of the project; and availability of space for contractor's plant, equipment, and storage at the construction site. Also, a limit is set on the permissible deviation of actual quantities from estimated quantities of the proposal without change in contract unit price.

Control of the Work ▪ This section deals with the authority of the engineer, plans, specifications, shop and working drawings, construction stakes, lines, and grades; inspection procedures; relations with other contractors at or adjacent to the site; provision of a field office and other facilities for the engineer needed in administration of the contract and control of the work; materials inspection, sampling, and testing; handling of unauthorized or defective work; contractor's claims for additional compensation or extension of time; delivery of spare parts, record documents; acceptance of work upon completion of project; and warranty maintenance.

Legal and Public Relations ▪ This section of the general provisions deals with legal aspects that determine the relations between the contractor and the owner agency and between the contractor and the general public. It sets up the requirements to be observed and protective measures to be taken by the contractor so that the liabilities for actions arising out of the prosecution of the work are properly oriented and provided for. Topics included are the disclaimer of any personal liability upon the contracting officer or the agency, the engineer, and their respective authorized representatives in carrying out the provisions of the contract or in exercising any power or authority granted them by virtue of their position; in such matters, they act as agents and representatives of the owner agency, such as federal government, state department, municipality, or authority.

Other features of legal and public relations that control contractors' procedures are damage claims; laws, ordinances, and regulations; responsibility for work; explosives; sanitary provisions; public safety and convenience; accident prevention; property damage; public utilities.

Damage Claims. Indemnification and save-harmless provisions are invoked to protect owners and their agents. The protection extends to suits and costs of every kind and description and all damages to which they may be subjected by reason of injury to person or property of others resulting from the performance of the contract work or through negligence of the contractor, use of improper or defective machinery, implements, or appliances, or any act or omission on the part of the contractor or contractor's agents or employees. These provisions are made to apply to subcontractors, material suppliers, and laborers performing work on the project.

These requirements are often implemented by requiring the contractor to provide insurance of specified character and in specified amounts as will provide adequate protection for the contractor, the

owners, their successors, officers, agents, or assigns and for others lawfully on the site of the work against all claims, liabilities, damages, and accidents. Insurance types and amounts are generally specified in the special provisions. However, neither approval nor failure to disapprove insurance furnished by the contractor releases the contractor of full responsibility for all liability inherent in the indemnification and save-harmless provisions. Generally included in the insurance to be carried by the contractor and in required minimum amounts of coverage established on the basis of loss in any one occurrence are:

Workmen's Compensation Insurance, statutory, as applicable. It should be extended where warranted to include obligations under the Longshoremen's and Harborworkers' Compensation Act and Admiralty law.

Contractor's Comprehensive General Liability, including Contractual Liability, with Bodily Injury Liability and Property Damage Liability. It should be augmented, by the prime contractor when there are subcontractors concerned, by *Contractor's Protective Liability Insurance* on the prime contractor's behalf and *Comprehensive General Liability* on behalf of each subcontractor. Policies should provide coverages for explosion, collapse, and other underground hazards (XCU coverage) when such hazards are incident to the work. To cover a lapse of time between the contractors' completion of the work and the owner's acceptance, the policies should bear endorsement for completed operations coverages. Also, Contractual Liability Insurance policies should bear endorsements noting acceptance by the underwriters of the indemnification and save-harmless clauses.

Comprehensive Automobile Liability providing coverage of all owned or rented vehicles and automotive construction equipment and with coverages of Bodily Injury Liability and Property Damage Liability.

Builder's Risk providing coverage of loss due to damage to a structure from fire, wind, etc.

Owner's Protective Public Liability and Property Damage Insurance, a separate original Public Liability, and Property Damage Insurance (Owner's Protective) should be provided by the contractor, designating the owner, successors, officers, agents, and employees as the named insured with respect to all operations performed by the contractor. Some specifications require the owner to maintain property insurance to cover full value of the project in addition to property insurance provided by the contractor. This owner-provided insurance will protect the owner from damage, by someone other than the contractor, to property that has been accepted and paid for prior to final acceptance.

Protection and Indemnity Insurance, or comparable coverage, should be carried by contractors, where applicable, with respect to all watercraft used by or operated for them, chartered for otherwise, covering bodily injury liability and property-damage liability. (See also Art. 4.16.)

Insurance is a specialized field. Hence, the specifying of insurance coverage should be left to those experienced in that field.

Laws, Ordinances, and Regulations. The pertinent federal and state laws, rules, and regulations, and local ordinances that affect those engaged or employed on the project, the materials or equipment used, or the conduct of the work are cited. All necessary permits and licenses for the conduct of the work are often specified to be procured by the contractors at their expense. Frequently the engineer prepares construction permits for the owner when those permits affect the final design of the project.

Responsibility for Work. Contractors are required to assume full responsibility for materials and equipment employed in the construction of the project. They are prohibited from making claim against the owner for damages to such materials or equipment from any cause whatsoever. Until final acceptance, the contractor is responsible for damage to or destruction of the project or any part thereof due to any cause, except for damage caused by owner-operated equipment. The contractor is required to make good all work damaged or destroyed, except that caused by others, before final acceptance of the project and to include all costs thereof in the prices bid for the various scheduled items in the proposal.

Explosives. The use, handling, and storage of explosives are required to conform to regulations of government agencies controlling these features of the work. Proper means are required to be used to avoid blasting damage to public and private property and construction personnel.

Sanitary Provisions. The contractor is required to provide and maintain suitable sanitary facilities for personnel in accordance with the requirements of federal, state, and local agencies having jurisdiction.

Public Safety and Convenience. This article provides that the contractor conduct the work so as to inconvenience as little as possible the public and residents adjacent to the project and provide protection for persons and property. The contractor must install temporary crossings to give access to private property. Also, measures must be taken to prevent deposits of earth or other materials on roads and thoroughfares on which hauling equipment is operated and to remove promptly such deposits, if they occur, and thoroughly clean the surfaces. The contractor must employ construction methods and means to keep flying dust to a minimum.

Accident Prevention. This article provides for observance of safety provisions outlined in the rules and regulations of public agencies functioning in this field (e.g. OSHA). It is the contractor's responsibility to provide safe working conditions on the project. The contractor is held fully responsible for the safe prosecution of the work at all times.

Property Damage. This article defines the contractor's obligations when entering upon or using private property in carrying out the work and in connection with any damage to that property.

Public Utilities. Through this article the contractor's attention is directed to the possibility of encountering public and private utility installations that either are obstructions to the prosecution of the work and need to be moved out of the way or, if not, must be properly protected during construction. It sets up the procedures to be followed and establishes costs to be absorbed by the contractor as well as the utility companies and the public agency in accordance with agency policy and laws dealing with such situations.

Abatement of Soil Erosion, Water Pollution, and Air Pollution. Through this article, the contractors are reminded of their responsibility for minimizing erosion of soils and preventing silting and muddying of streams, irrigation systems, impoundments, and adjacent lands. Pollutants such as fuels, lubricants, and other harmful materials are not to be discharged into soils or near streams, impoundments, or channels. No burning of any material is permitted.

Prosecution and Progress ▪ This section of the general provisions deals with such pertinent considerations as commencement and prosecution of the work, time of completion of the contract, suspension of the work, unavoidable delays, annulment and default of contract, liquidated damages, and extension of time.

Commencement and Prosecution of the Work. This article establishes the date on which work is to start and from which contract time is to run. It requires that construction proceed in a manner and sequence ensuring completion established by the contractor's progress schedule previously reviewed and accepted by the engineer. It describes whatever limitations of operations there may be at the site of work, including traffic, work by others, and schedule of stage completion. It also requires that the ability, adequacy, and character of workers, construction methods, and equipment be suitable for full prosecution of the work to completion in the time and manner specified.

Time of Completion. It is advantageous to specify time of completion in calendar days from date of commencement of work rather than working days because the actual determination of a working day is often a cause of contention. Herein may be specified stage completion when it is to the owner's advantage to have occupancy of a part of the work prior to completion of the entire contract or where a priority of construction of a particular feature of the work is essential to subsequent procedures.

Suspension of Work. This article covers the usual conditions under which the owner may suspend work, in whole or in part, for such period of time as may be deemed necessary, without breach of contract, and the period of time that suspension may be effected without allowance of compensation. These conditions may include weather, owner's or adjacent owner's operations, or other conditions unfavorable for prosecution of the work and the contractor's failure to perform in accordance with provisions of the contract or to correct conditions unsafe for workers or the general public.

Unavoidable Delays. For delays for any reason beyond the contractor's control, other than those caused by suspension of the work, the contractor may be granted an extension of the contract time. This citation, however, gives the contractor no right or claim to additional compensation unless the contract specifically provides for such compensation.

Annulment and Default of Contract. Provision is made for terminating the contract as follows:

For annulment: A public officeholder acting in the public interest or a national or state agency ordering a work stoppage may result in the owner's annulment of a contract. With a contractor not in default, settlement is usually made for work completed and proper costs of work in progress and for moving from the site, with no allowances for anticipated profit. Also, the owner may annul a contract when a contractor is found to have compensated others for soliciting a public contract, thus violating the warranty of noncollusion with others.

For default: When a project or any part of it has been abandoned, is unnecessarily delayed, or cannot be completed by the contractor within the time specified, or on which the contractor willfully violates terms of the contract or carries out the contract in bad faith, the owner usually has just cause to declare the contractor in default on the contract and notify the contractor to discontinue work on the project. When a contractor is in default, the owner may make use of contractor-furnished material and equipment to complete the project through the contractor's surety or by other means considered necessary for completion of the contract in an acceptable manner. All costs, over and above contract costs, for completing the project are recoverable from the contractor or the contractor's surety.

Liquidated Damages. Provision is made for the contractor to pay the owner a sum of money for each day of delay in completing specified stages or the complete contract beyond the dates due. This agreement on damages prior to breach of contract avoids litigation and dispute over almost undeterminable actual damage while providing an incentive to the contractor to complete work on time. When the specified sum of money is unsupportable as representative of the actual damage suffered by the owner in added costs, it becomes, in fact, a penalty for delayed completion and unenforceable in the courts.

Extension of Time. This article establishes certain conditions that will be considered just cause for an extension of the time stipulated in the contract for completion of the project. These conditions may include change orders adding to the work of the contract, suspension of work, or delay of work for other than normal weather conditions.

Measurement and Payment ▪ This section of the general provisions provides for measurement of quantities of the completed work; scope of payment; change of plans and consequent methods of payment; procedures for partial and final payments; termination of contractor's responsibility; and guaranty against defective work.

Measurement of Quantities. This article stipulates that all completed work of the contract will be measured for payment by the engineer according to United States, international or other standard measures.

Scope of Payment. This article establishes that payment for a measured quantity at the unit-price bid will constitute full compensation for performing and completing the work and for furnishing all labor, materials, tools, equipment, and all else necessary and incidental thereto.

Change of Plans. Provision is made for payments pertinent to changes in the work; i.e., the measured quantities of work completed or materials furnished which are greater than or less than the corresponding estimated quantities listed in the proposal and the quantitative limits of such changes permitted by change orders; the context of the change order, inclusive of kind and character of work, materials to be furnished, and changes in contract time of completion; supplementary agreement for changes in contract prices of scheduled items and the performance of work not identified with any scheduled item in the proposal.

Payment. This article establishes the procedure by which payment will be made for the actual quantity of authorized work completed and accepted under each item listed in the proposal either at the unit-price bid or at the unit price stipulated in the supplementary agreement.

The procedure usually provides for partial payments to be made periodically. These payments are based on approximate quantities of work completed during the preceding period, as measured by the engineer and attested to by certificates for payment. The owner may retain a percentage of the amount of each certificate, pending completion of the contract. Upon completion and acceptance of the contract, a final certificate of cost prepared by the engineer and approved by the owner determines the total amount of money due the contractor and from which previous payments on account will be deducted. Final payment is made upon satisfactory representation by the contractor that there are no outstanding claims against the contractor filed with the owner, that the contractor has satisfied or arranged for payment of all due obligations incurred personally and by subcontractors in carrying out the project, as evidenced by final releases of liens, and that whatever guaranty bond may be required has been posted.

Termination of Contractor Responsibility. This article establishes that upon completion and acceptance of all work included in the contract and payment of final certificate, the project is considered complete and the contractor is released from further obligation and requirements.

Guaranty against Defective Work. A guaranty period is established for all or portions of the work, together with an amount of guaranty, usually calculated as a percentage of the contract cost. A guaranty bond is furnished by the contractor and conditioned to replace all work and all materials that were not performed or furnished according to the terms and performance requirements of the contract and to make good defects that become apparent before the end of the guaranty period.

Dispute Resolution. Some specifications stipulate that disputes are to be handled by binding arbitration. Other specifications require disputes to go directly to court with the location of venue usually stipulated to be in the county of the owner's business location.

3.7 Technical Specifications

These specifications, which are described briefly in Art. 3.1, may take several forms. One or more of these forms may be selected to serve best the purpose for which the specifications are prepared. Types of technical specifications in common use are:

Materials and workmanship specifications, commonly called descriptive specifications

Material procurement specifications

Performance specifications (procurement)

Materials and Workmanship Specifications ▪ This type of specification is almost universally used on construction contracts. It is comprehensive in its coverage of the principal factors entering into the prosecution and completion of the work covered by the contract. These factors include the general and special conditions affecting the performance of the work, material requirements, construction details, measurement of quantities under the scheduled items of work, and basis of payment for these items.

Material Procurement Specifications ▪ These specifications are used on projects of considerable magnitude requiring many separate general construction contracts, usually in simultaneous operation and under which the types of construction are similar. For example, material procurement specifications may be desirable for a considerably long highway involving the construction of grade-crossing structures of structural steel or precast and prestressed concrete items. In such cases, it has often been found advantageous to separate contracts for the structural steel or prestressed concrete from the general contracts for the overall project. This procedure ensures uniformity and availability of the materials. It facilitates construction by scheduling deliveries to coincide with the general contractors' needs for these items at any particular location throughout the entire project. A similar procedure may also be used for the procurement of other construction materials in quantity.

The specifications for contracts of this nature contain, besides fabrication processes, all the elements of materials and workmanship specifications, except for the field construction details. If erection of the items is to be included in the procurement specifications, the procedure is the same as for materials and workmanship specifications.

Performance Specifications ▪ These specifications are used to a great extent in procurement contracts for machinery and plant operating equipment, as distinct from material procurement contracts. Contracts for machinery and equipment may be let separately by the owner prior to a construction contract under which installation will be made, to ensure delivery to the job in time for installation within the scheduled construction sequence. Advance letting of procurement contracts is usually necessary because of the great amount of time consumed in the manufacture of such items. In general, performance specifications, in addition to defining the materials entering into the manufacture of equipment, with all the pertinent physical and chemical properties, prescribe those characteristics that evidence equipment capability under actual operating conditions. Thus, the specifications must completely define quality, function, and other requirements that must be met. Since a performance specification requires samples, tests, affidavits, and other supporting evidence of compliance, it tends to increase contractor's costs for furnishing the items and engineer's costs for checking submitted data. It also adds to the designer's responsibility for an unsatisfactory or inadequate product.

Requirements for tests and certification of the results are set up in the specifications in accordance with test procedures established by the appropriate industry associations.

When not critical from the standpoint of manufacture and delivery schedules, machinery and equipment may be covered by the construction specifications. For a typical technical specification, see Art. 3.12.

3.7.1 Materials Specifications

Under this division of standard specifications are prescribed the various materials of construction to be used in the work and their properties. The principal properties to be considered in the preparation of specifications of materials for construction are:

1. Physical properties, such as strength, durability, hardness, and elasticity

2. Chemical composition

3. Electrical, thermal, and acoustical properties

4. Appearance, including color, texture, pattern, and finishes

Materials specifications should also include procedures and requirements to be met in inspections, tests, and analyses made by the manufacturer during manufacture and processing of the material and later by the owner. Note should be made as to whether a material is to be inspected at the shop or mill during manufacture and the number of test specimens, identified with the material proposed to be furnished, that will be furnished to the owner for test.

In addition, the specifications should cover the protection necessary in the interval between manufacture and processing of the materials and their incorporation into the work. Some materials are subject to deterioration or damage, under certain conditions of exposure, during stages of transportation, handling, and storage.

See also Art. 3.7.3.

3.7.2 Reference Standards

Standards published as reference specifications for construction materials and processes by professional engineering societies, government agencies, and industry associations are widely followed for construction work. The recommendations of these organizations are the bases of current construction practice, particularly with regard to quality of materials and, in some cases, fabrication practices, construction methods, and testing requirements.

3.7.3 Arrangement and Composition of Technical Specifications

The general provisions, as Division 1 of the specifications are followed by the various divisions of the technical specifications in numerical order and in sequence generally based on a logical order of construction stages for progressing the work. For example, in the Construction Specifications Institute 16-division MASTERFORMAT, successive divisions are:

Division 2 • *Underground, Pavement, and Site Work:* Section 02010—Subsurface Exploration; Section 02100—Clearing and Grubbing; Section 02110—Removal of Structures and Obstructions; Section 02200—Excavation and Backfill; Section 02552—Precast Concrete Structures; Section 02600—Pavements, Curbs, and Walks; Section 02710—Fencing; Section 02800—Sodding, Seeding, and Mulching; Section 02900—Landscaping

Division 3 • *Concrete:* Section 03100—Waterstop; Section 03200—Concrete Reinforcement; Section 03300—Cast-in-Place Concrete; Section 03350—Concrete Tank Bottoms; Section 03400—Precast Concrete Structures

Division 4 • *Masonry:* Section 04200—Masonry

Division 5 • *Metals:* Section 05100—Miscellaneous and Structural Steel; Section 05120—Aluminum Plates and Covers; Section 05200—Steel Joists; Section 05300—Metal Decking; Section 05530—Metal Floor Grating; Section 05540—Iron Castings; Section 05550—Stair Nosings; Section 05560—Steel Stairs and Platforms; Section 05700—Steel Storage Tanks

Division 6 • *Wood and Plastics:* Section 06100—Rough Carpentry; Section 06110—Stop Planks; Section 06200—Finish Carpentry; Section 06610—Fiberglass Gratings; Section 06615—Fiberglass Ceiling Panels; Section 06620—Fiberglass Handrailing; Section 06640—Fiberglass Cover Plates

Division 7 • *Thermal and Moisture Protections:* Section 07110—Expansion Joints; Section 07120—Mastic and Asphalt Joints; Section 07150—Waterproofing and Dampproofing; Section 07200—Wall Insulation; Section 07250—Roof Insulation; Section 07400—Preformed Metal Siding; Section 07500—Membrane Roofing; Section 07600—Sheet Metal and Flashing; Section 07800—Roof Accessories; Section 07900—Sealants and Caulking

Division 8 • *Doors and Windows:* Section 08100—Steel Doors and Frames; Section 08200—Aluminum Doors and Frames; Section 08320—Rolling Metal Doors; Section 08350—Folding Doors; Section 08500—Aluminum Windows; Section 08700—Finish Hardware; Section 08800—Glazing

Division 9 • *Finishes:* Section 09200—Lath and Plaster; Section 09300—Tile; Section 09500—Acoustical Ceilings; Section 09800—Concrete Coatings; Section 09650—Resilient Flooring; Section 09900—Painting and Coatings

Division 10 • *Specialties:* Section 10200—Rolling Stock; Section 10310—Portable Radios; Section 10320—Weigh Scale; Section 10400—Food Service Equipment; Section 10500—Shop Equipment; Section 10520—Fire Extinguisher; Section 10600—Movable Partitions; Section 10610—Toilet Partitions; 10700—Plaques and Signs; Section 10800—Toilet Room Accessories

Division 11 • *Equipment and Systems:* Section 11000—Air Diffusion Equipment; Section 11120—Air Blowers; Section 11230—Chlorination System; Section 11260—Effluent Filter; Section 11430—Scum System; Section 11480—Incineration Systems; Section 11600—Mixing Equipment; Section

11700—Pumping Equipment; Section 11800—Sampler Equipment; Section 11810—Rotary Fine Screens; Section 11820—Sludge Degritting Equipment; Section 11830—Gravity Sludge Thickeners; Section 11831—Odor Control Systems; Section 11950—Fiberglass Weirs and Troughs

Division 12 • *Furnishings:* Section 12100—Interior Furnishings

Division 13 • *Special Construction:* Not used.

Division 14 • *Conveying Systems:* Section 14300—Hoists and Cranes; Section 14500—Belt Conveyors; Section 14600—Screw Conveyors

Division 15 • *Mechanical:* Section 15100—General Mechanical Requirements; Section 15200—Piping; Section 15210—Valves; Section 15250—Sluice and Slide Gates; Section 15400—Plumbing; Section 15600—Heating, Ventilating and Air Conditioning (HVAC); Section 15700—Fuel System

Division 16 • *Electrical, Instrumentation and Controls:* Section 16000—Electrical; Section 16500—Instrumentation and Controls; Section 16600—Supervisory Data and Control Acquisition (SCADA) System; Section 16720—Fire Detection System

As indicated above, each division is composed of sections. The detailed specifications for each section (for example, Section 04200, "Masonry," under Division 4) are generally arranged under the following headings:

1. Description
2. Materials
3. Construction Requirements
4. Method of Measurement
5. Basis of Payment

Items 4 and 5 may be combined under a single heading, "Measurement and Payment."

Description ▪ Under this heading, a concise statement is made of the nature and extent of the work included in the section and its pertinent features, including the general requirement that work conform to the plans and specifications.

Materials ▪ This article presents the requirements for the various materials involved in the performance of the work of the section. If a separate division on materials has been included as a part of the technical specifications, simple references to specific articles that detail required material properties are made (see also Art. 3.7.1). If such a division is not included, reference to standard specifications of the professional engineering societies, government agencies, and industry associations are appropriate. When manufactured products are not listed in available reference standards, it is customary to name several of proven quality and performance. Usually, three are specified by name and manufacture, any one of which is considered acceptable for use on the work.

Sometimes, owners prefer to limit the purchase of items from one manufacturer to minimize their spare parts requirements. This sole source procurement may require specific justification for public owners.

"Or Equal" ▪ When a given construction material or piece of equipment does not lend itself readily to standard-specification designation or easily describable specifications most public bodies require the names of at least two or three suppliers or the name of one supplier with the added phrases "or equal," "or approved equal," "or equal as approved by the engineer." This requirement promotes fair competition and complies with the law for public bids in many states. In many instances, the procedure originates in the office of an attorney general or other public official and is based on a ruling that competition is a requirement of most public-works laws. In private-ownership practice, the main reason for use of this procedure is to obtain the best product for a client at the most economical price.

The "or equal" clause has often been a source of contention among engineers and contractors. However, careful use of the "or equal" clause promotes competition and can lower the delivered cost of work items. Allowing substitutes lets contractors bring their valuable experience with materials, equipment, and suppliers to projects.

Use of the "or equal" clause requires the engineer and the owner to be prepared and to budget time to investigate and evaluate substitutions offered by the contractor. The salient features of the originally specified item should be carefully documented and recorded for use during evaluation of proposed substitutes.

Some specifications stipulate that the contractor shall reimburse the engineer for the costs of such investigations and evaluations, including costs to redesign affected project items, e.g., foundations, electrical, and piping.

The specifications should require the contractor to assume full responsibility for compliance with all applicable provisions of the specifications on approval of a substitution. An exception to this occurs when the owner waives the requirements of the specifications to take advantage of the lower cost of a substitute, thereby relieving the engineer of responsibility. Approval of substitutions should always be given in writing.

Some specifications required bidders to offer substitutions for major work items with their bids. Under this scheme, the specifications prescribe the exact items required. Bidders must describe substitutions in detail with accompanying product specifications, drawings, catalog cut sheets, etc. Also, the contractor must stipulate the amount to deduct or add to the base bid for acceptance of the offered substitution. This method allows the engineer to review the proposed substitution along with the rest of the bid, free from the pressures that exist after contract award.

Construction Requirements ▪ The primary purpose of this article in the detailed specifications for each work item is to prescribe the requirements for its construction without relieving the contractor of responsibility for the satisfactory accomplishment of the end result. Among the principal features to be stressed are workmanship and finish, with consideration given to practical limitations in tolerances, clearances, and other limiting factors. Necessary precautions should be given for the protection of the work and adjacent property. Methods of inspection and tests applicable to the work, with particulars as to off-site inspection at mill or shop, as well as inspection at the site, should be specified.

Specifications for workmanship should indicate the results to be attained insofar as practicable. Thereby, the contractor obtains latitude in selection of construction procedures. In some instances, however, it may be necessary to designate methods to ensure satisfactory completion of the work, for example, compaction of earth embankments or shop and field welding procedures on steel structures. It may also be necessary to specify precautions and restrictions for purposes of protection and coordination of the work as a whole or when a definite sequence in construction operations

is made necessary by design conditions or to meet conditions contemplated by the owner.

Measurement and Payment ▪ This heading combines method of measurement and basis of payment. Every contract, regardless of type, must include provisions for payment. For a unit-price contract, the quantity of work completed under each bid item listed in the proposal must be measured by applying an appropriate unit of measurement. Some items, such as assembled units, are measured by the number required; others are measured by linear foot, square yard, cubic yard, pound, or gallon, as applicable.

The quantity to be considered for payment should be clearly defined so as to cover all deductions to be made for deficiencies and unauthorized work performed beyond the limits delineated on the plans or ordered by the engineer. Partial and final payments for the actual quantity of work completed and accepted can then be computed. To determine the payment due, each such quantity is multiplied by the corresponding unit price bid by the contractor and the extended subtotals for all items are totaled.

It is essential for payment purposes that the specifications define precisely each bid item per unit of measurement (cubic yard, linear foot, cubic metre, etc.). The specifications should clearly and fully state all the work and incidentals that should be included by the bidder in the item for which the unit price is to be submitted. When there are operations closely associated with a particular item of work for which separate payment is provided, the specifications should make this clear to avoid controversy or double payment for the work.

It is not uncommon in a unit-price contract to include items for which lump-sum prices are required. These are subject to all the conditions governing unit-price items, except measurement for payment and the right of the owner to vary the quantity of work without change order. The cost of all work and materials necessary to complete the construction of the lump-sum items, as delineated on the drawings and required by the specifications, must be included in the lump-sum bid. Work associated with construction of lump-sum items but not made a part thereof must be indicated as being included for payment under other bid items.

To facilitate partial payments for work performed on lump-sum items as well as for contract lump-sum bids, the contractor should be required to submit a breakdown for the components of the work. This breakdown is referred to as the schedule of values. The breakdown should include quantities for the different types of work or trades involved and unit prices applicable to each. When extended and summarized, the prices should equal the lump-sum bid for the completed item or contract. The specifications should require submittal of the schedule of values prior to the preconstruction conference. The schedule must be approved by the engineer before it becomes effective.

See also Art. 3.11 & 3.12.

3.8 Bidding and Award of Contracts

It is standard practice for government and public agencies to provide for the public letting of contracts for public works. Sealed bids are invited by advertising in newspapers and engineering publications for legally required periods. The advertisement should contain the following information: issuing office, date of issue, date for receipt of bids, location for receipt of bids and time of opening of bids, brief description of work (identification of project), location of project, quantities of major items of work, office where plans and specifications can be obtained and charges for them, proposal security, and rights reserved to the owner. For private work, an invitation for bids is issued by the owner to a selected group of contractors. The invitation conveys much of the information that would be included in an advertisement that may apply to the particular project.

3.8.1 Bidding Requirements for Public Works

Bidding requirements for public-works contracts are usually defined in the general provisions of the standard specifications for the particular agency. The object of these requirements is to advise prospective bidders of the routine to be followed for submitting a bid and their eligibility to do so. The principal points covered are:

Prequalification or Qualification ▪ For a bid to be acceptable, the bidder must have been either prequalified with the contracting agency for

capability and financial standing, by submission of documents furnishing required information (updated to reflect the situation at bid time), or otherwise qualified along the same lines by furnishing evidence thereof with the bid. Some states require that contractors be licensed, in which case a record of the contractor's license is filed with the contracting agency.

Preparation and Delivery of Proposal ▪ Instructions for preparing a proposal on forms furnished by the contracting agency are given to avoid irregularities, which could nullify the bid. Proposals must be signed and signatures legally acknowledged before being placed in envelopes (sometimes furnished for the purpose) and then sealed. Receipt of all addenda issued during the bidding period must be acknowledged on the proposal form, where provision is made for this purpose. Information requested of the bidder on the exterior of the envelope (when one is provided) must be entered in the spaces provided. A bid may be delivered by mail or messenger but must be received before the time set for opening; otherwise, it may not be accepted. (See also Art. 4.3).

Proposal Guaranty ▪ Public agencies always require a guaranty that the bidder will execute the contract agreement if awarded the contract. The guaranty may be in the form of a surety bond or certified check for a stated percentage of the bid. Usually this is 5 or 10%, with maximum limit of a fixed amount, but this could vary to serve the interest of the particular agency. Sometimes both a surety bond and certified check are required. The amount of the surety bond may vary from 100% of the bid price down to 5% at the discretion of the contracting agency. (See also Art. 4.3.)

Proposal guaranties must accompany the proposal. Bid securities are returned to all but the lowest three bidders within a short time after bids have been opened. Those of the lowest three bidders are returned after a contract has been executed.

Noncollusion Affidavit ▪ A noncollusion affidavit is generally required by public agencies by law.

3.8.2 Bidding Requirements for Private Works

For private owners, the procedures for submitting, receiving, and opening bids are more informal since they are not subject to the laws governing such procedures for public-works contracts. The manner in which these steps are handled is entirely at the discretion of the owner or engineer. Bid securities are not required. Advertisement for bids is not usually employed. Instead, a **Notice to Contractors** is issued to a selected group of contractors, known to the owner to be qualified. This notice is accompanied by Instructions to Bidders and a proposal form when competitive bids are required. The **Instructions to Bidders** generally include the information necessary for preparing and delivering the proposal. Noncollusion affidavits are not required. Tabulation and evaluation of bids and award and execution of contract usually follow the procedure for public-works contracts, modified to suit the owner's particular needs.

3.8.3 Evaluation and Comparison of Bids

Following the opening of bids, a public announcement is made of the prices bid for the various items listed in the proposal. These data then are tabulated, the totals for each item verified, and their summation, establishing the total amounts of bids, is checked for each bid submitted. Comparison of the total amounts of the bids establishes the lowest bid and those that follow in the order of increasing amounts.

3.8.4 Award and Execution of Contract

Having verified all specified submissions, such as licensing, prequalification statements, and noncollusion affidavits, and having established the low bidder, the owner officially notifies the successful bidder of the award of the contract; the bidder is then expected to execute the contract agreement within a specified time. It is a requisite for this final step in the contracting procedure that the successful bidder furnish performance and payment bonds acceptable to the contracting agency. The amount of these bonds equals the total amount of the bid. The two bonds often are combined into a single performance and payment bond, a guaranty to the owner that all the work required to be performed will be faithfully carried out according to the terms of the contract. Also, it guarantees that the contractor will pay all lawful claims for payment to subcontractors, material suppliers, and labor for all work done and materials supplied in the performance of the work under the contract.

The bond must also provide that the owner be saved harmless, defended, and indemnified against and from all suits and costs of any kind and damages to which the owner may be put by reason of injury to the person or property of others resulting from performance of the work or through negligence of the contractor. In addition, the owner must be shielded from all suits and actions that may be brought or instituted by subcontractors, material suppliers, or laborers who have performed work or furnished material on the project and on account of any claims, or amount recovered, by infringement of patents or copyrights. The requirement of the contractor to indemnify and save harmless the owner may be implemented by insurance, by retaining a percentage of the contract amount until final acceptance of the work, and by the contract bonds. (See also Art. 4.17.)

3.9 Specifications Writing: Style and Form

Preparation of the specifications for a construction contract starts with an overall analysis of requirements based on a survey of the proposed work, conditions under which it must be accomplished, materials, details of construction, and owner's administrative procedures. The analysis provides the various items for appropriate distribution among the contract documents. Also, a close study of the contract drawings will reveal that which is insufficiently shown and needs to be supplemented in the specifications. A descriptive outline of such a distribution or proposed contents with subheadings facilitates and expedites the work of the specifications writer when assembling the documents.

Design/build projects are increasingly used to expedite project delivery. This form of project delivery requires additional considerations of risk. Courts have used the distinction between design specifications and performance specifications to assign liability for design defects on design/build projects. A descriptive definition of these two types of specifications follows. Design specifications are those that tightly circumscribe the contractor's latitude in choosing products that achieve the specified standard of performance. Whereas performance specifications prescribe an objective or standard to be achieved and leave it to the ingenuity of the contractor to select the methods and materials to achieve the specified results.

One example of a design specification is to specify a brand name product without allowance for possible substitution. Conversely, courts have ruled that specifying a brand-name product with an "or equal" clause allowing substitution is an example of a technical specification. A contract due date has been determined to be a performance specification. Courts have decided that a due date is a warranty by the bidding contractor that it can do the work in the specified time and thus is a performance specification.

Design specification should be used when the project owner has strong preferences such as using one brand and type of motor actuated valve for all valves serving a specific duty. This allows owners to minimize the need for warehouse space and the use of maintenance staff. However, there is more risk on the part of the writer when design specifications are used. Performance specifications should be used when the owner is unfamiliar with a process of mechanism and it wishes to employ the knowledge and expertise of the contractor to accomplish the end goal.

3.9.1 Specifications Format

A basic format for specifications may be oriented for a particular project and its sponsor. There should be a title page identifying the documents and a table of contents listing the various sections of general provisions and technical specifications by section number, title, and page. Cross references in a section should be made by title only. Otherwise, unnecessary cross checking of references becomes unmanageable. This results from numerous revisions of specifications until their release for bidding.

Specifications should be organized in divisions and the divisions into sections (Art. 3.7.3). Each technical section usually begins with a brief description of the work included in it. Work contingent upon but not included in the work specified under

a particular section may be referenced as "Related work specified under other sections." Each section should be complete, with description of materials, workmanship, and requirements for testing clearly defined. All payment items must be mentioned, with methods of measurement and basis of payment specified for each item.

3.9.2 Precedence of Contract Documents

Of major importance in coordination and interpretation of contract documents is the establishment of an order of precedence. It is usual to provide that the contract drawings govern over the standard specifications and that the special provisions govern over the standard specifications and the contract drawings. Thus, in the preparation of special provisions, care must be exercised to avoid conflict with the other contract documents and to ensure a definite and clear description of the required work. Care must also be taken to avoid duplication of information in the special provisions or in both the drawings and special provisions to preclude conflict and errors, especially in the event of changes. It is advisable not to specify both the method to be used and the desired results thereof because a conflict may relieve the contractor of responsibility.

3.9.3 References to Standard Specifications

When preparing specifications for a project for which there are owner's standard specifications, for example, for a project of a public agency, the specifications writer is obliged to incorporate these specifications either directly or by reference and to identify and establish this standard in the special provisions. It is not unusual to cite sections of the standard specifications by reference at the beginning of each applicable section of the special provisions, with a paragraph similar to the following:

All work shall be in accordance with Standard Specifications (list section number and title), as amended herein.

However, in the text of a section of the special provisions, references may be made to other sections of the standard specifications or to standards other than the owner's, in whole or in part.

Special provisions therefore modify, restrict, or add to the standard specifications, where necessary, and admit such options and alternatives as may be permitted. Do not repeat portions of the standard specifications in the special provisions, and avoid repeated references in special provisions to a standard specifications section. *Redundancy leads to error!*

3.9.4 Basic Principles of Good Specification Writing

Specifications usually are written in the traditional style of composition, grammatically correct. They should go into as much qualitative and quantitative detail as necessary, to convey that which is required and therefore agreed to. Chances for misunderstandings and disputes, which frequently result in expensive litigation, should be kept to a minimum. Ambiguity and verbosity should be avoided. A good specification is clear, concise, complete and easily understood. It gives little cause for doubt of the intentions of the parties concerned and leaves nothing to be taken for granted. The courts have traditionally interpreted ambiguous requirements against the party who prepared them.

Inasmuch as the specifications, in conjunction with the drawings, are the means employed to guide the contractor in producing the desired end product, it is essential that they be correlated to avoid conflicts and misunderstandings of the requirements. Instructions more readily described in words belong in the technical specifications, whereas information that can be more effectively portrayed graphically should appear on the drawings. *Information on the drawings should not be duplicated in the specifications* or vice versa because there may be a discrepancy between the information provided in the two documents that may cause trouble.

Since specifications complement the drawings, the special provisions and standard specifications together should leave no doubt as to the quality and quantity of the required work. The function of the drawings is to show location, dimensions, scope, configuration, and detail of the required work. The function of the specifications is to define the minimum requirements of quality of material

and workmanship, prescribe tests by which these must be established, and describe methods of measurement and payment.

The contract documents should be fair to owner, bidders, contractor, engineer/architect, and others concerned. Any aspect of the work not clearly defined in the specifications or on the drawings will result in time and effort wasted during bidding or during construction, higher contract prices including, "contingencies," and in all probability arguments over extras, with ensuing delays.

Following are some general suggestions for writing specifications: Be specific, not indefinite. Be brief; avoid unnecessary words or phrases. Give all facts necessary. Avoid repetition. Specify in the positive form. Use correct grammar. Direct rather than suggest. Use short rather than long sentences. Do not specify both methods and results. Do not specify requirements in conflict with each other. Do not justify a requirement. Avoid sentences that require other than the simplest punctuation. Also, avoid words that are likely to be unfamiliar to users of the specifications, especially if the words have more than one meaning.

Be particularly careful when requiring approval by the engineer. Specific approval by the engineer of the contractor's equipment, methods, temporary construction, or safety standards, in certain situations, can relieve the construction contractor of responsibility under the terms of the contract. It is best, and usually the general provisions of specifications require, that the contractor be responsible for means, methods, and scheduling of construction.

When preparing the Construction Details of a specification, arrange the material in the sequence in which the work will be done. For example, specify the curing of concrete after specifying formwork, concrete mixing, and concrete placing. When inserting a reference to a national standard, such as a standard ASTM specification, read the standard first to assure yourself that it contains nothing that conflicts with job requirements.

The measurement and payment portion of a specification is most important to both the contractor and owner. Every item of work to be done by the contractor must be accounted for, whether it be measured and paid for separately or included for payment in another item.

Refer only to the principals to the contract: the owner, as represented by the engineer, or the contractor. Do not refer to other contractors, subcontractors, bidders, etc.

Refer to "these" specifications rather than "this" specification; use the plural.

Workmanship should be in accordance with, and materials should conform to, a reference specification.

Use the phrase "at no additional cost to the owner" only when there is a definite possibility of the contractor's not understanding that he or she is to bear a certain expense. Liberal use of the phrase might imply that other work specified is not at the contractor's expense.

Use the word "shall" for requirements placed on the contractor and the word "will" for expressions of intent on the part of the owner.

Do not confuse the meaning of words; proper word usage is of utmost importance.

Do not use indefinite words when more exact words may be substituted.

Avoid repeated use of stock phrases and stereotyped expressions. Specifications should not be encumbered with legal phrases that obscure their meaning or subordinate their function to that of a legal document.

Streamlined Specifications ▪ An alternative to the traditional style of specifications is a streamlined form. This is accomplished by shortening sentence structure wherever practicable. Properly employed, streamlining may be a major improvement. In general, streamlining omits from the specifications, without a change in meaning, those words having no legal significance. Only necessary provisions are retained. A good long-form specification can be streamlined without the slightest adulteration and yet reduce its bulk by one-third or more.

The technique of streamlining specifications may be adopted as a simplification of style, productive of a distinctive form of writing specifications, whereas the general format remains the same. However, note that this style is more readily adaptable to building construction contracts, wherein each section of the technical specifications relates directly to a particular construction trade.

Some aspects and considerations in streamlining specifications advocated by Ben John Small ("The Case for Streamlined Specifications," *The Construction Specifier*, July 1949) are:

The term "streamlining" should not be interpreted to mean that it refers to a specification lacking thoroughness or that streamlining is synonymous with specifications devoid of the three C's (Clarity-Conciseness-Comprehensiveness). Any specification long or short must be equipped with the requisite C's if it is to associate properly with its other relatives, which constitute the family of Contract Documents, such as the Agreement, General Conditions, the Drawings, etc.

Streamlining offers no cure for ineptitude in writing specifications, such as conflicting repetitions, giving contradictory instructions, etc. What it does, affirmatively, is to translate the writer's knowledge of construction and materials into simple, readable expressions subject to less misinterpretation. The most important part of streamlining is a statement that not only explains the use of the streamlined specification format but states only once in the entire specifications the requisite mandatory provisions that are usually repeated ad nauseam in traditional specifications. By requisite mandatory provisions we mean expressions such as "The Contractor shall...," "The Contractor must...," "The Contractor may...." These expressions tell the contractor to do something in different ways, which in a dispute could bring as many interpretations. The explanatory statement of streamlined specifications should be included as an article in the General Conditions, such as:

ART. 64—SPECIFICATIONS EXPLANATION

(a) The Specifications are of the abbreviated, simplified or streamlined type and include incomplete sentences. Omissions of words or phrases, such as "The Contractor shall," "in conformity therewith," "shall be," "as noted on the Drawings," "according to the plans," "a," "an," "the," and "all" are intentional. Omitted words or phrases shall be supplied by inference in the same manner as they are when a "note" occurs on the Drawings.

(b) The Contractor shall provide all items, articles, materials, operations, or methods listed, mentioned, or scheduled either on the Drawings or specified herein, or both, including all labor, materials, equipment, and incidentals necessary and required for their completion.

(c) Whenever the words "approved," "satisfactory," "directed," "submitted," "inspected," or similar words or phrases are used, it shall be assumed that the words "Engineer or his or her representative" follow the verb as the object of the clause, such as "approved by the Engineer or his or her representative."

(d) All references to standard specifications or manufacturer's installation directions shall mean the latest edition at the time of advertisement, unless specifically noted otherwise.

References. "Standard Specifications for Highway Bridges," and "Guide Specifications for Highway Construction," American Association of State Highway and Transportation Officials, 444 N. Capitol St., N.W., Washington, DC 20001. www.transportation.org

"Manual of Practice," including such documents as CSI MASTERFORMAT, Section Format, Construction Documents and the Project Manual, Bidding Requirements, Methods of Specifying, Performance Specifications, Specification Language, and Specification Writing and Production, Construction Specifications Institute, 99 Canal Center Plaza, Suite 300 Alexandria, VA 22314. www.csinet.org

Federal Highway Administration, "Standard Specifications for Construction of Roads and Bridges on Federal Highway Projects," FP-92, U.S. Government Printing Office, Washington, DC 20402. www.gpo.gov

J. Goldbloom, "Engineering Construction Specifications," Van Nostrand Reinhold, New York.

3.10 Word Processing of Specifications

Use of personal computers and word-processing software simplifies, speeds, and lowers cost of specification writing. The information is stored in a manner that enables it to be easily modified and reproduced accurately and efficiently.

A word processor produces normal finished pages (hard copy) of text and concurrently stores the text as files on the computer's hard disk, central server, diskettes, tape, CDs, etc. Diskettes and CDs allow easy transport and sharing of master specifications documents. Diskettes and CDs can be reused many times, but the stored document files should be restored every other year or so to ensure integrity of the stored specification. Document files stored on hard drives, diskettes, and CDs can be retrieved and printed to provide hard copies of the specifications in their latest version.

A first step in establishing a system is preparation of master specifications for storage (Art. 3.5). The stored master specifications are used by specifications writers as a basis for preparing hard copies of project specifications. Using word-processing software, a specification writer edits the master and deletes inapplicable sections.

To facilitate editing, much current word-processing software contains editing assistance called *strikeout* and *underline*. The word processor edits the standard specification document per the specification writer's editing markups. Then, using the word-processing software, the writer compares the edited version with the standard specification. Any deleted information is designated by a strikeout: for example, ~~strikeout~~. Any added information is designated by underline: for example, <u>underline</u>. These features allow the writer to review quickly only those portions that have been modified. Once the editing is completed, the writer can simply eliminate the underlines and the strikeout text to provide a finished specification.

A primary task of the specifications writer when using a specifications system is to constantly upgrade and update the master specifications. The use of the Internet, makes continuous improvement of the quality of specifications a relatively easy task for the specifications writer.

Master specifications are becoming increasingly available from specifications authoring entities via the Internet. Some sites with master specifications available by subscription are located at: www.csinet.org, www.4specs.com, and www.spectext.com.

3.11 Example of a Standard Specification in CSI Format

The following example of a CSI-format standard specification, *Section 02113, Site Preparation*, and modification by special provision is taken from Baltimore Region Rapid Transit System Standard Specifications, Mass Transit Administration, State of Maryland Department of Transportation. (See Art. 3.7.3.)

SECTION 02113—SITE PREPARATION

Part 1: General

1.01 Description:

    A. This Section includes specifications for removal, salvage, demolition in place, or other disposition of basement walls, slabs and footings; existing pavement, curbs and gutters, sidewalks, headwalls, walls, and steps;

utility service facilities; guardrail and posts, highway and street signs and fences; and other miscellaneous structures which interfere with construction, as indicated on the Contract Drawings or as required by the Engineer.

B. Maintenance, support, protection, relocation, reconstruction and adjusting-to-grade, restoration, and abandonment of existing utilities are specified in Section 02550.

C. Subsurface extraction of the items listed in paragraph 1.01.A herein, and salvaging of topsoil, are specified in Section 02200.

Part 2: Products (not used)

Part 3: Execution

3.01 *Removal:*

A. Remove entirely all existing miscellaneous facilities which interfere with construction as shown on the Contract Drawings or designated by the Engineer to be removed.

B. Remove walls and masonry construction to a minimum depth of 12 inches below existing ground level in areas where such items do not interfere with construction.

C. Abandoned Rail and Track Materials: Take possession of, remove, and dispose of, off site all materials between boundaries located two feet outside of the rails including the space between double tracks.

3.02 *Salvage:*

A. Salvage all items designated to be salvaged or determined by the Engineer to be suitable for use in reconstruction, including: grates, frames, other metal castings, and miscellaneous parts of inlets and manholes; hydrants, fire alarm posts and boxes; metal light poles; sound pipe; metal fencing and guard rail; highway and street signs and posts.

B. Protect metallic coatings on salvaged items. Remove adhering concrete from salvaged items.

C. Repair, or replace with new materials, any salvage material damaged or destroyed due to the Contractor's negligence.

3.03 *Demolition in Place:* Slabs may be broken up for drainage and left in place where such method of disposal is determined by the Engineer not to be detrimental to the structural integrity of the fill or structure to be placed above.

3.04 *Backfill:* Backfill trenches and excavations resulting from work under this section in accordance with Section 02200.

3.05 *Disposal of Materials:* Dispose of materials not salvaged or suitable for reuse outside the work site at no additional expense to the Administration.

Part 4: Measurement and Payment

4.01 *Measurement:*

A. Work performed under this Section will be measured by the linear dimension, by areas, by volumes, per each, or by other units appropriate to the item of work, as designated in the Proposal Form.

B. Excavating and backfilling incidental to work under this section will not be separately measured for payment. Subsurface extraction will be measured and paid for under Section 02200.

4.02 *Payment:* Payment for site preparation will be made at the Contract unit prices as indicated above.

The preceding standard specification was modified by a special provision, with the same section number and title, to meet the particular requirements of a specific contract. The following example of a special provision is taken from the Contract Specifications Book, Contract No. NW-02-06, for construction of the Lexington Market Station Structure, part of the Baltimore Region Rapid Transit System.

SECTION 02113—SITE PREPARATION (STATION)

Part 1: General

1.01 *Description:*

A. This Section includes specifications for removal, salvage, demolition in place, or other disposition of existing surface facilities including pavement, streetcar tracks, granite curb, concrete curbs and gutters, sidewalks, walls, street signs, fences, trees and shrubs, and other miscellaneous surface facilities which interfere with construction of the station, as indicated on the Contract Drawings or as required by the Engineer, and not specified elsewhere in other sections of the Specifications. Except as modified herein, the work shall be in accordance with Standard Specifications Section 02113.

B. Streetcar Tracks: Streetcar tracks include any streetcar rail facilities, concrete cable conduit, remnants of cast iron yokes, and concrete between yokes.

Part 2: Products (not used)

Part 3: Execution

3.01 Removal:

A. The requirements specified apply to those existing miscellaneous surface facilities not required to be removed under other sections of the Specifications.

D. Do not use a ball, weight or ram for breaking pavement within five feet of a pavement joint or within three feet of any structure or other pavement that is to remain in place. Protect existing underground utilities. Delineate removal limits of concrete base pavement by saw cutting two inches deep.

E. Stripping: Strip bituminous surfacing materials from existing rigid base pavement where shown on the Contract Drawings.

3.02 Salvage:

D. Maintain and have available for inspection by the Engineer, a detailed record of salvaged items.

E. Salvage granite curb removed during sidewalk and roadway pavement removal and deliver to City of Baltimore Department of Public Works, Special Services Yard, 6400 Pulaski Highway, Baltimore, Maryland.

Part 4: Measurement and Payment

4.01 Measurement:

A. The third line is revised to read: the Unit Price Schedule.

C. Removal of streetcar tracks and removal, salvage and delivery of granite curb will not be separately measured for payment; all work in connection therewith shall be considered incidental to the item of work, Removal of Roadway Pavement.

4.02 Payment: The first and second lines are revised to read: unit prices for the quantities as determined above.

A. Removal of concrete driveways and alleyways will be paid for as Removal of Sidewalk.

B. All work not otherwise paid for will be included for payment in the Contract lump sum price for Site Preparation.

3.12 Example of a Technical Specification Not in CSI Format

The following example illustrates a technical specification (not in CSI format) that was part of the project specifications prepared for the construction of a wharf and approach trestles in the Caribbean area.

SECTION T3—STEEL PIPE PILES

1. **Description.** The work specified in this Section includes the furnishing and driving of closed-end steel pipe piles, including protective coating, test piles, load tests and concrete fill, all as shown on the plans and as specified herein.

2. **Materials.**

   a. *Pipe* for piles shall be new, seamless, steel pipe conforming to the requirements of ASTM Designation A252, Grade 2. Pipe shall be eighteen inches outside diameter with a wall thickness of one-half inch, ordered in double random lengths. Ends of pipe sections shall be perpendicular to the longitudinal axis and shall be beveled as shown on the plans, where required for welded splices. Mill certificates for chemical composition and two certified copies of the records of the physical tests performed on the newly manufactured pipe in accordance with the above ASTM requirements shall be furnished before any driving is started.

b. *Steel Points* for pile tips shall be of cast steel conforming to the requirements of ASTM Designation A27, Grade 65-35. They shall be a standard 60° point with inside flange and two interior cross ribs. Each point shall be marked with the manufacturer's name or identification number. The Contractor shall submit to the Engineer for approval, details of the point he proposes to use.

c. *Splice Rings* as shown on the plans shall be of structural steel conforming to the requirements of ASTM Designation A36.

d. *Concrete* for piles shall be 3,500 psi conforming to the requirements of Section T5, Concrete.

e. *Reinforcement* for cages in the top of piles shall conform to the requirements of Section T5, Concrete.

f. *Welding Electrodes* shall conform to the requirements of the American Welding Society "Specifications for Mild-Steel Covered-Arc Welding Electrodes."

g. *Protective Coatings* shall consist of the following:

(1) Inorganic zinc-rich paint (1 coat), self-curing, with zinc pigment packaged separately, to be mixed at time of application. Zinc dust content to be 75% by weight of total non-volatile content. Acceptable products are Mobilzinc No. 7 by Mobil Chemical Co., No. 92 Tneme-Zinc by Tnemec Co., or Zinc-Rich 220 by USS Chemicals, Div. of U.S. Steel Corp.

(2) Coal-tar epoxy coating (2 coats), to be a two-component amine or polyamide-epoxy coal-tar product, black in color. Acceptable products are Amercoat No. 78 Ameron Corrosion Control Div., Tar-Coat No. 78-J-2 Val-Chem by Mobil Chemical Co., or Tarset No. C-200 by USS Chemicals.

(3) Both the zinc-rich paint and coal-tar epoxy shall conform to the applicable requirements of Federal Spec. MIL-P-23236.

3. Construction Details.

a. *Protective Coatings.* Zinc-rich paint and coal-tar epoxy shall be applied to exterior surfaces of pipe piles, including splice areas, within the respective limits shown on the plans. The Contractor shall apply the protective coatings to a sufficient length of pile sections to insure that the pile when driven to its required resistance, will be protected within the required limits.

Prior to the application of the zinc-rich paint and coal-tar epoxy, bare surfaces shall be blast cleaned to white metal in accordance with the Steel Structures Painting Council Specification No. SP-5.

The zinc-rich paint shall be applied in the shop to a dry-film thickness of 2 mils. The coal-tar epoxy may be applied in the shop or in the field and shall have a total dry film thickness of 16 mils. Coated pile sections shall not be stored in direct sunlight longer than one month without a tarpaulin covering.

Care shall be taken while handling coated pile sections during loading, transporting, unloading and placing, so that the protective coating is not penetrated or removed. Coated pile sections shall be inspected before placing in the leads and any damaged surfaces shall be repaired and recoated to the satisfaction of the Engineer.

The Contractor's attention is directed to the "Hazardous Warning Label" on the coal-tar epoxy products and the manufacturer's literature regarding the use of protective clothing, gloves, creams and goggles during mixing, application and cleanup.

The cured coal-tar epoxy coating will be tested by the Engineer to determine resistance to film removal by a mechanical force, as follows:

(1) Lay a sharp wood chisel almost flat on the coating surface in line with the pipe length.

(2) Drive the chisel using a hammer, through the coating and along the substrate.

(3) If the coating film is acceptably bonded to the surface, considerable force will be required to lift a layer of the film.

(4) Portions of the coating should remain in the valleys of the blast pattern adhering to surface for an acceptable test.

(5) The tested area shall be repaired as per these specifications by the Contractor.

(6) The number of tests will be limited to two acceptable tests for each shipment or for each day's field application of coating.

b. *Preparation for Driving*

(1) Piles shall not be driven in any area until all necessary excavation or grading has been completed.

(2) *Pile Points:* The tip of every pile shall be closed with an approved pile point, welded in place to produce a watertight joint.

(3) *Splices:* The number of splices shall be kept to the practical minimum. The number and location of splices will be subject to the approval of the Engineer. Splices shall be made with full strength butt welds utilizing an internal steel back-up splice ring as shown on the plans. Should the Contractor desire to use an alternate splice design, he shall submit full details of his proposed splice to the Engineer for approval. All splices shall be watertight.

(4) *Welding:* Welding shall conform to the applicable requirements of the current edition of the American Welding Society "Specifications for Welded Highway and Railway Bridges." Welders shall be qualified for the work, as prescribed in the AWS Specifications.

c. *Equipment for Driving:* All equipment shall be subject to the approval of the Engineer. Piles shall be driven with a single-acting hammer which shall develop a manufacturer's rated energy per blow at full stroke of not less than 30,000 foot-pounds. The striking weight shall be not less than 10,000 pounds.

Sufficient boiler or compressor capacity must be provided at all times to maintain the rated speed of the hammer during the full time of driving a pile. The valve mechanism and other parts of the hammer shall be maintained in first-class condition so that the length of stroke for which the hammer is designed will be obtained.

Piles shall be driven with leads constructed in such a manner as to afford freedom of movement of the hammer. Leads shall be held in position by guys or stiff braces to give the required support to the pile during driving. Inclined leads shall be used for driving batter piles. Leads shall be of sufficient length, as the use of a follower will not be permitted.

Water jets shall not be used for pile penetration unless authorized by the Engineer. When water jets are authorized, the Contractor shall submit to the Engineer for approval full details of his proposed jetting operation. In no event shall a pile be jetted within ten feet of its anticipated final tip elevation.

d. *Accuracy of Driving:* Completed piles at the cut-off elevation shall not vary from the plan locations by more than three inches. Piles shall be driven with a variation of not more than one-eighth inch per foot from the vertical or from the batter shown on the plans or as directed by the Engineer.

Piles shall not be subjected to force in order to place them in correct alignment or horizontal position. Piles exceeding the allowable tolerances will be considered unacceptable unless the Contractor submits a satisfactory working plan showing the corrective work he proposes. Such work shall not proceed until the working plan has been approved by the Engineer.

e. *Defective Piles:* Piles damaged by reason of internal defects or by improper handling or driving will be rejected. Corrective measures shall be submitted by the Contractor to the Engineer for approval. Approved corrective measures undertaken by the Contractor shall be at no additional cost to the owner.

f. *Limitations of Driving:* The Contractor's attention is directed to the existence of cement-waste fill material in the proposed work area, as indicated in the boring logs. All piles shall penetrate this layer. The Contractor shall take the necessary measures to accomplish this penetration subject to the approval of the Engineer.

g. *Lengths of Piles:* The lengths of piles indicated in the Proposal are for estimating purposes only. The actual lengths of piles necessary will be determined in the field by driving the pile sections to the required resistance established by the test piles and pile load tests.

h. *Pile Cut-offs:* Pile cut-offs may be used in other piles. However, useable cut-offs must be at least ten feet in length and only one cut-off length will be permitted in any one pile.

i. *Driving:* Driving of a pile shall be continuous as far as practicable. When driving is resumed after an interruption, the blow count shall not be taken into consideration until the temporary set of the pile resulting from the interruption has been broken.

Piles shall not be driven within 60 feet of concrete that is less than 7 days old.

Piles shall be driven for the last six inches to the resistance determined from the test piles and pile load tests and as established by the Engineer.

All piles forced up by any cause shall be driven down again as directed by the Engineer and any such costs shall be included in the unit price bid for the piles.

j. *Inspection.* The Contractor shall have available at all times a suitable drop-light for the inspection of each pile throughout its entire length.

k. *Concrete:* No concrete shall be placed in a pile until it has been inspected and accepted by the Engineer. Accumulations of water in the pile shall be removed before concrete placement. Concrete, 3,500 psi, shall be mixed and conveyed as specified in Section T5, *Concrete.* Concrete shall be placed continuously in each pile to the extent that there will be no cold joints. The slump shall not exceed 3 inches. Special care shall be exercised in filling the piles to prevent honeycomb and air pockets from forming in the concrete. Internal vibration and other means shall be used to the maximum depth practicable, to consolidate the concrete.

Should the Contractor be unable to remove water from within the pile to enable the concrete to be placed in "the dry," he shall submit details of his proposed tremie operation for filling the pile.

l. *Cutting off:* The tops of piles shall be cut off at the elevations shown on the plans.

m. *Reinforcement:* The tops of piles shall be reinforced as shown on the plans. The reinforcing steel shall be secured in such a manner as to insure its proper location in the finished piles.

n. *Test Piles:* Test piles shall be driven at the locations shown on the plans or directed by the Engineer, for determining approximate pile lengths. In addition, test piles will be load tested to verify the bearing value of the driven pile.

o. *Pile Load Tests:* Load tests shall be performed in accordance with the requirements of ASTM Designation D1143, "Load-Settlement Relationship for Individual Vertical Piles Under Static Axial Load" as modified herein:

(1) Pretest Information specified in Section 2 will not be required.

(2) Under Section 5, Procedure:

   (a) A time period of at least 7 days shall elapse between driving and loading the test pile.

   (b) The test pile shall be filled with concrete at least 3 days before loading.

   (c) No further loading beyond 200% of the design working load of 150 tons will be required.

   (d) Intermediate loads shall not be removed.

   (e) The full test load shall remain in place a minimum of 24 hours, as determined by the Engineer.

   (f) A final rebound reading shall be recorded 24 hours after the entire test load has been removed.

   (g) The increase in loading shall be applied at a uniform rate with no sudden load impact. Reducing the load shall be handled in the same manner.

   The Contractor shall submit to the Engineer full details of his proposed method of performing the load tests, including arrangement of equipment.

The safe bearing capacity of the test pile will be considered as one-half that test load which produces a permanent settlement of the top of the pile of not more than one-quarter inch.

4. Method of Measurement.

a. The quantity of 18-in. steel pipe piles to be paid for will be the number of linear feet of piles, including test piles in the completed structure, installed in accordance with the plans and specifications, measured from the point of the pile to cut-off.

b. The quantity of Pile Load Tests to be paid for will be the number of completed tests performed in accordance with the plans and specifications.

5. Basis of Payment.

a. The unit price bid per linear foot of 18-in. steel pipe piles shall include the cost of furnishing all labor, materials and equipment necessary to complete the work, including protective coatings, pile points, splices, concrete, reinforcement, jetting when authorized, corrective measures, unused pile cut-offs and test piles.

b. The unit price bid per each Pile Load Test shall include the cost of furnishing all labor, materials and equipment necessary to complete the work including the removal of all temporary materials and equipment.

3.13 Qualifications for Specifications Engineers

A review of the character and function of specifications bears witness to the knowledge specifications engineers must have of the proposed work and the conditions under which it must be accomplished, the materials and methods of construction that may be used, and the owner's prescribed procedures for administering the contract. In addition to technical skill, a major requisite of a specifications engineer is ability to convey full understanding of the contract to others: engineers, constructors, workers, lawyers, financiers, and the general public. Writing ability is an important element because specifications are of little value unless they can be clearly understood.

Specifications writers for civil construction should be graduate civil engineers with design and broad field experience. Mechanical and electrical engineers and architects should prepare the technical input to the specifications for their respective fields.

A specifications engineer should have a minimum of 10 years' exposure to construction practices, preferably as a representative of the owner. At least 3 to 5 years should have been served as a resident engineer, interpreting, enforcing, and defending the project specifications. The specifications engineer will thus have acquired an appreciation of the part that specifications play in the development, construction, and successful completion of projects.

Basically, contractors want to know what they are required to do under the terms of the contract and how they are to be paid for it. The more clearly and simply this information can be presented in the contract documents, the less likelihood of problems, delays, and claims developing on the job.

The Construction Committee of the U.S. Committee on Large Dams stated in Paper 8781, published by the American Society of Civil Engineers:

The proper framing of a set of construction specifications is not easy. Engineering specialists called specifications writers are employed for that purpose, and their work requires good judgment, a broad knowledge of the technical aspects of the job, and appreciation of the construction problems plus the ability to express clearly and concisely all of the terms, conditions, and provisions necessary to present an accurate picture to the constructor. It is a very large order.

4

Jonathan T. Ricketts
Consulting Engineer
Palm Beach Gardens, Florida

CONSTRUCTION MANAGEMENT

Construction is the mobilization and utilization of capital and specialized personnel, materials, and equipment to assemble materials and equipment on a specific site in accordance with drawings, specifications, and contract documents prepared to serve the purposes of a client. The organizations that perform construction usually specialize in one of four categories into which construction is usually divided: **housing**, including single-family homes and apartment buildings; **nonresidential building**, such as structures erected for institutional, educational, commercial, light-industry, and recreational purposes; **engineering construction**, which involves works designed by engineers and may be classified as **highway construction** or **heavy construction** for bridges, tunnels, railroad, waterways, marine structures, etc.; and **industrial construction**, such as power plants, steel mills, chemical plants, factories, and other highly technical structures. The reason for such specialization is that construction methods, supervisory skills, labor, and equipment are considerably different for each of the categories.

Construction involves a combination of specialized organizations, engineering science, studied guesses, and calculated risks. It is complex and diversified and the end product typically is nonstandard. Since operations must be performed at the site of the project, often affected by local codes and legal regulations, every project is unique. Furthermore, because of exposure to the outdoors, construction is affected by both daily and seasonal weather variations. It is also often influenced significantly by the availability of local construction financing, labor, materials, and equipment.

Construction Management can be performed by construction contractors, construction consultants also known as construction managers, or design build contractors. All of these individuals or entities have as their goal the most efficient, cost effective completion of a given construction project. Construction contractors typically employ supervisory and administrative personnel, labor, materials and equipment to perform construction in accordance with the terms of a contract with a client, or owner. Construction managers may provide guidance to an owner from inception of the project to completion, including oversight of design, approvals, and construction, or just provide construction advisory services to an owner. A construction manager may also act as an agent for the owner, contracting with others for performance of the work and provide administrative and supervisory services during construction. A design build entity can provide all of the above-mentioned activities providing a completed project for the owner with a single contract through one entity.

4.1 Tasks of Construction Management

Construction management can involve the planning, execution, and control of construction operations for any of the aforementioned types of construction.

Planning requires determination of financing methods, estimating of construction costs, scheduling of the work, and selection of construction methods and equipment to be used. Initially, a detailed study of the contract documents is required, leading to compilation of all items of work to be performed and grouping of related items in a master schedule. This is followed by the establishment of a sequence of construction operations. Also, time for execution is allotted for each work item. Subsequent planning steps involve selection of construction methods and equipment to be used for each work item to meet the schedule and minimize construction costs; preparation of a master, or general, construction schedule; development of schedules for procurement of labor, materials, and equipment; and forecasts of expenditures and income for the project.

In planning for execution, it is important to recognize that not only construction cost but also the total project cost increases with duration of construction. Hence, fast execution of the work is essential. To achieve this end, construction management must ensure that labor, materials, and equipment are available when needed for the work. Construction management may have the general responsibility for purchasing of materials and equipment and expediting their delivery not only to the job but also to utilization locations. For materials requiring fabrication by a supplier, arrangements should be made for preparation and checking of fabrication drawings and inspection of fabrication, if necessary. Also, essential for execution of construction are layout surveys, inspection of construction to check conformance with contract documents, and establishment of measures to ensure job safety and that operations meet Occupational Safety and Health Act (OSHA) regulations and environmental concerns. In addition, successful execution of the work requires provision of temporary construction facilities. These include field offices, access roads, cofferdams, drainage, utilities and sanitation, and design of formwork for concrete.

Control of construction requires up-to-date information on progress of the work, construction costs, income, and application of measures to correct any of these not meeting forecasts. Progress control typically is based on comparisons of actual performance of construction with forecast performance indicated on master or detailed schedules. Lagging operations generally are speeded by overtime work or addition of more crews and equipment and expedited delivery of materials and equipment to be installed. Cost and income control usually is based on comparisons of actual costs and income with those budgeted at the start of the project. Such comparisons enable discovery of the sources of cost overruns and income shortfalls so that corrective measures can be instituted.

Role of Contractors ■ The client, or owner, seeking construction of a project, contracts with an individual or construction company for performance of all the work and delivery of the finished project within a specific period of time and usually without exceeding estimated cost. This individual or company is referred to as a general contractor.

The general contractor primarily provides construction management for the entire construction process. This contractor may supply forces to perform all of the work, but usually most of the work is subcontracted to others. Nevertheless, the contractor is responsible for all of it. Completely in charge of all field operations, including procurement of construction personnel, materials, and equipment, the contractor marshals and allocates these to achieve project completion in the shortest time and at the lowest cost.

The contractor should have two prime objectives: (1) provision to the owner of a service that is satisfactory and on time; (2) making a profit.

Construction Manager ■ This is a general contractor or construction consultant who performs construction management under a professional service contract with the owner. When engaged at the start of a project, the construction manager will be available to assist the owner and designers by providing information and recommendations on construction technology and economics. The construction manager can also prepare cost estimates during the preliminary design and design development phases, as well as the final cost estimate after completion of the contract documents. Additional tasks include recommending procurement of long-lead-time materials and equipment to ensure delivery when needed; review of plans and specifications to avoid conflicts and overlapping in the work of subcontractors; preparing a progress schedule for all project activities of the owner, designers, general contractor, subcontractors, and construction manager; and providing all concerned with periodic reports of the status of the job relative to the project schedules. Also, the construction manager, utilizing knowledge of such factors as local labor availability and overlapping trade jurisdictions, can offer recommendations concerning the division of work in the specifications that will facilitate bidding and awarding of competitive trade contracts. Furthermore, on behalf of the owner, the manager can take and analyze competitive bids on the work and award or recommend to the owner award of contracts.

During construction, the construction manager may serve as the general contractor or act as an agent of the owner to ensure that the project meets the requirements of the contract documents, legal regulations, and financial obligations. As an agent of the owner, the construction manager assumes the duties of the owner for construction and organizes a staff for the purpose. Other functions of construction management are to provide a resident engineer, or clerk of the works; act as liaison with the prime design professional, general contractor, and owner; keep job records; check and report on job progress; direct the general contractor to bring behind-schedule items, if any, up to date; take steps to correct cost overruns, if any; record and authorize with the owner's approval, expenditures and payments; process requests for changes in the work and issue change orders; expedite checking of shop drawings; inspect construction for conformance with contract documents; schedule and conduct job meetings; and perform such other tasks for which an owner would normally be responsible.

(D. Barry and B. C. Paulson, Jr., "Professional Construction Management," 2nd ed., G. J. Ritz, "Total Project Management," and S. M. Levy, "Project Management in Construction," 3rd ed., McGraw-Hill, Inc., New York (books.mcgraw-hill.com).)

4.2 Organization of Construction Firms

The type of organization employed to carry out construction is influenced by considerations peculiar to that industry, many of which are unlike those affecting manufacturing, merchandising, or distribution of goods. This is due largely to the degree of mobility required, type of risk inherent in the particular type of construction, and geographic area to be served.

4.2.1 Contractor Organization as a Business

These contracting entities employ the usual business forms. Perhaps the greater number are sole proprietorships, where one person owns or controls the enterprise. Many others are partnerships, where two or more individuals form a voluntary association to carry on a business for profit. The corporate form has a particular appeal to both large- and small-scale enterprises operating in the construction field. To the large enterprise, corporate structure is an easier way to finance itself by dividing ownership into many small units that can be sold to a wide economic range of purchasers, including those with only small amount of capital to invest. In addition to assisting financing operations, the corporate device brings a limited liability to the persons interested in the enterprise and a perpetual succession not affected by the death of any particular owner or by the transfer of any owner's interest. Because of these features, the corporate vehicle is also used by numerous small contractors.

4.2.2 Special Considerations in Organization for Construction

Each facility that a construction team produces, it produces only once; the next time its work will be done at a new location, to a new pattern, and under new, although often similar, specifications. Furthermore, from the very inception of each construction project, contractors are wholly devoted to completion of the undertaking as quickly and economically as possible and then moving out.

The problems of construction differ from those of industrial-type businesses. The solutions can best be developed within the construction industry itself, recognizing the unique character of the construction business, which calls for extreme flexibility in its operations. Based on foundations resting within the industry itself, the construction industry has erected organizational structures under which most successful contractors find it necessary to operate. They tend to take executives away from the conference table and put them in close touch with the field. This avoids the type of organizational bureaucracy that hinders rapid communication between office and field and delays vital decisions by management.

Contractor workforces usually are organized by crafts or specialty work classifications. Each unit is directed by a supervisor who reports to a general construction superintendent (Fig. 4.1).

The general construction superintendent is in charge of all actual construction, including direction of the production forces, recommendation of construction methods, and selection of personnel, equipment, and materials needed to accomplish the work. This superintendent supervises and coordinates the work of the various craft superintendents and foremen. The general construction superintendent reports to management, or in cases where the magnitude or complexity of the project warrants, to a project manager, who in turn reports to management. To enable the general construction superintendent and project manager to achieve efficient on-the-job production of completed physical facilities, they must be backed up by others not in the direct line of production.

Figure 4.1 is representative of the operation of a small contracting business where the sole proprietor or owner serves as general construction superintendent. Such owners operate their businesses with limited office help for payroll preparation. They may do their own estimating and make commitments for major purchases, but often they use outside accounting and legal services.

As business expands and the owner undertakes larger and more complex jobs, more crafts, functions, or work classifications are involved than can be properly supervised by one person. Accordingly, additional crews with their supervisors may be grouped under as many craft superintendents as required. The latter report to the general construction superintendent, who in turn reports to the project manager, who still may be the owner (Fig. 4.2).

Along with this expansion of field forces, the owner of a one-person business next finds that the volume and complexities of the growing business require specialized support personnel who have to perform such services as:

1. Purchasing, receiving, and warehousing permanent materials to be incorporated into the completed project, as well as purchasing, receiving, and warehousing goods and supplies consumed or required by the contractor in doing the work

2. Timekeeping and payroll, with all the ramifications arising out of federal income tax and Social Security legislation, and detail involved in contracts with organized labor

3. Accounting and auditing, financing, and tax reporting

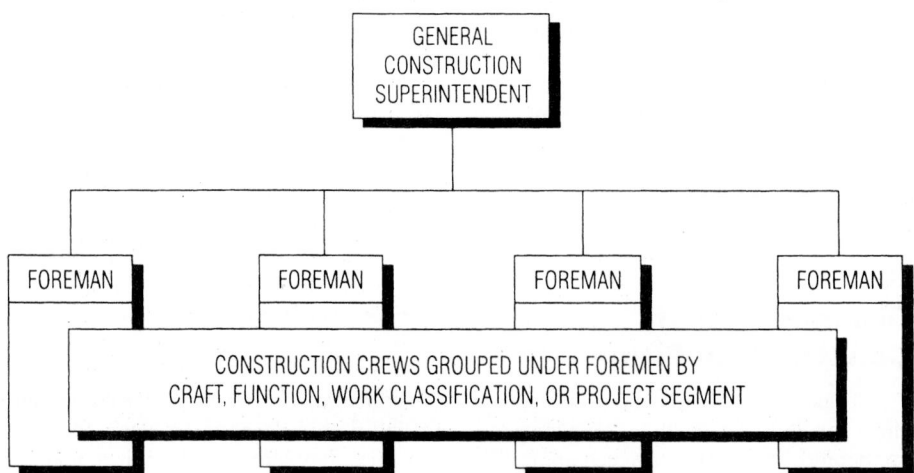

Fig. 4.1   Basic work-performing unit and organization for a small construction company.

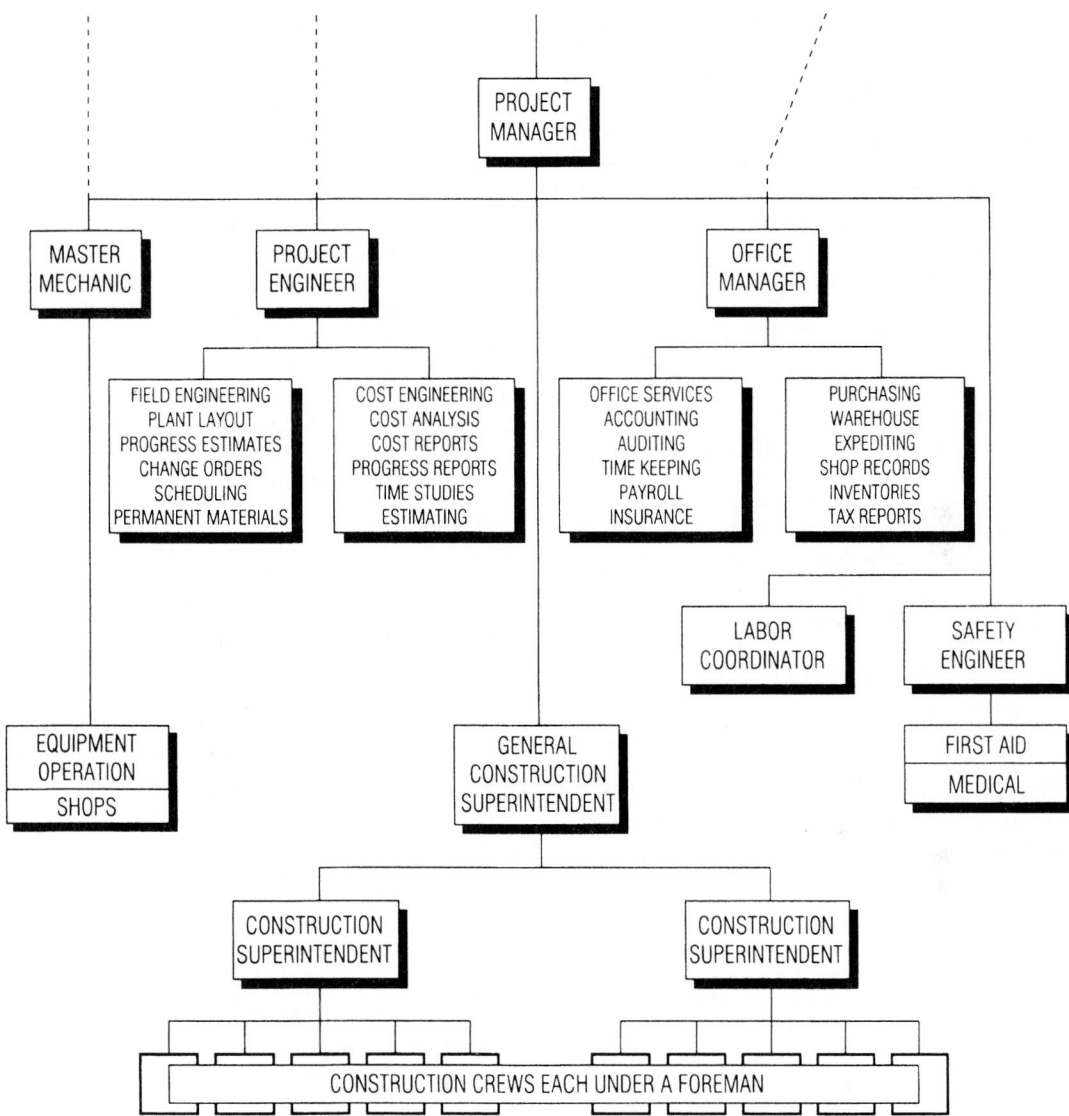

Fig. 4.2 Project organization, with that for the smallest unit as shown in Fig. 4.1.

4. Engineering estimating, cost control, plant layout, etc.

5. Accident prevention, labor relations, human resources etc.

To coordinate the operation of support staff required for general administration of the business and servicing of its field forces, the head of the organization needs freedom from the direct demands of on-the-job supervision of construction operations. This problem may be solved by employing a general construction superintendent or project manager or by entering into a partnership with an outside person capable of filling that position, with the owner taking the overall management position.

Further growth may find the company operating construction jobs simultaneously at a number of locations. Arrangements for the operation of this type of business take the form of an expanded headquarters organization to administer and control the jobs and service the general construction superintendent or project manager at each location. This concept contemplates, in general, delegation to the field of those duties and responsibilities that cannot best be executed by the headquarters function.

Accordingly, the various jobs usually have a project manager in charge (Fig. 4.2). On small jobs, or in those cases where the general construction superintendent is in direct charge, the project manager is accompanied by service personnel to perform the functions that must be conducted in the field, such as timekeeping, warehousing, and engineering layout.

Some large construction firms, whose operations are regional, nationwide, or worldwide in scope, delegate considerable authority to operate the business to districts or divisions formed on a geographical or functional basis (Fig. 4.3). District managers, themselves frequently corporate officers, are responsible to the general management of the home office for their actions. But they are free to conduct the business within their jurisdiction with less detailed supervision, although within definite confines of well-established company policies. The headquarters office maintains overall administrative control and close communication but constructs projects by and through its district organizations (Fig. 4.4).

4.2.3 Joint Ventures

Since risk is an important factor in construction, it is only prudent to spread it as widely as possible. One safeguard is a joint venture with other contractors whenever the financial hazard of any par-ticular project makes such action expedient. In brief, a joint venture is a short-term partnership arrangement wherein each of two or more participating construction companies is committed to a predetermined percentage of a contract and each shares proportionately in the final profit or loss. One of the participating companies acts as the manager or sponsor of the project.

4.2.4 Business Consultants

Contractors often employ experts from various disciplines to advise them on conduct of their business. For example, in addition to the usual architectural and engineering consultants, contractors consult the following:

Accountant ▪ Preferably one experienced in construction contracting, the accountant should be familiar with the generally accepted principles of accounting applicable to construction projects, such as costs, actual earnings, and estimated earnings on projects still in progress. Also, the accountant should be able to help formulate the financial status of the contractor, including estimates of the probable earnings from jobs in progress and the amounts of reserves that should be provided for contingencies on projects that have been completed but for which final settlements have not been made with all the subcontractors and suppliers.

Attorneys ▪ More than one attorney may be needed to handle a contractor's legal affairs. For example, the contractor may require an attorney for most routine matters of corporate business, such as formation of the corporation, registration of the corporation in other states, routine contract advice, and legal aid in general affairs. In addition, the company may need different attorneys to handle claims, personal affairs, estate work, real estate matters, taxes, and dealings with various government bodies.

Insurance and Bonding Brokers ▪ Contractors would be well advised to select an insurance broker who manages a relatively large volume of general insurance. This type of broker can be expected to have large leverage with insurance companies when conditions are encountered involving claims for losses or when influence is needed in establishment of premiums at policy renewal time.

For bonding matters, however, contractors will find it advisable to select a broker who specializes in bonding of general contractors and would be helpful in solving their bonding problems. Bonding and general insurance involve entirely different principles. A broker who provides many clients with performance and payment bonds should be able to recommend bonding and insurance companies best suited for the contractor's needs. Also, the broker should be able to assist the contractor and the contractor's accountant in preparation of financial statements with the objective of showing the contractor's position most favorably for bonding purposes.

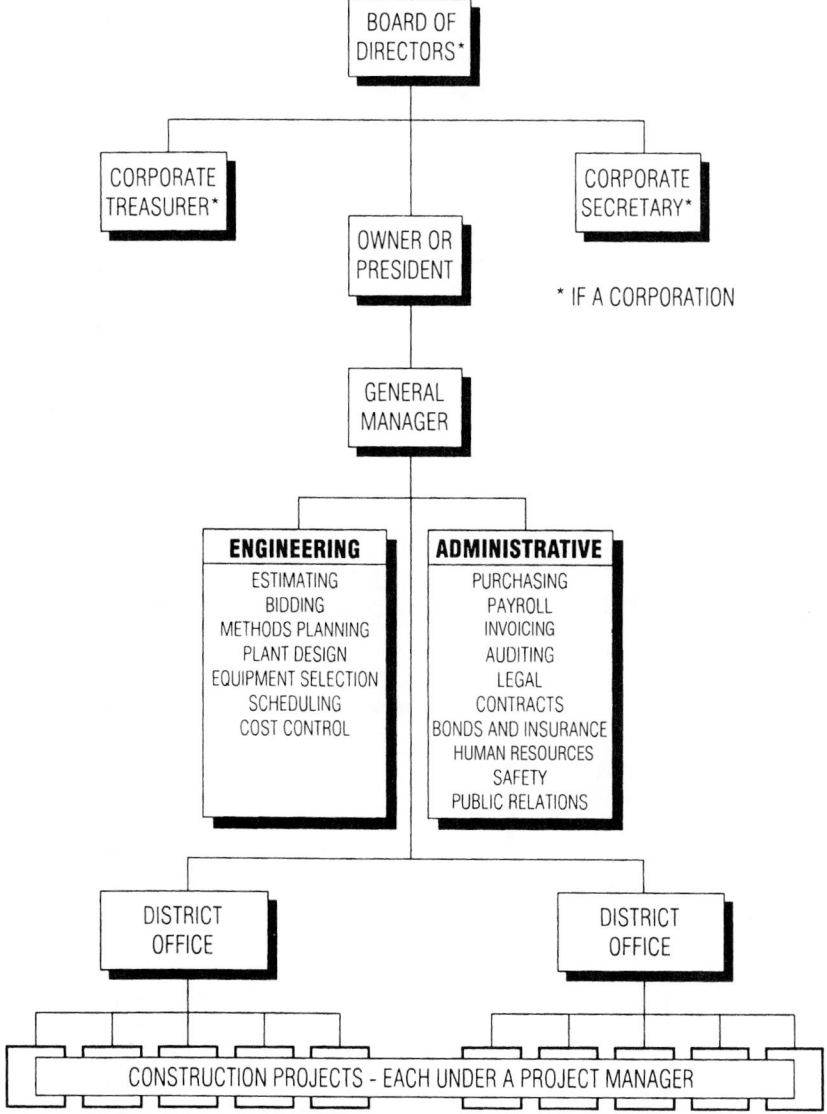

Fig. 4.3 District-type organization, with district offices organized as shown in Fig. 4.4 and projects as indicated in Fig. 4.2.

4.3 Nature and Significance of a Proposal

Contractors obtain most of their business from offers submitted in response to invitations to bid issued by owners, both public and private (Art. 3.8). Inasmuch as award is usually made to the "lowest bidder" or "lowest responsible bidder," the contractor is constantly faced with the likelihood of failing to secure the business if a bid is too high. On the other hand, the contractor risks financial loss in executing the work if a bid is low enough so that the contract is awarded. Therefore, the submission of a proposal is a commitment of far-reaching

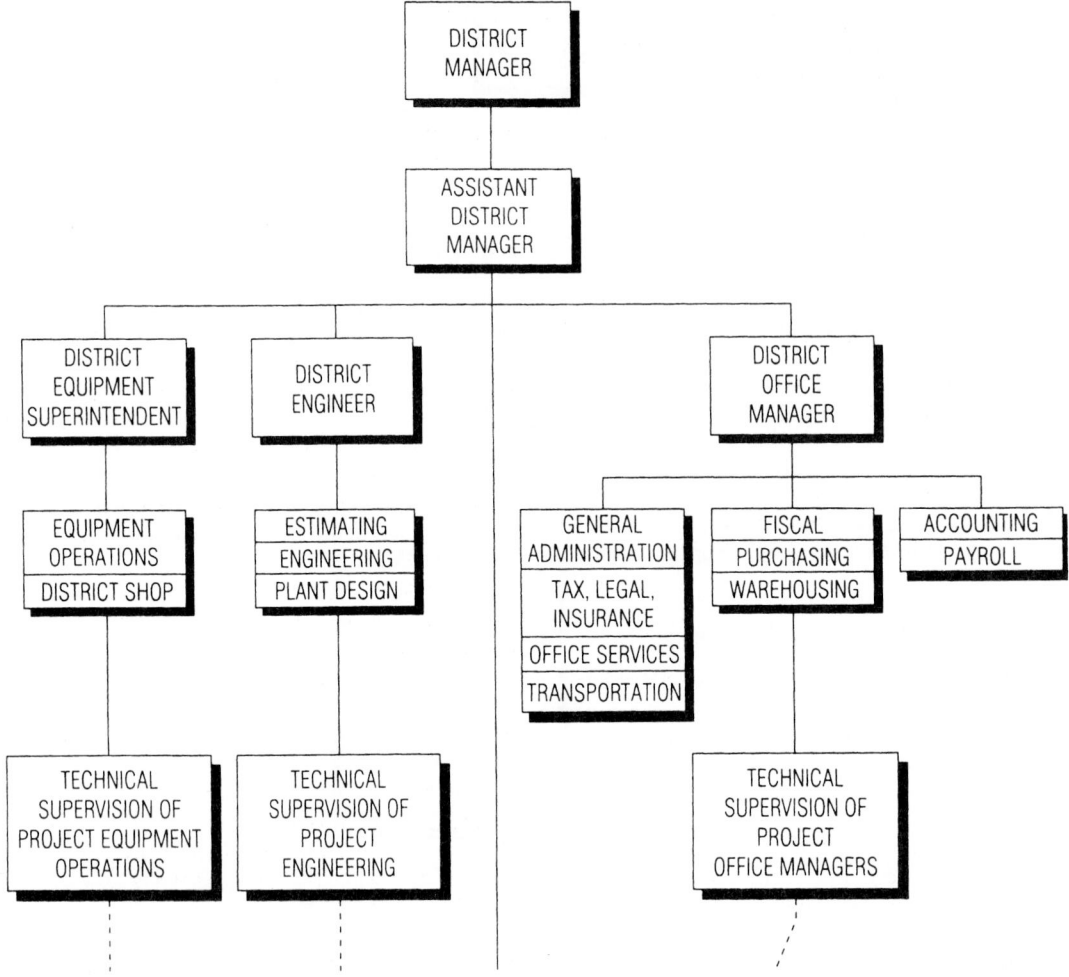

Fig. 4.4 District-type organization for a construction company.

significance. The contractor is responsible for the consequences of such mistakes as may be made as well as those risks inherent in construction over which the contractor may have no control.

A proposal is an offer made by the contractor to the owner to perform the work required by the contract documents for a stated sum of money. Furthermore, the proposal is a promise by the contractor that upon acceptance of the proposal by the owner, the contractor will enter into a contract and perform the work for the stated remuneration. Note that the proposal and its acceptance, together with the monetary consideration, constitute the essential elements of a contract between competent

parties. Ordinarily, a proposal is effective until it is rejected by the owner. Most owners, however, provide in their invitations for bid that award of contract will be made within a stipulated period of time, such as 30 days after the opening date.

By furnishing the form of proposal to be used by contractors in submitting bids and stipulating how it must be completed, the owner intends to put all bids on the same basis, thereby permitting equitable comparison and selection for award of contract. Although the time allotted for preparation of the estimate and submission of bid is seldom regarded as sufficient by the contractor, it is nonetheless incumbent upon the contractor to prepare the proposal in strict conformity with instructions in the invitation to bidders and other documents. Failure to do so may result in disqualification of the bid on the grounds of irregularity, with a resulting loss of the time and money expended in the preparation of the bid.

Bid Alternatives ▪ In addition to the basic bid, the owner may call for prices on alternative materials, equipment, or work items. These prices may be either added to or deducted from the base bid. This device is generally employed as a means of ensuring that an award can be made within the amount of the owner's available funds. It serves also as an aid to selection by the owner after having the benefit of firm prices on the various alternatives. Accordingly, figures quoted by the contractor on alternatives should be complete within themselves, including overhead and profit.

4.4 Prime Contracts

A construction contract is an agreement to construct a definite project in accordance with plans and specifications for an agreed sum and to complete it, ready for use and occupancy, within a certain time. Although contracts may be expressed or implied, oral or written, agreements between owners and contractors are almost universally reduced to writing. Their forms may vary from the simple acceptance of an offer to the usual fully documented contracts in which the complete plans, specifications, and other instruments used in bidding, including the contractor's proposal, are made a part of the contract by reference.

Recognizing that there are advantages to standardization and simplification of construction contracts, the Joint Conference on Standard Construction Contracts prepared standard documents for construction contracts intended to be fair to both parties. The American Institute of Architects also has developed standard contract documents. And the Contract Committees of the American Society of Municipal Engineers and the Associated General Contractors of America have proposed and approved a Standard Code for Municipal Construction.

Contractors generally secure business by submitting proposals in response to invitations to bid or by negotiations initiated by either party without

formal invitation or competitive bidding. Agencies and instrumentalities of the federal government and most state and municipal governments, however, are generally required by law to let construction contracts only on the basis of competitive bidding. However, certain federal agencies, for security reasons or in an emergency, may restrict bidders to a selected list, and, in these cases, may not open bids in public.

Normally, competitive bidding leads to fixed-price contracts. These may set either a lump-sum price for the job as a whole or unit prices to be paid for the number of prescribed units of work actually performed. Although negotiated contracts may be on a lump-sum or unit-price basis, they often take other forms embodying devices for making possible start of construction in the absence of complete plans and specifications, for early-completion bonus, or for profit-sharing arrangements as incentives to the contractor (see also Art. 3.3).

One alternative often used is a cost-plus-fixed-fee contract. When this is used, the contractor is reimbursed for the cost plus a fixed amount, the fee for accomplishment of the work. After the scope of the work has been clearly defined and both parties have agreed on the estimated cost, the amount of the contractor's fee is determined in relation to character and volume of work involved and the duration of the project. Thereafter the fee remains fixed, regardless of any fluctuation in actual cost of the project. There is no incentive for the contractor to inflate the cost under this type of contract since the contractor's fee is unaffected thereby. But maximum motivation toward efficiency and quick completion inherent in fixed-price contracts may be lacking.

A profit-sharing clause is sometimes written into the cost-plus-fixed-fee contracts as an incentive for the contractor to keep cost at a minimum, allowing the contractor a share of the savings if the actual cost, upon completion, underruns the estimated cost. This provision may also be accompanied by a penalty to be assessed against the contractor's fee in case the actual cost exceeds the agreed estimated cost.

A fundamental requirement for all cost-plus-fixed-fee contract agreements is a definition of cost. A clear distinction should be made between reimbursable costs and costs that make up the contractor's general expense, payable out of the contractor's fee. Some contracts, which would otherwise run smoothly, become difficult because of failure to define cost clearly. Usually, only the cost directly and solely assignable to the project is reimbursed to the contractor. Therefore, the contractor's central office overhead and general expense, salaries of principals and headquarters staff, and interest on capital attributable to the project frequently come out of the fee, although a fixed allowance in cost for contractor's home-office expense may be allowed.

Cost-plus-fixed-fee contracts do not guarantee a profit to the contractor. They may also result, particularly in government cost-plus-fixed-fee contracts, in unusually high on-job overhead occasioned by frequent government requirements for onerous and cumbersome procedures in accountability and accounting.

(B. M. Jervis and P. Levin, "Construction Law: Principles and Practice," M. Millman, "General Contracting: Winning Techniques for Starting and Operating a Successful Business," and M. Stokes, "Construction Law in Contractor's Language," 2nd ed., McGraw-Hill, Inc., New York (books.mcgraw-hill.com).)

4.5 Subcontracts

General contractors generally obtain subcontract and material-price bids before submitting a bid for a project to the owner. Usually, these bids are incorporated into the subcontracts. (Sometimes, general contractors continue shopping for subcontract bids after the award of the general contract, to attain budget goals that may have been exceeded by the initial bids.)

For every project, the contractor should keep records of everything to be purchased for the job and prepare a budget for each of the items. As each subcontract is awarded, the contractor should enter the subcontractor's name and the amount of the subcontract. Later, the profit or loss on the purchase should be entered in the record, thus maintaining a continuous tabulation of the status of the purchase. For convenience, priority numbers may be assigned to the various items, in order of preference in purchasing. Examination of the numbers enables a contractor to concentrate efforts on the subcontracts that must be awarded first.

Contractors typically solicit bids from subcontractors employed previously with satisfactory results and through notices in trade publications, such as *The Dodge Bulletin*. If the owner or the law requires use of specific categories of

subcontractors, the contractor must obtain bids from qualified members of such categories. After receipt of subcontractor bids, the contractor should analyze and tabulate them for fair comparison. To make such a comparison, the contractor should ensure that the bidders for a trade are including the same items. For this purpose, the contractor should question each of the bidders, when necessary, and from the answers received tabulate the exact items that are included in or excluded from each bid. Although this may seem obvious, it should be reiterated that a good construction manager may alter the division of work among subcontractors to receive the most cost-efficient completion of work. If a subcontractor's proposal indicates that a portion of the work is being omitted, the contractor should cross-check the specifications and other trades to be purchased to determine if the missing items are the province of other subcontractors.

Various forms are available for use as subcontract agreements. The standard form, "Contractor-Subcontractor Agreement," A401, American Institute of Architects, is commonly used. A subcontract rider tailored for each job usually is desirable and should be initialed by both parties to the contract and attached to all copies of the subcontract. The rider should take into account modifications required to adapt the standard form to the job. It should cover such items as start and completion dates, options, alternatives, insurance and bonding requirements, and special requirements of the owner or leading agency.

To achieve a fair distribution of risks and provide protective techniques for the benefit of both parties, it is necessary for subcontracts to be carefully drawn. The prime contractor wishes to be assured that the subcontractor will perform in a timely and efficient manner. On the other hand, the subcontractor wishes to be assured of being promptly and fairly compensated and that no onerous burdens of performance or administration will be imposed.

Basic problems arise where parties fail to agree with respect at least to the essentials of the transaction, including the scope of work to be performed, price to be paid, and performance. The subcontract must include the regulatory requirements of the prime contract and appropriate arrangements for price, delivery, and specifications. It is insufficient to assume that writing a subcontractor a purchase order binds that subcontractor to the terms of the prime contractor's agreement. Subcontracts should be explicit with respect to observance of the prime contract. Also, subcontractors should be fully informed by being furnished with the prime-contract plans, specifications, and other construction documents necessary for a complete understanding of the obligations to which they are bound.

Although prime contracts often provide for approval of subcontractors as to fitness and responsibility, the making of a subcontract establishes only indirect relationships between owner and subcontractor. The basis upon which subcontract agreements are drawn on fixed-price work is of no concern to the owner because the prime contractor, by terms of the agreement with the owner, assumes complete responsibility. Under cost-plus prime contracts, however, subcontracts are items of reimbursable cost. As such, their terms, particularly the monetary considerations involved, are properly subject to the owner's approval.

Subcontract agreements customarily define the sequence in which the work is to be done. They also set time limits on the performance of the work. Nevertheless, prime contractors are reluctant to delegate by means of subcontracts portions of a project where failure to perform might have serious consequences on completion of the whole project—for example, the construction of a tunnel for diversion of water in dam construction.

In the heavy-construction industry, the greater the risk of loss from failure to perform, the less work is subcontracted. Such damages as may be recovered under subcontract agreements for lack of performance are usually small recompense for the overall losses arising out of the detrimental effect on related operations and upon execution of the construction project as a whole.

This situation has given rise to a common trade practice in the heavy-construction industry: The prime contractor builds up a following of subcontractors known for their ability to complete commitments properly and on time and generally to cooperate with and fit into the contractor's job operating team. The prime contractor often negotiates subcontracts or limits bidding to a few such firms. As a result, the same subcontractors may follow the prime contractor from job to job.

Retainage ▪ Prime contracts require, as a rule, that a percentage—usually 10%—of the contractor's earnings be retained by the owner until final completion of the job and acceptance by the owner. Unless otherwise arranged, the provisions of the prime contract regarding payment and retainage pass into the subcontract. This is done with the usual stipulation in the prime contract that makes the subcontract subject to all the requirements of the prime contract.

Subcontractors whose work, such as site clearing, access-road building, or excavation, is performed in the early construction stages of a project may be severely impacted financially by this retainage. The standard retainage provisions may result in their having to wait a long time after completion of their work to collect the retained percentage. So the retainage on the general run of subcontracts, particularly those for work in the early phases of a project, often is reduced to a nominal amount after completion of the subcontractor's work. Justification for waiting until final completion of the job and acceptance by the owner may exist, however, under subcontracts for installed equipment carrying performance guarantees or for other items with vital characteristics.

An agreement may be negotiated, however, for early release or reduction of retainage. The subcontract should be specific in the matter of payment and release of retained earnings.

4.6 Prebid Site Investigations and Observations

A contractor should never bid a job without first thoroughly examining the site. This should be done early enough for the owner to have sufficient time to issue addenda to the plans and specifications, if required, to clarify questionable items.

Before visiting the site, the contractor should prepare a checklist of items to be investigated. The checklist should include, where applicable, the following: transportation facilities, electric power supply, water supply, source of construction materials, type of material to be encountered in required excavation or borrow pits, possible property damage from blasting and other operations of the contractor, interference from traffic, available labor supply (number and length of shifts per week being worked in the vicinity), areas available for construction of special plant, location of waste-disposal areas and access thereto, and weather records if not otherwise available.

It is sometimes helpful to take pictures of critical areas of the site at the time the investigation is made. Frequently, questionable items that were not covered on the original visit can be cleared up by referring to the photographs. They are sometimes of great value to the estimators doing the takeoff work and can help explain the job to others reviewing the estimate who have not visited the site.

4.7 Estimating Construction Costs

The two most important requisites for success in the construction business are efficient management of work in progress and correct estimating. Costs cannot be forecast exactly. But the contractor who can approach most nearly an accurate forecast of cost will bid intelligently a high percentage of the time and will be most successful over a period of years.

Construction estimates are prepared to determine the probable cost of constructing a project. Such estimates are almost universally prepared by contractors prior to submitting bids or entering into contracts for important projects. To be of value, an estimate must be based on a detailed mental picture of the entire operation; that is, it is necessary to plan the job and picture just how it is going to be done. Accordingly, it is wise to have the general construction superintendent or project manager who will be in charge of the job take part in the preparation of the job estimate.

4.7.1 Relationship of Estimating to Cost Accounting

Estimating and cost accounting should be very closely tied together. The estimate should be prepared in such a way that if the bid is successful, the estimate can be used as the framework for the cost accounts.

Estimating should be based on cost records to whatever extent may be reasonable in the particular case. But, prominently in the picture, there should also be a continuous study of new equipment, methods, and cost-cutting possibilities. The data most valuable, when used with due consideration of surrounding conditions and possible improvement, are cost records of the details of operations rather than of operations as a whole. Cost records and estimated costs for the labor portion of an operation should be expressed in both man-hours and dollars. A clear and complete narrative des-

cription of all the circumstances affecting the work should be made a part of the cost records prepared for use in future bidding. Otherwise, the usefulness of the data is greatly reduced.

The need for good production and cost records is emphasized by an increasing reluctance of some engineers and owners to make decisions and adjustments on the job. The resulting tendency is to throw the settlement of ordinary business items into arbitration or into court, where basic information is a fundamental requirement.

Normally, cost records in full detail are not available with sufficient promptness to be of substantial value on the job on which the costs are incurred. It is very desirable, however, that a current check on operating costs be maintained. This may be done by less formal procedures and still be adequate to provide timely information on undesirable deviations in progress and cost.

4.7.2 Forms for Estimating

Preparation of estimates is facilitated by standardization of forms. These are used for recording construction methods, equipment, and procedures that the estimator proposes as best adapted to the various items of work; to record calculations of the estimated cost of performing the work; and to summarize the estimated cost of the project. It is unnecessary and impractical to provide detailed printed forms for all types of work. A few simple forms are all that are needed. The mechanical makeup of an estimate must be simple, because conditions usually require that it be prepared in a short time—sometimes only two or three days when the estimator would like to have a month. These conditions do not change; it will always be necessary to make estimates quickly.

4.7.3 Steps in Preparation of an Estimate

It is advisable to have the routine to be followed in preparing cost estimates and submitting bids well established in a contractor's organization. For example:

1. Examine the contract documents for completeness of plans and specifications, and for the probable accuracy that an estimate will yield from the information being furnished.

2. Prepare a tentative progress schedule (Art. 4.9.1).

3. Prepare a top sheet based on an examination of the specifications table of contents. If there are no specifications, then the contractor should use as a guide top sheets (summary sheets showing each trade) from previous estimates for jobs of a similar natures or checklists.

4. Decide on which trades subcontractor bids will be obtained, and calculate prices on work of those trades where the work will be done by the contractor's own forces. Then, prepare a detailed estimate of labor and material for those trades.

5. Use unit prices arrived at from the contractor's own past records, from estimates made by the members of the contractor's organization, or various reference books that list typical unit prices. It is advantageous to maintain a computerized database of unit prices derived from previously completed work. The data can be updated with new wage and material costs, depending on the software used, so that prices can be adjusted nearly automatically.

6. Carefully examine the general conditions of the contract and visit the site, so as to have a full knowledge of all the possible hidden costs, such as special insurance requirements, portions of site not yet available, and complicated logistics.

7. Receive and record prices for materials and subcontracts. Compute the total price (see Art. 4.7.4).

8. Review the estimate and carefully note exclusions and exceptions in each subcontract bid and in material quotations. Fill in with allowances or budgets those items or trades for which no prices are available.

9. Decide on the markup, weighing factors such as the amount of extras that may be expected, the reputation of the owner, the need for work on the part of the contractor, and the contractor's overhead.

10. Submit the estimate to the owner in the form requested by the owner. It should be filled in completely, without any qualifying language or exceptions and submitted at the time and place specified in the invitation to bid.

4.7.4 Constituents of a Cost Estimate

The total price of a construction project is the sum of direct costs, contingency costs, and margin.

Direct costs are the labor, material, and equipment costs of project construction.

Contingency costs are those that should be added to the costs initially calculated to take into account events, such as rain or snow, or a probable increase in the cost of material or labor if the job duration is lengthy.

Margin (sometimes called **markup**) has three components: indirect, or distributable, costs; companywide, or general and administrative, costs; and profit.

Indirect costs are project-specific costs that are not associated with a specific physical item. They include such items as the cost of project management, payroll preparation, receiving, accounts payable, waste disposal, and building permits.

Companywide costs include the following: (1) costs that are incurred during the course of a project but are not project related—for example, costs of some portions of company salaries and rentals; (2) costs that are incurred before or after a project—for example, cost of proposal preparation and cost of outside auditing.

Profit is the amount of money that remains from the funds collected from the client after all costs have been paid.

4.7.5 Types of Estimates

Typical types of estimates are as follows: feasibility, order of magnitude, preliminary, baseline, definitive, fixed price, and claims and changes. There is some overlap from one type to another.

Feasibility estimates provide rough approximations of the cost of the project. They usually enable the owner to determine whether to proceed with construction. The estimate is made before design starts and may not be based on a specific design for the project under consideration. Such estimates are not very accurate.

Order-of-magnitude estimates are more detailed than feasibility estimates, because more information is available. For example, a site for the project may have been selected and a schematic design may have been developed. Generally made by the designer, these estimates are prepared after about 1% of the design has been completed.

Preliminary estimates reflect the basic design parameters. For the purpose, a site plan and a schematic design are required. Preliminary estimates can reflect solutions, identify unique construction conditions, and take into account construction alternatives. Usually, this type of estimate does not reveal design interferences. Generally prepared by the designer, preliminary estimates are made after about 5 to 10% of the design has been completed. Several preliminary estimates may be made for a project as the design progresses.

Baseline estimates and preliminary estimates. Identifying all cost components, the estimate provides enough detail to permit price comparisons of material options and is sufficiently detailed to allow equipment quotations to be obtained. The baseline estimate, generally prepared by the designer, is made after about 10 to 50% of the design has been completed.

Definitive estimates enable the owner to learn what the total project cost should be. The estimate is based on plan views, elevations, section, and outline specifications. It identifies all costs. It is sufficiently detailed to allow quotes to be obtained for materials, to order equipment and to commit to material prices for approximate quantities. Generally prepared by the designer, it represents the end of the designer's responsibility for cost estimates. It is made after about 30 to 100% of the design has been completed.

Fixed-price estimates, or **bids**, are prepared by a general contractor and represent a firm commitment by the contractor to build the project. A bid is based on the contractor's interpretation of the contract documents. To be accurate, it should be in sufficient detail to enable the contractor to obtain quotes from suppliers and to identify possible substitutes for specific items. It is made after 90% to 100% of the design has been completed.

Claims and changes estimates are prepared when a difference arises between actual construction and the requirements of the contract. This type of estimate should identify the changes clearly and concisely. It should specify, whenever possible, the additional costs that will be incurred and provide strong support for the price adjustments required.

4.7.6 Estimating Techniques

In preparing an estimate of the construction cost of a project, an estimator may use the parametric, unit-price, or crew-development technique. During the course of a project, any combination of these may be used. In general, the parametric technique is the least expensive, least time consuming, and least accurate. The crew-development technique is the most expensive, most time consuming, and most accurate. Of the three techniques, the parametric requires the most experience, and the unit-price technique the least.

Parametric estimating takes into account the strong correlation of project cost and project components that because of size, quantity, installation expense, or purchase price represent a very large portion of project cost. A parameter need not pertain to a specific design or to an item incorporated in the drawings; for example, it could be the number of barrels to be processed in a refinery project to be estimated. For an office building, the parameter could be floor area. For a warehouse, the parameter could be the size and number of items to be stored and the expected length of time each item is expected to be stored. The parametric technique obtains data from experience with completed work, standard tables, or proprietary tables that compile data from many projects of different types and are updated at frequent intervals.

Unit-price estimating is based on data contained in the contract documents. The project cost estimate is obtained by adding the products obtained by multiplying the unit cost of each item by the quantity required; for example, cubic yards of concrete, tons of structural steel, number of electric fans. The information needed is obtained from databases of quantities per work item and unit prices.

Crew-development estimating is based on the costs for personnel and equipment required for each item during each construction phase. Employment of these resources varies with project status, site conditions, and availability of labor, materials, and equipment. For example, for a tight completion schedule, the estimate might be based on a large crew and multiple shifts or overtime. For a site with limited access or storage area for construction materials and equipment, the estimate may assume that a small crew will be used. Furthermore, utilization of personnel and equipment may have to be varied as the work progresses. Data for the estimate may be obtained from production handbooks, which usually are organized by trades or in accordance with the use of a facility. Since it is based on the sequence of construction for the project, crew-development estimating is the most accurate of the estimating techniques.

Indirect Costs. When parametric estimating is used, indirect costs may be determined as a percentage of the direct cost of the project or as a percentage of the labor cost, or they may be based on the distance and the volume of materials that must be moved from source to site. For the other two methods of estimating, the estimator determines the various project activities, such as accounting, project management, staff overhead, and provision of temporary site offices, that are not associated with a specific physical item. In unit-price estimating, these activities are expressed in some unit of measurement, such as linear feet or cubic yards, and multiplied by an appropriate unit price to obtain the activity cost. The total indirect cost is the sum of the costs of all the activities. In crew-development estimating, the estimator determines the starting and ending dates and salaries for the personnel needed for those activities, such as project engineer, project manager, and payroll clerks. From these data, the estimator computes the total cost of personnel. Also, the estimator determines the length of time and cost of each facility and service needed for the project. These costs are added to the personnel costs to obtain the total indirect cost.

Margin, or Profit. The amount that a contractor includes for profit in the cost estimate for a project depends on many factors. These include capital required and capital risks involved, anticipated troublesome conditions during construction, locale, state of the industry, estimated competition for the job, general economic conditions, need of the firm for additional work, and disciplines required, such as structural, mechanical, and electrical. When a contractor is very anxious to obtain the job, the bid submitted based on the cost estimate may not include much, if any, margin. This may be done because of the prestige associated with the project or the expectation of profits from changes during construction.

Normally, to establish margin for an estimate, the estimator consults handbooks that express gross margin as a percent of project cost for various geographic regions and industries. Also, the estimator consults periodicals to obtain the current price for specific work. These data, adjusted for the effects of other considerations, form the basis for the margin to be included in the estimate.

Quantity Surveys. A quantity survey is a listing of all the materials and items of work required for a construction project by the contract documents. Together with prices for these components, the quantities taken off from these documents are the basis for calculation of the direct cost of the project. In the United States, it is customary, except for some public works, for contractors to make quantity surveys at their own expense. The contractors may prepare the surveys with their own forces or contract with professional quantity surveyors for the task. Often, a contractor's estimator will take off the quantities and price them either simultaneously with or after completion of the quantity survey.

Preparation of a quantity survey requires that the project be resolved into its components, work classifications, and trades. Because of the large number of items involved, professional quantity surveyors and estimators generally use checklists to minimize the chance of overlooking items. When each item in a checklist is assigned a code number, the list serves the additional purpose of being a code of accounts against which all expenses are charged to the benefiting item. It is good practice in recording an item on a quantity survey sheet or estimate form to indicate this step with a check mark on the checklist next to the item and to place items in the same sequence as they appear on the checklist. This will help ensure that items are not overlooked. Furthermore, when a search has to be made for an item, it will always appear in the same place.

Computer Estimating. Several types of computer software are available for facilitating construction cost estimating. The most common may be classified as utilities, databases, and expert systems (artificial intelligence).

Utilities compile information and perform arithmetic on the data, for example, in spreadsheet programs. Enabling quick extraction and presentation of needed information in convenient form for analysis and reporting, utilities supplement the expertise of estimators.

Databases are listings of unit prices for materials, equipment, fixtures, and work items. They are usually designed for use with a specific utility and may be limited to a specific type of estimate or estimating technique.

Expert systems should ideally, when fed complete, appropriate data, prepare an estimate automatically, with a minimum of assistance from a human estimator. In practice, they question the estimator and use the answers to produce the estimate.

(N. Foster et al., "Construction Estimates from Take-Off to Bid," 3rd ed., G. E. Deatherage, "Construction Estimating and Job Preplanning," McGraw-Hill, Inc., New York (books.mcgraw-hill.com); J. P. Frein, "Handbook of Construction Management and Organization," Van Nostrand Reinhold, New York.)

4.8 Bookkeeping and Accounting

Contractors must maintain financial records for many purposes. These include tax reporting, meeting requirements of government agencies, providing source data for indispensable support services, serving the purposes of company management, and submission of financial statements and reports to bankers, sureties, insurance companies, clients, public agencies, and others. Company management is especially concerned with financial accounts. Without complete, accurate records, management would find it impracticable to, among other things, estimate construction costs accurately, keep the firm in a fluid cash position, make sound decisions regarding acquisition of equipment, or control costs of projects under way.

4.8.1 Bookkeeping

Bookkeeping is the art of recording business transactions in a regular, systematic manner so as to show their relationship and the state of the business in which they occur. General practice in contractor bookkeeping is to divide every transaction into two entries of equal amount.

One entry, called a debit, indicates the income, materials, and services received by the contractor. The other entry, called a credit, is entered in a column on the right. Balancing and checking the first entry, it records outflow, such as payments.

Usually, bookkeepers maintain at least two sets of books, a journal and a ledger, both with debit and credit entries. In the journal, transactions are posted chronologically as they occur. For each transaction, the date, nature or source of transaction, purpose, and amount involved are recorded in successive entries. The amount received by the contractor (debit) is recorded one line above the outgoing amount (credit).

A second book, a ledger, is used to group transactions by type. It allots a page or two for each

kind of transaction posted in the journal, such as salaries, or taxes, or rent. Every debit entry in the journal is recorded as a debit entry in the ledger. Every credit entry in the journal is posted as a credit entry in the ledger. Consequently if no mistakes are made, the two books must balance: The sum of the money recorded in the ledger must equal the sum of the money posted in the journal.

4.8.2 Accounting Methods

Accounting includes bookkeeping but also other services that provide more detail and explanations affecting the financial health of a business. The main objective is job costing or the determination of income and expense from each construction project. The cost estimate for each project serves as a budget for it. Costs, as reported, are charged against the project that incurs them.

General practice for contractors is to use an accounting procedure known as the accrual method. (It differs from the alternative cash method in which income is recognized as received, not when billed. Expense is posted as incurred.) For the accrual method, income is recorded in the fiscal period during which it is earned, even though payment may not have been received. Also, expenses are posted in the period in which they incur.

A procedure known as the straight accrual method is used for accounting for short-term contracts (projects completed within a single accounting period). For long-term contracts (projects started in one taxable year and completed in another), contractors usually use the completed-contract or percentage-of-completion methods, which are variations of the accrual method.

Percentage-of-Completion Method ▪ In this procedure, income and expenses are reported as a project progresses, thus on a current basis rather than at irregular intervals when projects are completed. The method also reflects the status of ongoing projects through current estimates of percent completion of projects or of costs to complete. Profit is distributed over the fiscal year in which the project is under construction. The percentage of the total anticipated profit earned to the end of any period is generally estimated as the percentage that incurred costs to that date are of the anticipated total cost, with allowances for revised estimates of costs to complete.

Completed-Contract Method ▪ In this procedure, income and expenses are reported only when the project is completed. This method offers the advantage that income is reported after final financial results are known rather than being dependent on estimates of costs to complete the project. It has several drawbacks, however, one of which is the inability to indicate the performance to date of long-term contracts. Also, it may result in irregular reporting of income and expenses and hence, sometimes, in larger income taxes.

Because the percentage-of-completion and completed-contract methods have advantages and disadvantages, particularly with respect to income taxes, a contractor may elect to use percentage-of-completion method for financial statements and the completed-contract method for reporting income taxes. Or, a contractor may use one method for some projects and the other method for other projects. But once a method has been adopted for tax-reporting purposes, approval of the Internal Revenue Service is needed before the contractor can change it.

Financial Reports ▪ Several types of financial reports are derived from business records. Two of the most important are the income statement and the balance sheet.

Income, or profit and loss, statements summarize the nature and amounts of income and expense over a specific period. A statement expresses profit or loss as the difference between income received and expenses paid out during the period.

Balance sheets, also known as financial statements or statements of assets and liabilities, summarize assets, liabilities, and net worth as of a specific date, such as the end of a fiscal year. These statements are intended to indicate the financial condition of a business on that date. Balance sheets derive their name from the requirement that total assets equal total liabilities plus net worth. Assets include anything of value accruing to the business, such as all property owned by the business (less depreciations), cash on hand or in the bank, receivables, and prepaid expenses. Liabilities include financial obligations, such as notes and accounts payable; accrued expenses, including wages and interest accrued; deferred taxes; and long-term debt. Net worth represents the contractor's equity in the business.

(G. E. Deatherage, "Construction Office Administration," W. E. Coombs and W. J. Palmer, "Construction Accounting and Financial Management," 5th ed., and M. Millman, "General Contracting: Winning Techniques for Starting and Operating a Successful Business," McGraw-Hill, Inc., New York (books.mcgraw-hill.com); "Construction Cost Control," ASCE Manuals and Reports of Engineering Practice No. 65, American Society of Civil Engineers (www.asce.org).)

4.9 Project Scheduling

One of the first things to be done by a contractor when beginning the preparation of an estimate is to make a time schedule of the proposed operation and set up a tentative plan for doing the work. It is necessary for the contractor to study the plans and specifications in detail before visiting the site of the project. This study should proceed far enough to establish a tentative progress schedule for the more important or governing items of work.

4.9.1 Job Progress Schedule

This schedule should show all items affecting the progress of the work and consider the length of the construction season (if applicable) or seasonal weather influence at the particular site. Where applicable, the schedule should note the most advantageous date or the required date for early-stage work, such as river diversion for a dam; when deliveries of new or specialized construction plant or equipment can be obtained; possible delivery dates for critical items of contractor-furnished permanent materials; delivery dates of major items of permanent equipment to be furnished by the owner; and other controlling factors. Based on the preceding dates, production rates for the controlling items of work should be determined. Also, the type, number, and size of the various units of construction plant and equipment needed to complete the work, as required by this schedule, should be tentatively decided upon. Progress schedules can be prepared in several forms. Figure 4.5 shows a form that can be adapted to fit most conditions.

Based on the progress schedule, a brief narrative description of the job should be written. The description should call attention to indefinite, hazardous, and uncertain features as well as to items likely to increase and decrease in quantity. Also, the description should include a statement of

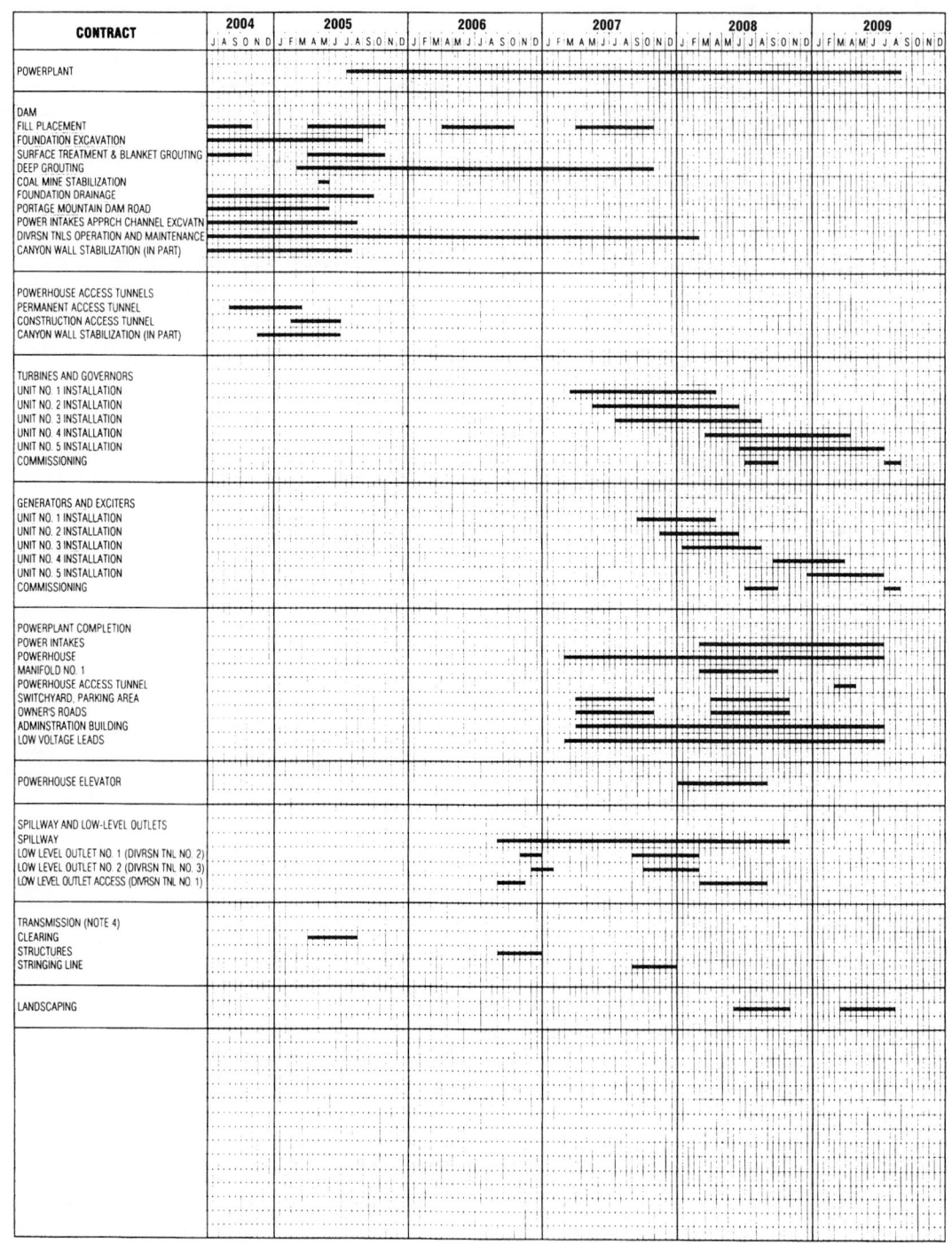

Fig. 4.5 Bar-chart progress schedule. Start and end of a horizontal line indicate, respectively, start and finish of an activity.

the total man-hours of labor and the total machine-hours for important equipment estimated as required for doing the work. In addition, the description should include peak labor requirements and controlling delivery requirements for important material and equipment items. Finally, the description should contain a statement of cash requirements derived from scheduled income and expenditures.

4.9.2 Scheduling to Save Money

Time is less tangible an ingredient of construction than labor or material but nonetheless real and important. Money and time are related in many ways.

For the owner of revenue-producing facilities, such as electric generating installations, processing plants, and rental buildings, reduction in time required for completion of construction results in less interest expense on investment over the period of construction. Also, increased income accrues to the extent that completion time is shortened, thereby permitting earnings to begin at an earlier date.

To the contractor, reduction in time for completing the job means, likewise, a reduction in interest charges on cash invested during construction. Also, the shorter the time to complete the job, the less the supervisory, administrative, and overhead expense. In addition, benefits accrue from shortened time because it permits earlier release of equipment for use on other work.

Construction scheduling consists essentially of arranging the several operations involved in the construction of a project in the sequence required to accomplish completion in the minimum period of time consistent with economy. To ensure completion within the contract time limit and to attempt to reduce the time required to do the job, it is necessary to program each unit of the project within itself and properly relate each unit to all the others.

4.9.3 Scheduling with a Rectangular-Bar Chart

Progress schedules show starting and completion dates for the various elements of a project. For unit-price work, the bid-item breakdown is normally used. On lump-sum contracts, subdivision according to that used in estimating the work is common. Schedules may be prepared in either tabular or graphical form, although the graphical form is generally used because of ease in visualization.

The most widely used graphical representation of the work schedule is the rectangular-bar chart (Fig. 4.5). It shows starting and completion dates for each item of work. Also, it indicates the items on which work must proceed concurrently, the items that overlap others and by how much, and the items that must be completed before work on others can begin.

Progress schedules should be prepared at the outset of the job as an aid in coordinating work by all departments of the contractor's organization (Art. 4.9.1). For instance, the progress schedule is a convenient way to advise the purchasing agent of critical material delivery dates.

Construction contracts often require the contractor to submit a progress schedule to the owner for approval within a specified time after award of the contract and before construction is started. The importance of this requirement often is emphasized in the contract by provisions to the effect that failure to submit a satisfactory schedule shall be just cause for annulment of the award and forfeiture of the proposal guarantee.

For comparing performance of work with that scheduled, a bar is often placed above the schedule bar to show actual start and completion dates. The chart in Fig. 4.6 indicates that excavation started on the date programmed and was completed ahead of time, whereas formwork began late. At the close of December, formwork was 60% complete. This method has the advantage of simplicity. It fails, however, to disclose the rate of progress required by the schedule or whether actual performance is ahead of or behind schedule.

4.9.4 Triangular-Bar Chart

The concept of rate of progress is introduced in Fig. 4.7, which deals with the same items charted in Fig. 4.6. In Fig. 4.7, horizontal distances represent time allotted for doing the work and vertical distances represent percentage of completion. Therefore, the sloping lines indicate the rate of progress.

For example, Fig. 4.7 indicates that excavation was scheduled to proceed from start to finish at a uniform rate (straight sloping line). Work started on time, progressed slowly at first, and tapered off at the end (crosshatched area). Greater production scheduled midway in the operation was sufficient,

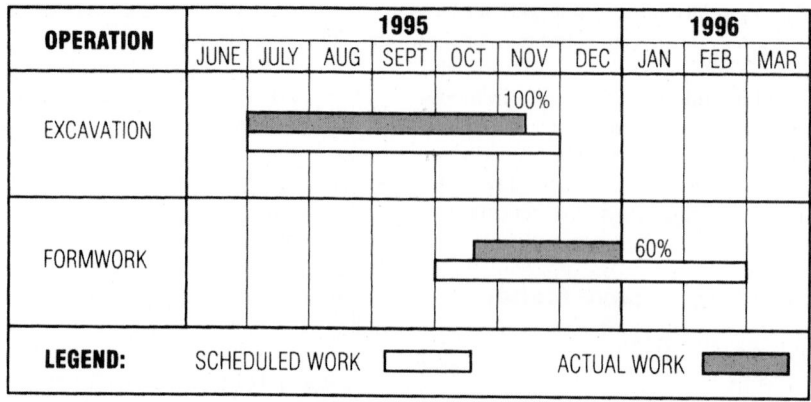

Fig. 4.6 Rectangular-bar progress schedule.

however, to bring the item to completion 15 days early. The date on which formwork would have begun was advanced by reason of the accelerated rate of excavation from October 1 to September 15 (dashed lines).

Instead of being stepped up to take advantage of the time gained on excavation, formwork was late in getting started and progressed slowly until December 1. Then it was speeded up, but the 60% completion reached at the close of December falls short of scheduled requirements. (In practice, the time gained on excavation would doubtless have been captured and put to beneficial use by arranging start of formwork on September 15, half a month ahead of schedule.)

Time gained or lost on any one work item affects many others. As a result, frequent revision is necessary to keep progress schedules currently accurate in all respects. Formalized revision of the overall progress schedule, however, is often rendered unnecessary because contractor dependency on it is gradually supplanted by such intimate acquaintance with the operations that controlling factors become common knowledge and all concerned know what must be done and when.

Critical items often are subjected to detailed analysis and scheduling. This may take the form of three-dimensional schematics, expanded views, stage-construction drawings, concrete-pouring diagrams, and similar devices as aids to visualization. After that, further scheduling, such as concrete-pouring programs, earthwork-quantity movement schedules, or programming of piping runs may be devised and utilized as required.

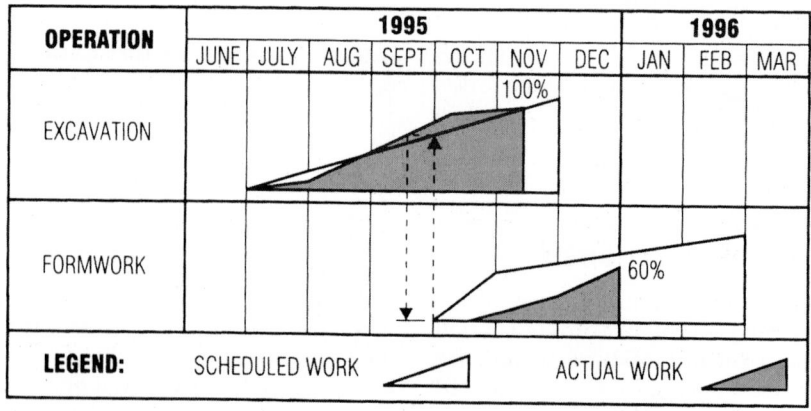

Fig. 4.7 Triangular-bar progress schedule.

4.9.5 Critical-Path Method of Scheduling (CPM)

The critical-path method has been developed as a tool of management useful in specialized situations. It is required by several federal and state agencies on some contracts. CPM is based on planning and job analysis going far beyond that necessary for bidding a job. In addition to the step-by-step breakdown of the job into its component tasks and subtasks, and the plotting of sequential relationships, the planner must know how long each task will take. For instance, the construction and installation of a large air handler inside the mechanical room requires shop drawings to be developed by the HVAC subcontractor. The HVAC contractor must calculate the time needed to prepare the shop drawings, have them reviewed by the engineer, allow time for any subsequent revisions after the review and then time for re-review and approval. All of these subtasks would need to be completed prior to manufacturing of the air ducts. All of the lead time for any other additional equipment needed inside the ducts, such as smoke detectors or dampers, would also need to be known. Some projects such as clean rooms or drug manufacturing facilities require lengthy testing periods of the HVAC equipment prior to acceptance. Even on the simplest of construction projects each task can have many subtasks. Most computer programs will allow a large number of subtasks to be shown, but for ease of reading, the subtasks can be hidden or represented by a task line.

After the project has been broken down into all its activities, the activities are listed or plotted in such a way that all sequential relationships are shown. Activities may be represented by arrows (Fig. 4.8a) or by circles, or nodes, connected by sequence lines (Fig. 4.8b). Analysis, by examination or computer, should guide establishment of a realistic time schedule and pinpointing of the operations whose completion times are responsible for establishing the overall project duration. Also, the analysis should facilitate settling change orders by determining the operations affected and the effect on project duration. In addition, it should help in establishing the proper sequence of work operations and determining the status of work in progress in relation to the number of days behind or ahead of schedule.

An **arrow diagram** (Fig. 4.8a) is drawn by setting the tail of an arrow representing an activity, such as placing concrete, at the tip of an arrow representing the immediately preceding activity, such as placing electrical conduit and junction boxes. The nodes (tips and tails) are assigned unique numbers to identify the activities (1–2, 2–3, etc.). Each node represents the completion of the preceding activities and the start of the following activities. Sometimes, a dummy arrow is needed to complete the network.

A **precedence (PERT) diagram** (Fig. 4.8b) is drawn by setting the node for an activity to the right of the node representing an immediately preceding activity. Each node is assigned a number greater than that of any preceding activity. The nodes are connected by lines to indicate the sequence of the work. Precedence diagrams are simpler to draw and analyze than arrow diagrams.

In either type of diagram, the **critical path** is the sequence of operations requiring the most time to complete. The critical path determines the duration of the project. To shorten the project, it is necessary to decrease the time required for one or more activities on the critical path (critical activities). These activities have zero total float.

Total float is the difference between time required and time available to execute an activity. It is equivalent to the difference between earliest and latest start (or finish) times for an activity. Table 4.1 shows the calculation of float for the simple network in Fig. 4.8. Float is determined in two steps: a forward and a backward pass over the network.

The **forward pass** starts with the early start (or scheduled) date for the first activity, Erect Forms. In this case, the date is 0. Addition of the duration of this activity, 2 days, to the early start time yields the early finish date, 2, which is also the early start date for the next activity, Place Reinforcing. The early finish date for this activity is obtained by adding its duration, 1 day, to the early start date. The forward pass continues with computation of early start and finish times for all subsequent activities. Where one activity follows several others, its early start date is the largest of the early finish dates of those activities.

The **backward pass** determines late start and finish dates. It begins with the late finish date of the final activity, Place Concrete, which is set equal to the early finish date, 6, of that activity. Subtraction of the duration, 1 day from the late finish date yields the late start date, 5, which is also the late finish date of preceding activities, Place Mechanical

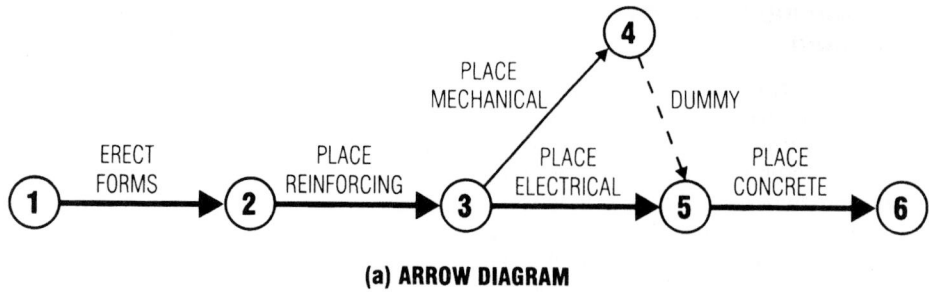

(a) ARROW DIAGRAM

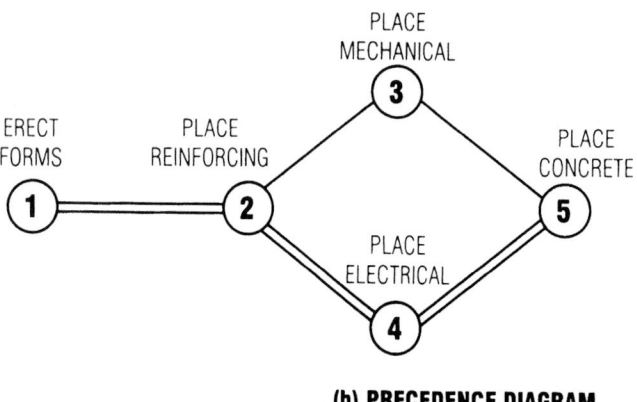

(b) PRECEDENCE DIAGRAM

Fig. 4.8 Representation of activities in a CPM network: (*a*) arrows; (*b*) nodes.

and Place Electrical, and their late start dates are found by subtracting their durations from the late finish dates. Where one activity precedes several others, its late finish date is the smallest of the late start dates of those activities. The backward pass continues until late start and finish dates are computed for all activities. Then, the float can be found for each activity as the difference between early and late start times. Critical activities (those with zero total float) are connected by heavy arrows in Fig. 4.8*a* and by double lines in Fig. 4.8*b* to indicate the critical path.

Table 4.1 Float Calculations for Critical-Path Method

Activity Number		Duration, days	Early Start Date	Early Finish Date	Late Start Date	Late Finish Date	Total Float, days
Arrow Diagram	Precedence Diagram						
1–2	1	2	0	2	0	2	0
2–3	2	1	2	3	2	3	0
3–4	3	1	3	4	4	5	1
3–5	4	2	3	5	3	5	0
5–6	5	1	5	6	5	6	0
4–5	—	0	4	4	5	5	1

4.9.6 Scheduling for Fast Tracking

CPM, described for application to construction of a project in Art. 4.9.5, can also be used for design, which usually is completed before the start of construction. In addition, CPM is useful for integrated scheduling of fast tracking, a procedure in which design and construction proceed simultaneously. When CPM is used for this purpose, it requires input from both design and construction personnel.

When a project is fast-tracked, final design and construction begin shortly after groundbreaking. Field work on components of the project proceeds as soon as applicable portions of the design have been completed. Thus, what would be the normal duration of the project is shortened by setting design and construction on separate but parallel tracks instead of in sequence, as is traditional.

One disadvantage of fast tracking is less control over costs than with projects where design has been completed before bids are taken. This disadvantage, however, can be partly overcome if a professional construction manager is employed for construction management or a cost-plus-fixed-fee or cost-plus-percentage-of-cost contract is awarded to a reputable general contractor. Another disadvantage of fast tracking is that coordination of the work is more difficult and the input from various consultants may be lacking. As a result, some work in place may have to be removed or redone. Because of the lower efficiency of fast tracking and the necessity of redoing work, construction costs may be larger than they would be when construction starts after completion of design. Despite this, the total cost of the project to the owner may be less, because of savings in interest on construction loans, revenues from earlier use of the project, and decreased effects of monetary inflation.

4.10 Role of Project Manager

A project manager, in brief, has responsibility for all construction functions for a project, including coordination of the work of job superintendents, crew supervisors, and subcontractors. For a small organization, the proprietor may serve as project manager. For a large firm, an experienced project manager may be assigned responsibility for one large project or several smaller ones.

Success of a construction project depends heavily on the abilities of the project manager. This individual should have administrative and managerial skills and be familiar with all details of the contract documents. Knowledge of all phases of construction is essential. From daily inspection of projects assigned, the construction manager should keep abreast of the current job status.

4.10.1 Duties of a Project Manager

Among the duties of a project manager are the following:

Coordination of contact with clients

Allocation of workforce to projects and organization of units for project operation

Coordination of the work of all units and divisions

Periodic review and analysis of project costs, schedule, progress, and other construction data

Insure timely submittal of pay requests to owner

Purchasing

Arranging for surveys and construction layout

Instituting and supervising job safety programs and compliance with all environmental programs

Securing permits from government agencies

Maintenance of files of labor agreements

Representing the contractor in jurisdictional disputes

Dealing with changes and extras

Submitting and obtaining approval of shop drawings and samples, and material certifications

Conducting conferences and job meetings with key personnel and following up on decisions made

After construction starts, the project manager should continually compare field performance with the established schedule. When the schedule is not being met, the corrective actions taken and rescheduling phases are known as **project time management**.

The monitoring phase of time management involves periodic measurement of actual job progress and comparison with the planned objectives. This should be done by determining the work quantities put into place and reporting this information for comparison with work quantities anticipated in the job schedule. Then, a determination can be made of the effect of the current status of the job on the completion date for the project. Any corrective actions necessary can then be planned and implemented. After that, the schedule can be updated.

CPM provides a convenient basis for measuring progress and for issuance of reports (Art. 4.9.5). The network diagram should be corrected as needed so that the current job schedule reflects actual job status.

A variety of software is available and can be used to produce reports that will assist project managers. Following are descriptions of some reports that contractors have found helpful:

Purchasing/Cost Report ▪ This report lists the various items to be procured and sets target dates for bidding and award of contracts. It keeps track of the budget and actual cost for each item. A summary prepared for top management provides totals in each category and indicates the status of the purchasing.

Expediting/Traffic Report ▪ This report lists the items when they are purchased. It also gives a continual update of delivery dates, shop drawing and approval status, shipping information, and location of the material when stored either on or off the site.

Furniture, Fixture, and Equipment List ▪ This report, which is normally used when the job involves a process or refinery, can also be used for lists of equipment in a complex building, such as a hospital or hotel. The report describes all the utility information for each piece of equipment, its size, functions, intent, characteristics, manufacturer, part number, location in the finished job, and guarantees. The report also provides information relating to the item's source, procurement, price, and location or drawing number of the plan it appears on.

Accounting System ▪ The system consists of a comprehensive series of accounting reports, including a register for each supplier, and shows all disbursements. This information is used in preparing requisitions for progress payments. It also can be used to report costs of the job to date and to make predictions of probable costs to complete.

4.10.2 Computerized Project Management Control System

This system combines project scheduling with cost controls, resource allocation controls, and a contract-progress statistical reporting system. The objective is to provide total control over time, cost, resources, and statistics.

Time ▪ The time aspect of the system is designed to produce, through project scheduling, a set of time objectives, a visual means of presenting these objectives, and the devising and enforcing of a corrective method of adhering to the objectives so that the desired results will be achieved.

Cost ▪ These are summary costs monitored by budget reports, produced monthly and distributed to the owner. In addition, detailed reports for construction company management list costs under each class of construction activity. These reports are used by project managers and field, purchasing, and top-management personnel. A report on probable total cost to complete the project is intended for all levels of construction company personnel but is used primarily by those responsible for corrective action.

Resource Allocation ▪ For the purpose of resource allocation, a graphical summary should be prepared of projected monthly use of personnel for individual activities and also of the estimated quantities of work to be in place for all trades on a cumulative basis. An update of these charts monthly will indicate which trades have low work quantities in place. With this information, the manager can ensure that lagging trades are augmented with the proper number of workers to permit them to catch up with and adhere to the schedule.

Statistics ▪ From the information received from the preceding reports, an accurate forecast can be made of the probable construction completion date and total cost of the project.

(F. S. Merritt and J. T. Ricketts, "Building Design and Construction Handbook," 5th ed., McGraw-Hill, Inc., New York; J. P. Frein, "Handbook of Construction Management and Organization," Van Nostrand Reinhold, New York.)

4.11 Role of Field Superintendent

A field superintendent has a wide variety of duties. Responsibilities include the following: field office (establishment and maintenance); fencing and security; watchmen; familiarity with contract documents; ordering out, receiving, storing, and installing materials; ordering out and operation of equipment and hoists; daily reports; assisting in preparation of the schedule for the project; maintenance of the schedule; accident reports; monitoring extra work; drafting of backcharges; dealing with inspectors, subcontractors, and field labor; punchlist work; quality control; and safety. Familiarity with contract documents and ability to interpret the plans and specifications are essential for performance of many of these duties.

Daily reports from the superintendent provide essential information on the construction. From these daily reports, the following information is derived: names of persons working and hours worked; cost code amounts; subcontractor operations and description of work being performed; materials received; equipment received or sent; visitors to the job site; summaries of discussions with key subcontractors and personnel; other remarks; temperature and weather; accidents or other unusual occurrences.

4.12 Purchase Orders

Issuance of a purchase order differs from award of a subcontract (Art 4.5). A purchase order is issued for material on which no labor is expected to be performed in the field. A subcontract, in contrast, is an agreement by a subcontractor not only to furnish materials but also to perform labor in the field. A purchase order notes the date, names of issuer and supplier, description, price, terms of payment, and signatures of the parties.

For the specific project, a purchase-order rider and list of contract drawings should be appended to the standard purchase-order form. The rider describes special conditions pertaining to the job, options or alternates, information pertaining to shop drawings, or sample submissions, and other particular requirements of the job.

Material price solicitations are handled in much the same manner as subcontract price solicitations. Material bids should be analyzed for complicated trades in the same manner as for subcontracts.

To properly administer both the subcontract and the purchase orders, it is necessary to have a purchasing log in which is entered every subcontract and purchase order after it has been sent to the subcontractor or vendor. The log serves as a ready cross reference, not only to names of subcontractors and vendors but also to the amounts of their orders and the dates the orders were sent.

A variety of software is available to keep track of all equipment and materials and related purchase information, such as specifications, quotations, final orders, shipment, and delivery dates. Software typically is based on the concept of critical-path items. The various tasks that must be performed are assigned due dates. For example, a report could be by project and show all open purchase-order items for one project, or by buyer name, with all open purchase-order items for each buyer, including all projects.

In negotiating and awarding either a subcontract or a material purchase, the contractor should take into account the scope of the work, list inclusions properly, note exceptions or exclusions, and, where practicable, record unit prices for added or deleted work. Consideration should be given to the time of performance of units of work and availability of workers and materials, or equipment for performing the work. Purchase orders should contain a provision for field measurements by the vendor, if this is required. In addition, purchase orders should indicate whether delivery and transportation charges and sales taxes are included in the prices.

4.13 Job Safety and Environmental Control

Accidents on a construction project, whether involving employees or the public, can impose an enormous burden on the construction contractor and others associated with the project. Consequently, it is of great importance to all concerned with the job to ensure that an appropriate job safety program is instituted. Although the owner of the construction firm or the company executives are legally responsible if an accident should occur, the project manager generally is responsible for establishing and supervising the safety program.

The federal government in 1970 passed the Occupational Safety and Health Act (OSHA) (Title 20—Labor Code of Federal Regulations, chap. XVII, part 1926, U.S. Government Printing Office). Compared with state safety laws, the federal law has much stricter requirements. For example, a state agency has to take the contractor to court for illegal practices. The Occupational Safety and Health Administration, however, can impose fines on the spot for violations, despite the fact that inspectors ask the employers to correct their deficiencies.

Construction accidents result from an unsafe act or an unsafe condition. Company policy should aim at preventing these through education, training, persuasion, and constant vigilance. On every project, the project manager should remind superintendents and supervisors of safety requirements. On visits to job sites, the manager should be constantly alert for violations of safety measures. The safety engineer or manager should ascertain that the construction superintendent holds weekly "toolbox" safety meetings with all supervisors and is writing accident reports and submitting them to the contractor's insurance administrator. In addition, the safety supervisor should maintain a file containing all the necessary records relative to government regulations and be familiar with record-keeping requirements under the Occupational Safety and Health Act (Occupational Safety and Health Administration, U.S. Department of Labor, Washington, D.C.). Management should hold frequent conferences with the project manager and with the insurance company to review the safety record of the firm and to obtain advice for improving this safety record.

("Manual of Accident Prevention in Construction," Associated General Contractors of America, Washington, DC 20006; "The 100 Most Frequently Cited OSHA Construction Standards in 1991," U.S. Government Printing Office, Washington, DC 20402.)

Another federal regulation that the construction manager must deal with is the Clean Water Act. The amendment in 1987 to this act required the Environmental Protection Agency to regulate storm water discharges from construction sites over five acres under the National Pollutant Discharge Elimination System. Subsequent amendments require almost all construction projects (those over one acre in size) to submit a notice of intent (NOI) to qualify the site for storm water discharges under the EPA's general permit. This permit requires the development of a pollution prevention plan that shows the devices that will be used during construction to prevent the discharge of sediment laden water from the site and minimize the amount of erosion generated within the site. These pollution prevention devices must be maintained throughout the course of construction and their effectiveness should be monitored by the construction manager. Additional information can be obtained at Environmental Protection Agency's website at www.epa.gov or from "Storm Water Management for Construction Activities: Developing Pollution Prevention Plans and Best Management Practices," USEPA, 1992.

4.14 Change Orders

Contract documents specify in detail the work the contractor is required to perform. Often, however, changes or extra work are found necessary after the award of the contract, especially after construction is under way. The contract documents generally contain provisions that allow the contractor or the owner to make changes if both parties agree to the change. If the change decreases construction costs, the owner receives a credit. If costs increase, the owner pays for the added costs. Cost of changes may be based on a negotiated lump sum; cost of labor and materials, plus markup; or unit prices.

The owner may issue a change order for any of several reasons. These include a change in the scope of the work from that described in the specifications; change in material or installed equipment; change to correct omissions; and change in expected conditions, such as subsurface rock not disclosed in plans and specifications, abnormal weather, or labor strikes. To provide for the occurrence of unexpected conditions, the construction contract should contain a changed-conditions clause in the general conditions. (See "General Conditions of the Contract for Construction," AIA A201, American Institute of Architects, 1735 New York Ave., N.W., Washington, DC 20006.) The American Society of Civil Engineers Committee on Contract Administration drafted the following recommended changed-conditions clause:

The contract documents indicating the design of the portions of the work below the surface are based upon available data and the judgment of the Engineer. The quantities, dimensions, and classes of work shown in the contract documents are agreed upon by the parties as embodying the assumptions from which the contract price was determined.

As the various portions of the subsurface are penetrated during the work, the Contractor shall promptly, and before such conditions are disturbed, notify the Engineer and Owner, in writing, if the actual conditions differ substantially from those which were assumed. The Engineer shall promptly submit to Owner and Contractor a plan or description of the modifications which he or she proposes should be made in the contract documents. The resulting increase or decrease in the contract price, or the time allowed for the completion of the contract, shall be estimated by the Contractor and submitted to the Engineer in the form of a proposal. If approved by the Engineer, he or she shall certify the proposal and forward it to the Owner with recommendation for approval. If no agreement can be reached between the Contractor and the Engineer, the question shall be submitted to arbitration or alternate dispute resolution as provided elsewhere herein. Upon the Owner's approval of the Engineer's recommendation, or receipt of the ruling of the arbitration board, the contract price and time of completion shall be adjusted by the issuance of a change order in accordance with the provision of the sections entitled, "Changes in the Work" and "Extensions of Time."

4.15 Claims and Disputes

During construction of a project, the contractor may claim that work ordered by the owner, or owner's representative, is not included in the contract and that there is no obligation to perform that work without adequate compensation. The contractor therefore may submit a change-order proposal before performing the work. (Sometimes, the contractor may proceed with the work before the order is issued so as not to delay the job.) If the owner disputes the claim, the contractor may continue the work or press for a decision on the claim through mediation, arbitration, or other remedy available under the contract or at law.

When a dispute between owner and contractor arises during construction, the first step is an effort to resolve it by negotiation. An optional procedure is to recognize before construction starts the possibility that disagreements may arise and make provisions for facilitating negotiations. One way is to appoint at that time a **dispute resolutions board** (DRB), consisting of three qualified persons, to assist in negotiation of a settlement. If an agreement

cannot be reached, the DRB should issue recommendations for a settlement. These, however, are nonbinding on the parties.

Another method of resolving disputes is arbitration, which may be required by the construction contract. If arbitration is agreed on or required, the parties involved submit the facts of the dispute to impartial third parties who examine the claims and render a decision, which is legally binding on the parties. (See "Construction Contract Disputes—How They May Be Resolved under the Construction Industry Arbitration Rules," American Arbitration Association, 140 W. 51st St. New York, NY 10020.) The American Arbitration Association can provide assistance for arbitration and also for mediation. The latter differs from arbitration in that mediation is entered into by the parties voluntarily and in addition the recommendations are not legally binding. In mediation, one or more impartial mediators consult with the parties with the objective of reaching an agreement that the parties find acceptable. Mediation is desirable as a less time-consuming and less costly step before recourse to arbitration or a judicial forum.

4.16 Insurance

Contractors should establish a sound insurance program for protection against financial losses due to unforeseen contingencies. For this purpose, insurance companies whose financial strength is beyond doubt should be selected. A competent insurance agent or broker experienced in construction work can be helpful in making such choices. The one selected should be capable of preparing a program that provides complete coverage of the hazards peculiar to the construction industry and of the more common perils. Also, the agent or broker should be able to obtain insurance contracts from qualified insurance companies that are in a position to render on-the-job service when needed. In addition, the contractor will need competent advice to be certain that all insurance policies protect all parties and provide adequate coverage limits.

4.16.1 Liability Insurance

Law, contracts, and common sense require that responsible contractors be adequately protected by liability insurance in all phases of their operations.

Required by Law ▪ Most states require users of highways to furnish evidence of automobile-bodily-injury and property-damage liability insurance in basic minimum limits. This is particularly true of businesses that have trucks or other heavy equipment on the highways or public roads. Special permits to move heavy equipment on the highways generally require somewhat higher limits of protection.

A contractor who operates in foreign nations generally finds that the liability-insurance requirements are even more stringent than those in the United States and that automobile liability insurance must be procured from an insurance company headquartered in the nation in which the contractor operates.

Required by Contract ▪ Almost without exception, construction contracts require the contractor to carry comprehensive liability insurance whose purpose is to protect the contractor, owner, and owner's engineers against all liability for bodily injuries or third-party property damage arising out of or in connection with the performance of the contract. Occasionally, the contract requires a separate Owner's Protective Liability Insurance policy. Also, when a contractor operates alongside or across the property of a railroad company, a Railroad Protective Liability Insurance policy is generally required.

Required by Common Sense ▪ Regardless of the coverages required by law or contract, the prudent contractor should carry liability-insurance protection in substantial amounts. The very nature of the construction industry subjects a contractor to the possibility of substantial risk of liability to third parties. In certain situations, particularly where the contractor is using explosives, risk may approach absolute liability.

4.16.2 Property Insurance

In addition to liability insurance, contractors must protect themselves against damage or loss to their own property and to the projects on which they are working.

Contractor's Equipment, Plant, Temporary Buildings, Materials, and Supplies Insurance ▪ Almost all assets of the typical construction contractor consist of contractor's equipment, construction plant, temporary buildings, materials, and supplies. Common sense dictates that contractors keep their property insured. Ordinarily, the contractor's heavy equipment and vehicles are purchased on conditional sales contracts or leased under agreements that require the contractor to maintain insurance against physical damage to the equipment and vehicles. Losses are payable to the contractor and the secured owners as their respective interests may appear at time of loss.

Contractors can maintain separate fire and theft coverage for heavy equipment and automobiles. They can also obtain collision coverage on their highway trucks and automobiles, and fire and extended-coverage insurance on plant and temporary buildings. However, "piecemeal" coverages do not provide sound all-risk protection on all property. Furthermore, the premiums in the aggregate often add up to more than the cost of a single all-risk blanket coverage on all property. Obviously too, the risks to which a contractor's property is subject stem from different and more varied sources than the risks of a merchant or manufacturer. For example, a contractor engaged in the construction of a dam may have little risk from fire or the usual extended perils, but risk from flood may be great. Yet, flood is generally a standard excepted peril in most property coverages.

Contractors' property insurance should be in an amount sufficient to cover the total values of property subject to any conceivable risk at one location. A contractor who has a normal recurrence of property losses may reduce the cost of insurance by arranging a deductible in an amount that approaches a normal loss recurrence. Ordinarily, the deductibles are based on the value of equipment at risk. A deductible of $1000 on equipment valued in excess of $5000 may be adequate to protect the ordinary contractor against calamitous loss and still be sufficient to provide coverage at the most reasonable premium cost. On equipment valued in excess of $10,000, a deductible of $2500 is reasonable. Generally, small tools, materials, and supplies can be covered in the same policy at a more reasonable premium than would be charged for a separate policy covering the contractor's inventory of these items.

Builder's All-Risk Insurance ▪ Invariably, the construction contract places full responsibility (and liability) on the contractor for protection of the project work and for repair or replacement of damage until the completed project work has been accepted by the owner. Occasionally, the owner carries "course of construction" insurance in which the contractor is an additional insured. In these situations, contractors should make sure they will be relieved of the responsibility for repair or replacement of damaged work. A contractor who assumes such responsibility, as is usual, should carry Builder's All-Risk Insurance.

Perhaps the most serious risk of damage to the work arises from the contractor's operations, such as failure of hoisting machinery or negligent operation of heavy equipment. Contractors' liability insurance would not be protection in such a situation because risks arising from their negligence or failure of machinery used by them are excluded under the standard "care, custody, and control" exclusion in the liability-insurance policy. Likewise, fire and extended-coverage insurance, being restricted to the specific perils named, would not insure contractors against loss resulting from operation of equipment, blasting, or other causes of risk usual to their operations.

Builder's All-Risk Insurance generally protects against any natural occurrence, act of God, or damage caused by human error. The possible loss can be substantial in amount. Hence, the policy limit should be adequate to cover the largest conceivable loss. Inasmuch as the contractor's main concern is protection against catastrophic loss, the contractor should require a high limit but permit a substantial deductible that will permit the purchase of this important coverage at the most reasonable cost.

4.16.3 Workmen's Compensation and Employee Benefits Insurance

In all states of the United States, Canada, and most foreign nations, Workmen's Compensation Insurance is required by law. The construction industry is regarded as "extra-hazardous" in the terminology employed in workers' compensation laws. Premiums are based on the classifications of work in which each craft of construction worker is engaged. Cost of Workmen's Compensation Insurance is an important factor in preparation of a bid.

Employer's Liability Insurance is automatically included in most Workmen's Compensation Insurance policies. A Workmen's Compensation Insurance claim is, without exception, the sole remedy of an injured worker or of the family of one who dies as the result of an industrial injury. Nevertheless, there may be occasions on which, because of liability assumed by contract or otherwise, a contractor may be required to defend an action at law or pay a judgment based on injuries to an employee or a subcontractor's employee.

In several states of the United States, commonly called the monopolistic-fund states, and in all provinces of Canada, Workmen's Compensation Insurance is required to be carried with the state or provincial fund. In these states and provinces, Employer's Liability Insurance is generally neither required by law nor furnished by the funds. The prudent contractor carries a special Employer's Liability Insurance policy with a private carrier when operating in these states and provinces.

Also, the contractor engaged in work bordering on a waterway or navigable stream should carry insurance for protection against liability under the Longshoremen's and Harbor Workers' Compensation Act and the Jones Act. These coverages can generally be provided by endorsement to the standard Workmen's Compensation Insurance policy at little or no additional premium.

Other coverages that the contractor may wish to consider but that are generally elective are group hospital, surgical and medical plans, and group term life and accidental death and dismemberment coverages. Often, these coverages will be provided by jointly administered employer-union benefit plans created by collective bargaining in the construction industry. The union plans, of course, are limited solely to the contractor's employees covered by a collective-bargaining agreement. It is up to the contractor to decide whether to provide similar coverage for salaried, managerial, engineering, and clerical personnel.

4.16.4 Miscellaneous Insurance Coverages

Contractors' miscellaneous insurance needs vary with the type and scope of their operations. Among those considered essential, however, are consequential loss insurance, fidelity and forgery insurance, and money and securities insurance.

Consequential Loss Insurance ▪ Contractors soon discover that physical-damage protection on the construction work in progress or on contractor's equipment will pay only a portion of their out-of-pocket financial losses. On permanent project work, Builder's All-Risk Insurance reimburses the actual cost of restoring the work. This recovery, of course, is limited to the original value of the work, and the deductible, which is generally substantial, is applied. No allowance is made for extra overhead incurred for the time required to repair or replace the damaged work, or for overtime expense. These are almost always excluded under the terms of the builder's-risk coverage. A contractor may procure a form of "business interruption" insurance that will pay the contractor any extra expense for extended overhead and overtime expense resulting from a builder's-risk type of loss.

The contractor who loses the use of equipment through physical damage must provide substitute equipment for the time during which damaged equipment is being repaired. Often, the contractor can obtain insurance along with the contractor's equipment coverage that covers rental expense of replacement equipment.

Fidelity and Forgery Insurance ▪ A contractor who has delegated authority with respect to the firm's business and financial affairs to one or more employees should carry fidelity insurance up to a limit adequate to cover such sums as the employees may deal with. Likewise, the prudent contractor carries depositor's forgery insurance to protect against financial loss caused by forgery of checks against banking accounts.

Money and Securities Insurance ▪ Ordinarily, the contractor keeps only small sums of cash in the office or at job offices. However, there may be some situations that require cash to be kept at the jobsite or office. In such situations, it is advisable to carry money and securities coverage, which protects the contractor against loss by outside theft, including burglary and robbery. This coverage should carry a limit equal to the largest sum of cash on hand at any one location.

4.16.5 "Coverage Boosters" and "Cost Savers"

A prudent selection of insurance plans, coupled with an active safety program, materially lessens the contractor's overall insurance costs.

Blanket Coverages and Package Plans ▪ One basic concept of insurance is "risk spreading." The more a risk can be spread, geographically or otherwise, the more economical will be the premium. Therefore, a contractor who insures all operations under a single policy against a common risk, whether it be a liability, physical damage, or fidelity, will enjoy the broadest protection at the lowest cost. Consider builder's-risk insurance, for instance: Some of a contractor's operations may be quite hazardous; others may be virtually risk-free. In such a situation, the contractor is able to maintain builder's-risk coverage at a reasonable rate on a hazardous project by charging all operations at the same premium rate, simply because the contractor's low-risk work contributes to the overall cost. The same reasoning holds with respect to other coverages.

Contractor's Safety Program ▪ Contractors should always be aware of one of the best cost savers available to them, namely, a good safety program. The largest insurance expenditure by far is the workers' compensation premium. Almost every underwriter of Workmen's Compensation Insurance offers substantial discounts, dividends, or retrospective premium-return plans that are based on low-accident experience. A contractor often can maintain a safety program at a cost much lower than the amount of the dividends earned from such premium returns. For the small contractor, almost every Workmen's Compensation Insurance carrier provides regular safety inspection and safety education materials and services.

On large projects with substantial payrolls, contractors can generally avail themselves of a retrospective premium-return plan, which essentially is a "cost-plus" insurance program. With a retrospective plan, the contractor pays the cost of injuries plus a modest amount to cover the insurance carrier's administrative expense and premium against a catastrophe or multiple-injury accident.

4.17 Bonds

Bonds are not insurance. A surety bond is equivalent to a cosigned promissory note. The principal on a surety bond, as on a promissory note, is primarily liable to the obligee. The surety, as is a cosigner, is liable only in the event that the principal fails to discharge the obligation undertaken.

The obligation undertaken in a contractor's surety bond runs in favor of the owner. And the owner, alone, is protected. The contractor, as principal, has no protection under a bond. On the contrary, the contractor is ultimately liable and fully obligated, not only to the owner, but to the surety company that issued the bond.

Contractors should read in full the applications they sign for bids, performance, or payment bonds. They will discover that they have pledged, transferred, and conveyed their entire assets and all contract revenues to the surety as security against the surety having to pay any amount or discharge any obligation under the bond. The smaller contractor pledges not only business but home and personal assets. If the contractor is an incorporated firm and its assets and income are insufficient to afford adequate security, the surety company will insist that the individual stockholders of the contracting company pledge sufficient personal assets to indemnify the surety adequately against loss.

The contractor pays a premium for a bond similar to interest on a promissory note. The premium charged depends on the type of construction to be performed, the time that the bond will be in effect, and the amount or contract price of the project to be built.

Almost all public construction and most larger private projects will require bid, performance, and payment bonds. Prudent contractors, intending to submit a bid, inquire of their surety companies whether they will write bid bonds for them. Generally, surety companies will not write a bid bond on a project without being satisfied as to the contractor's financial capacity. Once so satisfied, the surety, by issuing its bid bond, indicates its intention to write the performance and payment bonds if the contractor's bid is successful and a contract is awarded.

Bid bonds are generally based on the amount of the bid. For the most part, they run from 5 to 20% of the amount of the accompanying bid. This amount represents the damages or costs that the owner may incur. These include the losses if the bidder fails to enter into a contract and the work has to be readvertised for bids. Also, losses may be due to the difference in cost between the low bid submitted by a defaulting bidder and the next responsible bid, in the cases where the work must be awarded to the next lowest bidder.

Performance and payment bonds are usually in the full contract amount, or at least 50% of the contract amount. If, during the course of a project, the contractor defaults or becomes insolvent and is financially unable to carry on the work, the owner will require the surety to complete the work and to pay for labor, materials, and supplies. In such event, the surety, in discharging its obligations under the bond, has first claim as a secured creditor against the contractor's assets. Ultimately, the surety company's loss is the cost of completing the work less the recovery it can make from the contractor's assets.

5

Ruth T. Brantley
Senior Lecturer
University of Hawaii
Honolulu, Hawaii

L. Reed Brantley
Emeritus Professor
University of Hawaii
Honolulu, Hawaii

CONSTRUCTION MATERIALS*

This section describes the basic properties of materials commonly used in construction. For convenience, materials are grouped in the following categories: cementitious materials, metals, organic materials, and composites. Application of these materials is discussed in following sections. In these sections also, environmental degradation on the materials are described.

Cementitious Materials

Any substance that bonds materials may be considered a cement. There are many types of cements. In construction, however, the term cement generally refers to bonding agents that are mixed with water or other liquid, or both, to produce a cementing paste. Initially, a mass of particles coated with the paste is in a plastic state and may be formed, or molded, into various shapes. Such a mixture may be considered a cementitious material because it can bond other materials together. After a time, due to chemical reactions, the paste sets and the mass hardens. When the particles consist of fine aggregate (sand), mortar is formed. When the particles consist of fine and coarse aggregates, concrete results.

5.1 Types of Cementitious Materials

Cementitious materials may be classified in several different ways. One way often used is by the chemical constituent responsible for setting or hardening the cement. Silicate and aluminate cements, in which the setting agents are calcium silicates and aluminates, are the most widely used types.

Limes, wherein the hardening is due to the conversion of hydroxides to carbonates, were formerly widely used as the sole cementitious material, but their slow setting and hardening are not compatible with modern requirements. Hence, their principal function today is to plasticize the otherwise harsh cements and add resilience to mortars and stuccoes. Use of limes is beneficial in that their slow setting promotes healing, the re-cementing of hairline cracks.

Another class of cements is composed of calcined gypsum and its related products. The gypsum cements are widely used in interior plaster and for fabrication of boards and blocks; but the solubility of gypsum prevents its use in construction exposed to any but extremely dry climates.

Oxychloride cements constitute a class of specialty cements of unusual properties. Their cost prohibits their general use in competition with the cheaper cements; but for special uses, such as the

---

*With excerpts from F. S. Merritt and J. T. Ricketts, "Building Design and Construction Handbook," Sec. 4, "Building Materials," by F. S. Merritt and D. J. Akers, McGraw-Hill, Inc., New York.

production of sparkproof floors, they cannot be equaled.

Masonry cements or mortar cements are widely used because of their convenience. While they are, in general, mixtures of one or more of the above-mentioned cements with some admixtures, they deserve special consideration because of their economies.

Other cementitious materials, such as polymers, fly ash, and silica fume, may be used as a cement replacement in concrete. Polymers are plastics with long-chain molecules. Concretes made with them have many qualities much superior to those of ordinary concrete.

Silica fume, also known as microsilica, is a waste product of electric-arc furnaces. The silica reacts with lime in concrete to form a cementitious material. A fume particle has a diameter only 1% of that of a cement particle.

5.2 Portland Cements

Particles that become a bonding agent when mixed with water are referred to as hydraulic cements. The most widely used cements in construction are portland cements, which are made by blending a mixture of calcareous (lime-containing) materials and argillaceous (clayey) materials. (See Art. 5.3 for descriptions of other types of hydraulic cements.) The raw materials are carefully proportioned to provide the desired amounts of lime, silica, aluminum oxide, and iron oxide. After grinding to facilitate burning, the raw materials are fed into a long rotary kiln, which is maintained at a temperature around 2700 °F. The raw materials, burned together, react chemically to form hard, walnut-sized pellets of a new material, clinker.

The clinker, after discharge from the kiln and cooling, is ground to a fine powder (not less than 1600 cm^2/g specific surface). During this grinding process, a retarder (usually a few percent of gypsum) is added to control the rate of setting when the cement is eventually hydrated. The resulting fine powder is portland cement.

Four compounds, however, make up more than 90% of portland cement, by weight; tricalcium silicate (C_3S), dicalcium silicate (C_2S), tricalcium aluminate (C_3A), and tetracalcium aluminoferrite (C_4AF). Each of these four compounds is identifiable in the highly magnified microstructure of portland cement clinker, and each has characteristic properties that it contributes to the final mixture.

5.2.1 Hydration of Cement

When water is added to portland cement, the basic compounds present are transformed to new compounds by chemical reactions [Eq. (5.2)].

$$\begin{aligned}
&\text{Tricalcium silicate} + \text{water} \\
&\qquad \rightarrow \text{tobermorite gel} + \text{calcium hydroxide} \\
&\text{Dicalcium silicate} + \text{water} \\
&\qquad \rightarrow \text{tobermorite gel} + \text{calcium hydroxide} \\
&\text{Tetracalcium aluminoferrite} + \text{water} \\
&\quad + \text{calcium hydroxide} \\
&\qquad \rightarrow \text{calcium aluminoferrite hydrate} \qquad (5.1) \\
&\text{Tricalcium aluminate} + \text{water} \\
&\quad + \text{calcium hydroxide} \\
&\qquad \rightarrow \text{tetracalcium aluminate hydrate} \\
&\text{Tricalcium aluminate} + \text{water} + \text{gypsum} \\
&\qquad \rightarrow \text{calcium monosulfoaluminate}
\end{aligned}$$

Two calcium silicates, which constitute about 75% of portland cement by weight, react with the water to produce two new compounds: tobermorite gel, which is not crystalline, and calcium hydroxide, which is crystalline. In fully hydrated portland cement paste, the calcium hydroxide accounts for 25% of the weight and the tobermorite gel makes up about 50%. The third and fourth reactions in Eq. (5.1) show how the other two major compounds in portland cement combine with water to form reaction products. The final reaction involves gypsum, the compound added to portland cement during grinding of the clinker to control set.

Each product of the hydration reaction plays a role in the mechanical behavior of the hardened paste. The most important of these, by far, is the *tobermorite gel*, which is the main cementing component of cement paste. This gel has a composition and structure similar to those of a naturally occurring mineral, called tobermorite, named for the area where it was discovered, Tobermory in Scotland. The gel is an extremely finely divided substance with a coherent structure.

The average diameter of a grain of portland cement as ground from the clinker is about 10 μm. The particles of the hydration product, tobermorite gel, are on the order of a thousandth of that size. Particles of such small size can be observed only by using the magnification available in an electron microscope. The enormous surface area of the gel (about 3 million cm^2/g) results in attractive forces between particles since atoms on each surface are attempting to complete their unsaturated bonds by adsorption. These forces cause particles of tobermorite gel to adhere to each other and to other particles introduced into the cement paste. Thus, tobermorite gel forms the heart of hardened cement paste and concrete in that it cements everything together.

5.2.2 Effects of Portland Cement Compounds

Each of the four major compounds of portland cement contributes to the behavior of the cement as it proceeds from the plastic to the hardened state after hydration. Knowledge of the behavior of each major compound upon hydration permits the amounts of each to be adjusted during manufacture to produce desired properties in the cement.

Tricalcium silicate (C_3S) is primarily responsible for the high early strength of hydrated portland cement. It undergoes initial and final set within a few hours. The reaction of C_3S with water gives off a large quantity of heat (heat of hydration). The rate of hardening of cement paste is directly related to the heat of hydration; the faster the set, the greater the exotherm. Hydrated C_3S compound attains most of its strength in 7 days.

Dicalcium silicate (C_2S) is found in three different forms, designated alpha, beta, and gamma. Since the alpha phase is unstable at room temperature and the gamma phase shows no hardening when hydrated, only the beta phase is important in portland cement.

Beta C_2S takes several days to set. It is primarily responsible for the later-developing strength of portland cement paste. Since the hydration reaction proceeds slowly, the heat of hydration is low. The beta C_2S compound in portland cement generally produces little strength until after 28 days, but the final strength of this compound is equivalent to that of the C_3S.

Tricalcium aluminate (C_3A) exhibits an instantaneous or flash set when hydrated. It is primarily responsible for the initial set of portland cement and gives off large amounts of heat upon hydration. The gypsum added to the portland cement

during grinding in the manufacturing process combines with the C_3A to control the time to set. The C_3A compound shows little strength increase after 1 day. Although hydrated C_3A alone develops a very low strength, its presence in hydrated portland cement produces more desirable effects. An increased amount of C_3A in portland cement results in faster sets and also decreases the resistance of the final product to sulfate attack.

Tetracalcium aluminoferrite (C_4AF), is similar to C_3A in that it hydrates rapidly and develops only low strength. Unlike C_3A, however, it does not exhibit a flash set.

In addition to composition, speed of hydration is affected by fineness of grinding, amount of water added, and temperatures of the constituents at the time of mixing. To achieve faster hydration, cements are ground finer. Increased initial temperature and the presence of a sufficient amount of water also speed the reaction rate.

5.2.3 Specifications for Portland Cements

Portland cements are normally made in five types, the properties of these types being standardized on the basis of the ASTM Standard Specification for Portland Cement (C150). Distinction between the types is based on both chemical and physical requirements. Some requirements, extracted from ASTM C150, are shown in Table 5.1. Most cements exceed the strength requirements of the specification by a comfortable margin.

Type I, general-purpose cement, is the one commonly used for structural purposes when the special properties specified for the other four types of cement are not required.

Type II, modified general-purpose cement, is used where a moderate exposure to sulfate attack is anticipated or a moderate heat of hydration is required. These characteristics are attained by placing limitations on the C_3A and C_3S content of the cement. Type II cement gains strength a little more slowly than Type I but ultimately reaches equal strength. Type II cement, when optional chemical requirements, as indicated in Table 5.2, are met, may be used as a low-alkali cement where alkali-reactive aggregates are present in concrete.

Type III, high-early-strength cement, is designed for use when early strength is needed in a

Table 5.1 Chemical and Physical Requirements for Portland Cement*

Type:	I and IA	II and IIA	III and IIIA	IV	V
Name:	General-Purpose	Modified	High-Early-Strength	Low-Heat	Sulfate-Resisting
C_3S, max %				35	
C_2S, min %				40	
C_3A, max %		8	15	7	5
SiO_2, min %		20			
Al_2O_3, max %		6			
Fe_2O_3, max %		6		6.5	
MgO, max %	6	6	6	6	6
SO_3, max %:					
When $C_3A \leq 8\%$	3	3	3.5	2.3	2.3
When $C_3A > 8\%$	3.5		4.5		
$C_4AF + 2(C_3A)$, max %					25
Fineness, specific surface, m^2/kg					
Average min, by turbidimeter	160	160		160	160
Average min, by air permeability test	280	280		280	280
Compressive strength, psi, mortar cubes of 1 part cement and 2.75 parts graded standard sand after:					
1 day min					
Standard			1800		
Air-entraining			1450		
3 days min					
Standard	1800	1500	3500		1200
Air-entraining	1450	1200	2800		
7 days min					
Standard	2800	2500		1000	2200
Air-entraining	2250	2000			
28 days min					
Standard				2500	3000

*Based on requirements in "Standard Specification for Portland Cement," ASTM C150. See current edition of C150 for exceptions, alternatives, and changes in requirements.

particular construction situation. Concrete made with Type III cement develops in 7 days the same strength that it takes 28 days to develop in concretes made with Types I or II cement. This high early strength is achieved by increasing the C_3S and C_3A content of the cement and by finer grinding. No minimum is placed upon the fineness by specification, but a practical limit occurs when the particles are so small that minute amounts of moisture will prehydrate the cement during handling and storage. Since it has high heat evolution, Type III cement should not be used in large masses. With 15% C_3A, it has poor sulfate resistance. The C_3A content may be limited to 8% to obtain moderate sulfate resistance or to 5% when high sulfate resistance is required.

Type IV, low-heat-of-hydration cement, has been developed for mass-concrete applications. If Type I cement is used in large masses that cannot lose heat by radiation, it liberates enough heat

Table 5.2 Optional Chemical Requirements for Portland Cement*

Cement type	I and IA	II and IIA	III and IIIA	IV	V
Tricalcium aluminate (C_3A), max %					
For moderate sulfate resistance			8		
For high sulfate resistance			5		
Sum of tricalcium silicate and tricalcium aluminate, max %[†]		58			
Alkalies ($Na_2O + 0.685K_2O$), max %[‡]	0.60	0.60	0.60	0.60	0.60

*These optional requirements apply only if specifically requested. Availability should be verified.

[†]For use when moderate heat of hydration is required.

[‡]Low-alkali cement. This limit may be specified when cement is to be used in concrete with aggregates that may be deleteriously reactive. See "Standard Specification for Concrete Aggregates," ASTM C33.

during hydration to raise the temperature of the concrete as much as 50 or 60 °F. This results in a relatively large increase in dimensions while the concrete is still plastic, and later differential cooling after hardening causes shrinkage cracks to develop. Low heat of hydration in Type IV cement is achieved by limiting the compounds that make the greatest contribution to heat of hydration, C_3A and C_3S. Since these compounds also produce the early strength of cement paste, their limitation results in a paste that gains strength relatively slowly. The heat of hydration of Type IV cement usually is about 80% of that of Type II, 65% of that of Type I, and 55% of that of Type III after the first week of hydration. The percentages are slightly higher after about 1 year.

Type V, sulfate-resisting cement, is specified where there is extensive exposure to sulfates. Typical applications include hydraulic structures exposed to water with high alkali content and structures subjected to seawater exposure. The sulfate resistance of Type V cement is achieved by reducing the C_3A content to a minimum since that compound is most susceptible to sulfate attack.

Types IV and V are specialty cements not normally carried in dealer's stocks. They are usually obtainable for use on a large project if advance arrangements are made with a cement manufacturer.

Air-entraining portland cements (ASTM C226) are available for the manufacture of concrete for exposure to severe frost action. These cements are available in Types I, II, and III but not in Types IV and V. When an air-entraining agent has been added to the cement by the manufacturer, the cement is designated Type IA, IIA, or IIIA.

5.3 Other Types of Hydraulic Cements

Although portland cements (Art. 5.2) are the most common modern hydraulic cements, several other kinds are in everyday use.

5.3.1 Aluminous Cements

These are prepared by fusing a mixture of aluminous and calcareous materials (usually bauxite and limestone) and grinding the resultant product to a fine powder. These cements are characterized by their rapid-hardening properties and the high strength developed at early ages. Table 5.3 shows the relative strengths of 4-in cubes of 1:2:4 concrete made with normal portland, high-early-strength portland, and aluminous cements.

Since a large amount of heat is liberated with rapidity by aluminous cement during hydration, care must be taken not to use the cement in places where this heat cannot be dissipated. It is usually not desirable to place aluminous-cement concretes in lifts of over 12 in; otherwise the temperature rise may cause serious weakening of the concrete.

Aluminous cements are much more resistant to the action of sulfate waters than are portland cements. They also appear to be much more resistant to attack by water containing aggressive carbon dioxide or weak mineral acids than the silicate cements. Their principal use is in concretes where advantage may be taken of their very high early strength or of their sulfate resistance, and where the extra cost of the cement is not an important factor.

Table 5.3 Relative Strengths of Concrete Made from Portland Aluminous Cements*

	Compressive Strength, psi		
Days	Normal Portland	High-Early Portland	Aluminous
1	460	790	5710
3	1640	2260	7330
7	2680	3300	7670
28	4150	4920	8520
56	4570	5410	8950

*Adapted from F. M. Lea, "Chemistry of Cement and Concrete," St. Martin's Press, Inc., New York.

Another use of aluminous cements is in combination with firebrick to make refractory concrete. As temperatures are increased, dehydration of the hydration products occurs. Ultimately, these compounds create a ceramic bond with the aggregates.

5.3.2 White Portland Cement

These produce mortars of brilliant white color for use in architectural applications. To obtain this white color in the cement, it is necessary to use raw materials with a low iron-oxide content, to use fuel free of pyrite, and to burn at a temperature above that for normal portland cement. The physical properties generally conform to the requirements of a Type I portland cement.

5.3.3 Natural Cements

Natural cements are formed by calcining a naturally occurring mixture of calcareous and argillaceous substances at a temperature below that at which sintering takes place. The "Specification for Natural Cement," ASTM C10, requires that the temperature be no higher than necessary to drive off the carbonic acid gas. Since natural cements are derived from naturally occurring materials and no particular effort is made to adjust the composition, both the composition and properties vary rather widely. Some natural cements may be almost the equivalent of portland cement in properties; others are much weaker. Natural cements are principally used in masonry mortars and as an admixture in portland-cement concretes.

5.3.4 Limes

These are made principally of calcium oxide (CaO), occurring naturally in limestone, marble, chalk, coral, and shell. For building purposes, they are used chiefly in mortars. Limes are produced by driving out water from the natural materials. Their cementing properties are caused by the reabsorption of the expelled water and the formation of the same chemical compounds of which the original raw material was composed.

Hydraulic lime is made by calcining a limestone containing silica and alumina to a temperature short of incipient fusion. In slaking (hydration), just sufficient water is provided to hydrate the free lime so as to form sufficient free lime (CaO) to permit hydration and to leave unhydrated sufficient calcium silicates to give the dry powder its hydraulic properties. Because of the low silicate and high lime contents, hydraulic limes are relatively weak. They are principally used in masonry mortars.

Quicklime is the product of calcination (making powdery by heating) of limestone containing large proportions of calcium carbonate ($CaCO_3$) and some magnesium carbonate ($MgCO_3$). The calcination evaporates the water in the stone, heats the limestone to a high enough temperature for chemical dissociation, and drives off carbon dioxide as a gas, leaving the oxides of calcium and magnesium. The resulting calcium oxide (CaO), called quicklime, has a great affinity for water.

Quicklime intended for use in construction must first be combined with the proper amount of water to form a lime paste, a process called *slaking*. When quicklime is mixed with from two to three times its weight of water, the calcium oxide combines with the water to form calcium hydroxide, and sufficient heat is evolved to bring the entire mass to a boil.

The resulting product is a suspension of finely divided calcium hydroxide (and magnesium oxide) which, upon cooling, stiffens to a putty. This putty, after a period of seasoning, is used principally in masonry mortar, to which it imparts workability. It may also be used as an admixture in concrete to improve workability.

Hydrated limes are prepared from quicklimes by the addition of a limited amount of water during the manufacturing process. Hydrated lime was developed so that greater control could be exercised over the slaking operation by having it carried out during manufacture rather than on the construction job. After the hydration process ceases to evolve heat, a fine, dry powder is left as the resulting product.

Hydrated lime can be used in the field in the same manner as quicklime, as a putty or paste, but it does not require a long seasoning period. It can also be mixed with sand while dry, before water is added. Hydrated lime can be handled more easily than quicklime because it is not so sensitive to moisture. The plasticity of mortars made with hydrated limes, although better than that obtained with most cements, is not nearly so high as that of mortars made with an equivalent amount of slaked quicklime putty.

5.3.5 Gypsum Cements

Mineral gypsum, when pure, consists of crystalline calcium sulfate dihydrate ($CaSO_4 \cdot 2H_2O$). When it is heated to temperatures above 212 °F but not exceeding 374 °F, three-fourths of the water of crystallization is driven off. The resulting product, $CaSO_4 \cdot \frac{1}{2}H_2O$, called **plaster of paris**, is a fine, white powder. When recombined with water, it sets rapidly and attains strength on drying by reforming the original calcium sulfate dihydrate. Plaster of paris is used as a molding or gaging plaster or is combined with fiber or sand to form a "cement" plaster. Gypsum plasters have a strong set and gain their full strength when dry.

5.3.6 Oxychloride Cements

Magnesium oxychloride cements are formed by a reaction between lightly calcined magnesium oxide (MgO) and a strong aqueous solution of magnesium chloride ($MgCl_2$). The resulting product is a dense, hard cementing material with a crystalline structure. This oxychloride cement, or Sorel cement, develops better bonding with aggregate than portland cement. It is often mixed with colored aggregate in making flooring compositions or used to bond wood shavings or sawdust in making partition block or tile. It is moderately resistant to water but should not be used under continuously wet conditions. A similar oxychloride cement is made by mixing zinc oxide and zinc chloride.

5.3.7 Masonry Cements

Masonry cements, or mortar cements, are intended to be mixed with sand and used for setting unit masonry, such as brick, tile, and stone. They may be any one of the hydraulic cements already discussed or mixtures of them in any proportion.

Many commercial masonry cements are mixtures of portland cement and pulverized limestone, often containing as much as 50 or 60% limestone. They are sold in bags containing from 70 to 80 lb, each bag nominally containing a cubic foot. Price per bag is commonly less than that of portland cement, but because of the use of the lighter bag, cost per ton is higher than that of portland cement.

Since there are no limits on chemical content and physical requirements, masonry cement specifications are quite liberal. Some manufacturers vary the composition widely, depending on competition, weather conditions, or availability of materials. Resulting mortars may vary widely in properties.

5.3.8 Fly Ashes

Fly ash meeting the requirements of ASTM C618, "Specification for Fly Ash and Raw or Calcined Natural Pozzolan for Use as a Mineral Admixture in Portland Cement Concrete," is generally used as a cementitious material as well as an admixture.

Natural pozzolans are derived from some diatomaceous earths, opaline cherts and shales, and other materials. While part of a common ASTM designation with fly ash, they are not as readily available as fly ashes and thus do not generate the same level of interest or research.

Fly ashes are produced by coal combustion, generally in an electrical generating station. The ash that would normally be released through the chimney is captured by various means, such as electrostatic precipitators. The fly ash may be sized prior to shipment to concrete suppliers.

All fly ashes possess pozzolanic properties, the ability to react with calcium hydroxide at ordinary temperatures to form compounds with cementitious properties. When cement is mixed with water, a chemical reaction (hydration) occurs. The product of this reaction is calcium silicate hydrate (CSH) and calcium hydroxide [$Ca(OH)_2$]. Fly ashes have high percentages of silicon dioxide (SiO_2). In the presence of moisture, the $Ca(OH)_2$, will react with the SiO_2 to form another CSH.

Type F ashes are the result of burning anthracite or bituminous coals and possess pozzolanic properties. They have been shown by research and practice to provide usually increased sulfate resistance and to reduce alkali-aggregate expansions. Type C fly ashes result from burning lignite or subbituminous coals. Because of the chemical properties of the coal, the Type C fly ashes have some cementitious properties in addition to their pozzolanic properties. Type C fly ashes may reduce the durability of concretes into which they are incorporated.

5.3.9 Silica Fume (Microsilica)

Silica fume, or microsilica, is a condensed gas, the by-product of metallic silicon or ferrosilicon alloys produced by electric arc furnaces. [While both terms are correct, microsilica (MS) is a less confusing name.] The Canadian standard CAN/CSA-A23.5-M86, "Supplementary Cementing Materials," limits amorphous SiO_2, to a maximum of 85% and oversize to 10%. Many microsilicas contain more than 90% SiO_2.

MS has an average diameter of 0.1 to 0.2 μm, a particle size of about 1% that of portland cement. Because of this small size, it is not possible to utilize MS in its raw form. Manufacturers supply it either densified, in a slurry (with or without water-reducing admixtures), or pelletized. Either the densified or slurried MS can be utilized in concrete. The pelletized material is densified to the point that it will not break down during mixing.

Because of its extremely small size, MS imparts several useful properties to concrete. It greatly increases long-term strength. It very efficiently reacts with the $Ca(OH)_2$ and creates a beneficial material in place of a waste product. MS is generally used in concrete with a design strength in excess of 12,000 psi. It provides increased sulfate resistance to concrete, and it significantly reduces the permeability of concrete. Also, its small size allows MS to physically plug microcracks and tiny openings.

5.4 Mortars and Grouts

Mortars are composed of a cement, fine aggregate (sand), and water. They are used for bedding unit masonry, for plasters and stuccoes, and with the addition of coarse aggregate, for concrete. Properties of mortars vary greatly, being dependent on the properties of the cement used, ratio of cement to sand, characteristics and grading of the sand, and ratio of water to solids.

Grouts are similar in composition to mortars but mixes are proportioned to produce, before setting, a flowable consistency without segregation of the components.

5.4.1 Packaging and Proportioning of Mortar

Mortars are usually proportioned by volume. A common specification is that not more than 3 ft^3 of sand be used with 1 ft^3 of cementitious material. Difficulty is sometimes encountered, however, in determining just how much material constitutes a cubic foot: a bag of cement (94 lb) by agreement is called a cubic foot in proportioning mortars or concretes, but an actual cubic foot of lime putty may be used in proportioning mortars. Since hydrated limes are sold in 50-lb bags (Art. 5.3.4), each of which makes somewhat more than a cubic foot of putty, weights of 40, 42, and 45 lb of hydrated lime have been used as a cubic foot in laboratory studies, but on the job, a bag is frequently used as a cubic foot. Masonry cements are sold in bags containing 70 to 80 lb (Art. 5.3.7), and a bag is considered a cubic foot.

5.4.2 Properties of Mortars

Table 5.4 lists types of mortars as a guide in selection for unit masonry.

Workability is an important property of mortars, particularly of those used in conjunction with unit masonry of high absorption. Workability is controlled by the character of the cement and amount of sand. For example, a mortar made from 3 parts sand and 1 part slaked-lime putty will be more workable than one made from 2 parts sand

Table 5.4   Types of Mortar

Mortar Type	Parts by Volume				Min Avg Compressive Strength of Three 2-in Cubes at 28 Days, psi
	Portland Cement	Masonry Cement	Hydrated Lime or Lime Putty	Aggregate Measured in Damp, Loose Condition	
M	1	1			2500
	1		¼		
S	½	1			1800
	1		Over ¼ to ½	Not less than 2¼ and not more than 3 times the sum of the volumes of the cements and limes used	
N	1	1			750
	1		Over ½ to 1¼		
O	1	1			350
	1		Over 1¼ to 2½		
K	1		Over 2¼ to 4		75
PL	1		¼ to ½		2500
PM	1	1			2500

and 1 part portland cement. But the 3 : 1 mortar has lower strength. By proper selection or mixing of cementitious materials, a satisfactory compromise may usually be obtained, producing a mortar of adequate strength and workability.

Water retention—the ratio of flow after 1-min standard suction to the flow before suction—is used as an index of the workability of mortars. A high value of water retention is considered desirable for most purposes. There is, however, a wide variation in water retention of mortars made with varying proportions of cement and lime and with varying limes. The "Standard Specification for Mortar for Unit Masonry," ASTM C270, requires mortar mixed to an initial flow of 100 to 115, as determined by the test method of ASTM C109, to have a flow after suction of at least 75%.

Strength of mortar is frequently used as a specification requirement, even though it has little relation to the strength of masonry. (See, for example, ASTM C270, C780, and C476.) The strength of mortar is affected primarily by the amount of cement in the matrix. Other factors of importance are the ratio of sand to cementing material, curing conditions, and age when tested.

Volume change of mortars constitutes another important property. Normal volume change (as distinguished from unsoundness) may be considered as the shrinkage during early hardening, shrinkage on drying, expansion on wetting, and changes due to temperature.

After drying, mortars expand again when wetted. Alternate wetting and drying produces alternate expansion and contraction which apparently continues indefinitely with portland-cement mortars.

Coefficients of thermal expansion of several mortars, reported in "Volume Changes in Brick Masonry Materials," Journal of Research of the National Bureau of Standards, vol. 6, p. 1003, ranged from 0.38×10^{-5} to 0.60×10^{-5} for masonry-cement mortars; from 0.41×10^{-5} to 0.53×10^{-5} for lime mortars, and from 0.42×10^{-5} to 0.61×10^{-5} for cement mortars. Composition of the cementitious material apparently has little effect on the coefficient of thermal expansion of a mortar.

5.4.3   High-Bond Mortars

When polymeric materials, such as styrene-butadiene and polyvinylidene chloride, are added to mortar, greatly increased bonding, compressive, and shear strengths result. To obtain high strength, the other materials, including sand, water, Type I or III portland cement, and a workability additive, such as pulverized ground limestone or marble dust, must be of quality equal to that of the ingredients of standard mortar. The high strength of the mortar enables masonry to withstand appreciable bending and tensile stresses. This makes possible thinner walls and prelaying of single-wythe panels that can be hoisted into place.

5.5 Types of Concrete

A concrete may be any of several manufactured, stone-like materials composed of particles, called aggregates, that are selected and graded into specified sizes for construction purposes, usually with a substantial portion retained on a No. 4 (4.75 mm) sieve, and that are bonded together by one or more cementitious materials into a solid mass.

The term concrete, when used without a modifying adjective, ordinarily is intended to indicate the product formed from a mix of portland cement, sand, gravel or crushed stone, and water. There are, however, many different types of concrete. Some are distinguished by the types, sizes, and densities of aggregates; for example, wood-fiber, lightweight, normal-weight, or heavy-weight concrete. The names of others may indicate the type of binder used; for example, blended-hydraulic-cement, natural-cement, polymer, or bituminous (asphaltic) concrete.

Concretes are similar in composition to mortars (Art. 5.4), which are used to bond unit masonry. Mortars, however, are normally made with sand as the sole aggregate, whereas concretes contain both fine aggregates and much larger size aggregates and thus usually have greater strength. Concretes therefore have a much wider range of structural applications, including pavements, footings, pipes, unit masonry, floor slabs, beams, columns, walls, dams, and tanks.

For design of a concrete mix, ingredients are specified to achieve specific objectives, such as strength, durability, abrasion resistance, low volume change, and minimum cost. The ingredients are mixed together so as to ensure that coarse, or large-size, aggregates are uniformly dispersed, that fine aggregates fill the gaps between the larger ones, and that all aggregates are coated with cement. Before the cementing action commences, the mix is plastic and can be rolled or molded in forms into desired shapes. Recommended practices for measuring, mixing, transporting, placing, and testing concretes are promulgated by such organizations as the American Concrete Institute (ACI) and the American Association of State Transportation and Highway Officials (AASHTO).

Concretes may be classified as flexible or rigid. These characteristics are determined mainly by the cementitious materials used to bond the aggregates.

5.5.1 Flexible Concretes

Usually, bituminous, or asphaltic, concretes are used when a flexible concrete is desired. Flexible concretes tend to deform plastically under heavy loads or when heated. The main use of such concretes is for pavements.

The aggregates generally used are sand, gravel, or crushed stone, and mineral dust, and the binder is asphalt cement, an asphalt specifically refined for the purpose. A semisolid at normal temperatures, the asphalt cement may be heated until liquefied for binding of the aggregates. Ingredients usually are mixed mechanically in a "pug mill," which has pairs of blades revolving in opposite directions. While the mix is still hot and plastic, it can be spread to a specified thickness and shaped with a paving machine and compacted with a roller or by tamping to a desired density. When the mix cools, it hardens sufficiently to withstand heavy loads.

Sulfur, rubber, or hydrated lime may be added to an asphaltic-concrete mix to improve the performance of the product.

5.5.2 Rigid Concretes

Ordinary rigid concretes are made with portland cement, sand, and stone or crushed gravel. The mixes incorporate water to hydrate the cement to bond the aggregates into a solid mass. These concretes meet the requirements of such standard specifications as ASTM C685, "Concrete Made by Volumetric Batching and Continuous Mixing," or C94, "Ready-Mixed Concrete." Substances called admixtures may be added to the mix to achieve specific properties both of the mix and the hardened concrete. ACI published a recommended practice for measuring, mixing, transporting, and placing concrete.

Other types of rigid concretes include nailable concretes; insulating concretes; heavyweight concretes; lightweight concretes; fiber-reinforced concretes, embedding short steel or glass fibers for resistance to tensile stresses; polymer and pozzolan concretes, to improve several concrete properties; and silica-fume concretes, for high strength. Air-entrained concretes, which contain tiny, deliberately created, air bubbles, may be considered variations of ordinary concrete if in conformance with ASTM C685 or C94. (See also Art. 5.6.)

Because ordinary concrete is much weaker in tension than in compression, it is usually reinforced or prestressed with a much stronger material, such as steel, to resist tension. Use of plain, or unreinforced, concrete is restricted to structures in which tensile stresses will be small, such as massive dams, heavy foundations, and unit-masonry walls.

5.6 Portland Cement Concretes

This mixture of portland cement (Art. 5.2) fine aggregate, coarse aggregate, air, and water is a temporarily plastic material, which can be cast or molded, but is later converted to a solid mass by chemical reaction. The user of concrete desires adequate strength, placeability, and durability at minimum cost. The concrete designer may vary the proportions of the five constituents of concrete over wide limits to attain these aims. The principal variables are the water-cement ratio, cement-aggregate ratio, size of coarse aggregate, ratio of fine aggregate to coarse aggregate, type of cement, and use of admixtures.

Established basic relationships and laboratory tests provide guidelines for approaching optimum combinations. ACI 211. 1, "Recommended Practice for Selecting Proportions for Normal and Heavy-weight Concrete," and ACI 211.2, "Recommended Practice for Selecting Proportions for Structural Lightweight Concrete," American Concrete Institute, P.O. Box 19150, Redford Station, Detroit, Mich. 48219, provide data for mix design under a wide variety of specified conditions.

5.6.1 Aggregates for Portland Cement Concretes

Aggregate is a broad term encompassing boulders, cobbles, crushed stone, gravel, air-cooled blast furnace slag, native and manufactured sands, and manufactured and natural lightweight aggregates. Aggregates may be further described by their respective sizes.

Normal-Weight Aggregates ▪ These typically have specific gravities between 2.0 and 3.0. They are usually distinguished by size as follows:

Boulders	Larger than 6 in
Cobbles	6 to 3 in
Coarse aggregate	3 in to No. 4 sieve
Fine aggregate	No. 4 sieve to No. 200 sieve
Mineral filler	Material passing No. 200 sieve

Used in most concrete construction, normal-weight aggregates are obtained by dredging riverbeds or mining and crushing formational material. Concrete made with normal-weight fine and coarse aggregates generally weighs about 144 lb/ft^3.

Boulders and cobbles are generally not used in their as-mined size but are crushed to make various sizes of coarse aggregate and manufactured sand and mineral filler. Gravels and naturally occurring sand are produced by the action of water and weathering on glacial and river deposits. These materials have round, smooth surfaces and particle-size distributions that require minimal processing. These materials can be supplied in either coarse or fine-aggregate sizes.

Fine aggregates have 100% of their material passing the ⅜-in sieve. Coarse aggregates have the bulk of the material retained on the No. 4 sieve.

Aggregates comprise about 75%, by volume, of a typical concrete mix. Cleanliness, soundness, strength, and particle shape are important in any aggregate. Aggregates are considered clean if they are free of excess clay, silt, mica, organic matter, chemical salts, and coated grains. An aggregate is physically sound if it retains dimensional stability under temperature or moisture change and resists weathering without decomposition. To be considered adequate in strength, an aggregate should be able to develop the full strength of the cementing matrix. When wear resistance is important, the aggregate should be hard and tough.

Several processes have been developed for improving the quality of aggregates that do not meet desired specifications. Washing may be used to remove particle coatings or change aggregate gradation. Heavy-media separation, using a variable-specific-gravity liquid, such as a suspension of water and finely ground magnetite and ferrosilicon, can be used to improve coarse aggregates. Deleterious lightweight material is removed by flotation, and heavyweight particles settle out. Hydraulic jigging, where lighter particles are carried upward by pulsations caused by air or rubber diaphragms, is also a means for separation of lighter particles. Soft, friable particles can be separated from hard, elastic particles by a process called *elastic fractionation*. Aggregates are dropped onto an inclined hardened-steel surface, and their quality is measured by the distance they bounce.

Aggregates that contain certain forms of silicas or carbonates may react with the alkalies present in portland cement (sodium oxide and potassium oxide). The reaction product cracks the concrete or may create pop-outs at the concrete surface. The reaction is more pronounced when the concrete is in a warm, damp environment.

The potential reactivity of an aggregate with alkalies can be determined either by a chemical test (ASTM C289) or by a mortar-bar method (ASTM C227). The mortar-bar method is the more rigorous test and provides more reliable results but it requires a much longer time to perform.

Hardness of coarse aggregate is measured by the Los Angeles Abrasion Test, ASTM C131 or C595. These tests break the aggregate down by impacting it with steel balls in a steel tumbler. The resulting breakdown is not directly related to the abrasion an aggregate receives in service, but the results can be empirically related.

Soundness of aggregate is measured by ASTM C88, "Test Method for Soundness of Aggregates by Use of Sodium Sulfate or Magnesium Sulfate." This test measures the amount of aggregate degradation when exposed to alternating cycles of wetting and drying in a sulfate solution.

Particle shape has a significant effect on properties of concrete. Natural sand and gravel have a round, smooth particle shape. Crushed aggregate (coarse and fine) may have shapes that are flat and elongated, angular, cubical, disk, or rodlike. These shapes result from the crushing equipment employed and the aggregate mineralogy. Extreme angularity and elongation increase the amount of cement required to give strength, difficulty in finishing, and effort required to pump the concrete. Flat and elongated particles also increase the amount of required mixing water.

The bond between angular particles is greater than that between smooth particles. Properly graded angular particles can take advantage of this property and offset the increase in water required to produce concrete with cement content and strength equal to that of a smooth-stone mix.

Resistance to freezing and thawing is affected by aggregate pore structure, absorption, porosity, and permeability. Aggregates that become critically saturated and then freeze cannot accommodate the expansion of the frozen water. Empirical data show that freeze-thaw deterioration of concrete is caused by coarse aggregates, not fine. A method prescribed in "Test Method for Resistance of Concrete to Rapid Freezing and Thawing," ASTM C666, measures concrete performance by weight changes, a reduction in the dynamic modulus of elasticity, and increases in sample length.

Erratic setting times and rates of hardening may be caused by organic impurities in the aggregates, primarily the sand. The presence of these impurities can be investigated by a method given in "Test Method for Organic Impurities in Fine Aggregates for Concrete," ASTM C40.

Pop-outs and reduced durability can be caused by soft particles, chert clay lumps and other friable particles, coal, lignite, or other lightweight materials in the aggregates. Coal and lignite may also cause staining of exposed concrete surfaces.

Volume stability refers to susceptibility of aggregate to expansion when heated or to cyclic expansions and contractions when saturated and dried. Aggregates that are susceptible to volume change due to moisture should be avoided.

The grading and maximum size of aggregate are important because of the effect on relative proportions, workability, economy, porosity, and shrinkage. The particle-size distribution is determined by separation with a series of standard screens. The standard sieve used are Nos. 4, 8, 16, 30, 50, and 100 for fine aggregate and 6, 3, $1\frac{1}{2}$, $\frac{3}{4}$, and $\frac{3}{8}$ in, and 4 for coarse aggregate.

Fineness modulus (*F.M.*) is an index used to describe the fineness or coarseness of aggregate. The *F.M.* of a sand is computed by adding the cumulative percentages retained on the six standard sieves and dividing the sum by 100. For example, Table 5.5 shows a typical sand analysis.

The *F.M.* is not an indication of grading since an infinite number of gradings will give the same value for fineness modulus. It does, however, give a measure of the coarseness or finery of the material. Values of *F.M.* from 2.50 to 3.00 are normal.

Table 5.5 Computation of Fineness Modulus

Screen No.	Individual Percentages Retained	Cumulative Percentages Retained
4	1	1
8	18	19
16	20	39
30	19	58
50	18	76
100	16	92
Pan	8	
	100	285

F.M. = 285/100 = 2.85.

ASTM C33 provides ranges of fine- and coarse-aggregate grading limits. The latter are listed from Size 1(3½ to 1½ in) to Size 8 (⅜ to No. 8). The National Stone Association specifies a gradation for manufactured sand that differs from that for fine aggregate in C33 principally for the No. 100 and 200 sieves. The NSA gradation is noticeably finer (greater percentages passing each sieve). The fine materials, composed of angular particles, are rock fines, as opposed to silts and clays in natural sand, and contribute to concrete workability.

The various gradations provide standard sizes for aggregate production and quality-control testing. They are conducive to production of concrete with acceptable properties. Caution should be exercised, however, when standard individual grading limits are used. If the number of aggregate sizes are limited or there is not sufficient overlap between aggregate sizes, an acceptable or economical concrete may not be attainable with acceptably graded aggregates. The reason for this is that the combined gradation is gap graded. The ideal situation is a dense or well-graded size distribution that optimizes the void content of the combined aggregates. It is possible, however, to produce acceptable concrete with individual aggregates that do not comply with the standard limits but that can be combined to produce a dense gradation.

The material passing the No. 200 sieve is clay, silt, or a combination of the two. It increases the water demand of the aggregate. Large amounts of materials smaller than No. 200 may also indicate the presence of clay coatings on the coarse aggregate that would decrease bond of the aggregate to the cement matrix. A test method is given in ASTM C117, "Materials Finer than 75 μm Sieve in Mineral Aggregates by Washing."

Changes in sand grading over an extreme range have little effect on the compressive strengths of mortars and concretes when water-cement ratio and slump are held constant. Such changes in sand grading, however, do cause the cement content to vary inversely with the F.M. of the sand. Although this cement-content change is small, the grading of sand has a large influence on the workability and finishing quality of concrete.

Coarse aggregate is usually graded up to the largest size practical for a job, with a normal upper limit of 6 in. As shown in Fig. 5.1, the larger the maximum size of coarse aggregate, the less the water and cement required to produce concrete of a given quality.

A grading chart is useful for depicting the size distribution of aggregate particles in both the fine and coarse ranges. Figure 5.2 illustrates grading curves for sand, gravel, and combined aggregate, showing recommended limits and typical size distributions.

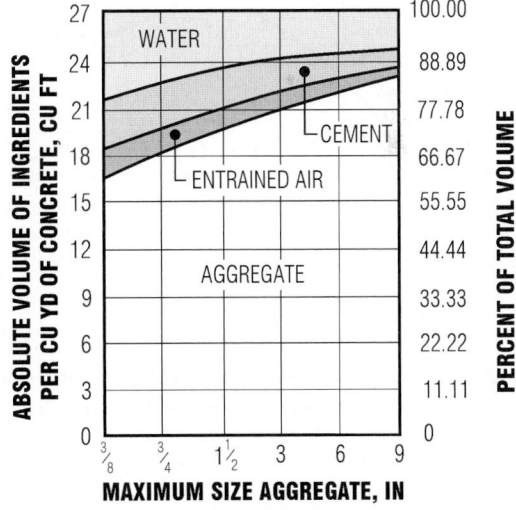

Fig. 5.1 Variations in amounts of water, cement, and entrained air in concrete mixes with maximum sizes of aggregates. The chart is based on natural aggregates of average grading in mixes with a water-cement ratio of 0.54 by weight, 3-in slump, and recommended air contents. (*From "Concrete Manual," 8th ed., U.S. Bureau of Reclamation.*)

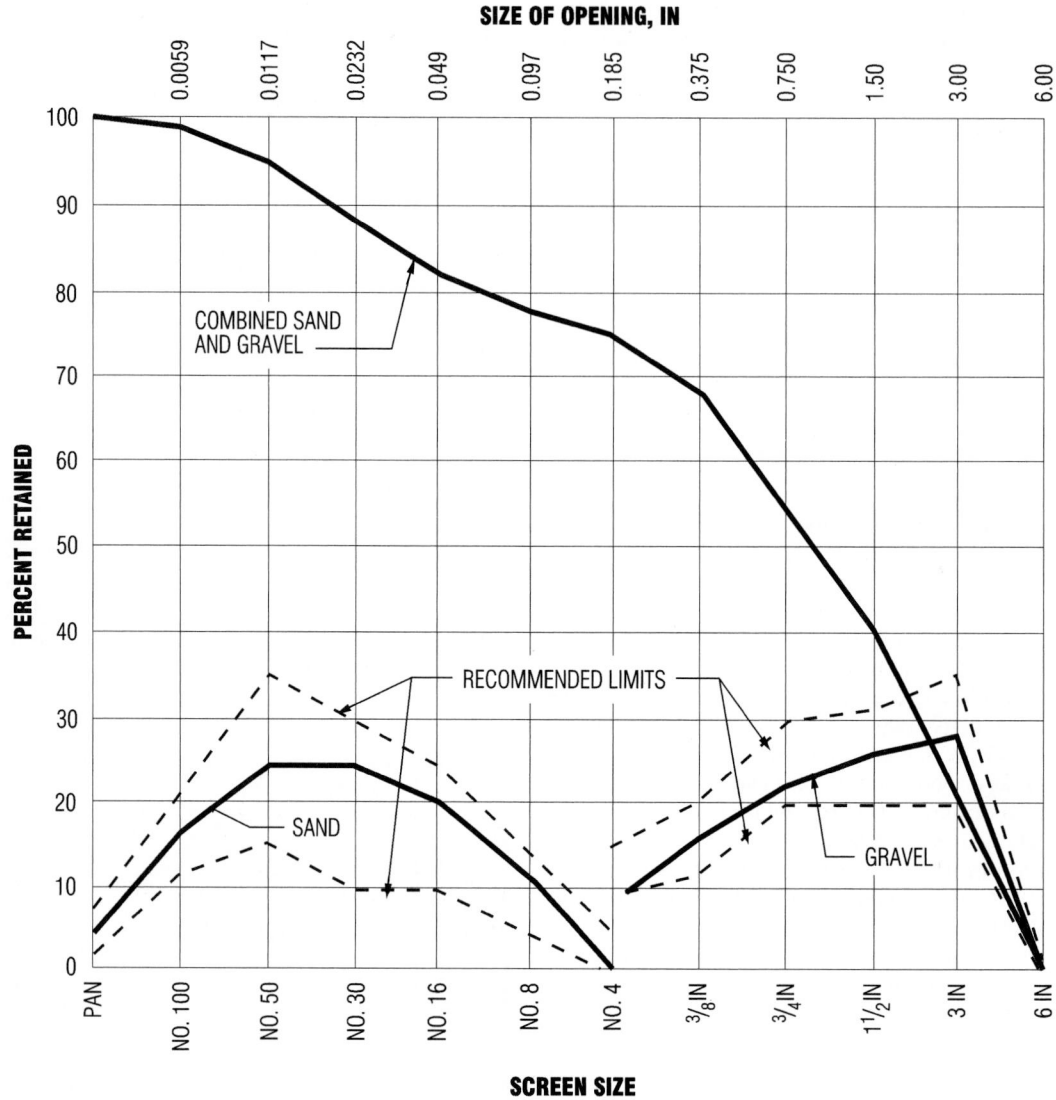

Fig. 5.2 Recommended and typical size distributions of natural aggregates for concrete mixes. Note that if No. 16 is 20% or less, No. 8 may be increased to 20%. (*From "Concrete Manual," 8th ed., U.S. Bureau of Reclamation.*)

Lightweight Aggregates ■ Lightweight aggregates are produced by expanding clay, shale, slate, perlite, obsidian, and vermiculite with heat; by expanding blast-furnace slag through special cooling processes; from natural deposits of pumice, scoria, volcanic cinders, tuff, and diatomite; and from industrial cinders. The strength of concrete made with lightweight aggregates is roughly proportional to its weight, which may vary from 35 to 115 lb/ft^3.

Lightweight aggregates can be divided into two categories: structural and nonstructural. The structural lightweight aggregates are defined by ASTM C330 and C331. They are either manufactured (expanded clay, shale, or slate, or blast-furnace slag) or natural (scoria and pumice). These aggregates produce concretes generally in the strength range of 3000 to 4000 psi; higher strengths are attainable.

The common nonstructural lightweight aggregates (ASTM C332) are vermiculite and perlite, although scoria and pumice can also be used. These materials are used in insulating concrete for soundproofing and nonstructural floor toppings.

Lightweight concrete has better fire resistance and heat- and sound-insulation properties than ordinary concrete, and it offers savings in structural supports and decreased foundations due to decreased dead loads. Structural concrete with lightweight aggregates costs 30 to 50% more, however, than that made with ordinary aggregates and has greater porosity and more drying shrinkage. Resistance to weathering is about the same for both types of concrete. Lightweight concrete can also be made with foaming agents, such as aluminum powder, which generates a gas while the concrete is still plastic and may be expanded.

Heavy Aggregates ▪ In the construction of atomic reactors, large amounts of heavyweight concrete are used for shielding and structural purposes. Heavy aggregates are used in shielding concrete because gamma-ray absorption is proportional to density. Heavy concrete may vary between the 150 lb/ft³ weight of conventional sand-and-gravel concrete and the theoretical maximum of 384 lb/ft³ where steel shot is used as fine aggregate and steel punchings as coarse aggregate. In addition to manufactured aggregates from iron products, various quarry products and ores, such as barite, limonite, hematite, ilemnite, and magnetite, have been used as heavy aggregates.

Table 5.6 shows the specific gravity of several heavy aggregates and the unit weights of concrete made with these aggregates. Since the introduction of high-density aggregates causes difficulty in mixing and placing operations due to segregation, grouting techniques are usually used in place of conventional methods.

5.6.2 Normal-Weight Concrete

The nominal weight of normal concrete is 144 lb/ft³ for non-air-entrained concrete but is less for air-entrained concrete. (The weight of concrete plus steel reinforcement is often assumed as 150 lb/ft³.)

Strength for normal-weight concrete ranges from 2000 to 20,000 psi. It is generally measured using a standard test cylinder 6 in in diameter by 12 in high. The strength of a concrete is defined as the average strength of two cylinders taken from the same load and tested at the same age. Flexural beams $6 \times 6 \times 20$ in may be used for concrete paving mixes.

Water-cement (W/C) ratio is the prime factor affecting the strength of concrete. Figure 5.3 shows how W/C, expressed as a ratio by weight, affects the compressive strength for both air-entrained and non-air-entrained concrete. Strength decreases with an increase in W/C in both cases.

Cement content itself affects the strength of concrete, with strength decreasing as cement content is decreased. In air-entrained concrete, this strength decrease can be partly overcome by taking advantage of the increased workability due to air entrainment, which permits a reduction in the amount of water. Strength vs. cement-content curves for two air-entrained concretes and non-air-entrained concretes are shown in Fig. 5.4.

Table 5.6 Heavy Aggregates for High-Density Concrete

Aggregate	Specific Gravity	Unit Weight of Concrete, lb per ft³	
		Conventional Placement	Grouting
Sand and stone		150	
Magnetite	4.30–4.34	220	346
Barite	4.20–4.31		232
Limonite	3.75–3.80		263
Ferrophosphorus	6.28–6.30	300	
Steel shot or punchings	7.50–7.78		384

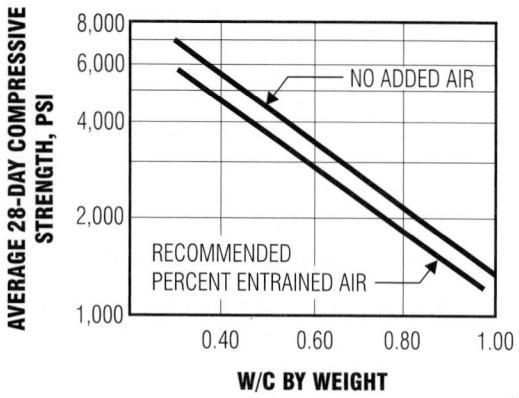

Fig. 5.3 Concrete strength decreases with increase in water-cement ratio for concrete with and without entrained air. (*From "Concrete Manual," 8th ed., U.S. Bureau of Reclamation.*)

Because of the water reduction possibility, the strengths of air-entrained concrete do not fall as far below those for non-air-entrained concrete as those previously indicated in Fig. 5.3.

Type of cement affects the rate at which strength develops and the final strength. Figure 5.5 shows how concretes made with each of the five types of portland cement compare when made and cured under similar conditions.

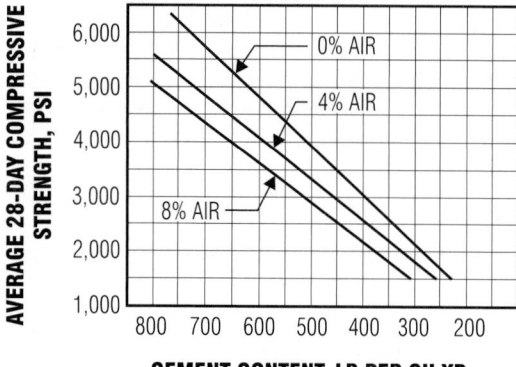

Fig. 5.4 Concrete strength increases with cement content but decreases with addition of air. Chart was drawn for concretes with ¾-in maximum size aggregates, 43% sand, and a 3-in maximum slump. (*From "Concrete Manual," 8th ed., U.S. Bureau of Reclamation.*)

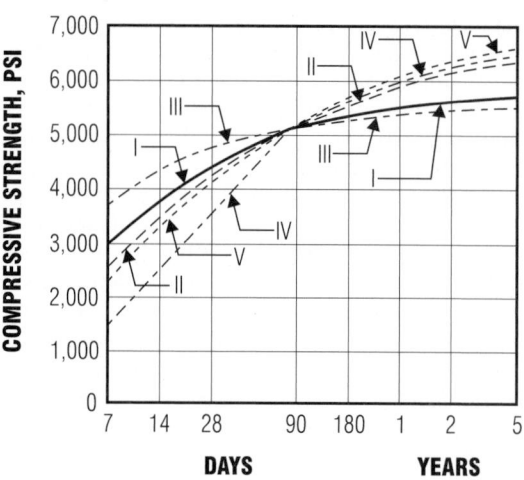

Fig. 5.5 Rates of strength development vary for concretes made with different types of cement. Tests were made on 6 × 12-in cylinders, fog cured at 70 °F. The cylinders were made from comparable concretes containing 1½-in maximum size aggregates and six bags of cement per cubic yard. (*From "Concrete Manual," 8th ed., U.S. Bureau of Reclamation.*)

Curing conditions are vital in the development of concrete strength. Since cement-hydration reactions proceed only in the presence of an adequate amount of water, moisture must be maintained in the concrete during the curing period. Curing temperature also affects concrete strength. Longer periods of moist curing are required at lower temperatures to develop a given strength. Although continued curing at elevated temperatures results in faster strength development up to 28 days, at later ages the trend is reversed; concrete cured at lower temperatures develops higher strengths.

Note that concrete can be frozen and will not gain strength in this state. Note also that, at low temperatures, strength gain of nonfrozen concrete is minimal and environmental factors, especially temperature and curing, are extremely important in development of concrete strength.

Stress-Strain Relations ▪ Concrete is not a linearly elastic material; the stress-strain relation for continuously increasing loading plots as a curved line. For concrete that has hardened thoroughly and has been moderately preloaded, however, the stress-strain curve is practically a straight line within the range of usual working stresses. As shown in Fig. 5.6, a modulus of elasticity can be determined from this portion of the curve. The elastic modulus for ordinary concretes at 28 days ranges from 2000 to 6000 ksi.

In addition to the elastic deformation that results immediately upon application of a load to concrete, deformation continues to increase with time under a sustained load. This plastic flow, or creep, continues for an indefinite time. It proceeds at a continuously diminishing rate and approaches a limiting value which may be one to three times the initial elastic deformation. Although increasing creep-deformation measurements have been recorded for periods in excess of 10 years, more than half of the ultimate creep usually takes place within the first 3 months after loading. Typical creep curves are shown in Fig. 5.7, where the effects of water-cement ratio and load intensity are illustrated. Upon unloading, an immediate elastic

recovery takes place, followed by a plastic recovery of lesser amount than the creep on the first loading.

Volume changes play an important part in the durability of concrete. Excessive or differential volume changes can cause cracking as a result of shrinkage and insufficient tensile strength, or spalling at joints due to expansion. Swelling and shrinkage of concrete occur with changes in moisture within the cement paste.

Hardened cement paste contains minute pores of molecular dimensions between particles of tobermorite gel and larger pores between aggregations of gel particles. The volume of pore space in a cement paste depends on the initial amount of water mixed with the cement; any excess water gives rise to additional pores, which weaken the structure of the cement paste. Movements of moisture into and out of this pore system cause volume changes. The drying shrinkage of concrete is about ½ in/100 ft. There is a direct relationship between mix-water content and drying shrinkage.

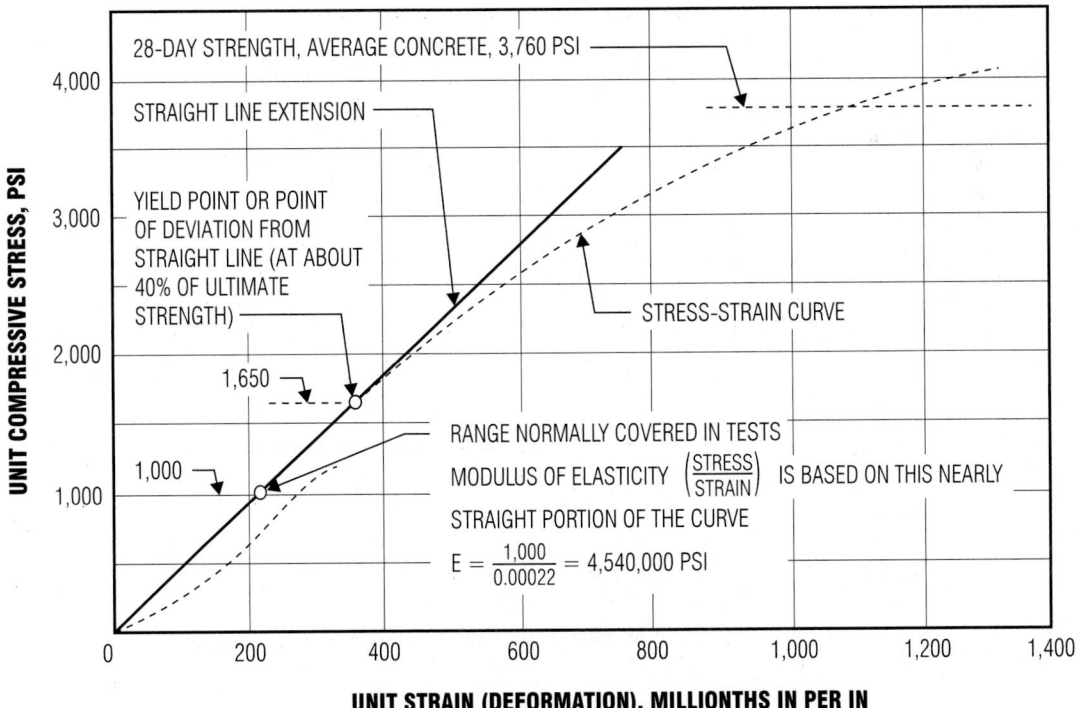

Fig. 5.6 Typical stress-strain diagram for cured concrete that has been moderately preloaded. (*From "Concrete Manual," 8th ed., U.S. Bureau of Reclamation.*)

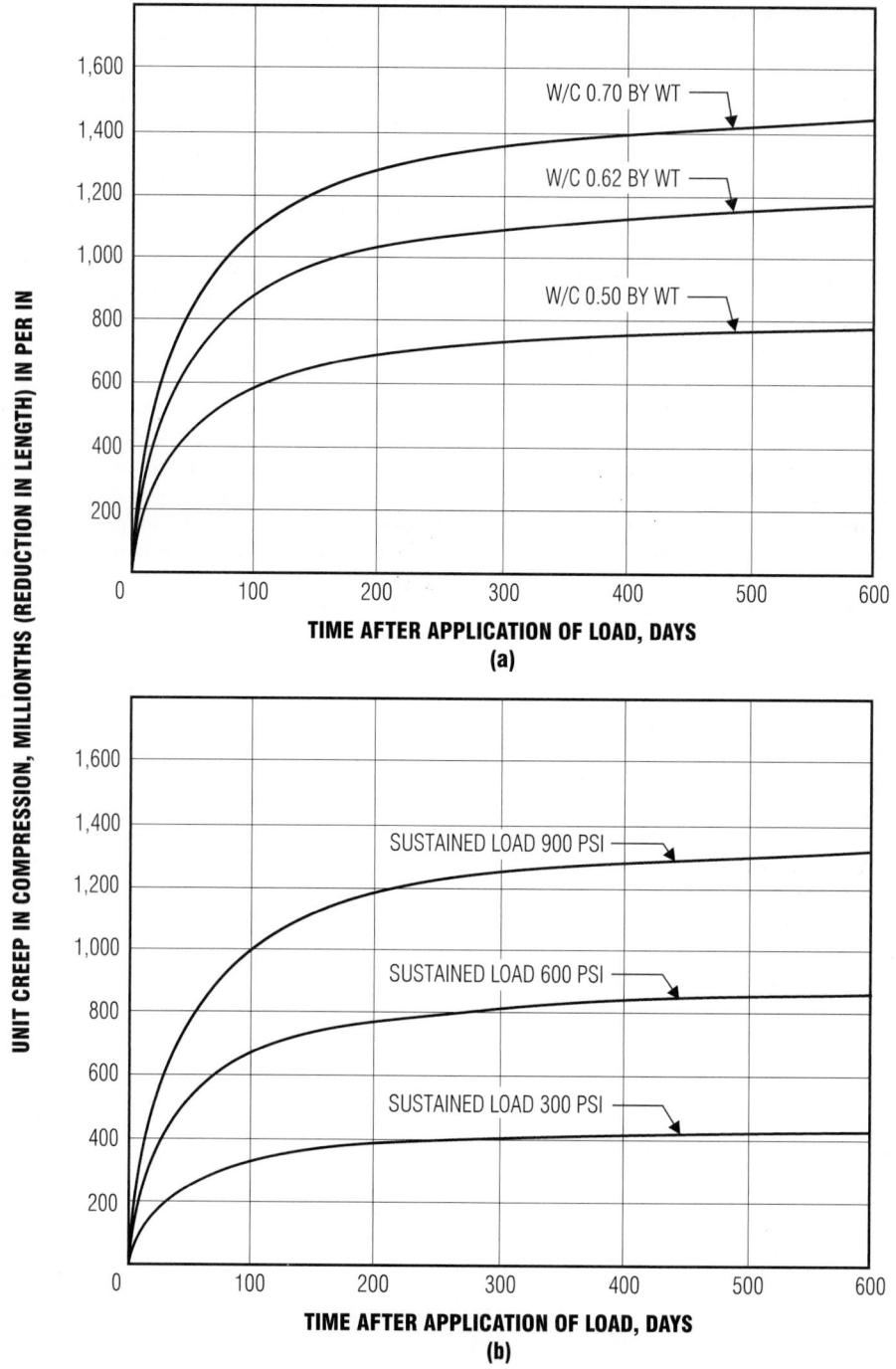

Fig. 5.7 Creep of concrete increases with increase in water-cement ratio or sustained load. (*a*) Effect of water-cement ratio on creep (applied-load constant). (*b*) Effect of intensity of applied load on creep (concretes identical). (*From "Concrete Manual," 8th ed., U.S. Bureau of Reclamation.*)

The cement content is of secondary importance in shrinkage considerations.

The thermal coefficient of expansion of concrete varies mainly with the type and amount of coarse aggregate used. The cement paste has a minor effect. An average value used for estimating is 5.5×10^{-6} in/(in · °F).

5.6.3 Admixtures for Concrete

Admixtures are anything other than portland cement, water, and aggregates that is added to a concrete mix to modify its properties. Included in this definition are chemical admixtures (ASTM C494 and C260), mineral admixtures such as fly ash (C618) and silica fume, corrosion inhibitors, colors, fibers, and miscellaneous (pumping aids, dampproofing, gas-forming, permeability-reducing agents). Many concrete admixtures are available to modify, improve, or give special properties to concrete mixtures. Admixtures should be used only when they offer a needed improvement not economically attainable by adjusting the basic mixture. Since improvement of one characteristic often results in an adverse effect on other characteristics, admixtures must be used with care.

Chemical admixtures used in concrete generally serve as water reducers, accelerators, set retarders, or a combination. ASTM C494, "Standard Specification for Chemical Admixtures for Concrete," contains the following classifications shown in Table 5.7. High-range admixtures reduce the amount of water needed to produce a concrete of a specific consistency by 12% or more.

Water-reducing admixtures decrease water requirements for a concrete mix by chemically

Table 5.7 Admixture Classification

Type	Property
A	Water reducer
B	Set retarder
C	Set accelerator
D	Water reducer and set retarder
E	Water reducer and set accelerator
F	High-range water reducer
G	High-range water reducer and set retarder

reacting with early hydration products to form a monomolecular layer at the cement-water interface that lubricates the mix and exposes more cement particles for hydration. The Type A admixture allows the amount of mixing water to be reduced while maintaining the same mix slump. If the amount of water is not reduced, the admixture will increase the slump of the mix and also strength of the concrete because more of the cement surface area will be exposed for hydration. Similar effects occur for Type D and E admixtures. Typically, a reduction in mixing water of 5 to 10% can be expected. Type F and G admixtures are used to achieve high-workability. A mix without an admixture typically has a slump of 2 to 3 in. After the admixture is added, the slump may be in the range of 8 to 10 in without segregation of mix components. These admixtures are especially useful for mixes with a low water-cement ratio. Their 12 to 30% reduction in water allows a corresponding reduction in cement.

The water-reducing admixtures are commonly manufactured from lignosulfonic acids and their salts, hydroxylated carboxylic acids and their salts, or polymers of derivatives of melamines or naphthalenes or sulfonated hydrocarbons. The combination of admixtures used in a concrete mix should be carefully evaluated and tested to ensure that the desired properties are achieved.

Superplasticizers are high-range water-reducing admixtures that meet the requirements of ASTM C494 Type F or G. They are often used to achieve high-strength concrete from mixes with a low water-cement ratio with good workability and low segregation. They also may be used to produce concrete of specified strengths with less cement at constant water-cement ratio. And they may be used to produce self-compacting, self-leveling flowing concretes, for such applications as long-distance pumping of concrete from mixer to formwork or placing concrete in forms congested with reinforcing steel. For these concretes, the cement content or water-cement ratio is not reduced, but the slump is increased substantially without causing segregation. For example, an initial slump of 3 to 4 in for an ordinary concrete mix may be increased to 7 to 8 in without addition of water and decrease in strength.

Superplasticizers may be classified as sulfonated melamine-formaldehyde condensates, sulfonated naphthaline-formaldehyde condensates, modified lignosulfonates, or synthetic polymers.

Air-entraining agents increase the resistance of concrete to frost action by introducing numerous tiny air bubbles into the hardened cement paste. These bubbles act as stress relievers for stresses induced by freezing and thawing. Air-entraining agents are usually composed of detergents. In addition to increasing durability of the hardened cement, they also decrease the amount of water required and increase the workability of the mix. Air contents are usually controlled to between 2 and 6%.

Because air-entrained concrete bleeds less than non-air-entrained concrete, fewer capillaries extend from the concrete matrix to the surface. Therefore, there are fewer avenues available for ingress of aggressive chemicals into the concrete.

The "Standard Specification for Air-Entraining Admixtures for Concrete," ASTM C260, covers materials for use of air-entraining admixtures to be added to concrete in the field. Air entrainment may also be achieved by use of Types IIA and IIIA portland cements. (See **air-entraining portland cements** in Art. 5.2.3.)

Set-accelerating admixtures are used to decrease the time from the start of addition of water to cement to initial set and to increase the rate of strength gain of concrete. The most commonly used set-accelerating admixture is calcium chloride. Calcium chloride offers advantages in cold-weather concreting by speeding the set at low temperature and reducing the time that protection is necessary. When used in usual amounts (less than 2% by weight of cement), however, it does not act as an antifreeze agent by lowering the freezing point. When 2% calcium chloride is used under normal conditions, it reduces the initial set time from 3 to 1 h and the final set time from 6 to 2 h, and at 70 °F it doubles the 1-day strength. Use of calcium chloride as an admixture improves workability, reduces bleeding, and results in a more durable concrete surface. Problems in its use may arise from impairment of volume stability (drying shrinkage may be increased as much as 50%) and an increase in the rate of heat liberation. Chloride ions can also contribute to corrosion of steel embedded in concrete. Limits on chloride ion concentration may be as low as 0.04% of the weight of the concrete.

Retarding admixtures are used to retard the initial set of concrete. A Type B or D admixture will allow transport of concrete for a longer time before initial set occurs. Final set also is delayed. Hence,

precautions should be taken if retarded concrete is to be used in walls.

Depending on the dosage and type of base chemicals in the admixture, initial set can be retarded for several hours to several days. A beneficial side effect of retardation of initial and final sets is an increase in the compressive strength of the concrete. A commonly used Type D admixture provides higher 7- and 28-day strengths than a Type A when used in the same mix design.

Mineral admixtures include fly ashes, pozzolans, and microsilicates (Arts. 5.3.8 and 5.3.9). Natural cement (Art. 5.3.3) is sometimes used as an admixture.

Corrosion inhibitors are sometimes added to a concrete mix to protect reinforcing steel. The steel usually is protected against corrosion by the high alkalinity of the concrete, which creates a passivating layer at the steel surface. This layer is composed of ferric oxide, a stable compound. Within and at the surface of the ferric oxide, however, are ferrous-oxide compounds, which are more reactive. When the ferrous-oxide compounds come into contact with aggressive substances, such as chloride ions, they react with oxygen to form solid, iron oxide corrosion products. These produce a fourfold increase in volume and create an expansion force greater than the concrete tensile strength. The result is deterioration of the concrete.

To inhibit corrosion, calcium nitrite admixtures may be added to the concrete mix. They do not create a physical barrier to chloride ion ingress. Instead, they modify the chemistry at the steel surface. The nitrite ions oxidize ferrous oxide present, converting it to ferric oxide. The nitrite is also absorbed at the steel surface and fortifies the ferric oxide passivating layer. For a calcium nitrite admixture to be effective, the dosage should be adjusted in accordance with the exposure of the concrete to corrosive agents. The greater the exposure, the larger should be the dosage.

Internal-barrier admixtures may be a waterproofing or a dampproofing compound or an agent that creates an organic film around the reinforcing steel, supplementing the passivating layer. The latter type of admixture may be added at a fixed rate regardless of expected chloride exposure.

Dampproofing admixtures include soaps, stearates, and other petroleum products. They are intended to reduce passage of water and water vapor through concrete. Caution should be exercised when using these materials inasmuch as they may increase water demand for the mix, thus increasing the permeability of the concrete. If dense, low-permeability concrete is desired, the water-cement ratio should be kept to a maximum of 0.50 and the concrete should be well vibrated and damp cured.

Permeability of concrete can be decreased by the use of fly ash and silica fume (Arts. 5.3.8 and 5.3.9) as admixtures. Also, use of a high-range water-reducing admixture and a water-cement ratio less than 0.50 will greatly reduce permeability.

Gas-forming admixtures are used to form lightweight concrete. They are also used in masonry grout where it is desirable for the grout to expand and bond to the concrete masonry unit. They are typically an aluminum powder.

Pumping aids are used to decrease the viscosity of harsh or marginally pumpable mixes. Organic and synthetic polymers, fly ash, bentonite, or hydrated lime may be used for this purpose. Results depend on concrete mix, including the effects of increased water demand and the potential for lower strength resulting from the increased water-cement ratio. If sand makes the mix marginally pumpable, fly ash is the preferred pumping additive. It generally will not increase the water demand and it will react with the calcium hydroxide in cement to provide some strength increase.

Coloring admixtures may be mineral oxides or manufactured pigments. Coloring requires careful control of materials, batching, and water addition in order to maintain a consistent color at the jobsite. Note that raw carbon black, commonly used for black color, greatly reduces the amount of entrained air in a mix. Therefore, if black concrete is desired for concrete requiring air-entrainment (for freeze-thaw or aggressive chemical exposure), either the carbon black should be modified to entrain air or an additional air-entraining agent may be incorporated in the mix. The mix design should be tested under field conditions prior to its use in construction.

5.7 Fiber Reinforcing for Concrete

Fibrous materials may be added to a concrete mix to improve strength, resilience, and crack control. Fiber lengths are small, and fibers may be de-scribed by their aspect ratio, the ratio of length to equivalent diameter.

The most commonly used types of fibers in concrete are synthetics, which include polypropylene, nylon, polyester, and polyethylene materials. Specialty synthetics include aramid, carbon, and acrylic fibers. Glass-fiber-reinforced concrete is made using E-glass and alkali-resistant (AR) glass fibers. Steel fibers are chopped high-tensile-strength or stainless steel.

Fibers should be dispersed uniformly throughout a mix. Orientation of the fibers in concrete generally is random. Conventional reinforcement, in contrast, typically is oriented in one or two directions, generally in planes parallel to the surface. Further, welded-wire fabric or reinforcing steel bars must be held in position as concrete is placed. Regardless of the type, fibers are effective in crack control because they provide omnidirectional reinforcement to the concrete matrix. With steel fibers, impact strength and toughness of concrete may be greatly improved and flexural and fatigue strengths enhanced.

Synthetic fibers are typically used to replace welded-wire fabric as secondary reinforcing for crack control in concrete flatwork. Depending on the fiber length, the fiber can limit the size and spread of plastic shrinkage cracks or both plastic and drying shrinkage cracks. Although synthetic fibers are not designed to provide structural properties, slabs tested in accordance with ASTM E72, "Standard Methods of Conducting Strength Tests of Panels for Building Construction," showed that test slabs reinforced with synthetic fibers carried greater uniform loads than slabs containing welded wire fabric. While much of the research for synthetic fibers has used reinforcement ratios greater than 2%, the common field practice is to use 0.1% (1.5 lb/yd^3). This dosage provides more cross-sectional area than 10-gage welded-wire fabric. The empirical results indicate that cracking is significantly reduced and is controlled. A further benefit of fibers is that after the initial cracking, the fibers tend to hold the concrete together.

Aramid, carbon, and acrylic fibers may be used for structural applications, such as wrapping concrete columns to provide additional strength. Other possible uses are for corrosion-resistance structures. The higher costs of the specialty synthetics limit their use in general construction.

Glass-fiber-reinforced concrete (GFRC) is used to construct many types of building elements, including architectural wall panels, roofing tiles, and water tanks. The full potential of GFRC has not been attained because the E-glass fibers are alkali reactive and the AR glass fibers are subject to embrittlement, possibly from infiltration of calcium hydroxide particles.

Steel fibers can be used as a replacement for conventional reinforcing steel. The volume of steel fiber in a mix ranges from 0.5 to 2%. American Concrete Institute Committee 544 states in "Guide for Specifying, Mixing, Placing, and Finishing Steel Fiber Reinforced Concrete," ACI 544.3R, that, in structural members such as beams, columns, and floors not on grade, reinforcing steel should be provided to support the total tensile load. In other cases, fibers can be used to reduce section thickness or improve performance. See also ACI 344.1R and 344.2R.

5.8 Polymer Concrete

When portland cement is replaced by a polymer, the resulting concrete has a lower rate of water absorption, higher resistance to cycles of freezing and thawing, better resistance to chemicals, greater strength, and excellent adhesion qualities compared to most other cementitious materials.

The most commonly used resins (polyesters and acrylics) are mixed with aggregates as a monomer, with a cross-linking agent (hardener) and a catalyst, to reach full polymerization. Polymer concretes are usually reinforced with metal fibers, glass fibers, or mats of glass fiber.

Polymer-impregnated concrete (PIC) is cured portland cement concrete that is impregnated with a monomer using pressure or a vacuum process. The monomer (most often an acrylic) is polymerized by a catalyst, heat, or ultraviolet radiation. A continuous surface layer is formed that waterproofs and strengthens and fills the voids.

5.9 Bituminous Concrete and Other Asphalt Composites

Mixtures of asphalt, serving as a binder, fine and coarse aggregates, and often fillers and admixtures are widely used as flexible pavements, dam facings, and canal linings. The aggregates, such as sand, gravel, and crushed stone, are similar to those used for portland cement concrete (Art. 5.6.1).

The American Association of State Highway and Transportation Officials (AASHTO), The Asphalt Institute, and ASTM publish specifications for asphalt. These generally are the basis for specifications of governmental departments of highways and transportation.

Asphalts are viscoelastic. Properties may range from brittle to rubbery. The hardness, or viscosity, depends on the temperature of the asphalts. The variation with temperature, however, depends on the shear susceptibility of the material, which indicates the state of its colloidal structure.

Asphalt which is a black or dark brown petroleum derivative, is distinct from tar, the residue from destructive distillation of coal. Asphalt consists of hydrocarbons and their derivatives and is completely soluble in carbon disulfide (CS_2). It is the residue of petroleums after the evaporation, by natural or artificial means, of their most volatile components.

Asphalt cements (ACs) are used as binders for almost all high-grade flexible pavements. They are mixtures of hard asphalts and nonvolatile oils that are brought to a usable consistency by heating, without being softened with a fluxing or emulsifying agent. They may be graded in accordance with their viscosity or penetration (distance to which the material is penetrated by a needle in a standard test) at a specified temperature.

Slow-curing (SC) road oils are liquid petroleum products that set slowly and are suitable for use where nearly the same consistency of cement is required both at the time of processing and the end of curing. They may be the product remaining after distillation of petroleum or the result of cutting back asphalt cements with a heavy distillate. More viscous than light grades of lubricating oil, SC binders are more fluid than asphalt cements.

Medium-curing (MC) cutback asphalts are asphalt cements that have been mixed (fluxed or cut back) with distillates of the kerosene or light diesel-oil type for greater fluidity. They evaporate relatively slowly. After an MC asphalt is applied, the flux evaporates from the cutbacks, leaving the semisolid asphalt cement as the binding agent. MC asphalts are used where greater fluidity is required at the time of processing than at the end of curing.

Rapid-curing (RC) cutback asphalts are asphalt cements that have been cut back with a heavier distillate, such as gasoline or naphtha, than that used for MC asphalts. RC asphalts evaporate rapidly. They are used where a speedy change, via evaporation, from applied liquid to semisolid asphalt-cement binder is required.

Emulsified asphalts are mixtures in which colloidal-size asphalt particles are dispersed in water in the presence of an emulsifying agent. Because the asphalt particles have like electrical charges, they do not coalesce until the water evaporates or the emulsion breaks. Asphalt content of the mixture may range from 55 to 70% by weight. The emulsions are applied unheated. They have low viscosity and can penetrate deeply into an aggregate matrix. When the water evaporates or flows away, the asphalt binder remains. Emulsions are available with fast (RS), medium (MS), and slow (SS) breaking times and thus are suitable for a wide variety of purposes. Emulsifying agents may be tallow derivates, soap of fatty and resinous acids, glue, or gelatin.

Bituminous concrete for pavements may be improved by addition of sulfur, lime, or rubber to the asphalt-aggregate mix (Sec. 16).

Asphalt Building Products ■ Because of its water-resistant qualities and durability, asphalt is used for many building applications. For damp-proofing (mopped-on coating only) and water-proofing (built-up coating of one or more plies), three types of asphalt are used: Type A, an easy-flowing, soft, adhesive material for use underground or in other moderate-temperature applications; Type B, a less susceptible asphalt for use aboveground where temperatures do not exceed 125 °F; and Type C, for use aboveground where exposed on vertical surfaces to direct sunlight or in other areas where temperatures exceed 125 °F.

Asphalt and asphalt products are also used extensively in roofing applications. Asphalt is used as a binder between layers in built-up roofing and as the impregnating agent in roofing felts, roll roofing, and shingles. Care should be taken not to mix asphalt and tar together, that is, to place asphalt layers on a tar-saturated felt or vice versa, unless their compatibility has been checked.

5.10 Cementitious Materials References

Brantley, L. R., and R. T. Brantley, "Building Materials Technology: Structural Performance and Environmental Impact," McGraw-Hill, Inc., New York.

Cowan, H. J., and P. R. Smith, "The Science and Technology of Building Materials," Van Nostrand Reinhold Company, New York.

Dikeou, J. T., and D. W. Powler, "Polymer Concrete: Uses, Materials, and Properties," American Concrete Institute, Farmington Hills, MI. www.aci-int.org

"McGraw-Hill Encyclopedia of Chemistry," 2nd ed., McGraw-Hill Book Company, New York. www.mcgraw-hill.com

"McGraw-Hill Dictionary of Science and Technology Terms," 4th ed., McGraw-Hill Book Company, New York. www.mcgraw-hill.com

Mendis, P., and C. McClaskey, "Polymers in Concrete: Advances and Applications," publication SP-116, American Concrete Institute, Farmington Hills, MI. www.aci-int.org

Swamy, R. N., and B. Barr, "Fibre-Reinforced Cements and Concretes: Recent Developments," Elsevier Applied Science, London.

Waddell, J. J., and J. A. Dobrowolski, "Concrete Construction Handbook," 3rd ed., McGraw-Hill Book Company, New York. www.mcgraw-hill.com

Metallic Materials

Regularity of atomic-level structure has made possible better understanding of the microscopic and atomic-level foundations of the mechanical properties of metals than of other kinds of materials. Attempts to explain macroscopic behavior on the basis of micromechanisms are relatively successful for metallic materials.

5.11 Deformation of Metals

Metals consist of atoms bonded together in large, regular aggregations. Metallic bonds between the atoms are due to the sharing of electrons in unsaturated covalent bonds. The elastic behavior of metallic materials under limited loadings can be explained in terms of interatomic bonding. The deformation of materials under applied load is elastic if the change in shape is entirely recovered when the material is returned to its original stress state. Elastic load-deformation relationships may or may not be linear, as shown in Fig. 5.8, but many metals behave linearly.

At a separation of a few atomic diameters, the repulsive forces between the like charges of the atomic nuclei start to assert themselves when a

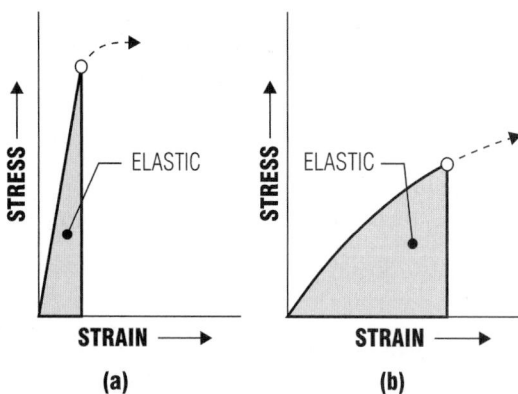

Fig. 5.8 Stress-strain diagram for metals may be (a) linear or (b) nonlinear elastic. Metals recover shape when returned to the original stress state when stressed within the elastic range.

compressive load is applied. At equilibrium separation, the forces of attraction just equal the forces of repulsion, and the potential energy is at a minimum. If the atoms try to move closer, the repulsive force increases much more rapidly than the attractive force as the electron clouds begin to overlap. If the atoms are pulled apart slightly, when released, they tend to go back to the equilibrium spacing, at which the potential energy is a minimum. The macroscopic modulus of elasticity thus has its basis in the limited stretching of the atomic bonds when the force vs. interatomic spacing curve is essentially linear near the equilibrium atomic spacing. Strongly bonded materials exhibit higher elastic moduli than do weakly bonded materials.

Ductile crystalline materials often fail by the slip of adjacent planes of atoms over each other. This mode of failure occurs when the resolved shear stress on some slip plane reaches a critical value before any possible brittle-fracture mode has been activated. If the shear stress to move one plane of atoms past another plane could be computed from atomic-bonding considerations, the strength of a material under a given external loading system could be predicted.

Slip on atomic planes actually proceeds in a stepwise manner, not by the gross slipping of whole atomic planes over each other. This stepwise slip is described in terms of *dislocations*, which are imperfections in the crystalline lattice at the atomic

scale. A *pure edge dislocation* is the discontinuity at the end of an extra half plane of atoms inserted in the crystal lattice. Under applied loading, an edge dislocation moves across the slip plane in a stepwise manner, breaking and reforming bonds as it moves. This movement results, in plastic deformation equivalent to the sliding of one whole plane of atoms across another by one atomic dimension. This dislocation mechanism is the one by which yield begins in metals and by which plastic deformation continues.

A second type of pure dislocation, known as a screw dislocation, is associated with shear deformations in crystalline structures. In general, dislocation in real crystalline lattices, which are usually in the form of loops, are mixed dislocations with both edge and screw components.

The elastic portion of a stress-strain curve, based on bond stretching at the atomic scale, ends with the onset of plastic deformation at the yield point. Yielding is associated with the irreversible movement of dislocations with which plastic straining begins. Beyond the yield point the material no longer returns to exactly its initial state with load removal; some plastic deformation remains.

A dislocation is surrounded by an elastic stress field that results in forces between dislocations and in interactions with other irregularities in the crystalline structure. The general effect of the interaction of dislocations with each other and with other obstacles after yielding is a work hardening of the material, that is, an increase in the stress required to continue plastic deformation. This arises from the increased difficulty of moving dislocations, with their surrounding stress fields, through the stress fields of other irregularites in the crystalline lattice.

Metals can be strengthened if ways can be found to keep dislocations from beginning to move or if obstacles to the movement can slow or stop them once the dislocations have begun to move. In addition to the strain hardening that results from interactions of moving dislocations, other means may be used to strengthen metals at the atomic level. See Art. 5.12.

5.12 Mechanisms for Strengthening Metals

Plastic deformation in metals is characterized by a phenomenon known as strain hardening (Art. 5.11).

When metals are deformed beyond the elastic limit, a permanent change in shape occurs. If a metal is loaded beyond its yield point, unloaded, then loaded again, the elastic limit is raised. This phenomenon, represented in Fig. 5.9, indicates that a metal can be strengthened by deformation previous to its loading in an engineering application. Its ductility, however, is decreased.

Dislocations piling up at obstacles on the slip plane cause strain hardening due to a back stress opposing the applied stress. The obstacles at which dislocations may be blocked during plastic deformation include foreign atoms in the lattice, precipitate particles, intersection of slip planes where dislocations combine to block each other, and grain boundaries.

Cold Working ▪ Plastic deformation that is carried out in a temperature range and over a time interval such that the strain hardening is not relieved is called *cold work*. Cold working is employed to harden and strengthen metals and alloys that do not respond to heat treatment.

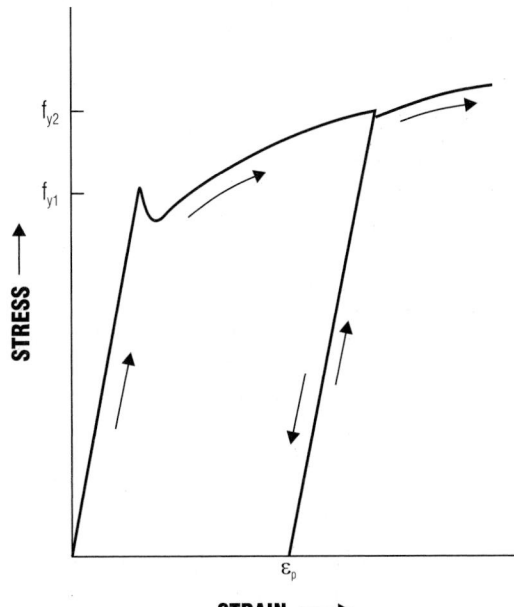

Fig. 5.9 Stress-strain curve for metal stressed beyond the elastic limit, unloaded, then reloaded. The yield stress on the second loading is higher than that on the first.

Although strength increases considerably, ductility, as measured by elongation, decreases greatly.

Cold work is often followed by annealing. This is a reheating process in which the metal is heated until it softens and reverts to a strain-free condition. Then, it is cooled slowly, usually in a furnace, to obtain the softest, most ductile state. Partial annealing may precede cold working to relieve internal stresses that might cause cracking during the cold working.

Solid-Solution Hardening ▪ Strengthening produced by dispersed, atomic-size lattice defects in a metal is referred to as solid-solution hardening. Substitutional and interstitial impurity atoms are the most common varieties of such defects. Whenever a dislocation (Art. 5.11) encounters an irregularity within a crystal lattice, hardening occurs.

Solute atoms introduced into solid solution in a pure metal produce an alloy stronger than the original metal. If the solute and solvent atoms are roughly similar, the solute atoms occupy lattice points in the crystal lattice of the solvent atoms. This forms a substitutional solid solution. If the solute atoms are considerably smaller than the solvent atoms, they occupy interstitial positions in the solvent lattice. Such elements as carbon, nitrogen, oxygen, hydrogen, and boron commonly form such interstitial solid solutions.

Precipitation Hardening ▪ Dispersion hardening is the strengthening produced by a finely dispersed insoluble second phase in a matrix of metal atoms. These second-phase particles act as obstacles to the movement of dislocations (Art. 5.11). Thus, higher stresses are required to cause plastic deformation when dislocations must overcome these obstacles to move across slip planes. The basic technique is to make the second phase as finely dispersed as possible. This can be achieved by supercooling.

One method of producing this type of strengthening, precipitation hardening, or age hardening is by a heat-treatment process. In any alloy such as copper-aluminum, a greater amount of the alloying element can be put into solid solution at an elevated temperature than at room temperature. If the temperature is reduced, a supersaturation of alloying atoms results. If the solid solution is cooled slowly, the excess solute atoms leave the solution by migrating to areas of disorder, such as grain boundaries, and forming large precipitates. Because of slow cooling, enough diffusion takes place that large precipitates that are not spaced closely enough to be effective in strengthening are formed. If rapid cooling follows the solutionizing treatment, however, the excess alloying atoms are retained in solid solution. In such a rapid quench, there is no time for diffusion to the grain boundaries to occur. Once the supersaturated solid solution exists at room temperature, it may be aged at room temperature or some slightly elevated temperature to allow precipitates to form on a very fine scale throughout the host metal. These fine precipitate particles effectively block dislocation movement and thus strengthen and harden the metal. Figure 5.10 shows how the properties of an aluminum alloy change during a precipitation heat treatment.

A continuation of the process of local segregation of alloying atoms over a long time leads to overaging, or softening. The continued growth of precipitates, in which small, closely spaced areas combine through diffusion to produce large precipitates, leaves a structure with less resistance to dislocation movement.

Grain Size ▪ Although single crystals of metals are specially grown for research investigations, commercial grades of metals are polycrystalline materials. Each grain in a polycrystalline metal is a small volume of atoms stacked in such a way that the atomic planes are essentially parallel. Each grain has an orientation quite different from that of neighboring grains. The interfaces between individual grains, called grain boundaries, are areas of great atomic misfit. Because of changes in orientation and the disruption of regular atomic structure at grain boundaries, dislocations are greatly inhibited in their motion at these areas. The more numerous the grain boundaries, the higher the strength of the metal.

Decreasing the average size of the grains in a polycrystalline metal increases the strength by increasing the number of grain-boundary obstacles to dislocation movement. Grain size can be controlled by the heating and rolling operations in the production of structural metals.

5.13 Structural Steels

High-strength steels are used in many civil engineering projects. New steels are generally introduced under trademarks by their producers, but a brief check into their composition, heat treatment, and properties will normally allow them to be related to other existing materials. Following are some working classifications that allow comparison of new products with standardized ones.

5.13.1 Classifications of Structural Steel

General classifications allow the currently available structural steels to be grouped into four major categories, some of which have further subcategories. The steels that rely on carbon as the main alloying element are called *structural carbon steels*. The older grades in this category were the workhorse steels of the construction industry for many years, and the newer, improved carbon steels still account for the bulk of structural tonnage.

Two subcategories can be grouped in the general classification *low-alloy carbon steels*. To develop higher strengths than ordinary carbon steels, the low-alloy steels contain moderate proportions of one or more alloying elements in addition to carbon. The *columbium-vanadium-bearing steels* are higher-yield-strength metals produced by addition of small amounts of these two elements to low-carbon steels.

Two kinds of *heat-treated steels* are on the market for construction applications. *Heat-treated carbon*

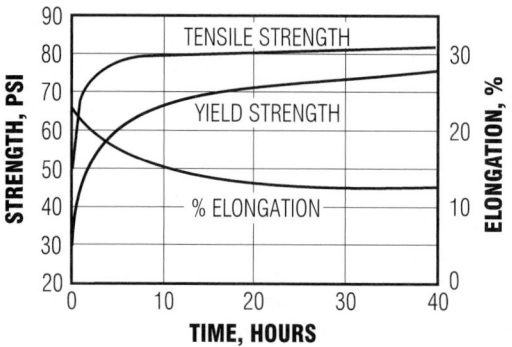

Fig. 5.10 Changes in mechanical properties during precipitation heat treatment of 7076 aluminum alloy at 250 °F.

steels are available in either normalized or quenched-and-tempered condition, both relying essentially on carbon alone for strengthening. *Heat-treated constructional alloy steels* are quenched-and-tempered steels containing moderate amounts of alloying elements in addition to carbon.

Another general category, *maraging steels*, consists of high-nickel alloys containing little carbon. These alloys are heat-treated to age the iron-nickel martensite. Maraging steels are unique in that they are construction-grade steels that are essentially carbon-free. They rely entirely on other alloying elements to develop their high strength. This class of steels probably represents the opening of a door to the development of a whole field of carbon-free alloys.

ASTM specification designations are usually used to classify the structural steels that have been in use long enough to be codified (Table 9.1). The "AASHTO Standard Specifications for Highway Bridges" (American Association of State Highway and Transportation Officials) contain similar specifications. These specifications cover production variables, such as process, chemical content, and heat treatment, as well as performance minima in tensile and hardness properties.

Chemical-content comparison of carbon and other alloying elements can be used to distinguish one structural steel from another. Most structural steels, except for the maraging steels, contain carbon in amounts between 0.10 and 0.28%. The older steels have few alloying elements and are usually classified as carbon steels. Steels containing moderate amounts of alloying elements, with less than about 2% of any one constituent element, are called low-alloy steels. Steels containing larger percentages of alloying elements, such as the 18% nickel maraging steels, are designated high-alloy steels. Specified chemical compositions of the codified structural steels are listed in ASTM specifications; typical chemical compositions of other structural steels are available from steel producers.

A basic numbering system sometimes is used to describe the carbon and alloy content of steels. In the American Iron and Steel Institute numbering system for low-alloy steels, the first two numbers indicate the alloy content and the last two numbers indicate the nominal carbon content in units of 0.01%. Complete listings of AISI steels, with composition limits and hardenability bands are in vol. 1 of "Metals Handbook" (American Society for Metals).

Heat treatment can be used as another means of classification. The older structural carbon steels and high-strength low-alloy steels are not specially heat-treated, but their properties are controlled by the hot-rolling process. The heat-treated, constructional alloy and carbon steels rely on a quenching and tempering process for development of their high-strength properties. The ASTM A514 steels are heat-treated by quenching in water or oil from not less than 1650 °F and then tempering at not less than 1100 °F. The heat-treated carbon steels are subjected to a similar quenching and tempering sequence: austenizing, water quenching, and then tempering at temperatures between 1000 and 1300 °F. The typical heat treatment of the maraging steels involves annealing at 1500 °F for 1 h, air cooling to room temperature, and then aging at 900 °F for 3 h. The aging treatment in the maraging steels may be varied to obtain different strength levels.

5.13.2 Effects of Steel Microstructure

Mechanical properties observed and measured at the macroscopic scale are based on the constituent microstructure of the steel. Although there are variations in the details of microstructure of a particular type of steel as the chemical composition and heat treatment vary within allowable limits, general characteristics of microstructure can be described for each of the broad classifications of structural steels.

If steel is cooled very slowly from its high temperature or molten condition to room temperature, it takes a characteristic form depending on the percentage of carbon present in the iron matrix. The forms present at any temperature and composition are readily displayed on the iron-carbon diagram shown in Fig. 5.11. This is a quasi-equilibrium diagram; it represents the situation for a given temperature and composition only if sufficient time has elapsed for the material to reach thermodynamic equilibrium. In many structural steels, nonequilibrium structures are purposely produced to obtain desired mechanical properties.

The structure of iron is different in each of its phases, just as ice, water, and steam have different structures in their respective stable temperature ranges. Ferrite, or alpha iron, is the body-centered-cubic structure iron found at room temperature.

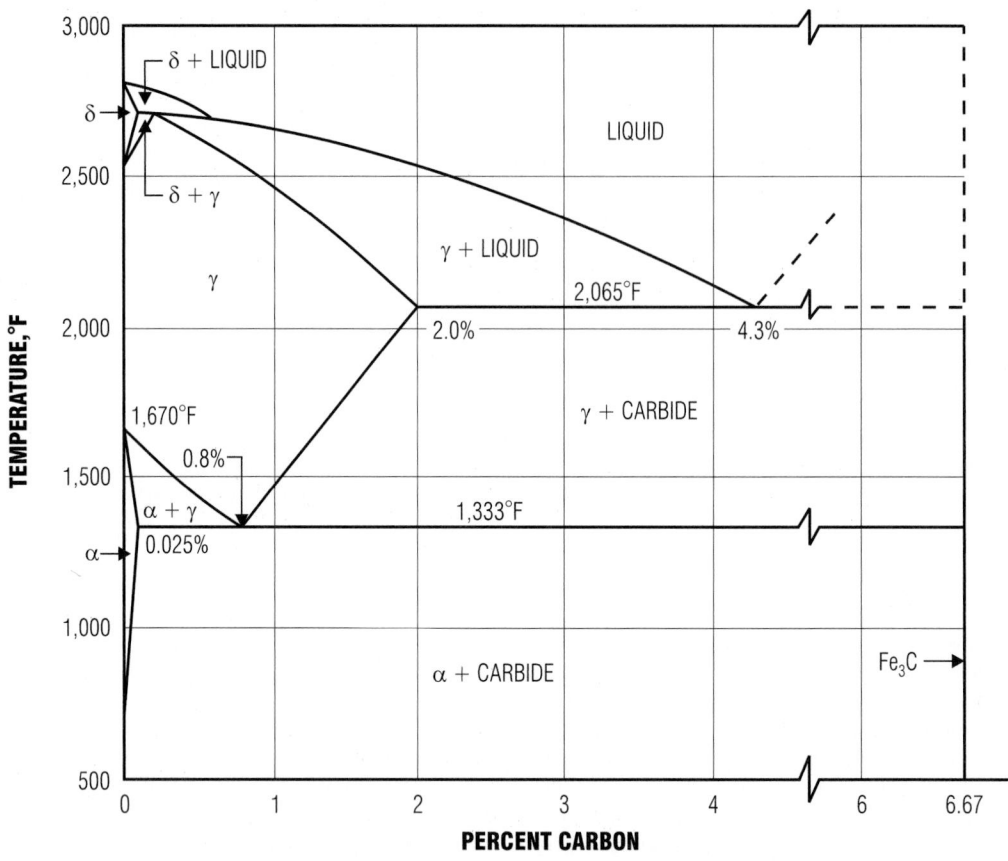

Fig. 5.11 Iron-carbon equilibrium diagram.

Ferrite has a low solubility of carbon since the carbon atom is too small for a substitutional solid solution and too large for an extensive interstitial solid solution (see Art. 5.12). Austenite, or gamma iron, is the face-centered-cubic form of iron that is stable between 1670 and 2550 °F. (These temperatures are for pure iron. see Fig. 5.11 for the entire range of stability of the gamma phase.) The face-centered-cubic structure has larger interstices than the ferrite and hence can have more carbon in the structure. The maximum solubility is 2% carbon by weight. Delta iron is the body-centered-cubic form of iron that is stable above 2550 °F. The relative solubilities of carbon in the iron matrix play an important part in the nonequilibrium structures that result from certain heat treatments of steel.

The combination of iron and carbon represented by the vertical line at 6.67% carbon content in

Fig. 5.11 is called **cementite** (or Fe_3C, iron carbide). Carbon in excess of the solubility limit in iron forms this second phase, in which the crystal lattice contains iron and carbon atoms in a 3:1 ratio. The iron-carbon eutectoid reaction, occurring as a dip in Fig. 5.11 at 0.8% carbon, involves the simultaneous formation of ferrite and carbide from austenite of eutectoid composition. Since the ferrite and Fe_3C form simultaneously, they are intimately mixed. The mixture, called **pearlite**, is a lamellar structure composed of alternate layers of ferrite and carbide.

The nonequilibrium structures produced by heat treatment can be represented on a time-temperature-transformation (TTT) plot. A typical TTT curve for a 1080 steel is shown in Fig. 5.12. When the temperature is decreased below the point where the gamma phase (**austenite**) is stable, there is a driving force for transformation to the

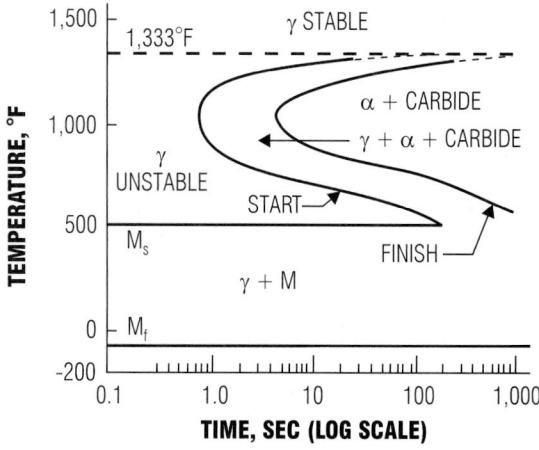

Fig. 5.12 Isothermal transformation curve for a plain carbon (0.80%) steel.

body-centered-cubic alpha phase (**ferrite**). This transformation takes some time, as shown on the TTT curve, and the time and temperature path followed determines the kind of structure formed.

If the temperature is maintained just below the transformation temperature, a coarse pearlite is formed because of high diffusion rates, which allow the excess carbon atoms to combine into large areas of Fe_3C. At somewhat lower temperatures, where diffusion rates are not so high, a fine pearlite is formed. If the unstable austenite is cooled quickly enough to prevent diffusion, the carbon present remains in solution instead of segregating out as a carbide. The resulting body-centered structure is tetragonal rather than cubic because of the strain in the lattice due to the excess carbon atoms. Since no diffusion occurs in the formation of this structure, which is called **martensite** (M in Fig. 5.12), there is essentially no time lag for this reaction.

The start of the martensitic transformation is labeled M_s and the finish M_f. Martensite is metastable, and its existence does not alter the validity of the iron-carbon equilibrium diagram. With sufficient time at temperatures below the eutectoid temperature, the supersaturated solution of carbon in iron transforms to an alpha-plus-carbide mixture called tempered martensite. The resulting microstructure is not lamellar like that of pearlite.

The rapid quenching of austenite to miss the "nose" on the TTT curve to form martensite is an important step in the heat treatment of steels. The

ensuing tempering at somewhat elevated temperatures produces steels of good toughness and high strength for construction applications.

TTT curves are also called *isothermal transformation* (IT) curves because of the way they are produced: by heating small samples into the austenite temperature range long enough for complete transformation, then quenching to various lower temperatures and holding. Samples are then quenched to room temperature at various times and the stages of transformation noted. Although the IT diagram is produced by observation of isothermal transformations, it is often used as an indication of results to be expected from non-isothermal transformations. The "Atlas of Isothermal Transformation Diagrams" (U.S. Steel Corp.) is a useful compilation of IT diagrams for a wide variety of steels.

Structural carbon steels contain about 0.2% carbon, an amount greater than that which can be dissolved in body-centered-cubic ferrite at room temperature. Little heat treatment is used with these steels, with control over the microstructure achieved by chemical composition and hot-rolling practice. Structural shapes are usually subjected to a low-temperature hot-rolling process, which results in a small, uniform grain size. Upon cooling, the final product is a fine ferrite plus pearlite (a lamellar aggregate of ferrite and iron carbide) structure.

High-strength low-alloy steels derive their strength increase from a finer microstructure and from solid-solution strengthening (Art. 5.12). Alloying elements delay the transformation of the austenite to pearlite and contribute elements that go into solution in the ferrite. This solid solutioning strengthens the ferrite.

Heat-treated carbon steels are subjected to a water quench from the austenite phase. The resulting low-temperature transformation products (martensite) are high in strength but very brittle. Tempering at about 1200 °F leads to improved toughness and ductility, with little loss in yield strength. This tempering results in the formation of a uniform structure consisting of a dense dispersion of carbides in a ferrite matrix.

Heat-treated constructional alloys are usually tempered martensitic structures. The M_s (martensitic transformation temperature) is about 700 °F for these steels. The presence of alloying elements pushes back the nose of the IT curve, thus allowing for more complete hardening. These steels are tempered at about 1200 °F, at which temperature the carbide-forming elements present (Cr, V, Mo) assist in the formation of various stable alloy carbides. The alloy carbides form a fine dispersion, strengthening the steel by dispersion hardening (Art. 5.12).

Maraging steels may owe their increased strength to formation of a finely dispersed nickel-based precipitate. During the aging process in 18% nickel maraging steels, extremely fine particles form on dislocation sites. These precipitates are responsible for the extremely high strength of the maraging steels. The difference in mechanical behavior between these nickel-based precipitates and the carbide precipitates found in heat-treated carbon steels seems to account for the superior toughness of the maraging steels.

Effects of Grain Size ▪ When a low-carbon steel is heated to hot-rolling and forging temperatures, about 1300 to 1600 °F, the steel may grow coarse grains. For some applications, this structure may be desirable; for example, it permits relatively deep hardening, and if the steel is to be used in elevated-temperature service, it will have higher load-carrying capacity and higher creep strength than if the steel had fine grains.

Fine grains, however, enhance many steel properties: notch toughness, bendability, and ductility. In quenched and tempered steels, higher yield strengths are obtained. Furthermore, fine-grain, heat-treated steels have less distortion, less quench cracking, and smaller internal stresses.

During the production of a steel, grain growth may be inhibited by an appropriate dispersion of nonmetallic inclusions or by carbides that dissolve slowly or remain undissolved during cooling. The usual method of making fine-grain steel employs aluminum deoxidation. In such steels, the inhibiting agent may be a submicroscopic dispersion of aluminum nitride or aluminum oxide. Fine grains also may be produced by hot working rolled or forged products, which otherwise would have a coarse-grain structure. The temperature at the final stage of hot working determines the final grain size. If the finishing temperature is relatively high and the grains after air cooling are coarse, the size may be reduced by normalizing. This requires heating of steel to about 1400 to 1800 °F. Then, the steel is allowed to cool in still air. (The rate of cooling is much more rapid than that used in annealing.) Fine- or coarse-grain steels may be heat treated to be coarse- or fine-grain.

5.13.3 Steel Alloys

Plain carbon steels can be given a great range of properties by heat treatment and by working; but addition of alloying elements greatly extends those properties or makes the heat-treating operations easier and simpler. For example, combined high tensile strength and toughness, corrosion resistance, high-speed cutting, and many other specialized purposes require alloy steels. However, the most important effect of alloying is the influence on hardenability.

Aluminum restricts grain growth during heat treatment and promotes surface hardening by nitriding.

Chromium is a hardener, promotes corrosion resistance, and promotes wear resistance.

Copper promotes resistance to atmospheric corrosion and is sometimes combined with molybdenum for this purpose in low-carbon steels and irons. It strengthens steel and increases the yield point without unduly changing elongation or reduction of area.

Manganese in low concentrations promotes hardenability and nondeforming, nonshrinking characteristics for tool steels. In high concentrations, the steel is austenitic under ordinary conditions, is extremely tough, and work-hardens readily. It is therefore used for teeth of power-shovel dippers, railroad frogs, rock crushers, and similar applications.

Molybdenum is usually associated with other elements, especially chromium and nickel. It increases corrosion resistance, raises tensile strength and elastic limit without reducing ductility, promotes casehardening, and improves impact resistance.

Nickel boosts tensile strength and yield point without reducing ductility; increases low-temperature toughness, whereas ordinary carbon steels become brittle; promotes casehardening; and in high concentrations improves corrosion resistance under severe conditions. It is often used with chromium. **Invar** contains 36% nickel.

Silicon strengthens low-alloy steels; improves oxidation resistance; with low carbon yields transformer steel, because of low hysteresis loss and high permeability; in high concentrations provides hard, brittle castings, resistant to corrosive chemicals, useful in plumbing lines for chemical laboratories.

Sulfur promotes free machining, especially in mild steels.

Titanium prevents intergranular corrosion of stainless steels by preventing grainboundary depletion of chromium during such operations as welding and heat treatment.

Tungsten, vanadium, and **cobalt** are all used in high-speed tool steels, because they promote hardness and abrasion resistance. Tungsten and cobalt also increase high-temperature hardness.

Stainless steels of primary interest in construction are the wrought stainless steels of the austenitic type. The austenitic stainless steels contain both chromium and nickel. Total content of alloy metals is not less than 23%, with chromium not less than 16% and nickel not less than 7%. Commonly used stainless steels have a tensile strength of 75 ksi and yield point of 30 ksi when annealed. Cold-finished steels may have a tensile strength as high as 125 ksi with a yield point of 100 ksi.

Austenitic stainless steels are tough, strong, and shock-resistant, but work-harden readily; so some difficulty on this score may be experienced with cold working and machining. These steels can be welded readily but may have to be stabilized (e.g., AISI Types 321 and 347) against carbide precipitation and intergranular corrosion due to welding unless special precautions are taken. These steels have the best high-temperature strength and resistance to scaling of all the stainless steels.

Types 303 and 304 are the familiar 18-8 stainless steels widely used for building applications. These and Types 302 and 316 are the most commonly employed stainless steels. Where maximum resistance to corrosion is required, such as resistance to pitting by seawater and chemicals, the molybdenum-containing Types 316 and 317 are best.

For resistance to ordinary atmospheric corrosion, some of the martensitic and ferritic stainless steels, containing 15 to 20% chromium and no nickel, are employed. The martensitic steels, in general, range from about 12 to 18% chromium and from 0.08 to 1.10% carbon. Their response to heat treatment is similar to that of the plain carbon steels. When chromium content ranges from 15 to 30% and carbon content is below 0.35%, the steels are ferritic and nonhardenable. The high-chromium steels are resistant to oxidizing corrosion and are useful in chemical plants.

5.13.4 Tubing for Structural Applications

Structural tubing is preferred to other steel members when resistance to torsion is required and when a smooth, closed section is esthetically desirable. In addition, structural tubing often may be the economical choice for compression members subjected to moderate to light loads. Square and rectangular tubing is manufactured either by cold or hot forming welded or seamless round tubing in a continuous process. A500 cold-formed carbon-steel tubing (Table 5.8) is produced in four strength grades in each of two product forms, shaped (square or rectangular) or round. A minimum yield point of up to 46 ksi is available for shaped tubes and up to 50 ksi for round tubes.

A501 tubing is a hot-formed carbon-steel product. It provides a yield point equal to that of A36 steel in tubing having a wall thickness of 1 in or less.

A618 tubing is a hot-formed HSLA product. It provides a minimum yield point of 33 to 50 ksi depending on grade and wall thickness. The three grades all have enhanced resistance to atmospheric corrosion. Grades Ia and Ib can be used in the bare condition for many applications when properly exposed to the astmosphere.

Table 5.8 Specified Minimum Mechanical Properties of Structural Tubing

ASTM Designation	Product Form	Yield Point, ksi	Tensile Strength, ksi	Elongation in 2 in, %
A500	Shaped			
Grade A		33	45	25
Grade B		42	58	23
Grade C		46	62	21
Grade D		36	58	23
A500	Round			
Grade A		39	45	25
Grade B		46	58	23
Grade C		50	62	21
Grade D		36	58	23
A501	Round or shaped	36	58	23
A618	Round or shaped			
Grades Ia, Ib, II				
Walls ≤ ¾ in		50	70	22
Walls > ¾ to 1½ in		46	67	22
Grade III		50	65	20

5.13.5 Mechanical Properties of Structural Steels

The tensile properties of steel are generally determined from tension tests on small specimens or coupons in accordance with standard ASTM procedures. The behavior of steels in these tests is closely related to the behavior of structural-steel members under static loads. Because, for structural steels, the yield points and moduli of elasticity determined in tension and compression are nearly the same, compression tests are seldom necessary.

Tensile strength of structural steels generally lies between about 60 and 80 ksi for the carbon and low-alloy grades and between 105 and 135 ksi for the quenched-and-tempered alloy steels (A514). Yield strengths are listed in Table 9.1. Elongation in 2 in, a measure of ductility, generally exceeds 20%, except for A514 steels. Modulus of elasticity usually is close to 29,000 ksi.

Typical stress-strain curves for several types of steels are shown in Fig. 5.13. The initial portion of the curves is shown to a magnified scale in Fig. 5.14. It indicates that there is an initial elastic range for the structural steels in which there is no permanent deformation on removal of the load. The modulus of elasticity E, which is given by the slope of the curves, is nearly a constant 29,000 ksi for all the steels. For carbon and high-strength, low-alloy steels, the inelastic range, where strains exceed those in the elastic range, consists of two parts: Initially, a plastic range occurs in which the steels yield; that is, strain increases with no increase in stress. Then follows a strain-hardening range in which increase in strain is accompanied by a significant increase in stress.

The curves in Fig. 5.14 also show an upper and lower yield point for the carbon and high-strength, low-alloy steels. The upper yield point is the one specified in standard specifications for the steels. In contrast, the curves do not indicate a yield point for the heat-treated steels. For these steels, ASTM 370, "Mechanical Testing of Steel Products," recognizes two ways of indicating the stress at which there is a significant deviation from the proportionality of stress to strain. One way, applicable to steels with a specified yield point of 80 ksi or less, is to define the yield point as the stress at which a test specimen reaches a 0.5% extension under load (0.5% EUL). The second way is to define the yield strength as the stress at which a test specimen reaches a strain (offset) 0.2% greater than that for elastic behavior. Yield point and yield strength are often referred to as yield stress.

Ductility is measured in tension tests by percent elongation over a given gage length—usually 2 or

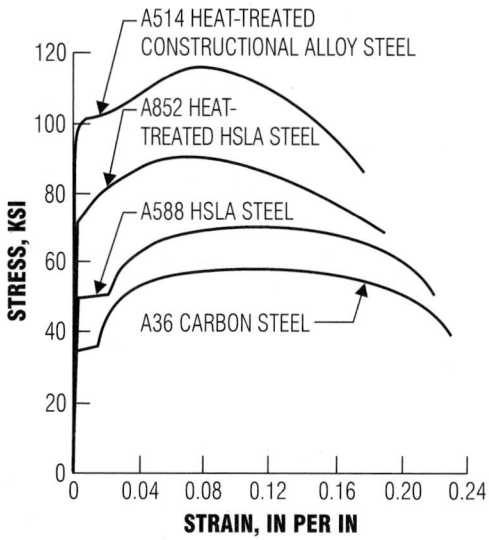

Fig. 5.13 Typical stress-strain curves for structural steels.

8 in—or percent reduction of cross-sectional area. Ductility is an important property because it permits redistribution of stresses in continuous members and at points of high local stresses.

Toughness is defined as the capacity of a steel to absorb energy; the greater the capacity, the greater the toughness. Determined by the area under the stress-strain curve, toughness depends on both strength and ductility of the metal. Notch toughness is the toughness in the region of notches or other stress concentrations. A quantitative measure of notch toughness is fracture toughness, which is determined by fracture mechanics from relationships between stress and flaw size.

Poisson's ratio, the ratio of transverse to axial strain, also is measured in tension tests. It may be taken as 0.30 in the elastic range and 0.50 in the plastic range for structural steels.

The high-strength low-alloy steels are as important in construction as the carbon steels. The A242 series, in addition to having a yield strength considerably higher than the structural carbon steels, also have four to six times the corrosion resistance of A36 carbon steel without copper. A441 is a manganese-vanadium steel with 0.20% minimum copper content and is intended primarily for welded construction. It has about twice the corrosion resistance of carbon steels. A588 steels have similar properties, but different chemistry

makes possible a 50-ksi yield strength in thicknesses up to 4 in, whereas the yield strength of A441 steels decreases from 50 to 46 ksi for thicknesses greater than $\frac{3}{4}$ in and to 42 ksi for thicknesses over $1\frac{1}{2}$ in.

The foremost property of the A514 steels is their high yield strength, which is almost three times that of A36. The heat-treated constructional alloy steels also exhibit good toughness over a wide range of temperatures and excellent resistance to atmospheric corrosion.

ASTM has also prepared a general specification, A709, for structural steel for bridges, encompassing previously generally used grades.

Cold working of structural steels, that is, forming plates or structural shapes into other shapes at room temperature, changes several properties of the steels. The resulting strains are in the strain-hardening range. Yield strength increases but ductility decreases. (Some steels are cold rolled to obtain higher strengths.) If a steel element is strained into the strain-hardening range, then unloaded and allowed to age at room or moderately elevated temperatures (a process called **strain aging**), yield and tensile strengths are increased, whereas ductility is decreased. Heat treatment can be used to modify the effects of cold working and strain aging.

Carbon-free iron-nickel martensite, the base material for maraging, is relatively soft and ductile compared with carbon-containing martensite. But iron-nickel martensite becomes hard, strong, and tough when aged. Thus maraging steels can be fabricated while they are in a comparatively ductile martensitic condition and later strengthened by a simple aging treatment.

Strain rate also changes the tensile properties of structural steels. In the ordinary tensile test, load is applied slowly. The resulting data are appropriate for design of structures for static loads. For design for rapid application of loads, such as impact loads, data from rapid tension tests are needed. Such tests indicate that yield and tensile strengths increase but ductility and the ratio of tensile strength to yield strength decrease.

High temperatures too affect properties of structural steels. As temperatures increase, the stress-strain curve typically becomes more rounded and tensile and yield strengths, under the action of strain aging, decrease. Poisson's ratio is not significantly affected but the modulus of elasticity decreases. Ductility is lowered until a minimum

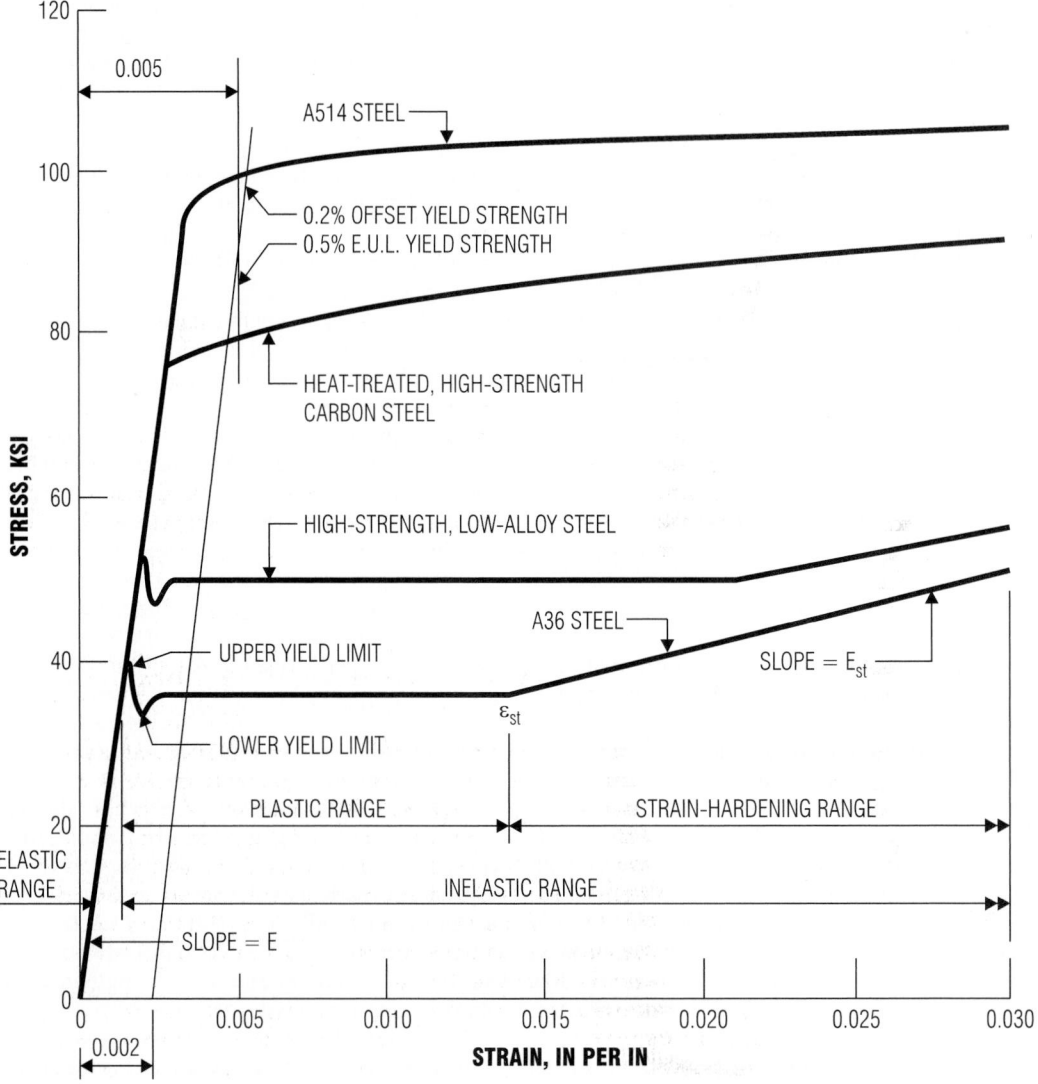

Fig. 5.14 Magnification of the initial portion of the stress-strain curves for structural steels shown in Fig. 5.13.

value is reached. Then, it rises with increase in temperature and becomes larger than the ductility at room temperature.

Low temperatures in combination with tensile stress and especially with geometric discontinuities, such as notches, bolt holes, and welds, may cause a brittle failure. This is a failure that occurs by cleavage, with little indication of plastic deformation. A ductile failure, in contrast, occurs mainly

by shear, usually preceded by large plastic deformation. One of the most commonly used tests for rating steels on their resistance to brittle fracture is the Charpy V-notch test. It evaluates notch toughness at specific temperatures.

Hardness is used in production of steels to estimate tensile strength and to check the uniformity of tensile strength in various products. Hardness is determined as a number related to resistance to indentation. Any of several tests may be used, the resulting hardness numbers being dependent on the type of penetrator and load. These should be indicated when a hardness number is given. Commonly used hardness tests are the Brinell, Rockwell, Knoop, and Vickers. ASTM A370, "Mechanical Testing of Steel Products," contains tables that relate hardness numbers from the different tests to each other and to the corresponding approximate tensile strength.

Creep, a gradual change in strain under constant stress, is usually not significant for structural steel framing, except in fires. Creep usually occurs under high temperatures or relatively high stresses, or both.

Relaxation, a gradual decrease in load or stress under a constant strain, is a significant concern in the application of steel tendons to prestressing. With steel wire or strand, relaxation can occur at room temperature. To reduce relaxation substantially, stabilized, or low-relaxation, strand may be used. This is produced by pretensioning strand at a temperature of about 600 °F. A permanent elongation of about 1% remains and yield strength increases to about 5% over stress-relieved (heat-treated but not tensioned) strand.

Residual stresses remain in structural elements after they are rolled or fabricated. They also result from uneven cooling after rolling. In a welded member, tensile residual stresses develop near the weld and compressive stresses elsewhere. Plates with rolled edges have compressive residual stresses at the edges, whereas flame-cut edges have tensile residual stresses. When loads are applied to such members, some yielding may take place where the residual stresses occur. Because of the ductility of steel, however, the effect on tensile strength is not significant but the buckling strength of columns may be lowered.

5.13.6 Fatigue of Structural Steels

Under cyclic loading, especially when stress reversal occurs, a structural member may eventually fail because cracks form and propagate. Known as a fatigue failure, this can take place at stress levels well below the yield stress. Fatigue resistance may be determined by a rotating-beam test, flexure test, or axial-load test. In these tests, specimens are subjected to stresses that vary, usually in a constant stress range between maximum and minimum stresses until failure occurs. Results of the tests are plotted on an S-N diagram, where S is the maximum stress (fatigue strength) and N is the number of cycles to failure (fatigue life). Such diagrams indicate that the fatigue strength of a structural steel decreases with increase in the number of cycles until a minimum value is reached, the **fatigue limit**. Presumably, if the maximum stress does not exceed the fatigue limit, an unlimited number of cycles of that ratio of maximum to minimum stress can be applied without failure. With tension considered positive and compression, negative, tests also show that as the ratio of maximum to minimum stress is decreased, fatigue strength is lowered significantly.

Since the tests are made on polished specimens and steel received from mills has a rough surface, fatigue data for design should be obtained from tests made on as-received material.

Tests further indicate that steels with about the same tensile strength have about the same fatigue strength. Hence the S-N diagram obtained for one steel may be used for other steels with about the same tensile strength.

5.13.7 Shear Properties of Structural Steels

The shear modulus of elasticity G is the ratio of shear stress to shear strain during initial elastic behavior. It can be computed from Eq. (6.5) from values of modulus of elasticity and Poisson's ratio developed in tension stress-strain tests. Thus G for structural steels is generally taken as 11,000 ksi.

The shear strength, or shear stress at failure in pure shear, ranges from $0.67F_t$ to $0.75F_t$ for structural steels, where F_t is the tensile strength. The yield strength in shear is about $0.57F_t$.

5.13.8 Effects of Steel Production Methods

The processing of steels after conversion of pig iron to steel in a furnace has an important influence on the characteristics of the final products. The general procedure is as follows: The molten steel at about 2900 °F is fed into a steel ladle, a refractory-lined open-top vessel. Alloying materials and deoxidizers may be added during the tapping of the heat or to the ladle. From the ladle, the liquid steel is poured into molds, where it solidifies. These castings, called ingots, then are placed in special furnaces, called soaking pits. There, the ingots are held at the desired temperature for rolling until the temperature is uniform throughout each casting.

Steel cools unevenly in a mold, because the liquid at the mold walls solidifies first and cools more rapidly than metal in the interior of the ingot. Gases, chiefly oxygen, dissolved in the liquid, are released as the liquid cools. Four types of ingot may result—killed, semikilled, capped, and rimmed—depending on the amount of gases dissolved in the liquid, the carbon content of the steel, and the amount of deoxidizers added to the steel.

A fully killed ingot develops no gas; the molten steel lies dead in the mold. The top surface solidifies relatively fast. Pipe, an intermittently bridged shrinkage cavity, forms below the top. Fully killed steels usually are poured in big-end-up molds with "hot tops" to confine the pipe to the hot top, which is later discarded. A semikilled ingot develops a slight amount of gas. The gas, trapped when the metal solidifies, forms blowholes in the upper portion of the ingot. A **capped ingot** develops rimming action, a boiling caused by evolution of gas, forcing the steel to rise. The action is stopped by a metal cap secured to the mold. Strong upward currents along the sides of the mold sweep away bubbles that otherwise would form blowholes in the upper portion of the ingot. Blowholes do form, however, in the lower portion, separated by a thick solid skin from the mold walls. A **rimmed ingot** develops a violent rimming action, confining blowholes to only the bottom quarter of the ingot. In rimmed steels, the effects of segregation are so marked that interior and outer regions differ enough in chemical composition to appear to be different steels. The boundary between these regions is sharp. Rimmed steels are preferred where surface finish is important and the effects of segregation will not be harmful.

Killed and semikilled steels require additional costs for deoxidizers if carbon content is low, and the deoxidation products form nonmetallic inclusions in the ingot. Hence, it often is advantageous for steel producers to make low-carbon steels by rimmed or capped practice, and high-carbon steels by killed or semikilled practice.

Pipe, or shrinkage cavities, generally is small enough in most steels to be eliminated by rolling. **Blowholes** in the interior of an ingot, small voids formed by entrapped gases, also usually are eliminated during rolling. If they extend to the surface, they may be oxidized and form seams when the ingot is rolled, because the oxidized metal cannot be welded together. Properly made ingots have a thick enough skin over blowholes to prevent oxidation.

Capped steels are made much like rimmed steels but with less rimming action. Capped steels have less segregation. They are used to make sheet, strip, skelp, tinplate, wire, and bars.

Semikilled steel is deoxidized less than killed steel. Most deoxidation is accomplished with additions of a deoxidizer to the ladle. Semikilled steels are used in structural shapes and plates.

Killed steels usually are deoxidized by additions to both furnace and ladle. Generally, silicon compounds are added to the furnace to lower the oxygen content of the liquid metal and stop oxidation of carbon (block the heat). This also permits addition of alloying elements that are susceptible to oxidation. Silicon or other deoxidizers, such as aluminum, vanadium, and titanium, may be added to the ladle to complete deoxidation. Aluminum, vanadium, and titanium have the additional beneficial effect of inhibiting grain growth when the steel is normalized. (In the hot-rolled conditions, such steels have about the same ferrite grain size as semikilled steels.) Killed steels deoxidized with aluminum and silicon (made to fine-grain practice) often are specified for construction applications because of better notch toughness and lower transition temperatures than semikilled steels of the same composition.

5.13.9 Effects of Hot Rolling

Plates and shapes for construction applications may be produced by casting and rolling of ingots or by continuous casting. Most plates and shapes are made by hot-rolling ingots. But usually, the final products are not rolled directly from ingots. First, the ingots are generally reduced in cross section by rolling into billets, slabs, and blooms. These forms permit correction of defects before finish rolling, shearing into convenient lengths for final rolling, reheating for further rolling, and transfer to other mills, if desired, for that processing.

ASTM A6 requires that material for delivery "shall be free from injurious defects and shall have a workmanlike finish." The specification permits manufacturers to condition plates and shapes "for the removal of injurious surface imperfections or surface depressions by grinding, or chipping and grinding. . . ."

Plates produced from slabs or directly from ingots, are distinguished from sheet, strip, and flat bars by size limitations in ASTM A6. Generally, plates are heavier, per linear foot, than these other products. Sheared plates, or sheared mill plates, are made with straight horizontal rolls and later trimmed on all edges. Universal plates, or universal mill plates, are formed between vertical and horizontal rolls and are trimmed on the ends only.

Some of the plates may be heat-treated, depending on grade of steel and intended use. For carbon steel, the treatment may be annealing, normalizing, or stress relieving. Plates of high-strength, low-alloy constructional steels may be quenched and tempered.

Shapes are rolled from blooms that first are reheated to 2250 °F. Rolls gradually reduce the plastic blooms to the desired shapes and sizes. The shapes then are cut to length for convenient handling with a hot saw.

Internal structure and many properties of plates and shapes are determined largely by the chemistry of the steel, rolling practice, cooling conditions after rolling, and heat treatment, where used. As a result of hot rolling, ductility and bendability are much better in the longitudinal direction than in the transverse, and these properties are poorest in the thickness direction. The cooling rate after rolling determines the distribution of ferrite and the grain size of the ferrite. Rolling, however, may induce residual stresses in plates and shapes. Still other effects are a consequence of the final thickness of the hot-rolled material.

Thicker material requires less rolling, the finish rolling temperature is higher, and the cooling rate is slower than for thin material. As a consequence, thin material has a superior microstructure. Furthermore, thicker material can have a more unfavorable state of stress because of stress raisers, such as tiny cracks and inclusions, and residual stresses. Consequently, thin material develops higher tensile and yield strengths than thick material of the same steel. ASTM specifications for structural steels recognize this usually by setting lower yield points for thicker material. A36 steel, however, has the same yield point for all thicknesses. To achieve this, the chemistry is varied for plates and shapes and for thin and thick plates. Thicker plates contain more carbon and manganese

to raise the yield point. This cannot be done for high-strength steels because of the adverse effect on notch toughness, ductility, and weldability.

Thin material has greater ductility than thick material of the same steel. Since normalizing refines the grain structure, thick material improves relatively more with normalizing than does thin material. The improvement is even greater with silicon-aluminum-killed steels.

5.13.10 Effects of Punching and Shearing

Punching holes and shearing during fabrication are cold-working operations that can cause brittle failure. Bolt holes, for example, may be formed by drilling, punching, or punching followed by reaming. Punching drastically cold-works the material at the edge of a hole. This makes the steel less ductile. Furthermore, there is a possibility that punching can produce short cracks extending radially from the hole. Hence, brittle failure can be initiated at the hole when the member is stressed.

Reaming a hole after punching can eliminate the short radial cracks and the risks of embrittlement. For the purpose, the hole diameter should be increased by $\frac{1}{16}$ to $\frac{1}{4}$ in by reaming, depending on material thickness and hole diameter.

Shearing has about the same effects as punching. If sheared edges are to be left exposed, $\frac{1}{16}$ in or more material, depending on thickness, should be trimmed by gas cutting. Note also that rough machining, for example, with edge planers making a deep cut, can produce the same effects as shearing or punching.

5.13.11 Welding

Fusion welding is a process for joining metals either by melting them together or by fusing them while a filler metal is deposited in the joint between them. During welding, the part of the base metal near the joint and all the filler metal are molten. Because of the good thermal conductivity of metal, a temperature gradient is developed, varying from the melting point at the fusion zone to the ambient temperature at some distance from the weld zone.

General welding characteristics of the various types of ferrous metals are as follows:

Wrought iron is ideally forged but may be welded by other methods if the base metal is thoroughly fused. Slag melts first and may confuse unwary operators.

Low-carbon iron and steels (0.30%C or less) are readily welded and require no preheating or subsequent annealing unless residual stresses are to be removed.

Medium-carbon steels (0.30 to 0.50%C) can be welded by the various fusion processes. In some cases, especially in steel with more than 0.40% carbon, preheating and subsequent heat treatment may be necessary.

High-carbon steels (0.50 to 0.90%C) are more difficult to weld and, especially in arc welding, may have to be preheated to at least 500 °F and subsequently heated between 1200 and 1450 °F. For gas welding, a carburizing flame is often used. Care must be taken not to destroy the heat treatment to which high-carbon steels may have been subjected.

Tool steels (0.80 to 1.50%C) are difficult to weld. Preheating, postannealing, heat treatment, special welding rods, and great care are necessary for successful welding.

Welding of structural steels is governed by the American Welding Society "Structural Welding Code," AWS D1.1, the American Institute of Steel Construction "Specification for the Design, Fabrication and Erection of Structural Steel for Buildings," or a local building code. AWS D1.1 specifies tests to be used in qualifying welders and types of welds. The AISC Specification and many building codes require, in general, that only qualified welds be used and that they be made only by qualified welders.

The heat required for fusion welding can be produced by burning together such gases as oxygen and acetylene in a welding torch but is more usually supplied by an electric arc. The arc may be struck either between the work and a consumable electrode, which also serves as the filler material, or between the work and a nonconsumable electrode, with external filler metal added.

A protective environment is usually provided to ensure weld soundness. This inert atmosphere may be formed by the decomposition of coatings on the welding electrodes or provided by separate means. Several welding processes are in common use today. Shielded metal-arc welding may employ coated electrodes or have bare electrodes passing through a separately maintained flux pool (submerged arc welding). Consumable metal-arc inert-gas welding is done under the protection of an inert shielding gas coming from a nozzle. Tungsten-arc inert-gas welding also employs inert shielding gas but uses a virtually nonconsumed tungsten electrode. On joints where filler metals are required with a tungsten arc, a filler rod is fed into the weld zone and melted with the base metal, as in the oxyacetylene process. These processes can be used manually or in semiautomatic or automatic equipment where the electrode may be fed continuously.

Stud welding is used to fuse metal studs or similar parts to other steel parts by the heat of an electric arc. A welding gun is usually used to establish and control the arc, and to apply pressure to the parts to be joined. At the end to be welded, the stud is equipped with a ceramic ferrule, which contains flux and which also partly shields the weld when molten.

Preheating before welding reduces the risk of brittle failure. Initially, its main effect is to lower the temperature gradient between the weld and adjoining base metal. This makes cracking during cooling less likely and gives entrapped hydrogen, a possible source of embrittlement, a chance to escape. A later effect of preheating is improved ductility and notch toughness of base and weld metals and lower transition temperature of weld. When, however, welding processes that deposit weld metal low in hydrogen are used and suitable moisture control is maintained, the need for preheat can be eliminated. Such processes include use of low-hydrogen electrodes and inert-arc and submerged arc welding.

Rapid cooling of a weld can have an adverse effect. One reason that arc strikes that do not deposit weld metal are dangerous is that the heated metal cools very fast. This causes severe embrittlement. Such arc strikes should be completely removed. The material should be preheated, to prevent local hardening, and weld metal should be deposited to fill the depression.

Weldability of structural steels is influenced by their chemical content. Carbon, manganese, silicon, nickel, chromium, and copper, for example, tend to have an adverse effect, whereas molybdenum and vanadium may be beneficial. To relate the influence of chemical content on structural steel properties to weldability, the use of a carbon equivalent has been proposed. One formula suggested is

$$C_{eq} = C + \frac{Mn}{4} + \frac{Si}{4} \qquad (5.2)$$

where C = carbon content, %

Mn = manganese content, %

Si = silicon content, %

Another proposed formula includes more elements:

$$C_{eq} = C + \frac{Mn}{6} + \frac{Ni}{20} + \frac{Cr}{10} - \frac{Mo}{50} - \frac{V}{10} + \frac{Cu}{40} \quad (5.3)$$

where Ni = nickel content, %

Cr = chromium content, %

Mo = molybdenum content, %

V = vanadium content, %

Cu = copper content, %

Carbon equivalent appears to be related to the maximum rate at which a weld and adjacent base metal may be cooled after welding without underbead cracking occurring. The higher the carbon equivalent, the lower will be the allowable cooling rate. Also, the higher the carbon equivalent, the more important use of low-hydrogen electrodes and preheating becomes.

Precautions are required to minimize pickup of hydrogen by the weld metal and heat-affected zone. Hydrogen tends to embrittle the steel and cause cracking underneath the deposited weld bead. In addition to providing a shielding atmosphere, it may be necessary to bake the electrodes to insure that moisture content is low at time of use.

5.14 Steel Sheet and Strip for Structural Applications

Steel sheet and strip are used for many structural applications, including cold-formed members in building construction and the stressed skin of transportation equipment. Mechanical properties of several of the more frequently used sheet steels are presented in Table 5.9.

ASTM A570 covers seven strength grades of uncoated, hot-rolled, carbon-steel sheets and strip. (See ASTM A611 for cold-rolled carbon-steel sheet.) A446 covers several grades of galvanized, carbon-steel sheets. The various weights of zinc coating available for A446 sheets afford excellent corrosion protection in many applications.

A607, available in six strength levels, covers high-strength, low-alloy columbium or vanad-

ium, or both, hot- and cold-rolled steel sheet and strip. The material may be in either cut lengths or coils. It is intended for structural or miscellaneous uses where greater strength and weight savings are important. A607 is available in two classes, each with six similar strength levels, but class 2 offers better formability and weldability than class 1. Without addition of copper, these steels are equivalent in resistance to atmospheric corrosion to plain carbon steel. With copper, however, resistance is twice that of carbon steel.

A606 covers high-strength, low-alloy, hot- and cold-rolled steel sheet and strip with enhanced corrosion resistance. This material is intended for structural or miscellaneous uses where weight savings or high durability are important. It is available, in cut lengths or coils, in either type 2 or type 4, with corrosion resistance two or four times, respectively, that of plain carbon steel.

5.15 Steel Cable for Structural Applications

Steel cables have been used for many years in bridge construction and are occasionally used in building construction for the support of roofs and floors. The types of cables used for these applications are referred to as **bridge strand** or **bridge rope**. In this use, **bridge** is a generic term that denotes a specific type of high-quality strand or rope.

A **strand** is an arrangement of wires laid helically about a center wire to produce a symmetrical section. A **rope** is a group of strands laid helically around a core composed of either a strand or another wire rope. The term **cable** is often used indiscriminately in referring to wires, strands, or ropes. Strand may be specified under ASTM A586; wire rope, under A603.

During manufacture, the individual wires in bridge strand and rope are generally galvanized to provide resistance to corrosion. Also, the finished cable is prestretched. In this process, the strand or rope is subjected to a predetermined load of not more than 55% of the breaking strength for a sufficient length of time to remove the "structural stretch" caused primarily by radial and axial adjustment of the wires or strands to the load. Thus, under normal design loadings, the elongation that occurs is essentially elastic and may be

Table 5.9 Specified Minimum Mechanical Properties for Steel Sheet and Strip for Structural Applications

ASTM Designation	Final Condition	Yield Point, ksi	Tensile Strength, ksi	Elongation, % In 2 in*	In 8 in
A446	Galvanized				
Grade A		33	45	20	
Grade B		37	52	18	
Grade C		40	55	16	
Grade D		50	65	12	
Grade E		80	82	—	
Grade F		50	70	12	
A570	Hot-rolled				
Grade 30		30	49	25	19
Grade 33		33	52	23	18
Grade 36		36	53	22	17
Grade 40		40	55	21	16
Grade 45		45	60	19	14
Grade 50		50	65	17	12
Grade 55		55	70	15	10
A606	Hot-rolled, cut length	50	70	22	
	Hot-rolled, coils	45	65	22	
	Cold-rolled	45	65	22	
A607	Hot- or cold-rolled				
Grade 45		45	$60^\dagger$	25–23	
Grade 50		50	$65^\dagger$	22–20	
Grade 55		55	$70^\dagger$	20–18	
Grade 60		60	$75^\dagger$	18–16	
Grade 65		65	$80^\dagger$	16–14	
Grade 70		70	$85^\dagger$	14–12	

*Modified for some thicknesses in accordance with the specification. Where two values are given, the first is for hot-rolled, the second for cold-rolled steel.

†For class 1 product. Reduce tabulated strengths 5 ksi for class 2.

calculated from the elastic-modulus values given in Table 5.10.

Strands and ropes are manufactured from cold-drawn wire and do not have a definite yield point. Therefore, a working load or design load is determined by dividing the specified minimum breaking strength for a specific size by a suitable safety factor. The breaking strengths for selected sizes of bridge strand and rope are listed in Table 5.10.

5.16 Aluminum Alloys

Aluminum alloys are generally harder and stronger but usually not as corrosion resistant as the pure metal. The alloys may be classified as (1) cast and wrought and (2) heat-treatable and non-heat-treatable. Wrought alloys can be worked

mechanically by such processes as rolling, extrusion, drawing, or forging.

5.16.1 Aluminum Alloy Designations

Wrought-aluminum alloys are designated by a four-digit index. The first digit identifies the alloy type, according to the following code:

Pure aluminum, 99.00% min and greater	1xxx
Copper	2xxx
Manganese	3xxx
Silicon	4xxx
Magnesium and silicon	6xxx
Zinc	7xxx
Other elements	8xxx

Table 5.10 Mechanical Properties of Steel Cables

Minimum Breaking Strength, ksi,* of Selected Cable Sizes			Minimum Modulus of Elasticity, ksi,* of Indicated Diameter Range	
Nominal Diameter, in	Zinc-Coated Strand	Zinc-Coated Rope	Nominal Diameter Range, in	Minimum Modulus, ksi
$\frac{1}{2}$	30	23	Prestretched	
$\frac{3}{4}$	68	52	Zinc-Coated Strand	
1	122	91.4	$\frac{1}{2}$ to $2\frac{9}{16}$	24,000
$1\frac{1}{2}$	276	208	$2\frac{5}{8}$ and over	23,000
2	490	372	Prestretched	
3	1076	824	Zinc-Coated Rope	
4	1850	1460	$\frac{3}{8}$ to 4	20,000

*Values are for cables with class A zinc coating on all wires. Class B or C can be specified where additional corrosion protection is required.

The second digit signifies specific alloy modifications, and the last two digits identify the specific aluminum alloy or indicate the aluminum purity. (EC is a special designation for electrical conductors.)

These wrought-aluminum alloys are heat-treatable if the dissolved alloying elements are less soluble in the solid state at ordinary temperatures than at elevated temperatures. This makes age hardening possible. Cold working or other forms of strain hardening may also be employed to strengthen aluminum alloys (Art. 5.12). The temper of an alloy is indicated by adding a symbol to the alloy designation, as follows:

-F   As fabricated, no control of temper

-O   Annealed (recrystallized)

-H   Strain-hardened

-T   Heat-treated to produce stable tempers other than F, O, or H

-N   Solution heat-treated

The letters H and T are usually followed by additional numbers indicating more details of the treatment. H1 designates an alloy that has been strain-hardened only, while H2 designates one that has been strain-hardened and then partially annealed. A second number following the H indicates increasing amounts of strain hardening on a scale from 2 to 9. H3 indicates

an alloy that has been strain-hardened and stabilized by suitable annealing. The various tempers produced by heat treatment are indicated by T followed by a number, as follows:

-T1   Naturally aged after an elevated-temperature fabrication process

-T2   Cold worked and then naturally aged after an elevated-temperature fabrication process

-T3   Solution heat treatment followed by strain hardening; different amounts of strain hardening are indicated by a second digit

-T4   Solution heat treatment followed by natural aging at room temperature

-T5   Artificial aging after an elevated-temperature fabrication process

-T6   Solution heat treatment followed by artificial aging

-T7   Solution heat treatment followed by stabilization with an overaging heat treatment

-T8   Solution heat treatment, strain hardening, and then artificial aging

-T9   Solution heat treatment, artificial aging, and then strain hardening

-T10 Cold worked and than artificially aged after an elevated-temperature fabricating process

As an example of the application of this system, consider alloy 7075. Its nominal composition is 5.6% zinc, 1.6% copper, 2.5% magnesium, 0.3% chromium, and the remainder aluminum and impurity traces. If it is designated 7075-O, it is in a soft condition produced by annealing at 775 °F for a few hours. If it is designated in a hard temper, 7075-T6, it has been solution heat-treated at 870 °F and aged to precipitation-harden it at 250 °F for about 25 h.

A similar designation system is used for cast alloys. Casting alloys may be sand or permanent-mold alloys.

5.16.2 Finishes for Aluminum

Almost all finishes used on aluminum may be divided into three major categories in the system recommended by the The Aluminum Association: mechanical finishes, chemical finishes, and coatings. The last may be subdivided into anodic coatings, resinous and other organic coatings, vitreous coatings, electroplated and other metallic coatings, and laminated coatings.

In The Aluminum Association system, mechanical and chemical finishes are designated by M and C, respectively, and each of the five classes of coating is also designated by a letter. The various finishes in each category are designated by two-digit numbers after a letter. The principal finishes are summarized in Table 5.11.

5.16.3 Structural Aluminum

Aluminum alloys are used in structural applications because the strength-to-weight ratio is often more favorable than that of other materials. Aluminum structures also need a minimum of maintenance since aluminum stabilizes in most atmospheres.

Wrought-aluminum alloys for structural applications are usually precipitation-hardened to strengthen them. Typical properties of some aluminum alloys frequently used in structural applications are in Table 5.12; the range of properties from the soft to the hardest available condition is shown.

Structural aluminum shapes are produced by extrusion. Angles, I beams, and channels are available in standard sizes and in lengths up to 85 ft. Plates up to 6 in thick and 200 in wide also may be obtained.

Table 5.11 Finishes for Aluminum and Aluminum Alloys

Type of Finish	Designation
Mechanical finishes:	
As fabricated	M1Y
Buffed	M2Y
Directional textured	M3Y
Nondirectional textured	M4Y
Chemical finishes:	
Nonetched cleaned	C1Y
Etched	C2Y
Brightened	C3Y
Chemical conversion coatings	C4Y
Coatings	
Anodic	
General	A1Y
Protective and decorative (less than 0.4 mil thick)	A2Y
Architectural Class II (0.4–0.7 mil thick)	A3Y
Architectural Class I (0.7 mil thick or more thick)	A4Y
Resinous and other organic coatings	R1Y
Vitreous coatings	V1Y
Electroplated and other metallic coatings	E1Y
Laminated coatings	L1Y

*Y represents digits (0, 1, 2, . . . , 9) or X (to be specified) that describe the surface, such as specular, satin, matte, degreased, clear anodizing or type of coating.

There are economic advantages in selecting structural aluminum shapes more efficient for specific purposes than the customary ones. For example, sections such as hollow tubes, shapes with stiffening lips on outstanding flanges, and stiffened panels can be formed by extrusion.

Aluminum alloys generally weigh about 170 lb/ft³, about one-third that of structural steel. The modulus of elasticity in tension is about 10,000 ksi, compared with 29,000 ksi for structural steel. Poisson's ratio may be taken as 0.50. The coefficient of thermal expansion in the 68 to 212 °F range is about 0.000013 in/in · °F, about double that of structural steel.

Table 5.12 Properties of Selected Structural Aluminum Alloys

Alloy Designation	Principal Alloying Elements	Hardening Process	Range of Properties (Soft to Hard Conditions)		
			Tensile Strength, ksi	Yield Strength, ksi	Elongation in 2 in, %
2014	4.4% Cu, 0.8% Si, 0.8% Mn, 0.4% Mg	Precipitation	27–70	14–60	18–13
2024	4.5% Cu, 1.5% Mg, 0.6% Mn	Precipitation	27–72	11–57	20–13
5456	5.0% Mg, 0.7% Mn, 0.15% Cu, 0.15% Cr	Cold working	45–51	23–37	24–16
6061	1.0% Mg, 0.6% Si, 0.25% Cu, 0.25% Cr	Precipitation	18–45	8–40	25–12
7075	5.5% Zn, 2.5% Mg, 1.5% Cu, 0.3% Cr	Precipitation	33–83	15–73	17–11
Clad 7075	Layer of pure aluminum bonded to surface of alloy to increase corrosion	Precipitation	32–76	14–67	17–11

Alloy 6061-T6 is often used for structural shapes and plates. ASTM B308 specifies a minimum tensile strength of 38 ksi, minimum tensile yield strength of 35 ksi, and minimum elongation in 2 in of 10%, but 8% when the thickness is less than $\frac{1}{4}$ in.

The preceding data indicate that, because of the low modulus of elasticity, aluminum members have good energy absorption. Where stiffness is important, however, the effect of the low modulus should be taken into account. Specific data for an application should be obtained from the producers.

5.16.4 Connections for Aluminum

Aluminum connections may be welded, brazed, bolted or riveted. Bolted connections are bearing type. Slip-critical connections, which depend on the frictional resistance of joined parts created by bolt tension, are not usually employed because of the relatively low friction and the potential relaxation of the bolt tension over time.

Bolts may be aluminum or steel. Bolts made of aluminum alloy 7075-T73 have a minimum expected shear strength of 40 ksi. Cost per bolt, however, is higher than that of 2024-T4 or 6061-T6, with tensile strengths of 37 and 27 ksi, respectively. Steel bolts may be used if the bolt material is selected to prevent galvanic corrosion or the steel is insulated from the aluminum. One option is use of stainless steel. Another alternative is to galvanize, aluminize, or cadmium plate the steel bolts.

Rivets typically are made of aluminum alloys. They are usually driven cold by squeeze-type riveters. Alloy 6053-T61, with a shear strength of 20 ksi, is preferred for joining relatively soft alloys, such as 6063-T5, Alloy 6061-T6, with a shear strength of 26 ksi, is usually used for joining 6061-T6 and other relatively hard alloys.

Brazing, a process similar to soldering, is done by furnace, torch, or dip methods. Successful brazing is done with special fluxes.

Welding of Aluminum ▪ All wrought-aluminum alloys are weldable but with different degrees of care required. The entire class of wrought alloys that are not heat-treatable can be welded with little difficulty.

Welds should be made to meet the requirements of the American Welding Society, "Structural Welding Code—Aluminum," AWS D1.2.

Inert-gas shielded-arc welding is usually used for welding aluminum alloys. The inert gas, argon or helium, inhibits oxide formation during welding. The electrode used may be consumable metal or tungsten. The gas metal arc is generally preferred for structural welding, because of the higher speeds that can be used. The gas tungsten arc is preferred for thicknesses less than $\frac{1}{2}$ in.

Butt-welded joints of annealed aluminum alloys and non-heat-treatable alloys have nearly the same strength as the parent metal. This is not true for strain-hardened or heat-tempered alloys. In these conditions, the heat of welding weakens the metal in the vicinity of the weld. The tensile strength of a butt weld of alloy 6061-T6 may be reduced to 24 ksi, about two-thirds that of the parent metal. Tensile yield strength of such butt welds may be only 15 to 20 ksi, depending on metal thickness and type of filler wire used in welding.

Fillet welds similarly weaken heat-treated alloys. The shear strength of alloy 6061-T6 decreases from about 27 ksi to 17 ksi or less for a fillet weld.

For annealed alloys that are not heat-treatable, joints can always be made to fail in the base metal as long as the thicker weld bead is left in place. For hard-rolled tempers, the base metal in the heat-affected zone is softened by the welding heat, so joint efficiency is less than 100%. With heat-treatable alloys in the 6000 series, 100% efficiency can be obtained if the welded structure can be solution and precipitation heat-treated after welding. Nearly 100% efficiency can also be obtained without the solution heat treatment if a high-speed welding technique (such as inert-gas shielded-metal arc) is used to limit heat flow into the base metal, and a precipitation heat treatment is used after welding. In the 2000 and 7000 series, such practices produce less improvement. Weld strengths in general range from about 60 to 100% of the strength of the alloy being welded.

5.17 Copper-Based Alloys

Copper and its alloys are widely used in construction for a large variety of purposes, particularly applications requiring corrosion resistance, high electrical conductivity, strength, ductility, impact resistance, fatigue resistance, or other special characteristics possessed by copper or its alloys. Some of the special characteristics of importance to construction are ability to be formed into complex shapes, appearance, and high thermal conductivity, although many of the alloys have low thermal conductivity and low electrical conductivity as compared with the pure metal. When copper is exposed to the air and oxidizes, a green patina forms on the surface that is sometimes objectionable when it is washed down over adjacent surfaces, such as ornamental stone. The patina is formed particularly in industrial atmospheres. In rural atmospheres, where industrial gases are absent, the copper normally turns to a deep brown color.

Principal types of copper and typical uses are:

Electrolytic tough pitch (99.90% copper) is used for electrical conductors—bus bars, commutators, etc.; building products—roofing, gutters, etc.; process equipment—kettles, vats, distillery equipment; forgings. General properties are high electrical conductivity, high thermal conductivity, and excellent working ability.

Deoxidized (99.90% copper and 0.025% phosphorus) is used, in tube form, for water and refrigeration service, oil burners, etc.; in sheet and plate form, for welded construction. General properties include higher forming and bending qualities than electrolytic copper. They are preferred for coppersmithing and welding (because of resistance to embrittlement at high temperatures).

5.17.1 Brass

A considerable range of brasses is obtainable for a large variety of end uses. The high ductility and malleability of the copper-zinc alloys, or brasses, make them suitable for operations like deep drawing, bending, and swaging. They have a wide range of colors. They are generally less expensive than the high-copper alloys.

Grain size of the metal has a marked effect upon its mechanical properties. For deep drawing and other heavy working operations, a large grain size is required, but for highly finished polished surfaces, the grain size must be small.

Like copper, brass is hardened by cold working. Hardnesses are sometimes expressed as quarter hard, half hard, hard, extra hard, spring, and extra spring, corresponding to reductions in cross section during cold working ranging from approximately 11 to 69%. Hardness is strongly influenced by alloy composition, original grain size, and form (strip, rod, tube, wire).

The principal plain brasses, with compositions ranging from high copper content to zinc contents of 40% or more, are the following: commercial bronze, employed in forgings, screws, stamped hardware, and weatherstripping; red brass, used for hardware and tubing and piping for plumbing;

cartridge brass, used in fabricating processes, pins, rivets, heating units, electrical sockets; Muntz metal, used in architectural work, condenser tubes, valve stems, brazing rods.

Leaded Brass ▪ Lead is added to brass to improve its machinability, particularly in such applications as automatic screw machines where a freely chipping metal is required. Leaded brasses cannot easily be cold-worked by such operations as flaring, upsetting, or cold heading. Several leaded brasses of importance in construction are the following: High-leaded brass, for keys, lock parts, scientific instruments; forging brass, used in hardware and plumbing; architectural bronze, for handrails, decorative molding, grilles, hinges.

Tin Brass ▪ Tin is added to a variety of basic brasses to obtain hardness, strength, and other properties that would otherwise not be available. Two important alloys are (1) admiralty, used for condenser and heat-exchanger plates and tubes, steam-power-plant equipment, chemical and process equipment, and marine applications; (2) manganese bronze, used for forgings, condenser plates, valve stems, coal screens.

5.17.2 Nickel Silvers

These are alloys of copper, nickel, and zinc. Depending on the composition, they range in color from a definite to slight pink cast through yellow, green, whitish green, whitish blue, to blue. A wide range of nickel silvers is made, of which only one typical composition will be described. Those that fall in the combined alpha-beta phase of metals are readily hot-worked and therefore are fabricated without difficulty into such intricate shapes as plumbing fixtures, stair rails, architectural shapes, and escalator parts. Lead may be added to improve machining.

5.17.3 Cupronickel

Copper and nickel are alloyed in a variety of compositions of which the high-copper alloys are called the cupronickels. Typical commercial types of cupronickel contain 10 or 30% nickel:

Cupronickel, 10% (88.5% copper, 10% nickel, 1.5% iron). Recommended for applications requiring corrosion resistance, especially to salt water, as in tubing for condensers, heat exchangers, and formed sheets.

Cupronickel, 30% (70.0% copper, 30.0% nickel). Typical uses are condenser tubes and plates, tanks, vats, vessels, process equipment, automotive parts, meters, refrigerator pump valves.

5.17.4 Bronzes

Originally, the bronzes were all alloys of copper and tin. Today, the term bronze is generally applied to engineering metals having high mechanical properties and the term brass to other metals. The commercial wrought bronzes do not usually contain more than 10% tin because the metal becomes extremely hard and brittle. When phosphorus is added as a deoxidizer, to obtain sound, dense castings, the alloys are known as phosphor bronzes. The two most commonly used tin bronzes contain 5 or 8% tin. Both have excellent cold-working properties. These are high-copper alloys containing percentages of silicon ranging from about 1% to slightly more than 3%. In addition, they generally contain one or more of the four elements, tin, manganese, zinc, and iron. A typical one is high-silicon bronze, type A, which is generally used for tanks, pressure vessels, vats; weatherstrips, forgings.

Aluminum bronzes, like aluminum, form an aluminum oxide skin on the surface, which materially improves resistance to corrosion, particularly under acid conditions. Since the color of the 5% aluminum bronze is similar to that of 18-carat gold, it is used for costume jewelry and other decorative purposes. Aluminum-silicon bronzes are used in applications requiring high tensile properties in combination with good corrosion resistance in such parts as valves, stems, air pumps, condenser bolts, and similar applications. Their wear-resisting properties are good; consequently, they are used in slide liners and bushings.

5.18 High-Performance Metal Composites

Additional strength can be obtained for an alloy by converting it into a high-performance, fiber-reinforced composite. Fibers of such materials as graphite, silicon carbide, silicon nitride, boron nitride, and alumna, may be used for the purpose.

Difficulties are frequently encountered, however, in forming a fiber composite in a molten, metallic matrix due to mechanical and chemical incompatibility.

To obtain desired mechanical properties, such as improved strength, toughness, and creep resistance, a thorough understanding of the transverse and shear fiber-matrix properties is required. Mismatch results in matrix cracking and breakdown of the fiber-matrix interface. For high-performance composites with relatively brittle metallic and ceramic matrices, a chemical reaction between the fiber and the matrix that forms an alloy can seriously deplete and weaken the fiber when the alloy has mechanical properties incompatible with the matrix.

When silicon-carbide-fiber reinforcement is incorporated in an aluminum alloy, the aluminum extracts silicon from the fiber to form aluminum silicide (Al_4Si_3). When the silicon concentration in the matrix is kept above a critical level, however, the need of the matrix to leach additional silicon from the fiber is relieved.

A more general method is to prevent an element in the fiber from forming an alloy with the matrix by giving the fiber a protective coating. For example, to provide a "sacrificial" coating on the fiber, it can be coated with silicon carbide, which is slowly sacrificed by a reaction with the aluminum alloy matrix to form aluminum silicide. Another technique is to coat the fiber with alumina, which is chemically inert. Proprietary processes are available, such as the Duralcan molten-metal mixing method, that produce low-cost composites. The Duralcan process permits use of conventional casting and fabrication practices.

5.19 Metal References

"Alcoa Structural Handbook" and "Welding Alcoa Aluminum," Aluminum Company of America, Pittsburgh, PA. www.alcoa.com

"Alloy Data," Copper Development Association, New York. www.copper.org

"Aluminum Standards and Data," "Aluminum Finishes," and "Specifications for Aluminum Structures," Aluminum Association, 900 19th street N.W., Washington, DC 20006. www.aluminum.org

Brady, G. S., and H. R. Clauser, "Materials Handbook," 13th ed., McGraw-Hill, Inc., New York. www.mcgraw-hill.com

Brantley, L. R., and R. T. Brantley, "Building Materials Technology: Structural Performance and Environmental Impact," McGraw-Hill, Inc., New York. www.mcgraw-hill.com

Callender, J. J., "Time Saver Standards for Architectural Design," 6th ed., McGraw-Hill, Inc., New York. www.mcgraw-hill.com

"Carbon Steels, Chemical Composition Limits, Constructional Alloys, Chemical Composition Limits," and "Steel Products Manual," American Iron and Steel Institute, 1101 17th St., Suite 1300, N.W.; Washington, DC 20036. www.steel.or

Merritt, F. S., and R. L. Brockenbrough, "Structural Steel Designers Handbook," 3rd ed., McGraw-Hill, Inc., New York. www.mcgraw-hill.com

"Metals Handbook," American Society for Metals, Materials Park, OH 44073. www.asminternational.org

"Welding Handbook," American Welding Society, 550 NW LeJone Rd, Miami, FL 33126. www.aws.org

Masonry Units and Tiles

A wide variety of manufactured products are produced from concrete and used in construction. These include concrete brick, block, or tile; floor and roof slabs; wall panels; cast stone; and precast beams and columns. Also, a wide variety of burned-clay units are manufactured for construction purposes. These products include common and face clay brick, hollow clay tile, ceramic tile, and architectural terra cotta. Various types of stone also are used as masonry.

Properties of concrete masonry units depend on the ingredients and proportioning of the mix and the method of manufacture and curing. Properties of burned-clay units vary with the type of clay or shale used as raw material, method of fabrication of the units, and temperature of burning. As a consequence, some units, such as salmon brick, are underburned, highly porous, and of poor strength. But others are almost glass hard, have been pressed and burned to almost eliminate porosity, and are very strong. Between these extremes lie most of the units used for construction.

5.20 Concrete Masonry Units

These are made both from normal, dense concrete mixes and from mixes with lightweight aggregates.

Concrete blocks are made with holes through them to reduce their weight and to enable masons to grip them.

Nominal size (actual dimensions plus width of mortar joint) of hollow concrete block generally is 8 × 8 × 16 in. Solid blocks often are available with nominal size of 4 × 8 × 16 in or 4 × 2½ × 8 in. For a list of modular sizes, see "Standard Sizes of Clay and Concrete Modular Units," ANSI A62.3.

Properties of the units vary widely—from strong, dense load-bearing units used under exposed conditions to light, relatively weak, insulating units used for roof and fire-resistant construction.

Requirements for strength and absorption of concrete brick and block established by ASTM for Type I, Grades N-I and S-I (moisture-controlled), and Type II, Grades N-II and S-II (non-moisture-controlled), units are summarized in Table 5.13.

Manufactured concrete units have the advantage (or sometimes disadvantage) that curing is under the control of the manufacturer. Many methods of curing are used, from simply stacking the units in a more or less exposed location to curing under high-pressure steam. The latter method appears to have considerable merit in reducing ultimate shrinkage of the block. Shrinkage may be as small as ¼ to ⅜ in per 100 ft for concrete units cured with high-pressure steam. These values are about one-half as great as those obtained with normal atmospheric curing. Tests for moisture movement in blocks cured with high-pressure and high-temperature steam indicate expansions of from ¼ to ½ in per 100 ft after saturation of previously dried specimens.

5.21 Bricks—Clay or Shale

These are burned-clay or shale products often used in wall and chimney construction and for refractory linings. Common nominal sizes of bricks in the United States are 4 or 6 in thick by 2⅔ or 4 in high by 8 or 12 in long. For a list of modular sizes, see "Standard Sizes of Clay and Concrete Modular Masonry Units," ANSI A62.3. Actual dimensions are smaller, usually by the amount of the width of the mortar joint. Current specification requirements for strength and absorption of building brick are given in Table 5.14 (see ASTM C652, C62, and C216). Strength and absorption of brick from different producers vary widely.

Thermal expansion of brick may range from 0.0000017 per °F for fire-clay brick to 0.0000069 per °F for surface-clay brick. Wetting tests of brick indicated expansions varying from 0.0005 to 0.025%.

The thermal conductivity of dry brick as measured by several investigators ranges from 1.29 to 3.79 Btu/(h)(ft³)(°F)(in). The values are increased by wetting.

5.22 Structural Clay Tiles

Structural clay tiles are hollow burned-clay masonry units with parallel cells. Such units have a multitude of uses: as a facing tile for interior and exterior unplastered walls, partitions, or columns; as load-bearing tile in masonry constructions designed to carry superimposed loads; as partition tile for interior partitions carrying no superimposed load; as fireproofing tile for protection of structural members against fire; as furring tile for lining the inside of exterior walls; as floor tile in floor and roof construction; and as header tiles, which are designed to provide recesses for header units in brick or stone-faced walls. Units are available with the following ranges in nominal dimensions: 8 to 16 in in length, 4 in for facing tile to 12 in for load-bearing tile in height, and 2 in for facing tile to 12 in for load-bearing tile in thickness.

Two general types of tile are available—side-construction tile, designed to receive its principal stress at right angles to the axis of the cells, and end-construction tile designed to receive its principal stress parallel to the axis of the cells.

Tiles are also available in a number of surface finishes, such as opaque glazed tile, clear ceramic-glazed tile, nonlustrous glazed tile, and scored, combed, or roughened finishes designed to receive mortar, plaster, or stucco.

Requirements of the appropriate ASTM specifications for absorption and strength of several types of tile are given in Table 5.15 (see ASTM C34, C56, C57, C212, and C126 for details pertaining to size, color, texture, defects, etc.). Strength and absorption of tile made from similar clays but from different sources and manufacturers vary widely. The modulus of elasticity of tile may range from 1,620,000 to 6,059,000 psi.

5.23 Ceramic Tiles

Ceramic tile is a burned-clay product used primarily for decorative and sanitary effects. It is

Table 5.13 Summary of ASTM Specification Requirements for Concrete Masonry Units

	Compressive Strength, min, psi		Moisture Content for Type I Units, max, % of Total Absorption (Average of 5 Units)			Moisture Absorption, max, lb/ft³ (Average of 5 Units)		
			Avg Annual Humidity, %			Oven-Dry Weight of Concrete, lb/ft³		
	Avg of 5 Units	Individual, min	Over 75	75 to 50	Under 50	125 or more	105 to 125	Under 105
Concrete building brick, ASTM C55:								
N-I, N-II (high strength severe exposures)	3500	3000				10	13	15
S-I, S-II (general use, moderate exposures) Linear shrinkage, %	2500	2000				13	15	18
0.03 or less			45	40	35			
0.03 to 0.045			40	35	30			
Over 0.045			35	30	25			
Solid, load-bearing units, ASTM C145:								
N-I, N-II (unprotected exterior walls below grade or above grade exposed to frost)	1800	1500				13	15	18
S-I, S-II (protected exterior walls below grade or above grade exposed to frost) Linear shrinkage, % (same as for brick)	1200	1000						20*
Hollow, load-bearing units, ASTM C90:								
N-I, N-II (general use)	1000	800				13	15	18
S-I, S-II (above grade, weather protected) Linear shrinkage, % (same as for brick)	700	600						20*
Hollow, non-load-bearing units, ASTM C129: Linear shrinkage, % (same as for brick)	600	500						

*For units weighing less than 85 lb/ft³.

composed of a clay body on which is superimposed a decorative glaze.

The tiles are usually flat but vary in size from about ½ in square to more than 6 in. Their shape is also widely variable—squares, rectangles, and hexagons are the predominating forms, to which must be added coved moldings and other decorative forms. These tiles are not dependent

Table 5.14 Physical Requirements for Clay or Shale Solid Brick

Grade	Compressive Strength, Flat, Min, psi		Water Absorption, 5-hr Boil, Max—%		Saturation* Coefficient, Max—%	
	Avg of 5	Individual	Avg of 5	Individual	Avg of 5	Individual
SW—Severe weathering	3000	2500	17.0	20.0	0.78	0.80
MW—Moderate weathering	2500	2200	22.0	25.0	0.88	0.90
NW—No exposure	1500	1250	No limit	No limit	No limit	No limit

*Ratio of 24-hr cold absorption to 5-hr boil absorption.

on the color of the clay for their final color, since they are usually glazed. Hence, they are available in a complete color gradation from pure whites through pastels of varying hue to deep solid colors and jet blacks.

Properties of the base vary somewhat. In particular, absorption ranges from almost zero to about 15%. The glaze is required to be impervious to liquids and should not stain, crack, or craze.

5.24 Architectural Terra Cotta

The term "terra cotta" has been applied for centuries to decorative molded-clay objects whose properties are similar to brick. The molded shapes are fired in a manner similar to brick.

Terra cotta is frequently glazed to produce a desired color or finish. This introduces the problem of cracking or crazing of the glaze, particularly over large areas.

Structural properties of terra cotta are similar to those of clay or shale brick.

5.25 Stone Masonry

Principal stones generally used in the United States as masonry are limestones, marbles, granites, and sandstones. Other stones such as serpentine and quartzite are used locally but to a much lesser extent. Stone, in general, makes an excellent building material, if properly selected on the basis of experience; but the cost may be relatively high.

Properties of stone depend on what nature has provided. Therefore, the designer does not have the choice of properties and color available in manufactured masonry units. The most stone producers

can do for purchasers is to provide stone that has been proved by experience to have good strength and durability.

Data on the strength of building stones are presented in Table 5.16, summarized from *U.S. National Bureau of Standards Technical Papers*, no. 123, B. S. vol. 12; no. 305, vol. 20, p. 191; no. 349, vol. 21, p. 497; *Journal of Research of the National Bureau of Standards*, vol. 11, p. 635; vol. 25, p. 161). The data in Table 5.16 pertain to dried specimens. Strength of saturated specimens may be either greater or less than that of completely dry specimens.

The modulus of rupture of dry slate is given in Table 5.16 as ranging from 6000 to 15,000 psi. Similar slates, tested wet, gave moduli ranging from 4700 to 12,300 psi. The ratio of wet modulus to dry modulus varied from 0.42 to 1.12 and averaged 0.73.

Permeability of stones varies with types of stone, thickness, and driving pressure that forces water through the stone. Following are some common building stones, listed in order of increasing permeability: slate, granite, marble, limestone, and sandstone.

Data on thermal expansion of building stones as given in Table 5.17 show that limestones have a wide range of expansion as compared with granites and slates.

Marble loses strength after repeated heating and cooling. A marble that had an original strength of 9174 psi had a strength after 50 heatings to 150 °C of 8998 psi—a loss of 1.9%. After 100 heatings to 150 °C, the strength was only 8507 psi, or a loss of 7.3%. The latter loss in strength was identical with that obtained on freezing and thawing the same marble for 30 cycles. Also, marble retains a permanent expansion after repeated heating.

Table 5.15 Physical Requirement Specification for Structural Clay Tile

Type and Grade	Absorption, % (1 h Boiling)		Compressive Strength, psi (Based on Gross Area)			
			End-Construction Tile		Side-Construction Tile	
	Avg of 5 Tests	Individual Max	Min, Avg 5 Tests	Individual Min	Min, Avg 5 Tests	Individual Min
Load-bearing (ASTM C34):						
LBX	16	19	1400	1000	700	500
LB	25	28	1000	700	700	500
Non-load-bearing (ASTM C56):						
NB		28				
Floor tile (ASTM C57):						
FT1		25	3200	2250	1600	1100
FT2		25	2000	1400	1200	850
Facing tile (ASTM C212):						
FTX	9 (max)	11				
FTS	16 (max)	19				
Standard			1400	1000	700	500
Special duty			2500	2000	1200	1000
Glazed units (ASTM C126)			3000	2500	2000	1500

LBX. Tile suitable for general use in masonry construction and adapted for use in masonry exposed to weathering. They may also be considered suitable for direct application of stucco.

LB. Tile suitable for general use in masonry where not exposed to frost action, or in exposed masonry where protected with a facing of 3 in or more of stone, brick, terra cotta, or other masonry.

NB. Non-load-bearing tile made from surface clay, shale, or fired clay.

FT 1 and FT 2. Tile suitable for use in flat or segmental panels or in combination tile and concrete ribbed-slab construction.

FTX. Smooth-face tile suitable for general use in exposed exterior and interior masonry walls and partitions, and adapted for use where tiles low in absorption, easily cleaned, and resistant to staining are required and where a high degree of mechanical perfection, narrow color range, and minimum variation in face dimensions are required.

FTS. Smooth or rough-texture face tile suitable for general use in exposed exterior and interior masonry walls and partitions and adapted for use where tile of moderate absorption, moderate variation in face dimensions, and medium color range may be used, and where minor defects in surface finish, including small handling chips, are not objectionable.

Standard. Tile suitable for general use in exterior or interior masonry walls and partitions.

Special duty. Tile suitable for general use in exterior or interior masonry walls and partitions and designed to have superior resistance to impact and moisture transmission, and to support greater lateral and compressive loads than standard tile construction.

Glazed units. Ceramic-glazed structural clay tile with a glossy or satin-mat finish of either an opaque or clear glaze, produced by the application of a coating prior to firing and subsequently made vitreous by firing.

Organic Materials

Through many generations of use, people have found ways of getting around some of the limitations of naturally occurring organic construction materials. Plywood, for instance, has overcome the problem of the highly directional properties of wood. In addition to improving natural materials, technologists have developed many synthe- tic polymers (plastics), which are important in construction.

5.26 Wood

Wood is a natural polymer composed of cells in the shape of long thin tubes with tapered ends. The cell wall consists of crystalline cellulose aligned

Table 5.16 Characteristics of Commercial Building Stones

Stone	Unit Weight, lb/ft³	Compressive Strength, psi, Range	Modulus of Rupture, psi, Range	Shear Strength, psi, Range	Tensile Strength, psi, Range	Elastic Modulus, psi, Range	Toughness Range	Toughness Avg	Wear Resistance Range	Wear Resistance Avg
Granite	157–187	7,700–60,000	1,430–5,190	2,000–4,800	600–1,000	5,700,000–8,200,000	8–27	13	43.9–87.9	60.8
Marble	165–179	8,000–50,000	600–4,900	1,300–6,500	150–2,300	7,200,000–14,500,000	2–23	6	6.7–41.7	18.9
Limestone	117–175	2,600–28,000	500–2,000	800–4,580	280–890	1,500,000–12,400,000	5–20	7	1.3–24.1	8.4
Sandstone	119–168	5,000–20,000	700–2,300	300–3,000	280–500	1,900,000–7,700,000	2–35	10	1.6–29.0	13.3
Quartzite	165–170	16,000–45,000					5–30	15		
Serpentine	158–183	11,000–28,000	1,300–11,000		800–1,600	4,800,000–9,600,000			13.3–111.4	46.9
Basalt	180–200	28,000–67,000					5–40	20		
Diorite		16,000–35,000					6–38	23		
Syenite		14,000–28,000								
Slate	168–180	6,000–15,000		2,000–3,600	3,000–4,300	9,800,000–18,000,000	10–56	19	5.6–11.7	7.7
Diabase							6–50			

Table 5.17 Coefficient of Thermal Expansion of Commercial Building Stones

Stone	Range of Coefficient
Limestone	$(4.2–22) \times 10^{-6}$
Marble	$(3.6–16) \times 10^{-6}$
Sandstone	$(5.0–12) \times 10^{-6}$
Slate	$(9.4–12) \times 10^{-6}$
Granite	$(6.3–9) \times 10^{-6}$

parallel to the axis of the cell. The cellulose crystals are bonded together by a complex amorphous lignin composed of carbohydrate compounds. Wood substance is 50 to 60% cellulose and 20 to 35% lignin, the remainder being other carbohydrates and mineral matter.

Most of the cells in trees are oriented vertically, but some are radially oriented to serve as reinforcement against spreading of the vertical fibers under the natural compressive loading of the tree trunk. Because of its directed cell structure, wood has greater strength and stiffness in the longitudinal direction than in other directions.

Specific gravity of the wood substance is about the same for all species: 1.56. The bulk density of the gross wood is much lower, however, because of voids (cavity cells) and accidental cracks in the cell structure. For common woods, specific gravity varies from the 0.12 of balsa to the 0.74 of oak. The various properties of wood, such as strength, can be correlated with density.

5.26.1 Moisture Effects on Woods

The cell wall has a high affinity for moisture because cellulose contains numerous hydroxyl groups, which are strongly hydrophilic. When exposed to moisture, often in the form of air with a high relative humidity, the cell walls in the wood absorb large amounts of water and swell. This process causes the intermolecular forces between the cellulose macromolecules to be neutralized by the absorbed water, thus reducing the strength and rigidity of the wood.

The moisture present in green wood consists of water absorbed in the cell walls and water contained in the cell cavities. As the wood dries,

water is first removed from the cell cavities. At the fiber-saturation point, the cavities are empty, while the cell walls are still fully saturated with water. On further drying in normal air, this moisture decrease continues until an equilibrium moisture content is reached. At an atmosphere of 60% relative humidity in 70 °F air, the moisture content of wood stabilizes at about 11%. Although kiln drying can lower the moisture content of the wood 2 to 6% more, the decrease is not permanent and the moisture content will go back to about 11% when returned to normal air.

Dimensional changes due to swelling and shrinking resulting from atmospheric moisture changes occur only at moisture contents below the fiber-saturation point. Additional moisture fills cell cavities but causes no appreciable dimensional changes. When dimensional changes occur, they take place in radial and tangential directions, transverse to the long axis of the wood, because the cell walls swell or shrink in the direction perpendicular to the long dimension of the fibers. Wood is seasoned before it is put into service, so that it comes to equilibrium under atmospheric conditions. See also Art. 11.1.

5.26.2 Properties of Wood

Wood has three mutually perpendicular axes of symmetry: longitudinal, or parallel to the grain; tangential; and radial. Strength and elastic properties differ in these three directions because of the directional cell structure of the wood. Values of modulus of elasticity in the two directions perpendicular to the grain are only one-twentieth to one-twelfth the value parallel to the grain. Table 5.18 compares the elastic and shear moduli of some typical woods in the longitudinal, tangential, and radial directions. These perpendicular moduli are important in the design of composite materials containing wood.

The principal mechanical properties of some woods commonly used in structural applications are discussed in section 11. Note that increasing moisture content reduces all the strength and stiffness properties except impact.

Table 5.19 lists weights and specific gravity of several commercial lumber species.

Table 5.18 Moduli of Various Woods*

Species	Longitudinal Modulus E_L, 10^3 psi	Young's Modulus Ratios		Modulus of Rigidity Ratios		
		E_T/E_L[†]	E_R/E_L[†]	G_{LR}/E_L[†]	G_{LT}/E_L[†]	G_{RT}/E_L[†]
Ash	2,180	0.064	0.109	0.057	0.041	0.017
Balsa	550	0.015	0.046	0.054	0.037	0.005
Birch, yellow	2,075	0.050	0.078	0.074	0.067	0.017
Douglas fir	2,280	0.050	0.068	0.064	0.078	0.007
Poplar, yellow	1,407	0.043	0.092	0.075	0.069	0.011
Walnut	1,630	0.056	0.106	0.085	0.062	0.021

*These data are for specific values of each wood species. From U.S. Forest Products Laboratory, "Wood Handbook."
[†]E_T = modulus of elasticity, psi, in tangential direction, E_R = modulus in radial direction, G_{LR} = shear modulus in a plane normal to the tangential direction, G_{LT} = shear modulus in a plane normal to the radial direction, and G_{RT} = shear modulus in a plane normal to the longitudinal direction.

5.26.3 Resistance of Wood to Chemical Attack

Wood is superior to many building materials in resistance to mild acids, particularly at ordinary temperatures. It has excellent resistance to most organic acids, notably acetic. However, wood is seldom used in contact with solutions that are more than weakly alkaline. Oxidizing chemicals and solutions of iron salts, in combination with damp conditions, should be avoided.

Wood is composed of roughly 50 to 70% cellulose, 25 to 30% lignin, and 5% extractives with less than 2% protein. Acids such as acetic, formic, lactic, and boric do not ionize sufficiently at room temperature to attack cellulose and thus do not harm wood.

When the pH of aqueous solutions of weak acids is 2 or more, the rate of hydrolysis of cellulose is small and is dependent on the temperature. A rough approximation of this temperature effect is that for every 20 °F increase, the rate of hydrolysis doubles. Acids with pH values above 2 or bases with pH below 10 have little weakening effect on wood at room temperature if the duration of exposure is moderate.

5.26.4 Commercial Grades of Wood

Lumber is graded to enable a user to buy the quality that best suits a particular purpose. The grade of a piece of lumber is based on the number, character, and location of strength-reducing features and factors affecting durability and utility. The best grades are virtually free of blemishes, but the other grades, which comprise the great bulk of lumber, contain numerous knots and other features that affect quality to varying degrees. Various associations of lumber manufacturers assume jurisdiction over the grading of certain species. Two principal sets of grading rules are employed for hardwood and softwood.

Hardwood is graded according to rules adopted by the National Hardwood Lumber Association. Since most hardwood boards are cut into smaller pieces to make a fabricated product, the grading rules are based on the proportion of a given piece that can be cut into smaller pieces. Usable material must have one clear face, and the reverse face must be sound.

Softwood is classified and graded under rules adopted by a number of regional lumber manufacturers' associations. American lumber standards for softwood lumber were formulated as a result of conferences organized by the U.S. Department of Commerce to improve and simplify the grading rules. These standards, issued in pamphlet form by the Department of Commerce, have resulted in more uniform practices throughout the country. Softwood lumber is classified according to use, size, and process of manufacture.

Use classifications include: (1) yard lumber, intended for general building purposes; (2) structural lumber, which is limited to the larger sizes and intended for use where minimum

Table 5.19 Weights and Specific Gravities of Commercial Lumber Species

Species	Specific Gravity Based on Oven-Dry Weight and Volume at 12% Moisture Content	Weight, lb/ft³			Moisture Content When Green (Avg) %	Specific Gravity Based On Oven-Dry Weight and Volume When Green	Weight When Green, lb/ft³
		At 12% Moisture content	At 20% Moisture Content	Adjusting Factor for Each 1% Change in Moisture Content			
Softwoods:							
Cedar							
Alaska	0.44	31.1	32.4	0.170	38	0.42	35.5
Incense	0.37	25.0	26.4	0.183	108	0.35	42.5
Port Orford	0.42	29.6	31.0	0.175	43	0.40	35.0
Western red	0.33	23.0	24.1	0.137	37	0.31	26.4
Cypress, southern	0.46	32.1	33.4	0.167	91	0.42	45.3
Douglas fir							
Coast region	0.48	33.8	35.2	0.170	38	0.45	38.2
Inland region	0.44	31.4	32.5	0.137	48	0.41	36.3
Rocky Mountain	0.43	30.0	31.4	0.179	38	0.40	34.6
Fir, white	0.37	26.3	27.3	0.129	115	0.35	39.6
Hemlock							
Eastern	0.40	28.6	29.8	0.150	111	0.38	43.4
Western	0.42	29.2	30.2	0.129	74	0.38	37.2
Larch, western	0.55	38.9	40.2	0.170	58	0.51	46.7
Pine							
Eastern white	0.35	24.9	26.2	0.167	73	0.34	35.1
Lodgepole	0.41	28.8	29.9	0.142	65	0.38	36.3
Norway	0.44	31.0	32.1	0.142	92	0.41	42.3
Ponderosa	0.40	28.1	29.4	0.162	91	0.38	40.9
Southern shortleaf	0.51	35.2	36.5	0.154	81	0.46	45.9
Southern longleaf	0.58	41.1	42.5	0.179	63	0.54	50.2
Sugar	0.36	25.5	26.8	0.162	137	0.35	45.8
Western white	0.38	27.6	28.6	0.129	54	0.36	33.0
Redwood	0.40	28.1	29.5	0.175	112	0.38	45.6
Spruce							
Engelmann	0.34	23.7	24.7	0.129	80	0.32	32.5
Sitka	0.40	27.7	28.8	0.145	42	0.37	32.0
White	0.40	29.1	29.9	0.104	50	0.37	33.0
Hardwoods:							
Ash, white	0.60	42.2	43.6	0.175	42	0.55	47.4
Beech, American	0.64	43.8	45.1	0.162	54	0.56	50.6
Birch							
Sweet	0.65	46.7	48.1	0.175	53	0.60	53.8
Yellow	0.62	43.0	44.1	0.142	67	0.55	50.8
Elm, rock	0.63	43.6	45.2	0.208	48	0.57	50.9
Gum	0.52	36.0	37.1	0.133	115	0.46	49.7
Hickory							
Pecan	0.66	45.9	47.6	0.212	63	0.60	56.7
Shagbark	0.72	50.8	51.8	0.129	60	0.64	57.0
Maple, sugar	0.63	44.0	45.3	0.154	58	0.56	51.1
Oak							
Red	0.63	43.2	44.7	0.187	80	0.56	56.0
White	0.68	46.3	47.6	0.167	68	0.60	55.6
Poplar, yellow	0.42	29.8	31.0	0.150	83	0.40	40.5

working stresses are required; and (3) factory and shop lumber, intended to be cut up for use in further manufacture.

Lumber classified according to manufacture includes: (1) rough lumber, which is in the undressed condition after sawing; (2) surfaced lumber, which is surface-finished by running through a planer; and (3) worked lumber, which has been matched or molded.

All softwood lumber is graded into two general categories, select and common, on the basis of appearance and characteristics. Structural lumber is graded according to strength for each species.

5.26.5 Improvement of Wood Properties

Because of its high anisotropy and hygroscopic properties, wood has limitations in use as a structural material. Various techniques are employed to improve the strength or dimensional stability of wood in service atmospheres. Preservatives may be applied to combat decay and attack by animal organisms. Thin sheets of wood may be bonded together to build up a modified wood structure. The sheets can be effectively impregnated to fill the cell cavities. As a further modification, the thin-sheet structure may be compressed during the period of bond curing to increase the density and strength. Such treatments improve the chemical resistance, decay resistance, and dimensional stability of the wood.

See also Art. 11.2.4.

5.27 Plastics

The synonymous terms *plastics* and *synthetic resins* denote synthetic organic high polymers. Polymers are compounds in which the basic molecular-level subunits are long-chain molecules. The word *plastic* has been adopted as a general name for this group of materials because all are capable of being molded at some stage in their manufacture.

5.27.1 Structure of Plastics

In polymerization, the simultaneous polymerization of two or more chemically different monomers can be employed to form a polymer containing both monomers in one chain. Such *copolymers* frequently have more desirable physical and mechanical properties than either of the polymers that have been combined. The range of properties available through copolymerization means that the engineer can have plastics tailor-made to specific requirements.

Polymers may be formed in either an amorphous or crystalline state, depending on the relative arrangements of the long-chain molecules. An amorphous (without form) state is characterized by a completely random arrangement of molecules. A crystalline state in a polymer consists of crystalline regions, called *crystallites*, embedded in an amorphous matrix.

Plasticizers and fillers may be added to polymers to change their basic properties. Plasticizers are low-molecular-weight (short-chain) substances added to reduce the average molecular weight of a polymer and thus make it more flexible. Fillers may be added, particularly to the softer plastics, to stiffen them, increase their strength and impact properties, or improve their resistance to heat. Wood flour, mica, asbestos fibers, and chopped fibers or fabric may be used as filler material for polymers.

Crystallization causes a denser packing of polymer molecules, thus increasing the intermolecular forces. The resulting polymers have greater strength and stiffness and a higher softening point than amorphous polymers of the same chemical structure and molecular weight. A typical example of this is high-density polyethylene.

Cross-linking, a common variation in the growth of polymers, ties the chains of molecules together at intervals by primary bonds. For effective cross-linking, there must be normally unsaturated carbon atoms present within the polymer chain since cross-linking takes place through such connecting points. Cross-linking greatly restricts the movement between adjacent polymer chains and thus alters the mechanical properties of the material. A cross-linked polymer has higher tensile strength, more recoverable deformation (elasticity), and less elongation at failure. The vulcanization of natural rubber with sulfur is the classic example of the kind of transformation that cross-linking can effect—from tire treads to battery cases.

Three-dimensional structures can also be formed from chain polymers by branching, where main chains are bifurcated into two chains. The extent of branching can be controlled in the production process. If branching is extensive enough, it restricts the movement between adjacent chains by causing intertangling.

5.27.2 Deformation of Polymers

The elastic moduli of plastics generally range from 10^4 to 10^6 psi, considerably lower than for metals. The greater strains observed when plastics are loaded result from the fact that there is chain straightening in polymers as well as bond lengthening. Network polymer structures are more rigid than linear structures and thus show higher moduli.

Deformation of a plastic favors crystallization since the molecular chains are pulled into closer alignment and proximity. Thus, the properties of polymers may be changed by large deformation. This phenomenon of orientation is employed to produce plastics with different properties in one direction than in others. Drawing, which orients the molecular chains in the direction of drawing, produces strength in the longitudinal direction that is several times that of undrawn material.

Polymers are viscoelastic in that they are subject to time-dependent phenomena. Polymeric materials, subjected to a steady load, creep to greater strains than under short-time loading. If the material is instead stretched to a given elongation, the stress necessary to maintain the elongation will diminish with time. Both creep and stress relaxation are accelerated at higher temperatures, where the molecular chains have more thermal energy to assist in reorientation or slipping. Since the properties are time-dependent, the rate of loading of a polymer can affect the observed behavior. Increased loading rates produce steeper stress-strain curves, indicating that the material is stiffer when the time for molecular readjustments is decreased.

Amorphous polymers have a characteristic temperature at which the properties make a drastic change, called the **glass transition temperature**. The transition from glassy behavior to rubbery behavior may occur at any temperature. On the high-temperature side of this transition, the molecular segments are free to move past each other; on the low-temperature side, they are rigidly confined. Thus, the temperature at which the polymer becomes glassy and brittle and no longer behaves as a rubbery polymer is a cause for concern in the use of any polymer system.

5.27.3 Thermosetting Plastics

This type of plastic is either originally soft or softens at once under a little heating, but upon further heating such plastics harden permanently. The final, continuous framework structure of thermosetting resins may develop from the condensation polymerization mechanism or may harden by the formation of primary bonds between molecular chains as thermal energy is applied. The completion of polymerization, which is accelerated at higher temperatures, provides a permanent set to the thermosetting resins. In general, thermosetting plastics are stronger than thermoplastic resins, particularly at elevated temperatures.

The principal varieties of thermosets are described briefly and their main applications noted in the following. (For detailed data on the properties of these plastics, see the latest Encyclopedia issue of *Modern Plastics*.)

Phenol formaldehydes provide the greatest variety of thermosetting molded plastic articles. They are used for chemical, decorative, electrical, mechanical, and thermal applications of all kinds. Hard and rigid, they change slightly, if at all, on aging indoors but on outdoor exposure lose their bright surface gloss. However, the outdoor-exposure characteristics of the more durable formulations are otherwise generally good. Phenol formaldehydes have good electrical properties, do not burn readily, and do not support combustion. They are strong, lightweight, and generally pleasant to the eye and touch. Light colors normally are not obtainable because of the dark brown basic color of the resin. They have low water absorption and good resistance to attack by most commonly found chemicals.

Epoxy and polyester resins are used for a variety of purposes. For example, electronic parts with delicate components are sometimes cast completely in these materials to give them complete and continuous support and resistance to thermal and mechanical shock. Some varieties must be cured at elevated temperatures; others can be formulated to be cured at room temperatures. One of the outstanding attributes of the epoxies is their excellent adhesion to a variety of materials, including such metals as copper, brass, steel, and aluminum.

Polyester molding materials, when compounded with fibers (particularly glass fibers) or with various mineral fillers (including clay), can be formulated into putties or premixes that are easily compression- or transfer-molded into parts having high impact resistance.

Melamine formaldehyde materials are unaffected by common organic solvents, greases, and oils and most weak acids and alkalies. Their water absorption is low. They are insensitive to heat and are highly flame-resistant, depending on the filler. Electrical properties are particularly good, especially resistance to arcing. Unfilled materials are highly translucent and have unlimited color possibilities. Principal fillers are alpha cellulose for general-purpose compounding; minerals to improve electrical properties, particularly at elevated temperatures; chopped fabric to afford high shock resistance and flexural strength; and cellulose, used mainly for electrical purposes.

Polyurethane is used several ways in construction. As thermal insulation, it is used in the form of foam, either prefoamed or foamed in place. The latter is particularly useful in irregular spaces. When blown with fluorocarbons, the foam has exceptionally low heat transmission and is therefore widely used in thin-walled refrigerators. Other uses include field-applied or baked-on clear or colored coatings and finishes for floors, walls, furniture, and casework generally. The rubbery form is employed for sprayed or troweled-on roofing and for gaskets and calking compounds.

Urea formaldehydes, like the melamines, offer unlimited translucent to opaque color possibilities, light fastness, good mechanical and electrical properties, and resistance to organic solvents and mild acids and alkalies. Although there is no swelling or change in appearance, the water absorption of urea formaldehyde is relatively high, and therefore it is not recommended for applications involving long exposure to water. Occasional exposure to water has no deleterious effect. Strength properties are good.

Silicones, unlike other plastics, are based on silicon rather than carbon. As a consequence, their inertness and durability under a wide variety of conditions are outstanding. As compared with the phenolics, their mechanical properties are poor, and consequently glass fibers are added. Molding is more difficult than with other thermosetting materials. Unlike most other resins, they may be used in continuous operations at 400 °F; they have very low water absorption; their dielectric properties are excellent over an extremely wide variety of chemical attack; and under outdoor conditions their durability is particularly outstanding. In liquid solutions, silicones are used to impart moisture resistance to masonry walls and to fabrics. They also form the basis for a variety of paints and other coatings capable of maintaining flexibility and inertness to attack at high temperatures in the presence of ultraviolet sunlight and ozone. Silicone rubbers maintain their flexibility at much lower temperatures than other rubbers.

5.27.4 Thermoplastic Resins

These plastics become easily deformable, at elevated temperatures. They become hard again on cooling. They can be so softened by heating and hardened by cooling any number of times. Thermoplastic resins deform easily under applied pressure, particularly at elevated temperatures, and so are used to make molded products.

The main varieties of thermoplastics are described briefly in the following paragraphs. (For detailed data on the properties of these plastics, see the latest Encyclopedia issue of *Modern Plastics*.)

Acrylics in the form of large transparent sheets are used in aircraft enclosures and building construction. Although not so hard as glass, acrylics have perfect clarity and transparency. They are the most resistant of the transparent plastics to sunlight and outdoor weathering, and they have an optimum combination of flexibility and rigidity, with resistance to shattering. A wide variety of transparent, translucent, and opaque colors can be produced. Sheets of acrylic are readily formed to complex shapes. They are used for such applications as transparent windows, outdoor and indoor signs, parts of lighting equipment, decorative and functional automotive parts, reflectors, household-appliance parts, and similar applications. Acrylics can be used as large sheets, molded from molding powders, or cast from the liquid monomer.

Acrylonitrile-butadiene-styrene (ABS) is a three-way copolymer that provides a family of tough, hard, chemically resistant resins. The greatest use is for pipes and fittings.

Polycarbonate has excellent transparency, high resistance to impact, and good resistance to weathering. It is used for safety glazing, general illumination, and hard hats.

Polyethylene, in its unmodified form, is a flexible, waxy, translucent plastic maintaining flexibility at very low temperatures, in contrast with many other thermoplastic materials. The heat-distortion point of the older, low-density polyethylenes is low; these plastics are not recommended for uses at temperatures above 150 °F. Newer, high-density materials have higher heat-distortion temperatures; some may be heated to temperatures above 212 °F. The heat-distortion point may rise well above 250 °F for plastics irradiated with high-energy beams or for polyethylene with ultrahigh molecular weight. Unlike most plastics, polyethylene is partly crystalline. It is highly inert to solvents and corrosive chemicals of all kinds at ordinary temperatures. Usually, low moisture permeability and absorption are combined with excellent electrical properties. Its density is lower than that of any other commercially available nonporous plastic. When compounded with black pigment, its weathering properties are good. Polyethylene is widely used as a primary insulating material on wire and cable and has been used as a replacement for the lead jacket on communication cables and other cables. It is widely used also as thin flexible film for packaging, particularly of food, and as corrosion-proof lining for tanks and other chemical equipment.

Polypropylene, a polyolefin, is similar in many ways to its counterpart, polyethylene, but is generally harder, stronger, and more temperature-resistant. It has a great many uses, among them for complete water cisterns for water closets in plumbing systems abroad.

Polytetrafluoroethylene, with the very active element fluorine in its structure, is a highly crystalline linear-type polymer, unique among organic compounds in its chemical inertness and resistance to change at high and low temperatures. It has an extremely low dielectric-loss factor. In addition, its other electrical properties are excellent. Its outstanding property is extreme resistance to attack by corrosive agents and solvents of all kinds. At temperatures well above 500 °F, polytetrafluoroethylene can be held for long periods with practically no change in properties except loss in tensile strength. Service temperatures are generally maintained below 480 °F. This material is not embrittled at low temperatures, and its films remain flexible at temperatures below −100 °F. It is used in bridges as beam seats or bearings and in buildings calling for resistance to extreme conditions, or for applications requiring low friction. In steam lines, for example, supporting pads of

polytetrafluoroethylene permit the line to slide easily over the pad as expansion and contraction with changes in temperature cause the line to lengthen and shorten. The temperatures involved have little or no effect. Mechanical properties are only moderately high, and reinforcement may be necessary to prevent creep and squeeze-out under heavy loads.

Polyvinyl fluoride has much of the superior inertness to chemical and weathering attack typical of the fluorocarbons. Among other uses, it is used as thin-film overlays for building boards to be exposed outdoors.

Polyvinyl formal resins are used principally as a base for tough, water-resistant insulating enamel for electric wire.

Polyvinyl butyral is the tough interlayer in safety glass. In its cross-linked and plasticized form, polyvinyl butyral is used extensively in coating fabrics for raincoats, upholstery, and other heavy-duty moisture-resistant applications.

Vinyl chloride polymers and copolymers vary from hard and rigid to highly flexible. Polyvinyl chloride is naturally hard and rigid but can be plasticized to any required degree of flexibility, as in raincoats and shower curtains. Copolymers, including vinyl chloride plus vinyl acetate, are naturally flexible without plasticizers. Nonrigid vinyl plastics are widely used as insulation and jacketing for electric wire and cable because of their electrical properties and resistance to oil and water. Thin films are used for rainwear and similar applications, whereas heavy-gage films and sheets are used widely for upholstery. Vinyl chlorides are used for floor coverings in the form of tile and sheet because of their abrasion resistance and relatively low water absorption. The rigid materials are used for tubing, pipe, and many other applications which require resistance to corrosion and action of many chemicals, especially acids and alkalies; they are attacked by a variety of organic solvents, however. Like all thermoplastics, vinyl chlorides soften at elevated temperatures; their maximum recommended temperature is about 140 °F. although at low loads they may be used at temperatures as high as 180 °F.

Vinylidene chloride is highly resistant to most inorganic chemicals and organic solvents generally. It is impervious to water on prolonged immersion, and its films are highly resistant to moisture-vapor transmission. It can be sterilized, if not under load, in boiling water, and its mechanical properties are good. Vinylidene chloride is not recommended for uses involving high-speed impact, shock resistance, or flexibility at subfreezing temperatures. It should not be used in applications requiring continuous exposure to temperatures in excess of 170 °F.

Polystyrene formulations constitute a large and important segment of the entire field of thermoplastic materials. Numerous modified polystyrenes provide a relatively wide range of properties. Polystyrene is one of the lightest of the presently available commercial plastics. It is relatively inexpensive and easily molded and has good dimensional stability and good stability at low temperatures. It is brilliantly clear when transparent but can be produced in an infinite range of colors. Water absorption is negligible even after long immersion. Electrical characteristics are excellent. It is resistant to most corrosive chemicals, such as acids, and a variety of organic solvents, although it is attacked by others. Polystyrenes, as a class, are considerably more brittle and less extendable than many other thermoplastic materials, but these properties are markedly improved by copolymerization. Under some circumstances, they tend to develop fine cracks, known as *craze marks*, on exposure, particularly outdoors. This is true of many other thermoplastics, especially when highly stressed.

Polyimide, in molded form, is used in increasing quantities for impact and high resistance to abrasion. It is employed in small gears, cams, and other machine parts because even when unlubricated, polyimide is highly resistant to wear. Its chemical resistance, except to phenols and mineral acids, is excellent. Extruded polyimide is coated onto electric wire, cable, and rope for abrasion resistance. Applications like hammerheads indicate its impact resistance.

Cellulose Derivatives ▪ Cellulose is a naturally occurring high polymer found in all woody plant tissue and in such materials as cotton. It can be modified by chemical processes into a variety of thermoplastic materials, which in turn may be still further modified with plasticizers, fillers, and other additives to provide a wide variety of properties. The oldest of all plastics is cellulose nitrate.

Cellulose acetate is the basis of safety film, developed to overcome the highly flammable nature of cellulose nitrate. Starting as film, sheet, or molding powder, it is made into a variety of items, such as transparent packages and a large variety of general-purpose items. Depending on the plasticizer content, it may be hard and rigid or soft and flexible. Moisture absorption of this and all other cellulosics is relatively high, and they are therefore not recommended for long-continued outdoor exposure. But cellulose acetate film, reinforced with metal mesh, is widely used for temporary enclosures of buildings during construction.

Cellulose acetate butyrate, a butyrate copolymer, is inherently softer and more flexible than cellulose acetate and requires less plasticizer to achieve a given degree of softness and flexibility. It is made in the form of clear transparent sheet and film, or in the form of molding powders which can be molded by standard injection-molding procedures into a wide variety of products. Like the other cellulosics, this material is inherently tough and has good impact resistance. It has infinite colorability, like the other cellulosics. Cellulose acetate butyrate tubing is used for such applications as irrigation and gas lines.

Ethyl cellulose is similar to cellulose acetate and acetate butyrate in its general properties. Two varieties, general-purpose and high-impact, are common; high-impact ethyl cellulose is made for better-than-average toughness at normal and low temperatures.

Cellulose nitrate, one of the toughest plastics, is widely used for tool handles and similar applications requiring high-impact strength. Its high flammability requires great caution, particularly when the plastic is in the form of film. Most commercial photographic film is made of cellulose nitrates rather than safety film. Cellulose nitrate is the basis of most of the widely used commercial lacquers for furniture and similar items.

5.27.5 PVC Siding

Palliside Weatherboard cladding system comprises extruded foamed PVC board with a co-extruded ultraviolet protection PVC exterior layer. There is an interlocking weatherseal between boards, and rigid PVC trims and flashings. Boards are prefinished, no need to prime or paint, will not rot or corrode, impervious to moisture, and from attack by termites or vermin. Used properly, PVC presents no greater fire risk than other natural or synthetic organic materials. The product is not easily ignited, shrinks, melts, and flows away from a heat source. Can be cleaned by washing with a hose. When used as external wall claddings, they provide good looks, insulation, easy care, years of service and a warranty of 25 years.

5.28 Elastomers, or Synthetic Rubbers

Rubber for construction purposes is both natural and synthetic. Natural rubber, often called crude rubber in its unvulcanized form, is composed of large complex molecules of isoprene. Synthetic rubbers, also known as elastomers, are generally rubberlike only in their high elasticity. The principal synthetic rubbers are the following:

GR-S is the one most nearly like crude rubber and is the product of styrene and butadiene copolymerization. It is the most widely used of the synthetic rubbers. It is not oil-resistant but is widely used for tires and similar applications.

Nitril is a copolymer of acrylonitrile and butadiene. Its excellent resistance to oils and solvents makes it useful for fuel and solvent hoses, hydraulic-equipment parts, and similar applications.

Butyl is made by the copolymerization of isobutylene with a small proportion of isoprene or butadiene. It has the lowest gas permeability of all the rubbers and consequently is widely used for making inner tubes for tires and other applications in which gases must be held with a minimum of diffusion. It is used for gaskets in buildings.

Neoprene is made by the polymerization of chloroprene. It has very good mechanical properties and is particularly resistant to sunlight, heat, aging, and oil; it is therefore used for making machine belts, gaskets, oil hose, insulation on wire cable, and other applications to be used for outdoor exposure, such as roofing, and gaskets for building and glazing.

Sulfide rubbers—the polysulfides of high molecular weight—have rubbery properties, and articles made from them, such as hose and tank linings and glazing compounds, exhibit good resistance to solvents, oils, ozone, low temperature, and outdoor exposure.

Silicone rubber, which also is discussed in Art. 5.27.3, when made in rubbery consistency forms a material exhibiting exceptional inertness and temperature resistance. It is therefore used in making gaskets, electrical insulation, and similar products that maintain their properties at both high and low temperatures.

Additional elastomers include polyethylene, cyclized rubber, plasticized polyvinyl chloride, and polybutene. A great variety of materials enters into various rubber compounds and therefore provide a wide range of properties. In addition, many elastomeric products are laminated structures of rubberlike compounds combined with materials like fabric and metals.

5.29 Geosynthetics

These are fabrics made of plastics, primarily polymers, but sometimes rubber, glass fibers, or other materials, that are incorporated in soils to improve certain geotechnical characteristics. The roles served by geosynthetics may be grouped into five main categories: separation of materials, reinforcement of soil, filtration, drainage within soil masses and barrier to moisture movement. There are several types of geosynthetics:

Geotextiles are flexible, porous fabrics made of synthetic fibers by standard weaving machines or by matting or knitting (nonwoven). They offer the advantages for geotechnical purposes of resistance to biodegration and porosity, permitting flow across and within the fabric.

Geogrids consist of rods or ribs made of plastics and formed into a net or grid. They are used mainly for reinforcement of and anchorage in soils. Aperture sizes for geogrids range from about 1 to 6 in in longitudinal and transverse directions, depending on the manufacturer.

Geonets are netlike fabrics similar to geogrids but with apertures of only about 0.25 in. The ribs generally are extruded polyethylene. Geonets are used as drainage media.

Geomembranes are relatively impervious, polymeric fabrics that are usually fabricated into continuous, flexible sheets. They are used primarily as a liquid or vapor barrier. They can serve as liners for landfills and covers for storage facilities. Some geomembranes are made by impregnating geotextiles with asphalt or elastomerics.

Geocomposites consist of a combination of other types of geosynthetics, formulated to fulfill specific functions.

Design of geosynthetic filters, or earth reinforcement, or an impervious membrane landfill liner requires a clear statement of the geotechnical characteristics to be achieved with geosynthetic application, a thorough understanding of geosynthetic properties, and a knowledge of materials currently available and their properties.

Specifications for Geosynthetics ▪ A joint committee of the American Association of State Highway and Transportation Officials (AASHTO), Associated General Contractors (AGC), and the American Road and Transportation Builders Association (ARTBA) has developed specifications and test procedures for geosynthetics intended for specific applications. ASTM has promulgated specifications for test methods for index properties, such as grab tensile strength (D4632), strip tensile strength (D1682), hydraulic (Mullen) bursting strength (D3786), trapezoid tearing strength (D4533), apparent opening size (D4751), degradation from exposure to ultraviolet light (D4355), temperature stability (D4594), permittivity (D4491), crush strength (D1621, and puncture strength (D4833). ASTM also publishes specifications for test methods for performance properties of geotextiles, geogrids, and geocomposites, such as tensile strength determined by the wide-width-strip method (D4595), sewn-seam strength (D4884), in-plane flow, or transmissivity (D4716).

In specification of a geosynthetic, consideration should be given not only to the type of application, such as soil reinforcement, drainage, or erosion control, but also to the function to be served by the material in that application and the required properties. Some properties that are of importance for other types of materials may not be significant for geosynthetics or lead to misleading or exclusionary specifications. For example, for geotextiles, thickness may not be relevant. Different manufacturing processes produce comparable fabrics with differing thicknesses. Furthermore, thickness may change during shipping and handling. Similarly, density, oz/yd^2 or g/m^2, may be useful only for estimating the weight of the geotextile. As another example, permeability, which is the product of permittivity and thickness, may be different for two fabrics with the same permittivity. The difference is a consequence of the fabrics differing in thickness. Hence, evaluation in terms of their coefficient of permeability can be misleading. Comparisons should be based on permittivity, which is the measure of the quantity of water that would pass through a unit thickness of a geotextile under a given head (Art. 7.39.2).

Specifications should be based on the specific properties required for the functions to be served. A geosynthetic may have secondary functions as well as a primary function. Consideration should be given to the following properties in specification of a geosynthetic:

Geotextiles

General: Fabric structure (woven, nonwoven, combination), polymer composition (polyester, polypropylene, polyethylene, combination), width and length of rolls, survivability. Fabrics may be formed from fibers or yarns. Fibers may be continuous filaments or staple fibers or produced by slitting an extruded plastic sheet to form thin, flat tapes. See Art. 7.39.2 for definitions of geotextile terms.

Storage and Handling: Protection against ultraviolet exposure, dust, mud, or other elements that may have a deleterious effect on performance.

Filtration and Hydraulic Properties: Percent of open area for woven fabrics, apparent opening size, permittivity.

Mechanical Properties: Sampling and testing requirements, puncture resistance, Mullen burst strength, trapezoid tear strength, tensile strength and elongation, wide-width-strip tensile strength and elongation in machine direction and cross direction, ultraviolet light resistance after 150 h, soil-fabric interface friction angle for reinforcement applications.

Seams and Overlaps: Overlaps dependent on application but minimum of 1 ft for all applications. Sewing of seams may be required. Seam thread should be polymeric and should be at least as durable as the main material. Seams should be placed directed upward. Factory-made sewn-seam strengths should be equal to or greater than the main material. Field-made sewn seams are weaker than the main material.

Placement: Grading and ground clearing, aggregates, cover thickness and lifts, equipment.

Repair: Procedures for repairing rips, tears, and other damages, including overlap, seam, and replacement requirements.

Geomembranes General: Polymer composition (polyvinyl chloride, hypalon, polyethylene, high density, very low density, or linear density and textured or nontextured), roll width and length, thickness, density, carbon-black content.

Mechanical Properties: Tensile strength (yield and break), elongation (at yield and break), tear resistance, low-temperature brittleness, seam shear strength and peel strength (fusion and extrusion), environmental-stress-crack resistance.

Other: under "Geotextiles" above.

Geosynthetic Clay Liners General: Roll width and length; average roll weight; bentonite density (exclusive of glue weight, if applicable); upper geosynthetic weight, thickness, and structure (woven, nonwoven layer in scrim-reinforced, nonwoven needle-punched); lower geosynthetic weight, thickness, and structure (woven, nonwoven, nonwoven needle-punched).

Mechanical Properties: Tensile strength and elongation.

Hydraulic Properties: Permeability.

Base Bentonite Properties: Moisture content, swell index, fluid loss.

Other: See under "Geotextiles" above.

Geonets General: Structure (geonet, single-or double-cuspated core, single-or-double-dimpled core, solid-or hollow-column core, entangled mesh), polymer composition (polyethylene, polypropylene, polystyrene), type of geotextile attached, roll width and length, core, net, and mesh thicknesses.

Mechanical Properties: Yield strength in compression.

Hydraulic Properties: In-plane flow rate.

Geogrids General: Manufacturing process (woven, punched, sheet drawn, extrusion), type of coating, polymer composition (polyester, polypropylene, polystyrene), roll width and length, density, aperture size.

Mechanical properties: Wide-width-strip tensile strength, long-term design strength.

Information on specific geosynthetics, including recommended applications, may be obtained from the manufacturers. Product data for a number of geosynthetics are presented in "Specifiers Guide," Geotechnical Fabrics Report, Industrial Fabrics Association International, 345 Cedar St., Suite 800, St. Paul, MN 55101-1088.

See also Art. 7.39.

("A Design Primer: Geotextiles and Related Materials," Industrial Fabrics Association International, St. Paul, Minn., R. M. Koerner, "Designing with Geosynthetics," Prentice-Hall, Englewood Cliffs, N.J.)

5.30 Organic Materials References

Brantley, L. R., and R. T. Brantley, "Building Materials Technology: Structural Performance and Environmental Impact," McGraw-Hill, Inc., New York. www.mcgraw-hill.com

Erikkson, K. L., et al., "Microbial and Enzymatic Degradation of Wood and Wood Components," Springer Verlag.

Faherty, K. F., and T. G. Williamson, "Wood Engineering and Construction Handbook," 2nd ed., McGraw-Hill, Inc., New York. www.mcgraw-hill.com

Harper, C. A., "Handbook of Plastics, Elastomers, and Composites," 2nd ed., McGraw-Hill, Inc., New York. www.mcgraw-hill.com

Koerner, R. M., "Designing with Geosynthetics," 2nd ed., Prentice-Hall, Upper Saddle River, N.J. www.prenhall.com

"Modern Plastics Encyclopedia," Plastics Catalog Corp., New York.

"Polymer Modified Concrete," SP-99; "Guide for the Use of Polymers in Concrete," ACI 548.1, and "Polymers in Concrete," ACI 548, American Concrete Institute, Farmington Hills, MI. www.asce.org

Skeist, I., "Plastics in Building," Van Nostrand Reinhold, New York.

"Structural Plastics Design Manual," American Society of Civil Engineers, 1801 Alexander Bell Drive Reston, VA. www.asce.org

Wilcox, W., et al., "Wood as a Building Material," John Wiley & Sons, Inc., Hoboken, N.J. www.wiley.com

Zabel, R. A., and J. J. Morell, "Wood Microbiology: Decay and Its Prevention," Academic Press, Inc., San Diego, Calif.

Joint Seals

Calking compounds, sealants, and gaskets are employed to seal the points of contact between similar and dissimilar building materials that cannot otherwise be made completely tight. Such points include glazing, the joints between windows and walls, the many joints occurring in the increasing use of panelized construction, the copings of parapets, and similar spots.

The requirements of a good joint seal are: (1) good adhesion to or tight contact with the surrounding materials, (2) good cohesive strength, (3) elasticity to allow for compression and extension as surrounding materials retract or approach each other because of changes in moisture content or temperature, (4) good durability or the ability to maintain their properties over a long period of time without marked deterioration, and (5) no staining of surrounding materials such as stone.

5.31 Calking Compounds

These sealers are used mostly with traditional materials such as masonry, with relatively small windows, and at other points where motion of building components is relatively small. They are typically composed of elastomeric polymers or bodied linseed or soy oil, or both, combined with calcium carbonate (ground marble or limestone), tinting pigments, a gelling agent, drier, and mineral spirits (thinners).

Two types are commonly employed, gun grade and knife grade. Gun grades are viscous semi-liquids suitable for application by hand or air-operated calking guns. Knife grades are stiffer and are applied by knife, spatula, or mason's pointing tools.

Because calking compounds are based on drying oils that eventually harden in contact with the air, the best joints are generally thick and deep, with a relatively small portion exposed to the air. The exposed surface is expected to form a tough protective skin for the soft mass underneath, which in turn provides the cohesiveness, adhesiveness, and elasticity required. Thin shallow beads cannot be expected to have the durability of thick joints with small exposed surface areas.

5.32 Sealants

For joints and other points where large movements of building components are expected, elastomeric materials may be used as sealants. Whereas traditional calking compounds should not be used where movements of more than 5% of the joint width or at most 10% are expected, larger movements, typically 10 to 25%, can be accommodated by the rubbery sealants.

Some elastomeric sealants consist of two components, mixed just before application. Polymerization occurs, leading to conversion of the viscous material to a rubbery consistency. The working time or pot life before this occurs varies, depending upon formulation and temperatures from a fraction of an hour to several hours or a day. Other formulations are single-component and require no mixing. They harden upon exposure to moisture in the air.

Various curing agents, accelerators, plasticizers, fillers, thickeners, and other agents may be added, depending on the basic material and the end-use requirements.

The proper choice of materials depends upon the application. A sealant with the appropriate hardness, extensibility, useful temperature ranges, expected life, dirt pickup, staining, colorability, rate of cure to tack-free condition, toxicity, resistance to ultraviolet light, and other attributes should be chosen for the specific end use.

In many joints, such as those between building panels, it is necessary to provide backup; that is, a foundation against which the compound can be applied. This serves to limit the thickness of the joint, to provide the proper ratio of thickness to width, and to force the compound into intimate contact with the substrate, thereby promoting adhesion. For the purpose, any of various compressible materials, such as polyethylene or polyurethane rope, or oakum, may be employed.

To promote adhesion to the substrate, various primers may be needed. (To prevent adhesion of the compound to parts of the substrate where adhesion is not wanted, any of various liquid and tape bond-breakers may be employed.) Generally, good adhesion requires dry, clean surfaces free of grease and other deleterious materials.

5.33 Gaskets

Joint seals described in Arts. 5.31 and 5.32 are formed in place; that is, soft masses are put into the joints and conform to their geometry. A gasket, on the other hand, is preformed and placed into a joint whose geometry must conform with the gasket in such a way as to seal the joint by compression of the gasket. Gaskets, however, are cured under shop-controlled conditions, whereas sealants cure under variable and not always favorable field conditions.

Rubbery materials most commonly employed for gaskets are cellular or noncellular (dense) Neoprene, EPDM (ethylene-propylene polymers and terpolymers), and polyvinylchloride polymers.

Gaskets are generally compression types or lock-strip (zipper) types. The former are forced into the joint and remain tight by being kept under compression. With lock-strip gaskets, a groove in the gasket permits a lip to be opened and admit glass or other panel, after which a strip is forced into the groove, tightening the gasket in place. If the strip is separable from the gasket, its composition is often harder than the gasket itself.

For setting large sheets of glass and similar units, setting or supporting spacer blocks of rubber are often combined with gaskets of materials such as vulcanized synthetic rubber and are finally sealed with the elastomeric rubber-based sealants or glazing compounds.

5.34 Joint Seals References

Brantley, L. R., and R. T. Brantley, "Building Materials Technology: Structural Performance and Environmental Impact," McGraw-Hill, Inc., New York. www.mcgraw-hill.com

"Building Seals and Sealants," STP 606, ASTM, Philadelphia, Pa.

Damusis, A., "Sealants," Van Nostrand Reinhold Company, New York.

Panek, J. A. and J. P. Cook, "Construction Sealants and Adhesives," 3rd ed., John Wiley & Sons, Inc., Hoboken, NJ. www.wiley.com

Paints and Other Coatings

These are widely used in construction for decoration, weatherproofing, and protection against fire, decay, and corrosion. They include such diverse coatings as paint, lacquer, varnish, baked-on finishes, and specialty systems.

5.35 Paints

Paint is a fluid comprising a pigment, vehicle or binder, a solvent or thinner, and dryer. Viscosity, drying time, and flowing properties are determined by the formulation. The fluid may be applied as one or more relatively thin coats, each coat usually changing to a solid before application of a successive coat. The change may be a result of chemical reaction or evaporation of the solvent or both.

Architectural paints are coatings that are applied by brush or spray to architectural and structural surfaces and dry when exposed to the air. They usually are solvent- or water-thinned.

Solvent-thinned paints that normally dry by evaporation of the solvent generally incorporate as a vehicle a hard resin, such as shellac or lacquer. (Shellac may be dissolved in alcohol and used as a varnish.) This classification also includes bitumens (asphalt or coal tar), which are used for roofing and waterproofing. Solvent-thinned paints that normally dry by oxidation generally use as a vehicle an oil or oil-based varnish. For exterior applications, polyvinyl-acetate and acrylic emulsion types of paint are often used. For interior surfaces, an alkyd enamel made from a drying oil, glycerin, and phthalit anhydride or water-thinned latexes made from polyvinyl-acetate or acrylic resins may be selected.

Water-thinned paints may have the vehicle dissolved in water or dispersed in an emulsion. The latter type are more widely used. They incorporate latexes; materials formed by copolymerization, such as butadiene-styrene; or polyvinyl-acetate or acrylic resins.

5.36 Commercial Finishes

These include coatings that are applied by brushing, spraying, or magnetic agglomeration and dry on exposure to the air or are cured by baking. Applications include highway marking and coatings on appliances and machinery.

Air-drying coatings for machinery include epoxy, urethane, or polyester resins that dry at room temperature. For use for highway markings and other areas painted for traffic control, latexes or solvent-thinned paints are specially formulated from alkyds, modified rubbers, or other resins.

Baked-on coatings include urea, acrylic, melamine, and some phenolic resins. They are generally used where hardness, chemical resistance, and color retention are required.

Porcelain enamel, also known as vitreous enamel, is an aluminum-silicate glass, which is fused to metal under high heat. Porcelain-enameled metal is used for indoor and outdoor applications because of its hardness, durability, washability, and color possibilities. For building purposes porcelain enamel is applied to sheet metal and cast iron, the former for a variety of purposes including trim, plumbing, and kitchen fixtures, and the latter almost entirely for plumbing fixtures. Most sheet metal used for porcelain enameling is steel—low in carbon, manganese, and other elements. Aluminum is also used for vitreous enamel.

Most enameling consists of a ground coat and one or two cover coats fired on at slightly lower temperatures; but one-coat enameling of somewhat inferior quality can be accomplished by first treating the metal surface with soluble nickel salts.

The usual high-soda glasses used to obtain low-temperature softening enamels are not highly acid-resistant and therefore stain readily and deeply when iron-containing water drips on them. Enamels highly resistant to severe staining conditions must be considerably harder; i.e., have higher softening temperatures and therefore require special techniques to avoid warping and distorting of the metal base.

5.37 Industrial Coatings

Materials in this category are intended for applications where resistance to high temperature or corrosion, or both, is desired. They typically require a base coat or primer, one or more intermediate coats, and a finish coat, or top coat.

Coatings for high-temperature applications include (1) inorganic zinc dispersed in an appropriate vehicle that permits use in temperatures up to 400 °C and (2) a phosphate bonding system with ceramic fillers in an aqueous solution of mono-aluminum phosphate that is cured at 400 °C and is serviceable in temperatures up to about 1500 °C. Silicone rubbers or resins, polyamide, or polytetrafluoroethylene polymers are used in ablative formulations that absorb heat through melting, sublimation decomposition, or vaporization or that expand when heated and form a foamlike insulation. They usually provide only short-term protection in the 150° to 500 °C range.

Corrosion-resistant coatings are used as a protective layer over metals or other substrate subject to attack by acids, alkalies, or other corrosive substances. The base coat should be applied to dry, clean, rough surfaces, after preparation by abrasive blasting, if necessary. This coat must provide adhesion to the substrate for the entire coating system. For steel, the primer often used is zinc dispersed in a suitable vehicle. Intermediate coats may not be necessary, but when used, they usually are layers of the same generic type as that specified for the top coat. The purpose is to build up the protective coat where corrosive attack is likely to be frequent. Vehicles in the top coat may be phenolic or polyamide resins, elastomers, polyesters, polyurethanes, chlorinated rubber, vinyl resin in solvent solution, epoxy resin cured from a solvent solution with polyfunctional amines, or a combination of coal tar and epoxy.

A variety of corrosion-resistant coatings also are available for protecting pipelines, hoppers, and other types of containers against attack by corrosive fluids or pellets or against abrasion. Coatings for such service include epoxy-furans, rubber, resinous cements, Neoprene, polyurethanes, unsaturated polyesters, baked-on phenolics, polyethylene, amine-cured epoxies, fluorocarbons, and asphalt.

Rubber-Lined Pipes, Tanks, and Similar Equipment ▪ The lining materials include all the natural and synthetic rubbers in various degrees of hardness, depending on the application. Frequently, latex rubber is deposited directly from the latex solution onto the metal surface to be covered. The deposited layer is subsequently vulcanized. Rubber linings can be bonded to ordinary steel, stainless steel, brass, aluminum, concrete, and wood. Adhesion to aluminum is inferior to adhesion to steel. Covering for brass must be compounded according to the composition of the metal.

5.38 Dryers, Thinners, and Pigments for Paints

Dryers. These are catalysts that hasten the hardening of drying oils. Most dryers are salts of heavy metals, especially cobalt, manganese, and lead, to which salts of zinc and calcium may be added. Iron salts, usable only in dark coatings, accelerate hardening at high temperatures. Dryers are normally added to paints to hasten hardening, but they must not be used too liberally or they cause rapid deterioration of the oil by overoxidation.

Thinners. These are volatile constituents added to coatings to promote their spreading qualities by reducing viscosity. They should not react with the other constituents and should evaporate completely. Commonly used thinners are turpentine and mineral spirits, i.e., derivatives of petroleum and coal tar.

Pigments may be classified as white and colored, or as opaque and extender pigments. The hiding power of pigments depends on the difference in index of refraction of the pigment and the surrounding medium—usually the vehicle of a protective coating. In opaque pigments, these indexes are markedly different from those of the vehicles (oil or other); in extender pigments, they are nearly the same. The comparative hiding efficiencies of various pigments must be evaluated on the basis of hiding power per pound and cost per pound.

Principal white pigments, in descending order of relative hiding power per pound, are approximately as follows: rutile titanium dioxide, anatase titanium dioxide, zinc sulfide, titanium-calcium, titanium-barium, zinc sulfide-barium, titanated lithopone, lithopone, antimony oxide, zinc oxide.

Zinc oxide is widely used by itself or in combination with other pigments. Its color is unaffected by many industrial and chemical atmospheres. It imparts gloss and reduces chalking but tends to crack and alligator instead.

Zinc sulfide is a highly opaque pigment widely used in combination with other pigments.

Titanium dioxide and extended titanium pigments have high opacity and generally excellent properties. Various forms of the pigments have different properties. For example, anatase titanium dioxide promotes chalking, whereas rutile inhibits it.

Colored pigments for building use are largely inorganic materials, especially for outdoor use, where the brilliant but fugitive organic pigments soon fade. The principal inorganic colored pigments are:

Metallic. Aluminum flake or ground particle, copper bronze, gold leaf, zinc dust

Black. Carbon black, lampblack, graphite, vegetable black, and animal blacks

Earth colors. Yellow ocher, raw and burnt umber, raw and burnt sienna; reds and maroons

Blue. Ultramarine, iron ferrocyanide (Prussian, Chinese, Milori)

Brown. Mixed ferrous and ferric oxide

Green. Chromium oxide, hydrated chromium oxide, chrome greens

Orange. Molybdated chrome orange

Red. Iron oxide, cadmium red, vermilion

Yellow. Zinc chromate, cadmium yellows, hydrated iron oxide

Extender pigments are added to extend dle opaque pigments, increase durability, provide better spreading characteristics, and reduce cost. The principal extender pigments are silica, china clay, talc, mica, barium sulfate, calcium sulfate, calcium carbonate, and such materials as magnesium oxide, magnesium carbonate, barium carbonate, and others used for specific purposes.

5.39 Paints and Coatings References

Banov, A., "Paints and Coatings Handbook," Structures Publishing Company, Farmington, Mich.

Brantley, L. R., and R. T. Brantley, "Building Materials Technology: Structural Performance and Environmental Impact," McGraw-Hill, Inc., New York. www.mcgraw.hill.com

Burns, R. M., and W. Bradley, "Protective Coatings for Metals," Van Nostrand Reinhold Company, New York.

Martens, C. R., "The Technology of Paints, Varnishes and Lacquers," Van Nostrand Reinhold Company, New York.

Weismantel, G. E., "Paint Handbook," 2nd ed., McGraw-Hill, Inc., New York. www.mcgraw-hill.com

Composite Materials

Well-known products such as plywood, reinforced concrete, and pneumatic tires are evidence that the concept of composite materials has been applied for many years. But new families of composites with expanding ranges and a variety of properties are continually being created. Composite materials for structural applications are particularly important where higher strength-to-weight and stiffness-to-weight ratios are desired than can be had with basic materials.

5.40 Types of Composites

Composites can be classified in seven basic material combinations and three primary forms. The materials categories are permutations of combinations of the three basic kinds of materials metal-metal, metal-inorganic, metal-organic, inorganic-inorganic, inorganic-organic, organic-organic, metal-inorganic-organic. Here inorganic applies to nonmetallic, inorganic materials such as ceramics, glasses, and minerals. No limitation on the number of phases embodied in a composite is intended by these designations. Thus metal-organic includes composites with two metallic phases and one organic phase or four-phase composites having two metallic and two organic components.

The three primary forms of composite structures are shown in Fig. 5.15. *Matrix systems* are characterized by a discontinuous phase, such as particles, flakes, or fibers, or combinations of these, in a continuous phase or matrix. *Laminates* are characterized by two or more layers bonded together. As a rule, strengthening is less an objective than other functional requirements in the design of laminated composites. *Sandwich structures* are characterized by a single, low-density core, such as honeycomb or foamed material, between two faces of comparatively higher density. A sandwich may have several cores or an open face. One primary form of composite may contain another. The faces of a sandwich, for example, might consist of a laminate or matrix system.

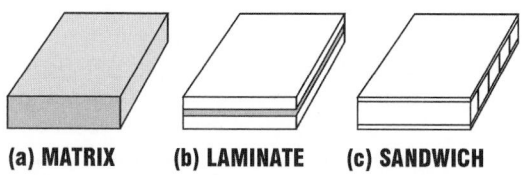

(a) MATRIX   **(b) LAMINATE**   **(c) SANDWICH**

Fig. 5.15 Primary forms of composite materials.

5.41 Matrix Systems

In construction the most important among the matrix systems are steel-reinforced concrete and those containing fibers or fiberlike material, such as whiskers, that enhance strength. Here advantage is taken of the high strengths available in some materials, especially when produced in the form of fine filaments a few micrometers in diameter.

Fiber-based structural composites are usually based on continuous filaments, (glass-reinforced plastics are typical of this group), or are whisker composites. The latter owe their useful properties to the extremely high strength available from materials in fine fibrous form. Alumina whiskers can now be made with strengths consistently ranging from 1000 to 3000 ksi. Silver has been strengthened from its normal level of 25 to 230 ksi, with a 24% (by volume) addition of these whiskers. Similarly, a 50% gain has been obtained with a 12% addition to an 80-20 nickel-chromium alloy.

See also Art. 5.43.

5.42 Sandwich Systems

A primary objective of most sandwich composites is superior structural performance. To this end, the core separates and stabilizes the faces against buckling under edgewise compression, torsion, or bending. Other considerations, such as heat resistance and electrical requirements, dictate the choice of materials. Cores are usually lightweight materials. Typical forms of core material are honeycomb structures (metal, glass-reinforced plastic, or resin-impregnated paper) and foams (generally plastic, but they may be ceramic). Synthetic organic adhesives (e.g., epoxies, phenolics, polyesters) are employed to assemble sandwich components, except when thermal considerations preclude them.

Vibration Insulators ▪ These usually consist of a layer of soft rubber bonded between two layers of metal. Another type of insulator consists of a rubber tube or cylinder vulcanized to two concentric metal tubes, the rubber being deflected in shear. A variant of this consists of a cylinder of soft rubber vulcanized to a tubular or solid steel core and a steel outer shell, the entire combination being placed in torsion to act as a spring. Heavy-duty mounts of this type are employed on trucks, buses, and other applications calling for rugged construction.

5.43 Continuous-Filament Composites

For matrix systems, fibers may be converted into yarns, rovings, and woven fabrics in a variety of configurations. Matrix materials employed with glass fibers generally have been synthetic resins, largely the polyester, phenolic, and epoxy families.

A variety of filaments can be used to obtain various composite properties and efficiencies: E-glass, Al_2O_3 glass, silica, beryllium, boron, and steel. Filament geometry presents still another degree of freedom. One example is hollow filament, which offers more stiffness than solid filament for the same weight. Also, matrix-filament ratios can be adjusted. And filament-alignment possibilities are infinite. E-glass 10 mm in diameter has a strength of 500 ksi, an elastic modulus of 10,500 ksi, and a density of 0.092 lb/in^3.

The attributes of glass-fiber-reinforced plastic make it an important structural material. Its mechanical properties are competitive with metals, considering density. It exhibits great freedom from corrosion, although it is not wholly immune to deterioration. The dielectric properties are very good. It may be fabricated in complex shapes, in limited quantities, with comparatively inexpensive tooling. In buildings, reinforced plastics have been rather widely used in the form of corrugated sheet for skylights and side lighting of buildings, and as molded shells, concrete forms, sandwiches, and similar applications.

Fabrics for Air-Supported Roofs ▪ Principal requirements for fabrics and coatings for air-supported structures are high strip tensile strength in both warp and fill directions, high tear resistance, good coating adhesion, maximum weathering resistance, maximum joint strength, good flexing resistance, and good flame resistance. Translucency may or may not be important, depending on the application. The most commonly used fabrics are nylon, polyester, and glass. Neoprene and Hypalon have commonly been employed for military and other applications where opacity is desired. For translucent application, vinyl chloride and fluorocarbon polymers are more

common. Careful analysis of loads and stresses, especially dynamic wind loads, and means of joining sections and attaching to anchorage is required.

Glass Composites ▪ Phase separation in which a solid phase precipitates to intermingle with the remaining liquid phase, is basic to glass ceramics. Combining glass and ceramics yields some of the best properties of each. By the use of a nucleating agent, such as finely divided titanium dioxide, and by controlled heat treatment, a 90% microcrystalline glass with tiny ceramic crystals embedded in the glass matrix results. One of the main differences between this material and the usual ceramic is the improved properties of the glass ceramic.

Glass ceramics are not as porous to stains and moisture as ceramics. In addition, the glass-ceramic composite is more shock-resistant because the cracks that would normally start at a grain boundary or an imperfection in a ceramic surface are arrested by the microcrystalline network of the glass structure. The thermal and mechanical shock resistance are further improved by use of aluminum-lithium-silicon-oxide glass. Failure by deformation and creep that occurs in metal does not occur in glass ceramics. Even the tendency of ceramics to fail in tension is offset by the glass matrix. These unique features account for the extensive use of glass ceramics in applications from oven cookware to the nose cones of rockets.

The procedure for making ceramic glass consists of melting the glass ingredients with a nucleating agent and then cooling the glass in the shape of the finished article. Reheating and controlled cooling produces nucleation and the desired amount of microcrystallization for the glass ceramic. A small amount of this microcrystalline phase is invisible to the eye, but serves as a reinforcing filler to strengthen the glass structure. In larger amounts, this microcrystalline phase gives an attractive milky appearance due to the multiple reflections of light from the tiny crystal surfaces.

With the wide variety of types of glasses available and the range of controlled-nucleation agents possible, the thermal expansion coefficient of glass ceramics can be varied over a wide range, particularly to match the coefficient of the metal to which they are to be attached.

5.44 High-Pressure Laminates

Laminated thermosetting products consist of fibrous sheet materials combined with a thermosetting resin, usually phenol formaldehyde or melamine formaldehyde. The commonly used sheet materials are paper, cotton fabric, asbestos paper or fabric, nylon fabric, and glass fabric. The usual form is flat sheet, but a variety of rolled tubes and rods is made.

Decorative laminates consist of a base of phenolic resin-impregnated kraft paper over which a decorative overlay, such as printed paper, is applied. Over all this is laid a thin sheet of melamine resin. When the entire assemblage is pressed in a hot-plate press at elevated temperatures and pressures, the various layers are fused together and the melamine provides a completely transparent finish, resistant to alcohol, water, and common solvents. This material is widely used for tabletops, counter fronts, wainscots, and similar building applications. It is customarily bonded to a core of plywood to develop the necessary thickness and strength. In this case, a backup sheet consisting of phenolic resin and paper alone, without the decorative surface, is employed to provide balance to the entire sandwich.

5.45 Laminated Rubber

Rubber is often combined with various textiles, fabrics, filaments, and metal wire to obtain strength, stability, abrasion resistance, and flexibility. Among the laminated materials are the following:

V Belts ▪ These consist of a combination of fabric and rubber, frequently combined with reinforcing grommets of cotton, rayon, steel, or other high-strength material extending around the central portion.

Flat Rubber Belting ▪ This laminate is a combination of several plies of cotton fabric or cord, all bonded together by a soft-rubber compound.

Conveyer Belts ▪ These, in effect, are moving highways used for transporting such material as crushed rock, dirt, sand, gravel, slag, and similar materials. When the belt operates at a steep angle, it is equipped with buckets or similar devices and becomes an elevator belt. A typical conveyor belt consists of cotton duck plies alternated with thin rubber plies; the assembly is wrapped in a rubber cover, and all elements are united into a single structure by vulcanization. A conveyor belt to withstand extreme conditions is made with some textile or metal cords instead of the woven fabric. Some conveyor belts are especially arranged to assume a trough form and made to stretch less than similar all-fabric belts.

Rubber Hose ▪ Nearly all rubber hose is laminated and composed of layers of rubber combined with reinforcing materials like cotton duck, textile cords, and metal wire. Typical hose consists of an inner rubber lining, a number of intermediate layers consisting of braided cord or cotton duck impregnated with rubber, and outside that, several more layers of fabric, spirally wound cord, spirally wound metal, or in some cases, spirally wound flat steel ribbon. Outside of all this is another layer of rubber to provide resistance to abrasion. Hose for transporting oil, water, wet concrete under pressure, and for dredging purposes is made of heavy-duty laminated rubber.

5.46 Composite Materials References

Brantley, L. R., and R. T. Brantley, "Building Materials Technology: Structural Performance and Environmental Impact," McGraw-Hill, Inc., New York. www.mcgraw-hill.com

Broughman, L. J., and R. H. Krock, "Modern Composite Materials," Addison-Wesley Publishing Company, Boston, Mass. www.aw.com

Das, S. K., et al., "High-Performance Composites for the 1990's," TMS, Warrendale, Pa.

Ishida, M., "Characterization of Composite Materials," Butterworth-Heinemann, Burlington, Mass. www.bhusa.com

Environmental Influences

Materials are usually subjected to atmospheres other than ideal inert conditions. They may encounter low or elevated temperatures, corrosion or oxidation, or irradiation by nuclear particles. Exposure to such environmental influences can affect the mechanical properties of the materials to such an extent that they do not meet service requirements.

5.47 Thermal Effects

Variations in temperature are often divided into two classifications: *elevated temperatures* (above room temperature) and *low temperatures* (below room temperature). This can be misleading because critical temperatures for the material itself may be high or low compared with room temperature. The lower limit of interest for all materials is absolute zero. The upper limit is the melting point for ceramics and metals, or melting or disintegration points for polymers and woods. Other critical temperatures include those for recrystallization in metals, softening and flow in thermosets, glass transition in thermoplastics, ductile-brittle transitions, and fictive temperature in glass. These temperatures mark the dividing lines between ranges in which materials behave in certain characteristic ways.

The immediate effect of thermal changes on materials is reflected in their mechanical properties, such as yield strength, viscous flow, and ultimate strength. For most materials there is a general downward trend of both yield and ultimate strength with increasing temperature. Sometimes, however, behavior irregularities in such materials are caused by structural changes (e g., polymorphic transformations). Low-temperature behavior is usually defined on the basis of transition from ductile to brittle behavior. This phenomenon is particularly important in body-centered-cubic metals which show well-defined transition temperatures.

Porous materials exhibit a special low-temperature effect: freezing and thawing. Concrete, for example, almost always contains water in its pores. Below 32 °F, the water is transformed to ice, which has a larger volume. The resulting swelling causes cracking. Thus, repeated thawing and freezing have a weakening effect on concrete. Brick is another, similar example.

Refractory Materials ▪ Materials whose melting points are very high relative to room temperature are called refractories. They may be either metallic or nonmetallic (ceramic) but are usually the latter. Generally, refractories are defined as those materials having melting points above 3000 °F. Their absolute maximum service temperatures may be as high as 90% of their absolute melting temperatures.

5.48 Metallic Corrosion

The simplest corrosion is by means of chemical solution, where an engineering material is dissolved by a strong solvent (e.g., when a rubber hose through which gasoline flows is in contact with hydrocarbon solvents).

Wet corrosion occurs by mechanisms essentially electrochemical in nature. This process requires that the liquid in contact with the metallic material be an electrolyte. Also, there must exist a difference of potential either between two dissimilar metals or between different areas on the surface of a metal. Many variables modify the course and extent of the electrochemical reactions, but it is usually possible to explain the various forms of corrosion by referring to basic electrochemical mechanisms.

Corrosion of metals is well understood. Corrosion as a chemical reaction is a characteristic of metals associated with the freedom of their valence electrons. It is this very freedom that produces the metallic bond that makes metals useful by allowing electric conduction. Being loosely bound to their atoms, the electrons in metals are easily removed in chemical reactions. In the presence of nonmetals, such as oxygen, sulfur, or chlorine, with their incomplete valence shells, there is a tendency for metals to form a compound, thus corroding the metal.

Each kind of corrosion involves the transfer of valence electrons from one metallic alloy to another. In the process, the metallic substance that is corroded provides the electrons that result in the other electrochemical reaction. One reaction cannot occur without the other taking place.

There are two kinds of corrosion. One is known as galvanic corrosion, or uniform corrosion. The other or more common kind is known as local corrosion, or nonuniform corrosion such as pitting, crevice, intergranular corrosion, and stress or mechanical corrosion. These differ chiefly in the method and location of occurrence. Galvanic corrosion should be avoided, or if this is not possible, be prevented by the insertion of a plastic electrical insulation barrier between the two dissimilar metals.

Galvanic corrosion occurs when two dissimilar metals are in electrical contact with each other and exposed to an electrolyte. A less noble metal will dissolve and form the anode, whereas the more noble metal will act as the cathode. The corrosion current flows from electrons at the anode metal, which is corroded, whereas the cathode metal is protected from the attack. A galvanic series lists metals in their order of corroding tendencies in a given environment and enables the probable corroding element to be identified. In seawater, for example, magnesium and zinc corrode more easily than steels, and lead, copper, and nickel corrode less than steels. Thus, in a galvanic cell of steel and magnesium in seawater, the steel would be anodic (corroded) and the magnesium cathodic (protected).

Several types of local corrosion are accelerated by the presence of some mechanical action. For example, if a local disorder is produced on a surface, the local energy is increased and the distorted material tends to become more anodic. The result is a localized decrease in resistance to corrosion. Examples of this stress corrosion include localized attack of cold-worked areas, such as sharp bends and punched holes; slip bands, which act as paths for internal corrosion across crystals, and stress-corrosion cracking, in which a metal under constant stress fails in tension after a time.

Pits and other surface irregularities produced by corrosion have the same effect on fatigue as other stress raisers, thus leading to corrosion fatigue. The constant reversal of strain has the effect of breaking any passivating film that may form on the surface. Thus, the corrosion fatigue strength of stainless steel may be as low as that of plain-carbon steels. With the formation of fatigue cracks at corrosion pits, the stress concentration at the crack tip further increases corrosion rate. Corrosion products fill the crack, exerting wedging action.

Other forms of corrosion include fretting corrosion due to mechanical wear in a corrosive atmosphere, cavitation damage serving to accelerate corrosion due to surface roughening, underground corrosion resulting from soil acidity, microbiological corrosion due to the metabolic activity of various microorganisms, and selective corrosion leading to the deterioration of alloys.

Concrete deterioration is generally attributable in part to chemical reactions between alkalies in the cement and mineral constituents of the concrete aggregates. Deterioration of concrete also results from contact with various chemical agents, which attack it in one of three forms: (1) corrosion resulting from the formation of soluble products that are removed by leaching; (2) chemical reactions producing products that disrupt the concrete because their volume is greater than that of the cement paste from which they were formed; and (3) surface deterioration by the crystallizing of salts in the pores of the concrete under alternate wetting and drying. The salts create pressures that can cause internal disruption.

5.49 Corrosion Control and Prevention

Proper selection of materials and sound engineering design are the best means of controlling and preventing degradation. For example, avoid use of dissimilar metals in contact where galvanic corrosion may result. Also, alloying can be used to improve chemical resistance.

Modifying the environment may also control corrosion. Such techniques as dehumidification and purification of the atmosphere or the addition of alkalies to neutralize the acidic character of a corrosive environment are typical of this approach. Inhibitors that effectively decrease the corrosion rate when added in small amounts to a corrosive environment may be used to prevent or control the anodic and cathodic reactions in electrochemical cells.

In corrosion, galvanic cells are formed in which certain areas become anodes and others cathodes. Ionic current flows through the electrolyte, and metal at the anode is dissolved or corroded. Cathodic protection reverses these currents and thereby makes cathodic all the metal to be protected.

Another procedure is to insert a new anode in the system, whose potential overcomes the potential of the original anode plus the resistance of the electrical elements. In this way, corrosion is concentrated in the new anode, which can be periodically replaced.

Application of protective coatings also furthers corrosion prevention and control. Three types of coatings are often employed: mechanical protection, separating the electrode from electrolyte (paints, grease, fired enamels); galvanic protection by being anodic to the base metal (zinc coating on galvanized iron); and passivators, which shift the base metal toward the cathodic end of the electromotive series.

5.49.1 Protection of Wood

Several types of preservatives are used to combat deterioration in woods: oily preservatives, such as coal-tar creosote; water-soluble salts, such as zinc chloride, sodium fluoride, copper salts, and mercuric salts; and solvent-soluble organic materials, such as pentachlorophenol. These preservatives may be applied by brushing, dipping, or pressure injection. Pressure treatments, by far the most effective, may be classified as either full- or empty-cell. In the full-cell treatment, a partial vacuum is first drawn to remove the air from the wood cells; then the preservative is pumped in under pressure. In the empty-cell treatment, air pressure in the cells restricts the pressure-applied preservative to the cell walls.

5.49.2 Corrosion Prevention for Steels

Corrosion of ferrous metals is caused by the tendency of iron (anode) to go into solution in water as ferrous hydroxide and displace hydrogen, which in turn combines with dissolved oxygen to form more water. At the same time, the dissolved ferrous hydroxide is converted by more oxygen to the insoluble ferric hydroxide, thereby allowing more iron to go into solution. Corrosion, therefore, requires liquid water (as in damp air) and oxygen (which is normally present dissolved in the water). Alloying elements can increase the resistance of steel considerably. For example, addition of copper to structural steels A36 and A529 can about double their corrosion resistance. Other steels, such as A242 and A588, are called weathering steels, because they have three to four times the resistance of A36 steel (Arts. 5.13.4, 9.1, and 9.4).

Protection against corrosion takes a variety of forms:

Deaeration ■ If oxygen is removed from water, corrosion stops. In hot-water heating systems, therefore, no fresh water should be added. Boiler feedwater is sometimes deaerated to retard corrosion.

Coatings ■

1. **Paints.** Some paints are based on oxidizing oil and a variety of pigments of which oxides of iron, zinc sulfate, graphite, aluminum, and various hydrocarbons are a few. No one paint is best for all applications. Other paints are coatings of asphalt and tar. The AISC "Specification for Structural Steel Buildings" (ASD and LRFD) states that, in general, steelwork to be concealed within a building need not be painted and that steel to be encased in concrete should not be painted. Inspections of old buildings have revealed that concealed steelwork withstands corrosion virtually to the same degree whether or not it is painted. (See also Art. 9.3.5.)

2. **Metallic.** Zinc is applied by hot dipping (**galvanizing**) or powder (**sherardizing**), hot tin dip, hot aluminum dip, and electrolytic plates of tin, copper, nickel, chromium, cadmium, and zinc. A mixture of lead and tin is called **terneplate**. Zinc is anodic to iron and protects, even after the coating is broken, by sacrificial protection. Tin and copper are cathodic and protect as long as the coating is unbroken but may hasten corrosion by pitting and other localized action once the coating is pierced.

3. **Chemical.** Insoluble phosphates, such as iron or zinc phosphate, are formed on the surface of the metal by treatment with phosphate solutions. These have some protective action and also form good bases for paints. Black oxide coatings are formed by treating the surface with various strong salt solutions. These coatings are good for indoor use but have limited life outdoors. They provide a good base for rust-inhibiting oils.

Cathodic Protection ■ As corrosion proceeds, electric currents are produced as the metal at the anode goes into solution. If a sufficient countercurrent is produced, the metal at the anode will not dissolve. This is accomplished in various ways, such as connecting the iron to a more active metal like magnesium (rods suspended in domestic water heaters) or connecting the part to be protected to buried scrap iron and providing an external current source such as a battery or rectified current from a power line (protection of buried pipe lines).

Reinforcing Steel Protection ▪ For chloride corrosion to occur in reinforcing steels in concrete, chloride in the range of 1.0 to 1.5 lb/yd³ must be present. If there is a possibility that chlorides may be introduced from outside the concrete matrix, for example, by de-icing salts, the steel can be protected by galvanizing, coating with epoxy, lowering the water-cement ratio, increasing the amount of cover over the reinforcing steel, adding a calcium nitrate admixture, adding an internal-barrier admixture, or cathodic protection, or a combination of these methods.

5.49.3 Corrosion Prevention for Aluminum

Although aluminum ranks high in the electromotive series of the metals it is highly corrosion resistant because of the tough, transparent, tenacious film of aluminum oxide that rapidly forms on any exposed surface. It is this corrosion resistance that recommends aluminum for construction applications. For most exposures, including industrial and seacoast atmospheres, the alloys normally recommended are adequate, particularly if used in usual thicknesses and if mild pitting is not objectionable.

Certain precautions should be taken in building. Aluminum is subject to attack by alkalies, and it should therefore be protected from contact with wet concrete, mortar, and plaster. Clear methacrylate lacquers or strippable plastic coatings are recommended for interiors and methacrylate lacquer for exterior protection during construction. Strong alkaline and acid cleaners should be avoided and muriatic acid should not be used on masonry surfaces adjacent to aluminum. If aluminum must be contiguous to concrete and mortar outdoors, or where it will be wet, it should be insulated from direct contact by asphalts, bitumens, felts, or other means. As is true of other metals, atmospheric-deposited dirt must be removed to maintain good appearance.

Electrolytic action between aluminum and less active metals should be avoided, because the aluminum then becomes anodic. If aluminum must be in touch with other metals, the faying surfaces should be insulated by painting with asphaltic or similar paints, or by gasketing. Steel rivets and bolts, for example, should be insulated. Drainage from copper-alloy surfaces onto aluminum must be avoided. Frequently, steel surfaces can be galvanized or cadmium-coated where contact is expected with aluminum. The zinc or cadmium coating is anodic to the aluminum and helps to protect it.

5.50 Irradiation

Radiation affects materials in many ways because of the diverse types of radiation and the differences in materials.

Radiation may be divided into two general groups:

1. Electromagnetic radiation, which is considered wavelike in nature (e.g., radio, heat, light, x-ray, gamma rays). These waves can also be considered as energy packets, called *photons.*

2. Radiation that is particulate in nature [e.g., accelerated protons (H+), neutrons, electrons (beta rays), and helium nuclei (alpha rays)]. These rays, although particulate, have many of the characteristics of waves.

Effects of Radiation ▪ The principal effect of radiation on materials arises from the extra energy it supplies, which helps break existing bonds and rearranges the atoms into new structures. In metals, heavy particles with sufficient radiant energy, such as fission fragments and fast neutrons, may displace atoms from the lattice, resulting in vacancies, interstitial atoms, and dislocations. These imperfections affect the physical and mechanical properties of metals. The general effect is similar to that brought about by precipitation hardening or by cold work.

The hardening effects, like strain hardening, can be removed by annealing, which allows vacancies and interstitials to become mobile enough to recombine. In some metals, if the metal is held at high enough temperature while being irradiated (common in reactors), little hardening will actually occur. A disturbing development is that radiation embrittlement of steels cannot be depended on to anneal out at ordinary reactor operating temperatures. Consequently, other materials (aluminum, titanium, and zirconium) are used for structural components in reactors.

In polymers, radiation damage seems to be a function of the actual radiation energy absorbed by the material regardless of the nature of the radiation. The energy imparted causes excitation and ionization of the molecules, which produce free radicals and ions. These molecular fragments may recombine with each other or with displaced electrons and oxygen from the air, causing either an increase or decrease in the molecular weight of the polymer. Thus, some polymers show an increased hardness, a higher softening point, and brittleness when irradiated, whereas others become soft. Most polymers lose strength through radiation damage.

5.51 Environmental Friendly Composites

Most lumber for outdoor decks is treated with compounds containing arsenic to ward off mold and insects. Finishes contain chemicals to prevent mold, mildew, and water penetration. These volatile organic compounds, unhealthy for individuals, may add to our greenhouse problems.

To avoid chemicals harmful to individuals and to the environment, a composite (TREX) is made from equal parts reclaimed hardwood sawdust and reclaimed/recycled polyethylene plastic and contains no preservatives or toxic chemicals. It withstands harsh conditions and heavy use, is undamaged by rot, mold, or termites, has low thermal expansion, is UV resistant, and is available at lumberyards. It is suitable for a variety of non-load bearing applications including decks, walkways, and stair treads.

A structural plastic lumber (TRIMAX) requiring superior strength and exceptional durability is a foamed polyolefin resin made from recycled plastic and reinforced by fiberglass. Available at lumberyards, this high-performance product is suited for outdoor structural applications such as pilings, posts, beams, joists, docks, and bridge fenders. Containing no harmful chemicals, it is resistant to marine borers, salt, spray, termites, corrosive substances, oil, fuels, and fungus.

5.52 Environmental Influences References

Brantley, L. R., and R. T. Brantley, "Building Materials Technology: Structural Performance and Environmental Impact," McGraw-Hill, Inc., New York. www.mcgraw-hill.com

Clauss, F. J., "Engineer's Guide to High-Temperature Materials," Addison-Wesley Publishing Company, Inc., Boston, Mass. www.aw.com

Fontana, M. G., "Corrosion Engineering," 3rd ed., McGraw-Hill Book Company, New York. www.mcgraw-hill.com

Kircher, J. F., and R. E. Bowman, "Effects of Radiation on Materials and Components," Van Nostrand Reinhold Company, New York.

Lane, R. W., "Control of Scale and Corrosion in Building Water Systems," McGraw-Hill, Inc., New York. www.mcgraw-hill.com

Uhlig, H. H., "Corrosion and Corrosion Control," John Wiley & Sons, Inc., Hoboken, NY. www.wiley.com

Akbar Tamboli, Mohsin Ahmed, Michael Xing
Thornton-Tomasetti Engineers
Newark, New Jersey

6

STRUCTURAL THEORY

Structural Theory Application to Model Structure to Predict Its Behavior

Structure design is the application of structural theory to ensure that buildings and other structures are built to support all loads and resist all constraining forces that may be reasonably expected to be imposed on them during their expected service life, without hazard to occupants or users and preferably without dangerous deformations, excessive sidesway (drift), or annoying vibrations. In addition, good design requires that this objective be achieved economically.

Applying structural theory to mathematic models is an essential and important tool in structural engineering. Over the past 200 years, many of the most significant contributions to the understanding of the structures have been made by scientist engineers while working on mathematical models, which were used for real structures.

Application of mathematical models of any sort to any real structural system must be idealized in some fashion; that is, an analytical model must be developed. There has never been an analytical model which is a precise representation of the physical system. While the performance of the structure is the result of natural effects, the development and thus the performance of the model is entirely under the control of the analyst. The validity of the results obtained from applying mathematical theory to the study of the model therefore rests on the accuracy of the model. While this is true, it does not mean that all analytical models must be elaborate, conceptually

sophisticated devices. In some cases very simple models give surprisingly accurate results. While in some other cases they may yield answers, which deviate markedly from the true physical behavior of the model, yet be completely satisfactory for the problem at hand.

Provision should be made in the application of structural theory to design for abnormal as well as normal service conditions. Abnormal conditions may arise as a result of accidents, fire, explosions, tornadoes, severer-than-anticipated earthquakes, floods, and inadvertent or even deliberate overloading of building components. Under such conditions, parts of a building may be damaged. The structural system, however, should be so designed that the damage will be limited in extent and undamaged portions of the building will remain stable. For the purpose, structural elements should be proportioned and arranged to form a stable system under normal service conditions. In addition, the system should have sufficient continuity and ductility, or energy-absorption capacity, so that if any small portion of it should sustain damage, other parts will transfer loads (at least until repairs can be made) to remaining structural components capable of transmitting the loads to the ground.

("Steel Design Handbook, LRFD Method", Akbar R. Tamboli Ed., McGraw-Hill 1997. "Design Methods for Reducing the Risk of Progressive Collapse in Buildings", NBS Buildings Science Series 98, National Institute of Standards and

Technology, 1977. "Handbook of Structural Steel Connection Design and Details," Akbar R. Tamboli Ed., McGraw-Hill 1999.)

6.1 Structural Integrity

Provision should be made in application of structural theory to design for abnormal as well as normal service conditions. Abnormal conditions may arise as a result of accidents, fire, explosions, tornadoes, severer-than-anticipated earthquakes, floods, and inadvertent or even deliberate overloading of building components. Under such conditions, parts of a building may be damaged. The structural system, however, should be so designed that the damage will be limited in extent and undamaged portions of the building will remain stable. For the purpose, structural elements should be proportioned and arranged to form a stable system under normal service conditions. In addition, the system should have sufficient continuity, redundancy and ductility, or energy-absorption capacity, so that if any small portion of it should sustain damage, other parts will transfer loads (at least until repairs can be made) to remaining structural components capable of transmitting the loads to the ground.

If a structure does not possess this capability, failure of a single component can lead, through progressive collapse of adjoining components, to collapse of a major part or all of the structure. For example, if the corner column of a multistory building should be removed in a mishap and the floor it supports should drop to the floor below, the lower floor and the column supporting it may collapse, throwing the debris to the next lower floor. This action may progress all the way to the ground. One way of avoiding this catastrophe is to design the structure so that when a column fails all components that had been supported by it will cantilever from other parts of the building, although perhaps with deformations normally considered unacceptable.

This example indicates that resistance to progressive collapse may be provided by inclusion in design of alternate load paths capable of absorbing the load from damaged or failed components. An alternative is to provide, in design, reserve strength against mishaps. In both methods, connections of components should provide continuity, redundancy and ductility.

(D. M. Schultz, F. F. P. Burnett, and M. Fintel, "A Design Approach to General Structural Integrity," in "Design and Construction of Large-Panel Concrete Structures," U.S. Department of Housing and Urban Development, 1977; E. V. Leyendecker and B. R. Ellingwood, "Design Methods for Reducing the Risk of Progressive Collapse in Buildings," NBS Buildings Science Series 98, National Institute of Standards and Technology, 1977.)

Equilibrium

6.2 Types of Load

Loads are the external forces acting on a structure. Stresses are the internal forces that resist the loads.

Tensile forces tend to stretch a component, **compressive forces** tend to shorten it, and **shearing forces** tend to slide parts of it past each other.

Loads also may be classified as static or dynamic. **Static loads** are forces that are applied slowly and then remain nearly constant, such as the weight, or dead load, of a floor system. **Dynamic loads** vary with time. They include repeated loads, such as alternating forces from oscillating machinery; moving loads, such as trucks or trains on bridges; impact loads, such as that from a falling weight striking a floor or the shock wave from an explosion impinging on a wall; and seismic loads or other forces created in a structure by rapid movements of supports.

Loads may be considered distributed or concentrated. **Uniformly distributed loads** are forces that are, or for practical purposes may be considered, constant over a surface of the supporting member; dead weight of a rolled-steel beam is a good example. **Concentrated loads** are forces that have such a small contact area as to be negligible compared with the entire surface area of the supporting member. For example, a beam supported on a girder, may, for all practical purposes, be considered a concentrated load on the girder.

In addition, loads may be axial, eccentric, or torsional. An **axial load** is a force whose resultant passes through the centroid of a section under consideration and is perpendicular to the plane of the section. An **eccentric load** is a force perpendicular to the plane of the section under consideration but not passing through the centroid of the section, thus bending the supporting member. **Torsional loads** are forces that are offset from the shear center of the section under consideration and are inclined to or in the plane of the section, thus twisting the supporting member.

Also, loads are classified according to the nature of the source. For example: **Dead loads** include materials, equipment, constructions, or other elements of weight supported in, on, or by a structural element, including its own weight, that are intended to remain permanently in place. **Live loads** include all occupants, materials, equipment, constructions, or other elements of weight supported in, on, or by a structural element that will or are likely to be moved or relocated during the expected life of the structure. **Impact loads** are a fraction of the live loads used to account for additional stresses and deflections resulting from movement of the live loads. **Wind loads** are maximum forces that may be applied to a structural element by wind in a mean recurrence interval, or a set of forces that will produce equivalent stresses. Mean recurrence intervals generally used are 25 years for structures with no occupants or offering negligible risk to life, 50 years for ordinary permanent structures, and 100 years for permanent structures with a high degree of sensitivity to wind and an unusually high degree of hazard to life and property in case of failure. **Snow loads** are maximum forces that may be applied by snow accumulation in a mean recurrence interval. **Seismic loads** are forces that produce maximum stresses or deformations in a structural element during an earthquake, or equivalent forces.

Probable maximum loads should be used in design. For buildings, minimum design load should be that specified for expected conditions in the local building code or, in the absence of an applicable local code, in "Minimum Design Loads for Buildings and Other Structures," ASCE 7-93, American Society of Civil Engineers, Reston, VA, (www.asce.org). For highways and highway bridges, minimum design loads should be those given in "Standard Specifications for Highway Bridges," American Association of State Highway and Transportation Officials, Washington, D.C. (www.transportation.org). For railways and railroad bridges, minimum design loads should be those given in "Manual for Railway Engineering," American Railway Engineering and Maintenance-of-Way Association, Chicago (www.arema.org).

6.3 Static Equilibrium

If a structure and its components are so supported that after a small deformation occurs no further motion is possible, they are said to be in equilibrium. Under such circumstances, external forces are in balance and internal forces, or stresses, exactly counteract the loads.

Since there is no translatory motion, the vector sum of the external forces must be zero. Since there is no rotation, the sum of the moments of the external forces about any point must be zero. For the same reason, if we consider any portion of the structure and the loads on it, the sum of the external and internal forces on the boundaries of that section must be zero. Also, the sum of the moments of these forces must be zero.

In Fig. 6.1, for example, the sum of the forces R_L and R_R needed to support the truss is equal to the 20-kip load on the truss (1 **kip** = 1 kilopound = 1000 lb = 0.5 ton). Also, the sum of the moments of the external forces is zero about any point; about the right end, for instance, it is $40 \times 15 - 30 \times 20 = 600 - 600$.

Figure 6.2 shows the portion of the truss to the left of section AA. The internal forces at the cut members balance the external load and hold this piece of the truss in equilibrium.

When the forces act in several directions, it generally is convenient to resolve them into components parallel to a set of perpendicular axes that will simplify computations. For example, for forces in a single plane, the most useful technique is to resolve them into horizontal and vertical components. Then, for a structure in equilibrium, if H represents the horizontal components, V the

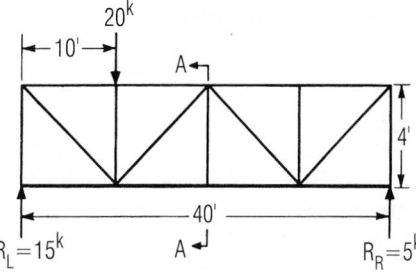

Fig. 6.1 Truss in equilibrium under load. Upward-acting forces, or reactions, R_L and R_R, equal the 20-kip downward-acting force.

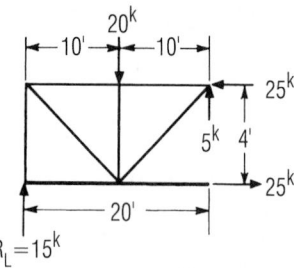

Fig. 6.2 Section of the truss shown in Fig. 6.1 is kept in equilibrium by stresses in the components.

vertical components, and M the moments of the components about any point in the plane,

$$\Sigma H = 0 \quad \Sigma V = 0 \quad \text{and} \quad \Sigma M = 0 \qquad (6.1)$$

These three equations may be used to determine three unknowns in any nonconcurrent coplanar force system, such as the truss in Figs. 6.1 and 6.2. They may determine the magnitude of three forces for which the direction and point of application already are known, or the magnitude, direction, and point of application of a single force. Suppose, for the truss in Fig. 6.1, the reactions at the supports are to be computed. Take the sum of the moments about the right support and equate them to zero to find the left reaction: $40R_L - 30 \times 20 = 0$, from which $R_L = 600/40 = 15$ kips. To find the right reaction, take moments about the left support and equate the sum to zero: $10 \times 20 - 40R_R = 0$, from which $R_R = 5$ kips. As an alternative, equate the sum of the vertical forces to zero to obtain R_R after finding R_L: $20 - 15 - R_R = 0$, from which $R_R = 5$ kips.

Stress and Strain

6.4   Unit Stress and Strain

It is customary to give the strength of a material in terms of unit stress, or internal force per unit of area. Also, the point at which yielding starts generally is expressed as a *unit stress*. Then, in some design methods, a safety factor is applied to either of these stresses to determine a unit stress that should not be exceeded when the member carries design loads. That unit stress is known as the *allowable stress*, or *working stress*.

In working-stress design, to determine whether a structural member has adequate load-carrying capacity, the designer generally has to compute the maximum unit stress produced by design loads in the member for each type of internal force—tensile, compressive, or shearing—and compare it with the corresponding allowable unit stress.

When the loading is such that the unit stress is constant over a section under consideration, the stress may be computed by dividing the force by the area of the section. But, generally, the unit stress varies from point to point. In those cases, the unit stress at any point in the section is the limiting value of the ratio of the internal force on any small area to that area, as the area is taken smaller and smaller.

Unit Strain ▪ Sometimes in the design of a structure, the designer may be more concerned with limiting deformation or strain than with strength. Deformation in any direction is the total change in the dimension of a member in that direction. *Unit strain* in any direction is the deformation per unit of length in that direction.

When the loading is such that the unit strain is constant over the length of a member, it may be computed by dividing the deformation by the original length of the member. In general, however, unit strain varies from point to point in a member. Like a varying unit stress, it represents the limiting value of a ratio.

6.5   Stress-Strain Relations

When a material is subjected to external forces, it develops one or more of the following types of strain: linear elastic, nonlinear elastic, viscoelastic, plastic, and anelastic. Many structural materials exhibit linear elastic strains under design loads. For these materials, unit strain is proportional to unit stress until a certain stress, the proportional limit, is exceeded (point A in Fig. 6.3a to c). This relationship is known as **Hooke's law**.

For axial tensile or compressive loading, this relationship may be written

$$f = E\varepsilon \quad \text{or} \quad \varepsilon = \frac{f}{E} \qquad (6.2)$$

where f = unit stress

e = unit strain

E = Young's modulus of elasticity

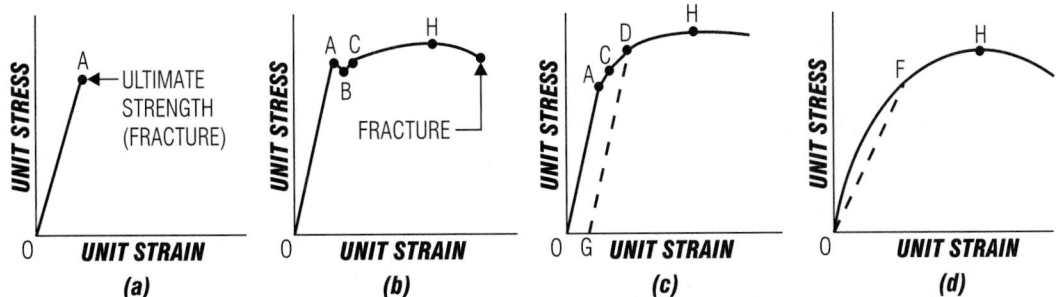

Fig. 6.3 Relationship of unit stress and unit strain for various materials. (*a*) Brittle. (*b*) Linear elastic with distinct proportional limit. (*c*) Linear elastic with an indistinct proportional limit. (*d*) Nonlinear.

Within the elastic limit, there is no permanent residual deformation when the load is removed. Structural steels have this property.

In nonlinear elastic behavior, stress is not proportional to strain, but there is no permanent residual deformation when the load is removed. The relation between stress and strain may take the form

$$\varepsilon = \left(\frac{f}{K}\right)^{n} \tag{6.3}$$

where K = pseudoelastic modulus determined by test

n = constant determined by test

Viscoelastic behavior resembles linear elasticity. The major difference is that in linear elastic behavior, the strain stops increasing if the load does; but in viscoelastic behavior, the strain continues to increase although the load becomes constant and a residual strain remains when the load is removed. This is characteristic of many plastics.

Anelastic deformation is time-dependent and completely recoverable. Strain at any time is proportional to change in stress. Behavior at any given instant depends on all prior stress changes. The combined effect of several stress changes is the sum of the effects of the several stress changes taken individually.

Plastic strain is not proportional to stress, and a permanent deformation remains on removal of the load. In contrast with anelastic behavior, plastic deformation depends primarily on the stress and is largely independent of prior stress changes.

When materials are tested in axial tension and corresponding stresses and strains are plotted, stress-strain curves similar to those in Fig. 6.3 result. Figure 6.3*a* is typical of a brittle material, which deforms in accordance with Hooke's law up to fracture. The other curves in Fig. 6.3 are characteristic of ductile materials; because strains increase rapidly near fracture with little increase in stress, they warn of imminent failure, whereas brittle materials fail suddenly.

Figure 6.3*b* is typical of materials with a marked proportional limit *A*. When this is exceeded, there is a sudden drop in stress, then gradual stress increase with large increases in strain to a maximum before fracture. Figure 6.3*c* is characteristic of materials that are linearly elastic over a substantial range but have no definite proportional limit. And Fig. 6.3*d* is a representative curve for materials that do not behave linearly at all.

Modulus of Elasticity ▪ E is given by the slope of the straight-line portion of the curves in Fig. 6.3*a* to *c*. It is a measure of the inherent rigidity or stiffness of a material. For a given geometric configuration, a material with a larger E deforms less under the same stress.

At the termination of the linear portion of the stress-strain curve, some materials, such as low-carbon steel, develop an upper and lower **yield point** (*A* and *B* in Fig. 6.3*b*). These points mark a range in which there appears to be an increase in strain with no increase or a small decrease in stress. This behavior may be a consequence of inertia effects in the testing machine and the deformation characteristics of the test specimen. Because of the location of the yield points, the yield stress sometimes is used erroneously as a synonym for proportional limit and elastic limit.

The **proportional limit** is the maximum unit stress for which Hooke's law is valid. The **elastic limit** is the largest unit stress that can be developed without a permanent set remaining after removal of the load (C in Fig. 6.3). Since the elastic limit is always difficult to determine and many materials do not have a well-defined proportional limit, or even have one at all, the offset yield strength is used as a measure of the beginning of plastic deformation.

The **offset yield strength** is defined as the stress corresponding to a permanent deformation, usually 0.01% (0.0001 in/in) or 0.20% (0.002 in/in). In Fig. 6.3c the yield strength is the stress at D, the intersection of the stress-strain curve and a line GD parallel to the straight-line portion and starting at the given unit strain. This stress sometimes is called the **proof stress**.

For materials with a stress-strain curve similar to that in Fig. 6.3d, with no linear portion, a **secant modulus**, represented by the slope of a line, such as OF, from the origin to a specified point on the curve, may be used as a measure of stiffness. An alternative measure is the **tangent modulus**, the slope of the stress-strain curve at a specified point.

Ultimate tensile strength is the maximum axial load observed in a tension test divided by the original cross-sectional area. Characterized by the beginning of necking down, a decrease in cross-sectional area of the specimen, or local instability, this stress is indicated by H in Fig. 6.3.

Ductility is the ability of a material to undergo large deformations without fracture. It is measured by elongation and reduction of area in a tension test and expressed as a percentage. Ductility depends on temperature and internal stresses as well as the characteristics of the material; a material that may be ductile under one set of conditions may have a brittle failure at lower temperatures or under tensile stresses in two or three perpendicular directions.

Modulus of rigidity, or shearing modulus of elasticity, is defined by

$$G = \frac{v}{\gamma} \qquad (6.4)$$

where G = modulus of rigidity

v = unit shearing stress

γ = unit shearing strain

It is related to the modulus of elasticity in tension and compression E by the equation

$$G = \frac{E}{2(1 + \mu)} \qquad (6.5)$$

where μ is a constant known as Poisson's ratio (Art. 6.7).

Toughness is the ability of a material to absorb large amounts of energy. Related to the area under the stress-strain curve, it depends on both strength and ductility. Because of the difficulty of determining toughness analytically, often toughness is measured by the energy required to fracture a specimen, usually notched and sometimes at low temperatures, in impact tests. Charpy and Izod, both applying a dynamic load by pendulum, are the tests most commonly used.

Hardness is a measure of the resistance a material offers to scratching and indention. A relative numerical value usually is determined for this property in such tests as Brinell, Rockwell, and Vickers. The numbers depend on the size of an indentation made under a standard load. Scratch resistance is measured on the Mohs scale by comparison with the scratch resistance of 10 minerals arranged in order of increasing hardness from talc to diamond.

Creep is a property of certain materials like concrete that deforms with time under constant load. Shrinkage for concrete is the volume reduction with time. It is unrelated to load application. **Relaxation** is a decrease in load or stress under a sustained constant deformation.

If stresses and strains are plotted in an axial tension test as a specimen enters the inelastic range and then is unloaded, the curve during unloading, if the material was elastic, descends parallel to the straight portion of the curve (for example, DG in Fig. 6.3c). Completely unloaded, the specimen has a permanent set (OG). This also will occur in compression tests.

If the specimen now is reloaded, strains are proportional to stresses (the curve will practically follow DG) until the curve rejoins the original curve at D. Under increasing load, the reloading curve coincides with that for a single loading. Thus, loading the specimen into the inelastic range, but not to ultimate strength, increases the apparent elastic range. The phenomenon, called **strain**

hardening, or work hardening, appears to increase the yield strength. Usually, when the yield strength of a material is increased through strain hardening, the ductility of the material is reduced.

But if the reloading is reversed in compression, the compressive yield strength is decreased, which is called the **Bauschinger effect.**

6.6 Constant Unit Stress

The simplest cases of stress and strain are those in which the unit stress and strain are constant. Stresses caused by an axial tension or compression load, a centrally applied shear, or a bearing load are examples. These conditions are illustrated in Figs. 6.4 to 6.7.

For constant unit stress, the equation of equilibrium may be written

$$P = Af \qquad (6.6)$$

where P = load, lb

A = cross-sectional area (normal to load) for tensile or compressive forces, or area on which sliding may occur for shearing forces, or contact area for bearing loads, in^2

f = tensile, compressive, shearing, or bearing unit stress, psi

For torsional stresses, see Art. 6.18.

Unit strain for the axial tensile and compressive loads is given by

$$\varepsilon = \frac{e}{L} \qquad (6.7)$$

where ε = unit strain, in/in

e = total lengthening or shortening of member, in

L = original length of the member, in

Application of Hooke's law and Eq. (6.6) to Eq. (6.7) yields a convenient formula for the deformation:

$$e = \frac{PL}{AE} \qquad (6.8)$$

where P = load on member, lb

A = its cross-sectional area, in^2

E = modulus of elasticity of material, psi

[Since long compression members tend to buckle, Eqs. (6.6) to (6.8) are applicable only to short members. See Arts. 6.39 to 6.41.]

Although tension and compression strains represent a simple stretching or shortening of a member, shearing strain is a distortion due to a small rotation. The load on the small rectangular portion of the member in Fig. 6.6 tends to distort it into a parallelogram. The unit shearing strain is the change in the right angle, measured in radians. (See also Art. 6.5.)

6.7 Poisson's Ratio

When a material is subjected to axial tensile or compressive loads, it deforms not only in the

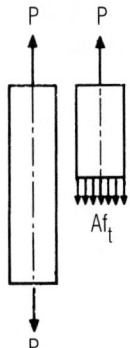

Fig. 6.4
Tension member axially loaded.

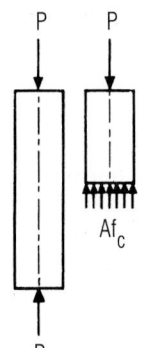

Fig. 6.5
Compression member axially loaded.

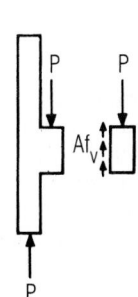

Fig. 6.6
Bracket in shear.

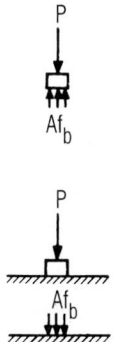

Fig. 6.7
Bearing load.

direction of the loads but normal to them. Under tension, the cross section of a member decreases, and under compression, it increases. The ratio of the unit lateral strain to the unit longitudinal strain is called *Poisson's ratio*.

Within the elastic range, Poisson's ratio is a constant for a material. For materials such as concrete, glass, and ceramics, it may be taken as 0.25; for structural steel, 0.3. It gradually increases beyond the proportional limit and tends to approach a value of 0.5.

Assume, for example, that a steel hanger with an area of 2 in^2 carries a 40-kip (40,000-lb) load. The unit stress is 40/2, or 20 ksi. The unit tensile strain, with modulus of elasticity of steel $E = 30,000$ ksi, is 20/30,000, or 0.00067 in/in. With Poisson's ratio as 0.3, the unit lateral strain is -0.3×0.00067, or a shortening of 0.00020 in/in.

6.8 Thermal Stresses

When the temperature of a body changes, its dimensions also change. Forces are required to prevent such dimensional changes, and stresses are set up in the body by these forces.

If α is the coefficient of expansion of the material and T the change in temperature, the unit strain in a bar restrained by external forces from expanding or contracting is

$$\varepsilon = \alpha T \tag{6.9}$$

According to Hooke's law, the stress f in the bar is

$$f = E\alpha T \tag{6.10}$$

where E = modulus of elasticity.

When a circular ring, or hoop, is heated and then slipped over a cylinder of slightly larger diameter d than d_n the original hoop diameter, the hoop will develop a tensile stress on cooling. If the diameter is very large compared with the hoop thickness, so that radial stresses can be neglected, the unit tensile stresses may be assumed constant. The unit strain will be

$$\varepsilon = \frac{\pi d - \pi d_1}{\pi d_1} = \frac{d - d_1}{d_1}$$

and the hoop stress will be

$$f = \frac{(d - d_1)E}{d_1} \tag{6.11}$$

6.9 Axial Stresses in Composite Members

In a homogeneous material, the centroid of a cross section lies at the intersection of two perpendicular axes so located that the moments of the areas on opposite sides of an axis about that axis are zero. To find the centroid of a cross section containing two or more materials, the moments of the products of the area A of each material and its modulus of elasticity E should be used, in the elastic range.

Consider now a prism composed of two materials, with modulus of elasticity E_1 and E_2, extending the length of the prism. If the prism is subjected to a load acting along the centroidal axis, then the unit strain ε in each material will be the same. From the equation of equilibrium and Eq. (6.8), noting that the length L is the same for both materials,

$$\varepsilon = \frac{P}{A_1 E_1 + A_2 E_2} = \frac{P}{\Sigma AE} \tag{6.12}$$

where A_1 and A_2 are the cross-sectional areas of each material and P the axial load. The unit stresses in each material are the products of the unit strain and its modulus of elasticity:

$$f_1 = \frac{PE_1}{\Sigma AE} \quad f_2 = \frac{PE_2}{\Sigma AE} \tag{6.13}$$

6.10 Stresses in Pipes and Pressure Vessels

In a cylindrical pipe under internal radial pressure, the circumferential unit stresses may be assumed constant over the pipe thickness t, in, if the diameter is relatively large compared with the thickness (at least 15 times as large). Then, the circumferential unit stress, in pounds per square inch, is given by

$$f = \frac{pR}{t} \tag{6.14}$$

where p = internal pressure, psi

R = average radius of pipe, in (see also Art. 21.14)

In a closed cylinder, the pressure against the ends will be resisted by longitudinal stresses in the cylinder. If the cylinder is thin, these stresses, psi, are given by

$$f_2 = \frac{pR}{2t} \qquad (6.15)$$

Equation (6.15) also holds for the stress in a thin spherical tank under internal pressure p with R the average radius.

In a thick-walled cylinder, the effect of radial stresses f_r becomes important. Both radial and circumferential stresses may be computed from Lamé's formulas:

$$f_r = p \frac{r_i^2}{r_o^2 - r_i^2} \left(1 - \frac{r_o^2}{r^2}\right) \qquad (6.16)$$

$$f = p \frac{r_i^2}{r_o^2 - r_o^2} \left(1 + \frac{r_o^2}{r^2}\right) \qquad (6.17)$$

where r_i = internal radius of cylinder, in

r_o = outside radius of cylinder, in

r = radius to point where stress is to be determined, in

The equations show that if the pressure p acts outward, the circumferential stress f will be tensile (positive) and the radial stress compressive (negative). The greatest stresses occur at the inner surface of the cylinder ($r = r_i$):

$$\text{Max } f_r = -p \qquad (6.18)$$

$$\text{Max } f = \frac{k^2 + 1}{k^2 - 1} p \qquad (6.19)$$

where $k = r_o/r_i$. Maximum shear stress is given by

$$\text{Max } f_v = \frac{k^2}{k^2 - 1} p \qquad (6.20)$$

For a closed cylinder with thick walls, the longitudinal stress is approximately

$$f_z = \frac{p}{r_i(k^2 - 1)} \qquad (6.21)$$

But because of end restraints, this stress will not be correct near the ends.

(S. Timoshenko and J. N. Goodier, "Theory of Elasticity," McGraw-Hill Book Company, New York.)

6.11 Strain Energy

Stressing a bar stores energy in it. For an axial load P and a deformation e, the energy stored called strain energy is

$$U = \frac{1}{2} Pe \qquad (6.22a)$$

assuming the load is applied gradually and the bar is not stressed beyond the proportional limit. The equation represents the area under the load-deformation curve up to the load P. Application of Eqs. (6.2) and (6.6) to Eq. (6.22a) yields another useful equation for energy, in-lb:

$$U = \frac{f^2}{2E} AL \qquad (6.22b)$$

where f = unit stress, psi

E = modulus of elasticity of material, psi

A = cross-sectional area, in^2

L = length of bar, in

Since AL is the volume of the bar, the term $f^2/2E$ gives the energy stored per unit of volume. It represents the area under the stress-strain curve up to the stress f.

Modulus of resilience is the energy stored per unit of volume in a bar stressed by a gradually applied axial load up to the proportional limit. This modulus is a measure of the capacity of the material to absorb energy without danger of being permanently deformed. It is important in designing members to resist energy loads.

Equation (6.22a) is a general equation that holds true when the **principle of superposition** applies (the total deformation produced at a point by a system of forces is equal to the sum of the deformations produced by each force). In the general sense, P in Eq. (6.22a) represents any group of statically interdependent forces that can be completely defined by one symbol, and e is the corresponding deformation.

The strain-energy equation can be written as a function of either the load or the deformation. For axial tension or compression, strain energy, in inch-pounds, is given by

$$U = \frac{P^2 L}{2AE} \qquad U = \frac{AEe^2}{2L} \qquad (6.23a)$$

where P = axial load, lb

e = total elongation or shortening, in

L = length of member, in

A = cross-sectional area, in^2

E = modulus of elasticity, psi

For pure shear:

$$U = \frac{V^2 L}{2AG} \quad U = \frac{AGe^2}{2L} \tag{6.23b}$$

where V = shearing load, lb

e = shearing deformation, in

L = length over which deformation takes place, in

A = shearing area, in^2

G = shearing modulus, psi

For torsion:

$$U = \frac{T^2 L}{2JG} \quad U = \frac{JG\phi^2}{2L} \tag{6.23c}$$

where T = torque, in-lb

ϕ = angle of twist, rad

L = length of shaft, in

J = polar moment of inertia of cross section, in^4

G = shearing modulus, psi

For pure bending (constant moment):

$$U = \frac{M^2 L}{2EI} \quad U = \frac{EI\theta^2}{2L} \tag{6.23d}$$

where M = bending moment, in-lb

θ = angle of rotation of one end of beam with respect to other, rad

L = length of beam, in

I = moment of inertia of cross section, in^4

E = modulus of elasticity, psi

For beams carrying transverse loads, the total strain energy is the sum of the energy for bending and that for shear. (See also Art. 6.54.)

Stresses at a Point

Tensile and compressive stresses sometimes are referred to as *normal stresses* because they act normal to the cross section. Under this concept, tensile stresses are considered positive normal stresses and compressive stresses negative.

6.12 Stress Notation

Consider a small cube extracted from a stressed member and placed with three edges along a set of x, y, z coordinate axes. The notations used for the components of stress acting on the sides of this element and the direction assumed as positive are shown in Fig. 6.8.

For example, for the sides of the element perpendicular to the z axis, the normal component of stress is denoted by f_z. The shearing stress v is resolved into two components and requires two subscript letters for a complete description. The first letter indicates the direction of the normal to the plane under consideration; the second letter gives the direction of the component of stress. Thus, for the sides perpendicular to the z axis, the shear component in the x direction is labeled v_{zx} and that in the y direction v_{zy}.

6.13 Stress Components

If, for the small cube in Fig. 6.8, moments of the forces acting on it are taken about the x axis, and assuming the lengths of the edges as dx, dy, and dz, the equation of equilibrium requires that

$$(v_{zy}\, dx\, dy)\, dz = (v_{yz}\, dx\, dz)\, dy$$

(Forces are taken equal to the product of the area of the face and the stress at the center.) Two similar equations can be written for moments taken about the y and z axes. These equations show that

$$v_{xy} = v_{yx} \quad v_{zx} = v_{xz} \quad v_{zy} = v_{yz} \tag{6.24}$$

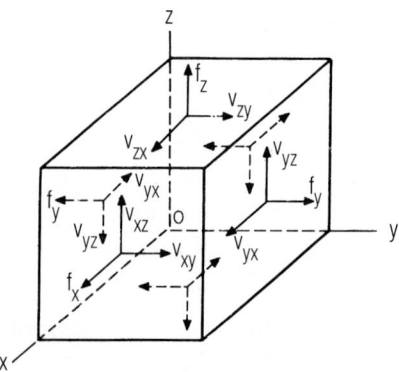

Fig. 6.8 Stresses at a point in a rectangular coordinate system.

Thus, components of shearing stress on two perpendicular planes and acting normal to the intersection of the planes are equal. Consequently, to describe the stresses acting on the coordinate planes through a point, only six quantities need be known: the three normal stresses f_x, f_y, f_z and three shearing components $v_{xy} = v_{yx}, v_{zx} = v_{xz}$, and $v_{zy} = v_{yz}$.

If only the normal stresses are acting, the unit strains in the x, y, and z directions are

$$\varepsilon_x = \frac{1}{E}[f_x - \mu(f_y + f_z)]$$

$$\varepsilon_y = \frac{1}{E}[f_y - \mu(f_x + f_z)] \qquad (6.25)$$

$$\varepsilon_z = \frac{1}{E}[f_z - \mu(f_x + f_y)]$$

where μ = Poisson's ratio. If only shearing stresses are acting, the distortion of the angle between edges parallel to any two coordinate axes depends only on shearing-stress components parallel to those axes. Thus, the unit shearing strains are (see Art. 6.5)

$$\gamma_{xy} = \frac{1}{G}v_{xy} \quad \gamma_{yz} = \frac{1}{G}v_{yz} \quad \gamma_{zx} = \frac{1}{G}v_{zx} \qquad (6.26)$$

6.14 Two-Dimensional Stress

When the six components of stress necessary to describe the stresses at a point are known (Art. 6.13), the stresses on any inclined plane through the same point can be determined. For two-dimensional stress, only three stress components need be known.

Assume, for example, that at a point O in a stressed plate, the components f_x, f_y, and v_{xy} are known (Fig. 6.9). To find the stresses on any other plane through the z axis, take a plane parallel to it close to O, so that this plane and the coordinate planes form a tiny triangular prism. Then, if α is the angle the normal to the plane makes with the x axis, the normal and shearing stresses on the inclined plane, to maintain equilibrium, are

$$f = f_x \cos^2 \alpha + f_y \sin^2 \alpha + 2v_{xy} \sin \alpha \cos \alpha \qquad (6.27)$$

$$v = v_{xy}(\cos^2 \alpha - \sin^2 \alpha) + (f_y - f_x)\sin \alpha \cos \alpha \qquad (6.28)$$

(See also Art. 6.17.)

Note: All structural members are three-dimensional. While two-dimensional stress calculations may be sufficiently accurate for most practical purposes, this is not always the case. For

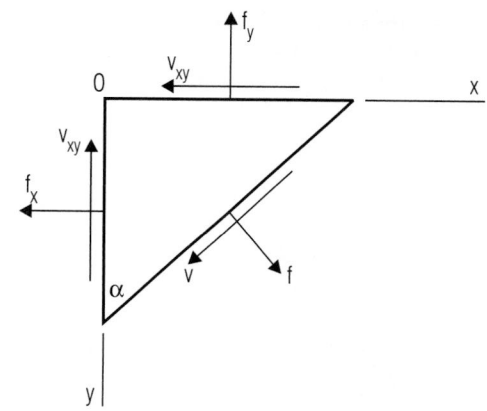

Fig. 6.9 Stresses at a point on a plane inclined to the axes.

example, although loads may create normal stresses on two perpendicular planes, a third normal stress also exists, as computed with Poisson's ratio. [See Eq. (6.25).]

6.15 Principal Stresses

If a plane at a point O in a stressed plate is rotated, it reaches a position for which the normal stress on it is a maximum or a minimum. The directions of maximum and minimum normal stress are perpendicular to each other, and on the planes in those directions, there are no shearing stresses.

The directions in which the normal stresses become maximum or minimum are called *principal directions*, and the corresponding normal stresses are called *principal stresses*. To find the principal directions, set the value of v given by Eq. (6.28) equal to zero. Then, the normals to the principal planes make an angle with the x axis given by

$$\tan 2\alpha = \frac{2v_{xy}}{f_x - f_y} \qquad (6.29)$$

If the x and y axes are taken in the principal directions, $v_{xy} = 0$. In that case, Eqs. (6.27) and (6.28) simplify to

$$f = f_x \cos^2 \alpha + f_y \sin^2 \alpha \qquad (6.30)$$

$$v = \frac{1}{2}(f_y - f_x)\sin 2\alpha \qquad (6.31)$$

where f_x and f_y are the principal stresses at the point, and f and v are, respectively, the normal and shearing stress on a plane whose normal makes an angle α with the x axis.

If only shearing stresses act on any two perpendicular planes, the state of stress at the point is said to be one of pure shear or simple shear. Under such conditions, the principal directions bisect the angles between the planes on which these shearing stresses act. The principal stresses are equal in magnitude to the pure shears.

6.16 Maximum Shearing Stress at a Point

The maximum unit shearing stress occurs on each of two planes that bisect the angles between the planes on which the principal stresses at a point act. The maximum shear equals half the algebraic difference of the principal stresses:

$$\text{Max } v = \frac{f_1 - f_2}{2} \tag{6.32}$$

where f_1 is the maximum principal stress and f_2 the minimum.

6.17 Mohr's Circle

As explained in Art. 6.14, if the stresses on any plane through a point in a stressed plate are known, the stresses on any other plane through the point can be computed. This relationship between the stresses may be represented conveniently on Mohr's circle (Fig. 6.10). In this diagram, normal stress f and shear stress v are taken as rectangular coordinates. Then, for each plane through the point there will correspond a point on the circle, the coordinates of which are the values of f and v for the plane.

Given the principal stresses f_1 and f_2 (Art. 6.15), to find the stresses on a plane making an angle α with the plane on which f_1 acts: Mark off the principal stresses on the f axis (points A and B in Fig. 6.10). Measure tensile stresses to the right of the v axis and compressive stresses to the left. Construct a circle passing through A and B and having its center on the f axis. This is the Mohr's circle for the given stresses at the point under consideration. Draw a radius making an angle 2α with the f axis, as indicated in Fig. 6.10. The coordinates of the intersection with the circle represent the normal and shearing stresses f and v acting on the plane.

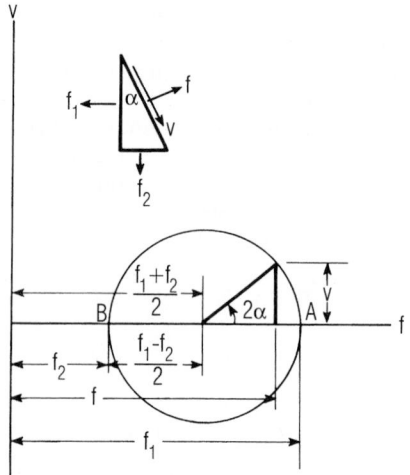

Fig. 6.10 Mohr's circle for stresses at a point—constructed from known principal stresses f_1 and f_2 in a plane.

Given the stresses on any two perpendicular planes f_x, f_y, and v_{xy}, but not the principal stresses f_1 and f_2, to draw the Mohr's circle: Plot the two points representing the known stresses with respect to the f and v axes (points C and D in Fig. 6.11). The line joining these points is a

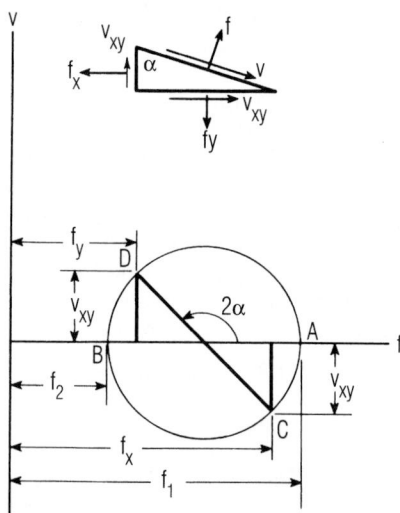

Fig. 6.11 Stress circle constructed from two known normal positive stresses f_x and f_y and a known shear v_{xy}.

diameter of the circle, so bisect CD to find the center of the circle and draw the circle. Its intersections with the f axis determine f_1 and f_2.

(S. Timoshenko and J. N. Goodier, "Theory of Elasticity," McGraw-Hill Book Company, New York, books.mcgraw-hill.com.)

6.18 Torsion

Forces that cause a member to twist about a longitudinal axis are called torsional loads. Simple torsion is produced only by a couple, or moment, in a plane perpendicular to the axis.

If a couple lies in a nonperpendicular plane, it can be resolved into a torsional moment, in a plane perpendicular to the axis, and bending moments, in planes through the axis.

Shear Center ▪ The point in each normal section of a member through which the axis passes and about which the section twists is called the shear center. If the loads on a beam, for example, do not pass through the shear center, they cause the beam to twist. See also Art. 6.36.

If a beam has an axis of symmetry, the shear center lies on it. In doubly symmetrical beams, the shear center lies at the intersection of two axes of symmetry and hence coincides with the centroid.

For any section composed of two narrow rectangles, such as a T beam or an angle, the shear center may be taken as the intersection of the longitudinal center lines of the rectangles.

For a channel section with one axis of symmetry, the shear center is outside the section at a distance from the centroid equal to $e(1 + h^2A/4I)$, where e is the distance from the centroid to the center of the web, h is the depth of the channel, A the cross-sectional area, and I the moment of inertia about the axis of symmetry. (The web lies between the centroid and the shear center.)

Locations of shear centers for several other sections are given in Freidrich Bleich, "Buckling Strength of Metal Structures," chap. 3, McGraw-Hill Publishing Company, New York, 1952, books. mcgraw-hill.com.

Stresses Due to Torsion ▪ Simple torsion is resisted by internal shearing stresses. These can be resolved into radial and tangential shearing stresses, which being normal to each other also are equal (see Art. 6.13). Furthermore, on planes that bisect the angles between the planes on which the

shearing stresses act, there also occur compressive and tensile stresses. The magnitude of these normal stresses is equal to that of the shear. Therefore, when torsional loading is combined with other types of loading, the maximum stresses occur on inclined planes and can be computed by the methods of Arts. 6.14 and 6.17.

Circular Sections ▪ If a circular shaft (hollow or solid) is twisted, a section that is plane before twisting remains plane after twisting. Within the proportional limit, the shearing stress at any point in a transverse section varies with the distance from the center of the section. The maximum shear, psi, occurs at the circumference and is given by

$$v = \frac{Tr}{J} \qquad (6.33)$$

where T = torsional moment, in-lb

r = radius of section, in

J = polar moment of inertia, in^4

Polar moment of inertia of a cross section is defined by

$$J = \int \rho^2 dA \qquad (6.34)$$

where ρ = radius from shear center to any point in section

dA = differential area at point

In general, J equals the sum of the moments of inertia about any two perpendicular axes through the shear center. For a solid circular section, $J = \pi r^4/2$. For a hollow circular section with diameters D and d, $J = \pi(D^4 - d^4)/32$.

Within the proportional limit, the angular twist between two points L inches apart along the axis of a circular bar is, in radians (1 rad = 57.3°):

$$\theta = \frac{TL}{GJ} \qquad (6.35)$$

where G is the shearing modulus of elasticity (see Art. 6.5).

Noncircular Sections ▪ If a shaft is not circular, a plane transverse section before twisting does not remain plane after twisting. The resulting warping increases the shearing stresses in some parts of the section and decreases them in others, compared with the shearing stresses that would occur if the section remained plane. Consequently, shearing stresses in a noncircular section are not proportional to distances from the shear center. In elliptical and rectangular sections, for example, maximum shear occurs on the circumference at a point nearest the shear center.

For a solid rectangular section, this maximum shear stress may be expressed in the following form:

$$v = \frac{T}{kb^2d} \qquad (6.36)$$

where b = short side of rectangle, in

d = long side, in

k = constant depending on ratio of these sides:

$d/b = 1.0 \quad 1.5 \quad 2.0 \quad 2.5 \quad 3 \quad 4 \quad 5 \quad 10 \quad \infty$

$k = 0.208 \quad 0.231 \quad 0.246 \quad 0.258 \quad 0.267$
  $0.282 \quad 0.291 \quad 0.312 \quad 0.333$

(S. Timoshenko and J. N. Goodier, "Theory of Elasticity," McGraw-Hill Publishing Company, New York, books.mcgraw-hill.com.)

Hollow Tubes ▪ If a thin-shell hollow tube is twisted, the shearing force per unit of length on a cross section (**shear flow**) is given approximately by

$$H = \frac{T}{2A} \qquad (6.37)$$

where A is the area enclosed by the mean perimeter of the tube, in². And the unit shearing stress is given approximately by

$$v = \frac{H}{t} = \frac{T}{2At} \qquad (6.38)$$

where t is the thickness of the tube, in. For a rectangular tube with sides of unequal thickness, the total shear flow can be computed from Eq. (6.37) and the shearing stress along each side from Eq. (6.38), except at the corners, where there may be appreciable stress concentration.

Channels and I Beams ▪ For a narrow rectangular section, the maximum shear is very nearly equal to

$$v = \frac{T}{1/3b^2d} \qquad (6.39)$$

This formula also can be used to find the maximum shearing stress due to torsion in members,

such as I beams and channels, made up of thin rectangular components. Let $J = 1/3\Sigma b^3d$, where b is the thickness of each rectangular component and d the corresponding length. Then, the maximum shear is given approximately by

$$v = \frac{Tb'}{J} \qquad (6.40)$$

where b' is the thickness of the web or the flange of the member. Maximum shear will occur at the center of one of the long sides of the rectangular part that has the greatest thickness.

(A. P. Boresi, O. Sidebottom, F. B. Seely, and J. O. Smith, "Advanced Mechanics of Materials," John Wiley & Sons, Inc., New York, www.wiley.com.)

Straight Beams

6.19 Types of Beams

Bridge decks and building floors and roofs frequently are supported on a rectangular grid of flexural members. Different names often are given to the components of the grid, depending on the type of structure and the part of the structure supported on the grid. In general, though, the members spanning between main supports are called **girders** and those they support are called **beams** (Fig. 6.12). Hence, this type of framing is known as beam-and-girder framing.

In bridges, the smaller structural members parallel to the direction in which traffic moves may be called **stringers** and the transverse members **floor beams**. In building roofs, the grid components may be referred to as **purlins** and **rafters;** and in floors, they may be called **joists** and **girders**.

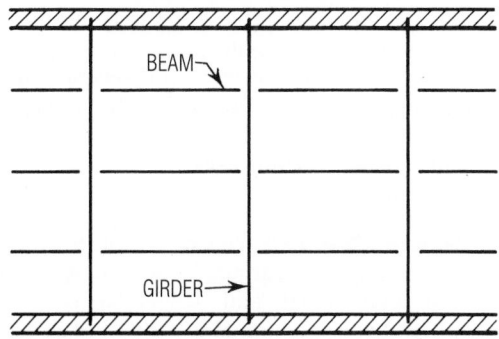

Fig. 6.12 Beam-and-girder framing.

Fig. 6.13 Simple beam, both ends free to rotate.

Fig. 6.14 Cantilever beam.

Fig. 6.15 Beam with one end fixed.

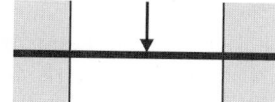

Fig. 6.16 Fixed-end beam.

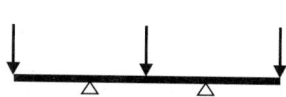

Fig. 6.17 Beam with over-hangs.

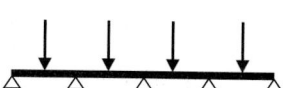

Fig. 6.18 Continuous beam.

Beam-and-girder framing usually is used for relatively short spans and where shallow members are desired to provide ample headroom underneath.

Beams and trusses are similar in behavior as flexural members. The term beam, however, usually is applied to members with top continuously connected to bottom throughout their length, while those with top and bottom connected at intervals are called trusses.

There are many ways in which beams may be supported. Some of the most common methods are shown in Figs. 6.13 to 6.19. The beam in Fig. 6.13 is called a simply supported beam, or **simple beam**. It has supports near its ends that restrain it only against vertical movement. The ends of the beam are free to rotate. When the loads have a horizontal component, or when change in length of the beam due to temperature may be important, the supports may also have to prevent horizontal motion, in which case horizontal restraint at one support generally is sufficient. The distance between the supports is called the **span**. The load carried by each support is called a **reaction**.

The beam in Fig. 6.14 is a **cantilever**. It has a support only at one end. The support provides restraint against rotation and horizontal and vertical movement. Such support is called a **fixed end**. Placing a support under the free end of the cantilever produces the beam in Fig. 6.15. Fixing the free end yields a **fixed-end beam** (Fig. 6.16); no rotation or vertical movement can occur at either

end. In actual practice, however, a fully fixed end can seldom be obtained. Most support conditions are intermediate between those for a simple beam and those for a fixed-end beam.

Figure 6.17 shows a beam that overhangs both its simple supports. The overhangs have a free end, like a cantilever, but the supports permit rotation.

Two types of beams that extend over several supports are illustrated in Figs. 6.18 and 6.19. Figure 6.18 shows a **continuous beam**. The one in Fig. 6.19 has one or two hinges in certain spans; it is called **hung-span**, or suspended-span, construction. In effect, it is a combination of simple beams and beams with overhangs.

Reactions for the beams in Figs. 6.13, 6.14, and 6.17 and the type of beam in Fig. 6.19 with hinges suitably located may be found from the equations of equilibrium, which is why they are classified as **statically determinate beams**.

The equations of equilibrium, however, are not sufficient to determine the reactions of the beams in Figs. 6.15, 6.16, and 6.18. For those beams, there are more unknowns than equations. Additional equations must be obtained based on a knowledge of the deformations, for example, that a fixed end permits no rotation. Such beams are classified as **statically indeterminate**. Methods for finding the stresses in that type of beam are given in Arts. 6.51 to 6.63.

6.20 Reactions

As pointed out in Art. 6.19, the loads imposed by a simple beam on its supports can be found by application of the equations of equilibrium [Eq. (6.1)]. Consider, for example, the 60-ft-long beam with overhangs in Fig. 6.20. This beam carries a uniform

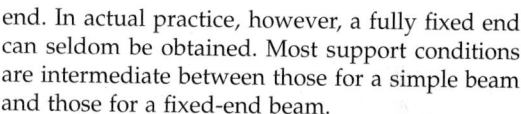

Fig. 6.19 Hung-span (suspended-span) construction.

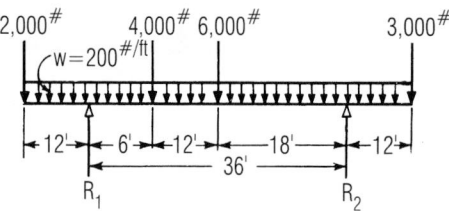

Fig. 6.20 Beam with overhangs loaded with both uniform and concentrated loads.

load of 200 lb/lin ft over its entire length and several concentrated loads. The span is 36 ft.

To find reaction R_1, take moments about R_2 and equate the sum of the moments to zero (assume clockwise rotation to be positive, counterclockwise, negative):

$$-2000 \times 48 + 36R_1 - 4000 \times 30 - 6000$$
$$\times 18 + 3000 \times 12 - 200 \times 60 \times 18 = 0$$
$$R_1 = 14,000 \, \text{lb}$$

In this calculation, the moment of the uniform load was found by taking the moment of its resultant, 200×60, which acts at the center of the beam.

To find R_2, proceed in a similar manner by taking moments about R_1 and equating the sum to zero, or equate the sum of the vertical forces to zero. Generally it is preferable to use the moment equation and apply the other equation as a check.

As an alternative procedure, find the reactions caused by uniform and concentrated loads separately and sum the results. Use the fact that the reactions due to symmetrical loading are equal, to simplify the calculation. To find R_2 by this procedure, take half the total uniform load

$$0.5 \times 200 \times 60 = 6000 \, \text{lb}$$

and add it to the reaction caused by the concentrated loads, found by taking moments about R_1, dividing by the span, and summing:

$$-2000 \times \frac{12}{36} + 4000 \times \frac{6}{36} + 6000 \times \frac{18}{36} + 3000$$
$$\times \frac{48}{36} = 7000 \, \text{lb}$$

$$R_2 = 6000 + 7000 = 13,000 \, \text{lb}$$

Check to see that the sum of the reactions equals the total applied load:

$$14,000 + 13,000 = 2000 + 4000 + 6000$$
$$+ 3000 + 200 \times 60$$
$$27,000 = 27,000$$

Reactions for simple beams with various loads are given in Figs. 6.33 to 6.38.

To find the reactions of a continuous beam, first determine the end moments and shears (Arts. 6.58 to 6.63); then if the continuous beam is considered as a series of simple beams with these applied as external loads, the beam will be statically determinate and the reactions can be determined from the equations of equilibrium. (For an alternative method, see Art. 6.57.)

6.21 Internal Forces

At every section of a beam in equilibrium, internal forces act to prevent motion. For example, assume the beam in Fig. 6.20 cut vertically just to the right of its center. Adding the external forces, including the reaction, to the left of this cut (see Fig. 6.21a) yields an unbalanced downward load of 4000 lb. Evidently, at the cut section, an upward-acting internal force of 4000 lb must be present to maintain equilibrium. Also, taking moments of the external forces about the section yields an unbalanced moment of 54,000 ft-lb. To maintain

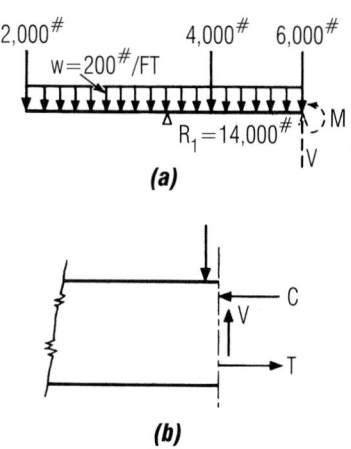

Fig. 6.21 Sections of beam kept in equilibrium by internal stresses.

equilibrium, there must be an internal moment of 54,000 ft-lb resisting it.

This internal, or resisting, moment is produced by a couple consisting of a force C acting on the top part of the beam and an equal but opposite force T acting on the bottom part (Fig. 6.21b). For this type of beam and loading, the top force is the resultant of compressive stresses acting over the upper portion of the beam, and the bottom force is the resultant of tensile stresses acting over the bottom part. The surface at which the stresses change from compression to tension—where the stress is zero—is called the **neutral surface**.

6.22 Shear Diagrams

As explained in Art. 6.21, at a vertical section through a beam in equilibrium, external forces on one side of the section are balanced by internal forces. The unbalanced external vertical force at the section is called the shear. It equals the algebraic sum of the forces that lie on either side of the section. For forces on the left of the section, those acting upward are considered positive and those acting downward negative. For forces on the right of the section, signs are reversed.

A shear diagram represents graphically the shear at every point along the length of a beam. The shear diagram for the beam in Fig. 6.20 is shown in Fig. 6.22b. The beam is drawn to scale and the loads and reactions are located at the points at which they act. Then, a convenient zero axis is drawn horizontally from which to plot the shears to scale. Start at the left end of the beam, and directly under the 2000-lb load there, scale off -2000 from the zero axis. Next, determine the shear just to the left of the next concentrated load, the left support: $-2000 - 200 \times 12 = -4400$ lb. Plot this downward under R_1. Note that in passing from just to the left of the support to just to the right, the shear changes by the magnitude of the reaction, from -4400 to $-4400 + 14,000$ or 9600 lb, so plot this value also under R_1. Under the 4000-lb load, plot the shear just to the left of it, $9600 - 200 \times 6$, or 8400 lb, and the shear just to the right, $8400 - 4000$, or 4400 lb. Proceed in this manner to the right end, where the shear is 3000 lb, equal to the load on the free end.

To complete the diagram, the points must be connected. Straight lines can be used because shear varies uniformly for a uniform load (see Fig. 6.24b)

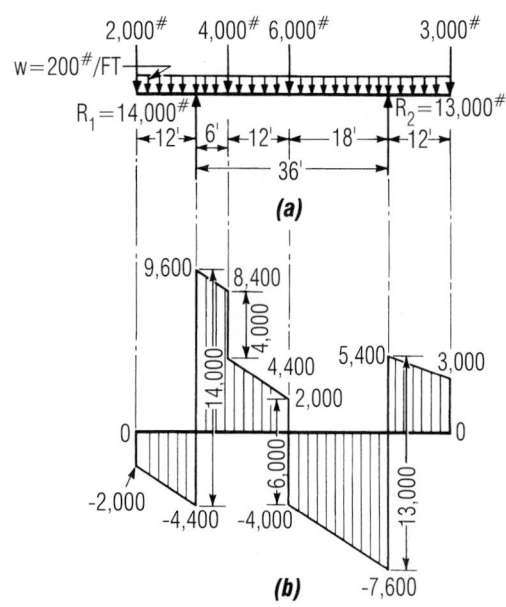

Fig. 6.22 Shear diagram for beam in Fig. 6.20.

6.23 Bending-Moment Diagrams

About a vertical section through a beam in equilibrium, there is an unbalanced moment due to external forces, called *bending moment*. For forces on the left of the section, clockwise moments are considered positive and counterclockwise moments negative. For forces on the right of the section, the signs are reversed. Thus, when the bending moment is positive, the bottom of a simple beam is in tension and the top is in compression.

A bending-moment diagram represents graphically the bending moment of every point along the length of the beam. Figure 6.23c is the bending-moment diagram for the beam with concentrated loads in Fig. 6.23a. The beam is drawn to scale, and the loads and reactions are located at the points at which they act. Then, a horizontal line is drawn to represent the zero axis from which to plot the bending moments to scale. Note that the bending moment at both supports for this simple beam is zero. Between the supports and the first load, the bending moment is proportional to the distance from the support since the bending moment in that region equals the reaction times the distance from the support. Hence, the bending-moment diagram for this portion of the beam is a sloping straight line.

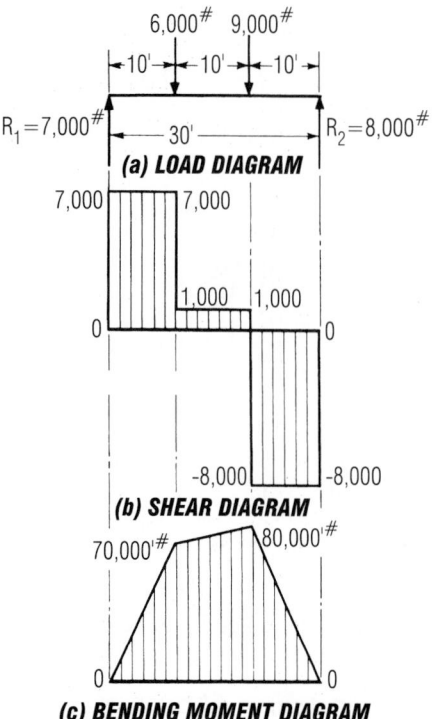

Fig. 6.23 Shear and moment diagrams for beam with concentrated loads.

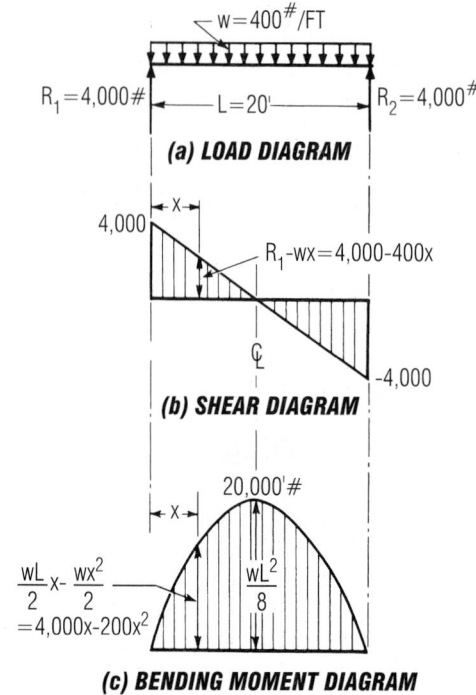

Fig. 6.24 Shear and moment diagrams for uniformly loaded beam.

To find the bending moment under the 6000-lb load, consider only the forces to the left of it, in this case only the reaction R_1. Its moment about the 6000-lb load is 7000 × 10, or 70,000 ft-lb. The bending-moment diagram, then, between the left support and the first concentrated load is a straight line rising from zero at the left end of the beam to 70,000, plotted, to a convenient scale, under the 6000-lb load.

To find the bending moment under the 9000-lb load, add algebraically the moments of the forces to its left: 7000 × 20 − 6000 × 10 = 80,000 ft-lb. (This result could have been obtained more easily by considering only the portion of the beam on the right, where the only force present is R_2, and reversing the sign convention: 8000 × 10 = 80,000 ft-lb.) Since there are no other loads between the 6000- and 9000-lb loads, the bending-moment diagram between them is a straight line.

If the bending moment and shear are known at any section, the bending moment at any other section can be computed if there are no un-known forces between the sections. The rule is:

The bending moment at any section of a beam equals the bending moment at any section to the left, plus the shear at that section times the distance between sections, minus the moments of intervening loads. If the section with known moment and shear is on the right, the sign convention must be reversed.

For example, the bending moment under the 9000-lb load in Fig. 6.23a also could have been determined from the moment under the 6000-lb load and the shear just to the right of that load. As indicated in the shear diagram (Fig. 6.23b), that shear is 1000 lb. Thus, the moment is given by 70,000 + 1000 × 10 = 80,000 ft-lb.

Bending-moment diagrams for simple beams with various loadings are shown in Figs. 6.33 to 6.38. To obtain bending-moment diagrams for loading conditions that can be represented as a sum of the loadings shown, sum the bending moments at corresponding locations on the beam as given on the diagram for the component loads.

For a simple beam carrying a uniform load, the bending-moment diagram is a parabola (Fig. 6.24c). The maximum moment occurs at the center and equals $wL^2/8$ or $WL/8$, where w is the load per linear foot and $W = wL$ is the total load on the beam.

The bending moment at any section of a simply supported, uniformly loaded beam equals one-half the product of the load per linear foot and the distances to the section from both supports:

$$M = \frac{w}{2}x(L - x) \qquad (6.41)$$

6.24 Shear-Moment Relationship

The slope of the bending-moment curve at any point on a beam equals the shear at that point. If V is the shear, M the moment, and x the distance along the beam,

$$V = \frac{dM}{dx} \qquad (6.42)$$

Since maximum bending moment occurs when the slope changes sign, or passes through zero, maximum moment (positive or negative) occurs at the point of zero shear.

Integration of Eq. (6.42) yields

$$M_1 - M_2 = \int_{x2}^{x1} V dx \qquad (6.43)$$

Thus, the change in bending moment between any two sections of a beam equals the area of the shear diagram between ordinates at the two sections.

6.25 Moving Loads and Influence Lines

Influence lines are a useful device for solving problems involving moving loads. An influence line indicates the effect at a given section of a unit load placed at any point on the structure.

For example, to plot the influence line for bending moment at a point on a beam, compute the moments produced at that point as a unit load moves along the beam and plot these moments under the corresponding positions of the unit load. Actually, the unit load need not be placed at every point along the beam. The equation of the influence line can be determined in many cases by placing

the load at an arbitrary point and computing the bending moment in general terms. (See also Art. 6.55.)

To draw the influence line for reaction at A for a simple beam AB (Fig. 6.25a), place a unit load at an arbitrary distance xL from B. The reaction at A due to this load is $1 \cdot xL/L = x$. Then, $R_A = x$ is the equation of the influence line. It represents a straight line sloping downward from unity at A, when the unit load is at that end of the beam, to zero at B, when the load is at B (Fig. 6.25a).

Figure 6.25b shows the influence line for bending moment at the center of a beam. It resembles in appearance the bending-moment diagram for a load at the center of the beam, but its significance is entirely different. Each ordinate gives the moment at midspan for a load at the location of the ordinate. The diagram indicates that if a unit load is placed at a distance xL from one end, it produces a bending moment of $xL/2$ at the center of the span.

Figure 6.25c shows the influence line for shear at the quarter point of a beam. When the load is to the right of the quarter point, the shear is positive and equal to the left reaction. When the load is to the left, the shear is negative and equals the right reaction. Thus, to produce maximum shear at the quarter point, loads should be placed only to the right of the quarter point, with the largest load at the quarter point, if possible. For a uniform load, maximum shear results when the load extends from the right end of the beam to the quarter point.

Suppose, for example, that a 60-ft crane girder is to carry wheel loads of 20 and 10 kips, 5 ft apart. For maximum shear at the quarter point, place the 20-kip wheel there and the 10-kip wheel 5 ft to the right. The corresponding ordinates of the influence line (Fig. 6.25c) are $\frac{3}{4}$ and $40/45 \times \frac{3}{4}$. Hence, the maximum shear is $20 \times \frac{3}{4} + 10 \times 40/45 \times \frac{3}{4} = 21.7$ kips.

Figure 6.25d shows influence lines for bending moment at several points on a beam. The apexes of the triangular diagrams fall on a parabola, as indicated by the dashed line. From the diagram, it can be concluded that the maximum moment produced at any section by a single concentrated load moving along a beam occurs when the load is at that section. And the magnitude of the maximum moment increases when the section is moved toward midspan, in accordance with the equation for the parabola given in Fig. 6.25d.

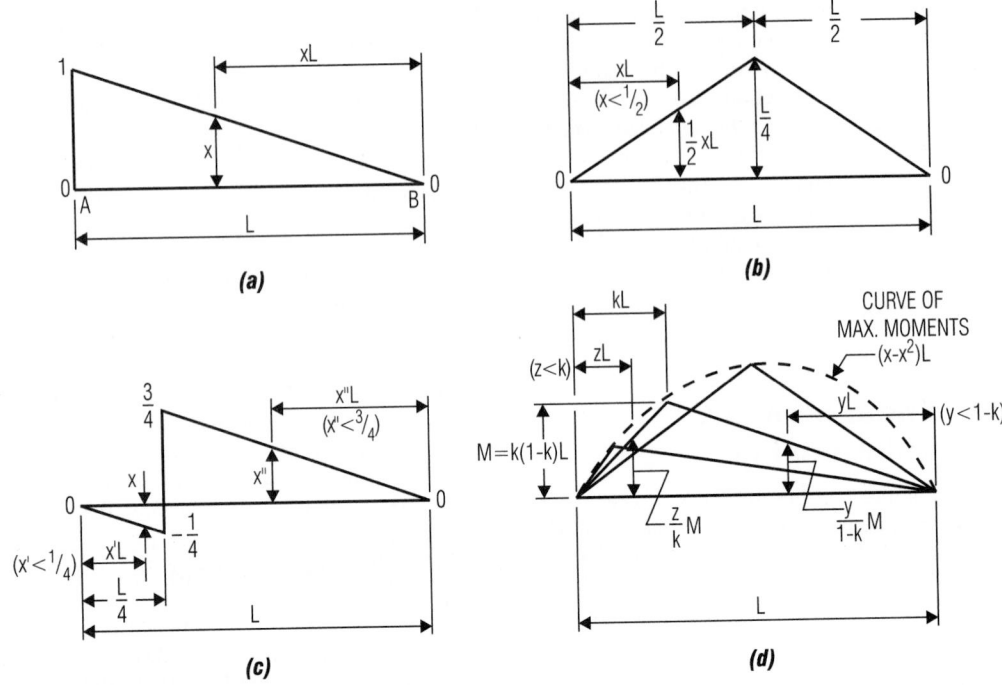

Fig. 6.25 Influence lines for (a) reaction at A, (b) midspan bending moment, (c) quarter-point shear, and (d) bending moments at several points in a beam.

6.26 Maximum Bending Moment

When a span is to carry several moving concentrated loads, an influence line is useful when determining the position of the loads for which bending moment is a maximum at a given section (see Art. 6.25). For a simple beam, maximum bending moment will occur at a section C as loads move across the beam when one of the loads is at C. The load to place at C is the one for which the expression $W_a/a - W_b/b$ (Fig. 6.26) changes sign as that load passes from one side of C to the other. (W_a is the sum of the loads on one side of C and W_b the sum of the loads on the other side of C.)

When several concentrated loads move along a simple beam, the maximum moment they produce in the beam may be near but not necessarily at midspan. To find the maximum moment, first determine the position of the loads for maximum moment at midspan. Then, shift the loads until the load P_2 (Fig. 6.27) that was at the center of the beam is as far from midspan as

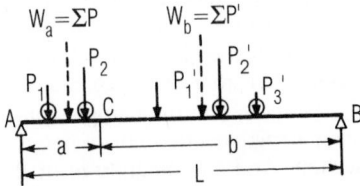

Fig. 6.26 Moving loads on simple beam AB placed for maximum moment at C.

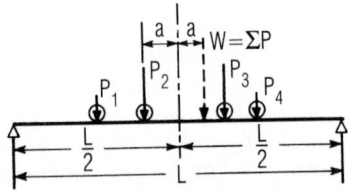

Fig. 6.27 Moving loads placed for maximum moment in a simple beam.

the resultant of all the loads on the span is on the other side of midspan. Maximum moment will occur under P_2. When other loads move on or off the span during the shift of P_2 away from midspan, it may be necessary to investigate the moment under one of the other loads when it and the new resultant are equidistant from midspan.

6.27 Bending Stresses in a Beam

The commonly used flexure formula for computing bending stresses in a beam is based on the following assumptions:

1. The unit stress parallel to the bending axis at any point of a beam is proportional to the unit strain in the same direction at the point. Hence, the formula holds only within the proportional limit.

2. The modulus of elasticity in tension is the same as that in compression.

3. The total and unit axial strain at any point are both proportional to the distance of that point from the neutral surface. (Cross sections that are plane before bending remain plane after bending. This requires that all fibers have the same length before bending, thus that the beam be straight.)

4. The loads act in a plane containing the centroidal axis of the beam and are perpendicular to that axis. Furthermore, the neutral surface is perpendicular to the plane of the loads. Thus, the plane of the loads must contain an axis of symmetry of each cross section of the beam. (The flexure formula does not apply to a beam with cross sections loaded unsymmetrically.)

5. The beam is proportioned to preclude prior failure or serious deformation by torsion, local buckling, shear, or any cause other than bending.

Equating the bending moment to the resisting moment due to the internal stresses at any section of a beam yields the **flexure formula**:

$$M = \frac{fI}{c} \qquad (6.44)$$

where M = bending moment at section, in-lb

f = normal unit stress at distance c, in, from the neutral axis (Fig. 6.28), psi

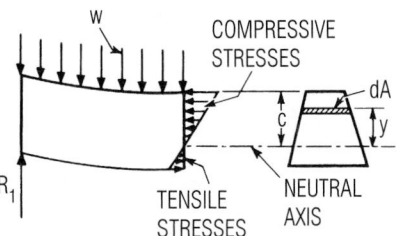

Fig. 6.28 Unit stresses on a beam section produced by bending.

I = moment of inertia of cross section with respect to neutral axis, in^4

Generally, c is taken as the distance to the outermost fiber to determine maximum f.

6.28 Moment of Inertia

The neutral axis in a symmetrical beam coincides with the centroidal axis; that is, at any section the neutral axis is so located that

$$\int y \, dA = 0 \qquad (6.45)$$

where dA is a differential area parallel to the axis (Fig. 6.28), y is its distance from the axis, and the summation is taken over the entire cross section.

Moment of inertia with respect to the neutral axis is given by

$$I = \int y^2 \, dA \qquad (6.46)$$

Values for I for several common cross sections are given in Fig. 6.29. Values for standard structural-steel sections are listed in manuals of the American Institute of Steel Construction. When the moments of inertia of other types of sections are needed, they can be computed directly by applying Eq. (6.46) or by breaking the section up into components for which the moment of inertia is known.

With the following formula, the moment of inertia of a section can be determined from that of its components:

$$I' = I + Ad^2 \qquad (6.47)$$

where I = moment of inertia of component about its centroidal axis, in^4

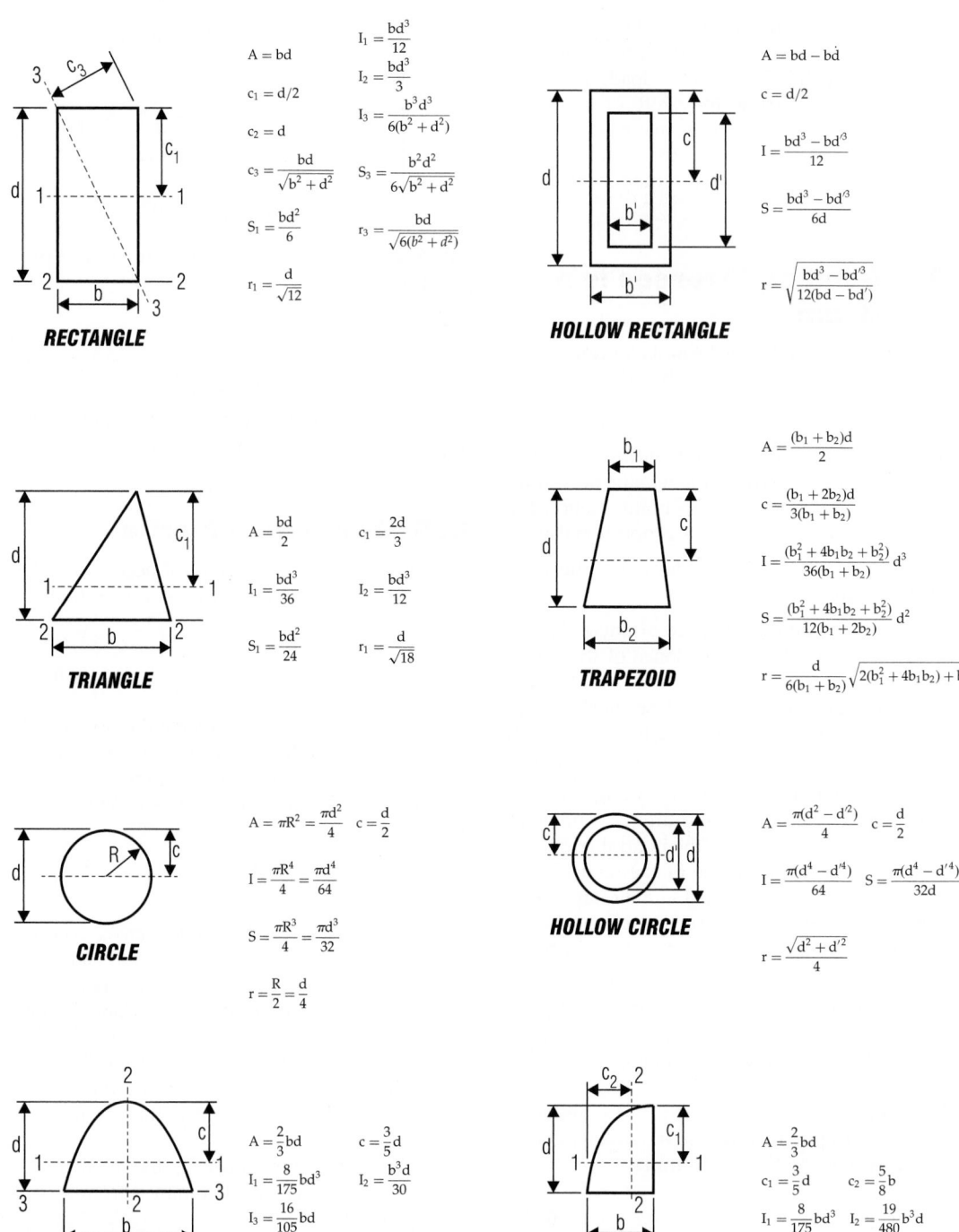

RECTANGLE

$$A = bd$$
$$c_1 = d/2$$
$$c_2 = d$$
$$c_3 = \frac{bd}{\sqrt{b^2 + d^2}}$$
$$S_1 = \frac{bd^2}{6}$$
$$r_1 = \frac{d}{\sqrt{12}}$$

$$I_1 = \frac{bd^3}{12}$$
$$I_2 = \frac{bd^3}{3}$$
$$I_3 = \frac{b^3 d^3}{6(b^2 + d^2)}$$
$$S_3 = \frac{b^2 d^2}{6\sqrt{b^2 + d^2}}$$
$$r_3 = \frac{bd}{\sqrt{6(b^2 + d^2)}}$$

HOLLOW RECTANGLE

$$A = bd - b'd'$$
$$c = d/2$$
$$I = \frac{bd^3 - b'd'^3}{12}$$
$$S = \frac{bd^3 - b'd'^3}{6d}$$
$$r = \sqrt{\frac{bd^3 - b'd'^3}{12(bd - b'd')}}$$

TRIANGLE

$$A = \frac{bd}{2} \qquad c_1 = \frac{2d}{3}$$
$$I_1 = \frac{bd^3}{36} \qquad I_2 = \frac{bd^3}{12}$$
$$S_1 = \frac{bd^2}{24} \qquad r_1 = \frac{d}{\sqrt{18}}$$

TRAPEZOID

$$A = \frac{(b_1 + b_2)d}{2}$$
$$c = \frac{(b_1 + 2b_2)d}{3(b_1 + b_2)}$$
$$I = \frac{(b_1^2 + 4b_1 b_2 + b_2^2)}{36(b_1 + b_2)} d^3$$
$$S = \frac{(b_1^2 + 4b_1 b_2 + b_2^2)}{12(b_1 + 2b_2)} d^2$$
$$r = \frac{d}{6(b_1 + b_2)}\sqrt{2(b_1^2 + 4b_1 b_2) + b_2^2}$$

CIRCLE

$$A = \pi R^2 = \frac{\pi d^2}{4} \qquad c = \frac{d}{2}$$
$$I = \frac{\pi R^4}{4} = \frac{\pi d^4}{64}$$
$$S = \frac{\pi R^3}{4} = \frac{\pi d^3}{32}$$
$$r = \frac{R}{2} = \frac{d}{4}$$

HOLLOW CIRCLE

$$A = \frac{\pi(d^2 - d'^2)}{4} \qquad c = \frac{d}{2}$$
$$I = \frac{\pi(d^4 - d'^4)}{64} \qquad S = \frac{\pi(d^4 - d'^4)}{32d}$$
$$r = \frac{\sqrt{d^2 + d'^2}}{4}$$

PARABOLA

$$A = \frac{2}{3}bd \qquad c = \frac{3}{5}d$$
$$I_1 = \frac{8}{175}bd^3 \qquad I_2 = \frac{b^3 d}{30}$$
$$I_3 = \frac{16}{105}bd$$

HALF PARABOLA

$$A = \frac{2}{3}bd$$
$$c_1 = \frac{3}{5}d \qquad c_2 = \frac{5}{8}b$$
$$I_1 = \frac{8}{175}bd^3 \qquad I_2 = \frac{19}{480}b^3 d$$

Fig. 6.29 Geometric properties of sections.

I' = moment of inertia of component about parallel axis, in^4

A = cross-sectional area of component, in^2

d = distance between centroidal and parallel axes, in

The formula enables computation of the moment of inertia of a component about the centroidal axis of a section from the moment of inertia about the component's centroidal axis, usually obtainable from Fig. 6.29 or the AISC manual. By summing up the transferred moments of inertia for all the components, the moment of inertia of the section is obtained.

When the moments of inertia of an area with respect to any two perpendicular axes are known, the moment of inertia with respect to any other axis passing through the point of intersection of the two axes may be obtained by using Mohr's circle as for stresses (Fig. 6.11). In this analog, I_x corresponds with f_x, I_y with f_y, and the **product of inertia** I_{xy} with v_{xy} (Art. 6.17)

$$I_{xy} = \int xy\, dA \qquad (6.48)$$

The two perpendicular axes through a point about which the moments of inertia are a maximum or a minimum are called the principal axes. The product of inertia is zero for the principal axes.

6.29 Section Modulus

The ratio $S = I/c$, relating bending moment and maximum bending stresses within the elastic range in a beam [Eq. (6.44)], is called the *section modulus*. I is the moment of inertia of the cross section about the neutral axis and c the distance from the neutral axis to the outermost fiber. Values of S for common types of sections are given in Fig. 6.29. Values for standard structural-steel sections are listed in manuals of the American Institute of Steel Construction.

6.30 Shearing Stresses in a Beam

Vertical shear at any section in a beam is resisted by nonuniformly distributed, vertical unit stresses (Fig. 6.30). At every point in the section, there also is a horizontal unit stress, which is equal in magnitude to the vertical unit shearing stress there [see Eq. (6.24)].

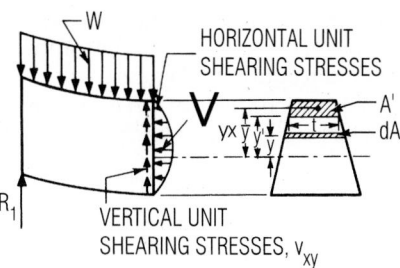

Fig. 6.30 Unit shearing stresses on a beam section.

At any distance y' from the neutral axis, both the horizontal and vertical shearing unit stresses are equal to

$$v = \frac{V}{It}A'\bar{y} \qquad (6.49)$$

where V = vertical shear at cross section, lb

t = thickness of beam at distance y' from neutral axis, in

I = moment of inertia of section about neutral axis, in^4

A' = area between outermost surface and surface for which shearing stress is being computed, in^2

$\bar{y}$ = distance of center of gravity of this area from neutral axis, in

For a rectangular beam, with width $t = b$ and depth d, the maximum shearing stress occurs at middepth. Its magnitude is

$$v = \frac{V}{(bd^3/12)b}\frac{bd}{2}\frac{d}{4} = \frac{3}{2}\frac{V}{bd}$$

That is, the maximum shear stress is 50% greater than the average shear stress on the section. Similarly, for a circular beam, the maximum is one-third greater than the average. For an I or wide-flange beam, however, the maximum shear stress in the web is not appreciably greater than the average for the web section alone, assuming that the flanges take no shear.

6.31 Combined Shear and Bending Stress

For deep beams on short spans and beams with low tensile strength, it sometimes is necessary to determine the maximum normal stress f' due to a combination of shear stress v and bending stress f. This maximum or principal stress (Art. 6.15) occurs on a plane inclined to that of v and of f. From Mohr's circle (Fig. 6.11) with $f = f_x, f_y = 0$, and $v = v_{xy}$,

$$f' = \frac{f}{2} + \sqrt{v^2 + \left(\frac{f}{2}\right)^2} \qquad (6.50)$$

6.32 Beam Deflections

The **elastic curve** is the position taken by the longitudinal centroidal axis of a beam when it deflects under load. The radius of curvature at any point of this curve is

$$R = \frac{EI}{M} \qquad (6.51)$$

where M = bending moment at point

E = modulus of elasticity

I = moment of inertia of cross section about neutral axis

Since the slope of the elastic curve is very small, $1/R$ is approximately d^2y/dx^2, where y is the deflection of the beam at a distance x from the origin of coordinates. Hence, Eq. (6.51) may be rewritten

$$M = EI\frac{d^2y}{dx^2} \qquad (6.52)$$

To obtain the slope and deflection of a beam, this equation may be integrated, with M expressed as a function of x. Constants introduced during the integration must be evaluated in terms of known points and slopes of the elastic curve.

After integration, Eq. (6.52) yields

$$\theta_B - \theta_A = \int_A^B \frac{M}{EI}\,dx \qquad (6.53)$$

in which θ_A and θ_B are the slopes of the elastic curve at any two points A and B. If the slope is zero at one of the points, the integral in Eq. (6.53) gives the slope of the elastic curve at the other. The integral represents the area of the bending-moment diagram between A and B with each ordinate divided by EI.

The **tangential deviation** t of a point on the elastic curve is the distance of this point, measured

in a direction perpendicular to the original position of the beam, from a tangent drawn at some other point on the curve.

$$t_B - t_A = \int_A^B \frac{Mx}{EI}\,dx \qquad (6.54)$$

Equation (6.54) indicates that the tangential deviation of any point with respect to a second point on the elastic curve equals the moment about the first point of the area of the M/EI diagram between the two points. The moment-area method for determining beam deflections is a technique employing Eqs. (6.53) and (6.54).

Moment-Area Method ■ Suppose, for example, the deflection at midspan is to be computed for a beam of uniform cross section with a concentrated load at the center (Fig. 6.31). Since the deflection at midspan for this loading is the maximum for the span, the slope of the elastic curve at midspan is zero; that is, the tangent is parallel to the undeflected position of the beam. Hence, the deviation of either support from the midspan tangent equals the deflection at the center of the beam. Then, by the moment-area theorem [Eq. (6.54)], the deflection y_c is given by the moment about either support of the area of the M/EI diagram included between an ordinate at the center of the beam and that support

$$y_c = \left(\frac{1}{2}\frac{PL}{4EI}\frac{L}{2}\right)\frac{L}{3} = \frac{PL^3}{48EI}$$

Suppose now that the deflection y at any point D at a distance xL from the left support (Fig. 6.31) is to be determined. Note that from similar triangles, $xL/L = DE/t_{AB}$, where DE is the distance from the undeflected position of D to the tangent to the elastic curve at support A, and t_{AB} is the tangential deviation of B from that tangent. But DE also equals $y + t_{AD}$, where t_{AD} is the tangential deviation of D from the tangent at A. Hence,

$$y + t_{AD} = xt_{AB}$$

This equation is perfectly general for the deflection of any point of a simple beam, no matter how loaded. It may be rewritten to give the deflection directly:

$$y = xt_{AB} - t_{AD} \qquad (6.55)$$

But t_{AB} is the moment of the area of the M/EI diagram for the whole beam about support B, and

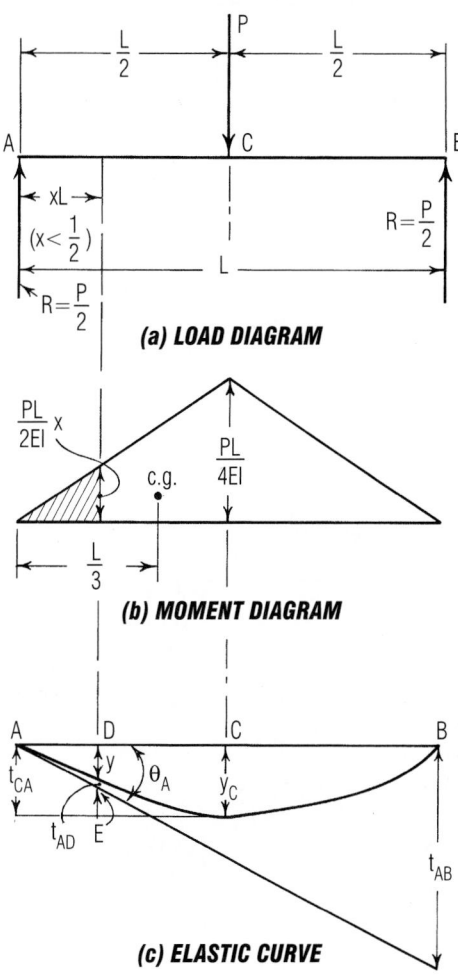

(a) LOAD DIAGRAM

(b) MOMENT DIAGRAM

(c) ELASTIC CURVE

Fig. 6.31 Elastic curve for a simple beam and tangential deviations at ends.

t_{AD} is the moment about D of the area of the M/EI diagram included between ordinates at A and D. So at any point x of the beam in Fig. 6.31, the deflection is

$$y = x \left[\frac{1}{2} \frac{PL}{4EI} \frac{L}{2} \left(\frac{L}{3} + \frac{2L}{3} \right) \right]$$

$$- \frac{1}{2} \frac{PLx}{2EI} (xL) \frac{xL}{3} = \frac{PL^3}{48EI} x(3 - 4x^2)$$

It also is noteworthy that, since the tangential deviations are very small distances, the slope of the elastic curve at A is given by

$$\theta_A = \frac{t_{AB}}{L} \tag{6.56}$$

This holds, in general, for all simple beams regardless of the type of loading.

Conjugate-Beam Method ▪ The procedure followed in applying Eq. (6.55) to the deflection of the loaded beam in Fig. 6.31 is equivalent to finding the bending moment at D with the M/EI diagram serving as the load diagram. The technique of applying the M/EI diagram as a load and determining the deflection as a bending moment is known as the conjugate-beam method.

The conjugate beam must have the same length as the given beam; it must be in equilibrium with the M/EI load and the reactions produced by the load; and the bending moment at any section must be equal to the deflection of the given beam at the corresponding section. The last requirement is equivalent to specifying that the shear at any section of the conjugate beam with the M/EI load be equal to the slope of the elastic curve at the corresponding section of the given beam. Figure 6.32 shows the conjugates for various types of beams.

Deflection Computations ▪ Deflections for several types of loading on simple beams are given in Figs. 6.33 and 6.35 to 6.38 and for cantilevers and beams with overhangs in Figs. 6.39 to 6.44.

When a beam carries several different types of loading, the most convenient method of computing its deflection usually is to find the deflections separately for the uniform and concentrated loads and add them.

For several concentrated loads, the easiest method of obtaining the deflection at a point on a beam is to apply the reciprocal theorem (Art. 6.55). According to this theorem, if a concentrated load is applied to a beam at a point A, the deflection the load produces at point B equals the deflection at A for the same load applied at B ($d_{AB} = d_{BA}$). So place the loads one at a time at the point for which the deflection is to be found, and from the equation of the elastic curve determine the deflections at the actual location of the loads. Then, sum these deflections.

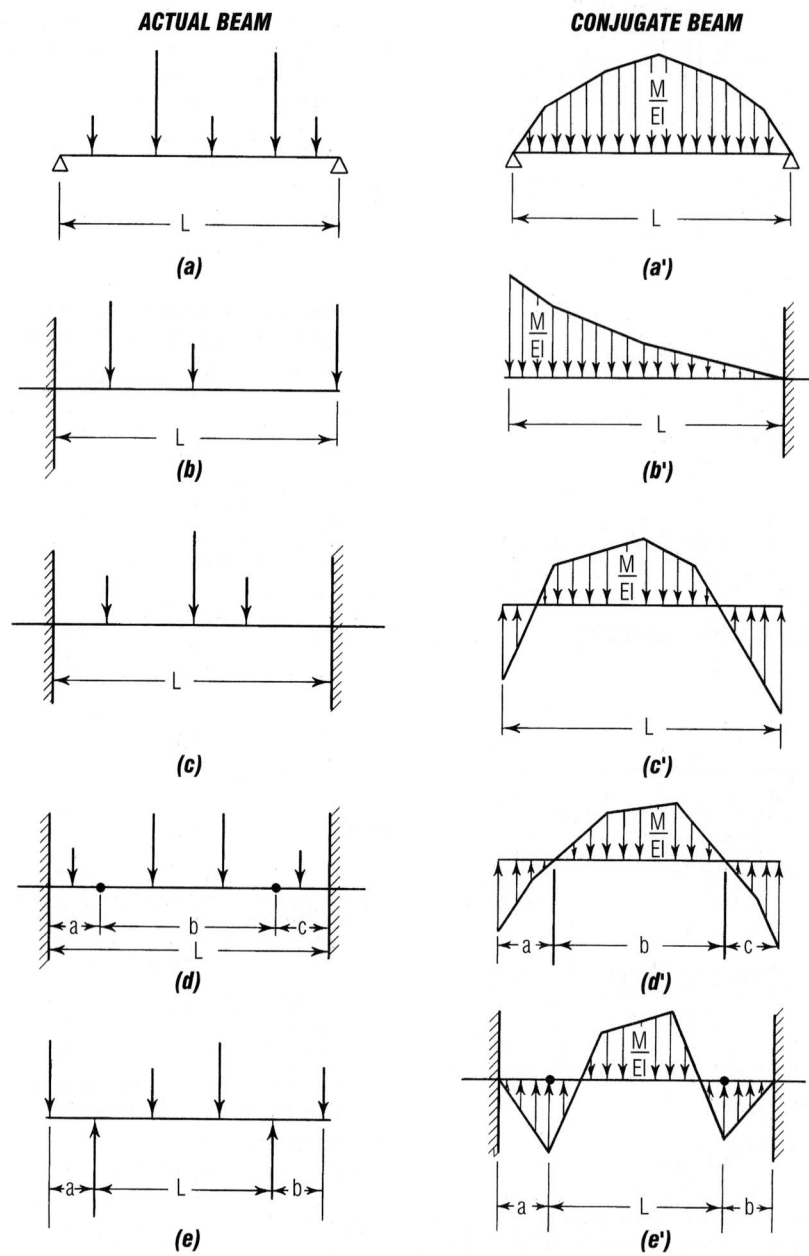

Fig. 6.32 Conjugate beams.

Suppose, for example, the midspan deflection is to be computed. Assume each load in turn applied at the center of the beam and compute the deflection at the point where it originally was applied from the equation of the elastic curve given in Fig. 6.36. The sum of these deflections is the total midspan deflection.

Another method for computing deflections is presented in Art. 6.54. This method also may be used to determine the deflection of a beam due to shear.

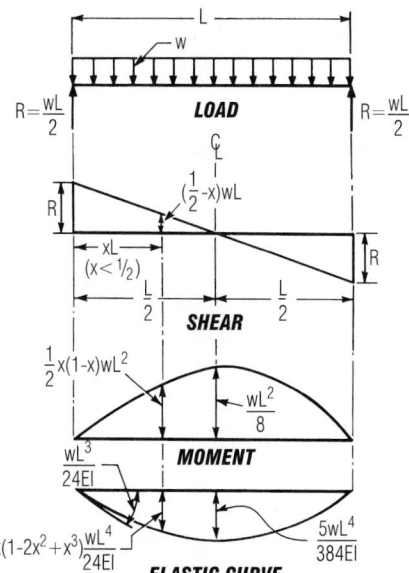

Fig. 6.33 Shears, moments, and deflections for full uniform load on a simply supported, prismatic beam.

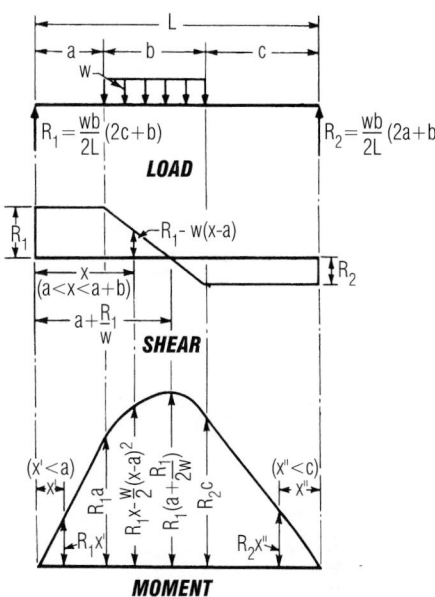

Fig. 6.34 Shears and moments for a uniformly distributed load over part of a simply supported beam.

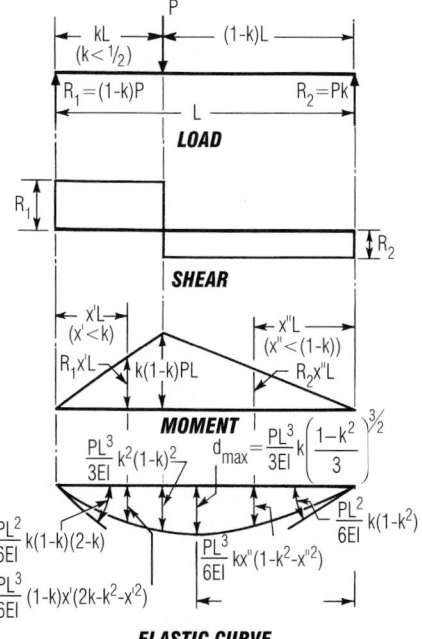

Fig. 6.35 Shears, moments, and deflections for a concentrated load at any point of a simply supported, prismatic beam.

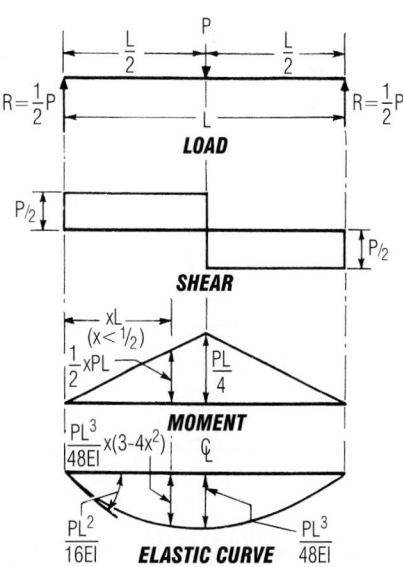

Fig. 6.36 Shears, moments, and deflections for a concentrated load at midspan of a simply supported, prismatic beam.

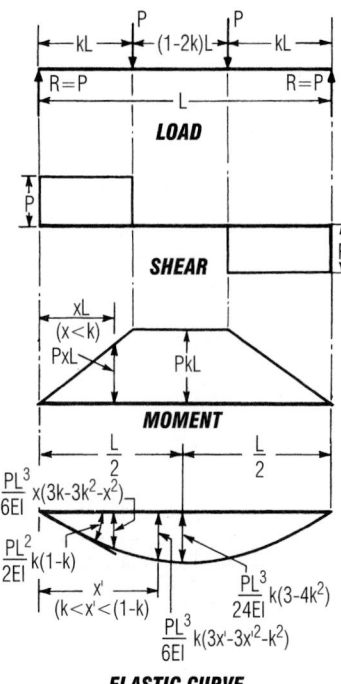

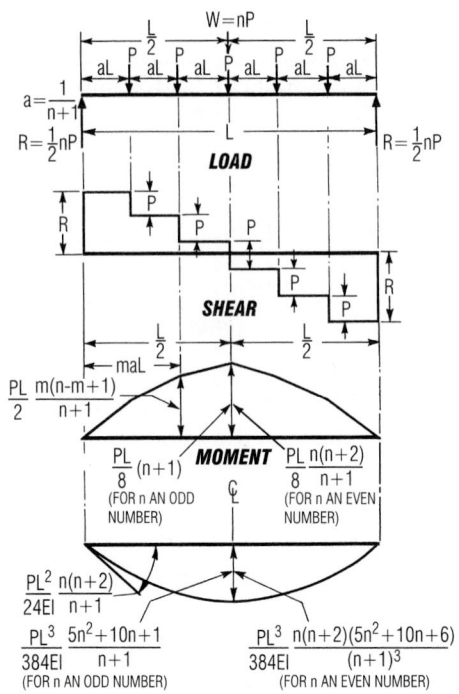

Fig. 6.37 Shears, moments, and deflections for two equal concentrated loads on a simply supported, prismatic beam.

Fig. 6.38 Shears, moments, and deflections for several equal loads equally spaced on a simply supported, prismatic beam.

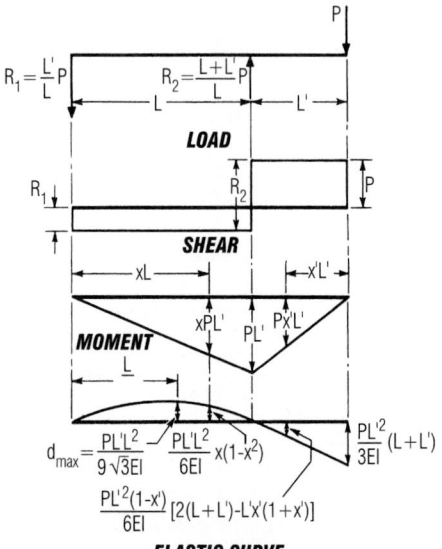

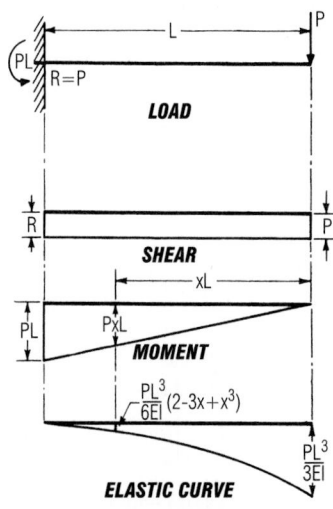

Fig. 6.39 Shears, moments, and deflections for a concentrated load on a beam overhang.

Fig. 6.40 Shears, moments, and deflections for a concentrated load on the end of a prismatic cantilever.

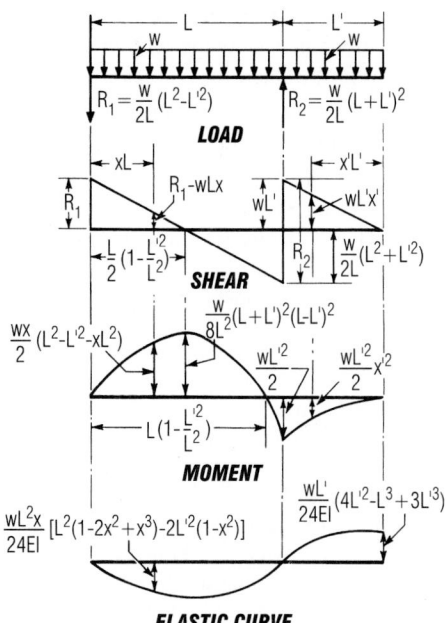

Fig. 6.41 Shears, moments, and deflections for a uniform load over a beam with overhang.

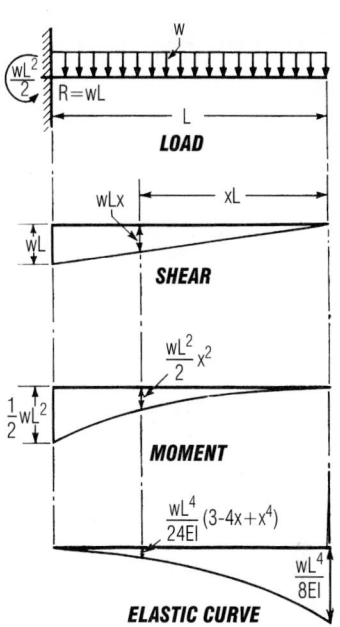

Fig. 6.42 Shears, moments, and deflections for a uniform load over the length of a cantilever.

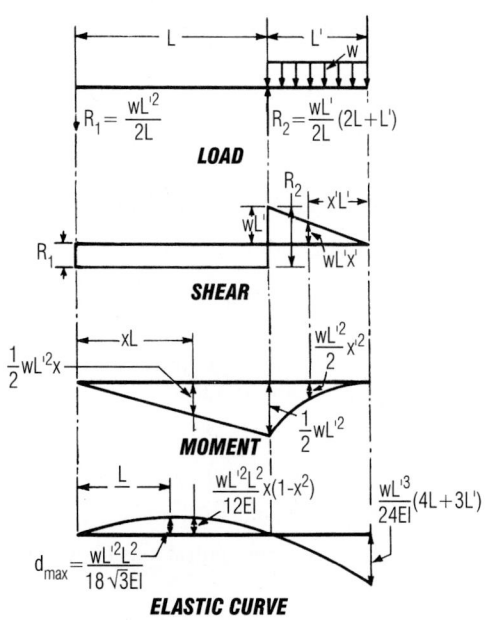

Fig. 6.43 Shears, moments, and deflections for a uniform load on a beam overhang.

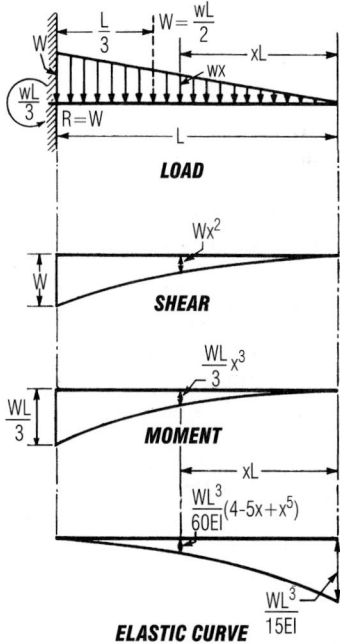

Fig. 6.44 Shears, moments, and deflections for a triangular loading on a prismatic cantilever.

6.33 Unsymmetrical Bending

When a beam is subjected to loads that do not lie in a plane containing a principal axis of each cross section, unsymmetrical bending occurs. Assuming that the bending axis of the beam lies in the plane of the loads, to preclude torsion (see Art. 6.36), and that the loads are perpendicular to the bending axis, to preclude axial components, the stress, psi, at any point in a cross section is

$$f = \frac{M_x y}{I_x} \pm \frac{M_y x}{I_y} \qquad (6.57)$$

where M_x = bending moment about principal axis XX, in-lb

M_y = bending moment about principal axis YY, in-lb

x = distance from point where stress is to be computed to YY axis, in

y = distance from point to XX, in

I_x = moment of inertia of cross section about XX, in^4

I_y = moment of inertia about YY, in^4

If the plane of the loads makes an angle θ with a principal plane, the neutral surface will form an angle α with the other principal plane such that

$$\tan \alpha = \frac{I_x}{I_y} \tan \theta$$

6.34 Combined Axial and Bending Loads

For short beams, subjected to both transverse and axial loads, the stresses are given by the principle of superposition if the deflection due to bending may be neglected without serious error. That is, the total stress is given with sufficient accuracy at any section by the sum of the axial stress and the bending stresses. The maximum stress, psi, equals

$$f = \frac{P}{A} + \frac{Mc}{I} \qquad (6.58a)$$

where P = axial load, lb

A = cross-sectional area, in^2

M = maximum bending moment, in-lb

c = distance from neutral axis to outermost fiber at section where maximum moment occurs, in

I = moment of inertia about neutral axis at that section, in^4

When the deflection due to bending is large and the axial load produces bending stresses that cannot be neglected, the maximum stress is given by

$$f = \frac{P}{A} + (M + Pd)\frac{c}{I} \qquad (6.58b)$$

where d is the deflection of the beam. For axial compression, the moment Pd should be given the same sign as M, and for tension, the opposite sign, but the minimum value of $M + Pd$ is zero. The deflection d for axial compression and bending can be obtained by applying Eq. (6.52).

(S. Timoshenko and J. M. Gere, "Theory of Elastic Stability," McGraw-Hill Book Company, New York, books.mcgraw-hill.com; Friedrich Bleich, "Buckling Strength of Metal Structures," McGraw-Hill Book Company, New York, books. mcgraw-hill.com.) But it may be closely approximated by

$$d = \frac{d_o}{1 - (P/P_c)} \qquad (6.59)$$

where d_o = deflection for transverse loading alone, in

P_c = critical buckling load, $\pi^2 EI/L^2$ (see Art. 6.39), lb

6.35 Eccentric Loading

If an eccentric longitudinal load is applied to a bar in the plane of symmetry, it produces a bending moment Pe, where e is the distance, in, of the load P from the centroidal axis. The total unit stress is the sum of the stress due to this moment and the stress due to P applied as an axial load:

$$f = \frac{P}{A} \pm \frac{Pec}{I} = \frac{P}{A}\left(1 \pm \frac{ec}{r^2}\right) \qquad (6.60)$$

where A = cross-sectional area, in^2

c = distance from neutral axis to outermost fiber, in

I = moment of inertia of cross section about neutral axis, in^4

r = **radius of gyration** = $\sqrt{I/A}$, in

Figure 6.29 gives values of the radius of gyration for several cross sections.

If there is to be no tension on the cross section under a compressive load, e should not exceed r^2/c. For a rectangular section with width b and depth d, the eccentricity, therefore, should be less than $b/6$ and $d/6$; i.e., the load should not be applied outside the middle third. For a circular cross section with diameter D, the eccentricity should not exceed $D/8$.

When the eccentric longitudinal load produces a deflection too large to be neglected in computing the bending stress, account must be taken of the additional bending moment Pd, where d is the deflection, in. This deflection may be computed by using Eq. (6.52) or closely approximated by

$$d = \frac{4eP/P_c}{\pi(1 - P/P_c)} \qquad (6.61)$$

P_c is the critical buckling load $\pi^2 EI/L^2$ (see Art. 6.39), lb.

If the load P does not lie in a plane containing an axis of symmetry, it produces bending about the two principal axes through the centroid of the section. The stresses, psi, are given by

$$f = \frac{P}{A} \pm \frac{Pe_x c_x}{I_y} \pm \frac{Pe_y c_y}{I_x} \qquad (6.62)$$

where A = cross-sectional area in^2

e_x = eccentricity with respect to principal axis YY, in

e_y = eccentricity with respect to principal axis XX, in

c_x = distance from YY to outermost fiber, in

c_y = distance from XX to outermost fiber, in

I_x = moment of inertia about XX, in^4

I_y = moment of inertia about YY, in^4

The principal axes are the two perpendicular axes through the centroid for which the moments of inertia are a maximum or a minimum and for which the products of inertia are zero.

6.36 Beams with Unsymmetrical Sections

The derivation of the flexure formula $f = Mc/I$ (Art. 6.27) assumes that a beam bends, without twisting, in the plane of the loads and that the

neutral surface is perpendicular to the plane of the loads. These assumptions are correct for beams with cross sections symmetrical about two axes when the plane of the loads contains one of these axes. They are not necessarily true for beams that are not doubly symmetrical because in beams that are doubly symmetrical, the bending axis coincides with the centroidal axis, whereas in unsymmetrical sections the two axes may be separate. In the latter case, if the plane of the loads contains the centroidal axis but not the bending axis, the beam will be subjected to both bending and torsion.

The **bending axis** is the longitudinal line in a beam through which transverse loads must pass to preclude the beam's twisting as it bends. The point in each section through which the bending axis passes is called the **shear center**, or center of twist. The shear center also is the center of rotation of the section in pure torsion (Art. 6.18). Its location depends on the dimensions of the section.

Computation of stresses and strains in members subjected to both bending and torsion is complicated, because warping of the cross section and buckling may occur and should be taken into account. Such computations may not be necessary if twisting is prevented by use of bracing or avoided by selecting appropriate shapes for the members and by locating and directing loads to pass through the bending axis.

(F. Bleich, "Buckling Strength of Metal Structures," McGraw-Hill Book Company, New York, books.mcgraw-hill.com.)

Curved Beams

Structural members, such as arches, crane hooks, chain links, and frames of some machines, that have considerable initial curvature in the plane of loading are called curved beams. The flexure formula of Art. 6.27, $f = Mc/I$, cannot be applied to them with any reasonable degree of accuracy unless the depth of the beam is small compared with the radius of curvature.

Unlike the condition in straight beams, unit strains in curved beams are not proportional to the distance from the neutral surface, and the centroidal axis does not coincide with the neutral axis. Hence the stress distribution on a section is not linear but more like the distribution shown in Fig. 6.45c.

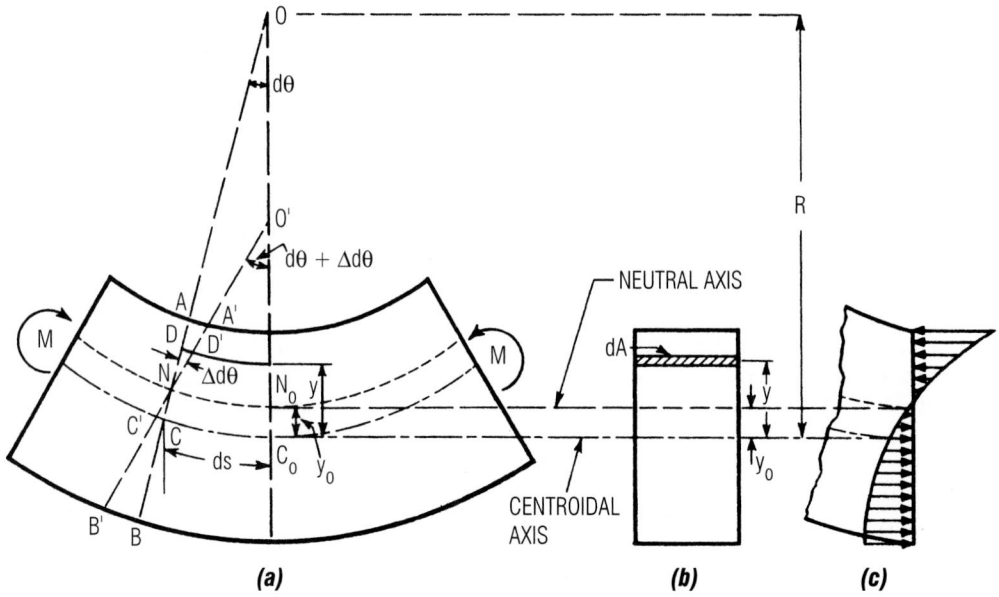

Fig. 6.45 Bending stresses in a curved beam.

6.37 Stresses in Curved Beams

Just as for straight beams, the assumption that plane sections before bending remain plane after bending generally holds for curved beams. So the total strains are proportional to the distance from the neutral axis. But since the fibers are initially of unequal length, the unit strains are a more complex function of this distance. In Fig. 6.45a, for example, the bending couples have rotated section AB of the curved beam into section $A'B'$ through an angle $\Delta d\theta$. If ε_o is the unit strain at the centroidal axis and ω is the angular unit strain $\Delta d\theta/d\theta$, then if M is the bending moment:

$$\varepsilon_o = \frac{M}{ARE} \quad \text{and} \quad \omega = \frac{M}{ARE}\left(1 + \frac{AR^2}{I'}\right) \quad (6.63)$$

where A is the cross-sectional area, E the modulus of elasticity, and

$$I' = \int \frac{y^2\, dA}{1 - y/R} = \int y^2\left(1 + \frac{y}{R} + \frac{y^2}{R^2} + \cdots\right) dA \quad (6.64)$$

It should be noted that I' is very nearly equal to the moment of inertia I about the centroidal axis when the depth of the section is small compared with R, so that the maximum ratio of y to R is small compared with unity. M is positive when it decreases the radius of curvature.

The stresses in the curved beam can be obtained from Fig. 6.45a with the use of ε_o and ω from Eq. (6.63):

$$f = \frac{M}{A\,R} - \frac{My}{I'}\frac{1}{1 - y/R} \quad (6.65)$$

Equation (6.65) for bending stresses in curved beams subjected to end moments in the plane of curvature can be expressed for the inside and outside beam faces in the form:

$$f = \frac{Mc}{I}K \quad (6.66)$$

where c = distance from the centroidal axis to the inner or outer surface. Table 6.1 gives values of K calculated from Eq. (6.66) for circular, elliptical, and rectangular cross sections.

Table 6.1 Values of K for Curved Beams

Section	R/c	K Inside face	K Outside face	y_o/R
CIRCLE / ELLIPSE	1.2	3.41	0.54	0.224
	1.4	2.40	0.60	0.151
	1.6	1.96	0.65	0.108
	1.8	1.75	0.68	0.084
	2.0	1.62	0.71	0.069
	3.0	1.33	0.79	0.030
	4.0	1.23	0.84	0.016
	6.0	1.14	0.89	0.0070
	8.0	1.10	0.91	0.0039
	10.0	1.08	0.93	0.0025
	1.2	3.28	0.58	0.269
	1.4	2.31	0.64	0.182
	1.6	1.89	0.68	0.134
	1.8	1.70	0.71	0.104
	2.0	1.57	0.73	0.083
	3.0	1.31	0.81	0.038
	4.0	1.21	0.85	0.020
	6.0	1.13	0.90	0.0087
	8.0	1.10	0.92	0.0049
	10.0	1.07	0.93	0.0031
	1.2	2.89	0.57	0.305
	1.4	2.13	0.63	0.204
	1.6	1.79	0.67	0.149
	1.8	1.63	0.70	0.112
	2.0	1.52	0.73	0.090
	3.0	1.30	0.81	0.041
	4.0	1.20	0.85	0.021
	6.0	1.12	0.90	0.0093
	8.0	1.09	0.92	0.0052
	10.0	1.07	0.94	0.0033

If Eq. (6.65) is applied to I or T beams or tubular members, it may indicate circumferential flange stresses that are much lower than will actually occur. The error is due to the fact that the outer edges of the flanges deflect radially. The effect is equivalent to having only part of the flanges active in resisting bending stresses. Also, accompanying the flange deflections, there are transverse bending stresses in the flanges. At the junction with the web, these reach a maximum, which may be greater than the maximum circumferential stress. Furthermore, there are radial stresses (normal stresses acting in the direction of the radius of curvature) in the web that also may have maximum values greater than the maximum circumferential stress.

If a curved beam carries an axial load P as well as bending loads, the maximum unit stress is

$$f = \frac{P}{A} \pm \frac{Mc}{I} K \qquad (6.67)$$

M is taken positive in this equation when it increases the curvature, and P is positive when it is a tensile force, negative when compressive.

6.38 Slope and Deflection of Curved Beams

If we consider two sections of a curved beam separated by a differential distance ds (Fig. 6.45a), the change in angle $\Delta d\theta$ between the sections caused by a bending moment M and an axial load P may be obtained from Eq. (6.63), noting that $d\theta = ds/R$.

$$\Delta d\theta = \frac{M\,ds}{EI'}\left(1 + \frac{I'}{A\,R^2}\right) + \frac{P\,ds}{A\,RE} \qquad (6.68)$$

where E is the modulus of elasticity, A the cross-sectional area, R the radius of curvature of the centroidal axis, and I' is defined by Eq. (6.64).

If P is a tensile force, the length of the centroidal axis increases by

$$\Delta ds = \frac{P\,ds}{A\,E} + \frac{M\,ds}{A\,RE} \qquad (6.69)$$

The effect of curvature on shearing deformations for most practical applications is negligible.

For shallow sections (depth of section less than about one-tenth the span), the effect of axial forces on deformations may be neglected. Also, unless the radius of curvature is very small compared with the depth, the effect of curvature may be ignored. Hence, for most practical applications, Eq. (6.68) may be used in the simplified form:

$$\Delta d\theta = \frac{M\,ds}{EI} \qquad (6.70)$$

For deeper beams, the action of axial forces, as well as bending moments, should be taken into account; but unless the curvature is sharp, its effect on deformations may be neglected. So only Eq. (6.70) and the first term in Eq. (6.69) need be used.

(S. Timoshenko and D. H. Young, "Theory of Structures," McGraw-Hill Publishing Company, New York, books.mcgraw-hill.com.) See also Arts. 6.69 and 6.70.

Buckling of Columns

Columns are compression members whose cross-sectional dimensions are small compared with their length in the direction of the compressive force. Failure of such members occurs because of instability when a certain load (called the critical or **Euler load**) is equaled or exceeded. The member may bend, or buckle, suddenly and collapse.

Hence, the strength of a column is determined not by the unit stress in Eq. (6.6) ($P = Af$) but by the maximum load it can carry without becoming unstable. The condition of instability is characterized by disproportionately large increases in lateral deformation with slight increase in load. It may occur in slender columns before the unit stress reaches the elastic limit.

6.39 Equilibrium of Columns

Figure 6.46 represents an axially loaded column with ends unrestrained against rotation. If the member is initially perfectly straight, it will remain straight as long as the load P is less than the critical load P_c (also called Euler load). If a small transverse force is applied, it will deflect, but it will return to

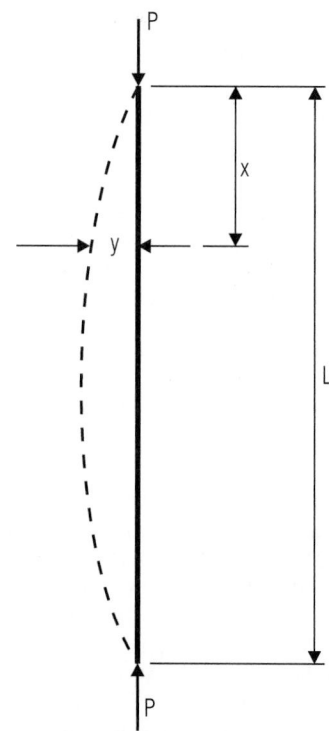

Fig. 6.46 Buckling of a column.

the straight position when this force is removed. Thus, when P is less than P_c, internal and external forces are in stable equilibrium.

If $P = P_c$ and a small transverse force is applied, the column again will deflect, but this time, when the force is removed, the column will remain in the bent position (dashed line in Fig. 6.46).

The equation of this elastic curve can be obtained from Eq. (6.52):

$$EI\frac{d^2y}{dx^2} = -P_c y \tag{6.71}$$

in which E = modulus of elasticity, psi

I = least moment of inertia of cross section, in^4

y = deflection of bent member from straight position at distance x from one end, in

This assumes that the stresses are within the elastic limit.

Solution of Eq. (6.71) gives the smallest value of the Euler load as

$$P_c = \frac{\pi^2 EI}{L^2} \tag{6.72}$$

Equation (6.72) indicates that there is a definite magnitude of an axial load that will hold a column in equilibrium in the bent position when the stresses are below the elastic limit. Repeated application and removal of small transverse forces or small increases in axial load above this critical load will cause the member to fail by buckling. Internal and external forces are in a state of unstable equilibrium.

It is noteworthy that the Euler load, which determines the load-carrying capacity of a column, depends on the stiffness of the member, as expressed by the modulus of elasticity, rather than on the strength of the material of which it is made.

By dividing both sides of Eq. (6.72) by the cross-sectional area A, in^2, and substituting r^2 for I/A (r is the radius of gyration of the section), we can write the solution of Eq. (6.71) in terms of the average unit stress on the cross section:

$$\frac{P_c}{A} = \frac{\pi^2 E}{(L/r)^2} \tag{6.73}$$

This holds only for the elastic range of buckling, that is, for values of the **slenderness ratio** L/r

above a certain limiting value that depends on the properties of the material.

Effects of End Conditions ▪ Equation (6.73) was derived on the assumption that the ends of the columns are free to rotate. It can be generalized, however, to take into account the effect of end conditions:

$$\frac{P_c}{A} = \frac{\pi^2 E}{(kL/r)^2} \tag{6.74}$$

where k is a factor that depends on the end conditions. For a pin-ended column, $k = 1$; for a column with both ends fixed, $k = \frac{1}{2}$; for a column with one end fixed and one end pinned, k is about 0.7; and for a column with one end fixed and one end free from all restraint, $k = 2$. When a column has different restraints or different radii of gyration about its principal axes, the largest value of kL/r for a principal axis should be used in Eq. (6.74).

Inelastic Buckling ▪ Equations (6.72) to (6.74), having been derived from Eq. (6.71), the differential equation for the elastic curve, are based on the assumption that the critical average stress is below the elastic limit when the state of unstable equilibrium is reached. In members with slenderness ratio L/r below a certain limiting value, however, the elastic limit is exceeded before the column buckles. As the axial load approaches the critical load, the modulus of elasticity varies with the stress. Hence, Eqs. (6.72) to (6.74), based on the assumption that E is a constant, do not hold for these short columns.

After extensive testing and analysis, prevalent engineering opinion favors the Engesser equation for metals in the inelastic range:

$$\frac{P_t}{A} = \frac{\pi^2 E_t}{(kL/r)^2} \tag{6.75}$$

This differs from Eq. (6.74) only in that the tangent modulus E_t (the actual slope of the stress-strain curve for the stress P_t/A) replaces E, the modulus of elasticity in the elastic range. P_t is the smallest axial load for which two equilibrium positions are possible, the straight position and a deflected position.

Another solution to the inelastic-buckling problem is called the double modulus method, in which the bending stiffness of the cross section is expressed in terms of E_t and E, representing the loading and unloading portions of materials on the cross section respectively. The critical stress obtained is higher than that of the Engesser equation.

Eccentric Loading ▪ Under eccentric loading, the maximum unit stress in short compression members is given by Eqs. (6.60) and (6.62), with the eccentricity e increased by the deflection given by Eq. (6.61). For columns, the stress within the elastic range is given by the **secant formula**:

$$f = \frac{P}{A}\left(1 + \frac{ec}{r^2}\sec\frac{kL}{2r}\sqrt{\frac{P}{AE}}\right) \qquad (6.76)$$

When the slenderness ratio L/r is small, the formula approximates Eq. (6.60).

6.40 Column Curves

The result of plotting the critical stress in columns for various values of slenderness ratios (Art. 6.39) is called a column curve. For axially loaded, initially straight columns, it consists of two parts: the Euler critical values [Eq. (6.73)] and the Engresser, or tangent-modulus, critical values [Eq. (6.75)], with $k = 1$.

The second part of the curve is greatly affected by the shape of the stress-strain curve for the material of which the column is made, as indicated in Fig. 6.47. The stress-strain curve for a material, such as an aluminum alloy or high-strength steel, which does not have a sharply defined yield point, is shown in Fig. 6.47a. The corresponding column curve is plotted in Fig. 6.47b. In contrast, Fig. 6.47c presents the stress-strain curve for structural steel,

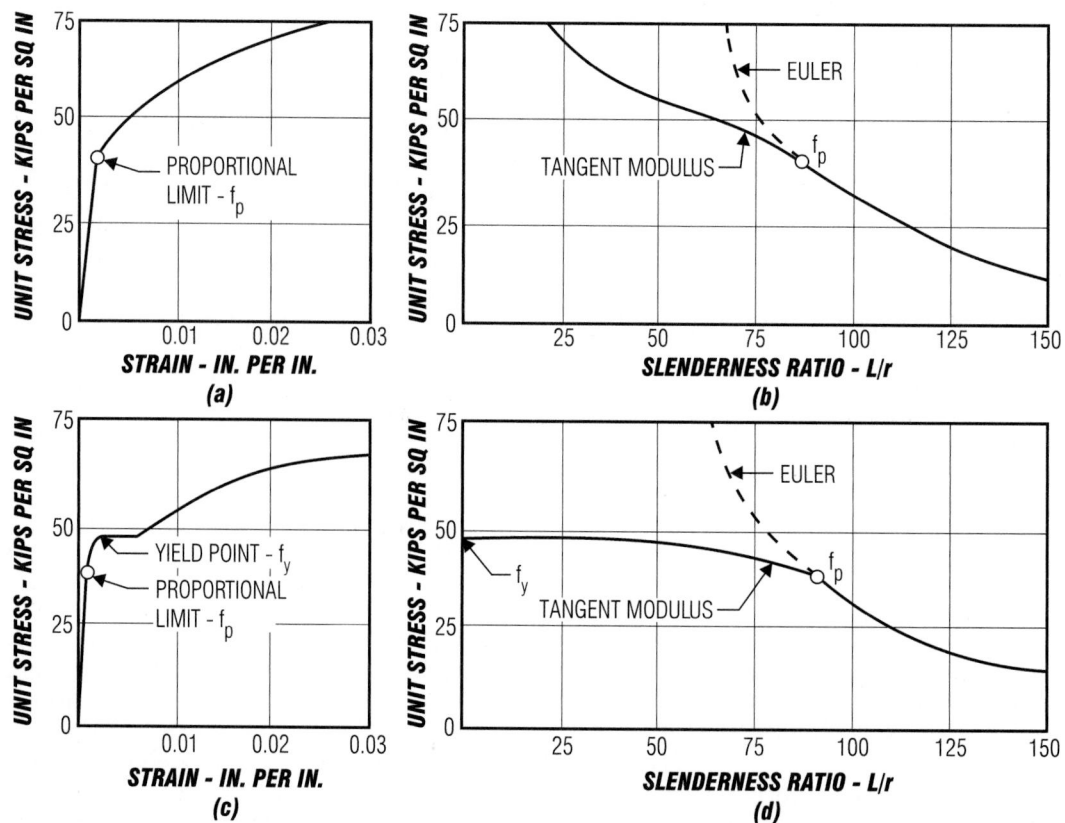

Fig. 6.47 Column curves: (a) Stress-strain curve for a material without a sharply defined yield point; (b) column curve for the material in (a); (c) stress-strain curve for a material with a sharply defined yield point; (d) column curve for the material in (c).

with a sharply defined yield point, and Fig. 6.47*d* the related column curve. This curve becomes horizontal as the critical stress approaches the yield strength of the material and the tangent modulus becomes zero, whereas the column curve in Fig. 6.47*b* continues to rise with decreasing values of the slenderness ratio.

Examination of Fig. 6.47*d* also indicates that slender columns, which fall in the elastic range, where the column curve has a large slope, are very sensitive to variations in the factor *k*, which represents the effect of end conditions. On the other hand, in the inelastic range, where the column curve is relatively flat, the critical stress is relatively insensitive to changes in *k*. Hence, the effect of end conditions is of much greater significance for long columns than for short columns.

6.41 Behavior of Actual Columns

For many reasons, columns in structures behave differently from the ideal column assumed in deriving Eqs. (6.72) to (6.76). A major consideration is the effect of accidental imperfections, such as nonhomogeneity of materials, initial crookedness, and unintentional eccentricities of the axial load. These effects can be taken into account by a proper choice of safety factor.

There are, however, other significant conditions that must be considered in any design procedure: continuity in framed structures and eccentricity of the load. Continuity affects column action two ways: The restraint and sidesway at column ends determine the value of *k*, and bending moments are transmitted to the columns by adjoining structural members.

Because of the deviation of the behavior of actual columns from the ideal, columns generally are designed by empirical formulas. Separate equations usually are given for short columns, intermediate columns, and long columns, and still other equations for combinations of axial load and bending moment.

Furthermore, a column may fail not by buckling of the member as a whole but, as an alternative, by buckling of one of its components. Hence, when members like I beams, channels, and angles are used as columns, or when sections are built up of plates, the possibility that the critical load on a component (leg, half flange, web, lattice bar) will be

less than the critical load on the column as a whole should be investigated.

Similarly, the possibility of buckling of the compression flange or the web of a beam should be investigated.

Local buckling, however, does not always result in a reduction in the load-carrying capacity of a column; sometimes it results in a redistribution of the stresses, which enables the member to carry additional load.

For more details on column action, see S. Timoshenko and J. M. Gere, "Theory of Elastic Stability," McGraw-Hill Book Company, New York, books.mcgraw-hill.com; B. G. Johnston, "Guide to Stability Design Criteria for Metal Structures," John Wiley & Sons, Inc., New Jersey, www.wiley.com; F. Bleich, "Buckling Strength of Metal Structures," McGraw-Hill Book Company, New York, books. mcgraw-hill.com; and T. V. Galambos, "Guide to Stability Design Criteria for Metal Structures," John Wiley & Sons, Inc., Hoboken, N.J., 1988, www. wiley.com.

Graphic-Statics Fundamentals

Since a force is completely determined when it is known in magnitude, direction, and point of application, any force may be represented by the length, direction, and position of a straight line. The length of line to a given scale represents the magnitude of the force. The position of the line parallels the line of action of the force, and an arrowhead on the line indicates the direction in which the force acts.

6.42 Force Polygons

Graphically represented, a force may be designated by a letter, sometimes followed by a subscript, such as P_1 and P_2 in Fig. 6.48. Or each extremity of the line may be indicated by a letter and the force referred to by means of these letters (Fig. 6.48*a*). The order of the letters indicates the direction of the force; in Fig. 6.48*a*, referring to P_1 as *OA* indicates it acts from *O* toward *A*.

Forces are concurrent when their lines of action meet. If they lie in the same plane, they are coplanar.

Parallelogram of Forces ▪ The **resultant** of several forces is a single force that would produce

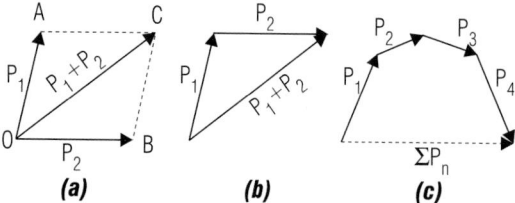

Fig. 6.48 Addition of forces by (*a*) parallelogram law, (*b*) triangle construction, and (*c*) polygon construction.

the same effect on a rigid body. The resultant of two concurrent forces is determined by the **parallelogram law**:

If a parallelogram is constructed with two forces as sides, the diagonal represents the resultant of the forces (Fig. 6.48*a*).

The resultant is said to be equal to the sum of the forces, sum here meaning vectorial sum, or addition by the parallelogram law. Subtraction is carried out in the same manner as addition, but the direction of the force to be subtracted is reversed.

If the direction of the resultant is reversed, it becomes the **equilibrant**, a single force that will hold the two given forces in equilibrium.

Resolution of Forces ■ Any force may be resolved into two components acting in any given direction. To resolve a force into two components, draw a parallelogram with the force as a diagonal and sides parallel to the given directions. The sides then represent the components.

The procedure is: (1) Draw the given force. (2) From both ends of the force draw lines parallel to the directions in which the components act. (3) Draw the components along the parallels through the origin of the given force to the intersections with the parallels through the other end. Thus, in Fig. 6.48*a*, P_1 and P_2 are the components in directions OA and OB of the force represented by OC.

Force Triangles and Polygons ■ Examination of Fig. 6.48*a* indicates that a step can be saved in adding forces P_1 and P_2. The same resultant could be obtained by drawing only the upper half of the parallelogram. Hence, to add two forces, draw the first force; then draw the second force at the end of the first one. The resultant is the force drawn from the origin of the first force to the end of the second force, as shown in Fig. 6.48*b*.

This diagram is called a force triangle. Again, the equilibrant is the resultant with direction reversed. If it is drawn instead of the resultant, the arrows representing the direction of the forces will all point in the same direction around the triangle. From the force triangle, an important conclusion can be drawn:

If three forces meeting at a point are in equilibrium, they form a closed force triangle.

To add several forces $P_1, P_2, P_3, \ldots, P_n$, draw P_2 from the end of P_1, P_3 from the end of P_2, and so on. The force required to complete the force polygon is the resultant (Fig. 6.48*c*).

If a group of concurrent forces is in equilibrium, they form a closed force polygon.

6.43 Equilibrium Polygons

When forces are coplanar but not concurrent, the force polygon will yield the magnitude and direction of the resultant but not its point of application. To complete the solution, the easiest method generally is to employ an auxiliary force polygon, called an equilibrium, or funicular (string), polygon. Sides of this polygon represent the lines of action of certain components of the given forces; more specifically, they take the configuration of a weightless string holding the forces in equilibrium.

In Fig. 6.49*a*, the forces P_1, P_2, P_3, and P_4 acting on the given body are not in equilibrium. The magnitude and direction of their resultant R are obtained from the force polygon *abcde* (Fig. 6.49*b*). The line of action may be obtained as follows:

From any point O in the force polygon, draw a line to each vertex of the polygon. Since the lines Oa and Ob form a closed triangle with force P_1, they represent two forces S_5 and S_1 that hold P_1 in equilibrium—two forces that may replace P_1 in a force diagram. So, as in Fig. 6.49*a*, at any point m on the line of action of P_1, draw lines mn and mv parallel to S_1 and S_5, respectively, to represent the lines of action of these forces. Similarly, S_1 and S_2 represent two forces that may replace P_2. The line of action of S_1 already is indicated by the line mn, and it intersects P_2 at n. So through n draw a line parallel to S_2, intersecting P_3 at r. Through r, draw rs parallel to S_3, and through s, draw st parallel to S_4. Lines mv and st, parallel to S_5 and S_4, respectively, represent the lines of action of S_5 and S_4. But these two forces form a closed force triangle with the resultant ae (Fig. 6.49*b*), and therefore the three

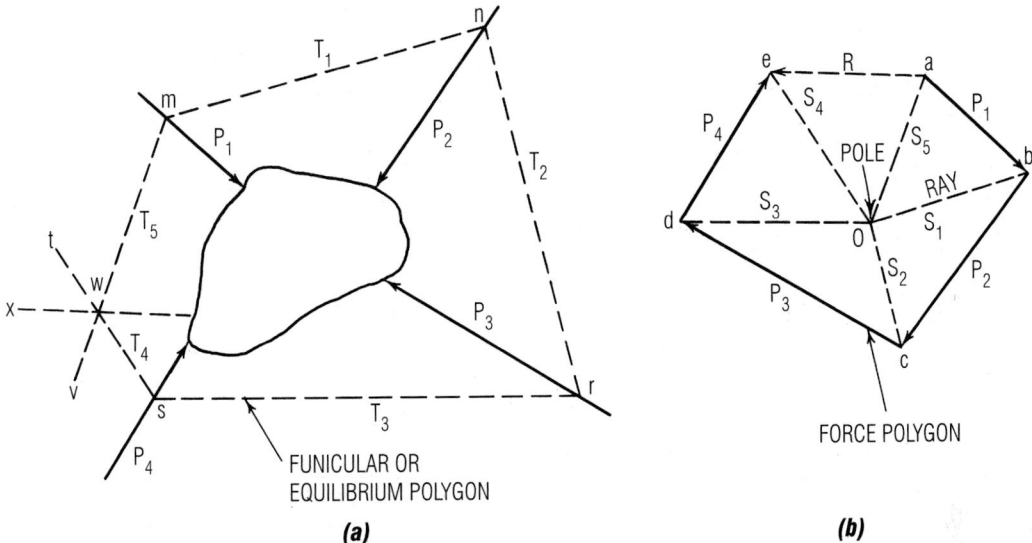

Fig. 6.49 Force and equilibrium polygons for a system of forces in equilibrium.

forces must be concurrent. Hence, the line of action of the resultant must pass through the intersection w of the lines mv and st. The resultant of the four given forces is thus fully determined. A force of equal magnitude but acting in the opposite direction, from e to a, will hold P_1, P_2, P_3, and P_4 in equilibrium.

The polygon $mnrsw$ is called an *equilibrium polygon*. Point O is called the *pole*, and $S_1 \ldots S_5$ are called the *rays of the force polygon.*

Stresses in Trusses

A truss is a coplanar system of structural members joined at their ends to form a stable framework. Usually, analysis of a truss is based on the assumption that the joints are hinged. Neglecting small changes in the lengths of the members due to loads, the relative positions of the joints cannot change. Stresses due to joint rigidity or deformations of the members are called **secondary stresses**.

6.44 Truss Characteristics

Three bars pinned together to form a triangle represent the simplest type of truss. Some of the

more common types of trusses are shown in Fig. 6.50.

The top members are called the **upper chord**, the bottom members the **lower chord**, and the verticals and diagonals **web members**.

Trusses act like long, deep girders with cutout webs. Roof trusses have to carry not only their own weight and the weight of roof framing but wind loads, snow loads, suspended ceilings and equipment, and a live load to take care of construction, maintenance, and repair loading. Bridge trusses have to support their own weight and that of deck framing and deck, live loads imposed by traffic (automobiles, trucks, railroad trains, pedestrians, and so on) and impact caused by live load, plus wind on structural members and vehicles. **Deck trusses** carry the live load on the upper chord and **through trusses** on the lower chord.

Loads generally are applied at the intersection of members, or panel points, so that the members will be subjected principally to direct stresses—tension or compression. To simplify stress analysis, the weight of the truss members is apportioned to upper- and lower-chord panel points. The members are assumed to be pinned at their ends, even though this may actually not be the case. However, if the joints are of such nature as to restrict relative rotation substantially, then

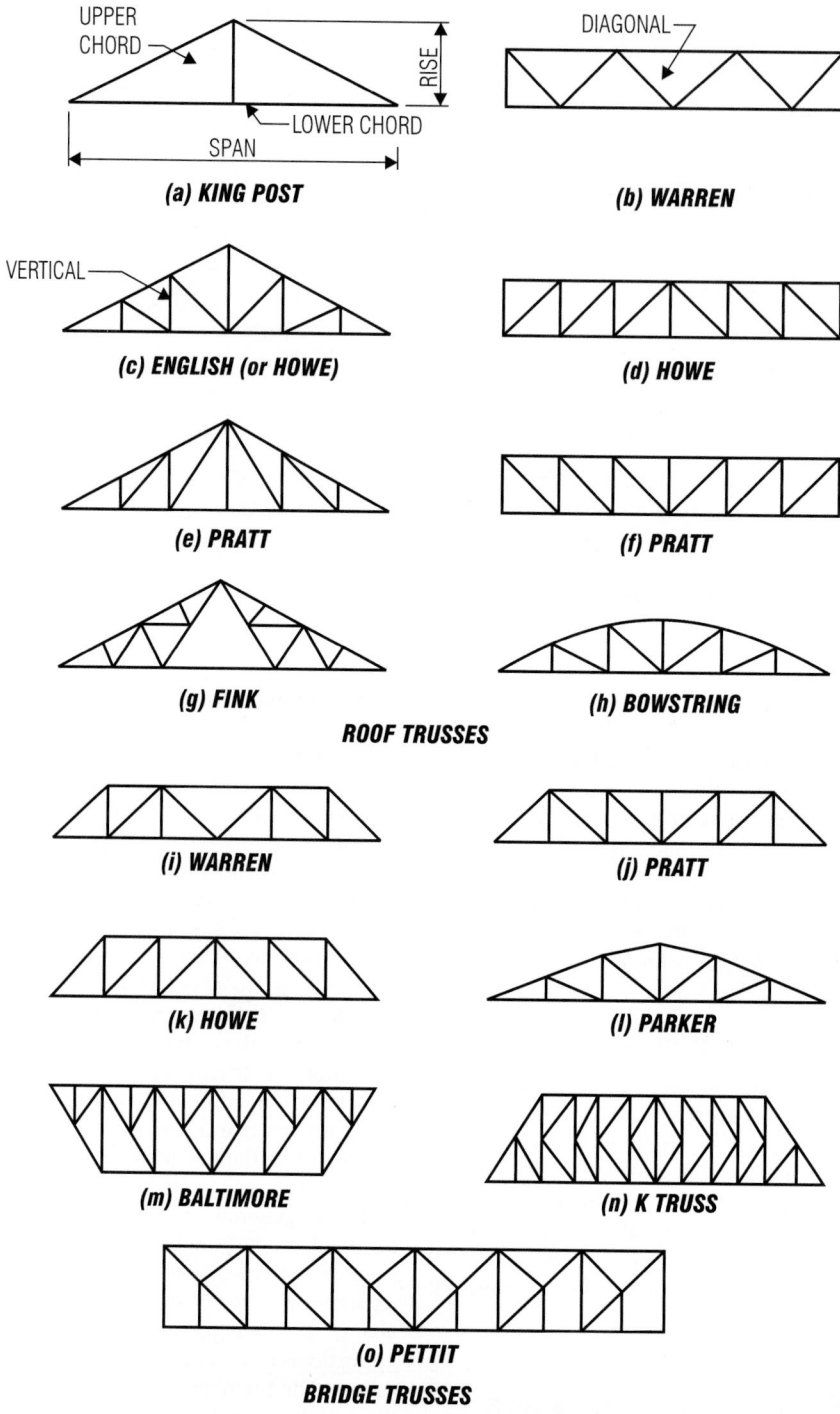

Fig. 6.50 Common types of trusses.

the "secondary" stresses set up as a result should be computed and superimposed on the stresses obtained with the assumption of pin ends.

6.45 Bow's Notation

In analysis of trusses, especially in graphical analysis, Bow's notation is useful for identifying truss members, loads, and stresses. Capital letters are placed in the spaces between truss members and between forces; each member and load is then designated by the letters on opposite sides of it. For example, in Fig. 6.51a, the upper-chord members are AF, BH, CJ, and DL. The loads are AB, BC, and CD, and the reactions are EA and DE. Stresses in the members generally are designated by the same letters but in lowercase.

6.46 Method of Sections for Truss Stresses

A convenient method of computing the stresses in truss members is to isolate a portion of the truss by a section so chosen as to cut only as many members with unknown stresses as can be evaluated by the laws of equilibrium applied to that portion of the truss. The stresses in the members cut by the section are treated as external forces and must hold the loads on that portion of the truss in equilibrium. Compressive forces act toward each joint or panel point, and tensile forces away from the joint.

Joint Isolation ▪ A choice of section that often is convenient is one that isolates a joint with only two unknown stresses. Since the stresses and load at the joint must be in equilibrium, the sum of the

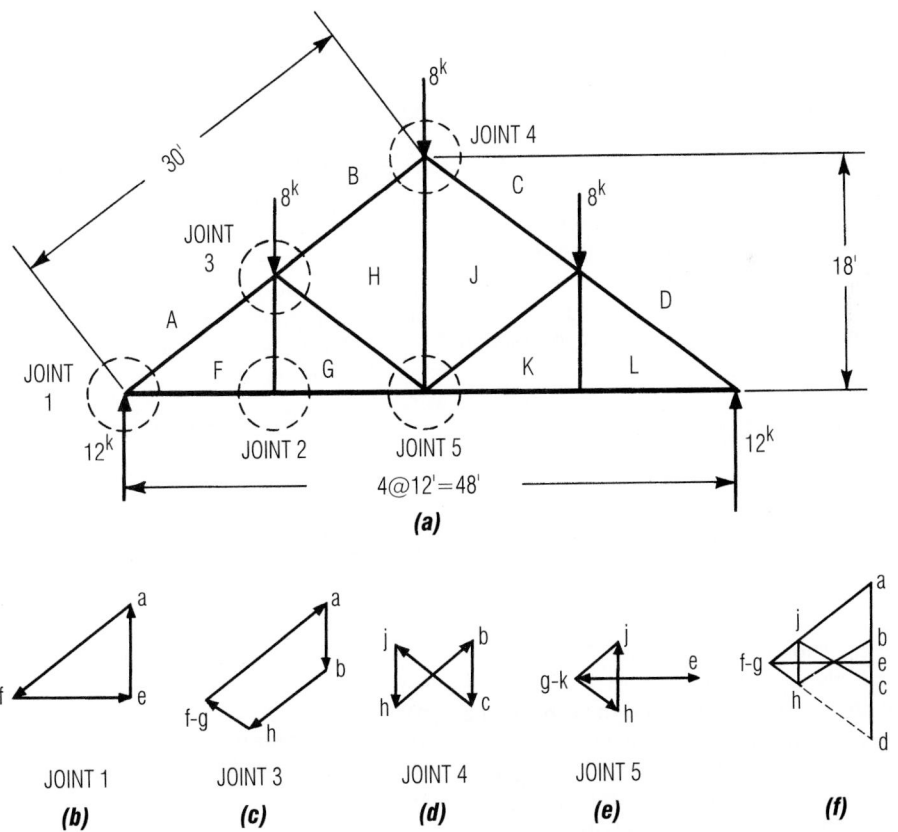

Fig. 6.51 Graphical determination of stresses at each joint of the truss in (a) may be expedited by constructing the single Maxwell diagram in (f).

horizontal components of the forces must be zero, and so must be the sum of the vertical components. Since the lines of action of all the forces are known (the stresses act along the longitudinal axes of the truss members), we can therefore compute two unknown magnitudes of stresses at each joint by this method.

To apply it to joint 1 of the truss in Fig. 6.51a, first equate the sum of the vertical components to zero. This equation shows that the vertical component af of the top chord must be equal and opposite to the reaction, 12 kips (see Fig. 6.51b and Bow's notation, Art. 6.45). The stress in the top chord ea at this joint, then, must be a compression equal to $12 \times 30/18 = 20$ kips. Next, equate the sum of the horizontal components to zero. This equation indicates that the stress in the bottom chord fe at the joint must be equal and opposite to the horizontal component of the top chord. Hence, the stress in the bottom chord must be a tension equal to $20 \times 24/30 = 16$ kips.

Taking a section around joint 2 in Fig. 6.51a reveals that the stress in the vertical fg is zero since there are no loads at the joint and the bottom chord is perpendicular to the vertical. Also, the stress must be the same in both bottom-chord members at the joint since the sum of the horizontal components must be zero.

After joints 1 and 2 have been solved, a section around joint 3 cuts only two unknown stresses: S_{BH} in top chord BH and S_{HG} in diagonal HG. Application of the laws of equilibrium to this joint yields the following two equations, one for the vertical components and the second for the horizontal components:

$$\Sigma V = 0.6S_{FA} - 8 - 0.6S_{BH}$$

$$+ 0.6S_{HG} = 0 \qquad (6.77)$$

$$\Sigma H = 0.8S_{FA} - 0.8S_{BH}$$

$$- 0.8S_{HG} = 0 \qquad (6.78)$$

Both unknown stresses are assumed to be compressive, i.e., acting toward the joint. The stress in the vertical does not appear in these equations because it already was determined to be zero. The stress in FA, S_{FA}, was found from analysis of joint 1 to be 20 kips. Simultaneous solution of the two equations yields $S_{HG} = 6.7$ kips and $S_{BH} = 13.3$ kips. (If these stresses had come out with a negative sign, it would have indicated that the original assumption of their directions was incorrect; they would, in that case, be tensile forces instead of compressive forces.)

Examination of the force polygons in Fig. 6.51 indicates that each stress occurs in two force polygons. Hence, the graphical solution can be shortened by combining the polygons. The combination of the various polygons for all the joints into one stress diagram is known as a **Maxwell diagram** (Fig. 6.51f).

Wind loads on a roof truss with a sloping top chord are assumed to act normal to the roof, in which case the load polygon will be an inclined line or a true polygon. The reactions are computed generally on the assumption either that both are parallel to the resultant of the wind loads or that one end of the truss is free to move horizontally and therefore will not resist the horizontal components of the loads. The stress diagram is plotted in the same manner as for vertical loads after the reactions have been found.

Some trusses are complex and require special methods of analysis. (C. H. Norris et al., "Elementary Structural Analysis," McGraw-Hill Book Company, New York, 1976, books.mcgraw-hill.com.)

Parallel-Chord Trusses ▪ A convenient section for determining the stresses in diagonals of parallel-chord trusses is a vertical one, such as N-N in Fig. 6.52a. The sum of the forces acting on that portion of the truss to the left of N-N equals the vertical component of the stress in diagonal cD (see Fig. 6.52b). Thus, if θ is the acute angle between cD and the vertical,

$$R_1 - P_1 - P_2 + S \cos \theta = 0 \qquad (6.79)$$

But $R_1 - P_1 - P_2$ is the algebraic sum of all the external vertical forces on the left of the section and is the vertical shear in the section. It may be designated as V. Therefore,

$$V + S \cos \theta = 0 \quad \text{or} \quad S = -V \sec \theta \qquad (6.80)$$

From this it follows that for trusses with horizontal chords and single-web systems, the stress in any web member, other than the subverticals, equals the vertical shear in the member multiplied by the secant of the angle that the member makes with the vertical.

Nonparallel Chords ▪ A vertical section also can be used to determine the stress in diagonals

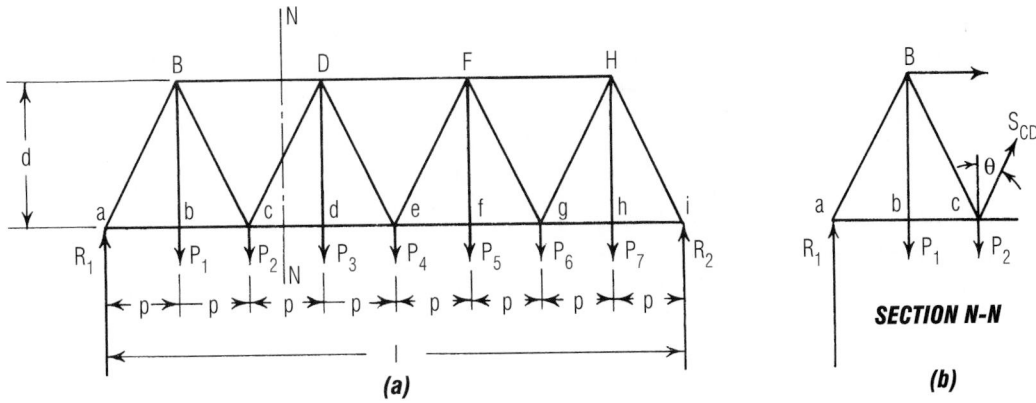

Fig. 6.52 Vertical section through the truss in (a) enables determination of stress in the diagonal (b).

when the chords are not parallel, but the previously described procedure must be modified. Suppose, for example, that the stress in the diagonal Bc of the Parker truss in Fig. 6.53 is to be found. Take a vertical section to the left of joint c. This section cuts BC, the top chord, and Bc, both of which have vertical components, as well as the horizontal bottom chord bc. Now, extend BC and bc until they intersect, at O. If O is used as the center for taking moments of all the forces, the moments of the stresses in BC and bc will be zero since the lines of action pass through O. Since Bc remains the only stress with a moment about O, Bc can be computed from the fact that the sum of the moments about O must equal zero, for equilibrium.

Generally, the calculation can be simplified by determining first the vertical component of the

diagonal and from it the stress. So resolve Bc into its horizontal and vertical components Bc_H and Bc_V, at c, so that the line of action of the horizontal component passes through O. Taking moments about O yields

$$(Bc_v \times Oc) - (R \times Oa) + (P_1 \times Ob) = 0 \qquad (6.81)$$

from which Bc_V may be determined. The actual stress in Bc is Bc_v multiplied by the secant of the angle that Bc makes with the vertical.

The stress in verticals, such as Cc, can be found in a similar manner. But take the section on a slope so as not to cut the diagonal but only the vertical and the chords. The moment equation about the intersection of the chords yields the stress in the vertical directly since it has no horizontal component.

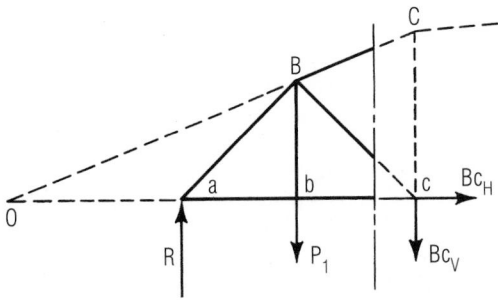

Fig. 6.53 Stress in a truss diagonal is determined by taking a vertical section and computing moments about the intersection of top and bottom chords.

Subdivided Panels ▪ In a truss with parallel chords and subdivided panels, such as the one in Fig. 6.54a, the subdiagonals may be either tension or compression. In Fig. 6.54a, the subdiagonal Bc is in compression and $d'E$ is in tension. The vertical component of the stress in any subdiagonal, such as $d'E$, equals half the stress in the vertical $d'd$ at the intersection of the subdiagonal and main diagonal. See Fig. 6.54b.

For a truss with inclined chords and subdivided panels, this is not the case. For example, the stress in $d'E$ for a truss with nonparallel chords is $d'd \times l/h$, where l is the length of $d'E$ and h is the length of Ee.

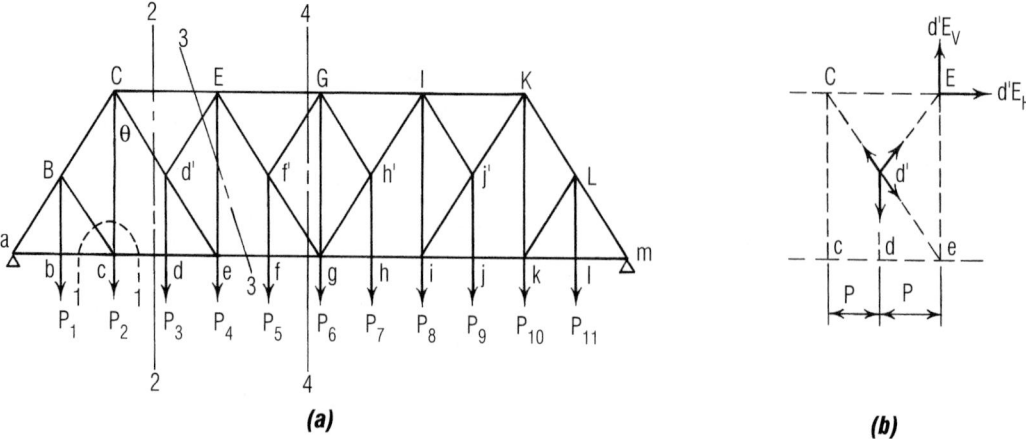

Fig. 6.54 Sections taken through truss with subdivided panels for finding stresses in web members.

6.47 Moving Loads on Trusses and Girders

To minimize bending stresses in truss members, framing is arranged to transmit loads to panel points. Usually, in bridges, loads are transmitted from a slab to stringers parallel to the trusses, and the stringers carry the load to transverse floor beams, which bring it to truss panel points. Similar framing generally is used for bridge girders.

In many respects, analysis of trusses and girders is similar to that for beams—determination of maximum end reaction for moving loads, for example, and use of influence lines (Art. 6.25). For girders, maximum bending moments and shears at various sections must be determined for moving loads, as for beams; and as indicated in Art. 6.46, stresses in truss members may be determined by taking moments about convenient points or from the shear in a panel. But girders and trusses differ from beams in that analysis must take into account the effect at critical sections of loads between panel points since such loads are distributed to the nearest panel points; hence, in some cases, influence lines differ from those for beams.

Stresses in Verticals ▪ The maximum total stress in a load-bearing stiffener of a girder or in a truss vertical, such as Bb in Fig. 6.55a, equals the maximum reaction of the floor beam at the panel point. The influence line for the reaction at b is shown in Fig. 6.55b and indicates that for maxi-

mum reaction, a uniform load of w lb/lin ft should extend a distance of $2p$, from a to c, where p is the length of a panel. In that case, the stress in Bb equals wp.

Maximum floor beam reaction for concentrated moving loads occurs when the total load between a and c, W_1 (Fig. 6.55c), equals twice the load between a and b. Then, the maximum live-load stress in Bb is

$$r_b = \frac{W_1 g - 2Pg'}{p} = \frac{W_1(g - g')}{p} \qquad (6.82)$$

where g is the distance of W_1 from c, and g' is the distance of P from b.

Stresses in Diagonals ▪ For a truss with parallel chords and single-web system, stress in a diagonal, such as Bc in Fig. 6.55a, equals the shear in the panel multiplied by the secant of the angle θ the diagonal makes with the vertical. The influence diagram for stresses in Bc, then, is the shear influence diagram for the panel multiplied by sec θ, as indicated in Fig. 6.55d. For maximum tension in Bc, loads should be placed only in the portion of the span for which the influence diagram is positive (crosshatched in Fig. 6.55d). For maximum compression, the loads should be placed where the diagram is negative (minimum shear).

A uniform load, however, cannot be placed over the full positive or negative portions of the span to get a true maximum or minimum. Any load in the panel is transmitted to the panel points at both

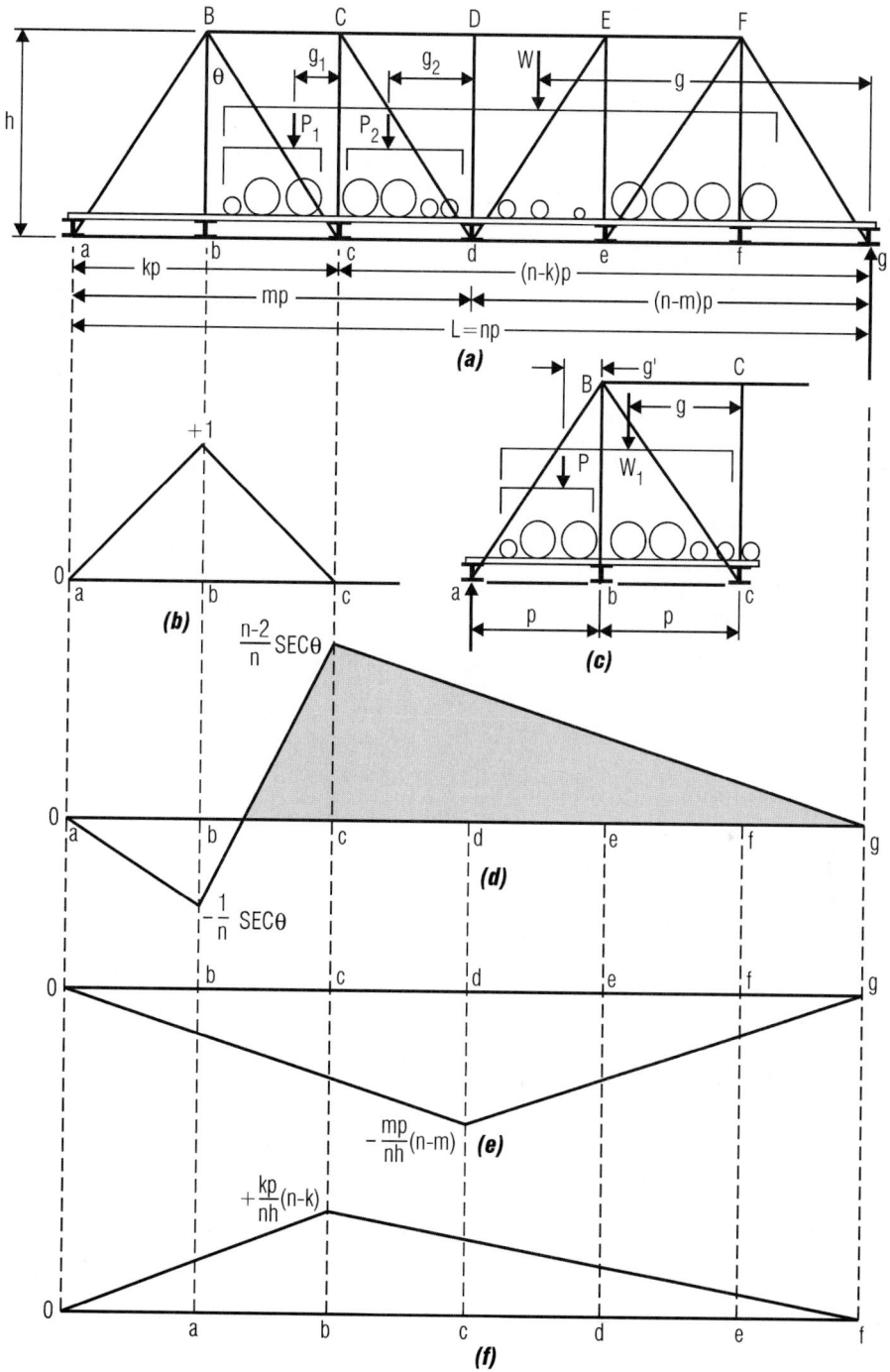

Fig. 6.55 Stresses produced in a truss by moving loads are determined with influence lines.

ends of the panel and decreases the shear. True maximum shear occurs for Bc when the uniform load extends into the panel a distance x from c equal to $(n - k)p/(n - 1)$, where n is the number of panels in the truss and k the number of panels from the left end of the truss to c.

For maximum stress in Bc caused by moving concentrated loads, the loads must be placed to produce maximum shear in the panel, and this may require several trials with different wheels placed at c (or, for minimum shear, at b). When the wheel producing maximum shear is at c, the loading will satisfy the following criterion: When the wheel is just to the right of c, W/n is greater than P_1, where W is the total load on the span and P_1 the load in the panel (Fig. 6.55a); when the wheel is just to the left of c, W/n is less than P_1.

Stresses in Chords ▪ Stresses in truss chords, in general, can be determined from the bending moment at a panel point, so the influence diagram for chord stress has the same shape as that for bending moment at an appropriate panel point. For example, Fig. 6.55e shows the influence line for stress in upper chord CD (minus signifies compression). The ordinates are proportional to the bending moment at d since the stress in CD can be computed by considering the

portion of the truss just to the left of d and taking moments about d. Figure 6.55f similarly shows the influence line for stress in bottom chord cd.

For maximum stress in a truss chord under uniform load, the load should extend the full length of the truss.

For maximum chord stress caused by moving concentrated loads, the loads must be placed to produce maximum bending moment at the appropriate panel point, and this may require several trials with different wheels placed at the panel point. Usually, maximum moment will be produced with the heaviest grouping of wheels about the panel point.

In all trusses with verticals, the loading producing maximum chord stress will satisfy the following criterion: When the critical wheel is just to the right of the panel point, Wm/n is greater than P, where mp is the distance of the panel point from the left end of the truss with span np and P is the sum of the loads to the left of the panel point; when the wheel is just to the left of the panel point, Wm/n is less than P.

In a truss without verticals, the maximum stress in the loaded chord is determined by a different criterion. For example, the moment center for the lower chord bc (Fig. 6.56) is panel point C, at a

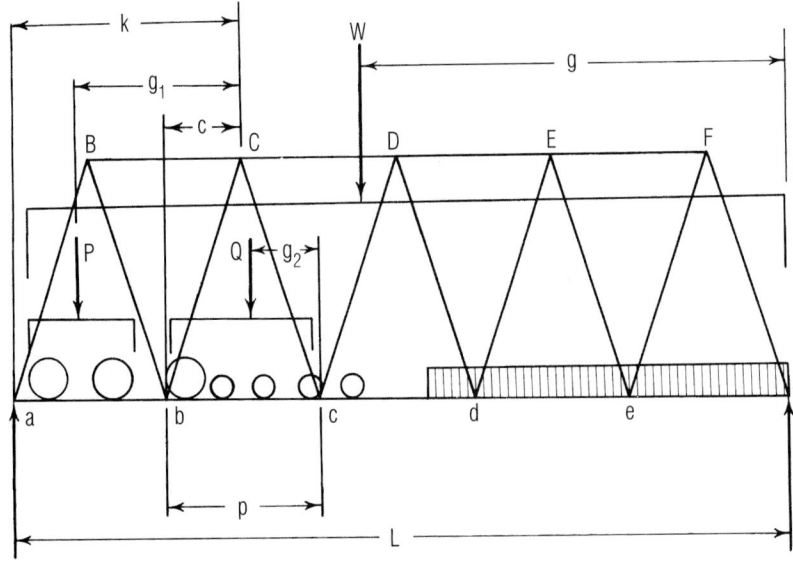

Fig. 6.56 Moving loads on a truss without verticals.

distance c from b. When the critical load is at b or c, the following criterion will be satisfied: When the wheel is just to the right of b or c, Wk/L is greater than $P + Qc/p$; when the wheel is just to the left of b or c, Wk/L is less than $P + Qc/p$, where W is the total load on the span, Q the load in panel bc, P the load to the left of bc, and k the distance of the center of moments C from the left support. The moment at C is $Wgk/L - Pg_1 - Qcg_2/p$, where g is the distance of the center of gravity of the loads W from the right support, g_1 the distance of the center of gravity of the loads P from C, and g_2 the distance of the center of gravity of the loads Q from c, the right end of the panel.

6.48 Counters

For long-span bridges, it often is economical to design the diagonals of trusses for tension only. But in the panels near the center of a truss, maximum shear due to live loads plus impact may exceed and be opposite in sign to the dead-load shear, thus inducing compression in the diagonal. If the tension diagonal is flexible, it will buckle. Hence, it becomes necessary to place in such panels another diagonal crossing the main diagonal (Fig. 6.57). Such diagonals are called *counters.*

Designed only for tension, a counter is assumed to carry no stress under dead load because it would buckle slightly. It comes into action only when the main diagonal is subjected to compression. Hence, the two diagonals never act together.

Although the maximum stresses in the main members of a truss are the same whether or not counters are used, the minimum stresses in the verticals are affected by the presence of counters. In most trusses, however, the minimum stresses in the verticals where counters are used are of the same sign as the maximum stresses and hence have no significance.

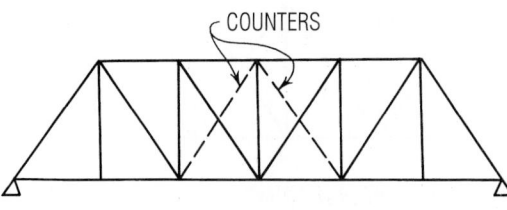

Fig. 6.57 Truss with counters.

6.49 Stresses in Trusses Due to Lateral Forces

To resist lateral forces on bridge trusses, trussed systems are placed in the planes of the top and bottom chords, and the ends, or **portals**, also are braced as low down as possible without impinging on headroom needed for traffic (Fig. 6.58). In stress analysis of lateral trusses, wind loads may be assumed as all applied on the windward chord or as applied equally on the two chords. In the former case, the stresses in the lateral struts are one-half panel load greater than if the latter assumption were made, but this is of no practical consequence.

Where the diagonals are considered as tension members only, counter stresses need not be computed since reversal of wind direction gives greater stresses in the members concerned than any partial loading from the opposite direction. When a rigid system of diagonals is used, the two diagonals of a panel may be assumed equally stressed. Stresses in the chords of the lateral truss should be combined with those in the chords of the main trusses due to dead and live loads.

In computation of stresses in the lateral system for the loaded chords of the main trusses, the wind on the live load should be added to the wind on the trusses. Hence, the wind on the live load should be positioned for maximum stress on the lateral truss. Methods described in Art. 6.46 can be used to compute the stresses on the assumption that each diagonal takes half the shear in each panel.

When the main trusses have inclined chords, the lateral systems between the sloping chords lie in several planes, and the exact determination of all the wind stresses is rather difficult. The stresses in the lateral members, however, may be determined without significant error by considering the lateral truss flattened into one plane. Panel lengths will vary, but the panel loads will be equal and may be determined from the horizontal panel length.

Since some of the lateral forces are applied considerably above the horizontal plane of the end supports of the bridge, these forces tend to overturn the structure (Fig. 6.58e). The lateral forces of the upper lateral system (Fig. 6.58a) are carried to the portal struts, and the horizontal loads at these points produce an overturning moment about the horizontal plane of the supports. In Fig. 6.58e, P represents the horizontal load brought to each portal strut by the upper lateral bracing, h the depth of the truss,

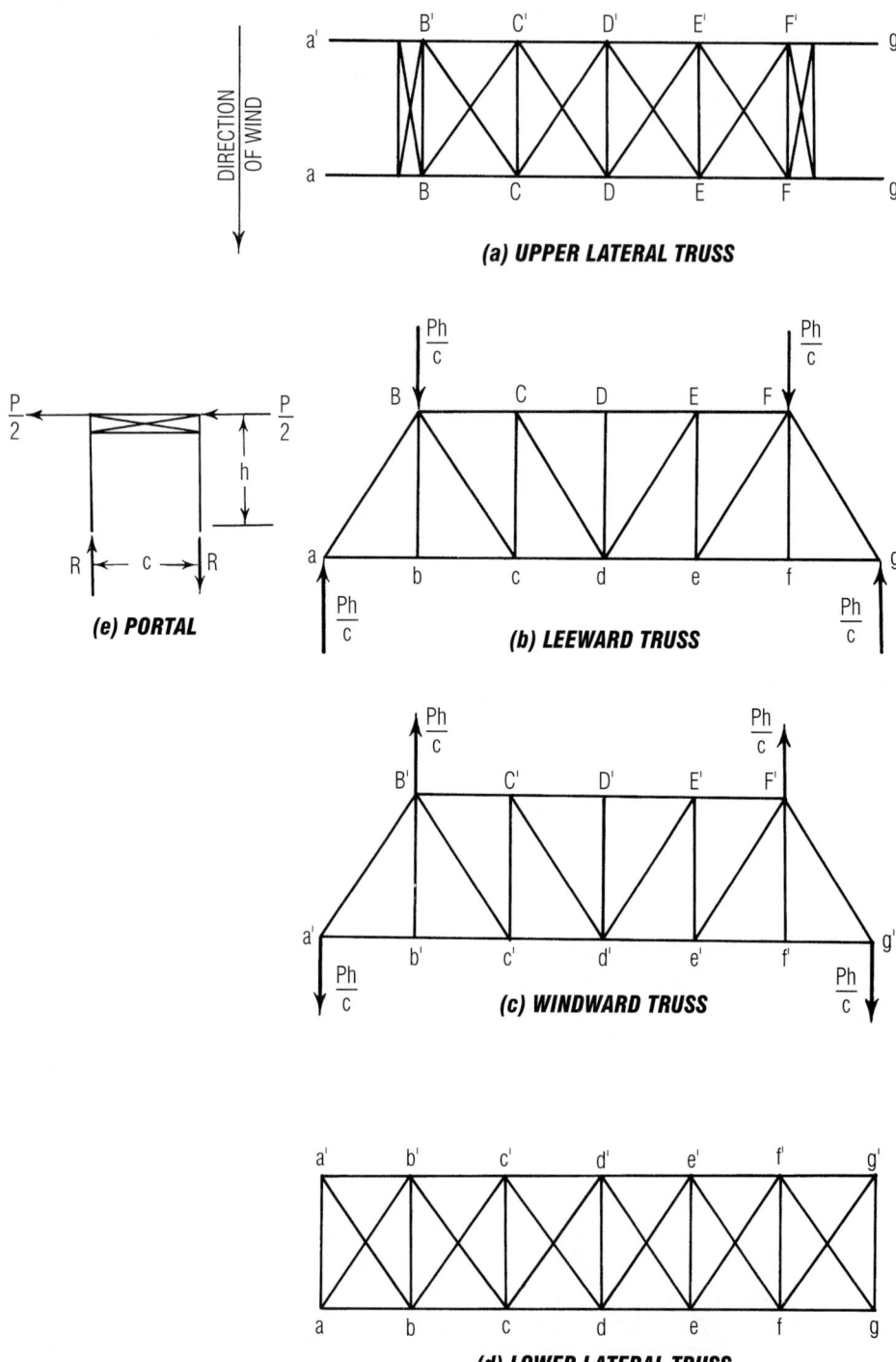

Fig. 6.58 Lateral trusses for bracing top and bottom chords of bridge trusses.

and c the distance between trusses. The overturning moment produced at each end of the structure is Ph, which is balanced by a reaction couple Rc. The value of the reaction R is then Ph/c. An equivalent effect is achieved on the main trusses if loads equal to Ph/c are applied at B and F and at B' and F', as shown in Fig. 6.58b and c. These loads produce stresses in the end posts and in the lower-chord members, but the web members are not stressed.

The lateral force on the live load also causes an overturning moment, which may be treated in a similar manner. But there is a difference as far as the web members of the main truss are concerned. Since the lateral force on the live load produces an effect corresponding to the position of the live load on the bridge, equivalent panel loads, rather than equivalent reactions, must be computed. If the distance from the resultant of the wind force to the plane of the loaded chord is h', the equivalent vertical panel load is Ph'/c, where P is the horizontal panel load due to the lateral force.

6.50 Complex Trusses

The method of sections may not provide a direct solution for some trusses with inclined chords and multiple-web systems. But if the truss is stable and statically determinate, a solution can be obtained by applying the equations of equilibrium to a section taken around each joint. The stresses in the truss members are obtained by solution of the simultaneous equations.

Since two equations of equilibrium can be written for the forces acting at a joint (Art. 6.46), the total number of equations available for a truss is $2n$, where n is the number of joints. If r is the number of horizontal and vertical components of the reactions, and s the number of stresses, $r + s$ is the number of unknowns.

If $r + s = 2n$, the unknowns can be obtained from solution of the simultaneous equations. If $r + s$ is less than $2n$, the structure is unstable (but the structure may be unstable even if $r + s$ exceeds $2n$). If $r + s$ is greater than $2n$, there are too many unknowns; the structure is statically indeterminate.

General Tools for Structural Analysis

For some types of structures, the equilibrium equations are not sufficient to determine the reactions or the internal stresses. These structures are called *statically indeterminate*.

For the analysis of such structures, additional equations must be written based on a knowledge of the elastic deformations. Hence, methods of analysis that enable deformations to be evaluated for unknown forces or stresses are important for the solution of problems involving statically indeterminate structures. Some of these methods, like the method of virtual work, also are useful in solving complicated problems involving statically determinate systems.

6.51 Virtual Work

A virtual displacement is an imaginary, small displacement of a particle consistent with the constraints upon it. Thus, at one support of a simply supported beam, the virtual displacement could be an infinitesimal rotation $d\theta$ of that end, but not a vertical movement. However, if the support is replaced by a force, then a vertical virtual displacement may be applied to the beam at that end.

Virtual work is the product of the distance a particle moves during a virtual displacement and the component in the direction of the displacement of a force acting on the particle. If the displacement and the force are in opposite directions, the virtual work is negative. When the displacement is normal to the force, no work is done.

Suppose a rigid body is acted on by a system of forces with a resultant R. Given a virtual displacement ds at an angle α with R, the body will have virtual work done on it equal to $R \cos \alpha \, ds$. (No work is done by internal forces. They act in pairs of equal magnitude but opposite direction, and the virtual work done by one force of a pair is equal and opposite in sign to the work done by the other force.) If the body is in equilibrium under the action of the forces, then $R = 0$, and the virtual work also is zero.

Thus, the principle of virtual work may be stated:

If a rigid body in equilibrium is given a virtual displacement, the sum of the virtual work of the forces acting on it must be zero.

As an example of how the principle may be used, let us apply it to the determination of the reaction R of the simple beam in Fig. 6.59a. First, replace the support by an unknown force R. Next,

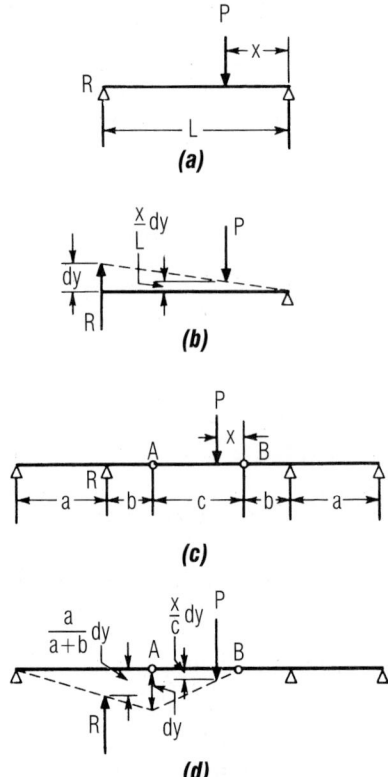

(a)

(b)

(c)

(d)

Fig. 6.59 Virtual work applied to determination of a simple-beam reaction (*a*) and (*b*) and the reaction of a beam with suspended span (*c*) and (*d*).

6.52 Strain Energy

When an elastic body is deformed, the virtual work done by the internal forces equals the corresponding increment of the strain energy dU, in accordance with the principle of virtual work.

Assume a constrained elastic body acted on by forces $P_1, P_2, \ldots$, for which the corresponding deformations are $e_1, e_2, \ldots$ Then, $\Sigma P_n de_n = dU$. The increment of the strain energy due to the increments of the deformations is given by

$$dU = \frac{\partial U}{\partial e_1} de_1 + \frac{\partial U}{\partial e_2} de_2 + \cdots$$

When solving a specific problem, a virtual displacement that is most convenient in simplifying the solution should be chosen. Suppose, for example, a virtual displacement is selected that affects only the deformation e_n corresponding to the load P_n, other deformations being unchanged. Then, the principle of virtual work requires that

$$P_n de_n = \frac{\partial U}{\partial e_n} de_n$$

This is equivalent to

$$\frac{\partial U}{\partial e_n} = P_n \qquad (6.83)$$

which states that the partial derivative of the strain energy with respect to a specific deformation gives the corresponding force.

Suppose, for example, the stress in the vertical bar in Fig. 6.60 is to be determined. All bars are

move the end of the beam upward a small amount dy as in Fig. 6.59*b*. The displacement under the load P will be $x\,dy/L$, upward. Then, the virtual work is $R\,dy - Px\,dy/L = 0$, from which $R = Px/L$.

The principle also may be used to find the reaction R of the more complex beam in Fig. 6.59*c*. Again, the first step is to replace one support by an unknown force R. Next, apply a virtual downward displacement dy at hinge A (Fig. 6.59*d*). The displacement under the load P will be $x\,dy/c$ and at the reaction R will be $a\,dy/(a+b)$. According to the principle of virtual work, $-Ra\,dy/(a+b) + Px\,dy/c = 0$; thus, $R = Px(a+b)/ac$. In this type of problem, the method has the advantage that only one reaction need be considered at a time and internal forces are not involved.

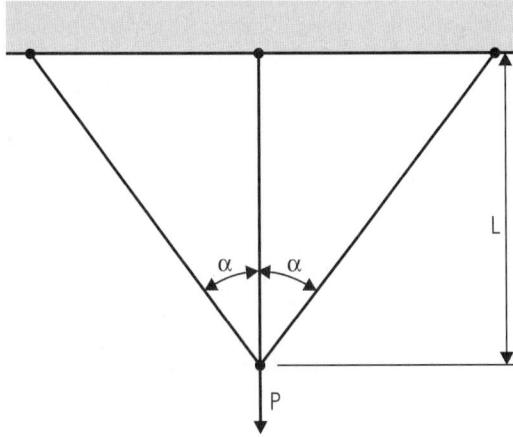

Fig. 6.60 Indeterminate truss.

made of the same material and have the same cross section A. If the vertical bar stretches an amount e under the load P, the inclined bars will each stretch an amount $e \cos \alpha$. The strain energy in the system is [from Eq. (6.23a)]

$$U = \frac{AE}{2L}(e^2 + 2e^2 \cos^3 \alpha)$$

and the partial derivative of this with respect to e must be equal to P; that is,

$$P = \frac{AE}{2L}(2e + 4e \cos^3 \alpha) = \frac{AEe}{L}(1 + 2\cos^3 \alpha)$$

Noting that the force in the vertical bar equals AEe/L, we find from the above equation that the required stress equals $P/(1 + 2\cos^3 \alpha)$.

Castigliano's Theorems ▪ If strain energy is expressed as a function of statically independent forces, the partial derivative of the strain energy with respect to a force gives the deformation corresponding to that force:

$$\frac{\partial U}{\partial P_n} = e_n \tag{6.84}$$

This is known as Castigliano's first theorem. (His second theorem is the principle of least work.)

6.53 Method of Least Work

Castigliano's second theorem, also known as the principle of least work, states:

The strain energy in a statically indeterminate structure is the minimum consistent with equilibrium.

As an example of the use of the method of least work, an alternative solution will be given for the stress in the vertical bar in Fig. 6.60 (see Art. 5.52). Calling this stress X, we note that the stress in each of the inclined bars must be $(P - X)/2 \cos \alpha$. Using Eq. (6.23a), we can express the strain energy in the system in terms of X:

$$U = \frac{X^2 L}{2AE} + \frac{(P - X)^2 L}{4AE \cos^3 \alpha}$$

Hence, the internal work in the system will be a minimum when

$$\frac{\partial U}{\partial X} = \frac{XL}{AE} - \frac{(P - X)L}{2AE \cos^3 \alpha} = 0$$

Solving for X gives the stress in the vertical bar as $P/(1 + 2\cos^3 \alpha)$, as in Art. 5.52.

6.54 Dummy Unit-Load Method for Displacements

The strain energy for pure bending is $U = M^2 L/2EI$ [see Eq. (6.23d)]. To find the strain energy due to bending stress in a beam, we can apply this equation to a differential length dx of the beam and integrate over the entire span. Thus,

$$U = \int_0^L \frac{M^2 dx}{2EI} \tag{6.85}$$

If we let M represent the bending moment due to a generalized force P, the partial derivative of the strain energy with respect to P is the deformation d corresponding to P. Differentiating Eq. (6.85) gives

$$d = \int_0^L \frac{M}{EI} \frac{\partial M}{\partial P} dx \tag{6.86}$$

The partial derivative in this equation is the rate of change of bending moment with the load P. It equals the bending moment m produced by a unit generalized load applied at the point where the deformation is to be measured and in the direction of the deformation. Hence, Eq. (6.86) can also be written as

$$d = \int_0^L \frac{Mm}{EI} dx \tag{6.87}$$

To find the vertical deflection of a beam, we apply a dummy unit load vertically at the point where the deflection is to be measured and substitute the bending moments due to this load and the actual loading in Eq. (6.87). Similarly, to compute a rotation, we apply a dummy unit moment.

Beam Deflections ▪ As a simple example, let us apply the dummy unit-load method to the determination of the deflection at the center of a simply supported, uniformly loaded beam of constant moment of inertia (Fig. 6.61a). As indicated in Fig. 6.61b, the bending moment at a distance x from one end is $(wL/2)x - (w/2)x^2$. If we apply a dummy

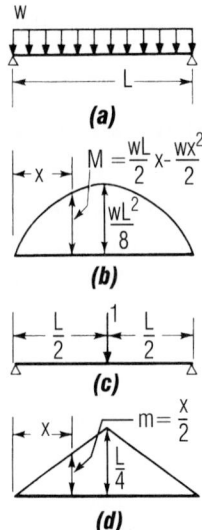

Fig. 6.61 Dummy unit-load method applied to a uniformly loaded beam (*a*) to find the midspan deflection; (*b*) moment diagram for the uniform load; (*c*) unit load at midspan; (*d*) moment diagram for the unit load.

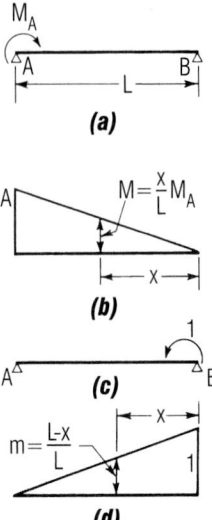

Fig. 6.62 End rotation at *B* in beam *AB* (*a*) caused by end moment at *A* is determined by dummy unit-load method; (*b*) moment diagram for the end moment; (*c*) unit moment applied at beam end; (*d*) moment diagram for that moment.

unit load vertically at the center of the beam (Fig. 6.61*c*), where the vertical deflection is to be determined, the moment at x is $x/2$, as indicated in Fig. 6.61*d*. Substituting in Eq. (6.87) and taking advantage of symmetry of loading gives

$$d = 2 \int_0^{L/2} \left(\frac{wL}{2} x - \frac{w}{2} x^2 \right) \frac{x \, dx}{2 \, EI} = \frac{SwL^4}{384EI}$$

Beam-End Rotations ▪ As another example, let us apply the method to finding the end rotation at one end of a simply supported, prismatic beam produced by a moment applied at the other end. In other words, the problem is to find the end rotation at B, θ_B in Fig. 6.62*a*, due to M_A. As indicated in Fig. 6.62*b*, the bending moment at a distance x from B due to M_A is $M_A x/L$. If we apply a dummy unit moment at B (Fig. 6.62*c*), it will produce a moment at x of $(L - x)/L$ (Fig. 6.62*d*).

Substituting in Eq. (6.87) gives

$$\theta_B = \int_0^L M_A \frac{x}{L} \frac{L - x}{L} \frac{dx}{EI} = \frac{M_A L}{6EI} \qquad (6.88)$$

Shear Deflections ▪ To determine the deflection of a beam due to shear, Castigliano's first

theorem can be applied to the strain energy in shear:

$$U = \int \int \frac{v^2}{2G} dA \, dx \qquad (6.89)$$

where v = shearing unit stress

G = modulus of rigidity

A = cross-sectional area

Truss Deflections ▪ The dummy unit-load method also may be adapted to computation of truss deformations. The strain energy in a truss is given by

$$U = \sum \frac{S^2 L}{2AE} \qquad (6.90)$$

which represents the sum of the strain energy for all the members of the truss. S is the stress in each member due to the loads, L the length of each, A the cross-sectional area, and E the modulus of elasticity. Application of Castigliano's first theorem (Art. 6.52) and differentiation inside the summation sign yield the deformation:

$$d = \sum \frac{SL}{AE} \frac{\partial S}{\partial P} \qquad (6.91)$$

where, as in Art. 6.54, P represents a generalized load. The partial derivative in this equation is the rate of change of axial stress with P. It equals the axial stress u produced in each member of the truss by a unit load applied at the point where the deformation is to be measured and in the direction of the deformation. Consequently, Eq. (6.91) also can be written

$$d = \sum \frac{SuL}{AE} \qquad (6.92)$$

To find the vertical deflection at any point of a truss, apply a dummy unit vertical load at the panel point where the deflection is to be measured. Substitute in Eq. (6.92) the stresses in each member of the truss due to this load and the actual loading. Similarly, to find the rotation of any joint, apply a dummy unit moment at the joint, compute the stresses in each member of the truss, and substitute in Eq. (6.92).

When it is necessary to determine the relative movement of two panel points in the direction of a member connecting them, apply dummy unit loads in opposite directions at those points.

Note that members not stressed by the actual loads or the dummy loads do not enter into the calculation of a deformation.

As an example of the application of Eq. (6.92), let us compute the midspan deflection of the truss in Fig. 6.63a. The stresses in kips due to the 20-kip load at every lower-chord panel point are given in Fig. 6.63a and Table 6.2. Also, the ratios of length of members in inches to their cross-sectional areas in square inches are given in Table 6.2. We apply a dummy unit vertical load at L_2, where the deflection is required. Stresses u due to this load are shown in Fig. 6.63b and Table 6.2.

Table 6.2 also contains the computations for the deflection. Members not stressed by the 20-kip loads or the dummy unit loads are not included. Taking advantage of the symmetry of the truss, the values are tabulated for only half the truss and the sum is doubled. Also, to reduce the amount of calculation, the modulus of elasticity E, which is equal to 30,000 is not included until the very last step since it is the same for all members.

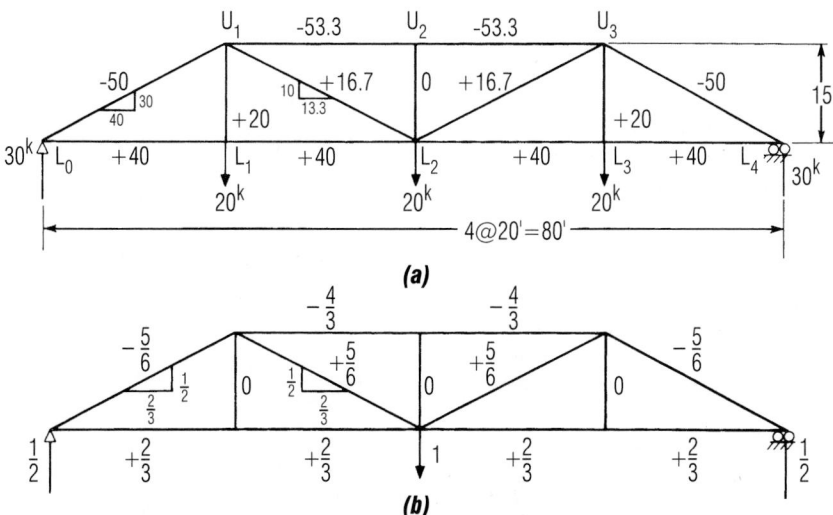

Fig. 6.63 Dummy unit-load method applied to a loaded truss to find (a) midspan deflection; (b) stresses produced by a unit load applied at midspan.

Table 6.2 Midspan Deflection of Truss of Fig. 6.63

Member	L/A	S	u	SuL/A
L_0L_2	160	$+40$	$+2/3$	4,267
L_0U_1	75	-50	$-5/6$	3,125
U_1U_2	60	-53.3	$-4/3$	4,267
U_1L_2	150	$+16.7$	$+5/6$	2,083
				13,742

Division of the summation of the last column by the modulus of elasticity $E = 30,000$ ksi yields the midspan deflection.

$$d = \sum \frac{SuL}{AE} = \frac{2 \times 13,742}{30,000} = 0.916 \text{ in}$$

6.55 Reciprocal Theorem and Influence Lines

Consider a structure loaded by a group of independent forces A, and suppose that a second group of forces B is added. The work done by the forces A acting over the displacements due to B will be W_{AB}.

Now, suppose the forces B had been on the structure first and then load A had been applied. The work done by the forces B acting over the displacements due to A will be W_{BA}.

The reciprocal theorem states that $W_{AB} = W_{BA}$.

Some very useful conclusions can be drawn from this equation. For example, there is the reciprocal deflection relationship:

The deflection at a point A due to a load at B equals the deflection at B due to the same load applied at A. Also, the rotation at A due to load (or moment) at B equals the rotation at B due to the same load (or moment) applied to A.

Another consequence is that deflection curves also may be influence lines, to some scale, for reactions, shears, moments, or deflections (**Mueller-Breslau principle**). For example, suppose the influence line for a reaction is to be found; that is, we wish to plot the reaction R as a unit load moves over the structure, which may be statically indeterminate. For loading condition A, we analyze the structure with a unit load on it at a distance x from some reference point. For loading condition B, we apply a dummy unit vertical load upward at the place where the reaction is to be determined,

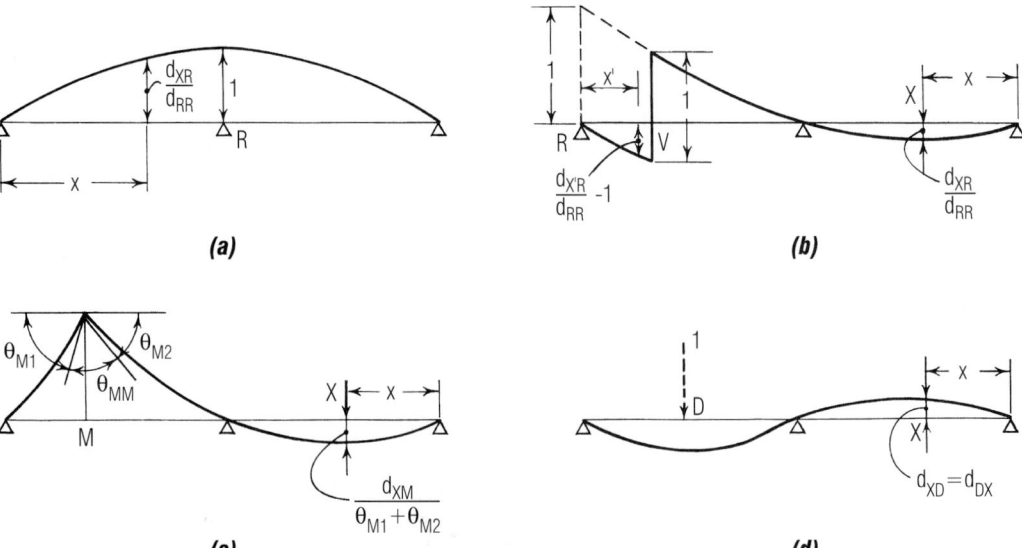

Fig. 6.64 Influence lines for a continuous beam are obtained from deflection curves. (*a*) Reaction at V; (*b*) shear at V; (*c*) bending moment at M; (*d*) deflection at D.

deflecting the structure off the support. At a distance x from the reference point, the displacement is d_{xR}, and over the support the displacement is d_{RR}. Hence, $W_{AB} = -1d_{xR} + Rd_{RR}$. On the other hand, W_{BA} is zero since loading condition A provides no displacement for the dummy unit load at the support in condition B. Consequently, from the reciprocal theorem, $W_{AB} = W_{BA} = 0$; hence

$$R = \frac{d_{xR}}{d_{RR}} \qquad (6.93)$$

Since d_{RR}, the deflection at the support due to a unit load applied there, is a constant, R is proportional to d_{xR}. So the influence line for a reaction can be obtained from the deflection curve resulting from a displacement of the support (Fig. 6.64a). The magnitude of the reaction is obtained by dividing each ordinate of the deflection curve by d_{RR}.

Similarly, the influence line for shear can be obtained from the deflection curve produced by cutting the structure and shifting the cut ends vertically at the point for which the influence line is desired (Fig. 6.64b).

The influence line for bending moment can be obtained from the deflection curve produced by cutting the structure and rotating the cut ends at the point for which the influence line is desired (Fig. 6.64c).

Finally, it may be noted that the deflection curve for a load of unity is also the influence line for deflection at that point (Fig. 6.64d).

6.56 Superposition Methods

The principle of superposition states that, if several loads are applied to a linearly elastic structure, the displacement at each point of the structure equals the sum of the displacements induced at the point when the loads are applied individually in any sequence. Furthermore, the bending moment (or shear) at each point equals the sum of the bending moments (or shears) induced at the point by the loads applied individually in any sequence.

The principle holds only when the displacement (deflection or rotation) at every point of the structure is directly proportional to applied loads. Also, it is required that unit stresses be proportional to unit strains and that displacements be very small so that calculations can be based on the undeformed configuration of the structure without significant error.

As a simple example, consider a bar with length L and cross-sectional area A loaded with n axial loads $P_1, P_2, \ldots, P_n$. Let F equal the sum of the loads. From Eq. (6.8), F causes an elongation $\delta = FL/AE$, where E is the modulus of elasticity of the bar. According to the principle of superposition, if e_1 is the elongation caused by P_1 alone, e_2 by P_2 alone, ... and e_n by P_n alone, then regardless of the sequence in which the loads are applied, when all the loads are on the bar,

$$\delta = e_1 + e_2 + \cdots + e_n$$

This simple case can be easily verified by substituting $e_1 = P_1L/AE$, $e_2 = P_2L/AE, \ldots$, and $e_n = P_nL/AE$ in this equation and noting that $F = P_1 + P_2 + \cdots + P_n$:

$$\delta = \frac{P_1L}{AE} + \frac{P_2L}{AE} + \cdots + \frac{P_nL}{AE}$$

$$= (P_1 + P_2 + \cdots + P_n)\frac{L}{AE} = \frac{FL}{AE}$$

In the preceding equations, L/AE represents the elongation induced by a unit load and is called the **flexibility** of the bar.

The reciprocal, AE/L, represents the force that causes a unit elongation and is called the **stiffness** of the bar.

Analogous properties of beams, columns, and other structural members and the principle of superposition are useful in analysis of many types of structures. Calculation of stresses and displacements of statically indeterminate structures, for example, often can be simplified by resolution of bending moments, shears, and displacements into components chosen to supply sufficient equations for the solution from requirements for equilibrium of forces and compatibility of displacements.

Consider the continuous beam $ALRBC$ shown in Fig. 6.65a. Under the loads shown, member LR is subjected to end moments M_L and M_R (Fig. 6.65b) that are initially unknown. The bending-moment diagram for LR for these end moments is shown at the left in Fig. 6.65c. If these end moments were known, LR would be statically determinate; that is, LR could be treated as a simply supported beam subjected to known end moments, M_L and M_R. The analysis can be further simplified by resolution of the bending-moment diagram into the three components shown to the right of the equals sign

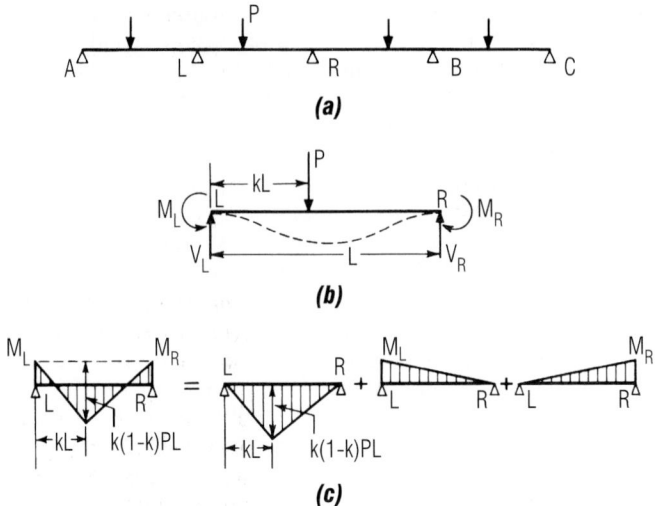

Fig. 6.65 Any span of a continuous beam (*a*) can be treated as a simple beam, as shown in (*b*) and (*c*). In (*c*), the moment diagram is resolved into basic components.

in Fig. 6.65*c*. This example leads to the following conclusion.

The bending moment at any section of a span *LR* of a continuous beam or frame equals the simple-beam moment due to the applied loads, plus the simple-beam moment due to the end moment at *L*, plus the simple-beam moment due to the end moment at *R*.

When the moment diagrams for all the spans of *ALRBC* in Fig. 6.65 have been resolved into components so that the spans may be treated as simple beams, all the end moments (moments at supports) can be determined from two basic requirements:

1. The sum of the moments at every support equals zero.

2. The end rotation (angular change at the support) of each member rigidly connected at the support is the same.

6.57 Influence-Coefficient Matrices

A matrix is a rectangular array of numbers in rows and columns that obeys certain mathematical rules known generally as matrix algebra and matrix calculus. A matrix consisting of only a single column is called a **vector**. In this book, matrices and vectors are represented by boldface letters, and

their elements by lightface symbols, with appropriate subscripts. It often is convenient to use numbers for the subscripts to indicate the position of an element in the matrix. Generally, the first digit indicates the row, and the second digit, the column. Thus, in matrix **A**, A_{23} represents the element in the second row and third column:

$$\mathbf{A} = \begin{bmatrix} A_{11} & A_{12} & A_{13} \\ A_{21} & A_{22} & A_{23} \\ A_{31} & A_{32} & A_{33} \end{bmatrix} \qquad (6.94)$$

Methods based on matrix representations often are advantageous for structural analysis and design of complex structures. One reason is that matrices provide a compact means of representing and manipulating large quantities of numbers. Another reason is that computers can perform matrix operations automatically and speedily. Computer programs are widely available for the purpose.

Matrix Equations ■ Matrix notation is especially convenient in representing the solution of simultaneous linear equations, which arise frequently in structural analysis. For example, suppose a set of equations is represented in matrix notation by $\mathbf{AX} = \mathbf{B}$, where **X** is the vector of variables $X_1, X_2, \ldots, X_n$, **B** is the vector of the constants on the right-hand side of the equations, and **A** is a matrix of the coefficients of the variables. Multiplication of both sides of the equation by $\mathbf{A}^{-1}$, the inverse of $\mathbf{A}$, yields $\mathbf{A}^{-1}\mathbf{A}\mathbf{X} = \mathbf{A}^{-1}\mathbf{B}$.

Since $\mathbf{A}^{-1}\mathbf{A} = \mathbf{I}$, the identity matrix, and $\mathbf{I}\mathbf{X} = \mathbf{X}$, the solution of the equations is represented by $\mathbf{X} = \mathbf{A}^{-1}\mathbf{B}$. The matrix inversion $\mathbf{A}^{-1}$ can be readily performed by computers. For large matrices, however, it often is more practical to solve the equations; for example, by the Gaussian procedure of eliminating one unknown at a time.

In the application of matrices to structural analysis, loads and displacements are considered applied at the intersection of members (joints, or nodes). The loads may be resolved into moments, torques, and horizontal and vertical components. These may be assembled for each node into a vector and then all the node vectors may be combined into a force vector $\mathbf{P}$ for the whole structure.

$$\mathbf{P} = \begin{bmatrix} P_1 \\ P_2 \\ \vdots \\ P_n \end{bmatrix} \quad (6.95)$$

Similarly, displacements corresponding to those forces may be resolved into rotations, twists, and horizontal and vertical components and assembled for the whole structure into a vector $\mathbf{\Delta}$.

$$\mathbf{\Delta} = \begin{bmatrix} \Delta_1 \\ \Delta_2 \\ \vdots \\ \Delta_n \end{bmatrix} \quad (6.96)$$

If the structure meets requirements for application of the principle of superposition (Art. 6.56) and forces and displacements are arranged in the proper sequence, the vectors of forces and displacements are related by

$$\mathbf{P} = \mathbf{K}_\Delta \quad (6.97a)$$
$$\mathbf{\Delta} = \mathbf{F}\mathbf{P} \quad (6.97b)$$

where $\mathbf{K}$ = stiffness matrix of the whole structure

$\mathbf{F}$ = flexibility matrix of the whole structure = $\mathbf{K}^{-1}$

The stiffness matrix $\mathbf{K}$ transforms displacements into loads. The flexibility matrix $\mathbf{F}$ transforms loads into displacements. The elements of $\mathbf{K}$ and $\mathbf{F}$ are functions of material properties, such as the modulus of elasticity; geometry of the structure; and sectional properties of members of the structure, such as area and moment of inertia. $\mathbf{K}$

and $\mathbf{F}$ are square matrices; that is, the number of rows in each equals the number of columns. In addition, both matrices are symmetrical; that is, in each matrix, the columns and rows may be interchanged without changing the matrix. Thus, $K_{ij} = K_{ji}$, and $F_{ij} = F_{ji}$, where i indicates the row in which an element is located, and j, the column.

Influence Coefficients ▪ Elements of the stiffness and flexibility matrices are influence coefficients. Each element is derived by computing the displacements (or forces) occurring at nodes when a unit displacement (or force) is imposed at one node, while all other displacements (or forces) are taken as zero.

Let Δ_i be the ith element of matrix $\mathbf{\Delta}$. Then, a typical element F_{ij} of $\mathbf{F}$ gives the displacement of a node i in the direction of Δ_i when a unit force acts at a node j in the direction of force P_j and no other forces are acting on the structure. The jth column of $\mathbf{F}$, therefore, contains all the nodal displacements induced by a unit force acting at node j in the direction of P_j.

Similarly, let P_i be the ith element of matrix $\mathbf{P}$. Then, a typical element K_{ij} of $\mathbf{K}$ gives the force at a node i in the direction of P_i when a node j is given a unit displacement in the direction of displacement Δ_j and no other displacements are permitted. The jth column of $\mathbf{K}$, therefore, contains all the nodal forces caused by a unit displacement of node j in the direction of Δ_j.

Application to a Beam ▪ A general method for determining the forces and moments in a continuous beam is as follows: Remove as many supports or members as necessary to make the structure statically determinate. (Such supports and members are often referred to as redundant.) Compute for the actual loads the deflections or rotations of the statically determinate structure in the direction of the unknown forces and couples exerted by the removed supports or members. Then, in terms of these forces and couples, treated as variables, compute the corresponding deflections or rotations the forces and couples produce in the statically determinate structure (see Arts. 6.32 and 6.54). Finally, for each redundant support or member, write equations that give the known rotations and deflections of the original structure in terms of the deformations of the statically determinate structure.

For example, one method of finding the reactions of the continuous beam AC in Fig. 6.66a is to remove supports 1, 2, and 3 temporarily. The beam is now simply supported between A and C. Hence, the reactions and the bending moments throughout can be computed from the laws of equilibrium. Beam AC deflects at points 1, 2, and 3, whereas we know that the continuous beam is prevented from deflecting at those points by the supports there. This information enables us to write three equations in terms of the three unknown reactions.

To determine the equations, assume that nodes exist at the location of the supports 1, 2, and 3. Then,

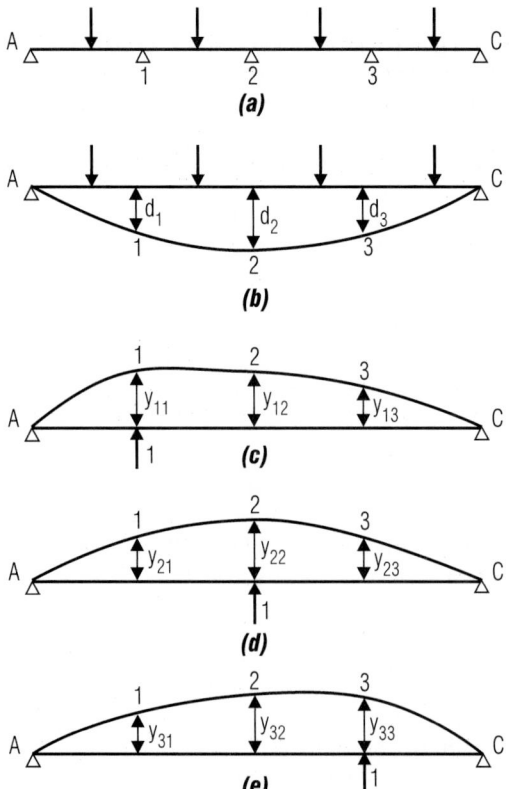

(a)

(b)

(c)

(d)

(e)

Fig. 6.66 Continuous beam (a) is converted into a simple beam (b) by temporary removal of interior supports. Reactions are then computed by equating the deflections due to the actual loads (b) to the sum of the deflections produced by the unknown reactions and the deflections due to the unit loads (c), (d), and (e).

for the actual loads, compute the vertical deflections d_1, d_2, and d_3 of simple beam AC at nodes 1, 2, and 3, respectively (Fig. 6.66b). Next, form two vectors, **d** with elements d_1, d_2, d_3, and **R** with the unknown reactions R_1 at node 1, R_2 at node 2, and R_3 at node 3 as elements. Since the beam may be assumed to be linearly elastic, set $\mathbf{d} = \mathbf{FR}$, where **F** is the flexibility matrix for simple beam AC. The elements y_{ij} of **F** are influence coefficients. To determine them, calculate column 1 of **F** as the deflections y_{11}, y_{21}, and y_{31} at nodes 1, 2, and 3, respectively, when a unit force is applied at node 1 (Fig. 6.66c). Similarly, compute column 2 of **F** for a unit force at node 2 (Fig. 6.66d) and column 3 for a unit force at node 3 (Fig. 6.66e). The three equations then are given by

$$\begin{bmatrix} y_{11} & y_{12} & y_{13} \\ y_{21} & y_{22} & y_{23} \\ y_{31} & y_{32} & y_{33} \end{bmatrix} \begin{bmatrix} R_1 \\ R_2 \\ R_3 \end{bmatrix} = \begin{bmatrix} d_1 \\ d_2 \\ d_3 \end{bmatrix} \qquad (6.98)$$

The solution may be represented by $\mathbf{R} = \mathbf{F}^{-1}\mathbf{d}$ and obtained by matrix or algebraic methods. See also Art. 6.66.

Continuous Beams and Frames

Continuous beams and frames are statically indeterminate. Bending moments in them are functions of the geometry, moments of inertia, and modulus of elasticity of individual members as well as of loads and spans. Although these moments can be determined by the methods described in Arts. 6.51 to 6.55, there are methods specially developed for beams and frames that often make analysis simpler. The following articles describe some of these methods.

6.58 Carry-Over and Fixed-End Moments

When a member of a continuous beam or frame is loaded, bending moments are induced at the ends of the member as well as between the ends. The magnitude of the end moments in the member depends on the magnitude and location of the loads, the geometry of the member, and the amount of restraint offered to end rotation of the member by other members connected to it. Connections are assumed to be rigid; that is, all members at a joint rotate through the same angle. As a result, end moments are induced in the connecting members, in addition to end moments that may be induced by loads on those spans.

Computation of end moments in a continuous beam or frame requires that the geometry and elastic properties of the members be known or assumed. (If these characteristics have to be assumed, computations may have to be repeated when they become known.)

Loads on any span, as well as the displacement of any joint, induce moments at the ends of the other members of the structure. As a result, an originating end moment may be considered distributed to the other members. The ratio of the end moment in an unloaded span to the originating end moment in the loaded span is a constant.

Sign Convention ▪ For computation of end moments, the following sign convention is most convenient: A moment acting at an end of a member or at a joint is positive if it tends to rotate the end or joint clockwise; it is negative, if it tends to rotate the end or joint counterclockwise.

Similarly, the angular rotation at the end of a member is positive if in a clockwise direction, negative if counterclockwise. Thus, a positive end moment produces a positive end rotation in a simple beam.

For ease in visualizing the shape of the elastic curve under the action of loads and end moments, plot bending-moment diagrams on the tension side of each member. Hence, if an end moment is represented by a curved arrow, the arrow will point in the direction in which the moment is to be plotted.

Carry-over Moments ▪ If a span of a continuous beam is loaded and if the far end of a connecting member is restrained by support conditions against rotation, a resisting moment is induced at the far end. That moment is called a carry-over moment. The ratio of the carry-over moment to the other end moment in the span is called carry-over factor. It is a constant for the member, independent of the magnitude and sign of the moments to be carried over. Every beam has two carry-over factors, one directed toward each end.

As pointed out in Art. 6.56, analysis of a span of a continuous beam or frame can be simplified by treating it as a simple beam subjected to applied

end moments. Thus, it is convenient to express the equations for carry-over factors in terms of the end rotations of simple beams: Convert a continuous member LR to a simple beam with the same span L. Apply a unit moment to one end (Fig. 6.67). The end rotation at the support where the moment is applied is α, and at the far end, the rotation is β. By the dummy-load method (Art. 6.54), if x is measured from the β end,

$$\alpha = \frac{1}{L^2} \int_0^L \frac{x^2}{EI_x} dx \qquad (6.99)$$

$$\beta = \frac{1}{L^2} \int_0^L \frac{x(L-x)}{EI_x} \qquad (6.100)$$

in which I_x is the moment of inertia at a section a distance of x from the β end, and E is the modulus of elasticity. In accordance with the reciprocal theorem (Art. 6.55), β has the same value regardless of the beam end to which the unit moment is applied (Fig. 6.67). For prismatic beams

$$\alpha_L = \alpha_R = \frac{L}{3EI} \qquad (6.101)$$

$$\beta = \frac{L}{6EI} \qquad (6.102)$$

The preceding equations can be used to determine carry-over factors for any magnitude of end restraint. The carry-over factors toward ends fixed against rotation, however, are of special importance

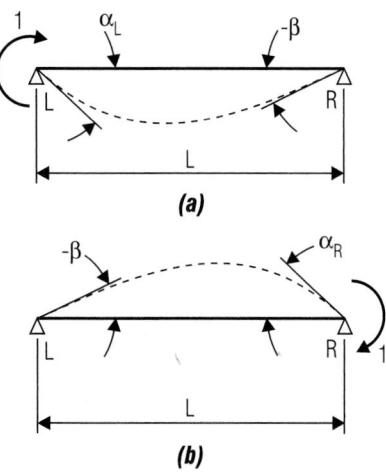

Fig. 6.67 End rotations of simple beam LR produced by a unit end moment (a) at L; (b) at R.

for moment distribution by converging approximations. For a span LR with ends L and R assumed to be fixed, the carry-over factor toward R is given by

$$C_R = \frac{\beta}{\alpha_R} \qquad (6.103)$$

Similarly, the carry-over factor toward support L is given by

$$C_L = \frac{\beta}{\alpha_L} \qquad (6.104)$$

If an end of a beam is free to rotate, the carry-over factor toward that end is zero.

Since the carry-over factors are positive, the moment carried over has the same sign as the applied moment.

Carry-over Factors for Prismatic Beams ■

For prismatic beams, $\beta = L/6EI$ and $\alpha = L/3EI$. Hence,

$$C_L = C_R = \frac{L}{6EI} \cdot \frac{3EI}{L} = \frac{1}{2} \qquad (6.105)$$

For beams with variable moment of inertia, β and α can be determined from Eqs. (6.99) and (6.100) and the carry-over factors from Eqs. (6.103) and (6.104).

Fixed-End Stiffness ■

The fixed-end stiffness of a beam is defined as the moment that is required to induce a unit rotation at the support where it is applied while the other end of the beam is fixed against rotation. Stiffness is important because it determines the proportion of the total moment applied at a joint, or intersection of members, that is distributed to each member of the joint.

In Fig. 6.68a, the fixed-end stiffness of beam LR at end R is represented by K_R. When K_R is applied to beam LR at R, a moment $M_L = C_L K_R$ is carried over to end L, where C_L is the carry-over factor toward L. K_R induces an angle change α_R at R, where α_R is given by Eq. (6.99). The carry-over moment induces at R an angle change $-C_L K_R \beta$, where β is given by Eq. (6.100). Since, by the definition of stiffness, the total angle change at R is unity, $K_R \alpha_R - C_L K_R \beta = 1$, from which

$$K_R = \frac{I/\alpha_R}{1 - C_R C_L} \qquad (6.106)$$

when C_R is substituted for β/α_R [see Eq. (6.103)].

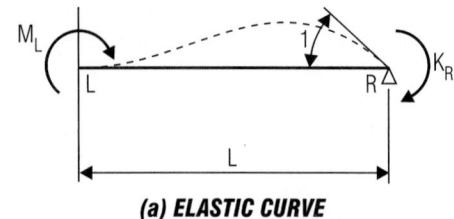

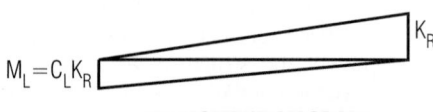

(a) ELASTIC CURVE

(b) MOMENT DIAGRAM

Fig. 6.68 Fixed-end stiffness.

In a similar manner, the stiffness at L is found to be

$$K_L = \frac{1/\alpha_L}{1 - C_R C_L} \qquad (6.107)$$

Stiffness of Prismatic Beams ■

With the use of Eqs. (6.101) and (6.105), the stiffness of a beam with constant moment of inertia is given by

$$K_L = K_R = \frac{3EI/L}{1 - \frac{1}{2} \times \frac{1}{2}} = \frac{4EI}{L} \qquad (6.108)$$

where L = span of the beam

E = modulus of elasticity

I = moment of inertia of beam cross section

Beam with Hinge ■

The stiffness of one end of a beam when the other end is free to rotate can be obtained from Eq. (6.106) or (6.107) by setting the carry-over factor toward the hinged end equal to zero. Thus, for a prismatic beam with one end hinged, the stiffness of the beam at the other end is given by

$$K = \frac{3EI}{L} \qquad (6.109)$$

This equation indicates that a prismatic beam hinged at only one end has three-fourths the stiffness, or resistance to end rotation, as a beam fixed at both ends.

Fixed-End Moments ■

A beam so restrained at its ends that no rotation is produced there by the loads is called a fixed-end beam, and the end moments are called fixed-end moments. Actually, it would be very difficult to construct a beam with ends that are truly fixed. The concept of fixed ends, however, is useful in determining the moments in continuous beams and frames.

Fixed-end moments may be expressed as the product of a coefficient and WL, where W is the total load on the span L. The coefficient is independent of the properties of other members of the structure. Thus, any member of a continuous beam or frame can be isolated from the rest of the structure and its fixed-end moments computed. Then, the actual moments in the beam can be found by applying a correction to each fixed-end moment.

Assume, for example, that the fixed-end moments for the loaded beam in Fig. 6.69a are to be determined. Let M_L^F be the moment at the left end L, and M_R^F the moment at the right end R of the beam. Based on the condition that no rotation is permitted at either end and that the reactions at the supports are in equilibrium with the applied loads, two equations can be written for the end moments in terms of the simple-beam end rotations, θ_L at L and θ_R at R for the specific loading.

Let K_L be the fixed-end stiffness at L and K_R the fixed-end stiffness at R, as given by Eqs. (6.106) and (6.107). Then, by resolution of the moment diagram into simple-beam components, as indicated in Figs. 6.69f to h, and application of the superposition principle (Art. 6.56), the fixed-end moments are found to be

$$M_L^F = -K_L(\theta_L + C_R\theta_R) \qquad (6.110)$$

$$M_R^F = -K_R(\theta_R + C_L\theta_L) \qquad (6.111)$$

where C_L and C_R are the carry-over factors to L and R, respectively [Eqs. (6.103) and (6.104)]. The end rotations θ_L and θ_R can be computed by a method described in Art. 6.32 or 6.54.

Moments for Prismatic Beams ▪ The fixed-end moments for beams with constant moment of inertia can be derived from the equations given above with the use of Eqs. (6.105) and (6.108):

$$M_L^R = -\frac{4EI}{L}\left(\theta_L + \frac{1}{2}\theta_R\right) \qquad (6.112)$$

$$M_R^F = -\frac{4EI}{L}\left(\theta_R + \frac{1}{2}\theta_L\right) \qquad (6.113)$$

where L = span of the beam

E = modulus of elasticity

I = moment of inertia

For horizontal beams with gravity loads only, θ_R is negative. As a result, M_L^F is negative and M_R^F positive.

For propped beams (one end fixed, one end hinged) with variable moment of inertia, the fixed-end moments are given by

$$M_L^F = -\frac{\theta_L}{\alpha_L} \quad \text{or} \quad M_R^F = -\frac{\theta_R}{\alpha_R} \qquad (6.114)$$

where α_L and α_R are given by Eq. (6.99). For prismatic propped beams, the fixed-end moments are

$$M_L^F = -\frac{3EI\theta_L}{L} \quad \text{or} \quad M_R^F = -\frac{3EI\theta_R}{L} \qquad (6.115)$$

Deflection of Supports ▪ Fixed-end moments for loaded beams when one support is displaced vertically with respect to the other support may be computed with the use of Eqs. (6.110) to (6.115) and the principle of superposition: Compute the fixed-end moments induced by the deflection of the beam when not loaded and add them to the fixed-end moments for the loaded condition with immovable supports.

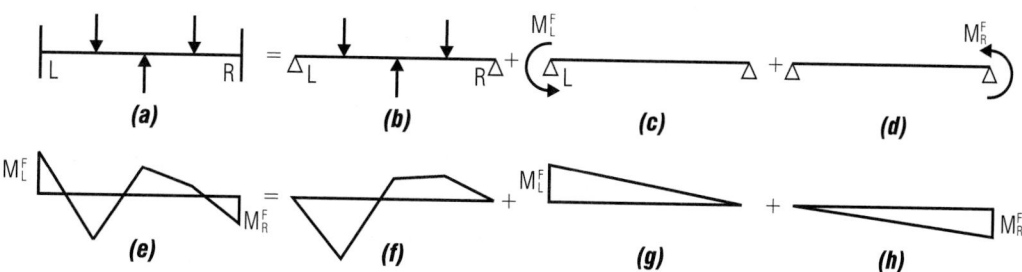

Fig. 6.69 Loads on fixed-end beam LR shown in (a) are resolved into component loads on a simple beam (b), (c), and (d). The corresponding moment diagrams are shown in (e) to (h).

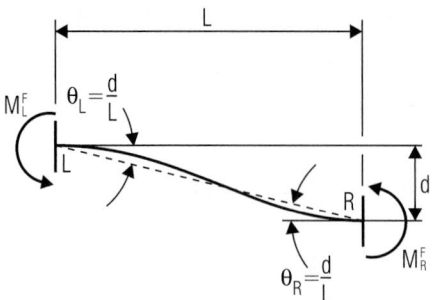

Fig. 6.70 End moments caused in a fixed-end beam by displacement d of one end.

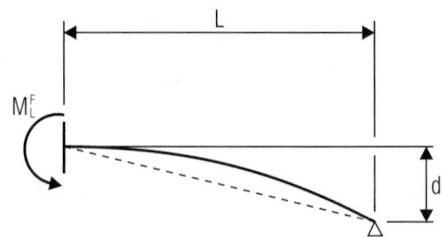

Fig. 6.71 End moment caused in a propped beam by displacement d of an end.

The fixed-end moments for the unloaded condition can be determined directly from Eqs. (6.110) and (6.111). Consider beam LR in Fig. 6.70, with span L and support R deflected a distance d vertically below its original position. If the beam were simply supported, the angle change caused by the displacement of R would be very nearly d/L. Hence, to obtain the fixed-end moments for the deflected condition, set $\theta_L = \theta_R = d/L$ and substitute these simple-beam end rotations in Eqs. (6.110) and (6.111):

$$M_L^F = -K_L(1 + C_R)\frac{d}{L} \qquad (6.116)$$

$$M_R^F = -K_R(1 + C_L)\frac{d}{L} \qquad (6.117)$$

If end L is displaced downward with respect to R, d/L would be negative and the fixed-end moments positive.

For beams with constant moment of inertia, the fixed-end moments are given by

$$M_L^F = M_R^F = -\frac{6EI}{L} \cdot \frac{d}{L} \qquad (6.118)$$

The fixed-end moments for a propped beam, such as beam LR shown in Fig. 6.71, can be obtained similarly from Eq. (6.114). For variable moment of inertia,

$$M^F = -\left(\frac{d}{L}\right)\left(\frac{1}{\alpha_L}\right) \qquad (6.119)$$

For a prismatic propped beam,

$$M^F = -\frac{3EI}{L} \cdot \frac{d}{L} \qquad (6.120)$$

Reverse signs for downward displacement of end L.

Computation Aids for Prismatic Beams ■ Fixed-end moments for several common types of loading on beams of constant moment of inertia (prismatic beams) are given in Fig. 6.72. Also, the curves in Fig. 6.74 enable fixed-end moments to be computed easily for any type of loading on a prismatic beam. Before the curves can be entered, however, certain characteristics of the loading must be calculated. These include $\tilde{x}L$, the location of the center of gravity of the loading with respect to one of the loads; $G^2 = \Sigma b_n^2 P_n/W$, where $b_n L$ is the distance from each load P_n to the center of gravity of the loading (taken positive to the right); and $S^3 = \Sigma b_n^3 P_n/W$ (see case 8, Fig. 6.73). These values are given in Fig. 6.73 for some common types of loading.

The curves in Fig. 6.74 are entered at the bottom with the location a of the center of gravity of the loading with respect to the left end of the span. At the intersection with the proper G curve, proceed horizontally to the left to the intersection with the proper S line, then vertically to the horizontal scale indicating the coefficient m by which to multiply WL to obtain the fixed-end moment. The curves solve the equations:

$$m_L = \frac{M_L^F}{WL} = G^2[1 - 3(1 - a)] + a(1 - a)^2 + S^3 \qquad (6.121)$$

$$m_R = \frac{M_R^F}{WL} = G^2(1 - 3a) + a^2(1 - a) - S^3 \qquad (6.122)$$

where M_L^F is the fixed-end moment at the left support and M_R^F at the right support.

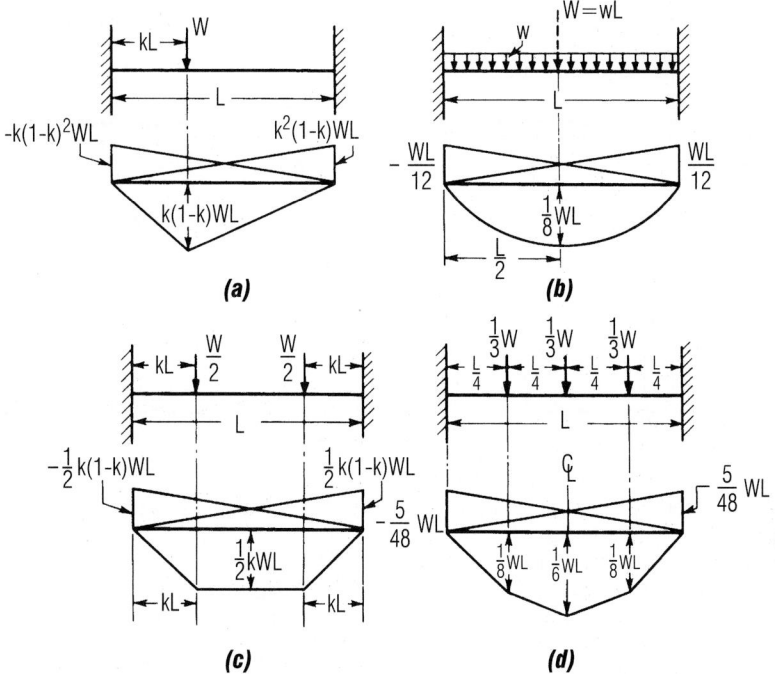

Fig. 6.72 Fixed-end moments for a prismatic beam: (*a*) for concentrated load; (*b*) for a uniform load; (*c*) for two equal concentrated loads; (*d*) for three equal concentrated loads.

As an example of the use of the curves, find the fixed-end moments in a prismatic beam of 20-ft span carrying a triangular loading of 100 kips, similar to the loading shown in case 4, Fig. 6.73, distributed over the entire span, with the maximum intensity at the right support.

Case 4 gives the characteristics of the loading: $y = 1$; the center of gravity is $L/3$ from the right support; so $a = 0.67$, $G^2 = 1/18 = 0.056$, and $S^3 = -1/135 = -0.007$. To find M_R^F, we enter Fig. 6.74 at the bottom with $a = 0.67$ on the upper scale and proceed vertically to the estimated location of the intersection of the coordinate with the $G^2 = 0.06$ curve. Then we move horizontally to the intersection with the line for $S^3 = -0.007$, as indicated by the dashed line in Fig. 6.74. Referring to the scale at the top of the diagram, we find the coefficient m_R to be 0.10. Similarly, with = 0.67 on the lowest scale, we find the coefficient m_L to be 0.07. Hence, the fixed-end moment at the right support is $0.10 \times 100 \times 20 = 200$ ft-kips, and at the left support $-0.07 \times 100 \times 20 = -140$ ft-kips.

6.59 Slope-Deflection Equations

In Arts. 6.56 and 6.58, moments and displacements in a member of a continuous beam or frame are obtained by addition of their simple-beam components. Similarly, moments and displacements can be determined by superposition of fixed-end-beam components. This method, for example, can be used to derive relationships between end moments and end rotations of a beam known as slope-deflection equations. These equations can be used to compute end moments in continuous beams.

Consider a member LR of a continuous beam or frame (Fig. 6.75). LR may have a moment of inertia that varies along its length. The support R is displaced vertically downward a distance d from its original position. Because of this and the loads on the member and adjacent members, LR is subjected to end moments M_L at L and M_R at R. The total end rotation at L is θ_L, and at R, θ_R. All displacements are so small that the member can be considered to

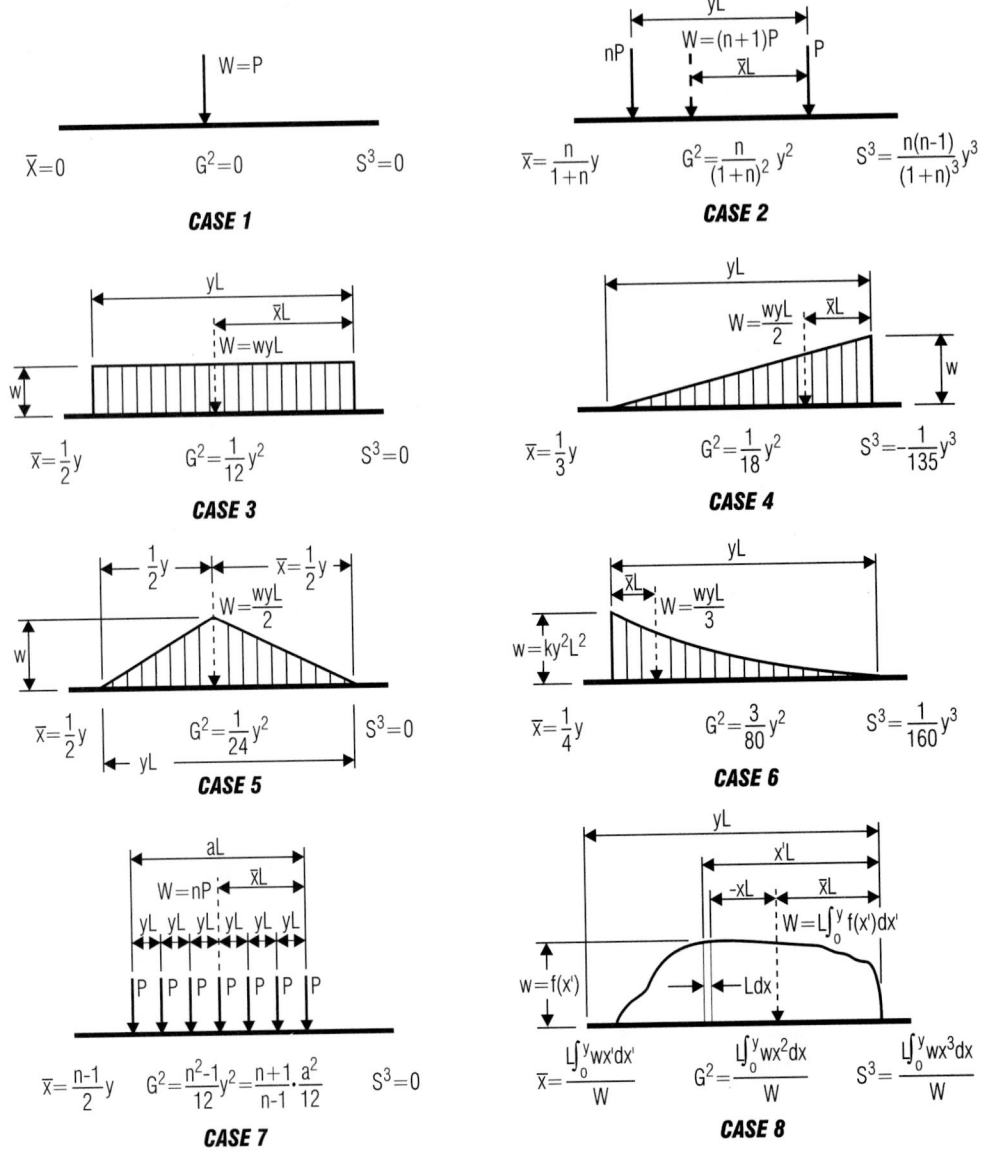

Fig. 6.73 Characteristics of loadings.

rotate clockwise through an angle nearly equal to d/L, where L is the span of the beam.

Assume that rotation is prevented at ends L and R by end moments m_L at L and m_R at R. Then, by application of the principle of superposition (Art. 6.56) and Eqs. (6.116) and (6.117),

$$m_L = M_L^F - K_L(1 + C_R)\frac{d}{L} \qquad (6.123)$$

$$m_R = M_R^F - K_R(1 + C_L)\frac{d}{L} \qquad (6.124)$$

where M_L^F = fixed-end moment at L due to the load on LR

M_R^F = fixed-end moment at R due to load on LR

K_L = fixed-end stiffness at end L

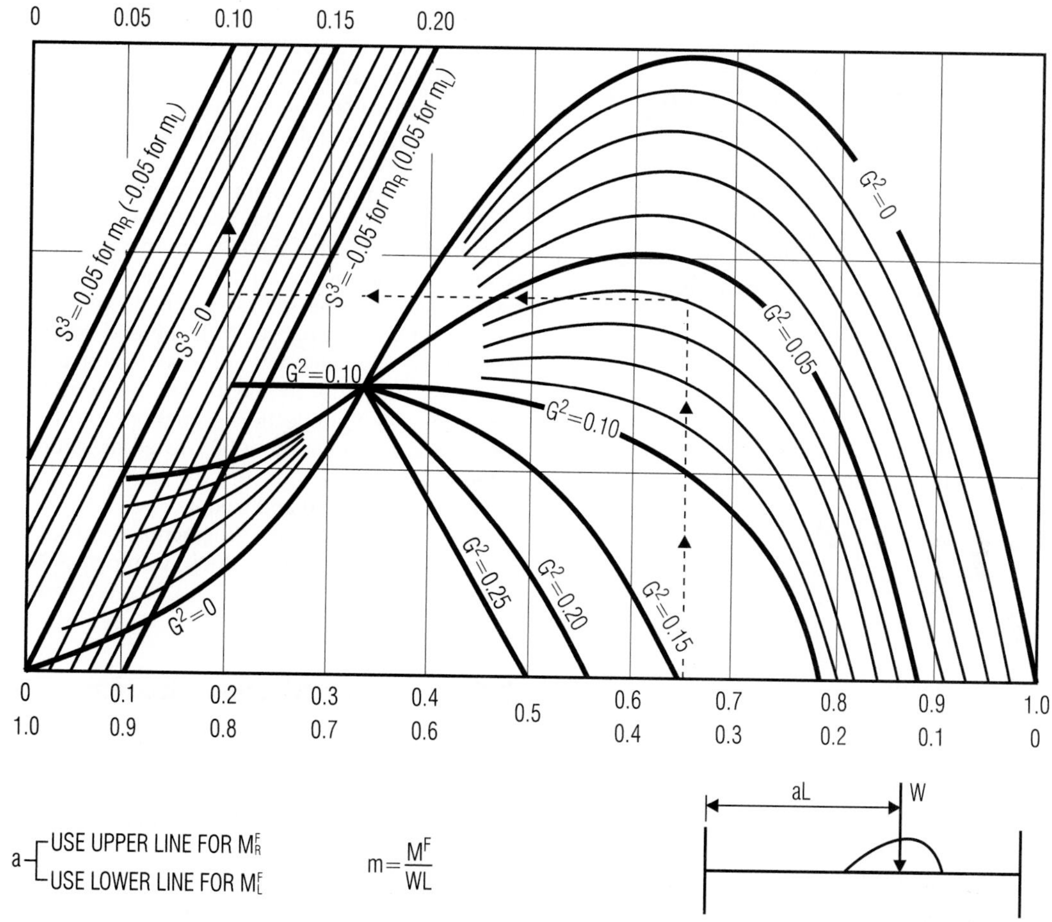

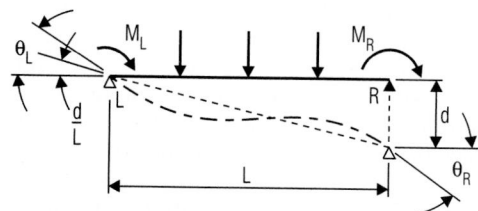

Fig. 6.74 Chart for fixed-end moments caused by any type of loading.

K_R = fixed-end stiffness at end R

C_L = carry-over factor toward end L

C_R = carry-over factor toward end R

Fig. 6.75 End moments M_L and M_R restrain against rotation the ends of loaded span LR of a continuous beam when one end is displaced.

Since ends L and R are not fixed but actually undergo angle changes θ_L and θ_R at L and R, respectively, the joints must now be permitted to rotate while an end moment M'_L is applied at L and an end moment M'_R, at R to produce those angle changes (Fig. 6.76). With the use of the definitions of carry-over factor and fixed-end stiffness (Art. 6.58), these moments are found to be

$$M'_L = K_L(\theta_L + C_R\theta_R) \qquad (6.125)$$
$$M'_R = K_R(\theta_R + C_L\theta_L) \qquad (6.126)$$

The slope-deflection equations for LR then result from addition of M'_L to m_L, which yields M_L, and of M'_R to m_R, which yields M_R,

$$M_L = K_L(\theta_L + C_R\theta_R) + M^F_L - K_L(1 + C_R)\frac{d}{L} \qquad (6.127)$$

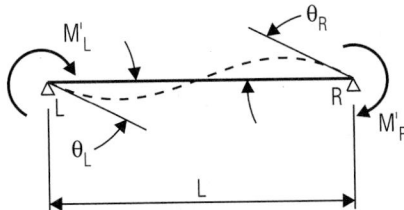

Fig. 6.76 Moments applied to the ends of a simple beam produce end rotations there.

$$M_R = K_R(\theta_R + C_L\theta_L) + M_R^F - K_R(1 + C_L)\frac{d}{L} \quad (6.128)$$

For beams with constant moment of inertia, the slope-deflection equations become

$$M_L = \frac{4EI}{L}\left(\theta_L + \frac{1}{2}\theta_R\right) + M_L^F - \frac{6EI}{L}\cdot\frac{d}{L} \quad (6.129)$$

$$M_R = \frac{4EI}{L}\left(\theta_R + \frac{1}{2}\theta_L\right) + M_R^F - \frac{6EI}{L}\frac{d}{L} \quad (6.130)$$

where E = modulus of elasticity

I = moment of inertia of the cross section

Note that if end L moves downward with respect to R, the sign for d in the preceding equations is changed.

If the end moments M_L and M_R are known and the end rotations are to be determined, Eqs. (6.125) to (6.128) can be solved for θ_L and θ_R or derived by superposition of simple-beam components, as is done in Art. 6.58. For beams with moment of inertia varying along the span:

$$\theta_L = (M_L - M_L^F)\alpha L - (M_R - M_R^F)\beta + \frac{d}{L} \quad (6.131)$$

$$\theta_R = (M_R - M_R^F)\alpha_R - (M_L - M_L^F)\beta + \frac{d}{L} \quad (6.132)$$

where α is given by Eq. (6.99) and β by Eq. (6.100). For beams with constant moment of inertia:

$$\theta_L = \frac{L}{3EI}(M_L - M_L^F) - \frac{L}{6EI}(M_R - M_R^F) + \frac{d}{L} \quad (6.133)$$

$$\theta_R = \frac{L}{3EI}(M_R - M_R^F) - \frac{L}{6EI}(M_L - M_L^F) + \frac{d}{L} \quad (6.134)$$

The slope-deflection equations can be used to determine end moments and rotations of the spans of continuous beams by writing compatibility and equilibrium equations for the conditions at each support. For example, the sum of the moments at each support must be zero. Also, because of continuity, the ends of all members at a support must rotate through the same angle. Hence, M_L for one span, given by Eq. (6.127) or (6.129), must be equal to $-M_R$ for the adjoining span, given by Eq. (6.128) or (6.130), and the end rotation θ at that support must be the same on both sides of the equation. One such equation, with the end rotations at the supports as the unknowns can be written for each support. With the end rotations determined by solution of the simultaneous equations, the end moments can be computed from the slope-deflection equations and the continuous beam can now be treated as statically determinate.

See also Arts. 6.60 and 6.66.

(C. H. Norris et al., "Elementary Structural Analysis,", McGraw-Hill Book Company, New York.)

6.60 Moment Distribution

The properties of fixed-end beams presented in Art. 6.58 enable the computation of end moments in continuous beams and frames by moment distribution, in which end moments induced by loads or displacements of joints are distributed to all the spans. The distribution is based on the assumption that translation is prevented at all joints and supports, rotation of the ends of all members of a joint is the same, and the sum of the end moments at every joint is zero.

The frame in Fig. 6.77 consists of four prismatic members rigidly connected together at O and fixed at ends A, B, C, and D. If an external moment U is applied at O, the sum of the end moments in each member at O must be equal to U. Furthermore, all members must rotate at O through the same angle θ since they are assumed to be rigidly connected there. Hence, by the definition of fixed-end stiffness (Art. 6.58), the proportion of U induced in or "distributed" to the end of each member at O equals the ratio of the stiffness of that member to the sum of the stiffnesses of all the members at O. This ratio is called the **distribution factor** at O for the member.

Suppose a moment of 100 ft-kips is applied at O, as indicated in Fig. 6.77b. The relative stiffness (or I/L) is assumed as shown in the circle on each

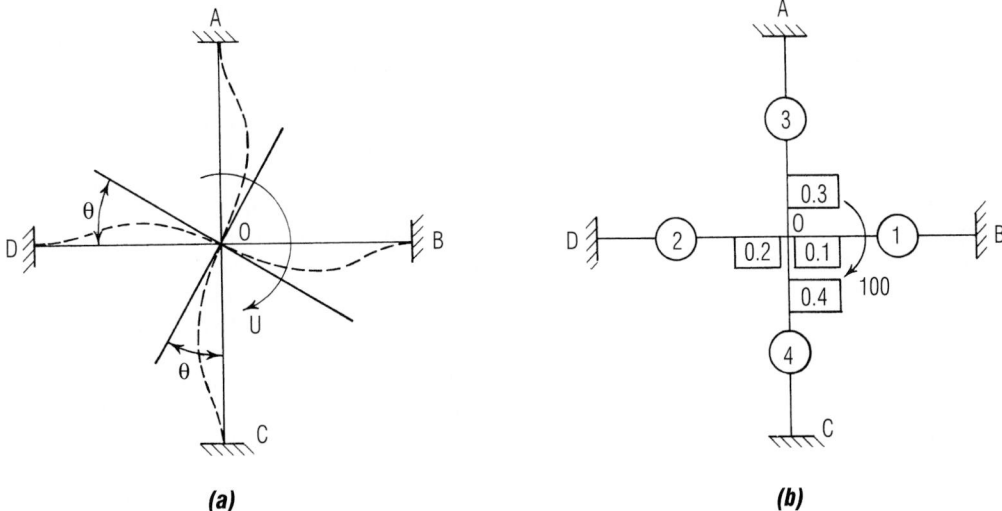

Fig. 6.77 Joint between four members of a simple frame is rotated by an applied moment. (*a*) Elastic curve; (*b*) stiffness and moment distribution factors.

member. The distribution factors for the moment at O are computed from the stiffnesses and shown in the boxes. For example, the distribution factor for OA equals its stiffness divided by the sum of the stiffnesses of all the members at the joint: $3/(3 + 1 + 4 + 2) = 0.3$. Hence, the moment induced in OA at O is $0.3 \times 100 = 30$ ft-kips. Similarly, OB gets 10 ft-kips, OC 40 ft-kips, and OD 20 ft-kips.

Because the far ends of these members are fixed, one-half of these moments are carried over to them (Art. 6.58). Thus $M_{AO} = 0.5 \times 30 = 15$; $M_{BO} = 0.5 \times 10 = 5$; $M_{CO} = 0.5 \times 40 = 20$; and $M_{DO} = 0.5 \times 20 = 10$.

Most structures consist of frames similar to the one in Fig. 6.77, or even simpler, joined together. Although the ends of the members may not be fixed, the technique employed for the frame in Fig. 6.77 can be applied to find end moments in such continuous structures.

Span with Simple Support ▪ Before the general method is presented, one shortcut is worth noting. Advantage can be taken when a member has a hinged end to reduce the work in distributing moments. This is done by using the true stiffness of the member instead of the fixed-end stiffness. (For a prismatic beam, the stiffness of a member with a hinged end is three-fourths the fixed-end stiffness; for a beam with variable moment of inertia, it is equal to the fixed-end stiffness times $1 - C_L C_R$, where C_L and C_R are the fixed-end carry-over factors to each end of the beam.) Naturally, the carry-over factor toward the hinge is zero.

Moment Release and Distribution ▪ When beam ends are neither fixed nor pinned but restrained by elastic members, moments can be distributed by a series of converging approximations. At first, all joints are locked against rotation. As a result, the loads will create fixed-end moments at the ends of every loaded member (Art. 6.58). At each joint, the unbalanced moment, a moment equal to the algebraic sum of the fixed-end moments at the joint, is required to hold it fixed. But if the joint actually is not fixed, the unbalanced moment does not exist. It must be removed by applying an equal but opposite moment. One joint at a time is unlocked by applying a moment equal but opposite in sign to the unbalanced moment. The unlocking moment must be distributed to the members at the joint in proportion to their fixed-end stiffnesses. As a result, the far end of each member should receive a "carry-over" moment equal to the distributed moment times a carry-over factor (Art. 6.58).

After all joints have been released at least once, it generally will be necessary to repeat the process—sometimes several times—before the corrections to the fixed-end moments become negligible. To reduce the number of cycles, start the unlocking of joints with those having the greatest unbalanced moments. Also, include carry-over moments with fixed-end moments in computing unbalanced moments.

Example ▪ Suppose the end moments are to be found for the continuous beam *ABCD* in Fig. 6.78, given the fixed-end moments on the first line of the figure. The *I/L* values for all spans are given equal; therefore, the relative fixed-end stiffness for all members is unity. But since *A* is a hinged end, the computation can be shortened by using the actual relative stiffness, which is ¾. Relative stiffnesses for all members are shown in the circle on each member. The distribution factors are shown in the boxes at each joint.

Begin the computation with removal of the unbalance in fixed-end moments (first line in Fig. 6.78). The greatest unbalanced moment, by inspection, occurs at hinged end *A* and is −400, so unlock this joint first. Since there are no other members at the joint, distribute the full unlocking moment of +400 to *AB* at *A* and carry over one-half to *B*. The unbalance at *B* now is +400−480 plus the carry-over of +200 from *A*, or a total of +120. Hence, a moment of −120

must be applied and distributed to the members at *B* by multiplying by the distribution factors in the corresponding boxes.

The net moment at *B* could be found now by adding the fixed-end and distributed moments at the joint. But it generally is more convenient to delay the summation until the last cycle of distribution has been completed.

After *B* is unlocked, the moment distributed to *BA* need not be carried over to *A* because the carry-over factor toward the hinged end is zero. But half the moment distributed to *BC* is carried over to *C*. Similarly, unlock joint *C* and carry over half the distributed moments to *B* and *D*, respectively. Joint *D* should not be unlocked since it actually is a fixed end. Thus, the first cycle of moment distribution has been completed.

Carry out the second cycle in the same manner. Release joint *B*, and carry over to *C* half the distributed moment in *BC*. Finally, unlock *C* to complete the cycle. Add the fixed-end and distributed moments to obtain the final moments.

6.61 Maximum Moments in Continuous Frames

In continuous frames, maximum end moments and maximum interior moments are produced by

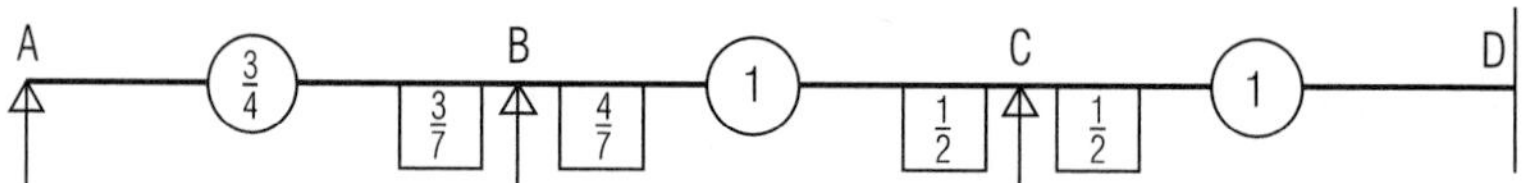

	A		B			C		D
		③⁄₄	③⁄₇	④⁄₇	①	½	½	①
1ST CYCLE								
FIXED-END MOMENTS	-400		+400	-480		+480	-540	+540
DISTRIBUTION AT A	+400	→	+200					
DISTRIBUTION AT B			-51	-69	→	-34	+47	→ +24
DISTRIBUTION AT C			+24	←	+47	-493		+564
MOMENTS	0		+549	-525		+493		
2ND CYCLE								
DISTRIBUTION AT B			-10	-14	→	-7		
DISTRIBUTION AT C			+2	←	+4	+3	→	+2
FINAL MOMENTS	0		+539	-537		+490	-490	+566

Fig. 6.78 Moment distribution in a beam by converging approximations.

different combinations of loadings. For maximum end moment in a beam, live load should be placed on that beam and on the beam adjoining the end for which the moment is to be computed. Spans adjoining these two should be assumed to be carrying only dead load.

For maximum midspan moments, the beam under consideration should be fully loaded, but adjoining spans should be assumed to be carrying only dead load.

The work involved in distributing moments due to dead and live loads in continuous frames in buildings can be greatly simplified by isolating each floor. The tops of the upper columns and the bottoms of the lower columns can be assumed fixed. Furthermore, the computations can be condensed considerably by following the procedure recommended in "Continuity in Concrete Building Frames," EB033D Portland Cement Association, Skokie, Ill. 60077 (www.portcement.org), and illustrated in Fig. 6.79.

Figure 6.79 presents the complete calculation for maximum end and midspan moments in four floor beams *AB*, *BC*, *CD*, and *DE*. Columns are assumed to be fixed at the story above and below. None of the beam or column sections is known to begin with; so as a start, all members will be assumed to have a fixed-end stiffness of unity, as indicated on the first line of the calculation.

Column Moments ▪ The second line gives the distribution factors (Art. 6.60) for each end of the beams; column moments will not be computed until moment distribution to the beams has been completed. Then, the sum of the column moments at each joint may be easily computed since they are the moments needed to make the sum of the end moments at the joint equal to zero. The sum of the column moments at each joint can then be distributed to each column there in proportion to its stiffness. In this example, each column will get one-half the sum of the column moments.

Fixed-end moments at each beam end for dead load are shown on the third line, just above the horizontal line, and fixed-end moments for live plus dead loads on the fourth line. Corresponding midspan moments for the fixed-end condition also are shown on the fourth line, and like the end moments will be corrected to yield actual midspan moments.

Maximum End Moments ▪ For maximum end moment at *A*, beam *AB* must be fully loaded, but *BC* should carry dead load only. Holding *A* fixed, we first unlock joint *B*, which has a total-load fixed-end moment of +172 in *BA* and a dead-load fixed-end moment of −37 in *BC*. The releasing

	A		B			C			D		E
1. STIFFNESS		1			1			1			1	
2. DISTRIBUTION FACTOR	0.33		0.25	0.25		0.25	0.25		0.25	0.25		0.33
3. F.E.M. DEAD LOAD	—		+91	-37		+37	-70		+70	-59		—
4. F.E.M. TOTAL LOAD	-172	+99	+172	-78	+73	+78	-147	+85	+147	-126	+63	+126
5. CARRY-OVER	-17	+11	+29	-1	+1	-2	-11	+7	+14	-21	+13	+7
6. ADDITION	-189	+18	+201	-79	-1	+76	-158	+9	+161	-147	+5	+133
7. DISTRIBUTION	+63		-30	-30		+21	+21		-4	-4		-44
8. MAX. MOMENTS	-126	+128	+171	-109	+73	+97	-137	+101	+157	-151	+81	+89

Fig. 6.79 Moment distribution in a continuous frame by converging approximations.

moment required, therefore, is $-(172-37)$, or -135. When B is released, a moment of -135×0.25 is distributed to BA. One-half of this is carried over to A, or $-135 \times 0.25 \times 0.5 = -17$. This value is entered as the carry-over at A on the fifth line in Fig. 6.79. Joint B then is relocked.

At A, for which we are computing the maximum moment, we have a total-load fixed-end moment of -172 and a carry-over of -17, making the total -189, shown on the sixth line. To release A, a moment of $+189$ must be applied to the joint. Of this, 189×0.33, or 63, is distributed to AB, as indicated on the seventh line. Finally, the maximum moment at A is found by adding lines 6 and 7: $-189 + 63 = -126$.

For maximum moment at B, both AB and BC must be fully loaded, but CD should carry only dead load. We begin the determination of the maximum moment at B by first releasing joints A and C, for which the corresponding carry-over moments at BA and BC are $+29$ and $-(+78-70) \times 0.25 \times 0.5 = -1$, shown on the fifth line in Fig. 6.79. These bring the total fixed-end moments in BA and BC to $+201$ and -79, respectively. The releasing moment required is $-(201-79) = -122$. Multiplying this by the distribution factors for BA and BC when joint B is released, we find the distributed moments, -30, entered on line 7. The maximum end moments finally are obtained by adding lines 6 and 7: $+171$ at BA and -109 at BC. Maximum moments at C, D, and E are computed and entered in Fig. 6.79 in a similar manner. This procedure is equivalent to two cycles of moment distribution.

Maximum Midspan Moments ■ The computation of maximum midspan moments in Fig. 6.79 is based on the assumption that in each beam the midspan moment is the sum of the simple-beam midspan moment and one-half the algebraic difference of the final end moments (the span carries full load but adjacent spans only dead load). Instead of starting with the simple-beam moment, however, we begin, for convenience, with the midspan moment for the fixed-end condition and then apply two corrections. In each span, these corrections equal the carry-over moments entered on line 5 for the two ends of the beam multiplied by a factor.

For beams with variable moment of inertia, the factor is $\pm\frac{1}{2}(1/C + D - 1)$, where C is the fixed-end

carry-over factor toward the end for which the correction factor is being computed and D the distribution factor for that end. The plus sign is used for correcting the carry-over at the right end of a beam and the minus sign for the carry-over at the left end. For prismatic beams, the correction factor becomes $\pm\frac{1}{2}(1 + D)$.

For example, to find the corrections to the midspan moment in AB, we first multiply the carry-over at A on line 5, -17, by $-\frac{1}{2}(1 + 0.33)$. The correction, $+11$, also is entered on the fifth line. Then we multiply the carry-over at B, $+29$, by $+\frac{1}{2}(1 + 0.25)$ and enter the correction, $+18$, on line 6. The final midspan moment is the sum of lines 4, 5, and 6: $+99 + 11 + 18 = +128$. Other midspan moments in Fig. 6.79 are obtained in a similar manner.

Approximate methods for determining wind and seismic stresses in tall buildings are given in Arts. 15.4, 15.9, and 15.10.

6.62 Moment-Influence Factors

For certain types of structures, particularly those for which different types of loading conditions must be investigated, it may be more convenient to find maximum end moments from a table of moment-influence factors. This table is made up by listing for the end of each member in a structure the moment induced in that end when a moment (for convenience, $+1000$) is applied to each joint successively. Once this table has been prepared, no additional moment distribution is necessary for computing the end moments due to any loading condition.

For a specific loading pattern, the moment at any beam end M_{AB} may be obtained from the moment-influence table by multiplying the entries under AB for the various joints by the actual

Table 6.3 Moment-Influence Factors for Fig. 6.80

Member	$+1,000$ at B	$+1,000$ at C
AB	351	-105
BA	702	-210
BC	298	210
CB	70	579
CD	-70	421
DC	-35	210

Table 6.4 Moment Collection Table for Fig. 6.80

Remarks	AB	BA	BC	CB	CD	DC
1. Sidesway *FEM*	−3,000*M*	−3,000*M*			−1,000*M*	−1,000*M*
2. Distribution for *B*	+1,053*M*	−2,106*M*	+894*M*	+210*M*	−210*M*	−105*M*
3. Distribution for *C*	−105*M*	−210*M*	+210*M*	+579*M*	+421*M*	+210*M*
4. Final sidesway *M*	−2,052*M*	−1,104*M*	+1,104*M*	+789*M*	−789*M*	−895*M*
5. For 2,000-lb horizontal	−17,000	−9,100	+9,100	+6,500	−6,500	−7,400
6. 4,000-lb vertical FEM			−12,800	+3,200		
7. Distribution for *B*	+4,490	+8,980	+3,820	+897	−897	−448
8. Distribution for *C*	+336	+672	−672	−1,853	−1,347	−673
9. Moments with no sidesway	+4,826	+9,652	−9,652	+2,244	−2,244	−1,121
10. Sidesway *M*	−4,710	−2,540	+2,540	+1,810	−1,810	−2,060
11. For 4,000-lb vertical	+116	+7,112	−7,112	+4,054	−4,054	−3,181

unbalanced moments at those joints divided by 1000 and summing. (See also Art. 6.64 and Tables 6.3 and 6.4.)

6.63 Procedure for Sidesway

For some structures, it is convenient to know the effect of a movement of a support normal to the original position. But the moment-distribution method is based on the assumption that such movement of a support does not occur. The method, however, can be modified to evaluate end moments resulting from a support movement.

The procedure is to distribute moments as usual, assuming no deflection at the supports. This implies that additional external forces are exerted at the supports to prevent movement. These forces can be computed. Then, equal and opposite forces are applied to the structure to produce the final configuration, and the moments that they induce are distributed as usual. These moments added to those obtained with undeflected supports yield the final moments.

Example—Horizontal Axial Load ▪ Suppose the rigid frame in Fig. 6.80 is subjected to a 2000-lb horizontal load acting to the right at the level of beam *BC*. The first step is to compute the moment-influence factors by applying moments of

+1000 at joints *B* and *C* (Art. 6.62), assuming sidesway is prevented, and enter the distributed moments in Table 6.3.

Since there are no intermediate loads on the beam and columns, the only fixed-end moments that need be considered are those in the columns due to lateral deflection of the frame.

This deflection, however, is not known initially. So we assume an arbitrary deflection, which produces a fixed-end moment of −1000*M* at the

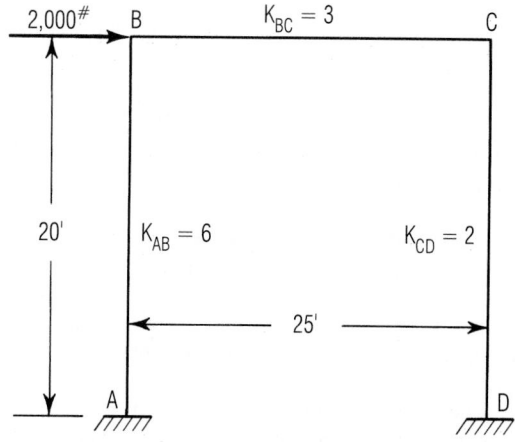

Fig. 6.80 Laterally loaded rigid frame.

top of column CD. M is an unknown constant to be determined from the fact that the sum of the shears in the deflected columns must equal the 2000-lb load. The deflection also produces a moment of $-1000M$ at the bottom of CD [see Eq. (6.118)].

From the geometry of the structure, we furthermore note that the deflection of B relative to A equals the deflection of C relative to D. Then, according to Eq. 6.118, the fixed-end moments of the columns of this frame are proportional to the stiffnesses of the columns and hence are equal in AB to $-1000M \times \frac{6}{2} = -3000M$. The column fixed-end moments are entered in the first line of Table 6.4, the moment-collection table for Fig. 6.80.

In the deflected position of the frame, joints B and C are unlocked in succession. First, we apply a releasing moment of $+3000M$ at B. We distribute it by multiplying by 3 the entries in the columns marked "+1000 at B" in Table 6.3. Similarly, a releasing moment of $+1000M$ is applied at C and distributed with the aid of the moment-influence factors. The distributed moments are entered in the second and third lines of the moment-collection table. The final moments are the sum of the fixed-end moments and the distributed moments and are given in the fourth line of Table 6.4, in terms of M.

Isolating each column and taking moments about one end, we find that the overturning moment due to the shear equals the sum of the end moments. We have one such equation for each column. Adding these equations, noting that the sum of the shears equals 2000 lb, we obtain

$$-M(2052 + 1104 + 789 + 895) = -2000 \times 20$$

from which we find $M = 8.26$. This value is substituted in the sidesway totals (line 4) in the moment-collection table to yield the end moments for the 2000-lb horizontal load (line 5).

Example—Vertical Load on Beam ■

Suppose a vertical load of 4000 lb is applied to BC of the rigid frame in Fig. 6.80, 5 ft from B. The same moment-influence factors and moment-collection table can again be used to determine the end moments with a minimum of labor.

The fixed-end moment at B, with sidesway prevented, is $-12,800$, and at $C +3200$ (Fig. 6.72a). With the joints still locked, the frame is permitted to

move laterally an arbitrary amount, so that in addition to the fixed-end moments due to the 4000-lb load, column fixed-end moments of $-3000M$ at A and B and $-1000M$ at C and D are induced. The moment-collection table already indicates in line 4 the effect of relieving these column moments by unlocking joints B and C. We now have to superimpose the effect of releasing joints B and C to relieve the fixed-end moments for the vertical load. This we can do with the aid of the moment-influence factors. The distribution is shown in lines 7 and 8 of Table 6.4, the moment-collection table. The sums of the fixed-end moments and distributed moments for the 4000-lb load are shown in line 9.

The unknown M can be evaluated from the fact that the sum of the horizontal forces acting on the columns must be zero. This is equivalent to requiring that the sum of the column end moments equal zero:

$$-M(2052 + 1104 + 789 + 895)$$
$$+ 4826 + 9652 - 2244 - 1121 = 0$$

from which $M = 2.30$. This value is substituted in line 4 of Table 6.4 to yield the sidesway moments for the 4000-lb load (line 10). Addition of these moments to the totals for no sidesway (line 9) gives the final moments (line 11).

Multistory Frames ■ This procedure permits analysis of one-story bents with straight beams by solution of one equation with one unknown, regardless of the number of bays. If the frame is multistory, the procedure can be applied to each story. Since an arbitrary horizontal deflection is introduced at each floor or roof level, there are as many unknowns and equations as there are stories. (For approximate methods for determining wind and seismic stresses in tall buildings, see Arts. 15.9 and 15.10.)

Arched Bents ■ The procedure is more difficult to apply to bents with curved or polygonal members between the columns. The effect of the change in the horizontal projection of the curved or polygonal portion of the bent must be included in the calculations. In many cases, it may be easier to analyze the bent as a curved beam (arch).

6.64 Load Distribution to Bents and Shear Walls

Provision should be made for all structures to transmit lateral loads, such as those from wind, earthquakes, and traction and braking of vehicles, to foundations and their supports that have high resistance to displacement. For the purpose, various types of bracing may be used, including struts, tension ties, diaphragms, trusses, and shear walls.

The various bracing members are usually designed to interact as a system. Structural analysis then is necessary to determine the distribution of the lateral loads on the system to the bracing members. The analysis may be based on the principles presented in the preceding articles but it requires a knowledge or assumption of the structural characteristics of the system components. For example, suppose a horizontal diaphragm, such as a concrete floor, is to be used to distribute horizontal forces to several parallel vertical trusses. In this case, the distribution would depend not only on the relative resistance of the trusses to the horizontal forces but also on the rigidity (or flexibility) of the diaphragm.

In tall buildings, bents or shear walls, which act as vertical cantilevers and generally are often also used to support some of the gravity loads, usually are spaced at appropriate intervals to transmit lateral loads to the foundations. A **bent** consists of vertical trusses or continuous rigid frames located in a plane (Fig. 6.81*a*). The trusses usually are an assemblage of columns, horizontal girders, and diagonal bracing (Fig. 6.81*b* to *e*). The rigid frames are composed of girders and columns, with so-called wind connections between them to establish continuity (Fig. 6.81*f*). **Shear walls** are thin cantilevers, usually constructed of concrete but sometimes of masonry or steel plates (Fig. 6.81*g*). They require bracing normal to their plane.

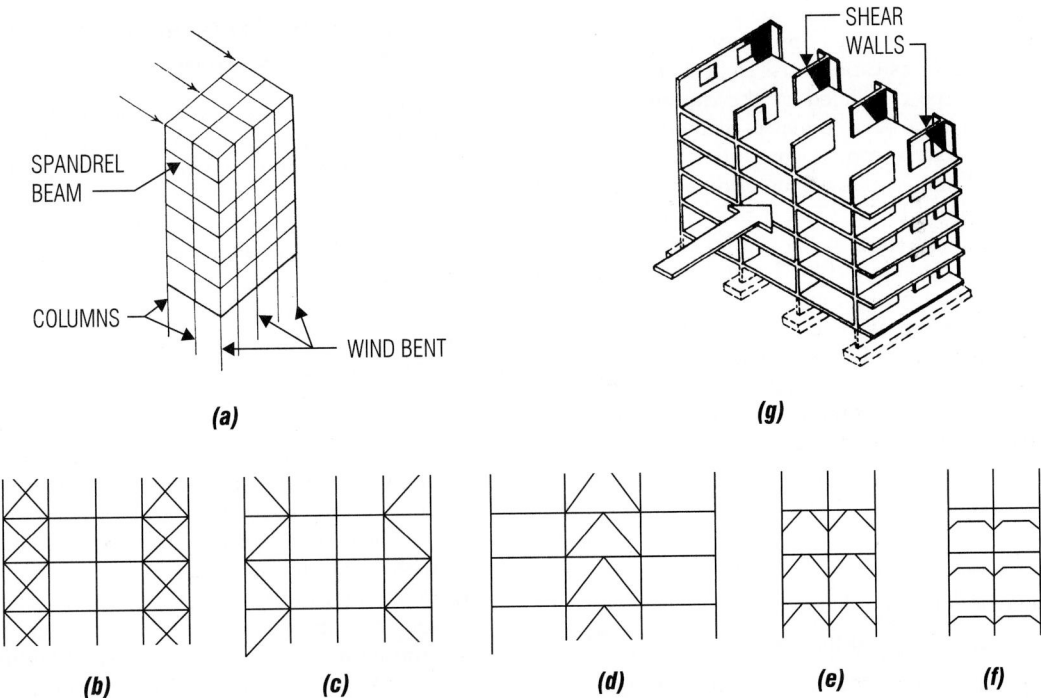

Fig. 6.81 Building frame resists lateral forces with (*a*) wind bents or (*g*) shear walls or a combination of the two. Bents may be braced in any of several ways, including (*b*) X bracing, (*c*) K bracing, (*d*) inverted V bracing, (*e*) knee bracing, and (*f*) rigid connections.

Where bents or shear walls are connected by rigid diaphragms so that they must deflect equally under horizontal loads, the proportion of the total horizontal load at any level carried by a bent or shear wall that is parallel to the load depends on the relative rigidity, or stiffness, of the bent or wall. Rigidity of this bracing is inversely proportional to its deflection under a unit horizontal load.

When the line of action of the resultant of the lateral forces does not pass through the center of rigidity of the vertical, lateral-force-resisting system, distribution of rotational forces must be considered as well as distribution of the translational forces. If relatively rigid diaphragms are used, the torsional forces may be distributed to the bents or shear walls in proportion to their relative rigidities and their distance from the center of rigidity. A flexible diaphragm should not be considered capable of distributing torsional forces.

Deflections of Bents and Shear Walls ■ Horizontal deflections in the planes of bents and shear walls can be computed on the assumption that they act as cantilevers. Deflections of braced bents can be calculated by the dummy-unit-load method (Art. 6.54) or a matrix method. Deflections of rigid frames can be computed by adding the drifts of the stories, as determined by moment distribution (Art. 6.60) or a matrix method. And deflections of shear walls can be calculated from formulas given in Art. 6.32, the dummy-unit-load method, or a matrix method.

For a shear wall, the deflection in its plane induced by a load in its plane is the sum of the flexural deflection as a cantilever and the deflection due to shear. Thus, for a wall with solid rectangular cross section, the deflection at the top due to uniform load is

$$\delta = \frac{1.5wH}{Et}\left[\left(\frac{H}{L}\right)^3+\frac{H}{L}\right] \qquad (6.135)$$

where w = uniform lateral load

H = height of the wall

E = modulus of elasticity of the wall material

t = wall thickness

L = length of wall

For a shear wall with a concentrated load P at the top, the deflection at the top is

$$\delta_c = \frac{4P}{Et}\left[\left(\frac{H}{L}\right)^3+0.75\frac{H}{L}\right] \qquad (6.136)$$

If the wall is fixed against rotation at the top, however, the deflection is

$$\delta_f = \frac{P}{Et}\left[\left(\frac{H}{L}\right)^3+3\frac{H}{L}\right] \qquad (6.137)$$

Where shear walls contain openings, such as those for doors, corridors, or windows, computations for deflection and rigidity are more complicated. Approximate methods, however, may be used.

(F. S. Merritt and Jonathan T. Ricketts, "Building Design and Construction Handbook," 5th ed., McGraw-Hill Publishing Co., New York, books. mcgraw-hill.com.)

6.65 Beams Stressed into the Plastic Range

When an elastic material, such as structural steel, is loaded with a gradually increasing load, stresses are proportional to strains nearly to the yield point. If the material, like steel, also is ductile, then it continues to carry load beyond the yield point, although strains increase rapidly with little increase in load (Fig. 6.82a).

Similarly, a beam made of a ductile material continues to carry more load after the stresses in the outer surfaces reach the yield point. The stresses, however, will no longer vary with distance from the neutral axis; so the flexure formula [Eq. (6.44)] no longer holds. But if simplifying assumptions are made, approximating the stress-strain relationship beyond the elastic limit, the load-carrying capacity of the beam can be computed with satisfactory accuracy.

Modulus of rupture is defined as the stress computed from the flexure formula for the maximum bending moment a beam sustains at failure. This is not a true stress but it is sometimes used to compare the strength of beams.

For a ductile material, the idealized stress-strain relationship in Fig. 6.82b may be assumed. Stress is proportional to strain until the yield-point stress

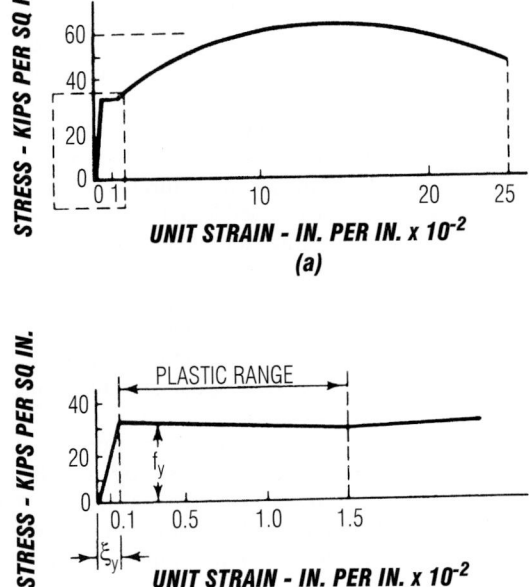

Fig. 6.82 Stress-strain relationship for a ductile material generally is similar to the curve in (*a*). To simplify plastic analysis, the portion of (*a*) enclosed by the dashed lines is approximated by the curve in (*b*), which extends to the range where strain hardening begins.

f_y is reached, after which strain increases at a constant stress.

For a beam of this material, it is also assumed that:

1. Plane sections remain plane, strains thus being proportional to distance from the neutral axis.

2. Properties of this material in tension are the same as those in compression.

3. Its fibers behave the same in flexure as in tension.

4. Deformations remain small.

Strain distribution across the cross section of a rectangular beam, based on these assumptions, is shown in Fig. 6.83*a*. At the yield point, the unit strain is ε_y and the curvature ϕ_y, as indicated in (1). In (2), the strain has increased several times, but the section still remains plane. Finally, at failure, (3),

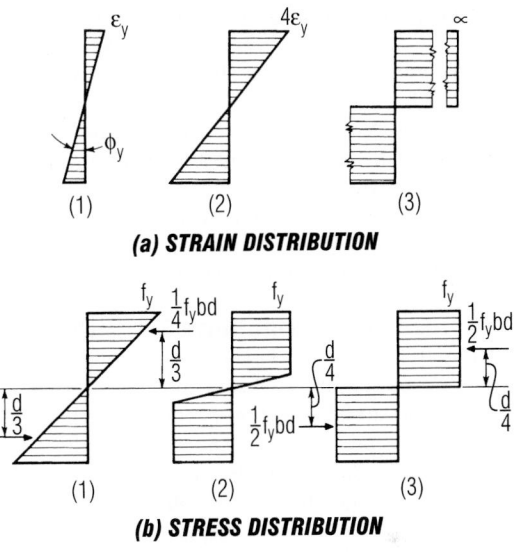

(a) STRAIN DISTRIBUTION

(b) STRESS DISTRIBUTION

Fig. 6.83 Strain distribution is shown in (*a*) and stress distribution in (*b*) for a cross section of a rectangular beam as it is loaded beyond the yield point, assuming the idealized stress-strain relationship in Fig. 6.82*b*. Stage (1) shows the conditions at the yield point of the outer surfaces; (2) after yielding starts; and (3) at ultimate load.

the strains are very large and nearly constant across the lower and upper halves of the section.

Corresponding stress distributions are shown in Fig. 6.83*b*. At the yield point (1), stresses vary linearly and the maximum is f_y. With increase in load, more and more fibers reach the yield point, and the stress distribution becomes nearly constant, as indicated in (2). Finally, at failure (3), the stresses are constant across the top and bottom parts of the section and equal to the yield-point stress.

The resisting moment at failure for a rectangular beam can be computed from the stress diagram for stage 3. If b is the width of the member and d its depth, then the ultimate moment for a rectangular beam is

$$M_p = \frac{bd^2}{4} f_y \tag{6.138}$$

Since the resisting moment at stage 1 is $M_y = f_y bd^2/6$, the beam carries 50% more moment before failure than when the yield-point stress is first reached in the outer fibers ($M_P/M_y = 1.5$).

A circular section has an M_P/M_y ratio of about 1.7, while a diamond section has a ratio of 2. The average wide-flange rolled-steel beam has a ratio of about 1.14.

The relationship between moment and curvature in a beam can be assumed to be similar to the stress-strain relationship in Fig. 6.82b. Curvature ϕ varies linearly with moment until $M_y = M_P$ is reached, after which ϕ increases indefinitely at constant moment. That is, a plastic hinge forms.

Moment Redistribution ▪ This ability of a ductile beam to form plastic hinges enables a fixed-end or continuous beam to carry more load after M_P occurs at a section because a redistribution of moments takes place. Consider, for example, a uniformly loaded fixed-end beam. In the elastic range, the end moments are $M_L = M_R = WL/12$, while the midspan moment M_C is $WL/24$. The load when the yield point is reached in the outer fibers is $W_y = 12M_y/L$. Under this load, the moment capacity of the ends of the beam is nearly exhausted; plastic hinges form there when the moment equals M_P. As load is increased, the ends then rotate under constant moment and the beam deflects as a simply supported beam. The moment at midspan increases until the moment capacity at that section is exhausted and a plastic hinge forms. The load causing that condition is the ultimate load W_u since, with three hinges in the span, a link mechanism is formed and the member continues to deform at constant load. At the time the third hinge is formed, the moments at ends and center are all equal to M_P. Therefore, for equilibrium, $2M_P = W_uL/8$, from which $W_u = 16M_P/L$. Since, for the idealized moment-curvature relationship, M_P was assumed equal to M_y, the carrying capacity due to redistribution of moments is 33% greater.

Finite-Element Methods

From the basic principles given in preceding articles, systematic procedures have been developed for determining the behavior of a structure from a knowledge of the behavior under load of its components. In these methods, called finite-element methods, a structural system is considered an assembly of a finite number of finite-size components, or elements. These are assumed to be connected to each other only at discrete points, called nodes. From the characteristics of the elements, such as their stiffness or flexibility, the characteristics of the whole system can be derived. With these known, internal stresses and strains throughout can be computed.

Choice of elements to be used depends on the type of structure. For example, for a truss with joints considered hinged, a natural choice of element would be a bar, subjected only to axial forces. For a rigid frame, the elements might be beams subjected to bending and axial forces, or to bending, axial forces, and torsion. For a thin plate or shell, elements might be triangles or rectangles, connected at vertices. For three-dimensional structures, elements might be beams, bars, tetrahedrons, cubes, or rings.

For many structures, because of the number of finite elements and nodes, analysis by a finite-element method requires mathematical treatment of large amounts of data and solution of numerous simultaneous equations. For this purpose, use of computers is advisable. The mathematics of such analyses is usually simpler and more compact when the data are handled in matrix form. (See also Art. 6.57.)

6.66 Force and Displacement Methods

The methods used for analyzing structures generally may be classified as force (flexibility) or displacement (stiffness) methods.

In analysis of statically indeterminate structures by force methods, forces are chosen as redundants, or unknowns. The choice is made in such a way that equilibrium is satisfied. These forces are then determined from the solution of equations that insure compatibility of all displacements of elements at each node. After the redundants have been computed, stresses and strains throughout the structure can be found from equilibrium equations and stress-strain relations.

In displacement methods, displacements are chosen as unknowns. The choice is made in such a way that geometric compatibility is satisfied. These displacements are then determined from the solution of equations that insure that forces acting at each node are in equilibrium. After the unknowns have been computed, stresses and strains throughout the structure can be found from equilibrium equations and stress-strain relations.

When choosing a method, the following should be kept in mind: In force methods, the number of unknowns equals the degree of indeterminancy. In displacement methods, the number of unknowns equals the degrees of freedom of displacement at nodes. The fewer the unknowns, the fewer the calculations required.

Both methods are based on the force-displacement relations and utilize the stiffness and flexibility matrices described in Art. 6.57. In these methods, displacements and external forces are resolved into components—usually horizontal, vertical, and rotational—at nodes, or points of connection of finite elements. In accordance with Eq. (6.97a), the stiffness matrix transforms displacements into forces. Similarly, in accordance with Eq. (6.97b), the flexibility matrix transforms forces into displacements. To accomplish the transformation, the nodal forces and displacements must be assembled into correspondingly positioned elements of force and displacement vectors. Depending on whether the displacement or the force method is chosen, stiffness or flexibility matrices are then established for each of the finite elements and these matrices are assembled to form a square matrix, from which the stiffness or flexibility matrix for the structure as a whole is derived. With that matrix known and substituted into equilibrium and compatibility equations for the structure, all nodal forces and displacements of the finite elements can be determined from the solution of the equations. Internal stresses and strains in the elements can be computed from the now known nodal forces and displacements.

6.67 Element Flexibility and Stiffness Matrices

The relationship between *independent* forces and displacements at nodes of finite elements in a structure is determined by flexibility matrices **f** or stiffness matrices **k** of the elements. In some cases, the components of these matrices can be developed from the defining equations:

The jth column of a flexibility matrix of a finite element contains all the nodal displacements of the element when one force S_j is set equal to unity and all other independent forces are set equal to zero.

The jth column of a stiffness matrix of a finite element consists of the forces acting at the nodes of the element to produce a unit displacement of the node at which displacement δ_j occurs and in the direction of δ_j but no other nodal displacements of the element.

Bars with Axial Stress Only ■ As an example of the use of the definitions of flexibility and stiffness, consider the simple case of an elastic bar under tension applied by axial forces P_i and P_j at nodes i and j, respectively (Fig. 6.84). The bar might be the finite element of a truss, such as a diagonal or a hanger. Connections to other members are made at nodes i and j, which can transmit only forces in the directions i to j or j to i.

For equilibrium, $P_i = P_j = P$. Displacement of node j relative to node i is e. From Eq. (6.8), $e = PL/AE$, where L is the initial length of the bar, A the bar cross-sectional area, and E the modulus of elasticity. Setting $P = 1$ yields the flexibility of the bar,

$$f = \frac{L}{AE} \qquad (6.139)$$

Setting $e = 1$ gives the stiffness of the bar,

$$k = \frac{AE}{L} \qquad (6.140)$$

Beams with Bending Only ■ As another example of the use of the definition to determine element flexibility and stiffness matrices, consider the simple case of an elastic prismatic beam in bending applied by moments M_i and M_j at nodes i and j, respectively (Fig. 6.85). The beam might be a finite element of a rigid frame. Connecztions to other members are made at nodes i and j, which can transmit moments and forces normal to the beam.

Nodal displacements of the element can be sufficiently described by rotations θ_i and θ_j relative to the straight line between nodes i and j. For equilibrium, forces $V_j = -V_i$ normal to the beam are required at nodes j and i, respectively, and $V_j = (M_i + M_j)/L$, where L is the span of the beam. Thus, M_i and M_j are the only independent forces

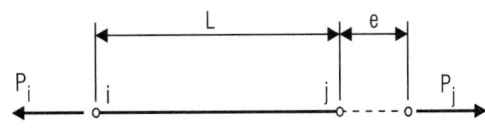

Fig. 6.84 Elastic bar in tension.

acting. Hence, the force-displacement relationship can be written for this element as

$$\boldsymbol{\theta} = \begin{bmatrix} \theta_i \\ \theta_j \end{bmatrix} = \mathbf{f} \begin{bmatrix} M_i \\ M_j \end{bmatrix} = \mathbf{fM} \qquad (6.141)$$

$$\mathbf{M} = \begin{bmatrix} M_i \\ M_j \end{bmatrix} = \mathbf{k} \begin{bmatrix} \theta_i \\ \theta_j \end{bmatrix} = \mathbf{k}\boldsymbol{\theta} \qquad (6.142)$$

The flexibility matrix $\mathbf{f}$ then will be a 2×2 matrix. The first column can be obtained by setting $M_i = 1$ and $M_j = 0$ (Fig. 6.85b). The resulting angular rotations are given by Eqs. (6.101) and (6.102). For a beam with constant moment of inertia I and modulus of elasticity E, the rotations are $\alpha = L/3EI$ and $\beta = -L/6EI$. Similarly, the second column can be developed by setting $M_i = 0$ and $M_j = 1$.

The flexibility matrix for a beam in bending then is

$$\mathbf{f} = \begin{bmatrix} \dfrac{L}{3EI} & -\dfrac{L}{6EI} \\ -\dfrac{L}{6EI} & \dfrac{L}{3EI} \end{bmatrix} = \dfrac{L}{6EI} \begin{bmatrix} 2 & -1 \\ -1 & 2 \end{bmatrix} \qquad (6.143)$$

The stiffness matrix, obtained in a similar manner or by inversion of $\mathbf{f}$, is

$$\mathbf{k} = \begin{bmatrix} \dfrac{4EI}{L} & \dfrac{2EI}{L} \\ \dfrac{2EI}{L} & \dfrac{4EI}{L} \end{bmatrix} = \dfrac{2EI}{L} \begin{bmatrix} 2 & 1 \\ 1 & 2 \end{bmatrix} \qquad (6.144)$$

Beams Subjected to Bending and Axial Forces ▪ For a beam subjected to nodal moments M_i and M_j and axial forces P, flexibility and stiffness are represented by 3×3 matrices. The load-displacement relations for a beam of span L, constant moment of inertia I, modulus of elasticity E, and cross-sectional area A are given by

$$\begin{bmatrix} \theta_i \\ \theta_j \\ e \end{bmatrix} = \mathbf{f} \begin{bmatrix} M_i \\ M_j \\ P \end{bmatrix} \qquad \begin{bmatrix} M_i \\ M_j \\ P \end{bmatrix} = \mathbf{k} \begin{bmatrix} \theta_i \\ \theta_j \\ e \end{bmatrix} \qquad (6.145)$$

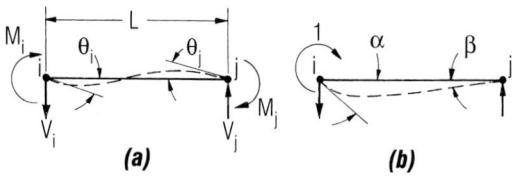

(a)   **(b)**

Fig. 6.85 Beam subjected to end moments and shears.

where e = axial displacement. In this case, the flexibility matrix is

$$\mathbf{f} = \dfrac{L}{6EI} \begin{bmatrix} 2 & -1 & 0 \\ -1 & 2 & 0 \\ 0 & 0 & \eta \end{bmatrix} \qquad (6.146)$$

where $\eta = 6I/A$, and the stiffness matrix, with $\psi = A/I$, is

$$\mathbf{k} = \dfrac{EI}{L} \begin{bmatrix} 4 & 2 & 0 \\ 2 & 4 & 0 \\ 0 & 0 & \psi \end{bmatrix} \qquad (6.147)$$

6.68 Displacement (Stiffness) Method

With the stiffness or flexibility matrix of each finite element of a structure known, the stiffness or flexibility matrix for the whole structure can be determined, and with that matrix, forces and displacements throughout the structure can be computed (Art. 6.67). To illustrate the procedure, the steps in the displacement, or stiffness, method are described in the following. The steps in the flexibility method are similar. For the stiffness method:

Step 1. Divide the structure into interconnected elements and assign a number, for identification purposes, to every node (intersection and terminal of elements). It may also be useful to assign an identifying number to each element.

Step 2. Assume a right-handed cartesian coordinate system, with axes x, y, z. Assume also at each node of a structure to be analyzed a system of base unit vectors, $\mathbf{e}_1$ in the direction of the x axis, $\mathbf{e}_2$ in the direction of the y axis, and $\mathbf{e}_3$ in the direction of the z axis. Forces and moments acting at a node are resolved into components in the directions of the base vectors. Then, the forces and moments at the node may be represented by the vector $P_i\mathbf{e}_i$, where P_i is the magnitude of the force or moment acting in the direction of $\mathbf{e}_i$. This vector, in turn, may be conveniently represented by a column matrix $\mathbf{P}$. Similarly, the displacements—translations and rotation—of the node may be represented by the vector $\Delta_i\mathbf{e}_i$, where Δ_i is the magnitude of the displacement acting in the direction of $\mathbf{e}_i$. This vector, in turn, may be represented by a column matrix Δ.

For compactness, and because, in structural analysis, similar operations are performed on all nodal forces, all the loads, including moments, acting on all the nodes may be combined into a single column matrix **P**. Similarly, all the nodal displacements may be represented by a single column matrix **Δ**.

When loads act along a beam, they could be replaced by equivalent forces at the nodes— simple-beam reactions and fixed-end moments, both with signs reversed from those induced by the loads. The final element forces are then determined by adding these moments and reactions to those obtained from the solution with only the nodal forces.

Step 3. Develop a stiffness matrix k_i for each element i of the structure (see Art. 6.67). By definition of stiffness matrix, nodal displacements and forces for the ith element are related by

$$S_i = k_i \delta_i \qquad i = 1, 2, \ldots, n \qquad (6.148)$$

where S_i = matrix of forces, including moments and torques acting at the nodes of the ith element

δ_i = matrix of displacements of the nodes of the ith element

Step 4. For compactness, combine this relationship between nodal displacements and forces for each element into a single matrix equation applicable to all the elements:

$$S = k\delta \qquad (6.149)$$

where **S** = matrix of all forces acting at the nodes of all elements

δ = matrix of all nodal displacements for all elements

$$k = \begin{bmatrix} k_1 & 0 & \cdots & 0 \\ 0 & k_2 & \cdots & 0 \\ \cdots & \cdots & \cdots & \cdots \\ 0 & 0 & \cdots & k_n \end{bmatrix} \qquad (6.150)$$

Step 5. Develop a matrix b_o that will transform the displacements **Δ** of the nodes of the structure into the displacement vector **δ** while maintaining geometric compatibility:

$$\delta = b_o \Delta \qquad (6.151)$$

b_o is a matrix of influence coefficients. The jth column of b_o contains the element nodal displacements when the node where Δ_j occurs is given a unit displacement in the direction of Δ_j, and no other nodes are displaced.

Step 6. Compute the stiffness matrix **K** for the whole structure from

$$K = b_o^T k b_o \qquad (6.152)$$

where b_o^T = transpose of b_o = matrix b_o with rows and columns interchanged.

This equation may be derived as follows: From energy relationships, $P = b_o^T S$. Substitution of $k\delta$ for **S** [Eq. (6.149)] and then substitution of $b_o \Delta$ for **δ** [Eq. (6.151)] yields $P = b_o^T k b_o \Delta$. Comparison of this with Eq. (6.97a), $P = k\Delta$, leads to Eq. (6.152).

Step 7. With the stiffness matrix **K** now known, solve the simultaneous equations

$$\Delta = K^{-1} P \qquad (6.153)$$

for the nodal displacements **Δ**. With these determined, calculate the member forces from

$$S = k b_o \Delta \qquad (6.154)$$

(N. M. Baran, "Finite Element Analysis on Microcomputers," and H. Kardestuncer and D. H. Norris, "Finite Element Handbook," McGraw-Hill Publishing Company, New York, books.mcgraw-hill.com; K. Bathe, "Finite Element Procedures in Engineering Analysis," T. R. Hughes, "The Finite Element Method," and H. T. Y. Yang, "Finite Element Structural Analysis," Prentice-Hall, Upper Saddle River, N.J.; W. Weaver, Jr., and J. M. Gere, "Matrix Analysis of Framed Structures," Van Nostrand Reinhold, New York.)

First- And Second-Order Analysis of Frames ■ The deformation of the brace frame due to the lateral forces is usually small and is not normally taken into account. Consequently, moments in the columns are amplified only by the moment produced by the axial force acting through the deflections along the member. These moments are called Pδ moments, where P is the column axial load and δ is the lateral deflection of the member with respect to the chord connecting its end points.

Unbraced frames subjected to unsymmetrical loads and/or lateral forces undergo lateral displacements. As a result of these displacements, columns in the frame are subjected to additional moments PΔ, where Δ is the lateral displacement of one end of a column with respect to the other end. In multistory structures, the PΔ moment for the columns in any one story is $(\Sigma P)\Delta$, where ΣP is the total vertical load on the story and Δ is the lateral deflection of the story with respect to the one below.

Analyses based on the dimensions of an undeformed frame are called first-order analyses, while those based on the deformed frame, taking into account both the Pδ and the PΔ effects, are called second-order analyses. Second-order analyses are basically geometrical nonlinear problems that required the use of computer programs. But not all programs that are advertised as for second-order analysis consider the Pδ moments. In the practical design, the PΔ effect is generally taken into account in the structural analysis through the iteration algorithm of computer programs, while the Pδ effect is considered during member design and only one step approximation is used through the magnification factor $1/(1 - P/Pc)$, as in Eq. (6.59).

(E. H. Gaylord, Jr. et al., "Design of Steel Structures" 3rd edition, McGraw-Hill, books: mcgraw-hill.com.)

Stresses in Arches

An arch is a curved beam, the radius of curvature of which is very large relative to the depth of section. It differs from a straight beam in that: (1) loads induce both bending and direct compressive stress in an arch; (2) arch reactions have horizontal components even though all loads are vertical, and (3) deflections have horizontal as well as vertical components. Names of arch parts are given in Fig. 6.86.

The necessity of resisting the horizontal components of the reactions is an important consideration in arch design. Sometimes these forces are taken by tie rods between the supports, sometimes by heavy abutments or buttresses.

Arches may be built with fixed ends, as can straight beams, or with hinges at the supports. They may also be built with an internal hinge, usually located at the uppermost point, or crown.

6.69 Three-Hinged Arches

An arch with an internal hinge and hinges at both supports (Fig. 6.87) is statically determinate. There are four unknowns—two horizontal and two vertical components of the reactions—but four equations based on the laws of equilibrium are available: (1) The sum of the horizontal forces must be zero. (In Fig. 6.87, $H_L = H_R = H$.) (2) The sum of the moments about the left support must be zero. ($V_R = Pk$.) (3) The sum of the moments about the right support must be zero. [$V_L = P(1 - k)$.] (4) The bending moment at the crown hinge must be zero (not to be confused with the sum of the moments about the crown, which also must be equal to zero but which would not lead to an independent equation for the solution of the reactions). Hence, for the right half of the arch in Fig. 6.87a, $Hh - V_R b = 0$, and $H = V_R b/h$. The influence line for H is a straight line, varying from zero for loads over the supports to the maximum of Pab/Lh for a load at C.

Reactions and stresses in three-hinged arches can be determined graphically by taking advantage of the fact that the bending moment at the crown hinge is zero. For example, in Fig. 6.87a, the load P is applied to segment AC of the arch. Then, since the bending moment at C must be zero, the line of action of the reaction R_R at B must pass through the crown hinge. It intersects the line of action of P at X. The line of action of the reaction R_L at A also must pass through X since P and the two reactions are in equilibrium. By constructing a force triangle with the load P and the lines of action of the reactions thus determined, you can obtain the magnitude of the reactions (Fig. 6.87b). After the reactions have been found, the stresses can be computed from the laws of statics or, in the case of a trussed arch, determined graphically.

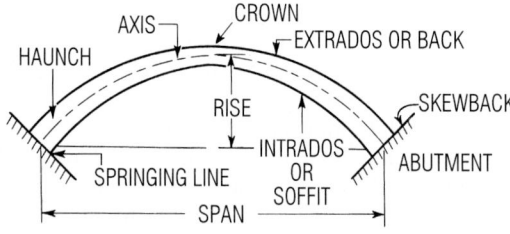

Fig. 6.86 Names of parts of a fixed arch.

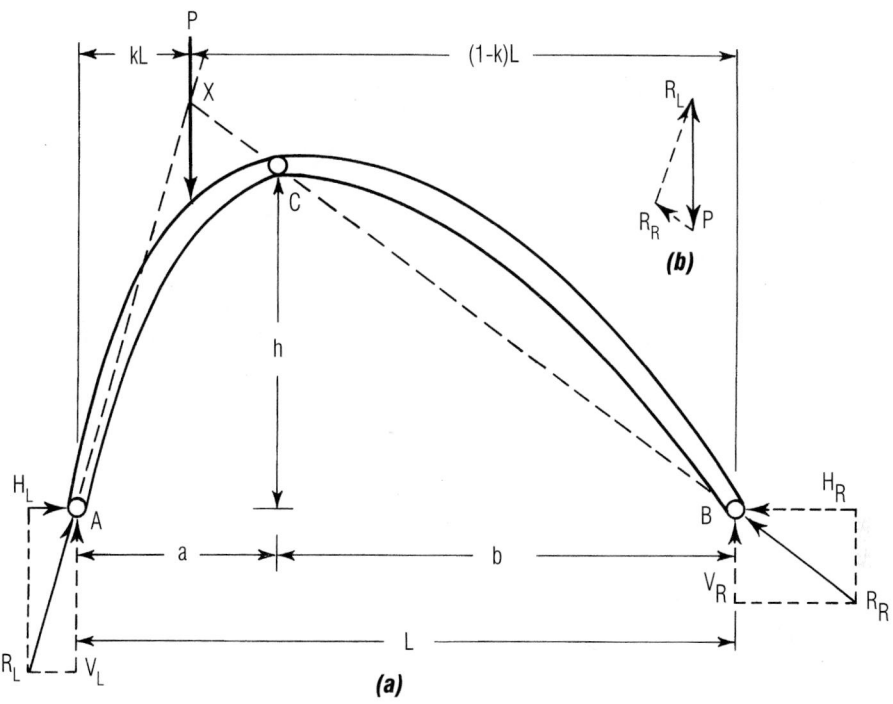

Fig. 6.87 Three-hinged arch.

6.70 Two-Hinged Arches

When an arch has hinges at the supports only (Fig. 6.88a), it is statically indeterminate; there is one more unknown reaction component than can be determined by the three equations of equilibrium. Another equation can be written from knowledge of the elastic behavior of the arch. One procedure is to assume that one of the supports is on rollers. The arch then is statically determinate, and the reactions and horizontal movement of the support can be computed for this condition (Fig. 6.88b). Next, the horizontal force required to return the movable support to its original position can be calculated (Fig. 6.88c). Finally, the reactions for the two-hinged arch (Fig. 6.88d) are obtained by superimposing the first set of reactions on the second.

For example, if δx is the horizontal movement of the support due to the loads on the arch, and if $\delta x'$ is the horizontal movement of the support

due to a unit horizontal force applied to the support, then

$$\delta x + H\delta x' = 0 \qquad (6.155)$$

$$H = -\frac{\delta x}{\delta x'} \qquad (6.156)$$

where H is the unknown horizontal reaction. (When a tie rod is used to take the thrust, the right-hand side of Eq. (6.155) is not zero but the elongation of the rod HL/A_sE_s, where L is the length of the rod, A_s its cross-sectional area, and E_s its modulus of elasticity. To account for the effect of an increase in temperature t, add to the left-hand side $EctL$, where E is the modulus of elasticity of the arch, c the coefficient of expansion.)

The dummy-unit-load method can be used to compute δx and $\delta x'$ (Art. 6.54):

$$\delta x = \int_A^B \frac{My\ ds}{EI} - \int_A^B \frac{N\ dx}{AE} \qquad (6.157)$$

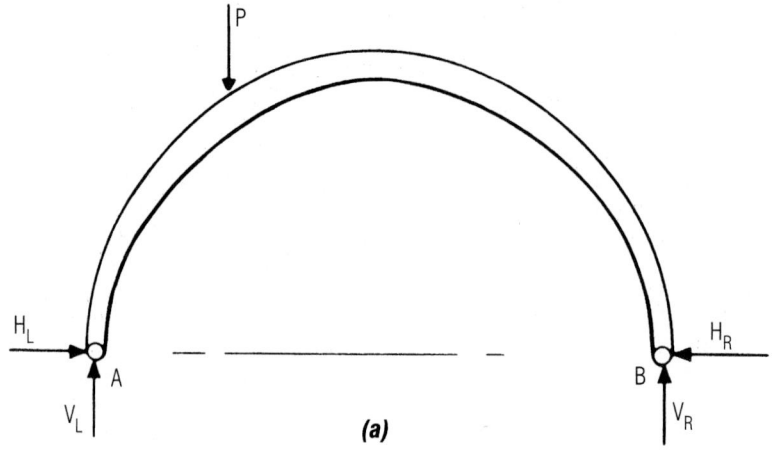

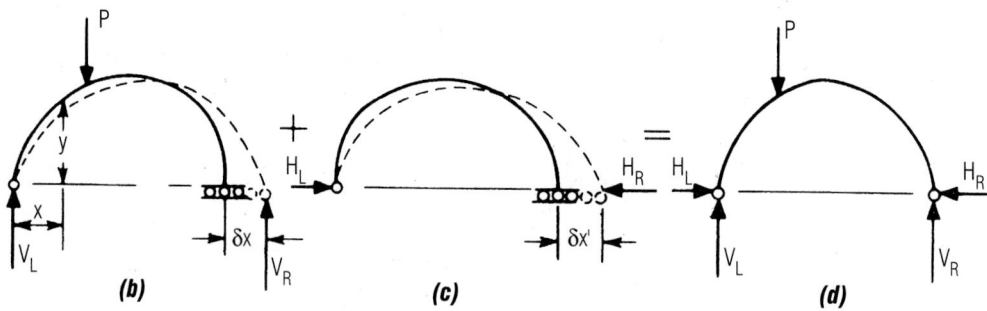

Fig. 6.88 Two-hinged arch.

where M = bending moment at any section due to loads

y = ordinate of section measured from immovable end of arch

I = moment of inertia of arch cross section

A = cross-sectional area of arch

ds = differential length along arch axis

dx = differential length along the horizontal

N = normal thrust on cross section due to loads

$$\delta x' = -\int_A^B \frac{y^2 \, ds}{EI} - \int_A^B \frac{\cos^2 \alpha \, dx}{AE} \quad (6.158)$$

where α = the angle the tangent to the axis at the section makes with the horizontal.

Equations (6.157) and (6.158) do not include the effects of shear deformation and curvature, which usually are negligible. Unless the thrust is very large, the second term on the right-hand side of Eq. (6.157) also can be dropped.

In most cases, integration is impracticable. The integrals generally must be evaluated by approximate methods. The arch axis is divided into a convenient number of elements of length Δs, and the functions under the integral sign are evaluated for each element. The sum of these terms is approximately equal to the integral. Thus, for the usual two-hinged arch

$$H = \frac{\sum_A^B (My \, \Delta s / EI)}{\sum^B (y^2 \, \Delta s / EI) + \sum^B (\cos^2 \alpha \, \Delta x / AE)} \quad (6.159)$$

(S. Timoshenko and D. H. Young, "Theory of Structures," McGraw-Hill Book Company,

New York, books.mcgraw-hill.com; S. F. Borg and J. J. Gennaro, "Modern Structural Analysis," Van Nostrand Reinhold Company, New York.)

6.71 Stresses in Arch Ribs

When the reactions have been found for an arch (Arts. 6.69 to 6.70), the principal forces acting on any cross section can be found by applying the equations of equilibrium. For example, consider the portion of an arch in Fig. 6.89, where the forces acting at an interior section X are to be found. The load P, H_L (or H_R), and V_L (or V_R) may be resolved into components parallel to the axial thrust N and the shear S at X, as indicated in Fig. 6.89. Then, by equating the sum of the forces in each direction to zero, we get

$$N = V_L \sin \theta_x + H_L \cos \theta_x + P \sin(\theta_x - \theta) \tag{6.160}$$

$$S = V_L \cos \theta_x - H_L \sin \theta_x + P \cos(\theta_x - \theta) \tag{6.161}$$

And the bending moment at X is

$$M = V_L x - H_L y - Pa \cos \theta - Pb \sin \theta \tag{6.162}$$

The shearing unit stress on the arch cross section at X can be determined from S with the aid of Eq. (6.49). The normal unit stresses can be calculated from N and M with the aid of Eq. (6.57).

When designing an arch, it may be necessary to compute certain secondary stresses, in addition to those caused by live, dead, wind, and snow loads. Among the secondary stresses to be considered are those due to temperature changes, rib shortening due to thrust or shrinkage, deformation of tie rods, and unequal settlement of footings. The procedure is the same as for loads on the arch, with the deformations producing the secondary stresses substituted for or treated the same as the deformations due to loads.

Also, the stability of arches should be considered. According to the mode of failure, there exist in-plane stability and out-of-plane buckling issues. (Theodore V. Galambos, "Guide to Stability Design Criteria for Metal Structures", 5th edition, John Wiley & Sons. Inc., www.wiley.com)

Thin-Shell Structures

A structural shell is a curved surface structure. Usually, it is capable of transmitting loads in more than two directions to supports. It is highly efficient structurally when it is so shaped, proportioned, and supported that it transmits the loads without bending or twisting.

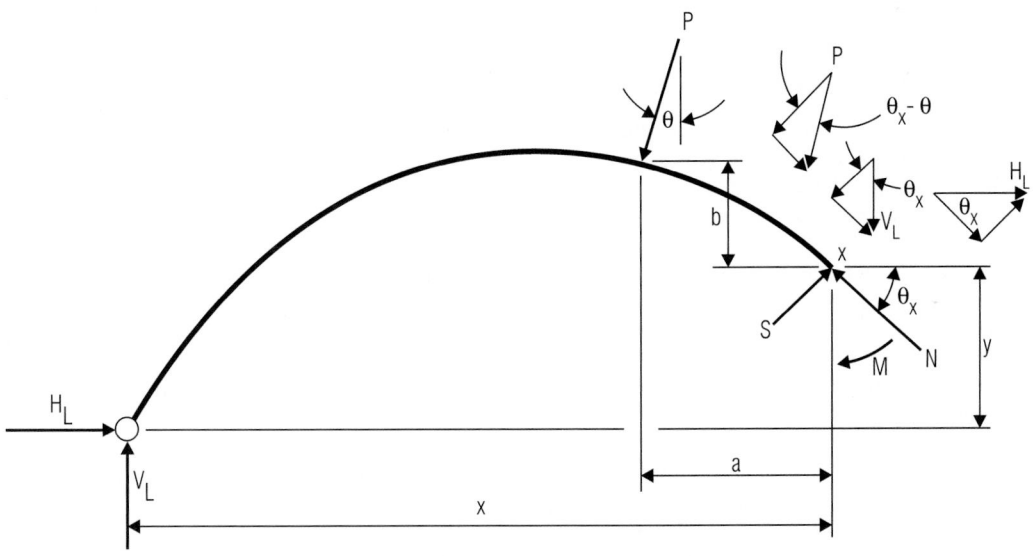

Fig. 6.89 Stresses in an arch rib.

A shell is defined by its middle surface, halfway between its extrados, or outer surface, and intrados, or inner surface. Thus, depending on the geometry of the middle surface, it might be a type of dome, barrel arch, cone, or hyperbolic paraboloid. Its thickness is the distance, normal to the middle surface, between extrados and intrados.

6.72 Thin-Shell Analysis

A thin shell is a shell with a thickness relatively small compared with its other dimensions. But it should not be so thin that deformations would be large compared with the thickness.

The shell should also satisfy the following conditions: Shearing stresses normal to the middle surface are negligible. Points on a normal to the middle surface before it is deformed lie on a straight line after deformation. And this line is normal to the deformed middle surface.

Calculation of the stresses in a thin shell generally is carried out in two major steps, both usually involving the solution of differential equations. In the first, bending and torsion are neglected (membrane theory, Art. 6.73). In the second step, corrections are made to the previous solution by superimposing the bending and shear stresses that are necessary to satisfy boundary conditions (bending theory, Art. 6.74).

6.73 Membrane Theory for Thin Shells

Thin shells usually are designed so that normal shears, bending moments, and torsion are very small, except in relatively small portions of the shells. In the membrane theory, these stresses are ignored.

Despite the neglected stresses, the remaining stresses are in equilibrium, except possibly at boundaries, supports, and discontinuities. At any interior point, the number of equilibrium conditions equals the number of unknowns. Thus, in the membrane theory, a thin shell is statically determinate.

The membrane theory does not hold for concentrated loads normal to the middle surface, except possibly at a peak or valley. The theory does not apply where boundary conditions are incompatible with equilibrium; and it is inexact where

there is geometric incompatibility at the boundaries. The last is a common condition, but the error is very small if the shell is not very flat. Usually, disturbances of membrane equilibrium due to incompatibility with deformations at boundaries, supports, or discontinuities are appreciable only in a narrow region about each source of disturbance. Much larger disturbances result from incompatibility with equilibrium conditions.

To secure the high structural efficiency of a thin shell, select a shape, proportions, and supports for the specific design conditions that come as close as possible to satisfying the membrane theory. Keep the thickness constant; if it must change, use a gradual taper. Avoid concentrated and abruptly changing loads. Change curvature gradually. Keep discontinuities to a minimum. Provide reactions that are tangent to the middle surface. At boundaries, insure, to the extent possible, compatibility of shell deformations with deformations of adjoining members, or at least keep restraints to a minimum. Make certain that reactions along boundaries are equal in magnitude and direction to the shell forces there.

Means usually adopted to satisfy these requirements at boundaries and supports are illustrated in Fig. 6.90. In Fig. 6.90a, the slope of the support and provision for movement normal to the middle surface insure a reaction tangent to the middle surface. In Fig. 6.90b, a stiff rib, or ring girder, resists unbalanced shears and transmits normal forces to columns below. The enlarged view of the ring girder in Fig. 6.90c shows gradual thickening of the shell to reduce the abruptness of the change in section. The stiffening ring at the lantern in Fig. 6.90d, extending around the opening at the crown, projects above the middle surface, for compatibility of strains, and connects through a transition curve with the shell; often, the rim need merely be thickened when the edge is upturned, and the ring can be omitted. In Fig. 6.90e, the boundary of the shell is a thickened edge. In Fig. 6.90f, a scalloped shell provides gradual tapering for transmitting the loads to the supports, at the same time providing access to the shell enclosure. And in Fig. 6.90g, a column is flared widely at the top to support a thin shell at an interior point.

Even when the conditions for geometric compatibility are not satisfactory, the membrane theory is a useful approximation. Furthermore, it yields a particular solution to the differential equations of the bending theory.

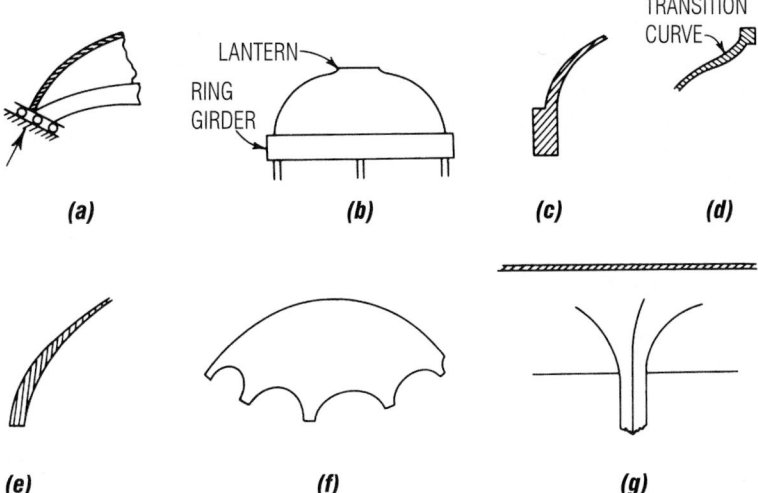

Fig. 6.90 Special provisions made at supports and boundaries of thin shells to meet requirements of the membrane theory include (*a*) a device to ensure a reaction tangent to the middle surface, (*b*) stiffened edges, such as the ring girder at the base of a dome, (*c*) gradually increased shell thickness at a stiffening member, (*d*) a transition curve at changes in section, (*e*) a stiffened edge obtained by thickening the shell, (*f*) scalloped edges, and (*g*) a flared support.

(D. P. Billington, "Thin-Shell Concrete Structures," 2nd ed., and S. Timoshenko and S. Woinowsky-Krieger, "Theory of Plates and Shells," McGraw-Hill Publishing Company, New York, books.mcgraw-hill.com; V. S. Kelkar and R. T. Sewell, "Fundamentals of the Analysis and Design of Shell Structures," Prentice-Hall, Englewood Cliffs, N.J., www.prenhall.com.)

6.74 Bending Theory for Thin Shells

When equilibrium conditions are not satisfied or incompatible deformations exist at boundaries, bending and torsion stresses arise in the shell. Sometimes, the design of the shell and its supports can be modified to reduce or eliminate these stresses (Art. 6.73). When the design cannot eliminate them, provision must be made for the shell to resist them.

But even for the simplest types of shells and loading, the stresses are difficult to compute. In bending theory, a thin shell is statically indeterminate; deformation conditions must supplement equilibrium conditions in setting up differential equations for determining the unknown forces and moments. Solution of the resulting equations may be tedious and time-consuming, if indeed solution is possible.

In practice, therefore, shell design relies heavily on the designer's experience and judgment. The designer should consider the type of shell, material of which it is made, and support and boundary conditions, and then decide whether to apply a bending theory in full, use an approximate bending theory, or make a rough estimate of the effects of bending and torsion. (Note that where the effects of a disturbance are large, these change the normal forces and shears computed by the membrane theory.) For domes, for example, the usual procedure is to use as a support a deep, thick girder or a heavily reinforced or prestressed tension ring, and the shell is gradually thickened in the vicinity of this support (Fig. 6.90*c*).

Circular barrel arches, with ratio of radius to distance between supporting arch ribs less than 0.25, may be designed as beams with curved cross section. Secondary stresses, however, must be taken into account. These include stresses due to volume change of rib and shell, rib shortening, unequal settlement of footings, and temperature differentials between surfaces.

Bending theory for cylinders and domes is given in W. Flügge, "Stresses in Shells," Springer-Verlag,

New York, www.springer-ny.com; S. Timoshenko and S. Woinowsky-Krieger, "Theory of Plates and Shells," McGraw-Hill Book Company, New York, books.mcgraw-hill.com; "Design of Cylindrical Concrete Shell Roofs," Manual of Practice no. 31, American Society of Civil Engineers. www.asce.org.

6.75 Stresses in Thin Shells

The results of the membrane and bending theories are expressed in terms of unit forces and unit moments, acting per unit of length over the thickness of the shell. To compute the unit stresses from these forces and moments, usual practice is to assume normal forces and shears to be uniformly distributed over the shell thickness and bending stresses to be linearly distributed.

Then, normal stresses can be computed from equations of the form

$$f_x = \frac{N_x}{t} + \frac{M_x}{t^3/12} z \qquad (6.163)$$

where z = distance from middle surface

t = shell thickness

M_x = unit bending moment about an axis parallel to direction of unit normal force N_x

Similarly, shearing stresses produced by central shears T and twisting moments D may be calculated from equations of the form

$$v_{xy} = \frac{T}{t} \pm \frac{D}{t^3/12} z \qquad (6.164)$$

Normal shearing stresses may be computed on the assumption of a parabolic stress distribution over the shell thickness:

$$v_{xz} = \frac{V}{t^3/6} \left(\frac{t^2}{4} - z^2 \right) \qquad (6.165)$$

where V = unit shear force normal to middle surface.

For axes rotated with respect to those used in the thin-shell analysis, use Eqs. (6.27) and (6.28) to transform stresses or unit forces and moments from the given to the new axes.

Folded Plates

A folded-plate structure consists of a series of thin planar elements, or flat plates, connected to

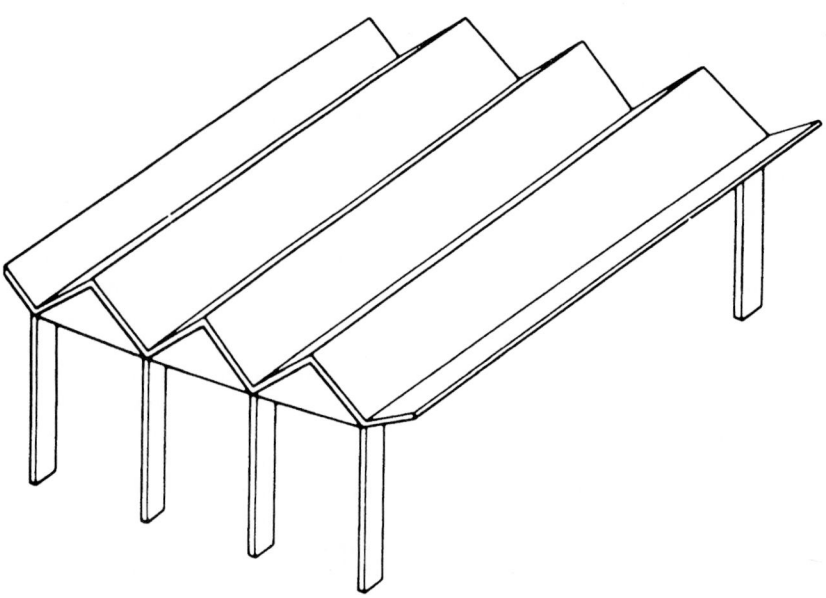

Fig. 6.91   Folded-plate structure.

one another along their edges. Usually used on long spans, especially for roofs, folded plates derive their economy from the girder action of the plates and the mutual support they give one another.

Longitudinally, the plates may be continuous over their supports. Transversely, there may be several plates in each bay (Fig. 6.91). At the edges, or folds, they may be capable of transmitting both moment and shear or only shear.

6.76 Folded-Plate Theory

A folded-plate structure has a two-way action in transmitting loads to its supports. Transversely, the elements act as slabs spanning between plates on either side. The plates then act as girders in carrying the load from the slabs longitudinally to supports, which must be capable of resisting both horizontal and vertical forces.

If the plates are hinged along their edges, the design of the structure is relatively simple. Some simplification also is possible if the plates, though having integral edges, are steeply sloped or if the span is sufficiently long with respect to other dimensions that beam theory applies. But there are no criteria for determining when such simplification is possible with acceptable accuracy. In general, a reasonably accurate analysis of folded-plate stresses is advisable.

Several good methods are available (D. Yitzhaki, "The Design of Prismatic and Cylindrical Shell Roofs," North Holland Publishing Company, Amsterdam, "Phase I Report on Folded-Plate Construction," *Proceedings Paper* 3741, *Journal of the Structural Division, ASCE,* December 1963; and A. L. Parme and J. A. Sbarounis, "Direct Solution of Folded Plate Concrete Roofs," EB021D Portland Cement Association, Skokie, IL. 60077). They all take into account the effects of plate deflections on the slabs and usually make the following assumptions:

The material is elastic, isotropic, and homogeneous. The longitudinal distribution of all loads on all plates is the same. The plates carry loads transversely only by bending normal to their planes and longitudinally only by bending within their planes. Longitudinal stresses vary linearly over the depth of each plate. Supporting members, such as diaphragms, frames, and beams, are infinitely stiff in their own planes and completely flexible normal to their own planes. Plates have no torsional stiffness normal to their own planes. Displacements due to forces other than bending moments are negligible.

Regardless of the method selected, the computations are rather involved; so it is wise to carry out the work in a well-organized table. The Yitzhaki method (Art. 6.77) offers some advantages over others in that the calculations can be tabulated, it is relatively simple, it requires the solution of no more simultaneous equations than one for each edge for simply supported plates, it is flexible, and it can easily be generalized to cover a variety of conditions.

6.77 Yitzhaki Method for Folded Plates

Based on the assumptions and general procedure given in Art. 6.76, the Yitzhaki method deals in two ways with the slab and plate systems that comprise a folded-plate structure. In the first, a unit width of slab is considered continuous over supports immovable in the direction of the load (Fig. 6.92b). The strip usually is taken where the longitudinal plate stresses are a maximum. Secondly, the slab reactions are taken as loads on the plates, which now are assumed to be hinged along the edges (Fig. 6.92c). Thus, the slab reactions cause angle changes in the plates at each fold. Continuity is restored by applying an unknown moment to the plates at each edge. The moments can be determined from the fact that at each edge the sum of the angle changes due to the loads and to the unknown moments must equal zero.

The angle changes due to the unknown moments have two components. One is the angle change at each slab end, now hinged to an adjoining slab, in the transverse strip of unit width. The second is the angle change due to deflection of the plates. The method assumes that the angle change at each fold varies in the same way longitudinally as the angle changes along the other folds.

For more details, see D. Yitzhaki and Max Reiss, "Analysis of Folded Plates," *Proceedings Paper* 3303, *Journal of the Structural Division, ASCE,* October 1962; F. S. Merritt and Jonathan T. Ricketts, "Building Design and Construction Handbook," 5th ed., McGraw-Hill Book Company, New York.

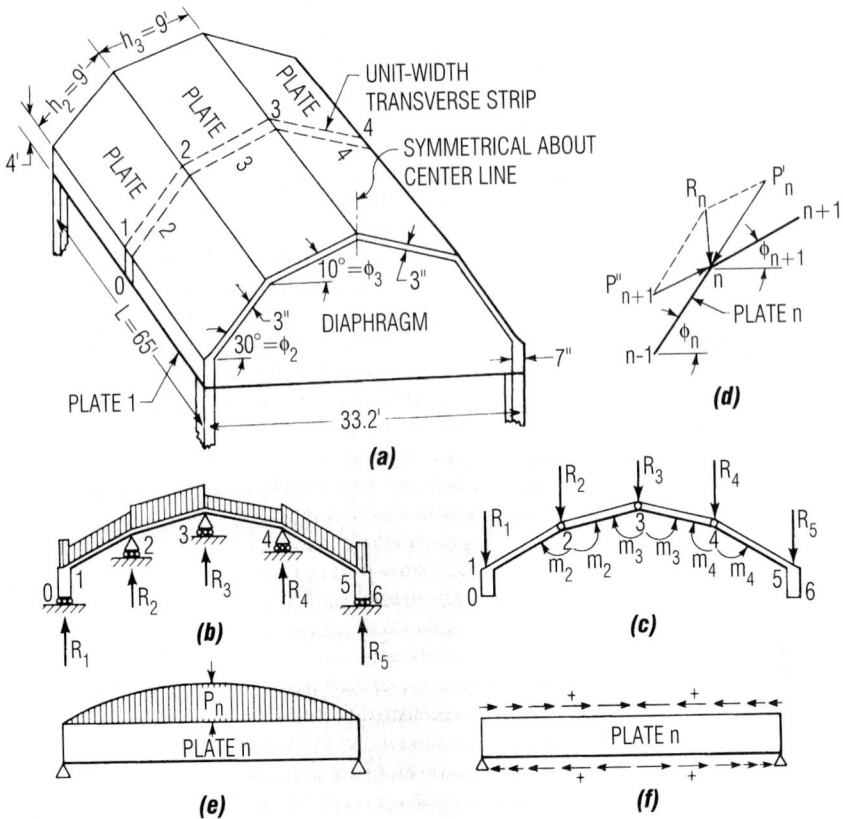

Fig. 6.92 Folded-plate structure is analyzed by first considering a transverse strip (*a*) as a continuous slab on supports that do not deflect (*b*). Then, the slabs are assumed hinged (*c*) and acted upon by the reactions computed in the first step and unknown moments to correct for this assumption. In the longitudinal direction, the plates act as deep girders (*e*) with shears along the edges, positive directions shown in (*f*). Slab reactions are resolved into plate forces, parallel to the planes of the plates (*d*).

Cable-Supported Structures*

A cable is a linear structural member, like a bar of a truss. The cross-sectional dimensions of a cable relative to its length, however, are so small that it cannot withstand bending or compression. Consequently, under loads at an angle to its longitudinal axis, a cable sags and assumes a shape that enables it to develop tensile stresses that resist the loads.

Structural efficiency results from two cable characteristics: (1) Uniformity of tensile stresses over the cable cross section, and (2) usually, small variation of tension along the longitudinal axis. Hence, it is economical to use materials with very high tensile strength for cables.

Cables sometimes are used in building construction as an alternative to such tension members as hangers, ties, or tension chords of trusses. For example, cables are used in a form of long-span cantilever-truss construction in which a horizontal roof girder is supported at one end by a column and near the other end by a cable that extends diagonally upward to the top of a vertical mast above the column support (cable-stayed-girder construction, Fig. 6.93).

---

*Reprinted with permission from F. S. Merritt, "Structural Steel Designers Handbook," McGraw-Hill Book Company, New York.

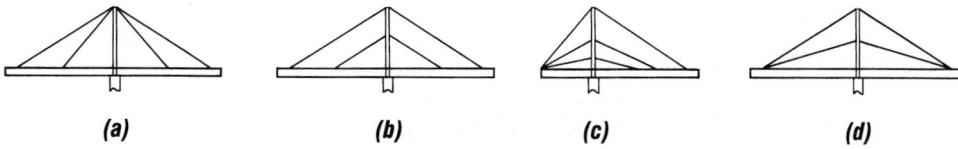

Fig. 6.93 Types of stayed girders: (*a*) Bundles (converging); (*b*) harp; (*c*) fan; (*d*) star.

Cable stress can be computed for this case from the laws of equilibrium. Similarly, cable-stayed girders are used to support bridge decks.

Cables also may be used instead of or with girders, trusses, or membranes to support roofs or bridge decks. For the purpose, cables may be arranged in numerous ways. It is consequently impractical to treat in detail in this book any but the simplest types of such applications of cables. Instead, general procedures for analyzing cable-supported structures are presented in the following. (See also Arts. 17.15 and 17.17).

6.78 Simple Cables

An ideal cable has no resistance to bending. Thus, in analysis of a cable in equilibrium, not only is the sum of the moments about any point equal to zero, but so is the bending moment at any point. Consequently, the equilibrium shape of the cable corresponds to the funicular, or bending-moment, diagram for the loading (Fig. 6.94*a*). As a result, the tensile force at any point of the cable is tangent there to the cable curve.

The point of maximum sag of a cable coincides with the point of zero shear. (Sag in this case should be measured parallel to the direction of the shear forces.)

Stresses in a cable are a function of the deformed shape. Equations needed for analysis, therefore, usually are nonlinear. Also, in general, stresses and deformations cannot be obtained accurately by superimposition of loads. A common procedure in analysis is to obtain a solution in steps by using linear equations to approximate the nonlinear ones and by starting with the initial geometry to obtain better estimates of the final geometry.

For convenience in analysis, the cable tension, directed along the cable curve, usually is resolved into two components. Often, it is advantageous to resolve the tension T into a horizontal component H and a vertical component V (Fig. 6.94*b*). Under vertical loading then, the horizontal component is constant along the cable. Maximum tension occurs at the support. V is zero at the point of maximum sag.

For a general, distributed vertical load q, the cable must satisfy the second-order linear differential equation

$$Hy'' = q \qquad (6.166)$$

where y = rise of cable at distance x from low point (Fig. 6.94*b*)

$$y'' = d^2y/dx^2$$

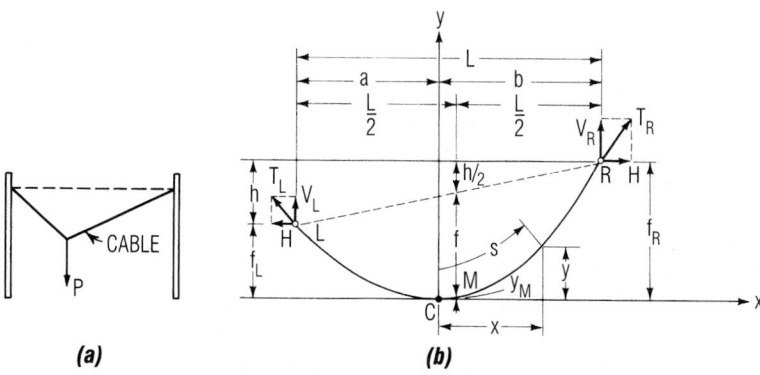

Fig. 6.94 Simple cables: (*a*) Shape of cable with a concentrated load; (*b*) shape of cable with supports at different levels.

6.78.1 Catenary

Weight of a cable of constant cross section represents a vertical loading that is uniformly distributed along the length of cable. Under such a loading, a cable takes the shape of a catenary.

Take the origin of coordinates at the low point C and measure distance s along the cable from C (Fig. 6.94b). If q_o is the load per unit length of cable, Eq. (6.166) becomes

$$Hy'' = \frac{q_o \, ds}{dx} = q_o\sqrt{1 + y'^2} \qquad (6.167)$$

where $y' = dy/dx$. Solving for y' gives the slope at any point of the cable

$$y' = \sinh\frac{q_o x}{H} = \frac{q_o x}{H} + \frac{1}{3!}\left(\frac{q_o x}{H}\right)^3 + \cdots \qquad (6.168)$$

A second integration then yields the equation for the cable shape, which is called a catenary.

$$y = \frac{H}{q_o}\left(\cosh\frac{q_o x}{H} - 1\right)$$
$$= \frac{q_o}{H}\frac{x^2}{2!} + \left(\frac{q_o}{H}\right)^3\frac{x^4}{4!} + \cdots \qquad (6.169)$$

If only the first term of the series expansion is used, the cable equation represents a parabola. Because the parabolic equation usually is easier to handle, a catenary often is approximated by a parabola.

For a catenary, length of arc measured from the low point is

$$s = \frac{H}{q_o}\sinh\frac{q_o x}{H} = x + \frac{1}{3!}\left(\frac{q_o}{H}\right)^2 x^3 + \cdots \qquad (6.170)$$

Tension at any point is

$$T = \sqrt{H^2 + q_o^2 s^2} = H + q_o y \qquad (6.171)$$

The distance from the low point C to the left support L is

$$a = \frac{H}{q_o}\cosh^{-1}\left(\frac{q_o}{H}f_L + 1\right) \qquad (6.172)$$

where f_L = vertical distance from C to L. The distance from C to the right support R is

$$b = \frac{H}{q_o}\cosh^{-1}\left(\frac{q_o}{H}f_R + 1\right) \qquad (6.173)$$

where f_R = vertical distance from C to R.

Given the sags of a catenary f_L and f_R under a distributed vertical load q_o, the horizontal component of cable tension H may be computed from

$$\frac{q_o l}{H} = \cosh^{-1}\left(\frac{q_o f_L}{H} + 1\right)$$
$$+ \cosh^{-1}\left(\frac{q_o f_R}{H} + 1\right) \qquad (6.174)$$

where l = span, or horizontal distance between supports L and $R = a + b$. This equation usually is solved by trial. A first estimate of H for substitution in the right-hand side of the equation may be obtained by approximating the catenary by a parabola. Vertical components of the reactions at the supports can be computed from

$$R_L = H\sinh\frac{q_o a}{H} \qquad R_R = H\sinh\frac{q_o b}{H} \qquad (6.175)$$

6.78.2 Parabola

Uniform vertical live loads and uniform vertical dead loads other than cable weight generally may be treated as distributed uniformly over the horizontal projection of the cable. Under such loadings, a cable takes the shape of a parabola.

Take the origin of coordinates at the low point C (Fig. 6.94b). If w_o is the load per foot horizontally, Eq. (6.166) becomes

$$Hy'' = w_o \qquad (6.176)$$

Integration gives the slope at any point of the cable

$$y' = \frac{w_o x}{H} \qquad (6.177)$$

A second integration then yields the parabolic equation for the cable shape

$$y = \frac{w_o x^2}{2H} \qquad (6.178)$$

The distance from the low point C to the left support L is

$$a = \frac{1}{2} - \frac{Hh}{w_o l} \qquad (6.179)$$

where l = span, or horizontal distance between supports L and $R = a + b$

h = vertical distance between supports

The distance from the low point C to the right support R is

$$b = \frac{1}{2} + \frac{Hh}{w_o l} \qquad (6.180)$$

Supports at Different Levels ■ The horizontal component of cable tension H may be computed from

$$H = \frac{w_o l^2}{h^2} \left(f_R - \frac{h}{2} \pm \sqrt{f_L f_R} \right) = \frac{w_o l^2}{8f} \qquad (6.181)$$

where f_L = vertical distance from C to L

f_R = vertical distance from C to R

f = sag of cable measured vertically from chord LR midway between supports (at $x = Hh/w_o l$)

As indicated in Fig. 6.94b,

$$f = f_L + \frac{h}{2} - y_M \qquad (6.182)$$

where $y_M = Hh^2/2w_o l^2$. The minus sign should be used in Eq. (6.181) when low point C is between supports. If the vertex of the parabola is not between L and R, the plus sign should be used.

The vertical components of the reactions at the supports can be computed from

$$V_L = w_o a = \frac{w_o l}{2} - \frac{Hh}{l}$$
$$\qquad (6.183)$$
$$V_r = w_o b = \frac{w_o l}{2} + \frac{Hh}{l}$$

Tension at any point is

$$T = \sqrt{H^2 + w_o^2 x^2} \qquad (6.184)$$

Length of parabolic arc RC is

$$L_{RC} = \frac{b}{2} \sqrt{1 + \left(\frac{w_o b}{H} \right)^2} + \frac{H}{2w_o} \sinh \frac{w_o b}{H}$$
$$\qquad (6.185)$$
$$= b + \frac{1}{6} \left(\frac{w_o}{H} \right)^2 b^3 + \cdots$$

Length of parabolic arc LC is

$$L_{LC} = \frac{a}{2} \sqrt{1 + \left(\frac{w_o a}{H} \right)^2} + \frac{H}{2w_o} \sinh \frac{w_o a}{H}$$
$$\qquad (6.186)$$
$$= a + \frac{1}{6} \left(\frac{w_o}{H} \right)^2 a^3 + \cdots$$

Supports at Same Level ■ In this case, $f_L = f_R = f$, $h = 0$, and $a = b = l/2$. The horizontal component of cable tension H may be computed from

$$H = \frac{w_o l^2}{8f} \qquad (6.187)$$

The vertical components of the reactions at the supports are

$$V_L = V_R = \frac{w_o l}{2} \qquad (6.188)$$

Maximum tension occurs at the supports and equals

$$T_L = T_R = \frac{w_o l}{2} \sqrt{1 + \frac{l^2}{16f^2}} \qquad (6.189)$$

Length of cable between supports is

$$L = \frac{1}{2} \sqrt{1 + \left(\frac{w_o l}{2H} \right)^2} + \frac{H}{w_o} \sinh \frac{w_o l}{2H}$$
$$\qquad (6.190)$$
$$= l \left(1 + \frac{8}{3} \frac{f^2}{l^2} - \frac{32}{5} \frac{f^4}{l^4} + \frac{256}{7} \frac{f^6}{l^6} + \cdots \right)$$

If additional uniformly distributed load is applied to a parabolic cable, the change in sag is approximately

$$\Delta f = \frac{15}{16f} \frac{l}{5 - 24f^2/l^2} \Delta L \qquad (6.191)$$

For a rise in temperature t, the change in sag is about

$$\Delta f = \frac{15}{16f(5 - 24f^2/l^2)} l^2 ct \left(1 + \frac{8}{3} \frac{f^2}{l^2} \right) \qquad (6.192)$$

where c = coefficient of thermal expansion.

Elastic elongation of a parabolic cable is approximately

$$\Delta L = \frac{Hl}{AE} \left(1 + \frac{16}{3} \frac{f^2}{l^2} \right) \qquad (6.193)$$

where A = cross-sectional area of cable

E = modulus of elasticity of cable steel

H = horizontal component of tension in cable

If the corresponding change in sag is small, so that the effect on H is negligible, this change may be computed from

$$\Delta f = \frac{15}{16} \frac{Hl^2}{AEf} \frac{1 + 16f^2/3l^2}{5 - 24f^2/l^2} \qquad (6.194)$$

For the general case of vertical dead load on a cable, the initial shape of the cable is given by

$$y_D = \frac{M_D}{H_D} \quad (6.195)$$

where M_D = dead-load bending moment that would be produced by load in a simple beam

H_D = horizontal component of tension due to dead load

For the general case of vertical live load on the cable, the final shape of the cable is given by

$$y_D + \delta = \frac{M_D + M_L}{H_D + H_L} \quad (6.196)$$

where δ = vertical deflection of cable due to live load

M_L = live-load bending moment that would be produced by live load in simple beam

H_L = increment in horizontal component of tension due to live load

Subtraction of Eq. (6.195) from Eq. (6.196) yields

$$\delta = \frac{M_L - H_L y_D}{H_D + H_L} \quad (6.197)$$

If the cable is assumed to take a parabolic shape, a close approximation to H_L may be obtained from

$$\frac{H_L}{AE} K = \frac{w_D}{H_D} \int_0^l \delta \, dx - \frac{1}{2} \int_0^l \delta'' \delta \, dx \quad (6.198)$$

$$K = l\left[\frac{1}{4}\left(\frac{5}{2} + \frac{16f^2}{l^2}\right)\sqrt{1 + \frac{16f^2}{l^2}} \right.$$
$$\left. + \frac{3l}{32f}\log_e\left(\frac{4f}{l} + \sqrt{1 + \frac{16f^2}{l^2}}\right) \right] \quad (6.199)$$

where $\delta'' = d^2\delta/dx^2$.

If elastic elongation and δ'' can be ignored, Eq. (6.198) simplifies to

$$H_L = \frac{\int_0^l M_L \, dx}{\int_0^l y_D \, dx} = \frac{3}{2fl}\int_0^l M_L \, dx \quad (6.200)$$

Thus, for a load uniformly distributed horizontally $w_{L'}$

$$\int_0^l M_L \, dx = \frac{w_L l^3}{12} \quad (6.201)$$

and the increase in the horizontal component of tension due to live load is

$$H_L = \frac{3}{2fl}\frac{w_L l^3}{12} = \frac{w_L l^2}{8f} = \frac{w_L l^2}{8}\frac{8H_D}{w_D l^2}$$

$$= \frac{w_L}{w_D} H_D \quad (6.202)$$

When a more accurate solution is desired, the value of H_L that is obtained from Eq. (6.202) can be used for an initial trial in solving Eqs. (6.197) and (6.198).

(S. P. Timoshenko and D. H. Young, "Theory of Structures," McGraw-Hill Book Company, New York, books.mcgraw-hill.com; W. T. O'Brien and A. J. Francis, "Cable Movements under Two-Dimensional Loads," *Journal of the Structural Division, ASCE*, vol. 90, no. ST3, *Proceedings Paper* 3929, June 1964, pp. 98–123; W. T. O'Brien, "General Solution of Suspended Cable Problems," *Journal of the Structural Division, ASCE*, vol. 93, no. ST1, *Proceedings Paper* 5085, February 1967, pp. 1–26, www.asce.org; W. T. O'Brien, "Behavior of Loaded Cable Systems," *Journal of the Structural Division, ASCE*, vol. 94, no. ST10, *Proceedings Paper* 6162, October 1968, pp. 2281–2302, www.asce.org; G. R. Buchanan, "Two-Dimensional Cable Analysis," *Journal of the Structural Division, ASCE*, vol. 96, no. ST7, *Proceedings Paper* 7436, July 1970, pp. 1581–1587, www.asce.org.)

6.79 Cable Systems

Analysis of simple cables is described in Art. 6.77. Cables, however, may be assembled into many types of systems. One important reason for such systems is that roofs to be supported are two-or three-dimensional. Consequently, three-dimensional cable arrangements often are advantageous. Another important reason is that cable systems can be designed to offer much higher resistance to vibrations than simple cables do.

Like simple cables, cable systems behave nonlinearly. Thus, accurate analysis is difficult, tedious, and time-consuming. As a result, many designers use approximate methods that appear to have successfully withstood the test of time. Because of the numerous types of cable systems and the complexity of analysis, only general procedures are outlined here.

Cable systems may be stiffened or unstiffened. Stiffened systems are usually used for suspension bridges. Our discussion here deals only with unstiffened systems, that is, systems where loads are carried to supports only by cables. Stiffened cable systems are discussed in Art. 17.15.

Often, unstiffened systems may be classified as a network or as a cable truss, or double-layered plane system.

Networks consist of two or three sets of parallel cables intersecting at an angle (Fig. 6.95). The cables are fastened together at their intersections.

Cable trusses consist of pairs of cables, generally in a vertical plane. One cable of each pair is concave downward, the other concave upward (Fig. 6.96).

Cable Trusses ▪ Both cables of a cable truss are initially tensioned, or prestressed, to a predetermined shape, usually parabolic. The prestress is made large enough that any compression that may be induced in a cable by loads only reduces the tension in the cable; thus, compressive stresses

cannot occur. The relative vertical position of the cables is maintained by verticals, or spreaders, or by diagonals. Diagonals in the truss plane do not appear to increase significantly the stiffness of a cable truss.

Figure 6.96 shows four different arrangements of the cables, with spreaders, in a cable truss. The intersecting types (Fig. 6.96*b* and *c*) usually are stiffer than the others, for a given size of cables and given sag and rise.

For supporting roofs, cable trusses often are placed radially at regular intervals. Around the perimeter of the roof, the horizontal component of the tension usually is resisted by a circular or elliptical compression ring. To avoid a joint with a jumble of cables at the center, the cables usually are also connected to a tension ring circumscribing the center.

Properly prestressed, such double-layer cable systems offer high resistance to vibrations. Wind or other dynamic forces difficult or impossible to anticipate may cause resonance to occur in a single cable, unless damping is provided. The probability of resonance occurring may be reduced by increasing the dead load on a single cable. But this is not economical because the size of cable and supports usually must be increased as well. Besides, the tactic may not succeed, because future loads may be outside the design range. Damping, however, may be achieved economically with interconnected cables under different tensions, for example, with cable trusses or networks.

The cable that is concave downward (Fig. 6.96) usually is considered the load-carrying cable. If the prestress in that cable exceeds that in the other cable, the natural frequencies of vibration of both cables will always differ for any value of live load. To avoid resonance, the difference between the frequencies of the cables should increase with increase in load. Thus, the two cables will tend to assume different shapes under specific dynamic loads. As a consequence, the resulting flow of energy from one cable to the other will dampen the vibrations of both cables.

Natural frequency, cycles per second, of each cable may be estimated from

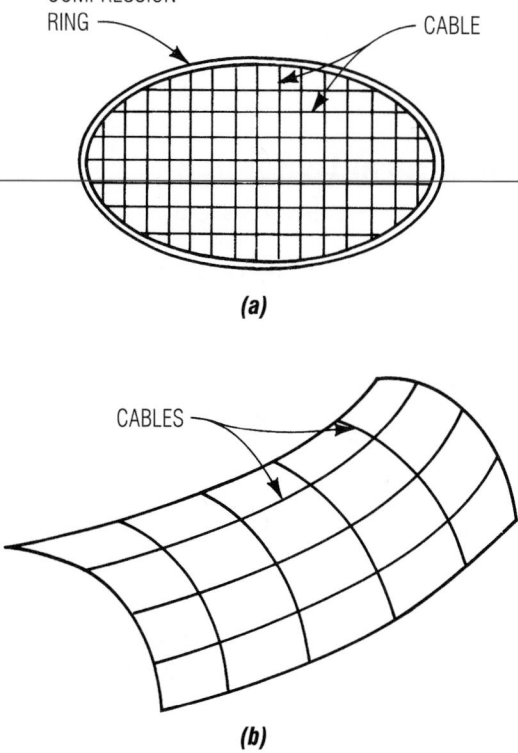

(a)

(b)

Fig. 6.95 Cable networks: (*a*) Cables forming a dish-shaped surface; (*b*) cables forming a saddle-shaped surface.

$$\omega_n = \frac{n\pi}{l}\sqrt{\frac{Tg}{w}} \qquad (6.203)$$

where n = integer, 1 for fundamental mode of vibration, 2 for second mode, ...

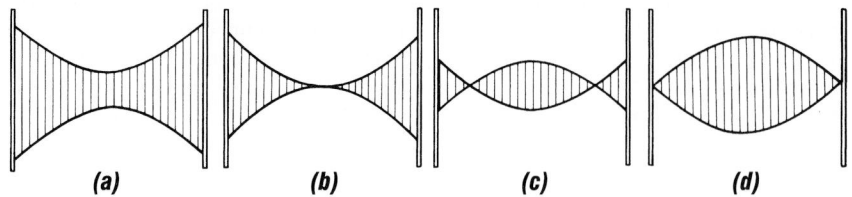

Fig. 6.96 Planar cable systems: (*a*) Completely separated cables; (*b*) cables intersecting at midspan; (*c*) crossing cables; (*d*) cables meeting at supports.

l = span of cable, ft

w = load on cable, kips/ft

g = acceleration due to gravity = 32.2 ft/s²

T = cable tension, kips

The spreaders of a cable truss impose the condition that under a given load the change in sag of the cables must be equal. But the changes in tension of the two cables may not be equal. If the ratio of sag to span f/l is small (less than about 0.1), Eq. (6.194) indicates that, for a parabolic cable, the change in tension is given approximately by

$$\Delta H = \frac{16}{3}\frac{AEf}{l^2}\Delta f \qquad (6.204)$$

where Δf = change in sag

A = cross-sectional area of cable

E = modulus of elasticity of cable steel

Double cables interconnected with struts may be analyzed as discrete or continuous systems. For a discrete system, the spreaders are treated as individual members. For a continuous system, the spreaders are replaced by a continuous diaphragm that insures that the changes in sag and rise of cables remain equal under changes in load. Similarly, for analysis of a cable network, the cables, when treated as a continuous system, may be replaced by a continuous membrane.

(H. Mollman, "Analysis of Plane Prestressed Cable Structures," *Journal of the Structural Division*, ASCE, vol. 96, no. ST10, *Proceedings Paper* 7598, October 1970, pp. 2059–2082; D. P. Greenberg, "Inelastic Analysis of Suspension Roof Structures," *Journal of the Structural Division*, ASCE, vol. 96, no. ST5, *Proceedings Paper* 7284, May 1970, pp. 905–930; H. Tottenham and P. G. Williams, "Cable Net: Continuous System Analysis," *Journal of the Engineering Mechanics Division*, ASCE, vol. 96, no. EM3, *Proceedings Paper* 7347, June 1970, pp. 277–293, www.asce.org; A. Siev, "A General Analysis of Prestressed Nets," *Publications, International Association for Bridge and Structural Engineering*, vol. 23, pp. 283–292, Zurich, Switzerland, 1963; A. Siev, "Stress Analysis of Prestressed Suspended Roofs," *Journal of the Structural Division*, ASCE, vol. 90, no. ST4, *Proceedings Paper* 4008, August 1964, pp. 103–121; C. H. Thornton and C. Birnstiel, "Three-Dimensional Suspension Structures," *Journal of the Structural Division*, ASCE, vol. 93, no. ST2, *Proceedings Paper* 5196, April 1967, pp. 247–270, www.asce.org.)

Structural Dynamics

Article 6.2 noted that loads can be classified as static or dynamic and that the distinguishing characteristic is the rate of application of load. If a load is applied slowly, it may be considered static. Since dynamic loads may produce stresses and deformations considerably larger than those caused by static loads of the same magnitude, it is important to know reasonably accurately what is meant by slowly.

A useful definition can be given in terms of the natural period of vibration of the structure or member to which the load is applied. If the time in which a load rises from zero to its maximum value is more than double the natural period, the load may be treated as static. Loads applied more rapidly may be dynamic. Structural analysis and design for such loads are considerably different from and more complex than those for static loads.

In general, exact dynamic analysis is possible only for relatively simple structures, and only when both the variation of load and resistance with time are a convenient mathematical function. Therefore, in practice, adoption of approximate methods that permit rapid analysis and design is advisable. And usually, because of uncertainties in loads and structural resistance, computations need not be carried out with more than a few significant figures, to be consistent with known conditions.

6.80 Material Properties Under Dynamic Loading

In general, mechanical properties of structural materials improve with increasing rate of load application. For low-carbon steel, for example, yield strength, ultimate strength, and ductility rise with increasing rate of strain. Modulus of elasticity in the elastic range, however, is unchanged. For concrete, the dynamic ultimate strength in compression may be much greater than the static strength.

Since the improvement depends on the material and the rate of strain, values to use in dynamic analysis and design should be determined by tests approximating the loading conditions anticipated.

Under many repetitions of loading, though, a member or connection between members may fail because of "fatigue" at a stress smaller than the yield point of the material. In general, there is little apparent deformation at the start of a fatigue failure. A crack forms at a point of high stress concentration. As the stress is repeated, the crack slowly spreads, until the member ruptures without measurable yielding. Although the material may be ductile, the fracture looks brittle.

Endurance Limit ▪ Some materials (generally those with a well-defined yield point) have what is known as an **endurance limit**. This is the maximum unit stress that can be repeated, through a definite range, an indefinite number of times without causing structural damage. Generally, when no range is specified, the endurance limit is intended for a cycle in which the stress is varied between tension and compression stresses of equal value.

A range of stress may be resolved into two components: a steady, or mean, stress and an alternating stress. The endurance limit sometimes is defined as the maximum value of the alternating stress that can be superimposed on the steady stress an indefinitely large number of times without causing fracture.

Improvement of Fatigue Strength ▪ Design of members to resist repeated loading cannot

be executed with the certainty with which members can be designed to resist static loading. Stress concentrations may be present for a wide variety of reasons, and it is not practicable to calculate their intensities. But sometimes it is possible to improve the fatigue strength of a material or to reduce the magnitude of a stress concentration below the minimum value that will cause fatigue failure.

In general, avoid design details that cause severe stress concentrations or poor stress distribution. Provide gradual changes in section. Eliminate sharp corners and notches. Do not use details that create high localized constraint. Locate unavoidable stress raisers at points where fatigue conditions are the least severe. Place connections at points where stress is low and fatigue conditions are not severe. Provide structures with multiple load paths or redundant members, so that a fatigue crack in any one of the several primary members is not likely to cause collapse of the entire structure.

Fatigue strength of a material may be improved by cold working the material in the region of stress concentration, by thermal processes, or by prestressing it in such a way as to introduce favorable internal stresses. Where fatigue stresses are unusually severe, special materials may have to be selected with high energy absorption and notch toughness.

(C. H. Norris et al., "Structural Design for Dynamic Loads," McGraw-Hill Book Company, New York, books.mcgraw-hill.com; W. H. Munse, "Fatigue of Welded Steel Structures," Welding Research Council, 3 Park Avenue 27th floor, New York, NY 10016.)

6.81 Natural Period of Vibration

A preliminary step in dynamic analysis and design is determination of this period. It can be computed in many ways, including application of the laws of conservation of energy and momentum or Newton's second law of motion, $F = M(dv/dt)$, where F is force, M mass, v velocity, and t time. But in general, an exact solution is possible only for simple structures. Therefore, it is general practice to seek an approximate—but not necessarily inexact—solution by analyzing an idealized representation of the actual member or structure. Setting up this model and interpreting the solution requires judgment of a high order.

Natural period of vibration is the time required for a structure to go through one cycle of free vibration, that is, vibration after the disturbance causing the motion has ceased.

To compute the natural period, the actual structure may be conveniently represented by a system of masses and massless springs, with additional resistances provided to account for energy losses due to friction, hysteresis, and other forms of damping. In simple cases, the masses may be set equal to the actual masses; otherwise, equivalent masses may have to be computed (Art. 6.84). The spring constants are the ratios of forces to deflections.

For example, a single mass on a spring (Fig. 6.97b) may represent a simply supported beam with mass that may be considered negligible compared with the load W at midspan (Fig. 6.97a). The spring constant k should be set equal to the load that produces a unit deflection at midspan; thus, $k = 48EI/L^3$, where E is the modulus of elasticity, psi; I the moment of inertia, in^4; and L the span, in, of the beam. The idealized mass equals W/g, where W is the weight of the load, lb, and g is the acceleration due to gravity, 386 in/s^2.

Also, a single mass on a spring (Fig. 6.97d) may represent the rigid frame in Fig. 6.97c. In that case, $k = 2 \times 12EI/h^3$, where I is the moment of inertia,

in^4, of each column and h the column height, in. The idealized mass equals the sum of the masses on the girder and the girder mass. (Weight of columns and walls is assumed negligible.)

6.81.1 Degree of a System

The spring and mass in Fig. 6.97b and d form a one-degree system. The degree of freedom of a system is determined by the least number of coordinates needed to define the positions of its components. In Fig. 6.97, only the coordinate y is needed to locate the mass and determine the state of the spring. In a two-degree system, such as one comprising two masses connected to each other and to the ground by springs and capable of movement in only one direction, two coordinates are required to locate the masses.

One-Degree System ■ If the mass with weight W, lb, in Fig. 6.97 is isolated, as shown in Fig. 6.97e, it will be in dynamic equilibrium under the action of the spring force $-ky$ and the inertia force $(d^2y/dt^2)(W/g)$.

Hence, the equation of motion is

$$\frac{W}{g}\frac{d^2y}{dt^2} + ky = 0 \qquad (6.205)$$

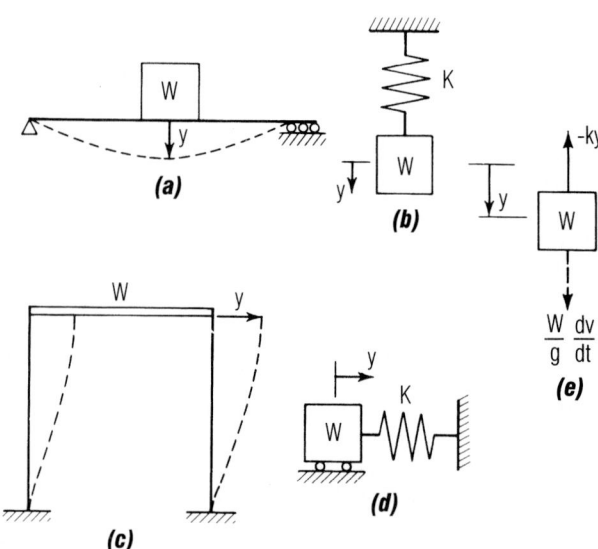

Fig. 6.97 Mass on weightless spring (b) or (d) may represent the motion of a beam (a) or a rigid frame (c) in free vibration.

This may be written in the more convenient form

$$\frac{d^2y}{dt^2} + \frac{kg}{W}y = \frac{d^2y}{dt^2} + \omega^2 y = 0 \tag{6.206}$$

The solution is

$$y = A \sin \omega t + B \cos \omega t \tag{6.207}$$

where A and B are constants to be determined from initial conditions of the system, and

$$\omega = \sqrt{\frac{kg}{W}} \tag{6.208}$$

is the natural circular frequency, radians per second.

The motion defined by Eq. (6.207) is harmonic. Its natural period in seconds is

$$T = \frac{2\pi}{\omega} = 2\pi\sqrt{\frac{W}{gk}} \tag{6.209}$$

Its natural frequency in cycles per second is

$$f = \frac{1}{T} = \frac{1}{2\pi}\sqrt{\frac{kg}{W}} \tag{6.210}$$

If, at time $t = 0$, the mass has an initial displacement y_0 and velocity v_0, substitution in Eq. (6.207) yields $A = v_0/\omega$ and $B = y_0$. Hence, at any time t, the mass is completely located by

$$y = \frac{v_0}{\omega}\sin \omega t + y_0 \cos \omega t \tag{6.211}$$

The stress in the spring can be computed from the displacement y, because the spring force equals $-ky$.

Multidegree Systems ▪ In multiple-degree systems, an independent differential equation of motion can be written for each degree of freedom. Thus, in an N-degree system with N masses, weighing $W_1, W_2, \ldots, W_N$, lb, and N^2 springs with constants k_{rj} ($r = 1, 2, \ldots, N$; $j = 1, 2, \ldots, N$), there are N equations of the form

$$\frac{W_r}{g}\frac{d^2y_r}{dt^2} + \sum_{j=1}^{N} k_{rj}y_j = 0 \quad r = 1, 2, \ldots, N \tag{6.212}$$

Simultaneous solution of these equations reveals that the motion of each mass can be resolved into N harmonic components. They are called the fundamental, second, third, and so on harmonics. Each set of harmonics for all the masses is called a normal mode of vibration.

There are as many normal modes in a system as degrees of freedom. Under certain circumstances, the system could vibrate freely in any one of these modes. During any such vibration, the ratio of displacement of any two of the masses remains constant. Hence, the solutions of Eqs. (6.212) take the form

$$y_r = \sum_{n=1}^{N} a_{rn} \sin \omega_n(t + \tau_n) \tag{6.213}$$

where a_{rn} and τ_n are constants to be determined from the initial conditions of the system and ω_n is the natural circular frequency for each normal mode.

6.81.2 Natural Periods

To determine ω_n set $y_1 = A_1 \sin \omega t$; $y_2 = A_2 \sin \omega_t \ldots$ Then, substitute these and their second derivatives in Eqs. (6.212). After dividing each equation by $\sin \omega t$, the following N equations result:

$$\left(k_{11} - \frac{W_1}{g}\omega^2\right)A_1 + K_{12}A_2 + \cdots + k_{1N}A_N = 0$$

$$k_{21}A_1 + \left(k_{22} - \frac{W_2}{g}\omega^2\right)A_2 + \cdots + k_{2N}A_N = 0$$

$$\cdots\cdots\cdots\cdots\cdots\cdots\cdots\cdots$$

$$k_{N1}A_1 + k_{N2}A_2 + \cdots + \left(k_{NN} - \frac{W_N}{g}\omega^2\right)A_N = 0 \tag{6.214}$$

If there are to be nontrivial solutions for the amplitudes $A_1, A_2, \ldots, A_N$, the determinant of their co-efficients must be zero. Thus,

$$\begin{vmatrix} k_{11} - \dfrac{W}{g}\omega^2 & k_{12} & \cdots & k_{1N} \\ k_{21} & k_{22} - \dfrac{W}{g}\omega^2 & \cdots & k_{2N} \\ \cdots\cdots\cdots\cdots\cdots \\ k_{N1} & k_{N2} & \cdots & k_{NN} - \dfrac{W_N}{g}\omega^2 \end{vmatrix} = 0 \tag{6.215}$$

Solution of this equation for ω yields one real root for each normal mode. And the natural period for each normal mode can be obtained from Eq. (6.209).

6.81.3 Modal Amplitudes

If ω for a normal mode now is substituted in Eqs. (6.214), the amplitudes $A_1, A_2, \ldots, A_N$ for that mode can be computed in terms of an arbitrary value, usually unity, assigned to one of them. The resulting set of modal amplitudes defines the **characteristic shape** for that mode.

The normal modes are mutually orthogonal; that is,

$$\sum_{r=1}^{N} W_r A_{rn} A_{rm} = 0 \qquad (6.216)$$

where W_r is the rth mass out of a total of N, A represents the characteristic amplitude of a normal mode, and n and m identify any two normal modes. Also, for a total of S springs

$$\sum_{s=1}^{S} k_s y_{sn} y_{sm} = 0 \qquad (6.217)$$

where k_s is the constant for the sth spring and y represents the spring distortion.

6.81.4 Stodola-Vianello Method

When there are many degrees of freedom, the preceding procedure for free vibration becomes very lengthy. In such cases, it may be preferable to solve Eqs. (6.214) by numerical, trial-and-error procedures, such as the Stodola-Vianello method, in which the solution converges first on the highest or lowest mode. Then, the other modes are determined by the same procedure after elimination of one of the equations by use of Eq. (6.216). The procedure requires assumption of a characteristic shape, a set of amplitudes A_{r1}. These are substituted in one of Eqs. (6.214) to obtain a first approximation of ω^2. With this value and with $A_{N1} = 1$, the remaining $(N-1)$ equations are solved to obtain a new set of A_{r1}. Then, the procedure is repeated until assumed and final characteristic amplitudes agree.

6.81.5 Rayleigh Method

Because even the Stodola-Vianello method is lengthy for many degrees of freedom, the Rayleigh approximate method may be used to compute the fundamental mode. The frequency obtained by this method, however, may be a little on the high side.

The Rayleigh method also starts with an assumed set of characteristic amplitudes A_{r1} and depends for its success on the small error in natural frequency produced by a relatively larger error in the shape assumption. Next, relative inertia forces acting at each mass are computed: $F_r = W_r A_{r1}/A_{N1}$, where A_{N1} is the assumed displacement at one of the masses. These forces are applied to the system as a static load and displacements B_{r1} due to them calculated. Then, the natural frequency can be obtained from

$$\omega^2 = \frac{g \sum_{r=1}^{N} F_r B_{r1}}{\sum_{r=1}^{N} W_r B_{r1}^2} \qquad (6.218)$$

where g is the acceleration due to gravity, 386 in/s². For greater accuracy, the computation can be repeated with B_{r1} as the assumed characteristic amplitudes.

When the Rayleigh method is applied to beams, the characteristic shape assumed initially may be chosen conveniently as the deflection curve for static loading.

The Rayleigh method may be extended to determination of higher modes by the Schmidt orthogonalization procedure, which adjusts assumed deflection curves to satisfy Eq. (6.216). The procedure is to assume a shape, remove components associated with lower modes, then use the Rayleigh method for the residual deflection curve. The computation will converge on the next higher mode. The method is shorter than the Stodola-Vianello procedure when only a few modes are needed.

For example, suppose the characteristic amplitudes A_{r1} for the fundamental mode have been obtained, and the natural frequency for the second mode is to be computed. Assume a value for the relative deflection of the rth mass A_{r2}. Then the shape with the fundamental mode removed will be defined by the displacements

$$a_{r2} = A_{r2} - c_1 A_{r1} \qquad (6.219)$$

where c_1 is the participation factor for the first mode.

$$c_1 = \frac{\sum_{r=1}^{N} W_r A_{r2} A_{r1}}{\sum_{r=1}^{N} W_r A_{r1}^2} \qquad (6.220)$$

Substitute a_{r2} for B_{r1} in Eq. (6.218) to find the second-mode frequency and, from deflections produced by $F_r = W_r a_{r2}$, an improved shape. (For more rapid convergence, A_{r2} should be selected to make c_1 small.) The procedure should be repeated, starting with the new shape.

For the third mode, assume deflections A_{r3} and remove the first two modes:

$$a_{r3} = A_{r3} - c_1 A_{r1} - c_2 A_{r2} \qquad (6.221)$$

The participation factors are determined from

$$c_1 = \frac{\sum_{r=1}^{N} W_r A_{r3} A_{r1}}{\sum_{r=1}^{N} W_r A_{r1}^2} \qquad c_2 = \frac{\sum_{r=1}^{N} W_r A_{r3} A_{r2}}{\sum_{r=1}^{N} W_r A_{r2}^2} \qquad (6.222)$$

Use a_{r3} to find an improved shape and the third-mode frequency.

6.81.6 Distributed Mass

For some structures with mass distributed throughout, it sometimes is easier to solve the dynamic equations based on distributed mass than the equations based on equivalent lumped masses. A distributed mass has an infinite number of degrees of freedom and normal modes. Every particle in it can be considered a lumped mass on springs connected to other particles. Usually, however, only the fundamental mode is significant, although sometimes the second and third modes must be taken into account.

For example, suppose a beam weighs w lb/lin ft and has a modulus of elasticity E, psi, and moment of inertia I, in^4. Let y be the deflection at a distance x from one end. Then, the equation of motion is

$$EI \frac{\partial^4 y}{\partial x^4} + \frac{w}{g} \frac{\partial^2 y}{\partial t^2} = 0 \qquad (6.223)$$

(This equation ignores the effects of shear and rotational inertia.) The deflection y_n for each mode, to satisfy the equation, must be the product of a harmonic function of time $f_n(t)$ and of the char-

acteristic shape $Y_n(x)$, a function of x with undetermined amplitude. The solution is

$$f_n(t) = c_1 \sin \omega_n t + c_2 \cos \omega_n t \qquad (6.224)$$

where ω_n is the natural circular frequency and n indicates the mode, and

$$Y_n(x) = A_n \sin \beta_n x + B_n \cos \beta_n x + C_n \sin h\beta_n x + D_n \cos h\beta_n x \qquad (6.225)$$

where

$$\beta_n = \sqrt[4]{\frac{w \omega_n^2}{EIg}} \qquad (6.226)$$

Equations (6.224) to (6.226) apply to spans with any type of end restraints. Figure 6.98 shows the characteristic shape and gives constants for determination of natural circular frequency ω and natural period T for the first four modes of cantilever, simply supported, fixed-end, and fixed-hinged beams. To obtain ω, select the appropriate constant from Fig. 6.98 and multiply it by $\sqrt{EI/wL^4}$. To get T, divide the appropriate constant by $\sqrt{EI/wL^4}$.

6.81.7 Simple Beam

For a simple beam, the boundary (support) conditions for all values of time t are $y = 0$ and bending moment $M = EI \, \partial^2 y/\partial x^2 = 0$. Hence, at $x = 0$ and $x = L$, the span length, $Y_n(x) = 0$ and $d^2 Y_n/dx^2 = 0$. These conditions require that $B_n = C_n = D_n = 0$ and $\beta_n = n\pi/L$, to satisfy Eq. (6.225). Hence, according to Eq. (6.226), the natural circular frequency for a simply supported beam is

$$\omega_n = \frac{n^2 \pi^2}{L^2} \sqrt{\frac{EIg}{w}} \qquad (6.227)$$

The characteristic shape is defined by

$$Y_n(x) = \sin \frac{n\pi x}{L} \qquad (6.228)$$

The constants c_1 and c_2 in Eq. (6.224) are determined by the initial conditions of the disturbance. Thus, the total deflection, by superposition of modes, is

$$y = \sum_{n=1}^{\infty} A_n(t) \sin \frac{n\pi x}{L} \qquad (6.229)$$

where $A_n(t)$ is determined by the load (see Art. 6.83).

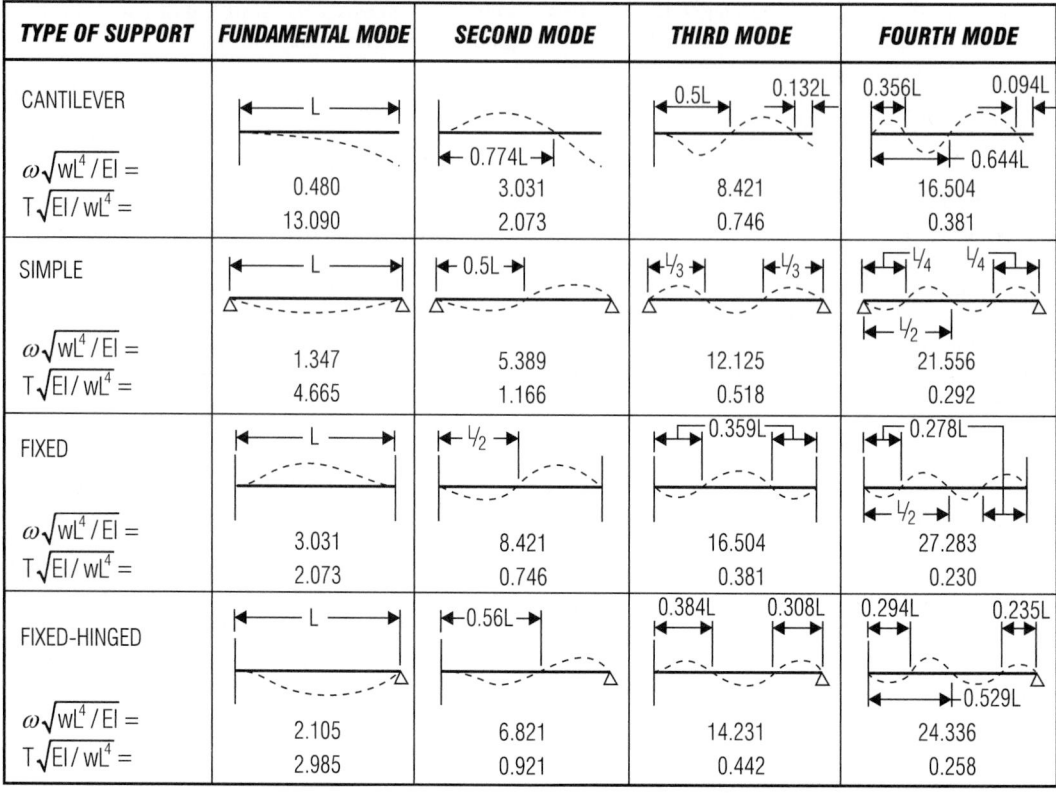

TYPE OF SUPPORT	FUNDAMENTAL MODE	SECOND MODE	THIRD MODE	FOURTH MODE
CANTILEVER $\omega\sqrt{wL^4/EI} =$ $T\sqrt{EI/wL^4} =$	0.480 13.090	3.031 2.073	8.421 0.746	16.504 0.381
SIMPLE $\omega\sqrt{wL^4/EI} =$ $T\sqrt{EI/wL^4} =$	1.347 4.665	5.389 1.166	12.125 0.518	21.556 0.292
FIXED $\omega\sqrt{wL^4/EI} =$ $T\sqrt{EI/wL^4} =$	3.031 2.073	8.421 0.746	16.504 0.381	27.283 0.230
FIXED-HINGED $\omega\sqrt{wL^4/EI} =$ $T\sqrt{EI/wL^4} =$	2.105 2.985	6.821 0.921	14.231 0.442	24.336 0.258

Fig. 6.98 Coefficients for computing natural circular frequencies ω and natural periods of vibration T, seconds, for prismatic beams: w = weight of beam, lb/lin ft; L = beam span, ft; E = modulus of elasticity, psi; I = moment of inertia, in^4.

To determine the characteristic shapes and natural periods for beams with variable cross section and mass, use the Rayleigh method. Convert the beam into a lumped-mass system by dividing the span into elements and assuming the mass of each element to be concentrated at its center. Also, compute all quantities, such as deflection and bending moment, at the center of each element. Start with an assumed characteristic shape and apply Eq. (6.218).

Methods are available for dynamic analysis of continuous beams. (R. Clough and J. Penzien, "Dynamics of Structures," McGraw-Hill Book Company, New York, books.mcgraw-hill.com; D. G. Fertis and E. C. Zobel, "Transverse Vibration Theory," The Ronald Press Company). But even for beams with constant cross section, these procedures are very lengthy. Generally, approximate solutions are preferable.

(J. M. Biggs, "Introduction to Structural Dynamics," McGraw-Hill Book Company, New York, books.mcgraw-hill.com; N. M. Newmark and E. Rosenblueth, "Fundamentals of Earthquake Engineering," Prentice-Hall, Inc., Englewood Cliffs, N.J., www.prenhall.com.)

6.82 Impact and Sudden Loads

Under impact, there is an abrupt exchange or absorption of energy and drastic change in velocity. Stresses caused in the colliding members may be several times larger than stresses produced by the same weights applied statically.

An approximation of impact stresses in the elastic range can be made by neglecting the inertia of the body struck and the effect of wave propagation and assuming that the kinetic energy is converted completely into strain energy in that body. Consider a prismatic bar subjected to an axial impact load in tension. The energy absorbed per unit of volume when the bar is stressed to the proportional limit is called the **modulus of resilience**. It is given by $f_y^2/2E$, where f_y is the yield stress and E the modulus of elasticity, both in psi. Below the proportional limit, the stress, psi, due to an axial load U, in-lb, is

$$f = \sqrt{\frac{2UE}{AL}} \qquad (6.230)$$

where A is the cross-sectional area, in^2, and L the length of bar, in.

This equation indicates that energy absorption of a member may be improved by increasing its length or area. Sharp changes in cross section should be avoided, however, because of associated high stress concentrations. Also, uneven distribution of stress in a member due to changes in section should be avoided. Energy absorption is larger with a uniform stress distribution throughout the length of the member.

If a static axial load W would produce a tensile stress f' in the bar and an elongation e', in, then the axial stress produced when W falls a distance h, in, is

$$f = f' + f'\sqrt{1 + \frac{2h}{e'}} \qquad (6.231)$$

if f is within the proportional limit. The elongation due to this impact load is

$$e = e' + e'\sqrt{1 + \frac{2h}{e'}} \qquad (6.232)$$

These equations indicate that the stress and deformation due to an energy load may be considerably larger than those produced by the same weight applied gradually.

The same equations hold for a beam with constant cross section struck by a weight at midspan, except that f and f' represent stresses at midspan and e and e', midspan deflections.

According to Eqs. (6.231) and (6.232), a sudden load ($h = 0$) causes twice the stress and twice the deflection as the same load applied gradually.

6.82.1 Impact on Long Members

For very long members, the effect of wave propagation should be taken into account. Impact is not transmitted instantly to all parts of the struck body. At first, remote parts remain undisturbed, while particles struck accelerate rapidly to the velocity of the colliding body. The deformations produced move through the struck body in the form of elastic waves. The waves travel with a constant velocity, ft/s,

$$c = 68.1\sqrt{\frac{E}{\rho}} \qquad (6.233)$$

where E = modulus of elasticity, psi

ρ = density of the struck body, lb/ft^3

6.82.2 Impact Waves

If an impact imparts a velocity v, ft/s, to the particles at one end of a prismatic bar, the stress, psi, at that end is

$$f = 0.0147v\sqrt{E\rho} \qquad (6.234)$$

if f is in the elastic range. In a compression wave, the velocity of the particles is in the direction of the wave. In a tension wave, the velocity of the particles is in the opposite direction to the wave.

In the plastic range, Eqs. (6.233) and (6.234) hold, but with E as the tangent modulus of elasticity. Hence, c is not a constant and the shape of the stress wave changes as it moves. The elastic portion of the stress wave moves faster than the wave in the plastic range. Where they overlap, the stress and irrecoverable strain are constant.

(The impact theory is based on an assumption difficult to realize in practice—that contact takes place simultaneously over the entire end of the bar.)

At a free end of a bar, a compressive stress wave is reflected as an equal tension wave, and a tension wave as an equal compression wave. The velocity of the particles at the free end equals $2v$.

At a fixed end of a bar, a stress wave is reflected unchanged. The velocity of the particles at the fixed end is zero, but the stress is doubled because of the superposition of the two equal stresses on reflection.

For a bar with a fixed end struck at the other end by a moving mass weighing W_m lb, the initial compressive stress, psi, is, from Eq. (6.234),

$$f_0 = 0.0147 v_0 \sqrt{E\rho} \qquad (6.235)$$

where v_0 is the initial velocity of the particles, ft/s, at the impacted end of the bar and E and ρ the modulus of elasticity, psi, and density, lb/ft³, of the bar. As the velocity of W_m decreases, so does the pressure on the bar. Hence, decreasing compressive stresses follow the wave front. At any time $t < 2L/c$, where L is the length of the bar, in, the stress at the struck end is

$$f = f_0 e^{-2\alpha t/\tau} \qquad (6.236)$$

where $e = 2.71828$; α is the ratio of W_b, the weight of the bar, to W_m; and $\tau = 2L/c$.

When $t = \tau$, the wave front with stress f_0 arrives back at the struck end, assumed still to be in contact with the mass. Since the velocity of the mass cannot change suddenly, the wave will be reflected as from a fixed end. During the second interval, $\tau < t < 2\tau$, the compressive stress is the sum of two waves moving away from the struck end and one moving toward this end.

Maximum stress from impact occurs at the fixed end. For α greater than 0.2, this stress is

$$f = 2f_0(1 + e^{-2\alpha}) \qquad (6.237)$$

For smaller values of α, it is given approximately by

$$f = f_0 \left(1 + \sqrt{\frac{1}{\alpha}}\right) \qquad (6.238)$$

Duration of impact, time it takes for the stress at the struck end to drop to zero, is approximately

$$T = \frac{\pi L}{c\sqrt{\alpha}} \qquad (6.239)$$

for small values of α.

When W_m is the weight of a falling body, velocity at impact is $\sqrt{2gh}$, when it falls a distance h, in. Substitution in Eq. (6.235) yields

$$f_0 = \sqrt{2EhW_b/AL}$$

since $W_b = \rho AL$ is the weight of the bar. Putting $W_b = \alpha W_m$; $W_m/A = f'$, the stress produced by W_m when applied gradually, and $E = f'L/e'$, where e' is

the elongation for the static load, gives

$$f_0 = f' \sqrt{2h\alpha/e'}$$

Then, for values of α smaller than 0.2, the maximum stress, from Eq. (6.238), is

$$f = f' \left(\sqrt{\frac{2h\alpha}{e'}} + \sqrt{\frac{2h}{e'}}\right) \qquad (6.240)$$

For larger values of α, the stress wave due to gravity acting on W_m during impact should be added to Eq. (6.237). Thus, for α larger than 0.2,

$$f = 2f'(1 - e^{-2\alpha}) + 2f' \sqrt{\frac{2h\alpha}{e'}} (1 + e^{-2\alpha}) \qquad (6.241)$$

Equations (6.250) and (6.251) correspond to Eq. (6.231), which was developed without taking wave effects into account. For a sudden load, $h = 0$, Eq. (6.241) gives for the maximum stress $2f'(1 - e^{-2\alpha})$, not quite double the static stress, the result indicated by Eq. (6.231). (See also Art. 6.83.)

(S. Timoshenko and J. N. Goodier, "Theory of Elasticity," S. Timoshenko and D. H. Young, "Engineering Mechanics," and D. D. Barkan, "Dynamics of Bases and Foundations," McGraw-Hill Book Company, New York, books.mcgraw-hill.com.)

6.83 Dynamic Analysis of Simple Structures

Articles 6.81 and 6.82 present a theoretical basis for analysis of structures under dynamic loads. As noted in Art. 6.81, an approximate solution based on an idealized representation of an actual member or structure is advisable for dynamic analysis and design. Generally, the actual structure may be conveniently represented by a system of masses and massless springs, with additional resistances to account for damping. In simple cases, the masses may be set equal to the actual masses; otherwise, equivalent masses may be substituted for the actual masses (Art. 6.85). The spring constants are the ratios of forces to deflections (see Art. 6.81).

Usually, for structural purposes, the data sought are the maximum stresses in the springs and their maximum displacements and the time of occurrence of the maximums. This time generally is computed in terms of the natural period of vibration of the member or structure or in terms of the duration of the load. Maximum displacement may be calculated in terms of the deflection that would result if the load were applied gradually.

The term D by which the static deflection e', spring forces, and stresses are multiplied to obtain the dynamic effects is called the **dynamic load factor**. Thus, the dynamic displacement is

$$y = De' \qquad (6.242)$$

and the maximum displacement y_m is determined by the maximum dynamic load factor D_m, which occurs at time t_m.

6.83.1 One-Degree System

Consider the one-degree-of-freedom system in Fig. 6.99a. It may represent a weightless beam with a mass weighing W lb applied at midspan and subjected to a varying force $F_o f(t)$, or a rigid frame with a mass weighing W lb at girder level and subjected to this force. The force is represented by an arbitrarily chosen constant force F_o times $f(t)$, a function of time.

If the system is not damped, the equation of motion in the elastic range is

$$\frac{W}{g}\frac{d^2y}{dt^2} + ky = F_o f(t) \qquad (6.243)$$

where k is the spring constant and g the acceleration due to gravity, 386 in/s^2. The solution consists of two parts. The first, called the complementary solution, is obtained by setting $f(t) = 0$. This solution is given by Eq. (6.211). To it must be added the second part, the particular solution, which satisfies Eq. (6.243).

The general solution of Eq. (6.243), arrived at by treating an element of the force-time curve (Fig. 6.99b) as an impulse, is

$$y = y_o \cos \omega t + \frac{v_o}{\omega}\sin \omega t$$

$$+ e'\omega \int_0^t f(\tau)\sin \omega(t-\tau)\,d\tau \qquad (6.244)$$

where $y =$ displacement of mass from equilibrium position, in

$y_o =$ initial displacement of mass $(t = 0)$, in

$\omega = \sqrt{kg/W} =$ natural circular frequency of free vibration

$k =$ spring constant = force producing unit deflection, lb/in

$v_o =$ initial velocity of mass, in/s

$e' = F_o/k =$ displacement under static load, in

A closed solution is possible if the integral can be evaluated.

Assume, for example, the mass is subjected to a suddenly applied force F_o that remains constant (Fig. 6.100a). If y_o and v_o are initially zero, the displacement y of the mass at any time t can be obtained from the integral in Eq. (6.244) by setting $f(\tau) = 1$:

$$y = e'\omega \int_0^t \sin \omega(t-\tau)\,d\tau = e'(1 - \cos \omega t) \qquad (6.245)$$

The dynamic load factor $D = 1 - \cos \omega t$. It has a maximum value $D_m = 2$ when $t = \pi/\omega$. Figure

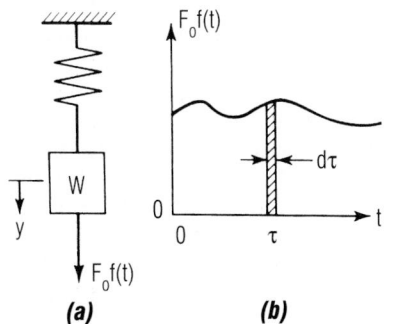

(a)   **(b)**

Fig. 6.99 One-degree system acted on by a varying force.

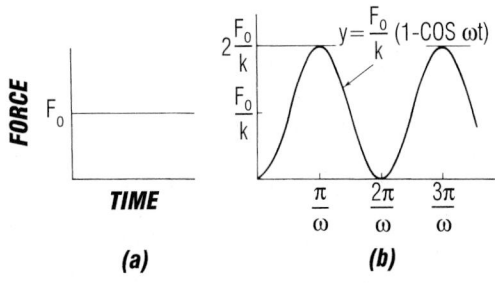

(a)   **(b)**

Fig. 6.100 Harmonic vibrations (b) result when a constant force (a) is applied to an undamped one-degree system like the one in Fig. 6.99a.

6.100b shows the variation of displacement with time.

6.83.2 Multidegree Systems

A multidegree lumped-mass system may be analyzed by the modal method after the natural frequencies of the normal modes have been determined (Art. 6.80). This method is restricted to linearly elastic systems in which the forces applied to the masses have the same variation with time. For other cases, numerical analysis must be used.

In the modal method, each normal mode is treated as an independent one-degree system. For each degree of the system, there is one normal mode. A natural frequency and a characteristic shape are associated with each mode. In each mode, the ratio of the displacements of any two masses is constant with time. These ratios define the characteristic shape. The modal equation of motion for each mode is

$$\frac{d^2 A_n}{dt^2} + \omega_n^2 A_n = \frac{g f(t) \sum_{r=1}^{j} F_r \phi_{rn}}{\sum_{r=1}^{j} W_r \phi_{rn}^2} \qquad (6.246)$$

where A_n = displacement in nth mode of arbitrarily selected mass

ω_n = natural frequency of nth mode

$F_r f(t)$ = varying force applied to rth mass

W_r = weight of rth mass

j = number of masses in system

ϕ_{rn} = ratio of displacement in nth mode of rth mass to A_n

g = acceleration due to gravity

We define the modal static deflection as

$$A_n' = \frac{g \sum_{r=1}^{j} F_r \phi_{rn}}{\omega_n^2 \sum_{r=1}^{j} W_r \phi_{rn}^2} \qquad (6.247)$$

Then, the response for each mode is given by

$$A_n = D_n A_n' \qquad (6.248)$$

where D_n is the dynamic load factor. Since D_n depends only on ω_n and $f(t)$, the variation of force with time, solutions for D_n obtained for one-degree systems also apply to multidegree systems. The total deflection at any point is the sum of the displacements for each mode, $\Sigma A_n \phi_{rn}$, at that point.

6.83.3 Response of Beams

The response of beams to dynamic forces can be determined in a similar way. The modal static deflection is defined by

$$A_n' = \frac{\int_0^L p(x)\phi_n(x)dx}{\omega_n^2(w/g)\int_0^L \phi_n^2(x)dx} \qquad (6.249)$$

where $p(x)$ = load distribution on span [$p(x)f(t)$ is varying force]

$\phi_n(x)$ = characteristic shape of nth mode (see Art. 6.81)

L = span length

w = uniformly distributed weight on span

The response of the beam then is given by Eq. (6.248) and the dynamic deflection is the sum of the modal components, $\Sigma A_n \phi_n(x)$.

Nonlinear Responses ■ When the structure does not react linearly to loads, the equations of motion can be solved by numerical analysis if resistance is a unique function of displacement. Sometimes, the behavior of the structure can be represented by an idealized resistance displacement diagram that makes possible a solution in closed form. Figure 6.101a shows such a diagram.

6.83.4 Elastic-Plastic Response

Resistance is assumed linear ($R = ky$) until a maximum R_m is reached. After that, R remains equal to R_m for increases in y substantially larger than the displacement y_e at the elastic limit. Thus, some portions of the structure deform into the plastic range. Figure 6.101a, therefore, may be used for ductile structures only rarely subjected to severe dynamic loads. When this diagram can be used for designing such structures, more economical designs can be produced than for structures limited to the elastic range because of the high energy-absorption capacity of structures in the plastic range.

For a one-degree system, Eq. (6.243) can be used as the equation of motion for the initial sloping part

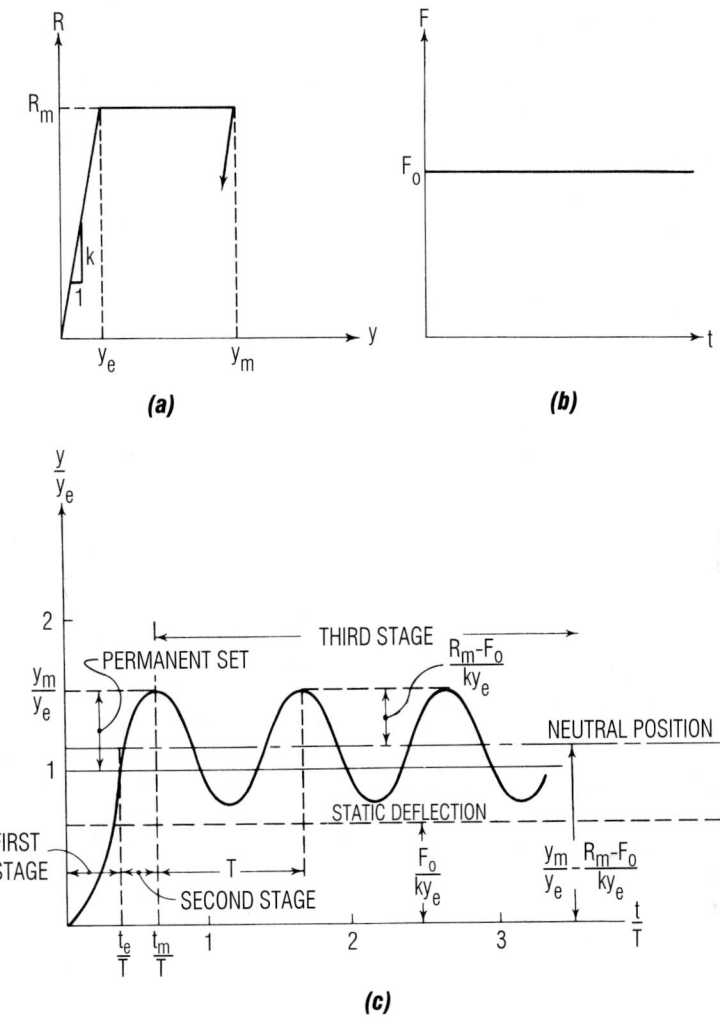

Fig. 6.101 Response in the elastic range of a one-degree system with resistance characteristics plotted in (*a*) to a constant force (*b*) is shown in (*c*).

of the diagram (elastic range). For the second stage, $y_e < y < y_m$ where y_m is the maximum displacement, the equation is

$$\frac{W}{g}\frac{d^2y}{dt^2} + R_m = F_o f(t) \qquad (6.250)$$

For the unloading stage, $y < y_m$, the equation is

$$\frac{W}{g}\frac{d^2y}{dt^2} + R_m - k(y_m - y) = F_o f(t) \qquad (6.251)$$

Suppose, for example, the one-degree undamped system in Fig. 6.99*a* behaves in accordance with the bilinear resistance function of Fig. 6.101*a* and is subjected to a suddenly applied constant load (Fig. 6.101*b*). With zero initial displacement and velocity, the response in the first stage ($y < y_e$), according to Eq. (6.245), is

$$y = e'(1 - \cos \omega t_1)$$

$$\frac{dy}{dt} = e' \omega \sin \omega t_1 \qquad (6.252)$$

Equation (6.245) also indicates that the displacement y_e will be reached at a time t_e such that $\cos \omega t_e = y_e/e'$.

For convenience, let $t_2 = t - t_e$ be the time in the second stage; thus, $t_2 = 0$ at the start of that stage. Since the condition of the system at that time is the same as the condition at the end of the first stage, the initial displacement is y_e and the initial velocity $e' \omega \sin \omega t_e$. The equation of motion is

$$\frac{W}{g}\frac{d^2y}{dt^2} + R_m = F_o \qquad (6.253)$$

The solution, taking into account initial conditions after integrating, for $y_e < y < y_m$ is

$$y = \frac{g}{2W}(F_o - R_m)t_2^2 + e' \omega t_2 \sin \omega t_e + y_e \qquad (6.254)$$

Maximum displacement occurs at the time

$$t_m = \frac{W \omega e'}{g(R_m - F_o)}\sin \omega t_e \qquad (6.255)$$

and can be obtained by substituting t_m in Eq. (6.254).

The third stage, unloading after y_m has been reached, can be determined from Eq. (6.251) and conditions at the end of the second stage. The response, however, is more easily found by noting that the third stage consists of an elastic, harmonic residual vibration. In this stage, the amplitude of vibration is $(R_m - F_o)/k$ since this is the distance between the neutral position and maximum displacement, and in the neutral position the spring force equals F_o. Hence, the response, obtained directly from Eq. (6.245), is $y_m - (R_m - F_o)/k$ for e' because the neutral position, $y = y_m - (R_m - F_o)/k$, occurs when $\omega t_3 = \pi/2$. The solution is

$$y = y_m - \frac{R_m - F_o}{k} + \frac{R_m - F_o}{k}\cos \omega t_3 \qquad (6.256)$$

where $t_3 = t - t_e - t_m$.

Response in the three stages is shown in Fig. 6.101c. In that diagram, however, to represent a typical case, the coordinates have been made nondimensional by expressing y in terms of y_e and the time in terms of T, the natural period of vibration.

(J. M. Biggs, "Introduction to Structural Dynamics" and R. Clough and J. Penzien, "Dynamics of Structures," McGraw-Hill Book Company, New York, books.mcgraw-hill.com; D. G. Fertis and E. C. Zobel, "Transverse Vibration Theory," The Ronald Press Company, New York; N. M. Newmark and E. Rosenbleuth, "Fundamentals of Earthquake, Engineering," Prentice-Hall, Inc., Englewood Cliffs, N.J., www.prenhall.com.)

6.84 Resonance and Damping

Damping in structures, due to friction and other causes, resists motion imposed by dynamic loads. Generally, the effect is to decrease the amplitude and lengthen the period of vibrations. If damping is large enough, vibration may be eliminated.

When maximum stress and displacement are the prime concern, damping may not be of great significance for short-time loads. These maximums usually occur under such loads at the first peak of response, and damping, unless unusually large, has little effect in a short period of time. But under conditions close to resonance, damping has considerable effect.

Resonance is the condition of a vibrating system under a varying load such that the amplitude of successive vibrations increases. Unless limited by damping or changes in the condition of the system, amplitudes may become very large.

Two forms of damping generally are assumed in structural analysis, viscous and constant (Coulomb). For viscous damping, the damping force is taken proportional to the velocity but opposite in direction. For Coulomb damping, the damping force is assumed constant and opposed in direction to the velocity.

6.84.1 Viscous Damping

For a one-degree system (Arts. 6.81 to 6.83), the equation of motion for a mass weighing W lb and subjected to a force F varying with time but opposed by viscous damping is

$$\frac{W}{g}\frac{d^2y}{dt^2} + ky = F - c\frac{dy}{dt} \qquad (6.257)$$

where y = displacement of mass from equilibrium position, in

k = spring constant, lb/in

t = time, s

c = coefficient of viscous damping

g = acceleration due to gravity = 386 in/s^2

Let us set $\beta = cg/2W$ and consider those cases in which $\beta < \omega$, the natural circular frequency [Eq. (6.208)], to eliminate unusually high damping (overdamping). Then, for initial displacement y_o and velocity v_o, the solution of Eq. (6.257) with $F = 0$ is

$$y = e^{-\beta t}\left(\frac{v_o + \beta y_o}{\omega_d}\sin\omega_d t + y_o\cos\omega_d t\right) \qquad (6.258)$$

where $\omega_d = \sqrt{\omega^2 - \beta^2}$ and $e = 2.71828$. Equation (6.258) represents a decaying harmonic motion with β controlling the rate of decay and ω_d the natural frequency of the damped system.

When $\beta = \omega$

$$y = e^{-\omega t}[v_o t + (1 + \omega t)y_o] \qquad (6.259)$$

which indicates that the motion is not vibratory. Damping producing this condition is called critical, and the critical coefficient is

$$c_d = \frac{2W\beta}{g} = \frac{2W\omega}{g} = 2\sqrt{\frac{kW}{g}} \qquad (6.260)$$

Damping sometimes is expressed as a percent of critical (β as a percent of ω).

For small amounts of viscous damping, the damped natural frequency is approximately equal to the undamped natural frequency minus $\frac{1}{2}\beta^2/\omega$. For example, for 10% critical damping ($\beta = 0.1\omega$), $\omega_d = \omega[1 - \frac{1}{2}(0.1)^2] = 0.995\omega$. Hence, the decrease in natural frequency due to small amount of damping generally can be ignored.

Damping sometimes is measured by **logarithmic decrement**, the logarithm of the ratio of two consecutive peak amplitudes during free vibration.

$$\text{Logarithmic decrement} = \frac{2\pi\beta}{\omega} \qquad (6.261)$$

For example, for 10% critical damping, the logarithmic decrement equals 0.2π. Hence, the ratio of a peak to the following peak amplitude is $e^{0.2\pi} = 1.87$.

The complete solution of Eq. (6.257) with initial displacement y_o and velocity v_o is

$$y = e^{-\beta t}\left(\frac{v_o + \beta y_o}{\omega_d}\sin\omega_d t + y_o\cos\omega_d t\right)$$
$$+ e'\frac{\omega^2}{\omega_d}\int_0^t f(\tau)e^{-\beta(t-\tau)}\sin\omega_d(t-\tau)\,d\tau \qquad (6.262)$$

where e' is the deflection that the applied force would produce under static loading. Equation (6.262) is identical to Eq. (6.244) when $\beta = 0$.

Unbalanced rotating parts of machines produce pulsating forces that may be represented by functions of the form $F_o\sin\alpha t$. If such a force is applied to an undamped one-degree system, Eq. (6.244) indicates that if the system starts at rest the response will be

$$y = \frac{F_o g}{W}\left(\frac{1/\omega^2}{1 - \alpha^2/\omega^2}\right)\left(\sin\alpha t - \frac{\alpha}{\omega}\sin\omega t\right) \qquad (6.263)$$

And since the static deflection would be $F_o/k = F_o g/W\omega^2$, the dynamic load factor is

$$D = \frac{1}{1 - \alpha^2/\omega^2}\left(\sin\alpha t - \frac{\alpha}{\omega}\sin\omega t\right) \qquad (6.264)$$

If α is small relative to ω, maximum D is nearly unity; thus, the system is practically statically loaded. If α is very large compared with ω, D is very small; thus, the mass cannot follow the rapid fluctuations in load and remains practically stationary. Therefore, when α differs appreciably from ω, the effects of unbalanced rotating parts are not too serious. But if $\alpha = \omega$, resonance occurs; D increases with time. Hence, to prevent structural damage, measures must be taken to correct the unbalanced parts to change α, or to change the natural frequency of the vibrating mass, or damping must be provided.

The response as given by Eq. (6.263) consists of two parts, the free vibration and the forced part. When damping is present, the free vibration is of the form of Eq. (6.268) and is rapidly damped out. Hence, the free part is called the **transient response**, and the forced part, the **steady-state response**. The maximum value of the dynamic load factor for the steady-state response D_m is called the **dynamic magnification factor**. It is given by

$$D_m = \frac{1}{\sqrt{(1 - \alpha^2/\omega^2)^2 + (2\beta\alpha/\omega^2)^2}} \qquad (6.265)$$

With damping, then, the peak values of D_m occur when

$$\alpha = \omega\sqrt{\frac{1 - \beta^2}{\omega^2}}$$

and are approximately equal to $\omega/2\beta$. For example, for 10% critical damping,

$$D_m = \frac{\omega}{0.2\omega} = 5$$

So even small amounts of damping significantly limit the response at resonance.

6.84.2 Coulomb Damping

For a one-degree system with Coulomb damping the equation of motion for free vibration is

$$\frac{W}{g}\frac{d^2y}{dt^2} + ky = \pm F_f \qquad (6.266)$$

where F_f is the constant friction force and the positive sign applies when the velocity is negative. If initial displacement is y_o and initial velocity is zero, the response in the first half cycle, with negative velocity, is

$$y = \left(y_o - \frac{F_f}{k}\right)\cos\omega t + \frac{F_f}{k} \qquad (6.267)$$

equivalent to a system with a suddenly applied constant force. For the second half cycle, with positive velocity, the response is

$$y = \left(-y_o + 3\frac{F_f}{k}\right)\cos\omega\left(t - \frac{\pi}{\omega}\right) - \frac{F_f}{k}$$

If the solution is continued with the sign of F_f changing in each half cycle, the results will indicate that the amplitude of positive peaks is given by $y_o - 4nF_f/k$, where n is the number of complete cycles, and the response will be completely damped out when $t = ky_oT/4F_f$, where T is the natural period of vibration, or $2\pi/\omega$.

Analysis of the steady-state response with Coulomb damping is complicated by the possibility of frequent cessation of motion.

(S. Timoshenko, D. H. Young, and W. Weaver, "Vibration Problems in Engineering," 4th ed., John Wiley & Sons, Inc., New York, www.wiley.com; D. D. Barkan, "Dynamics of Bases and Foundations," McGraw-Hill Book Company, New York; W. C. Hurty and M. F. Rubinstein, "Dynamics of Structures," Prentice-Hall, Inc., Englewood Cliffs, N.J., www.prenhall.com.)

6.85 Approximate Design for Dynamic Loading

Complex analysis and design methods seldom are justified for structures subjected to dynamic loading because of lack of sufficient information on loading, damping, resistance to deformation, and other factors. In general, it is advisable to represent the actual structure and loading by idealized systems that permit a solution in closed form. (See Arts. 6.80 to 6.83.)

Whenever possible, represent the actual structure by a one-degree system consisting of an equivalent mass with massless spring. For structures with distributed mass, simplify the analysis in the elastic range by computing the response only for one or a few of the normal modes. In the plastic range, treat each stage—elastic, elastic-plastic, and plastic—as completely independent; for example, a fixed-end beam may be treated, when in the elastic-plastic stage, as a simply supported beam.

Choose the parameters of the equivalent system to make the deflection at a critical point, such as the location of the concentrated mass, the same as it would be in the actual structure. Stresses in the actual structure should be computed from the deflection in the equivalent system.

Compute an assumed shape factor ϕ for the system from the shape taken by the actual structure under static application of the loads. For example, for a simple beam in the elastic range with concentrated load at midspan, ϕ may be chosen, for $x < L/2$, as $(Cx/L^3)(3L^2 - 4x^2)$, the shape under static loading, and C may be set equal to 1 to make ϕ equal to 1 when $x = L/2$. For plastic conditions (hinge at midspan), ϕ may be taken as Cx/L, and C set equal to 2, to make $\phi = 1$ when $x = L/2$.

For a structure with concentrated forces, let W_r be the weight of the rth mass, ϕ_r the value of ϕ at the location of that mass, and F_r the dynamic force acting on W_r. Then, the equivalent weight of the idealized system is

$$W_e = \sum_{r=1}^{j} W_r\phi_r^2 \qquad (6.268)$$

where j is the number of masses. The equivalent force is

$$F_e = \sum_{r=1}^{j} F_r\phi_r \qquad (6.269)$$

For a structure with continuous mass, the equivalent weight is

$$W_e = \int w\phi^2 \, dx \qquad (6.270)$$

where w is the weight in lb/lin ft. The equivalent force is

$$F_e = \int q\phi \, dx \qquad (6.271)$$

for a distributed load q, lb/lin ft.

The resistance of a member or structure is the internal force tending to restore it to its unloaded static position. For most structures, a bilinear resistance function, with slope k up to the elastic limit and zero slope in the plastic range (Fig. 6.101a), may be assumed. For a given distribution of dynamic load, maximum resistance of the idealized system may be taken as the total load with that distribution that the structure can support statically. Similarly, stiffness is numerically equal to the total load with the given distribution that would cause a unit deflection at the point where the deflections in the actual structure and idealized system are equal. Hence, the equivalent resistance and stiffness are in the same ratio to the actual as the equivalent forces to the actual forces.

Let k be the actual spring constant, g the acceleration due to gravity, 386 in/s^2, and

$$W' = \frac{W_e}{F_e} \Sigma F \qquad (6.272)$$

where ΣF represents the actual total load. Then, the equation of motion of an equivalent one-degree system is

$$\frac{d^2 y}{dt^2} + \omega^2 y = g \frac{\Sigma F}{W'} \qquad (2.273)$$

and the natural circular frequency is

$$\omega = \sqrt{\frac{kg}{W'}} \qquad (6.274)$$

The natural period of vibration equals $2\pi/\omega$. Equations (6.273) and (6.274) have the same form as Eqs. (6.206), (6.208), and (6.243). Consequently, the response can be computed as indicated in Arts. 6.80 to 6.82.

Whenever possible, select a load-time function for ΣF to permit use of a known solution.

For preliminary design of a one-degree system loaded into the plastic range by a suddenly applied force that remains substantially constant up to the time of maximum response, the following approximation may be used for that response:

$$y_m = \frac{y_e}{2(1 - F_o/R_m)} \qquad (6.275)$$

where y_e is the displacement at the elastic limit, F_o the average value of the force, and R_m the maximum resistance of the system. This equation indicates that for purely elastic response, R_m must be twice F_o; whereas, if y_m is permitted to be large, R_m may be made nearly equal to F_o, with greater economy of material.

For preliminary design of a one-degree system subjected to a sudden load with duration t_d less than 20% of the natural period of the system, the following approximation can be used for the maximum response:

$$y_m = \frac{1}{2} y_e \left[\left(\frac{F_o}{R_m} \omega t_d \right)^2 + 1 \right] \qquad (6.276)$$

where F_o is the maximum value of the load and ω the natural frequency. This equation also indicates that the larger y_m is permitted to be, the smaller R_m need be.

For a beam, the spring force of the equivalent system is not the actual force, or reaction, at the supports. The real reactions should be determined from the dynamic equilibrium of the complete beam. This calculation should include the inertia force, with distribution identical with the assumed deflected shape of the beam. For example, for a simply supported beam with uniform load, the dynamic reaction in the elastic range is $0.39R + 0.11F$, where R is the resistance, which varies with time, and $F = qL$ is the load. For a concentrated load F at midspan, the dynamic reaction is $0.78R - 0.28F$. And for concentrated loads $F/2$ at each third point, it is $0.62R - 0.12F$. (Note that the sum of the coefficients equals 0.50, since the dynamic-reaction equations must hold for static loading, when $R = F$.) These expressions also can be used for fixed-end beams without significant error. If high accuracy is not required, they also can be used for the plastic range.

Structures usually are designed to resist the dynamic forces of earthquakes by use of equivalent static loads. (See Arts. 15.4 and 17.3.)

6.85.1 Basics of Structural Dynamics

The basic element in structural dynamics is the single-degree-of-freedom system. Many of the available vibration criteria utilize a strategy to simplify a complex floor system into this basic element. The single-degree-of-freedom system is represented by a single mass m, spring k, and damper c, as shown in Fig. 6.102. The governing differential equation of motion for this system follows.

Equation of motion for single degree of freedom:

$$m\ddot{y}(t) + c\dot{y}(t) + ky(t) = F(t)$$

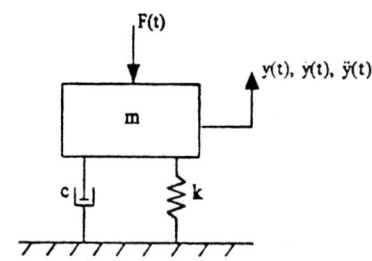

Single-Degree-of-Freedom System

Sinusoidal Force Input

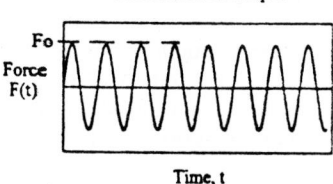

Ramp Force Input

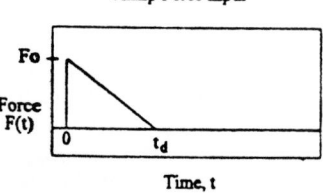

Fig. 6.102 Single-degree-of-freedom system and two common input forces.

When the mass m is subjected to a time-dependent input force $F(t)$, the result is a vibration response which can be described by the displacement $y(t)$, the velocity, $\dot{y}(t)$, and the acceleration, $\ddot{y}(t)$. The equation of motion for a single-degree-of-freedom system can also be formulated in terms of the natural frequency of the free vibration and ratio of critical damping.

$$\ddot{y}(t) + 2\zeta\omega_0\dot{y}(t) + \omega_0^2 y(t) = \frac{F(t)}{m} \qquad (6.277)$$

where ω_0 = circular natural frequency, radians/s
$\quad = \sqrt{k/m} = 2\pi f_0$
$\quad f_0$ = natural frequency, Hz
$\quad \zeta$ = ratio of critical damping
$\quad = c/c_{cr}$
$\quad c_{cr}$ = critical damping, the value of damping for which the roots of the characteristic equation are equal

$$c_{cr} = 2\sqrt{km} \qquad (6.278)$$

Also shown in Fig. 6.102 are two input forces commonly used to represent different sources of floor excitations. A sinusoidal force input function is often used to predict floor response due to rhythmic excitations. The ramp force input function is often used to assess the floor system response to transient excitations such as walking. The closed-form solutions for the response of a single-degree-of-freedom system subjected to these input forces can be found in most structural dynamics textbooks and will not be presented here.

Continuous systems, such as beams or plate-like structures, contain an infinite number of free vibration modes. Each of these modes can be characterized by its mode shape and its associated natural frequency. Figure 6.103 illustrates the first three modes of vibration for a simply supported beam with a uniform mass distribution. The vibration response at any point on a beam can be approximated by the sum of the individual modal contributions, truncated at some finite mode, at that point in space and time.

The fundamental natural frequency f_1 of a simply supported beam with a uniform mass distribution, as shown in Fig. 6.103, can also be conveniently expressed in terms of the static

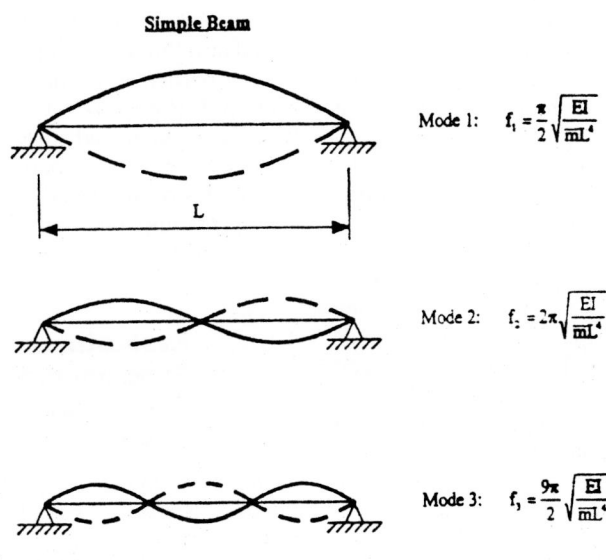

Simple Beam

Mode 1: $f_1 = \frac{\pi}{2}\sqrt{\frac{EI}{\bar{m}L^4}}$

Mode 2: $f_2 = 2\pi\sqrt{\frac{EI}{\bar{m}L^4}}$

Mode 3: $f_3 = \frac{9\pi}{2}\sqrt{\frac{EI}{\bar{m}L^4}}$

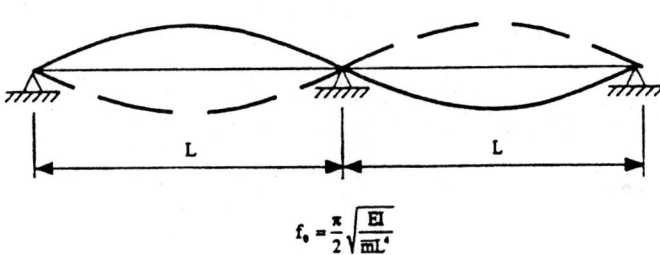

Continuous Beam

$f_0 = \frac{\pi}{2}\sqrt{\frac{EI}{\bar{m}L^4}}$

Fig. 6.103 Modes of vibration for beams with uniformly distributed mass.

deflection due to distributed weight. The derivation of this expression is as follows:

$$f_1 = \frac{\pi}{2}\sqrt{\frac{EI}{\bar{m}L^4}} = \frac{\pi}{2}\sqrt{\frac{5g}{384\Delta}} = 0.18\sqrt{\frac{g}{\Delta}} \quad (6.279)$$

$$\Delta = \frac{5wL^4}{384EI} = \frac{5g}{384}\frac{\bar{m}L^4}{EI} \quad (6.280)$$

$$\frac{EI}{\bar{m}L^4} = \frac{5g}{384\Delta} \quad (6.281)$$

where w = uniformly distributed load on a beam

$= \bar{m} \cdot g$

$\bar{m}$ = uniformly distributed mass on beam

g = acceleration of gravity = 386.4 in/s^2 or 9800 mm/s^2

L = beam length

E = modulus of elasticity

I = moment of inertia for the beam cross section

The expression for the fundamental natural frequency, in terms of static deflection, is often misused in determining the natural frequency for other beam configurations. In particular, the expression f_1 above cannot be used for continuous beams. There is a common misconception that providing continuity of beams over a support will raise the fundamental frequency of the system. While it is true that continuity reduces the maximum static deflection, the fundamental natural frequency remains the same. This concept is illustrated in Fig. 6.103.

Platelike structures, such as beam and girder systems, also possess an infinite number of natural frequencies and mode shapes. Figure 6.104 illustrates the natural frequencies and mode shapes, for the first four modes, for a one-bay floor system comprised of a slab, joists, and girders. In addition to the mass distribution, the frequencies and mode shapes are affected by the slab, joist (or beam), and girder properties. This concept is explored in the following subsection. Close inspection of Fig. 6.104 and some intuition reveals that an activity like jumping at the center of the floor would cause dynamic amplitudes consisting of the superposi-

tion of modes 1, 4 and higher-order modes with a modal amplitude at that point.

One particular phenomenon to carefully consider and, if possible, avoid is that of resonance. Resonance occurs when a component of a harmonic excitation corresponds to one of the natural frequencies of the structure. Vibration amplitudes are greatly amplified in lightly damped structures such as steel floor systems.

6.85.2 Evaluation of Fundamental Natural Frequency for a Floor System

As illustrated in Fig. 6.104, the dynamic behavior of a floor system is very complex. There are, however,

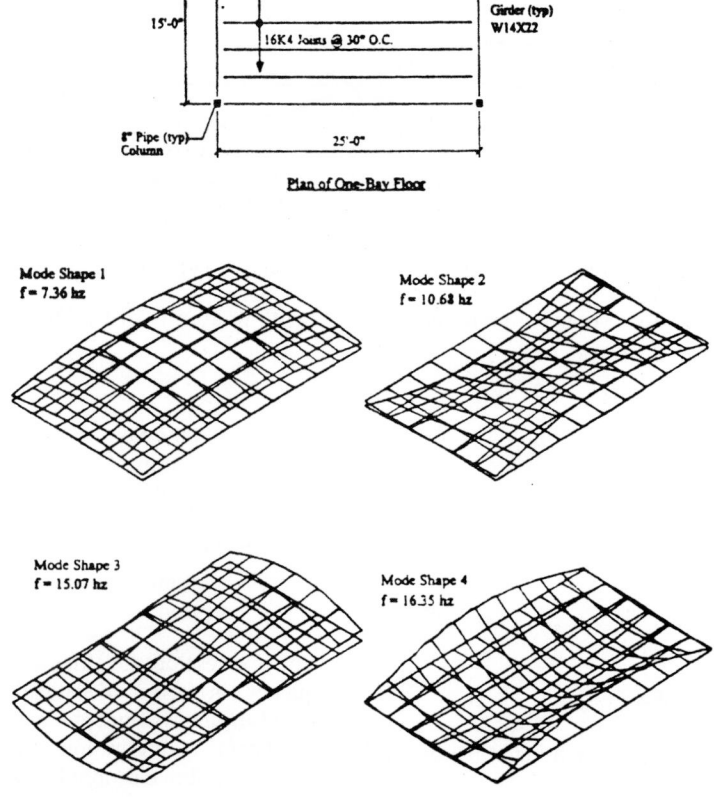

Fig. 6.104 Mode shapes and natural frequencies from a computer analysis.

commonly accepted procedures to determine dynamic characteristics of floor systems. The following discussion provides necessary information and a procedure to estimate the frequency of the first mode of free vibration for a steel floor system. A close approximation of the fundamental natural frequency of a floor system can be achieved by considering the frequencies of the major components of the floor system independently and then combining them as outlined in the procedure below.

Estimated System Frequency

$$\frac{1}{f_s^2} = \frac{1}{f_b^2} + \frac{1}{f_g^2} + \frac{1}{f_c^2} \qquad (6.282)$$

where f_s = first natural frequency of the floor system, Hz

f_b = frequency of the beam or joist member, Hz; see equations below

f_g = frequency of the girder members; the lowest girder frequency should be used if the girder frequencies differ; the girder term in the system expression above can be neglected if the beams or joists are supported by a rigid support such as a wall

f_c = frequency of the column, Hz; except in unusual circumstances, this term is generally neglected; the movement of the columns is usually insignificant relative to the beam and girder motion

Beam or Joist Frequency

$$f_b = K\sqrt{\frac{gEI_t}{wL^4}} \qquad (6.283)$$

where $K = 1.57$ for simply supported beams; 0.56 for cantilevered beams; refer to Murray and Hendrick for overhanging beams

g = acceleration of gravity; 386.4 in/s² or 9800 mm/s²

E = modulus of elasticity for transformed section, 29,000 ksi for steel

I_t = transformed moment of inertia; when the steel deck supporting the concrete rests directly on the beam or joist (connected by welds, screws, mechanical shear connectors, etc.), assume composite action between the steel member and the concrete slab; see Sec. 9.3.4 for more information on the computation of composite member properties

w = floor weight per unit length of beam; value should be the actual expected service load on the beam; overestimating this value can result in a nonconservative prediction of acceptability; 10 percent to 25 percent of the live load used in strength calculations is suggested for design

L = beam or joist span

Girder Frequency

$$f_g = K\sqrt{\frac{gEI_t}{wL^4}} \qquad (6.284)$$

where I_t = transformed moment of inertia

w = floor weight per unit length of girder; value should be the actual expected service load on the girder; loads from the beams or joists framing into the girder can usually be treated as continuous regardless of the spacing

Note: All other variables are as defined for the beam or joist frequency above.

Column Frequency

$$f_c = \frac{1}{2\pi}\sqrt{\frac{gAE}{PL}} \qquad (6.285)$$

where A = area of the column section

P = load on the column; value should be the actual expected service load

L = length of column

Note: All other variables are as defined previously.

Mohamad H. Hussein
Vice President
GRL Engineers, Inc., Orlando, Florida
Partner, Pile Dynamics, Inc., Cleveland, Ohio

Jerry A. DiMaggio
Senior Geotechnical Engineer
Federal Highway Administration
Washington, DC

7

GEOTECHNICAL ENGINEERING

In a broad sense, geotechnical engineering is that branch of civil engineering that employs scientific methods to determine, evaluate, and apply the interrelationship between the geologic environment and engineered works. In a practical context, geotechnical engineering encompasses evaluation, design, and construction involving earth materials.

The broad nature of this branch of civil engineering is demonstrated by the large number of technical committees comprising the Geo-Institute of the American Society of Civil Engineers (ASCE). In addition, the International Society for Soil Mechanics and Geotechnical Engineering (ISSMGE) includes the following 31 Technical Committees: Calcareous Sediments, Centrifuge and Physical Model Testing, Coastal Geotechnical Engineering, Deformation of Earth Materials, Earthquake Geotechnical Engineering, Education in Geotechnical Engineering, Environmental Geotechnics, Frost, Geophysical Site Characterization, Geosynthetics and Earth Reinforcement, Ground Improvement, Ground Property Characterization from In-situ Testing, Indurated Soils and Soft Rocks, Instrumentation for Geotechnical Monitoring, Landslides, Limit State Design in Geotechnical Engineering, Micro-geomechanics, Offshore Geotechnical Engineering, Peat and Organic Soils, Pile Foundations, Preservation of Historic Sites, Professional Practice, Risk Assessment and Management, Scour of Foundations, Soil Sampling Evaluation and Interpretation, Stress-Strain Testing of Geomaterials in the Laboratory, Tailings Dams, Tropical and Residual Soils, Underground Construction in Soft Ground, Unsaturated Soils, and Validation of Computer Simulations.

Unlike other civil engineering disciplines, which typically deal with materials whose properties are well defined, geotechnical engineering is concerned with subsurface materials whose properties, in general, cannot be specified. Pioneers of geotechnical engineering relied on the "observational approach" to develop an understanding of soil and rock mechanics and behavior of earth materials under loads. This approach was enhanced by the advent of electronic field instrumentation, wide availability of powerful personal computers, and development of sophisticated numerical techniques. These now make it possible to determine with greater accuracy the nonhomogeneous, nonlinear, anisotropic nature and behavior of earth materials for application to engineering works.

Geotechnical engineers should be proficient in the determination of soil and rock properties, engineering mechanics, subsurface investigation methods and laboratory testing techniques. They should have a thorough knowledge of design methods, construction methods, monitoring/inspection procedures, and specifications and contracting practices. Geotechnical engineers should have broad practical experience, in as much as the practice of geotechnical engineering involves art as much as science. This requirement was clearly expressed by

Karl Terzaghi, who made considerable contributions to the development of soil mechanics: "The magnitude of the difference between the performance of real soils under field conditions and the performance predicted on the basis of theory can only be ascertained by field experience."

Geotechnical engineering is the engineering science of selecting, designing, and constructing features constructed of or upon soils and rock. Shallow foundations, deep foundations, earth retaining structures, soil and rock embankments and cuts are all specialty areas of geotechnical engineering.

Foundation engineering is the art of selecting, designing, and constructing for engineering works structural support systems based on scientific principles of soil and engineering mechanics and earth-structure interaction theories, and incorporating accumulated experience with such applications.

7.1 Lessons from Construction Claims and Failures

Unanticipated subsurface conditions encountered during construction are by far the largest source of construction-related claims for additional payment by contractors and of cost overruns. Failures of structures as a result of foundation deficiencies can entail even greater costs, and moreover jeopardize public safety. A large body of experience has identified consistently recurring factors contributing to these occurrences. It is important for the engineer to be aware of the causes of cost overruns, claims, and failures and to use these lessons to help minimize similar future occurrences.

Unanticipated conditions (changed conditions) are the result of a variety of factors. The most frequent cause is the lack of definition of the constituents of rock and soil deposits and their variation throughout the construction site. Related claims are for unanticipated or excessive quantities of soil and rock excavation, misrepresentation of the quality and depth of bearing levels, unsuitable or insufficient on-site borrow materials, and unanticipated obstructions to pile driving or shaft drilling. Misrepresentation of groundwater condition is another common contributor to work *extras* as well as to costly construction delays and emergency redesigns. Significant claims have also been generated by the failure of geotechnical investigations to identify natural hazards, such as swelling soils and rock minerals, unstable natural and cut slopes, and old fill deposits.

Failures of structures during construction are usually related to undesirable subsurface conditions not detected before or during construction, faulty design, or poor quality of work. Examples are foundations supported on expansive or collapsing soils, on solutioned rock, or over undetected weak or compressible subsoils; foundation designs too difficult to construct properly; foundations that do not perform as anticipated; and deficient construction techniques or materials. Another important design-related cause of failure is underestimation or lack of recognition of extreme loads associated with natural events, such as earthquakes, hurricanes, floods, and prolonged precipitation. Related failures include soil liquefaction during earthquakes, hydrostatic uplift or water damage to structures because of a rise in groundwater level, undermining of foundations by scour and overtopping, or wave erosion of earth dikes and dams.

It is unlikely that conditions leading to construction claims and failures can ever be completely precluded, inasmuch as discontinuities and extreme variation in subsurface conditions occur frequently in many types of soil deposits and rock formations. An equally important constraint that must be appreciated by both engineers and clients is the limitations of the current state of geotechnical engineering practice.

Mitigation of claims and failures, however, can be achieved by fully integrated geotechnical investigation, design, and construction quality assurance conducted by especially qualified professionals. Integration, rather than departmentalization of these services, ensures a continuity of purpose and philosophy that effectively reduces the risks associated with unanticipated subsurface conditions and design and construction deficiencies. It is also extremely important that owners and prime design professionals recognize that cost savings that reduce the quality of geotechnical services may purchase liabilities several orders of magnitude greater than their initial "savings."

7.2 Soil and Rock Classifications

All soils are initially the products of chemical alteration or mechanical disintegration of bedrock that has been exposed to weathering processes. Soil constituents may have been subsequently modified by transportation processes such as water, wind, and ice and by inclusion and decomposition of organic matter. Consequently, soil deposits may be given a geologic as well as a constitutive classification.

Rock types are broadly classified by their mode of formation into igneous, metamorphic, and sedimentary deposits. The supporting ability (quality) assigned to rock for design or analysis should reflect the degree of alteration of the rock minerals due to weathering, the frequency of discontinuities within the rock mass, and the susceptibility of the rock to deterioration upon exposure.

7.2.1 Geologic Classification of Soils

The classification of a soil deposit with respect to its mode of deposition and geologic history is an important step in understanding the variation in soil type and the maximum stresses imposed on the deposit since deposition. (A geologic classification that identifies the mode of deposition of soil deposits is shown in Table 7.1.) The geologic history of a soil deposit may also provide valuable information on the rate of deposition, the amount of erosion, and the tectonic forces that may have acted on the deposit subsequent to deposition.

Geological and agronomic soil maps and detailed reports are issued by the U.S. Department of Agriculture (www.usda.gov), U.S. Geological Survey (www.usgs.gov), and corresponding state offices. Old surveys are useful for locating original shore lines, stream courses, and surface-grade changes.

7.2.2 Unified Soil Classification System

This is the most widely used of the various constitutive classification systems and correlates soil type with generalized soil behavior. All soils are classified as coarse-grained (50% of the particles >0.074 mm), fine-grained (50% of the particles <0.074 mm), or predominantly organic (see Table 7.2).

Coarse-grained soils are categorized by their particle size into boulders (particles larger than 8 in),

Table 7.1 Geologic Classification of Soil Deposits

Classification	Mode of Formation
Aeolian	
Dune	Wind deposition (coastal and desert)
Loess	Deposition during glacial periods
Alluvial	
Alluvium	River and stream deposition
Lacustrine	Lake waters, including glacial lakes
Floodplain	Floodwaters
Colluvial	
Colluvium	Downslope soil movement
Talus	Downslope movement of rock debris
Glacial	
Ground moraine	Deposited and consolidated by glaciers
Terminal moraine	Scour and transport at ice front
Outwash	Glacier melt waters
Marine	
Beach or bar	Wave deposition
Estuarine	River estuary deposition
Lagoonal	Deposition in lagoons
Salt marsh	Deposition in sheltered tidal zones
Residual	
Residual soil	Complete alteration by in situ weathering
Saprolite	Incomplete but intense alteration and leaching
Laterite	Complex alteration in tropical environment
Decomposed rock	Advanced alteration within parent rock

cobbles (3 to 8 in), gravel, and sand. For sands (S) and gravels (G), grain-size distribution is identified as either poorly graded (P) or well-graded (W), as indicated by the group symbol in Table 7.2. The presence of fine-grained soil fractions (under 50%), such as silt and clay, is indicated by the symbols M and C, respectively. Sands may also be classified as coarse (larger than No. 10 sieve), medium (smaller than No. 10 but larger than No. 40), or fine

Table 7.2 Unified Soil Classification Including Identification and Description[a]

Major Division	Group Symbol	Typical Name	Field Identification Procedures[b]	Laboratory Classification Criteria[c]
A. Coarse-grained soils (more than half of material larger than No. 200 sieve)[d]				
1. Gravels (more than half of coarse fraction larger than No. 4 sieve)[e]				
Clean gravels (little or no fines)	GW	Well-graded gravels, gravel-sand mixtures, little or no fines	Wide range in grain sizes and substantial amounts of all intermediate particle sizes	$D_{60}/D_{10} > 4$ $1 < D_{30}/D_{10}D_{60} < 3$ D_{10}, D_{30}, D_{60} = sizes corresponding to 10, 30, and 60% on grain-size curve
	GP	Poorly graded gravels or gravel-sand mixtures, little or no fines	Predominantly one size, or a range of sizes with some intermediate sizes missing	Not meeting all gradation requirements for GW
Gravels with fines (appreciable amount of fines)	GM	Silty gravels, gravel-sand-silt mixtures	Nonplastic fines or fines with low plasticity (see ML soils)	Atterberg limits below A line or PI < 4 — Soils above A line with 4 < PI < 7 are borderline cases, require use of dual symbols
	GC	Clayey gravels, gravel-sand-clay mixtures	Plastic fines (see CL soils)	Atterberg limits above A line with PI > 7
2. Sands (more than half of coarse fraction smaller than No. 4 sieve)[e]				
Clean sands (little or no fines)	SW	Well-graded sands, gravelly sands, little or no fines	Wide range in grain sizes and substantial amounts of all intermediate particle sizes	$D_{60}/D_{10} > 6$ $1 < D_{30}/D_{10}D_{60} < 3$
	SP	Poorly graded sands or gravel-sands, little or no fines	Predominantly one size, or a range of sizes with some intermediate sizes missing	Not meeting all gradation requirements for SW
Sands with fines (appreciable amount of fines)	SM	Silty sands, sand-silt mixtures	Nonplastic fines or fines with low plasticity (see ML soils)	Atterberg limits below A line or PI < 4 — Soils with Atterberg limits above A line while 4 < PI < 7 are borderline cases; require use of dual symbols
	SC	Clayey sands, sand-clay mixtures	Plastic fines (see CL soils)	Atterberg limits above A line with PI > 7

Information required for describing coarse-grained soils:

For undisturbed soils, add information on stratification, degree of compactness, cementation, moisture conditions, and drainage characteristics. Give typical name; indicate approximate percentage of sand and gravel; maximum size, angularity, surface condition, and hardness of the coarse grains; local or geological name and other pertinent descriptive information; and symbol in parentheses. Example: *Salty sand*, gravelly; about 20% hard, angular gravel particles, ½ in maximum size; rounded and subangular sand grains, coarse to fine; about 15% nonplastic fines with low dry strength; well compacted and moist in place; alluvial sand; (SM).

B. Fine-grained soils (more than half of material larger than No. 200 sieve)[d]

Major Division	Group Symbol	Typical Name	Identification Procedures[f]			Laboratory Classification Criteria[c]
			Dry Strength (Crushing Characteristics)	Dilatancy (Reaction to Shaking)	Toughness (Consistency Near PL)	
Silts and clays with liquid limit less than 50	ML	Inorganic silts and very fine sands, rock flour, silty or clayey fine sands, or clayey silts with slight plasticity	None to slight	Quick to slow	None	
	CL	Inorganic clays of low to medium plasticity, gravelly clays, sandy clays, silty clays, lean clays	Medium to high	None to very slow	Medium	
	OL	Organic silts and organic silty clays of low plasticity	Slight to medium	Slow	Slight	
Silts and clays with liquid limit more than 50	MH	Inorganic silts, micaceous or diatomaceous fine sandy or silty soils, elastic silts	Slight to medium	Slow to none	Slight to medium	Plasticity chart laboratory classifications of fine-grained soils compares them at equal liquid limit. Toughness and dry strength increase with increasing plasticity index (PI)
	CH	Inorganic clays of high plasticity, fat clays	None to very high	None	High	
	CH	Organic clays of medium to high plasticity	Medium to high	None to very slow	Slight to medium	

(Table continued)

Table 7.2 (Continued)

C. Highly organic soils

Pt	Peat and other highly organic soils	Readily identified by color, odor, spongy feel, and often by fibrous texture

Field identification procedures for fine-grained soils or fractions:[a]

Dilatancy (Reaction to Shaking)	Dry Strength (Crushing Characteristics)	Toughness (Consistency Near PL)
After removing particles larger than No. 40 sieve, prepare a pat of moist soil with a volume of about $\frac{1}{2}$ in³. Add enough water if necessary to make the soil soft but not sticky. Place the pat in the open palm of one hand and shake horizontally, striking vigorously against the other hand several times. A positive reaction consists of the appearance of water on the surface of the pat, which changes to a livery consistency and becomes glossy. When the sample is squeezed between the fingers, the water and gloss disappear from the surface, the pat stiffens, and finally it cracks or crumbles. The rapidity of appearance of water during shaking and of its disappearance during squeezing assist in identifying the character of the fines in a soil. Very fine clean sands give the quickest and most distinct reaction, whereas a plastic clay has no reaction. Inorganic silts, such as a typical rock flour, show a moderately quick reaction.	After removing particles larger than No. 40 sieve, mold a pat of soil to the consistency of putty, adding water if necessary. Allow the pat to dry completely by oven, sun, or air drying, then test its strength by breaking and crumbling between the fingers. This strength is a measure of character and quantity of the colloidal fraction contained in the soil. The dry strength increases with increasing plasticity. High dry strength is characteristic of clays of the CH group. A typical inorganic silt possesses only very slight dry strength. Silty fine sands and silts have about the same slight dry strength but can be distinguished by the feel when powdering the dried specimen. Fine sand feels gritty, whereas a typical silt has the smooth feel of flour.	After particles larger than the No. 40 sieve are removed, a specimen of soil about $\frac{1}{2}$ in³ in size is molded to the consistency of putty. If it is too dry, water must be added. If it is too sticky, the specimen should be spread out in a thin layer and allowed to lose some moisture by evaporation. Then, the specimen is rolled out by hand on a smooth surface or between the palms into a thread about $\frac{1}{8}$ in in diameter. The thread is then folded and rerolled repeatedly. During this manipulation, the moisture content is gradually reduced and the specimen stiffens, finally loses its plasticity, and crumbles when the plastic limit (PL) is reached. After the thread crumbles, the pieces should be lumped together and a slight kneading action continued until the lump crumbles. The tougher the thread near the PL and the stiffer the lump when it finally crumbles, the more potent is the colloidal clay fraction in the soil. Weakness of the thread at the PL and quick loss of coherence of the lump below the PL indicate either organic clay of low plasticity or materials such as kaolin-type clays and organic clays that occur below the A line. Highly organic clays have a very weak and spongy feel at PL.

Information required for describing fine-grained soils:

For undisturbed soils, add information on structure, stratification, consistency in undisturbed and remolded states, moisture, and drainage conditions. Give typical name; indicate degree and character of plasticity; amount and maximum size of coarse grains; color in wet conditions, odor, if any; local or geological name and other pertinent descriptive information; and symbol in parentheses. Example: *Clayey silt*, brown; slightly plastic; small percentage of fine sand; numerous vertical root holes; firm and dry in place; loess; (ML)

[a]Adapted from recommendations of U.S. Army Corps of Engineers and U.S. Bureau of Reclamation. All sieve sizes United States standard.

[b]Excluding particles larger than 3 in and basing fractions on estimated weights.

[c]Use grain-size curve in identifying the fractions as given under field identification.

For coarse-grained soils, determine percentage of gravel and sand from grain-size curve. Depending on percentage of fines (fractions smaller than No. 200 sieve), coarse-grained soils are classified as follows:

Less than 5% fines:	GW, GP, SW, SP
More than 12% fines:	GM, GC, SM, SC
5% to 12% fines:	Borderline cases requiring use of dual symbols

Soils possessing characteristics of two groups are designated by cobinations of group symbols; for example, GW-GC indicates a well-graded, gravel-sand mixture with clay binder.

[d]The No. 200 sieve size is about the smallest particle visible to the naked eye.

[e]For visual classification, the $\frac{1}{4}$ in size may be used as equivalent to the No. 4 sieve size.

[f]Applicable to fractions smaller than No. 40 sieve.

[g]These procedures are to be performed on the minus 40-sieve-size particles (about $\frac{1}{64}$ in). Simply remove by hand the coarse particles that interfere with the tests.

For field classification purposes, screening is not intended.

(smaller than No. 40). Because properties of these soils are usually significantly influenced by relative density D_r, rating of the in situ density and D_r is an important consideration (see Art. 7.4).

Fine-grained soils are classified by their liquid limit and plasticity index as organic clays OH or silts OL, inorganic clays CH or CL, or silts or sandy silts MH or ML, as shown in Table 7.2. For the silts and organic soils, the symbols H and L denote a high and low potential compressibility rating; for clays, they denote a high and low plasticity. Typically, the consistency of cohesive soils is classified from pocket penetrometer or Torvane tests on soil samples. These index tests are convenient for relative comparisons but do not provide design strength values and should not be used as property values for design or analysis. The consistency ratings are expressed as follows:

Soft—under 0.25 tons/ft^2

Firm—0.25 to 0.50 tons/ft^2

Stiff—0.50 to 1.0 tons/ft^2

Very stiff—1.0 to 2.0 tons/ft^2

Hard—more than 2.0 tons/ft^2

7.2.3 Rock Classification

Rock, obtained from core samples, is commonly characterized by its type, degree of alteration (weathering), and continuity of the core. (Where outcrop observations are possible, rock structure may be mapped.) Rock-quality classifications are typically based on the results of compressive strength tests or the condition of the core samples, or both. Rock types typical of igneous deposits include basalt, granite, diorite, rhyolite, and andesite. Typical metamorphic rocks include schist, gneiss, quartzite, slate, and marble. Rocks typical of sedimentary deposits include shale, sandstone, conglomerate, and limestone.

Rock structure and degree of fracturing usually control the behavior of a rock mass that has been significantly altered by weathering processes. It is necessary to characterize both regional and local structural features that may influence design of foundations, excavations, and underground openings in rock. Information from geologic publications and maps are useful for defining regional

trends relative to the orientation of bedding, major joint systems, faults, and so on.

Rock-quality indices determined from inspection of rock cores include the fracture frequency (*FF*) and rock-quality designation (*RQD*). *FF* is the number of naturally occurring fractures per foot of core run, whereas *RQD* is the cumulative length of naturally separated core pieces, 4 in or more in dimension, expressed as a percentage of the length of core run. The rock-quality rating also may be based on the velocity index obtained from laboratory and in situ seismic-wave-propagation tests. The velocity index is given by $(V_s/V_l)^2$, where V_s and V_l represent seismic-wave velocities from in situ and laboratory core measurements, respectively. Proposed *RQD* and velocity index rock-quality classifications and in situ deformability correlations are in Table 7.3. A relative-strength rating of the quality rock cores representative of the intact elements of the rock mass, proposed by Deere and Miller, is based on the uniaxial compressive (*UC*) strength of the core and its tangent modulus at one-half of the *UC*.

(D. U. Deere and R. P. Miller, "Classification and Index Properties for Intact Rock," Technical Report AFWL-TR-65-116, Airforce Special Weapons Center, Kirtland Airforce Base, New Mexico, 1966.)

Inasmuch as some rocks tend to disintegrate rapidly (slake) upon exposure to the atmosphere, the potential for slaking should be rated from laboratory tests. These tests include emersion in water, Los Angeles abrasion, repeated wetting and drying, and other special tests, such as a

Table 7.3 Rock-Quality Classification and Deformability Correlation

Classification	RQD	Velocity Index	Deformability E_d/E_t*
Very poor	0–25	0–0.20	Under 0.20
Poor	25–50	0.20–0.40	Under 0.20
Fair	50–75	0.40–0.60	0.20–0.50
Good	75–90	0.60–0.80	0.50–0.80
Excellent	90–100	0.80–1.00	0.80–1.00

*E_d = in situ deformation modulus of rock mass; E_t = tangent modulus at 50% of *UC* strength of core specimens.

Source: Deere, Patton and Cording, "Breakage of Rock," *Proceedings*, 8th Symposium on Rock Mechanics, American Institute of Mining and Metallurgical Engineers, Minneapolis, Minn.

slaking-durability test. Alteration of rock minerals due to weathering processes is often associated with reduction in rock hardness and increase in porosity and discoloration. In an advanced stage of weathering, the rock may contain soil-like seams, be easily abraded (friable), readily broken, and may (but will not necessarily) exhibit a reduced *RQD* or *FF*. Rating of the degree of rock alteration when logging core specimens is a valuable aid in assessing rock quality.

7.3 Physical Properties of Soils

Basic soil properties and parameters can be subdivided into physical, index, and engineering categories. Physical soil properties include density, particle size and distribution, specific gravity, and water content.

The **water content** w of a soil sample represents the weight of free water contained in the sample expressed as a percentage of its dry weight.

The **degree of saturation** S of the sample is the ratio, expressed as percentage, of the volume of free water contained in a sample to its total volume of voids V_v.

Porosity n, which is a measure of the relative amount of voids, is the ratio of void volume to the total volume V of soil:

$$n = \frac{V_v}{V} \tag{7.1}$$

The ratio of V_v to the volume occupied by the soil particles V_s defines the **void ratio** e. Given e, the degree of saturation may be computed from

$$S = \frac{wG_s}{e} \tag{7.2}$$

where G_s represents the specific gravity of the soil particles. For most inorganic soils, G_s is usually in the range of 2.67 ± 0.05.

The dry unit weight γ_d of a soil specimen with any degree of saturation may be calculated from

$$\gamma_d = \frac{\gamma_w G_s S}{1 + wG_s} \tag{7.3}$$

where γ_w is the unit weight of water and is usually taken as 62.4 lb/ft^3 for fresh water and 64.0 lb/ft^3 for seawater.

The particle-size distribution (gradation) of soils can be determined by mechanical (sieve) analysis and combined with hydrometer analysis if the sample contains a significant amount of particles finer than 0.074 mm (No. 200 sieve). The soil particle gradation in combination with the maximum, minimum, and in situ density of cohesionless soils can provide useful correlations with engineering properties (see Arts. 7.4 and 7.52).

7.4 Index Parameters for Soils

Index parameters of cohesive soils include liquid limit, plastic limit, shrinkage limits, and activity. Such parameters are useful for classifying cohesive soils and providing correlations with engineering soil properties.

The **liquid limit** of cohesive soils represents a near liquid state, that is, an undrained shear strength about 0.01 lb/ft^2. The water content at which the soil ceases to exhibit plastic behavior is termed the **plastic limit**. The **shrinkage limit** represents the water content at which no further volume change occurs with a reduction in water content. The most useful classification and correlation parameters are the plasticity index I_p, the liquidity index I_l, the shrinkage index I_s, and the activity A_c. These parameters are defined in Table 7.4.

Relative density D_r of cohesionless soils may be expressed in terms of void ratio e or unit dry weight γ_d:

$$D_r = \frac{e_{max} - e_o}{e_{max} - e_{min}} \tag{7.4a}$$

$$D_r = \frac{1/\gamma_{min} - 1/\gamma_d}{1/\gamma_{min} - 1/\gamma_{max}} \tag{7.4b}$$

D_r provides cohesionless soil property and parameter correlations, including friction angle, permeability, compressibility, small-strain shear modulus, cyclic shear strength, and so on.

In situ field tests such as the Standard Penetration Test (SPT), static cone penetrometer, pressuremeter, and dilatometer can also be used to determine the index properties of cohensionless and cohensive soils.

(H. Y. Fang, "Foundation Engineering Handbook," 2nd ed., Van Nostrand Reinhold, New York.)

Table 7.4 Soil Indices

Index	Definition*	Correlation
Plasticity	$I_p = W_l - W_p$	Strength, compressibility, compactibility, and so forth
Liquidity	$I_l = \dfrac{W_n - W_p}{I_p}$	Compressibility and stress rate
Shrinkage	$I_s = W_p - W_s$	Shrinkage potential
Activity	$A_c = \dfrac{I_p}{\mu}$	Swell potential, and so on

*W_l = liquid limit; W_p = plastic limit; W_n = moisture content, %; W_s = shrinkage limit; μ = percent of soil finer than 0.002 mm (clay size).

7.5 Engineering Properties of Soils

Engineering soil properties and parameters describe the behavior of soil under induced stress and environmental changes. Of interest to most geotechnical applications are the strength, deformability, and permeability of in situ and compacted soils. ASTM promulgates standard test procedures for soil properties and parameters.

7.5.1 Shear Strength of Cohesive Soils

The undrained shear strength c_u of cohesive soils under static loading can be determined by several types of laboratory tests, including uniaxial compression, triaxial compression (TC) or extension (TE), simple shear, direct shear, and torsion shear. The objective of soil laboratory testing is to replicate the field stress, loading and drainage conditions with regard to magnitude, rate and orientation. All laboratory strength testing require extreme care in securing, transporting and preparing the test sample. The triaxial test is the most versatile but yet the most complex strength test to perform. Triaxial tests involve application of a controlled confining pressure σ_3 and axial stress σ_1 to a soil specimen. σ_3 may be held constant and σ_1 increased to failure (TC tests), or σ_1 may be held constant while σ_3 is decreased to failure (TE tests). Specimens may be sheared in a drained or undrained condition.

The unconsolidated-undrained (UU) triaxial compression test is appropriate and commonly used for determining the c_u of relatively good-quality samples. For soils that do not exhibit changes in soil structure under elevated consolidation pressures, consolidated-undrained (CU) tests following the SHANSEP testing approach mitigate the effects of sample disturbance.

(C. C. Ladd and R. Foott, "New Design Procedures for Stability of Soft Clays," *ASCE Journal of Geotechnical Engineering Division*, vol. 99, no. GT7, 1974, www.asce.org.)

For cohesive soils exhibiting a normal clay behavior, a relationship between the normalized undrained shear strength c_u/σ'_{vo} and the overconsolidation ratio OCR can be defined independently of the water content of the test specimen by

$$\frac{c_u}{\sigma'_{vo}} = K(OCR)^n \qquad (7.5)$$

where c_u is normalized by the preshear vertical effective stress, the effective overburden pressure σ'_{vo}, or the consolidation pressure σ'_{lc} triaxial test conditions. OCR is the ratio of preconsolidation pressure to overburden pressure. The parameter K represents the c_u/σ'_{vo} of the soil in a normally consolidated state, and n primarily depends on the type of shear test. For CU triaxial compression tests, K is approximately 0.32 ± 0.02 and is lowest for low plasticity soils and is a maximum for soils with plasticity index I_p over 40%. The exponent n is usually within the range of 0.70 ± 0.05 and tends to be highest for OCR less than about 4.

In situ vane shear tests also are often used to provide c_u measurements in soft to firm clays. Tests are commonly made on both the undisturbed and remolded soil to investigate the **sensitivity**, the ratio of the undisturbed to remolded soil strength. This test is not applicable in sand or silts or where hard inclusions (nodules, shell, gravel, and so forth) may be present. (See also Art. 7.6.3.)

The Standard Penetration Test, cone penetrometer, pressuremeter, and dilatometer provide guidance on engineering properties of soils. Similar to laboratory tests ASTM procedures have been developed for several of these tests. Each test, similar to the vane shear test, has advantages and disadvantages, as well as being limited to certain soil types.

Drained shear strength of cohesive soils is important in design and control of construction embankments on soft ground as well as in other evaluations involving effective-stress analyses. Conventionally, drained shear strength τ_f is expressed by the Mohr-Coulomb failure criteria as:

$$\tau_f = c' + \sigma'_n \tan \phi' \qquad (7.6)$$

The c' and ϕ' parameters represent the effective cohesion and effective friction angle of the soil, respectively. σ'_n is the effective stress normal to the plane of shear failure and can be expressed in terms of total stress σ_n as $(\sigma_n - u_e)$, where u_e is the excess pore-water pressure at failure. u_e is induced by changes in the principal stresses $(\Delta\sigma_1, \Delta\sigma_3)$. For saturated soils, it is expressed in terms of the pore-water-pressure parameter A_f at failure as:

$$u = \Delta\sigma_3 + A_f(\Delta\sigma_1 - \Delta\sigma_3)_f \qquad (7.7)$$

The effective-stress parameters c', ϕ', and A_f are readily determined by CU triaxial shear tests employing pore-water-pressure measurements or, excepting A_f, by consolidated-drained (CD) tests.

After large movements along preformed failure planes, cohesive soils exhibit a significantly reduced (residual) shear strength. The corresponding effective friction angle ϕ'_r is dependent on I_p. For many cohesive soils, ϕ'_r is also a function of σ'_n. The ϕ'_r parameter is applied in analysis of the stability of soils where prior movements (slides) have occurred.

Cyclic loading with complete stress reversals decreases the shearing resistance of saturated cohesive soils by inducing a progressive buildup in pore-water pressure. The amount of degradation depends primarily on the intensity of the cyclic shear stress, the number of load cycles, the stress history of the soil, and the type of cyclic test used. The strength degradation potential can be determined by postcyclic, UU tests.

7.5.2 Shear Strength of Cohesionless Soils

The shear strength of cohesionless soils under static loading can be interpreted from results of drained or undrained TC tests incorporating pore-pressure measurements. The effective angle of internal friction ϕ' can also be expressed by Eq. (7.6), except that c' is usually interpreted as zero. For cohesionless soils, ϕ' is dependent on density or void ratio, gradation, grain shape, and grain mineralogy. Markedly stress-dependent, ϕ' decreases with increasing σ'_n, the effective stress normal to the plane of shear failure.

In situ cone penetration tests in sands may be used to estimate ϕ' from cone resistance q_c records. One approach relates the limiting q_c values directly to ϕ'. Where q_c increases approximately linearly with depth, ϕ' can also be interpreted from the slope of the curve for $q_c - \sigma_{vo}$ vs. ϕ'_{vo}, where σ_{vo} = total vertical stress, $\sigma'_{vo} = \sigma_{vo} - u$, and u = pore-water pressure. The third approach is to interpret the relative density D_r from q_c and then relate ϕ' to D_r as a function of the gradation and grain shape of the sand.

Relative density provides good correlation with ϕ' for a given gradation, grain shape, and normal stress range. A widely used correlation is shown in Fig. 7.1. D_r can be interpreted from standard penetration resistance tests (Fig. 7.12) and cone penetration resistance tests (see Arts. 7.6.2 and 7.6.3) or calculated from the results of in situ or maximum and minimum density tests. The most difficult property to determine in the relative density equation is e_0, the in-situ void ratio.

Dense sands typically exhibit a reduction in shearing resistance at strains greater than those required to develop the peak resistance. At relatively large strains, the stress-strain curves of loose and dense sands converge. The void ratio at which there is no volume change during shear is called the critical void ratio. A volume increase during shear (dilatancy) of saturated, dense, cohesionless soils produces negative pore-water pressures and a temporary increase in shearing resistance. Subsequent dissipation of negative pore-water pressure accounts for the "relaxation effect" sometimes observed after piles have been driven into dense, fine sands.

Saturated, cohesionless soils subject to cyclic loads exhibit a significant reduction in strength if cyclic loading is applied at periods smaller than the time required to achieve significant dissipation of pore pressure. Should the number of load cycles N_c

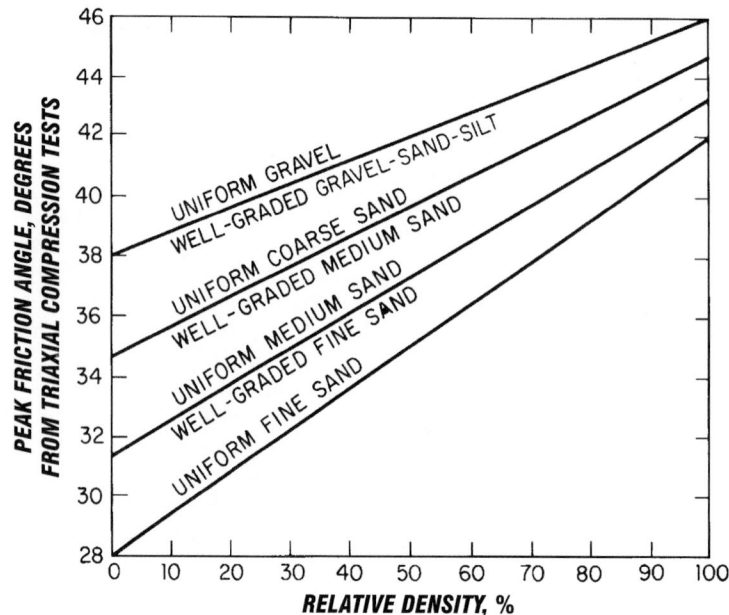

Fig. 7.1 Chart for determining friction angles for sands. (*After J. H. Schmertmann.*)

be sufficient to generate pore pressures that approach the confining pressure within a soil zone, excessive deformations and eventually failure (liquefaction) is induced. For a given confining pressure and cyclic stress level, the number of cycles required to induce initial liquefaction N_{c1} increases with an increase in relative density D_r. Cyclic shear strength is commonly investigated by cyclic triaxial tests and occasionally by cyclic, direct, simple-shear tests.

7.5.3 State of Stress of Soils

Assessment of the vertical σ'_{vo} and horizontal σ'_{ho} effective stresses within a soil deposit and the maximum effective stresses imposed on the deposit since deposition σ'_{vm} is a general requirement for characterization of soil behavior. The ratio $\sigma'_{vm}/\sigma'_{vo}$ is termed the **overconsolidation ratio** (OCR). Another useful parameter is the ratio of $\sigma'_{ho}/\sigma'_{vo}$, which is called the **coefficient of earth pressure at rest** (K_o).

For a simple gravitation piezometric profile, the effective overburden stress σ'_{vo} is directly related to the depth of groundwater below the surface and the effective unit weight of the soil strata.

Groundwater conditions, however, may be characterized by irregular piezometric profiles that cannot be modeled by a simple gravitational system. For these conditions, **sealed piezometer measurements** are required to assess σ'_{vo}.

Maximum Past Consolidation Stress ■ The maximum past consolidation stress σ'_{vm} of a soil deposit may reflect stresses imposed prior to geologic erosion or during periods of significantly lower groundwater, as well as desiccation effects and effects of human activity (excavations). The maximum past consolidation stress is conventionally interpreted from consolidation (oedometer) tests on undisturbed samples.

Normalized-shear-strength concepts provide an alternate method for estimating OCR from good-quality UU compression tests. In the absence of site-specific data relating c_u/σ'_{vo} and OCR, a form of Eq. (7.5) may be applied to estimate OCR. In this interpretation, σ'_{vo} represents the effective over-burden pressure at the depth of the UU-test sample. A very approximate estimate of σ'_{vm} can also be obtained for cohesive soils from relationships proposed between liquidity index and effective vertical stress ("Design Manual—Soil

Mechanics, Foundations, and Earth Structures," NAVDOCKS DM-7, U.S. Navy). For coarse-grained soil deposits, it is difficult to characterize σ'_{vm} reliably from either in situ or laboratory tests because of an extreme sensitivity to disturbance.

The coefficient of earth pressure at rest K_o can be determined in the laboratory from "no-lateral strain" TC tests on undisturbed soil samples or from consolidation tests conducted in specially constructed oedometers. Interpretation of K_o from in situ CPT, PMT, and dilatometer tests has also been proposed. In view of the significant impact of sample disturbance on laboratory results and the empirical nature of in situ test interpretations, the following correlations of K_o with friction angle ϕ' and OCR are useful. For both coarse- and fine-grained soils:

$$K_o = (1 - \sin\phi')OCR^m \qquad (7.8)$$

A value for m of 0.5 has been proposed for overconsolidated cohesionless soils, whereas for cohesive soils it is proposed that m be estimated in terms of the plasticity index I_p as $0.581I_p^{-0.12}$.

7.5.4 Deformability of Fine-Grained Soils

Deformations of fine-grained soils can be classified as those that result from volume change, (elastic) distortion without volume change, or a combination of these causes. Volume change may be a one-dimensional or, in the presence of imposed shear stresses, a three-dimensional mechanism and may occur immediately or be time-dependent. **Immediate deformations** are realized without volume change during undrained loading of saturated soils and as a reduction of air voids (volume change) within unsaturated soils.

The rate of volume change of saturated, fine-grained soils during loading or unloading is controlled by the rate of pore-fluid drainage from or into the stressed soil zone. The compression phase of delayed volume change associated with pore-pressure dissipation under a constant load is termed **primary consolidation**. Upon completion of primary consolidation, some soils (particularly those with a significant organic content) continue to decrease in volume at a decreasing rate. This response is usually approximated as a straight line for a plot of log time vs. compression and is termed **secondary compression**.

As the imposed shear stresses become a substantial fraction of the undrained shear strength of the soil, time-dependent deformations may occur under constant load and volume conditions. This phenomenon is termed **creep deformation**. Failure by creep may occur if safety factors are insufficient to maintain imposed shear stresses below the creep threshold of the soil. (Also see Art. 7.10.)

One-dimensional volume-change parameters are conveniently interpreted from consolidation (oedometer) tests. A typical curve for log consolidation pressure vs. volumetric strain ε_v (Fig. 7.2) demonstrates interpretation of the strain-referenced compression index C'_c, recompression index C'_r, and swelling index C'_s. The secondary compression index C'_a represents the slope of the near-linear portion of the volumetric strain vs. log-time curve following primary consolidation (Fig. 7.2b). The parameters C'_c, C'_r, and C'_a be roughly estimated from soil-index properties.

Deformation moduli representing three-dimensional deformation can be interpreted from the stress-strain curves of laboratory shear tests for application to either volume change or elastic deformation problems.

("Design Manual—Soil Mechanics, Foundations, and Earth Structures," NAVDOCKS DM-7, U.S. Navy; T. W. Lambe and R. V. Whitman, "Soil Mechanics," John Wiley & Sons, Inc., New York, www.wiley.com.)

7.5.5 Deformability of Coarse-Grained Soils

Deformation of most coarse-grained soils occurs almost exclusively by volume change at a rate essentially equivalent to the rate of stress change. Deformation moduli are markedly nonlinear with respect to stress change and dependent on the initial state of soil stress. Some coarse-grained soils exhibit a delayed volume-change phenomenon known as **friction lag**. This response is analogous to the secondary compression of fine-grained soils and can account for a significant amount of the compression of coarse-grained soils composed of weak or sharp-grained particles.

The laboratory approach previously described for derivation of drained deformation parameters for fine-grained soils has a limited application for coarse-grained soils because of the difficulty in

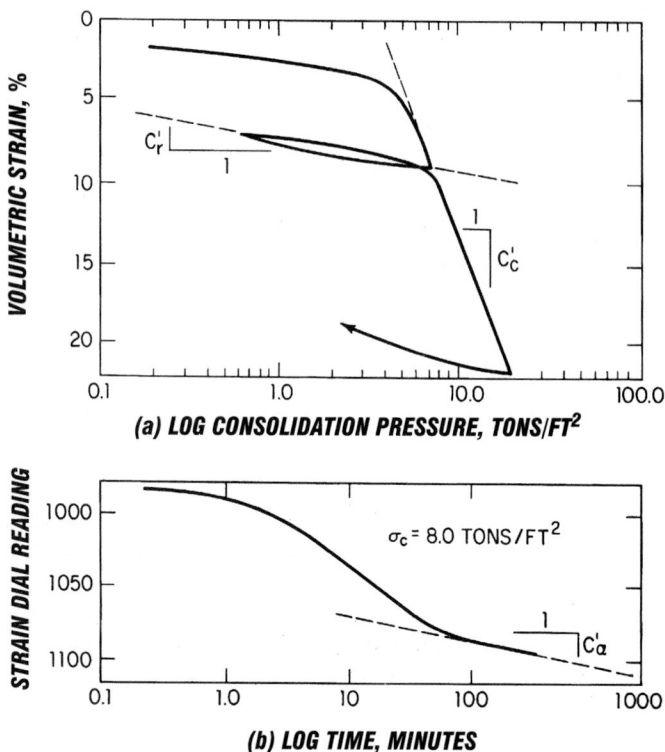

VOLUMETRIC STRAIN, %

C_r'

C_c'

(a) LOG CONSOLIDATION PRESSURE, TONS/FT^2

STRAIN DIAL READING

$\sigma_c = 8.0$ TONS/FT^2

C_α'

(b) LOG TIME, MINUTES

Fig. 7.2 Typical curves plotted from data obtained in consolidation tests.

obtaining reasonably undisturbed samples. Tests may be carried out on reinstituted samples but should be used with caution since the soil fabric, *aging*, and stress history cannot be adequately simulated in the laboratory. As a consequence, in situ testing techniques are often the preferred investigation and testing approach to the characterization of cohesionless soil properties (see Art. 7.6.3).

The Static Cone Penetration Test (CPT) ■ The CPT is one of the most useful in situ tests for investigating the deformability of cohesionless soils. The secant modulus E_s', tons/ft^2, of sands has been related to cone resistance q_c by correlations of small-scale plate load tests and load tests on footings. The relationship is given by Eq. (7.9a). The empirical correlation coefficient α in Eq. (7.9a) is influenced by the relative density, grain characteristics, and stress history of the soil (see Art. 7.6.3).

$$E_s' = \alpha q_c \qquad (7.9a)$$

The α parameter has been reported to range between 1.5 and 3 for sands and can be expressed in terms of relative density D_r as $2(1 + D_r^2)$. α may also be derived from correlations between q_c and standard penetration resistance N by assuming that q_c/N for mechanical cones or $q_c/N + 1$ for electronic type cone tips is about 6 for sandy gravel, 5 for gravelly sand, 4 for clean sand, and 3 for sandy silt. However, it should be recognized that E_s' characterizations from q_c or N are empirical and can provide erroneous characterizations. Therefore, the validity of these relationships should be confirmed by local correlations. Cone penetration test soundings should be conducted in accordance with ASTM D-3441 (Briaud, J. L. and Miran J. (1992). "The Cone Penetrometer Test", Federal Highway Administration, FHWA Report No. SA-91-043, Washington D.C; See also Art. 7.13)

Load-Bearing Test ■ One of the earliest methods for evaluating the in situ deformability of coarse-grained soils is the small-scale *load-bearing test*. Data developed from these tests have been used to provide a scaling factor to express the settlement ρ of a full-size footing from the settlement ρ_1 of a 1-ft^2 plate. This factor ρ/ρ_1 is given as a function of the width B of the full-size bearing plate as:

$$\frac{\rho}{\rho_1} = \left(\frac{2B}{1+B}\right)^2 \qquad (7.10)$$

From an elastic half-space solution, E'_s can be expressed from results of a plate load test in terms of the ratio of bearing pressure to plate settlement k_v as:

$$E'_s = \frac{k_v(1-\mu^2)\pi/4}{4B/(1+B)^2} \qquad (7.9b)$$

μ represents Poisson's ratio, usually considered to range between 0.30 and 0.40. Equation (7.9b) assumes that ρ_1 is derived from a rigid, 1-ft-diameter circular plate and that B is the equivalent diameter of the bearing area of a full-scale footing. Empirical formulations such as Eq. (7.10) may be significantly in error because of the limited footing-size range used and the large scatter of the data base. Furthermore, consideration is not given to variations in the characteristics and stress history of the bearing soils.

Pressuremeter tests (PMTs) in soils and soft rocks have been used to characterize E'_s from radial pressure vs. volume-change data developed by expanding a cylindrical probe in a drill hole (see Art. 7.6.3). Because cohesionless soils are sensitive to comparatively small degrees of soil disturbance, proper access-hole preparation is critical.

(K. Terzaghi and R. B. Peck, "Soil Mechanics and Engineering Practice," John Wiley & Sons, Inc., New York; H. Y. Fang, "Foundation Engineering Handbook," 2nd ed., Van Nostrand Reinhold, New York; Briaud, J. L. (1989). "The Pressuremeter Test For Highway Applications," Federal Highway Administration Publication No. FHWA-IP-89-008, Washington D.C.)

7.5.6 California Bearing Ratio (CBR)

This ratio is often used as a measure of the quality or strength of a soil that will underlie a pavement, for determining the thickness of the pavement, its base, and other layers.

$$CBR = \frac{F}{F_0} \qquad (7.11)$$

where F = force per unit area required to penetrate a soil mass with a 3-in^2 circular piston (about 2 in in diameter) at the rate of 0.05 in/min

F_0 = force per unit area required for corresponding penetration of a standard material

Typically, the ratio is determined at 0.10 in penetration, although other penetrations sometimes are used. An excellent base course has a CBR of 100%. A compacted soil may have a CBR of 50%, whereas a weaker soil may have a CBR of 10.

Tests to determine CBR may be performed in the laboratory or the field. ASTM standard tests are available for each case: "Standard Test Method for CBR (California Bearing Ratio) for Laboratory Compacted Soils," D1883, and "Standard Test Method for CBR (California Bearing Ratio) of Soils in Place," D4429 (www.astm.org).

One criticism of the method is that it does not simulate the shearing forces that develop in supporting materials underlying a flexible pavement.

7.5.7 Soil Permeability

The coefficient of permeability k is a measure of the rate of flow of water through saturated soil under a given hydraulic gradient i, cm/cm, and is defined in accordance with Darcy's law as:

$$V = kiA \qquad (7.12)$$

where V = rate of flow, cm^3/s.

A = cross-sectional area of soil conveying flow, cm^2

k is dependent on the grain-size distribution, void ratio, and soil fabric and typically may vary from as much as 10 cm/s for gravel to less than 10^{-7} cm/s for clays. For typical soil deposits, k for horizontal flow is greater than k for vertical flow, often by an order of magnitude.

Soil-permeability measurements can be conducted in tests under falling or constant head, either in the laboratory or the field. Large-scale pumping (drawdown) tests also may be conducted in the field to provide a significantly larger scale measurement of formation permeability. Correlations of k with soil gradation and relative density or void ratio have been developed for a variety of coarse-grained materials. General correlations of k with soil index and physical properties are less reliable for fine-grained soils because other factors than porosity may control.

(T. W. Lambe and R. V. Whitman, "Soil Mechanics," John Wiley & Sons, Inc., New York, www.wiley.com.)

7.6 Site Investigations

The objective of most geotechnical site investigations is to obtain information on the site and subsurface conditions that is required for design and construction of engineered facilities and for evaluation and mitigation of geologic hazards, such as landslides, subsidence, and liquefaction. The site investigation is part of a fully integrated process that includes:

1. Synthesis of available data

2. Field and laboratory investigations

3. Characterization of site stratigraphy and soil properties

4. Engineering analyses

5. Formulation of design and construction criteria or engineering evaluations

7.6.1 Planning and Scope

In the planning stage of a site investigation, all pertinent topographical, geologic, and geotechnical information available should be reviewed and assessed. In urban areas, the development history of the site should be studied and evaluated. It is particularly important to provide or require that a qualified engineer direct and witness all field operations.

The scope of the geotechnical site investigation varies with the type of project but typically includes topographic and location surveys, exploratory "drilling and sampling, in situ testing and groundwater monitoring. Frequently the investigation is supplemented by test pits, geophysical tests, air photos and remote sensing".

7.6.2 Exploratory Borings

Typical boring methods employed for geotechnical exploration consist of rotary drilling, auger drilling, percussion drilling, or any combination of these. Deep soil borings (greater than about 100 ft) are usually conducted by rotary-drilling techniques recirculating a weighted drilling fluid to maintain borehole stability. Auger drilling, with hollow-stem augers to facilitate sampling, is a widely used and economical method for conducting short- to intermediate-length borings. Most of the drill rigs are truck-mounted and have a rockcoring capability. A wide variety of drilling machines are available to provide access to the most difficult projects.

With percussion drilling, a casing is usually driven to advance the boring. Water circulation or driven, clean-out spoons are often used to remove the soil (cuttings) in the casing. This method is employed for difficult-access locations where relatively light and portable drilling equipment is required. A rotary drill designed for rock coring is often included.

Soil Samples ▪ These are usually obtained by driving a split-barrel sampler or by hydraulically or mechanically advancing a thin-wall (Shelby) tube sampler. Driven samplers, usually 2 in outside diameter (OD), are advanced 18 in by a 140-lb hammer dropped 30 in (ASTM D1586). The number of blows required to drive the last 12 in of penetration constitutes the **standard penetration resistance** (SPT) value. The **Shelby tube sampler**, used for undisturbed sampling, is typically a 12- to 16-gage seamless steel tube and is nominally 3 in OD (ASTM D1587). In soils that are soft or otherwise difficult to sample, a stationary piston sampler is used to advance a Shelby tube either hydraulically (pump pressure) or by the down-crowd system of the drill.

Rotary core drilling is typically used to obtain core samples of rock and hard, cohesive soils that cannot be penetrated by conventional sampling techniques. Typically, rock cores are obtained with diamond bits that yield core-sample diameters from $\frac{7}{8}$ (AX) to $2\frac{1}{8}$ (NX). For hard clays and soft rocks, a 3- to 6-in OD undisturbed sample can also be obtained by rotary drilling with a **Dennison** or **Pitcher sampler**.

Test Boring Records (Logs) ▪ These typically identify the depths and material classification of the various strata encountered, the sample location and penetration resistance, rock-core interval and recovery, groundwater levels encountered during and after drilling. Special subsurface conditions should be noted on the log, for example, changes in drilling resistance, hole caving, voids, and obstructions. General information required includes the location of the boring, surface elevation, drilling procedures, sampler and core barrel types, and other information relevant to interpretation of the boring log.

Groundwater Monitoring ■ Monitoring groundwater levels is an integral part of boring and sampling operations. Groundwater measurements during and at least 12 h after drilling are usually required. Standpipes are often installed in test borings to provide longer-term observations; they are typically small-diameter pipes perforated in the bottom few feet of casing.

If irregular piezometric profiles are suspected, piezometers may be set and sealed so as to measure hydrostatic heads within selected strata. Piezometers may consist of watertight $\frac{1}{2}$ to $\frac{3}{4}$-in OD standpipes or plastic tubing attached to porous ceramic or plastic tips. Piezometers with electronic or pneumatic pressure sensors have the advant-age of quick response and automated data acquisition. However, it is not possible to conduct in situ permeability tests with these closed-system piezometers.

7.6.3 In Situ Testing Soils

In situ tests can be used under a variety of circumstances to enhance profile definition, to provide data on soil properties, and to obtain parameters for empirical analysis and design applications.

Quasi-static and dynamic cone penetration tests (**CPTs**) quite effectively enhance profile definition by providing a continuous record of penetration resistance. Quasi-static cone penetration resistance is also correlated with the relative density, *OCR*, friction angle, and compressibility of coarse-grained soils and the undrained shear strength of cohesive soils. Empirical foundation design parameters are also provided by the CPT.

The standard CPT in the United States consists of advancing a 10-cm^2, 60° cone at a rate between 1.5 and 2.5 cm/s and recording the resistance to cone penetration (ASTM D3441). A friction sleeve may also be incorporated to measure frictional resistance during penetration. The cone may be incrementally (mechanical penetrometer) or continuously (electronic penetrometer) advanced.

Dynamic cones are available in a variety of sizes, but in the United States, they typically have a 2-in upset diameter with a 60° apex. They are driven by blows of a 140-lb hammer dropped 30 in. Automatically driven cone penetrometers are widely used in western Europe and are portable and easy to operate.

Pressuremeter tests (**PMTs**) provide an in situ interpretation of soil compressibility and undrained shear strength. Pressuremeters have also been used to provide parameters for foundation design.

The PMT is conducted by inserting a probe containing an expandable membrane into a drill hole and then applying a hydraulic pressure to radially expand the membrane against the soil, to measure its volume change under pressure. The resulting curve for volume change vs. pressure is the basis for interpretation of soil properties.

Vane shear tests provide in situ measurements of the undrained shear strength of soft to firm clays, usually by rotating a four-bladed vane and measuring the torsional resistance *T*. Undrained shear strength is then calculated by dividing *T* by the cylindrical side and end areas inscribing the vane. Account must be taken of torque rod friction (if unsleeved), which can be determined by calibration tests (ASTM D2573). Vane tests are typically run in conjunction with borings, but in soft clays the vane may be advanced without a predrilled hole.

Other in situ tests occasionally used to provide soil-property data include plate load tests (PLTs), borehole shear (BHS) tests, and dilatometer tests. The PLT technique may be useful for providing data on the compressibility of soils and rocks. The BHS may be useful for characterizing effective-shear-strength parameters for relatively free-draining soils as well as total-stress (undrained) shear-strength parameters for fine-grained soils. Dilatometer tests provide a technique for investigating the horizontal effective stress σ'_{ho} and soil compressibility. Some tests use small-diameter probes to measure pore-pressure response, acoustical emissions, bulk density, and moisture content during penetration.

Prototype load testing as part of the geotechnical investigation represents a variation of in situ testing. It may include pile load tests, earth load tests to investigate settlement and stability, and tests on small-scale or full-size shallow foundation elements. Feasibility of construction can also be evaluated at this time by test excavations, indicator pile driving, drilled shaft excavation, rock rippability trials, dewatering tests, and so on.

(H. Y. Fang, "Foundation Engineering Handbook," 2nd ed., Van Nostrand Reinhold, New York.)

7.6.4 Geophysical Investigations

Geophysical measurements are often valuable in evaluation of continuity of soil and rock strata between boring locations. Under some circumstances, such data can reduce the number of borings required. Certain of these measurements can also provide data for interpreting soil and rock properties. The techniques often used in engineering applications are as follows:

Seismic-wave-propagation techniques include seismic refraction, seismic rejection, and direct wave-transmission measurements. Refraction techniques measure the travel time of seismic waves generated from a single-pulse energy source to detectors (geophones) located at various distances from the source. The principle of seismic refraction surveying is based on refraction of the seismic waves at boundaries of layers with different acoustical impedances. This technique is illustrated in Fig. 7.3.

Compression P wave velocities are interpreted to define velocity profiles that may be correlated with stratigraphy and the depth to rock. The P wave velocity may also help identify type of soil. However, in saturated soils the velocity measured represents wave transmission through water-filled voids. This velocity is about 4800 ft/s regardless of the soil type. Low-cost single- and dual-channel seismographs are available for routine engineering applications.

Seismic reflection involves measuring the times required for a seismic wave induced at the surface to return to the surface after reflection from the interfaces of strata that have different acoustical impedances. Unlike refraction techniques, which usually record only the first arrivals of the seismic waves, wave trains are concurrently recorded by several detectors at different positions so as to

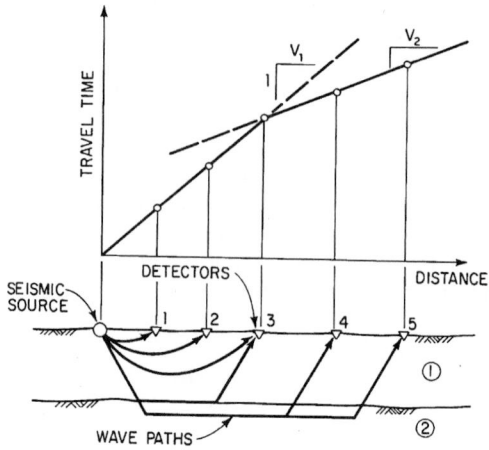

Fig. 7.3 Illustration of seismic refraction concept.

provide a pictorial representation of formation structure. This type of survey can be conducted in both marine and terrestrial environments and usually incorporates comparatively expensive multiple-channel recording systems.

Direct seismic-wave-transmission techniques include measurements of the arrival times of P waves and shear S waves after they have traveled between a seismic source and geophones placed at similar elevations in adjacent drill holes. By measuring the precise distances between source and detectors, both S and P wave velocities can be measured for a given soil or rock interval if the hole spacing is chosen to ensure a direct wave-transmission path.

Alternatively, geophones can be placed at different depths in a drill hole to measure seismic waves propagated down from a surface source near the drill hole. The detectors and source locations can also be reversed to provide up-hole instead of down-hole wave propagation. Although this method does not provide as precise a measure of interval velocity as the cross-hole technique, it is substantially less costly.

Direct wave-transmission techniques are usually conducted so as to maximize S wave energy generation and recognition by polarization of the energy input. S wave interpretations allow calculation of the small-strain shear modulus G_{max} required for dynamic response analysis. Poisson's ratio can also be determined if both P and S wave velocities can be recorded.

Resistivity and conductance investigation techniques relate to the proposition that stratigraphic details can be derived from differences in the electrical resistance or conductivity of individual strata. Resistivity techniques for engineering purposes usually apply the Wenner method of investigation, which involves four equally spaced steel electrodes (pins). The current is introduced through the two end pins, and the associated potential drop is measured across the two center pins. The apparent resistivity ρ is then calculated as a function of current I, potential difference V, and pin spacing a as:

$$\rho = \frac{2\pi a V}{I} \qquad (7.13)$$

To investigate stratigraphic changes, tests are run at successively greater pin spacings. Interpretations are made by analyzing accumulative or discrete-interval resistivity profiles or by theoretical curve-matching procedures.

A conductivity technique for identifying subsurface anomalies and stratigraphy involves measuring the transient decay of a magnetic field with the source (dipole transmitter) in contact with the surface. The depth of apparent conductivity measurement depends on the spacing and orientation of the transmitter and receiver loops.

Both resistivity and conductivity interpretations are influenced by groundwater chemistry. This characteristic has been utilized to map the extent of some groundwater pollutant plumes by conductivity techniques.

Other geophysical methods with more limited engineering applications include gravity and magnetic field measurements. Surveys using these techniques can be airborne, shipborne, or ground-based. Microgravity surveys have been useful in detecting subsurface solution features in carbonate rocks.

Aerial surveys are appropriate where large areas are to be explored. Analyses of conventional aerial stereoscopic photographs; thermal and false-color, infrared imagery; multispectral satellite imagery; or side-looking aerial radar can disclose the surface topography and drainage, linear features that reflect geologic structure, type of surface soil and often the type of underlying rock. These techniques are particularly useful in locating filled-in sinkholes in karst regions, which are often characterized by closely spaced, slight surface depressions.

(M. B. Dobrin, "Introduction to Geophysical Prospecting," McGraw-Hill Book Company, New York, books.mcgraw-hill.com.)

7.7 Hazardous Site and Foundation Conditions

There are a variety of natural hazards of potential concern in site development and foundation design. Frequently, these hazards are overlooked or not given proper attention, particularly in areas where associated failures have been infrequent.

7.7.1 Solution-Prone Formations

Significant areas in the eastern and midwestern United States are underlain by formations (carbonate and evaporate rocks) susceptible to dissolution. Subsurface voids created by dissolution range from open jointing to huge caverns. These features have caused catastrophic failures and detrimental settlements of structures as a result of ground loss or surface subsidence.

Special investigations designed to identify rock-solution hazards include geologic reconnaissance, air photo interpretation, and geophysical (resistivity, microgravity, and so on) surveys. To mitigate these hazards, careful attention should be given to:

1. Site drainage to minimize infiltration of surface waters near structures

2. Limitation of excavations to maximize the thickness of soil overburden

3. Continuous foundation systems designed to accommodate a partial loss of support beneath the foundation system

4. Deep foundations socketed into rock and designed solely for socket bond resistance

It is prudent to conduct special **proof testing** of the bearing materials during construction in solution-prone formations. Proof testing often consists of soundings continuously recording the penetration resistance through the overburden and the rate of percussion drilling in the rock. Suspect zones are thus identified and can be improved by excavation and replacement or by in situ grouting.

7.7.2 Expansive Soils

Soils with a medium to high potential for causing structural damage on expansion or shrinkage are found primarily throughout the Great Plains and Gulf Coastal Plain Physiographic Provinces. Heave or settlement of *active* soils occurs because of a change in soil moisture in response to climatic changes, construction conditions, changes in surface cover, and other conditions that influence the groundwater and evapotransportation regimes. Differential foundation movements are brought about by differential moisture changes in the bearing soils. Figure 7.4 presents a method for classifying the volume-change potential of clay as a function of activity.

Investigations in areas containing potentially expansive soils typically include laboratory swell tests. Infrequently, soil suction measurements are made to provide quantitative evaluations of volume-change potential. Special attention during the field investigation should be given to evaluation of the groundwater regime and to delineation of the depth of active moisture changes.

Common design procedures for preventing structural damage include mitigation of moisture changes, removal or modification of expansive material and deep foundation support. Horizontal

and vertical moisture barriers have been utilized to minimize moisture losses due to evaporation or infiltration and to cut off subsurface groundwater flow into the area of construction. Excavation of potentially active materials and replacement with inert material or with excavated soil modified by the addition of lime have proved feasible where excavation quantities are not excessive.

Deep foundations (typically drilled shafts) have been used to bypass the active zone and to resist or minimize uplift forces that may develop on the shaft. Associated grade beams are usually constructed to prevent development of uplift forces.

("Engineering and Design of Foundations on Expansive Soils," U.S. Department of the Army, 1981. L. D. Johnson, "Predicting Potential Heave and Heave with Time and Swelling Foundation Soils," Technical Report S-78-7, U.S. Army Engineers Waterways Experiment Station, Vicksburg, Miss., 1978.)

7.7.3 Landslide Hazards

Landslides are usually associated with areas of significant topographic relief that are characterized by relatively weak sedimentary rocks (shales, siltstones, and so forth) or by relatively impervious soil deposits containing interbedded water-bearing strata. Under these circumstances, slides that have occurred in the geologic past, whether or not currently active, represent a significant risk for hillside site development. In general hillside development in potential landslide areas is a most hazardous undertaking. If there are alternatives, one of those should be adopted.

Detailed geologic studies are required to evaluate slide potential and should emphasize detection of old slide areas. Procedures that tend to stabilize an active slide or to provide for the continued stability of an old slide zone include:

1. Excavation at the head of the sliding mass to reduce the driving force

2. Subsurface drainage to depress piezometric levels along potential sliding surface

3. Buttressing at the toe of the potential sliding mass to provide a force-resisting slide movement

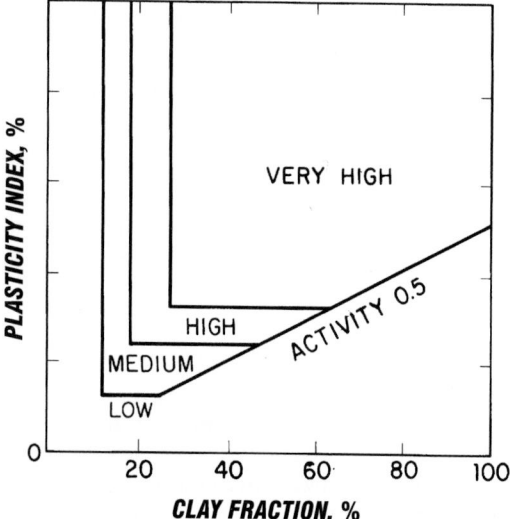

Fig. 7.4 Chart for rating volume-change potential of expansive soils.

Within the realm of economic feasibility, the reliability of these or any other procedures to stabilize active or old slides involving a significant sliding mass are generally of a relatively low order.

On hillsides where prior slides have not been identified, care should be taken to reduce the sliding potential of superimposed fills by removing weak or potentially unstable surficial materials, benching and keying the fill into competent materials, and (most importantly) installing effective subsurface drainage systems. Excavations that result in steepening of existing slopes are potentially detrimental and should not be employed. Direction and collection of surface runoff so as to prevent slope erosion and infiltration are recommended. ("Landslides Investigation and Mitigation," Transportation Research Board Special Report 247 National Academy Press 1996)

7.7.4 Liquefaction of Soils

Relatively loose saturated cohesionless soils may become unstable under cyclic shear loading such as that imposed by earthquake motions. A simplified method of analysis of the *liquefaction potential* of cohesionless soils has been proposed for predicting the ratio of the horizontal shear stress τ_{av} to the effective overburden pressure σ'_{vo} imposed by an earthquake. (τ_{av} represents a uniform cyclic-stress representation of the irregular time history of shear stress induced by the design earthquake.) This field stress ratio is a function of the maximum horizontal ground-surface acceleration a_{max}, the acceleration of gravity g, a stress-reduction factor r_d, and total vertical stress σ_{vo} and approximated as

$$\frac{\tau_{av}}{\sigma'_{vo}} = 0.65 \frac{a_{max}}{g} \frac{\sigma_{vo}}{\sigma'_{vo}} r_d \qquad (7.14)$$

r_d varies from 1.0 at the ground surface to 0.9 for a depth of 30 ft. (H. B. Seed and I. M. Idriss, "A Simplified Procedure for Evaluating Soil Liquefaction Potential," Report EERC 70-9, Earthquake Engineering Research Center, University of California, Berkeley, 1970.)

Stress ratios that produce liquefaction may be characterized from correlations with field observations (Fig. 7.5). The relevant soil properties are represented by their corrected penetration resistance

$$N_1 = (1 - 1.25 \log \sigma'_{vo})N \qquad (7.15)$$

where σ'_{vo} is in units of tons/ft^2. The stress ratio causing liquefaction should be increased about 25% for earthquakes with Richter magnitude 6 or lower. (H. B. Seed, "Evaluation of Soil Liquefaction Effects on Level Ground during Earthquakes," *Symposium on Liquefaction Problems and Geotechnical Engineering,*

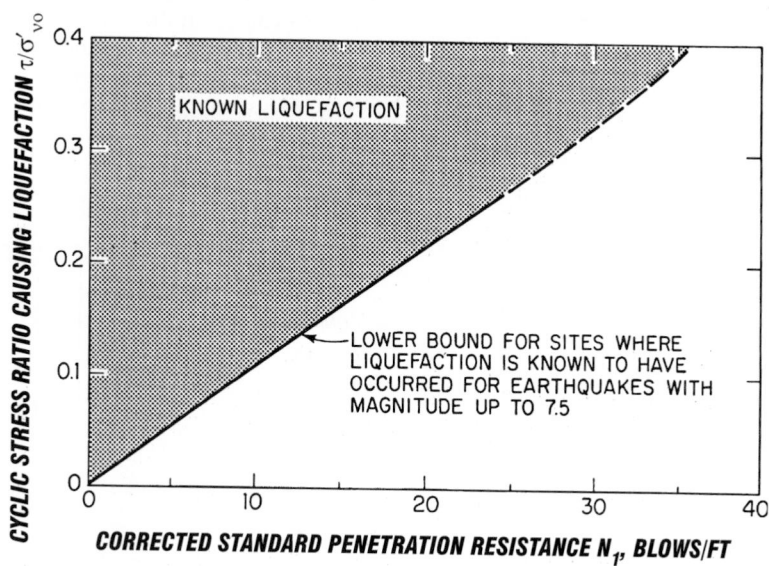

Fig. 7.5 Chart correlates cyclic-stress ratios that produce soil liquefaction with standard penetration resistance. (*After H. B. Seed.*)

ASCE National Convention, Philadelphia, Pa., 1976. ("Design Guidance: Geotechnical Earthquake Engineering For Highways," Federal Highway Administration, Publication No. FHWA-SA-97-076, May 1997)

Significantly, more elaborate, dynamic finite-element procedures have been proposed to evaluate soil liquefaction and degradation of undrained shear strength as well as generation and dissipation of pore-water pressure in soils as a result of cyclic loading. Since stress increases accompany dissipation of pore-water pressures, settlements due to cyclic loading can also be predicted. Such residual settlements can be important even though liquefaction has not been induced.

(P. B. Schnabel, J. Lysmer, and H. B. Seed, "A Computer Program for Earthquake Response Analysis of Horizontally Layered Sites," Report EERC 72-12, Earthquake Engineering Research Center, University of California, Berkeley, 1972; H. B. Seed, P. P. Martin, and J. Lysmer, "Pore-Water Pressure Changes During Soil Liquefaction," *ASCE Journal of Geotechnical Engineering Division*, vol. 102, no. GT4, 1975; K. L. Lee and A. Albaisa, "Earthquake-Induced Settlements in Saturated Sands," *ASCE Journal of Geotechnical Engineering Division*, vol. 100, no. GT4, 1974, www.asce.org.)

7.8 Types of Footings

Spread (individual) footings (Fig. 7.6) are the most economical shallow foundation types but are more susceptible to differential settlement. They usually support single concentrated loads, such as those imposed by columns.

Shallow Foundations

Shallow foundation systems can be classified as spread footings, wall and continuous (strip) footings, and mat (raft) foundations. Variations are combined footings, cantilevered (strapped) footings, two-way strip (grid) footings, and discontinuous (punched) mat foundations.

Combined footings (Fig. 7.7) are used where the bearing areas of closely spaced columns overlap. Cantilever footings (Fig. 7.8) are designed to accommodate eccentric loads.

Continuous wall and strip footings (Fig. 7.9) can be designed to redistribute bearing-stress concentrations and associated differential settlements in the event of variable bearing conditions or localized ground loss beneath footings.

Mat foundations have the greatest facility for load distribution and for redistribution of subgrade stress concentrations caused by localized anomalous bearing conditions. Mats may be constant section, ribbed, waffled, or arched. Buoyancy mats are used on compressible soil sites in combination with basements or subbasements, to create a permanent unloading effect, thereby reducing the net stress change in the foundation soils.

(M. J. Tomlinson, "Foundation Design and Construction," John Wiley & Sons, Inc., New York (www.wiley.com); J. E. Bowles, "Foundation

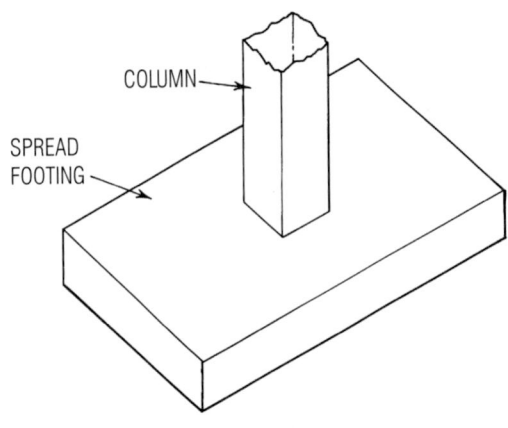

Fig. 7.6 Spread footing.

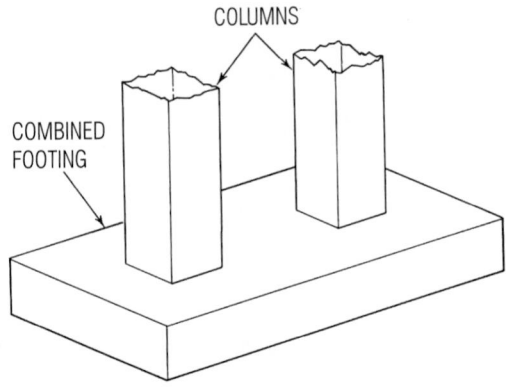

Fig. 7.7 Combined footing.

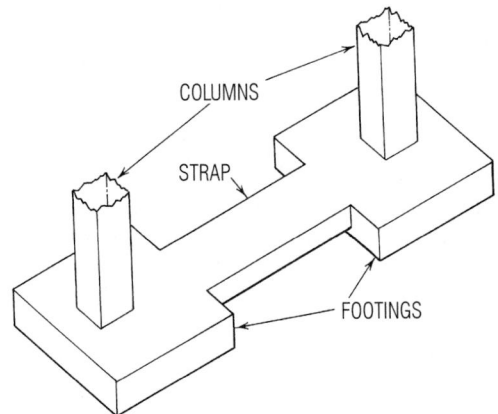

Fig. 7.8 Cantilever footing.

Analysis and Design," McGraw-Hill Book Company, New York. (books.mcgraw-hill.com) "Spread Footings for Highway Bridges," Federal Highway Administration, Publication No. FHWA-RD-86-185 October, 1987)

7.9 Approach to Foundation Analysis

Shallow-foundation analysis and formulation of geotechnical design provisions are generally approached in the following steps:

1. Establish project objectives and design or evaluation conditions.

2. Characterize site stratigraphy and soil rock properties.

3. Evaluate load-bearing fill support or subsoil-improvement techniques, if applicable.

4. Identify bearing levels; select and proportion candidate foundation systems.

5. Conduct performance, constructibility, and economic feasibility analyses.

6. Repeat steps 3 through 5 as required to satisfy the design objectives and conditions.

The scope and detail of the analyses vary according to the project objectives.

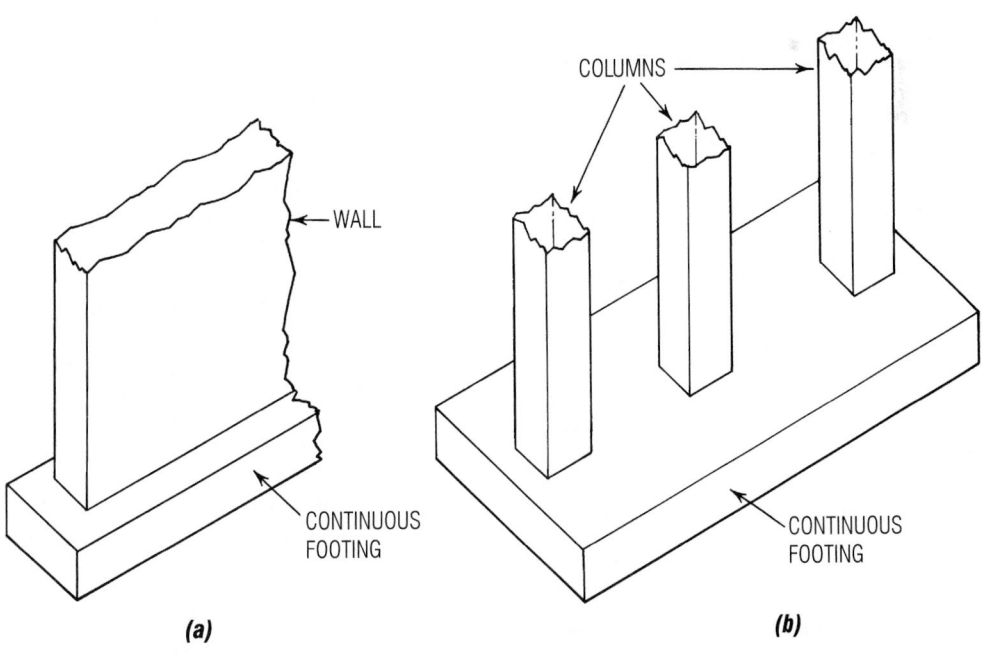

(a)   *(b)*

Fig. 7.9 Continuous footings for (*a*) a wall, (*b*) several columns.

Project objectives to be quantified are essentially the intent of the project assignment and the specific scope of associated work. The conditions that control geotechnical evaluation or design work include criteria for loads and grades, facility operating requirements and tolerances, construction schedules, and economic and environmental constraints. Failure to provide a clear definition of relevant objectives and design conditions can result in significant delays, extra costs, and, under some circumstances, unsafe designs.

During development of design conditions for structural foundations, tolerances for total and differential settlements are commonly established as a function of the ability of a structure to tolerate movement. Suggested structure tolerances in terms of angular distortion are in Table 7.5. Angular distortion represents the differential vertical movement between two points divided by the horizontal distance between the points.

Development of *design profiles* for foundation analysis ideally involves a synthesis of geologic and geotechnical data relevant to site stratigraphy and soil and rock properties. This usually requires site investigations (see Arts. 7.6.1 to 7.6.4) and in situ or laboratory testing, or both, of representative soil and rock samples (see Arts. 7.3 to 7.5.6).

Table 7.5 Limiting Angular Distortions*

Structural Response	Angular Distortion
Cracking of panel and brick walls	1/100
Structural damage to columns and beams	1/150
Impaired overhead crane operation	1/300
First cracking of panel walls	1/300
Limit for reinforced concrete frame	1/400
Limit for wall cracking	1/500
Limit for diagonally braced frames	1/600
Limit for settlement-sensitive machines	1/750

* Limits represent the maximum distortions that can be safely accommodated.
Source: After L. Bjerrum, European Conference on Soil Mechanics and Foundation Engineering, Wiesbaden, Germany, vol. 2, 1963.

To establish and proportion candidate foundation systems, consideration must first be given to identification of feasible bearing levels. The *depth of the foundation* must also be sufficient to protect exposed elements against frost heave and to provide sufficient confinement to produce a factor of safety not less than 2.5 (preferably 3.0) against shear failure of the bearing soils. Frost penetration has been correlated with a **freezing index**, which equals the number of days with temperature below 32° F multiplied by $T - 32$, where $T =$ average daily temperature. Such correlations can be applied in the absence of local codes or experience. Generally, footing depths below final grade should be a minimum of 2.0 to 2.5 ft.

For marginal bearing conditions, consideration should be given to improvement of the quality of potential bearing strata. Soil-improvement techniques include excavation and replacement or overlaying of unsuitable subsoils by *load-bearing fills*, *preloading* of compressible subsoils, *soil densification*, *soil reinforcement*, and *grouting techniques*. Densification methods include high energy surface impact (dynamic compaction), on grade vibratory compaction, and subsurface vibratory compaction by vibro-compaction techniques. Soil reinforcement methods include: stone columns, soil mixing, mechanically stabilized earth and soil nailing.

"Ground Improvement Technical Summaries," Federal Highway Administration, Publication No. FHWA-SA-98-086, December 1999.

The choice of an appropriate soil improvement technique is highly dependent on performance requirements (settlement), site and subsurface conditions, time and space constraints and cost.

Assessment of the suitability of candidate foundation systems requires evaluation of the safety factor against both catastrophic failure and excessive deformation under sustained and transient design loads. Catastrophic-failure assessment must consider overstress and creep of the bearing soils as well as lateral displacement of the foundation. Evaluation of the probable settlement behavior requires analysis of the stresses imposed within the soil and, with the use of appropriate soil parameters, prediction of foundation settlements. Typically, *settlement analyses* provide estimates of total and differential settlement at strategic locations within the foundation area and may include time-rate predictions of settlement. Usually, the suitability of shallow foundation systems is governed by the systems' load-settlement response rather than bearing capacity.

7.10 Foundation-Stability Analysis

The maximum load that can be sustained by shallow foundation elements at incipient failure (*bearing capacity*) is a function of the cohesion and friction angle of bearing soils as well as the width B and shape of the foundation. The **net bearing capacity** per unit area q_u of a long footing is conventionally expressed as:

$$q_u = \alpha_f c_u N_c + \sigma'_{vo} N_q + \beta_f \gamma B N_\gamma \qquad (7.16)$$

where $\alpha_f = 1.0$ for strip footings and 1.3 for circular and square footings

c_u = undrained shear strength of soil

σ'_{vo} = effective vertical shear stress in soil at level of bottom of footing

$\beta_f = 0.5$ for strip footings, 0.4 for square footings, and 0.6 for circular footings

γ = unit weight of soil

B = width of footing for square and rectangular footings and radius of footing for circular footings

N_c, N_q, N_γ = bearing-capacity factors, functions of angle of internal friction ϕ (Fig. 7.10)

For undrained (rapid) loading of cohesive soils, $\phi = 0$ and Eq. (7.16) reduces to

$$q_u = N'_c c_u \qquad (7.17)$$

where $N'_c = \alpha_f N_c$. For drained (slow) loading of cohesive soils, ϕ and c_u are defined in terms of effective friction angle ϕ' and effective stress c'_u.

Modifications of Eq. (7.16) are also available to predict the bearing capacity of layered soil and for eccentric loading.

Rarely, however, does q_u control foundation design when the safety factor is within the range of 2.5 to 3. (Should creep or local yield be induced, excessive settlements may occur. This consideration is particularly important when selecting a safety factor for foundations on soft to firm clays with medium to high plasticity.)

Equation (7.16) is based on an infinitely long strip footing and should be corrected for other shapes. Correction factors by which the bearing-capacity factors should be multiplied are given in Table 7.6, in which L = footing length.

The derivation of Eq. (7.16) presumes the soils to be homogeneous throughout the stressed zone, which is seldom the case. Consequently, adjustments may be required for departures from homo-

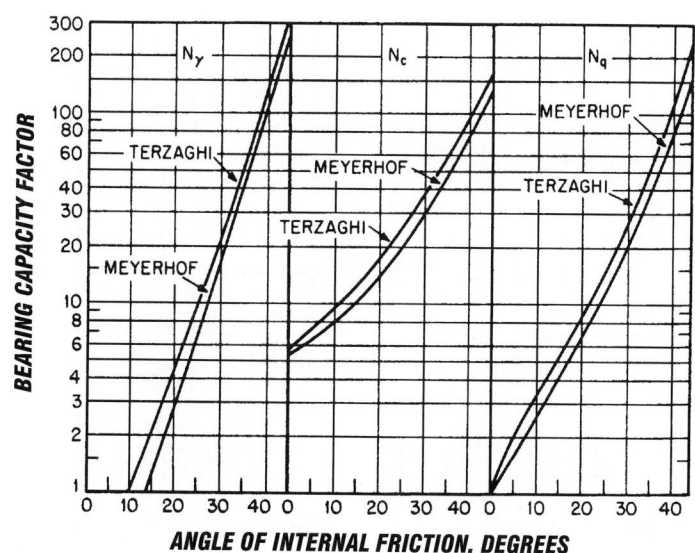

Fig. 7.10 Bearing capacity factors for use in Eq. (7.16) as determined by Terzaghi and Meyerhof.

Table 7.6 Shape Corrections for Bearing-Capacity Factors of Shallow Foundations*

Shape of Foundation	Correction Factor		
	N_c	N_q	N_γ
Rectangle[†]	$1 + (B/L)(N_q/N_c)$	$1 + (B/L)\tan\phi$	$1 - 0.4(B/L)$
Circle and square	$1 + (N_q/N_c)$	$1 + \tan\phi$	0.60

* After E. E. De Beer, as modified by A. S. Vesic. See H. Y. Fang, "Foundation Engineering Handbook," Van Nostrand Reinhold, 2d ed., New York.

[†] No correction factor is needed for long strip foundations.

geneity. In sands, if there is a moderate variation in strength, it is safe to use Eq. (7.16), but with bearing-capacity factors representing a weighted average strength.

For strongly varied soil profiles or interlayered sands and clays, the bearing capacity of each layer should be determined. This should be done by assuming the foundation bears on each layer successively but at the contact pressure for the depth below the bottom of the foundation of the top of the layer.

Eccentric loading can have a significant impact on selection of the bearing value for foundation design. The conventional approach is to proportion the foundation to maintain the resultant force within its middle third. The footing is assumed to be rigid and the bearing pressure is assumed to vary linearly as shown by Fig. 7.11b. If the resultant lies outside the middle third of the footing, it is assumed that there is bearing over only a portion of the footing, as shown in Fig. 7.11d. For the conventional case (Fig. 7.11a), the maximum and minimum bearing pressures are:

$$q_m = \frac{P}{BL}\left(1 \pm \frac{6e}{B}\right) \qquad (7.18)$$

where B = width of rectangular footing

L = length of rectangular footing

e = eccentricity of loading

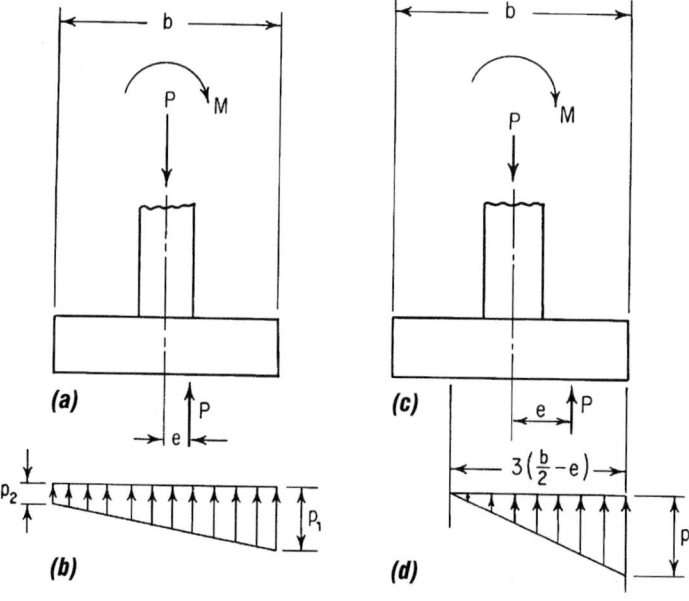

Fig. 7.11 Footings subjected to overturning.

For the other case (Fig. 7.11c), the soil pressure ranges from 0 to a maximum of:

$$q_m = \frac{2P}{3L(B/2 - e)} \quad (7.19)$$

For square or rectangular footings subject to overturning about two principal axes and for unsymmetrical footings, the loading eccentricities e_1 and e_2 are determined about the two principal axes. For the case where the full bearing area of the footings is engaged, q_m is given in terms of the distances from the principal axes c_1 and c_2, the radius of gyration of the footing area about the principal axes r_1 and r_2, and the area of the footing A as:

$$q_m = \frac{P}{A}\left(1 + \frac{e_1 c_1}{r_1^2} + \frac{e_2 c_2}{r_2^2}\right) \quad (7.20)$$

For the case where only a portion of the footing is bearing, the maximum pressure may be approximated by trial and error.

For all cases of *sustained eccentric loading*, the maximum (edge) pressures should not exceed the shear strength of the soil and also the factor of safety should be at least 1.5 (preferably 2.0) against overturning.

The foregoing analyses, except for completely rigid foundation elements, are a very conservative approximation. Because most mat foundations and large footings are not completely rigid, their deformation during eccentric loading acts to produce a more uniform distribution of bearing pressures than would occur under a rigid footing and to reduce maximum contact stresses.

In the event of *transient eccentric loading*, experience has shown that footings can sustain maximum edge pressures significantly greater than the shear strength of the soil. Consequently, some building codes conservatively allow increases in sustained-load bearing values of 30% for transient loads. Reduced safety factors have also been used in conjunction with transient loading. For cases where significant cost savings can be realized, finite-element analyses that model soil-structure interaction can provide a more realistic evaluation of an eccentrically loaded foundation.

Allowable Bearing Pressures ▪ Approximate allowable soil bearing pressures, without tests, for various soil and rocks are given in Table 7.7 for normal conditions. These basic bearing pressures should be used for preliminary

Table 7.7 Allowable Bearing Pressures for Soils

Soil Material	Pressure, tons/ft^2	Notes
Unweathered sound rock	60	No adverse seam structure
Medium rock	40	
Intermediate rock	20	
Weathered, seamy, or porous rock	2 to 8	
Hardpan	12	Well cemented
Hardpan	8	Poorly cemented
Gravel soils	10	Compact, well graded
Gravel soils	8	Compact with more than 10% gravel
Gravel soils	6	Loose, poorly graded
Gravel soils	4	Loose, mostly sand
Sand soils	3 to 6	Dense
Fine sand	2 to 4	Dense
Clay soils	5	Hard
Clay soils	2	Medium stiff
Silt soils	3	Dense
Silt soils	1½	Medium dense
Compacted fills		Compacted to 90% to 95% of maximum density (ASTM D1557)
Fills and soft soils	2 to 4	By field or laboratory test only

design only. Final design values should be based on the results of a thorough subsurface investigation and the results of engineering analysis of potential failure and deformation limit states.

Resistance to Horizontal Forces ▪ The horizontal resistance of shallow foundations is mobilized by a combination of the passive soil resistance on the vertical projection of the embedded foundation and the friction between the foundation base and the bearing soil and rock. The soil pressure mobilized at full passive resistance, however, requires lateral movements greater than can be sustained by some foundations. Consequently, a soil resistance between the at-rest and passive-pressure cases should be determined on the basis of the allowable lateral deformations of the foundation.

The frictional resistance f to horizontal translation is conventionally estimated as a function of the sustained, real, load-bearing stresses q_d from

$$f = q_d \tan \delta \qquad (7.21)$$

where δ is the friction angle between the foundation and bearing soils. δ may be taken as equivalent to the internal-friction angle ϕ' of the subgrade soils. In the case of cohesive soils, $f = c_u$. Again, some relative movement must be realized to develop f, but this movement is less than that required for passive-pressure development.

If a factor of safety against translation of at least 1.5 is not realized with friction and passive soil/rock pressure, footings should be keyed to increase sliding resistance or tied to engage additional resistance. Building basement and shear walls are also commonly used to sustain horizontal loading.

7.11 Stress Distribution Under Footings

Stress changes imposed in bearing soils by earth and foundation loads or by excavations are conventionally predicted from elastic half-space theory as a function of the foundation shape and the position of the desired stress profile. Elastic solutions available may take into account foundation rigidity, depth of the compressible zone, superposition of stress from adjacent loads, layered profiles, and moduli that increase linearly with depth.

For most applications, stresses may be computed by the pressure-bulb concept with the methods of either Boussinesq or Westergaard. For thick deposits, use the Boussinesq distribution shown in Fig. 7.12a; for thinly stratified soils, use the Westergaard approach shown in Fig. 7.12b. These charts indicate the stresses q beneath a single foundation unit that applies a pressure at its base of q_o.

Most facilities, however, involve not only multiple foundation units of different sizes, but also floor slabs, perhaps fills, and other elements that contribute to the induced stresses. The stresses used for settlement calculation should include the overlapping and contributory stresses that may arise from these multiple loads.

7.12 Settlement Analyses of Cohesive Soils

Settlement of foundations supported on cohesive soils is usually represented as the sum of the primary one-dimensional consolidation ρ_c, immediate ρ_i, and secondary ρ_s settlement components. Settlement due to primary consolidation is conventionally predicted for n soil layers by Eq. (7.22) and (7.23). For normally consolidated soils,

$$\rho_c = \sum_{i=1}^{n} H_i \left(C_c' \log \frac{\sigma_v}{\sigma_{vo}'} \right) \qquad (7.22)$$

where H_i = thickness of ith soil layer

C_c' = strain referenced compression index for ith soil layer (Art. 7.5.4)

σ_v = sum of average σ_{vo}' and average imposed vertical-stress change $\Delta\sigma_v$ in ith soil layer

σ_{vo}' = initial effective overburden pressure at middle of ith layer (Art 7.5.3)

For overconsolidated soils with $\sigma_v > \sigma_{vm}'$,

$$\rho_c = \sum_{i=1}^{n} H_i \left(C_r' \log \frac{\sigma_{vm}'}{\sigma_{vo}'} + C_c' \log \frac{\sigma_v}{\sigma_{vm}'} \right) \qquad (7.23)$$

where C_r' = strain referenced recompression index of ith soil layer (Art. 7.5.4)

σ_{vm}' = preconsolidation (maximum past consolidation) pressure at middle of ith layer (Art. 7.5.3)

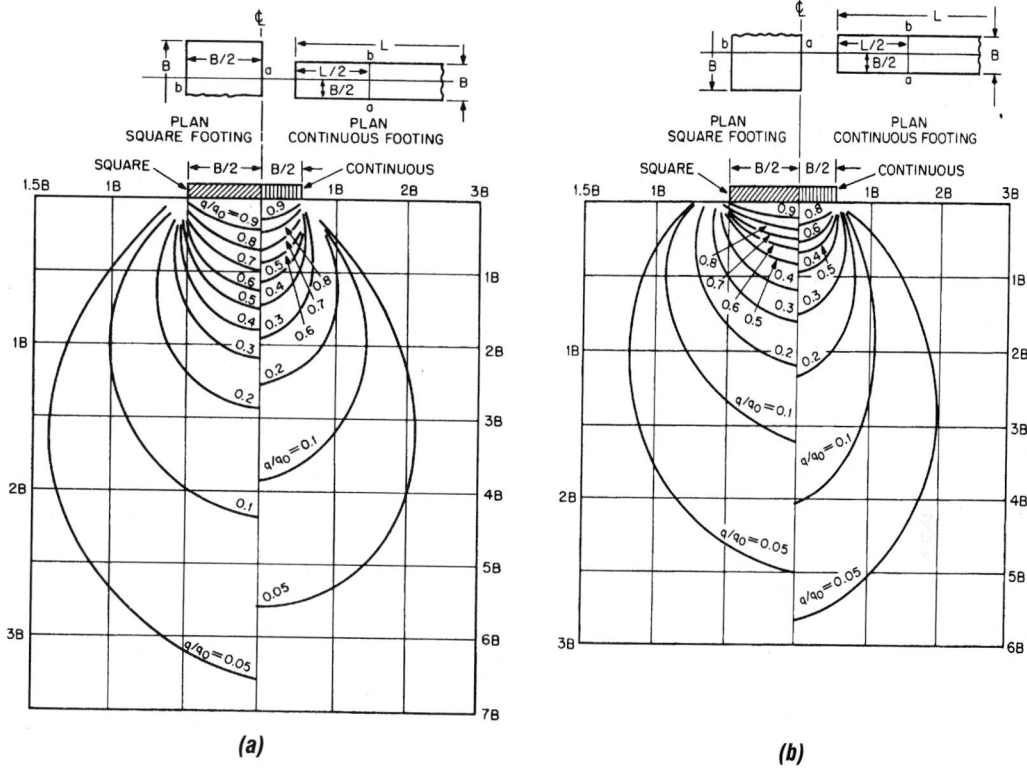

Fig. 7.12 Stress distribution under a square footing with side B and under a continuous footing with width B, as determined by equations of (a) Boussinesq and (b) Westergaard.

The maximum thickness of the compressible soil zone contributing significant settlement can be taken to be equivalent to the depth where $\Delta\sigma_v = 0.1\sigma'_{vo}$.

Equation (7.22) can also be applied to overconsolidated soils if σ_v is less than σ'_{vm} and C'_r is substituted for C'_c.

Inasmuch as Eqs. (7.22) and (7.23) represent one-dimensional compression, they may provide rather poor predictions for cases of three-dimensional loading. Consequently, corrections to ρ_c have been derived for cases of three-dimensional loading. These corrections are approximate but represent an improved approach when loading conditions deviate significantly from the one-dimensional case. (A. W. Skempton and L. Bjerrum, "A Contribution to Settlement Analysis of Foundations on Clay," *Geotechnique*, vol. 7, 1957.)

The stress-path method of settlement analysis attempts to simulate field loading conditions by conducting triaxial tests so as to track the sequential stress changes of an average point or points beneath the foundation. The strains associated with each drained and undrained load increment are summed and directly applied to the settlement calculation. Deformation moduli can also be derived from stress-path tests and used in three-dimensional deformation analysis.

Three-dimensional settlement analyses using elastic solutions have been applied to both drained and undrained conditions. Immediate (elastic) foundation settlements ρ_i, representing the undrained deformation of saturated cohesive soils, can be calculated by discrete analysis [Eq. (7.25)].

$$\rho_i = \sum_{i=1}^{n} H_i \frac{\sigma_1 - \sigma_3}{E_i} \qquad (7.24)$$

where $\sigma_1 - \sigma_3 =$ change in average deviator stress within each layer influenced by applied load. Note that Eq. (7.24) is strictly applicable only for axisymmetrical loading. Drained three-dimensional deformation can be estimated from Eq. (7.24) by substituting the secant modulus E'_s for E (see Art. 7.5.5).

The rate of one-dimensional consolidation can be evaluated with Eq. (7.26) in terms of the degree of consolidation U and the nondimensional time factor T_v. U is defined by

$$U = \frac{\rho_t}{\rho_c} = 1 - \frac{u_t}{u_i} \qquad (7.25)$$

where ρ_t = settlement at time t after instantaneous loading

ρ_c = ultimate consolidation settlement

u_t = excess pore-water pressure at time t

u_i = initial pore-water pressure ($t = 0$)

To correct approximately for the assumed instantaneous load application, ρ_t at the end of the loading period can be taken as the settlement calculated for one-half of the load application time. The time t required to achieve U is evaluated as a function of the shortest drainage path within the compressible zone h, the coefficient of consolidation C_v, and the dimensionless time factor T_v from

$$t = T_v \frac{h^2}{C_v} \qquad (7.26)$$

Closed-form solutions T_v vs. U are available for a variety of initial pore-pressure distributions. (H. Y. Fang, "Foundation Engineering Handbook," 2nd ed., Van Nostrand Reinhold Company, New York.)

Solutions for constant and linearly increasing u_i are shown in Fig. 7.13. Equation (7.27) presents an approximate solution that can be applied to the constant initial u_i distribution case for $T_v > 0.2$.

$$U = 1 - \frac{8}{\pi^2} e^{-\pi^2 T_v/4} \qquad (7.27)$$

where $e = 2.71828$. Numerical solutions for any u_i configuration in a single compressible layer as well as solutions for contiguous clay layers may be derived with finite-difference techniques.

(R. F. Scott, "Principles of Soil Mechanics," Addison-Wesley Publishing Company, Inc., Reading, Mass.)

The coefficient of consolidation C_v should be established based on experience and from site

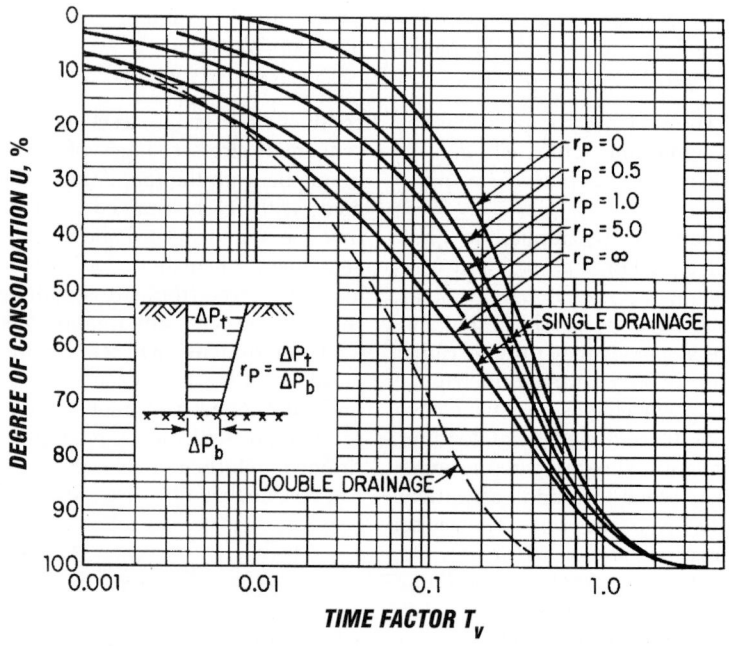

Fig. 7.13 Curves relate degree of consolidation and time factor T_v.

specific conventional consolidation tests by fitting the curve for time vs. deformation (for an appropriate load increment) to the theoretical solution for constant u_i. For tests of samples drained at top and bottom, C_v may be interpreted from the curve for log time or square root of time vs. strain (or dial reading) as

$$C_v = \frac{T_v H^2}{4t} \qquad (7.28)$$

where H = height of sample, in

t = time for 90% consolidation ($\sqrt{t}$ curve) or 50% consolidation (log t curve), days

T_v = 0.197 for 90% consolidation or 0.848 for 50% consolidation

(See T. W. Lambe and R. V. Whitman, "Soil Mechanics," John Wiley & Sons, Inc., New York, www.wiley.com), for curve-fitting procedures.) Larger values of C_v are usually obtained with the $\sqrt{t}$ method and appear to be more representative of field conditions.

Secondary compression settlement ρ_s is assumed, for simplicity, to begin on completion of primary consolidation, at time t_{100} corresponding to 100% primary consolidation. ρ_s is then calculated from Eq. (7.29) for a given period t after t_{100}.

$$\rho_s = \sum_{i=1}^{n} H_i C_\alpha \log \frac{t}{t_{100}} \qquad (7.29)$$

H_i represents the thickness of compressible layers and C_α is the coefficient of secondary compression given in terms of volumetric strain (Art. 7.5.4).

The ratio C_α to compression index C_c is nearly constant for a given soil type and is generally within the range of 0.045 ± 0.015. C_α, as determined from consolidation tests (Fig. 7.2), is extremely sensitive to pressure-increment ratios of less than about 0.5 (standard is 1.0). The effect of overconsolidation, either from natural or construction preload sources, is to significantly reduce C_α. This is an important consideration in the application of preloading for soil improvement.

The rate of *consolidation due to radial drainage* is important for the design of vertical *wick drains*. As a rule, drains are installed in compressible soils to reduce the time required for consolidation and to accelerate the associated gain in soil strength. Vertical drains are typically used in conjunction with preloading as a means of improving the supporting ability and stability of the subsoils.

(S. J. Johnson, "Precompression for Improving Foundation Soils," *ASCE Journal of Soil Mechanics and Foundation Engineering Division*, vol. 96, no. SM1, 1970 (www.asce.org); R. D. Holtz and W. D. Kovacs, "An Introduction to Geotechnical Engineering," Prentice-Hall, Inc., Englewood Cliffs, N.J. (www.prenhall.com) "Prefabricated Vertical Drains" Federal Highway Administration, Publication No. FHWA-RD-86-168, 1986)

7.13 Settlement Analysis of Sands

The methods most frequently used to estimate the settlement of foundations supported by relatively free-draining cohesionless soils generally employ empirical correlations between field observations and in situ tests. The primary correlative tests are plate bearing (PLT), cone penetration resistance (CPT), and standard penetration resistance (SPT) (see Art 7.6.3). These methods, however, are developed from data bases that contain a number of variables not considered in the correlations and, therefore, should be applied with caution.

Time rate of settlement in coarse grained soils is extremely rapid. This behavior may be used to the advantage of the designer where both dead and live loads are applied to the foundation. Often vertical deformations which occur during the construction process will have a minimal effect on the completed facility.

Plate Bearing Tests ▪ The most common approach is to scale the results of PLTs to full-size footings in accordance with Eq. (7.10). A less conservative modification of this equation proposed by A. R. S. S. Barazaa is

$$\rho = \left[\frac{2.5B}{1.5 + B} \right]^2 \rho_1 \qquad (7.30)$$

where B = footing width, ft

ρ = settlement of full-size bearing plate

ρ_1 = settlement of 1-ft square bearing plate

These equations are not sensitive to the relative density, gradation, and OCR of the soil or to the effects of depth and shape of the footing.

Use of *large-scale load tests* or, ideally, full-scale load tests mitigates many of the difficulties of the preceding approach but is often precluded by costs and schedule considerations. Unless relatively uniform soil deposits are encountered, this approach also requires a number of tests, significantly increasing the cost and time requirements. (See J. K. Mitchell and W. S. Gardner, "In-Situ Measurement of Volume-Change Characteristics," ASCE Specialty Conference on In-Situ Measurement of Soil Properties, Raleigh, N.C., 1975.)

Cone Penetrometer Methods ■ Correlations between quasi-static penetration resistance q_c and observation of the settlement of bearing plates and small footings form the basis of foundation settlement estimates using CPT data. The Buisman-DeBeer method utilizes a one-dimensional compression formulation. A recommended modification of this approach that considers the influence of the relative density of the soil D_r and increased secant modulus E_s' is

$$\rho = \sum_{i=1}^{n} H_i \frac{1.15\sigma_{vo}'}{(1 + D_r^2)q_c} \log \frac{\sigma_{vo}' + \Delta\sigma_v}{\sigma_{vo}'} \qquad (7.31)$$

where ρ = estimated footing settlement. The σ_{vo}' and $\Delta\sigma_v$ parameters represent the average effective overburden pressure and vertical stress change for each layer considered below the base of the foundation (see Art 7.12). Equation (7.31) has limitations because no consideration is given to: (1) soil stress history, (2) soil gradation, and (3) three-dimensional compression. Also, Eq. (7.31) incorporates an empirical representation of E_s', given by Eq. (7.32), and has all the limitations thereof (see Art. 7.5.5).

$$E_s' = 2(1 + D_r^2)q_c \qquad (7.32)$$

The foregoing procedures are not applicable to large footings and foundation mats. From field observations relating foundation width B, meters, to ρ/B, the upper limit for ρ/B, for $B > 13.5$ m, is given, in percent, approximately by

$$\frac{\rho}{B} = 0.194 - 0.115 \log \frac{B}{10} \qquad (7.33)$$

For the same data base, the best fit of the average ρ/B measurements ranges from about 0.09% ($B = 20$ m) to 0.06% ($B = 80$ m).

Standard Penetration Resistance Methods ■ A variety of methods have been proposed to relate foundation settlement to standard penetration resistance N. An approach proposed by I. Alpan and G. G. Meyerhof appears reasonable and has the advantage of simplicity. Settlement S, in, is computed for $B < 4$ ft from

$$S = \frac{8q}{N'} \qquad (7.34a)$$

and for $B \geq 4$ ft from

$$S = \frac{12q}{N'}\left(\frac{2B}{1 + B}\right)^2 \qquad (7.34b)$$

where q = bearing capacity of soil, tons/ft^2

B = footing width, ft

N' is given approximately by Eq. (7.34c) for $\sigma_{vo}' \leq 40$ psi.

$$N' = \frac{50N}{\sigma_{vo}' + 10} \qquad (7.34c)$$

and represents N (blows per foot) normalized for $\sigma_{vo}' = 40$ psi (see Fig. 7.14).

(G. G Meyerhof, "Shallow Foundations," *ASCE Journal of Soil Mechanics and Foundation Engineering Division*, vol. 91, no. SM92, 1965; W. G. Holtz and H. J. Gibbs, "Shear Strength of Pervious Gravelly Soils," *Proceedings ASCE*, paper 867, 1956 (www.asce.org); R. B. Peck, W. E. Hanson, and T. H. Thornburn, "Foundation Engineering," John Wiley & Sons, Inc., New York, www.wiley.com.)

Laboratory Test Methods ■ The limitations in developing representative deformation parameters from reconstituted samples were described in Art. 7.5.5. A possible exception may be for the settlement analyses of foundations supported by compacted fill. Under these circumstances, consolidation tests and stress-path, triaxial shear tests on the fill materials may be appropriate for providing the parameters for application of settlement analyses described for cohesive soils.

(D. J. D'Appolonia, E. D'Appolonia, and R. F. Brisette, "Settlement of Spread Footings on Sand," *ASCE Journal of Soil Mechanics and Foundation Engineering Division*, vol. 94, no. SM3, 1968, www.asce.org.)

Deep Foundations

Subsurface conditions, structural requirements, site location and features, and economics generally dictate the type of foundation to be employed for a given structure.

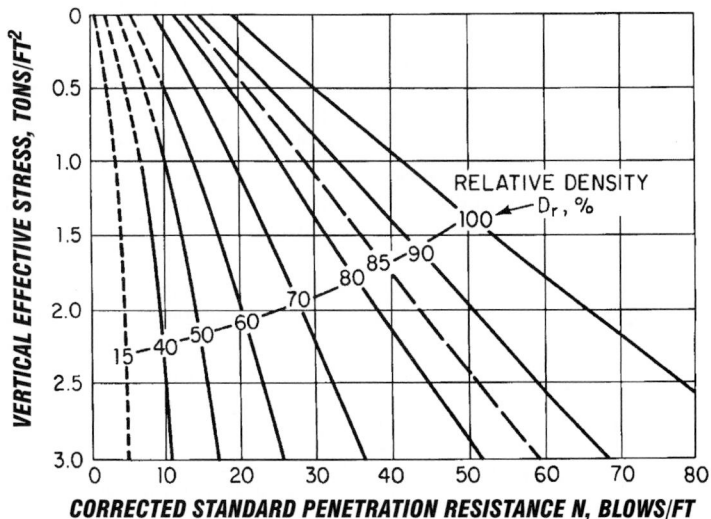

Fig. 7.14 Curves relate relative density to standard penetration resistance and effective vertical stress.

Deep foundations, such as piles, drilled shafts, and caissons, should be considered when:

Shallow foundations are inadequate and structural loads need to be transmitted to deeper, more competent soil or rock

Loads exert uplift or lateral forces on the foundations

Structures are required to be supported over water

Functionality of the structure does not allow for differential settlements

Future adjacent excavations are expected

7.14 Application of Piles

Pile foundations are commonly installed for bridges, buildings, towers, tanks, and offshore structures. Piles are of two major types: prefabricated and installed with a pile-driving hammer, or cast-in-place. In some cases, a pile may incorporate both prefabricated and cast-in-place elements. Driven piles may be made of wood, concrete, steel, or a combination of these materials. Cast-in-place piles are made of concrete that is placed into an auger-drilled hole in the ground. When the diameter of a drilled or augured cast-in-place pile exceeds about 24 in, it is then generally classified as

a drilled shaft, bored pile, or caisson (Arts. 7.15.2, 7.2, and 7.23).

The load-carrying capacity and behavior of a single pile is governed by the lesser of the structural strength of the pile shaft and the strength and deformation properties of the supporting soils. When the latter governs, piles derive their capacity from soil resistance along their shaft and under their toe. The contribution of each of these two components is largely dependent on subsurface conditions and pile type, shape, and method of installation. Piles in sand or clay deposits with shaft resistance predominant are commonly known as **friction piles**. Piles with toe resistance primary are known as **end-bearing piles**. In reality, however, most piles have both shaft and toe resistance, albeit to varying degrees. The sum of the ultimate resistance values of both shaft and toe is termed the pile **capacity**, which when divided by an appropriate safety factor yields the **allowable load** at the pile head.

The capacity of a laterally loaded pile is usually defined in terms of a limiting lateral deflection of the pile head. The ratio of the ultimate lateral load defining structural or soil failure to the associated lateral design load represents the safety factor of the pile under lateral load.

Piles are rarely utilized singly, but are typically installed in groups. The behavior of a pile in a group differs from that of the single pile. Often, the group effect dictates the overall behavior of the pile foundation system.

The following articles provide a general knowledge of pile design, analysis, construction, and testing methods. For major projects, it is advisable that the expertise of a geotechnical engineer with substantial experience with deep-foundations design, construction, and verification methods be employed.

7.15 Pile Types

Piles that cause a significant displacement of soil during installation are termed **displacement piles**. For example, closed-end steel pipes and precast concrete piles are displacement piles, whereas open-end pipes and H piles are generally **limited-displacement piles**. They may plug during driving and cause significant soil displacement. Auger-cast piles are generally considered **nondisplacement piles** since the soil is removed and replaced with concrete during pile installation.

Piles are usually classified according to their method of installation and type of material. Preformed driven piles may be made of concrete, steel, timber, or a combination of these materials.

7.15.1 Precast Concrete Piles

Reinforced or prestressed to resist handling and pile-driving stresses, precast concrete piles are usually constructed in a casting yard and transported to the jobsite. Pretensioned piles (commonly known as **prestressed piles**) are formed in very long casting beds, with dividers inserted to produce individual pile sections. Precast piles come in a variety of cross sections; for example, square, round, octagonal. They may be manufactured full length or in sections that are spliced during installation. They are suitable for use as friction piles for driving in sand or clay or as end-bearing piles for driving through soft soils to firm strata.

Prestressed concrete piles usually have solid sections between 10 and 30 in square. Frequently, piles larger than 24 in square and more than 100 ft long are cast with a hollow core to reduce pile weight and facilitate handling.

Splicing of precast concrete piles should generally be avoided. When it is necessary to extend pile length, however, any of several splicing methods may be used. Splicing can be accomplished, for instance, by installing dowel bars of sufficient length and then injecting grout or epoxy to bond them and the upper and lower pile sections. Oversize grouted sleeves may also be used. Alternatives to these bonding processes include welding of steel plates or pipes cast at pile ends. Some specialized systems employ mechanical jointing techniques using pins to make the connection. These mechanical splices reduce field splice time, but the connector must be incorporated in the pile sections at the time of casting.

All of the preceding methods transfer some tension through the splice. There are, however, systems, usually involving external sleeves (or *cans*), that do not transfer tensile forces; this is a possible advantage for long piles in which tension stresses would not be high, but these systems are not applicable to piles subject to uplift loading. For prestressed piles, since the tendons require bond-development length, the jointed ends of the pile sections should also be reinforced with steel bars to transfer the tensile forces across the spliced area.

Prestressed piles also may be posttensioned. Such piles are mostly cylindrical (typically up to 66-in diameter and 6-in wall thickness) and are centrifugally cast in sections and assembled to form the required length before driving. Stressing is achieved with the pile sections placed end to end by threading steel cables through precast ducts and then applying tension to the cables with hydraulic devices. Piles up to 200 fit long have been thus assembled and driven.

Advantages of precast concrete piles include their ability to carry high axial and inclined loads and to resist large bending moments. Also, concrete piles can be used as structural columns when extended above ground level. Disadvantages include the extra care required during handling and installation, difficulties in extending and cutting off piles to required lengths, and possible transportation difficulties. Special machines, however, are available for pile cutting, such as saws and hydraulic crushing systems. Care, however, is necessary during all stages of pile casting, handling, transportation, and installation to avoid damaging the piles.

Precast concrete piles are generally installed with pile-driving hammers. For this purpose, pile heads should always be protected with cushioning material. Usually, sheets of plywood are used. Other precautions should also be taken to protect piles during and after driving. When driving is expected to be through hard soil layers or into rock, pile toes should generally be fitted with steel shoes for reinforcement and protection from damage. When piles are driven into soils and groundwater containing destructive chemicals, special cement additives or coatings should be used to protect concrete piles. Seawater may also cause damage to concrete piles by chemical reactions or mechanical forces.

("Recommended Practice for Design, Manufacture, and Installation of Prestressed Concrete Piling," Prestressed Concrete Institute, 209 W. Jackson Blvd., Chicago, IL 60604, www.pci.org.)

7.15.2 Cast-in-Place Concrete Piles

These are produced by forming holes in the ground and then filling them with concrete. A steel cage may be used for reinforcement. There are many methods for forming the holes, such as driving of a closed-end steel pipe, with or without a mandrel. Alternatively, holes may be formed with drills or continuous-flight augers. Two common methods of construction are (1) a hole is excavated by drilling before placement of concrete to form a **bored pile**, and (2) a hole is formed with a continuous-flight auger (CFA) and grout is injected into the hole under pressure through the toe of the hollow auger stem during auger withdrawal. A modification of the CFA method is used to create a mixed-in-place concrete pile in clean granular sand. There are numerous other procedures used in constructing cast-in-place concrete piles, most of which are proprietary systems.

Advantages of cast-in-place concrete piles include: relatively low cost, fast execution, ease of adaptation to different lengths, capability for soil sampling during construction at each pile location, possibility of penetrating undesirable hard layers, high load-carrying capacity of large-size piles, and low vibration and noise levels during installation. Construction time is less than that needed for precast piles inasmuch as cast-in-place piles can be formed in place to required lengths and without having to wait for curing time before installation.

Pile foundations are normally employed where subsurface conditions are likely to be unfavorable for spread footings or mats. If cast-in-place concrete piles are used, such conditions may create concerns about the structural integrity, bearing capacity, and general performance of the pile foundation. The reason for this is that the constructed shape and structural integrity of such piles depend on subsurface conditions, concrete quality and method of placement, quality of work, and design and construction practices, all of which require tight control. Structural deficiencies may result from degraded or debonded concrete, necking, or inclusions or voids. Unlike pile driving, where the installation process itself constitutes a crude qualitative pile-capacity test and hammer-pile-soil behavior may be evaluated from measurements made during driving, methods for evaluating cast-in-place piles during construction are generally not available. Good installation procedures and inspection are critical to the success of uncased augured or drilled piles.

("Drilled Shafts: Construction Procedures and Design Methods," 1999, by Michael W. O'Neil and Lymon C. Reese, Report No. FHWA-IF-99-025, Federal Highway Administration, 400 7th Street SW, Washington, D.C. 20590 (www.fhwa.gov) various publications of The International Association of Foundation Drilling (ADSC), P.O. Box 280379, Dallas, TX 75228.)

7.15.3 Steel Piles

Structural steel H and pipe sections are often used as piles. Pipe piles may be driven open- or closed-end. After being driven, they may be filled with concrete. Common sizes of pipe piles range from 8 to 48 in in diameter. A special type of pipe pile is the Monotube, which has a longitudinally fluted wall, may be of constant section or tapered, and may be filled with concrete after being driven. Closed-end pipes have the advantage that they can be visually inspected after driving. Open-end pipes have the advantage that penetration of hard layers can be assisted by drilling through the open end.

H-piles may be rolled or built-up steel sections with wide flanges. Pile toes may be reinforced with special shoes for driving through soils with obstructions, such as boulders, or for driving to rock. If splicing is necessary, steel pile lengths may be connected with complete-penetration welds or commercially available special fittings. H piles, being low-displacement piles, are advantageous in situations where ground heave and lateral movement must be kept to a minimum.

Steel piles have the advantages of being rugged, strong, and easy to handle. They can be driven through hard layers. They can carry high compressive loads and withstand tensile loading. Because of the relative ease of splicing and cutting to length, steel piles are advantageous for use in sites where the depth of the bearing layer varies. Disadvantages of steel piles include small cross-sectional area and susceptibility to corrosion, which can cause a significant reduction in load-carrying capacity. Measures that may be taken when pile corrosion is anticipated include the use of larger pile sections than otherwise needed, use of surface-coating materials, or cathodic protection. In these cases, the pipes are usually encased in or filled with concrete.

Specifications pertaining to steel pipe piles are given in "Specification for Welded and Seamless Steel Pipe Piles," ASTM A252 (www.astm.org). For dimensions and section properties of H piles, see "HP Shapes" in "Manual of Steel Construction," American Institute of Steel Construction, One East Wacker Drive, Suite 3100, Chicago, IL 60601-2001.

7.15.4 Timber Piles

Any of a variety of wood species but usually southern pine or douglas fir, and occasionally red or white oak, can be used as piles. Kept below the groundwater table, timber piles can serve in a preserved state for a long time. Untreated piles that extend above the water table, however, may be exposed to damaging marine organisms and decay. Such damage may be prevented or delayed and service life prolonged by treating timber piles with preservatives. Preservative treatment should match the type of wood.

Timber piles are commonly available in lengths of up to 75 ft. They should be as straight as possible and should have a relatively uniform taper. Timber piles are usually used to carry light to moderate loads or in marine construction as dolphins and fender systems.

Advantages of timber piles include their relatively low cost, high strength-to-weight ratio, and ease of handling. They can be cut to length after driving relatively easily. Their naturally tapered shape (about 1 in in diameter per 10 ft of length) is advantageous in situations where pile capacities derive mostly from shaft resistance. Disadvantages include their susceptibility to damage during hard driving and difficulty in splicing.

Timber piles should be driven with care to avoid damage. Hammers with high impact velocities should not be used. Protective accessories should

be utilized, when hard driving is expected, especially at the head and toe of the pile.

Specifications relevant to timber piles are contained in "Standard Specifications for Round Timber Piles," ASTM D25; "Establishing Design Stresses for Round Timber Piles," ASTM D2899 (www.astm.org); and "Preservative Treatment by Pressure Processes," AWPA C3, American Wood Preservers Association (www.awpa.com). Information on timber piles also may be obtained from the National Timber Piling Council, Inc., 446 Park Ave., Rye, NY 10580.

7.15.5 Composite Piles

This type of pile includes those made of more than one major material or pile type, such as thick-walled, concrete-filled, steel pipe piles, precast concrete piles with steel (pipe or H section) extensions, and timber piles with cast-in-place concrete extensions.

7.15.6 Selection of Pile Type

The choice of an appropriate pile type for a particular application is essential for satisfactory foundation functioning. Factors that must be considered in the selection process include subsurface conditions, nature and magnitude of loads, local experience, availability of materials and experienced labor, applicable codes, and cost. Pile drivability, strength, and serviceability should also be taken into account. Figure 7.15 presents general

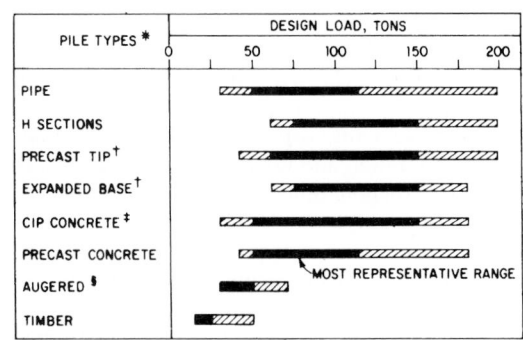

Fig. 7.15 Approximate ranges of design loads for vertical piles in axial compression.

*For shaft diameters not exceeding 18 in.
† Primary end bearing.
‡ Permanent shells only.
§ Uncased only.

guidelines and approximate ranges of design loads for vertical piles in axial compression. Actual loads that can be carried by a given pile in a particular situation should be assessed in accordance with the general methods and procedures presented in the preceding and those described in more specialized geotechnical engineering books.

7.16 Pile-Driving Equipment

Installation of piles by driving is a specialized field of construction usually performed by experienced contractors with dedicated equipment. The basic components of a pile-driving system are shown in Figure 7.16 and described in the following. All components of the driving system have some effect on the pile-driving process. The overall stability and capacity of the pile-driving crane should be assessed for all stages of loading conditions, including pile pickup and driving.

Lead ▪ The functions of the lead (also known as leader or guide) are to guide the hammer, maintain pile alignment, and preserve axial alignment between the hammer and pile. For proper functioning, leads should have sufficient strength and

be straight and well greased to allow free hammer travel.

There are four main types of leads: swinging, fixed, semifixed, and offshore. Depending on the relative positions of the crane and the pile, pile size, and other factors, a specific type of lead may have to be employed. Swinging leads are the simplest, lightest in weight, and most versatile. They do not, however, provide much fixity for prevention of lateral pile movement during driving. Fixed leads maintain the position of the pile during driving and facilitate driving a pile at an inclined angle. However, they are the most expensive type of lead. Semifixed leads have some of the advantages and disadvantages of the swinging and fixed leads. Offshore leads are used mostly in offshore construction to drive large-size steel piles and on land or near shore when a template is used to hold the pile in place. Their use for inclined piles is limited by the pile flexural strength.

Pile Cap (Helmet) ▪ The pile cap (also referred to as helmet) is a boxlike steel element inserted in the lead between the hammer and pile (Fig. 7.17). The function of the cap is to house both hammer and pile cushions and maintain axial alignment between hammer and pile. The size of the cap needed depends on the pile size and the jaw-opening size of the lead. In some cases, an adaptor is inserted under the cap to accommodate various pile sizes, assuring that the hammer and

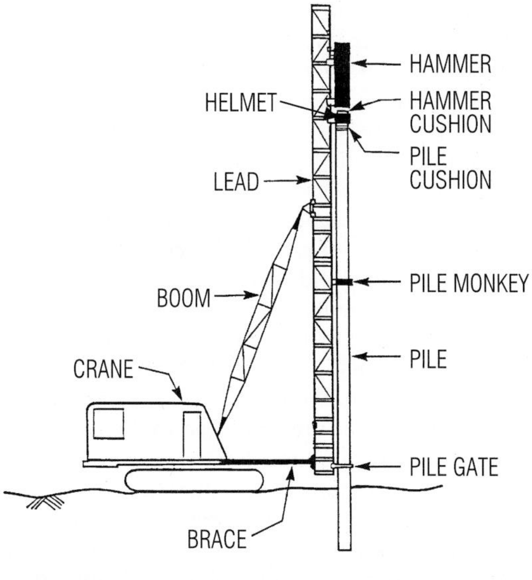

Fig. 7.16 Basic components of a pile-driving rig. (*From "The Performance of Pile Driving Systems—Inspection Manual," FHWA/RD-86/160, Federal High-way Administration.*)

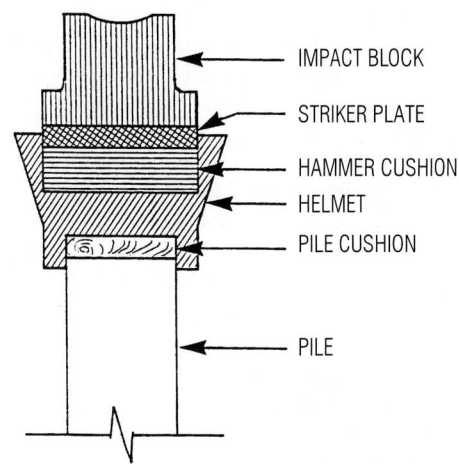

Fig. 7.17 Pile helmet and adjoining parts.

pile are concentrically aligned. A poor seating of the pile in the cap can cause pile damage and buckling due to localized stresses and eccentric loading at the pile top.

Cushions ▪ Hammers, except for some hydraulic hammers, include a cushion in the hammer (Fig. 7.17). The function of the hammer cushion is to attenuate the hammer impact forces and protect both the pile and hammer from damaging driving stresses. Normally, a steel striker plate, typically 3 in thick, is placed on top of the cushion to insure uniform cushion compression. Most cushions are produced by specialized manufacturers and consist of materials such as phenolic or nylon laminate sheets.

For driving precast concrete piles, a pile cushion is also placed at the pile top (Fig. 7.17). The most common material is plywood. It is placed in layers with total thickness between 4 and 12 in. In some cases, hardwood boards may be used (with the grain perpendicular to the pile axis) as pile cushions. Specifications often require that a fresh pile cushion be used at the start of the driving of a pile. The wood used should be dry. The pile cushion should be changed when significantly compressed or when signs of burning are evident. Cushions (hammer and pile) should be durable and reasonably able to maintain their properties. When a change is necessary during driving, the driving log should record this. The measured resistance to driving immediately thereafter should be discounted, especially if the pile is being driven close to its capacity, inasmuch as a fresh cushion will compress significantly more from a hammer blow than would an already compressed one. Hence, measurements of pile movement per blow will be different.

With the aid of a computer analytical program, such as one based on the wave equation, it is possible to design a cushion system for a particular hammer and pile that allows maximum energy transfer with minimum risk of pile damage.

Hammer ▪ This provides the energy needed for pile installation. Basically, an impact pile-driving hammer consists of a striking part, called the ram, and a means of imposing impacts in rapid succession to the pile.

Hammers are commonly rated by the amount of potential energy per blow. This energy basically is the product of ram weight and drop height (stroke). To a contractor, a hammer is a mass-production machine; hammers with higher efficiency are generally more productive and can achieve higher pile capacity. To an engineer, a hammer is an instrument that is used to measure the quality of the end product, the driven pile. Implicit assumptions regarding hammer performance are included in common pile evaluation procedures. Hammers with low energy transfer are the source of poor installation. Hence, pile designers, constructors, and inspectors should be familiar with operating principles and performance characteristics of the various hammer types. Following are brief discussions of the major types of impact pile-driving hammers.

Impact pile-driving hammers rely on a falling mass to create forces much greater than their weight. Usually, strokes range between 3 and 10 ft.

These hammers are classified by the mode used in operating the hammer; that is, the means used to raise the ram after impact for a new blow. There are two major modes: external combustion and internal combustion. Hammers of each type may be single or double acting. For single-acting hammers (Fig. 7.18), power is only needed to raise the ram, whereas the fall is entirely by gravity. Double-acting hammers also apply power to assist the ram during downward travel. Thus these hammers deliver more blows per minute than single-acting hammers; however, their efficiency may be lower, since the power source supplies part of the impact energy.

External-combustion hammers (ECHs) rely on a power source external to the hammer for their operation. One type is the drop hammer, which is raised by a hoist line from the crane supporting the pile and leads and then dropped to fall under the action of gravity to impact the pile. Main advantages of drop hammers are relatively low cost and maintenance and the ability to vary the stroke easily. Disadvantages include reduction of the effectiveness of the drop due to the cable and winch assembly required for the operation, slow operation, and hammer efficiency dependence on operator's skills. (The operator must allow the cable to go slack when the hammer is raised to drop height.) Consequently, use of drop hammers is generally limited to small projects involving lightly loaded piles or sheetpiles.

For some pile drivers, hydraulic pressure is used to raise the ram. The hammers, known as air/steam

Fig. 7.18 Single-acting, external-combustion hammer driving a precast concrete pile.

or hydraulic hammers, may be single or double acting. Action starts with introduction of the motive fluid (steam, compressed air, or hydraulic fluid) under the piston in the hammer chamber to lift the ram. When the ram attains a prescribed height, flow of the motive fluid is discontinued and the ram "coasts" against gravity up to the full stroke. At top of stroke, the pressure is vented and the ram falls under gravity. For double-acting hammers, the pressure is redirected to act on top of the piston and push the ram downward during its fall. Many hydraulic hammers are equipped with two stroke heights for more flexibility.

The next cycle starts after impact, and the start should be carefully controlled. If pressure is introduced against the ram too early, it will slow down the ram excessively and reduce the energy available to the pile. Known as **preadmission**, this is not desirable due to its adverse effect on energy transfer. For some hammers, the ram, immediately preceding impact, activates a valve to allow the

motive fluid to enter the cylinder to start the next cycle. For most hydraulic hammers, the ram position is detected by proximity switches and the next cycle is electronically controlled.

Main advantages of external combustion hammers include their higher rate of operation than drop hammers, long track record of performance and reliability, and their relatively simple design. Disadvantages include the need to have additional equipment on site, such as boilers and compressors, that would not be needed with another type of hammer. Also disadvantageous is their relatively high weight, which requires equipment with large lifting capacity.

Diesel hammers are internal-combustion hammers (ICHs). The power needed for hammer operation comes from fuel combustion inside the hammer, therefore eliminating the need for an outside power source. Basic components of a diesel hammer include the ram, cylinder, impact block, and fuel distribution system. Hammer operation is started by lifting the ram with one of the hoist lines from the crane or a hydraulic jack to a preset height. A tripping mechanism then releases the ram, allowing it to fall under gravity. During its descent, the ram closes cylinder exhaust ports, as a result of which gases in the combustion chamber are compressed. At some point before impact, the ram activates a fuel pump to introduce into the combustion chamber a prescribed amount of fuel in either liquid or atomized form. The amount of fuel depends on the fuel pump setting.

For liquid-injection hammers, the impact of the ram on the impact block atomizes the fuel. Under the high pressure, ignition and combustion result. For atomized-fuel-injection hammers, ignition occurs when the pressure reaches a certain threshold before impact. The ram impact and the explosive force of the fuel drive the pile into the ground while the explosion and pile reaction throw the ram upward past the exhaust ports, exhausting the combustion gases and drawing in fresh air for the next cycle. With an open-end diesel (OED), shown in Fig. 7.19, the ram continues to travel upward until arrested by gravity. Then the next cycle starts. The distance that the ram travels upward (stroke) depends on the amount of fuel introduced into the chamber (fuel-pump setting), cushions, pile stiffness, and soil resistance. In the case of closed-end diesel hammers (CEDs), the top of the cylinder is closed, creating an air-pressure, or bounce, chamber. Upward movement of the ram compresses the

Fig. 7.19 Open-end, single-acting diesel hammer driving a pile.

air in the bounce chamber and thus stores energy. The pressure shortens ram stroke and the stored energy accelerates the ram downward.

The energy rating of diesel hammers is commonly appraised by observing the ram stroke (or bounce-chamber pressure for closed-end hammers). This is an important indication but can be misleading; for example, when the hammer becomes very hot during prolonged driving. Because the fuel then ignites too early, the ram expends more energy to compress the gases and less energy is available for transmission into the pile. The high pressure still causes a relatively high stroke. This condition is commonly referred to as preignition. In contrast, short ram strokes may be caused by lack of fuel or improper fuel type, lack of compression in the chamber due to worn piston rings, excessive ram friction, pile stiffness, or lack of soil resistance.

Internal-combustion hammers are advantageous because they are entirely self-contained. They are relatively lightweight and thus permit use of smaller cranes than those required for external-combustion hammers. Also, stroke adjustment to soil resistance with internal-combustion hammers is advantageous in controlling dynamic stresses during driving of concrete piles. Among disadvantages is stroke dependence on the hammer-pile-soil system, relatively low blow rate, and potential cessation of operation when easy driving is encountered.

Table 7.8 presents the characteristics of impact pile-driving hammers. The hammers are listed by rated energy in ascending order. The table indicates for each model the type of hammer: ECH, external combustion hammer, or OED, open-end diesel hammer; manufacturer; model number; ram weight; and equivalent stroke. Note, however, that new models of hammers become available at frequent intervals.

Vibratory hammers drive or extract piles by applying rapidly alternating forces to the pile. The forces are created by eccentric weights (eccentrics) rotating around horizontal axes. The weights are placed in pairs so that horizontal centrifugal forces cancel each other, leaving only vertical-force components. These vertical forces shake piles up and down and cause vertical pile penetration under the weight of the hammer. The vibration may be either low frequency (less than 50 Hz) or high frequency (more than 100 Hz).

The main parameters that define the characteristics of a vibratory hammer are amplitude produced, power consumption, frequency (vibrations per minute), and driving force (resultant vertical force of the rotating eccentrics). Vibratory hammers offer the advantages of fast penetration, limited noise, minimal shock waves induced in the ground, and usually high penetration efficiency in cohesionless soils. A disadvantage is limited penetration capacity under hard driving conditions and in clay soils. Also, there is limited experience in correlating pile capacity with driving energy and penetration rate. This type of hammer is often used to install non-load-bearing piles, such as retaining sheetpiles.

("Vibratory Pile Driving," J. D. Smart, Ph.D. Thesis, University of Illinois, Urbana, 1969; "Driveability and Load Transfer Characteristics of Vibro-Driven Piles," D. Wong, Ph.D Thesis, University of Houston, Texas, 1988; various publications of the

Table 7.8 Impact Pile-Driving Hammers

Rated Energy kip-ft	Manufacturer	Hammer Model	Ram Weight, kips	Equivalent Stroke, ft	Hammer Type
1.00	MKT	No. 5	0.20	5.00	ECH
2.50	MKT	No. 6	0.40	6.25	ECH
4.15	MKT	No. 7	0.80	5.19	ECH
7.26	VULCAN	VUL 30C	3.00	2.42	ECH
7.26	VULCAN	VUL 02	3.00	2.42	ECH
8.10	LINKBELT	LB 180	1.73	4.68	CED
8.13	ICE	180	1.73	4.70	CED
8.23	DELMAG	D 5	1.10	7.48	OED
8.65	DAWSON	HPH 1200	2.29	3.77	ECH
8.75	MKT	9B3	1.60	5.47	ECH
10.50	DELMAG	D 6-32	1.32	7.94	OED
13.11	MKT	10B3	3.00	4.37	ECH
14.20	MKT	C5-Air	5.00	2.84	ECH
15.00	CONMACO	C 50	5.00	3.00	ECH
15.00	RAYMOND	R 1	5.00	3.00	ECH
15.00	VULCAN	VUL 01	5.00	3.00	ECH
15.02	LINKBELT	LB 312	3.86	3.89	CED
15.10	VULCAN	VUL 50C	5.00	3.02	ECH
16.00	MKT	DE 20	2.00	8.00	OED
16.20	MKT	C5-Steam	5.00	3.24	ECH
16.25	MKT	S-5	5.00	3.25	ECH
17.32	DAWSON	HPH 2400	4.19	4.13	ECH
17.34	BANUT	3 Tonnes	6.61	2.62	ECH
17.60	DELMAG	D 8-22	1.76	10.00	OED
18.00	BERMINGH	B200	2.00	9.00	OED
18.20	LINKBELT	LB 440	4.00	4.55	CED
18.56	ICE	440	4.00	4.64	CED
19.15	MKT	11B3	5.00	3.83	ECH
19.18	VULCAN	VUL 65C	6.50	2.95	ECH
19.50	CONMACO	C 65	6.50	3.00	ECH
19.50	RAYMOND	R 65C	6.50	3.00	ECH
19.50	RAYMOND	R 1S	6.50	3.00	ECH
19.50	RAYMOND	R 65CH	6.50	3.00	ECH
19.50	VULCAN	VUL 06	6.50	3.00	ECH
19.57	VULCAN	VUL 65CA	6.50	3.01	ECH
20.00	MKT 20	DE333020	2.00	10.00	OED
21.00	MKT	DA 35B	2.80	7.50	CED
21.20	MKT	C826 Air	8.00	2.65	ECH

Rated Energy kip-ft	Manufacturer	Hammer Model	Ram Weight, kips	Equivalent Stroke, ft	Hammer Type
22.13	IHC Hydh	SC 30	3.64	6.08	ECH
22.40	MKT	DE 30	2.80	8.00	OED
22.50	FEC	FEC 1200	2.75	8.18	OED
22.50	ICE	30 S	3.00	7.50	OED
22.99	BERMINGH	B23	2.80	8.21	CED
22.99	BERMINGH	B23 5	2.80	8.21	CED
23.12	ICE	422	4.00	5.78	CED
23.14	BANUT	4 Tonnes	8.82	2.62	ECH
23.59	DELMAG	D 12	2.75	8.58	OED
23.80	MKT	DA35B SA	2.80	8.50	OED
23.80	MKT	DE30B	2.80	8.50	OED
24.38	RAYMOND	R 0	7.50	3.25	ECH
24.40	MKT	C826 Stm	8.00	3.05	ECH
24.48	RAYMOND	R 80CH	8.00	3.06	ECH
24.48	RAYMOND	R 80C	8.00	3.06	ECH
24.48	VULCAN	VUL 80C	8.00	3.06	ECH
24.76	MENCK	MHF3-3	7.05	3.51	CED
24.88	UDDCOMB	H3H	6.60	3.77	ECH
26.00	ICE	32S	3.00	8.67	OED
28.14	MITSUB.	MH 15	3.31	8.50	OED
28.31	DELMAG	D 15	3.30	8.58	OED
28.92	BANUT	5 Tonnes	11.02	2.62	ECH
29.25	BERMINGH	B225	3.00	9.75	OED
29.48	IHC Hydh	SC 40	5.51	5.35	ECH
30.36	IHC Hydh	S 40	5.51	5.51	ECH
30.37	ICE	520	5.07	5.99	CED
30.41	HERA	1500	3.37	9.02	OED
30.72	MKT	DA 45	4.00	7.68	CED
30.80	MKT	MS-350	7.72	3.99	ECH
30.96	MENCK	MHF3-4	8.82	3.51	ECH
31.33	DELMAG	D 12-32	2.82	11.11	OED
32.00	MKT	DE 40	4.00	8.00	OED
32.50	CONMACO	C 100	10.00	3.25	ECH
32.50	CONMACO	C565	6.50	5.00	ECH
32.50	HPSI	650	6.50	5.00	ECH
32.50	MKT	S 10	10.00	3.25	ECH
32.50	RAYMOND	R 2/0	10.00	3.25	CED
32.50	VULCAN	VUL 506	6.50	5.00	ECH

Rated Energy kip-ft	Manufacturer	Hammer Model	Ram Weight, kips	Equivalent Stroke, ft	Hammer Type
32.50	VULCAN	VUL 010	10.00	3.25	ECH
32.55	FAIRCHLD	F-32	10.85	3.00	ECH
32.90	VULCAN	VUL 100C	10.00	3.29	ECH
33.00	MKT 33	DE333020	3.30	10.00	OED
33.18	UDDCOMB	H4H	8.80	3.77	ECH
34.72	BANUT	6 Tonnes	13.23	2.62	ECH
34.72	Pilemer	DKH-4	8.82	3.95	ECH
35.37	JUNTTAN	HHK 4	8.82	4.01	ECH
35.40	BERMINGH	B250 5	3.00	11.80	OED
35.98	VULCAN	VUL 140C	14.00	2.57	ECH
37.38	CONMACO	C 115	11.50	3.25	ECH
37.52	MKT	S 14	14.00	2.68	ECH
37.72	ICE	115	11.50	3.28	ECH
38.20	MKT	DA 55B	5.00	7.64	CED
38.69	MENCK	MHF3-5	11.02	3.51	ECH
38.69	MENCK	MHF5-5	11.02	3.51	ECH
39.00	VULCAN	VUL 012	12.00	3.25	ECH
39.25	DELMAG	D 16-32	3.52	11.15	OED
40.00	CONMACO	C 80E5	8.00	5.00	ECH
40.00	ICE	40-S	4.00	10.00	OED
40.00	MKT	DA55B SA	5.00	8.00	OED
40.00	MKT 40	DE333020	4.00	10.00	OED
40.00	VULCAN	VUL 508	8.00	5.00	ECH
40.31	BERMINGH	B300	3.75	10.75	OED
40.31	BERMINGH	B300 M	3.75	10.75	OED
40.49	BANUT	7 Tonnes	15.43	2.62	ECH
40.61	DELMAG	D 22	4.91	8.27	OED
40.62	ICE	640	6.00	6.77	CED
40.63	RAYMOND	R 3/0	12.50	3.25	ECH
41.47	UDDCOMB	H5H	11.00	3.77	ECH
42.00	CONMACO	C 140	14.00	3.00	ECH
42.00	ICE	42-S	4.09	10.27	OED
42.00	VULCAN	VUL 014	14.00	3.00	ECH
42.40	DELMAG	D 19-32	4.00	10.60	OED
42.50	MKT	DE 50B	5.00	8.50	OED
43.01	MITSUB.	M 23	5.06	8.50	OED
43.20	BERMINGH	B400 4.8	4.80	9.00	OED
43.37	BSP	HH 5	11.02	3.94	ECH

(Table continued)

Table 7.8 (Continued)

Rated Energy kip-ft	Manufacturer	Hammer Model	Ram Weight, kips	Equivalent Stroke, ft	Hammer Type
43.40	Pilemer	DKH-5	11.02	3.95	ECH
44.00	HPSI	110	11.00	4.00	ECH
44.00	MKT	MS 500	11.00	4.00	ECH
44.23	JUNTTAN	HHK 5	11.03	4.01	ECH
44.24	IHC Hydh	SC60	7.72	5.73	ECH
44.26	IHC Hydh	S. 60	13.23	3.35	ECH
45.00	BERMINGH	B400 5.0	5.00	9.00	OED
45.00	FAIRCHLD	F-45	15.00	3.00	ECH
45.07	MENCK	MRBS 500	11.02	4.09	ECH
45.35	KOBE	K22-Est	4.85	9.35	OED
46.43	MENCK	MHF5-6	13.23	3.51	ECH
46.43	MENCK	MHF3-6	13.23	3.51	ECH
46.84	MITSUB.	MH 25	5.51	8.50	OED
47.20	BERMINGH	B350 5	4.00	11.80	OED
48.50	DELMAG	D 22-13	4.85	10.00	OED
48.50	DELMAG	D 22-02	4.85	10.00	OED
48.75	CONMACO	C 160	16.25	3.00	ECH
48.75	RAYMOND	R 4/0	15.00	3.25	ECH
48.75	RAYMOND	R 150C	15.00	3.25	ECH
48.75	VULCAN	VUL 016	16.25	3.00	ECH
49.18	MENCK	MH 68	7.72	6.37	ECH
49.76	UDDCOMB	H6H	13.20	3.77	ECH
50.00	CONMACO	C 100E5	10.00	5.00	ECH
50.00	FEC	FEC 2500	5.50	9.09	OED
50.00	HPSI	1000	10.00	5.00	ECH
50.00	MKT 50	DE70/50B	5.00	10.00	OED
50.00	VULCAN	VUL 510	10.00	5.00	ECH
50.20	VULCAN	VUL 200C	20.00	2.51	ECH
50.69	HERA	2500	5.62	9.02	OED
51.26	DELMAG	D 22-23	4.85	10.57	OED
51.52	KOBE	K 25	5.51	9.35	CED
51.63	ICE	660	7.57	6.82	CED
51.63	LINKBELT	LB 660	7.57	6.82	ECH
51.65	IHC Hydh	S70	7.72	6.69	ECH
51.78	CONMACO	160 **	17.26	3.00	ECH
53.05	JUNTTAN	HHK 6	13.23	4.01	ECH
53.75	BERMINGH	B400	5.00	10.75	OED
54.17	MENCK	MHF3-7	15.43	3.51	ECH
54.17	MENCK	MHF5-7	15.43	3.51	ECH
55.99	FEC	FEC 2800	6.16	9.09	OED
56.77	HERA	2800	6.29	9.02	OED
56.88	RAYMOND	R 5/0	17.50	3.25	ECH
57.50	CONMACO	C 115E5	11.50	5.00	ECH
58.90	IHC Hydh	SC 80	11.24	5.24	ECH
59.00	BERMINGH	B400 5	5.00	11.80	OED
59.50	MKT	DE 70B	7.00	8.50	OED
59.60	DELMAG	D 30	6.60	9.03	OED
60.00	CONMACO	C 200	20.00	3.00	OED
60.00	HPSI	1500	15.00	4.00	ECH
60.00	ICE	60S	7.00	8.57	OED
60.00	MKT	S 20	20.00	3.00	ECH
60.00	VULCAN	VUL 320	20.00	3.00	ECH
60.00	VULCAN	VUL 512	12.00	5.00	ECH
60.00	VULCAN	VUL 020	20.00	3.00	ECH
60.76	Pilemer	DKH-7	15.43	3.95	ECH
60.78	BSP	HH 7	15.43	3.94	ECH
61.49	DELMAG	D 25-32	5.51	11.16	OED
61.71	MITSUB.	M 33	7.26	8.50	OED
61.91	JUNTTAN	HHK 7	15.44	4.01	ECH
61.91	MENCK	MHF5-8	17.64	3.51	ECH
62.50	CONMACO	C 125E5	12.50	5.00	ECH
63.03	FEC	FEC 3000	6.60	9.55	OED
64.00	ICE	160	16.00	4.00	ECH
65.62	MITSUB.	MH 35	7.72	8.50	OED
66.00	DELMAG	D 30-02	6.60	10.00	OED
66.00	DELMAG	D 30-13	6.60	10.00	OED
66.36	IHC Hydh	S90	9.92	6.69	ECH
67.77	MENCK	MRBS 750	16.53	4.10	ECH
69.34	UDDCOMB	H8H	17.60	3.94	ECH
69.43	MENCK	MH 96	11.02	6.30	ECH
69.50	BSP	HH 8	17.64	3.94	ECH
69.65	MENCK	MHF5-9	19.84	3.51	ECH
70.00	MKT 70	DE70/50B	7.00	10.00	OED
70.96	HERA	3500	7.87	9.02	OED
72.18	KOBE	K 35	7.72	9.35	OED
72.60	ICE	1070	10.00	7.26	CED
73.00	FEC	FEC 3400	7.48	9.76	OED
73.66	DELMAG	D 30-32	6.60	11.16	OED
73.66	DELMAG	D 30-23	6.60	11.16	OED
75.00	RAYMOND	R 30X	30.00	2.50	ECH
77.39	MENCK	MHF5-10	22.04	3.51	ECH
77.42	IHC Hydh	SC 110	15.21	5.09	ECH
77.88	BERMINGH	B450 5	6.60	11.80	OED
78.17	BSP	HH 9	19.84	3.94	ECH
80.00	HPSI	2000	20.00	4.00	ECH
80.00	ICE	80S	8.00	10.00	OED
80.41	MITSU8.	M 43	9.46	8.50	OED
81.25	RAYMOND	R 8/0	25.00	3.25	ECH
83.82	DELMAG	D 36-13	7.93	10.57	OED
83.82	DELMAG	D 36-02	7.93	10.57	OED
83.82	DELMAG	D 36	7.93	10.57	OED
85.13	MENCK	MHF5-11	24.25	3.51	ECH
85.43	MITSUB.	MH 45	10.05	8.50	OED
86.80	Pilemer	DKH-10	22.04	3.95	ECH
86.88	UDDCOMB	H10H	22.05	3.94	ECH
88.00	BERMINGH	B550 C	11.00	8.00	OED
88.42	JUNTTAN	HHK 10	22.05	4.01	ECH
88.50	DELMAG	D 36-32	7.93	11.16	OED
88.50	DELMAG	D 36-23	7.93	11.16	OED
90.00	HPSI	225	22.50	4.00	ECH
90.00	ICE	90-S	9.00	10.00	OED
90.00	VULCAN	VUL 330	30.00	3.00	ECH
90.00	VULCAN	VUL 030	30.00	3.00	ECH
90.44	DELMAG	D 44	9.50	9.52	OED
92.04	BERMINGH	B500 5	7.80	11.80	OED

92.75	KOBE	K 45	9.92	9.35	OED
92.87	MENCK	MHF5-12	26.45	3.51	ECH
93.28	MENCK	MRBS 850	18.96	4.92	ECH
93.28	MENCK	MRBS 800	18.98	4.92	ECH
95.54	BSP	HH 11	24.25	3.94	ECH
96.53	DELMAG	D 46-13	10.14	9.52	OED
100.00	CONMACO	C 5200	20.00	5.00	ECH
100.00	ICE	100	10.00	10.00	OED
100.00	RAYMOND	R 40X	40.00	2.50	ECH
100.00	VULCAN	VUL 520	20.00	5.00	ECH
101.37	HERA	5000	11.24	9.02	OED
103.31	IHC Hydh	SC 150	24.25	4.26	ECH
104.80	MENCK	MH 145	16.53	6.34	ECH
106.10	JUNTTAN	HHK 12	26.46	4.01	ECH
106.20	BERMINGH	B550 5	9.00	11.80	OED
107.18	DELMAG	D 46-02	10.14	10.57	OED
107.18	DELMAG	D 46	10.14	10.57	OED
107.18	DELMAG	D 46-23	10.14	10.57	OED
110.00	MKT 110	DE110150	11.00	10.00	OED
112.84	Pilemer	DKH-13	28.66	3.95	ECH
113.16	DELMAG	D 46-32	10.14	11.16	OED
113.60	VULCAN	VUL 400C	40.00	2.84	ECH
115.57	HERA	5700	12.81	9.02	OED
116.04	MENCK	MHF10-15	33.06	3.51	ECH
120.00	ICE	120S	12.00	10.00	OED
120.00	VULCAN	VUL 340	40.00	3.00	ECH
120.00	VULCAN	VUL 040	40.00	3.00	ECH
121.59	BSP	HH 14	30.86	3.94	ECH
123.43	MENCK	MRBS110	24.25	5.09	ECH
123.79	JUNTTAN	HHK 14	30.87	4.01	ECH
124.53	DELMAG	D 55	11.86	10.50	OED
125.70	HERA	6200	13.93	9.02	OED
130.18	KOBE	KB 60	13.23	9.84	OED
132.50	ICE	120S-15	15.00	8.83	OED
135.15	MITSUB.	MH 72B	15.90	8.50	OED
135.59	MENCK	MRBS150	33.07	4.10	ECH
138.87	BSP	HH 16	35.27	3.94	ECH
138.87	Pilemer	DKH-16	35.27	3.95	ECH
141.12	MENCK	MH 195	22.05	6.40	ECH
147.18	IHC Hydh	S 200	22.00	6.69	ECH
147.38	MENCK	MHUT 200	26.46	5.57	ECH
149.60	MITSUB.	MH 80B	17.60	8.50	OED
150.00	CONMACO	C 5300	30.00	5.00	ECH
150.00	MKT 150	DE110150	15.00	10.00	OED
150.00	RAYMOND	R 60X	60.00	2.50	ECH
150.00	VULCAN	VUL 530	30.00	5.00	ECH
151.00	IHC Hydh	SC 200	30.20	5.00	ECH
152.06	HERA	7500	16.85	9.02	OED
152.45	DELMAG	D 62-02	13.66	11.16	ECH
152.45	DELMAG	D 62-22	13.66	11.16	ECH
152.45	DELMAG	D 62-12	13.66	11.16	OED
154.32	HPSI	3005	30.86	5.00	OED
154.69	MENCK	MHF10-20	44.07	3.51	OED
170.00	ICE	205S	20.00	8.50	OED
173.58	BSP	HH 20S	44.09	3.94	OED
173.60	Pilemer	DKH-20	44.10	3.95	OED
173.58	KOBE	KB 80	17.64	9.84	ECH
176.32	HPSI	3505	35.26	5.00	OED
176.97	IHC Hydh	SC 250	39.24	4.51	ECH
178.42	HERA	8800	19.78	9.02	OED
179.16	VULCAN	VUL 600C	60.00	2.99	ECH
180.00	VULCAN	VUL 360	60.00	3.00	OED
180.00	VULCAN	VUL 060	60.00	3.00	ECH
184.64	IHC Hydh	S 250	27.60	6.69	ECH
186.24	DELMAG	D 80-12	17.62	10.57	ECH
189.81	MENCK	MRBS180	38.58	4.92	ECH
196.64	DELMAG	D 80-23	17.62	11.16	ECH
200.00	VULCAN	VUL 540	40.90	4.89	OED
206.51	IHC Hydh	S 280	29.80	6.93	OED
225.00	CONMACO	C 5450	45.00	5.00	ECH
225.95	MENCK	MRBS250	55.11	4.10	ECH
225.95	MENCK	MRBS250	55.11	4.10	OED
245.85	DELMAG	D100-13	22.03	11.16	ECH
260.37	BSP	HA 30	66.13	3.94	ECH
262.11	MENCK	MRBS250	63.93	4.10	ECH
289.55	MENCK	MHU 400	50.71	5.71	ECH
294.60	IHC Hydh	S 400	44.30	6.65	ECH
295.12	MENCK	MHUT 400	52.70	5.60	ECH
300.00	VULCAN	VUL 3100	100.00	3.00	ECH
300.00	VULCAN	VUL 560	62.50	4.80	ECH
325.36	MENCK	MRBS300	66.13	4.92	ECH
347.16	BSP	HA 40	88.18	3.94	ECH
350.00	CONMACO	C 5700	70.00	5.00	ECH
368.30	IHC Hydh	S 500	55.30	6.66	ECH
368.55	MENCK	MHUT 500	59.54	6.19	ECH
433.64	MENCK	MHU 600	77.16	5.62	ECH
498.94	MENCK	MRBS460	101.41	4.92	ECH
500.00	VULCAN	VUL 5100	100.00	5.00	ECH
510.00	CONMACO	C 6850	85.00	6.00	ECH
513.34	MENCK	MRBS390	86.86	5.91	ECH
516.13	MENCK	MHUT700	92.83	5.56	ECH
542.33	MENCK	MRBS500	110.23	4.92	ECH
589.85	IHC Hydh	S 800	81.57	7.23	ECH
619.18	MENCK	MHUT700	92.83	6.67	ECH
631.40	MENCK	MRBS700	154.00	4.10	ECH
736.91	IHCHydh	MHUT100	132.30	5.57	ECH
737.26	IHCHydh	S 1000	101.41	7.27	ECH
737.70	MENCK	MHU 1000	126.97	5.81	ECH
750.00	VULCAN	VUL 5150	150.00	5.00	ECH
759.23	MENCK	MRBS600	132.27	5.74	ECH
867.74	MENCK	MRBS800	176.37	4.92	ECH
954.53	MENCK	MRBS880	194.01	4.92	ECH
1177.80	IHC Hydh	S 1600	156.00	7.55	ECH
1228.87	MENCK	MHU 1700	207.23	5.93	ECH
1547.59	MENCK	MHU 2100	255.80	6.05	ECH
1581.83	MENCK	MBS12500	275.58	5.74	ECH
1694.97	IHC Hydh	S 2300	226.60	7.48	ECH
1800.00	VULCAN	VUL 6300	300.00	6.00	ECH
2171.65	MENCK	MHU 3000	363.76	5.97	ECH
2210.12	IHC Hydh	S 3000	332.00	6.66	ECH

Deep Foundations Institute, 120 Charolette Place, Englewood Cliffs, NJ 07632).

Other Pile Driving Accessories ▪ In addition to the basic equipment discussed in the preceding, some pile driving requires employment of special accessories, such as an adaptor, follower, mandrel, auger, or water jet.

An **adaptor** is inserted between a pile helmet and pile head to make it possible for one helmet to accommodate different pile sizes.

A **follower** is usually a steel member used to extend a pile temporarily in cases where it is necessary to drive the pile when the top is below ground level or under water. For efficiency in transmitting hammer energy to the pile, stiffness of the follower should be nearly equal to that of the pile. The follower should be integrated into the driving system so that it maintains axial alignment between hammer and pile.

Mandrels are typically used to drive steel shells or thin-wall pipes that are later filled with concrete. A mandrel is a uniform or tapered, round steel device that is inserted into a hollow pile to serve as a rigid core during pile driving.

Water jets or **augers** are sometimes needed to advance a pile tip through some intermediate soil layers. Jet pipes may be integrated into the pile shaft or may be external to the pile. Although possibly advantageous in assisting in pile penetration, jetting may have undesirable effects on pile capacity (compression and particularly uplift) that should be considered by the engineer.

(Department of Transportation Federal Highway Administration, "The Performance of Pile Driving Systems: Inspection Manual," FHWA Report No. FHWA/RD-86/160, National Technical Information Service, Springfield, VA 22161, www.ntis.gov.)

7.17 Vibration and Noise

With every hammer blow, pile driving produces vibration and noise effects that may extend a long distance from the source. Hundreds, or even thousands, of hammer blows are typically needed to drive a single pile. These environmental factors are increasingly becoming an issue concerning pile driving activities, particularly in urban areas.

Careful planning and execution of pile driving can limit the potential for real damage, and also for litigation based on human perception.

A significant portion of the pile driving hammer energy radiates from the pile into the surrounding ground and propagates away as seismic stress waves. The nature (transient or steady state) and characteristics (frequency, amplitude, velocity, attenuation, etc.) of these traveling waves depend on the type and size of hammer (impact or vibratory), pile (displacement or non-displacement, impedance, length etc.), and subsurface soil conditions. The resulting motion can adversely affect structures above or below ground surface, underground utilities, sensitive equipment and processes, and annoy the public. Structural damage may range from superficial plaster cracking to failure of structural elements. Experience has shown that damage to structures is not likely to occur at a distance greater than the pile embedded depth, or 50 feet minimum (Wood, 1997). In situations where liquefaction or shakedown settlement of loose granular materials may occur, pile driving vibration effects may extend to more than 1000 feet. The following information should be recorded as part of a survey of the pile driving site and surrounding area: distance to the nearest structure or underground utility, function and condition of nearby structures and facilities, and ground conditions of site and vicinity. The Florida Department of Transportation, for example, requires the monitoring of structures for settlement by recording elevations to 0.001 foot within a distance, in feet, of pile driving operations equal to 0.5 times the square root of the hammer energy, in foot-pounds.

People are much more sensitive to ground vibrations than structures. Since they can become annoyed with vibrations that are only 1/100th of those that might be harmful to most structures, human sensitivity should not be used as a measure of vibration for engineering purposes. Vibration limits to prevent damage and human discomfort are not clearly and firmly established. Pile driving vibrations are typically measured by monitoring ground peak particle velocity (ppv) using specialty equipment. Published limits range from 0.2 inch/second (for historical buildings) to 2 inch/second (for industrial structures); the Florida Department of Transportation typically uses 0.5 inch/second as a limiting value. The prediction of the vibration level which may be induced for a particular combination of hammer, pile, and soil is fraught with difficulties, nevertheless, the literature contains several equations for computing predicted peak particle velocity (Wiss, 1981). Vibration mitigation measures include: active isolation screening by means of a wave barrier near the pile driving location, passive isolation screening by means of a wave barrier near the target affected structure, and pile driving operation controls (jetting, predrilling, change of piling system, hammer type, pile top cushions, and driving sequence). Wave barriers (trenches or sheet pile wall) are attractive, but they are expensive and difficult to design and implement. Design parameters are available in the technical literature (Wood, 1968).

Noise annoys, frustrates and angers people. During the last 30 years there has been increasing concern with the quality of the environment. Through the Noise Control Act of 1972, the United States Congress directed the Environmental Protection Agency (EPA) to publish information about all identifiable effects of noise and to define acceptable levels which would protect the public health and welfare. Impact pile driving is inherently noisy, perhaps the noisiest of all construction operations. Noise is an environmental issue in populated areas, and should be of concern to those involved in pile driving activities. The usual unit of sound measurement is the decibel (dB), and one decibel is the lowest value a normal human ear can detect. The ear also registers pitch in addition to loudness. It is sensitive to frequencies of 0.01 to 16 kHz and most responsive at about 1 to 5 kHz. Measuring devices take this into account, by filtering components outside the most responsive range, and express results in dB(A). Thus, sound intensity, noise, is typically recorded in dB(A). The scale is logarithmic. Apparent loudness doubles for each 10 decibel increment. Typical values are: noisy factory 90 dB(A), busy street 85 dB(A), radio at full volume 70 dB(A), and normal speech 35 dB(A). Most people become annoyed by steady levels above 55 dB(A). Levels below 80 dB(A) would not cause hearing loss, sustained exposure to levels above 90 dB(A) cause physical and mental discomfort and result in permanent hearing damage.

Sources of noise in pile driving operations include hammer (or related equipment) exhaust, impact of hammer, and noise radiating from the pile itself. At distances of 10 to 100 feet, pile driving generally produce levels between 75 to 115 dB(A). Typically, a distance of about 300 feet would be needed for the noise level to be below the OSHA allowed 8-hour exposure (90 dB(A)), and the sound from the noisiest hammer/pile would have to travel miles before decreasing to below moderately annoying levels. Sound level drops by about 6 dB(A) as the distance doubles from the source.

Acoustic shrouds or curtain enclosures have been successfully employed to reduce the pile driving noise, by 15 to 30 dB(A), to below annoying levels. A reduction of 30 dB(A) would make the noisiest pile driving operation acceptable to most people located farther than 500 feet from the pile driving operation.

Smoke is another environmental factor to be considered in planning and executing a pile driving project, especially in metropolitan areas. Sources of smoke are the exhaust from the hammer itself (diesel hammers), the external combustion equipment such as boiler (steam hammers), compressor (air hammers), or power pak (hydraulic hammers). Some modern internal combustion hammers use environmentally friendly fuels.

(Wood, R. D., "Dynamic Effects of Pile Installations on Adjacent Structures", Synthesis of Highway Practice No. 253, National Cooperative Highway Research Program, National Academy Press, Washington, D.C., 1997, 86 pages; Wood, R. D. (1968), "Screening of elastic waves by trenches", ASCE Journal of the Soil Mechanics and Foundations Division, vol. 94, pp 951–979; Wiss, J. F. (1981), "Construction vibrations; state-of-the-art", ASCE Journal of Geotechnical Engineering, vol. 107, No. GT2, pp 167–181, www. asce.org.)

7.18 Pile-Design Concepts

Methods for evaluating load-carrying capacity and general behavior of piles in a foundation range from simple empirical to techniques that incorporate state-of-the-art analytical and field verification methods. Approaches to pile engineering include (1) precedence, (2) static-load analysis (3) static-load testing, and (4) dynamic-load analytical and testing methods. Regardless of the method selected, the foundation designer should possess full knowledge of the site subsurface conditions. This requires consultations with a geotechnical engineer and possibly a geologist familiar with the local area to ensure that a sufficient number of borings and relevant soil and rock tests are performed.

Design by precedent includes application of building code criteria, relevant published data, performance of similar nearby structures, and experience with pile design and construction. Under some circumstances this approach may be acceptable, but it is not highly recommended. Favorable situations include those involving minor and temporary structures where failure would not result in appreciable loss of property or any loss of life and construction sites where long-term experience has been accumulated and documented for a well-defined set of subsurface and loading conditions.

Static-load analysis for design and prediction of pile behavior is widely used by designers who are practitioners of geotechnical engineering. This approach is based on soil-mechanics principles, geotechnical engineering theories, pile characteristics, and assumptions regarding pile-soil interaction. Analysis generally involves evaluations of the load-carrying capacity of a single pile, of pile-group behavior, and of foundation settlement under service conditions. Designs based on this approach alone usually incorporate relatively large factors of safety in determination of allowable working loads. Safety factors are based on the engineer's confidence in parameters obtained from soil exploration and their representation of the whole site, anticipated loads, importance of the structure, and the designer's experience and subjective preferences. Static-load analysis methods are used in preliminary design to calculate required pile lengths for cost estimation and bidding purposes. Final pile design and acceptability are based on additional methods of verification. Pile-group behavior and settlement predictions are, however, usually based entirely on static analyses due to the lack of economical and efficient routine field verification techniques.

Field testing should be performed on a sufficient number of piles to confirm or revise initial design assumptions, verify adequacy of installation equipment and procedures, evaluate the effect of subsurface profile variations, and form the basis of final acceptance. Traditionally, piles were tested with a **static-load test** (by loading in axial compression, uplift, or laterally). The number of such tests that will be performed on a site is limited, however, due to the expense and time required when large number of piles are to be installed.

See Art. 7.19 for a description of static-load analysis and pile testing.

Dynamic-load pile testing and analysis is often used in conjunction with or as an alternative to static testing. Analytical methods utilizing computers and numerical modeling and based on one-dimensional elastic-wave-propagation theories are helpful in selecting driving equipment, assessing pile drivability, estimating pile bearing capacity, and determining required driving criteria, that is, blow count. Dynamic analysis is commonly known as *wave equation analysis of pile driving*. Field dynamic testing yields information on driving-system performance of piles: static axial capacity, driving stresses, structural integrity, pile-soil interaction, and load-movement behavior.

See Art. 7.20 for a description of dynamic analysis and pile testing.

(Manual for "Design and Construction of Driven Pile Foundations," 1996, Federal Highway Administration, Report No. FHWA-H1-96-033, Federal Highway Administration, 400 7th Street SW, Washington, D.C. 20590)

7.19 Static-Analysis and Pile Testing

Static analysis of piles and pile design based on it commonly employ global factors of safety. Use of the load-and-resistance-factors approach is growing, however. Steps involved in static analysis include calculation of the static load-carrying capacity of single piles, evaluation of group behavior, and assessment of foundation settlement. Normally, capacity and settlement are treated separately and either may control the design. Pile drivability is usually treated as a separate item and is not considered in static-load analysis. See also Art. 7.18.

7.19.1 Axial-Load Capacity of Single Piles

Pile capacity Q_u may be taken as the sum of the shaft and toe resistances, Q_{su} and Q_{bu} respectively. The allowable load Q_a may then be determined from either Eq. (7.35) or (7.36)

$$Q_a = \frac{Q_{su} + Q_{bu}}{F} \qquad (7.35)$$

$$Q_a = \frac{Q_{su}}{F_1} + \frac{Q_{bu}}{F_2} \qquad (7.36)$$

where F, F_1, F_2, are safety factors. Typically F for permanent structures is between 2 and 3 but may be larger, depending on the perceived reliability of the analysis and construction as well as the consequences of failure. Equation (7.36) recognizes that the deformations required to fully mobilize Q_{su} and Q_{bu} are not compatible. For example, Q_{su} may be developed at displacements less than 0.25 in, whereas Q_{bu} may be realized at a toe displacement equivalent to 5% to 10% of the pile diameter. Consequently, F_1 may be taken as 1.5 and F_2 as 3.0, if the equivalent single safety factor equals F or larger. (If $Q_{su}/Q_{bu} < 1.0$, F is less than the 2.0 usually considered as a minimum safety factor for permanent structures.)

7.19.2 Shaft Resistance in Cohesive Soils

The ultimate stress $\bar{f}_s$ of axially loaded piles in cohesive soils under compressive loads is conven-

tionally evaluated from the ultimate frictional resistance

$$Q_{su} = A_s \bar{f}_s = A_s \alpha \bar{c}_u \qquad (7.37)$$

where c_u = average undrained shear strength of soil in contact with shaft surface

A_s = shaft surface area

α = shear-strength (adhesion) reduction factor

One relationship for selection of α is shown in Fig. 7.20. This and similar relationships are empirical and are derived from correlations of load-test data with the c_u of soil samples tested in the laboratory. Some engineers suggest that $\bar{f}_s$ is influenced by pile length and that a limiting value of 1 ton/ft² be set for displacement piles less than 50 ft long and reduced 15% for each 50 ft of additional length. This suggestion is rejected by other engineers on the presumption that it neglects the effects of pile residual stresses in evaluation of the results of static-load tests on piles.

The shaft resistance stress $\bar{f}_s$ for cohesive soils may be evaluated from effective-stress concepts:

$$\bar{f}_s = \beta \sigma'_{vo} \qquad (7.38)$$

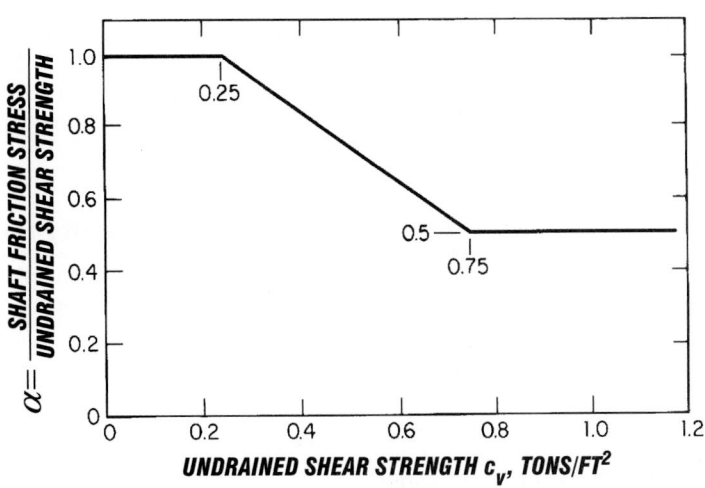

Fig. 7.20 Variation of shear-strength (adhesion) reduction factor α with undrained shear strength. (*After "Recommended Practice for Planning, Designing, and Constructing Fixed Off-Shore Platforms," American Petroleum Institute, Dallas.*)

where σ'_{vo} = effective overburden pressure of soil

β = function of effective friction angle, stress history, length of pile, and amount of soil displacement induced by pile installation

β usually ranges between 0.22 and 0.35 for intermediate-length displacement piles driven in normally consolidated soils, whereas for piles significantly longer than 100 ft, β may be as small as 0.15. Derivations of β are given by G. G. Meyerhof, "Bearing Capacity and Settlement of Pile Foundations," *ASCE Journal of Geotechnical Engineering Division*, vol. 102, no. GT3, 1976; J. B. Burland, "Shaft Friction of Piles in Clay," *Ground Engineering*, vol. 6, 1973; "Soil Capacity for Supporting Deep Foundation Members in Clay," STP 670, ASTM.

Both the α and β methods have been applied in analysis of n discrete soil layers:

$$Q_{su} = \sum_{i=1}^{n} A_{si} \bar{f}_{si} \qquad (7.39)$$

The capacity of friction piles driven in cohesive soils may be significantly influenced by the elapsed time after pile driving and the rate of load application. The frictional capacity Q_s of displacement piles driven in cohesive soils increases with time after driving. For example, the pile capacity after substantial dissipation of pore pressures induced during driving (a typical design assumption) may be three times the capacity measured soon after driving. This behavior must be considered if piles are to be rapidly loaded shortly after driving and when load tests are interpreted.

Some research indicates that frictional capacity for tensile load Q_{ut} may be less than the shaft friction under compression loading Q_{su}. In the absence of load-test data, it is therefore appropriate to take Q_{ut}, as $0.80Q_{su}$ and ignore the weight of the pile. Also, Q_{ut} is fully developed at average pile deformations of about 0.10 to 0.15 in, about one-half those developed in compression. Expanded-base piles develop additional base resistance and can be used to substantially increase uplift resistances.

(V. A. Sowa, "Cast-In-Situ Bored Piles," *Canadian Geotechnical Journal*, vol. 7, 1970; G. G. Meyerhof and J. I. Adams, "The Ultimate Uplift Capacity of Foundations," *Canadian Geotechnical Journal*, vol. 5, no. 4, 1968.)

7.19.3 Shaft Resistance in Cohesionless Soils

The shaft resistance stress $\bar{f}_s$ is a function of the soil-shaft friction angle δ, deg, and an empirical lateral earth-pressure coefficient K:

$$\bar{f}_s = K\bar{\sigma}'_{vo} \tan \delta \leq f_l \qquad (7.40)$$

At displacement-pile penetrations of 10 to 20 pile diameters (loose to dense sand), the average skin friction reaches a limiting value f_l. Primarily depending on the relative density and texture of the soil, f_l has been approximated conservatively by using Eq. (7.40) to calculate $\bar{f}_s$. This approach employs the same principles and involves the same limitations discussed in Art. 7.8.2.

For relatively long piles in sand, K is typically taken in the range of 0.7 to 1.0 and δ is taken to be about $\phi - 5$, where ϕ' is the angle of internal friction, deg. For piles less than 50 ft long, K is more likely to be in the range of 1.0 to 2.0 but can be greater than 3.0 for tapered piles.

Empirical procedures have also been used to evaluate $\bar{f}_s$ from in situ tests, such as cone penetration, standard penetration, and relative density tests. Equation (7.41), based on standard penetration tests, as proposed by Meyerhof, is generally conservative and has the advantage of simplicity.

$$\bar{f}_s = \frac{\bar{N}}{50} \qquad (7.41)$$

where $\bar{N}$ = average average standard penetration resistance within the embedded length of pile and $\bar{f}_s$ is given in tons/ft². (G. G. Meyerhof, "Bearing Capacity and Settlement of Pile Foundations," *ASCE Journal of Geotechnical Engineering Division*, vol. 102, no. GT3, 1976.)

7.19.4 Toe Capacity Load

For piles installed in cohesive soils, the ultimate toe load may be computed from

$$Q_{bu} = A_b q = A_b N_c c_u \qquad (7.42)$$

where A_b = end-bearing area of pile

q = bearing capacity of soil

N_c = bearing-capacity factor

c_u = undrained shear strength of soil within zone 1 pile diameter above and 2 diameters below pile tip

Although theoretical conditions suggest that N_c may vary between about 8 and 12, N_c is usually taken as 9.

For cohesionless soils, the toe resistance stress q is conventionally expressed by Eq. (7.43) in terms of a bearing-capacity factor N_q and the effective overburden pressure at the pile tip σ'_{vo}.

$$q = N_q \sigma'_{vo} \leq q_l \qquad (7.43)$$

Some research indicates that, for piles in sands, q, like $\bar{f}_s$, reaches a quasi-constant value q_l after penetrations of the bearing stratum in the range of 10 to 20 pile diameters. Approximately,

$$q_l = 0.5 N_q \tan \phi \qquad (7.44)$$

where ϕ is the friction angle of the bearing soils below the critical depth. Values of N_q applicable to piles are given in Fig. 7.21. Empirical correlations of CPT data with q and q_l have also been applied to predict successfully end-bearing capacity of piles in sand. (G. G. Meyerhof, "Bearing Capacity and Settlement of Pile Foundations," *ASCE Journal of Geotechnical Engineering Division*, vol. 102, no. GT3, 1976.)

7.19.5 Pile Settlement

Prediction of pile settlement to confirm allowable loads requires separation of the pile load into shaft friction and end-bearing components. Since q and $\bar{f}_s$ at working loads and at ultimate loads are different, this separation can only be qualitatively evaluated from ultimate-load analyses. A variety of methods for settlement analysis of single piles have been proposed, many of which are empirical or semiempirical and incorporate elements of elastic solutions.

(H. Y. Fang, "Foundation Engineering Handbook," 2nd ed., Van Nostrand Reinhold, New York.)

7.19.6 Groups of Piles

A pile group may consist of a cluster of piles or several piles in a row. The response of an individual pile in a pile group, where the piles are situated close to one another, may be influenced by the response and geometry of neighboring piles. Piles in such groups interact with one another through the surrounding soil, resulting in what is called the pile-soil-pile interaction, or group effect. The efficiency of a pile group (η_g) is defined as the ratio of the actual capacity of the group to the summation of the capacities of the individual piles in the group when tested as single piles. The pile-soil-pile interaction has two components: pile installation, and loading effects. Analytical models developed to analyze the pile-soil-pile interaction by considering strain superposition in the soil mass neglect the effect of installation and the alteration of the failure zone around an individual pile by those of neighboring piles.

In loose sand, the group efficiency in compression exceeds unity, with the highest values occurring at a pile center to center spacing(s) to diameter or width (d) ratio (s/d) of 2. Generally, higher efficiencies occur with an increase in the number of piles in the group. However, in dense sand, efficiency may be either greater or less than unity, although the trend is toward $\eta_g > 1$. An efficiency smaller than one is probably due to dilatancy and would generally be expected for bored or partially jetted piles. Conventional practice for the analysis of pile groups in sand has been based on assigning a conservative upper bound for η_g of unity for driven piles and 0.67 for bored piles.

Piles in clay always yield values of group efficiencies less than unity with a distinctive trend toward block failure in square groups with an (s/d) ratio of less than 2. Historically, the geotechnical practice was based on a value of η_g of unity for pile

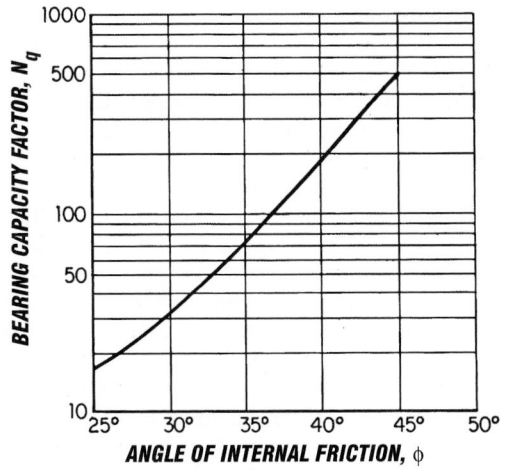

Fig. 7.21 Bearing-capacity factor for granular soils related to angle of internal friction.

groups in clay, provided that block failure does not occur and that sufficient time has elapsed between installation and the first application of load to permit excess pore pressure to dissipate.

Several efficiency formulas have been published in the literature. These formulas are mostly based on relating the group efficiency to the spacing between the piles and generally yield efficiency values of less than unity, regardless of the pile/soil conditions. A major apparent shortcoming in most of the efficiency formulas is that they do not account for the characteristics of the soil in contact with the pile group. Comparison of different efficiency formulas show considerable difference in their results. There is no comprehensive mathematical model available for computing the efficiency of a pile group. Any general group efficiency formula that only considers the planar geometry of the pile group should be considered with caution. Soil characteristics, time-dependent effects, cap contact, order of pile driving, and the increase of lateral pressure influence the efficiency of a pile group.

The pile group efficiency formula developed by Sayed and Bakeer (1992) accounts for the three-dimensional geometry of the pile group, soil strength and time-dependent change, and type of embedding soil (cohesive and cohensionless). For a typical configuration of pile group, the group efficiency η_g is expressed as:

$$\eta_g = 1 - (1 - \eta_s' \cdot K) \cdot \rho \qquad (7.45)$$

where $\eta_s' =$ geometric efficiency

$\qquad K =$ group interaction factor, and

$\qquad \rho =$ friction factor

This equation is particularly applicable for computing the efficiency of pile groups where a considerable percentage of the load is carried through shaft resistance. For an end bearing pile group, the term ρ becomes practically equal to zero, and accordingly, the formula yields a value of η_g of one. The formula does not account for the contribution of pile cap resistance to the overall bearing capacity of the pile group (neglected due to the potential of erosion or loss of support from settlement of the soil).

For a pile group arranged in a rectangular or square array, the geometric efficiency η_s' is defined as $\eta_s' = P_g / \Sigma P_p$, where $P_g =$ the perimeter of the pile group; and ΣP_p is the summation of the

perimeters of the individual piles in the group. Generally, η_s' increases with an increase in the pile spacing-to-diameter (width) ratio s/d, and its typical values are between 0.6 and 2.5.

The factor K is a function of the method of pile installation, pile spacing, and soil type. It is also used to model the change in soil strength due to pile driving (e.g., compaction in cohesionless soils or remolding in cohesive soils). The value of K may range from 0.4 to 9, where higher values are expected in dense cohesionless or stiff cohesive soils and smaller values are expected in loose or soft soils. A value of 1 is obtained for piles driven in soft clay and a value greater than 1 is expected for sands. The appropriate value of K is determined according to the relative density of the sand or the consistency of the clay. For example, a value of 2 to 3 is appropriate in medium-dense sand. These values were back-calculated from the results of several load tests on pile groups.

The friction factor ρ is defined as Q_{su}/Q_u, where Q_{su} and Q_u are the ultimate shaft resistance and total capacity of a single pile, respectively. This factor can be used to introduce the effect of time into the analysis when it is important to assess the short-term as well as the long-term efficiencies of a pile group. This is achieved by considering the gain or loss in the shear strength of the soil in the calculation of Q_{su} and Q_u. Typical values of ρ may range from zero for end-bearing piles to one for friction piles, with typical values of greater than 0.60 for friction or floating type foundations.

Pile dynamic measurements and related analysis (i.e., PDA and CAPWAP) made at the End Of Initial Driving (EOID) and the Beginning Of Restrike (BOR) can provide estimates of the friction factor ρ for the short-term and/or long-term conditions, respectively. For bored piles in cohesive soils, some remolding and possibly lateral stress relief usually occur during construction. It is suggested that the friction factor ρ be determined from a total stress analysis to calculate the short-term efficiency. The long-term efficiency should be based on an effective stress approach. Two types of triaxial tests, unconsolidated-undrained (UU) and consolidated-undrained with pore-pressure measurement (CU), can be used to provide the required strength parameters for the analysis. Moreover, high-strain dynamic tests can be performed on bored piles to provide similar information on the friction factor ρ, analogous to that obtained for driven piles. Some geotechnical engineers may prefer to use a total stress approach to compute the ultimate load capacity for both the short and long-term capacities. One can still compute the friction factor ρ for the short-term and/or long-term conditions provided that adequate information is available regarding the thixotropic gain of strength with time.

The three-dimensional modeling of pile groups incorporating the efficiency formula can be performed using the Florida-Pier (FLPIER) computer program developed by the Florida Department of Transportation (FDOT) at the University of Florida. FLPIER considers both axial and lateral pile-soil interaction and group effects. Pile-soil-pile interaction effects are considered through p-y multipliers which are assigned for each row within the group for lateral loading and group efficiency η_g for axial loading. FHWA's computer program COM624 is also available for modeling the lateral pile-soil interaction.

(S. M. Sayed and R. M. Bakeer (1992), "Efficiency Formula For Pile Groups," *Journal of Geotechnical Engineering*, ASCE, 118 No. 2, pages 278–299; S. T. Wang and L. C. Reese, L. C. (1993) "COM624P–Laterally Loaded Pile Analysis Program for the Microcomputer, Version 2.0," *FHWA Office of Technology Applications*, Publication No. FHWA-SA-91-048, Washington, D.C. 20590; Florida Department of Transportation FDOT (1995) "User's Manual for Florida Pier Program", www.dot.state.fl.us)

A very approximate analysis of group settlement, applicable to friction piles, models the pile group as a raft of equivalent plan dimensions situated at a depth below the surface equal to two-thirds the pile length. Subsequently, conventional settlement analyses are employed (see Arts. 7.12 and 7.13).

Negative Skin Friction, Dragload, and Downdrag ▪ Influenced by consolidation induced by placement of fill and/or lowering of the water table, soils along the upper portion of a pile will tend to compress and move down relative to the pile. In the process, load is transferred to the pile through negative skin friction. The permanent load (dead load) on the pile and the dragload imposed by the negative skin friction are transferred to the lower portion of the pile and resisted by means of positive shaft resistance and by toe resistance. A point of equilibrium, called the neutral plane,

exists where the negative skin friction changes over into positive shaft resistance. This is where there is no relative movement between the pile and the soil, which means that if the neutral plane is located in non-settling soil, then, the pile does not settle. If on the other hand, the soil experiences settlement at the level of the neutral plane, the pile will settle the same amount, i.e., be subjected to downdrag. Dragload is of concern if the sum of the dragload and the dead load exceeds the structural strength of the pile. The following must be considered:

1. Sum of dead plus live loads is smaller than the pile capacity divided by an appropriate factor of safety. The dragload is not included with these loads.

2. Sum of dead load and dragload is smaller than the structural strength with an appropriate factor of safety. The live load is not included because live load and drag load can not coexist.

3. The settlement of the pile (pile group) is smaller than a limiting value. The live load and dragload are not included in this analysis.

A procedure for construing the neutral plane and determining pile allowable load is illustrated in Fig. 7.22. The diagrams assume that above the neutral plane, the unit negative skin friction, q_n, and positive shaft resistance, r_s, are equal, which is an assumption on the safe side. A key factor is the estimate of the pile toe resistance, R_t. If the pile toe resistance is small, the neutral plane lies higher than when the toe resistance is large. Further, if the pile toe is located in a non-settling soil, the pile settlement will be negligible and only a function of the pile toe penetration necessary to mobilize the pile and bearing resistance. The maximum negative skin friction that can be developed on a single pile can be calculated with Eq. (7.38) with β factors for clay of 0.20 to 0.25, for silt of 0.25 to 0.35, and for sand of 0.35 to 0.50.

A very approximate method of pile-group analysis calculates the upper limit of group drag load Q_{gd} from

$$Q_{gd} = A_F \gamma_F H_F + PH c_u \qquad (7.46)$$

H_F, γ_F, and A_F represent the thickness, unit weight, and area of fill contained within the group. P, H, and c_u are the circumference of the group, the thickness of the consolidating soil layers penetrated by the piles, and their undrained shear strength, respect-

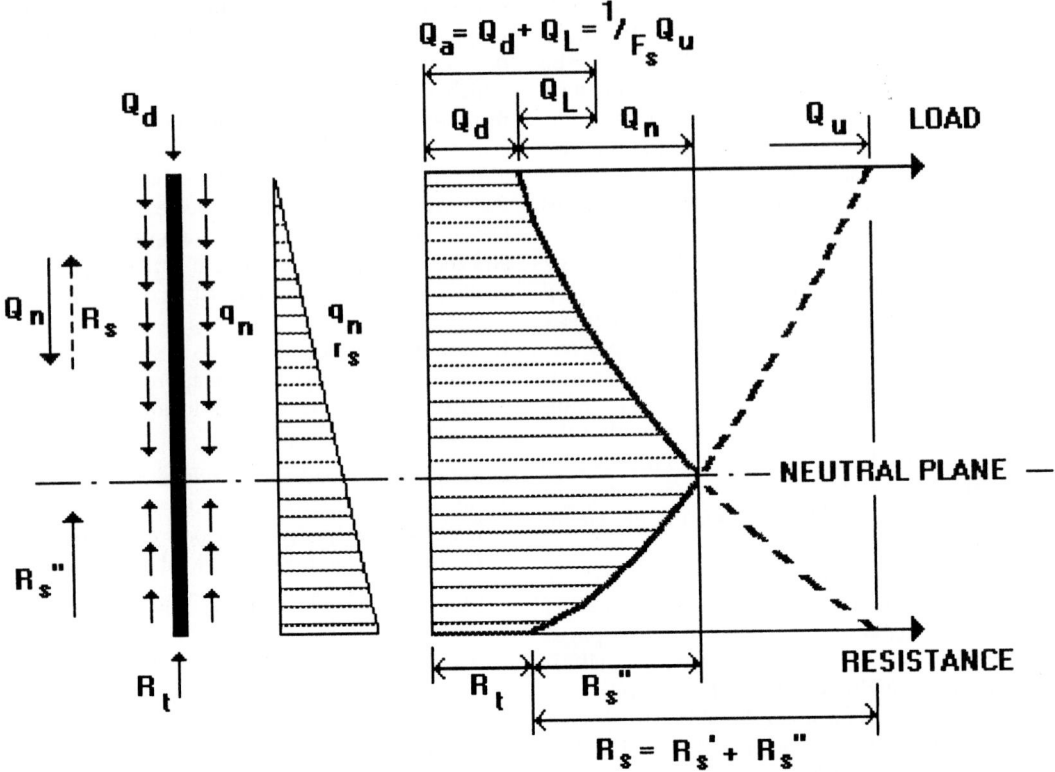

Fig. 7.22 Construing the Neutral Plane and determining allowable load (Guidelines for Static Pile Design, – A Continuing Education Short Course Text, B. H. Fellenius, Deep Foundations Institute, 1991 (www.dfi.org)).

ively. Such forces as Q_{gd} could only be approached for the case of piles driven to rock through heavily surcharged, highly compressible subsoils.

(H. G. Poulos and E. H. Davis, "Elastic Solutions for Soil and Rock Mechanics," and K. Terzaghi and R. B. Peck, "Soil Mechanics and Engineering Practice," John Wiley & Sons, Inc., New York; J. E. Garlanger and W. T. Lambe, *Symposium on Downdrag of Piles*, Research Report 73–56, Soils Publication no. 331, Massachusetts Institute of Technology, Cambridge, 1973; B. H. Fellenius, "Basics of Foundation Design," BiTech Publishers, Richmond, BC, Canada, 1999).

7.19.7 Design of Piles for Lateral Loads

Piles and pile groups are typically designed to sustain lateral loads by the resistance of vertical piles, by inclined, or batter, piles, or by a combination. Tieback systems employing ground anchor or deadmen reactions are used in conjunction with laterally loaded sheetpiles (rarely with foundation piles).

Lateral loads or eccentric loading produce overturning moments and uplift forces on a group of piles. Under these circumstances, a pile may have to be designed for a combination of both lateral and tensile load.

Inclined Piles ▪ Depending on the degree of inclination, piles driven at an angle with the vertical can have a much higher lateral-load capacity than vertical piles since a large part of the lateral load can be carried in axial compression. To minimize construction problems, however, pile batters (rake) should be less than 1 horizontal to 2 vertical.

Evaluation of the load distribution in a pile group consisting of inclined piles or combined vertical and batter piles is extremely complex because of the three-dimensional nature and indeterminacy of the system. A variety of computer solutions have become available and allow a rational evaluation of the load distribution to inclined group piles. The same methods of axial-capacity evaluation developed for vertical piles are applied to inclined-pile design, although higher driving energy losses during construction suggest that inclined piles would have a somewhat reduced axial-load capacity for the same terminal resistance. (A. Hrennikoff, "Analysis of Pile Foundations with Batter Piles," *ASCE Transactions*, vol. 115, 1950.)

Laterally Loaded Vertical Piles ▪ Vertical-pile resistance to lateral loads is a function of both the flexural stiffness of the pile, the stiffness of the bearing soil in the upper $4D$ to $6D$ length of pile, where D = pile diameter, and the degree of pile-head fixity. The lateral-load design capacity is also related to the amount of lateral deflection permitted and, except under very exceptional circumstances, the tolerable-lateral-deflection criteria will control the lateral-load design capacity.

Design loads for laterally loaded piles are usually evaluated by beam theory for both an elastic and nonlinear soil reaction, although elastic and elastoplastic continuum solutions are available. *Nonlinear solutions* require characterization of the soil reaction p versus lateral deflection y along the shaft. In obtaining these solutions, degradation of the soil stiffness by cyclic loading is an important consideration.

The lateral-load vs. pile-head deflection relationship is readily developed from charted non-dimensional solutions of Reese and Matlock. The solution assumes the soil modulus K to increase linearly with depth z; that is, $K = n_h z$, where n_h = coefficient of horizontal subgrade reaction. A characteristic pile length T is calculated from

$$T = \sqrt{\frac{EI}{n_h}} \qquad (7.47)$$

where EI = pile stiffness. The lateral deflection y of a pile with head free to move and subject to a lateral load P_t and moment M_t applied at the ground line is given by

$$y = A_y P_t \frac{T^3}{EI} + B_y M_t \frac{T^2}{EI} \qquad (7.48)$$

where A_y and B_y are nondimensional coefficients. Nondimensional coefficients are also available for evaluation of pile slope, moment, shear, and soil reaction along the shaft.

For positive moment,

$$M = A_m P_t T + B_m M_t \qquad (7.49)$$

Positive M_t and P_t values are represented by clockwise moment and loads directed to the right on the pile head at the ground line. The coefficients applicable to evaluation of pile-head deflection and to the maximum positive moment and its approximate position on the shaft z/T, where z = distance below the ground line, are listed in Table 7.9.

The negative moment imposed at the pile head by pile-cap or other structural restraint can be evaluated as a function of the head slope (rotation) from

$$-M_t = \frac{A_\theta P_t T}{B_\theta} - \frac{\theta_s EI}{B_\theta T} \qquad (7.50)$$

Table 7.9 Deflection, Moment, and Slope Coefficients

z_{max}	A_y	B_y	A_θ	B_θ	$A_m{}^*$	$B_m{}^*$	z/T^*
2	4.70	3.39	−3.40	−3.21	0.51	0.84	0.85
3	2.65	1.77	−1.75	−1.85	0.71	0.60	1.49
4	2.44	1.63	−1.65	−1.78	0.78	0.70	1.32
>5	2.43	1.62	−1.62	−1.75	0.77	0.69	1.32

*Coefficients for maximum positive moment are located at about the values given in the table for z/T.

Source: L. C. Reese and H. Matlock, "Non-Dimensional Solutions for Laterally Loaded Piles with Soil Modulus Assumed Proportional to Depth," *8th Texas Conference of Soil Mechanics and Foundation Engineering,* University of Texas, 1956.

where θ_s, rad, represents the counterclockwise $(+)$ rotation of the pile head and A_θ and B_θ are coefficients (see Table 7.9). The influence of the degrees of fixity of the pile head on y and M can be evaluated by substituting the value of $-M_t$ from Eq. (7.50) in Eqs. (7.48) and (7.49). Note that for the fixed-head case:

$$y_f = \frac{P_t T^3}{EI}\left(A_y - \frac{A_\theta B_y}{B_\theta}\right) \tag{7.51}$$

Improvement of Lateral Resistance ■ The lateral-load capacity of a specific pile type can be most effectively increased by increasing the diameter, i.e., the stiffness and lateral-bearing area. Other steps are to improve the quality of the surficial bearing soils by excavation and replacement or in-place densification, to add reinforcement, and to increase the pile-head fixity condition.

Typical lateral-load design criteria for buildings limit lateral pile-head deformations to about $\frac{1}{4}$ in. Associated design loads for foundation piles driven in medium dense sands or medium clays are typically in the range of 2 to 4 tons, although significantly higher values have been justified by load testing or detailed analyses or a combination.

Resistance of pile groups to lateral loads is not well-documented by field observations. Results of model testing and elastic analysis, however, indicate that pile spacings less than about 8 pile diameters D in the direction of loading reduce the soil modulus K. The reduction factors are assumed to vary linearly from 1.0 at $8D$ to 0.25 at a $3D$ spacing if the number of piles in the group is 5 or more and the passive resistance of the pile cap is ignored. The effect of this reduction is to "soften" the soil reaction and produce smaller lateral resistance for a given group deflection. Elastic analyses also confirm the long-held judgment that batter piles in the center of a pile group are largely ineffective in resisting lateral load.

(B. B. Broms, "Design of Laterally Loaded Piles," *ASCE Journal of Soil Mechanics and Foundation Engineering Division*, vol. 91, no. SM3, 1965. H. Y. Fang, "Foundation Engineering Handbook," Van Nostrand Reinhold, New York. B. H. Fellenius, "Guidelines for Static Pile Design," Deep Foundation Institute, 120 Charolette Place, Englewood Cliffs, NJ 07632 (www.dfi.org). H. G. Poulos and E. H. Davis, "Elastic Solutions for Soil and Rock Mechanics," John Wiley & Sons, Inc., New York. L. C. Reese and R. C. Welch, "Lateral Loading of Deep Foundations in Stiff Clay,"

ASCE Journal of Geotechnical Engineering, vol. 101, no. GT7, 1975.)

7.19.8 Static-Load Pile Testing

Because of the inherent uncertainty in static pile-design methods and the influence of construction procedures on the behavior of piles, static-load tests are desirable or may be required. Static-load tests are almost always conducted on single piles; testing of pile groups is very rare.

Engineers use static-load tests to determine the response of a pile under applied loads. Axial compression testing is the most common, although when other design considerations control, uplift or lateral loading tests are also performed. In some special cases, testing is performed with cyclic loadings or with combined loads; for example, axial and lateral loading. Pile testing may be performed during the design or construction phase of a project so that foundation design data and installation criteria can be developed or verified, or to prove the adequacy of a pile to carry design load.

Use of static-load pile testing is limited by expense and time required for the tests and analyses. For small projects, when testing costs add significantly to the foundation cost, the increased cost often results in elimination of pile testing. For projects involving a large number of piles, static-load pile tests usually are performed, but only a few piles are tested. (A typical recommendation is that of the total number of piles to be installed in normal practice 1% be tested, but the percentage of piles tested in actual practice may be much lower.) The number and location of test piles should be determined by the foundation design engineer after evaluation of the variability of subsurface conditions, pile loadings, type of pile, and installation techniques. Waiting time between pile installation and testing generally ranges from several days to several weeks, depending on pile type and soil conditions.

The foundation contractor generally is responsible for providing the physical setup for conducting a static-load test. The foundation designer should supervise the testing.

Standards detailing procedures on how to arrange and conduct static-load pile tests include "Standard Test Method for Piles under Static Axial Compression Load," ASTM D1143 (www.astm. org); "Standard Method of Testing Individual Piles under Static Axial Tension Load," ASTM D3689;

and "Standard Method of Testing Piles under Lateral Loads," ASTM D3966. See also "Static Testing of Deep Foundations," U. S. Federal Highway Administration, Report No. FHWA-SA-91-042 (www.fhwa.gov), 1992; "Axial Pile Loading Test—Part 1: Static Loading," International Society for Soil Mechanics and Foundation Engineering, 1985; and "Canadian Foundation Engineering Manual," 2nd ed., Canadian Geotechnical Society, 1985.

Load Application ▪ In a static-load pile test, a hydraulic jack, acting against a reaction, applies load at the pile head. The reaction may be provided by a kentledge, or platform loaded with weights (Fig. 7.23), or by a steel frame supported by reaction piles (Fig. 7.24), or by ground anchors. The distance to be used between the test pile and reaction-system supports depends on the soil conditions and the level of loading but is generally three pile diameters or 8 ft, whichever is greater. It may be necessary to have the test configuration evaluated by a structural engineer.

Hydraulic jacks including their operation should conform to "Safety Code for Jacks," ANSI B30. 1, American National Standards Institute. The jacking system should be calibrated (with load cells, gages, or machines having an accuracy of at least 2%) within a 6-month period prior to pile testing. The available jack extension should be at least 6 in. The jack should apply the load at the center of the pile (Fig. 7.25). When more than one jack is needed for the test, all jacks should be pressurized by a common device.

Loads should be measured by a calibrated pressure gage and also by a load cell placed between the jack and the pile. Internal forces in the pile may be measured with strain gages installed along the pile. Two types of test-loading procedures are used: the *maintained load* (ML) and the *constant rate of penetration* (CRP) methods.

In the ML method, load is applied in increments of 25% of the anticipated pile capacity until failure occurs or the load totals 200% of the design load. Each increment is maintained until pile movement is less than 0.01 in/h or for 2 h, whichever occurs first. The final load is maintained for 24 h. Then, the test load is removed in decrements of 25% of the total test load, with 1 h between decrements. This procedure may require from 1 to 3 days to complete. According to some practices, the ML method is changed to the CRP procedure as soon as the rate exceeds 0.8 in/h.

Tests that consist of numerous load increments (25 to 40 increments) applied at constant time intervals (5 to 15 min) are termed *quick tests*.

In the CRP procedure, the pile is continuously loaded so as to maintain a constant rate of penetration into the ground (typically between 0.01 and 0.10 in/min for granular soils and 0.01 to 0.05 in/min for cohesive soils). Loading is continued until no further increase is necessary for continuous pile penetration at the specified rate. As long as pile penetration continues, the load

Fig. 7.23 Static-load test on a pile with dead weight as the reaction load.

Fig. 7.24 Reaction piles used in static-load test on a pile.

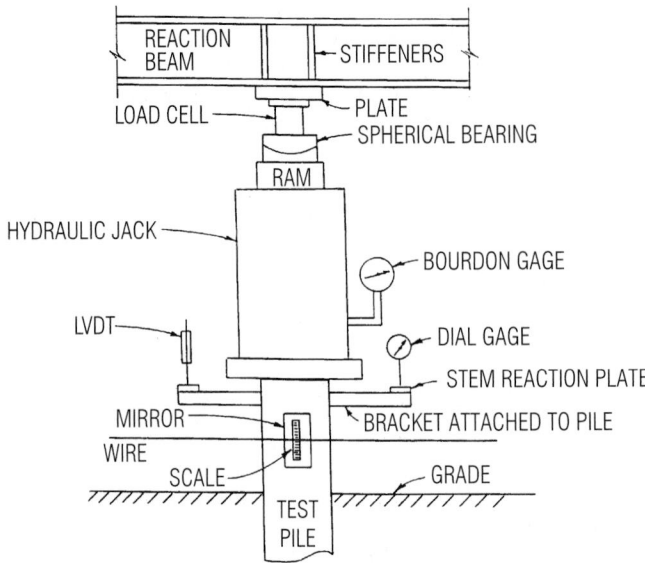

Fig. 7.25 Typical rearrangement of loading equipment and instrumentation at pile head for a compression static-load test. (*From "Static Testing of Deep Foundations," FHWA SA-91-042, Federal Highway Administration.*)

inducing the specified penetration rate is maintained until the total pile penetration is at least 15% of the average pile diameter or diagonal dimension, at which time the load is released. Also, if, under the maximum applied load, penetration ceases, the load is released.

Alternatively, for axial-compression static-load tests, sacrificial jacks or other equipment, such as the Osterberg Cell, may be placed at the bottom of the pile to load it (J. O. Osterberg, "New Load Cell Testing Device," Deep Foundations Institute (www.dfi.org)). One advantage is automatic separation of data on shaft and toe resistance. Another is elimination of the expense and time required for constructing a reaction system, inasmuch as soil resistance serves as a reaction. A disadvantage is that random pile testing is not possible since the loading apparatus and pile installations must be concurrent.

Penetration Measurements ▪ The axial movement of the pile head under applied load may be measured by mechanical dial gages or electromechanical devices mounted on an independently supported (and protected) reference beam. Figure 7.25 shows a typical arrangement of equipment and instruments at the pile head. The gages should have at least 2 in of travel (extendable to 6 in) and typically a precision of at least 0.001 in. For redundancy, measurements may also be taken with a surveyor's rod and precise level and referenced to fixed benchmarks. Another alternative is a tightly stretched piano wire positioned against a mirror and scale that are attached to the side of the pile. Movements at locations along the pile length and at the pile toe may be determined with the use of telltales.

For the ML or quick-test procedures, pile movements are recorded before and after the application of each load increment. For the CRP method, readings of pile movement should be taken at least every 30 s.

Pile-head transverse displacements should be monitored and controlled during the test. For safety and proper evaluation of test results, movements of the reaction supports should also be monitored during the test.

Interpretation of Test Results ▪ A considerable amount of data is generated during a static-load test, particularly with instrumented piles. The most widely used procedure for presenting test results is the plot of pile-head load vs.

movement. Other results that may be plotted include pile-head time vs. movement and load transfer (from instrumentation along the pile shaft). Shapes of load vs. movement plots vary considerably; so do the procedures for evaluating them for calculation of limit load (often mistakenly referred to as *failure load*).

Problems in data interpretation arise from the lack of a universally recognized definition of *failure*. For a pile that has a load-carrying capacity greater than that of the soil, failure may be considered to occur when pile movement continues under sustained or slightly increasing load (pile plunging). In general, the term *failure load* should be replaced with *interpreted failure load* for evaluations from plots of pile load vs. movement. The definition of interpreted failure load should be based on mathematical rules that produce repeatable results without being influenced by the subjective interpretation of the engineer. In the **offset limit method**, interpreted failure load is defined as the value of the load ordinate of the load vs. movement curve at $\rho + 0.15 + D/120$, where ρ is the movement, in, at the termination of elastic compression and D is the nominal pile diameter, in. One advantage of this technique is the ability to take pile stiffness into consideration. Another advantage is that maximum allowable pile movement for a specific allowable load can be calculated prior to proof testing of a pile. Interpretation methods that rely on extrapolation of the load-movement curve should be avoided. ("Guidelines for the Interpretation and Analysis of the Static Loading Test," 1990, B. H. Fellenius, Deep Foundation Institute, www.dfi.com).

The test report should include the following as well as other relevant data:

1. Information on general site subsurface conditions, emphasizing soil data obtained from exploration near the test pile

2. Descriptions of the pile and pile installation procedure

3. Dates and times of pile installation and static testing

4. Descriptions of testing apparatus and testing procedure

5. Calibration certificates

6. Photographs of test setup

7. Plots of test results

8. Description of interpretation methods

9. Name of testing supervisor

The cost, time, and effort required for a static-load test should be carefully weighed against the many potential benefits. A static-load test on a single pile, however, does not account for the effects of long-term settlement, downdrag loads, time-dependent soil behavior, or pile group action, nor does the test eliminate the need for an adequate foundation design.

7.20 Dynamic Pile Testing and Analysis

Simple observations made during impact pile driving are an important and integral part of the pile installation process. In its most basic form, dynamic-load pile testing encompasses visual observations of hammer operation and pile penetration during pile driving. Some engineers apply equations based on the Newtonian physics of rigid bodies to the pile movements recorded during pile driving to estimate the load-carrying capacity of the pile. The basic premise is that the harder it is to drive the pile into the ground, the more load it will be able to carry. The equations, generally known as energy formulas, typically relate hammer energy and work done on the pile to soil resistance. More than 400 formulas have been proposed, including the widely used and simple *Engineering News formula.*

This method of estimating load capacity, however, has several shortcomings. These include incomplete, crude, and oversimplified representation of pile driving, pile and soil properties, and pile-soil interaction. Often, the method has been found to be grossly inaccurate and unreliable to the extent that many engineers believe that it should be eliminated from contemporary practice.

Modern rational dynamic testing and analysis incorporates pile dynamic measurements analyzed with one-dimensional elastic stress wave propagation principles and theories. Such testing methods have become routine procedures in contemporary foundation engineering practice worldwide. They are covered in many codes and specifications. (ASTM D 4945-96: Standard Test Method for High-Strain Dynamic Testing of Piles; "Application of Stress Wave Theory to Piles: Quality Assurance on Land and Offshore Piling,"

Proceedings of 6th International Conference, San Paulo, Brazil, 2000, A. A. Balkema Publishers, www.balkema.nl).

7.20.1 Wave Equation

In contrast to the deficiencies of the energy formulas, analysis of pile-driving blow count or penetration per blow yields more accurate estimates of the load-carrying capacity of a pile, if based on accurate modeling and rational principles. One such type of analysis employs the wave equation based on a concept developed by E. A. Smith (*ASCE Journal of Geotechnical Engineering Division*, August 1960). Analysis is facilitated by use of computer programs such as GRLWEAP (Goble Rausche Likins and Associates, Inc., Cleveland, Ohio) that simulate and analyze impact pile driving. Sophisticated numerical modeling, advanced analytical techniques, and one-dimensional elastic-wave-propagation principles are required. Computations can be performed with personal computers. A substantial improvement that the wave equation offers over the energy approach is the ability to model all hammer, cushion, pile-cap, and pile and soil components realistically.

Figure 7.26 illustrates the lumped-mass model used in wave equation analyses. All components that generate, transmit, or dissipate energy are represented by a spring, mass, or dashpot. These permit representation of mass, stiffness, and viscosity.

A series of masses and springs represent the mass and stiffness of the pile. Elastic springs and linear-viscous dashpots model soil-resistance forces along the pile shaft and under the toe. The springs represent the displacement-dependent static-loaded components, and the dashpots the loading-dependent dynamic components. Springs model stiffness and coefficient of restitution (to account for energy dissipation) of hammer and pile cushions. A single mass represents the pile-cap. For external-combustion hammers, the representation is straightforward: a stocky ram, by a single mass; the hammer assembly (cylinder, columns, etc.), by masses and springs. For internal-combustion hammers, the modeling is more involved. The slender ram is divided into several segments. The gas pressure of the diesel combustion cycle is calculated according to the thermodynamic gas law for either liquid or atomized fuel injection.

The parameters needed for execution of a wave equation analysis with the GRLWEAP computer program are:

Hammer: Model and efficiency

Hammer and Pile Cushions: Area, thickness, elastic modulus, and coefficient of restitution

Pile Cap: Weight, including all cushions and any inserts

Pile: Area, elastic modulus, and density, all as a function of length

Soil: Total static capacity, percent shaft resistance and its distribution, quake and damping constants along the shaft and under the toe

In practice, wave equation analysis is employed to deal with the following questions:

1. If the input to the computer program provides a complete description of hammer, cushions, pile cap, pile, and soil, can the pile be driven safely and economically to the required static capacity?

2. If the input provides measurements of pile penetration during pile driving or restriking blow count, what is the static-load capacity of the pile?

For case 1, pile design and proper selection of hammer and driving system can be verified to ensure that expected pile-driving stresses are below allowable limits and reasonable blow count is attainable before actual field work starts. For case 2, given field observations made during pile driving, the analysis is used as a quality-control tool to evaluate pile capacity.

Generally, wave equation analysis is applied to a pile for the cases of several static-load resistances covering a wide range of values (at a constant pile penetration corresponding to the expected final pile-toe depth). Analysis results are then plotted as a *bearing graph* relating static pile capacity and driving stresses to blow counts.

Figure 7.27 presents a bearing graph from an analysis of a single-acting external-combustion hammer (Vulcan 012) and a precast concrete pile (18 in square, 95 ft long).

For a diesel hammer, the stroke or bounce-chamber pressure is also included in the plot. Alternatively, for an open-end diesel hammer (or any hammer with

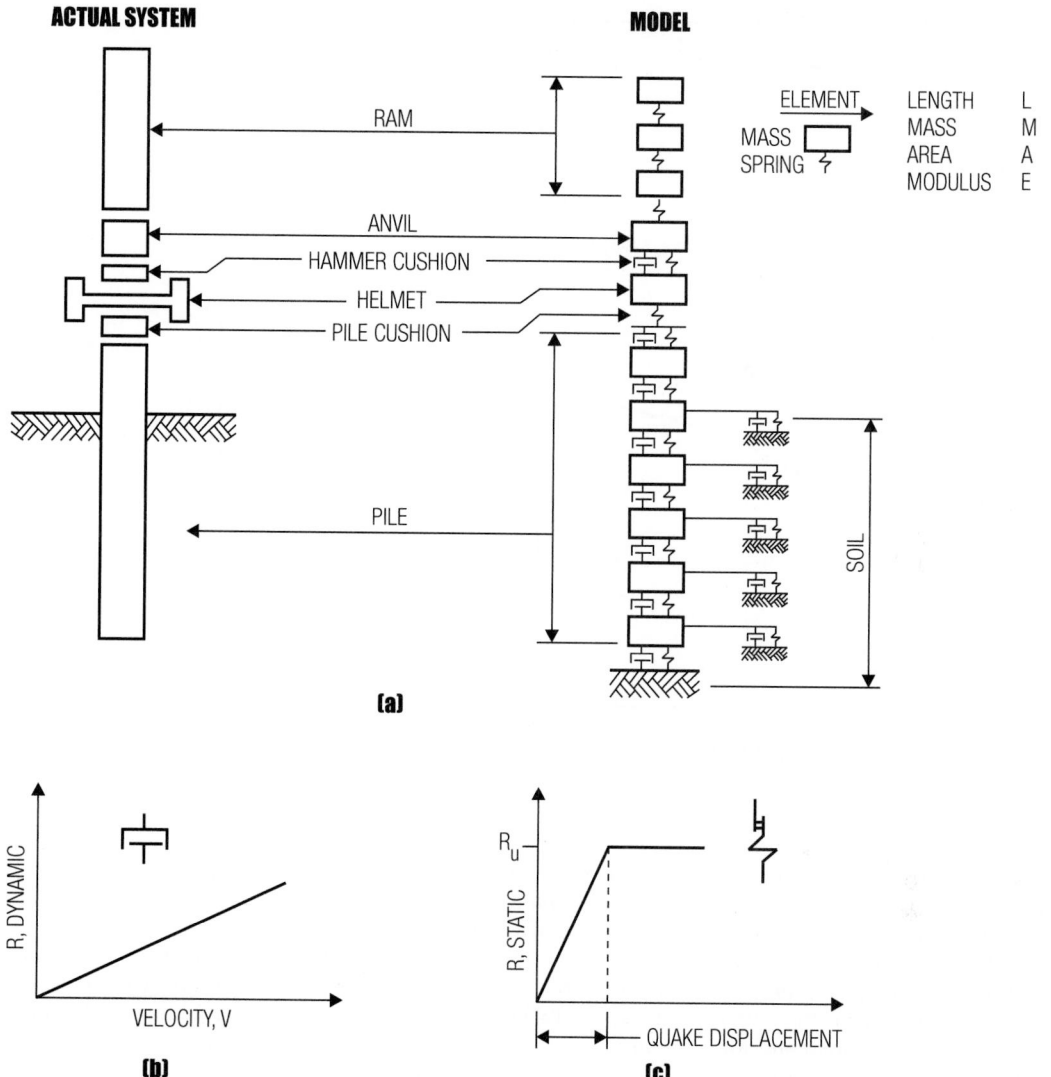

ACTUAL SYSTEM

MODEL

RAM

ANVIL
HAMMER CUSHION
HELMET
PILE CUSHION

PILE

SOIL

ELEMENT
MASS
SPRING

LENGTH   L
MASS   M
AREA   A
MODULUS   E

(a)

R, DYNAMIC

VELOCITY, V

(b)

R, STATIC

R_u

QUAKE DISPLACEMENT

(c)

Fig. 7.26 Lumped-mass model of a pile used in wave equation analysis. (*a*) Rectangular block with spring represents mass and stiffness; dashpot, the loading-dependent dynamic components; small square block with spring, the soil resistance forces along the pile shaft. (*b*) Variation of dynamic soil resistance with pile velocity. (*c*) Variation of static soil resistance with pile displacement.

variable stroke), the analysis may be performed with a constant pile static capacity and various strokes. In this way, the required blow count can be obtained as a function of the actual stroke.

Wave equation analysis may also be based on pile penetration (commonly termed pile drivability). In this way, variations of soil resistance with depth can be taken into account. Analysis results are obtained as a function of pile penetration.

Pile specifications prescribe use of wave equation analysis to determine suitability of a pile-driving system. Although it is an excellent tool for analysis of impact pile driving, the wave equation approach has some limitations. These are mainly

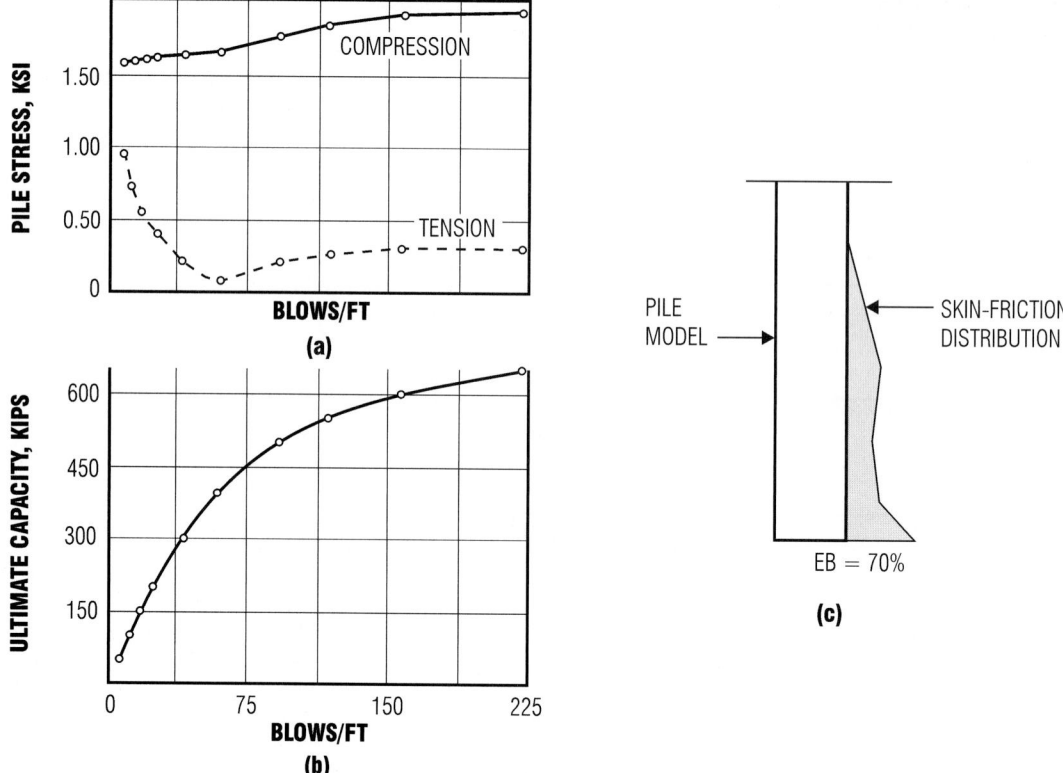

Fig. 7.27 Bearing graph derived from a wave equation analysis. (*a*) Variation of pile tensile and compressive stresses with blows per foot. (*b*) Ultimate capacity of pile indicated by blows per foot. (*c*) Skin friction distribution along the tested pile. Driving was done with a Vulcan hammer, model 012, with 67% efficiency. Helmet weighed 2.22 kips (k). Stiffness of hammer cushion was 5765 k/in and of pile cushion, 1620 k/in. Pile was 95 ft long and had a top area of 324 in². Other input parameters were quake (soil maximum elastic deformation), 0.100 in for shaft resistance and 0.150 in for toe resistance; soil damping factor, 0.150 s/ft for shaft and toe resistance.

due to uncertainties in quantifying some of the required inputs, such as hammer performance and soil parameters. The hammer efficiency value needed in the analysis is usually taken as the average value observed in many similar situations. Also, soil damping and quake values (maximum elastic soil deformation) needed in modeling soil behavior cannot be readily obtained from standard field or laboratory soil tests or related to other conventional engineering soil properties.

Dynamic-load pile testing and data analysis yield information regarding the hammer, driving system, and pile and soil behavior that can be used to confirm the assumptions of wave equation analysis. Dynamic-load pile tests are routinely performed on projects around the world for the purposes of monitoring and improving pile installation and as construction control procedures. Many professional organizations have established standards and guidelines for the performance and use of this type of testing; for example, ASTM (D4945), Federal Highway Administration ("Manual on Design and Construction of Driven Pile Foundation"). Dynamic-load testing methods are also effectively employed for evaluating cast-in-place piles ("Dynamic Load Testing of Drilled

Shaft—Final Report," Department of Civil Engineering, University of Florida, Gainesville 1991).

Main objectives of dynamic-load testing include evaluation of driving resistance and static-load capacity; determination of pile axial stresses during driving; assessment of pile structural integrity; and investigation of hammer and driving system performance.

(ASTM D 4945-96: Standard Test Method for High-Strain Dynamic Testing of Piles; "Dynamic Testing of Pile Foundations During Construction," M. Hussein and G. Likins, Proceedings of ASCE Structures Congress XIII, Boston, MA, 1995).

7.20.2 Case Method

A procedure developed at Case Institute of Technology (now Case Western Reserve University), Cleveland, Ohio, by a research team headed by G. G. Goble, enables calculation of pile static capacity from measurements of pile force and acceleration under hammer impacts during pile driving. The necessary equipment and analytical methods developed have been expanded to evaluate other aspects of the pile-driving process. These procedures are routinely applied in the field using a device called the **Pile Driving Analyzer (PDA)**. As an extension of the original work the researchers developed a computer program known as the CAse Pile Wave Analysis Program (CAPWAP), which is described later.

Measurements of pile force and velocity records under hammer impacts are the basis for modern dynamic pile testing. Data are obtained with the use of reusable strain transducers and accelerometers. Strain gages are bolted on the pile shaft, usually at a distance of about two pile diameters below the pile head. The PDA serves as a data acquisition system and field computer that provides signal conditioning, processing, and calibration of measurement signals. It converts measurements of pile strains and acceleration to pile force and velocity records. Dynamic records and testing results are available in real time following each hammer impact and are permanently stored in digital form. Using wave propagation theory and some assumptions regarding pile and soil, the PDA applies Case method equations and computes in a closed-form solution some 40 variables that fully describe the condition of the hammer-pile-soil system in real time following each hammer impact.

When a hammer or drop weight strikes the pile head, a compressive-stress wave travels down the pile shaft at a speed c, which is a function of the pile elastic modulus and mass density (Art. 6.82.1). The impact induces at the pile head a force F and a particle velocity v. As long as the wave travels in one direction, force and velocity are proportional; that is, $F = Zv$, where Z is the pile impedance, and $Z = EA/c$, where A is the cross-sectional area of the pile and E is its elastic modulus. Changes in impedance in the pile shaft and pile toe, and soil-resistance forces, produce wave reflections. The reflected waves arrive at the pile head after impact at a time proportional to the distance of their location from the toe. Soil-resistance forces or increase in pile impedance cause compressive-wave reflections that increase pile force and decrease velocity. Decrease in pile impedance has the opposite effect.

For a pile of length L, impedance Z, and stress-wave velocity c, the PDA computes total soil resistance from measured force and velocity records during the first stress-wave cycle; that is, when $0 < t \leq 2L/c$, where t is time measured from start of hammer impact. This soil resistance includes both static and viscous components. In the computation of pile bearing capacity under static load RS at the time of testing, effects of soil damping must be considered. Damping is associated with velocity. By definition, the Case method damping force is equal to $ZJ_c v_b$, where J_c is the dimensionless Case damping factor, and v_b the pile toe velocity, which can be computed from measured data at the pile head by applying wave mechanics principles. The static capacity of a pile can be calculated from:

$$RS = \frac{1}{2}[(1 - J_c)(Ft_1 + Zvt_1) + (1 + J_c)(Ft_2 - Zvt_2)]$$

$$(7.51a)$$

where $t_2 = t_1 + 2L/c$ and t_1 is normally the time of the first relative velocity peak. The damping constant J_c is related to soil grain size and may be taken for clean sands as 0.10 to 0.15, for silty sands as 0.15 to 0.25, for silts as 0.25 to 0.40, for silty clays as 0.4 to 0.7, and for clays as 0.7 to 1.0.

The computed RS value is the pile static capacity at the time of testing. Time-dependent effects can be evaluated by testing during pile restrikes. For this purpose, the pile must have sufficient penetration under the hammer impact to achieve full mobilization of soil-resistance forces. (F. Rausche, G. Goble, and G. Likins, "Dynamic Determination of Pile Capacity," *ASCE Journal of Geotechnical Engineering Division*, vol. 111, no. 3, 1985.)

The impact of a hammer subjects piles to a complex combination of compression, tension, torsional, and bending forces. Maximum pile compressive stress at the transducers' location is directly obtained from the measured data as the maximum recorded force divided by the pile area. For piles with mainly toe soil resistance, the compressive force at the pile toe is calculated from pile-head measurements and one-dimensional wave propagation considerations. Maximum tension force in the pile shaft can be computed from measurements near the pile head by considering the magnitude of both upward- and downward-traveling force components. Pile damage occurs if driving stresses exceed the strength of the pile material.

For a pile with a uniform cross-sectional area initially, damage after driving can be indicated by a change in area. Since pile impedance is proportional to pile area, a change in impedance would indicate pile damage. Hence, a driven pile can be tested for underground damage by measuring changes in pile impedance. Changes in pile impedance cause wave reflections and changes in the upward-traveling wave measured at the pile head. From the magnitude and time after impact of the relative wave changes, the extent and location of impedance change and hence of pile damage can be determined. Determination of pile damage can be assisted with the use of the PDA, which computes a relative integrity factor (unity for uniform piles and zero for a pile end) based on measured data near the pile head. (F. Rausche and G. G. Goble, "Determination of Pile Damage by Top Measurements," ASTM STP-670; "Structured Failure of Pile Foundations During Installation," M. J. Hussein and G. G. Goble, ASCE, Proceedings of the Construction Congress VI, Orlando, FL, 2000.)

The PDA also is helpful in determining the energy actually received by a pile from a hammer blow. Whereas hammers are assigned an energy rating by manufacturers, only the energy reaching the pile is of significance in effecting pile penetration. Due to many factors related to the hammer mechanical condition, driving-system behavior, and general dynamic hammer-cushions-pile-soil incompatibility, the percentage of potential hammer energy that actually reaches the pile is quite variable and often less than 50%. ("The Performance of Pile Driving Systems—Main Report," vol. 1–4, FHWADTFH 61-82-1-00059, Federal Highway Administration.) Figure 7.28 presents a summary of data obtained on hundreds of sites to indicate the percentage of all hammers of a specific type with an energy-transfer efficiency less than a specific percentage. Given records of pile force and velocity, the PDA calculates the transferred energy as the time integral of the product of force and velocity. The maximum transferred energy value for each blow represents the single most important parameter for an overall evaluation of driving-system performance.

7.20.3 CAPWAP Method

The CAse Pile Wave Analysis Program (CAPWAP) combines field-measured dynamic-load data and wave-equation-type analytical procedures to predict pile static-load capacity, soil-resistance distribution, soil-damping and quake values, pile load vs. movement plots, and pile-soil load-transfer characteristics. CAPWAP is a signal-matching or system identification method; that is, its results are based on a best possible match between a computed variable and its measured equivalent.

The pile is modeled with segments about 3 ft long with linearly elastic properties. Piles with nonuniform cross sections or composite construction can be accurately modeled. Static and dynamic forces along the pile shaft and under its toe represent soil resistance. Generally, the soil model follows that of the Smith approach (Art. 7.20.1) with modifications to account for full pile-penetration and rebound effects, including radiation damping. At the start of the analysis, an accurate pile model (incorporating splices, if present) is established and a complete set of soil constants is assumed. The hammer model used for the wave equation method is replaced by the measured velocity imposed as a boundary condition. The program calculates the force necessary to induce the imposed velocity. Measured and calculated forces are compared. If they do not agree, the soil model is adjusted and the analysis repeated. This iterative process is continued until no further improvement in the match can be obtained. The total number of unknowns to be evaluated during the analysis is $N_s + 18$, where N_s is the number of soil elements. Typically, one soil

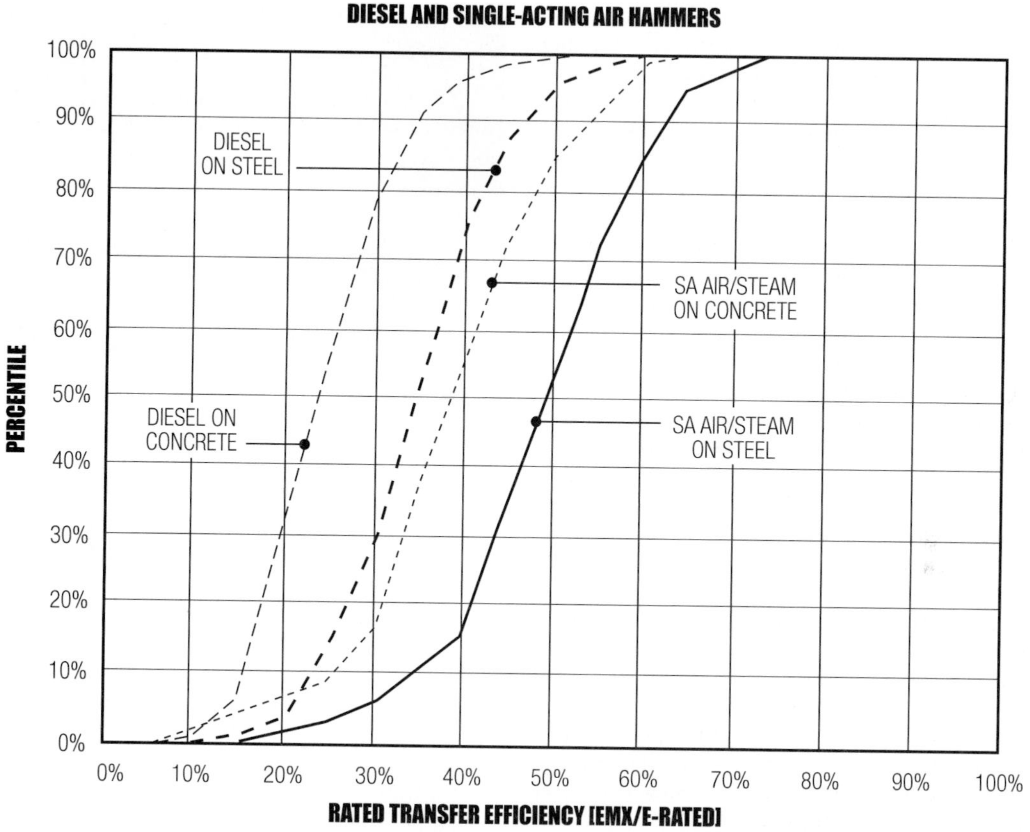

DIESEL AND SINGLE-ACTING AIR HAMMERS

Fig. 7.28 Comparison of performance of two types of hammers when driving steel or concrete piles. The percentile indicates the percentage of all hammers in each case with a rated transfer efficiency less than a specific percentage.

element is placed at every 6 ft of pile penetration plus an additional one under the toe.

Results that can be obtained from a CAPWAP analysis include the following:

Comparisons of measured values with corresponding computed values

Soil-resistance forces and their distribution for static loads

Soil-stiffness and soil-damping parameters along the pile shaft and under its toe

Forces, velocities, displacements, and energies as a function of time for all pile segments

Simulation of the relationship between static loads and movements of pile head and pile toe

Pile forces at ultimate soil resistance

Correlations between CAPWAP predicted values and results from static-load tests indicate very good agreement. (ASCE Geotechnical Special Publication No. 40, 1994.)

7.20.4 Low-Strain Dynamic Integrity Testing

The structural integrity of driven or cast-in-place concrete piles may be compromised during installation. Piles may also be damaged after installation by large lateral movements from impacts of heavy equipment or from slope or retaining-wall failures. Procedures such as excavation around a suspect pile or drilling and coring through its shaft are crude methods for investigating possible pile damage. Several testing techniques are available, however, for evaluation of the structural integrity of deep foundation elements in a more sophisticated manner (W. G. Fleming, A. J. Weltmen, M. F. Randolph, and W. K. Elson, "Piling Engineering," Surrey University Press, London). Some of these tests, though, require that the pile be prepared or instrumented before or during installation. These requirements make random application prohibitively expensive, if not impossible. A convenient and economical method is the low-strain pulse-echo technique, which requires relatively little instrumentation and testing effort and which is employed in low-strain, dynamic-load integrity testing. This method is based on one-dimensional wave mechanics principles and the measurement of dynamic-loading effects at the pile head under impacts of a small hand-held hammer. (ASTM D 5882-00: Standard Test Method for Low-Strain Integrity Testing of Piles.)

The following principle is utilized: When impacted at the top, a compressive-stress wave travels down the pile shaft at a constant speed c and is reflected back to the pile head from the toe. Changes in pile impedance Z change wave characteristics and indicate changes in pile cross-sectional size and quality and thus possible pile damage (Art. 7.20.2). Low-strain integrity testing is based on the premise that changes in pile impedance and soil-resistance forces produce predictable wave reflections at the pile head. The time after impact that the reflected wave is recorded at the pile head can be used to calculate the location on the pile of changes in area or soil resistance.

Field equipment consists of an accelerometer, a hand-held hammer (instrumented or without instrumentation), dedicated software, and a Pile Integrity Tester (Fig. 7.29), a data acquisition system capable of converting analog signals to digital form, data processing, and data storage. Pile preparation involves smoothing and leveling of a small area of the pile top. The accelerometer is affixed to the pile top with a jell-type material, and hammer blows are applied to the pile head. Typically, pile-head data resulting from several hammer blows are averaged and analyzed.

Data interpretation may be based on records of pile-top velocity (integral of measured acceleration), data in time or frequency domains, or more rigorous dynamic analysis. For a specific stress-wave speed (typically 13,000 ft/s), records of velocity at the pile head can be interpreted for pile

Fig. 7.29 Pile Integrity Tester. (*Courtesy of Pile Dynamics, Inc., Cleveland, Ohio.*)

nonuniformities and length. As an example, Fig. 7.30 shows a plot in which the abscissa is time, measured starting from impact, and the ordinate is depth below the pile top. The times that

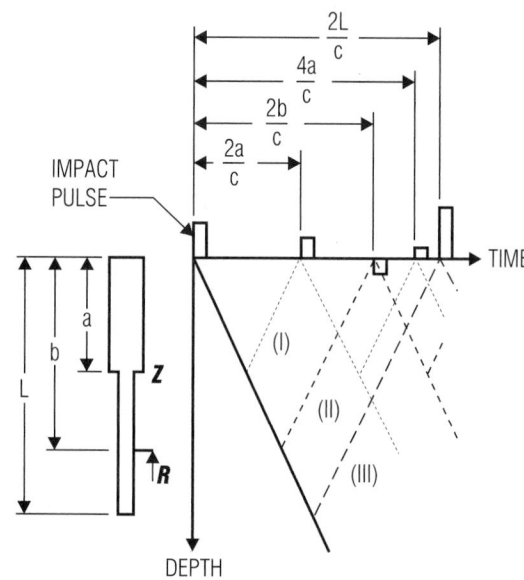

Fig. 7.30 Graph relates distance from a pile head to a depth where a change in the pile cross section or soil resistance occurs to the time it takes an impact pulse applied at the pile head and traveling at a velocity c to reach and then be reflected from the change back to the pile head. Line I indicates the reflection due to impedance, II the reflection due to passive resistance R (modeled velocity proportional), and III the reflection from the pile toe.

changes in wave characteristics due to pile impedance or soil resistance are recorded at the pile head are represented along the time axis by small rectangles. The line from the origin extending downward to the right presents the position of the wave traveling with velocity c after impact. Where a change in pile impedance Z occurs, at depth a and time a/c, a line (I) extends diagonally upward to the right and indicates that the wave reaches the pile top at time $2a/c$. Hence, with the time and wave velocity known, the distance a can be calculated. Similarly, from time $2b/c$, as indicated by line II, the distance b from the pile top of the change in soil resistance R can be computed. Line III indicates that the wave from the toe at distance L from the pile head reaches the head at time $2L/c$.

Dynamic analysis may be done in a signal-matching process or by a method that generates a pile impedance profile from the measured pile-top data. (F. Rausche et al., "A Formalized Procedure for Quality Assessment of Cast-in-Place Shafts Using Sonic Pulse Echo Methods," Transportation Research Board, Washington, D.C. 1994.)

The low-strain integrity method is applicable to concrete (cast-in-place and driven) and wood piles. Usually, piles are tested shortly after installation so that deficiencies may be detected early and corrective measures taken during foundation construction and before erection of the superstructure. As for other nondestructive testing methods, the results of measurements recorded may be divided into four main categories: (1) clear indication of a sound pile, (2) clear indication of a serious defect, (3) indication of a somewhat defective pile, and (4) records do not support any conclusions. The foundation engineer, taking into consideration structural, geotechnical, and other relevant factors, should decide between pile acceptability or rejection.

The low-strain integrity method may be used to determine the length and condition of piles under existing structures. (M. Hussein, G. Likins, and G. Goble, "Determination of Pile Lengths under Existing Structures," Deep Foundations Institute, 1992, www.dfi.org.)

The method has some limitations. For example, wave reflections coming from locations greater than about 35 pile diameters may be too weak to be detected at the pile head with instruments currently available. Also, gradual changes in pile impedance may escape detection. Furthermore, the method may not yield reliable results for steel piles.

Concrete-filled steel pipe piles may be evaluated with this method.

7.21 Specification Notes

Specifications for pile installation should provide realistic criteria for pile location, alignment, and minimum penetration or termination driving resistance. Particular attention should be given to provisions for identification of pile heave and relaxation and for associated remedial measures. Corrective actions for damaged or out-of-position piles should also be identified. Material quality and quality control should be addressed, especially for cast-in-place concrete piles. Tip protection of piles is an important consideration for some types of high-capacity, end-bearing piles or piles driven through obstructions. Other items that may be important are criteria for driving sequence in pile groups, preexcavation procedures, protection against corrosive subsoils, and control of pile driving in proximity to open or recently concreted pile shells. Guidelines for selected specification items are in Table 7.10.

The following is a list of documents containing sample guidelines, standards, and specifications related to deep foundations design and construction. Clear, comprehensive, reasonable, and fair project specifications greatly reduce the potential for disputes and costly delays.

"Guidelines for writing Construction Specifications for Piling" — Deep Foundations Institute, www.dfi.org.

"Recommended Design Specifications for Driven Bearing Piles" — Pile Driving Contractors Association, www.piledrivers.org.

"Standard Guidelines for the Design and Installation of Pile Foundations" — American Society of Civil Engineers, www.asce.org.

"Standards and Specifications for the Foundation Drilling Insustry" — International Association of Foundation Drilling, www.adsc-iafd.com.

"Standard Specifications for Road and Bridge Construction, Section 455: Structures Foundations" — Florida Department of Transportation, www.dot.state.fl.us.

"Standard Specification for the Construction of Drilled Piers" — Americal Concrete Institute, www.aci-int.org.

Table 7.10 Guide to Selected Specification Provisions

Position	. . . within 6 in of plan location (3 in for pile groups with less than 5 piles)
Plumbness	. . . deviation from vertical shall not exceed 2% in any interval (4% from axis for batter piles)
Pile-hammer assembly	Verification of the suitability of the proposed hammer assembly to drive the designated piles shall be provided by wave equation or equivalent analysis subject to the engineer's approval.
Pile-driving leaders	All piles shall be driven with fixed leaders sufficiently rigid to maintain pile position and axial alignment during driving.
Driving criteria	. . . to an elevation of at least ___ and/or to a terminal driving resistance of __ blows/__in
Indicator piles	Before start of production driving, indicator piles should be driven at locations determined by the engineer. Continuous driving resistance records shall be maintained for each indicator pile.
Preboring	Preboring immediately preceding pile installation shall extend to elevation ___. The bore diameter for friction piles shall not be less than 1 or more than 2 in smaller than the pile diameter.
Heave*	The elevation of pile butts or tips of CIPC pile shells shall be established immediately after driving and shall be resurveyed on completion of the pile group. Should heave in excess of $\frac{1}{4}$ in be detected, the piles shall be redriven to their initial elevation or as directed by the engineer.
Relaxation or setup[†]	Terminal driving resistance of piles in-place for at least 24 h shall be redriven as directed by the engineer.

* Can be initially conducted on a limited number of pile groups and subsequently extended to all piles if required.

[†] May be specified as part of initial driving operations.

"Recommended Practice for Design, Manufacturing, and Installation of Prestressed Concrete Piling" — Precast/Prestressed Concrete Institute, www.pci.org.

"International Building Code — Chapter 18: Soils and Foundations" — International Code Council, 5203 Leesburg Pike, Falls Church, Virginia 22041.

7.22 Drilled Shafts

Drilled shafts are commonly used to transfer large axial and lateral loads to competent bearing materials by shaft or base resistance or both. Also known as drilled piers, drilled-in caissons, or large-diameter bored piles, drilled shafts are cylindrical, cast-in-place concrete shafts installed by large-diameter, auger drilling equipment. Shaft diameters commonly range from 2.5 to 10 ft and lengths from 10 to 150 ft, although shafts with dimensions well outside these ranges can be installed. Shafts may be of constant diameter (straight shafts, Fig. 7.31a) or may be underreamed (belled Fig. 7.31b) or socketed into rock (Fig. 7.31c). Depending on load

requirements, shafts may be concreted with or without steel reinforcement.

Under appropriate foundation conditions, a single drilled shaft is well-suited for support of very heavy concentrated loads; 2000 tons with rock bearing is not unusual.

Subsurface conditions favoring drilled shafts are characterized by materials and groundwater

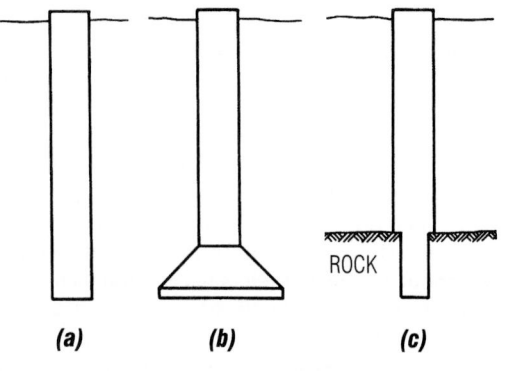

Fig. 7.31 Types of drilled shafts.

conditions that do not induce caving or squeezing of subsoils during drilling and concrete placement. High-capacity bearing levels at moderate depths and the absence of drilling obstructions, such as boulders or rubble, are also favorable conditions. Current construction techniques allow drilled shafts to be installed in almost any subsurface condition, although the cost-effectiveness or reliability of the system will vary significantly.

7.22.1 Construction Methods for Drilled Shafts

In stable soil deposits, such as stiff clays, concrete with or without reinforcement may be placed in uncased shafts. Temporary casing, however, may be employed during inspection of bearing conditions. Temporary casing may also be installed in the shaft during or immediately after drilling to prevent soil intrusion into the concrete during placement. During this process, the height of concrete in the casing should at all times be sufficient so that the weight more than counterbalances hydrostatic heads imposed by groundwater or by fluid trapped in the annular space between the soil and the casing. The lack of attention to this requirement is perhaps the greatest contributor to drilled-shaft failures.

Unstable soil conditions encountered within a limited interval of shaft penetration can be handled by advancing a casing into stable subsoils below the caving zone via vibratory driving or by screwing down the casing with a torque-bar attachment. The hole is continued by augering through the casing, which may subsequently be retracted during concrete placement or left in place. A shaft through unstable soils may also be advanced without a casing if a weighted drilling fluid (slurry) is used to prevent caving.

For a limited unstable zone, underlain by relatively impervious soils, a casing can be screwed into these soils so as to form a water seal. This allows the drilling slurry to be removed and the shaft continued through the casing and completed by normal concreting techniques.

The shaft may also be advanced entirely by slurry drilling techniques. With this method, concrete is tremied in place so as to completely displace the slurry.

Reinforcing steel must be carefully designed to be stable under the downward force exerted by the concrete during placement. Utilization of reinforcement that is not full length is not generally recommended where temporary casing is used to facilitate concrete placement or slurry methods of construction.

Concrete can be placed in shafts containing not more than about 4 in of water (less for belled shafts). Free-fall placement can be used if an unobstructed flow is achieved. Bottom-discharge hoppers centered on the shaft facilitate unrestricted flow, whereas flexible conduits (elephant trunks) attached to the hopper can be used to guide concrete fall in heavily reinforced shafts. Rigid tremie pipes are employed to place concrete in water or in slurry-filled shafts.

Equipment and Tools ▪ Large-diameter drills are crane- and truck-mounted, depending on their size and weight. The capacity of the drill is rated by its maximum continuous torque, ft-lb, and the force exerted on the drilling tool. This force is the weight of the **Kelly bar** (drill stem) plus the force applied with some drills by their Kelly-bar down-crowd mechanism.

Downward force on the auger is a function of the Kelly-bar length and cross section. Telescoping Kelly bars with cross sections up to 12 in square have been used to drill 10-ft-diameter shafts in earth to depths over 220 ft. Solid pin-connected Kelly sections up to 8 in square have also been effectively used for drilling deep holes. Additional down-crowd forces exerted by some drills are on the order of 20 to 30 kips (crane-mounts) and 15 to 50 kips (truck-mounts).

Drilling tools consisting of open helix (single-flight) and bucket augers are typically used for earth drilling and may be interchanged during construction operations. To drill hard soil and soft and weathered rock more efficiently, flight augers are fitted with hard-surfaced teeth. This type of auger can significantly increase the rate of advance in some materials and provides a more equitable definition of "rock excavation" when compared to the refusal of conventional earth augers. Flight augers allow a somewhat faster operation and in some circumstances have a superior penetration capability. Bucket augers are usually more efficient for excavating soft soils or running sands and provide a superior bottom cleanout.

Belled shafts in soils and soft rocks are constructed with special underreaming tools. These are usually limited in size to a diameter three times the diameter of the shaft. Hand mining techniques may be required where hard seams or other obstructions limit machine belling.

Cutting tools consisting of roller bits or core barrels are typically used to extend shafts into harder rock and to form rock sockets. Multiroller bits are often used with a *reverse circulation* type of rotary drilling rig. This technique, together with percussion bits activated by pneumatically powered drills, usually provide the most rapid advance in rock but have the disadvantage of requiring special drill types that may not be efficient in earth drilling.

7.22.2 Construction Quality Control and Assurance for Drilled Shafts

During the preparation of drilled-shaft design and construction specifications, special attention should be given to construction-related design features, including shaft types, shaft-diameter variations, site trafficability, ground-loss potential, and protection of adjacent facilities. Proper technical specifications and contract provisions for such items as payment for rock excavation and drilling obstructions are instrumental in preventing significant cost overruns and associated claims.

Some cost-reduction or quality-related precautions in preparation of drilled-shaft designs and specifications are:

1. Minimize the number of different shaft sizes; extra concrete quantities for diameters larger than actually needed are usually far less costly than use of a multiplicity of drilling tools and casings.

2. Delete a requirement for concrete vibration and use concrete slumps not less than 6 ± 1 in.

3. Do not leave a casing in place in oversized holes unless pressure grouting is used to prevent ground loss.

4. Shaft diameters should be at least 2.5 ft, preferably 3 ft.

Tolerances for location of drilled shafts should not exceed 3 in or $\frac{1}{24}$ of the shaft diameter, whichever is less. Vertical deviation should not be more than 2% of the shaft length or 12.5% of the diameter, whichever controls, except for special conditions.

Provisions for proof testing are extremely important for shafts designed for high-capacity end bearing. This is particularly true for bearing materials that may contain discontinuities or have random variations in quality. Small percussion drills (jackhammers) are often used for proof testing and may be supplemented by diamond coring, if appropriate.

Because many drilled-shaft projects involve variations in bearing levels that cannot be quantified during the design stage, the limitations of the bearing level and shaft quantities estimated for bidding purposes must be clearly identified. Variations in bearing level and quality are best accommodated by specifications and contract provisions that facilitate field changes. The continuous presence of a qualified engineer inspector, experienced in drilled-shaft construction, is required to ensure the quality and cost-effectiveness of the construction.

7.22.3 Drilled-Shaft Design

Much of the design methodology for drilled shafts is similar to that applied to pile foundations and usually differs only in the manner in which the design parameters are characterized. Consequently, drilled-shaft design may be based on precedent (experience), load testing, or static analyses. The ultimate-load design approach is currently the most common form of static analysis applied, although load-deformation compatibility methods are being increasingly used (see Art. 7.18).

7.22.4 Skin Friction in Cohesive Soils

Ultimate skin friction for axially loaded shafts drilled into cohesive soils is usually evaluated by application of an empirically derived reduction (adhesion) factor to the undrained shear strength of the soil in contact with the shaft [see Eq. (7.37)]. For conventionally drilled shafts in stiff clays ($c_u \geq 0.50$ tons/ft^2), the adhesion factor α has been observed to range usually between 0.3 and 0.6.

Based on analysis of high-quality load-test results, primarily in stiff, fissured Beaumont and London clays, α factors of 0.5 and 0.45 have been recommended. Unlike the criteria applied to pile design, these factors are independent of c_u, but are largely dependent on construction methods and practices. Reese has recommended that the shaft length assumed to be effective in transferring load should be reduced by 5 ft to account for base interaction effects and that a similar reduction be applied to account for surface effects such as soil shrinkage.

Table 7.11 lists recommended α factors for straight shafts as a function of the normalized shear strength c_u/σ'_{vo} (Art. 7.5.1) and plasticity index I_p (Art. 7.4). These factors reflect conventional dry-hole construction methods and the influence of the stress history and plasticity of the soil in contact with the shaft. The α factors in Table 7.11 may be linearly interpreted for specific values of c_u/σ'_{vo} and I_p. To account for tip effects, the part of the shaft located 1 diameter above the base should be ignored in evaluation of Q_{su} (see Art. 7.17).

Where shafts are drilled to bearing on relatively incompressible materials, the amount of relative movement between the soil and the shaft may be insufficient to develop a substantial portion of the ultimate skin friction, particularly for short, very stiff shafts. Under these circumstances, Q_s should be ignored in design or analyzed with load-displacement compatibility procedures.

Because belled shafts usually require larger deformations than straight shafts to develop design loads, Q_{su} may be reduced for such shafts as a result of a progressive degradation at relative deformations greater than that required to develop peak values. Limited data on belled vs. straight shafts suggest a reduction in the α factor on the order of 15% to account for the reduced shaft friction of the belled shafts. It is also conservative to assume that there is no significant load transfer by friction in that portion of the shaft located about 1 diameter above the top of the bell.

There is some evidence that when shaft drilling is facilitated by the use of a weighted drilling fluid (mud slurry), there may be a substantial reduction in Q_{su}, presumably as a result of entrapment of the slurry between the soil and shaft concrete. Where this potential exists, it has been suggested that the α factor be reduced about 40%.

(A. W. Skempton, "Summation, Symposium on Large Bored Piles," Institute of Civil Engineers, London; L. C. Reese, F. T. Toma, and M. W. O'Neill, "Behavior of Drilled Piers under Axial Loading," *ASCE Journal of Geotechnical Engineering Division,* vol. 102, no. GT5, 1976; W. S. Gardner, "Investigation of the Effects of Skin Friction on the Performance of Drilled Shafts in Cohesive Soils," Report to U.S. Army Engineers Waterways Experiment Station (Contract no. DACA 39-80-C-0001), vol. 3, Vicksburg, Miss., 1981.)

7.22.5 Skin Friction in Cohesionless Soils

Ultimate skin friction in cohesionless soils can be evaluated approximately with Eq. (7.40). In the absence of more definitive data, K in Eq. (7.40) may be taken as 0.6 for loose sands and 0.7 for medium dense to dense sands, on the assumption that the soil-shaft interface friction angle is taken as $\phi' - 5°$. As for piles, limited test data indicate that the average friction stress f_{su} is independent of overburden pressure for shafts drilled below a critical depth z_c of from 10 (loose sand) to 20 (dense sand) shaft diameters. The limiting skin friction f_i for shafts with ratio of length to diameter $L/D \leq 25$ should appreciably not exceed 1.0 ton/ft^2. Average f_{su} may be less than 1.0 ton/ft^2 for shafts longer than about 80 ft.

Equation (7.52) approximately represents a correlation between f_{su} and the average standard penetration test blow count $\bar{N}$ within the embedded pile length recommended for shafts in sand with effective $L/D \leq 10$. The stress f_{su} so computed, however, is less conservative than the foregoing design approach, particularly for $\bar{N} \geq 30$ blows per foot.

$$f_{su} = 0.03\bar{N} \leq 1.6 \text{ tons/ft}^2 \qquad (7.52)$$

7.22.6 End Bearing on Soils

End bearing of drilled shafts in cohesive soils is typically evaluated as described for driven piles [Eq. (7.42)]. The shear-strength term in this equation represents the average c_u within a zone of 2

Table 7.11 Adhesion Factors α for Drilled Shafts

Plasticity Index	Normalized Shear Strength c_u/σ'_{vo} *		
	0.3 or Less	1.0	2.5 or More
20	0.33	0.44	0.55
30	0.36	0.48	0.60
60	0.40	0.52	0.65

* Based on *UU* tests on good-quality samples selected so that c_u is not significantly influenced by the presence of fissures.

diameters below the shaft space. For smaller shafts, the suggested reduction factor is 0.8.

End bearing in cohesionless soils can be estimated in accordance with Eq. (7.43) with the same critical-depth limitations described for pile foundations. N_q for drilled shafts, however, has been observed to be significantly smaller than that applied to piles (see Fig. 7.19). Meyerhof has suggested that N_q should be reduced by 50%. Alternatively, q_u can be expressed in terms of the average SPT blow count $\bar{N}$ as:

$$q_u = 0.67\bar{N} \leq 40 \, \text{tons/ft}^2 \qquad (7.53)$$

where q_u = ultimate base resistance at a settlement equivalent to 5% of the base diameter. (G. G. Meyerhof, "Bearing Capacity and Settlement of Pile Foundations," *ASCE Journal of Geotechnical Engineering Division*, vol. 102, no. GT3, 1976.)

7.22.7 Shaft Settlement

Drilled-shaft settlements can be estimated by load-deformation compatibility analyses (see Art. 7.18). Other methods used to estimate settlement of drilled shafts, singly or in groups, are identical to those used for piles (Art. 7.19.5). These include elastic, semiempirical elastic, and load-transfer solutions for single shafts drilled in cohesive or cohesionless soils. (H. G. Poulos and E. H. Davis, "Elastic Solutions for Soil and Rock Mechanics," John Wiley & Sons, Inc., New York; A. S. Vesic, "Principles of Pile Foundation Design," Soil Mechanics Series no. 38, Duke University, Durham, N.C., 1975; H. Y. Fang, "Foundation Engineering Handbook," 2nd ed., Van Nostrand Reinhold, New York.)

Resistance to tensile and lateral loads by straight-shaft drilled shafts should be evaluated as described for pile foundations (see Art. 7.19).

7.22.8 Rock-Supported Shafts

Drilled shafts may be designed to be supported on rock or to be socketed into rock. Except for long, relatively small-diameter (comparatively compressible) shafts, conventional design ignores the skin friction of belled or straight shafts founded on relatively incompressible materials. Where shafts are socketed in rock, the design capacity is considered a combination of the sidewall shearing resistance (bond) and the end bearing of the socket.

In practice, both end-bearing and rock-socket designs are based on local experience, presumptive values in codes or semi empirical design methods. Latter methods based on field load test results are preferred for design efficiency.

Bearing values on rock given in design codes typically range from 50 to 100 tons/ft² for massive crystalline rock, 20 to 50 tons/ft² for sound foliated rock, 15 to 25 tons/ft² for sound sedimentary rock, 8 to 10 tons/ft² for soft and fractured rock, and 4 to 8 tons/ft² for soft shales.

The supporting ability of a specific rock type is primarily dependent on the frequency, orientation, and size of the discontinuities within the rock mass and the degree of weathering of the rock minerals. Consequently, application of presumptive bearing values is not recommended without specific local performance correlations. (R. W. Woodward, W. S. Gardner, and D. M. Greer, "Drilled Pier Foundations," McGraw-Hill Book Company, New York (books.mcgraw-hill.com).)

Some analyses relate the bearing values q_u in jointed rock to the uniaxial compressive (UC) strength of representative rock cores. These analyses indicate that q_u should not be significantly less than UC, possibly excluding weak sedimentary rocks such as compacted shales and siltstones. With a safety factor of 3, the maximum allowable bearing value q_a can be taken as: $q_a \leq 0.3UC$. In most instances, however, the compressibility of the rock mass rather than rock strength governs. Elastic solutions can be used to evaluate the settlement of shafts bearing on rock if appropriate deformation moduli of the rock mass E_r can be determined. (H. G. Poulos and E. H. Davis, "Elastic Solutions for Soil and Rock Mechanics," John Wiley & Sons, Inc., New York (www.wiley.com); D. U. Deere, A. J. Hendron, F. D. Patton, and E. J. Cording, "Breakage of Rock," *Eighth Symposium on Rock Mechanics*, American Institute of Mining and Metallurgical Engineers, Minneapolis, Minn., 1967; F. H. Kulhawy, "Geotechnical Model for Rock Foundation Settlement," *ASCE Journal of Geotechnical Engineering Division*, vol. 104, no. GT2, 1978.)

Concrete-rock bond stresses f_R used for the design of rock sockets have been empirically established from a limited number of load tests. Typical values range from 70 to 200 psi, increasing with rock quality. For good-quality rock, f_R may be related to the 28-day concrete strength f'_c and to the uniaxial compressive (UC) strength of rock cores. For rock with $RQD \geq 50\%$ (Table 7.3), f_R can be estimated as $0.05f'_c$ or $0.05UC$ strength, whichever is smaller, except that f_R should not exceed 250 psi. As shown by Fig. 7.31, the ultimate rock-concrete bond f_{Ru} is significantly higher than f_R except for very high UC values. (P. Rosenberg and N. L. Journeaux, "Friction and End-Bearing Tests on Bedrock for High-Capacity Socket Design," *Canadian Geotechnical Journal*, vol. 13, no. 3, 1976.)

Design of rock sockets is conventionally based on

$$Q_d = \pi d_s L_s f_R + \frac{\pi}{4} d_s^2 q_a \qquad (7.54)$$

where Q_d = allowable design load on rock socket

   d_s = socket diameter

   L_s = socket length

   f_R = allowable concrete-rock bond stress

   q_a = allowable bearing pressure on rock

Load-distribution measurements show, however, that much less of the load goes to the base than is indicated by Eq. (7.54). This behavior is demonstrated by the data in Table 7.12, where L_s/d_s is the ratio of the shaft length to shaft diameter and E_r/E_p is the ratio of rock modulus to shaft modulus. The finite-element solution summarized in Table 7.12 probably reflects a realistic trend if the average socket-wall shearing resistance does not exceed the ultimate f_R value; that is, slip along the socket sidewall does not occur.

A simplified design approach, taking into account approximately the compatibility of the socket and base resistance, is applied as follows:

1. Proportion the rock socket for design load Q_d with Eq. (7.54) on the assumption that the end-bearing stress is less than q_a [say $q_a/4$, which

Table 7.12 Percent of Base Load Transmitted to Rock Socket

L_s/d_s	E_r/E_p		
	0.25	1.0	4.0
0.5	54*	48	44
1.0	31	23	18
1.5	17*	12	8*
2.0	13*	8	4

*Estimated by interpretation of finite-element solution; for Poisson's ratio = 0.26.

is equivalent to assuming that the base load $Q_b = (\pi/4)d_s^2 q_a/4$].

2. Calculate $Q_b = RQ_d$, where R is the base-load ratio interpreted from Table 7.12.

3. If RQ_d does not equal the assumed Q_b, repeat the procedure with a new q_a value until an approximate convergence is achieved and $q \le q_a$.

The final design should be checked against the established settlement tolerance of the drilled shaft. (B. Ladanyi, discussion of "Friction and End-Bearing Tests on Bedrock," *Canadian Geotechnical Journal*, vol. 14, no. 1, 1977; H. G. Poulos and E. H. Davis, "Elastic Solutions for Rock and Soil Mechanics," John Wiley & Sons, Inc., New York (www.wiley.com).)

Following the recommendations of Rosenberg and Journeaux, a more realistic solution by the above method is obtained if f_{Ru} is substituted for f_R. Ideally, f_{Ru} should be determined from load tests. If this parameter is selected from Fig. 7.32 or from other data that are not site-specific, a safety factor of at least 1.5 should be applied to f_{Ru} in recognition of the uncertainties associated with the UC strength correlations. (P. Rosenberg and N. L. Journeaux, "Friction and End-Bearing Tests on Bedrock for High-Capacity Socket Design," *Canadian Geotechnical Journal*, vol. 13, no. 3, 1976.)

7.22.9 Testing of Drilled Shafts

Static-load capacity of drilled shafts may be verified by either static-load or dynamic-load testing (Arts. 7.19 and 7.20). Testing by applying static loads on the shaft head (conventional static-load test) or against the toe (Osterberg cell) provides information on shaft capacity and general behavior. Dynamic-load testing in which pile-head force and velocity under the impact of a falling weight are measured with a Pile Driving Analyzer and subsequent analysis with the CAPWAP method (Art. 7.20.3) provide information on the static-load capacity and shaft-movement and shaft-soil load-transfer relationships of the shaft.

Structural integrity of a drilled shaft may be assessed after excavation or coring through the shaft. Low-strain dynamic-load testing with a Pile Integrity Tester (Art. 7.20.4), offers many advantages, however. Alternative integrity evaluation methods are parallel seismic or cross-hole sonic logging.

For parallel seismic testing, a small casing is inserted into the ground near the tested shaft and

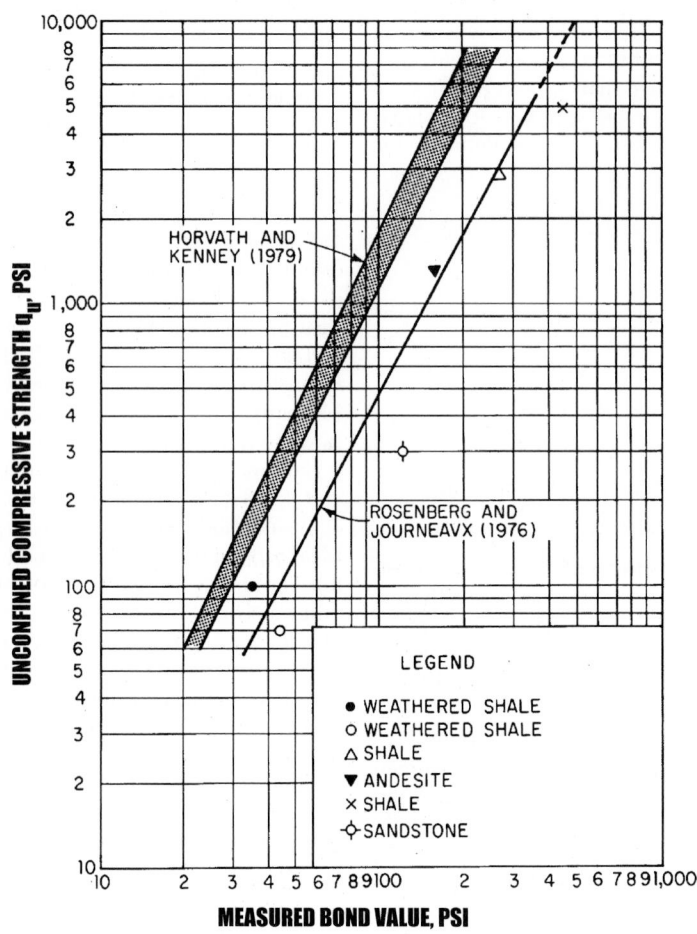

Fig. 7.32 Chart relates the bond of a rock socket to the unconfined compressive strength of cores.

to a greater depth than the shaft length. A hydrophone is lowered into the casing to pick up the signals resulting from blows on the shaft head from a small hand-held hammer. Inasmuch as wave velocity in the soil and shaft are different, the unknown length of the pile can be discerned from a series of measurements. One limitation of this method is the need to bore a hole adjacent to the shaft to be tested.

Cross-hole testing requires two full-length longitudinal access tubes in the shaft. A transmitter is lowered through one of the tubes to send a signal to a receiver lowered into the other tube. The arrival time and magnitude of the received signal are interpreted to assess the integrity of the shaft

between the two tubes. For large-diameter shafts, more than two tubes may be needed for thorough shaft evaluation. A disadvantage of this method is the need to form two or more access tubes in the shaft during construction. Furthermore, random testing or evaluations of existing shafts may not be possible with this method.

(C. L. Crowther, "Load Testing of Deep Foundations," John Wiley & Sons, Inc., New York (www.wiley.com); "New Failure Load Criterion for Large Diameter Bored Piles in Weathered Geometerials," by Charles W. W. Ng, et al., *ASCE Journal of Geotechnical and Geoenvironmental Engineering*, vol. 127, no. 6, June 2001.)

Retaining Methods For Excavation

The simplest method of retaining the sides of an excavation in soil is to permit the soil to form a natural slope that will be stable even in the presence of water. When there is insufficient space for this inside the excavation or when the excavation sides must be vertical, construction such as that described in the following must be used.

7.23 Caissons

Load-bearing enclosures known as caissons are formed in the ground, usually to protect excavation for a foundation, aid construction of the substructure, and serve as part of the permanent structure. Sometimes, a caisson is used to enclose a subsurface space to be used for such purposes as a pump well, machinery pit, or access to a deeper shaft or tunnel. Several caissons may be aligned to form a bridge pier, bulkhead, seawall, foundation wall for a building, or impervious core wall for an earth dam.

For foundations, caissons are used to facilitate construction of shafts or piers extending from near the surface of land or water to a bearing stratum. This type of construction can carry heavy loads to great depths. Built of common structural materials, they may have any shape in cross section. They range in size from about that of a pile to over 100 ft in length and width. Some small ones are considered drilled shafts. Previously described construction methods for drilled shafts are employed based on subsurface conditions and drilled shaft depth and diameter.

Caissons often are installed by sinking them under their own weight or with a surcharge. The operation is assisted by jacking, jetting, excavating, and undercutting. Care must be taken during this operation to maintain alignment. The caissons may be built up as they sink, to permit construction to be carried out at the surface, or they may be completely prefabricated. Types of caissons used for foundation work are as follows:

Chicago caissons are used for constructing foundation shafts through a thick layer of clay to hardpan or rock. The method is useful where the soil is sufficiently stiff to permit excavation for short distances without caving. A circular pit about 5 ft deep is dug and lined with wood staves. This vertical lagging is braced with two rings made with steel channels. Then, 5 ft of soil is removed, and the operation is repeated. If the ground is poor, shorter lengths are dug until the bearing stratum is reached. If necessary, the caissons can be belled at the bottom to carry large loads. Finally, the hole is filled with concrete. Minimum economical diameter for hand digging is 4 ft.

Sheeted piers or caissons are similarly constructed, but the vertical lagging of wood or steel is driven down during or before excavation. This system usually is used for shallow depths in wet ground.

In dry ground, horizontal wood sheeting may be used. This is economical and necessary where there is inadequate vertical clearance. Louvered construction should be used to provide drainage and to permit packing behind the wood sheeting where soil will not maintain a vertical face long enough to permit insertion of the next sheet. This type of construction requires over-excavating so that the wood sheets can be placed. Openings must be wide enough between sheets to allow backfilling and tamping, to correct the excavation irregularities and equalize pressure on all sides. Small blocks may be inserted between successive sheets to leave packing gaps. If the excavation is large, **soldier beams**, vertical cantilevers, can be driven to break up the long sheeting spans.

Benoto caissons up to 39 in in diameter may be sunk through water-bearing sands, hardpan, and boulders to depths of 150 ft. Excavation is done with a hammer grab, a single-line orange-peel bucket, inside a temporary, cylindrical steel casing. The hammer grab is dropped to cut into or break up the soil. After impact, the blades close around the soil. Then, the bucket is lifted out and discharged. Boulders are broken up with heavy, percussion-type drills. Rock is drilled out by churn drills. To line the excavation, a casing is bolted together in 20-ft-deep sections, starting with a cutting edge. A hydraulic attachment oscillates the casing continuously to ease sinking and withdrawal, while jacks force the casing into the ground. As concrete is placed, the jacks withdraw the casing in a way that allows concreting of the caisson. Benoto caissons are slower to place and more expensive than drilled shafts, except in wet granular material and where soil conditions are too tough for augers or rotating-bucket diggers.

Open caissons (Fig. 7.33) are enclosures without top and bottom during the lowering process. When

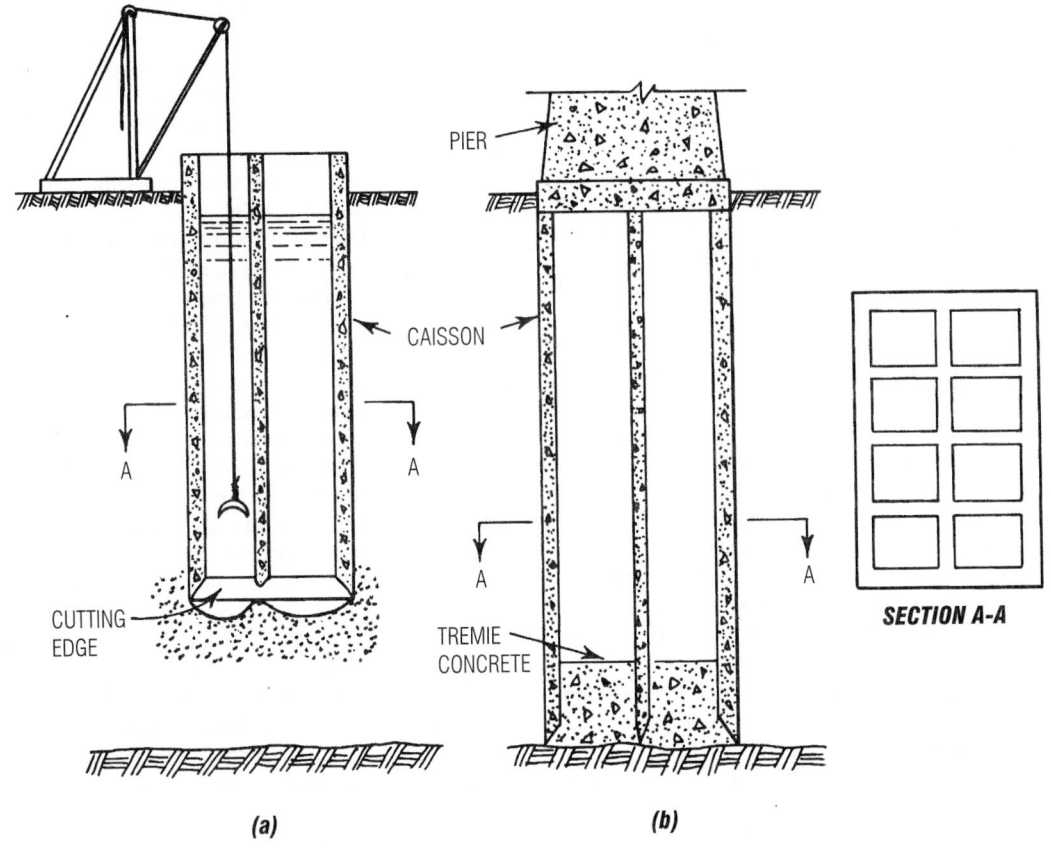

Fig. 7.33 Construction with an open concrete caisson.

used for pump wells and shafts, they often are cylindrical. For bridge piers, these caissons usually are rectangular and compartmented. The compartments serve as dredging wells, pipe passages, and access shafts. Dredging wells usually have 12- to 16-ft clear openings to facilitate excavation with clamshell or orange-peel buckets.

An open caisson may be a braced steel shell that is filled with concrete, except for the wells, as it is sunk into place. Or a caisson may be constructed entirely of concrete.

Friction along the caisson sides may range from 300 to over 1000 lb/ft². So despite steel cutting edges at the wall bottoms, the caisson may not sink. Water and compressed-air jets may be used to lubricate the soil to decrease the friction. For that purpose, vertical jetting pipes should be embedded in the outer walls.

If the caisson does not sink under its own weight with the aid of jets when soil within has been removed down to the cutting edge, the caisson must be weighted. One way is to build it higher, to its final height, if necessary. Otherwise, a platform may have to be built on top and weights piled on it, a measure that can be expensive.

Care must be taken to undercut the edges evenly, or the caisson will tip. Obstructions and variations in the soil also can cause uneven sinking.

When the caisson reaches the bearing strata, the bottom is plugged with concrete (Fig. 7.41b). The plug may be placed by tremie or made by injecting grout into the voids of coarse aggregate.

When a caisson must be placed through water, marine work sometimes may be converted to a land job by construction of a sand island. Fill is placed until it projects above the water surface.

Then, the caisson is constructed and sunk as usual on land.

Pneumatic caissons contain at the base a working chamber with compressed air at a pressure equal to the hydrostatic pressure of the water in the soil. Without the balancing pressure, the water would force soil from below up into a caisson. A working chamber clear of water also permits hand work to remove obstructions that buckets, air lifts, jets, and divers cannot. Thus, the downward course of the caisson can be better controlled. But sinking may be slower and more expensive, and compressed-air work requires precautions against safety and health hazards.

Access to the working chamber for workers, materials, and equipment is through air locks, usually placed at the top of the caisson (Fig. 7.34). Steel access cylinders 3 ft in diameter connect the air locks with the working chamber in large caissons.

Entrance to the working chamber requires only a short stay for a worker in an air lock. But the return stop may be lengthy, depending on the pressure in the chamber, to avoid the bends, or caisson disease, which is caused by air bubbles in muscles, joints, and the blood. Slow decompression gives the body time to eliminate the excess air. In addition to slow decompression, it is necessary to restrict the hours worked at various pressures and limit the maximum pressure to 50 psi above atmospheric or less. The restriction on pressure limits the maximum depth at which compressed-air work can be done to about 115 ft. A medical, or recompression, lock is also required on the site for treatment of workers attacked by the bends.

Floating caissons are used when it is desirable to fabricate caissons on land, tow them into position, and sink them through water. They are constructed much like open or pneumatic caissons but with a "false" bottom, "false" top, or buoyant cells. When floated into position, a caisson must be kept in alignment as it is lowered. A number of means may be used for the purpose, including anchors, templates supported on temporary piles, anchored barges, and cofferdams. Sinking generally is accomplished by adding concrete to the walls. When the cutting edges reach the bottom, the temporary bulkheads at the base, or false bottoms, are removed since buoyancy no longer is necessary. With false tops, buoyancy is controlled with compressed air, which can be released when the caisson sits on the bottom. With buoyant cells, buoyancy is gradually lost as the cells are filled with concrete.

Closed-box caissons are similar to floating caissons, except the top and bottom are permanent. Constructed on land, of steel or reinforced concrete, they are towed into position. Sometimes, the site can be dredged in advance to expose soil that can safely support the caisson and loads that will be imposed on it. Where loads are heavy, however, this may not be practicable; then, the box caisson may have to be supported on piles, but allowance can be made for its buoyancy. This type of caisson has been used for breakwaters, seawalls, and bridge-pier foundations.

Potomac caissons have been used in wide tidal rivers with deep water underlain by deep, soft deposits of sand and silt. Large timber mats are placed on the river bottom, to serve as a template for piles and to retain tremie concrete. Long, steel pipe or H piles are driven in clusters, vertical and battered, as required. Prefabricated steel or concrete caissons are set on the mat over the pile clusters, to serve as permanent forms for concrete shafts to be supported on the piles. Then, concrete is tremied into the caissons. Since the caissons are

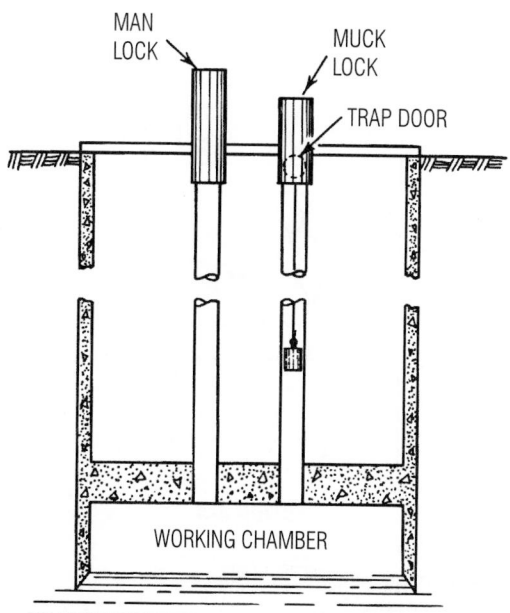

MAN LOCK

MUCK LOCK

TRAP DOOR

WORKING CHAMBER

Fig. 7.34 Pneumatic caisson. Pressure in working chamber is above atmospheric.

used only as forms, construction need not be so heavy as for conventional construction, where they must withstand launching and sinking stresses, and cutting edges are not required.

(H. Y. Fang, "Foundation Engineering Handbook," 2nd ed., Van Nostrand Reinhold Company, New York.)

7.24 Dikes and Cribs

Earth dikes, when fill is available, are likely to be the least expensive for keeping water out of an excavation. If impervious material is not easily obtained, however, a steel sheetpile cutoff wall may have to be driven along the dike, to permit pumps to handle the leakage. With an impervious core in the dike, wellpoints, deep-well pumps, or sumps and ditches may be able to keep the excavation unwatered.

Timber cribs are relatively inexpensive excavation enclosures. Built on shore, they can be floated to the site and sunk by filling with rock. The water side may be faced with wood boards for watertightness (Fig. 7.35). For greater watertightness, two lines of cribs may be used to support two lines of wood sheeting between which clay is tamped to form a "puddle" wall. Design of timber cribs

should provide ample safety against overturning and sliding.

7.25 Cofferdams

Temporary walls or enclosures for protecting an excavation are called cofferdams. Generally, one of the most important functions is to permit work to be carried out on a nearly dry site.

Cofferdams should be planned so that they can be easily dismantled for reuse. Since they are temporary, safety factors can be small, 1.25 to 1.5, when all probable loads are accounted for in the design. But design stresses should be kept low when stresses, unit pressure, and bracing reactions are uncertain. Design should allow for construction loads and the possibility of damage from construction equipment. For cofferdams in water, the design should provide for dynamic effect of flowing water and impact of waves. The height of the cofferdam should be adequate to keep out floods that occur frequently.

7.25.1 Double-Wall Cofferdams

These may be erected in water to enclose large areas. Double-wall cofferdams consist of two lines of sheetpiles tied to each other; the space between

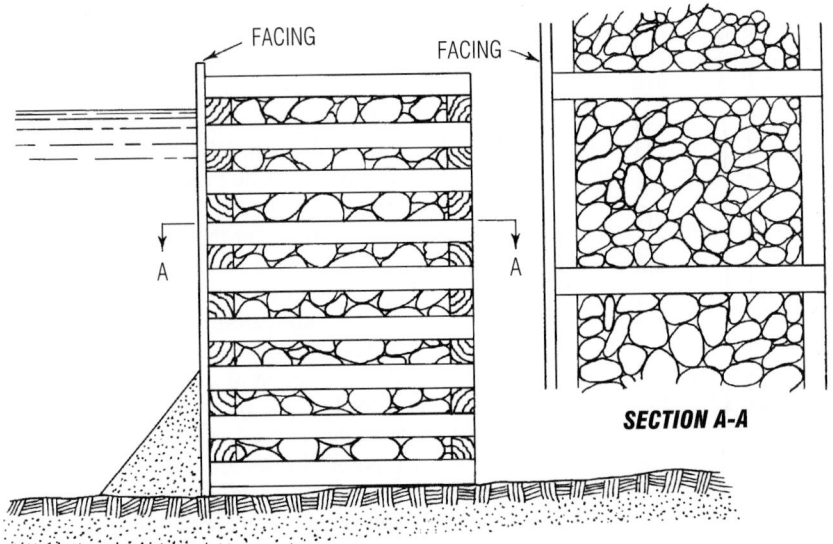

ELEVATION

Fig. 7.35 Timber crib with stone filling.

is filled with sand (Fig. 7.36). For sheetpiles driven to irregular rock, or gravel, or onto boulders, the bottom of the space between walls may be plugged with a thick layer of tremie concrete to seal gaps below the tips of the sheeting. Double-wall cofferdams are likely to be more watertight than single-wall ones and can be used to greater depths.

A berm may be placed against the outside face of a cofferdam for stability. If so, it should be protected against erosion. For this purpose, riprap, woven mattresses, streamline fins or jetties, or groins may be used. If the cofferdam rests on rock, a berm needs to be placed on the inside only if required to resist sliding, overturning, or shearing. On sand, an ample berm must be provided so that water has a long path to travel to enter the cofferdam (Fig. 7.36). (The amount of percolation is proportional to the length of path and the head.) Otherwise, the inside face of the cofferdam may settle, and the cofferdam may overturn as water percolates under the cofferdam and causes a quick, or boiling, excavation bottom. An alternative to a wide berm is wider spacing of the cofferdam walls. This is more expensive but has the added advantage that the top of the fill can be used by construction equipment and for construction plant.

7.25.2 Cellular Cofferdams

Used in construction of dams, locks, wharves, and bridge piers, cellular cofferdams are suitable for enclosing large areas in deep water. These enclosures are composed of relatively wide units.

Average width of a cellular cofferdam on rock should be 0.70 to 0.85 times the head of water against the outside. When constructed on sand, a cellular cofferdam should have an ample berm on the inside to prevent the excavation bottom from becoming quick (Fig. 7.37d).

Steel sheetpiles interlocked form the cells. One type of cell consists of circular arcs connected by straight diaphragms (Fig. 7.37a). Another type comprises circular cells connected by circular arcs (Fig. 7.37b). Still another type is the cloverleaf, composed of large circular cells subdivided by straight diaphragms (Fig. 7.37c). The cells are filled with sand. The internal shearing resistance of the sand contributes substantially to the strength of the cofferdam. For this reason, it is unwise to fill a cofferdam with clay or silt. Weepholes on the inside sheetpiles drain the fill, thus relieving the hydrostatic pressure on those sheets and increasing the shear strength of the fill.

In circular cells, lateral pressure of the fill causes only ring tension in the sheetpiles. Maximum stress in the pile interlocks usually is limited to 8000 lb/lin in. This in turn limits the maximum diameter of the circular cells. Because of numerous uncertainties, this maximum generally is set at 60 ft. When larger-size cells are needed, the cloverleaf type may be used.

Circular cells are preferred to the diaphragm type because each circular cell is a self-supporting unit. It may be filled completely to the top before construction of the next cell starts. (Unbalanced fills in a cell may distort straight diaphragms.)

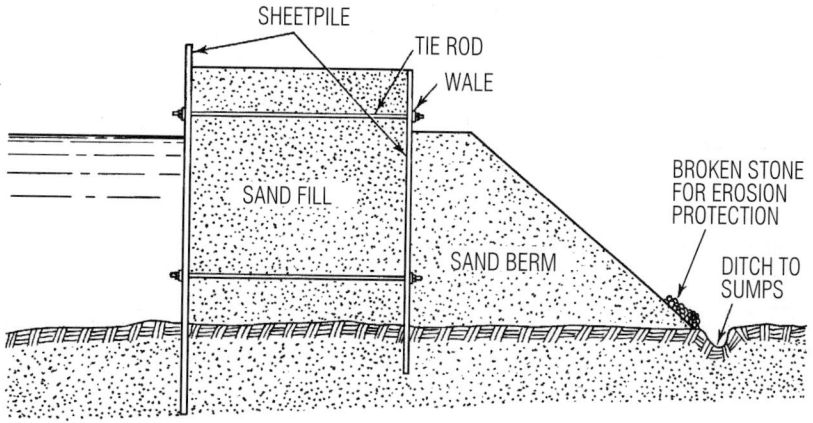

Fig. 7.36 Double-wall cofferdam.

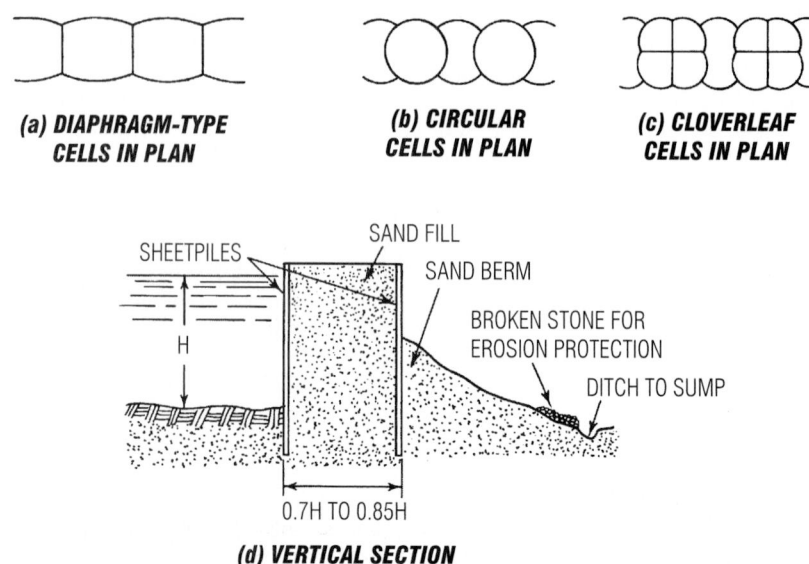

Fig. 7.37 Cellular sheetpile cofferdam.

When a circular cell has been filled, the top may be used as a platform for construction of the next cell. Also, circular cells require less steel per linear foot of cofferdam. The diaphragm type, however, may be made as wide as desired.

When the sheetpiles are being driven, care must be taken to avoid breaking the interlocks. The sheetpiles should be accurately set and plumbed against a structurally sound template. They should be driven in short increments, so that when uneven bedrock or boulders are encountered, driving can be stopped before the cells or interlocks are damaged. Also, all the piles in a cell should be started until the cell is ringed. This can reduce jamming troubles with the last piles to be installed for the cell.

7.25.3 Single-Wall Cofferdams

These form an enclosure with only one line of sheeting. If there will be no water pressure on the sheeting, they may be built with **soldier beams** (piles extended to the top of the enclosure) and horizontal wood lagging (Fig. 7.38). If there will be water pressure, the cofferdam may be constructed of sheetpiles. Although they require less wall material than double-wall or cellular cofferdams, single-wall cofferdams generally require bracing on the inside. Also, unless the bottom is driven into a thick, impervious layer, they may leak excessively

at the bottom. There may also be leakage at interlocks. Furthermore, there is danger of flooding and collapse due to hydrostatic forces when these cofferdams are unwatered.

For marine applications, therefore, it is advantageous to excavate, drive piles, and place a seal of tremie concrete without unwatering single-wall sheetpile cofferdams. Often, it is advisable to predredge the area before the cofferdam is constructed, to facilitate placing of bracing and to remove obstructions to pile driving. Also, if blasting is necessary, it would severely stress the sheeting and bracing if done after they were installed.

For buildings, single-wall cofferdams must be carefully installed. Small movements and consequent loss of ground usually must be prevented to avoid damaging neighboring structures, streets, and utilities. Therefore, the cofferdams must be amply braced. Sheeting close to an existing structure should not be a substitute for underpinning.

Bracing ▪ Cantilevered sheetpiles may be used for shallow single-wall cofferdams in water or on land where small lateral movement will not be troublesome. Embedment of the piles in the bottom must be deep enough to insure stability. Design usually is based on the assumptions that lateral passive resistance varies linearly with depth and the point of inflection is about two-thirds the

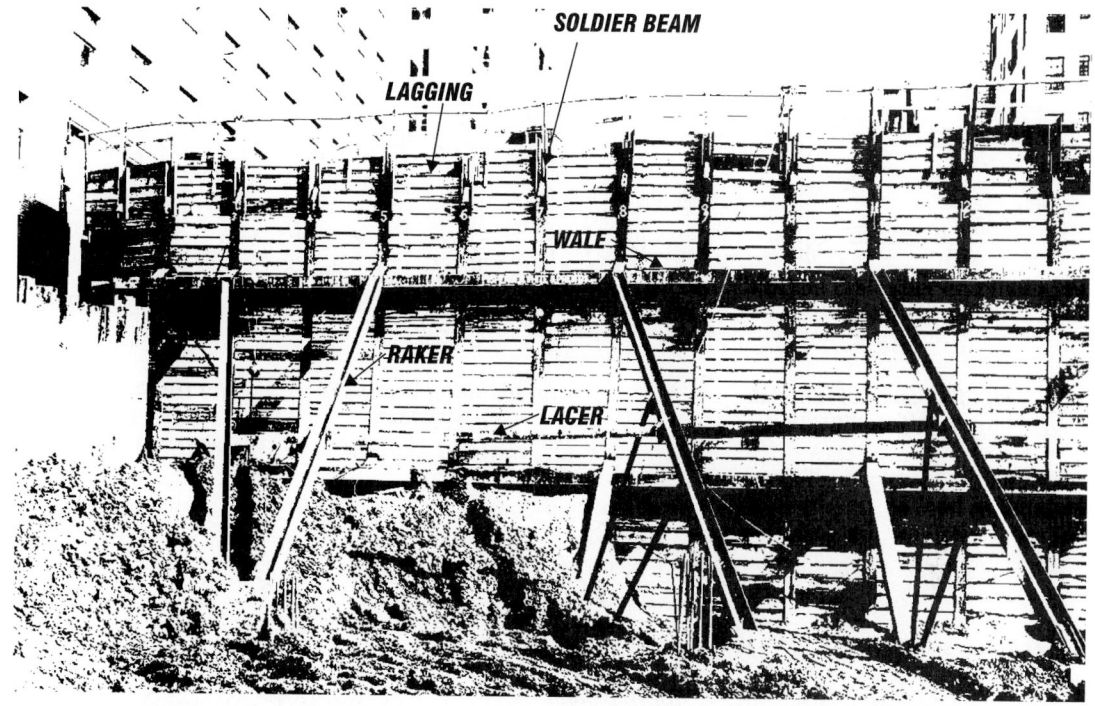

Fig. 7.38 Soldier beams and wood lagging retain the sides of an excavation.

embedded length below the surface. In general, however, cofferdams require bracing.

Cofferdams may be braced in many ways. Figure 7.39 shows some commonly used methods. Circular cofferdams may be braced with horizontal rings (Fig. 7.39a). For small rectangular cofferdams, horizontal braces, or wales, along sidewalls and end walls may be connected to serve only as struts. For larger cofferdams, diagonal bracing (Fig. 7.39b) or cross-lot bracing (Fig. 7.39d and e) is necessary. When space is available at the top of an excavation, pile tops can be anchored with concrete dead men (Fig. 7.39c). Where rock is close, the wall can be tied back with tensioned wires or bars that are anchored in grouted sockets in the rock (Fig. 7.40). See also Art. 7.40.4.

Horizontal cross braces should be spaced to minimize interference with excavation, form construction, concreting, and pile driving. Spacing of 12 and 18 ft is common. Piles and wales selected should be strong enough as beams to permit such spacing. In marine applications, divers often have to install the wales and braces underwater. To reduce the amount of such work, tiers of bracing

may be prefabricated and lowered into the cofferdam from falsework or from the top set of wales and braces, which is installed above the water surface. In some cases, it may be advantageous to prefabricate and erect the whole cage of bracing before the sheetpiles are driven. Then, the cage, supported on piles, can serve also as a template for driving the sheetpiles.

All wales and braces should be forced into bearing with the sheeting by wedges and jacks.

When pumping cannot control leakage into a cofferdam, excavation may have to be carried out in compressed air. This requires a sealed working chamber, access shafts, and air locks, as for pneumatic caissons (Art. 7.23). Other techniques, such as use of a tremie concrete seal or chemical solidification or freezing of the soil, if practicable, however, will be more economical.

Braced sheetpiles may be designed as continuous beams subjected to uniform loading for earth and to loading varying linearly with depth for water (Art. 7.27). (Actually, earth pressure depends on the flexibility of the sheeting and relative stiffness of supports.) Wales may be designed for

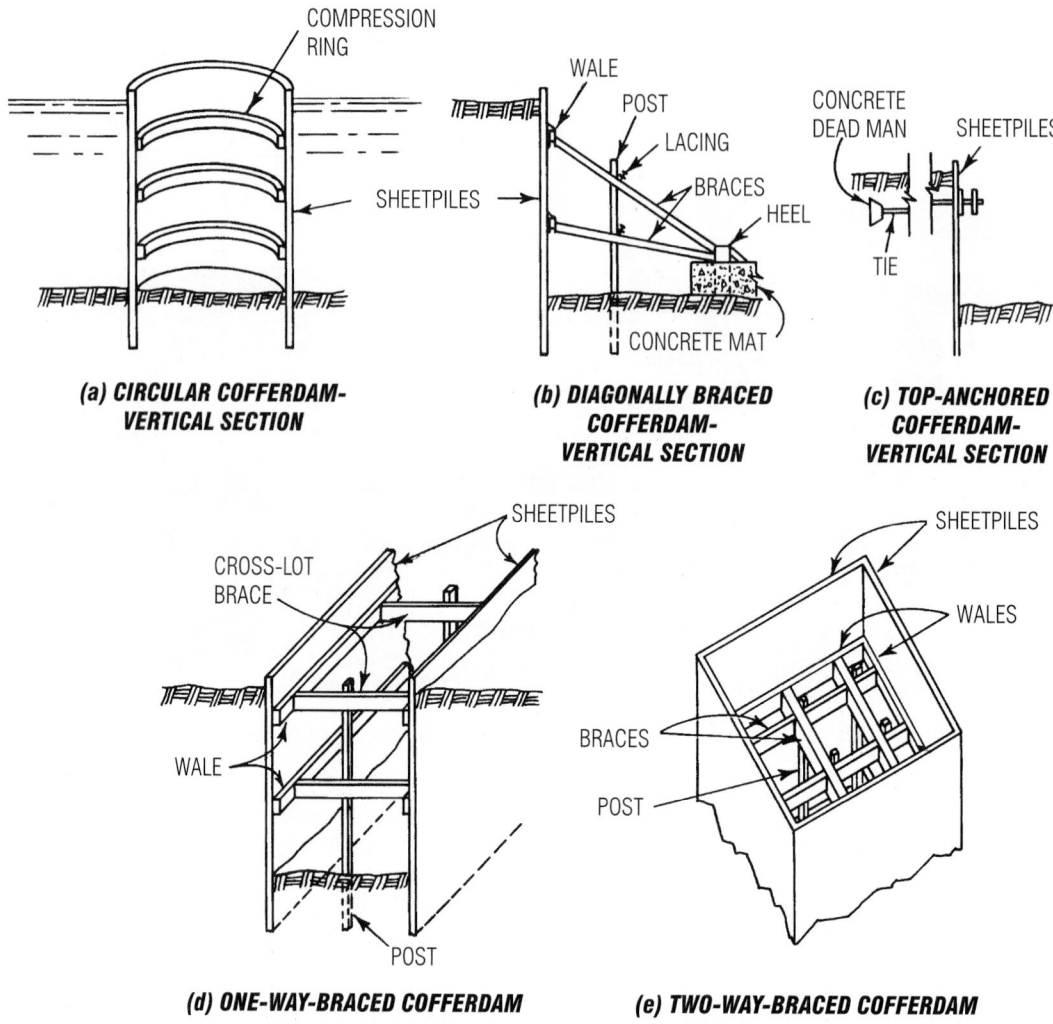

Fig. 7.39 Types of cofferdam bracing include compression rings; bracing, diagonal (rakers) or cross lot; wales, and tiebacks.

uniform loading. Allowable unit stresses in the wales, struts, and ties may be taken at half the elastic limit for the materials because the construction is temporary and the members are exposed to view. Distress in a member can easily be detected and remedial steps taken quickly.

Soldier beams and horizontal wood sheeting are a variation of single-wall cofferdams often used where impermeability is not required. The soldier beams, or piles, are driven vertically into the ground to below the bottom of the proposed

excavation. Spacing usually ranges from 5 to 10 ft (Table 7.13). (The wood lagging can be used in the thicknesses shown in Table 7.13 because of arching of the earth between successive soldier beams.)

As excavation proceeds, the wood boards are placed horizontally between the soldiers (Fig. 7.38). Louvers or packing spaces, 1 to 2 in high, are left between the boards so that earth can be tamped behind them to hold them in place. Hay may also be stuffed behind the boards to keep the ground from running through the gaps. The louvers permit

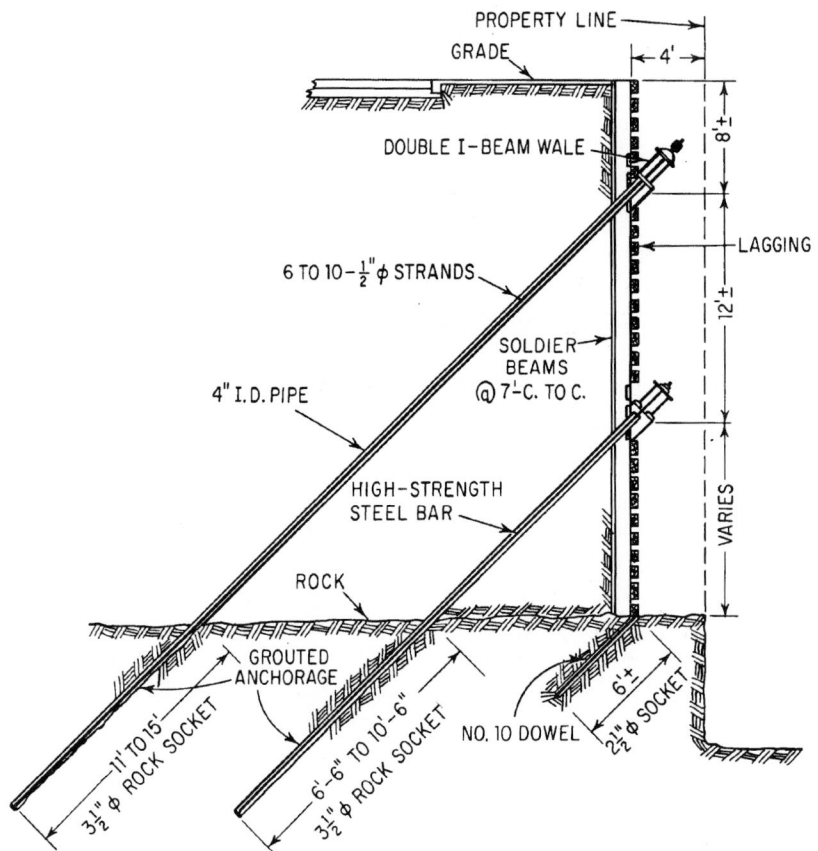

Fig. 7.40 Vertical section shows prestressed tiebacks for soldier beams.

drainage of water, to relieve hydrostatic pressure on the sheeting and thus allow use of a lighter bracing system. The soldiers may be braced directly with horizontal or inclined struts; or wales and braces may be used.

Advantages of soldier-beam construction include fewer piles; the sheeting does not have to

Table 7.13 Usual Maximum Spans of Horizontal Sheeting with Soldier Piles, ft

Nominal Thickness of Sheeting, in	In Well-Drained Soils	In Cohesive Soils with Low Shear Resistance
2	5	4.5
3	8.5	6
4	10	8

extend below the excavation bottom, as do sheetpiles; and the soldiers can be driven more easily in hard ground than can sheetpiles. Varying the spacing of the soldiers permits avoidance of underground utilities. Use of heavy sections for the piles allows wide spacing of wales and braces. But the soldiers and lagging, as well as sheetpiles, are no substitute for underpinning; it is necessary to support and underpin even light adjoining structures.

Liner-plate cofferdams may be used for excavating circular shafts. The plates are placed in horizontal rings as excavation proceeds. Stamped from steel plate, usually about 16 in high and 3 ft long, light enough to be carried by one person, liner plates have inward-turned flanges along all edges. Top and bottom flanges provide a seat for successive rings. End flanges permit easy bolting

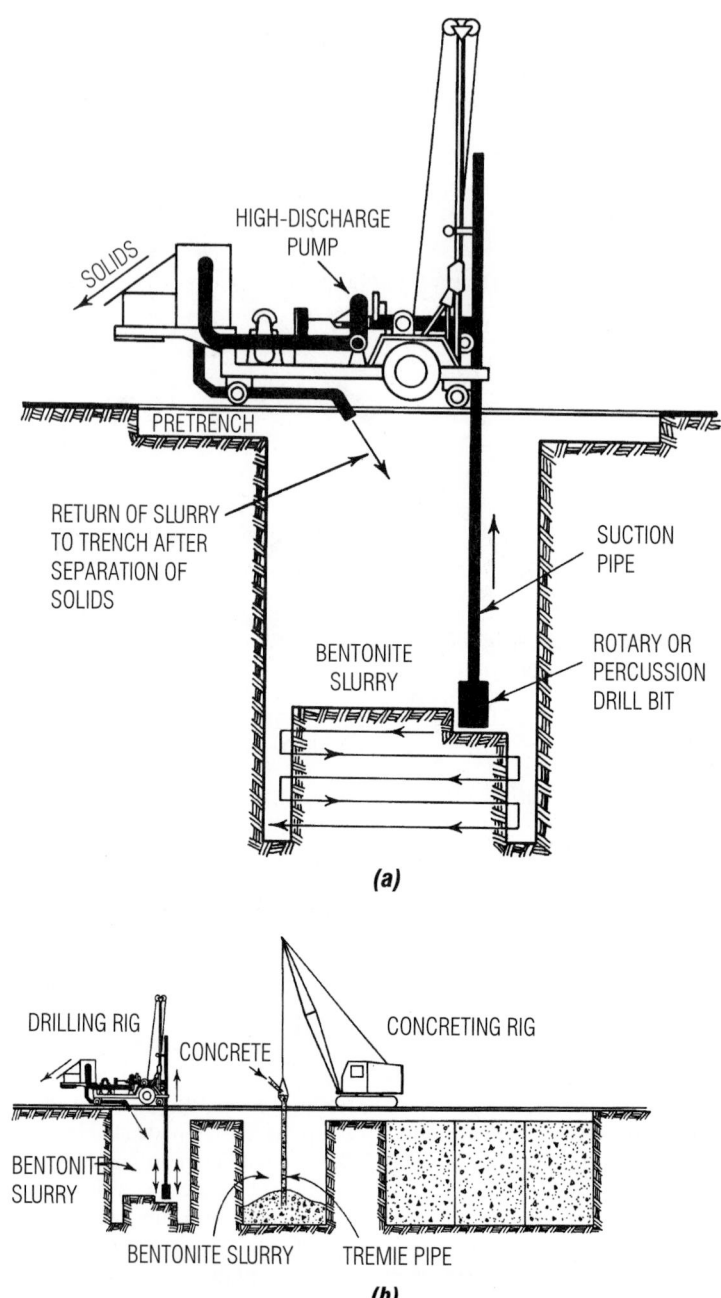

Fig. 7.41 Slurry-trench method for constructing a continuous concrete wall: (*a*) Excavating one section; (*b*) concreting one section while another is being excavated.

of adjoining plates in a ring. The plates also are corrugated for added stiffness. Large-diameter cofferdams may be constructed by bracing the liner plates with steel beam rings.

Vertical-lagging cofferdams, with horizontal-ring bracing, also may be used for excavating circular shafts. The method is similar to that used for Chicago caissons (Art. 7.23). It is similarly restricted to soils that can stand without support in depths of 3 to 5 ft for a short time.

Slurry trenches may be used for constructing concrete walls. The method permits building a wall in a trench without the earth sides collapsing. While excavation proceeds for a 24- to 36-in-wide trench, the hole is filled with a bentonite slurry with a specific gravity of 1.05 to 1.10 (Fig. 7.41*a*). The fluid pressure against the sides and caking of bentonite on the sides prevent the earth walls of the trench from collapsing. Excavation is carried out a section at a time. A section may be 20 ft long and as much as 100 ft deep. When the bottom of the wall is reached in a section, reinforcing is placed in that section. (Tests have shown that the bond of the reinforcing to concrete is not materially reduced by the bentonite.) Then, concrete is tremied into the trench, replacing the slurry, which may flow into the next section to be excavated or be pumped into tanks for reuse in the next section (Fig. 7.41*b*). The method has been used to construct cutoffs for dams, cofferdams, foundations, walls of buildings, and shafts.

7.26 Soil Solidification

Grouting is the injection of cement or chemicals into soil or rock to enhance engineering properties. During the past 20 years significant developments in materials and equipment have transformed grouting from art to science. See Federal Highway Administration publication "Ground Improvement Technical Summaries" Publication No. FHWA-SA-98-086 December 1999 for an extensive primer on grouting techniques, applications and procedures.

Freezing is another means of solidifying water-bearing soils where obstructions or depth preclude pile driving. It can be used for deep shaft excavations and requires little material for temporary construction; the refrigeration plant has high salvage value. But freezing the soil may take a very long time. Also, holes have to be drilled below

the bottom of the proposed excavation for insertion of refrigeration pipes.

(L. White and E. A. Prentis, "Cofferdams," Columbia University Press, New York; H. Y. Fang, "Foundation Engineering Handbook," 2nd ed., Van Nostrand Reinhold Company, New York.)

7.27 Lateral Active Pressures on Retaining Walls

Water exerts against a vertical surface a horizontal pressure equal to the vertical pressure. At any level, the vertical pressure equals the weight of a 1-ft² column of water above that level. Hence, the horizontal pressure p, lb/ft³ at any level is

$$p = wh \tag{7.55}$$

where w = unit weight of water, lb/ft³

h = depth of water, ft

The pressure diagram is triangular (Fig. 7.42). Equation (7.55) also can be written

$$p = Kwh \tag{7.56}$$

where K = pressure coefficient = 1.00.

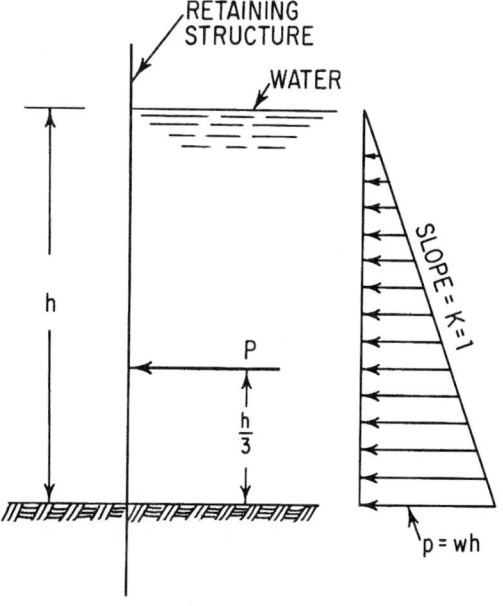

Fig. 7.42 Pressure diagram for water.

The resultant, or total, pressure, lb/lin ft, represented by the area of the hydrostatic-pressure diagram, is

$$P = K\frac{wh^2}{2} \tag{7.57}$$

It acts at a distance $h/3$ above the base of the triangle.

Soil also exerts lateral pressure. But the amount of this pressure depends on the type of soil, its compaction or consistency, and its degree of saturation, and on the resistance of the structure to the pressure. Also, the magnitude of passive pressure differs from that of active pressure.

Active pressure tends to move a structure in the direction in which the pressure acts. **Passive pressure** opposes motion of a structure.

Retaining walls backfilled with cohesionless soils (sands and gravel) tend to rotate slightly around the base. Behind such a wall, a wedge of sand ABC (Fig. 7.43a) tends to shear along plane AC. C. A. Coulomb determined that the ratio of sliding resistance to sliding force is a minimum when AC makes an angle of $45° + \phi/2$ with the horizontal,

where ϕ is the angle of internal friction of the soil, deg.

For triangular pressure distribution (Fig. 7.43b), the active lateral pressure of a cohesionless soil at a depth h, ft, is

$$p = K_a wh \tag{7.58}$$

where K_a = coefficient of active earth pressure

w = unit weight of soil, lb/ft^3

The total active pressure, lb/lin ft, is

$$E_a = K_a \frac{wh^2}{2} \tag{7.59}$$

Because of frictional resistance to sliding at the face of the wall, E_a is inclined at an angle δ with the normal to the wall, where δ is the angle of wall friction, deg (Fig. 7.43a). If the face of the wall is vertical, the horizontal active pressure equals $E_a \cos \delta$. If the face makes an angle β with the vertical (Fig. 7.43a), the pressure equals $E_a \cos (\delta + \beta)$. The resultant acts at a distance of $h/3$ above the base of the wall.

If the ground slopes upward from the top of the wall at an angle α, deg, with the horizontal, then for cohesionless soils

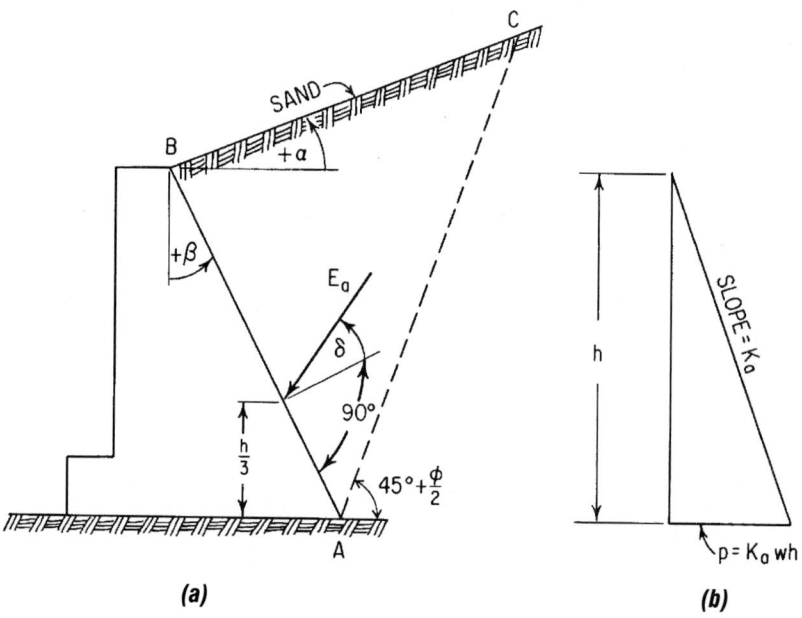

(a)   **(b)**

Fig. 7.43 Free-standing wall with sand backfill (a) is subjected to (b) triangular pressure distribution.

Table 7.14 Active-Lateral-Pressure Coefficients K_a

		ϕ						
		10°	15°	20°	25°	30°	35°	40°
$\beta = 0$	$\alpha = 0$	0.70	0.59	0.49	0.41	0.33	0.27	0.22
	$\alpha = 10°$	0.97	0.70	0.57	0.47	0.37	0.30	0.24
	$\alpha = 20°$	—	—	0.88	0.57	0.44	0.34	0.27
	$\alpha = 30°$	—	—	—	—	0.75	0.43	0.32
	$\alpha = \phi$	0.97	0.93	0.88	0.82	0.75	0.67	0.59
$\beta = 10°$	$\alpha = 0$	0.76	0.65	0.55	0.48	0.41	0.43	0.29
	$\alpha = 10°$	1.05	0.78	0.64	0.55	0.47	0.38	0.32
	$\alpha = 20°$	—	—	1.02	0.69	0.55	0.45	0.36
	$\alpha = 30°$	—	—	—	—	0.92	0.56	0.43
	$\alpha = \phi$	1.05	1.04	1.02	0.98	0.92	0.86	0.79
$\beta = 20°$	$\alpha = 0$	0.83	0.74	0.65	0.57	0.50	0.43	0.38
	$\alpha = 10°$	1.17	0.90	0.77	0.66	0.57	0.49	0.43
	$\alpha = 20°$	—	—	1.21	0.83	0.69	0.57	0.49
	$\alpha = 30°$	—	—	—	—	1.17	0.73	0.59
	$\alpha = \phi$	1.17	1.20	1.21	1.20	1.17	1.12	1.06
$\beta = 30°$	$\alpha = 0$	0.94	0.86	0.78	0.70	0.62	0.56	0.49
	$\alpha = 10°$	1.37	1.06	0.94	0.83	0.74	0.65	0.56
	$\alpha = 20°$	—	—	1.51	1.06	0.89	0.77	0.66
	$\alpha = 30°$	—	—	—	—	1.55	0.99	0.79
	$\alpha = \phi$	1.37	1.45	1.51	1.54	1.55	1.54	1.51

Table 7.15 Angles of Internal Friction and Unit Weights of Soils

Types of Soil	Density or Consistency	Angle of Internal Friction ϕ, deg	Unit Weight w, lb/ft^3
Coarse sand or sand and gravel	Compact	40	140
	Loose	35	90
Medium sand	Compact	40	130
	Loose	30	90
Fine silty sand or sandy silt	Compact	30	130
	Loose	25	85
Uniform silt	Compact	30	135
	Loose	25	85
Clay-silt	Soft to medium	20	90–120
Silty clay	Soft to medium	15	90–120
Clay	Soft to medium	0–10	90–120

$$K_a = \cfrac{\cos^2(\phi - \beta)}{\cos^2\beta\cos(\delta+\beta)\left[1+\sqrt{\cfrac{\sin(\phi+\delta)\sin(\phi-\alpha)}{\cos(\delta+\beta)\cos(\alpha-\beta)}}\right]^{-2}}$$

(7.60)

The effect of wall friction on K_a is small and usually is neglected. For $\delta = 0$,

$$K_a = \cfrac{\cos^2(\phi - \beta)}{\cos^3\beta\left[1+\sqrt{\cfrac{\sin\phi\sin(\phi-\alpha)}{\cos\beta\cos(\alpha-\beta)}}\right]^2}$$

(7.61)

Table 7.14 lists values of K_a determined from Eq. (7.61). Approximate values of ϕ and unit weights for various soils are given in Table 7.15.

For level ground at the top of the wall ($\alpha = 0$):

$$K_a = \cfrac{\cos^2(\phi - \beta)}{\cos^3\beta\left(1+\cfrac{\sin\phi}{\cos\beta}\right)^2}$$

(7.62)

If in addition, the back face of the wall is vertical ($\beta = 0$), Rankine's equation is obtained:

$$K_a = \frac{1 - \sin\phi}{1 + \sin\phi}$$

(7.63)

Coulomb derived the trigonometric equivalent:

$$K_a = \tan^2\left(45° - \frac{\phi}{2}\right)$$

(7.64)

The selection of the wall friction angle should be carefully applied as it has a significant effect on the resulting earth pressure force.

Unyielding walls retaining walls backfilled with sand and gravel, such as the abutment walls of a rigid-frame concrete bridge or foundation walls braced by floors, do not allow shearing resistance to develop in the sand along planes that can be determined analytically. For such walls, triangular pressure diagrams may be assumed, and K_a may be taken equal to 0.5.

Braced walls retaining cuts in sand (Fig. 7.44a) are subjected to earth pressure gradually and

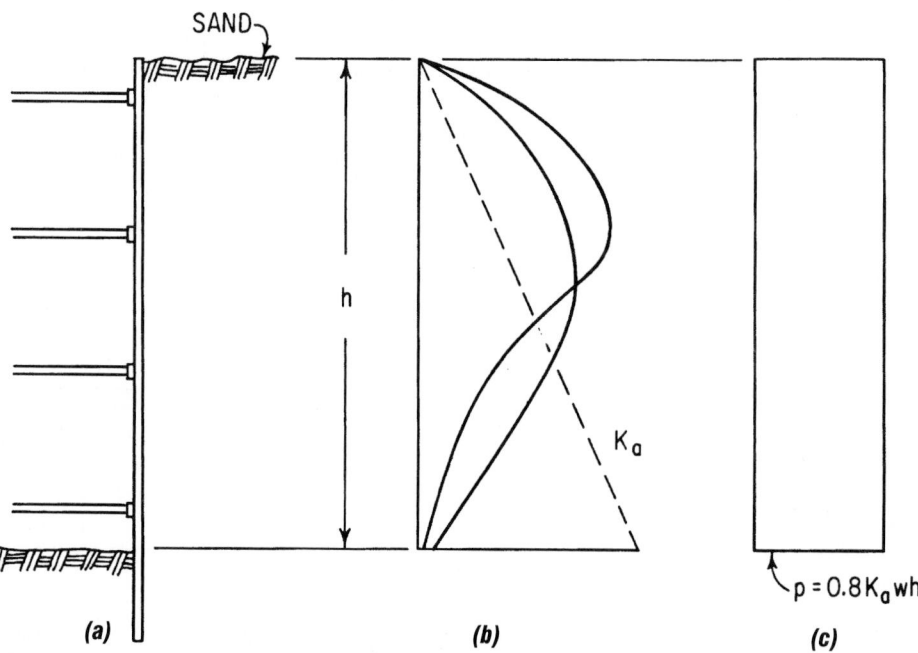

Fig. 7.44 Braced wall retaining sand (a) may have to resist pressure distributions of the type shown in (b). (c) Uniform pressure distribution may be assumed for design.

develop resistance in increments as excavation proceeds and braces are installed. Such walls tend to rotate about a point in the upper portion. Hence, the active pressures do not vary linearly with depth. Field measurements have yielded a variety of curves for the pressure diagram, of which two types are shown in Fig. 7.44b. Consequently, some authorities have recommended a trapezoidal pressure diagram, with a maximum ordinate

$$p = 0.8K_a wh \qquad (7.65)$$

K_a may be obtained from Table 7.14. The total pressure exceeds that for a triangular distribution.

Figure 7.45 shows earth-pressure diagrams developed for a sandy soil and a clayey soil. In both cases, the braced wall is subjected to a 3-ft-deep surcharge, and height of wall is 34 ft. For the sandy soil (Fig. 7.45a), Fig. 7.45b shows the pressure diagram assumed. The maximum pressure can be

obtained from Eq. (7.65), with $h = 34 + 3 = 37$ ft and K_a assumed as 0.30 and w as 110 lb/ft³.

$$p_1 = 0.8 \times 0.3 \times 110 \times 37 = 975 \, \text{lb/ft}^2$$

The total pressure is estimated as

$$P = 0.8 \times 975 \times 37 = 28,900 \, \text{lb/lin ft}$$

The equivalent maximum pressure for a trapezoidal diagram for the 34-ft height of the wall then is

$$p = \frac{28,900}{0.8 \times 34} = 1060 \, \text{lb/ft}^2$$

Assumption of a uniform distribution (Fig. 7.44c), however, simplifies the calculations and has little or no effect on the design of the sheeting and braces, which should be substantial to withstand construction abuses. Furthermore, trapezoidal loading terminating at the level of the excavation may not apply if piles are driven inside the completed

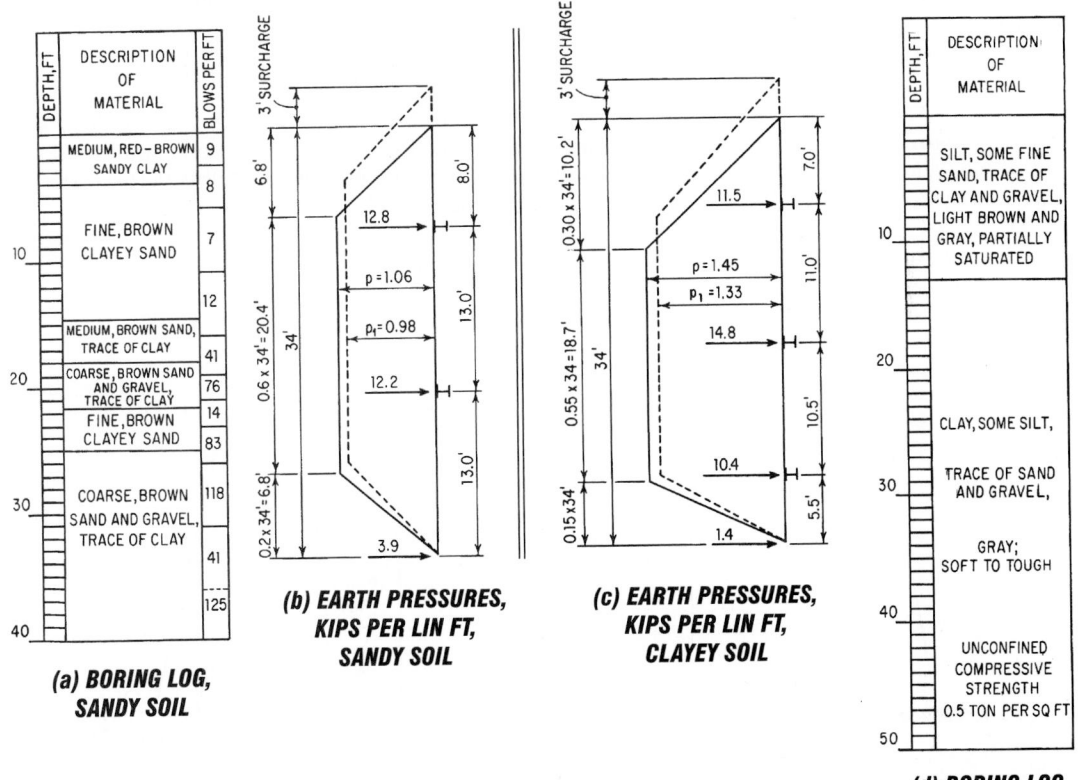

(a) BORING LOG, SANDY SOIL

(b) EARTH PRESSURES, KIPS PER LIN FT, SANDY SOIL

(c) EARTH PRESSURES, KIPS PER LIN FT, CLAYEY SOIL

(d) BORING LOG, CLAYEY SOIL

Fig. 7.45 Assumed trapezoidal diagrams for a braced wall in soils described by boring logs in (a) and (d).

excavation. The shocks may temporarily decrease the passive resistance of the sand in which the wall is embedded and lower the inflection point. This would increase the span between the inflection point and the lowest brace and increase the pressure on that brace. Hence, uniform pressure distribution may be more applicable than trapezoidal for such conditions.

See Note for free-standing walls.

Flexible retaining walls in sand cuts are subjected to pressures that depend on the fixity of the anchorage. If the anchor moves sufficiently or the tie from the anchor to the upper portion of the bulkhead stretches enough, the bulkhead may rotate slightly about a point near the bottom. In that case, the sliding-wedge theory may apply. The pressure distribution may be taken as triangular, and Eqs. (7.58) to (7.64) may be used. But if the anchor does not yield, then pressure distributions much like those in Fig. 7.44b for a braced cut may occur. Either a trapezoidal or uniform pressure distribution may be assumed, with maximum pressure given by Eq. (7.65). Stresses in the tie should be kept low because it may have to resist unanticipated pressures, especially those resulting from a redistribution of forces from soil arching. (Federal Highway Administration Geotechnical Engineering Circular No.4 "Ground Anchors and Anchored Systems" FHWA-IF-99015, June 1999).

Free-standing walls retaining plastic-clay cuts (Fig. 7.46a) may have to resist two types of active lateral pressure, both with triangular distribution. In the short term the shearing resistance is due to cohesion only, a clay bank may be expected to stand with a vertical face without support for a height, ft, of

$$h' = \frac{2c}{w} \qquad (7.66)$$

where $2c$ = unconfined compressive strength of clay, lb/ft^2

w = unit weight of clay, lb/ft^3

So if there is a slight rotation of the wall about its base, the upper portion of the clay cut will stand vertically without support for a depth h'. Below that, the pressure will increase linearly with depth as if the clay were a heavy liquid (Fig. 7.46b):

$$p = wh - 2c$$

The total pressure, lb/lin ft, then is

$$E_a = \frac{w}{2}\left(h - \frac{2c}{w}\right)^2 \qquad (7.67)$$

It acts at a distance $(h - 2c/w)/3$ above the base of the wall. These equations assume wall friction is zero, the back face of the wall is vertical, and the ground is level.

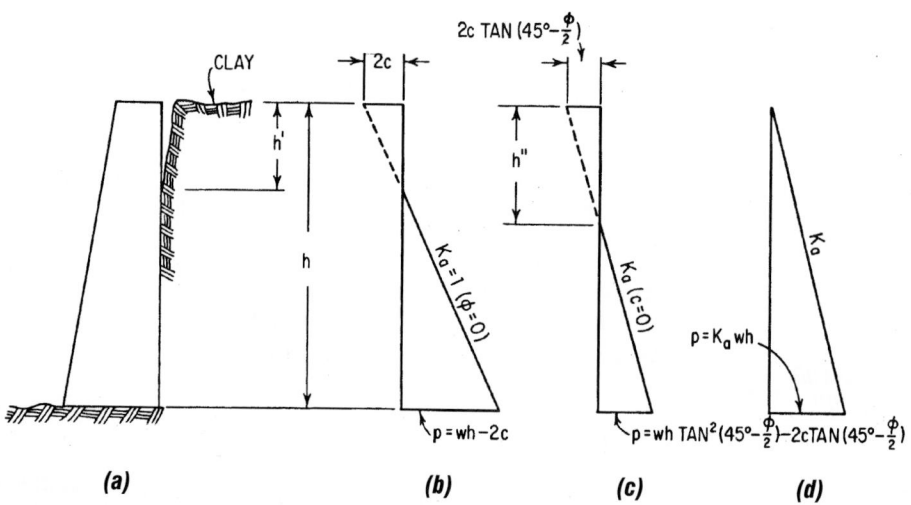

Fig. 7.46 Free-standing wall retaining clay (a) may have to resist the pressure distribution shown in (b) or (d). For mixed soils, the distribution may approximate that shown in (c).

In time the clay will reach its long term strength and the pressure distribution may become approximately triangular (Fig. 7.46d) from the top of the wall to the base. The pressures then may be calculated from Eqs. (7.58) to (7.64) with an apparent angle of internal friction for the soil (for example, see the values of ϕ in Table 7.15). The wall should be designed for the pressures producing the highest stresses and overturning moments.

Note: The finer the backfill material, the more likely it is that pressures greater than active will develop, because of plastic deformations, water-level fluctuation, temperature changes, and other effects. As a result, it would be advisable to use in design at least the **coefficient for earth pressure at rest:**

$$K_o = 1 - \sin\phi \qquad (7.68)$$

The safety factor should be at least 2.5.

Clay should not be used behind retaining walls, where other economical alternatives are available. The swelling type especially should be avoided because it can cause high pressures and progressive shifting or rotation of the wall.

For a mixture of cohesive and cohesionless soils, the pressure distribution may temporarily be as shown in Fig. 7.46c. The height, ft, of the unsupported vertical face of the clay is

$$h'' = \frac{2c}{w\tan(45° - \phi/2)} \qquad (7.69)$$

The pressure at the base is

$$p = wh\tan^2\left(45° - \frac{\phi}{2}\right) - 2c\tan\left(45° - \frac{\phi}{2}\right) \qquad (7.70)$$

The total pressure, lb/lin ft, is

$$E_a = \frac{w}{2}\left[h\tan\left(45° - \frac{\phi}{2}\right) - \frac{2c}{w}\right]^2 \qquad (7.71)$$

It acts at a distance $(h - h'')/3$ above the base of the wall.

Braced walls retaining clay cuts (Fig. 7.47a) also may have to resist two types of active lateral pressure. As for sand, the pressure distribution may temporarily be approximated by a trapezoidal diagram (Fig. 7.47b). On the basis of field observations, R. B. Peck has recommended a maximum pressure of

$$p = wh - 4c \qquad (7.72)$$

and a total pressure, lb/lin ft, of

$$E_a = \frac{1.55h}{2}(wh - 4c) \qquad (7.73)$$

[R. B. Peck, "Earth Pressure Measurements in Open Cuts, Chicago (Ill.) Subway," *Transactions, American Society of Civil Engineers,* 1943, pp. 1008–1036.]

Figure 7.47c shows a trapezoidal earth-pressure diagram determined for the clayey-soil condition of Fig. 7.47d. The weight of the soil is taken as 120 lb/ft³; c is assumed as zero and the active-

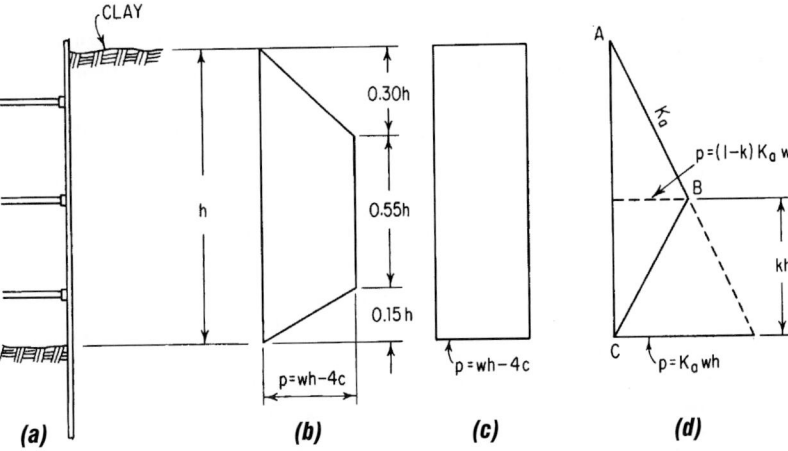

Fig. 7.47 Braced wall retaining clay (a) may have to resist pressures approximated by the pressure distribution in (b) or (d). Uniform distribution (c) may be assumed in design.

lateral-pressure coefficient as 0.3. Height of the wall is 34 ft, surcharge 3 ft. Then, the maximum pressure, obtained from Eq. (7.58) since the soil is clayey, not pure clay, is

$$p_1 = 0.3 \times 120 \times 37 = 1330 \, \text{lb/ft}^2$$

From Eq. (7.73) with the above assumptions, the total pressure is

$$P = \frac{1.55}{2} \times 37 \times 1330 = 38{,}100 \, \text{lb/lin ft}$$

The equivalent maximum pressure for a trapezoidal diagram for the 34-ft height of wall is

$$p = \frac{38{,}100}{34} \times \frac{2}{1.55} = 1450 \, \text{lb/ft}^2$$

To simplify calculations, a uniform pressure distribution may be used instead (Fig. 7.47c).

If after a time the clay should attain a consolidated equilibrium state, the pressure distribution may be better represented by a triangular diagram *ABC* (Fig. 7.47d), as suggested by G. P. Tschebotarioff. The peak pressure may be assumed at a distance of $kh = 0.4h$ above the excavation level for a stiff clay; that is, $k = 0.4$. For a medium clay, k may be taken as 0.25, and for a soft clay, as 0. For computing the pressures, K_a may be estimated from Table 7.14 with an apparent angle of friction obtained from laboratory tests or approximated from Table 7.15. The wall should be designed for the pressures producing the highest stresses and overturning moments.

See also Note for free-standing walls.

Flexible retaining walls in clay cuts and anchored near the top similarly should be checked for two types of pressures. When the anchor is likely to yield slightly or the tie to stretch, the pressure distribution in Fig. 7.47d with $k = 0$ may be applicable. For an unyielding anchor, any of the pressure distributions in Fig. 7.47 may be assumed, as for a braced wall. The safety factor for design of ties and anchorages should be at least twice that used in conventional design. See also Note for free-standing walls.

Backfill placed against a retaining wall should preferably be sand or gravel (free draining) to facilitate drainage. Also, weepholes should be provided through the wall near the bottom and a drain installed along the footing, to conduct water from the back of the wall and prevent buildup of hydrostatic pressures.

Saturated or submerged soil imposes substantially greater pressure on a retaining wall than dry or moist soil. The active lateral pressure for a soil-fluid backfill is the sum of the hydrostatic pressure and the lateral soil pressure based on the buoyed unit weight of the soil. This weight roughly may be 60% of the dry weight.

Surcharge, or loading imposed on a backfill, increases the active lateral pressure on a wall and raises the line of action of the total, or resultant, pressure. A surcharge w_s, lb/ft², uniformly distributed over the entire ground surface may be taken as equivalent to a layer of soil of the same unit weight w as the backfill and with a thickness of w_s/w. The active lateral pressure, lb/ft², due to the surcharge, from the backfill surface down, then will be $K_a w_s$. This should be added to the lateral pressures that would exist without the surcharge. K_a may be obtained from Table 7.14.

(A. Caquot and J. Kérisel, "Tables for Calculation of Passive Pressure, Active Pressure, and Bearing Capacity of Foundations," Gauthier-Villars, Paris.)

7.28 Passive Lateral Pressure on Retaining Walls and Anchors

As defined in Art. 7.27, active pressure tends to move a structure in the direction in which pressure acts, whereas passive pressure opposes motion of a structure.

Passive pressures of cohesionless soils, resisting movement of a wall or anchor, develop because of internal friction in the soils. Because of friction between soil and wall, the failure surface is curved, not plane as assumed in the Coulomb sliding-wedge theory (Art. 7.27). Use of the Coulomb theory yields unsafe values of passive pressure when the effects of wall friction are included.

Total passive pressure, lb/lin ft, on a wall or anchor extending to the ground surface (Fig. 7.48a) may be expressed for sand in the form

$$P = K_p \frac{wh^2}{2} \tag{7.74}$$

where K_p = coefficient of passive lateral pressure

w = unit weight of soil, lb/ft³

h = height of wall or anchor to ground surface, ft

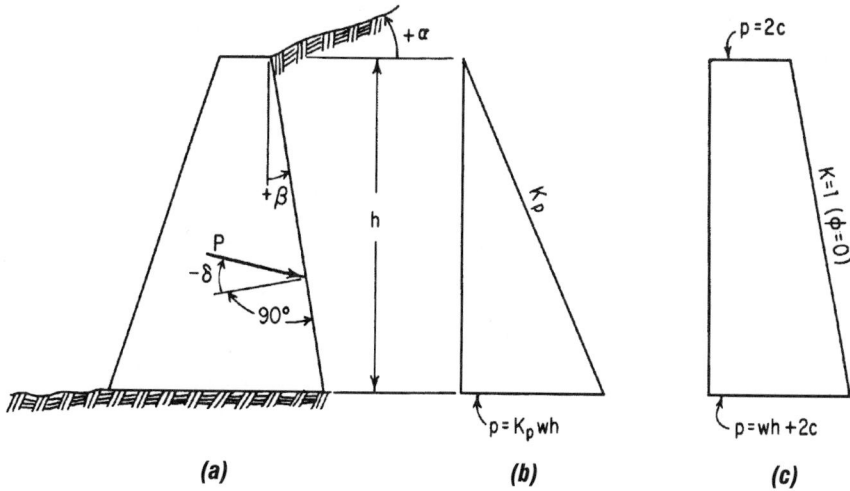

Fig. 7.48 Passive pressure on a wall (*a*) may vary as shown in (*b*) for sand or as shown in (*c*) for clay.

The pressure distribution usually assumed for sand is shown in Fig. 7.48*b*. Table 7.16 lists values of K_p for a vertical wall face ($\beta = 0$) and horizontal ground surface ($\alpha = 0$), for curved surfaces of failure. (Many tables and diagrams for determining passive pressures are given in A. Caquot and J. Kérisel, "Tables for Calculation of Passive Pressure, Active Pressure, and Bearing Capacity of Foundations," Gauthier-Villars, Paris.)

Since a wall usually transmits a downward shearing force to the soil, the angle of wall friction δ correspondingly is negative (Fig. 7.48*a*). For embedded portions of structures, such as anchored sheetpile bulkheads, δ and the angle of internal friction ϕ of the soil reach their peak values simultaneously in dense sand. For those conditions, if specific information is not available, δ may be assumed as $-\frac{2}{3}\phi$ (for $\phi > 30°$). For such structures as a heavy anchor block subjected to a horizontal pull or thrust, δ may be taken as $-\phi/2$ for dense sand. For those cases, the wall friction develops as the sand is pushed upward by the anchor and is unlikely to reach its maximum

value before the internal resistance of the sand is exceeded.

When wall friction is zero ($\delta = 0$), the failure surface is a plane inclined at an angle of $45° - \phi/2$ with the horizontal. The sliding-wedge theory then yields

$$K_p = \frac{\cos^2(\phi + \beta)}{\cos^3 \beta \left[1 - \sqrt{\dfrac{\sin \phi \sin(\phi + \alpha)}{\cos \beta \cos(\alpha - \beta)}} \right]^2} \qquad (7.75)$$

When the ground is horizontal ($\alpha = 0$):

$$K_p = \frac{\cos^2(\phi + \beta)}{\cos^3 \beta (1 - \sin \phi / \cos \beta)^2} \qquad (7.76)$$

If, in addition, the back face of the wall is vertical ($\beta = 0$):

$$K_p = \frac{1 + \sin \phi}{1 - \sin \phi} = \tan^2\left(45° + \frac{\phi}{2}\right) = \frac{1}{K_a} \qquad (7.77)$$

Table 7.16 Passive Lateral-Pressure Coefficients K_p*

	$\phi = 10°$	$\phi = 15°$	$\phi = 20°$	$\phi = 25°$	$\phi = 30°$	$\phi = 35°$	$\phi = 40°$
$\delta = 0$	1.42	1.70	2.04	2.56	3.00	3.70	4.60
$\delta = -\phi/2$	1.56	1.98	2.59	3.46	4.78	6.88	10.38
$\delta = -\phi$	1.65	2.19	3.01	4.29	6.42	10.20	17.50

* For vertical wall face ($\beta = 0$) and horizontal ground surface ($\alpha = 0$).

The first line of Table 7.16 lists values obtained from Eq. (7.77).

Continuous anchors in sand ($\phi = 33°$), when subjected to horizontal pull or thrust, develop passive pressures, lb/lin ft, of about

$$P = 1.5wh^2 \qquad (7.78)$$

where h = distance from bottom of anchor to the surface, ft.

This relationship holds for ratios of h to height d, ft, of anchor of 1.5 to 5.5, and assumes a horizontal ground surface and vertical anchor face.

Square anchors within the same range of h/d develop about

$$P = \left(2.50 + \frac{h}{8d}\right)^2 d\frac{wh^2}{2} \qquad (7.79)$$

where P = passive lateral pressure, lb

d = length and height of anchor, ft

Passive pressures of cohesive soils, resisting movement of a wall or anchor extending to the ground surface, depend on the unit weight of the soil w and its unconfined compressive strength $2c$, psf. At a distance h, ft, below the surface, the passive lateral pressure, psf, is

$$p = wh + 2c \qquad (7.80)$$

The total pressure, lb/lin ft, is

$$P = \frac{wh^2}{2} + 2ch \qquad (7.81)$$

and acts at a distance, ft, above the bottom of the wall or anchor of

$$\bar{x} = \frac{h(wh + 6c)}{3(wh + 4c)}$$

The pressure distribution for plastic clay is shown in Fig. 7.48c.

Continuous anchors in plastic clay, when subjected to horizontal pull or thrust, develop passive pressures, lb/lin ft, of about

$$P = cd\left[8.7 - \frac{11,600}{(h/d + 11)^3}\right] \qquad (7.82)$$

where h = distance from bottom of anchor to surface, ft

d = height of anchor, ft

Equation (7.82) is based on tests made with horizontal ground surface and vertical anchor face.

Safety factors should be applied to the passive pressures computed from Eqs. (7.74) to (7.82) for design use. Experience indicates that a safety factor of 2 is satisfactory for clean sands and gravels. For clay, a safety factor of 3 may be desirable because of uncertainties as to effective shearing strength.

(G. P. Tschebotarioff, "Soil Mechanics, Foundations, and Earth Structures," McGraw-Hill Book Company, New York (books.mcgraw-hill.com); K. Terzaghi and R. B. Peck, "Soil Mechanics in Engineering Practice," John Wiley & Sons, Inc., New York (www.wiley.com); Leo Casagrande, "Comments on Conventional Design of Retaining Structures," *ASCE Journal of Soil Mechanics and Foundations Engineering Division*, 1973, pp. 181–198; H. Y. Fang, "Foundation Engineering Handbook," 2nd ed., Van Nostrand Reinhold Company, New York.)

7.29 Vertical Earth Pressure on Conduit

The vertical load on an underground conduit depends principally on the weight of the prism of soil directly above it. But the load also is affected by vertical shearing forces along the sides of this prism. Caused by differential settlement of the prism and adjoining soil, the shearing forces may be directed up or down. Hence, the load on the conduit may be less or greater than the weight of the soil prism directly above it.

Conduits are classified as ditch or projecting, depending on installation conditions that affect the shears. A ditch conduit is a pipe set in a relatively narrow trench dug in undisturbed soil (Fig. 7.49). Backfill then is placed in the trench up to the original ground surface. A projecting conduit is a pipe over which an embankment is placed.

A projecting conduit may be positive or negative, depending on the extent of the embankment vertically. A positive projecting conduit is installed in a shallow bed with the pipe top above the surface of the ground. Then, the embankment is placed over the pipe (Fig. 7.50a). A negative projecting conduit is set in a narrow, shallow trench with the pipe top below the original ground surface (Fig. 7.50b). Then, the ditch is backfilled, after which the embankment is placed. The load on the conduit is less when the backfill is not compacted.

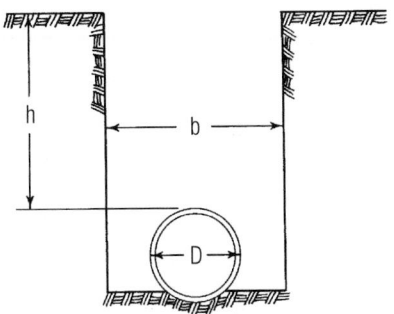

Fig. 7.49 Ditch conduit.

Load on underground pipe also may be reduced by the imperfect-ditch method of construction. This starts out as for a positive projecting conduit, with the pipe at the original ground surface. The embankment is placed and compacted for a few feet above the pipe. But then, a trench as wide as the conduit is dug down to it through the compacted soil. The trench is backfilled with a loose, compressible soil (Fig. 7.50c). After that, the embankment is completed.

The load, lb/lin ft, on a rigid ditch conduit may be computed from

$$W = C_D whb \qquad (7.83)$$

and on a flexible ditch conduit from

$$W = C_D whD \qquad (7.84)$$

where C_D = load coefficient for ditch conduit

w = unit weight of fill, lb/ft³

h = height of fill above top of conduit, ft

b = width of ditch at top of conduit, ft

D = outside diameter of conduit, ft

From the equilibrium of vertical forces, including shears, acting on the backfill above the conduit, C_D may be determined:

$$C_D = \frac{1 - e^{-kh/b}}{k} \frac{b}{h} \qquad (7.85)$$

where $e = 2.718$

$k = 2K_a \tan \theta$

K_a = coefficient of active earth pressure [Eq. (7.64) and Table 7.14]

θ = angle of friction between fill and adjacent soil ($\theta \leq \phi$, angle of internal friction of fill)

Table 7.17 gives values of C_D for $k = 0.33$ for cohesionless soils, $k = 0.30$ for saturated topsoil, and $k = 0.26$ and 0.22 for clay (usual maximum and saturated).

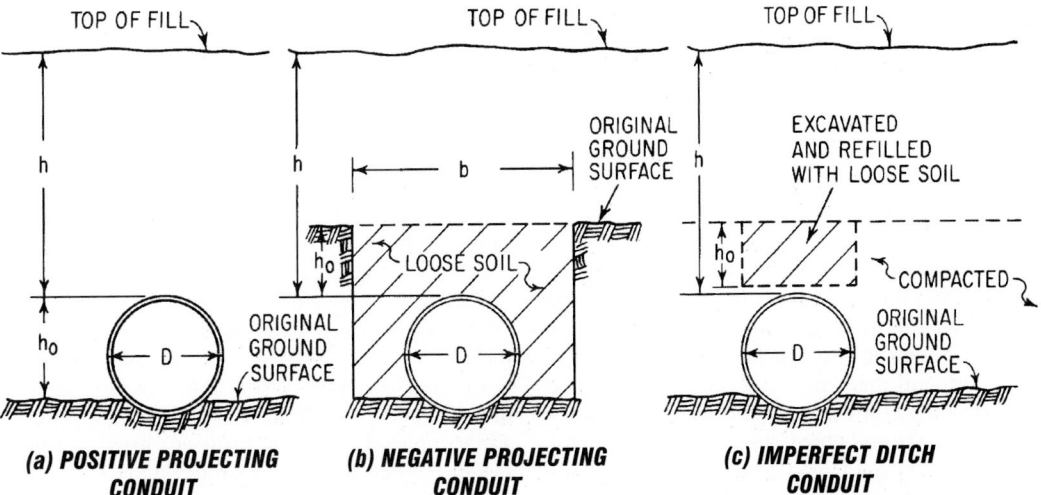

Fig. 7.50 Type of projecting conduit depends on method of backfilling.

Table 7.17 Load Coefficients C_D for Ditch Conduit

h/b	Cohesionless Soils	Saturated Topsoil	Clay $k = 0.26$	Clay $k = 0.22$
1	0.85	0.86	0.88	0.89
2	0.75	0.75	0.78	0.80
3	0.63	0.67	0.69	0.73
4	0.55	0.58	0.62	0.67
5	0.50	0.52	0.56	0.60
6	0.44	0.47	0.51	0.55
7	0.39	0.42	0.46	0.51
8	0.35	0.38	0.42	0.47
9	0.32	0.34	0.39	0.43
10	0.30	0.32	0.36	0.40
11	0.27	0.29	0.33	0.37
12	0.25	0.27	0.31	0.35
Over 12	$3.0b/h$	$3.3b/h$	$3.9b/h$	$4.5b/h$

Vertical load, lb/lin ft, on conduit installed by tunneling may be estimated from

$$W = C_D b(wh - 2c) \qquad (7.86)$$

where c = cohesion of the soil, or half the unconfined compressive strength of the soil, psf. The load coefficient C_D may be computed from Eq. (7.85) or obtained from Table 7.17 with b = maximum width of tunnel excavation, ft, and h = distance from tunnel top to ground surface, ft.

For a ditch conduit, shearing forces extend from the pipe top to the ground surface. For a projecting conduit, however, if the embankment is sufficiently high, the shear may become zero at a horizontal plane below grade, the plane of equal settlement. Load on a projecting conduit is affected by the location of this plane.

Vertical load, lb/lin ft, on a positive projecting conduit may be computed from

$$W = C_P whD \qquad (7.87)$$

where C_P = load coefficient for positive projecting conduit. Formulas have been derived for C_P and the depth of the plane of equal settlement. These formulas, however, are too lengthy for practical application, and the computation does not appear to be justified by the uncertainties in actual relative settlement of the soil above the conduit. Tests may be made in the field to determine C_P. If so, the possibility of an increase in earth pressure with time should be considered. For a rough estimate,

C_P may be assumed as 1 for flexible conduit and 1.5 for rigid conduit.

The vertical load, lb/lin ft, on negative projecting conduit may be computed from

$$W = C_N whb \qquad (7.88)$$

where C_N = load coefficient for negative projecting conduit

h = height of fill above top of conduit, ft

b = horizontal width of trench at top of conduit, ft

The load on an imperfect ditch conduit may be obtained from

$$W = C_N whD \qquad (7.89)$$

where D = outside diameter of conduit, ft.

Formulas have been derived for C_N, but they are complex, and insufficient values are available for the parameters involved. As a rough guide, C_N may be taken as 0.9 when depth of cover exceeds conduit diameter. (See also Art. 10.31.)

Superimposed surface loads increase the load on an underground conduit. The magnitude of the increase depends on the depth of the pipe below grade and the type of soil. For moving loads, an impact factor of about 2 should be applied. A superimposed uniform load w', lb/ft^2, of large extent may be treated for projecting conduit as an equivalent layer of embankment with a thickness, ft, of w'/w. For ditch conduit, the load due to the soil should be increased by $bw'e^{-kh/b}$, where $k = 2K_a\tan\theta$, as in Eq. (7.85). The increase due to concentrated loads can be estimated by assuming the loads to spread out linearly with depth, at an angle of about 30° with the vertical. (See also Art. 7.11).

(M. G. Spangler, "Soil Engineering," International Textbook Company, Scranton, Pa.; "Handbook of Steel Drainage and Highway Construction Products," American Iron and Steel Institute, Washington, D. C. (www.steel.org).)

7.30 Dewatering Methods for Excavations

The main purpose of dewatering is to enable construction to be carried out under relatively dry conditions. Good drainage stabilizes excavated slopes, reduces lateral loads on sheeting and bracing, and reduces required air pressure in tunneling. Dewatering makes excavated material lighter and easier to handle. It also prevents loss of soil below slopes or from the bottom of the excavation, and prevents a "quick" or "boiling" bottom. In addition, permanent lowering of the groundwater table or relief of artesian pressure may allow a less expensive design for the structure, especially when the soil consolidates or becomes compact. If lowering of the water level or pressure relief is temporary, however, the improvement of the soil should not be considered in foundation design. Increases in strength and bearing capacity may be lost when the soil again becomes saturated.

To keep an excavation reasonably dry, the groundwater table should be kept at least 2 ft, and preferably 5 ft, below the bottom in most soils.

Site investigations should yield information useful for deciding on the most suitable and economical dewatering method. Important is a knowledge of the types of soil in and below the site, probable groundwater levels during construction, permeability of the soils, and quantities of water to be handled. A pumping test may be desirable for estimating capacity of pumps needed and drainage characteristics of the ground.

Many methods have been used for dewatering excavations. Those used most often are listed in Table 7.18 with conditions for which they generally are most suitable. (See also Art. 7.37.)

In many small excavations, or where there are dense or cemented soils, water may be collected in ditches or sumps at the bottom and pumped out. This is the most economical method of dewatering, and the sumps do not interfere with future construction as does a comprehensive wellpoint system. But the seepage may slough the slopes, unless they are stabilized with gravel, and may hold up excavation while the soil drains. Also, springs may develop in fine sand or silt and cause underground erosion and subsidence of the ground surface.

For sheetpile-enclosed excavations in pervious soils, it is advisable to intercept water before it enters the enclosure. Otherwise, the water will put high pressures on the sheeting. Seepage also can cause the excavation bottom to become quick, overloading the bottom bracing, or create piping, undermining the sheeting. Furthermore, pumping from the inside of the cofferdam is likely to leave the soil to be excavated wet and tough to handle.

Wellpoints often are used for lowering the water table in pervious soils. They are not suitable, however, in soils that are so fine that they will flow with the water or in soils with low permeability. Also, other methods may be more economical for deep excavations, very heavy flows, or considerable lowering of the water table (Table 7.18).

Wellpoints are metal well screens about 2 to 3 in in diameter and up to about 4 ft long. A pipe connects each wellpoint to a header, from which water is pumped to discharge (Fig. 7.51). Each pump usually is a combined vacuum and centrifugal pump. Spacing of wellpoints generally ranges from 3 to 12 ft c to c.

A wellpoint may be jetted into position or set in a hole made with a hole puncher or heavy steel casing. Accordingly, wellpoints may be self-jetting or plain-tip. To insure good drainage in fine and dirty sands or in layers of silt or clay, the wellpoint and riser should be surrounded by sand to just below the water table. The space above the filter should be sealed with silt or clay to keep air from getting into the wellpoint through the filter.

Wellpoints generally are relied on to lower the water table 15 to 20 ft. Deep excavations may be dewatered with multistage wellpoints, with one row of wellpoints for every 15 ft of depth. Or when the flow is less than about 15 gal/min per wellpoint, a single-stage system of wellpoints may be installed above the water table and operated with jet-eductor pumps atop the wellpoints. These pumps can lower the water table up to about 100 ft, but they have an efficiency of only about 30%.

Deep wells may be used in pervious soils for deep excavations, large lowering of the water table, and heavy water flows. They may be placed along the top of an excavation to drain it, to intercept seepage before pressure makes slopes unstable, and to relieve artesian pressure before it heaves the excavation bottom.

Usual spacing of wells ranges from 20 to 250 ft. Diameter generally ranges from 6 to 20 in. Well screens may be 20 to 75 ft long, and they are surrounded with a sand-gravel filter. Generally, pumping is done with a submersible or vertical turbine pump installed near the bottom of each well.

Figure 7.52 shows a deep-well installation used for a 300-ft-wide by 600-ft-long excavation for a building of the Smithsonian Institution, Washington, D. C. Two deep-well pumps lowered the general water level in the excavation 20 ft. The well installation proceeded as follows: (1) Excavation to water level (elevation 0.0). (2) Driving of sheetpiles around the well area (Fig. 7.52a). (3) Excavation underwater inside the sheetpile enclosure to elevation -37.0 ft (Fig. 7.52b). Bracing installed as digging progressed. (4) Installation of a wire-mesh-wrapped timber frame, extending from elevation 0.0 to -37.0 (Fig. 7.52c). Weights added to sink the frame. (5) Backfilling of space between sheetpiles and mesh with $\frac{3}{16}$- to $\frac{3}{8}$-in gravel. (6) Removal of sheetpiles. (7) Installation of pump and start of pumping.

Vacuum well or wellpoint systems may be used to drain silts with low permeability (coefficient between 0.01 and 0.0001 mm/s). In these systems,

Table 7.18 Methods for Dewatering Excavations

Saturated-Soil Conditions	Dewatering Method Probably Suitable
Surface water	Ditches; dikes; sheetpiles and pumps or underwater excavation and concrete tremie seal
Gravel	Underwater excavation, grout curtain; gravity drainage with large sumps with gravel filters
Sand (except very fine sand)	Gravity drainage
Waterbearing strata near surface; water table does not have to be lowered more than 15 ft	Wellpoints with vacuum and centrifugal pumps
Waterbearing strata near surface; water table to be lowered more than 15 ft, low pumping rate	Wellpoints with jet-eductor pumps
Excavations 30 ft or more below water table; artesian pressure; high pumping rate; large lowering of water table—all where adequate depth of pervious soil is available for submergence of well screen and pump	Deep wells, plus, if necessary, wellpoints
Sand underlain by rock near excavation bottom	Wellpoints to rock, plus ditches, drains, automatic "mops"
Sand underlain by clay	Wellpoints in holes 3 or 4 into the clay, backfilled with sand
Silt; very find sand (permeability coefficient between 0.01 and 0.0001 mm/s)	For lifts up to 15 ft, wellpoints with vacuum; for greater lifts, wells with vacuum; sumps
Silt or silty sand underlain by pervious soil	At top of excavation, and extending to the pervious soil, vertical sand drains plus well points or wells
Clay-silts, silts	Electro-osmosis
Clay underlain by pervious soil	At top of excavation, wellpoints or deep wells extending to pervious soil
Dense or cemented soils: small excavations	Ditches and sumps

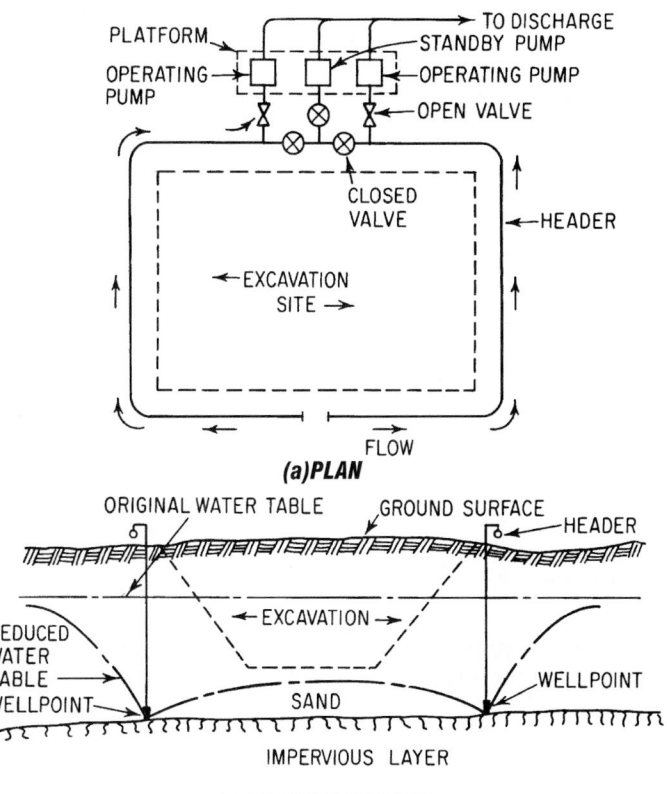

Fig. 7.51 Wellpoint system for dewatering an excavation.

wells or wellpoints are closely spaced, and a vacuum is held with vacuum pumps in the well screens and sand filters. At the top, the filter, well, and risers should be sealed to a depth of 5 ft with bentonite or an impervious soil to prevent loss of the vacuum. Water drawn to the well screens is pumped out with submersible or centrifugal pumps.

Where a pervious soil underlies silts or silty sands, vertical sand drains and deep wells can team up to dewater an excavation. Installed at the top, and extending to the pervious soil, the sand piles intercept seepage and allow it to drain down to the pervious soil. Pumping from the deep wells relieves the pressure in that deep soil layer.

For some silts and clay-silts, electrical drainage with wells or wellpoints may work, whereas gravity methods may not (Art. 7.37). In saturated clays, thermal or chemical stabilization may be necessary (Arts. 7.38 and 7.39).

Small amounts of surface water may be removed from excavations with "mops." Surrounded with gravel to prevent clogging, these drains are connected to a header with suction hose or pipe. For automatic operation, each mop should be opened and closed by a float and float valve.

When structures on silt or soft material are located near an excavation to be dewatered, care should be taken that lowering of the water table does not cause them to settle. It may be necessary to underpin the structures or to pump discharge water into recharge wells near the structures to maintain the water table around them.

(L. Zeevaert, "Foundation Engineering for Difficult Subsoil Conditions," H. Y. Fang, "Foundation Engineering Handbook," 2nd ed., Van Nostrand Reinhold Company, New York.)

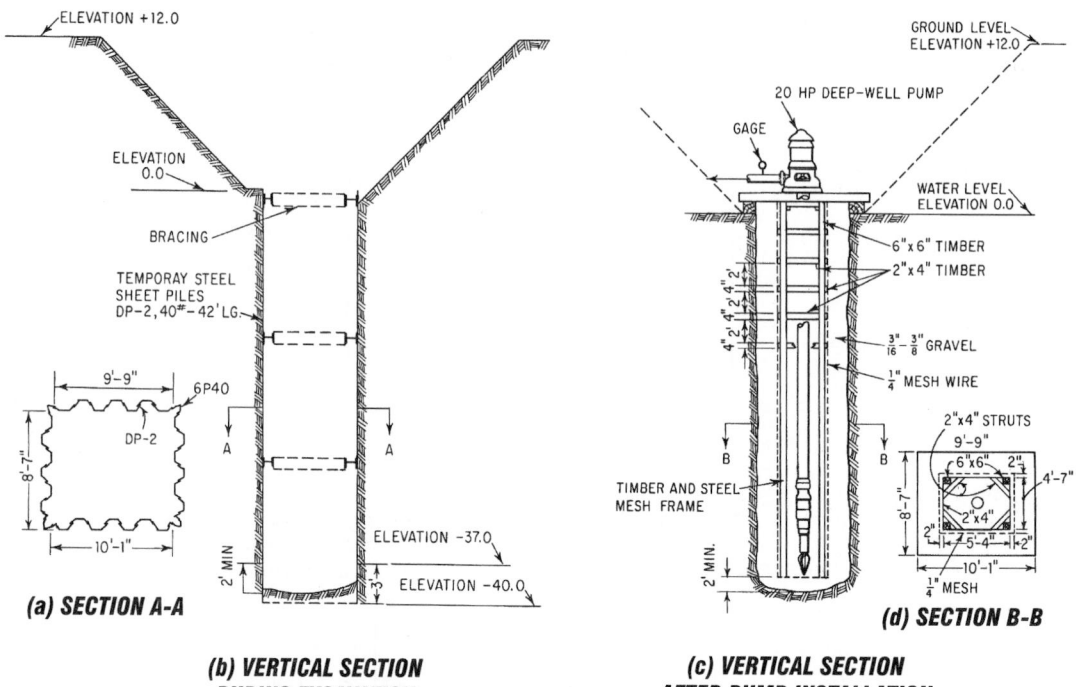

(a) SECTION A-A

**(b) VERTICAL SECTION
DURING EXCAVATION**

**(c) VERTICAL SECTION
AFTER PUMP INSTALLATION**

(d) SECTION B-B

Fig. 7.52 Deep-well installation used at Smithsonian Institution, Washington, D.C. (*Spencer, White & Prentis, Inc.*)

Underpinning

The general methods and main materials used to give additional support at or below grade to structures are called underpinning. Usually, the added support is applied at or near the footings.

7.31 Underpinning Procedures

Underpinning may be remedial or precautionary. Remedial underpinning adds foundation capacity to an inadequately supported structure. Precautionary underpinning is provided to obtain adequate foundation capacity to sustain higher loads, as a safeguard against possible settlement during adjacent excavating, or to accommodate changes in ground conditions. Usually, this type of underpinning is required for the foundations of a structure when deeper foundations are to be constructed nearby for an addition or another structure. Loss of ground, even though small, into

an adjoining excavation may cause excessive settlement of existing foundations.

Presumably, an excavation influences an existing substructure when a plane through the outermost foundations, on a 1-on-1 slope for sand or a 1-on-2 slope for unconsolidated silt or soft clay, penetrates the excavation. For a cohesionless soil, underpinning exterior walls within a 1-on-1 slope usually suffices; interior columns are not likely to be affected if farther from the edge of the excavation than half the depth of the cut.

The commonly accepted procedures of structural and foundation design should be used for underpinning. Data for computing dead loads may be obtained from plans of the structure or a field survey. Since underpinning is applied to existing structures, some of which may be old, engineers in charge of underpinning design and construction should be familiar with older types of construction as well as the most modern.

Before underpinning starts, the engineers should investigate and record existing defects in the structure. Preferably, the engineers should be accompanied in this investigation by a representative of the owner. The structure should be thoroughly inspected, from top to bottom, inside (if possible) and out. The report should include names of inspectors, dates of inspection, and description and location of defects. Photographs are useful in verifying written descriptions of damaged areas. The engineers should mark existing cracks in such a way that future observations would indicate whether they are continuing to open or spread.

Underpinning generally is accompanied by some settlement. If design and field work are good, the settlement may be limited to about $\frac{1}{4}$ to $\frac{3}{8}$ in. But as long as settlement is uniform in a structure, damage is unlikely. Differential settlement should be avoided. To check on settlement, elevations of critical points, especially columns and walls, should be measured frequently during underpinning. Since movement may also occur laterally, the plumbness of walls and columns also should be checked.

One of the first steps in underpinning usually is digging under a foundation, which decreases its load-carrying capacity temporarily. Hence, preliminary support may be necessary until underpinning is installed. This support may be provided by shores, needles, grillages, and piles. Sometimes, it is desirable to leave them in place as permanent supports.

Generally, it is advisable to keep preliminary supports at a minimum, for economy and to avoid interference with other operations. For the purpose, advantage may be taken of arching action and of the ability of a structure to withstand moderate overloads. Also, columns centrally supported on large spread footings need not be shored when digging is along an edge and involves only a small percent of the total footing area. A large part of the column load is supported by the soil directly under the column.

When necessary, weak portions of a structure, especially masonry, should be repaired or strengthened before underpinning starts.

7.32 Shores

Installed vertically or on a slight incline, shores are used to support walls or piers while underpinning pits are dug (Fig. 7.53a). Good bearing should be provided at top and bottom of the shores. One way

of providing bearing at the top is to cut a niche and mortar a steel bearing plate against the upper face. An alternate to the plate is a Z shape, made by removing diagonally opposite half flanges from an H beam. When the top of the shore is cut to fit between flange and web of the Z, movement of the shore is restrained. For a weak masonry wall, the load may have to be distributed over a larger area. One way of doing this is to insert a few lintel angles about 12 in apart vertically and bolt them to a vertical, heavy timber or steel distributing beam. The horizontal leg of an angle on the beam then transmits the load to a shore.

Inclined shores on only one side of a wall require support at the base for horizontal as well as vertical forces. One way is to brace the shores against an opposite wall at the floor. Preferably, the base of each shore should sit on a footing perpendicular to the axis of the shore. Sized to provide sufficient bearing on the soil, the footing may be made of heavy timbers, steel beams, or reinforced concrete, depending on the load on the shore.

Loads may be transferred to a shore by wedges or jacks. Oak wedges are suitable for light loads; forged steel wedges and bearing plates are desirable for heavy loads. Jacks, however, offer greater flexibility in length adjustments and allow corrections during underpinning for settlement of shore footings.

(H. A. Prentis and L. White, "Underpinning," Columbia University Press, New York; M. J. Tomlinson, "Foundation Design and Construction," Halsted Press, New York.)

7.33 Needles and Grillages

Needles are beams installed horizontally to transfer the load of a wall or column to either or both sides of its foundation, to permit digging of underpinning pits (Fig. 7.53b). These beams are more expensive than shores, which transmit the load directly into the ground. Needles usually are steel wide-flange beams, sometimes plate girders, used in pairs, with bolts and pipe spreaders between the beams. This arrangement provides resistance to lateral buckling and torsion. The needles may be prestressed with jacks to eliminate settlement when the load is applied.

The load from steel columns may be transmitted through brackets to the needles. For masonry walls, the needles may be inserted through niches. The

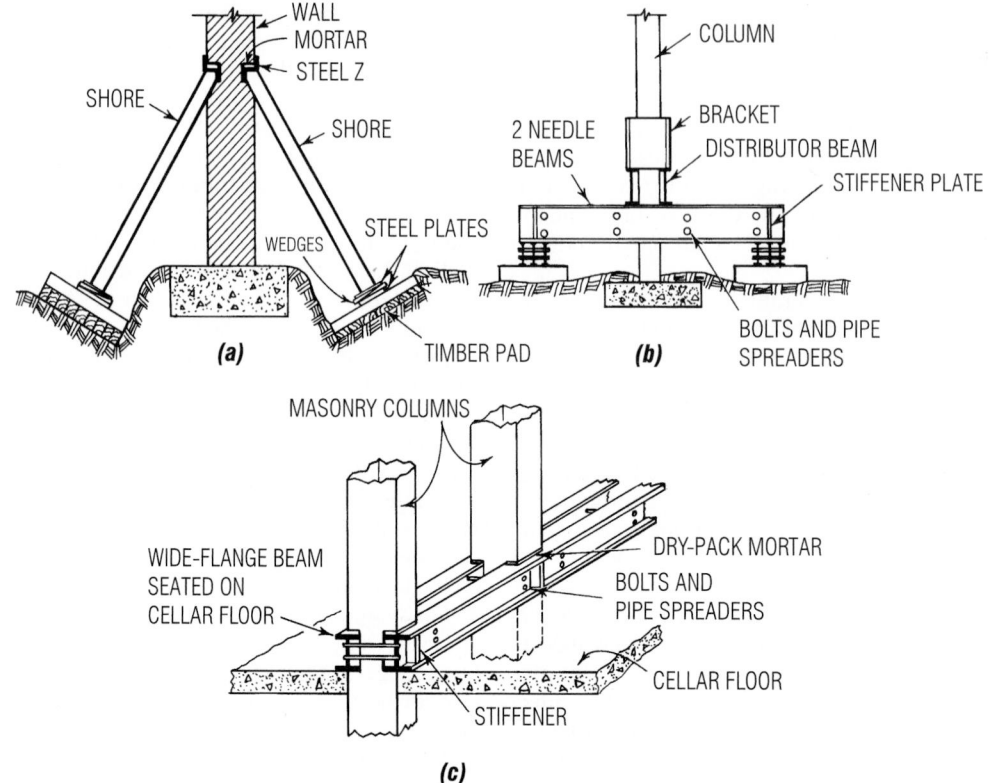

Fig. 7.53 Temporary supports used in underpinning: (*a*) shores; (*b*) needle beams; (*c*) grillage.

load should be transferred from the masonry to the needles through thin wood fillers that crush when the needles deflect and maintain nearly uniform bearing.

Wedges may be placed under the ends of the needles to shift the load from the member to be supported to those beams. The beam ends may be carried on timber pads, which distribute the load over the soil.

Grillages, which have considerably more bearing on the ground than needles, often are used as an alternative to needles and shores for closely spaced columns. A grillage may be installed horizontally on soil at foundation level to support and tie together two or more column footings, or it may rest on a cellar floor (Fig. 7.53*c*). These preliminary supports may consist of two or more steel beams, tied together with bolts and pipe spreaders, or of a steel-concrete composite. Also, grillages sometimes are used to strengthen or

repair existing footings by reinforcing them and increasing their bearing area. The grillages may take the form of dowels or of encircling concrete or steel-concrete beams. They should be adequately cross-braced against buckling and torsion. Holes should be made in steel beams to be embedded in concrete, to improve bond.

7.34 Pit Underpinning

After preliminary supports have been installed and weak construction strengthened or repaired, underpinning may start. The most common method of underpinning a foundation is to construct concrete piers down to deeper levels with adequate bearing capacity and to transfer the load to the piers by wedging up with dry packing. To build the piers, pits must be dug under the foundation. Because of the danger of loss of ground and consequent settlement where soils are saturated, the method usually is restricted to dry subsoil.

When piers have to be placed close together, a continuous wall may be constructed instead. But the underpinning wall should be built in short sections, usually about 5 ft long, to avoid undermining the existing foundation. Alternate sections are built first, and then the gaps are filled in.

Underpinning pits rarely are larger than about 5 ft square in cross section. Minimum size for adequate working room is 3 × 4 ft. Access to the pit

is provided by an approach pit started alongside the foundation and extending down about 6 ft. The pits must be carefully sheeted and adequately braced to prevent loss of ground, which can cause settlement of the structure.

In soils other than soft clay, 2-in-thick wood planks installed horizontally may be used to sheet pits up to 5 ft square, regardless of depth. Sides of the pits should be trimmed back no more than absolutely necessary. The boards, usually 2 × 8's, are installed one at a time with 2-in spaces between

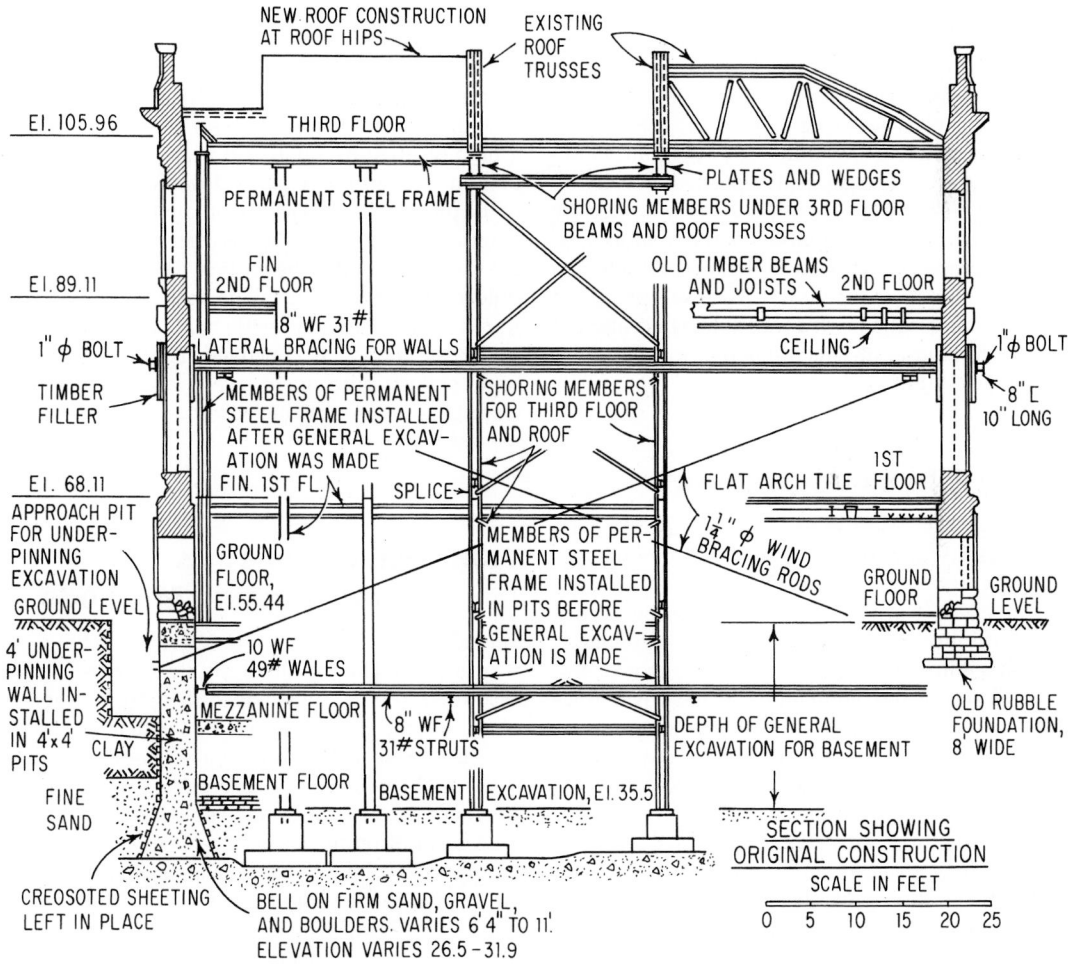

Fig. 7.54 Vertical section through White House, Washington, D.C., during restoration. Pit underpinning was used for the walls. (*Spencer, White & Prentis, Inc.*)

them vertically. Soil is repacked through these louvers to fill the voids behind the boards. In running sands, hay may be stuffed behind the boards to block the flow. Corners of the sheeting often are nailed with vertical 2 × 4 in wood cleats.

In soft clay, sheeting must be tight and braced against earth pressure. Chicago caissons, or a similar type, may be used (Art. 7.23).

In water-bearing soil with a depth not exceeding about 5 ft, vertical sheeting can sometimes be driven to cut off the water. For the purpose, light steel or tongue-and-groove wood sheeting may be used. The sheeting should be driven below the bottom of the pit a sufficient distance to prevent boiling of the bottom due to hydrostatic pressure. With water cut off, the pit can be pumped dry and excavation continued.

After a pit has been dug to the desired level, it is filled with concrete to within 3 in of the foundation to be supported. The gap is dry-packed, usually by ramming stiff mortar in with a 2 × 4 pounded by an 8-lb hammer. The completed piers should be laterally braced if soil is excavated on one side to a depth of more than about 6 ft. An example of pit underpinning is the work done in the restoration of the White House, in which a cellar and subcellar were created (Fig. 7.54).

7.35 Pile Underpinning

If water-bearing soil more than about 5 ft deep underlies a foundation, the structure may have to be underpinned with piles. Driven piles generally are preferred to jacked piles because of lower cost. The feasibility of driven piles, however, depends on availability of at least 12 ft of headroom and space alongside the foundations. Thus, driven piles often can be used to underpin interior building columns when headroom is available. But they are hard to install for exterior walls unless there is ample space alongside the walls. For very lightly loaded structures, brackets may be attached to underpinning piles to support the structure. But such construction puts bending into the piles, reducing their load-carrying capacity.

Driven piles usually are 12- to 14-in diameter steel pipe, $\frac{3}{8}$ in thick. They are driven open-ended, to reduce vibration, and in lengths determined by available headroom. Joints may be made with cast-steel sleeves. After soil has been removed from the pipe interior, it is filled with concrete.

Jacked piles require less headroom and may be placed under a footing. Also made of steel pipe installed open-ended, these piles are forced down by hydraulic jacks reacting against the footing. The operation requires an approach pit under the footing to obtain about 6 ft of headroom.

Pretest piles, originally patented by Spencer, White & Prentis, New York City, are used to prevent the rebound of piles when jacking stops and subsequent settlement when the load of the structure is transferred to the piles. A pipe pile is jacked down, in 4-ft lengths, to the desired depth. The hydraulic jack reacts against a steel plate mortared to the underside of the footing to be supported. After the pile has been driven to the required depth and cleaned out, it is filled with concrete and capped with a steel bearing plate. Then two hydraulic jacks atop the pile overload it 50%. As the load is applied, a bulb of pressure builds up in the soil at the pile bottom. This pressure stops downward movement of the pile. While the jacks maintain the load, a short length of beam is wedged between the pile top and the steel plate under the footing. Then, the jacks are unloaded and removed. The load, thus, is transferred without further settlement. Later the space under the footing is concreted. Figure 7.55 shows how pretest piles were used for underpinning existing structures during construction of a subway in New York City.

7.36 Miscellaneous Underpinning Methods

Spread footings may be pretested in much the same way as piles. The weight of the structure is used to jack down the footings, which then are wedged in place, and the gap is concreted. The method may be resorted to for unconsolidated soils where a high water table makes digging under a footing unsafe or where a firm stratum is deep down.

A form of underpinning may be used for slabs on ground. When concrete slabs settle, they may be restored to the proper elevation by mud jacking. In this method, which will not prevent future settlement, a fluid grout is pumped under the slab through holes in it, raising it. Pressure is maintained until the grout sets. The method also may be used to fill voids under a slab.

Chemical or thermal stabilization (Arts. 7.38 and 7.39) sometimes may be used as underpinning.

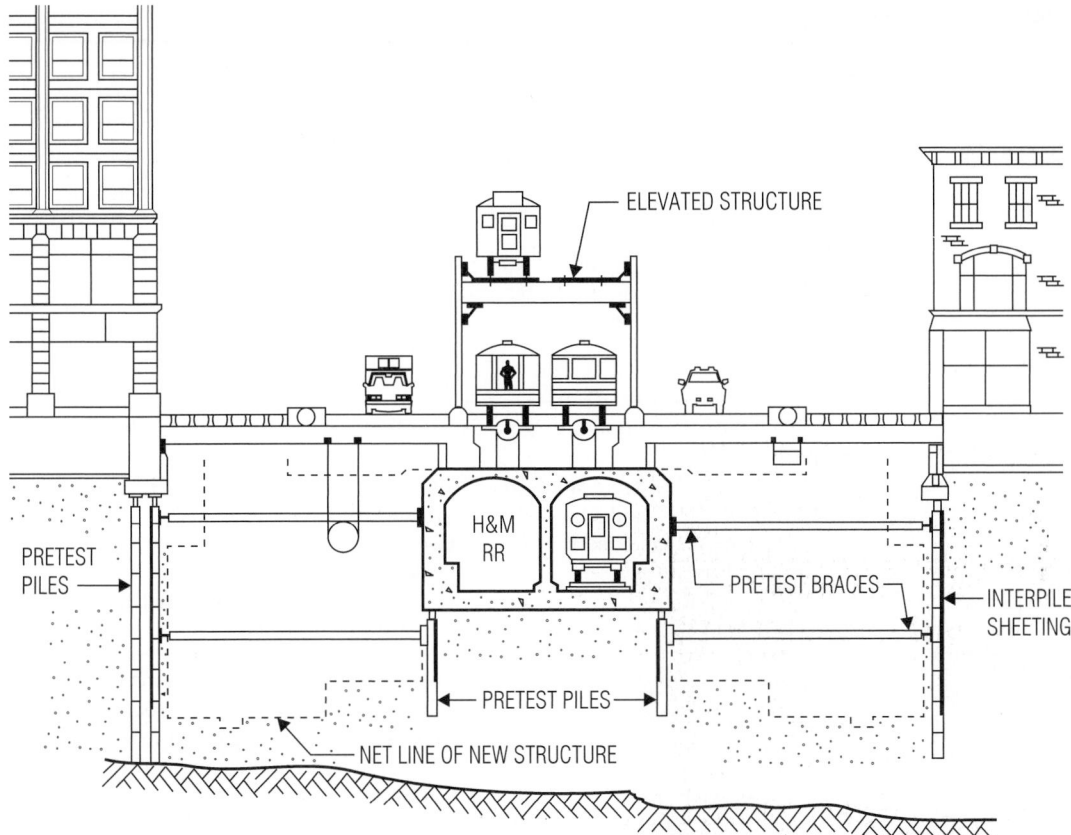

Fig. 7.55 Pretest piles support existing buildings, old elevated railway, and a tunnel during construction of a subway. (*Spencer, White & Prentis, Inc.*)

Ground Improvement

Soil for foundations can be altered to conform to desired characteristics. Whether this should be done depends on the cost of alternatives.

Investigations of soil and groundwater conditions on a site should indicate whether soil improvement, or stabilization, is needed. Tests may be necessary to determine which of several applicable techniques may be feasible and economical. Table 7.19 lists some conditions for which soil improvement should be considered and the methods that may be used.

As indicated in the table, ground improvement may increase strength, increase or decrease permeability, reduce compressibility, improve stability, or decrease heave due to frost or swelling. The main techniques used are constructed fills, replacement of unsuitable soils, surcharges, reinforcement, mechanical stabilization, thermal stabilization, and chemical stabilization. (Federal Highway Administration "Ground Improvement Technical Summaries" Publication No. FHWA-SA-98-086, December 1999).

7.37 Mechanical Stabilization of Soils

This comprises a variety of techniques for rearranging, adding, or removing soil particles. The objective usually is to increase soil density, decrease water content, or improve gradation. Particles may be rearranged by blending the layers of a stratified soil, remolding an undisturbed soil, or densifying a soil. Sometimes, the desired improvement can be obtained by drainage alone.

Table 7.19 Where Soil Improvement May Be Economical

Soil Deficiency	Probable Type of Failure	Probable Cause	Possible Remedies
Slope instability	Slides on slope	Pore-water pressure	Drain; flatten slope; freeze
		Loose granular soil	Compact
		Weak soil	Mix or replace with select material
	Mud flow	Excessive water	Exclude water
	Slides—movement at toe	Toe instability	Place toe fill, and drain
Low bearing capacity	Excessive settlement	Saturated clay	Consolidate with surcharge, and drain
		Loose granular soil	Compact; drain; increase footing depth; mix with chemicals
		Weak soil	Superimpose thick fill; mix or replace with select material; inject or mix with chemicals; freeze (if saturated); fuse with heat (if unsaturated)
Heave	Excessive rise	Frost	For buildings: place foundations below frost line; insulate refrigeration-room floors; refrigerate to keep ground frozen. For roads: Remove fines from gravel; replace with nonsusceptible soil
		Expansion of clay	Exclude water; replace with granular soil
Excessive Permeability	Seepage	Pervious soil or fissured rock	Mix or replace soil with select material; inject or mix soil with chemicals; construct cutoff wall with grout; enclose with sheetpiles and drain
"Quick" bottom	Loss of strength	Flow under	Add berm against cofferdam inner face; increase width of cofferdam between lines of sheeting; drain with wellpoints outside the cofferdam

Often, however, compactive effort plus water control is needed.

7.37.1 Embankments

Earth often has to be placed over the existing ground surface to level or raise it. Such constructed fills may create undesirable conditions because of improper compaction, volume changes, and unexpected settlement under the weight of the fill. To prevent such conditions, fill materials and their gradation, placement, degree of compaction, and thickness should be suitable for properly supporting the expected loads.

Fills may be either placed dry with conventional earthmoving equipment and techniques or wet by hydraulic dredges. Wet fills are used mainly for filling behind bulkheads or for large fills.

A variety of soils and grain sizes are suitable for topping fills for most purposes. Inclusion of organic matter or refuse should, however, be prohibited. Economics usually require that the source of fill material be as close as possible to the site. For most fills, soil particles in the 18 in below foundations, slabs, or the ground surface should not be larger than 3 in in any dimension.

For determining the suitability of a soil as fill and for providing a standard for compaction, the moisture-density relationship test, or Proctor test (ASTM D698 and D1557), is often used. Several of these laboratory tests should be performed on the borrowed material, to establish moisture-density curves. The peak of a curve indicates the maximum density achievable in the laboratory by the test method as well as the optimum moisture content. ASTM D1557 should be used as the standard when high bearing capacity and low compressibility are required; ASTM D698 should be used when requirements are lower, for example, for fills under parking lots.

The two ASTM tests represent different levels of compactive effort. But much higher compactive effort may be employed in the field than that used in the laboratory. Thus, a different moisture-density relationship may be produced on the site. Proctor test results, therefore, should not be considered an inherent property of the soil. Nevertheless, the test results indicate the proposed fill material's sensitivity to moisture content and the degree of field control that may be required to obtain the specified density.

See also Art. 7.40.

7.37.2 Fill Compaction

The degree of compaction required for a fill is usually specified as a minimum percentage of the maximum dry density obtained in the laboratory tests. This compaction is required to be accomplished within a specific moisture range. Minimum densities of 90 to 95% of the maximum density are suitable for most fills. Under roadways, footings, or other highly loaded areas, however, 100% compaction is often required. In addition, moisture content within 2 to 4% of the optimum moisture content usually is specified.

Field densities can be greater than 100% of the maximum density obtained in the laboratory test. Also, with greater compactive effort, such densities can be achieved with moisture contents that do not lie on the curves plotted from laboratory results. (Fine-grained soils should not be overcompacted on the dry side of optimum because when they get wet, they may swell and soften significantly.)

For most projects, lift thickness should be restricted to 8 to 12 in, each lift being compacted before the next lift is placed. On large projects where heavy compaction equipment is used, a lift thickness of 18 to 24 in is appropriate.

Compaction achieved in the field should be determined by performing field density tests on each lift. For that purpose, wet density and moisture content should be measured and the dry density computed. Field densities may be ascertained by the sand-cone (ASTM D1556) or balloon volume-meter (ASTM D2167) method, from an undisturbed sample, or with a nuclear moisture-density meter. Generally, one field density test for each 4000 to 10,000 ft^2 of lift surface is adequate.

Hydraulically placed fills composed of dredged soils normally need not be compacted during placement. Although segregation of the silt and clay fractions of the soils may occur, it usually is not harmful. But accumulation of the fine-grained material in pockets at bulkheads or under structures should be prevented. For the purpose, internal dikes, weirs, or decanting techniques may be used.

7.37.3 Soil Replacement or Blending

When materials at or near grade are unsuitable, it may be economical to remove them and substitute a fill of suitable soil, as described in Art. 7.37.1. When this is not economical, consideration should be given to improving the soil by other methods, such as densification or addition or removal of soil particles.

Mixing an existing soil with select materials or removing selected sizes of particles from an existing soil can change its properties considerably. Adding clay to a cohesionless soil in a nonfrost region, for example, may make the soil suitable as a base course for a road (if drainage is not too greatly impaired). Adding clay to a pervious soil may reduce its permeability sufficiently to permit its use as a reservoir bottom. Washing particles finer than 0.02 mm from gravel makes the soil less susceptible to frost heave (desirable upper limit for this fraction is 3%).

7.37.4 Surcharges

Where good soils are underlain by soft, compressible clays that would permit unacceptable settlement, the site often can be made usable by surcharging, or preloading, the surface. The objective is to use the weight of the surcharge to consolidate the underlying clays. This offsets the settlement of the completed structure that would otherwise occur. A concurrent objective may be to increase the strength of the underlying clays.

If the soft clay is overlain by soils with adequate bearing capacity, the area to be improved may be loaded with loose, dumped earth, until the weight of the surcharge is equivalent to the load that later will be imposed by the completed structure. (If highly plastic clays or thick layers with little internal drainage are present, it may be necessary to insert vertical drains to achieve consolidation within a reasonable time.) During and after placement of the surcharge, settlement of the original ground surface and the clay layer should be closely monitored. The surcharge may be removed after little or no settlement is observed. If surcharging has been properly executed, the completed structure should experience no further settlement due to primary consolidation. Potential settlement due to secondary compression, however, should be evaluated, especially if the soft soils are highly organic.

7.37.5 Densification

Any of a variety of techniques, most involving some form of vibration, may be used for soil densification. The density achieved with a specific technique, however, depends on the grain size of the soil. Consequently, grain-size distribution must be taken into account when selecting a densification method.

Compaction of clean sands to depths of about 6 ft usually can be achieved by rolling the surface with a heavy, vibratory, steel-drum roller. Although the vibration frequency is to some extent adjustable, the frequencies most effective are in the range of 25 to 30 Hz. Bear in mind, however, that little densification will be achieved below a depth of 6 ft, and the soil within about 1 ft of the surface may actually be loosened. Compactive effort in the field may be measured by the number of passes made with a specific machine of given weight and at a given speed. For a given compactive effort, density varies with moisture content. For a given moisture content, increasing the compactive effort increases the soil density and reduces the permeability.

Compaction piles also may be used to densify sands. For that purpose, the piles usually are made of wood. Densification of the surrounding soils results from soil displacement during driving of the pile or shell and from the vibration produced during pile driving. The foundations to be constructed need not bear directly on the compaction piles but may be seated anywhere on the densified mass.

Vibroflotation and Terra-Probe are alternative methods that increase sand density by multiple insertions of vibrating probes. These form cylindrical voids, which are then filled with offsite sand, stone, or blast-furnace slag. The probes usually are inserted in clusters, with typical spacing of about $4\frac{1}{2}$ ft, where footings will be placed. Relative densities of 85% or more can be achieved throughout the depth of insertion, which may exceed 40 ft. Use of vibrating probes may not be effective, however, if the fines content of the soil exceeds about 15% or if organic matter is present in colloidal form in quantities exceeding about 5% by weight.

Another technique for densification is dynamic compaction, which in effect subjects the site to numerous mini-earthquakes. In saturated soils, densification by this method also results from partial liquefaction, and the elevated pore pressures produced must be dissipated between each application of compactive energy if the following application is to be effective. As developed by Techniques Louis Menard, dynamic compaction is achieved by dropping weights ranging from 10 to 40 tons from heights up to 100 ft onto the ground surface. Impact spacings range up to 60 ft. Multiple drops are made at each location to be densified. This technique is applicable to densification of large areas and a wide range of grain sizes and materials.

7.37.6 Drainage

This is effective in soil stabilization because strength of a soil generally decreases with an increase in amount and pressure of pore water. Drainage may be accomplished by gravity, pumping, compression with an external load on the soil, electro-osmosis, heating, or freezing.

Pumping often is used for draining the bottom of excavations (Art. 7.30). For slopes, however, advantage must be taken of gravity flow to attain permanent stabilization. Vertical wells may be used to relieve artesian pressures. Usually, intercepting drains, laid approximately along contours, suffice.

Where mud flows may occur, water must be excluded from the area. Surface and subsurface flow must be intercepted and conducted away at the top of the area. Also, cover, such as heavy mulching and planting, should be placed over the entire surface to prevent water from percolating downward into the soil. See also Art. 7.40.

Electrical drainage adapts the principle that water flows to the cathode when a direct current passes through saturated soil. The water may be pumped out at the cathode. Electro-osmosis is relatively expensive and therefore usually is limited to special conditions, such as drainage of silts, which ordinarily are hard to drain by other methods.

Vertical drains, or piles, may be used to compact loose, saturated cohesionless soils or to consolidate saturated cohesive soils. They provide an escape channel for water squeezed out of the soil by an external load. A surcharge of pervious material placed over the ground surface also serves as part of the drainage system as well as part of the fill, or external load. Usually, the surcharge is placed before the vertical drains are installed, to support equipment, such as pile drivers, over the soft soil. Fill should be placed in thin layers to avoid formation of mud flows, which might shear the sand drains and cause mud waves. Analyses should be made of embankment stability at various stages of construction.

7.38 Thermal Stabilization of Soils

Thermal stabilization generally is costly and is restricted to conditions for which other methods are not suitable. Heat may be used to strengthen nonsaturated loess and to decrease the compressibility of cohesive soils. One technique is to burn liquid or gas fuel in a borehole.

Freezing a wet soil converts it into a rigid material with considerable strength, but it must be kept frozen. The method is excellent for a limited excavation area, for example, freezing the ground to sink a shaft. For the purpose, a network of pipes is placed in the ground and a liquid, usually brine, at low temperature is circulated through the pipes. Care must be taken that the freezing does not spread beyond the area to be stabilized and cause heaving damage.

7.39 Chemical Stabilization of Soils

Utilizing, portland cement, bitumens, or other cementitious materials, chemical stabilization meets many needs. In surface treatments, it supplements mechanical stabilization to make the effects more lasting. In subsurface treatments, chemicals may be used to improve bearing capacity or decrease permeability.

Soil-cement, a mixture of portland cement and soil, is suitable for subgrades, base courses, and pavements of roads not carrying heavy traffic ("Essentials of Soil-Cement Construction," Portland Cement Association). Bitumen-soil mixtures are extensively used in road and airfield construction and sometimes as a seal for earth dikes ("Guide Specifications for Highway Construction," American Association of State Highway and Transportation Officials, 444 North Capitol St., N.W., Washington, DC 20001 (www.ashto.org)). Hydrated, or slaked, lime may be used alone as a soil stabilizer, or with fly ash, portland cement, or bitumen ("Lime Stabilization of Roads," National Lime Association, 200 North Globe Road, Suite 800 Arlington, VA 22203 (www.lime.org)). Calcium or sodium chloride is used as a dust palliative and an additive in construction of granular base and wearing courses for roads ("Calcium Chloride for Stabilization of Bases and Wearing Courses," Calcium Chloride Institute).

Grouting, with portland cement or other chemicals, often is used to fill rock fissures, decrease soil permeability, form underground cutoff walls to eliminate seepage, and stabilize soils at considerable depth. The chemicals may be used to fill the voids in the soil, to cement the particles, or to form a rocklike material.

(K. Terzaghi and R. B. Peck, "Soil Mechanics in Engineering Practice," John Wiley & Sons, Inc., New York (www.wiley.com); G. P. Tschebotarioff, "Soil Mechanics, Foundations, and Earth Structures," McGraw-Hill Book Company, New York (books.mcgraw-hill.com); H. Y. Fang, "Foundation Engineering Handbook," 2nd ed., Van Nostrand Reinhold Company, New York.)

7.40 Geosynthetics

In the past, many different materials have been used for soil separation or reinforcement, including grasses, rushes, wood logs, wood boards, metal mats, cotton, and jute. Because they deteriorated in a relatively short time, required maintenance frequently, or were costly, however, use of more efficient, more permanent materials was desirable. Synthetic fabrics, grids, nets, and other structures are now used as an alternative.

Types of geosynthetics, polymer compositions generally used, and properties important for specifying materials to achieve desired performance are described in Art. 5.29. Principal applications of geosynthetics, functions of geosynthetics in those applications, recommended structures for each case, and design methods are discussed in the following. Table 7.20 summarizes the primary functions of geosynthetics in applications often used and indicates the type of geosynthetic generally recommended by the manufacturers of these materials for the applications. ("Geosynthetic Design and Construction Guidelines" Federal Highway Administration Publication No. FHWA-HI-95-038, May, 1995).

7.40.1 Design Methods for Geosynthetics

The most commonly used design methods for geosynthetics in geotechnical applications are the empirical (design by experience), specification, and rational (design by function) methods.

The empirical design process employs a selection process based on the experience of the project geo-technical engineer or of others, such as designers of projects reported in engineering literature, manufacturers of geosynthetics, and professional associations.

Design by specification is often used for routine applications. Standard specifications for specific applications may be available from geosynthetics manufacturers or developed by an engineering organization or a government agency for its own use or by an association or group of associations, such as the joint committee established by the American Association of State Highway and Transportation Officials, Associated General Contractors, and American Road and Transportation Builders Association (Art. 5.29).

When using the rational design method, designers evaluate the performance, construction methods required, and durability under service conditions of geosynthetics that are suitable for the planned application. This method can be used for all site conditions to augment the preceding methods. It is necessary for applications not covered by standard specifications. It also is required for projects of such nature that large property damage or personal injury would result if a failure should occur. The method requires the following:

A decision as to the primary function of a geosynthetic in the application under consideration

Table 7.20 Primary Function of Geosynthetics in Geotechnical Applications

Application	Function	Geosynthetic
Subgrade stabilization	Reinforcement, separation, filtration	Geotextile or geogrid
Railway trackbed stabilization	Drainage, separation, filtration	Geotextile
Sedimentation-control silt fence	Sediment retention, filtration, separation	Geotextile
Asphalt overlays	Stress-relieving layer and waterproofing	Geotextile
Soil reinforcement:		
Embankments	Reinforcement	Geotextile or geogrid
Steep slopes	Reinforcement	Geotextile or geogrid
Retaining walls	Reinforcement	Geotextile or geogrid
Erosion control:		
Reinforcement	Reinforcement, separation	Geocomposite
Riprap	Filtration and separation	Geotextile
Mats	Filtration and separation	Geotextile
Subsurface drainage filter	Filtration	Geotextile
Geomembrane protection	Protection and cushion	Geotextile
Subsurface drainage	Fluid transmission and filtration	Prefabricated drainage composite

Estimates or calculations to establish the required properties (design values) of the material for the primary function

Determination of the allowable properties, such as minimum tensile or tear strength or permittivity, of the material by tests or other reliable means

Calculation of the safety factor as the ratio of allowable to design values

Assessment of this result to ascertain that it is sufficiently high for site conditions

("A Design Primer: Geotextiles and Related Materials," Industrial Fabric Association International, 345 Cedar St., Suite 800, St. Paul, MN 55101; "Geotextile Testing and the Design Engineer," STP 952, ASTM; R. M. Koerner, "Designing with Geosynthetics," 2nd ed., Prentice-Hall, Englewood Cliffs, N. J.)

7.40.2 Geosynthetics Nomenclature

Following are some of the terms generally used in design and construction with geosynthetics:

Apparent Opening Size (AOS). A property designated O_{95} applicable to a specific geotextile that indicates the approximate diameter of the largest particle that would pass through the geotextile. A minimum of 95% of the openings are the same size or smaller than that particle, as measured by the dry sieve test specified in ASTM D4751.

Blinding. Blocking by soil particles of openings in a geotextile, as a result of which its hydraulic conductivity is reduced.

Chemical Stability. Resistance of a geosynthetic to degradation from chemicals and chemical reactions, including those catalyzed by light.

Clogging. Retention of soil particles in the voids of a geotextile, with consequent reduction in the hydraulic conductivity of the fabric.

Cross-Machine Direction. The direction within the plane of a fabric perpendicular to the direction of manufacture. Generally, tensile strength of the fabric is lower in this direction than in the machine direction.

Denier. Mass, g, of a 9000-m length of yarn.

Fabric. Polymer fibers or yarn formed into a sheet with thickness so small relative to dimensions in the plane of the sheet that it cannot resist compressive forces acting in the plane. A needle-punched fabric has staple fibers or filaments mechanically bonded with the use of barbed needles to form a compact structure. A spun-bonded fabric is formed with continuous filaments that have been spun (extruded), drawn, laid into a web, and bonded together in a continuous process, chemically, mechanically, or thermally. A woven fabric is produced by interlacing orthogonally two or more sets of elements, such as yarns, fibers, rovings, or filaments, with one set of elements in the machine direction. A monofilament woven fabric is made with single continuous filaments, whereas a multifilament woven fabric is composed of bundles of continuous filaments. A split-film woven fabric is constructed of yarns formed by splitting longitudinally a polymeric film to form a slit-tape yarn. A nonwoven fabric is produced by bonding or interlocking of fibers, or both.

Fiber. Basic element of a woven or knitted fabric with a length-diameter or length-width ratio of at least 100 and that can be spun into yarn or otherwise made into a fabric.

Filament. Variety of fiber of extreme length, not readily measured.

Filtration. Removal of particles from a fluid or retention of soil particles in place by a geosynthetic, which allows water or other fluids to pass through.

Geocomposite. Manufactured laminated or composite material composed of geotextiles, geomembranes, or geogrids, and sometimes also natural materials, or a combination.

Geogrid. Orthogonally arranged fibers, strands, or rods connected at intersections, intended for use primarily as tensile reinforcement of soil or rock.

Geomembrane. Geosynthetic, impermeable or nearly so, intended for use in geotechnical applications.

Geosynthetics. Materials composed of polymers used in geotechnical applications.

Geotextile. Fabric composed of a polymer and used in geotechnical applications.

Grab Tensile Strength. Tensile strength determined in accordance with ASTM D4632 and typically found from a test on a 4-in-wide strip of fabric, with the tensile load applied at the midpoint of the fabric width through 1-in-wide jaw faces.

Gradient Ratio. As measured in a constant-head permittivity test on a geotextile, the ratio of the average hydraulic gradient across the fabric plus 1 in of soil adjoining the fabric to the average hydraulic gradient of the 2 in of soil between 1 and 3 in above the fabric.

Machine Direction. The direction in the plane of the fabric parallel to the direction of manufacture. Generally, the tensile strength of the fabric is largest in this direction.

Monofilament. Single filament, usually of a denier higher than 15.

Mullen Burst Strength. Hydraulic bursting strength of a geotextile as determined in accordance with ASTM D3786.

Permeability (Hydraulic Conductivity). A measure of the capacity of a geosynthetic to allow a fluid to move through its voids or interstices, as represented by the amount of fluid that passes through the material in a unit time per unit surface area under a unit pressure gradient. Accordingly, permeability is directly proportional to thickness of the geosynthetic.

Permittivity. Like permeability, a measure of the capacity of a geosynthetic to allow a fluid to move through its voids or interstices, as represented by the amount of fluid that passes through a unit surface area of the material in a unit time per unit thickness under a unit pressure gradient, with laminar flow in the direction of the thickness of the material. For evaluation of geotextiles, use of permittivity, being independent of thickness, is preferred to permeability.

Puncture Strength. Ability of a geotextile to resist puncture as measured in accordance with ASTM D3787.

Separation. Function of a geosynthetic to prevent mixing of two adjoining materials.

Soil-Fabric Friction. Resistance of soil by friction to sliding of a fabric embedded in it, exclusive of resistance from cohesion. It is usually expressed as a friction angle.

Staple Fibers. As usually used in geotextiles, very short fibers, typically 1 to 3 in long.

Survivability. Ability of geosynthetics to perform intended functions without impairment.

Tearing Strength. Force required either to start or continue propagation of a tear in a fabric as determined in accordance with ASTM D4533.

Tenacity. Fiber strength, grams per denier.

Tex. Denier divided by 9.

Transmissivity. Amount of fluid that passes in unit time under unit pressure gradient with laminar flow per unit thickness through a geosynthetic in the in-plane direction.

Yarn. Continuous strand composed of textile fibers, filaments, or material in a form suitable for knitting, weaving, or otherwise intertwining to form a geotextile.

7.40.3 Geosynthetic Reinforcement of Steep Slopes

Geotextiles and geogrids are used to reinforce soils to permit slopes much steeper than the shearing resistance of the soils will permit. (**Angle of repose**, the angle between the horizontal and the maximum slope that a soil assumes through natural processes, is sometimes used as a measure of the limiting slopes for unconfined or unreinforced cuts and fills, but it is not always relevant. For dry, cohesionless soils, the effect of height of slope on this angle is negligible. For cohesive soils, in contrast, the height effect is so large that angle of repose is meaningless.) When geosynthetic reinforcement is used, it is placed in the fill in horizontal layers. Vertical spacing, embedment length, and tensile strength of the geosynthetic are critical in establishing a stable soil mass.

For evaluation of slope stability, potential failure surfaces are assumed, usually circular or wedge-shaped but other shapes also are possible. Figure 7.56a shows a slope for which a circular failure surface starting at the bottom of the slope and extending to the ground surface at the top is assumed. An additional circular failure surface is indicated in Fig. 7.56b. Fig. 7.56c shows a wedge-shaped failure surface. An infinite number of such failure surfaces are possible. For design of the reinforcement, the surfaces are assumed to pass through a layer of reinforcement at various levels and apply tensile forces to the reinforcement, which must have sufficient tensile strength to resist them. Sufficient reinforcement

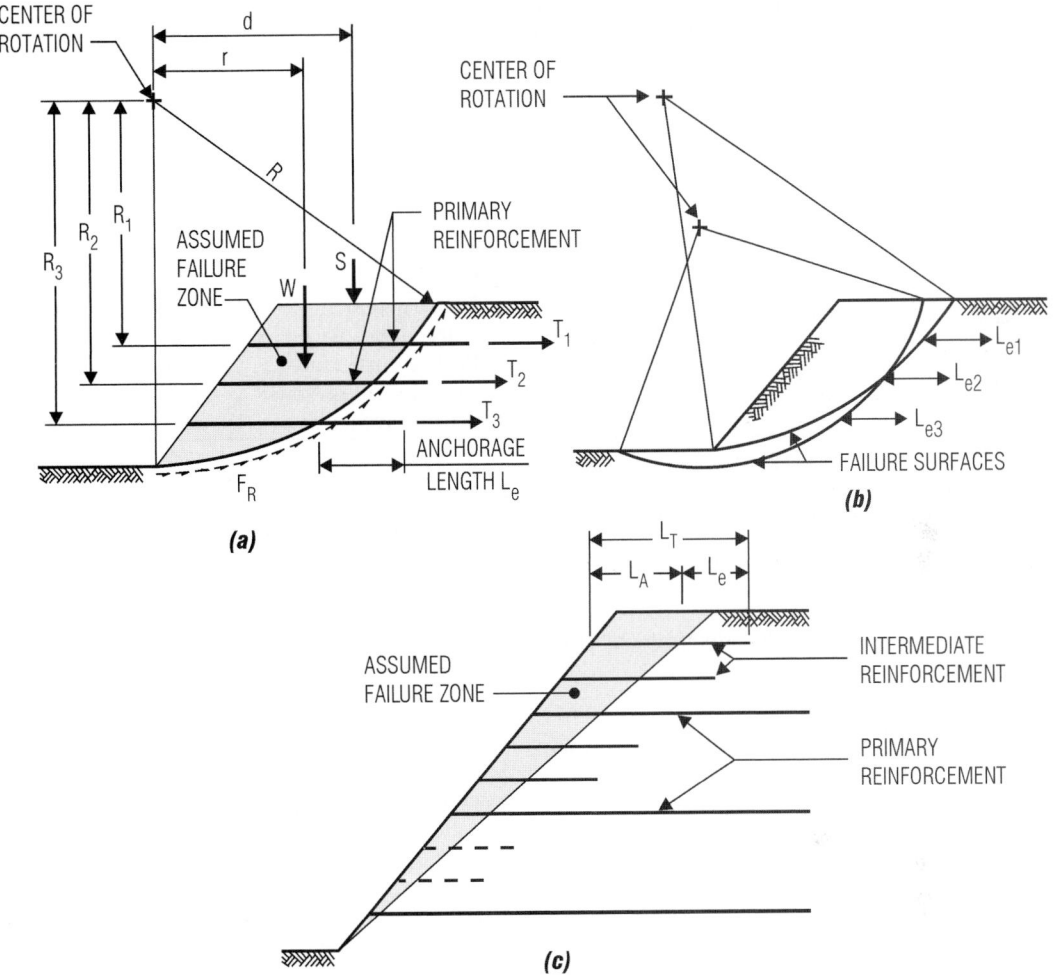

Fig. 7.56 Stabilization of a steep slope with horizontal layers of geosynthetic reinforcement. (*a*) Primary reinforcement for a circular failure surface. (*b*) Embedment lengths of reinforcement extend from critical failure surfaces into the backfill. (*c*) Intermediate reinforcement for shallow failure surfaces.

embedment lengths extending into stable soil behind the surfaces must be provided to ensure that the geosynthetic will not pull out at design loads.

Pull-out is resisted by geotextiles mainly by friction or adhesion—and by geogrids, which have significant open areas, also by soil-particle *strike-through*. The soil-fabric interaction is determined in laboratory pull-out tests on site-specific soils and the geosynthetic to be used, but long-term load-transfer effects may have to be estimated. Design of the reinforcement requires calculation of the

embedment required to develop the reinforcement fully and of the total resisting force (number of layers and design strength) to be provided by the reinforcement. The design should be based on safety factors equal to or greater than those required by local design codes. In the absence of local code requirements, the values given in Table 7.21 may be used. A stability analysis should be performed to investigate, at a minimum, circular and wedge-shaped failure surfaces through the toe (Fig. 7.56*a*), face (Fig. 7.56*c*), and deep seated below the toe (Fig. 7.56*b*). The total resisting moment for a

Table 7.21 Minimum Safety Factors K for Slope Reinforcement

External Stability		Internal Stability	
Condition	K	Condition	K
Sliding	1.5	Slope stability	1.3
Deep seated (overall stability)	1.3	Design tensile strength T_d	*
Dynamic loading	1.1	Allowable geosynthetic strength T_a[†]	
		Creep	4
		Construction	1.1 to 1.3
		Durability	1.1 to 1.2
		Pull-out resistance	
		Cohesionless soils	1.5[‡]
		Cohesive soils	2

[*] T_d at 5% strain should be less than T_a.

[†] $T_a = T_L/K_c K_d$, where T_L is the creep limit strength, K_c is the safety factor for construction, and K_d is the safety factor for durability. In the absence of creep tests or other pertinent data, the following may be used: $T_a = T_u/10.4$ or $T_u \geq 10.4T_d$, where T_u is the ultimate tensile strength of the geosynthetic.

[‡] For a 3-ft minimum embedment.

circular slip surface can be determined from Fig. 7.56b as

$$M_R = RF_r + \sum_{i=1}^{i=n} R_i T_i \qquad (7.90)$$

where R = radius of failure circle

F_R = soil shearing resistance along the slip surface = $\tau_f L_{sp}$

τ_f = soil shear strength

L_{sp} = length of slip surface

R_i = radius of slip surface at layer i

T_i = strength of the reinforcing required for layer i

The driving moment, or moment of the forces causing slip, is

$$M_D = Wr + Sd \qquad (7.91)$$

where W = weight of soil included in assumed failure zone (Fig. 7.56a)

r = moment arm of W with respect to the center of rotation (Fig. 7.56a)

S = surcharge

d = moment arm of S with respect to the center of rotation (Fig. 7.56a)

The safety factor for the assumed circular failure surface then is

$$K_D = \frac{M_R}{M_D} \qquad (7.92)$$

A safety factor should be computed for each potential failure surface. If a safety factor is less than the required minimum safety factor for prevention of failure of the soil unreinforced, either a stronger reinforcement is required or the number of layers of reinforcement should be increased. This procedure may also be used to determine the reinforcement needed at any level to prevent failure above that layer.

The next step is calculation of length L_e of reinforcement required for anchorage to prevent pullout.

$$L_e = \frac{KF_D}{2\sigma_o \tan \phi_{sr}} \qquad (7.93)$$

where F_D = required pull-out strength

K = minimum safety factor: 1.5 for cohesionless soils; 2 for cohesive soils

σ_o = overburden pressure above the reinforcing level = wh

w = density of the soil

h = depth of overburden

ϕ_{sr} = soil-reinforcement interaction angle, determined from pull-out tests

Embedment length L_e should be at least 3 ft. The total length of a reinforcement layer then is L_e plus the distance from the face of the slope to the failure circle (Fig. 7.56b). The total length of the reinforcement at the toe should be checked to ascertain that it is sufficient to resist sliding of the soil mass above the base of the slope.

Among the family of potential failure surfaces that should be investigated is the wedge shape, such as the one shown in Fig. 7.56c. To reinforce the failure zones close to the face of the slope, layers of reinforcement are required in addition to those provided for the deep failure zones, as indicated in Fig. 7.56c. Such face reinforcement should have a maximum vertical spacing of 18 in and a minimum length of 4 ft. Inasmuch as the tension in this reinforcement is limited by the short embedment, a geosynthetic with a smaller design allowable tension than that required for deep-failure reinforcement may be used. In construction of the reinforced slope, fill materials should be placed so that at least 4 in of cover will be between the geosynthetic reinforcement and vehicles or equipment operating on a lift. Backfill should not incorporate particles larger than 3 in. Turning of vehicles on the first lift above the geosynthetic should not be permitted. Also, end dumping of fill directly on the geosynthetic should not be allowed.

7.40.4 Geosynthetics in Retaining-Wall Construction

Geotextiles and geogrids are used to form retaining walls (Fig. 7.57a) or to reinforce the backfill of a retaining wall to create a stable soil mass (Fig. 7.57b). In the latter application, the reinforcement reduces the potential for lateral displacement of the wall under the horizontal pressure of the backfill.

As in the reinforcement of steep slopes discussed in Art. 7.40.3, the reinforcement layers must intersect all critical failure surfaces. For cohesionless backfills, the failure surface may be assumed to be wedge shaped, as indicated in Fig. 7.56c, with the sloping plane of the wedge at an angle of $45° + \phi/2$ with the horizontal. If the backfill is not homogeneous, a general stability analysis should be carried out as described in Art. 7.40.3.

The design process for cohesionless soils can be simplified by use of a constant vertical spacing s_v for the reinforcement layers. This spacing would be approximately

$$S_v = \frac{T_a}{KK_awH} \tag{7.94}$$

where T_a = allowable tension in the reinforcement

K = safety factor as specified in a local code or as given in Table 7.21

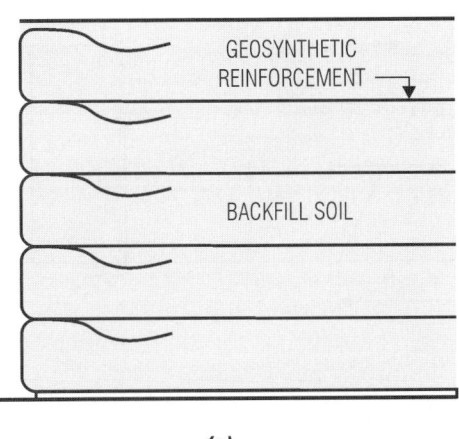

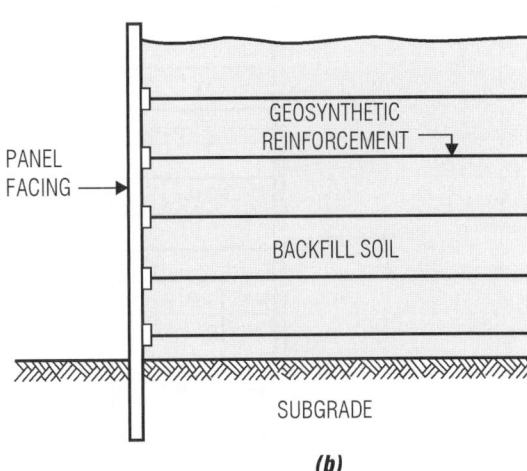

(a)   *(b)*

Fig. 7.57 Geosynthetic applications with retaining walls: (a) Reinforced earth forms a retaining wall. (b) Retaining wall anchored into backfill.

K_a = coefficient of active earth pressure (Art. 7.27)

w = density of backfill

H = average height of embankment

If Eq. (7.94) yields a value for s_v less than the minimum thickness of a lift in placement of the backfill, a stronger geosynthetic should be chosen. The minimum embedment length L_e may be computed from Eq. (7.93). Although the total reinforcement length thus computed may vary from layer to layer, a constant reinforcement length would be convenient in construction.

When the earth adjoining the backfill is a random soil with lower strength than that of the backfill, the random soil exerts a horizontal pressure on the backfill that is transmitted to the wall (Fig. 7.58). This may lead to a sliding failure of the reinforced zone. The reinforcement at the base should be sufficiently long to prevent this type of failure. The total horizontal sliding force on the base is, from Fig. 7.58,

$$P = P_b + P_s + P_v \qquad (7.95)$$

where $P_b = K_a w_b H^2 / 2$

wb = density of soil adjoining the reinforcement zone

$P_s = K_a w_s h H$

$w_s h$ = weight of uniform surcharge

P_v = force due to live load V as determined by the Boussinesq method (Art. 7.11)

The horizontal resisting force is

$$F_H = [(w_s h + w_r H) \tan \phi_{sr} + c] L \qquad (7.96)$$

where $w_r H$ = weight of soil in the reinforcement zone

ϕ_{sr} = soil-reinforcement interaction angle

c = undrained shear strength of the backfill

L = length of the reinforcement zone base

The safety factor for sliding resistance then is

$$K_{sl} = \frac{F_H}{P} \qquad (7.97)$$

and should be 1.5 or larger. A reinforcement length about 0.8H generally will provide base resistance sufficient to provide a safety factor of about 1.5.

The most economical retaining wall is one in which the reinforcement is turned upward and backward at the face of the wall and also serves as the face (Fig. 7.58a) The backward embedment should be at least 4 ft. If desired for esthetic reasons

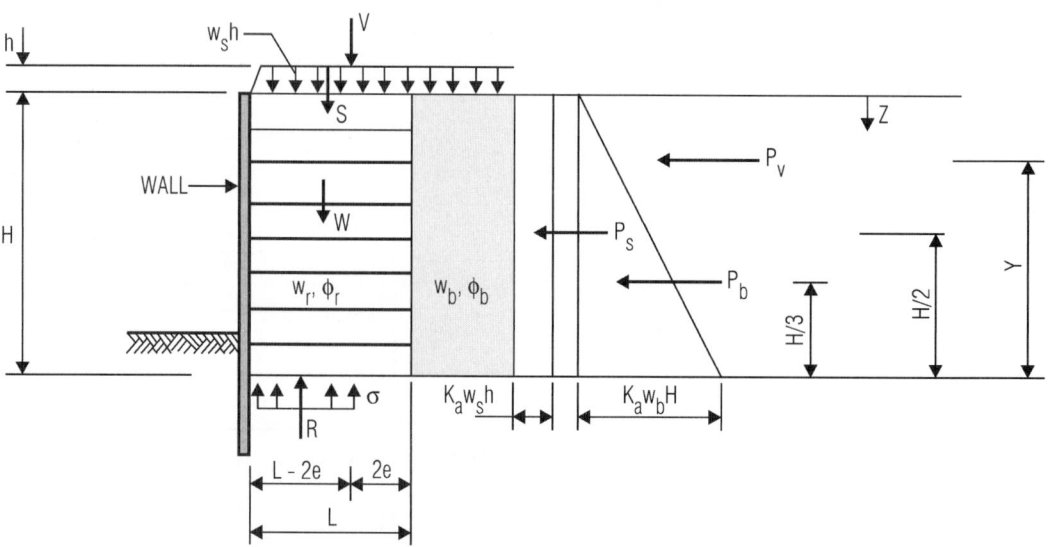

Fig. 7.58 Retaining wall anchored with geosynthetic reinforcement is subjected to pressure from random-soil backfill, sand backfill, surcharge, and live load. Assumed pressure distribution diagrams are rectangular and triangular.

or to protect the geosynthetic from damage or deterioration from exposure to ultraviolet light, sprayed concrete may be applied to the wall face.

As an alternative, the wall may be composed of concrete block or precast concrete panels that are anchored to soil reinforcement. The reinforcement should be installed taut to limit lateral movement of the wall during construction.

See Art. 7.40.3 for other precautions to be taken during construction.

7.40.5 Geosynthetic Reinforcement for Embankments

Geosynthetics placed in horizontal layers may be used to reinforce embankments in a manner similar to that used to reinforce steep slopes (Art. 7.40.3). The reinforcement may permit greater embankment height and a larger safety factor in embankment design than would an unreinforced embankment. Also, displacements during construction may be smaller, thus reducing fill requirements. Furthermore, reinforcement properly designed and installed can prevent excessive horizontal displacements along the base that can cause an embankment failure when the underlying soil is weak. Moreover, reinforcement may decrease horizontal and vertical displacements of the underlying soil and thus limit differential settlement. Reinforcement, however, will reduce neither long-term consolidation of the underlying weak soil nor secondary settlement.

Either geotextiles or geogrids may be used as reinforcement. If the soils have very low bearing capacity, it may be necessary to use a geotextile separator with geogrids for filtration purposes and to prevent the movement of the underlying soil into the embankment fill.

Figure 7.59 illustrates reinforcement of an embankment completely underlain by a weak soil. Without reinforcement, horizontal earth pressure within the embankment would cause it to spread laterally and lead to embankment failure, in the absence of sufficient resistance from the soil. Reinforcement is usually laid horizontally in the direction of major stress; that is, with strong axis normal to the longitudinal axis of the embankment. Reinforcement with strong axis placed parallel to the longitudinal axis of the embankment may also be required at the ends of the embankment. Seams should be avoided in the high-stress direction.

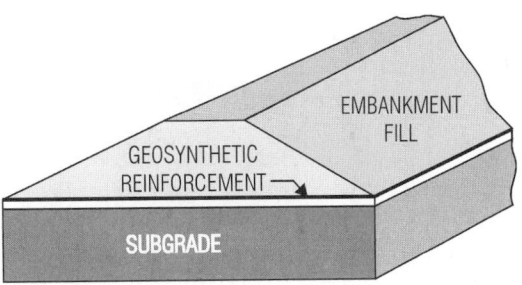

Fig. 7.59 Geosynthetic reinforcement for an embankment on weak soil is placed directly on the subgrade.

Design of the reinforcement is similar to that required for steep slopes (Art. 7.40.3).

For an embankment underlain by locally weak areas of soil or voids, reinforcement may be incorporated at the base of the embankment to bridge them.

7.40.6 Soil Stabilization with Geosynthetics

Woven or nonwoven geotextiles are used to improve the load-carrying capacity of roads over weak soils and to reduce rutting. Acting primarily as a separation barrier, the geosynthetic prevents the subgrade and aggregate base from mixing. The geosynthetic may also serve secondary functions. Acting as a filter, it prevents fines from migrating into the aggregate due to high water pressure. Also, the geotextile may facilitate drainage by allowing pore water to pass through and dissipate into the underlying soil. In addition, acting as reinforcement, the geotextile can serve as a membrane support of wheel loads and provide lateral restraint of the base and subgrade through friction between the fabric, aggregate, and soil.

Installation techniques to be used depend on the application. Usually, geosynthetics are laid directly on the subgrade (Fig. 7.60a). Aggregate then is placed on top to desired depth and compacted.

Design of permanent roads and highways consists of the following steps: If the CBR ≤ 3, need for a geotextile is indicated. The pavement is designed by usual methods with no allowance for structural support from the geotextile. If a thicker subbase than that required for structural support would have to be specified because of the sus-

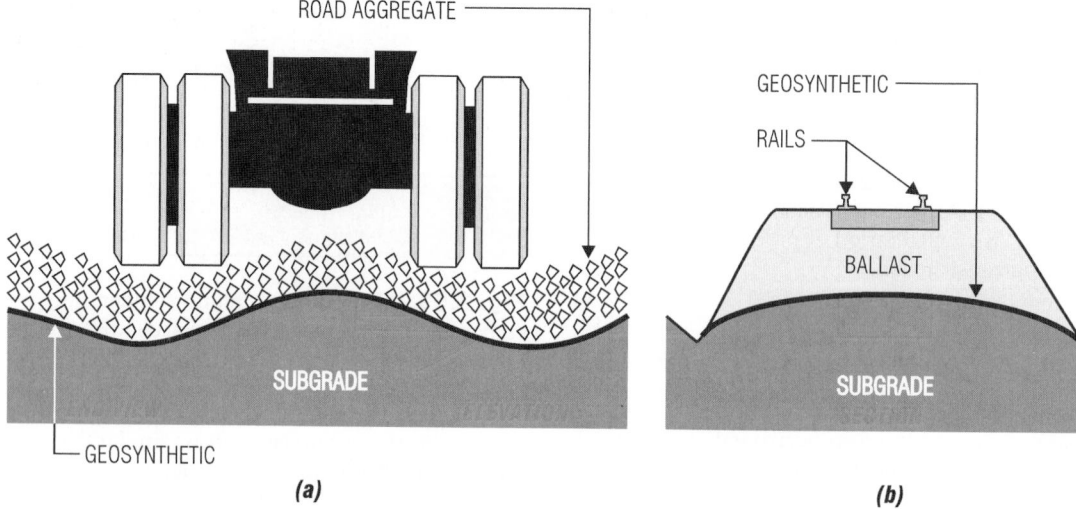

Fig. 7.60 Geosynthetic is used (*a*) to reinforce a road, (*b*) to reinforce a railway roadbed.

ceptibility of the underlying soil to pumping and subbase intrusion, the subbase may be reduced 50% and a geotextile selected for installation at the subbase-subgrade interface. For stabilization of the subgrade during construction, an additional determination of thickness of subbase assisted by a geotextile is made by conventional methods (bearing capacity N_c about 3.0 without geotextiles and about 5.5 with) to limit rutting to a maximum of 3 in under construction vehicle loads. The thicker subbase thus computed is selected. Then, the geotextile strength requirements for survivability and filtration characteristics are checked. (Details for this are given in B. R. Christopher and R. D. Holtz, "Geotextile Design and Construction Guidelines," FHWA DTFH61-86-C-00102, National Highway Institute, Federal Highway Administration, Washington, DC 20590 (www.fhwa.dot.gov).)

Geosynthetics are also used under railway tracks for separation of subgrade and subballast or subballast and ballast (Fig. 7.60*b*). They also are used for roadbed filtration, lateral permeability, and strength and modulus improvement.

7.40.7 Geosynthetics in Erosion Control

For purposes of erosion control, geosynthetics are used as turf reinforcement, as separators and filters

under riprap, or armor stone, and as an alternative to riprap. Different types of geosynthetics are used for each of these applications.

Turf Control ■ To establish a reinforced turf in ditches and water channels and on slopes, three-dimensional erosion-control mats often are used. Entangling with the root and stem network of vegetation, they greatly increase resistance to flow of water down slopes and thus retard erosion.

Mats used for turf reinforcement should have a strong, stable structure. They should be capable of retaining the underlying soil but have sufficient porosity to allow roots and stem to grow through them. Installation generally requires pinning the mat to the ground and burying mat edges and ends. Topsoil cover may be used to reduce erosion even more and promote rapid growth of vegetation.

When a geosynthetic is placed on a slope, it should be rolled in the direction of the slope. Horizontal joints should not be permitted. Vertical joints should be shingled downstream. Ditch and channel bottoms should be lined by rolling the geosynthetic longitudinally. Joints transverse to water flow should have a 3-ft overlap and be shingled downstream. Roll edges should be overlapped 2 to 4 in. They should be staked at intervals not exceeding 5 ft to prevent relative movement.

Where highly erodible soils are encountered, a geotextile filter should be installed under the turf reinforcement and staked or otherwise bonded to the mats. For stability and seeding purposes, wood chips may be used to infill the turf reinforcement.

Use of Geosynthetics with Riprap ■ Large armor stones are often used to protect soil against erosion and wave attack. Graded-aggregate filter generally is placed between the soil and the riprap to prevent erosion of the soil through the armoring layer. As a more economical alternative, geotextiles may be used instead of aggregate. They also offer greater control during construction, especially in underwater applications. Geosynthetics typically used are nonwoven fabrics, monofilament nonwoven geotextiles, and multifilament or fibrillated woven fabrics.

The geosynthetics should have sufficient permeability to permit passage of water to relieve hydrostatic pressure behind the riprap. Also, the geosynthetic should be capable of retaining the underlying soil. Conventional filter criteria can be used for design of.the geosynthetic, although some modifications may be required to compensate for properties of the geosynthetic.

Installation precautions that should be observed include the following: Riprap should be installed with care to avoid tearing the geosynthetic inasmuch as holes would decrease its strength. Stone placement, including drop heights, should be tested in field trials to develop techniques that will not damage the geosynthetic. As a general guide, for material protected by a sand cushion and material with properties exceeding that required for unprotected applications, drop height for stones weighing less than 250 lb should not exceed 3 ft; without a cushion, 1 ft. Stone weighing more than 250 lb should be placed without free fall, unless field tests determine a safe drop height. Stone weighing more than 100 lb should not be permitted to roll along the geosynthetic. Installation of the armor layer should begin at the base of slopes and at the center of the zone covered by the geosynthetic. After the stones have been placed, they should not be graded.

Special construction procedures are required for slopes greater than 2.5:1. These include increase in overlap, slope benching, elimination of pins at overlaps, toe berms for reaction against slippage, and laying of the geosynthetic suffi-ciently loose to allow for downstream movement, but folds and wrinkles should not be permitted.

The geosynthetic should be rolled out with its strong direction (machine direction for geotextiles) up and down the slope. Adjoining rolls should be seamed or shingle overlapped in the downslope or downstream direction. Joints should be stapled or pinned to the ground. Recommended pin spacing is 2 ft for slopes up to 3:1, 3 ft for slopes between 3:1 and 4:1, 5 ft for 4:1 slopes, and 6 ft for slopes steeper than 4:1. For streambanks and slopes exposed to wave action, the geosynthetic should be anchored at the base of the slope by burial around the perimeter of a stone-filled key trench. It should also be keyed at the top of the slope if the armor-geosynthetic system does not extend several feet above high water.

Riprap Replacement ■ Instead of the riprap generally used for erosion control, concrete mats may be used. For this purpose, the concrete conventionally has been cast in wood or steel forms. Use of expandable fabric forms, however, may be more economical. Such forms are made by joining two fabric sheets at discrete points. After the sheets are placed over the area to be protected, grout is pumped into the space between the sheets to form a mattress that initially will conform to the shape of the ground and later harden. Thickness of the mattress is controlled by internal spacer threads. Filter points and bands are formed in the mattress to dissipate pore water from the subsoil. The fabric forms may be grouted underwater, even in flowing water, and in hazardous-liquid conditions. The fabric usually used is a woven geotextile.

7.40.8 Uses of Geosynthetics in Subsurface Drainage

Subsurface drainage is required for many construction projects and geotextiles find many uses in such applications. Their primary function is to serve, with graded granular filter media, as a permeable separator to exclude soil from the drainage media but to permit water to pass freely. Nonwoven geotextiles are usually used for this purpose because of their high flow capacity and small pore size. Generally, fabric strength is not a primary consideration for subsurface drainage applications, except during installation.

Following are brief descriptions of typical applications of geotextiles in subsurface drainage:

Permeable separators wrapped around trench or edge drains

Drains for retaining walls and bridge abutments with the geotextile enclosing the backfill

Encirclement of slotted or jointed drains and wall pipes to prevent filter particles from entering the drains while permitting passage of water

Wraps for interceptor, toe, and surface drains on slopes to assist stabilization by dissipating excess pore-water pressures and retarding erosion

Seepage control with chimney and toe drains for earth dams and levees with the geotextile laid along the upstream face and anchored by a berm.

7.40.9 Geosynthetics as Pond Liners

Geomembranes, being impermeable, appear to be an ideal material for lining the bottom of a pond to retain water or other fluid. Used alone, however, they have some disadvantages. In particular, they are susceptible to damage from many sources and require a protective soil cover of at least 12 in. Also, for several reasons, it is advisable to lay a geotextile under the geomembrane. The geotextile provides a clean working area for making seams. It adds puncture resistance to the liner. It increases friction resistance at the interface with the soil, thus permitting steeper side slopes. And it permits lateral and upward escape of gases emitted from the soil. For this purpose, needle-punched non-woven geotextiles, geonets, or drainage composites with adequate transmissivity for passing the gases are needed. In addition, it is advantageous to cover the top surface of the geomembrane with another geotextile. Its purpose is to maintain stability of cover soil on side slopes and to prevent sharp stones that may be present in the cover soil from puncturing the liner. This type of construction is also applicable to secondary containment of underground storage tanks for prevention of leakage into groundwater.

In selection of a geosynthetic for use as a pond liner, consideration should be given to its chemical resistance with regard to the fluid to be contained and reactive chemicals in the soil. For determination of liner thickness, assumptions have to be made as to loads from equipment during installation and basin cleaning as well as to pressure from fluid to be contained.

7.40.10 Geosynthetics as Landfill Liners

Liners are used along the bottom and sides of landfills to prevent leachate formed by reaction of moisture with landfill materials from contaminating adjacent property and groundwater. Clay liners have been traditional for this purpose (Fig. 7.61a). They have the disadvantage of being thick, often in the range of 2 to 6 ft, and being subject to piping under certain circumstances, permitting leakage of leachate. Geomembranes, geotextiles, geonets, and geocomposites offer an alternative that prevents rather than just minimizes leachate migration from landfills.

The U. S. Environmental Protection Agency (EPA) requires that all new hazardous-waste landfills, surface impoundments, and waste piles have two or more liners with a leachate-collection system between the liners. This requirement may be satisfied by installation of a top liner constructed of materials that prevent migration of any constituent into the liner during the period such facility remains in operation and a lower liner with the same properties. In addition, primary leachate-collection and leak-detection systems must be installed with the double liners to satisfy the following criteria:

The primary leachate-collection system should be capable of keeping the leachate head from exceeding 12 in.

Collection and leak-detection systems should incorporate granular drainage layers at least 12 in thick that are chemically resistant to the waste and leachate. Hydraulic conductivity should be at least 0.02 ft/min. An equivalent drainage geosynthetic, such as a geonet, may be used instead of granular layers. Bottom slope should be at least 2%.

A granular filter or a geotextile filter should be installed in the primary system above the drainage layer to prevent clogging.

When gravel is used as a filter, pipe drains resistant to chemicals should be installed to collect leachate efficiently (Fig. 7.61a).

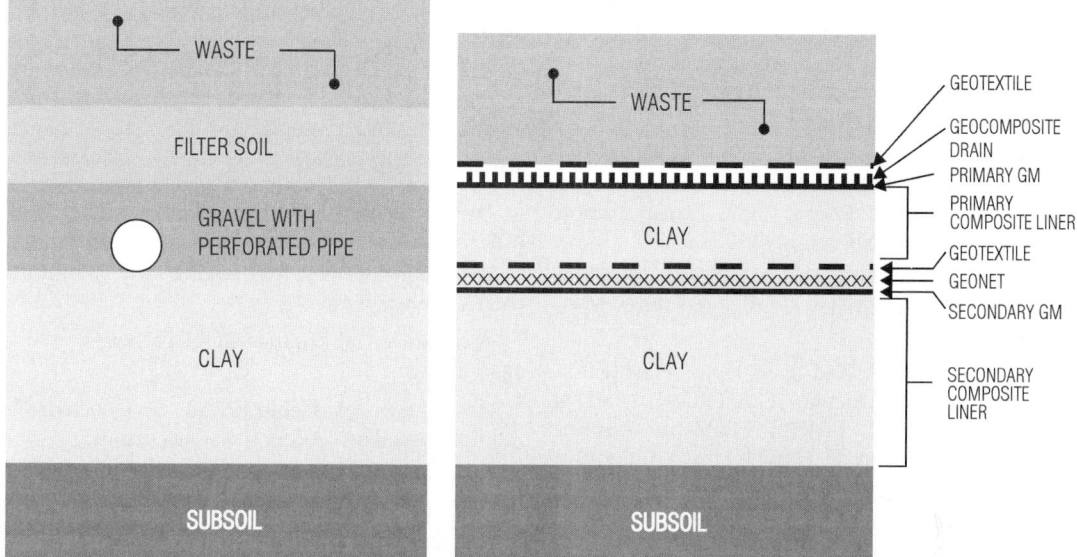

Fig. 7.61 Landfill liner systems: (*a*) with filter soil, gravel, drain pipe, and clay liner; (*b*) with geotextile separator-filter, geocomposite leachate drain, primary geomembrane and clay liner, geotextile filter, geonet for leak detection, and secondary geomembrane and clay liner.

Figure 7.61*b* illustrates a lining system that meets these criteria. Immediately underlying the wastes is a geotextile that functions as a filter. It overlies the geocomposite primary leachate drain. Below is the primary liner consisting of a geomembrane above a clay blanket. Next comes a geotextile filter and separator, followed underneath by a geonet that functions as a leak-detection drain. These are underlain by the secondary liner consisting of another geomembrane and clay blanket, which rests on the subsoil.

The EPA requires the thickness of a geomembrane liner for containment of hazardous materials to be at least 30 mils (0.75 mm) with timely cover or 45 mils (1.2 mm) without such cover. The secondary geomembrane liner should be the same thickness as the primary liner. Actual thickness required depends on pressures from the landfill and loads from construction equipment during installation of the liner system.

Terminals of the geosynthetics atop the side slopes generally consist of a short runout and a drop into an anchor trench, which, after insertion of the geosynthetics, is backfilled with soil and compacted. Side-slope stability of liner system and wastes needs special attention in design.

7.40.11 Geosynthetics Bibliography

"A Design Primer: Geotextiles and Related Materials," Industrial Fabrics Association International, 1801 County Road BW Roseville, MN 55113, (www.ifai.com).

"Construction over Soft Soils," The Tensar Corporation, 1210 Citizens Parkway, Morrow, GA 30260 (www.tensarcorp.com).

"Guidelines for Selection and Installation of Exxon Geotextiles" and "Exxon Geotextile Design Manual for Paved and Unpaved Roads," Exxon Chemical Americas, 380 Interstate North, Suite 375, Atlanta, GA 30339.

"Specifiers Guide,"(annual) *Geotechnical Fabrics Report*, Industrial Fabrics Association International, St. Paul, Minn. (www.ifai.com)

"Woven Geotextiles" and "Nonwoven Geotextiles," Amoco Fabrics and Fibers Company, 900 Circle 75 Parkway, Suite 300, Atlanta GA 30339.

Ahlrich, R. C., "Evaluation of Asphalt Rubber and Engineering Fabrics as Pavement Interlayers," GL-86-34, U.S. Army Corps of Engineers, Vicksburg, Miss.

Berg, R. R., "Guidelines for Design, Specification, and Contracting of Geosynthetic Mechanically Stabilized Earth Slopes on Firm Foundations," FHWA-SA-93-025, Federal Highway Administration, Washington, D. C.

Bonaparte, R., and B. Christopher, "Design and Construction of Reinforced Embankments over Weak Foundations," Record 1153, Transportation Research Board, Washington, D. C.

Cedergren, H. R., "Seepage, Drainage, and Flow Nets," 3rd ed., John Wiley & Sons, Inc., New York (www.wiley.com).

Christopher, B. B., and R. D. Holtz, "Geotextile Design and Construction Guidelines," FHWA DTFH61-86-C-00102, Federal Highway Administration, Washington, D. C (www.fhwa.dot.gov).

Ingold, T. S., and and K. S. Miller, "Geotextiles Handbook," Thomas Telford, London.

Kays, W. B., "Construction of Linings for Reservoirs, Tanks, and Pollution Control Facilities, "John Wiley & Sons, Inc., New York (www.wiley.com).

Koerner, R. M., "Designing with Geosynthetics," 2nd ed., Prentice-Hall, Englewood Cliffs, N. J (www.prenhall.com).

Matrecon, Inc., "Lining of Waste Impoundment and Disposal Facilities," EPA SW-870, U.S. Environmental Protection Agency, Washington, D.C.

Mitchell, J. K., et al., "Reinforcement of Earth Slopes and Embankments," Report 290, Transportation Research Board, Washington, D. C.

Richardson, G. N., and R. M. Koerner, "Geosynthetic Design Guidance for Hazardous Waste Landfill Cells and Surface Impoundments," EPA 68-03-3338, U.S. Environmental Protection Agency, Washington, D. C.

Rixner, J. J., et al., "Prefabricated Vertical Drains," FHWA/RD-86/168, Federal Highway Administration, Washington, D. C (www.fhwa.dot.gov).

7.41 Geotechnical Engineering Sites on the World Wide Web

American Association of State Highway and Transportation Officials (AASHTO), www.aashto.org

American Geological Institute, www.agiweb.org

American Geophysical Union (AGU), www.agu.org

American Rock Mechanics Association, www.armarocks.org

American Society of Civil Engineers, www.asce.org

American Society for Testing and Materials (ASTM), www.astm.org

Association of Engineering Geologists, www.aegweb.org

Association of Geotechnical & Geoenvironmental Specialists (AGS), www.ags.org.uk

British Geotechnical Society (BGS), www.geo.org.uk

Canadian Geotechnical Society, www.cgs.ca

Deep Foundations Institute, www.dfi.org

Federal Highway Administration (FHWA), Bridge Technology, Geotechnical Section, www.fhwa.dot.gov/bridge/geo.htm

Geo-Institute of the American Society of Civil Engineers, www.geoinstitute.org

Geological Society of America, www.geosociety.org

Geosynthetic Institute (GSI), www.geosynthetic-institute.org

Geosynthetic Materials Association (GMA), www.gmanow.com

Geotechnical & Geoenvironmental Software Directory, www.ggsd.com

International Association of Foundation Drilling (ADSC), www.adsc-iafd.com

International Geosynthetic Society (IGS), http://mergrb.rmc.ca/

International Society for Soil Mechanics and Geotechnical Engineering (ISSMGE), www.issmge.org

Mexican Society of Soil Mechanics, www.smms.org.mx

National Council for Geo-Engineering and Construction (Geo Council), www.geocouncil.org

National Geophysical Data Center (NGDC), www.ngdc.noaa.gov

National Geotechnical Experimentation Sites (NGES) www.geocouncil.org/nges/nges.html

National Science Foundation Directorate for Engineering (NSF), www.eng.nsf.gov

Pile Driving Contractors Association (PDCA), www.piledrivers.org

Professional Firms Practicing in the Geosciences (ASFE), www.asfe.org

United States Universities Council for Geotechnical Engineering Research (USUCGER), www.usucger.org

U.S. Corps of Engineers Waterways Experiment Station (WES), www.wes.army.mil

U.S. Geological Survey, www.usgs.gov

World Wide Web Virtual Library of Geotechnical Engineering, http://geotech.civen.okstate.edu/wwwvl/

8

Anne M. Ellis
Earth Tech., Inc.
Alexandria, VA

S.K. Ghosh
President
S.K. Ghosh Associates Inc.
Northbrook, IL

David A. Fanella
Dir. of Engineering
S.K. Ghosh Associates Inc.
Northbrook, IL

CONCRETE DESIGN AND CONSTRUCTION

Concrete made with portland cement is widely used as a construction material because of its many favorable characteristics. One of the most important is a large strength-cost ratio in many applications. Another is that concrete, while plastic, may be cast in forms easily at ordinary temperatures to produce almost any desired shape. The exposed face may be developed into a smooth or rough hard surface, capable of withstanding the wear of truck or airplane traffic, or it may be treated to create desired architectural effects. In addition, concrete has high resistance to fire and penetration of water.

But concrete also has disadvantages. An important one is that quality control sometimes is not so good as for other construction materials because concrete often is manufactured in the field under conditions where responsibility for its production cannot be pinpointed. Another disadvantage is that concrete is a relatively brittle material—its tensile strength is small compared with its compressive strength. This disadvantage, however, can be offset by reinforcing or prestressing concrete with steel. The combination of the two materials, reinforced concrete, possesses many of the best properties of each and finds use in a wide variety of constructions, including building frames, floors, roofs, and walls; bridges; pavements; piles; dams; and tanks.

8.1 Important Properties of Concrete

Characteristics of portland cement concrete can be varied to a considerable extent by controlling its ingredients. Thus, for a specific structure, it is economical to use a concrete that has exactly the characteristics needed, though weak in others. For example, concrete for a building frame should have high compressive strength, whereas concrete for a dam should be durable and watertight, and strength can be relatively small. Performance of concrete in service depends on both properties in the plastic state and properties in the hardened state.

8.1.1 Properties in the Plastic State

Workability is an important property for many applications of concrete. Difficult to evaluate, workability is essentially the ease with which the ingredients can be mixed and the resulting mix handled, transported, and placed with little loss in homogeneity. One characteristic of workability that engineers frequently try to measure is consistency, or fluidity. For this purpose, they often make a slump test.

In the slump test, a specimen of the mix is placed in a mold shaped as the frustum of a cone, 12 in high, with 8-in-diameter base and 4-in-diameter top (ASTM Specification C143). When the mold is removed, the change in height of the specimen is measured. When the test is made in accordance with the ASTM Specification, the change in height may be taken as the slump. (As measured by this test, slump decreases as temperature increases; thus the temperature of the mix at time of test should be specified, to avoid erroneous conclusions.)

Tapping the slumped specimen gently on one side with a tamping rod after completing the test may give additional information on the cohesiveness, workability, and placeability of the mix ("Concrete Manual," Bureau of Reclamation, Government Printing Office, Washington, DC 20402 (www.gpo.gov)). A well-proportioned, workable mix settles slowly, retaining its original identity. A poor mix crumbles, segregates, and falls apart.

Slump of a given mix may be increased by adding water, increasing the percentage of fines (cement or aggregate), entraining air, or incorporating an admixture that reduces water requirements. But these changes affect other properties of the concrete, sometimes adversely. In general, the slump specified should yield the desired consistency with the least amount of water and cement.

8.1.2 Properties in the Hardened State

Strength is a property of concrete that nearly always is of concern. Usually, it is determined by the ultimate strength of a specimen in compression, but sometimes flexural or tensile capacity is the criterion. Since concrete usually gains strength over a long period of time, the compressive strength at 28 days is commonly used as a measure of this property. In the United States, it is general practice to determine the compressive strength of concrete by testing specimens in the form of standard cylinders made in accordance with ASTM Specification C192 or C31. C192 is intended for research testing or for selecting a mix (laboratory specimens). C31 applies to work in progress (field specimens). The tests should be made as recommended in ASTM C39. Sometimes, however, it is necessary to determine the strength of concrete by taking drilled cores; in that case, ASTM C42 should be adopted. (See also American Concrete Institute Standard 214, "Recommended Practice for Evaluation of Strength Test Results of Concrete." (www.aci-int.org))

The 28-day compressive strength of concrete can be estimated from the 7-day strength by a formula proposed by W. A. Slater (*Proceedings of the American Concrete Institute,* 1926):

$$S_{28} = S_7 + 30\sqrt{S_7} \qquad (8.1)$$

where S_{28} = 28-day compressive strength, psi

S_7 = 7-day strength, psi

Concrete may increase significantly in strength after 28 days, particularly when cement is mixed

with fly ash. Therefore, specification of strengths at 56 or 90 days is appropriate in design.

Concrete strength is influenced chiefly by the water-cement ratio; the higher this ratio, the lower the strength. In fact, the relationship is approximately linear when expressed in terms of the variable C/W, the ratio of cement to water by weight: For a workable mix, without the use of water reducing admixtures

$$S_{28} = 2700\frac{C}{W} - 760 \qquad (8.2)$$

Strength may be increased by decreasing water-cement ratio, using higher-strength aggregates, grading the aggregates to produce a smaller percentage of voids in the concrete, moist curing the concrete after it has set, adding a pozzolan, such as fly ash, incorporating a superplasticizer admixture, vibrating the concrete in the forms, and sucking out excess water with a vacuum from the concrete in the forms. The short-time strength may be increased by using Type III (high-early-strength) portland cement (Art. 5.6) and accelerating admixtures, and by increasing curing temperatures, but long-time strengths may not be affected. Strength-increasing admixtures generally accomplish their objective by reducing water requirements for the desired workability. (See also Art. 5.6.)

Availability of such admixtures has stimulated the trend toward use of high-strength concretes. Compressive strengths in the range of 20,000 psi have been used in cast-in-place concrete buildings.

Tensile Strength, f_{ct}, of concrete is much lower than compressive strength. For members subjected to bending, the modulus of rupture f_r is used in design rather than the concrete tensile strength. For normal weight, normal-strength concrete, ACI specifies $f_r = 7.5\sqrt{f'_c}$.

The stress-strain diagram for concrete of a specified compressive strength is a curved line (Fig. 8.1). Maximum stress is reached at a strain of 0.002 in/in, after which the curve descends.

Modulus of elasticity E_c generally used in design for concrete is a secant modulus. In ACI 318, "Building Code Requirements for Reinforced Concrete," it is determined by

$$E_c = w^{1.5}33\sqrt{f'_c}, \text{ psi} \qquad (8.3a)$$

where w_c = density of concrete lb/ft³

f'_c = specified compressive strength at 28 days, psi

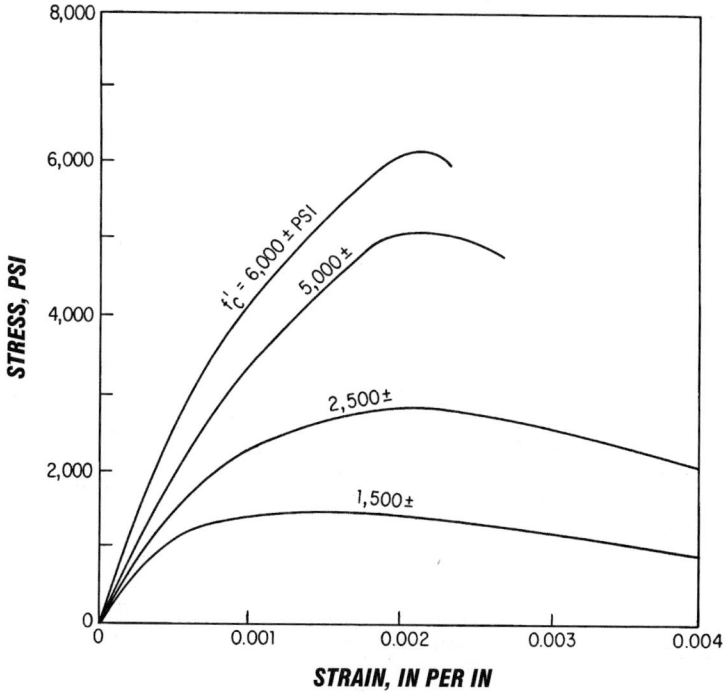

Fig. 8.1 Stress-strain curves for concrete.

This equation applies when 90 pcf $< w_c <$ 155 pcf. For normal-weight concrete, with $w = 145 \, \text{lb/ft}^3$,

$$E_c = 57,000\sqrt{f_c'}, \, \text{psi} \qquad (8.3b)$$

The modulus increases with age, as does the strength. (See also Art. 5.6)

Durability is another important property of concrete. Concrete should be capable of withstanding the weathering, chemical action, and wear to which it will be subjected in service. Much of the weather damage sustained by concrete is attributable to freezing and thawing cycles. Resistance of concrete to such damage can be improved by using appropriate cement types, lowering w/c ratio, providing proper curing, using alkali-resistant aggregates, using suitable admixtures, using an air-entraining agent, or applying a protective coating to the surface.

Chemical agents, such as inorganic acids, acetic and carbonic acids, and sulfates of calcium, sodium, magnesium, potassium, aluminum, and iron, disintegrate or damage concrete. When contact between these agents and concrete may occur, the concrete should be protected with a resistant coating. For resistance to sulfates, Type V portland cement may be used (Art. 5.6). Resistance to wear usually is achieved by use of a high-strength, dense concrete made with hard aggregates.

Watertightness is an important property of concrete that can often be improved by reducing the amount of water in the mix. Excess water leaves voids and cavities after evaporation, and if they are interconnected, water can penetrate or pass through the concrete. Entrained air (minute bubbles) usually increases watertightness, as does prolonged thorough curing.

Volume change is another characteristic of concrete that should be taken into account. Expansion due to chemical reactions between the ingredients of concrete may cause buckling and drying shrinkage may cause cracking.

Expansion due to alkali-aggregate reaction can be avoided by selecting nonreactive aggregates. If reactive aggregates must be used, expansion may be reduced or eliminated by adding pozzolanic material, such as fly ash, to the mix. Expansion due to heat of hydration of cement can be reduced by keeping cement content as low as possible, using Type IV cement (Art. 5.6), and chilling the aggregates, water, and concrete in the forms. Expansion due to increases in air temperature may be decreased by producing concrete with a lower coefficient of expansion, usually by using coarse aggregates with a lower coefficient of expansion.

Drying shrinkage can be reduced principally by cutting down on water in the mix. But less cement also will reduce shrinkage, as will adequate moist curing. Addition of pozzolans, however, unless enabling a reduction in water, may increase drying shrinkage.

Autogenous volume change, a result of chemical reaction and aging within the concrete and usually shrinkage rather than expansion, is relatively independent of water content. This type of shrinkage may be decreased by using less cement, and sometimes by using a different cement.

Whether volume change will damage the concrete often depends on the restraint present. For example, a highway slab that cannot slide on the subgrade while shrinking may crack; a building floor that cannot contract because it is anchored to relatively stiff girders also may crack. Hence, consideration should always be given to eliminating restraints or resisting the stresses they may cause.

Creep is strain that occurs under a sustained load. The concrete continues to deform, but at a rate that diminishes with time. It is approximately proportional to the stress at working loads and increases with increasing water-cement ratio. It decreases with increase in relative humidity. Creep increases the deflection of concrete beams and scabs and causes loss of prestress.

Density of ordinary sand-and-gravel concrete usually is about 145 lb/ft^3. It may be slightly lower if the maximum size of coarse aggregate is less than $1\frac{1}{2}$ in. It can be increased by using denser aggregate, and it can be decreased by using lightweight aggregate, increasing the air content, or incorporating a foaming, or expanding, admixture.

(J. G. MacGregor, "Reinforced Concrete," McGraw-Hill Book Company, New York (books.mcgraw-hill.com); M. Fintel, "Handbook of Concrete Engineering," 2nd ed., Van Nostrand Reinhold, New York.)

8.2 Lightweight Concretes

Concrete lighter in weight than ordinary sand-and-gravel concrete is used principally to reduce dead load, or for thermal insulation, nailability, or fill. Structural lightweight concrete must be of sufficient density to satisfy fire ratings.

Lightweight concrete generally is made by using lightweight aggregates or using gas-forming or foaming agents, such as aluminum powder, which are added to the mix. The lightweight aggregates are produced by expanding clay, shale, slate, diatomaceous shale, perlite obsidian, and vermiculite with heat and by special cooling of blast-furnace slag. They also are obtained from natural deposits of pumice, scoria, volcanic cinders, tuff, and diatomite, and from industrial cinders. Usual ranges of weights obtained with some lightweight aggregates are listed in Table 8.1.

Production of lightweight-aggregate concretes is more difficult than that of ordinary concrete because aggregates vary in absorption of water, specific gravity, moisture content, and amount and grading of undersize. Frequent unit-weight and slump tests are necessary so that cement and water content of the mix can be adjusted, if uniform results are to be obtained. Also, the concretes usually tend to be harsh and difficult to place and finish because of the porosity and angularity of the aggregates. Sometimes, the aggregates may float to the surface. Workability can be improved by increasing the percentage of fine aggregates or by using an air-entraining admixture to incorporate from 4 to 6% air. (See also ACI 211.2, "Recommended Practice for Selecting Proportions for Structural Lightweight Concrete," American Concrete Institute (www.aci-int.org).)

To improve uniformity of moisture content of aggregates and reduce segregation during stockpiling and transportation, lightweight aggregate

Table 8.1 Approximate Weights of Lightweight Concretes

Aggregate	Concrete Weight, lb/ft^3
Cinders:	
Without sand	85
With sand	110–115
Shale or clay	90–110
Pumice	90–100
Scoria	90–110
Perlite	50–80
Vermiculite	35–75

should be wetted 24 h before use. Dry aggregate should not be put into the mixer because the aggregate will continue to absorb moisture after it leaves the mixer and thus cause the concrete to segregate and stiffen before placement is completed. Continuous water curing is especially important with lightweight concrete.

Other types of lightweight concretes may be made with organic aggregates, or by omission of fines, or gap grading, or replacing all or part of the aggregates with air or gas. Nailing concrete usually is made with sawdust, although expanded slag, pumice, perlite, and volcanic scoria also are suitable. A good nailing concrete can be made with equal parts by volume of portland cement, sand, and pine sawdust, and sufficient water to produce a slump of 1 to 2 in. The sawdust should be fine enough to pass through a $\frac{1}{4}$-in screen and coarse enough to be retained on a No. 16 screen. (Bark in the sawdust may retard setting and weaken the concrete.) The behavior of this type of concrete depends on the type of tree from which the sawdust came. Hickory, oak, or birch may not give good results ("Concrete Manual," U.S. Bureau of Reclamation, Government Printing Office, Washington, DC, 20402 (www.gpo.gov)). Some insulating lightweight concretes are made with wood chips as aggregate.

For no-fines concrete, 20 to 30% entrained air replaces the sand. Pea gravel serves as the coarse aggregate. This type of concrete is used where low dead weight and insulation are desired and strength is not important. No-fines concrete may weigh from 105 to 118 lb/ft^3 and have a compressive strength from 200 to 1000 psi.

A porous concrete may be made by gap grading or single-size aggregate grading. It is used where drainage is desired or for light weight and low conductivity. For example, drain tile may be made with a No. 4 to $\frac{3}{8}$- or $\frac{1}{2}$-in aggregate and a low water-cement ratio. Just enough cement is used to bind the aggregates into a mass resembling popcorn.

Gas and foam concretes usually are made with admixtures. Foaming agents include sodium lauryl sulfate, alkyl aryl sulfonate, certain soaps, and resins. In another process, the foam is produced by the type of foaming agents used to extinguish fires, such as hydrolyzed waste protein. Foam concretes range in weight from 20 to 110 lb/ft^3.

Aluminum powder, when used as an admixture, expands concrete by producing hydrogen bubbles. Generally, about $\frac{1}{4}$ lb of the powder per bag of cement is added to the mix, sometimes with an alkali, such as sodium hydroxide or trisodium phosphate, to speed the reaction.

The heavier cellular concretes have sufficient strength for structural purposes, such as floor slabs and roofs. The lighter ones are weak but provide good thermal and acoustic insulation or are useful as fill; for example, they are used over structural floor slabs to embed electrical conduit.

(ACI 213R, "Guide for Structural Lightweight-Aggregate Concrete," and 211.2 "Recommended Practice for Selecting Proportions for Structural Lightweight Concrete," American Concrete Institute, 38800 Country Club Drive Farmington Hills, MI, 48331 (www.aci-int.org).)

8.3 Heavyweight Concretes

Concrete weighing up to about 385 lb/ft^3 can be produced by using heavier-than-ordinary aggregate. Theoretically, the upper limit can be achieved with steel shot as fine aggregate and steel punchings as coarse aggregate. (See also Art. 5.6.) The heavy concretes are used principally in radiation shields and counterweights.

Concrete made with barite develops an optimum density of 232 lb/ft^3 and compressive strength of 6000 psi; with limonite and magnetite, densities from 210 to 224 lb/ft^3 and strengths of 3200 to 5700 psi; with steel punchings and sheared bars as coarse aggregate and steel shot for fine aggregate, densities from 250 to 288 lb/ft^3 and strengths of about 5600 psi. Gradings and mix proportions are similar to those used for conventional concrete. These concretes usually do not have good resistance to weathering or abrasion.

Structural Concrete
8.4 Proportioning and Mixing Concrete

Components of a mix should be selected to produce a concrete with the desired characteristics for the service conditions and adequate workability at the lowest cost. For economy, the amount of cement should be kept to a minimum. Generally, this objective is facilitated by selecting the largest-size coarse aggregate consistent with job requirements and good gradation, to keep the volume of voids small. The smaller this volume, the less cement paste needed to fill the voids.

The water-cement ratio, for workability, should be as large as feasible to yield a concrete with the desired compressive strength, durability, and watertightness and without excessive shrinkage. Water added to a stiff mix improves workability, but an excess of water has deleterious effects (Art. 8.1).

8.4.1 Proportioning Concrete Mixes

A concrete mix is specified by indicating the weight, in pounds, of water, cement, sand, coarse aggregate, and admixture to be used per cubic yard of mixed concrete. In addition, type of cement, fineness modulus of the aggregates, and maximum sizes of aggregates should be specified. (In the past, one method of specifying a concrete mix was to give the ratio, by weight, of cement to sand to coarse aggregate; for example, 1:2:4; plus the minimum cement content per cubic yard of concrete.)

Because of the large number of variables involved, it usually is advisable to proportion concrete mixes by making and testing trial batches. A start is made with the selection of the water-cement ratio. Then, several trial batches are made with varying ratios of aggregates to obtain the desired workability with the least cement. The aggregates used in the trial batches should have the same moisture content as the aggregates to be used on the job. The amount of mixing water to be used must include water that will be absorbed by dry aggregates or must be reduced by the free water in wet aggregates. The batches should be mixed by machine, if possible, to obtain results close to those that would be obtained in the field. Observations should be made of the slump of the mix and

Table 8.2 Estimated Compressive Strength of Concrete for Various Water-Cement Ratios*

Water-Cement Ratio by Weight	28-day Compressive Strength	
	Air-Entrained Concrete	Non-Air-Entrained Concrete
0.40	4,300	5,400
0.45	3,900	4,900
0.50	3,500	4,300
0.55	3,100	3,800
0.60	2,700	3,400
0.65	2,400	3,000
0.70	2,200	2,700

* "Concrete Manual," U.S. Bureau of Reclamation.

appearance of the concrete. Also, tests should be made to evaluate compressive strength and other desired characteristics. After a mix has been selected, some changes may have to be made after some field experience with it.

Table 8.2 estimates the 28-day compressive strength that may be attained with various water-cement ratios, with and without air entrainment. Note that air entrainment permits a reduction of water, so a lower water-cement ratio for a given workability is feasible with air entrainment.

Table 8.3 lists recommended maximum sizes of aggregate for various types of construction. These tables may be used with Table 8.4 for proportioning concrete mixes for small jobs where time or other conditions do not permit proportioning by the trial-batch method. Start with mix B in Table 8.4 corresponding to the selected maximum size of aggregate. Add just enough water for the desired

Table 8.3 Recommended Maximum Sizes of Aggregate*

Minimum Dimension of Section, in	Maximum Size, in, of Aggregate for		
	Reinforced-Concrete Beams, Columns, Walls	Heavily Reinforced Slabs	Lightly Reinforced or Unreinforced Slabs
5 or less	—	$\frac{3}{4}-\frac{1}{2}$	$\frac{3}{4}-\frac{1}{2}$
6–11	$\frac{3}{4}-1\frac{1}{2}$	$1\frac{1}{2}$	$1\frac{1}{2}-3$
12–29	$1\frac{1}{2}-3$	3	3–6
30 or more	$1\frac{1}{2}-3$	3	6

* "Concrete Manual," U.S. Bureau of Reclamation.

Table 8.4 Typical Concrete Mixes*

Maximum Size of Aggregate, in	Mix Designation	Bags of Cement per yd³ of Concrete	Aggregate, lb per Bag of Cement		
			Sand		Gravel or Crushed Stone
			Air-Entrained Concrete	Concrete without Air	
½	A	7.0	235	245	170
	B	6.9	225	235	190
	C	6.8	225	235	205
¾	A	6.6	225	235	225
	B	6.4	225	235	245
	C	6.3	215	225	265
1	A	6.4	225	235	245
	B	6.2	215	225	275
	C	6.1	205	215	290
1½	A	6.0	225	235	290
	B	5.8	215	225	320
	C	5.7	205	215	345
2	A	5.7	225	235	330
	B	5.6	215	225	360
	C	5.4	205	215	380

* "Concrete Manual," U.S. Bureau of Reclamation.

workability. If the mix is undersanded, change to mix A; if oversanded, change to mix C. Weights are given for dry sand. For damp sand, increase the weight of sand 10 lb, and for very wet sand, 20 lb, per bag of cement.

8.4.2 Admixtures

These may be used to modify and control specific characteristics of concrete. Major types of admixtures include set accelerators, water reducers, air entrainers, and waterproofing compounds. In general, admixtures are helpful in improving concrete workability. Some admixtures, if not administered properly, could have undesirable side effects. Hence, every engineer should be familiar with admixtures and their chemical components as well as their advantages and limitations. Moreover, admixtures should be used in accordance with manufacturers' recommendations and, if possible, under the supervision of a manufacturer's representative. Many admixtures are covered by ASTM specifications.

Accelerating admixtures are used to reduce the time of setting and accelerating early strength

development and are often used in cold weather, when it takes too long for concrete to set naturally. The best-known accelerator is calcium chloride, but it is not recommended for use in prestressed concrete, in reinforced concrete containing embedded dissimilar metals, or where progressive corrosion of steel reinforcement can occur. Nonchloride, noncorrosive accelerating admixtures, although more expensive than calcium chloride, may be used instead.

Water reducers lubricate the mix. Most of the water in a normal concrete mix is needed for workability of the concrete. Reduction in the water content of a mix may result in either a reduction in the water-cement ratio (w/c) for a given slump and cement content or an increased slump for the same w/c and cement content. With the same cement content but less water, the concrete attains greater strength. As an alternative, reduction of the quantity of water permits a proportionate decrease in cement and thus reduces shrinkage of the hardened concrete. An additional advantage of a water-reducing admixture is easier placement of concrete. This, in turn, helps the workers and reduces the possibility of honeycombed concrete. Some water-reducing admixtures also act as retarders of concrete set, which is helpful in hot weather and in integrating consecutive pours of concrete.

High-range water-reducing admixtures, also known as superplasticizers, behave much like conventional water-reducing admixtures. They help the concrete achieve high strength and water reduction without loss of workability. Superplasticizers reduce the interparticle forces that exist between cement grains in the fresh paste, thereby increasing the paste fluidity. However, they differ from conventional admixtures in that superplasticizers do not affect the surface tension of water significantly, as a result of which, they can be used at higher dosages without excessive air entrainment.

Air-entraining agents entrain minute bubbles of air in concrete. This increases resistance of concrete to freezing and thawing. Therefore, air-entraining agents are extensively used in exposed concrete. Air entrainment also affects properties of fresh concrete by increasing workability.

Waterproofing chemicals may be added to a concrete mix, but often they are applied as surface treatments. Silicones, for example, are used on hardened concrete as a water repellent. If applied properly and uniformly over a concrete surface, they can effectively prevent rainwater from penetrating the surface. (Some silicone coatings discolor with age. Most lose their effectiveness after a number of years. When that happens, the surface should be covered with a new coat of silicone for continued protection.) Epoxies also may be used as water repellents. They are much more durable, but they also may be much more costly. Epoxies have many other uses in concrete, such as protection of wearing surfaces, patching compounds for cavities and cracks, and glue for connecting pieces of hardened concrete.

Miscellaneous types of admixtures are available to improve properties of concrete either in the plastic or the hardened state. These include polymer-bonding admixtures used to produce modified concrete, which has better abrasion resistance, better resistance to freezing and thawing, and reduced permeability; dampproofing admixtures; permeability-reducing admixtures; and corrosion-inhibiting admixtures.

8.4.3 Mixing Concrete Mixes

Components for concrete generally are stored in batching plants before being fed to a mixer. These plants consist of weighing and control equipment and hoppers, or bins, for storing cement and aggregates. Proportions are controlled by manually operated or automatic scales. Mixing water is measured out from measuring tanks or with the aid of water meters.

Machine mixing is used wherever possible to achieve uniform consistency of each batch. Good results are obtained with the revolving-drum-type mixer, commonly used in the United States, and countercurrent mixers, with mixing blades rotating in the direction opposite to that of the drum.

Mixing time, measured from the time the ingredients, including water, are in the drum, should be at least 1.5 min for a 1-yd^3 mixer, plus 0.5 min for each cubic yard of capacity over 1 yd^3. But overmixing may remove entrained air and increase fines, thus requiring more water to maintain workability, so it is advisable also to set a maximum on mixing time. As a guide, use three times the minimum mixing time.

Ready-mixed concrete is batched in central plants and delivered to various job-sites in trucks, usually in mixers mounted on the trucks. The concrete may be mixed en route or after arrival at the site. Though concrete may be kept plastic and workable for as long as $1\frac{1}{2}$ h by slow revolving of the mixer, better control of mixing time can be maintained if water is added and mixing started after arrival of the truck at the job, where the operation can be inspected.

(ACI 212.2, "Guide for Use of Admixtures in Concrete," ACI 211.1, "Recommended Practice for Selecting Proportion for Normal and Heavyweight Concrete," ACI 213R, "Recommended Practice for Selecting Proportions for Structural Lightweight Concrete," and ACI 304, "Recommended Practice for Measuring, Mixing, Transporting, and Placing Concrete," American Concrete Institute, 38800 Country Club Drive Farmington Hills, MI 48331; G. E. Troxell, H. E. Davis, and J. W. Kelly, "Composition and Properties of Concrete," McGraw-Hill Book Company, New York (books.mcgraw-hill.com); D. F. Orchard, "Concrete Technology," John Wiley & Sons, Inc., New York; M. Fintel, "Handbook of Concrete Engineering," 2nd ed., Van Nostrand Reinhold, New York.)

8.5 Concrete Placement

When concrete is discharged from the mixer, precautions should be taken to prevent segregation because of uncontrolled chuting as it drops into buckets, hoppers, carts, or forms. Such segregation is less likely to occur with tilting mixers than with nontilting mixers with discharge chutes that let the concrete pass in relatively small streams. To prevent segregation, a baffle, or better still, a section of downpipe should be inserted at the end of the chutes so that the concrete will fall vertically into the center of the receptacle.

8.5.1 Concrete Transport and Placement Equipment

Steel buckets, when selected for the job conditions and properly operated, handle and place concrete very well. But they should not be used if they have to be hauled so far that there will be noticeable separation, bleeding, or loss of slump exceeding 1 in. The discharge should be controllable in amount and direction.

Rail cars and trucks sometimes are used to transport concrete after it is mixed. But there is a risk of stratification, with a layer of water on top, coarse aggregate on the bottom. Most effective prevention is use of dry mixes and air entrainment. If stratification occurs, the concrete should be remixed either as it passes through the discharge gates or by passing small quantities of compressed air through the concrete en route.

Chutes frequently are used for concrete placement. But the operation must be carefully controlled to avoid segregation and objectionable loss of slump. The slope must be constant under varying loads and sufficiently steep to handle the stiffest concrete to be placed. Long chutes should be shielded from sun and wind to prevent evaporation of mixing water. Control at the discharge end is of utmost importance to prevent segregation. Discharge should be vertical, preferably through a short length of downpipe.

Tremies, or elephant trunks, deposit concrete under water. Tremies are tubes about 1 ft or more in diameter at the top, flaring slightly at the bottom. They should be long enough to reach the bottom. When concrete is being placed, the tremie is always kept full of concrete, with the lower end immersed in the concrete just deposited. The tremie is raised as the level of concrete rises. Concrete should never be deposited through water unless confined.

Belt conveyors for placing concrete also present segregation and loss-of-slump problems. These may be reduced by adopting the same precautions as for transportation by trucks and placement with chutes.

Sprayed concrete (shotcrete or gunite) is applied directly onto a form by an air jet. A "gun," or mechanical feeder, mixer, and compressor comprise the principal equipment for this method of placement. Compressed air and the dry mix are fed to the gun, which jets them out through a nozzle equipped with a perforated manifold. Water flowing through the perforations is mixed with the dry mix before it is ejected. Because sprayed concrete can be placed with a low water-cement ratio, it usually has high compressive strength. The method is especially useful for building up shapes without a form on one side.

Pumping is a suitable method for placing concrete, but it seldom offers advantages over other methods. Curves, lifts, and harsh concrete reduce substantially maximum pumping distance. For best performance, an agitator should be installed in the pump feed hopper to prevent segregation.

Barrows are used for transporting concrete very short distances, usually from a hopper to the forms. In the ordinary wheelbarrow, a worker can move $1\frac{1}{2}$ to 2 ft^3 of concrete 25 ft in 3 min.

Concrete carts serve the same purpose as wheelbarrows but put less load on the transporter. Heavier and wider, the carts can handle 4.5 ft^3. Motorized carts with $\frac{1}{2}$-yd^3 capacity also are available.

Regardless of the method of transportation or equipment used, the concrete should be deposited as nearly as possible in its final position. Concrete should not be allowed to flow into position but should be placed in horizontal layers because then less durable mortar concentrates in ends and corners where durability is most important.

8.5.2 Vibration of Concrete in Forms

This is desirable because it eliminates voids. The resulting consolidation also ensures close contact of the concrete with the forms, reinforcement, and other embedded items. It usually is accomplished with electric or pneumatic vibrators.

For consolidation of structural concrete and tunnel-invert concrete, immersion vibrators are recommended. Oscillation should be at least 7000 vibrations per minute when the vibrator head is immersed in concrete. Precast concrete of relatively small dimensions and concrete in tunnel arch and sidewalls may be vibrated with vibrators rigidly attached to the forms and operating at 8000 vibrations per minute or more. Concrete in canal and lateral linings should be vibrated at more than 4000 vibrations per minute, with the immersion type, though external vibration may be used for linings less than 3 in thick. For mass concrete, with 3- and 6-in coarse aggregate, vibrating heads should be at least 4 in in diameter and operate at frequencies of at least 6000 vibrations per minute when immersed. Each cubic yard should be vibrated for at least 1 min. A good small vibrator can handle from 5 to 10 yd^3/h and a large two-person, heavy-duty type, about 50 yd^3/h in uncramped areas. Over vibration can be detrimental as it can cause segregation of the aggregate and bleeding of the concrete.

8.5.3 Construction Joints

A construction joint is formed when unhardened concrete is placed against concrete that has become so rigid that the new concrete cannot be incorporated into the old by vibration. Generally, steps must be taken to ensure bond between the two.

Method of preparation of surfaces at construction joints vary depending on the orientation of the surface.

("Concrete Manual," U.S. Bureau of Reclamation, Government Printing Office, Washington, DC, 20402 (www.gpo.gov); ACI 311 "Recommended Practice for Concrete Inspection"; ACI 304, "Recommended Practice for Measuring, Mixing, Transporting, and Placing Concrete"; and ACI 506 "Recommended Practice for Shotcreting"; also, ACI 304.2R, "Placing Concrete by Pumping Methods," ACI 304.1R, "Preplaced Aggregate Concrete for Structural and Mass Concrete," and "ACI Manual of Concrete Inspection," SP-2, American Concrete Institute (www.aci-int.org).)

8.6 Finishing of Unformed Concrete Surfaces

After concrete has been consolidated, screeding, floating, and the first troweling should be performed with as little working and manipulation of

the surface as possible. Excessive manipulation draws inferior fines and water to the top and can cause checking, crazing, and dusting.

To avoid bringing fines and water to the top in the rest of the finishing operations, each step should be delayed as long as possible. If water accumulates, it should be removed by blotting with mats or draining, or it should be pulled off with a loop of hose, and the next finishing operation should be delayed until the water sheen disappears. Do not work neat cement into wet areas to dry them.

Screeds are guides for a straightedge to bring a concrete surface to a desired elevation or for a template to produce a desired curved shape. The screeds must be sufficiently rigid to resist distortion as the concrete is spread. They may be made of lumber or steel pipe.

For floors, screeding is followed by hand floating with wood floats or power floating. Permitting a stiffer mix with a higher percentage of large-size aggregate, power-driven floats with revolving disks and vibrators produce a sounder, more durable surface than wood floats. Floating may begin as soon as the concrete surface has hardened sufficiently to bear a person's weight without leaving an indentation. The operation continues until hollows and humps are removed or, if the surface is to be troweled, until a small amount of mortar is brought to the top.

If a finer finish is desired, the surface may be steel-troweled, by hand or by powered equipment. This is done as soon as the floated surface has hardened enough so that excess fine material will not be drawn to the top. Heavy pressure during troweling will produce a dense, smooth, watertight surface. Do not permit sprinkling of cement or cement and sand on the surface to absorb excess water or facilitate troweling. If an extra hard finish is desired, the floor should be troweled again when it has nearly hardened.

Concrete surfaces dust to some extent and may benefit from treatment with certain chemicals. They penetrate the pores to form crystalline or gummy deposits. Thus, they make the surface less pervious and reduce dusting by acting as plastic binders or by making the surface harder. Poor-quality concrete floors may be improved more by such treatments than high-quality concrete, but the improvement is likely to be temporary and the treatment will have to be repeated periodically.

("Concrete Manual," U.S. Bureau of Reclamation, U.S. Government Printing Office, Washington, DC 20402 (www.gpo.gov).)

8.7 Forms for Concrete

Formwork retains concrete until it has set and produces the desired shapes and, sometimes, desired surface finishes. Forms must be supported on falsework of adequate strength and sufficient rigidity to keep deflections within acceptable limits. The forms too must be strong and rigid, to meet dimensional tolerances. But they also must be tight, or mortar will leak out during vibration and cause unsightly sand streaks and rock pockets. Yet they must be low-cost and often easily demountable to permit reuse. These requirements are met by steel, reinforced plastic, and plain or coated lumber and plywood.

Unsightly bulges and offsets at horizontal joints should be avoided. This can be done by resetting forms with only 1 in of form lining overlapping the existing concrete below the line made by a grade strip. Also, the forms should be tied and bolted close to the joint to keep the lining snug against existing concrete (Fig. 8.2). If a groove along a joint will not be esthetically objectionable, forming of a groove along the joint will obscure unsightliness often associated with construction joints (Art. 8.5.3).

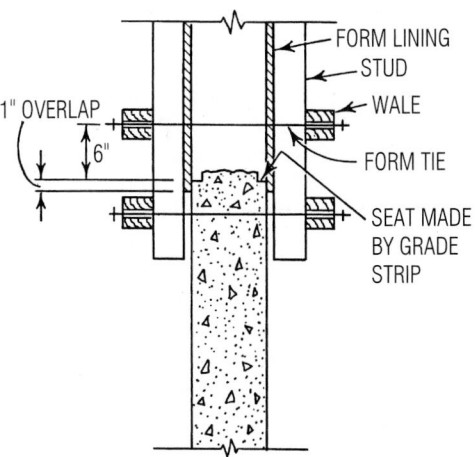

Fig. 8.2 Form set to avoid bulges at a horizontal joint in a concrete wall.

Where form ties have to pass through the concrete, they should be as small in cross section as possible. (The holes they form sometimes have to be plugged to stop leaks.) Ends of form ties should be removed without spalling adjacent concrete.

Plastic coatings, proper oiling, or effective wetting can protect forms from deterioration, weather, and shrinkage before concreting. Form surfaces should be clean. They should be treated with a suitable form-release oil or other coating that will prevent the concrete from sticking to them. A straight, refined, pale, paraffin-base mineral oil usually is acceptable for wood forms. Synthetic castor oil and some marine-engine oils are examples of compounded oils that give good results on steel forms. The oil or coating should be brushed or sprayed evenly over the forms. It should not be permitted to get on construction joint surfaces or reinforcing bars because it will interfere with bond.

Forms should provide ready access for placement and vibration of concrete for inspection. Formed areas should be clean of debris prior to concrete placement.

Generally, forms are stationary. But for some applications, such as highway pavements, precast-concrete slabs, silos, and service cores of buildings, use of continuous moving forms—sliding forms or slip forms—is advantageous.

8.7.1 Slip Forms

A slip form for vertical structures consists principally of a form lining or sheathing about 4 ft high, wales or ribs, yokes, working platforms, suspended scaffolds, jacks, climbing rods, and control equipment (Fig. 8.3). Spacing of the sheathing is slightly larger at the top to permit easy upward movement. The wales hold the sheathing in alignment, support the working platforms and scaffolds, and transmit lifting forces from yokes to sheathing. Each yoke has a horizontal cross member perpendicular to the wall and connected to a jack. From each end of the member, vertical legs extend downward on opposite sides of and outside the wall. The lower end of each leg is attached to a bottom wale. The jack pulls the slip form upward by climbing a vertical steel rod, usually about 1 in in diameter, embedded in the concrete. The suspended scaffolds provide access for finishers to the wall. Slip-form climbing rates range upward from about 2 to about 12 in/h.

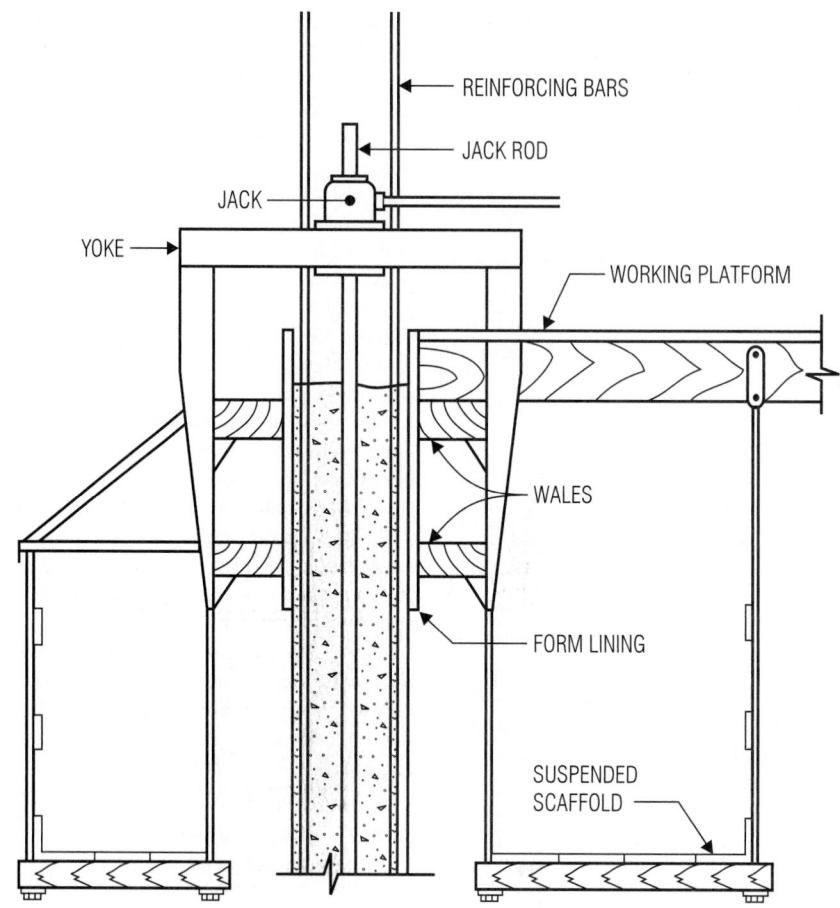

Fig. 8.3 Slip form for a concrete wall.

8.7.2 Form Removal

Stationary forms should be removed only after the concrete has attained sufficient strength so that there will be no noticeable deformation or damage to the concrete. If supports are removed before beams or floors are capable of carrying superimposed loads, they should be reshored until they have gained sufficient strength.

Early removal of forms generally is desirable to permit quick reuse, start curing as soon as possible, and allow repairs and surface treatment while the concrete is still green and conditions are favorable for good bond. In cold weather, however, forms should not be removed while the concrete is still warm. Rapid cooling of the surface will cause checking and surface cracks. For this reason also, curing water applied to newly stripped surfaces should not be much cooler than the concrete.

(R. L. Peurifoy, "Formwork for Concrete Structures," 2nd ed., McGraw-Hill Book Company, New York (books.mcgraw-hill.com); "Concrete Manual," U.S. Bureau of Reclamation, Government Printing Office, Washington, DC, 20402 (www.gpo. gov); ACI 347 "Recommended Practice for Concrete Formwork," "ACI Manual of Concrete Inspection," SP-2, and "Formwork for Concrete," SP-4, American Concrete Institute (www.aci-int.org).)

8.8 Curing Concrete

While more than enough mixing water for hydration is incorporated into normal concrete mixes, drying of the concrete after initial set may delay or prevent complete hydration. Curing includes all operations after concrete has set that improve hydration. Properly done for a sufficiently long period, curing produces stronger, more watertight concrete.

Methods may be classified as one of the following: maintenance of a moist environment by addition of water, sealing in the water in the concrete, and those hastening hydration.

8.8.1 Curing by Surface Moistening

Maintenance of a moist environment by addition of water is the most common field procedure. Generally, exposed concrete surfaces are kept continuously moist by spraying or ponding or by a covering of earth, sand, or burlap kept moist. Concrete made with ordinary and sulfate-resistant cements (Types I, II, and V) should be cured this way for 7 to 14 days, that made with low-heat cement (Type IV) for at least 21 days. Concrete made with high-early-strength cement should be kept moist until sufficient strength has been attained, as indicated by test cylinders.

8.8.2 Steam Curing

Precast concrete and concrete placed in cold weather often are steam-cured in enclosures. Although this is a form of moist curing, hydration is speeded by the higher-than-normal temperature, and the concrete attains a high early strength. Temperatures maintained usually range between 100 and 165 °F. Higher temperatures produce greater strengths shortly after steam curing commences, but there are severe losses in strength after 2 days. A delay of 1 to 6 h before steam curing will produce concrete with higher 24-h strength than if the curing starts immediately after the concrete is cast. This "preset" period allows early cement reactions to occur and development of sufficient hardness to withstand the more rapid temperature curing to follow. Length of the preset period depends on the type of aggregate and temperature. The period should be longer for ordinary aggregate than for lightweight and for higher temperatures. Duration of steam curing depends on the concrete mix, temperature, and desired results.

Autoclaving, or high-pressure steam curing, maintains concrete in a saturated atmosphere at temperatures above the boiling point of water.

Generally, temperatures range from 325 to 375 °F at pressures from 80 to 170 psig. Main application is for concrete masonry. Advantages claimed are high early strength, reduced volume change in drying, better chemical resistance, and lower susceptibility to efflorescence. For steam curing, a preset period of 1 to 6 h is desirable, followed by single- or two-stage curing. Single-curing consists of a pressure buildup of at least 3 h, 8 h at maximum pressure, and rapid pressure release (20 to 30 min). The rapid release vaporizes moisture from the block. In two-stage curing, the concrete products are placed in kilns for the duration of the preset period. Saturated steam then is introduced into the kiln. After the concrete has developed sufficient strength to permit handling, the products are removed from the kiln, set in a compact arrangement, and placed in the autoclave.

8.8.3 Curing by Surface Sealing

Curing concrete by sealing the water in can be accomplished by either covering the concrete or coating it with a waterproof membrane. When coverings, such as heavy building paper or plastic sheets, are used, care must be taken that the sheets are sealed airtight and corners and edges are adequately protected against loss of moisture. Coverings can be placed as soon as the concrete has been finished.

Coating concrete with a sealing compound generally is done by spraying to ensure a continuous membrane. Brushing may damage the concrete surface. Sealing compound may be applied after the surface has stiffened so that it will no longer respond to float finishing. But in hot climates, it may be desirable, before spraying, to moist cure for 1 day surfaces exposed to the sun. Surfaces from which forms have been removed should be saturated with water before spraying with compound. But the compound should not be applied to either formed or unformed surfaces until the moisture film on them has disappeared. Spraying should be started as soon as the surfaces assume a dull appearance. The coating should be protected against damage. Continuity must be maintained for at least 28 days.

White or gray pigmented compound often is used for sealing because it facilitates inspection and reflects heat from the sun. Temperatures with white pigments may be decreased as much as 40 °F, reducing cracking caused by thermal changes.

Surfaces of ceilings and walls inside buildings require no curing other than that provided by forms left in place at least 4 days. But wood forms are not acceptable for moist curing outdoor concrete. Water should be applied at the top, for example, by a soil-soaker hose and allowed to drip down between the forms and the concrete.

("Concrete Manual," U.S. Bureau of Reclamation, Government Printing Office, Washington, DC, 20402 (www.gpo.gov); ACI 517, "Recommended Practice for Atmospheric Pressure Steam Curing of Concrete," ACI 517.1R, "Low-Pressure Steam Curing," and ACI 516R, "High-Pressure Steam Curing: Modern Practice, and Properties of Autoclaved Products," American Concrete Institute (www.aci-int.org).)

8.9 Cold-Weather Concreting

Hydration of cement takes place in the presence of moisture at temperatures above 50 °F. Methods used during cold weather should prevent damage to concrete from freezing and thawing at an early age. (Concrete that is protected from freezing until it has attained a compressive strength of at least 500 psi will not be damaged by exposure to a single freezing cycle.) Neglect of protection against freezing can cause immediate destruction or permanent weakening of concrete. Therefore, if concreting is performed in cold weather, protection from low temperatures and proper curing are essential. Except within heated protective enclosures, little or no external supply of moisture is required for curing during cold weather. Under such conditions, the temperature of concrete placed in the forms should not be lower than the values listed in Table 8.5. Protection against freezing should be provided until concrete has gained sufficient strength to withstand exposure to low temperatures, anticipated environment, and construction and service loads.

The time needed for concrete to attain the strength required for safe removal of shores is influenced by the initial concrete temperature at placement, temperatures after placement, type of cement, type and amount of accelerating admixture, and the conditions of protection and curing. The use of high-early-strength cement or the addition of accelerating admixtures may be an economic solution when schedule considerations are critical. The use of such admixtures does not justify a reduction in the amount of protective cover, heat, or other winter protection.

Although freezing is a danger to concrete, so is overheating the concrete to prevent it. By accelerating chemical action, overheating can cause excessive loss of slump, raise the water requirement for a given slump, and increase thermal shrinkage. Rarely will mass concrete leaving the mixer have to be at more than 55 °F and thin-section concrete at more than 75 °F.

Table 8.5 Recommended Concrete Temperatures for Cold-Weather Construction—Air Entrained Concrete

	Minimum Cross-Sectional Dimension, in			
	less than 12	12 to 36	36 to 72	72 or more
	(a) Minimum Temperature of Concrete as Placed or Maintained, °F			
	55	50	45	40
	(b) Maximum Allowable Gradual Temperature Drop of Concrete in First 24 h after Protection Is Discounted, °F			
	50	40	30	20
Temperature of air, °F	(c) Minimum Temperature of Concrete as Mixed, °F			
30 or higher	60	55	50	45
0 to 30	65	60	55	50
0 or lower	70	65	60	55

To obtain the minimum temperatures for concrete mixes in cold weather, the water and, if necessary, the aggregates should be heated. The proper mixing water temperature for the required concrete temperature is based upon the temperature and weight of the materials in the concrete and the free moisture on aggregates. To avoid flash set of cement and loss of entrained air due to the heated water, aggregates and water should be placed in the mixer before the cement and air-entraining agent so that the colder aggregates will reduce the water temperature to below 80 °F.

When heating of aggregates is necessary, it is best done with steam or hot water in pipes. Use of steam jets is objectionable because of resulting variations in moisture content of the aggregates. For small jobs, aggregates may be heated over culvert pipe in which fires are maintained, but care must be taken not to overheat.

Before concrete is placed in the forms, the interior should be cleared of ice, snow, and frost. This may be done with steam under canvas or plastic covers.

Concrete should not be placed on frozen earth. It would lower the concrete temperature below the minimum and may cause settlement on thawing. The subgrade may be protected from freezing by a covering of straw and tarpaulins or other insulating blankets. If it does freeze, the subgrade must be thawed deep enough so that it will not freeze back up to the concrete during the required protection period.

The usual method of protecting concrete after it has been cast is to enclose the structure with tarpaulins or plastic and heat the interior. Since corners and edges are especially vulnerable to low temperatures, the enclosure should enclose corners and edges, not rest on them. The enclosure must be not only strong but windproof. If wind can penetrate it, required concrete temperatures may not be maintained despite high fuel consumption. Heat may be supplied by live or piped steam, salamanders, stoves, or warm air blown in through ducts from heaters outside the enclosure. But strict fire-prevention measures should be enforced. When dry heat is used, the concrete should be kept moist to prevent it from drying.

Concrete also may be protected with insulation. For example, pavements may be covered with layers of straw, shavings, or dry earth. For structures, forms may be insulated.

When protection is discontinued or when forms are removed, precautions should be taken that the drop in temperature of the concrete will be gradual. Otherwise, the concrete may crack and deteriorate. Table 8.5 lists recommended limitations on temperature drop in the first 24 hours. Special care shall be taken with concrete test specimens used for acceptance of concrete. Cylinders shall be properly stored and protected in insulated boxes with a thermometer to maintain temperature records.

("Concrete Manual," U.S. Bureau of Reclamation, Government Printing Office, Washington, DC 20402 (www.gpo.gov); ACI 306R "Cold-Weather Concreting," American Concrete Institute (www.aci-int.org).)

8.10 Hot-Weather Concreting

Hot weather is defined as any combination of the following: high ambient air temperature, high concrete temperature, low relative humidity, high wind velocity, and intense solar radiation. Such weather may lead to conditions in mixing, placing, and curing concrete that can adversely affect the properties and serviceability of the concrete.

The higher the temperature, the more rapid the hydration of cement, the faster the evaporation of mixing water, the lower the concrete strength and the larger the volume change. Unless precautions are taken, setting and rate of hardening will accelerate, shortening the available time for placing and finishing the concrete. Quick stiffening encourages undesirable additions of mixing water, or retempering, and may also result in inadequate consolidation and cold joints. The tendency to crack is increased because of rapid evaporation of water, increased drying shrinkage, or rapid cooling of the concrete from its high initial temperature. If an air-entrained concrete is specified, control of the air content is more difficult. And curing becomes more critical. Precautionary measures required on a calm, humid day will be less restrictive than those required on a dry, windy, sunny day, even if the air temperatures are identical.

Placement of concrete in hot weather is too complex to be dealt with adequately by simply setting a maximum temperature at which concrete may be placed. A rule of thumb, however, has been that concrete temperature during placement should be maintained as much below 90 °F as is economically feasible.

The following measures are advisable in hot weather: The concrete should have ingredients and proportions with satisfactory records in field use in hot weather. To keep the concrete temperature within a safe range, the concrete should be cooled with iced water or cooled aggregate, or both. Also, to minimize slump loss and other temperature effects, the concrete should be transported, placed, consolidated, and finished as speedily as possible. Materials and facilities not otherwise protected from the heat should be shaded. Mixing drums should be insulated or cooled with water sprays or wet burlap coverings. Also, water-supply lines and tanks should be insulated or at least painted white. Cement with a temperature exceeding 170 °F should not be used. Forms, reinforcing steel, and the subgrade should be sprinklered with cool water. If necessary, work should be done only at night. Futhermore, the concrete should be protected against moisture loss at all times during placing and curing.

Self-retarding admixtures counteract the accelerating effects of high temperature and lessen the need for increase in mixing water. Their use should be considered when the weather is so hot that the temperature of concrete being placed is consistently above 75 °F.

Continuous water curing gives best results in hot weather. Curing should be started as soon as the concrete has hardened sufficiently to withstand surface damage. Water should be applied to formed surfaces while forms are still in place. Surfaces without forms should be kept moist by wet curing for at least 24 h. Moist coverings are effective in eliminating evaporation loss from concrete, by protecting it from sun and wind. If moist curing is discontinued after the first day, the surface should be protected with a curing compound (Art. 8.8).

(ACI 305R, "Hot-Weather Concreting," American Concrete Institute (www.aci-int.org).)

8.11 Contraction and Expansion Joints

Contraction joints are used mainly to control locations of cracks caused by shrinkage of concrete after it has hardened. If the concrete, while shrinking, is restrained from moving, by friction or attachment to more rigid construction, cracks are likely to occur at points of weakness. Contraction joints, in effect, are deliberately made weakness planes. They are formed in the expectation that if a crack occurs it will be along the neat geometric pattern of a joint, and thus irregular, unsightly cracking will be prevented. Such joints are used principally in floors, roofs, pavements, and walls.

A contraction joint is an indentation in the concrete. Width may be $\frac{1}{4}$ or $\frac{3}{8}$ in and depth one-fourth the thickness of the slab. The indentation may be made with a saw cut while the concrete still is green but before appreciable shrinkage stress develops. Or the joint may be formed by insertion of a strip of joint material before the concrete sets or by grooving the surface during finishing. Spacing of joints depends on the mix, strength and thickness of the concrete, and the restraint to shrinkage. The indentation in highway and airport pavements usually is filled with a sealing compound.

Construction joints occur where two successive placements of concrete meet. They may be designed to permit movement and/or to transfer load.

Expansion or isolation joints are used to help prevent cracking due to thermal dimension changes in concrete. They usually are placed where there are abrupt changes in thickness, offsets, or changes in types of construction, for example, between a bridge pavement and a highway pavement. Expansion joints provide a complete separation between two parts of a slab. The opening must be large enough to prevent buckling or other undesirable deformation due to expansion of the concrete.

To prevent the joint from being jammed with dirt and becoming ineffective, the opening is sealed with a compressible material. For watertightness, a flexible water stop should be placed across the joint. And if load transfer is desired, dowels should be embedded between the parts separated by the joint. The sliding ends of the dowels should be enclosed in a close-fitting metal cap or thimble, to provide space for movement of the dowel during expansion of the concrete. This space should be at least $\frac{1}{4}$ in longer than the width of the joint.

(ACI 504R, "Guide to Joint Sealants for Concrete Structures," American Concrete Institute (www.aci-int.org).)

8.12 Steel Reinforcement in Concrete

Because of the low tensile strength of concrete, steel reinforcement is embedded in it to resist tensile stresses. Steel, however, also is used to take compression, in beams and columns, to permit use of smaller members. It serves other purposes too: It controls strains due to temperature and shrinkage and distributes load to the concrete and other reinforcing steel; it can be used to prestress the concrete; and it ties other reinforcing together for easy placement or to resist lateral stresses.

Most reinforcing is in the form of bars or wires whose surfaces may be smooth or deformed. The latter type is generally used because it produces better bond with the concrete because of the raised pattern on the steel.

Bars range in diameter from $\frac{1}{4}$ to $2\frac{1}{4}$ in (Table 8.11, p 8.36). Sizes are designated by numbers, which are approximately eight times the nominal diameters. (See the latest edition of ASTM "Specifications for Steel Bars for Concrete Reinforcement." These also list the minimum yield points and tensile strengths for each type of steel.) Use of bars with yield points over 60 ksi for flexural reinforcement is limited because special measures are required to control cracking and deflection.

Wires usually are used for reinforcing concrete pipe and, in the form of welded-wire fabric, for slab reinforcement. The latter consists of a rectangular grid of uniformly spaced wires, welded at all intersections, and meeting the minimum requirements of ASTM A185 and A497. Fabric offers the advantages of easy, fast placement of both longitudinal and transverse reinforcement and excellent crack control because of high mechanical bond with the concrete. (Deformed wires are designated by D followed by a number equal to the nominal area, in^2, times 100.) Bars and rods also may be prefabricated into grids, by clipping or welding (ASTM A184).

Sometimes, metal lath is used for reinforcing concrete, for example, in thin shells. It may serve as both form and reinforcing when concrete is applied by spray (gunite or shotcrete.)

8.12.1 Bending and Placing Reinforcing Steel

Bars are shipped by a mill to a fabricator in uniform long lengths and in bundles of 5 or more tons. The fabricator transports them to the job straight and cut to length or cut and bent.

Bends may be required for beam-and-girder reinforcing, longitudinal reinforcement of columns where they change size, stirrups, column ties and spirals, and slab reinforcing. Dimensions of standard hooks and typical bends and tolerances for cutting and bending are given in ACI 315, "Manual of Standard Practice for Detailing Reinforced Concrete Structures," American Concrete Institute (www.aci-int.org).

Some preassembling of reinforcing steel is done in the fabricating shop or on the job. Beam, girder, and column steel often is wired into frames before placement in the forms. Slab reinforcing may be clipped or welded into grids, or mats, if not supplied as welded-wire fabric.

Some rust is permissible on reinforcing if it is not loose and there is no appreciable loss of cross-sectional area. In fact, rust, by creating a rough surface, will improve bond between the steel and concrete. But the bars should be free of loose rust, scale, grease, oil, or other coatings that would impair bond.

Bars should not be bent or straightened in any way that will damage them. All reinforcement shall be bent cold unless permitted by the engineer. If heat is necessary for bending, the temperature should not be higher than that indicated by a cherry-red color (1200 °F), and the steel should be allowed to cool slowly, not quenched, to 600 °F.

Reinforcing should be supported and tied in the locations and positions called for in the plans. The steel should be inspected before concrete is placed. Neither the reinforcing nor other parts to be embedded should be moved out of position before or during the casting of the concrete.

Bars and wire fabric should not be kinked or have unspecified curvatures when positioned. Kinked and curved bars, including those misshaped by workers walking on them, may cause the hardened concrete to crack when the bars are tensioned by service loads.

Usually, reinforcing is set on wire bar supports, preferably galvanized for exposed surfaces. Lower-layer bars in slabs usually are supported on bolsters consisting of a horizontal wire welded to two legs about 5 in apart. The upper layer generally is supported on bolsters with runner wires on the bottom so that they can rest on bars already in place. Or individual or continuous high chairs can be used to hold up a support bar, often a No. 5, at appropriate intervals, usually 5 ft. An individual high chair is a bar seat that looks roughly like an inverted U braced transversely by another inverted U in a perpendicular plane. A continuous high chair consists of a horizontal wire welded to two inverted-U legs 8 or 12 in apart. Beam and joist chairs have notches to receive the reinforcing. These chairs usually are placed at 5-ft intervals.

Although it is essential that reinforcement be placed exactly where called for in the plans, some tolerances are necessary. Reinforcement in beams and slabs, walls and compression members should be within $\pm \frac{3}{8}''$ for members where $d \le 8''$, $\pm \frac{1}{2}''$ for members where $d > 8''$ of the specified distance from the tension or compression face. Lengthwise, a cutting tolerance of ± 1 in and a placement tolerance of ± 2 in are normally acceptable. If length of embedment is critical, the designer should specify bars 3 in longer than the computed minimum to allow for accumulation of tolerances. Spacing of reinforcing in wide slabs and tall walls may be permitted to vary $\pm \frac{1}{2}$ in or slightly more if necessary to clear obstructions, so long as the required number of bars are present.

Lateral spacing of bars in beams and columns, spacing between multiple reinforcement layers, and concrete cover over stirrups, ties, and spirals in beams and columns should never be less than that specified but may exceed it by $\frac{1}{4}$ in. A variation in setting of an individual stirrup or column hoop of 1 in may be acceptable, but the error should not be permitted to accumulate.

("CRSI Recommended Practice for Placing Reinforcing Bars," and "Manual of Standard Practice," Concrete Reinforcing Steel Institute, 180 North La Salle St., Chicago, IL 60601 (www.crsi.org).)

8.12.2 Minimum Spacing of Reinforcement

In buildings, the minimum clear distance between parallel bars should be 1 in for bars up to No. 8 and the nominal bar diameter for larger bars. For columns, however, the clear distance between longitudinal bars should be at least 1.5 in for bars up to No. 8 and 1.5 times the nominal bar diameter for larger bars. And the clear distance between multiple layers of reinforcement in building beams and girders should be at least 1 in. Upper-layer bars should be directly above corresponding bars below. These minimum-distance requirements also apply to the clear distance between a contact splice and adjacent splices or bars.

A common requirement for minimum clear distance between parallel bars in highway bridges is 1.5 times the diameter of the bars, and spacing center to center should be at least 1.5 times the maximum size of coarse aggregate.

Many codes and specifications relate the minimum bar spacing to maximum size of coarse aggregate. This is done with the intention of providing enough space for all of the concrete mix to pass between the reinforcing. But if there is a space to place concrete between layers of steel and between the layers and the forms, and the concrete is effectively vibrated, experience has shown that bar spacing or form clearance does not have to exceed the maximum size of coarse aggregate to ensure good filling and consolidation. That portion of the mix which is molded by vibration around bars, and between bars and forms, is not inferior to that which would have filled those parts had a larger bar spacing been used. The remainder of the mix in the interior, if consolidated layer after layer, is superior because of its reduced mortar and water content ("Concrete Manual," U.S. Bureau of Reclamation, Government Printing Office, Washington, D.C. 20402 (www.gpo.gov)).

Bundled Bars ■ Groups of parallel reinforcing bars bundled in contact to act as a unit may be used only when they are enclosed by ties or stirrups. Four bars are the maximum permitted in a bundle, and all must be deformed bars. If full-length bars cannot be used between supports, then there should be a stagger of at least 40 bar diameters between any discontinuities. Also, the length of lap should be increased 20% for a three-bar bundle and 33% for a four-bar bundle. In determining minimum clear distance between a bundle and parallel reinforcing, the bundle should be treated as a single bar of equivalent area.

8.12.3 Maximum Spacing

In walls and slabs in buildings, except for concrete-joist construction, maximum spacing, center to center, of principal reinforcement should be 18 in, or three times the wall or slab thickness, whichever is smaller.

8.12.4 Concept of Development Length

Bond of steel reinforcement to the concrete in a reinforced concrete member must be sufficient so that the steel will yield before it is freed from the concrete. Furthermore, the length of embedment must be adequate to prevent highly stressed reinforcement from splitting relatively thin sections of restraining concrete. Hence, design codes specify a required length of embedment, called development length, for reinforcing steel. The concept of development length is based on the attainable average bond stress over the embedment length of the reinforcement.

Each reinforcing bar at a section of a member must develop on each side of the section the calculated tension or compression in the bar through development length l_d or end anchorage, or both. Development of tension bars can be assisted by hooks.

8.12.5 Tension Development Lengths

For bars and deformed wire in tension, basic development length is defined by Eqs. (8.4). For No. 11 and smaller bars,

$$l_d = \left[\frac{3}{40} \frac{f_y}{\sqrt{f'_c}} \frac{\alpha\beta\gamma\lambda}{(c + k_{tr})/d_b} \right] d_b \qquad (8.4)$$

Where α = traditional reinforcement location factor

β = coating factor

γ = reinforcement size factor

λ = lightweight aggregate factor

c = spacing or cover dimension

k_{tr} = transverse reinforcement index

d_b = bar diameter

8.12.6 Compression Development Lengths

For bars in compression, the basic development length l_d is defined as

$$l_d = \frac{0.02 f_y d_b}{\sqrt{f'_c}} \geq 0.0003 d_b f_y \qquad (8.5)$$

but l_d not be less than 8 in. See Table 8.6.

For f_y greater than 60 ksi or concrete strengths less than 3000 psi, the required development length in Table 8.6 should be increased as indicated by Eq. (8.5). The values in Table 8.6 may be multiplied by the applicable factors:

a) reinforcement in excess of that required by analyses: $\frac{As \; required}{As \; provided}$

b) reinforcement enclosed within spiral reinforcement not less than $\frac{1}{4}''$ diameter and not more than $4''$ pitch or within #4 ties spaced not more than $4''$ on center.

8.12.7 Bar Lap Splices

Because of the difficulty of transporting very long bars, reinforcement cannot always be continuous. When splices are necessary, it is advisable that they

Table 8.6 Compression Development in Normal-Weight Concrete for Grade 60 Bars

Bar Size No.	f'_c (Normal-Weight Concrete)			
	3000 psi	3750 psi	4000 psi	Over 4444 psi*
3	8	8	8	8
4	11	10	10	9
5	14	12	12	11
6	17	15	15	14
7	19	17	17	16
8	22	20	19	18
9	25	22	22	20
10	28	25	24	23
11	31	27	27	25
14	38	34	34	32
18	50	44	43	41

* For $f'_c > 4444$ psi, minimum embedment = $18 d_b$.

should be made where the tensile stress is less than half the permissible stress.

Bars up to No. 11 in size may be spliced by overlapping them and wiring them together.

Bars spliced by noncontact lap splices in flexural members should not be spaced transversely farther apart than one-fifth the required lap length or 6 in.

8.12.8 Welded or Mechanical Splices

These other positive connections should be used for bars larger than No. 11 and are an acceptable alternative for smaller bars. Welding should conform to AWS D12.1, "Reinforcing Steel Welding Code," American Welding Society, 550 N.W. LeJeune Road, Miami, FL 33126 (www.aws.org). Bars to be spliced by welding should be butted and welded so that the splice develops in tension at least 125% of their specified yield strength. Mechanical coupling devices should be equivalent in strength.

8.12.9 Tension Lap Splices

The length of lap for bars in tension should conform to the following, with l_d taken as the tensile development length for the full yield strength f_y of the reinforcing steel [Eq. (8.4)]:

Class A splices (lap of l_d) are permitted where both conditions 1 and 2 occur.

1. The area of reinforcement provided is at least twice that required by analysis over the entire lengths of splices.

2. No more than one-half of the total reinforcement is spliced within the required lap length.

Class B splices (lap of 1.3 l_d) are required where either 1 or 2 does not apply.

Bars in tension splices should lap at least 12 in. Splices for **tension tie members** should be fully welded or made with full mechanical connections and should be staggered at least 30 in. Where feasible, splices in regions of high stress also should be staggered.[7]

8.12.10 Compression Lap Splices

For a bar in compression, the minimum length of a lap splice should be the largest of 12 in, or $0.0005f_y d_b$, for f_c' of 3000 psi or larger and steel yield strength f_y of 60 ksi or less, where d_b is the bar diameter.

For tied compression members where the ties have an area, in², of at least $0.0015hs$ in the vicinity of the lap, the lap length may be reduced to 83% of the preceding requirements but not to less than 12 in (h is the overall thickness of the member, in, and s is the tie spacing, in).

For spirally reinforced compression members, the lap length may be reduced to 75% of the basic required lap but not to less than 12 in.

In columns where reinforcing bars are offset and one bar of a splice has to be bent to lap and contact the other one, the slope of the bent bar should not exceed 1 in 6. Portions of the bent bar above and below the offset should be parallel to the column axis. The design should account for a horizontal thrust at the bend taken equal to at least 1.5 times the horizontal component of the nominal stress in the inclined part of the bar. This thrust should be resisted by steel ties, or spirals, or members framing into the column. This resistance should be provided within a distance of 6 in of the point of the bend.

Where column faces are offset 3 in or more, vertical bars should be lapped by separate dowels.

In columns, a minimum tensile strength at each face equal to one-fourth the area of vertical reinforcement multiplied by f_y should be provided at horizontal cross sections where splices are located. In columns with substantial bending, full tensile splices equal to double the factored tensile stress in the bar are required.

8.12.11 Splices of Welded-Wire Fabric

Wire reinforcing normally is spliced by lapping. For plainwire fabric in tension, when the area of reinforcing provided is more than twice that required, the overlap measured between outermost cross wires should be at least 2 in or $1.5l_d$. Otherwise, the overlap should equal the spacing of the cross wires but not less than $1.5l_d$ nor 6 in. For deformed wire fabric, the overlap measured between outermost cross wires should be at least 2 in. The overlap should be at least 8″ or $1.3l_d$.

8.12.12 Slab Reinforcement

Structural floor and roof slabs with principal reinforcement in only one direction should be reinforced for shrinkage and temperature stresses in a perpendicular direction. The crossbars may be spaced at a maximum of 18 in or five times the slab thickness. The ratio of reinforcement area of these bars to gross concrete area should be at least 0.0020 for deformed bars with less than 60 ksi yield strength, 0.0018 for deformed bars with 60 ksi yield strength and welded-wire fabric with welded intersections in the direction of stress not more than 12 in apart, and 0.0018 $(60/f_y)$ for bars with f_y greater than 60 ksi.

8.12.13 Concrete Cover

To protect reinforcement against fire and corrosion, thickness of concrete cover over the outermost steel should be at least that given in Table 8.7.

(ACI 318, "Building Code Requirements for Reinforced Concrete," American Concrete Institute; "Standard Specifications for Highway Bridges," American Association of State Highway and Transportation Officials, 444 N. Capitol St., N.W., Washington, DC 20001 (www.aashto.org).)

8.13 Tendons

High-strength steel is required for prestressing concrete to make the stress loss due to creep and shrinkage of concrete and to other factors a small percentage of the applied stress (Art. 8.37). This type of loss does not increase as fast as increase in strength in the prestressing steel, or tendons.

Tendons should have specific characteristics in addition to high strength to meet the requirements of prestressed concrete. They should elongate uniformly

Table 8.7 Cast-in-Place Concrete Cover for Steel Reinforcement (Non-prestressed)

1. Concrete deposited against and permanently exposed to the ground, 3 in.

2. Concrete exposed to seawater, 4 in; except precast-concrete piles, 3 in.

3. Concrete exposed to the weather or in contact with the ground after form removal, 2 in for bars larger than No. 5 and 1½ in for No. 5 or smaller.

4. Unexposed concrete slabs, walls, or joists, ¾ in for No. 11 and smaller, 1½ in for No. 14 and No. 18 bars. Beams, girders, and columns, 1½ in. Shells and folded-plate members, ¾ in for bars larger than No. 5, and ½ inch for No. 5 and smaller.

up to initial tension for accuracy in applying the prestressing force. After the yield strength has been reached, the steel should continue to stretch as stress increases, before failure occurs. ASTM Specifications for prestressing wire and strands, A421 and A416, set the yield strength at 80 to 85% of the tensile strength. Furthermore, the tendons should exhibit little or no creep, or relaxation, at the high stresses used.

ASTM A421 covers two types of uncoated, stress-relieved, high-carbon-steel wire commonly used for linear prestressed-concrete construction. Type BA wire is used for applications in which cold-end deformation is used for end anchorages, such as buttonheads. Type WA wire is intended for end anchorages by wedges and where no cold-end deformation of the wire is involved. The wire is required to be stress-relieved by a continuous-strand heat treatment after it has been cold-drawn to size. Type BA usually is furnished 0.196 and 0.250 in in diameter, with an ultimate strength of 240 ksi and yield strength (at 1% extension) of 192 ksi. Type WA is available in those sizes and also 0.192 and 0.276 in in diameter, with ultimate strengths ranging from 250 for the smaller diameters to 235 ksi for the largest. Yield strengths range from 200 for the smallest to 188 ksi for the largest (Table 8.8).

For pretensioning, where the steel is tensioned before the concrete is cast, wires usually are used individually, as is common for reinforced concrete. For posttensioning, where the tendons are tensioned and anchored to the concrete after it has attained sufficient strength, the wires generally are placed parallel to each other in groups, or cables, sheathed or ducted to prevent bond with the concrete.

A seven-wire strand consists of a straight center wire and six wires of slightly smaller diameter winding helically around and gripping it. High friction between the center and outer wires is important where stress is transferred between the strand and concrete through bond. ASTM A416 covers strand with ultimate strengths of 250 and 270 ksi (Table 8.8).

Galvanized strands sometimes are used for posttensioning, particularly when the tendons may not be embedded in grout. Sizes normally available range from a 0.5-in-diameter seven-wire strand, with 41.3-kip breaking strength, to $1^{11}\!/_{16}$-in-diameter strand, with 352-kip breaking strength. The cold-drawn wire comprising the strand is stress-relieved when galvanized, and stresses due to stranding are offset by prestretching the strand to about 70% of

Table 8.8 Properties of Tendons

Diameter, in	Area, in^2	Weight per ft-kip	Ultimate Strength
Uncoated Type WA Wire			
0.276	0.05983	203.2	235 ksi
0.250	0.04909	166.7	240 ksi
0.196	0.03017	102.5	250 ksi
0.192	0.02895	98.3	250 ksi
Uncoated Type BA Wire			
0.250	0.04909	166.7	240 ksi
0.196	0.03017	102.5	240 ksi
Uncoated Seven-Wire Strands, 250 Grade			
$\frac{1}{4}$	0.04	122	9 kips
$\frac{5}{16}$	0.058	197	14.5 kips
$\frac{3}{8}$	0.080	272	20 kips
$\frac{7}{16}$	0.108	367	27 kips
$\frac{1}{2}$	0.144	490	36 kips
270 Grade			
$\frac{3}{8}$	0.085	290	23 kips
$\frac{7}{16}$	0.115	390	31 kips
$\frac{1}{2}$	0.153	520	41.3 kips

its ultimate strength. Tendons 0.5 and 0.6 in in diameter are typically used sheathed and unbonded.

Hot-rolled alloy-steel bars used for prestressing concrete generally are not so strong as wire or strands. The bars usually are stress-relieved, then cold-stretched to at least 90% of ultimate strength to raise the yield point. The cold stretching also serves as proof stressing, eliminating bars with defects.

(H. K. Preston and N. J. Sollenberger, "Modern Prestressed Concrete," McGraw-Hill Book Company, New York (books.mcgraw-hill.com); J. R. Libby, "Modern Prestressed Concrete," Van Nostrand Reinhold Company, New York.)

8.14 Fabrication of Prestressed-Concrete Members

Prestressed concrete may be produced much like high-strength reinforced concrete, either cast in place or precast. Prestressing offers several advantages for precast members, which have to be transported from casting bed to final position and handled several times. Prestressed members are lighter than reinforced members of the same capacity, both because higher-strength concrete generally is used and because the full cross section is effective. In addition, prestressing of precast members normally counteracts handling stresses. And, if a prestressed, precast member survives the full prestress and handling, the probability of its failing under service loads is very small.

Two general methods of prestressing are commonly used—pretensioning and posttensioning—and both may be used for the same member. See also Art. 8.37.

Pretensioning, where the tendons are tensioned before embedment in the concrete and stress transfer from steel to concrete usually is by bond, is especially useful for mass production of precast elements. Often, elements may be fabricated in long lines, by stretching the tendons (Art. 8.13) between abutments at the ends of the lines. By use of tiedowns and struts, the tendons may be draped in a vertical plane to develop upward and downward components on release. After the tendons have been jacked to their full stress, they are anchored to the abutments.

The casting bed over which the tendons are stretched usually is made of a smooth-surface concrete slab with easily stripped side forms of steel. (Forms for pretensioned members must permit them to move on release of the tendons.) Separators are placed in the forms to divide the long line into members of required length and provide space for cutting the tendons. After the concrete has been cast and has attained its specified strength, generally after a preset period and steam curing, side forms are removed. Then, the tendons are detached from the anchorages at the ends of the line and relieved of their stress. Restrained from shortening by bond with the concrete, the tendons compress it. At this time, it is safe to cut the tendons between the members and remove the members from the forms.

In pretensioning, the tendons may be tensioned one at a time to permit the use of relatively light jacks, in groups, or all simultaneously. A typical stressing arrangement consists of a stationary anchor post, against which jacks act, and a moving crosshead, which is pushed by the jacks and to which the tendons are attached. Usually, the tendons are anchored to a thick steel plate that serves as a combination anchor plate and template. It has holes through which the tendons pass to place them in the desired pattern. Various patented grips are available for anchoring the tendons to the plate. Generally, they are a wedge or chuck type capable of developing the full strength of the tendons.

Posttensioning frequently is used for cast-in-place members and long-span flexural members. Cables or bars (Art. 8.13) are placed in the forms in flexible ducts to prevent bond with the concrete. They may be draped in a vertical plane to develop upward and downward forces when tensioned. After the concrete has been placed and has attained sufficient strength, the tendons are tensioned by jacking against the member and then are anchored to it. Grout may be pumped into the duct to establish bond with the concrete and protect the tendons against corrosion. Applied at pressures of 75 to 100 psi, a typical grout consists of 1 part portland cement, 0.75 parts sand (capable of passing through a No. 30 sieve), and 0.75 parts water, by volume.

Concrete with higher strengths than ordinarily used for reinforced concrete offers economic advantages for prestressed concrete. In reinforced concrete, much of the concrete in a slab or beam is assumed to be ineffective because it is in tension and likely to crack under service loads. In prestressed concrete, the full section is effective because it is always under either compression or very low tension. Furthermore, high-strength concrete develops higher bond stresses with the tendons, greater bearing strength to withstand the pressure of anchorages, and a higher modulus of elasticity. The last indicates reductions in initial strain and camber when prestress is applied initially and in creep strain. The reduction in creep strain reduces the loss of prestress with time. Generally, concrete with a 28-day strength of 5000 psi or more is advantageous for prestressed concrete.

Concrete cover over prestressing steel, ducts, and nonprestressed steel should be at least 3 in for concrete surfaces in contact with the ground; $1\frac{1}{2}$ in for prestressing steel and main reinforcing bars, and 1 in for stirrups and ties in beams and girders, 1 in in slabs and joists exposed to the weather; and $\frac{3}{4}$ in for unexposed slabs and joists. In extremely corrosive atmospheres or other severe exposures, the amount of protective cover should be increased.

Minimum clear spacing between pretensioning steel at the ends of a member should be four times the diameter of individual wires and three times the diameter of strands. Some codes also require that the spacing be at least $1\frac{1}{3}$ times the maximum size of aggregate. (See also Art. 8.12.2.) Away from the ends of a member, prestressing steel or ducts may be bundled. Concentrations of steel or ducts, however, should be reinforced to control cracking.

Prestressing force may be determined by measuring tendon elongation, by checking jack pressure on a recently calibrated gage, or by using a recently calibrated dynamometer. If several wires or strands are stretched simultaneously, the method used should be such as to induce approximately equal stress in each.

Splices should not be used in parallel-wire cables, especially if a splice has to be made by welding, which would weaken the wire. Failure is likely to occur during tensioning of the tendon.

Strands may be spliced, if necessary, when the coupling will develop the full strength of the tendon, not cause it to fail under fatigue loading, and does not displace sufficient concrete to weaken the member.

High-strength bars are generally spliced mechanically. The couplers should be capable of developing the full strength of the bars without decreasing resistance to fatigue and without replacing an excessive amount of concrete.

Posttensioning End Anchorages ▪ Anchor fittings are different for pretensioned and posttensioned members. For pretensioned members, the fittings hold the tendons temporarily against anchors outside the members and therefore can be reused. In posttensioning, the fittings usually anchor the tendons permanently to the members. In unbonded tendons, the sheathing is typically plastic or impregnated paper.

A variety of patented fittings are available for anchoring in posttensioned members. Such fittings should be capable of developing the full strength of the tendons under static and fatigue loadings. The fittings also should spread the prestressing force over the concrete or transmit it to a bearing plate. Sufficient space must be provided for the fittings in the anchor zone.

Generally, all the wires of a parallel-wire cable are anchored with a single fitting (Figs. 8.4 and 8.5). The type shown in Fig. 8.5 requires that the wires

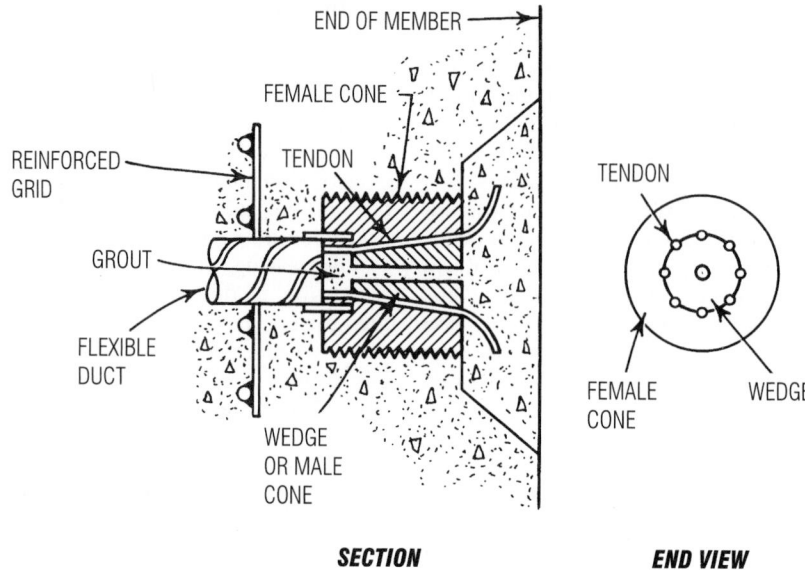

Fig. 8.4 Conical wedge anchorage for prestressing wires.

be cut to exact length and a buttonhead be cold-formed on the ends for anchoring.

The wedge type in Fig. 8.4 requires a double-acting jack. One piston, with the wires wedged to it, stresses them, and a second piston forces the male cone into the female cone to grip the tendons. Normally, a hole is provided in the male cone for grouting the wires. After final stress is applied, the anchorage may be embedded in concrete to prevent corrosion and improve appearance.

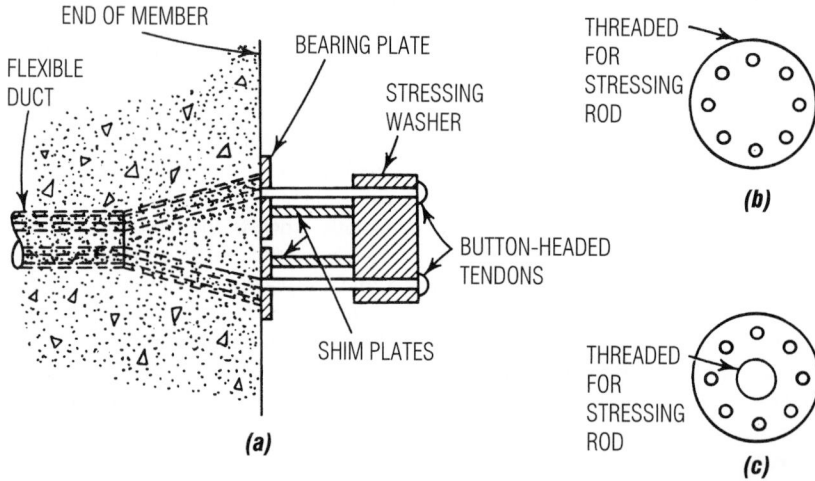

Fig. 8.5 Detail at end of prestressed concrete member. (*a*) End anchorage for button-headed wires. (*b*) Externally threaded stressing head. (*c*) Internally threaded stressing head. Heads are used for attachment to stressing jack.

With the buttonhead type, a stressing rod may be screwed over threads on the circumference of a thick, steel stressing washer (Fig. 8.5*b*) or into a center hole in the washer (Fig. 8.5*c*). The rod then is bolted to a jack. When the tendons have been stressed, the washer is held in position by steel shims inserted between it and a bearing plate embedded in the member. The jack pressure then can be released and the jack and stressing rod removed. Finally, the anchorage is embedded in concrete.

Posttensioning bars may be anchored individually with steel wedges (Fig. 8.6*a*) or by tightening a nut against a bearing plate (Fig. 8.6*b*). The former has the advantage that the bars do not have to be threaded.

Posttensioning strands normally are shop-fabricated in complete assemblies, cut to length, anchor fittings attached, and sheathed in flexible duct. Swaged to the strands, the anchor fittings have a threaded steel stud projecting from the end. The threaded stud is used for jacking the stress into the strand and for anchoring by tightening a nut against a bearing plate in the member (Fig. 8.7).

To avoid overstressing and failure in the anchorage zone, the anchorage assembly must be

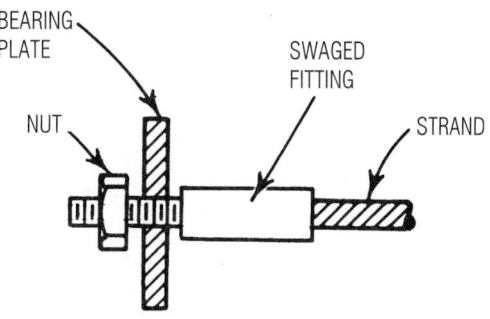

Fig. 8.7 Swaged fitting for strands. Prestress is maintained by tightening the nut against the bearing plate.

placed with care. Bearing plates should be placed perpendicular to the tendons to prevent eccentric loading. Jacks should be centered for the same reason and so as not to scrape the tendons against the plates. The entire area of the plates should bear against the concrete.

Prestress normally is applied with hydraulic jacks. The amount of prestressing force is determined by measuring tendon elongation and comparing with an average load-elongation curve for the steel used. In addition, the force thus determined should be checked against the jack pressure registered on a recently calibrated gage or by use of a recently calibrated dynamometer. Discrepancies of less than 5% may be ignored.

When prestressed-concrete beams do not have a solid rectangular cross section in the anchorage zone, an enlarged end section, called an end block, may be necessary to transmit the prestress from the tendons to the full concrete cross section a short distance from the anchor zone. End blocks also are desirable for transmitting vertical and lateral forces to supports and to provide adequate space for the anchor fittings for the tendons.

The transition from end block to main cross section should be gradual (Fig. 8.8). Length of end block, from beginning of anchorage area to the start of the main cross section, should be at least 24 in. The length normally ranges from three-fourths the depth of the member for deep beams to the full depth for shallow beams. The end block should be reinforced vertically and horizontally to resist tensile bursting and spalling forces induced by the concentrated loads of the tendons. In particular, a grid of reinforcing should be placed directly behind the anchorages to resist spalling.

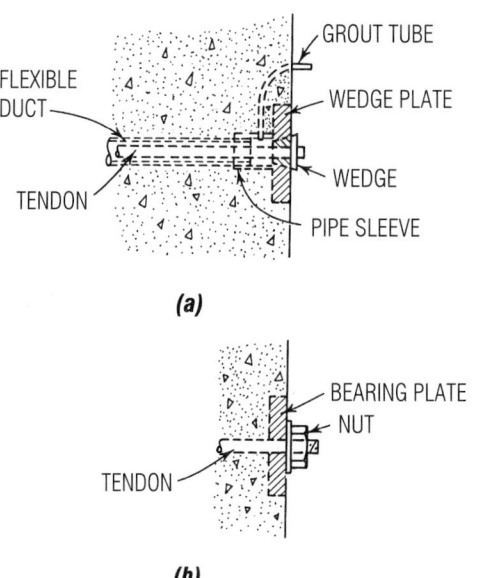

Fig. 8.6 End anchorages for bars. (*a*) Conical wedge. (*b*) Nut and washer acting against a bearing plate at a threaded end of tendon.

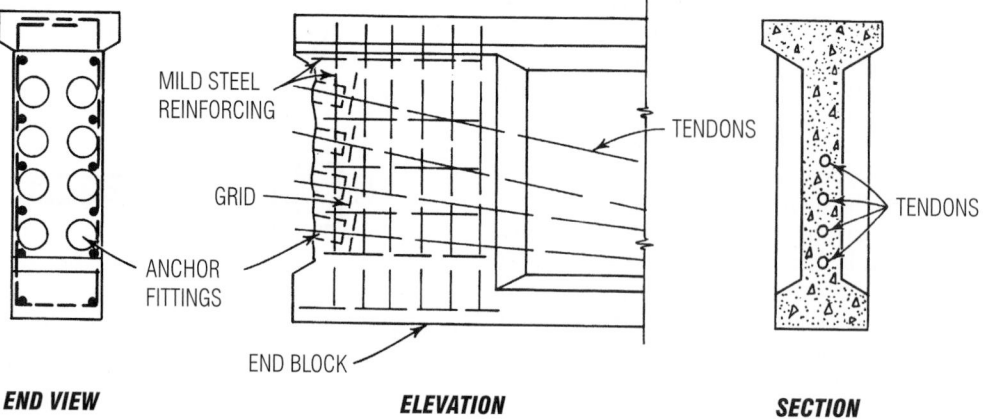

END VIEW   **ELEVATION**   **SECTION**

Fig. 8.8 Transition from cross section of the end block of a prestressed concrete beam to the main cross section.

Ends of pretensioned beams should be reinforced with vertical stirrups over a distance equal to one-fourth the beam depth. The stirrups should be capable of resisting in tension a force equal to at least 4% of the prestressing force.

Camber ▪ Control of camber is important for prestressed members. Camber tends to increase with time because of creep. If a prestressed beam or slab has an upward camber under prestress and long-time loading, the camber will tend to increase upward. Excessive camber should be avoided, and for deck-type structures, such as highway bridges and building floors and roofs, the camber of all beams and girders of the same span should be the same.

Computation of camber with great accuracy is difficult, mainly because of the difficulty of ascertaining with accuracy the modulus of elasticity of the concrete, which varies with time. Other difficult-to-evaluate factors also influence camber: departure of the actual prestressing force from that calculated, effects of long-time loading, influence of length of time between prestressing and application of full service loads, methods of supporting members after removal from the forms, and influence of composite construction.

When camber is excessive, it may be necessary to use concrete with higher strength and modulus of elasticity, for example, change from lightweight to ordinary concrete; increase the moment of inertia of the section; use partial prestressing, that is,

decrease the prestressing force and add reinforcing steel to resist the tensile stresses; or use a larger prestressing force with less eccentricity.

To ensure uniformity of camber, a combination of pretensioning and posttensioning can be provided for precast members. Sufficient prestress may be applied initially to permit removal of the member from the forms and transportation to a storage yard. After the member has increased in strength but before erection, additional prestress is applied by posttensioning to bring the camber to the desired value. During storage, the member should be supported in the same manner as it will be in the structure.

(H. K. Preston and N. J. Sollenberger, "Modern Prestressed Concrete," McGraw-Hill Book Company, New York (books.mcgraw-hill.com); J. B. Libby, "Modern Prestressed Concrete," Van Nostrand Reinhold Company, New York.)

8.15 Precast Concrete

When concrete products are made in other than their final position, they are considered precast. They may be unreinforced, reinforced, or prestressed. They include in their number a wide range of products: block, brick, pipe, plank, slabs, conduit, joists, beams and girders, trusses and truss components, curbs, lintels, sills, piles, pile caps, and walls.

Precasting often is chosen because it permits efficient mass production of concrete units. With precasting, it usually is easier to maintain quality control and produce higher-strength concrete than with field concreting. Formwork is simpler, and a good deal of falsework can be eliminated. Also, since precasting normally is done at ground level, workers can move about more freely. But sometimes these advantages are more than offset by the cost of handling, transporting, and erecting the precast units. Also, joints may be troublesome and costly.

Design of precast products follows the same rules, in general, as for cast-in-place units. However, ACI 318, "Building Code Requirements for Reinforced Concrete" (American Concrete Institute (www.ACI-int.org)), permits the concrete cover over reinforcing steel to be as low as $\frac{5}{8}$ in for slabs, walls, or joists not exposed to weather. Also, ACI Standard 525, "Minimum Requirements for Thin-Section Precast Concrete Construction," permits the cover for units not exposed to weather to be only $\frac{3}{8}$ in for bars smaller than #6.

Precast units must be designed for handling and erection stresses, which may be more severe than those they will be subjected to in service. Normally, inserts are embedded in the concrete for picking up the units. They should be picked up by these inserts, and when set down, they should be supported right side up, in such a manner as not to induce stresses higher than the units would have to resist in service.

For precast beams, girders, joists, columns, slabs, and walls, joints usually are made with cast-in-place concrete. Often, in addition, steel reinforcing projecting from the units to be joined is welded together. (ACI 512.1R, "Suggested Design of Joints and Connections in Precast Structural Concrete," American Concrete Institute (www.aci-int.org).)

8.16 Lift-Slab Construction

A type of precasting used in building construction involves casting floor and roof slabs at or near ground level and lifting them to their final position, hence the name lift-slab construction. It offers many of the advantages of precasting (Art. 8.15) and eliminates many of the storing, handling, and transporting disadvantages. It normally requires fewer joints than other types of precast building systems.

Typically, columns are erected first, but not necessarily for the full height of the building. Near the base of the columns, floor slabs are cast in succession, one atop another, with a parting compound between them to prevent bond. The roof slab is cast last, on top. Usually, the construction is flat plate, and the slabs have uniform thickness; waffle slabs or other types also can be used. Openings are left around the columns, and a steel collar is slid down each column for embedment in every slab. The collar is used for lifting the slab, connecting it to the column, and reinforcing the slab against shear.

To raise the slabs, jacks are set atop the columns and turn threaded rods that pass through the collars and do the lifting. As each slab reaches its final position, it is wedged in place and the collars are welded to the columns.

Design of Concrete Flexural Members

ACI 318, "Building Code Requirements for Reinforced Concrete," specifies that the span of members not integral with supports should be taken as the clear span plus the depth of the member but not greater than the distance center to center of supports. For analysis of continuous frames, spans should be taken center-to-center of supports for determination of bending moments in beams and girders, but moments at the faces of supports may be used in the design of the members. Solid or ribbed slabs integral with supports and with clear spans up to 10 ft may be designed for the clear span.

"Standard Specifications for Highway Bridges" (American Association of State Highway and Transportation Officials) has the same requirements as the ACI Code for spans of simply supported beams and slabs. For slabs continuous over more than two supports, the effective span is the clear span for slabs monolithic with beams or walls (without haunches); the distance between stringer-flange edges plus half the stringer-flange width for slabs supported on steel stringers; clear span plus half the stringer thickness for slabs supported on timber stringers. For rigid frames, the span should be taken as the distance between centers of bearings at the top of the footings. The span of continuous beams should be the clear distance between faces of supports.

Where fillets or haunches make an angle of 45° or more with the axis of a continuous or restrained slab and are built integral with the slab and support, AASHTO requires that the span be measured from the section where the combined depth of the slab and fillet is at least 1.5 times the thickness of slab. The moments at the ends of this span should be used in the slab design, but no portion of the fillet should be considered as adding to the effective depth of the slab.

8.17 Ultimate-Strength Theory for Reinforced-Concrete Beams

For consistent, safe, economical design of beams, their actual load-carrying capacity should be known. The safe load then can be determined by dividing this capacity by a safety factor. Or the design load can be multiplied by the safety factor to indicate what the capacity of the beams should be. It should be noted, however, that under service loads, stresses and deflections may be computed with good approximation on the assumption of a linear stress-strain diagram and a cracked cross section.

ACI 318, "Building Code Requirements for Reinforced Concrete" (American Concrete Institute), provides for design by ultimate-strength theory. Bending moments in members are determined as if the structure were elastic. Ultimate-strength theory is used to design critical sections, those with the largest bending moments, shear, torsion, etc. The ultimate strength of each section is computed, and the section is designed for this capacity.

8.17.1 Stress Redistribution

The ACI Code recognizes that, below ultimate load, a redistribution of stress occurs in continuous beams, frames, and arches. This allows the structure to carry loads higher than those indicated by elastic analysis. The code permits an increase or decrease of up to 10% in the negative moments calculated by elastic theory at the supports of continuous flexural members. But these modified moments must also be used for determining the moments at other sections for the same loading conditions. [The modifications, however, are permissible only for relatively small steel ratios at each support. The steel ratios ρ or $\rho-\rho'$ (see Arts. 8.20, 8.21, and 8.24 to 8.27) should be less than half ρ_b, the steel ratio for balanced conditions (concrete strength equal to steel strength) at ultimate load.]

For example, suppose elastic analysis of a continuous beam indicates a maximum negative moment at a support of $wL^2/12$ and maximum positive moment at midspan of $wL^2/8 - wL^2/12$, or $wL^2/24$. Then, the code permits the negative moment to be decreased to $0.9wL^2/12$, if the positive moment is increased to $wL^2/8 - 0.9wL^2/12$, or $1.2wL^2/24$.

8.17.2 Design Assumptions for Ultimate-Strength Design

Ultimate strength of any section of a reinforced-concrete beam may computed assuming the following:

1. Strain in the concrete is directly proportional to the distance from the neutral axis (Fig. 8.9b).

2. Except in anchorage zones, strain in reinforcing steel equals strain in adjoining concrete.

3. At ultimate strength, maximum strain at the extreme compression surface equals 0.003 in/in.

4. When the reinforcing steel is not stressed to its yield strength f_y, the steel stress is 29,000 ksi times the steel strain, in/in. After the yield strength has been reached, the stress remains constant at f_y, though the strain increases.

5. Tensile strength of the concrete is negligible.

At ultimate strength, concrete stress is not proportional to strain. The actual stress distribution may be represented by an equivalent rectangle, known as the Whitney rectangular stress block, that yields ultimate strengths in agreement with numerous, comprehensive tests (Fig. 8.9c).

The ACI Code recommends that the compressive stress for the equivalent rectangle be taken as $0.85f'_c$, where f'_c is the 28-day compressive strength of the concrete. The stress is assumed constant from the surface of maximum compressive strain over a depth $a = \beta_1 c$, where c is the distance to the neutral axis (Fig. 8.9c). For $f'_c \leq 4000$ psi, $\beta_1 = 0.85$; for greater concrete strengths, β_1 is reduced 0.05 for each 1000 psi in excess of 4000.

Formulas in the ACI Code based on these assumptions usually contain a factor ϕ which is applied to the theoretical ultimate strength of a section, to provide for the possibility that small adverse variations in materials, quality of work, and dimensions, while individually within acceptable tolerances, occasionally may combine, and actual

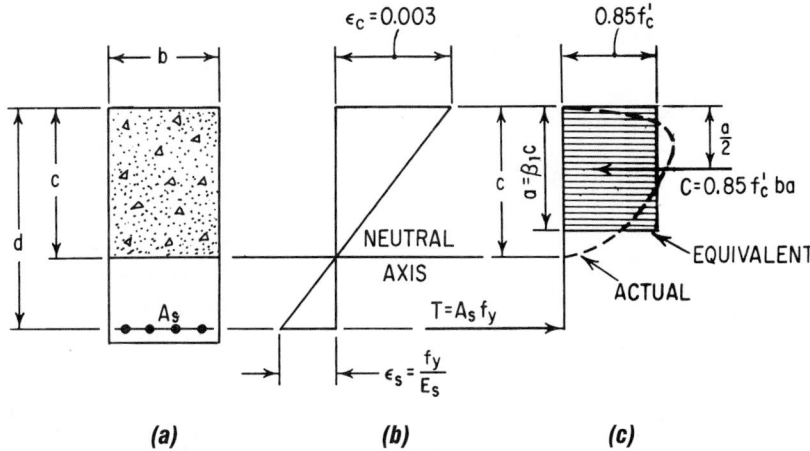

Fig. 8.9 Stresses and strains on a reinforced-concrete beam section: (*a*) At ultimate load, after the section has cracked and only the steel carries tension. (*b*) Strain diagram. (*c*) Actual and assumed compression-stress block.

capacity may be less than that computed. The coefficient ϕ is taken as 0.90 for flexure, 0.85 for shear and torsion, 0.75 for spirally reinforced compression members, and 0.70 for tied compression members. Under certain conditions of load (as the value of the axial load approaches zero) and geometry, the ϕ value for compression members may increase linearly to a maximum value of 0.90.

8.17.3 Crack Control of Flexural Members

Because of the risk of large cracks opening up when reinforcement is subjected to high stresses, the ACI Code recommends specific provisions on crack control through reinforcement distribution limits on spacings:

$$s = \frac{540}{f_s} - 2.5C_c \qquad (8.6)$$

where s = center to center spacing of flexural tension reinforcement (in),

$f_s = 0.6f_y$(ksi),

C_c = clear cover from nearest surface in tension to flexural tension reinforcement (in). These provisions apply to reinforced concrete beams and one-way slabs subject to normal environmental condition.

8.17.4 Required Strength

For combinations of loads, the ACI Code requires that a structure and its members should have the following ultimate strengths (capacities to resist design loads and their related internal moments and forces):

$$U = 1.4D + 1.7L \qquad (8.7a)$$
$$U = 1.4(D + F) \qquad (8.7b)$$
$$U = 1.2(D + F + T) + 1.6(L + H)$$
$$+ 0.5(L_r \text{ or } S \text{ or } R) \qquad (8.7c)$$
$$U = 1.2D + 1.6(L_r \text{ or } S \text{ or } R)$$
$$+ (1.0L \text{ or } 0.8W) \qquad (8.7d)$$
$$U = 1.2D + 1.6W + 0.5L$$
$$+ 1.0(L_r \text{ or } S \text{ or } R) \qquad (8.7e)$$
$$U = 1.2D + 1.0E + 1.0L + 0.2S \qquad (8.7f)$$
$$U = 0.9D + 1.6W + 1.6H \qquad (8.7g)$$
$$U = 0.9D + 1.0E + 1.6H \qquad (8.7h)$$

where D = dead load; E = earthquake load; F = lateral fluid pressure load and maximum height;

H = load due to the weight and lateral pressure of soil and water in soil;

L = live load; L_r = roof load; R = rain load; S = snow load;

T = self-straining force such as creep, shrinkage, and temperature effects;

W = wind load

For ultimate-strength loads (load-factor method) for bridges, see Art. 17.4.

Although structures may be designed by ultimate-strength theory, it is not anticipated that service loads will be substantially exceeded. Hence, deflections that will be of concern to the designer are those that occur under service loads. These deflections may be computed by working-stress theory. (See Art. 8.18.)

8.17.5 Deep Members

Due to the nonlinearity of strain distribution and the possibility of lateral buckling, deep flexural members must be given special consideration. The ACI Code considers members with clear span, ln, equal to or less than 4 times the overall member depth as deep members. The ACI Code provides special shear design requirements and minimum requirements for both horizontal and vertical reinforcement for such members.

8.18 Working-Stress Theory for Reinforced-Concrete Beams

Stress distribution in a reinforced-concrete beam under service loads is different from that at ulti- mate strength (Art. 8.17). Knowledge of this stress distribution is desirable for many reasons, including the requirements of some design codes that specified working stresses in steel and concrete not be exceeded.

Working stresses in reinforced-concrete beams are computed from the following assumptions:

1. Longitudinal stresses and strains vary with distance from the neutral axis (Fig. 8.10c and d); that is, plane sections remain plane after bending. (Strains in longitudinal reinforcing steel and adjoining concrete are equal.)

2. The concrete does not develop any tension. (Concrete cracks under tension.)

3. Except in anchorage zones, strain in reinforcing steel equals strain in adjoining concrete. But because of creep, strain in compressive steel in beams may be taken as half that in the adjoining concrete.

4. The modular ratio $n = E_s/E_c$ is constant. E_s is the modulus of elasticity of the reinforcing steel and E_c of the concrete.

Table 8.9 lists allowable stresses that may be used for flexure. For other than the flexural stresses in Table 8.9a, allowable or maximum stresses to be used in design are stated as a percentage of the values given for ultimate-strength design. See, for example, service loads in Table 8.9b.

Allowable stresses may be increased one-third when wind or earthquake forces are combined with other loads, but the capacity of the resulting

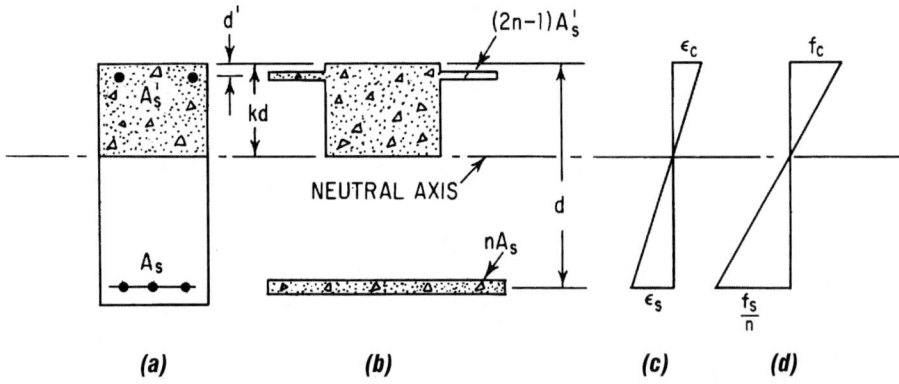

Fig. 8.10 Typical cracked cross section of a reinforced concrete beam: (a) Only the reinforcing steel is effective in tension. (b) Section treated as an all-concrete transformed section. In working-stress design, linear distribution is assumed for (c) strains and (d) stresses.

Table 8.9 Allowable Stresses for Concrete Flexural Members

(a)

Type of Stress	Buildings	Bridges
Compression in extreme compression surface	$0.45f_c^*$	$0.4f_c^*$
Tension in reinforcement		
Grade 40 or 50 steel	20 ksi	20 ksi
Grade 60 or higher yield strength	24 ksi	24 ksi

(b)

Type of Member and Stress	Allowable Stresses or Capacity, %, of Ultimate (Nominal)
Compression members, walls	40
Shear or tension in beams, joists, walls, one-way slabs	55
Shear or tension in two-way slabs, footings	50
Bearing in concrete	35

* f_c' is the 28-day compressive strength of the concrete.

section should not be less than that required for dead plus live loads.

Other equivalency factors are also given in terms of ultimate-strength values. Thus, the predominant design procedure is the ultimate-strength method, but for reasons of background and historical significance and because the working-stress design method is sometimes preferred for bridges and certain foundation and retaining-wall design, examples of working-stress design procedure are presented in Arts. 8.21, 8.25, and 8.27.

Transformed Section ▪ According to the working-stress theory for reinforced-concrete beams, strains in reinforcing steel and adjoining concrete are equal. Hence f_s, the stress in the steel, is n times f_c, the stress in the concrete, where n is the ratio of modulus of elasticity of the steel E_s to that of the concrete E_c. The total force acting on the steel then equals $(nA_s)f_c$. This indicates that the steel area can be replaced in stress calculations by a concrete area n times as large.

The transformed section of a concrete beam is one in which the reinforcing has been replaced by an equivalent area of concrete (Fig. 8.10b). (In doubly reinforced beams and slabs, an effective modular ratio of $2n$ should be used to transform the compression reinforcement, to account for the effects of creep and nonlinearity of the stress-strain diagram for concrete. But the computed stress should not exceed the allowable tensile stress.) Since stresses and strains are assumed to vary with distance from the neutral axis, conventional elastic theory for homogeneous beams holds for the transformed section. Section properties, such as location of neutral axis, moment of inertia, and section modulus S, can be computed in the usual way, and stresses can be found from the flexure formula $f = M/S$, where M is the bending moment.

8.19 Deflection Computations and Criteria for Concrete Beams

The assumptions of working-stress theory (Art. 8.18) may also be used for computing deflections under service loads; that is, elastic-theory deflection formulas may be used for reinforced-concrete beams (Art. 6.32). In these formulas, the **effective moment of inertia** I_e is given by Eq. (8.8).

$$I_e = \left(\frac{M_{cr}}{M_a}\right)^3 I_g + \left[1 - \left(\frac{M_{cr}}{M_a}\right)^3\right]I_{cr} \leq I_g \quad (8.8)$$

where I_g = moment of inertia of the gross concrete section

M_{cr} = cracking moment

M_a = moment for which deflection is being computed

I_{cr} = cracked concrete (transformed) section

If y_t is taken as the distance from the centroidal axis of the gross section, neglecting the reinforcement, to the extreme surface in tension, the cracking moment may be computed from

$$M_{cr} = \frac{f_r I_g}{y_t} \quad (8.9)$$

with the modulus of rupture of the concrete $f_r = 7.5\sqrt{f_c'}$ Eq. (8.8) takes into account the variation of the moment of inertia of a concrete section based on whether the section is cracked or uncracked. The modulus of elasticity of the concrete E_c may be computed from Eq. (8.3) in Art. 8.1.

The deflections thus calculated are those assumed to occur immediately on application of load. The total long-term deflection is

$$\Delta_{LT} = \Delta_L + \lambda_\infty \Delta_D + \lambda_t \Delta_{LS} \qquad (8.10)$$

where Δ_L = initial live load deflection,

Δ_D = initial dead load deflection,

Δ_{LS} = initial sustained live-load deflection,

λ_∞ = time dependent multiplier for infinite duration of sustained load,

λ_t = time dependent multiplier for limited load duration.

Deflection Limitations ▪ The ACI Code recommends the following limits on deflections in buildings:

For roofs not supporting and not attached to nonstructural elements likely to be damaged by large deflections, maximum immediate deflection under live load should not exceed $L/180$, where L is the span of beam or slab.

For floors not supporting partitions and not attached to nonstructural elements, the maximum immediate deflection under live load should not exceed $L/360$.

For a floor or roof construction intended to support or to be attached to partitions or other construction likely to be damaged by large deflections of the support, the allowable limit for the sum of immediate deflection due to live loads and the additional deflection due to shrinkage and creep under all sustained loads should not exceed $L/480$. If the construction is not likely to be damaged by large deflections, the deflection limitation may be increased to $L/240$. But tolerances should be established and adequate measures should be taken to prevent damage to supported or nonstructural elements resulting from the deflections of structural members.

8.20 Ultimate-Strength Design of Rectangular Beams with Tension Reinforcement Only

Generally, the area A_s of tension reinforcement in a reinforced-concrete beam is represented by the ratio $\rho = A_s/bd$, where b is the beam width and d the distance from extreme compression surface to the centroid of tension reinforcement (Fig. 8.11a). At ultimate strength, the steel at a critical section of the beam will be at its yield strength f_y if the concrete does not fail in compression first (Art. 8.17). Total tension in the steel then will be $A_s f_y = \rho f_y bd$. It will be opposed, according to Fig. 8.11c, by an equal compressive force, $0.85 f_c' ba = 0.85 f_c' b \beta_1 c$, where f_c' is the 28-day strength of the concrete, ksi, a the depth of the equivalent rectangular stress distribution, c the distance from the extreme compression surface to the neutral axis, and β_1 a constant (see Art. 8.17). Equating the compression and tension at the critical section yields

$$c = \frac{\rho f_y}{0.85 \beta_1 f_c'} d \qquad (8.11)$$

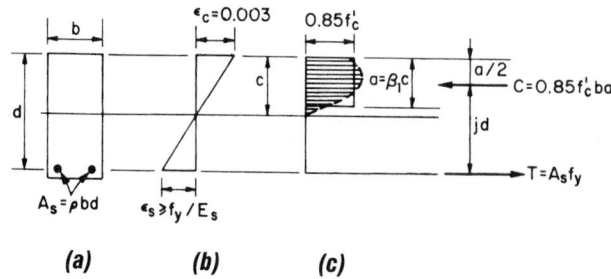

Fig. 8.11 Rectangular concrete beam reinforced for tension only: (*a*) Beam cross section. (*b*) Linear distribution assumed for strains at ultimate load. (*c*) Equivalent rectangular stress block assumed for compression stresses at ultimate load.

The criterion for compression failure is that the maximum strain in the concrete equals 0.003 in/in. In that case

$$c = \frac{0.003}{f_s/E_s + 0.003} d \qquad (8.12)$$

where f_s = steel stress, ksi

E_s = modulus of elasticity of steel

= 29,000 ksi

Table 8.10 lists the nominal diameters, weights, and cross-sectional areas of standard steel reinforcing bars.

8.20.1 Balanced Reinforcing

Under balanced conditions, the concrete will reach its maximum strain of 0.003 when the steel reaches its yield strength f_y. Then, c as given by Eq. (8.11) will equal c as given by Eq. (8.12) since c determines the location of the neutral axis. This determines the steel ratio for balanced conditions:

$$\rho_b = \frac{0.85\beta_1 f_c'}{f_y} \frac{87,000}{87,000 + f_y} \qquad (8.13)$$

8.20.2 Reinforcing Limitations

All structures are designed to collapse not suddenly but by gradual deformation when overloaded. This condition is referred to as a ductile mode of failure. To achieve this end in concrete, the reinforcement should yield before the concrete crushes. This will occur if the quantity of tensile reinforcement is less than the critical percentage determined by ultimate-strength theory [Eq. 8.13]. The ACI Code, to avoid compression failures, limits the steel ratio ρ to a maximum of $0.75\rho_b$. The Code also requires that ρ for positive-moment reinforcement be at least $200/f_y$.

8.20.3 Moment Capacity

For such underreinforced beams, the nominal moment strength is

$$M_n = [bd^2 f_c' w(1 - 0.59w)]$$

$$= \left[A_s f_y \left(d - \frac{a}{2} \right) \right] \qquad (8.14)$$

where $w = \rho f_y / f_c'$

$a = A_s f_y / 0.85 f_c' b$

The design moment strength, ϕM_n, must be equal to or greater than the external factored moment, M_u.

8.20.4 Shear Reinforcement

The nominal shear strength, V_n, of a section of a beam equals the sum of the nominal shear strength provided by the concrete, V_c, and the nominal shear strength provided by the reinforcement, V_s;

Table 8.10 Areas of Groups of Standard Bars, in^2

Bar No.	Diam, in	Weight, lb per ft	Number of Bars								
			1	2	3	4	5	6	7	8	9
2	0.250	0.167	0.05	0.10	0.15	0.20	0.25	0.30	0.35	0.40	0.45
3	0.375	0.376	0.11	0.22	0.33	0.44	0.55	0.66	0.77	0.88	0.99
4	0.500	0.668	0.20	0.39	0.58	0.78	0.98	1.18	1.37	1.57	1.77
5	0.625	1.043	0.31	0.61	0.91	1.23	1.53	1.84	2.15	2.45	2.76
6	0.750	1.502	0.44	0.88	1.32	1.77	2.21	2.65	3.09	3.53	3.98
7	0.875	2.044	0.60	1.20	1.80	2.41	3.01	3.61	4.21	4.81	5.41
8	1.000	2.670	0.79	1.57	2.35	3.14	3.93	4.71	5.50	6.28	7.07
9	1.128	3.400	1.00	2.00	3.00	4.00	5.00	6.00	7.00	8.00	9.00
10	1.270	4.303	1.27	2.53	3.79	5.06	6.33	7.59	8.86	10.12	11.39
11	1.410	5.313	1.56	3.12	4.68	6.25	7.81	9.37	10.94	12.50	14.06
14	1.693	7.650	2.25	4.50	6.75	9.00	11.25	13.50	15.75	18.00	20.25
18	2.257	13.600	4.00	8.00	12.00	16.00	20.00	24.00	28.00	32.00	36.00

that is, $V_n = V_c + V_s$. The factored shear force, V_n, on a section should not exceed

$$\phi V_n = \phi(V_c + V_s) \qquad (8.15)$$

where ϕ = strength reduction factor (0.75 for shear and torsion). Except for brackets and other short cantilevers, the section for maximum shear may be taken at a distance equal to d from the face of the support.

The shear V_c carried by the concrete alone should not exceed $2\sqrt{f_c'}b_w d$ where b_w is the width of the beam web and d the depth from the extreme compression fiber to centroid of longitudinal tension reinforcement. (For members subject to shear and flexure only, the maximum for V_c may be taken as

$$V_c = \left(1.9\sqrt{f_c'} + 2500\rho_w \frac{V_u d}{M_u}\right)b_w d \qquad (8.16)$$

$$\leq 3.5\sqrt{f_c'}b_w d$$

where $\rho_w = A_s/b_w d$ and V_u and M_u are the shear and bending moment, respectively, at the section considered, but M_u should not be less than $V_u d$.)

When V_u is larger than ϕV_c, the excess shear will have to be resisted by web reinforcement. In general, this reinforcement should be stirrups perpendicular to the axis of the member (Fig. 8.12). Shear or torsion reinforcement should extend the full depth d of the member and should be adequately anchored at both ends to develop the design yield strength of the reinforcement. An alternative is to incorporate welded-wire fabric with wires perpendicular to the axis of the member. In members without prestressing, however, the stirrups may be inclined, as long as the angle is at least 45° with the axis of the member. As an alternative, longitudinal reinforcing bars may be bent up at an angle of 30° or more with the axis, or spirals may be used. Spacing should be such that

every 45° line, representing a potential crack and extending from middepth $d/2$ to the longitudinal tension bars, should be crossed by at least one line of reinforcing.

The area of steel required in vertical stirrups, in^2 per stirrup, with a spacing s, in, is

$$A_v = \frac{V_s s}{f_y d} \qquad (8.17)$$

where f_y = yield strength of the shear reinforcement. A_v is the area of the stirrups cut by a horizontal plane. V_s should not exceed $8\sqrt{f_c'}b_w d$ in sections with web reinforcement, nor should f_y exceed 60 ksi. Where shear reinforcement is required and is placed perpendicular to the axis of the member, it should not be spaced farther apart than $0.5d$, nor more than 24 in c to c. When V_s exceeds $4\sqrt{f_c'}b_w d$, however, the maximum spacing should be limited to $0.25d$.

Alternatively, for practical design, Eq. (8.17) can be transformed into Eq. (8.18) to indicate the stirrup spacing s for the design shear V_u, stirrup area A_v, and geometry of the member b_w and d:

$$s = \frac{A_v \phi f_y d}{V_u - 2\phi\sqrt{f_c'}b_w d} \qquad (8.18)$$

The area required when a single bar or a single group of parallel bars are all bent up at the same distance from the support at α angle with the longitudinal axis of the member is

$$A_v = \frac{V_s}{f_y \sin \alpha} \qquad (8.19)$$

in which V_s should not exceed $3\sqrt{f_c'}b_w d$. A_v is the area cut by a plane normal to the axis of the bars. The area required when a series of such bars are bent up at different distances from the support or when inclined stirrups are used is

$$A_v = \frac{V_s s}{(\sin \alpha + \cos \alpha)f_y d} \qquad (8.20)$$

A minimum area of shear reinforcement is required in all members, except slabs, footings, and joists or where V_u exceeds $0.5V_c$.

8.20.5 Torsion Reinforcement

Types of stresses induced by torsion and reinforcement requirements for members subjected to torsion are discussed in Art. 8.28.

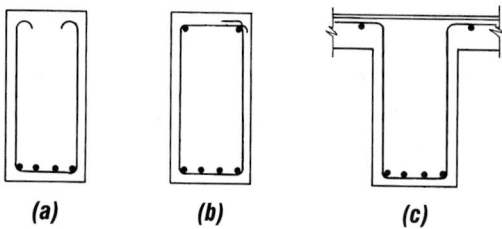

| (a) | (b) | (c) |

Fig. 8.12 Typical stirrups in a concrete beam.

8.20.6 Development of Tensile Reinforcement

To prevent bond failure or splitting, the calculated stress in any bar at any section must be developed on each side of the section by adequate embedment length, end anchorage, or hooks. The critical sections for development of reinforcement in flexural members are at points of maximum stress and at points within the span where adjacent reinforcement terminates. See Art. 8.22.

At least one-third of the positive-moment reinforcement in simple beams and one-fourth of the positive-moment reinforcement in continuous beams should extend along the same face of the member into the support, in both cases, at least 6 in into the support. At simple supports and at points of inflection, the diameter of the reinforcement should be limited to a diameter such that the development length l_d defined in Art. 8.12.5 satisfies

$$l_d \leq \frac{M_n}{V_u} + l_a \qquad (8.21)$$

where M_n = nominal moment strength with all reinforcing steel at section stressed to f_y

V_u = factored shear at section

l_a = additional embedment length beyond inflection point or center of support

At an inflection point, l_a is limited to a maximum of d, the depth of the centroid of the reinforcement, or 12 times the reinforcement diameter.

Negative-moment reinforcement should have an embedment length into the span to develop the calculated tension in the bar, or a length equal to the effective depth of the member, or 12 bar diameters, whichever is greatest. At least one-third of the total negative reinforcement should have an embedment length beyond the point of inflection not less than the effective depth of the member, or 12 bar diameters, or one-sixteenth of the clear span, whichever is greatest.

8.20.7 Hooks on Bars

When straight embedment of reinforcing bars in tension is inadequate to provide the required development lengths of the bars as specified in Art. 8.12.5, the bar ends may be bent into standard

90° and 180° hooks (Table 8.11) to provide additional development. The basic development length for a hooked bar with $f_y = 60$ ksi is defined as

$$l_{dh} = \left(\frac{0.02\beta\lambda f_y}{f_c'}\right) d_b \qquad (8.22)$$

where d_b is the bar diameter, in, and f_c' is the 28-day compressive strength of the concrete, psi. $\beta = 1.2$ for epoxy coated reinforcement and $\lambda = 1.3$ for lightweight aggregate concrete. For all other cases, β and λ shall be taken as 1.0. Figure 8.13 illustrates embedment lengths for standard hooks.

A footnote to Table 8.12 indicates some of the factors by which basic development length should be multiplied for values of f_y other than 60 ksi and for excess reinforcement. For bars sizes up to No. 11, side cover (normal to the plane of the hook) of at least $2\frac{1}{2}$ in, and for a 90° hook, cover on the bar extension of 2 in or more, the modification may be taken as 0.7. Also, for bars sizes up to No. 11 with the hook enclosed vertically or horizontally and enclosed within ties or stirrup-ties spaced along the full development length at $3d_b$ or less, the modification factor may be taken as 0.8.

Hooks should not be considered effective in adding to the compressive resistance of reinforcement. Thus, hooks should not be used on footing dowels. Instead, when depth of footing is less than that required by large-size bars, the designer should substitute smaller-diameter bars with equivalent area and lesser embedment length. It may be possible sometimes to increase the footing depth where large-diameter dowel reinforcement is used so that footing dowels can have the proper embedment length. Footing dowels need only transfer the excess load above that transmitted in bearing and therefore may be bars with areas different from those required for compression design for the first column lift.

(P. F. Rice and E. S. Hoffman, "Structural Design Guide to the ACI Building Code," Van Nostrand Reinhold Company, New York; "CRSI Handbook," Concrete Reinforcing Steel Institute, Chicago, Ill.; ACI SP-17, "Design Handbook in Accordance with the Strength Design Method of ACI 318-77 (www. aci-int.org)," American Concrete Institute; G. Winter and A. H. Nilson, "Design of Concrete Structures," McGraw-Hill Book Company, New York (books.mcgraw-hill.com).)

Table 8.11 Standard Hooks*

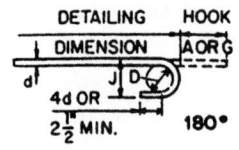

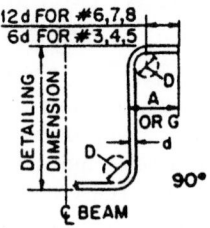

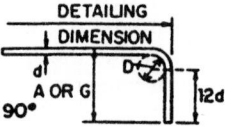

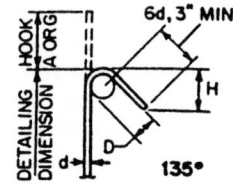

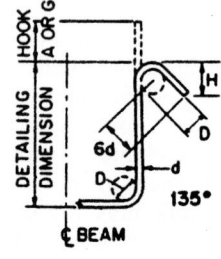

Stirrups (ties similar)

Recommended End Hooks—All Grades, in or ft-in

| Bar Size No. | 180° Hooks | | | 90° Hooks |
	$D^\dagger$	A or G	J	A or G
3	$2\frac{1}{4}$	5	3	6
4	3	6	4	8
5	$3\frac{3}{4}$	7	5	10
6	$4\frac{1}{2}$	8	6	1–0
7	$5\frac{1}{4}$	10	7	1–2
8	6	11	8	1–4
9	$9\frac{1}{2}$	1–3	$11\frac{3}{4}$	1–7
10	$10\frac{3}{4}$	1–5	$1–1\frac{1}{4}$	1–10
11	12	1–7	$1–2\frac{3}{4}$	2–0
14	$18\frac{1}{4}$	2–3	$1–9\frac{3}{4}$	2–7
18	24	3–0	$2–4\frac{1}{4}$	3–5

$^\dagger D$ = finished bend diameter, in.

Stirrup and Tie Hook Dimensions, in—Grades
40–50–60 ksi

| Bar Size No. | D, in | 90° Hook | 135° Hook | |
		Hook A or G	Hook A or G	H, Approx.
3	$1\frac{1}{2}$	4	4	$2\frac{1}{2}$
4	2	$4\frac{1}{2}$	$4\frac{1}{2}$	3
5	$2\frac{1}{2}$	6	$5\frac{1}{2}$	$3\frac{3}{4}$
6	$4\frac{1}{2}$	1–0	8	$4\frac{1}{2}$
7	$5\frac{1}{4}$	1–2	9	$5\frac{1}{4}$
8	6	1–4	$10\frac{1}{2}$	6

135° Seismic Stirrup/Tie Hook
Dimensions (Ties Similar) in—Grades
40–50–60 ksi

| Bar Size No. | D, in | 135° Hook | |
		Hook A or G	H, Approx.
3	$1\frac{1}{2}$	$4\frac{1}{4}$	3
4	2	$4\frac{1}{2}$	3
5	$2\frac{1}{2}$	$5\frac{1}{2}$	$3\frac{3}{4}$
6	$4\frac{1}{2}$	8	$4\frac{1}{2}$
7	$5\frac{1}{4}$	9	$5\frac{1}{4}$
8	6	$10\frac{1}{2}$	6

* Notes:

1. All specific sizes recommended by CRSI in this table meet minimum requirements of ACI 318.

2. 180° hook J dimension (sizes 10, 11, 14, and 18) and A or G dimension (Nos. 14 and 18) have been revised to reflect recent test research using ASTM/ACI bend-test criteria as a minimum.

3. Tables for stirrup and tie hook dimensions have been expanded to include sizes 6, 7, and 8, to reflect current design practices. Courtesy of the Concrete Reinforcing Steel Institute.

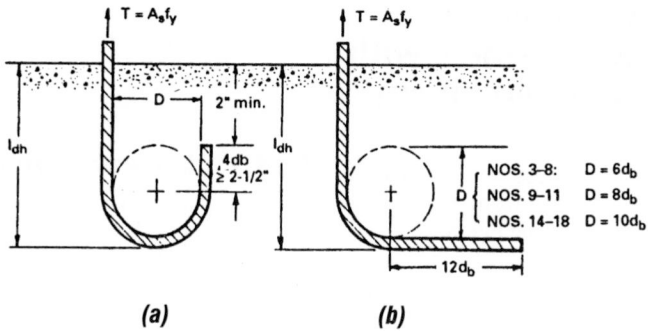

Fig. 8.13 Embedment lengths for 90° and 180° hooks.

Table 8.12 Minimum Embedment Lengths for Hooks on Steel Reinforcement in Tension

a. Embedment Lengths l_{dh}, in, for Standard End Hooks on Grade 60 Bars in Normal-Weight Concrete*

Bar Size No.	Concrete Compressive Strength f'_c, psi					
	3000	4000	5000	6000	7000	8000
3	6	6	6	6	6	6
4	8	7	6[†]	6[†]	6[†]	6[†]
5	10	9	8	7	7	6[†]
6	12	10	9	8	8	7[†]
7	14	12	11	10	9	9
8	16	14	12	11	10	10
9	18	15	14	13	12	11
10	20	17	15	14	13	12[†]
11	22	29	17	16	14	14[†]
14	37	32	29	27	25	23
18	50	43	39	35	33	31

b. Embedment Lengths, in, to Provide 2-in Concrete Cover over Tail of Standard 180° End Hooks

No. 3	No. 4	No. 5	No. 6	No. 7	No. 8	No. 9	No. 10	No. 11	No. 14	No. 18
6	7	7	8	9	10	12	14	15	20	25

* Embedment length for 90° and 180° standard hooks is illustrated in Fig. 8.13. Details of standard hooks are given in Table 8.11. Side cover required is a minimum of $2\frac{1}{2}$ in. End cover required for 90° hooks is a minimum of 2 in. To obtain embedment lengths for grades of steel different from Grade 60, multiply l_{dh} given in Table 8.12 by $f_y/60$. If reinforcement exceeds that required, multiply l_{dh} by the ratio of area required to that provided.

† For 180° hooks at right angles to exposed surfaces, obtain l_{dh} from Table 8.12b to provide 2-in minimum cover to tail (Fig. 8.13a).

8.21 Alternate Design of Rectangular Beams with Tension Reinforcement Only

From the assumption that stress varies across a beam section with the distance from the neutral axis (Art. 8.18), it follows that (see Fig. 8.14)

$$\frac{nf_c}{f_s} = \frac{k}{1-k} \qquad (8.23)$$

where n = modular ratio E_s/E_c

E_s = modulus of elasticity of steel reinforcement, ksi

E_c = modulus of elasticity of concrete, ksi

f_c = compressive stress in extreme surface of concrete, ksi

f_s = stress in steel, ksi

kd = distance from extreme compression surface to neutral axis, in

d = distance from extreme compression to centroid of reinforcement, in

When the steel ratio $\rho = A_s/bd$, where A_s = area of tension reinforcement, in², and b = beam width, in, is known, k can be computed from

$$k = \sqrt{2n\rho + (n\rho)^2} - n\rho \qquad (8.24)$$

Wherever positive-moment steel is required, ρ should be at least $200/f_y$, where f_y is the steel yield stress. The distance jd between the centroid of compression and the centroid of tension, in, can be obtained from Fig. 8.14:

$$j = 1 - \frac{k}{3} \qquad (8.25)$$

8.21.1 Allowable Bending Moment

The moment resistance of the concrete, in-kips, is

$$M_c = \frac{1}{2} f_c k j b d^2 = K_c b d^2 \qquad (8.26)$$

where $K_c = \frac{1}{2} f_c kj$. The moment resistance of the steel reinforcement is

$$M_s = f_s A_s j d = f_s \rho j b d^2 = K_s b d^2 \qquad (8.27)$$

where $K_s = f_s \rho j$. Allowable stresses are given in Art. 8.18. Table 8.10 lists nominal diameters, weights, and cross-sectional areas of standard steel reinforcing bars.

8.21.2 Allowable Shear

The nominal unit shear stress acting on a section with shear V is

$$v = \frac{V}{bd} \qquad (8.28)$$

Allowable shear stresses are 55% of those for ultimate-strength design (Art. 8.20.4). Otherwise, designs for shear by the working-stress and ultimate-strength methods are the same. Except for brackets and other short cantilevers, the section for maximum shear may be taken at a distance d from the face of the support. In working-stress design, the shear stress v_c carried by the concrete

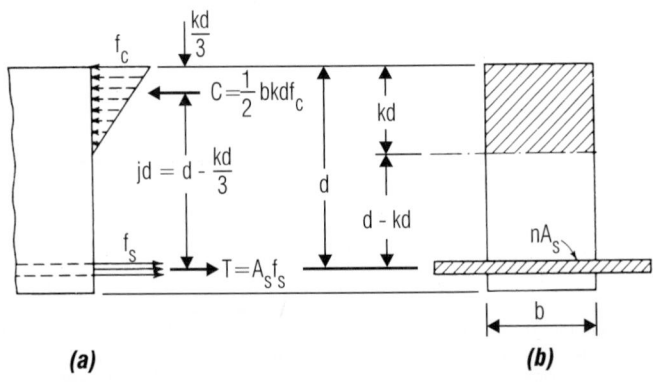

Fig. 8.14 Rectangular concrete beam reinforced for tension only: (*a*) In working-stress design, a linear distribution is assumed for compression stresses. (*b*) Transformed all-concrete section.

alone should not exceed $1.1\sqrt{f_c'}$. (As an alternative, the maximum for v_c may be taken as $\sqrt{f_c'} + 1300\rho Vd/M$, with a maximum of $1.9\sqrt{f_c'}$, f_c' is the 28-day compressive strength of the concrete, psi, and M is the bending moment at the section but should not be less than Vd.)

At cross sections where the torsional stress v_t exceeds $0.825\sqrt{f_c'}$, v_c should not exceed

$$v_c = \frac{1.1\sqrt{f_c'}}{\sqrt{1 + (v_t/1.2v)^2}} \tag{8.29}$$

The excess shear $v - v_c$ should not exceed $4.4\sqrt{f_c'}$ in sections with web reinforcement. Stirrups and bent bars should be capable of resisting the excess shear $V' = V - v_c bd$.

The area required in the legs of a vertical stirrup, in^2, is

$$A_v = \frac{V's}{f_v d} \tag{8.30}$$

where s = spacing of stirrups, in

f_v = allowable stress in stirrup steel, psi (see Art. 8.21)

For a single bent bar or a single group of parallel bars all bent at an angle α with the longitudinal axis at the same distance from the support, the required area is

$$A_v = \frac{V'}{f_v \sin \alpha} \tag{8.31}$$

For inclined stirrups and groups of bars bent up at different distances from the support, the required area is

$$A_v = \frac{V's}{f_v d(\sin \alpha + \cos \alpha)} \tag{8.32}$$

Where shear reinforcing is required and the torsional moment T exceeds the value calculated from Eq. (8.64), the minimum area of shear reinforcement provided should be that given by Eq. (8.60).

8.21.3 Allowable Torsion

Torsion effects should be considered whenever the torsion T due to service loads exceeds the torsion capacity of the concrete T_c given by Eq. (8.63). For working-stress design for torsion, see Art. 8.28.2.

8.21.4 Development of Reinforcement

To prevent bond failure or splitting, the calculated stress in reinforcement at any section should be developed on each side of that section by adequate embedment length, end anchorage, or, for tension only, hooks. Requirements are the same as those given for ultimate-strength design in Art. 8.20.6. Embedment length required at simple supports and inflection points can be computed from Eq. (8.25) by substituting double the computed shears for V_u. In computation of M_t, the moment arm, $d - a/2$ may be taken as $0.85d$ (Fig. 8.12). See also Art. 8.22.

8.22 Bar Cutoffs and Bend Points

It is common practice to stop or bend main reinforcement in beams and slabs where it is no longer required. But tensile steel should never be discontinued exactly at the theoretical cutoff or bend points. It is necessary to resist tensile forces in the reinforcement through embedment beyond those points.

All reinforcement should extend beyond the point at which it is no longer needed to resist flexure for a distance equal to the effective depth of the member or 12 bar diameters, whichever is greater except at supports of simple spans and at free end of a cantilever. Lesser extensions, however, may be used at supports of a simple span and at the free end of a cantilever. See Art. 8.20.6 for embedment requirements at simple supports and inflection points and for termination of negative-moment bars. Continuing reinforcement should have an embedment length beyond the point where bent or terminated reinforcement is no longer required to resist flexure. The embedment should be at least as long as the development length l_d defined in Art. 8.12.5.

Flexural reinforcement should not be terminated in a tension zone unless one of the following conditions is satisfied:

1. Shear is less than two-thirds that normally permitted, including allowance for shear reinforcement, if any.

2. Continuing bars provide double the area required for flexure at the cutoff, and the shear

does not exceed three-quarters of that permitted (No. 11 bar or smaller).

3. Stirrups in excess of those normally required are provided each way from the cutoff for a distance equal to 75% of the effective depth of the member. Area and spacing of the excess stirrups should be such that

$$A_v \geq 60 \frac{b_w s}{f_y} \tag{8.33}$$

where A_v = stirrup cross-sectional area, in^2

b_w = web width, in

s = stirrup spacing, in

f_y = yield strength of stirrup steel, psi

Stirrup spacing s should not exceed $d/8\beta_b$, where β_b is the ratio of the area of bars cut off to the total area of tension bars at the section and d is the effective depth of the member.

The location of theoretical cutoffs or bend points may usually be determined from bending moments since the steel stresses are approximately proportional to them. The bars generally are discontinued in groups or pairs. So, for example, if one-third the bars are to be bent up, the theoretical bend-up point lies at the section where

the bending moment is two-thirds the maximum moment. The point may be found analytically or graphically.

(G. Winter and A. H. Nilson, "Design of Concrete Structures," McGraw-Hill Book Company, New York (books.mcgraw-hill.com); P. F. Rice and E. S. Hoffman, "Structural Design Guide to the ACI Building Code," Van Nostrand Reinhold Company, New York, ACI 315, "Manual of Standard Practice for Detailing Reinforced Concrete Structures," American Concrete Institute (www.aci-int.org).)

8.23 One-Way Slabs

If a slab supported on beams or walls spans a distance in one direction more than twice that in the perpendicular direction, so much of the load is carried on the short span that the slab may reasonably be assumed to be carrying all the load in that direction. Such a slab is called a one-way slab.

Generally, a one-way slab is designed by selecting a 12-in-wide strip parallel to the short direction and treating it as a rectangular beam. Reinforcing steel usually is spaced uniformly in both directions (Table 8.13). In addition to the main reinforcing in the short span, steel should be provided in the long direction to distribute

Table 8.13 Areas of Bars in Slabs, in^2/ft of Slab

Spacing, in	Bar No.								
	3	4	5	6	7	8	9	10	11
3	0.44	0.78	1.23	1.77	2.40	3.14	4.00	5.06	6.25
3½	0.38	0.67	1.05	1.51	2.06	2.69	3.43	4.34	5.36
4	0.33	0.59	0.92	1.32	1.80	2.36	3.00	3.80	4.68
4½	0.29	0.52	0.82	1.18	1.60	2.09	2.67	3.37	4.17
5	0.26	0.47	0.74	1.06	1.44	1.88	2.40	3.04	3.75
5½	0.24	0.43	0.67	0.96	1.31	1.71	2.18	2.76	3.41
6	0.22	0.39	0.61	0.88	1.20	1.57	2.00	2.53	3.12
6½	0.20	0.36	0.57	0.82	1.11	1.45	1.85	2.34	2.89
7	0.19	0.34	0.53	0.76	1.03	1.35	1.71	2.17	2.68
7½	0.18	0.31	0.49	0.71	0.96	1.26	1.60	2.02	2.50
8	0.17	0.29	0.46	0.66	0.90	1.18	1.50	1.89	2.34
9	0.15	0.26	0.41	0.59	0.80	1.05	1.33	1.69	2.08
10	0.13	0.24	0.37	0.53	0.72	0.94	1.20	1.52	1.87
12	0.11	0.20	0.31	0.44	0.60	0.79	1.00	1.27	1.56

concentrated loads and resist shrinkage and thermal stresses. The bars or wires should not be spaced farther apart than five times the slab thickness for shrinkage and temperature steel and three times the slab thickness for main reinforcing. Spacing in either direction should not exceed 18 in.

For shrinkage and temperature stresses, ACI 318, "Building Code Requirements for Reinforced Concrete," requires the following ratio of reinforcement to gross concrete areas, in²/ft: deformed bars with yield strength less than 60 ksi, 0.0020; deformed bars with 60 ksi yield strength or welded-wire fabric with wires not more than 12 in apart, 0.0018. For highway bridge slabs, "Standard Specifications for Highway Bridges" (American Association of State Highway and Transportation Officials) requires reinforcing steel in the bottoms of all slabs transverse to the main reinforcement for lateral distribution of wheel loads. The area of the distribution steel should be at least the following percentages of the main steel required for positive moment, where S is the effective span, ft. When main steel is parallel to traffic, $100/\sqrt{S}$ with a maximum of 50%; when the main steel is perpendicular to traffic, $200/\sqrt{S}$, with a maximum of 67%.

To control deflections, the ACI Code sets limitations on slab thickness unless deflections are computed and determined to be acceptable (Art. 8.19). Otherwise, thickness of one-way slabs must be at least $L/20$ for simply supported slabs; $L/24$ for slabs with one end continuous; $L/28$ for slabs with both ends continuous; and $L/10$ for cantilevers; where L is the span, in.

8.24 Ultimate-Strength Design of Rectangular Beams with Compression Bars

The steel ratio ρ_b for balanced conditions at ultimate strength of a rectangular beam is given by Eq. (8.13) in Art. 8.20.1. When the tensile steel ratio ρ exceeds $0.75\rho_b$, compression reinforcement should be used. When ρ is equal to or less than $0.75\rho_b$, the strength of the beam may be approximated by Eq. (8.14), disregarding any

compression bars that may be present, since the strength of the beam will usually be controlled by yielding of the tensile steel.

The bending-moment capacity of a rectangular beam with both tension and compression steel is

$$M_u = 0.90\left[(A_s - A'_s)f_y\left(d - \frac{a}{2}\right)\right.$$
$$\left. + A'_s f_y(d - d')\right] \quad (8.34)$$

where a = depth of equivalent rectangular compressive stress distribution

$= (A_s - A'_s)f_y/f'_c b$

b = width of beam, in

d = distance from extreme compression surface to centroid of tensile steel, in

d' = distance from extreme compression surface to centroid of compressive steel, in

A_s = area of tensile steel, in²

A'_s = area of compressive steel, in²

f_y = yield strength of steel, ksi

f'_c = 28-day strength of concrete, ksi

Equation (8.35) is valid only when the compressive steel reaches f_y. This occurs when

$$(\rho - \rho') \geq 0.85\beta_1 \frac{f'_c d'}{f_y d} \frac{87,000}{87,000 - f_y} \quad (8.35)$$

where $\rho = A_s/bd$, $\rho' = A'_s/bd$, and β_1 is a constant defined in Art. 8.17. When $\rho - \rho'$ is less than the right-hand side of Eq. (8.36), calculate the moment capacity from Eq. (8.15) or from an analysis based on the assumptions of Art. 8.17. ACI 318, "Building Code Requirements for Reinforced Concrete," requires also that $\rho - \rho'$ not exceed $0.75\rho_b$ to avoid brittle failure of the concrete.

Compressive steel should be anchored by ties or stirrups at least ⅜ in in diameter and spaced no more than 16 bar diameters or 48 tie diameters apart. Tie reinforcement requirements are the same as those for columns.

Design for shear and development lengths of reinforcement is the same as for beams with tension reinforcement only (Art. 8.20.4 and 8.20.6).

8.25 Alternate Design of Rectangular Beams with Compression Bars

The following formulas, based on the linear variation of stress and strain with distance from the neutral axis (Fig. 8.15) may be used in design:

$$k = \frac{1}{1 + f_s/nf_c} \qquad (8.36)$$

where f_s = stress in tensile steel, ksi

f_c = stress in extreme compression surface, ksi

n = modular ratio, E_s/E_c

$$f_s' = \frac{kd - d'}{d - kd} 2f_s \qquad (8.37)$$

where f_s' = stress in compressive steel, ksi

d = distance from extreme compression surface to centroid of tensile steel, in

d' = distance from extreme compression surface to centroid of compressive steel, in

The factor 2 is incorporated into Eq. (8.37) in accordance with ACI 318, "Building Code Requirements for Reinforced Concrete," to account for the effects of creep and nonlinearity of the stress-strain diagram for concrete. But f_s' should not exceed the allowable tensile stress for the steel.

Since total compressive force equals total tensile force on a section,

$$C = C_c + C_s' = T \qquad (8.38)$$

where C = total compression on beam cross section, kips

C_c = total compression on concrete, kips, at section

C_s' = force acting on compressive steel, kips

T = force acting on tensile steel, kips

$$\frac{f_s}{f_c} = \frac{k}{2[\rho - \rho'(kd - d')/(d - kd)]} \qquad (8.39)$$

where $\rho = A_s/bd$ and $\rho' = A_s'/bd$.

For reviewing a design, the following formulas may be used:

$$k = \sqrt{2n\left(\rho + \rho'\frac{d'}{d}\right) + n^2(\rho + \rho')^2} - n(\rho + \rho') \quad (8.40)$$

$$\bar{z} = \frac{(k^3 d/3) + 4n\rho' d'[k - (d'/d)]}{k^2 + 4n\rho'[k - (d'/d)]} \qquad (8.41)$$

$$jd = d - \bar{z} \qquad (8.42)$$

where jd is the distance between the centroid of compression and the centroid of the tensile steel. The moment resistance of the tensile steel is

$$M_s = Tjd = A_s f_s jd \qquad (8.43)$$

$$f_s = \frac{M}{A_s jd} \qquad (8.44)$$

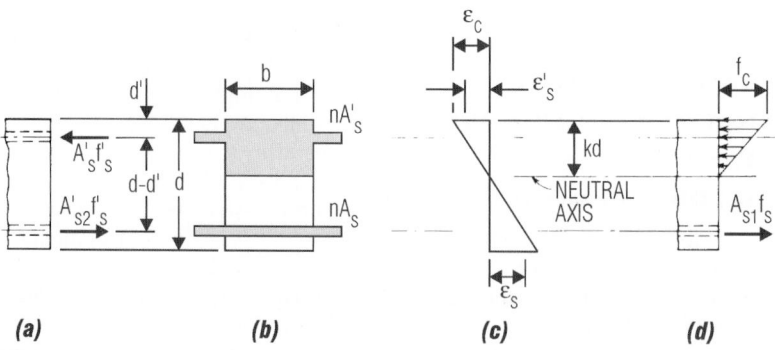

Fig. 8.15 Rectangular concrete beam: (*a*) Reinforced for both tension and compression. (*b*) Transformed all-concrete section. (*c*) Strain distribution. (*d*) Stresses.

where M is the bending moment at the section of beam under consideration. The moment resistance in compression is

$$M_c = \frac{1}{2}f_c jbd^2\left[k + 2n\rho'\left(1 - \frac{d'}{kd}\right)\right] \quad (8.45)$$

$$f_c = \frac{2M}{jbd^2\{k + 2n\rho'[1 - (d'/kd)]\}} \quad (8.46)$$

Computer software is available for the preceding calculations. Many designers, however, prefer the following approximate formulas:

$$M_1 = \frac{1}{2}f_c bkd\left(d - \frac{kd}{3}\right) \quad (8.47)$$

$$M'_s = M - M_1 = 2f'_s A'_s(d - d') \quad (8.48)$$

where M = bending moment

M'_s = moment-resisting capacity of compressive steel

M_1 = moment-resisting capacity of concrete

For determination of shear, see Art. 8.21. Compressive steel should be anchored by ties or stirrups at least No. 3 in size and spaced not more than 16 bar diameters or 48 tie diameters apart. At least one tie within the required spacing, throughout the length of the beam where compressive reinforcement is required, should extend completely around all longitudinal bars.

8.26 Ultimate-Strength Design of I and T Beams

A reinforced-concrete beam may be shaped in cross section like a T, or it may be composed of a slab and integral rectangular beam that, in effect, act as a T beam. According to ACI 318, "Building Code Requirements for Reinforced Concrete" (American Concrete Institute), and "Standard Specifications for Highway Bridges" (American Association of State Highway and Transportation Officials), when the slab forms the compression flange, its effective width b may be assumed to be not larger than one-fourth the beam span and not greater than the distance center to center of beams. In addition, the ACI Code requires that the overhanging width on either side of the beam web should not be assumed to be larger than

eight times the slab thickness nor one-half the clear distance to the next web. The AASHTO Specifications more conservatively limit the effective width to 12 times the slab thickness plus the beam width. For beams with a flange on only one side, the effective overhanging flange width should not exceed one-twelfth the beam span, or six times the slab thickness, or half the clear distance to the next beam.

Two cases may occur in the design of T and I beams: The neutral axis lies in the compression flange (Fig. 8.16a and b) or in the web (Fig. 8.16c and d). For negative moment, a T beam should be designed as a rectangular beam with width b equal to that of the stem. (See Arts. 8.17 and 8.20.)

When the neutral axis lies in the flange, the member may be designed as a rectangular beam, with effective width b and depth d, by Eq. (8.14). For that condition, the flange thickness t will be greater than the distance c from the extreme compression surface to the neutral axis.

$$c = \frac{1.18\omega d}{\beta_1} \quad (8.49)$$

where β_1 = constant defined in Art. 8.17

$\omega = A_s f_y/bdf'_c$

A_s = area of tensile steel, in^2

f_y = yield strength of steel, ksi

f'_c = 28-day strength of concrete, ksi

When the neutral axis lies in the web, the ultimate moment should not exceed

$$M_u = 0.90\left[(A_s - A_{sf})f_y\left(d - \frac{a}{2}\right) + A_{sf}f_y\left(d - \frac{t}{2}\right)\right] \quad (8.50)$$

where A_{sf} = area of tensile steel required to develop compressive strength of overhanging flange, in^2 = $0.85(b - b_w)tf'_c/f_y$

b_w = width of beam web or stem, in

a = depth of equivalent rectangular compressive stress distribution, in = $(A_s - A_{sf})f_y/0.85f'_c b_w$

The quantity $\rho_w - \rho_f$ should not exceed $0.75\rho_b$, where ρ_b is the steel ratio for balanced conditions [Eq. (8.13)], $\rho_w = A_s/b_w d$, and $\rho_f = A_{sf}/b_w d$.

For determination of ultimate shear, see Art. 8.20.4. Note, however, that the web or stem

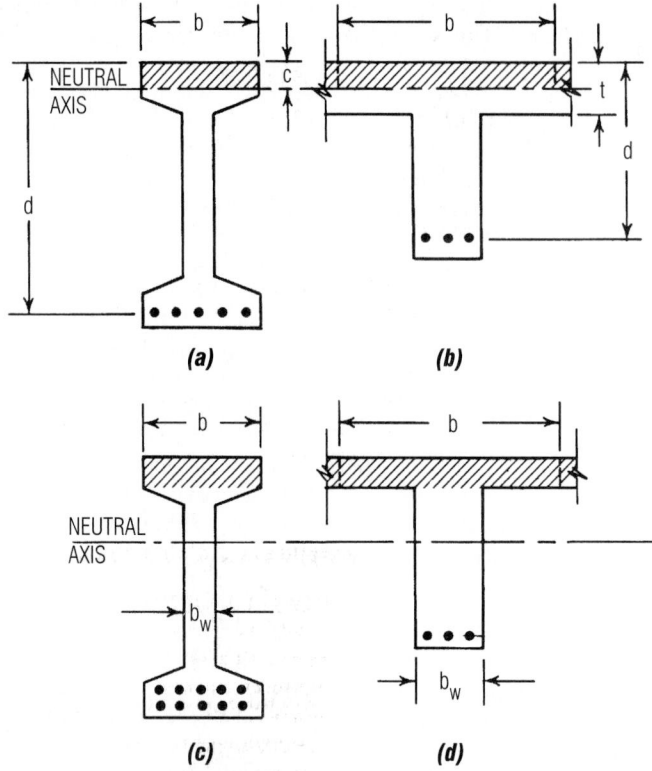

Fig. 8.16 I and T beams: (*a*) and (*b*) Neutral axis in the flange. (*c*) and (*d*) Neutral axis in the web.

width b_w should be used instead of b in these calculations.

8.27 Working-Stress Design of I and T Beams

For T beams, effective width of compression flange is determined by the same rules as for ultimate-strength design (Art. 8.26). Also, for working-stress design, two cases may occur: The neutral axis may lie in the flange (Fig. 8.16*a* and *b*) or in the web (Fig. 8.16*c* and *d*). (For negative moment, a T beam should be designed as a rectangular beam with width b equal to that of the stem.) See Art. 8.21.

If the neutral axis lies in the flange, a T or I beam may be designed as a rectangular beam with effective width b. If the neutral axis lies in the web or stem, an I or T beam may be designed by the following formulas, which

ignore the compression in the stem, as is customary:

$$k = \frac{I}{1 + f_s/nf_c} \qquad (8.51)$$

where kd = distance from extreme compression surface to neutral axis, in

d = distance from extreme compression surface to centroid of tensile steel, in

f_s = stress in tensile steel, ksi

f_c = stress in concrete at extreme compression surface, ksi

n = modular ratio = E_s/E_c

Since the total compressive force C equals the total tension T,

$$C = \frac{1}{2}f_c(2kd - t)\frac{bt}{kd} = T = A_s f_s \qquad (8.52)$$

$$kd = \frac{2ndA_s + bt^2}{2nA_s + 2bt} \tag{8.53}$$

where A_s = area of tensile steel, in^2

t = flange thickness, in

The distance between the centroid of the area in compression and the centroid of the tensile steel is

$$jd = d - \bar{z} \tag{8.54}$$

$$\bar{z} = \frac{t(3kd - 2t)}{3(2kd - t)} \tag{8.55}$$

The moment resistance of the steel is

$$M_s = Tjd = A_s f_s jd \tag{8.56}$$

The moment resistance of the concrete is

$$M_c = Cjd = \frac{f_c btjd}{2kd}(2kd - t) \tag{8.57}$$

In design, M_s and M_c can be approximated by

$$M_s = A_s f_s \left(d - \frac{t}{2}\right) \tag{8.58}$$

$$M_c = \frac{1}{2} f_c bt \left(d - \frac{t}{2}\right) \tag{8.59}$$

derived by substituting $d - t/2$ for jd and $f_c/2$ for $f_c(1 - t/2kd)$, the average compressive stress on the section.

For determination of shear, see Art. 8.21. Note, however, that the web or stem width b_w should be used instead of b in these calculations.

8.28 Torsion in Reinforced-Concrete Members

Under twisting or torsional loads, a member develops normal (warping) and shear stresses. The warping, normal stresses help greatly in resisting torsion. But there are no accurate ways of computing this added resistance.

The maximum shears at any point are accompanied by equal tensile stresses on planes bisecting the angles between the planes of maximum shears.

As for ordinary shear, reinforcement should be incorporated to resist the diagonal tension in excess of the tensile capacity of the concrete. If web reinforcement is required for vertical shear in a horizontal beam subjected to both flexure and torsion, additional web reinforcement should be included to take care of the full torsional shear.

8.28.1 Ultimate-Strength Design for Torsion

When the factored torsion T_u is less than the value calculated from Eq. (8.63), the area A_v of shear reinforcement should be at least

$$A_v = 50 \frac{b_w s}{f_y} \tag{8.60}$$

But when the ultimate torsion exceeds T_u calculated from Eq. (8.63) and where web reinforcement is required, either nominally or by calculation, the minimum area of closed stirrups required is

$$A_v + 2A_t = \frac{50 b_w s}{f_y} \tag{8.61}$$

where A_t is the area of one leg of a closed stirrup resisting torsion within a distance s.

While shear reinforcement may consist of stirrups (Fig. 8.12), bent-up longitudinal bars, spirals, or welded-wire fabric (Art. 8.20.4), torsion reinforcement should consist of closed ties, closed stirrups, or spirals—all combined with longitudinal bars. Closed ties or stirrups may be formed either in one piece by overlapping standard tie end hooks around a longitudinal bar (Fig. 8.12b), or in two pieces spliced as a Class B splice or adequately embedded. Pairs of U stirrups placed so as to form a closed unit should be lapped at least $1.3l_d$, where l_d is the tensile development length (Art. 8.12.5).

Torsion effects should be considered whenever the ultimate torsion exceeds

$$T_u = \phi \sqrt{f_c'} \left(\frac{A_{cp}^2}{p_{cp}}\right) \tag{8.62}$$

where A_{cp} = area enclosed by the outside perimeter of concrete cross section

p_{cp} = outside perimeter of the concrete cross section

The design torsional strength should be equal to or greater than the required torsional strength:

$$\phi T_n \geq T_u \tag{8.63}$$

The nominal torsional moment strength in terms of stirrup yield strength was derived above.

$$T_n = \frac{2A_o A_t f_{yv}}{s} \cot \theta \tag{8.64}$$

where $A_o = 0.85A_{oh}$ (this is an assumption for simplicity)

A_{oh} = area enclosed by centerline of the outermost closed transverse torsional reinforcement

θ = angle of compression diagonal, ranges between 30° and 60°. It is suggested in 11.6.3.6 to use 45° for nonprestressed members and 37.5° for prestressed members with prestress force greater than 40 percent of tensile strength of the longitudinal reinforcement.

The spacing of closed stirrups, however, should not exceed $p_h/8$ nor 12 in, where p_h is the perimeter of centerline of outermost closed transverse torsional reinforcement, in.

At least one longitudinal bar should be placed in each corner of the stirrups. Size of longitudinal bars should be at least No. 3, and their spacing around the perimeters of the stirrups should not exceed 12 in. Longitudinal bars larger than No. 3 are required if indicated by the larger of the values of Al computed from Eq. (8.65).

$$Al = \left(\frac{A_t}{s}\right)p_h\left(\frac{f_{yv}}{f_{yl}}\right)\cot^2\theta \qquad (8.65)$$

8.29 Two-Way Slabs

When a rectangular reinforced-concrete slab is supported on all four sides, reinforcement placed perpendicular to the sides may be assumed to be effective in the two directions if the ratio of the long sides to the short sides is less than about 2:1. "Standard Specifications for Highway Bridges" (American Association of State Highway and Transportation Officials) requires that the slab be designed as a one-way slab if the ratio is more than 1.5:1. In effect, a two-way slab distributes part of the load on it in the long direction and usually a much larger part in the short direction. For a symmetrically supported square slab, however, distribution is the same in the two directions for symmetrical loading.

Because precise determination of reactions and moments for two-way slabs with various edge conditions is complex and tedious, most codes offer empirical formulas to simplify the calculation.

The reactions of the slab on supporting beams and walls are not constant along the sides, which should be taken into account in the design of the supports. (One method is to use a triangular distribution on the short sides and a trapezoidal distribution on the long sides. The legs of the triangles and the trapezoids usually are assumed to make a 45° angle with the slab edges.)

8.29.1 Flat-Slab Construction

Two-way slab systems supported directly on columns, without beams or girders, and thickened locally around the columns creating drop panels, are classified as flat slabs. Generally, the columns flare out at the top in capitals (Fig. 8.17a). But only the portion of the inverted truncated cone thus formed that lies inside a 90° vertex angle is considered effective in resisting stress. Sometimes, the capital for an exterior column is a bracket on the inner face.

To reduce the shear stresses in the region of the columns and the amount of steel needed for negative bending moments, especially when the live load exceeds 150 psf, a rectangular drop panel, or thicker slab, is formed over the columns (Fig. 8.17a). For similar spans and loads, use of a drop panel permits a reduced slab thickness between panels. For the full effective depth of the drop to be used in determination of negative-moment reinforcement, ACI 318, "Building Code Requirements for Reinforced Concrete" (American Concrete Institute), specifies that a drop panel should extend in each direction from the center of support a distance equal to at least one-sixth the span in that direction. The difference in thickness between the drop panel and slab should be at least one-fourth the slab thickness but, for determining reinforcement, should not be taken as more than one-fourth the distance from the edge of the drop panel to the edge of the column or capital.

To control deflection, the ACI Code establishes minimum thicknesses for slabs, without interior beams as a ratio of the length of the clear space in the long direction. The minimum slab thickness for slabs with drop panels is 4 inches.

In general, flat slabs are more economical than beam-and-girder construction. They yield a lower building for the same number of stories. Formwork is simpler. Fire resistance is greater because of fewer sharp corners where spalling may occur. And there is less obstruction to light with flat slabs. The design procedure is similar to that for flat plates and is described in Art. 8.29.2.

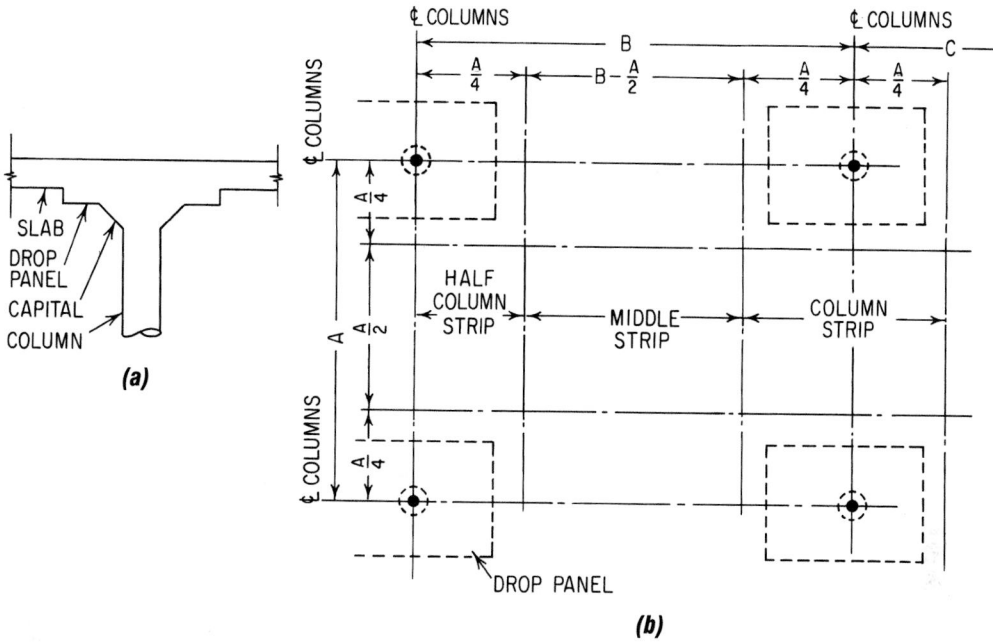

Fig. 8.17 Concrete flat slab: (*a*) Vertical section through drop panel and column at a support. (*b*) Plan view indicates division of slab into column and middle strips.

8.29.2 Flat-Plate Construction

Flat slabs with constant thickness between supports are called flat plates. Generally, capitals are omitted from the columns.

Exact analysis or design of flat slabs or flat plates is very complex. It is common practice to use approximate methods. The ACI Code presents two such methods: direct design and equivalent-frame.

In both methods, a flat slab is considered to consist of strips parallel to column lines in two perpendicular directions. In each direction, a **column strip** spans between columns and has a width of one-fourth the shorter of the two perpendicular spans on each side of the column centerline. The portion of a slab between parallel column strips in each panel is called the **middle strip** (Fig. 8.17).

Direct Design Method ▪ This may be used when all the following conditions exist:

The slab has three or more bays in each direction.

Ratio of length to width of panel is 2 or less.

Loads are uniformly distributed over the panel.

Ratio of live to dead load is 2 or less.

Columns form an approximately rectangular grid (10% maximum offset).

Successive spans in each direction do not differ by more than one-third of the longer span.

Moment redistribution shall not be applied.

When a panel is supported by beams on all sides, the relative stiffness of the beams satisfies

$$0.2 \le \frac{\alpha_1}{\alpha_2} \left(\frac{l_2}{l_1}\right)^2 \le 5 \qquad (8.66)$$

where $\alpha_1 = \alpha$ in direction of l_1

$\alpha_2 = \alpha$ in direction of l_2

α = ratio of flexural stiffness $E_{cb}I_b$ of the beam section to flexural stiffness $E_{cs}I_s$ of width of slab bounded laterally by centerline of adjacent panel, if any, on each side of beam.

$l_1 =$ span in the direction in which moments are being determined, c to c of supports

$l_2 =$ span perpendicular to l_1, c to c of supports

The basic equation used in direct design is the total static design moment in a strip bounded laterally by the centerline of the panel on each side of the centerline of the supports:

$$M_o = \frac{wl_2 l_n^2}{8} \qquad (8.67)$$

where $w =$ uniform design load per unit of slab area

$l_n =$ clear span in direction moments are being determined

The strip, with width l_2, should be designed for bending moments for which the sum in each span of the absolute values of the positive and average negative moments equals or exceeds M_o.

Interior Panels. Following is the procedure for direct design of an interior panel of a flat slab (or flat plate or two-way beam-and-slab construction):

Step 1. Determine the minimum allowable and practical slab thickness to control deflections and verify adequacy for shear strength.

Step 2. Determine the ultimate design load from Eq. (8.7a), $U = 1.4D + 1.7L$, where D represents the moments and shears caused by dead load and L those caused by live load. (This assumes that horizontal loads are taken by shear walls or other vertical elements.)

Step 3. Determine M_o from Eq. (8.67).

Step 4. For an interior span, distribute M_o as follows:

Negative design moment $= 0.65 M_o$

Positive design moment $= 0.35 M_o$

The negative-moment section should be designed to resist the larger of the two interior negative design moments determined for the spans framing into a common support.

Step 5. Proportion design moments and shears in column and middle strips as follows:

1. Column Strip. The interior negative moment should be determined in accordance with Table

8.14. Values not given may be obtained by linear interpolation.

The positive design moment should be determined in accordance with Table 8.15. Values not given may be obtained by linear interpolation.

When there is a beam between columns in the direction of the span in which moments are being considered, the beam should be proportioned to resist 85% of the column strip moment if $\alpha_1 l_2/l_1$ is greater than 1.0. For values of $\alpha_1 l_2/l_1$ between 1.0 and zero, the proportion of moment resisted by the beam may be obtained by linear interpolation between 85 and 0%. The slab in the column strip should be proportioned to resist that portion of the design moment not resisted by the beam.

2. Middle Strip. The interior negative or positive design moment assigned to a middle strip is that portion of the design moments not resisted by the column strips bounding it. Thus, each middle strip should be proportioned to resist the sum of the negative moment not taken by the column strip along one side and the negative moment not resisted by the column strip on the other side and, similarly, the sum of the positive moments.

3. Moment Modification. A design moment may be modified by 10% if the total static design moment for the panel in the direction considered is not less than that required by Eq. (8.67).

Step 6. Walls and columns built integrally with the slab should be designed to resist the moments due to loads on the slab system.

Exterior Panels. The ACI Code lists design criteria for exterior panels for a wide range of support conditions. These criteria require determination of the relative flexural stiffness of supports at edges, including torsional resistance.

Equivalent-Frame Method ▪ The equivalent-frame method typically is used when all

Table 8.14 Percent of Interior Negative Design Moment in Column Strips

$\alpha_1 l_2/l_1$	Span Ratio l_2/l_1		
	0.5	1.0	2.0
0	75	75	75
1 or more	90	75	45

Table 8.15 Percent of Positive Design Moment in Column Strips

$\alpha_1 l_2/l_1$	Span Ratio l_2/l_1		
	0.5	1.0	2.0
0	60	60	60
1.0 or more	90	75	45

the conditions required for the direct design method are not satisfied. The slab is initially divided into a series of bents, or equivalent frames, on column lines taken longitudinally and transversely through the building. Each frame consists of a row of equivalent columns and slab-beam strips, bounded laterally by the centerline of the panel on each side of the column line under investigation. Each such frame may be analyzed in its entirety. Or for vertical loads, each floor may be analyzed, with columns, above and below, assumed fixed at floors above and below. For purposes of computation, the slab-beam may be assumed fixed at any support two panels away from the support where the bending moment is being determined. The moments thus determined may be distributed to the column strips, middle strips, and beams as previously described for the direct design method if Eq. (8.66) is satisfied.

The critical section for negative moment in both the column and middle strips should be taken at the face of supports, but for interior supports not farther than $0.175l_1$ from the center of the column, where l_1 is the span center to center of supports.

Note that where slabs designed by the equivalent-frame method meet the criteria of the direct design method, the computed moments in any span may be reduced in a proportion such that the sum of the absolute values of the positive and average negative bending moments used in design does not exceed M_o given by Eq. (8.67).

Determination of reinforcement, based on the bending moments at critical sections, is the same as described for rectangular beams (Art. 8.20 or 8.21). Requirements for minimum reinforcement should be respected.

The equivalent-frame method attempts to represent the effects of torsional stiffness of the three-dimensional slab system by defining and using the flexural stiffness of the slab-beam-column system in geometric terms applicable to a two-dimensional analysis. The ACI Code assigns a finite moment of inertia to the slab-beam from center to face of column equal to the moment of inertia of the slab-beam at the face of the column divided by $(1 - c_2/l_2)^2$, where c_2 is the dimension of column, capital, or bracket in the direction of l_2. This assigned I represents the flexibility of the slab on the sides of the column. This simulates additional stiffness in the area of the slab-column and is reflected by the change in the coefficients used to determine fixed-end moments, stiffness factors, and carry-over factors for slabs. The ACI Code also modifies the column flexural stiffness to account for the torsional flexibility of the slab. The part of the slab providing the torsional restraint is transverse to the direction in which moments are being determined for the width of the column and extends to the bounding lateral panel centerlines on each side of the column.

8.29.3 Shear in Slabs

Slabs should also be investigated for shear, both beam-type and punching shear. For beam-type shear, the slab is considered as a thin, wide rectangular beam. The critical section for diagonal tension should be taken at a distance from the face of the column or capital equal to the effective depth d of the slab. The critical section extends across the full width b of the slab. Across this section, the nominal shear stress v_u on the unreinforced concrete should not exceed the ultimate capacity $2\sqrt{f_c'}$.

Punching shear may occur along several sections extending completely around the support, for example, around the face of the column or column capital or around the drop panel. These critical sections occur at a distance $d/2$ from the faces of the supports, where d is the effective depth of the slab or drop panel. Design for punching shear should be based on Eq. (8.16), with shear strength V_n taken not larger than the concrete strength V_c. V_c shall be the smallest of (a), (b) and (c).

(a)
$$V_c = \left(2 + \frac{4}{\beta_c}\right)\sqrt{f_c'}b_o d \tag{8.68}$$

where b_o = perimeter of critical section

β_c = ratio of long side to short side of critical section

(b)
$$V_c = \left(\frac{\alpha_s d}{b_o} + 2\right)\sqrt{f_c'}\,b_o d \qquad (8.69)$$

where $\alpha_s = 40$ for interior columns, 30 for edge columns, 20 for corner columns

(c)
$$V_c 2 4\sqrt{f_c'}\,b_o d \qquad (8.70)$$

Shear reinforcement for slabs generally consists of bent bars and is designed in accordance with the provisions for beams (Art. 8.20.4), with the shear strength of the concrete at critical sections taken as $2\sqrt{f_c'}\,b_o d$ at ultimate strength and $V_n \leq 6\sqrt{f_c'}\,b_o d$. Extreme care should be taken to ensure that shear reinforcement is accurately placed and properly anchored, especially in thin slabs.

The ACI Code also includes instructions for design of steel shear heads. Because of the cost of steel shear-head reinforcement, however, it is preferable to either thicken the slab or design concrete beams to support heavy loads.

8.29.4 Column Moments

Another important consideration in design of two-way slab systems is the transfer of moments to columns. This is generally a critical condition at edge columns, where the unbalanced slab moment is very high due to the one-sided panel.

The unbalanced slab moment is considered to be transferred to the column partly by flexure across a critical section, which is $d/2$ from the periphery of the column, and partly by eccentric shear forces acting about the centroid of the critical section.

That portion of unbalanced slab moment M_u transferred by the eccentricity of the shear is given by $\gamma_v M_u$.

$$\gamma_v = 1 - \gamma_f \qquad (8.71)$$

$$= 1 - \frac{1}{1 + (2/3)\sqrt{b_1/b_2}} \qquad (8.72)$$

where b_1 = width, in, of critical section in the span direction for which moments are being computed

b_2 = width, in, of critical section in the span direction perpendicular to b_1

For that portion of the unbalanced moment transferred to the column by flexure, it is accepted practice to concentrate or add reinforcement across the critical slab width, determined as the sum of the column width plus the thickness of the slab.

(G. Winter and A. H. Nilson, "Design of Concrete Structures," McGraw-Hill Book Company, New York (books.mcgraw-hill.com); P. F. Rice and E. S. Hoffman, "Structural Design Guide to the ACI Building Code," Van Nostrand Reinhold Company, New York: "CRSI Handbook," and "Two-Way Slab Design Supplements," Concrete Reinforcing Steel Institute, Chicago, Ill (www.crsi.org).)

8.30 Brackets and Corbels

Brackets and corbels are members having a ratio of shear span to depth a/d of 1 or less. The shear span a is the distance from the point of load to the face of support (Fig. 8.18).

The depth of a bracket or corbel at its outer edge should not be less than one-half of the required depth d at the support. Reinforcement should consist of main tension bars with area A_s and shear reinforcement with area A_h. The shear reinforcement should consist of closed ties parallel to the main tension reinforcement (Fig. 8.18). The area of shear reinforcing should not be less than $0.5(A_s - A_n)$ where A_n is the area of reinforcement to resist the tensile force and should be uniformly distributed within two-thirds of the depth of the bracket adjacent to the main tension bars. Also, the ratio $\rho = A_s/bd$ should not be less than $0.04f_c'/f_y$, where f_c' is the 28-day concrete strength and f_y the steel yield point.

It is good practice to anchor main tension reinforcement bars as close as possible to the outer edge by welding a crossbar or steel angle to them. Also, the bearing area should be kept at least 2 in from the outer edge, and the bearing plate should be welded to the main tension reinforcement if horizontal forces are present.

Tension Reinforcement ▪ A_s should be adequate at the face of the support to resist the moments due to the vertical load and any horizontal forces. This reinforcement must be properly developed to prevent pull-out, by proper anchorage within the support and by a crossbar welded to the bars at the end of the bracket.

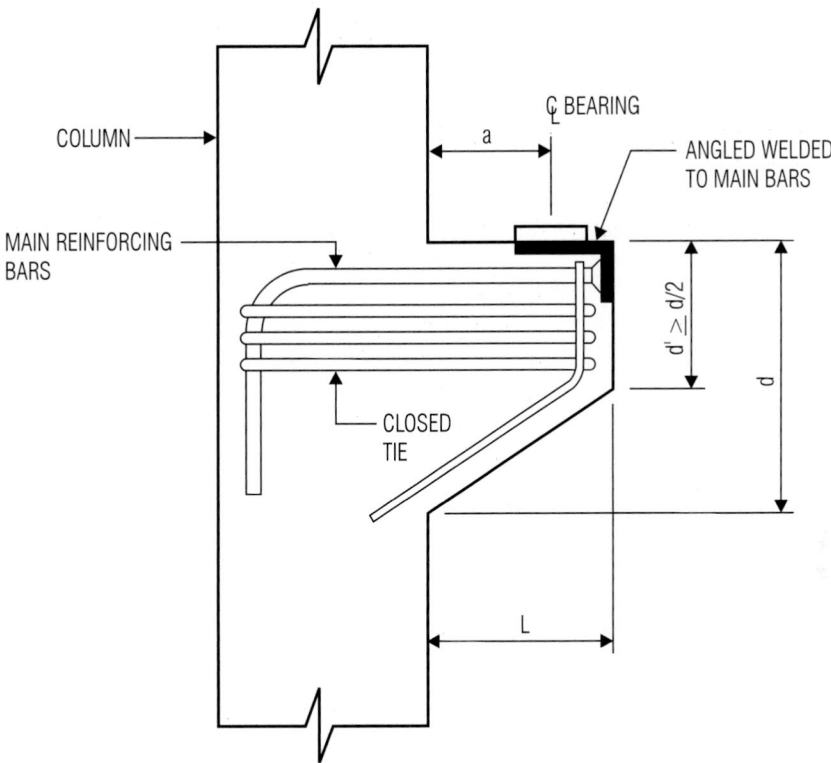

Fig. 8.18 Steel reinforcement of concrete corbel.

Concrete Compression Members

ACI 318, "Building Code Requirements for Reinforced Concrete," American Concrete Institute, sets limitations on column geometry and reinforcement. Following are some of the more important.

8.31 Column Reinforcement

In reinforced concrete columns, longitudinal steel bars help the concrete carry the load. Steel ties or spiral wrapping around those bars prevent the bars from buckling outward and spalling the outer concrete shell. Since spirals are more effective, columns with closely spaced spirals are allowed to carry greater loads than comparable columns with ties.

Reinforcement Cover ▪ For cast-in-place columns, spirals and ties should be protected with a monolithic concrete cover of at least $1\frac{1}{2}$ in. But for severe exposures, the amount of cover should be increased.

Minimum Reinforcement ▪ Columns should be reinforced with at least six longitudinal bars in a circular arrangement or with four longitudinal bars in a rectangular arrangement, of at least No. 5 bar size. Area of column reinforcement should not be less than 1% or more than 8% of the gross cross-sectional area of a column.

Excess Concrete ▪ In a column that has a larger cross section than that required by load, the effective area A_g used to determine minimum reinforcement area and load capacity may be reduced proportionally, but not to less than half the total area.

8.31.1 Spirals

This type of transverse reinforcement should be at least $\frac{3}{8}$ in in diameter. A spiral may be anchored at each of its ends by $1\frac{1}{2}$ extra turns of the spiral. Splices may be made by welding or by a lap of 48 bar diameters (but at least 12 in). Spacing (pitch) of spirals should not exceed 3 in or be less than 1 in. Clear spacing should be at least $1\frac{1}{3}$ times the maximum size of coarse aggregate.

A spiral should extend to the level of the lowest horizontal reinforcement in the slab, beam, or drop panel above. Where beams are of different depth or are not present on all sides of a column, ties should extend above the termination of the spiral to the bottom of the shallowest member. In a column with a capital, the spiral should extend to a plane at which the diameter or width of the capital is twice that of the column.

The ratio of the volume of spiral reinforcement to volume of concrete core (out to out of spiral) should be at least

$$\rho_s = 0.45\left(\frac{A_g}{A_c} - 1\right)\frac{f_c'}{f_y} \tag{8.73}$$

where A_g = gross area of column

A_c = area of core of column measured to outside of spiral

f_y = yield strength of spiral reinforcement

f_c' = 28-day compressive strength of concrete

8.31.2 Column Ties

Lateral ties should be at least $\frac{3}{8}$ in in diameter for No. 10 or smaller bars and $\frac{1}{2}$ in diameter for No. 11 and larger bars. Spacing should not exceed 16 bar diameters, 48 tie diameters, or the least dimension of the column. The ties should be so arranged that every corner bar and alternate longitudinal bars will have lateral support provided by the corner of a tie having an included angle of not more than 135° (Fig. 8.19). No bar should be more than 6 in from such a laterally supported bar. Where bars are located around a circle, a complete circular tie may be used. (For more details, see ACI 315, "Manual of Standard Practice for Detailing Reinforced Concrete Structures," American Concrete Institute (www.aci-int.org).)

8.32 Effects of Column Slenderness

Building columns generally are relatively short. Thus, an approximate evaluation of slenderness effects can usually be used in design. Slenderness, which is a function of column geometry and bracing, can reduce the load-carrying capacity of compression members by introducing bending stresses and can lead to a buckling failure.

Load-carrying capacity of a column decreases with increase in unsupported length l_u, beyond a certain length. In buildings, l_u should be taken as the clear distance between floor slabs, girders, or other members capable of providing lateral support to the column or as the distance from a floor to a column capital or a haunch, if one is present.

In contrast, load-carrying capacity increases with increase in radius of gyration r of the column cross section. For rectangular columns, r may be taken as 30% of the overall dimension in the direction in which stability is being considered and for circular members as 25% of the diameter.

8.32.1 Effective Column Length

Also, the greater the resistance offered by a column to sidesway, or drift, because of lateral bracing or

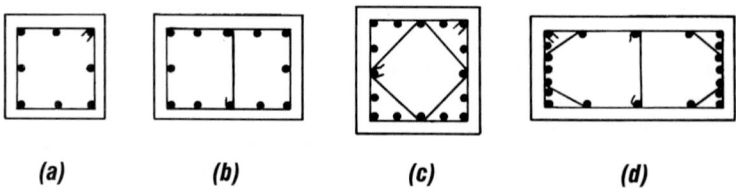

Fig. 8.19 Column ties provide lateral support at corners and to alternate reinforcing bars at a horizontal section. (*a*) Square column with single tie. (*b*) Rectangular column with a pair of ties. (*c*) Square column with a pair of ties. (*d*) Rectangular column with inclined ties.

restraint against end rotations, the higher the load-carrying capacity. This resistance is represented by application of a factor k to the unsupported length of the column, and kl_u is referred to as the **effective length** of the column.

The combination of these factors, which is a measure of the slenderness of a column, kl_u/r, is called the **slenderness ratio** of the column.

The effective-length factor k can be determined by analysis. If an analysis is not made, for compression members in nonsway frames, k should be taken as unity. For columns not braced against sidesway, k will be greater than unity; analysis should take into account the effects of cracking and reinforcement on relative stiffness. See also Art. 8.32.3.

ACI Committee 441 has proposed that k should be obtained from the Jackson and Moreland alignment chart, reproduced as Fig. 8.20. For determination of k with this chart, a parameter ψ_A must be computed for end A of column AB, and a similar parameter ψ_B must be computed for end B. Each parameter equals the ratio at that end of the column of the sum of EI/l_u for the compression members meeting there to the sum of EI/l for the flexural members meet-

ing there, where EI is the flexural stiffness of a member.

8.32.2 Non-Sway and Sway Frames

As a guide in judging whether a frame is non-sway or sway, ACI 318 indicates that a column in a structure can be considered non-sway if the column end moments due to second-order effects do not exceed 5% of the first-order end moments. It is also permitted to assume a story within a structure is non-sway if:

$$Q = \frac{\sum P_u \Delta_o}{V_u \ell_c} \leq 0.05 \qquad (8.74)$$

where Q = stability index for a story

$\sum P_u$ = total factored vertical load in the story corresponding to the lateral loading case for which $\sum P_u$ is greatest

V_u = total story shear

Δ_o = first-order relative deflection between the top and bottom of the story due to V_u

ℓ_c = column length, measured from center-to-center of the joints in the frame

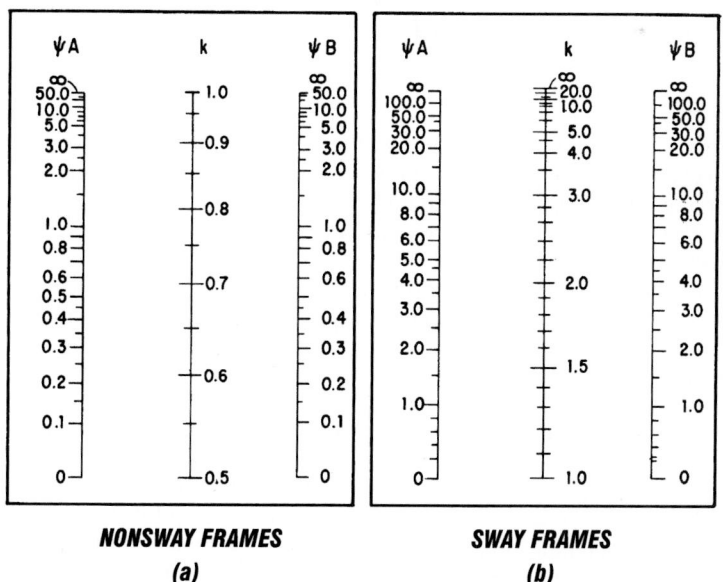

NONSWAY FRAMES

(a)

SWAY FRAMES

(b)

Fig. 8.20 Alignment charts for determination of effective-length factor k for columns. ψ is the ratio for each end of a column of $\Sigma EI/l_u$ for the compression members to $\Sigma EI/l$ for the girders.

For compression members in non-sway frames, the slenderness effect may be neglected under the following conditions:

For columns braced against sideway, when

$$\frac{kl_u}{r} < 34 - 12\frac{M_1}{M_2} \qquad (8.75)$$

where M_1 = smaller of two end moments on column as determined by conventional elastic frame analysis, with positive sign if column is bent in single curvature and negative sign if column is bent in double curvature

M_2 = absolute value of larger of the two end moments on column as determined by conventional elastic frame analysis

For columns not braced against sidesway, when

$$\frac{kl_u}{r} < 22 \qquad (8.76)$$

8.32.3 Column Design Loads

Analysis taking into account the influence of axial loads and variable moment of inertia on member stiffness and fixed-end moments, the effects of deflections on moments and forces, and the effects of duration of loads is required for all columns when

$$\frac{kl_u}{r} > 100 \qquad (8.77)$$

For columns for which the slenderness ratio lies between 22 and 100, and therefore the slenderness effect on load-carrying capacity must be taken into account, either an elastic analysis can be performed to evaluate the effects of lateral deflections and other effects producing secondary stresses, or an approximate method based on moment magnification may be used. In the approximate method, the compression member in a non-sway frame is designed for the factored axial load P_u and the moment amplified for the effects of member curvature M_c defined by

$$M_c = \delta_{ns}M_2 \qquad (8.78)$$

where δ_{ns} is the moment magnification factor for non-sway frames and may be determined from:

$$\delta_{ns} = \frac{C_m}{1 - P_u/0.75P_c} \geq 1 \qquad (8.79)$$

where C_m = factor relating actual moment diagram to that for equivalent uniform moment

P_c = critical load for column

$$= \frac{\pi^2 EI}{(kl_u)^2} \qquad (8.80)$$

$$EI = \frac{(0.2E_cI_g + E_sI_{se})}{1 + \beta_d} \qquad (8.81)$$

Or

$$EI = \frac{0.4E_cI_g}{1 + \beta_d} \qquad (8.82)$$

For members without transverse loads between supports,

$$C_m = 0.6 + 0.4\frac{M_1}{M_2} \geq 0.4 \qquad (8.83)$$

For members with transverse loads between supports $C_m = 1$.

The critical load is given by

$$P_c = \frac{\pi^2 EI}{(kl_u)^2} \qquad (8.84)$$

where EI is the flexural stiffness of the column.

The flexural stiffness EI may be computed approximately from

$$EI = \frac{E_cI_g/2.5}{1 + \beta_d} \qquad (8.85)$$

where E_c = modulus of elasticity of concrete, psi

I_g = moment of inertia about centroidal axis of gross concrete section, neglecting load reinforcement, in^4

E_s = modulus of elasticity of reinforcement, psi

I_{se} = moment of inertia of reinforcement, in^4

β_d = ratio of maximum design dead load to total load moment (always taken positive)

Because a column has different properties, such as stiffness, slenderness ratio, and δ, in different directions, it is necessary to check the strength of a column in each of its two principal directions.

For design of compression members in sway frames for slenderness, the magnified sway moment may be computed using a second-order elastic analysis, or an approximate method in the ACI 318 code.

8.33 Unified Design Provisions of ACI 318-02

The Unified Design Provisions, which were introduced in Appendix B of the 1995 edition of ACI 318 "Building Code Requirements for Structural Concrete (American Concrete Institute), are incorporated in the body of the 2002 edition. A version of this design method was initially introduced in a paper by Robert Mast in the *ACI Structural Journal*. (Robert Mast, "Unified Design Provisions for Reinforced and Prestressed Concrete Flexural and Compression Members," *ACI Structural Journal*, vol. 89, no. 2, March–April 1992, pp. 185–199)

Before describing the Unified Design Provisions, a brief review of the Strength Design Method that has been utilized for many years to design reinforced concrete members may be helpful. According to this method, the design strength of a member at any section must be greater than or equal to the required strength that is calculated by the load combinations specified in Chapter 9 (see also Art. 8.17.4) of the code:

Design Strength ≥ Required Strength

where

Design Strength
= Strength Reduction Factor (ϕ)
× Nominal Strength
Required Strength
= Load Factors × Service Load Effects

Strength reduction factors (ϕ-factors) account for the probability of understrength of a member due to variations in material strengths and member dimensions, inaccuracies in design equations, the degree of ductility (range of deformations beyond the stage of elastic response, over which full gravity loads can be sustained), the probable quality control achievable, and the importance of a member in a structure.

The nominal strength of a member or cross-section is determined using the assumptions given in Chapter 10 of the code and the design equations given in various chapters throughout the code.

The Unified Design Provisions modify the Strength Design Method for nonprestressed and prestressed members subjected to flexure and axial loads. Affected are strength reduction factors, reinforcement limits, and moment redistribution.

Like the Strength Design Method, members are proportioned by the Unified Design Provisions using factored loads and strength reduction factors. It is important to recognize that these provisions do not alter nominal strength calculations; the nominal strength of a section is computed in the same way as before. What is modified is the design strength of a section via the strength reduction factors. According to the Unified Design Provisions, ϕ-factors are determined based on the strain conditions in the reinforcement farthest from the extreme compression face. Prior to this, ϕ-factors depended only on the type of loading (axial load, flexure, or both) on the section. The Unified Design Provisions provide a rational means for designing nonprestressed and prestressed concrete members subjected to flexural and axial loads, and eliminate many of the inconsistencies in the previous design requirements. This method produces results similar to those from the Strength Design Method. The Unified Design Provisions apply to:

- Flexural and compression members
- Nonprestressed members, prestressed members, and members with a combination of nonprestressed and prestressed reinforcement
- Sections with reinforcement at various depths
- Sections of any shape
- Composite (precast and cast-in-place) concrete sections

These provisions, as they appear in the body of the 2002 ACI code, are described below.

The following definitions are relevant to the Unified Design Provisions. They can be found in Chapter 2 of the code.

- **Net tensile strain, ε_t:** the tensile strain at nominal strength, exclusive of strains due to effective prestress, creep, shrinkage, and temperature.

The net tensile strain is caused by external axial loads and/or bending moments at a section due to the loads applied on the member at the time when the concrete strain at the extreme compression fiber reaches its assumed limit of 0.003. Generally speaking, the net tensile strain can be used as a measure of excessive cracking or excessive deflection.

- **Extreme tension steel:** the reinforcement (prestressed or nonprestressed) that is the farthest from the extreme compression fiber.

Figure 8.21 depicts the location of the extreme tension steel for two sections with different reinforcement arrangements where the top fiber of the section is the extreme compression fiber. The distance from the extreme compression fiber to the centroid of the extreme tension steel is denoted in the figure as d_t. The net tensile strain ε_t in the extreme tension steel due to the external loads can be determined from a strain compatibility analysis for sections with multiple layers of reinforcement. For sections with one layer of reinforcement, it can easily be determined from the strain diagram by similar triangles.

- **Compression-controlled strain limit:** The net tensile strain at balanced conditions.

The definition of a balanced strain condition, which is given in ACI Section 10.3.2, is unchanged from previous editions of the code: a balanced strain condition exists at a cross-section when tension reinforcement reaches the strain corresponding to its specified yield strength just as the concrete strain in the extreme compression fiber reaches its assumed limit of 0.003.

For Grade 60 reinforcement and all prestressed reinforcement, ACI Section 10.3.3 permits the compression-controlled strain limit to be taken equal to 0.002. For Grade 60 bars, this limit is actually equal to $f_y/E_s = 60{,}000/29{,}000{,}000 = 0.00207$ where f_y and E_s are the specified yield strength and modulus of elasticity of the nonprestressed reinforcement, respectively. For other grades of nonprestressed steel, this limit is computed from the ratio f_y/E_s.

- **Compression-controlled section:** a cross-section in which the net tensile strain in the extreme tension steel at nominal strength is less than or equal the compression-controlled strain limit.

When the net tensile strain in the extreme tension steel is small, a brittle failure condition is expected. In such cases, there is little warning of impending failure. Cross-sections of compression members such as columns, subject to significant axial compression, are usually compression-controlled.

- **Tension-controlled section:** a cross-section in which the net tensile strain in the extreme tension steel at nominal strength is greater than or equal to 0.005.

The net tensile strain limit of 0.005 applies to both nonprestressed and prestressed reinforcement and provides ductile behavior for most designs. When the net tensile strain in the extreme tension steel is greater than or equal to 0.005, the section is expected to have sufficient ductility so that ample warning of failure in the form of visible cracking and deflection should be available. Cross-sections of flexural members such as beams, if not heavily reinforced, are usually tension-controlled.

Some sections have a net tensile strain in the extreme tension steel between the limits for compression-controlled and tension-controlled sec-

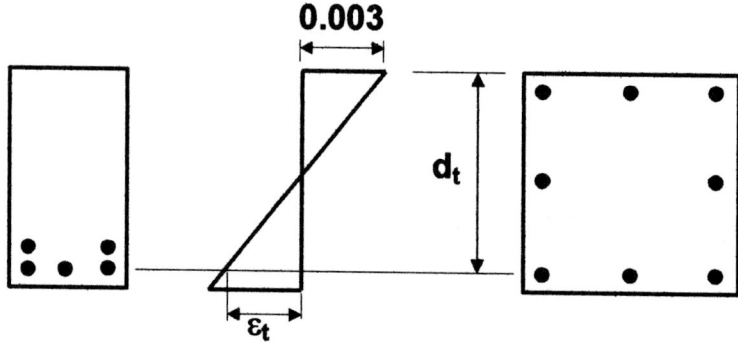

Fig. 8.21 Location of extreme tension steel and net tensile strain at nominal strength.

tions. An example of this is a section subjected to a small axial load and a large bending moment. These members are in a transition region, which is described below.

All of these definitions are utilized when determining strength reduction factors and design strengths.

8.33.1 Strength Reduction Factors, ϕ—Unified Design

In previous editions of the code, the appropriate ϕ-factor to use in design depended on the type of loading that the member was subjected to. For example, for members subjected to flexure without axial load, ϕ was equal to 0.90.

According to the Unified Design Provisions, strength reduction factors are a function of the net tensile strain ε_t in the extreme tension steel. ACI Section 9.3.2 contains ϕ-factors for tension-controlled sections, compression-controlled sections, and sections in which the net tensile strain in the extreme tension steel is between the limits for tension-controlled and compression-controlled sections. Variation of ϕ with respect to ε_t is depicted in ACI Fig. R9.3.2, which is reproduced here as Fig. 8.22.

For compression-controlled sections, ϕ is equal to 0.70 for members with spiral reinforcement conforming to ACI Section 10.9.3 and is equal to 0.65 for other members. For tension-controlled

sections, ϕ is equal to 0.90. For sections that fall between these two limits, it is permitted to linearly increase ϕ from the applicable value for compression-controlled sections to 0.90.

The following equations can be used to determine ϕ in the transition region:

• For sections with spiral reinforcement:

$$\phi = 0.57 + 67\varepsilon_t \qquad (8.86)$$

• For other sections:

$$\phi = 0.48 + 83\varepsilon_t \qquad (8.87)$$

ACI Fig. R9.3.2 also contains equations to determine ϕ as a function of the ratio c/d_t where c is the distance from the extreme compression fiber to the neutral axis at nominal strength.

Once ε_t has been computed, Fig. 8.22 can be used to determine the appropriate ϕ-factor.

8.33.2 Nominal Flexural Strength—Unified Design

As noted previously, nominal strength calculations for members subjected to flexure and/or axial loads have not been changed in the Unified Design Provisions. Nominal strength of any cross-section with any amount and arrangement of reinforcement is determined by satisfying force

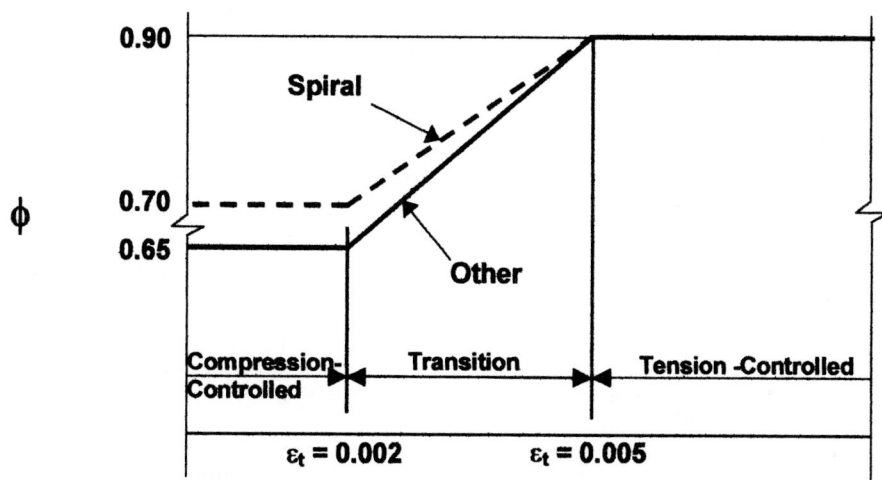

Fig. 8.22 Variation of ϕ with net tensile strain ε_t.

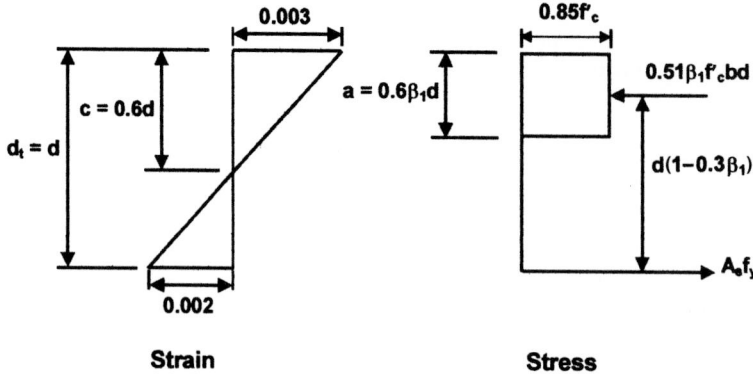

Fig. 8.23 Strains and stresses at compression-controlled limit.

and moment equilibrium, based on the design assumptions given in Section 10.2 of the code, which include compatibility of strain.

The following equations can be used to determine the nominal flexural strength of compression-controlled sections and tension-controlled sections.

Compression-Controlled Sections. For a rectangular section with one layer of Grade 60 reinforcement or prestressed reinforcement, when it is at the compression-controlled strain limit of $\varepsilon_t = 0.002$, the nominal flexural strength is obtained by summing moments about any point on the stress diagram shown in Fig. 8.23:

$$M_{nc} = f'_c bd^2 (0.51\beta_1 - 0.153\beta_1^2) \qquad (8.88)$$

Since the compression-controlled strain limit is not equal to 0.002 for other grades of nonprestressed reinforcement, similar equations to determine the nominal flexural strength at the compression-controlled strain limit for those grades of steel can easily be derived.

Tension-Controlled Sections. The nominal flexural strength of a rectangular section with one layer of Grade 60 or prestressed reinforcement, when it is at the tension-controlled strain limit of $\varepsilon_t = 0.005$, is determined in a similar fashion (see Fig. 8.24):

$$M_{nt} = f'_c bd^2 (0.319\beta_1 - 0.06\beta_1^2) \qquad (8.89)$$

By equating the tension force in the reinforcing steel to the compression force in the concrete,

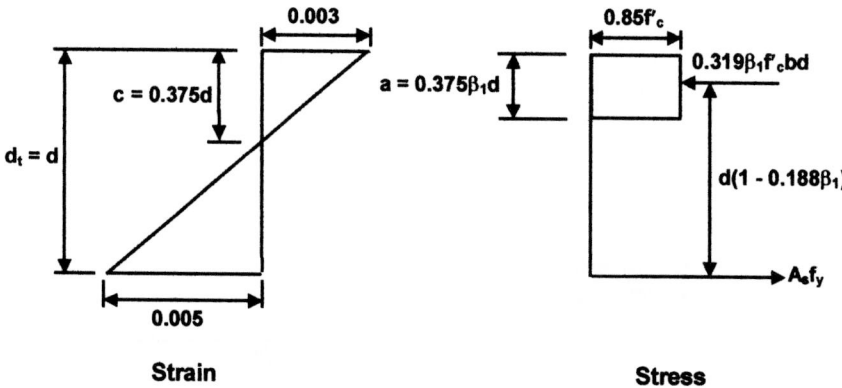

Fig. 8.24 Strains and stresses at tension-controlled limit.

the area of steel at the tension-controlled limit is:

$$A_s = \frac{0.319\beta_1 f_c' bd}{f_y} \qquad (8.90)$$

or

$$\rho_t = \frac{A_s}{bd} = \frac{0.319\beta_1 f_c'}{f_y} \qquad (8.91)$$

Define the reinforcement index ω_t at the tension-controlled strain limit as $\rho_t f_y / f_c'$ and substitute this into the above equation for M_{nt}:

$$M_{nt} = \omega_t(1 - 0.59\omega_t)f_c' bd^2 \qquad (8.92)$$

or

$$R_{nt} = \frac{M_{nt}}{bd^2} = \omega_t(1 - 0.59\omega_t)f_c' \qquad (8.93)$$

where R_{nt} is the nominal strength coefficient of resistance at the tension-controlled strain limit. Values of R_{nt} can readily be determined for any concrete strength and reinforcement. For example, for a section with 4000 psi concrete and Grade 60

reinforcing bars:

$$\rho_t = \frac{0.319 \times 0.85 \times 4}{60} = 0.0181 \qquad (8.94)$$

$$\omega_t = \frac{0.1081 \times 60}{4} = 0.2715 \qquad (8.95)$$

$$R_{nt} = \frac{M_{nt}}{bd^2} = 0.2715[1 - (0.59 \times 0.2715)] \times 4000$$

$$= 912\,\text{psi} \qquad (8.96)$$

The nominal strength of tension-controlled sections is controlled by the strength of the reinforcement, which is less variable than that of the concrete.

General Case. The following equation can be used to determine the nominal flexural strength of a rectangular section with tension reinforcement only:

$$M_n = \omega(1 - 0.59\omega)f_c' bd^2 \qquad (8.97)$$

where

$$\omega = \frac{\rho f_y}{f_c'} \qquad (8.98)$$

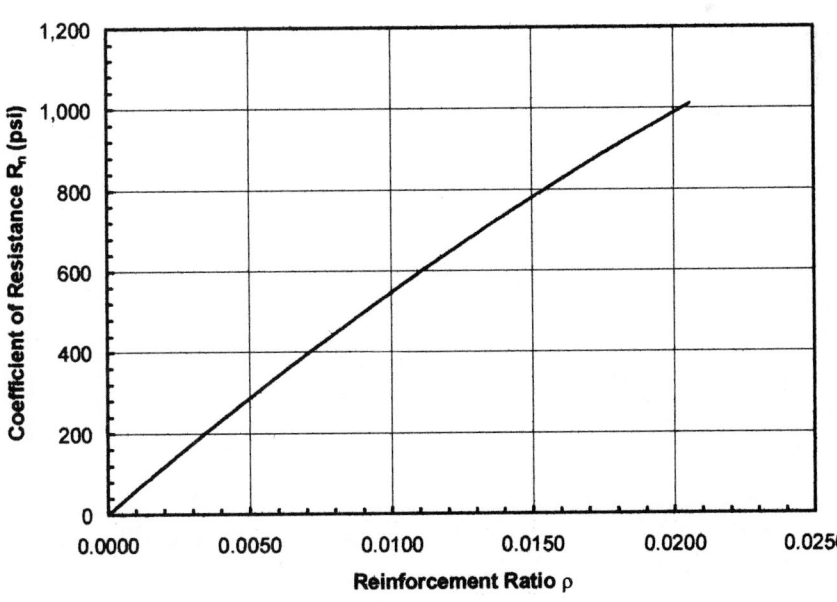

Fig. 8.25 Strength curve for 4000 psi concrete and Grade 60 reinforcement.

and

$$\rho = \frac{A_s}{bd}. \tag{8.99}$$

The nominal strength coefficient of resistance R_n is:

$$R_n = \frac{M_n}{bd^2} = \omega(1 - 0.59\omega)f_c' \tag{8.100}$$

$$= \rho f_y \left(1 - 0.59\rho\frac{f_y}{f_c'}\right) \tag{8.101}$$

Figure 8.25 shows the strength curve for 4000 psi concrete and Grade 60 reinforcement.

It is important to note that ACI Section 10.3.5 limits the amount of flexural reinforcement in nonprestressed flexural members that are subjected to an axial load less than $0.10f_c'A_g$ by requiring that the net tensile strain ε_t must be greater than or equal to 0.004. This limit is slightly more conservative than the reinforcement limit of $0.75\rho_b$ that was imposed in previous editions of the code, since the net tensile strain at nominal strength is 0.00376 when $\rho = 0.75\rho_b$.

8.33.3 Design Flexural Strength—Unified Design

Once the nominal flexural strength has been determined, the design flexural strength is computed by multiplying the nominal flexural strength by the strength reduction factor ϕ, which, as shown above, is determined based on the magnitude of the net tensile strain in the extreme tension steel. The design flexural strength must be greater than or equal to the required strength M_u due to the factored loads:

$$\phi M_n \geq M_u \tag{8.102}$$

The following equation can be used to determine the required area of reinforcement for a factored bending moment M_u. Substitute $M_n = M_u/\phi$ into the equation for R_n:

$$R_n = \frac{M_u}{\phi bd^2} = \rho f_y \left(1 - 0.59\rho\frac{f_y}{f_c'}\right) \tag{8.103}$$

$$\phi R_n = \frac{M_u}{bd^2} = \phi\rho f_y \left(1 - 0.59\rho\frac{f_y}{f_c'}\right) \tag{8.104}$$

The design strength curve for 4000 psi concrete and Grade 60 reinforcement is depicted in Fig. 8.26.

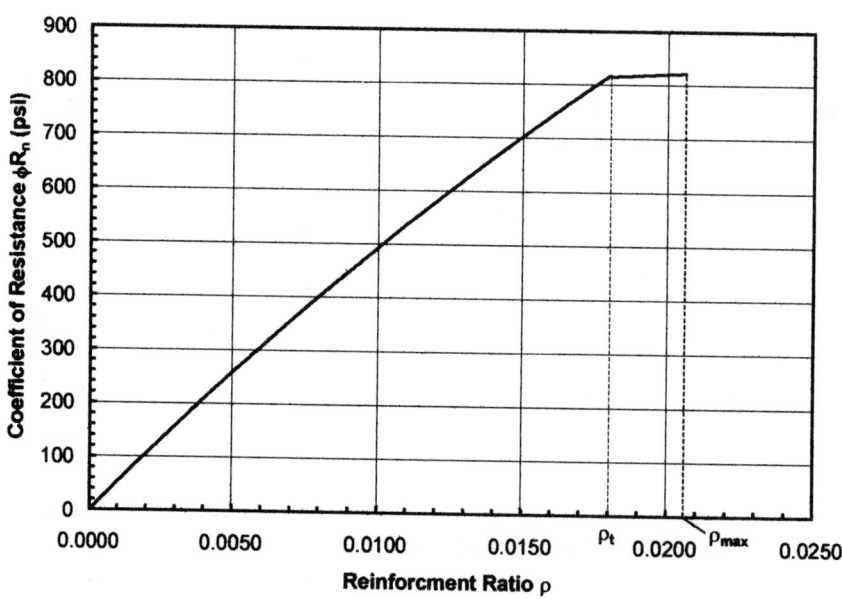

Fig. 8.26 Design strength curve for 4000 psi concrete and Grade 60 reinforcement.

The required reinforcement ratio ρ can be obtained from the figure for values of $\phi R_n = M_u/bd^2$.

Figure 8.26 shows the effect that ϕ has on the design strength. The relationship between the design strength and the reinforcement ratio is approximately linear up to $\rho = \rho_t$, which is the reinforcement ratio corresponding to the net tensile strain ε_t at the tension-controlled limit of 0.005. For the portion of the design strength curve up to $\rho = \rho_t$, $\phi = 0.90$. For reinforcement ratios greater than ρ_t, the net tensile strain is less than 0.005, and ϕ is less than 0.90. The maximum reinforcement ratio ρ_{max} corresponds to a net tensile strain of 0.004.

It is clear from the figure that there is no benefit in designing a flexural member with a reinforcement ratio greater than ρ_t, since any gain in strength with greater amounts of tension reinforcement is cancelled by the reduction in ϕ. Therefore, whenever possible, flexural members should be designed as tension-controlled sections. In cases where member size is limited, it is advisable to increase member strength by adding compression reinforcement instead of additional tension reinforcement so that the section remains tension-controlled.

8.33.4 Nominal Strength for Combined Flexure and Axial Load—Unified Design

The nominal strength of a member subjected to combined flexure and axial load must satisfy equilibrium and compatibility of strains, which are the same two conditions required for members subjected to flexure only. Fig. 8.27 depicts the general condition of strain and stress at nominal strength for a member subjected to combined flexure and axial compression.

The nominal axial load strength is computed by summing forces on the section, while the nominal flexure strength is obtained by summing moments about any point on the stress diagram. The forces in the reinforcement depend on the corresponding magnitude of the strains, which are determined from a strain compatibility analysis.

Once the net tensile strain is determined in the extreme tension steel for a given combination of axial load and bending moment, ϕ is determined as described above. The design axial load strength and design flexural strength of the section are obtained by multiplying the nominal axial load strength and nominal flexural strength by the strength reduction factor, respectively.

A design axial load-bending moment interaction diagram can be constructed for a section in the same manner as before. The only difference in using the Unified Design Provisions is that the ϕ-factors are computed at the various points along the strength curve as a function of the net tensile strain in the extreme tensile steel, as shown above.

8.33.5 Redistribution of Negative Moments— Unified Design

Prior to the 2002 code, the permissible percentage of redistribution of negative moments in continuous

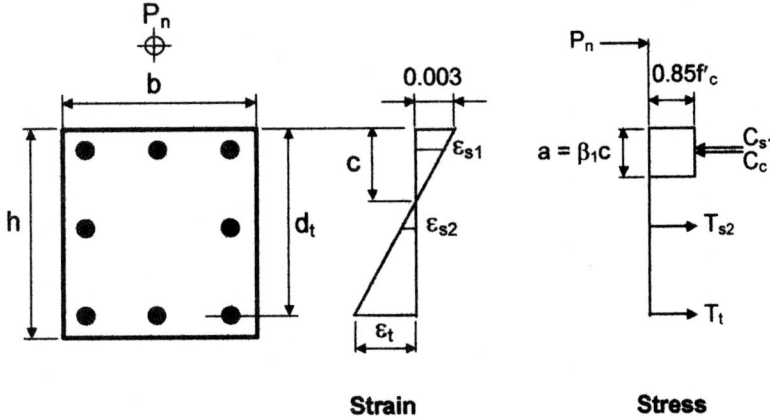

Strain   **Stress**

Fig. 8.27 Strains and stresses for section subjected to combined flexure and axial load.

nonprestressed flexural members was determined as a function of the negative and positive reinforcement in the section and the balanced reinforcement ratio. Similar provisions for continuous prestressed flexural members were also given in a different section of the code. Moment redistribution requirements based on the Unified Design Provisions are given in Section 8.4 of the 2002 code, and are applicable to both continuous nonprestressed and prestressed flexural members. They may not be applied to cases where approximate values for bending moments are used, as in the Direct Design Method of two-way slab design.

According to ACI 318-02 requirements, negative moments may be increased or decreased by not more than $1000\varepsilon_t$ percent, with a maximum of 20 percent, at sections where ε_t is greater than or equal to 0.0075. The lower limit on ε_t is required since moment redistribution depends on adequate ductility. The lower limit in prior codes used to be $\rho \leq 0.5\rho_b$.

These new requirements for moment redistribution are much simpler to apply than those in previous editions of the code. Figure R8.4 in the commentary of the 2002 ACI code shows a comparison between the permissible moment redistribution in the 2002 and earlier codes.

8.34 Planar Walls

These are vertical or near vertical members with length exceeding three times the thickness. Concrete walls may be classified as non-load bearing, load-bearing, or shear walls, which may be either load-bearing or non-load-bearing. Retaining walls are discussed in Arts. 8.41 to 8.43.

8.34.1 Non-Load-Bearing Walls

These are generally basement, retaining, or facade-type walls that support only their own weight and also resist lateral loads. Such walls are principally designed for flexure. By the ACI Code, design requirements include:

1. Ratio of vertical reinforcement to gross concrete area should be at least 0.0012 for deformed bars No. 5 or smaller, 0.0015 for deformed bars No. 6 and larger, and 0.0012 for welded-wire fabric not larger than $\frac{5}{8}$ in in diameter.

2. Spacing of vertical bars should not exceed three times the wall thickness or 18 in.

3. Lateral or cross ties are not required if the vertical reinforcement is 1% or less of the concrete area, or where the vertical reinforcement is not required as compression reinforcement.

4. Ratio of horizontal reinforcement to gross concrete area should be at least 0.0020 for deformed bars No. 5 or smaller, 0.0025 for deformed bars No. 6 and larger, and 0.0020 for welded-wire fabric not larger than $\frac{5}{8}$ in in diameter.

5. Spacing of horizontal bars should not exceed 3 times the wall thickness or 18 in.

Note that walls more than 10″ thick shall have nominal reinforcement for each direction place within two layers parallel with faces of wall.

8.34.2 Load-Bearing Walls

These are subject to axial compression loads in addition to their own weight and, where there is eccentricity of load or lateral loads, to flexure. Load-bearing walls may be designed in a manner similar to that for columns, but including the preceding design requirements for non-load-bearing walls.

As an alternative, load-bearing walls may be designed by an empirical procedure given in the ACI Code when the resultant of all factored loads is located within the middle third of the overall wall thickness.

Load-bearing walls designed by either method should meet the minimum reinforcing requirements for non-load-bearing walls.

In the empirical method the axial capacity, kips, of the wall is

$$\phi P_{nw} = 0.55\phi f_c' A_g \left[1 - \left(\frac{kl_c}{32h} \right)^2 \right] \qquad (8.105)$$

where $f_c' =$ 28-day compressive strength of concrete, ksi

$A_g =$ gross area of wall section, in^2

$\phi =$ strength reduction factor $= 0.70$

$l_c =$ vertical distance between supports, in

$h =$ overall thickness of wall, in

$k =$ effective-length factor

For a wall supporting a concentrated load, the length of wall effective for the support of that concentrated load should be taken as the smaller of the distance center to center between loads and the bearing width plus $4h$.

Reinforced bearing walls designed using Eq. (8.86) should have a thickness of at least $\frac{1}{25}$ of the unsupported height or width, whichever is shorter, but not less than 4 in. Thickness of exterior basement walls and foundation walls must be $7\frac{1}{2}$ in or greater. Also, walls more than 10 in thick, except for basement walls, should have two layers of reinforcement in each direction, with between one-half and two-thirds of the total steel area in the layer near the exterior face of the wall. This layer should be placed at least 2 in but not more than one-third the wall thickness from the face. Walls should be anchored to the floors, or to the columns, pilasters, or intersecting walls.

Walls designed as grade beams should have top and bottom reinforcement as required by the ACI Code for beam design.

8.34.3 Shear Walls

Walls subject to horizontal shear forces in the plane of the wall should, in addition to satisfying flexural requirements, be capable of resisting the shear. The nominal shear stress can be computed from

$$v_u = \frac{V_u}{\phi h d} \qquad (8.106)$$

where V_u = total design shear force

ϕ = capacity reduction factor = 0.75

$d = 0.8l_w$

h = overall thickness of wall

l_w = horizontal length of wall

The shear V_c carried by the concrete depends on whether N_u, the design axial load, lb, normal to the wall horizontal cross section and occurring simultaneously with V_u at the section, is a compression or tension force. When N_u is a compression force, V_c may be taken as $2\sqrt{f_c'}hd$, where f_c' is the 28-day strength of concrete, psi. When N_u is a tension force, V_c may be taken as:

$$V_c = 2\left(1 + \frac{N_u}{500A_g}\right)\sqrt{f_c'}b_w d > 0 \qquad (8.107)$$

where N_u = negative for tension

A_g = gross area of section

N_u/A_g = expressed in psi

Alternatively, more detailed calculations may be made for V_c when V_c is the smallest of:

$$V_c = 3.3\sqrt{f_c'}hd + \frac{N_u d}{4l_w} \qquad (8.108)$$

$$V_c = hd\left[0.6\sqrt{f_c'} + \frac{l_w(1.25\sqrt{f_c'} + 0.2N_u/l_w h)}{M_u/V_u - l_w/2}\right] \qquad (8.108a)$$

where N_u is negative for tension.

Equation (8.90) does not apply, however, when $M_u/V_u - l_w/2$ is negative.

When the factored shear V_u is less than $0.5\phi V_c$, reinforcement should be provided as required by the empirical method for bearing walls.

When V_u exceeds $0.5\phi V_c$, horizontal reinforcement should be provided in accordance with Eq. (8.18), with $V_s = A_v f_y d/s_2$, where s_2 = spacing of horizontal reinforcement and A_v = reinforcement area. Also, the ratio ρ_h of horizontal shear reinforcement to the gross concrete area of the vertical section of the wall should be at least 0.0025. Spacing of horizontal shear bars should not exceed $l_w/5$, $3h$, or 18 in. In addition, the ratio of vertical shear reinforcement area to gross concrete area of the horizontal section of wall need not be greater than that required for horizontal reinforcement but should not be less than

$$\rho_h = 0.0025 + 0.5\left(2.5 - \frac{h_w}{l_w}\right) \qquad (8.109)$$

$$(\rho_h - 0.0025) \le 0.0025$$

where h_w = total height of wall. Spacing of vertical shear reinforcement should not exceed $l_w/3$, $3h$, or 18 in.

In no case should the shear strength V_n be taken greater than $10\sqrt{f_c'}hd$ at any section.

8.35 Composite Columns

A composite column consists of a structural-steel shape, pipe, or tube compression member completely encased in concrete, with or without longitudinal reinforcement.

Composite compression members should be designed in accordance with the provisions applicable to ordinary reinforced concrete columns. Loads assigned to the concrete portion of a member must be transferred by direct bearing on the concrete through brackets, plates, reinforcing bars, or other structural shapes that have been welded to the central structural-steel compression members prior to placement of the perimeter concrete. The balance of the load should be assigned to the structural-steel shape and should be developed by direct connection to the structural shape.

8.35.1 Concrete-Filled Steel Columns

When the composite member consists of a steel-encased concrete core, the required thickness of metal face of width b of a rectangular section is not less than

$$t = b\sqrt{\frac{f_y}{3E_s}} \qquad (8.110)$$

for each face of width, b for circular sections of diameter h,

$$t = h\sqrt{\frac{f_y}{8E_s}} \qquad (8.111)$$

where f_y is the yield strength and E_s the modulus of elasticity of the steel.

8.35.2 Steel-Core Columns

When the composite member consists of a spiral-bound concrete encasement around a structural-steel core, the concrete should have a minimum strength of 2500 psi, and spiral reinforcement should conform to the requirements of Art. 8.31.

When the composite member consists of a laterally tied concrete encasement around a steel core, the concrete should have a minimum strength of 2500 psi. The lateral ties should completely encase the core. Ties should be No. 3 bars or larger but should have a diameter of at least 1/50 the longest side of the cross section. Ties need not be larger than No. 5. Vertical spacing should not exceed one-half of the least width of the cross section, or 48 tie bar diameters, or 16 longitudinal bar diameters. The area of vertical reinforcing bars within the ties should not be less than 1% or more than 8% of the net concrete section. In rectangular sections, one longitudinal bar should be placed in

each corner and other bars, if needed, spaced no farther apart than half the least side dimension of the section.

The design yield strength of the structural core should not be taken greater than 50 ksi, even though a larger yield strength may be specified.

Prestressed Concrete

Prestressing is the application of permanent forces to a member or structure to counteract the effects of subsequent loading. Applied to concrete, prestressing takes the form of precompression, usually to eliminate disadvantages stemming from the weakness of concrete in tension.

8.36 Basic Principles of Prestressed Concrete

The usual prestressing procedure is to stretch high-strength steel (Art. 8.13) and anchor it to the concrete, which resists the tendency of the stretched steel to shorten and thus is compressed. The amount of prestress used generally is sufficient to prevent cracking or sometimes to avoid tension entirely, under service loads. As a result, the whole concrete cross section is available to resist tension and bending, whereas in reinforced-concrete construction, concrete in tension is considered ineffective. Hence, it is particularly advantageous with prestressed concrete to use high-strength concrete. (See also Art. 8.14.)

Prestressed-concrete pipe and tanks are made by wrapping steel wire under high tension around concrete cylinders. Domes are prestressed by wrapping tensioned steel wire around the ring girders. Beams and slabs are prestressed linearly with steel tendons anchored at their ends or bonded to the concrete (Art. 8.14). Piles also are prestressed linearly, usually to counteract handling stresses.

Prestressed concrete may be either pretensioned or posttensioned. For pretensioned concrete, the steel is stretched before the concrete is placed around it and the forces are transferred to the concrete by bond. For posttensioned concrete, bars or tendons are sheathed in ducts within the concrete forms and are tensioned after the concrete attains sufficient strength.

The final precompression of the concrete is not equal to the initial tension applied to the tendons.

There are both immediate and long-time losses, (Art. 8.37), which should be deducted from the initial prestress to determine the effective prestress to be used in design. One reason high-tensioned tendons are used for prestressing is to maintain the sum of these losses at a small percentage of the applied prestress.

In determining stresses in prestressed members, the prestressing forces may be treated the same way as other external loads. If the prestress is large enough to prevent cracking under design loads, elastic theory may be applied to the entire concrete cross section.

For example, consider the simple beam in Fig. 8.28a. Prestress P is applied by a straight tendon at a distance e_1 below the neutral axis. The resulting prestress in the extreme surfaces throughout equals $P/A \pm Pe_1c/I$, where P/A is average stress on a cross section and Pe_1c/I, the bending stress ($+$ represents compression, $-$ represents tension), as indicated in Fig. 8.28c. If, now, stresses $+ Mc/I$ due to downward-acting loads are superimposed at midspan, the net stresses in the extreme surfaces may become zero at the bottom and compressive at the top (Fig. 8.28c). Since the stresses due to loads at the beam ends are zero, however, the prestress is the final stress there. Hence, the top of the beam at the ends will be in tension.

If this is objectionable, the tendons may be draped, or harped, in a vertical curve, as shown in Fig. 8.28b. Stresses at midspan will be substantially the same as before (assuming the horizontal component of P approximately equal to P), and the stress at the ends will be a compression, P/A, since P passes through the centroid of the section there. Between midspan and the ends, the cross sections also are in compression (Fig. 8.28d).

8.37 Losses in Prestress

As pointed out in Art. 8.36, the prestressing force acting on the concrete differs from the initial tension on the tendons by losses that occur immediately and over a long time.

8.37.1 Elastic Shortening of Concrete

In pretensioned members (Art. 8.14), when the tendons are released from fixed abutments and the steel stress is transferred to the concrete by bond, the concrete shortens because of the compressive stress. For axial prestress, the decrease in inches per inch of length may be taken as P_i/AE_c, where P_i is the initial prestress, kips; A the concrete area, in^2; and E_c the modulus of elasticity of the concrete, ksi. Hence, the decrease in unit stress in the tendons equals $P_iE_s/AE_c = nf_c$, where E_s is the modulus of elasticity of the steel, ksi; n the modular ratio; and f_c the stress in the concrete, ksi.

In posttensioned members, if tendons or cables are stretched individually, the stress loss in each due to compression of the concrete depends on the order of stretching. The loss will be greatest for the first tendon or cable stretched and least for the last one. The total loss may be approximated by assigning half the loss in the first cable to all. As an alternative, the tendons may be brought to the final prestress in steps.

8.37.2 Frictional Losses

In posttensioned members, there may be a loss of prestress where curved tendons rub against their enclosure. For harped tendons, the loss may be computed in terms of a curvature-friction coefficient μ. Losses due to unintentional misalignment may be calculated from a wobble-friction coefficient K (per lin ft). Since the coefficients vary considerably with duct material and construction methods, they should, if possible, be determined experimentally or obtained from the tendon manufacturer. Table 8.16 lists values of K and μ suggested in the Commentary to the 2002 Edition "Building Code Requirements for Structural Concrete" (American Concrete Institute (www.aci-int.org)) for posttensioned tendons.

With K and μ known or estimated, the friction loss can be computed from

$$P_s = P_x e^{Kl_x + \mu\alpha} \tag{8.112}$$

where P_s = force in tendon at prestressing jack, lb

P_x = force in tendon at any point x, lb

$e = 2.718$

l_x = length of tendon from jacking point to point x, ft

α = total angular change of tendon profile from jacking end to point x, rad

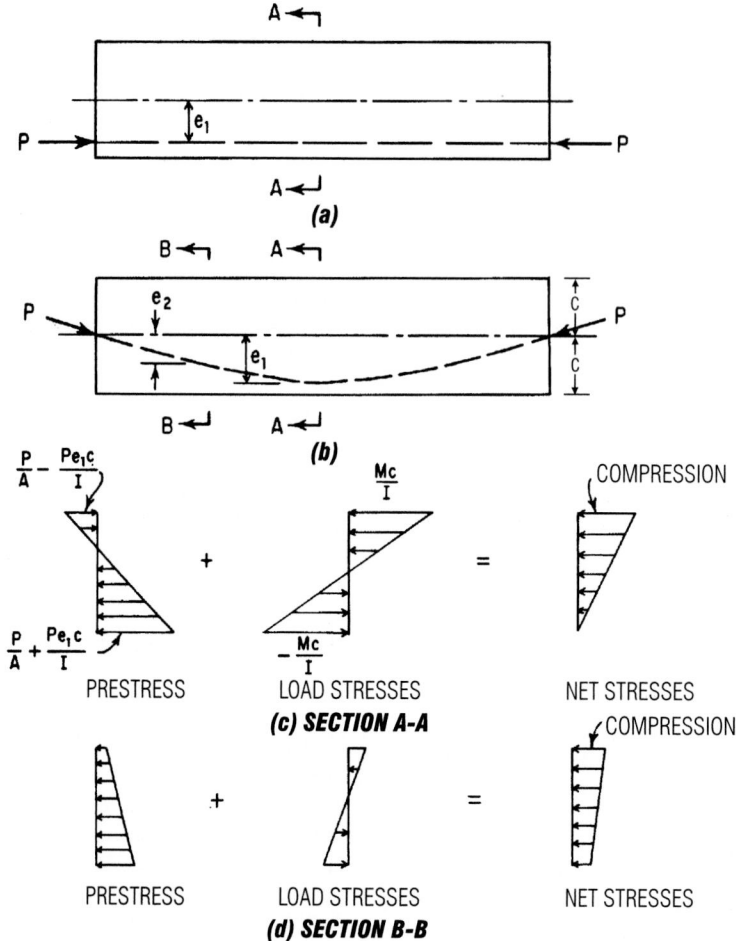

Fig. 8.28 Concrete beams: (*a*) Prestressed with straight tendons. (*b*) Prestressed with draped tendons. (*c*) Stress distribution at midspan. (*d*) Stress distribution for draped tendons at section between support and midspan. For straight tendons, net stress may be tensile near the supports.

When $Kl_x + \mu\alpha$ does not exceed 0.3, P_s may be obtained from

$$P_s = P_x(1 + Kl_x + \mu\alpha) \qquad (8.113)$$

8.37.3 Slip at Anchorages

For posttensioned members, prestress loss may occur at the anchorages during the anchoring. For example, seating of wedges may permit some shortening of the tendons. If tests of a specific anchorage device indicate a shortening δl, the

decrease in unit stress in the steel is $E_s \delta l / l$, where l is the length of the tendon.

8.37.4 Shrinkage of Concrete

Change in length of a member due to concrete shrinkage results over time in prestress loss. This should be determined from test or experience. Generally, the loss is greater for pretensioned members than for posttensioned members, which are prestressed after much of the shrinkage has occurred. Assuming a shrinkage of 0.0002 in/in for

Table 8.16   Friction Coefficients for Post-tensioned Tendons

		Wobble Coefficient, K	Curvature coefficient, μ
Grouted tendons in metal sheathing	Wire tendons	0.0010–0.0015	0.15–0.25
	High-strength bars	0.0001–0.0006	0.08–0.30
	7-wire strand	0.0005–0.0020	0.15–0.25
Unbonded tendons	Mastic coated	Wire tendons	0.0010–0.0020	0.05–0.15
		7-wire strand	0.0010–0.0020	0.05–0.15
	Pre-greased	Wire tendons	0.0003–0.0020	0.05–0.15
		7-wire strand	0.0003–0.0020	0.05–0.15

a pretensioned member, the loss in tension in the tendons will be

$$0.0002E_s = 0.0002 \times 30,000 = 6\,\text{ksi}$$

8.37.5   Creep of Concrete

Change in length of concrete under sustained load induces a prestress loss over time. This loss may be several times the elastic shortening. An estimate of the loss may be made with a creep coefficient C_c, equal to the ratio of additional long-time deformation to initial elastic deformation, determined by test. Hence, for axial prestress, the loss in tension in the steel is $C_c n f_c$, where n is the modular ratio and f_c is the prestressing force divided by the concrete area. (Values ranging from 1.5 to 2.0 have been recommended for C_c.)

8.37.6   Relaxation of Steel

Decrease in stress under constant high strain occurs with some steels. For example, for steel tensioned to 60% of ultimate strength, relaxation loss may be 3%. This type of loss may be reduced by temporary overstressing, stabilizing the tendons by artificially accelerating relaxation and thus reducing the loss that will occur later at lower stresses.

Actual losses should be computed based on the actual initial stress level, type of steel (stress-relieved or low relaxation; wire, strand or bar), and prestressing method (pretensioned or posttensioned).

8.38   Allowable Stresses in Prestressed Concrete — AASHTO

The "Standard Specifications for Highway Bridges" (American Association of State Highway and Transportation Officials) require that the design of precast prestressed members ordinarily must be based on $f'_c = 5000\,\text{psi}$. An increase to 6000 psi is permissible where, in the Engineer's judgment, it is reasonable to expect that this strength will be obtained consistently. Still higher concrete strengths may be considered on an individual area basis. In such cases, the Engineer must satisfy himself completely that the controls over materials and fabrication procedures will provide the required strengths.

In setting allowable stresses for prestressed concrete, design codes recognize two loading stages: application of initial stress and loading under service conditions. The codes permit higher stresses for the temporary loads during the initial stage.

Stresses due to the jacking force and those produced in the concrete and steel immediately after prestress transfer or tendon anchorage, before losses due to creep and shrinkage, are considered temporary. Permissible temporary stresses in the concrete are specified as a percentage of f'_{ci}, the compressive strength of the concrete psi, at time of initial prestress, instead of the usual f'_c, 28-day strength of the concrete. This is done because prestress usually is applied only a few days after casting the concrete.

Table 8.17 Allowable Stresses in Prestressed-Concrete Flexural Members—AASHTO Standard Specifications

---

Prestressing Steel
  Pretensioned members:
    Stress immediately prior to transfer—
    Low relaxation strands $\qquad$ $0.75f_s'$
    Stress-relieved strands $\qquad$ $0.70f_s'$
  Post-tensioned members:
    Stress immediately after seating—
    At anchorage $\qquad$ $0.70f_s'$
    At the end of the seating loss zone $\qquad$ $0.83f_y*$
    Tensioning to $0.90f_y*$ for short periods of time prior to seating
    may be permitted to offset seating and friction losses provided
    the stress at the anchorage does not exceed the above value.
  Stress at service load after losses $\qquad$ $0.80f_y*$

Concrete
  Temporary Stresses Before Losses Due to Creep and Shrinkage
  Compresion:
    Pretensioned members $\qquad$ $0.60f_{ci}'$
    Post-tensioned members $\qquad$ $0.55f_{ci}'$
  Tension:
    Precompressed tensile zone $\qquad$ No temporary allowable stresses are specified

    Other Areas
    In tension areas with no bonded reinforcement $\qquad$ 200 psi or $3\sqrt{f_{ci}'}$
    Where the calculated tensile stress exceeds this value, bonded
    reinforcement shall be provided to resist the total tension force in
    the concrete computed on the assumption of an uncracked
    section. The maximum tensile stress shall not exceed $\qquad$ $7.5\sqrt{f_{ci}'}$

  Stress at Service Load After Losses Have Occurred
  Compression:
    (a) The compression stresses under all load combinations, except
        as stated in (b) and (c), shall not exceed $0.60f_c'$.
    (b) The compressive stresses due to effective prestress plus
        permanent (dead) loads shall not exceed $0.40f_c'$.
    (c) The compressive stress due to live loads plus one-half of the
        sum of the compressive stresses due to prestress and
        permanent (dead) loads shall not exceed $0.40f_c'$.
  Tension in the precompressed tensile zone:
    (a) For members with bonded reinforcement (including bonded $\qquad$ $6\sqrt{f_{ci}'}$
        prestressed strands)
    (b) For severe corrosive exposure conditions, such as coastal areas $\qquad$ $3\sqrt{f_{ci}'}$
    (c) For members without bonded reinforcement $\qquad$ 0
  Tension in other areas is limited by allowable temporary
  stresses specified above.

Anchorage Bearing Stress
  Post-tensioned anchorage at service load $\qquad$ 3,000 psi (but not to exceed $0.9f_{ci}'$)

The allowable stresses for prestressed concrete, in accordance with the ASSHTO Standard Specifications, are given in Table 8.17. In the table, f'_s is the ultimate stress (tensile strength) of prestressing steel, and f^*_y is the yield stress (strength) of prestressing steel.

$$f^*_y = 0.90\,f'_s \text{ for low-relaxation wire or strand}$$
$$= 0.85\,f'_s \text{ for stress-relieved wire or strand}$$
$$= 0.85\,f'_s \text{ for Type I (smooth) high-strength bar}$$
$$= 0.80\,f'_s \text{ for Type II (deformed)}$$
high-strength bar

8.39 Allowable Stresses in Prestressed Concrete— ACI 318

The 2002 edition of ACI 318 "Building Code Requirements for Structural Concrete" (American Concrete Institute) requires prestressed flexural members to be classified as Class U, Class T, or Class C based on the computed extreme fiber stress f_t at service loads in the precompressed tensile zone, as follows:

$$\text{Class U: } f_t \le 7.5\sqrt{f'_c}$$

$$\text{Class T: } 7.5\sqrt{f'_c} \le f_t \le 12\sqrt{f'_c}$$

$$\text{Class C: } f_t > 12\sqrt{f'_c}$$

Class U members are assumed to behave as uncracked members. Class C members are assumed to behave as cracked members. The behavior of Class T members is assumed to be in transition between uncracked and cracked. The serviceability requirements for each class are summarized in Table R18.3.3, in the Commentary to the 2002 edition of ACI 318. For comparison, the table also shows the corresponding requirements for nonprestressed members.

The classes apply to both bonded and unbonded prestressed flexural members, but prestressed two-way slab systems are required to be designed as Class U.

For Class U and Class T flexural members, stresses at service loads are permited to be calculated using the uncracked section. For Class C flexural members, stresses at service loads must be calculated using the cracked transformed section.

Table 8.18 Allowable Stresses in Prestressed-Concrete Flexural Members—ACI 318

Stresses at transfer or anchoring:	
Compression in concrete	$0.60f'_{ci}$
Tension in concrete without auxiliary reinforcement in the tension zone*	$3\sqrt{f'_{ci}}$
Prestress in tendons due to jacking force†	$0.94f_{py} \le 0.80f_{pu}$
Prestress in tendons immediately after transfer or anchoring	$0.82f_{py} \le 0.74f_{pu}$
Stresses under service loads:	
Compression in concrete‡	
• due to prestress plus sustained load	$0.45f'_c$
• due to prestress plus total load	$0.60f'_c$

* Where the calculated tension stress exceeds this value, bonded additional reinforcement (prestressed or nonprestressed) should be provided to resist the total tension force on the concrete computed on the assumption of an uncracked section. At ends of simply supported beams, the allowable stress may be taken as $6\sqrt{f'_{ci}}$

† But not greater than the maximum value recommended by the manufacturer of the steel or anchorages. f_{py} = yield strength of tendons.

‡ For Class U and class T prestressed flexural members only.

The allowable stresses for prestressed concrete, in accordance with ACI 318, are listed in Table 8.18. In the table, f_{pu} is the specified tensile strength of prestressing steel, and f_{py} is the specified yield strength of prestressing steel.

Class C prestressed members are, for all practical purposes, treated like nonstressed members. As such, there is no upper limit on the tensile stress that may develop at a section under service loads. There is no explicit limit on the compressive stresses that can develop under service-level loads either. However, crack control and deflection control requirements become applicable for Class C prestressed flexural members. The introduction of the Class C flexural member enables a user of ACI 318 to design a partially prestressed concrete flexural member (combining mild reinforcement with prestressing steel) for the first time under the 2002 edition of the code.

For prestressed concrete members exposed to corrosive environments or other severe exposure conditions, and which are classified as Class T or C, minimum cover to the prestressed reinforcement (see Sections 7.7.2 and 7.7.3 of ACI 318-02) is required to be increased 50 percent. This requirement is permitted to be waived if the precompressed tensile zone is not in tension under sustained loads. The engineer may also consider reducing tensile stresses in the concrete to eliminate possible cracking at service loads.

Comprehensive requirements for post-tensioned anchorage zone design have been included in ACI 318 since its 1999 edition (see Section 18.13).

8.40 Design of Prestressed-Concrete Beams

This involves selection of shape and dimensions of the concrete portion, type and positioning of tendons, and amount of prestress. After a concrete shape and dimensions have been assumed, determine geometric properties: cross-sectional area, center of gravity, distances of extreme surfaces from the centroid, section moduli, and dead load of member per unit of length. Treat prestressing forces as a system of external forces acting on the concrete (see Art. 8.37).

Compute bending stresses due to dead and live loads. From these, determine the magnitude and location of the prestressing force at points of maximum moment. This force must provide sufficient compression to offset the tensile stresses caused by the bending moments due to loads (Fig. 8.21). But at the same time, it must not create any stresses exceeding the allowable values listed in Art. 8.38 or Art. 8.39. Investigation of other sections will guide selection of tendons to be used and determine their position in the beam.

After establishing the tendon profile, prestressing forces, and tendon areas, check critical points along the beam under initial and final conditions, on removal from the forms, and during erection. Check ultimate strength in flexure and shear and the percentage of prestressing steel. Design anchorages, if required, and diagonal-tension steel. Finally, check camber.

The design may be based on the following assumptions. Strains vary linearly with depth. At cracked sections, the concrete cannot resist tension. Before cracking, stress is proportional to strain. The transformed area of bonded tendons may be included in pretensioned members and in posten-

sioned members after the tendons have been grouted. Areas of open ducts should be deducted in calculations of section properties before bonding of tendons. The modulus of rupture should be determined from tests, or the cracking stress may be assumed as $7.5\sqrt{f_c'}$, where f_c' is the 28-day strength of the concrete, psi.

Prestressed beams should be checked by the strength theory (Art. 8.17). Beams for building should be capable of supporting the factored loads given in Chap. 9 of ACI 318-02. For bridge beams, the nominal strength should not be less than

$$\frac{U}{\phi} = \frac{1.30}{\phi}\left[D + \frac{5}{3}(L+I)\right] \qquad (8.114)$$

where D = effect of dead load

L = effect of design live load

I = effect of impact

ϕ = 1.0 for factory-produced precast, prestressed members

= 0.95 for posttensioned, cast-in-place members

= 0.90 for shear

The "Standard Specifications for Highway Bridges" (American Association of State Highway and Transportation Officials) recommend that prestressed-concrete flexural members be assumed to act as uncracked members subjected to combined axial and bending stresses under specified service loads. In pretensioned members and in posttensioned members after tendons have been grouted, the transformed area of bonded reinforcement may be taken into account in computations of section properties. For calculations of section properties before bonding of tendons, areas of open ducts should be deducted.

8.40.1 Steel Stress—AASHTO

The following definitions will be used for the formulas that follow:

A_s = area of non-prestressed tension reinforcement, in^2

A_s' = area of compression reinforcement, in^2

A_s^* = area of prestressing steel, in^2

A_{sf} = steel area required to develop the compressive strength of the overhanging portions of the flange

A_{sr} = steel area required to develop the compressive strength of the web of a flanged section, in²

b = width of a rectangular member or flange of a flanged member, in

b' = width of a web of a flanged member, in

d = distance, in, from extreme compression fiber to the centroid of the prestressing force

d_t = distance, in, from extreme compression fiber to the centroid of the non-prestressed tension reinforcement

f_c' = 28-day compressive strength of the concrete, psi

f_s' = ultimate strength of prestressing steel, psi

f_{su}^* = average stress in prestressing steel at ultimate load, psi

f_{sy} = yield strength of non-prestressed tension reinforcement, psi

f_y' = yield strength of non-prestressed compression reinforcement, psi

$p = A_s/bd_t$

$p^* = A_s^*/bd$

The value of the average stress in prestressing steel at ultimate load f_{su}^* may be determined by analysis. It may be estimated, however, from the following if the effective prestress after losses is at least half the ultimate strength of the prestressing steel:

For prestressed beams with unbonded tendons:

$$f_{su}^* = f_{se} + 900\{(d - y_u)/l_e\} \leq f_y^* \tag{8.115}$$

where f_{se} = effective prestress after losses

y_u = distance from extreme compression fiber to the neutral axis assuming the prestressing steel has yielded

$l_e = l_i/(1 + 0.5N_s)$, effective tendon length

l_i = tendon length between anchorages (in)

N_s = number of support hinges crossed by the tendon between anchorages or discretely bonded points.

f_y^* = average yield stress as defined in Art 8.37.

For prestressed beams with bonded prestressing steel and no non-prestressed tension reinforcement:

$$f_{su}^* = f_s'\left(1 - \frac{\gamma^*}{\beta_1}\frac{p^*f_s'}{f_c'}\right) \tag{8.116}$$

and with non-prestressed tension reinforcement:

$$f_{su}^* = f_s'\left[1 - \frac{\gamma^*}{\beta_1}\left(\frac{p^*f_s'}{f_c'} + \frac{d_t}{d}\frac{pf_{sy}}{f_c'}\right)\right] \tag{8.117}$$

where f_s' = ultimate strength of prestressing steel, psi

γ = factor for type of tendon used

= 0.28 for low-relaxation steel

= 0.40 for stress-relieved steel

= 0.55 for high-strength steel bars

β_1 = factor a/c defined in Art. 8.17.2

The design strength of prestressed beams depends on whether the reinforcement indexes $(p^*f_{su}^*)/f_c'$ for rectangular sections and $A_{sr}f_{su}^*/(b'df_c')$ for flanged sections are less than $36\beta_1$.

8.40.2 Steel Stress—ACI

As an alternative to a more accurate determination of f_{ps} (stress in prestressed reinforcement at nominal strength) based on strain compatibility, the following approximate values of f_{ps} may be used if f_{se} is not less than $0.5f_{pu}$.

For members with bonded tendons:

$$f_{ps} = f_{pu}\left\{1 - \frac{\gamma_p}{\beta_1}\left[\rho_p\frac{f_{pu}}{f_c'} + \frac{d}{d_p}(\omega - \omega')\right]\right\} \tag{8.118}$$

If any compression reinforcement is taken into account when calculating f_{ps} by the above equation, the term

$$\left[\rho_p\frac{f_{pu}}{f_c'} + \frac{d}{d_p}(\omega - \omega')\right]$$

is to be taken not less than 0.17 and d' is to be taken no greater than $0.15d_p$.

For members with unbonded tendons and with a span-to-depth ratio of 35 or less:

$$f_{ps} = f_{se} + 10,000 + \frac{f'_c}{100\rho_p} \qquad (8.119)$$

but no greater than f_{py}, nor greater than $(f_{se} + 60,000)$.

For members with unbonded tendons and with a span-to-depth ratio greater than 35:

$$f_{ps} = f_{se} + 10,000 + \frac{f'_c}{300\rho_p} \qquad (8.120)$$

but no greater than f_{py}, nor greater than $(f_{se} + 30,000)$

The notation used above is explained below:

A_{ps} = area of prestressed reinforcement in tension zone, in^2

A_s = area of nonstressed tension reinforcement, in^2

A'_s = area of compression reinforcement, in^2

b = width of compression face of member, in

d = distance from extreme compression fiber to centroid of nonprestressed tension reinforcement, in

d' = distance from extreme compression fiber to centroid of compression reinforcement, in

d_p = distance from extreme compression fiber to centroid of prestressed reinforcement, in

f'_c = specified compressive strength of concrete, psi

f_{ps} = stress in prestressed reinforcement at nominal strength, psi

f_{pu} = specified tensile strength of prestressing steel, psi

f_{py} = specified yield strength of prestressing steel, psi

f_{se} = effective stress in prestressed reinforcement (after allowance for all prestress losses), psi

f_y = specified yield strength of nonstressed reinforcement, psi

β_1 = stress block depth parameter

= 0.85 for $f'_c \leq 4000$ psi

= $0.85 - 0.05(f'_c - 4000/1000) \geq 0.65$ for $f'_c > 4000$ psi

γ_p = factor for type of prestressing steel

= 0.55 for f_{py}/f_{pu} not less than 0.80

= 0.40 for f_{py}/f_{pu} not less than 0.85

= 0.28 for f_{py}/f_{pu} not less than 0.90

ρ = ratio of nonprestressed tension reinforcement = A_s/bd

ρ' = ratio of compression reinforcement = A'_s/bd

ρ_p = ratio of prestressed reinforcement = A_{ps}/bd_p

$\omega = \rho f_y/f'_c$

$\omega' = \rho' f_y/f'_c$

8.40.3 Design Strength When Indexes Are 36β_1 or Less—AASHTO

AASHTO Specifications state that prestressed concrete members should be designed so that the steel is yielding as ultimate strength is approached. This requires, in general, that reinforcement indexes not exceed $36\beta_1$. When this requirement is met, design flexural strength ϕM_n, in-kips, with ϕ as given for Eq. (8.114), is determined as follows:

For rectangular sections with prestressing steel only and for flanged sections with prestressing steel only, when the depth of the equivalent rectangular stress block, $(A^*_s f^*_{su}/0.85f'_c b)$, does not exceed the compression flange thickness t,

$$\phi M_n = \phi\left[A^*_s f^*_{su} d\left(1 - 0.6\frac{p^* f^*_{su}}{f'_c}\right)\right] \qquad (8.121)$$

When non-prestressed tension reinforcement with a yield strength f_{sy} is used and the depth of the equivalent rectangular stress block, $[(A^*_s f^*_{su} + A_s f_y)/0.85f'_c b]$ does not exceed the compression flange thickness t,

$$\phi M_n = \phi\left[A^*_s f^*_{su} d\left(1 - 0.6\frac{p^* f^*_{su}}{f'_c} + \frac{d_t}{d}\frac{p f_{sy}}{f'_c}\right)\right. $$
$$\left. + A_s f_{sy} d_t\left[1 - 0.6\left(\frac{d}{d_t}\frac{p^* f^*_{su}}{f'_c} + \frac{p f_{sy}}{f'_c}\right)\right]\right] \qquad (8.122)$$

For flanged sections with prestressing steel only but with a deeper stress block than that specified for Eq. (8.121),

$$\phi M_n = \phi \left[A_{sr} f_{su}^* d \left(1 - 0.6 \frac{A_{sr} f_{su}^*}{b' d f_c'} \right) \right. \\ \left. + 0.85 f_c' t (b - b')(d - 0.5t) \right] \quad (8.123)$$

where $A_{sr} = A_s^* - A_{sf}$

A_{sf} = steel area required to develop the ultimate compressive strength of the overhanging portion of the flange

$= 0.85 f_c'(b - b')t/f_{su}^*$

For flanged sections with non-prestressed tension reinforcement but with a deeper stress block than that specified for Eq. (8.122),

$$\phi M_n = \phi \left[A_{sr} f_{su}^* d \left(1 - 0.6 \frac{A_{sr} f_{su}^*}{b' d f_c'} \right) \right. \\ + A_s f_{sy}(d_t - d) \\ \left. + 0.85 f_c' t (b - b')(d - 0.5t) \right] \quad (8.124)$$

where $A_{sr} = A_s^* + A_s f_{sy}/f_{su}^* - A_{sf}$.

8.40.4 Design Strength When Indexes Are $36\beta_1$ or More—AASHTO

The design flexural strength ϕM_n, in-kips, for prestressed beams with reinforcement indexes larger than $36\beta_1$ may be determined as follows: For rectangular sections, with ϕ given by Eq. (8.114),

$$\phi M_n = \phi[(0.36\beta_1 - 0.08\beta_1^2) f_c' bd^2] \quad (8.125)$$

For flanged sections,

$$\phi M_n = \phi[(0.36\beta_1 - 0.08\beta_1^2) f_c' b' d^2 \\ + 0.85 f_c' t(b - b')(d - 0.5t)] \quad (8.126)$$

8.40.5 Design Strength—ACI

Design moment strength of flexural members is to be computed by the strength design method of ACI 318. For prestressing steel, f_{ps} is to be substituted for f_y is strength computations. Irrespective of whether the traditional ACI load combination of Appendix C or the ASCE 7-98 load combinations from Chapter 9 of ACI 318-02 are used, the strength

reduction factor, ϕ is as follows (see Article 8.33 on Unified Design Procedure)

Tension-controlled sections	0.90
Compression-controlled sections, spirally reinforced	0.75
Compression-controlled sections, other	0.70

For sections in which the net tensile strain is between the limits for compression-controlled and tension-controlled sections, ϕ may be increased linearly from that for compression-controlled sections to 0.90 as the net tensile strain increases from the compression-controlled strain limit (which may be taken as 0.002 for Grade 60 reinforcement and for all prestressed reinforcement) to the tension-controlled strain limit (of 0.005).

If the traditional design approach in Appendix B of ACI 318-02 is used, the indexes ω_p, $[\omega_p + d/d_p(\omega - \omega')]$, or $[\omega_{pw} + d/d_p(\omega_w - \omega_w')]$ are restricted to $0.36\beta_1$,

where $\omega_p = \rho_p f_{ps}/f_c' = A_{ps} f_{ps}/bd_p f_c'$

$\omega_w, \omega_{pw}, \omega_w'$ = reinforcement indices for flanged sections computed as for ω, ω_p, and ω' except that b is to be the web width, and reinforcement area must be that required to develop compressive strength of web only.

Design moment strength of over-reinforced sections may be computed using strength equations similar to those for nonprestressed concrete members. The 1983 edition of ACI 318 provided strength equations for over-reinforced rectangular and flanged sections.

If the Unified Design Procedure in Chapter 18 of ACI 318-02 (see Article 8.33) is used, the net tensile strain, ϵ_t is restricted to an upper limit of 0.004.

8.40.6 Minimum Steel Required—AASHTO

The AASHTO Specifications require that the total amount of tendons and non-prestressed reinforcement be adequate to develop an ultimate strength ϕM_n that is at least 20% larger than the cracking moment M_{cr}^*. For a composite section,

$$M_{cr}^* = (f_r + f_{pe})S_c - M_{d/nc}\left(\frac{S_c}{S_b} - 1\right) \quad (8.127)$$

where f_r = modulus of rupture of the concrete

$= 7.5\sqrt{f_c'}$ for normal-weight concrete

f_{pe} = compressive stress in the concrete due to effective prestress forces only, after allowing for prestress losses, at the extreme surface of the section where tensile stress is caused by externally applied loads

S_b, S_c = noncomposite and composite section modulus, respectively, for the extreme surface of the section where tensile stress is caused by externally applied loads

$M_{d/nc}$ = noncomposite dead-load moment at the section

For a noncomposite section,

$$M^*_{cr} = (f_r + f_{pe})S_b \tag{8.128}$$

The above requirement may be varied if the area of prestressed and non-prestressed reinforcement provided at a section is at least one-third greater than that required by analysis based on the factored load combinations of AASHTO. The minimum amount of non-prestressed longitudinal reinforcement provided in the cast-in-place position of slabs utilizing precast prestressed deck panels must be 0.25 in² per foot of slab width.

8.40.7 Minimum Steel Required—ACI

The total amount of prestressed and nonprestressed reinforcement must be adequate to develop a factored load at least 1.2 times the cracking load computed on the basis of the modulars of rupture f_r specified in ACI 318 ($7.5\sqrt{f'_c}$ for normal-weight concrete). This provision may be waived for flexural members with shear and flexural strength at least twice those required by the factored loads.

Part or all of the bonded reinforcement consisting of bars or tendons are to be provided as close as practicable to the extreme tension fiber in all prestressed flexural members. In members prestressed with unbonded tendons, the minimum bonded reinforcement consisting of bars or wires must satisfy specific code requirements (See Article 8.40.8).

8.40.8 Shear in Prestressed Beams—AASHTO

"Standard Specifications for Highway Bridges" (American Association of State Highway and Transportation Officials) require that prestressed members be designed to resist diagonal tension by the strength theory.

Shear reinforcement should consist of stirrups or welded-wire fabric. The area of shear reinforcement, in², set perpendicular to the beam axis, should not be less than

$$A_v = \frac{50b's}{f_{sy}} \tag{8.129}$$

where s is the reinforcement spacing, except when the factored shear force V_u is less than one-half ϕV_c. The capacity reduction factor ϕ should be taken as 0.85.

The yield strength of shear reinforcement, f_{sy}, used in design calculations should not exceed 60,000 psi.

Where shear reinforcement is required, it should be placed perpendicular to the axis of the member and should not be spaced farther apart than $0.75h$, where h is the overall depth of the member, or 24 in. Web reinforcement between the face of support and the section at a distance $h/2$ from it should be the same as the reinforcement required at that section.

When V_u exceeds the design shear strength ϕV_c of the concrete, shear reinforcement must be provided. The shear strength provided by concrete, V_c, must be taken as the lesser of the value V_{ci} or V_{cw}.

The shear strength, V_{ci}, is to be computed by

$$V_{ci} = 0.6\sqrt{f'_c}b'd + V_d + \frac{V_i M_{cr}}{M_{max}} \tag{8.130}$$

$$\geq 1.7\sqrt{f'_c}b'd$$

and d need not be taken less than $0.8h$. V_d = shear force at section due to unfactored dead load.

The moment causing flexural cracking at the section due to externally applied loads, M_{cr}, is to be computed by

$$M_{cr} = \frac{1}{Y_t}(6\sqrt{f'_c} + f_{pe} - f_d) \tag{8.131}$$

where Y_t = distance from centroidal axis of gross section, neglecting reinforcement, to extreme fiber in tension

f_d = stress due to unfactored dead load, at extreme fiber of section where tensile stress is caused by externally applied loads

f_{pe} = compressive stress in concrete due to effective prestress forces only (after allowance for all prestress losses) at extreme fiber of section where tensile stress is caused by externally applied loads.

The shear strength, V_{cw}, is to be computed by

$$V_{cw} = (3.5\sqrt{f'_c} + 0.3f_{pc})b'd + V_p \qquad (8.132)$$

but d need to be taken less than $0.8h$. f_{pc} = compressive stress in concrete (after allowance for all prestress losses) at centroid of cross section resisting externally applied loads or at junction of web and flange when the centroid lies within the flange. V_p = vertical component of effective prestress force at section.

For a pretensioned member in which the section at a distance $h/2$ from the face of support is closer to the end of the member than the transfer length of the prestressing steel, the reduced prestress should be considered when computing V_{cw}. The prestress force may be assumed to vary linearly from zero at the end of the prestressing steel to a maximum at a distance from the end of the prestressing steel equal to the transfer length, assumed to be 50 diameters for strand and 100 diameters for single wire.

When $V_u - \phi V_c$ exceeds $4\sqrt{f'_c}b_wd$, the maximum spacing of stirrups must be reduced to $0.375h$, but not more than 12 in. But $V_u - \phi V_c$ must not exceed $8\sqrt{f'_c}b_wd$.

The shear strength provided by web reinforcement is to be taken as

$$V_s = \frac{A_v f_{sy} d}{s} \qquad (8.133)$$

where A_v is the area of web reinforcement within a distance s. V_s must not be taken greater than $8\sqrt{f'_c}b'd$ and d need not be taken less than $0.8h$.

8.40.9 Shear in Prestressed Beams—ACI

ACI 318, "Building Code Requirements for Reinforced Concrete" (American Concrete Institute) also requires that prestressed members be designed to resist diagonal tension by the strength theory.

Shear reinforcement should consist of stirrups or welded-wire fabric. The area of shear reinfor-

cement, in^2, set perpendicular to the beam axis, should not be less than

$$A_v = 0.75\sqrt{f'_c}\frac{b_w s}{f_y} \geq 50\frac{b_w s}{f_y} \qquad (8.134)$$

where s is the reinforcement spacing, in, except when the factored shear force V_u is less than one-half ϕV_c; or when the depth of the member h is less than 10 in or 2.5 times the thickness of the compression flange, or one-half the width of the web, whichever is largest. The capacity reduction factor ϕ should be taken as 0.85, if the traditional ACI load combinations from Appendix C of ACI 318-02 are used, or 0.75 if the ASCE 7-98 (American Society of Civil Engineers) load combinations adopted in Chapter 9 of ACI 318-02 are used.

Alternatively, a minimum area

$$A_v = \frac{A_{ps}f_{pu}s}{80f_yd}\sqrt{\frac{d}{b_w}} \qquad (8.135)$$

may be used if the effective prestress force is at least equal to 40% of the tensile strength of the flexural reinforcement.

The yield strength of shear reinforcement, f_y used in design calculations should not exceed 60,000 psi.

Where shear reinforcement is required, it should be placed perpendicular to the axis of the member and should not be spaced farther apart than $0.75h$, where h is the overall depth of the member, or 24 in. Web reinforcement between the face of support and the section at a distance $h/2$ from it should be the same as the reinforcement required at that section.

When V_u exceeds the nominal shear strength ϕV_c of the concrete, shear reinforcement must be provided. V_c may be computed from Eq. (8.136) when the effective prestress force is 40% or more of the tensile strength of the flexural reinforcement, but this shear stress must not exceed $5\sqrt{f'_c}b_wd$.

$$V_c = \left(0.6\sqrt{f'_c} + 700\frac{V_ud}{M_u}\right)b_wd \geq 2\sqrt{f'_c}b_wd \quad (8.136)$$

where M_u = factored moment at section occurring simultaneously with shear V_u at section

b_w = web width

d = distance from extreme compression surface to centroid of prestressing steel or $0.80h$, whichever is larger

V_ud/M_u should not be taken greater than 1. For some sections, such as medium- and long-span I-shaped members, Eq. (8.136) may be overconservative, and the following more detailed analysis would be preferable.

The ACI Code requires a more detailed analysis when the effective prestress force is less than 40% of the tensile strength of the flexural reinforcement. The governing shear stress is the smaller of the values computed for inclined flexure-shear cracking V_{ci} from Eq. (8.137) and web-shear cracking V_{cw} from Eq. (8.138).

$$V_{ci} = 0.6\sqrt{f_c'}b_wd + V_d + \frac{V_iM_{cr}}{M_{max}} \quad (8.137)$$

$$\geq 1.7\sqrt{f_c'}b_wd$$

$$V_{cw} = (3.5\sqrt{f_c'}b_wd + 0.3f_{pc})b_wd + V_p \quad (8.138)$$

where V_d = shear force at section due to unfactored dead load

V_i = factored shear force at section due to externally applied loads occurring simultaneously with M_{max} and produced by external loads

M_{cr} = moment causing flexural cracking at section due to externally applied loads [see Eq. (8.139)]

M_{max} = maximum factored moment at section due to externally applied loads

b_w = web width or diameter of circular section

d = distance from extreme compression fiber to centroid of longitudinal tension reinforcement or 80% of overall depth of beam, whichever is larger

f_{pc} = compressive stress in concrete occurring, after all prestress losses have taken place, at centroid of cross section resisting applied loads or at junction of web and flange when centroid lies in flange

V_p = vertical component of effective prestress force at section considered

The cracking moment is given by

$$M_{cr} = \frac{I}{y_t}(6\sqrt{f_c'} + f_{pe} - f_d) \quad (8.139)$$

where I = moment of inertia of section resisting externally applied factored loads, m^4

y_t = distance from centroidal axis of gross section, neglecting reinforcement, to extreme fiber in tension, in

f_{pe} = compressive stress in concrete due to effective prestress forces only, after all losses, occurring at extreme fiber of section at which tension is produced by externally applied loads, psi

f_d = stress due to unfactored dead load at extreme fiber of section at which tension is produced by externally applied loads, psi

Alternatively, V_{cw} may be taken as the shear force corresponding to dead load plus live load that results in a principal tensile stress of $4\sqrt{f_c'}$ at the centroidal axis of the member or, when the centroidal axis is in the flange, induces this tensile stress at the intersection of flange and web.

The values of M_{max} and V_i used in Eq. (8.137) should be those resulting from the load combination causing maximum moment to occur at the section.

In a pretensioned beam in which the section at a distance of half the overall beam depth $h/2$ from face of support is closer to the end of the beam than the transfer length of the tendon, the reduced prestress in the concrete at sections falling within the transfer length should be considered when calculating V_{cw}. The prestress may be assumed to vary linearly along the centroidal axis from zero at the beam end to a maximum at a distance from the beam end equal to the transfer length. This distance may be assumed to be 50 diameters for strand and 100 diameters for single wire.

When $V_u - \phi V_c$ exceeds $4\sqrt{f_c'}b_wd$, the maximum spacing of stirrups must be reduced to $0.375h$ but not to more than 12 in. But $V_u - \phi V_c$ must not exceed $8\sqrt{f_c'}b_wd$.

8.40.10 Bonded Reinforcement in Prestressed Beams—ACI

When prestressing steel is not bonded to the concrete, some bonded reinforcement should be provided in the precompressed tension zone of flexural members. The bonded reinforcement should be distributed uniformly over the tension zone near the extreme tension surface in beams and one-way slabs and should have an area of at least

$$A_s = 0.004A \qquad (8.140)$$

where A = area, in^2, of that part of cross section between flexural tension face and center of gravity of gross section.

In positive-moment regions of two-way slabs where the tensile stress under service loads exceeds $2\sqrt{f_c'}$, the area of bonded reinforcement should be at least

$$A_s = \frac{N_c}{0.5f_y} \qquad (8.141)$$

where N_c = tensile force in concrete due to unfactored dead plus live loads, lb

f_y = yield strength, psi, of bonded reinforcement ≤ 60 ksi

At column supports in negative-moment regions of two-way slabs, at least four bonded reinforcing bars should be placed in each direction and provide a minimum steel area

$$A_s = 0.00075hl \qquad (8.142)$$

where l = span of slab in direction parallel to that of reinforcement being determined, in

h = overall thickness of slab, in

The bonded reinforcement should be distributed, with a spacing not exceeding 12 in, over the slab width between lines that are $1.5h$ outside opposite faces of the columns.

8.40.11 Prestressed Compression Members

Prestressed concrete members subject to combined flexure and axial load, with or without nonprestressed reinforcement, must be proportioned by the strength design method, including effects of prestress, shrinkage, and creep. Reinforcement in columns with an average prestress less than 225 psi should have an area equal to at least 1% of the gross concrete area A_c. For walls subject to an average prestress greater than 225 psi and for which structural analysis shows adequate strength, the minimum reinforcement requirements given in Art. 8.35 may be waived.

Tendons in columns with average prestress f_{pc} equal to or greater than 225 psi should be enclosed in spirals or closed lateral ties. The spiral should comply with the requirements given in Art. 8.31.1. Ties should be at least No. 3 bar size and spacing should not exceed 48 tie diameters or the least dimension of the column.

8.40.12 Ducts for Posttensioning

Tendons for posttensioned members generally are sheathed in ducts before prestress is applied so that the tendons are free to move when tensioned. The tendons may be grouted in the ducts after transfer of prestress to the concrete and thus bonded to the concrete.

Ducts for grouting bonded bars or strand should be at least $\frac{1}{4}$ in larger than the diameter of the posttensioning bars or strand or large enough to produce an internal area at least twice the gross area of the prestressing steel. The temperature of members at time of grouting should be above 50 °F, and members must be maintained at this temperature for at least 48 h.

Unbonded prestressing steel should be completely coated with suitable material to ensure corrosion protection and protect the tendons against infiltration of cement during casting operations.

8.40.13 Deflections of Prestressed Beams

The immediate deflection of prestressed members may be computed by the usual formulas for elastic deflections. If cracking may occur, however, the effective moment of inertia (Art. 8.19) should be used. The PCI Design Handbook (Precast/Prestressed Concrete Institute) contains a deflection calculation method using bilinear moment-deflection relationships, which has been widely used in practice. Long-time deflection computations should include effects of the sustained load and effects of creep and shrinkage and relaxation of the steel (Art. 8.19).

("PCI Design Handbook," Precast/Prestressed Concrete Institute, 209 West Jackson Boulevard, Chicago, IL 60606)

Retaining Walls

8.41 Concrete Gravity Walls

Generally economical for walls up to about 15 ft high, gravity walls use their own weight to resist lateral forces from earth or other materials (Fig. 8.29*a*). Such walls usually are sufficiently massive to be unreinforced.

Forces acting on gravity walls include the walls' own weight, the weight of the earth on the sloping back and heel, lateral earth pressure, and resultant soil pressure on the base. It is advisable to include a force at the top of the wall to account for frost action, perhaps 700 lb/lin ft. A wall, consequently, may fail by overturning or sliding, overstressing of the concrete, or settlement due to crushing of the soil.

Design usually starts with selection of a trial shape and dimensions, and this configuration is checked for stability. For convenience, when the wall is of constant height, a 1-ft-long section may be analyzed. Moments are taken about the toe. The

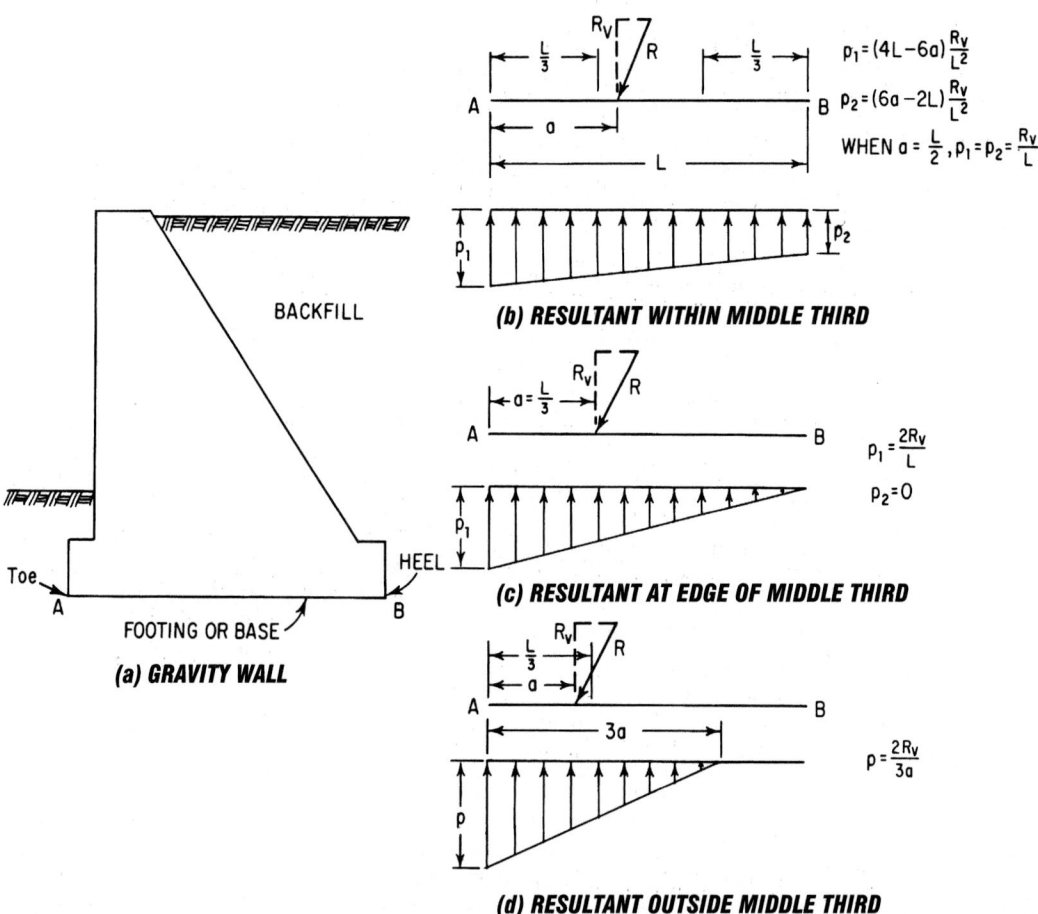

Fig. 8.29 Diagrams for pressure of the base of a concrete gravity wall on the soil below. (*a*) Vertical section through the wall. (*b*) Significant compression under the entire base. (*c*) No compression along one edge of the base. (*d*) Compression only under part of the base. No support from the soil under the rest of the beam.

sum of the righting moments should be at least 1.5 times the sum of the overturning moments. To prevent sliding

$$\mu R_v \geq 1.5 P_h \qquad (8.143)$$

where μ = coefficient of sliding friction

R_v = total downward force on soil, lb

P_h = horizontal component of earth thrust, lb

Next, the location of the vertical resultant R_v should be found at various sections of the wall by taking moments about the toe and dividing the sum by R_v. The resultant should act within the middle third of each section if there is to be no tension in the wall.

Finally, the pressure exerted by the base on the soil should be computed to ensure that the allowable pressure will not be exceeded. When the resultant is within the middle third, the pressures, psf, under the ends of the base are given by

$$p = \frac{R_v}{A} \pm \frac{Mc}{I} = \frac{R_v}{A}\left(1 \pm \frac{6e}{L}\right) \qquad (8.144)$$

where A = area of base, ft^2

L = width of base, ft

e = distance, parallel to L, from centroid of base to R_v, ft

Figure 8.29b shows the pressure distribution under a 1-ft strip of wall for $e = L/2 - a$, where a is the distance of R_v from the toe. When R_v is exactly $L/3$ from the toe, the pressure at the heel becomes zero (Fig. 8.29c). When R_v falls outside the middle third, the pressure vanishes under a zone around the heel, and pressure at the toe is much larger than for the other cases (Fig. 8.29d). "Standard Specifications for Highway Bridges" (American Association of State Highway and Transportation Officials) requires that contraction joints be provided at intervals not exceeding 30 ft. Alternate horizontal bars should be cut at these joints for crack control. Expansion joints should be located at intervals of up to 90 ft.

8.42 Cantilever Retaining Walls

This type of wall resists the lateral thrust of earth pressure through cantilever action of a vertical stem and horizontal base (Fig. 8.30a). Cantilever walls generally are economical for heights from 10 to 20 ft. For lower walls, gravity walls may be less costly; for taller walls, counterforts may be less expensive.

Usually, the force acting on the stem is the lateral earth pressure, including the effect of frost action, perhaps 700 lb/lin ft. The base is loaded by the moment and shear from the stem, upward soil pressure, its own weight, and that of the earth above. The weight of the soil over the toe, however, may be ignored in computing stresses in the toe since the earth may not be in place when the wall is first loaded or may erode. For walls of constant height, it is convenient to design and analyze a 1-ft-long strip.

The stem is designed to resist the bending moments and shear due to the earth thrust. Then, the size of the base slab is selected to meet requirements for resisting overturning and sliding and to keep pressure on the soil within the allowable. If the flat bottom of the slab does not provide sufficient friction [Eq. (8.143)], a key, or lengthwise projection, may be added on the bottom for that purpose. The key may be reinforced by extending and bending up the dowels between stem and base.

To provide an adequate safety factor against overturning, the sum of the righting moments about the toe should be at least 1.5 times the sum of the overturning moments. The pressure under the base can be computed, as for gravity walls, from Eq. (8.144). (See also Fig. 8.30b to d.)

Generally, the stem is made thicker at the bottom than required for shear and balanced design for moment because of the saving in steel. Since the moment decreases from bottom to top, the earth side of the wall usually is tapered, and the top is made as thin as convenient concreting will permit (8 to 12 in). The main reinforcement is set, in vertical planes, parallel to the sloping face and 3 in away. The area of this steel at the bottom can be computed from Eq. (8.27). Some of the steel may be cut off where it no longer is needed. Cutoff points may be determined graphically (Fig. 8.30b). The bending-moment diagram is plotted and the resisting moment of steel not cut off is superimposed. The intersection of the two curves determines the theoretical cutoff point. The bars should extend upward beyond this point a distance equal to d or 12 bar diameters.

In addition to the main steel, vertical steel is set in the front face of the wall and horizontal steel in both faces to resist thermal and shrinkage stresses

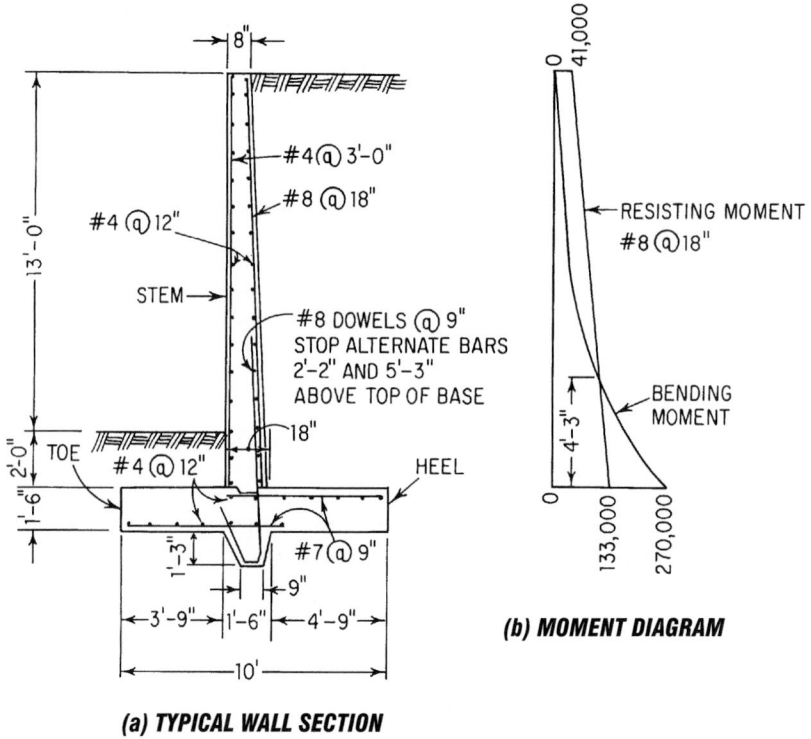

(a) TYPICAL WALL SECTION

(b) MOMENT DIAGRAM

Fig. 8.30 Cantilever retaining wall. (*a*) Vertical section shows main reinforcing steel placed vertically in the stem. (*b*) Moment diagram.

(Art. 8.23). "Standard Specifications for Highway Bridges" (American Association of State Highway and Transportation Officials) requires at least $\frac{1}{8}$ in^2 of horizontal reinforcement per foot of height.

The heel and toe portions of the base are both designed as cantilevers supported by the stem. The weight of the backfill tends to bend the heel down against relatively small resistance from soil pressure under the base. In contrast, the upward soil pressure tends to bend the toe up. So for the heel, main steel is placed near the top, and for the toe, near the bottom. Also, temperature steel is set lengthwise in the bottom. The area of the main steel may be computed from Eq. (8.27), but the bars should be checked for development length because of the relatively high shear.

To eliminate the need for diagonal-tension reinforcing, the thickness of the base should be sufficient to hold the shear stress, $v_c = V/bd$, below $1.1\sqrt{f_c'}$ where f_c' is the 28 day strength of the concrete, psi, as computed by the working-stress method. The critical section for shear is at a distance d from the face of the stem, where d is the distance from the extreme compression surface to the tensile steel.

The stem is constructed after the base. A key usually is formed at the top of the base to prevent the stem from sliding. Also, dowels are left projecting from the base to tie the stem to it, one dowel per stem bar. The dowels may be extended to serve also as stem reinforcing (Fig. 8.30*a*).

The AASHTO Specifications require that contraction joints be provided at intervals not exceeding 30 ft. Expansion joints should be located at intervals up to 90 ft.

To relieve the wall of water pressure, weep holes should be formed near the bottom of the stem. Also, porous pipe and backfill may be set behind the wall to conduct water to the weep holes.

(M. Fintel, "Handbook of Concrete Engineering," Van Nostrand Reinhold Company, New York; "CRSI Handbook," Concrete Reinforcing Steel

Institute, 180 North La Salle St., Chicago, IL 60601 (www.crsl.org).)

8.43 Counterfort Retaining Walls

Counterforts are ties between the vertical stem of a wall and its base (Fig. 8.31*a*). Placed on the earth side of the stem, they are essentially wedge-shaped cantilevers. (Walls with supports on the opposite side are called buttressed retaining walls.) Counterfort walls are economical for heights for which gravity and cantilever walls are not suitable.

Stability design is the same as for gravity walls (Art. 8.41) and cantilever walls (Art. 8.42). But the

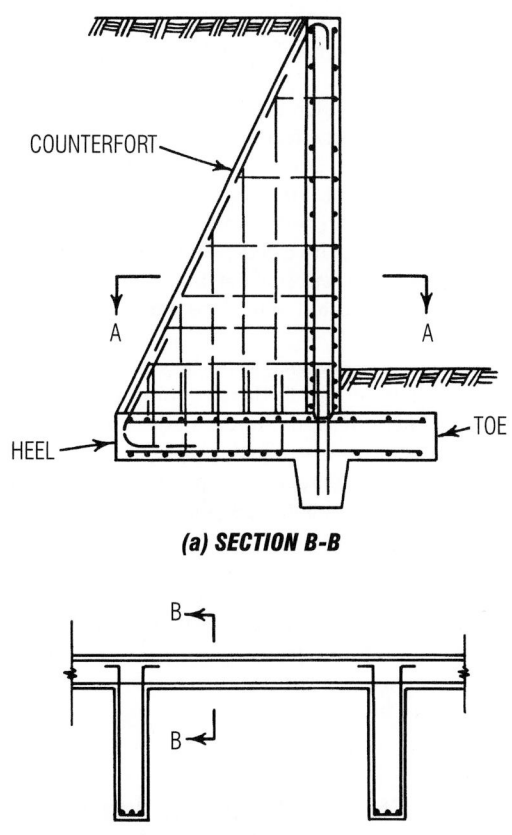

(a) SECTION B-B

(b) PLAN A-A

Fig. 8.31 Counterfort retaining wall. (*a*) Vertical section. (*b*) Horizontal section.

design is applied to a section of wall center to center of counterforts.

The vertical face resists lateral earth pressure as a continuous slab supported by the counterforts. It also is supported by the base, but an exact analysis of the effects of the three-sided supports would not be worthwhile except for very long walls. Similarly, the heel portion of the base is designed as a continuous slab supported by the counterforts. In turn, the counterforts are subjected to lateral earth pressure on the sloping face and the pull of the vertical stem and base. The toe of the base acts as a cantilever; as in a cantilever wall.

Main reinforcing in the vertical face is horizontal. Since the earth pressure increases with depth, reinforcing area needed also varies with depth. It is customary to design a 1-ft-wide strip of slab spanning between counterforts at the bottom of the wall and at several higher levels. The steel area and spacing for each strip then are held constant between strips. Negative-moment steel should be placed near the backfill face of the wall at the counterforts, and positive-moment steel near the opposite face between counterforts (Fig. 8.31*b*). Concrete cover should be 3 in over reinforcing throughout the wall. Design requirements are substantially the same as for rectangular beams and one-way slabs, except the thickness is made large enough to eliminate the need for shear reinforcing (Arts. 8.20 to 8.23). The vertical face also incorporates vertical steel, equal to about 0.3 to 1% of the concrete area, for placement purposes and to resist temperature and shrinkage stresses.

In the base, main reinforcing in the heel portion extends lengthwise, whereas that in the toe runs across the width. The heel is subjected to the downward weight of the backfill above and its own weight and to the upward pressure of the soil below and the pull of the counterforts. So longitudinal steel should be placed in the top face at the counterforts and near the bottom between counterforts. Main transverse steel should be set near the bottom to resist the cantilever action of the toe.

The counterforts, resisting the lateral earth pressure on the sloping face and the pull of the vertical stem, are designed as T beams. Maximum moment occurs at the bottom. It is resisted by main reinforcing along the sloping face. (The effective depth should be taken as the distance from the outer face of the wall to the steel along a perpendicular to the steel.) At upper levels, main steel not required may be cut off. Some of the steel, however, should be extended and bent down into the vertical face. Also, dowels equal in area to the main steel at the bottom should be hooked into the base to provide anchorage.

Shear unit stress on a horizontal section of a counterfort may be computed from $v_c = V_1/bd$, where b is the thickness of the counterfort and d is the horizontal distance from face of wall to main steel.

$$V_1 = V - \frac{M}{d}(\tan\theta + \tan\phi) \qquad (8.145)$$

where V = shear on section

$\quad M$ = bending moment at section

$\quad \theta$ = angle earth face of counterfort makes with vertical

$\quad \phi$ = angle wall face makes with vertical

For a vertical wall face, $\phi = 0$ and $V_1 = V - (M/d)$ $\tan\theta$. The critical section for shear may be taken conservatively at a distance up from the base equal to $d'\sin\theta\cos\theta$, where d' is the depth of counterfort along the top of the base.

Whether or not horizontal web reinforcing is needed to resist the shear, horizontal bars are required to dowel the counterfort to the vertical face (Fig. 8.25b). They should be designed for the full wall reaction. Also, vertical bars are needed in the counterfort to resist the pull of the base. They should be doweled to the base.

The base is concreted first. Vertical bars are left projecting from it to dowel the counterforts and the vertical face. Then, the counterforts and vertical stem are cast together.

Footings

Footings should be designed to satisfy two objectives: limit total settlement to an acceptable small amount and eliminate differential settlement between parts of a structure as nearly as possible. To limit the amount of settlement, a footing should be constructed on soil with sufficient resistance to deformation, and the load should be spread over a large soil area. The load may be spread horizontally, as is done with spread footings, or vertically, as with friction-pile foundations.

8.44 Types of Footings

There are a wide variety of spread footings. The most commonly used ones are illustrated in Fig. 8.32a to g. A simple pile footing is shown in Fig. 8.32h.

For walls, a spread footing is a slab wider than the wall and extending the length of the wall (Fig. 8.32a). Square or rectangular slabs are used under single columns (Fig. 8.32b to d). When two columns are so close that their footings would merge or nearly touch, a combined footing (Fig. 8.32e) extending under the two should be constructed. When a column footing cannot project in one direction, perhaps because of the proximity of a property line, the footing may be helped out by an adjacent footing with more space. Either a combined footing or a strap (cantilever) footing (Fig. 8.32f) may be used under the two columns.

For structures with heavy loads relative to soil capacity, a mat or raft foundation (Fig. 8.32g) may prove economical. A simple form is a thick, two-way-reinforced-concrete slab extending under the entire structure. In effect, it enables the structure to float on the soil, and because of its rigidity, it permits negligible differential settlement. Even greater rigidity can be obtained by building the raft foundation as an inverted beam-and-girder floor, with the girders supporting the columns. Sometimes, also, inverted flat slabs are used as mat foundations.

In general, footings should be so located under walls or columns as to develop uniform pressure below. The pressure under adjacent footings should be as nearly equal as possible, to avoid differential settlement. In the computation of stresses in spread footings, the upward reaction of the soil may be assumed to vary linearly. For pile-cap stresses, the reaction from each pile may be assumed to act at the pile center.

Simple footings act as cantilevers under the downward column or wall loads and upward soil or pile reactions. Therefore, they can be designed as rectangular beams.

8.45 Stress Transfer from Columns to Footings

For a footing to serve its purpose, column stresses must be distributed to it and spread over the soil or to piles, with a safety factor against failure of the

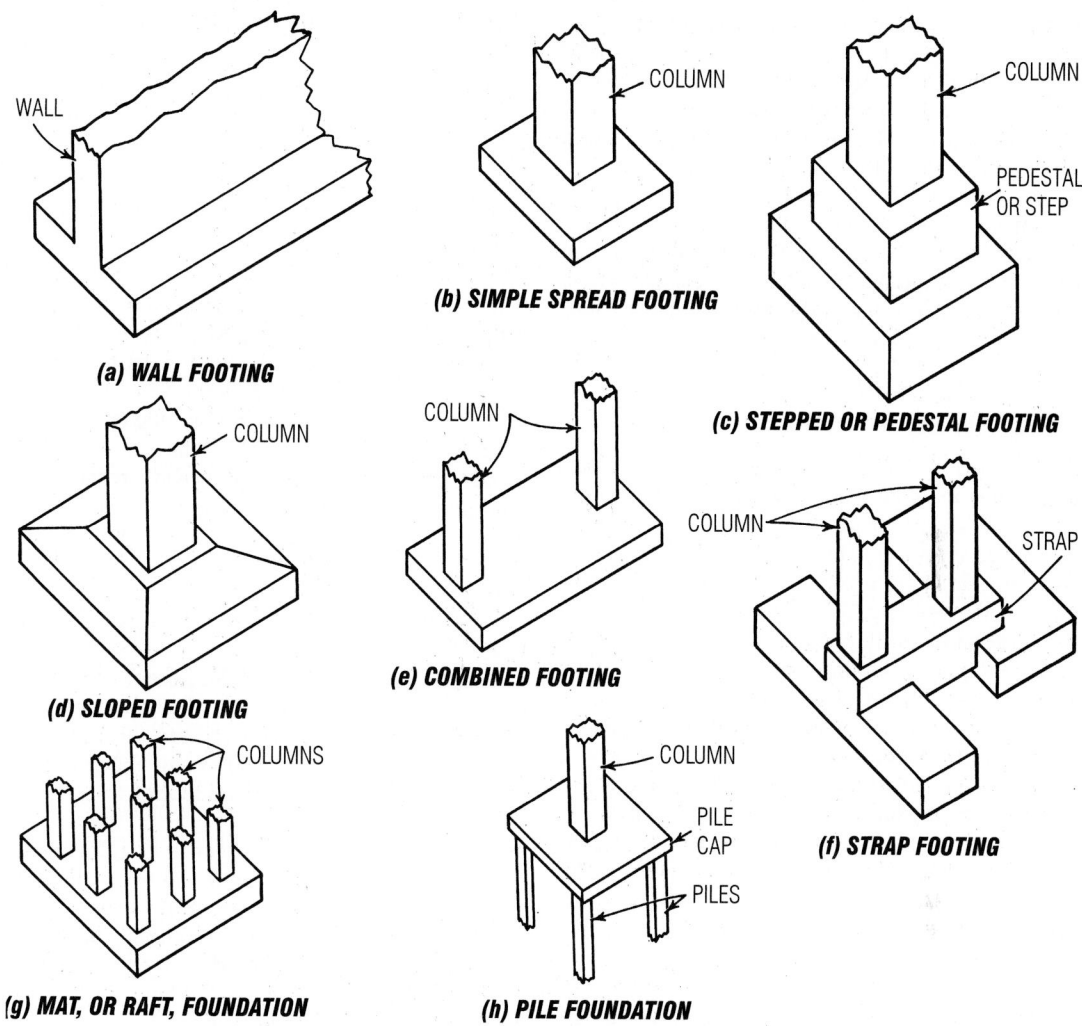

(a) WALL FOOTING

(b) SIMPLE SPREAD FOOTING

(c) STEPPED OR PEDESTAL FOOTING

(d) SLOPED FOOTING

(e) COMBINED FOOTING

(f) STRAP FOOTING

(g) MAT, OR RAFT, FOUNDATION

(h) PILE FOUNDATION

Fig. 8.32 Common types of concrete footings for walls and columns.

footing. Stress in the longitudinal reinforcement of a column should be transferred to its pedestal or footing either by extending the longitudinal steel into the support or by dowels. At least four bars should be extended or four dowels used. In any case, a minimum steel area of 0.5% of the column area should be supplied for load transfer. The stress-transfer bars should project into the base a sufficient compression-embedment distance to transfer the stress in the column bars to the base concrete. Where dowels are used, their total area should be adequate to transfer the compression in excess of that transmitted by the column concrete to the footing

in bearing, and the dowels should not be larger than #11 bars. If the required dowel length is larger than the footing depth less 3 in, either smaller-diameter bars with equivalent area should be used or a monolithic concrete cap should be added to increase the concrete depth. The dowels, in addition, should provide at least one-quarter of the tension capacity of the column bars on each column face. The dowels should extend into the column a distance equal to that required for compression lapping of column bars (Art. 8.12.6).

Stress in the column concrete should be considered transferred to the top of the pedestal or footing by bearing. ACI 318, "Building Code Requirements for Reinforced Concrete" (American Concrete Institute), specifies two bearing stresses:

For a fully loaded area, such as the base of a pedestal, allowable bearing stress is $0.85\phi f_c'$, where f_c' is the strength of the concrete and $\phi = 0.65$.

If the area A_1, the loaded portion at the top of a pedestal or footing, is less than the area of the top, the allowable pressure may be multiplied by $\sqrt{A_2/A_1}$ but not more than 2, where A_2 is the area of the top that is geometrically similar to and concentric with the loaded area A_1.

8.46 Wall Footings

The spread footing under a wall (Fig. 8.32a) distributes the wall load horizontally to preclude excessive settlement. (For retaining-wall footings, see Arts. 8.41 to 8.43.) The wall should be so located on the footing as to produce uniform bearing pressure on the soil (Fig. 8.33), ignoring the variation due to bending of the footing. The pressure, lb/ft², is determined by dividing the load per foot by the footing width, ft.

The footing acts as a cantilever on opposite sides of the wall under downward wall loads and upward soil pressure. For footings supporting concrete walls, the critical section for bending moment is at the face of the wall; for footings under masonry walls, halfway between the middle and edge of the wall. Hence, for a 1-ft-long strip of symmetrical concrete-wall footing, symmetrically loaded, the maximum moment, ft-lb, is

$$M = \frac{p}{8}(L - a)^2 \qquad (8.146)$$

where p = uniform pressure on soil, psf

L = width of footing, ft

a = wall thickness, ft

If the footing is sufficiently deep that the tensile bending stress at the bottom, $6M/t^2$, where M is the factored moment and t is the footing depth, in, does not exceed $5\phi\sqrt{f_c'}$, where f_c' is the 28-day concrete strength, psi, and $\phi = 0.90$, the footing need not be reinforced. If the tensile stress is larger, the footing should be designed as a 12-in-wide rectangular, reinforced beam. Bars should be placed across the width of the footing, 3 in from the bottom. Bar development length is measured from the point at which the critical section for moment occurs. Wall footings also may be designed by ultimate-strength theory.

ACI 318, "Building Code Requirements for Reinforced Concrete" (American Concrete Institute), requires at least 6 in of cover over the reinforcement at the edges. Hence, allowing about 1 in for the bar diameter, the minimum footing thickness is 10 in.

The critical section for shear is at a distance d from the face of the wall, where d is the distance from the top of the footing to the tensile reinforcement, in. Since diagonal-tension reinforcement is undesirable, d should be large enough to keep the shear unit stress $V/12d$ below $1.1\sqrt{f_c'}$, as computed by the working-stress method, or below $2\sqrt{f_c'}b_w d$ for factored shear loads. V is the shear at the critical section per foot of wall.

In addition to the main steel, some longitudinal steel also should be placed parallel to the wall to resist shrinkage stresses and facilitate placement of the main steel. (See also Art. 8.45.)

(G. Winter and A. H. Nilson, "Design of Concrete Structures," McGraw-Hill Book Company, New York (books.mcgraw-hill.com).)

8.47 Single-Column Spread Footings

The spread footing under a column (Fig. 8.32b to d) distributes the column load horizontally to prevent

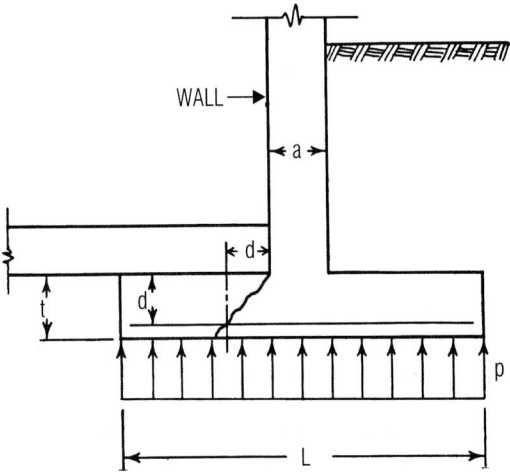

Fig. 8.33 Reinforced concrete wall footing.

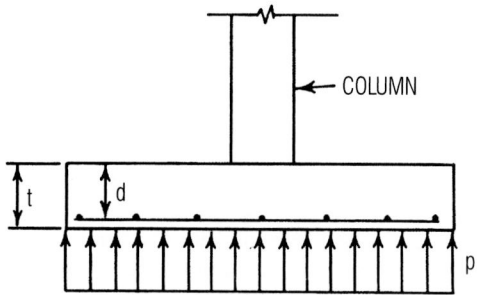

Fig. 8.34 Spread footing for column.

excessive total and differential settlement. The column should be located on the footing so as to produce uniform bearing pressure on the soil (Fig. 8.34), exclusive of the variation due to bending of the footing. The pressure equals the load divided by the footing area.

Single-column footings usually are square, but they may be made rectangular to satisfy space restrictions or to support elongated columns.

Under the downward load of the column and the upward soil pressure, a footing acts as a cantilever in two perpendicular directions. For rectangular concrete columns and pedestals, the critical section for bending moment is at the face of the loaded member (*ab* in Fig. 8.35*a*). (For round or octagonal columns or pedestals, the face may be taken as the side of a square with the same area.)

For steel baseplates, the critical section for moment is halfway between the face of the column and the edge of the plate.

The bending moment on *ab* is produced by the upward pressure of the soil on the area *abcd*. That part of the footing is designed as a rectangular beam to resist the moment. Another critical section lies along a perpendicular column face and should be similarly designed. If the footing is sufficiently deep that the factored tensile bending stress at the bottom does not exceed $5\phi\sqrt{f_c'}$, where $\phi = 0.90$ and f_c' is the 28-day strength of the concrete, psi, the footing need not be reinforced. If the tensile stress is larger, reinforcement should be placed parallel to both sides of the footing, with the lower layer 3 in above the bottom of the footing and the upper layer a bar diameter higher. The critical section for anchorage (or bar embedment length) is the same as for moment.

In square footings, the steel should be uniformly spaced in each layer. Although the effective depth *d* is less for the upper layer, thus requiring more steel, it is general practice to compute the required area and spacing for the upper level and repeat them for the lower layer.

In rectangular footings, reinforcement parallel to the long side, with length *A*, ft, should be uniformly distributed over the width of the footing, *B*, ft. Bars parallel to the short side should be more closely spaced under the column than near the edges. ACI 318, "Building Code Requirements for Reinforced Concrete" (American Concrete Institute), recommends that the short bars should be

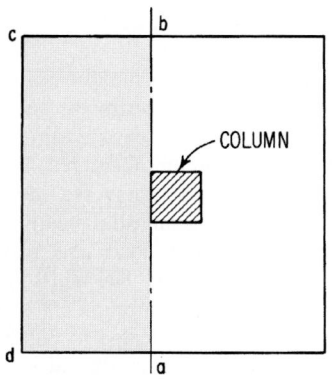

(a) MOMENT AND ANCHORAGE

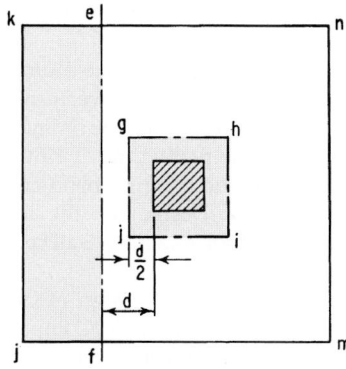

(b) SHEAR

Fig. 8.35 Critical section in a column footing as viewed in plan.

given a constant but closer spacing over a width B centered under the column. The area of steel in this band should equal twice the total steel area required in the short direction divided by $A/B + 1$. The remainder of the reinforcement should be uniformly distributed on opposite sides of the band. (See also Art. 8.45.)

Two types of shear should be investigated: two-way action (punching shear) and beam-type shear. The critical section for beam-type shear lies at a distance d from the face of column or pedestal (ef in Fig. 8.35b). The shear equals the total upward pressure on area $efjk$. To eliminate the need for diagonal-tension reinforcing, d should be made large enough that the unit shear stress does not exceed $1.1\sqrt{f_c'}$ ($2\sqrt{f_c'}$ for ultimate-strength design).

The critical section for two-way action is concentric with the column or pedestal. It lies at a distance $d/2$ from the face of the loaded member ($ghij$ in Fig. 8.35b). The shear equals the column load less the upward soil pressure on area $ghij$. In this case, d should be large enough that the factored shear on the concrete does not exceed

$$V_c = \left(2 + \frac{4}{\beta_c}\right)\sqrt{f_c' b_o d} \qquad (8.147)$$

where β_c = ratio of long side to short side of critical shear section

b_o = perimeter of critical section, in

d = depth of centroid of reinforcement, in

Shearhead reinforcement (steel shapes), although generally uneconomical, may be used to obtain a shallow footing.

Footings for columns designed to take moment at the base should be designed against overturning and nonuniform soil pressures. When the moments are about only one axis, the footing may be made rectangular with the long direction perpendicular to that axis, for economy. Design for the long direction is similar to that for retaining-wall bases (Art. 8.45 to 8.47).

(G. Winter and A. H. Nilson, "Design of Concrete Structures," McGraw-Hill Book Company, New York (books.mcgraw-hill.com); M. Fintel, "Handbook of Concrete Engineering," Van Nostrand Reinhold Company, New York; "CRSI Handbook," Concrete Reinforcing Steel Institute, Chicago, Ill. (www.crsi.org); ACI SP-17, "Design Handbook," American Concrete Institute, Detroit, MI (www.aci-int.org).)

8.48 Combined Footings

These are spread footings extended under more than one column (Fig. 8.32e). They may be necessary when two or more columns are so closely spaced that individual footings would interfere with each other. Or they may be desirable when space is restricted for a column footing, such as an exterior member so close to a property line that an individual footing would be so short that it would have excessive eccentric loading. In that case, the footing may be extended under a rear column. If the footing can be continued past that column a sufficient distance, and the exterior column has a lighter load, the combined footing may be made rectangular (Fig. 8.36a). If not, it may be made trapezoidal.

If possible, the columns should be so placed on the combined footing as to produce a uniform pressure on the soil. Hence, the resultant of the column loads should coincide with the centroid of the footing in plan. This requirement usually determines the length of the footing. The width is computed from the area required to keep the pressure on the soil within the allowable.

In the longitudinal direction, the footing should be designed as a rectangular beam with overhangs. This beam is subjected to the upward pressure of the soil. Hence, the main steel consists of top bars between the columns and bottom bars at the columns where there are overhangs (Fig. 8.36b). Depth of footing may be determined by moment or shear (see Art. 8.41).

The column loads may be assumed distributed to the longitudinal beam by beams of the same depth as the footing but extending in the narrow, or transverse, direction. Centered, if possible, under each column, the transverse member should be designed as a rectangular beam subjected to the downward column load and upward soil pressure under the beam. The width of the beam may be estimated by assuming a 60° distribution of the column load, as indicated in Fig. 8.36c. Main steel in the transverse beam should be placed near the bottom.

Design procedure for a trapezoidal combined footing is similar. But the reinforcing steel in the longitudinal direction is placed fanwise, and

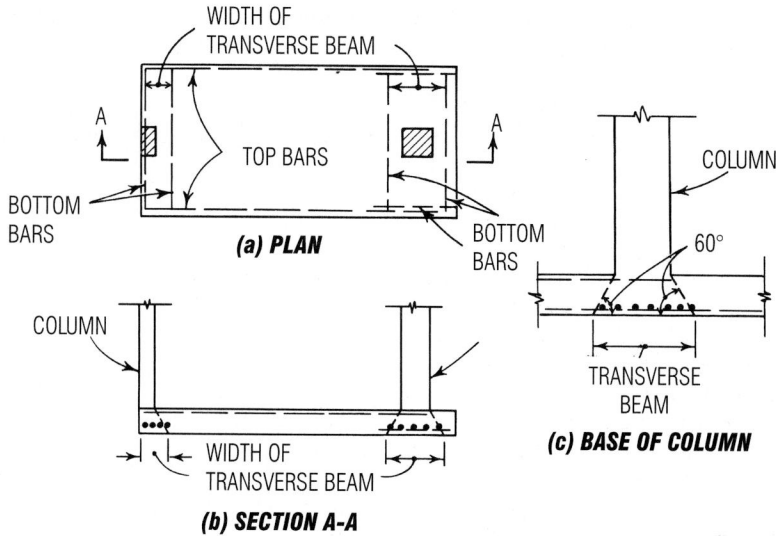

Fig. 8.36 Combined footing. (*a*) Plan view. (*b*) Vertical section. (*c*) Detail at base of interior column.

alternate bars are cut off as the narrow end is approached. (See also Art. 8.45.)

(G. Winter and A. H. Nilson, "Design of Concrete Structures," McGraw-Hill Book Company, New York (books.mcgraw-hill.com); M. Fintel, "Handbook of Concrete Engineering," Van Nostrand Reinhold Company, New York.)

8.49 Strap or Cantilever Footings

In Art. 8.48, the design of a combined footing is explained for a column footing in restricted space, such as an exterior column at a property line. As the distance between such a column and a column with adequate space around it increases, the cost of a combined footing rises rapidly. For column spacing more than about 15 ft, a strap footing (Fig. 8.32*f*) may be more economical. It consists of a separate footing under each column connected by a beam or strap to distribute the column loads (Fig. 8.37*a*).

The footings are sized to produce the same, constant pressure under each (Fig. 8.37*c*). This requires that the centroid of their areas coincide with the resultant of the column loads. Usually, the strap is raised above the bottom of the footings so as not to bear on the soil. The sum of the footing

areas, therefore, must be large enough for the allowable bearing capacity of the soil not to be exceeded. When these requirements are satisfied, the total net pressure under a footing does not necessarily equal the column design load on the footing.

The strap should be designed as a rectangular beam spanning between the columns. The loads on it include its own weight (when it does not rest on the soil) and the upward pressure from the footings. Width of the strap usually is selected arbitrarily as equal to that of the largest column plus 4 to 8 in so that column forms can be supported on top of the strap. Depth is determined by the maximum bending moment.

The main reinforcing in the strap is placed near the top. Some of the steel can be cut off where not needed. For diagonal tension, stirrups normally will be needed near the columns (Fig. 8.37*b*). In addition, longitudinal placement steel is set near the bottom of the strap, plus reinforcement to guard against settlement stresses.

The footing under the exterior column may be designed as a wall footing (Art. 8.46). The portions on opposite sides of the strap act as cantilevers under the constant upward pressure of the soil.

The interior footing should be designed as a single-column footing (Art. 8.47). The critical section for punching shear, however, differs from that for

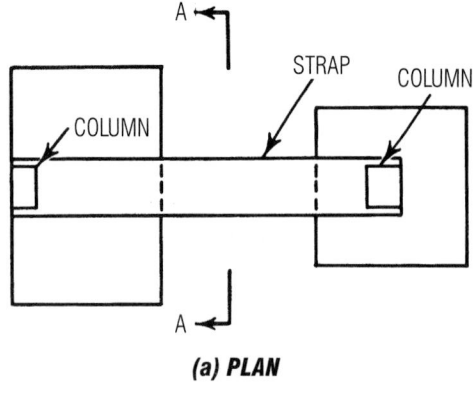

(a) PLAN

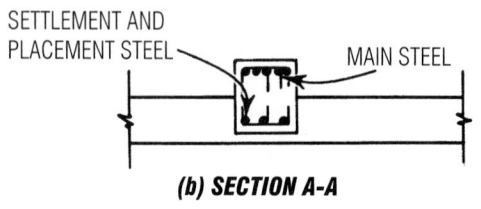

(b) SECTION A-A

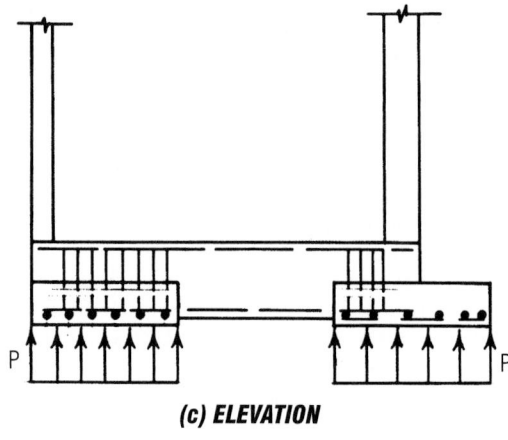

(c) ELEVATION

Fig. 8.37 Strap (cantilever) footing.

a conventional footing. This shear should be computed on a section parallel to the strap and at a distance $d/2$ from the sides and extending around the column at a distance $d/2$ from its faces; d is the effective depth of the footing, the distance from the bottom steel to the top of the footing.

(G. Winter and A. H. Nilson, "Design of Concrete Structures," McGraw-Hill Book Company, New York (books.mcgraw-hill.com); M. Fintel,

"Handbook of Concrete Engineering," Van Nostrand Reinhold Company, New York.)

8.50 Footings on Piles

When piles are required to support a structure, they are capped with a thick concrete slab, on which the structure rests. The pile cap should be reinforced. ACI 318, "Building Code Requirements for Reinforced Concrete" (American Concrete Institute), requires that the thickness above the tops of the piles be at least 12 in. The piles should be embedded from 6 to 9 in, preferably the larger amount, into the footing. They should be cut to required elevation before the footing is cast.

Like spread footings, pile footings for walls are continuous, the piles being driven in line under the wall. For a single column or pier, piles are driven in a cluster. "Standard Specifications for Highway Bridges" (American Association of State Highway and Transportation Officials) requires that piles be spaced at least 2 ft 6 in center to center. And the distance from the side of a pile to the nearest edge of the footing should be 9 in or more.

Whenever possible, the piles should be located so as to place their centroid under the resultant of the column load. If this is done, each pile will carry the same load. If the load is eccentric, then the load on a pile may be assumed to vary linearly with distance from an axis through the centroid.

The critical section for bending moment in the footing and embedment length of the reinforcing should be taken as follows:

At the face of the column, pedestal, or wall, for footings supporting a concrete column, pedestal, or wall

Halfway between the middle and edge of the wall, for footings under masonry walls

Halfway between the face of the column or pedestal and the edge of the metallic base, for footings under steel baseplates

The moment is produced at the critical section by the upward forces from all the piles lying between the section and the edge of the footing.

For diagonal tension, two types of shear should be investigated—punching shear and beamlike shear—as for single-column spread footings (Art. 8.47). The ACI Code requires that in computing the external shear on any section through a footing supported on piles, the entire reaction from any pile whose center is located half the pile diameter or more outside the section shall be assumed as producing shear on the section. The reaction from any pile whose center is located half the pile diameter or more inside the section shall be assumed as producing no shear on the section. For intermediate positions of the pile center, the portion of the pile reaction to be assumed as producing shear on the section should be based on straight-line interpolation between the full value at half the pile diameter outside the section and zero value at that distance inside the section.

(G. Winter and A. H. Nilson, "Design of Concrete Structures," McGraw-Hill Book Company, New York (books.mcgraw-hill.com); M. Fintel, "Handbook of Concrete Engineering," Van Nostrand Reinhold Company, New York).

Frames and Shells
8.51 Structural Analysis of Frames and Shells

Analysis of structural frames yields values of internal forces and moments at various sections. Results include bending moments (about two principal axes of each section), concentric normal forces (axial tension or compression), tangential forces (shear), and torsion (bending moment parallel to the section). In design, critical cross sections are selected and designed to resist the internal forces and moments acting on them.

Geometry of a structural frame and its components has a great bearing on distribution of internal forces and moments and their magnitude. Thus, the geometry affects economy and esthetics of a structural system and its components. Rigid frames, arches, folded plates, and shells are examples of the use of geometry for support of loads at relatively low cost.

Once any of these structures has been analyzed and internal forces and moments on critical cross sections have been determined, design becomes nearly identical with that of cross sections covered in previous articles in this section. Additional consideration, however, should be given to secondary stresses in detailing the reinforcement.

In practice, most structures and their components are analyzed only for the primary stresses caused by external loads. But most structural components, including beams, columns, and slabs

discussed previously, are subjected to secondary stresses. They could be due to many causes:

External loads normally not considered during the design, for example, when one side of a building is heated by sun more than the others

Nonhomogeneity of material, such as concrete

Geometry of structural members, for example, deep rather than shallow cross sections

Additional forces and moments due to deformations

Most of the formulas used in everyday structural design are simplified versions of more accurate but complicated mathematical expressions. The simplified formulas give results only for an approximate stress distribution. To provide for the difference between approximate and accurate analyses, design of members, including secondary stresses, should incorporate a margin of safety. Stress concentration, for example, is a secondary stress. In general, there are no set rules or formulas for predicting secondary stresses and designing for them.

In conventional reinforced-concrete structures, secondary stresses are relatively small compared with the primary stresses. But if secondary stresses are not provided for in design, cracks may develop in the structure. Usually, these cracks are not serious and are acceptable. In view of the difficulty, perhaps impossibility, of predicting the location and magnitude of secondary stresses in most cases, normal practice does not include analysis of structures for secondary stresses.

To protect structures against unpredictable stresses, ACI 318, "Building Code Requirements for Reinforced Concrete" (American Concrete Institute), specifies minimum reinforcement for beams, columns, and slabs. Spacing and size of this reinforcement take care of the secondary stresses. These provisions and some additional reinforcement requirements apply to design of rigid frames, arches, folded plates, and shells. But these types of structures often have larger secondary stresses than conventional structures, and these stresses are distributed differently from those in beams and columns. There are no code provisions for designing against these secondary stresses other than the general requirements of elastic behavior, equilibrium checks, and accounting for effects of large deflections, creep, and possible construction defects. But observations of the behavior of rigid frames, arches, folded plates, and shells, along with more accurate mathematical treatment and analysis, do help to design against secondary stresses.

The following articles point out the more salient considerations in designing these reinforced-concrete structures. Engineers, however, should have sufficient experience in design of such structures to take steps to avoid undue cracking of concrete.

One of the most important duties of a structural engineer is to choose an appropriate structural system, for example, to decide whether to span with a simply supported beam, a rigid frame, an arch, a folded plate, or a shell. The engineer must know the advantages of these structural systems to be able to select a proper structure for a project.

In indeterminate structures such as rigid frames, arches, folded plates, and shells, the sizes and thicknesses of the components of these structures affect the magnitude and distribution of the bending moments and, hence, shears and axial forces. For example, if the horizontal member of the rigid frame of Fig. 8.38a is made much deeper than the width of the vertical member, that is, the beam is much stiffer than the column, the maximum moment in the beam would be relatively large and that in the column small. Conversely, if the vertical member is made much wider than the depth of the horizontal member, that is, the column is much stiffer than the beam, the maximum bending moment in the column would be relatively large.

Similarly, deepening the haunches in the horizontal member of Fig. 8.38b would increase the negative bending moment at the haunches and decrease the positive bending moment at midspan, where the beam is shallow.

Because of the properties described, indeterminate structures are analyzed by first assuming sizes and shapes of components. After internal forces

and moments have been determined, the assumed sections are checked for adequacy. If the assumed sizes must be adjusted, another analysis is performed with the adjusted sizes. Then, these are checked for adequacy. If necessary, the cycle is repeated.

8.52 Concrete Rigid Frames

Rigid frame implies a plane structural system consisting of straight members meeting each other at an angle and rigidly connected at the junction. A rigid connection keeps unchanged the angle between members as the entire frame distorts under load.

Rigid frames may be one bay long and one tier high (Fig. 8.38a and b), or they may have multiple bays and multiple tiers (Fig. 8.39a and b). They may be built of reinforced concrete or prestressed concrete, cast in place or precast.

Because of continuity between columns and beams, columns in rigid frames participate with the beams in bending and thus in resisting external loads. This participation results in both smaller bending moments and different moment distribution along the beam than in a simply supported beam with the same span and loads. But for these advantages in bending-moment distribution along the beam, the column is penalized. Under vertical loading, for example, it is subjected to bending moments in addition to axial force. (See also Arts. 6.61 to 6.63 and 8.57.)

Since the bases of most rigid frames develop horizontal reactions, the beams usually are subjected to a small axial force. Also, the beams and columns are subjected to shear forces.

It is not advisable, in general, to differentiate between beams and columns in a rigid frame, but to consider each as a member subjected to axial loads and flexure. Find bending moments, shear, and axial forces in each, and design for these.

Because of continuity between members in a rigid frame, this type of structure is particularly advantageous in resisting wind and seismic loads. It does not necessarily have to be subjected to vertical loads only or consist of vertical and horizontal members. Figures 8.40 and 8.41 show examples of rigid frames with sloping members subjected to vertical and lateral loads.

Dimensions of cross sections and the amount of reinforcement in concrete rigid frames are

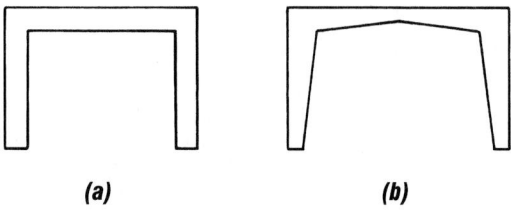

(a)   **(b)**

Fig. 8.38 Rigid frames: (a) with prismatic members; (b) with haunched beam.

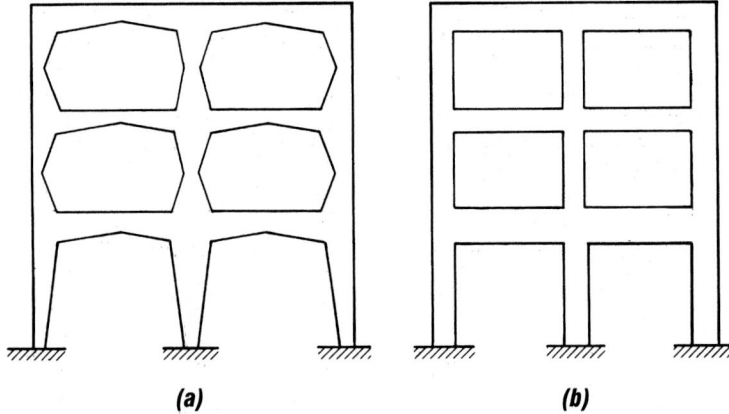

Fig. 8.39 Multistory rigid frames: (*a*) with haunched members; (*b*) with prismatic members.

determined by primary stresses due to bending moments, axial forces, and shears, as in beams and columns. In addition, the following require special attention:

Rigid joints, where members meet, particularly at reentrant corners

Toes of legs at the foundations

Exceptionally deep members (Art. 8.17.5)

Typical details of rigid joints in a reinforced-concrete frame are shown in Fig. 8.42*a* and *b*. Ample embedment of bars at supports should be provided

at all corners, as well as at overlaps (Art. 8.20.6). No interior or exterior face of a rigid joint should be left without reinforcement.

Note that in Fig. 8.42 reinforcing bars extend without bends past the reentrant corners. Reinforcing never should be bent around a reentrant corner. When the reinforcement is in tension, it tends to tear concrete at the corner away from the joint. Furthermore, sufficient stirrups should be provided around all bars that cross a joint. The amount of stirrups may be computed from the

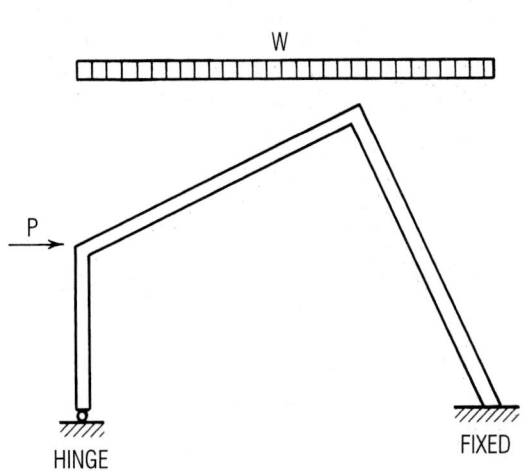

Fig. 8.40 Rigid frame with sloping beam, one vertical column, and one sloping column.

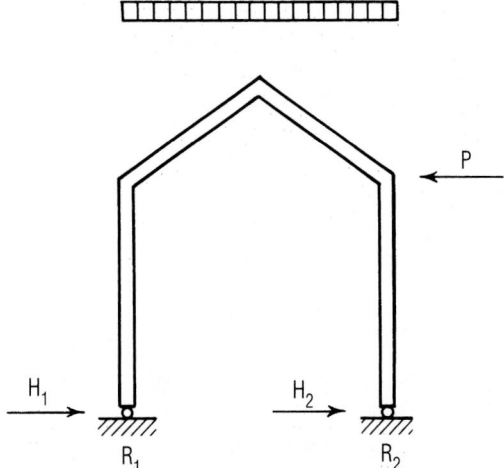

Fig. 8.41 Gable frame with vertical columns and two sloping beams.

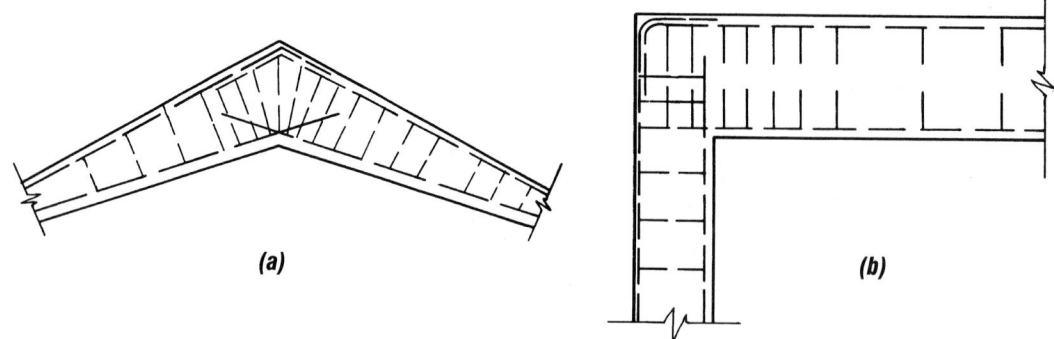

(a)

(b)

Fig. 8.42 Reinforcing arrangements at right-frame joints.

component of tensile force in the reinforcement, but preferably a lower limit should be the minimum size and number of ties required for columns.

All toes of rigid frames are subjected to horizontal forces, or thrust. In a hinged rigid frame, an additional axial force (compression or tension) acts on the base, while in a fixed rigid frame, an additional axial force and a bending moment act.

Usually, analysis assumes that the toes of rigid frames do not move relative to each other. The designer should check this assumption in the design. If the toes do spread under load, the horizontal thrust, as well as all the internal forces and moments within the frame, will change. The actual internal forces due to movement of the toes should be computed and the frame designed accordingly. Similarly, if the base is not truly hinged or fixed, but only partly so, the effect of partial fixity on the frame should be taken into account.

The thrust may be resisted by a footing pressing against rock (Fig. 8.43), by friction of the footing against the soil (Fig. 8.44), or by a tie (Fig. 8.45). In the cases illustrated by Figs. 8.44 and 8.45, the likelihood of the toes spreading apart is considerable.

If the toe is hinged, the hinge detail could be provided in the field (Fig. 8.46). Or it could be a prefabricated steel hinge (Fig. 8.47).

In a fixed rigid frame, the connection of the toe to the footing (Fig. 8.48) should be strong enough to develop the computed bending moment. Since this moment is to be transferred to the ground, it is usual to construct a heavy eccentric footing that counterbalances this moment by its weight, as shown in Fig. 8.48.

To obtain an advantageous moment distribution in a frame, a designer might find it desirable to increase the sizes of some members of the frame.

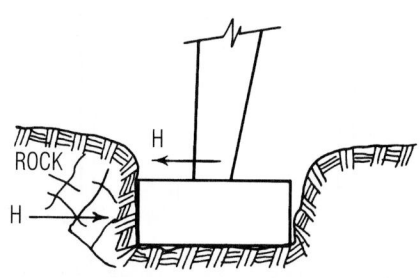

Fig. 8.43 Footing thrust resisted by side bearing.

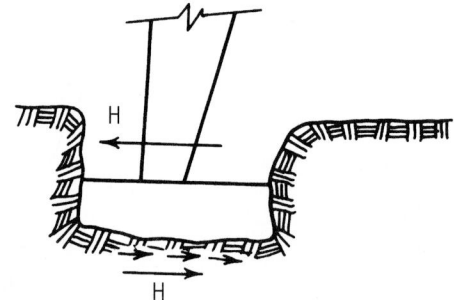

Fig. 8.44 Footing thrust resisted by base friction.

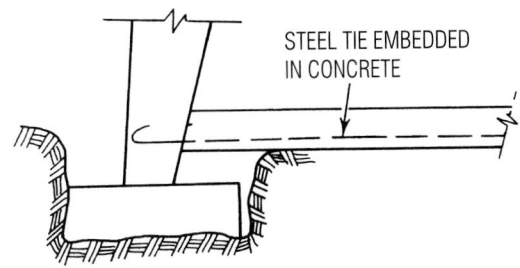

Fig. 8.45 Ties between footings take thrust at the base of a rigid frame.

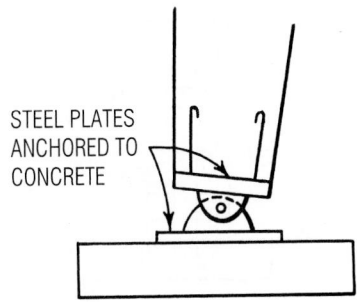

Fig. 8.47 Base with steel hinge.

For example, for a long-span, low, rigid frame, increasing the width of the vertical legs would reduce positive bending moments in the horizontal members and increase moments in the vertical members. The vertical members could become stubby, as in Fig. 8.49. According to the ACI Code, when the ratio of depth d to length L of a continuous member exceeds 0.4, the member becomes a "deep" beam; the bending stresses and resistance to them do not follow the patterns described previously in this section. The designer should provide more than the usual stirrups and distribute reinforcement along the faces of the deep members, as in Fig. 8.49 (Art 8.17.5).

Design of precast-concrete rigid frames is identical to that of cast-in-place frames, except for connections. It is quite common to precast parts of frames between points of counterflexure, or sec-

tions where bending moment is small, as shown in Fig. 8.50a. This eliminates the need for a moment connection (often referred to as a continuity connection) at a joint. Only a shear connection is required (Fig. 8.50b). Since some bending moment might occur at the joint due to live, wind, seismic, and other loads, moment resistance should be provided by grouting longitudinal bars (Fig. 8.50b) or welding steel plates embedded in the precast concrete (Fig. 8.50c). When this type of connection is used, however, bending moments in the structure should be determined for continuity at the joint to verify the adequacy of the joint.

Rigid frames also may be prestressed and cast in place or precast. Prestressed, cast-in-place frames are posttensioned. Usually, the prestress is applied to each member with tendons anchored within the

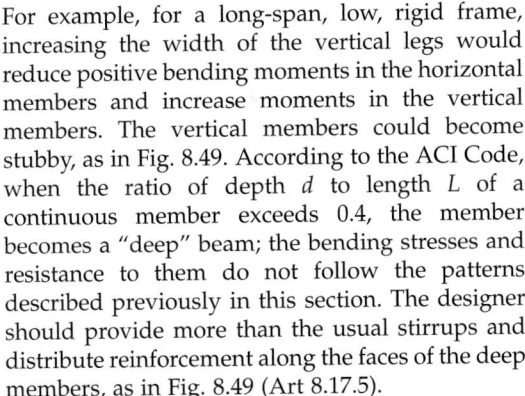

Fig. 8.46 Hinge built with reinforcing bars at the top of a footing.

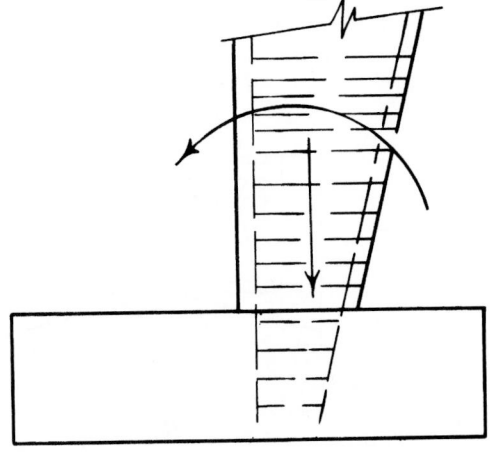

Fig. 8.48 Base with moment resistance.

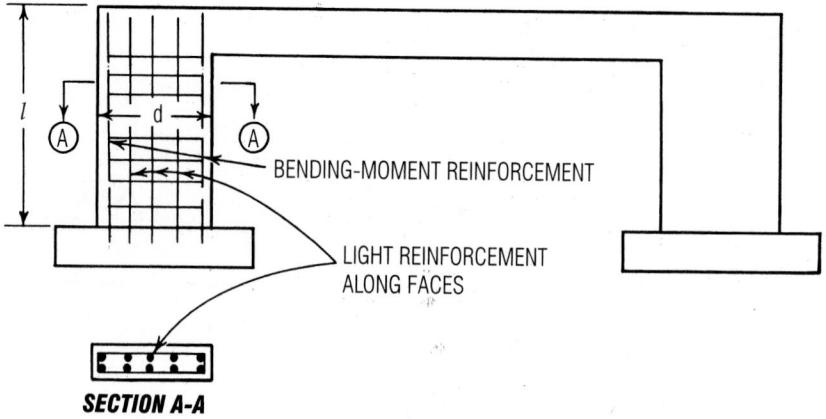

Fig. 8.49 Rigid frame with stubby columns.

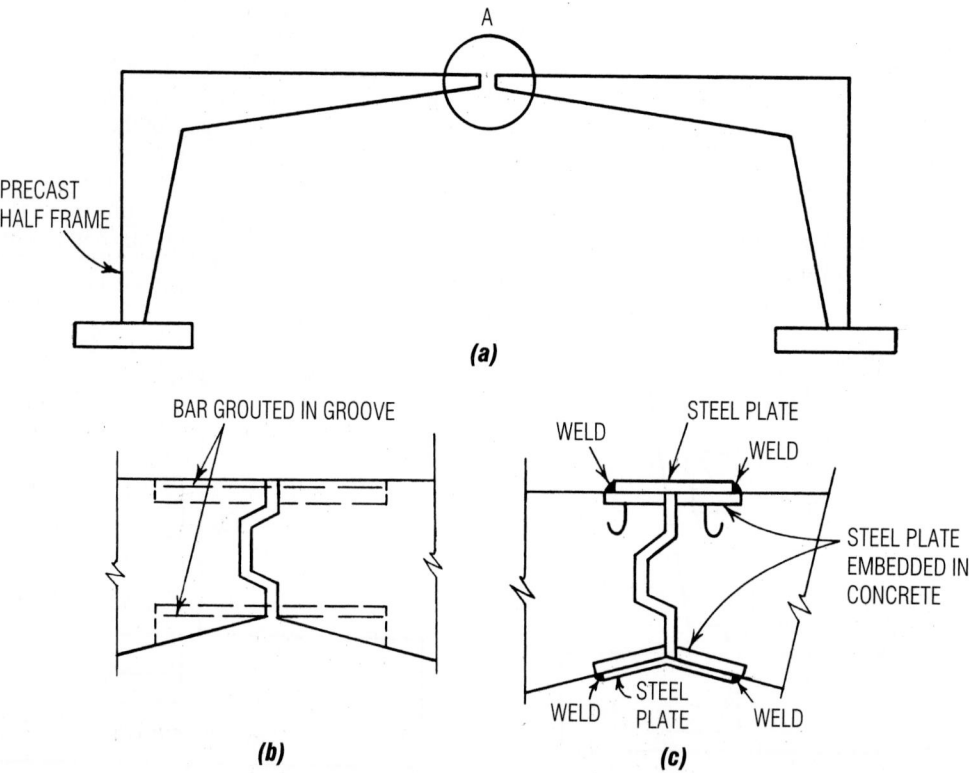

Fig. 8.50 Precast-concrete rigid frame. (*a*) Halves connected at midspan. (*b*) Midspan joint with grouted longitudinal reinforcing bars. (*c*) Welded connection at midspan.

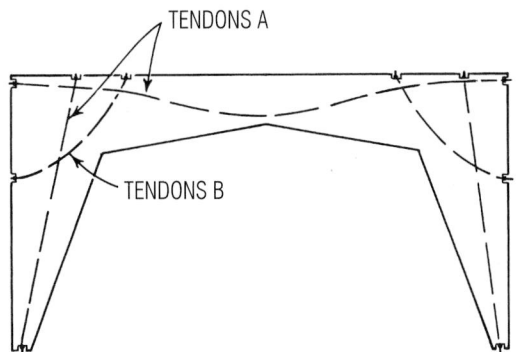

Fig. 8.51 Prestressed-concrete rigid frame.

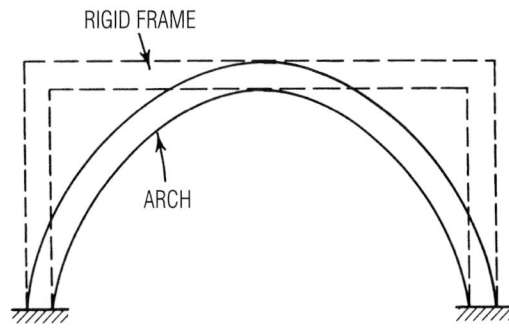

Fig. 8.52 Arch replacement for rigid frame.

member (Fig. 8.51). Although continuous tendons may be more efficient structurally, friction losses due to bending the tendons make application of prestress in the field as intended by the design difficult. Such losses cannot be estimated. Hence, the magnitude of the prestress imparted is uncertain. The rigid joints, though, may be prestressed by individual straight or slightly bent tendons anchored in adjacent members (tendons B in Fig. 8.51).

When selecting the magnitude of the prestressing force in each member, the designer should ascertain that the bending moments at the ends of members meeting at a joint are in equilibrium and that the end rotation there is the same for each member.

Precast rigid frames may be pretensioned, posttensioned, or both. In prestressed, precast rigid frames, it is common to fabricate the individual members between joints, rather than between points of counterflexure, and connect them rigidly at the joints. The members are connected at the rigid joints by grouting reinforcing bars, welding steel inserts, or posttensioning. In all cases, the designer should make sure that the rotations of the ends of all members meeting at a joint are equal.

8.53 Concrete Arches

Structurally, arches are, in many respects, similar to rigid frames (Arts. 8.51 and 8.52). An arch may be considered a rigid frame with one curved member instead of a number of straight members (Fig. 8.52). The internal forces in the two structural

systems are of the same nature: bending moments, axial forces, and shears. The difference is that bending moments predominate in rigid frames, while arches may be shaped so that axial (compression) force predominates. Nevertheless, general design procedures for arches and rigid frames are identical.

Design of details, however, differs since arches have no rigid joints above the abutments, and arches, being predominantly subjected to compression, must be provided with more resistance against buckling. Also, because arches are dependent on development of thrust resistance for their strength, all the requirements for rigid frames for thrust resistance are even more critical for arches.

Precasting of arches is not common because the curvature makes stacking for transportation difficult. Some small-span site-precast arches, however, have been successfully erected.

Prestressing of arch ribs is not very common because the arches are subjected to large compressive forces; thus, prestressing rarely offers advantages. But prestressing of abutments and of connections of a fixed-end arch to abutments, where bending moments are large, could be beneficial in resisting these moments.

See also Arts. 6.69 to 6.71.

(G. Winter and A. H. Nilson, "Design of Concrete Structures," McGraw-Hill Book Company, New York.)

8.54 Concrete Folded Plates

The basic structural advantage of a folded-plate structure (Fig. 8.53) over beams and slabs for a given span is that more material in a folded plate

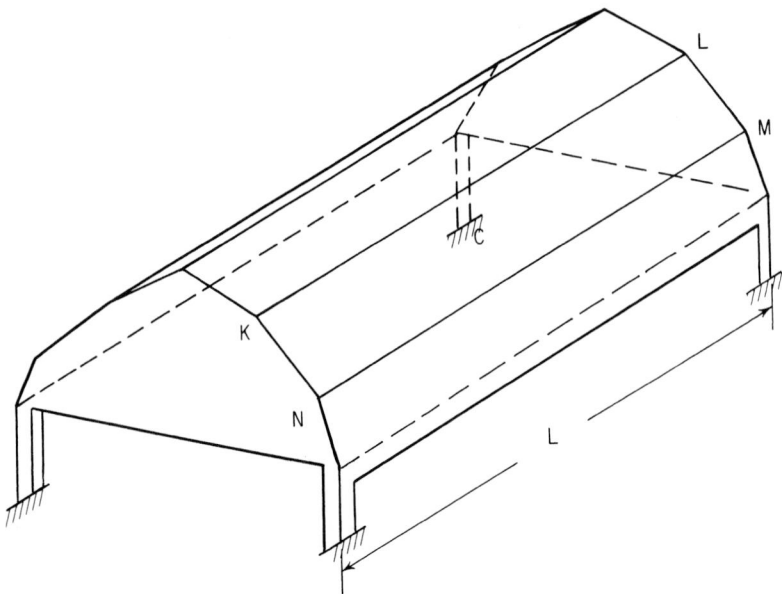

Fig. 8.53   Folded-plate roof.

carries stresses, and stress distribution may be more uniform. For example, Fig. 8.54a shows cross sections of alternative structural systems of the same span and depth superimposed. One section is for a folded plate, the other for a system with two solid beams. The stress distribution in the solid beams is shown in Fig. 8.54b. Only the extreme fibers are stressed to the maximum allowable, while the remainder, the largest part of the cross section, is subjected to much smaller stresses. The stresses in the folded plate, as shown in Fig. 8.54c, are more uniformly distributed through the depth D of the structure. Furthermore, folded plates inherently enclose a space, whereas, for the same function, beams require a deck to span between them. Hence, a folded-plate structure needs less material than solid beams and may therefore be more economical.

It should be noted, however, that longitudinal-stress distribution in a folded-plate structure spanning a distance L (Fig. 8.53) is not given accurately by simple-beam theory; that is, the longitudinal normal stresses are not as shown in Fig. 8.54b. Under vertical loads, one cannot compute the moment of inertia of the folded-plate section in Fig. 8.54a about the centroidal axis and

find the stresses from Mc/I. The cross section distorts under load, invalidating the elementary bending theory. Hence, the result may be more nearly the stress distribution shown in Fig. 8.54c. See also Arts. 6.76 and 6.77.

These normal stresses are perpendicular to the plane of the folded-plate section (Fig. 8.54a). They and the shear stresses parallel to the section may be assumed uniformly distributed over the thickness of the plates. The same is true of membrane stresses in shell structures.

Reinforcement in each plate, such as KLMN (Fig. 8.53), in the transverse and longitudinal directions, is determined from stresses obtained from analysis. Typical reinforcement is shown in Fig. 8.55. The quantity of longitudinal reinforcement is determined by the tensile stresses in each plate. But reinforcement should not be less than that indicated in Art. 8.23 for minimum quantity in slabs. In addition, a minimum of temperature reinforcement as required for slabs should be distributed uniformly throughout each plate. (See also Art. 8.51.)

Transverse reinforcement is determined by the transverse bending in each plate between support points A, B, C, D, . . . (Fig. 8.55). But reinforcement

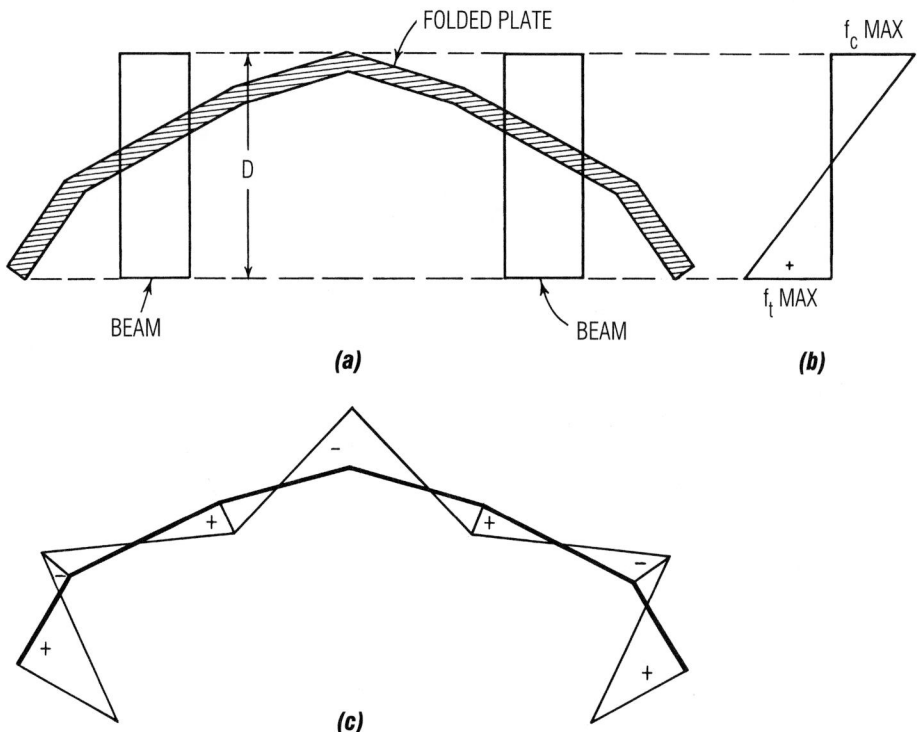

Fig. 8.54 Comparison of folded-plate with beams. (*a*) Vertical section through a folded-plate roof with superimposition of two solid, rectangular beams that could replace it as roof supports. (*b*) Stress distribution at midspan of a beam. (*c*) Longitudinal stress distribution at midspan of the folded-plate roof.

should not be less than the temperature reinforcement indicated in Art. 8.23. Because the regions around plate intersections, such as *B* and *C*, are subjected to negative transverse bending moments, negative (top) reinforcement is required there. This reinforcement, as well as the bottom bars, should be carried far enough past the corner for proper embedment. Because of the distortions of the section and the uncertainty of the extent of transverse negative moments, it is good practice to carry reinforcement along the top of all plates, as shown for plate *CD* (Fig. 8.55). Such top reinforcement also is efficient in resisting shear.

Essentially, Fig. 8.55 represents a cross section of a rigid frame. The joints between plates have to be maintained rigid to correspond to assumptions made in the analysis. Thus, these joints should be reinforced as in rigid frames. When the angle between two plates is large, it is desirable to tie top and bottom reinforcement with ties, as indicated in Fig. 8.55.

If the concrete alone is not sufficient to resist diagonal tension due to shear, reinforcement should be provided for the excess diagonal tension. Such reinforcement may be inclined, as at *A* in Fig. 8.56, or a grid of longitudinal and transverse bars may be used, as at *B*. In the latter case, the reinforcement will have the pattern indicated in Fig. 8.55. The quantity needed to resist diagonal tension, then, should be added to that required for bending. Both the transverse and longitudinal reinforcement inserted for this purpose preferably should be distributed evenly between the top and bottom faces of the plates.

Elementary analysis of folded plates usually assumes that the cross sections at the supports do not distort. Therefore, it is common practice to provide rigid diaphragms at the ends of folded plates in planes of supports (Fig. 8.57). The diaphragms act as transverse beams, as well as ties, between supports. Hence, they usually have

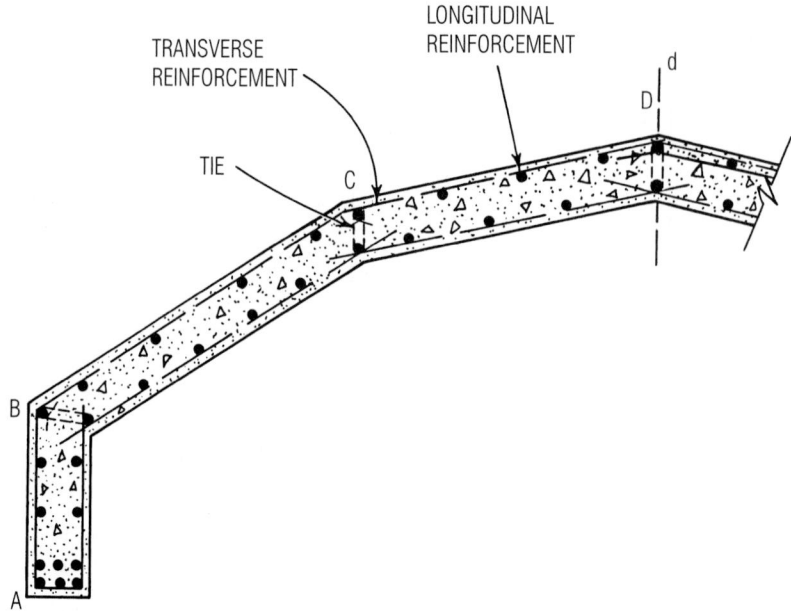

Fig. 8.55 Typical reinforcement at a section of a folded plate.

relatively heavy bottom reinforcement. The strains in the end diaphragms should be kept small, to keep the end sections of the folded-plate structure from distorting. It is advisable, therefore, that the reinforcement in the diaphragm be evenly distributed throughout each face.

8.55 Concrete Shells

Thin shells are curved or folded slabs whose thicknesses are small compared with their other dimensions. In addition, shells are characterized by their three-dimensional load-carrying behavior, which is determined by their geometric shape, their boundary conditions, and the nature of the applied load. Many forms of concrete shells are used. To be amenable to theoretical analysis, these forms have geometrically expressible surfaces.

8.55.1 Stress Analysis of Shells

Elastic behavior is usually assumed for shell structural analysis, with suitable assumptions to approximate the three-dimensional behavior of

shells. The ACI Building Code includes special provisions for shells. It suggests model studies for complex or unusual shapes, prescribes minimum reinforcement, and specifies design by the ultimate-strength method with the same load factors as for design of other elements.

Stresses usually are determined by membrane theory and are assumed constant across the shell thickness. The membrane theory for shells, however, neglects bending stresses. Yet, every shell is subjected to bending moments, not only under unsymmetrical loads but under uniform and symmetrical loads. Stress analysis of shells, however, by bending theory is more complex than by membrane theory but with the use of computers and finite-element, boundary-element, or numerical integration methods, it can be readily executed. See also Arts. 6.72 to 6.75.

Although unsymmetrical loads cause bending moments throughout a whole shell, symmetrical loads cause moments mainly at edges and supports. These edge and support moments may be very large. Provision should be made to resist them. If they are not properly provided for, not only would unsightly cracks occur, but the shell may distort, progressively increasing the size of

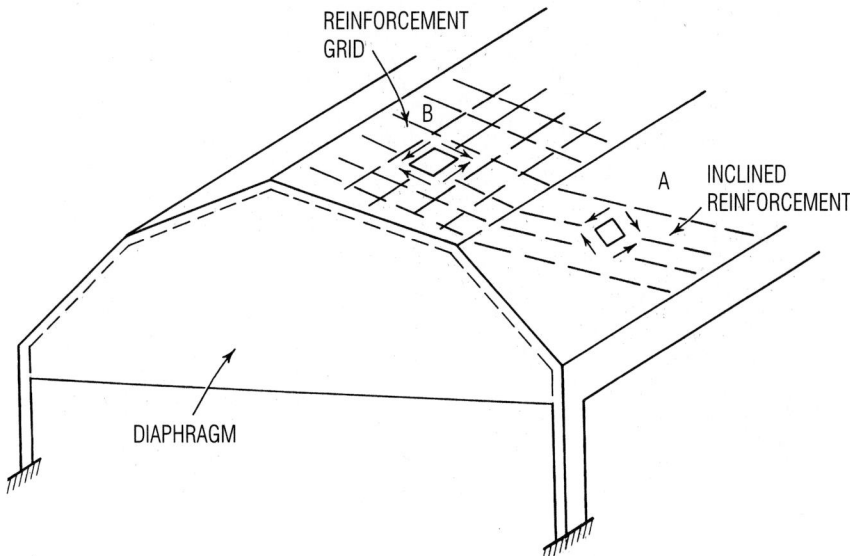

Fig. 8.56   Reinforcement patterns in the plates of a folded-plate roof.

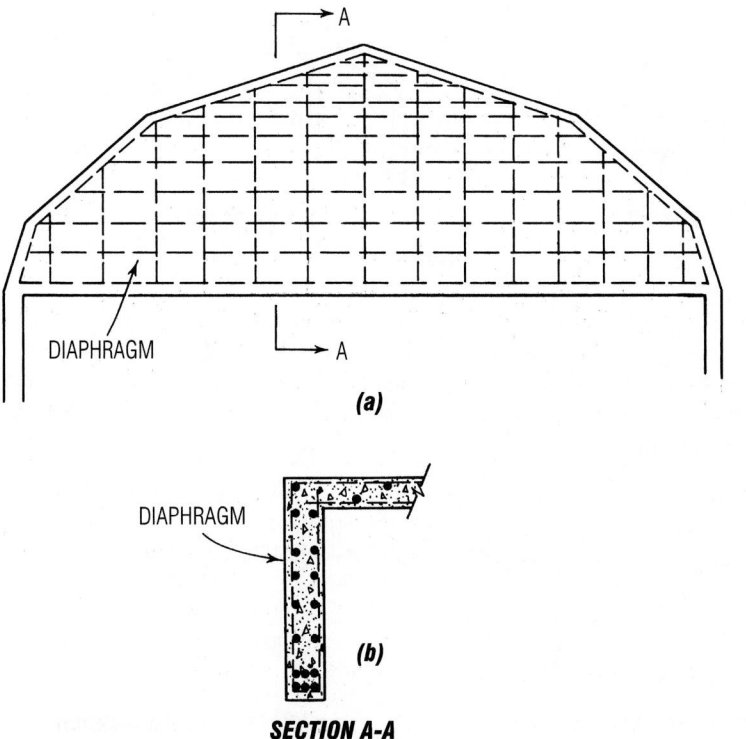

Fig. 8.57   Reinforcement in the diaphragm of a folded-plate roof.

the cracks and causing large deflections, rendering the shell unusable. Therefore, past experience in design, field observations, and knowledge of results of tests on shells are a necessity for design of shell structures, to insure the proper quantity of reinforcement in critical locations, even though the reinforcement is not predicted by theory. Model testing is a helpful tool for shell design, but small-scale models may not predict all the possible stresses in a prototype.

Because of the difficulties in determining stresses accurately, only those forms of shell that have been successfully constructed and tested in the past are usually undertaken for commercial uses. These forms include barrel arches, domes, and hyperbolic paraboloids (Fig. 8.58).

8.55.2 Cylindrical Shells

Also known as **barrel shells**, cylindrical shells may consist of single transverse spans (Fig. 8.58a) or multiple spans (Fig. 8.59). Analysis yields a different stress distribution for a single barrel shell from that for a multiple one. But design considerations are the same.

Usually, the design stresses in a shell are quite small, requiring little reinforcement. The reinforcement, both circumferential and longitudinal, however, should not be less than the minimum reinforcement required for slabs (Art. 8.23).

Barrel shells usually are relatively thin. Thickness varies from 4 to 6 in for most parts of shells with spans up to 300 ft transversely and longitudinally. But the shells generally are thickened at

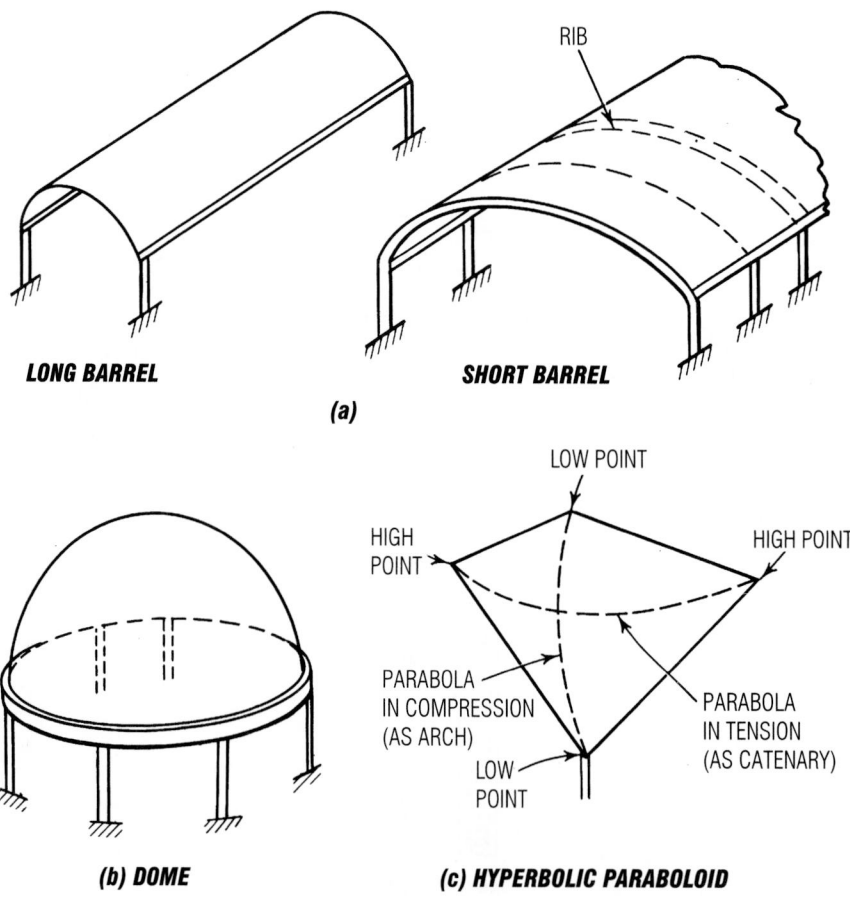

Fig. 8.58 Common types of concrete shells.

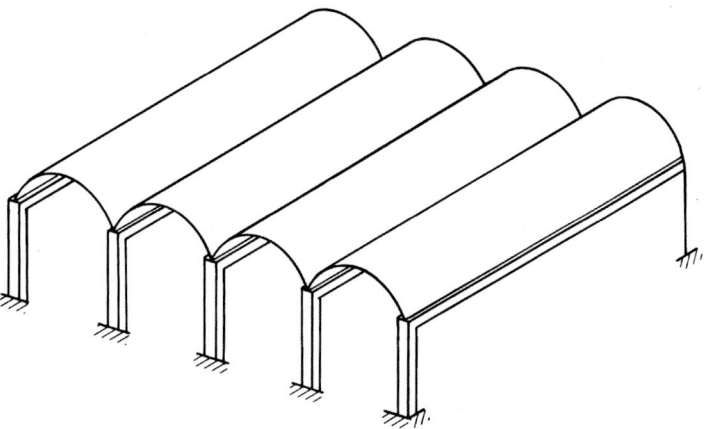

Fig. 8.59 Multiple barrel-arch roof.

edges and supports and stiffened by edge beams. With analysis, including model testing, it is possible to design barrel shells of uniform thickness throughout, without stiffening edge members. But if the more simplified method of analysis (membrane theory) is employed, which is more usual and practical, stiffening edge members should be provided, as shown in Fig. 8.60. These consist of

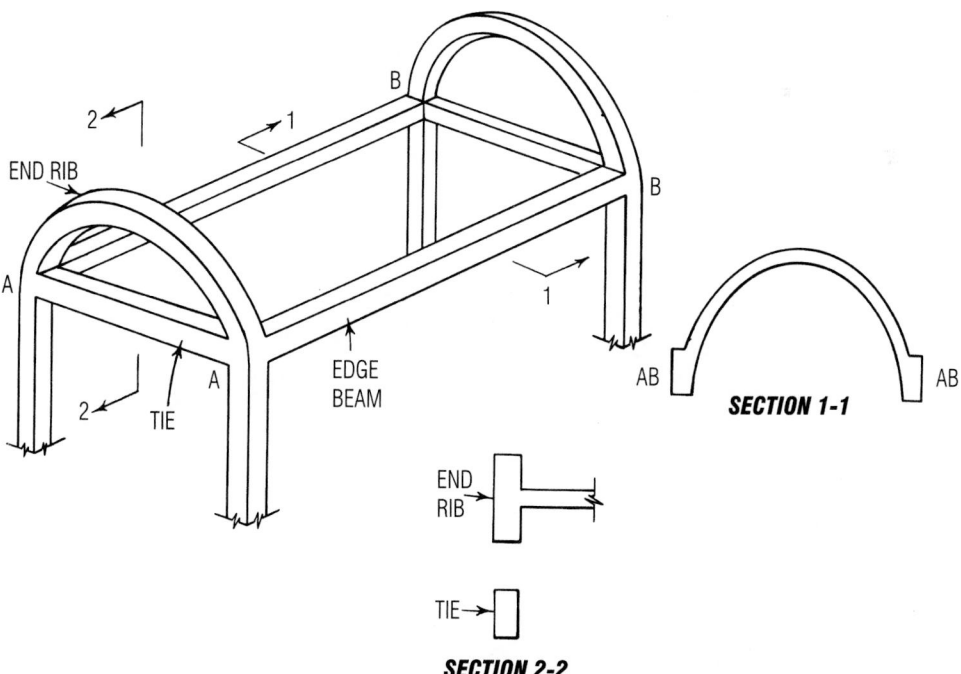

Fig. 8.60 Stiffening members in thin-shell arch roof.

edge beams *AB* and end arch ribs *AA* and *BB*. Instead of an end arch rib, an end diaphragm may be employed as indicated in Fig. 8.57 for a folded-plate roof.

Stresses determined from analysis may be combined to give the principal stresses, or maximum tension and compression, at each point in the shell. If these are plotted on a projection of the shell, the lines of constant stress, or stress trajectories, will be curved. The tensile-stress trajectories generally follow a diagonal pattern near supports and are nearly horizontal around midspan. Reinforcing bars to resist these stresses, therefore, may be draped along the lines of principal stress. This, however, makes fieldwork difficult because large-diameter bars may have to be bent and extra care is needed in placing them. Hence, main steel usually is placed in a grid pattern, with the greatest concentration along longitudinal edges or valleys. To control temperature and shrinkage cracks, minimum reinforcement should be provided.

Reinforcement may be placed in the shell in one layer (Fig. 8.61a) or two layers (Fig. 8.61b), depending on the stresses; that is, the span and design loads. (Very thin shells, for example, those 3 to $4\frac{1}{2}$ in thick, may offer space for only a single layer.) Shells with one layer of reinforcement are more likely to crack because of local deformations. Although such cracks may not be structurally detrimental, they could permit rainwater leakage. Hence, shells with one layer of reinforcement should have built-up roofing or other water-proofing applied to the outer surface. In reinforcing small-span shells, two-way wire fabric may be used instead of individual bars.

The area of reinforcement, in^2/ft width of shell, should not exceed $7.2f_c'/f_y$ or $29,000h/f_y$, where h is

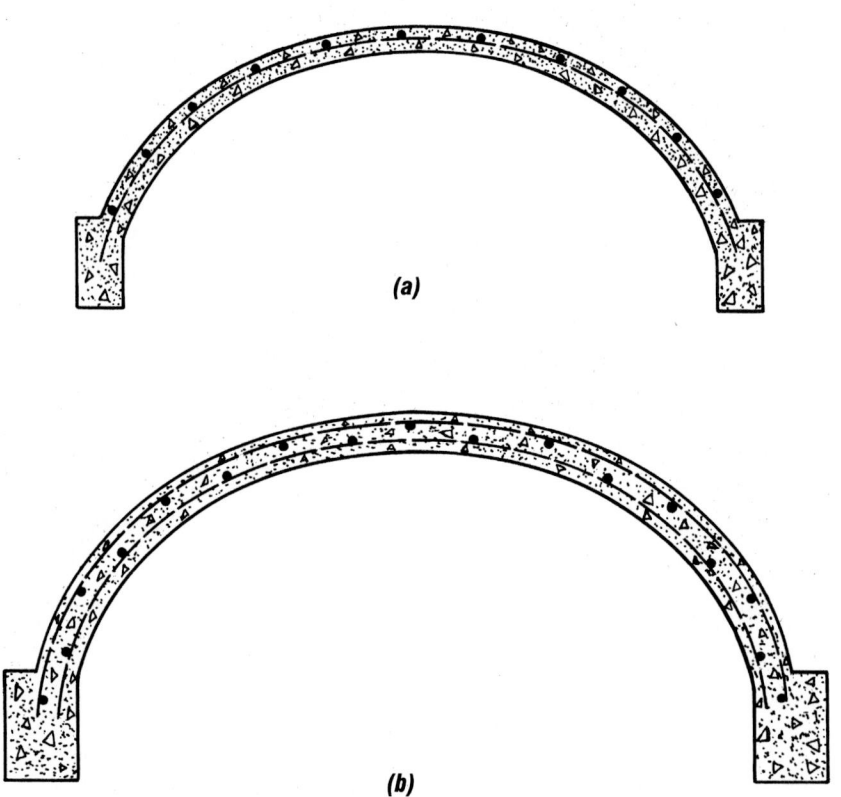

Fig. 8.61 Arch reinforcement: (*a*) single layer; (*b*) double layer.

the overall thickness of the shell, in; f_y the yield strength, psi, of the reinforcement; and f_c' the compressive strength, psi, of the concrete. Reinforcement should not be spaced farther apart than five times the shell thickness or 18 in. Where the computed factored principal tensile stress exceeds $4\sqrt{f_c'}$, the reinforcement should not be spaced farther apart than three times the shell thickness.

Minimum specified compressive strength of concrete f_c' should not be less than 3000 psi, while specified yield strength of reinforcement f_y should not exceed 60,000 psi.

Edge beams of barrel arches behave like ordinary beams under vertical loads, except that additional horizontal shear is applied at the top face at the junction with the shell. (If these shear stresses are high, reinforcement should be provided to resist them.) Also, a portion of the shell equal to the flange width permitted for T beams may be assumed to act with the supporting members. Furthermore, transverse reinforcement from the shell equal to that required for the flange of a T beam should be provided and should be adequately anchored into the edge beam. A typical detail of an edge beam is shown in Fig. 8.62.

Computed stresses in the end arch ribs or diaphragms usually are small. The minimum amount of reinforcement in a rib should be the minimum specified by the ACI Code for a beam and, in a diaphragm, the minimum specified for a slab. Longitudinal reinforcement from the shell should be adequately embedded in the ribs. Because of shear transmission between shell and ribs, the shear stresses should be checked and adequate shear reinforcement provided, if necessary. Typical reinforcement in end ribs and diaphragms is shown in Fig. 8.63.

High tensile stresses and considerable distortions, particularly in long barrels, usually occur near supports. If the stresses in those areas are not computed accurately, reinforcement should be increased there substantially over that required by simplified analysis. The increased quantity of reinforcement should form a grid. In arches with very long spans and where stresses are computed more accurately, prestressing of critical areas may be efficient and economical. But the ratio of steel to concrete in any portion of the tensile zone should be at least 0.0035.

When barrel shells are subjected to heavy concentrated loads, such as in factory roofs or bridges, economy may be achieved by providing interior ribs (Fig. 8.64), rather than increasing the thickness throughout the whole shell. Such ribs increase both the strength and stiffness of the shell without increasing the weight very much.

In many cases, only part of a barrel shell may be used. This could occur in end bays of multiple barrels or in interior barrels where large openings are to be provided for windows. Stress distribution in such portions of shells is different from that in whole barrels, but design considerations for edge members and reinforcement placement are the same.

8.55.3 Domes

These are shells curved in two directions. One of the oldest types of construction, domes were often built of large stone pieces. Having a high ratio of thickness to span, this type of construction is excluded from the family of thin shells.

Concrete domes are built relatively thin. Domes spanning 300 ft have been constructed only 6 in thick. Ratio of rise to span usually is in the range of 0.10 to 0.25.

A dome of revolution is subjected mostly to pure membrane stresses under symmetrical, uniform live load. These stresses are compressive in most of the dome and tensile in some other por-

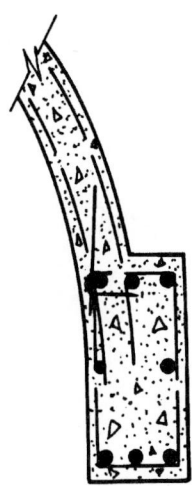

Fig. 8.62 Edge beam for arch.

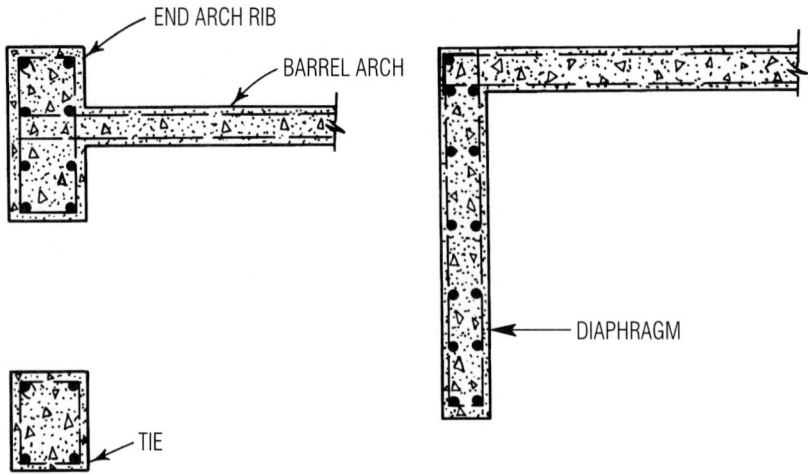

Fig. 8.63 Reinforcing in end ribs, tie, and diaphragm of an arch.

tions, mainly in the circumferential direction. Under unsymmetrical loading, bending moments may occur. Hence, it is common to place reinforcement both in the circumferential direction and perpendicular to it (Fig. 8.65). The reinforcement may be welded-wire fabric or individual bars. It may be placed in one layer (Fig. 8.65*b*), depending on stresses. Concrete for domes may be cast in

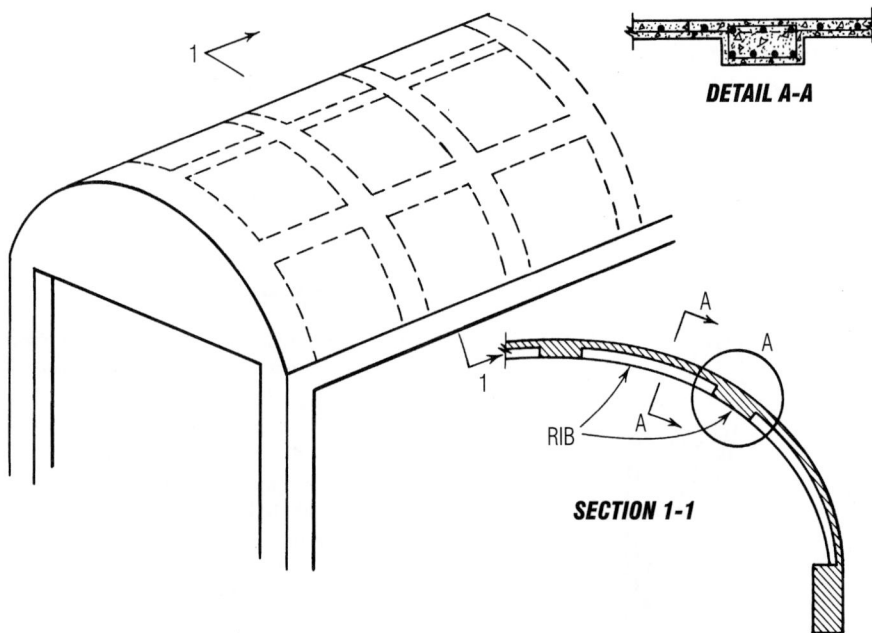

Fig. 8.64 Arch with ribs in longitudinal and transverse directions.

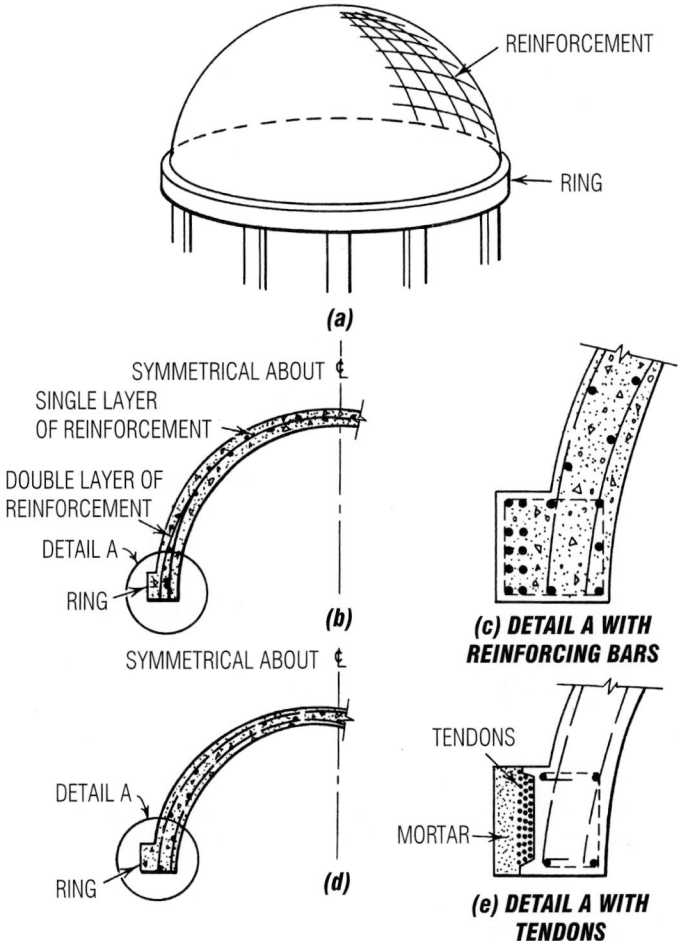

Fig. 8.65 Reinforcing arrangements for domes.

forms, as are other more conventional structures, or sprayed.

The critical portion of a dome is its base. Whether the dome is supported continuously there, for example, on a continuous footing, or on isolated supports (Fig. 8.65a), relatively large bending moments and distortions occur in the shell close to the supports. These regions should be designed to resist the resulting stresses. In domes reinforced with one layer of bars or mesh, it is advisable to provide in the vicinity of the base a double layer of reinforcement (Fig. 8.65b). It also is advisable to thicken the dome close to its base.

The base is subjected to a very large outward-acting radial force, causing large circumferential tension. To resist this force, a concrete ring is constructed at the base (Fig. 8.65). The ring and thickening of the concrete shell in the vicinity of the ring help reduce distortions and cracking of the dome at its base.

Reinforcement of the shell should be properly embedded in the ring (detail A, Fig. 8.65c). The ring should be reinforced or prestressed to resist the circumferential tension. Prestressing is efficient and hence often used. One method of applying prestress is shown in detail A, Fig. 8.65d and e.

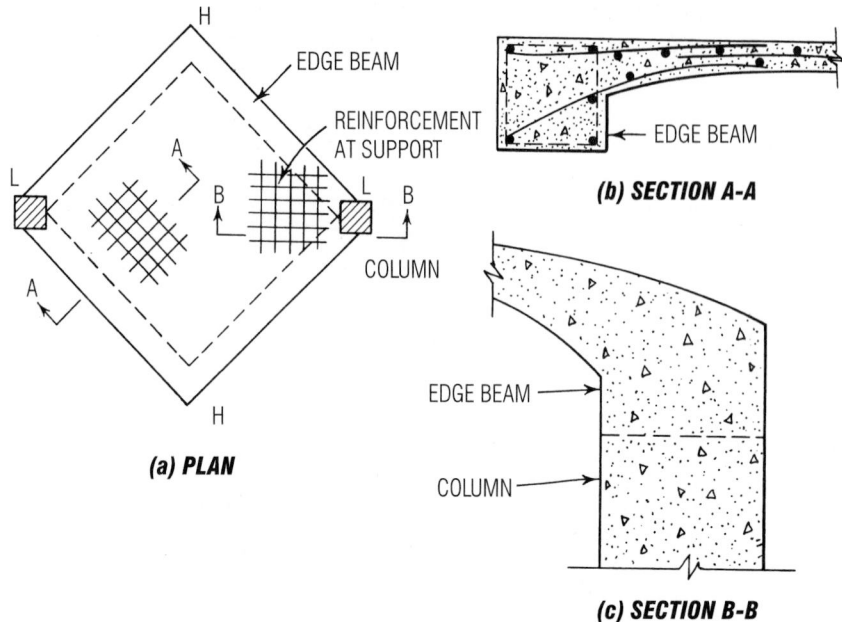

Fig. 8.66 Hyperbolic-paraboloid shell. *H* indicates high point, *L* low point.

Wires are wrapped under tension around the ring and then covered with mortar, for protection against rust and fire. Stirrups should be provided throughout the ring.

8.55.4 Hyperbolic-Paraboloid Shells

Also referred to as a **hypar**, this type of shell, like a dome, is double-curved, but it can be formed with straight boards. Furthermore, since the principal stresses throughout the shell interior consist of equal tension and compression in two perpendicular, constant directions, placement of reinforcement is simple.

Figure 8.66*a* shows a plan of a hypar supported by two columns at the low points *L*. The other corners *H* are the two highest points of the shell. Although strips parallel to *LL* are in compression and strips parallel to *HH* in tension, it is customary to place reinforcement in two perpendicular directions parallel to the generatrices of the shell, as shown at section *A-A*, Fig. 8.66*a*. The reinforcement should be designed for diagonal tension parallel to the generatrices. Since considerable

bending moments may occur in the shell at the columns, this region of the shell usually is made thicker than other portions and requires more reinforcement. The added reinforcement may be placed in the *HH* and *LL* directions, as shown at section *B-B*, Fig. 8.66*a*.

Shell reinforcement may be placed in one or two layers, depending on the intensity of stresses and distribution of superimposed load. If the superimposed load is irregular and can cause significant bending moments, it is advisable to place the reinforcement in two layers.

As for other types of shells, edges of a hypar are subjected to larger distortions and bending moments than its interior. Therefore, it is desirable to construct edge beams and thicken the shell in the vicinity of these beams (Fig. 8.66*b*). A double layer of reinforcement at the edge beams helps reduce cracking of the shell in the vicinity of the beams.

The edge beams are designed as compression or tension members, depending on whether the hypar is supported at the low points or high points. Prestress in the shell is most efficient in the vicinity of supports. It also is efficient along the edge beams if supports are at the high points.

8.55.5 Shells with Complex Shapes

Curved shells also may be built with more complex shapes. For example, they may be undulating or have elliptical or irregular boundaries. In some cases, they may be derived by inverting structures in pure tension, such as bubbles or fabric hung from posts.

(D. P. Billington "Thin-Shell Concrete Structures," 2nd ed., and A. H. Nilson and G. Winter, "Design of Concrete Structures," 11th ed., McGraw-Hill, Inc., New York.)

9

Roger L. Brockenbrough
President
R. L. Brockenbrough & Associates, Inc.
Pittsburgh, Pennsylvania

STRUCTURAL STEEL DESIGN AND CONSTRUCTION

The many desirable characteristics of structural steels has led to their widespread use in a large variety of applications. Structural steels are available in many product forms and offer an inherently high strength. They have a very high modulus of elasticity, so deformations under load are very small. Structural steels also possess high ductility. They have a linear or nearly linear stress-strain relationship up to relatively large stresses, and the modulus of elasticity is the same in tension and compression. Hence, structural steels' behavior under working loads can be accurately predicted by elastic theory. Structural steels are made under controlled conditions, so purchasers are assured of uniformly high quality.

Standardization of sections has facilitated design and kept down the cost of structural steels. For tables of properties of these sections, see "Manual of Steel Construction," American Institute of Steel Construction, One East Wacker Dr., Chicago, IL 60601-2001 www.aisc.org.

This section provides general information on structural-steel design and construction. Any use of this material for a specific application should be based on a determination of its suitability for the application by professionally qualified personnel.

9.1 Properties of Structural Steels

The term *structural steels* includes a large number of steels that, because of their economy, strength, ductility, and other properties, are suitable for load-carrying members in a wide variety of fabricated structures. Steel plates and shapes intended for use in bridges, buildings, transportation equipment, construction equipment, and similar applications are generally ordered to a specific specification of ASTM and furnished in "Structural Quality" according to the requirements (tolerances, frequency of testing, and so on) of ASTM A6. Plate steels for pressure vessels are furnished in "Pressure Vessel Quality" according to the requirements of ASTM A20.

Each structural steel is produced to specified minimum mechanical properties as required by the specific ASTM designation under which it is ordered. Generally, the structural steels include steels with yield points ranging from about 30 to 100 ksi. The various strength levels are obtained by varying the chemical composition and by heat treatment. Other factors that may affect mechanical properties include product thickness, finishing temperature, rate of cooling, and residual elements.

The following definitions aid in understanding the properties of steel.

Yield point F_y is that unit stress, ksi, at which the stress-strain curve exhibits a well-defined increase in strain without an increase in stress. Many design rules are based on yield point.

Tensile strength, or ultimate strength, is the largest unit stress, ksi, the material can achieve in a tensile test.

Modulus of elasticity E is the slope of the stress-strain curve in the elastic range, computed by dividing the unit stress, ksi, by the unit strain, in/in. For all structural steels, it is usually taken as 29,000 ksi for design calculations.

Ductility is the ability of the material to undergo large inelastic deformations without fracture. It is generally measured by the percent elongation for a specified gage length (usually 2 or 8 in). Structural steel has considerable ductility, which is recognized in many design rules.

Weldability is the ability of steel to be welded without changing its basic mechanical properties. However, the welding materials, procedures, and techniques employed must be in accordance with the approved methods for each steel. Generally, weldability decreases with increase in carbon and manganese.

Notch toughness is an index of the propensity for brittle failure as measured by the impact energy necessary to fracture a notched specimen, such as a Charpy V-notch specimen.

Toughness reflects the ability of a smooth specimen to absorb energy as characterized by the area under a stress-strain curve.

Corrosion resistance has no specific index. However, relative corrosion-resistance ratings are based on the slopes of curves of corrosion loss (reduction in thickness) vs. time. The reference of comparison is usually the corrosion resistance of carbon steel without copper. Some high-strength structural steels are alloyed with copper and other elements to produce high resistance to atmospheric deterioration. These steels develop a tight oxide that inhibits further atmospheric corrosion. Figure 9.1 compares the rate of reduction of thickness of typical proprietary "corrosion-resistant" steels with that of ordinary structural steel. For standard methods of estimating the atmospheric corrosion resistance of low-alloy steels, see ASTM Guide G101, American Society of Testing and Materials, 100 Barr Harbor Drive West Conshchoken, PA, 19428-2959, www.astm.org.

(R. L. Brockenbrough and B. G. Johnston, "USS Steel Design Manual," R. L. Brockenbrough & Associates, Inc., Pittsburgh, PA 15243.)

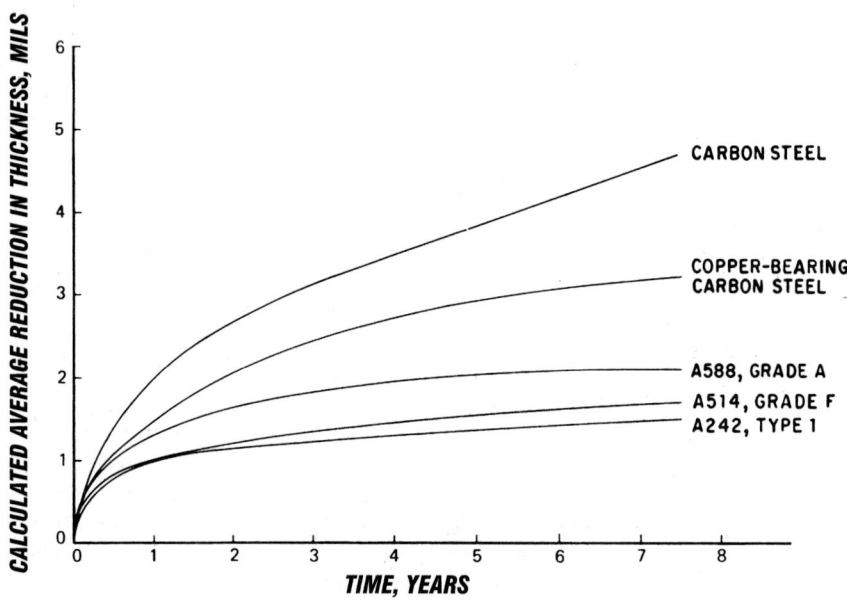

Fig. 9.1 Curves show corrosion rates for steels in an industrial atmosphere.

9.2 Summary of Available Structural Steels

The specified mechanical properties of typical structural steels are presented in Table 9.1. These steels may be considered in four general categories, depending on chemical composition and heat treatment, as indicated below. The tensile properties for structural shapes are related to the size groupings indicated in Table 9.2.

Carbon steels are those steels for which (1) the maximum content specified for any of the following elements does not exceed the percentages noted: manganese—1.65%, silicon—0.60%, and copper—0.60%, and (2) no minimum content is specified for the elements added to obtain a desired alloying effect.

The first carbon steel listed in Table 9.1—A36— is a weldable steel available as plates, bars, and structural shapes. The last steel listed in the table. A992, which is available only for W shapes (rolled wide flange shapes), was introduced in 1998 and has rapidly become the preferred steel for building construction. It is unique in that the steel has a maximum ratio specified for yield to tensile strength, which is 0.85. The specification also includes a maximum carbon equivalent of 0.47 percent to enhance weldability. A minimum average Charpy V-notch toughness of 20 ft-lb at 70 °F can be specified as a supplementary requirement. The other carbon steels listed in Table 9.1 are available only as plates. Although each steel is available in three or more strength levels, only one strength level is listed in the table for A283 and A285 plates.

A283 plates are furnished as structural-quality steel in four strength levels—designated as Grades A, B, C, and D—having specified minimum yield points of 24, 27, 30, and 33 ksi. This plate steel is of structural quality and has been used primarily for oil- and water-storage vessels. A573 steel, which is available in three strength levels, is a structural-quality steel intended for service at atmospheric temperatures at which improved notch toughness is important. The other plate steels—A285, A515, and A516—are all furnished in pressure-vessel quality only and are intended for welded construction in more critical applications, such as pressure vessels. A516 is furnished in four strength levels— designated as Grades 55, 60, 65, and 70 (denoting their tensile strength)—having specified minimum yield points of 30, 32, 35, and 38 ksi. A515 has similar grades except there is no Grade 55. A515 steel is for "intermediate and higher temperature service," whereas A516 is for "moderate and lower temperature service."

Carbon steel pipe used for structural purposes is usually A53 Grade B with a specified minimum yield point of 35 ksi. Structural carbon-steel hot-formed tubing, round and rectangular, is furnished to the requirements of A501 with a yield point of 36 ksi. Cold-formed tubing is also available in several grades with a yield point from 33 to 50 ksi.

High-strength, low-alloy steels have specified minimum yield points above about 40 ksi in the hot-rolled condition and achieve their strength by small alloying additions rather than through heat treatment. A588 steel, available in plates, shapes, and bars, provides a yield point of 50 ksi in plate thicknesses through 4 in and in all structural shapes and is the predominant steel used in structural applications in which durability is important. Its resistance to atmospheric corrosion is about four times that of carbon steel. A242 steel also provides enhanced atmospheric-corrosion resistance. Because of this superior atmospheric-corrosion resistance, A588 and A242 steels provide a longer paint life than other structural steels. In addition, if suitable precautions are taken, these steels can be used in the bare, uncoated condition in many applications in which the members are exposed to the atmosphere because a tight oxide is formed that substantially reduces further corrosion. Bolted joints in bare steel require special considerations as discussed in Art. 9.36.

A572 high-strength, low-alloy steel is used extensively to reduce weight and cost. It is produced in several grades that provide a yield point of 42 to 65 ksi. Its corrosion resistance is the same as that of carbon steel.

Heat-Treated Carbon and High-Strength, Low-Alloy Steels ■ This group is comprised of carbon and high-strength, low-alloy steels that have been heat-treated to obtain more desirable mechanical properties.

A633, Grades A through E, are weldable plate steels furnished in the normalized condition to provide an excellent combination of strength (42 to 60 ksi minimum yield point) and toughness (up to 15 ft-lb at − 75 °F).

Table 9.1 Specified Mechanical Properties of Steel*

ASTM Designation	Plate Thickness, in	ANSI/ASTM Group or Weight/ft for Structural Shapes	Yield Point or Yield Strength, ksi	Tensile Strength, ksi
Carbon Steels				
A36	To 8, incl	To 426 lb/ft, incl.	36	58–80
	Not applicable	Over 426 lb/ft	36	58
	over 8	Not applicable	32	58–80
A283, Grade C	None specified	Not applicable	30	55–70
A285, Grade C	To 2, incl	Not applicable	30	55–75
A516, Grade 55	To 12, incl	Not applicable	30	55–75
A516, Grade 60	To 8, incl	Not applicable	32	60–80
A516, Grade 65	To 8, incl	Not applicable	35	65–85
A516, Grade 70	To 8, incl	Not applicable	38	70–90
A573, Grade 58	To $1\frac{1}{2}$, incl	Not applicable	32	58–71
A573, Grade 65	To $1\frac{1}{2}$, incl	Not applicable	35	65–77
A573, Grade 70	To $1\frac{1}{2}$, incl	Not applicable	42	70–90
A992	Not Applicable	All w shapes	50–65	65
High-Strength, Low-Alloy Steels				
A242	To $\frac{3}{4}$, incl	Groups 1 and 2	50	70
	Over $\frac{3}{4}$ to $1\frac{1}{2}$, incl	Group 3	46	67
	Over $1\frac{1}{2}$ to 4, incl	Groups 4 and 5	42	63
A588	To 4, incl	Groups 1–5	50	70
	Over 4 to 5, incl		46	67
	Over 5 to 8, incl		42	63
A572, Grade 42	To 6, incl	Groups 1–5	42	60
A572, Grade 50	To 4, incl	Groups 1–5	50	65
A572, Grade 60	To $1\frac{1}{4}$, incl	Groups 1 and 3	60	75
A572, Grade 65	To $1\frac{1}{4}$, incl	Groups 1 and 3	65	80
Heat-Treated Carbon and High-Strength, Low-Alloy Steels				
A633, Grade C and D	To $2\frac{1}{2}$, incl	Not applicable	50	70–90
	Over $2\frac{1}{2}$ to 4, incl		46	65–85
A633, Grade E	To 4, incl		60	80–100
	Over 4 to 6, incl		55	75–95
A678, Grade C	To $\frac{3}{4}$, incl	Not applicable	75	95–115
	Over $\frac{3}{4}$ to $1\frac{1}{2}$, incl		70	90–110
	Over $1\frac{1}{2}$ to 2, incl		65	85–105
A852	To 4, incl	Not applicable	70	90–110
A913, Grade 50	Not applicable	Groups 1–5	50	65
A913, Grade 60	Not applicable	Groups 1–5	60	75
A913, Grade 65	Not applicable	Groups 1–5	65	80
A913, Grade 70	Not applicable	Groups 1–5	70	90

(Table continued)

Table 9.1 (*Continued*)

ASTM Designation	Plate Thickness, in	ANSI/ASTM Group or Weight/ft for Structural Shapes	Yield Point or Yield Strength, ksi	Tensile Strength, ksi
		Heat-Treated Constructional Alloy Steel		
A514	To 2½ incl	Not applicable	100	110–130
	Over 2½ to 6, incl		90	100–130

* Mechanical properties listed are specified minimum values except where a specified range of values (minimum to maximum) is given. The following properties are approximate values for all the structural steels: modulus of elasticity—29,000 ksi; shear modulus—11,000 ksi; Poisson's ratio—0.30; yield stress in shear—0.57 times yield stress in tension; ultimate strength in shear—$\frac{2}{3}$ to $\frac{3}{4}$ times tensile strength; coefficient of thermal expansion—6.5×10^{-6} in/in/°F for temperature range -50 to $+150$ °F.

A678, Grades A through D, are weldable plate steels furnished in the quenched and tempered condition to provide a minimum yield point of 50 to 75 ksi.

A852 is a quenched and tempered, weathering, plate steel with corrosion resistance similar to that of A588 steel. It has been used for bridges and construction equipment.

A913 is a high-strength low-alloy steel for structural shapes, produced by the quenching and self-tempering process, and intended for buildings, bridges, and other structures. Four grades provide a minimum yield point of 50 to 70 ksi. Maximum carbon equivalents range from 0.38 to 0.45 percent, and the minimum average Charpy V-notch toughness is 40 ft-lb at 70 °F.

Heat-Treated, Constructional-Alloy Steels ▪ Heat-treated steels that contain alloying elements and are suitable for structural applications are called heat-treated, constructional-alloy steels. A514 (Grades A through Q) covers quenched and tempered alloy-steel plates with a minimum yield strength of 90 or 100 ksi.

Bridge Steels ▪ Steels for application in bridges are covered by A709, which includes steel in several of the categories mentioned above. Under this specification, Grades 36, 50, 70, and 100 are steels with yield strengths of 36, 50, 70, and 100 ksi, respectively. The grade designation is followed by the letter W, indicating whether ordinary or high atmospheric-corrosion resistance is required. An

Table 9.2 Wide-Flange Size Groupings for Tensile-Property Classification

Group 1	Group 2	Group 3	Group 4	Group 5
W24 × 55, 62	W40 × 149, 268	W40 × 277–328	W40 × 362–655	W36 × 920
W21 × 44–57	W36 × 135–210	W36 × 230–300	W36 × 328–798	W14 × 605–873
W18 × 35–71	W33 × 118–152	W33 × 201–291	W33 × 318–619	
W16 × 26–57	W30 × 99–211	W30 × 235–261	W30 × 292–581	
W14 × 22–53	W27 × 84–178	W27 × 194–258	W27 × 281–539	
W12 × 14–58	W24 × 68–162	W24 × 176–229	W24 × 250–492	
W10 × 12–45	W21 × 62–147	W21 × 166–223	W21 × 248–402	
W8 × 10–48	W18 × 76–143	W18 × 158–192	W18 × 211–311	
W6 × 9–25	W16 × 67–100	W14 × 145–211	W14 × 233–550	
W5 × 16, 19	W14 × 61–132	W12 × 120–190	W12 × 210–336	
W4 × 13	W12 × 65–106			
	W10 × 49–112			
	W8 × 58, 67			

additional letter, T or F, indicates that Charpy V-notch impact tests must be conducted on the steel. The T designation indicates the material is to be used in a nonfracture-critical application as defined by the American Association of State Highway and Transportation Officials (AASHTO). The F indicates use in a fracture-critical application. A trailing numeral, 1, 2, or 3, indicates the testing zone, which relates to the lowest ambient temperature expected at the bridge site. See Table 9.3. As indicated by the first footnote in the table, the service temperature for each zone is considerably

less than the Charpy V-notch impact-test temperature. This accounts for the fact that the dynamic loading rate in the impact test is severer than that to which the structure is subjected. The toughness requirements depend on fracture criticality, grade, thickness, and method of connection. Additionally, A709-HPS70W, designated as a High Performance Steel (HPS), is also available for highway bridge construction. This is a weathering plate steel, designated HPS because it possesses superior weldability and notch toughness as compared to conventional steels of similar strength.

Table 9.3 Charpy V-Notch Toughness for A709 Bridge Steels*

Grade	Max Thickness, in, Inclusive	Joining/ Fastening Method	Min Avg Energy, ft-lb	Test Temp, °F		
				Zone 1	Zone 2	Zone 3
Non-Fracture-Critical Members						
36T	4	Mech/Weld	15	70	40	10
50T,† 50WT†	2	Mech/Weld	15	70	40	10
	2 to 4	Mechanical	15			
	2 to 4	Welded	20			
70WT‡	2½	Mech/Weld	20	50	20	−10
	2½ to 4	Mechanical	20			
	2½ to 4	Welded	25			
100T, 100WT	2½	Mech/Weld	25	30	0	−30
	2½ to 4	Mechanical	25			
	2½ to 4	Welded	35			
Fracture-Critical Members						
36F	4	Mech/Weld	25	70	40	10
50F,† 50WF†	2	Mech/Weld	25	70	40	10
	2 to 4	Mechanical	25			−10
	2 to 4	Welded	30			−10
70WF‡	2½	Mech/Weld	30	50	20	−10
	2½ to 4	Mechanical	30			−10
	2½ to 4	Welded	35			−10
100F, 100WF	2½	Mech/Weld	35	30	0	−30
	2½ to 4	Mechanical	35			−30
	2½ to 4	Welded	45			NA

* Minimum service temperatures: Zone 1, 0 °F; Zone 2, <0 to −30 °F; Zone 3, < −30 to −60 °F.
† If yield strength exceeds 65 ksi, reduce test temperature by 15 °F for each 10 ksi above 65 ksi.
‡ If yield strength exceeds 85 ksi, reduce test temperature by 15 °F for each 10 ksi above 85 ksi.

Lamellar Tearing ▪ The information on strength and ductility presented generally pertains to loadings applied in the planar direction (longitudinal or transverse orientation) of the steel plate or shape. Note that elongation and area-reduction values may well be significantly lower in the through-thickness direction than in the planar direction. This inherent directionality is of small consequence in many applications, but it does become important in the design and fabrication of structures containing massive members with highly restrained welded joints.

With the increasing trend toward heavy welded-plate construction, there has been a broader recognition of occurrences of lamellar tearing in some highly restrained joints of welded structures, especially those in which thick plates and heavy structural shapes are used. The restraint induced by some joint designs in resisting weld-deposit shrinkage can impose tensile strain high enough to cause separation or tearing on planes parallel to the rolled surface of the structural member being joined.

The incidence of this phenomenon can be reduced or eliminated through use of techniques based on greater understanding by designers, detailers, and fabricators of the (1) inherent directionality of constructional forms of steel, (2) high restraint developed in certain types of connections, and (3) need to adopt appropriate weld details and welding procedures with proper weld metal for through-thickness connections. Furthermore, steels can be specified to be produced by special practices or processes to enhance through-thickness ductility and thus assist in reducing the incidence of lamellar tearing.

However, unless precautions are taken in both design and fabrication, lamellar tearing may still occur in thick plates and heavy shapes of such steels at restrained through-thickness connections. Some guidelines for minimizing potential problems have been developed by the American Institute of Steel Construction (AISC). (See "The Design, Fabrication, and Erection of Highly Restrained Connections to Minimize Lamellar Tearing," *AISC Engineering Journal*, vol. 10, no. 3, 1973, www.aisc.org.)

Welded Splices in Heavy Sections ▪ Shrinkage during solidification of large welds causes strains in adjacent restrained material that can exceed the yield-point strain. In thick material, triaxial stresses may develop because there is restraint in the thickness direction as well as the planar directions. Such conditions inhibit the ability of the steel to act in a ductile manner and increase the possibility of brittle fracture. Therefore, for building construction, AISC imposes special requirements when splicing either Group 4 or Group 5 rolled shapes, or shapes built up by welding plates more than 2 in thick, if the cross section is subject to primary tensile stresses due to axial tension or flexure. Included are notch toughness requirements, the removal of weld tabs and backing bars (ground smooth), generous-sized weld access holes, preheating for thermal cutting, and grinding and inspecting cut edges. Even when the section is used as a primary compression member, the same precautions must be taken for sizing the weld access holes, preheating, grinding, and inspection. See the AISC Specification for further details.

Cracking ▪ An occasional problem known as "k-area cracking" has been identified. Wide flange sections are typically straightened as part of the mill production process. Often a rotary straightening process is used, although some heavier members may be straightened in a gag press. Some reports in recent years have indicated a potential for crack initiation at or near connections in the "k" area of wide flange sections that have been rotary straightened. The k area is the region extending from approximately the midpoint of the web-to-flange fillet, into the web for a distance approximately 1 to 1-$\frac{1}{2}$ in. beyond the point of tangency. Apparently, in some cases, this limited region had a reduced notch toughness due to cold working and strain hardening. Most of the incidents reported occurred at highly restrained joints with welds in the "k" area. However, the number of examples reported has been limited and these have occurred during construction or laboratory tests, with no evidence of difficulties with steel members in service. Research has confirmed the need to avoid welding in the "k" area. AISC issued the following recommendations concerning fabrication and design practices for rolled wide flange shapes:

• Welds should be stopped short of the "k" area for transverse stiffeners (continuity plates).

- For continuity plates, fillet welds and/or partial joint penetration welds, proportioned to transfer the calculated stresses to the column web, should be considered instead of complete joint penetration welds. Weld volume should be minimized.

- Residual stresses in highly restrained joints may be decreased by increased preheat and proper weld sequencing.

- Magnetic particle or dye penetrant inspection should be considered for weld areas in or near the "k" area of highly restrained connections after the final welding has completely cooled.

- When possible, eliminate the need for column web doubler plates by increasing column size.

Good fabrication and quality control practices, such as inspection for cracks, gouges, etc., at flame-cut access holes or copes, should continue to be followed and any defects repaired and ground smooth. All structural wide flange members for normal service use in building construction should continue to be designed per AISC Specifications and the material furnished per ASTM standards.

(AISC Advisory Statement, *Modern Steel Construction*, February 1997.)

Fasteners ▪ Steels for structural bolts are covered by A307, A325, and A490 Specifications. A307 covers carbon-steel bolts for general applications, such as low-stress connections and secondary members. Specification A325 includes two type of quenched and tempered high-strength bolts for structural steel joints: Type 1—medium-carbon, carbon-boron, or medium-carbon alloy steel, and Type 3—weathering steel with atmospheric corrosion resistance similar to that of A588 steel. A previous Type 2 was withdrawn in 1991.

Specification A490 includes three types of quenched and tempered high-strength steel bolts for structural-steel joints: Type 1—bolts made of alloy steel; Type 2—bolts made from low-carbon martensite steel, and Type 3—bolts having atmospheric-corrosion resistance and weathering characteristics comparable to that of A588, A242, and A709 (W) steels. Type 3 bolts should be specified when atmospheric-corrosion resistance is required. Hot-dip galvanized A490 bolts should not be used.

Bolts having diameters greater than $1\frac{1}{2}$ in are available under Specification A449.

Rivets for structural fabrication were included under Specification A502 but this designation has been discontinued.

9.3 Structural-Steel Shapes

Most structural steel used in building construction is fabricated from rolled shapes. In bridges, greater use is made of plates since girders spanning over about 90 ft are usually built-up sections.

Many different rolled shapes are available: W shapes (wide-flange shapes), M shapes (miscellaneous shapes), S shapes (standard I sections), angles, channels, and bars. The "Manual of Steel Construction," American Institute of Steel Construction, lists properties of these shapes.

Wide-flange shapes range from a W4 × 13 (4 in deep weighing 13 lb/lin ft) to a W36 × 920 (36 in deep weighing 920 lb/lin ft). "Jumbo" column sections range up to W14 × 873.

In general, wide-flange shapes are the most efficient beam section. They have a high proportion of the cross-sectional area in the flanges and thus a high ratio of section modulus to weight. The 14-in W series includes shapes proportioned for use as column sections; the relatively thick web results in a large area-to-depth ratio.

Since the flange and web of a wide-flange beam do not have the same thickness, their yield points may differ slightly. In accordance with design rules for structural steel based on yield point, it is therefore necessary to establish a "design yield point" for each section. In practice, all beams rolled from A36 steel (Art. 9.2) are considered to have a yield point of 36 ksi. Wide-flange shapes, plates, and bars rolled from higher-strength steels are required to have the minimum yield and tensile strength shown in Table 9.1.

Square, rectangular, and round structural tubular members are available with a variety of yield strengths. Suitable for columns because of their symmetry, these members are particularly useful in low buildings and where they are exposed for architectural effect.

Connection Material ▪ Connections are normally made with A36 steel. If, however, higher-strength steels are used, the structural size groupings for angles and bars are:

Group 1:   Thicknesses of $\frac{1}{2}$ in or less

Group 2: Thicknesses exceeding $\frac{1}{2}$ in but not more than $\frac{3}{4}$ in

Group 3: Thicknesses exceeding $\frac{3}{4}$ in

Structural tees fall into the same group as the wide-flange or standard sections from which they are cut. (A WT7 × 13, for example, designates a tee formed by cutting in half a W 14 × 26 and therefore is considered a Group 1 shape, as is a W 14 × 26.)

9.4 Selecting Structural Steels

The following guidelines aid in choosing between the various structural steels. When possible, a more detailed study that includes fabrication and erection cost estimates is advisable.

A basic index for cost analysis is the cost-strength ratio, p/F_y, which is the material cost, cents per pound, divided by the yield point, ksi. For tension members, the relative material cost of two members, C_2/C_1, is directly proportional to the cost-strength ratios; that is,

$$\frac{C_2}{C_1} = \frac{p_2/F_{y2}}{p_1/F_{y1}} \qquad (9.1a)$$

For bending members, the relationship depends on the ratio of the web area to the flange area and the web depth-to-thickness ratios. For fabricated girders of optimum proportions (half the total cross-sectional area is the web area),

$$\frac{C_2}{C_1} = \frac{p_2}{p_1}\left(\frac{F_{y1}}{F_{y2}}\right)^{1/2} \qquad (9.1b)$$

For hot-rolled beams,

$$\frac{C_2}{C_1} = \frac{p_2}{p_1}\left(\frac{F_{y1}}{F_{y2}}\right)^{2/3} \qquad (9.1c)$$

For compression members, the relation depends on the allowable buckling stress F_c, which is a function of the yield point directly; that is,

$$\frac{C_2}{C_1} = \frac{F_{c1}/p_1}{F_{c2}/p_2} \qquad (9.1d)$$

Thus, for short columns, the relationship approaches that for tension members. Table 9.4 gives ratios of F_c that can be used, along with typical material prices p from producing mills, to calculate relative member costs.

Higher strength steels are often used for columns in buildings, particularly for the lower floors when the slenderness ratios is less than 100. When bending is dominant, higher strength steels are economical where sufficient lateral bracing is present. However, if deflection limits control, there is no advantage over A36 steel.

On a piece-for-piece basis, there is substantially no difference in the cost of fabricating and erecting the different grades. Higher-strength steels, however, may afford an opportunity to reduce the number of members, thus reducing both fabrication and erection costs.

9.5 Tolerances for Structural Shapes

ASTM Specification A6 lists mill tolerances for rolled-steel plates, shapes, sheet piles, and bars. Included are tolerances for rolling, cutting, section

Table 9.4 Ratio of Allowable Stress in Columns of High-Strength Steel to That of A36 Steel

Specified Yield Strength F_y, ksi	Slenderness Ratio Kl/r											
	5	15	25	35	45	55	65	75	85	95	105	115
65	1.80	1.78	1.75	1.72	1.67	1.62	1.55	1.46	1.35	1.22	1.10	1.03
60	1.66	1.65	1.63	1.60	1.56	1.52	1.47	1.40	1.32	1.21	1.10	1.03
55	1.52	1.51	1.50	1.48	1.45	1.42	1.38	1.33	1.27	1.20	1.10	1.03
50	1.39	1.38	1.37	1.35	1.34	1.32	1.29	1.26	1.22	1.17	1.10	1.03
45	1.25	1.24	1.24	1.23	1.22	1.21	1.19	1.17	1.15	1.12	1.08	1.03
42	1.17	1.16	1.16	1.15	1.15	1 14	1.13	1.12	1.10	1.08	1.06	1.03

area, and weight, ends out of square, camber, and sweep. The "Manual of Steel Construction" contains tables for applying these tolerances.

The AISC "Code of Standard Practice" gives fabrication and erection tolerances for structural steel for buildings. Figures 9.2 and 9.3 show permissible tolerances for column erection for a multistory building. In these diagrams, a working point for a column is the actual center of the member at each end of a shipping piece. The working line is a straight line between the member's working points.

Both mill and fabrication tolerances should be considered in designing and detailing structural steel. A column section, for instance, may have an actual depth greater or less than the nominal depth. An accumulation of dimensional variations, therefore, would cause serious trouble in erection of a building with many bays. Provision should be made to avoid such a possibility.

Tolerances for fabrication and erection of bridge girders are usually specified by highway departments.

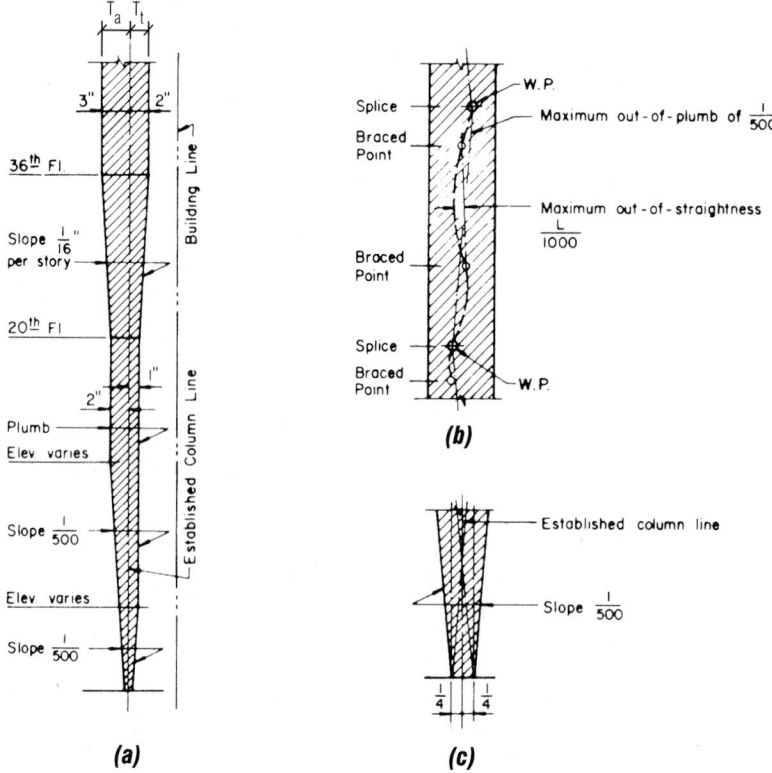

Fig. 9.2 Tolerances permitted for exterior columns for plumbness normal to the building line. (*a*) Envelope within which all working points must fall. (*b*) For individual column sections lying within the envelope shown in (*a*), maximum out-of-plumb of an individual shipping piece, as defined by a straight line between working points, is 1/500 and the maximum out-of-straightness between braced points is *L*/1000, where *L* is the distance between braced points. (*c*) Tolerance for the location of a working point at a column base. The plumb line through that point is not necessarily the precise plan location, inasmuch as the 2000 AISC "Code of Standard Practice" deals only with plumbness tolerance and does not include inaccuracies in location of established column lines, foundations, and anchor bolts beyond the erector's control.

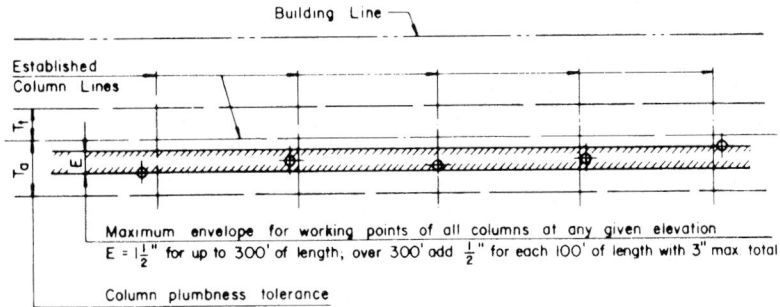

Fig. 9.3 Tolerance in plan permitted for exterior columns at any splice level. Circles indicate column working points. At any splice level, the horizontal envelope defined by E lies within the distances T_a and T_t from the established column line (Fig. 9.2a). Also, the envelope E may be offset from the corresponding envelope at the adjacent splice levels, above and below, by a distance not more than $L/500$, where L is the column length. Maximum E is $1\frac{1}{2}$ in for buildings up to 300 ft long. E may be increased by $\frac{1}{2}$ in for each additional 100 ft of length but not to more than 3 in.

9.6 Structural-Steel Design Specifications

The design of practically all structural steel for buildings in the United States is based on two specifications of the American Institute of Steel Construction. AISC has long maintained a traditional allowable-stress design (ASD) specification, including a comprehensive revised specification issued in 1989, "Specification for Structural Steel for Buildings—Allowable Stress Design and Plastic Design." AISC also publishes an LRFD specification, "Load and Resistance Factor Design Specification for Structural Steel for Buildings." Other important design specifications published by AISC include "Seismic Provisions for Structural Steel Buildings," "Specification for the Design of Steel Hollow Structural Sections," "Specification for the Design, Fabrication and Erection of Steel Safety Related Structures for Nuclear Facilities," and "Specification for Load and Resistance Factor Design of Single-Angle members."

Design rules for bridges are given in "Standard Specifications for Highway Bridges," (American Association of State Highway and Transportation Officials, N. Capitol St, Suite 249 N.W., Washington, DC 20001, www.ashto.org). They are somewhat more conservative than the AISC Specifications. AASHTO gives both an allowable-stress method and a load-factor method. However, the most recent developments in bridge design are reflected in the AASHTO publication. "LRFD Bridge Design Specifications."

Other important specifications for the design of steel structures include the following:

The design of structural members cold-formed from steel not more than 1 in thick follows the rules of AISI "Specification for the Design of Cold-Formed Steel Structural Members" (American Iron and Steel Institute, 1101 17th St., N.W., Washington, DC 20036-4700, www.aisc.org. See Sec. 10).

Codes applicable to welding steel for bridges, buildings, and tubular members are offered by AWS (American Welding Society, 550 N.W. LeJone Road, Miami, FL 33126).

Rules for the design, fabrication, and erection of steel railway bridges are developed by AREMA (American Railway Engineering and Maintenance-of-Way Association, 8201 Corporate Drive, Suite 1125, Landover, Md., 20785-2230). See Sec. 17.

Specifications covering design, manufacture, and use of open-web steel joists are available from SJI (Steel Joist Institute, www.steeljoist.org). See Sec. 10.

9.7 Structural-Steel Design Methods

Structural steel for buildings may be designed by either the allowable-stress design (ASD) or load-and-resistance-factor design (LRFD) method

(Art. 9.6). The ASD Specification of the American Institute of Steel Construction follows the usual method of specifying allowable stresses that represent a "failure" stress (yield stress, buckling stress, etc.) divided by a safety factor. In the AISC-LRFD Specification, both the applied loads and the calculated strength or resistance of members are multiplied by factors. The load factors reflect uncertainties inherent in load determination and the likelihood of various load combinations. The resistance factors reflect variations in determining strength of members such as uncertainty in theory and variations in material properties and dimensions. The factors are based on probabilistic determinations, with the intent of providing a more rational approach and a design with a more uniform reliability. In general, the LRFD method can be expected to yield some savings in material requirements but may require more design time.

Factors to be applied to service loads for various loading combinations are given in Art. 15.5. Rules for "plastic design" are included in both specifications. This method may be applied for steels with yield points of 65 ksi or less used in braced and unbraced planar frames and simple and continuous beams. It is based on the ability of structural steel to deform plastically when strained past the yield point, thereby developing plastic hinges and redistributing loads (Art. 6.65). The hinges are not anticipated to form at service loads but at the higher factored loads.

Steel bridge structures may be designed by ASD, LFD, or LRFD methods in accordance with the specifications of the American Association of State Highway and Transportation Officials (AASHTO). With the load-factor design (LFD) method, only the loads are factored, but with the load-and-resistance-factor (LRFD) method, factors are applied to both loads and resistances. For load factors for highway bridges, see Art. 17.3. Railroad bridges are generally designed by the ASD method.

9.8 Dimensional Limitations on Steel Members

Design specifications, such as the American Institute of Steel Construction "Specification for Structural Steel Buildings—Allowable Stress Design and Plastic Design" and "Load and Resistance Factor Design for Structural Steel Buildings" and the American Association of State Highway and Transportation Officials "Standard Specifications for Highway Bridges" and "LRFD Bridge Design Specifications" set limits, maximum and minimum, on the dimensions and geometry of structural-steel members and their parts. The limitations generally depend on the types and magnitudes of stress imposed on the members and may be different for allowable-stress design (ASD) and load-and-resistance-factor design (LRFD).

These specifications require that the structure as a whole and every element subject to compression be constructed to be stable under all possible combinations of loads. The effects of loads on all parts of the structure when members or their components deform under loads or environmental conditions should be taken into account in design and erection.

(T. V. Galambos, "Guide to Stability Design Criteria for Metal Structures," 5th ed., John Wiley & Sons, Inc., New York.)

Vibration Considerations ▪ In large open areas of buildings, where there are few partitions or other sources of damping, transient vibrations caused by pedestrian traffic may become annoying. Beams and slender members supporting such areas should be designed with due regard for stiffness and damping. Special attention to vibration control should be given in design of bridges because of their exposure to wind, significant temperature changes, and variable, repeated, impact and dynamic loads. Some of the restrictions on member dimensions in standard building and bridge design specifications are intended to limit amplitudes of vibrations to acceptable levels.

Minimum Thickness ▪ Floor plates in buildings may have a nominal thickness as small as $\frac{1}{8}$ in. Generally, minimum thickness available for structural-steel bars 6 in or less wide is 0.203 in and for bars 6 to 8 in wide, 0.230 in. Minimum thickness for plates 8 to 48 in wide is 0.230 in and for plates over 48 in wide, 0.180 in.

The AASHTO Specification requires that, except for webs of certain rolled shapes, closed ribs in orthotropic-plate decks, fillers, and railings, structural-steel elements be at least $\frac{5}{16}$ in thick. Web thickness of rolled beams may be as small as 0.23 in. Thickness of closed ribs in orthotropic-plate decks should be at least $\frac{3}{16}$ in. No minimum is established for fillers. The American Railway

Engineering and Maintenance-of-Way Association "Manual for Railway Engineering" requires that bridge steel, except for fillers, be at least 0.335 in thick. Gusset plates connecting chords and web members of trusses should be at least ½ in thick. In any case, where the steel will be exposed to a substantial corrosive environment, the minimum thicknesses should be increased or the metal should be protected.

Maximum Slenderness Ratios ▪ The AISC Specifications require that the slenderness ratio, the ratio of effective length to radius of gyration of the cross section, should not exceed 200 for members subjected to compression in buildings. For steel highway bridges the AASHTO Specification limits slenderness ratios for compression members to a maximum of 120 for main members and 140 for secondary members and bracing. The AREMA Manual lists the following maximum values for slenderness ratios for compression members in bridges: 100 for main members, 120 for wind and sway bracing, 140 for single lacing, and 200 for double lacing.

For members in tension, the AISC Specifications limit slenderness ratio to a maximum of 300 in buildings. For tension members other than rods, eyebars, cables, and plates, AASHTO specifies for bridges a maximum ratio of unbraced length to radius of gyration of 200 for main tension members, 240 for bracing, and 140 for main-members subject to stress reversal. The AREMA Manual limits the ratio for tension members to 200 for bridges.

Compact Sections ▪ The AISC and AASHTO specifications classify structural-steel sections as compact, noncompact, slender, or hybrid. Slender members have elements that exceed the limits on width-thickness ratios for compact and noncompact sections and are designed with formulas that depend on the difference between actual width-thickness ratios and the maximum ratios permitted for noncompact sections. Hybrid beams or girders have flanges made of steel with yield strength different from that for the webs.

For a specific cross-sectional area, a compact section generally is permitted to carry heavier loads than a noncompact one of similar shape. Under loads stressing the steel into the plastic range, compact sections should be capable of forming plastic hinges with a capacity for inelastic rotation at least three times the elastic rotation corresponding to the plastic moment. To qualify as compact, a section must have flanges continuously connected to the webs, and thickness of its elements subject to compression must be large enough to prevent local buckling while developing a fully plastic stress distribution.

Tables 9.5 and 9.6 present, respectively, maximum width-thickness ratios for structural-steel compression elements in buildings and highway bridges. See also Arts. 9.12 and 9.13.

9.9 Allowable Tension in Steel

For buildings, AISC specifies a basic allowable unit tensile stress, ksi, $F_t = 0.60F_y$, on the gross cross section area, where F_y is the yield strength of the steel, ksi (Table 9.7). F_t is subjected to the further limitation that it should not exceed on the net cross section area, one-half the specified minimum tensile strength F_u of the material. On the net section through pinholes in eyebars, pin-connected plates, or built-up members, $F_t = 0.45F_y$.

For bridges, AASHTO specifies allowable tensile stresses as the smaller of $0.55F_y$ on the gross section, or $0.50F_u$ on the net section ($0.46F_y$ for 100 ksi yield strength steels), where F_u = tensile strength (Table 9.7). In determining gross area, area of holes for bolts and rivets must be deducted if over 15 percent of the gross area. Also, open holes larger than $1\frac{1}{4}$ in, such as perforations, must be deducted.

Table 9.7 and subsequent tables apply to two strength levels, $F_y = 36$ ksi and $F_y = 50$ ksi, the ones generally used for construction.

The net section for a tension member with a chain of holes extending across a part in any diagonal or zigzag line is defined in the AISC Specification as follows: The net width of the part shall be obtained by deducting from the gross width the sum of the diameters of all the holes in the chain and adding, for each gage space in the chain, the quantity $s^2/4g$, where s = longitudinal spacing (pitch), in, of any two consecutive holes and g = transverse spacing (gage), in, of the same two holes. The critical net section of the part is obtained from the chain that gives the least net width.

Table 9.5 Maximum Width-Thickness Ratios b/t [a] for Compression Elements for Buildings[b]

Description of Element	ASD and LRFD[c] Compact—λ_p	ASD[c] Noncompact [d]	LRFD[c] Noncompact—λ_r
Projecting flange element of I-shaped rolled beams and channels in flexure	$65/\sqrt{F_y}$	$95/\sqrt{F_y}$	$141/\sqrt{F_L}$ [g]
Projecting flange element of I-shaped hybrid or welded beams in flexure	$65/\sqrt{F_y}$	$95/\sqrt{F_{yt}/k_c}$ [e]	$162/\sqrt{F_L/k_{cc}}$ [h]
Projecting flange element of I-shaped sections in pure compression, plates projecting from compression elements; outstanding legs of pairs of angles in continuous contact; flanges of channels in pure compression	Not specified	$95/\sqrt{F_y}$	$95/\sqrt{F_y}$
Flanges of square and rectangular box and hollow structural sections of uniform thickness subject to bending or compression; flange cover plates and diaphragm plates between lines of fasteners or welds	$190/\sqrt{F_y}$ uniform comp. $160/\sqrt{F_y}$ plastic anal.	$238/\sqrt{F_y}$	$238/\sqrt{F_y}$
Unsupported width of cover plates perforated with a succession of access holes	Not specified	$317/\sqrt{F_y}$	$317/\sqrt{F_y}$
Legs of single-angle struts; legs of double-angle struts with separators; unstiffened elements; i.e., supported along one edge	Not specified	$76/\sqrt{F_y}$	$76/\sqrt{F_y}$
Stems of tees	Not specified	$127/\sqrt{F_y}$	$127/\sqrt{F_y}$
All other uniformly compressed stiffened elements; i.e., supported along two edges	Not specified	$253/\sqrt{F_y}$	$253/\sqrt{F_y}$
Webs in flexural compression[a]	$640/\sqrt{F_y}$	$760/\sqrt{F_y}$	$970/\sqrt{F_y}$
D/t for circular hollow sections[f] In axial compression for ASD In flexure for ASD	$3,300/F_y$ $3,300/F_y$		

(Table continued)

Table 9.5 *(Continued)*

Description of Element	ASD and LRFD[c] Compact—λ_p	ASD[c] Noncompact[d]	LRFD[c] Noncompact—λ_r
In axial compression for LRFD		Not specified	$3,190/F_y$
In flexure for LRFD	$2,030/F_y$		$8,990/F_y$
In plastic design for LRFD	$1,300/F_y$		

[a]b = width of element or projection (half the nominal width of rolled beams and tees; full width of angle legs and Z and channel flanges). For webs in flexural compression, b should be taken as h, the clear distance between flanges (less fillets for rolled shapes) or distance between adjacent lines of fasteners; t should be taken as t_w, web thickness.

[b]As required in AISC Specifications for ASD and LRFD. These specifications also set specific limitations on plate-girder components.

[c]F_y = specified minimum yield stress of the steel, ksi, but for hybrid beams, use F_{yt}, the yield strength, ksi, of flanges; F_b = allowable bending stress, ksi, in the absence of axial force; F_r = compressive residual stress in flange, ksi (10 ksi for rolled shapes, 16.5 ksi for welded shapes).

[d]Elements with width-thickness ratios that exceed the noncompact limits should be designed as slender sections.

[e]$k_c = 4.05/(h/t)^{0.46}$ for $h/t > 70$; otherwise $k_c = 1$.

[f]D = outside diameter; t = section thickness.

[g]F_L = smaller of $(F_{yf} - F_r)$ or F_{yw}, ksi; F_{yf} = yield strength, ksi, of flanges and F_{yw} = yield strength, ksi, of web.

[h]$k_{cc} = 4/(h/t)^{0.46}$ and $0.35 \le k_{cc} \le 0.763$.

For splice and gusset plates and other connection fittings, the design area for the net section taken through a hole should not exceed 85% of the gross area. When the load is transmitted through some but not all of the cross-sectional elements—for example, only through the flanges of a W shape—an effective net area should be used (75 to 90% of the calculated net area).

LRFD for Tension in Buildings ▪ The limit states for yielding of the gross section and fracture in the net section should be investigated. For yielding, the design tensile strength P_u, ksi, is given by

$$P_u = 0.90F_yA_g \qquad (9.2)$$

where F_y = specified minimum yield stress, ksi

A_g = gross area of tension member, in²

For fracture,

$$P_u = 0.75F_uA_e \qquad (9.3)$$

where F_u = specified minimum tensile strength, ksi

A_e = effective net area, in²

In determining A_e for members without holes, when the tension load is transmitted by fasteners or welds through some but not all of the cross-sectional elements of the member, a reduction factor U is applied to account for shear lag. The factor ranges from 0.75 to 1.00.

9.10 Allowable Shear in Steel

The AASHTO "Standard Specification for Highway Bridges" (Art. 9.6) specifies an allowable shear stress of $0.33F_y$, where F_y is the specified minimum yield stress of the web. Also see Art. 9.10.2. For buildings, the AISC Specification for ASD (Art. 9.6.) relates the allowable shear stress in flexural members to the depth-thickness ratio, h/t_w, where t_w is the web thickness and h is the clear distance between flanges or between adjacent lines of fasteners for built-up sections. In design of girders, other than hybrid girders, larger shears may be allowed when intermediate stiffeners are used. The stiffeners permit tension-field action; that is, a strip of web acting as a tension diagonal resisted by the transverse stiffeners acting as struts, thus enabling the web to carry greater shear.

9.10.1 ASD for Shear in Buildings

The AISC Specification for ASD specifies the following allowable shear stresses F_v, ksi:

$$F_v = 0.40F_y \quad h/t_w \le 380/\sqrt{F_y} \qquad (9.4)$$

$$F_v = C_vF_y/2.89 \le 0.40F_y \quad h/t_w > 380/\sqrt{F_y} \quad (9.5)$$

Table 9.6 Maximum Width-Thickness Ratios b/t^a for Compression Elements for Highway Bridges[b]

	Load-and-Resistance-Factor Design[c]	
Description of Element	Compact	Noncompact [d]
Flange projection of rolled or fabricated I-shaped beams	$65/\sqrt{F_y}$	$235\sqrt{\dfrac{1}{f_c\sqrt{\frac{2D_c}{t_w}}}}^{-h}$
Webs in flexural compression without longitudinal stiffeners	$640/\sqrt{F_y}$	$\dfrac{2D_c}{t_w} = \dfrac{1150}{\sqrt{f_c}}$

	Allowable-Stress Design[e]		
Description of Element (Compression Members)	$f_a < 0.44\,F_y$	$f_a = 0.44\,F_y$	
		$F_y = 36$ ksi	$F_y = 50$ ksi
Plates supported on one side and outstanding legs of angles			
In main members	$51/\sqrt{f_a} \le 12$	12	11
In bracing and other secondary members	$51/\sqrt{f_a} \le 16$	12	11
Plates supported on two edges or webs of box shapes [f]	$126/\sqrt{f_a} \le 45$	32	27
Solid cover plates supported on two edges or solid webs[g]	$158/\sqrt{f_a} \le 50$	40	34
Perforated cover plates supported on two edges for box shapes	$190/\sqrt{f_a} \le 55$	48	41

[a]b = width of element or projection; t = thickness. The point of support is the inner line of fasteners or fillet welds connecting a plate to the main segment or the root of the flange of rolled shapes. In LRFD, for webs of compact sections, $b = d$, the beam depth, and for noncompact sections, $b = D$, the unsupported distance between flange components.

[b]As required in AASHTO "Standard Specification for Highway Bridges." The specifications also provide special limitations on plate-girder elements.

[c]F_y = specified minimum yield stress, ksi, of the steel.

[d]Elements with width-thickness ratios that exceed the noncompact limits should be designed as slender elements.

[e]f_a = computed axial compression stress, ksi.

[f]For box shapes consisting of main plates, rolled sections, or component segments with cover plates.

[g]For webs connecting main members or segments for H or box shapes.

[h]D_c = depth of web in compression, in; f_c = stress in compression flange, ksi, due to factored loads; t_w = web thickness, in.

where $C_v = 45{,}000k_v/F_y(h/t_w)^2$ for $C_v < 0.8$

$\quad\quad = \sqrt{36{,}000k_v/F_y(h/t_w)^2}$ for $C_v > 0.8$

$\quad k_v = 4.00 + 5.34/(a/h)^2$ for $a/h < 1.0$

$\quad\quad = 5.34 + 4.00/(a/h)^2$ for $a/h > 1.0$

$\quad a$ = clear distance between transverse stiffeners

The allowable shear stress with tension-field action is

$$F_v = \frac{F_y}{289}\left[C_v + \frac{1 - C_v}{1.15\sqrt{1 + (a/h)^2}}\right] \le 0.40F_y \quad (9.6)$$

$C_v \le 1$

Table 9.7 Allowable Tensile Stresses in Steel for Buildings and Bridges, ksi

| Yield Strength | Buildings | | Bridges | |
	On Gross Section	On Net Section*	On Gross Section	On Net Section*
36	22.0	29.0	20.0	29.0
50	30.0	32.5	27.5	32.5

* Based on A36 and A572 Grade 50 steels with $F_u = 58$ ksi and 65 ksi, respectively.

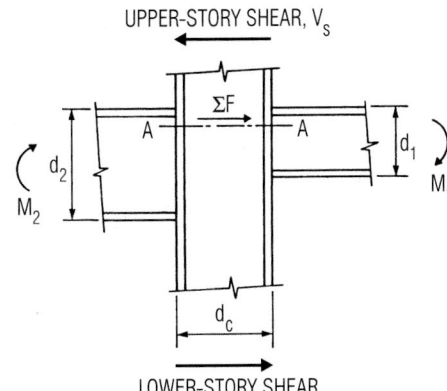

UPPER-STORY SHEAR, V_s

LOWER-STORY SHEAR

Fig. 9.4 Rigid connection of steel members with webs in a common plane.

When the shear in the web exceeds F_v, stiffeners are required. See also Art. 9.13.

The area used to compute shear stress in a rolled beam is defined as the product of the web thickness and the overall beam depth. The webs of all rolled structural shapes are of such thickness that shear is seldom the criterion for design.

At beam-end connections where the top flange is coped, and in similar situations in which failure might occur by shear along a plane through the fasteners or by a combination of shear along a plane through the fasteners and tension along a perpendicular plane, AISC employs the **block shear** concept. The load is assumed to be resisted by a shear stress of $0.30F_u$ along a plane through the net shear area and a tensile stress of $0.50F_u$ on the net tension area, where F_u is the minimum specified tensile strength of the steel.

Within the boundaries of a rigid connection of two or more members with webs lying in a common plane, shear stresses in the webs generally are high. The Commentary on the AISC Specification for buildings states that such webs should be reinforced when the calculated shear stresses, such as those along plane AA in Fig. 9.4, exceed F_v; that is, when ΣF is larger than $d_c t_w F_v$, where d_c is the depth and t_w is the web thickness of the member resisting ΣF. The shear may be calculated from

$$\Sigma F = \frac{M_1}{0.95d_1} + \frac{M_2}{0.95d_2} - V_s \qquad (9.7)$$

where V_s = shear on the section

$M_1 = M_{1L} + M_{1G}$

M_{1L} = moment due to the gravity load on the leeward side of the connection

M_{1G} = moment due to the lateral load on the leeward side of the connection

$M_2 = M_{2L} - M_{2G}$

M_{2L} = moment due to the lateral load on the windward side of the connection

M_{2G} = moment due to the gravity load on the windward side of the connection

9.10.2 ASD for Shear in Bridges

Based on the AASHTO Specification for Highway Bridges, transverse stiffeners are required where h/t_w exceeds 150 and must not exceed a spacing, a, of $3h$, where h is the clear unsupported distance between flange components, t_w is the web thickness, and all dimensions are in inches. Where transverse stiffeners are required, the allowable shear stress, ksi, may be computed from

$$F_v = \frac{F_y}{3}\left[C + \frac{0.87(1 - C)}{\sqrt{1 - (a/h)^2}}\right] \qquad (9.8)$$

where $C = 1.0$ when $\dfrac{h}{t_w} < \dfrac{190\sqrt{k}}{\sqrt{F_y}}$

$C = \dfrac{190\sqrt{k}}{(h/t_w)\sqrt{F_y}}$ when $\dfrac{190\sqrt{k}}{\sqrt{F_y}} \le \dfrac{h}{t_w} \le \dfrac{237\sqrt{k}}{\sqrt{F_y}}$

$C = \dfrac{45,000\sqrt{k}}{(h/t_w)^2\sqrt{F_y}}$ when $\dfrac{h}{t_w} > \dfrac{237\sqrt{k}}{\sqrt{F_y}}$

See also Art. 9.13.

9.10.3 LRFD for Shear in Buildings

Based on the AISC Specifications for LRFD for buildings, the shear capacity V_u, kips, of flexural members with unstiffened webs may be computed from the following:

$$V_u = 0.54 F_{yw} A_w \quad \text{when } h/t_w = 417\sqrt{1/F_{yw}} \quad (9.9)$$

$$V_u = 0.54 F_{yw} A_w \left(\frac{417\sqrt{1/F_{yw}}}{h/t_w} \right)$$

$$\text{when } 417\sqrt{1/F_{yw}} < h/t_w \leq 523\sqrt{1/F_{yw}} \quad (9.10)$$

$$V_u = A_w \left[\frac{131,000}{(h/t_w)^2} \right] \quad (9.11)$$

$$\text{when } 523\sqrt{1/F_{yw}} < h/t_w \leq 260$$

where F_{yw} = specified minimum yield stress of web, ksi

A_w = web area, in^2 = dt_w

Stiffeners are required when the shear exceeds V_u (Art. 9.13). In unstiffened girders, h/t_w may not exceed 260. For shear capacity with tension-field action, see the AISC Specification for LRFD.

9.10.4 LFD Shear Strength Design for Bridges

Based on the AASHTO Specifications for load-factor design, the shear capacity, kips, may be computed from:

$$V_u = 0.58 F_y h t_w C \quad (9.12a)$$

for flexural members with unstiffened webs with $h/t_w < 150$ or for girders with stiffened webs but a/h exceeding 3 or $67,600(h/t_w)^2$.

$$C = 1.0 \quad \text{when } \frac{k}{t_w} < \beta$$

$$= \frac{\beta}{h/t_w} \quad \text{when } \beta \leq \frac{h}{t_w} \leq 1.25\beta$$

$$= \frac{45,000k}{F_y(h/t_w)^2} \quad \text{when } \frac{h}{t_w} > 1.25\beta$$

where $\beta = 190\sqrt{k/F_y}$

$k = 5$ for unstiffened webs

$k = 5 + \lfloor 5/(a/h)^2 \rfloor$ for stiffened webs

For girders with transverse stiffeners and a/h less than 3 and $67,600(h/t_w)^2$, the shear capacity is given by

$$V_u = 0.58 F_y d t_w \left[C + \frac{1-C}{1.15\sqrt{1+(a/h)^2}} \right] \quad (9.12b)$$

Stiffeners are required when the shear exceeds V_u (Art. 9.13).

9.11 Allowable Compression in Steel

The allowable compressive load or unit stress for a column is a function of its slenderness ratio. The slenderness ratio is defined as Kl/r, where K = effective-length factor, which depends on restraints at top and bottom of the column; l = length of column between supports, in; and r = radius of gyration of the column section, in. For combined compression and bending, see Art. 9.17. For maximum permissible slenderness ratios, see Art. 9.8. Columns may be designed by allowable-stress design (ASD) or load-and-resistance-factor design (LRFD).

9.11.1 ASD for Building Columns

The AISC Specification for ASD for buildings (Art. 9.7) provides two formulas for computing allowable compressive stress F_a, ksi, for main members. The formula to use depends on the relationship of the largest effective slenderness ratio Kl/r of the cross section of any unbraced segment to a factor C_c defined by Eq. (9.13a). See Table 9.8a.

$$C_c = \sqrt{\frac{2\pi^2 E}{F_y}} = \frac{756.6}{\sqrt{F_y}} \quad (9.13a)$$

where E = modulus of elasticity of steel

= 29,000 ksi

F_y = yield stress of steel, ksi

Table 9.8a Values of C_c

F_y	C_c
36	126.1
50	107.0

Table 9.8b Allowable Stresses F_a, ksi, in Steel Building Columns for $Kl/r \leq 120$

Kl/r	Yield Strength of Steel F_y, ksi	
	36	50
10	21.16	29.26
20	20.60	28.30
30	19.94	27.15
40	19.19	25.83
50	18.35	24.35
60	17.43	22.72
70	16.43	20.94
80	15.36	19.01
90	14.20	16.94
100	12.98	14.71
110	11.67	12.34*
120	10.28	10.37*

* From Eq. (9.13c) because $Kl/r > C_c$.

Table 9.8c Allowable Stresses, ksi, in Steel Building Columns for $Kl/r > 120$

Kl/r	F_a
130	8.84
140	7.62
150	6.64
160	5.83
170	5.17
180	4.61
190	4.14
200	3.73

When Kl/r is less than C_c,

$$F_a = \frac{[1 - (Kl/r)^2/2C_c^2]F_y}{\text{F.S.}} \qquad (9.13b)$$

where F.S. = safety factor = $5/3 + 3(Kl/r)/8C_c - (Kl/r^3)/8C_c^3$
(See Table 9.8b).

When Kl/r exceeds C_c,

$$F_a = \frac{12\pi^2 E}{23(Kl/r)^2} = \frac{150,000}{(Kl/r)^2} \qquad (9.13c)$$

(See Table 9.8c.)

The effective-length factor K, equal to the ratio of effective-column length to actual unbraced length, may be greater or less than 1.0. Theoretical K values for six idealized conditions, in which joint rotation and translation are either fully realized or nonexistent, are tabulated in Fig. 9.5.

An alternative and more precise method of calculating K for an unbraced column uses a nomograph given in the "Commentary" on the AISC Specification for ASD. This method requires calculation of "end-restraint factors" for the top and bottom of the column, to permit K to be determined from the chart.

9.11.2 ASD for Bridge Columns

In the AASHTO bridge-design Specifications, allowable stresses in concentrically loaded columns are determined from Eq. (9.14a) or (9.14b). When Kl/r is less than C_c,

$$F_a = \frac{F_y}{2.12}\left[1 - \frac{(Kl/r)^2}{2C_c^2}\right] \qquad (9.14a)$$

When Kl/r is equal to or greater than C_c,

$$F_a = \frac{\pi^2 E}{2.12(Kl/r^2)} = \frac{135,000}{(Kl/r)^2} \qquad (9.14b)$$

See Table 9.9.

9.11.3 LRFD for Building Columns

For axially loaded members with $b/t < \lambda_r$ given in Table 9.5, the maximum load P_u, ksi, may be computed from

$$P_u = 0.85A_g F_y \qquad (9.15)$$

where A_g = gross cross-sectional area of the member

$$F_{cr} = (0.658^{\lambda_c^2})F_y \quad \text{for } \lambda \leq 1.5$$

$$F_{cr} = \left[\frac{0.877}{\lambda_c^2}\right]F_y \quad \text{for } \lambda > 1.5$$

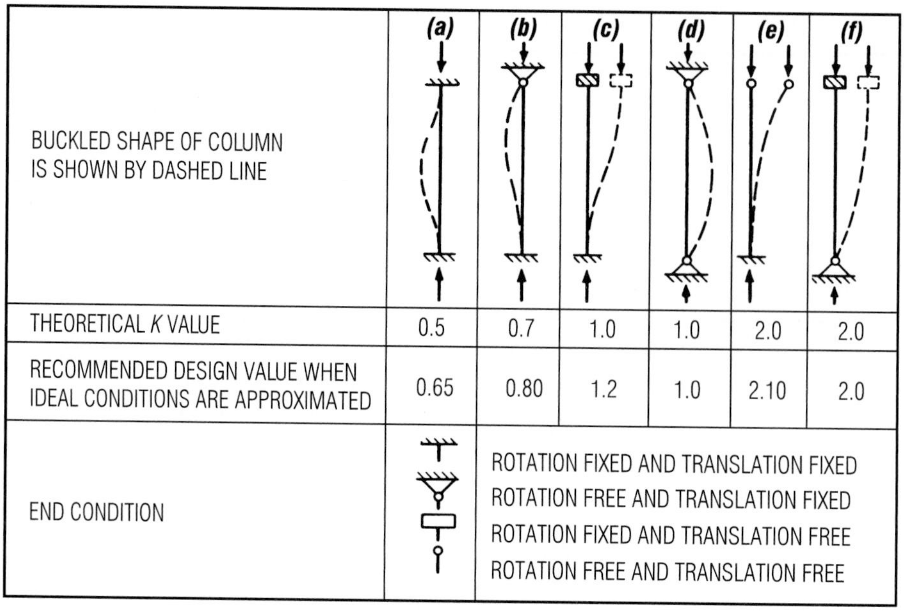

Fig. 9.5 Values of effective-length factor K for columns.

$$\lambda = \frac{Kl}{r\pi}\sqrt{\frac{F_y}{E}}$$

The AISC Specification for LRFD also presents formulas for designing members with slender elements.

9.11.4 LFD for Bridge Columns

Compression members designed by load-factor design should have a maximum strength, kips,

$$P_u = 0.85 A_s F_{cr} \qquad (9.16)$$

Table 9.9 Column Formulas for Bridge Design

Yield Strength, ksi		Allowable Stress, ksi	
	C_c	$Kl/r < C_c$	$Kl/r \geq C_c$
36	126.1	$16.98 - 0.00053\,(Kl/r)^2$	
50	107.0	$23.58 - 0.00103\,(Kl/r)^2$	
90	79.8	$42.45 - 0.00333\,(Kl/r)^2$	$135,000/(Kl/r)^2$
100	75.7	$47.17 - 0.00412\,(Kl/r)^2$	

where A_s = gross effective area of column cross section, in.^{2}.

For $KL_c/r \leq \sqrt{2\pi^2 E/F_y}$,

$$F_{cr} = F_y\left[1 - \frac{F_y}{4\pi^2 E}\left(\frac{KL_c}{r}\right)^2\right] \qquad (9.17a)$$

For $KL_c/r > \sqrt{2\pi^2 E/F_y}$,

$$F_{cr} = \frac{\pi^2 E}{(KL_c/r)^2} = \frac{286,220}{(KL_c/r)^2} \qquad (9.17b)$$

where F_{cr} = buckling stress, ksi

F_y = yield strength of the steel, ksi

K = effective-length factor in plane of buckling

L_c = length of member between supports, in

r = radius of gyration in plane of buckling, in

E = modulus of elasticity of the steel, ksi

Equations (9.17a) and (9.17b) can be simplified by introducing a Q factor:

$$Q = \left(\frac{KL_c}{r}\right)^2 \frac{F_y}{2\pi^2 E} \qquad (9.18)$$

Then, Eqs. (9.17a) and (9.17b) can be rewritten as follows:
For $Q < 1.0$:

$$F_{cr} = \left(1 - \frac{Q}{2}\right)F_y \qquad (9.19a)$$

For $Q > 1.0$:

$$F_{cr} = \frac{F_y}{2Q} \qquad (9.19b)$$

9.12 Allowable Stresses and Loads in Bending

In allowable-stress design (ASD), bending stresses may be computed by elastic theory. The allowable stress in the compression flange usually governs the load-carrying capacity of steel beams and girders.

(T. V. Galambos, "Guide to Design Criteria for Metal Compression Members," 5th ed., John Wiley & Sons, Inc., New York.)

9.12.1 ASD for Building Beams

The maximum fiber stress in bending for laterally supported beams and girders is $F_b = 0.66F_y$ if they are compact (Art. 9.8), except for hybrid girders and members with yield points exceeding 65 ksi. $F_b = 0.60F_y$ for noncompact sections. F_y is the minimum specified yield strength of the steel, ksi. Table 9.10 lists values of F_b for two grades of steel.

Because continuous steel beams have considerable reserve strength beyond the yield point, a redistribution of moments may be assumed when compact sections are continuous over supports

Table 9.10 Allowable Bending Stresses in Braced Beams for Buildings, ksi

Yield Strength, ksi	Compact $(0.66F_y)$	Noncompact $(0.60F_y)$
36	24	22
50	33	30

or rigidly framed to columns. In that case, negative gravity-load moments over the supports may be reduced 10%. If this is done, the maximum positive moment in each span should be increased by 10% of the average negative moments at the span ends.

The allowable extreme-fiber stress of $0.60F_y$ applies to laterally supported, unsymmetrical members, except channels, and to noncompact-box sections. Compression on outer surfaces of channels bent about their major axis should not exceed $0.60F_y$ or the value given by Eq. (9.22).

The allowable stress of $0.66F_y$ for compact members should be reduced to $0.60F_y$ when the compression flange is unsupported for a length, in, exceeding the smaller of

$$l_{max} = \frac{76.0b_f}{\sqrt{F_y}} \qquad (9.20a)$$

$$l_{max} = \frac{20,000}{F_y d/A_f} \qquad (9.20b)$$

where b_f = width of compression flange, in

d = beam depth, in

A_f = area of compression flange, in^2

The allowable stress should be reduced even more when l/r_T exceeds certain limits, where l is the unbraced length, in, of the compression flange and r_T is the radius of gyration, in, of a portion of the beam consisting of the compression flange and one-third of the part of the web in compression.

For $\sqrt{102,000C_b/F_y} \le l/r_T \le \sqrt{510,000C_b/F_y}$, use

$$F_b = \left[\frac{2}{3} - \frac{F_y(l/r_T)^2}{1,530,000C_b}\right]F_y \qquad (9.21a)$$

For $l/r_T > \sqrt{510,000C_b/F_y}$, use

$$F_b = \frac{170,000C_b}{(l/r_T)^2} \qquad (9.21b)$$

where C_b = modifier for moment gradient [Eq. (9.23)].

When, however, the compression flange is solid and nearly rectangular in cross section and its area is not less than that of the tension flange, the allowable stress may be taken as

$$F_b = \frac{12,000C_b}{ld/A_f} \qquad (9.22)$$

When Eq. (9.22) applies (except for channels), F_b should be taken as the larger of the values computed from Eqs. (9.22) and (9.21a) or (9.21b) but not more than $0.60F_y$.

The moment-gradient factor C_b in Eqs. (9.20) to (9.22) may be computed from

$$C_b = 1.75 + 1.05\frac{M_1}{M_2} + 0.3\left(\frac{M_1}{M_2}\right)^2 \leq 2.3 \quad (9.23)$$

where M_1 = smaller beam end moment

M_2 = larger beam end moment

The algebraic sign of M_1/M_2 is positive for double-curvature bending and negative for single-curvature bending. When the bending moment at any point within an unbraced length is larger than that at both ends, the value of C_b should be taken as unity. For braced frames, C_b should be taken as unity for computation of F_{bx} and F_{by} with Eq. (9.65).

Equations (9.21a) and (9.21b) can be simplified by introduction of a new term:

$$Q = \frac{(l/r_T)^2 F_y}{510,000C_b} \quad (9.24)$$

Now, for $0.2 \leq Q \leq 1$,

$$F_b = \frac{(2 - Q)F_y}{3} \quad (9.25)$$

For $Q > 1$,

$$F_b = \frac{F_y}{3Q} \quad (9.26)$$

As for the preceding equations, when Eq. (9.22) applies (except for channels), F_b should be taken as the largest of the values given by Eqs. (9.22) and (9.25) or (9.26), but not more than $0.60F_y$.

9.12.2 ASD for Bridge Beams

AASHTO (Art. 9.6) gives the allowable unit (tensile) stress in bending as $F_b = 0.55F_y$ (Table 9.11). The same stress is permitted for compression when the compression flange is supported laterally for its full length by embedment in concrete or by other means.

When the compression flange is partly supported or unsupported in a bridge, the allowable bending stress, ksi, is

$$F_b = (5 \times 10^7 C_b/S_{xc})(I_{yc}/L)$$
$$\times \sqrt{0.772J/I_{yc} + 9.87(d/L)^2} \leq 0.55F_y \quad (9.27)$$

where L = length, in, of unsupported flange between connections of lateral supports, including knee braces

S_{xc} = section modulus, in^3, with respect to the compression flange

I_{yc} = moment of inertia, in^4, of the compression flange about the vertical axis in the plane of the web

$J = \frac{1}{3}(b_c t_c^3 + b_t t_t^3 + Dt_w^3)$

b_c = width, in, of compression flange

b_t = width, in, of tension flange

t_c = thickness, in, of compression flange

t_t = thickness, in, of tension flange

t_w = thickness, in, of web

D = depth, in, of web

d = depth, in, of flexural member

In general, the moment-gradient factor C_b may be computed from Eq. (9.23). It should be taken as unity, however, for unbraced cantilevers and members in which the moment within a significant portion of the unbraced length is equal to or greater than the larger of the segment end moments. If cover plates are used, the allowable static stress at the point of cutoff should be computed from Eq. (9.27).

The allowable compressive stress for bridge beams may be roughly estimated from the expressions given in Table 9.12, which are based on a formula used prior to 1992.

Table 9.11 Allowable Bending Stress in Braced Bridge Beams, ksi

F_y	F_b
36	20
50	27

Table 9.12 Allowable Compressive Stress in Flanges of Bridge Beams, ksi

F_y	Max l/b	F_b
36	36	$20 - 0.0075\,(l/b)^2$
50	30	$27 - 0.0144\,(l/b)^2$

9.12.3 LRFD for Building Beams

The AISC Specification for LRFD (Art. 9.6) permits use of elastic analysis as described previously for ASD. Thus, negative moments produced by gravity loading may be reduced 10% for compact beams, if the positive moments are increased by 10% of the average negative moments. The reduction is not permitted for hybrid beams, members of A514 steel, or moments produced by loading on cantilevers.

For more accurate plastic design of multistory frames, plastic hinges are assumed to form at points of maximum bending moment. Girders are designed as three-hinged mechanisms. The columns are designed for girder plastic moments distributed to the attached columns plus the moments due to girder shears at the column faces. Additional consideration should be given to moment-end rotation characteristics of the column above and the column below each joint.

For a compact section bent about the major axis, however, the unbraced length L_b of the compression flange where plastic hinges may form at failure may not exceed L_{pd} given by Eqs. (9.28) and (9.29). For beams bent about the minor axis and square and circular beams, L_b is not restricted for plastic analysis.

For I-shaped beams, symmetric about both the major and the minor axis or symmetric about the minor axis but with the compression flange larger than the tension flange, including hybrid girders, loaded in the plane of the web,

$$L_{pd} = \frac{3480 + 2200(M_1/M_2)}{F_{yc}} r_y \qquad (9.28)$$

where F_{yc} = minimum yield stress of compression flange, ksi

M_1 = smaller of the moments, in-kips, at the ends of the unbraced length of beam

M_2 = larger of the moments in-kips, at the ends of the unbraced length of beam

r_y = radius of gyration, in, about minor axis

The plastic moment M_p equals $F_y Z$ for homogenous sections, where Z = plastic modulus, in³ (Art. 6.65), and for hybrid girders, it may be computed from the fully plastic distribution. M_1/M_2 is positive for beams with reverse curvature.

For solid rectangular bars and symmetric box beams,

$$L_{pd} = \frac{4930 + 2900(M_1/M_2)}{F_y} r_y \geq 2900 \frac{r_y}{F_y} \qquad (9.29)$$

The **flexural design strength** is limited to $0.90M_p$ or $0.90M_n$, whichever is less. M_n is determined by the limit state of lateral-torsional buckling and should be calculated for the region of the last hinge to form and for regions not adjacent to a plastic hinge. The Specification gives formulas for M_n that depend on the geometry of the section and the bracing provided for the compression flange.

For compact sections bent about the major axis, for example, M_n depends on the following unbraced lengths:

L_b = the distance, in, between points braced against lateral displacement of the compression flange or between points braced to prevent twist

L_p = limiting laterally unbraced length, in, for full plastic bending capacity

= $300 r_y / \sqrt{F_{yf}}$ for I shapes and channels, $L_b \leq L_r$

= $3750(r_y/M_p)/\sqrt{JA}$ for solid rectangular bars and box beams, $L_p \leq L_r$

F_{yf} = flange yield stress, ksi

J = torsional constant, in⁴ (see AISC "Manual of Steel Construction" on LRFD)

A = cross-sectional area, in²

L_r = limiting laterally unbraced length, in, for inelastic lateral buckling

For doubly symmetric I-shaped beams and channels

$$L_r = \frac{r_y X_1}{F_L} \sqrt{1 + \sqrt{1 + X_2 F_L^2}} \qquad (9.30)$$

where F_L = smaller of $F_{yf} - F_r$ or F_{yw}

F_{yf} = specified minimum yield stress of flange, ksi

F_{yw} = specified minimum yield stress of web, ksi

F_r = compressive residual stress in flange

= 10 ksi for rolled shapes, 16.5 ksi for welded sections

$$X_1 = (\pi/S_x)\sqrt{EGJA/2}$$

$$X_2 = (4C_w/I_y)(S_x/GJ)^2$$

E = elastic modulus of the steel

G = shear modulus of elasticity

S_x = section modulus about major axis, in³
(with respect to the compression flange if that flange is larger than the tension flange)

C_w = warping constant, in⁶ (see AISC Manual—LRFD)

I_y = moment of inertia about minor axis, in⁴

For the aforementioned shapes, the limiting buckling moment M_r, ksi, may be computed from,

$$M_r = F_L S_x \tag{9.31}$$

For doubly symmetric shapes and channels with $L_b \leq L_r$, bent about the major axis

$$M_n = C_b\left[M_p - (M_p - M_r)\frac{L_b - L_p}{L_r - L_p}\right] \leq M_p \tag{9.32}$$

where $C_b = \dfrac{12.5 M_{max}}{2.5 M_{max} + 3M_A + 4M_B + 3M_C}$

M_{max} = absolute value of maximum moment in the unbraced segment, kip-in

M_A = absolute value of moment at quarter point of the unbraced segment, kip-in

M_B = absolute value of moment at centerline of the unbraced segment, kip-in

M_C = absolute value of moment at three-quarter point of the unbraced segment, kip-in

Also, C_b is permitted to be conservatively taken as 1.0 for all cases.

(See T. V. Galambos, "Guide to Stability Design Criteria for Metal Structures," 5th ed., John Wiley & Sons, Inc., New York, for use of larger values of C_b.)

For solid rectangular bars and box section bent about the major axis,

$$L_r = 58,000\left(\frac{r_y}{M_r}\right)\sqrt{JA} \tag{9.33}$$

and the limiting buckling moment is given by

$$M_r = F_y S_x \tag{9.34}$$

For doubly symmetric shapes and channels with $L_b > L_r$, bent about the major axis,

$$M_n = M_{cr} \leq C_b M_r \tag{9.35}$$

where M_{cr} = critical elastic moment, kip-in.

For shapes to which Eq. (9.30) applies,

$$M_{cr} = C_b\frac{\pi}{L_b}\sqrt{EI_yGJ + I_yC_w\left(\frac{\pi E}{L_b}\right)^2} \tag{9.36a}$$

For solid rectangular bars and symmetric box sections,

$$M_{cr} = \frac{57,000 C_b \sqrt{JA}}{L_b/r_y} \tag{9.36b}$$

For determination of the flexural strength of noncompact plate girders and other shapes not covered by the preceding requirements, see the AISC Manual on LRFD.

9.12.4 LFD for Bridge Beams

For load-factor design of symmetrical beams, there are three general types of members to consider: compact, braced noncompact, and unbraced sections. The maximum strength of each (moment, in-kips) depends on member dimensions and unbraced length as well as on applied shear and axial load (Table 9.13).

The maximum strengths given by the formulas in Table 9.13 apply only when the maximum axial stress does not exceed $0.15 F_y A$, where A is the area of the member. Symbols used in Table 9.13 are defined as follows:

D_c = depth of web in compression

F_y = steel yield strength, ksi

Z = plastic section modulus, in³ (See Art. 6.65.)

S = section modulus, in³

b' = width of projection of flange, in

d = depth of section, in

h = unsupported distance between flanges, in

M_1 = smaller moment, in-kips, at ends of unbraced length of member

$M_u = F_y Z$

M_1/M_u is positive for single-curvature bending.

Table 9.13   Design Criteria for Symmetrical Flexural Sections for Load-Factor Design of Bridges

Type of Section	Maximum Bending Strength M_u, in-kips	Flange Minimum Thickness t_f, in**	Web Minimum Thickness t_w, in**	Maximum Unbraced Length L_b, in
Compact*	$F_y Z$	$(b'\sqrt{F_y})/65.0$	$(d\sqrt{F_y})/608$	$([3600 - 2200(M_1/M_u)]r_y)/F_y$
Braced noncompact*	$F_y S$	$(b'\sqrt{F_y})/69.6$	$(D_c\sqrt{F_y})/487$	$(20,000A_f)/(F_y d)$
Unbraced	See AASHTO Specification			

* Straight-line interpolation between compact and braced noncompact moments may be used for intermediate criteria, except that $t_w \le d\sqrt{F_y}/608$ should be maintained.

** For compact sections, when both b'/t_f and d/t_w exceed 75% of the limits for these ratios, the following interaction equation applies:

$$\frac{d}{t_w} + 9.35\frac{b'}{t_f} \le \frac{1064}{\sqrt{F_{yf}}}$$

where F_{yf} is the yield strength of the flange, ksi; t_w is the web thickness, in; and t_f = flange thickness, in.

9.13   Plate Girders

Flexural members built up of plates that form horizontal flanges at top and bottom and joined to vertical or near vertical webs are called plate girders. They differ from beams primarily in that their web depth-to-thickness ratio is larger, for example, exceeds $760/\sqrt{F_b}$ in buildings, where F_b is the allowable bending stress, ksi, in the compression flange.

The webs generally are braced by perpendicular plates called stiffeners, to control local buckling or withstand excessive web shear. Plate girders are most often used to carry heavy loads or for long spans for which rolled shapes are not economical.

9.13.1   Allowable-Stress Design

In computation of stresses in plate girders, the moment of inertia I, in^4, of the gross cross section generally is used. Bending stress f_b due to bending moment M is computed from $f_b = Mc/I$, where c is the distance, in, from the neutral axis to the extreme fiber. For determination of stresses in bolted or riveted girders for bridges, no deduction need be made for rivet or bolt holes unless the reduction in flange area, calculated as indicated in Art. 9.9, exceeds 15%; then the excess should be deducted. For girders for buildings, no deduction need be made provided that

$$0.5F_u A_{fn} \ge 0.6F_y A_{fg} \qquad (9.37a)$$

where F_y is the yield stress, ksi; F_u is the tensile strength, ksi; A_{fg} is the gross flange area, in^2; and A_{fn} is the net flange area, in^2, calculated as indicated in Art. 9.9. If this condition is not met, member flexural properties must be based on an effective tension flange area, A_{fe}, given by

$$A_{fe} = \frac{5F_u A_{fn}}{6F_y} \qquad (9.37b)$$

In welded-plate girders, each flange should consist of a single plate. It may, however, comprise a series of shorter plates of different thickness joined end to end by full-penetration groove welds. Flange thickness may be increased or decreased at a slope of not more than 1 in 2.5 at transition points. In bridges, the ratio of compression-flange width to thickness should not exceed 24 or $103/\sqrt{f_b}$, where f_b = computed maximum bending stress, ksi.

The web depth-to-thickness ratio is defined as h/t, where h is the clear distance between flanges, in, and t is the web thickness, in. Several design rules for plate girders depend on this ratio.

9.13.2   Load-and-Resistance-Factor Design

The AISC and AASHTO specifications (Art. 9.6) provide rules for LRFD for plate girders. These are not given in the following.

9.13.3 Plate Girders in Buildings

For greatest resistance to bending, as much of a plate girder cross section as practicable should be concentrated in the flanges, at the greatest distance from the neutral axis. This might require, however, a web so thin that the girder would fail by web buckling before it reached its bending capacity. To preclude this, the AISC Specification (Art. 9.6) limits h/t. (See also Art. 9.8).

For an unstiffened web, this ratio should not exceed

$$\frac{h}{t} = \frac{14{,}000}{\sqrt{F_y(F_y + 16.5)}} \qquad (9.38)$$

where F_y = yield strength of compression flange, ksi.

Larger values of h/t may be used, however, if the web is stiffened at appropriate intervals.

For this purpose, vertical plates may be welded to it. These transverse stiffeners are not required, though, when h/t is less than the value computed from Eq. (9.38) or given in Table 9.14.

With transverse stiffeners spaced not more than 1.5 times the girder depth apart, the web clear-depth-to-thickness ratio may be as large as

$$\frac{h}{t} = \frac{2000}{\sqrt{F_y}} \qquad (9.39)$$

(See Table 9.14.) If, however, the web depth-to-thickness ratio h/t exceeds $760/\sqrt{F_b}$ where F_b, ksi, is the allowable bending stress in the compression flange that would ordinarily apply, this stress should be reduced to F_b', given by Eqs. (9.40) and (9.41).

$$F_b' = R_{PG}R_eF_b \qquad (9.40)$$

$$R_{PG} = \left[1 - 0.0005\frac{A_w}{A_f}\left(\frac{h}{t} - \frac{760}{\sqrt{F_b}}\right)\right] \le 1.0 \qquad (9.41a)$$

$$R_e = \left[\frac{12 + (A_w/A_f)(3\alpha - \alpha^3)}{12 + 2(A_w/A_f)}\right] \le 1.0 \qquad (9.41b)$$

Table 9.14 Critical h/t for Plate Girders in Buildings

F_y, ksi	$14{,}000/\sqrt{F_y(F_y + 16.5)}$	$2000/\sqrt{F_y}$
36	322	333
50	243	283

where A_w = web area, in^2

A_f = area of compression flange, in^2

$\alpha = 0.6F_{yw}/F_b \le 1.0$

F_{yw} = minimum specified yield stress, ksi, of web steel

In a hybrid girder, where the flange steel has a higher yield strength than the web, Eq. (9.41b) protects against excessive yielding of the lower-strength web in the vicinity of the higher-strength flanges. For nonhybrid girders, $R_e = 1.0$.

Stiffeners on Building Girders ▪ The shear and allowable shear stress may determine required web area and stiffener spacing. Equations (9.5) and (9.6) give the allowable web shear F_v, ksi, for any panel of a building girder between transverse stiffeners.

The average shear stress f_v, ksi, in a panel of a plate girder (web between successive stiffeners) is defined as the largest shear, kips, in the panel divided by the web cross-sectional area, in^2. As f_v approaches F_v given by Eq. (9.6), combined shear and tension become important. In that case, the tensile stress in the web due to bending in its plane should not exceed $0.6F_y$ or $(0.825 - 0.375f_v/F_v)F_y$, where F_v is given by Eq. (9.6).

The spacing between stiffeners at end panels, at panels containing large holes, and at panels adjacent to panels containing large holes, should be such that f_v does not exceed the value given by Eq. (9.5).

Intermediate stiffeners, when required, should be spaced so that a/h is less than 3 and less than $[260/(h/t)]^2$, where a is the clear distance, in, between stiffeners. Such stiffeners are not required when h/t is less than 260 and f_v is less than F_v computed from Eq. (9.5).

An infinite combination of web thicknesses and stiffener spacings is possible with a particular girder. Figure 9.6, developed for A36 steel, facilitates the trial-and-error process of selecting a suitable combination. Similar charts can be developed for other steels.

The required area of intermediate stiffeners is determined by

$$A_{st} = \frac{1 - C_v}{2}\left[\frac{a}{h} - \frac{(a/h)^2}{\sqrt{1 + (a/h)^2}}\right]YDht \qquad (9.42)$$

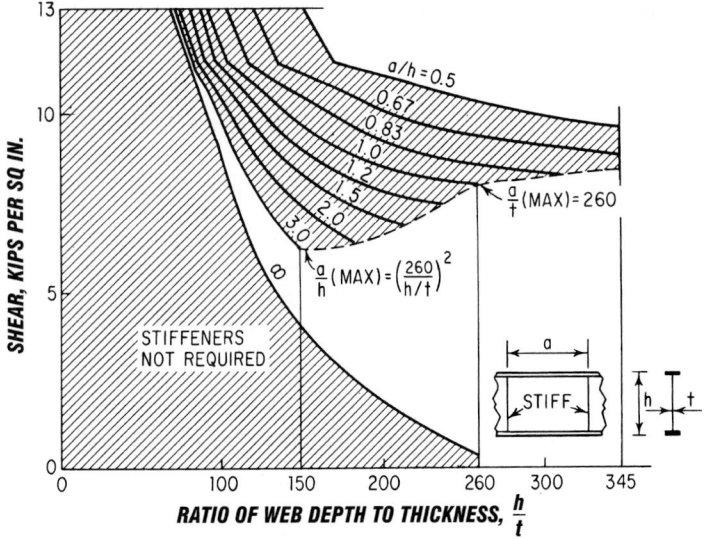

Fig. 9.6 Chart for determining spacing of girder stiffeners of A36 steel.

where A_{st} = gross stiffener area, in^2 (total area, if in pairs)

Y = ratio of yield point of web steel to yield point of stiffener steel

D = 1.0 for stiffeners in pairs

= 1.8 for single-angle stiffeners

= 2.4 for single-plate stiffeners

If the computed web-shear stress f_v is less than F_v computed from Eq. (9.6), A_{st} may be reduced by the ratio f_v/F_v.

The moment of inertia of a stiffener or pair of stiffeners should be at least $(h/50)^4$.

The stiffener-to-web connection should be designed for a shear, kips/lin in of single stiffener, or pair of stiffeners, of at least

$$f_{vs} = h\sqrt{\left(\frac{F_y}{340}\right)^3} \qquad (9.43)$$

This shear may also be reduced by the ratio f_v/F_v.

Spacing of fasteners connecting stiffeners to the girder web should not exceed 12 in c to c. If intermittent fillet welds are used, the clear distance between welds should not exceed 10 in or 16 times the web thickness.

Bearing stiffeners are required on webs where ends of plate girders do not frame into columns or other girders. They may also be needed under

concentrated loads and at reaction points. Bearing stiffeners should be designed as columns, assisted by a strip of web. The width of this strip may be taken as $25t$ at interior stiffeners and $12t$ at the end of the web. Effective length for l/r (slenderness ratio) should be 0.75 of the stiffener length. See Art. 9.18 for prevention of web crippling.

Butt-welded splices should be complete-penetration groove welds and should develop the full strength of the smaller spliced section. Other types of splices in cross sections of plate girders should develop the strength required by the stresses at the point of splice but not less than 50% of the effective strength of the material spliced.

Flange connections may be made with rivets, high-strength bolts, or welds connecting flange to web, or cover plate to flange. They should be proportioned to resist the total horizontal shear from bending. The longitudinal spacing of the fasteners, in, may be determined from

$$P = \frac{R}{q} \qquad (9.44)$$

where R = allowable force, kips, on rivets, bolts, or welds that serve length p

q = horizontal shear, kips/in

For a rivet or bolt, $R = A_v F_v$, where A_v is the cross-sectional area, in^2, of the fastener and F_v

the allowable shear stress, ksi. For a weld, R is the product of the length of weld, in, and allowable unit force, kips/in. Horizontal shear may be computed from

$$q = \frac{VQ}{I} \qquad (9.45a)$$

where V = shear, kips, at point where pitch is to be determined

I = moment of inertia of section, in^4

Q = static moment about neutral axis of flange cross-sectional area between outermost surface and surface at which horizontal shear is being computed, in^3

Approximately,

$$q = \frac{V}{d}\frac{A}{A_f + A_w/6} \qquad (9.45b)$$

where d = depth of web, in, for welds between flange and web; distance between centers of gravity of tension and compression flanges, in, for bolts between flange and web; distance back to back of angles, in, for bolts between cover plates and angles

A = area of flange, in^2, for welds, rivets, and bolts between flange and web; area of cover plates only, in^2, for bolts and rivets between cover plates and angles

A_f = flange area, in^2

A_w = web area, in^2

If the girder supports a uniformly distributed load w, kips/in, on the top flange, the pitch should be determined from

$$p = \frac{R}{\sqrt{q^2 + w^2}} \qquad (9.46)$$

(See also Art. 9.16.)

Maximum longitudinal spacing permitted in the compression-flange cover plates is 12 in or the thickness of the thinnest plate times $127\sqrt{F_y}$ when fasteners are provided on all gage lines at each section or when intermittent welds are provided along the edges of the components. When rivets or bolts are staggered, the maximum spacing on each gage line should not exceed 18 in or the thickness of the thinnest plate times $190\sqrt{F_y}$. Maximum spacing in tension-flange cover plates is 12 in or 24 times the thickness of the thinnest plate. Maximum spacing for connectors between flange angles and web is 24 in.

9.13.4 Girders in Bridges

For highway bridges, Table 9.15 gives critical web thicknesses t, in, for two grades of steel as a fraction of h, the clear distance, in, between flanges. When t is larger than the value in column 1, intermediate transverse (vertical) stiffeners are not required. If shear stress is less than the allowable, the web may be thinner. Thus, stiffeners may be omitted if $t \geq h\sqrt{f_n}/271$, where f_v = average unit shear, ksi (vertical shear at section, kips, divided by web cross-sectional area). But t should not be less than $h/150$.

When t lies between the values in columns 1 and 2, transverse intermediate stiffeners are required. Webs thinner than the values in column 2 are permissible if they are reinforced by a longitudinal (horizontal) stiffener. If the computed maximum compressive bending stress f_b, ksi, at a section is less than the allowable bending stress, a longitudinal stiffener is not required if $t \geq h\sqrt{f_b}/727$; but t should not be less than $h/170$. When used, a plate longitudinal stiffener should be attached to the web at a distance $h/5$ below the inner surface of the compression flange. [See also Eq. (9.49).]

Webs thinner than the values in column 3 are not permitted, even with transverse stiffeners and one longitudinal stiffener, unless the computed

Table 9.15 Minimum Web Thickness, in, for Highway-Bridge Plate Girders*

Yield, Strength, ksi	Without Intermediate Stiffeners (1)	Transverse Stiffeners, No Longitudinal Stiffeners (2)	Longitudinal Stiffener, Transverse Stiffeners (3)
36	$h/78$	$h/165$	$h/327$
50	$h/66$	$h/140$	$h/278$

* "Standard Specifications for Highway Bridges," American Association of State Highway and Transportation Officials.

compressive bending stress is less than the allowable. When it is, t may be reduced in accord with AASHTO formulas, but it should not be less than $h/340$.

Stiffeners on Bridge Girders ▪ The shear and allowable shear stress may determine required web area and stiffener spacing. Equation (9.8) gives the allowable web shear F_v, ksi, for panels between intermediate transverse stiffeners. Maximum spacing a, in, for such panel is $3h$ but not more than $67,600h(h/t_w)^2$. The first intermediate stiffener from a simple support should be located not more than $1.5h$ from the support and the shear in the end panel should not exceed F_v given by Eq. (9.8) nor $F_y/3$.

Intermediate stiffeners may be a single angle fastened to the web or a single plate welded to the web. But preferably they should be attached in pairs, one on each side of the web. Stiffeners on only one side of the web should be attached to the outstanding leg of the compression flange. At points of concentrated loading, stiffeners should be placed on both sides of the web and designed as bearing stiffeners.

The minimum moment of inertia, in^4, of a transverse stiffener should be at least

$$I = a_o t^3 J \qquad (9.47)$$

where $J = 2.5h^2/a_o^2 - 2 \geq 0.5$

h = clear distance between flanges, in

a_o = actual stiffener spacing, in

t = web thickness, in

For paired stiffeners, the moment of inertia should be taken about the centerline of the web; for single stiffeners, about the face in contact with the web.

The gross cross-sectional area of intermediate stiffeners should be at least

$$A = \left[0.15BDt_w(1-C)\frac{V}{V_u} - 18t_w^2\right]Y \qquad (9.48)$$

where Y is the ratio of web-plate yield strength to stiffener-plate yield strength: $B = 1.0$ for stiffener pairs, 1.8 for single angles, and 2.4 for single plates; and C is defined in Eq. (9.8). V_u should be computed from Eq. (9.12a) or (9.12b).

The width of an intermediate transverse stiffener, plate or outstanding leg of an angle, should be at least 2 in plus $\frac{1}{30}$ of the depth of the girder and preferably not less than one-fourth the width of the flange. Minimum thickness is $\frac{1}{16}$ of the width.

Transverse intermediate stiffeners should have a tight fit against the compression flange but need not be in bearing with the tension flange. The distance between the end of the stiffener weld and the near edge of the web-to-flange fillet weld should not be less than $4t$ or more than $6t$. However, if bracing or diaphragms are connected to an intermediate stiffener, care should be taken in design to avoid web flexing, which can cause premature fatigue failures.

Bearing stiffeners are required at all concentrated loads, including supports. Such stiffeners should be attached to the web in pairs, one on each side, and they should extend as nearly as practicable to the outer edges of the flanges. If angles are used, they should be proportioned for bearing on the outstanding legs of the flange angles or plates. (No allowance should be made for the portion of the legs fitted to the fillets of flange angles.) The stiffener angles should not be crimped.

Bearing stiffeners should be designed as columns. The allowable unit stress is given in Table 9.9, with $L = h$. For plate stiffeners, the column section should be assumed to consist of the plates and a strip of web. The width of the strip may be taken as 18 times the web thickness t for a pair of plates. For stiffeners consisting of four or more plates, the strip may be taken as the portion of the web enclosed by the plates plus a width of not more than $18t$. Minimum bearing stiffener thickness is $(b'/12)\sqrt{F_y/33}$, where b' = stiffener width, in.

Bearing stiffeners must be ground to fit against the flange through which they receive their load or attached to the flange with full-penetration groove welds. But welding transversely across the tension flanges should be avoided to prevent creation of a severe fatigue condition.

Termination of Top Flange ▪ Upper corners of through-plate girders, where exposed, should be rounded to a radius consistent with the size of the flange plates and angles and the vertical height of the girder above the roadway. The first flange plate, or a plate of the same width, should be bent around the curve and continued to the bottom of the girder. In a bridge consisting of two or more spans, only the corners at the extreme ends of the bridge need to be rounded, unless the spans have girders of different heights. In such a case, the higher girders should have the top flanges curved down at the ends to meet the top corners of the girders in adjacent spans.

Seating at Supports ▪ Sole plates should be at least $\frac{3}{4}$ in thick. Ends of girders on masonry should be supported on pedestals so that the bottom flanges will be at least 6 in above the bridge seat. Elastomeric bearings often are cost-effective.

Longitudinal Stiffeners ▪ These should be placed with the center of gravity of the fasteners $h/5$ from the toe, or inner face, of the compression flange. Moment of inertia, in^4, should be at least

$$I = ht^3 \left(2.4 \frac{a_o^2}{h^2} - 0.13 \right) \qquad (9.49)$$

where a_o = actual distance between transverse stiffeners, in

t = web thickness, in

Thickness of stiffener, in, should be at least $b\sqrt{F_y}/96$ where b is the stiffener width, in, and F_y is the yield strength of the compression flange, ksi. The bending stress in the stiffener should not exceed the allowable for the material.

Longitudinal stiffeners usually are placed on one side of the web. They need not be continuous. They may be cut at their intersections with transverse stiffeners.

Splices ▪ These should develop the strength required by the stresses at the splices but not less than 75% of the effective strength of the material spliced. Splices in riveted flanges usually are avoided. In general, not more than one part of a girder should be spliced at the same cross section. Bolted web splices should have plates placed symmetrically on opposite sides of the web. Splice plates for shear should extend the full depth of the girder between flanges. At least two rows of bolts on each side of the joint should fasten the plates to the web.

Rivets, high-strength bolts, or welds connecting flange to web, or cover plate to flange, should be proportioned to resist the total horizontal shear from bending, as described for plate girders in buildings. In riveted bridge girders, legs of angles 6 in or more wide connected to webs should have two lines of rivets. Cover plates over 14 in wide should have four lines of rivets.

Hybrid Bridge Girders ▪ These may have flanges with larger yield strength than the web and may be composite or noncomposite with a concrete slab, or they may utilize an orthotropic-plate deck as the top flange. At any cross section where the bending stress in either flange exceeds 55 percent of the minimum specified yield strength of the web steel, the compression-flange area must not be less than the tension-flange area. The top-flange area includes the transformed area of any portion of the slab or reinforcing steel that acts compositely with the girder.

Computation of bending stresses and allowable stresses is generally the same as for girders with uniform yield strength. The bending stress in the web, however, may exceed the allowable bending stress if the computed flange bending stress does not exceed the allowable stress multiplied by a factor R.

$$R = 1 - \frac{\beta\psi(1-\alpha)^2(3-\psi+\psi\alpha)}{6+\beta\psi(3-\psi)} \qquad (9.50)$$

where α = ratio of web yield strength to flange yield strength

ψ = distance from outer edge of tension flange or bottom flange of orthotropic deck to neutral axis divided by depth of steel section

β = ratio of web area to area of tension flange or bottom flange of orthotropic-plate bridge

The rules for shear stresses are as previously described, except that for transversely stiffened girders, the allowable shear stress (throughout the length of the girder) is given by the following instead of Eq. (9.8): $F_v = CF_y/3 \le F_y/3$.

9.14 Deflection Limitations

For buildings, beams and girders supporting plastered ceilings should not deflect under live load more than 1/360 of the span. To control deflection, fully stressed floor beams and girders should have a minimum depth of $F_y/800$ times the span, where F_y is the steel yield strength, ksi. Depth of fully stressed roof purlins should be at least $F_y/1000$ times the span, except for flat roofs, for which ponding conditions should be considered (Art. 9.15).

For bridges, simple-span or continuous girders should be designed so that deflection due to live load plus impact should not exceed $\frac{1}{800}$ of the span.

For bridges located in urban areas and used in part by pedestrians, however, deflection preferably should not exceed $\frac{1}{1000}$ of the span. To control deflections, depth of noncomposite girders should be at least $\frac{1}{25}$ of the span. For composite girders, overall depth, including slab thickness, should be at least $\frac{1}{25}$ of the span, and depth of steel girder alone, at least $\frac{1}{30}$ of the span. For continuous girders, the span for these ratios should be taken as the distance between inflection points.

9.15 Ponding Considerations in Buildings

Flat roofs on which water may accumulate may require analysis to ensure that they are stable under ponding conditions. A flat roof may be considered stable and an analysis need not be made if both Eqs. (9.51) and (9.52) are satisfied.

$$C_p + 0.9C_s \leq 0.25 \qquad (9.51)$$

$$I_d \geq 25S^4/10^6 \qquad (9.52)$$

where $C_p = 32L_sL_p^4/10^7I_p$

$C_s = 32SL_s^4/10^7I_s$

L_p = length, ft, of primary member or girder

L_s = length, ft, of secondary member or purlin

S = spacing, ft, of secondary members

I_p = moment of inertia of primary member, in^4

I_s = moment of inertia of secondary member, in^4

I_d = moment of inertia of steel deck supported on secondary members, in^4/ft

For trusses and other open-web members, I_s should be decreased 15%. The total bending stress due to dead loads, gravity live loads, and ponding should not exceed $0.80F_y$, where F_y is the minimum specified yield stress for the steel.

9.16 Allowable Bearing Stresses and Loads

Load transfer between steel members and their supports may be designed by the allowable-stress or load-and-resistance-factor method (Art. 9.7).

The AISC and AASHTO Specifications provide rules for these methods, but the following covers only ASD.

The Specifications require that provision be made for safe transfer of loads in bearing between steel components and between steel members and supports of different materials. In the latter case, base plates are generally set under columns and bearing plates are placed under beams to transfer loads between the steel members and their supports. When the supports are rigid, such as concrete or masonry, axial loads may be assumed to be uniformly distributed over the bearing areas. It is essential that the load be spread over an area such that the average pressure on the concrete or masonry does not exceed the allowable stress for the material. In the absence of building code or other governing regulations, the allowable bearing stresses in Table 9.16 may be used.

Bearing on Fasteners ▪ See Art. 9.24.

Bearing Plates ▪ To resist a beam reaction, the minimum bearing length N in the direction of the beam span for a bearing plate is determined by equations for prevention of local web yielding and web crippling (Art. 9.18). A larger N is generally desirable but is limited by the available wall thickness.

When the plate covers the full area of a concrete support, the area, in^2, required by the bearing plate is

$$A_1 = \frac{R}{0.35f_c'} \qquad (9.53)$$

where R = beam reaction, kips

f_c' = specified compressive strength of the concrete, ksi

Table 9.16 Allowable Bearing Stress, F_p, on Concrete and Masonry, ksi

Full area of concrete support	$0.35f_c'$
Less than full area of concrete support	$0.35f_c'\sqrt{A_2/A_1} \leq 0.70f_c'$
Sandstone and limestone	0.40
Brick in cement mortar	0.25

When the plate covers less than the full area of the concrete support, then, as determined from Table 9.16,

$$A_1 = \left(\frac{R}{0.35f_c'\sqrt{A_2}}\right)^2 \tag{9.54}$$

where A_2 = full cross-sectional area of concrete support, in^2

With N established, usually rounded to full inches, the minimum width of plate B, in, may be calculated by dividing A_1 by N and then rounded off to full inches so that $BN \geq A_1$. Actual bearing pressure f_p, ksi, under the plate then is

$$f_p = \frac{R}{BN} \tag{9.55}$$

The plate thickness usually is determined with the assumption of cantilever bending of the plate.

$$t = \left(\frac{1}{2}B - k\right)\sqrt{\frac{3f_p}{F_b}} \tag{9.56}$$

where t = minimum plate thickness, in

  k = distance, in, from beam bottom to top of web fillet (Fig. 9.7)

  F_b = allowable bending stress of plate, ksi

Column Base Plates ■ The area A_1, in^2, required for a base plate under a column supported by concrete should be taken as the larger of the

values calculated from Eq. (9.54), with R taken as the total column load, kips, or

$$A_1 = \frac{R}{0.70f_c'} \tag{9.57}$$

Unless the projections of the plate beyond the column are small, the plate may be designed as a cantilever assumed to be fixed at the edges of a rectangle with sides equal to $0.80b$ and $0.95d$, where b is the column flange width, in, and d the column depth, in.

To minimize material requirements, the plate projections should be nearly equal. For this purpose, the plate length N, in (in the direction of d), may be taken as

$$N = \sqrt{A_1} + 0.5(0.95d - 0.80b) \tag{9.58}$$

The width B, in, of the plate then may be calculated by dividing A_1 by N. Both B and N may be selected in full inches so that $BN \geq A_1$. In that case, the bearing pressure f_p, ksi, may be determined from Eq. (9.55). Thickness of plate, determined by cantilever bending, is given by

$$t = 2p\sqrt{\frac{f_p}{F_y}} \tag{9.59}$$

where F_y = minimum specified yield strength, ksi, of plate

  p = larger of $0.5(N - 0.95d)$ and $0.5(B - 0.80b)$

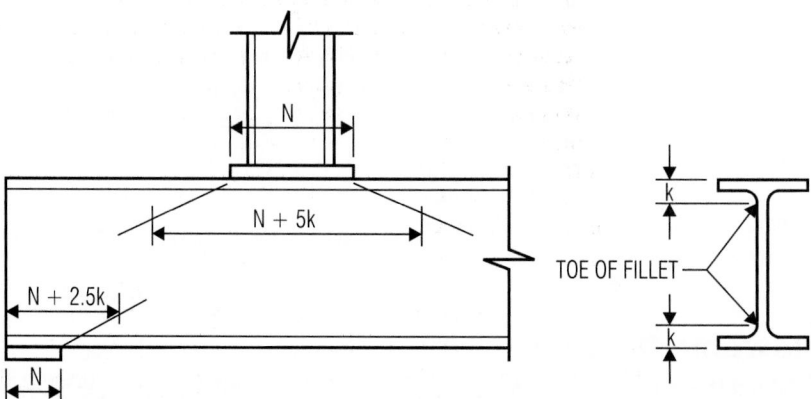

Fig. 9.7 For investigating web yielding, stresses are assumed to be distributed over lengths of web indicated at the bearings, where N is the length of bearing plates and k is the distance from outer surface of beam to the toe of the fillet.

When the plate projections are small, the area A_2 should be taken as the maximum area of the portion of the supporting surface that is geometrically similar to and concentric with the loaded area. Thus, for an H-shaped column, the column load may be assumed distributed to the concrete over an H-shaped area with flange thickness L, in, and web thickness $2L$.

$$L = \frac{1}{4}(d+b) - \frac{1}{4}\sqrt{(d+b)^2 - \frac{4R}{F_p}} \quad (9.60)$$

where F_p = allowable bearing pressure, ksi, on support. (If L is an imaginary number, the loaded portion of the supporting surface may be assumed rectangular as discussed above.) Thickness of the base plate should be taken as the larger of the values calculated from Eq. (9.59) and

$$t = L\sqrt{\frac{3f_p}{F_b}} \quad (9.61)$$

Bearing on Milled Surfaces ▪ In building construction, allowable bearing stress for milled surfaces, including bearing stiffeners, and pins in reamed, drilled, or bored holes, is $F_p = 0.90F_y$, where F_y is the yield strength of the steel, ksi.

For expansion rollers and rockers, the allowable bearing stress, kips/lin in, is

$$F_p = \frac{F_y - 13}{20}0.66d \quad (9.62)$$

where d is the diameter, in, of the roller or rocker. When parts in contact have different yield strengths, F_y is the smaller value.

For highway design, AASHTO limits the allowable bearing stress on milled stiffeners and other steel parts in contact to $F_p = 0.80F_y$. Allowable bearing stresses on pins are in Table 9.17.

The allowable bearing stress for expansion rollers and rockers used in bridges depends on the yield point in tension F_y of the steel in the roller or the base, whichever is smaller. For diameters up to 25 in, the allowable stress, kips/lin in, is

$$p = \frac{F_y - 13}{20}0.6d \quad (9.63)$$

For diameters from 25 to 125 in,

$$p = \frac{F_y - 13}{20}3\sqrt{d} \quad (9.64)$$

where d = diameter of roller or rocker, in.

Table 9.17 Allowable Bearing Stresses on Pins, ksi

		Bridges	
F_y	Buildings $F_p = 0.90F_y$	Pins Subject to Rotation $F_p = 0.40F_y$	Pins Not Subject to Rotation $F_p = 0.80F_y$
36	33	14	29
50	45	20	40

9.17 Combined Axial Compression or Tension and Bending

The AISC Specification for allowable stress design for buildings (Art. 9.6) includes three interaction formulas for combined axial compression and bending:

When the ratio of computed axial stress to allowable axial stress f_a/F_a exceeds 0.15, both Eqs. (9.65) and (9.66) must be satisfied.

$$\frac{f_a}{F_a} + \frac{C_{mx}f_{bx}}{(1 - f_a/F'_{ex})F_{bx}} + \frac{C_{my}f_{by}}{(1 - f_a/F'_{ey})F_{by}} \leq 1 \quad (9.65)$$

$$\frac{f_a}{0.60F_y} + \frac{f_{bx}}{F_{bx}} + \frac{f_{by}}{f_{by}} \leq 1 \quad (9.66)$$

When $f_a/F_a \leq 0.15$, Eq. (9.67) may be used instead of Eqs. (9.65) and (9.66).

$$\frac{f_a}{F_a} + \frac{f_{bx}}{F_{bx}} + \frac{f_{by}}{F_{by}} \leq 1 \quad (9.67)$$

In the preceding equations, subscripts x and y indicate the axis of bending about which the stress occurs, and

F_a = axial stress that would be permitted if axial force alone existed, ksi (see Arts. 9.9 and 9.11)

F_b = compressive bending stress that would be permitted if bending moment alone existed, ksi (see Art. 9.12)

$F'_e = 149,000/(Kl_b/r_b)^2$, ksi; as for F_a, F_b, and $0.6F_y$, F'_e may be increased one-third for wind and seismic loads where permitted by codes

l_b = actual unbraced length in plane of bending, in

r_b = radius of gyration about bending axis, in

K = effective-length factor in plane of bending

f_a = computed axial stress, ksi

f_b = computed compressive bending stress at point under consideration, ksi

C_m = adjustment coefficient

For compression members in frames subject to joint translation (sidesway), $C_m = 0.85$ in Eq. (9.65). For restrained compression members in frames braced against joint translation and not subject to transverse loading between supports in the plane of bending, $C_m = 0.6 - 0.4M_1/M_2$. M_1/M_2 is the ratio of the smaller to larger moment at the ends of that portion of the member unbraced in the plane of bending under consideration. M_1/M_2 is positive when the member is bent in reverse curvature and negative when it is bent in single curvature. For compression members in frames braced against joint translation in the plane of loading and subject to transverse loading between supports, the value of C_m may be determined by rational analysis. But in lieu of such analysis, the following values may be used: For members whose ends are restrained, $C_m = 0.85$. For members whose ends are unrestrained, $C_m = 1.0$.

Building members subject to combined axial tension and bending should satisfy Eq. (9.67), with f_b and F_b, respectively, as the computed and permitted bending tensile stress. But the computed bending compressive stress is limited by Eqs. (9.22) and (9.21a) or (9.21b).

Combined compression and bending stresses in bridge design are covered by equations similar to Eqs. (9.65) and (9.66) but adjusted to reflect the lower allowable stresses of AASHTO.

9.18 Webs under Concentrated Loads

Yielding or crippling of webs of rolled beams and plate girders at points of application of concentrated loads should be investigated.

Criteria for Buildings ▪ The AISC Specification for allowable-stress design for buildings (Art. 9.6) places a limit on compressive stress in webs to prevent local web yielding. For a rolled beam, bearing stiffeners are required at a concentrated

load if the stress f_a, ksi, at the toe of the web fillet exceeds $F_a = 0.66F_{yw}$, where F_{yw} is the minimum specified yield stress of the web steel, ksi. In the calculation of the stressed area, the load may be assumed distributed over the distance indicated in Fig. 9.7.

For a concentrated load applied at a distance larger than the depth of the beam from the end of the beam,

$$f_a = \frac{R}{t_w(N + 5k)} \tag{9.68}$$

where R = concentrated load of reaction, kips

t_w = web thickness, in

N = length of bearing, in (for end reaction, not less than k)

k = distance, in, from outer face of flange to web toe of fillet (Fig. 9.7)

For a concentrated load applied close to the beam end,

$$f_a = \frac{R}{t_w(N + 2.5k)} \tag{9.69}$$

To prevent web crippling, the AISC Specification requires that bearing stiffeners be provided on webs where concentrated loads occur when the compressive force exceeds R, kips, computed from the following:

For a concentrated load applied at a distance from the beam end of at least $d/2$, where d is the depth of beam

$$R = 67.5t_w^2\left[1 + 3\left(\frac{N}{d}\right)\left(\frac{t_w}{t_f}\right)^{1.5}\right]\sqrt{F_{yw}t_f/t_w} \tag{9.70}$$

where t_f = flange thickness, in.

For a concentrated load applied closer than $d/2$ from the beam end,

$$R = 34t_w^2\left[1 + 3\left(\frac{N}{d}\right)\left(\frac{t_w}{t_f}\right)^{1.5}\right]\sqrt{F_{yw}t_f/t_w} \tag{9.71}$$

If stiffeners are provided and extend at least one-half the web, R need not be computed.

Another consideration is prevention of sidesway web buckling. The AISC Specification requires bearing stiffeners when the compressive force from a concentrated load exceeds limits that depend on the relative slenderness of web and flange r_{wf} and whether or not the loaded flange is restrained against rotation.

$$r_{wf} = \frac{d_c/t_w}{l/b_f} \qquad (9.72)$$

where l = largest unbraced length, in, along either top or bottom flange at point of application of load

b_f = flange width, in

d_c = web depth clear of fillets = $d - 2k$

Stiffeners are required if the concentrated load exceeds R, kips, computed from

$$R = \frac{6800t_w^2}{h}(1 + 0.4r_{wf}^3) \qquad (9.73a)$$

where h = clear distance, in, between flanges, and r_{wf} is less than 2.3 when the loaded flange is restrained against rotation. If the loaded flange is not restrained and r_{wf} is less than 1.7,

$$R = 0.4r_{wf}^3 \frac{6800t_w^3}{h} \qquad (9.73b)$$

R need not be computed for larger values of r_{wf}.

Criteria for Bridges ▪ Rolled beams used as flexural members in bridges should be provided with stiffeners at bearings when the unit shear in the web exceeds $0.25F_{yw}$.

For plate-girder bridges, bearing stiffeners should always be installed over the end bearings and over the intermediate bearings of continuous girders. See Art. 9.13.

9.19 Design of Stiffeners under Loads

AISC requires that fasteners or welds for end connections of beams, girders, and trusses be designed for the combined effect of forces resulting from moment and shear induced by the rigidity of the connection. When flanges or moment-connection plates for end connections of beams and girders are welded to the flange of an I- or H-shape column, a pair of column-web stiffeners having a combined cross-sectional area A_{st} not less than that calculated from Eq. (9.74) must be provided whenever the calculated value of A_{st} is positive.

$$A_{st} = \frac{P_{bf} - F_{yc}t_{wc}(t_b + 5k)}{F_{yst}} \qquad (9.74)$$

where F_{yc} = column yield stress, ksi

F_{yst} = stiffener yield stress, ksi

k = distance, in, between outer face of column flange and web toe of its fillet, if column is rolled shape, or equivalent distance if column is welded shape

P_{bf} = computed force, kips, delivered by flange of moment-connection plate multiplied by $\frac{5}{3}$, when computed force is due to live and dead load only, or by $\frac{4}{3}$, when computed force is due to live and dead load in conjunction with wind or earthquake forces

t_{wc} = thickness of column web, in

t_b = thickness of flange or moment-connection plate delivering concentrated force, in

Notwithstanding the above requirements, a stiffener or a pair of stiffeners must be provided opposite the beam compression flange when the column-web depth clear of fillets d_c is greater than

$$d_c = \frac{4100t_{wc}^3\sqrt{F_{yc}}}{P_{bf}} \qquad (9.75)$$

and a pair of stiffeners should be provided opposite the tension flange when the thickness of the column flange t_f is less than

$$t_f = 0.4\sqrt{\frac{P_{bf}}{F_{yc}}} \qquad (9.76)$$

When the preceding conditions for Eqs. (9.74) to (9.76) or for prevention of web yielding, crippling, or sidesway buckling (Art. 9.18) require stiffeners, they should be applied in pairs. Stiffeners required to keep within the limits for Eqs. (9.68) (9.69), (9.74), and (9.76) need not extend more than half the depth of the web.

Each stiffener pair required by Eqs. (9.70), (9.71), and (9.74) should be designed as a column consisting of the stiffener pair and an effective portion of the beam or column web having a width of 25 times the web thickness at interior points and 12 times the web thickness at the column ends. Effective length of the column should be taken as 75% of the clear depth of the web.

Stiffeners required by Eqs. (9.74) to (9.76) should comply with the following additional criteria:

1. The width of each stiffener plus half the thickness of the column web should not be less than one-third the width of the flange or moment-connection plate delivering the concentrated force.

2. The thickness of stiffeners should not be less than $t_b/2$.

3. The weld joining stiffeners to the column web must be sized to carry the force in the stiffener caused by unbalanced moments on opposite sides of the column.

Connections having high shear in the column web should be investigated as described by AISC. Equation (9.7) gives the condition for investigating high shear in the column web within the boundaries of the connection.

Stiffeners opposite the beam compression flange may be fitted to bear on the inside of the column flange. Stiffeners opposite the tension flange should be welded, and the welds must be designed for the forces involved.

See also Art. 9.13 and Art. 9.2 (cracking).

9.20 Design of Beam Sections for Torsion

Torsional stresses may be induced in steel beams either by unsymmetrical loading or by symmetrical loading on unsymmetrical shapes, such as channels or angles. In most applications, they are much smaller than the concurrent axial or bending stresses, but the resultant of the combined stresses should not exceed the allowable stress. In bridge design, torsional effects are important in design of curved girders.

9.21 Wind and Seismic Stresses

In allowable-stress design for buildings, allowable stresses may be increased one-third under wind and seismic forces acting alone or with gravity loads, where permitted by codes. The resulting design, however, should not be less than that required for dead and live loads without the increase in allowable stress. The increased stress is permitted because of the short duration of the load. Its validity has been justified by many years of satisfactory performance.

For allowable stresses, including wind and seismic effects on bridges, see Art. 17.4.

Successful wind or seismic design is dependent on close attention to connection details. It is good practice to provide as much ductility as practical in such connections so that the fasteners are not overstressed.

In load-and-resistance-factor design, load factors are applied to adjust for wind and seismic effects.

9.22 Fatigue Strength of Structural Components

Extensive research programs have been conducted to determine the fatigue strength of structural members and connections. These programs included large-scale beam specimens with various details such as flange-to-web fillet welds, flange cover plates, flange attachments, and web stiffeners. These studies showed that the stress range (algebraic difference between maximum and minimum stress) and the notch severity of the details were the dominant variables. For design purposes, the effects of steel yield point and stress ratio are not considered significant.

Allowable stress ranges based on this research have been adopted by the American Institute of Steel Construction (AISC), the American Association of State Highway and Transportation Officials (AASHTO), and the American Welding Society (AWS) as indicated in Table 9.18. Plain material and various details have been grouped in categories of increasing severity, A through F. The allowable-stress ranges are given for various numbers of cycles, from 20,000 to over 2 million. The over-2 million-cycles life corresponds to the fatigue limit; the detail is considered to have infinite life if the allowable stress range listed for over 2 million cycles is not exceeded. The allowable fatigue-stress range is applicable to any of the structural steels, but the maximum stress cannot exceed the maximum permitted under static loadings.

Under AASHTO allowable-stress design, the allowable stress ranges given in Table 9.18 are for redundant load path structures and more conservative values are given for non-redundant load

Table 9.18 Allowable Range of Stress under Fatigue Loading, ksi*

Stress. Category[†]	Number of Cycles			
	From 20,000–100,000	From 100,000–500,000	From 500,000–2,000,000	Over 2,000,000
A	63	37	24	24
B	49	29	18	16
B′	39	23	15	12
C	35	21	13	10[‡]
D	28	16	10	7
E	22	13	8	5
E′	16	9	6	3
F	15	12	9	8

* Based on the requirements of AISC and AWS. Allowable ranges F_{sr} are for tension or reversal, except as noted. Values given represent 95% confidence limits for 95% survival. See AISC and AWS for description of stress categories. AASHTO requirements are similar, except as noted below, for structures with redundant load paths but are more severe for structures with nonredundant load paths.

[†] Typical details included in each category are as follows (see specifications for complete descriptions): A—base metal of plain material; B—base metal and weld metal at full-penetration groove welds, reinforcement ground off; B′—base metal in full-penetration groove welds in built-up members, backing bars not removed; C—base metal and weld metal at full-penetration groove welds, reinforcement not removed; D—base metal at certain attachment details; E—base metal at end of cover plate; E′—base metal at end of cover plate over 0.8 in thick; F—shear in weld metal of fillet welds.

[‡] Flexural stress range of 12 ksi permitted at toe of stiffener welds on webs or flanges.

path structures. LRFD specifications of AISC and AASHTO use a somewhat different approach to fatigue design. Under that method, the stress range is a design consideration only when it exceeds a fatigue threshold stress given for the various details. In such cases, a design stress range for various stress categories is calculated from an equation that includes a fatigue constant for each detail. For plain material away from any welding, non-coated weathering steel is classified as stress Category B (16 ksi stress threshold) because of its rougher surface, rather than Category A (24 ksi stress threshold) for other base metal. However, the prevailing structural detail category, such as groove welds or attachments, usually controls and these are unaffected by the weathered surface.

Note that the AISC, AASHTO, and AWS Specifications do not require fatigue checks in elements of members where calculated stresses are always in compression because although a crack may initiate in a region of tensile residual stress, the crack will generally not propagate beyond.

In design of a structural member to resist fatigue, each detail should be checked for the stress conditions that exist at that location. When a severe detail cannot be avoided, it is often advantageous to locate it in a region where the stress range is low so that the member can withstand the desired number of cycles.

9.23 Load Transfer and Stresses at Welds

Various coated stick electrodes for shielded metal-arc welding and various wire electrodes and flux or gas combinations for other processes may be selected to produce weld metals that provide a wide range of specified minimum-strength levels. AWS Specifications give the electrode classes and welding processes that can be used to obtain matching weld metal, that is, weld metal that has a minimum tensile strength similar to that of various groups of steel. As indicated in Tables 9.19 and 9.20, however, matching weld metal is not always required, particularly in the case of fillet welds.

The differential cooling that accompanies welding causes residual stresses in the weld and the material joined. Although these stresses have an important effect on the strength of compression members, which is included in the design

Table 9.19 Allowable Stresses for Welds in Building Construction

Stresses in Weld*	Allowable Stress	Required Weld Strength Level[†]
Complete-Penetration Groove Welds		
Tension normal to effective area	Same as base metal	Matching weld metal must be used
Compression normal to effective area	Same as base metal	Weld metal with strength level equal to or less than matching weld metal may be used.
Tension or compression parallel to axis of weld	Same as base metal	Weld metal with strength level equal to or less than matching weld metal may be used.
Shear on effective area	0.30 nominal tensile strength of weld metal, ksi	Weld metal with strength level equal to or less than matching weld metal may be used.
Partial-Penetration Groove Welds[‡]		
Compression normal to effective area	Same as base metal	Weld metal with strength level equal to or less than matching weld metal may be used.
Tension or compression parallel to axis of weld[§]	Same as base metal	Weld metal with strength level equal to or less than matching weld metal may be used.
Shear parallel to axis of weld	0.30 nominal tensile strength of weld metal, ksi	Weld metal with strength level equal to or less than matching weld metal may be used.
Tension normal to effective area	0.30 nominal tensile strength of weld metal, ksi, except tensile stress on base metal shall not exceed 0.60 yield strength of base metal	Weld metal with strength level equal to or less than matching weld metal may be used.
Fillet Welds		
Shear on effective area	0.30 nominal tensile strength of weld metal, ksi	Weld metal with strength level equal to or less than matching weld metal may be used.
Tension or compression parallel to axis of weld[§]	Same as base metal	Weld metal with strength level equal to or less than matching weld metal may be used.
Plug and Slot Welds		
Shear parallel to faying surfaces (on effective area)	0.30 nominal tensile strength of weld metal, ksi	Weld metal with strength level equal to or less than matching weld metal may be used.

* For definition of effective area, see AWS, D1.1.
† For matching weld metal, see AWS, D1.1. Weld metal one strength level stronger than matching weld metal may be used.
‡ See the AISC Specification for allowable-stress design for buildings for limitations on use of partial-penetration groove welds.
§ Fillet welds and partial-penetration groove welds joining the component elements of built-up members, such as flange-to-web connections, may be designed without regard to the tensile or compressive stress in those elements parallel to the axis of the welds.

Table 9.20 Allowable Stresses for Welds in Bridge Construction[†]

Stresses in Weld*	Allowable Stress
Complete-Penetration Groove Welds	
Tension normal to effective area	Same as base metal
Compression normal to the effective area	Same as base metal
Tension or compression parallel to axis of weld	Same as base metal
Shear on effective area	0.27 nominal tensile strength of weld metal, ksi, but not greater than the tensile strength of the connected part.
Partial-Penetration Groove Welds	
Compression normal to effective area	Same as base metal
Tension or compression parallel to axis of weld[‡]	Same as base metal
Shear parallel to axis of weld	0.27 nominal tensile strength of weld metal, ksi, except shear stress on base metal shall not exceed 0.36 yield strength of base metal
Fillet Welds	
Shear on effective area	0.27 nominal tensile strength of weld metal, ksi, except shear stress on base metal shall not exceed 0.36 yield strength of base metal
Tension or compression parallel to axis of weld[‡]	Same as base metal
Plug and Slot Welds	
Shear parallel to faying surfaces (on effective area)	0.27 nominal tensile strength of weld metal, ksi, except shear stress on base metal shall not exceed 0.36 yield strength of base metal

* For definition of effective area, see AWS, D1.5.
[†] Matching weld metal is usually specified. However, designers may use an electrode classification with strength less than the base metal in the case of quenched and tempered steels. For matching weld metal, see AWS, D1.5
[‡] Fillet welds and partial-penetration groove welds joining the component elements of built-up members, such as flange-to-web connections, may be designed without regard to the tensile or compressive stress in those elements parallel to the axis of the welds.

equations, they do not usually have a significant effect on the strength of welded connections.

In groove welds, the loads are transferred directly across the weld by tensile or compressive stresses. For complete-penetration groove welds, the welding grade or electrode class is selected so that the resulting weld is as strong as the steel-joined. Partial-penetration groove welds, in which only part of the metal thickness is welded, are sometimes used when stresses are low and there is no need to develop the complete strength of the material. The stress area of such a weld is the product of the length of the weld and an effective throat thickness. In single J- or U-type joints, the effective throat thickness is equal to the depth of the groove, and in bevel- or V-type joints, it is equal to the depth of the chamfer or the depth of the chamfer minus $\frac{1}{8}$ in, depending on the included angle and the welding process. AWS does not permit partial-penetration groove welds to be used for cyclic tension normal to the weld axis; also, if the weld is made from one side only, it must be restrained from rotation. AISC permits such welds to be used for cyclic loading, but the allowable stress range is only one-third to one-half that of a complete-penetration groove weld. Details of recommended types of joints are given by AWS.

In fillet welds, the load is transferred between the connected plates by shear stresses in the welds. The shear stress in a fillet weld is computed on an area equal to the product of the length of the weld by the effective throat thickness.

The **effective throat thickness** is the shortest distance from the root to the face of the weld, a flat face being assumed, and is 0.707 times the nominal size or leg of a fillet weld with equal legs. AISC specifies that the effective throat for submerged-arc fillet welds be taken equal to the leg size for welds $\frac{3}{8}$ in or less and to the theoretical throat plus 0.11 in for larger welds.

Plug welds and slot welds are occasionally used to transfer shear stresses between plates. The shear area for the weld is the nominal cross-sectional area of the hole or slot. This type of connection should be avoided because of the difficulty in inspecting to ensure a satisfactory weld and the severe stress concentration created.

The basic allowable stresses for welds in buildings and bridges are shown in Tables 9.19 and 9.20. As indicated in the tables, complete-penetration groove welds in building or bridge construction and certain other welds in building construction have the same allowable stress as the steel that is joined. The allowable stresses shown for fillet welds provide a safety factor against ultimate weld shear failure of about 3 for building construction and about 10% higher for bridge construction.

9.24 Stresses for Bolts

In bolted connections, shear is transmitted between connected parts by friction until slip occurs. Then, the load is resisted by shear on the bolts, bearing on the connected parts, and residual friction between the faying surfaces of those parts. Where slippage is undesirable, for example, when a joint would be subjected to frequent reversals of load direction, slip-critical joints, formerly called friction-type, may be specified. To prevent slippage, the parts are squeezed together by pretensioning the bolts during installation to create enough friction to resist service loads without slippage (Art. 9.27). High-strength bolts are required. A325 and A490 bolts are usually tightened to a minimum tension of at least 70% of the tensile strength.

In bearing-type connections, load is transferred between parts by shear on a bolt and bearing on the parts, and bolts may be tightened to a *snug-tight* condition (Art. 9.27). Higher shear stresses are permitted for high-strength bolts in such joints than in slip critical joints. Also, lower-cost, but lower-strength A307 bolts may be used.

Fasteners in Buildings ■ The AISC Specification for allowable stresses for buildings (Art. 9.6) specifies allowable unit tension and shear stresses on the cross-sectional area on the unthreaded body area of bolts and threaded parts, as given in Table 9.21. (Generally, rivets should not be used in direct tension.) When wind or seismic load are combined with gravity loads, the allowable stresses may be increased one-third, where permitted by code.

Most building construction is done with bearing-type connections. Allowable bearing stresses apply to both bearing-type and slip-critical connections. In buildings the allowable bearing stress F_p, ksi, on projected area of fasteners with two or more bolts in the line of force is

$$F_p = 1.2F_u \qquad (9.77)$$

where F_u is the tensile strength of the connected part, ksi. Distance measured in the line of force to the nearest edge of the connected part (end distance) should be at least $1.5d$, where d is the fastener diameter. The center-to-center spacing of fasteners should be at least $3d$.

Bridge Fasteners ■ For bridges, AASHTO (Art. 9.6) specifies the working stresses for bolts listed in Table 9.22. Bearing-type connections with high-strength bolts are limited to members in compression and secondary members. The allowable bearing stress, ksi, in standard holes for high-strength bolts is

$$\frac{0.5L_cF_u}{d} \le F_u \qquad (9.78)$$

where L_c is the clear distance between holes or between hole and edge in the direction of the load, in; d is the bolt diameter, in; and F_u is the tensile strength of the connected material, ksi. For other than high-strength bolts, the bearing stress is limited by allowable bearing on the

Table 9.21 Allowable Stresses for Bolts in Buildings*

| | | Shear Stress, ksi | |
Type of Fastener	Tensile Stress, ksi	Slip-Critical Connections	Bearing-Type Connections
A307 bolts	20	Not applicable	10
Threaded parts of suitable steels, threads not excluded from shear plane†	$0.33F_u$	Not applicable	$0.17F_u$
Threaded parts of suitable steels, threads excluded from shear plane	$0.33F_u$	Not applicable	$0.22F_u$
A325 bolts, threads not excluded from shear plane	44	17	21
A325 bolts, threads excluded from shear plane	44	17	30
A490 bolts, threads not excluded from shear plane	54	21	28
A490 bolts, threads excluded from shear plane	54	21	40

* Stresses are for nominal bolt areas, except as noted, and are based on AISC Specifications. F_u is tensile strength, ksi. F_y is yield point, ksi. Allowable stresses are decreased for slip-critical joints with oversized or slotted holes.
   Allowable stresses in tension are for static loads only, except for A325 and A490 bolts. For fatigue loadings, see AISC Specifications. Allowable shear stresses for friction-type connections are for clean mill scale on faying surfaces. See AISC for allowable stresses for other surface conditions.
   When bearing-type connections used to splice tension members have a fastener pattern whose length, measured parallel to the line of force, exceeds 50 in, allowable shear stresses must be reduced by 20%.
   † In addition, the tensile capacity of the threaded portion of an upset rod, based on the cross-sectional area at its major thread diameter, must be larger than the nominal body area of the rod before upsetting times $0.60F_y$.

fasteners. The allowable bearing stress on A307 bolts is 20 ksi, and on structural-steel rivets 40 ksi.

Combined Stresses in Fasteners ▪ The AISC and AASHTO Specifications for allowable-stress design provide formulas that limit unit stresses in bolts subjected to a combination of tension and shear.

For buildings, the allowable tension stress is based on the calculated shear stress f_v, ksi, with an upper limit on allowable tension based on type and grade of fastener. The shear stress for bearing-type joints, however, should not exceed the allowable shear in Table 9.21. For slip-critical joints, the allowable shear stress for bolts is based on the calculated tensile stress in the bolts, f_t, ksi, and the specified pretension on the bolts, kips.

For highway bridges, the shear and tension stresses for bolts are required to satisfy an inter-action formula involving f_v, f_t, and the allowable shear shear stress given in Table 9.22. For slip-critical joints, the allowable shear stress is based on f_t.

9.25 Composite Construction

In composite construction, steel beams and a concrete slab are connected so that they act together to resist the load on the beam. The slab, in effect, serves as a cover plate. As a result, a lighter steel section may be used.

Construction in Buildings ▪ There are two basic methods of composite construction.

Method 1. The steel beam is entirely encased in the concrete. Composite action in this case depends on steel-concrete bond alone. Since the beam is completely braced laterally, the allowable stress in the flanges is $0.66F_y$, where F_y is the yield strength, ksi, of the steel. Assuming the steel to

Table 9.22 Allowable Stresses for Bolts in Bridges*

| | | Shear Stress, ksi | |
Type of Fastener	Tensile Stress, ksi	Slip-Critical Connections	Bearing-Type Connections
A307 bolts	18.0[†]	Not applicable	11.0
A325 bolts, threads not excluded from shear plane	38.0	15.0	19.0
A325 bolts, threads excluded from shear plane	38.0	15.0	23.8
A490 bolts, threads not excluded from shear plane	47.0	19.0	24.0
A490 bolts, threads excluded from shear plane	47.0	19.0	30.0

* Stresses are for nominal bolt area, except as noted, and are based on AASHTO Specifications. AASHTO specifies reduced shear stress values under certain conditions.

For fatigue loadings, see AASHTO Specifications.

Allowable shear stresses for friction-type connections are for clean mill scale on faying surfaces. See AASHTO for allowable stresses for other surface conditions.

In bearing-type connections whose length between extreme fasteners in each of the spliced parts measured parallel to the line of an axial force exceeds 50 in, allowable shear stresses must be reduced by 20%.

[†] Based on area at the root of thread.

carry the full dead load and the composite section to carry the live load, the maximum unit stress, ksi, in the steel is

$$f_s = \frac{M_D}{S_s} + \frac{M_L}{S_{tr}} \leq 0.66F_y \qquad (9.79)$$

where M_D = dead-load moment, in-kips

M_L = live-load moment, in-kips

S_s = section modulus of steel beam, in^3

S_{tr} = section modulus of transformed composite section, in^3

An alternative, shortcut method is permitted by the AISC Specification (Art. 9.6). It assumes the steel beam will carry both live and dead loads and compensates for this by permitting a higher stress in the steel:

$$f_s = \frac{M_D + M_L}{S_s} \leq 0.76F_y \qquad (9.80)$$

Method 2. The steel beam is connected to the concrete slab by shear connectors. Design is based on ultimate load and is independent of use of temporary shores to support the steel until the concrete hardens. The maximum stress in the bottom flange is

$$f_s = \frac{M_D + M_L}{S_{tr}} \leq 0.66F_y \qquad (9.81)$$

To obtain the transformed composite section, treat the concrete above the neutral axis as an equivalent steel area by dividing the concrete area by n, the ratio of modulus of elasticity of steel to that of the concrete. In determination of the transformed section, only a portion of the concrete slab over the beam may be considered effective in resisting compressive flexural stresses (positive-moment regions). None of the concrete is assumed capable of resisting tensile flexural stresses, although the longitudinal steel reinforcement in the effective width of slab may be included in the computation of the properties of composite beams if shear connectors are provided. The width of slab on either side of the beam centerline that may be considered effective should not exceed any of the following:

1. One-eighth of the beam span between centers of supports

2. Half the distance to the centerline of the adjacent beam

3. The distance from beam centerline to edge of slab (see Fig. 9.8)

When the steel beam is not shored during the casting of the concrete slab, the steel section alone should be considered to be carrying all loads until the concrete attains 75% of its required strength. Stresses in the steel should not exceed $0.90F_y$ for this condition in building construction. Later, the compressive flexural stress in the concrete should not exceed 45% of its specified compressive strength f_c'. The section modulus used for calculating that stress should be that of the transformed composite section.

9.25.1 Shear on Connectors

The total horizontal shear to be resisted by the shear connectors in building construction is taken as the smaller of the values given by Eqs. (9.82) and (9.83).

$$V_h = \frac{0.85f_c'A_c}{2} \qquad (9.82)$$

$$V_h = \frac{A_sF_y}{2} \qquad (9.83)$$

where V_h = total horizontal shear, kips, between maximum positive moment and each end of steel beams (or between point of maximum positive moment and point of contraflexure in continuous beam)

f_c' = specified compressive strength of concrete at 28 days, ksi

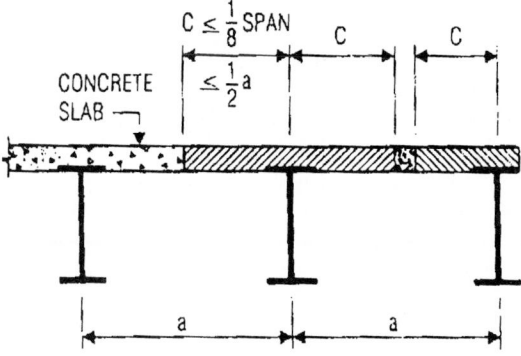

Fig. 9.8 Limitations on effective width of concrete slab in a composite steel-concrete beam.

A_c = actual area of effective concrete flange, in^2

A_s = area of steel beam, in^2

In continuous composite construction, longitudinal reinforcing steel may be considered to act compositely with the steel beam in negative-moment regions. In this case, the total horizontal shear, kips, between an interior support and each adjacent point of contraflexure should be taken as

$$V_h = \frac{A_{sr}F_{yr}}{2} \qquad (9.84)$$

where A_{sr} = area of longitudinal reinforcement at support within effective area, in^2

F_{yr} = specified minimum yield stress of longitudinal reinforcement, ksi

9.25.2 Number of Connectors Required for Building Construction

The total number of connectors to resist V_h is computed from V_h/q, where q is the allowable shear for one connector, kips. Values of q for connectors in buildings are in Table 9.23.

Table 9.23 is applicable only to composite construction with concrete made with stone aggregate conforming to ASTM C33. For lightweight concrete weighing at least 90 lb/ft^3 and made with rotary-kiln-produced aggregates conforming to ASTM C330, the allowable shears in Table 9.23 should be reduced by multiplying by the appropriate coefficient of Table 9.24.

The required number of shear connectors may be spaced uniformly between the sections of maximum and zero moment. Shear connectors should have at least 1 in of concrete cover in all directions, and unless studs are located directly over the web, stud diameters may not exceed 2.5 times the beam-flange thickness.

With heavy concentrated loads, the uniform spacing of shear connectors may not be sufficient between a concentrated load and the nearest point of zero moment. The number of shear connectors in this region should be at least

$$N_2 = \frac{N_1[(M\beta/M_{max}) - 1]}{\beta - 1} \qquad (9.85)$$

Table 9.23 Allowable Shear Loads on Connectors for Composite Construction in Buildings

Type of Connector	Allowable Horizontal Shear Load q, kips (Applicable Only to Concrete Made with ASTM C33 Aggregates)		
	f_c', ksi		
	3.0	3.5	4.0
$\frac{1}{2}$-in dia × 2-in hooked or headed stud	5.1	5.5	5.9
$\frac{5}{8}$-in dia × 2$\frac{1}{2}$-in hooked or headed stud	8.0	8.6	9.2
$\frac{3}{4}$-in dia × 3-in hooked or headed stud	11.5	12.5	13.3
$\frac{7}{8}$-in dia × 3$\frac{1}{2}$-in hooked or headed stud	15.6	16.8	18.0
3-in channel, 4.1 lb	4.3w*	4.7w	5.0w
4-in channel, 5.4 lb	4.6w	5.0w	5.3w
5-in channel, 6.7 lb	4.9w	5.3w	5.6w

* w = length of channel, in.

where M = moment at concentrated load, ft-kips

M_{max} = maximum moment in span, ft-kips

N_1 = number of shear connectors required between M_{max} and zero moment

$\beta = S_{tr}/S_s$ or S_{eff}/S_s, as applicable

S_{eff} = effective section modulus for partial composite action, in^3

9.25.3 Partial Composite Construction

This is used when the number N_1 of shear connectors required would provide a beam considerably stronger than necessary. In that case, the effective section modulus is used in stress computation instead of the transformed section modulus, and S_{eff} is calculated from Eq. (9.86).

$$S_{eff} = S_s + \sqrt{\frac{V_h'}{V_h}}(S_{tr} - S_s) \qquad (9.86)$$

where V_h' = number of shear connectors provided times allowable shear load q of Table 9.23 (times coefficient of Table 9.24, if applicable).

Composite construction of steel beams and concrete slab cast on a cold-formed-steel deck can also be designed with the information provided, but certain modifications are required as described in the AISC Specification for allowable stress. Various dimensional requirements must be met. Also, the allowable shear loads for the stud connectors must be multiplied by a reduction factor. The ribs in the steel deck may be oriented perpendicular to or parallel to the steel beam or girder. The studs are typically welded directly through the deck by following procedures recommended by stud manufacturers.

9.25.4 Composite Construction in Highway Bridges

Shear connectors between a steel girder and a concrete slab in composite construction in a highway

Table 9.24 Shear Coefficient for Lightweight Concrete with Aggregates Conforming to ASTM C330

	Air-Dry Unit Weight, lb/ft^3						
	90	95	100	105	110	115	120
When $f_c' \leq 4$ ksi	0.73	0.76	0.78	0.81	0.83	0.86	0.88
When $f_c' \geq 5$ ksi	0.82	0.85	0.87	0.91	0.93	0.96	0.99

bridge should be capable of resisting both horizontal and vertical movement between the concrete and steel. Maximum spacing for shear connectors generally is 24 in, but wider spacing may be used over interior supports, to avoid highly stressed portions of the tension flange (Fig. 9.9). Clear depth of concrete cover over shear connectors should be at least 2 in and they should extend at least 2 in above the bottom of the slab. In simple spans and positive-moment regions of continuous spans, composite sections generally should be designed to keep the neutral axis below the top of the steel girder. In negative-moment regions, the concrete is assumed to be incapable of resisting tensile stresses, but the longitudinal steel reinforcement may be considered to participate in the composite action if shear connectors are provided.

For composite action, stresses should be computed by the moment-of-inertia method for a transformed composite section, as for buildings, except that the AASHTO Specification (Art. 9.6) requires that the effect of creep be included in the computations. When shores are used and kept in place until the concrete has attained 75% of its specified 28-day strength, the stresses due to dead and live loads should be computed for the composite section.

Creep and Shrinkage ▪ AASHTO requires that the effects of creep be considered in the design of composite beams with dead loads acting on the composite section. For such beams, tension, compression, and horizontal shears produced by dead loads acting on the composite section should be computed for n (Table 9.25) or $3n$, whichever gives the higher stresses.

Shrinkage also should be considered. Resistance of a steel beam to longitudinal contraction of the

Table 9.25 Ratio of Moduli of Elasticity of Steel and Concrete for Bridges

f_c' for Concrete	$n = E_s/E_c$
2.0–2.3	11
2.4–2.8	10
2.9–3.5	9
3.6–4.5	8
4.6–5.9	7
6.0 and over	6

concrete slab produces shear stresses along the contact surface. Associated with this shear are tensile stresses in the slab and compressive stresses in the steel top flange. These stresses also affect the beam deflection. The magnitude of the shrinkage effect varies within wide limits. It can be qualitatively reduced by appropriate casting sequences, for example, by placing concrete in a checkerboard pattern.

The steel beams or girders should be investigated for strength and stability for the loading applied during the time the concrete is in place and before it has hardened. The casting or placing sequence specified for the deck must be considered when calculating moments and shears in the beams or girders.

Span-Depth Ratios ▪ In bridges, for composite beams, preferably the ratio of span to steel beam depth should not exceed 30 and the ratio of span to depth of steel beam plus slab should not exceed 25.

Effective Width of Slabs ▪ For a composite interior girder, the effective width assumed for

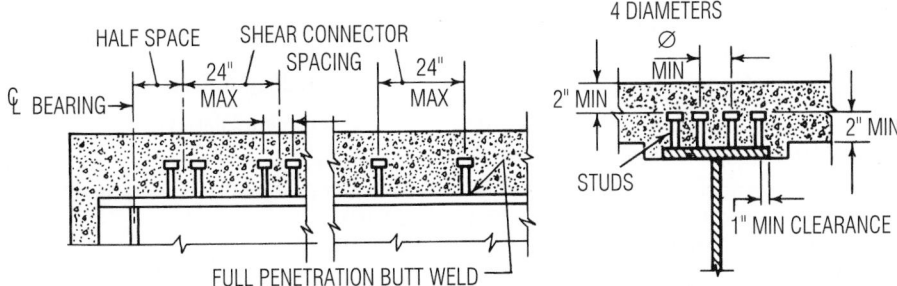

Fig. 9.9 Maximum pitch for stud shear connectors in composite beams.

the concrete flange should not exceed any of the following:

1. One-fourth the beam span between centers of supports

2. Distance between centerlines of adjacent girders

3. Twelve times the least thickness of the slab

For a girder with the slab on only one side, the effective width of slab should not exceed any of the following:

1. One-twelfth the beam span between centers of supports

2. Half the distance to the centerline of the adjacent girder

3. Six times the least thickness of the slab

Bending Stresses ▪ In composite beams in bridges, stresses depend on whether or not the members are shored; they are determined as for beams in buildings [see Eqs. (9.79) and (9.81)], except that the stresses in the steel may not exceed $0.55F_y$ [see Eqs. (9.87) and (9.88)].

Unshored:

$$f_s = \frac{M_D}{S_s} + \frac{M_L}{S_{tr}} \leq 0.55F_y \qquad (9.87)$$

Shored:

$$f_s = \frac{M_D + M_L}{S_{tr}} \leq 0.55F_y \qquad (9.88)$$

Shear Range ▪ Shear connectors in bridges are designed for fatigue and then are checked for ultimate strength. The horizontal-shear range for fatigue is computed from

$$S_r = \frac{V_r Q}{I} \qquad (9.89)$$

where S_r = horizontal-shear range at juncture of slab and beam at point under consideration, kips/lin in

V_r = shear range (difference between minimum and maximum shears at the point) due to live load and impact, kips

Q = static moment of transformed compressive concrete area about neutral axis of transformed section, in^3

I = moment of inertia of transformed section, in^4

The transformed area is the actual concrete area divided by n (Table 9.25).

The allowable range of horizontal shear Z_r, kips, for an individual connector is given by Eq. (9.90) or (9.91), depending on the connector used.

For channels (with a minimum of $\frac{3}{16}$-in fillet welds along heel and toe):

$$Z_r = Bw \qquad (9.90)$$

where w = channel length, in, in transverse direction on girder flange

B = cyclic variable = 4.0 for 100,000 cycles, 3.0 for 500,000 cycles, 2.4 for 2 million cycles, 2.1 for over 2 million cycles

For welded studs (with height-diameter ratio $H/d \geq 4$):

$$Z_r = \alpha d^2 \qquad (9.91)$$

where d = stud diameter, in

α = cyclic variable = 13.0 for 100,000 cycles, 10.6 for 500,000 cycles, 7.85 for 2 million cycles, 5.5 for over 2 million cycles

Required pitch of shear connectors is determined by dividing the allowable range of horizontal shear of all connectors at one section Z_r, kips, by the horizontal range of shear S_r, kips/lin in.

Number of Connectors in Bridges ▪ The ultimate strength of the shear connectors is checked by computation of the number of connectors required from

$$N = \frac{P}{\phi S_u} \qquad (9.92)$$

where N = number of shear connectors between maximum positive moment and end supports

S_u = ultimate shear connector strength, kips [see Eqs. (9.97) and (9.98) and Table 9.26]

ϕ = reduction factor = 0.85

P = force in slab, kips

At points of maximum positive moments, P is the smaller of P_1 and P_2, computed from Eqs. (9.93) and (9.94).

$$P_1 = A_s F_y \qquad (9.93)$$

$$P_2 = 0.85f'_c A_c \qquad (9.94)$$

Table 9.26 Ultimate Horizontal-Shear Load for Connectors in Composite Beams in Bridges*

Type of Connector	Ultimate Shear, kips, for Concrete Compressive Strength f_c', ksi		
	f_c', ksi		
	3.0	3.5	4.0
Welded stud			
¾-in dia × 3 in min high	21.8	24.4	27.0
⅞-in dia × 3.5 in min high	29.6	33.3	36.8
Rolled channel[†]			
3-in deep, 4.1 lb/ft	$10.78w$	$11.65w$	$12.45w$
4-in deep, 5.4 lb/ft	$11.69w$	$12.62w$	$13.50w$
5-in deep, 6.7 lb/ft	$12.50w$	$13.50w$	$14.43w$

* The values are based on the requirements of AASHTO and include no safety factor. Values are for concrete with unit weight of 144 lb/ft³.
[†] w is channel length, in.

where A_c = effective concrete area, in²

f_c' = 28-day compressive strength of concrete, ksi

A_s = total area of steel section, in²

F_y = steel yield strength, ksi

The number of connectors required between points of maximum positive moment and points of adjacent maximum negative moment should equal or exceed N_2, given by

$$N_2 = \frac{P + P_3}{\phi S_u} \qquad (9.95)$$

At points of maximum negative moments, the force in the slab P_3, is computed from

$$P_3 = A_{sr}F_{yr} \qquad (9.96)$$

where A_{sr} = area of longitudinal reinforcing within effective flange, in²

F_{yr} = reinforcing steel yield strength, ksi

Ultimate Shear Strength of Connectors, kips, in Bridges ▪ For channels:

$$S_u = 17.4\left(h + \frac{t}{2}\right)w\sqrt{f_c'} \qquad (9.97)$$

where h = average channel-flange thickness, in

t = channel-web thickness, in

w = channel length, in

For welded studs ($H/d \geq 4$ in):

$$S_u = 0.4d^2\sqrt{f_c'E_c} \qquad (9.98)$$

where E_c is the modulus of elasticity of the concrete, ksi, given by $E_c = 1.044w^{3/2}\sqrt{f_c'}$ and w is the unit weight of concrete, lb/ft³.

Table 9.26 gives the ultimate shear for connectors as computed from Eqs. (9.97) and (9.98) for some commonly used concrete strengths.

9.26 Bracing

It usually is necessary to provide bracing for the main members or secondary members in most buildings and bridges.

9.26.1 In Buildings

There are two general classifications of bracing for building construction: sway bracing for lateral loads and lateral bracing to increase the capacity of individual beams and columns.

Both low- and high-rise buildings require sway bracing to provide stability to the structure and to resist lateral loads from wind or seismic forces. This bracing can take the form of diagonal members or X bracing, knee braces, moment connections, and shear walls.

X bracing is probably the most efficient and economical bracing method. Fenestration or architectural considerations, however, often preclude it. This is especially true for high-rise structures.

Knee braces are often used in low-rise industrial buildings. They can provide local support to the column as well as stability for the overall structure.

Moment connections are frequently used in high-rise buildings. They can be welded, riveted, or bolted, or a combination of welds and bolts can be used. End-plate connections, with shop welding and field bolting, are an economical alternative. Figure 9.10 shows examples of various end-plate moment connections.

In many cases, moment connections may be used in steel frames to provide continuity and thus reduce the overall steel weight. This type of framing is especially suitable for welded construction; full moment connections made with bolts may be cumbersome and expensive.

In low buildings and the top stories of high buildings, moment connections may be designed to resist lateral forces alone. Although the overall steel weight is larger with this type of design, the connections are light and usually inexpensive.

Shear walls are also used to provide lateral bracing in steel-framed buildings. For this purpose, it often is convenient to reinforce the walls needed for other purposes, such as fire walls, elevator shafts, and divisional walls. Sometimes shear walls are used in conjunction with other forms of bracing. Steel plate shear walls have also proven effective.

For plastic-design of multistory frames under factored gravity loads (service loads times 1.7) or under factored gravity plus wind loads (service loads times 1.3), unbraced frames may be used if designed to preclude instability, including the effects of axial deformation of columns. The factored column axial loads should not exceed $0.75AF_y$. Otherwise, frames should incorporate a vertical bracing system to maintain lateral stability. This vertical system may be used in selected braced bents that must carry not only horizontal loads directly applied to them but also the horizontal loads of unbraced bents. The latter loads may be transmitted through diaphragm action of the floor system.

Lateral bracing of columns, arches, beams, and trusses in building construction is used to reduce their critical or effective length, especially of those portions in compression. In floor or roof systems, for instance, it may be economical to provide a strut at midspan of long members to obtain an increase in the allowable stress for the load-carrying members. (See also Arts. 9.11 and 9.12 for effects on allowable stresses of locations of lateral supports.)

Usually, normal floor and roof decks can be relied on to provide sufficient lateral support to compression chords or flanges to warrant use of the full allowable compressive stress. Examples of cases where it might be prudent to provide supplementary support include purlins framed into beams well below the compression flange or precast-concrete planks inadequately secured to the beams.

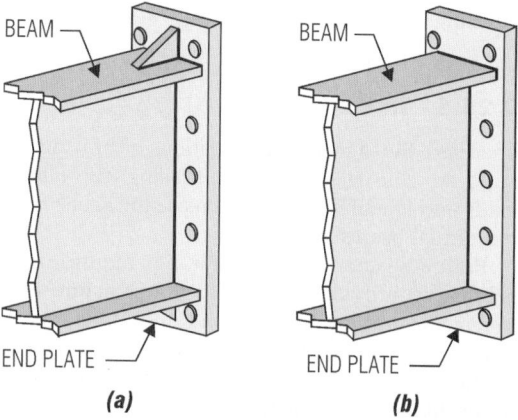

Fig. 9.10 End-plate connections for girders: (a) stiffened moment connection; (b) unstiffened moment connection.

BEAM

END PLATE

(a)

BEAM

END PLATE

(b)

9.26.2 In Bridges

Bracing requirements for highway bridges are given in detail in "Standard Specifications for Highway Bridges," American Association of State Highway and Transportation Officials.

Through trusses require top and bottom lateral bracing. Top lateral bracing should be at least as deep as the top chord. Portal bracing of the two-plane or box type is required at the end posts and should take the full end reaction of the top-chord lateral system. In addition, sway bracing at least 5 ft deep is required at each intermediate panel point.

Deck-truss spans and spandrel arches also require top and bottom lateral bracing. Sway bracing, extending the full depth of the trusses, is required in the plane of the end posts and at all intermediate panel points. The end sway bracing carries the entire upper lateral stress to the supports through the end posts of the truss.

A special case arises with a half-through truss because top lateral bracing is not possible. The main truss and the floor beams should be designed for a lateral force of 300 lb/lin ft, applied at the top-chord panel points. The top chord should be treated as a column with elastic lateral supports at each panel point. The critical buckling force should be at at least 50% greater than the maximum force from dead load, live load, and impact in any panel of the top chord.

Lateral bracing is not usually necessary for deck plate-girder or beam bridges. Most deck construction is adequate as top bracing, and substantial diaphragms (with depth preferably half the girder depth) or cross frames obviate the necessity of bottom lateral bracing. Cross frames are required at each end to resist lateral loads. The need for lateral bracing should be investigated with the use of equations and wind forces specified by AASHTO.

Through-plate girders should be stiffened against lateral deformation by gusset plates or knee braces attached to the floor beams. If the unsupported length of the inclined edge of a gusset plate exceeds $350/\sqrt{F_y}$ times the plate thickness, it should be stiffened with angles.

All highway bridges should be provided with cross frames or diaphragms spaced at a maximum of 25 ft.

("Detailing for Steel Construction," American Institute of Steel Construction, www.aisc.org.)

9.27 Mechanical Fasteners

Unfinished bolts are used mainly in building construction where slip and vibration are not a factor. Characterized by a square head and nut, they also are known as machine, common, ordinary, or rough bolts. They are covered by ASTM A307 and are available in diameters over a wide range (see also Art. 9.2).

A325 bolts are identified by the notation A325. Additionally, Type 1 A325 bolts may optionally be marked with three radial lines 120° apart; Type 2 A325 bolts, withdrawn in 1991, were marked with three radial lines 60° apart; and Type 3 A325 bolts must have the A325 notation underlined. Heavy hexagonal nuts of the grades designated in A325 are manufactured and marked according to specification A563.

A490 bolts are identified by the notation A490. Additionally, Type 2 A490 bolts must be marked with six radial lines 30° apart, and Type 3 A490 bolts must have the A490 notation underlined. Heavy hexagonal nuts of the grades designated in A490 are manufactured and marked according to specification A563.

9.27.1 Types of Bolted Connections

Two different types of bolted connections are recognized for bridges and buildings: bearing and slip critical. Bearing-type connections are allowed higher shear stresses and thus require fewer bolts. Slip-critical connections offer greater resistance to repeated loads and therefore are used when connections are subjected to stress reversal or where slippage would be undesirable. See Art. 9.24.

9.27.2 Symbols for Bolts and Rivets

These are used to denote the type and size of rivets and bolts on design drawings as well as on shop and erection drawings. The practice for buildings and bridges is similar.

Figure 9.11 shows the conventional signs for rivets and bolts.

9.27.3 Bolt Tightening

High-strength bolts for bearing-type connections can generally be installed in the *snug-tight* condition. This is the tightness that exists when all plies in the joint are in firm contact, and may be obtained by a few impacts of an impact wrench or by a full manual effort with a spud wrench. High-strength bolts in slip-critical connections and in connections that are subject to direct tension must be fully pretensioned. Such bolts can be tightened by a calibrated wrench or by the *turn-of-the-nut* method. Calibrated wrenches are powered and have an automatic cutoff set for a predetermined torque. With this method, a hardened washer must be used under the element turned.

The turn-of-the-nut method requires snugging the plies together and then turning the nut a specified amount. From one-third to one turn is specified; increasing amounts of turn are required for long bolts or for bolts connecting parts with slightly sloped surfaces. Alternatively, a direct tension indicator,

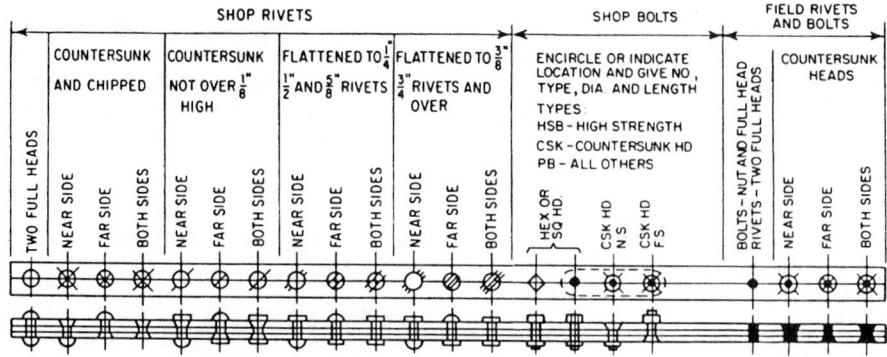

Fig. 9.11 Conventional symbols for bolts and rivets.

such as a load-indicating washer, may be used. This type of washer has on one side raised surfaces which when compressed to a predetermined height (0.005 in measured with a feeler gage) indicate attainment of required bolt tension. Another alternative is to use fasteners that automatically provide the required tension, such as by yielding of or twisting off of an element. The Research Council on Structural Connections "Specification for Structural Steel Joints Using A325 or A490 Bolts," gives detailed specifications for all tightening methods.

9.27.4 Holes

These generally should be $\frac{1}{16}$ in larger than the nominal fastener diameter. Oversize and slotted holes may be used subject to the limitations of Table 9.27.

("Detailing for Steel Construction," American Institute of Steel Construction.)

9.28 Welded Connections

Welding, a method of joining steel by fusion, is used extensively in both buildings and bridges. It usually requires less connection material than other methods. No general rules are possible regarding the economics of the various connection methods; each job must be individually analyzed.

Although there are many different welding processes, shielded-arc welding is used almost exclusively in construction. Shielding serves two purposes: It prevents the molten metal from oxidizing and it acts as a flux to cause impurities to float to the surface.

In manual arc welding, an operator maintains an electric arc between a coated electrode and the work. Its advantage lies in its versatility; a good operator can make almost any type of weld. It is used for fitting up as well as for finished work. The coating turns into a gaseous shield, protecting the weld and concentrating the arc for greater penetrative power.

Automatic welding, generally the submerged-arc process, is used in the shop, where long lengths

Table 9.27 Oversized- and Slotted-Hole Limitations for Structural Joints with A325 and A490 Bolts

Bolt Diameter, in	Maximum Hole Size, in*		
	Oversize Holes†	Short Slotted Holes‡	Long Slotted Holes‡
$\frac{1}{2}$	$\frac{5}{8}$	$\frac{9}{16} \times \frac{11}{16}$	$\frac{9}{16} \times 1\frac{1}{4}$
$\frac{5}{8}$	$\frac{13}{16}$	$\frac{11}{16} \times \frac{7}{8}$	$\frac{11}{16} \times 1\frac{9}{16}$
$\frac{3}{4}$	$\frac{15}{16}$	$\frac{13}{16} \times 1$	$\frac{13}{16} \times 1\frac{7}{8}$
$\frac{7}{8}$	$1\frac{1}{16}$	$\frac{15}{16} \times 1\frac{1}{8}$	$\frac{15}{16} \times 2\frac{3}{16}$
1	$1\frac{1}{4}$	$1\frac{1}{16} \times 1\frac{5}{16}$	$1\frac{1}{16} \times 2\frac{1}{2}$
$1\frac{1}{8}$	$1\frac{7}{16}$	$1\frac{3}{16} \times 1\frac{1}{2}$	$1\frac{3}{16} \times 2\frac{13}{16}$
$1\frac{1}{4}$	$1\frac{9}{16}$	$1\frac{5}{16} \times 1\frac{5}{8}$	$1\frac{5}{16} \times 3\frac{1}{8}$
$1\frac{3}{8}$	$1\frac{11}{16}$	$1\frac{7}{16} \times 1\frac{3}{4}$	$1\frac{7}{16} \times 3\frac{7}{16}$
$1\frac{1}{2}$	$1\frac{13}{16}$	$1\frac{9}{16} \times 1\frac{7}{8}$	$1\frac{9}{16} \times 3\frac{3}{4}$

* In slip-critical connections, a lower allowable shear stress, as given by AISC, should be used for the bolts.
† Not allowed in bearing-type connections.
‡ In bearing-type connections, slot must be perpendicular to direction of load application.

of welds in the flat position are required. In this method, the electrode is a base wire (coiled) and the arc is protected by a mound of granular flux fed to the work area by a separate flux tube. Most welded bridge girders are fabricated by this method, including the welding of transverse stiffeners.

Other processes, such as gas metal or flux-cored arc welding, are also used.

There are basically two types of welds: fillet and groove. Figure 9.12 shows conventional symbols for welds, and Figs. 9.13 to 9.15 illustrate typical fillet, complete-penetration groove, and partial-

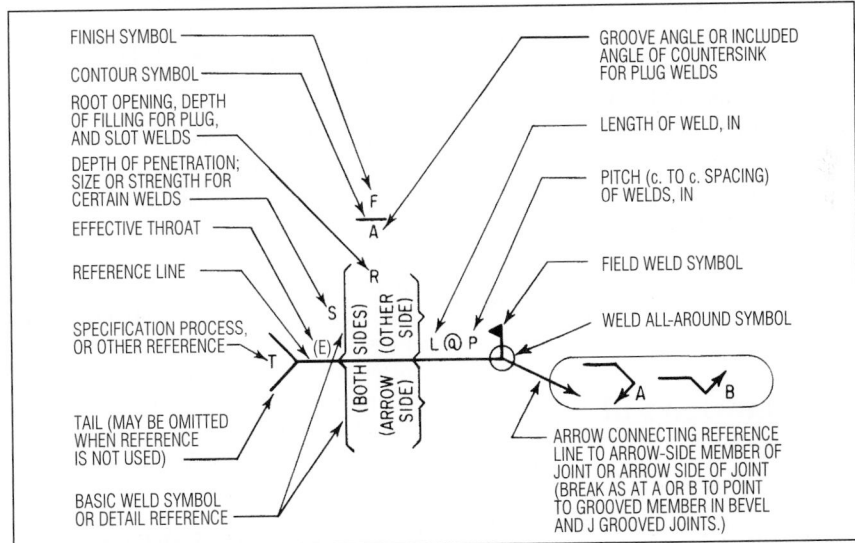

Fig. 9.12 Symbols recommended by the American Welding Society for welded joints. Size, weld symbol, length of weld, and spacing should read in that order from left to right along the reference line, regardless of its orientation or arrow location. The perpendicular leg of symbols for fillet, bevel, J, and flare-bevel-groove should be on the left. Arrow and Other Side welds should be the same size. Symbols apply between abrupt changes in direction of welding unless governed by the *all-around* symbol or otherwise dimensioned. When billing of detail material discloses the existence of a member on the far side (such as a stiffened web or a truss gusset), welding shown for the near side should also be duplicated on the far side.

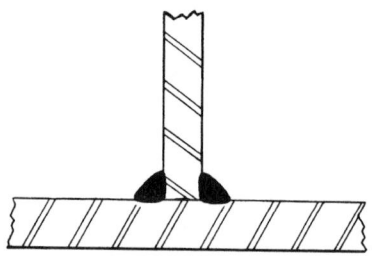

Fig. 9.13 Typical fillet welds.

Fig. 9.14 Typical complete-penetration groove weld.

Fig. 9.15 Typical partial-penetration groove weld.

penetration groove welds. AISC (Art. 9.6) permits partial-penetration groove welds with a reduction in allowable stress. AASHTO (Art. 9.6) does not allow partial-penetration groove welds for bridges where tension may be applied normal to the axis of the weld. Allowable stresses for welds in buildings and bridges are presented in Art. 9.19.

("Detailing for Steel Construction," American Institute of Steel Construction.)

9.29 Combinations of Fasteners

In new construction, different types of fasteners (bolts, or welds) are generally not combined to share the same load because varying amounts of deformation are required to load the different fasteners properly. AISC (Art. 9.6) permits one exception to this rule: Slip-critical bolted connections may be used with welds if the bolts are tightened prior to welding. When welding is used in alteration of existing building framing, existing rivets and existing high-strength bolts in slip-

critical connections may be assumed to resist stresses from loads present at the time of alteration, and the welding may be designed to carry only the additional stresses.

9.30 Column Splices

Connections between lengths of a compression member are often designed more as an erection device than as stress-carrying elements.

Building columns usually are spliced at every second or third story, about 2 ft above the floor. AISC (Art. 9.6) requires that the connectors and splice material be designed for 50% of the stress in the columns. In addition, they must be proportioned to resist tension that would be developed by lateral forces acting in conjunction with 75% of the calculated dead-load stress and without live load.

The AISC "Manual of Steel Construction" (ASD and LRFD) illustrates typical column splices for riveted, bolted, and welded buildings. Where joints depend on contact bearing as part of the splice capacity, the bearing surfaces may be prepared by milling, sawing or suitable means.

Bridge Splices ▪ AASHTO (Art. 9.6) requires splices (tension, compression, bending, or shear) to be designed for the average of the stress at the point of splice and the strength of the member but not less than 75% of the strength of the member. Splices in truss chords should be located as close as possible to panel points.

In bridges, if the ends of columns to be spliced are milled, the splice bolts can be designed for 50 percent of the lower allowable stress of the section spliced. In buildings, AISC permits other means of surfacing the end, such as sawing, if the end is accurately finished to a true plane.

("Detailing for Steel Construction," American Institute of Steel Construction.)

9.31 Beam Splices

Connections between lengths of a beam or girder are designed as either shear or moment connections (Fig. 9.16) depending on their location and function in the structure. In cantilever or hung-span construction in buildings, where beams are extended over the tops of columns and spliced, or connected by another beam, it is sometimes

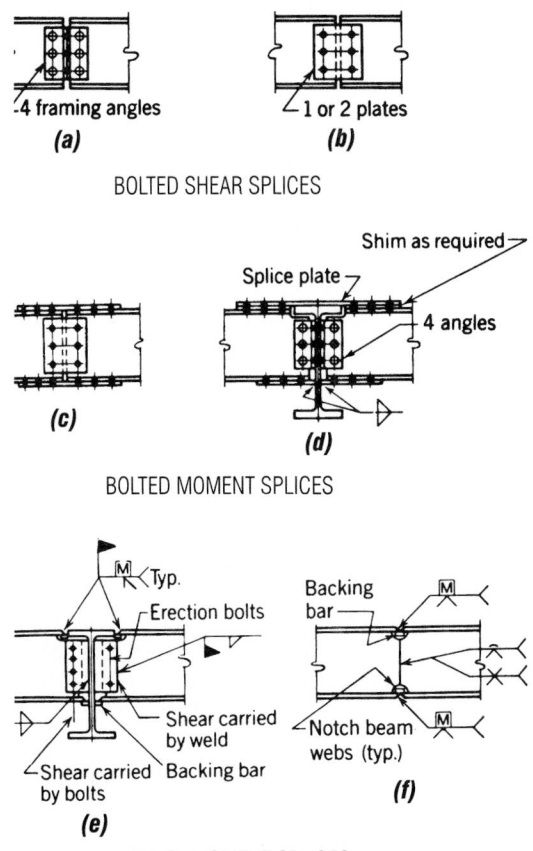

Fig. 9.16 Examples of beam splices used in building construction.

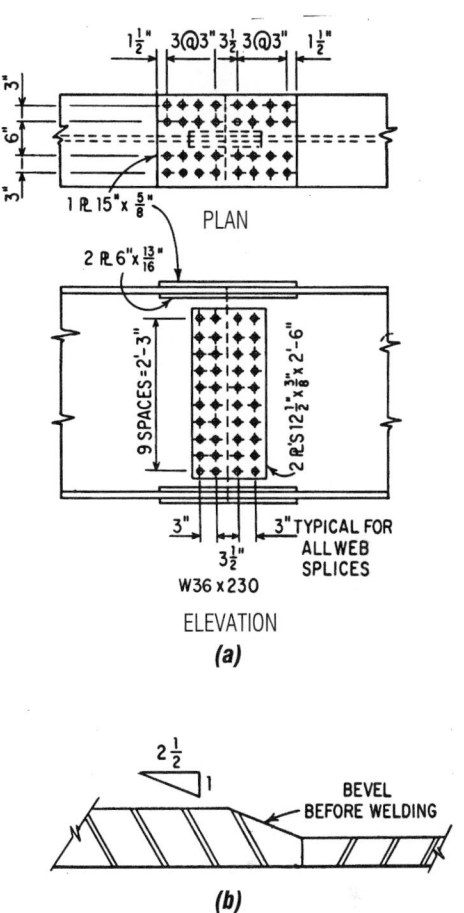

Fig. 9.17 Bridge-beam splices: (*a*) Bolted moment splice. (*b*) Welded flange splice.

possible to use only a shear splice (Fig. 9.16*a* and *b*) if no advantage is taken of continuity and alternate span loading is not likely. Otherwise, at least a partial moment splice is required, depending on the loading and span conditions. Splices may be welded or bolted. The AISC "Manual of Steel Construction" (ASD and LRFD) illustrates typical beam splices.

For continuous bridges, beam splices are designed for the full moment capacity of the beam or girder and are usually bolted (Fig. 9.17*a*). Field-welded splices, although not so common as field-bolted splices, may be an economical alternative.

Special flange splices are always required on welded girders where the flange thickness changes. Care must be taken to ensure that the stress

flow is uniform. Figure 9.17*b* shows a typical detail.

("Detailing for Steel Construction," American Institute of Steel Construction.)

9.32 Erecting Structural Steel

Structural steel is erected by either hand-hoisting or power-hoisting devices.

The simplest hand device is the **gin pole** (Fig. 9.18). The pole is usually a sound, straight-grained timber, although metal poles can also be used. The guys, made of steel strands, generally are set at an angle of 45° or less with the pole. The

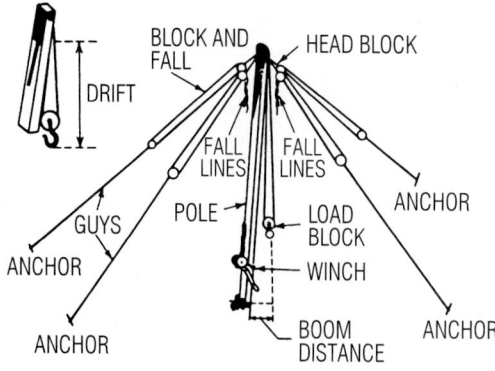

Fig. 9.18 Gin pole.

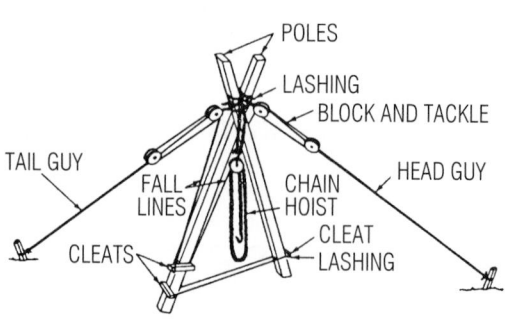

Fig. 9.19 A or shear-leg frame.

hoisting line may be manila or wire rope. The capacity of a gin pole is determined by the strength of the guys, hoist line, winch, hook, supporting structure, and the pole itself.

There are several variations of gin poles, such as the **A frame** (Fig. 9.19) and the **Dutchman** (Fig. 9.20).

A **stiffleg derrick** consists of a boom, vertical mast, and two inclined braces, or stifflegs (Fig. 9.21). It is provided with a special winch, which is furnished with hoisting drums to provide separate load and boom lines. After the structural frame of a high building has been completed, a stiffleg may be installed on the roof to hoist building materials, mechanical equipment, and so forth to various floors.

Guy derricks (Fig. 9.22) are advantageous in erecting multistory buildings. These derricks can jump themselves from one story to another. The boom temporarily serves as a gin pole to hoist the mast to a higher level. The mast is then secured in place and, acting as a gin pole, hoists the boom into its next position. Slewing (rotating) the derrick may be handled manually or by power.

A **Chicago boom** is a lifting device that uses the structure being erected to support the boom (Fig. 9.23).

Cranes are powered erection equipment consisting primarily of a rotating cab with a counterweight and a movable boom (Fig. 9.24). Sections of boom may be inserted and removed, and jibs may be added to increase the reach.

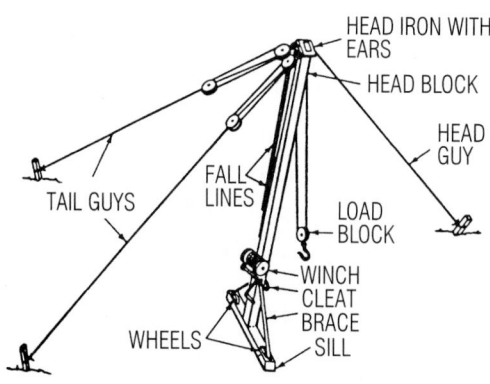

Fig. 9.20 Dutchman.

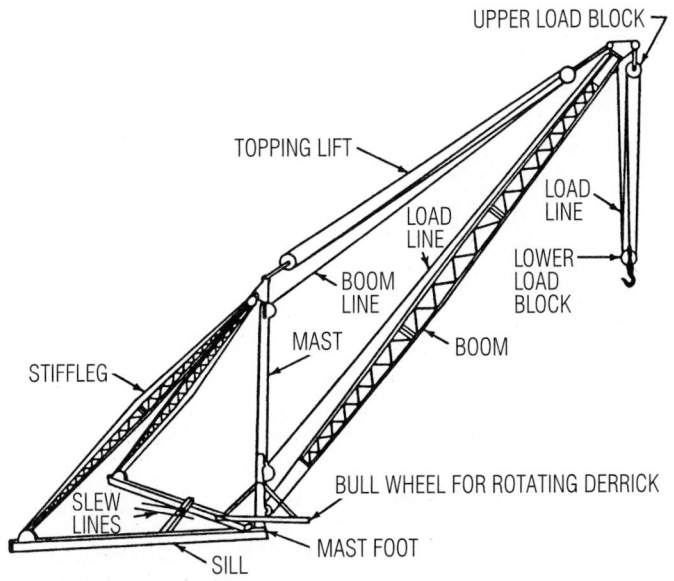

Fig. 9.21 Stiffleg derrick.

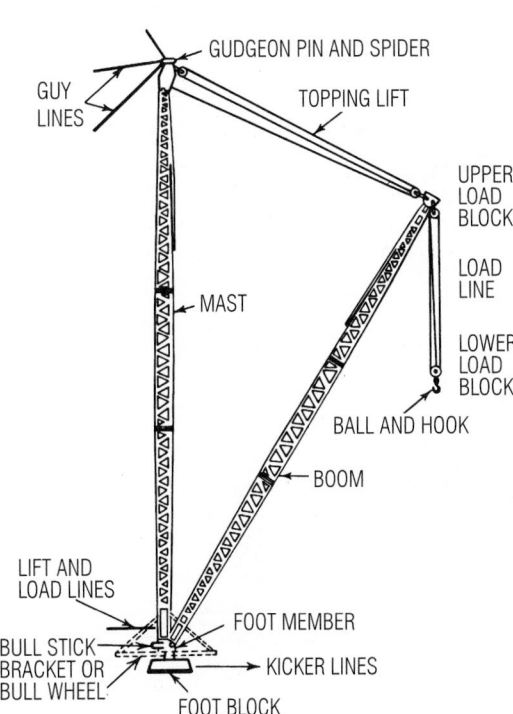

Fig. 9.22 Guy derrick.

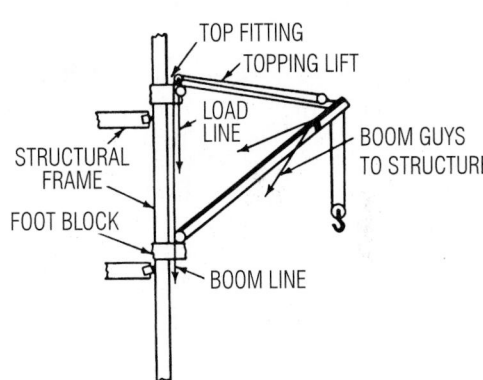

Fig. 9.23 Chicago boom.

Cranes may be mounted on a truck, crawler, or locomotive frame. The truck-mounted crane requires firm, level ground. It is useful on small jobs, where maneuverability and reach are required. Crawler cranes are more adaptable for use on soggy soil or where an irregular or pitched surface exists. Locomotive cranes are used for bridge erection or for jobs where railroad track exists or when it is economical to lay track.

The **tower crane** (Fig. 9.25) has important advantages. The control station can be located on

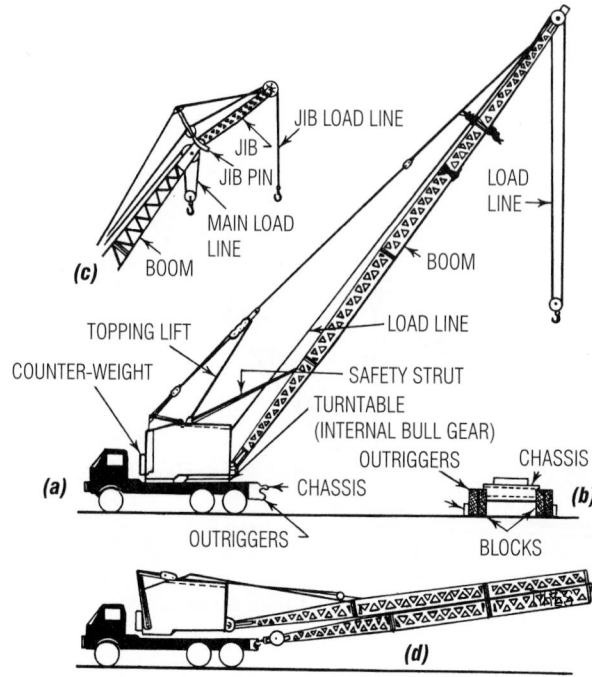

Fig. 9.24 Truck crane.

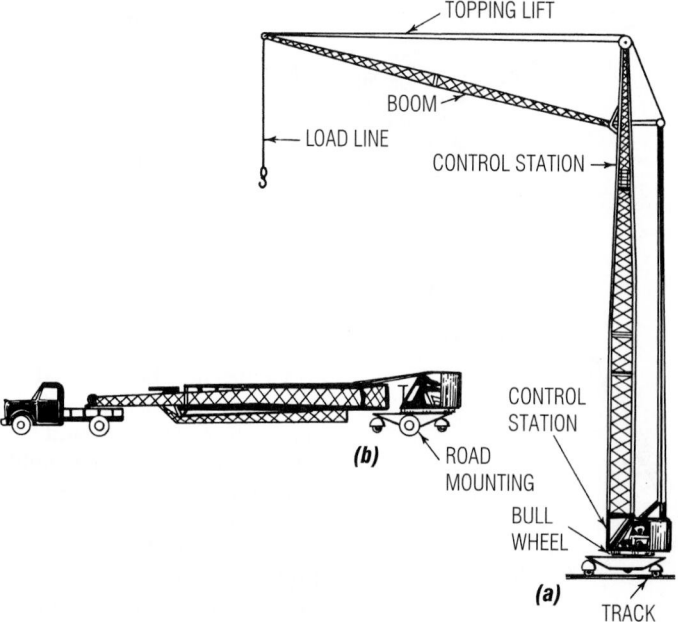

Fig. 9.25 Tower or slewing crane.

the crane or at a distant position that enables the operator to see the load at all times. Also, the equipment can be used to place concrete directly in the forms for floors and roofs, eliminating chutes, hoppers, and barrows.

Variations of the tower crane include the *kangaroo* (Fig. 9.26*a*) and the *hammerhead types* (Fig. 9.26*b*). The control station is located at the top of the tower and gives the operator a clear view of erection from above. A hydraulic jacking system is

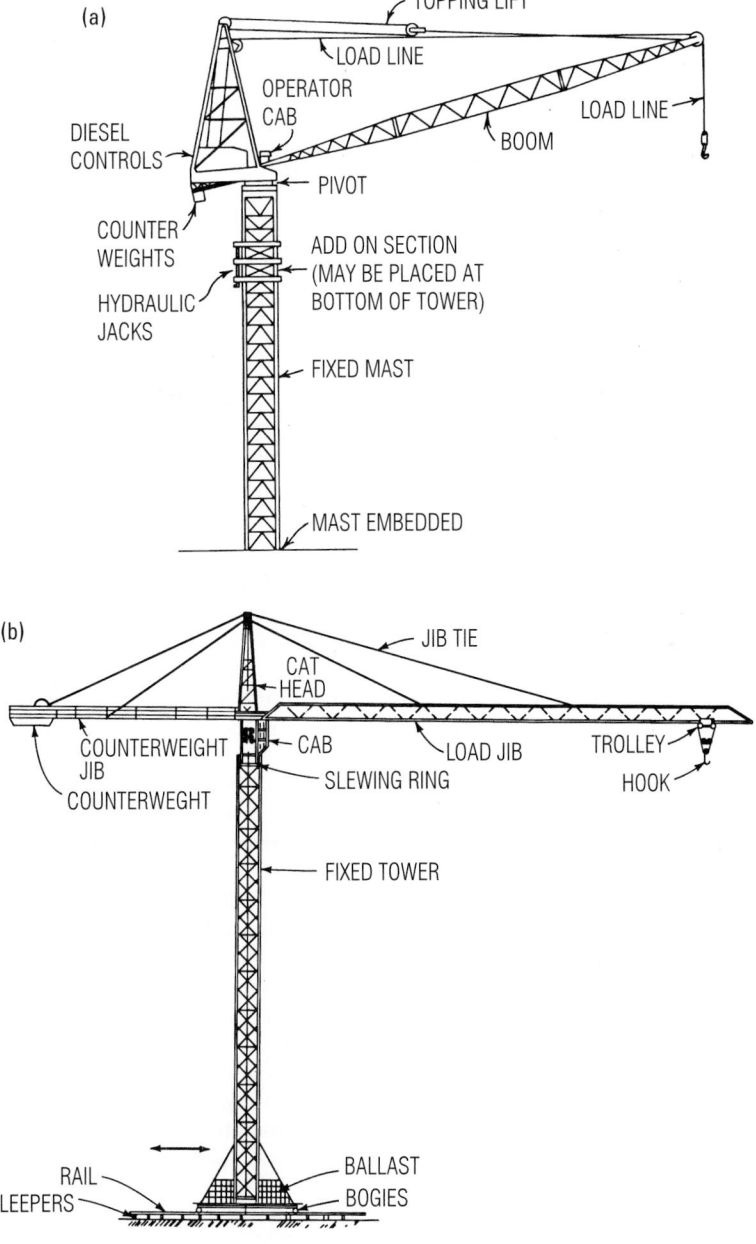

Fig. 9.26 Variations of the tower crane: (*a*) Kangaroo. (*b*) Hammerhead.

built into the fixed mast, and new mast sections are added to increase the height. As the tower gets higher, the mast must be tied into the structural framework for stability.

No general rules can be given regarding the choice of an erection device for a particular job. The main requirement is usually speed of erection, but other factors must be considered, such as the cost of the machine, labor, insurance, and cost of the power. Also, it is important to follow safety regulations set forth by the U.S. Office of Safety and Health Administration (OSHA).

9.33 Tolerances and Clearances for Erecting Beams

It is the duty of the structural-shop drafter to detail the steel so that each member may be swung into position without shifting members already in place.

Over the years, experience has resulted in "standard" practices in building work. The following are some examples:

In a framed connection, the total out-to-out distance of beam framing angles is usually $\frac{1}{8}$ in shorter than the face-to-face distance between the columns or other members to which the beam will be connected. Once the beam is in place, it is an easy matter to bend the outstanding legs of the angle, if necessary, to complete the connection. With a relatively short beam, the drafter may determine that it is impossible to swing the beam into place with only the $\frac{1}{8}$-in clearance. In such cases, it may be necessary to ship the connection angles "loose" for one end of the beam. Alternatively, it may be advantageous to connect one angle of each end connection to the supporting member and complete the connection after the beam is in place.

The common case of a beam framing into webs of columns must also be carefully considered. The usual practice is to place the beam in the "bosom" of the column by tilting it in the sling as shown in Fig. 9.27. It must, of course, clear any obstacle above. Also, the greatest diagonal distance G must be about $\frac{1}{8}$ in less than the distance between column webs. After the beam is seated, the top angle may be attached.

It is standard detailing practice to compensate for anticipated mill variations. The limits for mill tolerances are prescribed in ASTM A6, "General

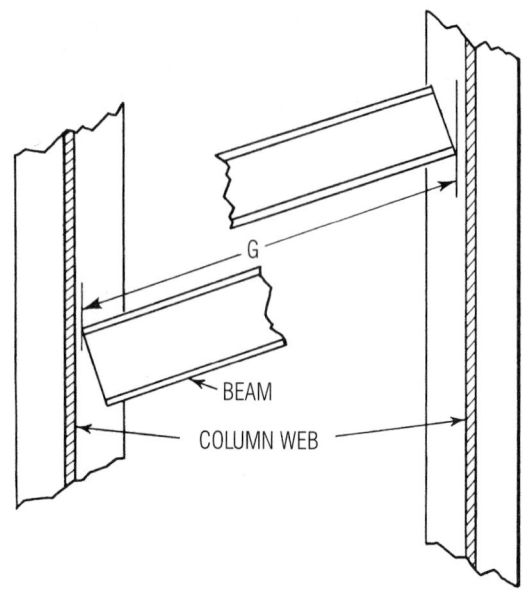

Fig. 9.27 Diagonal distance G for beam should be less than the clear distance between column webs, to provide erection clearance.

Requirements for Delivery of Rolled Steel Plates, Shapes, Sheet Piling, and Bars for Structural Use." For example, wide-flange beams are considered straight, vertically or laterally, if they are within $\frac{1}{8}$ in for each 10 ft of length. Similarly, columns are straight if the deviation is within $\frac{1}{8}$ in/10 ft, with a maximum deviation of $\frac{3}{8}$ in.

The "Code of Standard Practice" of the American Institute of Steel Construction gives permissible tolerances for the completed frame; Fig. 9.2 summarizes these. As shown, beams are considered level and aligned if the deviation does not exceed 1:500. With columns, the 1:500 limitation applies to individual pieces between splices. The total or cumulative displacement for multistory buildings is also given. The control is placed on exterior columns or those in elevator shafts.

There are no rules covering tolerances for milled ends of columns. It is seldom possible to achieve tight bearing over the cross section, and there is little reason for such a requirement. As the column receives its load, portions of the bearing area may quite possibly become plastic, which tends to redistribute stresses. Within practical limits, no harm is done to the load-carrying capacity of the member.

9.34 Fire Protection of Steel

Although structural steel does not support combustion and retains significant strength at elevated temperatures as subsequently discussed, the threat of sustained high-temperature fire, in certain types of construction and occupancies, requires that a steel frame be protected with fire-resistive materials.

In many buildings, no protection at all is required because they house little combustible material or they incorporate sprinkler systems. Therefore, "exposed" steel is often used for industrial-type buildings, hangars, auditoriums, stadiums, warehouses, parking garages, billboards, towers, and low stores, schools, and hospitals. Bridges require no fire protection.

The factors that determine fire-protection requirements, if any, are height, floor area, type of occupancy (a measure of combustible contents), availability of fire-fighting apparatus, sprinkler systems, and location in a community (fire zone), which is a measure of hazard to adjoining properties.

Fire Ratings ▪ Based on the above factors, building codes specify minimum fire-resistance requirements. The degree of fire resistance required for any structural component is expressed in terms of its ability to withstand fire exposure in accordance with the requirements of the ASTM standard time-temperature fire test, as shown in Fig. 9.28.

Under the standard fire-test ASTM Specification (E119), each tested assembly is subjected to the standard fire of controlled extent and severity. The fire-resistance rating is expressed as the time, in hours, that the assembly is able to withstand exposure to the standard fire before the criterion of failure is reached. These tests indicate the period of time during which the structural members, such as columns and beams, are capable of maintaining their strength and rigidity when subjected to the standard fire. They also establish the period of time during which floors, roofs, walls, or partitions will prevent fire spread by protecting against the passage of flame, hot gases, and excessive heat.

Strength Changes ▪ When evaluating fire-protection requirements for structural steel, it is useful to consider the effect of heat on its strength. In general, the yield point decreases linearly from its value at 70 °F to about 80% of that value at 800°F. At 1000 °F, the yield point is about 70% of its value

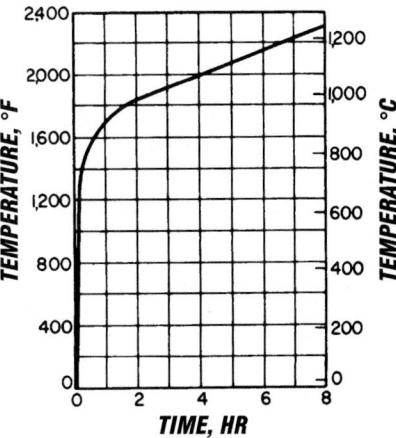

Fig. 9.28 ASTM time-temperature curve for fire test. Air temperature reaches 1000 °F in 5 min, 1700 °F in 1 h, and 2000 °F in 4 h.

at 70 °F and approaches the working stress of the structural members. Tension and compression members, therefore, are permitted to carry their maximum working stresses if the average temperature in the member does not exceed 1000 °F or the maximum at any one point does not exceed 1200 °F. (For steels other than carbon or low-alloy, other temperature limits may be necessary.)

Coefficient of Expansion ▪ The average coefficient of expansion for structural steel between temperatures of 100 and 1200 °F is given by the formula

$$c = (6.1 + 0.0019t) \times 10^{-6} \qquad (9.99)$$

where c = coefficient of expansion per °F

t = temperature, °F

Change in Modulus ▪ The modulus of elasticity is about 29,000 ksi at room temperature and decreases linearly to about 25,000 ksi at 900 °F. Above that, it decreases more rapidly.

Fire-Protection Methods ▪ Once the required rating has been established for a structural component, there are many ways in which the steel frame may be protected. For columns, one popular-fire-protection material is lightweight plaster (Fig. 9.29). Generally, a vermiculite or perlite

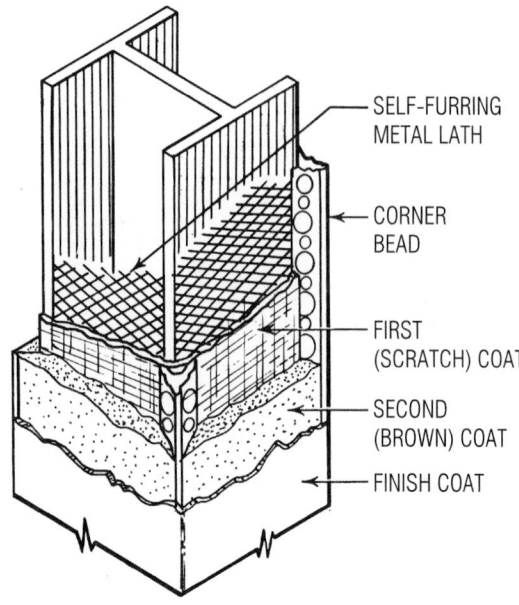

- SELF-FURRING METAL LATH
- CORNER BEAD
- FIRST (SCRATCH) COAT
- SECOND (BROWN) COAT
- FINISH COAT

Fig. 9.29 Column fireproofing with plaster on metal lath.

plaster thickness of 1 to $1\frac{3}{4}$ in affords protection of 3 to 4 h, depending on construction details.

Concrete, brick, or tile is sometimes used on columns where rough usage is expected. Ordinarily, however, these materials are inefficient because of the large dead weight they add to the structure. Lightweight aggregates would, of course, reduce this inefficiency.

Beams, girders, and trusses may be fireproofed individually or by a membrane ceiling. Lath and plaster, sprayed mineral fibers, or concrete encasement may be used. As with columns, concrete adds considerably to the weight. The sprayed systems usually require some type of finish for architectural reasons.

The membrane ceiling is used quite often to fireproof the entire structural floor system, including beams, girders, and floor deck. For many buildings, a finished ceiling is required for architectural reasons. It is therefore logical and economical to employ the ceiling for fire protection also. Figure 9.30 illustrates typical installations. As can be seen, the rating depends on the thickness and type of material.

Two alternative methods of fire protection are flame shielding and water-filled columns. These methods are usually used together and are employed where the exposed steel frame is used architecturally.

Another method of fire protection is by separation from a probable source of heat. If a structural member is placed far enough from the source of heat, its temperature will not exceed the critical limit. Mathematical procedures for determining the temperature of such members are available. (See, for example, "Fire-Safe Structural Steel—A Design Guide," American Iron and Steel Institute, 1001 17th St., Washington, D.C. 20036, www.aisc.org.)

Figure 9.31 illustrates the principle of flame shielding. The spandrel web is exposed on the exterior side and sprayed with fireproofing material on the inside. The shield in this case is the insulated bottom flange, and its extension protects the web from direct contact with the flame. The web is heated by radiation only and will achieve a maximum

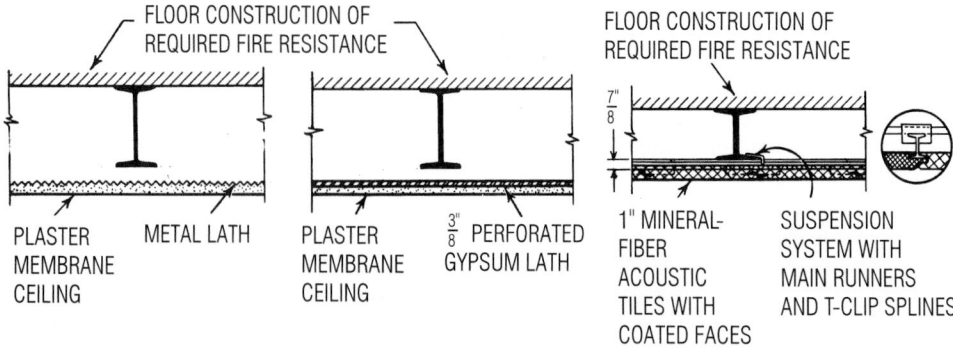

FLOOR CONSTRUCTION OF REQUIRED FIRE RESISTANCE

FLOOR CONSTRUCTION OF REQUIRED FIRE RESISTANCE

$\frac{7}{8}$"

PLASTER MEMBRANE CEILING   METAL LATH

PLASTER MEMBRANE CEILING $\frac{3}{8}$" PERFORATED GYPSUM LATH

1" MINERAL-FIBER ACOUSTIC TILES WITH COATED FACES

SUSPENSION SYSTEM WITH MAIN RUNNERS AND T-CLIP SPLINES

Fig. 9.30 Ceiling-membrane fireproofing applied below floor beams and girders.

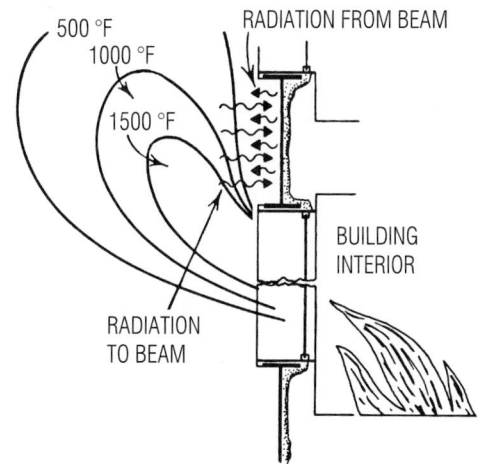

Fig. 9.31 Flame-shielded spandrel girder. (*From "Fire-Resistant Steel-Frame Construction," American Iron and Steel Institute, with permission.*)

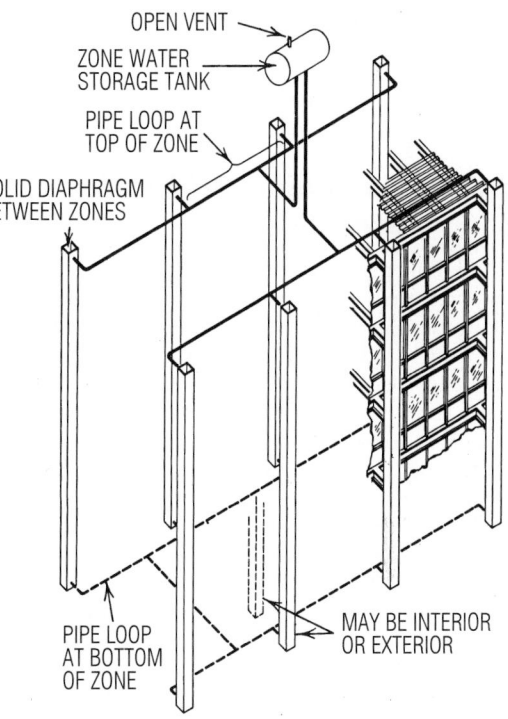

Fig. 9.32 Piping arrangement for liquid-filled-column fire-protection system. (*From "Fire-Resistant Steel-Frame Construction," American Iron and Steel Institute, with permission.*)

temperature well below the critical temperature associated with structural failure.

Water-filled columns can be used with flame-shielded spandrels and are an effective fire-resistance system. The hollow columns are filled with water plus antifreeze (in northern climates). The water is stationary until the columns are exposed to fire. Once exposed, heat that penetrates the column walls is absorbed by the water. The heated water rises, causing water in the entire system to circulate. This takes heated water away from the fire and brings cooler water to the fire-affected columns (Fig. 9.32).

Another alternative in fire protection is intumescent paint. Applied by spray or trowel, this material has achieved a 1-h rating and is very close to a 2-h rating. When subjected to heat, it puffs up to form an insulating blanket. It can be processed in many colors and has an excellent architectural finish.

In building construction, it is often necessary to pierce the ceiling for electrical fixtures and air-conditioning ducts. Tests have provided data for the effect of these openings. The rule that has resulted is that ceilings should be continuous, except that openings for noncombustible pipes, ducts, and electrical outlets are permissible if they do not exceed 100 in^2 in each 100 ft^2 of ceiling area. All duct openings must be protected with approved fusible-link dampers.

Summaries of established fire-resistance ratings are available from the following organizations:

American Insurance Association, 1130 Connecticut Ave NW, Washington, DC 20036.

National Institute of Standards and Technology, Washington, DC 20234

Gypsum Association, 810 First St., Washington, DC 20002.

Metal Lath/Steel Framing Association, 8 S. Michigan Ave, Chicago, IL 60603

Perlite Institute, 88 New Dorp Plaza, Staten Island, NY, 10306-2994

Vermiculite Association, Whitegate Acre, Metheringham, Fen, Lincoln, LN43AL, UK

American Iron and Steel Institute, 1140 Connecticut Ave., N.W., Washington, DC 20036

American Institute of Steel Construction, One East Wacker Dr., Chicago, IL 60601-2001

9.35 Corrosion Protection of Steel

The following discussion applies to all steels used in applications for which a coating is required for protection against atmospheric corrosion. As previously indicated (Art. 9.3), some high-strength, low-alloy steels can, with suitable precautions (including those in Art. 9.36), be used in the bare, uncoated condition for some applications in which a coating is otherwise required for protection against atmospheric corrosion.

Steel does not rust except when exposed to atmospheres above a critical relative humidity of about 70%. Serious corrosion occurs at normal temperature only in the presence of both oxygen and water, both of which must be replenished continually. In a hermetically sealed container, corrosion of steel will continue only until either the oxygen or water, or both, are exhausted.

To select a paint system for corrosion prevention, therefore, it is necessary to begin with the function of the structure, its environment, maintenance practices, and appearance requirements. For instance, painting steel that will be concealed by an interior building finish is usually not required. On the other hand, a bridge exposed to severe weather conditions would require a paint system specifically designed for that purpose.

The Society for Protective Coatings, SPC (Forty 24th St., Pittsburgh, PA 15222, www.sspc.org) issues specifications covering practical and economical methods of surface preparation and painting steel structures. The SPC also engages in research aimed at reducing or preventing steel corrosion. This material is published in two volumes: I, "Good Painting Practice," and II, "Systems and Specifications."

The SPC Specifications include numerous paint systems. By reference to a specific specification number, it is possible to designate an entire proved paint system, including a specific surface preparation, pretreatment, paint-application method, primer, and intermediate and top coat. Each specification includes a "scope" clause recommending the type of usage for which the system is intended.

In addition to the overall system specification, the SPC publishes individual specifications for surface preparation and paints. Surface preparations included are solvent, hand tool, power tool, pickling, flame, and several blast techniques.

When developing a paint system, it is extremely important to relate properly the type of paint to the surface preparation. For instance, a slow-drying paint containing oil and rust-inhibitive pigments and one possessing good wetting ability could be applied on steel nominally cleaned. On the other hand, a fast-drying paint with poor wetting characteristics requires exceptionally good surface cleaning, usually entailing complete removal of mill scale.

"Standard Specifications for Highway Bridges," (American Association of State Highway and Transportation Officials), gives detailed specifications and procedures for the various painting operations and for paint systems. AASHTO Specifications for surface preparation include hand cleaning, blast cleaning, and steam cleaning. Application procedures are given for brush, spray, or roller, as well as general requirements.

Concrete Protection ▪ In bridge and building construction, steel may be in contact with concrete. According to SPC vol. I, "Good Painting Practice":

1. Steel that is embedded in concrete for reinforcing should not be painted. Design considerations require strong bond between the reinforcing and the concrete so that the stress is distributed; painting of such steel does not supply sufficient bond. If the concrete is properly made and of sufficient thickness over the metal, the steel will not corrode.

2. Steel encased with exposed lightweight concrete that is porous should be painted with at least one coat of good-quality rust-inhibitive primer. When conditions are severe or humidity is high, two or more coats of paint should be applied since the concrete may accelerate corrosion.

3. When steel is enclosed in concrete of high density or low porosity and the concrete is at least 2 to 3 in thick, painting is not necessary since the concrete will protect the steel.

4. Steel in partial contact with concrete is generally not painted. This creates an undesirable condition, for water may seep into the crack between the steel and the concrete, causing corrosion. A sufficient volume of rust may build up, spalling the concrete. The only remedy is to chip or leave a groove in the concrete at the edge next to the steel and seal the crack with an alkali-resistant calking compound (such as bituminous cement).

5. Steel should not be encased in concrete that contains cinders since the acidic condition will cause corrosion of the steel.

9.36 Bolted Joints in Bare-Steel Structures

Special considerations are required for the design of joints in bare weathering steels. Atmospheric-corrosion-resistant, high-strength, low-alloy steels are used in the unpainted (bare) condition for such diverse applications as buildings, railroad hopper cars, bridges, light standards, transmission towers, plant structures, conveyor-belt systems, and hoppers because these steels are relatively inexpensive and require little maintenance. Under alternate wetting and drying conditions, a protective oxide coating that is resistant to further corrosion forms. But if such atmospheric-corrosion-resistant steels remain wet for prolonged periods, their corrosion resistance will not be any better than that of carbon steel. Thus, the design of the structure should minimize ledges, crevices, and other areas that can hold water or collect debris.

Experience with bolted joints in exposed frameworks of bare weathering steel indicates that if the stiffness of the joint is adequate and the joint is tight, the space between two faying surfaces of weathering-type steel seals itself with the formation of corrosion products around the periphery of the joint. However, if the joint design does not provide sufficient stiffness, continuing formation of corrosion products within the joint leads to expansive forces that can (1) deform the connected elements such as cover plates and (2) cause large tensile loads on the bolts.

Consequently, in the design of bolted joints in bare weathering steel, it is important to adhere to the following guidelines:

1. Limit pitch to 14 times the thickness of the thinnest part (7-in maximum).

2. Limit edge distance to 8 times the thickness of the thinnest part (5-in maximum).

3. Use fasteners such as ASTM A325, Type 3, installed in accordance with specifications approved by the Research Council on Structural Connections. (Nuts should also be of weathering steel; galvanized nuts may not provide adequate service if used with weathering steel.)

10

Don S. Wolford
Consulting Engineer
Middletown, Ohio

Wei-Wen Yu
University of Missouri-Rolla
Rolla, Missouri

COLD-FORMED-STEEL DESIGN AND CONSTRUCTION

The introduction of sheet rolling mills in England in 1784 by Henry Cort led to the first cold-formed-steel structural application, light-gage corrugated steel sheets for building sheathing. Continuous hot-rolling mills, developed in America in 1923 by John Tytus, led to the present fabricating industry based on coiled strip steel. This is now available in widths up to 90 in and in coil weights up to 40 tons, hot- or cold-rolled.

Formable, weldable, flat-rolled steel is available in a variety of strengths and in black, galvanized, or aluminum-coated. Thus, fabricators can choose from an assortment of raw materials for producing cold-formed-steel products. (In cold forming, bending operations are done at room temperature.) Large quantities of cold-formed sections are most economically produced on multistand roll-forming machines from slit coils of strip steel. Small quantities can still be produced to advantage in presses and bending brakes from sheared blanks of sheet and strip steel. Innumerable cold-formed-steel products are now made for building, drainage, road, and construction uses. Design and application of such lightweight-steel products are the principal concern of this section.

10.1 How Cold-Formed Shapes are Made

Cold-formed shapes are relatively thin sections made by bending sheet or strip steel in roll-forming machines, press brakes, or bending brakes. Because of the relative ease and simplicity of the bending operation and the comparatively low cost of forming rolls and dies, the cold-forming process also lends itself well to the manufacture of special shapes for specific architectural purposes and for maximum section stiffness.

Door and window frames, partitions, wall studs, floor joists, sheathing, and moldings are made by cold forming. There are no standard series of cold-formed structural sections, like those for hot-rolled structural shapes, although some dimensional requirements are specified in the American Iron and Steel Institute (AISI) Standards for cold-formed steel framing.

Cold-formed shapes cost a little more per pound than hot-rolled sections. They are nevertheless more economical under light loading.

10.2 Steel for Cold-Formed Shapes

Cold-formed shapes are made from sheet or strip steel, usually from 0.020 to 0.125 in thick. In thicknesses available (usually 0.060 to $\frac{1}{2}$ in), hot-rolled steel usually costs less to use. Cold-rolled steel is used in the thinner gages or where the surface finish, mechanical properties, or more uniform thickness resulting from cold reducing are desired. (The commercial distinction between steel plates, sheets, and strip is principally a matter of thickness and width of material.)

Cold-formed shapes may be either black (uncoated) or galvanized. Despite its higher cost, galvanized material is preferable where exposure conditions warrant paying for increased corrosion protection. Uncoated material to be used for structural purposes generally conforms to one of the standard ASTM Specifications for structural-quality sheet and strip (A1008, A1011 and others). ASTM A653 covers structural-quality galvanized sheets. Steel with a hot-dipped aluminized coating (A792 and A875) is also available.

The choice of grade of material usually depends on the severity of the forming operation required to make the desired shape. Low-carbon steel has wide usage. Most shapes used for structural purposes in buildings are made from material with yield points in the range of 33 to 50 ksi under ASTM Specifications A1008 and A1011. Steel conforming generally to ASTM A606, "High-Strength, Low-Alloy, Hot-Rolled and Cold-Rolled Steel Sheet and Strip with Improved Corrosion Resistance," A1008, "Steel, Sheet, Cold-Rolled, Carbon, Structural, High-Strength Low-Alloy and High-Strength Low-Alloy with Improved Formability," or A1011, "Steel, Sheet and Strip, Hot-Rolled, Carbon, Structural, High-Strength Low-Alloy and High-Strength Low-Alloy with Improved Formability," is often used to achieve lighter weight by designing at yield points from 45 to 70 ksi, although higher yield points are also being used.

Sheet and strip for cold-formed shapes are usually ordered and furnished in decimal or millimetre thicknesses. (The former practice of specifying thickness based on weight and gage number is no longer appropriate.)

For the use of steel plates for cold-formed shapes, see the AISI Specification.

10.3 Types of Cold-Formed Shapes

Some cold-formed shapes used for structural purposes are similar in general configuration to hot-rolled structural shapes. Channels (C-sections), angles, and Z's can be roll-formed in a single operation from one piece of material. I sections are usually made by welding two channels back to back, or by welding two angles to a channel. All such sections may be made with either plain flanges, as in Fig. 10.1a to d, j, and m, or with flanges stiffened by lips at outer edges, as in Fig. 10.1e to h, k, and n.

In addition to these sections, the flexibility of the forming process makes it relatively easy to obtain hat-shaped sections, open box sections, or inverted-U sections (Fig. 10.1o, p, and q). These sections are very stiff in a lateral direction.

The thickness of cold-formed shapes can be assumed to be uniform throughout in computing weights and section properties. The fact that cold-formed sections have corners rounded on both the inside and outside of the bend has only a slight effect on the section properties, and so computations may be based on sharp corners without serious error.

Cracking at 90° bends can be reduced by use of inside bend radii not smaller than values recommended for specific grades of the steels mentioned in Art. 10.2. For instance, A1008, SS Grade 33 steel, for which a minimum yield point of 33 ksi is specified, should be bent around a die with a radius equal to at least $1\frac{1}{2}$ times the steel thickness. See ASTM Specification grade for appropriate bend radius that can safely be used in making right angle bends.

10.4 Design Principles for Cold-Formed Sections

In 1939, the American Iron and Steel Institute (AISI) started sponsoring studies, which still continue, under the direction of structural specialists associated with the AISI Committees of Sheet and Strip

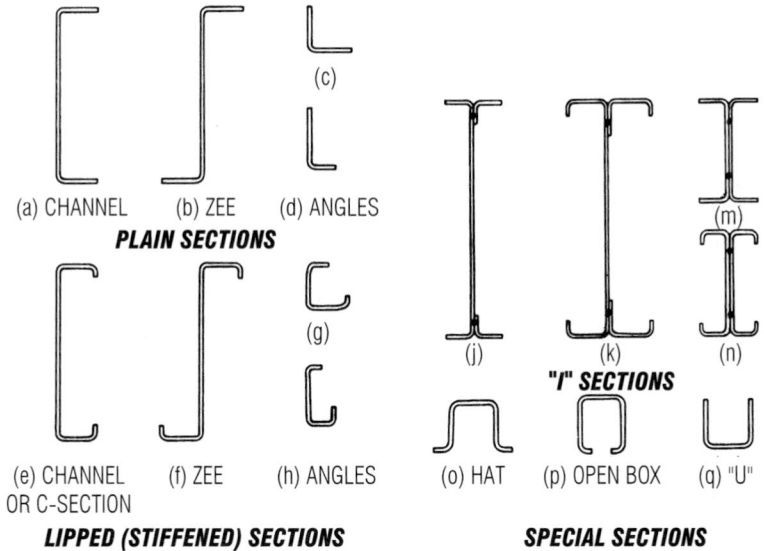

Fig. 10.1 Typical cold-formed-steel structural sections.

Steel Producers, that have yielded the AISI Specification for the Design of Cold-Formed Steel Structural Members. (American Iron and Steel Institute, 1140 Connecticut Ave., N.W., Washington, DC 20036.) The specification, which has been revised and amended repeatedly since its initial publication in 1946, has been adopted by the major building codes of the United States.

Structural behavior of cold-formed shapes conforms to classic principles of structural mechanics, as does the structural behavior of hot-rolled shapes and sections of built-up plates. However, local buckling of thin, wide elements, especially in cold-formed sections, must be prevented with special design procedures. Shear lag in wide elements remote from webs that causes nonuniform stress distribution and torsional instability that causes twisting in columns and beam of open sections also need special design treatment.

Uniform thickness of cold-formed sections and the relative remoteness from the neutral axis of their thin, wide flange elements make possible the assumption that, in computation of section properties, section components may be treated as line elements. (See "Section 3 of Part I of the AISI Cold-Formed Steel Design Manual," 2002.)

(Wei-Wen Yu, "Cold-Formed Steel Design," John Wiley & Sons, Inc., New York.)

Design Basis ▪ The Allowable Strength Design Method (ASD) is used currently in structural design of cold-formed steel structural members and described in the rest of this section using US customary units. In addition, the Load and Resistance Factor Design Method (LRFD) can also be used for design. Both methods are included in the 2001 edition of the AISI "North American Specification for the Design of Cold-Formed Steel Structural Members." However, these two methods cannot be mixed in designing the various cold-formed steel components of a structure.

In the allowable strength design method, the required strengths (bending moments, shear forces, axial loads, etc.) in structural members are computed by structural analysis for the working or service loads using the load combinations given in the AISI Specification. These required strengths are not to exceed the allowable design strengths as follows:

$$R \le \frac{R_n}{\Omega}$$

where R = required strength

R_n = nominal strength specified in the AISI Specification

Ω = safety factor specified in the AISI Specification

R_n / Ω = allowable design strength

Unlike the allowable strength design method, the LRFD method uses multiple load factors and resistance factors to provide a refinement in the design that can account for different degrees of the uncertainties and variabilities of analysis, design, loading, material properties and fabrication. In this method, the required strengths are not to exceed the design strengths as follows:

$$R_u \leq \phi R_n$$

where $R_u = \Sigma \gamma_i Q_i$ = required strength

R_n = nominal strength specified in the AISI Specification

ϕ = resistance factor specified in the AISI Specification

γ_i = load factors

Q_i = load effects

ϕR_n = design strength

The load factors and load combinations are also specified in the AISI North American Specification for the design of different type of cold-formed steel structural members and connections. For design examples, see AISI "Cold-Formed Steel Design Manual," 2002 edition.

The ASD and LRFD methods discussed above are used in the United States and Mexico. The AISI North American Specification also includes the Limit States Design Method (LSD) for use in Canada. The methodology for the LSD method is the same as the LRFD method, except that the load factors, load combinations, and some resistance factors are different. The North American Specification includes Appendixes A, B, and C, which are applicable in the United States, Canada, and Mexico, respectively.

10.5 Structural Behavior of Flat Compression Elements

For buckling of flat compression elements in beams and columns, the flat-width ratio w/t is an important factor. It is the ratio of width w of a single flat element, exclusive of edge fillets, to the thickness t of the element (Fig. 10.2).

Flat compression elements of cold-formed structural members are classified as stiffened and unstiffened. Stiffened compression elements have both edges parallel to the direction of stress stiffened by a web, flange, or stiffening lip. Unstiffened compression elements have only one edge parallel to the direction of stress stiffened. If the sections in Fig. 10.1a to n are used as compression members, the webs are considered stiffened compression elements. But the wide, lipless flange elements and the lips that stiffen the outer edges of the flanges are unstiffened elements. Any section composed of a number of plane elements can be broken down into a combination of stiffened and unstiffened elements.

The cold-formed structural cross sections shown in Fig. 10.3 illustrate how effective portions of stiffened compression elements are considered to be divided into two parts located next to the two edge stiffeners of that element. In beams, a stiffener may be a web, another stiffened element, or a lip.

In computing net section properties, only the effective portions of elements are considered and the ineffective portions are disregarded. For beams, flange elements subjected to uniform compression may not be fully effective. Accordingly, section properties, such as moments of inertia and section moduli, should be reduced from those for a fully effective section. (Effective widths of webs can be determined using Section B2.3 of the AISI North American Specification.) Effective areas of column cross sections needed for determination of column loads from Eq. (10.21) of Art. 10.12 are based on full cross-sectional areas less all ineffective portions.

Elastic Buckling ▪ Euler, in 1744, determined the critical load for an elastic prismatic bar end-

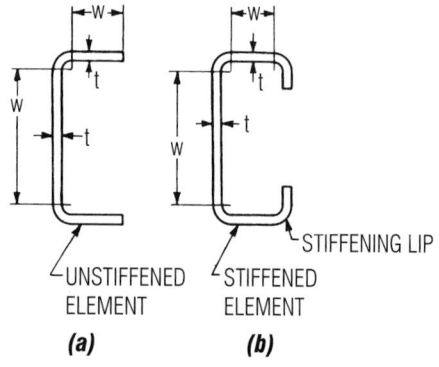

Fig. 10.2 Compression elements.

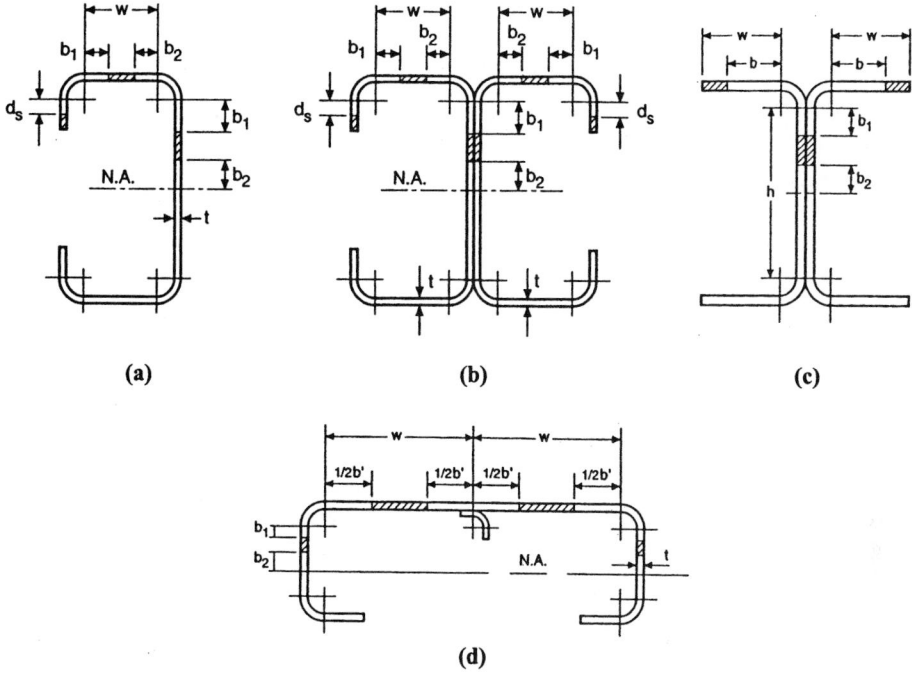

(a)   **(b)**   **(c)**

(d)

BEAMS – TOP FLANGE IN COMPRESSION

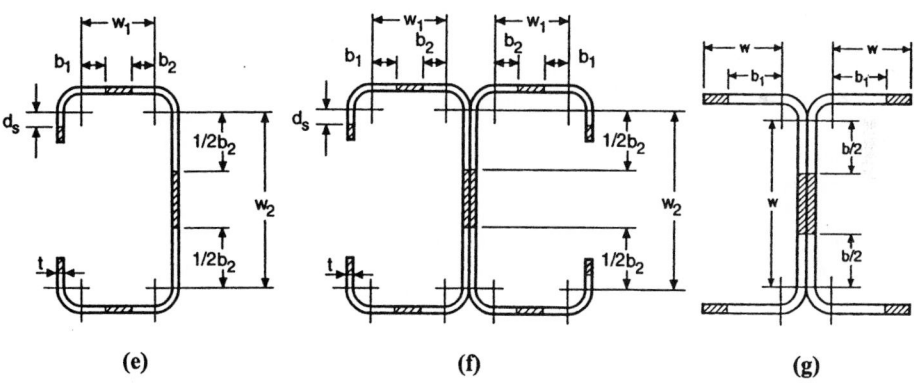

(e)   **(f)**   **(g)**

COLUMNS – EFFECTIVE AREA FOR COMPUTING COLUMN LOADS

Fig. 10.3 Effective width of compression elements.

loaded as a column from

$$P_{cr} = \frac{\pi^2 EI}{L^2} \qquad (10.1)$$

where P_{cr} = critical load at which bar buckles, kips

E = modulus of elasticity, 29,500 ksi for steel

I = moment of inertia of bar cross section, in^4

L = column length of bar, in

This equation is the basis for designing long columns of prismatic cross section subject to elastic buckling. It might be regarded as the precursor of formulas used in the design of thin rectangular plates in compression.

Bryan, in 1891, proposed for design of a thin rectangular plate compressed between two opposite edges with the other two edges supported:

$$f_{cr} = \frac{k\pi^2 E(t/w)^2}{12(1-\nu^2)} \qquad (10.2)$$

where f_{cr} = critical local buckling stress, ksi

k = a coefficient depending on edge-support restraint

w = width of plate, in

ν = Poisson's ratio

t = thickness, in

Until the 1986 edition, all AISI Specifications based strength of thin, flat elements stiffened along one edge on *buckling stress* rather than *effective width* as used for thin, flat elements stiffened along both edges. Although efforts were made by researchers to unify element design using a single concept, unification did not actually occur until Pekoz, in 1986, presented his unified approach using effective width as the basis of design for both stiffened and unstiffened elements and even for web elements subjected to stress gradients. Consequently, the AISI Specification uses the following equations to determine the effective width of uniformly compressed stiffened and unstiffened elements based on a slenderness factor λ:

$$\lambda = \sqrt{\frac{f}{f_{cr}}} = \frac{1.052(w/t)\sqrt{f/E}}{\sqrt{k}} \qquad (10.3)$$

where k = 4.00 for stiffened elements

= 0.43 for unstiffened elements

f = unit stress in the compression element of the section, computed on the basis of the design width, ksi

f_{cr} = Eq. (10.2)

w = flat width of the element exclusive of radii, in

t = base thickness of element, in

The effective width is given by

$$b = w \qquad \lambda \le 0.673 \qquad (10.4)$$
$$b = \rho w \qquad \lambda > 0.673 \qquad (10.5)$$

The reduction factor ρ is given by

$$\rho = \frac{(1-0.22/\lambda)}{\lambda} \qquad (10.6)$$

10.6 Unstiffened Elements Subject to Local Buckling

By definition, unstiffened cold-formed elements have only one edge in the compression-stress direction supported by a web or stiffened element, while the other edge has no auxiliary support (Fig. 10.1a). The coefficient k in Eq. (10.3) is 0.43 for such an element. When the ratio of flat width to thickness does not exceed $72/\sqrt{f}$, an unstiffened element with unit stress f is fully effective; that is, the effective width b equals flat width w. Generally, however, Eq. (10.3) becomes

$$\lambda = \frac{1.052}{\sqrt{0.43}} \frac{w}{t} \sqrt{\frac{f}{E}} = 0.0093 \frac{w}{t} \sqrt{f} \qquad (10.7)$$

where E = 29,500 ksi for steel

f = unit compressive stress, ksi, computed on the basis of effective widths, Eq. (10.3)

When λ is substituted in Eq. (10.6), the b/w ratio ρ results. The lower portion of Fig. 10.5 shows curves for determining the effective-width ratio b/t for unstiffened elements for w/t between 0 and 60, with f between 15 and 90 ksi.

In beam-deflection determinations requiring the use of the moment of inertia of the cross section, f is the allowable stress used to calculate the effective width of an unstiffened element in a cold-formed-steel beam. However, in beam-strength determinations requiring use of the section modulus of the cross section, f is the unit compression stress to be used in Eq. (10.7) to calculate the effective width of the unstiffened element and provide an adequate margin of safety. In determining safe column loads, effective width for the unstiffened element must be determined for a nominal column buckling stress to ensure adequate margin of safety for such elements.

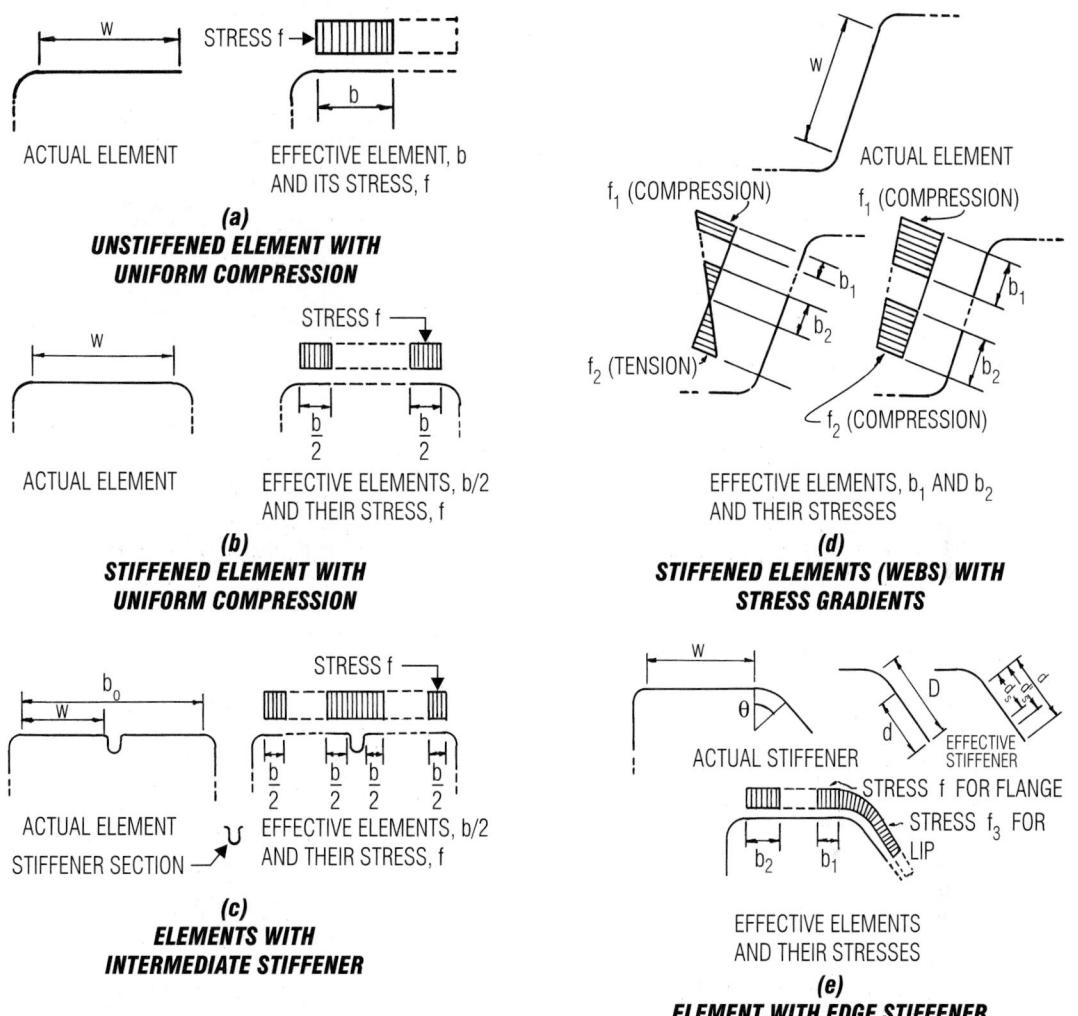

Fig. 10.4 Schematic diagrams showing effective widths for unstiffened and stiffened elements, intermediate stiffeners, beam webs, and edge stiffeners.

("Cold-Formed Steel Design Manual," American Iron and Steel Institute, Washington, D.C.)

10.7 Stiffened Elements Subject to Local Buckling

By definition, stiffened cold-formed elements have one edge in the compression-stress direction supported by a web or stiffened element and the other edge is also supported by a qualified stiffener (Fig.

10.4*b*). The coefficient *k* in Eq. (10.3) is 4.00 for such an element. When the ratio of flat width to thickness does not exceed $220/\sqrt{f}$, the stiffened element is fully effective, in which $f =$ unit stress, ksi, in the compression element of the structural section computed on the basis of effective widths, Eq. (10.3) becomes

$$\lambda = \frac{1.052}{\sqrt{4}}\frac{w}{t}\sqrt{\frac{f}{E}} = 0.0031\frac{w}{t}\sqrt{f} \qquad (10.8)$$

where $E = 29{,}500$ ksi for steel.

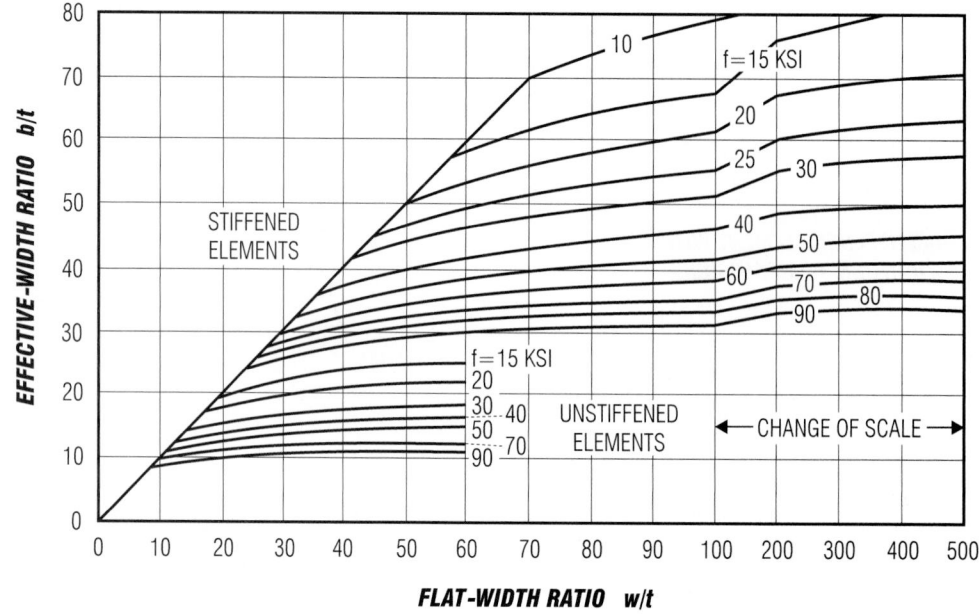

Fig. 10.5 Curves relate the effective-width ratio b/t to the flat-width ratio w/t for various stresses f for unstiffened and stiffened elements.

If λ is substituted in Eq. (10.6), the b/w ratio ρ results. Moreover, when $\lambda \leq 0.673$, $b = w$, and when $\lambda > 0.673$, $b = \rho w$. The upper portion of Fig. 10.5 shows curves for determining the effective-width ratio b/t for stiffened elements w/t between 0 and 500 with f between 10 and 90 ksi.

In beam-deflection determinations requiring the use of the moment of inertia of the cross section, f is the allowable stress used to calculate the effective width of a stiffened element in a cold-formed-steel member loaded as a beam. However, in beam-strength determinations requiring the use of the section modulus of the cross section, f is the unit compression stress to be used in Eq. (10.8) to calculate the width of a stiffened element in a cold-formed-steel beam. In determination of safe column loads, effective width for a stiffened element should be determined for a nominal column buckling stress to ensure an adequate margin of safety for such elements.

Note that the slenderness factor is $\sqrt{4.00/0.43} = 3.05$ times as great for unstiffened elements as for stiffened elements at applicable combinations of stress f and width-thickness ratio w/t. This emphasizes the greater effective width and economy of stiffened elements.

Single Intermediate Stiffener ■ For uniformly compressed stiffened elements with a single intermediate stiffener, as shown in Fig. 10.4c, the required moment of inertia I_a, in^4, is determined by a parameter $S = 1.28\sqrt{E/f}$:

For $b_o/t \leq S$, $I_a = 0$ and no intermediate stiffener is needed, $b = w$.

For $b_o/t > S$, the effective width of the compression flange can be determined by the following local buckling coefficient k:

$$k = 3(R_I)^n + 1 \qquad (10.9a)$$

where

$$n = \left[0.583 - \frac{b_o/t}{12S} \right] \geq \frac{1}{3} \qquad (10.9b)$$

$$R_I = I_s/I_a \leq 1 \qquad (10.9c)$$

For $S < b_o/t < 3S$:

$$I_a = t^4 \left[50\frac{b_o/t}{S} - 50 \right] \qquad (10.10a)$$

For $b_o/t \geq 3S$:

$$I_a = t^4 \left[128 \frac{b_o/t}{S} - 285 \right] \qquad (10.10b)$$

In the above equations,

b_o = flat width including the stiffener, in

I_s = moment of inertia of full section of stiffener about its own centroidal axis parallel to the element to be stiffened, in^4

Webs Subjected to Stress Gradients ■ Pekoz's unified approach using effective widths (Art. 10.5) also applies to stiffened elements subjected to stress gradients in compression, such as in webs of beams (Fig. 10.4d). The effective widths b_1 and b_2 are determined from the following, with $\psi = |f_2/f_1|$, where f_1 and f_2 are stresses shown in Fig. 10.4d calculated on the basis of the effective section. Stress f_1 is assumed to be in compression (positive) and f_2 can be either tension (negative) or compression. In case f_1 and f_2 are both in compression, f_1 is the larger of the two stresses.

$$b_1 = \frac{b_e}{3 + \psi} \qquad (10.11)$$

where b_e = effective width b determined from Eqs. (10.3) to (10.6), with f_1 substituted for f and with k calculated from

$$k = 4 + 2(1 + \psi)^3 + 2(1 + \psi) \qquad (10.12)$$

The value of b_2 is calculated as follows:

For $h_o/b_o \leq 4$:

$$b_2 = b_e/2, \text{ when } \psi > 0.236$$
$$b_2 = b_e - b_1, \text{ when } \psi \leq 0.236$$

For $h_o/b_o > 4$:

$$b_2 = b_e/(1 + \psi) - b_1$$

where b_o = out-to-out width of the compression flange, in

h_o = out-to-out depth of web, in

In addition, $b_1 + b_2$ should not exceed the compression portion of the web calculated on the basis of effective section.

Uniformly Compressed Elements with an Edge Stiffener ■ It is important to understand the capabilities of edge stiffeners (depicted in Fig. 10.4e for a slanted lip). However, due to the complexity of this subject, the following presentation is confined primarily to simple lip stiffeners.

Two ranges of w/t values are considered relative to a parameter 0.328 S. The limit value of w/t for full effectiveness of the flat width without auxiliary support is

$$0.328\,S = (0.328)(1.28)\sqrt{\frac{E}{f}} = 0.420\sqrt{\frac{E}{f}} \quad (10.13)$$

where E = modulus of the elasticity, ksi

f = unit compressive stress computed on the basis of effective widths, ksi

1. For the first case, where $w/t \leq 0.328S$, $b = w$, and no edge support is needed.

2. For the second case, where $w/t > 0.328S$, edge support is needed with the required moment of inertia I_a, in^4, determined from

$$I_a = 399t^4 \left[\frac{w/t}{S} - 0.328 \right]^3$$

$$\leq t^4 \left[115 \frac{w/t}{S} + 5 \right] \qquad (10.14)$$

For a slanted lip, as shown in Fig. 10.4e, the moment of inertia of full stiffener I_s, in^4, is

$$I_s = \frac{d^3 t}{12} \sin^2 \theta \qquad (10.15)$$

where d = flat width of lip, in

θ = angle between normals to stiffened element and its lip (90° for a right-angle lip) (Fig. 10.4e)

The effective width, b, of the compression flange can be determined from Eqs. (10.3) to (10.6) with k calculated from the following equations for single lip edge stiffener having $(140° \geq \theta \geq 40°)$:

For $D/w \leq 0.25$, $k = 3.57(R_I)^n + 0.43 \leq 4$

$$(10.16a)$$

For $0.25 < D/w \leq 0.8$,

$$k = \left(4.82 - \frac{5D}{w} \right)(R_I)^n + 0.43 \leq 4 \qquad (10.16b)$$

where $n = \left[0.582 - \frac{w/t}{4S} \right] \geq \frac{1}{3} \qquad (10.16c)$

$$R_I = I_s/I_a \leq 1 \qquad (10.16d)$$

The values of b_1 and b_2, as shown in Fig. 10.4e, can be computed as follows:

$$b_1 = \frac{b}{2}(R_l)$$

$$b_2 = b - b_1$$

The effective width b depends on the actual stress f, which, in turn, is determined by reduced section properties that are a function of effective width. Employment of successive approximations consequently may be necessary in using these equations. This can be avoided and the correct values of b/t obtained directly from the formulas when f is known or is held to a specified maximum value. This is true, though, only when the neutral axis of the section is closer to the tension flange than to the compression flange, so that compression controls. The latter condition holds for symmetrical channels, Z's, and I sections used as flexural members about their major axis, such as Fig. 10.1e, f, k, and n. For wide, inverted, pan-shaped sections, such as deck and panel sections, a somewhat more accurate determination, using successive approximations, is necessary.

For computation of moment of inertia for deflection or stiffness calculations, properties of the full unreduced section can be used without significant error when w/t of the compression elements does not exceed 60. For greater accuracy, use Eqs. (10.7) and (10.8) to obtain effective widths.

Example ▪ As an example of effective-width determination, consider the hat section in Fig. 10.6. The section is to be made of steel with a specified minimum yield point of $F_y = 33$ ksi. It is to be used as a simply supported beam with the top flange in compression. Safe load-carrying capacity is to be computed. Because the compression and tension flanges have the same width, $f = 33$ ksi is used to compute b/t.

The top flange is a stiffened compression element 3 in wide. If the thickness is $\frac{1}{16}$ in, then the flat-width ratio is 48 ($> 220/\sqrt{f}$) and Eq. (10.8) applies. For this value of w/t and $f = 33$ ksi, Eq. (10.8) or Fig. 10.5 gives b/t as 41. Thus, only 85% of the top-flange flat width can be considered effective in this case. The neutral axis of the section will lie below the horizontal center line, and compression will control. In this case, the assumption that $f = F_y = 33$ ksi, made at the start, controls maximum stress, and b/t

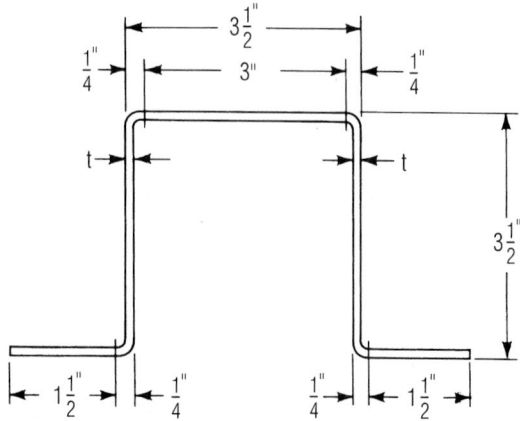

Fig. 10.6 Hat section.

can be determined directly from Eq. (10.8), without successive approximations.

For a wide hat section in which the horizontal centroidal axis is nearer the compression than the tension flange, the stress in the tension flange controls. So determination of unit stress and effective width of the compression flange requires successive approximations.

("Cold-Formed Steel Design Manual," American Iron and Steel Institute, Washington, D.C., 2002 Edition.)

10.8 Maximum Flat-Width Ratios for Cold-Formed Elements

When the flat-width ratio exceeds about 30 for an unstiffened element and 250 for a stiffened element, noticeable buckling of the element may develop at relatively low stresses. Present practice is to permit buckles to develop in the sheet and take advantage of what is known as the postbuckling strength of the section. The effective-width formulas [Eqs. (10.3), (10.6), (10.7), and (10.8)] are based on this practice of permitting some incipient buckling to occur at the allowable stress. To avoid intolerable deformations, however, overall flat-width ratios, disregarding intermediate stiffeners and based on the actual thickness of the element, should not exceed the following:

Stiffened compression element having one longitudinal edge connected to a web or flange, the other to a simple lip	60
Stiffened compression element having one longitudinal edge connected to a web or flange, the other stiffened by any other kind of stiffener	90
Stiffened compression element with both longitudinal edges connected to a web or flange element, such as in a hat, U, or box type of section	500
Unstiffened compression element	60

10.9 Beam Design Considerations

For the design of beams, considerations should be given to (a) bending strength and deflection, (b) web strength for shear, combined bending and shear, web crippling, and combined bending and web crippling, (c) bracing requirements, (d) shear lag, and (e) flange curling.

Based on the AISI ASD method, the required bending moment computed from working loads shall not exceed the allowable design moment determined by dividing the nominal bending moment by a factor of safety. For *laterally supported beams*, the nominal bending moment is based on the nominal section strength calculated on the basis of either (a) initiation of yielding in the effective section or (b) the inelastic reserve capacity in accordance with the AISI Specification. The factor of safety for bending is taken as 1.67.

10.10 Laterally Unsupported Cold-Formed Beams

In the relatively infrequent cases in which cold-formed sections used as beams are not laterally supported at frequent intervals, the strength must be reduced to avoid failure from lateral instability. The amount of reduction depends on the shape and proportions of the section and the spacing of lateral supports. This is not a difficult obstacle. (For details, see the AISI "North American Specification for the Design of Cold-Formed Steel Structural Members," 2001.)

Because of the torsional flexibility of cold-formed channel and Z sections, their use as beams without lateral support is not recommended. When one flange is connected to a deck or sheathing material, the nominal flexural strength of the member can be determined in accordance with the AISI specification.

When laterally unsupported beams must be used, or where lateral buckling of a flexural member is likely to be a problem, consideration should be given to the use of relatively bulky sections that have two webs, such as hat or box sections (Fig. 10.1o and p).

10.11 Allowable Shear Strength and Web Crippling Strength in Webs

The shear force at any section should not exceed the allowable shear V_a, kips, calculated as follows:

1. For $h/t \leq 1.510\sqrt{Ek_v/F_y}$,

$$V_a = 0.375t^2\sqrt{K_vF_yE} \leq 0.375F_yht \qquad (10.17a)$$

2. For $h/t > 1.510\sqrt{Ek_v/F_y}$,

$$V_a = 0.565\frac{Ek_vt^3}{h} \qquad (10.17b)$$

where t = web thickness, in

h = depth of the flat portion of the web measured along the plane of the web, in

k_v = shear buckling coefficient = 5.34 for unreinforced webs for which $(h/t)_{max}$ does not exceed 200

F_y = design yield stress, ksi

E = modulus of elasticity = 29,500 ksi

For design of reinforced webs, especially when h/t exceeds 200, see AISI "North American Specification for the Design of Cold-Formed Steel Structural Members," 2001.

For a web consisting of two or more sheets, each sheet should be considered a separate element carrying its share of the shear force.

For beams with unreinforced webs, the moment M, and the shear V, should satisfy the following interaction equation:

$$\left(\frac{M}{M_{axo}}\right)^2 + \left(\frac{V}{V_a}\right)^2 \le 1.0 \qquad (10.18)$$

where M_{axo} = allowable moment about the centroidal axis, in-kips

V_a = allowable shear force when shear alone exists, kips

M = applied bending moment, in-kips

V = actual shear load, kips

For beams with reinforced webs, the interaction equation for combined bending and shear is given in the AISI North American Specification.

In addition to the design for shear strength of beam webs, consideration should also be given to the web crippling strength and combined bending and web crippling strength as necessary. The web crippling strength depends on several parameters including $h/t, N/t, R/t, F_y, t$, and the angle between the plane of the web and the plane of the bearing surface. In the above ratios, N is the actual bearing length and R is the inside bend radius. Other symbols were defined previously.

The 2001 edition of the AISI North American Specification includes the following equation for determining the nominal web crippling strength of webs without holes:

$$P_n = Ct^2F_y \sin\theta\left(1 - C_R\sqrt{\frac{R}{t}}\right)$$
$$\times \left(1 + C_N\sqrt{\frac{N}{t}}\right)\left(1 - C_h\sqrt{\frac{h}{t}}\right) \qquad (10.19)$$

In the above equation, coefficients C, C_h, C_N, and C_R together with factors of safety are given in the Specification for built-up sections, single web channel and C-sections, single web Z-sections, single hat sections, and multi-web deck sections under different support and loading conditions. For beam webs with holes, the web crippling strength should be multiplied by the reduction factor, R_c. In addition, the AISI Specification provides interaction equations for combined bending and web crippling strength.

10.12 Concentrically Loaded Compression Members

The following applies to members in which the resultant of all loads acting on the member is an axial load passing though the centroid of the effective section calculated for the nominal buckling stress F_n, ksi. The axial load should not exceed P_a calculated as follows:

$$P_a = \frac{P_n}{\Omega_c} \qquad (10.20)$$

$$P_n = A_eF_n \qquad (10.21)$$

where P_a = allowable compression load, kips

P_n = ultimate compression load, kips

Ω_c = factor of safety for axial compression = 1.80

A_e = effective area at stress F_n, in^2

The magnitude of F_n is determined as follows, ksi:

$$\text{For } \lambda_c \le 1.5, \quad F_n = (0.658^{\lambda_c^2})F_y \qquad (10.22)$$

$$\text{For } \lambda_c > 1.5, \quad F_n = \left[\frac{0.877}{\lambda_c^2}\right]F_y \qquad (10.23)$$

where $\lambda_c = \sqrt{F_y/F_e}$

F_y = yield stress of the steel, ksi

F_e = the least of the elastic flexural, torsional and torsional-flexural buckling stress

Figure 10.7 shows the ratio between the column buckling stress F_n and the yield strength F_y.

For the elastic flexural mode,

$$F_e = \frac{\pi^2E}{(KL/r)^2} \qquad (10.24)$$

where K = effective-length factor

L = unbraced length of member, in

r = radius of gyration of full, unreduced cross section, in

E = modulus of elasticity, ksi

Moreover, non-compact angle sections should be designed for the applied axial load P acting simultaneously with a moment equal to $PL/1000$ applied about the minor principal axis causing compression in the tips of the angle legs.

The slenderness ratio KL/r of all compression members preferably should not exceed 200 except that, during construction only, KL/r preferably should not exceed 300.

For treatment of open cross sections which may be subject to torsional-flexural buckling, refer to

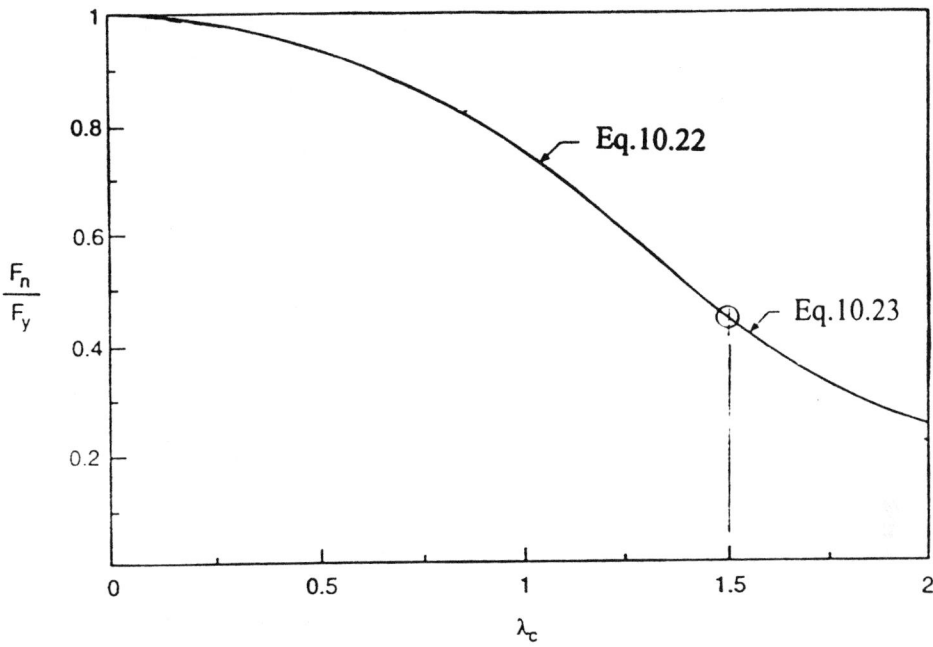

Fig. 10.7 Ratio of nominal column buckling stress to yield strength.

AISI "North American Specification for the Design of Cold-Formed Steel Structural Members," 2001.

10.13 Combined Axial and Bending Stresses

Combined axial and bending stresses in cold-formed sections can be handled in a similar way as for structural steel. The interaction criterion to be used is given in the AISI "North American Specification for the Design of Cold-Formed-Steel Structural Members," 2001.

10.14 Welding of Cold-Formed Steel

Welding offers important advantages to fabricators and erectors in joining metal structural components. Welded joints make possible continuous structures, with economy and speed in fabrication; 100% joint efficiencies are possible.

Conversion to welding of joints initially designed for mechanical fasteners is poor practice. Joints should be specifically designed for welding, to take full advantage of possible savings. Important considerations include the following: The overall assembly should be weldable, welds should be located so that notch effects are minimized, the final appearance of the structure should not suffer from unsightly welds, and welding should not be expected to correct poor fit-up.

Steels bearing protective coatings require special consideration. Surfaces precoated with paint or plastic are usually damaged by welding. And coatings may adversely affect weld quality. Metallically coated steels, such as galvanized (zinc-coated), aluminized, and terne-coated (lead-tin alloy), are now successfully welded using procedures tailored for the steel and its coating.

Generally, steel to be welded should be clean and free of oil, grease, paints, scale, and so on. Paint should be applied only after the welding operation.

("Welding Handbook," American Welding Society, 550 N.W. LeJeune Rd., Miami, FL 33135 www.aws.org; O. W. Blodgett, "Design of Weld-ments," James F. Lincoln Arc Welding Foundation, Cleveland, OH 44117 www.weldinginnovation. com.)

10.15 Arc Welding of Cold-Formed Steel

Arc welding may be done in the shop and in the field. The basic sheet-steel weld types are shown in Fig. 10.8. Factors favoring arc welding are porta-bility and versatility of equipment and freedom in joint design. (See also Art. 10.14.) Only one side of a joint need be accessible, and overlap of parts is not required if joint fit-up is good.

Distortion is a problem with lightweight steel weldments, but it can be minimized by avoiding overwelding. Weld sizes should be matched to service requirements.

Always design joints to minimize shrinking, warping, and twisting. Jigs and fixtures for holding lightweight work during welding should be used to control distortion. Directions and amounts of distortion can be predicted and sometimes counter-acted by preangling the parts. Discrete selection of welding sequence can also be used to control distortion.

Groove welds (made by butting the sheet edges together) can be designed for 100% joint efficiency. Calculations of design stress is usually unnecessary if the weld penetrates 100% of the section.

Stresses in fillet welds should be considered as shear on the throat for any direction of the applied stress. The dimension of the throat is calculated as 0.707 times the length of the shorter leg of the weld. For example, a 12-in-long, $\frac{1}{4}$-in fillet weld has a leg dimension of $\frac{1}{4}$ in, a throat of 0.177 in, and an equivalent area of 2.12 in^2. For all grades of steel, fillet and plug welds should be proportioned according to the AISI specification. For the allowable strength design method, the factors of safety for various weld types are given in the AISI North American Specification.

Shielded-metal-arc welding, also called man-ual stick electrode, is the most common arc welding process because of its versatility, but it calls for skilled operators. The welds can be made in any position. Vertical and overhead welding should be avoided when possible.

Gas-metal-arc welding uses special equipment to feed a continuous spool of bare or flux-cored wire into the arc. A shielding gas such as argon or carbon dioxide is used to protect the arc zone from the contaminating effects of the atmosphere. The process is relatively fast, and close control can be maintained over the deposit. The process is not

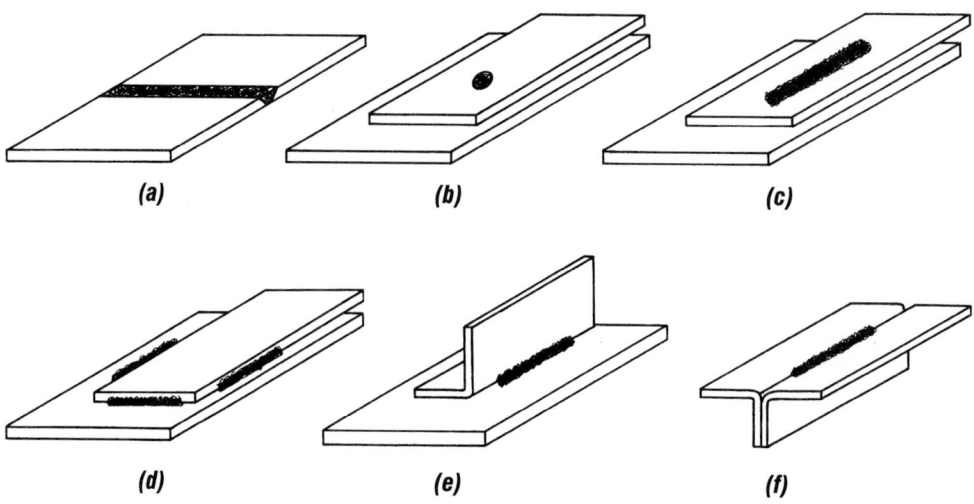

Fig. 10.8 Types of sheet-steel welds: (*a*) Square-groove weld; (*b*) arc spot weld (round puddle weld); (*c*) arc seam weld (oblong puddle weld); (*d*) fillet welds; (*e*) flare-bevel-groove weld; (*f*) flare-V-groove weld.

applicable to materials below $\frac{1}{32}$ in thick but is extensively used for thicker steels.

Gas-tungsten-arc welding operates by maintaining an arc between a nonconsumable tungsten electrode and the work. Filler metal may or may not be added. Close control over the weld can be maintained. This process is not widely used for high-production fabrication, except in specialized applications, because of higher cost.

One form of **arc spot welding** is an adaption of gas-metal-arc welding wherein a special welding torch and automatic timer are employed. The welding torch is positioned on the work and a weld is deposited by burning through the top component of the lap joint. The filler wire provides sufficient metal to fill the hole, thereby fusing together the two parts. Access to only one side of the joint is necessary. Field welding by unskilled operators often makes this process desirable.

Another form of arc spot welding utilizes gas-tungsten arc welding. The heat of the arc melts a spot through one of the sheets and partly through the second. When the arc is cut off, the pieces fuse. No filler metal is added. Design of arc-welded joints of sheet steel is fully treated in the American Welding Society "Structural Welding Code-Sheet Steel," AWS D1.3, www.aws.org. Allowable maximum-load capacities of arc-welded joints of sheet steel, including cold-formed members 0.180 in or less thick, are determined in the following ways.

Groove Welds in Butt Joints ▪ The maximum load for a groove weld in a butt joint, welded from one or both sides, is determined by the base steel with the lower strength in the connection, provided that an effective throat equal to or greater than the thickness of the material is consistently obtained.

Arc Spot Welds ▪ These are permitted for welding sheet steel to thicker supporting members in the flat position. Arc spot welds (puddle welds) may not be made on steel where the thinnest connected part is over 0.15 in thick, nor through a combination of steel sheets having a total thickness of over 0.15 in. Arc spot welds should be specified by minimum effective diameter of fused area d_e. Minimum effective allowable diameter is $\frac{3}{8}$ in. The nominal shear load P_n, on each arc spot weld between two or more sheets and a supporting member should not exceed the smaller of the

values calculated from Eq. (10.25) or, as appropriate, Eqs. (10.26), (10.27), (10.28).

$$P_n = 0.589 d_e^2 F_{xx} \qquad (10.25)$$

For $d_a/t \le 0.815\sqrt{E/F_u}$:

$$P_n = 2.20 t d_a F_u \qquad (10.26)$$

For $0.815\sqrt{E/F_u} < d_a/t < 1.397\sqrt{E/F_u}$:

$$P_n = 0.280\left[1 + 5.59\frac{t}{d_a}\sqrt{\frac{E}{F_u}}\right]t d_a F_u \qquad (10.27)$$

For $d_a/t \ge 1.397\sqrt{E/F_u}$:

$$P_n = 1.40 t d_a F_u \qquad (10.28)$$

where t = sum of thicknesses, in (exclusive of coatings), of all the sheets involved in shear transfer through the spot weld

d_a = average diameter, in, of spot weld at middepth of the shear transfer zone

= $d - t$ for a single sheet or multiple sheets (not more than four lapped sheets over a supporting member)

d = visible diameter, in, of outer surface of spot weld

d_e = effective diameter, in, of fused area

= $0.7d - 1.5t$ but not more than $0.55d$

F_{xx} = stress-level designation, ksi, in AWS electrode classification

F_u = tensile strength of base metal as specified, ksi

The distance measured in the line of force from the centerline of a weld to the nearest edge of an adjacent weld or to the end of the connected part toward which the force is directed should not be less than the value of e_{min} as given by

$$e_{min} = e\Omega_e \qquad (10.29)$$

where $e = P/(F_u t)$

Ω_e = factor of safety for sheet tearing

= 2.20 when $F_u/F_{sy} \ge 1.08$

= 2.55 when $F_u/F_{sy} < 1.08$

F_u = tensile strength of base metal as specified, ksi

P = force transmitted by weld, kips

t = total combined base steel thickness, in

In addition, the distance from the centerline of any weld to the end or boundary of the connected member may not be less than $1.5d$. In no case may the clear distance between welds and the end of the member be less than d.

The nominal tension load P_n on each arc spot weld between sheet and supporting member should be computed as the smaller of either:

$$P_n = 0.785d_e^2 F_{xx} \quad (10.30a)$$

or

$$P_n = 0.8\left(\frac{F_u}{F_y}\right)^2 td_a F_u \quad (10.30b)$$

and the following limitations also apply:

$$td_a F_u \le 3; \; e_{min} \ge d; \; F_{xx} \ge 60\,\text{ksi};$$
$$F_u \le 82\,\text{ksi}; \; F_{xx} > F_u$$

If it can be shown by measurement that a given weld procedure will consistently give a larger effective diameter d_e, or larger average diameter d_a, as applicable, this larger diameter may be used, if the welding procedure required for making those welds is followed.

Arc Seam Welds ▪ These apply to the following joints:

1. Sheet to thicker supporting member in the flat position

2. Sheet to sheet in the horizontal or flat position

The nominal shear load P_n on each arc seam weld should not exceed the values calculated from either Eq. (10.31) or (10.32).

$$P_n = \left[\frac{\pi d_e^2}{4} + Ld_e\right]0.75F_{xx} \quad (10.31)$$

$$P_n = 2.5tF_u(0.25L + 0.96d_a) \quad (10.32)$$

where d = width of arc seam weld, in

L = length of seam weld not including the circular ends, in (For computation purposes, L should not exceed $3d$)

d_a = average width of arc seam weld, in

= $d - t$ for a single sheet or double sheets

d_e = effective width of arc seam weld at fused surfaces, in

= $0.7d - 1.5t$

F_u and F_{xx} are strengths as previously defined for arc spot welds. Also, minimum edge distance is the same as that defined for arc spot welds. If it can be shown by measurement that a given weld procedure will consistently give a larger effective width d_e or larger average width, d_a, as applicable, this value may be used, if the welding procedure required for making the welds that were measured is followed.

Fillet Welds ▪ These may be used for welding of joints in any position, either sheet to sheet or sheet to thicker steel member. The nominal shear load P_n, kips, on a fillet weld in lap or T joints should not exceed the following:

For Longitudinal Loading
For $L/t < 25$:

$$P_n = \left(1 - \frac{0.01L}{t}\right)tLF_u \quad (10.33)$$

For $L/t \ge 25$:

$$P_n = 0.75tLF_u \quad (10.34)$$

For Transverse Loading

$$P_n = tLF_u \quad (10.35)$$

where t = least thickness of sheets being fillet welded, in

L = length of fillet weld, in

In addition, for $t > 0.10$ in, the nominal load for a fillet weld in lap and T joints should not exceed

$$P_n = 0.75t_w LF_{xx} \quad (10.36)$$

where t_w = effective throat, in, = lesser of $0.707w_1$ or $0.707w_2$; w_1 and w_2 are the width of the weld legs; and F_u and F_{xx} are strengths as previously defined.

Flare-Groove Welds ▪ These may be used for welding of joints in any position, either:

1. Sheet to sheet for flare-V-groove welds

2. Sheet to sheet for flare-bevel-groove welds

3. Sheet to thicker steel member for flare-bevel-groove welds

The nominal shear load, P_n, kips, on a weld is governed by the thickness, t, in, of the sheet steel adjacent to the weld.

For flare-bevel-groove welds, the transverse load should not exceed

$$P_n = 0.833tLF_u \qquad (10.37)$$

For flare-V-groove welds, when the effective throat t_w is equal to or greater than the least thickness t of the sheets being joined but less than $2t$, or if the lip height is less than the weld length L, in, the longitudinal loading should not exceed

$$P_n = 0.75tLF_u \qquad (10.38)$$

If t_w is equal to or greater than $2t$ and the lip height is equal to or greater than L,

$$P_n = 1.50tLF_u \qquad (10.39)$$

In addition, if $t > 0.10$ in

$$P_n = 0.75t_wLF_{xx} \qquad (10.40)$$

10.16 Resistance Welding of Cold-Formed Steel

Resistance welding comprises a group of welding processes wherein coalescence is produced by the heat obtained from resistance of the work to flow of electric current in a circuit of which the work is a part and by the application of pressure. Because of the size of the equipment required, resistance welding is essentially a shop process. Speed and low cost are factors favoring its selection.

Almost all resistance-welding processes require a lap-type joint. The amount of contacting overlap varies from $\frac{3}{8}$ to 1 in, depending on sheet thickness. Access to both sides of the joint is normally required. Adequate clearance for electrodes and welder arms must be provided.

Spot welding is the most common resistance-welding process. The work is held under pressure between two electrodes through which an electric current passes. A weld is formed at the interface between the pieces being joined and consists of a cast-steel nugget. The nugget has a diameter about equal to that of the electrode face and should penetrate about 60 to 80% of each sheet thickness.

For structural design purposes, spot welding can be treated the same way as rivets, except that no reduction in net section due to holes need be made. Table 10.1 gives the essential information for uncoated material based on "Recommended Practices for Resistance Welding," American Welding Society. Note that the thickest material

for plain spot welding is $\frac{1}{8}$ in. Thicker material can be resistance-welded by projection or by pulsation methods if high-capacity spot welders for material thicker than $\frac{1}{8}$ in are not available.

Projection welding is a form of spot welding in which the effects of current and pressure are intensified by concentrating them in small areas of projections embossed in the sheet to be welded. Thus, satisfactory resistance welds can be made on thicker material using spot welders ordinarily limited to thinner stocks.

Pulsation welding, or **multiple-impulse welding**, is the making of spot welds with more than one impulse of current, a maneuver that makes some spot welders useful for thicker materials. The trade-offs influencing choice between projection welding and impulse welding involve the work being produced, volume of output, and equipment available.

The spot welding of higher-strength steels than those contemplated under Table 10.1 may require special welding conditions to develop the higher shear strengths of which the higher-strength steels are capable. All steels used for spot welding should be free of scale; therefore, either hot-rolled and pickled or cold-rolled steels are usually specified. Steels containing more than 0.15% carbon are not as readily spot welded as lower-carbon steels, unless special techniques are used to ensure ductile welds. However, high-carbon steels such as ASTM A653, SS Grade 50 (formerly, Grade D), which can have a carbon content as high as 0.40% by heat analysis, are not recommended for resistance welding. Designers should resort to other means of joining such steels.

Maintenance of sufficient overlaps in detailing spot-welded joints is important to ensure consistent weld strengths and minimum distortions at joints. Minimum weld spacings specified in Table 10.1 should be observed, or shunting to previously made adjacent welds may reduce the electric current to a level below that needed for welds being made. Also, the joint design should provide sufficient clearance between electrodes and work to prevent short-circuiting of current needed to make satisfactory spot welds. For design purposes, the AISI North American Specification provides design equations and a factor of safety on the basis of "Recommended Practices for Resistance Welding of Coated Low-Carbon Steel," American Welding Society, 550 N.W. LeJeune Rd., Miami, FL 33135, www.aws.org.

Table 10.1 Test Data for Spot and Projection Welding

Thickness t of Thinnest Piece, in	Min OD of Electrode D, in	Min Contacting Overlap, in	Min Weld Spacing c to c, in	Approx Dia of Fused Zone, in	Min Shear Strength per Weld, lb	Dia of Projection D, in
	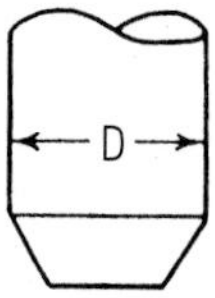	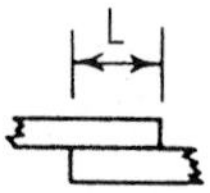		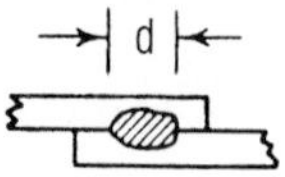		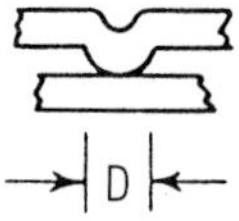
Spot Welding						
0.021	⅜	⁷⁄₁₆	⅜	0.13	320	
0.031	⅜	⁷⁄₁₆	½	0.16	570	
0.040	½	½	¾	0.19	920	
0.050	½	⁹⁄₁₆	⅞	0.22	1,350	
0.062	½	⅝	1	0.25	1,850	
0.078	⅝	¹¹⁄₁₆	1¼	0.29	2,700	
0.094	⅝	¾	1½	0.31	3,450	
0.109	⅝	¹³⁄₁₆	1⅝	0.32	4,150	
0.125	⅞	⅞	1¾	0.33	5,000	
Projection Welding						
0.125	¹¹⁄₁₆	⁹⁄₁₆	0.338	4,800	0.281	
0.140	¾	⅜	⁷⁄₁₆	6,000	0.312	
0.156	¹³⁄₁₆	¹¹⁄₁₆	½	7,500	0.343	
0.171	⅞	¾	⁹⁄₁₆	8,500	0.375	
0.187	¹⁵⁄₁₆	¹³⁄₁₆	⁹⁄₁₆	10,000	0.406	

10.17 Bolting of Cold-Formed-Steel Members

Bolting is convenient in cold-formed-steel construction. Bolts, nuts, and washers should generally conform to the requirements of the ASTM specifications listed in Table 10.2.

Maximum sizes permitted for bolt holes are given in Table 10.3. Holes for bolts may be standard or oversized round or slotted. Standard holes should be used in bolted connections when possible. The length of slotted holes should be normal to the direction of shear load. Washers should be installed over oversized or slotted holes.

Hole Locations ▪ The distance e, measured in the line of force from the center of a standard hole to the nearest edge of an adjacent hole or to the end of the connected part toward which the force is directed, should not be less than the value of e_{min}

Table 10.2 ASTM Bolt, Nut, and Washer Steels

A194	Carbon and Alloy Steel Nuts for High-Pressure and High-Temperature Service
A307	(Type A) Carbon Steel Bolts and Studs
A325	High Strength Bolts for Structural Steel Joints
A354	(Grade BD) Quenched and Tempered Alloy Steel Bolts, Studs, and Other Externally Threaded Fasteners (for diameter of bolt smaller than $\frac{1}{2}$ in)
A449	Quenched and Tempered Steel Bolts and Studs (for diameter of bolt smaller than $\frac{1}{2}$ in)
A490	Heat-Treated Steel Structural Bolts
A563	Carbon and Alloy Steel Nuts
F436	Hardened Steel Washers
F844	Washers, Steel, Plain (Flat), Unhardened for General Use
F959	Compressible Washer-Type Direct Tension Indicators for Use with Structural Fasteners

determined by Eq. (10.41),

$$e_{\min} = e\Omega_e \qquad (10.41)$$

where $e = \dfrac{P}{F_u t}$ (10.42)

Ω_e = factor of safety for sheet tearing

    = 2.00 when $F_u/F_{sy} \geq 1.08$

    = 2.22 when $F_u/F_{sy} < 1.08$

P = force transmitted by bolt, kips

t = thickness of thinnest connected part, in

F_u = tensile strength of connected part, ksi

F_{sy} = yield strength of connected part, ksi

In addition, the minimum distance between centers of bolt holes should provide sufficient clearance for bolt heads, nuts, washers, and the wrench but not less than three times the nominal bolt diameter d. The distance from the center of any standard hole to the end or boundary of the connecting member should not be less than $1\frac{1}{2} d$.

Allowable Tension ■ The tension force on the net sectional area A_n of a bolted connection should not exceed P_a calculated from Eq. (10.43).

$$P_a = \frac{P_n}{\Omega_t} \qquad (10.43)$$

where $P_n = A_n F_t$ (10.44)

    F_t = nominal limit for tension stress on net section, ksi

F_t and Ω_t are determined as follows:

1. When $t \geq \frac{3}{16}$ in, as required by the AISC Specification.

2. When $t < \frac{3}{16}$ in, the tensile capacity of a bolted member should be determined from Section C2 of the AISI North American Specification. For fracture in the effective net section of flat sheet connections having washers provided under the bolt head and the nut, the tensile stress F_t can be computed as follows:

   a. For a single bolt or a single row of bolts perpendicular to the force,

   $$F_t = \left(0.1 + \frac{3d}{s}\right)F_u \leq F_u \qquad (10.45a)$$

   b. For multiple bolts in the line parallel to the force,

   $$F_t = F_u \qquad (10.45b)$$

where Ω_t = factor of safety for tension on the net section

Table 10.3 Maximum Size of Bolt Holes, in

Nominal Bolt Dia, d, in	Standard Hole Dia, d, in	Oversized Hole Dia, d, in	Short-Slotted Hole Dimensions, in	Long-Slotted Hole Dimensions, in
$<\frac{1}{2}$	$d + \frac{1}{32}$	$d + \frac{1}{16}$	$(d + \frac{1}{32}) \times (d + \frac{1}{4})$	$(d + \frac{1}{32}) \times (2\frac{1}{2} d)$
$\geq\frac{1}{2}$	$d + \frac{1}{16}$	$d + \frac{1}{8}$	$(d + \frac{1}{16}) \times (d + \frac{1}{4})$	$(d + \frac{1}{16}) \times (2\frac{1}{2} d)$

= 2.22 for single shear and 2.00 for double shear

d = nominal bolt diameter, in

s = sheet width divided by number of bolt holes in cross section, in

F_u = tensile strength of the connected part, ksi

When washers are not provided under the bolt head and nut, see AISI Specification. The Specification also provides the design information for flat sheet connections having staggered hole patterns and structural members such as angles and channels.

Table 10.4b Modification Factor, m_f, for Type of Bearing Connection

Type of Bearing Connection	m_f
Single Shear and Outside Sheets of Double Shear Connection with Washers under Both Bolt Head and Nut	1.00
Single Shear and Outside Sheets of Double Shear Connection without Washers under Both Bolt Head and Nut, Or with only One Washer	0.75
Inside Sheet of Double Shear Connection with or without Washers	1.33

Allowable Bearing ■ The bearing force should not exceed P_a calculated from Eq. (10.46).

$$P_a = \frac{P_n}{\Omega_b} \qquad (10.46)$$

where $P_n = m_f C d t F_u$, kips $\qquad (10.47)$

Ω_b = factor of safety for bearing = 2.50

C = bearing factor determined from Table 10.4a

d = nominal bolt diameter, in

t = uncoated sheet thickness, in

F_u = tensile strength of sheet, ksi

m_f = modification factor determined from Table 10.4b

Allowable Bolt Sresses ■ Table 10.5 lists nominal shear and tension for various grades of bolts. The bolt force resulting in shear, tension, or combination of shear and tension should not exceed allowable bolt force P_a calculated from

Eq. (10.48).

$$P_a = \frac{A_b F}{\Omega} \qquad (10.48)$$

where A_b = gross cross-sectional area of bolt, in^2

F = nominal unit stress given by F_{nv}, F_{nt} or F'_{nt} in Tables 10.5 and 10.6

Factors of safety given in Tables 10.5 and 10.6 should be used to compute allowable loads on bolted joints.

Table 10.6 lists nominal tension stresses for bolts subject to the combination of shear and tension.

Example—Tension Joints with Two Bolts ■ Assume that the bolted tension joints of Fig. 10.9 comprise two sheets of $\frac{3}{16}$-in-thick, A1008 SS Grade 33 steel. For this steel, $F_y = 33$ ksi and $F_u = 48$ ksi. The sheets in each joint are 4 in wide and are connected to two $\frac{5}{8}$-in-diameter, A325 bolts, with washers under both bolt head and nut. Determine the allowable load based on the ASD method.

A. *Based on Tensile Strength of Steel Sheets*

Case 1 shows the two bolts arranged in a single transverse row. A force $T/2$ is applied to each bolt and the total force T has to be carried by the net section of each sheet through the bolts. So, in Eq. (10.45a), spacing of the bolts $s = 2$ in and $d/s = (\frac{5}{8})/2 = 0.312$. The tension stress in the net section,

Table 10.4a Bearing Factor, C

Thickness of Connected Part, t, in	Ratio of Fastener Diameter to Member Thickness, d/t	C
	$d/t < 10$	3.0
$0.024 \leq t < 0.1875$	$10 \leq d/t \leq 22$	$4 - 0.1(d/t)$
	$d/t > 22$	1.8

Table 10.5 Nominal Tensile and Shear Strength for Bolts

Description of Bolts	Tensile Strength		Shear Strength	
	Factor of Safety Ω	Nominal Stress F_{nt}, ksi	Factor of Safety Ω	Nominal Stress F_{nv}, ksi
A307 Bolts, Grade A, $\frac{1}{4}$ in $\leq d <\frac{1}{2}$ in.	2.25	40.5	2.4	24.0
A307 Bolts, Grade A, $d \geq\frac{1}{2}$ in	2.25	45.0		27.0
A325 bolt, when threads are not excluded from shear planes	2.0	90.0		54.0
A325 bolts, when threads are excluded from shear planes		90.0		72.0
A354 Grade BD Bolts $\frac{1}{4}$ in $\leq d <\frac{1}{2}$ in, when threads are not excluded from shear planes		101.0		59.0
A354 Grade BD Bolts $\frac{1}{4}$ in $\leq d <\frac{1}{2}$ in, when threads are excluded from shear planes		101.0		90.0
A449 Bolts, $\frac{1}{4}$ in $\leq d <\frac{1}{2}$ in, when threads are not excluded from shear planes		81.0		47.0
A449 Bolts, $\frac{1}{4}$ in $\leq d <\frac{1}{2}$ in, when threads are excluded from shear planes		81.0		72.0
A490 Bolts, when threads are not excluded from shear planes		112.5		67.5
A490 Bolts, when threads are excluded from shear planes		112.5		90.0

Table 10.6 Nominal Tension Stress, F'_{nt} (ksi), for Bolts Subject to the Combination of Shear and Tension

Description of Bolts	Threads Not Excluded from Shear Planes	Threads Excluded from Shear Planes	Factor of Safety Ω
A325 Bolts	$110-3.6f_v \leq 90$	$110-2.8f_v \leq 90$	2.0
A354 Grade BD Bolts	$122-3.6f_v \leq 101$	$122-2.8f_v \leq 101$	
A449 Bolts	$100-3.6f_v \leq 81$	$100-2.8f_v \leq 81$	
A490 Bolts	$136-3.6f_v \leq 112.5$	$136-2.8f_v \leq 112.5$	
A307 Bolts, Grade A			
When $\frac{1}{4}$ in $\leq d < \frac{1}{2}$ in	$52 - 4f_v \leq 40.5$		2.25
When $d \geq\frac{1}{2}$ in	$58.5 - 4f_v \leq 45$		

The shear stress, f_v, shall also satisfy Table 10.5.

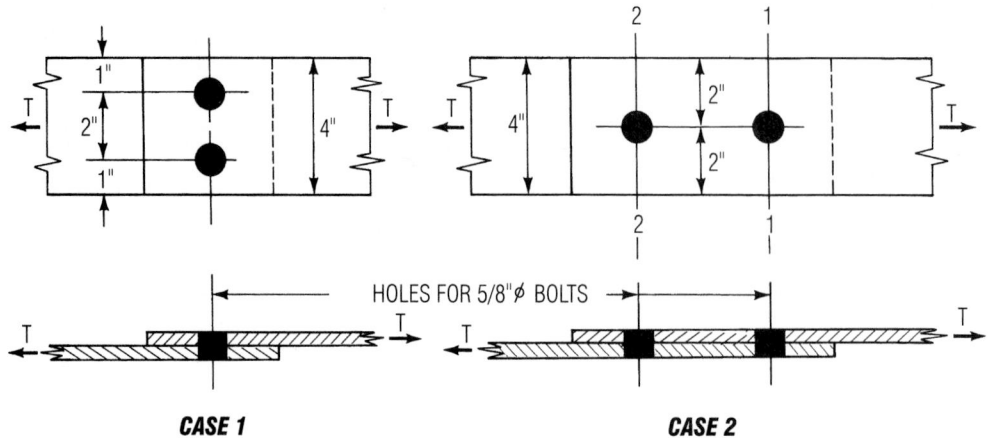

Fig. 10.9 Bolted connections with two bolts.

computed from Eq. (10.45a), is then

$$F_t = (0.1 + 3 \times 0.312)F_u = 1.04F_u > F_u$$

Use $F_t = F_u$.

Substitution in Eq. (10.44) with $F_u = 48$ ksi yields the nominal tension load on the net section:

$$P_n = [4 - (2 \times 11/16)] \times 3/16 \times 48 = 23.63 \text{ kips}$$

The allowable load is

$$P_a = \frac{P_n}{\Omega} = \frac{23.63}{2.22} = 10.64 \text{ kips}$$

This compares with the tensile strength of each sheet for tension member design according to Section C2 of Appendix A of the 2001 edition of the AISI North American Specification:

For yielding:

$$T_n = A_g F_y = (4 \times 3/16)(33) = 24.75 \text{ kips}$$

$$T_a = \frac{T_n}{\Omega} = \frac{24.75}{1.67} = 14.82 \text{ kips}$$

For fracture away from connection:

$$T_n = A_n F_u = [4 - (2 \times 11/16)] \times 3/16 \times 48$$
$$= 23.63 \text{ kips}$$

$$T_a = \frac{T_n}{\Omega} = \frac{23.63}{2.00} = 11.82 \text{ kips} < 14.82 \text{ kips}$$

Use $T_a = 11.82$ kips. Since $T_a > P_a$, use $P_a = 10.64$ kips for Case 1.

Case 2 shows the two bolts, with 4-in spacing, arranged in a single line along the direction of applied force. From Eq. (10.45b),

$$F_t = F_u \text{ and } P_n = A_n F_t$$

$$P_a = 10.64 \text{ kips (same as Case 1)}$$

Compare with the tensile strength for tension member design:

For yielding (same as Case 1):

$$T_a = 14.82 \text{ kips}$$

For fracture away from connection:

$$T_n = A_n F_u = (4 - 11/16) \times 3/16 \times 48 = 29.81 \text{ kips}$$
$$T_a = 29.81/2.00 = 14.91 \text{ kips} > 14.82 \text{ kips}$$

Use $T_a = 14.82$ kips. Since $T_a > P_a$, use $P_a = 10.64$ kips for Case 2.

B. *Check for Bearing Capacity*

From Eq. (10.47), the bearing strength P_n per bolt of the $\frac{3}{16}$-in-thick steel sheet is:

$$P_n = m_f C d t F_u$$

Since $d/t = (5/8)/(3/16) = 3.33 < 10$, $C = 3.0$. For single shear connection with washers under both bolt head and nut, $m_f = 1.00$. Therefore,

$$P_n = 1 \times 3 \times 5/8 \times 3/16 \times 48 = 16.88 \text{ kips}$$

The allowable bearing load for two bolts:

$$P_a = 2\frac{P_n}{\Omega} = 2 \times \frac{16.88}{2.50} = 13.50 \text{ kips}$$

$$> 10.64 \text{ kips O.K.}$$

C. *Check for Shear Strength of Bolts*

Using the A325 bolts with threads not excluded from the shear plane, the allowable shearing strength of each bolt is:

$$P_s = A_b \frac{F_{nv}}{\Omega} = (5/8)^2 \times 0.7854 \times \frac{54}{2.4} = 6.9 \text{ kips}$$

For two bolts, the allowable load is:

$$P_a = 2 \times 6.9 = 13.8 \text{ kips} > 10.64 \text{ kips O.K.}$$

D. *Bolt Spacing and Edge Distance*

From the above calculations, the allowable load for Cases 1 and 2 is 10.64 kips. The minimum distance between a bolt center and adjacent bolt edge or sheet edge in the direction of applied force for Cases 1 and 2 is:

$$e = \frac{P}{F_u t} = \frac{10.64/2}{(48)(3/16)} = 0.59 \text{ in}$$

$$e_{\min} = e\Omega = 0.59 \times 2 = 1.18 \text{ in}$$

The bolt spacing and edge distance should also be checked for other AISI dimensional requirements.

In addition to the above calculations, block shear rupture should also be considered according to the AISI North American Specification.

10.18 Tapping Screws for Joining Light-Gage Members

Tapping screws are often used for making field joints in lightweight construction, especially in connections that do not carry any calculated gravity load. Such screws are of several types (Fig. 10.10). Tapping screws used for fastening sheet-metal siding and roofing are generally preassembled with Neoprene washers for effective control of leaks, squeaks, cracks, or crazing, depending on the surface of the material. For best results, when Type A sheet-metal screws are specified, screws should be fully threaded to the head to assure maximum hold in sheet metal.

Tapping screws are made of steel so hardened that their threads form or cut mating threads in one or both relatively soft materials being joined. Slotted, hexagon, and plain heads are provided for installing them. The thread-forming types all

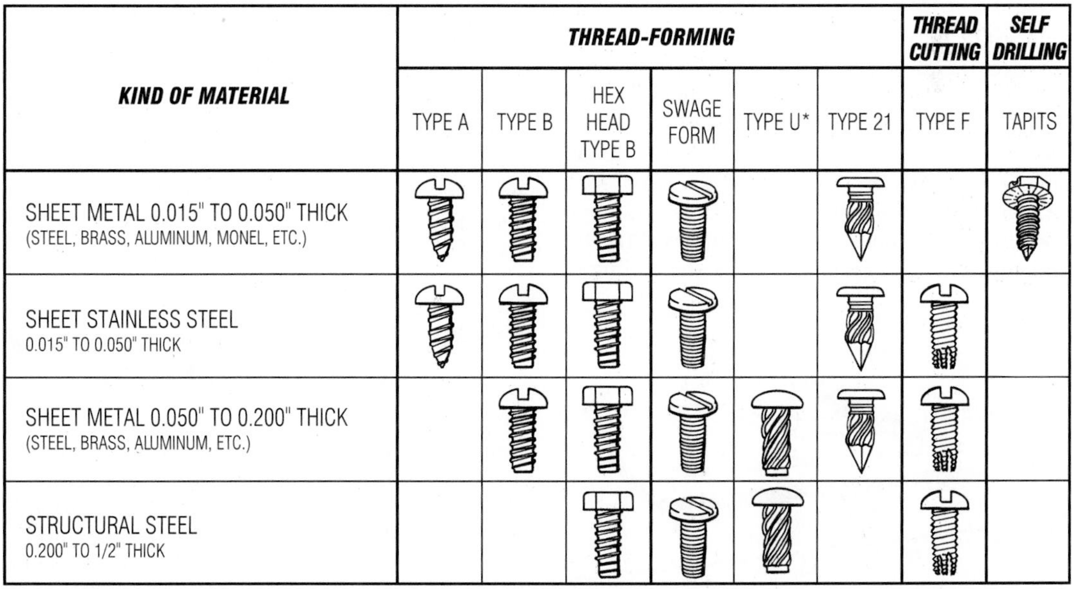

KIND OF MATERIAL	THREAD-FORMING						THREAD CUTTING	SELF DRILLING
	TYPE A	TYPE B	HEX HEAD TYPE B	SWAGE FORM	TYPE U*	TYPE 21	TYPE F	TAPITS
SHEET METAL 0.015" TO 0.050" THICK (STEEL, BRASS, ALUMINUM, MONEL, ETC.)								
SHEET STAINLESS STEEL 0.015" TO 0.050" THICK								
SHEET METAL 0.050" TO 0.200" THICK (STEEL, BRASS, ALUMINUM, ETC.)								
STRUCTURAL STEEL 0.200" TO 1/2" THICK								

Fig. 10.10 Tapping screws. Note: A blank space does not necessarily signify that the type of screw cannot be used for this purpose; it denotes that the type of self-tapping screw will not generally give the best results in the material. (*Parker-Kalon Corp., Emhart Corp.*)

require predrilled holes appropriate in diameter to the hardness and thickness of the material being joined. Types A and B are screwed, whereas types U and 21 are driven. Predrilled holes are required for thread-cutting Type F, but no hole is required for self-drilling TAPIT type.

Tapping screws may be used for light-duty connections, such as fastening bridging to sheet-metal joists and studs. Since 1996 the AISI Specification included design rules for determining nominal load for shear and tension. The factors of safety to be used for computing the allowable load is 3.0.

Steel Roof and Floor Deck

Steel roof deck consists of ribbed sheets with nesting or upstanding-seam joints designed for the support of roof loads between purlins or frames. A typical roof-deck assembly is shown in Fig. 10.11. The Steel Deck Institute, P.O. Box 25, Fox River Grove, IL 6002, www.sdi.org, has developed much useful information on steel roof-deck.

10.19 Types of Steel Roof Deck

As a result of the Steel Deck Institute's efforts to improve standardization, steel roof deck is now classified. All types consist of long, narrow sections with longitudinal ribs at least $1\frac{1}{2}$ in deep spaced about 6 in on centers. Other rib dimensions are shown in Fig. 10.12*a* to *c* for some standard styles. Such steel roof deck is commonly available in 24- and 30-in covering widths, but sometimes in 18- and 36-in widths, depending on

the manufacturer. Figure 10.12*d* and *e* shows full-width executions in cross section. Usual spans, which may be simple, two-span continuous, or three-span continuous, range from 4 to 10 ft. The SDI "Design Manual for Composite Decks, Form Decks, Roof Decks and Cellular Deck Floor Systems with Electrical Distribution" gives allowable total uniform loading (dead and live), lb/ft², for various gages, spans, and rib widths.

Some manufacturers make special long-span roof-deck sections, such as the 3-in-deep Type N roof deck shown in Fig. 10.13.

The weight of the steel roof deck shown in Fig. 10.12 varies, depending on rib dimensions and edge details. For structural design purposes, weights of 2.8, 2.1, and 1.7 lb/ft² can be used for the usual design thicknesses of 0.048, 0.036, and 0.030 in, respectively, for black steel in all rib widths, as commonly supplied.

Steel roof deck is usually made of structural-quality sheet or strip, either black or galvanized, ASTM A611, Grade C, D or E or A653 Structural Quality with a minimum yield strength of 33 ksi. Black steel is given a shop coat of priming paint by the roof-deck manufacturer. Galvanized steel may or may not be painted; if painted, it should first be bonderized to ensure paint adherence.

The thicknesses of steel commonly used are 0.048 and 0.036 in, although most building codes also permit 0.030-in-thick steel to be used.

SDI Design Manual includes "Recommendations for Site Storage and Erection," and also provides standard details for accessories. See also SDI "Manual of Construction with Steel Deck."

10.20 Load-Carrying Capacity of Steel Roof Deck

The Steel Deck Institute has adopted a set of basic design specifications, with limits on rib dimensions, as shown in Fig. 10.12*a* to *c*, to foster standardization of steel roof deck. This also has made possible publication by SDI of allowable uniform loading tables. These tables are based on section moduli and moments of inertia computed with effective-width procedures stipulated in the AISI "Specification for the Design of Cold-Formed Steel

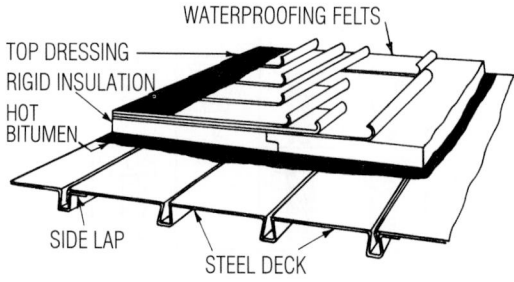

Fig. 10.11 Roof-deck assembly.

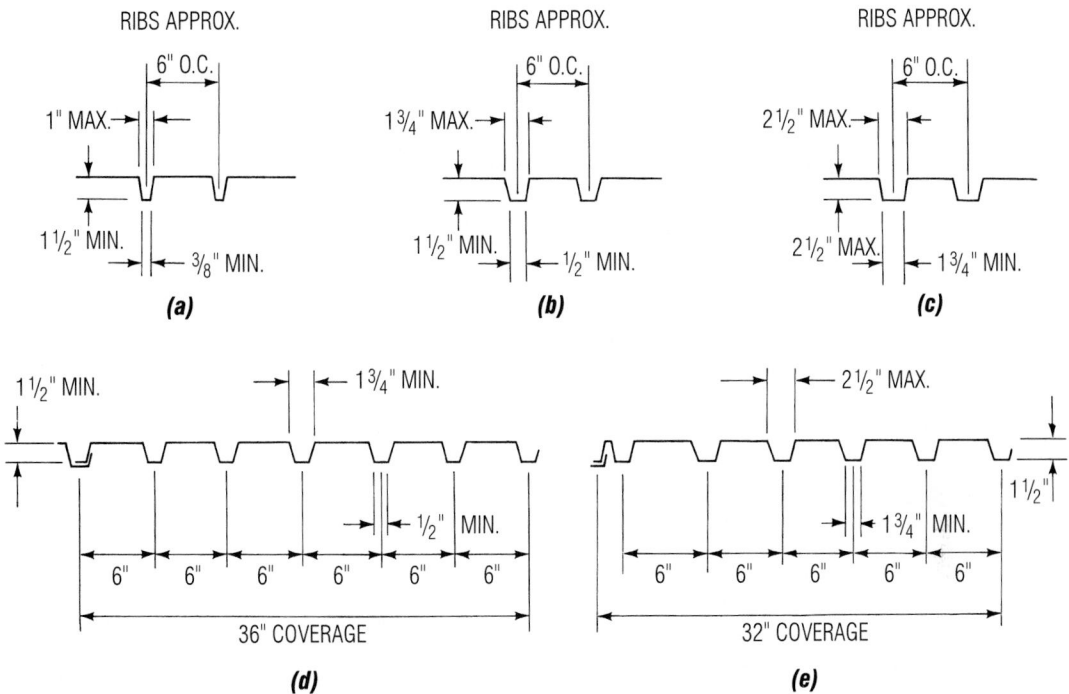

Fig. 10.12 Typical cold-formed-steel roof-deck sections: (*a*) Narrow-rib; (*b*) intermediate rib; (*c*) wide rib; (*d*) intermediate rib in 36-in-wide sheets with nested side laps; (*e*) wide rib in 32-in-wide sheets with upstanding seams.

Structural Members" (Art. 10.4). SDI has banned compression flange widths otherwise assumed to be effective. SDI "Basic Design Specifications" contain the following provisions:

Moment and Deflection Coefficients ■ Where steel roof decks are welded to the supports, a moment coefficient of $\frac{1}{10}$ (applied to WL) shall be used for three or more spans. Deflection coefficients of 0.0054 and 0.0069 (applied to WL^3/EI) shall be used for two span and three span, respectively. All other steel roof-deck installations shall be designed as simple spans, for which moment and deflection coefficients are $\frac{1}{8}$ and $\frac{5}{384}$, respectively.

Maximum Deflections ■ The deflection under live load shall not exceed $\frac{1}{240}$ of the clear span, center to center of supports. (Suspended ceiling, lighting fixtures, ducts, or other utilities shall not be supported by the roof deck.)

Anchorage ■ Steel roof deck shall be anchored to the supporting framework to resist the following gross uplifts:

45 lb/ft^2 for eave overhang

30 lb/ft^2 for all other roof areas

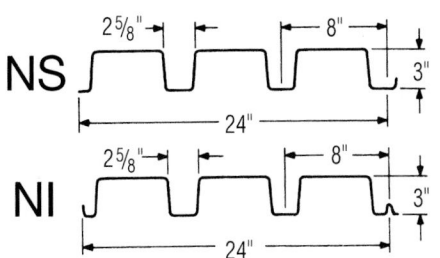

Fig. 10.13 Roof-deck cross sections types NS and NI of 9- to 15-ft spans.

The dead load of the roof-deck construction may be deducted from the above uplift forces.

Diaphragm Action ▪ Steel deck when properly attached to a structural frame becomes a diaphragm capable of resisting in-plane shear forces. A major SDI steel-deck diaphragm testing program at West Virginia University has led to tentative shear-design recommendations given in two publications that can be ordered from SDI. For design purposes, see SDI Diaphragm Design Manual.

10.21 Details and Accessories for Steel Roof Deck

In addition to the use of nesting or upstanding seams, most roof-deck sections are designed so that ends can be lapped shingle fashion.

Special ridge, valley, eave, and cant strips are provided by the roof-deck manufacturers.

Connections ▪ Roof decks are commonly arc welded to structural steel with puddle welds at least $\frac{1}{2}$ in in diameter or with elongated welds of equal perimeter. Electrodes should be selected and amperage adjusted to fuse all layers of roof deck to steel supporting members without creating blow-holes around the welds. Welding washers are recommended for thicknesses less than 0.030 in.

One-inch-long fillet welds should be used to connect lapped edges of roof deck.

Tapping screws are an alternative means of attaching steel roof deck to structural support members, which should be at least $\frac{1}{16}$ in thick. All edge ribs and a sufficient number of interior ribs should be connected to supporting frame members at intervals not exceeding 18 in. When standard steel roof-deck spans are 5 ft or more, adjacent sheets should be fastened together at midspan with either welds or screws. Details to be used depend on job circumstances and manufacturer's recommendations.

Insulation ▪ Although insulation is not ordinarily supplied by the roof-deck manufacturer, it is standard practice to install $\frac{3}{4}$- or 1-in-thick mineral fiberboard between roof deck and roofing. The Steel Deck Institute further recommends: All

steel decks shall be covered with a material of sufficient insulating value as to prevent condensation under normal occupancy conditions. Insulation shall be adequately attached to the steel deck by means of adhesives or mechanical fasteners. Insulation materials shall be protected from the elements at all times during storage and installation.

Fire Resistance ▪ The "Fire Resistance Directory," Underwriters' Laboratories Inc., 333 Pfingsten Rd., Northbrook, IL 60062, lists fire-resistance ratings for steel roof-deck construction. SDI Design Manual provides the UL Designs for 2-hour rating with directly-applied protection, 2-hour rating with metal lath and plaster ceiling, and 1-hour rating with suspended acoustical ceiling.

10.22 Composite Floor Deck

Research on the structural behavior of cold-formed-steel decks filled with concrete has demonstrated that composite action between these materials can be achieved in floors. Floor deck from one supplier is available in the thicknesses from 0.030 to 0.060 in and rib depths of $1\frac{1}{2}$, 2, and 3 in, with embossed surfaces for improved bonding with the concrete in-fill. Figure 10.14 shows three cross sections of composite floor deck.

10.23 Cellular Steel Floor and Roof Panels*

Several designs of cellular steel panels and fluted steel panels for floor and roof construction are shown in Fig. 10.15. One form of cellular steel floor for distribution of electrical wiring, telephone cables, and data cables is described in the following and illustrated in Fig. 10.16. This system is used in many kinds of structures, including massive high-rise buildings for institutional, business, and mercantile occupancies. It consists of profiled steel deck containing multiple wiring cells with structural concrete on top. The closely spaced, parallel, cellular raceways are connected

*Written by R. E. Albrecht.

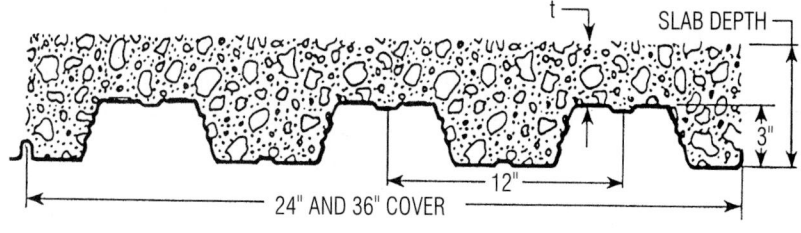

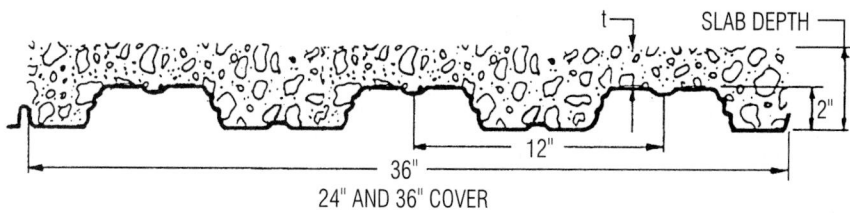

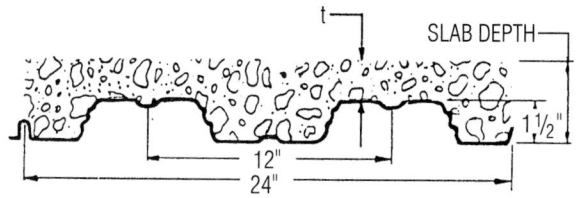

Fig. 10.14 Types of composite floor deck (LOK-FLOR, by United Steel Deck, Inc.).

to a header duct usually placed perpendicular to the cells. The header duct is equipped with a removable cover plate for lay-in wiring. On a repetitive module, the cellular raceways are assigned to electrical power, telephone, and data wiring. Preset inserts for activation of workstations may be installed at prescribed intervals, as close as 2 ft longitudinally and transversely.

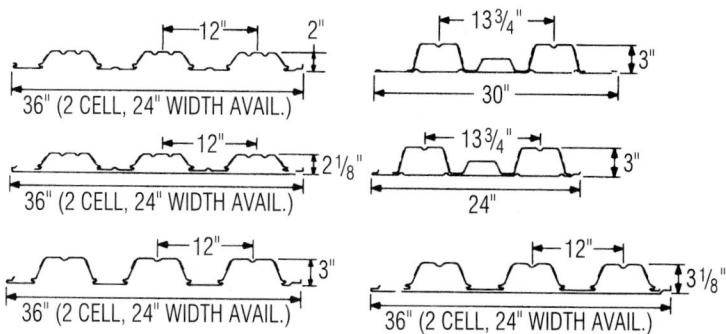

Fig. 10.15 Composite cellular and fluted steel floor sections. (*H. H. Robertson Co.*)

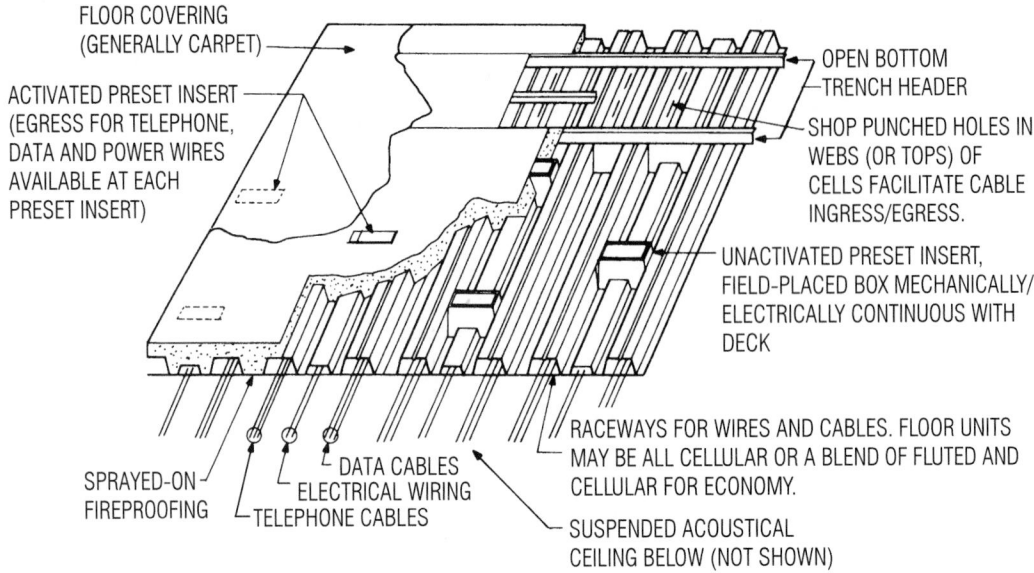

FLOOR COVERING
(GENERALLY CARPET)

ACTIVATED PRESET INSERT
(EGRESS FOR TELEPHONE,
DATA AND POWER WIRES
AVAILABLE AT EACH
PRESET INSERT)

OPEN BOTTOM
TRENCH HEADER

SHOP PUNCHED HOLES IN
WEBS (OR TOPS) OF
CELLS FACILITATE CABLE
INGRESS/EGRESS.

UNACTIVATED PRESET INSERT,
FIELD-PLACED BOX MECHANICALLY/
ELECTRICALLY CONTINUOUS WITH
DECK

RACEWAYS FOR WIRES AND CABLES. FLOOR UNITS
MAY BE ALL CELLULAR OR A BLEND OF FLUTED AND
CELLULAR FOR ECONOMY.

SPRAYED-ON
FIREPROOFING

DATA CABLES
ELECTRICAL WIRING
TELEPHONE CABLES

SUSPENDED ACOUSTICAL
CEILING BELOW (NOT SHOWN)

Fig. 10.16 Cellular steel floor raceway system. (*H. H. Robertson Co.*)

When an insert is activated at a workstation, connections for electrical power, telephone, and data are provided at one outlet.

Features ▪ During construction, the cellular steel floor decking serves as a working platform and as concrete forms. Afterward, the steel deck serves as the tensile reinforcement for the composite floor slab. The system also provides the required fire-resistive barrier between stories of the building.

Cellular steel floor raceway systems have many desirable features, including moderate first cost, flexibility in accommodating owners' needs (which lowers life-cycle costs), and minimal limitations on placement of outlets, which may be installed as close as 2 ft on centers in longitudinal and transverse directions. Physically, the wiring must penetrate the floor surface at outlet fittings. Therefore, the carpet (or other floor covering) has to be cut and a flap peeled back to expose each outlet. Use of carpet tiles rather than sheet carpet facilitates activation of preset inserts.

Where service outlets are not required to be as close as 2 ft, a blend of cellular and fluted floor sections may be used. For example, alternating 3-ft-wide fluted floor sections with 2-ft-wide cellular floor sections results in a module for service outlets

of 5 ft in the transverse direction and as close as 2 ft in the longitudinal direction. Other modules and spacings are also available.

Flexibility in meeting owners' requirements can be achieved with little or no change in required floor depth to accommodate the system. Service fittings may be flush with the floor or may project above the floor surface, depending on the owners' desires.

Specifications ▪ Cellular steel floor and roof sections (decking) usually are made of steel 0.030 in or more thick complying with requirements of ASTM A1008, SS Grade 33, for uncoated steel or ASTM A653, SS Grade 33, for galvanized steel, both having specified minimum yield strengths of 33 ksi. Steel for decking may be galvanized or painted.

Structural design of cold-formed-steel floor and roof panels is usually based on the American Iron and Steel Institute "Specification for the Design of Cold-Formed Steel Structural Members." Structural design of composite slabs incorporating cold-formed-steel floor and roof panels is usually based on the American Society of Civil Engineers "Standard for the Structural Design of Composite Slabs" and "Standard Practice for Construction and Inspection of Composite Slabs" (www.asce.org).

Details of design and installation vary with types of panels and manufacturers. For a specific installation, follow the manufacturer's recommendations.

Fire Resistance ■ Any desired degree of fire protection for cellular and fluted steel floor and roof assemblies can be obtained with concrete toppings and plaster ceilings or direct-application compounds (sprayed-on fireproofing). Fire-resistance ratings for many assemblies are available (Table 10.7). ("Fire-Resistant Steel-Frame Construction," American Institute of Steel Construction www.aisc.org; "Fire Resistance Directory," 1990, Underwriters' Laboratories, www.ul.com.)

Open-Web Steel Joists

As defined by the Steel Joist Institute, 3127 10th Avenue, North Ext., Myrtle Beach, SC 29577 (www.steeljoist.org), open-web steel joists are load-carrying members suitable for the direct support of floors and roof decks in buildings when these members are designed in accordance with SJI specifications and standard load tables.

As usually employed in floor construction, open-web steel joists support on top a slab of concrete, 2 to $2\frac{1}{2}$ in thick, placed on permanent forms (Fig. 10.17). In addition to light weight, one advantage claimed for open-web steel-joist construction is that the open-web system provides space for electrical work, ducts, and piping.

10.24 Joist Fabrication

Standardization under the specifications of the Steel Joist Institute consists of definition of product; specification of materials, design stresses, manufacturing features, accessories, and installation procedures; and handling and erection techniques. Most manufacturers have made uniform certain details, such as end depths, which are desirably standardized for interchangeability. Exact forms of the members, configuration of web systems, and methods of manufacture are matters for the individual manufacturers of these joists. A number of proprietary designs have been developed.

Open-web steel joists are different in one important respect from fabricated structural-steel framing members commonly used in building construction: The joists usually are manufactured by production line methods with special equipment designed to produce a uniform product. Components generally are joined by either resistance or electric-arc welding. Various joist designs are shown in Fig. 10.18.

K-series open-web joists are manufactured in standard depths from 8 to 30 in in 2-in increments and in different weights. The K series is designed with higher allowable stresses, for either high-strength, hot-rolled steel or cold-worked sections that utilize an increase in base-material yield point. Thus, such steel having a specified minimum yield point of 50 ksi can be designed at a basic allowable stress of 30 ksi. The K series is intended for spans from 8 to 60 ft.

LH-series, longspan joists have been standardized with depths from 18 to 48 in for clear spans from 25 to 96 ft. DLH-series, deep, longspan joists have been standardized with depths from 52 to 72 in for clear spans from 89 to 144 ft. Basic allowable design stress is taken at 0.6 times the specified minimum yield point for the LH and DLH series, values from 36 to 50 ksi being feasible.

Joist girders have been standardized with depths from 20 to 72 in for clear spans from 20 to 60 ft. Basic allowable design stress is taken at 0.6 times the specified minimum yield point for joist girders, values from 36 to 50 ksi being contemplated.

The safe load capacities of each series are listed in SJI "Standard Specifications, Load Tables, and Weight Tables for Steel Joists and Joist Girders," 1994.

10.25 Design of Open-Web Joist Floors

Open-web joists are designed primarily for use under uniformly distributed loading and at substantially uniform spacing. But they can safely carry concentrated loads if proper consideration is given to the effect of such loads. Good practice requires that heavy concentrated loads be applied at joist panel points. The weight of a partition running crosswise to the joists usually is considered satisfactorily distributed by the floor slab and assumed not to cause local bending in the top chords of the joists. Even so, joists must be selected to resist the bending moments, shears, and end reactions due to such loads.

The method of selecting joist sizes for any floor depends on whether or not the effect of any cross

Table 10.7 Fire-Resistance Ratings for Steel Floor and Roof Assemblies (2-h Ratings)*

Roof Construction	Insulation	Underside Protection	Authority
Min 1½-in-deep steel deck on steel joists or steel beams	Min 1¾-in-thick, listed mineral fiberboard	Min 1¾-in-thick, direct-applied, sprayed vermiculite plaster, UL Listed	UL Design P711[†]
Min 1½-in-deep steel deck on steel joists or steel beams	Min 1¹⁄₁₆-in-thick, listed mineral fiberboard	Min 1⁹⁄₁₆-in-thick, direct-applied, sprayed fiber protection, UL Listed	UL Design P818[†]

Floor Construction	Concrete	Underside Protection	Authority
1½-, 2-, or 3-in-deep, steel floor units on steel beams	2½-in-thick, normal-weight or light-weight concrete	Min. ³⁄₈-in-thick, direct-applied, sprayed vermiculite plaster, UL Listed	UL Design P739[†]
1½-, 2-, or 3-in-deep, steel floor units on steel beams	2½-in-thick, normal-weight or light-weight concrete	Min. ³⁄₈-in-thick, direct-applied, sprayed fiber protection, UL Listed	UL Design P858[†]

*For roof construction, 1½-h and 1-h ratings are also available. For floor construction, 2½-h, 3-h, and 4-h ratings are also available.
[†]"Fire Resistance Index," Underwriters' Laboratories, Inc., 1990.

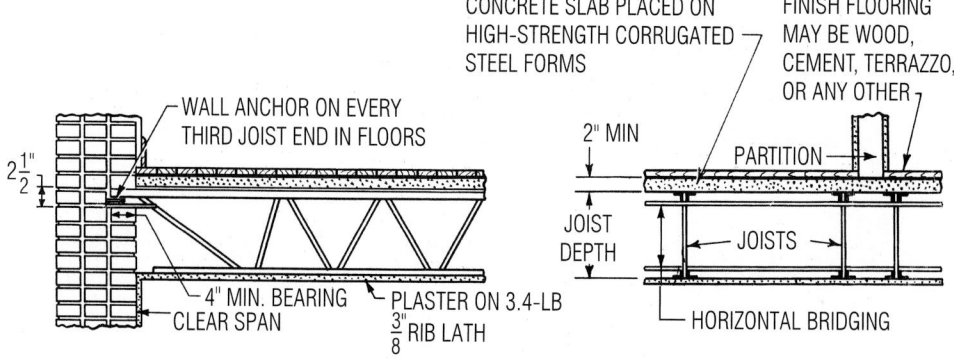

Fig. 10.17 Open-web steel joist construction.

partitions or other concentrated loads must be considered. Under uniform loading only, joist sizes and spacings are most conveniently selected from a table of safe loads. Where concentrated or nonuniform loads exist, calculate bending moments, end reactions, and shears, and select joists accordingly.

The chord section and web details are different for different joist designs made by different manufacturers. Information relating to the size and properties of the members may be obtained from manufacturers' catalogs.

Open-web steel-joist specifications require that the clear span not exceed 24 times the depth of the joist.

10.26 Construction Details for Open-Web Steel Joists

It is essential that bridging be installed between joists as soon as possible after the joists have been placed and before application of construction loads. The most commonly used type of bridging is a continuous horizontal bracing composed of rods fastened to the top and bottom chords of the joists. Diagonal bridging, however, also is permitted. The attachment of the floor or roof deck must provide lateral support for design loads.

It is important that masonry anchors be used on wall-bearing joists. Where the joists rest on steel beams, the joists should be welded, bolted, or clipped to the beams.

Fire resistance ratings of 1, 1½, 2 and 3 hours are possible using concrete floors above decks as thin as 2 in and as thick as 3½ in with various types of

ceiling protection systems. The Steel Joist Institute identifies such ceiling protection systems as exposed grids, concealed grids, gypsum board, cementitious or sprayed fiber.

When the usual cast-in-place concrete floor slab is used, it is customary to install reinforcing bars in two perpendicular directions or welded-wire fabric. Stirrups are not usually necessary. Forms for the concrete slab usually consist of corrugated steel sheets, expanded-metal rib lath, or welded-wire fabric. Corrugated sheets can be fastened with self-tapping screws or welded to the joists, with a bent washer to reinforce the weld and anchor the slab.

Pre-Engineered Steel Buildings and Housing

10.27 Characteristics of Pre-Engineered Steel Buildings

These structures may be selected from a catalog fully designed and supplied with all structural and covering material, with all components and fasteners. Such buildings eliminate the need for engineers and architects to design and detail both the structure and the required accessories and openings, as would be done for conventional buildings with components from many individual suppliers. Available with floor area of up to 1 million ft^2, pre-engineered buildings readily meet requirements for single-story structures, especially for industrial plants and commercial buildings (Fig. 10.19).

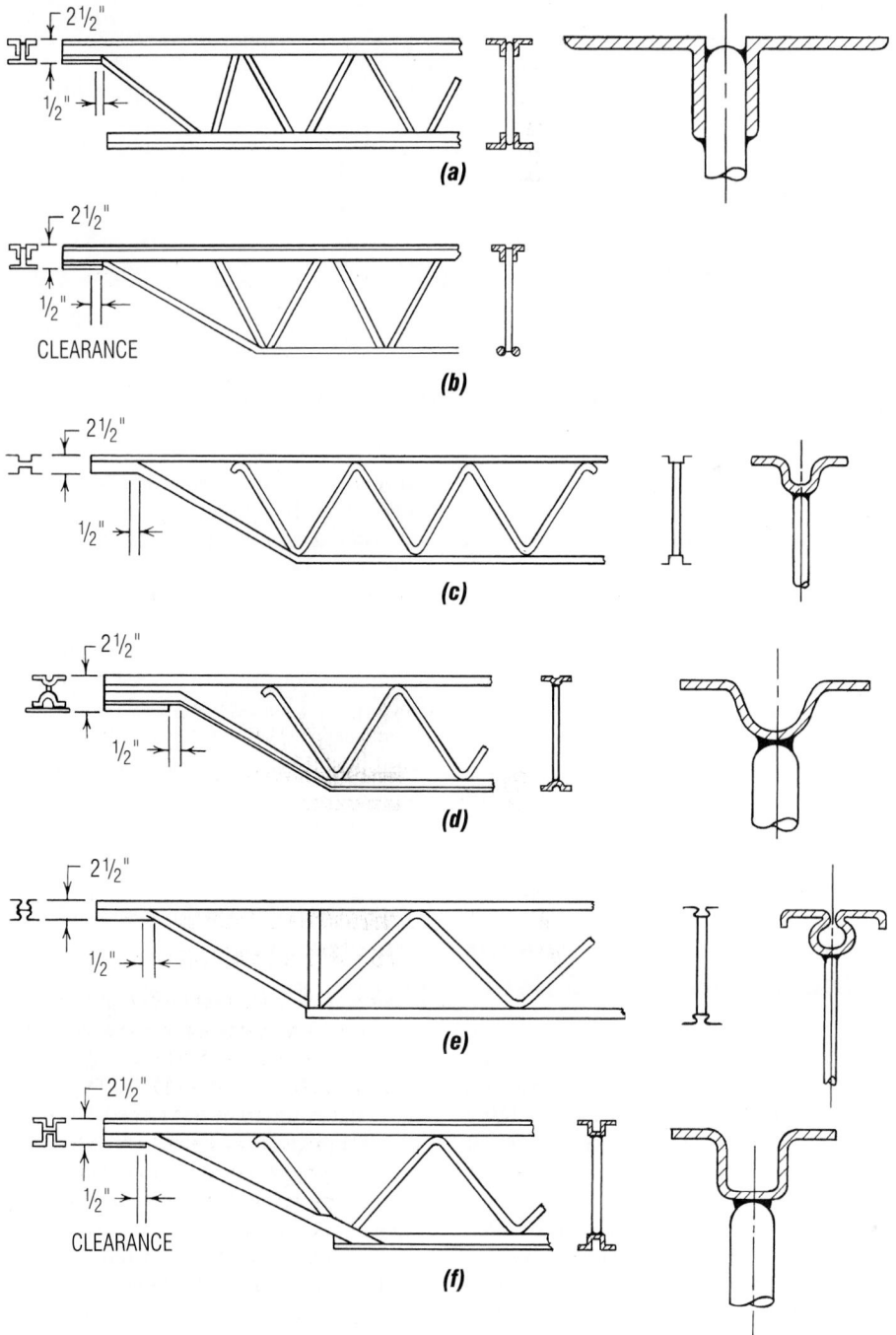

Fig. 10.18   Open-web steel joists.

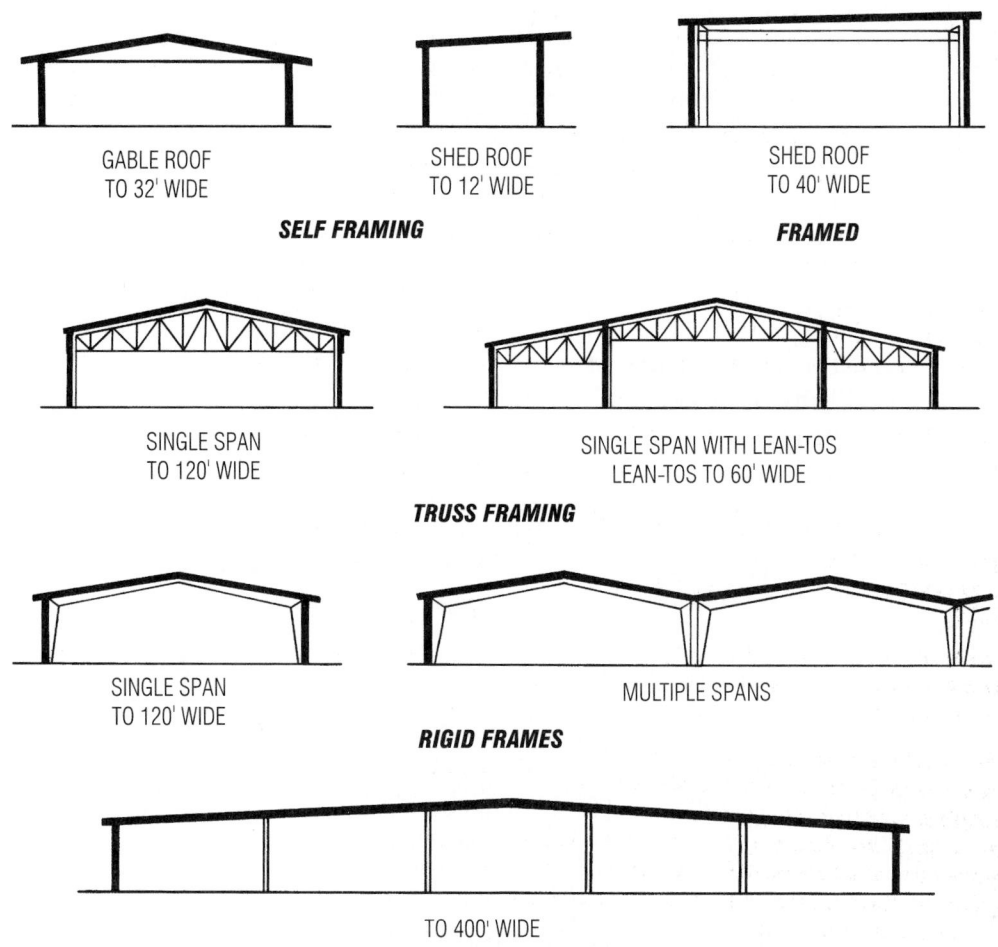

GABLE ROOF
TO 32' WIDE

SHED ROOF
TO 12' WIDE

SHED ROOF
TO 40' WIDE

SELF FRAMING

FRAMED

SINGLE SPAN
TO 120' WIDE

SINGLE SPAN WITH LEAN-TOS
LEAN-TOS TO 60' WIDE

TRUSS FRAMING

SINGLE SPAN
TO 120' WIDE

MULTIPLE SPANS

RIGID FRAMES

TO 400' WIDE
POST-AND-BEAM FRAMES

Fig. 10.19 Principal framing systems for pre-engineered buildings.

Pre-engineered buildings may be provided with custom architectural accents. Also, standard insulating techniques may be used and thermal accessories incorporated to provide energy efficiency. Exterior wall panels are available with durable factory-applied colors.

Many pre-engineered metal building suppliers are also able to modify structurally their standard designs, within certain limits, while retaining the efficiencies of predesign and automated volume fabrication. Examples of such modification include the addition of cranes; mezzanines; heating, ventilating, and air-conditioning equipment; sprinklers; lighting; and ceiling loads with special building dimensions.

Pre-engineered buildings make extensive use of cold-formed structural members. These parts lend themselves to mass production, and their design can be more accurately fitted to the specific structural requirement. For instance, a roof purlin can be designed with the depth, moment of inertia, section modulus, and thickness required to carry the load, as opposed to picking the next-higher-size standard hot-rolled shape, with more weight than required. Also, because this purlin is used on thousands of buildings, the quantity justifies investment in automated equipment for forming and punching. This equipment is flexible enough to permit a change of thickness or depth of section to produce similar purlins for other loadings.

The engineers designing a line of pre-engineered buildings can, because of the repeated use of the design, justify spending additional design time refining and optimizing the design. Most pre-engineered buildings are designed with the aid of electronic computers. Their programs are specifically tailored for the product. A rerun of a problem to eliminate a few pounds of steel is justified since the design will be reused many times during the life of that model.

10.28 Structural Design of Pre-Engineered Buildings

The buildings are designed for loading criteria in such a way that any building may be specified to meet the geographical requirements of any location. Combinations of dead load, snow load, live load, and wind load conform with requirements of several model building codes.

The Metal Building Dealers Association, 1406 Third National Building, Dayton, OH 45402, and the Metal Building Manufacturers Association, 1300 Summer Ave., Cleveland, OH 44115 (www.mbma.com), have established design standards (see MBMA, "Metal Building Systems Manual" and "Metal Roofing System Design Manual"). These standards discuss methods of load application and maximum loadings, for use where load requirements are not established by local building codes. Other standard design specifications include:

Structural Steel—"Specification for Structural Steel Buildings," American Institute of Steel Construction (www.asic.org).

Cold-Formed Steel—"Specification for the Design of Cold-Formed Steel Structural Members," American Iron and Steel Institute (www.steel.org).

Welding—"Structural Welding Code—Steel," D1.1, and "Structural Welding Code—Sheet Steel," D1.3, American Welding Society (www.aws.org).

Cold-formed steel structural members have been used for residential construction for many years. To satisfy the needs of design and construction information, the AISI Committee on Framing Standards has developed several "Standards for Cold-Formed Steel Framing," including General Provisions, Truss Design, Header Design, Prescriptive Method for One and Two-Family Dwellings, Wall Stud Design and Lateral Resistance Design. (American Iron and Steel Institute, 1140 Connecticut Ave., N.W., Washington, DC 20036.)

Structural Design of Corrugated Steel Pipe

10.29 Corrugated Steel Pipe

Corrugated steel pipe was first developed and used for culvert drainage in 1896. It is now produced in full-round diameters from 6 in in diameter and 0.064 in thick to 144 in in diameter and 0.168 in thick. Heights of cover up to 100 ft are permissible with highway or railway loadings.

Riveted corrugated pipe (Fig. 10.20a shows pipe-arch shape) is produced by riveting together circular corrugated sheets to form a tube. The corrugations are annular.

Helically corrugated pipe (Fig. 10.20b) is manufactured by spirally forming a continuously corrugated strip into a tube with a locked or welded seam

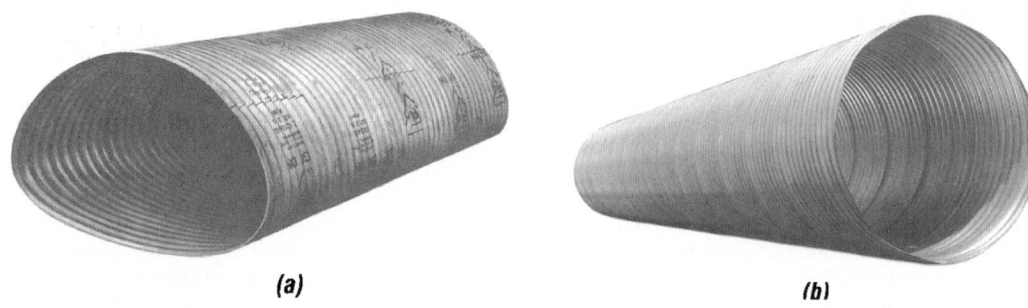

(a)   *(b)*

Fig. 10.20 Corrugated steel structures. (*a*) Riveted pipe arch. (*b*) Helical pipe.

joining abutting edges. This pipe is stronger in ring compression because of the elimination of the longitudinal riveted joints. Also, the seam is more watertight than the lap joints of riveted pipe.

Besides being supplied in full-round shapes, both types of pipe are also available in pipe-arch shape. This configuration, with a low and wide waterway area in the invert, is beneficial for headroom conditions. It provides adequate flow capacity without raising the grade.

Corrugated steel pipe and pipe arch are produced with a variety of coatings to resist corrosion and erosion.

The zinc coating provided on these structures is adequate protection under normal drainage conditions with no particular corrosion hazard. Additional coatings or pavings may be specified for placing over the galvanizing.

Asbestos-bonded steel has a coating in which a layer of asbestos fiber is embedded in molten zinc and then saturated with bituminous material. This provides protection for extreme corrosion conditions. Asbestos-bonded steel is available in riveted pipe only. Helical corrugated structures may be protected with a hot-dip coating of bituminous material for severe soil or effluent conditions.

For erosive hazards, a paved invert of bituminous material can be applied to give additional protection to the bottom of the pipe. And for improved flow, these drainage conduits may also be specified with a full interior paving of bituminous material.

Normally, pipe-arch structures are supplied in a choice of span-and-rise combinations that have a periphery equal to that available with full-round corrugated pipe.

10.30 Structural Plate Pipe

To extend the diameter or span-and-rise dimensions of corrugated steel structures beyond that (120 in) available with factory-fabricated drainage conduits, structural plate pipe and other shapes may be used. These are made of heavier gages of steel and are composed of curved and corrugated steel plates bolted together at the installation site. Their shapes include full-round, elliptical, pipe-arch, arch, and horseshoe or underpass shapes. Applications include storm drainage, stream enclosures, vehicular and pedestrian underpasses, and small bridges.

Such structures are field-assembled with curved and corrugated steel plates that may be 10 or 12 ft long (Fig. 10.21). The wall section of the standard structures has 2-in-deep corrugations, 6 in c to c. Thickness ranges from 0.109 to 0.380 in. Each plate is punched for field bolting and special high-strength bolts are supplied with each structure. The number of bolts used can be varied to meet the ring-compression stress.

Circular pipes are available in diameters ranging from 5 to 26 ft, with structures of other configurations available in a similar approximate size range. Special end plates can be supplied to fit a skew or bevel, or a combination of both.

Plates of all structures are hot-dip galvanized. They are normally shipped in bundles for handling convenience. Instructions for assembly are also provided.

10.31 Design of Culverts

Formerly, design of corrugated steel structures was based on observations of how such pipes performed structurally under service conditions. From these observations, data were tabulated and gage tables established. As larger pipes were built and installed and experience was gained, these gage tables were revised and enlarged.

Following is the design procedure for corrugated steel structures recommended in the "Handbook of Steel Drainage and Highway Construction Products" (American Iron and Steel Institute, 1140 Connecticut Ave., N.W., Washington, D.C. 20036).

1. Backfill Density ▪ Select a percent compaction of pipe backfill for design. The value chosen should reflect the importance and size of the structure and the quality that can reasonably be expected. The recommended value for routine use is 85%. This value will usually apply to ordinary installations for which most specifications will call for compaction to 90%. But for more important structures in higher-fill situations, consideration must be given to selecting higher-quality backfill and requiring this quality for construction.

2. Design Pressure ▪ When the height of cover is equal to or greater than the span or diameter of the structure, enter the load-factor chart (Fig. 10.22) to determine the percentage of the total load acting on the steel. For routine use, the

Fig. 10.21 Structural-plate pipe is shown being bolted together at right. Completely assembled structural-plate pipe arch is shown at left.

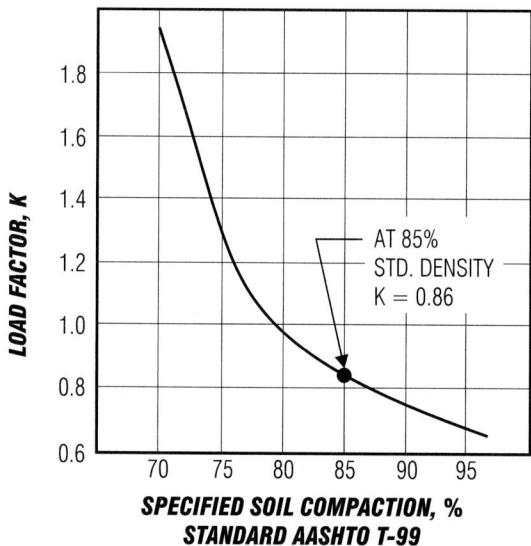

Fig. 10.22 Load factors for corrugated steel pipe are plotted as a function of specified compaction of backfill.

85% soil compaction will provide a load factor $K = 0.86$. The total load is multiplied by K to obtain the design pressure P_n acting on the steel. If the height of cover is less than one pipe diameter, the total load TL is assumed to act on the pipe, and $TL = P_n$; that is,

$$P_n = DL + LL + I \quad H < S \tag{10.49}$$

When the height of cover is equal to or greater than one pipe diameter,

$$P_n = K(DL + LL + I) \quad H \geq S \tag{10.50}$$

where P_n = design pressure, kips/ft²

K = load factor

DL = dead load, kips/ft²

LL = live load, kips/ft²

I = impact, kips/ft²

H = height of cover, ft

S = span or pipe diameter, ft

3. Ring Compression ▪ The compressive thrust C, kips/ft, on the conduit wall equals the radial pressure P_n kips/ft², acting on the wall multiplied by the wall radius R, ft; or $C = P_n R$. This thrust, called ring compression, is the force carried by the steel. The ring compression is an axial load acting tangentially to the conduit wall (Fig. 10.23). For conventional structures in which the top arc approaches a semicircle, it is convenient

to substitute half the span for the wall radius. Then

$$C = P_n \frac{S}{2} \tag{10.51}$$

4. Allowable Wall Stress ▪ The ultimate compression in the pipe wall is expressed by Eqs. (10.52) and (10.53). The ultimate wall stress is equal to the specified minimum yield point of the steel and applies to the zone of wall crushing or yielding. Equation (10.52) applies to the interaction zone of yielding and ring buckling; Eq. (10.53) applies to the ring-buckling zone.

When the ratio D/r of pipe diameter—or span D, in, to radius of gyration r, in, of the pipe cross section—does not exceed 294, the ultimate wall stress may be taken as equal to the steel yield strength:

$$F_b = F_y = 33 \, \text{ksi}$$

When D/r exceeds 294 but not 500, the ultimate wall stress, ksi, is given by

$$F_b = 40 - 0.000081 \left(\frac{D}{r}\right)^2 \tag{10.52}$$

When D/r is more than 500

$$F_b = \frac{4.93 \times 10^6}{(D/r)^2} \tag{10.53}$$

A safety factor of 2 is applied to the ultimate wall stress to obtain the design stress F_c, ksi,

$$F_c = \frac{F_b}{2} \tag{10.54}$$

5. Wall Thickness ▪ Required wall area A, in²/ft of width, is computed from the calculated compression C in the pipe wall and the allowable stress F_c.

$$A = \frac{C}{F_c} \tag{10.55}$$

From Table 10.8, select the wall thickness that provides the required area with the same corrugation used for selection of the allowable stress.

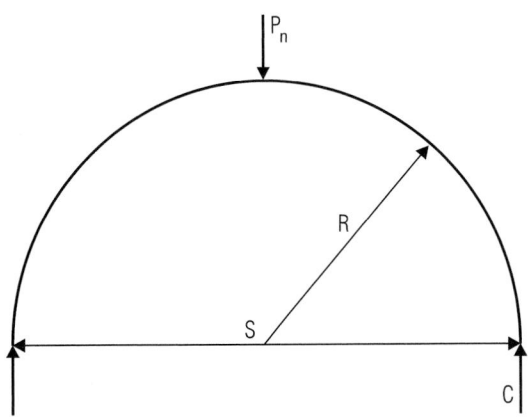

Fig. 10.23 Radical pressure P_n, on the wall of a curved conduit is resisted by compressive thrust, C.

6. Check Handling Stiffness ▪ Minimum pipe stiffness requirements for practical handling

Table 10.8 Moments of Inertia, Cross-Sectional Areas, and Radii of Gyration for Corrugated Steel Sheets and Plates for Underground Conduits*

Corrugation pitch × depth, in	Specified Thickness Including Galvanized Coating, in												
	0.040	0.052	0.064	0.079	0.109	0.138	0.168	0.188	0.218	0.249	0.280	0.310	0.380
	Base-Metal Thickness, in												
	0.0359	0.0478	0.0598	0.0747	0.1046	0.1345	0.1644	0.1838	0.2145	0.2451	0.2758	0.3125	0.3750
Moment of Inertia I, in^4/ft of Width													
1½ × ¼	0.0030	0.0041	0.0053	0.0068	0.0103	0.0145	0.0196						
2 × ½	0.0137	0.0184	0.0233	0.0295	0.0425	0.0566	0.0719						
2⅔ × ½	0.0135	0.0180	0.0227	0.0287	0.0411	0.0544	0.0687						
3 × 1	0.0618	0.0827	0.1039	0.1306	0.1855	0.2421	0.3010						
5 × 1			0.1062	0.1331	0.1878	0.2438	0.3011						
6 × 2					0.725	0.938	1.154	1.296	1.523	1.754	1.990	2.280	2.784
Cross-Sectional Wall Area A, in^2/ft of Width													
1½ × ¼	0.456	0.608	0.761	0.950	1.331	1.712	2.093						
2 × ½	0.489	0.652	0.815	1.019	1.428	1.838	2.249						
2⅔ × ½	0.465	0.619	0.775	0.968	1.356	1.744	2.133						
3 × 1	0.534	0.711	0.890	1.113	1.560	2.008	2.458						
5 × 1			0.794	0.992	1.390	1.788	2.186						
6 × 2					1.556	2.003	2.449	2.739	3.199	3.658	4.119	4.671	5.613
Radius of Gyration r, in													
1½ × ¼	0.0816	0.0824	0.0832	0.0846	0.0879	0.0919	0.0967						
2 × ½	0.1676	0.1682	0.1690	0.1700	0.1725	0.1754	0.1788						
2⅔ × ½	0.1702	0.1707	0.1712	0.1721	0.1741	0.1766	0.1795						
3 × 1	0.3403	0.3410	0.3417	0.3427	0.3448	0.3472	0.3499						
5 × 1			0.3657	0.3663	0.3677	0.3693	0.3711						
6 × 2					0.682	0.684	0.686	0.688	0.690	0.692	0.695	0.698	0.704

*Corrugation dimensions are nominal, subject to manufacturing tolerances. Section properties were calculated from base-metal thicknesses, that is, with galvanized-coating thickness excluded.

and installation, without undue care or bracing, have been established through experience. The resulting flexibility factor *FF* limits the size of each combination of corrugation pitch and metal thickness.

$$FF = \frac{D^2}{EI} \qquad (10.56)$$

where E = modulus of elasticity, ksi, of steel = 30,000 ksi

I = moment of inertia of wall, in^4/in

The following maximum values of *FF* are recommended for ordinary installations:

$FF = 0.0433$ for factory-made pipe with riveted, welded, or helical seams

$FF = 0.0200$ for field-assembled pipe with bolted seams

Higher values can be used with special care or where experience indicates. Trench condition, as in sewer design, can be one such case; use of aluminum pipe is another. For example, the flexibility factor permitted for aluminum pipe in some national specifications is more than twice that recommended here for steel because aluminum has only one-third the stiffness of steel, the modulus for aluminum being about 10,000 ksi vs. 30,000 ksi for steel. Where a high degree of flexibility is acceptable for aluminum, it will be equally acceptable for steel.

7. Check Longitudinal Seams ▪ Most pipe seams develop the full yield strength of the pipe wall. However, there are exceptions in standard pipe manufacture and these are identified here. Shown in Table 10.9 are those standard riveted and bolted seams which do not develop a strength equivalent to F_y = 33 ksi. To maintain a consistent factor of safety of 2.0 for these pipes, it is necessary to reduce the maximum ring compression to one half the indicated seam strength. Nonstandard, or new longitudinal seam details should be checked for this same possible condition.

Other Types of Lightweight-Steel Construction

10.32 Lightweight-Steel Bridge Decking

This trapezoidal-corrugated plank, welded to steel (Fig. 10.24) or lagged to wood stringers, gives a strong, secure base for a smooth bituminous traffic surface. It may be used for replacement of old wood decks and for new construction.

10.33 Beam-Type Guardrail

The beam-type guardrail in Fig. 10.25 has the flexibility necessary to absorb impact as well as the beam strength to prevent pocketing of a car against a post. Standard post spacing is 12½ ft. The rail is anchored with one bolt to each post, and there are eight bolts in the rail splice to assure continuous-

Table 10.9 Ultimate Longitudinal Seam Strengths (kips/ft)

Thickness, in		6 × 2 in 4 Bolts Per Ft	3 × 1 in	2⅔ × ½ in Rivet Seams		
Corrugated Steel Pipe	Structural Plate			5⁄16 in Single Rivet	3⁄8 in Single Rivet	3⁄8 in Double Rivet
0.064			28.7[1]	16.7		
0.079			35.7[1]	18.2		
0.109	0.111	42.0			23.4	
0.138	0.140	62.0	63.7[2]		24.5	49.0
0.168			70.7[2]		25.6	51.3

Standard seams not shown develop full yield strength of pipe wall.

[1]Double ⅜-in. rivets.

[2]Double 7⁄16-in. rivets.

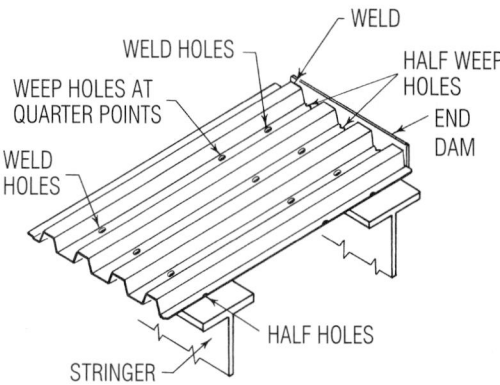

Fig. 10.24 Lightweight-steel bridge plank.

beam strength. Available lengths are $12\frac{1}{2}$ and 25 ft. Standard steel thickness is 0.109 in; heavy-duty is 0.138 in thick. The guardrail is furnished galvanized or as prime-painted steel. (See also Art. 16.17.)

10.34 Bin-Type Retaining Wall

A bin-type retaining wall (Fig. 10.26) is a series of closed-face bins, which when backfilled transform the soil mass into an economical retaining wall. The flexibility of steel allows for adjustments due to uneven ground settlement. There are standard designs for these walls with vertical or battered face, heights to 30 ft, and various conditions of surcharge.

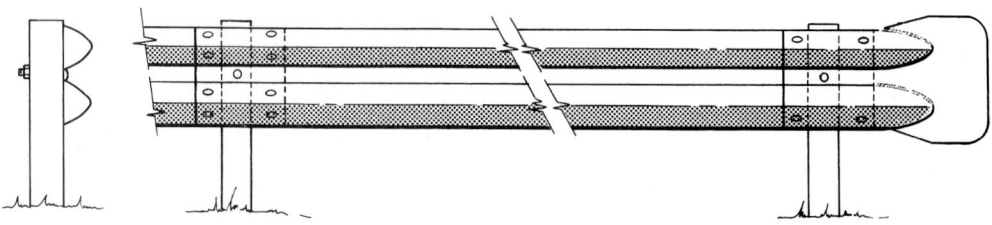

Fig. 10.25 Beam-type guardrail of steel.

Fig. 10.26 Bin-type retaining wall of cold-formed steel.

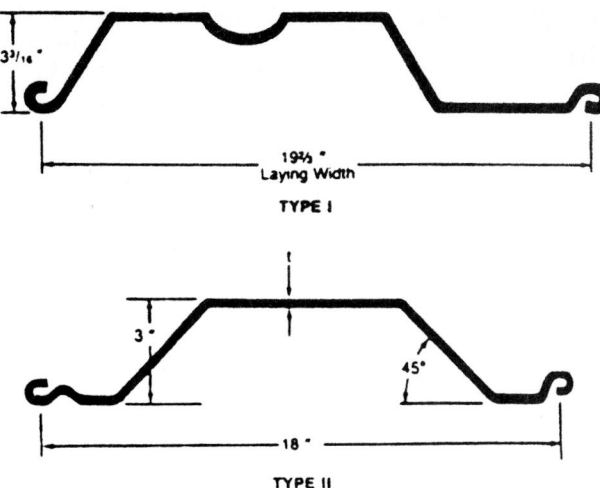

Fig. 10.27 Lightweight steel sheeting

Table 10.10(a) Physical Properties of Type I Lightweight Steel Sheeting

Gage	Uncoated Thickness in	Weight		Section Properties			
		lbs/ft of pile	lbs/ft² of wall	Section Modulus, in³		Moment of Inertia, in⁴	
				per section	per ft	per section	per ft
5	0.209	19.1	11.6	5.50	3.36	9.40	5.73
7	0.179	16.4	10.0	4.71	2.87	8.00	4.88
8	0.164	15.2	9.3	4.35	2.65	7.36	4.49
10	0.134	12.5	7.6	3.60	2.20	6.01	3.67
12	0.105	9.9	6.0	2.80	1.71	4.68	2.85

Based on AISI "Handbook of Steel Drainage & Highway Construction Products," 1994.

Table 10.10(b) Physical Properties of Type II Lightweight Steel Sheeting

Gage	Uncoated Thickness, in.	Weight		Section Modulus, in³/ft	Moment of Inertia in⁴/ft
		lbs/ft	lbs/ft²		
7	0.179	13.86	9.24	2.37	3.56
8	0.164	12.70	8.47	2.13	3.20
10	0.134	10.37	6.91	1.84	2.74

Based on AISI "Handbook of Steel Drainage & Highway Construction Products," 1994.

10.35 Lightweight-Steel Sheeting

Corrugated sheeting has beam strength to support earth pressure on walls of trenches and excavations, and column strength for driving. The sheeting presents a small end cross section for easy driving (Fig. 10.27). Physical properties of the sheeting shown in Fig. 10.27 are listed in Table 10.10.

11

Maurice J. Rhude
President
Sentinel Structures, Inc.
Peshtigo, Wisconsin

WOOD DESIGN AND CONSTRUCTION

Wood is remarkable for its beauty, versatility, strength, durability, and workability. It possesses a high strength-to-weight ratio. It has flexibility. It performs well at low temperatures. It withstands substantial overloads for short periods. It has low electrical and thermal conductance. It resists the deteriorating action of many chemicals that are extremely corrosive to other building materials. There are few materials that cost less per pound than wood.

As a consequence of its origin, wood as a building material has inherent characteristics with which users should be familiar. For example, although cut simultaneously from trees growing side by side in a forest, two boards of the same species and size most likely do not have the same strength. The task of describing this nonhomogeneous material, with its variable biological nature, is not easy, but it can be described accurately, and much better than was possible in the past because research has provided much useful information on wood properties and behavior in structures.

Research has shown, for example, that a compression grade cannot be used, without modification, for the tension side of a deep bending member. Also, a bending grade cannot be used, unless modified, for the tension side of a deep bending member or for a tension member. Experience indicates that typical growth characteristics are more detrimental to tensile strength than to compressive strength. Furthermore, research has made possible better estimates of

wood's engineering qualities. No longer is it necessary to use only visual inspection, keyed to averages, for estimating the engineering qualities of a piece of wood. With a better understanding of wood now possible, the availability of sound structural design criteria, and development of economical manufacturing processes, greater and more efficient use is being made of wood for structural purposes.

Improvements in adhesives also have contributed to the betterment of wood construction. In particular, the laminating process, employing adhesives to build up thin boards into deep timbers, improves nature. Not only are stronger structural members thus made available, but also higher grades of lumber can be placed in regions of greatest stress and lower grades in regions of lower stress, for overall economy. Despite variations in strength of wood, lumber can be transformed into glued-laminated timbers of predictable strength and with very little variability in strength.

11.1 Basic Characteristics of Wood

Wood differs in several significant ways from other building materials, mainly because of its cellular structure. Because of this structure, structural properties depend on orientation. Although most structural materials are essentially isotropic, with nearly equal properties in all directions, wood has three principal grain directions: longitudinal, radial, and tangential. (Loading in the longitudinal direction is referred to as parallel to the grain, whereas transverse loading is considered across the grain.) Parallel to the grain, wood possesses high strength and stiffness. Across the grain, strength is much lower. (In tension, wood stressed parallel to the grain is 25 to 40 times stronger than when stressed across the grain. In compression, wood loaded parallel to the grain is 6 to 10 times stronger than when loaded perpendicular to the grain.) Furthermore, a wood member has three moduli of elasticity, with a ratio of largest to smallest as large as 150:1.

Wood undergoes dimensional changes from causes different from those for dimensional changes in most other structural materials. For instance, thermal expansion of wood is so small as to be unimportant in ordinary usage. Significant dimensional changes, however, occur because of gain or loss in moisture. Swelling and shrinkage from this cause vary in the three grain directions; size changes about 6 to 16% tangentially, 3 to 7% radially, but only 0.1 to 0.3% longitudinally.

Wood offers numerous advantages nevertheless in construction applications—beauty, versatility, durability, workability, low cost per pound, high strength-to-weight ratio, good electrical insulation, low thermal conductance, and excellent strength at low temperatures. It is resistant to many chemicals that are highly corrosive to other materials. It has high shock-absorption capacity. It can withstand large overloads of short time duration. It has good wearing qualities, particularly on its end grain. It can be bent easily to sharp curvature. A wide range of finishes can be applied for decoration or protection. Wood can be used in both wet and dry applications. Preservative treatments are available for use when necessary, as are fire retardants. Also, there is a choice of a wide range of species with a wide range of properties.

In addition, many wood framing systems are available. The intended use of a structure, geographical location, configuration required, cost, and many other factors determine the framing system to be used for a particular project.

11.1.1 Moisture Content of Wood

Wood is unlike most structural materials in regard to the causes of its dimensional changes, which are primarily from gain or loss of moisture, not change in temperature. For this reason expansion joints are seldom required for wood structures to permit movement with temperature changes. It partly accounts for the fact that wood structures can withstand extreme temperatures without collapse.

A newly felled tree is green (contains moisture). When the greater part of this water is being removed, seasoning first allows free water to leave the cavities in the wood. A point is reached where these cavities contain only air, and the cell walls still are full of moisture. The moisture content at which this occurs, the fiber-saturation point, varies from 25 to 30% of the weight of the oven-dry wood.

During removal of the free water, the wood remains constant in size and in most properties (weight decreases). Once the fiber-saturation point has been passed, shrinkage of the wood begins as the cell walls lose water. Shrinkage continues nearly linearly down to zero moisture content (Table 11.1). (There are, however, complicating factors, such as the effects of timber size and relative rates of moisture movement in three directions: longitudinal, radial, and tangential to the growth rings.) Eventually, the wood assumes a condition of equilibrium, with the final moisture content dependent on the relative humidity and temperature of the ambient air. Wood swells when it absorbs moisture, up to the fiber-saturation point. The relationship of wood moisture content, temperature, and relative humidity can actually define an environment (Fig. 11.1).

This explanation has been simplified. Outdoors, rain, frost, wind, and sun can act directly on the wood. Within buildings, poor environmental conditions may be created for wood by localized heating, cooling, or ventilation. The conditions of service must be sufficiently well known to be specifiable. Then, the proper design value can be assigned to wood and the most suitable adhesive selected.

Dry Condition of Use ▪ Design values for dry conditions of use are applicable for normal loading when the wood moisture content in service is less than 16%, as in most covered structures.

Dry-use adhesives perform satisfactorily when the moisture content of wood does not exceed 16% for repeated or prolonged periods of service and are to be used only when these conditions exist.

Table 11.1 Shrinkage Values of Wood Based on Dimensions When Green

Species	Dried to 20% MC*			Dried to 6% MC†			Dried to 0% MC		
	Radial, %	Tangential, %	Volumetric, %	Radial, %	Tangential, %	Volumetric, %	Radial, %	Tangential, %	Volumetric, %
Softwoods:‡									
Cedar:									
Alaska	0.9	2.0	3.1	2.2	4.8	7.4	2.8	6.0	9.2
Incense	1.1	1.7	2.5	2.6	4.2	6.1	3.3	5.2	7.6
Port Orford	1.5	2.3	3.4	3.7	5.5	8.1	4.6	6.9	10.1
Western red	0.8	1.7	2.3	1.9	4.0	5.4	2.4	5.0	6.8
Cypress, southern	1.3	2.1	3.5	3.0	5.0	8.4	3.8	6.2	10.5
Douglas fir:									
Coast region	1.7	2.6	3.9	4.0	6.2	9.4	5.0	7.8	11.8
Inland region	1.4	2.5	3.6	3.3	6.1	8.7	4.1	7.6	10.9
Rocky Mountain	1.2	2.1	3.5	2.9	5.0	8.5	3.6	6.2	10.6
Fir, white	1.1	2.4	3.3	2.6	5.7	7.8	3.2	7.1	9.8
Hemlock:									
Eastern	1.0	2.3	3.2	2.4	5.4	7.8	3.0	6.8	9.7
Western	1.4	2.6	4.0	3.4	6.3	9.5	4.3	7.9	11.9
Larch, western	1.4	2.7	4.4	3.4	6.5	10.6	4.2	8.1	13.2
Pine:									
Eastern white	0.8	2.0	2.7	1.8	4.8	6.6	2.3	6.0	8.2
Lodgepole	1.5	2.2	3.8	3.6	5.4	9.2	4.5	6.7	11.5
Norway	1.5	2.4	3.8	3.7	5.8	9.2	4.6	7.2	11.5
Ponderosa	1.3	2.1	3.2	3.1	5.0	7.7	3.9	6.3	9.6
Southern (avg.)	1.6	2.6	4.1	4.0	6.1	9.8	5.0	7.6	12.2
Sugar	1.0	1.9	2.6	2.3	4.5	6.3	2.9	5.6	7.9
Western white	1.4	2.5	3.9	3.3	5.9	9.4	4.1	7.4	11.8
Redwood (old growth)	0.9	1.5	2.3	2.1	3.5	5.4	2.6	4.4	6.8
Spruce:									
Engelmann	1.1	2.2	3.5	2.7	5.3	8.3	3.4	6.6	10.4
Sitka	1.4	2.5	3.8	3.4	6.0	9.2	4.3	7.5	11.5
Hardwoods:‡									
Ash, white	1.6	2.6	4.5	3.8	6.2	10.7	4.8	7.8	13.4
Beech, American	1.7	3.7	5.4	4.1	8.8	13.0	5.1	11.0	16.3
Birch:									
Sweet	2.2	2.8	5.2	5.2	6.8	12.5	6.5	8.5	15.6
Yellow	2.4	3.1	5.6	5.8	7.4	13.4	7.2	9.2	16.7
Elm, rock	1.6	2.7	4.7	3.8	6.5	11.3	4.8	8.1	14.1
Gum, red	1.7	3.3	5.0	4.2	7.9	12.0	5.2	9.9	15.0
Hickory:									
Pecan§	1.6	3.0	4.5	3.9	7.1	10.9	4.9	8.9	13.6
True	2.5	3.8	6.0	6.0	9.0	14.3	7.5	11.3	17.9
Maple, hard	1.6	3.2	5.0	3.9	7.6	11.9	4.9	9.5	14.9
Oak:									
Red	1.3	2.7	4.5	3.2	6.6	10.8	4.0	8.2	13.5
White	1.8	3.0	5.3	4.2	7.2	12.6	5.3	9.0	15.8
Poplar, yellow	1.3	2.4	4.1	3.2	5.7	9.8	4.0	7.1	12.3

*MC = moisture content as a percent of weight of oven-dry wood. These shrinkage values have been taken as one-third the shrinkage to the oven-dry conditions as given in the last three columns.

† These shrinkage values have been taken as four-fifths of the shrinkage to the oven-dry condition as given in the last three columns.

‡ The total longitudinal shrinkage of normal species from fiber saturation to oven-dry condition is minor. It usually ranges from 0.17 to 0.3% of the green dimension.

§ Average of butternut hickory, nutmeg hickory, water hickory, and pecan.

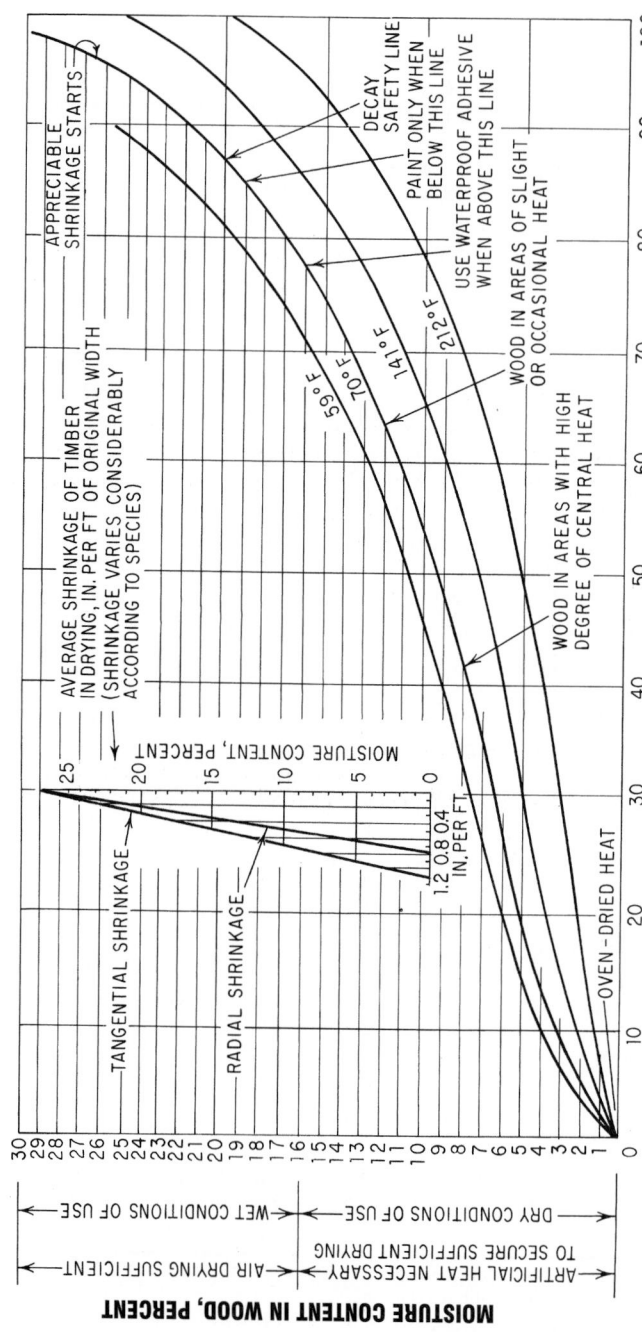

Fig. 11.1 Chart shows the approximate relationship for wood of equilibrium moisture content, temperature, and relative humidity. The triangular diagram indicates the effect of wood moisture content on the shrinkage of wood.

Wet Condition of Use ▪ Design values for wet condition of use are applicable for normal loading when the moisture content in service is 16% or more. This may occur in members not covered or in covered locations of high relative humidity.

Wet-use adhesives will perform satisfactorily for all conditions, including exposure to weather, marine use, and where pressure treatments are used, whether before or after gluing. Such adhesives are required when the moisture content exceeds 16% for repeated or prolonged periods of service.

11.1.2 Checking in Timbers

Separation of grain, or checking, is the result of rapid lowering of surface moisture content combined with a difference in moisture content between inner and outer portions of the piece. As wood loses moisture to the surrounding atmosphere, the outer cells of the member lose at a more rapid rate than the inner cells. As the outer cells try to shrink, they are restrained by the inner portion of the member. The more rapid the drying, the greater the differential in shrinkage between outer and inner fibers and the greater the shrinkage stresses. Splits may develop. Splits are cracks from separation of wood fibers across the thickness of a member that extend parallel to the grain.

Checks, radial cracks, affect the horizontal shear strength of timber. A large reduction factor is applied to test values in establishing design values, in recognition of stress concentrations at the ends of checks. Design values for horizontal shear are adjusted for the amount of checking permissible in the various stress grades at the time of the grading. Since strength properties of wood increase with dryness, checks may enlarge with increasing dryness after shipment without appreciably reducing shear strength.

Cross-grain checks and splits that tend to run out the side of a piece, or excessive checks and splits that tend to enter connection areas, may be serious and may require servicing. Provisions for controlling the effects of checking in connection areas may be incorporated into design details.

To avoid excessive splitting between rows of bolts due to shrinkage during seasoning of solid-sawn timbers, the rows should not be spaced more than 5 in apart, or a saw kerf, terminating in a bored hole, should be provided between the lines of bolts. Whenever possible, maximum end distances for connections should be specified to minimize the effect of checks running into the joint area. Some designers require stitch bolts in members, with multiple connections loaded at an angle to the grain. Stitch bolts, kept tight, will reinforce pieces where checking is excessive.

One principal advantage of glued-laminated timber construction is relative freedom from checking. Seasoning checks may however, occur in laminated members for the same reasons that they exist in solid-sawn members. When laminated members are glued within the range of moisture contents set in American National Standard, "Structural Glued Laminated Timber," ANSI/AITC A190.1, they will approximate the moisture content in normal-use conditions, thereby minimizing checking. Moisture content of the lumber at the time of gluing is thus of great importance to the control of checking in service. However, rapid changes in moisture content of large wood sections after gluing will result in shrinkage or swelling of the wood, and during shrinking, checking may develop in both glued joints and wood.

Differentials in shrinkage rates of individual laminations tend to concentrate shrinkage stresses at or near the glue line. For this reason, when checking occurs, it is usually at or near glue lines. The presence of wood-fiber separation indicates glue bonds and not delamination.

In general, checks have very little effect on the strength of glued-laminated members. Laminations in such members are thin enough to season readily in kiln drying without developing checks. Since checks lie in a radial plane, and the majority of laminations are essentially flat grain, checks are so positioned in horizontally laminated members that they will not materially affect shear strength. When members are designed with laminations vertical (with wide face parallel to the direction of load application), and when checks may affect the shear strength, the effect of checks may be evaluated in the same manner as for checks in solid-sawn members.

Seasoning checks in bending members affect only the horizontal shear strength. They are usually not of structural importance unless the checks are significant in depth and occur in the midheight of the member near the support, and then only if shear governs the design of the members. The reduction in shear strength is nearly directly proportional to the ratio of depth of check to width of beam. Checks in columns are not of structural importance unless the check develops into a split, thereby increasing the slenderness ratio of the columns.

Minor checking may be disregarded since there is an ample factor of safety in design values. The final decision as to whether shrinkage checks are detrimental to the strength requirements of any particular design or structural member should be made by a competent engineer experienced in timber construction.

11.1.3 Standard Sizes of Lumber and Timber

Details regarding dressed sizes of various species of wood are given in the grading rules of agencies that formulate and maintain such rules. Dressed sizes in Table 11.2 are from the American Softwood Lumber Standard, "Voluntary Product Standard PS20-70." These sizes are generally available, but it is good practice to consult suppliers before specifying sizes not commonly used to find out what sizes are on hand or can be readily secured.

11.1.4 Standard Sizes of Glued-Laminated Timber

Standard finished sizes of structural glued-laminated timber should be used to the extent that conditions permit. These standard finished sizes are based on lumber sizes given in "Voluntary Product Standard PS20-70." Other finished sizes may be used to meet the size requirements of a design or other special requirements.

Nominal 2-in-thick lumber, surfaced to $1\frac{3}{8}$ or $1\frac{1}{2}$ in before gluing, is used to laminate straight members and curved members with radii of curvature within the bending-radius limitations for the species. Nominal 1-in-thick lumber, surfaced to $\frac{5}{8}$ or $\frac{3}{4}$ in before gluing, may be used for laminating curved members when the bending radius is too short to permit use of nominal 2-in-thick laminations if the bending-radius limitations for the species are observed. Other lamination thicknesses may be used to meet special curving requirements.

11.1.5 Section Properties of Wood Members

Sectional properties of solid-sawn lumber and timber and glue-laminated timber members are shown on the web page for the American Institute of Timber Construction (AITC) and listed in AITC's "Timber Construction Manual," 4th ed., published by John Wiley & Sons (www.wiley.com).

11.2 Structural Grading of Wood

Strength properties of wood are intimately related to moisture content and specific gravity. Therefore, data on strength properties unaccompanied by corresponding data on these physical properties are of little value.

The strength of wood is actually affected by many other factors, such as rate of loading, duration of load, temperature, direction of grain, and position of growth rings. Strength is also influenced by such inherent growth characteristics as knots, cross grain, shakes, and checks.

Analysis and integration of available data have yielded a comprehensive set of simple principles for grading structural lumber.

The same characteristics, such as knots and cross grain, that reduce the strength of solid timber also affect the strength of laminated members. However, additional factors peculiar to laminated wood must be considered: Effect on strength of bending members is less from knots located at the neutral plane of the beam, a region of low stress. Strength of a bending member with low-grade laminations can be improved by substituting a few high-grade laminations at the top and bottom of the member. Dispersement of knots in laminated members has a beneficial effect on strength. With sufficient knowledge of the occurrence of knots within a grade, mathematical estimates of this effect may be established for members containing various numbers of laminations.

Design values taking these factors into account are higher than for solid timbers of comparable grade. But cross-grain limitations must be more restrictive than for solid timbers, to justify these higher design values.

Table 11.2 Nominal and Minimum Dressed Sizes of Boards, Dimension, and Timbers

	Thickness, in			Face Width, in		
		Minimum Dressed			Minimum Dressed	
Item	Nominal	Dry*	Green[†]	Nominal	Dry	Green[†]
Boards	1	$\frac{3}{4}$	$\frac{25}{32}$	2	$1\frac{1}{2}$	$1\frac{9}{16}$
	$1\frac{1}{4}$	1	$1\frac{1}{32}$	3	$2\frac{1}{2}$	$2\frac{9}{16}$
	$1\frac{1}{2}$	$1\frac{1}{4}$	$1\frac{9}{16}$	4	$3\frac{1}{2}$	$3\frac{9}{16}$
				5	$4\frac{1}{2}$	$4\frac{5}{8}$
				6	$5\frac{1}{2}$	$5\frac{5}{8}$
				7	$6\frac{1}{2}$	$6\frac{5}{8}$
				8	$7\frac{1}{4}$	$7\frac{1}{2}$
				9	$8\frac{1}{4}$	$8\frac{1}{2}$
				10	$9\frac{1}{4}$	$9\frac{1}{2}$
				11	$10\frac{1}{4}$	$10\frac{1}{2}$
				12	$11\frac{1}{4}$	$11\frac{1}{2}$
				14	$13\frac{1}{4}$	$13\frac{1}{2}$
				16	$15\frac{1}{4}$	$15\frac{1}{2}$
Dimension	2	$1\frac{1}{2}$	$1\frac{9}{16}$	2	$1\frac{1}{2}$	$1\frac{9}{16}$
	$2\frac{1}{2}$	2	$2\frac{1}{16}$	3	$2\frac{1}{2}$	$2\frac{9}{16}$
	3	$2\frac{1}{2}$	$2\frac{9}{16}$	4	$3\frac{1}{2}$	$3\frac{9}{16}$
	$3\frac{1}{2}$	3	$3\frac{1}{16}$	5	$4\frac{1}{2}$	$4\frac{5}{8}$
				6	$5\frac{1}{2}$	$5\frac{5}{8}$
				8	$7\frac{1}{4}$	$7\frac{1}{2}$
				10	$9\frac{1}{4}$	$9\frac{1}{2}$
				12	$11\frac{1}{4}$	$11\frac{1}{2}$
				14	$13\frac{1}{4}$	$13\frac{1}{2}$
				16	$15\frac{1}{4}$	$15\frac{1}{2}$
	4	$3\frac{1}{2}$	$3\frac{9}{16}$	2	$1\frac{1}{2}$	$1\frac{9}{16}$
	$4\frac{1}{2}$	4	$4\frac{1}{16}$	3	$2\frac{1}{2}$	$2\frac{9}{16}$
				4	$3\frac{1}{2}$	$3\frac{9}{16}$
				5	$4\frac{1}{2}$	$4\frac{5}{8}$
				6	$5\frac{1}{2}$	$5\frac{5}{8}$
				8	$7\frac{1}{4}$	$7\frac{1}{2}$
				10	$9\frac{1}{4}$	$9\frac{1}{2}$
				12	$11\frac{1}{4}$	$11\frac{1}{2}$
				14		$13\frac{1}{2}$
				16		$15\frac{1}{2}$
Timbers	5 and thicker	$\frac{1}{2}$ in less		5 and wider	$\frac{1}{2}$ in less	

* Dry lumber is defined as lumber seasoned to a moisture content of 19% or less.

[†] Green lumber is defined as lumber having a moisture content in excess of 19%.

Table 11.3 Standard Nominal and Finished Widths of Glued-Laminated Timber

Nominal width of stock, in	3	4	6	8	10	12	14	16
Net finished width, in (western softwoods)	$2\frac{1}{8}$	$3\frac{1}{8}$	$5\frac{1}{8}$	$6\frac{3}{4}$	$8\frac{3}{4}$	$10\frac{3}{4}$	$12\frac{1}{4}$	$14\frac{1}{4}$
Net finished width, in (southern pine)	$2\frac{1}{8}$	3 or $3\frac{1}{8}$	5 or $5\frac{1}{8}$	$6\frac{3}{4}$	$8\frac{1}{2}$	$10\frac{3}{4}$	$12\frac{1}{4}$	$14\frac{1}{4}$

11.3 Design Values for Lumber, Timber, and Structural Glued-Laminated Timber

Testing a species to determine average strength properties should be carried out from either of two viewpoints:

1. Tests should be made on specimens of large size containing defects. Practically all structural uses involve members of this character.

2. Tests should be made on small, clear specimens to provide fundamental data. Factors to account for the influence of various characteristics may be applied to establish the design values of structural members.

Tests made in accordance with the first viewpoint have the disadvantage that the results apply only to the particular combination of characteristics existing in the test specimens. To determine the strength corresponding to other combinations requires additional tests; thus, an endless testing program is necessary. The second viewpoint permits establishment of fundamental strength properties for each species and application of general rules to cover the specific conditions involved in a particular case.

This second viewpoint has been generally accepted. When a species has been adequately investigated under this concept, there should be no need for further tests on that species unless new conditions arise.

Basic stresses are essentially unit stresses applicable to clear and straight-grained defect-free material. These stresses, derived from the results of tests on small, clear specimens of green wood, include an adjustment for variability of material, length of loading period, and factor of safety. They

are considerably less than the average for the species. They require only an adjustment for grade to become allowable unit stresses.

Allowable unit stresses are computed for a particular grade by reducing the basic stress according to the limitations on defects for that grade. The basic stress is multiplied by a strength ratio to obtain an allowable stress. This strength ratio represents that proportion of the strength of a defect-free piece that remains after taking into account the effect of strength-reducing features.

The principal factors entering into the establishment of allowable unit stress for each species include inherent strength of wood, reduction in strength due to natural growth characteristics permitted in the grade, effect of long-time loading, variability of individual species, possibility of some slight overloading, characteristics of the species, size of member and related influence of seasoning, and factor of safety. The effect of these factors is a strength value for practical-use conditions lower than the average value taken from tests on small, clear specimens.

When moisture content in a member will be low throughout its service, a second set of higher basic stresses, based on the higher strength of dry material, may be used. Technical Bulletin 479, U.S. Department of Agriculture, "Strength and Related Properties of Woods Grown in the United States," presents tests results on small, clear, and straight-grained wood species in the green state and in the 12%-moisture-content, air-dry condition.

Design values for an extensive range of sawn lumber and timber are tabulated in "National Design Specification for Wood Construction," (NDS), American Forest and Paper Association (AFPA), 1111 19th St., N. W., Suite 800, Washington, DC 20036 (www.afandapa.org).

Lumber ▪ Design values for lumber are contained in grading rules established by the National

Lumber Grades Authority (Canadian), Northeastern Lumber Manufacturers Association, Northern Softwood Lumber Bureau, Redwood Inspection Service, Southern Pine Inspection Bureau, West Coast Lumber Inspection Bureau, and Western Wood Products Association. Design values for most species and grades of visually graded dimension lumber are based on provisions in "Establishing Allowable Properties for Visually Graded Dimension Lumber from In-Grade Tests of Full-Size Specimens," ASTM D1990. Design values for visually graded timbers, decking, and some species and grades of dimension lumber are based on provisions of "Establishing Structural Grades and Related Allowable Properties for Visually Graded Lumber," ASTM D245. This standard specifies adjustments to be made in the strength properties of small clear specimens of wood, as determined in accordance with "Establishing Clear Wood Strength Values," ASTM D2555, to obtain design values applicable to normal conditions of service. The adjustments account for the effects of knots, slope of grain, splits, checks, size, duration of load, moisture content, and other influencing factors. Lumber structures designed with working stresses derived from D245 procedures and standard design criteria have a long history of satisfactory performance.

Design values for machine stress-rated (MSR) lumber and machine-evaluated lumber (MEL) are based on nondestructive tests of individual wood pieces. Certain visual-grade requirements also apply to such lumber. The stress rating system used for MSR lumber and MEL is checked regularly by the responsible grading agency for conformance with established certification and quality-control procedures.

Glued-Laminated Timber ▪ Design values for glued-laminated timber, developed by the American Institute of Timber Construction (AITC) and published by American Wood Systems (AWS) in accordance with principles originally established by the U.S. Forest Products Laboratory, are included in the NDS. The principles are the basis for the "Standard Method for Establishing Stresses for Structural Glued-Laminated Timber (Glulam)," ASTM D3737. It requires determination of the strength properties of clear, straight-grained lumber in accordance with the methods of ASTM D2555 or as given in a table in D3737. The

ASTM test method also specifies procedures for obtaining design values by adjustments to those properties to account for the effects of knots, slope of grain, density, size of member, curvature, number of laminations, and other factors unique to laminating.

See also Art. 11.4.

11.4 Adjustment Factors for Design Values

Design values obtained by the methods described in Art. 11.2 should be multiplied by adjustment factors based on conditions of use, geometry, and stability. The adjustments are cumulative, unless specifically indicated in the following.

The adjusted design value F_b' for extreme-fiber bending is given by

$$F_b' = F_b C_D C_M C_t C_L C_F C_V C_r C_c \qquad (11.1)$$

where F_b = design value for extreme-fiber bending

C_D = load-duration factor (Art. 11.4.2)

C_M = wet-service factor (Art. 11.4.1)

C_t = temperature factor (Art. 11.4.3)

C_L = beam stability factor (Arts. 11.4.6 and 11.5)

C_F = size factor—applicable only to visually graded, sawn lumber and round timber flexural members (Art. 11.4.4)

C_V = volume factor—applicable only to glued-laminated beams (Art. 11.4.4)

C_r = repetitive-member factor—applicable only to dimension-lumber beams 2 to 4 in thick (Art. 11.4.9)

C_c = curvature factor—applicable only to curved portions of glued-laminated beams (Art. 11.4.8)

For glued-laminated beams, use either C_L or C_V, whichever is smaller, not both, in Eq. (11.1).

The adjusted design value for tension F_t' is given by

$$F_t' = F_t C_D C_M C_t C_F \qquad (11.2)$$

where F_t = design value for tension.

For shear, the adjusted design value F_V' is computed from

$$F_V' = F_V C_D C_M C_t C_H \qquad (11.3)$$

where F_V = design value for shear and C_H = shear stress factor ≥ 1—permitted for F_V parallel to the grain for sawn lumber members (Art. 11.4.12).

For compression perpendicular to the grain, the adjusted design value $F'_{c\perp}$ is obtained from

$$F'_{c\perp} = F_{c\perp}C_M C_t C_b \qquad (11.4)$$

where $F_{c\perp}$ = design value for compression perpendicular to the grain and C_b = bearing area factor (Art. 11.4.10).

For compression parallel to the grain, the adjusted design value F'_c is given by

$$F'_c = F_c C_D C_M C_t C_F C_P \qquad (11.5)$$

where F_c = design value for compression parallel to grain and C_P = column stability factor (Arts. 11.4.11 and 11.11).

For end grain in bearing parallel to the grain, the adjusted design value F'_g is computed from

$$F'_g = F_g C_D C_t \qquad (11.6)$$

where F_g = design value for end grain in bearing parallel to the grain. See also Art. 11.14.

The adjusted design value for modulus of elasticity E' is obtained from

$$E' = E C_M C_T C \ldots \qquad (11.7)$$

where E = design value for modulus of elasticity

C_T = buckling stiffness factor—applicable only to sawn-lumber truss compression chords 2 × 4 in or smaller, when subject to combined bending and axial compression and plywood sheathing $\frac{3}{8}$ in or more thick is nailed to the narrow face (Art. 11.4.11).

$C \ldots$ = other appropriate adjustment factors

11.4.1 Wet-Service Factor

As indicated in Art 11.1.1, design values should be adjusted for moisture content.

Sawn-lumber design values apply to lumber that will be used under dry-service conditions; that is, where moisture content (MC) of the wood will be a maximum of 19% of the oven-dry weight regardless of MC at time of manufacture. When the MC of structural members in service will exceed 19% for an extended period of time, design values should be multiplied by the appropriate wet-service factor listed in Table 11.4.

Table 11.4 Wet-Service Factors C_M

Design Value	C_M for Sawn Lumber*	C_M for Glulam Timber[†]
F_b	0.85[‡]	0.80
F_t	1.0	0.80
F_V	0.97	0.875
$F_{c\perp}$	0.67	0.53
F_c	0.80[§]	0.73
E	0.90	0.833

* For use where moisture content in service exceeds 19%.
[†] For use where moisture content in service exceeds 16%.
[‡] $C_M = 1.0$ when $F_b C_F \leq 1150$ psi.
[§] $C_M = 1.0$ when $F_c C_F \leq 750$ psi.

MC of 19% or less is generally maintained in covered structures or in members protected from the weather, including windborne moisture. Wall and floor framing and attached sheathing are usually considered to be such dry applications. These dry conditions are generally associated with an average relative humidity of 80% or less. Framing and sheathing in properly ventilated roof systems are assumed to meet MC criteria for dry conditions of use, even though they are exposed periodically to relative humidities exceeding 80%.

Glued-laminated design values apply when the MC in service is less than 16%, as in most covered structures. When MC is 16% or more, design values should be multiplied by the appropriate wet-service factor C_M in Table 11.4.

11.4.2 Load-Duration Factor

Wood can absorb overloads of considerable magnitude for short periods; thus, allowable unit stresses are adjusted accordingly. The elastic limit and ultimate strength are higher under short-time loading. Wood members under continuous loading for years will fail at loads one-half to three-fourths as great as would be required to produce failure in a static-bending test when the maximum load is reached in a few minutes.

Normal load duration contemplates fully stressing a member to the allowable unit stress by the application of the full design load for a duration of about 10 years (either continuously or cumulatively). When the cumulative duration of the full design load differs from 10 years, design values, except $F_{c\perp}$ for compression perpendicular to grain and modulus of elasticity E, should be multiplied by the appropriate load-duration factor C_D listed in Table 11.5.

When loads of different duration are applied to a member, C_D for the load of shortest duration should be applied to the total load. In some cases, a larger-size member may be required when one or more of the shorter-duration loads are omitted. Design of the member should be based on the critical load combination. If the permanent load is equal to or less than 90% of the total combined load, the normal load duration will control the design. Both C_D and the modification permitted in design values for load combinations may be used in design.

The duration factor for impact does not apply to connections or structural members pressure-treated with fire retardants or with waterborne preservatives to the heavy retention required for marine exposure.

Table 11.5 Frequently Used Load-Duration Factors C_D

Load Duration	C_D	Typical Design Loads
Permanent	0.9	Dead load
10 years	1.0	Occupancy live load
2 months	1.15	Snow load
7 days	1.25	Construction load
10 minutes	1.6	Wind or seismic load
Impact	2.0	Impact load

11.4.3 Temperature Factor

Tests show that wood increases in strength as temperature is lowered below normal. Tests conducted at about $-300\,°F$ indicate that the important strength properties of dry wood in bending and compression, including stiffness and shock resistance, are much higher at extremely low temperatures.

Some reduction of the design values for wood may be necessary for members subjected to elevated temperatures for repeated or prolonged periods. This adjustment is especially desirable where high temperature is associated with high moisture content.

Temperature effect on strength is immediate. Its magnitude depends on the moisture content of the wood and, when temperature is raised, the duration of exposure.

Between 0 and 70 °F, the static strength of dry wood (12% moisture content) roughly increases from its strength at 70 °F about $\frac{1}{3}$ to $\frac{1}{2}$% for each 1 °F decrease in temperature. Between 70 and 150 °F, the strength decreases at about the same rate for each 1 °F increase in temperature. The change is greater for higher wood moisture content.

After exposure to temperatures not much above normal for a short time under ordinary atmospheric conditions, the wood, when temperature is reduced to normal, may recover essentially all its original strength. Experiments indicate that air-dry wood can probably be exposed to temperatures up to nearly 150 °F for a year or more without a significant permanent loss in most strength properties. But its strength while at such temperatures will be temporarily lower than at normal temperature.

When wood is exposed to temperatures of 150 °F or more for extended periods of time, it will be permanently weakened. The nonrecoverable strength loss depends on a number of factors, including moisture content and temperature of the wood, heating medium, and time of exposure. To some extent, the loss depends on the species and size of the piece.

Design values for structural members that will experience sustained exposure to elevated temperatures up to 150 °F should be multiplied by the appropriate temperature factor C_t listed in Table 11.6.

Glued-laminated members are normally cured at temperatures of less than 150 °F. Therefore, no reduction in allowable unit stresses due to temperature effect is necessary for curing.

Adhesives used under standard specifications for structural glued-laminated members, for example, casein, resorcinol-resin, phenol-resin, and melamine-resin adhesives, are not affected substantially by temperatures up to those that char wood. Use of adhesives that deteriorate at high temperatures is not permitted by standard specifications for structural glued-laminated timber. Low temperatures appear to have no significant effect on the strength of glued joints.

Table 11.6 Temperature Factors C_t

Design Values and In-Service Moisture Conditions	$T \leq 100°F$	$100°F < T \leq 125°F$	$125°F < T \leq 150°F$
F_t and E, wet or dry	1.0	0.09	0.9
F_b, F_V, F_c, and $F_{c\perp}$			
Dry	1.0	0.8	0.7
Wet	1.0	0.7	0.5

Modifications for Pressure-Applied Treatments ▪ The design values given for wood also apply to wood treated with a preservative when this treatment is in accordance with American Wood Preservers Association (AWPA) standard specifications, which limit pressure and temperature. Investigations have indicated that, in general, any weakening of timber as a result of preservative treatment is caused almost entirely by subjecting the wood to temperatures and pressures above the AWPA limits.

The effects on strength of all treatments, preservative and fire-retardant, should be investigated, to ensure that adjustments in design values are made when required ("Manual of Recommended Practice," American Wood Preservers Association).

11.4.4 Size and Volume Factors

For visually graded dimension lumber, design values F_b, F_t, and F_c for all species and species combinations, except southern pine, should be multiplied by the appropriate size factor C_F given in Table 11.7 to account for the effects of member size. This factor and the factors used to develop size-specific values for southern pine are based on the adjustment equation given in ASTM D1990. This equation based on in-grade test data, accounts for differences in F_b, F_t, and F_c related to width and in F_b and F_t related to length (test span).

For visually graded timbers (5 × 5 in or larger), when the depth d of a stringer beam, post, or timber

Table 11.7 Size Factors C_F

		F_b			
		Thickness, in			
Grades	Width, in	2 and 3	4	F_t	F_c
Select Structural	2, 3, and 4	1.5	1.5	1.5	1.15
No. 1 and better	5	1.4	1.4	1.4	1.1
No. 1, No. 2,	6	1.3	1.3	1.3	1.1
No. 3	8	1.2	1.3	1.2	1.05
	10	1.1	1.2	1.1	1.0
	12	1.0	1.1	1.0	1.0
	14 and wider	0.9	1.0	0.9	0.9
Stud	2, 3, and 4	1.1	1.1	1.1	1.05
	5 and 6	1.0	1.0	1.0	1.0
Construction and Standard	2, 3, and 4	1.0	1.0	1.0	1.0
Utility	4	1.0	1.0	1.0	1.0
	2 and 3	0.4		0.4	0.6

exceeds 12 in, the design value for bending should be adjusted by the size factor

$$C_F = \left(\frac{12}{d}\right)^{1/9} \tag{11.8}$$

Design values for bending F_b for glued-laminated beams should be adjusted for the effects of volume by multiplying by

$$C_V = K_L \left[\left(\frac{5.125}{b}\right)^{1/x} \left(\frac{12}{d}\right)^{1/x} \left(\frac{21}{L}\right)^{1/x} \right] \tag{11.9}$$

where L = length of beam between inflection points, ft

d = depth, in, of beam

b = width, in, of beam

x = 20 for southern pine

= 10 for other species

K_L = loading condition coefficient (Table 11.8)

For glued-laminated beams, the smaller of C_V and the beam stability factor C_L should be used, not both.

Table 11.8 Loading-Condition Coefficient K_L for Glued-Laminated Beams

Single-Span Beams	
Loading condition	K_L
Concentrated load at midspan	1.09
Uniformly distributed load	1.0
Two equal concentrated loads at third points of span	0.96
Continuous Beams or Cantilevers	
All loading conditions	1.00

11.4.5 Beam Stability Factor

Design values F_b for bending should be adjusted by multiplying by the beam stability factor C_L specified in Art. 11.5. For glued-laminated beams, the smaller value of C_L and the volume factor C_V should be used, not both. See also Art. 11.4.4.

11.4.6 Form Factor

Design values for bending F_b for beams with a circular cross section may be multiplied by a form factor $C_f = 1.18$. For a flexural member with a square cross section loaded in the plane of the diagonal (diamond-shape cross section), C_f may be taken as 1.414.

These form factors ensure that a circular or diamond-shape flexural member has the same moment capacity as a square beam with the same cross-sectional area. If a circular member is tapered, it should be treated as a beam with variable cross section.

11.4.7 Curvature Factor

The radial stress induced by a bending moment in a member of constant cross section may be computed from

$$f_r = \frac{3M}{2Rbd} \tag{11.10}$$

where M = bending moment, in-lb

R = radius of curvature at centerline of member, in

b = width of cross section, in

d = depth of cross section, in

When M is in the direction tending to decrease curvature (increase the radius), tensile stresses occur across the grain. For this condition, the allowable tensile stress across the grain is limited to one-third the allowable unit stress in horizontal shear for southern pine for all load conditions, and for Douglas fir and larch for wind or earthquake loadings. The limit is 15 psi for Douglas fir and larch for other types of loading. These values are subject to modification for duration of load. If these values are exceeded, mechanical reinforcement sufficient to resist all radial tensile stresses is required.

When M is in the direction tending to increase curvature (decrease the radius), the stress is compressive across the grain. For this condition, the design value is limited to that for compression perpendicular to grain for all species.

For the curved portion of members, the design value for wood in bending should be modified by multiplication by the following curvature factor:

$$C_c = 1 - 2000 \left(\frac{t}{R}\right)^2 \tag{11.11}$$

where t = thickness of lamination, in

R = radius of curvature of lamination, in

t/R should not exceed $\frac{1}{100}$ for hardwoods and southern pine, or $\frac{1}{125}$ for softwoods other than southern pine. The curvature factor should not be applied to stress in the straight portion of an assembly, regardless of curvature elsewhere.

The recommended minimum radii of curvature for curved, structural glued-laminated members of Douglas fir are 9 ft 4 in for $\frac{3}{4}$-in laminations, and 27 ft 6 in for $1\frac{1}{2}$-in laminations. Other radii of curvature may be used with these thicknesses, and other radius-thickness combinations may be used.

Certain species can be bent to sharper radii, but the designer should determine the availability of such sharply curved members before specifying them.

11.4.8 Repetitive-Member Factor

Design values for bending F_b may be increased when three or more members are connected so that they act as a unit. The members may be in contact or spaced up to 24 in c to c if joined by transverse load-distributing elements that ensure action of the assembly as a unit. The members may be any piece of dimension lumber subjected to bending, including studs, rafters, truss chords, joists, and decking.

When the criteria are satisfied, the design value for bending of dimension lumber 2 to 4 in thick may be multiplied by the repetitive-member factor $C_r = 1.15$.

A transverse element attached to the underside of framing members and supporting no uniform load other than its own weight and other incidental light loads, such as insulation, qualifies as a load-distributing element only for bending moment associated with its own weight and that of the framing members to which it is attached. Qualifying construction includes subflooring, finish flooring, exterior and interior wall finish, and cold-formed metal siding with or without backing. Such elements should be fastened to the framing

members by approved means, such as nails, glue, staples, or snap-lock joints.

Individual members in a qualifying assembly made of different species or grades are each eligible for the repetitive-member increase in F_b if they satisfy all the preceding criteria.

11.4.9 Bearing Area Factor

Design values for compression perpendicular to the grain $F_{c\perp}$ apply to bearing surfaces of any length at the ends of a member and to all bearings 6 in or more long at other locations. For bearings less than 6 in long and at least 3 in from the end of a member, $F_{c\perp}$ may be multiplied by the bearing area factor

$$C_b = \frac{L_b + 0.375}{L_b} \tag{11.12}$$

where L_b = bearing length, in, measured parallel to grain. Equation (11.12) yields the values of C_b for elements with small areas, such as plates and washers, listed in Table 11.9. For round bearing areas, such as washers, L_b should be taken as the diameter.

11.4.10 Column Stability and Buckling Stiffness Factors

Design values for compression parallel to the grain F_c should be multiplied by the column stability factor C_P given by Eq. (11.13).

$$C_P = \frac{1 + (F_{cE}/F_c^*)}{2c}$$

$$- \sqrt{\left[\frac{1 + (F_{cE}/F_c^*)}{2c}\right]^2 - \frac{(F_{cE}/F_c^*)}{c}} \tag{11.13}$$

where F_c^* = design value for compression parallel to the grain multiplied by all applicable adjustment factors except C_P

$$F_{cE} = K_{cE}E'/(L_e/d)^2$$

Table 11.9 Bearing Area Factors C_b

Bearing length, in	0.50	1.00	1.50	2.00	3.00	4.00	6 or more
Bearing area factor	1.75	1.38	1.25	1.19	1.13	1.10	1.00

E' = modulus of elasticity multiplied by adjustment factors

K_{cE} = 0.3 for visually graded lumber and machine-evaluated lumber

= 0.418 for products with a coefficient of variation less than 0.11

c = 0.80 for solid-sawn lumber

= 0.85 for round timber piles

= 0.90 for glued-laminated timber

For a compression member braced in all directions throughout its length to prevent lateral displacement, $C_P = 1.0$. See also Art. 11.11.

The buckling stiffness of a truss compression chord of sawn lumber subjected to combined flexure and axial compression under dry service conditions may be increased if the chord is 2×4 in or smaller and has the narrow face braced by nailing to plywood sheathing at least $\frac{3}{8}$ in thick in accordance with good nailing practice. The increased stiffness may be accounted for by multiplying the design value of the modulus of elasticity E by the buckling stiffness factor C_T in column stability calculations. When the effective column length L_e, in, is 96 in or less, C_T may be computed from

$$C_T = 1 + \frac{K_M L_e}{K_T E} \qquad (11.14)$$

where K_M = 2300 for wood seasoned to a moisture content of 19% or less at time of sheathing attachment

= 1200 for unseasoned or partly seasoned wood at time of sheathing attachment

K_T = 0.59 for visually graded lumber and machine-evaluated lumber

= 0.82 for products with a coefficient of variation of 0.11 or less

When L_e is more than 96 in, C_T should be calculated from Eq. (11.14) with $L_e = 96$ in. For additional information on wood trusses with metal-plate connections, see design standards of the Truss Plate Institute, Madison, Wisconsin.

11.4.11 Shear Stress Factor

For dimension-lumber grades of most species or combinations of species, the design value for shear parallel to the grain F_V is based on the assumption that a split, check, or shake that will reduce shear strength 50% is present. Reductions exceeding 50% are not required inasmuch as a beam split lengthwise at the neutral axis will still resist half the bending moment of a comparable unsplit beam. Furthermore, each half of such a fully split beam will sustain half the shear load of the unsplit member. The design value F_V may be increased, however, when the length of split or size of check or shake is known and is less than the maximum length assumed in determination of F_V, if no increase in these dimensions is anticipated. In such cases, F_V may be multiplied by a shear stress factor C_H greater than unity.

In most design situations, C_H cannot be applied because information on length of split or size of check or shake is not available. The exceptions, when C_H can be used, include structural components and assemblies manufactured fully seasoned with control of splits, checks, and shakes when the products, in service, will not be exposed to the weather. C_H also may be used in evaluation of the strength of members in service. The "National Design Specification for Wood Construction," American Forest and Paper Association, lists values of C_H for lumber and timber of various species.

11.5 Lateral Support of Wood Framing

To prevent beams and compression members from buckling, they may have to be braced laterally. Need for such bracing and required spacing depend on the unsupported length and cross-sectional dimensions of members.

When buckling occurs, a member deflects in the direction of its least dimension b, unless prevented by bracing. (In a beam, b usually is taken as the width.) But if bracing precludes buckling in that direction, deflection can occur in the direction of the perpendicular dimension d. Thus, it is logical that unsupported length L, b, and d play important roles in rules for lateral support, or in formulas for reducing allowable stresses for buckling.

For flexural members, design for lateral stability is based on a function of Ld/b^2. For solid-sawn beams of rectangular cross section, maximum depth-width ratios should satisfy the approximate rules, based on nominal dimensions, summarized

Table 11.10 Approximate Lateral-Support Rules for Lumber Beams*

Depth-Width Ratio (Nominal Dimensions)	Rule
2 or less	No lateral support required
3	Hold ends in position
4	Hold ends in position and member in line, e.g., with purlins or sag rods
5	Hold ends in position and compression edge in line, e.g., with direct connection of sheathing, decking, or joists
6	Hold ends in position and compression edge in line, as for 5 to 1, and provide adequate bridging or blocking at intervals not exceeding 6 times the depth
7	Hold ends in position and both edges firmly in line.

If a beam is subject to both flexure and compression parallel to grain, the ratio may be as much as 5:1 if one edge is held firmly in line, e.g., by rafters (or roof joists) and diagonal sheathing. If the dead load is sufficient to induce tension on the underside of the rafters, the ratio for the beam may be 6:1.

* From "National Specification for Wood Construction," American Forest and Paper Association.

in Table 11.10. When the beams are adequately braced laterally, the depth of the member below the brace may be taken as the width.

No lateral support is required when the depth does not exceed the width. In that case also, the design value does not have to be adjusted for lateral instability. Similarly, if continuous support prevents lateral movement of the compression flange, lateral buckling cannot occur and the design value need not be reduced.

When the depth of a flexural member exceeds the width, bracing must be provided at supports. This bracing must be so placed as to prevent rotation of the beam in a plane perpendicular to its longitudinal axis. Unless the compression flange is braced at sufficiently close intervals between the supports, the design value should be adjusted for lateral buckling.

The slenderness ratio R_B for beams is defined by

$$R_B = \sqrt{\frac{L_e d}{b^2}} \qquad (11.15)$$

The slenderness ratio should not exceed 50.

The effective length L_e for Eq. (11.15) is given in terms of unsupported length of beam in Table 11.11. Unsupported length is the distance between supports or the length of a cantilever when the beam is laterally braced at the supports to prevent rotation and adequate bracing is not installed elsewhere in the span. When both rotational and lateral displacement are also prevented at intermediate points, the unsupported length may be taken as the distance between points of lateral support. If the compression edge is supported throughout the length of the beam and adequate bracing is installed at the supports, the unsupported length is zero.

Acceptable methods of providing adequate bracing at supports include anchoring the bottom of a beam to a pilaster and the top of the beam to a parapet; for a wall-bearing roof beam, fastening the roof diaphragm to the supporting wall or installing a girt between beams at the top of the wall; for beams on wood columns, providing rod bracing.

For continuous lateral support of a compression flange, composite action is essential between deck elements, so that sheathing or deck acts as a diaphragm. One example is a plywood deck with edge nailing. With plank decking, nails attaching the plank to the beams must form couples, to resist rotation. In addition, the planks must be nailed to each other, for diaphragm action. Adequate lateral support is not provided when only one nail is used per plank and no nails are used between planks.

The beam stability factor C_L may be calculated from

$$C_L = \frac{1 + (F_{bE}/F_b^*)}{1.9}$$

$$- \sqrt{\left[\frac{1 + (F_{bE}/F_b^*)}{1.9}\right]^2 - \frac{F_{bE}/F_b^*}{0.95}} \qquad (11.16)$$

Table 11.11 Effective Length L_e for Lateral Stability of Beams*

Loading	For Depth Greater than Width[†]	For Loads from Secondary Framing[‡]
Simple Beam[§]		
Uniformly distributed load	$1.63L_u + 3d$	
Load concentrated at midspan	$1.37L_u + 3d$	$1.11L_u$
Equal end moments	$1.84L_u$	
Equal concentrated loads at third points		$1.68L_u$
Equal concentrated loads at quarter points		$1.54L_u$
Equal concentrated loads at fifth points		$1.68L_u$
Cantilever[§]		
Uniformly distributed load	$0.90L_u + 3d$	
Concentrated load on the end	$1.44L_u + 3d$	

* As specified in the "National Design Specification for Wood Construction," American Forest and Paper Association.

[†] L_u = clear span when depth d exceeds width b and lateral support is provided to prevent rotational and lateral displacement at bearing points in a plane normal to the beam longitudinal axis and no lateral support is provided elsewhere.

[‡] L_u = maximum spacing of secondary framing, such as purlins, when lateral support is provided at bearing points and the framing members prevent lateral displacement of the compression edge of the beam at the connections.

[§] For a conservative value of L_e for any loading on simple beams or cantilevers, use $1.63L_u + 3d$ when $L_u/d > 14.3$ and $1.84L_u$ when $L_u/d > 14.3$.

where $F_b^* =$ design value for bending multiplied by all applicable adjustment factors except C_{fu}, C_V, and C_L (Art. 11.4)

$$F_{bE} = K_{bE}E'/R_B^2$$

$K_{bE} = 0.438$ for visually graded lumber and machine-evaluated lumber

$= 0.609$ for products with a coefficient of variation of 0.11 or less

$E' =$ design modulus of elasticity multiplied by applicable adjustment factors (Art. 11.4)

(American Institute of Timber Construction (www.aitc-glulam.org), "Timber Construction Manual," 4th ed., John Wiley & Sons, Inc., New York (www.wiley.com); "National Design Specification," American Forest and Paper Association (www.afandpa.org); "Western Woods Use Book," Western Wood Products Association, 522 S.W. Fifth Ave., Portland, OR 97204 (www.wwpa.org).)

11.6 Manufacture of Glued-Laminated Lumber

Structural glued-laminated lumber is made by bonding together layers of lumber with adhesive so that the grain direction of all laminations is essentially parallel. Narrow boards may be edge-glued; short boards, end-glued; and the resultant wide and long laminations then face-glued into large, shop-grown timbers.

Recommended practice calls for lumber of nominal 1- and 2-in thicknesses for laminating. The thinner laminations are generally used in curved members.

Depth of constant-depth members normally is a multiple of the thickness of the lamination stock used. Depths of variable-depth members, due to tapering or special assembly techniques, may not be exact multiples of these lamination thicknesses.

Industry-standard finished widths correspond to the nominal widths in Table 11.3 after allowance for drying and surfacing of nominal lumber widths. Standard widths are most economical since they represent the maximum width of board normally obtained from the lumber stock used in laminating.

When members wider than the stock available are required, laminations may consist of two boards side by side. These edge joints must be staggered, vertically in horizontally laminated beams (load acting normal to wide faces of laminations) and horizontally in vertically laminated beams (load acting normal to the edge of laminations). In horizontally laminated beams, edge joints need not be edge-glued. Edge gluing is required in vertically laminated beams. The objective when creating long laminations required from lumber of shorter lengths is to avoid butt joints at the lumber ends. Wood, being of a hollow tube structure, does not bond well end to end.

Edge and face gluings are the simplest to make, end gluings the most difficult. Ends are also the most difficult surfaces to machine. Scarfs or finger joints generally are used to avoid end gluing.

A plane sloping scarf (Fig. 11.2), in which the tapered surfaces of laminations are glued together, can develop 85 to 90% of the strength of an unscarfed, clear, straight-grained control specimen. A relatively flat slope on the plane scarf or on the individual slopes of the finger joint provide gluing surfaces that can give high shear resistance to a tension parallel to grain force along the lamination. Finger joints (Fig. 11.3) are less wasteful of lumber. Quality can be adequately controlled in machine

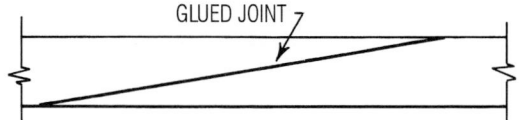

Fig. 11.2 Plane sloping scarf.

cutting and in high-frequency gluing. A combination of thin tip, flat slope on the side of the individual fingers, and a narrow pitch is desired. The length of fingers should be kept short for savings of lumber but long for maximum strength. In testing the quality of glued end joints, the objective is failure to occur in the wood as opposed to adhesive failure.

The usefulness of structural glued-laminated timbers is determined by the lumber used and glue joint produced. Certain combinations of adhesive, treatment, and wood species do not produce the same quality of glue bond as other combinations, although the same gluing procedures are used.

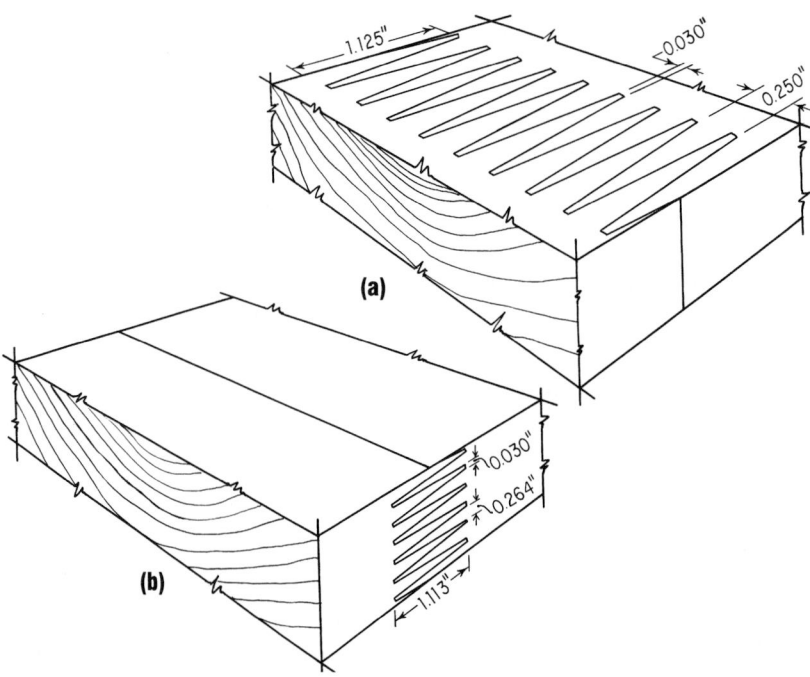

Fig. 11.3 Finger joint: (*a*) Fingers formed by cuts perpendicular to the wide face of the board; (*b*) fingers formed by cuts perpendicular to the edges.

Thus, a combination must be supported by adequate experience with a laminator's gluing procedure (see also Art. 11.25).

The only adhesives currently recommended for wet-use and preservative-treated lumber, whether gluing is done before or after treatment, are the resorcinol and phenol-resorcinol resins. Melamine and melamine-urea blends are used in smaller amounts for high-frequency curing of end gluings.

Glued joints are cured with heat by several methods. R. F. (high-frequency) curing of glue lines is used for end joints and for limited-size members where there are repetitive gluings of the same cross section. Low-voltage resistance heating, where current is passed through a strip of metal to raise the temperature of a glue line, is used for attaching thin facing pieces. The metal may be left in the glue line as an integral part of the completed member. Printed electric circuits, in conjunction with adhesive films, and adhesive films, impregnated on paper or on each side of a metal conductor placed in the glue line, are other alternatives.

Preheating the wood to ensure reactivity of the applied adhesive has limited application in structural laminating. The method requires adhesive application as a wet or dry film simultaneously to all laminations and then rapid handling of multiple laminations.

Curing the adhesive at room temperature has many advantages. Since wood is an excellent insulator, a long time is required for elevated ambient temperature to reach inner glue lines of a large assembly. With room-temperature curing, equipment needed to heat the glue line is not required, and the possibility of injury to the wood from high temperatures is avoided.

11.7 Fabrication of Structural Timber

Fabrication consists of boring, cutting, sawing, trimming, dapping, routing, planing, and otherwise shaping, framing, and furnishing wood units, sawn or laminated, including plywood, to fit them for particular places in a final structure. Whether fabrication is performed in shop or field, the product must exhibit a high quality of work.

Jigs, patterns, templates, stops, or other suitable means should be used for all complicated and multiple assemblies to insure accuracy, uniformity, and control of all dimensions. All tolerances in cutting, drilling, and framing must comply with good practice in the industry and applicable specifications and controls. At the time of fabrication, tolerances must not exceed those listed below unless they are not critical and not required for proper performance. Specific jobs, however, may require closer tolerances.

Location of Fastenings ▪ Spacing and location of all fastenings within a joint should be in accordance with the shop drawings and specifications with a maximum permissible tolerance of $\pm \frac{1}{16}$ in. The fabrication of members assembled at any joint should be such that the fastenings are properly fitted.

Bolt-Hole Sizes ▪ Bolt holes in all fabricated structural timber, when loaded as a structural joint, should be $\frac{1}{16}$ in larger in diameter than bolt diameter for $\frac{1}{2}$-in and larger-diameter bolts, and $\frac{1}{32}$ in larger for smaller-diameter bolts. Larger clearances may be required for other bolts, such as anchor bolts and tension rods.

Holes and Grooves ▪ Holes for stress-carrying bolts, connector grooves, and connector daps must be smooth and true within $\frac{1}{16}$ in per 12 in of depth. The width of a split-ring connector groove should be within $+0.02$ in of and not less than the thickness of the corresponding cross section of the ring. The shape of ring grooves must conform generally to the cross-sectional shape of the ring. Departure from these requirements may be allowed when supported by test data. Drills and other cutting tools should be set to conform to the size, shape, and depth of holes, grooves, daps, and so on specified in the "National Design Specification for Wood," American Forest and Paper Association.

Lengths ▪ Members should be cut within $\pm \frac{1}{16}$ in of the indicated dimension when they are up to 20 ft long and $\pm \frac{1}{16}$ in per 20 ft of specified length when they are over 20 ft long. Where length dimensions are not specified or critical, these tolerances may be waived.

End Cuts ▪ Unless otherwise specified, all trimmed square ends should be square within $\frac{1}{16}$ in/ft of depth and width. Square or sloped ends to be loaded in compression should be cut to provide contact over substantially the complete surface.

Shrinkage or Swelling Effects on Shape of Curved Members
■ Wood shrinks or swells across the grain but has practically no dimensional change along the grain. Radial swelling causes a decrease in the angle between the ends of a curved member; radial shrinkage causes an increase in this angle.

Such effects may be of great importance in three-hinged arches that become horizontal, or nearly so, at the crest of a roof. Shrinkage, increasing the relative end rotations, may cause a depression at the crest and create drainage problems. For such arches, therefore, consideration must be given to moisture content of the member at time of fabrication and in service and to the change in end angles that results from change in moisture content and shrinkage across the grain.

11.8   Timber Erection

Erection of timber framing requires experienced crews and adequate lifting equipment to protect life and property and to assure that the framing is properly assembled and not damaged during handling.

On receipt at the site, each shipment of timber should be checked for tally and evidence of damage. Before erection starts, plan dimensions should be verified in the field. The accuracy and adequacy of abutments, foundations, piers, and anchor bolts should be determined. And the erector must see that all supports and anchors are complete, accessible, and free from obstructions.

Jobsite Storage
■ If wood members must be stored at the site, they should be placed where they do not create a hazard to other trades or to the members themselves. All framing, and especially glued-laminated members, stored at the site should be set above the ground on appropriate blocking. The members should be separated with strips so that air may circulate around all sides of each member. The top and all sides of each storage pile should be covered with a moisture-resistant covering that provides protection from the elements, dirt, and jobsite debris. (Do not use clear polyethylene films since wood members may be bleached by sunlight.) Individual wrappings should be slit or punctured on the lower side to permit drainage of water that accumulates inside the wrapping.

Glued-laminated members of Premium and Architectural Appearance (and Industrial Appearance in some cases) are usually shipped with a protective wrapping of water-resistant paper. Although this paper does not provide complete freedom from contact with water, experience has shown that protective wrapping is necessary to ensure proper appearance after erection. Used specifically for protection in transit, the paper should remain in place until the roof covering is in place. It may be necessary, however, to remove the paper from isolated areas to make connections from one member to another. If temporarily removed, the paper should be replaced and should remain in position until all the wrapping may be removed.

At the site, to prevent surface marring and damage to wood members, the following precautions should be taken:

Lift members or roll them on dollies or rollers out of railroad cars. Unload trucks by hand or crane. Do not dump, drag, or drop members.

During unloading with lifting equipment, use fabric or plastic belts, or other slings that will not mar the wood. If chains or cables are used, provide protective blocking or padding.

Equipment
■ Adequate equipment of proper load-handling capacity, with control for moving and placing members, should be used for all operations. It should be of such nature as to ensure safe and expedient placement of the material. Cranes and other mechanical devices must have sufficient controls that beams, columns, arches, or other elements can be eased into position with precision. Slings, ropes, cables, or other securing devices must not damage the materials being placed.

The erector should determine the weights and balance points of the framing members before lifting begins so that proper equipment and lifting methods may be employed. When long-span timber trusses are raised from a flat to a vertical position preparatory to lifting, stresses entirely different from normal design stresses may be introduced. The magnitude and distribution of these stresses depend on such factors as weight, dimensions, and type of truss. A competent rigger will consider these factors in determining how much suspension and stiffening, if any, is required and where it should be located.

Accessibility ▪ Adequate space should be available at the site for temporary storage of materials from time of delivery to the site to time of erection. Material-handling equipment should have an unobstructed path from jobsite storage to point of erection. Whether erection must proceed from inside the building area or can be done from outside will determine the location of the area required for operation of the equipment. Other trades should leave the erection area clear until all members are in place and are either properly braced by temporary bracing or permanently braced in the building system.

Assembly and Subassembly ▪ Whether these are done in a shop or on the ground or in the air in the field depends on the structural system and the various connections involved.

Care should be taken with match marking on custom materials. Assembly must be in accordance with the approved shop drawings for the materials. Any additional drilling or dapping, as well as the installation of all field connections, must be done in a workmanlike manner.

Trusses are usually shipped partly or completely disassembled. They are assembled on the ground at the site before erection. Arches, which are generally shipped in half sections, may be assembled on the ground or connections may be made after the half arches are in position. When trusses and arches are assembled on the ground at the site, assembly should be on level blocking to permit connections to be properly fitted and securely tightened without damage. End compression joints should be brought into full bearing and compression plates installed where intended.

Prior to erection, the assembly should be checked for prescribed overall dimensions, prescribed camber, and accuracy of anchorage connections. Erection should be planned and executed in such a way that the close fit and neat appearance of joints and the structure as a whole will not be impaired.

Field Welding ▪ Where field welding is required, the work should be done by a qualified welder in accordance with job plans and specifica-tions, approved shop drawings, and specifications of the American Institute of Steel Construction and the American Welding Society.

Cutting and Fitting ▪ All connections should fit snugly in accordance with job plans and specifications and approved shop drawings. Any field cutting, dapping, or drilling should be done in a workmanlike manner with due consideration given to final use and appearance.

Bracing ▪ Structural elements should be placed to provide restraint or support, or both, to insure that the complete assembly will form a stable structure. This bracing may extend longitudinally and transversely. It may comprise sway, cross, vertical, diagonal, and like members that resist wind, earthquake, erection, acceleration, braking, and other forces. And it may consist of knee braces, cables, rods, struts, ties, shores, diaphragms, rigid frames, and other similar components in combinations.

Bracing may be temporary or permanent. Permanent bracing, required as an integral part of the completed structure, is shown on the architectural or engineering plans and usually is also referred to in the job specifications. Temporary construction bracing is required to stabilize or hold in place permanent structural elements during erection until other permanent members that will serve the purpose are fastened in place. This bracing is the responsibility of the erector, who normally furnishes and erects it. It should be attached so that children and other casual visitors cannot remove it or prevent it from serving as intended. Protective corners and other protective devices should be installed to prevent members from being damaged by the bracing.

In timber-truss construction, temporary bracing can be used to plumb trusses during erection and hold them in place until they receive the rafters and roof sheathing. The major portion of temporary bracing for trusses is left in place because it is designed to brace the completed structure against lateral forces.

Failures during erection occur occasionally and regardless of construction material used. The blame can usually be placed on insufficient or improperly located temporary erection guys or braces, overloading with construction materials, or an externally applied force sufficient to render temporary erection bracing ineffective.

Structural members of wood must be stiff as well as strong. They must also be properly guyed or laterally braced, both during erection and permanently in the completed structure. Large rectangular cross sections of glued-laminated timber have relatively high lateral strength and resistance to torsional stresses during erection. However, the erector must never assume that a wood arch, beam, or column cannot buckle during handling or erection.

Specifications often require that:

1. Temporary bracing shall be provided to hold members in position until the structure is complete.

2. Temporary bracing shall be provided to maintain alignment and prevent displacement of all structural members until completion of all walls and decks.

3. The erector should provide adequate temporary bracing and take care not to overload any part of the structure during erection.

The magnitude of the restraining force that should be provided by a cable guy or brace cannot be precisely determined, but general experience indicates that a brace is adequate if it supplies a restraining force equal to 2% of the applied load on a column or of the force in the compression flange of a beam. It does not take much force to hold a member in line, but once it gets out of alignment, the force then necessary to hold it is substantial.

11.9 Design Recommendations

The following recommendations aim at achieving economical designs with wood framing:

Use standard sizes and grades of lumber. Consider using standardized structural components, whether lumber, stock glued beams, or complex framing designed for structural adequacy, efficiency, and economy.

Use standard details wherever possible. Avoid specially designed and manufactured connecting hardware.

Use as simple and as few joints as possible. Place splices, when required, in areas of lowest stress. Do not locate splices where bending moments are large, and thus avoid design, erection, and fabrication difficulties.

Avoid unnecessary variations in cross section of members along their length.

Use identical member designs repeatedly throughout a structure, whenever practicable. Keep the number of different arrangements to a minimum.

Consider using roof profiles that favorably influence the type and amount of load on the structure.

Specify design values rather than the lumber grade or combination of grades to be used.

Select an adhesive suitable for the service conditions, but do not overspecify. For example, waterproof resin adhesives need not be used where less expensive water-resistant adhesives will do the job.

Use lumber treated with preservatives where service conditions dictate. Such treatment need not be used where decay hazards do not exist. Fire-retardant treatments may be used to meet a specific flame-spread rating for interior finish but are not necessary for large-cross-sectional members that are widely spaced and already a low fire risk.

Instead of long, simple spans, consider using continuous or suspended spans or simple spans with overhangs.

Select an appearance grade best suited to the project. Do not specify premium appearance grade for all members if it is not required.

Table 11.12 is a guide to economical span ranges for roof and floor framing in buildings.

Designing for Fire Safety ▪ Maximum protection of the occupants of a building and the property itself can be achieved in timber design by taking advantage of the fire-endurance properties of wood in large cross sections and by close attention to details that make a building fire-safe. Building materials alone, building features alone, or detection and fire-extinguishing equipment alone cannot provide maximum safety from fire in buildings. A proper combination of these three factors will provide the necessary degree of protection for the occupants and the property.

Table 11.12 Economical Span Range for Framing Members

Framing Member	Economical Span Range, ft	Usual Spacing, ft
Roof beams (generally used where a flat or low-pitched roof is desired):		
Simple span:		
Constant depth		
Solid-sawn	0–40	4–20
Glued-laminated	20–100	8–24
Tapered	25–100	8–24
Double tapered (pitched beams)	25–100	8–24
Curved beams	25–100	8–24
Simple beam with overhangs (usually more economical than simple span when span is over 40 ft):		
Solid-sawn	24	4–20
Glued-laminated	10–90	8–24
Continuous span:		
Solid-sawn	10–50	4–20
Glued-laminated	10–50	8–24
Arches (three-hinged for relatively high-rise applications and two-hinged for relatively low-rise applications):		
Three-hinged:		
Gothic	40–90	8–24
Tudor	30–120	8–24
A-frame	20–160	8–24
Three-centered	40–250	8–24
Parabolic	40–250	8–24
Radial	40–250	8–24
Two-hinged:		
Radial	50–200	8–24
Parabolic	50–200	8–24
Trusses (provide openings for passage of wires, piping, etc.)		
Flat or parallel chord	50–150	12–20
Triangular or pitched	50–90	12–20
Bowstring	50–200	14–24
Tied arches (where no ceiling is desired and where a long, clear span is desired with low rise):		
Tied segment	50–100	8–20
Buttressed segment	50–200	14–24
Domes	50–350	8–24
Simple-span floor beams:		
Solid-sawn	6–20	4–12
Glued-laminated	6–40	4–16
Continuous floor beams	25–40	4–16
Roof sheathing and decking		
1-in sheathing	1–4	
2-in sheathing	6–10	
3-in roof deck	8–15	
4-in roof deck	12–20	
Plywood sheathing	1–4	
Sheathing on roof joists	1.33–2	
Plank floor decking (floor and ceiling in one):		
Edge to edge	4–16	
Wide face to wide face	4–16	

The following should be investigated:

Degree of protection needed, as dictated by occupancy or operations taking place

Number, size, type (such as direct to the outside), and accessibility of exits (particularly stairways), and their distance from each other

Installation of automatic alarm and sprinkler systems

Separation of areas in which hazardous processes or operations take place, such as boiler rooms and workshops

Enclosure of stairwells and use of self-closing fire doors

Fire stopping and elimination, or proper protection of concealed spaces

Interior finishes to assure surfaces that will not spread flame at hazardous rates

Roof venting equipment or provision of draft curtains where walls might interfere with production operations

When exposed to fire, wood forms a self-insulating surface layer of char, which provides its own fire protection. Even though the surface chars, the undamaged wood beneath retains its strength and will support loads in accordance with the capacity of the uncharred section. Heavy-timber members have often retained their structural integrity through long periods of fire exposure and remained serviceable after the charred surfaces have been refinished. This fire endurance and excellent performance of heavy timber are attributable to the size of the wood members and to the slow rate at which the charring penetrates.

The structural framing of a building, which is the criterion for classifying a building as combustible or noncombustible, has little to do with the hazard from fire to the building occupants. Most fires start in the building contents and create conditions that render the inside of the structure uninhabitable long before the structural framing becomes involved in the fire. Thus, whether the building is classified as combustible or noncombustible has little bearing on the potential hazard to the occupants. However, once the fire starts in the contents, the material of which the building is constructed can significantly help facilitate evacuation, fire fighting, and property protection.

The most important protection factors for occupants, firefighters, and the property, as well as adjacent exposed property, are prompt detection of the fire, immediate alarm, and rapid extinguishment of the fire. Firefighters do not fear fires in buildings of heavy-timber construction as they do those in buildings of many other types of construction. They need not fear sudden collapse without warning; they usually have adequate time, because of the slow-burning characteristics of the timber, to ventilate the building and fight the fire from within the building or on top.

With size of member of particular importance to fire endurance of wood members, building codes specify minimum dimensions for structural members and classify buildings with wood framing as heavy-timber construction, ordinary construction, or wood-frame construction.

Heavy-timber construction is that type in which fire resistance is attained by placing limitations on the minimum size, thickness, or composition of all load-carrying wood members; by avoidance of concealed spaces under floors and roofs; by use of approved fastenings, construction details, and adhesives; and by providing the required degree of fire resistance in exterior and interior walls. (See AITC 108, "Heavy Timber Construction," American Institute of Timber Construction.)

Ordinary construction has exterior masonry walls and wood-framing members of sizes smaller than heavy-timber sizes.

Wood-frame construction has wood-framed walls and structural framing of sizes smaller than heavy-timber sizes.

Depending on the occupancy of a building or hazard of operations within it, a building of frame or ordinary construction may have its members covered with fire-resistive coverings. The interior finish on exposed surfaces of rooms, corridors, and stairways is important from the standpoint of its tendency to ignite, flame, and spread fire from one location to another. The fact that wood is combustible does not mean that it will spread flame at a hazardous rate. Most codes exclude the exposed wood surfaces of heavy-timber structural members from flame-spread requirements because such wood is difficult to ignite and, even with an external source of heat, such as burning contents, is resistant to spread of flame.

Fire-retardant chemicals may be impregnated in wood with recommended retentions to lower the rate of surface flame spread and make the wood self-extinguishing if the external source of heat is removed. After proper surface preparation, the surface is paintable. Such treatments are accepted under several specifications, including federal government and military. They are recommended only for interior or dry-use service conditions or locations protected against leaching. These treatments are sometimes used to meet a specific flame-spread rating for interior finish or as an alternate to noncombustible secondary members and decking meeting the requirements of Underwriters' Laboratories, Inc., NM 501 or NM 502, nonmetallic roof-deck assemblies in otherwise heavy-timber construction.

11.10 Wood Tension Members

The tensile stress f_t parallel to the grain should be computed from P/A_n, where P is the axial load and A_n is the net section area. This stress should not exceed the design value for tension parallel to grain f_t, adjusted as required by Eq. (11.2).

Tensile stress perpendicular to the grain should be avoided as there are no such allowable design values for this condition.

11.11 Wood Columns

Wood compression members may be a solid piece of lumber or timber (Fig. 11.4a), or spaced columns, connector-joined (Fig. 11.4b and c), or built-up (Fig. 11.4d).

Solid Columns ▪ These consist of a single piece of lumber or timber or of pieces glued together to act as a single member. In general,

$$f_c = \frac{P}{A_g} \leq F'_c \qquad (11.17)$$

where P = axial load on the column

A_g = gross area of column

F'_c = design value in compression parallel to grain multiplied by the applicable adjustment factors, including column stability factor C_P given by Eq. (11.13)

There is an exception, however, applicable when holes or other reductions in area are present in the critical part of the column length most susceptible to buckling; for instance, in the portion between supports that is not laterally braced. In that case, f_c should be based on the net section and should not

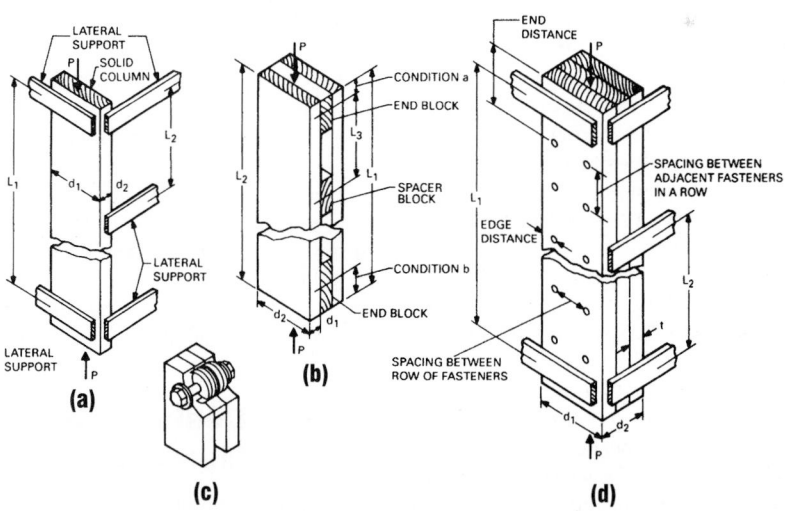

Fig. 11.4 Bracing of wood columns to control length-thickness and depth-thickness ratios: (a) For a solid wood column; (b) For a spaced column (the end distance for condition a should not exceed $L_1/20$ and for condition b should be between $L_1/20$ and $L_1/10$). (c) Shear plate connection in the end block of the spaced column. (d) Bracing for a built-up column. (*From F. S. Merritt and J. T. Ricketts, "Building Design and Construction Handbook," 5th ed., McGraw-Hill Publishing Company, New York.*)

exceed F_c, the design value for compression parallel to grain, multiplied by applicable adjustment factors, except C_P; that is,

$$f_c = \frac{P}{A_n} \le F_c \qquad (11.18)$$

where A_n = net cross-sectional area.

C_P represents the tendency of a column to buckle and is a function of the slenderness ratio. For a rectangular wood column, a modified slenderness ratio, L_e/d, is used, where L_e is the effective unbraced length of column, and d is the smallest dimension of the column cross section. The effective length L_e may be taken as the actual column length multiplied by the appropriate buckling-length coefficient K_e indicated in Fig. 9.5, p. 9.18. For the column in Fig. 11.4a, the slenderness ratio should be taken as the larger of the ratios L_{e1}/d_1 or L_{e2}/d_2, where each unbraced length is multiplied by the appropriate value of K_e. For solid columns, L_e/d should not exceed 50, except that during construction, L_e/d may be as large as 75.

The critical section of columns supporting trusses frequently exists at the connection of knee brace to column. Where no knee brace is used, or the column supports a beam, the critical section for moment usually occurs at the bottom of truss or beam. Then, a rigid connection must be provided to resist moment, or adequate diagonal bracing must be provided to carry wind loads into a support.

(American Institute of Timber Construction (www.aitc.org), "Timber Construction Manual," John Wiley & Sons, Inc., New York (www.wiley.com); "National Design Specification for Wood Construction," American Forest and Paper Association, 1111 19th St., N. W., Washington, DC 20036 (www.afandpa.org).)

Built-up Columns ■ These often are fabricated by joining together individual pieces of lumber with mechanical fasteners, such as nails, spikes, or bolts, to act as a single member (Fig. 11.4d). Strength and stiffness properties of a built-up column are less than those of a solid column with the same dimensions, end conditions, and material (equivalent solid column). Strength and stiffness properties of a built-up column, however, are much greater than those of an unconnected assembly in which individual pieces act as independent columns. Built-up columns obtain their efficiency

from the increase in the buckling resistance of the individual laminations provided by the fasteners. The more nearly the laminations of a built-up column deform together—that is, the smaller the slip between laminations, under compressive load—the greater is the relative capacity of the column compared with an equivalent solid column.

When built-up columns are nailed or bolted in accordance with provisions in the "National Design Specification for Wood Construction," American Forest and Paper Association, the capacity of nailed columns exceeds 60% and of bolted built-up columns, 75% of an equivalent solid column for all L/d ratios. The NDS contains criteria for design of built-up columns based on tests performed on built-up columns with various fastener schedules.

Spaced Columns ■ These consists of the following elements: (1) two or more individual, rectangular wood compression members with their wide faces parallel; (2) wood blocks that separate the members at their ends and one or more points between; and (3) steel bolts through the blocks to fasten the components, with split-ring or shear-plate connectors at the end blocks (Fig. 11.4b). The connectors should be capable of developing required shear resistance.

The advantage of a spaced column over an equivalent solid column is the increase permitted in the design value for buckling for the spaced-column members because of the partial end fixity of those members. The increased capacity may range from $2\frac{1}{2}$ to 3 times the capacity of a solid column. This advantage applies only to the direction perpendicular to the wide faces. Design of the individual members in the direction parallel to the wide faces is the same for each as for a solid column. The NDS gives design criteria, including end fixity coefficients, for spaced columns.

11.12 Design of Wood Flexural Members

Standard beam formulas for bending, shear, and deflection may be used to determine beam and joist sizes. Ordinarily, deflection governs design, but for short, heavily loaded beams, shear is likely to control. Bracing for beam stability is discussed in Art. 11.5. Bearing on beams is treated in Art. 11.14.

Joists are relatively narrow beams, usually spaced 12 to 24 in c to c. They generally are topped with sheathing and braced with diaphragms or cross bridging at intervals up to 10 ft. For joist spacings of 16 to 24 in c to c, 1-in sheathing usually is required. For spacings over 24 in, 2 in or more of wood decking is necessary.

Figure 11.5 shows the types of beams commonly produced in timber. Straight and single- and double-tapered straight beams can be furnished solid-sawn or glued-laminated. The curved surfaces can be furnished only glued-laminated. Beam names describe the top and bottom surfaces of the beam: The first part describes the top surface, the word following the hyphen the bottom. Sawn surfaces on the tension side of a beam should be avoided.

Table 11.13 gives the load-carrying capacity for various cross-sectional sizes of glued-laminated, simply supported beams.

Example ▪ Design a straight, glued-laminated beam, simply supported and uniformly loaded: span, 28 ft; spacing, 9 ft c to c; live load, 30 lb/ft^2; dead load, 5 lb/ft^2 for deck and 7.5 lb/ft^2 for roofing. Allowable bending stress of combination grade is 2400 psi, with modulus of elasticity $E = 1,800,000$ psi. Deflection limitation is $L/180$, where L is the span, ft. Assume the beam is laterally supported by the deck throughout its length and held in line at the ends.

With a 15% increase for short-duration loading, the allowable bending stress F_b becomes 2760 psi and the allowable horizontal shear F_v, 230 psi.

Assume the beam will weigh 22.5 lb/lin ft, averaging 2.5 lb/ft^2. Then, the total uniform load comes to 45 lb/ft^2. So the beam carries $w = 45 \times 9 = 405$ lb/lin ft.

The end shear $V = wL/2$ and the maximum shearing stress $= 3V/2 = 3wL/4$. Hence, the required area, in^2, for horizontal shear is

$$A = \frac{3wL}{4F_v} = \frac{wL}{306.7} = \frac{405 \times 28}{306.7} = 37.0$$

The required section modulus, in^3, is

$$S = \frac{1.5wL^2}{F_b} = \frac{1.5 \times 405 \times 28^2}{2760} = 172.6$$

If $D = 180$, the reciprocal of the deflection limitation, then the maximum deflection equals $5 \times 1728wL^4/384EI \leq 12L/D$, where I is the moment of inertia of the beam cross section, in^4. Hence, to control deflection, the moment of inertia must be at least

$$I = \frac{1.875DwL^3}{E}$$
$$= \frac{1.875 \times 180 \times 405 \times 28^3}{1,800,000} = 1688 \text{ in}^4$$

Assume that the beam will be fabricated with 1½-in laminations. The most economical section satisfying all three criteria is $5\frac{1}{8} \times 16\frac{1}{2}$, with $A = 84.6$, $S = 232.5$, and $I = 1918.5$. But it has a volume factor of 0.97, so the allowable bending stress must be reduced to $2760 \times 0.97 = 2677$ psi. And the required section modulus must be increased accordingly to $172.6/0.97 = 178$. Nevertheless, the selected section still is adequate.

Suspended-Span Construction ▪ Cantilever systems may comprise any of the various types and combinations of beam illustrated in Fig. 11.6. Cantilever systems permit longer spans or larger loads for a given size member than do simple-span systems if member size is not controlled by compression perpendicular to grain at the supports or by horizontal shear. Substantial design economies can be effected by decreasing the

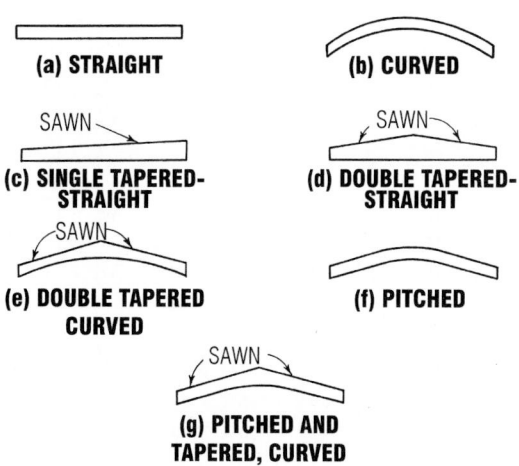

(a) STRAIGHT

(b) CURVED

SAWN
(c) SINGLE TAPERED-STRAIGHT

SAWN
(d) DOUBLE TAPERED-STRAIGHT

SAWN
(e) DOUBLE TAPERED CURVED

(f) PITCHED

SAWN
(g) PITCHED AND TAPERED, CURVED

Fig. 11.5 Types of timber beams.

Table 11.13 Load-Carrying Capacity of Simple-Span Laminated Beams*

Span, ft	Spacing, ft	Roof Beam Total-Load-Carrying Capacity						Floor Beams Total Load
		$30\,lb/ft^2$	$35\,lb/ft^2$	$40\,lb/ft^2$	$45\,lb/ft^2$	$50\,lb/ft^2$	$55\,lb/ft^2$	$50\,lb/ft^2$
8	4	$3\frac{1}{8} \times 4\frac{1}{2}$	$3\frac{1}{8} \times 4\frac{1}{2}$	$3\frac{1}{8} \times 6$	$3\frac{1}{8} \times 6$	$3\frac{1}{8} \times 6$	$3\frac{1}{8} \times 6$	$3\frac{1}{8} \times 6$
	6	$3\frac{1}{8} \times 4\frac{1}{2}$	$3\frac{1}{8} \times 4\frac{1}{2}$	$3\frac{1}{8} \times 6$	$3\frac{1}{8} \times 6$	$3\frac{1}{8} \times 6$	$3\frac{1}{8} \times 6$	$3\frac{1}{8} \times 6$
	8	$3\frac{1}{8} \times 4\frac{1}{2}$	$3\frac{1}{8} \times 4\frac{1}{2}$	$3\frac{1}{8} \times 6$	$3\frac{1}{8} \times 6$	$3\frac{1}{8} \times 6$	$3\frac{1}{8} \times 6$	$3\frac{1}{8} \times 7\frac{1}{2}$
10	4	$3\frac{1}{8} \times 4\frac{1}{2}$	$3\frac{1}{8} \times 4\frac{1}{2}$	$3\frac{1}{8} \times 6$	$3\frac{1}{8} \times 6$	$3\frac{1}{8} \times 6$	$3\frac{1}{8} \times 6$	$3\frac{1}{8} \times 7\frac{1}{2}$
	6	$3\frac{1}{8} \times 4\frac{1}{2}$	$3\frac{1}{8} \times 6$	$3\frac{1}{8} \times 6$	$3\frac{1}{8} \times 6$	$3\frac{1}{8} \times 6$	$3\frac{1}{8} \times 7\frac{1}{2}$	$3\frac{1}{8} \times 7\frac{1}{2}$
	8	$3\frac{1}{8} \times 6$	$3\frac{1}{8} \times 6$	$3\frac{1}{8} \times 7\frac{1}{2}$	$3\frac{1}{8} \times 7\frac{1}{2}$	$3\frac{1}{8} \times 7\frac{1}{2}$	$3\frac{1}{8} \times 7\frac{1}{2}$	$3\frac{1}{8} \times 9$
	10	$3\frac{1}{8} \times 6$	$3\frac{1}{8} \times 7\frac{1}{2}$	$3\frac{1}{8} \times 7\frac{1}{2}$	$3\frac{1}{8} \times 7\frac{1}{2}$	$3\frac{1}{8} \times 7\frac{1}{2}$	$3\frac{1}{8} \times 9$	$3\frac{1}{8} \times 9$
12	6	$3\frac{1}{8} \times 6$	$3\frac{1}{8} \times 6$	$3\frac{1}{8} \times 7\frac{1}{2}$	$3\frac{1}{8} \times 7\frac{1}{2}$	$3\frac{1}{8} \times 7\frac{1}{2}$	$3\frac{1}{8} \times 7\frac{1}{2}$	$3\frac{1}{8} \times 9$
	8	$3\frac{1}{8} \times 6$	$3\frac{1}{8} \times 7\frac{1}{2}$	$3\frac{1}{8} \times 9$	$3\frac{1}{8} \times 9$	$3\frac{1}{8} \times 9$	$3\frac{1}{8} \times 9$	$3\frac{1}{8} \times 10\frac{1}{2}$
	10	$3\frac{1}{8} \times 7\frac{1}{2}$	$3\frac{1}{8} \times 7\frac{1}{2}$	$3\frac{1}{8} \times 9$	$3\frac{1}{8} \times 9$	$3\frac{1}{8} \times 9$	$3\frac{1}{8} \times 10\frac{1}{2}$	$3\frac{1}{8} \times 10\frac{1}{2}$
	12	$3\frac{1}{8} \times 7\frac{1}{2}$	$3\frac{1}{8} \times 9$	$3\frac{1}{8} \times 9$	$3\frac{1}{8} \times 9$	$3\frac{1}{8} \times 10\frac{1}{2}$	$3\frac{1}{8} \times 10\frac{1}{2}$	$3\frac{1}{8} \times 12$
14	8	$3\frac{1}{8} \times 7\frac{1}{2}$	$3\frac{1}{8} \times 9$	$3\frac{1}{8} \times 9$	$3\frac{1}{8} \times 9$	$3\frac{1}{8} \times 10\frac{1}{2}$	$3\frac{1}{8} \times 10\frac{1}{2}$	$3\frac{1}{8} \times 12$
	10	$3\frac{1}{8} \times 9$	$3\frac{1}{8} \times 9$	$3\frac{1}{8} \times 10\frac{1}{2}$	$3\frac{1}{8} \times 10\frac{1}{2}$	$3\frac{1}{8} \times 10\frac{1}{2}$	$3\frac{1}{8} \times 12$	$3\frac{1}{8} \times 12$
	12	$3\frac{1}{8} \times 9$	$3\frac{1}{8} \times 10\frac{1}{2}$	$3\frac{1}{8} \times 10\frac{1}{2}$	$3\frac{1}{8} \times 10\frac{1}{2}$	$3\frac{1}{8} \times 12$	$3\frac{1}{8} \times 12$	$3\frac{1}{8} \times 13\frac{1}{2}$
	14	$3\frac{1}{8} \times 10\frac{1}{2}$	$3\frac{1}{8} \times 10\frac{1}{2}$	$3\frac{1}{8} \times 12$	$3\frac{1}{8} \times 12$	$3\frac{1}{8} \times 12$	$3\frac{1}{8} \times 13\frac{1}{2}$	$3\frac{1}{8} \times 13\frac{1}{2}$
16	8	$3\frac{1}{8} \times 9$	$3\frac{1}{8} \times 9$	$3\frac{1}{8} \times 10\frac{1}{2}$	$3\frac{1}{8} \times 10\frac{1}{2}$	$3\frac{1}{8} \times 12$	$3\frac{1}{8} \times 12$	$3\frac{1}{8} \times 13\frac{1}{2}$
	12	$3\frac{1}{8} \times 10\frac{1}{2}$	$3\frac{1}{8} \times 12$	$3\frac{1}{8} \times 12$	$3\frac{1}{8} \times 12$	$3\frac{1}{8} \times 13\frac{1}{2}$	$3\frac{1}{8} \times 13\frac{1}{2}$	$3\frac{1}{8} \times 15$
	14	$3\frac{1}{8} \times 12$	$3\frac{1}{8} \times 12$	$3\frac{1}{8} \times 13\frac{1}{2}$	$3\frac{1}{8} \times 13\frac{1}{2}$	$3\frac{1}{8} \times 15$	$3\frac{1}{8} \times 15$	$3\frac{1}{8} \times 15$
	16	$3\frac{1}{8} \times 12$	$3\frac{1}{8} \times 13\frac{1}{2}$	$3\frac{1}{8} \times 13\frac{1}{2}$	$3\frac{1}{8} \times 15$	$3\frac{1}{8} \times 15$	$3\frac{1}{8} \times 16\frac{1}{2}$	$3\frac{1}{8} \times 15$
18	8	$3\frac{1}{8} \times 9$	$3\frac{1}{8} \times 10\frac{1}{2}$	$3\frac{1}{8} \times 12$	$3\frac{1}{8} \times 12$	$3\frac{1}{8} \times 12$	$3\frac{1}{8} \times 13\frac{1}{2}$	$3\frac{1}{8} \times 15$
	12	$3\frac{1}{8} \times 12$	$3\frac{1}{8} \times 12$	$3\frac{1}{8} \times 13\frac{1}{2}$	$3\frac{1}{8} \times 13\frac{1}{2}$	$3\frac{1}{8} \times 15$	$3\frac{1}{8} \times 16\frac{1}{2}$	$3\frac{1}{8} \times 16\frac{1}{2}$
	16	$3\frac{1}{8} \times 13\frac{1}{2}$	$3\frac{1}{8} \times 15$	$3\frac{1}{8} \times 15$	$3\frac{1}{8} \times 16\frac{1}{2}$	$5\frac{1}{8} \times 13\frac{1}{2}$	$5\frac{1}{8} \times 13\frac{1}{2}$	$5\frac{1}{8} \times 15$
	18	$3\frac{1}{8} \times 15$	$3\frac{1}{8} \times 15$	$3\frac{1}{8} \times 16\frac{1}{2}$	$3\frac{1}{8} \times 18$	$5\frac{1}{8} \times 15$	$5\frac{1}{8} \times 15$	$5\frac{1}{8} \times 15$
20	8	$3\frac{1}{8} \times 12$	$3\frac{1}{8} \times 12$	$3\frac{1}{8} \times 13\frac{1}{2}$	$3\frac{1}{8} \times 13\frac{1}{2}$	$3\frac{1}{8} \times 13\frac{1}{2}$	$3\frac{1}{8} \times 15$	$3\frac{1}{8} \times 16\frac{1}{2}$
	12	$3\frac{1}{8} \times 13\frac{1}{2}$	$3\frac{1}{8} \times 13\frac{1}{2}$	$3\frac{1}{8} \times 15$	$3\frac{1}{8} \times 16\frac{1}{2}$	$3\frac{1}{8} \times 16\frac{1}{2}$	$5\frac{1}{8} \times 13\frac{1}{2}$	$5\frac{1}{8} \times 15$
	16	$3\frac{1}{8} \times 15$	$3\frac{1}{8} \times 16\frac{1}{2}$	$3\frac{1}{8} \times 18$	$3\frac{1}{8} \times 18$	$5\frac{1}{8} \times 15$	$5\frac{1}{8} \times 16\frac{1}{2}$	$5\frac{1}{8} \times 18$
	18	$3\frac{1}{8} \times 16\frac{1}{2}$	$3\frac{1}{8} \times 16\frac{1}{2}$	$3\frac{1}{8} \times 18$	$5\frac{1}{8} \times 15$	$5\frac{1}{8} \times 16\frac{1}{2}$	$5\frac{1}{8} \times 16\frac{1}{2}$	$5\frac{1}{8} \times 18$
22	8	$3\frac{1}{8} \times 13\frac{1}{2}$	$3\frac{1}{8} \times 13\frac{1}{2}$	$3\frac{1}{8} \times 13\frac{1}{2}$	$3\frac{1}{8} \times 15$	$3\frac{1}{8} \times 15$	$3\frac{1}{8} \times 16\frac{1}{2}$	$5\frac{1}{8} \times 15$
	12	$3\frac{1}{8} \times 15$	$3\frac{1}{8} \times 15$	$3\frac{1}{8} \times 16\frac{1}{2}$	$3\frac{1}{8} \times 18$	$3\frac{1}{8} \times 18$	$5\frac{1}{8} \times 15$	$5\frac{1}{8} \times 16\frac{1}{2}$
	16	$3\frac{1}{8} \times 16\frac{1}{2}$	$3\frac{1}{8} \times 18$	$5\frac{1}{8} \times 15$	$5\frac{1}{8} \times 16\frac{1}{2}$	$5\frac{1}{8} \times 16\frac{1}{2}$	$5\frac{1}{8} \times 18$	$5\frac{1}{8} \times 19\frac{1}{2}$
	18	$3\frac{1}{8} \times 18$	$5\frac{1}{8} \times 15$	$5\frac{1}{8} \times 16\frac{1}{2}$	$5\frac{1}{8} \times 16\frac{1}{2}$	$5\frac{1}{8} \times 18$	$5\frac{1}{8} \times 18$	$5\frac{1}{8} \times 19\frac{1}{2}$
24	8	$3\frac{1}{8} \times 13\frac{1}{2}$	$3\frac{1}{8} \times 15$	$3\frac{1}{8} \times 15$	$3\frac{1}{8} \times 16\frac{1}{2}$	$3\frac{1}{8} \times 16\frac{1}{2}$	$3\frac{1}{8} \times 18$	$5\frac{1}{8} \times 16\frac{1}{2}$
	12	$3\frac{1}{8} \times 16\frac{1}{2}$	$3\frac{1}{8} \times 16\frac{1}{2}$	$3\frac{1}{8} \times 18$	$5\frac{1}{8} \times 15$	$5\frac{1}{8} \times 16\frac{1}{2}$	$5\frac{1}{8} \times 16\frac{1}{2}$	$5\frac{1}{8} \times 18$
	16	$3\frac{1}{8} \times 18$	$5\frac{1}{8} \times 16\frac{1}{2}$	$5\frac{1}{8} \times 16\frac{1}{2}$	$5\frac{1}{8} \times 18$	$5\frac{1}{8} \times 18$	$5\frac{1}{8} \times 19\frac{1}{2}$	$5\frac{1}{8} \times 21$
	18	$5\frac{1}{8} \times 15$	$5\frac{1}{8} \times 16\frac{1}{2}$	$5\frac{1}{8} \times 18$	$5\frac{1}{8} \times 18$	$5\frac{1}{8} \times 19\frac{1}{2}$	$5\frac{1}{8} \times 21$	$5\frac{1}{8} \times 21$
26	8	$3\frac{1}{8} \times 15$	$3\frac{1}{8} \times 16\frac{1}{2}$	$3\frac{1}{8} \times 16\frac{1}{2}$	$3\frac{1}{8} \times 16\frac{1}{2}$	$3\frac{1}{8} \times 18$	$5\frac{1}{8} \times 16\frac{1}{2}$	$5\frac{1}{8} \times 18$
	12	$3\frac{1}{8} \times 18$	$3\frac{1}{8} \times 18$	$5\frac{1}{8} \times 16\frac{1}{2}$	$5\frac{1}{8} \times 16\frac{1}{2}$	$5\frac{1}{8} \times 18$	$5\frac{1}{8} \times 18$	$5\frac{1}{8} \times 19\frac{1}{2}$
	16	$5\frac{1}{8} \times 16\frac{1}{2}$	$5\frac{1}{8} \times 16\frac{1}{2}$	$5\frac{1}{8} \times 18$	$5\frac{1}{8} \times 18$	$5\frac{1}{8} \times 19\frac{1}{2}$	$5\frac{1}{8} \times 21$	$5\frac{1}{8} \times 22\frac{1}{2}$
	18	$5\frac{1}{8} \times 16\frac{1}{2}$	$5\frac{1}{8} \times 18$	$5\frac{1}{8} \times 18$	$5\frac{1}{8} \times 19\frac{1}{2}$	$5\frac{1}{8} \times 21$	$5\frac{1}{8} \times 21$	$5\frac{1}{8} \times 22\frac{1}{2}$

Table 11.13 (*Continued*)

Span, ft	Spacing, ft	Roof Beam Total-Load-Carrying Capacity						Floor Beams Total Load
		30 lb/ft²	35 lb/ft²	40 lb/ft²	45 lb/ft²	50 lb/ft²	55 lb/ft²	50 lb/ft²
28	8	3⅛ × 16½	3⅛ × 16½	3⅛ × 16½	3⅛ × 18	5⅛ × 16½	5⅛ × 16½	5⅛ × 19½
	12	3⅛ × 18	5⅛ × 16½	5⅛ × 18	5⅛ × 18	5⅛ × 18	5⅛ × 19½	5⅛ × 21
	16	5⅛ × 18	5⅛ × 18	5⅛ × 19½	5⅛ × 19½	5⅛ × 21	5⅛ × 22½	5⅛ × 24
	18	5⅛ × 18	5⅛ × 19½	5⅛ × 19½	5⅛ × 21	5⅛ × 22½	5⅛ × 24	5⅛ × 24
30	8	3⅛ × 18	3⅛ × 18	5⅛ × 16½	5⅛ × 16½	5⅛ × 18	5⅛ × 18	5⅛ × 21
	12	5⅛ × 16½	5⅛ × 18	5⅛ × 18	5⅛ × 19½	5⅛ × 19½	5⅛ × 21	5⅛ × 22½
	16	5⅛ × 18	5⅛ × 19½	5⅛ × 21	5⅛ × 21	5⅛ × 22½	5⅛ × 24	5⅛ × 25½
	18	5⅛ × 19½	5⅛ × 21	5⅛ × 21	5⅛ × 22½	5⅛ × 24	5⅛ × 25½	5⅛ × 27
32	8	3⅛ × 18	5⅛ × 16½	5⅛ × 18	5⅛ × 18	5⅛ × 18	5⅛ × 19½	5⅛ × 21
	12	5⅛ × 18	5⅛ × 19½	5⅛ × 19½	5⅛ × 21	5⅛ × 21	5⅛ × 22½	5⅛ × 24
	16	5⅛ × 19½	5⅛ × 21	5⅛ × 22½	5⅛ × 22½	5⅛ × 24	5⅛ × 25½	5⅛ × 27
	18	5⅛ × 21	5⅛ × 21	5⅛ × 22½	5⅛ × 24	5⅛ × 25½	5⅛ × 27	5⅛ × 28½
34	8	5⅛ × 16½	5⅛ × 18	5⅛ × 18	5⅛ × 19½	5⅛ × 19½	5⅛ × 21	5⅛ × 22½
	12	5⅛ × 19½	5⅛ × 19½	5⅛ × 21	5⅛ × 21	5⅛ × 22½	5⅛ × 24	5⅛ × 25½
	16	5⅛ × 21	5⅛ × 22½	5⅛ × 22½	5⅛ × 24	5⅛ × 25½	5⅛ × 27	5⅛ × 28½
	18	5⅛ × 22½	5⅛ × 22½	5⅛ × 24	5⅛ × 25½	5⅛ × 27	5⅛ × 28½	5⅛ × 28½
36	12	5⅛ × 19½	5⅛ × 21	5⅛ × 22½	5⅛ × 22½	5⅛ × 24	5⅛ × 25½	6¾ × 25½
	16	5⅛ × 22½	5⅛ × 24	5⅛ × 24	5⅛ × 25½	5⅛ × 27	5⅛ × 28½	6¾ × 27
	18	5⅛ × 22½	5⅛ × 24	5⅛ × 25½	5⅛ × 28½	5⅛ × 30	6¾ × 27	6¾ × 28½
	20	5⅛ × 24	5⅛ × 25½	5⅛ × 27	5⅛ × 30	6¾ × 27	6¾ × 28½	6¾ × 30
38	12	5⅛ × 21	5⅛ × 22½	5⅛ × 24	5⅛ × 24	5⅛ × 25½	5⅛ × 27	6¾ × 27
	16	5⅛ × 24	5⅛ × 24	5⅛ × 25½	5⅛ × 27	5⅛ × 28½	5⅛ × 30	6¾ × 28½
	18	5⅛ × 24	5⅛ × 25½	5⅛ × 27	5⅛ × 30	6¾ × 27	6¾ × 28½	6¾ × 30
	20	5⅛ × 25½	5⅛ × 27	5⅛ × 28½	6¾ × 27	6¾ × 28½	6¾ × 30	6¾ × 31½
40	12	5⅛ × 22½	5⅛ × 24	5⅛ × 24	5⅛ × 25½	5⅛ × 27	6¾ × 25½	6¾ × 28½
	16	5⅛ × 24	5⅛ × 25½	5⅛ × 27	5⅛ × 28½	6¾ × 27	6¾ × 28½	6¾ × 31½
	18	5⅛ × 25½	5⅛ × 27	5⅛ × 28½	6¾ × 27	6¾ × 28½	6¾ × 30	6¾ × 31½
	20	5⅛ × 27	5⅛ × 28½	6¾ × 27	6¾ × 28½	6¾ × 30	6¾ × 31½	6¾ × 33
42	12	5⅛ × 24	5⅛ × 24	5⅛ × 25½	5⅛ × 27	6¾ × 25½	6¾ × 25½	6¾ × 30
	16	5⅛ × 25½	5⅛ × 27	5⅛ × 28½	5⅛ × 30	6¾ × 28½	6¾ × 30	6¾ × 33
	18	5⅛ × 27	5⅛ × 28½	5⅛ × 30	6¾ × 28½	6¾ × 30	6¾ × 31½	6¾ × 33
	20	5⅛ × 28½	5⅛ × 30	6¾ × 28½	6¾ × 30	6¾ × 31½	6¾ × 33	6¾ × 34½
44	12	5⅛ × 24	5⅛ × 25½	5⅛ × 27	5⅛ × 27	6¾ × 25½	6¾ × 27	6¾ × 31½
	16	5⅛ × 27	5⅛ × 28½	5⅛ × 30	6¾ × 28½	6¾ × 30	6¾ × 31½	6¾ × 33
	18	5⅛ × 28½	5⅛ × 30	6¾ × 28½	6¾ × 30	6¾ × 31½	6¾ × 33	6¾ × 34½
	20	5⅛ × 30	6¾ × 27	6¾ × 30	6¾ × 30	6¾ × 33	6¾ × 34½	6¾ × 36
46	12	5⅛ × 25½	5⅛ × 27	5⅛ × 28½	6¾ × 25½	6¾ × 27	6¾ × 28½	6¾ × 31½
	16	5⅛ × 28½	5⅛ × 30	6¾ × 28½	6¾ × 28½	6¾ × 31½	6¾ × 33	6¾ × 36
	18	5⅛ × 28½	6¾ × 28½	6¾ × 30	6¾ × 31½	6¾ × 33	6¾ × 34½	6¾ × 36

(*Continued*)

Table 11.13 (*Continued*)

| Span, ft | Spacing, ft | Roof Beam Total-Load-Carrying Capacity | | | | | | Floor Beams Total Load |
		30 lb/ft²	35 lb/ft²	40 lb/ft²	45 lb/ft²	50 lb/ft²	55 lb/ft²	50 lb/ft²
	20	5⅛ × 30	6¾ × 28½	6¾ × 31½	6¾ × 33	6¾ × 34½	6¾ × 36	8¾ × 34½
48	12	5⅛ × 27	5⅛ × 28½	5⅛ × 30	5⅛ × 30	6¾ × 28½	6¾ × 30	6¾ × 33
	16	5⅛ × 30	6¾ × 28½	6¾ × 30	6¾ × 30	6¾ × 31½	6¾ × 34½	6¾ × 37½
	18	5⅛ × 30	6¾ × 30	6¾ × 30	6¾ × 33	6¾ × 34½	6¾ × 36	8¾ × 34½
	20	6¾ × 28½	6¾ × 30	6¾ × 31½	6¾ × 34½	6¾ × 36	6¾ × 37½	8¾ × 36
50	12	5⅛ × 28½	5⅛ × 28½	5⅛ × 30	6¾ × 28½	6¾ × 30	6¾ × 31½	6¾ × 34½
	16	5⅛ × 30	6¾ × 30	6¾ × 30	6¾ × 31½	6¾ × 33	6¾ × 36	8¾ × 34½
	18	6¾ × 28½	6¾ × 30	6¾ × 31½	6¾ × 34½	6¾ × 36	8¾ × 33	8¾ × 36
	20	6¾ × 30	6¾ × 31½	6¾ × 33	6¾ × 36	6¾ × 37½	8¾ × 34½	8¾ × 37½
52	12	5⅛ × 28½	5⅛ × 30	6¾ × 28½	6¾ × 30	6¾ × 31½	6¾ × 31½	6¾ × 36
	16	6¾ × 28½	6¾ × 30	6¾ × 31½	6¾ × 33	6¾ × 34½	6¾ × 37½	8¾ × 36
	18	6¾ × 30	6¾ × 31½	6¾ × 33	6¾ × 34½	6¾ × 37½	6¾ × 39	8¾ × 37½
	20	6¾ × 31½	6¾ × 33	6¾ × 34½	6¾ × 37½	6¾ × 39	8¾ × 36	8¾ × 39
54	12	5⅛ × 30	6¾ × 28½	6¾ × 30	6¾ × 31½	6¾ × 33	6¾ × 33	6¾ × 37½
	16	6¾ × 30	6¾ × 31½	6¾ × 33	6¾ × 34½	6¾ × 36	6¾ × 37½	8¾ × 37½
	18	6¾ × 31½	6¾ × 33	6¾ × 34½	6¾ × 36	6¾ × 39	8¾ × 36	8¾ × 39
	20	6¾ × 33	6¾ × 34½	6¾ × 36	6¾ × 39	8¾ × 36	8¾ × 37½	8¾ × 40½
56	12	6¾ × 28½	6¾ × 30	6¾ × 31½	6¾ × 33	6¾ × 33	6¾ × 34½	8¾ × 36
	16	6¾ × 31½	6¾ × 33	6¾ × 34½	6¾ × 36	6¾ × 37½	8¾ × 34½	8¾ × 39
	18	6¾ × 33	6¾ × 34½	6¾ × 36	6¾ × 37½	8¾ × 34½	8¾ × 37½	8¾ × 40½
	20	6¾ × 33	6¾ × 36	6¾ × 37½	8¾ × 34½	8¾ × 37½	8¾ × 39	8¾ × 42
58	12	6¾ × 30	6¾ × 31½	6¾ × 31½	6¾ × 33	6¾ × 34½	6¾ × 36	8¾ × 37½
	16	6¾ × 31½	6¾ × 34½	6¾ × 36	6¾ × 37½	6¾ × 39	8¾ × 36	8¾ × 40½
	18	6¾ × 33	6¾ × 34½	6¾ × 37½	6¾ × 39	8¾ × 36	8¾ × 39	8¾ × 42
	20	6¾ × 34½	6¾ × 36	6¾ × 39	8¾ × 36	8¾ × 39	8¾ × 40½	8¾ × 43½
60	12	6¾ × 30	6¾ × 31½	6¾ × 33	6¾ × 34½	6¾ × 36	6¾ × 37½	8¾ × 39
	16	6¾ × 33	6¾ × 34½	6¾ × 36	6¾ × 39	8¾ × 36	8¾ × 37½	8¾ × 42
	18	6¾ × 34½	6¾ × 36	6¾ × 39	8¾ × 36	8¾ × 37½	8¾ × 39	8¾ × 43½
	20	6¾ × 36	6¾ × 37½	8¾ × 36	8¾ × 37½	8¾ × 40½	8¾ × 42	8¾ × 45

* This table applies to straight, simply supported, laminated timber beams. Other beam support systems may be employed to meet varying design conditions.

1. Roofs should have a minimum slope of ¼ in/ft to eliminate water ponding.

2. Beam weight must be subtracted from total load-carrying capacity. Floor beams are designed for uniform loads of 40 lb/ft² live load and 10 lb/ft² dead load.

3. Allowable stresses: Bending stress, F_b = 2400 psi (reduced by the volume factor for southern pine). Shear stress F_v = 165 psi. Modulus of elasticity E = 1,800,000 psi. For roof beams, F_b and F_v were increased 15% for short duration of loading.

4. Deflection limits: Roof beams—1/180 span for total load. Floor beams—1/360 span for 40 lb/ft² live load only. For preliminary design purposes only. For more complete design information, see the AITC "Timber Construction Manual."

5. Maximum shear stress increased to 270 psi for southern pine and to 270 psi for western species. Shear will not govern for single span beams.

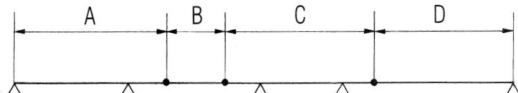

Fig. 11.6 Cantilevered-beam systems. *A* is a single cantilever; *B* is a suspended beam; *C* has a double cantilever; *D* is a beam with one end suspended.

depths of the members in the suspended portions of a cantilever system.

For economy, the negative bending moment at the supports of a cantilevered beam should be equal in magnitude to the positive moment.

Consideration should be given to deflection and camber in cantilevered multiple spans. When possible, roofs should be sloped the equivalent of $\frac{1}{4}$ in/ft of horizontal distance between the level of drains and the high point of the roof to eliminate water pockets, or provision should be made to ensure that accumulation of water does not produce greater deflection and live loads than anticipated. Unbalanced loading conditions should be investigated for maximum bending moment, deflection, and stability.

(American Institute of Timber Construction, "Timber Construction Manual," John Wiley & Sons, Inc., New York; "National Design Specification for Wood Construction," American Forest and Paper Association, 1111 19th St., N. W., Washington, DC 20036.)

11.13 Deflection and Camber of Wood Beams

The design of many structural systems, particularly those with long spans, is governed by deflection. Strength calculations based on allowable stresses alone may result in excessive deflection. Limitations on deflection increase member stiffness.

Table 11.14 gives recommended deflection limits, as a fraction of the beam span, for wood beams. The limitation applies to live load or total load, whichever governs.

Glued-laminated beams are cambered by fabricating them with a curvature opposite in direction to that corresponding to deflections under load. Camber does not, however, increase stiffness. Table 11.15 lists recommended minimum cambers for glued-laminated timber beams.

Table 11.14 Recommended Beam Deflection Limitations, in* (in terms of Span *l*, in)

Use Classification	Live Load Only	Dead Load Plus Live Load
Roof beams:		
Industrial	*l*/180	*l*/120
Commercial and institutional:		
Without plaster ceiling	*l*/240	*l*/180
With plaster ceiling	*l*/360	*l*/240
Floor beams:		
Ordinary usage†	*l*/360	*l*/240
Highway bridge stringers	*l*/200 to *l*/300	
Railway bridge stringers	*l*/300 to *l*/400	

* "Camber and Deflection," AITC 102, app. B, American Institute of Timber Construction.

† Ordinary usage classification is intended for construction in which walking comfort, minimized plaster cracking, and elimination of objectionable springiness are of prime importance. For special uses, such as beams supporting vibrating machinery or carrying moving loads, more severe limitations may be required.

Minimum Roof Slopes ■ Flat roofs have collapsed during rainstorms, although they were adequately designed on the basis of allowable stresses and definite deflection limitations. The reason for these collapses was the same, regardless

Table 11.15 Recommended Minimum Camber for Glued-Laminated Timber Beams*

Roof beams†	1½ times dead-load deflection
Floor beams‡	1½ times dead-load deflection
Bridge beams:§	
Long span	2 times dead-load deflection
Short span	2 times dead-load + ½ applied load-deflection

* "Camber and Deflection," AITC 102, app. B, American Institute of Timber Construction.

† The minimum camber of 1½ times dead-load deflection will produce a nearly level member under dead load alone after plastic deformation has occurred. Additional camber is usually provided to improve appearance or provide necessary roof drainage (see under "Minimum Roof Slopes").

‡ The minimum camber of 1½ times dead-load deflection will produce a nearly level member under dead load alone after plastic deformation has occurred. On long spans, a level ceiling may not be desirable because of the optical illusion that the ceiling sags. For warehouse or similar floors where live load may remain for long periods, additional camber should be provided to give a level floor under the permanently applied load.

§ Bridge members are normally cambered for dead load only on multiple spans to obtain acceptable riding qualities.

of the structural framing used—the failures were caused by ponding of water as increasing deflections permitted more and more water to collect.

Roof beams should have a continuous upward slope equivalent to $\frac{1}{4}$ in/ft between a drain and the high point of a roof, in addition to minimum recommended camber (Table 11.15), to avoid ponding. When flat roofs have insufficient slope for drainage (less than $\frac{1}{4}$ in/ft), the stiffness of supporting members should be such that a 5-lb/ft^2 load will cause no more than $\frac{1}{2}$-in deflection.

Because of ponding, snow loads or water trapped by gravel stops, parapet walls, or ice dams magnify stresses and deflections from existing roof loads by

$$C_p = \frac{1}{1 - W'L^3/\pi^4 EI} \qquad (11.19)$$

where C_p = factor for multiplying stresses and deflections under existing loads to determine stresses and deflections under existing loads plus ponding

W' = weight of 1 in of water on roof area supported by beam, lb

L = span of beam, in

E = modulus of elasticity of beam material, psi

I = moment of inertia of beam, in^4

(Kuenzi and Bohannan, "Increases in Deflection and Stresses Caused by Ponding of Water on Roofs," Forest Products Laboratory, Madison, Wis.)

11.14 Bearing on Wood Members

Bearing stresses, or compression stresses perpendicular to the grain, in a beam occur at the supports or at places where other framing members are supported on the beam. The compressive stress in the beam $f_{c\perp}$ is given by

$$f_{c\perp} = \frac{P}{A} \qquad (11.20)$$

where P = load transmitted to or from the beam and A = bearing area. This stress should be less than the design value for compression perpendicular to the grain $F_{c\perp}$ multiplied by applicable adjustment factors (Art. 11.4). (The duration-of-load factor does not apply to $F_{c\perp}$ for either solid-sawn lumber or glued laminated timber.)

Design values for $F_{c\perp}$ are averages based on a maximum deformation of 0.04 in in tests conforming with ASTM D143. Design values $F_{c\perp}$ for glued laminated beams are generally lower than for solid sawn lumber with the same deformation limit. This is due partly to use of larger-size sections for glued laminated beams, length of bearing and partly to the method used to derive the design values.

Where deformations are critical, the deformation limit may be decreased, with resulting reduction in $F_{c\perp}$. For example, for a deformation maximum of 0.02 in, the "National Design Specification for Wood Construction," (American Forest and Paper Association), recommends that $F_{c\perp}$, psi, be reduced to $0.73F_{c\perp} + 5.60$. For glued-laminated beams, $F_{c\perp}$ may be taken as $0.73F_{c\perp}$.

Bearing stress parallel to grain f_g on a wood member should be computed for the net bearing area. This stress may not exceed the design value for bearing parallel to grain F_g multiplied by load duration factor C_D and temperature factor C_t (Art. 11.4). The adjusted design value applies to end-to-end bearing of compression members if they have adequate lateral support and their end cuts are accurately squared and parallel to each other.

When f_g exceeds 75% of the adjusted design value, the member should bear on a metal plate, strap, or other durable, rigid, homogeneous material with adequate strength. In such cases, when a rigid insert is required, it should be a steel plate with a thickness of 20 ga or more or the equivalent, and it should be inserted with a snug fit between abutting ends.

Bearing perpendicular to grain is equivalent to compression perpendicular to grain. The compressive stress should not exceed the design value perpendicular to grain multiplied by applicable adjustment factors, including the bearing area factor (Art. 11.4.10). In the calculation of bearing area at the end of a beam, an allowance need not be made for the fact that, as the beam bends, it creates a pressure on the inner edge of the bearing that is greater than at the end of the beam.

Bearing at an angle to grain is assigned a design value that is a function of the design value F_g for bearing parallel to grain and the design value for bearing perpendicular to grain $F_{c\perp}$, which differ considerably. When load is applied at an angle θ with respect to the grain, where $0 \le \theta \le 90°$ (Fig. 11.7), the design value for bearing lies between F_g and $F_{c\perp}$. The "National Design Specification for Wood Construction," (American Forest and Paper Association) recommends that the design

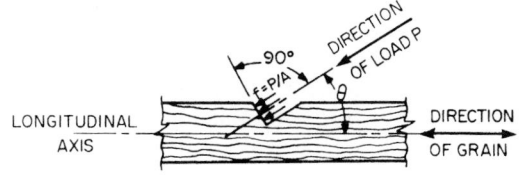

Fig. 11.7 Load applied to a member in bearing at an angle to the grain.

value for such loading be calculated from the Hankinson formula:

$$F'_n = \frac{F'_g F'_{c\perp}}{F'_g \sin^2\theta + F'_{c\perp} \cos^2\theta} \qquad (11.21)$$

where F'_n = adjusted design value for bearing at angle θ to the grain (longitudinal axis)

F'_g = design value for end bearing multiplied by applicable adjustment factors

$F'_{c\perp}$ = design value for compression perpendicular to grain multiplied by applicable adjustment factors

11.15 Combined Stresses in Wood Members

Design values given in the "National Design Specification for Wood Construction" apply directly to bending, horizontal shear, tension parallel to grain, and compression parallel or perpendicular to grain. When a bending moment and an axial force act on a section of a structural member, the effects of the combined stresses must be provided for in design of the member.

11.15.1 Bending and Axial Tension

Members subjected to combined bending and axial tension should be proportioned to satisfy the interaction equations, Eqs. (11.22) and (11.23).

$$\frac{f_t}{F'_t} + \frac{f_b}{F^*_b} \leq 1 \qquad (11.22)$$

$$\frac{(f_b - f_t)}{F^{**}_b} \leq 1 \qquad (11.23)$$

where f_t = tensile stress due to axial tension acting alone

f_b = bending stress due to bending moment alone

F'_t = design value for tension multiplied by applicable adjustment factors

F^*_b = design value for bending multiplied by applicable adjustment factors except C_L

F^{**}_b = design value for bending multiplied by applicable adjustment factors except C_V

Adjustment factors are discussed in Art. 11.4.

The load duration factor C_D associated with the load of shortest duration in a combination of loads with differing duration may be used to calculate F'_t and F^*_b. All applicable load combinations should be evaluated to determine the critical load combination.

11.15.2 Bending and Axial Compression

Members subjected to a combination of bending and axial compression (beam-columns) should be proportioned to satisfy the interaction equation, Eq. 11.24.

$$\left(\frac{f_c}{F'_c}\right)^2 + \frac{f_{b1}}{[1 - (f_c/F_{cE1})]F'_{b1}} \qquad (11.24)$$

$$+ \frac{f_{b2}}{[1 - (f_c/F_{cE2}) - (f_{b1}/F_{bE})^2]F'_{b2}} \leq 1$$

where f_c = compressive stress due to axial compression acting alone

F'_c = design value for compression parallel to grain multiplied by applicable adjustment factors, including the column stability factor

f_{b1} = bending stress for load applied to the narrow face of the member

f_{b2} = bending stress for load applied to the wide face of the member

F'_{b1} = design value for bending for load applied to the narrow face of the member multiplied by applicable adjustment factors, including the column stability factor

F'_{b2} = design value for bending for load applied to the wide face of the member multiplied by applicable adjustment factors, including the column stability factor

For either uniaxial or biaxial bending, f_c should not exceed

$$F_{cE1} = \frac{K_{cE}E'}{(L_{e1}/d_1)^2} \qquad (11.25)$$

where E' = modulus of elasticity multiplied by adjustment factors. Also, for biaxial bending, f_c should not exceed

$$F_{cE2} = \frac{K_{cE}E'}{(L_{e2}/d_2)^2} \qquad (11.26)$$

and f_{b1} should not be more than

$$F_{bE} = \frac{K_{bE}E'}{R_B^2} \qquad (11.27)$$

where d_1 = width of the wide face and d_2 = width of the narrow face. Slenderness ratio R_B for beams is given by Eq. (11.15). K_{bE} is defined for Eq. (11.16). The effective column lengths L_{e1} for buckling in the d_1 direction and L_{e2} for buckling in the d_2 direction, E', F_{cE1}, and F_{cE2} should be determined in accordance with Art. 11.11.

As for the case of combined bending and axial tension, F'_c, F'_{b1}, and F'_{b2} should be adjusted for duration of load by applying C_D. See Art. 11.4.

11.16 Characteristics of Mechanical Fastenings

Various kinds of mechanical fastenings are used in wood construction. The most common are nails, spikes, screws, lags, bolts, and timber connectors, such as shear plates and split rings (Art. 11.19). Joint-design data have been established by experience and tests because determination of stress distribution in wood and metal fasteners is complicated.

Design values and methods of design for bolts, connectors, and other fasteners used in one-piece sawn members also are applicable to laminated members.

Problems can arise, however, if a deep-arch base section is bolted to the shoe attached to the foundation by widely separated bolts. A decrease in wood moisture content and shrinkage will set up considerable tensile stress perpendicular to grain, and splitting may occur. If the moisture content at erection is the same as that to be reached in service, or if the bolt holes in the shoe are slotted to permit bolt movement, the tendency to split will be reduced.

Fasteners subject to corrosion or chemical attack should be protected by painting, galvanizing, or plating. In highly corrosive atmospheres, such as in chemical plants, metal fasteners and connections should be galvanized or made of stainless steel. Consideration may be given to covering connections, with hot tar or pitch. In such extreme conditions, lumber should be at or below equilibrium moisture content at fabrication, to reduce subsequent shrinkage, which could open avenues of attack for the corrosive atmosphere.

Iron salts are frequently very acidic and show hydrolytic action on wood in the presence of free water. This accounts for softening and discoloration of wood observed around corroded nails. This action is especially pronounced in acidic woods, such as oak, and in woods containing considerable tannin and related compounds, such as redwood. It can be eliminated, however, by using zinc-coated aluminum, or copper nails.

11.16.1 Nails and Spikes

Common wire nails and spikes conform to the minimum sizes in Table 11.16.

Hardened deformed-shank nails and spikes are made of high-carbon-steel wire and are headed, pointed, annularly or helically threaded, and heat-treated and tempered, to provide greater strength than common wire nails and spikes. But the same loads are given for common wire nails and spikes or the corresponding lengths are used with a few exceptions.

Nails should not be driven closer together than half their length, unless driven in prebored holes. Nor should nails be closer to an edge than one-quarter their length. When one structural member is joined to another, penetration of nails into the second or farther timber should be at least half the length of the nails. Holes for nails, when

Table 11.16   Nail and Spike Dimensions

Pennyweight	Length, in	Wire Dia, in
Nails:		
6d	2	0.113
8d	2½	0.131
10d	3	0.148
12d	3¼	0.148
16d	3½	0.162
20d	4	0.192
30d	4½	0.207
40d	5	0.225
50d	5½	0.244
60d	6	0.263
Spikes:		
10d	3	0.192
12d	3¼	0.192
16d	3½	0.207
20d	4	0.225
30d	4½	0.244
40d	5	0.263
50d	5½	0.283
60d	6	0.283
5/16	7	0.312
3/8	8½	0.375

necessary to prevent splitting, should be bored with a diameter less than that of the nail. If this is done, the same allowable load as for the same-size fastener with a bored hole applies in both withdrawal and lateral resistance.

Nails or spikes should not be loaded in withdrawal from the end grain of wood. Also, nails inserted parallel to the grain should not be used to resist tensile stresses parallel to the grain.

Design values for nails and spikes and adjustment factors are discussed in Art. 11.17.

11.16.2   Wood Screws

The common types of wood screws have flat, oval, or round heads. The flathead screw is commonly used if a flush surface is desired. Oval- and round-headed screws are used for appearance or when countersinking is objectionable.

Wood screws should not be loaded in withdrawal from end grain. They should be inserted perpendicular to the grain by turning into predrilled holes and should not be started or driven with a hammer. Spacings, end distances,

and side distances must be such as to prevent splitting.

For Douglas fir and southern pine, the lead hole for a screw loaded in withdrawal should have a diameter of about 70% of the root diameter of the screw. For lateral resistance, the part of the hole receiving the shank should be about seven-eighths the diameter of the screw at the root of the thread.

Design values for wood screws and adjustment factors are discussed in Art. 11.17.

11.16.3   Lag Screws

Also known as lag bolts, lag screws are large screws with a square or hexagonal bolt head. They range, usually, from about 0.2 to 1.0 in in diameter and from 1 to 16 in in length. The threaded portion ranges from ¾ in for 1- and 1¼-in-long lag screws to half the length for all lengths greater than 10 in.

As is the case with bolts and timber connectors, lag screws are used where relatively heavy loads have to be transmitted in a connection. They are used particularly where it would be difficult to fasten a bolt or where a nut on the surface would be objectionable. They also are used instead of bolts where the components of a joint are so thick that an excessively long bolt would be needed or where heavy withdrawal loads have to be resisted.

Lag screws are turned with a wrench into prebored holes with total length equal to the nominal screw length. Soap or other lubricant may be used to facilitate insertion and prevent damage to screws. Two holes are drilled for each lag screw. The first and deepest hole has a diameter, as specified in the NDS for various species, depending on the wood density, ranging from 40 to 85% of the shank diameter. The second hole should have the same diameter as the shank, or unthreaded portion of the lag screw, and the same depth as the unthreaded portion.

Lag screws loaded in withdrawal should be designed for allowable tensile strength in the net (root-of-thread) section as well as for resistance to withdrawal. For single-shear wood-to-wood connections, the lag screw should be inserted in the side grain of the main member with the screw axis perpendicular to the wood fibers. Penetration of the threaded portion to a distance of about 7 times the shank diameter in the denser species and 10 to 12 times the shank diameter in the less dense species will develop approximately the ultimate tensile strength of a lag screw.

Lag screws should preferably not be driven into end grain because splitting may develop under lateral load. The resistance of a lag screw to withdrawal from end grain is about three-quarters that from side grain.

Spacings, edge and end distances, and net section for lag-screw joints should be the same as those for joints with bolts of a diameter equal to the shank diameter of the lag screw.

For more than one lag screw, the total allowable load equals the sum of the loads permitted for each lag screw, provided that spacings, end distances, and edge distances are sufficient to develop the full strength of each lag screw.

Design values for lag screws and adjustment factors are discussed in Art. 11.17.

11.16.4 Bolts and Dowels

Machine bolts conforming to ANSI/ASME Standard B18.2.1, with square heads and nuts, are used extensively in wood construction. Spiral-shaped dowels are also used at times to hold two pieces of wood together; they are used to resist checking and splitting in railroad ties and other solid-sawn timbers.

Holes for bolts should always be prebored and have a diameter that permits the bolt to be driven easily (Art. 11.6). Careful centering of holes in main members and splice plates is necessary. The holes should have a diameter from $\frac{1}{32}$ to $\frac{1}{16}$ in larger than the bolt diameter. Tight fit of bolts in the holes, requiring forced insertion, is not recommended. A metal plate, strap, or washer (not smaller than a standard cut washer) should be placed between the wood and bolt head and between the wood and the nut. The length of bolt threads subject to bearing on the wood should be kept to a practical minimum.

Two or more bolts placed in a line parallel to the direction of the load constitute a **row**. **End distance** is the minimum distance from the end of a member to the center of the bolt hole that is nearest to the end. **Edge distance** is the minimum distance from the edge of a member to the center of the nearest bolt hole. Figure 11.8 illustrates these distances, the spacing between rows, and the spacing of bolts in a row. NDS requirements are listed for minimum end distance in Table 11.17, for minimum edge distance in Table 11.18, and for minimum spacing between rows and between bolts in a row in Table 11.19. The geometry factor C_Δ discussed in Art. 11.17 is applied to the design value for a bolted connection when the end distance or spacing between bolts is less than that given in these tables for full design value.

The critical section is that section at right angles to the direction of the load that gives maximum stress in the member over the net area remaining after bolt holes at the section are deducted. For parallel-to-grain loads, the net area at a critical section should be at least 100% for hardwoods and 80% for softwoods of the total area in bearing under all the bolts in the joint.

For parallel- or perpendicular-to-grain loads, spacing between rows paralleling a member should not exceed 5 in unless separate splice plates are used for each row.

Groups of Bolts ▪ When bolts are properly spaced and aligned, the allowable load on a group

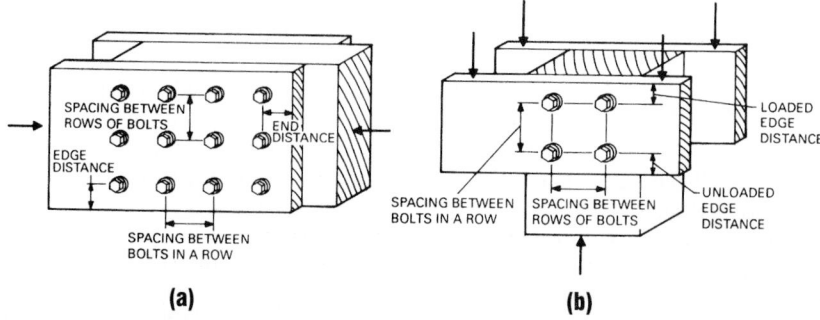

Fig. 11.8 Bolt spacing and edge distances in connections are defined with respect to load direction: (*a*) Parallel to grain; (*b*) perpendicular to grain. (*From F. S. Merritt and J. T. Ricketts, "Building Design and Construction Handbook," 5th ed., McGraw-Hill Publishing Company, New York.*)

Table 11.17 Minimum End Distance for Bolts*

Direction of Loading	For Reduced Design Value	For Full Design Value
Perpendicular to grain	2D	4D
Compression parallel to grain (bolt bearing away from member end)	2D	4D
Tension parallel to grain (bolt bearing toward member end):		
For softwoods	3.5D	7D
For hardwoods	2.5D	5D

* D = bolt diameter.

of bolts may be taken as the sum of the individual load capacities.

Design Values ▪ These and adjustment factors for bolts are discussed in Art. 11.17.

11.16.5 Timber Connectors

These are metal devices used with bolts for producing joints with fewer bolts without reduction in strength. Several types of connectors are available. Usually, they are either steel rings, called split rings, that are placed in grooves in adjoining members to prevent relative movement or metal plates, called shear plates, embedded in the faces of adjoining timbers. The bolts used with these connectors prevent the timbers from separating. The load is transmitted across the joint through the connectors.

Split rings, used for joining wood to wood, are placed in circular grooves cut by a hand tool in the

Table 11.18 Minimum Edge Distance for Bolts

Direction of loading*	Minimum edge distance
Parallel to grain:	
When $L/D \leq 6$	1.5D
When $L/D > 6$	1.5D or half the spacing between rows, whichever is greater
Perpendicular to grain:	
To loaded edge	4D
To unloaded edge	1.5D

* L = length of bolt in main member and D = bolt diameter.

contact surfaces. About half the depth of each ring is in each of the two members in contact (Fig. 11.9b). A bolt hole is drilled through the center of the core encircled by the groove. Split rings require greater accuracy for fabricating the wood members properly and the relative difficulty of installation make these connectors more costly than shear plates.

Shear plates are intended for wood-to-steel connections (Fig. 11.9c and d). But when used in pairs, they may be used for wood-to-wood connections (Fig. 11.9e), replacing split rings. Set with one plate in each member at the contact surface, they enable the members to slide easily into position during fabrication of the joint, thus reducing the labor needed to make the connection. Shear plates are placed in precut daps and are completely embedded in the timber, flush with the surface. As with split rings, the role of the bolt through each plate is to prevent the components of the joint from separating; loads are transmitted across the joint through the plates. They come in $2\frac{5}{8}$- and 4-in diameters.

Shear plates are useful in demountable structures. They may be installed in the members immediately after fabrication and held in position by nails.

Toothed rings and spike grids sometimes are used for special applications. Shear plates are the prime connectors for timber construction subject to heavy loads.

Tables in the NDS list the least thickness of member that should be used with the various sizes of connectors. The NDS also lists minimum end and edge distances and spacing for timber connectors (Table 11.20). Edge distance is the distance from the edge of a member to the center

Table 11.19 Minimum Spacing for Bolts*

(a) For Bolts in a Row		
Direction of Loading	For Reduced Design Value	For Full Design Value
Parallel to grain	3D	4D
Perpendicular to grain	3D	Required spacing for attached member(s)

(b) Between Bolts in a Row	
Direction of Loading	Minimum Spacing
Parallel to grain	1.5D

Perpendicular to grain
  When $L/D \leq 2$
  When $2 < L/D < 6$
  When $L/D \geq 6$
	2.5D
	$(5L + 10D)/8$
	5D

*L = length of bolt in main member and D = bolt diameter.

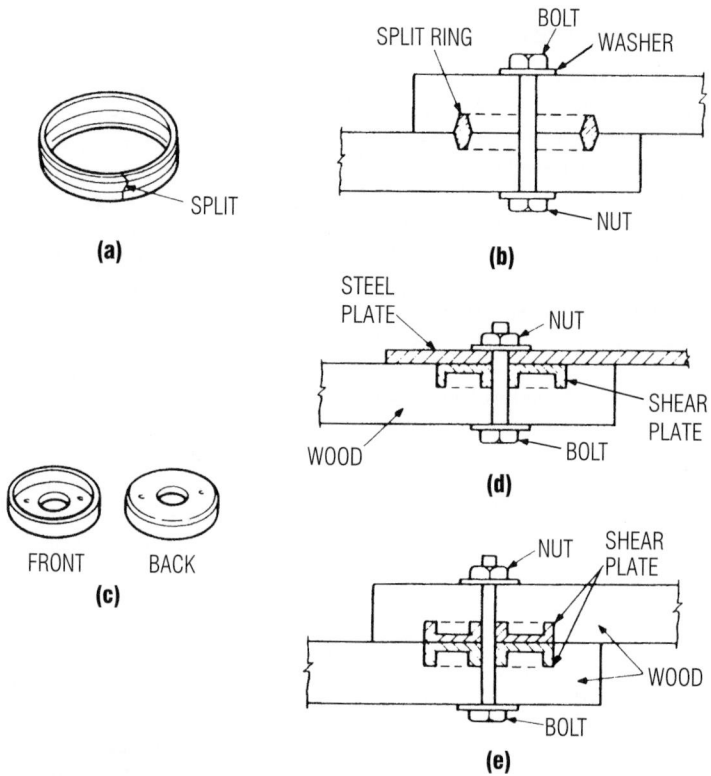

Fig. 11.9 Timber connectors: (a) Split ring; (b) wood members connected with split ring and bolt; (c) shear plate; (d) steel plate connected to a wood member with a shear plate and bolt; (e) wood members connected with a pair of shear plates and bolt.

Table 11.20 Minimum Edge and End Distances, Spacing, and Geometry Factors C_Δ for Shear-Plate Connectors.

	2⅝-in Shear-Plate Connectors				4-in Shear-Plate Connectors			
	Loads Parallel to Grain		Loads Perpendicular to Grain		Loads Parallel to Grain		Loads Perpendicular to Grain	
	For Reduced Design Value	For Full Design Value	For Reduced Design Value	For Full Design Value	For Reduced Design Value	For Full Design Value	For Reduced Design Value	For Full Design Value
Edge distance								
Unloaded edge, in	1¾	1¾	1¾	1¾	2¾	2¾	2¾	2¾
C_Δ	1.0	1.0	1.0	1.0	1.0	1.0	1.0	1.0
Loaded edge, in	1¾	1¾	1¾	2¾	2¾	2¾	2¾	3¾
C_Δ	1.0	1.0	0.83	1.0	1.0	1.0	0.83	1.0
End distance								
Tension member, in	2¾	5½	2¾	5½	3½	7	3½	7
C_Δ	0.625	1.0	0.625	1.0	0.625	1.0	0.625	1.0
Compression member, in	2½	4	2¾	5½	3½	5½	3½	7
C_Δ	0.625	1.0	0.625	1.0	0.625	1.0	0.625	1.0
Spacing								
Spacing parallel to grain, in	3½	6¾	3½	3½	5	9	5	5
C_Δ	0.5	1.0	1.0	1.0	0.5	1.0	1.0	1.0
Spacing perpendicular to grain, in	3½	3½	3½	4¼	5	5	5	6
C_Δ	1.0	1.0	0.5	1.0	1.0	1.0	0.5	1.0

of the connector closest to that edge and measured perpendicular to the edge. End distance is measured parallel to the grain from the center of the connector to the square-cut end of the member. If the end of the member is not cut normal to the longitudinal axis, the end distance, measured parallel to that axis from any point on the center half of the connector diameter that is perpendicular to the axis, should not be less than the minimum end distance required for a square-cut member. Spacing of connectors is measured between their centers along a line between centers.

Placement of connectors in joints with members at right angles to each other is subject to the limitations of either member. Since rules for alignment, spacing, and edge and end distance of connectors for all conceivable directions of applied load would be complicated, designers must rely on a sense of proportion and adequacy in applying the above rules to conditions of loading outside the specific limitations mentioned.

Design values for shear plates and adjustment factors are discussed in Art. 11.17.

11.16.6 Anchor Bolts

To attach columns or arch bases to concrete foundations, anchor bolts are embedded in the concrete, with sufficient projection to permit placement of angles or shoes bolted to the wood. Sometimes, instead of anchor bolts, steel straps are embedded in the concrete with a portion projecting above for bolt attachment to the wood members.

11.16.7 Washers

Bolt heads and nuts bearing on wood require metal washers to protect the wood and to distribute the pressure across the surface of the wood. Washers may be cast, malleable, cut, round-plate, or square-plate. When subjected to salt air or salt water, they should be galvanized or given some type of effective coating. Ordinarily, washers are dipped in red lead and oil prior to installation.

Setscrews should never be used against a wood surface. It may be possible, with the aid of proper washers, to spread the load of the setscrew over sufficient surface area of the wood that the compression strength perpendicular to grain is not exceeded.

11.16.8 Tie Rods

To resist the horizontal thrust of arches not buttressed, tie rods are required. The tie rods may be installed at ceiling height or below the floor.

11.16.9 Hangers

Standard and special hangers are used extensively in timber construction. Stock hangers are available from a number of manufacturers. But by far the greater number of hangers are of special design. Where appearance is of prime importance, concealed hangers are frequently selected.

11.17 Design Values and Adjustment Factors for Mechanical Fastenings

Determination of stress distribution in connections made with wood and metal is complicated. Consequently, information for design of joints has been developed from tests and experience. The data indicate that design values and methods of design for mechanical connections are applicable to both solid-sawn lumber and laminated members. The "National Design Specification for Wood Construction" (NDS), American Forest and Paper Association, lists design values for connections made with various types of fasteners. Design values for connections made with more than one type of fastener, however, should be based on tests or special analysis.

Design values for shear plates subject to loads at an angle between $0°$ (parallel to grain) and $90°$ (perpendicular to grain) may be computed from Eq. (11.21). In this case, F_n', F_g', and $F_{c\perp}'$ are, respectively, the adjusted design value at inclination θ with the direction of grain, parallel to grain, and perpendicular to grain.

Article 11.19 illustrates connections often used in wood structural framing.

Design values are based on the assumption that the wood at the joint is clear and relatively free from checks, shakes, and splits. If knots are present in the longitudinal projection of the net section within a distance from the critical section of half the diameter of the connector, the area of the knot should be subtracted from the area of the critical section. It is assumed that slope of the grain at the joint does not exceed 1 in 10.

The stress, whether tension or compression, in the net area, the area remaining at the critical section after subtracting the projected area of the connectors and the bolt from the full cross-sectional area of the member, should not exceed the design value of clear wood in compression parallel to the grain. The design values listed in the NDS for the greatest thickness of member with each type and size of connector unit are the maximums to be used for all thicker material. Design values for members with thicknesses between those listed may be obtained by interpolation.

11.17.1 Adjustment of Design Values for Connections with Fasteners

Nominal design values for connections or wood members with fasteners should be multiplied by applicable adjustment factors described in Art. 11.17.2 to obtain adjusted design values. The types of loading on the fasteners may be divided into four classes: lateral loading, withdrawal, loading parallel to grain, and loading perpendicular to grain. Adjusted design values are given in terms of nominal design values and adjustment factors in Eqs. (11.28) to (11.40),

where Z' = adjusted design value for lateral loading

Z = nominal design value for lateral loading

$W' =$ adjusted design value for withdrawal

$W =$ nominal design value for withdrawal

$P' =$ adjusted value for loading parallel to grain

$P =$ nominal value for loading parallel to grain

$Q' =$ adjusted value for loading normal to grain

$Q =$ nominal value for loading normal to grain

Bolts:

$$Z' = ZC_DC_MC_tC_gD_\Delta \qquad (11.28)$$

where $C_D =$ load-duration factor, not to exceed 1.6 for connections

$C_M =$ wet-service factor, not applicable to toenails loaded in withdrawal

$C_t =$ temperature factor

$C_R =$ group-action factor

$C_\Delta =$ geometry factor

Split-ring and shear-plate connectors:

$$P' = PC_DC_MC_tC_gC_\Delta C_dC_{st} \qquad (11.29)$$

$$Q' = QC_DC_MC_tC_gC_\Delta C_d \qquad (11.30)$$

where $C_d =$ penetration-depth-factor

$C_{st} =$ metal-side-plate factor

Nails and spikes:

$$W' = WC_DC_MC_tC_{tn} \qquad (11.31)$$

$$Z' = ZC_DC_MC_tC_dC_{eg}C_{di}C_{tn} \qquad (11.32)$$

where $C_{di} =$ diaphragm factor

$C_{tn} =$ toenail factor

Wood screws:

$$W' = WC_DC_MC_t \qquad (11.33)$$

$$Z' = ZC_DC_MC_tC_dC_{eg} \qquad (11.34)$$

where $C_{eg} =$ end-grain factor

Lag screws:

$$W' = WC_DC_MC_tC_{eg} \qquad (11.35)$$

$$Z' = ZC_DC_MC_tC_gC_\Delta C_dC_{eg} \qquad (11.36)$$

Metal plate connectors:

$$Z' = ZC_DC_MC_t \qquad (11.37)$$

Drift bolts and drift pins:

$$W' = WC_DC_MC_tC_{eg} \qquad (11.38)$$

$$Z' = ZC_DC_MC_tC_gC_\Delta C_dC_{eg} \qquad (11.39)$$

Spike grids:

$$Z' = ZC_DC_MC_tC_\Delta \qquad (11.40)$$

Adjustments for Fire-Retardant Treatment ▪ For connections made with lumber or structural glued-laminated timber pressure-treated with fire-retardant chemicals, design values should be obtained from the company providing the treatment and redrying service. The load-duration factor for impact does not apply to such connections.

11.17.2 Adjustment Factors for Connections with Fasteners

Design values for connections with fasteners should be adjusted as indicated in Art. 11.7.1. The adjustment factors are the following:

Load-Duration Factor ▪ Except when connection capacity is governed by the strength of metal, values of C_D may be taken from Table 11.5, Art. 11.4.2. For connections, C_D may not exceed 1.6.

Wet-Service Factor ▪ Nominal design values apply to wood that will be used where moisture content of the wood will be a maximum of 19% of the oven-dry weight, as would be the case in most covered structures. For connections in wood that is unseasoned or partly seasoned, or when connections will be exposed to wet-service conditions in use, nominal design values should be multiplied by the appropriate wet-service factor C_M in Table 11.21

Temperature Factor ▪ Values of C_t are listed in Table 11.22 for connections that will experience sustained exposure to elevated temperatures from 100 to 150 °F.

Group-Action Factor ▪ Values of C_g are given in Table 11.23. The NDS contains design criteria for determination of C_g for configurations

Table 11.21 Wet-Service Factors C_M for Connections

	Conditions of Wood*		
Fastener Type	At Time of Fabrication	In Service	C_M
Split-ring or shear-plate connectors[†]	Dry	Dry	1.0
	Partially seasoned	Dry	‡
	Wet	Dry	0.8
	Dry or wet	Partially seasoned or wet	0.67
Bolts or lag screws	Dry	Dry	1.0
	Partially seasoned or wet	Dry	§
	Dry or wet	Exposed to weather	0.75
	Dry or wet	Wet	0.67
Wood screws	Dry or wet	Dry	1.0
	Dry or wet	Exposed to weather	0.75
	Dry or wet	Wet	0.67
Common wire nails, box nails:			
For withdrawal loads	Dry	Dry	1.0
	Partially seasoned or wet	Wet	1.0
	Partially seasoned or wet	Dry	0.25
	Dry	Subject to wetting and drying	0.25
For lateral loads	Dry	Dry	1.0
	Partially seasoned or wet	Dry or wet	0.75
	Dry	Partially seasoned or wet	0.75

* Conditions of wood for determining wet-service factors for connections:
  Dry wood—moisture content up to 19%.
  Wet wood—moisture content at or above 30% (approximate fiber saturation point).
  Partly seasoned wood—moisture content between 19% and 30%.
  Exposed to weather—wood will vary in moisture content from dry to partly seasoned but is not expected to reach the fiber saturation point when the connection is supporting full design load. Subject to wetting and drying—wood will vary in moisture content from dry to partly seasoned or wet, or vice versa, with consequent effects on the tightness of the connection.

[†] For split-ring or shear-plate connectors, moisture-content limitations apply to a depth of $\frac{3}{4}$ in below the surface of the wood.

‡ When split-ring or shear-plate connectors are installed in wood that is partly seasoned at time of fabrication but will be dry before full design load is applied, proportional intermediate wet-service factors may be used.

§ When bolts or lag screws are installed in wood that is wet at the time of fabrication but will be dry before full design load is applied, the following wet service factors C_M apply:

For one fastener only or two or more fasteners placed in a single row parallel to grain or fasteners placed in two or more rows parallel to grain with separate splice plates for each row, $C_M = 1.0$.

When bolts or lag screws are installed in wood that is partly seasoned at the time of fabrication but will be dry before full design load is applied, proportional intermediate wet-service factors may be used.

not included in the table. For determination of C_g, a row of fasteners is defined as any of the following:

1. Two or more split-ring or shear-plate connectors aligned with the direction of the load.
2. Two or more bolts with the same diameter, loaded in shear, and aligned with the direction of the load.
3. Two or more lag screws of the same type and size loaded in single shear and aligned with the direction of the load.

The group-action factor is applied because the two end fasteners carry a larger load than the interior fasteners. With six or more fasteners in a row, the two end fasteners carry more than 50% of the load. With bolts, however, a small redistribution of load from the end bolts to the interior bolts occurs due to

Table 11.22 Temperature Factor C_t for Connections

In-Service Moisture Conditions*	$T \leq 100\,°F$	$100\,°F \leq T \leq 125\,°F$	$125\,°F < T \leq 150\,°F$
Dry	1.0	0.8	0.7
Wet	1.0	0.7	0.5

* Wet and dry service conditions are defined in a Table 11.21 footnote.

Table 11.23 Group-Action Factors

(a) For Bolt or Lag-Screw Connections with Wood Side Members*

		Number of Fasteners in a Row										
A_s/A_m†	A_{sr}‡ in²	2	3	4	5	6	7	8	9	10	11	12
0.5	5	0.98	0.92	0.84	0.75	0.68	0.61	0.55	0.50	0.45	0.41	0.38
	12	0.99	0.96	0.92	0.87	0.81	0.76	0.70	0.65	0.61	0.57	0.53
	20	0.99	0.98	0.95	0.91	0.87	0.83	0.78	0.74	0.70	0.66	0.62
	28	1.00	0.98	0.96	0.93	0.90	0.87	0.83	0.79	0.76	0.72	0.69
	40	1.00	0.99	0.97	0.95	0.93	0.90	0.87	0.84	0.81	0.78	0.75
	64	1.00	0.99	0.98	0.97	0.95	0.93	0.91	0.89	0.87	0.84	0.82
1	5	1.00	0.97	0.91	0.85	0.78	0.71	0.64	0.59	0.54	0.49	0.45
	12	1.00	0.99	0.96	0.93	0.88	0.84	0.79	0.74	0.70	0.65	0.61
	20	1.00	0.99	0.98	0.95	0.92	0.89	0.86	0.82	0.78	0.75	0.71
	28	1.00	0.99	0.98	0.97	0.94	0.92	0.89	0.86	0.83	0.80	0.77
	40	1.00	1.00	0.99	0.98	0.96	0.94	0 92	0.90	0.87	0.85	0.82
	64	1.00	1.00	0.99	0.98	0.97	0.96	0.95	0.93	0.91	0.90	0.88

(b) For 4-in Split-Ring or Shear-Plate Connectors with Wood Side Members§

		Number of Fasteners in a Row										
A_s/A_m†	A_{sr}‡ in²	2	3	4	5	6	7	8	9	10	11	12
0.5	5	0.90	0.73	0.59	0.48	0.41	0.35	0.31	0.27	0.25	0.22	0.20
	12	0.95	0.83	0.71	0.60	0.52	0.45	0.40	0.36	0.32	0.29	0.27
	20	0.97	0.88	0.78	0.69	0.60	0.53	0.47	0.43	0.39	0.35	0.32
	28	0.97	0.91	0.82	0.74	0.66	0.59	0.53	0.48	0.44	0.40	0.37
	40	0.98	0.93	0.86	0.79	0.72	0.65	0.59	0.54	0.49	0.45	0.42
	64	0.99	0.95	0.91	0.85	0.79	0.73	0.67	0.62	0.58	0.54	0.50
1	5	1.00	0.87	0.72	0.59	0.50	0.43	0.38	0.34	0.30	0.28	0.25
	12	1.00	0.93	0.83	0.72	0.63	0.55	0.48	0.43	0.39	0.36	0.33
	20	1.00	0.95	0.88	0.79	0.71	0.63	0.57	0.51	0.46	0.42	0.39
	28	1.00	0.97	0.91	0.83	0.76	0.69	0.62	0.57	0.52	0.47	0.44
	40	1.00	0.98	0.93	0.87	0.81	0.75	0.69	0.63	0.58	0.54	0.50
	64	1.00	0.98	0.95	0.91	0.87	0.82	0.77	0.72	0.67	0.62	0.58

* For fastener diameter $D = 1$ in and fastener spacing $s = 4$ in in bolt or lag-screw connections with modulus of elasticity for wood $E = 1,400,000$ psi. Tabulated values of C_g are conservative for $D < 1$ in, $s < 4$ in, or $E > 1,400,000$ psi.

† A_s = cross-sectional area of the main members before boring or grooving and A_m = sum of the cross-sectional areas of the side members before boring or grooving. When $A_s/A_m > 1$, use A_m/A_s.

‡ When $A_s/A_m > 1$, use A_m instead of A_s.

§ For spacing $s = 9$ in in connections made with 4-in split rings or shear plates with modulus of elasticity for wood $E = 1,400,000$ psi. Tabulated values of C_g are conservative for 2½-in split-ring connectors, 2⅝-in shear-plate connectors, $s < 9$ in, or $E > 1,400,000$ psi.

crushing of the wood at the end bolts. If failure is in shear, though, a partial failure occurs before substantial redistribution of load takes place.

When fasteners in adjacent rows are staggered but close together, they may have to be treated as a single row in determination of C_g. This occurs when the distance between adjacent rows is less than one-fourth of the spacing between the closest fasteners in adjacent rows.

Geometry Factor ▪ When the end distance or the spacing is less than the minimum required by the NDS for full design value but larger than the minimum required for reduced design value for bolts, lag screws, and split-ring and shear-plate connectors, nominal design values should be multiplied by the smallest applicable geometry factor C_Δ determined from the end distance and spacing requirements for the type of connector specified (Table 11.20). The smallest geometry factor for any connector in a group should be applied to all in the group. For multiple shear connections or for asymmetric three-member connections, the smallest geometry factor for any shear plane should be applied to all fasteners in the connection.

Penetration Factor ▪ For wood screws, lag screws, nails, and spikes, when the penetration is larger than the minimum required by the NDS (Table 11.24) but less than that assumed in the establishment of the full lateral design value, linear interpolation should be used in determination of C_d. This factor should not exceed unity. Table 11.24 lists values of C_d for the aforementioned fasteners.

End-Grain Factor ▪ Application of C_{eg} is necessary because connections are weaker when fasteners, such as screws and nails, are inserted in the end grain than when they are inserted in the side grain. Woods screws, nails, and spikes should not be loaded in withdrawal from end grain. Lag screws may be so loaded, but the nominal design value should be multiplied by $C_{eg} = 0.75$. Wood screws, lag screws, nails, and spikes may be permitted to carry lateral loading when inserted, parallel to grain, into end grain. In such cases, the nominal design value for lateral loads should be multiplied by $C_{eg} = 0.67$.

Metal-Side-Plate Factor ▪ When metal side plates are used in joints made with nails, spikes, or wood screws, the design value for wood side plates may be multiplied by the metal-side-plate factor $C_{st} = 1.25$. For 4-in shear-plate connectors, the nominal design value for load parallel to grain P should be multiplied by the appropriate C_{st} given in Table 11.25. The values depend on the species of wood used in the connection, such as group A, B, C, or D listed in the NDS.

Diaphragm Factor ▪ A diaphragm is a large, thin structural element that is loaded in its plane. When nails or spikes are used in a diaphragm connection, the nominal lateral design value should be multiplied by the diaphragm factor $C_{di} = 1.1$.

Toenail Factor ▪ For such connections as stud-to-plate, beam-to-plate, and blocking-to-plate nailing, the NDS recommends that toenails be driven at an angle of about 30° with the face of the stud, beam, or blocking and started about one-third the length of the nail from the end of the member. For toenailed connections the nominal lateral design values for connections with nails driven into side grain should be multiplied by the toenail factor $C_{tn} = 0.83$.

Table 11.24 Penetration and Penetration-Depth Factor*

Penetration p	Lag Screws	Wood Screws	Nails or Spikes
For full design value	8D	7D	12D
Minimum p	4D	4D	6D
C_d	p/8D	p/7D	p/12D

*D = bolt diameter.

Table 11.25 Metal-Side-Plate Factors for Shear-Plate Connectors*

Species Group[†]	C_{st}
A	1.18
B	1.11
C	1.05
D	1.00

* For 4-in shear plates loaded parallel to grain.
[†] For components of each species group, see the groupings in the NDS.

11.18 Glued Joints

Glued joints are generally made between two pieces of wood where the grain directions are parallel (as between the laminations of a beam or arch). Such joints may be between solid-sawn or laminated timber and plywood, where the face grain of the plywood is either parallel or at right angles to the grain direction of the timber.

Only in special cases may lumber be glued with the grain direction of adjacent pieces at an angle. When the angle is large, dimensional changes from changes in wood moisture content set up large stresses in the glued joint. Consequently, the strength of the joint may be considerably reduced over a period of time. Exact data are not available, however, on the magnitude of this expected strength reduction.

In joints connected with plywood gusset plates, this shrinkage differential is minimized because plywood swells and shrinks much less than does solid wood.

Glued joints can be made between end-grain surfaces, but they are seldom strong enough to meet the requirements of even ordinary service. Seldom is it possible to develop more than 25% of the tensile strength of the wood in such butt joints. For this reason plane sloping scarfs of relatively flat slope (Fig. 11.2) or finger joints with thin tips and flat slope on the sides of the individual fingers (Fig. 11.3) are used to develop a high proportion of the strength of the wood.

Joints of end grain to side grain are also difficult to glue properly. When subjected to severe stresses as a result of unequal dimensional changes in the members due to changes in moisture content, joints suffer from severely reduced strength.

For the above reasons, joints between end-grain surfaces and between end-grain and side-grain surfaces should not be used if the joints are expected to carry load.

For joints made with wood of different species, the allowable shear stress for parallel-grain bonding is equal to the allowable shear stress parallel to the grain for the weaker species in the joint. This assumes uniform stress distribution in the joint. When grain direction is not parallel, the allowable shear stress on the glued area between the two pieces may be computed Eq. (11.21).

[Military Specification MIL-A-397B, "Adhesive, Room-Temperature and Intermediate-Temperature Setting Resin (Phenol, Resorcinol, and Melamine Base)," and Military Specification MIL A-5534A, "Adhesive, High-Temperature Setting Resin (Phenol, Melamine, and Resorcinol Base)," U.S. Naval Supply Depot, Philadelphia, PA 19120.]

11.19 Wood Structural Framing Details

Wood structural frames are frequently used for single-family residences, apartment buildings, and commercial and industrial structures. The framing usually is wood beam-and-girder with wood columns, wood beam-and-post, wood joists with wood-stud bearing walls, or glued-laminated timber arches or rigid frames. Roofs may be supported on wood trusses or sloping wood rafters.

Timber bridges generally are the trestle, girder, truss, or arch types. If sawn timbers are used, they should be pressure-treated with a preservative after fabrication. For glued-laminated members, either individual laminations should be pressure-treated with a preservative before they are glued together or the member should be treated after the gluing, depending on the type of treatment specified. Some preservative treatments may not be suitable for use after gluing. (Consult a local laminator or "Standard for Preservative Treatment of Structural Glued Laminated Timber," AITC 109, American Institute of Timber Construction, or both.) See also Art. 11.25.

Connections in the structural framing in buildings and bridges are made with mechanical fastenings, such as nails, spikes, wood screws, lag screws, bolts, and timber connectors (see Arts. 11.16 and 11.17). Standard and special pre-engineered metal hangers are used extensively. Stock hangers are generally available and most manufacturers also provide hangers of special design. Where appearance is of prime importance, concealed hangers may be specified.

Figures 11.10 to 11.12 show structural framing details such as beam hangers and connectors and column anchors.

Wood Framing for Small Houses ▪ Although skeleton framing may be used for one- and two-family dwellings, such structures up to three stories high generally are built with load-bearing walls. When wood framing is used, the walls are conventionally built with slender studs spaced 16 or 24 in c to c. Similarly, joists and

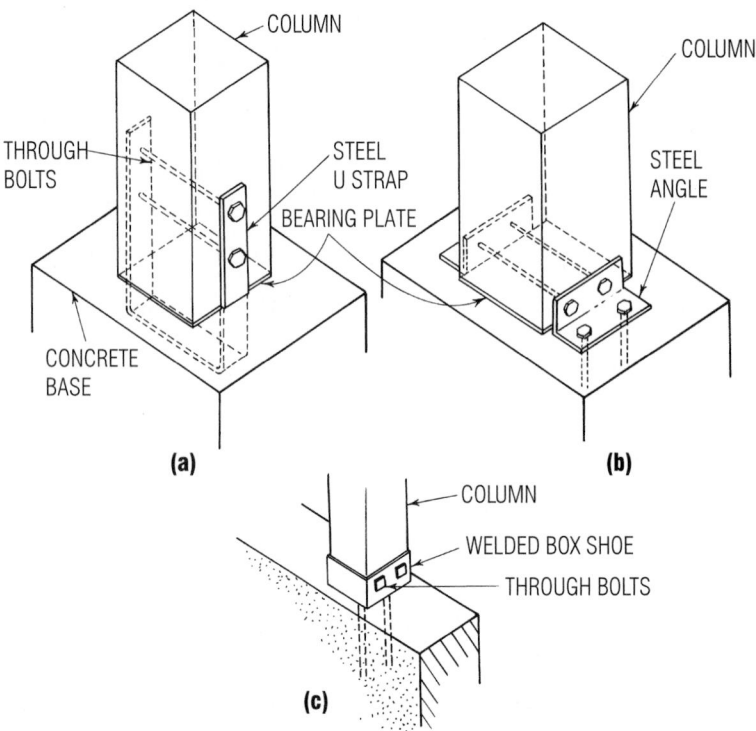

Fig. 11.10 Typical anchorages of wood column to base: (*a*) Wood column anchored to concrete base with U strap; (*b*) anchorage with steel angles; (*c*) with a welded box shoe.

rafters, which are supported on the walls and partitions, are usually also spaced 16 or 24 in c to c. Facings, such as sheathing and wallboard, and decking, floor underlayment, and roof sheathing are generally available in appropriate sizes for attachment to studs, joists, and rafters with that spacing.

Wood studs are usually set in walls and partitions with wide faces perpendicular to the face of the wall or partition. The studs are nailed at the bottom to bear on a horizontal plank, called the bottom or sole plate, and at the top to a pair of horizontal planks, called the top plate. These plates often are the same size as the studs. Joists or rafters may be supported on the top plate or on a header, called a ribband, supported in a cutout in the studs (Fig. 11.14).

Studs may be braced against racking by diagonals or horizontal blocking and facing materials, such as plywood or gypsum sheathing.

Three types of wood-frame construction are generally used: platform frame, balloon frame, and plank-and-beam frame.

In **platform framing**, first-floor joists are completely covered with subflooring to form a platform on which exterior walls and interior partitions are built (Fig. 11.13). This is the type of framing usually used for single-family dwellings.

Balloon framing is generally used for construction more than one story high. Wall studs are continuous from story to story. First-floor joists and exterior wall studs bear on an anchored sill plate (Fig. 11.14). Joists for the second and higher floors bear on a 1 × 4-in ribband let in to the inside edges of exterior wall studs. In two-story buildings, with brick or stone veneer exteriors, balloon framing minimizes variations in settlement of the framing and the masonry veneer.

Plank-and-beam framing (Fig. 11.15) requires fewer but larger-size piers, and wood components are spaced farther apart than in platform and balloon framing. In plank-and-beam framing, subfloors or roofs, typically composed of planks with a nominal thickness of 2 in, are supported on beams

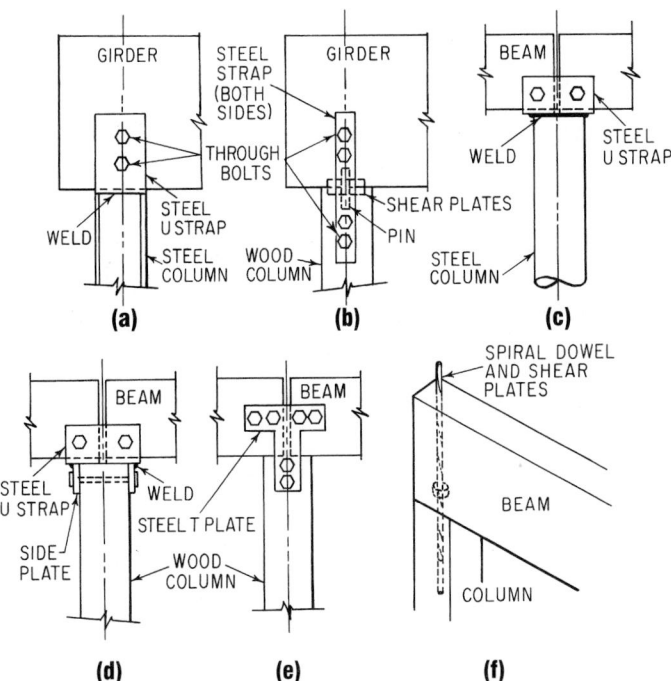

Fig. 11.11 Typical wood beam and girder connections to columns: (*a*) Wood girder to steel column; (*b*) girder to wood column; (*c*) beam to steel pipe column; (*d*) beam to wood column, with steel strap welded to steel side plates; (*e*) beam to wood column, with a steel T plate; (*f*) beam to wood column with a spiral dowel and shear plates.

spaced 8 ft c to c. Ends of the beams are supported on posts or concrete piers. Supplemental framing, set between posts for attachment of exterior and interior wall framing and finishes, also provides lateral support or bracing for the frame. Construction labor savings result from the use of larger and fewer framing members, which require less handling and fewer mechanical fasteners. Another advantage is that the need for cross bracing, which is often required in platform and balloon framing, is eliminated.

("Plank and Beam Framing for Residential Buildings," WCD No. 4, American Forest and Paper Association, Washington, D.C.)

11.20 Design of Wood Trusses

Wood trusses are used for long-span bridges and for support of roofs for buildings. For the latter, trusses offer the advantage that the type and arrangement of members may be chosen to suit the shape of the structure and the loads and stresses involved. Prefabricated, lightweight wood and wood-steel trusses are available and offer economy through use of repetitive design and mass production in truss assembly plants.

Joints are critical in truss design. Use of a specific truss type is often governed by joint considerations.

11.20.1 Lightweight Trusses

Chords and web members of lightweight trusses are generally made of dimension lumber, either visually graded or machine stress-rated. The trusses are usually installed 12 to 24 in c to c and are designed to take advantage of repetitive-member action (Art. 11.4.9). At a joint, members are connected by sheet-metal-gusset nail plates with projections, or teeth, that are pressed into the wood on opposite faces of the joint.

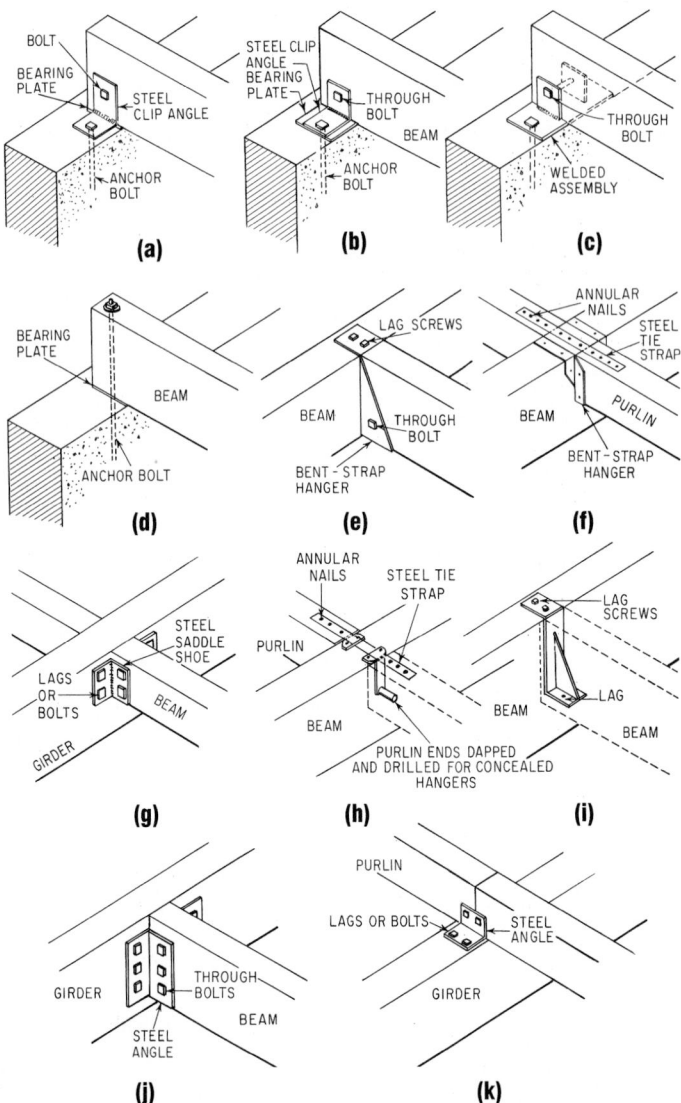

Fig. 11.12 Beam connections: (*a*) and (*b*) Wood beam anchored on a wall with steel angles; (*c*) with welded assembly; (*d*) beam anchored directly with a bolt; (*e*) beam supported on a bent-strap hanger on a girder; (*f*) similar support for purlins; (*g*) saddle connects beam to girder (suitable for one-sided connection); (*h*) and (*i*) connections with concealed hangers; (*j*) and (*k*) connections with steel angles.

Load-transfer capacity at a joint is based on an allowable load per unit of surface area of plate. Accordingly, a plate should be sized to cover all the members at the joint with an area sufficient to transfer loads from each member to the others. The allowable load depends on the number, size, and design of the steel teeth of the gusset plate. The load capacities of specific gusset plates should be obtained from the manufacturer. Additional information on this type of truss may be obtained from the Truss Plate Institute and the Wood Truss Council of America, both located in Madison, Wisconsin.

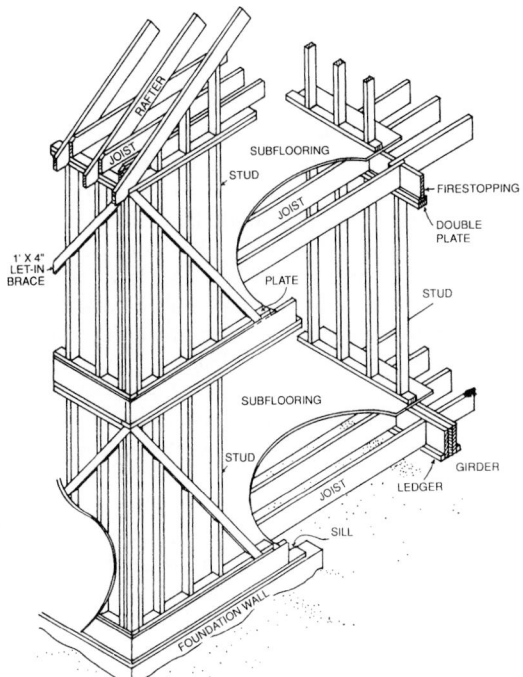

Fig. 11.13 Platform framing for two-story building.

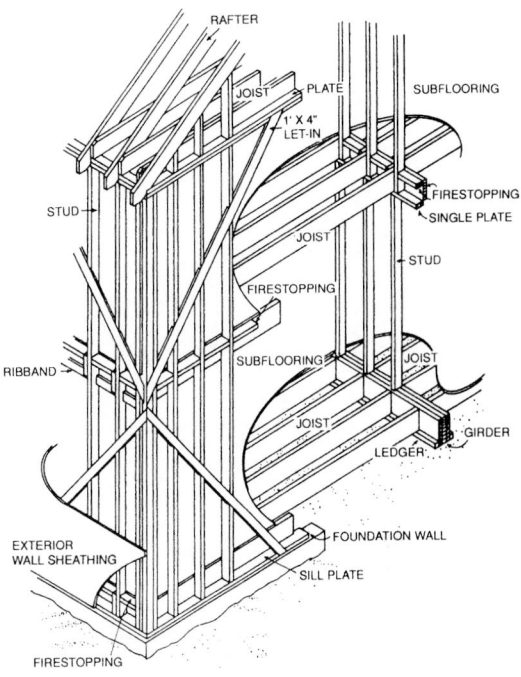

Fig. 11.14 Balloon framing for two-story building.

11.20.2 Timber Trusses

For long spans or large truss spacings, for example, 8 ft c to c or more, heavier wood chords and webs will be required. These members may have a nominal thickness of 3 or 4 in, or they may be glued-laminated timbers. At joints, the members will be connected with thicker metal-gusset plates than those required for lightweight trusses. As an alternative, composite wood-steel trusses with lumber chords and steel webs may be used.

Types of timber trusses generally used are bowstring, flat or parallel chord, and scissors (Fig. 11.16). For commercial buildings, trusses usually are spaced 8 to 24 ft apart.

Chords and webs may be single-leaf (or monochord), double-leaf, or multiple-leaf members Monochord trusses and trusses with double-leaf chords and single-leaf web system are the most common arrangements. Web members may be attached to the sides of the chords, or the web members may be in the same plane as the chords and attached with straps or gussets.

Individual truss members may be solid-sawn, glued-laminated, or mechanically laminated. Glued-laminated chords and solid-sawn web members are usually used. Steel rods or other steel shapes may be used as members of timber trusses if they meet design and service requirements.

The bowstring truss is by far the most popular. In building construction, spans of 100 to 200 ft are common, with single or two-piece top and bottom chords of glued-laminated timber, webs of solid-sawn timber, and metal heel plates, chord splice plates, and web-to-chord connections. This system is light in weight for the loads that it can carry; it can be shop- or field-assembled. Attention to the top chord, bottom chord, and heel connections is of prime importance since they are the major stress-carrying components. Since the top chord is nearly the shape of an ideal arch, stresses in chords are almost uniform throughout a bowstring truss; web stresses are low under uniformly distributed loads.

Parallel-chord trusses, with slightly sloping top chords and level bottom chords, are used less often because chord stresses are not uniform along their length and web stresses are high. Hence, different

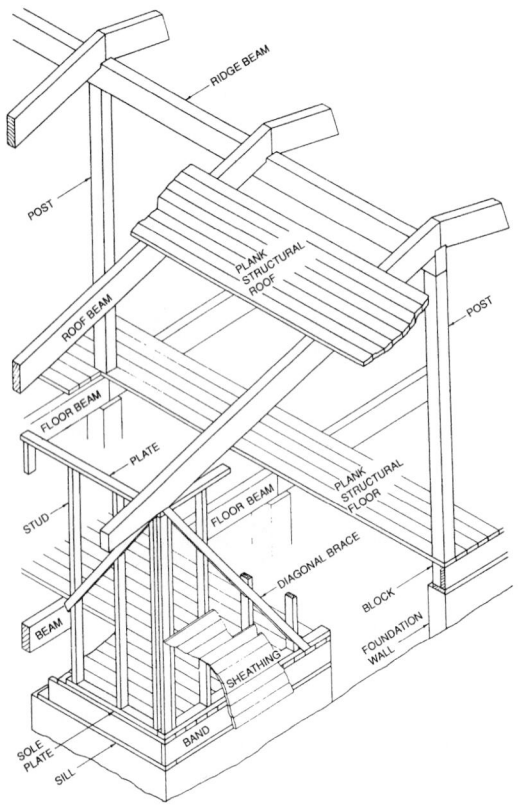

Fig. 11.15 Plank-and-beam framing for one-story building.

cross sections are required for successive chords, and web members and web-to-chord connections are heavy. Eccentric joints and tension stresses across the grain should be avoided in truss construction whenever possible, but particularly in parallel-chord trusses.

Triangular trusses and the more ornamental camelback and scissors trusses are used for shorter spans. They usually have solid-sawn members for both chords and webs where degree of seasoning of timbers, hardware, and connections are of considerable importance.

Truss Joints ▪ For joints, bolts, lag screws, metal-gusset nail plates (Art. 11.20.1), or shear-plate connectors are generally used. Sometimes, when small trusses are field-fabricated, only bolted joints are used. However, grooving tools for

connectors can also be used effectively in the field. Metal-gusset plates are usually installed in a truss assembly plant.

Framing between Trusses ▪ Longitudinal sway bracing perpendicular to the truss is usually provided by solid-sawn X bracing. Lateral wind bracing may be provided by end walls or intermediate walls, or both. The roof system and horizontal bracing should be capable of transferring the wind load to the walls. Knee braces between trusses and columns are often used to provide resistance to lateral loads.

Horizontal framing between trusses consists of struts between trusses at bottom-chord level and diagonal tie rods, often of steel with turnbuckles for adjustment.

("Design Manual for TECO Timber Connector Construction," Timber Engineering Co., Colliers, W. Va.; AITC 102, app. A, "Trusses and Bracing," American Institute of Timber Construction, Englewood, CO 80110; K. F. Faherty and T. G. Williamson, "Wood Engineering and Construction Handbook," 2nd ed., McGraw-Hill Publishing Company, New York.)

11.21 Design of Timber Arches

Arches may be two-hinged, with hinges at each base, or three-hinged, with a hinge at the crown. Figure 11.17 presents typical forms of arches.

Tudor arches are gabled rigid frames with curved haunches. Columns and pitched roof beam on each side of the crown usually are one piece of glued-laminated timber. This type of arch is frequently used in church construction with a high rise.

A-frame arches are generally used where steep pitches are required. They may spring from grade, or concrete abutments, or other suitably designed supports.

Radial arches are often used where long clear spans are required. They have been employed for clear spans up to 300 ft.

Gothic, parabolic, and three-centered arches are often selected for architectural and aesthetic considerations.

Timber arches may be tied or buttressed. If an arch is tied, the tie rods, which resist the horizontal thrust, may be above the ceiling or below grade,

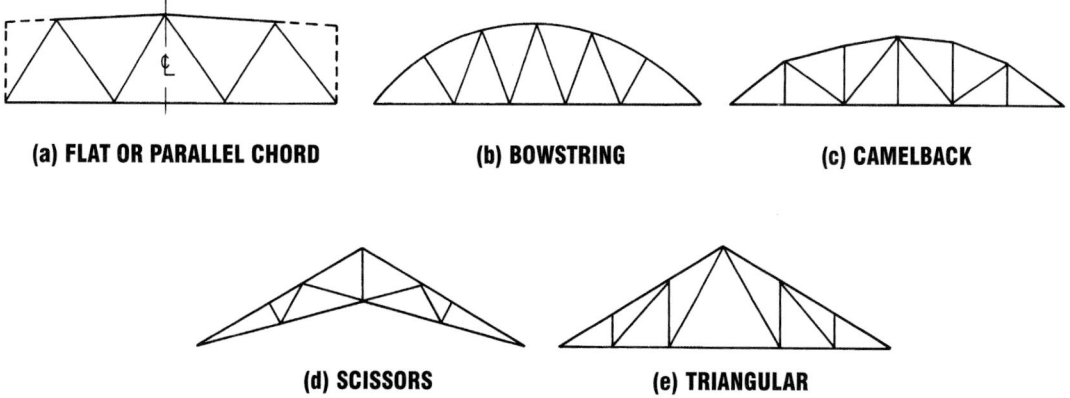

Fig. 11.16 Types of wood trusses.

and simple connections may be used where the arch is supported on masonry walls, concrete piers, or columns (Fig. 11.18).

Segmented arches are fabricated with overlapping lumber segments, nailed- or glued-laminated. They generally are three-hinged, and they may be tied or buttressed. They are economical because of the ease of fabrication and simplicity of field erection. Field splice joints are minimized;

generally there is only one simple connection, at the crown (Fig. 11.19c). Except for extremely long spans, they are shipped in only two pieces. Erected, they need not be concealed by false ceilings, as may be necessary with trusses. And the cross section is large enough for segmented arches to be classified as heavy-timber construction.

A long-span arch may require a splice or moment connection to segment the arch to facilitate transportation to the jobsite. Figure 11.20 shows typical moment connections for wood arches.

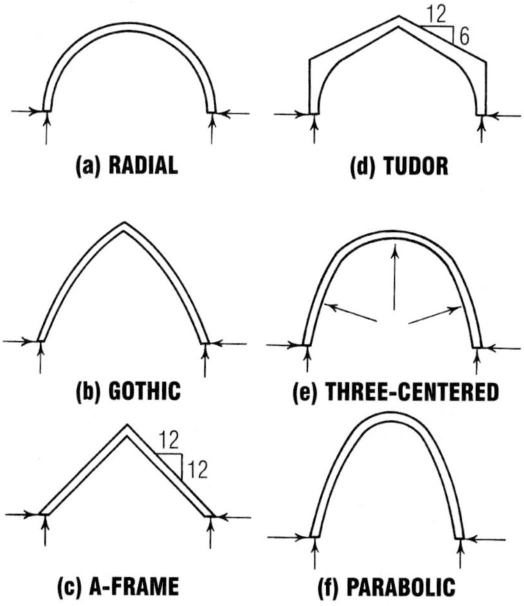

11.22 Timber Decking

Wood decking used for floor and roof construction may consist of solid-sawn planks with nominal thickness of 2, 3, or 4 in. Or it may be panelized or laminated. Panelized decking is made up of splined panels, usually about 2 ft wide.

For glued-laminated decking, two or more pieces of lumber are laminated into a single decking member, usually with 2- to 4-in nominal thickness.

Solid-sawn decking usually is fabricated with edges tongued and grooved or shiplapped for transfer of vertical load between pieces. The decking may be end matched, square end, or end grooved for splines. As indicated in Fig. 11.21, the decking may be arranged in various patterns over supports.

In Type 1, the pieces are simply supported. Type 2 has a controlled random layup. Type 3 contains intermixed cantilevers. Type 4 consists of a

(a) RADIAL

(d) TUDOR

(b) GOTHIC

(e) THREE-CENTERED

(c) A-FRAME

(f) PARABOLIC

Fig. 11.17 Types of wood arches.

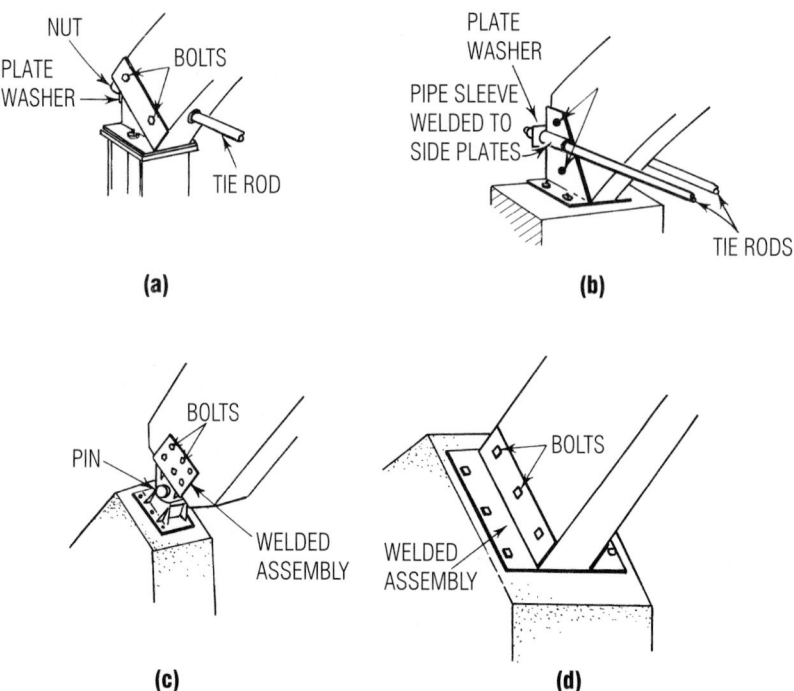

Fig. 11.18 Bases for segmented wood arches: (*a*) and (*b*) Tie rod anchored to arch shoe; (*c*) hinge anchorage for large arch; (*d*) welded arch shoe.

combination of simple-span and two-span continuous pieces. Type 5 is two-span continuous.

In Types 1, 4, and 5, end joints bear on supports. For this reason, these types are recommended for thin decking, such as 2-in.

Type 3, with intermixed cantilevers, and Type 2, with controlled random layup, are used for deck continuous over three or more spans. These types permit some of the end joints to be located between supports. Hence, provision must be made for stress transfer at those joints. Tongue-and-groove edges, wood splines on each edge of the course, horizontal spikes between courses, and end matching or metal end splines may be used to transfer shear and bending stresses.

In Type 2, the distance between end joints in adjacent courses should be at least 2 ft for 2-in deck and 4 ft for 3- and 4-in deck. Joints approximately lined up (within 6 in of being in line) should be separated by at least two courses. All pieces should rest on at least one support. And not more than one end joint should fall between supports in each course.

In Type 3, every third course is simple span. Pieces in other courses cantilever over supports, and end joints fall at alternate quarter or third points of the spans. Each piece rests on at least one support.

To restrain laterally supporting members of 2-in deck in Types 2 and 3, the pieces in the first and second courses and in every seventh course should bear on at least two supports. End joints in the first course should not occur on the same supports as end joints in the second course unless some construction, such as plywood overlayment, provides continuity. Nail end distance should be sufficient to develop the lateral nail strength required.

Heavy-timber decking is laid with wide faces bearing on the supports. Each piece must be nailed to each support. Each end at a support should be nailed to it. For 2-in decking, a $3\frac{1}{2}$-in (16d) toe and face nail should be used in each 6-in-wide piece at supports, and three nails for wider pieces. Tongue-and-groove decking generally is also toenailed through the tongue. For 3-in

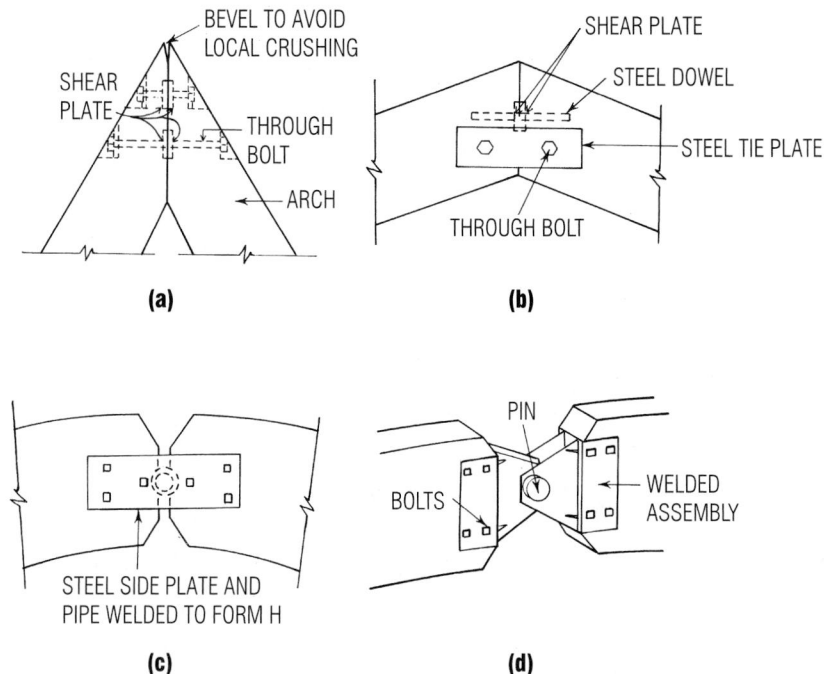

Fig. 11.19 Crown connections for arches: (*a*) For arches with slope 4:12 or greater, the connection consists of pairs of back-to-back shear plates with through bolts or threaded rods counterbored into the arch. (*b*) For arches with flatter slopes, shear plates centered on a dowel may be used in conjunction with the plates and through bolts. (*c*) and (*d*) Hinge at crown.

decking, each piece should be toenailed with one 4-in (20d) spike and face-nailed with 5-in (40d) spike at each support. For 4-in decking, each piece should be toenailed at each support with one 5-in (40d) nail and face-nailed there with one 6-in (60d) spike.

Courses of 3- and 4-in double tongue-and-groove decking should be spiked to each other with 8½-in spikes not more than 30 in apart. One spike should not be more than 10 in from each end of each piece. The spikes should be driven through predrilled holes. Two-inch decking is not fastened together horizontally with spikes.

Deck design usually is governed by maximum permissible deflection in end spans. But each design should be checked for bending stress.

(AITC 112, "Standard for Heavy Timber Roof Decking," and AITC 118, "Standard for 2-in. Nominal Thickness Lumber Roof Decking for Structural Applications," American Institute of Timber Construction, 7012 S. Revere Parkway, Englewood, Colo (www.AITC-glulam.org); AITC

"Timber Construction Manual," 4th ed., John Wiley & Sons, Inc., New York (www.wiley.com).)

11.23 Pole Construction

Wood poles are used for various types of construction, including flagpoles, utility poles, and framing for buildings. These employ preservative-treated round poles set into the ground as columns. The ground furnishes vertical and horizontal support and prevents rotation at the base.

For allowable foundation and lateral pressures, consult the local building code or a model code.

In buildings, a bracing system can be provided at the top of the poles to reduce bending moments at the base and to distribute loads. Design of buildings supported by poles without bracing requires good knowledge of soil conditions, to eliminate excessive deflection or sidesway.

Bearing values under the base of poles should be checked. For backfilling the holes, well-tamped

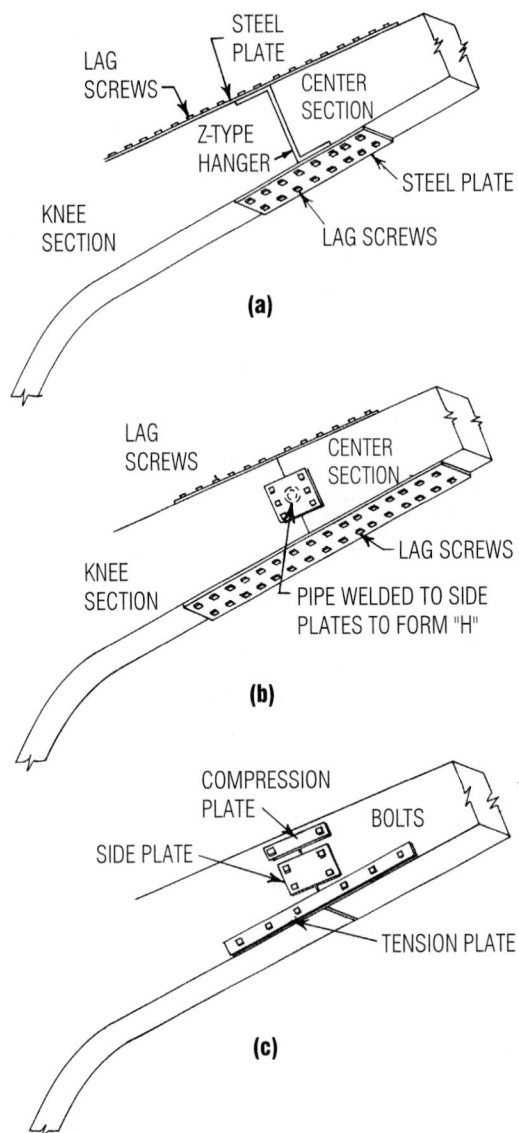

(a)

(b)

(c)

Fig. 11.20 Schematics of some moment connections for timber arches: (*a*) and (*b*) Connections with top and bottom steel plates; (*c*) connection with side plates.

native soil, sand, or gravel may be satisfactory. But concrete or soil cement is more effective. They can reduce the required depth of embedment and improve bearing capacity by increasing the skin-friction area of the pole. Skin friction is effective in reducing uplift due to wind.

To increase bearing capacity under the base end of poles for buildings, concrete footings often are used. They should be designed to withstand the punching shear of the poles and bending moments. Thickness of concrete footings should be at least 12 in. Consideration should be given to use of concrete footings even in firm soils, such as hard dry clay, coarse firm sand, or gravel.

Calculation of required depth of embedment in soil of poles subject to lateral loads generally is impractical without many simplifying assumptions. An approximate analysis can be made, but the depth of embedment should be checked by tests or at least against experience in the same type of soil. See "Post and Pole Foundation Design," ASAE Engineering Practice, EP486, American Society of Agricultural Engineers, St. Joseph, Mich.

("Design Properties of Round, Sawn and Laminated Preservatively Treated Construction Poles and Posts," ASAE Engineering Practice, EP388.2; "Standard Specifications and Dimensions for Wood Poles." ANSI 05.1, American National Standards Institute (www.ansi.org).)

11.24 Wood Structural Panels

Structural panels are composed of two or more materials with different structural characteristics formed into a thin, flat configuration and capable of resisting applied loads. The panels may be classified, in accordance with the manufacturing process, as plywood; mat-formed panels, such as oriented-strand board (OSB); and composite panels.

Plywood is a structural panel comprising wood veneer plies, united under pressure by adhesive. The bond between plies is at least as strong as solid wood. The panel is formed of an odd number of layers, with the grain of each layer perpendicular to the grain of adjoining layers. A layer may consist of a single ply or two or more plies laminated with grain parallel. Outer layers and all odd-numbered layers usually have the grain oriented parallel to the long dimension of the panel. The variation in grain direction, or cross lamination, makes the panel strong and stiff, equalizes strains under load, and limits panel dimensional changes, warping, and splitting.

Mat-formed panels are structural panels, such as particleboard, waferboard, and OSB, that do not contain wood veneer. Particleboard consists of a combination of wood particles and adhesive and is

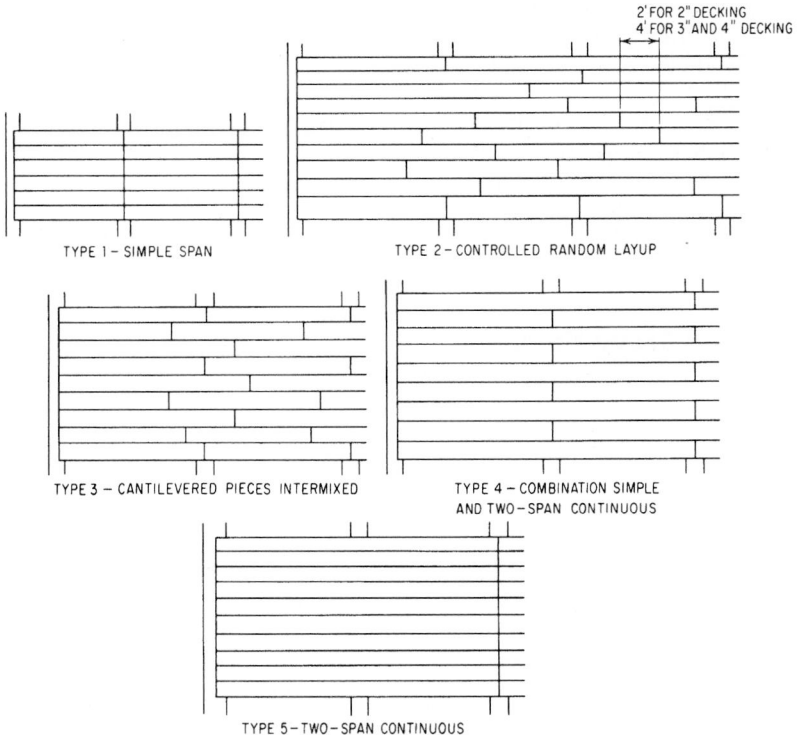

Fig. 11.21 Typical arrangements for heavy-timber decking.

widely used as floor underlayment in buildings. Waferboard is similar to particleboard but is made with flakes instead of particles. OSB is composed of compressed wood strands arranged in layers at right angles to one another and bonded with a waterproof adhesive. Like plywood, OSB has the strength and stiffness that result from cross lamination of layers.

Composite panels consist of combinations of veneer and other wood-based materials.

Structural wood panels may be used in construction as sheathing, decking, floor underlayment, siding, and concrete forms. Plywood, in addition, may serve as a component of stressed-skin panels and built-up (I- or box-shape) beams and columns.

To satisfy building code requirements, structural wood panels should meet the requirements of one or more of the following standards:

"U.S. Product Standard PS 1-83 for Construction and Industrial Plywood," applicable to plywood only.

"Voluntary Product Standard PS 2-92, Performance Standard for Wood-Based Structural-Use Panels," applicable to plywood, OSB, and composite panels.

"APA Performance Standards and Policies for Structural-Use Panels," PRP 108, which is similar to PS 2 but also incorporates performance-based qualifications procedures for siding panels.

11.24.1 Classification of Structural Panels

To meet building code requirements, structural wood panels should carry the trademark of a code-approved agency, such as the American Plywood Association (APA). Construction grades are generally fabricated with a waterproof adhesive and may be classified as Exterior or Exposure 1.

Exterior panels are suitable for permanent exposure to weather or moisture.

Exposure 1 panels may be used where they are not permanently exposed to the weather and where exposure durability is needed to resist the effects of moisture during construction delays, high humidity, water leakage, and other conditions of similar severity.

Exposure 2 panels are suitable for interior use where exposure durability to resist the effects of high humidity and water leakage is required.

Interior panels are intended for interior use where they will be exposed only to minor amounts of moisture and only temporarily.

11.24.2 Plywood Group Number

Plywood can be manufactured from over 70 species of wood. These species are divided on the basis of strength and stiffness into five groups under U.S. Product Standard PS 1-83.

Group 1. Douglas fir from Washington, Oregon, California, Idaho, Montana, Wyoming, British Columbia, and Alberta; western larch; southern pine (loblolly, longleaf, shortleaf, slash); yellow birch; tan oak

Group 2. Port Orford cedar; Douglas fir from Nevada, Utah, Colorado, Arizona, and New Mexico; fir (California red, grand, noble, Pacific silver, white); western hemlock; red and white lauan; western white pine; red pine; black maple; yellow poplar; red and Sitka spruce

Group 3. Red alder; Alaska cedar; jack, lodgepole, spruce and ponderosa pine; paper birch; subalpine fir; eastern hemlock; bigleaf maple; redwood; black, Engelmann, and white spruce

Group 4. Incense and western red cedar, sugar and eastern white pine, eastern and black (western poplar) cottonwood, cativo, paper birch, and bigtooth and quaking aspen

Group 5. Balsam fir, basswood, and balsam poplar

The strongest species are in Group 1; the next strongest in Group 2, etc. The group number that appears in the trademark on some APA trademarked panels, primarily sanded grades, is based on the species used for face and back veneers. Where face and back veneers are not from the same species group, the higher group number is used, except for sanded panels $\frac{3}{8}$ in thick or less and decorative panels of any thickness. These are identified by face species if C or D grade backs are at least $\frac{1}{8}$ in thick and are not more than one species group number larger.

11.24.3 Grades of Structural Wood Panels

Wood veneers are graded in accordance with appearance. Veneer grades define veneer appearance in terms of natural, unrepaired-growth characteristics and allowable number and size of repairs that may be made during manufacture (Table 11.26). The highest quality veneer grades are N and A. The minimum grade of veneer permitted in Exterior plywood is C grade. D-grade veneer is used in panels intended for interior use or applications protected from permanent exposure to weather.

Plywood is generally graded in accordance with the veneer grade used on the face and back of the panel; for example, A-B, B-C, . . . , or by a name suggesting the panel's intended end use, such as APA Rated Sheathing or APA Rated Sturd-I-Floor. Since OSB panels are composed of flakes or strands instead of veneers, they are graded without reference to veneers or species. Composite panels are graded on an OSB performance basis by end use and exposure durability. Typical panel trademarks for all three panel types and an explanation of how to read them are shown in Fig. 11.22.

Plywood panels with B-grade or better veneer faces are supplied in sanded-smooth condition to fulfill the requirements of their intended end use—applications such as cabinets, shelving, furniture, and built-ins. Rated sheathing panels are unsanded since a smooth surface is not a requirement of their intended end use. Still other panels, such as Underlayment, Rated Sturd-I-Floor, C-D Plugged, and C-C Plugged, require only touch sanding for "sizing" to make the panel thickness more uniform.

Standard panel dimensions are 4×8 ft, although some mills also produce plywood panels 9 or 10 ft long or longer. OSB panels may be ordered in lengths up to 28 ft.

Construction plywood is graded under the standard in accordance with two basic systems. One system covers engineered grades, the other appearance grades.

Engineered grades consist largely of unsanded sheathing panels designated C-D Interior or C-C Exterior. The latter is bonded with exterior glue. Either grade may be classified as Structural I or

Table 11.26 Veneer-Grade Designations

Grade N
Smooth surface "natural finish" veneer. Select, all heartwood or all sapwood. Free of open defects. Allows not more than six repairs, wood only, per 4 × 8-ft panel, made parallel to grain and well matched for grain and color.

Grade A
Smooth, paintable. Not more than 18 neatly made repairs, boat, sled, or router type, and parallel to grain, permitted. May be used for natural finish in less demanding applications. Synthetic repairs permitted.

Grade B
Solid surface. Shims, circular repair plugs, and tight knots up to 1 in across grain permitted. Some minor splits and synthetic repairs permitted.

Grade C—Plugged
Improved C veneer with splits limited to $\frac{1}{8}$-in width and knotholes and borer holes limited to $\frac{1}{4} \times \frac{1}{2}$ in. Admits some broken grain. Synthetic repairs permitted.

Grade C
Tight knots to $1\frac{1}{2}$ in. Knotholes up to 1 in across grain and some up to $1\frac{1}{2}$ in if total width of knots and knotholes is within specified limits. Synthetic or wood repairs and discoloration and sanding defects that do not impair strength permitted. Limited splits allowed. Stitching permitted.

Grade D
Knots and knotholes up to $2\frac{1}{2}$ in wide across grain and $\frac{1}{2}$ in larger within specified limits, limited splits, and stitching permitted. Limited to Exposure 1 or interior panels.

Structural II, both of which are made with exterior glue and subject to other requirements, such as limitations on knot size and repairs of defects. Structural I is made only of Group I species and is stiffer than other grades. Structural II may be made of Group 1, 2, or 3 or any combination of these species. Structural I and II are suitable for such applications as box beams, gusset plates, stressed-skin panels, and folded-plate roofs.

Appearance grades, except for Plyform, are designated by panel thickness, veneer classification of face and back veneers, and species group of the veneers. For Plyform, class designates a mix of species.

11.24.4 Plywood Applications

Table 11.27 describes the various grades of plywood and indicates how they are generally used.

PS 1-83 classifies plywood made for use as concrete forms in two grades. Plyform (B-B) Class I is limited to Group I species on face and back, with limitations on inner plies. Plyform (B-B) Class II permits Group 1, 2, or 3 for face and back, with limitations on inner plies. High-density overlay should be specified for both classes when highly smooth, grain-free concrete surfaces and maximum reuses are required. The bending strength of Plyform Class I is greater than that of Class II. Grades other than Plyform, however, may be used for forms.

Span-rated panels are available that are designed specifically for use in buildings in single-layer floor construction beneath carpet and pad. The maximum spacing of floor joists, or span rating, is stamped on each panel. Panels are manufactured with span ratings of 16, 20, 24, 32, and 48 in. These assume the panel continuous over two or more spans with the long dimension or

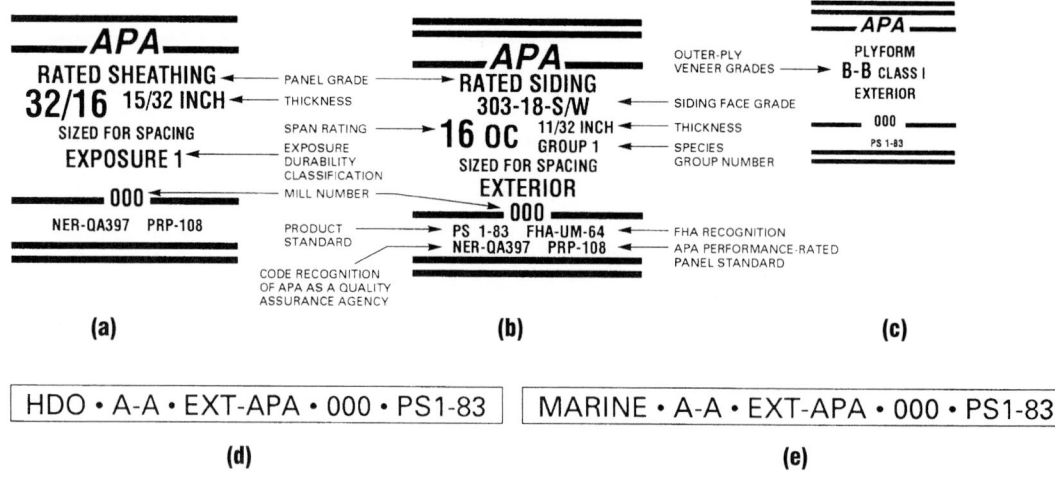

Fig. 11.22 Typical trademarks for structural panels. (*a*) APA Rated Sheathing with a thickness of $^{15}/_{32}$ in and a span rating $^{32}/_{16}$. The left-hand number denotes the recommended maximum spacing of supports when the panel is used for roof sheathing with the long dimension or strength axis of the panel across three or more supports. The right-hand number indicates the maximum recommended spacing of supports when the panel is used for subflooring with the long dimension or strength axis of the panel across three or more supports. (*b*) APA Rated Siding, grade 303-18-S/W, with a span rating of 16 in. (*c*) APA Ply-form, intended for use in formwork for concrete. (*d*) APA high-density overlay (HDO), abrasion resistant and suitable for exterior applications (used for concrete forms, cabinets, countertops, and signs). (*e*) APA Marine, used for boat hulls.

strength axis across supports (Fig. 11.23*a*). The span rating in the trademark applies when the long panel dimension is across supports unless the strength axis is otherwise identified. Glue-nailing is preferred, though panels may be nailed only. Figure 11.23*b* illustrates application of panel subflooring.

Rated siding (panel or lap) may be applied directly to studs or over nonstructural fiberboard, or gypsum or rigid-foam-insulation sheathing. (Nonstructural sheathing is defined as sheathing not recognized by building codes as meeting both bending and racking-strength requirements.) A single layer of panel siding, since it is strong and rack resistant, eliminates the cost of installing separate structural sheathing or diagonal wall bracing. Panel sidings are normally installed vertically, but most may also be placed horizontally (long dimension across supports) if horizontal joints are blocked.

Building paper is generally not required over wall sheathing, except under stucco or under brick

veneer where required by the local building code. Recommended wall sheathing spans with brick veneer and masonry are the same as those for nailable panel sheathing.

Rated sheathing meets building code wall-sheathing requirements for bending and racking strength without let-in corner bracing. Installation is illustrated in Fig. 11.24. Either rated sheathing or all-veneer plywood rated siding can be used in shear walls.

Publications of the American Plywood Association, P.O. Box 11700, Tacoma, WA 98411-0700 (www.apawood.org): "U.S. Product Standard PS 1-83 for Construction and Industrial Plywood," H850; "Voluntary Product Standard PS 2-92," S350; "Performance Standards and Policies for Structural-Use Panels," E445; "Nonresidential Roof Systems," A310; "APA Design Construction Guide, Residential & Commercial," E30; "Diaphragms," L350; "Concrete Forming," V345; "Plywood Design Specifications (PDS)", Y510; PDS "Supplements;" "House Building Basics," X461.)

Table 11.27 Applications of Plywood Grades

Plywood Grade	Description and use	Veneer Grade			Common Thicknesses, in
		Face	Back	Inner	
(a) Interior-Type Plywood					
C-D INT-APA	Unsanded sheathing grade for wall, roof subflooring, and industrial applications such as pallets and for engineering design, with proper stresses. Also available with intermediate and exterior glue.* For permanent exposure to weather or moisture only, exterior-type plywood is suitable.	C	D	D	$\frac{5}{16}$, $\frac{3}{8}$, $\frac{1}{2}$, $\frac{5}{8}$, $\frac{3}{4}$
Structural I C-D INT-APA or Structural II C-D INT-APA	Plywood grades to use where strength properties are of maximum importance, such as plywood-lumber components. Made with exterior glue only. Structural I is made from all Group 1 woods. Structural II allows Group 3 woods.	C	D	D	$\frac{5}{16}$, $\frac{3}{8}$, $\frac{1}{2}$, $\frac{5}{8}$, $\frac{3}{4}$
Underlayment INT-APA	For underlayment or combination subfloor-underlayment under resilient floor coverings. Available with exterior glue. Touch-sanded. Available with tongue and groove.	C Plugged	D	C & D	$\frac{1}{2}$, $\frac{19}{32}$, $\frac{5}{8}$, $\frac{23}{32}$, $\frac{3}{4}$
C-D Plugged INT-APA	For built-ins, wall and ceiling tile backing, not for underlayment. Available with exterior glue. Touch-sanded.	C Plugged	D	D	$\frac{1}{2}$, $\frac{19}{32}$, $\frac{5}{8}$, $\frac{23}{32}$, $\frac{3}{4}$
Structural I or II[†] Underlayment or C-D plugged	For higher strength requirements for underlayment or built-ins. Structural I constructed from all Group 1 woods. Made with exterior glue only.	C Plugged	D	C & D	$\frac{1}{2}$, $\frac{19}{32}$, $\frac{5}{8}$, $\frac{23}{32}$, $\frac{3}{4}$
2.4.1 INT-APA	Combination subfloor-underlayment. Quality floor base. Available with exterior glue, most often touch-sanded. Available with tongue and groove.	C Plugged	D	C & D	$1\frac{1}{8}$
Appearance grades	Generally applied where a high quality surface is required. Includes N-N, N-A, N-B, N-D, A-A, A-B, A-D, B-B, and B-D INT-APA grades.	B or better	D or better	D	$\frac{1}{4}$, $\frac{3}{8}$, $\frac{1}{2}$, $\frac{5}{8}$, $\frac{3}{4}$

(Continued)

Table 11.27 *(Continued)*

Plywood Grade	Description and use	Veneer Grade			Common Thicknesses, in
		Face	Back	Inner	
	(b) Exterior-Type Plywood				
C-D EXT-APA	Unsanded sheathing grade with waterproof glue bond for wall, roof, subfloor, and industrial applications such as pallet bins.	C	C	C	$\frac{5}{16}$, $\frac{3}{8}$, $\frac{1}{2}$, $\frac{5}{8}$, $\frac{3}{4}$
Structural I C-C EXT-APA or Structural II C-C EXT-APA[†]	"Structural" is a modifier for this unsanded sheathing grade. For engineering applications in construction and industry where full exterior-type panels are required. Structural I is made from Group 1 woods only.	C	C	C	$\frac{5}{16}$, $\frac{3}{8}$, $\frac{1}{2}$, $\frac{5}{8}$, $\frac{3}{4}$
Underlayment EXT-APA and C-C plugged EXT-APA	Underlayment for combination subfloor underlayment or two-layer floor under resilient floor coverings where severe moisture conditions may exist. Also for controlled-atmosphere rooms and many industrial applications. Touch-sanded. Available with tongue and groove.	C Plugged	C	C	$\frac{1}{2}$, $\frac{19}{32}$, $\frac{5}{8}$, $\frac{23}{32}$, $\frac{3}{4}$
Structural I or II[†] Underlayment EXT-APA or C-C plugged EXT-APA	For higher-strength underlayment where severe moisture conditions may exist. All Group 1 construction in Structural I. Structural II allows Group 3 woods.	C Plugged	C	C	$\frac{1}{2}$, $\frac{19}{32}$, $\frac{5}{8}$, $\frac{23}{32}$, $\frac{3}{4}$
B-B Plyform Class I or II[†]	Concrete-form grade with high reuse factor. Sanded both sides, mill-oiled unless otherwise specified. Available in HDO. For refined design information on this special-use panel see APA publication "Plywood for Concrete Forming" (form V345). Design using values from this specification will result in a conservative design.	B	B	C	$\frac{5}{8}$, $\frac{3}{4}$
Marine EXT-APA	Superior exterior-type plywood made only with Douglas fir or Western larch. Special solid-core construction. Available with MDO or HDO face. Ideal for boat hull construction.	A or B	A or B	B	$\frac{1}{4}$, $\frac{3}{8}$, $\frac{1}{2}$, $\frac{5}{8}$, $\frac{3}{4}$

Table 11.27 (*Continued*)

Plywood Grade	Description and use	Veneer Grade			Common Thicknesses, in
		Face	Back	Inner	
Appearance grades	Generally applied where a high quality surface is required. Includes A-A, A-B, A-C, B-B, B-C, HDO, and MDO EXT-APA. Appearance grades may be modified to Structural I. For such designation use Group 1 stresses and Table 11-33*b* (sanded) section properties.	B or better	C or better	C	$\frac{1}{4}$, $\frac{3}{8}$, $\frac{1}{2}$, $\frac{5}{8}$, $\frac{3}{4}$

* When exterior glue is specified, i.e., "interior with exterior glue glue," stress level 2 (S-2) should be used.
† Check local suppliers for availability of Structural II and Plyform Class II grades.
Source: "Plywood Design Specifications," American Plywood Association.

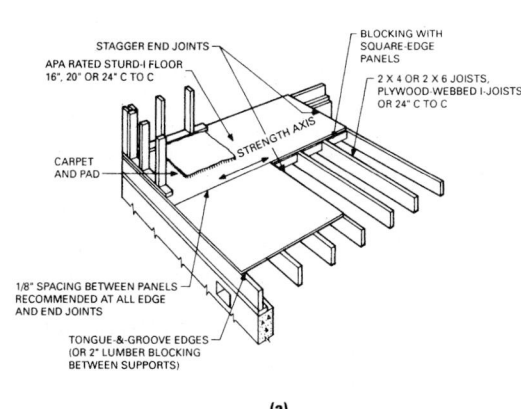

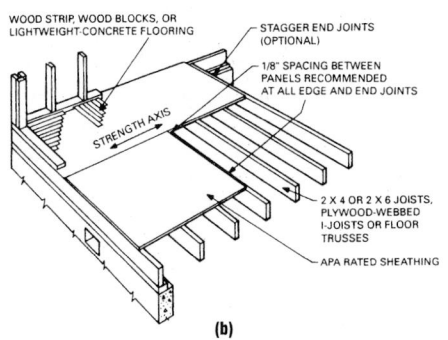

Fig. 11.23 Floor construction with structural wood panels: (*a*) Single-layer floor; (*b*) subfloor.

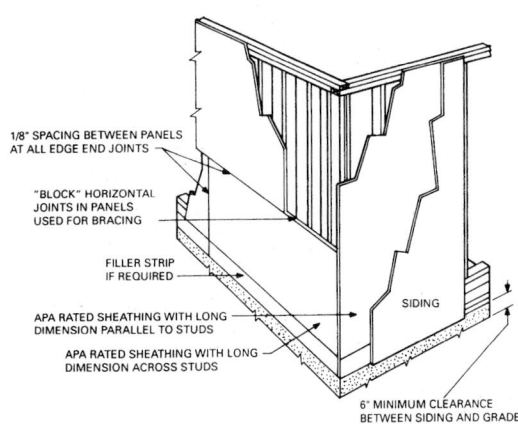

Fig. 11.24 Structural panel sheathing applied to studs.

11.25 Preservative Treatments for Wood

Wood-destroying fungi must have air, suitable moisture, and favorable temperatures to develop and grow in wood. Submerge wood permanently and totally in water to exclude air, keep the moisture content below 18 to 20%, or hold temperature below 40 °F or above 110 °F, and wood remains permanently sound. If wood moisture content is kept below the fiber-saturation point

(25 to 30%) when the wood is untreated, decay is greatly retarded. Below 18 to 20% moisture content, decay is completely inhibited.

If wood cannot be kept dry, a wood preservative, properly applied, must be used. The following can be a guide to determine if treatment is necessary.

Wood members are permanent without treatment if located in enclosed buildings where good roof coverage, proper roof maintenance, good joint details, adequate flashing, good ventilation, and a well-drained site assure moisture content of the wood continuously below 20%. Also, in arid or semiarid regions, where climatic conditions are such that the equilibrium moisture content seldom exceeds 20%, and then only for short periods, wood members are permanent without treatment.

Where wood is in contact with the ground or water, where there is air and the wood may be alternately wet and dry, a preservative treatment, applied by a pressure process, is necessary to obtain an adequate service life. In enclosed buildings where moisture given off by wet-process operations maintains equilibrium moisture contents in the wood above 20%, wood structural members must be treated with a preservative, as must wood exposed outdoors without protective roof covering and where the wood moisture content can go above 18 to 20% for repeated or prolonged periods.

Where wood structural members are subject to condensation by being in contact with masonry, preservative treatment is necessary.

Design values for wood structural members apply to products pressure-treated by an approved process and with an approved preservative. (The "AWPA Book of Standards," American Wood Preservers Association, Stevensville, Md., describes these approved processes.) Design values for pressure-preservative-treated lumber are modified with the usual adjustment factors described in Art. 11.4 with one exception. The load-duration factor for impact (Table 11.5) does not apply to structural members pressure-treated with waterborne preservatives to the heavy retentions required for "marine" exposure or to structural members treated with fire-retardant chemicals.

To obtain preservative-treated glued-laminated timber, lumber may be treated before gluing and the members then glued to the desired size and shape. The already glued and machined members may be treated with certain treatments. When laminated members do not lend themselves to treatment because of size and shape, gluing of treated laminations is the only method of producing adequately treated members.

There are problems in gluing some treated woods. Certain combinations of adhesive, treatment, and wood species are compatible; other combinations are not. All adhesives of the same type do not produce bonds of equal quality for a particular wood species and preservative. The bonding of treated wood depends on the concentration of preservative on the surface at the time of gluing and the chemical effects of the preservative on the adhesive. In general, longer curing times or higher curing temperatures, and modifications in assembly times, are needed for treated than for untreated wood to obtain comparable adhesive bonds (see also Art. 11.7).

Each type of preservative and method of treatment has certain advantages. The preservative to be used depends on the service expected of the member for the specific conditions of exposure. The minimum retentions shown in Table 11.28 may be increased where severe climatic or exposure conditions are involved.

Creosote and creosote solutions have low volatility. They are practically insoluble in water and thus are most suitable for severe exposure, contact with ground or water, and where painting is not a requirement or a creosote odor is not objectionable.

Oil-borne chemicals are organic compounds dissolved in a suitable petroleum carrier oil and are suitable for outdoor exposure or where leaching may be a factor, or where painting is not required. Depending on the type of oil used, they may result in relatively clean surfaces. There is a slight odor from such treatment, but it is usually not objectionable.

Waterborne inorganic salts are dissolved in water or aqua ammonia, which evaporates after treatment and leaves the chemicals in wood. The strength of solutions varies to provide net retention of dry salt required. These salts are suitable where clean and odorless surfaces are required. The surfaces are paintable after proper seasoning.

When treatment before gluing is required, waterborne salts, oil-borne chemicals in mineral spirits, or AWPA P9 volatile solvent are recommended. When treatment before gluing is not required or desired, creosote, creosote solutions, or oil-borne chemicals are recommended.

Table 11.28 Recommended Minimum Retentions of Preservatives, lb/ft*

Preservatives	Sawn and Laminated Timbers		Laminations		Sawn and Laminated Timbers		Laminations	
	Western Woods[†]	Southern Pine	Western Woods[†]	Southern Pine	Western Woods[†]	Southern Pine	Western Woods[†]	Southern Pine
Creosote or creosote solutions:								
Creosote	10	10	10	10	8	8	8	8
Creosote—coal-tar solution	10	10	NR[‡]	10	8	8	NR[‡]	8
Creosote-petroleum solution	12	NR[‡]	12	NR[‡]	6	NR[‡]	6	NR[‡]
Oil-borne chemicals:								
Pentachlorophenol (5% in specified petroleum oil)	0.6	0.6	0.6	0.6	0.3	0.3	0.3	0.3
Waterborne inorganic salts:								
Acid copper chromate (ACC)	NR[‡]	NR[‡]	0.50	0.50	0.25	0.25	0.25	0.25
Ammoniacal copper arsenite (ACA)	0.40	0.40	0.40	0.40	0.25	0.25	0.25	0.25
Chromated zinc chloride (CZC)	NR[‡]	NR[‡]	NR[‡]	NR[‡]	0.45	0.45	0.45	0.45
Chromated copper arsenate (CCA)	0.40	0.40	0.40	0.40	0.25	0.25	0.25	0.25
Ammoniacal copper zinc arsenate (ACZA)	0.40	0.40	0.40	0.40	0.25	0.25	0.25	0.25

* See latest edition of AITC 109, "Treating Standard for Structural Timber Framing," American Institute of Timber Construction or AWPA Standards C2 and C28, American Wood Preservers Association.

† Douglas fir, western hemlock, western larch.

‡ NR = not recommended.

("Design of Wood-Frame Structures for Permanence," WCD No. 6, American Forest and Paper Association, Washington, D.C.)

Fire-retardant treatment with approved chemicals can make wood highly resistant to the spread of fire. The fire retardant may be applied as a paint or by impregnation under pressure. The latter is more effective. It may be considered permanent if the wood is used where it will be protected from the weather.

Design values, including those for connections, for lumber and structural glued-laminated timber pressure-treated with fire-retardant chemicals should be obtained from the company providing the treatment and redrying service. The load-duration factor for impact (Table 11.5) should not be applied to structural members pressure-treated with fire-retardant chemicals.

12

Jonathan T. Ricketts
Consulting Engineer
Palm Beach Gardens, Florida

SURVEYING*

Surveying is the science and art of making the measurements necessary to determine the relative positions of points above, on, or beneath the surface of the earth or to establish such points. Surveying continues to undergo important changes.

12.1 Types of Surveys

Plane surveying neglects curvature of the earth and is suitable for small areas.

Geodetic surveying takes into account curvature of the earth. It is applicable for large areas, long lines, and precisely locating basic points suitable for controlling other surveys.

Land, boundary, and cadastral surveys usually are closed surveys that establish property lines and corners. The term *cadastral* is now generally reserved for surveys of the public lands. There are two major categories: **retracement surveys** and **subdivision surveys.**

Topographic surveys provide the location of natural and artificial features and elevations used in map making.

Route surveys normally start at a control point and progress to another control point in the most direct manner permitted by field conditions. They are used for surveys for railroads, highways, pipelines, etc.

Construction surveys are made while construction is in progress to control elevations, horizontal positions and dimensions, and configuration. Such surveys also are made to obtain essential data for computing construction pay quantities.

As-built surveys are postconstruction surveys that show the exact final location and layout of civil engineering works, to provide positional verification and records that include design changes.

Hydrographic surveys determine the shoreline and depths of lakes, streams, oceans, reservoirs, and other bodies of water.

Sea surveying covers surveys for port and offshore industries and the marine environment, including measurement and marine investigations by ship-borne personnel.

Solar surveying includes surveying and mapping of property boundaries, solar access easements, positions of obstructions and collectors, determination of minimum vertical sun angles, and other requirements of zoning boards and title insurance companies.

Satellite surveying provides positioning data and imagery, which is received by equipment, stored, and automatically verified in selected data coordinates with each satellite pass. Doppler and global positioning are used as standard practice in remote regions and on subdivided lands.

Global positioning system (GPS) utilizes a constellation of 24 high-altitude navigational satellites positioned in six orbital planes and spaced so that an operator of specialized equipment can receive signals from between five to eight satellites at all times.

Inertial surveying systems acquire coordinate data obtained by use of a helicopter or ground vehicle. Inertial equipment now coming into use has a dramatic impact on the installation of geodetic and cadastral control.

Photogrammetric surveys utilize terrestrial and aerial photographs or other sensors that provide data and can be a part of all the types of surveys listed in the preceding.

---

*Revised and updated from Sec. 12, Surveying, by Roy Minnick in the 3rd edition.

12.2 Surveying Sources and Organizations

Land and boundary surveying is a regulated activity; each state licenses those who practice land surveying. Boards are established to test prospective land surveyors and to ensure compliance with state laws. Rosters of licensees are usually maintained. There is no national licensing of land surveyors. Each state defines land surveying, who must be licensed, and the activities that are subject to regulation, and those that are exempt. Information about licensing and regulations may be obtained from the American Congress of Surveying and Mapping (ACSM), 6 Montgomery Village Avenue, Suite 403, Gaithersburg, MD 20879 (www.acsm.org). ACSM is also the national membership organization for all branches of surveying. It can provide information about survey education and licensing, state societies, and state registration boards.

The National Geodetic Survey (NGS), formerly called the U.S. Coast and Geodetic Survey, coordinates activities of the Federal Geodetic Control Committee, which develops standards and specifications for conducting Federal geodetic surveys. NGS is the source for geodetic control data, both historic and current. Information on products, programs, and services may be obtained from the National Geodetic Information Branch, 1315 East-West Highway, Silver Spring, MD 20910 (www.ngs.noaa.gov).

The Geological Survey's (USGS) National Mapping Program is responsible for commonly used 7.5-min quadrangle maps and other multipurpose maps. The Earth Science Information Office, in USGS, informs the public of sources of maps, aerial photographs, digital products, and other cartographic and earth science products. The U.S. Geological Survey and the Earth Science Information Center are located at National Center, John W. Powers Federal Building, 12201 Sunrise Valley Drive, Reston, VA 20192 (www.usgs.gov).

The Bureau of Land Management, Cadastral Survey (BLM) is the agency responsible for survey and resurvey of the public lands. The agency is the source for information about the public lands surveys. A starting place for seeking survey information is the Division of Cadastral Surveys, 1849 C Street NW, MS L302, Washington, DC 20240.

Surveying equipment, using the computer, satellites, and a wide array of other technologies, is evolving rapidly. Two magazines, sent free on request, contain topical articles and useful information about all types of surveying and surveying equipment: Professional Surveyor Magazine, 100 Tuscany Drive, Suite B1, Frederick, MD 21702 (www.profsurv.com), and P.O.B. Magazine, 755 West Big Beaver Rd, Ste 1000, Troy, MI 48084.

See also Art. 12.19.

12.3 Units of Measurement

Units of measurement used in past and present surveys are:

For construction work: feet, inches, fractions of inches

For most surveys: feet, tenths, hundredths, thousandths

For National Geodetic Survey control surveys: metres, 0.1, 0.01, 0.001 m

The most-used equivalents are:

1 metre = 39.37 in (exactly) = 3.2808 ft

1 rod = 1 pole = 1 perch = $16\frac{1}{2}$ ft

1 engineer's chain = 100 ft = 100 links

1 Gunter's chain = 66 ft = 100 Gunter's links (lk) = 4 rods = $\frac{1}{80}$ mi

1 acre = 100,000 sq (Gunter's) links = 43,560 ft^2 = 160 $rods^2$ = 10 sq (Gunter's) chains = 4046.87 m^2 = 0.4047 hectare

1 rood = $\frac{1}{4}$ acre = 40 $rods^2$ (also local unit = $5\frac{1}{2}$ to 8 yd)

1 hectare = 10,000 m^2 = 107,639.10 ft^2 = 2.471 acres

1 arpent = about 0.85 acre, or length of side of 1 square arpent (varies)

1 statute mile = 5280 ft = 1609.35 m

1 mi^2 = 640 acres

1 nautical mile (U.S.) = 6080.27 ft = 1853.248 m

1 fathom = 6 ft

1 cubit = 18 in

1 vara = 33 in (Calif.), $33\frac{1}{3}$ in (Texas), varies

1 degree = $\frac{1}{360}$ circle = 60 min = 3600 s = 0.01745 rad

sin $1° = 0.01745241$

1 rad $= 57°17'44.8''$ or about $57.30°$

1 grad (grade) $= \frac{1}{400}$ circle $= \frac{1}{100}$ quadrant $= 100$ centesimal min $= 10^4$ centesimals (French)

1 mil $= \frac{1}{6400}$ circle $= 0.05625°$

1 military pace (milpace) $= 2\frac{1}{2}$ ft

12.4 Theory of Errors

When a number of measurements of the same quantity have been made, they must be analyzed on the basis of probability and the theory of errors. After all systematic (cumulative) errors and mistakes have been eliminated, random (compensating) errors are investigated to determine the most probable value (**mean**) and other critical values. Formulas determined from statistical theory and the normal, or Gaussian, bell-shaped probability distribution curve, for the most common of these values follow:

Standard deviation of a series of observations is

$$\sigma_s = \pm\sqrt{\frac{\Sigma d^2}{n-1}} \qquad (12.1)$$

where d = residual (difference from mean) of single observation

n = number of observations

The **probable error** of a single observation is

$$PE_s = \pm 0.6745\sigma_s \qquad (12.2)$$

(The probability that an error within this range will occur is 0.50.)

The probability that an error will lie between two values is given by the ratio of the area of the probability curve included between the values to the total area. Inasmuch as the area under the entire probability curve is unity, there is a 100% probability that all measurements will lie within the range of the curve.

The area of the curve between $\pm \sigma_s$ is 0.683; that is, there is a 68.3% probability of an error between $\pm \sigma_s$ in a single measurement. This error range is also called the one-sigma or 68.3% confidence level. The area of the curve between $\pm 2\sigma_s$ is 0.955. Thus there is a 95.5% probability of an error between $\pm 2\sigma_s$, and $\pm 2\sigma_s$ represents the 95.5% error (two-sigma or 95.5% confidence level). Similarly, $\pm 3\sigma_s$ is referred to as the 99.7% error (three-sigma or 99.7% confidence level). For practical purposes,

a maximum tolerable level often is assumed to be the 99.9% error. Table 12.1 indicates the probability of occurrence of larger errors in a single measurement.

The probable error of the combined effects of accidental errors from different causes is

$$E_{sum} = \sqrt{E_1^2 + E_2^2 + E_3^2 + \cdots} \qquad (12.3)$$

where E_1, E_2, E_3, ... are probable errors of the separate measurements.

Error of the mean is

$$E_m = \frac{E_{sum}}{n} = \frac{E_s\sqrt{n}}{n} = \frac{E_s}{\sqrt{n}} \qquad (12.4)$$

where E_s = specified error of a single measurement.

Probable error of the mean is

$$PE_m = \frac{PE_s}{\sqrt{n}} = \pm 0.6745\sqrt{\frac{\Sigma d^2}{n(n-1)}} \qquad (12.5)$$

12.5 Significant Figures

These are the digits read directly from a measuring device plus one digit that must be estimated and therefore is questionable. For example, a reading of 654.32 ft from a steel tape graduated in tenths of a foot has five significant figures. In multiplying 798.16 by 37.1, the answer cannot have more significant figures than either number used; i.e., three in this case. The same rule applies in division. In addition or subtraction, for example, $73.148 + 6.93 + 482$, the answer will have three significant figures, all on the left side of the decimal point.

Table 12.1 Probability of Error in a Single Measurement

Error	Confidence level, percent	Probability of larger error
Probable ($0.6745\sigma_s$)	50	1 in 2
Standard deviation (σ_s)	68.3	1 in 3
90% ($1.6449\sigma_s$)	90	1 in 10
$2\sigma_s$ or 95.5%	95.5	1 in 20
$3\sigma_s$ or 97.7%	99.7	1 in 370
Maximum ($3.29\sigma_s$)	99.9+	1 in 1000

Small, hand-held and large computers now available provide 10 or more digits, but carrying computation results beyond justifiable significant figures leads to false impressions of precision.

12.6 Measurement of Distance with Tapes

Reasonable precisions for different methods of measuring distances are

Pacing (ordinary terrain): $\frac{1}{50}$ to $\frac{1}{100}$.

Taping (ordinary steel tape): $\frac{1}{1000}$ to $\frac{1}{10,000}$. (Results can be improved by use of tension apparatus, transit alignment, leveling.)

Base line (invar tape): $\frac{1}{50,000}$ to $\frac{1}{1,000,000}$.

Stadia: $\frac{1}{300}$ to $\frac{1}{500}$ (with special procedures).

Subtense bar: $\frac{1}{1000}$ to $\frac{1}{700}$ (for short distances, with a 1-s theodolite, averaging angles taken at both ends).

Electronic distance measurement (EDM) devices have been in use since the middle of the twentieth century and have now largely replaced steel tape measurements on large projects. The continued development, and the resulting drop in prices, are making their use widespread. A knowledge of steel taping errors and corrections remains important, however, because use of earlier survey data requires a knowledge of how the measurements were made, common sources for errors, and corrections that were typically required.

Slope Corrections ■ In slope measurements, the horizontal distance $H = L \cos x$, where $L =$ slope distance and $x =$ vertical angle, measured from the horizontal—a simple hand calculator operation. For slopes of 10% or less, the correction to be applied to L for a difference d in elevation between tape ends, or for a horizontal offset d between tape ends, may be computed from

$$C_s = \frac{d^2}{2L} \tag{12.6}$$

For a slope greater than 10%, C_s may be determined from

$$C_s = \frac{d^2}{2L} + \frac{d^4}{8L^3} \tag{12.7}$$

Temperature Corrections ■ Table 12.2 lists temperature corrections for steel tapes. Formulas for other tape corrections, ft, with L as the measured distance ft, are as follows: **For incorrect tape length,**

$$C_t = \frac{(\text{actual tape length} - \text{nominal tape length})L}{\text{nominal tape length}} \tag{12.8}$$

For nonstandard tension,

$$C_t = \frac{(\text{applied pull} - \text{standard tension})L}{AE} \tag{12.9}$$

where $A =$ cross-sectional area of tape, in^2

$E =$ modulus of elasticity $= 29,000,000$ psi for steel

For sag correction between points of support, ft,

$$C = -\frac{w^2 L_s^3}{24P^2} \tag{12.10}$$

where $w =$ weight of tape per foot, lb

$L_s =$ unsupported length of tape, ft

$P =$ pull on tape, lb

Sources and Types of Error ■ There are three sources of error in taping—instrumental, natural, and personal—and nine general types of errors. Table 12.3 lists the types of errors and their sources and classifies them as systematic or accidental.

All errors in Table 12.3 produce, in effect, an incorrect tape length. Therefore, only four basic tape problems exist: *measuring* a line between fixed points with a tape too long or too short, and *laying out* a line from one fixed point with a tape too long or too short. A simple oneline sketch (Fig. 12.1) with tick marks for nominal and actual tape lengths is a foolproof method for deciding whether to add or subtract the correction in any case.

In base-line measurements with steel or invar tapes (three or more tapes should be used on different sections of the line), corrections are applied for inclination; temperature; nonstandard length of tape, for both full and partial tape lengths; and reduction to sea level.

12.7 Leveling

A few definitions introduce the subject:

Vertical Line ■ A line to the center of the earth from any point. Commonly considered to coincide with a plumb line.

Table 12.2 Temperature Corrections for Steel Tapes*

Subtract corrections for these temperatures, °F	Length of line, ft				Add corrections for these temperatures, °F
	5000	1000	500	100	
68	0.00	0.00	0.00	0.00	68
66	0.06	0.01	0.01	0.00	70
64	0.13	0.03	0.01	0.00	72
62	0.20	0.04	0.02	0.00	74
60	0.26	0.05	0.03	0.01	76
58	0.32	0.06	0.03	0.01	78
56	0.39	0.08	0.04	0.01	80
54	0.46	0.09	0.04	0.01	82
52	0.52	0.10	0.05	0.01	84
50	0.58	0.12	0.06	0.01	86
48	0.65	0.13	0.06	0.01	88
46	0.72	0.14	0.07	0.01	90
44	0.78	0.16	0.08	0.02	92
42	0.84	0.17	0.08	0.02	94
40	0.91	0.18	0.09	0.02	96
38	0.98	0.20	0.10	0.02	98
36	1.04	0.21	0.10	0.02	100
34	1.10	0.22	0.11	0.02	102
32	1.17	0.23	0.12	0.02	104
30	1.24	0.25	0.12	0.02	106
28	1.30	0.26	0.13	0.03	108
26	1.36	0.27	0.14	0.03	110

Example: Given a recorded distance of 8785.32 ft for a line measured when the average temperature is 80°F. Correction to be added is $0.39 + 3(0.08) + 0.04 + 2(0.01) + 0.01 = 0.70$ ft. Because of rounding off in the table, the total correction of 0.70 is 0.01 ft larger than the value computed directly by formula, $C = 0.0000065(T - 68F)L$.

* By permission of Marvin C. May, University of New Mexico.

Table 12.3 Types, Sources, and Classification of Taping Errors

Type of Error	Source*	Classification[†]	Departure from standard to produce 0.01-ft error for a 100-ft tape
Tape length	I	S	0.01 ft
Temperature	N	S or A	15°F
Tension	P	S or A	15 lb
Sag	N, P	S	$7\frac{3}{8}$ in at center as compared with support throughout
Alignment	P	S	1.4 ft at one end, or $8\frac{1}{2}$ in at midpoint
Tape not level	P	S	1.4 ft
Interpolation	P	A	0.01 ft
Marking	P	A	0.01 ft
Plumbing	P	A	0.01 ft

* I = instrumental, N = natural, P = personal.
[†] S = systematic, A = accidental.

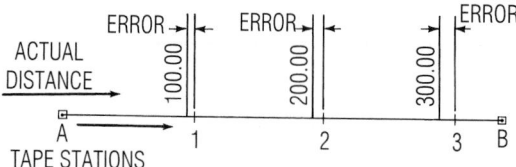

Fig. 12.1 Cumulative error from measuring with a tape that is too long.

Level Surface ▪ A curved surface that, at every point, is perpendicular to a plumb line through the point.

Level Line ▪ A line in a level surface, therefore a curved line.

Horizontal Plane ▪ A plane perpendicular to the plumb line.

Horizontal Line ▪ A straight line perpendicular to the vertical.

Datum ▪ Any level surface to which elevations are referred, such as mean sea level, which is most commonly used; also called datum plane, although not actually a plane.

Mean Sea Level (MSL) ▪ The average height of the surface of the sea. MSL was established originally over a 19-year period, for all tidal stages, at United States and Canadian coastal stations. The basic National Geodetic Vertical Datum net is being connected to all accessible primary tide and water-level stations.

Orthometric Correction ▪ This is a correction applied to preliminary elevations due to flattening of the earth in the polar direction. Its value is a function of the latitude and elevation of the level circuit.

Curvature of the earth causes a horizontal line to depart from a level surface. The departure, C_f, ft; or C_m, m, may be computed from

$$C_f = 0.667.4M^2 = 0.0239F^2 \qquad (12.11a)$$

$$C_m = 0.0785K^2 \qquad (12.11b)$$

where M, F, and K are distances in miles, thousands of feet, and kilometres, respectively, from the point of tangency to the earth.

Refraction causes light rays that pass through the earth's atmosphere to bend toward the earth's surface. For horizontal sights, the average angular displacement (like the sun's diameter) is about 32 min. The displacement, R_f, ft, or R_m, m, is given approximately by

$$R_f = 0.093M^2 = 0.0033F^2 \qquad (12.12a)$$

$$R_m = 0.011K^2 \qquad (12.12b)$$

To obtain the combined effect of refraction and curvature of the earth, subtract R_f from C_f or R_m from C_m.

Differential leveling is the process of determining the difference in elevation of two points. The procedure involves sighting with a level on a ruled rod set on a point of known elevation (backsight or plus sight), then on the rod set on points (or intermediate points) whose elevations are to be determined (foresights). These elevations equal the height of instrument minus the foresight reading. The height of instrument equals the known elevation plus the backsight reading. For accuracy, the sum of backsight and foresight distances should be kept nearly equal.

Elevations commonly are taken to 0.01 ft in engineering surveys and to 0.001 m in precise National Geodetic Survey work.

Table 12.4 shows a typical left-hand page of open-style notes. In closed-style (condensed) notes, B.S., H.I., F.S., and elevation values are placed on the same line, thereby saving space (which is cheap in a field book) but reducing the clarity of steps for beginners. The right-hand page contains bench-mark descriptions, sketches, date of survey, names of survey-party members, and information on the weather, equipment used, and other necessary remarks.

As noted in Brinker, Austin, and Minnick, "Note Forms for Surveying Measurements," Landmark Enterprises, Rancho Cordova, Cal.: The critical importance of field notes is sometimes neglected. If any of the five main features used in evaluating notes—accuracy, integrity, legibility, arrangement and clarity—is absent, delays, mistakes and increased costs in completing field work, computations, and mapping result.

Table 12.4 Differential Leveling Notes

Differential Leveling—BM Civil to BM Dorm					
Station	B.S.	H.I.*	F.S.	Elev.[†]	Dist.
BM Civil				100.00	
	4.08	104.08			175
TP 1			0.20	103.88	180
	6.09	109.97			160
BM Dorm			4.32	105.65	155
	___		___		___
	10.17		4.52		670
	4.52				
	___				
	5.65				
BM Dorm				105.65	
	4.37	110.02			165
TP 2			6.14	103.88	165
	0.93	104.81			170
BM Civil			4.80	100.01	175
	___		___		___
	5.30		10.94		675
			5.30		
			___		
			5.64		
			Elev Diff = 5.64 ft		
			Loop Closure = 0.01 ft		

* Height of instrument (H.I.) = elevation + backsight (B.S.).

[†] Elevation = H.I. − foreset (F.S.).

Digital (electronic) notes for field measurements of angles and distances, as well as reduction of slope distances and computation of coordinates, are now being recorded in various types of Data Collectors. They are displayed and the data automatically recorded. Reading and transcribing errors are thus eliminated, both in the field and office where the data collector automatically transfers the field notes to a computer for processing. The results then go to a printer, which makes working plots and convenient page-width printouts.

Data collectors should not completely replace the field book, which still is used to record backup information, including sketches and notes to show station identification for the permanent project. Actually, since only a small part of total field time is occupied in recording measurements in a field book, the important time-saving advantage of a data collector is gained in the office and plan production.

A usable tool for notekeepers is photography. With a reasonably priced, digital camera a *photographic record of monuments set or found, and other field evidence to the survey can be prepared.*

Profile leveling determines the elevations of points at known distances along a line. When these points are plotted, a vertical section through the earth's surface is shown. Elevations are taken at full stations (100 ft) or closer in irregular terrain, at breaks in the ground surface, and at critical points such as bridge abutments and road crossings. Profiles are generally plotted on special paper with an exaggeration of from 5:1 up to 20:1, or even more, so that elevation differentials will show up better. Profiles are needed for route surveys, to select grades and find earthwork quantities. Elevations are usually taken to 0.01 ft on bench marks and 0.1 ft on the ground.

Reciprocal leveling is used to cross streams, lakes, canyons, and other topographic barriers that prevent balancing of backsights and foresights. On each side of the obstruction to be crossed, a plus sight is taken on the near rod and several minus sights on the far rod. The resulting differences in elevation are averaged to eliminate the effects of curvature and refraction, and inadjustment of the instrument. Even though a number of minus sights are taken for averaging, their length may reduce the accuracy of results.

Borrow-pit or cross-section leveling produces elevations at the corners of squares or rectangles whose sides are dependent on the area to be covered, type of terrain, and accuracy desired. For example, sides may be 10, 20, 40, 50, or 100 ft. Contours can be located readily, topographic features not so well. Quantities of material to be excavated or filled are computed, in cubic yards, by selecting a grade elevation, or final ground elevation, computing elevation differences for the corners, and substituting in

$$Q = \frac{nxA}{108} \qquad (12.13)$$

where n = number of times a particular corner enters as part of a division block

x = difference in ground and grade elevation for each corner, ft

A = area of each block, ft^2

Cross-section leveling also is the term applied to the procedure for locating contours or taking elevations on lines at right angles to the center line in a route survey.

Three-wire leveling is a type of differential leveling with three horizontal sighting wires in the level. Upper, middle, and lower wires are read to obtain an average value for the sight, check the

precision of reading the individual wires, and secure stadia distances for checking lengths of backsights and foresights. The height of instrument is not needed or computed. The National Geodetic Survey has long used three-wire leveling for its control work, but more general use is now being made of the method.

Grade designates the elevation of the finished surface of an engineering project and also the rise or fall in 100 ft of horizontal distance, for example, a 4% grade (also called gradient). Note that since the common stadia interval factor is 100, the difference in readings between the middle and upper (or lower) wire represents ½ ft in 100 ft, or a ½% grade.

Types of levels in general use are listed in Table 12.5.

Special construction levels include the Blout & George Laser Tracking Level (which can search a 360° horizontal plane and lock on a pocket-size target), the Dietzgen Laser Swinger, Spectra-Physics Rotolite Building Laser, and AGL Construction Laser. Some laser instruments are available for shaft plumbing and setups inside large pipe lines.

12.8 Vertical Control

The National Geodetic Survey provides vertical control for all types of surveys. NGS furnishes descriptions and elevations of bench marks on request. As given in "Standards and Specifications for Geodetic Control Networks," Federal Geodetic Control Committee, the relative accuracy C, mm,

Table 12.5 Types of Levels

Type	Use
Hand level	Rough work. Sights on ordinary level rod limited to about 50 ft because of zero- to 2-power magnification
Engineer's level, Wye or Dumpy	Suitable for ordinary work (third- or fourth-order). Elevations to 0.01 ft without target
Tilting level	Faster, more accurate sighting. Good for third-, second-, or first order work depending upon refinement
Self-leveling, automatic levels	Fast, suitable for second-order and third-order work
Precise level	Very sensitive level vials, high magnification, tilting, other features

Note: Instruments are arranged in ascending order of cost.

required between directly connected bench marks for the three orders of leveling is:

First-order: $C = 0.5\sqrt{K}$ for Class I and $0.7\sqrt{K}$ for Class II

Second-order: $C = 1.0\sqrt{K}$ for Class I and $1.3\sqrt{K}$ for Class II

Third-order: $C = 2.0\sqrt{K}$

where K is the distance between bench marks, km.

12.9 Magnetic Compass

A magnetic compass consists of a magnetized needle mounted on a pivot at the center of a graduated circle. The compass is now used primarily for retracement purposes and checking, although some surveys not requiring precision are made with a compass, for example, in forestry and geology. American transits have traditionally come with a long compass needle, whereas on optical instruments the compass is merely an accessory, and therefore the instruments can be smaller and lighter.

A small weight is placed on the south end of the needle in the northern hemisphere to counteract the dip caused by magnetic lines of force. Since the magnetic poles are not located at the geographic poles, a horizontal angle (declination) results between the axis of the needle and a true meridian. East declination occurs if the needle points east of true (due) north, west declination if the needle points west of true north.

The National Geodetic Survey publishes a world chart every fifth year showing the positions of the agonic line, isogonic lines for each degree, and values for annual variation of the needle. The **agonic line** is a line of zero declination; i.e., a magnetic compass set up on points along this line would point to true north as well as magnetic north. For points along an isogonic line, declination should be constant, barring local attraction.

Table 12.6 lists the periodic variations in the declination of the needle that make it unreliable. In addition, local attraction resulting from power sources, metal objects, etc., may produce considerable error in bearings taken with a compass. If the source of local attraction is fixed and constant, however, angles between bearings are correct, even though the bearings are uniformly distorted.

The Brunton compass or pocket transit has some of the features of a sighting compass, a prismatic compass, a hand level, and a clinometer. It is suitable for some forest, geological, topographical, and preliminary surveys of various kinds.

A common problem today is the conversion of past magnetic bearings based upon the declination of a given date to present bearings with today's declination, or to true bearings. A sketch, such as Fig. 12.2, showing all data with pencils of different colors, will make the answer evident.

12.10 Bearings and Azimuths

The direction of a line is the angle measured from any reference line, such as a magnetic or true meridian. Bearings are angles measured from the north and the south, toward the east or the west. They can never be greater than 90° (Fig. 12.3).

Bearings read in the advancing direction are forward bearings; those in the opposite direction are back bearings. Computed bearings are obtained by using a bearing and applying a direct, deflection, or other angle. Bearings, either magnetic or true, are used in rerunning old surveys, in computations, on maps, and in deed descriptions.

An azimuth is a clockwise angle measured from some reference line, usually a meridian. Government surveys use geodetic south as the base of azimuths. Other surveys in the northern hemisphere may employ north. Azimuths are advantageous in topographic surveys, plotting, direction

Table 12.6 Periodic Variations in Declination of Magnetic Needle

Variation	Remarks
Secular	Largest and most important. Produces wide unpredictable swings over a period of years, but records permit comparison of past and present declinations
Daily (diurnal)	Swings about 8 min per day in the U.S. Relatively unimportant
Annual	Periodic swing amounting to less than 1 min of arc; it is unimportant
Irregular	From magnetic storms and other sources. Can pull needle off more than a degree

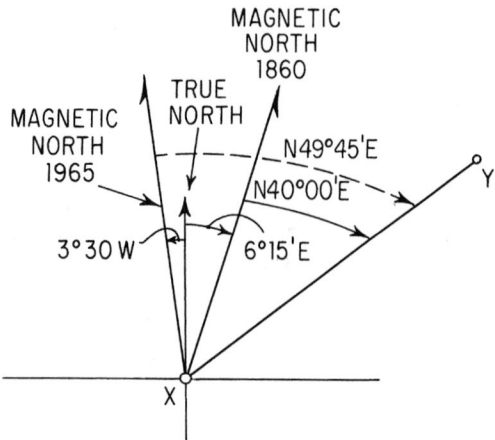

Fig. 12.2 Magnetic bearing of a line *XY* in a past year is found by plotting magnetic north for that year with respect to true north.

problems, and other work where omission of the quadrant letters and a range of angular values from 0 to 360° simplify the work.

12.11 Horizontal Control

All surveys require some kind of control, be it a base line or bench mark, or both. Horizontal control consists of points whose positions are established by traverse, triangulation, or trilateration. The National Geodetic Survey has established control monuments throughout the country and tabulated azimuths, latitude and longitude, statewide coordinates, and other data for them. Surveys on the statewide coordinate system have increased the number of control points available to all surveyors.

12.11.1 Traverses

For a traverse, the survey follows a line from point to point in succession. The lengths of lines between points and their directions are measured. If the traverse returns to the point of origin, it is called a closed traverse. The United States–Canada boundary, for example, was run by traverse. In contrast, the boundary of a construction site would be surveyed by a closed traverse. Permissible closures for traverses that make a closed loop or connect adjusted positions of equal-order or higher-order control surveys are given in Table 12.7.

Transit-tape traverses provide control for areas of limited size as well as for the final results on property surveys, route surveys, and other work. Stadia traverses are good enough for small-area topographic surveys when tied to higher control. Faster and more accurate traversing may be accomplished with electronic distance-measuring devices and with theodolites with direct readings to seconds and much lighter than the older-type, bulky transits.

As a result of developing technologies, the acceptable ratio of error to distance measured for

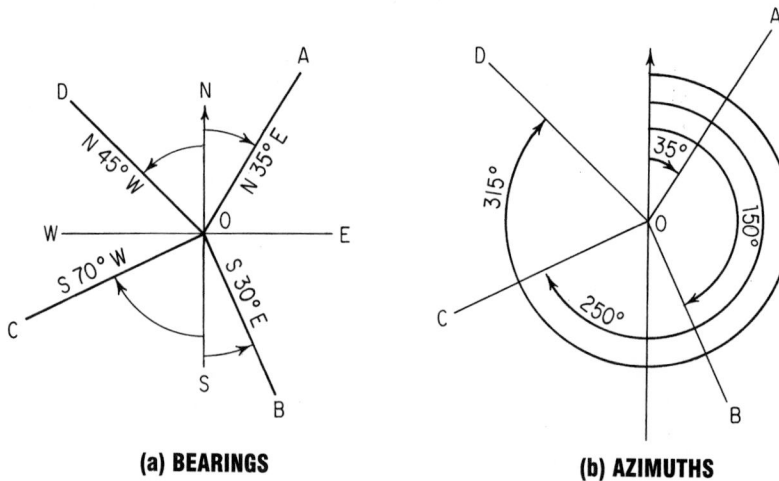

(a) BEARINGS   **(b) AZIMUTHS**

Fig. 12.3 Direction of lines may be specified by (*a*) bearings or (*b*) azimuth.

Table 12.7 Periodic Traverse Closures*

Traverse order	Max permissible closure after azimuth adjustment	Max azimuth closure at azimuth checkpoint	
		Sec per station	Sec[†]
First order	1:100,000	1.0	$2\sqrt{N}$
Second order			
Class I	1:50,000	1.5	$3\sqrt{N}$
Class II	1:20,000	2.0	$6\sqrt{N}$
Third order			
Class I	1:10,000	3.0	$10\sqrt{N}$
Class II	1:5,000	8.0	$30\sqrt{N}$

* National Geodetic Survey.

[†] N = number of stations.

various types of survey is being reviewed and is subject to change. To obtain the latest recommendations, contact the organizations mentioned in Art. 12.2. See also Art. 12.19.

12.11.2 Triangulation

In triangulation, points are located at the apexes of triangles, and all angles and one base line are measured. Additional base lines are used when a chain of triangles, quadrilaterals, or central-point figures is required (Fig. 12.4). All other sides are computed and adjustments carried from the fixed base lines forward and backward to minimize the corrections. Angles used in computation should exceed 15°, and preferably 30°, to avoid the rapid change in sines for small angles.

Chains of triangles are unsuitable for high-precision work since they do not permit the rigid adjustments available in quadrilaterals and more complicated figures. Quadrilaterals are advantageous for long, relatively narrow chains; polygons and central-point figures for wide systems and perhaps for large cities, where stations can be set on tops of buildings.

Strength of figure in triangulation is an expression of relative precision possible in the system based on the route of computation of a triangle side. It is independent of the accuracy of observations and utilizes the number of directions observed, conditions to be satisfied, and rates of changes for the sines of distance angles. Triangulation stations that cannot be occupied require additional computation for reduction to center in obtaining coordinates and other data.

Permissible triangulation closures for the three orders of triangulation specified by the National Geodetic Survey are given in Table 12.8 and specifications for base-line measurements in Table 12.9.

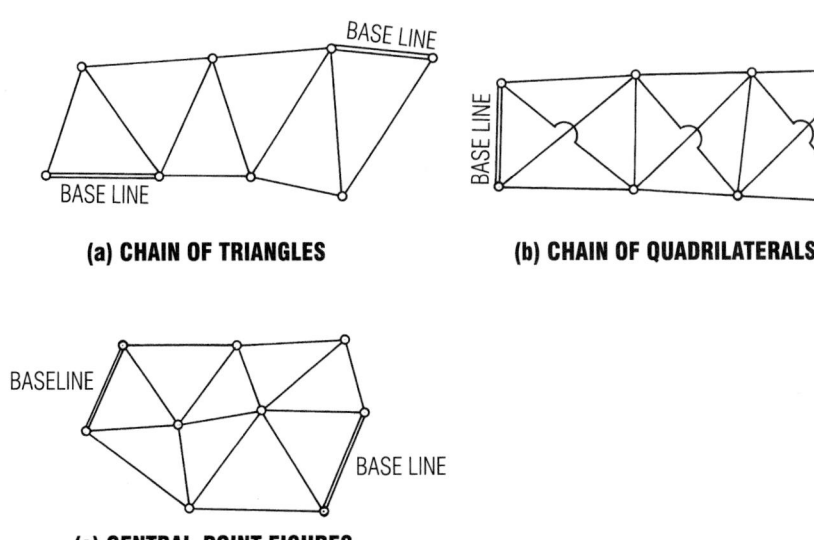

(a) CHAIN OF TRIANGLES

(b) CHAIN OF QUADRILATERALS

(c) CENTRAL-POINT FIGURES

Fig. 12.4 Triangular chains.

Table 12.8 Triangulation Closures

Specification, item	First order	Second order		Third order	
		Class I	Class II	Class I	Class II
Avg. triangle closure, s	1.0	1.2	2.0	3.0	2.0
Max triangle closure, s	3.0	3.0	5.0	5.0	10.0

12.11.3 Trilateration

This method has replaced triangulation for establishment of control in many cases, such as photogrammetry, since the development of electronic measuring devices. All distances are measured and the angles computed as needed.

12.11.4 Trilateration vs. Triangulation

In triangulation, one or more base lines and all angles are measured. Astronomical observations made at some monuments control directions. In trilateration, lengths of all lines to be used are measured, slope and atmospheric corrections applied, and astronomical observations taken at intervals. Reading some directions contributes additional strengthening.

Various field and office studies show that time and cost for triangulation or trilateration are about the same for some networks. A combination of observed directions and distances determined with electronic distance-measuring instruments may be best. Trilateration networks covering basically square blocks provide a better strength of figure than long narrow chains (where a number of angles should be read also).

Table 12.9 Specifications for Base-Line Measurements

Order	Max standard error of base
First	1/1,000,000
Second	
Class I	1/900,000
Class II	1/800,000
Third	
Class I	1/500,000
Class II	1/250,000

If one monument is fixed and one azimuth known, both trilateration and triangulation surveys can be extended through other points.

Usefulness of the trilateration method is not confined to high-order-precision large geodetic control networks. Field-proven (using only simple mathematics) satisfactory closures are obtained for small jobs with reasonably strong triangles.

12.12 Stadia

Stadia is a method of measuring distances by noting the length of a stadia or level rod intercepted between the upper and lower sighting wires of a transit, theodolite, or level. Most transits and levels have an interval between stadia wires that gives a vertical intercept of 1 ft on a rod 100 ft away. A stadia constant varying from about $\frac{3}{4}$ to $1\frac{1}{4}$ ft (usually assumed to be 1 ft) must be added for older-type external-focusing telescopes. The internal-focusing short-length telescopes common today have a stadia constant of only a few tenths of a foot, and so it can be neglected for normal readings taken to the nearest foot.

Figure 12.5 shows stadia relationships for a horizontal sight with the older-type external-focusing telescope. Relationships are comparable for the internal-focusing type.

For horizontal sights, the stadia distance, ft (from instrument spindle to rod), is

$$D = R\frac{f}{i} + C \qquad (12.14)$$

where R = intercept on rod between two sighting wires, ft

f = focal length of telescope, ft (constant for specific instrument)

i = distance between stadia wires, ft

$C = f + c$

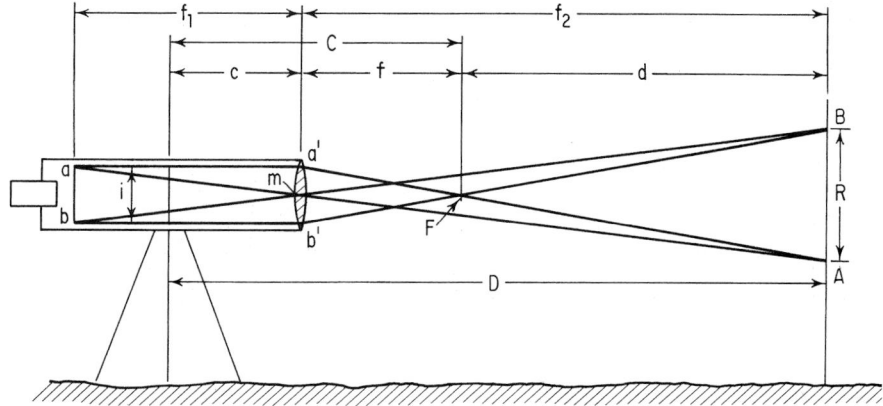

Fig. 12.5 Distance D is measured with an external-focusing telescope by determining interval R intercepted on a rod AB by two horizontal sighting wires a and b.

c = distance from center of spindle to center of objective lens, ft

C is called the stadia constant, although c and C vary slightly.

The value of f/i, the stadia factor, is set by the manufacturer to be about 100, but it is not necessarily 100.00. The value should be checked before use on important work, or when the wires or reticle are damaged and replaced.

For inclined sights (Fig. 12.6) the rod is held vertical, as indicated by a rod level or other means because it is difficult to assure perpendicularity to the sight line on sloping shots. Reduction to horizontal and vertical distances is made according to formulas, such as

$$H = 100R - 100R \sin^2 \alpha + C \qquad (12.15)$$

$$V = 100R(\tfrac{1}{2} \sin 2\alpha) \qquad (12.16)$$

where H = horizontal distance from instrument to rod, ft

V = vertical distance from instrument to rod, ft

α = vertical angle above or below level sight

A Beaman arc on transits and alidades simplifies reduction of slope sights. It consists of an H scale and a V scale, both graduated in percent, with spacing based on the stadia formulas. The H scale gives the correction per 100 ft of slope distance, which is subtracted from $100R + C$ to get the horizontal distance. A V-scale index of 50 for level sights eliminates minus values in determining vertical distance. Readings above 50 are angles of elevation; below 50, angles of depression. Each unit above or below 50 represents 1-ft difference in elevation per 100 ft of sight. By setting the V scale to a whole number, even though the middle wire does not fall on the height of the instrument, you need only mental arithmetic to compute vertical distance. The H scale is read by interpolation since the value generally is small and falls in the area of wide spaces.

As an illustration, to determine the elevation of a point X from a setup at point Y, compute elevation X = elevation Y + height of instrument + (arc reading − 50)(rod intercept) − reading of middle wire.

Some self-reducing tachymeters have curved stadia lines engraved on a glass plate, which turns and appears to make the lines move closer or farther apart. A fixed stadia factor of 100 is used for horizontal reduction, but several factors are required for elevation differences, depending on the slope.

Stadia traverses can be run with direct or azimuth angles. Distances and elevation differences should be averaged for the foresights and backsights. Elevation checks on bench marks are necessary at frequent intervals to maintain reasonable precision.

Poor closures in stadia work usually result from incorrect rod readings rather than errors in angles.

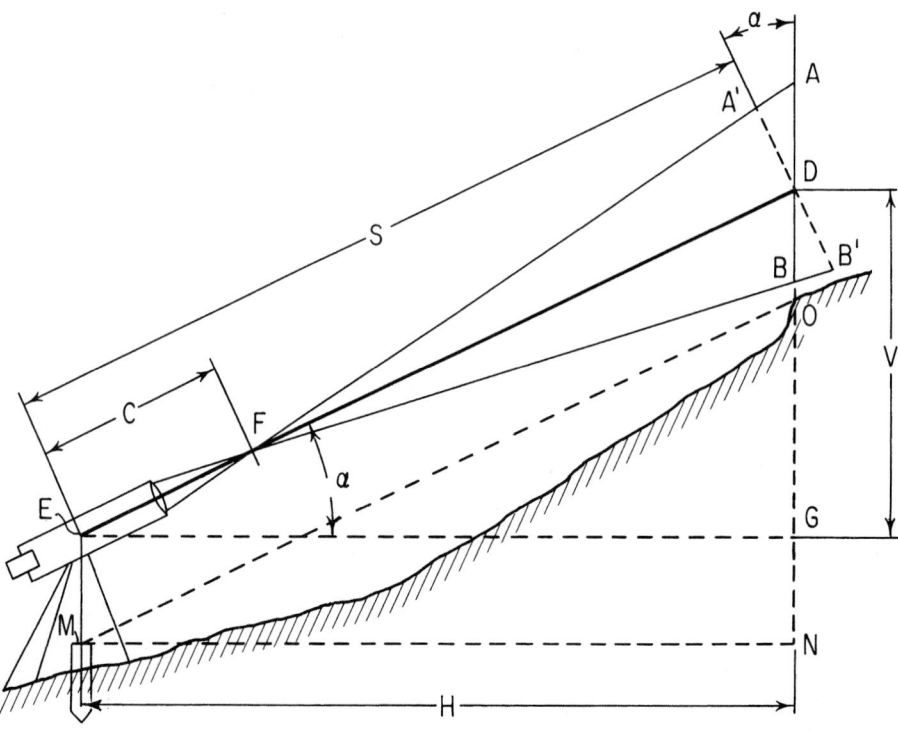

Fig. 12.6 Stadia measurement of vertical distance V and horizontal distance H by reading with a telescope the rod intercept AB and vertical angle α.

A difference of 1 min in vertical angle has little effect on horizontal distances and produces a correction smaller than 0.1 ft for sights up to 300 ft.

Stadia distances, normally read to the nearest foot, are assumed to be valid within about $\frac{1}{2}$ ft. For the same line and lateral error on a 300-ft shot, $\sin \alpha = \frac{1}{2}/300 = 0.00167$ and $\alpha = 5.7$ min. Thus for stadia sights up to 300 ft, comparable distance-angle precision is obtained by reading horizontal angles to the nearest 5 or 6 min. This can be done by estimation on the scale without using the vernier graduations. (See Fig. 12.7.)

Closely approximate answers to many problems in surveying, engineering, mechanics, and other

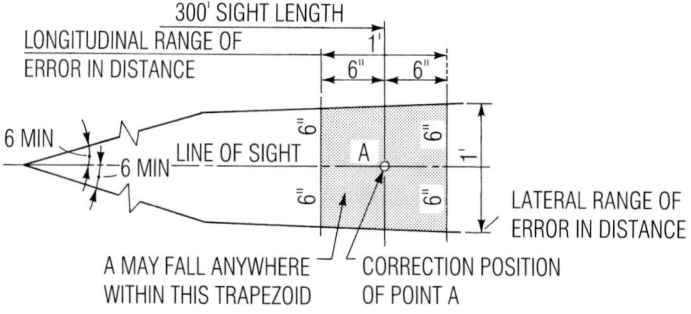

Fig. 12.7 Comparable precision of angles and stadia distances.

fields can be computed mentally by memorizing the sin of $1' = 0.00029$ (or roughly 0.0003), and $\sin 1° = 0.01745$ (about $0.01\frac{3}{4}$). For sines of angles from 0° to 10°, the numerical values increase almost linearly. The divergence from true value at 10° is only $\frac{1}{2}$%; at 30° just $4\frac{1}{2}$%—high for surveying but within, say, some design load estimates. Values of tangents found by multiplying the tangent of 1° by other angular sizes diverge more rapidly but still are off only 1% at 10°.

12.13 Topographic Surveys

Topographic surveys are made to locate natural and constructed features for mapping purposes. By means of conventional symbols, culture (bridges, buildings, boundary lines, etc.), relief, hydrography, vegetation, soil types, and other topographic details are shown for a portion of the earth's surface.

Planimetric (line) maps define natural and cultural features in plane only. **Hypsometric maps** give elevations by contours, or less definitely by means of hachures, shading, and tinting.

Horizontal and vertical control of a high order is necessary for accurate topographic work. Triangulation, trilateration, traversing, and photogrammetry furnish the skeleton on which the topographic details are hung. A level net must provide elevations with closures smaller than expected of the topographic traverse and side shots. For surveys near lake shores or slow-moving streams, the water surface on calm days is a continuous bench mark.

Seven methods are used to locate points in the field, as listed in Table 12.10. The first four require a *base line* of known length. An experienced instrument person selects the simplest method considering both fieldwork and office work involved.

A **contour** is a line connecting points of equal elevation. The shoreline of a lake not disturbed by wind, inlet, or outlet water forms a contour. The vertical distance (elevation) between successive contours is the contour interval. Intervals commonly used are 1, 2, 5, 10, 20, 25, 40, 50, 80, and 100 ft, depending on the map scale, type of terrain, purpose of the map, and other factors.

Methods of taking topography and pertinent points on the suitability of each for given conditions are given in Table 12.11.

12.14 Satellite Doppler Positioning

Satellite Doppler positioning is a three-dimensional measurement system based on the radio signals of the U.S. Navy Navigational Satellite System (NNSS), commonly referred to as the TRANSIT system. Satellite Doppler positioning is used primarily to establish horizontal control.

The Doppler observations are processed to determine station positions in Cartesian coordinates, which can be transformed to geodetic coordinates (geodetic latitude and longitude and height above reference ellipsoid). There are two methods by which station positions can be derived: point positioning and relative positioning.

Point positioning, for geodetic applications, requires that the processing of the Doppler data be performed with the precise ephemerides that are supplied by the Defense Mapping Agency. In this method, data from a single station are processed to yield the station coordinates.

Table 12.10 Methods for Locating Points in the Field

Method	Principal use
1. Two distances	Short taping, details close together, trilateration
2. Two angles	Graphical triangulation, phone table
3. One angle, adjacent distance	Transit and stadia
4. One angle, opposite distance	Special cases
5. One distance, right-angle offset	Route surveys, curved shorelines, or boundaries
6. String lines from straddle hubs	Referencing hubs for relocation
7. Two angles at point to be located	Three-point location for planetable, navigation

Table 12.11 Methods of Obtaining Topography

Method	Suitability
Transit and tape	Accurate, but slow and costly. Used where accuracy beyond plotting precision is desired
Transit and stadia	Fast, reasonably accurate for plotting purposes. Contours by direct (trace contour) method in gently rolling country, or by indirect (controlling-point) system where high, low, and break points are found in rugged terrain, or on uniform slopes, and contours interpolated
Planetable	Plotting and checking in field. Good in cluttered areas of many details. Contours by direct or indirect method. Now generally replaced by photogrammetry for large areas. Used to check photogrammetric maps
Coordinate squares	Better for contours than culture. Elevations at corners and slope changes interpolated for contours. Size of squares dependent on area covered, accuracy desired, and terrain. Best in level to gently rolling country
Offsets from center line or cross sectioning	On route surveys, right-angle offsets taken by eye or prism at full stations and critical points, along with elevations, to get a cross profile and topographic details. Contours by direct or indirect method. Elevations or contours recorded as numerator, and distance out as denominator
Photogrammetry	Fast, cheap, and now commonly used for large areas covering any terrain, where ground can be seen. Basic control by ground methods, some additional control from photographs

Relative positioning is possible when two or more receivers are operated together in the survey area. The processing of the Doppler data can be performed in four modes: simultaneous point positioning, translocation, semishort arc, and short arc. The specifications for relative positioning are valid only for data reduced by the semishort- or short-arc methods. The semishort-arc mode allows up to 5 degrees of freedom in the ephemerides; the short arc mode allows 6 or more degrees of freedom. These modes allow the use of the broadcast ephemerides in place of the precise ephemerides.

See also Arts. 12.2 and 12.18.

12.15 Global Positioning System (GPS)

This system utilizes radio signals from a worldwide set of navigational satellites, which broadcast continuously on two L-band carrier frequencies. These provide coded information, such as predicted satellite ephemeris, satellite identification, and time data. Each satellite provides strong radio signals that can be compared with the same signals arriving at other positions on earth for determination of relative earth positions (Fig. 12.8). For the most precise measurements, the surveyor should have three or more receivers simultaneously observing GPS satellites. When four satellites are observed simultaneously, it is possible to determine the timing and three-dimensional positioning of a ground receiver. In effect, the satellites serve as control points and ground positions are determined by distance–distance intersection. Compared with satellite Doppler positioning, GPS offers an order-of-magnitude increase in accuracy and shorter occupation time.

A variation of this system known as real-time kinematic (RTK) GPS offers advantages over other systems for boundary surveys. It enables a surveyor to determine the position of a corner and establish a corner without having to make

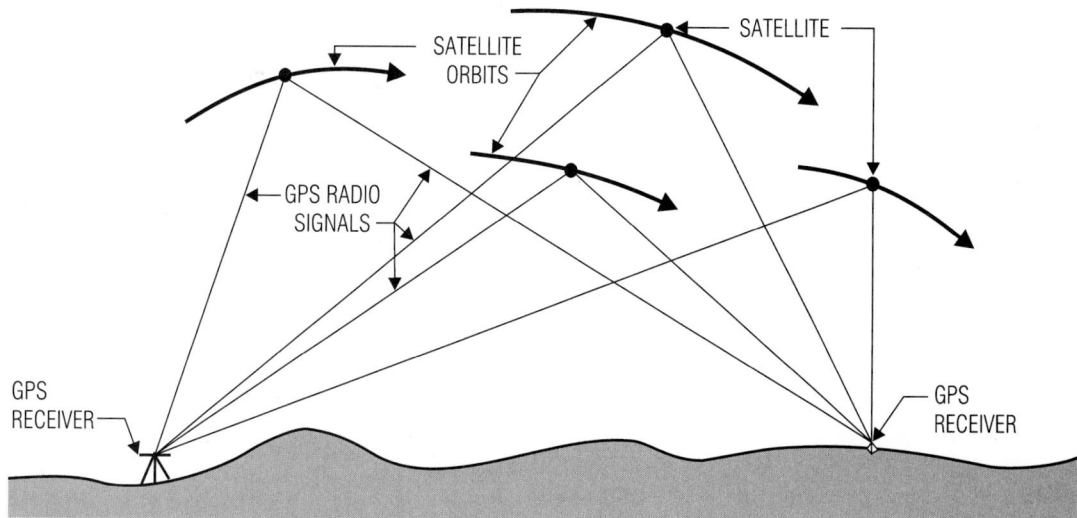

Fig. 12.8 Radio signals from orbiting GPS satellites determine the relative position of receivers on the earth's surface.

traditional corner moves with conventional surveying instruments and practices and without having to postprocess the data. An RTK GPS system generally comprises two or more GPS receivers, three or more radio modems, a fixed-plate initializer, and a hand-held data collector and portable computer. One receiver occupies a control point and broadcasts a correction message, or compact measurement record, to one or more roving receivers. These process the information to produce an accurate position relative to the control point. (C. W. Sumpter and G. W. Asher, "GPS Goes Real Time," *Civil Engineering*, September 1994, p. 64.)

See also Arts. 12.2 and 12.18.

12.16 Inertial Surveying

The inertial surveying system (ISS) is a relative positioning system, in which changes in position are determined from measurements of acceleration and time and by sensing the earth's rotation and the local vertical direction. Distance components are measured from an initial known reference location, used as a control point, and new positions are located relative to that point. Equipment required, which may be mounted on a light-duty truck or a helicopter, consists of accelerometers, stabilized by gyroscopes and mounted on an inertial platform, and control and data-handling components, including a computer. The system is self-contained and has no line-of-sight limitations. The equipment can be moved rapidly and produces three-dimensional geodetic positions with an accuracy acceptable for many purposes.

Inertial surveying is a measurement system composed of lines or a grid of ISS observations (Fig. 12.9). Specifications given in Table 12.12 cover use of ISS only for horizontal control.

Each inertial survey line is required to tie into a minimum of four horizontal network control points spaced well apart and should begin and end at network control points. These network control points must have horizontal datum values better than the intended order (and class) of the new survey. Whenever the shortest distance between two new unconnected survey points is less than 20% of the distance between those points traced along existing or new connections, then a direct connection should be made between those two survey points. In addition, the survey should connect to any sufficiently accurate network control points within the distance specified by the station spacing. The connections may be measured by electronic distance measurement or tape traverse, or by another ISS line. If an ISS line is used, then these lines should follow the same specifications as all other ISS lines in the survey.

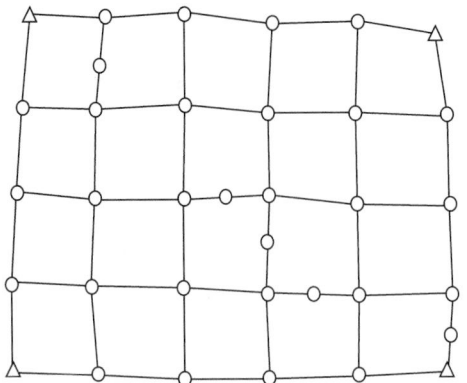

Fig. 12.9 Inertial grid configuration for an inertial traverse survey.

For extended area surveys by ISS, a grid of intersecting lines that satisfies the 20% rule stated above can be designed.

A grid of intersecting lines should contain a minimum of eight network points, and should have a network control point at each corner. The remaining network control points may be distributed about the interior or the periphery of the grid. However, there should be at least one network control point at an intersection of the grid lines near the center of the grid. If the required network points are not available, then they should be established by some other measurement system. See also Arts. 12.2 and 12.18.

12.17 Photogrammetry

Photogrammetry is the art and science of obtaining reliable measurements by photography (metric photogrammetry) and qualitative evaluation of image data (photo interpretation). It includes use of terrestrial, close-range, aerial, vertical, oblique, strip, and space photographs along with their interpretation. Remote sensing and side-looking radar are also used. Some advantages of mapping by aerial photographs are rapid coverage of large accessible or inaccessible areas and assurance of getting all visible detail. Note that an aerial photo is not a map, i.e., an orthographic projection; it is a perspective projection and may contain unnecessary details that overshadow the critical ones. However, orthophotos may be prepared from a pair of overlapping photos to eliminate the perspective factor and can serve as topographic maps.

Four of the five cameras most commonly used for topographic mapping have f5.6 lenses. They may have narrow-, normal-, wide-, or ultrawide-angle lenses. Most use roll film. Four fiducial marks printed on each photograph locate the geometric axes and principal point. Photographs are taken in strips with a side lap (strip overlap) of 25% and a forward overlap (advance) averaging 60% to ensure that images of ground points appear in at least two and preferably three or more pictures.

Because vertical photographs are perspective views, the scale is not uniform. Equal-length ground lines at higher elevations and near the edges of photographs will be longer than those at lower elevations and near the center. An average scale can be selected as an approximate value.

The basic photogrammetric formulas presented in the following paragraphs are used by the equipment and operators to make measurements and draw maps. Three types available use direct-optical, mechanical, or optical-mechanical projection systems. Among the various models are the Multiplex, Balplex, Kelsh, Zeiss Double Projection, Planimat and Stereoplanigraph, Wild Aviograph and Autograph A10, and Kern PG2.

Scale formulas are as follows (refer to Fig. 12.10)

$$\frac{\text{Photo scale}}{\text{Map scale}} = \frac{\text{photo distance}}{\text{map distance}} \quad (12.17)$$

$$\text{Photo scale} = \frac{ab}{AB} = \frac{f}{H - h_1} \quad (12.18)$$

Table 12.12 Network Geometry

Order class	Second I	Second II	Third I	Third II
Minimum station spacing, km	10	4	2	1
Maximum deviation from straight line connecting endpoints, deg	20	25	30	35

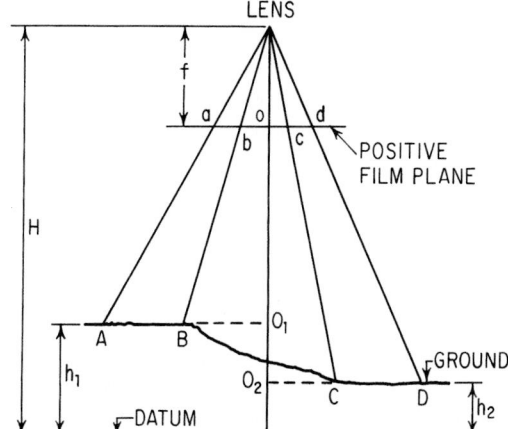

Fig. 12.10 Photographic scale depends on focal length of lens f and height H of airplane. *(Reprinted with permission from R. C. Brinker, "Elementary Surveying," Harper & Row, Publishers, New York.)*

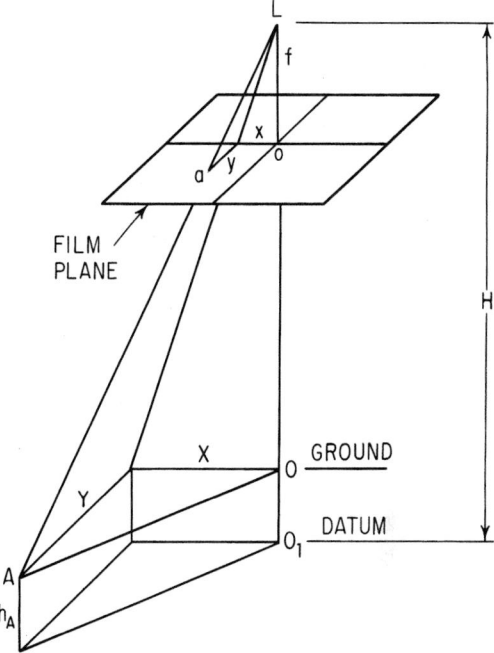

Fig. 12.11 Photograph coordinates x, y are proportional to ground coordinates X, Y when the optical axis is vertical. *(Reprinted with permission from R. C. Brinker, "Elementary Surveying," Harper & Row, Publishers, New York.)*

where f = focal length of lens, in

$\qquad H$ = flying height of airplane above datum (usually mean sea level), ft

$\qquad h_1$ = elevation of point, line, or area with respect to datum, ft

Ground distances can be found from measurements on a photograph by using photograph coordinates x, y and ground coordinates X, Y (Fig. 12.10 and 12.11). For a line AB with unequal elevations at A and B, length is determined by

$$AB = \sqrt{(X_A - X_B)^2 + (Y_A - Y_B)^2} \qquad (12.19)$$

where $X_A = x_a(H - h_A)/f$
$\qquad Y_A = y_a(H - h_A)/f$
$\qquad X_B = x_b(H - h_B)/f$
$\qquad Y_B = y_b(H - h_B)/f$

Average displacements caused by topographic relief on vertical aerial photographs always radiate from the principal point o (Fig. 12.12), which is directly above the nadir point O on the ground when the optical axis is vertical. The displacement d, in, is the distance on a photograph from the image of a ground point to its fictitious image projected on a datum plane (Fig. 12.12). Then,

$$d = r - r_1 \qquad r = \frac{Rf}{H - h_1} \qquad r_1 = \frac{Rf}{H} \qquad (12.20)$$

Substituting for r and r_1 in the first equation yields

$$d = \frac{Rf}{H - h_1} - \frac{Rf}{H} = \frac{Rfh_1}{H(H - h_1)} = \frac{rh_1}{H}$$

$$= \frac{r_1 h_1}{H - h_1} \qquad (12.21)$$

where r = radial distance on photograph from principal point to ground image of point P in (or mm)

$\qquad r_1$ = radial distance on photograph from principal point to P_1, the fictitious image position of point P projected to datum, in (or mm)

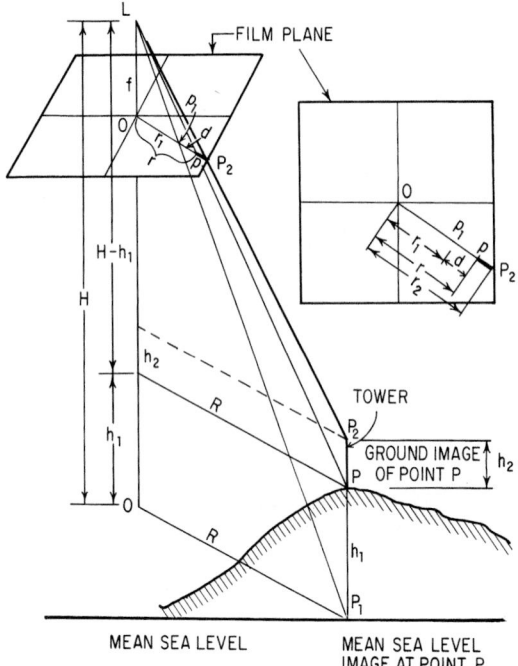

Fig. 12.12 Elevation differences cause topographic-relief displacements. (*Reprinted with permission from R. C. Brinker, "Elementary Surveying," Harper & Row, Publishers, New York.*)

h_1 = height of point P above datum, ft

H = height of airplane above datum, ft

As an example, find the height of a tower on an aerial photograph where the flight altitude above mean sea level is 5000 ft. ground elevation is 1000 ft. and measurements give $r_2 = 8.65$ mm and $r = 8.52$ mm (Fig. 12.12).

$$d' = r_2 - r = 0.13 \text{ mm}$$

$$h_2 = \frac{d'(H - h_1)}{r_2} = \frac{0.13(5000 - 1000)}{8.65} = 60 \text{ ft}$$

Stereoscopic vision is that particular application of binocular vision (simultaneous vision with both eyes) that enables an observer to view two different perspective photographs of an object (such as two photographs taken from different camera stations) and get the mental impression of three dimensions. Thus, a stereoscope permits each eye to see as one a pair of photographs that

shows an area from different exposure points and thereby produces a three-dimensional (stereoscopic) image (model).

Parallax is the apparent displacement of the position of a body with respect to a reference point or system caused by a shift in the point of observation. As a result of the forward movement of a camera in flight, positions of all images travel across the focal plane from one exposure to the next, with images of higher elevations moving farther than those at lower levels.

Absolute parallax of a point is the total movement of the image of a point in the focal plane between exposures and is found as follows: (1) Locating the principal points of adjacent photographs containing the images of the point (Fig. 12.13); (2) transferring each principal point to the other photograph; (3) connecting each principal and transferred principal point to define the flight line; (4) drawing a line on each photograph through the principal point perpendicular to the flight line; and (5) measuring the x coordinate (parallel to flight line) of the point under study on each photograph.

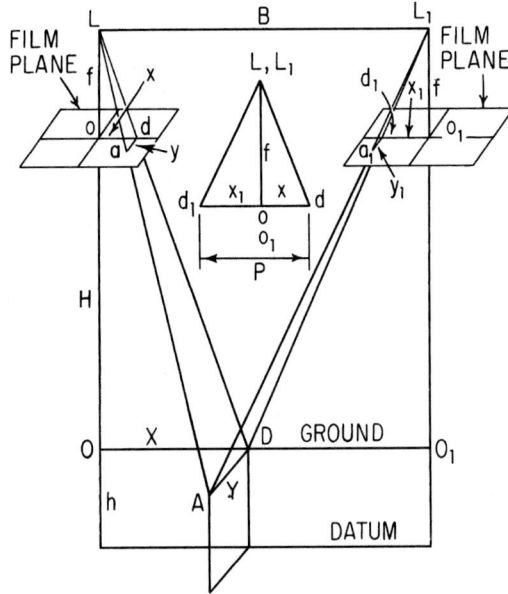

Fig. 12.13 Parallax shifts image of line *AD* on successive photographs. (*Reprinted with permission from R. C. Brinker, "Elementary Surveying," Harper & Row, Publishers, New York.*)

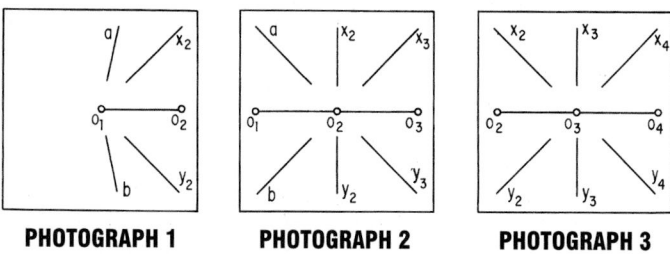

PHOTOGRAPH 1   **PHOTOGRAPH 2**   **PHOTOGRAPH 3**

Fig. 12.14   Radial-line plotting of horizontal control points on aerial photographs.

Absolute parallax of a point, in (or mm) (observing algebraic signs), is

$$p = x - x_1 \qquad (12.22)$$

Also,

$$X = \frac{xB}{p} \quad Y = \frac{yB}{p} \quad H - h = \frac{fB}{p} \qquad (12.23)$$

For untilted photographs,

$$Y = \frac{y_1 B}{p}$$

where X, Y = ground coordinates measured from plumb point, ft

B = air base = distance between exposure stations, ft

x, y = photograph coordinates, in (or mm)

H = altitude of airplane above datum, ft

h = elevation of object above datum, ft

f = focal length of lens, in (or mm)

Measuring stereoscopes, such as the stereo-comparator and contour finder, are satisfactory for small areas. The multiplex, Kelsh plotter, Wild Autograph, and other large plotters usually are preferred for extensive projects. The latter instruments measure differences in parallax by means of a floating dot—actually two dots superimposed on the photographs and mentally fused by the operator to produce the floating dot. The operator places it at apparent ground level in the photograph for contouring or finding spot elevations.

The accuracy of photogrammetric contouring depends on camera precision, type of terrain and ground cover, type of stereoscopic plotter, and experience of the operator.

$$C \text{ factor} = \frac{\text{flying height}}{\text{control interval}} \qquad (12.24)$$

is an empirical ratio used to express the efficiency of stereoscopic plotters. Photogrammetrists get C factors of 750 to 2500, and elevations to the nearest foot and half foot with present equipment.

Radial-line plotting is a graphical method of extending horizontal control between fixed ground points on aerial photographs. In Fig. 12.14, points o_1, o_2, and o_3, the principal points in photographs 1, 2, and 3, are located on adjacent photographs. Control points a and b are identified in photograph 1. Additional control points called pass points (x_2 and y_2 in photograph 1, a and b in photograph 2, x_3 and y_3 in photograph 3) are established and transferred to the other photographs. On a sheet of tracing paper or template

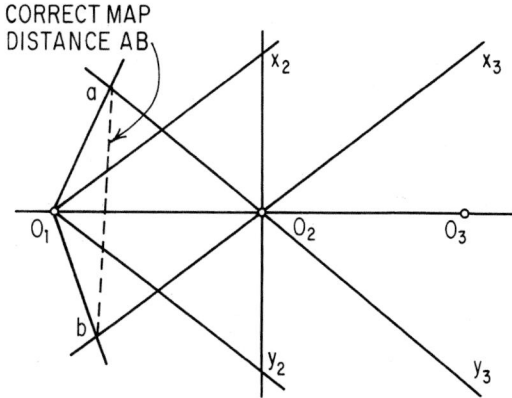

Fig. 12.15   Correct map scale and location of points are obtained with the aid of radial lines of Fig. 12.14.

placed over each photograph, a set of rays is drawn from the principal point, through each conjugate principal point, control point, and pass point. The templates are superimposed, as shown in Fig. 12.15, until all rays to each point, such as *a* or *b*, provide a single intersection. The map positions of the points then have been fixed.

The method is based upon two fundamental photogrammetric principles:

On truly vertical photographs, image displacements caused by topographic relief radiate from the principal point.

Angles between rays passing through the principal point are equal to the horizontal angles formed by the corresponding lines on the ground.

12.18 Bibliography

R. C. Brinker and R. Minnick, "The Surveying Handbook," Van Nostrand Reinhold, New York.

"Definitions of Surveying and Associated Terms," Manual No. 34, American Society of Civil Engineers.

"Handbook of Close-Range Photogrammetry and Surveying," American Society of Photogrammetry, Falls Church, Va.

J. A. Nathanson and P. C. Kissam, "Surveying Practice," 4th ed., McGraw-Hill, Inc., New York.

G. O. Stenstrom, Jr., "Surveying Ready-Reference Manual," McGraw-Hill, Inc., New York.

See also Art. 12.2.

13

Charles H. Sain
Consulting Engineer
Birmingham, Alabama

G. William Quinby
Consulting Engineer
Golden, Colorado

EARTHWORK

Earthwork involves movement of a portion of the earth's surface from one location to another and, in its new position, creation of a desired shape and physical condition. Occasionally, the material moved is disposed of as spoil. Because of the wide variety of soils encountered and jobs to be done on them, much equipment and many methods have been developed for the purpose. This section describes and analyzes the equipment and methods.

13.1 Types of Excavation

A common method of classifying excavation is by type of excavated material: topsoil, earth, rock, muck, and unclassified.

Topsoil excavation is removal of the exposed layer of the earth's surface, including vegetation. Since the topsoil, or mantle soil, supports growth of trees and other vegetation, this layer contains more moisture than that underneath. So that the lower layer will lose moisture and become easier to handle, it is advantageous to remove the topsoil as soon as possible. When removed, topsoil usually is stockpiled. Later, it is restored on the site for landscaping or to support growth of vegetation to control erosion.

Earth excavation is removal of the layer of soil immediately under the topsoil and on top of rock. Used to construct embankments and foundations, earth usually is easy to move with scrapers or other types of earthmoving equipment.

Rock excavation is removal of a formation that cannot be excavated without drilling and blasting. Any boulder larger than $\frac{1}{2}$ yd^3 generally is classified as rock. In contrast, earth is a formation that when plowed and ripped breaks down into small enough pieces to be easily moved, loaded in hauling units, and readily incorporated into an embankment or foundation in relatively thin layers. Rock, when deposited in an embankment, is placed in thick layers, usually exceeding 18 in.

Muck excavation is removal of material that contains an excessive amount of water and undesirable soil. Its consistency is determined by the percentage of water contained. Because of lack of stability under load, muck seldom can be used in an embankment. Removal of water can be accomplished by spreading muck over a large area and letting it dry, by changing soil characteristics, or by stabilizing muck with some other material, thereby reducing the water content.

Unclassified excavation is removal of any combination of topsoil, earth, rock, and muck. Contracting agencies frequently use this classification. It means that earthmoving must be done without regard to the materials encountered. Much excavation is performed on an unclassified basis because of the difficulty of distinguishing, legally or practically, between earth, muck, and rock. Unclassified excavation must be carried out to the lines and grades shown on the plans without regard to percentage of moisture and type of material found between the surface and final depth.

Excavation also may be classified in accordance with the purpose of the work, such as stripping, roadway, drainage, bridge, channel, footing, borrow. In this case, contracting agencies indicate the nature of the excavation for which materials are to be removed. Excavation designations differ with agencies and locality. Often, the only reason a certain type of excavation has a particular designation is local custom.

Stripping usually includes removal of all material between the original surface and the top of any material that is acceptable for permanent embankment.

Roadway excavation is that portion of a highway cut that begins where stripping was completed and terminates at the line of finished subgrade or bottom of base course. Often, however, stripping is made part of roadway excavation.

Drainage excavation or structure excavation is removal of material encountered during installation of drainage structures other than bridges. Those structures are sometimes referred to as minor drainage structures and include roadway pipe and culverts. A culvert is usually defined as any structure under a roadway with a clear span less than 20 ft, whereas a bridge is a structure spanning more than 20 ft. After a pipe or culvert has been installed, backfilling must be done with acceptable material. This material usually is obtained from some source other than drainage excavation, which generally is not acceptable or workable. Often, culvert excavation does not include material beyond a specified distance from the end of a culvert.

Bridge excavation is removal of material encountered in digging for footing and abutments. Often, bridge excavation is subdivided into wet, dry, and rock excavation. The dividing line between wet and dry excavation usually is denoted by specification of a ground elevation, above which material is classified as dry and below which as wet. A different elevation may be specified for each foundation.

Channel excavation is relocation of a creek or stream, usually because it flows through a right-of-way. A contracting agency will pay for any inlet or outlet ditch needed to route water through a pipe as channel excavation, to the line where culvert excavation starts.

Footing excavation is the digging of a column or wall foundation for a building. This work usually is done to as neat a line and grade as possible, so that concrete may be cast without forms. Although elimination of forms saves money, special equipment and more-than-normal handwork are usually required for this type of excavation.

Borrow excavation is the work done in obtaining material for embankments or fills from a source other than required excavation. In most instances, obtaining material behind slope lines is classified as borrow, although it commonly is considered as getting material from a source off the site. Most specifications prohibit borrow until all required excavation has been completed or the need for borrow has been established beyond a reasonable doubt. In some cases, need for a material not available in required excavation makes borrow necessary. A borrow pit usually has to be cleared of timber and debris and then stripped of topsoil before desired material can be excavated.

Dredge excavation is the removal of material from under water.

13.2 Basic Excavating Equipment

A tractor is the most widely used excavating tool. Essentially, it is a power source on wheels or tracks (crawler). Equipped on the front with a **bulldozer**, a steel blade that can be raised and lowered, a tractor can push earth from place to place and shape the ground. If a scraper is hooked to the drawbar and means of raising, lowering, and dumping are provided, a tractor-drawn scraper results. Addition of other attachments creates tools suitable for different applications (see also Art. 13.7).

Another basic machine is one that by attachment of different fronts may be converted into a shovel, dragline, clamshell, backhoe, crane, or pile driver. The basic machine made for a shovel, however, has shorter and narrower tracks than one made for a dragline or clamshell, and more counterweight has to be added to the back. A shovel attachment will fit the basic machine made for a dragline or clamshell, but the longer tracks will interfere with the shovel (see also Art. 13.4).

Scrapers may be tractor-drawn or self-propelled. More excavation is moved with self-propelled or rubber-tired scrapers than with scrapers towed and controlled by crawler tractors (see also Art. 13.8).

Trenchers, used for opening trenches and ditches, may be ladder or wheel type. They do most of the pipeline excavation in earth. The ladder type has chains to which are attached buckets that scoop up earth as the chains move. It is adaptable to deep excavation. The wheel type has digging buckets on the circumference of a rotating wheel. The buckets dump excavated material into a conveyor mounted in the center of the wheel. This type of trencher is used mainly for shallow trenches. Neither type is used to any great extent when rock is encountered in trench excavation.

Wheel excavators, used in constructing earth dams or in strip mining, excavate soft or granular materials at very high rates. For example, one excavator with a 28-ft wheel moves 1500 tons of iron ore per hour. A typical wheel excavator resembles a wheel-type trencher. Buckets mounted on a wheel 12 or more ft in diameter scoop up the earth. They may be 2 ft or more wide, with a capacity of $\frac{1}{3}$ yd^3 or more, and equipped with a straight cutting edge or teeth. The buckets dump into a hopper, which feeds the earth onto a conveyor belt. The belt moves along a boom, which may be 200 ft or more long, to dump the earth into another hopper. This hopper in turn feeds the earth to a stockpile or to earthmoving equipment.

13.3 Selecting Basic Equipment

Type of material to be excavated may determine the basic equipment to be used. But length and type of haul road must also be considered. For example, suppose excavation is in earth and best results could be obtained with rubber-tired scrapers, but the haul is over city streets. In this case, this type of equipment probably could not be used because of heavy wheel loads and interference with traffic.

For rock, a front-end loader, backhoe, or shovel would be the basic rig. For earth, when a haul road can be built, scrapers would be chosen. But if the earth has to be moved several miles over existing streets or highways, the choice would be a front-end loader, shovel, or backhoe that would load dump trucks. Whether a shovel or backhoe would be used depends on whether the excavation bottom can support a front-end loader or shovel and hauling units. If the bottom is too soft, a dragline or backhoe would be required. A dragline can sit outside the excavation and load a hauling unit at the same level (loading on top). But when a backhoe can be used, it is preferred to a dragline because of greater production.

Therefore, in selecting basic equipment, consider:

Types of material to be excavated

Types and size of hauling equipment to be used

Load-supporting ability of original ground

Load-supporting ability of material to be excavated

Volume of excavation to be moved

Volume to be moved per unit of time

Length of haul

Type of haul road

13.4 General Equipment for Excavation and Compaction

Clearing or Grubbing

Use *tractor* with bulldozer or root rake. *Bulldozer* can fell trees, uproot stumps. *Root rake* piles for burning, makes cleaner pile. *Brush hog* may be required for light brush.

Grubbing

Use low-strength *explosives*, slow detonation speed.

Clearing

Drag *chain* or *chain and heavy ball* between two tractors. Useful for trees that break easily. Tractors equipped with cutter blades can operate on any footing and cut any tree at ground level.

Stripping

Bulldozers are limited by length of push or haul but are useful for swampy conditions.

Scrapers are limited by terrain and support ability of ground; they may be tractor-drawn for short hauls.

Draglines are limited by depth of stripping, ability to service with hauling units, and space for casting the bucket. They are used where swampy conditions prevent other equipment from being used.

Graders are limited to use where stripping can be windrowed on final position. Material can be loaded from a windrow by a front-end loader.

Pipe Installation

Backhoes are used on firm soil where depth of trench is not excessive; they are good in rock.

Draglines are used for deep trenches if the sides can be flattened; they have difficulty digging vertical walls.

Clamshells are used where sheeting of sides is required and it is necessary to dig between braces and to great depths. They are inefficient in rock.

Bulldozers are limited to shallow excavation.

Trenching machines produce vertical or near-vertical walls and can maintain line and earth grade.

Earth Excavation

Tractor-drawn scrapers are limited by length of haul and supporting ability of the soil. Cost gets excessive if haul distance greatly exceeds 1000 ft.

Two-axle, rubber-tired, self-propelled scrapers are limited by length of haul, terrain, and supporting ability of the soil; they bounce on long hauls at top speed.

Three-axle, rubber-tired, self-propelled scrapers need maneuvering or working space and are limited by terrain and supporting ability of soil. They are most efficient on long hauls.

Twin-engine, rubber-tired scrapers have few limitations. They are useful in rough terrain and where traction is needed on all wheels.

Front-end loaders generally discharge into hauling units if the haul greatly exceeds 100 ft and they also are limited by digging and dumping ease of excavated material.

Shovels are also used to load into hauling units. Working room must be ample and distance to cast short. Shovels also have to dig from a face.

Draglines may be used where excavation is deep and the material has no supporting ability. Material should be easy to dig. Draglines usually load into hauling units.

Wheel excavators offer high excavation rate and loading into hauling units with soft or granular soils.

Mobile belt loaders (Fig. 13.1) give high-production loading into hauling units but are limited by working room and supporting capacity of excavation bottom. Belt loaders are limited to short, infrequent moves. A wide belt handles some rock excavation.

Dredges usually are used where transportation and digging costs are prohibitive if other than water-borne equipment is used. Water must be available for mixing with the excavated material for pumping through pipes. Distance to spoil area should not be too great.

Clamshells are low producers but are useful in small or deep spaces, where there is no overhead interference with swinging of the boom.

Gradall, not a high-production tool, is suitable for dressing or finishing where tolerances are close.

Scoopers, hydraulically operated, are high-production equipment, limited by dumping height and to easily dug material. Production is not so greatly influenced by height of face as for a shovel.

Rock Excavation

Shovels can dig any type of rock broken into pieces that can be easily dug. Limited to digging from a

Fig. 13.1 Mobile belt loader. (*Barber-Greene Co.*)

face, shovels are used for high-production loading into hauling units.

Bulldozers are limited to short movements and easily dug rock. Sometimes, they are used to dispose of boulders when drilling and blasting are not economical.

Front-end loaders are used instead of shovels because of their high production, lower cost of operation, and ease of moving from job to job.

Backhoes are used for foundation excavation, trenches, and high production in rough terrain. They must dig below their tracks.

Scrapers are suitable for short movements and rock broken down to small sizes, such as blasted shale, but tire wear may be greater than in other applications.

Scoopers may be used instead of shovels where working space is tight. They are limited by the height of hauling units and to easily dug rock.

Gradalls are used for trench and foundation excavation, but material must be well blasted.

Clamshells are most suitable for deep foundations or where the reach from machine position to excavation prohibits other equipment from being used. Rock must be well broken for maximum production.

Compaction

Sheepsfoot rollers, made with feet of various shapes, offer high-speed production. Compaction depends on unit pressure and speed of roller. They are not suitable for compacting sand. They also are limited by depth of layer to be compacted.

Rubber-tired rollers are used for granular soils, including shales and rock. Ranging from very light weight to 200 tons, they may be self-propelled or towed. Depth of lift compacted depends on weight.

Vibratory compactors, towed, self-propelled, or hand-held, are also used for granular soils. Compaction ability depends on frequency and energy of vibrations. Depth of lift is not so much a factor as for other types of compactors.

Grid rollers, useful in breaking down oversize particles, are limited to shallow lifts of nonsticky material. They can be towed at any safe, economical speed.

Air tamps are used to backfill pipe and foundations and for work in areas not accessible to power equipment. Usually hand-held, they are powered by compressed air imparting reciprocating blows.

They are limited to low production and shallow lifts.

Paddlefoot rollers, usually self-propelled, compact from the top of lift down. They are limited to an average depth (up to 8 in) of lift in all soils.

A *rubber-tired front-end loader* can be converted to this type of roller by a change of wheels.

Steel-wheel rollers, self-propelled, are used where a smooth, sealed surface is desired. They are limited to shallow depth of lift.

13.5 Power Shovels, Draglines, Clamshells, and Backhoes

These four machines are made by installing an attachment on a basic machine, which may be mounted on crawler tracks or a trucklike chassis (Art. 13.2). (See Figs. 13.2 to 13.5.) When mounted on a trucklike chassis, the machine usually is designed for use as a truck crane, but it also can be used as a shovel or backhoe if mobility is desired and low production is acceptable. Most backhoes, however, are hydraulic and cannot be converted.

There is not much difference between equipment used as a clamshell and that used as a dragline or crane. A boom used with a clamshell has two-point sheaves, so that two cables can attach to the bucket. One cable is used to open and close the bucket and the other to hoist or lift the bucket. Since the two cables should travel at the same

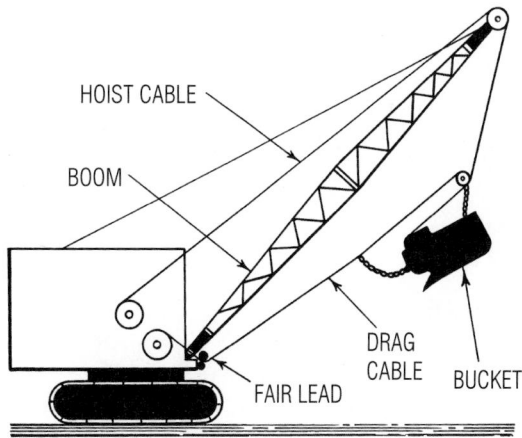

HOIST CABLE

BOOM

DRAG CABLE

BUCKET

FAIR LEAD

Fig. 13.2 Dragline.

Fig. 13.3 Hydraulic excavator (backhoe). (*Caterpillar Tractor Co.*)

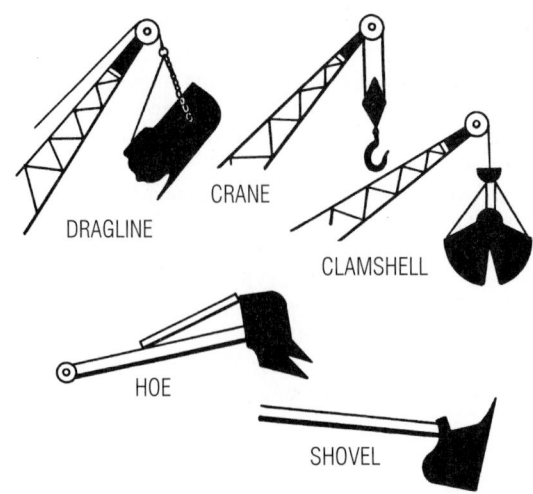

Fig. 13.5 Excavating and crane attachments.

speed, the drums on a clamshell are the same size. To keep the bucket from spinning and twisting the hoist and closing lines, a tagline extends between the bucket and a spring-loaded reel on the side of the boom (Fig. 13.4).

A dragline has a hoist cable that goes through a point sheave atop the boom and attaches to the bucket. Another line, the drag cable, goes through the fairlead and attaches to the bucket (Fig. 13.2). The drum that exerts pull on the drag cable is smaller than the hoist drum because more force is required on the drag cable than on the hoist lines. Typical performance factors for a dragline are given in Tables 13.1 and 13.6.

Power shovels are used primarily to load rock into hauling units. Production depends on type of material to be loaded, overall job efficiency, angle

of swing, height of bank or face the shovel digs against, ability of operator, swell of material, slope of ground machine is working on, and whether hauling units are of optimum size and adequate in number. For highest efficiency, the degree of swing should be held to a minimum. (Typical performance factors are given in Table 13.2.) Working the shovel so that a hauling unit can be loaded on each side is desirable so there is no lost time waiting for a hauling unit to get into position.

Table 13.3 gives estimated hourly production of power shovels. It is based on bank cubic yards measure, 90° swing, optimum digging depth, grade-level loading, 100% efficiency, 60-min hour, and bucket-fill factor of 1.00 (see Table 13.5).

Table 13.1 Typical Dragline Calculating Factors: Average Swing Cycle with 110° Swing

Bucket capacity, yd^3	$\frac{1}{2}$	$1\frac{1}{2}$	2
Time, s	24	30	33

Bucket Factors		

Type of digging	% of rated capacity (approx)
Easy	95–100
Medium	80–90
Medium hard	65–75
Hard	40–65

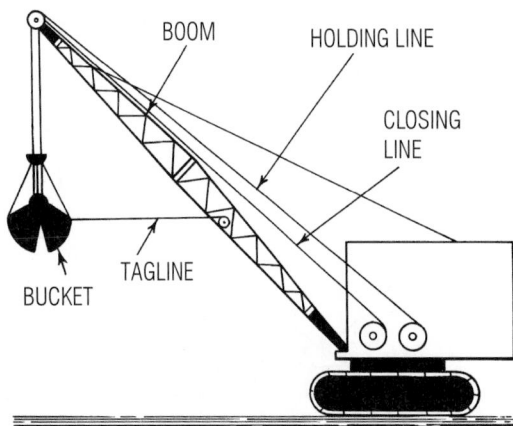

Fig. 13.4 Clamshell.

Table 13.2 Typical Shovel Calculating Factors: Average Swing Cycle with 90° Swing

Bucket capacity, yd^3	½	1	1½	2	2½
Time, s	20	21	22	23	24

Dipper Factors

Type of digging	% of rated capacity (approx)
Easy	95–100
Medium	85–90
Medium hard	70–80
Hard	50–70

Table 13.4 indicates the effect on production of depth of cut and angle of swing.

Optimum digging depth is the shortest distance a bucket must travel up a face or bank to obtain its load. This depth usually is the vertical distance from shipper shaft (dipper-stick pivot shaft) to ground level. Optimum depth varies with type of material to be excavated since a lower boom is needed for hard materials than for soft.

Work must be planned to load or move the maximum yardage each shift: Locate the shovel and hauling units for the shortest swing of the shovel. If it is necessary to work high, dig the upper portion first. Move up to the face while a hauling unit is getting into position. Make short moves frequently, instead of less frequent long moves. Stay close to the face; do not dig at the end of the stick. Lower the dipper only enough to get a full bucket; this cuts down on hoist time. Keep dipper teeth sharp. Have spare cables and dipper teeth readily available near the shovel. Hoist the load no more than necessary to clear the hauling-unit bed. Start the swing when the bucket is full and clear of the bank. Spot the hauling unit under the boom point so it is not necessary to crowd or retract to dump into the bed (Fig. 13.6). Break rock well for easier digging.

A dragline is more versatile than a shovel. With a dragline, load can be obtained from a greater distance from the machine (reach is greater). Excavation can be done below water and at a long distance above or below the dragline. A larger bucket than the machine's rated capacity can be used if a short boom is installed. It is not uncommon for a machine rated at 2½ yd^3 to be loading with a 4-yd^3 bucket into hauling units. But weight of bucket and load should not exceed 70% of the tipping load of the machine. (Lifting-crane capacity is based on 75% of actual tipping load. A dragline may approach this if it is on solid footing and is digging good-handling material.)

Since a dragline loads its bucket by pulling it toward the machine, the pit or face slopes from bottom to top toward the dragline. Best production is obtained by removing material in nearly horizontal layers and working from side to side of the excavation. A keyway, or slot, should be cut next to the slope. This keyway should always be slightly lower than the area being taken off in horizontal layers. A good operator fills the bucket

Table 13.3 Estimated Hourly Production of Dipper-Type Power Shovel*

Material class	Shovel dipper sizes, yd^3															
	½	¾	1	1¼	1½	2	2½	3	4	4½	5	6	7	8	9	10
Moist loam or sandy clay	115	165	205	250	285	355	405	454	580	635	685	795	895	990	1075	1160
Sand and gravel	110	155	200	230	270	330	390	450	555	600	645	740	835	925	1010	1100
Common earth	95	135	175	210	240	300	350	405	510	560	605	685	765	845	935	1025
Clay, tough hard	75	110	145	180	210	265	310	360	450	490	530	605	680	750	840	930
Rock, well blasted	60	95	125	155	180	230	275	320	410	455	500	575	650	720	785	860
Common with rock	50	80	105	130	155	200	245	290	380	420	460	540	615	685	750	820
Clay, wet and sticky	40	70	95	120	145	185	230	270	345	385	420	490	555	620	680	750
Rock, poorly blasted	25	50	75	95	115	160	195	235	305	340	375	440	505	570	630	695

* Caterpillar Tractor Co.

Table 13.4 Correction Factors for Effect of Depth of Cut and Angle of Swing on Power-Shovel Output*

Depth of cut, % of optimum	Angle of swing, deg						
	45	60	75	90	120	150	180
40	0.93	0.89	0.85	0.80	0.72	0.65	0.59
60	1.10	1.03	0.96	0.91	0.81	0.73	0.66
80	1.22	1.12	1.04	0.98	0.86	0.77	0.69
100	1.26	1.16	1.07	1.00	0.88	0.79	0.71
120	1.20	1.11	1.03	0.97	0.86	0.77	0.70
140	1.12	1.04	0.97	0.91	0.81	0.73	0.66
160	1.03	0.96	0.90	0.85	0.75	0.67	0.62

* "Earthmoving Data," Caterpillar Tractor Co.

as soon as possible, within a distance less than the bucket length. Digging on a slight incline helps fill the bucket. When the bucket is full, it should be nearly under the boom point and should be lifted as drag ceases.

As with shovels, a relatively shallow pit yields the greatest efficiency for draglines. The hauling units should be in the excavation or at the same elevation to which the dragline is digging. Thus, when the bucket is full, it will have a short lift to reach the top of the hauling units. If the pit bottom is soft or for some other reason hauling units cannot be spotted below the machine, then loading on top must be resorted to, with a loss in loading efficiency. Table 13.6 indicates dragline production in cubic yards bank measure per hour. The table is based on suitable depth of cut for maximum effect, no delays, 90° swing, and all materials loaded into hauling units (see also Table 13.1).

Production of clamshells, like that of draglines, depends on radius of operation and lifting capacity. It is general practice to limit the clamshell load, including bucket weight, to 50% of the full power-line pull at the short boom radius.

Table 13.5 Bucket-Fill Factor*

Material	Fill-factor range
Sand and gravel	0.90–1.00
Common earth	0.80–0.90
Hard clay	0.65–0.75
Wet clay	0.50–0.60
Rock, well blasted	0.60–0.75
Rock, poorly blasted	0.40–0.50

* "Earthmoving Data," Caterpillar Tractor Co.

Types of clamshell bucket are general-purpose, rehandling, and heavy excavating. The rehandling bucket is best for unloading materials from bins or railroad cars or loading materials from stockpiles. The heavy-excavation bucket is used for extreme service, such as placing riprap. It can be adjusted so that the operation is easy on components since a clamshell does not demand a tightly adjusted friction band. The general-purpose bucket is between the rehandling and excavating buckets and can be used with or without teeth.

13.6 Tractor Shovels

Also commonly known as front-end loaders, tractor shovels can be mounted on wheels (Fig. 13.7) or crawler tracks (Fig. 13.8). A crawler is desirable if moving it from one job to another is no problem, haul distance is short, and type of excavation bottom is not suitable for rubber tires. Most wheel loaders have four-wheel drive.

Capacity of a tractor shovel depends on unit weight of material to be handled, so there is a variety of buckets for each loader. These are of three basic types: hydraulically controlled, gravity dump, and overhead (overshot). Hydraulically controlled machines are preferable for most operations. The overhead is desirable where working room for turning is unavailable.

All loaders except overhead use a load, turn, dump cycle. For best efficiency and reduction of wear on tires or undercarriage, turning should be held to a minimum.

A loader should dig from a relatively low height of bank or face. Since most loaders are equipped with automatic bucket positions, the height of bank

Fig. 13.6   Hydraulic shovel loads off-highway dump truck. (*Caterpillar Tractor Co.*)

Table 13.6   Hourly Dragline Handling Capacity, yd^3

	Bucket capacity, yd^3								
Class of Material	$\frac{3}{8}$	$\frac{1}{2}$	$\frac{3}{4}$	1	$1\frac{1}{4}$	$1\frac{1}{2}$	$1\frac{3}{4}$	2	$2\frac{1}{2}$
Moist loam or sandy clay	70	95	130	160	195	220	245	265	305
Sand and gravel	65	90	125	155	185	210	235	255	295
Good, common earth	55	75	105	135	165	190	210	230	265
Clay, hard, tough	35	55	90	110	135	160	180	195	230
Clay, wet, sticky	20	30	55	75	95	110	130	145	175

should be adjusted so it is not higher than necessary to fill the bucket; this is about the same height as the push-arm hinges.

On an average construction job, a front-end loader is a versatile tool. Attachments are available so that it can be used as a bulldozer, rake, clamshell, log loader, crane, or loader.

13.7   Tractors and Tractor Accessories

Tractors are the prime movers on any construction job where earth or rock must be moved. They may be mounted on wheels or crawler tracks. Properly equipped, a tractor usually is the first item moved onto a job and one of the last to finish.

Crawlers are more widely used than wheel tractors. Crawlers will work on steep, rugged terrain; soft, marshy conditions; and solid rock. Rubber-tired tractors are suitable for specific projects or uses, such as excavation of earth or sand where track wear would be excessive. Tires and track system are the most expensive parts to maintain.

Basic components of a crawler tractor include engine, radiator, transmission, clutch, steering clutches, final drives, and undercarriage, consisting of tracks, rollers, sprockets, and idlers. Components

Fig. 13.7 Wheel loader loads off-highway dump truck. (*Caterpillar Tractor Co.*)

of a wheeled tractor include engine, radiator, transmission, clutch, tires, and rear end. A wheeled tractor may have two- or four-wheel drive. Its travel speed may range from a minimum of 3 mi/h to a maximum of over 40 mi/h. Travel speed of a crawler may range from less than 1 mi/h to not much more than 8 mi/h.

A crawler tractor can be equipped with accessories that enable it to perform a wide variety of tasks:

Rear Double-Drum Cable Control Unit ■ This is used for pulling a scraper; or cable control for a bulldozer by using only one drum.

Bulldozer ■ Either cable-controlled by rear or front unit, or hydraulically controlled (Fig. 13.9). Several different types of blade are available, such as angle, straight, U, root rake, rock rake, stump dozer, tree dozer, push dozer.

Ripper ■ Rear-mounted and hydraulically controlled to provide pressure up or down (Fig. 13.10).

Side Boom ■ A short, cable-operated boom mounted on one side with a counterweight on the opposite side of the tractor. The main use is laying cross-country pipelines (Fig. 13.11).

Tractor Crane ■ A boom with limited swinging radius.

Fig. 13.8 Track-type loader. (*Caterpillar Tractor Co.*)

Fig. 13.9 Tractor with bulldozer attachment. (*Caterpillar Tractor Co.*)

Fig. 13.10 Tractor (bulldozer) with ripper. (*Caterpillar Tractor Co.*)

Pusher Block or Blade ▪ Used for pushing scrapers, to assist and speed up loading (Fig. 13.12). A pusher block may be attached rigidly to the frame, mounted on an angle bulldozer C frame, or mounted in the center of a bulldozer. Or it can be mounted as a short bulldozer. Although designed especially as a tool for pushing, a pusher block can be used in a limited manner as a bulldozer. One

Fig. 13.11 Pipelayers. (*Caterpillar Tractor Co.*)

Fig. 13.12 Tractor pushing a scraper. (*Caterpillar Tractor Co.*)

Fig. 13.13 Tandem-powered, elevating scraper. (*Caterpillar Tractor Co.*)

way to absorb the shock when a pusher makes contact with a scraper is with springs; another is to use a hydraulic accumulator, which eliminates the need for stopping.

Welder ▪ Mounted on the tractor for mobility, welding machines are powered by the tractor engine.

Drills ▪ Often, a tractor serves as the prime mover for a rotary drill. During the drilling, the tractor engine powers the drill-steel rotation, hydraulic pumps, and air compressors. A percussion-type drill and air compressor also can be mounted on a tractor for mobility. Instead of a separate compressor, a piston-type compressor, powered by the tractor engine, can be mounted on the front or rear. Except for very large tractors, horsepower available is sufficient to furnish the required air for only one drill at a time.

13.8 Scrapers

Commonly used for earthmoving, a scraper may be self-propelled or powered by a crawler tractor. A self-propelled scraper may have two or three axles and may be single- or twin-engine. The single engine powers the front wheels (Fig. 13.13). With twin engines, one drives the front wheels; the second, the trailer wheels. Scrapers also can be worked in tandem, that is, two scrapers behind one power unit or tractor.

Essentially, a scraper acts like a scoop. A bowl hung from the frame tilts downward to permit its cutting edge to scrape off a thin layer of earth.

As the scraper moves forward, the bowl fills. When it is full, it is tilted up and an apron is dropped down over the open end to close the bowl. For discharge in thin layers, the bowl tilts down and an ejector pushes the earth out.

On most scrapers, the bowl and apron are hydraulically operated, with pressure applied to the cutting edge and the apron forceably closed to retain shale, rock, or lumpy material in the bowl. Bowl and apron also may be cable-operated, but with hydraulic pressure on the cutting edge, harder material can be loaded.

Tractor-drawn scrapers can be used for short hauls. Maximum economic haul is about 1000 ft. This type of scraper is useful for stripping topsoil and earthmoving in marshy conditions.

Twin-engine scrapers are suitable for steep grades and swampy conditions. Such a scraper can outperform a crawler tractor-scraper combination on adverse grades and marshy soil. For best performance, a twin-engine scraper should be equipped with the largest-size tires, to obtain the greatest flotation under difficult conditions. Although this machine can obtain a load without a pusher, the scraper can load faster with a pusher, tire wear will be less, and there will be other benefits to offset the higher cost of pushing.

Two-axle self-propelled scrapers are more maneuverable, adaptable to rougher terrain and more difficult conditions, and better for shorter haul distance than three-axle units. The latter are more efficient on long hauls because they move faster on a haul road. Two-axle units bounce at high speeds, even on smooth haul roads. A three-axle unit is suitable for short hauls where there are no adverse return grades and there is ample maneuvering room, such as sites for airports, railroad classification yards, or industrial buildings.

On jobs with a small amount of rock, scrapers may be competitive with a shovel and rock-type trucks. Scraper tire and cutting-edge costs may exceed normal, but properly evaluated, wear costs may not be excessive. To keep costs within economic limits, cuts must be laid out so that scrapers can load without difficulty. At least half a cut, but preferably the entire cut, should be blasted for its entire length before scrapers start excavation. Better results will be obtained if some earth remains at the end of the cut or in locations where scrapers can complete a load in earth. Broken rock and shale do not "boil" or roll into the scraper; more power is required to force them into the bowl than for earth. When loading is completed in earth, however, rock is forced into the bowl. Hydraulic scrapers can forceably close the apron, thus reducing spillage. But a heaped or full load is very difficult to obtain. So the amount of material moved per trip is less with rock than with earth.

For handling by scrapers, rock has to be broken into small particles efficiently. Blasting holes have to be spaced closer, and more explosives per cubic yard are needed than for shovel-and-truck excavation. Most shales and sandstones can be blasted so that the maximum size can be easily controlled and enough fines produced to facilitate scraper loading. But igneous and metamorphic rocks, when blasted, do not readily produce a material that can be scraper-loaded. They have cleavage planes that form oversize particles and few fines. Experience, observations of rock formation, and comparisons of unit and total cost are necessary to determine whether to use scrapers or a shovel and supporting equipment.

13.9 Formulas for Earthmoving

External forces offer *rolling resistance* to the motion of wheeled vehicles, such as tractors and scrapers. The engine has to supply power to overcome this resistance; the greater the resistance, the more power needed to move a load. Rolling resistance depends on the weight on the wheels and the tire penetration into the ground.

$$R = R_f W + R_p pW \qquad (13.1)$$

where R = rolling resistance, lb

R_f = rolling-resistance factor, lb/ton

W = weight on wheels, tons

R_p = tire-penetration factor, lb/ton · in penetration

p = tire penetration, in

R_f usually is taken as 40 lb/ton (or 2% lb/lb) and R_p as 30 lb/ton · in (1.5% lb/lb · in). Hence Eq. (13.1) can be written as

$$R = (2\% + 1.5\%p)W' = R'W' \qquad (13.2)$$

where W' = weight on wheels, lb

$R' = 2\% + 1.5\%p$ (see Table 13.7)

Table 13.7 Typical Total Rolling Resistances of Wheeled Vehicles

Surface	Lb per ton	Lb per lb
Hard, smooth, stabilized, surfaced roadway without penetration under load, watered, maintained	40	0.020
Firm, smooth roadway, with earth or light surfacing, flexing slightly under load, maintained fairly regularly, watered	65	0.033
Snow:		
Packed	50	0.025
Loose	90	0.045
Earth roadway, rutted, flexing under load, little if any maintenance, no water	100	0.50
Rutted earth roadway, soft under travel, no maintenance, no stabilization	150	0.75
Loose sand or gravel	200	1.00
Soft, muddy, rutted roadway, no maintenance	300–400	1.50–2.00

Additional power is required to overcome rolling resistance on a slope. Grade resistance also is proportional to weight.

$$G = R_g sW \qquad (13.3)$$

where G = grade resistance, lb

R_g = grade-resistance factor = 20 lb/ton = 1% lb/lb

s = percent grade, positive for uphill motion, negative for downhill

Thus, the total road resistance is the algebraic sum of the rolling and grade resistances, or the total pull, lb, required:

$$T = (R' + R_g s)W' = (2\% + 1.5\%p + 1\%s)W' \qquad (13.4)$$

In addition, an allowance may have to be made for loss of power with altitude. If so, allow 3% pull loss for each 1000 ft above 2500 ft.

Usable pull P depends on the weight W on the drivers:

$$P = fW \qquad (13.5)$$

where f = coefficient of traction (Table 13.8). (See also Art. 13.12.)

Table 13.8 Approximate Traction Factors*

Traction surface	Traction factors	
	Rubber tires	Tracks
Concrete	0.90	0.45
Clay loam, dry	0.55	0.90
Clay loam, wet	0.45	0.70
Rutted clay loam	0.40	0.70
Loose sand	0.30	0.30
Quarry pit	0.65	0.55
Gravel road (loose not hard)	0.36	0.50
Packed snow	0.20	0.25
Ice	0.12	0.12
Firm earth	0.55	0.90
Loose earth	0.45	0.60
Coal, stockpiled	0.45	0.60

* See also Art. 13.12.

Earth Quantities Hauled. When soils are excavated, they increase in volume, or swell, because of an increase in voids (Table 13.9).

$$V_b = V_L L = \frac{100}{100 + \% \text{ swell}} V_L \qquad (13.6)$$

Table 13.9 Load Factors for Earthmoving

Swell, %	Voids, %	Load factor	Swell, %	Voids, %	Load factor
5	4.8	0.952	5.3	5	0.95
10	9.1	0.909	11.1	10	0.90
15	13.0	0.870	17.6	15	0.85
20	16.7	0.833	25.0	20	0.80
25	20.0	0.800	33.3	25	0.75
30	23.1	0.769	42.9	30	0.70
35	25.9	0.741	53.8	35	0.65
40	28.6	0.714	66.7	40	0.60
45	31.0	0.690	81.8	45	0.55
50	33.3	0.667	100.0	50	0.50
55	35.5	0.645			
60	37.5	0.625			
65	39.4	0.606			
70	41.2	0.588			
75	42.9	0.571			
80	44.4	0.556			
85	45.9	0.541			
90	47.4	0.526			
95	48.7	0.513			
100	50.0	0.500			

where V_b = original volume, yd^3, or bank yards

V_L = loaded volume, yd^3, or loose yards

L = load factor (Tables 13.9 and 13.10)

When soils are compacted, they decrease in volume.

$$V_c = V_b S \qquad (13.7)$$

where V_c = compacted volume, yd^3

S = shrinkage factor

Bank yards moved by a hauling unit equals weight of load, lb, divided by density of the material in place, lb per bank yard.

Table 13.10 Percentage Swell and Load Factors of Materials

Material	Swell, %	Load Factor
Cinders	45	0.69
Clay:		
Dry	40	0.72
Wet	40	0.72
Clay and gravel:		
Dry	40	0.72
Wet	40	0.72
Coal, anthracite	35	0.74
Coal, bituminous	35	0.74
Earth, loam:		
Dry	25	0.80
Wet	25	0.80
Gravel:		
Dry	12	0.89
Wet	12	0.89
Gypsum	74	0.57
Hardpan	50	0.67
Limestone	67	0.60
Rock, well blasted	65	0.60
Sand:		
Dry	12	0.89
Wet	12	0.89
Sandstone	54	0.65
Shale and soft rock	65	0.60
Slag, bank	23	0.81
Slate	65	0.60
Traprock	65	0.61

13.10 Scraper Production

Production is measured in terms of tons or bank cubic yards of material a machine excavates and discharges, under given job conditions, in 1 hour.

$$\text{Production, bank yd}^3/\text{h} \qquad (13.8)$$

$$= \text{load, yd}^3 \times \text{trips per hour}$$

$$\text{Trips per hour} = \frac{\text{working min/h}}{\text{cycle time, min}} \qquad (13.9)$$

The load, or amount of material a machine carries, can be determined by weighing or estimating the volume. Payload estimating involves determination of the bank cubic yards being carried, whereas the excavated material expands when loaded into the machine. For determination of bank cubic yards from loose volume, the amount of swell or the load factor must be known (Tables 13.9 and 13.10); then the conversion can be made by use of Eq. (13.6).

Weighing is the most accurate method of determining the actual load. This is normally done by weighing one wheel or axle at a time with portable scales, adding the wheel or axle weights, and subtracting the weight empty. To reduce error, the machine should be relatively level. Enough loads should be weighed to provide a good average.

$$\text{Bank yd}^3 = \frac{\text{weight of load, lb}}{\text{density of material, lb/bank yd}^3}$$

$$(13.10)$$

For Eq. (13.9), cycle time, the time to complete one round trip, may be measured with a stopwatch. Usually, an average of several complete cycles is taken. Sometimes, additional information is desired, such as load time and wait time, which indicate loading ability and job efficiency, so the watch is kept running continuously and the times for beginning and ending certain phases are recorded. Table 13.11 is an example of a simple time-study form. It can easily be modified to include other segments of the cycle, such as haul time and dump time, if desired. Similar forms can be made for pushers, bulldozers, and other equipment.

Wait time is the time a unit must wait for another machine so that the two can work together, for example, a scraper waiting for a pusher. Delay time is any time other than wait time when a

Table 13.11 Cycle-Time Observations

Total cycle times (less delays), min	Arrive cut	Wait time	Begin load	Load time	End load	Begin delay	Delay time	End delay
	0.00	0.30	0.30	0.60	0.90			
3.50	3.50	0.30	3.80	0.65	4.45			
4.00	7.50	0.35	7.85	0.70	8.55	9.95	1.00	10.95
4.00	12.50	0.42	12.92	0.68	13.60			

machine is not working, for example, a scraper waiting to cross a road.

Since cycle time is involved in computation of production [Eq. (13.8)], different types of production may be measured, depending on whether cycle time includes wait or delay time. Measured production includes all waits and delays. Production without delays includes normal wait time but no delay time. For maximum production, wait time is minimized or eliminated and delay time is eliminated. Cycle time may be further altered by using an optimum load time, as determined from a load-growth study. (See also Art. 13.13.)

Example 13.1: A job study of wheeled scrapers yields the following data:

Weight of haul unit empty, 44,800 lb. Average of three weighings of haul unit loaded, 81,970 lb. Density of material to be excavated, 3140 lb/bank yd^3.

Average wait time, 0.28 min; average delay time, 0.25 min; average load time, 0.65 min; average total cycle time, less delays, 7.50 min.

What will be the production of the haul unit?

Average load will be $81,970 - 44,800$ lb $= 37,170$ lb. This load is equivalent to $37,170/3140 = 11.8$ bank yd^3. In 60 working min/h, the scraper will make $60/7.50 = 8.0$ trips per hour. Hence, production (less delays) will be $11.8 \times 8.0 = 94$ bank yd^3/h.

Equipment Required ▪ To determine the number of scrapers needed on a job, required production must first be computed.

Production required,

$$\frac{yd^3}{h} = \frac{quantity, \ bank \ yd^3}{working \ time, \ h} \quad (13.11)$$

No. of scrapers needed

$$= \frac{production \ required, \ yd^3/h}{production \ per \ unit, \ yd^3/h} \quad (13.12)$$

No. of scrapers a pusher will load

$$= \frac{scraper \ cycle \ time, \ min}{pusher \ cycle \ time, \ min} \quad (13.13)$$

For computation of rolling resistance, see Art. 13.9; for traction, see Art. 13.12.

("Earthmoving Data," Caterpillar Tractor Co.)

13.11 Bulldozer Production

Production normally is measured by bank cubic yards dozed per hour. Because of the number of variables involved, determination of bulldozer production is difficult. A simplified method can provide a satisfactory estimate:

Two workers, using a 50-ft tape measure, can determine bulldozer payloads on the job. The bulldozer pushes its load onto a level area, stops, raises the blade while moving slightly forward, then reverses to clear the pile. The workers measure height, width, and length of the pile. To determine the average height, one worker holds the tape vertically at the inside edge of each grouser mark. The second worker, on the other side of the pile, aligns the tape with the top of the pile. To measure average width and length, the two hold the tape horizontally and sight on each end of the pile. Sighting to the nearest tenth of a foot is sufficiently accurate. Multiplication of the dimensions yields the loose volume in cubic feet; division by 27, in cubic yards. Application of a load factor [Eq. (13.6) and Tables 13.9 and 13.10] gives the loads in bank cubic yards.

13.12 Traction

This normally is measured by the maximum drawbar pull or rim pull, lb, a tractor exerts before the tracks or driving wheels slip and spin. When computing pull requirements for track-type tractors, rolling resistance does not apply to the tractor, only to the trailed unit. Since track-type tractors move on steel wheels rolling on steel "roads," rolling resistance is relatively constant and accounted for in the drawbar-pull rating.

Traction depends on the weight on driving tracks or wheels, gripping action with the ground, and condition of the ground. The coefficient of traction (Table 13.8) is the ratio of the maximum pull, lb, a tractor exerts when on a specific surface to the total weight on the drivers.

Example 13.2: What usable drawbar pull can a 59,100-lb tractor exert while working on firm earth? On loose earth?

The solution can be obtained with Eq. (13.5) and Table 13.8:

Firm earth: $P = 0.90 \times 59,100 = 53,200$ lb

Loose earth: $P = 0.60 \times 59,100 = 35,500$ lb

If 48,000 lb were required to move a load, then this tractor could move it on firm earth, but on loose earth the tracks would spin.

Example 13.3: What usable rim pull can a wheeled tractor-scraper exert while working on firm earth? On loose earth? Assume the weight distribution for the loaded unit as 49,670 lb on the drive wheels and 40,630 lb on the scraper wheels.

The solution can be obtained with Eq. (13.5) and Table 13.8. Use the weight on the drivers only.

Firm earth: $P = 0.55 \times 49,670 = 27,320$ lb

Loose earth: $P = 0.45 \times 49,670 = 22,350$ lb

If 25,000 lb were required to move a load and the engine were sufficiently powerful, the tractor-scraper could move the load on firm earth, but the drivers would slip on loose earth.

Equipment specification sheets show how many pounds pull a machine can exert in a given gear at a given speed. But if the engine works at high altitudes, it cannot produce as much power as it is rated for at sea level because of the decrease in air density. Up to 2500 ft above sea level, the reduction is insignificant. For each 1000 ft above 2500 ft an engine loses about 3% of its horsepower. But some machines with turbocharged engines operate at altitudes above 2500 ft without loss of power, so consult service literature on a machine before de-rating for altitude.

("Earthmoving Data," Caterpillar Tractor Co.)

13.13 How to Estimate Cycle Time and Job Efficiency

Before production on an earthmoving job can be estimated, cycle time for the equipment must be known [Eqs. (13.8) and (13.9)]. Cycle time is the time required to complete one round trip in moving material. Different approaches are used in estimating cycle time for each type of machine.

Track-Type Tractor-Scrapers ▪ Cycle time is the sum of fixed times and variable times. Fixed times in scraper work are the number of minutes for loading, turning, dumping, and in bulldozing, for gear shifting. Variable times comprise haul and return times. Experience shows that the fixed times in Table 13.12 are satisfactory for estimating purposes.

Since speeds and distances may vary on haul and return, haul and return times are estimated separately.

Variable time, min   (13.14)

$$= \frac{\text{haul distance, ft}}{88 \times \text{speed, mi/h}} + \frac{\text{return distance, ft}}{88 \times \text{speed, min/h}}$$

Haul speed may be obtained from the equipment specification sheet when the drawbar pull required is known.

Wheel-Type Tractor-Scrapers ▪ The procedure for estimating cycle time for wheel- and track-type tractors is about the same. But for wheel-type tractors, time consumed in acceleration and deceleration must be included in the estimate of fixed time. The values given in Table 13.12 may be used for estimating.

To determine the haul speed of a wheel-type tractor-scraper, it is necessary to match the rim pull required (total road resistance) against rim pull available (obtained from equipment specifications) and select a reasonable operating gear (from the specifications). Equation (13.14) may be used to

Table 13.12 Fixed Times for Estimating Cycle Time, min

Track-type Tractor and Scraper			
	Self-loaded		Push loaded
	$15\ yd^3$ or more	$14\ yd^3$ or less	$15\ yd^3$ or more
Loading	1.5	1.0	1.0
Dumping, turning	1.0	1.0	1.0
Total fixed time	2.5	2.0	2.0

Track-type Tractor and Bulldozer	
Shuttle bulldozing using same gear and shifting only forward-reverse lever	0.1
Shuttle bulldozing shifting to higher reverse gear	0.2
Power-shift tractors	0

Track-type Tractor-Shovels (Fixed time for loading, turning, dumping)		
	Manual Shift	Power Shift
Bank or stockpile loading	0.35	0.25
Excavation	0.61	0.43

Wheel-type Tractor-Shovels	
Stockpile loading, power shift	0.20

Wheel-type Tractor-Scrapers with Pushers			
	5th-gear hauls	4th-gear hauls	3d-gear hauls
Loading	1.0	1.0	1.0
Maneuver and spread	0.5	0.5	0.5
Acceleration and deceleration	1.5	0.8	0.4
Total fixed time	3.0	2.3	1.9

figure variable time. The sum of fixed and variable times gives the estimated cycle time.

Power loss due to altitude is taken into account by dividing the total road-resistance factor [Eq.

Table 13.13 Efficiency Factors for Average Job Conditions*

	Min per working h	Efficiency factor
Day Operation		
Track-type tractor	50	0.83
Wheel-type tractor	45	0.75
Night Operation		
Track-type tractor	45	0.75
Wheel-type tractor	40	0.67

* These take into consideration only minor delays. No time is included for major overhauls and repairs. Machine availability and weather also should be taken into account.

(13.4)] by a correction factor k. The resulting effective resistance factor then is used to compute travel time.

$$k = 1 - 0.03\frac{H - 2500}{1000} \qquad (13.15)$$

where H = altitude above sea level, ft. Travel time can be determined from data supplied by the scraper manufacturer.

Job efficiency depends on many variables, including operator skill, minor repairs and adjustments, delays caused by personnel, and delays caused by job layout. Table 13.13 lists approximate efficiency factors for estimating when job data are unavailable. Production, cubic yards per working hour, then equals production, yd^3/h, times the efficiency factor.

13.14 Mass Diagram

This is a graph showing the accumulation of cut and fill with distance from a starting point, or origin. Cut usually is considered positive and fill negative. The volume of each is plotted in cubic yards. Distance normally is measured, along the center line of the construction, in stations 100 ft apart, starting with the origin as $0 + 00$. Swell factors are applied to the cuts and shrinkage factors to the embankments [Eqs. (13.6) and (13.7)] to obtain bank cubic yards excavated and compacted fill, respectively.

Figure 13.14b shows a mass diagram for the profile in Fig. 13.14a (shrinkage factor of 10% and

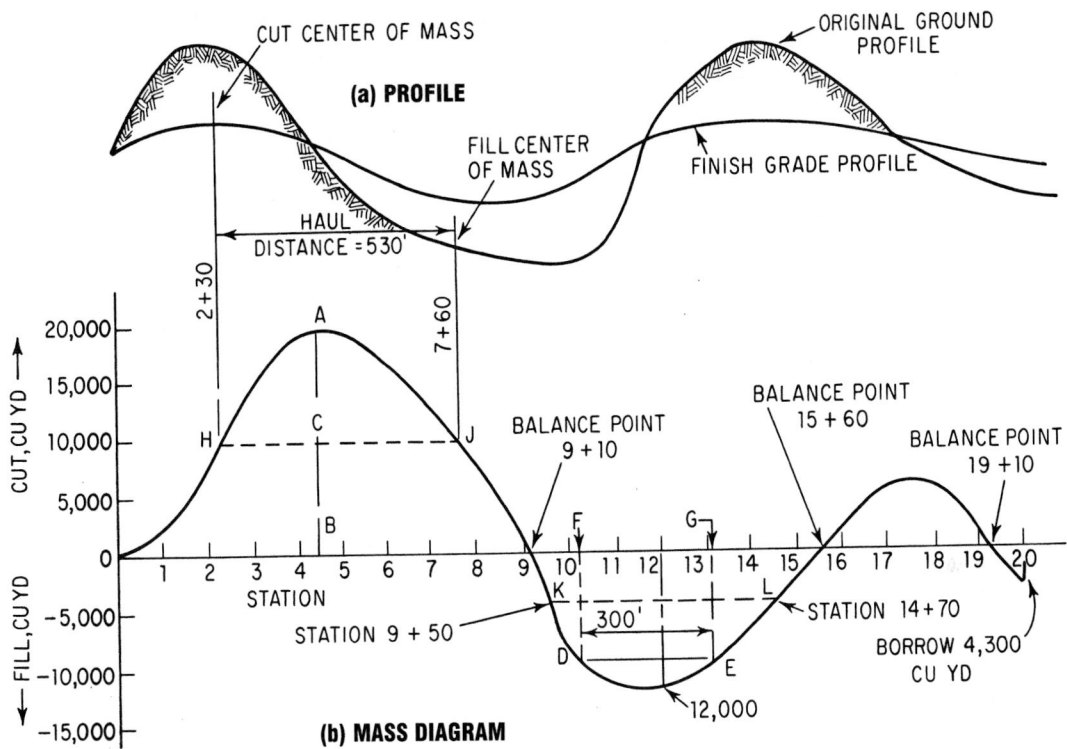

Fig. 13.14 Profile and mass diagram for cut and fill for grading a highway.

swell factor of 20% included). Between 0 + 00 and 1 + 00, there is a cut of 2000 yd³. This is plotted at 1 + 00. Between 1 + 00 and 2 + 00, there is a cut of 5000 yd³, making a total of 7000 yd³ between 0 + 00 and 2 + 00; 7000 is plotted at 2 + 00. At 4 + 00, there is a total accumulation of 18,000 yd³ of cut. Between 4 + 00 and 5 + 00, there are 1000 yd³ of cut and 550 yd³ of embankment (corrected for shrinkage), making a net accumulation of

$$18,000 + 1000 - 550 = 18,450 \, \text{yd}^3$$

From 6 + 00 to 12 + 00, there is mostly embankment, and the accumulation decreases to −12,000 yd³. Cut follows, then some more embankment. At the end of the construction, 20 + 00, there is a net of −4300 yd³ embankment that must be obtained from borrow.

If a mass curve is horizontal between stations, the implication is that no material has to be moved

in that stretch. Actually, there may be cuts and fills but they balance. If work consists of side-hill cuts and fills, the mass diagram tends to flatten because the cuts can be moved into the fills and not moved from one station to another. Moving excavation from one side of the center line to the other is called **cross haul**.

The slope of the mass curve increases with volume between stations. An ascending mass curve indicates cut; a descending diagram, fill. The curve reaches a maximum where cut ends and fill begins, and a minimum where fill ends and cut begins.

If a mass diagram is intersected by a horizontal line, cuts balance fills between the points of intersection. If the mass curve loops above the line, cuts will have to be hauled forward (in the direction of increasing stations) for the embankments; if the diagram lies below the horizontal line, the haul will have to be backward.

Haul, station-yards, for a section of earthwork is the product of the amount of excavation, cubic yards, and the distance it is moved, stations. Total haul is the product of total amount of excavation hauled and average haul distance. The area between the mass diagram and a balancing (horizontal) line equals the haul, station-yards, between the two points cut by that line. Average haul distance equals the area between the mass diagram and the balancing line divided by the total cut (maximum ordinate) between the points of intersection.

Center of mass of cut and fill can be determined from the mass diagram. Draw the maximum ordinate between a balancing line and the curve (for example, BA in Fig. 13.14b). Then, draw a horizontal line (HJ) through the midpoint of that ordinate, and note the stations at the points of intersection with the curve. The station (H) on the increasing portion of the diagram is the center of mass of cut; the station (J) on the decreasing portion, the center of fill. The distance between the stations is the haul distance.

If the mass curve terminates below the horizontal axis, borrow is required. If the curve ends above the axis, excavation must be wasted.

Free haul is the distance excavation may be moved without an increase in contract price; that is, the unit bid price for excavation applies only to haul distances less than free haul. Overhaul is haul distance exceeding free haul. The bid price for overhaul usually is given in terms of dollars per station-yard.

Example 13.4: For Fig. 13.14, if free haul is 300 ft. determine the overhaul between $9 + 10$ and $15 + 60$.

Draw horizontal line DE with length 300 ft between two points on the mass curve. Draw ordinates FD at D and GE at E. These vertical lines set the limits of free haul. Next, the center of mass of cut and fill outside these limits must be found. To do this, draw a horizontal line through the midpoints of FD and GE intersecting the mass curve at K and L. The center of mass of cut is at L, $14 + 70$, and of fill, at K, $9 + 50$. $KL = 5.2$ stations represents the average haul distance. Hence, the overhaul equals the product of $DF = 9500$ yd^3 and KL less the free-haul distance ($5.2 - 3.0$), or 20,900 station-yards.

(C. F. Allen, "Railroad Curves and Earthwork," McGraw-Hill Book Company, New York.)

13.15 Drilling for Rock Excavation

Usually, before rock can be excavated, it must be blasted into pieces small enough for efficient handling by available equipment. To place explosive charges for this purpose, holes have to be drilled into the rock. This is done with percussion or rotary drills. Percussion generally is used for hard rock and small-diameter holes. Maximum size of bit for percussion drills is about 6 in. Larger bits may be used on rotary drills (Fig. 13.15), but they rarely exceed 9 in in diameter.

Normally, percussion drills are mounted on self-propelled crawlers (Fig. 13.16). Drilling commonly is done with sectional drill steel and carbide-insert bits, both of which have to be rugged. A bit has first to crush its way into the rock. Next, the hole must be reamed. Finally, the cuttings are mixed and

Fig. 13.15 Track-mounted rotary drill with air compressor. (*Caterpillar Tractor Co.*)

Fig. 13.16 Percussion drill powered by an air compressor and mounted on a tractor. (*Caterpillar Tractor Co.*)

blown from the hole by compressed air fed through a hole in the center of the drill steel and discharged through holes in the bit. Hard rock requires a bit with good crushing or penetrating and reaming ability. Shales, usually soft, require a bit that mixes material fast. The bit does not need to have good crushing ability. For sandstone, the gage will usually be destroyed first or the bit will lose its reaming ability. A bit used in sandstone must have exceptional reaming ability plus good mixing features.

The best way to determine how well a bit is performing is to inspect the chips. They should be firm pieces of rock, not dust. When cuttings are dust, usually the chips are not being blown from the hole until they have been reground several times. This causes more than normal bit wear. Low air pressure also may produce excessive dust. Pressure at the drill should be at least 90 psi. When computing the pressure, take into account the drop in pressure due to friction in the hose.

Rotary drilling is more suitable for large holes. With low-cost ammonium nitrate and fuel oil as the explosive, economical production results. With large holes, spacing can be greater and more cubic yards can be produced per foot of hole. When determining whether to use large or small holes, the engineer should bear in mind that the amount of explosives is directly proportional to the area of the hole.

In rotary drilling, it is essential to maintain sufficient down pressure, rotation speed, and volume and pressure of air used to blow cuttings out of the hole or excessive bit wear and low production results. Down pressure should be at

least 5000 psi/in of bit diameter. Rotation speed should be the largest possible without regrinding chips before they are blown out of the hole. Therefore, rotation speed depends on air volume. Air is blown through the center of the drill steel and discharged through passages in the bit. Except in extremely deep holes, 40 psi usually is enough pressure to clean holes.

13.16 Explosives for Rock Excavation

Explosives are used to blast rock into pieces small enough to be handled efficiently by available equipment. The charges usually are set in holes drilled into the rock (Art. 13.15) and detonated.

If the reaction is instantaneous or extremely rapid over the entire mass of the explosive, detonation has occurred. Deflagration, however, takes place when the reaction particles move away from the unreacted particles or the material burns. The basic difference between these two reactions is that detonation produces a high-pressure shock wave that is self-propagating throughout the charge.

Several factors contribute to the effectiveness of an explosive charge: confinement, density, most efficient uniform propagation diameter, and critical mass.

Confinement helps the reacted products contribute to detonation of the unreacted products. If the reacted portions can escape, the reactions will cease. An air space can be very effective in dampening a reaction.

The denser the mass of the charge, the more effective it will be, up to a point. For every explosive, there is an optimum density. Since drilling costs more than explosives per cubic yard of excavation, it is desirable to use as many pounds of explosive per foot of borehole as possible.

The most efficient uniform propagation diameter is the width or length over which the explosive mass will be self-propagating after detonation starts. This length ranges from very small to about 9 in for ammonium nitrate.

The self-propagating diameter can be lowered by the overdrive method. Overdrive is the ability of an explosive to detonate at a rate greater than the self-propagating detonating rate. Suppose, for example, an explosive that detonates at 21,000 ft/s is set off in contact with another type of explosive that detonates at 12,000 ft/s. Then, the slower explosive will detonate at more than 12,000 ft/s but less than 21,000 ft/s for a given distance, usually less than 2 ft.

Sensitivity of an explosive is very important from a safety standpoint. An explosive should be easy to detonate by specific methods, but hard or impossible to set off with normal or careful handling during manufacture, shipment, storage, and preparation for detonation.

Critical mass is that amount of an explosive that must be present for the reaction to change from deflagration to detonation. This mass is very small for high-order explosives but about 123 tons for ammonium nitrate.

Explosive manufacturers generally balance the ingredients of their products to get maximum gas volume. This usually depends on the amount of oxygen available from an unstable oxidizer in the explosive. A combination of gas ratio and brisance (shattering effect) is called power factor. Explosive ingredients can be combined many ways to provide almost any power factor.

Rate of detonation is a rough measure of the shattering ability of an explosive. Mass formations of rock require a rate of at least 12,000 ft/s. Maximum detonating rate for commercial explosives is 26,000 ft/s.

Explosive strength generally is rated by the percent of nitroglycerin or equivalent in explosive power contained in an explosive. Straight dynamites contain only nitroglycerin and an inert ingredient. In an ammonia dynamite, some of the nitroglycerin is replaced by other ingredients, such as ammonium nitrate. Explosive power may be denoted by weight strength or bulk or cartridge strength. When weight strength is given, an ammonia dynamite will have the same explosive power as a straight dynamite of the same strength. Following are important features of explosives commonly used in construction:

Gelatin Dynamites ▪ Weight strength from 100 to 60%. Detonation rate from 26,200 to 19,700 ft/s, respectively. Suitable for submarine blasting or for use where considerable water pressure will be encountered. Inflammable. Has high shattering action.

Gelatin Extras ▪ Weight strength from 80 to 30%. Detonation rate from 24,000 to 15,000 ft/s, respectively. Ammonium nitrate replaces part of nitroglycerin. Gelatin extras have less water resistance than gelatins but can be used satisfactorily except under the most severe conditions.

Extra Dynamites ▪ Weight strength from 60 to 20%. Detonation rate from 12,450 to 8200 ft/s. Ammonium nitrate replaces part of nitroglycerin. Extra dynamites can be used in average water conditions if properly wrapped with waterproofing. They usually are called original ammonia dynamites.

Semigelatins ▪ Weight strength from 65 to 40%; bulk strength from 65 to 30%. Detonation speed from 17,700 to 9850 ft/s. Higher detonation speeds for larger-diameter cartridges. Can be used instead of gelatins in most blasting uses. Water resistance is adequate for average conditions.

High-Ammonium-Nitrate-Content Dynamites ▪ Weight strength from 68 to 46%; bulk strength from 50 to 20%. Detonation speed from 10,500 to 5250 ft/s. Has low water resistance but can be used if fired within a relatively short time of exposure.

Boosters or Primers ▪ Have high density. Detonation speed of 25,000 ft/s. Used to detonate ammonium nitrates and fuel oil or any non-cap-sensitive explosive because boosters and primers have a very high detonation pressure.

Detonating Cord ▪ Used as a fuse. Has high-explosive core that detonates at 21,000 ft/s with sufficient energy to detonate another, less sensitive explosive alongside in a borehole. When strung from top to bottom of a hole, detonating cord will act as a detonating agent throughout the length of the hole.

Ammonium nitrate, for best results, should be mixed with at least 6% fuel oil, by weight. The oil is added for oxygen balancing and to lower the self-propagating diameter. Quantities of fuel oil greatly in excess of 6% have a dampening effect on the explosion. By use of the overdrive method, the rate of detonation for ammonium nitrate and fuel oil will be sufficient to shatter any rock formation encountered. Ammonium nitrate plus 10% booster has a rate of 4500 to 10,000 ft/s; when fuel oil is added, the rate increases to 10,000 to 16,500 ft/s. For overdrive, best results are obtained with at least 5% of a primer with a high detonation rate. The primers should be properly spaced to ensure that critical propagation length will not be exceeded and detonation will occur throughout.

Special precautions should be observed when overdrive is used. If free fuel oil is available in the mixture, an ammonia dynamite should not be used as a primer. Fuel oil will desensitize ammonia dynamite, and a partial or complete failure will result. Fuel oil also has an adverse effect on the explosive contained in detonating cord. This, however, can be avoided by using a plastic coating on the cord.

Table 13.14 gives the approximate amount of ammonium nitrate to use per foot of borehole. The table assumes a density of 47 lb/ft³ for ammonium nitrate and fuel oil.

Ammonium nitrate is soluble in water. It develops some water resistance when mixed with fuel oil. But exposure to water results in loss of efficiency, and detonation becomes difficult.

Table 13.14 Amount of Ammonium Nitrate per Foot of Borehole

Hole dia., in	Approx. weight, lb per ft	Approx. volume, ft³ per ft
2	1.02	0.0218
2¼	1.29	0.0275
2½	1.59	0.034
3	2.30	0.049
3¼	2.67	0.057
3½	3.00	0.064
4	4.09	0.087
4½	5.17	0.110
5	6.39	0.136
5½	7.75	0.165
6	9.21	0.196
6¼	10.01	0.213
6½	10.81	0.230
6⅞	12.03	0.256
7	12.54	0.267
7¼	13.44	0.286
7⅞	15.79	0.336
8	16.40	0.349
8½	18.51	0.394
9	20.72	0.441
9½	23.12	0.492
10	25.61	0.545
10½	28.24	0.601
11	30.97	0.659
11½	33.88	0.721
12	36.89	0.785

13.17 Rock Excavation by Blasting

To secure the desired shape of rock surface after blasting, explosive charges must be placed in boreholes laid out in the proper pattern and of sufficient depth. (See also Arts. 13.15 and 13.16.) Before the pattern is chosen, an explosive factor must be selected (Table 13.15).

Next, drill size, burden, and spacing can be selected. Then, the amount of stemming can be determined. **Stemming** is the top portion of a borehole that contains a tightly tamped backfill, not explosive. Since an explosive exerts equal pressure in all directions, depth of stemming should not exceed the width of burden. **Burden** is the distance from the borehole to the rock face. Burden distance should be less than the hole spacing so that the blasted rock will be thrown in the direction of the burden.

Holes should be placed in lines parallel to the rock face because a rectangular pattern gives better breakage and vibration control. Depth of drill holes is determined by height of face desired and the distance it is necessary to drill below grade so that the bottom can be controlled.

A mathematical check should be made to determine that the explosive factor is correct for the burden and spacing selected. If properly blasted rock is not produced when a drill pattern is tried, a new spacing or burden width should be tried. It is best to vary only one dimension at a time until desired fragmentation is obtained.

Delay caps may be used on the explosive charges for better fragmentation and vibration control. Delay caps permit detonation of explosive charges in different holes at intervals of a few milliseconds. The result is better fragmentation,

Table 13.15 Explosive Factors

Types of rock	Explosive factor, lb/yd³
Shales	0.25–0.75
Sandstone	0.30–0.60
Limestone	0.40–1.00
Granite	1.00–1.50

controlled throw, and less back break since better displacement is obtained. Table 13.16 gives characteristics of short-period delay caps. Use of regular delays is not recommended because of "hole robbing" and uncontrolled throw.

Presplitting is a technique for producing a reasonably smooth, nonshattered wall, free from loose rock. An objective is to hold maintenance of slopes and ditches to a minimum. Presplit holes are drilled in a single line in a plane that will be the final slope or wall face. Line drilling also may be used, with holes spaced about two times the bit diameter. But for presplitting, the spacing is much greater. Dynamite, evenly spaced on detonating cord, is exploded to break the web between holes. Manufacturers can furnish explosives made for

presplitting. When this type of explosive is used, loading of holes is easier since no detonating cord is required. The resulting saving of labor will usually more than offset additional explosive costs.

Percussion drills commonly are used for drilling presplitting holes. An air track with hydraulic controls is very effective in enabling the driller to move from hole to hole and reset the drill in a minimum time. Number of drills required varies with capacity of loading shovel, width of cut, and spacing of presplit holes.

For presplitting, 40% extra gelatin works satisfactorily. This explosive has a detonation speed that can break the hardest rock formations and is adequate under the most adverse conditions. Speed of detonation should not be less than 15,000 ft/s for presplitting.

Figure 13.17a shows a presplit hole loaded with $1\frac{1}{4} \times 8$-in cartridges spaced 18 to 24 in apart on

Table 13.16 Characteristics of Millisecond Delay Caps*

Delay period	Nominal firing time, ms	Interval between delay periods, ms
0	12	
SP-1	25	13
SP-2	50	25
SP-3	75	25
SP-4	100	25
SP-5	135	35
SP-6	170	35
SP-7	205	35
SP-8	240	35
SP-9	280	40
SP-10	320	40
SP-11	360	40
SP-12	400	40
SP-13	450	50
SP-14	500	50
SP-15	550	50
SP-16	600	50
SP-17	700	100
SP-18	900	200
SP-19	1100	200
SP-20	1300	200
SP-21	1500	200
SP-22	1700	200
SP-23	1950	250
SP-24	2200	250
SP-25	2450	250
SP-26	2700	250
SP-27	2950	250

* Courtesy of Hercules Powder Co.

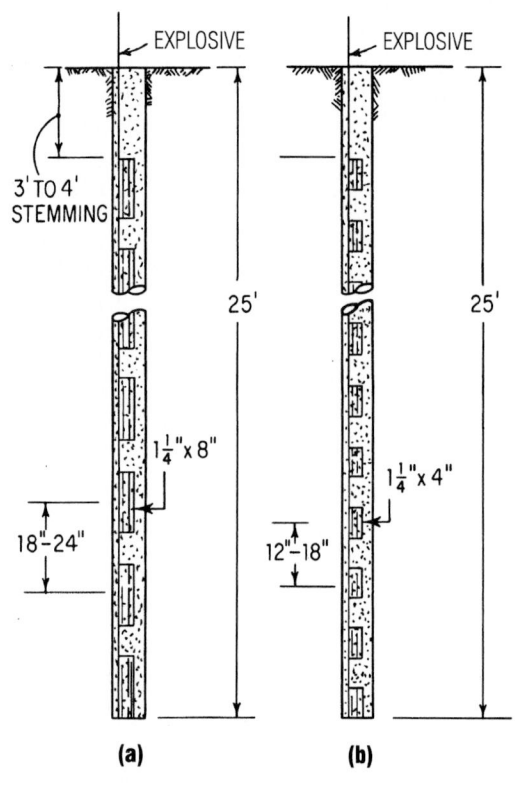

Fig. 13.17 Drill holes loaded with (a) $1\frac{1}{4} \times 8$-in catridges and (b) $1\frac{1}{4} \times 4$-in catridges on detonation cord for presplitting. Prepackaged explosives are available from explosive manufacturers.

Table 13.17 Pounds of 40% Gelatin Extra to Produce 2500 ft² of Wall by Presplitting

Hole Spacing	1¼ × 8-in cartridges		1¼ × 4-in cartridges	
	18 in c to c	24 in c to c	12 in c to c	18 in c to c
18	362	272	272	181
24	270	203	203	135
30	215	161	161	108
36	178	134	134	89
42	152	114	114	76
48	132	99	99	66
54	117	88	88	58
60	105	79	79	52
66	95	71	71	47
72	86	64	64	43

primacord; Fig. 13.17b shows 1¼ × 4-in cartridges on 12- to 18-in spacing. Table 13.17 indicates the number of pounds of 40% gelatin extra required to produce a wall 25 ft high by 100 ft long.

Presplitting should precede the primary blast. Some locations, however, preclude this; for example, a side hill where there would not be sufficient burden in front of the presplit holes. In such a case, presplitting will be accomplished, but the burden in front will be shifted, causing loss of primary blast holes or difficult drilling if the holes were not drilled previously. If a sidehill condition exists, delay caps should be used to ensure that presplitting is done before detonation of the primary blast.

Spacing of holes for presplitting varies considerably with material, location, and method of primary blasting. Spacings up to 6 ft have been found adequate where no restrictions are imposed on explosives and primary blasting can be adjusted to obtain correct balance for removal of material within the walls. Obtaining a good wall is the result of balancing primary blasting with as wide a spacing as possible for the type of rock. Use of close spacing of holes without consideration of other factors may be wasteful and not yield best overall results.

Spacing of presplit holes and charges for best results may be determined by trial. Vary only one variable at a time. For example, initially drill the holes for 25 ft of wall 18 in apart and detonate. Then, for the next 25 ft of wall, drill the holes 24 in on centers and detonate with the same loading. Continue increasing the spacing until a maximum is reached. Next, vary the charge. If too much dynamite is used, the resulting surface between

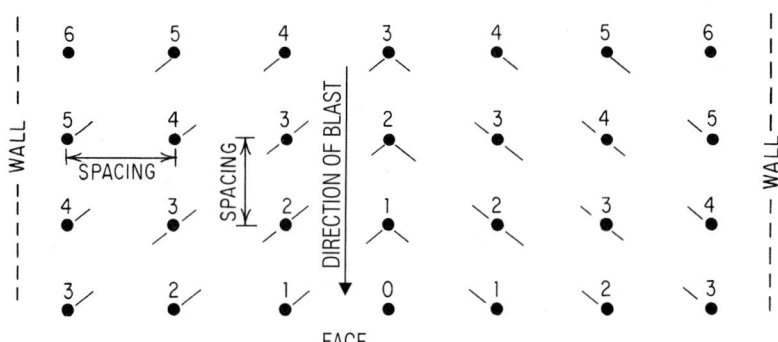

Fig. 13.18 Drill pattern for conventional blasting with holes all of the same diameter. Numbers indicate order of firing with delays.

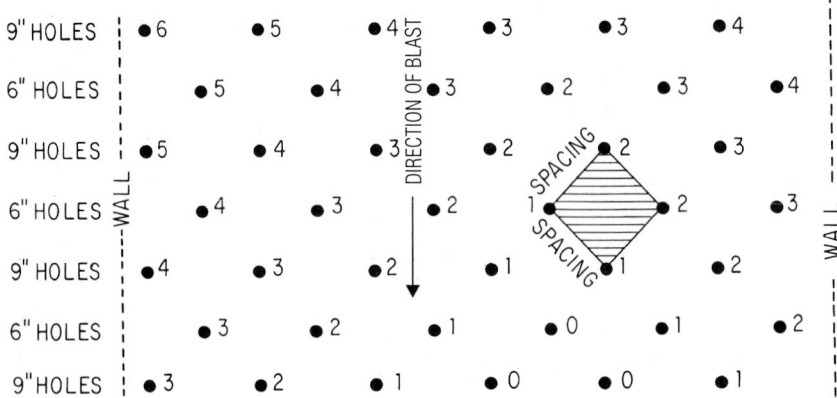

Fig. 13.19 Drill pattern for conventional blasting with two sizes of holes. Numbers indicate order of firing with delays.

holes will be concave. Conversely, with insufficient dynamite, the surface will be convex.

In **conventional blasting**, placing of delays in the primary blast is important. The more relief that can be given to holes near the wall, the less opportunity for damage to the wall (Figs. 13.18 and 13.19 and Tables 13.18 and 13.19).

The depth of each lift in presplitting is governed by the size of the shovel excavating equipment. Lifts generally average 20 to 25 ft. The last lift may be deeper, to reach grade in one setup. For efficiency, each lift should be presplit separately. Drilling speed diminishes rapidly as a 40-ft depth is approached.

When more than one lift is required, the drill has to be set up for successive lifts at least 1 ft away from the face, to provide clearance for drilling (Fig. 13.20).

Loading of deep holes, particularly if they contain water, can be very difficult. Stringing sticks of dynamite on a long detonating cord can exceed the structural strength of the cord, breaking it and causing a misfire.

After holes have been drilled, dynamite cartridges are fastened to a detonating cord, usually 50 grain, long enough to reach the bottom of the hole. Spacing of charges on the cord varies with rock formation and hole spacing. Charges may be attached with tape or rubber bands. With rubber bands, spacing is easier to maintain because the charges do not slip so easily. In a limestone formation with holes drilled at 4-ft intervals, $1\frac{1}{4} \times 8$-in charges spaced 18 in on centers have been

Table 13.18 Powder Factor for Drill Pattern of Fig. 13.18

Spacing of holes, ft	Burden, yd^3	Powder factor*
For 9-in-Dia Holes, 25 ft Deep, 10 ft Loaded, 207 lb of Ammonium Nitrate		
20 × 18	333	0.62
18 × 16	267	0.78
16 × 14	207	1.00
14 × 12	156	1.33
12 × 10	111	1.87
For 6-in-Dia Holes, 25 ft Deep, 16 ft Loaded, 147 lb of Ammonium Nitrate		
18 × 16	267	0.55
16 × 14	207	0.71
14 × 12	156	0.94
12 × 10	111	1.32
10 × 8	74	1.99
For 5-in-Dia Holes, 25 ft Deep, 17 ft Loaded, 109 lb of Ammonium Nitrate		
16 × 14	207	0.52
14 × 12	156	0.70
12 × 10	111	0.98
10 × 8	74	1.47
8 × 6	44	2.46

*Pounds of ammonium nitrate, density 47 lb/ft^3, per cubic yard of burden.

Table 13.19 Powder Factor for Drill Pattern of Fig. 13.19

Hole dia, in	Hole depth, ft	Load depth, ft	Charge	
			Lb	Lb per ft
5	25	17	108.63	6.39
6	25	16	147.36	9.21
9	25	10	207.20	20.72

Spacing, ft	Burden, yd^3	Powder factor*					
		9-in holes	9- and 6-in holes	9- and 5-in holes	6-in holes	6- and 5-in holes	5-in holes
8 × 8	59	3.51	3.00	2.68	2.50	2.17	1.84
10 × 10	93	2.23	1.91	1.70	1.58	1.38	1.17
12 × 12	133	1.56	1.33	1.19	1.11	0.96	0.82
14 × 14	194	1.07	0.91	0.81	0.76	0.66	0.56
16 × 16	237	0.87	0.75	0.67	0.62	0.54	0.46
18 × 18	300	0.69	0.59	0.53	0.49	0.43	0.36
20 × 20	370	0.56	0.48	0.43	0.40	0.35	0.29
22 × 22	448	0.46	0.40	0.35	0.33	0.29	0.24

* Pounds of ammonium nitrate, density 47 lb/ft^3, per cubic yard of rock.

found adequate, whereas good results have been obtained in soft shale with a 50% reduction in the charge, to $1\frac{1}{4} \times 4$ in, and the same hole spacing. Detonation cord from each hole is attached to a trunk line, which when fired causes each hole to detonate instantaneously.

Stemming can be done several ways. In one method, after the charge has been placed in a hole,

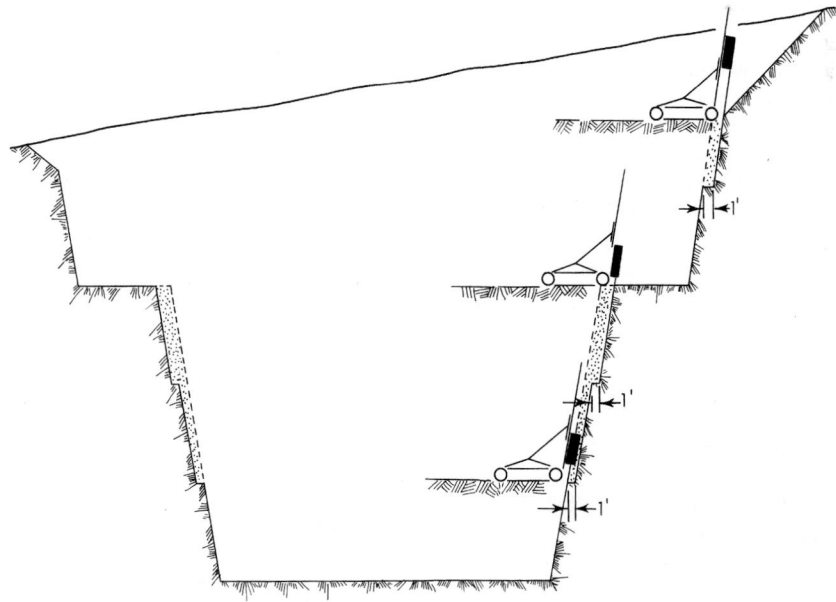

Fig. 13.20 Placement of a drill in a multilift cut.

clean stone chips or sand that will pass a $\frac{3}{8}$-in standard sieve is placed on top. For best results, the stemming should be worked around the charges by holding the end of the detonation cord in the center of the hole and working it up and down. Another stemming method is to push newspaper into the hole until it reaches the top charge. On top of the paper, the hole is stemmed with drill cuttings or other suitable available material.

In most blasting procedures, it is good practice to have as much confinement as possible. In presplitting, some means must be provided to allow excess gases to escape. Use of detonation cord and top firing produces best results. Most instant blasting caps have so much delay that breakage occurs in the wall if they are used. To reduce noise and vibration, delay connectors may be used between groups of two or more holes.

The cost of presplitting per cubic yard excavated depends on the distance between walls or volume to be removed per square foot of presplit wall. Presplitting eliminates the need for small-diameter holes for a primary blast, moving of material from behind a pay line, and scaling of slopes. If presplitting is not required and no pay will be received for material excavated behind a pay line set 18 in beyond the design slope, then, to control excess excavation, small-diameter blast holes would be drilled near the slope at a minimum spacing of 6 ft. These holes would be the same diameter as presplit holes in most cases. Usually, two rows of these holes would be required. The primary blast holes would be at a greater distance from the design slope than for presplitting. When presplitting is used, the spacing of the primary blast holes can be rearranged to produce well-broken rock that will load more easily at less cost.

A cost comparison between presplitting and conventional blasting should compare the cost of blasting the entire cut without presplitting with the cost of presplitting, rearranging the primary blast, and shooting. Generally, presplitting will cost less. For most formations, this will be true when the ratio of cubic yards excavated to square feet of presplit wall exceeds 1.5:1.

13.18 Vibration Control in Blasting

Explosive users should take steps to minimize vibration and noise from blasting and protect themselves against damage claims.

Before blasting, an explosive user should conduct a survey of nearby structures. Experienced, qualified personnel should make this survey. They should carefully inspect every structure within a preselected distance, at least 500 ft. for cracks, deformation from any cause, and other damage that could be claimed. They should make a written report of all observations, wall by wall, and take pictures of all previous damage. This is known as a preblast survey and should be well documented for future use in case of a claim.

Any rock excavation project is a part of some community and has an effect on the surrounding environment. The explosive user can be a good neighbor, enjoying that position, or an undesirable one and suffer the consequences. The decision as to whether the explosive user is an asset or a liability is not made by people familiar with blasting problems. Quarries and rock excavation projects, therefore, should be operated with the realization that any right to exist will have to be proved by behavior acceptable to the community.

To be a good neighbor, an explosive user must not make noise, create vibrations, or throw projectile rocks. The first and last factors are easy to control if proper supervision and good guidance are used. If a neighbor does not hear or see the blast, annoyance is greatly diminished.

Noise and throw are best controlled during drilling and loading cycles. No explosives should be loaded closer to the ground surface than the least dimension used for drill-hole spacing. In other words, put explosives in the bottom of holes and use as much stemming as possible; when noise occurs, energy has been wasted. Using larger holes, with resulting wide spacing, will usually produce oversize stone in the top of the shot. This can be controlled by the use of small (satellite) holes drilled to a shallow depth, below the top of the explosives, between large-diameter holes. This is one method used to get explosives evenly distributed.

Extreme care should be exercised with detonating cord. Nothing makes a sharper and more startling noise. When detonating cord is demanded, a low-noise-level cord should be used, and it should be covered with some material that will not contaminate the desired product. Considerable depth of covering is required to control noise: Experience dictates not less than 3 ft for $\frac{1}{4}$-cord.

Knowledge of human habits and how to use the surrounding environment will greatly help reduce complaints. Set off blasts while people are busy with their daily tasks. Bear in mind that weather conditions affect noise transmission. Blasting during cloudy, overcast weather is like shooting in a room that has a roof. Use other noise- and vibration-producing elements of the surrounding environment for dampening or overriding effects, for example, scheduling and performing blasts while a freight train passes or while a jet airplane is taking off.

Vibrations caused by blasting are propagated with a velocity V, ft/s, frequency f, Hz, and wavelength L, ft, related by

$$L = \frac{V}{f} \tag{13.16}$$

Velocity v, in/s, of the particles disturbed by the vibrations depends on the amplitude of the vibrations A, in.

$$v = 2\pi f A \tag{13.17}$$

If the velocity v_1 at a distance D_1 from the explosion is known, the velocity v_2 at a distance D_2 from the explosion may be estimated from

$$v_2 \approx v_1 \left(\frac{D_1}{D_2}\right)^{1.5} \tag{13.18}$$

The acceleration a, in/s^2, of the particles is given by

$$a = 4\pi^2 f^2 A \tag{13.19}$$

For a charge exploded on the ground surface, the overpressure P, psi, may be computed from

$$P = 226.62 \left(\frac{W^{1/3}}{D}\right)^{1.407} \tag{13.20}$$

where W = maximum weight of explosives, lb per delay

D = distance, ft, from explosion to exposure

The sound pressure level, decibels, may be computed from

$$dB = \left(\frac{P}{6.95 \times 10^{-28}}\right)^{0.084} \tag{13.21}$$

For vibration control, blasting should be controlled with the scaled-distance formula:

$$v = H \left(\frac{D}{\sqrt{W}}\right)^{-\beta} \tag{13.22}$$

where β = constant (varies for each site)

H = constant (varies for each site)

Distance to exposure, ft, divided by the square root of maximum pounds per delay (Fig. 13.21) is known as **scaled distance**.

Most courts have accepted the fact that a particle velocity not exceeding 2 in/s will not damage any part of any structure. This implies that, for this velocity, vibration damage is unlikely at scaled distances larger than 8 (see Fig. 13.22).

Without specific information about a particular blasting site, the maximum weight of explosives per delay should conform with explosive weight and distance limits to prevent vibration damage. This conforms with a scaled distance of 50 or greater without known facts (Fig. 13.21).

To control vibration, the scaled-distance formula should be applied for each blasting location. If formations vary around the site, each formation will have a different formula, which should be computed. The more blasts used in determining the formula constants, the more accurate the scaled-distance formula becomes. Only two easily determined factors must be known: distance from seismograph and maximum weight of explosive used with any delay. Once a safe scaled distance has been determined, the need to use a seismograph for vibration measurements of future blasts is unlikely. Particle velocity can be computed with actual measured distance and known maximum weight of explosives used with any delay.

There is a direct relationship between particle velocity (vibration) and number of complaints expected from families exposed. This is shown in Fig. 13.23.

When a complaint is received, it should be handled firmly and expeditiously. Following are some suggestions:

Assign one person the primary responsibility for handling complaints. This person should be mature and capable of communicating with complainants who are sincerely upset and afraid of not only property damage but bodily injury. A minimum of two employees should, preferably,

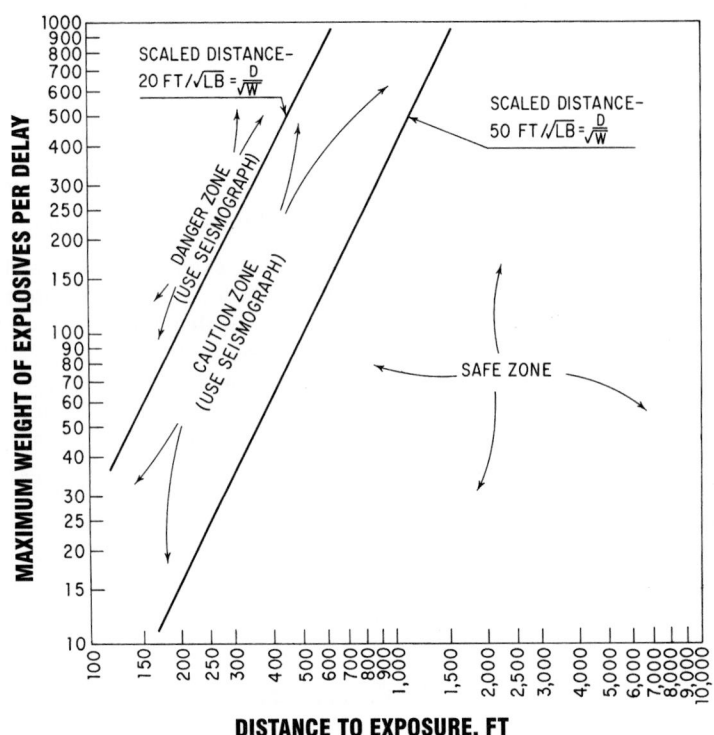

Fig. 13.21 Explosive weight and distance limits for prevention of damage from the vibrations of blasting.

be detailed because the primary employee may not always be available. The primary employee should always be held responsible and informed of all complaints.

Before blasting begins, the public should be advised about whom to contact for any information. When a complaint is received, record the complainant's name, address, and telephone number. Ask at what time the blast was felt and heard. Ask if blast was felt or heard first. Was the complainant's building included in the preblast survey?

Employees handling complaints should be courteous but firm, never apologizing or saying that fewer explosives will be used in the future. Never admit or imply any possible damage until your consultant has advised you of the findings. A completely informed public will want progress, and that is what your organization owes its success to.

Inform the complaints that a consulting engineer has been retained to design and control

your blasting, and this consultant is concerned with nothing but facts. This consultant has been retained to protect the public, to help you do a more efficient job, and to inform the blaster of any potential liability. An independent consultant will know if and where damage may have occurred, probably before the property owner does.

Emphasize that your organization does blasting as a normal operation and has enjoyed success for a considerable length of time, that you have very competent personnel with years of experience, and that you are producing work as efficiently as possible with the least inconvenience to everyone.

People are afraid of noise made by explosives. Noise can be controlled by proper drilling, loading, and stemming. If a shot cannot be seen or heard, your complaints will be few. Remember, it takes only one hole that is not properly tamped to blow out, and then everyone believes the entire shot was not controlled.

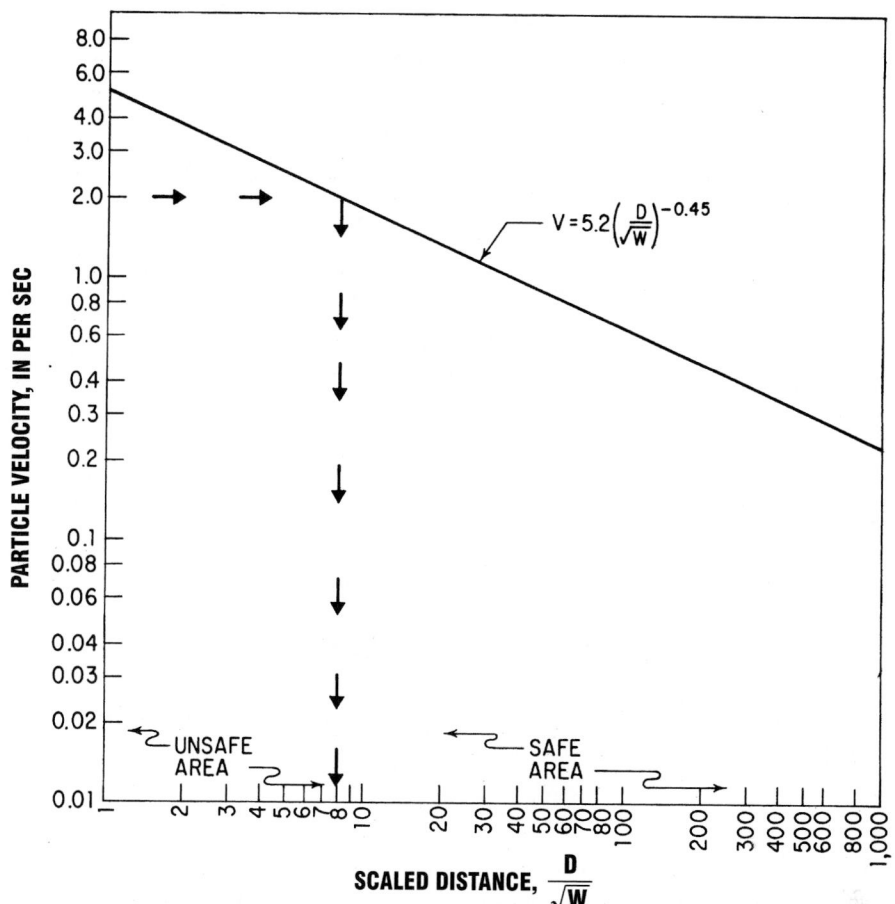

Fig. 13.22 Relationship between particle velocity (vibration) and scaled distance for a specific site for which $H = 5.2$ and $\beta = 0.45$ in Eq. (13.22). For a maximum particle velocity of 2 m/s, the scaled distance is 8. Hence, vibration damage is unlikely at scaled distances larger than 8.

Safe blasting is not only demanded and practical, but essential.

13.19 Compaction

This is the process by which soils are densified. It may be done by loading with static weight, striking with an object, vibration, explosives, or rolling. Compaction is used to help eliminate settlement and to make soil more impervious to water. Compaction is costly, and for some embankments, the results cannot be justified because reduced settlement and other desired benefits are not economical.

For a given soil and given compactive effort, there is an optimum moisture content, expressed in percent of soil dry weight, which gives the greatest degree of compaction. ASTM D698, AASHTO T99, and a modified AASHTO method are widely used for determining moisture content. The modified method may be specified if the soil engineer's investigation indicates that T99 will not yield the desired consolidation. In these tests, soil density of a compacted sample is plotted against percent of moisture in the sample. Maximum density and optimum moisture for the sample can be determined from the resulting curve (Fig. 13.24).

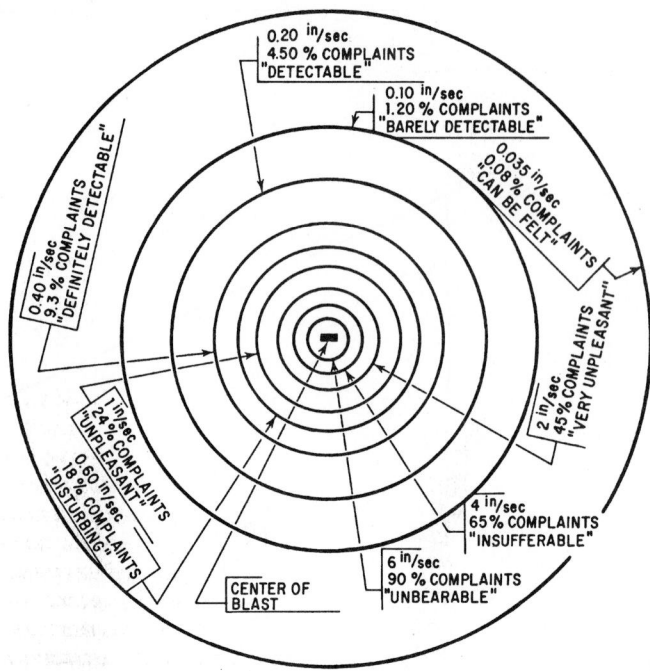

Fig. 13.23 Public reaction to blasting is indicated by the percentage of the total number of families exposed to a specific particle velocity that should be expected to complain, plotted to a logarithmic scale.

Compaction to be obtained on embankments is expressed as a percent of maximum density. For example, 90% compaction means that the soil in place in the field should have a density of 90% of the maximum obtained in the laboratory. Moisture content should not vary more than 3% above or below optimum. To obtain proper com-

paction in the field, moisture must be controlled and compactive effort applied to the entire lift.

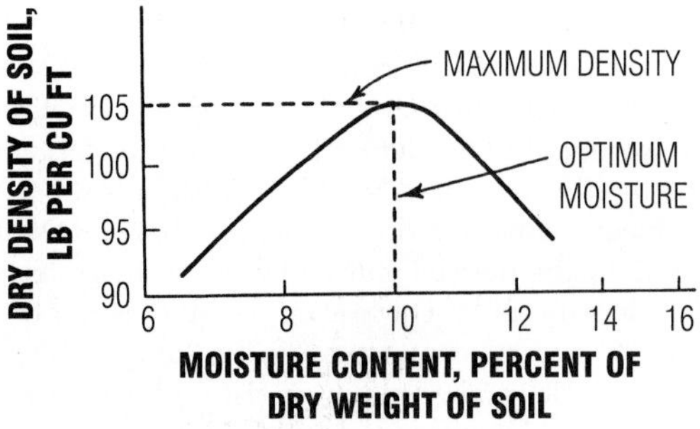

In-Place Density Tests ▪ Several ASTM standard test methods are available for determination of soil density in the field. The two types most frequently used are nuclear methods (ASTM D2992), applicable for shallow depths, and the sand-cone, or calibrated-sand, method (D1556).

Nuclear methods offer the advantage over the others in the relative ease with which the tests can be made. They eliminate the need for digging holes and collecting samples. More tests can be carried out per day than by the other methods. Also, they have the advantage of being more nearly non-destructive tests, thus permitting immediate detection of apparent erratic measurements. Since nuclear methods measure density of the soil near the surface, however, they preclude examination of the soil in depth.

Fig. 13.24 Maximum-density graph.

In these tests, a gamma-ray source and a gamma-ray detector placed on, into, or adjacent to the soil to be tested are used to determine the total or wet density of the soil. A counter or scaler capable of automatic and precise timing is generally used to report the rate at which gamma rays emitted by the source and modified by the soil arrive at the detector. This rate depends partly on the density of the underlying soil. The scaler reading is converted to measured wet density with the aid of a calibration curve that relates soil density to nuclear count rate as determined by correlation tests of soils with known average density. The nuclear methods are normally suitable for test depths of about 2 to 12 in.

The sand-cone method is used to determine in the field the density of compacted soils in earth embankments, road fill, and structure backfill, as well as the density of natural soil deposits, aggregates, soil mixtures, or other similar materials. It is not suitable, however, for soils that are saturated or soft or friable (crumble easily). The method requires that a small hole be dug in the soil to be tested. Hence, the soil should have sufficient cohesion to maintain stable sides. It should be firm enough to withstand, without deforming or sloughing, the pressures involved in forming the hole and placing the test apparatus over it. Furthermore, the hole should not be subjected to seepage of water into it.

All soil removed from the hole is weighed, and a sample is saved for moisture determination.

Then, the hole is filled with dry sand of known density. The weight of sand used to fill the hole is determined and used to compute the volume of the hole. Characteristics of the soil are computed from

$$\text{Volume of soil, ft}^3 \qquad (13.23)$$

$$= \frac{\text{weight of sand filling hole, lb}}{\text{density of sand, lb/ft}^3}$$

$$\% \text{ moisture} \qquad (13.24)$$

$$= \frac{100(\text{weight of moist soil} - \text{weight of dry soil})}{\text{weight of dry soil}}$$

$$\text{Field density, lb/ft}^3 = \frac{\text{weight of soil, lb}}{\text{volume of soil, ft}^3}$$

$$(13.25)$$

$$\text{Dry density} = \frac{\text{field density}}{1 + \% \text{ moisture}/100} \qquad (13.26)$$

$$\% \text{ compaction} = \frac{100(\text{dry density})}{\text{max dry density}} \qquad (13.27)$$

Maximum density is found by plotting a density-moisture curve, similar to Fig. 13.25, and corresponds to optimum moisture. Table 13.20 lists recommended compaction for fills.

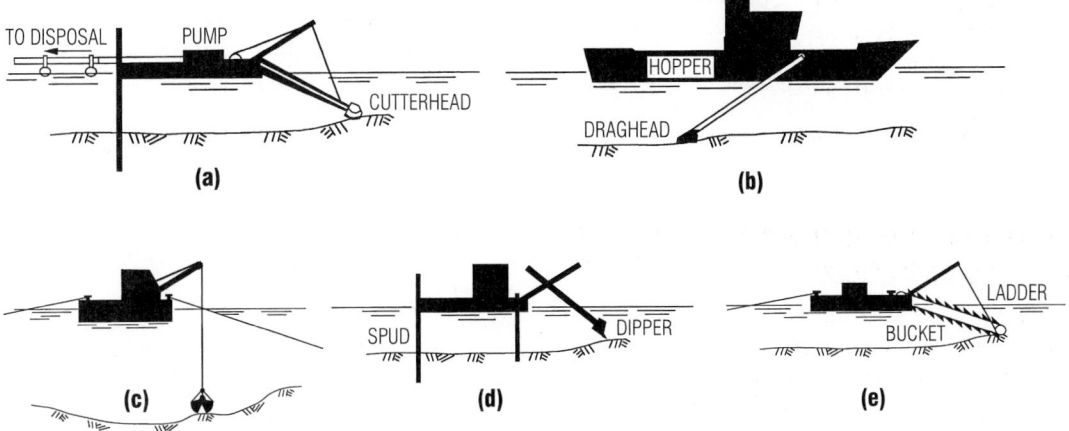

Fig. 13.25 Types of dredges, (*a*) cutterhead; (*b*) trailing hopper; (*c*) grab; (*d*) dipper; (*e*) ladder.

Table 13.20 Recommended Compaction of Fills

Dry density, lb per ft3	Recommended compaction, %
Less than 90	—
90–100	95–100
100–110	95–100
110–120	90–95
120–130	90–95
Over 130	90–95

A mistake commonly made in the field is application of compactive effort when either insufficient or excessive moisture is present in the soil. Under such conditions, it is impossible to obtain recommended compaction no matter how great the effort.

Compaction Equipment ▪ A wide variety of equipment is used to obtain compaction in the field (Table 13.21). Sheepsfoot rollers generally are used on soils that contain high percentages of clay. Vibrating rollers are used on more granular soils.

To determine maximum depth of lift, make a test fill. In the process, the most suitable equipment and pressure to be applied, psi of ground

Table 13.22 Average Speeds, Mi/h, of Rollers

Grid rollers	12
Sheepsfoot rollers	3
Tamping rollers	10
Pneumatic rollers	8

contact, also can be determined. Equipment selected should be able to produce desired compaction with four to eight passes. Desirable speed of rolling also can be determined. Average speeds, mi/h, under normal conditions are given in Table 13.22. Compaction production can be computed from

$$yd^3/h = \frac{16WSLFE}{P} \qquad (13.28)$$

where W = width of roller, ft

S = roller speed, mi/h

L = lift thickness, in

F = ratio of pay yd^3 to loose yd^3

E = efficiency factor (allows for time losses, such as those due to turns): 0.90, excellent; 0.80, average; 0.75, poor

P = number on passes

Table 13.21 Compaction Equipment

Compactor type	Soil best suited for	Max effect in loose lift, in	Density gained in lift*	Max weight, tons
Steel tandem 2–3 axle	Sandy silts, most granular materials, some clay binder	4–8	Average	16
Grid and tamping rollers	Clays, gravels, silts with clay binder	7–12	Nearly uniform	20
Pneumatic small tire	Sandy silts, sandy clays, gravelly sands and clays, few fines	4–8	Uniform to average	12
Pneumatic large tire	All (if economical)	To 24	Average	50
Sheepsfoot	Clays, clay silts, silty clays, gravels with clay binder	7–12	Nearly uniform	20
Vibratory	Sands, sandy silts, silty sands	3–6	Uniform	30
Combinations	All	3–6	Uniform	20

* Density diminishes with depth.

13.20 Dredging

Dredges are used for excavating in or under water. They may be classified by the method used for excavation and the method of transporting and disposing of the excavated material.

13.20.1 Methods of Excavation

Hydraulic, or suction, dredges are the most widely used type of dredge. They move material by suction and pumping through pipes.

Plain suction dredges often have the suction pipe mounted in the bow. They may use water jets to loosen the material to be moved. Plain suction dredges perform well in sand. They remain stationary and dredge a depression into which the surrounding sand flows. They can dredge to a depth of 85 m.

Cutterheads are often used on suction dredges to cut or loosen material to permit handling by the suction line and discharge pipes (Fig. 13.25a).

Trailing, or drag, dredges have the suction pipe mounted on the side and extending toward the stern (Fig. 13.25b). This type of dredge, often employing a draghead attachment and cutting a small bank with each pass, is widely used for maintenance dredging of shoaling in navigation channels.

Bucket, or mechanical, dredges excavate with grab buckets, dippers, and bucket ladders.

Grab dredges (Fig. 13.25c), also known as clamshells or orange peels, are often used close to obstructions, such as docks, piers, and other marine structures, and for the corner of cuts. These dredges can operate to large depths, limited only by the length of wire from the boom to the bucket. They perform well in silts and stiff mud. Performance is poor, however, in hard, consolidated materials, and this type of dredge is not suitable for hard clays.

Dipper dredges are used for excavating broken rock or hard material (Fig. 13.25d). As is the case for power shovels, operating depth is limited by the length of boom.

Ladder dredges employ a continuous chain of buckets to excavate material and transport it to the dredges (Fig. 13.25e). Commonly used for sand and gravel dredging and mining, they also work well in soft clays and rock. Disadvantages of ladder dredges include high maintenance costs, inability to operate in rough water, and the need for mooring lines and anchors, which may interfere with navigation traffic.

Bucket dredges can cause considerable turbidity due to material escaping from the buckets. Consequently, in some locations, bucket-dredge operation is limited during "environmental windows," such as fish migration periods.

13.20.2 Transportation and Disposal

Disposal of dredged material, often as difficult as the dredging itself, is a serious concern.

Bucket dredges typically discharge dredged material into a scow or barge, into a hopper in the dredge itself, or onto an onshore disposal area, if it is within reach.

Pipeline dredges transport dredged material by direct pumping through a floating pipeline to the disposal area. They are normally referred to by the size of their discharge pipeline.

Hopper dredges, floating counterparts of scrapers, transport dredged material in the dredge hoppers to a disposal area. The hopper dredges may be unloaded by opening the hoppers and bottom-dumping the material or by pulling alongside a mooring barge at the disposal area and connecting to a pipeline. Use of this type of dredge is indicated in situations where the distance to a disposal area is too large to permit pumping the full distance by pipeline. This type of dredge, however, has the disadvantage that it must stop excavating during transport.

A third form of disposal is **side casting** dredged materials in a direction that permits the current to carry them away from the project area. This method of disposal is used for dredging of navigation inlets to remove shoaling.

Water-injection dredging (WID), a newer dredging and disposal method, uses water injected through jets in a horizontal pipe to fluidize fine-grained materials, such as sand and silt. The fluidized sediment is carried away from the project site by a density current or natural currents. For projects characterized by fine sediments, favorable currents, and nearby deep water for reception of dredged material, WID is an alternative to conventional dredging and disposal. Advantages include low cost, no need for pipes to transport dredged material, and little disruption of navigation traffic as would occur with conventional pipelines. Also, turbidity is less with WID since the fluidized material stays within about 2 ft of the bottom.

13.20.3 Dredge Production Rates

Prediction of production rates for dredges is extremely complex. Production rates depend on many factors: soil type, uniformity, and grain size; digging depth, work-face height, tides and currents, pipeline length, and disposal elevation; nearby navigation traffic; and equipment maintenance and crew training.

Measurement of dredging quantities for progress and payment also can be difficult. The standard method used by those who require dredging work (government officials, shippers, marina owners) is the in situ volume based on pre- and postdredging surveys. Payment is made for dredging down to a design depth and width, plus a tolerance.

A method of measurement more favorable to dredge operators, however, is the volume or weight transported by the dredging equipment. Unless close controls are maintained, this method is rarely satisfactory to the paying authority, who does not want to pay for overexcavation beyond specified dimensions.

Another method is to measure the dredged material after disposal. This, however, is suitable only when the objective of the dredging is to create a fill.

13.20.4 Permits and Authorizations

A permit is needed for dredging in or over any navigable water in the United States, in accordance with requirements of Section 10 of the 1899 Rivers and Harbors Act. Also, Section 404 of the Clean Water Act requires authorization for practically all dredging discharges. These permits are administered by the U.S. Army Corps of Engineers.

13.21 Earthwork Bibliography

J. E. Clausner, "Dredging Research," vol. DRP-93-3, Nov 1993, U.S. Army Corps of Engineers, Waterways Experiment Station, Vicksburg, Miss.

J. F. Riddell, "Methods of Measurement for Payment Purposes," Proceedings, Maintenance Dredging Conference, May 1987, Institution of Civil Engineers, Bristol, U.K.

T. M. Turner, "Estimating Hydraulic Dredge Capacity," Proceedings of the XIIth World Dredging Conference, 1989.

R. L. Nichols, "Moving the Earth—The Workbook of Excavation," 3rd ed., McGraw-Hill, Inc., New York (books.mcgraw-hill.com).

14

William N. Lane
Dane County Regional Planning Commission
Madison, Wisconsin

COMMUNITY AND REGIONAL PLANNING

Planning for the physical development of communities and regions generally requires an interdisciplinary team approach. Although planning generalists may prepare general development plans for smaller communities, more complex and specialized studies require consultation with and guidance from experts and specialists in a diverse array of disciplines. Since civil engineering works are fundamental components in the physical development of cities and regions, the civil engineer plays a prominent role in community and regional planning efforts.

This section is only a brief overview of the most significant issues and approaches used in planning studies. The references at the end of this section contain additional information as well as more extensive bibliographies.

Basic Approach and Organization of Planning

14.1 Need and Justification for Planning

Planning is practiced routinely by individuals, corporations, and governments. Time and activities are scheduled. Efforts are directed toward achieving goals and objectives. Scarce material and time resources are allocated to various competing demands.

Planning is good management, allowing anticipation and preparation for future events, and is concerned with both avoiding future problems and correcting existing problems. Planning helps meet basic human needs, such as housing, transportation, and goods, and conserves and protects resources and maintains environmental quality. The planning approach, by rationally examining the range of available solutions to existing problems, can result in the selection of a solution that does not itself become a future problem.

Governmental actions and programs often require major expenditures for public works and public services. It is therefore understandable that a principal goal of planning is to achieve the most efficient allocation of scarce resources to competing demands.

Governmental decisions have long-term effects yet are made by officials elected to short terms of office. This can result in emphasizing short-term actions and programs at the expense of long-term programs and long-range objectives. Planning is a framework in which to undertake immediate action in the context of long-range goals and objectives.

Other important objectives of planning, particularly regionally, are to enhance intergovernmental coordination and cooperation and to manage intergovernmental conflicts. Many management decisions must be made regarding environmental quality and regional services, and they must transcend the boundaries of individual communities.

Regional planning can be a structure that addresses important intergovernmental and regional issues and finds solutions through cooperation between individual units of government and coordination of their efforts.

In searching for solutions to problems, attention may be too narrowly focused on a specific functional area or problem, and potential effects on other aspects of physical development may not be fully evaluated. The overall comprehensive approach of community and regional planning encourages examination of side effects and relationships between functional planning areas. In a world of increasing fragmentation and complexity, where emphasis may be on short-term results, planning satisfies the need for a rational decision-making framework based on long-term, comprehensive, and regional viewpoints.

Since many governmental decisions and actions provide guidance for private decision making, the influence of governmental planning extends beyond the sphere of public actions and decision.

14.2 Scale of Planning— Neighborhood, Community, Region

The scale of a particular planning program depends on the nature of the problem being addressed and the potential solutions to that problem. An air-quality problem cannot be addressed or corrected at the neighborhood level. On the other hand, it is not usually necessary to address the specific location and design of neighborhood facilities regionally. Generally, planning at the neighborhood scale is directed toward basic services and common issues, whereas planning at the regional level is directed toward more highly specialized services and facilities, or at those problems that require regional solutions.

Planning at each scale—neighborhood, community, region—should be in the context of the next larger scale: The neighborhood should be planned in relation to the community, and the community within the context of the region.

Neighborhood ▪ A neighborhood has from 2000 to 10,000 residents and is oriented around significant neighborhood facilities, such as an elementary school, neighborhood parkland, and neighborhood shopping area. A well-defined neighborhood has a distinct identity and is clearly separated from surrounding neighborhoods. It also has a cohesiveness and commonality of interests of residents. In many communities, neighborhoods have established citizen organizations that are active in politics and planning. It is useful for planners and elected officials to work with these organizations, if they are representative of the neighborhood.

Neighborhood planning, involving primarily housing and neighborhood facilities, is concerned with specific sites, and there is considerable emphasis on esthetic concerns, such as site design and character of public spaces, and issues like historic preservation. Neighborhood planning can also effectively address the level and quality of basic public services, such as public safety, solid-waste collection, and street maintenance.

In addition to addressing internal conditions and facilities serving the individual neighborhood, planners need to consider the interaction of the neighborhood with other neighborhoods and access of neighborhood residents to community and regional facilities and services, such as public transportation. Neighborhood planning is also important in addressing the neighborhood impacts of major community or regional projects, such as routing of a major highway through a neighborhood or the location of a major public facility within a neighborhood.

Community ▪ Although there is no precise definition, a community ordinarily consists of a number of neighborhoods and generally reflects a greater diversity of interests and concerns as well as a greater degree of economic self-sufficiency than an individual neighborhood. Community planning, often referred to as city planning, is concerned with providing the basic services and facilities of concern in neighborhood planning, but community planning is also directed toward more centralized and specialized facilities and services, such as the location and design of major industrial and shopping areas (including central business districts), middle and high schools, and cultural facilities like libraries, community centers, and other similar community-wide facilities. Community planning ordinarily deals with a single unit of government that has the capability of exercising the control and taking the actions recommended in the plan without the need for substantial coordination or cooperative efforts with other units of government.

Region ▪ Regional planning is concerned primarily with issues, problems, or services that overlap or transcend community boundaries. Typical examples include air quality, water quality, transportation systems, specialized cultural and higher educational facilities, regional shopping centers, and specialized industry. The regional approach is sometimes dictated by sheer size, as in the need to achieve economies of scale in wastewater treatment plants. The need for a regional approach may also be necessitated by the proximity of communities to each other.

14.3 Structure and Organization of Planning Agencies

The structure and organization of planning agencies varies from community to community, and there are differences between community planning and regional agencies.

Community Planning ▪ Most communities are engaged in both functional and comprehensive planning activities. Even small units of government have planning commissions engaged in land-use planning and zoning activities, and most operating agencies with significant budgets are engaged in functional planning (for example, a local sewerage or public-works department engaged in functional planning for providing wastewater collection and disposal). There are a number of advantages to functional planning, including the involvement of governmental staff directly responsible for providing the service and having the most intimate knowledge of service provision. The comprehensive viewpoint, on the other hand, permits an assessment of the impacts of a particular single-function plan on other areas of governmental concern, allows relative priorities to be established among separate governmental functions, and permits coordination of separate departmental efforts to achieve common goals and objectives. It is usually beneficial to combine the advantages of the two approaches by establishing a comprehensive planning framework within which detailed

functional planning can proceed. This way, functional planning can satisfy the needs of that particular governmental function while contributing to overall community goals and objectives.

Most communities have established comprehensive planning agencies that prepare and administer comprehensive plans for community development and provision of services and coordinate functional planning by individual agencies or departments. These comprehensive planning agencies range from small volunteer committees to large municipal agencies with sizable staffs and significant operating or regulatory responsibilities.

Community comprehensive planning organizations generally take the form of an independent planning commission, planning department, or community development agency. These various forms differ in their administration and relationship to the legislative and executive branches of local government. The independent planning commission is usually governed by a body of representatives appointed by the executive or legislative branch of government. Traditionally, a significant responsibility of independent planning commissions has been to administer and advise on zoning matters. Since zoning decisions are ordinarily a matter of legislative concern, independent planning commissions normally have closer ties with the legislative branch of government than do planning departments or community development agencies and somewhat less direct ties to the executive branch. Independent planning commissions have usually been more oriented toward long-term planning rather than direct involvement in short-term development proposals and projects, although the commissions may administer zoning and subdivision regulations.

A community planning department or development agency is a common fixture in larger units of government with significant executive and administrative programs. These agencies are more closely related to the executive branch and are usually under the authority and administration of the chief executive. A planning department is primarily concerned with planning, land-use decision making, and coordination of functional planning. A community development agency is usually more involved in initiating development proposals, managing specific development projects, and administering governmental operating and regulatory functions, including code enforcement and housing functions. Both planning departments and community development agencies tend to place greater emphasis on short-term or current planning and development activities than on long-term goal setting and decision making.

A dual approach, incorporating both an independent planning commission and a community development agency or planning department within the executive branch, can offer the advantages of both types of agencies if the relative roles of the two agencies are clearly defined to avoid unnecessary duplication, overlap, or conflict.

Regional Planning ▪ Regional planning agencies, as differentiated from community planning agencies, are not associated with a single governmental unit but perform planning functions for a region that contains many governmental units. Regional planning agencies are usually independent commissions or councils of governments, with only limited authority, and are primarily advisory in nature.

Regional planning agencies are ordinarily governed by a body of representatives known as a commission or council and commonly named from constituent units of government in the region. Because regional agencies are not associated with a single unit of government, their activities are heavily oriented to intergovernmental coordination and cooperation and to dealing with issues or problems that transcend the boundaries of individual local communities.

14.4 Basic Approach and Methodology in Planning

Figure 14.1 shows a work program for a typical land-use plan. Nearly all planning problems are approached with the same basic methodology, which includes the following key elements:

Identify Current Conditions and Problems ▪ One of the earliest and most time-consuming aspects of any planning process is collecting and analyzing data on current conditions. This analysis should include an evaluation of resources and constraints (physical or economic) that might affect or limit future opportunities, identification of existing problems and deficiencies, and identification of present assets and resources that need to be protected and maintained. It is important to avoid concentrating solely on deficiencies and problems since it is equally important to protect available resources and strengthen or reinforce a community's assets and strong points.

Forecast Trends and Needs ▪ Since the purpose of a plan is to direct and control future events, it is important to understand the changes

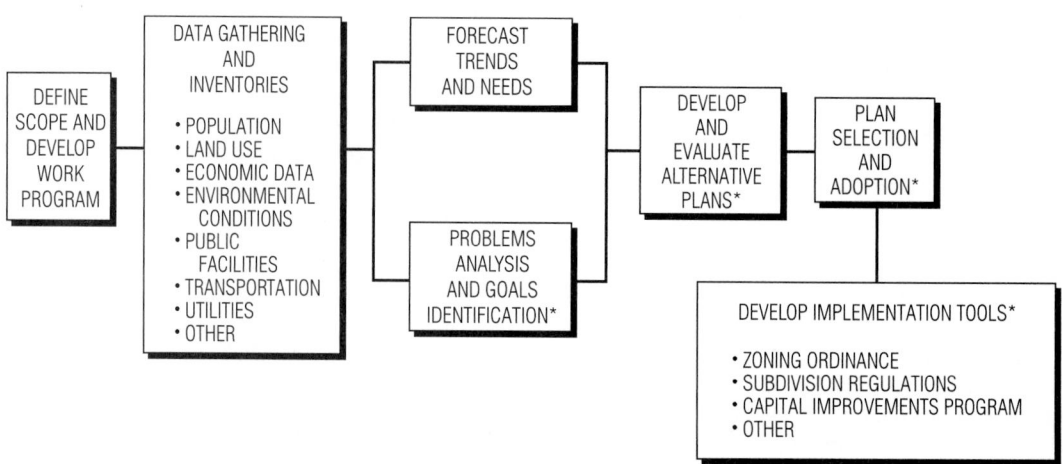

Fig. 14.1 Steps in preparation of a land-use plan. Asterisks indicate steps requiring significant review and participation by the community, elected officials, and concerned organizations.

that may result from continuation of current trends and programs. This process entails the identification of historical changes and trends and an analysis of whether the basic causes of historical trends and changes retain current validity. Trends are then projected into the future, and future needs and demands are forecast based on these trends. Finally, an evaluation is made of the trends to determine whether or not:

1. They represent future conflicts or problems
2. Projected needs and demands will exceed available resources
3. Projections and forecasts are realistic in light of current information and future changes that can be reasonably anticipated

Establish Goals and Objectives ▪ It is wise to state explicitly the plan's goals and objectives, to help ensure that the goals and objectives are those desired by the community or region and that any conflicts between goals are addressed in the planning process.

Goals and objectives are the ends the plan is designed to achieve. The terms are used interchangeably, but a goal usually represents a long-term target to aim at, without the implication that this target will actually be achieved. An objective is generally considered an end that can be achieved within the planning period. Goals and objectives, as ends, should be distinguished from policies, strategies, programs, and actions, which are means to achieve ends.

In some cases, an objective might be to alter a projected trend or future demand believed undesirable. As an example, a planning study of future water supply may have as an objective the satisfaction of future water needs. Although one approach would be to provide a basic water supply and facilities to satisfy future demand based on projections of existing trends, it may also be possible to alter future demand for water through water-conservation programs. Thus, in planning aimed at satisfying future needs, consideration of ways to alter future demand may be as important as consideration of ways to satisfy that demand.

Outline and Evaluate Alternative Plans ▪ Once goals and objectives have been established, the plan focuses on policies, strategies, programs, and actions designed to achieve the stated goals and objectives. Because there are usually alternative ways of achieving goals and objectives, it is common practice to evaluate a number of alternative plans to provide a range of choices for the general public and elected officials. Each alternative plan should be evaluated for satisfying each individual goal or objective.

Select Recommended Plan ▪ After each alternative plan has been evaluated, the recommended plan is selected from the array of alternatives because it best satisfies all the goals and objectives, although there are often conflicts between various goals and objectives and some goals and objectives are more important than others. A goals-achievement matrix chart, as in Table 14.1, may be used to display the various

Table 14.1   Goals-Achievement Matrix

Alternative plans	Objectives			Cumulative weighted Score	Rank
	Objective 1 (Weight =)	Objective 2 (Weight =)	Objective 3 (Weight =)		
	Score / Weighted Score	Score / Weighted Score	Score / Weighted Score		
A					
B					
C					

planning objectives, the ability of each alternative plan to satisfy each individual objective, and an evaluation of the overall satisfaction of all objectives by each alternative plan. A goals-achievement matrix can help make the final plan selection more explicit and understandable to citizens and elected officials.

Develop Detailed Implementation and Financing Techniques ▪ After the recommended plan has been selected, it is necessary to outline specific programs and actions necessary for carrying out the plan's policies and strategies. The specific programs and actions should include such important details as financing methods, scheduling, and staffing needs. Practically demonstrating how the plan can be achieved is as important as determining what should be achieved.

14.5 Public Information and Citizen Participation

An essential ingredient in any successful planning program is a public-information and citizen-participation program. Public information is a one-way process designed to inform the general public and elected officials about the planning program and proposals. On the other hand, citizen participation is a two-way mechanism designed to permit the general public to participate directly in formulating objectives and plans and to allow citizens to provide input and opinions to the planning process.

Need ▪ Successful adoption and implementation of any plan requires the support of elected officials and the general public. Hence, they should be extensively consulted during the planning process. In addition, public information and citizen participation improve democratic processes by keeping citizens informed of and involved in governmental decision making. Also, public information and citizen participation often improve the quality of the plan by incorporating additional specific knowledge of local conditions and concerns, as expressed by citizens and elected officials, and ensuring that plans address citizens' real concerns.

Table 14.2 illustrates the roles of citizens, planners, and elected officials at various stages of the planning process.

Timing ▪ Public information and citizen participation are usually continuous activities throughout the planning process, but several key stages of a planning program require significant efforts to inform and involve the public.

It is important to conduct a public-information and citizen-participation effort early in the planning process. At this stage, the purposes are:

1. Inform citizens and elected officials of the objectives and schedule.

2. Enlist elected officials and the public in identifying problems, concerns, assets, and other existing conditions that may affect the planning program.

Table 14.2 Public Participation in the Planning Process

Stage of Process	Role of Participants*		
	Elected Officials	Citizens	Planners
1. Define scope and develop work program	S		P
2. Data gathering and inventories			P
3. Forecast trends and needs			P
4. Problems' analysis and goals identification	P	P	S
5. Develop alternative plans			P
6. Evaluate alternative plans	P	P	S
7. Plan selection and adoption	P	P	S
8. Develop implementation tools	S		P

*P = primary role; S = supporting role.

3. Let elected officials and the general public participate in formulating goals and objectives.

The second stage of the planning process requiring extensive public information and involvement occurs when alternative plans have been formulated and a preliminary evaluation of those alternatives has been completed. Elected officials and citizens can help identify additional alternatives not considered or overlooked, evaluate the ability of alternatives to satisfy goals and objectives, and identify favored alternatives or features of alternatives.

Public information and citizen participation are also required after a recommended plan has been tentatively selected by the planning staff and sponsoring agency. This allows elected officials and citizens to comment on the final selected plan, identify potential modifications of the selected plan that would improve or increase support for the plan, and document support for or opposition to the recommended plan.

Mechanisms ▪ A variety of tools and techniques are available for providing public information and citizen participation. Since nearly all these tools and techniques have limitations, a good public-information and citizen-participation program includes many techniques.

The goal of a public-information program is to communicate efficiently key features of the plan to a wide audience. The tools and techniques most commonly used for disseminating information to the public and elected officials include advisory committees, direct mail efforts (including newsletters and project summaries), use of the print media (including news releases, articles and newspaper tabloids or supplements), use of broadcast media (including radio and TV shows), and public meetings. Internet computer access to planning information and for interactive feedback is becoming an important mechanism for public information and citizen participation.

The most common tools and techniques used for citizen participation include advisory committees or task forces, public meetings and public hearings, and public-opinion surveys. Although creation of citizen advisory committees limits participation to a small group of citizens, those citizens are well-informed on the issues and alternatives. Care must be taken to ensure that committees are representative of the community or targeted audience.

Public meetings and hearings, often legally required, are included in most planning programs. Meetings and hearings offer opportunities to present information and to obtain input and opinions from citizens and elected officials. Ordinarily, the format for such meetings includes an initial presentation designed to provide information to citizens and elected officials attending the meeting, followed by an opportunity for those citizens and elected officials to state opinions and participate in analysis or evaluation. Care should be taken to schedule public meetings and hearings at times and locations convenient for citizens and elected officials. In addition, the atmosphere and format should be as comfortable as possible, to encourage participation. Long and technical presentations by the planning staff or consultants should be avoided. An excessively legal or formal atmosphere at public meetings and hearings can be intimidating and discourage participation.

Public-opinion surveys and questionnaires are commonly used in citizen-participation programs. These may take the form of written questionnaires distributed by mail or at public meetings or of direct personal or phone interviews, where the interviewer records responses to standard questions.

Direct personal interviews and telephone surveys are very effective means of obtaining citizen participation. Since these techniques require an interviewer, they usually involve more effort and cost than mail questionnaires. It may be possible in some communities to have volunteers conduct the interviews. The use of an interviewer provides the opportunity to explain the questions and provide clarification if the citizen is confused.

Problems and Pitfalls ▪ The principal pitfalls to avoid in a public-information program are confusion, tedium, and limited distribution. Since the goal is to communicate key features of the plan as efficiently as possible and to as many people as possible, brevity and clarity are of the utmost importance. Language should be nontechnical. Interesting graphics (drawings, graphs, and photographs) should be used liberally to sustain interest and make it easier to understand important information. Lengthy technical documents are too bulky and cumbersome to distribute widely and usually provide more information than the average citizen desires or needs. Project summaries and brief articles and presentations can be widely distributed and more effectively communicate with large numbers of persons.

The principal problems and pitfalls in citizen participation include lack of interest and participation and failure to ensure representativeness and randomness in sampling procedures and participation. It can be assumed that voluntary participants in advisory committees and public meetings and hearings are not truly representative of the population as a whole. On the other hand, the involvement of citizens with high interest or who are likely to be directly affected by the plan is crucial to the success of any planning project, even though their opinions may not represent the entire population.

It is difficult to structure survey techniques and written questionnaires to get representative answers. The best results can be obtained by consulting experts on opinion research and survey design, to ensure that questionnaires and survey techniques are formulated to obtain unbiased results. Also, elected officials can usually be expected to represent their entire constituency. The involvement of elected officials in the planning process can be a good tool for ensuring that narrow or unrepresentative interest groups are not exerting undue influence over the outcome and results.

Each public-information and citizen-participation technique has limitations and drawbacks. An effective public-information and citizen-participation program therefore includes a variety of these techniques, used in combination to supplement and complement each other. Nearly every public-information and citizen-participation program should include advisory committees, public meetings and hearings, information dissemination through print and broadcast media, surveys of public opinion, and wide distribution of project or program summaries.

14.6 Projections and Forecasts

Projections and forecasts are made to determine future needs for land and resources and demand for public services. In most cases, projections and forecasts are based on historical data and trends, modified by future expectations and anticipated changes. Historical data are available from various Federal agencies, including the Bureau of the Census, U.S. Departments of Commerce, Health and Human Resources, and Housing and Urban Development. These data are also available from state, regional, and community planning agencies.

Because projections and forecasts deal with the future and the unknown, their validity should be viewed with suspicion. It is usually impossible to predict with any accuracy future effects of technological and societal changes.

Projections for larger and more heterogeneous regions are generally more accurate than those for small geographic areas or for specialized or homogeneous areas. Short-term projections are usually more accurate than projections into the distant future.

Projections based on extrapolation of only a few years of recent historical data risk being influenced by unusual short-term events or trends. Utilizing a longer historical base of record ensures that the future trend projected is not based on a short-term deviation from historical patterns. Since basic changes in growth rates and development patterns can occur rapidly, however, recent trends and events must be taken into account and evaluated to determine whether they will exert any lasting influence in future years.

One of the most common mistakes in making future projections and forecasts is to assume that historical patterns and relationships will remain constant. There have, in fact, been substantial changes within relatively short times in the relationships between population or employment and measures of demand for land or public services. In some localities, for example, in recent years there has been a rapid decline in the number of persons per dwelling unit. This has the effect of requiring substantial additional housing units, even in the absence of population growth, and results in a declining population density even though the density of dwelling units per acre remains the same. Other relationships and factors that have exhibited substantial change include the relative density of industrial employment, expressed as employees per acre; the ratio of total employment to population; the number of automobiles per household; and other similar and basic factors used in planning projections and forecasts. In all cases, it is wise to use local data wherever available, to compare these data with other similar communities, and to carefully evaluate potential future changes in trends and relationships.

Population ▪ Population projections and forecasts are fundamental to most planning studies since population is one of the most important measures of demand for land, goods, and public services. Historical population data are available from Census publications and can be obtained from state, regional, and community planning agencies. In addition, future population forecasts and projections are commonly made for communities and regions by state, regional, and community planning agencies.

The most common methodology used for making population projections is the cohort-survival technique, which estimates the natural increase in a resident population by subdividing the population into classifications of age and sex and applying specific birthrates and death rates to these classifications. Migration is then factored into the analysis by examining historical migration rates and estimating future migration.

Migration of population into or out of a region or community is the most difficult component of population change to accurately forecast. Migration is heavily influenced by employment and availability of job opportunities. It is therefore important to coordinate population forecasts with forecasts of future economic conditions and employment. Historical migration rates should not be automatically assumed to continue, and future migration forecasts should be based on anticipated rates of job creation and employment opportunities.

Simpler population projection techniques are often used where population projections are not available from planning agencies and where the time and effort involved in making a cohort-survival population forecast are not justified. These include simple graphical projection techniques or arithmetic or geometric projections based on historical growth rates. Projecting future population growth based on historical trends is often inaccurate, and these methods should preferably be used only as a check against other population forecasting techniques. If future employment projections are available for an area, it may be possible to project future population as a ratio between population and employment, taking into account potential changes in that ratio.

Because of uncertainties in population projections and forecasts, particularly for smaller communities, it is often desirable to use several different techniques and establish a potential range of reasonably expected future population forecasts.

Once a range has been established, the effects and impacts of using the higher or lower end of the range and the consequences of possible over-design or underdesign can be evaluated. In some cases, inaccuracies in the forecast may simply alter the useful design life of a facility by a few years; in other cases, the consequences of inaccurate forecasts can be quite serious.

Economic Factors ▪ In most planning studies, there is a need to identify the local economy's present strengths and weaknesses and future potential and needs for growth. The primary factors to be considered include employment, characteristics of the labor force, income, and retail market opportunities. The local economy is the driving force behind population growth because it is growth in the local economy that creates jobs and affects migration rates.

Economic studies are ordinarily restricted to large areas or regions. In most urban areas, the basic unit of study for economic projections is the Metropolitan Statistical Area (MSA), as defined by the Federal government. An MSA is more suitable for economic studies and projections than a smaller area because the MSA is relatively self-contained from an economic standpoint (as one measure, most people who live within an MSA also work within the MSA).

The most common methods of making economic projections and forecasts include economic base studies and input-output studies. Input-output studies are less commonly used for metropolitan areas and are somewhat more complicated than economic base studies. Economic base studies examine that portion of the local economy (the economic base) that exports goods and services from the region and generates income from outside the region. This outside income then generates other local economic activities through a multiplier effect. The economic base technique utilizes ratios to develop relationships between base export market activity, local market activity, and overall local economic activity.

Projections of employment and other economic factors are often broken down into classifications according to the Standard Industrial Classification Codes. This makes the economic projection data more useful for converting into related projections of land needs, waste-generation rates, and other data required in the planning process.

Land Requirements ▪ Forecasts of future land-conversion needs are usually based on population and employment forecasts. Population is commonly used as a basis for projecting needs for all land-use classifications for areas that are relatively self-contained from an economic standpoint. Where employment projections are available, they may be used to forecast land needs for commercial and industrial land. In cases where both population forecasts and employment forecasts are available, both methods should be used and compared with each other. Land-use data and projections are usually broken down into classifications. These classifications are often based on the Standard Land-Use Coding Manual, but they vary from community to community. Common classifications of land use include residential; manufacturing; transportation, communications, and utilities; wholesale and retail trade; cultural, entertainment, and recreational services; resource production and extraction; and undeveloped land and water areas. It is useful to use a land-use classification system that is compatible with other state, regional, and community planning agencies working in the area.

When making forecasts of future land-conversion needs, it is advisable to base density figures on existing land-use patterns and local conditions. For new development, it is preferable to develop figures reflecting recent typical development densities. Use of total figures that include older development may not reflect future development densities or types. Development densities should be examined and compared with figures from other typical communities to get representative values. Once representative values for the present densities of various land uses have been determined, it is possible to develop density standards for future development, based on historical data and trends as modified by anticipated future changes or plan policies. It is possible for a community to encourage higher or lower densities of development than have been historically experienced by adopting planning policies and taking actions to encourage changes in density.

Resources and Environmental Quality

14.7 Soils, Geology, and Land Characteristics

An analysis of the soils, geology, and land characteristics of the planning area allows the planner to obtain an understanding of surface and subsurface characteristics so that land uses can be located in compatible areas.

Concerns and Potential Problems ▪ Potential problems arising from surface and subsurface conditions can be subdivided into human problems and environmental problems. Human problems include hazards to health, potential injury and loss of life, and economic losses. Health hazards and potential injury or loss of life can result from earthquake damage, landslides, or pollution of surface or groundwater. Economic losses caused by inattention to surface and subsurface conditions can be staggering in magnitude. Examples include high construction costs due to high water tables, shallow bedrock or unstable or compressible soils; high maintenance costs due to expansive soils, compressible or unstable soils, or excessive erosion and deposition; and property damage due to earthquake damage, landslides, expansive soils, and compressible or unstable soils.

Environmental problems may include damage to resources, such as surface and groundwater pollution caused by erosion or poor location and design of waste-disposal sites. An important but neglected resource problem can be declining soil fertility and productivity from excessive erosion. Another factor, which may be important in some areas, is the subsequent inability to extract or utilize subsurface mineral resources after urban development has been allowed to occur on the land surface.

Besides damage to resources, potential environmental problems can include disturbance of natural ecosystems, alteration of wildlife habitat, and removal of vegetative cover. Esthetic problems, such as poor visual or scenic quality as a result of development or visual incompatibility with the landscape, can also occur.

Data and Pertinent Factors to Consider ▪ In most parts of the continental United States, there is a considerable amount of available data regarding surface and subsurface characteristics. Proper use and interpretation of these data require considerable skill and experience. As is the case with most other specialized areas of community and regional planning, it is recommended that qualified experts in soils and geology be consulted in the process of interpreting and evaluating data and arriving at conclusions.

Data most commonly used in evaluating surface and subsurface conditions include aerial photographs, topographic maps, geologic reports and mapping, soil surveys, and well data and drillers' logs. The U.S. Geological Survey and state geological surveys are sources of topographic maps and geologic reports and mapping. State geological surveys and water-supply or environmental agencies may be a source of well data and drillers' logs. Aerial photographs are generally available from a variety of Federal, state, and local agencies.

In most areas, modern soil surveys are available from the Soil Conservation Service of the U.S. Department of Agriculture. These detailed soil surveys are a primary tool used by community and regional planners to evaluate surface and subsurface conditions. USDA soil surveys, however, are limited to rather shallow surface deposits. These soil surveys are suitable for general locational planning; for specific site studies, this information needs to be supplemented by more detailed on-site information.

The specific factors of concern will depend on the particular issue or land use in question. Factors important in most planning studies include:

1. Topography (land slopes) and landforms

2. Existing vegetative cover

3. Surface-water features and areas subject to flooding

4. Depth to and type of bedrock

5. Depth to and quality and availability of groundwater

6. Groundwater-flow patterns

7. Soil types and characteristics (fertility and productivity, erodibility, strength and stability, compressibility, swelling characteristics, permeability)

8. Location of mineral resources, including sand and gravel

Mapping and Geographic Information Systems ▪ Community and regional planning practice relies heavily on the visual display of information in the form of maps, airphotos, charts and graphs, sectional and perspective views, and three-dimensional views. Most information displays are now computer-based, and the use of geographic information systems (GIS) for storage, analysis and display of spatial information is commonplace.

GIS systems are computer-based systems that provide a rapid and convenient approach to storing, processing and displaying all kinds of spatial or geographic information—virtually any information that can be mapped. GIS systems make possible the storage and retrieval of vast amounts of spatial data, allow categories of data to be separated into discrete layers of information, and greatly facilitate the processing, analysis and display of individual or combined information layers, or the relationship between information layers. In addition, GIS systems are very useful for communicating and transmitting spatial information, and allow widespread access to geographic information. GIS systems have proven to be such a valuable tool that they are being utilized at every level of government.

Land-Suitability Analysis ▪ One of the most important tools in making locational and siting decisions as a part of planning studies is the analysis of suitability of land for different purposes or land uses.

A common approach to land-suitability analysis is to use a multiple overlay mapping process, which involves four basic steps:

First, the factors of importance to the particular land-use or siting question are selected. For example, if the problem is to site a sanitary landfill, the important factors might include such characteristics as slope, soil type, depth to bedrock, depth to groundwater, and some measure of flooding hazard or distance to surface-water bodies.

In the second step, criteria are established by dividing each factor into relative degrees of suitability. Using the example of siting a sanitary landfill, the slope factor may be subdivided into slight limitations (flat slopes), moderate limitations (moderate slopes), and severe limitations (steep slopes).

In the third step, a suitability map for each factor is prepared for the entire area under consideration. The map indicates areas that have severe limitations, moderate limitations, or slight limitations.

The fourth and final step is to combine the information from the individual factor maps to produce a composite land-suitability map based on all factors. Two methods are commonly used to produce composite maps. In the first method, the individual factor maps are simply superimposed on top of each other, and the resulting visual composite image is used. If on each factor map the areas with most severe limitations are mapped as darker shades while areas with less severe limitations are mapped in lighter shades, superimposing these overlays on top of each other yields a composite map in which the darkest areas are the least suitable and the lightest areas the most suitable, taking into account all the factors included in the analysis.

Another approach to deriving the composite map is to divide the planning area into sufficiently small units, for example, by using a two-dimensional grid and then assigning numerical scores to each cell in the grid for each factor. A mathematical composite score is then derived for each cell based on the scores for each factor. The composite scores for each cell are then mapped to derive an overall composite map reflecting all the factors used in the analysis.

A number of examples of practical application of the land-suitability-analysis technique to regional planning problems is included in Ian McHarg, "Design with Nature," Natural History Press, Garden City, New York, and more recent updated information regarding the technique is included in Kaiser, Godschalk and Chapin, "Urban Land Use Planning," University of Illinois Press.

Land-suitability analyses have a number of advantages when making locational and siting decisions as a part of planning studies: The technique has the capability of analyzing the suitability of all land within the area to be considered, rather than being limited to a few selected alternative sites. The technique can generally be adapted to use data readily available in most areas. The technique, while suited to manual analysis and the capabilities of small planning agencies and consulting firms, is also adaptable to the use of computers and modern data-processing techniques and GIS systems. Furthermore, one of the most important advantages of land-suitability analysis is that it provides the capability of communicating and explaining a complex multifactor environmental analysis in a format easy for public officials and the public to understand. Thus, land-suitability analysis can often be instrumental in explaining and obtaining support for controversial locational and siting decisions.

14.8 Water Resources and Supply

One of the most fundamental aspects of planning for future urban development and land uses is the provision of adequate supplies of water necessary for domestic and commercial potable water needs, manufacturing needs, irrigation, cooling water needs, and power generation. The basic water-supply system includes sources of supply, treatment (if required), and distribution to the locations of use. Storage of untreated or treated water is an important element in all water-supply systems.

Water Use and Water Consumption ■ Water for domestic uses, most manufacturing and industrial cooling uses, power generation, and other purposes is usually available for nearby reuse. Water that is converted into water vapor through evaporation or transpiration from plants is considered consumed because it is not available for nearby reuse. Irrigation is usually the most significant aspect of water consumption. (See Sec. 21 for more information on water supply, use, and demand.)

There are variations over time in both supply of and demand for water. Variations in the quantity of water available are caused by long-term drought conditions or seasonal variations in precipitation, stream flow, and groundwater recharge. Variations in the demand for or use of water are caused by long-term trends (population increases, industrial growth, increasing or decreasing per capita use), seasonal variations (lawn watering), and daily and hourly fluctuations caused by living patterns or firefighting.

Impoundments or storage facilities are commonly used to accommodate fluctuations in either supply of or demand for water. Basic supply storage of untreated water in surface impoundments is commonly used to accommodate long-term droughts and seasonal variations. Storage of treated water close to the distribution system, on the other hand, is a common technique for accommodating short-term daily and hourly fluctuations in water demand. In all cases, sufficient storage must be provided to last through those periods when demand is greater than basic supply or distribution capabilities. Withdrawals from storage during periods when demand is higher than basic capacity are then compensated by replenishing storage during periods when demand is less than basic supply capacity. Where groundwater is used for water supplies, the aquifers themselves may act as storage facilities. On a long-term basis, withdrawal must be balanced by recharge to avoid depletion of groundwater supplies.

Overall demand for water and patterns of water use can be significantly affected by metering and pricing policies and by information and regulatory programs directed at water conservation.

Development and Protection of Water Supplies ▪ The basic water supply for a community or region is obtained from either surface-water or groundwater supplies. Surface water may be obtained by direct draft from rivers or by withdrawal from lakes and impoundments. Direct withdrawal from rivers, without storage, is normally feasible only where low flows during periods of prolonged drought are greater than peak water demand. Rivers are also subject to rapid variations in water quality and vulnerable to discharges or spills of toxic and hazardous substances.

Withdrawal from lakes and impoundments is a more common source of water for most metropolitan areas. Lakes and impoundments offer storage capacity to offset supply shortages during long-term droughts and accommodate short-term seasonal fluctuations. Lakes and impoundments have advantages and disadvantages from a water-quality standpoint. Because a substantial volume of water offers dilution, lakes and impoundments are less subject to significant water-quality changes from temporary or accidental pollution. On the other hand, because of the longer retention time of water in lakes and impoundments, pollution is retained longer rather than being transported downstream. Many lakes and impoundments are also subject to a natural aging process called eutrophication. Land-use changes and development activities in the watershed of a lake or impoundment can accelerate the discharge of sediment and nutrients to lakes and impoundments, speeding up the eutrophication process and creating nuisance conditions because of excessive aquatic weed and algae growth.

In many areas, part or all of a community's or region's basic water needs may be satisfied by groundwater supplies. Protection of groundwater recharge and discharge areas is important in managing the groundwater resource. Groundwater discharge areas, such as springs or wetlands, are important in providing base flow in surface streams. Protection of groundwater recharge areas is critical for protecting groundwater supplies. Discharge of pollution in the recharge area can result in transport of contaminants throughout the groundwater aquifer and elimination of the aquifer as a source of supply. In addition, land-use changes and surface activities in recharge areas can affect the rate of recharge and limit the aquifer's long-term supply capacity.

The quality of groundwaters may be adversely affected by waste-disposal practices or land application of chemicals. Waste-disposal practices of concern include septic tanks, livestock waste, sanitary landfills, and waste-storage or -treatment lagoons. Proper location and design of these installations can prevent groundwater pollution. Wide-scale land applications of fertilizers, pesticides, and herbicides and use of road deicing salts and other chemicals are more difficult to control and can have a widespread effect on groundwater quality. Waste-injection wells and leaking underground fuel tanks can also create problems.

14.9 Drainage and Flooding

Flooding problems and surface drainage, as concerns of community and regional planning studies, differ primarily in degree of severity. The principal concern with flooding is the desire to avoid injury and loss of life and reduce property damages caused by major floods (those having a recurrence interval of 25 to 100 years). Surface-drainage systems, on the other hand, are primarily concerned with convenience and providing access to property during relatively minor storms (those having a recurrence interval of 2 to 10 years).

Flood-Damage Prevention ▪ The most important means of reducing loss of life and damages from flooding include flood warning and evacuation systems, flood insurance, flood-proofing, watershed management, floodplain management, and flood-control structures, such as reservoirs, levees, floodwalls, and channel improvements.

Flood warning and evacuation systems are economical and effective means of avoiding injury and loss of life during major flooding events. They are most effective near large streams when there is adequate time to provide warning and allow evacuation. Flood insurance programs seek only to minimize the economic hardship associated with property damages and losses caused by flooding by compensating for those losses.

Floodplain management is a land-use management technique designed to avoid increases in flood damages by restricting new development and construction in areas subject to flooding. This is normally accomplished through zoning restrictions on the area subject to flooding, usually the 100-year floodplain. These restrictions commonly require that new construction be constructed on fill or elevated by other means above the level of the design flood. Land-use activities in the floodprone area may be restricted to uses that would not suffer extensive damage from flooding. In some cases, a two-zone system is adopted. A floodway (the area necessary to convey flood waters without substantial increases in upstream flood stages) is defined, and no filling or obstruction is allowed within the floodway. In the remainder of the floodprone area (called the flood fringe area), development and structures may be permitted if they are protected from flooding and adequate access is provided.

Floodproofing is a technique for designing buildings and structures to resist flood damages when it is necessary to locate these structures in the floodplain (as may be the case with wastewater-treatment plants). Floodproofing can be accomplished through provisions in zoning and building regulations. Floodproofing can be costly and yet does not ensure adequate access to structures. For these reasons, the primary use of floodproofing should be those instances where it is necessary to locate new buildings in the floodplain or where existing high-value structures need protection.

Watershed management can affect flooding by encouraging detention and infiltration of rainfall in upland areas, thereby decreasing runoff and resultant flooding. Watershed management is most effective in reducing flood damages from minor storms. It is very effective in reducing increases in flooding that result from land-use changes and urbanization.

Major flood-control works are the most frequently used method of reducing flooding and flood damages. Flood-control reservoirs, levees, floodwalls, and channel improvements are the most common elements of a structural flood-control program. These works usually involve large capital expenditures. They have significant direct effects on flooding conditions. They are most effective in protecting existing development and should be used in combination with nonstructural techniques, such as floodplain management, to be fully effective. They are, however, ineffective in avoiding greater future flood damage caused by increased downstream occupancy of the floodplain. In some instances, levees, floodwalls, and channel improvements can increase flood stages and damage downstream from the area protected by these improvements.

Urban Drainage Systems ■ Urban drainage systems are designed to remove surface water from land surfaces expeditiously, to avoid inconvenience and provide access to property during minor storms. Urban drainage systems are costly and involve a significant investment by both individual property owners and governments.

As urban areas are developed, the amount of impervious land surface increases as the land is covered with parking lots, streets, and rooftops. Increasing imperviousness results in dramatic increases in storm runoff, frequency and intensity of peak flows, and flooding resulting from storms. As a result, the flow capacity and erosion resistance of natural channels are overtaxed. Subsequent improvements to natural channels and enclosure of flow in conduits speeds runoff processes even more. Full development of a watershed, therefore, may cause peak rates of runoff several times greater than those experienced prior to development.

In lower density areas, it may be advantageous to retain the natural drainage system rather than invest in substantial improvements to open channels or conduits. Where development densities and land-use patterns permit, protection and preservation of natural drainage systems can provide the following:

1. Lower costs than conventional storm-sewer systems

2. Open space and recreation opportunities

3. Scenic beauty associated with streams and greenways incorporated into drainage corridors

4. Reduction of nonpoint-source pollution by providing vegetated areas that filter surface runoff entering the surface-drainage network

5. Decrease in streambank and streambed erosion by vegetative stabilization and streambank protection

6. Opportunities to offset increased stormwater runoff by providing for increased infiltration of storm runoff

7. Reduction in the effect of increased peak runoff and increased flooding by providing opportunities for incorporating temporary storage and detention of runoff and flood waters

Environmental corridors—linear systems of open-space lands developed around drainage and stream networks—offer the opportunity for protection and multipurpose use of natural drainage corridors in urban areas. Environmental corridors are discussed in more detail in Art. 14.12.

Use of natural channels and maintenance of natural drainage systems may not be possible or economical in high-density areas. In addition, upland watershed management techniques, including detention and infiltration of storm runoff, are necessary to offset any potential increases in flow volumes and peak rates of runoff over natural conditions. If these increases are not mitigated, the change from natural conditions can destabilize the natural channel and subsequently cause erosion and deposition.

In urbanized areas, streets and roadways are fundamental and integral parts of the urban drainage system. Local streets and associated channels may be the main conveyance of runoff from minor storms, and streets and associated drainage systems need to be located, designed, and maintained with their drainage function in mind.

14.10 Water Quality and Waste Disposal

Water-quality and waste-disposal issues often overlap or transcend municipal boundaries; hence they are frequently addressed at a regional level. The intergovernmental nature of water-quality and waste-disposal issues usually requires close cooperation and coordination between the planning agency, state and Federal environmental regulatory agencies, local units of general government, and local special-purpose sewerage and solid-waste agencies.

Water-Quality Objectives ▪ An initial step in the water-quality planning process is establishing water-quality objectives related to proposed or desired uses of surface water and groundwater. Water uses usually considered include use as a potable water supply, industrial use for cooling or process water, livestock watering, irrigation, recreation, power generation, and support and propagation of fish and aquatic life. In most cases there will be multiple uses of water.

After proposed or anticipated uses of a water body have been determined, specific water-quality standards can be established. These represent minimum or maximum limits for specific parameters or constituents that are consistent with proposed uses. Water-quality standards are usually established by state regulatory agencies.

When evaluating surface-water-quality conditions and objectives, it is important to distinguish between base flow and runoff or high-flow conditions. Base-flow water quality, or water quality during low-flow conditions, is of most concern for continuous uses of water and for recreation and support and propagation of fish and aquatic life. Base-flow water-quality conditions exist most of the time and are affected most by continuous municipal and industrial wastewater and cooling-water discharges. The quality of groundwater contributions to base flow can also be important in determining base-flow water quality. The water quality of storm runoff during high-flow conditions may be important in some instances, particularly if runoff water quality is worse than base-flow water quality. Runoff water-quality conditions are temporary and usually of most concern in determining loading of pollutants to downstream water bodies.

Discharge of organic and chemical materials from municipalities and industries causes pollution. Thermal pollution may result from discharges of cooling water from industries and power generation. These discharges often involve extremely large volumes of water with elevated temperatures, and the discharges can affect water uses and have significant effects on fish and aquatic life.

Wastewater Treatment and Discharge ▪ The most common approach in urban areas is to collect municipal and industrial wastewaters at a central wastewater- or sewage-treatment plant, treat the wastewaters, and discharge the treated effluent to surface waters. The principal alternatives to treatment and surface discharge include land application of treated or untreated wastewaters, underground injection of wastewaters, and recycling of treated wastewater for direct reuse. Although these alternative approaches are not widely used, they may be appropriate in specific circumstances.

Wastewater-treatment plants may utilize biological or physical-chemical-treatment processes, discussed in Sec. 22. The level or degree of treatment provided varies; however, secondary treatment using biological processes is quite common. This type of treatment removes about 80 to 90% of the solids and organic materials from the wastewater and is economical.

Advanced waste treatment (beyond secondary) is provided where necessary to meet water-quality objectives. It usually involves additional processes directed at reducing or removing specific substances or constituents in the wastewater.

In addition to removing solids, organic materials, and other substances from wastewater, it is necessary to develop environmentally sound means of disposing of the solids, sludge, and other materials removed in the treatment process.

The location of treatment plants and level of treatment are selected to be consistent with the assimilative capacity of the receiving waters. Small streams with low flows may be capable of absorbing and purifying only small quantities of wastewater. When the projected waste load will overtax a stream's assimilative capacity, alternatives include finding another discharge location where the receiving water has a greater assimilative capacity or providing a higher level of treatment. Regionalization of wastewater treatment (replacing several smaller, wastewater treatment plants with a large regional plant) may be cost-effective in some cases because of economies of large-scale construction and operations. Disadvantages of regionalization include concentrating wastewaters at one location, which may overtax the assimilative capacity of streams, and the additional cost of transporting the wastewaters to the regional treatment plant.

In rural or low-density areas, the provision of sanitary sewers and central wastewater treatment plants is ordinarily not economically feasible. Septic tanks are the most common waste-disposal method in these areas. If properly designed and maintained, septic tanks and tile-drainage fields can be an acceptable permanent method of disposal of domestic wastewaters in rural or low-density areas. This requires suitable soils to accommodate tile-drainage disposal systems. Also, sufficient land area is needed to replace the tile-drainage field in the event of failure. Special design and construction techniques can be used in situations involving poor soils, shallow bedrock, and high water tables, but these techniques usually mean high installation costs.

Nonpoint-Source Pollution ■ A significant element of regional water-quality planning is nonpoint-source pollution. This is usually defined as pollution from widespread and diffuse sources, such as storm runoff from rural and urban land surfaces, as well as pollution from activities such as mining and silviculture. The main concern of nonpoint-source pollution is pollutant loading to downstream water bodies, particularly sediment and plant nutrients.

Sediment affects water clarity and aquatic life, and deposition of sediment can adversely affect habitats such as spawning areas, clog drainage structures, reduce flow capacity, and create navigation problems. In addition, sediment often carries other pollutants, including nutrients, metals, and toxic substances. Plant nutrients (such as phosphorus and nitrogen) can create nuisance algae and aquatic weed conditions in downstream water bodies and accelerate the eutrophication (or aging) of downstream lakes and impoundments.

Phosphorus is the plant nutrient most directly related to peoples' activities, more subject to control, and often the limiting nutrient for plant growth. Thus control of nonpoint-source pollution is often directed at control of phosphorus. Organic materials, metals, and toxic substances may also be of concern in urban runoff, and pesticides and herbicides may be significant concerns in agricultural areas.

The main sources of urban nonpoint-source pollution include erosion from construction sites, runoff from developed urban lands, and runoff from streets and parking lots. Erosion from construction sites is often a major source of sediment and pollutants in runoff from developing urban areas because erosion from unprotected construction sites occurs at rates from 10 to 100 times the normal rate of erosion on agricultural cropland. Management practices directed at controlling urban nonpoint-source pollution include vegetative management and erosion control at construction sites; detention and infiltration of urban runoff; and improved housekeeping practices, such as street sweeping and removal of leaves and yard debris.

The main sources of agricultural nonpoint-source pollution are runoff from barnyards and livestock concentrations and cropland erosion and runoff. Irrigation return waters may also create significant water-quality problems in some areas. Management practices directed at controlling agricultural nonpoint-source pollution include soil conservation, barnyard-runoff control programs, management of livestock waste, and careful application of fertilizers and pesticides.

A number of other potential sources of nonpoint-source pollution include mining, silviculture, and logging practices. Management of vegetation and erosion-control practices are applicable to many of these nonpoint sources. In addition, other specific management practices are utilized to address particular problems.

Solid-Waste Disposal ▪ The collection and disposal of solid wastes is one of the more significant and costly public services in urban areas.

Resource recovery (recycling) and volume reduction of solid wastes are often utilized to reduce the quantity of waste for ultimate disposal. Resource recovery may be directed at recovering the energy value, the materials value, or the organic and biological value of solid waste. Energy recovery includes the production of steam or hot water through incineration or direct burning.

In addition to resource recovery, incineration has been commonly used as a volume-reduction technique for solid waste. Conversion of solid-waste materials into fuels for generating energy is also practiced in a number of areas. Facilities that convert solid wastes into fuel for power generation or that produce steam or electricity through direct incineration of wastes involve high initial construction costs and annual operating and maintenance costs. Significant savings can be realized if fuel prepared from solid wastes can be burned in existing boilers and power-generation facilities as a substitute for a portion of the normal fuel.

Organic materials can often be recycled by returning them to the land and processing and using these wastes as soil conditioners and natural fertilizers and for erosion control. This approach may be applicable for crop residues, animal manures, wastewater-treatment sludges, leaves and yard wastes, tree wastes, canning wastes, and other decomposable organic wastes.

Individual materials can be recovered and recycled from solid waste either before or after the waste has been collected. This approach is most applicable to materials of relatively high value or materials easily segregated. Newsprint, ferrous metals, office paper, scrap metal of high value (such as copper), used motor oil, and similar materials can often be economically recovered and recycled from solid wastes. The economics of recovering and recycling individual materials varies, and the success of these endeavors usually depends on the availability of markets for the recovered materials.

The ultimate disposal of solid wastes and residue from volume-reduction and resource-recovery processes is placement of these wastes and residues in open dumps and landfills. Open dumps and uncontrolled landfills are common, but they are not environmentally sound or acceptable means of solid-waste disposal. Sanitary landfills, in contrast, are an engineered method of land disposal, which can be acceptable from an environmental standpoint and which are flexible and economical means of solid-waste disposal. Such landfills should be located in environmentally suitable locations and designed to prevent pollution of groundwater. It is also desirable to locate disposal sites close to major waste generators, to minimize hauling costs. This is an important factor in overall solid-waste costs because hauling costs often exceed disposal costs.

14.11 Air Quality

Air-quality problems are usually not confined to a single local jurisdiction. Therefore, they are usually addressed on a metropolitan or regional basis. Air-quality problems are most serious in metropolitan areas where climate and physiography contribute to serious air-pollution episodes as a result of temporary temperature inversions and stagnation of air-circulation patterns.

Pollutants of most concern in air-quality studies include sulfur dioxide, suspended particulates; hydrocarbons, oxides of nitrogen, carbon monoxide, and ozone—all of which have serious potential ill effects on public health, particularly in exacerbating respiratory illnesses. Besides affecting public health, most of these pollutants have significant adverse effects on plants' growth and health. Sulfur dioxide and nitrogen oxides can also contribute to acidic precipitation, which can have important adverse effects on fish and aquatic life in some areas. Finally, air pollution often results in increased haze and reduced visibility.

The principal sources of sulfur dioxide and suspended particulate matter are stationary, such as industrial smokestacks, power generation, and combustion of fossil fuels. The principal means of reducing these emissions include controls on stack emissions and substitution of low-sulfur or cleaner fuels in fuel combustion. Forest fires, burning solid wastes, and fugitive dust from wind erosion and construction sites can also significantly contribute these pollutants in some areas.

The principal sources of hydrocarbons, oxides of nitrogen, and carbon monoxide are vehicle emissions. Ozone results from the action of sunlight on nitrogen oxides and hydrocarbons and is therefore also a result of vehicle emissions. The principal means of reducing vehicle emissions include emission controls and altering transportation systems to encourage reduced travel or to make travel more efficient by encouraging greater use of mass transit, ridesharing and carpooling, bicycling, and other modes of transportation.

14.12 Outdoor Recreation and Open Space

Outdoor recreation and open-space planning is conducted at both community and regional levels. The objective of a community or regional outdoor recreation and open-space plan is to provide opportunities, lands, and facilities to satisfy demand for such activities as walking, swimming, boating, bicycling, fishing, camping, sports and games, and outdoor cultural events.

Function and Classification of Open Space ▪ Open-space lands can be classified into five general categories, according to the primary function of the open space. These include lands devoted to:

1. Resource production, including such uses as agriculture, forestry, and mining

2. Environmental protection, including fish and wildlife refuges, wetlands and marshes, ground water recharge areas, and watershed and stream-corridor protection areas

3. Protection of public health and safety, including floodplains, landslide and earthquake-hazard areas, fire-hazard areas, steep slopes, and areas of high noise exposure (such as airport flight paths)

4. Satisfaction of community outdoor recreation needs, including parks, playgrounds, trails, and other recreation areas

5. Buffer areas and separation between adjacent communities or incompatible land uses, growth-management boundaries and barriers, and utility and transportation corridors—as a determinant of urban form

A thoughtful and well-designed open-space system includes elements of all five open-space functions, within an integrated multipurpose network of community and regional open spaces.

Environmental Corridors ▪ The discussion in Art. 14.9 of urban drainage systems indicated the potential advantages of protecting and utilizing natural drainage systems. The concept of environmental corridors is utilized in community and regional planning programs to address the multiple concerns of drainage, water quality, recreation, and open space. Environmental corridors are linear systems of open space that include environmentally sensitive lands and natural resources requiring protection from disturbance and development and lands needed for open space and recreational use. These linear features are developed around drainageways and stream channels, floodplains, wetlands, and other resource lands and features (Fig. 14.2).

Protection and preservation of environmental corridors directly relate to environmental protection in general, and specifically to water-quality enhancement through reduction of nonpoint-source pollution and protection of natural drainage systems. In addition to protecting natural drainage systems in urban areas, environmental corridors can protect and preserve sensitive environmental areas, such as wetlands, floodplains, woodlands, steep slopes, and other areas that would impair surface or groundwater quality if disturbed or developed.

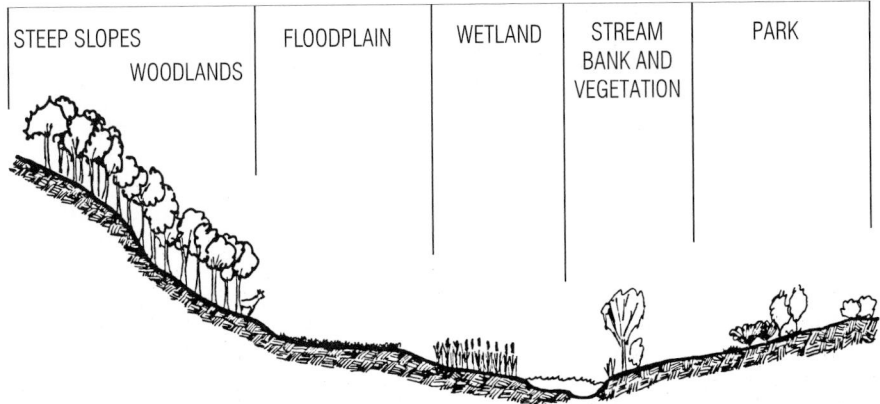

| STEEP SLOPES WOODLANDS | FLOODPLAIN | WETLAND | STREAM BANK AND VEGETATION | PARK |

Fig. 14.2 Components of an environmental corridor: woods on steep slopes to minimize erosion and provide habitat for wildlife, floodplains, wetlands, parks, and stream banks and vegetation, with buffer zones along the streams to filter out pollutants, prevent flood damage, and provide recreation areas.

Most open-space and recreational uses are compatible with these lands. Therefore, environmental corridors can be a major part of the needed open space for a community or region. In addition, the linear nature of environmental corridors is suited to increasingly popular recreational activities requiring trail development, such as hiking, biking, cross-country skiing, and nature walks. Finally, the linear aspect of environmental corridors provides continuity, which enhances the value of the corridors as wildlife habitat.

Multipurpose environmental corridors can be delineated and protected in both urban and rural areas through a combination of regulation and acquisition. Regulatory protection through floodplain zoning and zoning restrictions on shorelands, wetlands, and conservancy areas is appropriate where public access is not needed and allows lands to remain in private ownership. It is necessary to acquire lands through dedication or purchase where public access is required for recreation, for provision of structures such as detention basins, or where access is needed for public maintenance of stream channels and structures. In addition, public acquisition through dedication or purchase may be required to protect important resource areas vulnerable to development and not adequately protected through zoning or other regulatory means. Conservation easements may also be used in instances where fee-simple title is not needed.

Parks and Playgrounds ▪ Traditionally, planning at the community level has focused heavily on public park and playground development. More recent trends have placed greater emphasis on the satisfaction of recreation needs through multipurpose use of resource protection areas and environmental corridors. There remains, however, a need for provision of parks and playgrounds in the community or regional outdoor recreation and open-space plan.

There is a difference between intensive and active use areas for outdoor games and sports and passive or extensive uses, such as scenic enjoyment and picnicking. Playgrounds and playfields are examples of recreation areas designed for intensive or active use, whereas parks are usually oriented to satisfy needs for both active and passive uses.

It is extremely important to provide for coordination and joint use of school and community recreation facilities. A neighborhood playground provided in conjunction with a neighborhood school site can satisfy students' recreation needs as well as be a neighborhood playground. Community playfields, located in conjunction with middle or high schools, can provide similar benefits of joint and efficient use of land and facilities. Specific programs for coordination and joint provision of outdoor recreation facilities are necessitated in most areas because separate government agencies are responsible for recreation and schools.

Convenient location and access are important attributes of parks and playgrounds. Playgrounds and playfields have a high degree of locational flexibility since the principal requirement is for relatively level, well-drained land. They should be located in conjunction with or in proximity to neighborhood and community schools. A location that provides safe and convenient access is important for both the school and the recreational facility.

Resource protection areas and parks have less flexibility in location. The best sites are often dependent on the location of scenic areas, water features, and resources needing protection. Hence, greater effort must be made in providing adequate and convenient access to the sites by transportation systems, including mass transit, bicycles, and pedestrians.

14.13 Urban Design and Esthetics

Urban design and esthetics are concerned with shaping urban form and identity and improving the quality of the visual landscape. The subject is complex, involving the design skills of many disciplines—architects, landscape architects, urban designers and planners, and civil engineers.

Principles and Elements of Urban Design and Form ▪ Perhaps the most important principle in urban and regional design and form is that each element in the physical landscape be designed within the context of the next larger unit. Buildings should be designed to relate to the street or block, blocks designed within the context of the neighborhood, the neighborhood related to the community, and the community planned within the context of the region. Ideally, each element will have a clear and distinct identity but will be compatible with its surroundings and landscape and appropriate in the context of its larger setting. Elements should work together as a cohesive unit if properly planned and designed.

The key elements in urban form and identity that most people use to construct a mental image of a city include pathways (routes of circulation), districts (component areas or neighborhoods), edges (of districts), landmarks (prominent visual features helpful in orientation and identification), and nodes (centers of activity). A visual experience may be relatively static if experienced at one location. Most visual experiences, however, consist of a dynamic sequence of activities and views that occurs during movement from one location to another.

The dynamic and sequential nature of this experience should be taken into account and can be improved by carefully designing and locating urban elements and considering the relationship between elements. Transitions from one district or visual setting to another are extremely important—diversity and contrast add variety to experience—but extreme contrast or rapid transition may disturb or disorient the viewer.

Appropriate scale, also important, depends on the mode and speed of movement. Correct scale of a visual element is substantially different for a pedestrian than for a passenger in a fast-moving automobile.

Tools and Techniques ▪ Tools and techniques available to governmental bodies to foster and enhance urban design and esthetic considerations include education and information programs, regulatory programs, and design and location of public facilities.

Visual surveys of the community or region are an important element in an urban-design education and information program. A visual survey is an inventory of view experiences, scenic vistas, landmarks, approaches, and other visual aspects of the community or region. The purposes of a visual survey are to inform the public and elected officials of visual assets to be protected and to point out visual problems needing correction or improvement.

Regulatory programs are one of government's principal tools in attempting to improve the quality of the visual landscape and to shape urban form and identity. An important approach to reducing visual clutter in urban areas is to require that utilities be placed underground wherever feasible.

Zoning and subdivision regulations usually provide for the regulation of signs, fences, landscaping requirements, buffer areas, and other features that can have an important impact on the visual environment. In addition, building height, bulk, and site placement (setbacks) are regulated through zoning.

In some communities, a substantial degree of architectural control and design review is provided through the establishment of architectural review boards or urban design commissions. These groups, which commonly include professional designers as well as citizens, review development proposals and projects for visual quality and architectural compatibility.

Land-use planning and zoning regulations can also be used to protect significant views and landmark visibility and to maximize the benefits of significant views through zoning and building placement. Regulations can preserve lines of sight and prevent adverse development of areas or landmarks that are viewed. Zoning regulations can also be used to encourage higher densities or public uses in prime viewing areas, thus maximizing the opportunities for enjoyment of such views. Thoughtful urban design can maximize views by limiting building height, with higher buildings permitted farther away from prime views and landmarks and lower buildings placed closer to the views. This permits maximum exposure and enjoyment of views and landmarks and avoids interruption of lines of sight by nearby buildings and obstructions.

Design and location of public spaces and facilities are important opportunities for government to exert a positive influence on urban design and visual quality. Perhaps the most important role for communities is utilizing open-space acquisition and protection policies to achieve the goals of urban design and esthetics: shaping urban form and identity and improving the quality of the visual landscape. Open-space acquisition and protection policies can be used to protect scenic resources, to define district or community edges and provide buffer areas between districts, and to define and protect pathways or corridors for future uses.

Streets and roadways are public spaces and facilities with pervasive and substantial impacts on the quality of the visual experience for most residents of a community or region. By properly fitting streets and roadways to the landscape, paying careful attention to street furnishings and hardware (such as paving materials, street lighting, traffic signs and control devices, bus shelters, and other structures), and providing for sensitive landscaping of street and roadway corridors, a community can have a significant beneficial impact on the everyday visual experience of most residents. It is important to pay particular attention to main corridors and entrances to a community with high volumes of travel because these routes have high visual impact and often shape the image of a community for visitors.

In addition to streets and roadways, government can influence visual quality through design and location of other public buildings, spaces, and facilities. In all cases, key elements include good design in the context of the immediate surroundings and the larger unit, sensitive landscaping of public spaces, and provision of art in public spaces.

Historic Preservation ▪ Preservation and reuse of existing historic buildings and landmarks is an important planning issue in many communities. Historic preservation is significant to urban design and esthetics, in addition to the historic heritage being preserved.

The usual approach is to designate historic sites, landmarks, or districts. Once designated (under a variety of local, state, and Federal laws), changes in the exterior appearances of buildings are restricted, and tax incentives are often provided to encourage preservation of the building exteriors. In most cases, the building interiors may be adapted to modern use, although interiors may also be preserved in outstanding cases. Designation of individual buildings, sites, or landmarks may be limited to outstanding examples of architecture or historic interest. Designation of entire historic districts permits the inclusion of many buildings of lesser importance from a historic or architectural standpoint, to retain the historic flavor of the district and compatible architectural surroundings.

Land-Use Planning

14.14 Housing and Residential Land Use

Housing is a basic need in society and occupies most of the developed or urban land in a community or region. Hence, housing and residential land use is a central element in most community and regional land-use plans.

Housing is provided primarily by the private sector. However, government is directly involved in controlling location and type of housing and residential development as well as providing public services and facilities in residential areas. Federal, state, and local governments also provide financial assistance for low-income and elderly housing.

Housing Types ▪ The principal distinction in housing types is between single-family detached homes, which represent the largest proportion of housing provided in most areas, and multifamily units, including apartments and townhouses. Duplexes, or attached two-family homes, may be treated similarly to either single-family detached homes or multifamily units, depending on the specific area and circumstances.

Group quarters, such as dormitories and residential hotels, may be important in some circumstances involving military installations or institutions like universities.

Mobile homes and factory-built housing units are an important segment of the housing market. These prefabricated dwelling units are often treated separately from housing constructed on-site, primarily because of differences in construction, architecture, and permanence of siting.

Goals and Objectives ▪ A fundamental goal of planning is to provide for a sufficient supply of housing units to accommodate the needs of the population. These housing units should be provided and maintained in safe, healthy, and sound condition. They should be located for convenient access to neighborhood and community activities and facilities. There should be a broad choice of housing types to suit the differing needs and desires of various segments of the population. In addition, an important objective is to ensure that housing is available at costs compatible with residents' income levels.

Establishing Needs ▪ A critical initial step in establishing housing and residential land needs in a community or region is to conduct a survey of the existing housing stock. This survey should be concerned not only with the numbers of available housing units but with their condition. Existing housing can be described as being in sound condition (requiring only conservation of existing condition), deteriorating condition (requiring repairs and rehabilitation), or dilapidated condition (requiring clearance or reconstruction). The survey of existing housing stock should also note other problems and deficiencies and describe trends and indicators of potential problems. Vacancy rates can be an important indicator of surpluses or deficiencies in the housing market or for specific housing types. Conversion of single-family homes to multifamily use is also an indicator of change in housing demand and may lead to neighborhood destabilization.

Future dwelling-unit needs can be predicted on the basis of population forecasts divided by the average number of persons per dwelling unit. Because the average number of persons per dwelling unit may change substantially over time, it is important to examine historical trends and future potential changes in this ratio.

The difference between total future needs and that portion of the existing housing stock that can supply safe and sound housing units at a future date represents the additional housing units to be constructed. Table 14.3 illustrates the assumptions and steps needed to complete a housing-needs analysis.

The total additional acreage required for new residential construction is based on the number of additional dwelling units needed and assumed future types and densities of housing units. In rural or undeveloped areas lacking public water or sewerage facilities, residential densities are usually less than 2 units per acre. These low densities result from reliance on on-site waste disposal (septic tanks) and water supply and the need to avoid overtaxing public services and facilities. In urban areas with public water and sewer service, densities generally range from 2 to 8 dwelling units per acre for detached single-family housing (with most new construction in the 2- to 5-dwelling unit per acre range); 5 to 20 units per acre for duplexes, townhouses, and garden apartments; and 30 or more dwelling units per acre for high-rise multifamily residential development.

Location and Design ▪ Planning and design of residential land uses are normally formulated on the neighborhood concept. The neighborhood is a relatively self-contained area containing from about 2000 to 10,000 residents, oriented around significant neighborhood facilities, such as an elementary school, neighborhood park or playground, and neighborhood shopping area. Proper neighborhood layout and design include adequate access to community and regional transportation systems, utilities, and public facilities. A neighborhood should not be divided by major transportation corridors or incompatible land uses. It is desirable to have a gradual transition from high- to low-density areas and to provide buffers between different types and densities of land uses. In addition, higher densities may be most appropriately

Table 14.3 Illustrative Housing-Needs Analysis

Component	2000 (Census)	2020 (Forecast)
Population and households		
Total population	300,000	400,000
Group quarters population	15,000	20,000
Household population	285,000	380,000
Average household size, persons/unit	3.1	2.9
Total households	91,900	131,000
Housing units		
Total households	91,900	131,000
Vacancy rate, %	3.9	5.0*
Total housing units	95,600	137,900
Housing needs, 2000–2020		
Increased number of units needed	42,300	
Replacement of existing units[†]	5,000	
Total additional units needed	47,300	
Annual housing unit construction needs	2,365	

* Desired vacancy rate.

† Includes units lost through disaster, demolition, and conversion.

located near major transportation corridors and activity nodes or major facilities.

In many areas, instead of typical rectilinear or gridiron layout of streets and lots, cluster developments and planned-unit developments are used. These alternative layouts provide the opportunity for curvilinear arrangements and street layouts, which permits siting streets and homes in a manner more compatible with topography. The curvilinear layouts made possible through clustering and planned-unit development also provide greater opportunities for preserving existing trees and natural vegetation and can reduce overall land-grading requirements. By providing for smaller clustered lots and common open space, these approaches reduce the need for improvements and utilities. The result is more effective use of land and less wasteful provision of open space. Figure 14.3 illustrates some of the open-space and cost advantages of cluster developments over conventional subdivisions.

Although there are a number of advantages to planned-unit development and cluster development, there have been some abuses and disadvantages. The use of dead-end streets and cul-de-sacs presents disadvantages to public services. Dead-end water mains do not provide as much available fire flow as do looped mains, and dead-end streets require service and delivery vehicles to turn around and double back, which can raise the cost of public services. Generally, it is advisable to avoid excessively long dead-end streets or cul-de-sacs.

The design and location of streets is commonly provided for and specified in subdivision regulations. Moderate street grades are important for ensuring access (particularly for public-safety vehicles during adverse weather conditions) and can also be important in avoiding drainage problems. The width of residential streets is also usually specified in subdivision regulations and depends on the need for on-street parking in a particular area. Also, because of the importance of pedestrian movement and bicycle transportation in neighborhood planning and layout, sidewalks and bicycle routes should be provided to ensure safe access of children and adult residents to neighborhood schools and facilities.

Mixed-Use Neighborhood Planning ■
Land use planning from the 1950s through the 1980s followed the post-war trend toward low-density auto-dependent suburban development patterns, characterized by a high degree of segregation or separation of land uses, as well as income stratification in new residential neighborhoods.

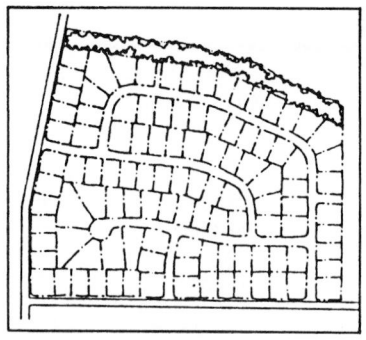 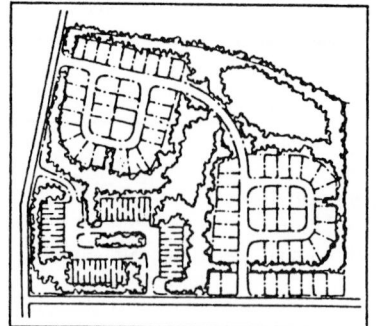

NUMBER OF LOTS: 108
OPEN SPACE: 10%
LINEAR FEET OF STREETS: 5,400
LINEAR FEET OF SEWER LINES: 5,400

NUMBER OF LOTS: 108
OPEN SPACE: 50%
LINEAR FEET OF STREETS: 4,900
LINEAR FEET OF SEWER LINES: 3,900

(a) CONVENTIONAL SUBDIVISION

(b) OPEN-SPACE SUBDIVISION

Fig. 14.3 Comparison of conventional and open-space subdivisions. (*Reprinted from "How will America Grow?" Citizens Advisory Committee on Environmental Quality, 1976.*)

As these suburban developments have aged, their problems and shortcomings have become both more evident and serious. In the 1990s, planners began to increasingly explore development patterns that would retain the desirable features and social functionality of the better urban neighborhoods built in the early 1900s, adapted to contemporary settings and circumstances.

This return to traditional neighborhood planning concepts is central to most current trends in planning thought, characterized by terms such as "traditional neighborhood development", "new urbanism" and "smart growth". Many of the concepts, and examples of this trend in neighborhood planning and development are presented in Peter Katz' 1994 book "The New Urbanism" (McGraw-Hill), included in the list of references at the end of this section.

The most basic principles and objectives of this approach to neighborhood planning and development (or redevelopment) are presented in Table 14.4.

14.15 Commercial Land and Shopping Areas

Planning of commercial land and shopping areas is basically concerned with provision of goods and services to consumers. In addition, commercial and service-related employment has a significant impact on the economy.

Classification and Types ▪ Standard land-use classifications for commercial land and shopping areas include wholesale and retail trade; finance, insurance, and real estate; services (personal, business, repair, and professional); and offices.

Commercial and shopping areas can be classified as regional centers, community areas, local or neighborhood areas, and limited-function or specialized areas. Regional centers include such major concentrations as the central business district of a metropolitan area and regional shopping centers. Community areas include the central business district of a community and community shopping centers. Local or neighborhood areas include neighborhood concentrations or shopping streets, neighborhood shopping centers, and rural concentrations.

Limited-function or specialized areas include highway service concentrations along major streets and highways and at highway intersections and freeway interchanges. Office space for professional, finance, insurance, real estate, and other similar services and entertainment areas, nightclubs, and specialized functions such as new and used car sales may be provided either as part of regional

Table 14.4 Desirable Neighborhood Characteristics

Principle or Characteristic	Purpose or Rationale
• Compact, higher density	• Provide population base for neighborhood facilities, services, shopping, transit, employment within a reasonable walking distance ($\frac{1}{4}$ – $\frac{1}{2}$ mile). Enhances social interaction, reduces auto dependence.
• Mix of land uses	• Provide opportunities for shopping, professional services, employment within neighborhood. Reduces auto dependence and travel.
• Clustered development with smaller lots, more common or public open space	• Allows easy access to expanded and more diverse recreational opportunities. More efficient use of land, improves community appearance, enhances social interaction.
• Diverse mix of housing types and costs	• Provide housing for full range of family sizes, ages and income levels. Reduces segregation and income stratification, enhances social diversity.
• Garages and parking lots behind structures, with alleys and service drives	• Improves community appearance.
• Neighborhood public facilities and spaces (schools, parks, libraries, neighborhood centers, etc.) and shopping areas coordinated as focal points at visible and accessible locations	• Improves visibility, access and use of neighborhood facilities and shopping areas.
• Interconnected or grid-style neighborhood street systems with narrower streets	• Fosters distribution rather than concentration of neighborhood auto traffic, reduces congestion, slows neighborhood auto traffic, better and safer pedestrian and bicycle travel, easier and more flexible navigation and movement within neighborhood.
• Transit-oriented development with transit stops within walking distance of all parts of the neighborhood and at activity centers	• Enhances mobility and access to the larger community and region, particularly important for those not able to drive. Increases ridership and support for the transit system, reduces auto dependence.
• Pedestrian-friendly design	• Improves access to neighborhood facilities, services, shopping, jobs; reduces neighborhood auto trips and auto dependence.

centers, community neighborhood areas, or as separate limited-function or specialized areas.

Goals and Objectives ▪ The basic goal of commercial land and shopping areas is to provide an adequate supply of goods and services. Also, different types of shopping areas and centers should be provided, to satisfy a variety of routine and specialized shopping needs. In addition, commercial land and shopping areas should be located and designed to maximize convenience to and safety of the individual consumer.

Establishing Needs ▪ Future needs for commercial land and shopping areas are usually based on population forecasts. If economic studies and forecasts are available, future needs based on population forecasts should be consistent with forecasts of employment, retail sales, or other economic factors. Economic studies and market surveys are often the basis for determining specific deficiencies and needs in the commercial sector. Market surveys can be used in conjunction with an inventory of existing commercial land and the present commercial economy and with forecasts of future needs to determine needed additions to commercial land and shopping areas.

Location and Design ▪ Since convenience and access are the fundamental objectives of planning for commercial land and shopping areas, location is the key attribute of a well-planned commercial land network. Regional or community centers should have direct access to major highways as well as mass transit (for employees as well as consumers). A highly visible location is important for regional and community centers.

Commercial lands and shopping areas usually attract and generate high volumes of traffic, so control of traffic and ingress and egress to the shopping area is extremely important. It is also essential, when designing the site, to provide for separation of pedestrian and vehicular traffic and to ensure that pedestrian movement is safe and convenient.

Since most consumers use automobiles for shopping trips, provision of adequate parking is a key element in design of a shopping area. New shopping centers ordinarily have adequate parking within a reasonable walking distance of a number of stores. The provision of adequate off-street parking is particularly important for existing downtown and community shopping areas if they are to compete with community and regional shopping centers.

The appearance of commercial areas and shopping centers can help attract consumers. Landscaping is an important element since many major shopping centers feature large barren parking areas. Control of signs is also important, particularly in commercial strips located along streets and highways. In addition, landscaped buffer areas are often important for protecting adjacent land uses.

Level, well-drained sites are desirable for shopping centers and commercial areas. Without good drainage, the volume and peak rates of storm runoff can be a problem in new shopping centers and commercial areas because of the relatively large areas of impervious paving and rooftops. Innovative approaches to incorporating treatment and infiltration of parking lot runoff have been developed to offset adverse impacts.

Shopping and commercial areas have some special service needs. These include truck deliveries, for which appropriately located loading and unloading areas of sufficient size for the purpose should be provided, and collection and removal of large quantities of refuse and paper wastes, for which solid-waste storage areas and efficient, economical collection practices are essential.

Pedestrian malls, which are open or enclosed pedestrian shopping ways or streets, are provided in many large shopping centers. This concept also has been adapted to central business districts and downtown community shopping areas, where pedestrian malls are created by eliminating vehicular traffic from former shopping streets. In some cases, however, limited vehicular access is retained for transit, delivery vehicles, and bicycles.

Undesirable Conditions for Commercial Development ▪ These often arise from strip commercial development, highway interchanges in rural or undeveloped areas, and commercial overzoning.

Inasmuch as commercial activities and shopping areas thrive on exposure and access, much commercial development has traditionally located along major highways and thoroughfares. This strip form of commercial development has created a number of undesirable conditions; as a result, most community and regional plans attempt to ensure that new shopping areas are not located and designed in this fashion.

Strip commercial development is inconvenient for the consumer who needs to make more than one shopping stop or who wants to comparison shop. Provision of ingress and egress directly from major traffic arteries to individual stores creates traffic-safety hazards and disrupts the flow of traffic, creating unnecessary congestion. Furthermore, there is evidence that it is more difficult to provide adequate levels of security, so strip commercial development may be somewhat more vulnerable to crime. For these reasons, most plans recommend the clustering of stores in commercial and shopping areas to permit consumers to satisfy their shopping needs at one location. Where it is necessary to develop commercial areas in a strip fashion, access can be provided from frontage roads paralleling the major traffic artery, rather than directly from the highway. Frontage roads can also be used in some circumstances to reduce the undesirable conditions associated with existing strip commercial development.

Highway interchanges in rural or undeveloped areas can make difficult the provision of adequate support services for commercial areas. In many instances, substantial high-value commercial facilities (often highway-service facilities oriented to motorists' needs) develop around rural intersections and interchanges, and services that are most important are often lacking, including sewage disposal, water supply, police and fire protection, and solid-waste collection and disposal. Generally, joint or cooperative provision of these services is better and more economical than individual on-site solutions. It may be necessary to devise or develop a joint management structure to provide these services cooperatively or to use special or existing governmental units to provide necessary services.

Another undesirable condition encountered in the planning of commercial land and shopping areas is overzoning. In many cases, communities attempt to attract and encourage commercial growth by zoning a great deal of land for commercial purposes in desirable locations. Overzoning commercial land or zoning all suitable areas should be avoided, to ensure that commercial development that does occur is in the proper location and that local governmental units retain control over the type and location of commercial areas. Zoning of commercial land should ordinarily be done only in response to specific proposals at the time of development, and the plan should specify only general locations for new shopping centers and commercial developments.

14.16 Industry

Planning for industry is primarily concerned with making suitable land and support services available for manufacturing and production of goods and materials. Agriculture, mining, and forestry are also industries but usually receive separate and special treatment in most planning studies.

Objectives and Needs ▪ The goals and objectives of planning for industry include:

1. Strengthening the economic base of the community. (Industry, by producing goods and materials for export from the community or region, brings in outside income that has a multiplier effect on the local economy.)

2. Encouraging development of various types of industrial and manufacturing concerns, to provide a diversified employment base. (A diversified employment base is usually more stable and less subject to sudden market fluctuations or economic downturns.)

3. Providing a sufficient number of industrial and manufacturing jobs tailored to the capabilities of the available workforce.

4. Providing adequate and suitable land for manufacturing and production purposes.

5. Locating industry to provide for economical and efficient transport of raw materials and finished goods and to be conveniently accessible to employees.

Future land needs for industry are usually based on employment forecasts. For industries utilizing multishift operations, peak shift employment should be utilized for determining land needs rather than total employment. For new industries in planned industrial parks, density may be as low as 5 to 10 employees per acre. Much higher densities are common for multistory plants and for industries and manufacturing plants in high-density areas. Additional land may be needed to accommodate the relocation of existing manufacturing enterprises to low-density or suburban areas.

In many communities, land well-suited for industry may be in short supply. Such land should be reserved and protected from development for other land uses (such as residential) that have less stringent locational and site requirements. Zoning land for industry and reserving land in planned industrial parks are common approaches to reserving future industrial land.

Location and Design ▪ Most new sites for industry and manufacturing plants are located in outlying or suburban areas, often in planned industrial parks. These suburban locations and industrial parks are characterized by low density and one-story buildings, to meet industrial needs for space for expansion, material storage, loading and unloading areas, and employee parking.

Manufacturing plants and industrial sites require access to major transportation facilities, both for importing raw materials and exporting finished products. Access to major highway and rail facilities is important in almost all cases. Roadways should be capable of accommodating large trucks. Airport industrial parks are sometimes provided for industries relying heavily on air transport. Access to water-transportation facilities, including docking and loading or unloading facilities, can be important and beneficial for industries receiving or shipping bulk materials.

Industrial sites should be level and well-drained and have suitable soils for foundation requirements for equipment, machinery, and structural loads. Land area should be sufficient to accommodate expansion needs, requirements for material storage and loading and unloading areas, and employee parking. For multishift operations, employee parking needs may have to take into account some overlap of shifts. Access to transit facilities for employees is desirable and can reduce employee parking needs. If public transit is not available, carpools or vanpools can be organized to reduce parking demands.

In most new industrial areas, substantial buffer areas are provided to protect adjacent land uses. Attractiveness and landscaping are considered important attributes, and substantial building setbacks are often required. Zoning and site regulations for industrial areas usually contain specific performance standards to limit emissions, pollution, noise, and other potential nuisances to acceptable levels.

14.17 Community Facilities and Institutions

Community facilities and institutions include private and public educational facilities, libraries and cultural facilities, hospitals and health-care facilities, public-safety facilities, governmental administrative facilities, water and wastewater-treatment plants, and solid-waste facilities.

Educational Facilities ■ Educational facilities include public elementary, middle (junior high) and high schools, private and parochial schools, universities, junior colleges, vocational and technical schools, and other specialized schools.

Public elementary schools are for students from kindergarten through sixth grade. Elementary schools are ordinarily designed to serve a neighborhood within a ½-mi service radius. They have an ideal enrollment of 400 to 600 students. As the discussion in Art. 14.12 indicated, the optimum land use results when elementary schools are provided in conjunction with a neighborhood park or playground. The ideal site size for a combined school-park incorporating an elementary school and a neighborhood playground is 10 to 15 acres. An elementary school-park should be located on a collector street and sited so that children do not have to cross major highways or traffic arterials to travel to and from school.

Middle or junior high schools are for grades 7 to 9. These schools are ordinarily designed to serve a community, with a desired service radius of 1 to 1¼ mi. The ideal enrollment of a middle school is from 500 to 1500 students. A middle school should be located in conjunction with a district park or community playfield. A middle-school site that includes sufficient open space and recreation area ranges from 25 to 35 acres. A middle school should be located near arterial streets, preferably with access to mass transit.

Senior high schools are for grades 10 to 12. The schools are usually designed to serve a community, preferably with a service radius of 1½ to 2 mi. The ideal enrollment for a high school is 1000 to 2000 students. A senior high school should be located in conjunction with a district park or community playfield. A site of 40 to 50 acres includes adequate lands for open space and recreation areas. A senior high school should be located near arterial streets, preferably with access to mass transit.

In many areas, parochial and private schools educate a significant proportion of the population. When projecting space and facility needs for public schools, it is necessary to account for anticipated enrollment in private and parochial schools. When planning for higher educational facilities, including junior colleges, universities, and vocational and technical schools, it is useful to distinguish between those facilities and campuses where most of the student body reside on campus and commuter campuses where most students travel daily to and from the campus. Location and access, including access to mass transit, is a particularly important consideration for commuter campuses.

Libraries and Cultural Facilities ▪ In larger communities, a headquarters or main library may be provided in addition to community branch libraries. An important attribute of a main library is accessibility, particularly access to mass transit and pedestrians. Community branch libraries serve a population of from 20,000 to 30,000 residents located within a 1- to 1½-mi service radius. Generally, community branch libraries serve about the same area as a middle or senior high school, and the ideal location for a community branch library is adjacent to or near a senior high school. This provides for joint use of the library facilities by students and residents and can allow the library joint use of the school's off-street parking facilities outside school hours. If a location near a middle or senior high school is not possible, a location in a community shopping center or community business district should be considered. Generally, branch libraries have a minimum site requirement of about 1 acre for a one-story building.

Community cultural facilities include museums, concert halls, civic centers, performing arts centers, and municipal stadiums. Many of these facilities are significant traffic generators and should be located close to major highways and accessible to mass transit.

Health Facilities and Hospitals ▪ Health facilities include general and special hospitals, nursing homes, and mental health institutions. The minimum desirable size for a full-service general hospital is about 200 beds, which requires a support population of 50,000 to 75,000. Major facilities, particularly general hospitals and emergency and trauma centers, should be easy for patients to find and accessible to major highways. Accessibility to mass transit is also important for serving the needs of the hospital staff.

Public-Safety Facilities ▪ Public-safety facilities include police stations, fire stations, and emergency medical and ambulance services.

Police stations are often centralized, but in large cities, district or precinct police stations may be provided. Because most responses to crimes in progress are from mobile patrol units, the location of police stations is not as critical as that for fire stations. The location of jails and detention facilities and convenient access to courthouses, however, can be important determinants in the location of police stations. The need for central communications may also be an important factor in location.

In selection of locations for fire stations, response time to fires and other emergencies is extremely important. A basic pumper company should be provided for a maximum service radius of 1½ mi, and a basic ladder company should be installed for a maximum radius of about 2 mi. High-value areas require backup response from other nearby stations. Fire stations should have direct access to the major street and highway network. Additional stations may be needed in areas that can be isolated, such as areas where at-grade railroad crossings or temporary flooding can prevent access.

Emergency medical service or ambulance service is increasingly a governmental responsibility. Response time is also important for emergency medical service, and it is often recommended that this service be operated out of local fire stations.

Other Governmental Facilities ▪ Governmental administrative facilities include city halls, courthouses, post offices, multipurpose community and neighborhood centers, and municipal garages and maintenance facilities. Because many governmental administrative facilities are used by local citizens, these facilities should be easy to find and have convenient access.

The location of water and wastewater-treatment plants is ordinarily constrained by the location of water sources or by location of the most suitable receiving water for treated wastewater. For wastewater-treatment plants, it is usually wise to provide buffer areas to protect surrounding areas from potential odor problems.

Solid-waste facilities include transfer stations, recycling centers, processing plants and incinerators, and disposal sites. For facilities used by residents, convenient access is an important factor. Other solid-waste facilities should be located to minimize transportation costs and impacts, by providing access to major streets and highways and avoiding truck traffic on local or neighborhood streets. Location of solid-waste facilities near compatible land uses is desirable but buffer areas must be provided where adjacent land uses are sensitive to truck traffic and other impacts.

Utility and Transportation Systems

14.18 Service-Area Planning

Delineating the most desirable forms and patterns of urban development partly depends on those patterns that allow public services and utilities to be provided in the most cost-effective and efficient manner. Conversely, governmental control over the extension and timing of extension of public services and utilities can influence growth patterns and be an important technique in guiding both the location and timing of urban development.

Urban Services ▪ In most governmental jurisdictions, basic or general governmental services are available to all residents regardless of location. These include courts, basic police and fire protection, hospitals and health-care facilities, public-health programs, and construction and maintenance of public streets and highways and solid-waste disposal facilities.

In cities, villages, and other incorporated municipalities where development takes place at urban densities, a higher level of governmental services is generally provided. These additional urban services include public water-supply and distribution systems, public sanitary sewerage systems, higher levels of police and fire protection, solid-waste collection services, urban mass transit, urban drainage facilities and streets with curbs and gutters, and neighborhood facilities such as parks and schools. Since a chief function of urban governments is to provide urban services, a primary focus of community and regional planning studies is the determination of those areas that are anticipated will develop to urban densities and require urban services.

Urban Service Areas as a Planning Tool ▪ Most regions have more than enough vacant developable land to accommodate anticipated urban development. In some communities, boundary restrictions may limit the amount of vacant developable land available, with the result that all vacant land will be developed within the

planning period. In most cases, however, only part of the available land area will be developed for urban uses within the planning period. The purpose of urban service areas is delineation of those areas that are proposed for development at urban densities within the planning period and where the governmental unit intends to provide urban services by the end of the planning period. Urban densities include most residential land uses with densities higher than 2 units per acre and nearly all commercial and industrial land uses.

Urban-service-area delineation allows the governmental unit to plan the orderly extension of utilities and public services. This permits efficient utilization of capacity of facilities and utilities and avoids premature extension of utilities and services to pockets of development scattered over wide areas. Orderly extension of public services and utilities is an additional tool for controlling location and timing of development, to complement zoning and other growth-management techniques. Urban-service-area delineation also provides guidance to the private sector as to which areas are to receive public services within the planning period.

Delineation of Urban Service Areas ▪ The procedure for delineation of an urban-service area is summarized in Table 14.5.

The initial step is to identify those lands that are environmentally sensitive or unsuitable for development at urban densities. Lands that might be excluded from consideration for urban development include floodplains, wetlands, areas of steep slopes, environmental corridors, and areas of unsuitable soils or geology, valuable or unique vegetation, mineral resources, or wildlife habitat.

The second step is to locate potential boundaries for the urban service area. These may include either natural or constructed barriers to development, such as limited-access highways, rail corridors, stream or environmental corridors, and floodplains or wetlands. Drainage area or watershed divides are important boundaries for potential urban service areas because of the efficiency and desirability of providing sanitary sewerage and drainage facilities on a drainage-area basis.

The third step is to determine the amount of vacant developable land required to accommodate anticipated urban development during the planning period. The total amount of additional land needed for urban development can be determined

Table 14.5   Steps in Delineation of an Urban Service Area

1.   Identification of environmentally sensitive or otherwise unsuitable land
2.   Location of potential boundaries for the urban service area
3.   Determination of vacant land needed for development
4.   Delineation of alternative service areas
5.   Recommendation of a specific urban service area

from the land-use needs and forecasts (based on density standards and population and economic forecasts, discussed in Art. 14.6).

In the fourth step, alternative service areas having sufficient developable land to accommodate needed development are delineated. Ordinarily, it is advisable to delineate alternative urban service areas somewhat larger than needed to accommodate development. This is necessary for recognition of typical patterns of development. (Urban services are normally provided to areas prior to complete development.) A slightly larger area also provides some flexibility in the location of urban development.

Finally, a recommended urban service area is delineated. Although it is desirable to use natural or constructed barriers to development as boundaries, arbitrary boundaries may sometimes be necessary to provide the proper size. Bear in mind that urban-service-area boundaries will change with time and with changing conditions and forecasts.

Ultimate or Design Service Areas ▪ Most community and regional plans are prepared for medium-term planning periods ranging up to 20 to 25 years in the future. This period is the longest term for which reasonably detailed forecasts can be made. The physical life of many facilities (such as underground water and sewer lines) is often much longer than 20 years, and some of these facilities may be designed for periods ranging up to 50 years or more. The designer is concerned therefore with growth and capacity needs substantially beyond normal planning periods. A common approach is to assume full or complete development of the potential or ultimate service area of the facility (which may be a drainage area for sanitary sewerage and drainage facilities, for example), if that assumption appears reasonable and consistent with medium-range (20-year) growth trends. It is also usual to assume distribution and types of land use consistent with the medium-term plan and

growth trends. It is advisable for designers to review these long-term design assumptions with the planning agency responsible for making medium-term growth forecasts, to ensure that the long-term assumptions are reasonable and realistic.

Long-term forecasts are extremely questionable. For some types of facilities (particularly underground pipes), it is usually economical to provide extra capacity to allow for future uncertainty. For other facilities, it may be best to assume the risk that capacity may be reached earlier or later than the specific design period. If capacity can be added easily, short design periods may be most appropriate and cost-effective.

Short-Term Staging Boundaries ▪ In addition to the medium-term (20 year) and long-term service areas used to forecast growth and plan and design for facility needs, it is useful to delineate areas where urban services and utilities will be extended within the short-term future (5 to 10 years). This approach guides private landowners and developers regarding the specific intentions of the governmental unit to provide services, directs short-term growth and development activities into specific areas, and avoids the inefficiency of serving, in the short term, individual areas or developments widely scattered throughout the 20-year urban service area. Short-term staging of public services and utilities can be extremely effective as a growth-management technique.

14.19   Utility Systems Planning

After areas proposed for urban development (urban service areas) have been delineated (Art. 14.18), plans can be prepared for providing these areas with urban services and utilities. Public water distribution and sanitary sewerage systems are significant governmental responsibilities and capital investments in newly developing areas. Natural gas, electricity, and telephone service are provided by the private sector in most areas but may be provided by government in some jurisdictions. In most communities, solid-waste collection is a governmental responsibility but may be provided by contracting with private collection firms.

Water-Distribution Systems ▪ Water-distribution systems provide an adequate supply of potable water for residential, commercial, and industrial use and adequate water for firefighting throughout the urban service area. The specific criteria for layout and design of water-supply and distribution facilities are covered in more detail in Sec. 21. Following is a brief overview of those aspects of water-supply and -distribution systems of concern in general planning studies.

Water-distribution systems provide potable water under pressure, either by gravity from high-level reservoirs or under pressure from pumping stations. Required flow rates for firefighting purposes are usually the determining factor in sizing water mains in distribution systems and establishing distribution-system storage requirements. Ordinarily, distribution mains are laid out in a grid or looped fashion to avoid dead-end mains. Distribution mains should be buried at sufficient depths to avoid freezing in winter.

Distribution storage may be provided in either ground or elevated storage tanks. The primary purposes for distribution storage include operating storage, fire-reserve storage, and emergency-reserve storage. Operating and fire-reserve storage capacity is provided to satisfy peak hourly demand rates or peak fire flow rates, which are greater than the basic supply and transmission capacity. Emergency-reserve storage provides water during an interruption in supply or a breakdown in basic supply or transmission facilities. Elevated storage also stabilizes pressures in the system and allows gravity distribution during power outages when pumping stations are out of service.

Sanitary Sewerage Systems ▪ Development patterns that conform to topography and result in the layout of sanitary sewerage systems that drain by gravity result in greater reliability and significant savings in operating costs. Lift or pump stations are required in many cases, however, and generally result in a system more costly to operate and more vulnerable to power outages or equipment breakdown.

Most areas developed in recent years have been provided with separate sanitary and stormwater sewers. There are in many cities, however, older areas that are served by combined sewers, which carry wastewater during periods of dry weather to a wastewater-treatment plant and carry storm runoff mixed with wastewater during periods of rainfall and discharge to surface-water bodies. Combined sewer systems can aggravate poor-water-quality conditions in receiving waters, and corrective solutions are difficult and expensive.

Sanitary sewers are sized to accommodate peak flows. Peak flows are based on average flow rates, determined from local flow data where available. Where local data are not available, a flow of 100 gal per capita per day is commonly assumed. This flow includes provision for usual residential and commercial wastewater flows plus an allowance for normal infiltration and inflow. Significant industrial contributions must be added to this figure. Peaking factors applied to the average daily flow rate commonly range from 1.5 to 2.0 (for sewers serving very large areas) to 4.0 (for small sewers serving relatively limited areas). The minimum diameter for sanitary sewers is 8 in, which, at minimum recommended slope, can serve a population of 1000 to 1500, or several hundred homes.

Where pumping or lift stations must be provided, it is desirable to have backup or standby facilities in the event of power outages or equipment breakdowns. More specific criteria and details regarding the design of elements of the sanitary sewerage system are in Sec. 22.

Solid-Waste Collection ▪ In urban residential areas, solid-waste collection is usually provided door to door by collection trucks and crews. It is one of the most costly public services provided to property and residents in urban areas. Collection is commonly provided as a governmental responsibility, utilizing either governmental employees and government-owned equipment or private contractors. Collection is usually provided once or twice weekly. Wastes are normally placed at the street curb for collection in residential areas. Some communities, however, collect at the rear of properties or provide a set-out and set-back service for residents.

Solid-waste collection is a very labor-intensive service. Efficiency and cost-effectiveness require good labor management and efficient layout of collection routes.

Waste containers and storage practices are extremely important for efficient and effective solid-waste collection. In residential areas, proper containers and storage practices improve the appearance of the neighborhood and reduce the potential for litter and health hazards. In addition, use of proper containers and locations for collection can improve the convenience and efficiency of the collection effort and yield substantial cost savings. In multifamily residential, commercial, and industrial areas, solid-waste storage needs and requirements are often neglected in site planning and layout, yet it is usually advisable to provide for containerized waste storage and collection in these high-density areas. In addition, on-site compaction of wastes often helps reduce storage-space requirements and enhances collection efforts.

The discussion of solid waste disposal in section 14.10 included resource recovery and recycling. Many communities have instituted or expanded curbside collection of recyclable materials. This often requires modifications or redesign of the solid waste collection system.

Private Utilities ■ Other significant utilities provided to individual homes and properties include natural gas, electricity, telephone, and cable television. In a few limited locations, steam or hot water may also be supplied and distributed for district heating purposes. These services are usually provided by private utilities, but in some cases they may be provided by government-owned utilities.

Underground installation of utilities is a significant issue in most communities. Underground installation of all utilities can substantially improve community appearance, as discussed in Art. 14.13, and it can reduce maintenance needs and increase reliability of utility service. Another major issue regarding utility service is the desirability of reserving or providing multipurpose utility corridors for joint use or installation of utilities. This can result in more efficient use of land and avoid problems such as disruption during construction and confusion as to location of utilities. The issue of joint use of multipurpose utility corridors has been thoroughly investigated by a number of organizations, including the American Society of Civil Engineers.

14.20 Transportation System Planning

Comprehensive multimodal transportation system planning is an extremely complex subject. This type of planning is a significant portion of the total planning effort of most regional and metropolitan planning agencies, working in concert with state highway departments. Following is only a general overview of the main principles and issues of transportation system planning.

Elements of Transportation System Planning ■ The principal modes of transportation include: pedestrian facilities and bicycles; street and highway vehicles (automobiles, buses, and trucks); rail transportation (intercity rail, commuter trains, subways); air transportation; and water transportation.

The principal elements for each mode of transportation include the vehicle (automobile, train, barge); the travel way (highway, rail line, waterway); and terminal or transfer facilities (parking lot, rail terminal, port). Because many trips involve the use of more than one mode of transportation, it is important for the plan to fully accommodate and facilitate transfer between modes. Terminal or transfer facilities should be provided for transfer from one mode to another and may also provide storage space for vehicles. Parking lots and garages are primary examples.

The most important goals and objectives of transportation planning include:

1. Enhancing the mobility of residents and accessibility to employment, shopping areas, education, health-care, and other public facilities

2. Increasing the convenience and safety of necessary travel, including consideration of different modes and transfer between modes

3. Avoiding detrimental impacts of transportation facilities on neighborhoods and communities

4. Reducing the monetary and time costs associated with travel and transportation facilities

Pedestrian Facilities and Bikeways ■ Walking and bicycles are important transportation modes in neighborhood travel, for short trips and for circulation in high-density areas, such as central business districts and university campuses. Bicycles also may be used extensively in some areas seasonally, for recreation and work trips.

Outdoor pedestrian facilities commonly include such elements as sidewalks, pedestrian malls, and pedestrian bridges or overpasses. Bike lanes may be provided as separately marked lanes on existing streets or as entirely separate paths. Separate bike paths may be reserved exclusively for bicycles or jointly used by cyclists and pedestrians. In areas of light travel, sidewalks may be used jointly as bikeways and pedestrian facilities if curbs are ramped. Adequate bicycle parking and storage facilities are critical elements at the terminus of trips or at transfer points. It is important to design pedestrian and bikeway facilities as integrated networks that tie into other modes of transportation and to design these networks to avoid conflicts between modes of transportation.

Street and Highway Planning ■ Street and highway planning is the most significant aspect of multimodal transportation system planning since most trips are made by automobile. For planning and designing the street and highway network, streets and highways are classified according to function; these classifications include freeways, expressways, arterial streets and highways, collector streets, and local streets. Section 16 contains detailed guidelines and design criteria for various classes of streets and highways.

Traffic-carrying capacity of streets and highways is based on the concept of level of service (see Sec. 16). The highest level of service is A, which represents free flow of traffic at design speeds. The lowest level is F, which represents unstable congested flow conditions at low speeds. The maximum carrying capacity of a street or highway generally corresponds most closely to level of service E, which is characterized by unstable traffic flow and average speeds of about 30 mi/h.

The primary function of **freeways and expressways** is to carry traffic. Ordinarily, no direct access is provided to adjacent land, although frontage roads paralleling the expressway or freeway may be utilized to provide access to abutting land. Expressways may have at-grade intersections with cross streets, and freeways usually have grade separations and interchanges to provide uninterrupted traffic flow. Freeways and expressways are commonly designed as 4- to 8-lane, divided highways capable of carrying average traffic volumes of 25,000 to over 40,000 vehicles per day. The primary use of freeways is for medium to long intraurban trips and for intercity travel.

The primary function of **arterial streets and highways** is to carry traffic for most major intraurban trips over 1 mi, particularly during peak hours. Arterial streets and highways ordinarily provide only limited direct access to adjacent land. This direct access is usually focused on major facilities, shopping centers, or other significant traffic generators. Arterial streets and highways are usually designed with 4 to 6 travel lanes, may have parking lanes, and may be either divided or undivided highways. These facilities are capable of carrying average traffic volumes up to about 25,000 vehicles per day.

Collector streets carry traffic and provide access to adjacent land. Collector streets are most appropriate for short to medium ($\frac{1}{2}$ to 1-mi) intraurban trips. Collector streets receive traffic from local streets and transmit it to arterial streets and highways, expressways, or freeways. Collector streets, spaced to relieve excessive traffic volumes on local streets, are commonly designed with 2 travel lanes and 2 parking lanes, with the capability of carrying up to 8000 to 10,000 vehicles per day.

Local streets provide access to adjacent land. They are normally used only for very short ($\frac{1}{2}$-mi or less) intraurban trips. Local streets commonly have 2 travel lanes and 1 parking lane. Traffic volumes are normally kept to minimum levels.

Right-of-way requirements for streets and highways vary considerably to accommodate space needs for utilities, sidewalks, and landscaping. Total public right-of-way for local streets is usually about 60 ft; for collector streets, 66 to 80 ft; and for arterial streets and highways, 80 to 120 ft. Special lanes for buses or bicycles are added in some cases, which results in additional right-of-way requirements. Specific cross-sectional details and geometry are discussed in more detail in Sec. 16.

Parking Requirements ■ Since most trips are by auto, parking needs are usually important considerations in land use, transportation, and site planning. An adequate supply of parking is necessary to ensure convenient access and smooth functioning of the transportation system, but parking often requires a large proportion of the land or site facilities at major activity centers or destinations. Table 14.6 indicates the parking requirements for major land uses, appropriate where most of the travel to and from the uses is by auto rather than transit or pedestrians.

Urban Mass Transit ▪ Urban mass transit includes bus systems (providing both local and express service) and rail systems, such as subways, elevated trains and tramways, and commuter railroads. Paratransit (taxicabs, airport limousines, and special services, such as elderly or handicapped vans) is important for serving special needs and areas.

Urban mass-transit systems are heavily oriented to serving trips to work or school and regular trips to large activity centers, such as central business districts. These systems can substantially reduce parking needs and congestion in central business districts.

Bus systems are the most commonly used form of urban mass transit. They have the advantage of flexibility, in that routes can be changed and new areas added quite easily, and the system can be adapted relatively easily and economically to changing conditions. Bus systems utilize existing streets and highways and do not require major investments in the travel way or terminal or transfer facilities. Bus systems are capable of providing transportation service to within convenient walking distance of residential areas.

Table 14.6 Parking Requirements

Office	3 spaces per 1000 ft² of gross leasable floor area (GLA)
Retail	Range from 3 spaces per 1000 ft² of GLA for convenience stores to 5 spaces per 1000 ft² for regional shopping centers
Restaurants	20 spaces per 1000 ft² of GLA
Hotels	1.25 spaces per room
Industrial	0.6 space per employee
Residential	Range from 1 to 1.5 spaces per multi-family unit to 2 spaces per single-family residence

Source: Institute of Transportation Engineers and Urban Land Institute.

Important factors in improving bus ridership and service include convenience, comfort, and speed. Local bus service can be convenient if stops are located relatively close together, routes are spaced to keep walking distances relatively short ($\frac{1}{4}$-mi maximum), and time intervals between buses are relatively short, particularly during peak hours. Comfort can be enhanced by providing modern equipment and bus shelters.

Local bus service can be quite slow compared to automobile or rail travel and is most suited for short to medium intraurban trips. Express bus service along major arterials or freeways, with limited stops, is more suitable for longer trips. Exclusive travel lanes and preferential treatment for buses have been utilized in some locations to significantly reduce travel times. Local buses can be feeders to express bus or rail transit for long intraurban trips. Besides serving local feeder buses, express buses can utilize park-and-ride facilities, where provision is made for all-day parking for commuters.

Rail facilities used in urban transit systems include subways and commuter railroads. Subways or sometimes elevated trains or tramways are used for circulation in very high density areas with significant travel volumes, such as central business districts of large cities. Commuter rail facilities are used for commuter travel along high-density corridors in major metropolitan areas.

Rail transit efficiently serves high volumes of travel with minimum use of land and facilities and effectively reduces congestion during peak hours. Rail transit requires a large population base generating high traffic volumes to be feasible. The disadvantages of rail transit include a fixed location and a high cost for the travel way and stations or terminals. Rail systems are inflexible in location and ordinarily not capable of providing convenient service within reasonable walking distance in residential areas. To overcome these disadvantages, the following guidelines are suggested:

1. Rail transit should be used where most appropriate: for medium to long intraurban trips along very high-density corridors, for travel to and between major activity centers in large urban areas, and for circulation within very high density areas, such as central business districts.

2. Rail transit facilities should be located in high-density corridors generating substantial travel demands.

3. Local feeder-bus service and park-and-ride facilities should be provided to maximize convenience.

4. Attempts should be made to reduce costs where possible by utilizing existing rail facilities and by locating rail routes in or adjacent to other transportation facilities, such as freeway rights-of-way.

Urban Transportation Modeling ▪ The urban transportation planning process in most metropolitan areas considers more than one major mode of travel. The process is sufficiently complex to require computer models. The models simulate travel patterns and volumes and allow evaluation of alternative land-use patterns and transportation-system network changes or adjustments needed to satisfy travel needs. Figure 14.4 illustrates the basic steps in the urban transportation modeling process. It includes feedback and reiteration designed to seek a match between travel demand and transportation facilities capable of satisfying that demand.

Intercity Rail, Air, and Water Transportation ▪ Rail, air, and water transportation facilities, like highways, are important for providing intercity movement of people and goods.

Intercity rail travel usually focuses on freight movement, with less emphasis on passenger travel. Because rail transportation is an important mode for movement of freight and bulk materials, availability of rail facilities can be a major factor in location of industries. Section 19 presents engineering and design aspects of rail facilities.

Air transportation is a significant mode for intercity passenger travel as well as high-value, low-bulk freight. The two major categories of air travel include commercial air carriers and general aviation, which uses small private and business planes. The needs of commercial air carriers and general aviation often conflict. In many areas, consideration is given to separating the two types of air travel and providing separate facilities for each.

Airports are the major terminal and transfer facilities for air travel of concern to the community and regional planner. (Specific engineering and design aspects of airports are discussed in Sec. 18.) Land needs for airports can range from as little as 50 to 100 acres, for a small airport serving light planes, up to 15,000 to 40,000 acres or more, for a major international airport. An important consideration is provision of sufficient spacing from other airports to avoid air-traffic conflicts.

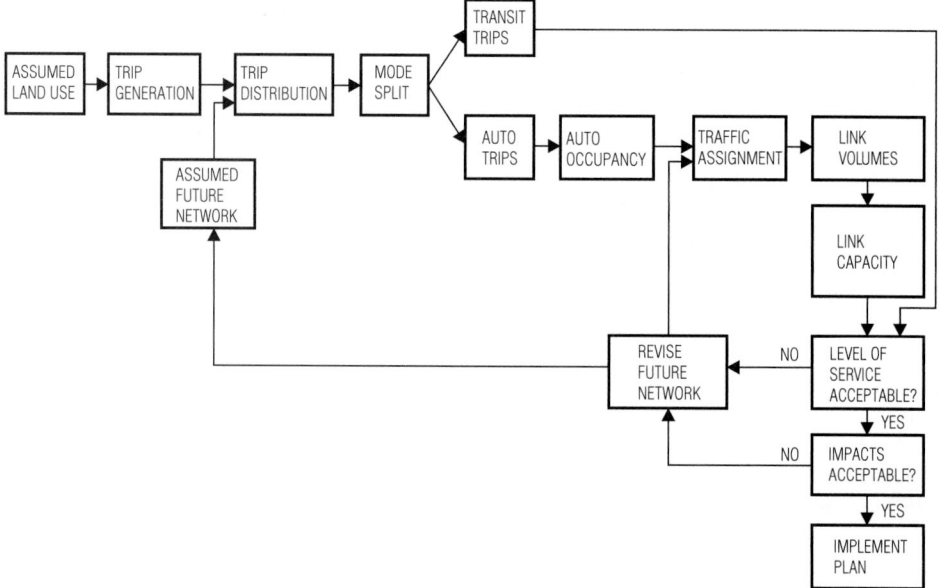

Fig. 14.4 Steps in urban transportation modeling.

Flat terrain in an elevated location, as well as absence of physical barriers or hazards, is important in airport siting. Good soils and drainage are key features, as is availability of utilities. Since trips rarely end at the airport, airports serve primarily as transfer facilities, so accessibility and interconnection with major regional transportation facilities are other critical elements in airport location. Compatibility with adjacent land uses also is a significant issue with most airports because of safety requirements and noise problems. Strict control of use of adjacent land, particularly land near approach and takeoff paths, is essential to avoid future problems.

Water transportation is an important mode of transportation for bulk materials. The main concern of community and regional planners is provision of terminal and transfer facilities (ports and harbors). Specific design and layout of ports and harbors is discussed in Sec. 23.

Implementation Tools and Techniques

14.21 Comprehensive and Functional Plans

The comprehensive plan for a community or region, sometimes called the master plan or city plan, is the most important and central document for management and control of an area's physical development and growth. The comprehensive plan addresses all aspects of the physical development of a community or region. The main subjects of most comprehensive plans include: resources and environmental quality, private and public uses of land, community facilities and utilities, and circulation.

The comprehensive plan serves a number of very important purposes:

1. As a statement of community goals and policies. (Thus, citizens and elected officials should be intimately involved in the preparation of a comprehensive plan.)

2. As a guide to governmental and private decision making. (To be effective and useful, the comprehensive plan should be designed to be used constantly by both the executive and legislative branches of government. The plan should be updated and revised frequently.)

3. As an overall framework to guide the preparation of more specific and detailed plans for individual services or facilities or plans for community subareas or neighborhoods and to coordinate those functional and subarea plans.

4. As a legal basis and foundation for land-use and growth-management tools, such as zoning, official mapping, subdivision regulation, annexation, public-utility and service extension policies, land acquisition, and capital-improvements programming.

Comprehensive plans may take the form of either a general or policy plan or a physical master plan that contains detailed locations of specific land uses and facilities. Each format has important advantages and disadvantages. The general or policy plan is generally more compact and easy to prepare, read, and use. The general plan addresses only the more important goals and policies, leaving specific details to individual cases and special studies. A general or policy plan is most useful as a guide to general governmental decision making and is the easiest form to update and keep current.

A plan that contains specific locations of land uses and facilities illustrates the practical results of applying the plan goals and policies, making it easier for citizens and elected officials to understand. A detailed physical plan addresses specific locational decisions and has the advantage of pointing out areas of potential controversy and conflict prior to implementation. Since detailed proposals are more likely to point out implementation and feasibility problems early in the process, detailed physical plans can be instrumental in moving expeditiously from plan policies to implementation. Detailed physical plans are more difficult to prepare and update and are much more cumbersome to use and read. A particular disadvantage of bulky and cumbersome detailed plans is that important or key issues tend to get lost in a maze of detail.

The most advisable approach is to combine the best elements or attributes of the general or policy plan with those of the detailed comprehensive plan. Plan goals and policies should be clearly highlighted as the overall framework. It should be recognized that goals and policies are not sufficient in themselves. Therefore, it is important that the plan include a sufficiently detailed illustration of the proposed physical development of the community or region to show the desired end results.

The format should allow for convenient and frequent revision and updating and contain text summaries of detailed functional or subarea plans. Detailed plans for specific functions, services, or subareas, such as neighborhoods, should be incorporated as separate documents or appendixes that can be individually prepared, changed, and updated.

Figure 14.5 illustrates a typical comprehensive growth and development plan for a small community; Fig. 14.6 shows one for a metropolitan region.

14.22 Zoning and Subdivision Regulations

Zoning ordinances and subdivision regulations are regulatory means of implementing the comprehensive plan and controlling land conversion and development processes. Both zoning and subdivision regulation are based on the community's exercise of police power to enact laws protecting its citizens' public health, safety, morals, and general welfare.

The zoning ordinance designates, with a map and text, land-use districts and outlines compatible land uses permitted within each district. Also specified are conditional uses or special uses that may be permitted within a land-use district under certain conditions after specific review in individual circumstances. The zoning ordinance normally includes provision for various residential, commercial, industrial, and open-land districts. Within each district, the zoning ordinance specifies (in addition to land use) density or area requirements for individual parcels; height, bulk, and placement of structures; and regulations regarding provision of off-street parking and loading areas, signs, landscaping and buffer areas, noise- and pollution-emission standards, and other requirements.

The zoning ordinance is based on the comprehensive plan, but it is not identical to the land-use plan element of the comprehensive plan. The zoning map is a short- to medium-term assignment of land uses rather than long-term. It often represents a compromise between existing land-use patterns and those in the comprehensive plan. The ordinance usually contains provisions for temporary nonconforming uses and procedures for variances from the ordinance requirements. In some communities, added flexibility is built into the zoning ordinance by providing more general requirements for density (such as Land-Use Intensity Standards) and performance standards rather than specific requirements for lot area and building placement. In many communities, the zoning ordinance contains a provision for planned-unit developments, which allows greater flexibility for large integrated developments but requires detailed review and approval of specific development proposals.

Subdivision regulations govern the process of dividing land and converting it to building sites. Subdivision regulations usually include procedures for submittal, review, and approval and recording of plats for land records.

A two-step planning process is normally specified. In the first step, a preliminary plat is prepared and submitted to obtain approval of the overall layout and design of the area prior to detailed design. In the second step, a final plat is prepared and submitted, to serve as the legal instrument of public record for land-recording purposes.

Subdivision regulations also specify requirements and design standards for layout of streets, blocks, lots and parcels, open spaces, and relationships with adjacent areas and land uses. Specifications for grading, streets, and other required improvements and facilities are often included. In addition to requiring installation of certain public improvements and utilities, many communities require dedication of land for public facilities or open spaces as part of the land-division procedure.

14.23 Capital-Improvements Programming and Financing

A capital-improvements program is a short-range (5- to 6-year) plan and schedule for financing and constructing major public facilities and physical improvements recommended in the comprehensive plan. A capital-improvements program is needed because many projects are too large to finance or complete in one year.

The program should be designed for use as a budgeting tool. It should be prepared for a specific governmental unit, with significant involvement of the legislative and executive branches of government, particularly governmental agency and department staff. Although usually prepared

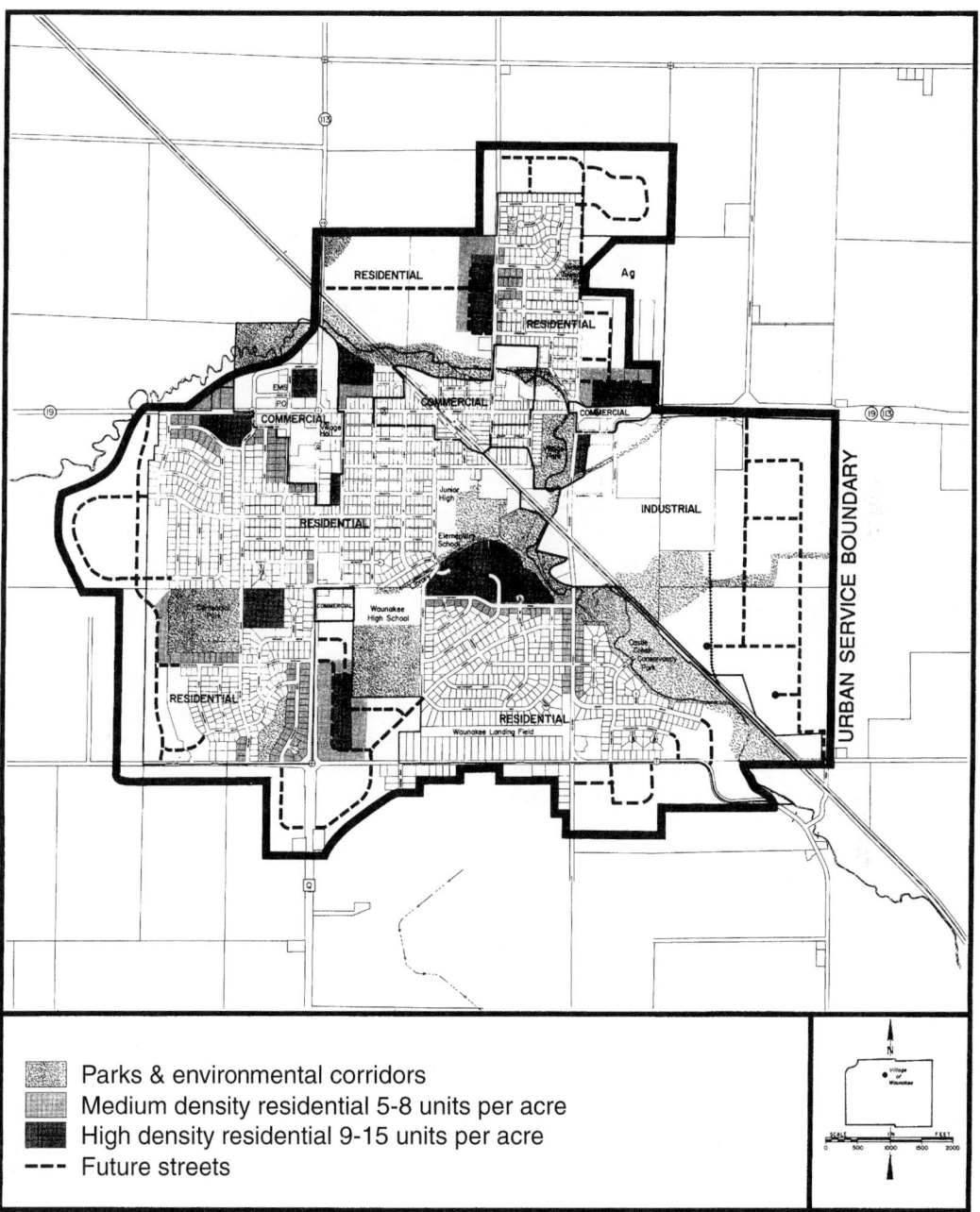

Parks & environmental corridors
Medium density residential 5-8 units per acre
High density residential 9-15 units per acre
--- Future streets

Fig. 14.5 Development plan for a small community, developed by the Dane County Regional Planning Commission, Madison, Wis., 1988, for the village of Waunakee.

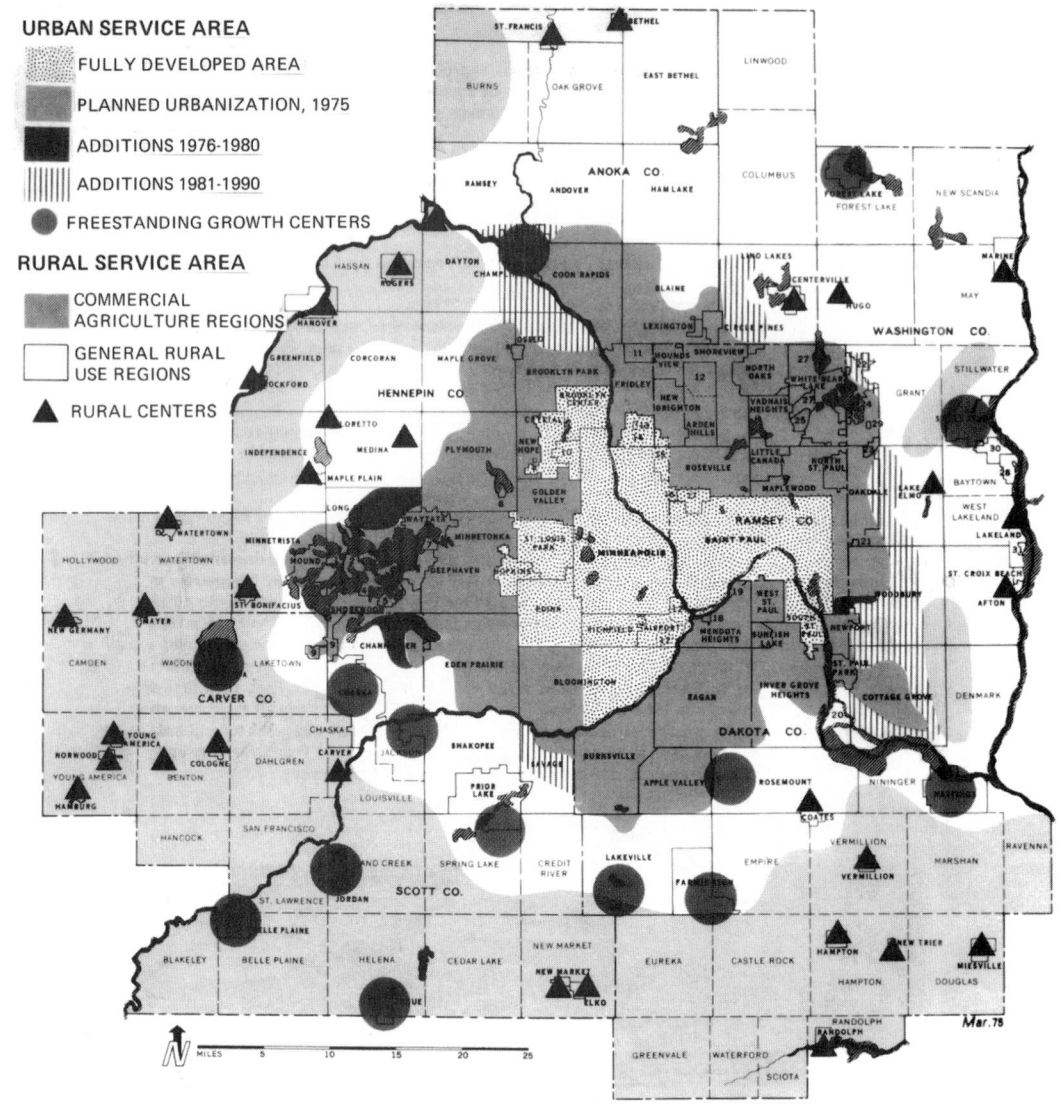

Fig. 14.6 Regional development plan. (*Reprinted with permission from "The Politics and Planning of a Metropolitan Growth Policy for the Twin Cities," The Metropolitan Council of the Twin Cities Area, St. Paul, Minn., 1976.*)

to cover a 5- to 6-year period, the program should be updated and revised annually as part of the budget process. The first year of the capital-improvements program should contain a detailed budget and description of activities; later years of the program should show an annual budget and activities in less detail.

The initial step in preparing a capital-improvements program is a financial analysis of the revenues, expenditures, and indebtedness of the governmental unit. Normal operating and maintenance expenditures, which must be paid out of current revenues, are evaluated and projected for future years. Projected future revenues in excess of operating and maintenance expenditures are available for financing capital improvements.

After the financial analysis has been completed, the recommended capital improvements from the comprehensive plan are evaluated and assigned priorities. Those improvements that are to be initiated within the short-range programming period are selected. Evaluation of and assignment of priorities to projects is a sensitive process, which requires extensive involvement from elected officials and governmental department heads.

Finally, based on available revenues and financing mechanisms, recommendations are made, projected budgets are prepared, and long-term financing arrangements are proposed.

Governmental expenditures are financed from a variety of general revenue sources and grants from other levels of government. Basic local governmental revenues include taxes on real and personal property, income taxes, sales taxes, direct user charges and fees, special assessments against benefited property, and grants from state and Federal governments.

Major capital expenditures may be financed over short- or long-term periods. Loans are commonly used for short-term financing periods of less than 10 years; bonds are normally used for long-term financing. Bonds may be revenue bonds (which are repaid from user fees and charges), general obligation bonds (repaid from general tax revenues), or assessment bonds (repaid from special assessments). Revenue bonds are most suitable for improvements that provide income in the form of user charges or fees (such as those from use of water and sewerage systems). Revenue bonds and assessment bonds often require somewhat higher interest payments than general obligation bonds. General obligation bonds, on the other hand, require voter approval in many states.

Equity is a principal concern in determining the appropriate method of financing. Although financing arrangements often try to assess costs against those who directly benefit from a service, many basic taxing and financial proposals also reflect other governmental objectives, such as lighter tax burdens for low-income individuals or heavier taxes on luxury items.

14.24 Other Implementation Tools

Other important tools that can significantly help implement comprehensive plans include phasing and extension of public services and utilities; official mapping; codes, permits, and impact statements; and rehabilitation and clearance.

The approach to and advantages of phasing and extension of public services and utilities is briefly outlined in Art. 14.18. By controlling the location and timing of the extension of public utilities and the provision of public services to newly developing areas, the governmental unit can exert considerable influence over the location and timing of development.

Many states permit communities to prepare and adopt an official map, which pinpoints the location of future streets and other public facilities. The map is an indication of the community's intent to acquire specified property for public purposes, and the adopting ordinance usually prohibits development of specified lands until the community is notified and has an opportunity to acquire them. Projects sufficiently defined to fall within the capital-improvements programming period are suitable for official mapping.

Codes, permits, and impact statements are also a means for allowing governmental examination and control of construction and activities having a significant impact on physical development. Codes, such as building and sanitary codes, contain detailed requirements and specifications to ensure adequate new construction and minimum acceptable conditions in existing facilities. Vigorous code enforcement is especially important for preventing further decline in areas showing signs of deteriorating housing and facilities.

Governmental agencies and departments require permits or impact statements for a variety of projects and activities. Permits, such as conditional-use permits, discharge permits, activity permits, and impact statements allow an examination of the impacts of particular projects or activities and may result in the placing of restrictions on the project or activity to ameliorate adverse impacts.

Where housing, structures, or facilities are deteriorating or dilapidated, rehabilitation or clearance may be the appropriate solution. Rehabilitation programs are oriented to deteriorating areas, structures, and facilities that can be restored to acceptable conditions. The principal elements of a successful rehabilitation program usually include vigorous code enforcement, repair or reconstruction of public facilities, and provision of technical and financial assistance in private upgrading efforts. Clearance and redevelopment may be necessary where structures and facilities are dilapidated and restoration to acceptable conditions is not cost-effective. Close cooperation between the public and private sectors in redevelopment projects is important to success.

14.25 References

The selected references listed below include more extensive bibliographies.

ASCE Manual and Report on Engineering Practice No. 49, *Urban Planning Guide*, Revised Edition. Reston, Virginia: American Society of Civil Engineers, 1986 (www.asce.org).

Corbitt, Robert A., ed. *Standard Handbook of Environmental Engineering*. New York: McGraw-Hill, 1990 (books.mcgraw-hill.com).

Dunne, Thomas; and Leopold, Luna B. *Water In Environmental Planning*. San Francisco: W.H. Freeman and Co., 1978.

Edwards, John D., ed. *Transportation Planning Handbook*, Second Edition. Washington, D.C.: Institute of Transportation Engineers, 1999 (www.ite.org).

The Practice of Local Government Planning, Third Edition, Washington, D.C.: International City/County Management Association, 2000 (www.icma.org).

The Practice of State and Regional Planning. Washington, D.C.: International City/County Management Association, 1986 (www.icma.org).

Katz, Peter. *The New Urbanism*. New York: McGraw-Hill, 1994 (books.mcgraw-hill.com).

Kaiser, Edward J.; Godschalk, David R.; and Chapin, F. Stuart, Jr. *Urban Land Use Planning*, Fourth Edition. Urbana, Illinois: University of Illinois Press, 1995 (www.press.uillinois.edu).

McHarg, Ian L. *Design With Nature*. Garden City, N.Y.: Natural History Press, 1969.

Schueler, Thomas R.; and Holland, Heather K. *The Practice of Watershed Protection*. Elliott City, MD: The Center for Watershed Protection, 2000 (www.cwp.org).

Stover, Vergil G.; and Koepke, Frank J. *Transportation and Land Development*. Institute of Transportation Engineers. Englewood Cliffs, N.J.: Prentice Hall, 1988 (www.prenhall.com).

15

M. Myint Lwin*
Structural Engineer

BUILDING ENGINEERING

Buildings include a wide range of construction intended for human occupancy or for sheltering machines or stored goods. Civil engineers play an important role in the design and construction of such structures. But sometimes, the civil engineer is only one of many design professionals participating in the planning and design of a building. Therefore, it is necessary that the engineer's design decisions take into consideration the objectives and needs of the other professionals. For this purpose, civil engineers must be well-informed on such subjects as architecture, building layout, lighting, electrical systems, elevators, plumbing, heating, and air conditioning, as well as structural design. To serve this need, this section summarizes briefly the design principles of those fields and lists references for more detailed study.

15.1 Influence of Zoning on Building Design

Localities use zoning to regulate use of land, control types of occupancies and size of buildings, and in various other ways safeguard the public health, safety, and welfare. Zoning regulations supplement building code requirements.

When selecting land for a building, the local zoning code should be checked to see if the type of occupancy planned—residential, commercial, industrial, school, church—is permitted. If it is not, the possibility of a variance or a code change should be investigated.

For some types of construction—housing, for example—lack of a zoning code may discourage selection of a parcel of land. Uncontrolled land use may permit undesirable neighbors—junkyards or odorous factories—with a resulting deflation in property values. When a zoning code exists, its control over land adjoining the parcels being considered should be examined, to determine whether neighboring occupancies will be desirable. See also Art. 14.22.

To ensure light and air to adjacent property, zoning codes restrict the height and bulk of buildings. Some codes limit the number of stories; some place a maximum on the height above the street. In some cases, codes permit unlimited height but require the buildings to be set back from the base after certain heights are reached, depending on the width of the street, measured between building lines. This type of requirement led in the past to "wedding cake" architecture, buildings that were made narrower and narrower in steps as they rose because of frequent setbacks. As an alternative, in which internal space is sacrificed in the interests of esthetics, a building may satisfy that type of regulation yet not have setbacks if it is erected as a sheer tower occupying only part of its site.

* Revised and updated from Sec. 15 "Building Engineering," Frederick S. Merritt "Standard Handbook for Civil Engineers," 4th ed., McGraw-Hill Book Company.

Some codes control height and bulk by establishing a ratio between total floor area permitted and the area of the site. Extra floor area sometimes is allowed if part of the site at or near street level is devoted to a plaza. Thus, designers may shape the building any way they please within the building lines; they may make it tall and thin or short and broad, so long as the total floor area does not exceed that permitted. Codes, however, sometimes also require buildings to be set back at the base minimum distances from lot lines; these regulations should be determined and observed in locating a building on its site.

In addition to being restricted by zoning, building height may be limited by Federal aviation authorities, especially in the vicinity of airports. These regulations should be considered before land is selected for a project and especially before building height is decided on in the early design stages.

15.2 Building Codes

Localities, and sometimes the states, exercise the police power to safeguard public health, safety, and welfare by controlling building design and construction through building codes. The control generally extends over all phases, including specification of permissible design and construction methods, as well as field inspection to ensure compliance.

Codes may be classified as specification or performance-standards type. Specification-type codes are characterized by requirements that list acceptable materials and their minimum sizes for specific applications. Performance-standards codes specify the end result to be obtained in terms of such characteristics as strength, stability, permeability, hardness, and fire resistance. In practice, this type of code generally is supplemented by a catalog of acceptable materials and constructions after tests show they meet code requirements.

Since most communities have their own codes, which may differ from those of adjoining localities, building designers should become familiar with the local building codes for the areas in which their projects will be erected. Even where state codes exist, communities may have the power to set more stringent requirements than the state.

For projects in areas not under the jurisdiction of a state or local building code, building designers should adopt the code of a nearby large city or a model code applicable to the region. Nationally recognized model codes include: National Building Code—National Conference of States on Building Codes and Standards; Uniform Building Code—International Conference of Building Officials; Standard Building Code—Southern Building Code Congress International, Inc.; Basic Building Code—Building Officials and Code Administrators International, Inc., and International Building Code—International Code Council.

There are also Federal regulations that must be complied with by building owners, designers, and contractors. For example, the Occupational Safety and Health Administration (OSHA) sets standards, regulations, and procedures for building construction and conditions in buildings after construction. In general, OSHA requires that contractors and subcontractors not permit personnel to work in surroundings or under working conditions that are unsanitary, hazardous, or dangerous to health and safety. Employers are responsible for initiating and maintaining accident prevention programs. Detailed requirements are given in "Construction Industry: OSHA Safety and Health Standards (29 CFR 1926/1910)," Superintendent of Documents, Government Printing Office, Washington, DC 20402. (http://bookstore.gpo.gov)

Another example of Federal rules that must be considered is the Americans with Disabilities Acts (ADA). This federal law requires that buildings be accessible to individuals with disabilities. A copy of the guidelines for conformance with this Federal law can be obtained from the Access Board, 1331 F Street, NW, Suite 1000 Washington, DC 20004-1111. (www.access-board.gov)

15.3 Fire Protection for Buildings

An important consideration in the design of nearly all buildings is the fire resistance required by building codes and insurance companies. This resistance may require use of incombustible materials, fire-protective coverings, and sprinkler systems, which generally cost more than constructions of lower fire resistance. Also, codes may prohibit use of hazardous materials, for example, materials that may explode or emit excessive smoke or poisonous gases.

Sometimes, the lowest long-run costs for a building are obtained with a higher fire resistance than required by the local building code because of reductions in fire insurance premiums.

Fire-resistant construction aims at withstanding a fire locally for a specific period of time and preventing it from spreading—throughout the level at which it starts, or from story to story, or to adjacent buildings. The objective of sprinkler systems is to extinguish the fire quickly.

Fire ratings are assigned to building components in accordance with their performance in standard fire tests (ASTM E119, "Standard Methods of Fire Tests of Building Construction and Material"). If a component meets requirements after 1 h exposure in a standard furnace test, it is given a 1-h rating; if it withstands the test for 2 h, it gets a 2-h rating; and so on.

Fire protection for a building and its occupants involves prevention, detection and warning, containment, and extinguishment of fires, and provision for life safety.

Prevention ▪ Buildings should be designed to minimize the possibility of fire other than in such authorized places as furnaces and fireplaces. Where possible, construction materials—roofing, flooring, ceilings, and sash, for example—and coatings, paints, and curtains should be incombustible. Also, the fuel load from furnishings should be kept small.

Detection and Warning ▪ Buildings should be equipped in every story with devices that can detect fire or smoke and sound an alarm. Such devices can also automatically instigate extinguishment procedures. There are five general types of detectors; each employs a different physical means of operation.

Fixed-temperature detectors indicate the presence of fire when the device reaches a predetermined temperature.

Rate-of-rise detectors function when there is a rapid increase in temperature.

Photoelectric detectors are sensitive to smoke.

Combustion-products detectors recognize combustion products and are designed for very early warning.

Flame detectors respond to light, infrared or ultraviolet, produced by combustion reactions.

Detection should immediately signal a warning so that building occupants who may be endangered, life-safety supervisory personnel, and firefighters may be alerted.

Large buildings, especially those with many occupants, should have an emergency control center, or fire command station, on the ground floor to which detection signals are communicated. The center should have and control two-way communication with every floor and be able to direct rescuers and firefighters and transmit instructions to occupants to guide them to safety. The center should also be able to control all electromechanical systems, such as elevators, air conditioning, and fans. To assist firefighters, controls should be capable of venting, pressurizing, or sealing any zone in a building.

Containment ▪ Buildings should be so designed that, if fire or smoke should occur, it would be extinguished almost immediately, but in any event, it could not spread much beyond the place of occurrence. Spread of fire or smoke can be prevented by fire barriers, venting of heat and gases, and dampers.

Barriers. Large floor areas should be partitioned by fire walls into smaller areas. Fire doors protecting wall openings should be kept closed. Plenums, such as the spaces between floor and ceiling or roof and ceiling, should be isolated at frequent intervals by fire stops. Spandrels should have a high fire rating and should be sufficiently deep at each floor level to prevent flames extending out the windows in one story from igniting materials in the story above. (The National Building Code recommends a minimum depth of 3 ft.)

Venting systems should be provided to cool fires and keep heat and smoke from escape routes and refuge areas. Areas adjacent to a fire should be pressurized to keep out smoke. To clear smoke, windows should be openable or have smoke ventilation panels. Alternatively or in addition, an automatically vented smoke shaft should be provided. Also, the tops of fire towers enclosing elevators or stairs should permit venting of hot gases and smoke. Emergency ventilation of stairwells and elevator shafts may be assisted by fans. Fresh makeup air should be provided to keep safe areas habitable.

Automatic fire dampers should be installed in ducts, along with fire or smoke detectors that sample all air passing through. The dampers should be controlled to seal control zones, prevent smoke from spreading to escape routes and refuge areas, and guide ventilation area air to points where it is needed.

Extinguishment ▪ Means of fire suppression range from hand-held extinguishers to high-pressure water flows from hoses and sprays from installed sprinkler systems (Art. 15.33). (For some types of fires, carbon dioxide or chemicals may be necessary instead of water.) In addition, firefighters have various types of equipment for firefighting. Regardless of the means used, life safety and property damage depend primarily on early detection of fire and smoke and rapid application of the appropriate extinguishment method.

To assist firefighters, water must be supplied to them in adequate quantities and at sufficient pressures for firefighting. If necessary, storage or pumping facilities must be provided. An elevated water tank may be used for this purpose (National Fire Protection Association standard 22). The supply may be augmented by a fire pump (NFPA standard 20). Pressure should be at least 15 psi at the highest level of sprinklers, while flow at the base of the risers is at least 250 gal/min for light-hazard occupancies and 500 gal/min for ordinary-hazard occupancies. (Local building codes usually specify minimum pressures.)

The usual means of manually applying water to interior building fires is with hoses receiving water from standpipes. These generally are required in buildings more than about 50 ft high and should be so located that any part of a floor is not more than 130 ft from a standpipe outlet valve. Risers up to 75 ft high may be 4 in in diameter, but for greater heights, 6 in. Hose valves usually are $2\frac{1}{2}$ in in diameter.

Life Safety ▪ Buildings should provide for safe, easy escape in emergencies, preferably but not necessarily to outdoors at ground level. In some cases, it may be advisable to instruct occupants to stay in place or to provide refuge areas within the buildings to which occupants may proceed when alerted. Doors, hallways, and stairs should be adequate in number, size, and location to accommodate the number of occupants that may have to be evacuated in emergencies. (Requirements are specified in local building codes and in "Life Safety Code," NFPA101, National Fire Protection Association.) In addition, firefighters should be provided with safe access to fires.

In buildings with elevators, the cars should be equipped with controls for emergency use by firefighters and move automatically to the lobby floor to be available to them if needed. Control wiring should be protected against accidental operation by high temperatures.

Elevators and stairs should be enclosed in fire towers with walls having a 4-h rating (*fire walls*) and fire-resistant doors that are kept closed. Building entrances and exits should be especially protected. (See also Art. 15.18.)

(F. S. Merritt, "Building Design and Construction Handbook," 6th ed., McGraw-Hill Publishing Company, New York (books.mcgraw-hill.com); F. S. Merritt, "Building Engineering and Systems Design," 2nd ed., Van Nostrand Reinhold Company, New York; "The SFPE Handbook of Fire Protection Engineering," "Automatic Sprinkler Systems Handbook," "Fire Protection Handbook," 17th ed., "Life Safety Code Handbook," and "National Fire Codes," National Fire Protection Association, Quincy, Mass (www.nfpa.org).)

15.4 Service Loads for Buildings

Service loads used in the design of a building should be the maximum probable loads to which the structure may be subjected. They should not be less, however, than the loads specified by the local building code. In the absence of a local code, the loads given in this article may be used, or the loads in a model building code, or the loads in "Building Code Requirements for Minimum Design Loads in Buildings and Other Structures," ASCE 7-02, American Society of Civil Engineers, New York (www.asce.org). See also Art. 15.5.

All structural components of a building must be designed for the full dead load as a minimum. When the dead load is not certain, for example, when the location of partitions is not known when the design is made, an allowance should be made. Some building codes require a uniformly distributed load of 20 psf to be added to the known dead load to allow for partitions not definitely located. Tables 15.1 and 15.2c list minimum design dead

Table 15.1 Minimum Design Dead Loads

Walls	lb/ft²	Insulation	lb/ft²
Clay brick		Cork, per in thick	1.0
High-absorption, per 4-in wythe	34	Foamed glass, per in thick	0.8
Medium-absorption, per 4-in wythe	39	Glass-fiber bats, per in thick	0.06
Low-absorption, per 4-in wythe	46	Polystyrene, per in thick	0.21
Sand-lime brick, per 4-in wythe	38	Urethane, per in thick	0.17
Concrete brick		Vermiculite, loose fill, per in thick	0.05
4-in, with heavy aggregate	46		
4-in, with light aggregate	33		

Wood joists, double wood floor, joist size	lb/ft² 12-in spacing	16-in spacing
2 × 6	6	5
2 × 8	6	6
2 × 10	7	6
2 × 12	8	7
3 × 6	7	6
3 × 8	8	7
3 × 10	9	8
3 × 12	11	9
3 × 14	12	10

Walls (continued):

Walls	lb/ft²
Concrete block, hollow	
8-in, with heavy aggregate	55
8-in, with light aggregate	35
12-in, with heavy aggregate	85
12-in, with light aggregate	55
Clay tile, load-bearing	
4-in	24
8-in	42
12-in	58
Clay tile, non-load-bearing	
2-in	11
4-in	18
8-in	34
Furring tile	
1½-in	8
2-in	10
Glass block, 4-in	18
Gypsum block, hollow	
2-in	9.5
4-in	12.5
6-in	18.5

Partitions	lb/ft²
Plaster on masonry	
Gypsum, with sand, per in thick	8.5
Gypsum, light aggregate, per in	4
Cement, with sand, per in thick	10
Cement, light aggregate, per in	5
Plaster, 2-in solid	20
Metal studs	
Plastered two sides	18
Gypsumboard each side	6
Wood studs, 2 × 4-in	
Unplastered	3
Plastered one side	11
Plastered two sides	19
Gypsumboard each side	7

Concrete Slabs	lb/ft²
Stone aggregate, reinforced, per in	12.5
Slag, reinforced, per in thick	11.5
Light aggregate, reinforced, per in	6–10

Floor Finishes	lb/ft²
Asphalt block, 2-in	24
Cement, 1-in	12
Ceramic or quarry tile, 1-in	12
Hardwood flooring, ⅞-in	4
Plywood subflooring, ½-in	1.5
Resilient flooring, such as asphalt tile	2
Slate, 1-in	15
Softwood subflooring per in thick	3
Terrazzo, 1-in	13
Wood block, 3-in	4

Floor Fill	lb/ft²
Cinders, no cement, per in thick	5
Cinders, with cement, per in thick	9
Sand, per in thick	8

Waterproofing	lb/ft²
Five-ply membrane	5

Glass	lb/ft²
Single-strength	1.2
Double-strength	1.6
Plate, ⅛-in	1.6

Ceilings	lb/ft²
Plaster (on tile or concrete)	5
Suspended metal lath, gypsum plaster	10
Suspended metal lath, cement plaster	15

(Table continued)

Table 15.1 (*Continued*)

	lb/ft^2		lb/ft^3
Roof and Wall Coverings		Masonry	
Clay tile	9–14	Cast-stone masonry	144
Asphalt shingles	2	Concrete, stone aggregate, reinforced	150
Composition:		Ashlar:	
3-ply ready roofing	1	Granite	165
4-ply felt and gravel	5.5	Limestone, crystalline	165
5-ply felt and gravel	6	Limestone, oölitic	135
Copper or tin	1	Marble	173
Corrugated steel	2	Sandstone	144
Sheathing (gypsum), $\frac{1}{4}$ in	2		
Sheathing (wood), per in thickness	3		
Slate, $\frac{1}{4}$ in	10		
Wood shingles	2		

loads for various materials. In computing dead load, the weight of the member being designed should be included, as well as the weight of the rest of the structure that it has to support.

Live loads for buildings generally are assumed uniformly distributed, except, of course, that the live load transmitted from a beam to a girder is a concentrated load on the girder. Some codes also require an additional concentrated load, applied at any point in a bay, for garages, machine rooms, and offices. But such loads as those from a moving crane on crane girders and columns should be treated as moving concentrated loads. Table 15.2 lists minimum design live loads for various occupancies.

Live loads should be placed on a structure to produce maximum stress and deformation in components being designed. For example, in design of a continuous beam for maximum positive moment at midspan, only alternate spans, including the one being designed, should carry full live load. Machine weight should be increased 25% and elevator loads 100% for impact.

When a very large area contributes live load to a member, most codes permit a reduction from that required for a member supporting a small loaded area. For floors, for example, some codes permit a reduction for members supporting 150 ft^2 or more of 0.08% per ft^2. But the reduction cannot exceed 60% or 23.1(1 + D/L)%, where D is the dead load per square foot of area supported and L the live load per square foot. And the reduction does not apply to one-way slabs, places of public assembly, or garages for trucks and buses. Where live loads exceed 100 lb/ft^2 and for

passenger-car garages, live loads on columns supporting more than one floor may be reduced 20%.

Snow loads on roofs should be treated as live load and placed to produce maximum stress and deformation. Ordinary roofs should be designed for a live load of at least 20 lb/ft^2 of horizontal projection to provide for sleet and minor snow loads and loads incidental to construction and repair. Where snow loads may exceed 20 lb/ft^2, the roof should be designed for the maximum anticipated or that required by the local building code or the loads given in ASCE 7-02. Roofs used for incidental promenade purpose should be designed for a minimum live load of 60 lb/ft^2. When used for roof gardens or assembly purposes, they should be designed for 100 lb/ft^2.

Wind loads vary with location and height of building. Buildings should be designed for wind coming from any direction. Wind loads and live loads may act simultaneously, but wind loads need not be combined with seismic loads.

Horizontal pressures produced by wind are assumed for design purposes to act normal to the faces of buildings and may be directed toward the interior of the buildings or outward. These forces are called velocity pressures because they are primarily a function of the velocity of the wind striking the buildings. Building codes usually permit wind pressures to be either calculated or determined by tests on models of buildings and terrain.

The basic wind speed used in design is the fastest-mile wind speed recorded at a height of 10 m (32.8 ft) above open, level terrain with a

Table 15.2 Minimum Design Live Loads

a. Uniformly distributed live loads, lb/ft², impact included*

Occupancy or use	Load	Occupancy or use	Load
Assembly spaces:		Marquees	75
Auditoriums[†] with fixed seats	60	Morgue	125
Auditoriums[†] with movable seats	100	Office buildings:	
Ballrooms and dance halls	100	Corridors above first floor	80
Bowling alleys, poolrooms, similar	75	Files	125
recreational areas		Offices	50
Conference and card rooms	50	Penal institutions:	
Dining rooms, restaurants	100	Cell blocks	40
Drill rooms	150	Corridors	
Grandstand and receiving-stand	100	Residential:	100
seating areas	100	Dormitories:	
Gymnasiums	100	Nonpartitioned	60
Lobbies, first-floor	100	Partitioned	40
Roof gardens, terraces	100	Dwellings, multifamily:	
Skating rinks	100	Apartments	40
Bakeries	150	Corridors	80
Balconies (exterior)	100	Hotels:	
Up to 100 ft³ on one- and two-family		Guest rooms, private corridors	40
houses	60	Public corridors	80
Bowling alleys, alleys only	40	Housing, one- and two-family:	
Broadcasting studios	100	First floor	40
Catwalks	30	Storage attics	80
Corridors:		Uninhabitable attics	20
Areas of public assembly, first-floor	100	Upper floors, habitable attics	30
lobbies		Schools:	
Other floors same as occupancy		Classrooms	40
served, except as indicated		Corridors	80
elsewhere in this table		Shops with light equipment	60
Fire escapes:		Stairs and exitways	100
Multifamily housing	40	Handrails, vertical and horizontal	
Others	100	thrust, lb/lin ft	50
Garages:		Storage warehouses:	
Passenger cars	50	Heavy	250
Trucks and buses	‡	Light	125
Hospitals:		Stores:	
Operating rooms, laboratories,	60	Retail:	
service areas		Basement and first floor	100
Patients' rooms, wards, personnel	40	Upper floors	75
areas		Wholesale:	100
Kitchens other than domestic	150	Telephone equipment rooms	80
Laboratories, scientific	100	Theaters:	
Libraries:		Aisles, corridors, lobbies	100
Corridors above first floor	80	Dressing rooms	40
Reading rooms	60	Projection rooms	100
Stack rooms, books and shelving at 65		Stage floors	150
lb/ft³ but at least	150	Toilet areas	40
Manufacturing and repair areas:			
Heavy	250		
Light	125		

(Table continued)

Table 15.2 (*Continued*)

b. Concentrated live loads[§]

Location	Load, lb
Elevator machine room grating (on 4-in^2 area)	300
Finish, light floor-plate construction (on 1-in^2 area)	200
Garages:	
Passenger cars:	
Manual parking (on 20-in^2 area)	2,000
Mechanical parking (no slab), per wheel	1,500
Trucks, buses (on 20-in^2 area), per wheel	16,000
Office floors (on area 2.5 ft square)	2,000
Roof-truss panel point over garage, manufacturing, or storage floors	2,000
Scuttles, skylight ribs, and accessible ceilings (on area 2.5 ft square)	200
Sidewalks (on area 2.5 ft square)	8,000
Stair treads (on 4-in^2 area at center of tread)	300

c. Minimum design loads for materials

Material	Load, lb/ft^3	Material	Load, lb/ft^3
Aluminum, cast	165	Gypsum, loose	70
Bituminous products:		Ice	57.2
Asphalt	81	Iron, cast	450
Petroleum, gasoline	42	Lead	710
Pitch	69	Lime, hydrated, loose	32
Tar	75	Lime, hydrated, compacted	45
Brass, cast	534	Magnesium alloys	112
Bronze, 8 to 14% tin	509	Mortar, hardened:	
Cement, portland, loose	90	Cement	130
Cement, portland, set	183	Lime	110
Charcoal	12	Riprap (not submerged):	
Coal, anthracite, piled	52	Limestone	83
Coal, bituminous or lignite, piled	47	Sandstone	90
Coal, peat, dry, piled	23	Sand, clean and dry	90
Copper	556	Sand, river; dry	106
Cork, compressed	14.4	Silver	656
Earth (not submerged):		Steel	490
Clay, dry	63	Stone, quarried, piled:	
Clay, damp	110	Basalt, granite, gneiss	96
Clay and gravel, dry	100	Limestone, marble, quartz	95
Sand and gravel, dry loose	100	Sandstone	82
Sand and gravel, dry, packed	110	Shale, slate	92
Sand and gravel, wet	120	Tin, cast	459
Silt, moist, loose	78	Water, fresh	62.4
Silt, moist, packed	96	Water, sea	64
Gold, solid	1,205	Zinc	450
Gravel, dry	104		

* See local building code for reductions permitted for members subjected to live loads from large loaded areas.

† Including churches, schools, theatres, courthouses, and lecture halls.

‡ Use American Association of State Highway and Transportation Officials highway lane loadings.

§ Use instead of uniformly distributed live load, except for roof trusses, if concentrated loads produce greater stresses or deflections. Add impact factor for machinery and moving loads: 100% for elevators, 20% for light machines, 50% for reciprocating machines, 33% for floor or balcony hangers. For craneways, add a vertical force equal to 25% of maximum wheel load; a lateral force equal to 10% of the weight of trolley and lifted load, at the top of each rail and a longitudinal force equal to 10% of maximum wheel loads, acting at top of rail.

50-year mean recurrence interval. In the absence of code specifications and reliable data, the basic wind speed may be approximated for preliminary design from the following:

Coastal areas, northwestern and southwestern	
United States and mountainous areas	110 mi/h
Northern and central United States	90 mi/h
Other parts of the contiguous states	80 mi/h

For design purposes, wind pressures should be determined in accordance with the degree to which terrain surrounding the proposed building exposes it to the wind. Exposures may be classified as follows:

Exposure A applies to centers of large cities, where for at least ½ mi upwind from the building the majority of structures are over 70 ft high and lower buildings extend at least one more mile upwind.

Exposure B applies to wooded or suburban terrain or to urban areas with closely spaced buildings mostly less than 70 ft high, where such conditions prevail upwind for a distance from the building of at least 1500 ft or 10 times the building height.

Exposure C exists for flat, open country or exposed terrain with obstructions less than 30 ft high.

Exposure D applies to flat, unobstructed areas exposed to wind blowing over a large expanse of water with a shoreline at a distance from the building of not more than 1500 ft or 10 times the building height.

For ordinary buildings not subject to hurricanes, the velocity pressure q_z, psf, at height z, ft, above grade may be calculated from

$$q_z = KV^2 \geq 10 \text{ psf} \qquad (15.1)$$

where V = basic wind speed, mi/h, but not less than 70 mi/h

K = pressure coefficient from Table 15.3

For important buildings, such as hospitals and communications buildings, for tall, slender structures, and for high-occupancy buildings, such as auditoriums, q_z computed from Eq. (15.1) should be increased 15%. To allow for hurricanes, q_z should be increased 5% for ordinary buildings and 20% for important, wind-sensitive, or high-risk buildings along hurricane coastlines, such as those along the Atlantic Ocean and Gulf of Mexico. These increases, however, may be assumed to reduce uniformly with distance from the shore to zero for ordinary buildings and 15% for the more important or sensitive structures at points 100 mi inland.

For design of the main wind-force resisting system of ordinary, rectangular, multistory buildings, wind pressures at any height z may be computed from

$$p_{zw} = G_o C_{pw} q_z \qquad (15.2)$$

where p_{zw} = design wind pressure, psf, on windward wall

G_o = gust response factor

C_{pw} = external pressure coefficient

For windward walls, C_{pw} may be taken as 0.8. For sidewalls, C_{pw} may be assumed as -0.7 (suction). For roofs and leeward walls, substitute in Eq. (15.2) an external pressure coefficient C_p for C_{pw}. Leeward walls are subject to suction and C_p depends on ratio of depth d to width b of the building. For d/b of 1 or less, $C_p = -0.5$; for $d/b = 2$, $C_p = -0.3$; and for d/b of 4 or more, $C_p = -0.2$. For roofs, q_z should be computed for z equal to the mean roof height. For flat roofs, C_p may be taken as -0.7. For sloping roofs, C_p depends on wind direction and roof slope

Table 15.3 Pressure Coefficients, K, for Computation of Wind Pressure

	$K \times 10^6$			
	Exposure A	Exposure B	Exposure C	Exposure D
Height z up to 15 ft	307	940	2046	3052
Height z over 15 ft	$50.45 q_z^{2/3}$	$282 q_z^{4/9}$	$943 q_z^{2/7}$	$1776 q_z^{1/5}$

(see "Minimum Design Loads for Buildings and Other Structures," ASCE 7-02, American Society of Civil Engineers, New York). The gust response factor may be taken approximately as

$$G_o = 0.65 + \frac{8.58D}{(h/30)^n} \geq 1 \qquad (15.3)$$

where D = 0.16 for Exposure A, 0.10 for Exposure B, 0.07 for Exposure C, and 0.05 for Exposure D

$n = \frac{1}{3}$ for Exposure A, $\frac{2}{9}$ for Exposure B, $\frac{1}{7}$ for Exposure C, and 0.1 for Exposure D

h = mean roof height, ft

For design of the main wind-force resisting system of rectangular, one-story buildings, wind pressures vary with relative areas of openings of windward and leeward walls. For windward walls, pressures computed from Eq. (15.2) should be increased by $C_{p1}q_z$, where $C_{p1} = 0.75$ if the percentage of openings in one wall exceeds that of other walls by 10% or more, and $C_{p1} = 0.25$ for all other cases. For roofs and leeward walls, $C_{p2}q_z$ should be subtracted from pressures computed from Eq. (15.2), where $C_{p2} = 0.75$ or -0.25 if the percentage of openings in one wall exceeds that of other walls by 10% or more, and $C_{p2} = \pm 0.25$ for all other cases.

For flexible buildings (those with a fundamental natural frequency less than 1 Hz or with a ratio of height to least horizontal dimension exceeding 5), see ASCE 7-02.

Design Seismic Forces ▪ The procedures and the limitations for determining seismic forces for the design of structures are dependent on site seismic hazard characteristics, site soil characteristics, site soil characteristics, occupancy, structure configuration, structural system and height. Current model building codes, such as, the 1999 National Building Code (NBC), the 1997 Uniform Building Code (UBC), the 2000 International Building Code (IBC), should be followed for determining seismic forces for the design and analysis of new structures and retrofit of existing structures. Building codes have been developed and updated based on better understanding of seismic risk and structural response through lessons learned from past earthquakes, and research in structural, geotechnical and earthquake engineering. The civil and structural engineers should follow the

provisions of the model building codes in seismic design and analysis. There are far fewer structural failures when modern building codes are used.

Some key aspects of the seismic design provisions of the UBC will be used to discuss the determination of seismic forces and other design and analysis requirements. The engineers should refer to the UBC or other controlling building codes for complete and detailed seismic design criteria, procedures and limitations.

Seismic hazard characteristics for the site are based on seismic zones, proximity of the site to active seismic sources, and site soil profile characteristics. Each site is assigned a seismic zone. A seismic zone map of the United States is shown in Fig. 15.1. National maps of seismic hazards have been produced by the U.S. Geological Survey (USGS) since 1948. These maps are revised as new earthquake studies provide better understanding of the seismic hazard. The national maps of seismic hazards provide information essential to creating and updating the seismic design provisions building codes used in the United States. These maps may be obtained from USGS, Central Region, Geologic Hazards Team, Golden, Colorado or from the USGS web site: http://geohazards.cr.usgs.gov/eg.

A typical seismic hazard map may have the title, "Ground motions having 10 percent probability of being exceeded in 50 years." The 10 percent is an "exceedance probability" and the 50 years is an "exposure time." This means that the ground motions shown in this map has a 10 percent probability of being exceeded in 50 years. An event with this probability has a return period of about 475 years. This means the same as saying that these ground motions have an annual probability of occurrence of $\frac{1}{475}$ per year. This is commonly used in building and bridge codes as the design earthquake and the ground motions associated with the design earthquake are termed design ground motions. The ground motions are generally expressed as peak ground accelerations (PGA). A site-specific procedure must be used for determining the design ground motions if the site conditions are unusual, such as, high seismicity zones, soft soils, site is near to an active fault, or a very important structure.

However, building codes have been replacing maps having numbered zones with maps showing contours of design ground motion. At the present, the 1997 Uniform Building Code is the only

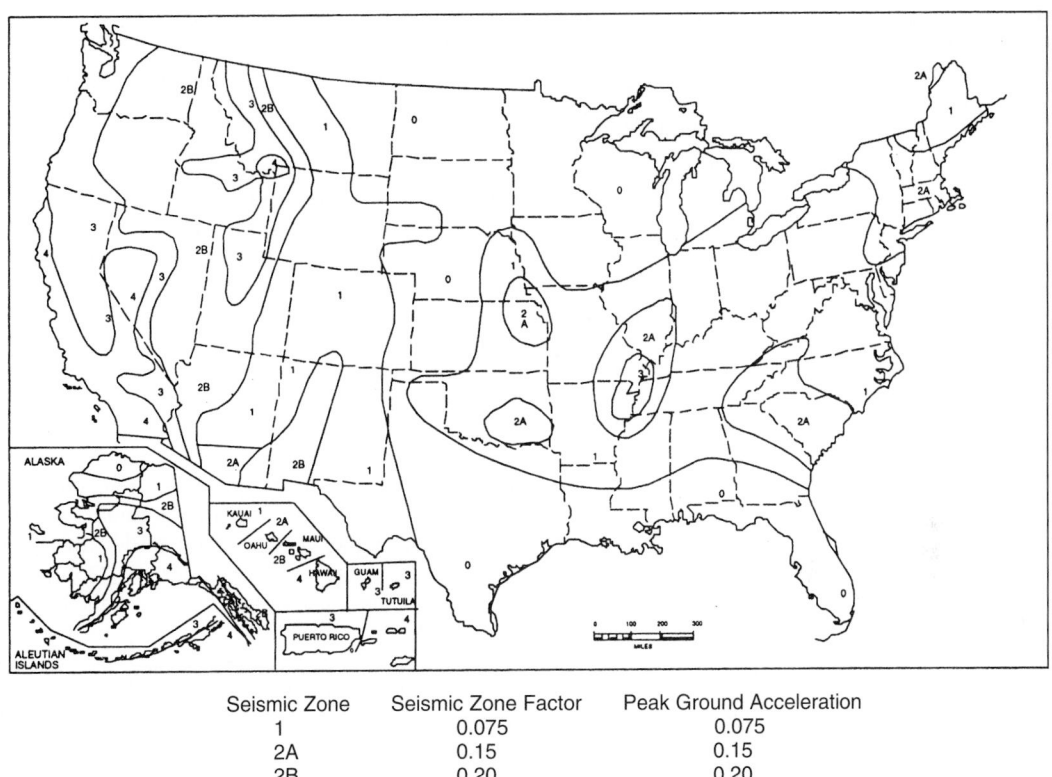

Seismic Zone	Seismic Zone Factor	Peak Ground Acceleration
1	0.075	0.075
2A	0.15	0.15
2B	0.20	0.20
3	0.30	0.30
4	0.40	0.40

Fig. 15.1 Seismic zone map of the United States [UBC].

building code that still uses seismic zones. It is anticipated that the new edition of the UBC will drop the zones and adopt the contours of design ground motion. The new format will avoid the need to revise zone boundaries by petition from various states.

Seismic Zones. Building codes traditionally divide a seismic hazard map into seismic zones in which a common level of seismic design is applied. For example, in the 1997 Uniform Building Code, the seismic hazard map of the United States is divided into six seismic zones of different levels of peak ground accelerations (PGA) as shown in Fig. 15.1.

For each structure, UBC assigns a seismic zone factor, Z, corresponding to the peak ground acceleration as shown in Fig. 15.1

Active Seismic Sources. Earthquakes occur on faults. A fault is a thin zone of crushed rock be-tween two tectonic plates. It can also be a fracture within a tectonic plate or in the crust of the earth where rocks have moved relatively to one another. A fault can be of any length, from inches to thousands of miles. Active faults move at average of a fraction of an inch to about 4 inches per year. For example, the Juan de Fuca plate is known to be subducting beneath the North American plate along the cascadia subduction zone off the Pacific coast of Washington State at a rate of about 1.2 to 1.6 inches per year. When the rock on one side of a fault suddenly slips with respect to the other, energy is abruptly released, causing ground mo-tions that rattle buildings. The larger slips correspond to larger energy release and larger ground motions. As expected, larger rupture length results in larger earthquake magnitude. The well known San Andreas Fault in California has a length of over 650 miles, extending to a depth of over 10 miles, and it has been the source of many

Table 15.4 Earthquake Magnitude vs Length of Slipped Fault

Magnitude	Length of slipped fault (miles)
8.0	190
7.0	25
6.0	5
5.0	2.1
4.0	0.83

large earthquakes, including the famous 1906 San Francisco Earthquake with magnitude 8.3. Table 15.4 gives an approximate relationship between earthquake magnitude and length of fault that has slipped.

The 1989 Loma Prieta Earthquake with magnitude 7.1 was reported to have a ruptured length of 25 to 30 miles.

Soil Characteristics. Each site is assigned a soil profile based on geotechnical data and the soil response to ground motions. For example, lessons from past earthquakes show that ground shaking is stronger in soft soil than in hard rock. There is amplification or deamplification in different soil types. Table 15.5 shows the 6 soil types used by UBC.

To account for the site effects of the soil profile types on structural response, seismic response coefficients are used by UBC to amplify the seismic zone factors Z for the seismic zones. The seismic response coefficients to be assigned to each structure are listed in Table 16-Q for C_a and Table 16-R for C_v of UBC. It may be noted that there is

deamplification for Soil Profile Type S_A which is hard rock, and no amplification for Soil Profile Types S_B which is rock. There are significant amplifications for the other Soil Profile Types S_C, S_D, S_E and S_F.

Occupancy. The UBC design ground motion is based on a 10% probability of being exceeded in 50 years, which is an earthquake having a return period of 475 years. Buildings designed in accordance with UBC are expected to perform without major structural failures and loss of life under this design earthquake. However, for essential facilities, such as hospitals, fire and police stations, emergency response centers (ERC), structures housing equipment for ERC, etc. and hazardous facilities housing or supporting toxic or explosive chemicals or substances, the UBC assigns higher Seismic Importance Factors I and I_p to provide higher seismic resistance. UBC Table 16-K contains the definitions for the Occupancy Categories and the assignments of Seismic Importance Factors.

Structure Configuration. For seismic design purposes, structures are designed as being structurally regular or irregular. Regular structures have no significant physical discontinuities in plan or vertical configuration or in their lateral-force-resisting system. Irregular structures have significant physical discontinuities in configuration or in their lateral-force-resisting systems. Irregular features include torsional irregularity, re-entrant corners, diaphragm discontinuity, out-of-plane offsets, and buildings with nonparallel systems.

Structural Systems. Structural systems are covered in detail in Sec. 15.7 of this handbook. For the

Table 15.5 Shear Wave Velocity

Profile type	Soil type	Shear wave velocity, ft/sec
S_A	Hard rock	>5,000
S_B	Rock	2,500 to 5,000
S_C	Very dense soil and soft rock	1,200 to 2,500
S_D	Stiff soil profile	600 to 1,200
S_E	Soft soil profile	<600
S_F	Soil requiring site-specific evaluation*	

* Soil Profile Type S_F is defined as soils requiring site-specific evaluation as follows:
1. Soils vulnerable to potential failure or collapse under seismic loading, such as liquefiable soils, quick and highly sensitive clays, and collapsible weakly cemented soils.
2. Peats and/or highly organic clays, where the thickness of peat or highly organic clay exceeds 10 ft.
3. Very high plasticity clays with a plasticity index, PI > 75, where the depth of clay exceeds 25 ft.
4. Very thick soft/medium stiff clays, where the depth of clay exceeds 120 ft.

purpose of seismic design, they may be classified as follows: **Bearing wall system** is a structural system for supporting vertical loads, such as dead loads. There is no vertical load-carrying space frame. Shear walls or braced frames are needed to resist lateral earthquake forces. **Building frame system** is a structural system using vertical load-carrying space frame to support vertical or gravity loads. Shear walls or braced frames are needed to resistance lateral earthquake forces. **Moment-resisting frame system** is a structural system using space frame to support vertical or gravity loads and the moment-resisting frames to provide lateral resistance to earthquake forces. **Dual system** is a structural system vertical load-carrying space frame to support vertical or gravity loads, and shear wall, braced frame and/or moment-resisting frame to resist the lateral seismic forces. The total base shear in a dual system is resisted proportionately in accordance with the relative rigidities of the individual systems. **Cantilevered column system** is a structural system relying on the cantilevered column elements for lateral resistance. UBC Table 16-N shows a list of lateral-force-resisting systems.

Height. UBC imposes height limits for the various structural systems in Seismic Zones 3 and 4. The limitations are given in UBC Table 16-N.

Near-Source Factor. When a fault ruptures, seismic waves are generated along the entire length of the fault. The direction in which the rupture propagates is a significant factor for most large earthquakes. This directivity effect causes significant amplification of shaking and velocity impulse to structures situated in the near-source or near-fault region, which is generally considered as within 6 to 7 miles of the active fault. The directivity effect was observed in the 1989 Loma Prieta earthquake in California, and the 1995 Kobe earthquake in Japan. To account for the near-source effect, UBC defines three Seismic Source Types A, B, C and assigns Near-Source Factors N_a and N_v to each Seismic Source Type, depending on the distance to known seismic source. Subduction sources are not included in these definitions and should be evaluated on a site-specific basis. The Near-Source Factors are given in Tables 16-S and 16-T and the Seismic Source Types are defined in Table 16-U of the UBC.

Seismic Factors. The UBC uses the Seismic Force Overstrength Factor Ω_o, to assure that the struc-

tures are designed to have minimum design strength over and above the seismic force determined by analysis. This may be considered as a seismic force amplification factor to obtain structural overstrength against earthquake forces higher than those anticipated in the design. The UBC introduces the Response Modification Factor, R, to recognize the ductility of a structure. Ductility is the ability of a structure to undergo inelastic deformation without collapse. It is not economical to design a structure to resist large earthquakes elastically. The Response Modification Factor approach takes advantage of the inherent energy dissipation capacity of a structural system as it undergoes inelastic deformation in the components or connections. This approach demands stringent detailing requirements to assure ductile behavior of the structural system. The values of the Seismic Force Overstrength Factor, Ω_o, and the Response Modification Factor, R, are given in Table 16-N of the UBC.

Redundancy Factor. The UBC introduces a Redundancy Factor, ρ, to recognize the importance of providing redundancy or multiple load paths in a structural system. The engineering profession has considered it good design practice to build in as much redundancy in a structural system as feasible. The building code provisions have begun to address redundancy after the 1994 Northridge earthquake in California. More stringent design provisions are imposed on nonredundant structures than on redundant structures.

Design and Analysis. The UBC identifies the design requirements that must be followed to assure adequate strength to withstand the lateral displacements induced by the Design Basis Ground Motion, considering the inelastic response of the structure and the inherent redundancy, overstrength and ductility of the lateral-force-resisting system. The design and analysis procedures and methods are outlined in the UBC for the structural systems. Diligently following the provisions in UBC and updating with new research findings and experience the civil and structural engineers will achieve seismic resistant structures consistent with the level of performance desired.

Depending on the structure configuration and height, structural systems, occupancy, seismic zones, and soil profile types, the static and/or dynamic lateral-force procedures may be used for seismic design and analysis. The static procedure is generally used for structures, regular under 240 feet in height or irregular not more than five stories or 65 feet in height, in Seismic Zone 1 and in Occupancy Categories 4 and 5 in Seismic Zone 2. The dynamic procedure is used for other major and more complex structures, and structures located in Soil Profile Type S_F, and have a period greater than 0.7 second. For the dynamic procedure, the ground motion, as a minimum, shall be one having a 10-percent probability of being exceeded in 50 years, and it must not be reduced by the response modification factor, R. The ground motion may be represented by a site-specific elastic design response spectrum using a damping ratio of 0.05 or an elastic design response spectrum constructed in accordance with Fig. 15.2 (see Figure 16-3 of UBC), using C_a and C_v consistent with the specific site.

For a time-history dynamic analysis, the ground motion time histories may be developed for the specific using actual earthquake motions, individually or in combination. The response spectra from time histories must approximate the site spectrum, using a damping ratio of 0.05 and considering geologic, tectonic, seismologic and soil characteristics of the site.

Earthquake Load Combinations. Structures are designed for seismic forces acting in any horizontal direction. UBC provides the following load combinations for earthquake loads:

$$E = \rho E_h + E_v \qquad (15.4a)$$
$$E_m = \Omega_o E_h \qquad (15.4b)$$

Where E = combined earthquake load on an element of the structure

E_h = earthquake load due to base shear, V or lateral force F_p

E_m = estimated maximum earthquake force that can be developed in the structure

E_v = load effect from vertical component of ground motion

Ω_o = seismic force amplification factor for overstrength

ρ = reliability/redundancy factor given by

$$\rho = 2 - \frac{20}{r_{max}\sqrt{A_B}} \geq 1.0 \qquad (15.4c)$$

But need not be greater than 1.5

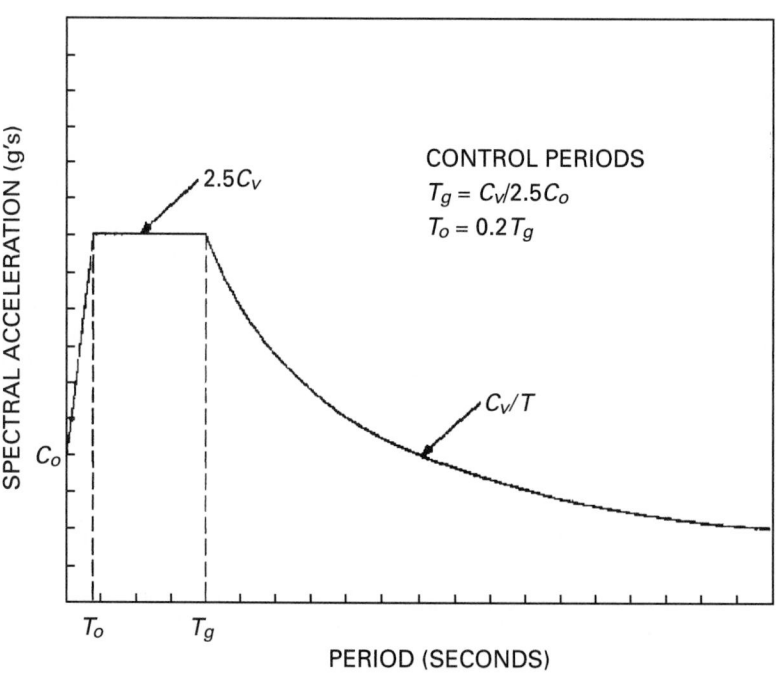

Fig. 15.2 Elastic design response spectrum (UBC).

where r_{max} = the maximum element-story shear ratio.

A_B = ground floor area of the structure in square feet.

Design Base Shear. The total design base shear in a given direction may be determined from the following formula based on the static force procedure:

$$V = \frac{C_v I}{RT} W \qquad (15.5)$$

The total design base shear need not exceed the following:

$$V = \frac{2.5 C_a I}{R} W \qquad (15.6)$$

The total design base shear shall not be less than the following:

$$V = 0.11 C_a I W \qquad (15.7)$$

where V = Total design base shear

C_a = Seismic coefficient, as given in UBC Table 16-Q

C_v = Seismic coefficient, as given in UBC Table 16-R

I = Importance factor, as given in UBC Table 16-K

R = Response modification factor, as given in UBC Table 16-N or 16-P

W = Total seismic dead load plus 25% of the floor live load for storage and warehouse, a minimum of 10 psf for a partition load, design snow load greater than 30 psf, and total weight of permanent equipment.

T = Fundamental period in seconds in the direction under consideration. T may be approximated from the following equation:

$$T = C_t(h_n)^{3/4} \qquad (15.8)$$

where C_t = 0.035 for steel moment-resisting frames

C_t = 0.030 for reinforced concrete moment-resisting frames and eccentrically braced frames.

C_t = 0.020 for all other buildings

h_n = height in feet above the base to the uppermost level.

The UBC requires that floor and roof diaphragms and horizontal bracing systems for seismic loads be designed to resist horizontal forces computed from

$$F_{px} = \frac{F_t + \sum_{i=x}^{n} F_i}{\sum_{i=x}^{n} w_i} w_{px} \leq 1.0 C_a I w_x \qquad (15.9)$$

But not less than $0.5 C_a I w_{px}$

where F_{px} = Design seismic force on a diaphragm

F_t = that portion of the shear force, V, considered concentrated at the top of the structure in addition to F_n

F_i = Design seismic force applied to Level i

n = number of levels in the structure

w_i = that portion of dead load located at or assigned to Level i

w_{px} = the weight of the diaphragm and the element tributary thereto at Level x, including portions of other loads defined in Section 1630.1.1 of UBC

When the diaphragm is required to transfer design seismic forces from the vertical-resisting elements above the diaphragm to other vertical-resisting elements below the diaphragm due to offset in the placement of the elements or to changes in stiffness in the vertical elements, these forces shall be added to those determined from Equation 15.9.

Distribution of Seismic Loads ■ Seismic forces are assumed to act at each floor level on vertical planar frames, or bents, or on shear walls extending in the direction of the loads. The seismic loads at each level should be distributed over the floor or roof area in accordance with the distribution of mass on that level.

The Uniform Building Code recommends that the seismic force F_x, to be assigned to any level at a height h_x, ft, above the ground be calculated from

$$F_x = (V - F_t) \frac{w_x h_x}{\sum_{i=1}^{n} w_i h_i} \qquad (15.10)$$

where w_x = portion of W located at or assigned to level x

h_x = height, ft, of level x above ground level

w_i = portion of W located at or assigned to level i

h_i = height, ft, of level i above ground level

n = number of levels in structure

V is the base shear computed from Eq. (15.4). F_t is an additional seismic force assigned to the top level of the structure and is calculated from

$$F_t = 0.07TV \qquad (15.11)$$

where T = fundamental natural period of vibration of the structure in the direction of the lateral force, s. F_t need not be more than 0.25 V and may be taken as zero when $T \leq 0.7$ s. Equation (15.12) recognizes the influence of higher modes of vibration as well as deviations from straight-line deflection patterns, particularly in tall buildings with relatively small dimensions in plan. Consequently, the design seismic shear at any level i is given by

$$V_i = F_t + \sum_{x=t}^{n} F_x \qquad (15.12)$$

This shear should be distributed to the bents or shear walls of the lateral-force system in proportion to their rigidities. The distribution-resisting should, however, take into account the rigidities of horizontal bracing and diaphragms (floors and roofs). In lightly loaded structures, for example, diaphragms may be sufficiently flexible to permit independent action of the lateral-force-resisting bents. A strong temblor could cause severe distress in frames and diaphragms if relative rigidities were not properly evaluated.

The design seismic force computed from Eq. (15.8) for an element of a structure or a nonstructural component supported by the structure should be distributed in proportion to the distribution of mass of the element or component.

Seismic force distribution for buildings or structural frames with irregular shapes should be determined by dynamic analysis.

Vertical Seismic Forces ■ Provision should be made in aseismic design for the possibility of uplift due to seismic loads. When design of a structure is based on allowable unit stresses, only 85% of the dead load and no live loads should be considered available to counteract the uplift. Furthermore, the UBC requires for structures in Seismic Zones 3 and 4 that horizontal cantilever components be designed for a net uplift force of $F_{u'}$, where

$$F_{u'} = 0.7C_a I W_p \qquad (15.13)$$

In addition to all other applicable load combinations, horizontal prestressed components must be designed using not more than 50% of the dead load for the gravity load, alone or in combination with the lateral force effects.

Horizontal Shear and Torsion ■ For calculating the effects of torsion due to seismic loads on a structure, the rigidity of diaphragms that distribute the seismic loads laterally to lateral-force-resisting framing should be considered. For the purpose, an inflexible diaphragm is defined as one for which the in-plane deflection of its midpoint due to the force F_{px}, computed from Eq. (15.8), is less than twice the average story drift of the stories above and below the diaphragm under the action of seismic forces V_i, calculated from Eq. (15.13). When diaphragms are inflexible, the Uniform Building Code requires that shears at any level i due to horizontal torsion be added to the direct horizontal shears. These are the shears at level i that result from the distribution of V_i, computed from Eq. (15.13), to the components of the vertical, lateral-force-resisting framing in proportion to their rigidities.

The design seismic torsion at any level i consists of two components: (1) The horizontal moment at level i due to eccentricities between the design seismic forces at upper levels and the vertical resisting components at level i. (2) An accidental torsion. This torsion is intended to account for uncertainties in location of seismic loads. For the purpose of computing eccentricities, the mass at each level is assumed to be displaced from the calculated center of mass a distance equal to 5% of the building dimension at that level and in the direction in which that dimension is measured. The displacements should be assumed to occur normal to the seismic load under consideration.

When a structure with inflexible diaphragms is torsionally irregular, the Uniform Building Code requires that the accidental torsion at each level i be multiplied by an amplification factor A_i. To determine whether a structure is torsionally irregular, locate the vertical, lateral-force-resisting bents (or shear walls) parallel to the design seismic loads and at or near the sides of the structure. Compute the maximum story drift due to the seismic shears, including accidental torsional shears, for each of those bents. (Story drift is the displacement of a level relative to the level above or below.) Let d_m be the larger of those drifts and d_a the average of the two. Then, if d_m exceeds $1.2d_a$, the structure is torsionally irregular. If it is, multiply the accidental torsion by A_i computed from

$$A_i = \left(\frac{d_m}{1.2d_a}\right)^2 < 3 \qquad (15.14)$$

Limitation on Story Drift ▪ To prevent damage to building components that could affect life safety, many building codes place limits on the amount of story drift permissible. For example, the UBC limits story drift to not more than 0.025 times the story height for structures having a fundamental period of less than 0.7 second, 0.02 times the story height for structures having a fundamental period of 0.7 second or greater.

Overturning ▪ The equivalent static lateral forces applied to a building at various levels induce overturning moments. At any level, the overturning moment equals the sum of the products of each force and its height above that level. The overturning moments acting on the base of the structure and in each story are resisted by axial forces in vertical elements and footings.

At any level, the increment in the design overturning moment should be distributed to the resisting elements in the same proportion as the distribution of shears to those elements. Where a vertical resisting element is discontinued, the overturning moment at that level should be carried down as loads to the foundation.

Importance of Proper Detailing. Proper detailing is of paramount importance in the design and construction of seismic-resistant structures. This fact is confirmed in every recent major earthquake. Experience in the 1994 Northridge Earthquake showed that new bridges designed and built to current design criteria and construction standards performed well. Existing bridges retrofitted to current retrofit standards also performed well.

With each major earthquake civil and structural engineers continue to learn and modify the seismic design criteria and construction practices to assure that new and retrofitted structures will perform well. Building and bridge codes have been undergoing progressive improvement based on research, experience and costly lessons from recent major earthquakes. Codes traditionally have been focusing on life safety. Modern codes, such as the UBC, are now paying attention to structural performance beyond the issues of life safety to more stringent performance based criteria. This shift is a direct reflection on the costly disruption of building use, commerce and communications. With the shift of emphasis to seismic performance criteria, the civil and structural engineers need to pay greater attention to proper detailing to assure adequate redundancy and ductility in the structures to meet the performance levels expected. For example, (1) for a low level earthquake there should be only minimal damage, and (2) for significant earthquake, collapse should be prevented but significant damage may occur. For critical structures, only repairable damage would be expected. The facility should be functional within a few days after the earthquake.

Proper detailing includes but not limited to the following:

1. Consider structural system reliability—a structure is a system of members, components and connections of the structure, including the foundation. Each structural element contributes to the integrity and safety of the bridge. Every member, component and connection must serve its function to resist and transmit seismic forces, and to accommodate displacements as expected in the design. Additionally, the structural elements should be designed to have reserved strength and ductility to absorb and dissipate energy of higher magnitude without fracture or collapse.

2. Provide at least one continuous viable load path to transmit inertial loads to the foundation—all members, components and connections along the load path must be capable of resisting the imposed load effects. Experience in past earthquake has shown that when one or more of the members, components or connections behaved in a ductile manner damage was much reduced.

3. Avoid irregularities in the structures as much as practicable—irregularities include geometric and stiffness irregularities, discontinuities in lateral force path, capacity and diaphragm, and large skews.

4. Consider commercially available and tested base isolation devices to limit the damaging seismic forces on the structure, and to maintain post-earthquake serviceability of critical existing and new structures.

5. Provide adequate anchorage between building and foundation—properly designed and detailed anchorage systems can have good ductility and absorb considerable energy without breaking.

6. Provide adequate reinforcing steel and confinement reinforcement in concrete and masonry members to assure ductile behavior under high seismic forces.

7. Design and detail steel members and connections to avoid local and global buckling, rupturing of welds and brittle fracture.

Good detailing practices are covered in the Building Codes, the provisions of Building Code requirements for Reinforced Concrete (ACI 318-95) and commentary, and the AISC LRFD Design Manual.

Risk Mitigation. Two most important lessons from recent major earthquakes in the United States and around the world are (1) thousands of nonductile buildings were damaged or collapsed, and (2) surface faulting ruptured lifelines, buildings, bridges and other critical facilities constructed over or across a fault. Nonductile structures, such as unreinforced masonry buildings, inadequately reinforced concrete buildings are prone to catastrophic failures. There is a large inventory of nonductile structures in the high seismicity regions in the United States. Surface faulting has caused some spectacular rupturing or fracturing of buildings, bridges and dams as evidenced in recent earthquakes in Turkey and Taiwan. The civil and structural engineers must work with governmental agencies to identify these failure-prone structures and take necessary actions to mitigate the risk.

In California the Alquist-Priolo Earthquake Fault Zoning Act was passed in 1972 to mitigate the hazard of surface faulting to structures for human occupancy. This state law was a direct result of the 1971 San Fernando Earthquake, which was associated with extensive surface fault ruptures that damaged numerous homes, commercial buildings, and other structures. The main purpose of the Act is to prevent the construction of structures used for human occupancy on the surface trace of active faults. Surface rupture is the most easily avoided seismic hazard for new construction. For existing structures, the only mitigation is to relocate the structures. In 1990, the

California Legisture passed the Seismic Hazards Mapping Act to address non-surface fault rupture earthquake hazards, including liquefaction and seismically induced landslides.

Utah is situated on the 240 mile-long Wasatch Fault, which has the potential of producing large earthquakes above magnitude 7.5. The highly populated areas of Salt Lake City, Ogden, and Provo lie on soft lake sediments that will shake violently during large earthquakes. The Wasatch Fault has not caused a powerful earthquake for the past 150 years. However, the people of Utah are aware of the threat of a catastrophic earthquake. They bond with the communities and public agencies to plan and take action to reduce loss of life and property in future earthquakes. For example, the historic Salt Lake City and County Building has been made safer from earthquakes by installing base isolation devices beneath this 101-year old unreinforced masonry structure. Utah has made major improvement in the public infrastructure to reduce seismic risk. At least 10 fire stations in Salt Lake City and 4 major hospitals have been strengthened or replaced with new earthquake-resistant structures. Over 400 public and private school buildings in the region have been evaluated for seismic resistance. Three high schools and one grade school have been strengthened or replaced.

Utah and California set the examples on what can and should be done to mitigate earthquake risk.

(J. M. Biggs, "Introduction to Structural Dynamics," and R. Clough and J. Penzien, "Dynamics of Structures," McGraw-Hill Publishing Company, New York (books.mcgraw-hill.com); E. Rosenblueth, "Design of Earthquake-Resistant Structures," John Wiley & Sons, Inc., New York (www.wiley.com); N. M. Newmark and E. Rosenblueth, "Fundamentals of Earthquake Engineering," Prentice-Hall, Inc., Englewood Cliffs, N.J. (www.prenhall.com); S. Okamoto, "Introduction to Earthquake Engineering," John Wiley & Sons, Inc., New York (www.wiley.com).)

15.5 Factored Loads

Structural members must be designed with sufficient capacity to sustain without excessive deformation or failure those combinations of service loads that will produce the most unfavorable effects. Also, the effects of such conditions as ponding of water on roofs, saturation of soils, settlement, and dimensional changes must be included. In determination of the structural capacity of a member or structure, a safety margin must be provided and the possibility of variations of material properties from assumed design values and of inexactness of capacity calculations must be taken into account.

Building codes may permit either of two methods, allowable-stress design or load-and-resistance factor design (also known as ultimate-strength design), to be used for a structural member. In both methods, design loads, which determine the required structural capacity, are calculated by multiplying combinations of service loads by factors. Different factors are applied to the various possible load combinations in accordance with the probability of occurrence of the loads.

In allowable-stress design, required capacity is usually determined by the load combination that causes severe cracking or excessive deformation. For the purpose, dead, live, wind, seismic, snow, and other loads that may be imposed simultaneously are added together, then multiplied by a factor equal to or less than 1. Load combinations usually considered in allowable-stress design are:

$$(1)\ D\ \ (2)\ D+G\ \ (3)\ D+(W\ \text{or}\ E)$$
$$(4)\ D+G+(W+E)$$

where D = dead load

$G = L + L$, or $L + S$ or $L + R$

L = live loads due to intended use of occupancy, including partitions

L_r = roof live loads

S = snow loads

R = rain loads

W = wind loads

E = effect of horizontal and vertical seismic loads = $\pm Q + 0.5ZD$

Q = effect of horizontal earthquake-induced forces

Z = seismic intensity coefficient defined for Eq. (15.4)

Building codes usually permit a smaller factor when the probability is small that combinations of extreme loads, such as dead load plus maximum live load plus maximum wind or seismic forces,

will occur. Generally, for example, a factor of 0.75 is applied to load-combination sums (3) and (4) and 0.66 when dimensional changes are added to (4). Such factors are equivalent to permitting higher allowable unit stresses for the applicable loading conditions than for load combinations (1) and (2), for which the allowable stress is obtained by dividing by a safety factor more than unity the unit stress causing excessive deformation or failure.

In ultimate-strength design, the various types of loads are each multiplied by a load factor, the value of which is selected in accordance with the probability of occurrence of each type of load. The factored loads are then added to obtain the total load a member or system must sustain. A structural member is selected to provide a load-carrying capacity exceeding that sum. This capacity is determined by multiplying the ultimate-load capacity by a resistance factor, the value of which reflects the reliability of the estimate of capacity. Load combinations that may be used in the absence of local building code revisions are as follows:

1. $1.4D$

2. $1.2D + 1.6L + 0.5(L_r\ \text{or}\ S\ \text{or}\ R)$

3. $1.2D + 1.6(L_r\ \text{or}\ S\ \text{or}\ R) + (0.5L\ \text{or}\ 0.8W)$

4. $1.2D + 1.3W + 0.5G$

5. $1.2D + 1.0E + 0.5L + 0.2S$

6. $0.9D - (1.3W\ \text{or}\ 1.0E')$

where E' = effect of horizontal and vertical seismic forces = $\pm Q - 0.5ZD$. For garages, places of public assembly, and areas for which live loads exceed 100 lb/ft^2, the load factor usually is taken as unity for L in combinations 3, 4, and 5. The load factor should be taken as 1.3 for liquid loads, 1.6 for loads from soils, and 1.2 for ponding loads, forces due to differential settlement, and restraining forces due to prevention of dimensional changes. The recommended load factors recognize the greater certainty of the magnitude of dead loads but provide a larger safety factor against overloads due to dead loads alone.

15.6 Modular Measure

This is a dimensioning system for building components and equipment to permit them to be field-assembled without cutting. The basic unit is a 4-in cube. Thus, buildings may be laid out around a continuous, three-dimensional rectangular grid with 4-in spacing (Fig. 15.3a).

Manufacturers make many building materials and some installed equipment to correspond to this module. The grid is a convenient tool for drawing assemblies of building products, be they modular or nonmodular.

Modular building products are assigned nominal dimensions corresponding to an even number of modules, although the actual dimensions may be slightly less to allow for joints. Nominal masonry dimensions, for example, equal the dimensions of a unit plus the thickness of one mortar joint. (Standard joint thickness is $\frac{3}{8}$ in for concrete block; $\frac{1}{2}$ in for clay backup and structural units; $\frac{3}{8}$ or $\frac{1}{2}$ in for brick; and $\frac{1}{4}$ in for salt-glazed, clear-glazed, and ceramic-glazed facing units.)

When preparing drawings, the designer can use the grid for both small-scale plans and large-scale details. At scales less than $\frac{3}{4}$ in = 1 ft, however, it is not practical to show grid lines at 4-in spacing. The designer should select a larger planning module that is a multiple of 4 in. For floor plans and

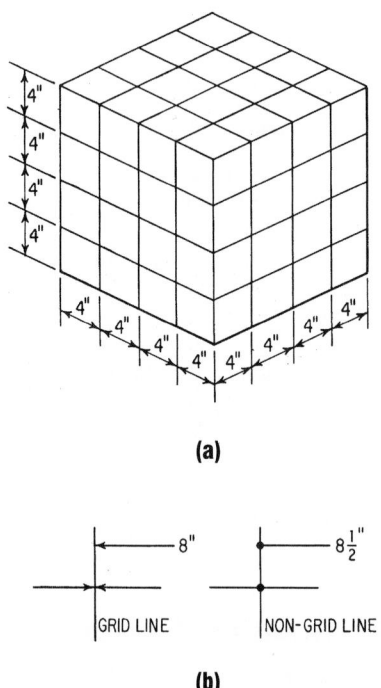

(a)

(b)

Fig. 15.3 Elements of modular measure.

elevations, for example, the module may be 2 ft 8 in, 4 ft, 5 ft, 6 ft 4 in, and so on. Materials should be shown actual size or to scale and located on or related to a grid line by a reference dimension. Dimensions on grid lines are shown by arrows; those not on grid lines, by dots (Fig. 15.3b).

15.7 Structural Systems

Foundations for buildings should be selected and designed in accordance with the principles given in Sec. 7. Basic principles for superstructure design are given in Secs. 6 and 8 to 11.

Buildings may have load-bearing-wall construction, skeleton framing, or a combination of the two. Generally, the engineer's responsibility is to select that type of construction that will serve the owner's total needs most economically. Thus, the most economical construction may not necessarily be the one that requires the least structural materials, or even the one that also has the lowest fabrication and erection costs. Architectural, mechanical, electrical, and other costs that may be affected by the structural system must be considered in any cost comparison.

Because of the large number of variables, which change with time and location, the superiority of one type of construction over the others is difficult to demonstrate, even for a specific building at a given location and time. Availability of materials and familiarity of contractors with required construction methods, or their willingness to take on a job, are important factors that complicate the selection of a structural system still more. Consequently, engineers should consider the specific conditions for each building in selecting the structural system.

Also, deciding on the spans to be used is no simple matter. Foundations, column or wall height, live load, bracing, and provisions for ducts and piping vary with each building and must be taken into account, along with the factors previously mentioned. It is possible, however, to standardize designs for simple buildings, such as one-story warehouses or factories, and determine the most economical arrangement and spans of structural components. But such designs should be reviewed and updated periodically because changing conditions, such as the introduction of new materials, new shapes, or new construction methods and price revisions, could change the economic balance.

Engineers also should bear in mind that the relative economics of a structural system can be improved if it can be made to serve more than just structural purposes. Money is saved if a facade also carries loads or if a structural slab is both floor and ceiling and also serves as air conditioning ducts.

Load-bearing wood walls frequently are used for one- and two-story houses. They usually consist of 2 × 4-in studs spaced 24 or 16 in c to c and set with wide faces perpendicular to the face of the wall. The walls have top and bottom plates, each consisting of two 2 × 4's. Unless supported laterally by adequate framing, maximum height of such a wall is 15 ft. Lumber or plywood sheathes the exterior; plaster or wallboard is placed on the interior. (N. L. Burbank and C. Phelps, "House Carpentry Simplified," McGraw-Hill Publishing Company, New York (books.mcgraw-hill.com).)

Load-bearing masonry walls have been used for buildings 10 or more stories high. But unless design is based on rational engineering analysis instead of empirical requirements, thickness required at the base is very large. Some building codes require plain masonry bearing walls to have a minimum thickness of 12 in for the top 35 ft and to increase in thickness 4 in for each successive 35 ft down. Thus, walls for a 20-story building would have to be about 3 ft thick at the bottom.

Since thickness must be increased from the top down, a natural shape in vertical cross section for load-bearing masonry walls is trapezoidal. With the widest section at the bottom, such a shape is good for resisting overturning. In practice, however, the exterior wall face usually is kept plumb and the inside face is stepped where thickness must be increased.

In low buildings, minimum wall thickness may be governed by the ratio of unsupported wall height or length to thickness, whichever ratio is smaller. (For cavity walls, thickness is the sum of the nominal thickness of inner and outer wythes.) Usually, bearing-wall thickness must be at least 6 in; check the local building code. (See also Art. 15.2.)

Much thinner walls can be used with steel-reinforced masonry designed in accordance with Building Code Requirements for "Masonry Structure," Brick Industry Association, Reston, Va.

("Recommended Practice for Engineered Brick Masonry," Brick Industry Association, Reston, Va.;

F. S. Merritt, "Building Design and Construction Handbook," 6th ed.; McGraw-Hill Publishing Company, New York (books.mcgraw-hill.com).)

Load-bearing reinforced concrete walls may be much thinner than masonry for a given height. The American Concrete Institute Building Code Requirements for Reinforced Concrete (ACI 318-86) sets for superstructure walls a minimum thickness of at least $\frac{1}{25}$ the unsupported height or length, but not less than 4 in. Thickness of exterior basement walls and foundations, however, should be at least $7\frac{1}{2}$ in.

Load-bearing walls may be used for the exterior, partitions, wind bracing, and service-core enclosure. For these purposes, masonry has the disadvantage when used in combinations with skeleton framing of being erected more slowly. Thus, there may be delays in erection of the framing while masonry is being placed to support it.

When **load-bearing partitions** can be placed at relatively short intervals across the width of a building, curtain walls can be used on the exterior along the length of the building. Such partitions, together with flat-plate reinforced concrete floors (Fig. 15.4), make an efficient structural system for certain types of buildings, such as multistory apartment houses. In such buildings also, concrete walls around closets can double as columns.

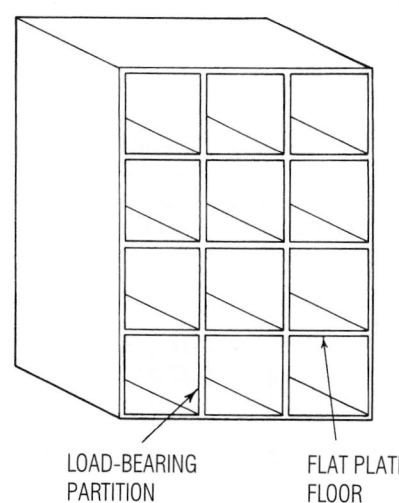

LOAD-BEARING   FLAT PLATE
PARTITION   FLOOR

Fig. 15.4 Load-bearing partitions support flat-plate floors in an apartment building.

Load-bearing walls may serve as shear walls. (But unless they are relatively long, bending stresses due to lateral forces acting parallel to the walls may be large.) Thus, the walls, if properly arranged, will resist wind and earthquake forces in shear and bending. For example, in Fig. 15.5a, shear walls placed at the ends of the building may be designed to resist lateral forces in the narrow direction. In Fig. 15.5b, perpendicular shear walls can take lateral forces from all directions since the forces can be resolved into components parallel to the walls. In Fig. 15.5c, walls enclosing stairs, elevators, toilets, and service rooms (**service core**) may serve as shear walls in perpendicular directions. For earthquake forces, however, it is desirable to supplement the shear walls with a ductile, moment-resisting space frame, to prevent sudden collapse if a shear wall should fail.

Load-bearing service-core walls can be designed, however, to carry all the loads in a building. In that case, the roof and floors cantilever from the walls (Fig. 15.6a). When spans are large,

cantilevers become uneconomical. Instead, columns may assist the service-core walls in carrying the vertical loads (Fig. 15.6b). As an alternative, the outer ends of the floors may be suspended from roof trusses, which are supported on but cantilever beyond the core walls (Fig. 15.6c). Other possibilities include service cores in pairs with floors supported between them, on girders, trusses, cables, or arches or combinations of these.

Architectural-structural walls represent a type of exterior construction somewhere between load-bearing walls and skeleton framing with curtain walls. The load-bearing elements in architectural-structural walls are linear, as in skeleton framing, rather than planar, as in load-bearing walls, and their function is clearly expressed architecturally. Spaces between the structural elements may be screens or curtains, or glass. The structural elements may lie on diagonal lines or verticals (Fig. 15.6d); they may be cross-shaped, combining columns and spandrels (Fig. 15.6e); they may be horizontal or vertical Vierendeel trusses (Fig. 15.6f); or they may be any other system that is structurally sound.

In **skeleton framing**, columns carry building loads to the foundations. Lateral forces are resisted by the columns and diagonal bracing or by rigid-frame action.

Floor and roof construction are much the same for skeleton and load-bearing construction. One principal component is a horizontal structural slab or deck. The deck underside may serve as or carry a ceiling. The upper surface may serve as or carry a wearing surface for traffic or weatherproofing. The deck may be solid, or it may be hollow to reduce weight, permit pipe and wiring to pass through, and serve as air ducts. When the deck does not transmit its loads directly to columns, as it does in flat-slab and flat-plate construction, other major components of floor and roof systems are trusses, beams, and girders (sometimes also called joists, purlins, or rafters, depending on arrangement and location). These support the deck and transmit the load to the columns.

Flat-plate construction employs a deck with constant thickness in each bay and transmits the load directly to columns. It generally is economical for residential and other lightly loaded structures, where spans are fairly short. It is used for *lift-slab construction*, in which the concrete slabs are cast on the ground, then raised to final position by jacks set on the columns. For longer spans, a waffle or two-way ribbed plate may be used.

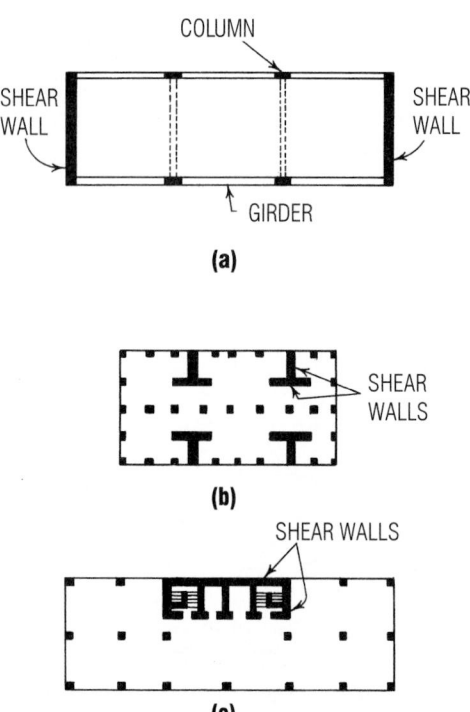

Fig. 15.5 Arrangements of shear walls for resisting lateral forces.

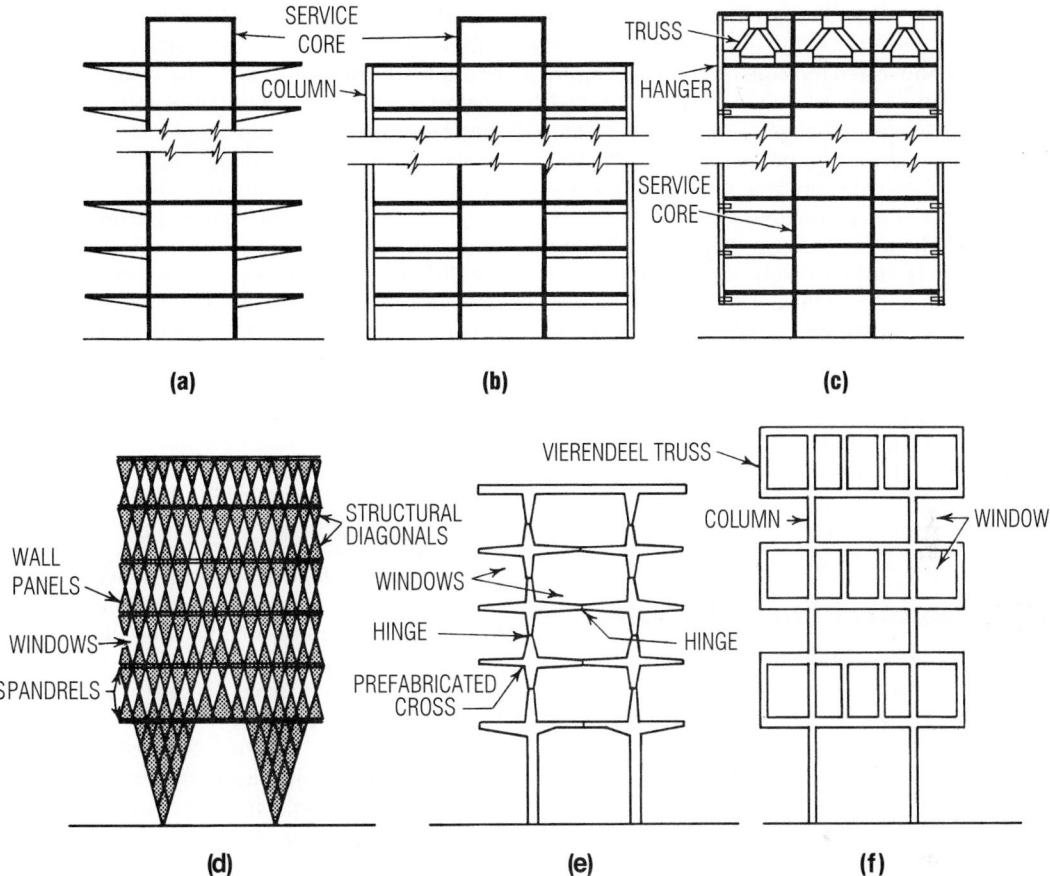

Fig. 15.6 (*a*), (*b*), (*c*)—Framing arranged to place all, or nearly all, loads on service-core walls; (*d*), (*e*), (*f*)—Examples of architectural-structural walls.

Flat-slab reinforced concrete construction may be more suitable for heavier loads. Also transmitting loads directly to columns, it differs from flat-plate in that the slab is thickened in the region around the columns (*drop panels*). Often too, the columns flare at the top (*capitals*). Waffle construction may be used for longer spans.

Slab-band construction is a variation of flat-plate and flat-slab in which wide, shallow beams are used to support the slab and transmit loads to the columns.

Two-way slabs are another variation; they are supported on girders spanning between columns along the border of each bay. Thus, longer spans and heavier loads can be supported more economically.

Beam-and-girder construction is economical for a wide range of conditions. In one- and two-story houses, wood joists or rafters spaced 16 or 24 in c to c generally are used on short spans in conjunction with lumber or plywood decking. For other lightly loaded structures, open-web steel joists, light, rolled-steel beams, or precast-concrete plank may be used, with wood or concrete floors. For heavier loads and longer spans, one-way ribbed-concrete slabs and girders (metal-pan construction); prestressed-concrete plank, tees, double tees, or girders; reinforced concrete beams and girders; laminated-wood girders; or structural-steel beams and girders, including steel-concrete composite construction, may be more suitable. For still longer spans, as usually is the case in industrial buildings, beams and trusses may be most economical.

Arches and catenary construction are appropriate for very long spans. Usually, they are used to support roofs of hangars, stadiums, auditoriums, railroad terminals, and exhibition halls. Their design must provide a means of resisting the horizontal thrust of their reactions.

Thin-shell construction is suitable for uniform loading where curved surfaces are permissible or desirable. It is economical for very long spans. **Folded-plate construction** often is an economic alternative.

(F. S. Merritt, "Building Design and Construction Handbook," 6th ed., McGraw-Hill Publishing Company, New York (books.mcgraw-hill.com); F. S. Merritt and J. Ambrose, "Building Engineering and Systems Design," Van Nostrand Reinhold Company, New York.)

15.8 Lateral-Force Bracing

No structural system is complete unless it transmits all forces acting on it into adequate support in the ground. Hence, provision must be made in both low and tall buildings to carry into the foundations not only vertical loads but lateral forces, such as those from wind and earthquake. Also, the possibilities of blast loading and collision with vehicles must be considered. Without adequate provision for resisting lateral forces, buildings may be so unstable that they may collapse during or after construction under loads considerably less than the full design wind or seismic loads.

Low Wood Buildings ▪ In wood-frame houses one and two stories high, plywood or diagonal lumber sheathing may provide adequate resistance to lateral forces if it is properly nailed and glued. With diagonal lumber, each board should be nailed with two nails to every stud it crosses. Plywood $\frac{5}{8}$ in thick should be nailed with 8d common nails, 6 in c to c; $\frac{1}{4}$ in thick, with 6d nails, 3 in c to c. With other types of sheathing, it is advisable to brace the frame with diagonal studs, especially at end corners of the outside walls and important intermediate corners.

Rigid Frames ▪ Buildings of reinforced concrete beam-and-girder construction generally are designed as rigid frames, capable of taking lateral forces. Except possibly for tall structures subjected to severe earthquakes, rarely does additional provision have to be made for bracing against lateral forces. Tall flat-plate buildings also may be designed as rigid frames to resist wind. If the height-width ratio is large, wind resistance can be improved at relatively low cost by placing wings perpendicular or nearly so to the main portion, so that there are rigid frames with several bays parallel to the directions in which wind-force components may be resolved. Thus, the buildings may be made T-shaped, H-shaped, or cross-shaped in plan, or may have V-shaped wings at the ends. Alternatively, buildings may be curved in plan to improve wind resistance.

Shear Walls ▪ When it is impractical to rely on a moment-resisting space frame to take 100% of the lateral forces, shear walls can be used to take all or part of them. Made of structural-steel plates or reinforced brick or concrete, such walls should be long enough parallel to the wind that bending stresses are within the allowable for the concrete and steel. As shown in Fig. 15.5, shear walls may be placed parallel to the narrow width of the building and rigid frames used in the longitudinal direction, or perpendicular shear walls may take lateral forces from any direction, or service-core walls may double as shear walls (Art. 15.7). The floors should be designed to act as diaphragms or adequate horizontal bracing should be provided to ensure transfer of horizontal forces to the walls. For wind loads, provision must be made to brace exterior walls and transmit the loads from them to the floors. Walls should be adequately anchored to floors and roofs to prevent separation by wind suction or seismic forces.

In areas subjected to severe earthquakes, it is advisable that shear walls be supplemented by ductile, moment-resisting space frames, to prevent sudden collapse if the walls should fail.

Braced Frames ▪ Another method of resisting lateral forces is to use diagonal bracing. Frames that are X-braced generally are stiffer than similar frames relying solely on rigid-frame action.

Roof trusses should be braced against horizontal forces since the spans usually are long and roof decks are made of light material. Additional horizontal and vertical trusses may be used for the purpose. Also, the framework in the plane of the trusses may be stiffened by inserting knee braces between the columns supporting the trusses and the bottom chord. Purlins carrying the roof deck should be securely fastened to the top chords, which are in compression, to brace them laterally.

Trussed roof bracing may be placed in the plane of the top or bottom chords. Putting it in the plane of the top chords offers the advantages of simpler details, shorter unsupported length of diagonals, and less sagging of bracing because it can be connected to the purlins at all intersections. Bracing both top and bottom chords with separate truss systems seldom is necessary. But the bottom chord should be braced at frequent intervals, even though it is a tension member, to reduce its unsupported length.

Figure 15.7 illustrates typical bracing for a mill-building roof. Diagonal bracing is placed in the plane of the top chord in three bays, assuming that the purlins will be sufficiently well-connected to the trusses to transmit longitudinal forces from the unbraced trusses to the braced bays. Not more than five unbraced bays should be permitted between

braced trusses. Struts are shown between lower chords at every panel point, but for a long truss, the struts may be placed at alternate panel points. At corresponding top-chord panel points, the purlins should be capable of carrying compressive forces in addition to vertical loading. The struts between the upper and lower chords should transmit longitudinal forces to the laterally braced bays, where cross frames are placed between the trusses in the plane of the struts, as indicated in Fig. 15.7, to prevent the trusses from tipping over.

Bracing the roof trusses, however, is not enough. The horizontal forces in the roof system must be brought to the ground. The designer must consider the building as a whole.

Figure 15.8 shows a simple bracing system to illustrate the principle. Wind forces on the windward long side of the building are transmitted to the leeward roof truss. This truss carries the loads to the ends of the building, where diagonals in the planes of the ends take the loads to the foundations. Wind on the ends is resisted by bracing in the side walls.

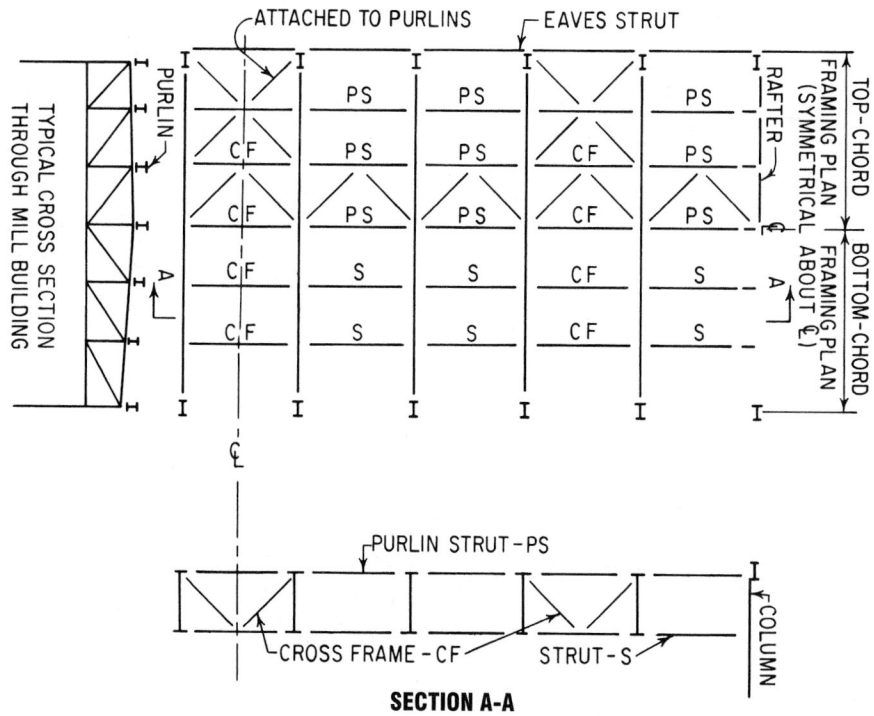

Fig. 15.7 Lateral bracing of roof trusses.

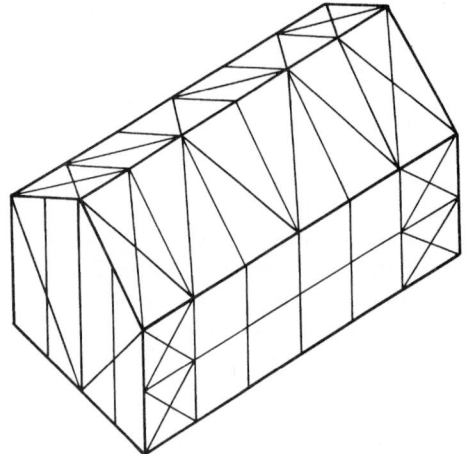

Fig. 15.8 X bracing carries lateral loads from roof to foundation.

Tall Buildings ▪ Similarly, when designing bracing for a tall building, the designer should consider the building as a whole. For example, lateral forces may be resisted by all the bents

(Fig. 15.9*a*) or only the outer bents (Fig. 15.9*b*). In the latter case, the building may be designed as a hollow-tube cantilever for the horizontal forces. The floor and roof systems then must be capable of distributing the loads from the windward wall to the side and leeward walls.

For the bents individually, X bracing (Fig. 15.9*c*) is both efficient and economical. But it usually is impractical because it interferes with doors, windows, and clearance between floors and ceilings above. Generally, the only places X bracing can be installed in tall buildings are in walls without openings, such as elevator-shaft and fire-tower walls. When X bracing cannot be used, bracing that does not interfere with openings should be placed in each bent.

There are many alternatives to X bracing. One is knee bracing between girders and columns (Fig. 15.9*d*); but the braces may interfere with windows in exterior bents or may be objectionable in interior bents because they are unsightly or reduce floor-to-ceiling clearance. Portal framing of several types, including haunched, solid-web spandrels (Fig. 15.9*e*) or trusses, are other alternatives. At the columns, these members provide sufficient

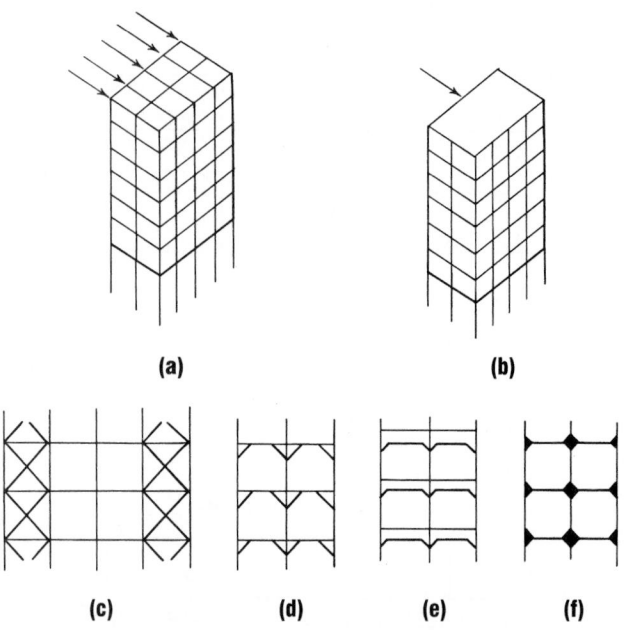

Fig. 15.9 Bracing for high-rise buildings: (*a*) All transverse bents resist lateral forces; (*b*) building acts as a vertical tube; (*c*) bent with X bracing; (*d*) knee bracing between columns and girders; (*e*) haunched spandrels; (*f*) moment-resisting connections between columns and girders.

depth for moment resistance, but at a short distance away from the columns, they become shallow enough to clear windows and doors. In exterior bents, the spandrels can extend from the window head in one story to the window sill in the story above. In interior bents, however, they may have the same disadvantages as knee braces.

Another alternative to diagonal bracing for tall buildings is moment-resisting or wind connections of the bracket type (Fig. 15.9f). Different types may be used, depending on size of members, magnitude of wind moment, and compactness needed to satisfy floor-to-ceiling clearance. In steel framing, the minimum type consists of angles attached to the columns and to top and bottom girder flanges. Plates welded to both girder flanges and butt-welded to the columns are an alternative. When greater moment resistance is needed, the angles may be replaced by tees (made by splitting a wide-flange beam at middepth). Also, the bottom flange may be seated on a beam-stub bracket.

The continuous rigid frames formed with these connections can be analyzed by the methods of Arts. 6.58 to 6.65. For preliminary design or to check computer programs, however, approximate methods may be used (Arts. 15.9 and 15.10).

It is noteworthy that for most buildings even the "exact" methods are not exact. First, the forces are not static loads but generally dynamic; they are uncertain in intensity, direction, and duration. Also, at the beginning of a design, the sizes of members are not known, so the exact resistance to lateral deformation cannot be calculated. Furthermore, floors, walls, and partitions help resist the lateral forces in a very uncertain way.

(F. S. Merritt, "Building Design and Construction Handbook," 6th ed., B. S. Taranath, "Structural Analysis and Design of Tall Buildings," McGraw-Hill Publishing Company, New York (books. mcgraw-hill.com).)

15.9   Portal Method

Since, as pointed out in Art. 15.8, an exact analysis of stresses due to lateral forces on a tall building is impractical, most designers prefer a wind-analysis method based on reasonable assumptions and requiring a minimum of calculations. One such method is the so-called portal method, which is based on the assumptions that points of inflection occur at the midpoints of all members and that

exterior columns take half as much shear as do interior columns. These assumptions enable all moments and shears to be computed by the laws of statics.

Consider, for example, the roof level (Fig. 15.10a) of a tall building. A wind load of 600 lb is assumed to act at the top line of girders. To apply the portal method, cut the building frame along a section through the inflection points of the top-story columns. These points of zero moment are assumed here to be at the column midpoints, 6 ft down from the top of the building. (Some designers prefer to take the top-story inflection points one-third the story height down from the roof girders because the sum of the stiffnesses of the members at each roof joint is likely to be much less than that at each joint in the story below. Similarly, they assume inflection points in the bottom story to be two-thirds the story height up from the base because the anchorage tends to fix the base.) Now, let us compute the stresses in the members above the section.

Since the exterior columns take only half as much shear as do the interior columns, 100 lb of the total 600-lb load is apportioned to each exterior column and 200 lb to each interior column. The moments at the top of the columns equal these shears times the distance from the top to the inflection point. The wall end of the end girders carries a moment equal to that in the exterior column. (At the floor below, as indicated in Fig. 15.10b, the end girder carries a moment equal to the sum of the column moments.) Since the inflection point in the girder is at the midpoint, the moment at the inner end of the girder must be the same as the outer end. The moment in the adjoining girder can be found by subtracting the end-girder moment from the column moment because the sum of the moments at the joint must be zero. (At the floor below, as shown in Fig. 15.10b, the moment in the interior girder is found by subtracting the moment in the end girder from the sum of the column moments.)

Girder shears then can be computed by dividing girder moments by the half span. When these shears have been found, column loads can be easily calculated from the fact that the sum of the vertical loads must be zero, by taking a section around each joint through column and girder inflection points. As a check, it should be noted that the column loads produce a moment that must be equal to the sum of the moments of the wind loads above the section for which the column loads were

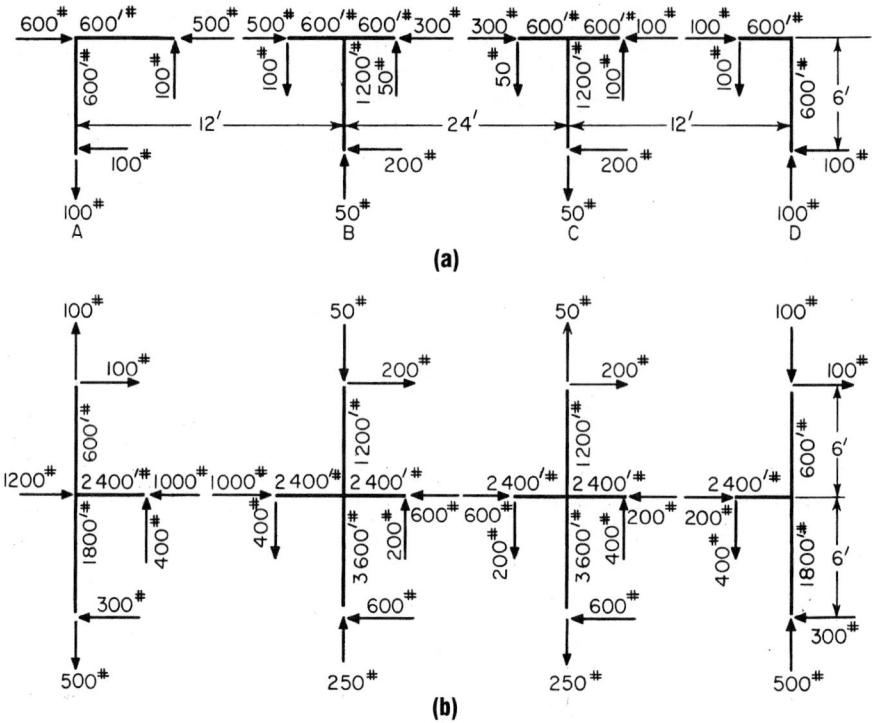

Fig. 15.10 Wind stresses in a tall building computed by the portal method.

computed. For the roof level section (Fig. 15.10a), for example, $-50 \times 24 + 100 \times 48 = 600 \times 6$.

See also Art. 15.10. (C. H. Norris et al., "Elementary Structural Analysis," 3rd ed., McGraw-Hill Publishing Company, New York (books.mcgraw-hill.com).)

15.10 Cantilever Method

This is an alternative to the portal method described in Art. 15.9 for determining stresses in tall buildings due to lateral forces. Basic assumptions are that inflection points are at the midpoints of all members and that direct stress in a column is proportional to its distance from the center of gravity of all the columns in the bent. The assumptions are sufficient to enable shears, axial forces in the columns, and moments in the frame to be determined from the laws of statics.

For multistory buildings with height-to-width ratio of 4 or more, the Spurr modification is recommended ("Welded Tier Buildings," U.S. Steel

Corp.). In this method, the moments of inertia of the girders at each level are made proportional to the girder shears when the spans are equal; otherwise, the moments of inertia must also be proportional to the square of the spans.

The results obtained from the cantilever method generally will be different from those obtained from the portal method. In general, neither solution is correct, but the answers provide a reasonable estimate of the resistance to be provided against lateral forces. In buildings over about 25 stories high, the effects of changes in column lengths should be considered in the analysis. (See also *Transactions of the American Society of Civil Engineers*, vol. 105, pp. 1713–1739, 1940; vol. 126, pp. 1124–1198, 1961.)

15.11 Floor Coverings

When concrete is used as the structural deck in a building, it may be left exposed to serve as a wearing surface, depending on the quality of surface and the type of occupancy. This is generally done in warehouses and industrial buildings with heavy moving loads. Some engineers, however, prefer to place a higher-quality topping on the structural concrete slab. The topping may be applied before or after the base slab has hardened. Usually, integral toppings are ½ in thick, independent toppings about 1 in ("Finishing Concrete Slabs with Color and Texture," SS391, Portland Cement Association, Skokie, Ill (www.portcement.org). In office buildings where electricity and telephone wiring are distributed above the structural slab, a lightweight concrete fill covers the conduit and a floor covering protects the fill.

Wood floors may be made of the hardwoods—maple, beech, birch, oak, or pecan; or of the softwoods—yellow pine, Douglas fir, or western hemlock. The hardwoods are more resistant to wear and indentation. Solid-unit wood blocks are made from two or more units of strip-wood flooring fastened together with metal splines or other suitable devices. The tongued-and-grooved blocks are held in place with nails or an asphalt adhesive. Also, a laminated block is formed with plywood. Average moisture content of wood flooring at time of installation should be 6% in the dry southern states, 10% in damp southern coastal states, and 7% in the rest of the United States ("Moisture Content of Wood in Use," *U.S. Forest Products Laboratory Publication* 1655, Madison, Wis.). Leave at least 1 in of expansion space at walls and columns.

Asphalt tiles, composed of reinforcing fibers, mineral coloring pigments, and inert fillers with asphalt as the binder, are intended for use on rigid subfloors. They may be used on below-grade concrete subject to slight moisture from the ground.

Cork tile is made by baking cork granules with phenolic or other resin binders under pressure. It yields a surface suitable for areas where quiet and comfort are of utmost importance. It is intended for use on rigid subfloors above grade and free of moisture. Cork tile with natural finish should be sanded to level, sealed, and waxed immediately after installation. All cork floors must be maintained with sealers and protective coatings to prevent soiling.

Unbacked vinyl flooring, for use on rigid subfloors above grade, is made of polyvinyl chloride resin as a binder, plasticizers, stabilizers, extenders, inert fillers, and coloring pigments. Resilient under foot, it can withstand heavy loads without indentation but is easily scuffed and scratched unless protected with a floor polish. It is practically unaffected by grease, fat, oils, household cleaners, or solvents.

Vinyl also may be laminated to various backing materials.

Rubber flooring generally is intended for use on rigid subfloors above grade. It is resilient and has excellent resistance to permanent deformation under load.

Linoleum is made from drying oils, such as linseed, natural and synthetic resins, a filler, and pigments similar to those used in paints. Usually, it is backed with burlap or rag felt. Since the backing is susceptible to moisture and fungus attack, linoleum should not be used for floors where moisture can reach the backing. Properly maintained, it performs outstandingly on rigid subfloors above grade in residential and commercial buildings.

Since protection from moisture is a prime consideration for most thin floor coverings, moisture within a concrete slab must be brought to a low level before installation of the flooring begins. Moisture barriers should be placed under concrete slabs at or below grade, and a minimum of 30 days drying time should be allowed after concrete placement before installing the flooring. A longer drying time should be allowed for lightweight concrete.

Adhesives ▪ Concrete surfaces to receive adhesive-applied thin flooring should be smooth. A troweled-on underlayment of rubber latex composition or asphalt mastic should be used over rough floors. Usually, the adhesive for asphalt and vinyl-asbestos tiles is an emulsion or cutback asphalt; for rubber and vinyl above grade, latex; for linoleum, cork, and vinyl backed with felt, linoleum paste; for laminated or solid wood block, hot-melt or cutback asphalt; for vinyl backed with asbestos felt, latex on concrete and linoleum paste on plywood and hardboard. Laminated wood blocks also may be set with a rubber-base adhesive.

Ceramic Tiles ▪ These generally are bonded to the subfloor with portland cement mortar (see American National Standards Institute "Standard Specifications for the installation of Ceramic Tile," ANSI A108, A118, A136.1-19). For areas not subject to heavy traffic, concentrated loads, or excessive amounts of water, organic-adhesive thin setting beds may be used instead. Appearance and resistance to wear make ceramic tiles suitable for use in kitchens and bathrooms.

Ceramic mosaic tile is less than 6 in^2 in area. Paver tiles are larger, usually 3×3 to 6×6 in. Quarry tile is a denser product, highly resistant to freezing, abrasion, and moisture.

Terrazzo is a mosaic topping composed of 2 parts marble chips to 1 part portland cement, sometimes with color pigments, applied to concrete or steel decks. Rubber latex, epoxy, and polyesters are alternative matrix materials. The topping may be precast or cast in place. Sand cushion (floating) terrazzo, at least 3 in thick, is used where structural movement that might injure the topping is anticipated. When terrazzo is bonded to the under slab, the topping usually is at least $1\frac{3}{4}$ in thick; a monolithic topping may be $\frac{5}{8}$ in thick.

(J. H. Callender, "Time-Saver Standards for Architectural Design Data," 6th ed., McGraw-Hill Publishing Company, New York (books.mcgraw-hill.com).)

15.12 Masonry Walls

Different design criteria are applied to masonry walls, depending on whether they are load-bearing or non-load-bearing. Minimum requirements for both types are given in "Building Code Requirements for Masonry," ACI 530-95/ASCE 5-95, and "Specifications for Masonry Structures," ACI 530.1-95/ASCE 6-95, American Concrete Institute and American Society of Civil Engineers.

Following are some terms most commonly encountered in masonry construction:

Architectural Terra Cotta ■ (See Ceramic Veneer.)

Ashlar Masonry ■ Masonry composed of rectangular units, usually larger in size than brick and properly bonded, having sawed, dressed, or squared beds. It is laid in mortar.

Bonder ■ (See Header.)

Brick ■ A rectangular masonry building unit, not less than 75% solid, made from burned clay, shale, or a mixture of these materials.

Buttress ■ A bonded masonry column built as an integral part of a wall and decreasing in thickness from base to top, although never thinner than the wall. It is used to provide lateral stability to the wall.

Ceramic Veneer ■ Hard-burned, non-load-bearing, clay building units, glazed or unglazed, plain or ornamental.

Chase ■ A continuous recess in a wall to receive pipes, ducts, and conduits.

Collar Joint ■ A vertical joint between wythes or a wythe and backup.

Column ■ A compression member with width not exceeding four times the thickness, and with height more than three times the least lateral dimension.

Concrete Block ■ A machine-formed masonry building unit composed of portland cement, aggregates, and water.

Coping ■ A cap or finish on top of a wall, pier, chimney, or pilaster to prevent penetration of water to masonry below.

Corbel ■ Successive courses of masonry projecting from the face of a wall to increase its thickness or to form a shelf or ledge.

Course ■ A continuous horizontal layer of masonry units bonded together.

Cross-Sectional Area ■ Net cross-sectional area of a masonry unit is the gross cross-sectional area minus the area of cores or cellular spaces. Gross cross-sectional area of scored units is determined to the outside of the scoring, but the cross-sectional area of the grooves is not deducted to obtain the net area.

Grout ■ A mixture of cementitious material, fine aggregate, and sufficient water to produce pouring consistency without segregation of the constituents.

Grouted Masonry ▪ Masonry in which the interior joints are filled by pouring grout into them as the work progresses.

Header (Bonder) ▪ A brick or other masonry unit laid flat across a wall with end surface exposed, to bond two wythes.

Height of Wall ▪ Vertical distance from top of wall to foundation wall or other intermediate support.

Hollow Masonry Unit ▪ Masonry with net cross-sectional area in any plane parallel to the bearing surface less than 75% of its gross cross-sectional area measured in the same plane.

Masonry ▪ A built-up construction or combination of masonry units, such as clay brick, concrete block, or stone, bonded together with mortar or other cementitious material.

Mortar ▪ A plastic mixture of cementitious materials, fine aggregates, and water.

Partition ▪ An interior wall one story or less in height.

Pier ▪ An isolated column of masonry. A bearing wall not bonded at the sides into associated masonry is considered a pier when its horizontal dimension measured at right angles to the thickness does not exceed four times its thickness.

Pilaster ▪ A bonded or keyed column of masonry built as part of a wall and of uniform thickness throughout its height. It serves as a vertical beam, column, or both.

Rubble:

Coursed Rubble. Masonry composed of roughly shaped stones fitting approximately on level beds, well bonded and brought at vertical intervals to continuous level beds or courses.

Random Rubble. Masonry composed of roughly shaped stones, well bonded and brought at irregular intervals vertically to discontinuous but approximately level beds or courses.

Rough or Ordinary Rubble. Masonry composed of irregularly shaped stones laid without regularity of coursing, but well bonded.

Solid Masonry Unit ▪ A masonry unit with net cross-sectional area in every plane parallel to the bearing surface 75% or more of its gross cross-sectional area measured in the same plane.

Veneer ▪ A wythe securely attached to a wall but not considered as sharing load with or adding strength to it.

Wall ▪ Vertical or near-vertical construction for enclosing space or retaining earth or stored materials.

Bearing Wall. A wall that supports any vertical load in addition to its own weight.

Cavity Wall. (See Hollow Wall.)

Curtain Wall. A non-load-bearing exterior wall.

Faced Wall. A wall in which the masonry facing and backing are of different materials and so bonded as to exert a common reaction under load.

Hollow Wall. A wall of masonry so arranged as to provide an air space within the wall between the inner and outer wythes. A cavity wall is built of masonry units or plain concrete, or a combination of these materials, arranged to provide an air space within the wall, which may be filled with insulation, and in which inner and outer wythes are tied together with metal ties.

Nonbearing Wall. A wall that supports no vertical load other than its own weight.

Party Wall. A wall on an interior lot line used or adapted for joint service between two buildings.

Shear Wall. A wall that resists horizontal forces applied in the plane of the wall.

Spandrel Wall. An exterior curtain wall at the level of the outside floor beams in multistory buildings. It may extend from the head of the window below the floor to the sill of the window above.

Veneered Wall. A wall having a facing of masonry or other material securely attached to a backing, but not so bonded as to exert a common reaction under load.

Wythe ▪ Each continuous vertical section of a wall one masonry unit in thickness.

Materials used in masonry construction should be capable of meeting the requirements of the applicable standard of ASTM. For unit masonry, mortar should meet the requirements of ASTM Specifications C270 and C476. Mortars containing lime generally are preferred because of greater workability. Commonly used:

For concrete block, 1 part cement, 1 part lime putty, 5 to 6 parts sand

For rubble, 1 part cement, 1 to 2 parts lime hydrate or putty, 5 to 7 parts sand

For brick, 1 part cement, 1 part lime, 6 parts sand

For setting tile, 1 part cement, $\frac{1}{2}$ part lime, 3 parts sand

Design of masonry structures should be based on elastic analysis, except that empirical design may be used for Seismic Zones 1 and 2 or where basic wind pressure is less than 25 psf, if no other lateral loads than wind or seismic loads are applied. Design should take into account the decrease in cross section and other weakening effects of embedding pipes and conduits in the masonry. Spacing should be at least three diameters center to center.

ACI 530-95/ASCE 5-95 sets the following requirements:

Masonry walls comprising two or more wythes, each wythe intended to resist individually the loads imposed on it (noncomposite action), should incorporate a cavity between the wythes, without headers, grout, or mortar. Width of the cavity should not exceed 4 in. The wythes should be connected by steel ties spaced not more than 16 in apart horizontally and vertically. Loads acting transversely to the plane of a wall should be distributed to each wythe in proportion to its relative stiffness.

Masonry walls designed for composite action of the wythes should have collar joints filled with mortar or grout or crossed by headers bonded to the wythes. For mortared collar joints, shear stresses between the wythes and collar joints or within headers should not exceed 5 psi; for grouted collar joints, these stresses should not exceed 10 psi. Headers should be embedded at least 3 in in each wythe and spaced uniformly over the wall. Total cross-sectional area of all the headers should be at least 4% of the wall surface area. Walls without headers should be bonded by steel ties, spaced not more than 36 in horizontally and 24 in vertically. At a minimum, one 9-gage tie should be used for every 2.67 ft^2 of wall surface, or one $\frac{3}{16}$-in-diameter tie for every 4.5 ft^2.

For resistance to wind and seismic loads, masonry walls should be anchored to floors and roofs that provide lateral support. Anchors should be embedded in reinforced bond beams or reinforced vertical cells and capable of resisting loads of at least 200 lb/lin ft of wall. Steel reinforcement should be incorporated both horizontally and vertically in the walls. Bearing walls should have a nominal thickness of at least 6 in.

Masonry columns should have a minimum nominal side dimension of at least 12 in for seismic resistance. The ratio of effective height to the smallest nominal side dimension should be 25 or less. Lateral ties at least $\frac{1}{4}$ in diameter should enclose longitudinal reinforcement bars in the columns. Spacing of the lateral ties should be less than 16 bar diameters, 48 tie diameters, and the smallest side dimension of the columns.

Empirical Design of Masonry Walls ▪ Where empirical design is permitted, bearing walls of one-story buildings may be only 6 in thick. Higher walls should be at least 8 in thick. Rubble stone walls, however, should be at least 16 in thick. Buildings using masonry walls to resist lateral loads should not be more than 35 ft high. If masonry shear walls are provided for lateral stability, they should be at least 8 in thick. Cumulative length of shear walls in any direction should be at least 40% of the long dimension of the building.

For lateral stability, solid or solid grouted, load-bearing masonry walls should have either a ratio of unsupported height to nominal thickness or a ratio of unsupported length to nominal thickness of 20 or less. For other types of load-bearing walls and exterior nonbearing walls, the ratio should not exceed 18. For interior nonbearing walls, the ratio should be 36 or less. Parapet walls should be at least 8 in thick. Their height should not exceed three times the thickness. Minimum thickness of foundation walls depends on the depth of unbalanced fill to be resisted. For example, walls 12 in thick are permitted a 6-ft depth for ungrouted hollow units, 7-ft depth for solid units, and 8-ft depth for fully grouted units.

Good Practice ▪ Backfill should not be placed against foundation walls until they have been braced to withstand horizontal pressure. Veneers should not be considered part of the wall when computing thickness for strength or stability.

When determining the unsupported length of walls, you may assume existing cross walls, piers, or buttresses as lateral supports if these members are well bonded or anchored to the walls and capable of transmitting the lateral forces to connected structural members or to the ground. When determining the unsupported height of walls, you may consider floors and roofs as lateral supports, if provision is made in the building to transmit the lateral forces to the ground. Ends of floor joists or beams bearing on masonry walls should be securely fastened to the walls. If lateral support of a partition depends on a ceiling, floor, or roof, the top of the partition should have adequate anchorage to transmit the forces. This anchorage may be accomplished with metal anchors or by keying the top of the partition to overhead work. Suspended ceilings may be considered as lateral support if ceilings and anchorages are capable of resisting a horizontal force of 200 lb/lin ft of wall.

Walls should not vary in thickness between lateral supports. When it is necessary to change thickness between floor levels to meet minimum thickness requirements, the greater thickness should be carried up to the next floor level. Where walls of hollow units or bonded hollow walls are decreased in thickness, a course of solid masonry should be interposed between the wall below and the thinner wall above, or else special units or construction should be used to transmit the loads between the walls of different thickness.

When two bearing walls intersect and the courses are built up together, the intersections should be bonded by laying in true bond at least half the units at the intersection. When the courses are carried up separately, the intersecting walls should be regularly toothed or blocked with 8-in maximum offsets. The joints should be provided with metal anchors having a minimum section of $\frac{1}{4} \times 1\frac{1}{2}$ in with ends bent up at least 2 in or with cross pins to form an anchorage. Such anchors should be at least 2 ft long and spaced not more than 4 ft apart.

(J. H. Matthys, "Masonry Designers' Guide," 3rd ed., The Masonry Society, 3970 Broadway, Ste 201D, Boulder, CO 80304-1135; F. S. Merritt, "Building Design and Construction Handbook," 6th ed., McGraw-Hill Publishing Company, New York (books.mcgraw-hill.com); J. Ambrose, "Simplified Design of Masonry Structures," John Wiley & Sons, Inc., New York (www.wiley.com).)

15.13 Glass Block

Masonry walls of glass block may be used to control light that enters a building and to obtain better thermal and acoustic insulation than with ordinary glass panes. These units are hollow, $3\frac{7}{8}$ in thick by 6, 8, or 12 in square (actual length and height $\frac{1}{4}$ in less, for modular dimensioning). Faces of the units may be cut into prisms to direct light, or the block may be treated to diffuse light.

Glass block may serve as nonbearing walls or to fill openings in walls. Block so used should have a minimum thickness of $3\frac{1}{2}$ in at the joint. Also, surfaces of the block should be treated to permit satisfactory mortar bonding.

For exterior walls, glass-block panels should not have an unsupported area exceeding 144 ft². They should not be more than 25 ft long or more than 20 ft high between supports.

For interior walls, glass-block panels should not have an unsupported area of more than 250 ft². Neither length nor height should exceed 25 ft.

Exterior panels should be held in place in the wall opening to resist both internal and external wind pressures. The panels should be set in recesses at the jambs so as to provide a bearing surface at least 1 in wide along the edges. Panels more than 10 ft long also should be recessed at the head. (Some building codes, however, permit anchoring small panels in low buildings with non-corrodible perforated metal strips.) The sides and top, kept free of mortar and filled with resilient material, should permit expansion.

Mortar joints should be from $\frac{1}{4}$ to $\frac{3}{8}$ in thick. Steel reinforcement should be placed in the horizontal mortar joints at vertical intervals of 2 ft or less and should extend the full length of the joints. When splices are necessary, the reinforcement should be lapped at least 6 in. It should consist of two parallel longitudinal galvanized-steel wires. They should be No. 9 gage or larger, spaced 2 in apart, and having welded to them No. 14 or heavier gage cross wires at intervals of up to 8 in.

15.14 Curtain Walls

With skeleton-frame construction, exterior walls need carry no load other than their own weight. Their principal function is to keep wind and weather out of the building—hence the name curtain wall. Nonbearing walls may be supported on the structural frame of the building or projections from it, on supplementary framing (girts, for example) in turn supported on the structural frame, or on the floors.

Curtain walls need not be any thicker than required to serve their principal function. Many industrial buildings are enclosed only with light-gage metal. For structures with certain types of occupancies and for buildings close to others, however, fire resistance is an important characteristic; fire-resistance requirements in local building codes often govern when determining the thickness and type of material used for curtain walls.

In many types of buildings, it is desirable to have an exterior wall with good insulating properties. Sometimes, a dead air space is used for this purpose; sometimes, insulating material is incorporated into the wall or erected as a backup.

The exterior surface of a curtain wall should be made of a durable material, capable of lasting as long as the building. Maintenance should be a minimum; initial cost of the wall is not so important as the annual cost (amortized initial cost plus annual maintenance and repair costs).

Wood walls are used on one- and two-story buildings, generally with a wood frame. The frame may be sheathed on its outer sides with gypsum, lumber, or plywood and then a finish applied, or sheathing and siding can be combined in one unit. The exterior finish may be in the form of wood shingles, siding, half timbers, or plywood sheets.

Drop, or novelty, siding—tongued-and-grooved boards—is not considered a good finish for permanent structures. **Lap siding** or **clapboards** are better. These are beveled boards, thinner along one edge than the opposite edge, which are nailed over sheathing and building paper. Usually, narrow boards lap each other 1 in, wide boards more than 2 in. Normally, clapboards are installed with edges horizontal.

Half timbers may be used to form a structural frame of heavy horizontal, vertical, and diagonal members, the spaces between being filled with brick. This type of construction sometimes is imitated by nailing boards in a pattern similar to an ordinary sheathed frame and filling the spaces between boards with stucco.

Plywood for exterior use should be an exterior grade, with plies bonded with permanent waterproof glue. The curtain wall may consist of a single sheet of plywood or a sandwich of which plywood is a component. Also, plywood may be laminated to another material, such as a light-gage metal, to give it stiffness.

Stucco is an exterior wall finish applied like plaster and made of sand, portland cement, lime, and water. Two coats are applied to masonry, three coats on metal lath. The lath should be heavily galvanized. It should weigh 3.4 lb/yd^2 when supports are 16 in c to c and at least 2.5 lb with closely spaced furring. For the first coat, a common mix is 1 part portland cement, 1 part lime putty, and 5 or 6 parts sand. The second, or brown, coat may be based on lime or portland cement. With lime, the mix may be 1 part quicklime putty or hydrated lime putty and 3 parts sand by volume. With cement, the mix may be 1 part portland cement to 3 parts sand, plus lime putty in an amount equal to 15 to 25% of the volume of cement. The finish coat may have the same proportions as the brown coat. The brown coat may be applied as soon as the first, or scratch, coat has hardened, usually in 7 to 10 days. For the finish coat, it may be wise to wait several months, to let the building settle and the base coats shrink.

Metal siding may be used as curtain walls with or without insulating backup. Precautions should be taken to prevent water from penetrating between sheets. With corrugated sheeting, horizontal splices should lap about 4 in. Vertical splices should lap at least 1½ corrugations. Flat sheets may be installed in sash like window glass, or the splices may be covered with battens. Edges of metal sheets may be flanged to interlock and exclude wind and rain. Provision should be made in all cases for expansion and contraction with temperature changes.

In contrast to siding, in which a single material may form the complete wall, metal or glass sometimes is used as a facing and backed up with insulation, fire-resistant material, and an interior finish. The glass usually is tinted and held in a light frame in the same manner as window glass. Metal panels may be fastened similarly in a light frame, attached to mullions or other secondary framing members, anchored to brackets at each floor level, or connected to the structural frame of the building. The panels may be small and light enough for one person to carry or as high as one or two stories, prefabricated with windows. Provisions should be made for expansion and contraction and for prevention of moisture penetration at joints. Flashing and other details should be so arranged that should water penetrate the facing, it will drain to the outside.

Curtain walls also may be built of prefabricated panels consisting of an insulation core sandwiched between a thin, lightweight facing and backing. Such panels may be fastened in place in much the same manner as metal or glass facings. And the same precautions regarding expansion, contraction, and water penetration should be taken.

Metal curtain walls may be custom, commercial, or industrial. Custom-type walls are designed for a specific project, generally multistory buildings. Commercial-type walls are built up of parts standardized by manufacturers. Industrial-type walls comprise ribbed, fluted, or otherwise preformed metal sheets in stock sizes, standard metal sash, and insulation.

Metal curtain walls may be classified according to the methods used for field installation:

Stick Systems (Fig. 15.11a). Walls installed piece by piece. Each principal framing member, with windows and panels, is assembled in place separately (Fig. 15.11b). This type of system involves more parts and field joints than other types and is not so widely used.

Mullion-and-Panel Systems (Fig. 15.11c). Walls in which vertical supporting members (mullions) are erected first, and then wall units, usually incorporating windows (generally unglazed), are placed between them (Fig. 15-11d). Often, a cover strip is added to cap the vertical joint between units.

Panel Systems (Fig. 15-11e). Walls composed of factory-assembled units (generally unglazed) and

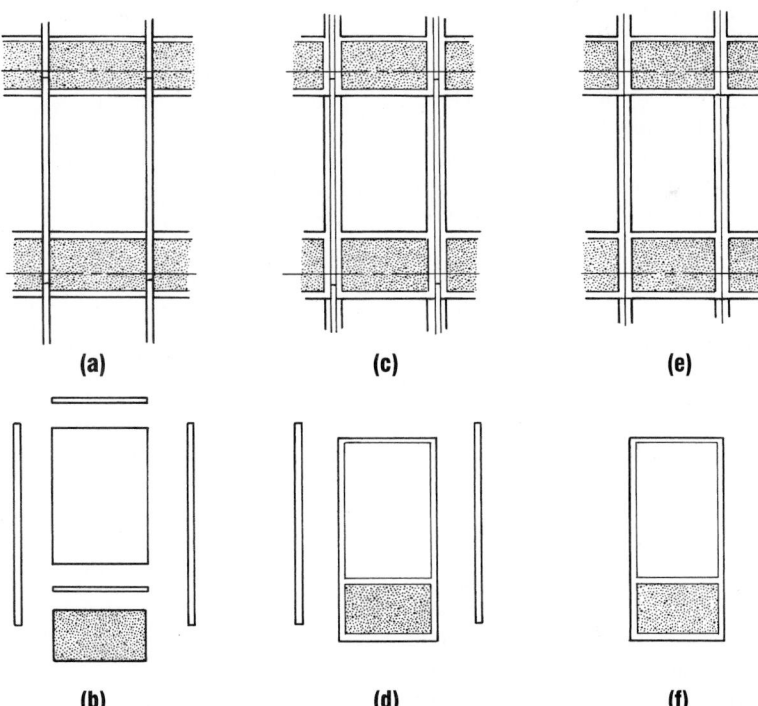

(a)   **(c)**   **(e)**

(b)   **(d)**   **(f)**

Fig. 15.11 Metal curtain walls: (*a*) Stick systems, which are installed piece by piece as shown in (*b*). Mullion-and-panel systems (*c*) have panels placed between verticals (*d*). Panel systems (*e*) come factory-assembled (*f*).

installed by connecting to anchors on the building frame and to each other (Fig. 15.11*f*). Units may be one or two stories high. This system requires fewer pieces and field joints than the other systems.

When mullions are used, it is customary to provide for horizontal movements of the wall at the mullions and, in multistory buildings, to accommodate vertical movement at each floor, or at alternate floors when two-story-high components are used. Common ways of providing for horizontal movements include use of split mullions, bellows mullions, batten mullions, and elastic structural gaskets. To permit vertical movement, mullions are spliced with a telescoping slip joint. When mullions are not used and wall panels are connected to each other along their vertical edges, the connection is generally made through deep flanges. With the bolts several inches from the face of the wall, movement is permitted by the flexibility of the flanges. Slotted holes are unreliable as a means of accommodating wall movement.

(W. F. Koppes, "Metal Curtain Wall Specifications Manual," National Association of Architectural Metal Manufacturers, 8 South Michigan A Suite 1000, Chicago, IL 60603 (www.naam.org); "Curtain Wall Handbook," U.S. Gypsum Co., Chicago, IL 60606; F. S. Merritt, "Building Design and Construction Handbook," 6th ed., and J. H. Callender, "Time-Saver Standards for Architectural Design Data," 6th ed., McGraw-Hill Publishing Company, New York (www.books. mcgraw-hill.com).)

15.15 Partitions

Partitions are walls one story or less in height used to subdivide the interior space in buildings. They may be bearing or nonbearing walls.

Bearing partitions may be built of masonry or wood or light-gage steel studs. The masonry or studs may be faced with plaster, wallboard, plywood, wood boards, plastic, or other materials that meet functional and architectural requirements.

Nonbearing partitions may be permanently fixed in place, or temporary (movable) so that the walls can be easily shifted when desired. The temporary type includes folding partitions. Since the principal function is to separate space, construction and materials used vary widely. Partitions may be opaque or transparent, louvered or hollow or solid, extend from floor to ceiling or

only partway, and may serve additionally as cabinets or closets or as a concealment for piping and electrical conduit. Fire resistance sometimes dictates the type of construction. Acoustic treatment may range from acoustic finishes on the surfaces to use of double walls separated completely by an air space or an insulating material.

Folding partitions, in a sense, are large doors. Depending on size and weight, they may be electrically or manually operated. They may be made of wood, light-gage metal, or synthetic fabric on a light collapsible frame. Provision should be made for framing and supporting them in a manner similar to that for large folding doors (Art. 15.18).

(F. S. Merritt, "Building Design and Construction Handbook," 6th ed., McGraw-Hill Publishing Company, New York (books.mcgraw-hill.com).)

15.16 Windows

Some building codes require that glass areas equal at least 10% of each room's floor area. But for many types of occupancy, it is good practice to provide glass areas in excess of 20%. Windows should be continuous and located as high as possible to lengthen the depth of light penetration. Continuous sash or one large window in a room gives better light distribution than separated narrow windows. Since windows also provide ventilation, the designer sometimes must compromise between locations, sizes, and arrangements of windows that give best lighting or best ventilation.

Window sash and frames generally are made of wood or metal. Fire-resistance requirements of building codes usually dictate use of metal.

White pine, sugar pine, ponderosa pine, fir, redwood, cedar, and cypress are used for exposed wood window parts because of their resistance to shrinkage and warping. A relatively hard wood should be used for the stiles against which a double-hung window slides. Inside parts of wood windows are usually made of the same material as trim.

Components of a typical window are shown in Fig. 15.12. Some commonly used terms are:

Sash ▪ A single assembly of stiles and rails made into a frame for holding glass, with or without dividing bars. It may be supplied glazed or unglazed.

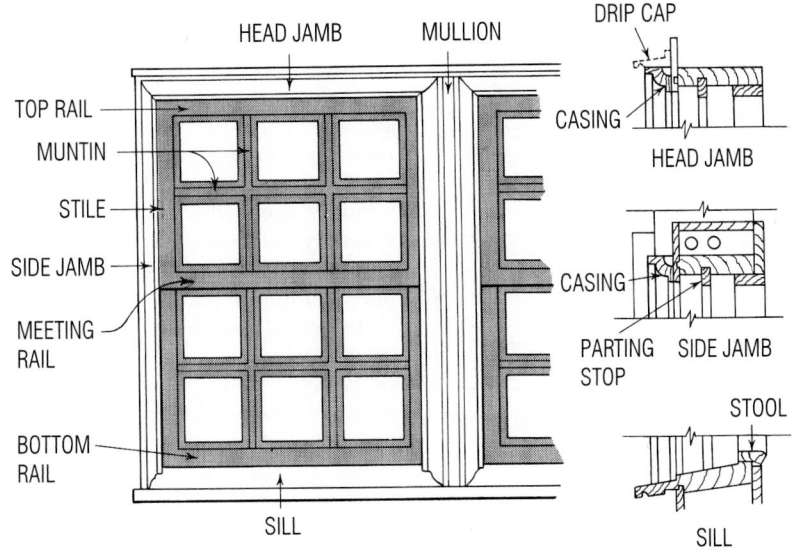

Fig. 15.12 Components of a window.

Window ▪ Sash and the glass that fill an opening.

Stiles ▪ Upright, or vertical, border pieces of a sash.

Rails ▪ Cross, or horizontal, members of a sash.

Check Rails ▪ Meeting rails sufficiently thicker than the window to fill the opening between the top and bottom sash made by the check strip or parting strip in the frame. They usually are beveled and rabbeted.

Bar ▪ Member that extends the height or width of the opening to be glazed.

Muntin ▪ A short, light bar.

Frame ▪ Wood parts machined and assembled to form an enclosure and support for a window.

Jamb ▪ Part of a frame that surrounds and contacts the window it supports. Side jambs form the vertical sides; head jamb, the top. A rabbeted jamb has a rectangular groove along its edges to receive a window.

Sill ▪ The horizontal bottom part of a frame.

Stool ▪ The part of the sill inside the building.

Pulley Stile ▪ A side jamb in which a pulley is installed and along which the sash slides.

Casing ▪ Molding of various widths and thicknesses used to trim window openings.

Drip Cap ▪ A molding placed on top of the head casing of a window frame to direct water away from it.

Blind Stop ▪ A thin strip of wood machined to fit the exterior vertical edge of the pulley stile or jamb and keep the sash in place.

Parting Stop ▪ A thin wood strip let into the jamb of a window frame to separate the sash.

Dado ▪ A rectangular groove cut across the grain of a frame member.

Jamb Liner ▪ A small strip of wood, either surfaced four sides or tongued on one edge, which,

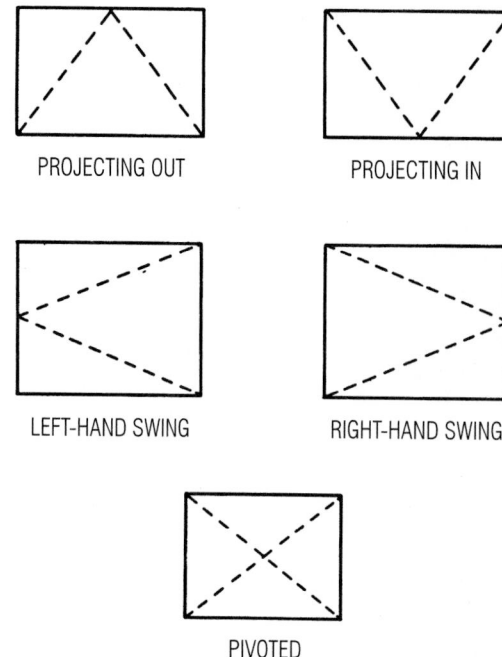

PROJECTING OUT   PROJECTING IN

LEFT-HAND SWING   RIGHT-HAND SWING

PIVOTED

Fig. 15.13 Symbols for common types of windows (viewed from outside).

when applied to the inside edge of a jamb, increases its width for use in thicker walls.

Generally, steel windows are made from hot-rolled structural-grade new billet steel. Dimensions usually conform to the specifications of The Steel Window Institute. Similar types of windows also

are available in aluminum but conforming to the specifications of the Aluminum Window Manufacturers Association.

Many types of windows are available (see symbols, Fig. 15.13). Among the more commonly used types are:

Pivoted windows (Fig. 15.14a), an industrial window used where very tight closure is not a necessity. Vents are pivoted about 2 in above the center. Top swings in. They may be mechanically operated in groups.

Projected windows (Fig. 15.14b), similar to pivoted windows except that the pivot is at the top or bottom. Commercial projected windows are used in commercial and industrial installations where initial cost is a prime consideration. Maximum opening is about 35°. Architectural projected windows are medium-quality, used for commercial, institutional, and industrial buildings. Intermediate projected windows are high-quality, used for schools, hospitals, commercial buildings, and many other structures. Basement and utility windows usually are projected, opening inward, generally with pivot at the bottom. Also, security windows, psychiatric projected, and detention windows generally are bottom-pivoted.

Double-hung windows (Fig. 15.14c), comprising a pair of vertically sliding sash. They are used for all types of buildings. Usually, the sash are balanced, to permit easy movement, by weights or other devices in the jambs. Horizontally sliding windows also are available.

Casement windows (Fig. 15.14d), consisting of a pair of vertically pivoted sash, generally opening

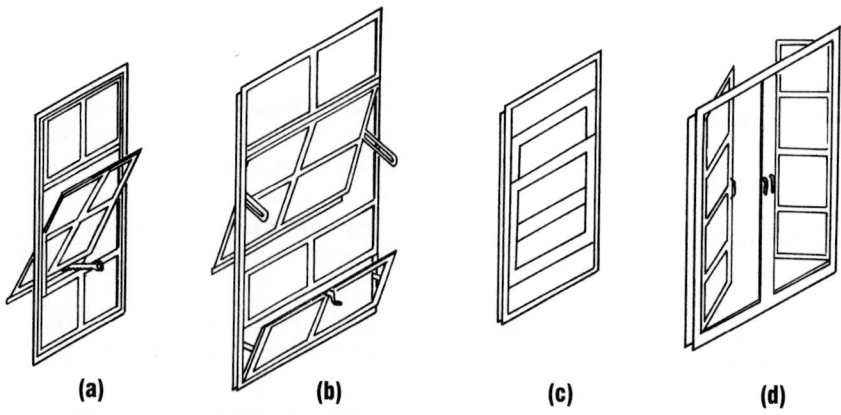

(a)   (b)   (c)   (d)

Fig. 15.14 Types of windows: (a) pivoted; (b) projected; (c) double-hung; (d) casement.

outward. Rotary or lever operators hold the vents at the desired position up to full opening. **Intermediate combination windows** come with casement windows above a vent that projects in.

Picture windows have fixed sash, and sometimes ventilating units.

Storm sash is another means of reducing heat loss. In effect, it is a second window installed outside the main window. The objective is to create a dead air space, which offers good thermal insulation, without decreasing visibility appreciably.

Windows may be supplied with or without screens. Storm sash and screens may be obtained as a single unit for some types of windows, for example, for double-hung and horizontally sliding windows.

Weather stripping is used on windows to reduce air leakage around joints. It is made of metal or a compressible resilient material.

(F. S. Merritt, "Building Design and Construction Handbook," 6th ed., and J. H. Callender, "Time-Saver Standards for Architectural Design Data," 6th ed., McGraw-Hill Publishing Company, New York (books.mcgraw-hill.com).)

15.17 Glazing

Window glass in common use includes:

Sheet Glass ▪ Used in all types of buildings. Classified by Federal standards according to defects. Grade A is used where appearance is important; Grade B for industrial buildings, low-cost housing, basements, garages, and so on. Sheet glass comes in single strength, $\frac{3}{32}$ in thick, up to 40×50 in; double strength, $\frac{1}{8}$ in thick, up to 60×80 in; and heavy sheet, $\frac{7}{32}$ in, up to 76×120 in. For appearance's sake, single and double strength should be limited in area to 7 ft^2.

Float Glass ▪ Used in store windows, picture windows, and better-grade buildings. Better appearance; no distortion of vision. Thickness ranges from $\frac{1}{8}$ to $\frac{7}{8}$ in.

Patterned (Figured or) Rolled Glass ▪ Types of obscure glass.

Obscure Wired Glass ▪ Used where resistance to fire or breakage is desired.

Polished Wired Glass ▪ More expensive than obscure wired glass. Used where clear vision is desired, such as in school or institutional doors.

There are also many special glasses for specific purposes:

Heat-absorbing glass reduces heat, glare, and a large percentage of ultraviolet rays, which bleach colored fabrics. It often is used for comfort and reduction of air conditioning loads where large areas of glass have a severe sun exposure. Because of differential temperature stresses and expansion induced by heat absorption under severe sun exposure, special attention must be given to edge conditions. Glass with clean-cut edges is particularly desirable because such cuts affect the edge strength, which in turn must resist the central-area expansion. A resilient glazing material should be used.

Corrugated glass, corrugated wire glass, and corrugated plastic panels are used for decorative treatments, diffusing light, or as translucent structural panels with color.

Laminated glass consists of two or more layers of glass laminated together by one or more coatings of a transparent plastic. This construction adds strength. Some types of laminated glass also provide a degree of security, sound isolation, heat absorption, and glare reduction. Where color and privacy are desired, fadeproof opaque colors can be included. When fractured, laminated glass tends to adhere to the inner layer of plastic and therefore shatters into small splinters, thus minimizing the hazard of flying glass.

Bullet-resisting glass is made of three or more layers of plate glass laminated under heat and pressure. Thicknesses of this glass vary from $\frac{3}{4}$ to 3 in. The more common thicknesses are $\frac{3}{16}$ in, to resist medium-powered small arms; $1\frac{1}{2}$ in, to resist high-powered small arms; and 2 in, to resist rifles and submachine guns. (Underwriters Laboratories lists materials having the required properties for various degrees of protection.) Greater thicknesses are used for protection against armor-piercing projectiles. Uses of bullet-resisting glass include cashier windows, bank teller cages, toll-bridge booths, peepholes, and many industrial and military applications. Transparent plastics also are used as bullet-resistant materials, and some of these materials have been tested by the Underwriters Laboratories. Thicknesses of $1\frac{1}{4}$ in or more have met UL standards for resisting medium-powered small arms.

Tempered glass is produced by a process of reheating and sudden cooling that greatly increases strength. All cutting and fabricating must be done before tempering. Doors of $\frac{1}{2}$ and $\frac{3}{4}$in-thick tempered glass are commonly used for commercial buildings. Other uses, with thicknesses from $\frac{1}{8}$ to $\frac{7}{8}$ in, include backboards for basketball, showcases, balustrades, sterilizing ovens, and windows, doors, and mirrors in institutions. Although tempered glass is $4\frac{1}{2}$ to 5 times as strong as annealed glass of the same thickness, it is breakable, and when broken, disrupts into innumerable small fragments more or less cube shaped.

Tinted and coated glasses are available in several types and for varied uses. As well as decor, these uses can provide for light and heat reflection, lower light transmission, greater safety, sound reduction, reduced glare, and increased privacy.

Transparent mirror glass appears as a mirror when viewed from a brightly lighted side but is transparent to a viewer on the darker opposite side. This one-way-vision glass is available as a laminate, plate or float, tinted, and in tempered quality.

Plastic window glazing, made of such plastics as acrylic or polycarbonate, is used for urban school buildings and in areas where high vandalism might be anticipated. These plastics have substantially higher impact strength than glass or tempered glass. Allowance should be made in the framing and installation for expansion and contraction of plastics, which may be about eight times as much as that of glass. Note also that the modulus of elasticity (stiffness) of plastics is about one-twentieth that of glass. Standard sash, however, usually will accommodate the additional thickness of plastic and have sufficient rabbet depth.

Suspended glazing utilizes metal clamps bonded to tempered plate glass at the top edge, with vertical glass supports at right angles for resistance to wind pressure. These vertical supports, called stabilizers, have their exposed edges polished. The joints between the large plates and the stabilizers are sealed with a bonding cement. The bottom edge or sill is held in position by a metal channel and sealed with resilient waterproofing. Suspended glazing offers much greater latitude in use of glass and virtually eliminates visual barriers.

Factory-sealed double glazing is an insulating-glass unit composed of two panes of glass separated by a dehydrated air space. This type of sash is also manufactured with three panes of glass and two air spaces, providing additional insulation against heat flow or sound transmission. Heat loss and heat gain can be substantially reduced by this insulated glass, permitting larger window areas and added indoor comfort. Heat-absorbing glass often is used for the outside pane and a clear sheet or float glass for the inside.

Special Glazing ▪ American National Standards Institute's Specification Z-97, adopted by many states, requires entrance-way doors and appurtenances glazed with tempered, laminated, or plastic material.

Glass Thickness for Wind ▪ Figure 15.15 can be used to determine the nominal thickness of float or sheet glass for a given glass area, or the maximum area for a given thickness to withstand a specified wind pressure. Based on minimum thickness permitted by Federal Specification DD-G-451c, the wind-load chart provides a safety factor of 2.5. It is intended for rectangular lights with four edges glazed in a stiff, weathertight rabbet. Deflection of a glass support should not exceed $\frac{1}{175}$ of the span.

For example, determine the thickness of a 108×120 in (90-ft^2) light of polished plate glass to withstand a 20-lb/ft^2 wind load. Since the 20-lb/ft^2 and 90-ft^2 ordinates intersect at the $\frac{3}{8}$in glass thickness line, the thickness to use is $\frac{3}{8}$ in.

The correction factors in Table 15.6 also allow Fig. 15.15 to be used to determine the thickness for certain types of fabricated glass products. However, the table makes no allowance for the weakening effect of such items as holes, notches, grooves, scratches, abrasion, and welding splatter.

The appropriate thickness for the fabricated glass product is obtained by multiplying the wind load, lb/ft^2, by the factor given in Table 15.6. The intersection of the vertical line drawn from the adjusted load and the horizontal line drawn from the glass area indicates the minimum recommended glass thickness.

Glazing Compounds ▪ Glass usually is held in place in sash by putty, glazing compound, rubber, plastic strips, metal clips (with metal sash), or glazing points (with wood sash). Commonly used glazing compounds include vegetable-oil

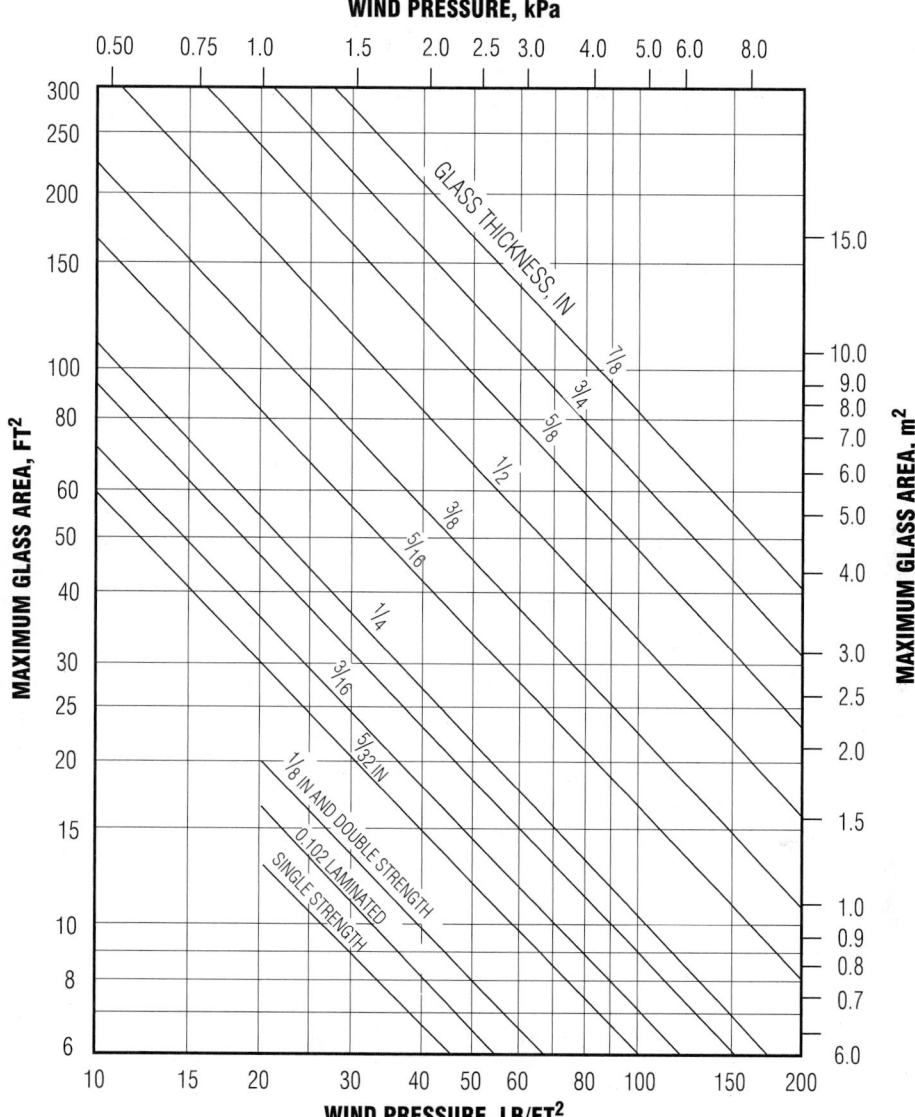

Fig. 15.15 Window chart indicates maximum glass area for various nominal thicknesses of sheet or float glass to withstand specified wind pressures, with design factor of 2.5. The chart is based on glazing firmly supported along all four edges and with length-to-width ratio not exceeding 5:1.

base (skin-forming type), vegetable-oil rubber or nondrying oil blends (polybutene), nondrying oil types—all of which may be applied by gun or knife—butyl rubber or polyisobutylene, applied as tacky tape; polysulfide rubber, applied by gun; Neoprene, applied by gun or as a preformed gasket; and vinyl chloride and copolymer, applied as preformed gaskets. Bedding of glass in glazing compound is desirable because it furnishes a smooth bearing surface for the glass, prevents rattling, and eliminates voids where moisture can collect. A thin layer of putty or bedding compound

Table 15.6 Relative Resistance of Glass to Wind Loads

Product	Factor*
Float sheet glass	1.0
Patterned glass	1.0
Sand blasted glass	2.5
Laminated glass[†]	1.33
Sealed double glazing[‡]	
2 panes	0.60
3 panes	0.40
Heat-strengthened glass	0.5
Fully tempered glass	0.25
Wired	2.0

  * Enter Fig. 15.15 with the product of the wind load in lb/ft^2 multiplied by the factor.

  [†] At 70 °F or above, for two lights of equal thickness laminated to 0.015-in-thick vinyl. At 0 °F, factor approaches 1.

  [‡] For thickness, use thinner of the two lights, not total thickness.

is first placed in the rabbet of the sash; the glass is then pressed into this bed, after which the sash is face-puttied and excess putty removed from the back.

Gaskets ■ Preformed structural gaskets can be used as an alternative to sash. Gaskets are extruded from rubberlike materials or plastics in a single strip, molded into the shape of the window perimeter, and installed against the glass and window frame. The gasket may fit into a groove or, H-shaped in cross section, grip the glass and a continuous metal fin on the frame. A continuous locking strip of the same material as the gasket is forced into one side of the gasket to make it grip.

(F. S. Merritt, "Building Design and Construction Handbook," 6th ed., and J. H. Callender, "Time-Saver Standards for Architectural Design Data," 6th ed., McGraw-Hill Publishing Company, New York (books.mcgraw-hill.com).)

15.18 Doors

Selection of a door depends on more than just its function as a barrier to trespass, weather, drafts, noise, drafts, fire, and smoke. Cost, psychological effect, fire resistance, architectural harmony, and ornamental considerations are only a few of the other factors that must be taken into account.

Traffic Flow and Safety ▪ Openings in walls and partitions must be sized for their primary function of providing entry to or exit from a building or its interior spaces. Doors must be sized and capable of operating so as to prevent or permit such passage, as required by the occupants of the building. In addition, openings must be adequately sized to serve as an exit under emergency conditions. In all cases, traffic must be able to flow smoothly through the openings.

To serve these needs, doors must be properly selected for their intended use and properly arranged for maximum efficiency. In addition, they must be equipped with suitable hardware for the application.

Safety. Exit doors and doors leading to exit passageways should be so designed and arranged as to be clearly recognizable as such and to be readily accessible at all times. A door from a room to an exit or to an exit passageway should be the swinging type, installed to swing in the direction of travel to the exit.

Code Limitations on Door Sizes. To ensure smooth, safe traffic flow, building codes generally place maximum and minimum limits on door sizes. Typical restrictions are as follows:

No single leaf in an exit door should be less than 32 in wide (28 in in an existing building) or more than 48 in wide.

Minimum clear width of opening should be at least:

36 in for single corridor or exit doors

32 in for each of a pair of corridor or exit doors with central mullion

48 in for a pair of doors with no central mullion

32 in for doors to all occupiable and habitable rooms

44 in for doors to rooms used by bedridden patients and single doors used by patients in such buildings as hospitals, sanitariums, and nursing homes

32 in for toilet-room doors

Jambs, stops, and door thickness when the door is open should not restrict the required clear width of opening.

Nominal opening height for exit and corridor doors should be at least 6 ft 8 in. Jambs, stops, sills, and closures should not reduce the clear opening to less than 6 ft 6 in.

Opening Width Determined by Required Capacity. The width of an opening used as an exit is a measure of the traffic flow that the opening is permitted to accommodate. Capacities of exits and access facilities generally are measured in units of width of 22 in, and the number of persons per unit of width is determined by the type of occupancy. Thus, the number of units of exit width for a doorway is found by dividing by 22 the clear width of the doorway when the door is in the open position. Fractions of a unit of width less than 12 in should not be credited to door capacity. If, however, 12 in or more is added to a multiple of 22 in, one-half unit of width can be credited. Local building codes list capacities in persons per unit of width that may be assumed for various types of occupancy.

Every floor of a building should be provided exit facilities for its occupant load. The number of occupants for whom exit facilities must be provided is determined by the actual number of occupants for which the space is designed, or by dividing the net floor area by the net floor area per person specified in the local building code.

Fire and Smokestop Doors ▪ Building codes require fire-resistant doors in critical locations to prevent passage of fire. Such doors are required to have a specific minimum fire-resistance rating and are usually referred to as fire doors. The codes also may specify that doors in other critical locations be capable of preventing passage of smoke. Such doors, known as smoke-stop doors, need not be fire-rated.

Fire protection of an opening in a wall or partition depends on the door frame and hardware as well as on the door. All these components must be "labeled" or "listed" as suitable for the specific application. Bear in mind that fire doors are tested as an assembly of these components, and hence only approved assemblies should be specified.

All fire doors should be self-closing or close automatically when a fire occurs. In addition, they should be self-latching, so they remain closed. Push-pull hardware should not be used. Exit doors for places of assembly for more than 100 persons usually must be equipped with fire-exit (panic) hardware capable of releasing the door latch when pressure of 15 lb, or less, is applied to the device in the direction of exit. Combustible materials, such as flammable carpeting, should not be permitted to pass under a fire door.

Fire door assemblies are rated, in hours, according to their ability to withstand a standard fire test, such as that specified in ASTM E152. They may be identified as products qualified by tests by a UL label, provided by Underwriters Laboratories, Inc.; an FM symbol of approval, authorized by Factory Mutual Research Corp.; or a self-certified label, provided by the manufacturer (not accepted by some code officials).

Openings in walls and partitions that are required to have a minimum fire-resistance rating must have protection with a corresponding fire-resistance rating. Typical requirements are in Table 15.7.

This table also gives typical requirements for fire resistance of doors to stairs and exit passageways, corridor doors, and smokestop doors.

In addition, some building codes also limit the size of openings in fire barriers. Typical maximum areas, maximum dimensions, and maximum percent of wall length occupied by openings are in Table 15.8.

Smokestop doors should be of the construction indicated in Table 15.7, last footnote. They should close openings completely, with only the amount of clearance necessary for proper operation.

["Standard for Fire Doors and Windows," NFPA No. 80; Life Safety code, NFPA No. 101; "Fire Tests of Door Assemblies,"NFPA No. 252, National Fire Protection Association, 1 Batterymarch Park, Quincy, MA 02269 (www.nfpa.org).

"Fire Tests of Door Assemblies," Standard UL 10(b); "Fire Door Frames," Standard UL 63; "Building Materials List" (annual, with bimonthly supplements), Underwriters' Laboratories, Inc., 333 Pfingsten Road, Northbrook, IL 60062 (www.ul.com).

"Factory Mutual Approval Guide," Factory Mutual Research Corp., 1151 Boston-Providence Turnpike, Norwood, MA 02062.]

Ordinary Doors ▪ Swinging doors are hung on butts or hinges that permit rotation about a vertical axis at an edge. The door is hinged to and closes against a door frame. The frame consists of two verticals, or jambs, and a horizontal member, the header (Fig. 15.16). *Single-acting doors* can swing 90° or more in only one direction; *double-acting*

Table 15.7 Typical Fire Ratings Required for Doors

Door use	Rating, h*
Doors in 3- or 4-h fire barriers	3[†]
Doors in 2- or 1½-h fire barriers	1½
Doors in 1-h fire barriers	¾
Doors in exterior walls	
Subject to severe fire exposure from outside the building	¾
Subject to moderate or light fire exposure from outside the building	⅓
Doors to stairs and exit passageways	1½
Doors in 1-h corridors	⅓
Other corridor doors	0[‡]
Smokestop doors	⅓[§]

   * Self-closing, swinging doors. Normally kept closed.

   [†] A door should be installed on each side of the wall.

   [‡] Should be noncombustible or 1¾-in solid-core wood doors. Some codes do not require self-closing for doors in hospitals, sanitariums, nursing homes, and similar occupancies.

   [§] May be metal, metal-covered, or 1¾-in solid-core wood doors (1⅜-in in buildings less than three stories high), with 600-in² or larger, clear, wire-glass panels in each door.

doors can swing 90° or more in each of two directions.

To stop drafts and passage of light, the jamb about which the door swings has a rebate or projection, extending the full height, against which the door closes. The projection may be integral with the frame, or formed by attaching a stop on the surface of the frame, or inset slightly. With single-acting doors, the opposite jamb also is provided with a stop, against which the door closes.

Door frames for swinging doors generally are fastened to bucks, rough construction members. Joints between the frame and wall are covered with casings, or trim. With metal construction, the trim often is integral with the frame and designed to grip the bucks. The sill, at the bottom of the door, forms a division between the finished floor on one side and that on the other side.

The sill for exterior doors generally serves also as a step since the door opening usually is raised above grade to prevent rain from entering. The top of the sill is sloped to drain water away from the interior. A raised section or separate threshold at the door is an additional barrier to water. A weather strip in the form of a hooked length of metal may be attached to the underside of the door. When the door is closed, the weather strip locks into the threshold to seal out water and reduce air leakage.

Selection of a swinging door involves consideration of the jamb to which the door is hinged and the direction in which it is to open. This relationship is called the *hand* of the door. Hand and hardware for swinging doors are discussed later.

Horizontally sliding doors roll on rails at top or bottom and slide in guides at the opposite edge. Some doors fold or collapse like an accordion, to occupy less space when open. A pocket should be provided in the walls on either or both sides to receive rigid doors; with folding or accordion types, a pocket is optional.

Vertically sliding doors may rise straight up, may rise up and swing in, or may pivot outward

Table 15.8 Maximum Sizes of Openings in Fire Barriers

Protection of adjoining areas	Max area, ft²	Max dimension, ft
Unsprinklered	120*	12[†]
Sprinklers on both sides	150*	15*
Building fully sprinklered	Unlimited*	Unlimited*

   * But not more than 25% of the wall length or 56 ft² per door if the fire barrier serves as a horizontal exit.

   [†] But not more than 25% of the wall length.

   Source: Based on New York City Building Code.

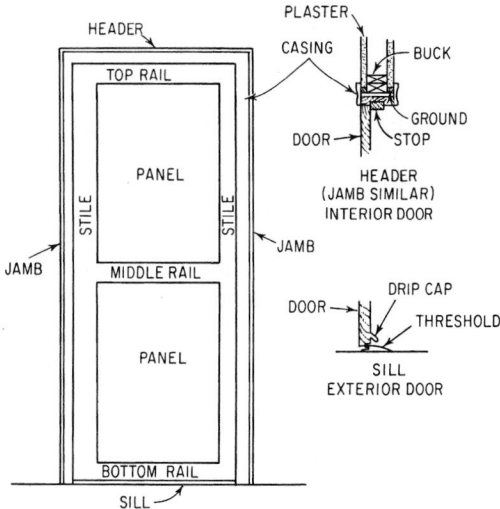

Fig. 15.16 Components of a door.

to form a canopy. Sometimes, the door may be in two sections, one rising up and the other dropping down. Generally, all types are counterweighted for ease of operation. To exclude weather, either the upper part is recessed into the wall above or the top part of the door extends slightly above the bottom of the wall on the inside. Similarly, door sides are recessed into the walls or lap them and are held firmly against the inside. Also, the finished floor is raised a little above outside grade.

Revolving doors are generally selected for entranceways carrying a continuous flow of traffic without a very high peak. They offer the advantage of keeping interchange of inside and outside air to a relatively small amount compared with other types of doors. Usually, they are used in combination with swinging doors because of the inability to handle large groups of people in a short time. Revolving doors consist of four leaves that rotate about a vertical axis inside a cylindrical enclosure with a 4- to 5-ft wide opening.

Special Doors ▪ Large-sized doors, such as those for hangars, garages, and craneway openings and for subdividing gymnasiums and auditoriums, often have to be designed individually, with special attention to their supports and controls. Manufacturers classify such special-purpose doors

as horizontal-sliding, vertical-sliding, swing, and top- or horizontal-hinge.

Telescoping doors, a horizontal-sliding type, are used for airplane hangars. Normally composed of 6 to 20 leaves, they generally are center-parting. When opened, the doors are stacked in pockets at each end of the opening. Operation is by electric motor installed in the end pockets and driving an endless chain attached to the tops of the center leaves.

Folding doors are commonly used for subdividing gymnasiums, auditoriums, and cafeterias and for hangars with very wide openings. This type of door is made up of a series of leaves hinged together in pairs. The leaves fold outward, and when the door is shut, they are held by automatic folding stays. Motors that operate the doors usually are installed in mullions adjacent to the center of the opening. The mullions are connected by cables to the ends of the opening, and when the door is to be opened, the mullions are drawn toward the ends, sweeping the leaves along. The chief advantage over telescoping types is that only two guide channels are required.

Vertical-sliding doors are advantageous when space is available above and below an opening into which door leaves can be moved. Usually, the doors are counterweighted, even when motor-operated.

Large swing doors are used when there is insufficient space around openings for sliding doors. Common applications have been for fire-stations, where width-of-building clearance is essential, and railway entrances, where doors are interlocked with the signal system. Common variations include single-swing (solid leaf with vertical hinge on one jamb), double-swing (hinges on both jambs), two-fold (hinge on one jamb and another between folds and leaves), and four-fold (hinges on both jambs and between each pair of folds). The more folds, the less time required for opening and the smaller the radius needed for swing.

Horizontal-hinge doors are used in craneway entrances to buildings. Sometimes, horizontal-sliding doors are installed below the crane doors to increase the opening. If so, the top guides are contained in the bottom of the crane door; the sliding door must be opened before the swing door.

Radiation-shielding doors are used as a barrier against harmful radiation and atomic particles across openings for access to "hot" cells, and against similar radioactive-isotope handling arrangements and radiation chambers of high-energy x-ray machines or accelerators. Usually, the doors must protect not only personnel but instruments even more sensitive to radiation than people. Shielding doors usually are much thicker and heavier than ordinary doors because density is an important factor in barring radiation. These special-purpose doors are made of steel plates, steel-sheathed lead, or concrete. To reduce thickness, concrete doors may be of medium-heavy ($240 \, \text{lb/ft}^3$) or heavy ($300 \, \text{lb/ft}^3$) concrete, often made with iron-ore aggregate. They usually are operated hydraulically or by electric motor.

Common types of shielding doors include hinged, plug, and overlap. The hinged type is similar to a bank vault door. The plug type, flush with the walls when closed, may roll on floor-mounted tracks or hang from rails. Overlap doors, surface-mounted, also may roll or hang from rails. In addition, vertical-lift doors sometimes are used.

Door Materials ▪ Doors are made of a wide variety of materials. Wood is used in several forms. Better-grade doors are made with panels set in a frame or with flush construction. Paneled doors consist of solid wood or plywood panels held in place by stiles and rails (Fig. 15.16). The joints permit expansion and contraction of the wood with atmospheric changes. If the rails and stiles are

made of a single piece of wood, the paneled door is called solid. When hardwood or better-quality woods are used, the doors generally are veneered; rails and stiles are made with cores of softwood sandwiched between the desired veneer. Tempered glass or plastic may be used instead of wood for panels. Flush doors also may be solid or veneered, or hollow-core.

Metal doors generally are constructed in one of three ways: cast as a single unit or separate frame and panel pieces; metal frame covered with sheet metal; and sheet metal over a wood or other type of insulating core. The heavier metal doors of the swinging type usually are pivoted at top and bottom. Metal-covered doors may be obtained with a wide variety of fire-resistant cores. A **Kalamein door** has a wood core (the wood will not burn as long as the sheet-metal cover prevents oxygen from reaching it).

Doors may be made wholly or partly transparent or translucent. Lights may be made of tempered glass or plastic. Doors made completely of glass are pivoted at top and bottom because the weight makes it difficult to support them with hinges or butts.

Sliding doors of the collapsible accordian type generally consist of wood slats or a light steel frame covered with textile. Plastic coverings frequently are used.

Hand of Doors ▪ Swinging doors are called left-hand doors if, when viewed from the outside, they are hinged to the left-hand jamb and open inward; they are left-hand reverse if they are hinged to the left-hand jamb and open outward. Similarly, they are right-hand and right-hand reverse, respectively, if they open inward and outward when hinged to the right-hand jamb. Since some butts and hinges are handed, the type of door may determine the type of hinge (Fig. 15.17).

Door Hardware ▪ The term **hinge** usually refers to the elongated strap type (Fig. 15.18a and b). It is suitable for mounting on the surface of a door. It consists of two leaves joined by a pin passing through knuckle joints where the leaves fit together.

When the device is to be mounted on the edge of a door, the length of the leaves must be shortened. The leaves thus retain only the portion near the pin,

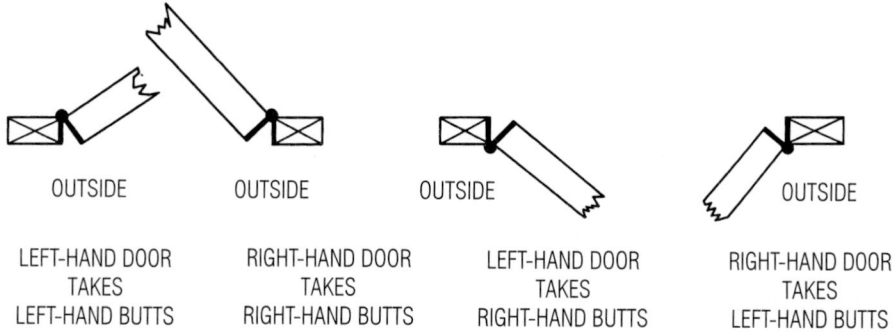

OUTSIDE   OUTSIDE   OUTSIDE   OUTSIDE

LEFT-HAND DOOR   RIGHT-HAND DOOR   LEFT-HAND DOOR   RIGHT-HAND DOOR
TAKES   TAKES   TAKES   TAKES
LEFT-HAND BUTTS   RIGHT-HAND BUTTS   RIGHT-HAND BUTTS LEFT-HAND BUTTS

Fig. 15.17 Direction of opening classifies swinging doors.

or the butt end of the hinge (Fig. 15.18c to g). Thus, hinges applied to the edge of a door are referred to as **butts** or **butt hinges**.

Butts usually are mortised into the door edge. The number of butts required per door depends on the size and weight of the door and the conditions of use. In general, use two butts on doors up to 68 in high and three butts on doors 68 to 90 in high. The weight and usage of the door also determine whether the butts should be two- or four-bearing. When the butts are the type that may be mounted on both left- and right-hand doors, only half the bearing units available participate in carrying the vertical load. This should be considered when selecting butts.

The upper butt may be attached with its top about 5 in below the rabbet of the head jamb. The lowest butt may be set with its bottom 10 in from the finished floor. A third butt may be installed about midway between the other two.

Bearing butts (Fig. 15.18e and f) or butts with Oilite bearings are used for doors requiring silent operation, subject to heavy usage, or equipped with a door closer. Plain bearings (Fig. 15.18g) usually are used for residential doors.

Template hardware, manufactured to close tolerances, is attached to metal jambs and doors with machine screws. In butts, holes that are template-drilled usually form a crescent pattern (Fig. 15.18c and e). When the holes are staggered, the butts are nontemplate (Fig. 15.18f and g).

Butts and hinges come with loose or fast pins. Loose pins are used wherever practicable because they simplify door hanging. The fast (or tight) pin is permanently set in the butt at the time of manufacture; thus, a locked door cannot be opened by

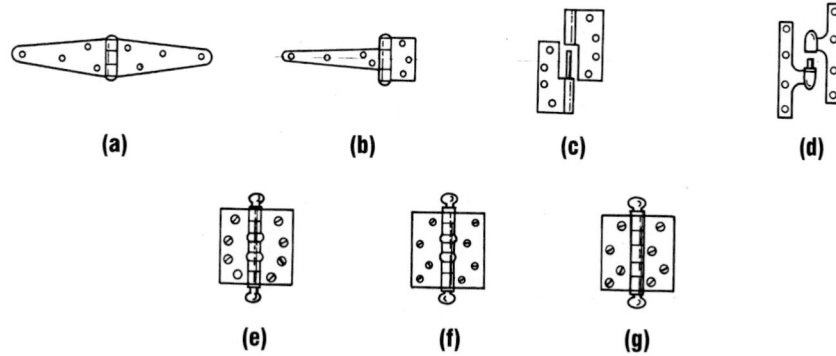

(a)   (b)   (c)   (d)

(e)   (f)   (g)

Fig. 15.18 Typical hinges and butts; (a) heavy strap hinge; (b) heavy T hinge; (c) loose-joint hinge; (d) olive knuckle hinge; (e) bearing template hinge; (f) bearing nontemplate hinge; (g) plain bearing nontemplate hinge.

removing the pins and separating the leaves of the butts. Also, nonremovable loose pins are available for the same purpose. Another type of loose pin is nonrising; it does not have the disadvantage of ordinary loose pins of working upward with repeated movements of the door.

Doors generally are equipped with locks or latches to hold them closed. **Rim locks or latches** are fastened on the surface of the door. Those mortised into the edge of the door are called **mortise locks or latches**. A latch has a beveled locking bolt, which slides into position automatically when the door is closed. Usually, it is operated by a knob or lever. When specifying a latch, the hand of the door should be given.

When the locking bolt is rectangular in shape and must be moved in and out by a thumb turn or key, the bolt is called a **dead bolt** and the lock a **dead lock**. A unit composed of latch bolts and dead bolts is known as a lock.

Unit locks are complete assemblies that can be installed in a standard notch. Bored-in locks similarly are complete assemblies, but installed in circular holes. Depending on the arrangement of holes, bored-in locks may be tubular or cylindrical lock sets.

Tubular locks have a horizontal tubular case perpendicular to the door edge. Another small hole is required normal to the first hole for the locking cylinder.

Cylinder locks need a relatively large hole perpendicular to the face of the door for the cylindrical case. Another small hole perpendicular to the door edge accommodates the bolt.

When selecting locks, choose a uniform size, if practicable, for the project. Then, standard-size cutouts or sinkages can be used throughout. This will reduce installation costs. In addition, if changes are made as the job progresses, hardware changes will be simple and special hardware avoided.

(F. S. Merritt, "Building Design and Construction Handbook," 6th ed., and J. H. Callender, "Time-Saver Standards for Architectural Design Data," 6th ed., McGraw-Hill Publishing Company, New York (books.mcgraw-hill.com); "Life Safety Code Handbook," National Fire Protection Association, 1 Batterymarch Park, Quincy, MA 02269 (www.nfpa.org); "Specifiers Guide to Windows and Doors;" Window & Doors Manufacturers Association, 1400 East Touhy Ave. Ste. 470, Des Plaines, IL 60018, (www.wdma.com); "Recom-

mended Standard Steel Doors, and Frame Details, Steel Door Institute, 712 Lakewood Center North, 30200 Detroit Ave., Cleveland, OH, 44145; "Entrance Manual," Door and Hardware Institute, 14150 Newbrook DR., Suite 200, Chantilly, VA 20151, (www.chi.org).)

15.19 Roof Coverings

Success of a roofing installation depends heavily on the roof deck. Roof framing should be sized and spaced to prevent significant deflection of the deck and consequent damage to the roofing. The deck itself should be smooth, dry, and clean. Many roof failures have resulted from application of roof coverings to damp decks. Pressures developed by the entrapped moisture caused blisters and rupture of the coverings.

Roofing may be single-unit or multiple-unit. The single-unit type, which includes built-up roofing of asphalt or coal-tar pitch, sprayed-on products, and flat-seam metal roofing, is suitable for flat roof decks, where water can collect before proceeding slowly to drainage outlets. Multiple-unit coverings, including shingles, tile, slate, and standing-seam metal panels, are used on steep roof decks, where water flows swiftly over each exposed unit to gutters and leaders.

A built-up roof consists of plies of felt mopped with asphalt or pitch. These form a seamless piece of flexible, waterproofed material, custom-built to conform to the roof deck and protect all angles formed by the roof deck and projecting surfaces.

Bitumen is a generic term used to indicate either asphalt or coal-tar pitch.

Asphalt is a by-product of the refining processes of petroleum oils.

Coal-tar pitch is a by-product of crude tars derived from the coking of coal. The crude tars are distilled to produce coal-tar pitch. It has a lower melting point than asphalt; hence pitch roofs must be protected by a covering of slag or gravel. Asphalt can be used on steeper slopes than pitch.

Felts, made of paper, wood pulp, rag, or glass fibers, aid the bitumens in water shedding and waterproofing. Felts are designated by a number that indicates their approximate weight in pounds per square (100 ft^2). The felts are saturated with bitumen and cemented to the deck and each other with bitumen. For roof decks that permit nailing, the first two plies of felt are nailed to the deck. For slopes steeper than 2 in on 12 in, where there is difficulty holding a slag or gravel covering, felt with minerals embedded in the surface may be used for the top, or cap, sheet. Minimum weight of the sheet should be 55 lb/square. For best results, the cap sheets should be applied over two 15 lb/square felts—the first one nailed, the second mopped.

Built-up roofing generally employs five plies of felt on a wood-sheathed or metal deck and four plies on a cast-in-place concrete deck. "Minimum" specifications require one or two fewer plies and layers of bitumen.

The bituminous roofing may be hot- or cold-process. For cold-process roofing, the bituminous materials, in some cases combined with chemicals, such as polyurethane, are thinned with solvents—for example, kerosene to cut back asphalt, toluene for tar—or emulsified with water. Felts and fabrics differ in the hot and cold processes. Since, in the cold process, cementing occurs on evaporation of solvents or emulsion water, the felts should be of a type that speeds the drying-out process, so an open weave is desirable. Cold-process materials are applied by brush or spray; hot bitumen is mopped in place.

Asphalt roofing materials may be used on slopes up to 4 in/ft:

1 in/ft	19-in selvage-edge roofing, built-up roofing
2 in/ft	Mineral-surfaced roll roofing, blind-nailed, square-butt strip shingles with two-ply felt underlay
3 in/ft	Mineral-surfaced roll roofing, exposed nailed
4 in/ft	Hexagonal, individual, three-tab square butt and lock-type shingles

Asphalt shingles are made of asphalt-saturated and coated felt in which is embedded a permanent mineral surfacing. Square-butt strip shingles are slotted at the butts to give the appearance of individual units. Other shapes, such as hexagonal, also are available. The shingles are installed over an underlayment of No. 15 asphalt felt fastened to the roof deck. A starter course at eaves should be cemented to the felt to prevent leakage through nail punctures. The starter strip may be a row of shingles turned upside down or, preferably, mineral-surfaced roll roofing, 18 in wide in normal-wind areas, 36 in wide in high-wind areas. The shingles should be nailed with nails at least $1\frac{1}{4}$ in long on new work and $1\frac{3}{4}$ in long on reroofing.

Wood shingles come in two varieties, machine-sawn and hand-split. Standard exposure is 5 in but may range from $3\frac{1}{2}$ to 12 in. The shingles are laid in alternate courses, with joints broken relative to courses above and below, starting preferably with a triple layer at the eaves.

Concrete tile, cured under controlled temperature and humidity, may be colored throughout or have an exposed surface of a cementitious material colored with synthetic oxides. The tile may be classified as roll or flat. Roll tiles are available pan shape, with or without interlocking edges; cover, or barrel, semicircular shape; pan and cover shape; and S shape. Flat tile may be supplied as flat shingles or with flat ribs, with interlocking edges. For one-piece barrel and flat tiles, the roof slope should be at least 4 in/ft; for flat shingles, a minimum of 5 in/ft.

Slate roofs may employ standard commercial slating or textural (random) slating. For commercial slating, the material is graded at the quarry; for textural slating, the slates are delivered to the job in random sizes. Longer and heavier slates are placed at the eaves, medium-sized at the center, and the smallest at the ridge. Application starts with an undereaves course fastened over a batten, which slopes the first course. Slating nails should be driven level with the slate.

Clay roofing tile comes in two varieties: roll and flat. Roll tiles may be semicircular, reverse curve, pan and cover, or flat shingle. Application is much the same as for slate.

Metal Roofing ▪ Principal metals used in roofing are galvanized iron, terneplate, Monel metal, aluminum, and copper. Sometimes, zinc, lead, cast iron, or stainless steel is used. In all cases, care should be taken to prevent corrosion, especially from galvanic action. For example, a copper roof should not be applied directly over a wood deck; the copper must be insulated from attack by the steel nails in the deck. Fasteners used with metal roofing should preferably be of the same metal as the roofing. Also, provision should be made for expansion and contraction.

Common types of metal roof installation that allow for thermal movements include batten-seam, standing-seam, flat-seam, and corrugated-metal. Batten-seam and standing-seam coverings consist of narrow strips with loose-locked seams that permit lateral motion. Flat-seam construction consists of small sheets soldered on all edges. Movement is absorbed by buckling in the center of the small sheets.

Except for corrugated metal, large roof areas should be broken into small units, and no metal sheets should be fastened directly to the deck. Clips or battens should be used for fastening. With corrugated-metal roofing, the corrugations absorb thermal movements. Fasteners for corrugated metal include nails, clips, straps, clinch rivets, hook bolts, and welded studs. Where the sheets are fastened directly to roof framing, lead or Neoprene washers should be used with the fasteners.

Plastics Roofing ▪ This may be applied as a roof covering in liquid, sheet, or rigid form. Fluid-applied *elastomeric coatings*, such as Neoprene and Hypalon, conform to any shape of surface and expand and contract with it. Manufacturer's specifications should be followed carefully in all applications. *Plastic roofing sheets* may be made of Neoprene, polyvinyl fluoride, polyisobutylene, EPDM membrane, or other suitable materials, alone or in combination with other materials, for example, bonded to felt. They are applied in the same way as conventional roofing sheets. *Rigid polyvinyl chloride* is available as flat, corrugated, and ribbed panels. They can be cut on the job with portable power saws with abrasive cutting wheels.

("The NRCA Roofing and Waterproofing Manual," National Roofing Contractors Association, 10255 W. Higgins Road, Suite 600, Rosemont, IL 60018 (www.nrca.net); F. S. Merritt, "Building Design and Construction Handbook," 4th ed., McGraw-Hill Publishing Company, New York (books.mcgraw-hill.com).)

15.20 Flashing

At all intersecting surfaces on a building exterior, flashing is necessary to prevent penetration of water through the joints or cracks that might form. Since thermal movements are likely to occur at the intersections, flashing should be elastic or shaped to permit motion.

Bituminous flashings have the ability to hug tight against building surfaces. Metal flashings require added protection, such as cap flashing, installed above and covering the top edge of the base flashing. Plastic flashing sheets have been particularly useful in sealing the junction of vents and pipes with roof decks.

At intersections of walls and flat roofs, at least 6 in of the base flashing should be fastened to the deck and 8 in to the wall. Counterflashing should overlap the base flashing from above at least 4 in and should penetrate at least $1\frac{1}{2}$ in into a raggle cut into the mortar line between the nearest row of bricks above the base flashing.

Step flashing should be used at intersections of walls and steep roofs. For this purpose, short pieces of metal are bent at right angles and one flange sandwiched between every roofing unit at the intersection and the other flange set in contact with the wall. Each flashing unit should lap the one below at least 2 in. Counterflashing also should be installed in steps.

Crickets or flashing saddles are needed between chimneys and sloping roofs to guide water away from the intersection. The saddle is a miniature roof, usually of metal, with a ridge and two valleys, secured to a base step flashing.

Flashing also is required at a number of places in exterior walls; for example, around spandrel beams, coping, sills, at grade, belt courses, water tables, cornices, roof valleys, gables, openings in roofs, and window and door heads.

("The NCRA Roofing and Waterproofing Manual," National Roofing Contractors Association 10255 W. Higgins Road, Suite 600, Rosemont, IL 60018 (www.nrca.net).)

15.21 Waterproofing

Properly built concrete and masonry walls, whether above or below grade, can keep water out of a building without protective coatings or integral waterproofing. Leakage through masonry walls usually occurs at the joints and results from failure to fill them with mortar and poor bond between masonry and mortar. Leakage through concrete walls usually occurs in porous material, or at wall ties or at intersections with other surfaces. Following are some terms most commonly encountered:

Permeability ▪ Quality or state of permitting passage of water and water vapor into, through, and from pores and interstices, without causing rupture or displacement.

Terms used in this section to describe the permeability of materials, coatings, structural elements, and structures follow in decreasing order of permeability:

Pervious or Leaky ▪ Cracks, crevices, leaks, or holes larger than capillary pores, which permit a flow or leakage of water, are present. The material may or may not contain capillary pores.

Water-Resistant ▪ Capillary pores exist that permit passage of water and water vapor, but there are few or no openings larger than capillaries that permit leakage of significant amounts of water.

Water-Repellent ▪ Not "wetted" by water; hence, not capable of transmitting water by capillary forces alone. However, the material may allow transmission of water under pressure and may be permeable to water vapor.

Waterproof ▪ No openings are present that permit leakage or passage of water and water vapor; the material is impervious to water and water vapor, whether or not under pressure.

These terms also describe the permeability of a surface coating or a treatment against water penetration, and they refer to the permeability of materials, structural members, and structures whether or not they have been coated or treated.

Permeability of Concrete and Masonry ▪ Concrete contains many interconnected voids and openings of various sizes and shapes, most of which are of capillary dimensions. If there are few larger voids and openings and they are not directly connected with each other, there will be little or no water penetration by leakage and the concrete may be considered water resistant.

Concrete in contact with water not under pressure ordinarily will absorb it. Resistance of concrete to penetration of water may be improved, however, by incorporating a water-repellent admixture into the mix during manufacture.

Water-repellent concrete is permeable to water vapor. The concrete is not made waterproof (in the full meaning of the term) by the use of an integral water repellent. Note also that water repellents may not make concrete impermeable to penetration of water under pressure.

Most masonry units also will absorb water. Some are highly pervious under pressure. The mortar commonly used in masonry will absorb water too but usually contains few openings permitting leakage.

Masonry walls may leak at the joints between the mortar and the units, however. Except in single-leaf walls of highly pervious units, leakage at the joints results from failure to fill them with mortar and poor bond between the masonry unit and mortar. As with concrete, rate of capillary penetration through masonry walls is small compared with the possible rate of leakage.

Capillary penetration of moisture through above-grade walls that resist leakage of wind-driven rain is usually of minor importance. Such penetration of moisture into well-ventilated subgrade structures may also be of minor importance if the moisture is readily evaporated. However, long-continued capillary penetration into some deep, confined subgrade interiors frequently results in an increase in relative humidity, a decrease in evaporation rate, and objectionable dampness.

Drainage for Subgrade Structures ▪ Subgrade structures located above groundwater level in drained soil may be in contact with water and wet soil for indefinite periods after long-continued rains and spring thaws. Drainage of surface and subsurface water, however, may greatly reduce the time during which the walls and floor of a structure are subjected to water, may prevent leakage through openings resulting from poor work, and reduce the capillary penetration of water into the structure. If subsurface water cannot be removed by drainage, the structure must be made waterproof or highly water-resistant.

Surface water may be diverted by grading the ground surface away from the walls and by carrying the runoff from roofs away from the building. The slope of the ground surface should be at least $\frac{1}{4}$ in/ft for a minimum distance of 10 ft from the walls. Runoff from high ground adjacent to the structure should also be diverted.

Proper subsurface drainage of groundwater away from basement walls and floors requires a drain of adequate size, sloped continuously, and, where necessary, carried around corners of the building without breaking continuity. The drain should lead to a storm sewer or to a lower elevation that will not be flooded and permit water to back up in the drain.

Drain tile should have a minimum diameter of 6 in and be laid in gravel or other kind of porous bed at least 6 in below the basement floor. The open joints between the tile should be covered with a wire screen or building paper to prevent clogging of the drain with fine material. Gravel should be laid above the tile, filling the excavation to an elevation well above the top of the footing. Where considerable water may be expected in heavy soil, the gravel fill should be carried up nearly to the ground surface and should extend from the wall a distance of at least 12 in (Fig. 15.19).

Concrete Floors on Ground ▪ These should preferably not be constructed in low-lying areas that are wet from groundwater or periodically flooded with surface water. The ground should slope away from the floor. The level of the finished floor should be at least 6 in above grade. Further protection against ground moisture and possible flooding of the slab from heavy surface runoffs may be obtained with subsurface drains located at the elevation of the wall footings.

All organic material and topsoil of poor bearing value should be removed in preparation of the subgrade, which should have a uniform bearing value to prevent unequal settlement of the floor slab. Backfill should be tamped and compacted in layers not exceeding 6 in in depth.

Where the subgrade is well-drained, as where subsurface drains are used or are unnecessary, floor slabs of residences should be insulated either by placing a granular fill over the subgrade or by using a lightweight-aggregate concrete slab covered with a wearing surface of gravel or stone concrete. The granular fill, if used, should have a minimum thickness of 5 in and may consist of coarse slag, gravel, or crushed stone, preferably of 1-in minimum size. A layer of 3-, 4-, or 6-in-thick hollow masonry building units is preferred to gravel fill for insulation and provides a smooth, level bearing surface.

Where a complete barrier against the rise of moisture from the ground is desired, a two-ply bituminous membrane or other waterproofing material should be placed beneath the slab and over the insulating concrete or granular fill. The top of the lightweight-aggregate concrete, if used, should be troweled or brushed to a smooth level surface for the membrane. The top of the granular fill should be covered with a grout coating, similarly finished. Where there is no possible danger of water reaching the underside of the floor, a single layer of 55-lb smooth-surface asphalt roll roofing or an equivalent waterproofing membrane may be used under the floor. Joints between the sheets should be lapped and sealed with bituminous mastic. Great care should be taken to prevent puncturing the roofing layer during concreting operations.

("A Guide to the Use of Waterproofing, Dampproofing, Protective and Decorative Barrier Systems for Concrete," ACI 515.1R-79, American Concrete Institute (www.concrete.org).)

Basement Floors ▪ Where a basement is to be used in drained soils as living quarters or for the storage of items that may be damaged by moisture, the floor should be insulated and preferably contain the membrane waterproofing previously described. In general, the design and construction of such basement floors are similar to those of floors on ground.

If passage of moisture from the ground into the basement is unimportant or can be satisfactorily controlled by air conditioning or ventilation, the

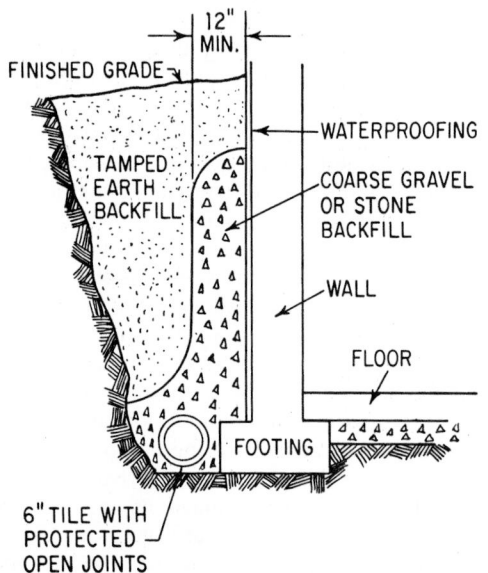

Fig. 15.19 Drainage of basement wall with drain tile along the footing and gravel fill.

waterproof membrane need not be used. The concrete slab should have a minimum thickness of 4 in and need not be reinforced, but it should be laid on a granular fill or other insulation placed on a carefully prepared subgrade. The concrete in the slab should have a minimum compressive strength of 2000 psi and may contain an integral water repellent.

A basement floor below the water table will be subjected to hydrostatic upward pressures. The floor should be made heavy enough to counteract the uplift.

An appropriate sealant in the joint between the basement walls and a floor over drained soil will prevent leakage into the basement of any water that may occasionally accumulate under the slab. Space for the joint may be provided by use of beveled siding strips, which are removed after the concrete has hardened. After the slab is properly cured, it and the wall surface should be in as dry a condition as practicable before the joint is filled to ensure a good bond of the filler and to reduce the effects of slab shrinkage on the joint's permeability.

("Guide to Joint Sealants for Concrete Structures," ACI 504R, American Concrete Institute.)

Monolithic Concrete Basement Walls ▪ These should have a minimum thickness of 6 in. Where insulation is desirable, as where the basement is used for living quarters, lightweight aggregate, such as those prepared by calcining or sintering blast-furnace slag, clay, or shale that meet the requirements of ASTM Standard C330, may be used in the concrete. The concrete should have a minimum compressive strength of 2000 psi.

For the forms in which concrete for basement walls is cast, form ties of an internal-disconnecting type are preferable to twisted-wire ties. Entrance holes for the form ties should be sealed with mortar after the forms are removed. If twisted-wire ties are used, they should be cut a minimum distance of $1\frac{1}{2}$ in inside the face of the wall and the holes filled with mortar.

The resistance of the wall to capillary penetration of water in temporary contact with the wall face may be increased by using a water-repellent admixture. The water repellent may also be used in the concrete at and just above grade to reduce the capillary rise of moisture from the ground into the superstructure walls.

Where it is desirable to make the wall resistant to passage of water vapor from the outside and to increase its resistance to capillary penetration of water, the exterior wall face may be treated with an impervious coating. The continuity and the resulting effectiveness in resisting moisture penetration of such a coating depend on the smoothness and regularity of the concrete surface and the skill and technique used in applying the coating to the dry concrete surface. Some bituminous coatings that can be used are listed in increasing order of their resistance to moisture penetration:

Spray- or brush-applied asphalt emulsions

Spray- or brush-applied bituminous cutbacks

Trowel coatings of bitumen with organic solvent, applied cold

Hot-applied asphalt or coal-tar pitch, preceded by application of a suitable primer

Cementitious brush-applied paints and grouts and trowel coatings of a mortar increase moisture resistance of monolithic concrete, especially if such coatings contain a water repellent. However, in properly drained soil, such coatings may not be justified unless needed to prevent leakage of water through openings in the concrete resulting from segregation of the aggregate and bad work in casting the walls. The trowel coatings may also be used to level irregular wall surfaces in preparation for the application of a bituminous coating. For information on other waterproofing materials, see "A Guide to the Use of Waterproofing, Dampproofing, Protective and Decorative Barrier Systems for Concrete," ACI 515.1R-79, American Concrete Institute.

Basement Walls of Masonry Units ▪ Water-resistant basement walls of masonry units should be carefully constructed of durable materials to prevent leakage and damage from frost and other weathering exposure. Frost action is most severe at the grade line and may cause structural damage and leakage of water.

The use of shotcrete or trowel-applied mortar coatings, $\frac{3}{4}$ in or more thick, to the outside faces of both monolithic concrete and unit-masonry walls greatly increases their resistance to penetration of moisture. Such plaster coatings cover and seal construction joints and other vulnerable joints in the walls against leakage. When applied in a thickness of 2 in or more, they may be reinforced with welded-wire fabric to reduce the incidence of large shrinkage cracks in the coating. However, the plaster does not protect the walls against leakage if the walls, and subsequently the coatings, are badly cracked as a result of unequal foundation settlement, excessive drying shrinkage, and thermal changes. ("Guide to Shotcrete," ACI 506, American Concrete Institute.)

Thin, impervious coatings may be applied to the plaster if resistance to penetration of water vapor is desired. (See ACI 515.1R.) The plaster should be dry and clean before the impervious coating is applied over the surface of the wall and the top of the footing.

Impervious Membranes ■ These are waterproof barriers providing protection against penetration of water under hydrostatic pressure and water vapor. To resist hydrostatic pressure, the membrane should be made continuous in a basement's walls and floor. It also should be protected from damage during building operations and should be laid by experienced workers under competent supervision. It usually consists of three or more alternate layers of hot, mopped-on asphalt or coal-tar pitch and plies of treated glass fabric or bituminous saturated cotton or woven burlap fabric. The number of moppings exceeds the number of plies by one.

Materials used in the hot-applied system should meet the requirements of the following current ASTM Standards:

Creosote primer for coal-tar pitch—D43

Primer for asphalt—D41

Coal-tar pitch—D450, Type II

Asphalt—D449, Type A

Cotton fabric, bituminous saturated—D173

Woven burlap fabric, bituminous saturated—D1327

Treated glass fabric—D1668

Coal-tar saturated organic felt—D227

Asphalt saturated organic felt—D226

The number of plies of saturated felt or fabric should be increased with increase in the hydrostatic head to which the membrane is to be subjected. Five plies is the maximum commonly used in building construction, but 10 or more plies have been recommended for pressure heads of 35 ft or greater. The thickness of the membrane crossing the wall footings at the base of the wall should be no greater than necessary, to keep very small the possible settlement of the wall due to plastic flow in the membrane materials.

The membrane should be built up ply by ply, the strips of fabric or felt being laid immediately after each bed has been hot-mopped. The lap of succeeding plies or strips over each other depends on the width of the roll and number of plies. In any membrane there should be a lap of the top or final ply over the first, initial ply of at least 2 in. End laps should be staggered at least 24 in, and the laps between succeeding rolls should be at least 12 in.

At least one ply of a membrane should be fabric. The minimum weight of felt should be 13 lb/100 ft^2; of fabric, 10 oz/yd^2. About 1 gal of primer per 100 ft^2 of wall should be applied to the walls before the first mopping of bitumen. Immediately after a membrane has been completed, it should be protected by a l-in mortar coat or by other facings.

Alternatives to a hot-applied membrane system are cold-applied bituminous systems, liquid-applied membranes, and sheet-applied membranes, similar to those used for roofing. See also "The NRCA Waterproofing Manual," National Roofing Manufacturers Association," and ACI 515.1R-79.

Bellows-type water stops should be placed in expansion joints in basement walls. Made of 16-oz copper sheets, they should extend at least 6 in on each side of a joint. Metal-bellows water stops should be placed in both expansion and contraction joints if there is hydrostatic pressure. The protective facing of the membrane should be disconnected at expansion joints and the line of the facing filled.

Above-Grade Walls ■ The rate of moisture penetration through capillaries in above-grade walls is low and usually of minor importance. However, such walls should not permit leakage of wind-driven rain through openings larger than those of capillary dimension.

Masonry walls above grade that leak may be made water-resistant with coatings of portland cement paints, grouts, stuccos, or pneumatically applied mortars. Pigmented organics, including conventional paints, if applied as a continuous coating without pinholes, also may be used. They are decorative but may not be so water-resistant, economical, or durable as cementitious coatings. Leakage through joints in masonry walls can be stopped by either repointing or grouting the joints. Repointing consists of cutting away and replacing the mortar from all joints to a depth of about $\frac{5}{8}$ in. Grouting consists of scrubbing a thin coating of grout over the joints. The grout may be composed of equal parts by volume of portland cement and sand passing a No. 30 sieve. Repointing is more effective but also more expensive.

(F. S. Merritt, "Building Design and Construction Handbook," 6th ed., McGraw-Hill Publishing Company, New York (books.mcgraw-hill.com); "The NRCA Roofing and Waterproofing Manual," National Roofing Contractors Association, 10255 W. Higgins Road, Suite 600, Rosemont, IL 60018, (www.nrca.net).)

15.22 Stairs

Principal components of a stairway are described below and some are illustrated in Fig. 15.20.

Flight ▪ A series of steps extending from floor to floor, or from a floor to an intermediate landing or platform. Landings are used where turns are necessary or to break up long climbs.

Rise ▪ Distance from floor to floor.

Run ▪ Total length of stairs in a horizontal plane, including landings.

Riser ▪ Vertical face of a step. Its height generally is taken as the vertical distance between treads.

Tread ▪ Horizontal face of a step. Its width usually is taken as the horizontal distance between risers.

Nosing ▪ Projection of a tread beyond the riser below.

Carriage ▪ Rough timber supporting the steps of wood stairs.

Stringers ▪ Inclined members along the sides of a stairway. The stringer along a wall is called a wall stringer. Open stringers are those cut to follow the lines of risers and treads. Closed stringers have parallel top and bottom, and treads and risers are supported along their sides or mortised into them. In wood stairs, stringers are placed outside the carriage to provide a finish.

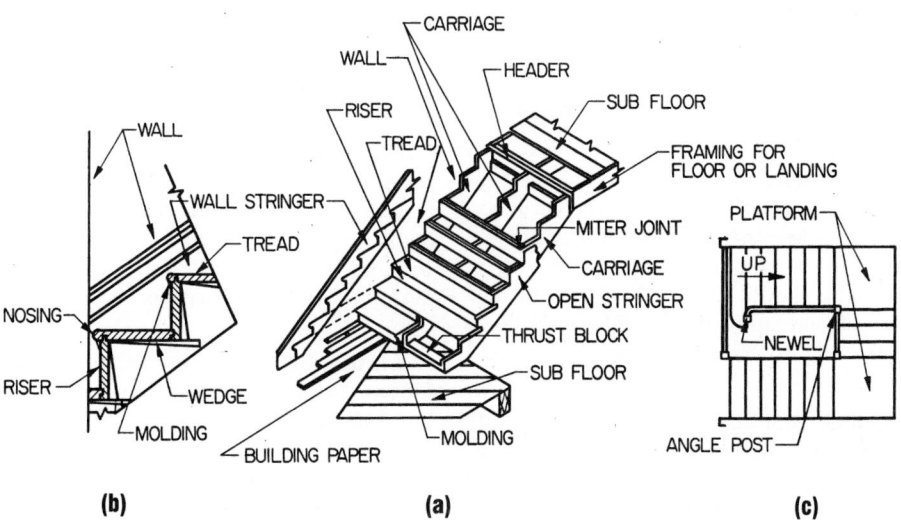

(b)   **(a)**   **(c)**

Fig. 15.20 Typical construction for wood stairs.

Railing ▪ Protective bar placed at a convenient distance above the stairs for a handhold.

Balustrade ▪ A railing composed of balusters capped by a handrail.

Handrail ▪ Protective bar placed at a convenient distance above the stairs for a handhold.

Baluster ▪ Vertical member supporting the railing.

Newel Post ▪ Post at which the railing terminates at each floor level.

Angel Post ▪ Railing support at landings or other breaks in the stairs. If an angle post projects beyond the bottom of the strings, the ornamental detail formed at the bottom of the post is called the **drop**.

Winders ▪ Steps with tapered treads in sharply curved stairs.

Headroom ▪ Minimum clear height from a tread to overhead construction, such as the ceiling of the next floor, ductwork, or piping.

Safety Rules ▪ Building codes restrict stair dimensions and also control the number of stairways. This control may be achieved by restricting the horizontal distance from any point on a floor to a stairway or the floor area contributory to a stairway. In addition, codes usually have special provisions for public buildings and the maximum capacity of a stairway.

Vertical Clearance. Minimum vertical distance from the nosing of a tread to overhead construction should never be less than 6 ft 8 in and preferably not less than 7 ft. But in general, a person of average height should be able to extend one hand forward and upward without touching an obstruction. Height between landings should not exceed 12 ft.

Stair Widths. Building codes usually specify minimum width of stairs for buildings of various types of occupancy; for example, 36 in for one- and two-family dwellings and 44 in for other types of occupancy. But the stairs should be wider than these minimums if necessary to accommodate the number of people who will use them in peak periods and emergencies. (See also "Life Safety Code," NFPA 101, National Fire Protection Association, Quincy, Mass.)

Step Sizes. The most comfortable height of risers is 7 to 7½ in. Risers less than 6 in and more than 8 in high should not be used. Treads should be 11 to 14 in wide, exclusive of nosing. Simple formulas sometimes used to proportion risers and treads include:

1. Product of riser and tread must be between 70 and 75.

2. Riser plus tread must equal 17 to 17.5.

3. Sum of the tread and twice the riser must lie between 24 and 25.5.

When designing stairs, account should be taken of the fact that there always is one less tread than riser per flight.

Guards. Barriers, called guards, at least 42 in high, should be placed along edges of stairs and landings, to prevent people from falling over the edges. The guards should be designed for a horizontal force of 50 lb/ft² applied 42 in above the floor or for the force transmitted by handrails on them.

Railings. Handrails, 1¼ to 2 in in diameter, should be set 2 ft 10 in to 3 ft 2 in above the intersections of tread and risers at the front of the steps on both sides of stairs. At turns of stairs, inside handrails at landings should be continuous between flights. Wide stairways should have intermediate handrails spaced not more than 60 in apart along the natural path of travel.

Emergency Use. In many types of buildings, exit stairs must be enclosed with walls having a high resistance to fire and self-closing fire-resistant doors, to prevent spread of smoke and flame (Arts. 15.3 and 15.18). In public buildings, there should be more than one fire tower, as far apart as possible.

Materials ▪ Stairs may be constructed of wood for wood-frame buildings, low nonfireproof buildings, and one- and two-story houses (Fig. 15.20). They may be built in place or shop-fabricated.

Cold-formed or plate steel stairs often are used in fire-resistant buildings. The sheets are formed into risers and subtreads or pans, into which one of several types of treads may be inserted. Treads may be made of stone, concrete, composition, or metal and usually have a nonslip surface. Stringers generally are channel-shaped.

Concrete stairs may be designed as cantilevered or inclined beams and slabs. The entire stairway may be cast in place as a single unit, or slabs or T beams formed first and the steps built up later. Concrete treads should have metal nosings to protect the edges.

(F. S. Merritt, "Building Design and Construction Handbook," 6th ed., and J. H. Callender, "Time-Saver Standards for Architectural Design Data," 6th ed., McGraw-Hill Publishing Company, New York (books.mcgraw-hill.com); "Metal Stairs Manual," National Association of Architectural Metal Manufacturers," 8 S. Michigan Ave., Chicago, IL 60603 (www.arcat.com); "Life Safety Code Handbook," National Fire Protection Association, 1 Batterymarch Park, Quincy, MA 02269 (www.nfpa.org).)

15.23 Escalators

Providing continuous operation without operators, escalators, or powered stairs, are used when it is necessary to move large numbers of people from floor to floor. They have large capacity with low power consumption. Large department stores provide vertical-transportation facilities for one person per hour for every 20 to 25 ft^2 of sales area above the entrance floor, and powered stairs generally carry 75 to 90% of the traffic, elevators the rest.

In effect, an escalator is an inclined bridge spanning between floors, with an endless belt to transport passengers. Main components are a steel trussed framework, handrails, and an endless belt with steps. At the upper end is a matching pair of motor-driven sprocket wheels and a worm-gear driving machine. At the lower end is a matching pair of sprocket wheels. Two precision-made roller chains travel over the sprockets pulling the endless belt. The steps move on an accurately made set of tracks attached to the trusses. Each step is mounted on resilient rollers.

Normally, escalators move at 90 or 120 ft/min and are reversible in direction. Slope is standardized at 30°.

For a given speed, width of step determines the capacity of the powered stairs. Standard widths are 32 and 48 in between handrails, with corresponding

capacities at 90 ft/min of 5000 and 8000 persons per hour. At 120 ft/min, a 48-in escalator can carry as many as 10,000 persons per hour.

Escalators usually are installed in pairs—one for carrying traffic up and the other for moving traffic down. The units may be placed parallel to each other in each story or crisscrossed; the latter placement generally is preferred for compactness. Fire-protection devices may be incorporated into the stairway installation.

A structural frame should be erected around the stairwell to carry the floor and wellway railing. The stairway should be independent of this frame.

("Life Safety Code Handbook," National Fire Protection Association, Quincy, Mass (www.nfpa.org); "Safety Code for Elevators, Dumbwaiters, Escalators, and Moving Walls," A 17.1, American National Standards Institute, New York (webstore.ansi.org); G. R. Strakosch, "Vertical Transportation: Elevators and Escalators," and B. Stein et al., "Mechanical and Electrical Equipment for Buildings," 7th ed., John Wiley & Sons, Inc., New York (www.wiley.com).)

15.24 Elevators

Electric traction elevators are used exclusively in tall buildings. Hydraulic elevators are usually used for low-rise freight, with lifts up to about 50 ft, but may be used for passenger service in buildings up to six stories high, where they cost less to operate.

Major components of an electric traction installation include the hoistway, car or cab, hoist wire ropes, driving machine, control equipment, counterweights, guide rails, safety devices, machine room, and pit (Fig. 15.21). The car is a cage of light metal supported on a structural frame, to the top of which the ropes are attached. The ropes raise and lower the car. They pass over a grooved motor-driven sheave and are fastened to the counterweights. The elevator machine that drives the sheave consists of an electric motor, brakes, and auxiliary equipment, which are mounted, with the sheave, on a heavy structural frame. The counterweights, consisting of blocks of cast iron in a frame, are needed to reduce power requirements.

The paths of the counterweights and the car are controlled by separate sets of T-shaped guide rails. The control and operating machinery may be placed in a penthouse above the shaft or in the basement. Safety springs or buffers are placed in

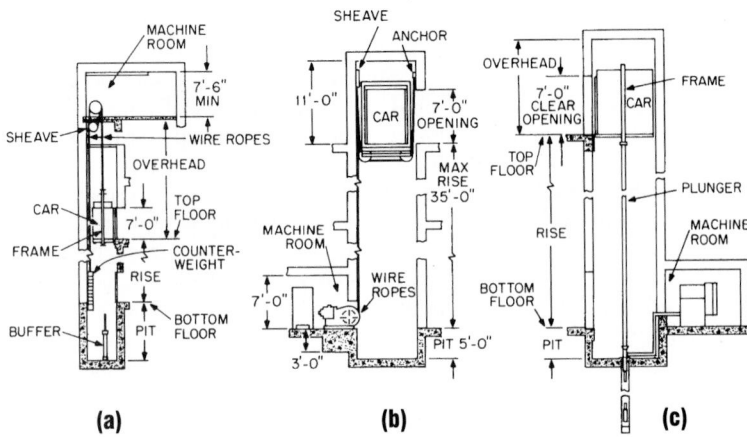

Fig. 15.21 Some lifting arrangements for elevators: (*a*) electric elevator with driving machine at the top of the hoistway; (*b*) electric elevator with driving machine in the basement; (*c*) hydraulic elevator moved by plunger.

the pit to bring the car and counterweights to a safe stop if either passes the bottom terminal at normal speed. Hoistways must be enclosed with non-combustible materials of high fire resistance (Art. 15.3).

Elevators and related equipment, such as machinery, signal systems, controls, ropes, and guide rails, are generally supplied and installed by the manufacturer. The general contractor has to guarantee the dimensions of the shaft and its freedom from encroachments. The owner's architect or engineer is responsible for the design and construction of components needed for supporting the plant, including buffer supports, machine-room floors, trolley beams, and guide-rail bracket supports. Magnitudes of loads generally are supplied by the manufacturer, with a 100% allowance for impact.

Driving machines may be winding-drum or traction type, depending on whether the ropes are wound on drums on the drive shaft or are powered by a drive sheave. The traction type is usually used; it may be double-wrap or single-wrap. For the double-wrap, to obtain sufficient traction between the ropes and the driving sheave, which has U-shaped or round-seat grooves, a secondary or idler sheave is used (Fig. 15.22*a*). In the single-wrap types, the ropes pass over the traction or driving sheave only once, so there is a single wrap, or less, of the ropes on the sheave (Fig. 15.22*d*). The traction sheave has wedge-shaped or undercut grooves for

gripping the ropes. For the same weight of car and counterweight, the sheave has half the loading of the double-wrap machine.

In most buildings, driving machines are installed in a penthouse. When heavy loads are to be handled and speed is not important, a 2:1 roping may be used (Fig. 15.22*c*), in which case the car speed is only half that of the rope. Ends of the rope are anchored to the overhead beams, instead of being attached to the car and counterweights, as for 1:1 roping. With this arrangement, the anchorages carry half the weight of car and counterweights, so the loading on the traction and secondary sheaves is only about half that for the 1:1 machine. Therefore, a less costly motor can be used. When a machine must be installed in the basement (Fig. 15.22*b*), the load on the overhead supports is increased, cable length is tripled, and additional sheaves are needed, adding substantially to the cost.

Passenger Elevators ▪ The number of passenger elevators needed to serve a building adequately depends on their capacity, volume of traffic, and interval between cars. Platform sizes should conform to the standards of the National Elevator Industry, Inc.

Traffic is measured by the number of persons handled in 5-min periods. Dividing the peak 5-min traffic flow by the 5-min handling capacity of an

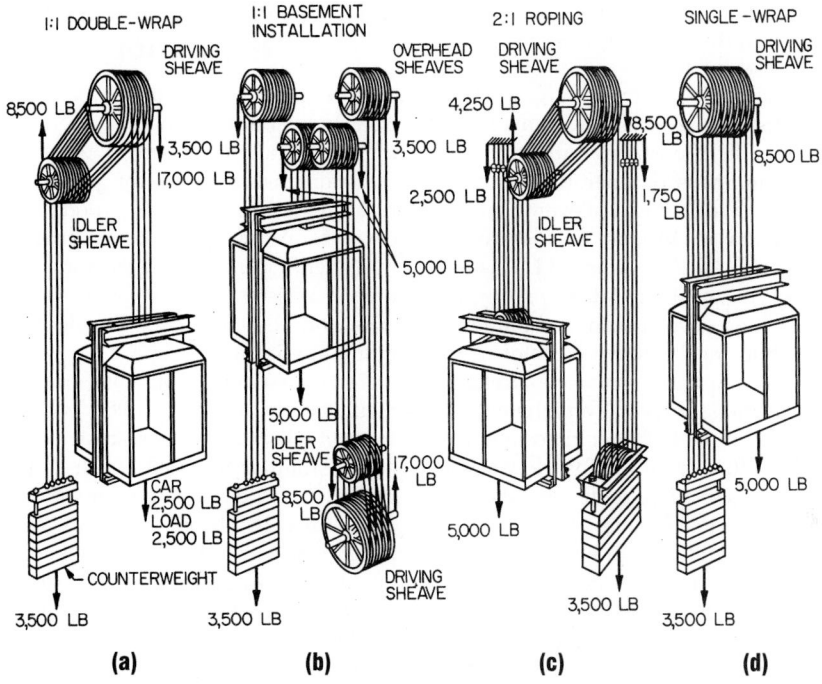

Fig. 15.22 Types of roping for elevators driven by electric traction machines.

elevator gives the minimum number of elevators required. The 5-min handling capacity of an elevator is determined from the round-trip time. Round-trip time is composed principally of the time for a full-speed round trip without stops, time for accelerating and decelerating per stop, time for leveling at each stop, time for opening or closing gates and doors, time for passengers to move in and out, reaction time of operator, lost time due to false stops, and standing time at top and bottom floors.

After the number of elevators has been computed on the basis of traffic flow, a check should be made on the **interval**, the average time between elevators leaving the ground floor. Interval is a significant measure of good service.

Fully automatic elevators are used in nearly all multistory buildings. These systems are capable of adjusting to varying traffic conditions. Since the elevators are operatorless, several safety devices are incorporated in addition to those commonly installed in manually operated systems—an automatic load weigher to prevent overcrowding, buttons in car and starter station to stop the doors from closing and hold them open, lights to indicate

floor stops pressed, two-way loudspeaker system for communication with the starter station, and auxiliary power systems if the primary power and supervisory systems fail. Safety devices also prevent the doors from closing when a passenger is standing in the doorway. The elevators cannot move when the doors are open.

Department stores should be served by a coordinated system of moving stairs and elevators (see Art. 15.23). The required capacity of the vertical-transportation system should be based on the transportation or merchandising area and the maximum density to which it is expected to be occupied by shoppers. The transportation area is all the floor space above or below the first floor to which shoppers and employees must be moved. The transportation capacity is the number of persons per hour that the vertical-transportation system can distribute from the main floor to the other merchandising floors. The ratio of the peak transportation capacity to the transportation area is called the density ratio, which is about 1:20 for a busy department store. So the required hourly handling capacity of a combined moving stairs and elevator system equals 5% of the transportation area. The elevator system generally is designed to handle about 10% of the total.

Multivoltage controls normally are used for passenger elevators. Freight elevators may have variable voltage or ac rheostatic. With multivoltage, the hoisting motor is dc operated. A motor-generator set is provided for each elevator, and the speed and direction of motion of the car are controlled by varying the generator field. This type of elevator permits the most accurate stops, the most rapid acceleration and deceleration, and minimum power consumption for an active elevator. Automatic leveling to compensate for rope stretch or other variations from floor level is an inherent part of multivoltage equipment. The ac rheostatic type generally is chosen to keep initial cost down when the elevator is to be used infrequently (less than five trips per hour on a normal business day).

For low-rise elevators, hydraulic equipment may be used to lift the car. It sits atop a plunger, or ram, which operates in a pressure cylinder (Fig. 15.21c). Oil is the pressure fluid and is supplied through a motor-driven positive-displacement pump, actuated by an electric-hydraulic control system. To raise the car, the pump is started, discharging oil into the pressure cylinder and forcing the ram up. When the car reaches the desired level, the pump stops. To lower the car, oil is released from the pressure cylinder and returns to a storage tank.

Capacity of electrohydraulic passenger elevators ranges from 1000 to 4000 lb at speeds from 40 to 125 ft/min. With gravity lowering, down speed may be 1.5 to 2 times up speed. So the average speed for a round trip can be considerably higher than the up speed. Capacity of standard electrohydraulic freight elevators ranges from 2000 to 20,000 lb at 20 to 85 ft/min, but they can be designed for much greater loads.

(G. R. Strakosch, "Vertical Transportation: Elevators and Escalators," 3rd ed. and B. Stein et al., "Mechanical and Electrical Equipment for Buildings," 9th ed., John Wiley & Sons, Inc., New York (www.wiley.com); "Safety Code for Elevators, Dumbwaiters, Escalators, and Moving Walks," ANSI A17.1, American National Standards Institute, New York (webstore.ansi.org); F. S. Merritt, "Building Design and Construction Handbook," 6th ed., McGraw-Hill Publishing Company, New York (books.mcgraw-hill.com).)

15.25 Heat Flow and Thermal Insulation

Heat transfer into and out of a building or its parts may be substantially decreased by use of materials that resist heat flow or a type of construction that achieves the purpose. Some structural materials, such as wood and lightweight concrete, also have good insulating properties. But in general, certain nonstructural materials offer greater resistance to heat flow for a given thickness and therefore may be more economical in many applications.

Most insulating materials employ still air as the insulator. Some, such as cork, cellular glass, and foamed plastics, enclose small particles of air in cells. Granular materials, such as pumice, vermiculite, and perlite, trap air in relatively large enclosures. In fibrous materials, thin films of air cling persistently to all surfaces and serve as the heat barrier. In cavity-wall construction, a dead air space is formed between the wythes.

Reflective insulation involves a different principle. Metal foil is combined with an air gap to reduce heat flow. The shiny metal reflects heat, conducts it rapidly away from a heat source, and radiates heat slowly. An air gap of $\frac{3}{4}$ to 2 in on at least one side of the foil acts as a barrier to heat transfer by conduction. So if heat is radiated to a bright aluminum foil, 95% will be reflected back. If it receives heat by conduction, it will lose only 5% by radiation from the opposite face. To prevent condensation troubles, use at least two reflective surfaces separated by a dead air space. But do not place a foil on the cold side of a construction unless a better vapor barrier is provided close to the warm side.

Heat is transmitted by conduction, convection, and radiation. All materials conduct heat; some, such as metals, are excellent conductors, but others, such as cork, are poor conductors. Convection occurs when heat is transmitted by air flow; heat is transferred by conduction from a warm surface to cooler air in contact with it and from warm air to a cooler surface. Since warm air tends to rise and cool air to fall, the air flow may carry heat from a warm area to a cooler one. Heat transmitted by conduction or convection is proportional to temperature difference. Radiation, in contrast, is the flow of heat between a warm and cool surface with no material contact.

Heat usually is measured in **British thermal units (Btu)**. For practical purposes, 1 Btu is the amount of heat required to raise the temperature of 1 lb of water 1 degree Fahrenheit. Heat flow is measured by **thermal conductivity** K, which is defined as the number of Btu that will flow in 1 h through a material 1 ft square and 1 in thick because of a temperature difference of 1 °F. Similarly, **thermal conductance** C is defined as the heat flow through a given thickness of 1-ft-square material with a 1 °F temperature differential. Note that these basic units do not include the insulating values of air films at the surfaces of the material, only flow from surface to surface. **Resistance** R is the reciprocal of conductance.

Since building components are built up of several materials, including air spaces and surface films, the **overall conductance** U of a construction is needed in heat-transfer calculations. This factor is defined as the number of Btu that will flow in 1 h through 1 ft^2 of the structure from air to air with a temperature differential of 1 °F. Values of K, C, and U or R have been determined experimentally for many materials and types of construction ("Handbook of Fundamentals," American Society of Heating, Refrigerating and Air Conditioning Engineers, 1791 Tullie Circle, N.E., Atlanta, Ga (www.ashrae.org).).

The thermal conductance of an outside air film in a 15-mi/h wind is 6.00 Btu/h, of an inside air film (still air), 1.65 Btu/h; and of an air space $\frac{3}{4}$ in or more wide, 1.10 Btu/h.

When the overall conductance of a construction is not given in a table, it may be computed from tabulated values of conductance of each component and air films. For example, consider a wall composed of 4 in of brick ($K = 9.2$) and $\frac{1}{2}$-in wallboard ($C = 1.00$) separated by an air space ($C = 1.10$). The calculations are shown in Table 15.9.

Suppose, now, 1 in of insulation ($K = 0.25$) were to be incorporated into this wall. The resistance R of the insulation ($1/K$) is 4. Thus, the resistance of the original wall is increased to $3.116 + 4$, or 7.116; the new overall conductance U becomes $1/7.116 = 0.14$.

15.26 Prevention of Condensation

Normally, air contains water vapor, which tends to move from a warm region to a cooler one. The lower the temperature, the less vapor the air can hold. If the air is saturated (100% relative humidity), a temperature drop will cause some of the vapor to condense. The temperature at which this occurs is called the **dew point**.

Since almost all building materials or the joints between them are porous, vapor will permeate them. If the dew point is reached between inner and outer surfaces, the vapor will condense, and the temperature differential will cause more vapor to penetrate, repeating the process. In cold weather, the dew point often occurs within insulation in walls and roofs. If vapor reaches it, the condensation may saturate the insulation, drastically reducing its insulating value. Furthermore, the moisture may rot or rust the structure or stain interior finishes. If temperatures are low, the water may freeze and in expanding, as ice always does, crack the structure.

A simple solution to condensation is to stop the flow of water vapor with a vapor barrier on the warm side. Since the dangers of condensation are greatest in the heating season, vapor barriers should be installed on the interior side of walls and roofs, next to the insulation.

Table 15.9 Calculation of Overall Conductance of a Wall

Item	K	Thickness, in	C	$R = 1/C$
Outside film			6	0.166
Brick	9.2	4	2.30	0.434
Air space			1.10	0.910
Wallboard		$\frac{1}{2}$	1.00	1.000
Inside film			1.65	0.606
Total resistance				3.116

Overall conductance $U = 1/3.116 = 0.32$.

Aluminum foil is a good, economical vapor barrier. Some insulations come equipped with it attached to one side. Other vapor barriers include aluminum paints, plastic paints and films, asphalt paints, rubber-base paints, asphalt, and foil-laminated papers.

The ability of a material to pass vapor is measured by the **perm**, defined as a vapor-transmission rate of 1 grain of water vapor through 1 ft^2 of material per hour when the vapor-pressure difference equals 1 in of mercury (7000 grains = 1 lb). A material with a vapor-transmission rate of 1 perm or less is considered a good vapor barrier. **Rep** is the reciprocal of perm; it measures resistance to vapor transmission.

Since vapor barriers are not likely to be perfect or installed perfectly, some vapor may penetrate to the insulation. Means should be provided to let this vapor escape. Hence, the exterior surface should be as porous as possible or vented and yet prevent rain from penetrating. Cold-side venting may be desirable, even though condensation does not occur in the insulation, because it may occur instead in back of the exterior facing. Whenever the dew point occurs within a material, condensation will not take place until the water vapor encounters the surface of another material with greater resistance to the vapor flow.

Vapor also tends to flow through insulated ceilings into attics and air spaces under a roof. If these spaces are not ventilated with air capable of removing the moisture, trouble can result. In general, vent area should total about $\frac{1}{300}$ of the horizontal projection of the roof area. If possible, both high and low vents should be installed to ensure air flow.

15.27 Heating

Required capacity of a heating plant is determined mainly by the total heat loss from a building through conduction, radiation, and infiltration. To allow for the temperature pickup usually required in the morning, however, the plant should have a capacity 20% larger than this heat loss. But do not choose too large a unit, because then operating efficiency suffers.

The heat loss depends on the design inside and outside temperatures. (See tables in the "ASHRAE Guide and Data Book," American Society of Heating, Refrigerating and Air Conditioning

Engineers, Atlanta, Ga. The design outdoor temperatures are not the lowest ever attained in the region, but a slightly higher recommended value.) The difference between inside and outside temperatures is the temperature gradient. When multiplied by the exposed surface area of a material or construction and its overall thermal conductance U (Art. 15.25), the gradient determines the hourly heat flow in Btu. The sum of these products for all exposed surfaces—walls, windows, roofs,—yields the total heat loss through them.

Heat loss through basement floors and walls may be determined from groundwater temperature, which ranges from about 40 to 60 °F in the northern sections of the United States and from 60 to 76 °F in the southern sections (Table 15.10). (For specific areas, see the "ASHRAE Guide.")

Heat loss from a floor on grade without edge insulation is about 75 Btu/h per linear foot of exposed edge in the cold northern sections of the United States, 65 in the temperate zones, and 60 in the warm south. With 1 in of insulation, these rates drop to 60, 55, and 50; with 2 in, to 50, 45, and 40.

To obtain the heat loss through unheated attics, the equilibrium attic temperature must first be computed by equating heat gain to the attic via the ceiling to heat loss through the roof. The same procedure should be used to obtain the temperature of other unheated spaces, such as cellars and attached garages.

To the heat load for exposed surfaces must be added the load due to cold air infiltrating and warm air leaking out. The amount of leakage depends on crack area, wind velocity, and number of exposures, among other things. To account for leakage, the assumption is made that cold outside air will be heated and pumped into the building to create a static pressure large enough to prevent

Table 15.10 Heat Losses Below Grade

Groundwater temp, °F	Basement-floor loss,* Btu/h ft^2	Below-grade wall loss, Btu/h ft^2
40	3.0	6.0
50	2.0	4.0
60	1.0	2.0

* Based on basement temperature of 70 °F.

cold air from infiltrating. The amount of heat q in Btu/h required to warm up this cold air is given by

$$q = 1.08QT \qquad (15.15)$$

where $Q = \text{ft}^3/\text{min}$ of air to be warmed $= VN/60$

$T = $ temperature rise of air, °F

$V = $ volume of room, ft^3

$N = $ number of air changes per hour

If the heating plant also will be used to produce hot water, the added capacity for this purpose should be determined and added to the heat load.

A **warm-air heating system** supplies heat to a room by bringing in a quantity of air above room temperature. The amount of heat added by the air must be at least equal to that required to counteract heat losses. Equation (15.15) gives this heat if T is taken as the difference between the temperature of the air leaving the grille and the room temperature and Q as the ft^3/min of air supplied to the room. In good systems, the discharge temperatures range from 135 to 140 °F. Supply grilles should be arranged to blow a curtain of air across exposed walls and windows. The best location is near the floor. Return-air grilles should be installed in the interior, preferably at the ceiling.

Ducts for warm-air systems generally are designed by the equal-friction method. Sizes are calculated to accommodate the design air quantity of the heater with a predetermined friction factor. The pressure loss due to friction should not exceed 0.15 in of water per 100 ft of duct. Also, starting velocity of the air in main ducts should be kept below 900 ft/min in residences; 1300 ft/min in schools, theaters, and public buildings; and 1800 ft/min in industrial buildings. Velocity

in branch ducts should be about two-thirds of these, and in branch risers, about one-half. But too low a velocity will require uneconomical, bulky ducts. (See the "ASHRAE Guide and Data Book.")

In a forced warm-air system, a thermostat calls for heat, starting a heat source. When the air chamber in the heater reaches about 120 °F, the fan starts. (If the discharge temperature exceeds 180 °F, a safety element in the air chamber shuts off the heat source.) The heat source stops when the indoor temperature reaches the value at which the thermostat is set. But the fan continues to operate until the air cools to below 120 °F.

In warm-air perimeter heating, often used with concrete floors on ground, the heater discharges warm air to two or more underfloor radial ducts feeding a perimeter duct. Floor grilles or baseboard grilles are located as in a conventional warm-air heating system, with collars connected to the perimeter duct.

A **hot-water heating system** consists of a heater or furnace, radiators, piping systems, and circulator. Normally, forced circulation systems are used because they can maintain higher water velocities and therefore require smaller pipes and provide more sensitive control.

Three types of piping systems are in general use. The one-pipe system (Fig. 15.23a) has many disadvantages and is not usually recommended. The two-pipe direct-return system (Fig. 15.23b) provides all radiators with the same supply-water temperature, but the last radiator has more pipe resistance than the first. This can be balanced out by installing orifices in the other radiators to add an equivalent resistance and by sizing the pump for the longest run. In the two-pipe reversed-return

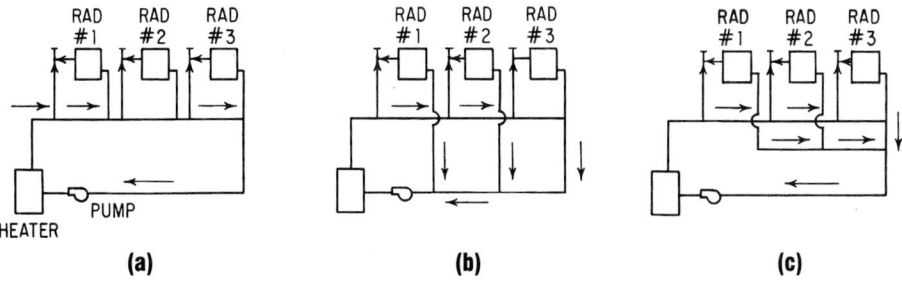

Fig. 15.23 Types of hot-water heating systems: (*a*) one pipe; (*b*) two-pipe direct-return; (*c*) two-pipe reversed-return.

system (Fig. 15.23c), the total pipe resistance is about the same for all radiators.

For a hot-water system, supply design temperatures usually are 180 °F, with a 20 °F drop assumed through the radiators. Thus, the temperature of the return riser would be 160 °F. The amount of heat required to offset the 20 °F drop, in Btu/h, is

$$q = 10,000Q \qquad (15.16)$$

where Q = flow of water, gal/min. Piping may be sized for the required water flow with the aid of friction flow charts and tables showing equivalent pipe lengths for fittings. (See, for example, the ASHRAE "Handbook of Fundamentals.") Water velocity should be limited to a maximum of 4 ft/s. Loss of pressure due to friction should be between 0.25 and 0.60 in of water per foot. The system must be provided with an expansion tank, located at least 3 ft above the highest radiator and in a location where the water will not freeze. The tank should be sized for 6% of the total volume of water in radiators, heaters, and piping. In very tall buildings, to avoid too high a static pressure on the boiler, heat exchangers should be provided in the upper levels.

In a hot-water system, an immersion thermostat in the heater controls the heat source to maintain design heater water temperature (usually about 180 °F). When the room thermostat calls for heat, the circulator starts. Thus, an immediate supply of hot water is available for the radiators. For 170 °F average water temperature, 1 ft^2 of radiation surface emits 150 Btu/h.

A **steam-heating system** consists of a boiler or steam generator and a piping system connecting to individual radiators or convectors. In a one-pipe system (Fig. 15.24a), the pipe supplying steam to the radiators also is used to return condensate to the boiler. On start-up, the steam must push air out of the pipe and radiators. For this purpose, the radiators are equipped with thermostatic air valves. Orifice size in the air vents must be varied to balance the system; otherwise, radiators at the far end of a pipe run may get steam much later than the near end. Valves in a one-pipe system must be fully open or closed. In a two-pipe system (Fig. 15.24b), steam is fed to the radiators through one pipe and the condensate returned through a second pipe. When condensate cools the radiator below 180 °F, a trap opens to allow the condensate to return to a collecting tank, from which it is pumped to the

boiler. The wet-return system (Fig. 15.24c) usually has a smaller pressure head available for pipe loss. It is a self-adjusting system depending on the load. When the condensate collects sufficiently in the return main above boiler level, the pressure will force the condensate into the boiler.

In all cases, the steam-supply pipes must be pitched to remove condensate from the pipe. Where condensate flows against the steam, the pipe may have to be oversized. Pipe capacities for supply risers, runouts, and radiator connections are given in the ASHRAE "Handbook of Fundamentals." Capacities are expressed in square feet of equivalent direct radiation (EDR).

$$1 \text{ ft}^2 \text{ EDR} = 240 \text{ Btu/h} \qquad (15.17)$$

Where capacities are in pounds per hour, 1 lb/h = 970 Btu/h.

A **vacuum-heating system** is similar to a steam pressure system with a condensate return pump. The vacuum pump pulls noncondensables from the piping and radiators for discharge to the atmosphere, whereas in a steam pressure system thermostatic vents are opened for this purpose.

Unit heaters often are used for large open areas, such as garages, showrooms, stores, and workshops. The units usually consist of a heat source or heat exchanger and an electrically operated fan. Heat may be supplied by steam or electricity or by burning gas. When gas-fired unit heaters are used, however, an outside flue must be provided to dispose of the products of combustion. Sizes of gas piping and burning rates for gas can be obtained from the ASHRAE handbook. Efficiency of most gas-fired equipment is between 70 and 80%.

Radiant heating, or panel heating, consists of a warm pipe or electric cables embedded in the floor, ceiling, or walls. Joints in ferrous pipe should be welded, whereas those in nonferrous pipe should be soldered, and return bends should be made with a pipe bender instead of with fittings, to avoid joints. All piping should be subjected to a hydrostatic test of at least three times the working pressure, with a minimum of 150 psig. Repairs are costly after construction has been completed. Piping and circuiting are similar to those for a hot-water system with radiators and convectors, except that cooler water is used. But a 20 °F temperature drop usually is assumed. Therefore, charts used for the design of hot-water piping systems may be used for panel heating too. Floor

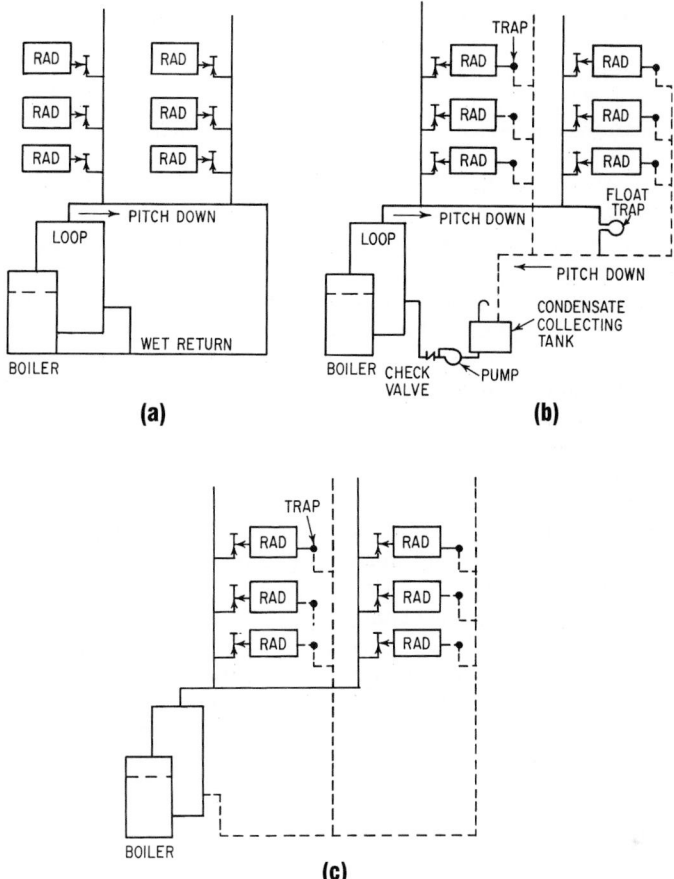

Fig. 15.24 Types of steam-heating systems: (*a*) one-pipe with condensate returning through the supply pipe; (*b*) two-pipe; (*c*) two-pipe wet-return.

panel temperatures generally are maintained about 85 °F or lower and ceiling panel temperatures at 100 °F or lower. It is possible with panel heating to maintain relatively low room air temperatures with comfort, but the system should be designed for standard room temperatures to prevent discomfort after the thermostat stops water circulation.

("ASHRAE Handbook of Fundamentals," American Society of Heating, Refrigerating and Air-Conditioning Engineers, Atlanta, Ga (www. ashrae.org); B. Stein et al., "Mechanical and Electrical Equipment for Buildings," John Wiley & Sons, Inc., New York (www.wiley.com); F. S. Merritt, "Building Design and Construction Handbook," 6th ed., McGraw-Hill Publishing Company, New York (books.mcgraw-hill.com).)

15.28 Air Conditioning

Required capacity of a cooling plant is determined by the heat transmitted to the conditioned space through the walls, glass, ceiling, floor, and so on and all the heat generated in the space. The total cooling load consists of sensible and latent heat. Sensible heat is the part that shows up in the form of a dry-bulb temperature rise. It includes heat transmitted through the building enclosure; radiation from the sun; and heat from lights, people, electrical and gas appliances, and outside air brought into the air-conditioned space. Latent heat is that needed to remove moisture from the air. Usually, the moisture is condensed out on the cooling coils in the cooling unit; 1050 Btu is required per pound of condensation.

Design conditions for comfort cooling usually are 80 °F dry bulb and 50% relative humidity. Design outdoor temperatures are not the highest ever recorded in a region but a slightly lower recommended value. (See tables in the "ASHRAE Handbook of Fundamentals.") The difference between indoor and outdoor temperatures multiplied by the area of walls, roofs, windows, and so forth and the respective overall coefficients of conductance U (Art. 15.24) yields the heat gain through each.

Radiation from the sun through glass and roofs adds substantially to the heat load. (The sun effect on walls, however, generally can be neglected.) Sun through unshaded window glass can add about 200 Btu/h·ft^2 through windows facing east and west; about three-fourths as much for windows facing northeast and northwest; and half as much for windows facing south. For most roofs, total equivalent temperature difference for calculating heat gain due to the sun is about 50 °F. Roof sprays sometimes are used to reduce this load. With a water spray, the equivalent temperature difference may be taken as 18 °F.

Heat from electric lights and other electrical appliances can be computed from

$$q = 3.42W \tag{15.18}$$

where q = Btu/h developed

W = watts of electricity used

For fluorescent lighting, add 25% of the lamp rating for the heat generated in the ballast.

Heat gain from people for various types of activities is given in tables in the ASHRAE handbook.

The sensible heat from outside air brought into a conditioned space can be computed from

$$q_s = 1.08Q(T_o - T_i) \tag{15.19}$$

where q_s = sensible load due to outside air, Btu/h

Q = ft^3/min of outside air brought into conditioned space

T_o = design dry-bulb temperature of outside air

T_i = design dry-bulb temperature of conditioned space

The latent load due to outside air in Btu/h is

$$q_l = 0.67Q(G_o - G_i) \tag{15.20}$$

where G_o = moisture content of outside air, grains/ lb of air

G_i = moisture content of inside air, grains/ lb of air

The moisture content of air at various conditions may be obtained from a psychrometric chart.

The total heat load for sizing a cooling plant also must include heat from fans in the air conditioning system, which usually ranges from $3\frac{1}{2}$ to 5% of the sensible load, and heat loss from ducts. The load can be converted to tons of refrigeration by

$$\text{Load in tons} = \frac{\text{load in Btu/h}}{12,000} \tag{15.21}$$

A **ton of refrigeration** is the amount of cooling that can be done by a ton of ice melting in 24 h.

Basic Cycle ▪ Figure 15.25a shows the basic air conditioning cycle of the direct-expansion type. The compressor takes refrigerant gas at a relatively low pressure and compresses it to a higher pressure. The hot gas is passed to a condenser where heat is removed and the refrigerant liquefied. This liquid then is piped to the cooling coil of an air-handling unit and allowed to expand to a lower pressure (suction pressure). The liquid vaporizes or is boiled off by the relatively warm air passing over the coil. The compressor pulls away the vaporized refrigerant to maintain the required low coil pressure with its accompanying low temperature. A system in which the refrigerant chills water, which is circulated to air-handling units for cooling air, is shown in Fig. 15.25b.

Air Quantity ▪ The amount of air, ft^3/min, to be handled can be computed from

$$Q = \frac{q_s}{1.08(T_i - T_d)} \tag{15.22}$$

where q_s = total sensible heat load, Btu/h

T_i = indoor temperature (dry bulb)

T_d = dry-bulb temperature of air discharged from air-handling unit

T_d should be about 3 °F higher than the room dew point, to avoid sweating ducts.

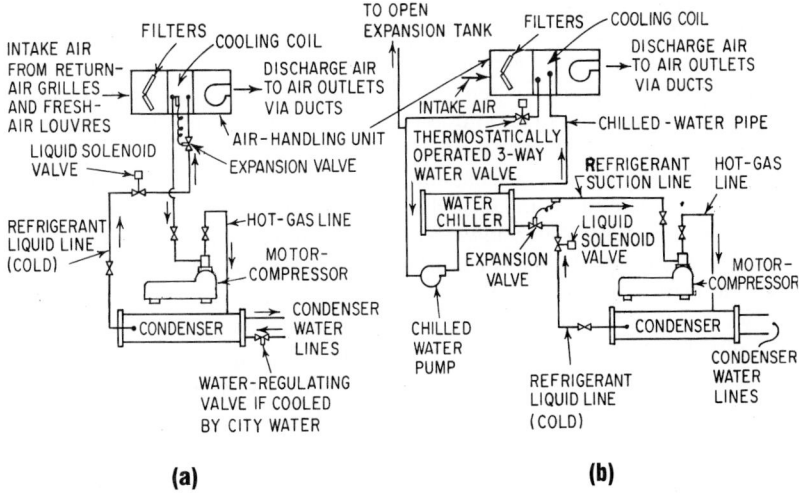

Fig. 15.25 Air conditioning cycles: (*a*) direct-expansion; (*b*) chilled water.

Condensers ▪ If a water-cooled condenser is used to remove heat from the refrigerant, it may be supplied with city water, and the warm water may be discharged to a sewer. Or a water tower may be used to cool condenser water, which then can be recirculated to the condenser. If the wet-bulb temperature is low enough, the condenser and water tower can be replaced by an evaporative condenser. The capacity of such water savers as towers and evaporative condensers decreases as the wet-bulb temperature increases. The amount of water in gallons per minute required for condensers is

$$Q = \frac{\text{tons of cooling} \times 30}{\text{water-temperature rise}} \quad (15.23)$$

Condensers for small cooling units can be cooled by a fan blowing air over the refrigerant coils.

Zoning ▪ Multizone air-handling units control the temperature of several zones in a building without a separate air-handling unit for each zone. When a zone thermostat calls for cooling, the damper motor for that zone opens the cold deck dampers and throttles the warm deck dampers. Thus, the same unit can provide cooling for one zone while it supplies heat for another zone.

Filters ▪ The area of the filters in the air-handling units should be large enough so that the air velocity does not exceed 350 ft/min for low-velocity filters and 550 ft/min for high-velocity filters. Minimum filter area in square feet equals air flow in ft^3/min divided by maximum air velocity across the filters, ft/min. Most filters are the throwaway or cleanable type. Electrostatic filters usually are used in industrial installations, where a higher percentage of dust removal must be obtained, in combination with regular throwaway or cleanable filters, which remove large particles.

Packaged Units ▪ For lower-cost air conditioning installations, "packaged" or preassembled units may be used. They generally operate on the complete cycle shown in Fig. 15.25*a*. For window units, the condenser, projecting outside the building, is air-cooled, and the same motor usually runs both the fan for the cooling coil and that for the condenser. Small floor-type units may be air-cooled; larger ones generally are water-cooled.

Built-Up Units ▪ These are air conditioning units assembled on the site. Usually limited to large units, with capacity of 50 tons or more, they provide cooling air in summer and heated air in winter. They are equipped with filters, return-air fan, compressor, condenser, dampers, and controls, as required. The units may be installed outdoors at grade or on roofs, or indoors as a central plant.

Absorption Chillers ▪ These use heat to regenerate the refrigerant. The compressor of the basic air conditioning cycle (Fig. 15.25a) is replaced by an absorber, pump, and generator. The refrigerant is regenerated by absorption in a weak solution of refrigerant and water, forming a strong solution, which is heated in the generator. The refrigerant vapor thus is driven out of the solution and brought to the condenser under pressure. When low-cost steam is available, absorption systems may be more economical to operate than the systems with compressors. In general, steam consumption is about 20 lb/h per ton of refrigeration.

Variable-Air-Volume (VAV) Systems ▪ In a VAV system, air is supplied at constant temperature, but the volume is varied to meet changing loads in interior spaces or zones (Fig. 15.26a). Zoning is important for good temperature control.

Any type of diffuser, or register, may be used with VAV control units. Generally, however, linear-slot-type diffusers (Fig. 15.26b) are preferred, because they discharge air into a room with a horizontal flow that hugs the ceiling and results in more uniform temperature within the room.

While variable air volume can be produced by modulating the supply-air fan, terminal control units usually give better results. Types of control units generally used include the following:

Shut-off control diffusers, which may provide shut-off operation, multiple slots, integral slot diffuser, and electric-pneumatic or system- powered controls.

Fan-powered control units, often used for perimeter and special areas, which may come with pressure-compensating controls with factory-installed hot-water or multistage electric coils that have pneumatic or electric controls.

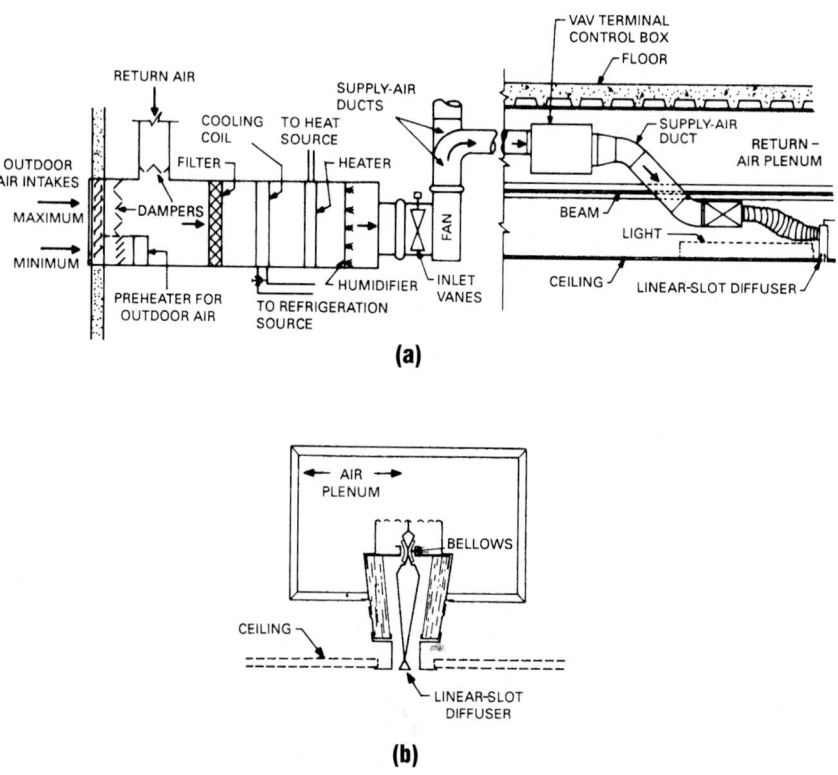

(a)

(b)

Fig. 15.26 Variable-air volume system. (a) Fan blows supply air at constant temperature but with variable flow through a terminal box to a diffuser. (b) Terminal box with linear-slot diffuser.

Dual-duct VAV control units, which are used for dual-duct systems that require VAV for perimeter areas. They feature pressured-compensated shut-off operation with pneumatic, electric, or system-powered controls, with a variable-volume cold deck and constant-volume or variable-volume hot deck.

Air-Water Systems ▪ An alternative to the all-air systems, air-water systems furnish chilled water from a remote chiller or central plant to the room terminal devices. These contain a cooling coil or a heating coil, or both. Room temperature is maintained by varying the flow of chilled water or heating fluid through the coils with valves that respond to the thermostat. Ventilation air is provided from a separate central plant directly to the room or the terminal device.

Two- or four-pipe systems are used for distribution of chilled water and hot water to the room terminals from a central plant. In a two-pipe system, the supply pipe may carry either chilled water or hot water and the second pipe is used as a return. The four-pipe system provides two pipes for chilled-water supply and return and two pipes for hot water supply and return. The installation cost of the two-pipe system is less than that of the four-pipe system, but the versatility is less. The major disadvantage of the two-pipe system is its inability to provide both heating and cooling with a common supply pipe on days for which both heating and cooling are desired. The four-pipe system has a major drawback in loss of temperature control whenever a changeover from cooling to heating is desired. To overcome this, thermostats are used that permit selection of either cooling or heating by a manual changeover at the thermostats.

Terminal devices for air-water systems are usually of the fan-coil or induction types.

A fan-coil terminal device consists of a fan or blower section, chilled-water coil, hot water heating coil or electric-resistance heating elements, filter, return-air connection, and a housing for these components with an opening for ventilation air. The electric-resistance heating coil is often used with two-pipe systems to provide the performance of a four-pipe system without the cost of the two extra pipes for hot water, insulation, pumps, and accessories. Fan-coil units may be floor-mounted, ceiling-mounted-exposed or ceiling-mounted-recessed, or ceiling-mounted-recessed with supply- and return-air ductwork. When furnished with

heating coils, the units are usually mounted on the outside wall or under a window, to neutralize the effects of perimeter heat losses.

Built-in centrifugal fans recirculate room air through the cooling coil. Chilled water circulating through the coil absorbs the room heat load. Ventilation air that is conditioned by another remote central plant is ducted throughout the building and supplied directly to the room or room terminal devices, such as a fan-coil unit. A room thermostat varies the amount of cooling water passing through the cooling coil, thus varying the discharge temperature from the terminal unit and satisfying the room thermostat.

Induction terminal units, frequently used in large office buildings, are served by a remote air-handling unit that provides high-pressure conditioned air, which may be heated or cooled and is referred to as primary air. At the terminal induction unit, which is mounted on the outside walls of each room or zone, a flow of the high-pressure primary air through several nozzles induces a flow of room air through the heating or cooling coil in the unit, to provide a mixed-air temperature that satisfies the thermal requirements of the space. The induction system is a large energy consumer because of the extra power required to maintain high air-distribution pressure and to operate simultaneously with heating and cooling.

Air-water systems generally have substantially lower installation and operating costs than all-air systems. They do not, however, provide as good control over room temperature, humidity, air quality, air movement, and noise. The best control of an air-water system is achieved with a fan-coil unit with supplemental ventilation air from a central, primary-air system that provides ventilation air.

A **heat pump** comprises equipment for heating a building by using the heat removed by the condenser in a refrigeration cycle. The heat absorbed by the refrigerant evaporator is taken from some other heat source, often outdoor air, instead of from the building. When the heat is exhausted outside the building, the heat pump also can be used for cooling. In general, heat pumps are economical for regions where the cooling season is substantially longer than the heating season and winter temperatures are not extreme. The colder the outdoor temperature, the lower the heat-pump capacity becomes, unless a constant-temperature heat source, such as warm water from a deep well, is available.

(H. E. Bovay, Jr., "Handbook of Mechanical and Electrical Systems for Buildings"; N. R. Grimm and R. C. Rosaler, "Handbook of HVAC Design"; R. W. Haines, "HVAC Systems Design Handbook"; F. S. Merritt, "Building Design and Construction Handbook," 6th ed.; R. Shuttleworth, "Mechanical and Electrical Systems for Construction," McGraw-Hill Publishing Company, New York (books. mcgraw-hill.com); F. C. McQuiston and J. D. Parker, "Heating, Ventilation, and Air Conditioning," 3rd ed.; and B. Stein et al., "Mechanical and Electrical Equipment for Buildings," 9th ed., John Wiley & Sons, Inc., New York (www.wiley.com).)

15.29 Ventilation

Natural air movement or air replacement in a room depends on prevailing winds, temperature difference between interior and exterior, height of structure, window openings, etc. For controlled ventilation, a mechanical method of air change is desirable.

Where people are working, the amount of ventilation air required will vary from one air change per hour where no heat or offensive odors are generated to about 60 air changes per hour. Table 15.11 gives the minimum amount of air recommended for various activities.

The number of air changes per hour equals $60Q/V$, where Q is the air supplied, ft^3/min, and V is the volume of ventilated space, ft^3. If there is less than one air change per hour, the ventilation system will take too long to produce a noticeable effect when first put into operation. Five air changes per hour generally is considered a practical minimum. Air changes above 60 per hour usually will create some

discomfort because air velocities will be too high. Toilet and locker-room ventilation generally are covered by local building codes; 50 ft^3/min per water closet and urinal is the usual minimum for toilets and six changes per hour minimum for both toilets and locker rooms.

Removal of heat by ventilation is best done by installing exhaust outlets close to the heat source. Where concentrated sources of heat are present, canopy hoods should be used. When heat is discharged into a room, the amount of ventilation air, ft^3/min, required to remove heat not lost by transmission through enclosures is

$$Q = \frac{q}{1.08(T_i - T_0)} \tag{15.24}$$

where q = heat, Btu/h, carried away by ventilation air

T_i = indoor temperature to be maintained

T_0 = temperature of fresh air (usually outdoor air)

If a gas or moisture in the air is to be diluted, the amount of ventilation air, ft^3/min, required is $Q = X/Y$, where the vapor or gas is formed at the rate of X lb/min and Y is the allowable concentration, lb/ft^3. See also Art. 15.28.

15.30 Electric Power for Buildings

Electrical design and construction for buildings are based usually on the National Electrical Code (National Fire Protection Association, 1 Battery--march Park, Quincy MA 02269). But local building codes may have some more restrictive require-

Table 15.11 Minimum Ventilation Air for Various Activities

Type of occupancy	Ventilation ft^3/min per person
Inactive, theaters	5
Light activity, offices	10
Light activity with some odor generation, restaurants	15
Light activity with moderate odor generation, bars	20
Active work, shipping rooms	30
Very active work, gymnasiums	50

ments and should be checked. These codes contain minimum safety standards. Use of these standards does not guarantee adequate performance of an electrical system.

A building's electrical systems operate on electric currents supplied at specified effective voltages. An electric current I, amperes (A), is the rate at which electric charges flow through a circuit. If the current always flows in the same direction, it is called a direct current (dc). The current is assumed to flow from a positive to a negative terminal. An alternating current (ac) reverses direction at regular intervals.

Electromotive force or potential difference E, volts, is the force that makes electrons move in the circuit. It is opposed by a resistance R. Ohm's law relates E, I, and R:

$$E = IR \qquad (15.25)$$

Electric power watts (W), or kilowatts (1 kW = 1000 W), is the rate of doing electrical work: 746 W = 0.746 kW = 1 hp. Direct-current power, W, is given by

$$W = EI = I^2R \qquad (15.26)$$

Phases ▪ In single-phase ac circuits, power is the product of voltage, current, and a power factor, which equals 100% only when current and voltage are in phase, that is, pass through zero, maximums, and minimums at the same time.

If current and voltage are represented by a sine curve, one may lead or lag the other by nearly 360°. If, for example, the maximum of a sinusoidal current occurs 60° before the maximum of the voltage, the current leads the voltage by 60° or lags by 300°. In a single-phase ac system, the power factor equals the cosine of the angle between the voltage and current phases. Hence, the closer the phase angle is to 90° or 270°, the smaller the power factor and the larger the equipment and conductors needed to deliver the required power. Low power factors often may be corrected by installing synchronous motors, or by connecting static condensers across the line.

Inductance L, henrys, makes current lag voltage. **Capacitance** C, farads, makes current lead voltage. Both inductive reactance X_L, ohms, and capacitive reactance X_c, ohms, impede the flow of current. **Impedance** Z, ohms, is the total opposition to the flow of current and equals the vector sum of resistance and reactance:

$$Z^2 = R^2 + (X_L - X_c)^2 \qquad (15.27)$$

Maximum voltage drop across an impedance equals maximum current times impedance.

Types of Circuits ▪ Basic circuits are either series or parallel types. A series circuit has components connected in sequence. If there is a break in a series circuit, current will not flow; hence, if one lamp goes out, all go out. Parallel (multiple or shunt) circuits, in contrast, have components with common terminals. Voltage across the components is the same, and the current divides among them, in accordance with Ohm's law [Eq. (15.25)]. Parallel circuits generally are used for electrical distribution in buildings, whereas series circuits often are used for street lighting.

Service equipment consists of a circuit breaker or switch and fuses, and their accessories, located near the point of entrance of supply conductors to a building and intended to constitute the main control and means of cutoff of the supply. **Feeders** are the conductors between the service equipment, or the generator switchboard of an isolated plant, and branch-circuit over-current-protective devices. A **branch circuit** is the part of the system between the feeder and the load, or current-consuming equipment. Branch circuits deliver current to outlets, points where current is taken for equipment. A **receptacle**, or convenience outlet, permits the circuit to be tapped with a plug and flexible cord.

Electrical Loads ▪ All conductors should be sized for the sum of the loads, in kilowatts, for lighting, motors, and appliances. Since all lights may not be on at the same time, codes generally permit for feeders a reduction in the lighting load by application of a demand factor. Codes also specify that feeders and branch circuits be sized for a minimum load, in W/ft² of floor area, that depends on type of occupancy. But usually, the actual load will exceed these minimums.

Conductors should not be smaller than No. 12 in branch circuits.

Small installations, such as dwellings, usually are supplied with three-wire service. This consists of a neutral and two power wires with current differing 180° in phase. Tapping across the phase wires yields a single-phase two-wire 230-volt (V) supply. Either phase wire and the neutral yield a single-phase two-wire 115-V supply. In addition, for safety reasons, a separate ground wire should be provided since the neutral, though grounded, carries current.

For larger installations, a 120/208-V three-phase, four-wire system usually is used. This consists of a neutral and three power wires carrying current differing 120° in phase. Tapping across any two phase wires yields a single-phase two-wire 208-V supply. Any phase wire and the neutral provide a single-phase two-wire 120-V supply. Other combinations yield two- or three-phase 120/208-V supplies.

No current flows in the neutral when the loads on the system's circuits are balanced. Hence, the system should be so designed that, under full load, the load on each phase leg will be nearly equal.

Current in a conductor may be computed from the following formulas, in which

I = conductor current, A

W = power, W

f = power factor, as decimal

E_p = voltage between any two phase legs

E_g = voltage between phase leg and neutral, or ground

Single-phase two-wire circuits:

$$I = \frac{W}{E_p f} \quad \text{or} \quad I = \frac{W}{E_g f} \qquad (15.28)$$

Single-phase three-wire (and balanced two-phase three-wire) circuits:

$$I = \frac{W}{2E_g f} \qquad (15.29)$$

Three-phase three-wire (and balanced three-phase four-wire) circuits:

$$I = \frac{W}{3E_g f} \qquad (15.30)$$

Voltage drop in a circuit may be computed from the following formulas, in which

V_d = voltage drop between any two phase legs, or between phase leg and neutral when only one phase wire is used in circuit

L = one-way run, ft

cmil = circular mils (1 cmil = area of a circle 0.001 in in diameter)

Single-phase two-wire (and balanced single-phase three-wire) circuits:

$$V_d = \frac{2RIL}{\text{cmil}} \qquad (15.31)$$

Balanced two-phase three-wire, three-phase three-wire, and balanced three-phase four-wire circuits:

$$V_d = \frac{\sqrt{3}\,RIL}{\text{cmil}} \qquad (15.32)$$

Equations (15.31) and (15.32) contain a factor R that represents the resistance to direct current, in ohms, of 1 mil-ft of wire. For wires smaller than No. 3, resistance is the same for alternating and direct currents. For wires larger than No. 3 carrying alternating current, a correction factor should be applied because of the higher resistance. The value for R may be taken as 10.7 for copper and 17.7 for aluminum.

Copper conductors may be more economical for small-diameter conductors, for which weight is not an important consideration. The smaller weight may be advantageous for large conductors. To avoid excessive heat and incendiary conditions at splices with aluminum, use conductors preferably No. 4 AWG or larger.

In design of feeders and branch conductors, voltage drops may range from 1 to 5%. Some codes limit the voltage drop to 2.5% for combined light and power circuits from the service equipment to branch panels. For economy, the greater part of the voltage drop, 1.5 to 2%, may be assigned to the smaller, more numerous feeders, and only 0.5 to 1% to the heavy main feeders. For motor loads only, the maximum voltage drop may be increased to 5%. Of this, 4% can be assigned to feeders.

The general procedure in sizing conductors is to start with the minimum-size wire permitted by code and test it for voltage drop. If this drop is excessive, test a larger size, and repeat until a wire is found for which the voltage drop is within the desired limit.

Fuses and circuit breakers should be incorporated into the circuits to protect motors from overcurrents of long duration yet permit high, short-duration starting currents to pass. The National Electrical Code allows such overcurrent protective devices to have a higher ampere rating than the allowable current-carrying capacity of the wire. In branch circuits with one motor, conductors should have an allowable current-carrying capacity of at least 125% of the motor full-load current. For feeders supplying several motors, the conductor capacity should be at least 125% of the full-load current of the largest motor plus the sum of the full-load currents of the other motors.

The part of the wiring system at service switches and main distribution panels connected near these switches consists of heavy cables or buses and large switches that have very low resistance. If a short circuit occurs, very high currents will flow, and ordinary fuses or circuit breakers will not be able to interrupt them before wiring or equipment is damaged. For this purpose, high-interrupting-capacity current-limiting fuses, such as Amp-Traps and Hi-Caps, are needed. Ask the utility company for the interrupting capacity required.

Codes generally require that incoming service in a multiple-occupancy building be controlled near the point of entry by not more than six switches or circuit breakers. Meters, furnished by the utility company, also must be installed near the point of entry. The service switch and metering equipment may be combined in one unit, or the switch may be connected by conduit to a separate meter trough.

(F. S. Merritt, "Building Design and Construction Handbook," 6th ed., D. G. Fink and H. W. Beaty, "Standard Handbook for Electrical Engineers," 14th ed., H. Richter and W. Schwan, "Practical Electrical Wiring," 15th ed., McGraw-Hill Publishing Company, New York (books.mcgraw-hill.com); "National Electrical Code Handbook," National Fire Protection Association, Quincy, MA 02269.)

15.31 Electric Lighting for Buildings

Artificial illumination is installed primarily for seeing, but it also may serve architectural purposes. With electric lighting, room illumination is not limited to window and skylight openings and by the vagaries of sunlight.

A basic lighting unit usually consists of a light source, or lamp, and a luminaire for housing the lamp, and accessory equipment, such as lenses and the ballasts required for fluorescent lighting. Either the lamp itself or, more commonly, both lamp and luminaire are designed to control brightness and light intensity in various directions. Generally, comfort in seeing is as important as ease in seeing.

Like design of other building systems, lighting design is significantly affected by building codes. These generally contain minimum requirements for illumination levels, for the safety and health of building occupants. In addition, electric-lighting equipment and electrical distribution must con-form to safety requirements in building codes and the National Electrical Code, which is promulgated by the National Fire Protection Association, and to standards of the Underwriters Laboratories, Inc. Also, the Illuminating Engineering Society has developed standards and recommended practices to promote good lighting design.

In the interests of energy conservation, Federal and state government agencies have set limits on the amount of energy that may be expended (energy budget) for operation of buildings. These limits may establish maximum levels of illumination for specific purposes in buildings.

Illumination level at any point is inversely proportional to the square of distance from the point light source.

This is known as the inverse square law for radiation of light. For large light sources, the law holds approximately at large distances (at least five times the largest dimension of the sources) from the sources.

Light Source Power ▪ Analogous to a pump in a water system or a battery in an electrical system, a light source emits luminous power. The unit used to measure this power is **candlepower** (cp), or **candela** (cd) (metric unit). (At one time, 1 cp was assumed equivalent to the luminous intensity of a wax candle, but now a more precise definition based on radiation from a heated blackbody is used.) The unit used to measure luminous power at a distance from the light source is the lumen (lm).

A **lumen** is the luminous power on an area of 1 ft^2 at a distance of 1 ft from a 1-cp light source.

Luminous efficacy is the unit used to measure the effectiveness of light sources. It is calculated by dividing the total lumen output of a light source by the total input, W, and thus is measured in lm/W.

Level of Illumination ▪ A major objective of lighting design is to provide a specified **illuminance**, or level of illumination, on a task. For design purposes, the task often is taken as a flat surface, called a **work plane**. If the task is uniformly illuminated, the level of illumination equals the lumens striking the surface divided by the area. The unit used to measure illuminance is the **footcandle** (fc). In accordance with the inverse square law, the illuminance on a work plane normal to the direction to a point light source is given by

$$fc = \frac{cp}{D^2} \qquad (15.33)$$

where D = distance, ft, from work plane to light source

cp = candlepower of light source

For a work plane at an angle θ to the direction of the light source,

$$fc = \frac{cp \sin \theta}{D^2} \qquad (15.34)$$

A **luminaire** is a lighting device that consists of one or more lamps, or light sources, a fixture that positions and shields them, components that distribute the light, and elements that connect the lamps to the power supply. In general, luminaires do not radiate light of equal intensity in all directions because of the characteristics of the lamps or the geometry of the fixtures. The actual illuminance around a single luminaire is an important design consideration. This environment may be characterized by the candlepower distribution curve of the luminaire. The curve indicates the variation in illuminance with direction from the light source.

Brightness ▪ An observer sees an object because of light reflected from it. The observer interprets the intensity of the sensation experienced as brightness. The sensation of brightness usually is partly attributable to the general luminous environment, which affects the state of adaptation of the eye, and partly attributable to the intensity of light emanating from the object. The latter component is called luminance, or photometric brightness.

Luminance is the luminous power emitted, transmitted, or reflected by a surface in a given direction, per unit of area of the surface projected on a plane normal to that direction. The unit of measurement of luminance is the footlambert (fL).

A **footlambert** of luminance in a given direction is produced by one lumen per square foot emanating from a surface in that direction. Thus, a self-luminous surface emitting $10 \, \text{lm/ft}^2$ has a luminance of 10 fL. For surfaces that reflect or transmit light, however, luminance depends on both the illuminance of light incident on the surface and characteristics of the surface.

For a reflecting surface, luminance is determined from

$$fL = fc \times \text{reflectance} \qquad (15.35)$$

where fc = footcandles of incident light. A mirror (specular reflector) may give almost 100% reflection, whereas a black surface absorbs light and therefore has negligible reflectance. Most materials have an intermediate value of reflectance.

For a transmitting surface, luminance is determined from

$$fL = fc \times \text{transmittance} \qquad (15.36)$$

Clear glass (transparent material) may have a transmittance of about 90%, whereas an opaque material has no transmittance. Transmittances of other transparent materials may be about the same as that for clear glass; transmittances of translucent materials may be 50% or less. Light incident on a surface and not reflected or transmitted is absorbed by it.

Generally, visibility improves with increase in brightness of a task. Because increase in brightness is usually accomplished at increase in operating cost caused by consumption of electric power, it is neither necessary nor desirable, however, to maintain levels of illumination higher than the minimum needed for satisfactory performance of the task. For example, tests show that speed of reading and comprehension are nearly independent of illuminance above a minimum level. This level depends on several factors, such as difficulty of the task, age of observers, duration of the task, and luminance relation between task and its surroundings. The more difficult the task, the older the occupants, and the longer the task, the higher the minimum level of illumination should be.

High brightness also is useful in attracting visual attention and accenting texture. For this reason, bright lights are played on merchandise and works of art.

Contrast ▪ This is created when the brightness of an object and its surroundings are different. The effects of contrast on visibility depend on several factors, especially the ratio of brightness of object to that of its background. Ideally, the brightness of a task should be the same as that of its background. A 3:1 brightness ratio, however, is not objectionable; it will be noticed but usually will not attract attention. A 10:1 brightness ratio will draw attention, and a brightness ratio of 50:1 or more will accent the object and detract attention from everything else in the field of vision.

High background brightness, or low brightness ratios, may have adverse or beneficial effects on visibility. Such high contrast is undesirable when it causes glare or draws attention from the task or creates discordant light and dark patterns (visual noise). On the other hand, high contrast is advantageous when it helps the observer detect task details, for example, read fine print. High contrast makes the object viewed appear dark so that its size and silhouette can be readily discerned. But under such circumstances, if surface detail on the object must be detected, object brightness must be increased at least to the level of that of the background.

Color Rendering ▪ A blackbody is colorless. When increasing heat is applied to such a body, it eventually develops a deep red glow, then cherry red, next orange, and finally blue-white. The color of the radiated light is thus related to the temperature of the heated body. This phenomenon is the basis for a temperature scale used for the comparison of the color of light from different sources. For example, the light from an incandescent lamp, which tends to be yellowish, may be designated in degrees as 2500 Kelvin (K), whereas a cool white fluorescent lamp may be designated 4500 K.

Light used for general illumination is mainly white, but white light is a combination of colors, and some colors are more predominant than others in light emitted from light sources commonly used. When light other than white is desired, it may be obtained by selection of a light source rich in the desired hue or through use of a filter that produces that hue by absorbing other colors.

Color rendering is the degree to which a light source affects the apparent color of objects. **Color rendering index** is a measure of this degree relative to the perceived color of the same objects when illuminated by a reference source for specified conditions. The index actually is a measure of how closely light approximates daylight of the same color temperature. The higher the index, the better the color rendering. The index for commonly used light sources ranges from about 20 to 99.

Quantity of Light ▪ Methods for selecting illuminance, or level of illumination, fc, have been developed by the Illuminating Engineering Society of North America (IES) and published in the "IES Lighting Handbook." These methods take into account

Luminance, or brightness, of the task

Luminance relation between task and surroundings

Color rendering of the light

Size of details to be detected

Contrast of the details with their background

Duration and frequency of occurrence of the task

Speed and accuracy required in performance of the task

Age of workers

Lighting Methods ▪ To meet specific lighting objectives, the following lighting methods may be used alone or in combination: general lighting, local or functional lighting, accent lighting, and decorative lighting.

Illumination may be classified as indirect, semiindirect, diffuse or direct-indirect, semidirect, or direct.

For indirect lighting, about 90 to 100% of the illumination provided in a space is directed at the ceiling and upper walls, and nearly all the light reaches the task by reflection from them. The resulting illumination is therefore diffuse and uniform, with little or no glare.

For semiindirect lighting, about 60 to 90% of the illumination is directed at the ceiling and upper walls, the remaining percentage in generally downward directions. When overhead luminaires are used, the downward components should be dispersed by passage through a diffusing or diffracting lens to reduce direct glare. The resulting illumination on a task is diffuse and nearly glarefree.

General diffuse or direct-indirect lighting is designed to provide nearly equal distribution of light upward and downward. General-diffuse luminaires enclose the light source in a translucent material to diffuse the light and produce light in all directions. Direct-indirect luminaires give little light near the horizontal. Quality of the resulting illumination from either type depends on the type of task and the layout of the luminaires.

For semidirect lighting, about 60 to 90% of the illumination is directed downward, the remaining percentage upward. Depending on the eye adaptation level, as determined by overall room luminance, the upward component may reduce glare. Diffuseness of the lighting depends on reflectance of room enclosures and furnishings.

For direct lighting, almost all the illumination is directed downward. If such luminaires are spread out, reflections from room enclosures and furnishings may diffuse the light sufficiently that it can be used for general lighting, for example, in large offices. A concentrated layout of these luminaires is suitable for accent, decorative, or local lighting. Because direct lighting provides little illumination on vertical surfaces, provision of supplementary perimeter lighting often is desirable.

Characteristics of Lamps ▪ Selection of the most suitable lamp consistent with design objectives is critical to performance and cost of a lighting system. This decision should be carefully made before selecting a fixture for the lamp. Luminaires are designed for specific lamps.

Lamps are constructed to operate at a specific voltage and wattage, or power consumption. Generally, the higher the wattage rating of a specific type of lamp, the greater its **efficacy**, or lumen output per watt.

Greatest economy will be secured for a lighting installation through use of a lamp with the highest lumen output per watt with good quality of illumination. In addition to lumen output, however, color rendering and other characteristics, such as lighting distribution, should also be considered in lamp selection. Information on these characteristics can be obtained from lamp manufacturers. Latest data should be requested because characteristics affecting lamp performance are changed periodically.

Lamps that are commonly used may be generally classified as incandescent, tungsten-halogen, fluorescent, or high-intensity-discharge (HID). HID lamps include mercury-vapor, metal-halide, low-pressure sodium, and high-pressure sodium lamps.

Consideration in Luminaire Selection ▪ Because luminaires are designed for specific types of lamps and for specific voltage and wattage ratings of the lamps, a prime consideration in choosing a luminaire is its compatibility with lamps to be used. Other factors to consider include:

Conformance with the chosen lighting method

Degree to which a luminaire helps meet objectives for quantity and quality of light through emission and distribution of light

Luminous efficiency of a luminaire, the ratio of lumens output by the luminaire to lumens produced by the lamps

Esthetics—in particular, coordination of size and shape of luminaires with room dimensions so that luminaires are not overly conspicuous

Durability

Ease of installation and maintenance

Light distribution from luminaires may be accomplished by means of transmission, reflection, refraction, absorption, and diffusion. Reflectors play an important role. Their reflectance, consequently, should be high—at least 85%. The shape of a reflector—spherical, parabolic, elliptical, hyperbolic—should be selected to meet design objectives; for example, to spot or spread light in a building space or to spread light over a luminaire lens that controls light distribution. (The need for a curved reflector, which affects the size of the luminaire, can be avoided by use of a Fresnel lens, which performs the same function as a reflector. With this type of lens, therefore, a smaller fixture is possible.) Light control also is affected by shielding, baffles, and louvers that are positioned on luminaires to prevent light from being emitted in undesirable directions.

A wide range of light control can be achieved with lenses. Flat or contoured lenses may be used to diffuse, diffract, polarize, or color light, as required. Lenses composed of prisms, cones, or spherical shapes may serve as refractors, producing uniform dispersion of light or concentration in specific directions.

Maintenance of Lamp Output ▪ The efficiency of a lighting system decreases with time because of dirt accumulation, decrease in lumen output as lamps age, lamp failures, and deteriorating luminaires. Depending on type of luminaires, cleanliness of the environment, and time between cleanings of lamps and luminaires, lumen losses due to dirt may range from 8 to 10% in a clean environment to more than 50% under severe conditions. Also, the longer lamps operate, the dimmer they become; for example, a fluorescent lamp at the end of its life will yield only 80 to 85% of its initial lumen output. And when one or more lamps fail and are not replaced, the space being illuminated may suffer a substantial loss in light. Furthermore, in the case of lights operating with ballasts, the lamps, before burning out, overload the ballast and may cause it to fail. Consequently, a poorly maintained lighting installation does not provide the illumination for which it was designed and wastes money on the power consumed.

("IES Lighting Handbook, 9th ed." Illuminating Engineering Society, 120 Wall Street, Floor 17, New York, NY 10017 (www.iesna.org); M. D. Egan, "Concepts in Lighting for Architecture," F. S. Merritt, "Building Design and Construction Handbook," 6th ed., and L. Watson, "Lighting Design Handbook," McGraw-Hill Publishing Company, New York (books.mcgraw-hill.com); B. Stein et al., "Mechanical and Electrical Equipment for Buildings," John Wiley & Sons, Inc., New York (www. wiley.com).)

15.32 Waste Piping

One function of a plumbing system in a building is to remove safely and quickly human, natural, and industrial wastes. The National Plumbing Code, ANSI Standard A40.8 (American Society of Mechanical Engineers, 345 E. 47th St., New York, NY 10017) (www.asme.org), contains minimum standards for the design of such systems. But local building codes may have more restrictive requirements and should be checked. Table 15.12, which lists the minimum number of fixtures required for various occupancies, indicates requirements of the New York City Building Code.

Associated with each fixture is a soil or waste stack, a vent or vent stack, and a trap. Soil stacks conduct wastes from one or more fixtures to a sloped house or building drain at the base of the building. Vents and vent stacks supply fresh air to the plumbing system to dilute gases and balance air pressure. Connected to each drainage pipe, vent stacks (vertical) must extend above the roof. They may have branch vents connected to them. Traps provide a water seal that prevents gases from discharging from the drainage pipes through the fixtures. The house or building drain, located

below the lowest fixture, conducts the waste to the house or building sewer, which starts 4 or 5 ft outside the foundation walls. That sewer, in turn, carries the wastes to a public sewer or other main sewer. Generally, a cleanout is required at the upper end of the house drain.

The piping generally is made of cast iron lined internally with cement or coal-tar enamel, copper, galvanized steel or ductile iron, or plastics. Codes specify the type of joint to be used with each material.

For convenience, the discharges from fixtures are measured in terms of fixture units, which are used to determine pipe sizes. Tables 15.13 and 15.14 list the number of fixture units assigned to various types of fixtures in the National Standard Plumbing Code, as well as the minimum trap size recommended. Table 15.15 notes the maximum number of fixture units (equivalent to maximum permissible discharge) that may be connected to stacks and horizontal fixture branches of various diameters. Also, the table gives the maximum number of fixture units that may be connected to building drains and sewers of various diameters. And Table 15.16 gives the diameter of vent and maximum length permitted with various sizes of soil or waste stacks and various fixture units.

The plumbing system also may be required to dispose of rain on roofs, yards, areaways, and exposed floors. Normally, exterior sheet-metal leaders and gutters are not included in the plumbing contract, but interior leaders and storm-water drains are. Although storm drains may sometimes be permitted to discharge into sanitary drains, codes generally prohibit use of storm drains for disposing of sewage. Recommended sizes of vertical leaders and horizontal storm drains are listed in Table 15.17. Table 15.18 lists recommended sizes of semicircular gutters.

Indirect-waste piping usually is required for the discharge from commercial food-handling equipment and dishwashers, rinse sinks, laundry washers, steam tables, refrigerators, egg boilers, iceboxes, coffee urns, stills, and sterilizers and from units that must be fitted with drip or drainage connections but are not ordinarily regarded as plumbing fixtures. An indirect-waste pipe is not connected directly to the building drains but discharges wastes into a plumbing fixture or receptacle, from which they flow to the drains. An air gap should separate the indirect-waste pipe from the drains. The length of the gap should be at

Table 15.12 Minimum Number of Plumbing Fixtures for Various Occupancies[a]

Type of building or occupancy	Water closets		Urinals		Lavatories		Bathtubs or showers	Drinking fountains[c]
Dwellings or apartment buildings[d]	1 for each dwelling or apartment unit				1 for each unit		1 for each unit	
Public buildings, offices, business mercantile, storage; warehouses, factories, and institutional employees[e]	No. of persons of each sex	No. of fixtures	Urinals may be provided in men's toilet rooms in lieu of water closets but for not more than ½ of required water closets when more than 35 persons.		No. of persons	No. of fixtures	1 shower for each 15 persons exposed to excessive heat or skin contamination	1 for each 75 persons
	1–15	1			1–20	1		
	16–35	2			21–40	2		
	36–55	3			41–60	3		
	56–80	4			61–90	4		
	81–110	5			91–125	5		
	111–150	6			1 fixture for each additional 45 persons[g, h]			
	1 fixture for each additional 40 persons							
Schools:[f]	Males	Females					For gyms or pools, one for every 3 pupils of largest class using pool at one time	1/50 persons but at least 1 per floor
Elementary	1/90	1/35	1/30 males		1/50 pupils			
Secondary	1/90	1/35	1/30 males		1/50 pupils			
					Over 300 pupils			
					1/100 pupils			
Assembly—	No. of persons	No. of fixtures	No. of persons	No. of fixtures	No. of persons	No. of fixtures		1 for each 1000 persons except that there shall be at least 1 fixture each assembly floor
Auditoriums, theaters, convention halls	1–100	1	1–200	1	1–200	1		
	101–200	2	201–400	2	201–400	2		
	201–400	3	401–600	3	401–750	3		
	Over 400, add 1 fixture for each additional 500 men and 1 for each 800 women		Over 600, add 1 fixture for each 300 men[e]		Over 750, add 1 fixture for each 500 persons			
Dormitories[i]	Men: 1 for each 10 persons Women: 1 for each 8 persons		1 for each 25 men. Over 150, add 1 fixture for each 50 men[e]		1 for each 12 persons		1/8 persons. For every 30 women, substitute 1 bathtub for 1 shower	1 for each 75 persons
Worker temporary facilities	1/30 workers		1/30 workers					At least 1 per floor equivalent for each 100 workers

[a] Figures shown are based on one fixture being the minimum required for the number of persons indicated or any fraction thereof. Population used in determining the number of fixtures required should be based on the number of persons to occupy the space but not less than 125 ft^2 of net floor area per person.

[b] Building categories not shown in this table will be considered separately by the administrative authority.

[c] Drinking fountains shall not be installed in toilet rooms.

[d] Laundry trays—one single compartment tray for each dwelling unit or 2 compartment trays for each 10 apartments. Kitchen sinks—1 for each dwelling or apartment unit.

[e] As required by the ANSI Standard Safety Code for Industrial Sanitation in Manufacturing Establishments (ANS Z4.1-1935).

[f] This schedule was adopted (1945) by the National Council on Schoolhouse Construction.

[g] Where there is exposure to skin contamination with poisonous, infectious, or irritating materials, provide 1 lavatory for each 5 persons.

[h] 24 lin in of wash sink or 18 in of a circular basin, when provided with water outlets for such space, shall be considered equivalent to 1 lavatory.

[i] Laundry trays, 1 for each 50 persons. Slop sinks, 1 for each 100 persons.

[j] Temporary work personnel facilities:
24-in urinal trough = 1 urinal. 48-in urinal trough = 2 urinals.
36-in urinal trough = 2 urinals. 60-in urinal trough = 3 urinals. 72-in urinal trough = 4 urinals.

General. In applying this schedule of facilities, consideration must be given to the fixtures. Conformity purely on a numerical basis may not result in an installation suited to the need of the individual establishment. For example, schools should be provided with toilet facilities on each floor having classrooms.

Table 15.13 Fixture Units per Fixture or Group

Fixture type	Fixture-unit value as load factors	Min size of trap, in
1 bathroom group consisting of water closet, lavatory, and bathtub or shower stall	Tank water closet, 6 Flush-valve water closet, 8	
Bathtub* (with or without overhead shower)	2	$1\frac{1}{2}$
Bidet	1	$1\frac{1}{4}$
Clothes washer, domestic, automatic	3	2
Drinking fountain	$\frac{1}{2}$	1
Dishwasher, domestic	2	$1\frac{1}{2}$
Floor drains†	1	2
Kitchen sink, domestic	2	$1\frac{1}{2}$
Kitchen sink, domestic, with food-waste grinder and dishwasher	2	2
Lavatory‡	1	$1\frac{1}{2}$
Laundry tray (1 or 2 compartments)	2	$1\frac{1}{2}$
Shower stall, domestic	2	2
Showers (group) per head	3	2
Urinal, pedestal, siphon jet, blowout	6	Nominal 3
Urinal, wall lip	4	$1\frac{1}{2}$
Urinal stall, washout	4	2
Urinal trough (each 2-ft section)	2	$1\frac{1}{2}$
Water closet, tank operated	4	2
Water closet, valve-operated	6	3

Source: National Standard Plumbing Code.

* A shower head over a bathtub does not increase the fixture value.

† Size of floor drain shall be determined by the area of surface water to be drained.

‡ Lavatories with $1\frac{1}{4}$- or $1\frac{1}{2}$-in trap have the same load value; larger P.O. (plumbing orifice) plugs have greater flow rate.

Table 15.14 Estimates of Other Fixture Units per Fixture or Group

Fixture drain or trap size, in	Fixture-unit value
$1\frac{1}{4}$ in and smaller	1
$1\frac{1}{2}$	2
2	3
$2\frac{1}{2}$	4
3	5
4	6

Source: National Standard Plumbing Code.

least twice the diameter of the drain served. This requirement is met by a pipe discharging into a vented or trapped floor drain, slop sink, or similar fixture not used for domestic or culinary purposes. The indirect-waste pipe should be terminated at least 2 in above the floor level of the fixture.

On completion of the plumbing, the system should be inspected or tested with either air or water. In a water test, all openings but the highest one are tightly sealed. The pipes then are filled with water, so that the minimum head is 10 ft, except for the top 10 ft of the system. In an air test, the system is sealed and subjected to 5-psi pressure. For a final test, either a strong-smelling smoke or peppermint is used. With smoke, a pressure of at least 1 in of water should be maintained on the sealed system

Table 15.15 Maximum Permissible Loads, Fixture Units, for Sanitary Drainage Piping

Dia of pipe, in	Max number of fixture units that may be connected to		More than 3 stories in height		Max number of fixture units that may be connected to any portion* of the building drain or the building sewer.			
	Any horizontal fixture branch*	One stack of 3 stories in height or 3 intervals	Total for stack	Total at one story or branch interval	Fall per ft			
					$\frac{1}{16}$ in	$\frac{1}{8}$ in	$\frac{1}{4}$ in	$\frac{1}{2}$ in
$1\frac{1}{4}$	1	2	2	1				
$1\frac{1}{2}$	3	4	8	2				
2	6	10	24	6			21	26
$2\frac{1}{2}$	12	20	42	9			24	31
3	20[†]	30[‡]	60[‡]	16[†]		20[‡]	27[†]	36[†]
4	160	240	500	90		180	216	250
5	360	540	1,100	200		390	480	575
6	620	960	1,900	350		700	840	1,000
8	1,400	2,200	3,600	600	1,400	1,600	1,920	2,300
10	2,500	3,800	5,600	1,000	2,500	2,900	3,500	4,200
12	3,900	6,000	8,400	1,500	3,900	4,600	5,600	6,700
15	7,000				7,000	8,300	10,000	12,000

Source: National Standard Plumbing Code.
* Include branches of the building drain.
[†] Not over two water closets.
[‡] Not over six water closets.

for 15 min before inspection begins. For the peppermint test, 2 oz of oil of peppermint is injected into each line or stack.

(H. E. Bovay, Jr., "Handbook of Mechanical and Electrical Design for Buildings," T. G. Hicks, "Plumbing Design and Installation Reference Guide," J. F. Mueller, "Plumbing Design and Installation Details," L. Nielsen, "Standard Plumbing Engineering Design," McGraw-Hill Publishing Company, New York (books.mcgraw-hill.com); B. Stein et al., "Mechanical and Electrical Equipment for Buildings," John Wiley & Sons, Inc., New York (www.wiley.com); "National Standard Plumbing Code," National Association of

Table 15.16 Size and Length of Vent Stacks and Branch Vents

Size, in, of soil or waste stack	Fixture units connected	Diameter of vent required, in								
		$1\frac{1}{4}$	$1\frac{1}{2}$	2	$2\frac{1}{2}$	3	4	5	6	8
		Maximum developed length of vent, ft								
$1\frac{1}{4}$	2	30								
$1\frac{1}{2}$	8		150							
2	24		50	150						
$2\frac{1}{2}$	42			100	300					
3	72				80	400				
4	500					180	700			
5	1100						200	700		
6	1900							200	700	
8	3600									800
10	5600									250

Table 15.17 Maximum Drainage Areas for Vertical Leaders and Horizontal Storm Drains for Rainfall of 1 in/h*

Vertical Leaders		Horizontal Storm Drains			
Size of leader or conductor,[†] in	Maximum projected area, ft²	Dia of Drain, in	Maximum projected roof area, ft², for drains of various slopes, in/ft		
			$\frac{1}{8}$	$\frac{1}{4}$	$\frac{1}{2}$
2	2,880	3	3,288	4,640	6,576
2½	5,200	4	7,520	10,600	15,040
3	8,800	5	13,360	18,880	26,720
4	18,400	6	21,400	30,200	42,800
5	34,600	8	46,000	65,200	92,000
6	54,000	10	82,800	116,800	165,600
8	116,000	12	133,300	188,000	266,400
		15	238,000	336,000	476,000

* Divide tabulated roof areas by design rainfall rate, in/h, if it is larger than 1 in/h.
[†] The equivalent diameter of square or rectangular leader may be taken as the diameter of a circle inscribed in leader area.

Plumbing-Heating-Cooling Contractors, Falls Church, VA 22046 (www.phccweb.org); "Uniform Plumbing Code," International Association of Plumbing and Mechanical Officials, Ontario, CA. (www.iapmo.org); "ASPE Data Book," American Society of Plumbing Engineers, Chicago, IL (www.aspe.org).)

15.33 Fire-Sprinkler Systems

Consisting essentially of parallel horizontal pipes installed near ceilings, sprinklers have been very effective in preventing spread of fires in buildings. The extinguishing agent usually is water, although for some hazards carbon dioxide is used. The agent, kept under pressure, is discharged from the pipes through sprinklers preset to open when air temperature rises rapidly or reaches a specified level, usually 135 to 160 °F.

Common types of sprinkler systems include wet-pipe, dry-pipe, preaction, and deluge. Pipes in a wet-pipe system contain water at all times and discharge immediately when the sprinklers open. In a dry-pipe system, air under pressure in the pipes is discharged when the sprinklers open, thus

Table 15.18 Maximum Drainage Areas for Roof Gutters for Rainfall of 1 in/h*

Dia of gutters,[†] in.	Maximum projected roof area, ft², for gutters of various slopes, in/ft			
	$\frac{1}{16}$	$\frac{1}{8}$	$\frac{1}{4}$	$\frac{1}{2}$
3	680	960	1,360	1,920
4	1,440	2,040	2,880	4,080
5	2,500	3,520	5,000	7,080
6	3,840	5,440	7,680	11,080
7	5,520	7,800	11,040	15,600
8	7,960	11,200	15,920	22,400
10	14,400	20,400	28,800	40,000

* Divide tabulated roof areas by design rainfall rate, in/h, if it is larger than 1 in/h.
[†] Gutters other than semicircular may be used if they have an equivalent cross-sectional area.

allowing water pressure to open a valve to allow water to flow to the sprinklers. Such systems are suitable for unheated areas in cold climates. Preaction systems have pipes containing air, which may not be under pressure. Heat-responsive devices near sprinklers open water valves when a fire occurs. The deluge system has sprinklers attached to a piping system. Heat responsive devices in the same area as the sprinklers open a valve permitting water to discharge from all sprinklers when a fire occurs in the area served.

The sprinkler system should have a water supply of adequate capacity and pressure. For a secondary supply, a motor-driven automatically controlled fire pump supplied from a water main or pressure-storage system of sufficient capacity may be acceptable.

Local fire-prevention authorities and fire underwriters have specific requirements for type, material, size, and spacing of pipes and sprinklers. They should be consulted before a system is designed.

(C. M. Harris, "Handbook of Utilities and Services for Buildings," and F. S. Merritt, "Building Design and Construction Handbook," 6th ed., McGraw-Hill Publishing Company, New York (books.mcgraw-hill.com); R. E. Solomon, "Automatic Sprinklers Handbook," "Life Safety Code Handbook," and "The SFPE Handbook of Fire Protection Engineering," National Fire Protection Association, Quincy, MA 02269 (www.nfpa.org).)

15.34 Hot- and Cold-Water Piping for Buildings

Local building codes contain minimum standards for hot- and cold-water piping. For example, the New York City Building Code requires a pressure at a faucet or water outlet with the outlet wide open of at least 8 psi. Maximum pressure should not exceed 85 psi with no flow. In addition, units may be used for which manufacturers have higher requirements, for example, for pressure. Care should be exercised using previously recommended criteria for fixture and faucet pressures and flows, because of advances in design since establishment of those criteria. Check with the manufacturers of the individual units for latest recommendations.

Sufficient allowance should be made for pressure loss in pipe and fittings between the supply source and the fixtures, so that required pressures will be present at the fixtures. If necessary to maintain pressure, booster pumps and gravity and pressure tanks may have to be used. If there is danger of excessive pressure in some parts of the system causing water hammer, an air chamber or approved device must be installed to protect the pipes from pressure surges and to reduce noise.

Pipes and tubes for water distribution may be made of copper, brass, cast iron, wrought iron, steel, or plastic. The local building code should be checked for approved materials and types of joints for each. Use of solder containing lead or other health hazards should be avoided.

The system should be designed so that there is no possibility of backflow at any time. Codes generally require that an air gap—space between fixture outlet and flood-level rim of the receptacle—of specified size for each type of fixture be maintained.

Hot water may be supplied by upfeed or downfeed systems, with unused water returned to the heater. In an upfeed system, fixtures are supplied with hot water by a riser directly from the heater. In a downfeed system, hot water is brought to the highest floor in a supply riser and vented at the top through a vent valve, and the fixtures are supplied by the return riser.

Generally, hot water is delivered at 130 to 140 °F. (See Table 15.19.) For floor cleaning, slop sinks may be fed 150 °F water.

Heaters ■ Domestic-water heaters may be direct-fired or unfired. Two types of unfired heaters are in general use: storage and instantaneous. Heat for storage heaters may be supplied by steam or hot water. The tank stores hot water for future use. For occupancies with uneven demand for hot water, such as industrial, office, and school buildings, a relatively large storage capacity is needed. But for occupancies with a nearly uniform demand, such as hotels, apartment buildings, and hospitals, storage capacity may be smaller, but capacity of the heating coil must be larger. With instantaneous heaters, however, water is heated as needed; there is no storage tank. These heaters have V-shaped or straight tubes through which the supply water passes to be heated.

Table 15.19 Hot-Water Demand per Fixture for Various Building Types (Based on average conditions for types of buildings listed, gallons of water per hour per fixture at 140 °F)

Type of fixture	Apartment	Hospital	Hotel	Industrial Plant	Office Building
Basins, private lavatories	2	2	2	2	2
Basins, public lavatories	4	6	8	12	6
Bathtubs	20	20	20		
Showers	30	75	75	225	30
Slop sinks	20	20	30	20	15
Dishwashers		50–150	50–200	20–100	
Pantry sinks	5	10	10		
Demand factor	0.30	0.25	0.25	0.40	0.30
Storage factor	1.25	0.60	0.80	1.00	2.00

Cold Water ▪ Required flow at cold-water fixtures usually is measured in terms of fixture units. The load for each type of fixture is determined by multiplying the number of each to be installed on a branch or a riser or in the building by the demand weight in fixture units (Table 15.20). Figure 15.27 relates the fixture units to the probable demand in gallons of water per minute. Figure 15.28 is an enlargement of Fig. 15.27 in the range of 0 to 250 units. Hot-water demand per fixture in gallons per hour at 140 °F for various building occupancies is given in Table 15.19.

Minimum sizes for hot- and cold-water fixture-supply pipes to keep flow velocity within code limits are listed in Table 15.21. For small buildings, the following diameters can be used for the mains supplying water to the fixture branches:

$\frac{1}{2}$ in for mains with up to three $\frac{3}{8}$-in branches

$\frac{3}{4}$ in for mains with up to three $\frac{1}{2}$-in or five $\frac{3}{8}$-in branches

1 in for mains with up to three $\frac{3}{8}$-in or eight $\frac{1}{2}$-in or fifteen $\frac{3}{8}$-in branches

The minimum pipe sizes in Table 15.21 may be satisfactory for short branches to individual fixtures. But required flow and pressure may make larger sizes necessary. Sizes of risers may be computed from the required flow in gallons per minute and the pressure drop permitted between the supply main and the highest fixtures.

When the potable-water piping has been completed and before it is put into use, it should be disinfected with chlorine by a procedure approved by the local code.

Bibliography ▪ See Art. 15.32, p. 15.72.

15.35 Acoustics

As applied to buildings, acoustics involves the creation of conditions for comfortable listening and means for noise control. At present, acoustics is both an art and a science, for what is comfortable and what is noise depends on judgment and the function of the room to be treated. A sound one person finds too loud may not bother someone else, what is comfortable in a factory may not be acceptable in a school; the music enjoyed by a high-fidelity fan may be noise to a neighbor trying to sleep. Noise is unwanted sound.

Sounds are characterized by their pitch, or frequency; intensity, or loudness; and spectral distribution of energy, or sound quality. An average person can hear from 20 Hz (Hertz or cycles, or vibrations, per second) to 20 kHz. High-frequency, or high-pitched, sounds are more annoying to most people than low-pitched sounds of the same intensity. But high-pitched sounds attenuate faster in air than low-pitched.

Loudness is a subjective evaluation of sound pressure or intensity. But because human response to loudness varies with frequency, any measure of loudness must, in some way, include frequency as well as pressure or intensity to be of significance in building acoustics. In addition, changes in human response to loudness depend on the ratio of the

Table 15.20 Fixture Units for Estimating Water Flow at Fixtures*

Fixture or group	Occupancy	Type of supply control	Fixture units Hot	Fixture units Cold	Fixture units Total
Water closet	Public	Flush valve		10	10
Water closet	Public	Flush tank		5	5
Pedestal urinal	Public	Flush valve		10	10
Stall or wall urinal	Public	Flush valve		5	5
Stall or wall urinal	Public	Flush tank		3	3
Lavatory	Public	Faucet	1.5	1.5	2
Bathtub	Public	Faucet	3	3	4
Shower head	Public	Mixing valve	3	3	4
Service sink	Office, etc.	Faucet	2	2	4
Kitchen sink	Hotel or restaurant	Faucet	3	3	4
Drinking fountain	Various	⅜-in valve	0.25		0.25
Water closet	Private	Flush valve		6	6
Water closet	Private	Flush tank		3	3
Lavatory	Private	Faucet	0.75	0.75	1
Bathtub	Private	Faucet	1.5	1.5	2
Shower head	Private	Mixing valve	1.5	1.5	2
Bathroom group	Private	Flush valve W.C.	2.25	6	8
Bathroom group	Private	Flush tank W.C.	2.25	4.5	6
Separate shower	Private	Mixing valve	1.5	1.5	2
Kitchen sink	Private	Faucet	1.5	1.5	2
Laundry tray	Private	Faucet	2	2	3
Dishwasher	Private	Automatic		1	1

* For calculating maximum probable demand, add fixture units; then determine flow from Fig. 15.26 or 15.27, or similar charts or tables. For example, to determine the probable water flow from two branches, add the fixture units assigned to each branch and use the sum to determine the flow, gal/min.

intensities of the sound. In acoustics, the ratio 10:1 is called a bel. For practical reasons, the unit frequently used, however, is the decibel (dB), equal to 0.1 bel.

Intensity level IL, dB, used as a measure of loudness, is defined by

$$IL = 10 \log_{10} \frac{I}{I_o} \qquad (15.37)$$

where I = intensity, measured, W/cm^2

I_o = reference intensity = 10^{-16} W/cm^2

Equation (15.37) indicates that zero level corresponds with $I = I_o$, the reference intensity, which, in turn, corresponds with the average threshold of human hearing of sound at about 1 kHz.

Sound pressure level SPL, dB, since intensity varies as the square of pressure, is accordingly defined by

$$SPL = 20 \log_{10} \frac{p}{p_o} \qquad (15.38)$$

where p = pressure, measured, pascals (Pa)

p_o = reference pressure = 0.00002 Pa

A change in sound level of less than about 3 dB is not likely to be perceptible, but a change of 5 dB will be noticeable. An increase of 10 dB will appear to be twice as great as an increase of 5 dB, and an increase of 20 dB much greater than an increase of 10 dB—not quite proportionately.

Sound levels usually are measured with electronic instruments that respond to sound pressures. Readings on the A scale of such instruments are often used because this scale is adjusted for frequencies that correspond somewhat with the

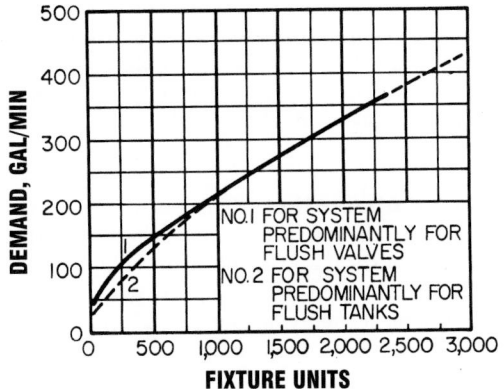

Fig. 15.27 Curves for estimating demand load for domestic water supply.

Table 15.22 Comparison of Intensity, Sound Pressure Level, and Common Sounds

Relative intensity	SPL, dBA*	Loudness
100,000,000,000,000	140	Jet aircraft and artillery fire
10,000,000,000,000	130	Threshold of pain
1,000,000,000,000	120	Threshold of feeling
100,000,000,000	110	Chainsaw, weed wacker
10,000,000,000	100	Loud factory
1,000,000,000	90	Full symphony or band
100,000,000	80	Busy restaurant
10,000,000	70	Conversation, face-to-face
1,000,000	60	Busy store
100,000	50	Inside general office
10,000	40	Inside private office
1,000	30	Inside bedroom
100	20	Inside empty theater
10	10	Normal breathing
1	0	Threshold of hearing

* SPL as measured on A scale of standard sound level meter.

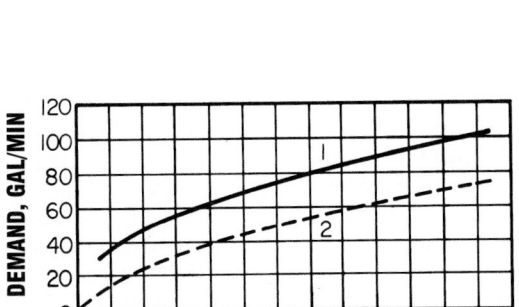

Fig. 15.28 Enlargement of the low-demand portion of Fig. 15.27.

response of the human ear. In such cases, the unit is indicated by dBA.

Table 15.22 presents a comparison of intensity, *SPL*, and common sounds.

Acoustical analysis and design aim at both sound and vibration control. Sound control is accomplished with barriers and enclosures, acoustically absorbent materials, and other materials properly shaped and assembled. Vibration control is accomplished with energy-absorptive construction, usually with resilient materials, or by damping, involving use of viscoelastic materials.

Table 15.21 Minimum Sizes for Fixture-Supply Pipes*

Type of fixture or device	Pipe size, in	Type of Fixture or device	Pipe size, in
Bathtubs	$\frac{1}{2}$	Shower (single head)	$\frac{1}{2}$
Combination sink and tray	$\frac{1}{2}$	Sinks(service, slop)	$\frac{1}{2}$
Drinking fountain	$\frac{3}{8}$	Sinks, Rushing rim	$\frac{3}{4}$
Dishwasher (domestic)	$\frac{1}{2}$	Urinal (flush tank)	$\frac{1}{2}$
Kitchen sink, residential	$\frac{1}{2}$	Urinal (direct flush valve)	$\frac{3}{4}$
Kitchen sink, commercial	$\frac{3}{4}$	Water closet (tank type)	$\frac{3}{8}$
Lavatory	$\frac{3}{8}$	Water closet (Rush valve type)	1
Laundry tray, 1, 2, or 3	$\frac{1}{2}$	Hose bibs	$\frac{1}{2}$
compartments		Wall hydrant	$\frac{1}{2}$

* Objective is to restrict the velocity of flow to not more than 8 or 10 ft/s, depending on the local building-code requirements.

Effectiveness of a barrier in stopping sound is measured by **sound transmission loss**, the loss in energy level as sound passes through a barrier. The greater the mass of a barrier, the greater the sound transmission loss and the more effective the barrier. Mass and transmission loss, however, are not related linearly. At low frequencies losses tend to be larger, and at other frequencies, smaller, than required for a linear relationship. Table 15.23 lists the performance of various barriers and rating systems for their performance.

The purpose of a barrier with a high sound transmission loss, however, can be defeated if sound can bypass the barrier through openings or by transmission through adjoining construction. Ducts, pipes, and almost any continuous, rigid component of a building can carry sound past a barrier. Therefore, precautions should be taken to prevent such bypassing. Carpet over a resilient pad, for example, effectively helps absorb such sounds as footfalls, heel clicks, and impact of dropped light objects. Openings should be plugged. Vibrations from machines and other equipment can be absorbed by supporting them on springs, elastomeric pads, or other resilient mounts.

Vibration of barriers resulting from impact of sound or transmission of vibrations from machines can be damped out by proper assembly in any of several ways. One way is to attach to a barrier a material with high internal friction or poor

connections between particles, or viscoelastic materials, such as asphaltic compounds that are neither completely elastic nor completely viscous. Also, components of the barrier may be connected with a viscoelastic adhesive.

Sound Absorption ▪ Reflection of sound from a surface can be curtailed by covering the surface with an acoustical absorbent, usually lightweight, porous boards, blankets, or panels, which convert the mechanical energy of sound into heat. Exposed surfaces may be smooth or textured, fissured or perforated, or decorated or etched in many ways. Selection of an absorbent usually is based on its absorptive efficiency, appearance, fire resistance, moisture resistance, strength, and maintenance requirements. An absorbent, however, may have poor resistance to sound transmission and should not be used in an attempt to improve the airborne sound isolation of a barrier.

Sound absorption coefficients are used as an indication of the absorptive efficiency of building products. The sound absorption coefficient of a product is the ratio of the energy it will absorb from a sound wave to the total energy impinging. A perfect absorber would be assigned a coefficient of 1. Sound absorption, however, depends on the frequency of the sound (Table 15.24). Consequently, coefficients for a product are given for specific frequencies, or sometimes as a composite for a group of frequencies, such as the noise reduction coefficient *NRC* in (Table 15.24).

Generally, absorbents are used not only to reduce undesirable sound reflection, such as echoes and flutter, but also to secure desirable reverberations. *Echoes* are distinct reflections. *Flutter* is produced by rapid, repeated, partly distinguishable echoes, such as those that occur between parallel sidewalls of a corridor, *Reverberation* results from very rapid, repeated, jumbled echoes, which produce a continuing indistinct sound that persists after the sound producing the echoes has ceased.

Reverberation within a room can garble speech or distort music. But, properly controlled, reverberation can enhance the sound of music. Good reverberation can be achieved with proper proportioning and shaping of rooms, echo control, and absorption of noise. Generally, acoustical absorbents on room surfaces are desirable to absorb acoustical energy to prevent buildup of undesirable sounds.

Table 15.23 Sound Transmission Class (STC) of Various Constructions

Construction	STC
$\frac{1}{4}$-in plate glass	26
Double glazing, $\frac{1}{4}$-in panes, 4-in air gap	48
$1\frac{1}{2}$-in hollow-core door	22
$1\frac{3}{4}$-in solid-wood door	30
$\frac{3}{4}$-in plywood	28
$\frac{1}{2}$-in gypsumboard, both sides of	
2×4 studs	33
$\frac{1}{4}$-in steel plate	36
6-in concrete block wall	42
8-in reinforced concrete wall	51
12-in concrete block wall	53
Cavity wall, 6-in concrete block,	
2-in air space, 6-in concrete block	56

Table 15.24   Noise Reduction and Sound Absorption Coefficients

Absorbent	Thickness, in	Density, lb per ft³	Noise reduction coefficient
Mineral or glass fiber blankets	½–4	½–6	0.45–0.95
Molded or felted tiles, panels, and boards	½–1⅛	8–25	0.45–0.90
Plasters (porous)	⅜–¾	20–30	0.25–0.40
Sprayed-on fibers and binders	⅜–1⅛	15–30	0.25–0.75
Foamed, open-cell plastics, elastomers, etc.	½–2	1–3	0.35–0.90
Carpets	Varies with weave, texture, backing pad, etc.		0.30–0.60
Draperies	Varies with weave, texture, weight, fullness		0.10–0.60

Absorbent	Absorption coefficient per ft² of floor area at frequencies, Hz					
	125	250	500	1000	2000	4000
Seated audience	0.60	0.75	0.85	0.95	0.95	0.85
Unoccupied upholstered (fabric) seats	0.50	0.65	0.80	0.90	0.80	0.70

Noise reduction NR, dB, achieved through addition of absorbents can be computed from

$$NR = 10 \log_{10} \frac{A_o + A_a}{A_o} \qquad (15.39)$$

where A_o = original acoustical absorption present

A_a = added acoustical absorption

Acoustical absorption equals the sum of the products of the absorption coefficient of each material forming the enclosure surface and the corresponding surface area.

Sometimes materials are rated with a **noise reduction coefficient** NRC, the arithmetic average of the sound absorption coefficients of a material at 250, 500, 1000, and 2000 Hz (Table 15.24).

Reverberation Time ▪ This is the time, in seconds, that a sound pulse within a room takes to decay 60 dB, to one-millionth of its original level. Reverberation time T can be computed from the Sabine formula:

$$T = \frac{0.49V}{A} \qquad (15.40)$$

where V = volume of room, ft³

A = total acoustical absorption in room

Reverberation times falling within the shaded area in Fig. 15.29 may be considered satisfactory under ordinary conditions. For critical spaces, such as concert halls, radio studios, and auditoriums,

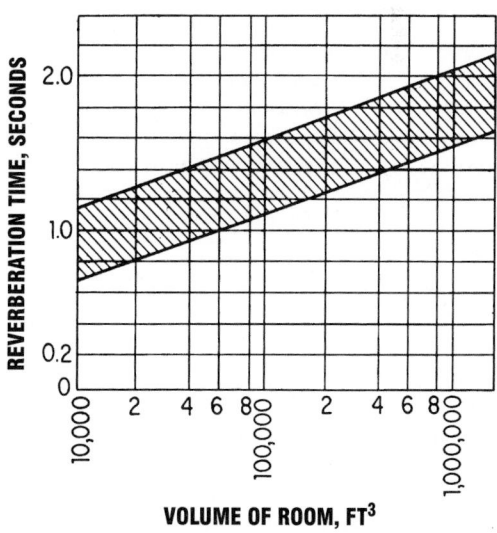

Fig. 15.29   Recommended reverberation time, indicated by shaded area, varies with size of room.

Table 15.25 Impact Insulation Class of Floor Constructions

Construction	IIC
Oak flooring on ½-in plywood subfloor, 2 × 10 joists, ½-in gypsumboard ceiling	23
With carpet and pad	48
8-in concrete slab	35
With carpet and pad	57
2½-in concrete on light metal forms, steel bar joists	27
With carpet and pad	50

advice of an acoustical consultant should be obtained.

Rating Systems ■ ASTM has adopted rating systems for evaluating acoustical performance of materials, such as:

Sound transmission class STC for indicating the insulation against airborne sound of partitions, floor-ceiling assemblies, and other barriers (ASTM E90 and E413). Table 15.23 lists some typical ratings.

Impact insulation class IIC for indicating the impact insulation of floor-ceiling assemblies (ASTM RM 14-4). See Table 15.25.

Table 15.26 Guide Criteria for Acoustical Design

Isolation requirements between rooms					
For sound			For impact		
Between Room and Adjacent area		Sound isolation requirement, STC	Between Room and Room below		Impact isolation requirement, IIC
Hotel bedroom	Hotel bedroom	47	Hotel bedroom	Hotel bedroom	55
Hotel bedroom	Corridor	47	Public spaces	Hotel bedroom	60
Hotel bedroom	Exterior	42	Classroom	Classroom	47
Normal office	Normal office	33	Music room	Classroom	55
Executive office	Executive office	42	Music room	Theater	62
Bedroom	Mechanical room	52	Office	Office	47
Classroom	Classroom	37			
Classroom	Corridor	33			
Theater	Classroom	52			
Theater	Music rehearsal	57			

Typical Sound Pressures		Acceptable background levels	
Source of sound	Pressure level, dBA*	Space	Background level, dBA*
Washing machine	60	Recording studio	25
TV (at 10 ft)	67	Suburban bedroom	30
Telephone (at 5 to 15 ft), ringing	61	Theater	30
Water closet, tank type, refilling	55	Church	35
Conversational speech, normal	73	Classroom	35
Rock-and-roll music (amplified)	102	Private office	40
Trucks (at 20 ft)	75	General office	50
Classrooms	73	Dining room	55
Computer room	77	Computer room	70

* Average of sound pressure at 250, 500, 1000, 2000 Hz.

Impact noise rating INR, an alternative measure of impact insulation of floor-ceiling assemblies. *IIC* ratings can be converted to *INR* by deduction of 51 points.

Sound absorption coefficients for indicating the absorptive efficiency of acoustical absorbents (ASTM C423). See Table 15.24.

Noise reduction coefficients, an alternative measure of absorptive efficiency (Table 15.24).

Acoustical Criteria ■ Although governmental agencies and others have adopted design criteria to achieve desired sound reduction and insulation, the criteria are based on subjective response to acoustical parameters. Accordingly, small deviations from such criteria may not necessarily result in unsatisfactory performance. Usually, a tolerance of ± 2.5 points from a numerical value is acceptable in practice. Table 15.26 presents some acoustical criteria that may be used as a guide.

("McGraw-Hill's Acoustics Source Book," M. D. Egan, "Architectural Acoustics," C. M. Harris, "Handbook of Acoustical Measurements and Noise Control," McGraw-Hill Publishing Company, New York (books.mcgraw-hill.com); L. F. Yerges, "Sound, Noise and Vibration Control," Van Nostrand Reinhold Company, New York.)

16

Richard Garrabrant, P.E.
Consulting Engineer
Buffalo, New York

HIGHWAY ENGINEERING

For the purposes of this section, a highway is considered a conduit that carries vehicular traffic from one location to another. Highway engineering deals with provisions for meeting public needs for highways; environmental impact of highways; budgeting, planning, design, construction, maintenance, and rehabilitation of highways; access to and exit from highways; economics and financing of highway construction; traffic control; and safety of those using or affected by the use of highways.

Highway engineering is continually evolving and improving. Many of the basic design techniques employed are the same today as they were fifty or more years ago. However, numerous incremental improvements have been made to provide safer, more efficient highway designs that prove less costly over the life cycle of a highway. Further, in today's culture of minimizing the highway's impact on the environment and society as a whole, a great deal of extra effort and time is expended to develop the final "highway design." Improved design methodologies and construction materials, enhanced analysis techniques and the use of computer-assisted design and drafting have significantly improved the overall quality and thoroughness of the resultant design, plans, and the roadway itself. Community involvement in the decision making process and associated environmental review processes help to ensure that a context sensitive design is developed and implemented that addresses the entire spectrum of

highway users and uses including access management and economic development. New concepts such as Intelligent Transportation Systems and Maintenance (ITSM), Pavement Management Systems (PMS) and Asset Management Systems (AMS) have significant effects on highway engineering and the final constructed product. These management systems and tools help to better address highway engineering design parameters, budgeting needs, and the maintenance and operation of the facility when it is constructed. This section presents these newer technique as well as fundamental principles of highway engineering and their practical applications based on long-time experience.

16.1 Classes of Highways

Highways can range in character from a dirt road in a rural setting to a multilane pavement in an urban environment. They are classified in accordance with functional characteristics. These characteristics are based on the location of the road, such as urban or rural; width of the road, such as single lane or multilane; and the type of service the road is to provide, such as local access or travel between cities.

Principal guidelines for classifying highways are provided in the American Association of State Highway and Transportation Officials (AASHTO) guide, "A Policy on Geometric Design of Highways and Streets" (Policy) (www.aashto.org). Highways are grouped in accordance with the type of service they provide; that is, the type of travel associated with the road. Travel is facilitated by a highway **network** that is comprised of various classes of highways. Figure 16.1 presents a schematic of a highway network composed of the three principal highway classes: arterials, collectors, and local roads.

Arterials are highways that provide direct service to major population centers. **Collectors** provide direct service to towns and link up with arterials. **Local roads** connect various regions of a municipality and tie into the system of collectors. Further subdivisions of these three principal categories can be made by defining principal and minor arterials and major and minor collectors.

A principal arterial provides for main movement whereas a minor arterial acts as a distributor. Major and minor collectors are subclassifications that can be used to define the types of population centers the collector serves and other impacting criteria such as spacing and population density.

Each functional class of highway is designed to meet specific needs. For instance, arterials are built to facilitate a high degree of mobility. This need for mobility typically is met by construction of multilane highways with strict control of access. Local roads, in contrast, are designed to facilitate access to various areas of a municipality; for example, commercial and residential areas.

Also associated with these basic functional classifications are the political classifications of highways. Criteria to which a roadway is constructed and maintained are related to the political entity, such as Federal, state, and county (local) government, that has jurisdiction over the highway. These criteria have a profound impact on the way a highway is designed. A local road in a rural setting, for example, may consist of asphalt and aggregate surfaces on gravel bases and would be financed by local taxes and some state funds. A toll road in the Interstate system, however, would have a higher-quality, durable pavement and would be funded by those who use the highway and perhaps by some Federal assistance. Thus, design of a highway is highly dependent on whether it will serve a rural or urban environment.

AASHTO defines an urban area as "those places within boundaries set by the responsible State and local officials having a population of 5,000 or more." Furthermore, the AASHTO Policy defines an *urbanized area* as one with a population of 50,000 and over and a *small urban area* as one with a population between 5000 and 50,000. *Rural areas* are defined as areas falling outside the definition of urban areas.

16.1.1 Rural Highway Systems

A **rural principal highway system** is comprised of those highways that offer corridor movement and are capable of supporting statewide or interstate travel on this class of highway. Rural principal arterial systems can be further subdivided into freeways and all other principal arterials.

A **freeway** is a divided highway with fully controlled access. Access to a freeway is made *without* use of at-grade intersections. Figure 16.2 shows a schematic of a rural highway network and its corresponding functional classifications. As illustrated in this figure, the arterials provide direct service between cities and large towns, which are the major traffic generators.

Rural minor arterial systems serve in conjunction with principal arterial systems to link together cities, large towns, and other traffic generators, for example, large resorts. The resulting network can also serve to integrate interstate and intercounty service. To provide for a consistent and high degree of mobility, maximum travel speeds are set as high as those on the associated principal arterial systems and therefore require a design that can accommodate such speeds.

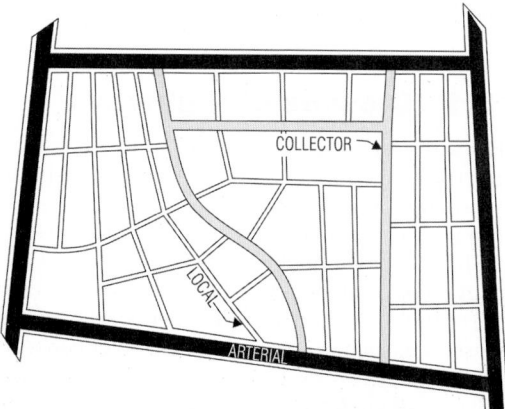

Fig. 16.1 Schematic of a suburban network with local, collector, and arterial roads.

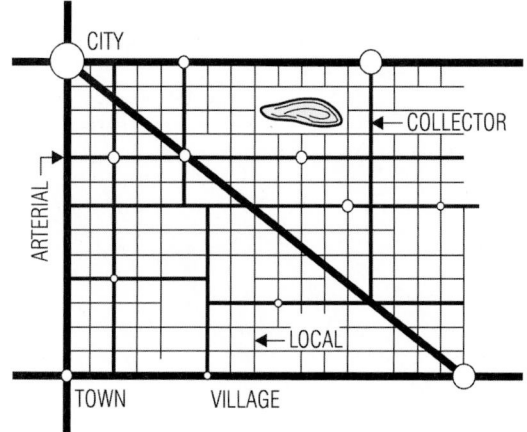

Fig. 16.2 Schematic of a rural highway network serving towns, villages, and cities.

Rural major collector systems contain routes that are intended to serve county seats and nearby large towns and cities that are not directly served by an arterial system. Other traffic generators that may be served by a major collector system are consolidated schools, shipping points, county parks, agricultural areas, and other locations of intracounty importance. In Fig. 16.2, the collectors are shown to collect traffic from local roads, which service specific land uses such as farms, and distribute this traffic to the arterials.

Rural minor collector systems contain routes that carry traffic from the system of local roads and other traffic generators of local importance to other facilities.

Rural local road systems include all roads in the rural system that do not fall into any of the preceding rural categories. These local roads carry traffic from land adjacent to the collector system and are useful for travel over relatively short distances.

16.1.2 Urban Highway Systems

An urban principal arterial system is designed to accommodate travel along heavily traveled corridors serving major centers of activity in urban areas. Although an urban principal arterial system may or may not be a controlled-access facility, all controlled-access facilities are classified as this type of system.

An urban principal arterial system is integrated with its rural counterpart that serves major centers. The urban system facilitates most of the trips that either enter or leave a population center in addition to serving most through traffic.

There are three major subclassifications of an urban principal arterial: interstate, other freeways, and other principal arterials. The last type may provide partial or no controlled access. Only this subclassification of primary arterial can be used to provide direct access to intersecting roads.

Urban minor arterial systems interconnect and augment an urban principal arterial system. In comparison to the primary arterials, a minor arterial system is intended more for use for direct access to intersecting roads and less for provision of travel mobility. While such arterials do not usually pass through identifiable neighborhoods, they may support local bus routes and provide some continuity between various communities in an urban area.

Urban collector street systems receive traffic from local roads in commercial, industrial, and residential areas for travel to an arterial system. Collector street systems may carry local bus routes and, in some instances, comprise the entire street grid of a central business district.

Urban local street systems include all roads in the urban system that do not fall into any of the preceding urban categories. These roads carry traffic from land adjacent to the collector system. Through traffic is generally discouraged.

Elements of Highway Transverse Cross Sections

The geometry of a typical highway comprises three basic components: cross-sectional geometry, horizontal geometry, and vertical geometry. The type, size, and number of elements used in a highway are directly related to its class (Sec. 16.1) and the corresponding function of the highway.

16.2 Travel Lanes

Travel lanes are that section of a roadway on which traffic moves. Figure 16.3 shows a typical two-lane highway and such cross-sectional components as travel lanes, shoulders, and side slopes.

From a geometric standpoint, the key parameters defining a travel lane are the number of

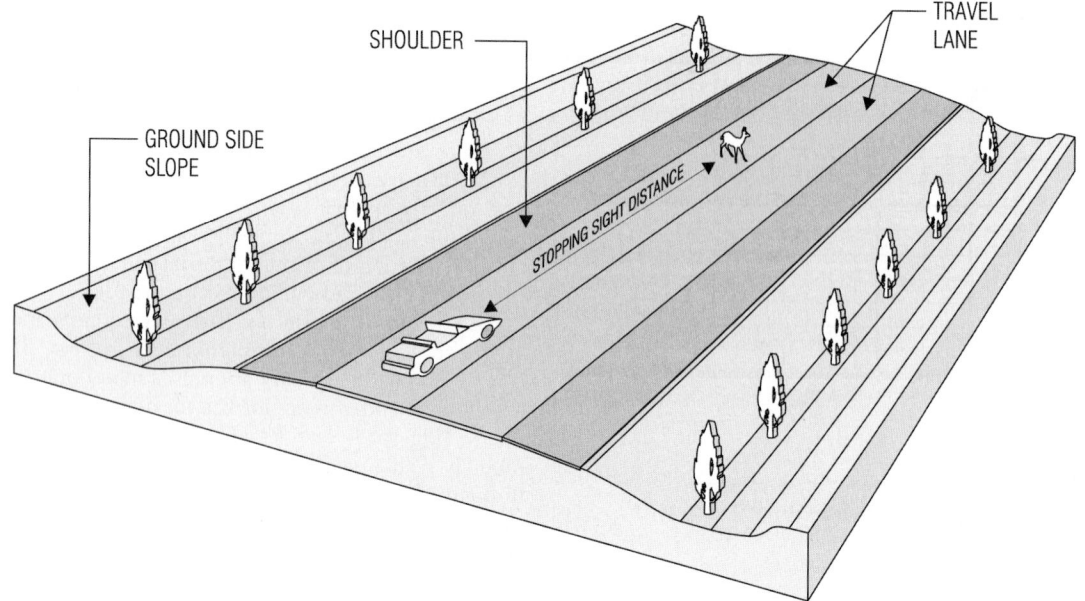

Fig. 16.3 Cross section of a typical two-lane highway.

lanes, their width, and cross slopes, all of which impact the level of service a highway can accommodate. Of equal importance are the characteristics of the pavement surface and its skid resistance. These features affect the overall ridability, safety, and future maintenance of a highway.

Travel-Lane Widths ▪ Travel lanes generally range in width from 10 to 13 ft. (Under extreme circumstances, a width of 9 ft may be used.) The width selected has a significant impact on highway capacity. A width of 12 ft predominates on high-type pavements since the cost differential for constructing a 12-ft lane width instead of a 10-ft width is usually offset by reduced maintenance costs for the shoulders and the edges of pavements. When the narrower, 10-ft width is used, the shoulders and edges of pavement undergo more wear and tear from wheel concentrations at these locations.

Number of Travel Lanes ▪ For most highways, three lanes in one direction usually is the maximum installed. In certain situations, four lanes in one direction may be provided. If more than three lanes are required, however, and sufficient

land is available, dual roadways should be constructed in each direction.

In general, the number of lanes selected should be based on the design traffic volume and other design-related considerations. For example, highways in steep or mountainous terrain may necessitate incorporation of a separate *climbing lane* for slow-moving trucks. Another example is the addition of an exclusive *bus lane* to the lanes that would otherwise be provided. As another example, a *reversible lane* may be used to expedite traffic flow on highways on which traffic flow fluctuates greatly from morning to night (owing to commuting patterns). A general rule of thumb is that lane changes should be avoided at intersections and interchanges.

16.3 Roadway Cross Slopes

For highways with two lanes or more, the roadway usually is sloped from a high point, or crown at the middle of the roadway downward toward the opposing edges. Figure 16.4 shows a typical two-lane highway with linear opposing cross slopes intersecting at the centerline of the total travel width.

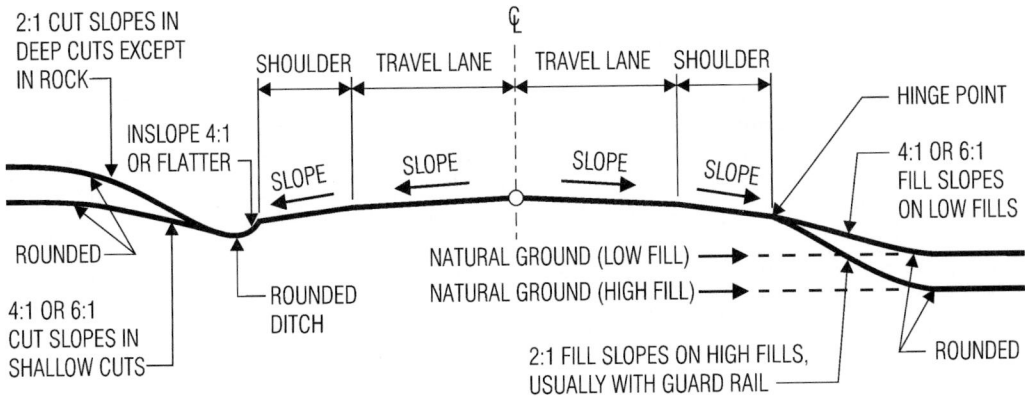

Fig. 16.4 Typical two-lane highway with linear cross slopes.

Alternatively, the roadway cross slope may be unidirectional. This roadway cross section is generally more pleasing to drivers since vehicles appear to be pulled in the same direction when changing lanes.

Opposing and unidirectional cross slopes have advantages and disadvantages for drainage of the highway. Opposing cross slopes have the advantage of being able to drain the roadway quickly during a heavy rainstorm. This layout, however, requires installation of drainage facilities on both sides of the roadway. Unidirectional cross slopes tend to drain more slowly, but they have the advantage of permitting drainage facilities to be consolidated along one edge of the roadway and thereby reduce construction and maintenance costs.

In some instances, a parabola is used in lieu of straight-line segments in forming the crown of the roadway. While the parabolic section provides good drainage of the roadway, it has relatively high construction costs, and difficulty may be encountered in grading at intersections.

When specifying travel-lane cross slopes, designers should consider the necessity of both adequate drainage and driver safety. A cross slope that is too flat will not drain properly and one that is too steep can cause vehicles to drift toward the edges of pavement, especially when the pavement is slick.

The slope selected generally depends on the type of pavement used. The American Association of State Highway and Transportation Officials (AASHTO) recommends a cross slope of 1.5 to 2%

for highways with high-type pavements, 1.5 to 3% for intermediate-type pavements, and 2 to 6% for low-type pavements (Art. 16.4).

From a safety standpoint, cross slopes greater than 2% should be avoided for high-type pavements on which high speeds are permitted and that have opposing cross slopes. Such steep slopes pose a hazard for drivers because, when passing another vehicle on two-lane highways with such pavements, a driver must cross and then recross the crown, where a total cross-slope change (rollover) of more than 4% occurs. In some instances, though, it may be necessary to use slightly steeper cross slopes to facilitate proper drainage. In doing this, however, designers should limit the total cross-slope change to minimize hazards to safe driving.

16.4 Types of Roadway Surfaces

The rate of cross slope specified generally depends on the type of pavement utilized. The American Association of State Highway and Transportation Officials (AASHTO) recognizes three major types of pavement: high-type, intermediate-type, and low-type.

Pavement classified as high-type possesses a wearing surface that can sustain heavy vehicles and high-volume and high-speed traffic over long periods of time without failure due to wear or fatigue. This type of pavement should keep non-routine maintenance to a minimum and sustain a consistent flow of traffic without repair-related interruptions. Intermediate-type pavements are similar to high-type pavements except that they are constructed in accordance with standards that are not as strict as those for high-type pavements. Low-type pavements, predominately used in low-cost roads, may be composed of surface-treated earth and stabilized materials or any of a variety of loose surfaces, such as earth, shell, crushed stone, or bank-run gravel.

The type of pavement to select depends on a variety of factors of which design speed is only one. For instance, skid resistance is an important factor. See also Arts. 16.18 to 16.24.

Skid Resistance ▪ The ability of a pavement to accommodate driver braking and steering maneuvers is a function of the pavement skid resistance, the ability of a pavement to prevent accidents due to skidding. While most pavements perform adequately in dry conditions, their ability to limit skidding may deteriorate under wet or icy conditions. The principal causes of skidding type accidents include: a) vehicular tire hydroplaning caused by water ponding in excessively rutted pavement surface; b) polishing of aggregates contained within the pavement structure; c) bleeding of asphalt emulsion on the surface of the pavement; and d) the presence of some form of lubrication on the pavement surface.

When excessive rutting occurs in a pavement surface, water accumulates (ponds) in the wheel tracks and during high travel speeds this can cause a vehicles tires to loose direct contact with the pavement surface, thus hydroplane and skid out of control. Polishing of the aggregates contained within the pavement surface create a smooth pavement surface that does not provide enough friction between the tires and the road surface and leads to skidding type accidents. Bleeding of a lubricating substance on the pavement can cover the pavement microtexture and diminish the pavement's textural features and reduce skid resistance characteristics. Loss of skid resistance due to any of the above mentioned causes are serious issues that require corrective maintenance action on the part of the maintaining agency.

16.5   Shoulders

A shoulder (also known as a **verge**) is that part of a roadway between the edge of the traveled way and the edge of an adjacent curb, ground side slope, or drainage feature, such as a ditch or gutter (Fig. 16.5). A shoulder is designed to accommodate stopping and temporary parking of vehicles, emergency use, and provide lateral support of base and surface courses. Shoulders should be capable of sustaining starting, stopping, and movement of vehicles without appreciable rutting or distortion.

Shoulder Widths. Shoulders usually range in width from 2 ft for minor local roads to 12 ft for major highways. Figure 16.5 shows various forms of graded and usable shoulder widths. Graded shoulder width is the distance from the edge of the traveled way to the intersection of the shoulder slope and the start of the ground side slope. Usable

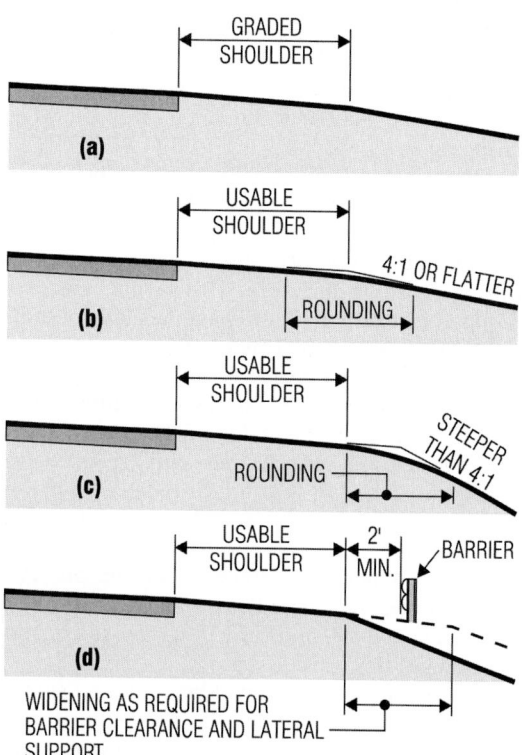

Fig. 16.5 Cross section of a highway with shoulders. (*a*) Graded shoulder. (*b*) Shoulder with side slope of 4:1 or less. (*c*) Shoulder with side slope exceeding 4:1. (*d*) Shoulder wide enough to permit installation of a guard rail, wall, or other barrier. Such vertical elements should be offset at least 2 ft from the outer edge of the usable pavement.

shoulder width is that part of the shoulder that drivers can use to stop and park a vehicle. If the ground side slope is 4:1 or flatter, then the usable width will be the same as the graded width.

A minimum distance of 2 ft should be maintained between the edge of the traveled way and a vehicle stopped on a shoulder. This results in shoulders with a width of at least 10 ft (preferably 12 ft) for heavily traveled roads. For minor roads, topography or other site-related constraints may necessitate use of smaller shoulder widths. In such conditions, a minimum width of 2 ft sometimes is used, but a range of 6 to 8 ft is preferable.

Usable shoulder widths used on the approach roadways to bridges should also be maintained on the bridge. Narrowing or elimination of the shoulder width on a bridge can create unsafe conditions when emergency stopping of vehicles on a bridge is required.

Shoulders are generally continuous along the length of the roadway. [In some European countries, intermittent shoulders (turnouts) are used on minor roads.] According to AASHTO, some shoulder is better than no shoulder at all. Even under the most severe topographic constraints, designers should endeavor to maximize the width and continuity of shoulders.

Shoulder Cross Slopes ▪ The cross slope to be used for a shoulder depends on the type of shoulder construction. Cross slopes ranging from 2 to 6% are generally used for bituminous and concrete surfaces, from 4 to 6% for gravel or crushed rock surfaces, and up to about 8% for turf shoulders. These values are presented as a guide and are neither maximum nor minimum values. It is noteworthy that the highway geometry can greatly impact the design of shoulder cross slopes. For example, long-radius, curved alignments or superelevated roadways can present drainage and other design conditions that require modification of the preceding slopes.

The region in the vicinity of the intersection of the shoulder and ground side slope may be rounded (Fig. 16.5b and c). When the side slope is 4:1 or flatter, the rounding may be 4 to 6 ft wide without adverse impact on the usable shoulder width.

If a barrier is installed outside a shoulder, there should be at least 2-ft clearance between the barrier and the usable shoulder, which should be widened as needed for barrier clearance and lateral support

(Fig. 16.5d). If curbs are placed on the outer side of shoulders, the design should ensure good drainage to prevent excessive ponding. In extreme conditions, ponding can encroach into the traveled way and hinder traffic or cause accidents.

Shoulder Stability ▪ Shoulders should be designed not only to support vehicle loading without appreciable rutting but also to be contiguous with the traveled way. They must be constructed flush with the paved surface of the traveled way if they are to function properly. In addition, shoulders should be stabilized so that they remain flush in service.

Shoulders that are not properly stabilized can settle enough to adversely affect a driver's control of a vehicle moving from the traveled way to the shoulder. This situation can also encourage drivers to avoid the pavement edge adjoining the shoulder intentionally, thereby increasing the chance of accidents.

Shoulder-Pavement Contrast ▪ It is desirable to vary the color and texture of a shoulder from that of the travel lanes. The resulting contrast serves the dual function of providing clear differentiation between travel lanes and shoulders and discouraging the use of shoulders as through lanes.

Bituminous, gravel, crushed rock, and turf shoulders offer excellent contrast with concrete lanes. For bituminous lanes, one method of enhancing contrast between the travel lanes and shoulders is to seal-coat the shoulders with lighter-color stone chips. A drawback to this method is that contrast may diminish with time. Additional contrast can be provided by installation of reflective striping at the edge of the traveled way.

16.6 Curbs

A curb is a raised element that is used, among other things, to denote the edge of a roadway. Curbs can be made of portland cement or bituminous concrete, granite, or some other hard durable material. In addition to pavement delineation, curbs provide drainage control, right-of-way reduction, enhanced appearance, delineation of pedestrian walkways, and reduction of maintenance operations. To facilitate drainage, curbs can be combined with a gutter to create a combined curb-gutter section.

There are two general classifications of curbs: barrier and mountable. Figure 16.6 illustrates various types of curbs.

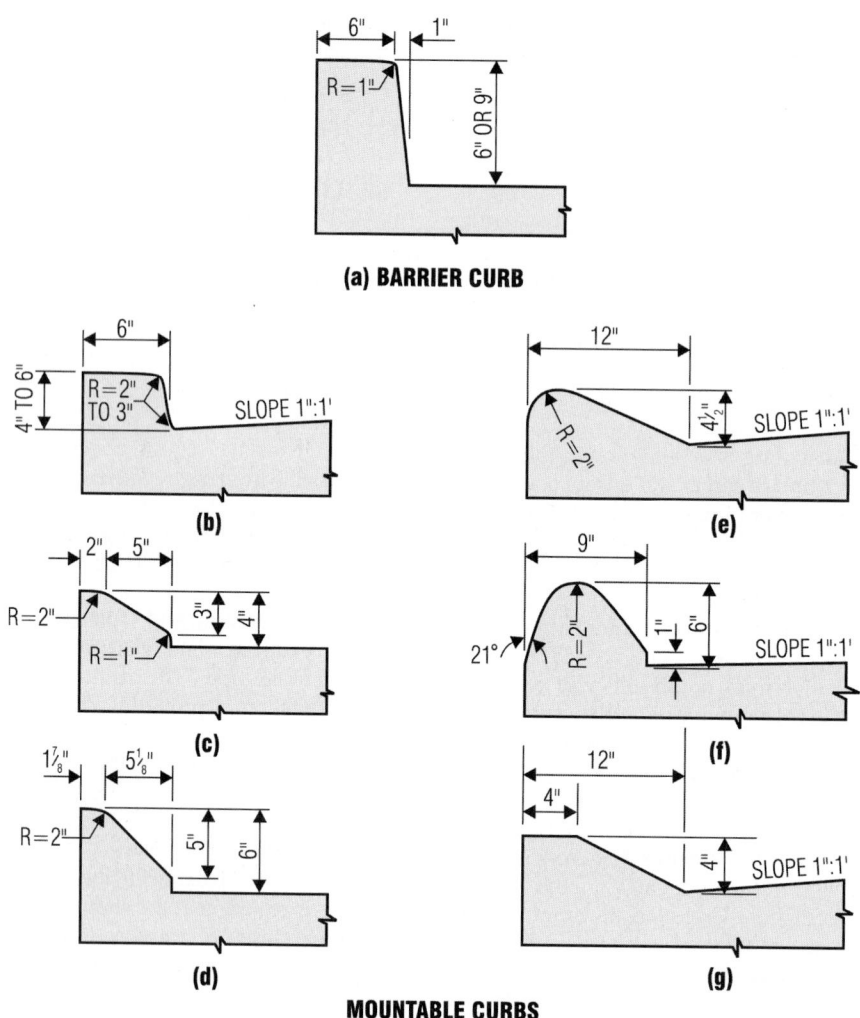

(a) BARRIER CURB

MOUNTABLE CURBS

Fig. 16.6 Typical highway curbs. (*a*) Barrier curb used to prevent vehicles from leaving the roadway. (*b*) to (*g*) Mountable curbs that permit vehicles to cross when necessary. Slopes of curb faces and rounding vary.

Barrier Curbs ▪ The purpose of a barrier curb is to prevent or limit the possibility of a vehicle's leaving the roadway. For this purpose, a barrier curb is made relatively high and given a steep face (Fig. 16.6*a*). Typical height is 6 to 9 in. When the roadway-side face is sloped, the batter should not exceed 1 in horizontal on 3 in vertical.

Barrier curbs are typically used along the faces of long walls and tunnels and along low-speed, low-volume roadways but rarely along major highways. Because of their height, these curbs can present a hazard to vehicles traveling at high speeds, inas-much as drivers can lose control of their vehicles on contacting the curbs. A general rule of thumb is that barrier curbs should not be used when the design speed is greater than 40 mi/h.

Mountable Curbs ▪ A mountable curb offers the advantage that a vehicle can cross it when necessary. Typical forms of mountable curbs are illustrated in Fig. 16.6*b* to *g*. In contrast to barrier curbs, mountable curbs are relatively low and have flat sloping faces.

To facilitate vehicle crossing of the curbs, the curb faces on the roadway side may be rounded. Curb height depends on the face slope. For face slopes steeper than 1:1, a height of 4 in or less is desirable. For face slopes that fall between 1:1 and 2:1, curb height is limited to a maximum of about 6 in.

Mountable curbs may be installed along median edges to delineate islands. Like barrier curbs, however, mountable curbs should not be used along the travel-way edges of high-speed, high-volume highways. Mountable curbs are often used along the outer edge of a shoulder for drainage control, reduction of erosion, and enhanced delineation.

Color and texture of mountable curbs should be contrasted with that of the adjacent roadway to enhance their visibility, especially at night and in adverse weather conditions. One method used to enhance visibility of curbs is to apply reflective surfaces. Another approach is to form on the curbs depressions and ribs that reflect headlight beams.

16.7 Sidewalks

Sidewalks are used predominately in urban environments, but they are also used in rural areas that are adjacent to schools or other regions, such as shopping centers, where pedestrian traffic is high and sidewalks can help minimize pedestrian-related accidents. Because of their expense, use of sidewalks must be warranted before they are incorporated in a highway cross section. A shoulder can sometimes fulfill the role of a sidewalk if it is constructed and maintained in a way that encourages pedestrian use. Sidewalks, when installed, however, should always be separated from a shoulder, preferably by a curb (Fig. 16.7).

Typical width of sidewalks is 4 to 8 ft. For areas with a large amount of pedestrian traffic, a sidewalk should be at least 6 ft wide.

Sidewalks should be constructed of weather-resistant materials. They should be maintained free from debris and vegetation growth. When allowed to deteriorate because of poor maintenance, sidewalks will go unused because pedestrians will choose to walk on the travel lanes rather than the sidewalks. Not only does this defeat the intended function of the sidewalks (and justification for the additional expense) but it also greatly increases the risk of pedestrian-related accidents.

16.8 Traffic Barriers

Roadside barriers are used to protect vehicles and their occupants from impact with natural or man-made features located at the side of the roadway. It is important to note that a vehicle striking the traffic barrier may cause damage to both the vehicle/occupants and the barrier itself. The damage caused by hitting the barrier generally should be less severe than if the vehicle hit the feature located behind the barrier, thus traffic barriers represent a compromise between these two concerns. In addition to protecting vehicles, traffic barriers are also utilized to shield pedestrians, maintenance/construction crews, or cyclists from errant traffic. In its most basic form, a traffic barrier is designed to prevent a vehicle leaving the traveled way from striking a fixed object. The barrier must first contain an errant vehicle and then redirect it. Because of the variable nature of vehicle impacts and the destructive effects at high speeds, extensive full-scale crash tests should be conducted to ensure the adequacy of the traffic barrier to be used.

Barriers are available in a large variety of sizes and shapes. Choice of type of barrier to use depends on a variety of factors, including the environment in which the highway is located and the speed and volume of traffic.

Traffic barriers may be classified as longitudinal barriers, bridge railings and barriers, and crash cushions.

16.8.1 Longitudinal Barriers

Longitudinal barriers can be classified as roadside barriers and median barriers. Whereas a roadside barrier may be placed on either side of a roadway, a median barrier is placed between lanes of traffic traveling in opposite directions.

Barriers differ in the amount of deflection they undergo when struck by a vehicle. The principal categories of longitudinal barriers, based on the amount of deflection allowed, are flexible, semirigid, and rigid systems. Table 16.1 presents some basic forms of roadside barriers as given in the "Roadside Design Guide," American Association of State Highway and Transportation Officials (AASHTO), which discusses selection and implementation of traffic barrier systems. Also, the National Cooperative Highway Research Program (NCHRP) has developed updated guidelines for

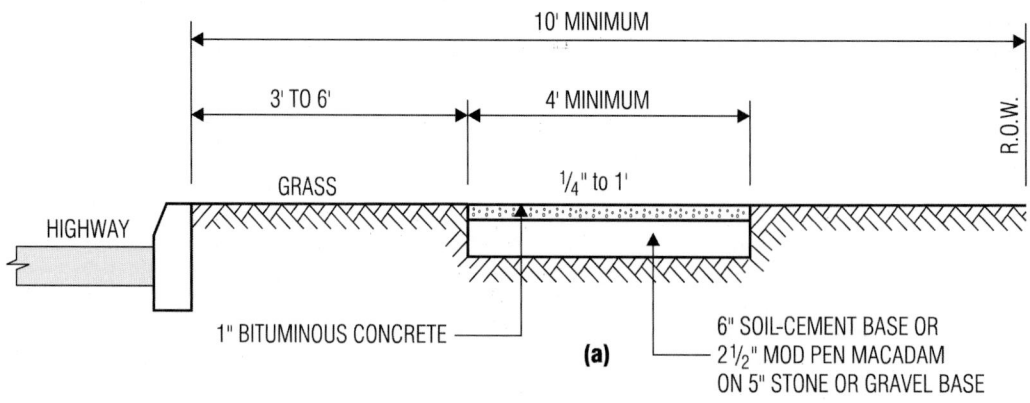

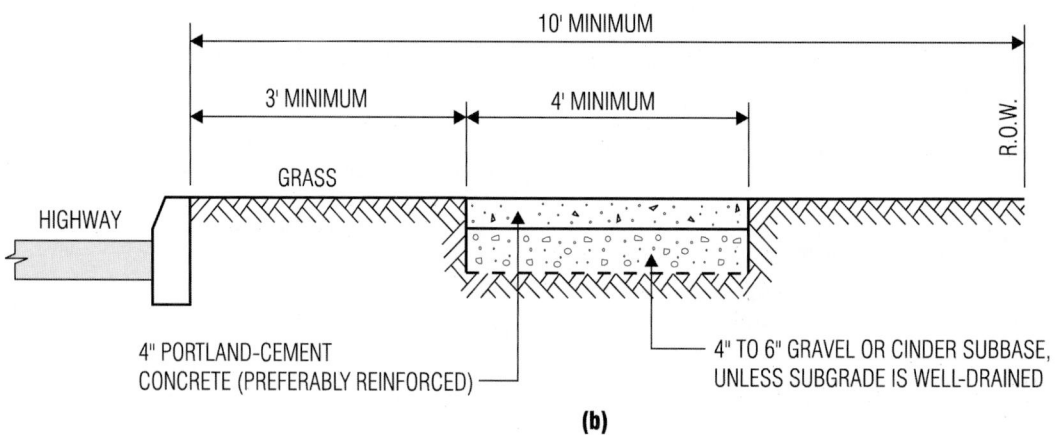

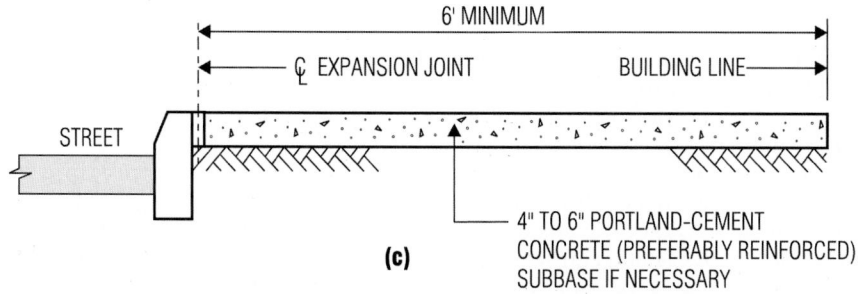

Fig. 16.7 Cross sections of sidewalks: (*a*) for rural or suburban areas; (*b*) for suburban or urban areas; and (*c*) for city streets in a business district.

Table 16.1 Standard Sections for Roadside Barriers

Barrier type	Description	Vehicle weight, lb	Maximum deflection, ft
	Flexible		
3-strand cable	¾-in-diameter steel cables 3 to 4 in apart, mounted on weak posts spaced 12 to 16 ft	1800–4500	11.5
W-beam weak post	Similar to cable guardrail except it uses a corrugated metal rail whose cross section resembles the letter w	1800–4000	7.3
Thrie beam* weak post	Same as the weak-post, W-beam except it uses a thrie beam rail	1800–4500	6.2
	Semirigid		
Box beam	Consists of a box rail mounted on steel posts (e.g., 6 in × 6-in box mounted on S3 × 5.7 steel posts on 6-ft centers)	1800–4000	4.8
Blocked-out W-beam (strong post)	Consists of wood or steel posts and a W-beam rail. Posts are set back or *blocked out* to minimize vehicle snagging	1800–4500	2.9
Blocked-out thrie beam* (strong post)	Same as blocked-out W-beam except with a thrie-beam rail. The added corrugation stiffens the system	1800–4000	3.3
Modified thrie beam*	Similar to a blocked-out W-beam with a triangular notch cut from the spacer block web. Minimizes vehicle rollover	Tested for 1800 lb, 20,000 lb (2.9-ft deflection), and 32,000 lb	
Self-restoring barrier (SERB) guardrail	Consists of tubular thrie beam rail supported from wood posts by steel pivot bars and cables. Classified experimental	1800–40,000	3.9
Steel-backed wood rail	Consists of wood rail backed with a steel plate and supported by timber posts	1800–4500	
	Rigid		
Concrete safety shape	Similar to a concrete median barrier but has a smaller section. Has sloped front face and vertical back face	1800–4500	
Stone masonry wall	A 2-ft-high barrier consisting of a reinforced concrete core faced and capped with stone and mortar	1800–4300	

* Cross section of a thrie beam looks like three vees (vvv).

the design of traffic barriers. These guidelines should be consulted when designing traffic barrier systems.

Flexible systems are designed for large deflections on impact. The primary objective is to contain rather than redirect an impacting vehicle. A flexible barrier generally consists of a weakly supported vertical post and a longitudinal member, such as a cable or a railing, designed to resist most of the tensile impact forces (Fig 16.8c). When subjected to an impact, the cable or beams separate from the post, offering little or no resistance in the area of impact.

Semirigid systems utilize the combined strength of the post and the longitudinal member (Fig. 16.8b). Posts at the point of impact help

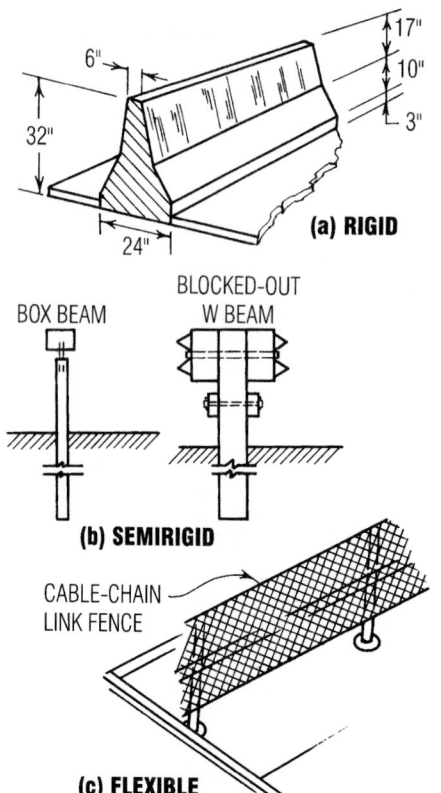

Fig. 16.8 Typical barriers for roadways.

While the main body of a longitudinal barrier is a safety device, an exposed end segment of barrier presents a significant hazard to oncoming traffic. Therefore, tapering or burying of the end section, or both, is a necessity. Another option is incorporation of some form of crash cushion or breakaway cable terminal.

16.8.2 Bridge Railings and Barriers

Bridge railings are installed on a highway bridge to prevent vehicular or pedestrian traffic from falling off the structure. They are an integral element of the bridge and thus must be designed to take into account the effects on the bridge superstructure of a vehicle impact.

"AASHTO Standard Specifications for Highway Bridges" presents guidelines for the design of highway bridge railings (See also Art. 17.3). The type of barriers provided on a bridge depends on the size of the structure, the volume of traffic passing over it, and the type of traffic, such as vehicular only or vehicular with pedestrians.

At each end of a bridge, a transition should be provided between the bridge railings and the approach railings. Since the two railings generally differ in stiffness, a sufficient length of transition railing should be provided to accomplish the change in stiffness smoothly so that snagging or pocketing of an impacting vehicle cannot occur.

16.8.3 Crash Cushions

Also known as an impact attenuator, a crash cushion protects against a head-on collision of an errant vehicle with a hazard by decelerating the vehicle to a safe stop or redirecting it from the hazard. The goal of crash cushions is to minimize the effects of accidents rather than to prevent them. In essence, a crash cushion limits the effects on a vehicle of a direct impact by absorbing the energy of the impact at a safe, controlled rate.

A crash cushion often is used at a critical location containing a fixed object. One such location is at a ramp gore (triangular area between an exit ramp and a roadway) where the highway and ramp railings join at a sharp angle. Another critical location is at obstacles, such as toll booths, that are installed directly in the flow of traffic.

Crash cushions usually are proprietary systems that are designed and tested by their manufacturers. Most of the systems are based on either

distribute impact forces to adjacent posts while posts outside the zone of impact help control the deflection of the railing. By limiting deflection, the outside posts assist in redirecting the impacting vehicle along the flow of traffic.

Rigid systems do not deflect appreciably when impacted by a vehicle. Instead, the impact forces are dissipated by raising and lowering the errant vehicle. Energy is also dissipated through deformation of the vehicle's sheet metal. One example of a rigid system is the concrete *Jersey barrier* used in construction zones (Fig. 16.8a). Rigid systems are primarily used in sections of highways where the angle of impact will be very shallow since little barrier deflection may occur. They also are used in front of bridge piers that are close to the flow of traffic, because, as a consequence of the limited deflection, they offer a high degree of protection to the hazard object.

absorption of kinetic energy or transfer of momentum to an inertial barrier.

To absorb kinetic energy, plastically deformable materials or hydraulic energy absorbers are placed in the front of the hazard. Dissipation of energy is also achieved through deformation of the front portion of an impacting vehicle. Rigid backup or support is provided to resist the impact force that causes deformation of the crash cushion. The goal of the system is primarily to protect the occupants of the impacting vehicle from injury and secondarily to preserve the integrity of the obstacle.

For transfer of momentum to an inertial barrier, an expendable mass of material is placed in the path of the vehicle to absorb impact. For example, containers filled with sand may be used as an inertial barrier (Fig. 16.9). If a vehicle were to impact such a crash cushion, the sand would absorb momentum from the vehicle. The momentum of the vehicle and the sand after impact would equal the momentum of the vehicle just before the impact occurred. While theoretically the vehicle would not come to a stop, the loss in momentum of

the vehicle would be sufficient to slow the vehicle to a speed of about 10 mi/h after impact with the last container. Design of crash cushions is typically accomplished through use of manufacturer-supplied design aids and charts.

16.9 Highway Medians

A median is a wide strip of a highway used to separate traffic traveling in opposite directions (Fig. 16.10). The width of the median in a two-lane highway is the distance between the inner edges of the lanes and includes the shoulders in the median. The width of the median for a highway with two or more lanes in each direction is the distance between the inner edges of the innermost lanes and includes the shoulders in the median.

In addition to separating the opposing flows of traffic, a median is designed to accomplish the following:

■ Offer a recovery area for errant vehicles

■ Provide an area for emergency stopping

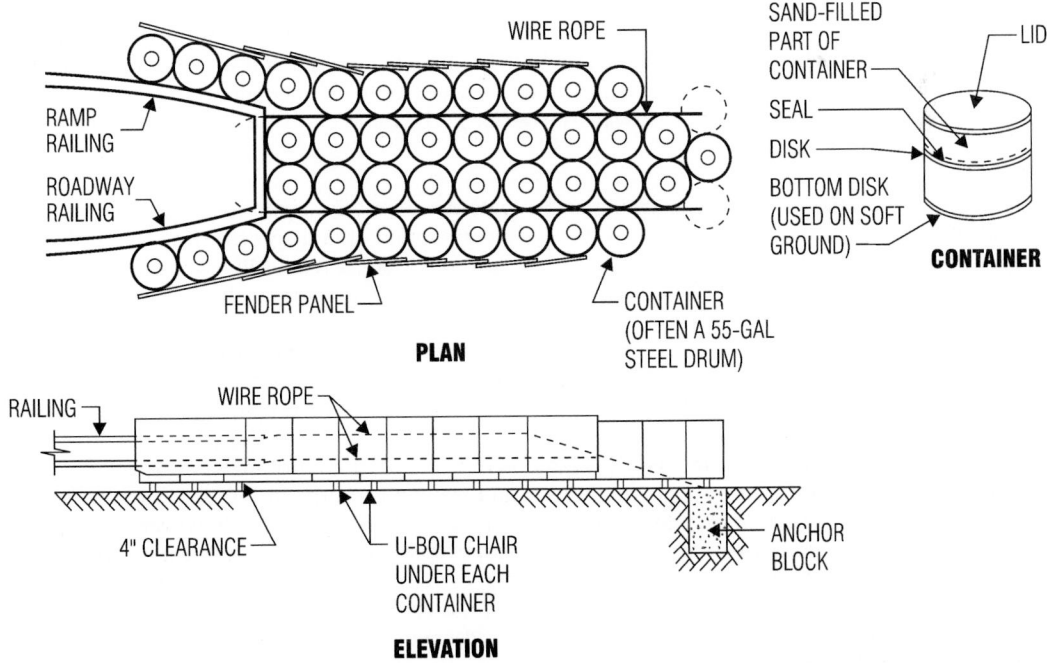

Fig. 16.9 Containers filled with sand used as an inertial barrier.

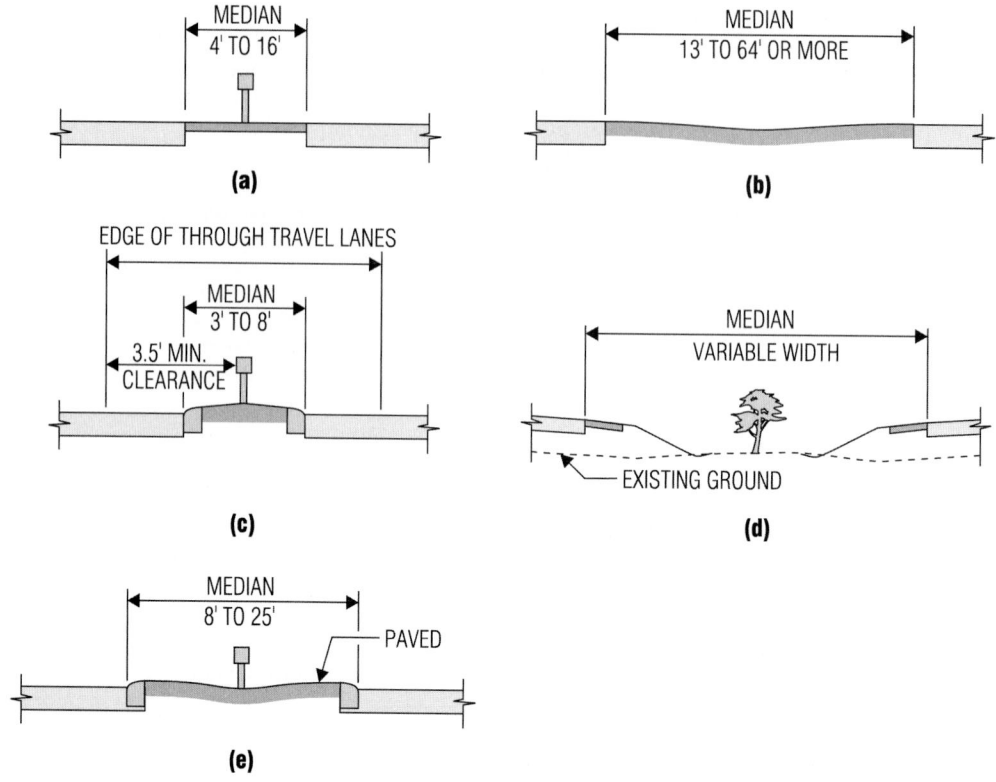

Fig. 16.10 Cross sections of a highway with medians (*a*) paved flush; (*b*) with swale and paved flush (maximum slope, 1:6) when median width exceeds 36 ft; otherwise, paved and incorporating a median barrier; (*c*) raised, curbed, and crowned, with 3-ft width when optional median barrier is installed; (*d*) natural ground between independent roadways; (*e*) raised, curbed, and depressed toward median barrier.

- Serve as a safe waiting area for left-turning and U-turning vehicles

- Decrease the amount of headlight glare

- Allow for expansion to future lanes

Medians may be flush, raised, or depressed. Figure 16.10 shows these basic forms in various configurations. Both flush and raised medians are generally used in urban environments whereas depressed medians are often used in high-speed freeways. Medians should be contrasted in color and texture with the roadways for maximum visibility.

Widths used for medians generally range from 4 to 80 ft. In general, the wider the median, the better. For one thing, median widths of 40 ft or more provide a distinct separation of noise and air

pressure from the opposing lanes. For another, incorporation of large green spaces with plantings can create an aesthetically pleasing appearance. Another consideration is that, depending on the width of the median, a traffic barrier may or may not be required. The larger the median width, the less the need for a barrier. Installation of a median barrier should be investigated for narrow medians (those less than 30 ft wide) and for medians that a vehicle out of control may be expected to cross and encounter traffic in the opposite direction. A balance should be struck, however, between the cost of increased median width and the overall cost of the project. In addition to economics, the psychology of drivers is also an important consideration in design of a median. Cross-over accidents, where a driver crosses the median and collides with opposing traffic, are typically severe accidents that result in severe vehicular damage, occupant injury or fatality. Federal, state, and local government agencies have different standards for placement of median traffic barrier systems. Consult with your applicable agency for their respective design standards/warrants for median barrier systems.

Design of medians should also take into account the possibility of their use to reduce glare from headlights in the opposing travel lanes. Visibility can be decreased by glare and shadows resulting from oncoming headlights. This condition may be especially acute when raised medians are used. It can be corrected through the incorporation of antiglare treatments in conjunction with a median barrier.

Regardless of the type of median chosen, drainage is an important design consideration. Flush and raised medians should be crowned or depressed for proper drainage. Depressed medians located on freeways should be designed to accommodate drainage and snow removal. For drainage, a ground slope of 6:1 is often used, but a slightly flatter slope may be adequate. Drainage inlets and culverts should be provided as necessary for runoff removal.

16.10 Highway Roadside

This is the area that adjoins a highway and can be used to accommodate drainage facilities and for recovery of errant vehicles (Fig. 16. 11). (Shoulders are not included in this area.) A roadside, however, can contain hazards to vehicles that leave the roadway, causing them to come in contact with obstacles or topography they cannot traverse.

A typical roadside that is not flat may contain one or more of the following elements: embankment or fill slope (negative slope), cut slope (positive slope), drainage channel or ditch (change in slope, usually negative to positive), clear zone, curb, sidewalk, berm, fence, traffic barrier, noise barrier, and highway light posts.

16.10.1 Clear Zone

Selection of width, slope, and other characteristics of roadside elements should provide for recovery of errant vehicles. To facilitate design of safe side slopes and related roadside elements, the American Association of State Highway and Transportation Officials (AASHTO) recommends establishment of a clear zone defined as that "area beyond the edge of the traveled way which is used for the recovery of errant vehicles." The traveled way does not include shoulders or auxiliary lanes.

The width to be used for a clear zone depends on traffic volume and speed, and embankment slopes. Rural local roads and collectors that carry low-speed traffic should have a minimum clear-zone width of 10 ft. For highways in an urban environment where space for clear zones is at a premium, a minimum clear-zone width of 1.5 ft should be maintained beyond the face of curbs.

16.10.2 Side Slopes

These provide stability for the roadway and give drivers of errant vehicles an opportunity to regain control. Composition to be used for side slopes depends on the geographic region and availability of materials. Rounding and blending of the slopes with the existing topography will enhance highway safety and aesthetics.

In Fig. 16.11, the hinge point is identified as the intersection of the extreme edge of the shoulder and the foreslope. From a safety standpoint, the

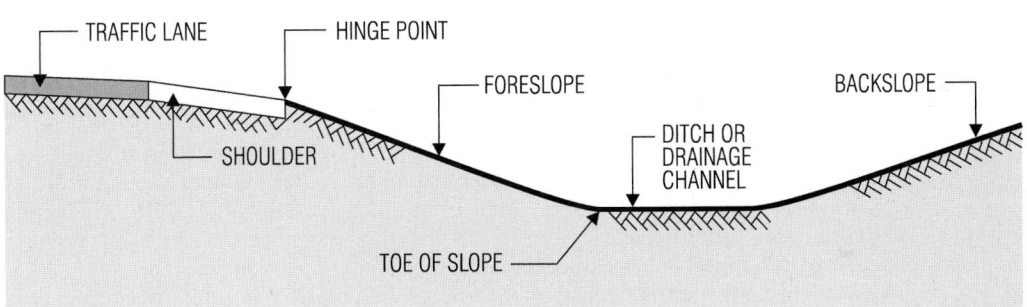

Fig. 16.11 Typical elements of a roadside.

hinge point is critical, since it is possible for drivers to lose control of their vehicles (and even become airborne) at this location. The foreslope and toe of slope also are critical because of potential safety hazards when vehicles attempt a recovery after leaving the roadway.

To help minimize these and other potential unsafe conditions, the hinge point and slopes are rounded, thus reducing the chance of an errant vehicle's becoming airborne. In addition, slopes should not be steeper than 3:1 and preferably not steeper than 4:1, especially for foreslopes, the region where vehicle recovery is likely to take place. When steeper slopes are demanded by specific site characteristics, a roadside barrier should be installed.

For backslopes, the slope should be 3:1 or flatter to facilitate operation of maintenance equipment, such as mowers. When site constraints mandate slopes steeper than 2:1, for example, in urban areas where real estate is at a premium, installation of retaining walls should be investigated.

16.10.3 Berms

These are used along rural highways on embankments or around islands to retain drainage in the shoulder and inhibit erosion of the side slope. A berm is a raised shelf that can be formed of plain earth and sodded or paved with road- or plant-mix bituminous material.

16.10.4 Fences

These are often installed along a highway to limit or control access to the highway right-of-way by pedestrians or vehicles. Fencing can also be used to prevent indiscriminate crossing of a median by vehicles, reduce headlight glare, and prevent animals from entering the highway. Important from a highway maintenance point of view, fences help to minimize right-of-way intrusions by adjacent property owners. Adjacent property owners typically construct garages, buildings, signs, store materials or install plantings on or near the right-of-way line. Fences also help to minimize routine intrusions by off-road vehicles (4-wheelers, snowmobiles, bicyclists) and prevent pedestrians from entering/crossing the highway. For these purposes, a chain-link fence 6 ft high is generally erected. In rural areas, however, a 4-ft-high farm fence is frequently used. In many instances, rural fencing is employed to prevent entry of livestock onto a highway. Fences also are installed on bridges to prevent vandals on the bridges from throwing objects down onto underpasses and causing accidents. When control of pedestrian access to a highway is the principal concern, a thick hedge may be planted to control access to the highway.

16.10.5 Noise Barriers

Incorporation of barriers to reduce the effects of noise on occupied areas adjacent to a highway, although often expensive, may be necessary. The noise generated by large volumes of traffic can severely impact residential and other properties where people live and work. Sources of highway-traffic noise include vehicle motors, vehicle exhaust, aerodynamic effects, and interaction of tires and roadway surface. For a major highway, design, beginning with the preliminary design stage, should take into account the anticipated noise levels and the type of noise barrier, if any, that will be required. Noise barriers are costly and many times do not reduce the noise concern to the degree that the adjacent effected parties believe that they will. The use of visual barrier, an aesthetically pleasing fence rather than a wall, may prove to be a better alternative from the effected parties point of view. Again, basic design parameters and community input should be thoroughly defined during the environmental and planning studies conducted prior to start of the actual highway design.

Noise barriers are sound-absorbing or sound-reflecting walls. They often are fabricated of concrete, wood, metal, or masonry. The type selected should be aesthetically pleasing and blend well with the surrounding topography. Local availability of materials or components and applicable standards often play a critical role in the selection of types of noise barriers.

Design and installation of noise barriers for a highway should conform with the general geometric design constraints of the highway. The barriers should be set as much as possible away from the highway and allow proper sight distance for drivers. When noise barriers are placed close to traffic, it may be necessary to erect protective barriers with the noise barriers.

As an alternative to employment of noise barriers, there are other ways to control the effects of noise on adjacent properties. One method is to depress the highway below the level of adjacent buildings. Another possibility is elevation of the highway on an embankment or bridge above the level of adjacent buildings. To further limit noise, shrubs and trees may be planted or ground covers placed between the highway and adjoining properties.

16.10.6 Roadside Drainage Channels

A drainage channel is often incorporated in a roadside to collect and convey surface water for drainage away from the roadbed. To perform this function, drainage channels should be sized for both design runoff and excessive storm water flows.

A drainage channel usually is a ditch formed by shaping the roadside ground surface (Fig. 16.11). From a hydraulic standpoint, the best drainage channel is the one with the steepest sides. Therefore, a balance between drainage needs and the need for flatter slopes must be achieved (Art. 16.10.2).

Drainage channels should be located to avoid creation of a hazard to errant vehicles. Maintenance crews should keep the channels free from debris, which can reduce the capacity of the channels. They should also ensure that the channels are not subjected to significant erosion, deposition of material, or other causes of channel deterioration.

16.11 Right-of-Way

This is the entire area needed for construction, drainage, and maintenance of a highway as well as for access to and exit from the highway. Achievement of many of the desirable design features discussed in Art. 16.10, such as flatter slopes and proper placement of drainage facilities, is facilitated by procurement of sufficient right-of-way. In addition, acquisition of large right-of-way allows future highway expansion to accommodate larger traffic volumes. As a minimum, however, the size of the right-of-way acquired for a highway should be at least that required for incorporation of all elements in the design cross section and the appropriate border areas.

For estimating right-of-way required for a typical ground-level freeway, for example, the cross section may be assumed to contain 12-ft lanes,

56-ft median, 50-ft outer roadsides, 30-ft frontage roads, and 15-ft borders. The American Association of State Highway and Transportation Officials (AASHTO) recommends a width of right-of-way of about 225 ft for such a freeway with no frontage roads and 300 to 350 ft with one-way frontage roads on both sides of the through pavement. For a ground-level freeway with restricted cross section, AASHTO recommends a width of 100 to 150 ft with no frontage road and 100 to 200 ft with a two-way frontage road on one side. Different sizes of right-of-way are recommended for other types of highways (AASHTO "A Policy on Geometric Design of Highways and Streets").

16.12 Superelevation

It is desirable to construct one edge of a roadway higher than the other along curves of highways to counteract centrifugal forces on passengers and vehicles, for the comfort of passengers and to prevent vehicles from overturning or sliding off the road if the centrifugal forces are not counteracted by friction between the roadway and tires. Because of the possibility of vehicle sliding when the curved road is covered with rain, snow, or ice, however, there are limitations on the amount of superelevation that can be used.

The maximum superelevation rate to use depends on local climate and whether the highway is classified as rural or urban. Table 16.2 presents typical limits for various design speeds, minimum radii, superelevation rates e, and transition spiral lengths L_s. The last is the distance over which the normal crown cross section changes to a fully banked section as the roadway alignment changes from tangent to start of a circular curve.

For the safety and comfort of drivers, provision usually is made for gradual change from a tangent to the start of a circular curve. One method for doing this is to insert a spiral curve between those sections of the roadway (Art. 16.13.3). A spiral provides a comfortable path for drivers since the radius of curvature of the spiral gradually decreases to that of the circular curve while the superelevation gradually increases from zero to full superelevation of the circular curve. A similar transition is inserted at the end of the circular curve. (An alternative is to utilize compound curves that closely approximate a spiral.) Over the length of the transition, the centerline of each

Table 16.2 Superelevation e, in/ft, of Pavement Width and Spiral Length, L_s, ft, for Horizontal Curves of Highway*

Degree D_c of curve	Radius of curve, ft	30 e	30 L_s 2	30 L_s 4	40 e	40 L_s 2	40 L_s 4	50 e	50 L_s 2	50 L_s 4	60 e	60 L_s 2	60 L_s 4	65 e	65 L_s 2	65 L_s 4	70 e	70 L_s 2	70 L_s 4	75 e	75 L_s 2	75 L_s 4
0° 15′	22,918	NC	0	0	NC	0	0	NC	0	0	NC	0	0	NC	0	0	NC	0	0	NC	0	0
0° 30′	11,459	NC	0	0	NC	0	0	NC	0	0	RC	175	175	RC	190	190	RC	200	200	0.022	220	220
0° 45′	7,639	NC	0	0	NC	0	0	RC	150	150	0.022	175	175	0.025	190	190	0.029	200	200	0.032	220	220
1° 00′	5,730	NC	0	0	RC	125	125	0.021	150	150	0.029	175	175	0.053	190	190	0.038	200	200	0.043	220	220
1° 30′	3,820	RC	100	100	0.021	125	125	0.030	150	150	0.040	175	175	0.046	190	200	0.053	200	240	0.080	220	290
2° 00′	2,865	RC	100	100	0.027	125	125	0.038	150	150	0.051	175	210	0.057	190	250	0.065	200	290	0.072	230	340
2° 30′	2,292	0.021	100	100	0.033	125	125	0.046	150	170	0.060	175	240	0.066	190	290	0.073	220	330	0.078	250	370
3° 00′	1,910	0.025	100	100	0.038	125	125	0.053	150	190	0.067	180	270	0.073	210	320	0.073	230	350	0.080	250	380
3° 30′	1,637	0.028	100	100	0.043	125	140	0.058	150	210	0.073	200	300	0.077	220	330	0.080	240	380	0.080	250	380
4° 00′	1,432	0.052	100	100	0.047	125	150	0.063	150	230	0.077	210	310	0.079	230	340	0.080	240	360	D_c max = 3.0°		
5° 00′	1,146	0.038	100	100	0.055	125	170	0.071	170	260	0.080	220	320	0.080	230	350	D_c max = 3.5°					
6° 00′	955	0.043	100	120	0.061	130	190	0.077	180	280	0.080	220	320	D_c max = 4.5°								
7° 00′	819	0.048	100	130	0.067	140	210	0.079	190	280	D_c max = 5.0°											
8° 00′	716	0.052	100	140	0.071	150	220	0.080	190	290												
9° 00′	637	0.056	100	150	0.075	160	240	D_c max = 7.5°														
10° 00′	573	0.059	110	160	0.077	160	240															
11° 00′	521	0.063	110	170	0.079	170	250															
12° 00′	477	0.066	120	180	0.080	170	230															
13° 00′	441	0.068	120	180	0.080	170	250															
14° 00′	409	0.070	130	190	D_c max = 12.5°																	
16° 00′	358	0.074	130	200																		
18° 00′	318	0.077	140	210																		
20° 00′	286	0.079	140	210																		
22° 00′	260	0.080	140	220																		
		0.080	140	220																		
		D_c max = 23.0°																				

* Adapted from "Highway Design Manual", New York State Department of Transportation.

roadway is maintained at profile grade while the outer edge of the roadway is raised and the inner edge is lowered to produce the required superelevation. As indicated in Fig. 16.12, typically the outer edge is raised first until the outer half of the cross section is level with the crown (point B). Then, the outer edge is raised further until the cross section is straight (point C). From there on, the entire cross section is rotated until the full superelevation is attained (point E). See also Art. 16.13.4.

Superelevated roadway cross sections are typically employed on curves of rural highways and urban freeways. Superelevation is rarely used on local streets in residential, commercial, or industrial areas.

Highway Alignments

Geometric design of a highway is concerned with horizontal and vertical alignment as well as the cross-sectional elements discussed in Arts. 16.2 to 16.12. Horizontal alignment of a highway defines its location and orientation in plan view. Vertical alignment of a highway deals with its shape in profile. For a roadway with contiguous travel lanes, alignment can be conveniently represented by the centerline of the roadway.

16.13 Horizontal Alignment

This comprises one or more of the following geometric elements: tangents (straight sections),

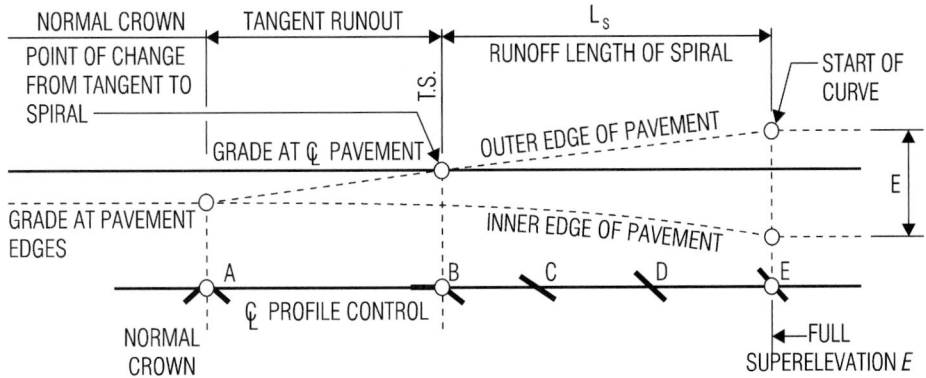

Fig. 16.12 Superelevation variations along a spiral transition curve.

circular curves (Art 16.13.2), and transition spirals (Arts. 16.12 and 16.13.3).

16.13.1 Stationing

Distance along a horizontal alignment is measured in terms of stations. A full station is defined as 100 ft and a half station as 50 ft. Station 100 + 50 is 150 ft from the start of the alignment, Station 0 + 00. A point 1492.27 ft from 0 + 00 is denoted as 14 + 92.27, indicating a location 14 stations (1400 ft) plus 92.27 ft from the starting point of the alignment. This distance is measured horizontally along the centerline of the roadway, whether it is a tangent, curve, or a combination of these.

16.13.2 Simple Curves

A simple horizontal curve consists of a part of a circle tangent to two straight sections on the horizontal alignment. The radius of a curve preferably should be large enough that drivers do not feel compelled to slow their vehicles. Such a radius, however, is not always feasible, inasmuch as the alignment should blend harmoniously with the existing topography as much as possible and balance other design considerations, such as overall project economy, sight distance, and side friction. Superelevation, usually employed on curves with sharp curvature, also should be taken into account (Art. 16.12).

A curve begins at a point designated *point of curvature* or *PC*. There, the curve is tangent to the straight section of the alignment, which is called a *tangent* (Fig. 16.13). The curve ends at the *point of* tangency *PT*, where the curve is tangent to another straight section of the alignment. The angle Δ formed at *PI*, *the point of intersection* of the two tangents, is called the *interior angle* or *intersection angle*.

The curvature of a horizontal alignment can be defined by the radius R of the curve or the degree of curve D. **One degree of curve** is the central angle that subtends a 100-ft arc (approximately a 100-ft chord). The degree of a curve is given by

$$D = \frac{5729.8}{R} \tag{16.1}$$

Values for the minimum design radius allowable for normal crown sections are presented in Table 16.3.

The length of the tangent T (distance from *PC* to *PI* or *PI* to *PT*) can be computed from

$$T = R \tan \frac{\Delta}{2} \tag{16.2}$$

The external distance E measured from *PI* to the curve on a radial line is given by

$$E = R\left(\sec \frac{\Delta}{2} - 1\right) \tag{16.3}$$

The middle ordinate distance M extends from the midpoint B of the chord to the midpoint A of the curve.

$$M = R\left(1 - \cos \frac{\Delta}{2}\right) \tag{16.4}$$

The length of the chord C from *PC* to *PT* is given by

$$C = 2R \sin \frac{\Delta}{2} = 2T \cos \frac{\Delta}{2} \tag{16.5}$$

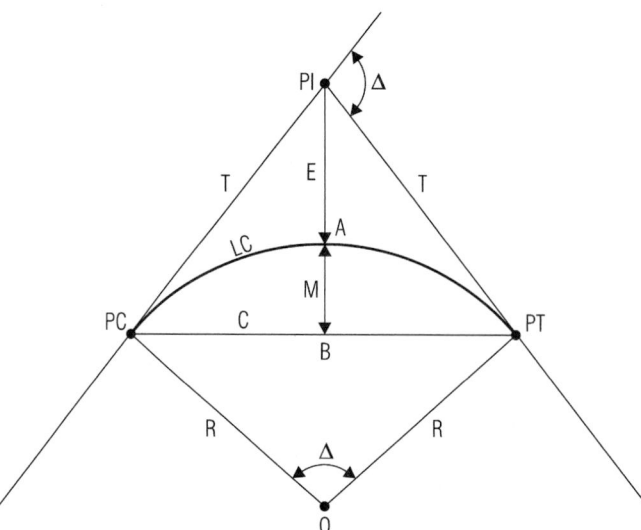

Fig. 16.13 Circular curve starting at point *PC* on one tangent and ending at *PT* on a second tangent that intersects the first one at *PI*. Curve radius is *R* and chord distance between *PC* and *PT* is *C*. Length of arc is *LC*. Tangent distance is *T*.

The length *L* of the curve can be computed from

$$L = \frac{\Delta \pi R}{180} = \frac{100\Delta}{D} \qquad (16.6)$$

where Δ = intersection angle, degrees.

16.13.3 Transition (Spiral) Curves

On starting around a horizontal circular curve, a vehicle and its contents are immediately subjected to centrifugal forces. The faster the vehicle enters the circle and the sharper the curvature, the greater the influence on vehicles and drivers of the change from tangent to curve. For example, depending on the friction between tires and road, vehicles may slide sideways, especially if the road is slick. Furthermore, drivers are uncomfortable because of the difficulty of achieving a position of equilibrium. A similar condition arises when a vehicle leaves a circular curve to enter a straight section of highway. To remedy these

Table 16.3 Maximum Curvature for Normal Crown Section*

Design speed, mi/h	Average running speed, mi/h	Maximum degree of curve	Minimum curve radius, ft
20	20	3° 23′	1,700
30	28	1° 43′	3,340
40	36	1° 02′	5,550
50	44	0° 41′	8,320
55	48	0° 35′	9,930
60	52	0° 29′	11,690
65	55	0° 26′	13,140
70	58	0° 23′	14,690

*Adapted from "A Policy on Geometric Design of Highways and Streets," American Association of State Highway and Transportation Officials.

conditions, especially where high-speed traffic must round sharp curves, a transition curve with a constantly changing radius should be inserted between the circular curve and the tangent. The radius of the transition curve should vary gradually from infinity at the tangent to that of the circular curve. Along the transition, superelevation should be applied gradually from zero to its full value at the circular curve.

An **Euler spiral** (also known as a **clothoid**) is typically used as the transition curve. The gradual change in radius results in a corresponding gradual development of centrifugal forces, thereby reducing the aforementioned adverse effects. In general, transition curves are used between tangents and sharp curves and between circular curves of substantially different radii. Transition curves also improve driving safety by making it easier for vehicles to stay in their own lanes on entering or leaving curves. When transition curves are not provided, drivers tend to create their own transition curves by moving laterally within their travel lane and sometimes the adjoining lane, a hazardous maneuver. In addition, transition curves provide a more aesthetically pleasing alignment, giving the highway a smooth appearance without noticeable breaks at the beginning and end of circular curves.

The minimum length L, ft, of a spiral may be computed from

$$L = \frac{3.15V^3}{RC} \qquad (16.7)$$

where V = vehicle velocity, mi/h

R = radius, ft, of the circular curve to which the spiral is joined

C = rate of increase of radial acceleration

An empirical value indicative of the comfort and safety involved, C values often used for highways range from 1 to 3. (For railroads, C is often taken as unity 1.) Another, more practical, method for calculating the minimum length of spiral required for use with circular curves is to base it on the required length for superelevation runoff (Art. 16.13.4).

16.13.4 Superelevation Runoff L_s

This is the length of highway required to alter the cross section of a roadway from normal crown to fully superelevated, or vice versa (Fig. 16.12).

Table 16.2 lists values of L_s for two- and four-lane pavements and for various design velocities. The table is based on the assumption that the centerline of each roadway is maintained at profile grade while the outer edge is raised and the inner edge lowered to create the required superelevation. The superelevation runoff is effected uniformly to provide both comfort and safety. AASHTO recommendations for superelevation runoff may differ somewhat from those given in Table 16.2. Also, high-type alignments may require longer runoffs, and while the runoff for wide pavements is greater than that for two-lane pavements, there are no generally accepted length ratios.

A certain amount of prudence should be exercised in design in the use of any of the preceding criteria. For example, if a highway is located in a cut with a relatively flat profile, lowering of the inner edge may result in a sag from which surface water cannot be properly drained. To prevent this condition, superelevation should be achieved through raising of the outer edge. This will require elevating this edge twice the distance needed when the inner edge is lowered. As another example, if superelevation is employed on a divided highway, an undesirable condition may arise if superelevation is applied by rotating about its centerline the pavement on each side of the median. The two sides of the median will end up at substantially different elevations. A better alternative is to rotate each pavement about the roadway edge adjoining the median.

16.13.5 Passing Sight Distance

On two-lane highways, drivers should be provided at intervals safe opportunities to pass slow-moving vehicles. Failure to do so increases the risk of head-on collisions and tends to decrease highway traffic capacity. To permit safe passing, a driver must be able to see far enough ahead to be certain that there is no danger of collision with an oncoming vehicle or an obstruction in the highway. Table 16.4 lists minimum sight distances that can serve as a guide in designing highway alignment.

16.14 Vertical Alignment

A vertical alignment defines the geometry of a highway in elevation, or profile. A vertical alignment can be represented by the highway centerline

Table 16.4 Minimum Passing Sight Distances for Design of Two-Lane Highways

Design speed, mi/h	Assumed passed-vehicle speed, mi/h*	Minimum passing sight distance, ft
30	26	1100
40	34	1500
50	41	1800
60	47	2100
65	50	2300
70	54	2500
75	56	2600
85	59	2700

* Assumed speed of passing vehicle 10 mi/h faster than that of the passed vehicle.

along a single tangent at a given grade, a vertical curve, or a combination of these.

16.14.1 Clearance for Bridges

When a highway is carried on a bridge over an obstruction, a minimum clearance should be maintained between the underside of the bridge superstructure and the feature crossed. AASHTO's Standard Specifications for Highway Bridges specifies an absolute minimum clearance of 14 ft and design clearance of 16 ft.

16.14.2 Vertical Curves

These are used as a transition where the vertical alignment changes grade, or slope. Vertical curves are designed to blend as best as possible with the existing topography, consideration being given to the specified design speed, economy, and safety. The tangents to a parabolic curve, known as grades, can affect traffic in many ways; for example, they can influence the speed of large tractor trailers and stopping sight distance.

Although a circular curve can be used for a vertical curve, common practice is to employ a para-bolic curve. It is linked to a corresponding horizontal alignment by common stationing. Figure 16.14 shows a typical vertical curve and its constituent elements.

A curve like the one shown in Fig. 16.14 is known as a crest vertical curve; that is, the curve crests like a hill. If the curve is concave, it is called a sag vertical curve; that is, the curve sags like a valley. As indicated in Fig. 16.14, the transition starts on a tangent at *PVC*, point of vertical curvature, and terminates on a second tangent at *PVT*, point of vertical tangency. The tangents, if extended, would meet at *PVI*.

The basic properties of a parabolic vertical curve are derived from an equation of the form $y = ax^2$. The rate of grade change r, percent per station of curve length, is

$$r = \frac{g_2 - g_1}{L} \qquad (16.8)$$

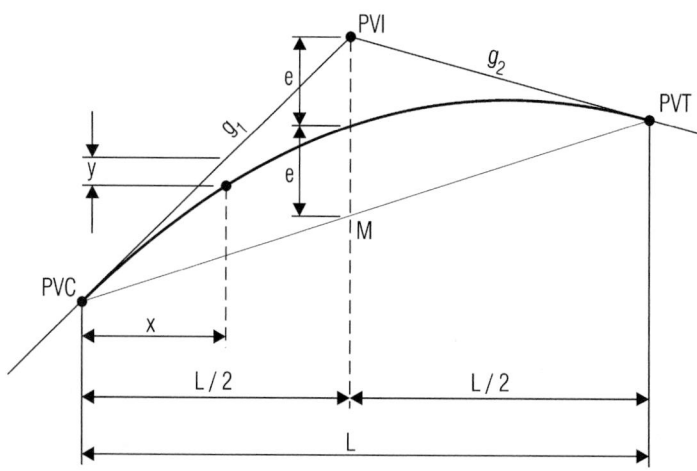

Fig. 16.14 Parabolic vertical curve starting at point *PVC* on one tangent and terminating at *PT* on a second tangent that intersects the first one at *PVI* at a distance *e* above the curve.

where g_1 = grade, percent, at PVC, shown positive (upward slope) in Fig. 16.14

g_2 = grade, percent, at PVT, shown negative (downward slope)

L = length, stations, of vertical curve

If a curve has a length of 700 ft, $L = 7$. If grade g_1 at PVC were 2.25% and grade g_2 at PVT were −1.25%, the rate of change would be $r = (−1.25 − 2.25)/7 = −0.50\%$ per station.

A key point on a vertical curve is the **turning point**, where the minimum or maximum elevation on a vertical curve occurs. The station at this point may be computed from

$$x_{TP} = \frac{-g_1}{r} \qquad (16.9)$$

The middle ordinate distance e, the vertical distance from the PVI to the vertical curve, is given by

$$e = \frac{(g_1 - g_2)L}{8} \qquad (16.10)$$

For the layout of a vertical curve in the field, it is necessary to know the elevations at points along the curve. From the equation of a parabola, the elevation H_x, ft, of the curve at a distance x, stations, from the PVC may be computed from

$$H_x = H_1 + g_1 x + \frac{rx^2}{2} \qquad (16.11)$$

where H_1 = elevation of the PVC. The last term of the equation $rx^2/2$ is the vertical offset of the curve from a point on the tangent to the curve at a distance x, stations, from PVC.

16.14.3 Stopping Sight Distance

This is the length of roadway needed between a vehicle and an arbitrary object (at some point down the road) to permit a driver to stop a vehicle safely before reaching the obstruction. This is not to be confused with passing sight distance, which AASHTO defines as the "length of roadway ahead visible to the driver." (Art. 16.13.5). Figure 16.15 shows the parameters governing stopping sight distance on a crest vertical curve.

The minimum stopping sight distance is computed for a height of eye (driver eye height) of 3.50 ft and a height of object (obstruction in roadway) of 6 in. The stopping distance on a level roadway comprises the distance over which a vehicle moves during the brake reaction time, the time it takes a driver to apply the brakes on sighting an obstruction, and the distance over which the vehicle travels before coming to a complete stop (braking distance). Table 16.5 lists approximate stopping sight distances on a level roadway for various design speeds and wet pavements. If the vehicle is traveling uphill, the

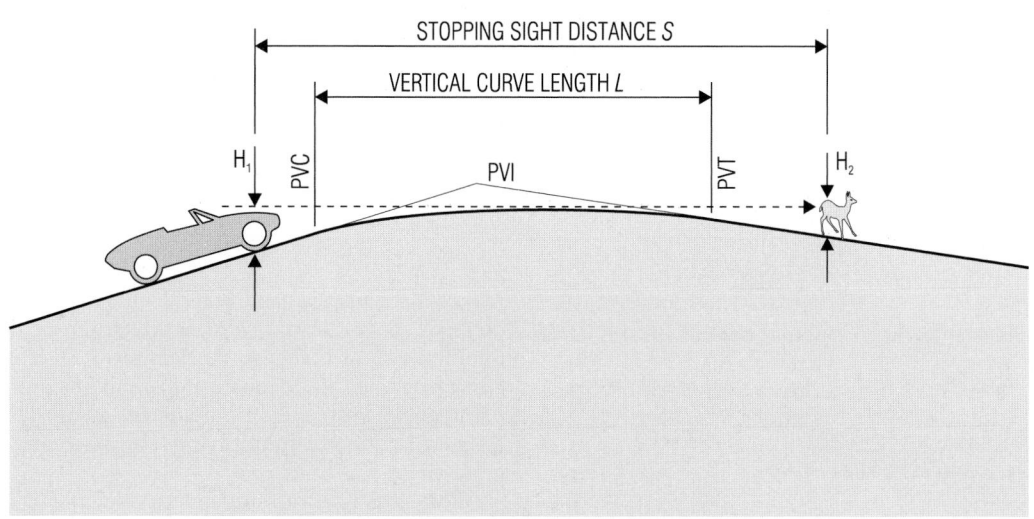

Fig. 16.15 Stopping sight distance on a crest vertical curve.

Table 16.5 Design Controls for Vertical Curves Based on Stopping Sight Distance*

Design speed, mi/h	Average speed for condition, mi/h	Coefficient of friction f	Stopping sight distance (rounded for design), ft	Rate of vertical curvature K, ft per percent of A			
				For crest curves		For sag curves	
				Computed	Rounded for design	Computed	Rounded for design
20	20–20	0.40	125–125	8.6–8.6	10–10	14.7–14.7	20–20
25	24–25	0.38	150–150	14.4–16.1	20–20	21.7–23.5	30–30
30	28–30	0.35	200–200	23.7–28.8	30–30	30.8–35.3	40–40
35	32–35	0.34	225–250	35.7–46.4	40–50	40.8–48.6	50–50
40	36–40	0.32	275–325	53.6–73.9	60–80	53.4–65.6	60–70
45	40–45	0.31	325–400	76.4–110.2	80–120	67.0–84.2	70–90
50	44–50	0.30	400–475	106.6–160.0	110–160	82.5–105.6	90–110
55	48–55	0.30	450–550	140.4–217.6	150–220	97.6–126.7	100–130
60	52–60	0.29	525–650	189.2–302.2	190–310	116.7–153.4	120–160
65	55–65	0.29	550–725	227.1–394.3	230–400	129.9–178.6	130–180
70	58–70	0.28	625–850	282.8–530.9	290–540	147.7–211.3	150–220

* Adapted from "A Policy on Geometric Design of Highways and Streets," American Association of State Highway and Transportation Officials.

braking distance is less, because gravity aids in slowing the vehicle. For downhill movement, braking distance is more.

A general rule of thumb is that the longer a vertical curve, the larger the safe stopping sight distance may be. Long curves, however, may be too costly to construct. Therefore, a balance should be reached between economy and safety without compromising safety.

For crest vertical curves AASHTO defines the minimum length L_{min}, ft, of crest vertical curves based on a required sight distance S, ft, as that given by Eqs. (16.12) to (16.15).

$$L_{min} = \frac{AS^2}{100(\sqrt{2H_1} + \sqrt{2H_2})^2} \quad S < L \quad (16.12)$$

When eye height is 3.5 ft and object height is 0.5 ft,

$$L_{min} = \frac{AS^2}{1329} \quad S < L \quad (16.13)$$

Also, for crest vertical curves,

$$L_{min} = 25 - \frac{200(\sqrt{H_1} + \sqrt{H_2})^2}{AS^2} \quad S > L \quad (16.14)$$

When eye height is 3.5 ft and object height 0.5 ft,

$$L_{min} = 25 - \frac{1329}{AS^2} \quad S > L \quad (16.15)$$

where A = algebraic difference in grades, percent, of the tangents to the vertical curve

H_1 = eye height, ft, above the pavement

H_2 = object height, ft, above the pavement

Design controls for vertical curves can be established in terms of the rate of vertical curvature K defined by

$$K = \frac{L}{A} \quad (16.16)$$

where L = length, ft, of vertical curve and A is defined above. K is useful in determining the minimum sight distance, the length of a vertical curve from the PVC to the turning point (maximum point on a crest and minimum on a sag). This distance is found by multiplying K by the approach gradient.

Table 16.5 lists recommended values of K for various design velocities and stopping sight distances for crest and sag vertical curves.

Highway Drainage

Proper drainage is a very important consideration in design of a highway. Inadequate drainage facilities can lead to premature deterioration of the highway and the development of adverse safety conditions such as hydroplaning. It is common, therefore, for a sizable portion of highway construction budgets to be devoted to drainage facilities.

In essence, the general function of a highway drainage system is to remove rainwater from the road and water from the highway right-of-way. The drainage system should provide for the drainage conditions described in Arts. 16.16 and 16.17.

16.15 Storm Frequency and Runoff

Storm frequency refers to the chance that a given intensity of rainfall will occur within a specific span of years. It is determined from historical data that indicate that a particular intensity of rainfall can be expected once in N years. A drainage system designed for such an intensity is intended to be capable of withstanding an N-year storm, runoff, or flood. A 25-year storm, for example, represents a 1 in 25 probability that the drainage system will have to accommodate such an intensity. This does not mean that every 25 years a certain storm of this magnitude will occur. It is possible that such a storm will not occur at all during any 25-year period. It is also possible, however, that two or more such storms will take place in a single year.

For highways, cross drains (small culverts) passed under major highways to carry the flow from defined watercourses are typically designed to accommodate a 25-year storm. Larger culverts and bridges on major highways are designed with capacity for 100-year storms. For nonmajor highways, the storm used for design can range from a 10- to 50-year storm, depending on the highway size and traffic volume expected.

Runoff Determination ▪ The amount of runoff to be used for design of surface drainage can be determined through physical stream-flow measurements or through the use of empirical formulas. A common approach is to utilize the rational method described in Art. 21.39 (also known as the Lloyd-Davies method in the United Kingdom). While this approach gives reasonable answers in most urban areas, care must be taken when applying the rational method in rural areas. Runoff for rural and large watershed areas is much more difficult to estimate accurately than the runoff in urban environments. Typically, for determination of runoff, a large watershed is divided into several smaller watershed areas, from which runoff flows to various inlets or waterways. In general, conservative design values of runoff can be determined for drainage areas of 100 acres or less. Some designers, however, have used 200-acre and even 500-acre maximum values.

16.16 Surface Drainage

Provision must be made for removal of water, from rain or melting snow, or both, that falls directly on a road or comes from the adjacent terrain. The road should be adequately sloped to drain the water away from the travel lanes and shoulders and then directed to drainage channels in the system, such as natural earth swales, concrete gutters, and ditches, for discharge to an adjacent body of water. The channels should be located and shaped to minimize the potential for traffic hazards and accommodate the anticipated storm-water flows. Drainage inlets should be provided as needed to prevent ponding and limit the spread of water into traffic lanes.

16.16.1 Surface Drainage Methods

For rural highways on embankments, runoff from the roadway should be allowed to flow evenly over the side slopes and then spread over the adjacent terrain. This method, however, can sometimes adversely impact surrounding land, such as farms. In such instances the drainage should be collected, for example, in longitudinal ditches and then conveyed to a nearby watercourse.

When a highway is located in a cut, runoff may be collected in shallow side ditches. These typically have a trapezoidal, triangular, or rounded cross section and should be deep enough to drain the pavement subbase and convey the design-storm flow to a discharge point. Care should be taken to design the ditches so that the toe of adjoining sloping fill does not suffer excessive erosion. For larger water flows than the capacity of a shallow ditch, paved gutters or drainpipes with larger capacities will have to be used.

In urban environments and built-up areas, use of roadside drainage channels may be severely limited by surrounding land uses. In most instances, the cost of acquiring the necessary right-of-way to implement such drainage facilities is prohibitive. For highways on embankments, a curb or an earth berm may be constructed along the outer edge of the roadway to intercept runoff and divert it to inlets placed at regular intervals. The inlets, in turn, should be connected to storm sewers that convey the water to points of disposal. In an urban area, it may be necessary to construct storm sewers of considerable length to reach the nearest body of water for discharge of the runoff.

16.16.2 Inlets

These are parts of a drainage system that receive runoff at grade and permit the water to flow downward into underground storm drains. Inlets should be capable of passing design floods without clogging with debris. The entrance to inlets should be protected with a grating set flush with the surface of gutters or medians, so as not to be a hazard to vehicles. There are several types of inlets.

A **drop inlet** is a box-type structure that is located in pipe segments of a storm-water collection system and into which storm water enters from the top. Most municipal agencies maintain design and construction standards for a wide variety of inlets, manholes, and other similar structures, but some large structures may require site-specific design.

A **curb inlet** consists of a vertical opening in a curb through which gutter flow passes. A **gutter inlet** is a horizontal opening in the gutter that is protected by a single grate or multiple grates through which the gutter flow passes. A **combination inlet** consists of both gutter and curb inlets with the gutter inlet placed in front of the curb inlet.

Inlet spacing depends on the quantity of water to be intercepted, shape of ditch or gutter conveying the water, and hydraulic capacity of the inlet.

16.16.3 Storm Sewers

These are underground pipes that receive the runoff from a roadside inlet for conveyance and discharge into a body of water away from the road. Storm sewers are often sized for anticipated runoff

and for pipe capacity determined from the Manning formula (Art. 21.9).

In general, changes in sewer direction are made at inlets, catch basins, or manholes.

The manholes should provide maintenance access to sewers at about every 500 ft.

A storm sewer system for a new highway should be connected to an existing drainage system, such as a stream or existing storm sewer system. If a storm sewer is to connect to a stream, the downstream conditions should be investigated to ensure that the waterway is adequate and that the new system will not have an adverse environmental impact. If the environmental impact is not acceptable, it will be necessary to study possible improvements to downstream outlets to accommodate the additional flow or to make the drainage scheme acceptable to local officials or adjacent property owners in some other fashion.

16.16.4 Open Channels

As indicated in Art. 16.16.1, side ditches may be used to collect runoff from a highway located in a cut. The ditches may be trapezoidal or V-shaped. The trapezoidal ditch has greater capacity for a given depth. Most roadway cross sections, however, include some form of V-shaped channel as part of their cross-sectional geometry. In most instances, it is not economical to vary the size of these channels. As a result, this type of channel generally has capacity to spare, since a normal depth must be maintained to drain the pavement subbase courses.

When steep grades are present, the possibility of ditch erosion becomes a serious consideration. Erosion can be limited by lining the channel with sod, stone, bituminous or concrete paving, or by providing small check dams at intervals that depend on velocity, type of soil, and depth of flows.

Linings for roadside channels are typically classified as either rigid or flexible. Paved and concrete linings are examples of rigid linings. Rock (riprap) and grass linings are examples of flexible linings. While rigid linings are better at limiting erosion, they often permit higher water velocities since they are smoother than flexible linings.

Roadside channels are often sized for anticipated runoff and for open-channel flow computed from the Manning equation (Art. 21.24). This equation includes a roughness coefficient n that may range from as low as 0.02 for concrete to 0.10 for thick grass. Flow in open channels is discussed in Arts. 21.23 to 21.33, which deal with hydraulic jump and normal, subcritical, and supercritical flow. Flow down gentle slopes is likely to be subcritical whereas flow down steep slopes may be supercritical. When the water depth is greater than the critical depth, subcritical flow occurs. Conversely, when the depth of water is less than the critical depth, supercritical flow occurs. The abrupt transition from subcritical to supercritical flow takes the form of a hydraulic jump.

Open channels should be designed to avoid supercritical flow. The reason for this is that water moving through a channel at high speeds can generate waves that travel downstream and cause water to overtop the sides of the channel and scour the downstream outlet. To limit the effects of scour at the outlet, energy dissipators may be incorporated in the channel. An energy dissipator may be a drop structure that alters the slope of the channel from steep to gentle. Alternatively, roughness elements, such as blocks and sills, can be placed in the channel to increase resistance to flow and decrease the probability of hydraulic jump's occurring.

16.16.5 Culverts

A culvert is a closed conduit for passage of runoff from one open channel to another. One example is a corrugated metal pipe under a roadway. Figure 16.16 shows various types of culvert cross sections and indicates material types used in highway design.

For small culverts, stock sizes of corrugated metal pipe may be used. For larger flows, however, a concrete box or multiple pipes may be needed. If the culvert foundation is not susceptible to erosion, a bridge may be constructed over the waterway (bridge culvert).

The section of a culvert passing under a highway should be capable of withstanding the loads induced by traffic passing over the culvert. Since corrugated metal pipes are flexible, they are assisted by surrounding soil in carrying gravity loads. Reinforced concrete culverts, however, have to support gravity loads without such assistance.

Empirical methods often are used for selecting and specifying culverts. With the use of data from previous experience, designers generally select

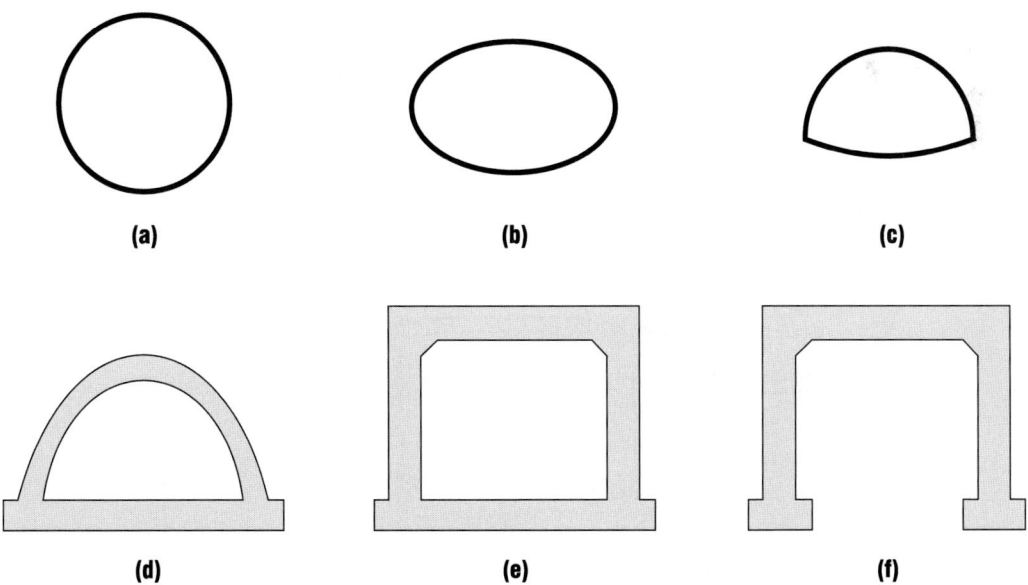

Fig. 16.16 Culvert cross sections: (*a*) circular pipe, usually concrete, corrugated metal, vitrified clay, or cast iron; (*b*) elliptical pipe, generally reinforced concrete or corrugated metal; (*c*) precast concrete pipe arch; (*d*) corrugated metal or reinforced concrete arch; (*e*) reinforced concrete box culvert; (*f*) reinforced concrete bridge culvert.

small-sized culverts from standards based on the characteristics of the project to be constructed. Larger concrete arch and box-type structures, however, are designed for the specific service loads.

Culverts are generally installed in an existing channel bed since this will result in the least amount of work in modifying existing drainage conditions. To avoid extremely long culvert lengths, however, it may be necessary to relocate an existing channel.

16.17 Subsurface Drainage

Water in underlying soil strata of a highway can move upward through capillary action and water can permeate downward to the underlying soil strata through cracks and joints in the pavement. In either case, the water can cause deterioration of the roadbed and pavement. To prevent this, subsurface drainage is used to remove water from the highway subgrade and intercept underground water before it flows to the subgrade. Although design of subsurface drainage systems depends on the specific geometry, topography, and subsurface conditions of the site to be drained, subsurface drainage facilities should be considered an integral component of the entire highway drainage system rather than treated as a separate component.

Failure to implement subsurface facilities that meet drainage requirements can lead to premature failure of major segments of the highway and to slope instability. Figures 16.17 to 16.19 illustrate some commonly used subgrade drainage methods.

Figure 16.17 shows an intercepting drain installed to cut off an underground flow of water to prevent it from seeping into the subgrade of a road. The top of the trench is sealed to prevent silting. In Fig. 16.18, drains are shown employed on both sides of a road to remove surface water that may be trapped when a pervious base is laid over a relatively impervious subgrade. When this detail is used, the longitudinal base drains should be outletted at convenient points, which may be 100 ft apart or more. On steep slopes, lateral drains may be added under the pavement.

Figure 16.19 shows a typical bedding and backfill detail for a pipe underdrain. It is constructed by digging a trench to a specified depth, placing a pipe in the trench, and then backfilling the trench with a porous, granular material. The pipes are generally fabricated of perforated corrugated metal pipe, vitrified clay, or porous concrete. Sizing of pipes is typically based on previous experience, but large projects may require site-specific design.

Road Surfaces

Roads may be paved with a durable material, such as portland cement concrete or bituminous

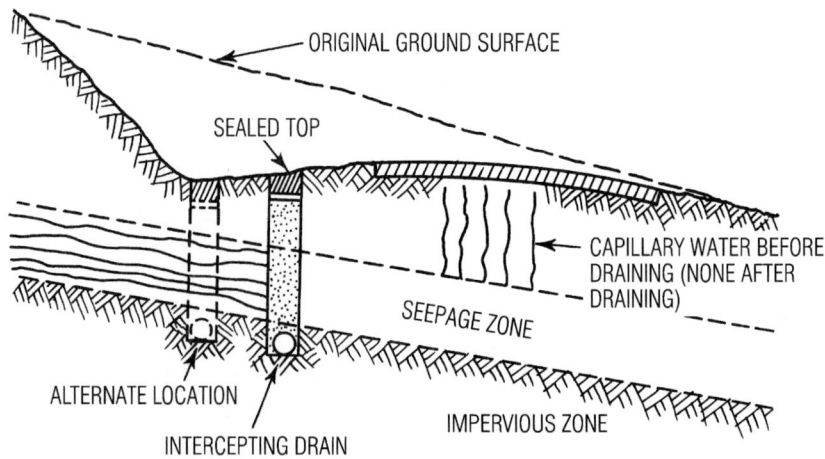

Fig. 16.17 Drain intercepts source of supply of harmful capillary and free water under a road. Top of trench is sealed to prevent silting. (*"Handbook of Drainage and Construction Products," Metal Products Division, Armco Steel Corp.*)

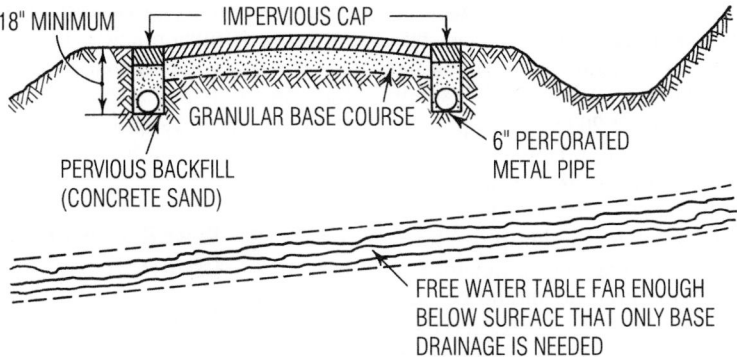

18" MINIMUM · IMPERVIOUS CAP

GRANULAR BASE COURSE

PERVIOUS BACKFILL
(CONCRETE SAND)

6" PERFORATED
METAL PIPE

FREE WATER TABLE FAR ENOUGH
BELOW SURFACE THAT ONLY BASE
DRAINAGE IS NEEDED

Fig. 16.18 Drains remove surface water that may be trapped when a pervious base is laid over a relatively impervious subgrade. On steep slopes, lateral may be added under the pavement. Longitudinal base drains should be outletted at convenient points, which may be 100 ft or more apart. (*"Handbook of Drainage and Construction Products," Metal Products Division, Armco Steel Corp.*)

concrete, or untreated. Pavement classifications and skid resistance are discussed in Art. 16.4. The economic feasibility of many types of road surfaces depends heavily on the costs and availability locally of suitable materials.

16.18 Untreated Road Surfaces

An untreated road surface is one that utilizes untreated soil mixtures composed of gravel,

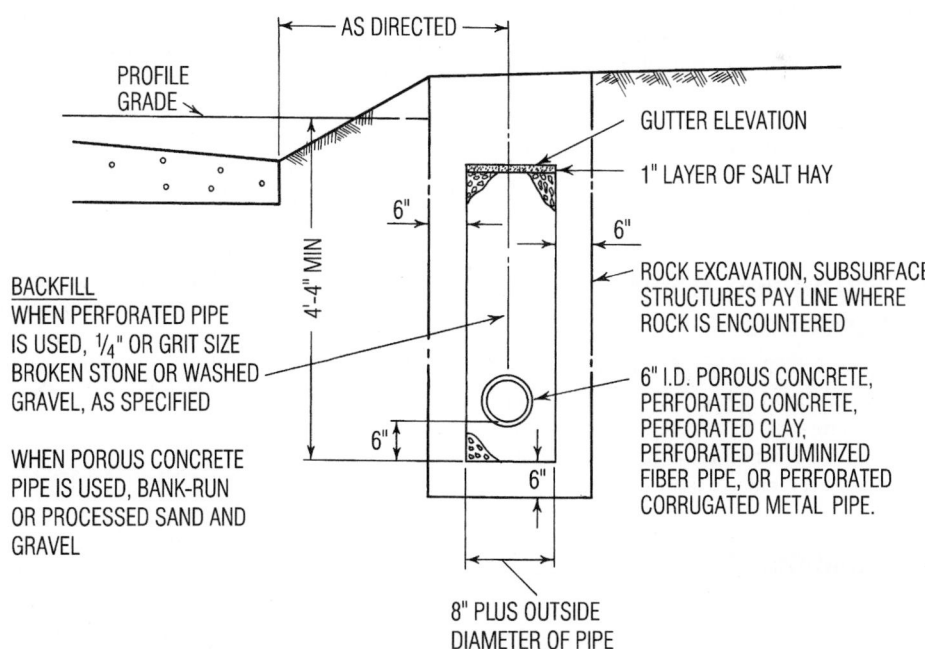

AS DIRECTED

PROFILE
GRADE

GUTTER ELEVATION

1" LAYER OF SALT HAY

6"

6"

4'-4" MIN

ROCK EXCAVATION, SUBSURFACE
STRUCTURES PAY LINE WHERE
ROCK IS ENCOUNTERED

BACKFILL
WHEN PERFORATED PIPE
IS USED, 1/4" OR GRIT SIZE
BROKEN STONE OR WASHED
GRAVEL, AS SPECIFIED

WHEN POROUS CONCRETE
PIPE IS USED, BANK-RUN
OR PROCESSED SAND AND
GRAVEL

6"

6"

6" I.D. POROUS CONCRETE,
PERFORATED CONCRETE,
PERFORATED CLAY,
PERFORATED BITUMINIZED
FIBER PIPE, OR PERFORATED
CORRUGATED METAL PIPE.

8" PLUS OUTSIDE
DIAMETER OF PIPE

Fig. 16.19 Underdrain detail.

crushed rock, or other locally available material, such as volcanic cinders, blast furnace slag, lime rock, chert, shells, or caliche. Such roads are sometimes used where traffic volume is low, usually no more than about 200 vehicles per day. Should larger traffic volumes develop in the future, the untreated road surface can be used as a subgrade for a higher class of pavement.

To withstand abrasion from superimposed traffic loads, a well-graded coarse aggregate (retained on No. 10 sieve) combined with sand should be used. This mixture provides a tight, water-resistant surface with interlocking aggregate that resists shearing forces. To limit deformation, sufficient binding material, such as clay, may be added to bind the aggregates. Excessive use of clay, however, can lead to surface dislocation brought on by expansion when high moisture is present.

Gravel roads are often used during staged highway construction. Staged construction allows for construction of a project in two or more phases. A dry gravel surface can serve as a temporary road for one phase while construction proceeds on another.

The initial cost of untreated surfaces is often very low compared with that of other types of surfaces. Long-term cost of the roadway may be high, however, because frequent maintenance of the surface may be required. The principal concern in maintenance of untreated road surfaces is providing a smooth surface. Smoothness may be accomplished by blading the surface of the road with a motor grader, drag, or similar device. The roadway cross slopes also need to be maintained; otherwise ponding and other associated drainage problems can occur.

16.19 Stabilized Road Surfaces

Controlled mixtures of native soil and an additive, such as asphalt, portland cement, calcium chloride, or sand-clay, can be used to form a stabilized road. Such roads can also serve as a base course for certain types of pavements.

16.19.1 Sand-Clay Roads

Sand-clay roads are composed of a mixture of clay, silt, fine and coarse sand, and, ideally, some fine gravel. This type of road is frequently used in areas where coarse gravel is not readily available. The thickness of this type of roadway is typically 8 in or more. Construction and maintenance of sand-clay roads are similar to that described in Art. 16.18 for untreated road surfaces. The economic feasibility of sand-clay roads is greatly dependent on the availability of suitable materials.

16.19.2 Stabilization with Calcium Chloride

Calcium chloride ($CaCl_2$) is a white salt with the ability to absorb moisture from the air and then dissolve in the moisture. These properties make it an excellent stabilizing agent and dust palliative. For the latter purpose, calcium chloride is most effective when the surface soil binder is more clayey than sandy.

When calcium chloride is used as a stabilizing agent on an existing surface course, the existing roadway surface should be scarified and mixed with about $\frac{1}{2}$ lb/yd^2 of calcium chloride per inch of depth. For this process to be successful, however, adequate moisture must be present.

The surface of calcium chloride-treated roads is maintained by blading with a motor grader, drag, or similar device. While, under normal conditions, calcium chloride-treated roads generally require less maintenance than untreated surfaces, they require blading immediately after rain. In dry periods, a thin layer of calcium chloride should be applied in order to maintain moisture. During extended dry periods, the road surface may require patching.

Calcium chloride often is used as a deicing agent on pavements and can cause corrosion of the metal bodies of vehicles. Similarly, when used in stabilized roads, calcium chloride can corrode the metal of vehicles, but it also can have adverse environmental effects, such as contamination of groundwater. Accordingly, calcium chloride should be used advisably as a stabilizing and dust control agent.

16.19.3 Stabilization with Portland Cement

Untreated road surfaces can be stabilized by mixing the existing road surface with portland cement if the clay content in the soil is favorable for this type of treatment. A general constraint to stabilization with portland cement is that the soils in the road surface contain less than 35% clay. The required rate of application of cement varies with soil classification and generally ranges from 6 to 12% by volume. The roadway surface to be treated should be scarified to accommodate a treated depth of about 6 in. The cement should be applied uniformly to the loose material, brought to the optimum moisture content, and then lightly rolled. The quality of soil-cement surfaces can be enhanced by mixing the soils, cement, and water in a central or traveling mixing plant, then rolling the mixture after it has been placed on the road.

16.19.4 Stabilization with Asphalt

Various asphalt surface treatments can be utilized to stabilize untreated road surfaces. The process consists of application of asphalt, then aggregate uniformly distributed, and rolling. For double, triple, or other multiple surface treatments, the process is repeated several times. This type of stabilization often is used for roads with low design speeds. Surface treatment with bituminous material should not be expected to accommodate high-speed traffic since vehicles traveling at high speeds tend to dislodge the loose aggregate.

For good results in stabilization with asphalt, at the time of application the temperature should be above 40 °F, there should be no rain, and the existing road surface should be dry and well compacted. Also, the quantity and viscosity of the asphalt should be in proper relationship with the temperature, size, and quantity of the aggregate used.

For use as a dust palliative, liquid asphalt may be applied at a rate of 0.1 to 0.5 gal/yd^2. This process is typically referred to as road oiling. This type of dust palliative treatment is often used as a preliminary to progressive improvement of low-type roadways.

16.20 Macadam Road Surfaces and Base Courses

Macadam pavements are derivatives of one of the oldest types of road surfaces. They were originally developed by Scottish road builder John Louden MacAdam (1756–1836). Used as both a road surface and base course, macadam pavements are usually classified as waterbound macadam or bituminous (penetration) macadam.

16.20.1 Waterbound Macadam

A waterbound macadam road is constructed with crushed stone, which is mechanically locked or keyed with stone screenings rolled into the voids and then set in place with water. For pavement thicknesses up to 9 in, a waterbound macadam pavement is typically constructed in two courses. Thicker pavements are generally constructed with three courses.

In two-course construction, the lower course is about 4 in thick and the upper course about 2 in thick. The stones in the lower course should pass a 3-in ring and be retained on a 2-in ring. The top-course stone should pass through a 2-in ring and be retained on a 1-in ring. In addition to the size requirements, the stones should also be of suitable hardness.

After the base course of stones has been put down, it is rolled with a roller weighing about 10 tons or compacted with vibratory compactors. The compaction generally shrinks the course depth by roughly one-third. Therefore, if a 4-in course were desired, stone would be spread to a depth of about 6 in prior to rolling.

After the lower course has been placed and rolled, the finer top course of stone is spread on top and compacted. Next, a layer of stone chips or stone dust is shoveled over the top course and broomed into the voids as a binder. The layer is then sprinkled with water to set it. Alternate applications of binder, water, and rolling follow until a wave of mortar appears ahead of the roller. With an experienced work crew, it is possible to obtain an excellent pavement that sheds water and is suitable for light traffic in rural areas.

Waterbound macadam has been generally superseded by asphalt concrete or portland cement-treated bases. This change occurred because of advances made in plant equipment and the time-consuming nature of waterbound macadam construction. In areas where labor is readily available and inexpensive, this type of pavement may prove feasible.

16.20.2 Bituminous (Penetration) Macadam

When a bituminous material is used as the binder material in macadam, bituminous macadam is formed. After the aggregate layer is compacted, the bituminous material is applied and penetrates into the voids, binding the stone particles together. This process has led to bituminous macadam also being referred to as penetration macadam.

When the bitumen is asphalt, it is heated to about 300 to 350 °F and applied as a liquid to the compacted aggregates. The air temperature should be 40 °F or higher at the time of application and for the preceding 24 h.

A penetration-macadam top course is usually 2 to 3 in thick. It is placed on a base course about 4 in thick, similar to the lower course of water-bound macadam in which the voids are filled with small stone (Art. 16.20.1). After the base course has been rolled, excess filler is removed by stiff brooming. Next, the large stones for the top course are spread on top and the bitumen is applied. Then, while the bitumen is still warm, the large stones are keyed or choked with small stone. Excess screenings are broomed off and the surface is rolled to ensure good keying. A second application of bitumen is made and followed by a covering of stone chips or pea gravel and rolling.

16.20.3 Inverted Penetration Macadam

For inverted penetration, the process described in Art. 16.20.2 for bituminous macadam is reversed. The asphalt binder is sprayed over a prepared surface first and then covered with aggregate. This approach can be utilized for dust control, prime coat or tack coat on which a new wearing surface will be constructed, surface treatment and armor coat for temporary protection of untreated surfaces, or seal coat for leveling, strengthening, or otherwise improving existing pavements.

16.21 Surface Treatments

Various types of surface treatment are available for improving the quality of an existing pavement. Typically, a surface treatment is a thin layer of material (about $\frac{1}{2}$ to $\frac{3}{4}$ in thick) applied to the surface of a road in single or multiple lifts. Surface treatments generally consist of a bituminous material applied to crushed stone by the inverted penetration method (Art. 16.20.3). Since the surface treatment is relatively thin, it is usually not intended to support loads by itself.

Surface treatments can be used to achieve a seal coat, armor coat, dust palliative, or prime or tack coat for a new wearing surface. A surface treatment

is applied to a granular-type base by a pressure distributor truck. This type of vehicle is equipped with a tank containing the surfacing material and a spray bar with nozzles that spread the binder over a given width of roadway.

16.21.1 Armor Coats

Named generically a surface treatment, an armor coat is applied in two or more lifts. It is generally used to provide protection to an untreated mineral surface. Armor coats are composed of a base consisting of gravel, waterbound macadam, earth, or similar material and a top course of bituminous binder covered by mineral aggregates.

16.21.2 Seal Coats

A seal coat is a coat of binder less than $\frac{1}{2}$ in thick that is applied to a pavement surface and covered with fine aggregates. Seal coats are used to waterproof (seal), protect, and enhance the skid resistance of an existing pavement. They may, however, be applied in multiple lifts in a fashion similar to that described for armor coats (Art. 16.21.1).

A seal coat comprised of fine sand, emulsified asphalt, and water is known as a **slurry seal**. This type of seal coat is used to fill cracks and otherwise rejuvenate the surface of deteriorated pavements.

16.21.3 Dust Palliatives

As a pavement deteriorates, dust and fine particles can be raised by traffic. At best, this can cause a severe hindrance to visibility and at worst, extremely hazardous conditions for vehicles traveling over the road. Dust palliative surface treatments, consisting of a small quantity of a light, slow-curing oil, such as SC-70 or SC-250, may be applied to the pavement surface to control dust. The oil penetrates the pavement surface, producing a film that surrounds individual particles and binds them together.

16.21.4 Prime Coats

Before a bituminous pavement is constructed over a base of earth, gravel, or waterbound macadam, the surface is sprayed with a bitumen. The purposes are to plug capillary voids to stop upward seepage of water from the subgrade, to coat and bind dust and loose mineral particles, and to enhance adhesion between the base and surface courses. The bitumen *primes* the surface by penetrating it until the bitumen is completely absorbed.

A liquid asphalt, such as MC-30 or MC-70, or a low-viscosity road tar, such as RT-1 to RT-3, usually is used as the bitumen. Its most important characteristic is penetrating capability. Before the prime coat is applied, the existing surface should be clean and dry.

16.21.5 Tack Coats

A tack coat is used to bind together two pavement surfaces, typically a new wearing surface to an existing base surface consisting of bitumen, portland cement concrete, or other road material. Before application of the tack coat, the existing surface should be properly prepared in order for a successful bond to be formed. It is important that the existing surface be dry and free from dirt and debris. The bitumen is applied by a pressure distributor. This type of vehicle is equipped with a tank containing the surfacing material and a spray bar with nozzles that spread the bitumen. Kept free of traffic, the tack coat should be allowed to dry until it reaches an appropriate degree of stickiness to allow for proper bonding between the two layers. Then the resurfacing layer may be applied.

16.22 Flexible Pavements

Bituminous pavements are classified as flexible, whereas portland cement-concrete pavements are considered rigid. Whereas under loads, a rigid pavement acts as a beam that can span across irregularities in an underlying layer, a flexible pavement stays in complete contact with the underlying layer. A rigid pavement is designed so that it can deflect like a beam and then return to the state that existed prior to loading. Flexible pavements, however, may deform and not entirely recover when subjected to repeated loading. The decision as to which type of pavement to use depends on local availability of materials, urban/rural location, traffic volume, costs, and future maintenance considerations.

16.22.1 Flexible-Pavement Courses

Figure 16.20 shows the constituent elements of a typical flexible pavement. The main components,

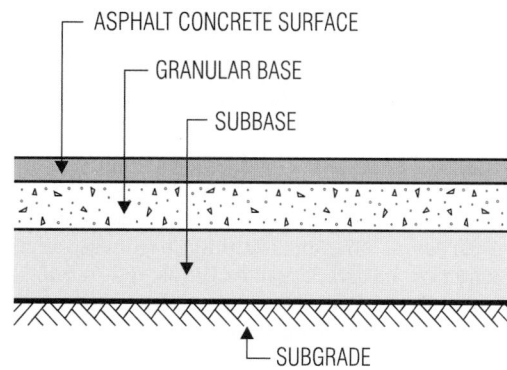

Fig. 16.20 Components of a flexible pavement.

from the bottom up, are the subgrade, subbase, granular base, and asphalt-concrete wearing surface. For Course thicknesses, see Art. 16.22.10.

Subgrade ▪ This is the underlying soil that serves as the foundation for a flexible pavement. It may be native soil or a layer of selected borrow materials that are compacted to a depth below the surface of the subbase.

Subbase ▪ As shown in Fig. 16.20, the subbase is the course between the subgrade and the base course. The subbase typically consists of a compacted layer of granular material, treated or untreated, or a layer of soil treated with a suitable admixture. It differs from the base course in that it has less stringent specifications for strength, aggregate types, and gradation. If the subgrade meets the requirements of a subbase course, the subbase course may be omitted. In addition to its major structural function as part of the pavement cross section, however, the subbase course can also serve many secondary functions, such as limiting damage due to frost, preventing accumulation of free water within or below the pavement structure, and preventing intrusion of fine-grain subgrade soils into the base courses. In rock cuts, the subbase course can also act as a working platform for construction equipment or for subsequent pavement courses. Performance of these secondary functions depends on the type of material selected for the subbase course.

Base Course ▪ This is the layer of material directly under the surface course. The base course rests on the subbase or, if no subbase is provided, on the subgrade. A structural portion of the pavement, the base course consists of aggregates such as crushed stone, crushed slag, gravel and sand, or a combination of these.

Specifications for base-course materials are much more stringent than those for subbase-course materials. This is especially the case for such properties as strength, stability, hardness, aggregate types, and gradation. Addition of a stabilizing admixture, such as portland cement, asphalt, or lime, can improve the characteristics of a wide variety of materials that, if untreated, would be unsuitable for use as a base course. From an economic standpoint, such treatment is especially beneficial when there is a limited supply of suitable untreated material.

Surface Course ▪ This is the uppermost layer of material in a flexible pavement. It is designed to support anticipated traffic, resist its abrasive forces, limit the amount of surface water that penetrates into the pavement, provide a skid-resistant surface, and offer a smooth riding surface. To serve these purposes, the surface course should be durable, regardless of weather conditions.

Surface courses typically consist of bituminous material and mineral aggregates that are well graded and have a maximum size of about ¾ to 1 in. Various other gradations ranging from sand (used in sheet asphalt) to coarse, open-graded mixtures of coarse and fine aggregates have been used with satisfactory results under specific conditions.

16.22.2 Flexible-Pavement Design Assumptions

Flexible pavements are designed as a multilayered elastic system. Each course of a pavement is a layer with specific material properties that differ from those of the other layers and that affect the overall performance of the pavement. All layers are assumed to be infinite in the horizontal plane. The subgrade, the bottom layer, is assumed to be infinite in the vertical plane as well.

As the wheel of a vehicle passes over the pavement, compressive stresses are imposed in the surface course directly under the wheel. The surface course distributes the stresses over the base course, which, in turn, transmits them to the lower courses. The stresses are greatest at the top of the surface course and decrease toward the subgrade. Horizontal stresses exist below the wheel load also. They vary from compression (above the neutral axis of the pavement cross section) to tension (below the neutral axis). In addition, the pavement is subjected to thermal stresses.

Flexible pavements usually are designed by a method promulgated by the American Association of State Highway and Transportation Officials (AASHTO), or the Asphalt Institute, or the California Department of Transportation (Caltrans). Article 16.22.3 presents an overview of the AASHTO method.

16.22.3 AASHTO Design Method for Flexible Pavements

The AASHTO "Guide for Design of Pavement Structures" takes into account pavement performance, traffic volume, subgrade soils, construction materials, environment, drainage, reliability, lifecycle costs, and shoulder design. In essence, the design procedure is to convert the varying axle loads to a single design load and to express the traffic volume as the number of repetitions of the design axle load (Arts. 16.22.4 to 16.22.10).

16.22.4 Flexible-Pavement Performance

Pavement performance includes both the structural and functional performance of the pavement structure. Structural performance describes the ability of the pavement to support traffic loading without excessive permanent deformations, cracking, faulting, raveling, etc. Functional performance addresses the ability of the pavement to fulfill its intended functions such as maintaining a smooth and uniform riding surface.

Pavement performance is also used to describe the ability of the pavement to provide for the safety of vehicles and their passengers. An important pavement feature that impacts safety is the friction between vehicle tires and the pavement.

The influence of pavement performance in the AASHTO design method is represented by the **present serviceability index (PSI)**, which takes into account pavement roughness and distress as indicated by the extent of cracking, patching, and rut depth present. The PSI is based on a scale from 0 to 5; the higher the number the better the condition; that is, the smoother the pavement. A pavement with a PSI of 4.5, for example, is smoother (less rough) than a pavement with a PSI of 4.0. The assumption is that a smooth pavement will have a longer life than a rough one.

Two serviceability indexes are used in design of a pavement structure. One is the initial serviceability index p_i, which represents the condition of the pavement when new. The second is the terminal serviceability index p_t, which represents the minimum acceptable roughness at which stage rehabilitation is needed. AASHTO suggests the following maximum values of p_t: 2.5 or 3.0 for major highways, 2.0 for lower classifications, and 1.5 for extreme situations for low-volume roads where costs must be kept low and then on a case-by-case basis.

While deterioration and the related loss of serviceability of a pavement are related to the age of the pavement, volume of traffic, and various environmental conditions, there is no direct relationship that incorporates the combined impact of these variables. Therefore, some degree of idealization is required; for example, age may be taken as a net negative factor that reduces serviceability.

Traffic predictions are made for a convenient period of time, typically 20 years. Any period, however, may be used with the AASHTO design method since traffic is expressed as daily or total ESAL applications. The total ESAL applications are the number of repetitions of the loading that the pavement is expected to carry, from opening of the road to the time when it reaches its terminal value, for example, when $p_t = 2.0$.

For design purposes, the traffic must be distributed by direction and by lanes. Directional distribution is generally made by assigning 50% of the traffic to each direction (if special conditions do not warrant some other distribution). Lane distribution is usually made by assigning 100% of the traffic in each direction to each lane. Some states, however, have developed lane distribution percentages for highways with more than one lane in a given direction. Depending on the total number of lanes present, these percentages typically range from 60 to 100% of the one-directional traffic.

Because of the importance of the traffic data design of a pavement, the design team should work closely with the personnel involved in the gathering of this information. Poor traffic estimates can adversely affect highway performance and economy.

16.22.5 Traffic Loads

The effects of traffic loads are determined by the use of an equivalent single 18-kip axle load (ESAL). The AASHTO method takes into account axle loads, axle configuration, and number of applications of the loads. The actual loading is related to ESAL by equivalence factors based on the terminal serviceability index p_t (Art. 16.22.4) and a parameter called structural number SN. The structural number is used to describe the overall thickness of the pavement (Art. 16.22.10).

Tables 16.6 and 16.7 list axle-load equivalence factors for single and tandem axles acting on flexible pavements with a p_t of 2.0 and 2.5, respectively. These tables can be used to convert mixed traffic loads to an equivalent number of 18-kip loads.

The accuracy of traffic estimates depends greatly on the accuracy of the following: load equivalence values, estimates of traffic volume and weight, prediction of ESAL over the design period, and interaction of age and traffic as it affects changes in PSI (Art. 16.22.4).

16.22.6 Subgrade Support for Flexible Pavements

A pavement is designed to distribute traffic loads to the subgrade, which must be capable of withstanding the resulting stresses. Hence the performance of the pavement depends greatly on the physical properties and condition of the subgrade soils. AASHTO characterizes the soil by its resilient modulus M_R, psi. The resilient modulus takes into account various nonlinear properties of the soil. (M_R replaces the soil support value S used in the past. The change was made because of the applicability of the resilient modulus to multilayered systems in general and pavement structures in particular.) Because some transportation agencies do not have the ability to perform the resilient modulus test (described in AASHTO Test Method T274), the AASHTO "Guide for Design of Pavement Structures" contains correlations that relate the frequently used California bearing ratios (CBR) and stabilometer R values to an equivalent M_R.

Table 16.6 Axle-Load Equivalence Factors for Flexible Pavement, $p_t = 2.0$*

	Single Axles					
Axle-Load, kips	Structural number SN					
	1	2	3	4	5	6
2	0.0002	0.0002	0.0002	0.0002	0.0002	0.0002
4	0.002	0.003	0.002	0.002	0.002	0.002
6	0.01	0.01	0.01	0.01	0.01	0.01
8	0.03	0.04	0.04	0.03	0.03	0.03
10	0.08	0.08	0.09	0.08	0.08	0.08
12	0.16	0.18	0.19	0.18	0.17	0.17
14	0.32	0.34	0.35	0.35	0.34	0.33
16	0.59	0.60	0.61	0.61	0.60	0.60
18	1.00	1.00	1.00	1.00	1.00	1.00
20	1.61	1.59	1.56	1.55	1.57	1.60
22	2.49	2.44	2.35	2.31	2.35	2.41
24	3.71	3.62	3.43	3.33	3.40	3.51
26	5.36	5.21	4.88	4.68	4.77	4.96
28	7.54	7.31	6.78	6.42	6.52	6.83
30	10.38	10.03	9.24	8.65	8.73	9.17
32	14.00	13.51	12.37	11.46	11.48	12.17
34	18.55	17.87	16.30	14.97	14.87	15.63
36	24.20	23.30	21.16	19.28	19.02	19.93
38	31.14	29.95	27.12	24.55	24.03	25.10
40	39.57	38.02	34.34	30.92	30.04	31.25

	Tandem Axles					
Axle-Load, kips	Structural number SN					
	1	2	3	4	5	8
10	0.01	0.01	0.01	0.01	0.01	0.01
12	0.01	0.02	0.02	0.01	0.01	0.01
14	0.02	0.03	0.03	0.03	0.02	0.02
16	0.04	0.05	0.05	0.05	0.04	0.04
18	0.07	0.08	0.08	0.08	0.07	0.07
20	0.10	0.12	0.12	0.12	0.11	0.10
22	0.16	0.17	0.18	0.17	0.16	0.16
24	0.23	0.24	0.26	0.25	0.24	0.23
26	0.32	0.34	0.36	0.35	0.34	0.33
28	0.45	0.46	0.49	0.48	0.47	0.46
30	0.61	0.62	0.65	0.64	0.63	0.62
32	0.81	0.82	0.84	0.84	0.83	0.82
34	1.06	1.07	1.08	1.08	1.08	1.07
36	1.38	1.38	1.38	1.38	1.38	1.38
38	1.76	1.75	1.73	1.72	1.73	1.74
40	2.22	2.19	2.15	2.13	2.16	2.18
42	2.77	2.73	2.64	2.62	2.66	2.70
44	3.42	3.36	3.23	3.18	3.24	3.31
46	4.20	4.11	3.92	3.83	3.91	4.02
48	5.10	4.98	4.72	4.58	4.68	4.83

* From "Guide for Design of Pavement Structures," American Association of State Highway and Transportation Officials.

Table 16.7   Axle-Load Equivalence Factors for Flexible Pavement, $p_t = 2.5$*

Axle-Load, kips	Single Axles Structural number SN					
	1	2	3	4	5	6
2	0.0004	0.0004	0.0003	0.0002	0.0002	0.0002
4	0.003	0.004	0.004	0.003	0.003	0.002
6	0.01	0.02	0.02	0.01	0.01	0.01
8	0.03	0.05	0.05	0.04	0.03	0.03
10	0.08	0.10	0.12	0.10	0.09	0.08
12	0.17	0.20	0.23	0.21	0.19	0.18
14	0.33	0.36	0.40	0.39	0.36	0.34
16	0.59	0.61	0.65	0.65	0.62	0.61
18	1.00	1.00	1.00	1.00	1.00	1.00
20	2.61	1.57	1.49	1.47	1.51	1.55
22	2.48	2.38	2.17	2.09	2.18	2.30
24	3.69	3.49	3.09	2.89	3.03	3.27
26	5.33	4.99	4.31	3.91	4.09	4.48
28	7.49	6.98	5.90	5.21	5.39	5.98
30	10.31	9.55	7.94	6.83	6.97	7.79
32	13.90	12.82	10.52	8.85	8.88	9.95
34	18.41	16.94	13.74	11.34	11.18	12.51
36	24.02	22.04	17.73	14.38	13.93	15.50
38	30.90	28.30	22.61	18.06	17.20	18.98
40	39.26	35.89	28.51	22.50	21.08	23.04

Axle-Load, kips	Tandem Axles Structural number SN					
	1	2	3	4	5	6
10	0.01	0.01	0.01	0.01	0.01	0.01
12	0.02	0.02	0.02	0.02	0.01	0.01
14	0.03	0.04	0.04	0.03	0.03	0.02
16	0.04	0.07	0.07	0.06	0.05	0.04
18	0.07	0.10	0.11	0.09	0.08	0.07
20	0.11	0.14	0.16	0.14	0.12	0.11
22	0.16	0.20	0.23	0.21	0.18	0.17
24	0.23	0.27	0.31	0.29	0.26	0.24
26	0.33	0.37	0.42	0.40	0.36	0.34
28	0.45	0.49	0.55	0.53	0.50	0.47
30	0.61	0.65	0.70	0.70	0.66	0.63
32	0.81	0.84	0.89	0.89	0.86	0.83
34	1.06	1.08	1.11	1.11	1.09	1.08
36	1.38	1.38	1.38	1.38	1.38	1.38
38	1.75	1.73	1.69	1.68	1.70	1.73
40	2.21	2.16	2.06	2.03	2.08	2.14
42	2.76	2.67	2.49	2.43	2.51	2.61
44	3.41	3.27	2.99	2.88	3.00	3.16
46	4.18	3.98	3.58	3.40	3.55	3.79
48	5.08	4.80	4.25	3.98	4.17	4.49

* From "Guide for Design of Pavement Structures," American Association of State Highway and Transportation Officials.

An equivalent M_R can be determined for the Corps of Engineers CBR value from

$$M_R = 1500\,\text{CBR} \qquad (16.17)$$

Equation (16.17) is valid for fine-grain soils with a soaked CBR of 10 or less. An equivalent value M_R based on an R value can be determined for fine-grain soils with an R value less than or equal to 20 from

$$M_R = 1000 + 555R \qquad (16.18)$$

The AASHTO "Guide" contains design curves for conversion to a structural number SN (Art. 16.22.9).

The resilient modulus is based on the properties of the compacted subgrade soils. It may be necessary, however, to include the properties of in situ materials in the uncompacted foundation if these materials are especially weak. Also, compaction of the subgrade is essential to ensure adequate performance and reliability.

16.22.7 Flexible-Pavement Material

For flexible pavements, materials used for subbase, base, and surface courses differ. Article 16.22.1 describes the properties and characteristics of these layers. For more detailed information, see AASHTO "Guide for the Design of Pavement Structures" and "Construction Manual for Highway Construction." (www.aashto.org)

In addition to the aforementioned three principal layers, the prepared roadbed is an important component of a flexible pavement, and a drainage layer also may be necessary. The prepared roadbed may be a layer of compacted roadbed soil or select borrow material that is compacted to a specified density. Examples of a drainage layer are given in Fig. 16.21. Figure 16.21a shows a base course that serves also as a drainage layer, whereas Fig. 16.21b shows a drainage layer between the subbase and the subgrade.

16.22.8 Flexible-Pavement Drainage

Rainfall is one of the principal environmental conditions that affects the design and performance of pavements. The major concern with rainwater is that it may penetrate through the pavement into the roadbed soil and weaken it. Proper drainage is an important element in preventing this. Experience has shown that pavements that are not properly drained deteriorate prematurely, especially when exposed to heavy traffic volumes and their related loads.

Articles 16.16 and 16.17 discuss the adverse effects when water penetrates a pavement and describes some methods employed to prevent this and to remove water from the surface of the roadway. The AASHTO design method for flexible pavements takes into account the impact of swelling, frost heave, and moisture on roadbed soil and base strength. This is done by multiplying the structural layer coefficients a_1 and a_2 (Art. 16.22.9) by a factor m_i that takes into account the quality of drainage and the percent of time the pavement is subjected to moisture levels approaching saturation. The quality of drainage is indicated by the amount of time needed to drain the base layer to 50% of saturation.

("Guide for Design of Pavement Structures," American Association of State Highway and Transportation Officials (www.aashto.org).)

16.22.9 Structural Numbers for Flexible Pavements

The design of a flexible pavement or surface treatment expected to carry more than 50,000 repetitions of ESAL (Art. 16.22.5) requires identification of a structural number SN that is used as a measure of the ability of the pavement to withstand anticipated axle loads. In the AASHTO design method, the structural number is defined by

$$SN = SN_1 + SN_2 + SN_3 \qquad (16.19)$$

where SN_1 = structural number for the surface course = $a_1 D_1$

a_1 = layer coefficient for the surface course

D_1 = actual thickness of the surface course, in

SN_2 = structural number for the base course = $a_2 D_2 m_2$

a_2 = layer coefficient for the base course

D_2 = actual thickness of the base course, in

m_2 = drainage coefficient for the base course

SN_3 = structural number for the subbase course = $a_3 D_3 m_3$

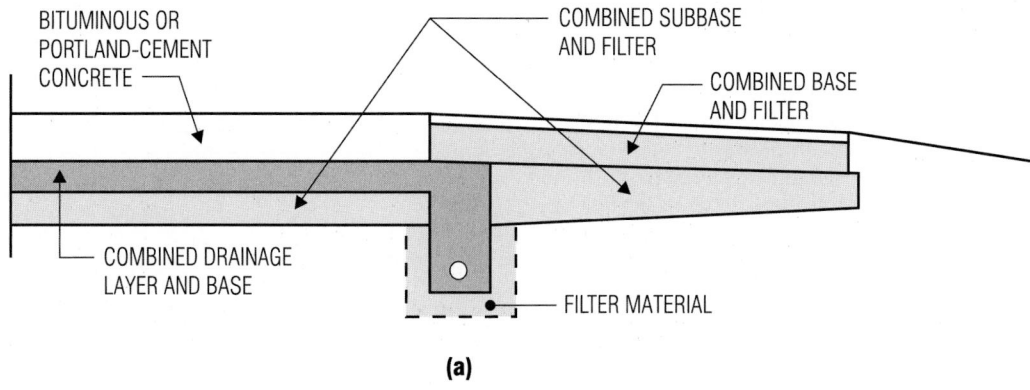

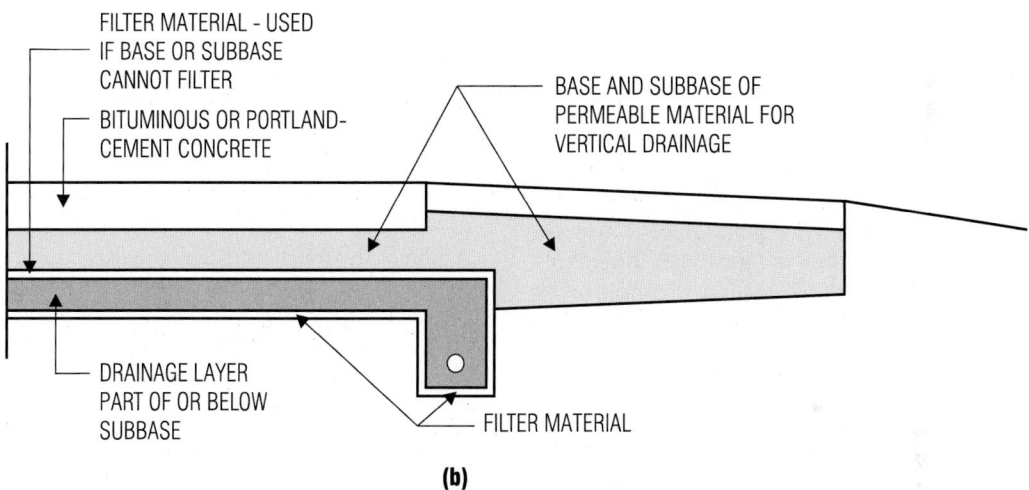

Fig. 16.21 Drainage layers under pavements: (*a*) base used as the drainage layer; (*b*) drainage layer as part of or below the subbase.

a_3 = layer coefficient for the subbase course

D_3 = actual thickness of the subbase course, in

m_3 = drainage coefficient for the subbase

The layer coefficients a_n are assigned to materials used in each layer to convert structural numbers to actual thickness. They are a measure of the relative ability of the materials to function as a structural component of the pavement. Many transportation agencies have their own values for these coefficients. As a guide, the layer coefficients may be 0.44 for asphaltic-concrete surface course, 0.14 for crushed-stone base course, and 0.11 for sandy-gravel subbase course. The drainage coefficient m_n is discussed in Art. 16.22.8.

The thicknesses D_1, D_2, and D_3 should be rounded to the nearest $\frac{1}{2}$ in. Selection of layer thicknesses usually is based on agency standards, maintainability of the pavement, and economic feasibility. See also Art. 16.22.10 and the AASHTO "Guide for Design of Pavement Structures."

16.22.10 Determination of Course Thicknesses

The thickness to be used for the various layers of a flexible pavement is, along with other parameters, a function of the material used and the load the pavement is expected to withstand. Minimum thickness for each layer depends on the size of aggregate used. With aggregate size as the controlling criterion, the following are minimum layer thicknesses: surface course, $1\frac{1}{2}$ in; base course, 3 in; and subbase course, 4 in. Table 16.8 lists minimum thicknesses recommended by AASHTO for various levels of ESAL. These are practical thicknesses and vary with local conditions and design practices.

A flexible pavement is essentially a composite of layers (Fig. 16.22) and is designed as such. The first step is to determine the structural number SN needed for the combination of layers above the subgrade with the use of the resilient modulus (see Art. 16.22.6). Next, the structural numbers needed for the combination of layers above the subbase and for the surface course are calculated. Taking into account the differences between these calculated structural numbers, a maximum allowable thickness for any layer can be found. Therefore, to determine the maximum allowable structural number for the subbase material, subtract the structural number required for the layers above the subbase from the structural number required for the subgrade. Repeat this process for the other layers in the pavement. After the structural numbers have been deter-

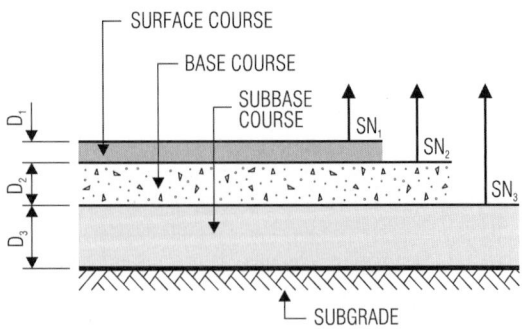

Fig. 16.22 Composite of layers forming a flexible pavement. SN indicates structural number of a layer.

mined, the respective layer thicknesses can be calculated as follows:

1. The thickness D_1 of the surface course is determined by dividing the structural number required SN_1 for the surface course by the layer coefficient a_1. Select a thickness D_1' by rounding the calculated value to the nearest larger $\frac{1}{2}$ in or a more practical dimension.

2. The structural number supplied then is $SN_1' = a_1 D_1'$, which is larger than SN_1.

3. The thickness D_2 to be used for the base course should be chosen selected equal to or larger than $(SN_2 - SN_1')/a_2 m_2$, where SN_2 is the required structural number for the base and surface layers. The sum of the structural numbers supplied for the base and surface courses then should be equal to or larger than SN_2.

4. The thickness D_3 to be used for the subbase course should be selected equal to or larger than $[SN_3 - (SN_1' + SN_2')]/a_3 m_3$.

The AASHTO "Guide" presents various charts and design aids for determining the structural numbers layer thicknesses required for a pavement. A limiting criterion of this method is that it cannot be used to determine the SN required above subbase or base materials possessing an elastic modulus greater than 40,000 psi. In such instances, the thickness of a layer above the high-modulus layer should be based on economic and practical minimum-thickness considerations.

Table 16.8 Minimum Layer Thickness, in, Based on ESAL*

Traffic, ESAL	Asphalt concrete, in	Aggregate base, in
Less than 50,000	1.0[†]	4
50,000–150,000	2.0	4
150,001–500,000	2.5	4
500,001–2,000,000	3.0	6
2,000,001–7,000,000	3.5	6
Greater than 7,000,000	4.0	6

* Adapted from AASHTO "Guide for Design of Pavement Structures."
† For surface treatment.

See also "Thickness Design—Full Depth Asphalt Pavement Structures for Highways and Streets," Manual MS-1, The Asphalt Institute, College Park, MD 10740 (www.asphaltinstitute.org).

16.23 Alternative Flexible Pavements

A variety of technologies are available as alternatives to the type discussed in Art. 16.22. Included in this category are porous pavements, sulfur-asphalt mixes, hydrated-lime additives, rubberized wearing surfaces, recycled asphalt pavements, and the Superpave mix design system.

16.23.1 Porous Pavements

These are essentially asphalt pavements without any fines (sand) in the mix. This type of pavement contains voids through which rainwater is allowed to seep into the subgrade. This characteristic offers several advantages: Removal of water from the pavement decreases the possibility of damage from trapped water, thus increasing pavement life. Also, if storm water from the pavement can percolate into the soil, a smaller highway drainage system is needed. If there is an existing storm sewer, the risk of overloading it is also greatly reduced. Furthermore, porous pavements enhance traffic safety by decreasing the risk of hydroplaning (wet skidding). In addition, driver visibility of pavement markings does not suffer in rain because the water percolates rapidly through the porous asphalt surface. From an aesthetic viewpoint, porous pavement is not objectionable since there is no basic visual difference between porous and conventional, nonpermeable pavements.

Porous pavements are generally used in highways, local streets, and parking lots. For parking lots, porous pavements are advantageous because rain seeping from them into the subgrade promotes healthy growth of trees, shrubs, ground cover, and other plantings and thus makes a parking area and associated landscaping more aesthetically pleasing. For a parking area, a typical porous pavement consists of a $2\frac{1}{2}$-in surface course of porous bituminous concrete over a 12-in graded, crushed-stone base. The base course is layered. Small stones form the top layer so that a paving machine can create a smooth surface for application of the surface course.

16.23.2 Sulfur-Asphalt Mixes

Sulfur is used in bituminous pavements in several ways. In one method, sulfur serves as a filler. It is added to a hot sand-asphalt mix after the asphalt and aggregate have been mixed. The sulfur fills the voids and locks the sand particles, stabilizing the mix. In another method, sulfur and asphalt are blended to form sulfur-extended asphalt (SEA). The hot sulfur is dispersed into the asphalt to create a binder that is then mixed with the aggregate. Production of SEA requires only a slight modification of the hot-mix plant. Otherwise, the construction operations and equipment for SEA are the same as that for asphalt concrete.

Sulfur is also used for roads in areas subject to permafrost. Conventional highway construction practice calls for gravel depths of 5 ft or more below the ground surface to provide a stable load-bearing surface. Also, thermal insulation is installed below the gravel to protect the underlying permafrost. When gravel is not available locally, as is often the case in many northern areas, it must be transported to the project site from elsewhere at considerable expense. Construction costs can be cut, however, by reducing significantly the amount of gravel required through use of sulfur in foam form with gravel. One test showed that 7 ft of gravel could be replaced by only 3 ft of gravel set atop 3 to 4 in of sulfur foam.

There are, however, health hazards associated with the use of sulfur in general. For instance, noxious gases, such as sulfur dioxide and hydrogen disulfide, can be generated at the plant and the construction site.

16.23.3 Hydrated Lime

This is widely used in hot mixes that contain marginally acceptable aggregates. The lime acts as a chemical additive rather than a void filler. It increases the strength and stability of an asphalt mix while making it more water resistant. Also by hardening mixes, it allows faster compaction and yields higher densities.

16.23.4 Rubber in Wearing Courses

Rubber is used to improve the paving qualities of hot mixes used in bituminous wearing courses. For this purpose, rubber may be added to an asphalt-concrete mix or applied to the pavement surface after placement and compaction. The rubber reduces temperature susceptibility, decreases raveling, offers better void control, and lessens the tendency to flow, improving flexibility and adhesion to aggregates.

16.23.5 Superpave Mix Design System

The Superior Performing Asphalt Pavements (Superpave) mix design system is a method of designing flexible-pavement mixes that are tailored to specific project characteristics. These include traffic, environment, pavement structural section, and reliability.

The Superpave mix design system assists in selection of combinations of asphalt binder, aggregate, and any necessary modifiers to obtain a desired level of pavement performance. The goal of the system is to create an *ideal* blend of asphalt binder and aggregate for production of the lowest-cost pavement for the anticipated level of service.

The Superpave system applies to three different levels of traffic, low, intermediate, and high, and employs laboratory and field testing techniques. Computer software based on the Superpave specifications is available to assist in the process. The software and associated specifications perform analysis and design of multiple layer pavements consisting of base, binder, and surface courses. For example, selection of the necessary materials used in the Superpave mix is based on, among other things, the design ESAL for the project (Art. 16.22.5). ESALs are used to ascertain whether the anticipated traffic level is low, intermediate, or high. Also taken into account are pavement environmental conditions influenced by climate. Based on these conditions, an asphalt binder, for example, can be chosen.

The Superpave system also allows for addition of modifiers, such as fibers or hydrated lime, to the mix to enhance the ability of the paving mixes to avoid pavement distresses. While the system does not offer a list of modifiers for correction of specific pavement distress, it does offer a guide based on AASHTO Practice PP5, "The Laboratory Evaluation of Modified Asphalt Systems," to assist in the selection of appropriate modifiers to enhance the performance of the pavement.

("The Superpave Mix Design Manual for New Construction and Overlays," SHRP-A-407, Strategic Highway Research Program, National Research Council, 2101 Constitution Ave., NW, Washington, DC 20418 (www.nas.edu/nrc/).)

16.23.6 Recycling of Asphalt Pavements

The materials in an asphalt pavement that is scheduled for replacement can be reused as ingredients of a new surface course or a new pavement, including underlying, untreated base material. The recycling may be performed in place or at a central plant.

When the asphalt pavement is to be recycled at the site, a process known as in-place, cold-mix, asphalt-pavement recycling is used. In this process, the existing pavement materials are ripped, broken, pulverized, and mixed in place with asphalt or other materials, such as aggregates or stabilizing agents. The other materials usually are required to provide a higher-strength base. The process requires that an asphalt surface course be placed on top of the recycled layer. One drawback to this process is that quality control is not so good in the field as it would be at a central plant. Another is that maintenance of traffic is difficult because of the necessity of avoiding interference with the recycling equipment. The cold-mix process, however, can also be conducted at a central plant where enhanced quality control provides higher mix efficiency and reliability. Central plants also offer higher production capacity and better uniformity and reliability.

An alternative to the cold-mix process is hot-mix, asphalt-pavement recycling, which is performed at a central plant. In this process, reclaimed asphalt-pavement materials are removed from an existing roadway in a fashion similar to that described for the cold-mix process and combined in the central plant with new asphalt or recycled agents. The hot-mix method also sometime utilizes uncoated aggregates from the base to produce the hot mix. For the hot-mix recycling process, one of the following types of plants is generally used: batch, drum mixer, or continuous mixer.

Several factors affect the feasibility of a recycling project. These include availability of recycling equipment, impact on traffic through the construction site, and the size and location of the project. In the right situation, however, recycling can offer many economic and environmental advantages.

16.24 Rigid Pavements

A rigid pavement typically consists of a portland cement-concrete slab resting on a subbase course. (The subbase course may be omitted when the subgrade material is granular.) The slab possesses beamlike characteristics that allow it to span across irregularities in the underlying material. When designed and constructed properly, rigid pavements provide many years of service with relatively low maintenance.

16.24.1 Subbase for a Rigid Pavement

This consists of one or more compacted layers of granular or stabilized material placed between the subgrade and the rigid slab. The subbase provides a uniform, stable, and permanent support for the concrete slab. It also can increase the modulus of subgrade reaction k, reduce or avert the adverse effects of frost, provide a working platform for equipment during construction, and prevent pumping of fine-grain soils at joints, cracks, and edges of the rigid slab.

In design and maintenance of a rigid pavement, a major concern is prevention of accumulation of water on or in the subbase or roadbed soils. AASHTO recommends that, if needed for drainage purposes, the subbase layer be carried 1 to 3 ft beyond the paved roadway width or to the inslope. Another concern is prevention of erosion, particularly at slab joints and pavement edges. To compensate for this, lean concrete or porous layers are sometimes used as the subbase material. This practice, however, requires close inspection by design and maintenance personnel.

16.24.2 Types of Concrete Pavements

A concrete pavement may be plain concrete, reinforced concrete, or prestressed concrete. Figure 16.23 shows a cross section of a reinforced concrete pavement. The half cross section in Fig. 16.23a is shown reinforced whereas that in Fig. 16.23b is unreinforced.

Reinforced concrete pavements may be jointed or continuously reinforced. Continuously reinforced pavements eliminate the need for transverse joints but do require construction joints or joints at physical interruptions of the highway, such as bridges. Plain-concrete pavements have no reinforcement except for steel tie bars used to hold longitudinal joints tightly closed.

Jointed Reinforced Concrete Pavement ▪ The main function of reinforcing steel in a jointed concrete pavement is to control cracking caused by thermal expansion and contraction, soil movement, and moisture. The amount and spacing of transverse and longitudinal reinforcing steel required for this purpose depend on slab length, type of steel used, and resistance between the bottom of the slab and the top of the underlying subgrade (or subbase) layer.

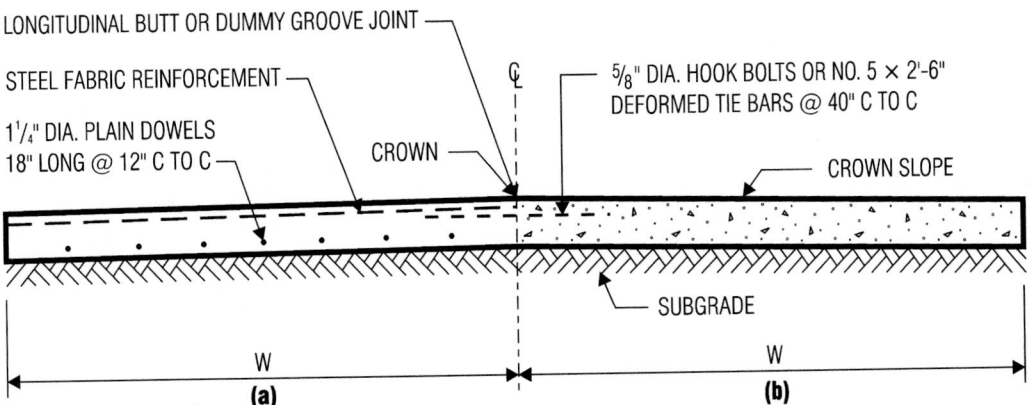

Fig. 16.23 Components of a reinforced concrete pavement.

Continuously Reinforced Concrete Pavement ▪ The principal reinforcement in a continuously reinforced pavement is longitudinal steel, which may be reinforcing bars or deformed wire fabric. It is used to control cracking caused by volume changes in the concrete.

In addition to the longitudinal steel, transverse reinforcement may be provided to control the width of longitudinal cracks. When longitudinal cracking is not expected to be troublesome, transverse reinforcement may not be required.

Design of continuously reinforced pavements should take into account the properties of the concrete used. Specifically, the concrete properties that affect the design of continuously reinforced pavements are tensile strength, shrinkage characteristics, and thermal coefficient. Pavement design should also take into account the anticipated drop in temperature, which for design purposes is the difference between the average concrete curing temperature and a design minimum temperature.

Longitudinal reinforcing steel used typically consists of No. 5 and No. 6 deformed bars. When wire fabric is used, the diameter should be of sufficient size that corrosion and deterioration will not significantly impair the cross-sectional properties of the fabric.

16.24.3 Reinforced Concrete Pavement Slabs

These are typically constructed of portland cement-concrete, reinforcing steel, load-transfer devices, and joint sealing materials. These materials should conform to the appropriate AASHTO or agency specifications to ensure that pavement distortion or disintegration is minimized.

The thickness of concrete pavement slabs generally is determined with the use of design charts or computer software. For details of a design method, see "Guide for Design of Pavement Structures," American Association of State Highway and Transportation Officials (AASHTO, www.aashto. org). In this method, the effects of traffic loads are determined by the use of an equivalent single 18-kip axle load (ESAL). See Arts. 16.22.4 and 16.22.5.

Concrete pavement slabs may be subject to surface deterioration caused by deicing agents or expansion and contraction due to temperature changes. To combat such forms of deterioration, air-entrained concrete is used. Air entrainment also improves the workability of the concrete mix. The design of the mix and its material specifications should be in accordance with the AASHTO "Guide Specifications for Highway Construction" and AASHTO "Standard Specifications for Transportation Materials" (www.aashto.org). Specifications for portland cement concrete are also promulgated by ASTM. The AASHTO and ASTM specifications contain requirements for the properties of the cement, coarse aggregates, and fine aggregates to be used in the mix.

16.24.4 Reinforcing Steel for Concrete Pavement

The purpose of reinforcing steel in a concrete pavement slab is to control cracking as well as tie together slab segments and act as a load-transfer mechanism at joints. The reinforcing steel, whether bars or wire fabric, is generally deformed; that is, the steel surfaces have a ribbed deformation that enhances bond between steel and concrete.

The reinforcing steel that is used primarily to control cracking is known as **temperature steel**. The steel used to tie two slabs together is known as **tie bars**. The steel bars that act as a load-transfer mechanism are called **dowels**.

Temperature steel may consist of deformed bars, a bar mat, or a wire mesh. The purpose of temperature steel is to control the width of cracks, not necessarily to prevent cracking. If a smooth wire mesh is used, the bond between steel and concrete is developed through the welded cross wires. When a deformed wire fabric is used, the bond is developed through the deformations on the steel and at the welded intersections. The steel mesh should be given adequate concrete cover on top, usually about 3 in. The amount of steel to be provided depends on slab thickness, length, and material properties, such as type of concrete and reinforcing steel used.

History has shown that the design of concrete pavements that include reinforcing steel have not performed as well as would be expected. Poor construction methods resulting in the installation of the reinforcing steel too high within the slab have resulted in premature pavement failure. Long term use of de-icing chemicals in northern climates combined with freeze-thaw cycles typically cause the concrete pavement to spall and expose the reinforcement steel when the reinforcing steel is placed too high in the slab. Over time progressive deterioration occurs within the pavement slab, spalls expand, further reinforcing steel is exposed which cause significant routine maintenance problems and reduces the intended life of the pavement.

16.24.5 Tie Bars

These are installed between abutting slabs to tie them together. For this purpose, the tie bars, which may be connectors or deformed bars, should have sufficient tensile strength to prevent the slabs from separating. (Tie bars are not intended to serve as load-transfer devices.) The tendency for the slabs to separate arises because they try to shorten when the temperature drops or moisture content of the concrete decreases (as is the case when the concrete cures). The resistance to movement provided by the tie bars induces tensile stresses in the concrete, which must have sufficient tensile strength to withstand these stresses, or reinforcing steel should be provided for this purpose. To facilitate bonding, the tie bars are usually equipped with a hook. It is often advantageous to provide the bars with a protective coating, especially when the pavement slab is exposed to deicing agents.

16.24.6 Load-Transfer Devices

Load-transfer devices are installed between the ends of abutting slabs to transfer traffic loads from one to the other yet offer little or no resistance to longitudinal movements of the slabs. The most common form of load-transfer device is a large-diameter dowel. Other mechanical devices, however, have been used successfully. It is also possible to achieve load transfer with the interlocking of aggregate alone.

A dowel provides flexural, shear, and bearing resistance. One end of the dowel is bonded to the concrete. The other end is allowed to move freely. For this purpose, half the dowel adjoining this end may be greased, painted, or coated with asphalt, thus preventing the dowel from bonding to the concrete. As a result, the dowel can slide freely in the concrete after embedment in the end of a slab. To ensure proper movement of the dowels, it is essential to maintain alignment between the abutting slabs.

Although offering little restraint to longitudinal movement of the slabs, load-transfer devices should also be mechanically stable under wheel loads and cyclical loading. It is often beneficial to provide the dowel, or other device, with a pro-

tective coating when the slab may be exposed to corrosive elements.

Dowel size depends on slab thickness and other site-specific criteria. A general rule of thumb is that the dowel diameter should be equal to one-eighth the slab thickness; for example, for a 9-in slab, a diameter of $1\frac{1}{8}$ in might be used. Dowel spacing is generally 12 in and dowel length 18 in.

16.24.7 Joints in Concrete Pavement

Joints are formed in concrete pavement to reduce the effects of expansion and contraction, facilitate concrete placement, and allow for bonding of abutting slabs. Joints may be perpendicular to the pavement centerline (transverse) and, depending on the intended function of the joints, longitudinal.

Transverse Expansion Joints ▪ The principal function of an expansion joint in concrete pavement is to allow movement of the slab due to changes in temperature. For example, when temperature increases, a slab increases in length, thereby creating compressive stresses in the concrete. If expansion joints are not provided, the slab, depending on its length, might buckle upward or blow up.

In concrete pavement, expansion joints are generally placed every 40 to 60 ft along the length of the pavement. The joints, which may range in thickness from $\frac{3}{4}$ to 1 in, should incorporate appropriate load-transfer devices (Art. 16.22.6). Fillers, such as rubber, bitumen, or cork, that permit expansion of the slab and exclude dirt should be placed in the joints.

Some transportation agencies specify expansion joints, but others do not and instead employ alternate methods to minimize the potential for blowups. One way this is done is to use cement and aggregates with properties that limit the amount that slabs increase in length with increases in temperature.

Transverse Contraction Joints ▪ Contraction joints are provided to limit the effects of tensile forces in a concrete slab caused by a drop in temperature. The objective is to weaken the slab so that if the tensile forces are large enough to crack it, the cracks will form at the joints. The depth of contraction joints generally is only one-quarter the thickness of the slab. When properly designed and spaced, however, they can also minimize cracking of the slab outside the joints.

Contraction joints can be formed by sawing into the hardened concrete, by inserting plastic inserts at joint locations before concrete is cast, or by tooling the concrete after placement but before the concrete has fully hardened. When aggregate interlocking is insufficient for transferring load between the pavement sections on either side of a joint, an appropriate load-transfer mechanism should be incorporated in the joint.

Longitudinal Joints ▪ These are formed parallel to the highway centerline to facilitate lane construction and prevent propagation of irregular, longitudinal cracks. The joints can be keyed, butted, mechanically formed, or saw grooved. To prevent adjacent lanes from separating or faulting, steel tie bars or connections should be embedded in the concrete, perpendicular to the joints.

Longitudinal joints are formed or sawed to a minimum depth of one-fourth the slab thickness. The maximum longitudinal joint spacing recommended by AASHTO is 16 ft.

Construction Joints ▪ When concrete placement for a concrete slab is interrupted, a construction joint is desirable at the *cold joint* between the two slab sections. In preparation for the interruption, a vertical face is formed with a header board at the end of the slab being cast. The header board is equipped with a protrusion that, when formed into the concrete, creates a key way for load transfer between the adjoining slab sections. It is also sometimes desirable to use deformed tie bars to hold the joint closed.

Joint Sealing ▪ Many highway agencies specify that all joints be cleaned and also sealed to exclude incompressibles (small stones and other small roadway debris contained within the joint itself), water, and dirt. Others seal only expansion joints. The basic types of sealants used are liquid seals, preformed elastomeric seals, and cork expansion-joint filler.

Liquid seals are poured into a joint where they are allowed to set. Types of liquid seals used include asphalt, hot-poured rubber, elastomeric compounds, silicone, and polymers.

Preformed elastomeric seals consists of extruded neoprene strips with internal webs that exert an outward force against the faces of the joint. The type of elastomeric seal to specify depends on the anticipated slab movement at the joint. The seals are provided with a coat of adhesive to fasten them to the faces of the joint.

Proper sealing and long-term bonding of asphalt cement concrete shoulders with portland cement concrete pavements requires different material type seals than those for bonding portland cement concrete pavements with portland cement concrete shoulders. Joint sealant manufacturers literature and transportation agency materials departments should be consulted for approved material types to ensure proper bonding of pavement materials and long term performance.

Highway Intersections and Interchanges

An intersection is the junction or crossing of two or more roads at the same or different elevations. When the roads are at the same level, the intersection is called an at-grade intersection. When the roads are at different elevations, the intersection is referred to as a grade separation when there is no connection between the intersecting roads or as an interchange when connecting roads, such as ramps or turning roadways, permit movement of vehicles between the intersecting roads.

Intersections should be kept simple so that necessary movements are obvious to drivers. Uniformity of intersections is important to avoid driver confusion. Factors to be considered for this purpose include design speed, intersection angles (90° is preferred), intersection curves, vehicle turning paths, roadway widths, and traffic control devices.

16.25 At-Grade Intersections

The junction or crossing of two or more highways at a point of common elevation is called an at-grade intersection.

Intersections of highways and railways at grade should be provided with protective and warning devices. Sight distance is an important design consideration when only advance warning of approaching trains and railway crossbuck signs are installed.

16.25.1 Geometric Design of At-Grade Intersections

Major influences on the geometric design of at-grade intersections include human factors, traffic considerations, physical elements, and economic factors. The goal is to reduce or eliminate the potential for accidents involving vehicular, bicycle, or pedestrian traffic through the intersection. Also, natural transitional paths for traffic must be provided.

Human Factors ▪ Design of an intersection is affected by human factors, such as driving habits, the ability of drivers to make decisions, adequate advance warning to drivers regarding the presence of an intersection, driver decision and reaction time, and the presence of pedestrians at the intersection.

Traffic Considerations ▪ Traffic volume and movement impact the design of an at-grade intersection. Both the design and actual capacity of the intersecting highways should be taken into account. Also of concern are the design-hour turning movement and other movements, such as diverging, merging, weaving, and crossing.

Other traffic criteria include vehicle size, speed and operating characteristics, transit involvement, and, if applicable, the history of accidents at the site. Storage requirements for traffic-signal-controlled approaches should also be taken into account.

Physical Elements ▪ Geometric and site-specific features whether natural such as topography and vegetation, or man-made, such as signs, presence of buildings, sidewalks, utilities have important influences on design of at-grade intersections. The horizontal and vertical alignment of the intersecting roadways also greatly affects the design. Both of these elements impact sight distance and angle of intersection. Other features affecting the design are traffic control devices, lighting equipment, and safety appurtenances.

Preexisting site conditions, including abutting property uses, such as shopping areas and industrial complexes, and the presence of sidewalks and their associated pedestrian traffic, should also be factored into the design process.

Economic Factors ▪ Design of an at-grade intersection should be both practicable and economically feasible. The cost of required improvements along with the impact on abutting residential or commercial properties should be taken into account.

16.25.2 Types of At-Grade Intersections

Each highway that radiates from an intersection and forms part of it is known as an **intersection leg**. The intersection of two highways generally results in four legs. Intersections with more than four legs are not recommended.

Three-Leg Intersections ▪ A three-leg intersection is formed when one highway starts or terminates at a junction with another highway (Fig. 16.24). Unchannelized T intersections (Fig. 16.24a) are usually employed at the intersection of minor roads with more important highways at an angle not exceeding 30° from the normal. At times, a right-turn lane is provided on one side of the through highway (Fig. 16.24b). This type of turn lane is used when right-turning traffic from the through highway is significant and left-turning traffic from the through highway is minor.

Four-Leg Intersections ▪ A four-leg intersection is formed when two highways cross at grade (Fig. 16.25). The design of four-leg intersections follows many of the general guidelines for three-leg intersections. As with T intersections, the roadway intersection angle typically should not be more than 30° from the normal. Figure 16.25c shows a four-leg intersection of a through highway and a minor highway. The through highway is flared to provide additional capacity for through and turning movements. The flaring is provided through incorporation of parallel auxiliary lanes that are required for major highways requiring an uninterrupted flow capacity. Flaring may also be needed where cross traffic is sufficiently high to warrant signal control.

Channelization at Intersections ▪ This is a method of creating defined paths for vehicle travel by installing traffic islands or pavement markings at at-grade intersections. These *defined*

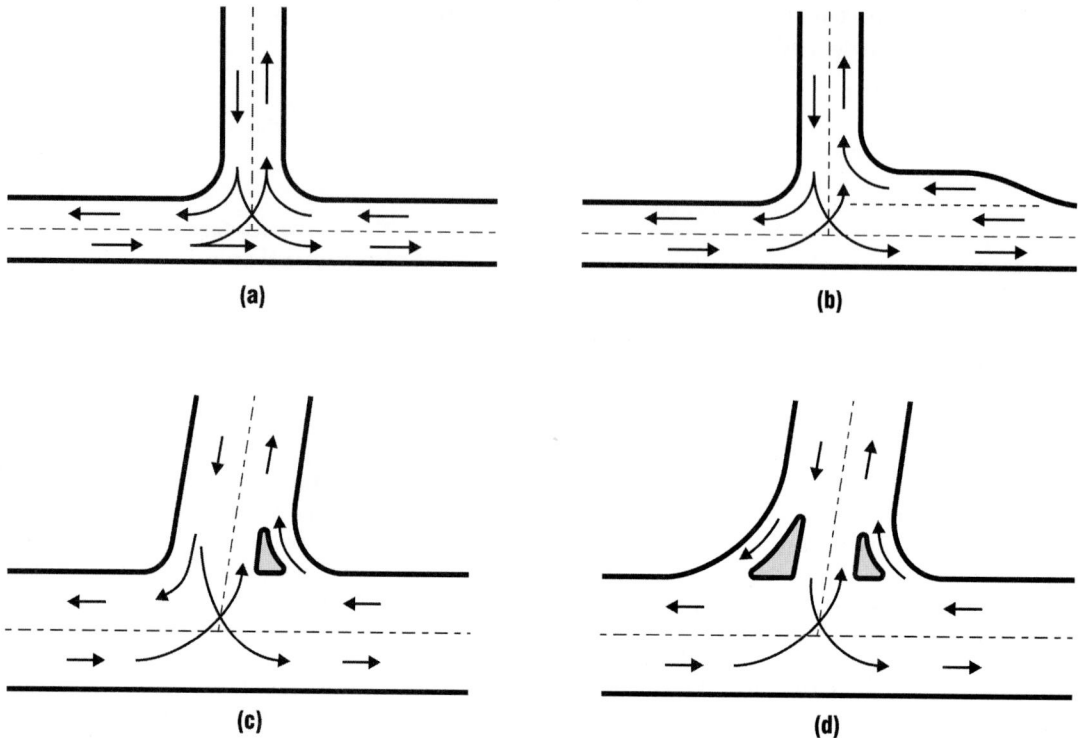

Fig. 16.24 Types of at-grade T (three-leg) intersections: (*a*) unchannelized; (*b*) intersection with a right-turn lane; (*c*) intersection with a single-turning roadway; (*d*) channelized intersection with a pair of turning roadways.

paths provide for the safe and orderly movement of both vehicles and pedestrians through the intersections. Channelized intersections are illustrated in Figs. 16.24 and 16.25.

Channelization should be used prudently; excessive use of channelization may worsen rather than improve conditions at an intersection. When properly implemented, channelization can dramatically reduce accidents at at-grade intersections. Factors that influence design of a channelized intersection include type of design vehicle, vehicle speed, cross sections of roadways, anticipated volumes of vehicle and pedestrian traffic, locations of bus stops, and type and location of traffic-control devices.

Figure 16.24*c* and *d* shows examples of channelization of three-leg intersections. In these intersections, the return radius between the intersecting roadways is larger than that used for the

unchannelized intersections in Fig. 16.24*a* and *b*. This is done to provide space for the channelizing traffic islands. Whether the approach roadway should have a turning lane for right-turning traffic depends on the volume of traffic that will make right turns. When speeds or turning paths are above a prescribed minimum, the incorporation of dual right-turning roadways, as shown in Fig. 16.24*d*, may be required.

Channelization of four-leg intersections is often incorporated for many of the reasons given above for three-leg intersections. In Fig. 16.25*b*, a four-leg intersection with right-turning lanes in all four quadrants is shown. This approach is taken when space is readily available and turning movements are critical. This type of channelized, four-leg intersection is used frequently in suburban areas where pedestrian traffic is present.

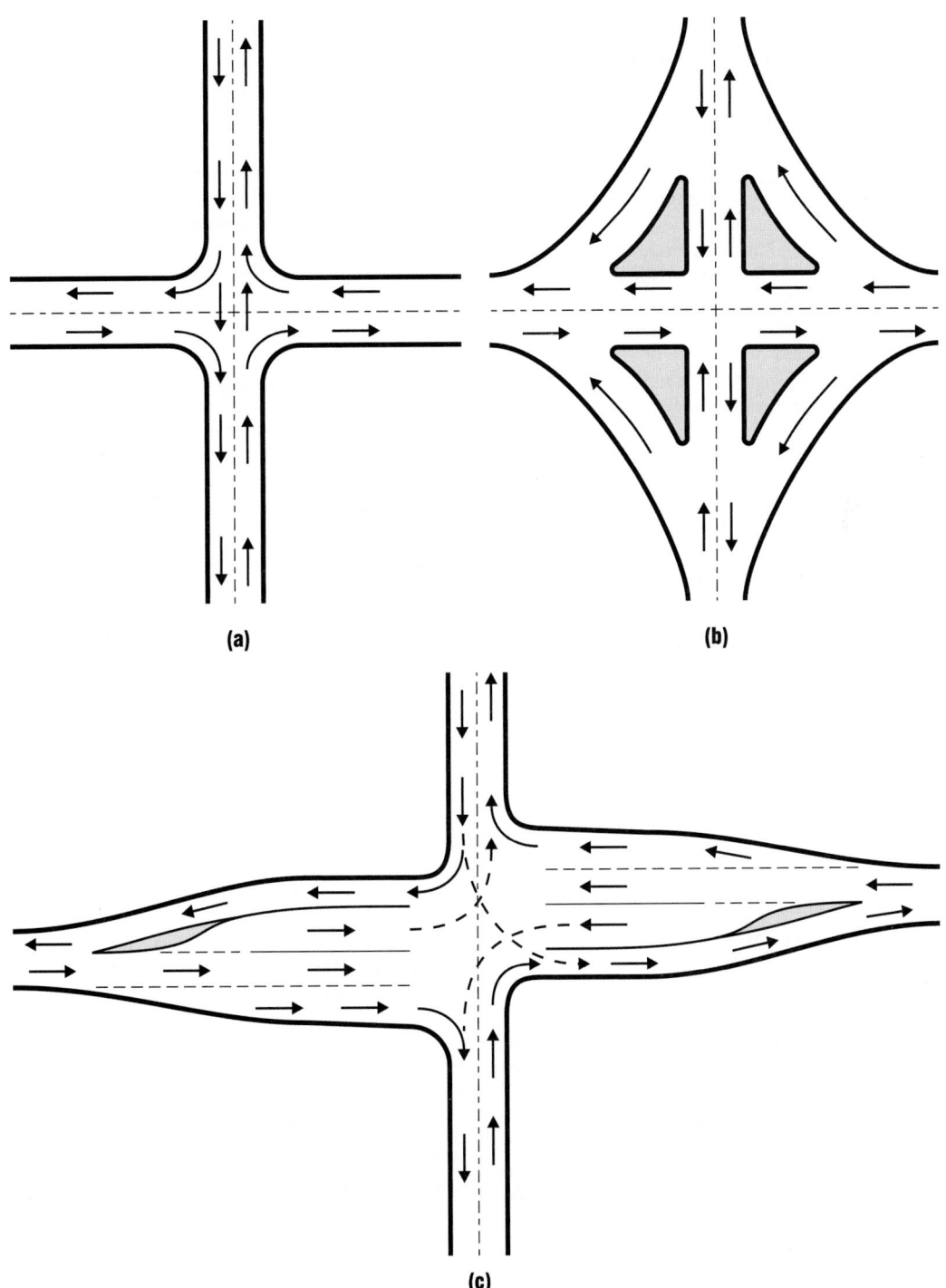

Fig. 16.25 Types of at-grade four-leg intersections: (*a*) unchannelized; (*b*) channelized; (*c*) flared.

16.25.3 Horizontal and Vertical Alignment at Intersections

Alignment geometries play a critical role in the design of an at-grade intersection. In the vertical plane, it is important that the profiles of the intersecting roadways be as flat as possible (preferably less than 3% through the intersection). Also, the horizontal alignment should be as straight as practical. Grade and curvature have considerable impact on sight distance at intersections, where it is desirable to have sight distances greater than specified minimum values. Adverse sight-distance conditions can be the source of accidents, because of driver inability to see other vehicles or discern the messages of traffic-control devices.

Horizontal Alignment ▪ A general rule of thumb when laying out the horizontal alignments of intersecting roadways is to minimize the deviation of the intersection angle from 90°. Excessively skewed intersections can result in poor driving conditions, especially for trucks. Where the acute angles are formed, visibility may be limited and the exposure time of vehicles dangerously large as they cross the main flow of traffic. Where the obtuse angles are formed, blind spots may occur on the right side of the vehicle. Therefore, it is desirable to have the intersection angle as close to 90° as possible.

Vertical Alignment ▪ Maximizing driver sight distance should be the goal of the vertical alignment. Proper sight distance should be provided along each highway and across the corners.

Whenever possible, major grade changes should be avoided at intersections. Generally, the profile of the major highway at an intersection should be extended through it and the profile of the minor crossing road adjusted to match that of the major highway at the intersection. This may require transitioning or warping the cross section of the minor road. For low-speed, unchannelized intersections where stop controls or signs are present, it may be desirable to warp the crown of each intersecting roadway. Any alteration of the roadway cross section should be gradual and take into account the effects on drainage.

16.25.4 Islands

An island is a defined area established between traffic lanes in channelized intersections to direct traffic into definite paths. It may consist of curbed medians or areas delineated by paint. In general, islands are provided in channelized intersections to separate and control the angle of conflicts in traffic movement, reduce excessive pavement areas, protect pedestrians and waiting areas for turning and crossing vehicles, and provide a location for traffic-control devices.

16.26 Highway Interchanges

An interchange is a system of interconnecting roadways used in conjunction with one or more grade separations of highways (Fig. 16.26). It accommodates movement of traffic between two or more roadways at different elevations. In so doing, it eliminates grade crossings, which may be unsafe and are inefficient in accommodating both turning and through traffic. When highways carrying high volumes of traffic intersect each other, the greatest degree of safety, efficiency, and capacity is achieved with grade separations of the highways.

There are in use numerous variations of the interchange types shown in Fig. 16.26. They vary in size and magnitude depending on the environment and scope of service for which they are intended. Design of an interchange is based on traffic volume, topography of the site, economic considerations, and environmental factors.

16.26.1 Justification of Interchanges

An interchange is not needed at every highway intersection. The American Association of State Highway and Transportation Officials (AASHTO) considers the following as warranting investigation of the advisability of selecting an interchange instead of an at-grade intersection: highway classifications, need for eliminating traffic bottlenecks and hazards, road-user benefits, and traffic volume.

Design Classification ▪ When a highway has been designated to serve as a freeway (Art. 16.1.1), the designer must decide whether each

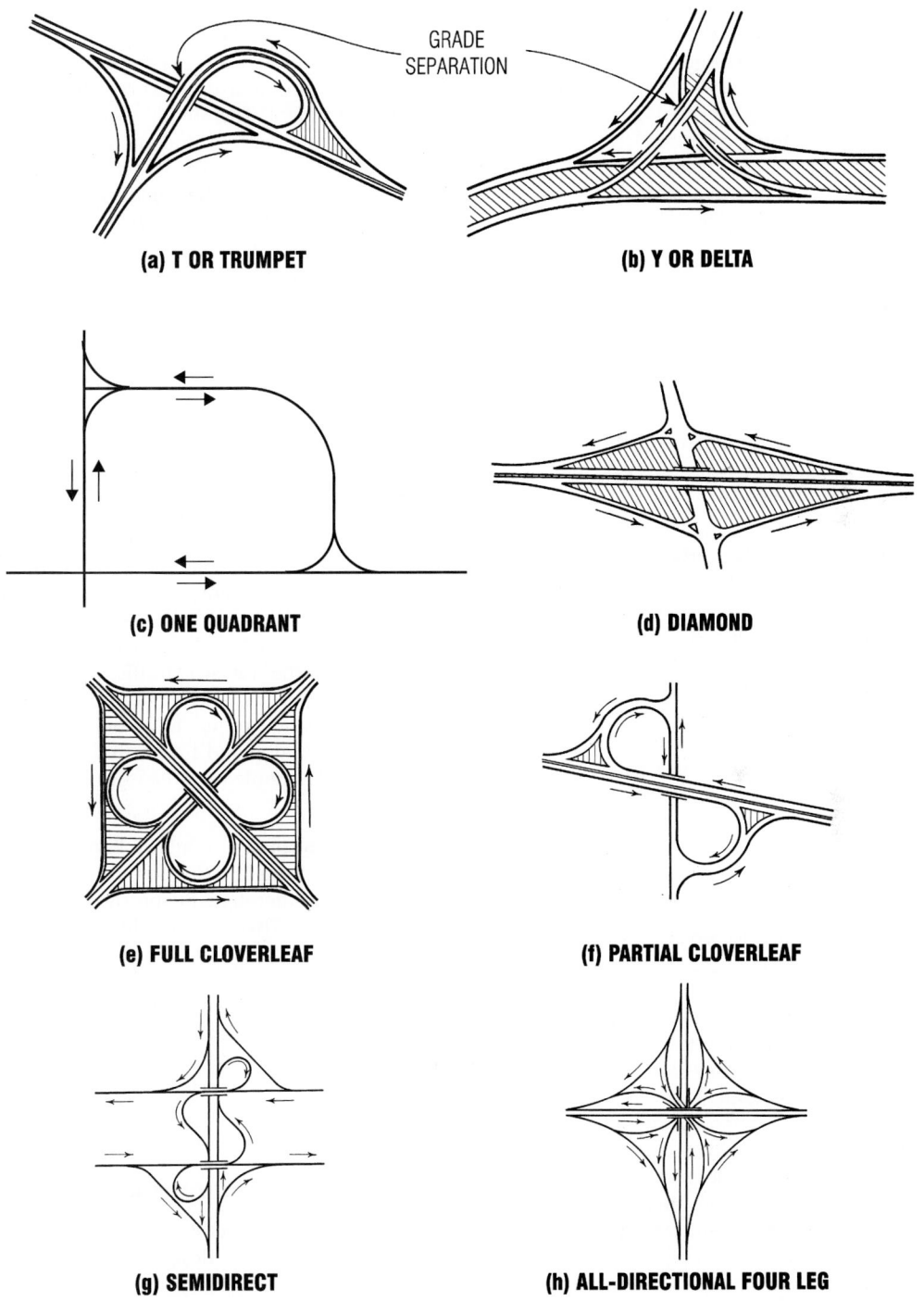

Fig. 16.26 Types of interchanges for intersecting grade-separated highways.

intersecting highway should be terminated, rerouted, or connected to the freeway with a grade separation or an interchange. The overriding goal should be maintenance of a safe and uninterrupted flow of traffic on the freeway. When traffic on an intersecting road must cross the freeway, a grade separation is necessary to eliminate interference with traffic flow on the freeway. When access from the other road to the freeway is required, an interchange is required.

Elimination of Bottlenecks ▪ A general drawback to at-grade intersections is the potential, due to high volume of traffic, for spot congestion or bottlenecks occurring at one or more of the approaches. If an at-grade interchange cannot satisfy the capacity requirements of the intersecting highways, then use of some form of interchange should be investigated.

Elimination of Hazards ▪ The occurrence of numerous accidents at an at-grade intersection may warrant construction of an interchange. Its feasibility, however, depends on the environment in which the intersection exists. An interchange necessitates acquisition of large amounts of right-of-way. Availability and cost of the needed land is an important consideration in deciding on an interchange. As a result, interchanges are more likely to be employed in rural areas than in urban areas for elimination of hazards.

Road-User Benefits ▪ Substitution of an interchange for an at-grade intersection can often lead to direct economic benefits for users. Delays and congestion at an at-grade intersection can be costly because of damage from accidents and the consumption of fuel, tires, oil, and time while waiting for an opportunity to cross or make turns or for signal changes. Time lost waiting at traffic signals can be extremely severe when traffic volumes are large. Although an interchange requires users to travel a longer distance than they would at a conventional at-grade intersection, this disadvantage is more than offset by the benefits from savings in time resulting from use of the interchange.

To determine whether road-user benefits justify construction of an interchange, the designer should compare the projected benefits with the cost of improvement required. This may be done with the use of the ratio of the annual user benefit to the annual capital cost of the improvement. The annual benefit is the difference between user cost for the existing condition and the cost for the condition after improvement. The annual capital cost is the sum of the annual interest and amortization for the cost of the improvement. The larger the benefit-cost ratio, the greater is the justification for the interchange based on road-user benefits. A general rule of thumb is that a ratio greater than 1 is the minimum required for economic justification. Another consideration is that an interchange can be implemented in stages, in which case incremental benefits can be realized that compare even more favorably with incremental costs.

Proper local land use planning, development activities, zoning, and access management are important considerations for a staged interchange implementation. Peripheral development and land use may preclude full implementation of the interchange if the above mentioned issues are not properly managed by both the local government and transportation agency. The full interchange design may involve relocation of facilities and redesign of the adjacent roadways to safely accommodate the revised traffic patterns. Full interchange implementation may be the optimal choice initially in order to avoid many time consuming planning and traffic studies necessary in order to fully implement the interchange.

Traffic Volume ▪ While a high volume of traffic is not sole justification for an interchange, it is a major consideration in the overall decision-making process. This is especially the case when traffic volumes exceed the capacity of an at-grade intersection, in which case use of an interchange is generally indicated. The unavailability or high costs of land for an interchange, however, may override the benefits accruing from elimination of the traffic conflicts associated with an at-grade intersection.

16.26.2 Types of Interchanges

After deciding to specify an interchange for a highway intersection, designers have a wide variety of interchange layouts from which to choose (Fig. 16.26). The type of interchange to use and its application at a given site depend on many factors including the number of intersection legs (parts of highways radiating from the intersection), anticipated volume of through and turning movements, topography of the site, culture, design controls, signing, and initiative of the designer.

Design of an interchange typically is custom fit to site conditions and constraints. It is desirable, however, to provide a certain degree of uniformity in interchanges to prevent driver confusion. Also, although interchanges offer greater safety than do at-grade intersections, there are safety issues of concern with interchanges, such as proper signing, lighting and placement of exits.

Three-Leg Interchanges ▪ These consist of one or more highway grade separations with three intersecting legs. All traffic moves over one-way roadways. In plan view, the roadway layout generally resembles a T or a Y, or delta. A T, or trumpet, interchange is a three-leg interchange in which two of the three legs form a through road and the angle of intersection with the third leg is about 90° (Fig. 16.26a). When all three intersection legs are through roads, or the intersection angle of two legs with the third leg is small, the interchange is called a Y, or delta, interchange (Fig. 16.26b). Any basic interchange pattern, regardless of through road characteristics or intersection angle, can be adapted to specific site conditions.

Four-Leg Interchanges ▪ These consist of one or more highway grade separations with four legs. General categories of four-leg interchanges include ramps in one quadrant, diamond, full cloverleaf, partial cloverleaf, and semidirect- and direct-connection interchanges. Partial cloverleafs include interchanges with ramps in two or three quadrants.

Interchanges with ramps in one quadrant (Fig. 16.26c) are generally used where low-volume roads intersect and topography necessitates incorporation of some form of interchange. With such interchanges, turning traffic is facilitated through the use of a single two-way ramp of near-minimum design. Since interchanges are rarely used in areas with a low volume of traffic, application of this type of interchange is somewhat limited. A possible use of a ramp in only one quadrant is for the intersection of a scenic parkway and a state or county two-lane highway. For such a setting, preservation of the existing topography, absence of truck traffic, and relatively small number of

turning movements would justify this type of interchange. To control turning movements, however, left-turn lanes must be provided on the through roads and a high degree of channelization is required at the terminals and the median.

Diamond Interchanges. One of the more common types of interchange used, diamond interchanges are generally employed where a highway with large traffic volume crosses but is separated by a bridge from a road carrying comparatively light or slow-speed traffic (Figs. 16.26d and 16.27). The diamond layout is the simplest form of all-movements interchange. The two highways are connected by four one-way ramps that may be straight or curved to suit the existing topography or site conditions. The ramps connect with one of the highways at a flat angle.

If the roads carry moderate to large traffic volumes, ramp traffic may be regulated through the use of signal-controlled ramp terminals. When this is the case, widening may be required at the ramp or at the cross street through the interchange area, or at both locations. Each ramp terminal at the minor road is formed with a T or Y at-grade intersection, which allows one left and one right turning movement. If the volume of traffic is large enough, the cross street may be divided and separate lanes provided for the left turns.

A diamond interchange has many advantages over a partial cloverleaf (Fig. 16.26f). Unlike a cloverleaf design where traffic typically slows when entering the ramp, diamond interchanges allow entry and exit at relatively high speeds. Also, they occupy a comparatively narrow band of right-of-way, which may be not more than that required for the highway proper.

Split-Diamond Interchanges. These consist of two pairs of parallel or nearly parallel streets connected by two pairs of ramps (Fig. 16.27). As indicated in Fig. 16.27a, which shows a split-diamond interchange for two-way streets, the parallel streets need not be consecutive. Figure 16.27b is an example of a split-diamond interchange for one-way streets. In the case illustrated, connecting (frontage) roads parallel to the elevated highway are provided between the cross streets to improve traffic movement.

A split-diamond interchange reduces traffic conflicts by accommodating the same amount of traffic at four rather than two crossroad intersections. This has the effect of reducing the left-turn

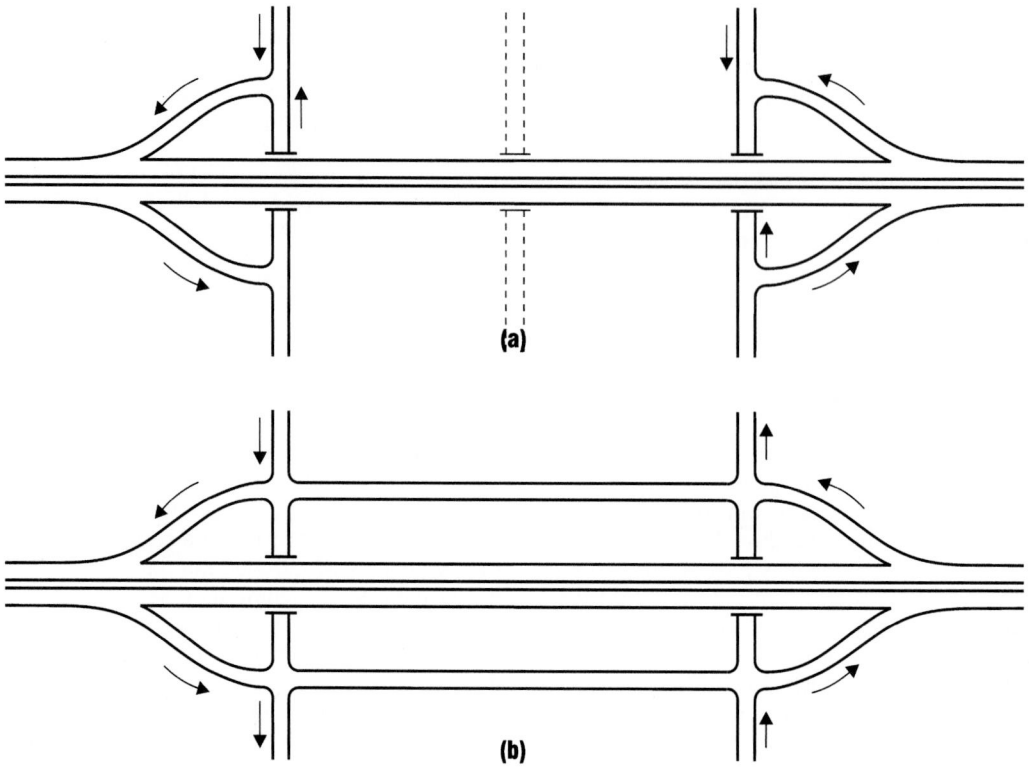

Fig. 16.27 Split-diamond interchanges: (*a*) with two-way streets; (*b*) with one-way streets.

movements at each intersection from two to one. One drawback to the split-diamond interchange, however, is that traffic leaving the elevated highway cannot return to the same interchange and continue in the same direction.

Cloverleaf Interchanges. A cloverleaf interchange provides direct connections for right turns between two highways but utilizes loop ramps to accommodate left turns. A full cloverleaf (Fig. 16.26*e*) has loops in four quadrants, whereas a partial cloverleaf (Fig. 16.26*f*) has loops in only two quadrants.

While a cloverleaf interchange greatly reduces accidents by eliminating all left turns, it does possess drawbacks. For example, high speed and large volume of traffic require large radii for the loop ramps and hence acquisition of very large areas of right-of-way. This has greatly limited use of cloverleafs in urban regions. Even a slight increase in design speed can require significantly

greater radii. For a design speed of 25 mi/h, for instance, design standards call for a loop radius of 150 ft. An increase of only 5 to 30 mi/h, an increase of 20%, requires a radius of 230 ft, an increase of 53%. Furthermore, the area required for right-of-way increases by about 130%.

Another disadvantage of cloverleafs is that left-turning traffic must travel a greater distance than otherwise would be required and significant weaving movement may be generated. For a loop designed for 20 mi/h and having a 90-ft radius, for example, the extra travel distance required is about 600 ft. In contrast, for a loop designed for 25 mi/h and having a 150-ft radius, the extra distance is roughly 1000 ft, and for 30 mi/h on a loop with a 230-ft radius, the extra distance is about 1500 ft. Moreover, since travel time on ramps varies almost directly with the length of ramp, the time that might be saved by increased speed is lost over the greater distance that must be traversed. In addition, weaving maneuvers associated with the use of a cloverleaf for left turns can result in serious vehicle interference and a corresponding slow-down of through traffic, especially when the flow exceeds 1000 vehicles per hour.

Since it is seldom practical to provide for more than a single lane on a loop, a ramp can be expected to accommodate no more than 800 vehicles per hour. If truck traffic is not anticipated and the design speed for the ramp is 30 mi/h or higher, a design capacity of 1200 vehicles per hour can be used. Thus loop-ramp traffic capacity is a major constraint and can limit the effectiveness of a cloverleaf interchange.

Partial Cloverleaf Ramp Arrangements. A partial cloverleaf interchange utilizes loop ramps in only two or three quadrants. This type of interchange is desirable when the anticipated traffic distribution does not require a full cloverleaf. A major design decision is selection of the quadrants in which ramps should be placed.

The arrangement of ramps in a partial cloverleaf interchange should facilitate major turning movements at right-turn exits and entrances and limit impediments to traffic flow on the major highway. If traffic on the major highway is much greater than that on the minor intersecting road, right-turn exits and entrances, in general, should be placed on the major highway. Such an arrangement, however, will require a direct left turn off the crossing road.

Directional Interchanges. These provide direct or semidirect connections between intersecting highways. They are often preferred to cloverleaf interchanges, which are composed only of loops and consequently may fail to meet the high speed and traffic volume demands of an expressway.

A **direct connection** is a one-way roadway that does not deviate greatly from the intended direction of travel. An interchange that utilizes direct connections for all major left turns is called a directional interchange (Fig. 16.26h). It may also incorporate loops for minor left turns. Loops in conjunction with direct connections are generally used in rural areas where direct connections in all quadrants cannot be justified.

A **semidirect connection** is a one-way roadway that deviates from the intended direction of travel but is more direct than a conventional loop (Fig. 16.26g) Therefore, a semidirectional interchange is similar to a directional interchange

except that it utilizes semidirect connections to accommodate major left turns.

Directional interchanges typically require several grade separations. Figure 16.26h shows a directional interchange with a four-level structure.

Directional interchanges are generally designed to accommodate many site-specific conditions, including topography, geometry, and traffic demands. The design of individual ramps should satisfy accepted standards for curvature, pavement widths, length of weaving sections, and entrance and exit design criteria.

16.26.3 Selection of Interchanges

The type of interchange to select is one that best meets the needs of the site and provides operational efficiency and safety, and adequate capacity for anticipated traffic volumes and turning patterns. It is advisable to choose the type of interchange before final route selection. This permits a determination that the interchange type selected can be adequately developed.

Interchanges generally fall into two basic categories: system interchanges and service interchanges. The former includes interchanges that connect one freeway to another. In contrast, service interchanges connect a freeway to a road with a lower-grade classification. In a rural environment, service demand is a principal issue.

When the intersecting roadways are freeways, all-directional interchanges may be advantageous to facilitate high turning volumes. When only some turning volumes are high, a combination of directional, semidirectional, and loop ramps may prove advantageous. Where directional or semidirectional interchanges are used in conjunction with loops, however, weaving sections should be avoided.

A cloverleaf interchange is the minimum type to use at the intersection of two highways both of which have partial or full control of access. Cloverleafs are also advantageous for intersections where left turns at grade are prohibited. A diamond intersection is appropriate where a major highway intersects a minor road. A partial cloverleaf may be advisable where traffic volumes or site conditions do not warrant or allow employment of a full cloverleaf interchange.

Interchanges for highways in urban areas should be selected on a systemwide rather than individual basis. Since interchanges are usually closely spaced in urban environments, an interchange may influence selection and design of preceding and subsequent interchanges. For example, additional lanes may be required to accommodate capacity, weaving, and lane-balance requirements.

A general rule of thumb is that a minimum interchange spacing of 1 mi should be used in urban areas, but a closer spacing may be used if grade-separated ramps are provided or collector-distributor roads are added. A minimum spacing of 2 mi should be used in rural regions.

16.26.4 Ramps in Interchanges

A ramp is a roadway that connects two or more legs of an interchange and is used for turning traffic (Fig. 16.28). The main elements of a ramp are a connecting roadway and a terminal at each end. The profile of the connecting roadway typically is sloped and the horizontal alignment is curved. In general, design criteria for horizontal and vertical alignments of ramps are less restrictive than those of the intersecting highways, but sometimes the design criteria are the same.

In design of a ramp, the designer has to balance several factors. For example, consider topography and costs of right-of-way, which influence selection and design of the ramp. To conserve land, it may be necessary to locate the ramp so close to the highway that a retaining wall must be constructed. The cost of the wall then has to be balanced against the cost of acquiring additional right-of-way to eliminate need for the wall.

The type of ramp to use depends on the type of interchange. A trumpet interchange, for example, utilizes one loop, one semidirectional ramp, and two right directional or diagonal ramps (Figs. 16.26a and 16.28). Usually, a ramp is a one-way roadway. Some ramps, such as a diagonal ramp, are one-way but allow both left and right turns at a terminal on a minor intersecting road.

Ramp Design Speeds ▪ The design speed for a ramp generally should be about the same as that for the intersecting highway with the lower traffic volume. Although lower ramp speeds may be necessary, the design ramp speed used should not be less than the values presented in Table 16.9. The table lists, as a guide, ramp design speeds to be used with various highway design speeds.

When a ramp connects a high-speed highway with a minor road or a city street, provision should be made for a considerable reduction in speed for traffic leaving the high-speed highway. An initial speed reduction can be accomplished through use of a deceleration lane on the main highway. To allow for continuing deceleration on the ramp, the radii of the curves on the ramp should be reduced in stages. At the ramp terminal at the minor road, it may be necessary to provide some form of signal or signing to stop or slow vehicles.

Ramp Curvature ▪ The principles governing horizontal curvature (Art. 16.13) are also applicable to the design of interchange ramps. For example, use of compound curves and spirals is often beneficial in adapting a ramp to site-specific conditions and providing a natural path for vehicles. Loops, except for their terminals, may be composed of circular arcs or some other curve that is formed with spiral transitions. It is very important for the highway designer to recognize the varied types and sizes of trucks that will utilize a particular facility. For interstate type highways, ramps turning radius and shoulders must be adequate to accommodate the wide turning radius of large, long trucks (tandem-trucks or double bottoms).

Ramp Sight Distance ▪ Safety demands provision for adequate sight distance along ramps and at the ramp terminals. Sight distance along ramps should be at least as long as the safe stopping sight distance. Sight distance for passing, however, is not required.

At the ramp terminals, a clear view of the entire exit terminal should be provided. The exit nose and a section of the ramp pavement beyond the gore, the area downstream from the shoulder intersection points, should be clearly visible

Ramp Vertical Curves ▪ In general, a ramp grade should be as flat as practical to limit the amount of driving effort needed to traverse from one road to another. A ramp profile typically resembles the letter S. It consists of a sag curve at the lower end and a crest curve at the upper end. When a ramp goes over or under another roadway, however, additional vertical curves may be required.

Ramp Terminal ▪ This is the portion of a ramp that adjoins the through traveled way. The terminal includes speed-change lanes, tapers, and islands. A ramp terminal may be an at-grade type, as is the case for a diamond interchange, or a

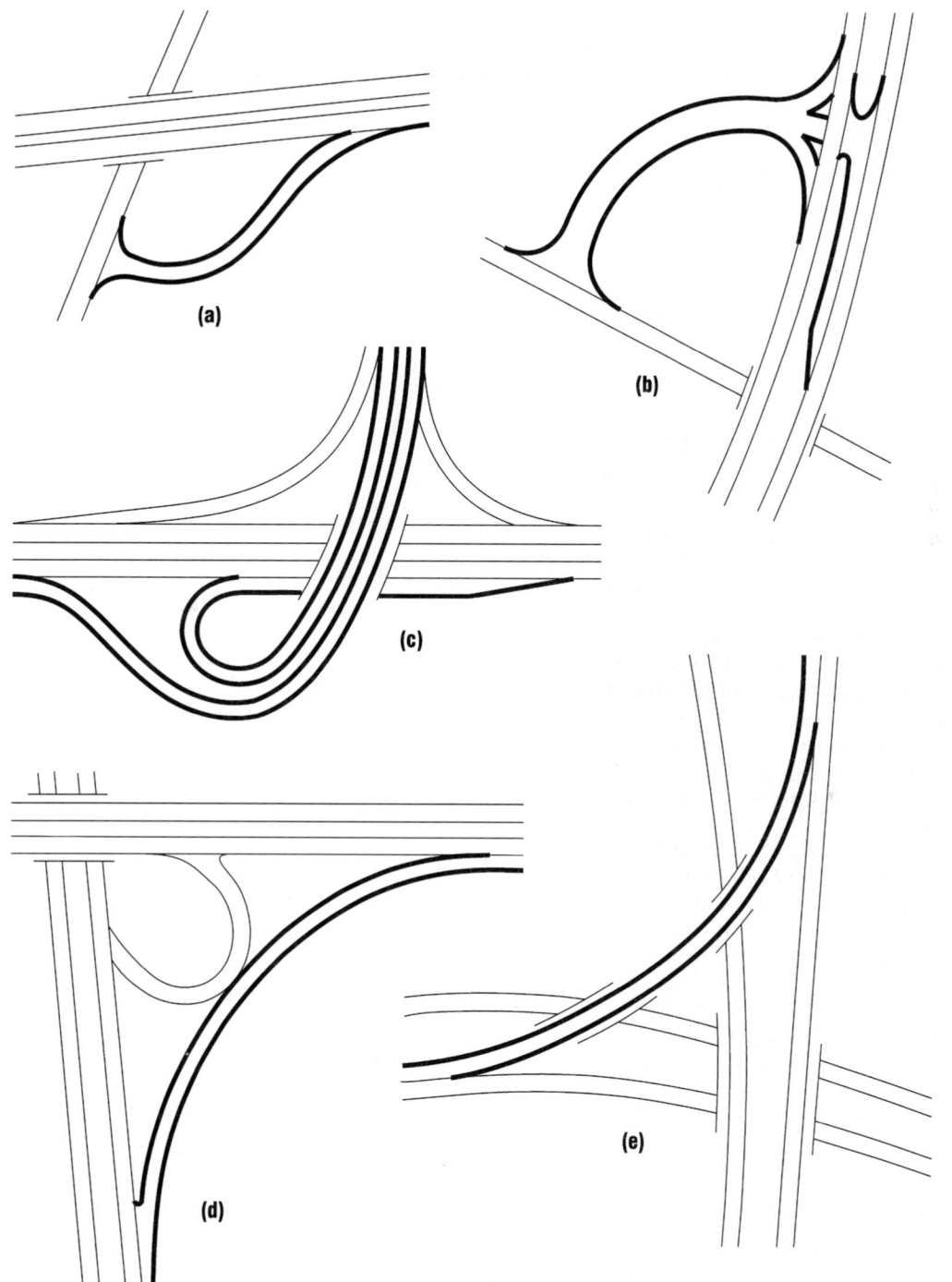

Fig. 16.28 Types of ramps: (*a*) diagonal; (*b*) one-quadrant; (*c*) loop and semidirect; (*d*) outer connection; (*e*) directional.

Table 16.9 Suggested Ramp Design Maximum Speeds, mi/h, Based on Design Speeds of Connected Highways*

Ramp design speed, mi/h	Highway design speed, mi/h					
	30	40	50	60	65	70
Upper range (85%)	25	35	45	50	55	60
Middle range (70%)	20	30	35	45	45	50
Lower range (85%)	15	20	25	30	30	35

* Adapted from "A Policy on Geometric Design of Highways and Streets," American Association of State Highway and Transportation Officials.

free-flow type that allows ramp traffic to merge with or diverge from high-speed through traffic. For the free-flow type, the intersection with the through traffic should be made at a relatively flat angle. The number of lanes on the ramp at the terminal and their configuration also influence the type of ramp terminal to be used and its associated design.

Traffic Control and Safety Provisions

To design roads that are safe and efficient in accommodating traffic flow, highway engineers should be familiar with the basic characteristics not only of roads but also of drivers and vehicles. In addition, these engineers should have knowledge of highway-related causes of accidents and means of avoiding them. To reduce the number of highway accidents, a multiple approach is necessary, including improvements in driver and pedestrian training and education, vehicle and traffic law enforcement, vehicle design, and highways. A very high percentage of highway accidents result from driver error, often associated with law violations. Good highway design, nevertheless, can help prevent accidents. Statistics indicate that the relative frequency of accidents associated with vehicle movements or maneuvers depends to a great extent on the type of highway and, in particular, on various design features that are intended to prevent traffic conflicts. Many features that are advantageous in accident avoidance are discussed in preceding articles. Other features, such as traffic control devices and highway lighting, are discussed in the following.

16.27 Traffic Control Devices

These provide for the safe and orderly movement of traffic on a highway by offering guidance and navigation information to drivers. Commonly used traffic control devices include traffic signals and signs that display regulatory warnings and route information. Other forms include pavement markings and delineators.

An effective traffic control device should command attention, convey a clear and simple meaning, acquire the respect of drivers, and allow adequate time for drivers to respond. Traffic control devices should be uniform, treating similar situations in the same fashion. Consistent use of symbols and location of signs and other traffic control devices helps to give drivers sufficient response time for reacting to traffic messages.

16.27.1 Traffic Signs

In general, traffic signs may be classified as regulatory, warning, or guide. Regulatory signs are used to indicate the required method of traffic movement. Examples of regulatory signs include STOP and YIELD signs used on intersecting roadways to establish the right of through movement. Warning signs, such as a FALLEN ROCK ZONE sign, are used to indicate potentially hazardous conditions. Variable message signs are used under remote control or automatic sensors, among other purposes, to convey emergency warnings. Guide signs, such as exit signs on a freeway, are used to direct traffic along a route toward a destination.

Placement and design of signs should be an integral part of highway design, especially in preparation of highway geometry. Such an approach will help ensure that future adverse operational conditions will be minimized or eliminated.

Signs are typically manufactured from high intensity light-reflective materials. In areas of high traffic and in construction zones, illuminated signs are often used.

Variable Message Signs (VMS)

Variable Message Signs (VMS) are used under remote control or automatic sensors, to convey emergency warnings, provide traffic conditions to motorists in order that they may avoid congestion problems ahead on the highway, and to advise motorists of up-coming maintenance/construction activities that will significantly effect travel time on the roadway. VMS can be either permanently fixed (overhead signs) or portable (trailer mounted) with remote programming capabilities. Remote programming capabilities built into the software of these signs save considerable time for traffic management personnel that do not have to drive through a congested highway to get to the VMS and then program messages on the VMS. When a VMS message is placed on the VMS in a timely manner, drivers can seek alternate routes to avoid traffic backups and minimize delays. VMS is an essential component of a traffic management system for daily commuting purposes, altering the traveling public of upcoming maintenance/construction activities that will effect commute time/travel plans, and incident management. Many transportation agencies in urban locations use these signs as an integral part of their traffic management planning, incident response, and traffic monitoring systems. Standardized messages and consistent usage of these signs by the transportation agency is essential in order to obtain the traveling publics reliance on VMS as a tool to help them minimize delays in their travel plans and to plan their trips accordingly.

16.27.2 Delineators

These are reflectors that are used to guide traffic, especially at night. They may be mounted above-ground or fixed to the pavement. In the latter case, delineators may act as a compliment to or replacement of conventional pavement markings (Art. 16.27.3) and are known as raised pavement delineators. Because they are subject to uprooting by snowplow blades, use of this type of delineator as a permanent marker is more predominant in warm climates than in cold ones. Raised pavement delineators, however, are commonly used in any environment as construction zone markers to delineate temporary travel lanes.

When mounted on a post, delineators are reflectors typically made of faceted plastic or glass. These units are installed at specific heights and spacings to delineate the horizontal alignment, typically in regions where alignment changes may be confusing or ill-defined. Delineators at interchanges are usually different in color and multiple-mounted to differentiate the interchange area from the roadway proper.

16.27.3 Pavement Markings

Pavement markings are markers that are on the roadway surface and that are used to regulate and guide movement of traffic in a safe, orderly, and efficient manner. The forms of pavement markings include centerline stripes, lane lines, no-passing barriers, and edge striping. Painting is the most common method of applying pavement markings. An alternative is plastic striping fixed to pavement with an adhesive. This method is often used for marking temporary lanes. For the pavement markings to fulfill their intended functions, they must be visible, and for this, they must be properly maintained by cleaning and renewal when required.

In northern climates, snow and ice removal operations typically abrade the pavement markings and necessitate a more frequent replacement cycle. Further, areas of high traffic volume and high amounts of weaving traffic at entrance/exit ramps require more frequent applications and maintenance.

16.27.4 Traffic Signals

Traffic signals are used to assign the right-of-way at intersections and thereby control the flow of vehicular and pedestrian traffic. Signals can also be used to emphasize a hazardous location, supplement conventional signs, and provide control at railroad-highway grade crossings.

Red, yellow (amber), and green signal lights are widely used. Depending on the type of intersection, the displays may have a circular or arrow configuration. Care should be taken in placement of traffic signals to ensure visibility, meet pedestrian requirements, and effect integration of the signals with the highway geometry. The design of a traffic signal system should also allow for future expansion.

Traffic signals may be pretimed, traffic-actuated, or pedestrian-activated. Pretimed signals operate on a predetermined, consistent, and regularly repeated sequence of intervals. Traffic-actuated controls utilize some form of vehicle and pedestrian detector that trips the signal to allow assigned movements. Typical uses of traffic-actuated signals are to control left turns and traffic flow from side roads, movements that are not permitted until a vehicle trips, or actuates, a time-delay detector. Pedestrian-actuated signals allow normal vehicular movement until a pedestrian presses a button that changes the signal light, halting traffic and allowing the pedestrian to cross safely.

Signal Systems ▪ These are used to coordinate the movement of traffic through intersections on major highways located in and on approaches to cities and large villages. In signal systems, a master controller resynchronizes various intersectional signal controllers to reduce the inconvenience and delays resulting from independent control of traffic signals in cities and large villages.

16.27.5 Colored Pavement

Another method of delineating pavement sections to guide and regulate traffic is to color sections of the pavement, such as shoulders (Art. 16.5). In order for colored pavement to serve successfully as a traffic control device, it should provide significant contrast with adjoining paved areas.

16.27.6 Ramp control

It is sometimes necessary to control entry of vehicles to limited-access highways from ramps. This purpose may be achieved in several ways. One way is to close the ramps. This involves complete diversion of ramp traffic. Another way is to apply ramp metering. This requires drivers to stop and wait before entering the highway when ramp traffic flow must be restricted. An alternative is merge control, which employs a ramp-metering system that stops vehicles at the ramp terminal when there is steady traffic on the highway and releases them when the system detects a gap in the highway traffic.

16.27.7 Traffic Surveillance and Control Systems

These utilize video and related equipment to monitor traffic manually. The objective is to keep traffic flowing as orderly and efficiently as possible. Traffic surveillance and control systems can range from limited types that use conventional detectors to elaborate systems that employ vehicle detection loops, helicopters, and video equipment.

The goal of traffic surveillance and control systems is to provide from remote locations observation of traffic movements and permit immediate identification of demands for service. These systems can also play an invaluable role in promoting highway safety by facilitating immediate recognition of emergencies. In such situations, the controllers can promptly notify the proper authorities who can take appropriate actions to dispatch aid to the scene. The controllers also can initiate management of traffic flow through the use of variable message signs (Art. 16.27.1).

16.28 Intelligent Transportation Systems Management

Intelligent Transportation Systems Management (ITSM) is a collection of systems that address a variety of important traffic management objectives. Integral components of ITSM include: Variable Message System (VMS) boards, Closed Caption Television (CCTV) cameras, Highway Advisory Radio (HAR), and working with the local media (television, radio, and newspapers) to keep the traveling public informed of roadway conditions, traffic backups, and planned maintenance/construction activities. The overall purpose of ITSM is to better coordinate traffic conditions and minimize delays to the traveling public. Many transportation agencies are aggressively pursuing ITSM to address significant traffic problems in major urban areas. Agency coordination and communication is an integral part of ITSM, in that planned construction activities by one agency may significantly impact other roadways and cause traffic backups. Through coordinated, planned lane closures and the use of VMS, CCTV, and HAR many traffic backups can be minimized and/or avoided thus reducing travel delays. Further, incident management, response by maintenance personnel and law enforcement to an accident, can be expedited through the use of these modern tools.

16.28.1 Advanced Traffic Management Systems (ATMS)

This is a component of ITSM that has the ability to detect accidents, construction work, and other causes of traffic backup and congestion. The ATMS also may offer alternate paths for vehicles in an effort to dispel congestion and provide optimal use of the open highway system.

An ATMS is essentially a form of traffic surveillance and control system (Art. 16.27.7). Two important aspects of an ATMS are the type and location of detection equipment used to identify points of incident. An ATMS requires robust detectors in order to provide necessary information. Loop detectors are often used in ITSM, since they generally are less expensive and more reliable than currently available video-image processing. Loop detectors alone, however, do not provide all the information needed for proper management of traffic congestion. They require supplementing by other technologies, such as closed-circuit television (CCTV), which can provide personnel located in a control center with more specific information concerning incidents. Typically, implementation of CCTV is confined to major highways and bottleneck points on minor roads.

16.28.2 Advanced Traveler Information Systems (ATIS)

These are components of an ITSM that provide drivers with in-vehicle navigational information and real-time data concerning the location of existing or potential traffic conflicts. An ATIS may also suggest alternative travel routes.

16.28.3 Advanced Vehicle Control Systems (AVCS)

These are components of an ITSM that are designed to give advance warning to drivers of potential collision with other vehicles. An AVCS may be able automatically to brake vehicles if a collision is imminent or direct vehicles away from a potential collision.

16.28.4 Commercial Vehicle Operation (CVO)

These are components of an ITSM that monitor movements of trucks, buses, vans, taxis, and emergency vehicles. Tracking of commercial vehicles has many benefits. One benefit is that CVO facilitates automation of toll collection, which greatly helps in reducing congestion at toll collection facilities. Another benefit is the ability to track movements of hazardous material and of large vehicles that exceed legal truck-load limitations.

16.28.5 Advanced Public Transportation Systems (APTS)

In addition to benefiting users of highway, an ITSM is also designed to benefit the users of mass transit or high-occupancy vehicles (HOV) through incorporation of an advanced public transportation system. An HOV may be a bus, van with more than one passenger, or car pool. The goals of an APTS are to reduce operating costs for transit systems and promote use of transit systems through increased efficiency. An APTS allows transit riders to prepay fares and receive in evidence of the payment a *smart* card that can be used to gain access to transit vehicles.

16.29 Highway Lighting

Nighttime illumination of a roadway is very important in promoting safety and operational efficiency. As with any highway appurtenance, however, there is an associated cost that should be balanced against the enhancement offered.

In general, lighting is used more extensively for urban rather than rural roadways. In addition to furthering highway safety, lighting in urban environments promotes safety to pedestrians. In rural areas, lighting is generally applied in critical areas, such as interchanges, intersections, railroad crossings at grade, narrow or long bridges, tunnels, sharp curves, and areas where roadside interference is a concern.

A typical highway lighting installation consists of an aluminum or steel standard (pole) on top of which is mounted a luminaire (Fig. 16.29). This lighting fixture comprises a lamp, its housing, and a lens.

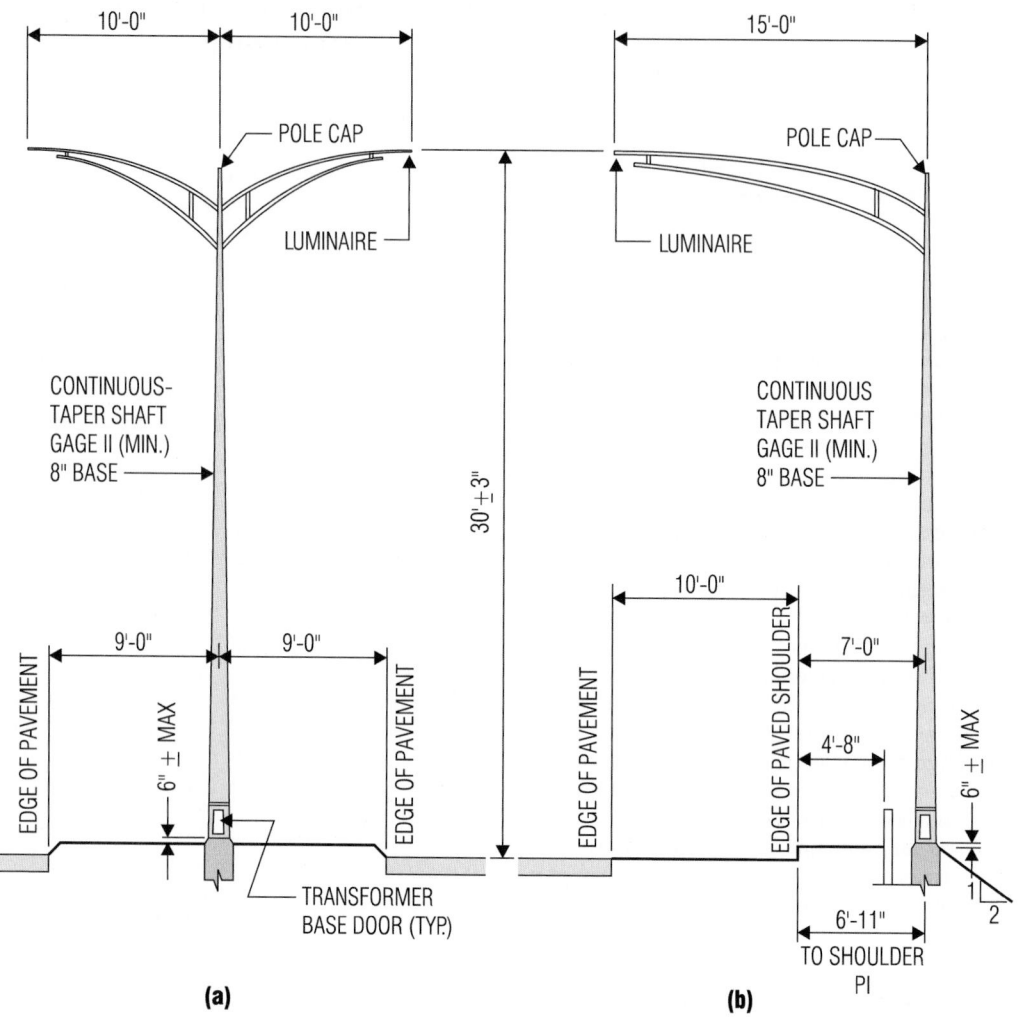

Fig. 16.29 Highway lighting installation with luminaires mounted on tall posts.

Like other roadside elements, lighting standards are susceptible to vehicle impact and therefore should be placed outside the roadway clear zone. If it is not possible or practical to locate the standards in a *safe* area, the standards should be equipped with some form of impact-attenuation feature. For this purpose, breakaway poles may be used. They should be installed along stretches of roadway where vehicles should be traveling at relatively low speeds at which damage to a vehicle striking a standard will not be severe. Breakaway poles should not be used, however, in heavily developed regions,

where there is the possibility of an impacted pole damaging adjacent buildings or pedestrians.

Installation of poles on the outside of curves on a ramp should also be avoided because in such locations they are likely to be struck by a vehicle. If lighting standards are placed behind a longitudinal traffic barrier, they should be offset to allow for deflection of the longitudinal barrier when it is impacted.

When installed for a divided highway, lighting standards may be placed either in the median (Fig. 16.29*a*) or on the right side of the road

(Fig. 16.29*b*). For high-speed lanes, it is generally preferable to place the standards, equipped with dual-mast arms, in the median since the cost is typically lower and the illumination provided greater than for standards on the right side of the roadways. Overhead lighting installations, such as those depicted in Fig. 16.29, generally extend 30 ft above grade and are equipped with mercury-vapor lamps.

For interchanges, traffic circles, and toll plaza areas, another form of overhead lighting, known as high-mast lighting, is used. In this case, luminaires are mounted on tapered steel poles or triangular steel towers that range in height from 50 to 150 ft. The luminaires can be lowered to within 3 ft of the ground for periodic inspection and maintenance. To further facilitate maintenance, hoisting and electric cables can be replaced at ground level, where electrical connections are made. The lamps typically are 1000-W mercury vapor, metal halide, or high-pressure sodium vapor.

Even if initially the design of a highway does not specify highway lighting, provision for future installation of lighting should at least be considered. If lighting should be required in the future, its installation will be greatly facilitated by provision of the necessary conduits under pavements and curbs during construction of the highway.

("An Informational Guide for Roadway Lighting," American Association of State Highway and Transportation Officials, Washington, D.C. (www.aashto.org).)

Highway Maintenance and Rehabilitation

Maintenance and rehabilitation of highway pavements are ongoing activities, critical for prolonging the life of highways. Methods for performing these tasks vary from region to region and depend on the type of pavement.

16.30 Maintenance of Asphalt Pavements

Deterioration of an asphalt pavement is evidenced by poor ride quality and surface pavement distresses: including rutting, shoving, distortion, and various forms of cracking.

16.30.1 Distortions of Asphalt Pavements

A distortion is a change of a pavement from its original shape. Asphalt pavements can suffer from a variety of distortions that can cause cracking and other adverse conditions. The principal forms of pavement distortions are channeling (rutting), corrugations, shoving, depressions, and upheaval.

Channeling is a lengthy depression formed in wheel tracks. **Corrugation** (washboarding) is the plastic movement of an asphalt surface that causes formation of ripples across the pavement. **Shoving** is plastic movement that results in a localized bulge in the pavement. **Upheaval** is the localized upward displacement of a pavement brought on by swelling of the subgrade or other portion of the pavement structure.

16.30.2 Cracking of Asphalt Pavements

This has many causes and takes a variety of forms, such as alligator, edge, joint, reflection, slippage, transverse, longitudinal, and diagonal cracking.

Alligator Cracking ▪ Cracks that form small interlocking rectangular shapes having the appearance of alligator skin are known as alligator cracks. These usually are initiated by failure of an untreated granular base or by a soft subgrade layer. Such conditions often arise from excessive saturation of pavement base, subgrade, or poor drainage.

Maintenance generally involves removal of all wet material and patching with a full-depth hot-mix asphalt. To prevent future occurrence of alligator cracking, new drainage facilities should be installed or existing drainage facilities should be improved (Arts. 16.16 and 16.17).

Edge Cracking ▪ Located at or near the edges of pavements, edge cracking extends longitudinally, nearly parallel to the centerline of the roadway. This type of cracking may be accompanied by transverse cracks, nearly perpendicular to the roadway centerline. Causes of edge cracking include settlement of the pavement, inadequate support for the pavement, inadequate drainage, and frost heave.

Repair of edge cracks requires filling the cracks with asphalt-emulsion slurry or cutback asphalt mixed with sand. If settlement has occurred, it may be necessary to bring the roadway surface to grade through use of hot-mix asphalt patching.

Joint Cracking ▪ This occurs at the interface between a pavement and adjacent shoulder. Joint cracking can be initiated by deformational loads due to thermal expansion and contraction or alternate wetting and drying. It also can be caused by intrusion of water as a result of inadequate drainage.

A crack between two adjacent paving lanes is known as a **lane-joint crack**. It typically is caused by inadequate bond or a poor seam between adjoining sections of pavement.

Repair of joint cracks requires filling of the cracks with an asphalt-emulsion slurry. In addition, poor drainage conditions should be corrected.

Reflection Cracking ▪ This is a crack that forms in an asphalt overlay and reflects the pattern of the underlying pavement surface. Reflection cracking can be induced by horizontal or vertical movements in the pavements beneath the overlay. These movements are generally caused by traffic loads, earth movement, or temperature. Reflection cracks typically occur in asphalt overlays placed on top of a portland cement concrete or cement-treated base.

Cracks less than $\frac{1}{8}$ in wide may either be ignored or, if intrusion of water is a concern, filled, with the use of a squeegee, with emulsified or cutback asphalt covered with sand. Cracks more than $\frac{1}{4}$ in wide should be filled with an asphalt-emulsion slurry or a light grade of cutback asphalt and fine sand.

Shrinkage Cracking ▪ This is evidenced by interconnected cracks that create a series of large blocks with sharp corners or angles. Shrinkage cracks are usually associated with a volume change in the pavement asphalt mix, base, or subgrade. They also may result from aging of the pavement. The constant exposure of the pavement materials to thermal expansion and contraction, may cause them to lose some of their elasticity or resiliency and bring about shrinkage cracking.

Slippage Cracking ▪ This cracking structure is represented by crescent shaped cracks generated by traffic-induced horizontal forces. Typically found on ramps and curved roadways sections, these cracks present a serious traffic safety concern that should be corrected as soon as possible by

maintenance forces. Slippage cracks are caused by insufficient bond between the surface layer and the underlying course. Dust, dirt, and oil atop the underlying course during placement of the surface course can contribute to this lack of bond. Also, omission of a tack coat during construction can lead to formation of slippage cracks. The pavement course are not properly bonded when a tack coat is eliminated between the pavement courses. This omission may lead to debonding of the courses and cause slippage cracks or complete delamination of the pavement course. Many times, delamination occurs after an extreme freeze-thaw cycle in the spring where water has entered between the layers and the freezing and thawing action debonds the pavement courses and causes premature pavement failure.

This type of cracking is repaired by removing the surface layer around the crack to locations where an adequate bond is present. The area from which the surface course was removed is then patched with a hot-mix asphalt.

("The Asphalt Handbook," The Asphalt Institute, College Park, Md. See also Art. 16.32.)

16.31 Maintenance of Portland Cement–Concrete Pavements

Deterioration of a Portland Cement–Concrete pavement (PCC) is evidenced by poor ride quality, faulting of pavement slabs and other forms of pavement distresses including: surface spalls, loss of sealant, and various forms of cracking. Further, cracking at the pavement-shoulder interface may indicate lack of proper drainage. Consequently, much maintenance work is concerned with filling of cracks and expansion joints. For this purpose, asphalt is often used. It is suitable for sealing joints and cracks, filling small cavities, and raising sunken slabs. A more extreme alternative in maintaining PCC pavements is to cover deteriorated pavement with a thin asphalt course (overlay).

16.31.1 Distortions of PCC Pavements

Major forms of distortion of PCC pavements are faulting and pumping at joints. A fault is a physical difference in elevation between two adjacent slabs; it may located at either a joint or a crack within the pavement slab. It is primarily caused by loss of load-transfer between the slabs where the load-transfer device has failed due to corrosion and/or structural failure. This distortion is a major concern for maintenance agencies, it results in poor ride quality and further progresses deterioration of the pavement structure.

Pumping is the upward and downward movement of a slab under traffic loads. Pumping can be typically observed by viewing the adjacent shoulders at the transverse joint or crack location. Fine silts and water can be seen pumping or squirting out of the joint when traffic passes over the joint or crack. Pumping may occur when pavements overlay very wet sand, clay or silt and are poorly drained. Pumping can be corrected, to some extent, by inserting asphalt or portland cement grout under the slab or better, by improving the subgrade drainage by installing edge drains.

16.31.2 Cracking of PCC Pavements

This has many causes and takes a variety of form, such as transverse, longitudinal, and diagonal cracking.

Transverse Cracking ▪ Extending roughly perpendicular to the pavement centerline, transverse cracking may be caused by the pavement slab length design, the slabs may have been designed too long for the loading condition and particular slab design type. Further, overloading of the pavement, faulting and pumping of slabs, settlement and failure of a soft foundation material, inadequate compaction of the subgrade materials, frozen load transfer joints, lack of joints, excessively shallow joints, or concrete shrinkage are additional causes of transverse concrete pavement cracking. Repair for tight cracks (small enough to insert a pencil) typically require that the cracks be cleaned of all loose material, routing the crack and then fill with a rubber-asphalt sealer. Open cracks and faulted cracks may require more substantial repairs involving full depth removal and replacement of the slab. Alternative methods include dowel inserts and stitching slabs together. Full depth slab replacements and other methods of slab repair are costly. The designer should familiarize themselves with applicable Portland Cement Concrete Association guidelines and manuals for repair of concrete pavements prior to implementing a repair technique. Proper guidelines and typical repair details are necessary to properly control the repair work in the field and avoid cost overruns.

Longitudinal Cracking ▪ This extends roughly parallel to the pavement centerline. Longitudinal cracking can be caused by shrinkage of the concrete, especially in wide pavements without a longitudinal joint. Other conditions that can create longitudinal cracking are pumping or an expansive subbase or subgrade.

Repair of longitudinal cracks in PCC pavements is the same as that for transverse cracking. For pumping-induced cracks, a high-softening-point asphalt may be used to fill the voids under the pavement slab.

Diagonal Cracking ▪ These run diagonally to the pavement centerline. They are induced by traffic loads on an unsupported end of the pavement slab.

Repair of the cracks is like that described for transverse and longitudinal cracking. As may be done for pumping-type cracks, an asphalt underseal may be applied to the slab and followed by filling of the crack with a rubber-asphalt sealing compound.

See also Art. 16.32.

16.32 Pavement Management Systems (PMS)

Constant exposure to the elements, combined with wear and tear from traffic, make highways extremely prone to deterioration. As a result, they must be repaired or replaced if they are to serve as intended. Highway maintenance and rehabilitation, however, is not limited just to application of the remedial measures described in Arts. 16.30 and 16.31.

Responsible for maintenance of immense lengths of roadway and associated appurtenance facilities, transportation agencies frequently have to decide which sections of highway need immediate attention and which can be deferred. A variety of factors influence this decision, and highway design is only one of these factors, albeit an important one. Economic, political, and a host of other factors must be evaluated before project selection may be made. A Pavement Management System (PMS) is an organized, systematic approach on the owner's part, to plan appropriate maintenance, rehabilitation, and reconstruction (MR&R) activities for their highway infrastructure. A PMS consists of data collection activities, computerized databases, computerized analysis tools, and decision supporting tools that help best allocate resources to maintain the highway pavement inventory subject to management goals, objectives, and funding and other resource constraints. The overriding theme in a PMS is to maximize the remaining service life of a pavement since pavements represent significant capital investments. The human component of a PMS is essential in the decision-making process, but proper use of computer software can play an important role in the decision-making process.

The basic components and associated products of a PMS are as follows: inventory database, maintenance database, budgetary information, project selection methods, and costing models.

The inventory database of a PMS details pavement conditions throughout the entire highway network. There are many ways to define the state of a section of pavement. One method is to rate the pavement in terms of various forms of pavement distress, such as edge cracking and rutting, as described in Arts. 16.30 and 16.31. The length of the pavement section to be rated depends on the detail desired. Use of small lengths, however, does not necessarily translate into a more accurate picture of pavement condition.

Typically, information on pavement condition is stored in a computerized database management system (DBMS) for both querying and modeling. The data may also be tied into a geographic information system (GIS), which allows excellent visualization of data. Historical data contained within the maintenance database describe what work has been performed on the pavement sections. The data are helpful in determining both the results of individual remedial methods and associated costs. Budgetary information may be derived from the inventory and maintenance databases. Based on the data thus made conveniently available, project selection and cost analysis methods can be applied to assist in selection of the sequence in which projects are to be implemented and formulation of highway maintenance and rehabilitation budgets.

16.32.1 Project and Network Level Analyses

A PMS can function using a project or network level analysis approach, or both. Project level analysis

deals with individual sections of pavement and the remedial measures to be taken to correct deficiencies. Associated cost estimating can be performed and ramifications of various remedial measures can be predicted with the objective of determining which method and level of repair will yield the best results in terms of both economy and safety.

Network level analysis is applicable to a group of projects comprising various sections of non-contiguous highway. This analysis permits formulation of alternatives based on the maintenance and rehabilitation needs not only of specific highway sections but also of the highway network as a whole. For example, one section of the network may require patching of alligator cracking and another may show evidence of subsurface drainage inadequacies. If funds are insufficient for correcting both conditions, the PMS could assist in the decision whether to correct the drainage condition, which if ignored could lead to failure of the pavement, or to patch the cracking, which should not be ignored but may be deferred for a short time without serious consequences. While this is a relatively simple example, it serves to illustrate the basic concepts behind network level analysis.

In addition to providing analysis, the PMS offers valuable support information in the form of cost and record data and ancillary backup information that can be used not only for formulating but also for justifying maintenance plans. Development of a PMS builds upon methods and information currently in use in an effort to create an integrated system for planning and performing pavement maintenance and rehabilitation.

16.32.2 Predicting Future Pavement Condition

In addition to assisting in selection of projects for repair, a PMS may be used to predict the future condition of pavement. Predictions are typically based on one of the following assumptions: no repair work is performed; partial, interim remedial measures are taken; or full repairs are made to correct all deficiencies.

The estimates of future pavement conditions provide maintenance planners with a more accurate picture of the ramifications of various options than could otherwise be obtained. This information is also useful in developing long-range plans and estimating future costs.

Thomas A. Ostrom
Chief
Office of Structures-North
California Department of Transportation
Sacramento, California

Steven L. Mellon
Senior Bridge Design Engineer
Quincy Engineering, Inc.
Sacramento, California

BRIDGE ENGINEERING

Bridge engineering covers the planning, design, construction, operation, and maintenance of structures that carry facilities for movement of humans, animals, or materials over natural or created obstacles.

Most of the diagrams used in this section were taken from the "Manual of Bridge Design Practice," State of California Department of Transportation and "Standard Specifications for Highway Bridges," American Association of State Highway and Transportation Officials. The authors express their appreciation for permission to use these illustrations from this comprehensive and authoritative publication.

General Design Considerations

17.1 Bridge Types

Bridges are of two general types: fixed and movable. They also can be grouped according to the following characteristics:

Supported facilities: Highway or railway bridges and viaducts, canal bridges and aqueducts, pedestrian or cattle crossings, material-handling bridges, pipeline bridges.

Bridge-over facilities or natural features: Bridges over highways and over railways; river bridges; bay, lake, slough and valley crossings.

Basic geometry: In plan—straight or curved, square or skewed bridges; in elevation—low-level bridges, including causeways and trestles, or high-level bridges.

Structural systems: Single-span or continuous-beam bridges, single- or multiple-arch bridges, suspension bridges, frame-type bridges.

Construction materials: Timber, masonry, concrete, and steel bridges.

17.2 Design Specifications

Designs of highway and railway bridges of concrete or steel often are based on the latest editions of the "Standard Specifications for Highway Bridges" or the "Load and Resistance Factor Design Specifications" (LRFD) of the American Association of State Highway and Transportation Officials (AASHTO) and the "Manual for Railway Engineering" of the American Railway Engineering and Maintenance-of-Way Association (AREMA). Also useful are standard plans issued by various highway administrations and railway companies.

Length, width, elevation, alignment, and angle of intersection of a bridge must satisfy the functional requirements of the supported facilities and the geometric or hydraulic requirements of the bridged-over facilities or natural features. Figure 17.1 shows typical highway clearance diagrams.

Selection of the structural system and of the construction material and detail dimensions is governed by requirements of structural safety; economy of fabrication, erection, operation, and maintenance; and aesthetic considerations.

Highway bridge decks should offer comfortable, well-drained riding surfaces. Longitudinal grades and cross sections are subject to standards similar to those for open highways (Sec. 16). Provisions for roadway lighting and emergency services should be made on long bridges.

Barrier railings should keep vehicles within the roadways and, if necessary, separate vehicular

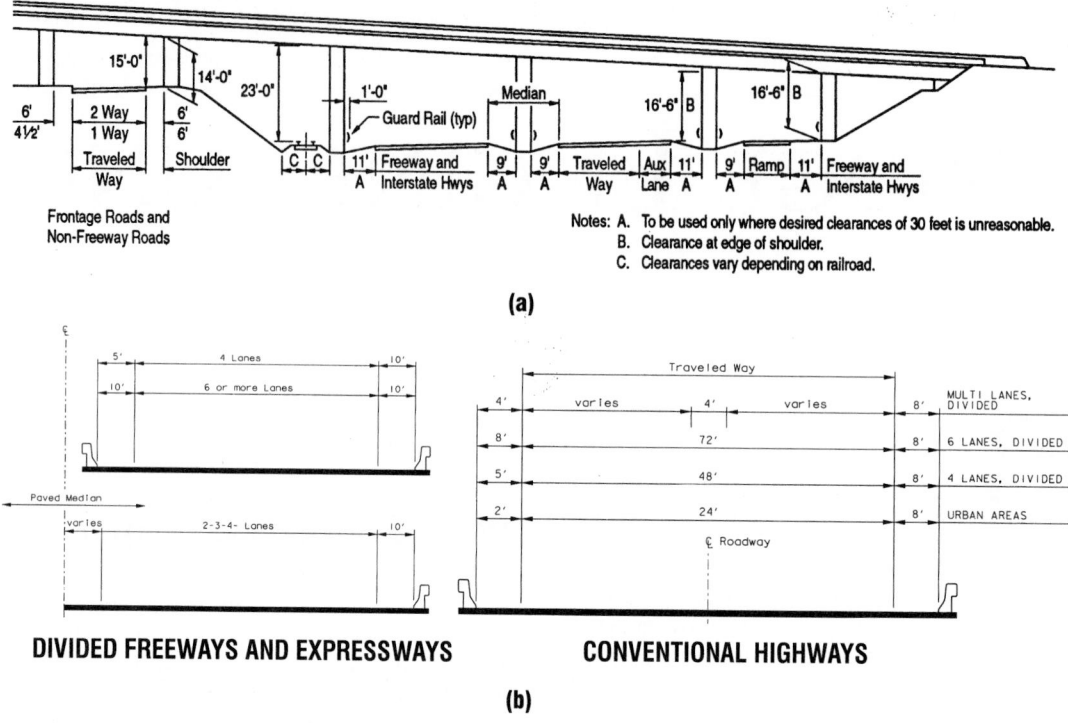

(a)

(b)

Fig. 17.1 Minimum clearances for highway structures. (*a*) Elevation of a highway bridge showing minimum vertical clearances below it. (*b*) Typical bridge cross sections indicating minimum horizontal clearances. Long-span bridges may have different details and requirements.

lanes from pedestrians and bicyclists. Utilities carried on or under bridges should be adequately protected and equipped to accommodate expansion or contraction of the structures.

Most railroads require that the ballast bed be continuous across bridges to facilitate vertical track adjustments. Long bridges should be equipped with service walkways.

17.3 Design Loads for Bridges

Bridges must support the following loads without exceeding permissible stresses and deflections:

Dead load D, including permanent utilities

Live load L and impact I

Longitudinal forces due to acceleration or deceleration LF and friction F

Centrifugal forces CF

Wind pressure acting on the structure W and the moving load WL

Earthquake forces EQ

Earth E, water and ice pressure ICE, stream flow SF, and uplift B acting on the substructure

Forces resulting from elastic deformations, including rib shortening R

Forces resulting from thermal deformations T, including shrinkage S, and secondary prestressing effects

17.3.1 Highway Bridge Loads

Vehicular live load of highway bridges is expressed in terms of design lanes and lane loadings. The number of design lanes depends on the width of the roadway.

In the standard specifications, each lane load is represented by a standard truck with trailer (Fig. 17.3) or, alternatively, as a 10-ft-wide uniform load in combination with a concentrated load (Fig. 17.2). As indicated in Fig. 17.3, there are two classes of loading: HS20 and HS15, which represent a truck and trailer with three loaded axles. These loading designations are followed by a 44, which indicates that the loading standard was adopted in 1944. The LRFD HL-93 vehicular live load consists of a combination of the HS20-44 design truck depicted in Fig. 17.3, or the LRFD design tandem, and the LRFD live load. The LRFD design tandem is defined as a pair of 25 kip axles spaced 4.0 ft apart. The LRFD live load consists of 0.64 k/lf applied uniformly in the longitudinal and transverse direction.

When proportioning any member, all lane loads should be assumed to occupy, within their respective lanes, the positions that produce maximum stress in that member. Table 17.1 gives maximum moments, shears, and reactions for one loaded lane. Effects resulting from the simultaneous loading of more than two lanes may be reduced

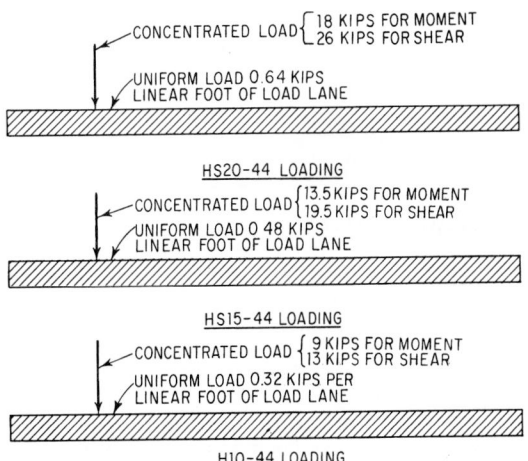

Fig. 17.2 HS loadings for simply supported spans. For maximum negative moment in continuous spans, an additional concentrated load of equal weight should be placed in one other span for maximum effect. For maximum positive moment, only one concentrated load should be used per lane, but combined with as many spans loaded uniformly as required for maximum effect.

by a loading factor, which is 0.90 for three lanes and 0.75 for four lanes.

In design of steel grid and timber floors for HS20 loading, one axle load of 24 kips or two axle loads of 16 kips each, spaced 4 ft apart, may be used, whichever produces the greater stress, instead of the 32-kip axle shown in Fig. 17.3. For slab design, the centerline of the wheel should be assumed to be 1 ft from the face of the curb.

Wind forces generally are considered as moving loads that may act horizontally in any direction. They apply pressure to the exposed area of the superstructure, as seen in side elevation; to traffic on the bridge, with the center of gravity 6 ft above the deck; and to the exposed areas of the substructure, as seen in lateral or front elevation. Wind loads in Tables 17.2 and 17.3 were taken from "Standard Specifications for Highway Bridges," American Association of State Highway and Transportation Officials. They are based on 100-mi/h wind velocity. They should be multiplied by $(V/100)^2$ for other design velocities except for Group III loading (Art. 17.4).

In investigation of overturning, add to horizontal wind forces acting normal to the longitudinal bridge axis an upward force of $20 \, \text{lb/ft}^2$ for the structure without live load or $6 \, \text{lb/ft}^2$ when the structure carries live load. This force should be applied to the deck and sidewalk area in plan at the windward quarter point of the transverse superstructure width.

Impact is expressed as a fraction of live-load stress and determined by the formula:

$$I = \frac{50}{125 + l} \quad 30\% \text{ maximum} \qquad (17.1)$$

where l = span, ft; or for truck loads on cantilevers, length from moment center to farthermost axle; or for shear due to truck load, length of loaded portion of span. For negative moments in continuous spans, use the average of two adjacent loaded spans. For cantilever shear, use $I = 30\%$. Impact is not figured for abutments, retaining walls, piers, piles (except for steel and concrete piles above ground rigidly framed into the superstructure), foundation pressures and footings, and sidewalk loads.

Longitudinal forces on highway bridges should be assumed at 5% of the lane load plus concentrated load for moment headed in one

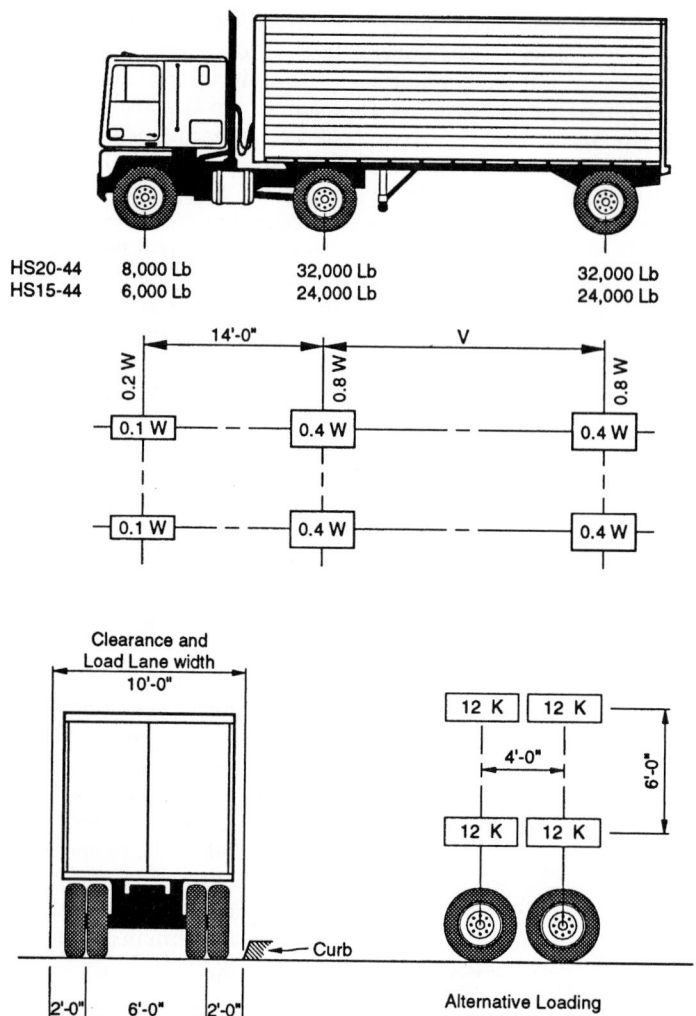

Fig. 17.3 Standard truck loading. HS trucks: W = combined weight on the first two axles, which is the same weight as for H trucks. V indicates a variable spacing from 14 to 30 ft that should be selected to produce maximum stress.

direction, plus forces resulting from friction in bridge expansion bearings.

Centrifugal forces should be computed as a percentage of design live load

$$C = \frac{6.68S^2}{R} \tag{17.2}$$

where S = design speed, mi/h

R = radius of curvature, ft

These forces are assumed to act horizontally 6 ft above deck level and perpendicular to the bridge centerline.

Restraint forces, generated by preventing rotations of deformations, must be considered in design.

Thermal forces, in particular, from restraint, may cause overstress, buckling, or cracking. Provision should be made for expansion and contraction due to temperature variations, and

Table 17.1 Maximum Moments, Shears, and Reactions for Truck Loads on One Lane, Simple Spans*

Span, ft	H15 Moment[†]	H15 End shear and end reaction[‡]	H20 Moment[†]	H20 End shear and end reaction[‡]	HS15 Moment[†]	HS15 End shear and end reaction[‡]	HS20 Moment[†]	HS20 End shear and end reaction[‡]
10	60.0[§]	24.0[§]	80.0[§]	32.0[§]	60.0[§]	24.0[§]	80.0[§]	32.0[§]
20	120.0[§]	25.8[§]	160.0[§]	34.4[§]	120.0[§]	32.2[§]	160.0[§]	41.6[§]
30	185.0[§]	27.2[§]	246.6[§]	36.3[§]	211.6[§]	37.2[§]	282.1[§]	49.6[§]
40	259.5[§]	29.1	346.0[§]	38.8	337.4[§]	41.4[§]	449.8[§]	55.2[§]
50	334.2[§]	31.5	445.6[§]	42.0	470.9[§]	43.9[§]	627.9[§]	58.5[§]
60	418.5	33.9	558.0	45.2	604.9[§]	45.6[§]	806.5[§]	60.8[§]
70	530.3	36.3	707.0	48.4	739.2[§]	46.8[§]	985.6[§]	62.4[§]
80	654.0	38.7	872.0	51.6	873.7[§]	47.7[§]	1164.9[§]	63.6[§]
90	789.8	41.1	1053.0	54.8	1008.3[§]	48.4[§]	1344.4[§]	64.5[§]
100	937.5	43.5	1250.0	58.0	1143.0[§]	49.0[§]	1524.0[§]	65.3[§]
110	1097.3	45.9	1463.0	61.2	1277.7[§]	49.4[§]	1703.6[§]	65.9[§]
120	1269.0	48.3	1692.0	64.4	1412.5[§]	49.8[§]	1883.3[§]	66.4[§]
130	1452.8	50.7	1937.0	67.6	1547.3[§]	50.7	2063.1[§]	67.6
140	1648.5	53.1	2198.0	70.8	1682.1[§]	53.1	2242.8[§]	70.8
150	1856.3	55.5	2475.0	74.0	1856.3	55.5	2475.1	74.0
160	2076.0	57.9	2768.0	77.2	2076.0	57.9	2768.0	77.2
170	2307.8	60.3	3077.0	80.4	2307.8	60.3	3077.0	80.4
180	2551.5	62.7	3402.0	83.6	2551.5	62.7	3402.0	83.6
190	2807.3	65.1	3743.0	86.8	2807.3	65.1	3743.0	86.8
200	3075.0	67.5	4100.0	90.0	3075.0	67.5	4100.0	90.0
220	3646.5	72.3	4862.0	96.4	3646.5	72.3	4862.0	96.4
240	4266.0	77.1	5688.0	102.8	4266.0	77.1	5688.0	102.8
260	4933.5	81.9	6578.0	109.2	4933.5	81.9	6578.0	109.2
280	5649.0	86.7	7532.0	115.6	5649.0	86.7	7532.0	115.6
300	6412.5	91.5	8550.0	122.0	6412.5	91.5	8550.0	122.0

* Based on "Standard Specifications for Highway Bridges," American Association of State Highway and Transportation Officials. Impact not included.
[†] Moments in thousands of ft-lb (ft-kips).
[‡] Shear and reaction in kips. Concentrated load is considered placed at the support. Loads used are those stipulated for shear.
[§] Maximum value determined by standard truck loading. Otherwise, standard lane loading governs.

Table 17.2 Wind Loads for Superstructure Design

	Trusses and arches	Beams and girders	Live Load
Wind load	75 lb/ft²	50 lb/ft²	100 lb/lin ft
Minimums:			
On loaded chord	300 lb/lin ft		
On unloaded chord	150 lb/lin ft		
On girders		300 lb/lin ft	

Table 17.3 Wind Loads for Substructure Design

---

a. Loads transmitted by superstructure to substructure slab and girder bridges (up to 125-ft span)

	Transverse	Longitudinal
Wind on superstructure when not carrying live load, lb/ft^2	50	12
Wind on superstructure when carrying live load, lb/ft^2	15	4
Wind on live load, lb/lin ft*	100	40

Major and unusual structures

	No live load on bridge				Live load on bridge					
	Wind on trusses, lb/ft^2		Wind on girders, lb/ft^2		Wind on trusses, lb/ft^2		Wind on girders, lb/ft^2		Wind on live load, lb/lin ft*	
Skew angle, or wind, deg	Lateral load	Longitudinal load	Lateral load	Longitudinal load	Lateral load	Longitudinal load	Lateral load	Longitudinal load	Lateral load	Longitudinal load
0	75	0	50	0	22.5	0	15	0	100	0
15	70	12	44	6	21	3.6	13.2	1.8	88	12
30	65	28	41	12	19.5	8.4	12.3	3.6	82	24
45	47	41	33	16	14.1	12.3	9.9	4.8	66	32
60	25	50	17	19	7.5	15	5.1	5.7	34	38

b. Loads from wind acting directly on the substructure[†]

Horizontal wind—no live load on bridge, lb/ft^2	40
Horizontal wind—live load on bridge, lb/ft^2	12

* Acting 6 ft above deck.

[†] Resolve wind forces acting at a skew into components perpendicular to side and front elevations of the substructure and apply at centers of gravity of exposed areas. These loads act simultaneously with wind loads from superstructure.

on concrete structures, also for shrinkage. For the continental United States, Table 17.4 covers temperature ranges of most locations and includes the effect of shrinkage on ordinary beam-type concrete structures. The coefficient of thermal expansion for both concrete and steel per °Fahrenheit is 0.0000065 (approximately $\frac{1}{150,000}$). The shrinkage coefficient for concrete arches and rigid frames should be assumed as 0.002, equivalent to a temperature drop of 31 °F.

Stream-flow pressure on a pier should be computed from

$$P = KV^2 \qquad (17.3)$$

where P = pressure, lb/ft^2

V = velocity of water, ft/s

$K = \frac{4}{3}$ for square ends, $\frac{1}{2}$ for angle ends when angle is 30° or less, and $\frac{2}{3}$ for circular piers

Ice pressure should be assumed as 400 psi. The design thickness should be determined locally.

Earth pressure on piers and abutments should be computed by recognized soil-mechanics formulas, but the equivalent fluid pressure should be at least 36 lb/ft^3 when it increases stresses and not more than 27 lb/ft^3 when it decreases stresses.

Sidewalks and their direct supports should be designed for a uniform live load of 85 lb/ft^2.

Table 17.4   Expansion and Contraction of Structures*

Air temp range	Steel		Concrete[†]	
	Temp rise and fall, °F	Movement per unit length	Temp rise and fall, °F	Movement per unit length
Extreme: 120 °F, certain mountain and desert locations	60	0.00039	40	0.00024
Moderate: 100 °F, interior valleys and most mountain locations	50	0.00033	35	0.00021
Mild: 80 °F, coastal areas, Los Angeles, and San Francisco Bay area	40	0.00026	30	0.00018

* This table was developed for California. For other parts of the United States, the temperature limits given by AASHTO "Standard Specifications for Highway Bridges" should be used.

[†] Includes shrinkage.

The effect of sidewalk live loading on main bridge members should be computed from

$$P = \left(30 + \frac{3000}{l}\right)\frac{55 - w}{50} \leq 60 \text{ lb/ft}^2 \qquad (17.4)$$

where P = sidewalk live load, lb/ft^2

l = loaded length of sidewalk, ft

w = sidewalk width, ft

Curbs should resist a force of 500 lb/lin ft acting 10 in above the floor. For design loads for *railings*, see Fig. 17.4.

17.3.2   Railway Bridge Loads

Live load is specified by axle-load diagrams or by the E number of a "Cooper's train," consisting of two locomotives and an indefinite number of freight cars. Figure 17.5 shows the typical axle spacing and axle loads for E80 loading.

Members receiving load from more than one track should be assumed to be carrying the following proportions of live load: For two tracks, full live load; for three tracks, full live load from two tracks and half from the third track; for four tracks, full live load from two, half from one, and one-fourth from the remaining one.

Impact loads, as a percentage of railroad live loads, may be computed from Table 17.5.

Longitudinal forces should be computed for braking and traction and **centrifugal forces** should be computed corresponding to each axle. See the AREMA "Manual for Railway Engineering" for more information (www.arema.org).

17.3.3   Proportioning of Bridge Members and Sections

The following groups represent various combinations of loads and forces to which a structure may be subjected. Each component of the structure, or the foundation on which it rests, should be proportioned to withstand safely all group combinations of these forces that are applicable to the particular site or type. Group loading combinations for service load design and load factor design are given by

$$\begin{aligned}
\text{Group } (N) = \gamma[&\beta_D D + \beta_L(L + I) + \beta_C CF \\
&+ \beta_E E + \beta_B B + \beta_S SF + \beta_W W \\
&+ \beta_{WL} WL + \beta_L LF \\
&+ \beta_R(R + S + T) + \beta_{EQ} EQ \\
&+ \beta_{ICE} ICE]
\end{aligned} \qquad (17.5)$$

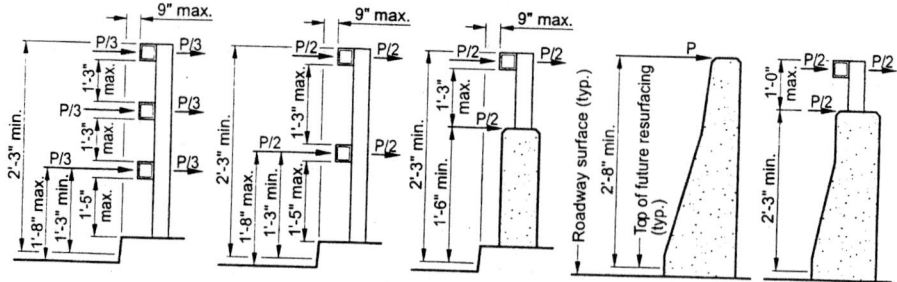

(To be used where there is no curb or curb project 9" or less from traffic face of railing.)

TRAFFIC RAILING

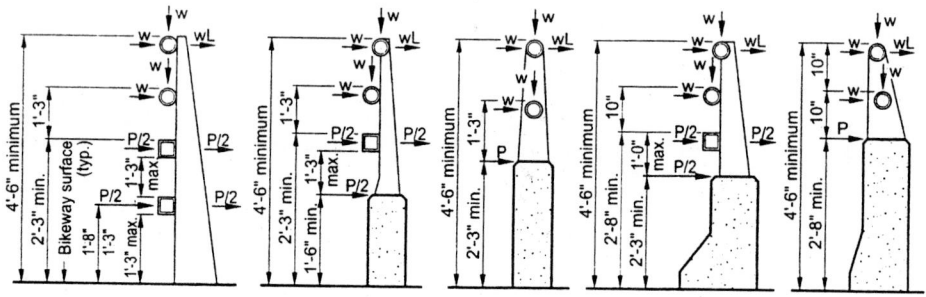

COMBINATION TRAFFIC AND PEDESTRIAN RAILING

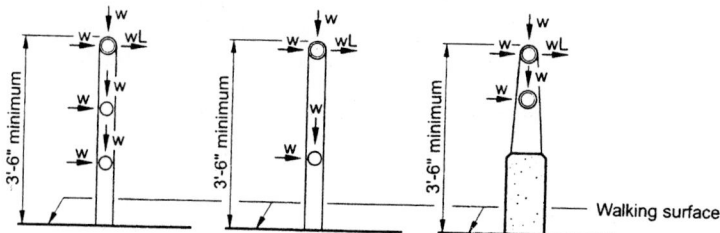

(To be used adjacent to a sidewalk when highway traffic is separated from pedestrian traffic by a traffic railing.)

PEDESTRIAN RAILING

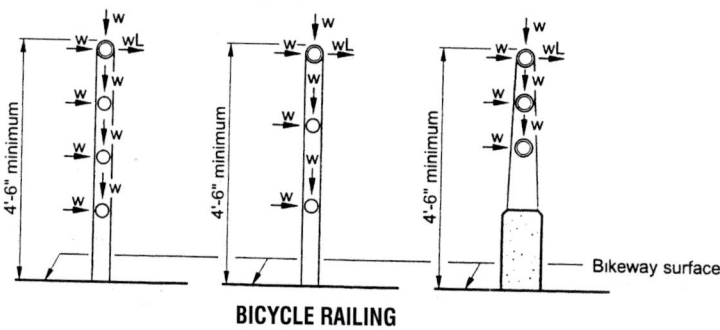

BICYCLE RAILING

Fig. 17.4 Service loads for railings: $P = 10$ kips, $L =$ post spacing, $w = 50$ lb/ft. Rail loads are shown on the left, post loads on the right. (Rail shapes are for illustrative purposes only.)

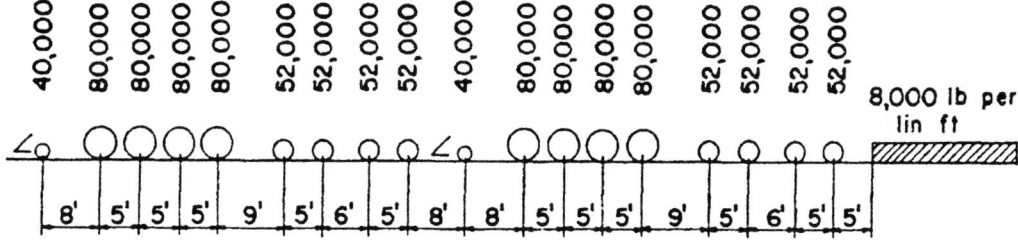

(a) Cooper E 80 Load

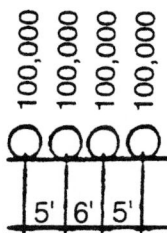

(b) Alternative Live Load on 4 Axles (Steel Railroad Bridges only)

Fig. 17.5 Axle Spacing and Axle Loads for E80 loading

where N = group number, or number assigned to a specific combination of loads

γ = capacity reduction factor to provide for small adverse variations in materials, workmanship, and dimensions within acceptable tolerances

β = load factor (subscript indicates applicable type of load)

See Table 17.6 for appropriate coefficients. See also Art. 17.3.1 and Secs. 8 and 9.

AASHTO LRFD associates load combinations with various limit states according to design objectives. The sum of the factored loads must be less than the sum of the factored resistance:

$$\sum \eta_i \gamma_i Q_i \le \varphi R_n \qquad (17.6)$$

where η_i = load modifier relating to ductility, redundancy, and operational importance

γ_i = load factor, a statistically based multiplier reflecting certainty in the value for force effect

Q_i = force effect i

φ = resistance factor, a statistically based multiplier reflecting certainty in value for particular material property

See Table 17.7 and 17.8 for design objectives, limit state load combinations and load factors. Resistance factors vary according to material and characteristic such as bending, shear, bearing, torsion, etc., and are not shown. In LRFD, both the γ's and φ's have been calibrated to achieve a uniform level of safety throughout the structure.

17.4 Seismic Design

Seismic forces are an important loading consideration that often controls the design of bridges in seismically active regions. All bridges should be designed to insure life safety under the demands imparted by the Maximum Considered Earthquake (MCE). Higher levels of performance may be required by the bridge owner to provide post earthquake access to emergency facilities or when the time required to restore service after an earthquake would create a major economic impact.

Table 17.5 Railroad Impact Factors

Structure type	Impact percent*
Prestressed concrete:	
$L < 60$	$35 - \dfrac{L^2}{500}$
$60 < L < 135$	$\dfrac{800}{L-2} + 14$
$L \geq 135$	20%
Reinforced concrete:	$\dfrac{100LL}{LL+DL}$
(80% max. for steam engines)	
(60% max. for diesel engines)	
Steel:**	
Non-hammerblow engine equipment	
$\quad L < 80$	$RE + 40 - \dfrac{3L^2}{1600}$
$\quad L \geq 80$	$RE + 16 + \dfrac{600}{L-30}$
Steam engine equipment with hammerblow	
$\quad L < 100$	$RE + 60 - \dfrac{L^2}{500}$
$\quad L \geq 100$	$RE + 10 + \dfrac{1800}{L-40}$
Truss spans	$RE + 15 + \dfrac{4000}{L+25}$

* For ballasted decks use 90% of calculated impact (steel bridges only)
L = span, ft; S = longitudinal beam spacing, ft; DL = applicable dead load; LL = applicable live load.
RE = the rocking effect consisting of the percentage of downward on one rail and upward on the other rail, increasing and decreasing, respectively, the loads otherwise specified. RE shall be expressed as a percentage; either 10% of the axle load or 20% of the wheel load.
** Impact is reduced for $L > 175$ ft or when load is received from more than two tracks.

All bridges should have a clearly identifiable system to resist forces and deformations imposed by seismic events. Experimental research and past performance has demonstrated that simple bridge features lead to more predictable seismic response. Irregular features lead to complex and less predictable seismic response and should be avoided in high seismic region whenever possible (See Table 17.9). Every effort should be made to balance the effective lateral stiffness between adjacent bents within a frame, adjacent columns within a bent, and adjacent frames. If irregular features or significant variations in lateral stiffness are unavoidable, they should be assessed with more rigorous analysis and designed for a higher level of seismic performance.

Seismic effects for box culverts and buried structures need not be considered, except when they cross active faults.

17.4.1 Seismic Design Approach

Ordinarily bridges are not designed to remain elastic during the MCE because of economic constraints and the uncertainties in predicting seismic demands. Design codes permit the designer to take advantage of ductility and post elastic strength as long as the expected deformations do not exceed the bridge's lateral displacement capacity. Ductile

Table 17.6 Capacity-Reduction and Load Factors

Group	γ	D	$(L+I)_n$	$(L+I)_P$	CF	E	B	SF	W	WL	LF	$R+S+T$	EQ	ICE	% of basic unit stresses
						Service-Load Design[†]									
I	1.0	1	1	0	1	β_E	1	1	0	0	0	0	0	0	100
IA	1.0	1	2	0	0	0	0	0	0	0	0	0	0	0	150
IB	1.0	1	0	1	1	β_E	1	1	0	0	0	0	0	0	‡
II	1.0	1	0	0	0	1	1	1	1	0	0	0	0	0	125
III	1.0	1	1	0	1	β_E	1	1	0.3	1	1	0	0	0	125
IV	1.0	1	1	0	1	β_E	1	1	0	0	0	1	0	0	125
V	1.0	1	0	0	0	1	1	1	1	0	0	1	0	0	140
VI	1.0	1	1	0	1	β_E	1	1	0.3	1	1	1	0	0	140
VII	1.0	1	0	0	0	1	1	1	0	0	0	0	1	0	133
VIII	1.0	1	1	0	1	1	1	1	0	0	0	0	0	1	140
IX	1.0	1	0	0	0	1	1	1	1	0	0	0	0	1	150
						Load-Factor Design[§]									
I	1.3	β_D	1.67[¶]	0	1.0	β_E	1	1	0	0	0	0	0	0	
IA	1.3	β_D	2.20	0	0	0	0	0	0	0	0	0	0	0	
IB	1.3	β_D	0	1	1.0	β_E	1	1	0	0	0	0	0	0	
II	1.3	β_D	0	0	0	β_E	1	1	1	0	0	0	0	0	
III	1.3	β_D	1	0	1	β_E	1	1	0.3	1	1	0	0	0	
IV	1.3	β_D	1	0	1	β_E	1	1	0	0	0	1	0	0	
V	1.25	β_D	0	0	0	β_E	1	1	1	0	0	1	0	0	
VI	1.25	β_D	1	0	1	β_E	1	1	0.3	1	1	1	0	0	
VII	1.3	β_D	0	0	0	β_E	1	1	0	0	0	0	1	0	
VIII	1.3	β_D	1	0	1	β_E	1	1	0	0	0	0	0	1	
IX	1.20	β_D	0	0	0	β_E	1	1	1	0	0	0	0	1	

* D = dead load
  L = live load
  I = live-load impact
  E = earth pressure
  B = buoyancy
  W = wind load on structure
WL = wind load on live load

LF = longitudinal force from live load
$(L+I)_n$ = live load plus impact for AASHTO highway loading
CF = centrifugal force
F = longitudinal force due to friction
R = rib shortening
S = shrinkage

T = temperature
EQ = earthquake
SF = stream flow pressure
ICE = ice pressure
$(L+I)_P$ = live load plus impact consistent with the overload criteria of the operating agency

[†] For service-load design: No increase in allowable unit stresses is permitted for members or connections carrying wind loads only.
  $\beta_E = 1.0$ for lateral loads on rigid frames subjected to full earth pressure
  $= 0.5$ when positive moment in beams and slabs is reduced by half the earth-pressure moment
Check both loadings to see which one governs.

[‡] $\% = \dfrac{\text{Maximum unit stress(operating rating)}}{\text{Allowable basic unit stress}} \times 100$

[§] For load factor design:
  $\beta_E = 1.3$ for lateral earth pressure for rigid frames excluding rigid culverts
  $= 0.5$ for lateral earth pressure when checking positive moments in rigid frames
  $= 1.0$ for vertical earth pressure
  $\beta_D = 0.75$ when checking member for minimum axial load and maximum moment or maximum eccentricity and column design
  $= 1.0$ when checking member for maximum axial load and minimum moment and column design
  $= 1.0$ for flexural and tension members

[¶] $\beta_D = 1.25$ for design of outer roadway beam for combination of sidewalk and roadway live load plus impact, if it governs the design, but section capacity should be at least that required for $\beta_D = 1.67$ for roadway live load alone
  $= 1.00$ for deck-slab design for $D + L + I$

Table 17.7 AASHTO LRFD Load Combinations and Load Factor

Design objective	Load combination / Limit state	DC DD DW EH EV ES EL	LL IM CE BR PL LS	WA	WS	WL	FR	TU CR SH	TG	SE	EQ	IC	CT	CV
											Use one of these at a time			
To have structural integrity for all statistically significant loads	STRENGTH I	γ_P	1.75	1.00	—	—	1.00	0.50/1.20	γ_{TG}	γ_{SE}	—	—	—	—
	STRENGTH II	γ_P	1.35	1.00	—	—	1.00	0.50/1.20	γ_{TG}	γ_{SE}	—	—	—	—
	STRENGTH III	γ_P	—	1.00	1.40	—	1.00	0.50/1.20	γ_{TG}	γ_{SE}	—	—	—	—
	STRENGTH IV	γ_P	—	1.00	—	—	1.00	0.50/1.20	—	—	—	—	—	—
	STRENGTH V	γ_P	1.35	1.00	0.40	1.0	1.00	0.50/1.20	γ_{TG}	γ_{SE}	—	—	—	—
To survive without collapsing during a flood, collision, or earthquake	EXTREME EVENT I	γ_P	γ_{EQ}	1.00	—	—	1.00	—	—	—	1.00	—	—	—
	EXTREME EVENT II	γ_P	0.50	1.00	—	—	1.00	—	—	—	—	1.00	1.00	1.00
To last 75 years	SERVICE-I	1.00	1.00	1.00	0.30	1.0	1.00	1.00/1.20	γ_{TG}	γ_{SE}	—	—	—	—
	SERVICE-II	1.00	1.30	1.00	—	—	1.00	1.00/1.20	—	—	—	—	—	—
	SERVICE-III	1.00	0.80	1.00	—	—	1.00	1.00/1.20	γ_{TG}	γ_{SE}	—	—	—	—
To withstand cyclic loading, especially at connections	FATIGUE	—	0.75	—	—	—	—	—	—	—	—	—	—	—

Table 17.8 Load Factors for Permanent Loads γ_p

Type of load	Load factor	
	Maximum	Minimum
DC: Component and Attachments	1.25	0.90
DD: Downdrag	1.80	0.45
DW: Wearing Surfaces and Utilities	1.50	0.65
EH: Horizontal Earth Pressure		
• Active	1.50	0.90
• At-Rest	1.35	0.90
EL: Locked in Erection Stresses	1.0	1.0
EV: Vertical Earth Pressure		
• Retaining Walls and Abutments	1.35	1.00
• Rigid Buried Structure	1.30	0.90
• Rigid Frames	1.35	0.90
• Flexible Buried Structures other than Metal Box Culverts	1.95	0.90
• Flexible Metal Box Culverts	1.50	0.90
ES: Earth Surcharge	1.50	0.75

response in bridge systems is typically achieved through sustained hysteric force-deformation cycles that dissipate energy. This dissipation occurs internally, within the structural members, by the formation of flexural plastic hinges, or externally with isolation bearings or external dampers.

Inelastic behavior should be limited to predetermined locations within the bridge that can be easily inspected and repaired following an earthquake. Preferable locations for inelastic behavior on most bridges include columns, pier walls, and abutment backwalls and wingwalls. Inelastic res-

Table 17.9 Examples of Irregular Bridge Features

Geometry
• Multiple superstructure levels
• Variable width, bifurcating, or highly curved superstructures
• Significant in-plane curvature
• Highly skewed supports

Framing
• Outrigger or C-bent supports
• Unbalanced mass and/or stiffness distribution
• Multiple superstructure types

Geologic Conditions
• Soft soil
• Moderate to high liquefaction potential
• Proximity to an earthquake fault

ponse in the superstructure is not desirable because it is difficult to inspect and repair and may prevent the bridge from being restored to a serviceable condition.

Members not participating in the primary energy dissipating system (i.e. column shear, joints, cap beams and foundations) should be capacity protected. This is achieved by ensuring that the maximum moment and shear from plastic hinges, isolation bearings and dampers can be dependably resisted by the adjoining elements.

17.4.2 Seismic Demands

The uniform load method can be used to determine the seismic loading for bridges that will respond principally in their fundamental mode of vibration. Equivalent static earthquake loads are calculated by multiplying the tributary permanent load by a response spectra coefficient:

$$P_e = \frac{C_{sm}W}{L} \qquad (17.7)$$

where p_e = Equivalent uniform static seismic load per unit length of bridge

C_{sm} = Elastic response coefficient see equation 17.8

W = Dead load of the bridge superstructure and tributary substructure

L = Total length of bridge in ft

$$C_{sm} = \frac{1.2AS}{T_m^{2/3}} \leq 2.5A \qquad (17.8)$$

where T_m = Period of vibration of the mth mode (seconds)

A = Acceleration coefficient from national ground motion maps.

S = Site coefficient specified in Table 17.10

$$T_m = 2\pi\sqrt{\frac{W}{gK}} \qquad (17.9)$$

where g = Acceleration of gravity

K = Bridge lateral stiffness

Single span bridges do not require seismic analysis. The minimum design force at the connections between the superstructure and substructure shall not be less than the product of the site coefficient S, the acceleration coefficient A, and the tributary permanent load.

The multimode spectral mode analysis method should be used if coupling between the longitudinal, transverse and/or vertical response is expected. A three dimensional linear dynamic model should be used to represent the bridge. The elastic seismic forces and displacement generated from multiple mode shapes are combined using acceptable methods such as the root-mean-square method or the complete quadratic combination method. The number of modes in the model should be at least three times the number of spans being modeled. Site-specific response spectra are often developed for multi-modal analysis that incorporates the seismic source, ground attenuation, and near fault phenomena.

When response spectra analysis is used, a maximum single seismic force is calculated by combining two horizontal orthogonal ground motion components. These components are applied along a

Table 17.10 Soil Coefficients

Soil profile type	Site coefficient S	Description
I	1.0	Rock of any description (shale-like or crystalline) or stiff soils (sands, gravels, stiff clays) less than 200 ft in depth overlying rock
II	1.2	Stiff cohesive or deep cohesionless soils more than 200 ft in depth overlying rock
III	1.5	Soft to medium-stiff clays and sands characterized by 30 ft or more of clay with or without intervening layers of sand.
IV	2.0	Soft clays or silts greater than 40 ft of depth.

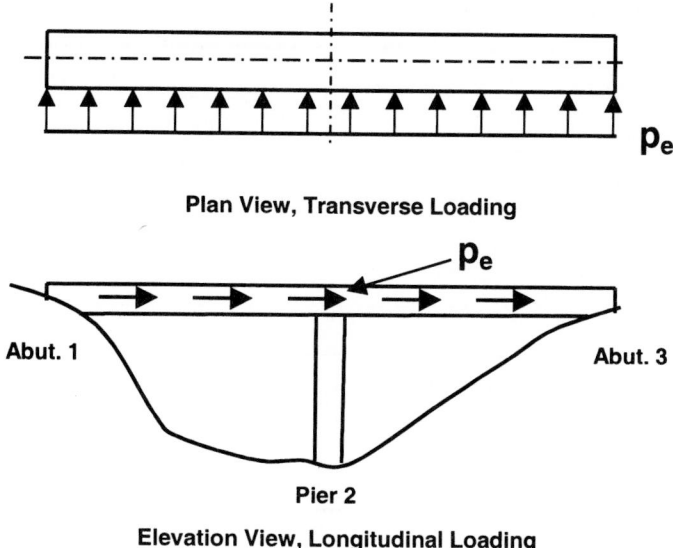

Fig. 17.6 Equivalent static earthquake loads.

longitudinal axis defined by a chord intersecting the centerline of the bridge at the abutments and a normal transverse axis (See Fig. 17.7).

It is uneconomical to design bridges to resist large earthquakes elastically. Columns are assumed to deform inelastically where seismic forces exceed design levels established by dividing the elastically computed moments by the appropriate response modification factors, R. (See Tables 17.11 and 17.12)

The AASHTO Bridge Design Specification defines three levels of response modification factors for critical bridges, essential bridges and other bridges. The bridge owner must determine the performance level required considering social/survival and security/defense requirements.

More rigorous analysis such as inelastic time history analysis should be used on geometrically complex bridges, critical bridges and bridges within close proximity of earthquake faults. The nonlinear analysis provides forces and deformations as a function of time for a specified earthquake motion.

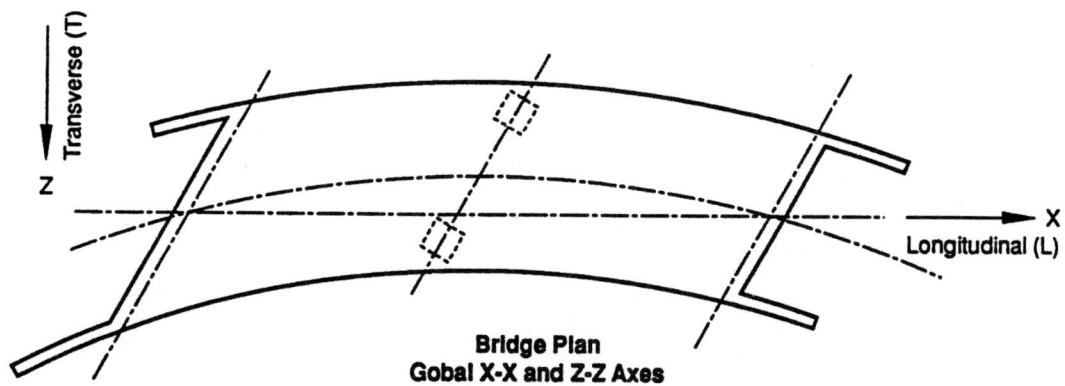

Fig. 17.7 Orthogonal bridge axis definition.

Table 17.11 AASHTO Substructure Response Modification Factors

Substructure	Importance Category		
	Critical	Essential	other
Wall-type piers—larger dimension	1.5	1.5	2.0
Reinforced concrete pile bents			
• Vertical piles only	1.5	2.0	3.0
• With batter piles	1.5	1.5	2.0
Single Columns	1.5	2.0	3.0
Steel or composite steel and concrete pile bents			
• Vertical pile only	1.5	3.5	5.0
• With batter piles	1.5	2.0	3.0
Multiple column bents	1.5	3.5	5.0

A minimum of three ground motions representing the design event should be used.

Nonlinear static analysis, commonly known as pushover analysis has recently been adopted by Caltrans. The inelastic displacement capacity of the piers is compared to the displacements from an elastic demand analysis that considers the bridge's cracked flexural stiffness. The inelastic deformation capacity of the earthquake resisting members is calculated using moment curvature analysis utilizing expected material properties and dependable materials strain limits. Displacement capacities are also limited by degradation of strength and P-Δ effects that occur under large inelastic deformations. If the P-Δ moments are less than 20% of the plastic moment capacity of the member, they are typically ignored.

17.4.3 Seismic Design of Concrete Bridge Columns

Cross sectional column dimensions should be limited to the depth of the superstructures or bent cap to reduce the potential for inelastic damage migrating into the superstructure. The longitudinal reinforcement for compression members should not exceed 4% of the columns gross cross sectional area to insure adequate ductility, avoid congestion and to permit adequate anchorage of the longitudinal reinforcement. Conversely not less than 1% of the columns gross cross sectional area to insure a reasonable level of strength. In the column potential plastic hinge regions, the transverse reinforcement ρ_s for circular columns shall not be less than:

$$\rho_s = 0.12 \frac{f'_c}{f_y} \qquad (17.10)$$

f'_c = specified compressive strength of concrete at 28 days (ksi)

f_y = yield strength of reinforcing bars (ksi)

Table 17.12 AASHTO Connection Response Modification Factors

Connection	All Importance Categories
Superstructure to abutment	0.8
Expansion joints within a span of the superstructure	0.8
Column, piers, or piles bent to cap beam or superstructure	1.0
Column or piers to foundations	1.0

For rectangular sections the total gross sectional area of rectangular hoop reinforcement shall not be less than either:

$$A_{sh} = 0.30 s h_c \frac{f_c'}{f_y}\left[\frac{A_g}{A_c} - 1\right] \text{ or } A_{sh} = 0.12 s h_c \frac{f_c'}{f_y}$$

$$(17.11)$$

where:

s = vertical spacing of hoops not to exceed 4.0 inches (in)

A_c = area of column core (in^2)

A_g = gross area of column (in^2)

A_{sh} = total cross sectional area of tie reinforcement, including supplementary cross ties having a vertical spacing "s" and cross section having core diameter of h_c (in^2)

h_c = core dimension of tied column in the direction under consideration (in^2)

The potential plastic hinge region is defined as the larger of 1.5 times the cross sectional dimension in the direction of bending or the region of column where the moment exceeds 75% of the maximum plastic moment.

The column design shear force should be calculated considering the flexural overstrength developed at the most probable location within the column with a rational combination of the most adverse end moments. The shear resisting mechanism is provided by a combination of truss action (V_s), concrete tensile contribution (V_c) and arch or strut action (V_p).

$$V_u < V_s + V_p + V_c \qquad (17.12)$$

The concrete contribution is significantly diminished under high ductilities and cyclic loading and is often ignored in the plastic moment regions. The flexural reinforcement in continuous or cantilever members needs to detailed to provide continuity of reinforcement at intersections with other members to develop nominal moment resistance of the joint can be developed to resist the shear depicted in Fig. 17.8. Several shear design models defining the shear resisting mechanisms for columns and joints can be found in the AASHTO Design Specifications or the Caltrans Seismic Design Criteria.

The unseating of girders and abutments must be avoided in all circumstances. The seat width needs to accommodate thermal movement, prestress shortening, creep, shrinkage and anticipated earthquake displacements. The seat width should not be less then 1.5 times the elastic seismic displacement

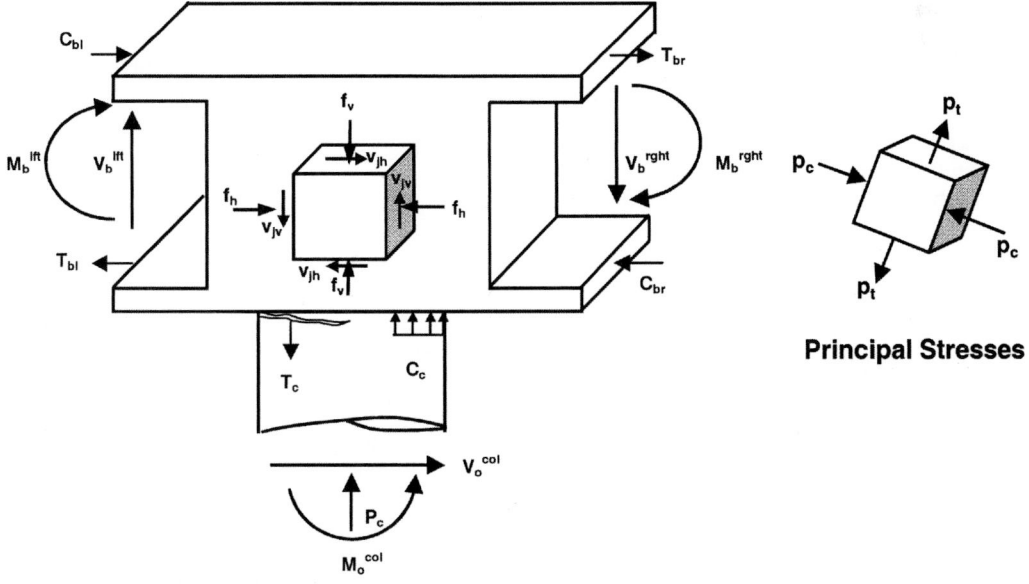

Principal Stresses

Fig. 17.8 Joint shear stresses in T-joints.

of the superstructure at the seat or:

$$N = \left[(8 + 0.02L + 0.08H)\left(1 + \frac{S^2}{8000}\right)\right] \quad (17.13)$$

    N = Support width normal to the centerline of bearing

    L = Length of the bridge deck to the adjacent expansion joint, or the end of the bridge (ft)

    H = Average height of the columns supporting the bridge deck to the next expansion joint (ft) ($H = 0$ for single span bridges)

    S = Skew of the support measured from line normal to span (deg)

(MCEER, "Recommended LRFD Guidelines for the Seismic Design of Highway Bridges," Caltrans Seismic Design Criteria, vol. 1.1 (www.dot.ca.gov); AASHTO LRFD Bridge Design Specification (www.aashto.org).)

Steel Bridges

Steel is competitive as a construction material for medium and long-span bridges for the following reasons: It has high strength in tension and compression. It behaves as a nearly perfect elastic material within the usual working ranges. It has strength reserves beyond the yield point. The high standards of the fabricating industry guarantee users uniformity of the controlling properties within narrow tolerances. Connection methods are reliable, and workers skilled in their application are available.

The principal disadvantage of steel in bridge construction, its susceptibility to corrosion, is being increasingly overcome by chemical additives or improved protective coatings.

17.5 Systems Used for Steel Bridges

The following are typical components of steel bridges. Each may be applied to any of the functional types and structural systems listed in Art. 17.1.

Main support: Rolled beams, plate girders, box girders, or trusses.

Connections: (See also Art. 17.7.) High-strength-bolted, welded, or combinations.

Materials for traffic-carrying deck: Timber stringers and planking, reinforced concrete slab or prestressed concrete slab, stiffened steel plate (orthotropic deck), or steel grid.

Timber decks are restricted to bridges on roads of minor importance. Plates of corrosion resistant steel should be used as ballast supports on through plate-girder bridges for railways. For roadway decks of stiffened steel plates, see Art. 17.13.

Deck framing: Deck resting directly on main members or supported by grids of stringers and floor beams.

Location of deck: On top of main members: deck spans (Fig. 17.9*a*); between main members, the underside of the deck framing being flush with that of the main members: through spans (Fig. 17.9*b*).

17.6 Grades and Design Criteria for Steel for Bridges

Preferred steel grades, permissible stresses, and standards of details, materials, and quality of work for steel bridges are covered in the AREMA and AASHTO specifications. Properties of the various grades of steel and the testing methods to be used to control them are regulated by specifications of ASTM. Properties of the structural steels presently preferred in bridge construction are Tabulated in Table 17.13.

Dimensions and geometric properties of commercially available rolled plates and shapes are tabulated in the "Steel Construction Manual," for allowable stress design and for load-and-resistance-factor design of the American Institute of Steel Construction (AISC), and in manuals issued by the major steel producers.

All members, connections, and parts of steel bridges should be designed by the load-factor design method, and then checked for fatigue at service-level loads. The fatigue check should assure that all connections are within allowable stress ranges (FSR).

The design strength of a beam or girder is based on the dimensional properties of the section and the spacing of compression flange bracing. The three types of member sections are (1) compact, (2) braced noncompact, and (3) partially braced. The AASHTO Flexural design formulas for the three types of I-Girder sections are shown in Table 17.14.

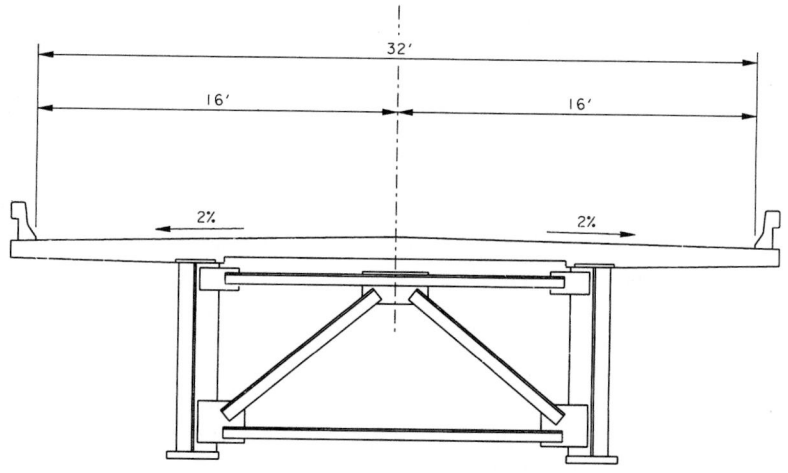

(a) TYPICAL SECTION - HIGHWAY BRIDGE

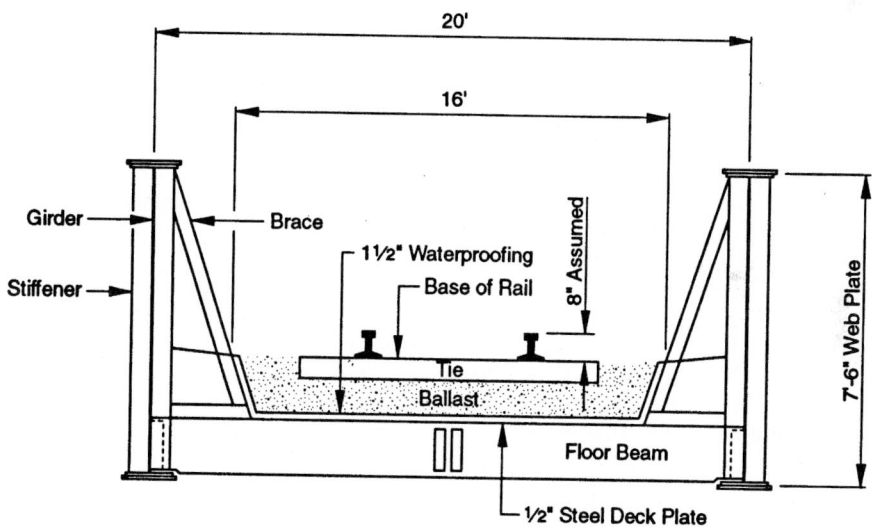

(b) TYPICAL SECTION - THROUGH GIRDER RAILROAD BRIDGE

Fig. 17.9 Two-lane deck-girder highway bridge

Design Limitations on Depth Ratios, Slenderness Ratios, Deflections ▪ AASHTO specifications restrict the depth-to-span ratios D/L of bridge structures and the slenderness ratios l/r of individual truss or bracing members to the values in Table 17.15.

where D = depth of construction, ft

L = span, ft, c to c bearings for simple spans or distance between points of contraflexure for continuous spans

l = unsupported length of member, in

r = radius of gyration, in

These are minimum values; preferred values are higher.

Table 17.13   Minimum Mechanical Properties of Structural Steel

Property	Structural steel	High-strength low-alloy steel		Quenched and tempered low-alloy steel	High-yield-strength, quenched and tempered alloy steel	
AASHTO designation	M270 Grade 36	M270 Grade 50	M270 Grade 50W	M270 Grade 70W	M270 Grades 100/100W	
Equivalent ASTM designations	A709 Grade 36	A709 Grade 50	A709 Grade 50W	A709 Grade 70W	A709 Grades 100/100W	
Thickness of plates, in	Up to 4 incl.	Up to 4 incl.	Up to 4 incl.	Up to 4 incl.	Up to 2.5 incl.	Over 2.5 to 4 incl.
Shapes	All groups	All groups	All groups	Not applicable	Not applicable	Not applicable
Minimum tensile strength, F_u, ksi	58	65	70	90	110	100
Minimum yield point or yield strength, F_y, ksi	36	50	50	70	100	90

Both specifications limit the elastic deflections of bridges under design live load plus impact to $\frac{1}{800}$ of the span, measured c to c bearings, except that $\frac{1}{1000}$ may be used for bridges used by pedestrians; $\frac{1}{300}$ of the length of cantilever arms. Deflection calculations should be based on the gross sections of girders or truss members. Anticipated dead-load deflections must be compensated by adequate camber in the fabrication of steel structures.

Splices ▪ Shop, assembly yard, or erection splices must be provided for units whose overall length exceeds available rolled lengths of plates and shapes or the clearances of available shipping facilities. Splices also must be provided when total weight exceeds the capacity of available erection equipment.

Accessibility ▪ All parts should be accessible and adequately spaced for fabrication, assembly, and maintenance. Closed box girders and box-type sections should be equipped with handholes or manholes.

On long and high bridges, installation of permanent maintenance travelers may be justified.

17.7   Steel Connections in Bridges

Connections of steel members to other steel members are usually made with high-strength bolts, welds, or pins. In composite construction, steel beams are joined to concrete decks by steel studs or channels welded to the top flange of the beams.

17.7.1   Connections with High-Strength Bolts

The parts may be clamped together by bolts of quenched and tempered steel, ASTM A325. The nuts are tightened to 70% of their specified tensile strength.

Details and quality of work are covered by the "Specifications for Structural Joints Using ASTM A325 and A490 Bolts," approved by the Research Council on Structural Connections of the Engineering Foundation. Maximum stresses for bearing type connections are given in Table 17.11.

Tensioned ASTM A325 bolts are the preferred bolt for all steel bridge connections. The nuts on

Table 17.14 ASSHTO Flexural Design Formulas for I-Girders

Section description	Compression flange slenderness	Web slenderness	Lateral Bracing of compression flange	M_u
Compact	$\dfrac{b}{t} \le \dfrac{4100}{\sqrt{F_y}}$ When b/t and D/t_w exceeds 75% of the above limits, the following interaction equation shall apply $\dfrac{D}{t_w} + 4.68\left(\dfrac{b}{t}\right) \le \dfrac{33,650}{\sqrt{F_{yf}}}$	$\dfrac{D}{t_w} \le \dfrac{19,230}{\sqrt{F_y}}$	$\dfrac{L_b}{r_y} \le \left[\dfrac{3.6 - 2.2\left(\dfrac{M_l}{M_u}\right)}{F_y}\right] \times 10^6$	$F_y Z$
Braced Non-Compact	$\dfrac{b}{t} \le 24$	With transverse stiffeners only: $\dfrac{D}{t_w} \le \dfrac{36,500}{\sqrt{F_y}}$ With transverse stiffeners and one longitudinal stiffeners: $\dfrac{D}{t_w} \le \dfrac{73,000}{\sqrt{F_y}}$	$L_b \le \dfrac{20,000,000}{F_y d}$	Lessor of: $F_y S_{xt}$ or $F_{cr} S_{xc} R_b$
Partially Braced	*No Max. Requirements*	*No Max. Requirements*	*No Max. Requirements*	See AASHTO Standard Specifications for Highway Bridges

A_f = flange area (sq in^2)

b = compression flange width

D = clear distance between flanges

d = depth of beam or girder

f = the lesser of $(f_b/R_b)\, F_y$ (psi)

f_b = factored bending stress in the compression flange (psi)

f_{cr} = critical stress of the compression flange (psi)

F_y = specified minimum field strength of the steel being used (psi)

F_{yf} = specified minimum field strength of the compression flange (psi)

L_b = distance between points of bracing of the compression flange (in)

M_l = smaller moment at the end of the unbraced length of the member (lb-in)

M_u = design strength equal $F_y{*}Z$ at the other end of the unbraced length:

   (M_l/M_u) is positive when moments cause single curvature between brace points

   (M_l/M_u) is negative when moments cause reverse curvature between brace points (lb-in)

R_b = Bending Capacity Reduction Factor, See AASHTO Standard Specifications for Highway Bridges

r_y = radius of gyration of the steel section with respect to the Y-Y axis (in)

S_{xc} = elastic section modulas with respect to compression flange (in^3)

S_{xt} = elastic section modulas with respect to tension flange (in^3)

t = flange thickness (in)

t_w = web thickness (in)

Z = plastic section modulas (in^3)

Table 17.15 Dimensional Limitations for Bridge Members

	AASHTO
Min depth-span ratios:	
For noncomposite beams or girders	1/25
For simple span composite girders*	1/22
For continuous composite girders*	1/25
For trusses	1/10
Max slenderness ratios:	
For main members in compression	120
For bracing members in compression	140
For main members in tension	200
For bracing members in tension	240

* For composite girders the depth shall include the concrete slab.

tensioned A325 and A490 bolts will not loosen under vibrations associated with bridge loadings. If ASTM A307 bolts or non tensioned high strength bolts are used, provisions should be made to prevent nut loosing by the use of thread locking adhesive, self-locking nuts, or double nuts. Bolted connections subject to tension, or combined tension and shear, or stress reversal, or severe vibration, or heavy impact loads, or any other condition where joint slippage would be detrimental, shall use tensioned High-Strength bolts and be designed as a slip-critical connection.

Slip-critical connections are designed to prevent slip at a specified overload condition in addition to meeting the strength requirements in bearing. The overload condition at which the connection is required work as a friction connection (no slip) is equal to Dead load + 1.67(Live load + Impact). The slip strength of the connection is based on the number of slip plains, the friction coefficient of the contact surfaces, the type of hole, and the bolt tension stress. The AASHTO specifications provide equations for determining the slip strength of connections.

17.7.2 Welded Connections

In welding, the parts to be connected are fused at high temperatures, usually with addition of suitable metallic material. The "Structural Welding Code," AWS D1.5, American Welding Society, regulates application of the various types and sizes of welds, permissible stresses in the weld and parent metal, permissible edge configurations, kinds and sizes of electrodes, details of quality of work, and qualification of welding procedures and welders. (For Maximum welding stresses, see Table 17.16.)

Many designers favor the combination of shop welding with high-strength-bolted field connections.

17.7.3 Pin Connections

Hinges between members subject to relative rotation are usually formed with pins, machined steel cylinders. They are held in either semicircular machined recesses or smoothly fitting holes in the connected members.

For fixation of the direction of the pin axis, pins up to 10-in diameter have threaded ends for recessed nuts, which bear against the connected members. Pins over 10-in diameter are held by recessed caps. These in turn are held by either tap bolts or a rod that runs axially through a hole in the pin itself and is threaded and secured by nuts at its ends.

Pins are designed for bending and shear and for bearing against the connected members. (For stresses, see Art. 9.6.)

17.8 Rolled-Beam Bridges

The simplest steel bridges consist of rolled wide-flange beams and a traffic-carrying deck. Rolled beams serve also as floor beams and stringers for decks of plate-girder and truss bridges.

Reductions in steel weight may be obtained, but with greater labor costs, by adding cover plates in

Table 17.16 Design Strength of Connectors

Type of fastener	Strength (ϕF)
Groove weld[a]	$1.00F_y$
Fillet weld[b]	$0.45F_u$
Low-carbon steel bolts ASTM A307	
Tension[g]	30 ksi
Shear on Bolt with threads in shear plane	18 ksi
Power-driven rivets ASTM A502	
Shear—Grade 1	25 ksi
Shear—Grade 2	30 ksi
High-Strength Bolts	
AASHTO M 164 (ASTM A325)	
Applied Static Tension[c,g]	68 ksi
Shear on bolt with threads in shear plane[c,d,e]	35 ksi
AASHTO M 253 (ASTM A490)	
Applied Static Tension[g]	85 ksi
Shear on bolt with threads in shear plane[d,e]	43 ksi
Bolt bearing on connected material[f]	$0.9L_c t F_u \leq 1.8 d t F_u$

[a]F_y = yield point of connected material
[b]F_u = minimum strength of the welding rod metal but not greater than the tensile strength of the connected parts.
[c]The tensile strength of M 164 (A 325) bolts decreases for diameters greater than 1 inch.
The design values listed are for bolts up to 1 inch diameter. The design values shall be multiplied by 0.875 for diameters greater than 1 inch.
[d]Tabulated values shall be reduced by 20 percent in bearing-type connections whose length between extreme fasteners in each of the spliced parts measured parallel to the line of axial force exceeds 50 inches.
[e]If material thickness or joint details preclude threads in the shear plane, multiply values by 1.25
[f]Bearing on connected material in standard oversized short slotted holes loaded in any direction or long slotted holes parallel to the applied bearing force. For long slotted holes perpendicular to the applied bearing force calculated values shall be reduced 20 percent. L_c is clear distance between the holes or between the hole and the edge of the material in the direction of the applied bearing force (in); d is the diameter of the bolt (in), t is thickness of connected material (in) and F_u is the specified minimum tensile strength of the connected part given in Table 17.9.
[g]For combined tension and shear when $f_v/F_v > 0.33$ the design tensile strength, F_t (in table) shall be reduced to F_t' where:

$$F_t' = F_t\sqrt{1 - (f_v/F_y)^2}$$
f_v = calculated bolt stress in shear
F_v = design shear strength of bolt (in table)

the area of maximum moments, by providing continuity over several spans, by utilizing the deck in composite action, or by a combination of these measures. The principles of design and details are essentially identical with those of plate girders (Art. 17.9).

17.9 Plate-Girder Bridges

The term plate girder applies to structural elements of I-shaped cross section that are welded from plates. Plate girders are used as primary supporting elements in many structural systems: as simple beams on abutments or, with overhanging ends, on piers; as continuous or hinged multispan beams; as stiffening girders of arches and suspension bridges, and in frame-type bridges. They also serve as floor beams and stringers on these other bridge systems.

Their prevalent application on highway and railway bridges is in the form of deck-plate girders in combination with concrete decks (Fig. 17.9). (For design of concrete deck slabs, see Art. 17.20. For girders with steel decks (orthotropic decks), see Art 17.13. Girders with track ties mounted directly on the top flanges, **open-deck girders**, are used on branch railways and industrial spurs. **Through plate girders** (Fig. 17.9b) are now practically restricted to railway bridges where allowable structure depth is limited.

The two or more girders supporting each span must be braced against each other to provide stability against overturning and flange buckling, to resist transverse forces (wind, earthquake, centrifugal), and to distribute concentrated heavy loads. On **deck girders**, this is done by transverse bracing in vertical planes. Transverse bracing should be installed over each bearing and at intermediate locations not over 25 ft apart. This bracing may consist either of full-depth cross frames or of solid diaphragms with depth at least half the web depth for rolled beams and preferably three-quarters the web depth for plate girders. End cross frames or diaphragms should be proportioned to transfer fully all vertical and lateral loads to the bearings. On **through-girder spans,** since top lateral and transverse bracing systems cannot be installed, the top flanges of the girders must be braced against the floor system. For the purpose, heavy gusset plates or knee braces may be used (Fig. 17.9b).

The most commonly used type of steel bridge girder is the welded plate girder. It is typically laterally braced, noncompact, and unsymmetrical, with top and bottom flanges of different sizes. Figure 17.10b shows a typical welded plate girder.

Variations in moment resistance are obtained by using flange plates of different thicknesses, widths, or steel grades, butt-welded to each other in succession. Web thickness too may be varied. Girder webs should be protected against buckling by transverse and, in the case of deep webs, longitudinal stiffeners. Transverse bearing stiffeners are required to transfer end reactions from the web into the bearings and to introduce concentrated loads into the web. Intermediate and longitudinal stiffeners are required if the girder depth-to-thickness ratios exceed critical values (see Art. 9.13.4).

Stiffeners may be plain plates, angles, or T sections. Transverse stiffeners can be in pairs or single elements. The AASHTO Specifications contain restrictions on width-to-thickness ratios and minimum widths of plate stiffeners (Art. 9.13.4).

Web-to-flange connections should be capable of carrying the stress flow from web to flange at every section of the girder. At an unloaded point, the stress flow equals the horizontal shear per linear inch. Where a wheel load may act, for example, at upper flange-to-web connections of deck girders, the stress flow is the vectorial sum of the horizontal shear per inch and the wheel load (assumed distributed over a web length equal to twice the deck thickness). Welds connecting bearing stiffeners to the web must be designed for the full bearing reaction.

Space restrictions in the shop, clearance restrictions in transportation, and erection considerations may require dividing long girders into shorter sections, which are then joined (spliced) in the field. Individual segments, plates or angles, must be spliced either in the shop or in the field if they exceed in length the sizes produced by the rolling mills or if shapes are changed in thickness to meet stress requirements.

Specifications require splices to be designed for the average between the stress due to design loads and the capacity of the unspliced segment, but for not less than 75% of the latter. In bolted design, material may have to be added at each splice to satisfy this requirement. Each splice element must be connected by a sufficient number of bolts to develop its full strength. Whenever it is possible to do so, splices of individual segments should be staggered. No splices should be located in the vicinity of the highest-stressed parts of the girder, for example, at midspan of simple-beam spans, or over the bearings on continuous beams. Figure 17.10a presents a design flow chart for welded plate girders.

(F. S. Merritt and R. L. Brockenbrough, "Structural Steel Designers' Handbook," 2nd ed., McGraw-Hill Inc., New York (books-mcgraw-hill.com).)

17.10 Composite-Girder Bridges

Installation of appropriately designed shear connectors between the top flange of girders or beams and the concrete deck allows use of the deck as part of the top flange (equivalent cover plate). The resulting increase in effective depth of the total section and possible reductions of the top-flange steel usually allow some savings in steel compared with the noncomposite steel section. The overall economy depends on the cost of the shear connectors and any other additions to the girder or the deck that may be required and on possible limitations in effectiveness of the composite section as such.

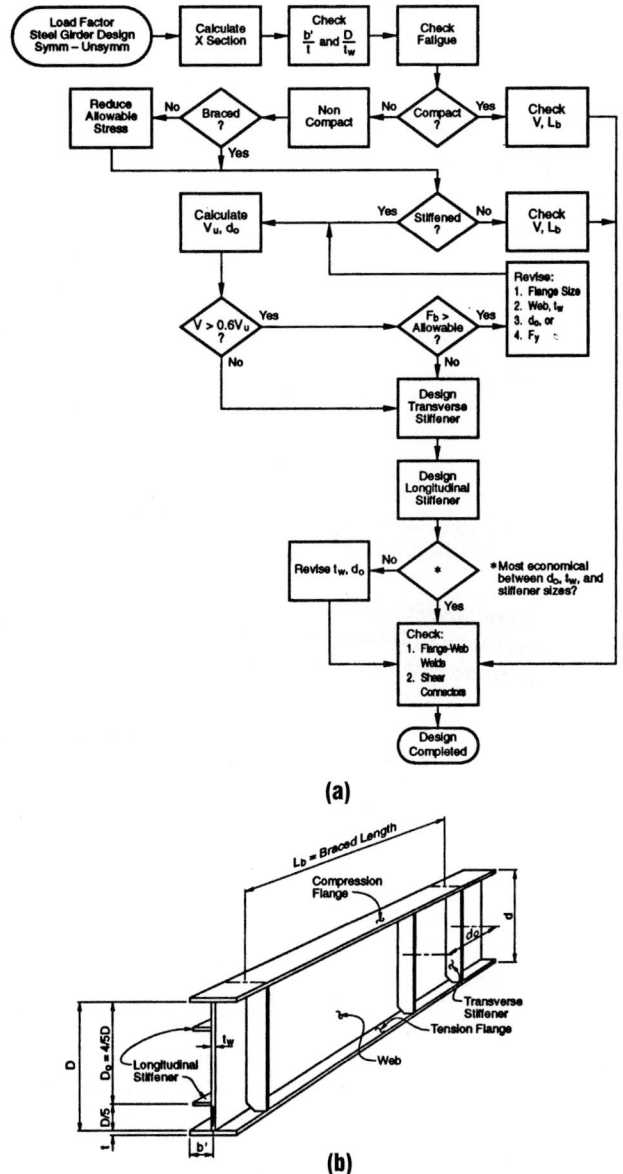

Fig. 17.10 Welded plate girder. (*a*) Flow chart gives steps in load-factor design. (*b*) Typical plate girder—stiffened, braced, and noncompact.

In areas of negative moment, composite effect may be assumed only if the calculated tensile stresses in the deck are either taken up fully by reinforcing steel or compensated by prestressing. The latter method requires special precautions to assure slipping of the deck on the girder during the prestressing operation but rigidity of connection after completion.

If the steel girder is not shored up while the deck concrete is placed, computation of dead-load stresses must be based on the steel section alone.

The effective flange width of the concrete slab that is used as a T-beam flange of a composite girder is the lesser of the following:

1. One-fourth of the span length of the girder

2. The center to center distance between adjacent girders

3. Twelve times the least thickness of the slab

Shear connectors should be capable of resisting all forces tending to separate the abutting concrete and steel surfaces, both horizontally and vertically. Connectors should not obstruct placement and thorough compaction of the concrete. Their installation should not harm the structural steel.

The types of shear connectors presently preferred are channels, or welded studs. Channels should be placed on beam flanges normal to the web and with the channel flanges pointing toward the girder bearings.

The modular ratio recommended for stress analysis of composite girders under live loads is given in Table 17.17.

For composite action under dead loads, the concrete section may be assumed to be subjected to constant compressive stress. This will cause the concrete to undergo plastic flow and thus will reduce its capacity to resist stress. This is taken into account in design of a composite girder for dead loads by multiplying by 3 the modular ratio n given in Table 17.17. Most composite girders, however, are designed for composite action only for live loads and dead loads (usually, curbs, railings, and utilities) that are added after the concrete deck has attained sufficient strength to support them.

Table 17.17 Modular Ratio for Composite Girders with Live Loads

Specified minimum compressive strength of concrete deck f'_c, psi	Modular ratio $n = E_s/E_c^*$
2000–2400	15
2500–2900	12
3000–3900	10
4000–4900	8
5000 or more	6

*E_s = elastic modulus of the steel
E_c = elastic modulus of the concrete

Example—Stress Calculations for a Composite Girder: The following illustrates the procedure for determining flexural stresses in a composite welded girder for factored loads. The girder is assumed to be fabricated of M270, Grade 50, steel, with yield point $F_y = 50$ ksi. It will not be shored during placement of the concrete deck. For the concrete, the 28-day compressive strength is assumed to be $f'_c = 3.25$ ksi, $n = 10$ for live loads, $n = 30$ for dead loads. Dimensions, section properties, and bending moments are given in the following:

The section properties of the steel girder alone are determined first. For the purpose, the moment of inertia I_{1-1} of the steel section (Fig. 17.11a) is calculated with respect to the bottom of the girder. Then, the moment of inertia I_{NA} with respect to the neutral axis is computed. Next, the section properties of the composite section (Fig. 17.11b) are calculated. Stresses in the concrete are small, since the steel girder carries the weight of the deck.

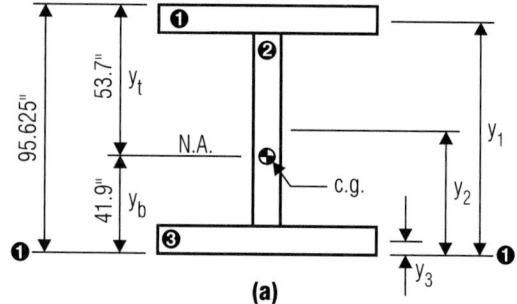

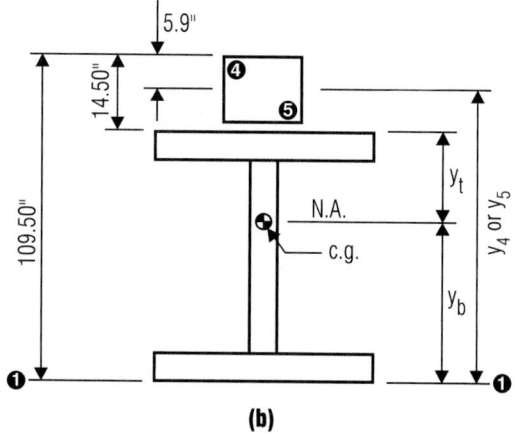

Fig. 17.11 Sections of composite plate girder: (a) steel section alone; (b) composite section.

Steel Section for Slab and Girder Loads (Fig. 17.11a)

Material	Size, in	Area	y	Ay	Ay^2
① Top flange	$12 \times \frac{5}{8}$	7.50	95.31	715	68,130
② Web	$94 \times \frac{5}{16}$	29.38	48.00	1410	67,680
③ Bottom flange	14×1	14.00	0.50	7	4

$$\Sigma A = 50.88 \qquad \Sigma Ay = 2132 \qquad 135,814$$

$$y_b = \frac{\Sigma Ay}{\Sigma A} = 41.9$$

Moment of inertia of web =	$+21,629$
$I_{1\text{-}1} =$	157,443
$-y_b^2 \Sigma A =$	$-89,325$
$I_{NA} =$	68,118

$$y_t = 95.6 - 41.9 = 53.7$$

Section for Curb Loads, Railing, Utilities (Composite) (Fig. 17.11b)

Material	Area	y	Ay	Ay^2
Steel section	50.88		2132	135,814
④ Concrete: A/n, $n = 30$	51.57	103.6	5342	553,463
	102.45		7474	689,277

$$y_b = \frac{7474}{102.45} = 72.9$$

Moment of inertia of girder web =	$+21,629$
$I_{1\text{-}1} =$	710,906
$-y_b^2 \Sigma A =$	$-544,461$
$I_{NA} =$	166,445

$$y_t = 95.6 - 72.9 = 22.7$$

Section for Live Loads (Composite) (Fig. 17.11b)

Material	Area	y	Ay	Ay^2
Steel section	50.88		2132	135,814
④ Concrete: A/n, $n = 10$	154.70	103.6	16,027	1,660,389
	205.58		18,159	1,796,203

$$y_b = \frac{18.159}{205.58} = 88.3$$

Moment of inertia of girder web =	$+21,629$
$I_{1\text{-}1} =$	1,817,832
$-y_b^2 \Sigma A =$	$-1,603,995$
$I_{NA} =$	213,837

$$y_t = 95.6 - 88.3 = 7.3$$

Moments

For slab and girder loads = 1825 kips-ft

For curb loads = 347 kips-ft

For live loads = 6295 kips-ft

Stresses in Steel Girder

Type of Load	Bottom f_s	Top f_s
Slab and girder loads	$\dfrac{1825 \times 12}{68,118} \times 41.9 = 13.5$	$\dfrac{1825 \times 12}{68,118} \times 53.7 = 17.3$
Curb loads	$\dfrac{347 \times 12}{166,445} \times 72.9 = 1.8$	$\dfrac{347 \times 12}{166,445} \times 22.7 = 0.6$
Live loads	$\dfrac{6295 \times 12}{213,937} \times 88.3 = 31.2$	$\dfrac{6295 \times 12}{213,937} \times 7.3 = 2.6$
	Bottom $f_s = 46.5$ kips/in^2	Top $f_s = 20.5$ kips/in^2

17.11 Fatigue Design of Bridge Members

All members and connections should be designed so that maximum stresses induced by loads are less than allowable stresses and also so that the range of stresses induced by variations in service loads is less than the allowable fatigue stress range. If a member is entirely in compression and never is subjected to tensile stresses, a fatigue check is not required.

Fatigue is an important consideration in design of all bridge components but may be especially critical for welded girders. Welding leaves residual stresses in welded regions due to heat input during the welding process and subsequent differential cooling.

The types of connections that are most commonly used with welded plate girders and that should be checked for fatigue are illustrated in Fig. 17.12, and the stress category assigned for each type is given in Table 17.18a. Table 17.19 gives the allowable stress ranges for various stress categories. Table 17.20 lists allowable stress cycles for various types of roads and bridge members.

("Economical and Fatigue-Resistant Steel Bridge Details," FHWA-HI-90-043, Federal Highway Administration; "Guide Specifications for Fatigue Critical Non-Redundant Steel Bridges," American Association of State Highway and Transportation Officials (www.aashto.org).)

17.12 Orthotropic-Deck Bridges

An orthotropic deck is, essentially, a continuous, flat steel plate, with stiffeners (ribs) welded to its underside in a parallel or rectangular pattern. The term *orthotropic* is shortened from *orthogonal anisotropic*, referring to the mathematical theory used for the flexural analysis of such decks.

When used on steel bridges, orthotropic decks are usually joined quasi-monolithically, by welding or high-strength bolting, to the main girders and floor beams. They then have a dual function as roadway and as structural top flange.

The combination of plate or box girders with orthotropic decks allows the design of bridges of considerable slenderness and of nearly twice the span reached by girders with concrete decks. The

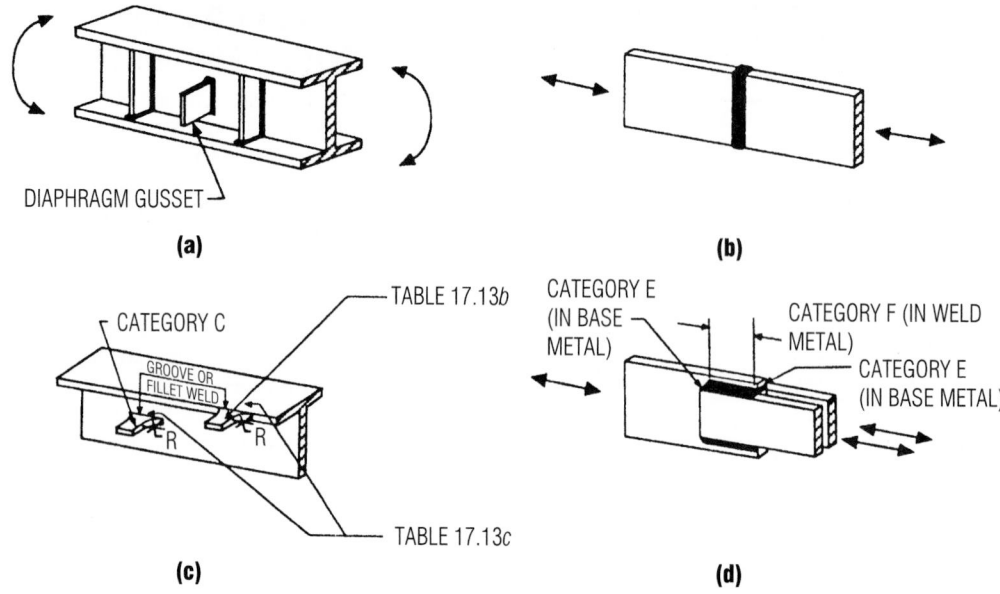

Fig. 17.12 Fatigue stress categories for some commonly used connections (see Table 17.13). In (c), category C applies also to transverse loading.

most widespread application of orthotropic decks is on continuous, two- to five-span girders on low-level river crossings in metropolitan areas, where approaches must be kept short and grades low. This construction has been used for main spans up to 1100 ft in cable-stayed bridges and up to 856 ft without cable stays. There also are some spectacular high-level orthotropic girder bridges and some arch and suspension bridges with orthotropic stiffening girders. On some of the latter, girders and deck have been combined in a single lens-shaped box section that has great stiffness and low aerodynamic resistance.

17.12.1 Box Girders

Single-web or box girders may be used for orthotropic bridges. Box girders are preferred if structure depth is restricted. Their inherent stiffness makes it possible to reduce, or to omit, unsightly transverse bracing systems. In cross section, they usually are rectangular, occasionally trapezoidal. Minimum dimensions of box girders are controlled by considerations of accessibility and ease of fabrication.

Wide decks are supported by either single box girders or twin boxes. Wide single boxes have been built with multiple webs or secondary interior trusses. Overhanging floor beams sometimes are supported by diagonal struts.

17.12.2 Depth-Span Ratios

Girder soffits are parallel to the deck, tapered, or curved. Parallel flanges, sometimes with tapered side spans, generally are used on unbraced girders with depth-to-main-span ratios as low as 1:70. Parallel-flange unbraced girders are practically restricted to high-level structures with unrestricted clearance. Unbraced low-level girders usually are designed with curved soffits with minimum depth-to-main-span ratios of about 1:25 over the main piers and 1:50 at the shallowest section.

17.12.3 Cable-Suspended Systems with Orthotropic Decks

Main spans of bridges may have girders suspended from or directly supported by cables that are hung from towers, or pylons. The cables are curved if the

Table 17.18 Fatigue Stress Categories for Bridge Members

(a) Stress Categories for Typical Connections			
Type of connection	Figure No.	Stress	Category
Toe of transverse stiffeners	17.13a	Tension or reversal	C
Butt weld at flanges	17.13b	Tension or reversal	B
Gusset for lateral bracing (assumed groove weld, R ≥ 24 in)	17.13c	Tension or reversal	B
Flange to web	17.13d	Shear	F

(b) Stress Categories for Weld Conditions in Fig. 17.13c	
Weld condition*	Category
Unequal thickness; reinforcement in place	E
Unequal thickness; reinforcement removed	D
Equal thickness; reinforcement in place	C
Equal thickness; reinforcement removed	B

(c) Stress Categories for Radii R in Fig. 17.13c		
	Category for welds	
R, in[†]	Fillet	Groove
24 or more	D	B
From 6 to 24	D	C
From 2 to 6	D	D
2 or less	E	E

* For transverse loading, check transition radius for possible assignment of lower category.

† Also applies to transverse loading.

girders are suspended at each floor beam (suspension bridges); otherwise they are straight (cable-stayed bridges). In cable-stayed bridges, the cables may extend from the pylons to the connections with the girders in tiers, parallel to each other (harped), or in a bundle pattern (radiating from the pylons). See Fig. 17.24.

Each cable stay adds one degree of statical indeterminancy to a system. To make the actual conditions conform to design assumptions, the cable length must be adjustable either at the anchorages to the girders or at the saddles on the towers. (See also Art. 17.16.)

17.12.4 Steel Grades

The steel commonly used for orthotropic plates is weldable high-strength, low-alloy structural steel

M270, Grade 50. Minimum thickness is seldom less than $\frac{7}{16}$ in (10 mm), to avoid excessive deflections under heavy wheel loads. The maximum thickness seldom exceeds $\frac{3}{4}$ in because of the decrease in permissible working stresses of high-strength low-alloy steel and the increase of fillet- and butt-weld sizes for plates of greater thickness.

17.12.5 Floor Beams

If, as in most practical cases, the deck spans transversely between main girders, transverse ribs are replaced by the floor beams, which are then built up of inverted T sections, with the deck plate acting as top flange. Floor-beam spacings are preferably kept constant on any given structure. They range from less than 5 ft to over 15 ft. Longer spacings have been suggested for greater economy.

Table 17.19 Allowable Fatigue Stress Range FSR*, ksi, for Bridge Members

	For Structures with Redundant Load Paths[†]			
Category	For 100,000 Cycles	For 500,000 Cycles	For 2,000,000 Cycles	For over 2,000,000 Cycles
A	63◇	37◇	24◇	24◇
B	49	29	18	16
C	35.5	21	13	10
				12[‡]
D	28	16	10	7
E	22	13	8	4.5
F	16	9.2	5.8	2.6
G	15	12	9	8

	Nonredundant-Load-Path Structures			
Category	For 100,000 Cycles	For 500,000 Cycles	For 2,000,000 Cycles	For over 2,000,000 Cycles
A	50◇	29◇	24◇	24◇
B	39	23	16	16
C	28	16	10	9
			12[‡]	11[‡]
D	22	13	8	5
E[§]	17	10	6	2.3
F	12	9	7	6

*The range of stress is defined as the algebraic difference between the maximum stress and the minimum stress. Tension stress is considered to have the opposite algebraic sign from compression stress.

[†]Structure types with multiload paths where a single fracture in a member cannot lead to the collapse. For example, a simple supported single-span multibeam bridge or a multielement eyebar truss member has redundant load path.

[‡]For transverse stiffener welds on girder webs or flanges.

[§]Partial-length welded cover plates should not be used on flanges more than 0.8 in thick for nonredundant-load path structures.

◇For unpainted weathering steel, A709, the category 'A' allowable FSR values are less then values shown, see AASHTO Specifications.

17.12.6 Ribs

These are either open (Fig. 17.13a) or closed (Fig. 17.13b). The spacing of open ribs is seldom less than 12 in or more than 15 in. The lower limit is determined by accessibility for fabrication and maintenance, the upper by considerations of deck-plate stiffness. To reduce deformations of the surfacing material under concentrated traffic loads, some specifications require the plate thickness to be not less than $\frac{1}{25}$ of the spacing between open ribs or between the weld lines of closed ribs.

Usually, the longitudinal ribs are made continuous through slots or cutouts in the floor-beam webs to avoid a multitude of butt welds. Rib splices can then be coordinated with the transverse deck splices.

Closed ribs, because of their greater torsional rigidity, give better load distribution and, other things being equal, require less steel and less welding than open ribs. Disadvantages of closed ribs are their inaccessibility for inspection and maintenance and more complicated splicing details. There have also been some difficulties in defining the weld between closed ribs and deck plate.

Table 17.20 Allowable Stress Cycles for Bridge Members

Main (longitudinal) load-carrying members				
Type of road	Case	ADTT*	Truck loading	Lane loading[†]
Freeways, expressways, major highways, and streets	I	2500 or more	2,000,000[‡]	500,000
Freeways, expressways, major highways, and streets	II	less than 2500	500,000	100,000
Other highways and streets not included in case I or II	III		100,000	100,000

Transverse members and details subjected to wheel loads			
Type of road	Case	ADTT*	Truck loading
Freeways, expressways, major highways, and streets	I	2500 or more	over 2,000,000
Freeways, expressways, major highways, and streets	II	less than 2500	2,000,000
Other highways and streets	III		500,000

* Average daily truck traffic (one direction).

[†] Longitudinal members should also be checked for truck loading.

[‡] Members should also be investigated for fatigue when over 2 million stress cycles are produced by a single truck on the bridge with load distributed to the girders as designated for traffic lane loading.

17.12.7 Fabrication

Orthotropic decks are fabricated in the shop in panels as large as transportation and erection facilities permit. Deck-plate panels are fabricated by butt-welding available rolled plates. Ribs and floor beams are fillet-welded to the deck plate in upside-down position. Then, the deck is welded to the girder webs.

It is important to schedule all welding sequences to minimize distortion and locked-up stresses. The most effective method has been to fit up all components of a panel—deck plate, ribs, and floor beams—before starting any welding, then to place the fillet welds from rib to rib and from floor beam to floor beam, starting from the panel center and uniformly proceeding toward the edges. Since this sequence practically requires manual welding throughout, American fabricators prefer to join the ribs to the deck by automatic fillet welding before assembly with the floor beams. After slipping the floor-beam webs over the ribs, the fabricators weld manually only the beam webs to the deck. This method requires careful preevaluation of rib distortions, wider floor-beam slots, and consequently more substantial or only one-sided rib-to-floor-beam welds.

17.12.8 Analysis

Stresses in orthotropic decks are considered as resulting from a superposition of four static systems:

System I consists of the deck plate considered as an isotropic plate elastically supported by the ribs

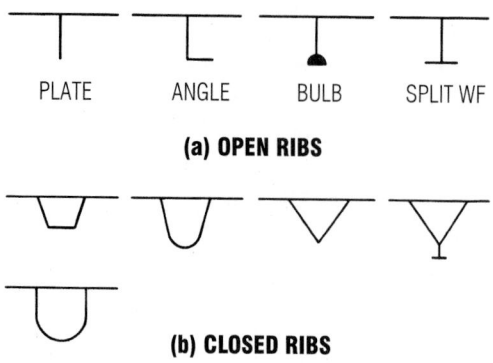

(a) OPEN RIBS

PLATE ANGLE BULB SPLIT WF

(b) CLOSED RIBS

Fig. 17.13 Rib shapes used in orthotropic-plate decks.

(Fig. 17.14a). The deck is subject to bending from wheel loads between the ribs.

System II combines the deck plate, as transverse element, and the ribs, as longitudinal elements. The ribs are continuous over, and elastically supported by, the floor beams (Fig. 17.14b). The orthotropic analysis furnishes the distribution of concentrated (wheel) loads to the ribs, their flexural and torsional stresses, and thereby the axial and torsional stresses of the deck plate as their top flange.

System III combines the ribs with the floor beams and is treated either as an orthotropic system or as a grid (Fig. 17.14c). Analysis of this system furnishes the flexural stresses of the floor beams,

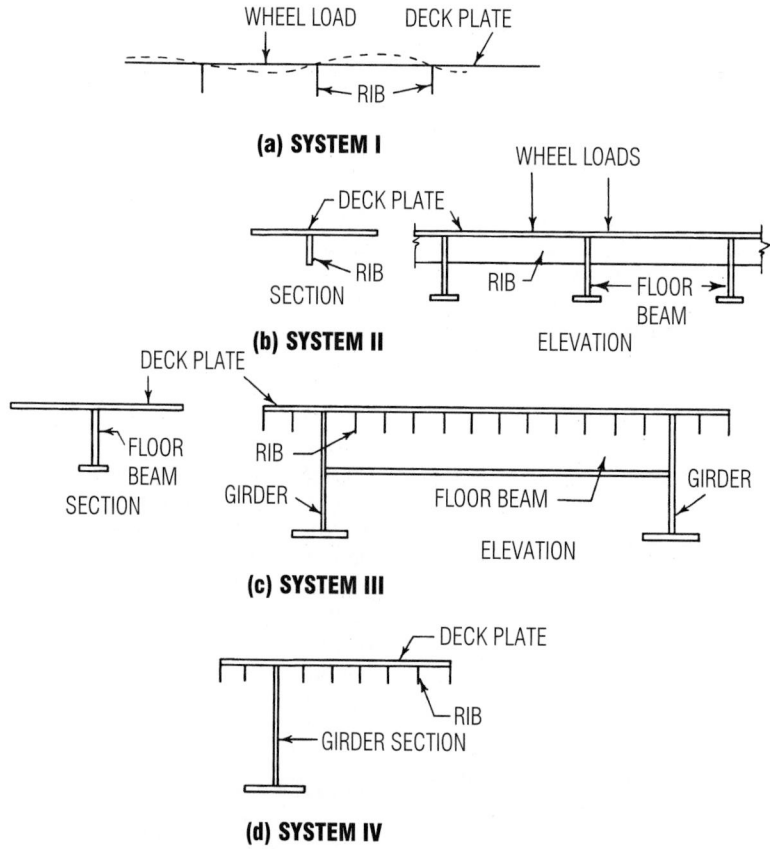

Fig. 17.14 Orthotropic-plate deck may be considered to consist of four systems: (*a*) Deck plate supported on ribs; (*b*) rib-deck T beams spanning between floor beams; (*c*) floor beam with deck plate as top flange, supported on girders; (*d*) girder with deck plate as top flange.

including the stresses the deck plate receives as their top flange.

System IV comprises the main girders with the orthotropic deck as top flange (Fig. 17.14*d*). Axial stresses in the deck plate and ribs and shear stresses in the deck plate are obtained from the flexural and torsional analysis of the main girders by conventional methods.

Theoretically, the deck plate should be designed for the maximum principal stresses that may result from the simultaneous effect of all four systems. Practically, because of the rare coincidence of the maxima from all systems and in view of the great inherent strength reserve of the deck as a membrane (second-order stresses), a design is generally satisfactory if the stresses from any one system do not exceed 100% of the ordinarily permissible working stresses and 125% from a combination of any two systems.

In the design of long-span girder bridges, special attention must be given to buckling stability of deep webs and of the deck. Also, consideration should be given to conditions that may arise at intermediate stages of construction.

17.12.9 Steel-Deck Surfacing

All traffic-carrying steel decks require a covering of some nonmetallic material to protect them from accidental damage, distribute wheel loads, compensate for surface irregularities, and provide a nonskid, plane riding surface. To be effective, the surfacing must adhere firmly to the base and resist wear and distortion from traffic under all conditions. Problems arise because of the elastic and thermal properties of the steel plate, its sensitivity to corrosion, the presence of bolted deck splices, and the difficulties of replacement or repair under traffic.

The surfacing material usually is asphaltic. Strength is provided by the asphalt itself (mastic-type pavements) or by mineral aggregate (asphalt-concrete pavement). The usefulness of mastic-type pavements is restricted to a limited temperature range, below which they become brittle and above which they may flow. The effectiveness of the mineral aggregate of asphalt concrete depends on careful grading and adequate compaction, which on steel decks sometimes is difficult to obtain. Asphalt properties may be improved by admixtures of highly adhesive or ductile chemicals of various plastics families.

("Design Manual for Orthotropic Steel Plate Deck Bridges," American Institute of Steel Construction, Chicago, Ill.; F. S. Merritt and R. L. Brockenbrough, "Structural Steel Designers' Handbook," 2nd ed., McGraw-Hill, Inc., New York.)

17.13 Truss Bridges

Trusses are lattices formed of straight members in triangular patterns. Although truss-type construction is applicable to practically every static system, the term is restricted here to beam-type structures: simple spans and continuous and hinged (cantilever) structures. For typical single-span bridge truss configurations, see Fig. 6.50. For the stress analysis of bridge trusses, see Arts. 6.46 through 6.50.

Truss bridges require more field labor than comparable plate girders. Also, trusses are more costly to maintain because of the more complicated makeup of members and poor accessibility of the exposed steel surfaces. For these reasons, and as a result of changing aesthetic preferences, use of trusses is increasingly restricted to long-span bridges for which the relatively low weight and consequent easier handling of the individual members are decisive advantages.

The superstructure of a typical truss bridge is composed of two main trusses, the floor system, a top lateral system, a bottom lateral system, cross frames, and bearing assemblies.

Decks for highway truss bridges are usually concrete slabs on steel framing. On long-span railway bridges, the tracks are sometimes mounted directly on steel stringers, although continuity of the track ballast across the deck is usually preferred. Orthotropic decks are rarely used on truss bridges.

Most truss bridges have the deck located between the main trusses, with the floor beams framed into the truss posts. As an alternative, the deck framing may be stacked on top of the top chord. **Deck trusses** have the deck at or above top-chord level (Fig. 17.15); **through trusses**, near the bottom chord (Fig. 17.16). Through trusses whose depth is insufficient for the installation of a top lateral system are referred to as **half through trusses** or **pony trusses**.

Figure 17.16 illustrates a typical cantilever truss bridge. The main span comprises a suspended

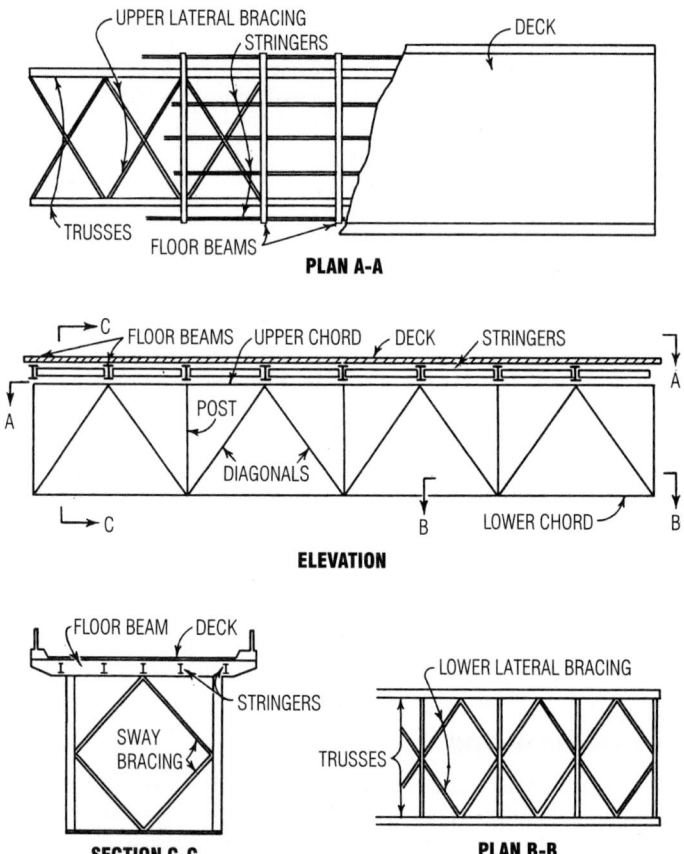

Fig. 17.15 Deck truss bridge.

span and two cantilever arms. The side, or anchor, arms counterbalance the cantilever arms.

Sections of truss members are selected to ensure effective use of material, simple details for connections, and accessibility in fabrication, erection, and maintenance. Preferably, they should be symmetrical.

In bolted design, the members are formed of channels or angles and plates, which are combined into open or half-open sections. Open sides are braced by lacing bars, stay plates, or perforated cover plates. Welded truss members are formed of plates. Figure 17.17 shows typical truss-member sections. For slenderness restrictions of truss members, see Art. 17.7.

The design strength of tensile members is controlled by their net section, that is, by the section area that remains after deduction of rivet

or bolt holes. In shop-welded field-bolted construction, it is sometimes economical to build up tensile members by butt-welding three sections of different thickness or steel grades. Thicker plates or higher-strength steel is used for the end sections to compensate for the section loss at the holes.

The permissible stress of compression members depends on the slenderness ratio (see Art. 9.11). Design specifications also impose restrictions on the width-to-thickness ratios of webs and cover plates to prevent local buckling.

The magnitude of stress variation is restricted for members subject to stress reversal during passage of a moving load (Art. 9.20).

All built-up members must be stiffened by diaphragms in strategic locations to secure their squareness. Accessibility of all members and con-

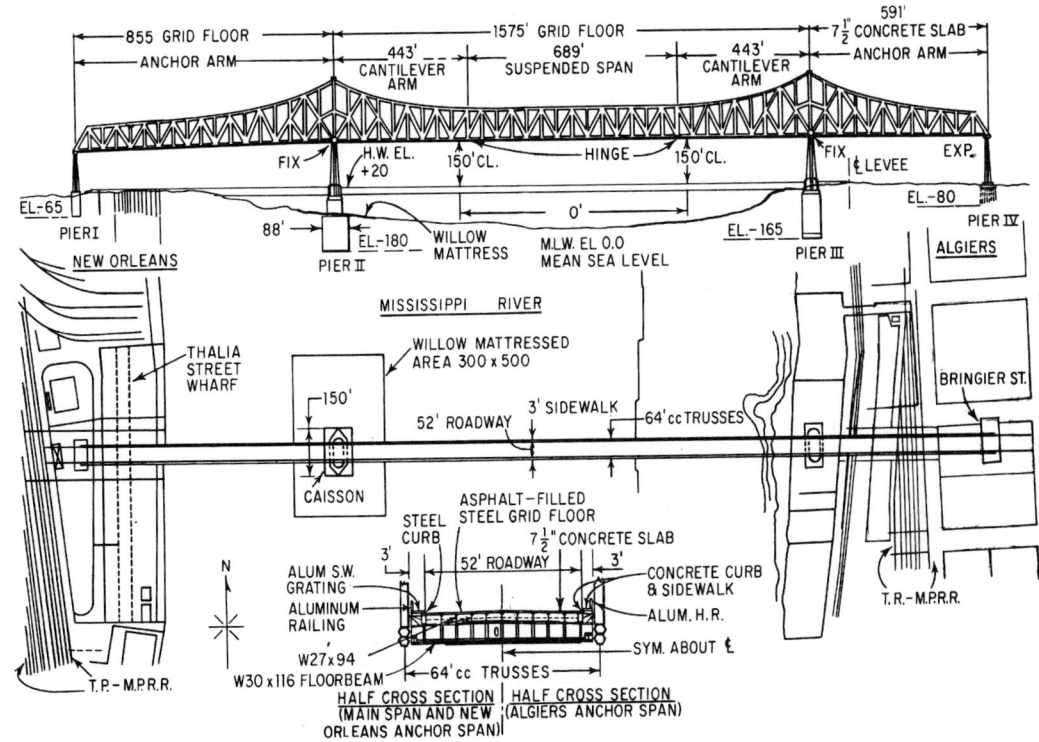

Fig. 17.16 Typical cantilever truss bridge.

nections for fabrication and maintenance should be a primary design consideration.

Whenever possible, each web member should be fabricated in one piece reaching from the top to the bottom chord. The shop length of chord members may extend over several panels. Chord splices should be located near joints and may be incorporated into the gusset plates of a joint.

In most trusses, members are joined by bolting or welding with gusset plates. Pin connections, which were used frequently in earlier truss bridges, are now the exception. As a rule, the centerlines or center-of-gravity lines of all members converging at a joint intersect in a single point (Fig. 17.19).

Stresses in truss members and connections are divided into primary and secondary stresses. Primary stresses are the axial stresses in the members of an idealized truss, all of whose joints are made with frictionless pins and all of whose loads are applied at pin centers. Secondary stresses are the stresses resulting from the incorrectness of these assumptions. Somewhat higher stresses are allowed when secondary stresses are considered.

(Some specifications require computation of the flexural stresses in compression members caused by their own weight as primary stresses.) Under ordinary conditions, secondary stresses must be computed only for members whose depth is more than one-tenth of their length.

(F. S. Merritt and R. L. Brockenbrough, "Structural Steel Designers' Handbook," 2nd ed., McGraw-Hill, Inc., New York (books.mcgraw-hill.com).)

17.14 Suspension Bridges

These are generally preferred for spans over 1800 ft, and they compete with other systems on shorter spans.

The basic structural system consists of flexible main cables and, suspended from them, stiffening girders or trusses (collectively referred to as "stiffening beams"), which carry the deck framing. The vehicular traffic lanes are as a rule accommodated between the main supporting systems.

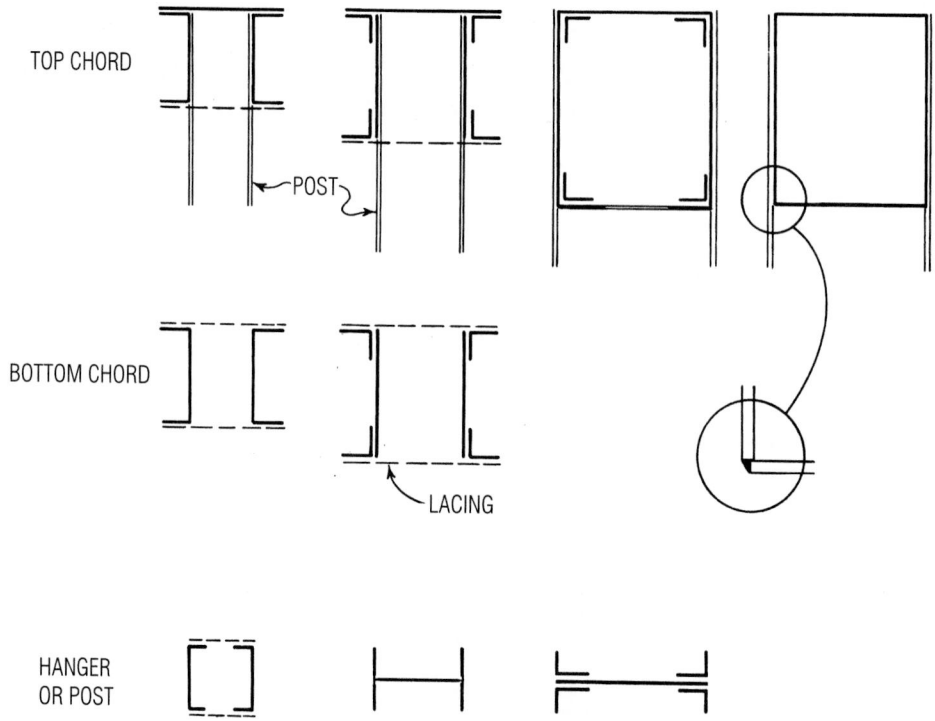

TOP CHORD

POST

BOTTOM CHORD

LACING

HANGER
OR POST

Fig. 17.17 Typical sections used in steel bridge trusses.

Sidewalks may lie between the main systems or cantilever out on both sides.

17.14.1 Stiffening Beams

Stiffening beams distribute concentrated loads, reduce local deflections, act as chords for the lateral system, and secure the aerodynamic stability of the structure. Spacing of the stiffening beams is controlled by the roadway width but is seldom less than $\frac{1}{50}$ the span.

Stiffening beams may be either plate girders, box girders, or trusses. On major bridges, their depth is at least $\frac{1}{180}$ of the main span.

17.14.2 Anchorages

The main cables are anchored in massive concrete blocks or, where rock subgrade is capable of resisting cable tension, in concrete-filled tunnels. Or the main cables are connected to the ends of the stiffening girders, which then are subjected to longitudinal compression equal to the horizontal component of the cable tension.

17.14.3 Continuity

Single-span suspension bridges are rare in engineering projects. They may occur in crossings of narrow gorges where the rock on both sides provides a reliable foundation for high-level cable anchorages.

The overwhelming majority of suspension bridges have main cables draped over two towers. Such bridges consist, thus, of a main span and two side spans. Preferred ratios of side span to main span are 1:4 to 1:2. Ratios of cable sag to main span are preferably in the range of 1:9 to 1:11, seldom less than 1:12.

If the side spans are short enough, the main cables may drop directly from the tower tops to the anchorages, in which case the deck is carried to the abutments on independent, single-span plate girders or trusses. Otherwise, the suspension system is extended over both side spans to the

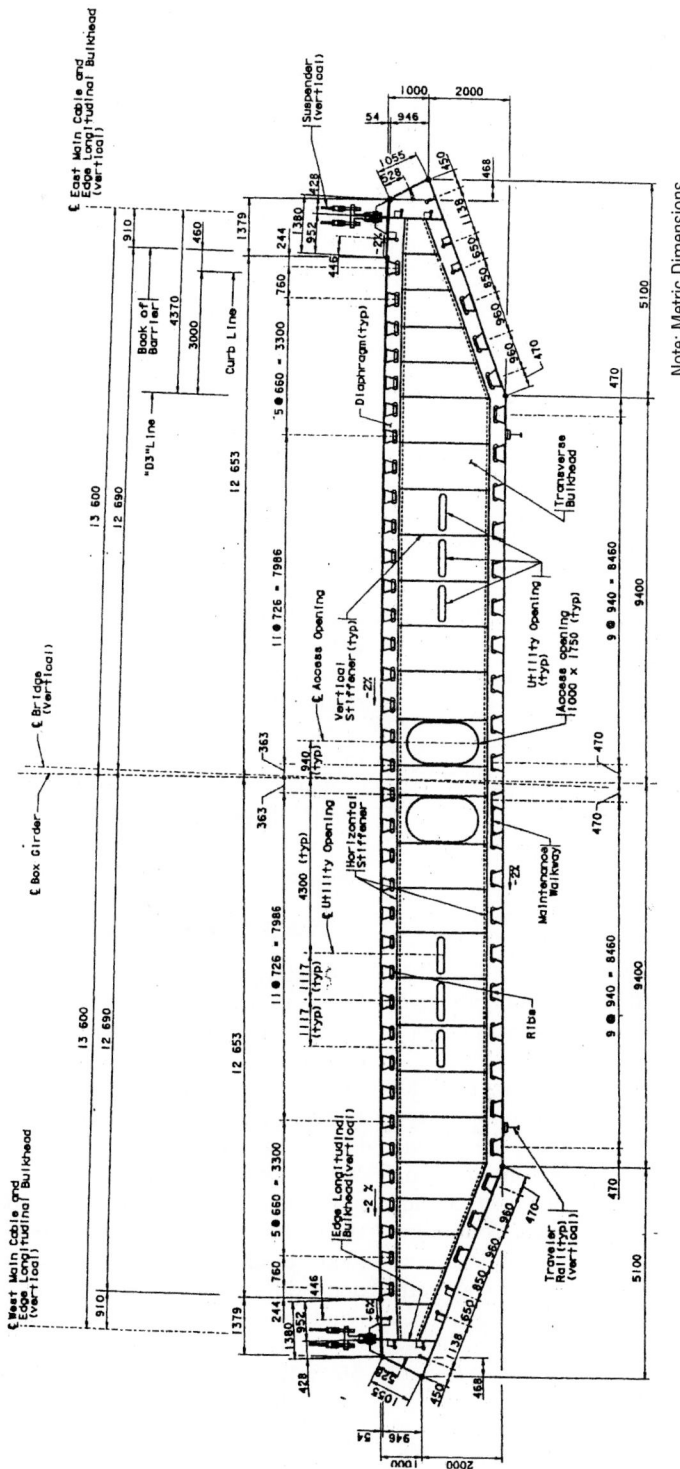

Fig. 17.18 Box girder stiffening beam—Carquinez Bridge.

Note: Metric Dimensions

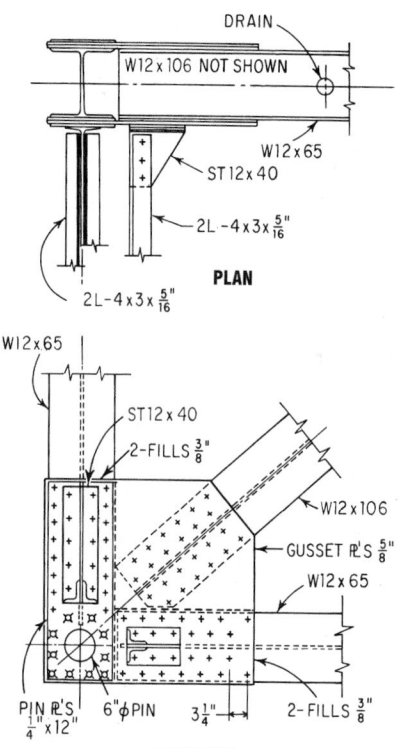

PLAN

ELEVATION

Fig. 17.19 Pin joint in the lower chord of a bridge truss at a support.

next piers. There, the cables are deflected to the anchorages. The first system allows the designer some latitude in alignment, for example, curved roadways. The second requires straight side spans, in line with the main span. It is the common system for suspension bridges that are links in a chain of multiple-span crossings.

When side spans are not suspended, the stiffening beam is of course restricted to the main span. When side spans are suspended, the stiffening beams of the three spans may be continuous or discontinuous at the towers. The spans are typically restrained to the tower at the ends. Continuity of stiffening beams is required in self-anchored suspension bridges, where the cable ends are anchored to the stiffening beams.

17.14.4 Cable Systems

The suspenders between main cables and stiffening beams are usually equally spaced and vertical.

Main cables, suspenders, and stiffening beams (girders or trusses) are usually arranged in vertical planes, symmetrical with the longitudinal bridge axis. Bridges with inward- or outward-sloping cables and suspenders and with offset stiffening beams are less common.

Three-dimensional stability is provided by top and bottom lateral systems and transverse frames, similar to those in ordinary girder and truss bridges. Rigid roadway decks may take the place of either or both lateral systems, especially in double decked trusses.

In the United States, the main cables are usually made up of 6-gage galvanized bridge wire of 220 to 225 ksi ultimate and 82 to maximum 90 ksi working stress. The wires are usually placed parallel but sometimes in strands and compacted and wrapped with No. 9 wire. In Europe, strands containing elaborately shaped heat-treated cast-steel wires are sometimes used. Strands must be prestretched. They have a lower and less reliable modulus of elasticity than parallel wires. The heaviest cables, those of the Golden Gate Bridge, are about 36 in in diameter. Twin cables are used if larger sections are required.

Suspenders may be eyebars, rods, single steel ropes, or pairs of ropes slung over the main cable. Connections to the main cable are made with cable bands. These are cast steel whose inner faces are molded to fit the main cable. The bands are clamped together with high-strength bolts.

17.14.5 Floor System

In the design of the floor system, reduction of dead load and resistance to vertical air currents should be the governing considerations. The deck is usually lightweight concrete or steel grating partly filled with concrete with the exception of box sections which usually have a wearing surface. Expansion joints should be provided every 100 to 120 ft to prevent mutual interference of deck and main structure. Stringers should be made composite with the deck for greater strength and stiffness. Floor beams may be plate girders or trusses, depending on available clearance. With trusses, wind resistance is less.

17.14.6 Towers

The towers may be portal type, multistory, or diagonally braced frames (Fig. 17.20). They may be of cellular construction, made of steel plates and

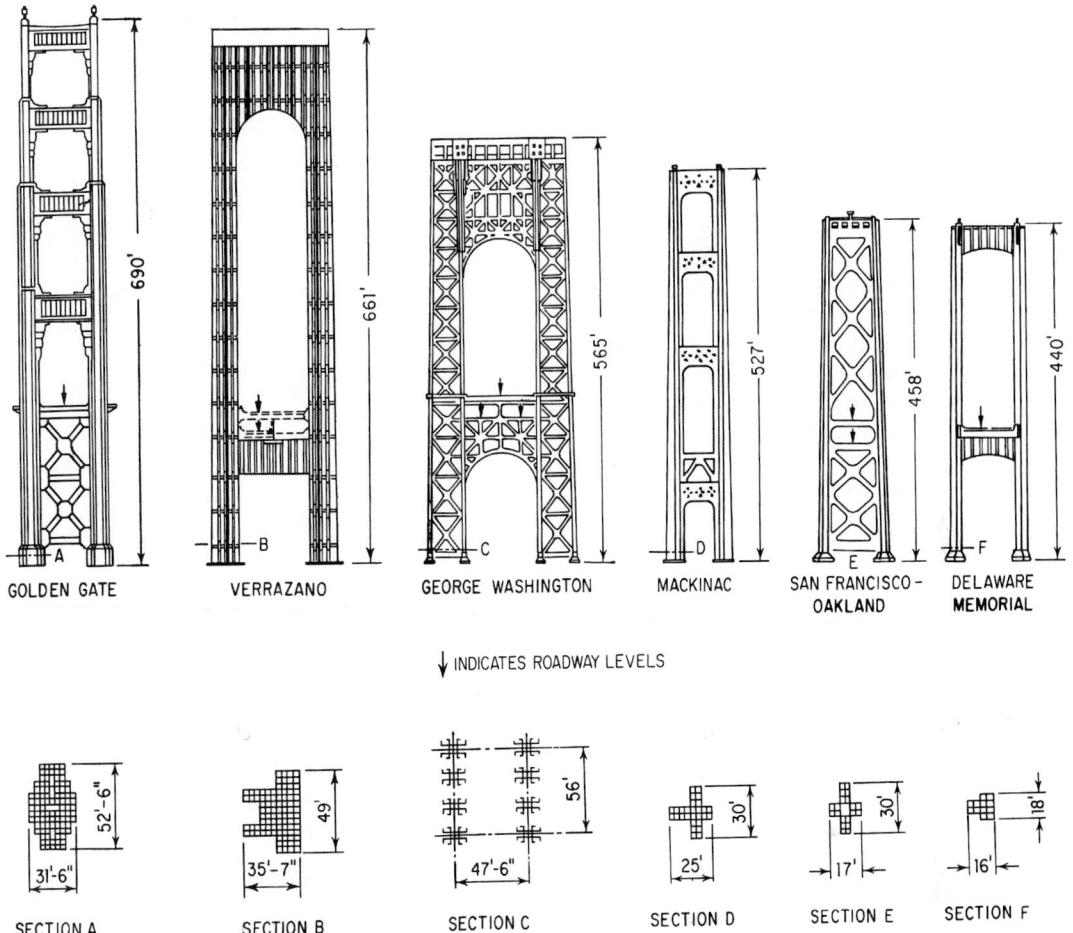

Fig. 17.20 Types of towers used for long suspension bridges.

shapes, or steel lattices, or of reinforced concrete. The substructure below the "spray" line is concrete. The base of steel towers is usually fixed, but it may be hinged. (Hinged towers, however, offer some erection difficulties.) The cable saddles at the top of fixed towers are sometimes placed on rollers to reduce the effect on the towers of unbalanced cable deflections. Cable bents can be considered as short towers, either fixed or hinged, whose axis coincides with the bisector of the angle formed by the cable.

17.14.7 Analysis

For gravity loads, the three elements of a suspension bridge in a vertical plane—the main cable or chain, the suspenders, and the stiffening beam—are considered as a single system. The system of discrete suspenders often is idealized as one of continuous suspension.

The stiffening beam is assumed stressless under dead load, a condition approximated by appropriate methods of erection. Moments and shears are produced by that part of the live load not taken up by the main cable through the suspenders. Also, moments and shears result from changes in cable length and sag due to temperature variations or unbalanced loadings of adjacent spans. Deflections of the stiffening beam are strictly elastic; that is, neglecting the effect of shear, the curvature at any section of the elastic line of the loaded beam is proportional to the bending moment divided by the moment of inertia of that section.

The suspenders are subject to tension only. Their elongation under live load is usually neglected in the analysis.

The main cable typically is assumed to have no flexural stiffness and to be subject to axial tension only. Its shape is that of a funicular polygon of the applied forces (which include the dead weight of the cable). The pole distance H, lb, which is the horizontal component of the cable tension, is constant for a given loading and a given sag. The shape of the cable under given loads, that is, its ordinate y, ft, and slope $\tan \alpha$ at any point with abscissa x, ft, can be expressed in terms of moment M_o, ft-kips, and shear V, kips, that a simple beam of the same span L, ft, as the cable would have under the same load (Fig. 17.21).

$$y = \frac{M_o}{H} \qquad \tan \alpha = \frac{V}{H} \qquad (17.14)$$

In the special case of a uniform load w, kips/lin ft,

$$H = \frac{wL^2}{8f} \qquad (17.15)$$

$$y = \frac{wx(L-x)}{2H} \text{ or } y = \frac{4fx(L-x)}{L^2} \qquad (17.16)$$

where f = cable sag, ft.

The shape of the cable under its own weight without suspended load would be a catenary; under full dead load, the cable shape is usually closer to a parabola. The difference is small.

Concentrated or sectionally uniform live load superimposed on the dead load subjects the cable to additional strain and causes it to adjust its shape to the changed load configuration. The resulting deformations are not exactly proportional to the additional loading; their magnitude is influenced by the already existing dead-load stresses.

If M_o is the bending moment of the stiffening beam under the applied load but without cooperation of the cable, the beam moment M with cooperation of the cable will be

$$M = M_o - Hy \qquad (17.17)$$

More specifically, using subscripts D and L, respectively, for dead and live load and considering that

$$y_L = y_D + \Delta y \qquad (17.18)$$

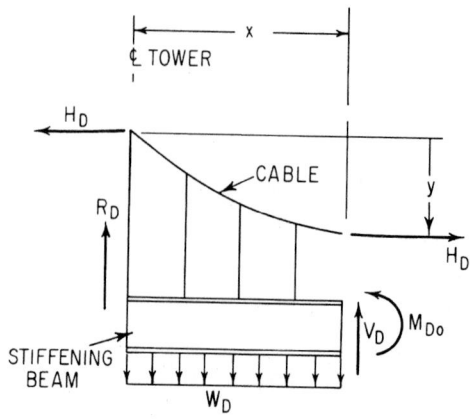

(a) **DEAD LOAD ON BRIDGE**

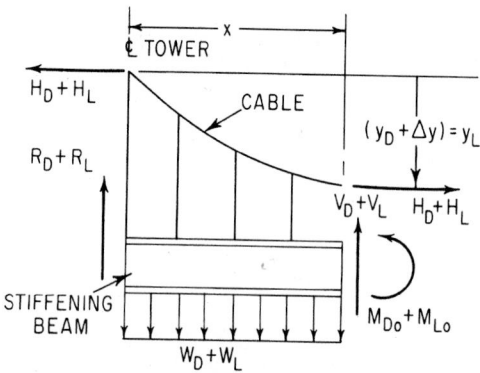

(b) **LIVE LOAD SUPERIMPOSED**

Fig. 17.21   Stresses in cable and stiffening beam of a suspension bridge.

one gets the following expression for the dead- plus live-load bending moment of the beam (see Fig. 17.21b):

$$M = M_D + M_L = M_{D0} + M_{L,0}$$
$$- (H_D + H_L)(y_D + \Delta y) \qquad (17.19)$$

But, since $M_D = M_{D0} - H_D y_D = 0$, because the stiffening beam has no bending moment under dead load (ideally),

$$M = M_{L0} - (H_D + H_L)\Delta y - H_L y_D \qquad (17.20)$$

This is the basic equation of the cable-beam system.

In this equation, M_{L0}, H_D, and y_D are given. H_L and Δy must be so determined that the conditions of static equilibrium of all forces and geometric compatibility of all deformations are satisfied throughout the system.

The mathematically exact solution of the problem is known as the deflection theory. A less exact, older theory is known as the elastic theory. Besides these, there are several approximate methods based on observed regularities in the behavior of suspension bridges, which are sufficiently accurate to serve for preliminary design.

17.14.8 Wind Resistance

Wind acting on the main cables and on part of the suspenders is carried to the towers by the cables. Wind acting on the deck, stiffening beams, and live load is resisted mainly by the lateral bracing system and slightly by the cables because of the gravity component resulting from any elastic lateral deflection of the main supporting system.

Oscillations of the structure may be generated by live load, earthquake, or wind. Live-load vibrations are insignificant in major bridges. (N. C. Raab and H. C. Wood, "Earthquake Stresses in the San Francisco–Oakland Bay Bridge," *Transactions of the American Society of Civil Engineers*, vol. 106, 1941). Oscillations due to wind, however, can become dangerous if excessive amplitudes build up; that is, if the exciting impulses approach the natural frequency of the structure. Oscillating wind forces are caused by eddies, which may be generated outside the structure or by the structure itself, especially on the lee side of large plates. Oscillations of the structure may be purely flexural, purely torsional, or coupled (flutter), the last two being the more dangerous.

Methods used to predict the aerodynamic behavior of suspension bridges include:

Mathematical analysis of the natural frequency of the structure in flexure and torsion [F. Bleich, C. B. McCullogh, R. Rosecrans, and G. S. Vincent, "Mathematical Theory of Vibration in Suspension Bridges," Government Printing Office, Washington, D.C.: A. G. Pugsley, "Theory of Suspension Bridges," Edward Arnold (Publishers) Ltd., London].

Wind-tunnel tests on scale models of the entire structure or of typical sections ("Aerodynamic Stability of Suspension Bridges with Special Reference to the Tacoma Narrows Bridge," *University of Washington Engineering Experiment Station Bulletin* 116).

Application of Steinman's criteria (these are controversial) (D. B. Steinman, "Rigidity and Aerodynamic Stability of Suspension Bridges," with discussion, *Transactions of the American Society of Civil Engineers*, vol. 110, 1945).

Tuned mass dampers and tuned liquid dampers have been used to decrease the amplitude of vortex oscillations.

17.14.9 Tower Stresses

The towers must resist the forces imposed on them by the main cables in addition to the gravity and wind loads acting directly.

The following forces must be considered: The vertical components of the main cables in main and side spans under dead load, live load, temperature change, seismic, and wind acting on the main cables, both parallel and transverse to the bridge axis; reactions to longitudinal cable movements due to unbalanced loading. These reactions will develop unless the movements are taken up by hinges or friction-free roller nests. Theoretically, the magnitude of these movements will be affected by the flexural resistance Q of the towers, but this effect, being comparatively small, is usually neglected.

Movement of the tower top generates bending moments. These increase from the top to the bottom at the rate of

$$M_x = Vy + Qx \tag{17.21}$$

where V = vertical cable reaction

x = distance below top

y = horizontal deflection at x

Q = horizontal resistance at top

The magnitude of Q is such that the total deflection equals the longitudinal cable movement. It is found by solving the differential equation for the elastic curve of the tower axis. Thus,

$$y = A \sin cx + B \cos cx - \frac{Q}{V}x \tag{17.22}$$

$$= \frac{Q}{V}\left(\frac{\sin cx}{c \cos cL} - x\right)$$

in which $c = \sqrt{V/EI}$, I = moment of inertia, and E = modulus of elasticity of tower, if the towers have constant cross sections. The bending moment at x is

$$M_x = \frac{Q \sin cx}{c \cos cL} \qquad (17.23)$$

where L = height of tower.

If, as is usual, the tower cross section varies in several steps, the coefficients A and B in Eq. (17.22) differ from section to section. They are found from the continuity conditions at each step.

Anchorages and footings should be designed for adequate safety against uplift, tipping, and sliding under any possible combination of acting forces.

(S. Hardesty and H. E. Wessman, "Preliminary Design of Suspension Bridges," *Transactions of the American Society of Civil Engineers*, vol. 104, 1939; R. J. Atkinson and R. V. Southard, "On the Problem of Stiffened Suspension Bridges and Its Treatment by Relaxation Methods," *Proceedings of the Institute of Civil Engineers*, 1939; C. D. Crosthwaite, "The Corrected Theory of the Stiffened Suspension Bridge," *Proceedings of the Institute of Civil Engineers*, 1946; Ling-Hi Tsien, "A Simplified Method of Analyzing Suspension Bridges," *Transactions of the American Society of Civil Engineers*, vol. 114, 1947; F. S. Merritt and R. L. Brockenbrough, "Structural Steel Designers' Handbook," 2nd ed., McGraw-Hill, Inc., New York (books.mcgraw-hill.com).)

17.15 Cable-Stayed Bridges*

The cable-stayed bridge, also called the stayed-girder (or truss), has come into wide use since about 1950 for medium- and long-span bridges because of its economy, stiffness, aesthetic qualities, and ease of erection without falsework. Design of cable-stayed bridges utilizes taut cables connecting pylons to a span to provide intermediate support for the span. This principle has been understood by bridge engineers for at least the last two centuries, as indicated by the bridge in Fig. 17.22. The Roeblings used cable stays as supplementary stiffening elements in the famous Brooklyn Bridge (1883). Many recently built and proposed suspension bridges also incorporate taut cable stays when dynamic (railroad) and long-span

---

* Extracted with permission from F. S. Merritt and R. L. Brockenbrough, "Structural Steel Designers Handbook," 2nd ed., McGraw-Hill, Inc., New York.

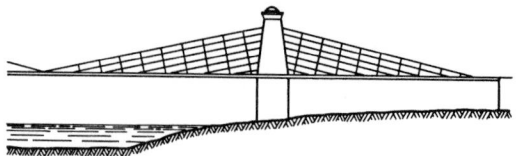

Fig. 17.22 Cable-stayed chain bridge (Hatley system, 1840).

effects have to be contended with, as in the Salazar Bridge.

17.15.1 Characteristics of Cable-Stayed Bridges

The cable-stayed bridge offers a proper and economical solution for bridge spans intermediate between those suited for deck girders (usually up to 600 to 800 ft but requiring extreme depths, up to 33 ft) and the longer-span suspension bridges (over 1000 ft). The cable-stayed bridge thus finds application in the general range of 600- to 1600-ft spans but may be competitive in cost for spans as long as 2900 ft.

A cable-stayed bridge has the advantage of greater stiffness over a suspension bridge. Use of single or multiple box girders gains large torsional and lateral rigidity. These factors make the structure stable against wind and aerodynamic effects.

The true action of a cable-stayed bridge (Fig. 17.23) is considerably different from that of a suspension bridge. As contrasted with the relatively flexible cable of the latter, the inclined, taut cables of the cable-stayed structure furnish relatively stable point supports in the main span. Deflections are thus reduced. The structure, in effect, becomes a continuous girder over the piers, with additional intermediate, elastic (yet relatively stiff) supports in the span. As a result, the girder may be shallow. Depths usually range from $\frac{1}{60}$ to $\frac{1}{80}$ the main span, sometimes even as small as $\frac{1}{100}$ the span.

Cable forces are usually balanced between the main and flanking spans, and the structure is internally anchored; that is, it requires no massive masonry anchorages. Analogous to the self-anchored suspension bridge, second-order effects of the type requiring analysis by a deflection theory are of relatively minor importance. Thus, static

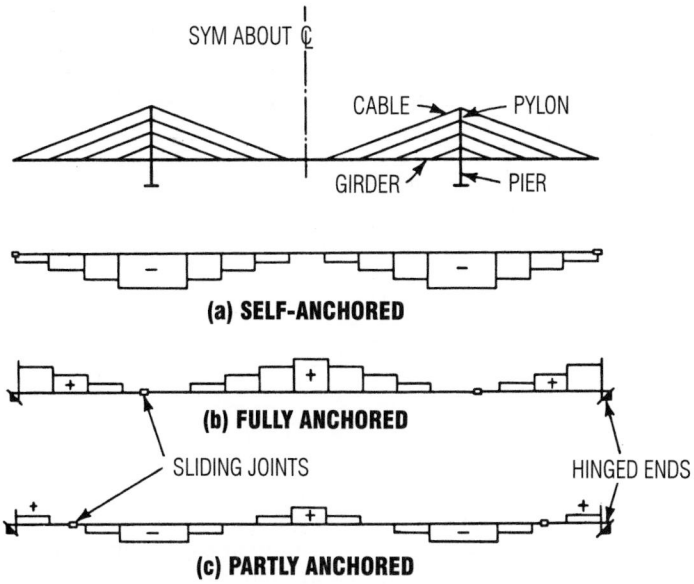

Fig. 17.23 Axial forces in a cable-stayed girder of (*a*) self-anchored, (*b*) fully anchored, (*c*) partly anchored cable-stayed bridges.

analysis is simpler, and the structural behavior may be more clearly understood.

The above remarks apply to the common, self-anchored type of cable-stayed bridges, characterized by compression in the main bridge girders (Fig. 17.23*a*). It is possible to conceive of the opposite extreme of a fully anchored (earth-anchored) cable bridge in which the main girders are in tension. This could be achieved by pinning the girders to the abutments and providing sliding joints in the side-span girders adjacent to the pylons (Fig. 17.23*b*). The fully anchored system is stiffer than the self-anchored system and may be advantageously analyzed by second-order deflection theory because (analogous to suspension

bridges) bending moments are reduced by the deformations.

A further increase in stiffness of the fully anchored system is possible by providing piers in the side spans at the cable attachments (Fig. 17.24). This is advantageous if the side spans are not used for boat traffic below, and if, as is often the case, the side spans cross over low water or land (Knieb-rücke at Düsseldorf, Fig. 17.27*i*).

A partly anchored cable-stayed system (Fig. 17.23*c*) has been proposed wherein some of the cables are self-anchored and some fully anchored. The axial forces in the girders are then partly compression and partly tension, but their magnitudes are considerably reduced.

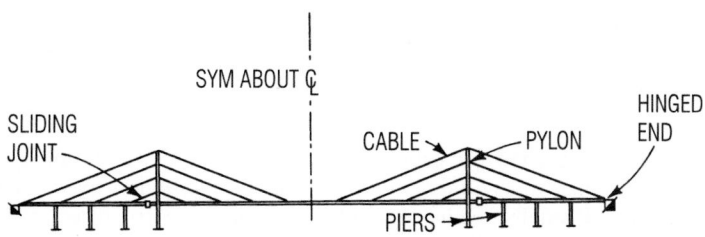

Fig. 17.24 Anchorage of side-span cables at piers and abutments increases stiffness of the center span.

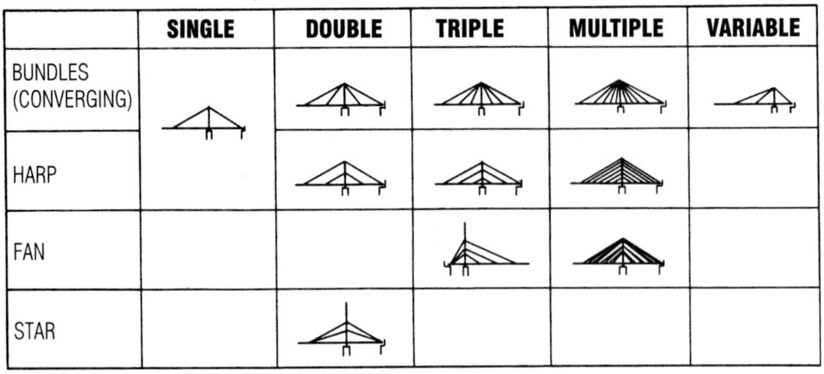

	SINGLE	DOUBLE	TRIPLE	MULTIPLE	VARIABLE
BUNDLES (CONVERGING)					
HARP					
FAN					
STAR					

Fig. 17.25 Classification of cable-stayed bridges by arrangement of cables. (*From A. Feige, "Evolution of German Cable-Stayed Bridges: An Overall Survey," Acier-Stahl-Steel, vol. 12, 1966.*)

17.15.2 Classification of Cable-Stayed Bridges

The relatively small diameter of the cables and the absolute minimum amount of overhead structure required are the principal features contributing to the excellent architectural appearance of cable-stayed bridges. The functional character of the structural design produces, as a by-product, a graceful and elegant solution for a bridge crossing. This is encouraged by the wide variety of possible types, using single or multiple cables, including the bundle, harp, fan, and star configurations, as seen in elevation (Fig. 17.25). These may be symmetrical or asymmetrical.

A wide latitude of choice of cross section of the bridge at the pylons is also possible (Fig. 17.26). The most significant distinction occurs between those with twin pylons (individual, portal, or A frame) and those with single pylons in the center of the roadway. The single pylons usually require a large box girder to resist the torsion of eccentric loadings, and the box is most frequently of steel with an anisotropic steel deck. The single-pylon type is advantageous in allowing a clear unobstructed view from cars passing over the bridge. The pylons may (as with suspension-bridge towers) be either fixed or pinned at their bases. In the case of fixity, this may be either with the girders or directly with the pier.

Some details of cable-stayed bridges are shown in the elevations and cross sections in Fig. 17.27.

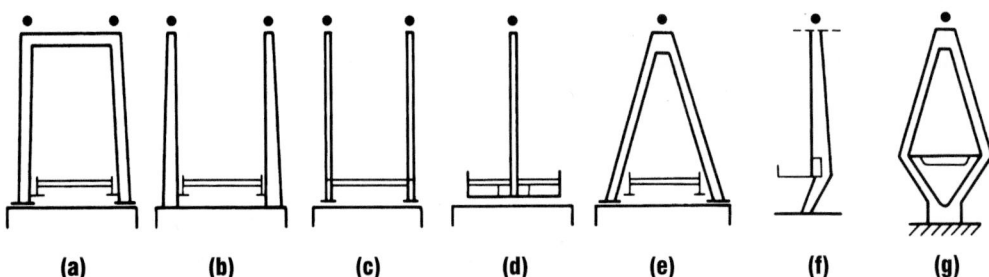

Fig. 17.26 Shapes of pylons used for cable-stayed bridges. (*a*) Portal-frame type with top cross member; (*b*) pylon fixed to pier and without top cross member; (*c*) pylon fixed to superstructure and without top cross member; (*d*) axially located pylon fixed to the superstructure; (*e*) A-shaped pylon; (*f*) laterally offset pylon fixed to a pier; (*g*) diamond-shaped pylon. (*From A. Feige, "Evolution of German Cable-Stayed Bridges: An Overall Survey," Acier-Stahl-Steel, vol. 12, 1966.*)

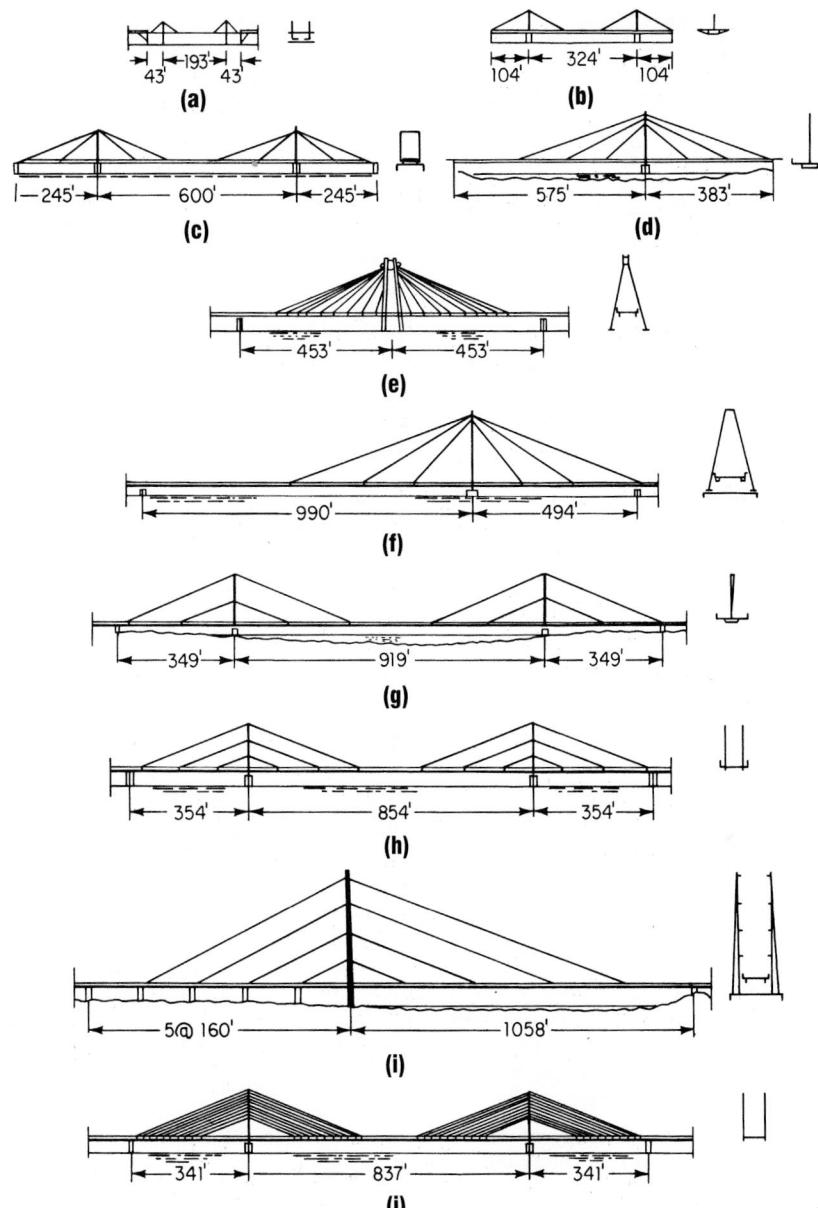

Fig. 17.27 Some cable-stayed bridges, with cross sections taken at pylons. (*a*) Büchenauer crossing at Bruchsal, 1956; (*b*) Jülicherstrasse crossing at Düsseldorf, 1964; (*c*) bridge over the Strömsund, Sweden, 1955; (*d*) bridge over the Rhine near Maxau, 1966; (*e*) bridge on the elevated highway at Ludswigshafen, 1969; (*f*) Severin Bridge, Cologne, 1959; (*g*) bridge over the Rhine near Levenkusen, 1965; (*h*) North Bridge at Düsseldorf, 1958; (*i*) Kniebrücke at Düsseldorf, 1969; (*j*) bridge over the Rhine at Rees, 1967.

17.15.3 Cable-Stayed Bridge Analysis

The static behavior of a cable-stayed girder can best be gaged from the simple, two-span example of Fig. 17.28. The girder is supported by one stay cable in each span, at E and F, and the pylon is fixed to the girder at the center support B. The static system has two internal cable redundants and one external support redundant.

If the cable and pylon were infinitely rigid, the structure would behave as a continuous four-span beam AC on five rigid supports A, E, B, F, and C. The cables are elastic, however, and correspond to springs. The pylon also is elastic but much stiffer because of its large cross section. If cable stiffness is reduced to zero, the girder assumes the shape of a deflected two-span beam ABC.

Cable-stayed bridges of the nineteenth century differed from those of the 1960s in that their tendons constituted relatively soft spring supports. Heavy and long, the tendons could not be stressed highly. Usually, the cables were installed with significant slack or sag. Consequently, large deflections occurred under live load as the sag was diminished. Modern cables have high-strength steel, are relatively short and taut, and their weight is low. Their elastic action may therefore be considered linear, and an equivalent modulus of elasticity may be used. The action of such cables then produces something more nearly like the four-span beam for a structure like the one in Fig. 17.28.

If the pylon were hinged at its base connection with the girder at B, the pylon would act as a pendulum column. This would have an important effect on the stiffness of the system, for the spring support at E would become more flexible. In magnitude, the effect might exceed that due to the elastic stretch of the cables. In contrast, the elastic shortening of the tower has no appreciable effect.

Relative girder stiffness plays a dominant role in the structural action. The girder tends to approach a beam on rigid supports A, E, B, F, C as girder stiffness decreases toward zero. With increasing girder stiffness, however, the action of the cables becomes minor and the bridge approaches a girder supported on its piers and abutments A, B, C.

In a three-span bridge, a side-span cable connected to the abutment furnishes more rigid support to the main span than does a cable attached to some point in the side span. In Fig. 17.28, for example, the support of the load P in the position shown would be improved if cable attachment at F were shifted to C. This explains why cables from the pylon top to the abutment are structurally more efficient, although not as aesthetically pleasing as other arrangements.

The stiffness of the system is also affected by whether the cables are fixed at the towers (at D, for example, in Fig. 17.28) or whether they run continuously over (or through) the towers. Most designs with more than one cable to a pylon from the main span require one of the cables to be fixed to the pylon and the others to be on movable saddle supports.

The curves of maximum-minimum girder moments for all load variations usually show a large range of stress. Designs providing for the corresponding normal forces in the girder may require large variation in cross sections. By prestressing the cables or by raising or lowering the support points, it is possible to achieve a more uniform and economical moment capacity. The amount of prestressing to use for this purpose may be calculated by successively applying a unit force in each cable and drawing the respective moment diagrams. Then, by trial, the proper multiples of each force are determined so that when their moments are superimposed on the maximum-minimum moment diagrams, an optimum balance results.

17.15.4 Static Analysis— Elastic Theory

Cable-stayed bridges may be analyzed by the general method of indeterminate analysis with the equations of virtual work.

The degree of internal redundancy of the system depends on the number of cables, types of

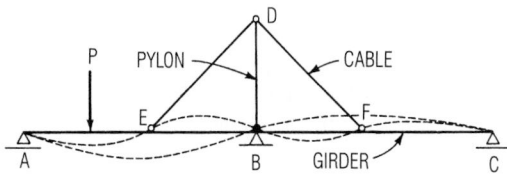

Fig. 17.28 Deflected positions (dash lines) of a cable-stayed bridge.

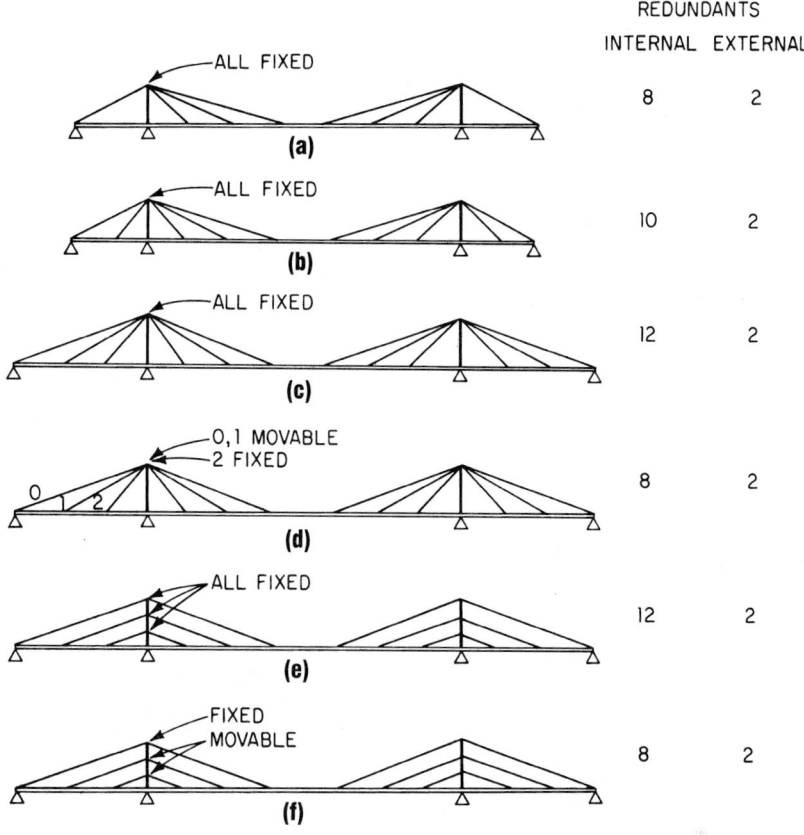

Fig. 17.29 Redundants in three-span cable-stayed bridges.

connections (fixed or movable) of cables with the pylons, and the nature of the pylon connection at its base with the girder or pier. The girder is usually made continuous over three spans. Figure 17.29 shows the order of redundancy for various single-plane systems of cables.

If the bridge has two planes of cables, two girders, and double pylons, it usually also must be provided with a number of intermediate cross diaphragms in the floor system, each of which is capable of transmitting moment and shear. The bridge may also have cross girders across the top of the pylons. Each cross member adds two redundants, to which must be added twice the internal redundancy of the single plane structure, and any additional reactions in excess of those needed for external equilibrium as a space structure. The redundancy of the space structure is very high, usually of the order of 40 to 60. Therefore, the methods of plane statics are normally used, except for larger structures.

It is convenient to select as redundants the bending moments in the girder at those points where the cables and pylons join the girder. When these redundants are set equal to zero, an articulated, statically determinate truss base system is obtained (Fig. 17.30). When the loads are applied to this choice of base system, the stresses in the cables do not differ greatly from their final values, so the cables may be dimensioned in a preliminary way.

Other approaches are also possible. One is to use the continuous girder itself as a statically indeterminate base system, with the cable forces as redundants. But computation is generally increased.

A third method involves imposition of hinges, for example at *a* and *b* (Fig. 17.31), so placed as to

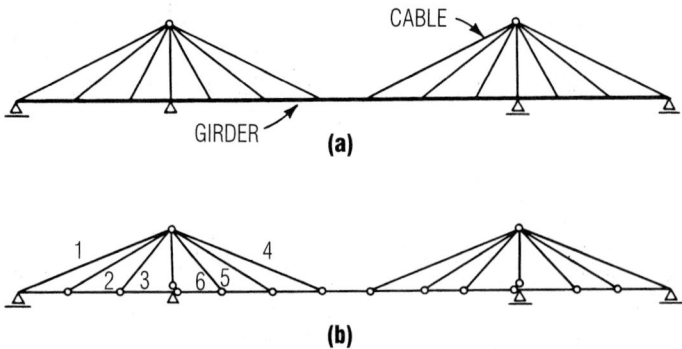

Fig. 17.30 Three-span cable-stayed bridge. (*a*) Girder is continuous over the three spans. (*b*) Insertion of hinges in the girder at cable-attachment points makes the structure statically determinate.

form two coupled symmetrical base systems, each statically indeterminate to the fourth degree. The influence lines for the four indeterminate cable forces of each partial base system are at the same time also the influence lines of the cable forces in the real system. The two redundant moments X_a and X_b are treated as symmetrical and antisymmetrical group loads, $Y = X_a + X_b$, and $Z = X_a - X_b$, to calculate influence lines for the 10-degree-indeterminate structure shown. Kern moments are plotted to determine maximum effects of combined bending and axial forces.

Note that the bundle system in Fig. 17.29*c* and *d* generally has more favorable bending moments for long spans than does the harp system of Fig. 17.29*e* and *f*. Cable stresses also are somewhat lower for the bundle system because the steeper cables are more effective. But the concentration of cable forces at the top of the pylon introduces detailing and construction difficulties. When viewed at an angle, the bundle system presents aesthetic problems because of the different intersection angles when the cables are in two planes. Furthermore, fixity of the cables at pylons with the bundle system in Fig. 17.29*c* and *d* produces a wider range of stress than does a movable arrangement. This can influence design for fatigue.

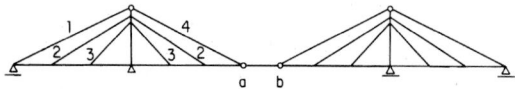

Fig. 17.31 Insertion of hinges at *a* and *b* in the center span of a three-span continuous girder reduces the degree of indeterminancy.

The secondary effect of creep of cables can be incorporated into the analysis. The analogy of a beam on elastic supports is changed thereby to a beam on linear viscoelastic supports.

17.15.5 Static Analysis— Deflection Theory

Distortion of the structural geometry of a cable-stayed bridge under action of loads is considerably less than in comparable suspension bridges. The influence on stresses of distortion is relatively small for cable-stayed bridges. In any case, the effect of distortion is to increase stresses, as in arches, rather than the reverse, as in suspension bridges. This effect for the Severin Bridge is 6% for the girder and less than 1% for the cables. Similarly for the Dusseldorf North Bridge, stress increase due to distortion amounts to 12% for the girders.

The calculations, therefore, most expeditiously take the form of a series of successive corrections to results from first-order theory. The magnitude of vertical and horizontal displacements of the girder and pylons can be calculated from the first-order theory results. If the cable stress is assumed constant, the vertical and horizontal cable components V and H change by magnitudes ΔV and ΔH by virtue of the new deformed geometry. The first approximate correction determines the effects of these ΔV and ΔH forces on the deformed system, as well as the effect of V and H due to the changed geometry. This process is repeated until convergence, which is fairly rapid.

17.15.6 Dynamic Analysis— Aerodynamic Stability

The aerodynamic action of cable-stayed bridges is less severe than that of suspension bridges because of increased stiffness due to the taut cables and the widespread use of torsion box decks.

17.15.7 Preliminary Design of Cable-Stayed Bridges

In general, height of a pylon in a cable-stayed bridge is about $\frac{1}{6}$ to $\frac{1}{8}$ the span. Depth of girder ranges from $\frac{1}{60}$ to $\frac{1}{80}$ the span and is usually 8 to 14 ft averaging 11 ft.

Wide box girders are mandatory for single-plane systems to resist the torsion of eccentric loads. Box girders, even narrow ones, are also desirable for double-plane systems to enable cable connections to be made without eccentricity. Single-web girders, however, are occasionally used.

To achieve symmetry of cables at pylons the ratio of side to main spans should be about 3:7 where three cables are used on each side of the pylons, and about 2:5 where two cables are used. A proper balance of side-span length to main-span length must be established if uplift at the abutments is to be avoided. Otherwise, movable (pendulum-type) tiedowns must be provided at the abutments.

The usual range of live-load deflections is from $\frac{1}{400}$ to $\frac{1}{500}$ the span.

Since elastic-theory calculations are relatively simple to program for a computer, a formal set is usually made for preliminary design after the general structure and components have been proportioned.

17.15.8 Design Details for Cable-Stayed Bridges

These structures differ from usual long-span girder bridges in only a few details.

Towers and Floor System ▪ The towers are composed basically of two parts: the pier (below the deck) and the pylon (above the deck). The pylons are frequently of steel box cross section, although concrete may also be used.

Bridge Deck ▪ Although cable-stiffened bridges usually incorporate an orthotropic steel deck with steel box girders, to reduce the dead load, other types of construction also are in use. For the Lower Yarra River Bridge in Australia, a concrete deck was specified to avoid site welding and to reduce the amount of shop fabrication. The Maracaibo Bridge likewise incorporates a concrete deck, and the Bridge of the Isles (Canada) has a concrete-slab deck supported on longitudinal and transverse steel box girders and steel floor beams. The Büchenauer Bridge also has a concrete deck. Use of a concrete deck in place of orthotropic-plate construction is largely a matter of local economics. The cost of structure to carry the added dead load should be compared with the lower cost per square foot of the concrete deck and other possible advantages, such as better durability and increased stability against wind.

(W. Podolny, Jr., and J. B. Scalzi, "Construction and Design of Cable-Stayed Bridges," 2nd ed., John Wiley & Sons, Inc., New York (www.wiley.com); "Guidelines for Design of Cable-Stayed Bridges," ASCE Committee on Cable-Stayed Bridges (www.asce.org).)

17.16 Steel Arch Bridges

A typical arch bridge consists of two or (rarely) more parallel arches or series of arches, plus necessary lateral bracing and end bearings, and columns or hangers for supporting the deck framing. Types of arches correspond roughly to positions of the deck relative to the arch ribs.

Bridges with decks above the arches and clear space underneath (Fig. 17.32a) are designed as open spandrel arches on thrust-resisting abutments. Given enough underclearance and adequate foundations, this type is usually the most economical. Often, it is competitive in cost with other bridge systems.

Bridges with decks near the level of the arch bearings (Fig. 17.32b) are usually designed as tied arches; that is, tie bars take the arch thrust. End bearings and abutments are similar to those for girder or truss bridges. Tied arches compete in cost with through trusses in locations where under-clearances are restricted. Arches sometimes are preferred for aesthetic reasons. Unsightly overhead laterals can be avoided by using arches with

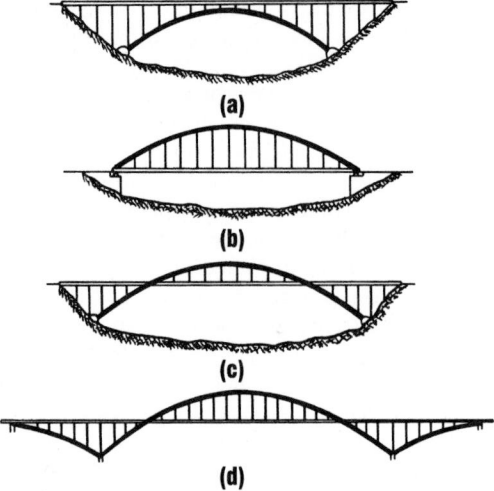

Fig. 17.32 Basic types of steel arch bridges: (*a*) Open spandrel arch; (*b*) tied arch; (*c*) arch with deck at an intermediate level; (*d*) multiple-arch bridge.

sufficiently high moment of inertia to resist buckling.

Bridges with decks at an intermediate level (Fig. 17.32*c*) may be tied, may rest on thrust-resisting abutments, or may be combined structurally with side spans that alleviate the thrust of the main span on the main piers (Fig. 17.32*d*). Intermediate deck positions are used for long, high-rising spans on low piers.

Spans of multiple-arch bridges are usually structurally separated at the piers. But such bridges may also be designed as continuous structures.

17.16.1 Hinges

Whether or not hinges are required for arch bridges depends on foundation conditions. Abutment movements may sharply increase rib stresses. Fully restrained arches are more sensitive to small abutment movements (and temperature variations) than hinged arches. Flat arches are more sensitive than high arches. If foundations are not fully reliable, hinged bearings should be used.

Complete independence from small abutment movements is achieved by installing a third hinge, usually at the crown. This hinge may be either permanent or temporary during erection, to be

locked after all dead-load deformations have been accounted for.

17.16.2 Arch Analysis

The elementary analysis of steel arches is based on the elastic, or first-order, theory, which assumes that the geometric shape of the center line remains constant, irrespective of the imposed load. This assumption is never mathematically correct. The effects of deviations caused by overall flattening of the arch due to the elastic rib shortening, elastic or inelastic displacements of the abutment, and local deformation due to live-load concentrations increase with initial flatness of the arch. An effort is usually made to eliminate the dead-load part of the effect of rib shortening and abutment yielding during erection by jacking the legs of an arch toward each other or the crown section apart before final closure. Arches subject to substantial deformation must be checked by the second-order, or deflection, theory.

For heavy moving loads, it is sometimes advantageous to assign the flexural resistance of the system to special stiffening girders or trusses, analogous to those of suspension bridges (Art. 17.15). The arches themselves are then subject, essentially, to axial stresses only and can be designed as slender as buckling considerations permit.

17.16.3 Arch Design

In general, steel arches must be designed for combined stresses due to axial loads and bending.

The height-to-span ratio used for steel arches varies within wide limits. Minimum values are around 1:10 for tied arches, 1:16 for open spandrel arches.

In cross section, steel arches may be I-shaped, box-shaped, or tubular. Or they may be designed as space trusses.

17.16.4 Deck Construction

The roadway deck of steel arch bridges is usually of reinforced concrete, often of lightweight concrete, on a framing of steel floor beams and stringers. To avoid undesirable cooperation with the primary steel structure, concrete decks either are provided with appropriately spaced expansion joints or prestressed. Orthotropic decks that combine the functions of traffic deck, tie bar, stiffening girder, and lateral diaphragm have been used on some major arch bridges.

(F. S. Merritt and R. L. Brockenbrough, "Structural Steel Designers' Handbook," 2nd ed., McGraw-Hill, Inc., New York (books.mcgraw-hill. com).)

17.17 Horizontally Curved Steel Girders

For bridges with curved steel girders, the effects of torsion must be taken into consideration by the designer. Also, careful attention should be given to cross frames—spacing, design, and connection details. The effects of torsion decrease the stresses in the inside girders (those nearer the center of curvature). But there is a corresponding increase in stresses in the outside girders. Although the differences are not large for multiple-girder systems, the differences in stress for two-girder systems with short-radius curves and long spans can be as high as 50%. The torsional forces translate into vertical and horizontal forces, which must be transferred from the outside to inside girders through the cross frames.

An approximate method for analysis of curved girder stresses is given in the U.S. Steel "Highway Structures Design Handbook." This approximate method has proven satisfactory for many structures, but for complex structures (those with long spans, short-radius curves, or with only two girders), it is recommended that a rigorous analysis using a computer program be used. For the structure in Fig. 17.33, the stress differentials in the two girders are 50% and the cross frames transfer up to 70 kips of vertical and horizontal forces between girders. The center of the main span rotated 4 in when the deck was placed. Such rotations should be anticipated and the girders erected "out of plumb" so that the final web position will be vertical.

Design of curved-girder bridges should consider the following:

1. Full-depth cross frames should be used to transfer the lateral forces from the flanges. (See Fig. 17.34.)

Fig. 17.33 Curved girders of Tuolomne River Bridge, California, were erected in pairs with their cross frames connected between them. (*California Department of Transportation.*)

2. The cross frames should be designed as primary stress-carrying members to transfer the loads.

3. Flange-plate width should be increased above the normal design minimums to provide stability during handling and erection.

4. Cross-frame connections at the web plates are critical. The web plate should be thickened to provide bending resistance

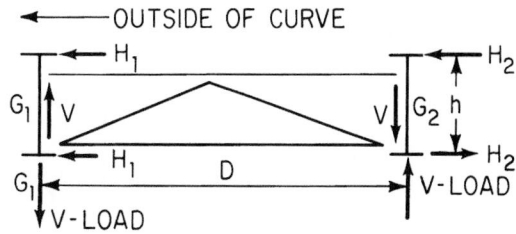

Fig. 17.34 Cross section of curved-girder bridge at cross frame, showing forces resulting from curvature.

17.18 Bridge Bearings

Bearings are structural assemblies installed to secure the safe transfer of all reactions from the superstructure to the substructure. They must fulfill two basic requirements: They must spread the reactions over adequate areas of the substructure, and they must be capable of adapting to elastic, thermal, and other deformations of the superstructure without generating harmful restraining forces.

Generally, bearings are classified as fixed, expansion, or elastomeric.

Fixed bearings adapt only to angular deflections. They must be designed to resist both vertical and horizontal components of reactions.

Expansion bearings adapt to both angular deflections and longitudinal movements of the superstructure. Except for friction, they resist only those components of the superstructure reactions perpendicular to these movements.

In both types of bearings, provision must be made for the safe transfer of all forces transverse to the direction of the span.

Elastomeric bearings are a very efficient bearing for short to medium span bridges. They are relatively maintenance free and are one of the safest bearings, under seismic loading. Elastomeric bearings generally consist of laminated layers of elastomer restrained at their surfaces by bonded laminas. The elastomer is a Neoprene rubber; the laminas consist of either glass-fiber fabrics or steel sheets. Steel-reinforced elastomeric bearings are usually used when anchor bolts are required through the bearing (Figs. 17.34 and 17.35).

The bearing pressure of elastomeric bearings should not exceed 800 psi under a service-load combination of dead load and live load, not including impact. For steel-reinforced bearing pads, the pressure should not exceed 1000 psi. The minimum pressure allowed on any pad due to dead load only is 200 psi.

The capacity of an elastomeric bearing to absorb angular deflections and longitudinal movements of the superstructure is a function of its thickness (or of the sum of the thicknesses of its rubber elements between steel laminas), its shape factor (area of the loaded face divided by the sum of the side areas free to bulge), and the properties of the elastomer.

AASHTO Specifications limit the overall thickness of a laminated bearing to one-third its length or width, whichever is smaller (or one-fourth of its diameter). The thickness should be at least twice the horizontal movement.

An alternative is a pot bearing, which supports the structure on a hydraulic cylinder with an elastomer as the liquid medium.

Concrete Bridges

Reinforced concrete is used extensively in highway bridges because of its economy in short and medium spans, durability, low maintenance costs, and easy adaptability to horizontal and vertical curvature. The principal types of cast-in-place supporting elements are the longitudinally reinforced slab, T beam or girder, and cellular or box girder. Precast construction, usually prestressed, often employs an I-beam or box-girder cross section. In long-span construction, posttensioned box girders often are used.

17.19 Slab Bridges

Concrete slab bridges, longitudinally reinforced, may be simply supported on piers and/or abutments, monolithic with wall supports, or continuous over intermediate supports.

17.19.1 Design Span

For simple spans, the design span is the distance center to center of supports but need not exceed the clear span plus slab thickness. For slabs monolithic with walls (without haunches), use the clear span. For slabs on steel or timber stringers, use the clear span plus half the stringer width.

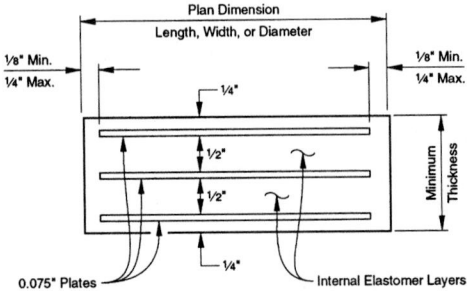

Fig. 17.35 Steel-laminated elastomeric bearing pad.

17.19.2 Load Distribution

In design, usually a 1-ft-wide longitudinal, typical strip is selected and its thickness and reinforcing determined for the appropriate HS loading. Wheel loads may be assumed distributed over a width, ft,

$$E = 4 + 0.06S \leq 7 \qquad (17.24)$$

where $S =$ span, ft. Lane loads should be distributed over a width of $2E$.

For simple spans, the maximum live-load moment, ft-kips, per foot width of slab, without impact, for HS20 loading is closely approximated by

$$M = 0.9S \quad S \leq 50 \, \text{ft} \qquad (17.25a)$$
$$M = 1.30S - 20 \quad 50 > S < 100 \qquad (17.25b)$$

For HS15 loading, use three-quarters of the value given by Eqs. (17.25).

For longitudinally reinforced cantilever slabs, wheel loads should be distributed over a width, ft,

$$E = 0.35X + 3.25 \leq 7 \, \text{ft} \qquad (17.26)$$

where $X =$ distance from load to point of support, ft. The moment, ft-kips per foot width of slab, is

$$M = \frac{P}{E}X \qquad (17.27)$$

where $P = 16$ kips for H20 loading and 12 kips for H15.

17.19.3 Reinforcement

Slabs should also be reinforced transversely to distribute the live loads laterally. The amount

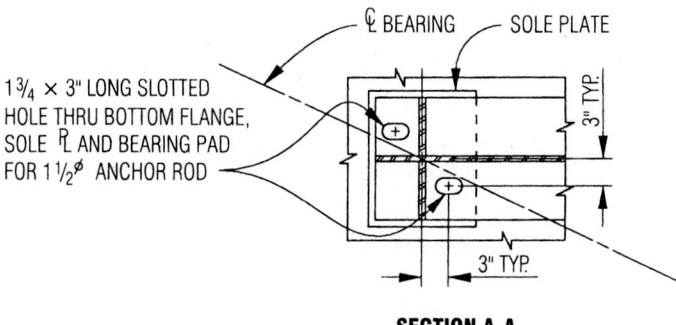

SECTION A-A

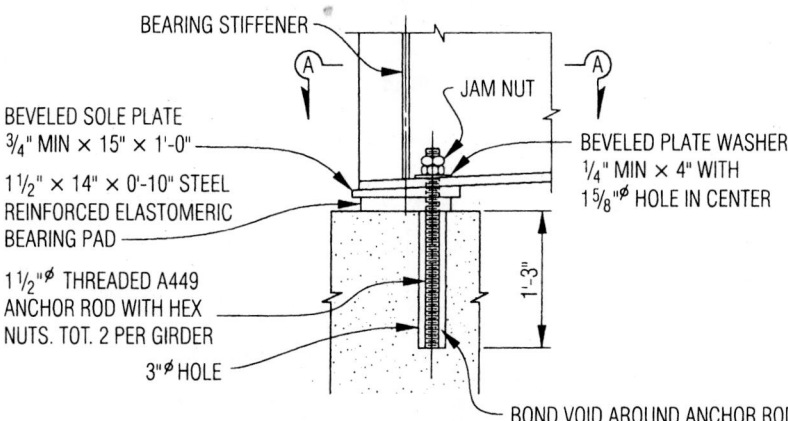

PART ELEVATION AT BEARINGS

Fig. 17.36 Example of an elastomeric bearing.

should be at least the following percentage of the main reinforcing steel required for positive moment: $100/\sqrt{S}$, but it need not exceed 50%.

The slab should be strengthened at all unsupported edges. In the longitudinal direction, strengthening may consist of a slab section additionally reinforced, a beam integral with and deeper than the slab, or an integral reinforced section of slab and curb. These should be designed to resist a live-load moment, ft-kips, of $1.6S$ for HS20 loading and $1.2S$ for HS15 loading on simply supported spans. Values for continuous spans may be reduced 20%. Greater reductions are permissible if justified by more exact analysis.

At bridge ends and intermediate points where continuity of the slabs is broken, the edges should be supported by diaphragms or other suitable means. The diaphragms should be designed to resist the full moment and shear produced by wheel loads that can pass over them.

17.19.4 Design Procedure

The following procedure may be used for design of a typical longitudinally reinforced concrete slab bridge (Fig. 17.37).

Step 1. Determine the live-load distribution (effective width). For the three-span, 90-ft-long

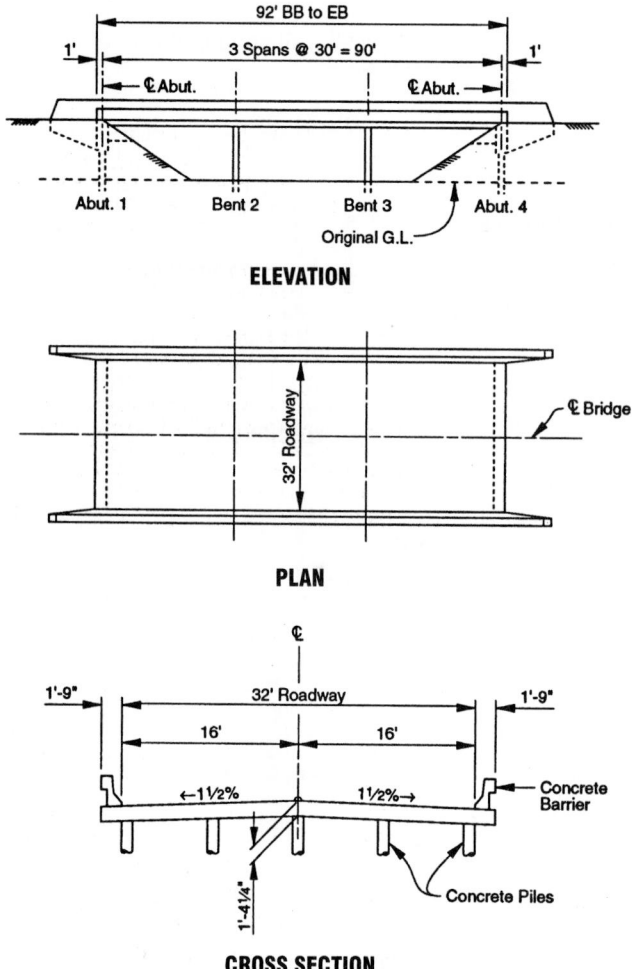

ELEVATION

PLAN

CROSS SECTION

Fig. 17.37 Three-span concrete-slab bridge.

bridge in Fig. 17.37, $S = 30$ ft and

$$E = 4 + 0.06 \times 30 = 5.8 \text{ ft}$$

The distributed load for a 4-kip front wheel then is 4/5.8, or 0.69 kips, and for a 16-kip rear or trailer wheel load, 16/5.8, or 2.76 kips, per foot of slab width. For an alternative 12-kip wheel load, the distributed load is 12/5.8, or 2.07 kips per foot of slab width (see Fig. 17.38).

Step 2. Assume a slab depth.

Step 3. Determine dead-load moments for the assumed slab depth.

Step 4. Determine live-load moment at point of maximum moment. (This is done at this stage to get a check on the assumed slab depth.)

Step 5. Combine dead-load, live-load, and impact moments at point of maximum moment. Compare the required slab depth with the assumed depth.

Step 6. Adjust the slab depth, if necessary. If the required depth differs from the assumed depth of step 2, the dead-load moments should be revised and step 5 repeated. Usually, the second assumption is sufficient to yield the proper slab depth. Steps 2 through 6 follow conventional structural theory.

Step 7. Place live loads for maximum moments at other points on the structure to obtain intermediate values for drawing envelope curves of maximum moment.

Step 8. Draw the envelope curves. Determine the sizes and points of cutoff for reinforcing bars.

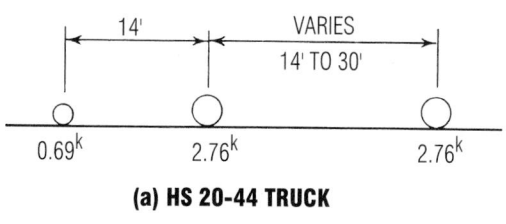

(a) HS 20-44 TRUCK

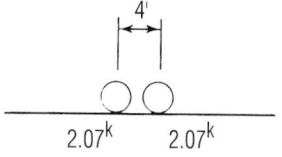

(b) ALTERNATIVE LOADING

Fig. 17.38 Wheel load per foot width of slab for bridge of Fig. 17.37.

Step 9. Determine distribution steel.

Step 10. Determine the number of piles required at each bent.

Figures 17.39 and 17.40 illustrate typical steel reinforcement patterns for a single-span and a two-span concrete-slab bridge, respectively, similar to Fig. 17.37, suitable for spans ranging from 16 to 44 ft and carrying HS20 or alternate loading. Reinforcement parallel to traffic in the single-span bridge is mainly in the bottom of the slab (Fig. 17.39*b*), rather than in the top (Fig. 17.39*a*). The two-span bridge has main steel reinforcement in the top of the slab (Fig. 17.40*b*) over the center bent, to resist negative moments and main steel reinforcement in the bottom of the slab (Fig. 17.40*a*) in positive-moment regions. Reinforcement in multi-span bridges is arranged similarly. Transverse distribution steel is spaced typically at 11 to 12 in. The thickness of the concrete slab and reinforcement sizes depend on the specified 28-day concrete compressive strength f_c' and yield point of the reinforcement steel.

For skews up to 20°, transverse reinforcement should be placed parallel to the bent. For larger skews, transverse reinforcement should be placed perpendicular to the center line of the bridge. Skews exceeding 50° require special design.

("Bridge Design Aids," Division of Structures, California Department of Transportation, Sacramento, Calif. (www.dot.ca.gov).)

17.20 Concrete T-Beam Bridges

Widely used in highway construction, this type of bridge consists of a concrete slab supported on, and integral with, girders (Fig. 17.41). It is especially economical in the 50- to 80-ft range. Where falsework is prohibited, because of traffic conditions or clearance limitations, precast construction of reinforced or prestressed concrete may be used.

17.20.1 Design of Transverse Slabs

Since the girders are parallel to traffic, main reinforcing in the slab is perpendicular to traffic. For simply supported slabs, the span should be the distance center to center of supports but need not exceed the clear distance plus thickness of slabs.

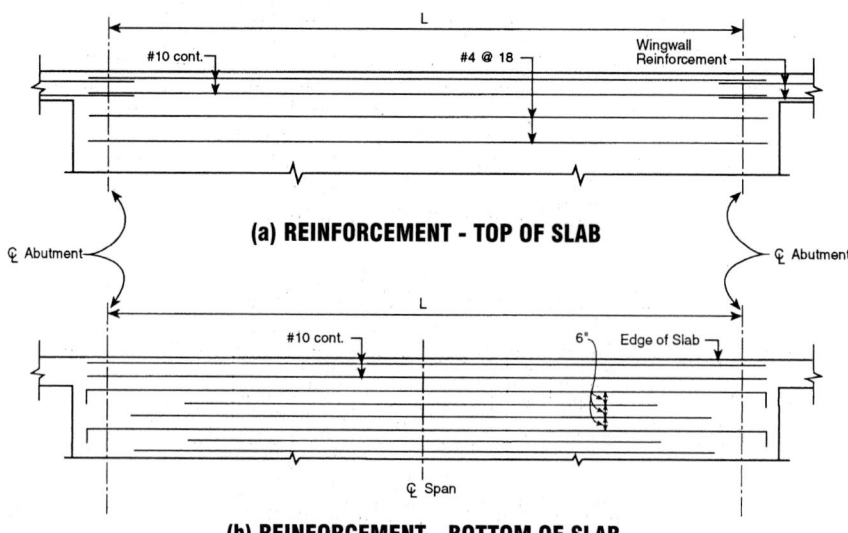

Fig. 17.39 Arrangement of slab reinforcement for a single-span bridge carrying HS20-44 or alternative loading.

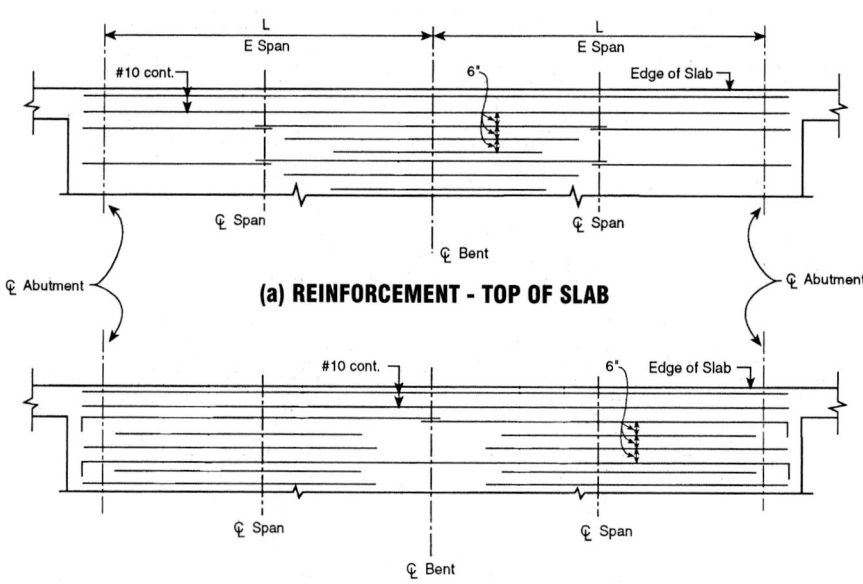

Fig. 17.40 Arrangement of slab reinforcement for a two-span bridge carrying HS20-44 or alternative loading.

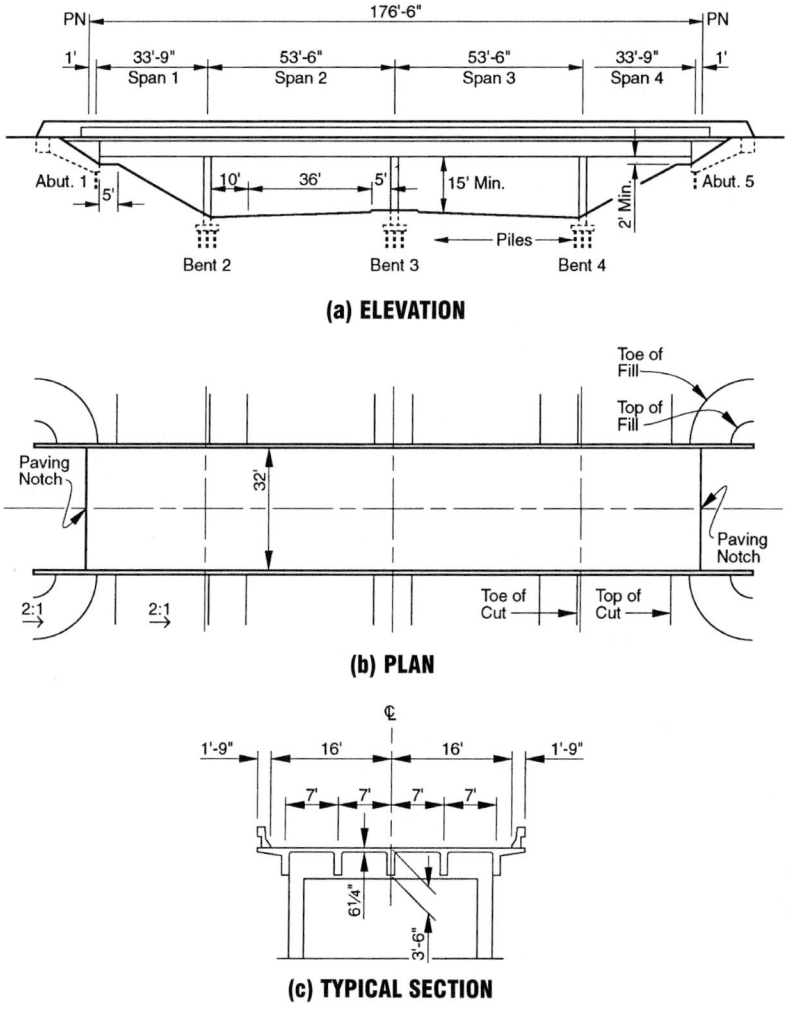

Fig. 17.41 Four-span bridge with concrete T beams.

For slabs continuous over more than two girders, the span may be taken as the clear distance between girders.

The live-load moment, ft-kips, for HS20 loading on simply supported slab spans is given by

$$M = 0.5(S + 2) \qquad (17.28)$$

where S = span, ft.

For slabs continuous over three or more supports, multiply M in Eq. (17.28) by 0.8 for both positive and negative moment. For HS15 loading, multiply M by $\frac{3}{4}$.

Reinforcement also should be placed in the slab parallel to traffic to distribute concentrated live loads. The amount should be the following percentage of the main reinforcing steel required for positive moment: $220/\sqrt{S}$, but need not exceed 67%.

Where a slab cantilevers over a girder, the wheel load should be distributed over a distance, ft, parallel to the girder of

$$E = 0.8X + 3.75 \qquad (17.29)$$

where X = distance, ft, from load to point of support. The moment, ft-kips per foot of slab parallel to girder, is

$$M = \frac{P}{E}X \qquad (17.30)$$

where $P = 16$ kips for HS20 loading and 12 kips for HS15. Equations (17.28) to (17.30) apply also to concrete slabs supported on steel girders, including composite construction.

In design of the slabs, a l-ft-wide strip is selected and its thickness and reinforcing determined. The dead-load moments, ft-kips, positive and negative, can be assumed to be $wS^2/10$, where w is the dead load, kips/ft². Live-load moments are given by Eq. (17.28) with a 20% reduction for continuity. Impact is a maximum of 30%. With these values, standard charts can be developed for design of slabs on steel and concrete girders. Figure 17.42 shows a typical slab-reinforcement layout.

17.20.2 T-Beam Design

The structure shown in Fig. 17.41 is a typical four-span grade-separation structure. The structural frame assumed for analysis is shown in Fig. 17.43. Columns with a pinned base are less stiff than fixed columns which minimizes shrinkage and temperature moments. In addition, foundation pressures in

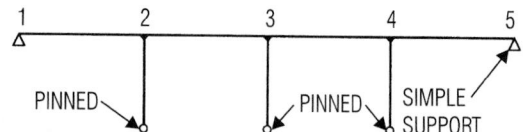

Fig. 17.43 Assumed support conditions for the bridge in Fig. 17.41.

pinned columns are considered fairly uniform, resulting in an economical footing size and design.

For concrete girder design, curves of maximum moments for dead load plus live load plus impact may be developed to determine reinforcement. For live-load moments, truck loadings are moved across the bridge. As they move, they generate changing moments, shears, and reactions. It is necessary to accumulate maximum combinations of moments to provide an adequate design. For heavy moving loads, extensive investigation is necessary to find the maximum stresses in continuous structures.

Figure 17.44 shows curves of maximum moments consisting of dead load plus live load plus impact combinations that are maximum along the span. From these curves, reinforcing steel amounts and lengths may be determined by plotting the moments developed. Figure 17.45 shows curves of maximum shears. Figure 17.46 shows the girder steel

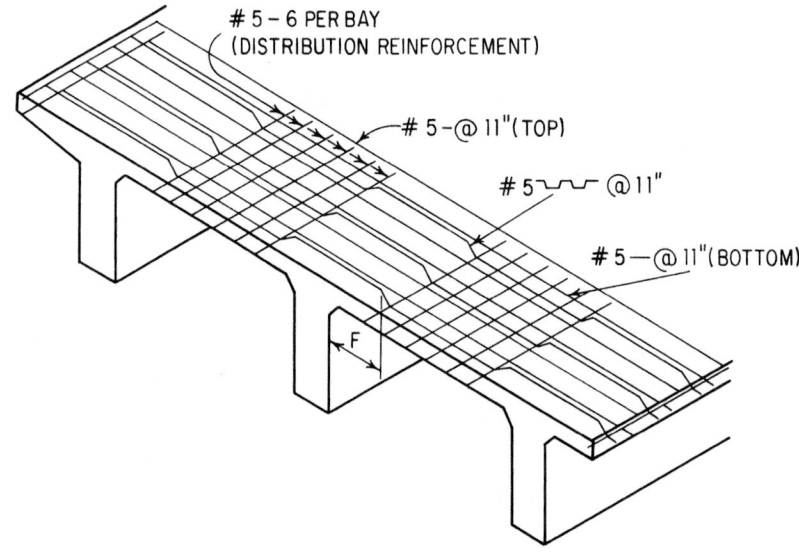

Fig. 17.42 Typical layout of reinforcement in the deck of a concrete T-beam bridge.

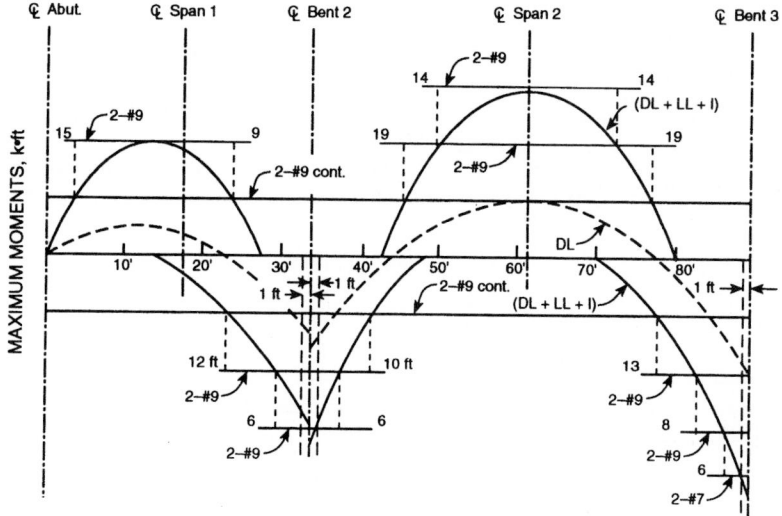

Fig. 17.44 Reinforcing for T beams of Fig. 17.41 is determined from curves of maximum bending moment. Numbers at the ends of the bars are distances, ft, from the center line of the span or bent.

reinforcement layout. Maximum-shear requirements are derived theoretically by a point-to-point study of variations. Usually, a straight line between center line and end maximums is adequate.

Girder spacing ranges from about 7 to 9 ft. Usually, a deck slab overhang of about 2 ft 6 in is economical.

When the slab is made integral with the girder, its effective width of compression flange in design may not exceed the distance center to center of girders, one-fourth the girder span, or girder web-width plus 12 times the least thickness of slab. For exterior girders, however, effective overhang width may not exceed half the clear distance to the next

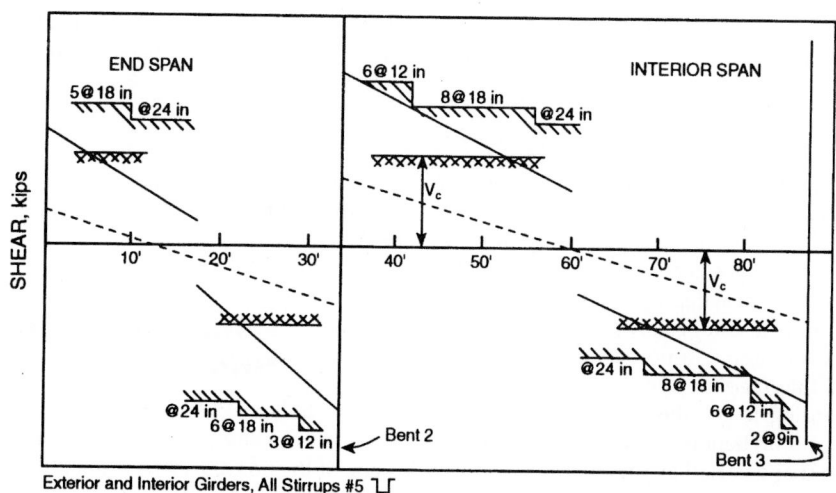

Fig. 17.45 Curves of maximum shear for T beams of Fig. 17.41.

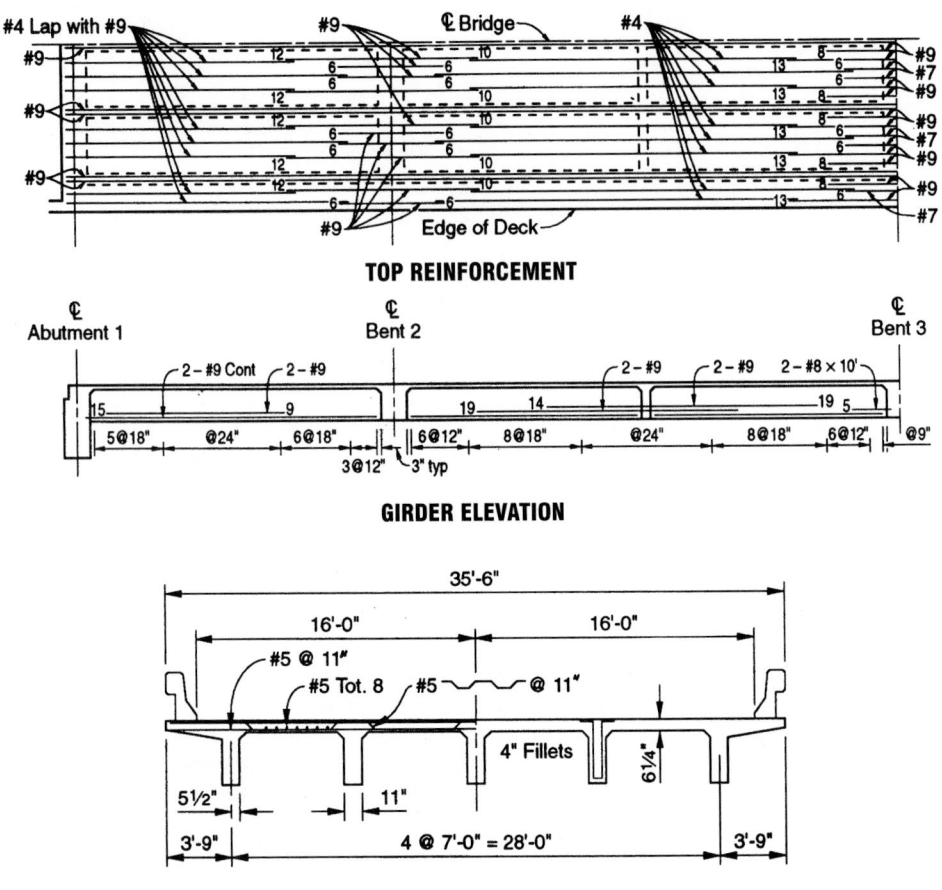

TOP REINFORCEMENT

GIRDER ELEVATION

TYPICAL SECTION

Fig. 17.46 Reinforcement layout for T beams of Fig. 17.41. Reinforcement is symmetrical about the center lines of the bridge and bent 3. Numbers at the ends of the bars indicate distances, ft, from the center line of bent or span.

girder, one-twelfth the girder span, or six times the slab thickness.

Ratios of beam depths to spans used in continuous T-beam bridges generally range from 0.065 to 0.075. An economical depth usually results when a small amount of compressive reinforcement is required at the interior supports.

Design of intermediate supports or bents varies widely, according to the designer's preference. A simple two-column bent is shown in Fig. 17.41. But considerable shape variations in column cross section and elevation are possible.

Abutments are usually seat type or a monolithic end diaphragm supported on piles.

("Bridge Design Aids," Division of Structures, California Department of Transportation, Sacramento, Calif. (www.dot.ca.gov).)

17.21 Concrete Box-Girder Bridges

Box or hollow concrete girders (Fig. 17.47) are favored by many designers because of the smooth plane of the bottom surface, uncluttered by lines of individual girders. Provision of space in the open cells for utilities is both a structural and an aesthetic advantage. Utilities are supported by the bottom

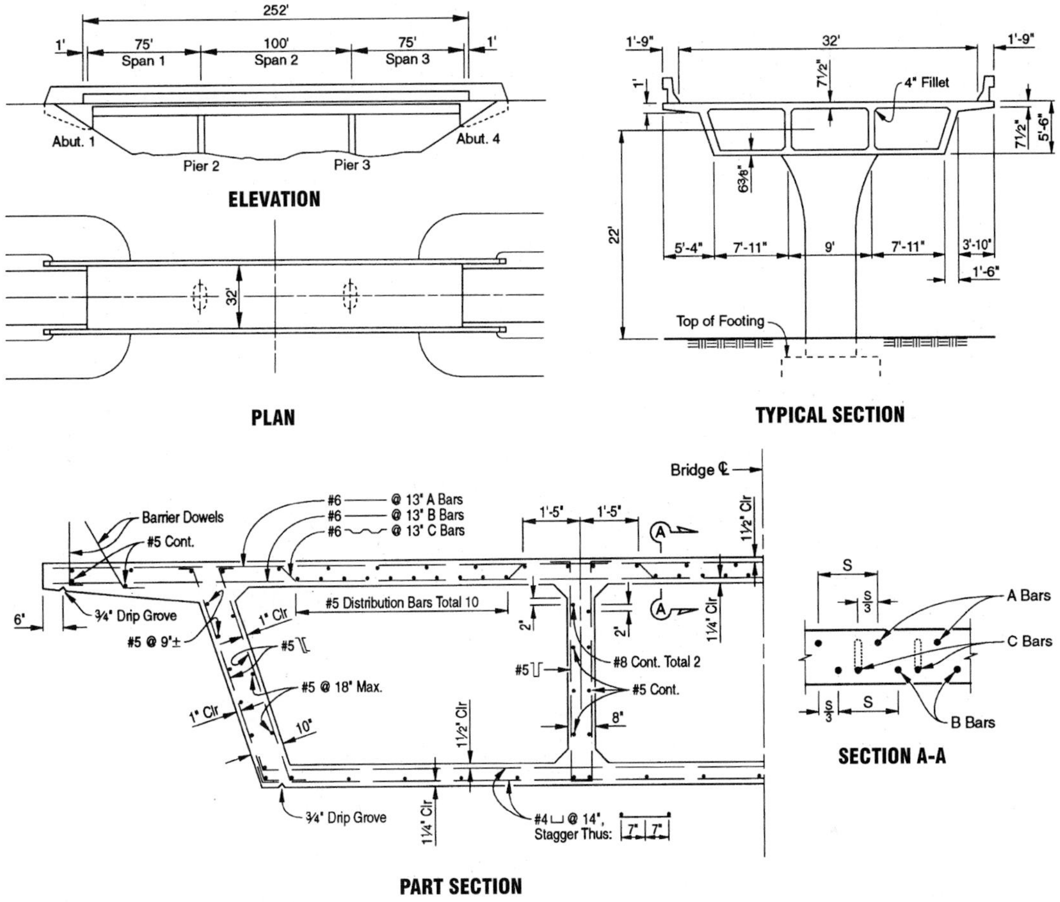

ELEVATION

PLAN

TYPICAL SECTION

PART SECTION

SECTION A-A

Fig. 17.47 Three-span, reinforced concrete box-girder bridge. For more details, see Fig. 17.51.

slab, and access can be made available for inspection and repair of utilities.

For sites where structure depth is not severely limited, box girders and T beams have been about equal in price in the 80-ft span range. For shorter spans, T beams usually are cheaper, and for longer spans, box girders. These cost relations hold in general, but box girders have, in some instances, been economical for spans as short as 50 ft when structure depth was restricted.

17.21.1 Girder Design

Structural analysis is usually based on two typical segments, interior and exterior girders (Fig. 17.48). An argument could be made for analyzing the

entire cross section as a unit because of its inherent transverse stiffness. Requirements in "Standard Specifications for Highway Bridges," American Association of State Highway and Transportation Officials, however, are based on live-load distributions for individual girders, and so design usu-

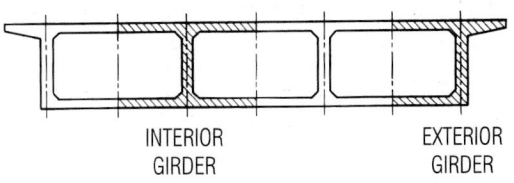

Fig. 17.48 Typical design sections (cross-hatched) for a box-girder bridge.

ally is based on the assumption that a box-girder bridge is composed of separate girders.

Effective width of slab as compression flange of an interior girder may be taken as the smallest of the distance center to center of girders, one-fourth the girder span, and girder-web width plus 12 times the least thickness of slab. Effective overhang width for an exterior girder may be taken as the smallest of half the clear distance to the next girder, one-twelfth the girder span, and six times the least thickness of the slab.

Usual depth-to-span ratio for continuous spans is 0.055. This may be reduced to about 0.048 with balanced spans, at some sacrifice in economy and increase in deflections. Simple spans usually require a minimum depth-to-span ratio of 0.06.

A typical concrete box-girder highway bridge is illustrated in Fig. 17.47. Girder spacing is approximately 1½ times the structure depth. Minimum girder web thickness is determined by shear but generally is at least 8 in. Changes should be gradual, spread over a distance at least 12 times the difference in web thickness.

Top-slab design follows the procedure described for T-beam bridges in Art. 17.20. Bottom-slab thickness and secondary reinforcement are usually controlled by specification minimums. AASHTO Specifications require that slab thickness be at least one-sixteenth the clear distance between girders but not less than 6 in for the top slab and 5½ in for the bottom slab. Fillets should be provided at the intersections of all surfaces within the cells.

Minimum flange reinforcement parallel to the girder should be 0.6% of the flange area. This steel may be distributed at top and bottom or placed in a single layer at the center of the slab. Spacing should not exceed 18 in. Minimum flange reinforcing normal to the girder should be 0.5% and similarly distributed. Bottom-flange bars should be bent up into the exterior-girder webs and anchored using a standard 90° hook or equivalent. At least one-third of the top flange tension reinforcement should extend to the exterior face of the outside girder and should be anchored with 90° bends or, where the flange projects beyond the girder sufficiently, extended far enough to develop bar strength in bond.

When the top slab is placed after the web walls have set, at least 10% of the negative-moment reinforcing should be placed in the web. The bars should extend a distance of at least one-fourth the span on each side of the intermediate supports of

continuous spans, one-fifth the span from the restrained ends of continuous spans, and the entire length of cantilevers. In any event, the web should have reinforcing placed horizontally in both faces, to prevent temperature and shrinkage cracks. The bars should be spaced not more than 12 inches c to c. Total area of this steel should be at least 10% of the area of flexural tension reinforcement.

Analysis of the structure in Fig. 17.47 for dead loads follows conventional moment-distribution procedure. Assumed end conditions are shown in Fig. 17.49a.

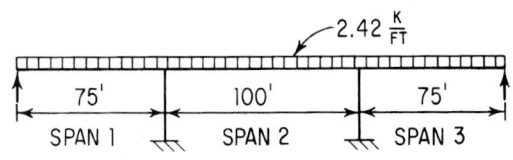

(a) DEAD LOAD ON BRIDGE

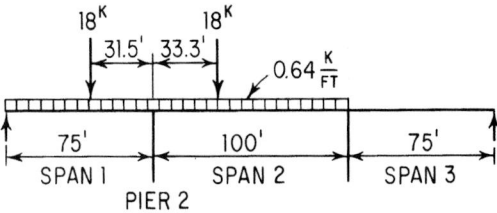

(b) LANE LOADING FOR MAXIMUM MOMENT OVER PIER 2

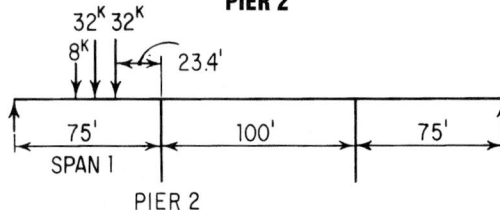

(c) TRUCK LOADING FOR MAXIMUM MOMENT OVER PIER 2

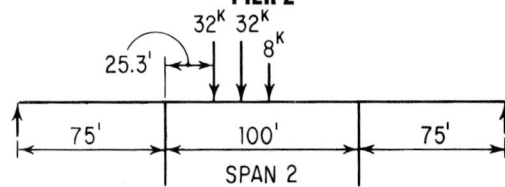

(d) ALTERNATE TRUCK LOADING FOR MAXIMUM MOMENT OVER PIER 2

Fig. 17.49 Loading patterns for maximum stresses in a box-girder bridge.

Live loads, positioned to produce maximum negative moments in the girders over Pier 2, are shown in Fig. 17.49*b* to *d*. Similar loadings should be applied to find maximum positive and negative moments at other critical points. Moments should be distributed and points plotted on a maximum-moment diagram (for dead load plus live load plus impact), as shown in Fig. 17.50. Layout of main girder reinforcement follows directly from this diagram. Figure 17.51 shows a typical layout.

("Bridge Design Details," Division of Structures, California Department of Transportation, Sacramento, Calif. (www.dot.ca.gov).)

17.22 Prestressed-Concrete Bridges

In prestressed-concrete construction, concrete is subjected to permanent compressive stresses of such magnitude that little or no tension develops when design loading is applied (Art. 8.42).

Prestressing allows considerably better utilization of concrete than conventional reinforcement. It results in an overall dead-load reduction, which

makes long spans possible with concrete, sometimes competitive in cost with those of steel. Prestressed concrete, however, requires greater sophistication in design, higher quality of materials (both concrete and steel), and greater refinement and controls in fabrication than does reinforced concrete.

Depending on the methods and sequence of fabrication, prestressed concrete may be precast, pretensioned; precast, posttensioned; cast-in-place posttentioned; composite; or partly prestressed.

In precast-beam bridges, the primary structure consists of precast-concrete units, usually I beams, channels, T beams, or box girders. They may be either pretensioned or posttensioned. Precast slabs may be solid or hollow. Precast I beams (Fig. 17.52) may be combined with fully or partly cast-in-place decks. This construction has the advantage that the deck can be shaped closely to the desired specifications. Precast slabs, incorporated into the deck, may be used in lieu of removable deck forms where accessibility is poor, for example, on over-water trestles or causeways. Precast T beams (Fig. 17.53) offer no advantage over the easier to fabricate, more compact I sections. Alignment of

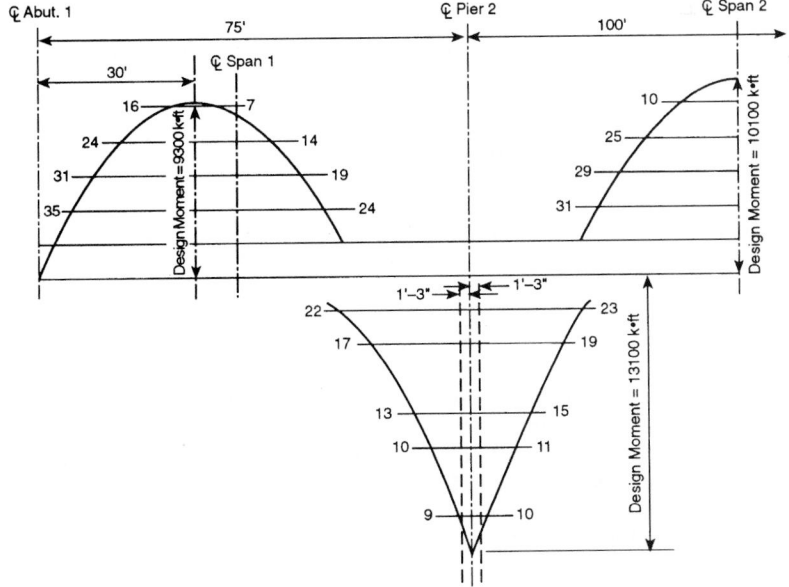

Fig. 17.50 Curves of maximum moment determine reinforcing for a box girder. Numbers at the ends of the bars indicate distances, ft, from the center line of piers or span.

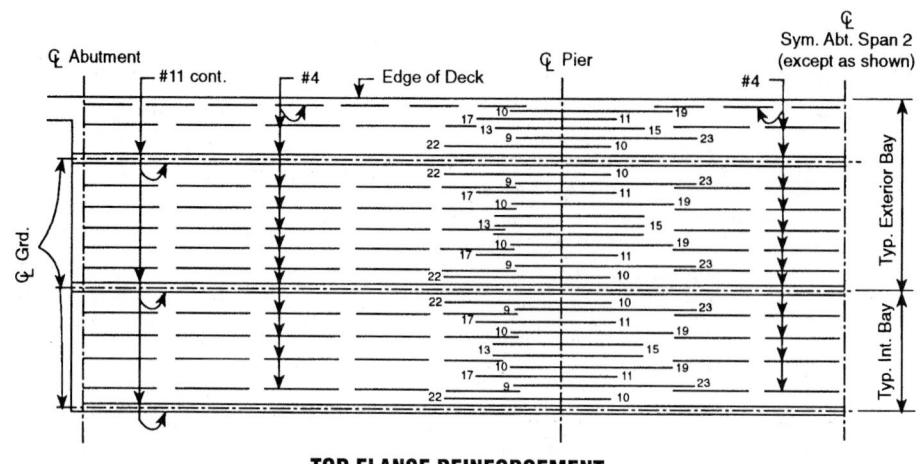

TOP FLANGE REINFORCEMENT

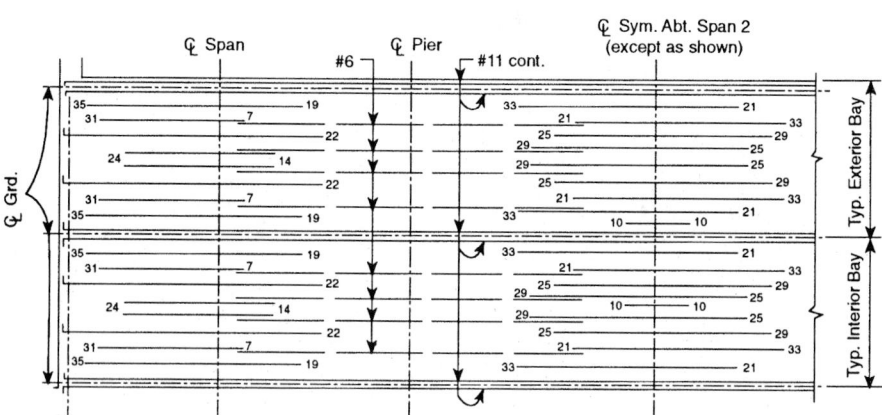

BOTTOM FLANGE REINFORCEMENT

Fig. 17.51 Reinforcing layout for the box-girder bridge Fig. 17.47 of and moment curves of Fig. 17.50. Design stresses for HS20 loading: $f_c' = 3500\,\text{psi}$, $f_y = 60\,\text{ksi}$.

the flanges of T sections often is difficult. And as with I beams, the flanges must be connected with cast-in-place concrete. Precast box sections may be placed side by side to form a bridge span. If desired, they may be posttensioned transversely.

Precast beams mainly are used for spans up to about 90 ft where erection of conventional falsework is not feasible or desirable. Such beams are particularly economical if conditions are favorable for mass fabrication, for example, in multispan viaducts or causeways or in the vicinity of centralized fabrication plants. Longer spans are

possible but require increasingly heavy handling equipment.

Standard designs for precast, prestressed girders have been developed by the Federal Highway Administration and state highway departments.

Cast-in-place prestressed concrete often is used for low-level bridges, where ground conditions favor erection of conventional falsework. Typical cross sections are essentially similar to those used for conventionally reinforced sections, except that, in general, prestressing permits structures with thinner depths.

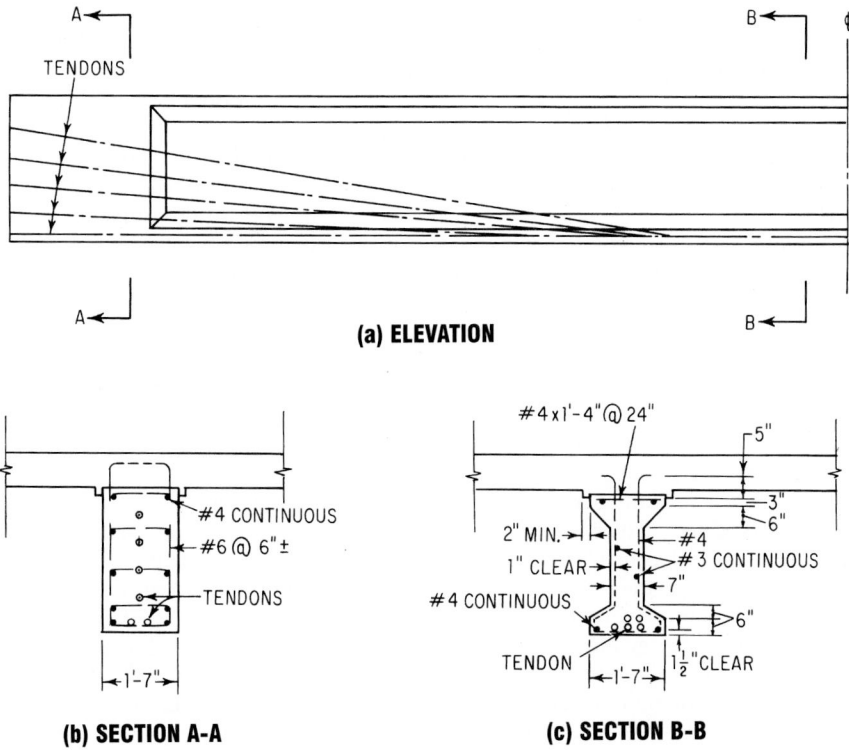

(a) ELEVATION

(b) SECTION A-A

(c) SECTION B-B

Fig. 17.52 Typical precast, prestressed I beam used in highway bridges.

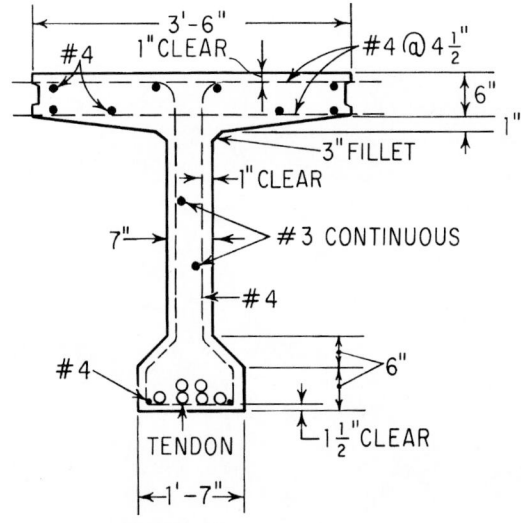

Fig. 17.53 Typical precast, prestressed T beam used in highway bridges.

For fully cast-in-place single-span bridges, post-tensioning differs only quantitatively from that for precast elements. In design of multispan continuous bridges, the following must be considered: Frictional prestress losses depend on the draping pattern of the ducts. To reduce potential losses and increase the reliability of effective prestress, avoid continuously waving tendon patterns. Instead, use discontinuous simple patterns. Another method is to place tendons, usually bundles of cables, in the hollows of box girders and to bend the tendons at lubricated, accessible bearings.

Prestressed concrete is competitive with other materials for spans of 150 to 250 ft or more. Construction techniques and improvements in prestressing hardware, such as smooth, lightweight conduits, which reduce friction losses, have brought prestressed concrete bridges into direct competition with structural steel, once preeminent in medium and long spans.

Segmental construction, both precast and cast-in-place, has eliminated the need for expensive falsework, which previously made concrete bridges uneconomical in locations requiring long spans over navigation channels or deep canyons. The two types of segmental construction used most in the United States are the cast-in-place and precast balanced cantilever types.

For cast-in-place construction, the movable formwork is supported by a structural framework, or traveler, which cantilevers from an adjacent completed section of the superstructure. As each section is cast, cured, and posttensioned, the framework is moved out and the process repeated. Figure 17.54 illustrates this type of construction. For precast construction the procedure is similar, except that the sections are prefabricated.

Other methods, such as full-span and incremental launching procedures, can be used to fit site conditions. In all segmental construction, special attention should be given in the erection plan to limitation of temporary stresses and to maintenance of balance during erection and prior to span closing. Also important are an accurate prediction of creep and accurate calculation of deflections to ensure attainment of the desired structure profile and deck grades in the completed structure.

Posttensioning makes possible widening or strengthening or other remodeling of existing concrete structures. For example, Fig. 17.55 shows a cross section through a double-deck viaduct. The row of columns under the upper deck had to be removed, and capacity had to be increased from H15 to HS20 loading. No interference with upper-deck traffic and a minimum of interference with lower-deck traffic were permitted. This objective was accomplished by reinforcing each floor beam with precast units incorporating preformed ducts for tendons. Then the entire upper deck was prestressed transversely. This permitted the beams to span the full width of the bridge and carry the heavier loading. Similar remodeling has been done with cast-in-place concrete.

Determination of stresses in prestressed bridges is similar to that for other structures. In analysis of statically indeterminate systems, however, the deformations caused by prestressing must be taken into account (see also Arts. 8.42 to 8.45).

[C. A. Ballinger and W. Podolny, Jr., "Segmental Bridge Construction in Western Europe," Transportation Research Board, Record 665, 1978; A. Grand, "Incremental Launching of Concrete Structures," *Journal of the American Concrete Institute*, August 1975; W. Baur, "Bridge Erection

Fig. 17.54 Segmental cast-in-place concrete construction in progress for the Pine Valley Bridge, California. (*California Department of Transportation.*)

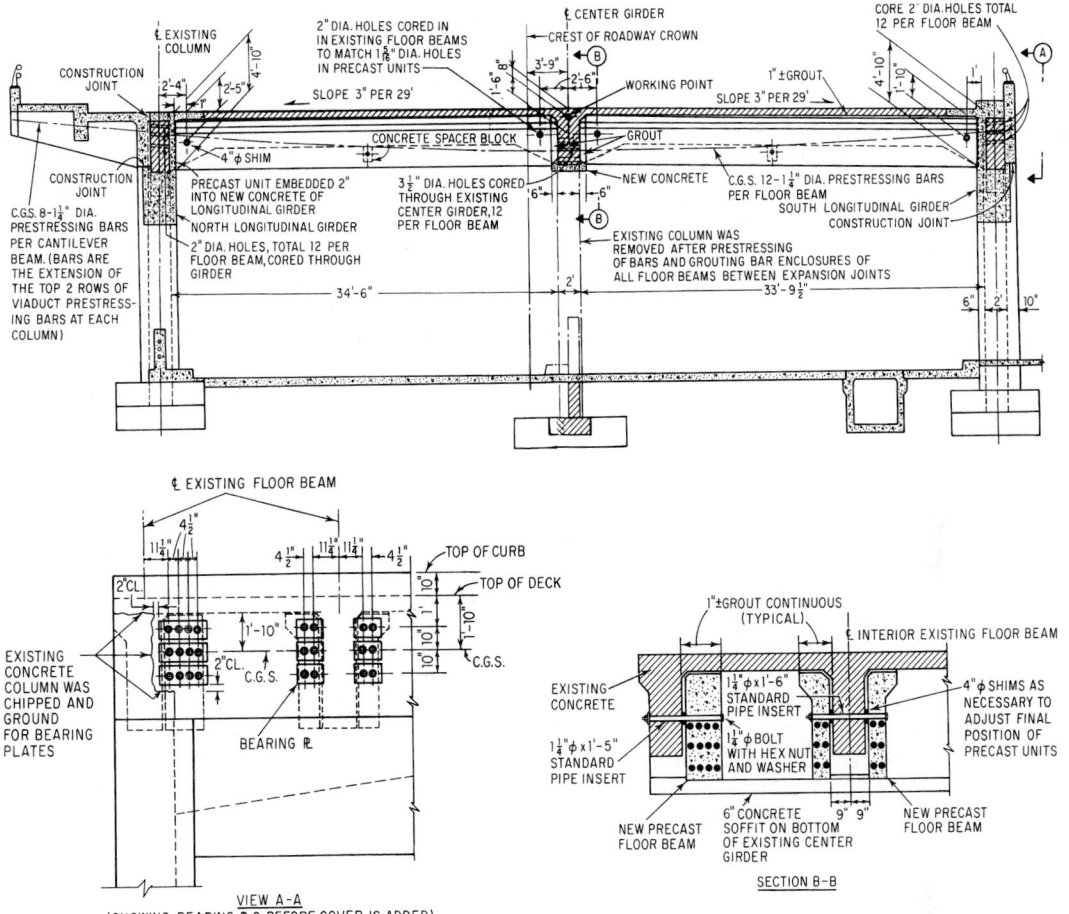

Fig. 17.55 Double-deck viaduct strengthened by prestressing to permit removal of column and passage of heavier trucks.

by Launching Is Fast, Safe, and Efficient," *Civil Engineering*, March 1977; F. Leonhardt, "New Trends in Design and Construction of Long-Span Bridges and Viaducts (Skew, Flat Slabs, Torsion Box)," Eighth Congress, International Association for Bridge and Structural Engineering, New York, Sept. 9 to 14, 1968.]

17.23 Concrete Bridge Piers and Abutments

Bridge piers are the intermediate supports of the superstructure of bridges with two or more openings. Abutments are the end supports and usually

have the additional function of retaining earth fill for the bridge approaches.

The minimum height of piers and abutments is governed by requirements of accessibility for maintenance of the superstructure, including bearings; of protection against spray for bridges over water; and of vertical clearance requirements for bridges over traveled ways. There is no upper limit for pier heights, except that imposed by economic considerations. One of the piers of the Europa Bridge, which carries an international freeway in Austria, for instance, soars to 492 ft above the ground surface of the valley.

The top surface of piers must have adequate length and width to accommodate the bridge bearings of the superstructure. On abutments, added width is required for the back wall (curtain wall or bulkhead), which retains approach fill and protects the end section of the superstructure. In designing the aboveground sections of piers, restrictions resulting from lateral-clearance requirements of adjacent traveled ways and visibility needs may have to be taken into account. Length and width at the base level are controlled by stability, stress limitations in the pier shaft, and foundation design.

For stress and stability analyses, the reactions from loadings (dead and live, but not impact) acting on the superstructure should be combined with those acting directly on the substructure. Longitudinal reactions depend on the type of bearing, whether fixed or expansion.

17.23.1 Piers

A number of basic pier shapes have been developed to meet the widely varying requirements. Enumerated below are some of the more common types and their preferred uses.

Trestle-type piers are preferred on low-level "causeways" carried over shallow waters or seasonally flooded land on concrete slab or beam-and-slab superstructures. Each pier or bent consists of two or more bearing piles, usually all driven in the same plane, and a thick concrete deck or a prismatic cap into which the piles are framed (Fig. 17.37). Both cap and piles may be of timber or, for more permanent construction, of precast conventionally reinforced or prestressed concrete.

Wall-type concrete piers on spread footings are generally used as supports for two-lane overcrossings over divided highways. Given adequate longitudinal support of the superstructure, these piers may be designed as pendulum walls, with joints at top and bottom; otherwise, as cantilever walls.

T-shaped piers on spread footings, with or without bearing piles, may be used to advantage as supports of twin girders. The girders are seated on bearings at both tips of the cross beam atop the pier stem. T-shaped piers have been built either entirely of reinforced concrete or of reinforced concrete in various combinations with structural steel.

Single-column piers of rectangular or circular cross section on spread footings may be used to support box girders, with built-in diaphragms acting as cross beams (Fig. 17.47).

Portal frames may be used as piers under heavy steel girders, with bearings located directly over the portal legs (columns). When more than two girders are to be supported, the designer may choose to strengthen the portal cap beam or to add more columns. Preferably, all legs of each portal frame should rest on a common base plate. If, instead, separate footings are used, as, for instance, on separate pile clusters, adequate tie bars must be used to prevent unintended spreading.

Massive masonry piers have been built since antiquity for multiple-arch river bridges, high-level aqueducts, and more recently, viaducts. In the twentieth century, their place has been taken by massive concrete construction, with or without natural stone facing. Where reduction of dead load is of the essence, hollow piers, often of heavily reinforced concrete, may be used.

Steel towers on concrete pedestals may be used for high bridge piers. They may be designed either as thin-membered, special trellis or as closed box portals, or combinations of these (Figs. 17.20 and 17.26).

Very tall piers, when used, are usually constructed of reinforced or prestressed concrete, either solid or cellular in design (Fig. 17.33).

Bridge abutments basically are piers with flanking (wing) walls. Abutments for short-span concrete bridges, such as T-beam or slab-type highway overcrossings, are frequently simple concrete trestles built monolithically with the superstructure (see Figs. 17.37 and 17.47). Abutments for steel bridges and for long-span concrete bridges that are subject to substantial end rotation and longitudinal movements should be designed as separate structures that provide a level area for the bridge bearings (bridge seat) and a back wall (curtain wall or bulkhead). The wall (stem) below the bridge seat of such abutments may be of solid concrete or thin-walled reinforced-concrete construction, with or without counterfort walls; but on rare occasions, masonry is used.

Sidewalls, which retain approach fill, should have adequate length to prevent erosion and undesired spill of the backfill. They may be built either monolithically with the abutment stem and backwall in which case they are designed as cantilevers subject to two-way bending, or as self-supporting retaining walls on independent footings. Sidewalls may be arranged in a straight line with the abutment face, parallel to the bridge axis, or at any intermediate angle to the abutment face that may suit local conditions. Given adequate foundation conditions, the parallel-to-bridge-axis arrangement (U-shaped abutment) is often preferred because of its inherent stability.

Abutments must be safe against overturning about the toe of the footing, against sliding on the footing, and against crushing of the underlying soil or overloading of piles. In earth-pressure computations, the vehicular load on highways may be taken into account in the form of an equivalent layer of soil 2 ft thick. Live loads from railroads may be assumed to be 0.5 kip/ft^2 over a 14-ft-wide strip for each track.

In computations of internal stresses and stability, the weight of the fill material over an inclined or stepped rear face and over reinforced concrete spread footings should be considered as fully effective. No earth pressures however, should be assumed from the earth prism in front of the wall. Buoyancy should be taken into account if it may occur.

18

Richard Harding
*Air Transportation &
Facilities Consultant
Harrisburg, Pennsylvania*

AIRPORT ENGINEERING

Airport engineering involves design and construction of a wide variety of facilities for the landing, takeoff, movement on the ground, and parking of aircraft; maintenance and repair of aircraft; fuel storage; and handling of passengers, baggage, and freight. Thus, at a typical airport, there are terminal buildings and hangars; pavements for aircraft runways, taxiways, and aprons; roads, bridges, and tunnels for automobiles and walks for pedestrians; automobile parking areas; drainage structures; and underground storage tanks. Aircraft include airplanes, helicopters, and the anticipated tilt rotor aircraft. Airport engineers have the responsibility of determining the size and arrangement of these facilities for safe, efficient, low-cost functioning of an airport.

18.1 Functions of Airport Components

A **runway**, the most essential component of an airport, enables landing and takeoff of airplanes. For all but the crudest airports, it is a paved strip. Many airports have more than one runway to accommodate aircraft landing and taking-off at locations where winds vary significantly in both direction and speed. Parallel runways are two runways laid out in the same direction to accommodate operations when the capacity of a single runway is exceeded.

Taxiways provide a convenient means for aircraft to enter and exit a runway. They are usually paved strips connecting runways with each other and with aircraft parking areas.

Parking aprons are typically paved areas adjacent to terminal buildings, storage hangers, aircraft maintenance hangers, and other buildings that pilots use as an approach to the building and to stop to permit passengers and crew to enter or exit the aircraft. Aprons at larger airports usually incorporate fuel systems, electrical power supply, and facilities for servicing aircraft.

A **terminal building usually** is incorporated in an airport layout to provide a transition for passengers and crew from ground to air and vice versa. It houses waiting rooms for passengers and at larger airports include facilities for baggage and cargo handling. Also, at larger airports it generally contains airline ticketing counters and offices. It is served by automobile access roads, and typically parking spaces for autos are provided nearby.

Control towers are built at many busy airports for air-traffic control. They provide a raised area from which traffic controllers can observe runways, taxiways, and aprons.

18.2 Classes of Airports

There are two categories of airports in the United States: civil and military. Civil airports serve the scheduled airlines and all phases of general aviation. They are developed through the local initiative of individual communities, with some assistance from state and Federal sources. Military airports serve as bases for Air Force,

Army, Navy, and Marine Corps aviation and are developed, as needed, through the Department of Defense.

Civil airports may be further classified as **air carrier airports** (those that serve the scheduled airlines) and **general aviation airports** (those that serve business and executive flying, air-taxi operations, commercial and industrial aviation, and student instruction). Although all air carrier airports accommodate considerable general aviation activity, the general aviation airports are usually not of a size sufficient to accommodate scheduled airlines. In each instance, the size and type of facility are determined by the existing and anticipated types and volume of air traffic that the facility will serve.

Military airports serve only the nation's defense needs. Only in rare instances is civil aviation activity permitted. There are, however, limited military facilities for Reserve and National Guard purposes at some civil airports. Military development is under the Corps of Engineers, United States Army, or the Facilities Engineering Command, United States Navy. Rigid adherence to standards and specifications for military airports is maintained.

18.3 National Airport Standards

The Federal Aviation Administration (FAA) publishes an advisory circular, AC 150/5300-13, 'Airport Design,' which replaces five previous advisory circulars dealing with site requirements for terminal navigation facilities, design of utility airports, aircraft data for airport design, and design of airport aprons. Airport design, under AC 150/5300-13, is guided by the Airport Reference Code (ARC), which correlates airport design criteria and operational and physical characteristics of the airplanes intended to operate at the airport.

The ARC consists of two components related to the design airplane selected for the airport. A letter depicts one component, the aircraft approach category. This is determined by the aircraft approach speed, which, in general, affects design of runways and runway-related facilities. A Roman numeral designates the second component, the airplane design group. This is related to airplane wing span, which primarily determines aircraft separation requirements and influences the design of taxiways and taxilanes.

The ARC specifies five aircraft approach categories, which are designated A through E, and six wing-span categories, which are labeled I through VI. Category A-I covers small and slow single-engine airplanes, whereas categories such as D-V and C-VI concern larger and faster airplanes (Table 18.1). Category E generally applies to high-speed military airplanes and is not referenced in the design of civil airports.

FAA advisory circulars contain standards for nationwide application to the design and construction of airports. These standards make possible the compatibility of local airports with each other and with the national system of airports. Although the standards are widely accepted, their use by communities is not mandatory unless Federal funds are involved in local airport development. Furthermore, regulations permit some latitude in deviation from the standards, where justified.

Table 18.2 summarizes physical characteristics set by national standards for airports. These are the minimum requirements that the FAA considers acceptable for safe operation. They can be used as a design guide in selection of physical characteristics of an airport to accommodate the aircraft anticipated to use its facilities. Information on how to obtain the standards is available at FAA district offices or one can obtain many of the advisory circulars for airport design by going to the FAA Web Site at http://www.faa.gov/arp/150acs.htm.

18.4 Airport Planning

Safety is the number one priority when designing or operating airports. All airport work must be carefully coordinated with the Federal Aviation Administration (FAA) and must be shown on an FAA (or designated representative) approved airport layout plan (ALP), and receive environmental clearance and air-space clearance to ensure its compatibility with the total airport and airspace system. The Airport Improvement Program (AIP), administered by the FAA, may provide funds for a major part of the development of landing areas including land acquisition. The FAA maintains national airport standards; offers advice on airport planning, design, and construction matters; maintains a national airport systems plan; certifies airports for operation; and conducts a compliance program to ensure adherence to regulations and requirements. The FAA operates

Table 18.1 Airplane Operational Characteristics for FAA Airport Reference Coding System

Approach Category (Speed, Knots)	Airplane Design Group (Wing Span, ft)					
	I Less than 49	II 49 to 78	III 79 to 117	IV 118 to 170	V 171 to 213	VI 214 to 261
A (less than 90)	A-I*	A-II*,†	A-III†	A-IV†		
B (91 to 120)	B-I*,†	B-II*,†	B-III†	B-IV†,‡		
C (121 to 140)	C-I†	C-II*,†	C-III†	C-IV†,‡	C-V‡	C-VI‡
D (141 to 165)	D-I†	D-II†	D-III†,‡	D-IV†,‡	D-V‡	
E (166 or more)		E-II†	E-III†,‡			

*Small airplanes (12,500 lb or less maximum takeoff weight). Examples:

A-I: Cessna 177 Cardinal   A-II: DHC-6 Twin Otter
B-I: Beech 100 King Air   B-II: Beech 200 Super King Air
C-II: Rockwell 980

†Large airplanes (more than 12,500 lb maximum takeoff weight). Examples:

A-II: Dassault 941   A-III: DHC-8 Dash 8-300   A-IV: Lockheed 1649 Constellation
B-I: Mitsubishi 300 Diamond   B-II: Cessna III Citation   B-III: BAe 146   B-IV: MDC DC-7
C-I: Gates 55 Learjet   C-II: Grumman III Gulfstream   C-III: Boeing 737 500   C-IV: Boeing 757
D-I: Gates 36A Learjet   D-II: Grumman IV Gulfstream   D-III: BAC III 500   D-IV: Boeing 707-200
E-II: Lockheed SR-71 Blackbird

‡Heavy airplanes (takeoff weight of 300,000 lb or more). Examples:

B-IV: Ilyushin Il-76
C-IV: Airbus A-300-B4   C-V: Boeing 747-SP   C-VI: Lockheed C-5B Galaxy
D-III: BAC/Aerospatiale Concord   D-IV: Boeing 777   D-V: Boeing 747-400
E-III: Tupolev TU-144

through conveniently located district offices. Liaison should be effected with the appropriate FAA office to ensure full consideration of FAA policies and procedures.

18.4.1 Airport Master Plans

In the event that a full master planning study has not been made for an existing or proposed airport, such a study might well precede the planning of an improvement to that airport. If a master plan study has been undertaken, it can be used as the basis for further planning, or it can be reappraised. The master plan presents the planner's conception of the ultimate development of a specific airport, together with priority phasing, cost estimates, and financial plan. The master plan should be reevaluated periodically to maintain its validity.

To be eligible for Federal funding, an airport must be included in the National Plan of Integrated Airport Systems (NPIAS), described in Art. 18.4.2. It also must have an FAA-approved airport layout plan (ALP). This is a scaled drawing of existing and proposed land and facilities necessary for airport operations and development. All airport development carried out with Federal financial assistance must be done in accordance with the FAA-approved ALP. To the extent practicable, this plan should conform to the FAA airport design standards existing at the time of its approval. See also Arts. 18.3 and 18.4.3.

Table 18.2 Federal Aviation Administration Standards for Airport Design[a]

Item		Airports Serving Aircraft Approach Categories[b] A and B — Airplane Design Group[c]					Airports Serving Aircraft Approach Categories[b] C and D — Airplane Design Group[c]					
		I[d]	I	II	III	IV	I	II	III	IV	V	VI
Length, ft.												
Runway[e]		2,800	3,200	4,370	5,360	6,370	5,490	6,370	7,290	9,580	10,700	12,000
Runway safety area	$<\frac{3}{4}$[f]	600	600	600	800	1,000	1,000	1,000	1,000	1,000	1,000	1,000
(beyond runway end)	$\geq\frac{3}{4}$[g]	240	240	300	600	1,000						
Runway object-free area	$<\frac{3}{4}$[f]	600	600	600	800	1,000	1,000	1,000	1,000	1,000	1,000	1,000
(beyond runway end)	$\geq\frac{3}{4}$[g]	240	240	300	600	1,000						
Width, ft.												
Runway	$<\frac{3}{4}$[f]	75	100	100	100	150	100	100	100	150	150	200
	$\geq\frac{3}{4}$[g]	60	60	75	100	150						
Runway safety area	$<\frac{3}{4}$[f]	300	300	300	400	500	500	500	500	500	500	500
	$\geq\frac{3}{4}$[g]	120	120	150	300	500						
Runway object-free area	$<\frac{3}{4}$[f]	800	800	800	800	800	800	800	800	800	800	800
	$\geq\frac{3}{4}$[g]	250	400	500	800	800						
Taxiway		25	25	35	50	75	25	35	50	75	75	100
Taxiway safety area		49	49	79	118	171	49	79	118	171	214	262
Taxiway object-free area		89	89	131	186	259	89	131	186	259	320	386
Taxilane object-free area		79	79	115	162	225	79	115	162	225	276	334
Minimum distance between, ft.:												
Center lines of parallel[a] runways[h]		See Advisory Circular 150/5300-13, Chapter 2										
Center lines of runway and	$<\frac{3}{4}$[f]	200	250	300	350	400	400	400	400	400	450	600
center line of taxiway	$\geq\frac{3}{4}$[g]	150	225	240	300	400	300	300	400	400	450	600
Center line of runway and	$<\frac{3}{4}$[f]	400	400	400	400	500	500	500	500	500	500	500
aircraft parking area	$\geq\frac{3}{4}$[g]	125	200	250	400	500	400	400	500	500	500	500
Center line of taxiway and aircraft parking apron		45	45	66	93	130	45	66	93	130	160	193
Center line of parallel taxiways		69	69	105	152	215	69	105	152	215	267	324
Center line of runway to	$<\frac{3}{4}$[f]	875	875	875	875	875	875	875	875	875	875	875
building line or obstruction[i]	$\geq\frac{3}{4}$[g]	600	600	600	600	600	713	713	713	713	713	713
Center line of taxiway to obstruction		45	45	66	93	130	45	66	93	130	160	193
Maximum runway grades[j], %:												
Longitudinal		2.0	2.0	2.0	2.0	2.0	1.5	1.5	1.5	1.5	1.5	1.5
Transverse[k]		2.0	2.0	2.0	2.0	2.0	1.5	1.5	1.5	1.5	1.5	1.5

[a]"Airport Design," FAA Advisory Circular 150/5300-13, Change 6.

[b]Aircraft Approach Categories are described in Art. 18.3.

[c]Airplane Design Group is described in Art. 18.3.

[d]Represents airports serving only small airplanes (an airplane of 12,500 lb or less maximum certificated takeoff weight).

[e]Runway lengths assume an airport elevation of 1000 ft above mean sea level (MSL) and a mean daily maximum temperature of 85 degrees in the hottest month. Actual runway lengths should be based on the selected design airplane adjusted for the local condition of elevation, temperature, and runway gradient. The lengths shown are representative of a runway that can accommodate selected airplanes found in the indicated Airport Reference Code (ARC). Runway length for airplanes over 60,000 lb is usually determined based on the amount of fuel needed to fly a certain distance or haul length and may need to be increased from that determined above.

[f]With approach visibility minimum less than $\frac{3}{4}$ mile.

[g]With approach visibility minimum greater than or equal to $\frac{3}{4}$ mile.

[h]Dual simultaneous precision instrument approaches normally require parallel runway center-line separartion of 4300 ft. A minimum distance of 3400 ft may be used if special radar and monitoring equipment is used. Simultaneous instrument flight regulation (IFR) operations to parallel runways are not authorized for nonprecision instrument approach procedures. Simultaneous precision instrument approach procedures serving parallel runways spaced 2500 ft require radar controlled approaches and departures. Consult with FAA.

[i]The numbers represent a building restriction line (BRL) that encompasses the runway protection zones and runway object-free area. The BRL should also encompass the runway visibility zone, NAVAID critical areas, areas required for terminal instrument procedures, and airport traffic control tower clear line of sight.

[j]Taxiway grades should be held to the same maximum grades as runways.

[k]Gradient shown is for pavement. To improve runoff, shoulder slopes may be increased to 5.0% for a distance of 10 ft from the edge of pavement, then continue at 5% maximum for approach categories A and B and 3% for approach categories C and D.

18.4.2 National Plan of Integrated Airport Systems (NPIAS)

Through constant research, the Federal Aviation Administration, Department of Transportation, has developed criteria for determining the aeronautical potential of a community and translating that potential into airport requirements. The overall airport needs of a community are summarized in the NPIAS, published by the FAA. For existing and proposed airports, the plan shows the type of activity forecast and the general facilities required to accommodate the activity. A brief text spells out broad items of recommended development.

In the past, communities to receive passenger service were certified by the Civil Aeronautics Board (CAB). The Airline Deregulation Act of 1978, however, called for the "sunset" of the CAB by the end of 1984. During the final years of the CAB and in the aftermath of deregulation, air carriers were permitted to change routes without government approval. As a result, the carriers dropped many unprofitable routes. With the end of the CAB, the few remaining essential functions performed by the CAB were transferred to the Department of Transportation (DOT). DOT oversees the Deregulation Act "Essential Air Service" provisions, which authorize subsidized passenger air service to some smaller communities. The historic trend of the number of enplaned airline passengers related to other factors can indicate a community's air carrier potential.

The number of based aircraft at an airport is an indication of the general aviation potential. At air carrier airports, the requirements for facilities to serve scheduled operations are greater than those for general aviation. Consequently, the overall needs of general aviation are usually met at airports that are developed to serve scheduled activity. Thus, the requirements of general aviation become a controlling factor only at airports that are not served by, or built for, the scheduled airline service.

The FAA changed the method of classifying airports in 1982. It now lists them in four major categories, which identify the broad functional mission of each airport in the NPIAS by relating the mission to the service level, including commercial service (primary and reliever) and general aviation airports (Table 18.3).

18.4.3 Airport Layout Plan

Every airport should have a layout plan showing ultimate development, even though construction is to be in stages. Such a plan is desirable to ensure an orderly development and an economical and functionally sound airport. All major components should be worked out in advance.

The airport layout plan is the basic element of the airport's master plan and shows all existing and proposed facilities, property lines, topography, utilities, airport approach surfaces, and runway protection zones, in addition to the ultimate runway and taxiway layout. The ultimate plan will provide a basis for acquiring ample land and for determining zoning required to protect future approaches. The plan should be flexible enough to permit modifications between stages of construction to meet the changing demands of air transportation.

18.4.4 Airport Zoning

In the planning of any airport, it is important that sponsors work closely with local communities and their planners to develop and implement sound land use compatibility plans. This also requires that all existing obstructions to air navigation be cleared or marked and lighted and that future obstructions be prevented. Where legally possible, steps should be taken to adopt appropriate airport zoning legislation to prevent the establishment of obstructions to air navigation. Ideally the zoning will be developed concurrently with the layout plan. If comprehensive zoning is in force or can be instituted, height restrictions and land use can both be incorporated.

18.4.5 Environmental Impact

Airport development is subject to state and Federal regulations that require careful consideration of environmental, ecological, and sociological matters in planning and construction. It is likely that preparation of an Environmental Impact Assessment Report will be required for airport development involving airport location, new runways, major runway extensions, runway strengthening if it might result in increased aircraft noise, adverse effects on the capacity of existing roads, certain land acquisitions, establishment or relocation of an instrument landing system or an approach lighting system. Such a statement should include, among other things, a description of the project and a

Table 18.3 National Airport Classification System—Aeronautical Activity Levels for Functional-Role Airport Classification System*

NPIAS[†] airport category	Classification description
Commercial service	Annual scheduled passenger service
Primary	Percent of total U.S. enplanements
Large hub (L)	1% or more
Medium hub (M)	0.25–1.00%
Small hub (S)	0.05–0.25%
Non-hub (N)	0.01–0.05%
Other	2500 or more but less than 0.01%
Reliever[‡]	Must have at least:
	50 based aircraft, or
	25,000 annual itinerant operations, or
	35,000 annual local operations
General aviation	All other civil airports

*Established by the Federal Aviation Administration.

[†]National Plan of Integrated Airport Systems.

[‡]Intended to reduce congestion at large commercial service airports by providing general aviation pilots with alternative landing areas and by providing more general aviation access to the overall community.

discussion of its purpose, impact on the natural environment, and impact on the human environment. Also, it should include alternatives to the proposed development, unavoidable adverse impact, short-term effects, long-term impact, irreversible or irretrievable commitments of resources, and long-term benefits. The statement should be carefully prepared, thorough and complete, unbiased, and clear, so that its review by many public bodies and agencies will not be unduly protracted.

18.4.6 Airport Construction Plans

Construction plans for airports should include a location and site plan, airport layout plan, grading and clearing plan, borings and soils-exploration plot, grading and drainage plan, runway and taxiway profiles, access-road plans and profiles, drainage-line profiles, pavement cross sections, drainage structures, lighting and conduit plan, landscaping plan, and summary of construction quantities. Plans are also required for development of terminal area and parking lots and for construction of terminal buildings.

18.5 Obstruction and Clearance Criteria for Airport Approaches

The FAA has established standards for determining obstructions to airports in Part 77 of the Federal Aviation Regulations. These standards set up **civil imaginary surfaces** (Fig. 18.1 and Table 18.4). Objects that extend above these surfaces are considered obstructions and should be removed or marked and lighted, depending on the nature of the obstruction and the feasibility of its removal.

The **airport reference point** is a centrally located point that defines the geographic location of the airport. The **primary surface** corresponds to a **landing surface**; it is a surface with a width of 250, 500, or 1000 feet depending upon the weight class of the airplane to be accommodated longitudinally centered on a runway and extending 200 ft beyond each end of the runway. The **horizontal surface** is a horizontal plane 150 ft above the established airport elevation (the highest point on the landing surfaces). It is bounded by a **conical surface**, which has a width of 4000 ft and rises on a 20:1 slope.

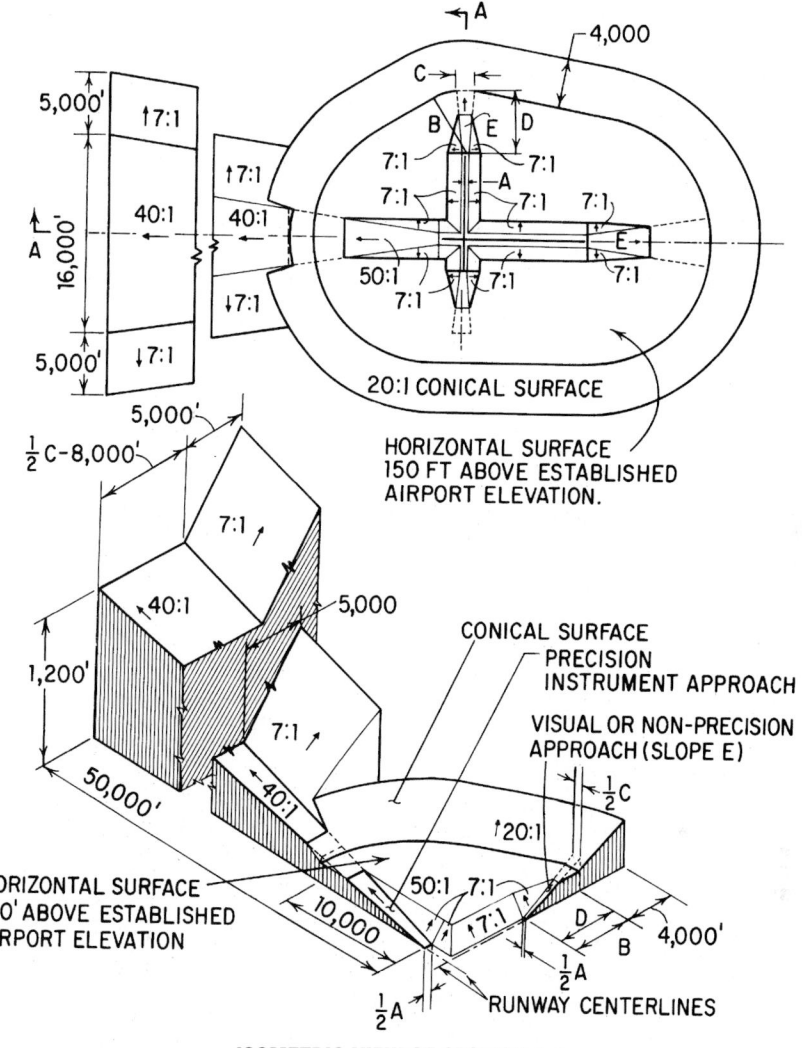

Fig. 18.1 Airport imaginary surfaces for determining obstructions. (*Federal Aviation Administration.*)

Approach surfaces are longitudinally centered on runway center lines extended outward from the primary surface. The dimensions and slopes vary, depending on the nature of the runway involved (Table 18.4). From the sides of the approach surfaces, transitional surfaces extend outward at 7:1 until they intersect the horizontal or conical surfaces. The transitional surface at each end of a precision instrument runway extends beyond the conical surface for the remaining length of the approach surface and has a width of 5000 ft.

All feasible steps should be taken to insure adequate protection of airports from obstructions above these imaginary surfaces.

18.5.1 Runway Protection Zones

These are land areas the function of which is to enhance protection of people or property on the ground from aircraft operation. Runway protec-

Table 18.4 Criteria for Airport Imaginary Surfaces for Determining Obstructions*

Dimension	Item	Dimensional standards, ft (see Fig. 18.1)					
		Visual runway		Nonprecision instrument runway			Precision instrument runway
					Other than utility runways		
		Utility runaways‡	Other than utility runways	Utility runways	Visibility minimums greater than ¾ mi	Visibility minimums as low as ¾ mi	
A	Width of primary surface and width of approach surface at inner end	250	500	500	500	1,000	1,000
B	Radius of horizontal surface	5,000	5,000	5,000	10,000	10,000	10,000
C	Approach surface width at end	1,250	1,500	2,000	3,500	4,000	16,000
D	Approach surface length	5,000	5,000	5,000	10,000	10,000	†
E	Approach slope	20:1	20:1	20:1	34:1	34:1	†

*Federal Aviation Administration.

†Precision instrument approach slope is 50:1 for inner 10,000 ft and 40:1 for an additional 40,000 ft.

‡Runways expected to serve propeller-driven airplanes with maximum certificated takeoff weight of 12,500 lb or less.

tion zones require elimination of objects and activities that are incompatible with airport operations. They also designate areas on which are prohibited land uses such as residences and places of public assembly, including churches, schools, hospitals, office buildings, shopping centers, and theaters.

Runway protection zones lie directly beneath the inner portions of runway approach surfaces (Fig. 18.2 and Table 18.5). The standard configurations of runway protection zones conform to the inner dimensions of approach surfaces. Zone length is a function of the type of aircraft and approach visibility minimums for the runway (Table 18.5).

Airport authorities should control sufficient property in the runway protection zone to provide for unobstructed passage of aircraft landing or taking off. All obstructions should be cleared and creation of future obstructions should be prohibited. Although protected areas should be completely cleared, grading of the areas is not necessary.

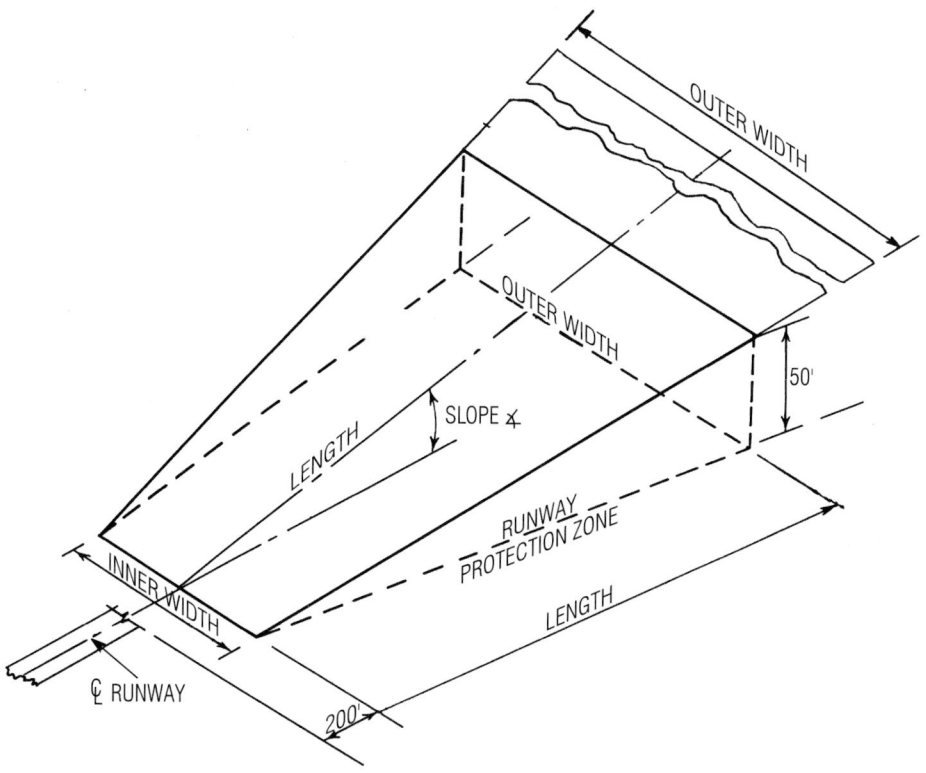

Fig. 18.2 Runway protection zones and approach surfaces. (*Federal Aviation Administration.*)

Also, although ownership of the areas is desirable, zoning or aviation easements give the necessary protection.

18.5.2 Clearance of Obstructions

To test approach zones for clearance of obstructions, a topographic map of the airport site and its environs is required for a radius of at least 4 mi from the airport boundary. A convenient test method is to prepare a transparent template showing the extension of the runway center line, the limits of the runway approach surface, and contour lines representing the elevations of the sloping runway approach surface and 7:1 transition surface. For an instrument-runway approach, the transparent template (Fig. 18.3) is fitted

to the end of each runway, and the ground-surface contours are compared with those of the runway approach surface. Any high places or created features on the ground that will protrude into the runway approach surface are noted. The runway layout is adjusted, if necessary, to avoid obstacles with a minimum sacrifice of wind coverage.

The horizontal surface clearances, 150 ft above the airport are examined in a similar manner. All obstructions above the horizontal surface are spotted. Measures should be taken to remove as many obstructions as possible and to mark and light those that cannot be removed.

Detailed plans should be made of critical areas in approach zones. The plans should show heights of trees, poles, buildings, etc. that come near the runway approach surface. Steps should then be taken to obtain control of these areas by easement

Table 18.5 Dimensions for Runway Protection Zones

Approach Visibility Minimums*	Facilities Expected to Serve	Length L, feet (meters)	Inner Width W_1, feet (meters)	Outer Width W_2, feet (meters)	RPZ, acres
Visual and not lower than 1-Mile (1,600 m)	Small aircraft exclusively	1,000 (300)	250 (75)	450 (135)	8.035
	Aircraft approach categories A & B	1,000 (300)	500 (150)	700 (210)	13.770
	Aircraft approach categories C & D	1,700 (510)	500 (150)	1,010 (303)	29.465
Not lower than 3/4 Mile (1,200 m)	All aircraft	1,700 (510)	1,000 (300)	1,510 (453)	48.978
Lower than 3/4 Mile (1,200 m)	All aircraft	2,500 (750)	1,000 (300)	1,750 (525)	78.914

*The RPZ dimensional standards are for the runway end with the specified approach visibility minimums. The departure RPZ dimensional standards are equal to or less than the approach RPZ dimensional standards. When a RPZ begins other than 200 feet (60 m) beyond the runway end, separate approach and departure RPZs should be provided. Refer to appendix 14 for approach and departure RPZs.

Small airplane: Airplanes of 12,500 lb or less maximum certificated takeoff weight.

Large airplane: Airplanes of more than 12,500 lb maximum certificated takeoff weight.

or purchase, so that the obstructions may be removed. Clearances for railroads and highways are shown in Fig. 18.4.

18.6 Airport Site Selection

Before investigating possible sites in detail for an airport, the engineer should assemble certain background data. These include U.S. Geological Survey topographic maps, aerial photographs in stereo-pairs for studying relief and culture, available soils maps and analyses, and overall development plans for the area. Data on winds and weather should be obtained from the most reliable sources possible. It is desirable to get complete weather information for a period of at least 10 years.

The engineer should establish liaison with appropriate representatives of the FAA, the state aviation agency, local and area planning groups, and the aeronautical interests that can be expected to use the airport. Finally, there must be evaluations, projections, and studies to develop forecasts of the volumes and types of anticipated activity and to establish the general size, character, and scope of the airport. With such information, a reconnaissance of the area can be made and the most likely sites identified for further study.

18.6.1 Physical Site Characteristics

Selection of an airport site is influenced by a number of physical factors. These can affect the utility of the airport and the economy of its development.

Adequate area must be provided to accommodate an airport of the type required and oriented for prevailing winds. The area is determined by the runway length and runway layout and by terminal-area requirements. A small airport may be located on 50 to 100 acres.

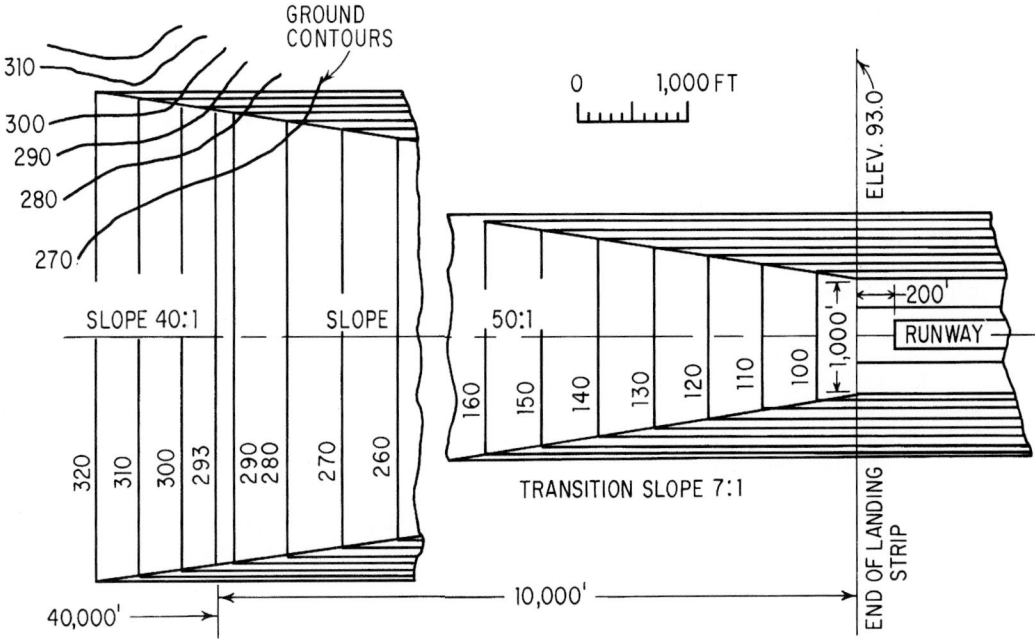

Fig. 18.3 Template for checking approach-zone clearance for instrument runways. Similar templates can be developed for noninstrument runways.

A large international airport may cover as much as 15,000 to 40,000 acres.

Possibility for expansion should be ensured by the selection of a site that is not constrained by built-up property, railroad yards, mountains, rivers, harbors, or other features that prohibit enlargement except at excessive cost. Although initial acquisition should include all land needed for ultimate development, there should be ample undeveloped land available adjacent to the site. This land should be protected by zoning against uncontrolled growth of industrial or residential

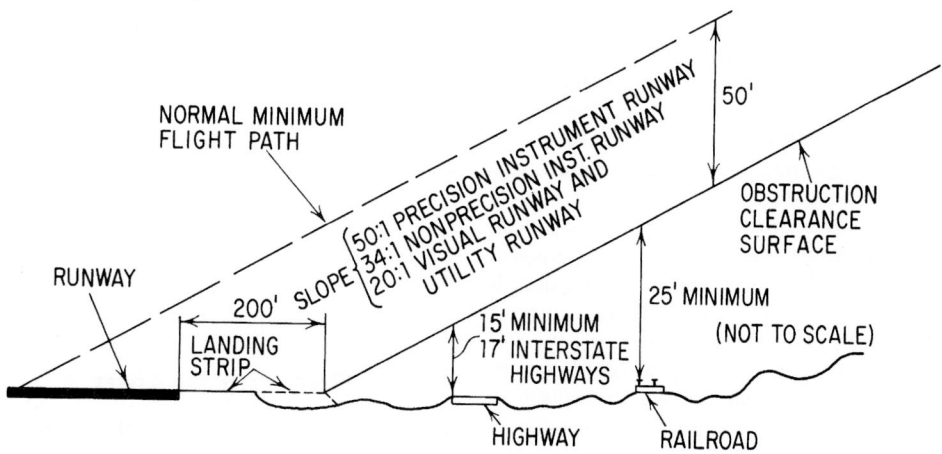

Fig. 18.4 Vertical profile along extended center line of runways shown minimum clearance required by the Federal Aviation Administration over highways and railways.

property that will block runway extensions or terminal-area expansion.

Terrain should be relatively flat to avoid excessive grading costs. Elevated sites are preferable to those in lowlands because they are usually free from obstructions in approach zones, less subject to fog and erratic winds, and easy to drain.

Soils should be studied and evaluated for their effect on grading, drainage, and pavements. The nature of the soil influences the cost of construction. Ideally, the site should be cleared ground that is easily drained and has sandy or gravelly soil that offers a satisfactory foundation for runway pavements without excessively thick subbases and costly subdrainage systems.

Drainage characteristics of the site should be investigated to ascertain the possibility of floods and the existence of high water tables. Natural drainage is most desirable. The ability to dispose of storm water should also be evaluated.

Air approaches to the proposed airport should be free of obstacles, such as mountains, hills, tall buildings, transmission lines, chimneys, and towers.

(A. T. Walls, "Airport Planning and Management," McGraw-Hill, Inc., New York.)

18.6.2 General Site Characteristics

In addition to the physical characteristics of an airport site, factors of a more general nature require consideration.

Accessibility to the community is essential, to preserve the speed advantage of air transportation. In general, accessibility is measured in time rather than distance. Sites near modern express highways are to be sought, and those bounded by traffic-congested streets avoided. On the other hand, the site should not be so remote from the community as to require excessive transportation time.

Availability of utilities, such as electric power, gas, telephone, water, sewers, and public transportation, is an important factor to be investigated. If these utilities are not available, the cost of providing them must be considered.

Control of the site and its surroundings by zoning should be investigated to ensure protection of aerial approaches and possibility of expansion. If the airport is located outside the community to be served, the means of guaranteeing proper control should be determined.

Compatibility with local and area planning is an important characteristic. It should be explored so that the airport and the area can develop without one interfering with the other. The effect on land values and tax assessments may be adverse or beneficial, depending on the nature of the site. If the airport is located near residential property, the value of that property could be affected because of the commercial nature of some types of airports. If located in an undeveloped area, the airport will increase the value of adjacent land for industrial sites and for other uses related to the airport. The possible impact of aircraft noise should be assessed.

Spacing of the airports is a consideration since airports should not be located so that air-traffic patterns interfere. Approval of the FAA is necessary to ensure air-space compatibility. This approval should be obtained before a final commitment is made for a specific airport site.

18.6.3 Site Evaluation

Having identified the most likely sites in an area, the engineer should review them on the basis of physical and general characteristics. It is not likely that any one site will possess all the desirable characteristics. Thus, it is necessary to evaluate the good and bad features of each site to make the best selection.

Preliminary runway patterns should be tested, approaches checked, real estate evaluated, and construction costs analyzed. The more promising sites can be evaluated in the field, and specific soil and topographic data developed. Before final selection is made, the engineer should ascertain that the most favored site will receive FAA airspace clearance, that an acceptable master plan can be developed for that site, and that it offers maximum compatibility with area planning.

18.7 Runway Design

Runways are the focal points of an airport. They must have a length and width adequate to accommodate the aircraft to be served. (See also Table 18.2.)

18.7.1 Runway Lengths

To determine the runway length required for a given airport location, the engineer should take into account the takeoff and landing performance of the most critical aircraft expected to make regular use of the airport. Aircraft performance decreases with increase in distance to be flown from the airport, airport elevation, runway gradient, and air temperature. The runway length chosen should be thoroughly reviewed and validated.

The Federal Aviation Administration (FAA) issues advisory circulars from time to time, giving performance data on aircraft that supplement its engineering data. The safe runway length for transport aircraft is based on Federal Aviation Regulations (Part 25), which specify three requirements for civil air transports, each of which must be met:

1. Runway lengths should be sufficient for airplanes to accelerate to the point of takeoff and then, in case of failure of a critical engine, to be braked and brought to a stop within the limits of the runway (or usable landing strip).

2. If failure of a critical engine occurs at point of takeoff, airplanes should be capable of takeoff with one or more operating engines. Aircraft powered by reciprocating engines should be able to clear the end of the runway at an elevation of 50 ft and those powered by turbine engines, at an elevation of 35 ft.

3. In landing, airplanes should clear the approach end of the runway by 50 ft and be able to touch down and stop within 60% of the available runway length.

Data published on runway requirements for transport aircraft usually incorporate the preceding so that no additional computation is required, except for effective gradient (Art. 18.7.2).

Normal requirements for landing of jet aircraft establish runway lengths that are valid only for normal instrument conditions. For jet airliners to land at lower weather minimums, runways should provide a landing length more than that normally required. Generally, the additional requirement will still be less than the required takeoff length.

Runway length requirements are established for instrument operations for runway visual ranges (RVR) from 2400 ft down to 1200 ft. The equivalent of a 100-ft ceiling and 0.25 mi visibility. With electronic and visual landing aids of greater integrity, weather minimums may be lowered. All-weather operations are the ultimate goal. The corrected landing length should be checked against required takeoff length to ascertain that an adequate length is provided if lower RVR operations can be forecast.

Future needs for a new runway at an existing airport or the need for an entirely new airport should be determined only after thorough study and review of the requirements to meet the anticipated demand. The study process should account for all factors that impact full use of a runway for the design airplane. These include length, width, and specially designated areas free of obstacles to provide an interconnected system of air space and land surfaces for the safe landing and takeoff of aircraft. To control areas off the airport from impacting the air space or approach surfaces, airport owners must have the authority to prohibit potential obstructions and incompatible land uses. The intent of airport planning, in this regard, is to maximize utilization and retention of paved areas on airports. Without proper planning, runways are subject to encroachment by obstructions or incompatible land uses that may restrict or preclude future use of the runways.

To meet obstacle clearance requirements for approaches to existing runways, where control of encroachments has not been possible, the threshold for an affected runway may be displaced or relocated if this is determined to be the only practical alternative. A **displaced threshold** reduces the length of runway available for landings. The portion of the runway behind a displaced threshold is available for takeoffs in either direction and for landings from the opposite direction. A **relocated threshold** is different in that the runway is not available for landings or takeoffs at that end of the runway.

Another way to increase runway utilization at constrained airports is to use *declared distances*. These provide an alternative design procedure in which distances are specified that satisfy requirements for aircraft takeoff run, takeoff distance, accelerate-stop distance, and landing distance. The application of declared distances at a specified location requires prior FAA approval, which is given on a case-by-case basis. Approval must be reflected on the FAA-approved Airport Layout Plan.

18.7.2 Runway Grades

Aircraft performance is influenced by the gradient of the runway. Ascending grades increase power required for takeoff. Descending grades increase braking distance. Not only is the gradient at any point along the runway of concern, but also the effective gradient of the overall runway. Other factors that influence grades are the sight distance and the transverse slopes of graded areas.

Longitudinal grades for airports serving aircraft in approach categories C and D should not exceed 1.50% at any point on the runway profile, but a 2.0% maximum is allowable for airports serving aircraft in categories A and B (Table 18.2).

Runway length determined for the critical aircraft at the elevation and mean temperature of the airport site is further increased at the rate of 20% for each 1.0% of effective gradient.

Longitudinal grade changes should be avoided. If changes are necessary, they should be in accordance with Table 18.6, which shows maximum grade changes and minimum length of vertical curves.

Minimum runway sight distances are necessary to permit safe visual aircraft operations. At noncontrolled airports, runway grade changes should be such that there will be an unobstructed line of sight from any point 5 ft above the center line of the runway to any other point 5 ft above the runway. If the airport has an operating control tower, adherence to longitudinal gradient standards for runways will provide an adequate line of sight from the tower to the runways.

A graded safety area 240 to 1000 ft long is required at each runway end, depending on aircraft approach category and design group (Table 18.2). The associated width required, also shown in

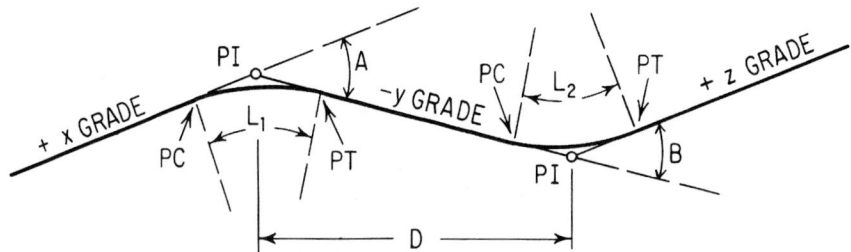

Fig. 18.5 Vertical profile around runway center line shows changes in longitudinal grades. (*Federal Aviation Administration.*)

Table 18.6 Vertical Curve Data and Maximum Grade Changes for Runways

	Runways serving categories A and B airplanes	Runways serving categories C and D airplanes
Maximum gradient at ends of runaway, such as x grade or z grade (Fig. 18.5)	0 to 2.0%	0 to 0.8%, first and last quarter of runway length
Maximum gradient in middle portion of runway, such as y grade (Fig. 18.5)	0 to 2.0%	0 to 1.5%
Maximum grade change, such as A or B (Fig. 18.5)	2.0%	1.5%
Minimum length of vertical curve L_1 or L_2 (Fig. 18.5) for each 1.0% grade change	300 ft*	1000 ft
Minimum distance between points of intersection for vertical curves, D (Fig. 18.5)	$250(A + B)$ ft[†]	$1000(A + B)$ ft[†]

*Vertical curves not required at utility airports for grade changes less than 0.4%.

[†]A% and B% are successive changes in grade.

Table 18.2, varies from 120 to 500 ft. For the first 200 ft, measured from the runway ends, the safety area should slope between 0 and 3% downward along the longitudinal axis of the runway. For the rest of the safety area, the maximum longitudinal grade should be selected to prevent any part of the runway area from penetrating the approach surface or clearway plane. The maximum downward slope permitted in the safety area is 5%. Longitudinal grade changes are limited to 2% per 100 ft up or down.

Transverse grades on runways should not exceed 2% for approach categories A and B and 1.5% for categories C and D (Table 18.2).

Unpaved shoulders may have a steeper slope to improve runoff. The first 10 ft of shoulder adjoining the pavement may be as steep as 5%. The transverse grade of shoulder beyond the 10-ft distance may be as steep as 5% for approach categories A and B and not more than 3% for categories C and D.

Graded shoulders should be set $1\frac{1}{2}$ in below the adjoining pavement edge to preclude future turf from developing a gutter that would impound water at the pavement.

18.7.3 Runway Numbering System

The runways at each airport are designated by numbers related to azimuth, measured clockwise from magnetic north. For simplicity, the numbers are expressed in 10° units of azimuth.

For example, if a runway has an azimuth measured from magnetic south of 32°, the southerly end is numbered 21 since $(32° + 180°)/10° = 21.2$. The other end is numbered 3 since $32°/10° = 3.2$. The runway is referred to as 3-21.

The object of the system is to have the number facing a landing airplane correspond (in 10° units) to the compass course of the airplane. Where there are parallel runways, the runway on the right of the landing airplane is designated with an R (right); the other is designated L (left). For example, if there were a runway parallel to 3-21, the runway would be 3R-21L or 3L-21R.

18.7.4 Runway Layout

Choice of runway pattern is influenced by the necessity of obtaining clear approaches, the desirability of providing maximum wind coverage,

and the necessity for fitting the layout to the topography so as to secure low grading and drainage costs. Shape and location of the terminal area also influence the layout. Furthermore, short and direct taxiing distances are desired between runways and the airport terminal.

The number of runways will depend on wind coverage and traffic volume to be handled. To increase capacity, the layout should permit simultaneous use of two or more runways.

Orientation of the runways depends on obstacle clearance requirements and prevailing wind directions. The instrument runway should, if possible, be aligned with the winds that prevail during instrument-flying conditions. Ideally, runway approaches should, if possible, be over sparsely settled or nonresidential areas where the public will be the least inconvenienced by aircraft operations.

18.7.5 Wind Coverage

The Federal Aviation Administration (FAA) specifies that runways be oriented with the prevailing winds. The intent is to ensure that aircraft may be landed at least 95% of the time without exceeding the cross-wind capability of aircraft forecast to use the airport on a regular basis. If the runway does not provide 95% wind coverage for the forecast aircraft, then a cross-wind runway may be necessary. Inasmuch as light airplanes are more susceptible to cross winds than heavy airplanes, allowable cross-wind components are specified for runways designed to serve the aircraft of different Airport Reference Codes, as indicated in Table 18.7.

The trend is toward one- or possibly two-directional layouts. In some localities, where the prevailing winds are consistently in one direction

Table 18.7 Allowable Cross-Wind Components for Aircrafts

Aircraft Reference Code	Cross-wind component, knots
A-I and B-I	10.5
A-II and B-II	13
A-III and B-III	16
C-I through D-III	16
A-IV through D-VI	20

or the reverse, a single runway will meet FAA requirements. One-runway layouts are sometimes adopted when wind-coverage requirements are not fully met but the approaches are excellent and other factors are satisfied.

18.7.6 Wind Rose

To determine the orientation of a runway that will offer the greatest wind coverage, a wind rose may be used. A simple type consists of bars radiating in several compass directions, each representing, to scale, the percentage of time that the wind blows from the direction in which that bar points.

For mathematical computation of wind coverage on the basis of cross-wind component, a wind rose similar to that shown in Fig. 18.6 is helpful.

This wind rose gives the percentage of time the wind blows in specified speed ranges as well as in specified directions. The small numbers on the diagram represent the percentages of time the wind blows from the several compass directions between specified velocities. For the wind rose in Fig. 18.6, the percentages of winds were known for velocity ranges of 0 to 3.5 knots (calm), 3.5 to 13 knots, 14 to 27 knots, 28 to 41 knots, and over 41 knots. Winds over 41 knots accounted for less than 0.1% and were neglected.

This wind rose may be used to determine the maximum wind coverage for a one-, two-, or three-directional runway layout. It may also be used to check the wind coverage for a layout adopted after a study of obstacles in approaches and other factors.

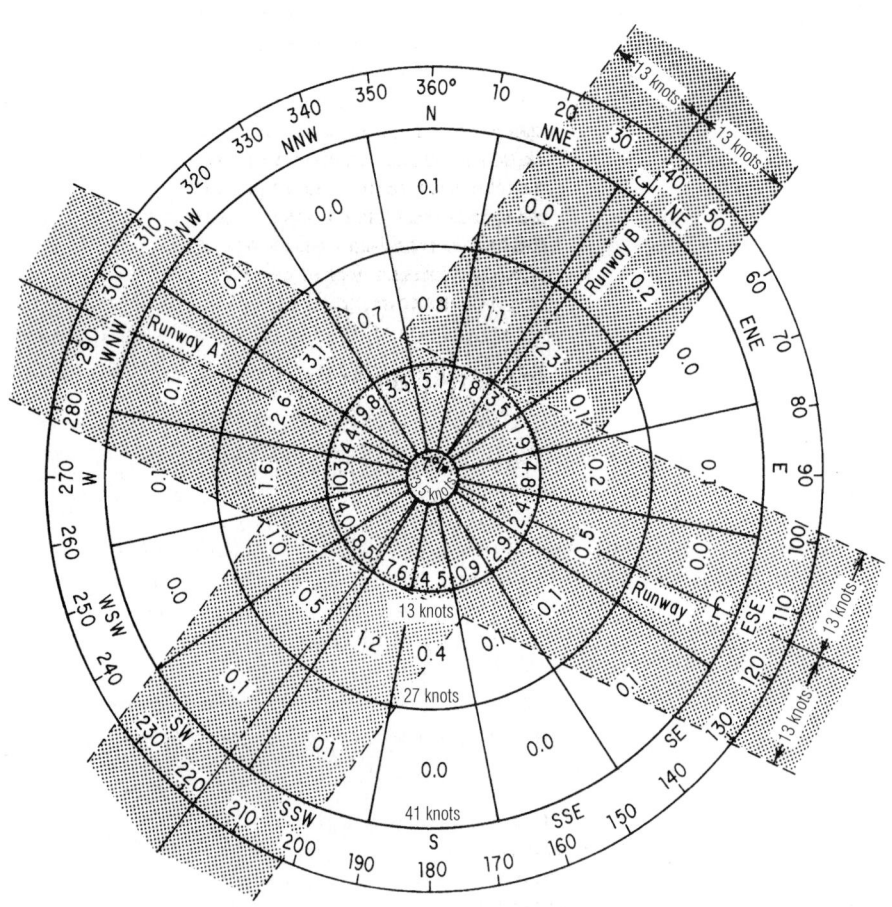

Fig. 18.6 Template aids determination of wind coverage for a cross-wind component of 13 knots.

For finding the maximum wind coverage possible for a given runway, a transparent template is made. On it are drawn the runway center line and parallel lines representing the limits of 13-knot cross-wind components on each side of the center line. This template is then superimposed on the wind rose, with the center line passing through the center of the rose. Next, the template is rotated until a direction is found in which the greatest percentage of wind is included within the 26-knot-wide band.

If the layout has more than one runway, templates are plotted for each runway and shifted about the center of the wind rose until the direction for each runway is found such that the total percentage of wind coverage by all runways is a maximum.

With Fig. 18.6, for example, a two-runway layout is to be checked for wind coverage; first for Runway A alone and then for both Runways A and B. The runway center lines are plotted on the wind rose in their proper compass directions. Lines are drawn parallel to each center line, to represent, to the scale of the wind rose, the limits of all cross-wind components of 13 knots. For simplicity, the percentage of winds not covered is computed and deducted from 100. The percentages and fractions of percentages outside the limits of coverage (dashed lines in Fig. 18.6) for Runway A are as follows: in directions NW to E, $0.4 \times 0.1 + 0.0 + 0.6 \times 0.7 + 0.1 \times 0.9 \times 0.8 + 0.0 + 1.1 + 0.2 + 2.3 + 0.0 + 0.8 \times 0.1 + 0.6 \times 0.1 + 0.1 \times 0.2$; from SE to W, $0.4 \times 0.1 + 0.0 + 0.5 \times 0.1 + 0.0 + 0.9 \times 0.4 + 0.1 + 1.2 + 0.1 + 0.9 \times 0.5 + 0.0 + 0.6 \times 1.0 + 0.6 \times 0.1 + 0.1 \times 1.6 = 8.16$ or 91.84% coverage. The addition of Runway B will add the following coverage: from N to ENE, $0.5 \times 0.8 + 0.0 + 1.1 + 0.2 + 2.3 + 0.0 + 0.6 \times 0.1$ and from S to WSW $0.5 \times 0.4 + 0.8 \times 0.1 + 1.2 + 0.1 + 0.9 \times 0.5 + 0.0 + 0.4 \times 1.0 = 6.49$, giving total coverage for two runways of 98.33%.

The analysis may be refined by using more wind-velocity groups if they are available. It may also be applied for other cross-wind components.

The wind rose usually employed for study purposes is plotted for annual data. In locations where the wind distribution varies during the year, roses should be plotted for the different seasons and the fluctuations taken into account in design, particularly if the airport is used mostly in certain seasons.

For selecting the instrument-runway orientation, a wind rose for low-visibility conditions is useful and can be developed from special studies undertaken by the U.S. Weather Bureau.

18.7.7　Runway Configurations

The simplest layout is a single runway with parallel taxiway and centrally located terminal area as shown by full lines in Fig. 18.7a. Two directions of operation are possible, 6-24 or 24-6 (Art. 18.7.3). Only one landing or takeoff can be made at a time.

Under these conditions, the capacity of the runway is about 50 movements per hour (including both landings and takeoffs). When more capacity is needed, a second parallel runway may be built as shown by dashed lines in Fig. 18.7a.

In this design, the original runway can be used for takeoffs, while the "future" runway is used for landings. The capacity under visual flight rules will be raised to about 70 movements per hour. Landing traffic will have to cross the takeoff runway under control from the tower.

Figure 18.7b shows parallel runways 5000 ft apart. The terminal area lies between the runways. This arrangement has definite operational advantages over the layout in Fig. 18.7a. Taxiways do not cross runways, the terminal area is centrally located with ample room for expansion, and the wide separation of runway approaches will increase capacity under conditions of low visibility since the 5000-ft separation is adequate for simultaneous operations. But the layout in Fig. 18.7b requires a larger area than the one in Fig. 18.7a. The two parallel runways, however, need not be opposite each other. Increasing the offset from the terminal area will decrease taxiing distance but may increase land and construction costs.

Taxiways may be extended to the runway ends to provide exits for incompleted takeoffs, to facilitate landings and takeoffs on the same runway, and to permit simultaneous use of both runways for takeoff or for landing. During peak-hour operations, arrivals and departures are not usually equal, so simultaneous use of both runways for the same type of operation is often desirable.

In Fig. 18.8 an open V-type layout is shown. This layout gives four directions of wind coverage and also allows simultaneous operation of runways in most directions when wind velocities are not unusually high. The traffic diagrams indicate a

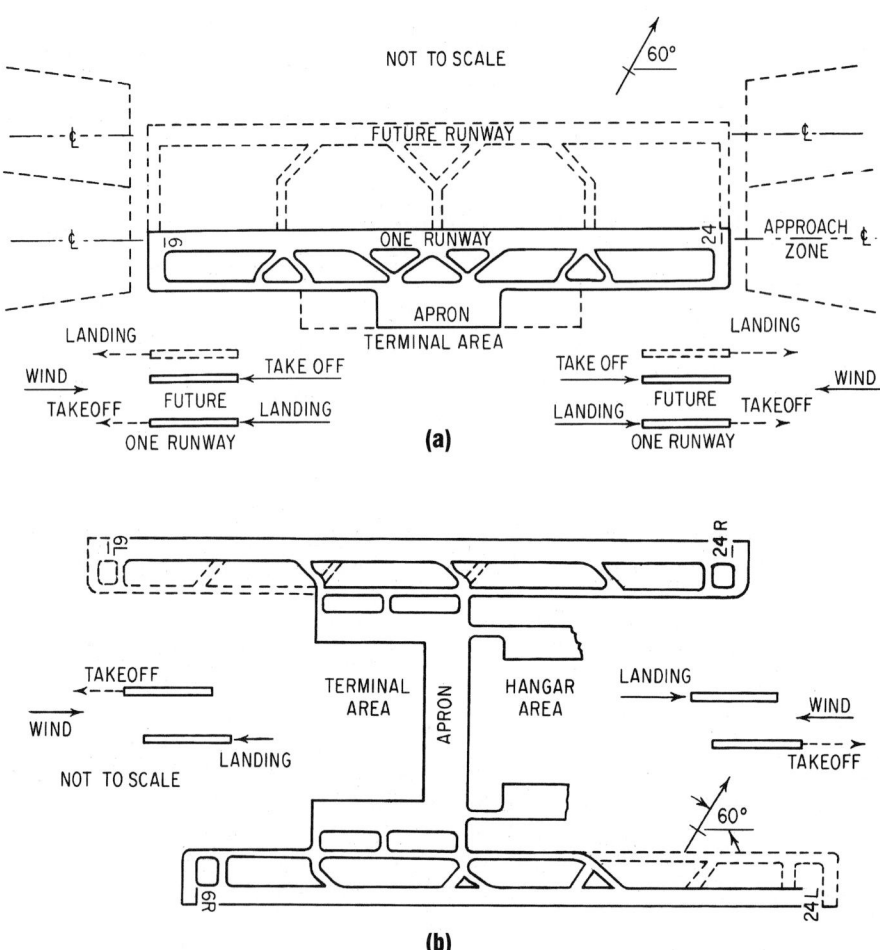

Fig. 18.7 Plan of simple runway layouts: (*a*) Single runway with future parallel runway. (*b*) Two parallel runways.

separation of landings and takeoffs in three or four wind directions. In the one situation where the landing go-around path intersects the takeoff path, the landings and takeoffs will have to be rigidly coordinated.

The V shape permits a centrally located terminal area with room for expansion. In some designs the angle of the V is made about 90°.

When additional capacity is required, the designs in Figs. 18.7*b* and 18.8 may be expanded by building a runway parallel to each of the original runways but 1000 to 3500 ft farther out.

Two runways would then be available for landings and two for takeoffs at all times for the layout in Fig. 18.7*b*, and for most of the time for the layout in Fig. 18.8. The greatest capacity can be obtained from the two sets of parallels for the configuration shown in Fig. 18.7*b*, with a third runway at a divergent angle on each side.

Most existing airports have intersecting runways. At some locations, it is impractical to build nonintersecting runways. When winds are not critical, the capacity of these designs can be improved over single-runway operations by using

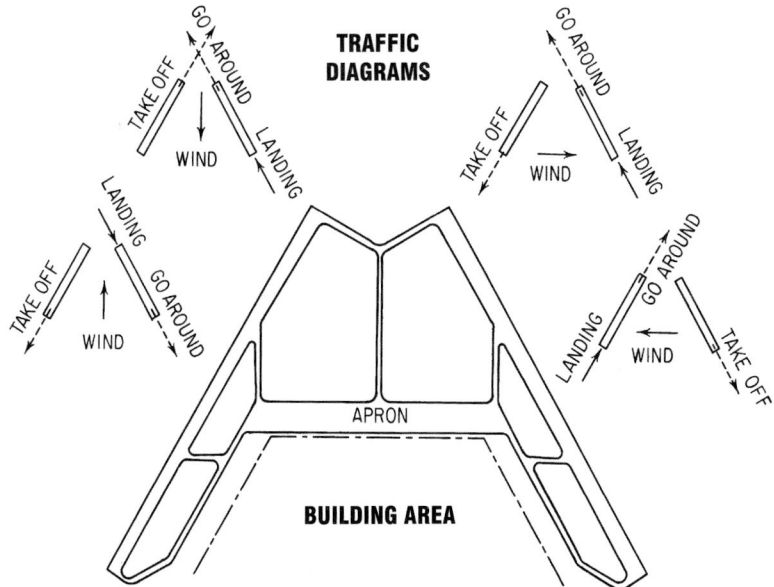

Fig. 18.8 V-type runway layout permits two-directional operation of aircraft. (*Federal Aviation Administration.*)

one runway for takeoffs and another for landings. The movements are alternated under rigid coordination from the air-traffic control tower.

Airport capacity is reduced under instrument-landing procedures, and delays to landings occur. Improvements in air-traffic control, however, have increased landing rates in overcast weather so they nearly equal those in good weather.

18.8 Taxiway Systems

Taxiways are laid out to connect the terminal area with ends of runways for takeoffs and to tap the runways at several points to provide exits for landing aircraft. Landings usually do not require the full length of the runway.

To clear a runway of landing planes as rapidly as possible, easy turns are introduced at exit taxiways (Fig. 18.7). Even faster aircraft exits are obtained when the runway is equipped with the taxiway illustrated in Fig. 18.9. These exit taxiways best serve a variety of aircraft when placed about 2500, 4000, and 6000 ft from the runway threshold.

Where there is a taxiway parallel to the runway, the exit taxiways can lead into the parallel taxiway with a reverse curve that permits the maintenance

of high-speed taxi operations. When applied bidirectionally to the same runway, the effect can be that in Fig. 18.7*a*. At the ends of a runway, the taxiways join the runway at about 90° to give the pilot a view of the runway and its extension in both directions. Additional pavement is added to make room for waiting airplanes and to allow one airplane to pass another in the takeoff sequence. Taxiway widths and clearances are given in Table 18.2. Figure 18.10 shows taxiway intersection details, and Table 18.8 lists standard dimensions.

18.9 Aprons

The apron or "ramp" adjacent to a terminal is used for loading and unloading airplanes, fueling, and minor servicing and checkup. The dimensions of the apron depend on the number of loading positions required and the size and turning characteristics of aircraft. The number of spaces depends on the time of occupancy per aircraft, the time being longer at terminal airports than at en-route stops. In most instances, airlines desire exclusive use of apron positions because of the complex equipment required to service transport aircraft. The resulting need is for a greater number

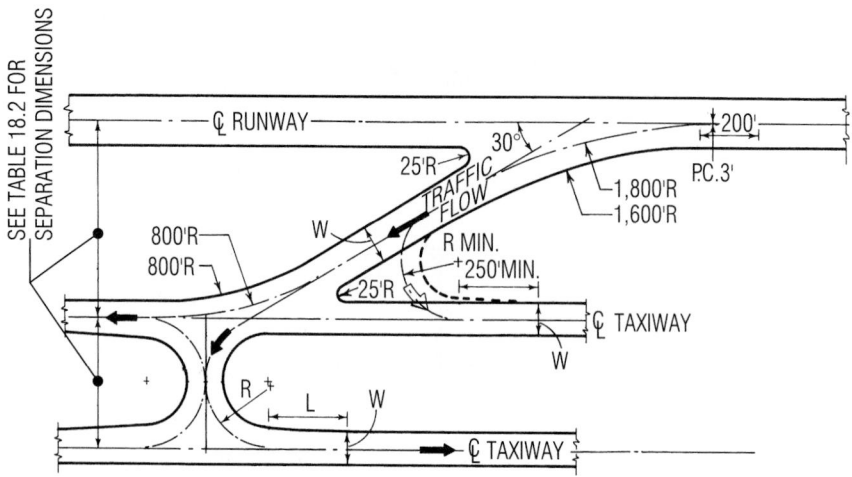

Fig. 18.9 Angled-exit taxiway design with dual parallel and crossover. (*Federal Aviation Administration.*)

of loading positions than would be required if positions were shared.

When determining area requirements for aprons, various methods of aircraft positioning should be explored. The size of airline loading aprons depends on the number and size of aircraft to be accommodated, as determined from a forecast of peak-hour aircraft movements. Aircraft loading positions are designated by circles of varying diameters, depending on wing span, length, and turning radius of the aircraft that will use the airport.

Provision of underground facilities in the apron is a requirement at some airports. At others, services such as fuel, air, power, and telephone are available at the edge of the apron or from the terminal building. Grounding connections should be provided.

18.10 Automobile Parking Areas

Ample parking facilities are required for airport patrons, passengers, employees, and spectators. Public parking should be developed as near the airline terminal as feasible, to minimize walking distance. Most visitors will come on Sundays and when special events occur at the airport.

The parking lot should be designed to handle overflow traffic, or a supplemental lot should

be developed for intermittent use. A design criterion of 150 parked automobiles per acre may be used when estimating the size of parking lot required.

To minimize walking distances, some airports have multilevel parking structures adjoining the terminal. Employee parking facilities are usually separate and more distant.

At busy terminals, temporary storage areas will be needed to park taxicabs, buses, and limousines waiting for turns or for scheduling. Parking might be required for service vehicles such as fuel trucks. There should be adequate truck-parking areas at the terminal for delivery of commodities and supplies.

18.11 Airport Grading and Drainage

A thorough analysis of the soils on an airport site is required for planning of grading and drainage and subdrainage systems and for designing pavements and base courses. Soils testing is also required for the control of compaction of fills and base courses so that there will be no detrimental settlement under heavy airplane loads.

Procedures for soil sampling and testing are much the same as for highways. Samples should be taken at 200-ft intervals along the center lines of planned runways and taxiways, one boring per

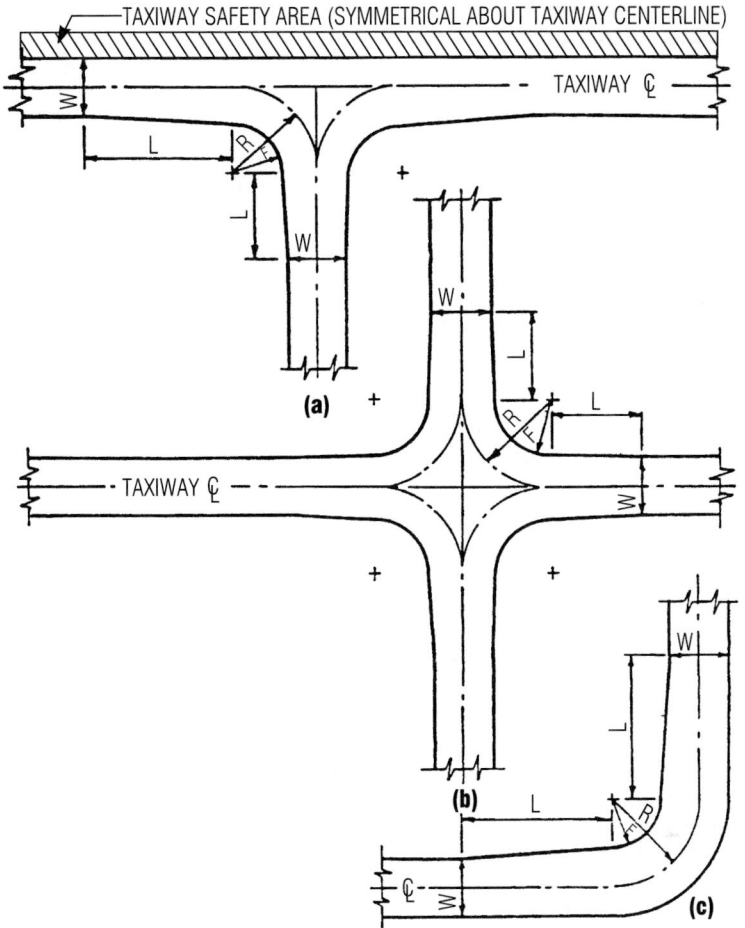

Fig. 18.10 Taxiway intersection details. The taxiway safety area shown in (*a*) has been omitted from (*b*) and (*c*) for clarity. Dimensions W, R, L, and F are given in Table 18.8. (*a*) T-shape intersection. (*b*) Crossover. (*c*) Turn.

10,000 ft^2 for other areas of pavement. Borrow areas should be tested sufficiently to define the borrow material clearly. Results of such tests are plotted on soils profiles or on a boring plan. This plan shows locations of borings with respect to proposed runway layout and individual profiles of the soil layers at each location, with a description of each soil type.

The FAA has adopted the Unified System of soil classification (ASTM D2487). Table 18.9 lists recommended spacings and depths for borings for soil investigations for airport construction. A typical graphic soil log is shown in Fig. 18.11.

FAA Advisory Circular "Airport Paving," AC 150/5230-6, discusses soils and paving topics.

18.11.1 Airport Grading

The surface of an airport should be relatively smooth but well-drained. Few natural sites provide these ideals; hence, proper grading is important. Grading plans and drainage plans must be carefully coordinated.

The grading plans should consist of runway and taxiway center-line profiles, cross sections showing areas of cut and fill, and a topographic map

Table 18.8 Taxiway Dimensional Standards

Design item	Symbol[a]	Airplane Design Group					
		I	II	III	IV	V	VI
Taxiway width	W	25	35	50[b]	75	75	100
Taxiway edge safety margin[c]		5	7.5	10[d]	15	15	20
Taxiway pavement fillet configuration:							
Radius of taxiway turn[e]	R	75	75	100[f]	150	150	170
Length of lead-in to fillet	L	50	50	150[f]	250	250	250
Fillet radius for center line	F	60	55	55[f]	85	85	85
Fillet radius for judgmental oversteering, symmetrical widening[g]	F	62.5	57.5	68[f]	105	105	110
Fillet radius for judgmental oversteering, symmetrical widening[h]	F	62.5	57.5	60[f]	97	97	100
Taxiway shoulder width		10	10	20	25	35[i]	40[i]
Taxiway safety area width		49	79	118	171	214	262
Taxiway object-free area width		89	131	186	259	320	386
Taxilane object-free area width		79	115	162	225	276	334

[a]Letters correspond to the dimensions in Fig. 18.10.

[b]For airplanes in Airplane Design Group III with wheelbase equal to or greater than 60 ft, the standard taxiway width is 60 ft.

[c]The taxiway edge safety margin is the minimum acceptable distance between the outside of the airplane wheels and the pavement edge.

[d]For airplanes in Airplane Design Group III with a wheelbase equal to or greater than 60 ft, the taxiway edge safety margin is 15 ft.

[e]Dimensions for taxiway fillet designs relate to the radius of taxiway turn specified. Additional design data can be found in "Airport Design," AC 150/5200-13.

[f]Airplanes in Airplane Design Group II with a wheelbase equal to or greater than 60 ft should use a fillet radius of 50 ft.

[g]Figure 18.10b displays pavement fillets with symmetrical taxiway widening.

[h]Figure 18.10c displays a pavement fillet with taxiway widening on one side.

[i]Airplanes in Airplane Design Groups V and VI normally require stabilized or paved taxiway shoulder surfaces.

Table 18.9 Recommended Spacings and Depths for Borings for Soil Investigations for Airport Construction

Area	Spacing	Depth
Runways and taxiways	Along center line, 200 ft c to c	Cut areas: 10 ft below finished grade Fill areas: 10 ft below existing ground surface*
Other areas of pavement	One boring per 10,000 ft² of area	Cut areas: 10 ft below finished grade Fill areas: 10 ft below existing ground surface*
Borrow areas	Sufficient tests to define borrow material clearly	To depth of proposed excavation of borrow

*For deep fills, boring depths should be used as necessary to determine the extent of consolidation and slippage that the fill to be placed may cause.

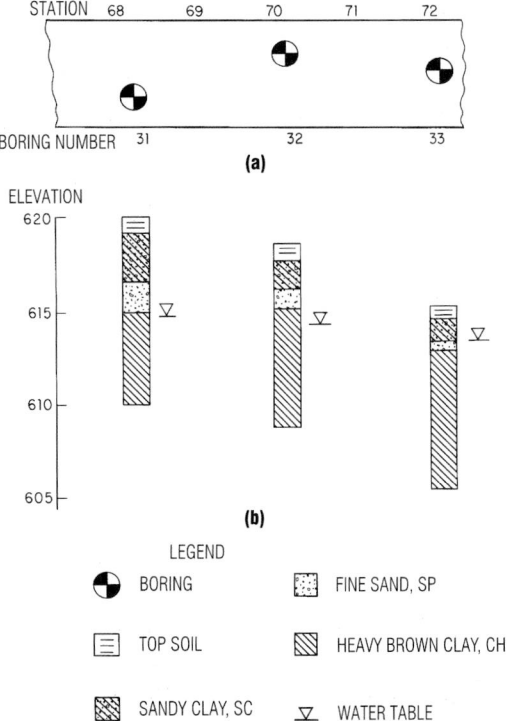

Fig. 18.11 Boring for subgrade investigation. (*a*) Plan of runway showing locations of borings. (*b*) Typical graphic boring log. (*Federal Aviation Administration.*)

showing initial and final contours. This latter map becomes the basis of the drainage-layout plan.

Cross sections of runways and taxiways should slope transversely each way from the center line to provide for surface drainage. Paved surfaces should slope 1 to 1½% for those serving approach categories C and D airplanes and 1 to 2% for those serving categories A and B.

Side slopes of cuts and fills should be as flat as possible. In cuts, the sides should not encroach on a lateral clearance ratio of 7:1 measured normal to the edge of the landing strip.

Properly designed grades can develop low areas that may be used for temporary ponding of storm runoff in the interest of a more economical storm-sewer system. Typical cross sections of runways are shown in Fig. 18.12.

(See "Airport Pavement Design and Evaluation" (www.faa.gov/arp/).)

18.11.2 Airport Drainage

With proper grading, the surface runoff is drained into collector sewers or ditches. Runoff is usually collected along the edges of runways in shallow ditches leading to inlets piped to storm sewers (Fig. 18.12*a*). At some airports in northern climates, where snowbanks along the edges of the runway block drainage across the runway, the surface water is also collected along the edges of the runway (Fig. 18.12*b*). Surface drainage inlets may be placed just outside the edges of runways, or they may be set in a shallow depression built in the outer edge of the pavement (Fig. 18.13). Inlets are usually spaced from 200 to 300 ft apart along the runway or taxiway.

Subsurface drainage is obtained by the use of interceptor drains and pervious base-course layers, in much the same way that highways are drained. Some of the smaller, turfed fields are drained by a network of subdrains covering the entire area. At airports with paved runways, subdrains are usually placed along the edges of the runways where soil conditions indicate that drainage is needed to lower the groundwater level. A combined interceptor and base drain is often used (Fig. 18.14).

Surface drainage is accomplished by the collection of surface water into inlets. A system of underground pipes is required to carry runoff from inlets and subdrains to outlets into waterways. In low areas, surface waters are sometimes drained into ditches or canals running around the perimeter of the airport.

For design of the drainage system, a topographic map is required, on which is plotted the proposed layout of runways, taxiways, aprons, and the terminal plan. The proposed surface grades of these features are shown by contours of small interval: 0.1 or 0.2 ft for paved areas and 0.5 or 1.0 ft for turfed areas. Inlet locations and subdrains are plotted, and storm-drain lines laid out to collect the discharge from them. The system should be as direct as possible to avoid excessive lengths of pipe; frequent changes in pipe size should also be avoided. Crossings of pipes under runways should be held to a minimum.

Figure 18.15 shows a portion of an airport drainage system. The pipe sizes are computed to accommodate the discharge from the design storm, which may be taken as that expected once in every 2 to 10 years, depending on how serious an effect

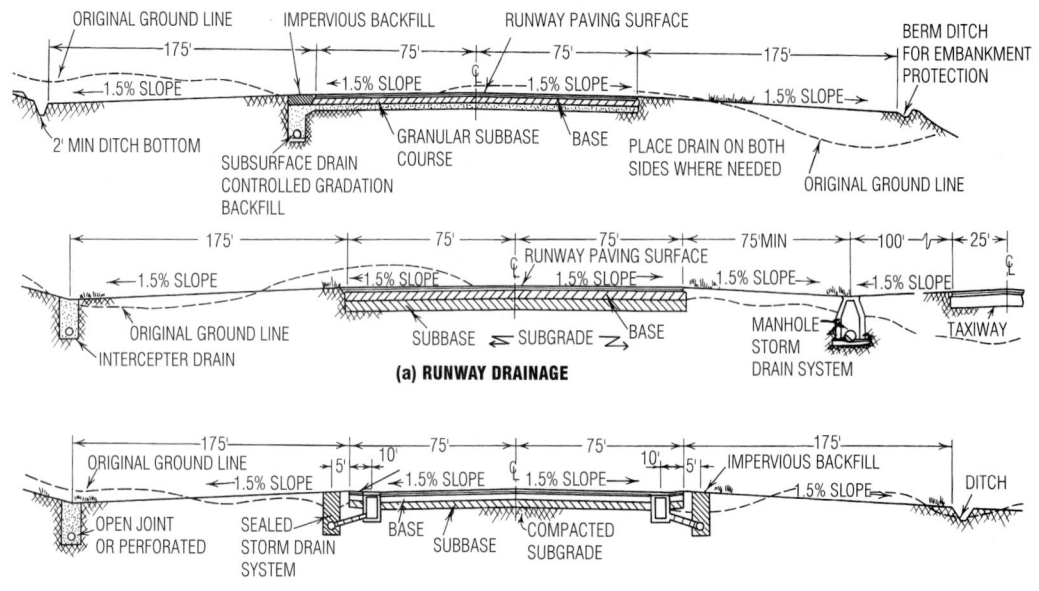

Fig. 18.12 Runway cross sections showing typical provisions for drainage.

an occasional flooding may have on airplane operations. In some designs, a certain amount of ponding is permitted in areas outside the runways.

The rational method (Art. 21.39) of calculating runoff is universally used in airport-drainage design.

The engineer should prepare studies of intersections to ensure good drainage. Center-line grades are held constant, and the grades of the outer portion of the runway or taxiway warped or adjusted so that there will be no abrupt changes in grade in the path of airplanes. The surface should have sufficient slope to drain properly. Intersection studies should be made at a scale of 1 in equals 50 ft. A contour interval of 0.10 ft will permit positive surface drainage to be designed. The

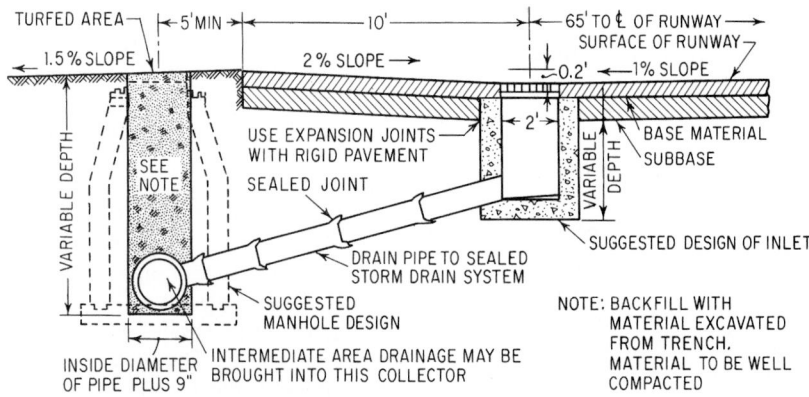

Fig. 18.13 Drainage inlet at outer edge of runway in northern climates. (*Federal Aviation Administration.*)

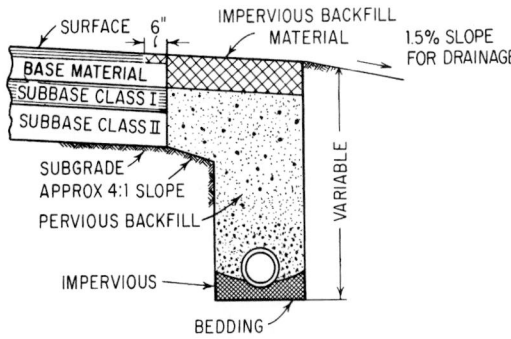

Fig. 18.14 Combined interceptor and base drain. (*Federal Aviation Administration.*)

studies will also be useful in establishing pavement grades.

("Airport Drainage," Federal Aviation Administration Advisory Circular AC150/5320-5.)

18.12 Airport Pavements

Airport pavements are constructed to support the loads imposed by aircraft using the airport and to produce a smooth, all-weather surface. Pavements are divided into two general types: *flexible* and *rigid*. Properly designed and constructed, either type will provide a satisfactory airport pavement. Specific types have, however, proved beneficial in

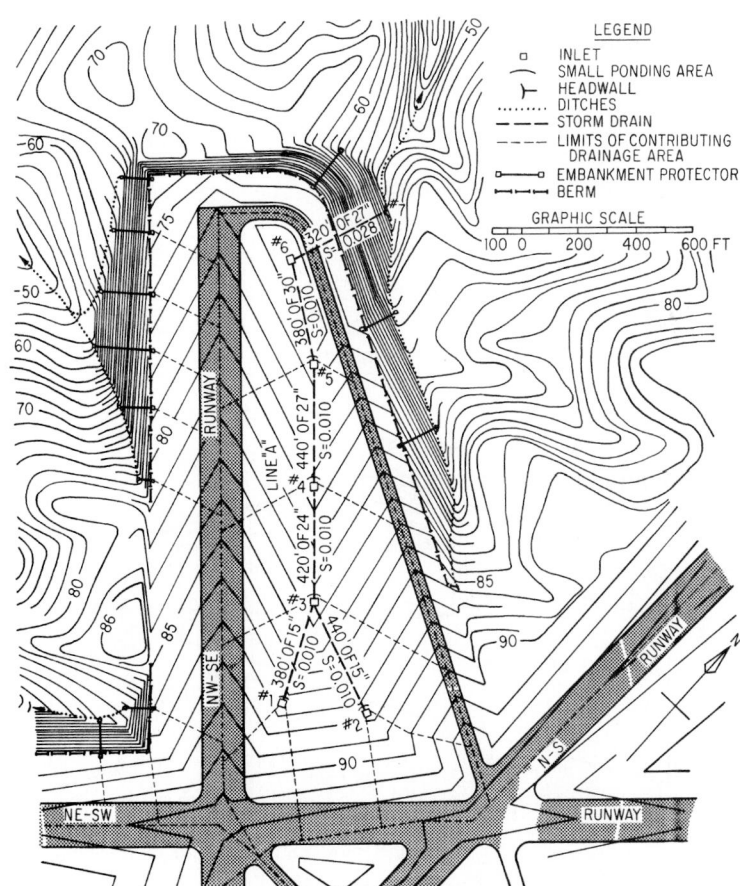

Fig. 18.15 Plan of a portion of an airport drainage system. (*Federal Aviation Administration Advisory Circular AC 150/5320-5.*)

specific applications: Rigid pavements are recommended for areas subjected to appreciable fuel spillage at aircraft gate positions or maintenance positions on the apron; a low-cost flexible pavement is adequate to stabilize an area subject to jet-blast erosion.

The "Airport Paving Manual," published by the Federal Aviation Administration, is the usually accepted guide for design of civil airport pavements. It contains methods and requirements to be used in designing projects involving Federal funds.

Subgrade is the foundation for airport pavements (Fig. 18.12). Its bearing capacity affects the thickness required in flexible and rigid pavements. Depth of frost penetration and influence of drainage conditions can affect the supporting value of the subgrade. Through selective grading, it might be economical to replace inferior subgrade material with superior material so as to reduce the subbase thickness requirement. Subgrades should be thoroughly compacted to provide the highest possible bearing capacity.

Subbase is a granular material placed on the compacted subgrade (Fig. 18.12). It usually is required under flexible and rigid pavements, except for the better soils groups. Thorough compaction is mandatory.

Figure 18.16 shows cross sections of typical runway pavements. The transverse slope of pavements usually is 1.50%, to minimize water ponding on the surface. The Federal Aviation Administration maintains "Standards for Specifying Construction of Airports," AC 150/5370-10, which cover most elements of airport development.

Critical areas are those requiring the thickest pavement. They include sections of runways, all taxiways, and aprons (Fig. 18.17). These are the areas subject to the most adverse aircraft loadings. Pavement thickness in noncritical areas may be reduced from the thickness required in critical areas (Fig. 18.16).

18.12.1 Flexible Pavements

These consist of a bituminous surface course, a base course of suitable material, and usually a granular subbase course (Fig. 18.16). Design of flexible pavements is based on the results of subgrade soils tests. The FAA has developed a relationship

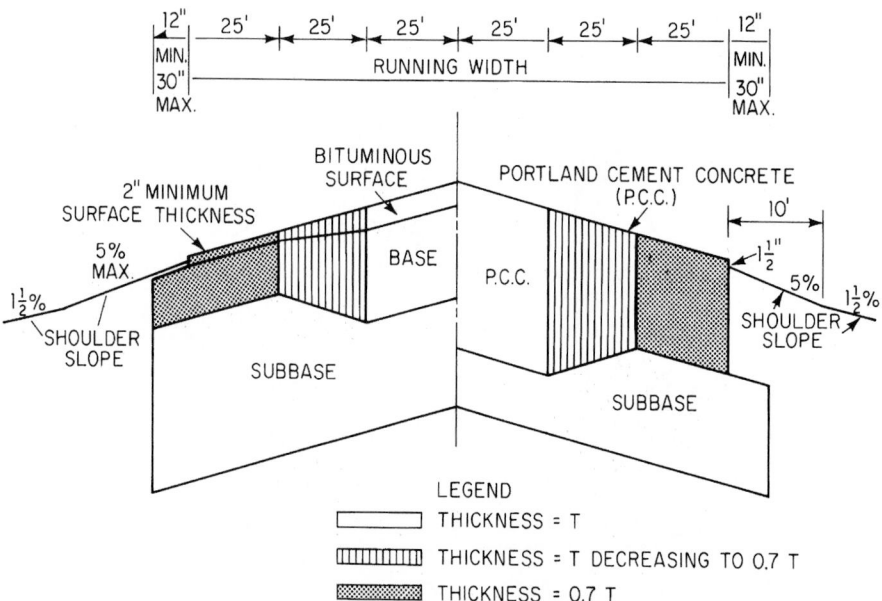

Fig. 18.16 Cross section shows typical bituminous pavement (left of center line) and portland cement–concrete pavement construction (right of center line) for critical areas of runways (Fig. 18.17).

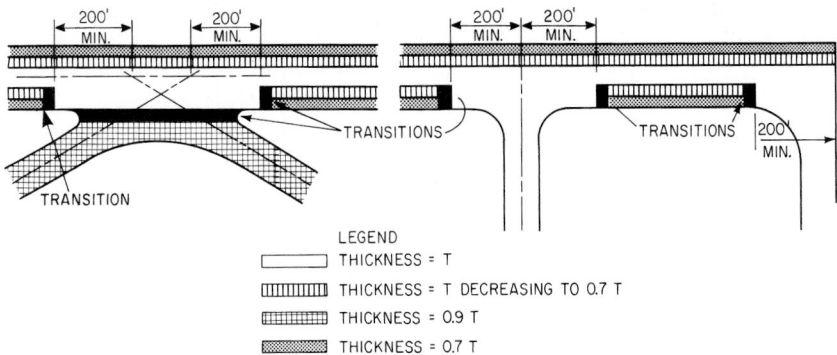

LEGEND
☐ THICKNESS = T
▦ THICKNESS = T DECREASING TO 0.7 T
▦ THICKNESS = 0.9 T
▦ THICKNESS = 0.7 T

Fig. 18.17 Critical areas of airport pavement. $T =$ total thickness of flexible pavement or concrete thickness for rigid pavement. See also Fig. 18.16.

between soil classes and thickness of surface course, base course, and subbase course required for various gross weights of aircraft, based on different conditions of drainage and frost action.

Design curves for flexible pavements are shown in Fig. 18.18, applicable to single-wheel landing gear, Fig. 18.19, applicable to dual-wheel landing gear; and Fig. 18.20, applicable to dual-tandem landing gear.

Curves are based on the assumption of a 20-year pavement life. Bituminous surfaces should be at least 4 in thick in critical areas, 3 in in noncritical areas, except for light aircraft. Use of the design curves for flexible pavements requires a California bearing ratio (CBR) value for the subgrade material and a CBR value for the subbase material. Also required are the gross weight of the design aircraft and the equivalent number of annual departures of the design aircraft. When the proper curve has been selected based on the landing-gear configuration of the design aircraft, the chart is entered at the appropriate CBR value at the top. A line is drawn vertically down to the gross weight of the design aircraft. From this point of intersection, a line is drawn horizontally to the number of annual departures. From this point of intersection, a vertical line is drawn to the base of the chart and the thickness read on the lower scale. The indicated thickness is for the surface and base courses combined, bearing on a subbase with the indicated CBR value.

Surface-course requirements are established to protect the base from surface water, provide a smooth running surface for aircraft, accommodate traffic loads, and resist skidding, traffic abrasion, and weathering. The surface course generally consists of two bituminous layers—a wearing course and a binder course. The binder course

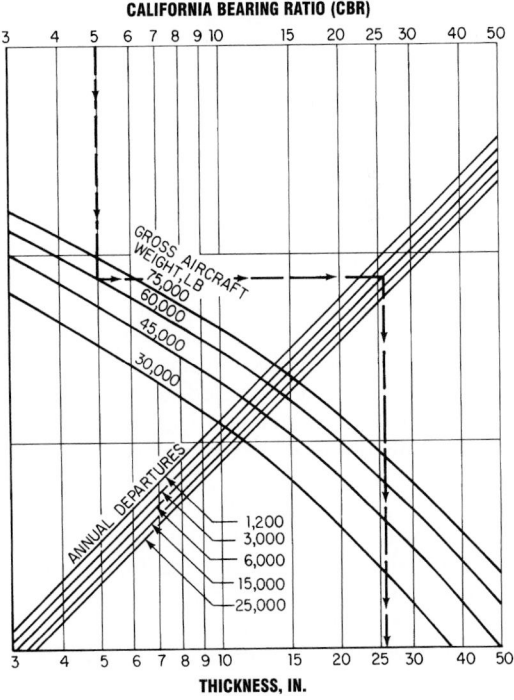

Fig. 18.18 Flexible-pavement design curves for critical areas—single-wheel gear. (*"Airport Pavement,"* *Federal Aviation Administration.*)

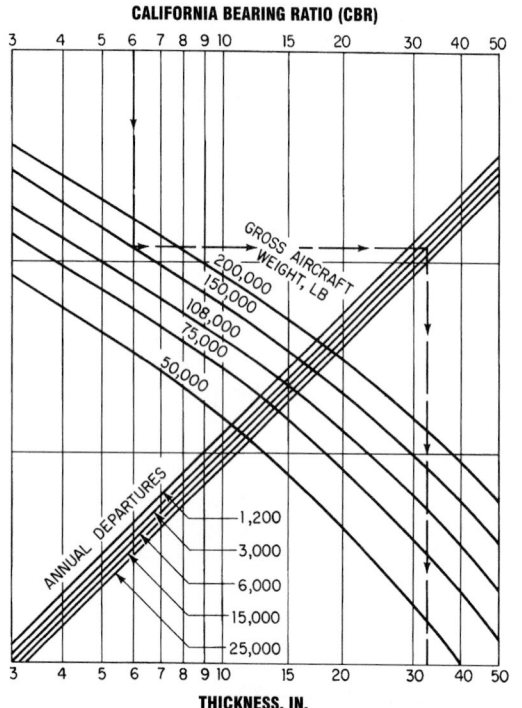

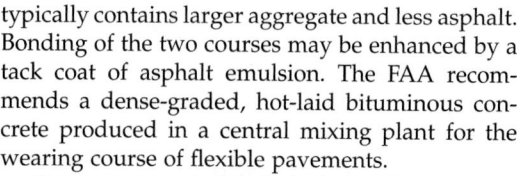

Fig. 18.19 Flexible-pavement design curves for critical areas—dual-wheel gear. (*"Airport Pavement," Federal Aviation Administration.*)

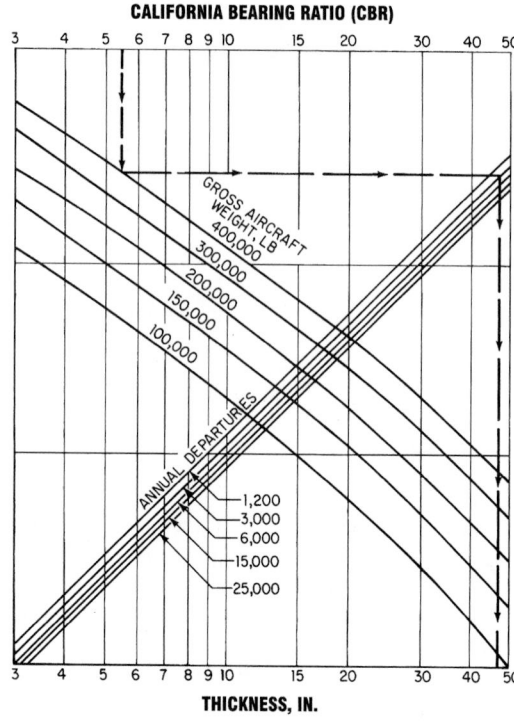

Fig. 18.20 Flexible-pavement design curves for critical areas—dual-tandem gear. (*"Airport Pavement," Federal Aviation Administration.*)

typically contains larger aggregate and less asphalt. Bonding of the two courses may be enhanced by a tack coat of asphalt emulsion. The FAA recommends a dense-graded, hot-laid bituminous concrete produced in a central mixing plant for the wearing course of flexible pavements.

Base-course materials include a wide variety, to take maximum advantage of local materials and construction practices. When high-quality aggregates are used, asphalt or portland cement treatments produce bases that are more effective than untreated bases. Accordingly, the FAA credits 1.0 in of certain treated base materials as being equivalent to 1.5 in of untreated base material.

Subbase is usually an integral part of the flexible-pavement structure. It is protected by the base and surface courses, and so the material requirements are not so strict as for the base course.

Pavements for light aircraft do not need to be so thick as for heavy aircraft. At airports that will not be required to accommodate aircraft in excess of

30,000 lb gross weight, the design curves in Fig. 18.21 should be used. The procedure is the same as with the design curves for heavier aircraft, except there is no reduction for noncritical areas.

18.12.2 Rigid Pavements

These are made of portland cement-concrete, usually placed on a suitable subbase course, which rests on a compacted subgrade (Fig. 18.16). Design curves for rigid pavements are shown in Fig. 18.22, applicable to single-wheel landing gear, Fig. 18.23, applicable to dual-wheel landing gear, and Fig. 18.24, applicable to dual-tandem landing gear.

Use of the design curves for rigid pavements requires the flexural strength of the concrete, the k value of the subbase, the gross weight of the design aircraft, and the equivalent number of annual departures of the design aircraft. When the proper curve has been selected based on the landing-gear configuration of the design aircraft, the chart

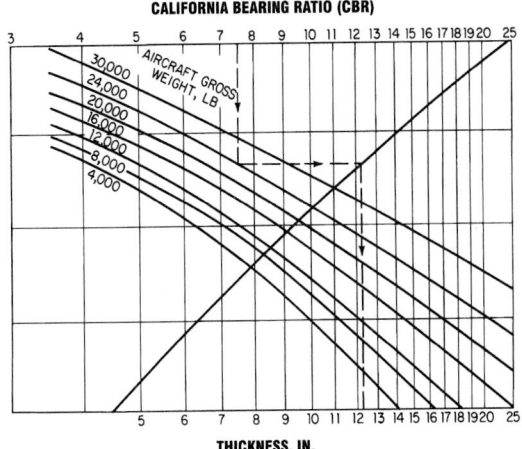

CALIFORNIA BEARING RATIO (CBR)

THICKNESS, IN.

Fig. 18.21 Flexible-pavement design curves—light aircraft. A minimum of 2 in is required for the surface course. (*Federal Aviation Administration.*)

thickness is read from that scale. The pavement thickness shown refers to the thickness of the concrete pavement only, exclusive of the subbase, and is that shown in Fig. 18.16 as *T*, referred to as the critical thickness.

The *k* value is based on the material directly beneath the concrete pavement. (See p. 16.42) A *k* value should be established for the subgrade and then corrected to account for the effects of the subbase.

Joints and reinforcing used in airport pavements are similar to those used in highway pavements, except that wider slabs and larger dowels are used for thick pavements. Longitudinal construction joints are doweled, keyed, or hinged (butt or keyed).

Longitudinal expansion joints are advisable at runway and taxiway intersections and next to structures. Where dowels are not suitable, a thickened edge may be introduced.

Transverse contraction joints are spaced 15 to 25 ft apart in unreinforced pavement. Transverse expansion joints are not generally used except at intersections.

Dowels are used across expansion joints and also across construction joints in some designs. The bars or bonded reinforcing are carried across certain longitudinal contraction joints and keyed construction joints to hold the slab faces in close contact.

Construction joints between runs of pavement are keyed, doweled, or hinged. The diameters of dowels vary from $\frac{3}{4}$ in for 6- to 7-in-thick slabs, to

containing the appropriate design curve is entered, from the left, with the concrete flexural strength. A horizontal projection is made until it intersects the appropriate foundation modulus (the *k* value). A vertical projection is made from this point of intersection to the appropriate gross weight of the design aircraft. A horizontal projection is then made to the scale on the right that corresponds with the annual departures. The pavement

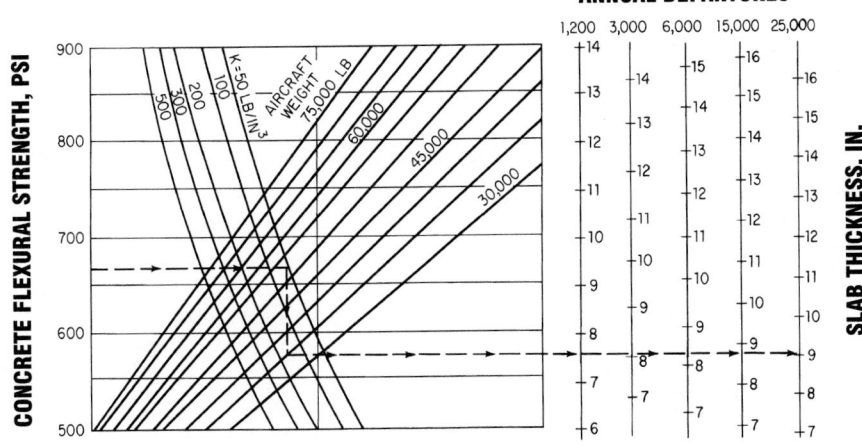

ANNUAL DEPARTURES

CONCRETE FLEXURAL STRENGTH, PSI

SLAB THICKNESS, IN.

Fig. 18.22 Rigid-pavement design curves—single-wheel gear. (*"Airport Pavement," Federal Aviation Administration.*)

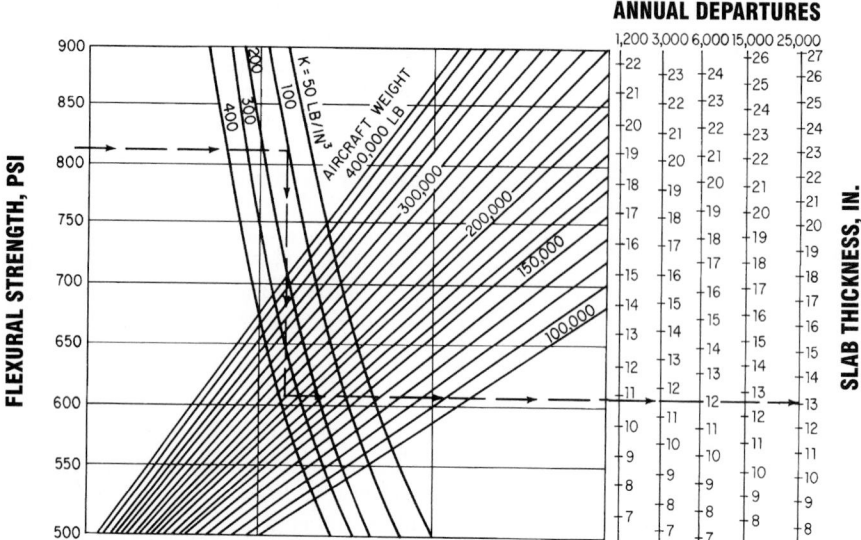

Fig. 18.23 Rigid-pavement design curves—dual-wheel gear. (*"Airport Pavement," Federal Aviation Administration.*)

2 in for 21- to 24-in slabs. Standard length of dowels is 18 to 24 in and spacing is 12 to 18 in c to c.

Reinforcing of the pavement is desirable to control cracks. Its installation should follow the latest design and construction practices.

Rigid pavement for small aircraft (weighing 12,500 lb or less) should be at least 5 in thick. For aircraft between 12,500 and 30,000 lb, the rigid pavement should be at least 6 in thick. All paved areas should be considered as critical areas (Fig. 18.16).

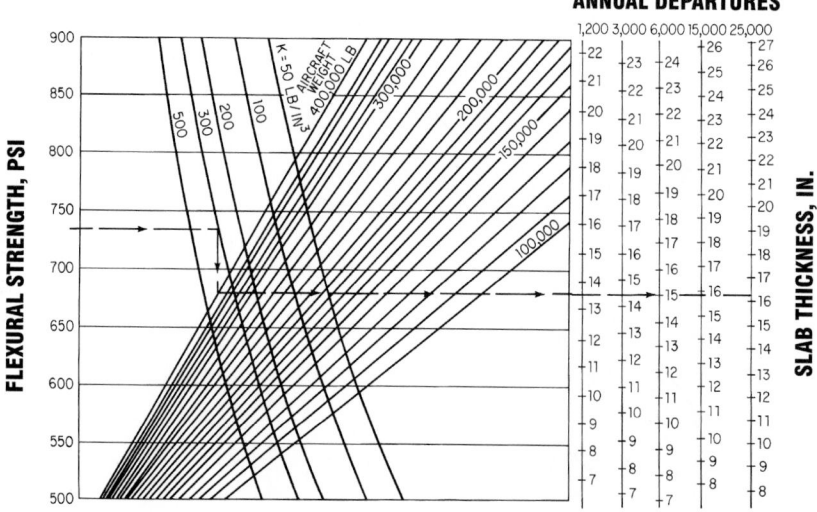

Fig. 18.24 Rigid-pavement design curves—dual-tandem-wheel gear. (*"Airport Pavement," Federal Aviation Administration.*)

18.12.3 Pavement Overlays

Pavement overlays restore a pavement. An overlay may be applied over a pavement that no longer can be maintained satisfactorily. Or an overlay may be used to increase the load-bearing qualities of a satisfactory pavement that must accommodate aircraft heavier than those for which the pavement was designed.

Flexible overlays involve a combination of a base course and a bituminous surface course. *Rigid overlays* involve the use of a layer of portland cement-concrete. *Bituminous overlays* consist entirely of bituminous concrete. In each instance, the qualities of the existing pavement must be fully ascertained and the overlay designed to make the resultant pavement capable of handling the required traffic, following procedures outlined in "Airport Paving," Federal Aviation Administration.

18.13 Unpaved Surfaces at Airports

Some airports do not require paved surfaces because of a low volume of traffic and use by only light aircraft. In some instances, turf surfaces are used for landings and takeoffs at small airports and on the unpaved areas of runways at larger airports.

A tough, thickly matted grass is required in these areas. The type of grass to use depends on soil characteristics and climate at the site. If tests show the soil deficient in nutrient elements, these may be supplied by appropriate fertilizers. When a fertile topsoil must be removed during grading operations, it should be stockpiled and later spread on the areas to be turfed. A vegetative cover is also desirable on embankment, cut slopes, and other interior areas of the airport to prevent dusting and erosion.

When turf is not adequate by itself, it may be possible to add to stability by adding coarse aggregate to the soil prior to the development of the turf. This will permit the soil to retain sufficient moisture to promote the growth of grass, yet provide a surface that will not become too soft in wet weather.

18.14 Soil Stabilization at Airports

Granular material, portland cement, tar, cut-back asphalt, or emulsified asphalt may be used to improve the qualities of a soil so that it can serve as a base or subbase. Such stabilized soils are not intended to serve as a surface course; a separate wearing surface must be provided. The same general procedures are followed in stabilizing airport soils as are followed in highway practice. (See Art. 16.19.)

18.15 Airport Terminal Buildings

Transition of passengers from ground to air occurs in the terminal area. Various methods are used to accommodate and transfer the public and its goods, arriving either by air or by ground, and to provide for parking, servicing, and storage of aircraft and vehicles used in ground transportation. The degree of development in the terminal area varies with the volume of airport operations, the type of traffic using the airport, the number of people to be served, and the manner in which they are to be accommodated.

18.15.1 Airplane Parking

The concept of a very small airport might involve only a hangar with simple office facilities, adequate for limited aeronautical activity. At larger airline terminals, demands are greater. The concept can involve bilevel terminal operations, auto parking in buildings, and elaborate passenger-loading devices. Various concepts of terminal systems are shown in Figs. 18.25 and 18.26.

Frontal layout of facilities is usual at airports of low activity. In the small-airport layout (Fig. 18.25a), the facilities required to serve a moderate volume of general aviation are in a row along the boundary road. At many small airports, the terminal is eliminated and its functions housed in a lean-to of a service hangar. At such an airport, the terminal (or lean-to) would usually have a waiting room, rest rooms, office for the airport manager or flight service operator, and perhaps a restaurant, snack bar, or vending machines.

At airline terminals of low activity, a frontal loading system as shown in Fig. 18.25b is usually preferred. Expansion possibilities are indicated. As the fingers are extended, however, the passenger walking distances increase. Likewise, the finger structure becomes less economical, since loading positions are on only one side.

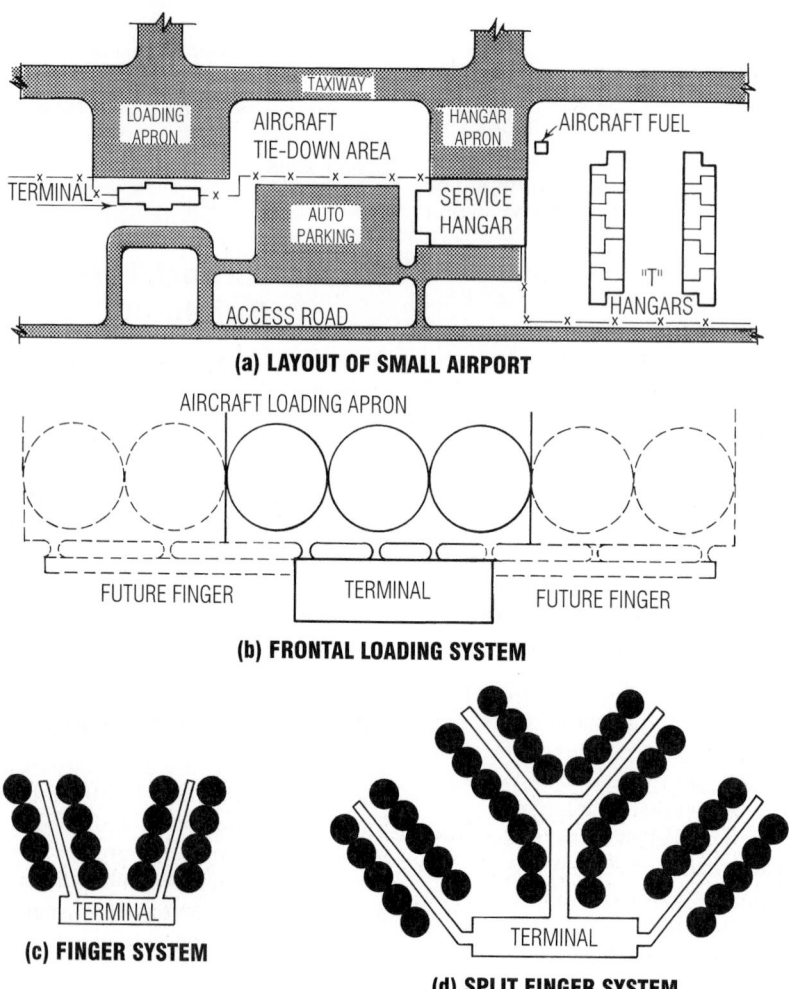

(a) LAYOUT OF SMALL AIRPORT

(b) FRONTAL LOADING SYSTEM

(c) FINGER SYSTEM

(d) SPLIT FINGER SYSTEM

Fig. 18.25 Simple terminal systems. Fingers are added to increase aircraft parking capacity.

Finger systems project onto the parking apron and permit aircraft to park closer to the terminal. This arrangement reduces structural cost since loading is accomplished on both sides.

The finger system shown in Fig. 18.25c is a simple solution for a hub airport. Walking distances to the extreme end positions, however, could be rather long.

A more elaborate finger layout is the split-finger system (Fig. 18.25d). Here the passenger walking distances become quite long. A passenger transferring from the end-loading position of one finger to the end-loading position of another finger would walk more than ½ mi, assuming the aircraft

parking positions to be 200 ft in diameter. Walking distances are inevitably long at centralized terminals that serve large numbers of gates, unless mechanical transfer of passengers is employed.

Unit terminals concentrate aircraft parking positions and minimize passenger walking distance except where interunit transfers are required. The movement from one unit terminal to another can involve excessive time and distance. Unit terminals are generally designed so that each unit is a self-contained entity.

Satellite terminals also concentrate aircraft parking positions in an effort to minimize walking distances. The satellites shown in Fig. 18.26a are fed

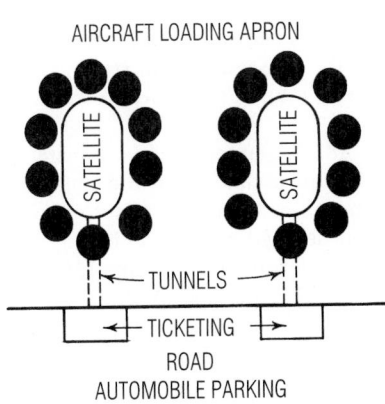

**(a) SATELLITE SYSTEM
(LOS ANGELES INTERNATIONAL AIRPORT)**

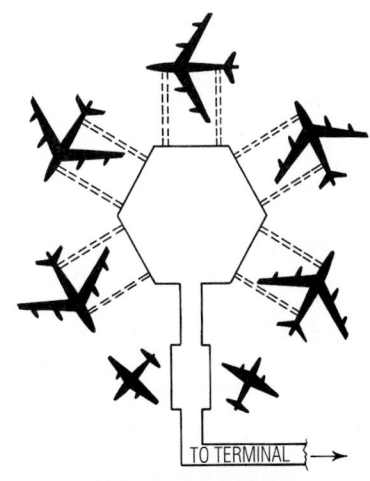

**(b) FINGER-PIER SYSTEM
(U.A.L.-SAN FRANCISCO INTERNATIONAL AIRPORT)**

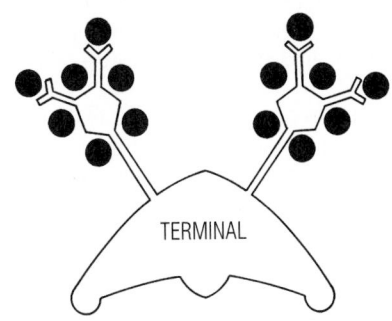

**(c) FINGER-PIER SYSTEM
(T.W.A.-J.F. KENNEDY INTERNATIONAL AIRPORT)**

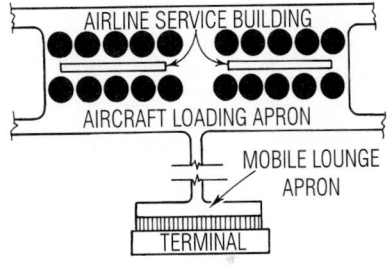

**(d) REMOTE PARKING SYSTEM
(DULLES INTERNATIONAL AIRPORT)**

Fig. 18.26 Terminal systems used at some international airports.

by tunnels from the ticketing area and provide a number of aircraft parking positions without excessive walking distances.

In Fig. 18.26*b* the satellite is a pier at the end of a finger and concentrates parking positions, with a resultant saving in walking distance. The terminal layout shown in Fig. 18.26*c* has two piers to serve 14 loading positions with relatively short walking distances. The pier-satellite approach offers minimum passenger walking distances for a large number of gate positions.

Remote parking of aircraft minimizes walking distances by using a vehicle to transport passengers from terminal to airplane. At some European airports, buses accomplish the transfer.

In an elaborate scheme under the remote concept, a mobile lounge moves passengers to and from aircraft parked some distance from the terminal (Fig. 18.26*d*). The mobile lounge is in use in several countries. At flight time, passengers are driven to the aircraft, or are met on arrival. Thus, the long walk between plane and terminal is eliminated.

18.15.2 Passenger Loading and Unloading

Passenger-loading devices permit weatherproof transfer from terminal to aircraft, usually with no change in level required. The *mobile lounge* is one type of loading device. Figure 18.27*a* shows the type of lounge vehicle used at Dulles Airport. Figure 18.27*b* shows passengers transferring from lounge to aircraft. Local-service airlines using small transport aircraft can be accommodated directly at the terminal.

The *telescoping gangplank*, shown in Fig. 18.27*c* and *d*, is the loading device in most general use. The telescoping passage has a swivel connection at the terminal. The aircraft end rides on a full-swivel gear that is electric-powered. Parallel parking is shown in Fig. 18.27*c*, with gangplanks serving front and rear doors of the aircraft. Airplanes that park at an angle may use a single gangplank, as shown in Fig. 18.27*d*. In both instances the aircraft can taxi into and out of gate positions. (Wide-body aircraft are normally towed away from parking positions.)

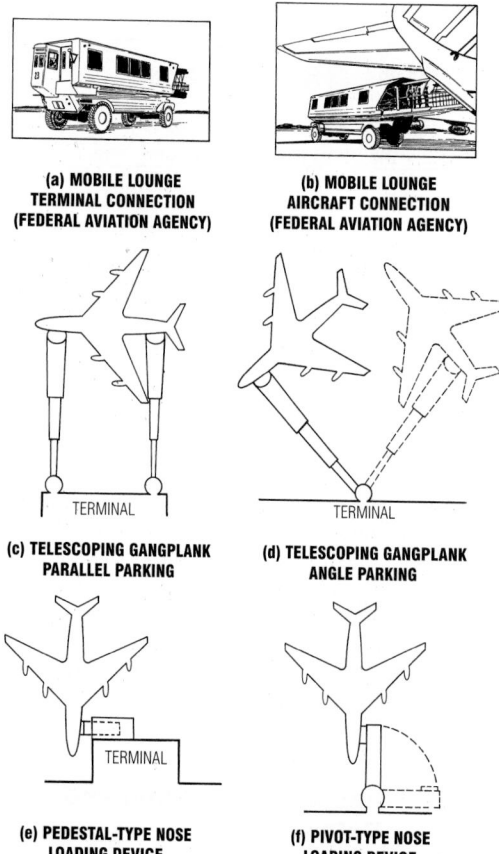

(a) MOBILE LOUNGE TERMINAL CONNECTION (FEDERAL AVIATION AGENCY)

(b) MOBILE LOUNGE AIRCRAFT CONNECTION (FEDERAL AVIATION AGENCY)

(c) TELESCOPING GANGPLANK PARALLEL PARKING

(d) TELESCOPING GANGPLANK ANGLE PARKING

(e) PEDESTAL-TYPE NOSE LOADING DEVICE

(f) PIVOT-TYPE NOSE LOADING DEVICE

Fig. 18.27 Devices used for passenger loading on and unloading from aircraft at airports.

The *nose devices* shown in Fig. 18.27*e* and *f* permit aircraft to taxi into the parking position, but aircraft must be towed away from the gate. Deplaning the passengers is faster with nose loading devices, but the departure from the gate is slower.

In Fig. 18.27*e*, an adjustable transfer device suspended from a canopy on the outside of the terminal moves only a few feet to the aircraft door. The nose loading device in Fig. 18.27*f* is pivoted at the terminal and supported on fixed, powered wheels at the aircraft end. When not in use, the device is stored against the wall of the terminal and swings into position to connect to the doorway of the aircraft. Experience has shown that aircraft can be precisely taxied into parking positions so that elaborate adjustments are not required.

The pedestal device (Fig. 18.27*e*) is the least expensive, but the swivel type (Fig. 18.27*f*) can serve a greater variety of aircraft since it can serve a wider span of aircraft heights because of the longer ramp.

Other types of passenger-transfer devices include moving sidewalks in fingers and other places where feasible, horizontal transportation systems that connect unit terminals and satellites, and similar systems that can serve individual loading positions, to keep walking distances to a minimum.

18.15.3 Terminal-Building Layout

The key feature of any terminal-area layout is the terminal building. In size, it can be small for airports with low activity, or large and complex at primary-system terminals.

The terminal should be planned to serve the number of peak-hour passengers forecast for 10 years in the future. Flexibility and expandability are paramount requirements.

The terminal building should provide a smooth flow of passengers from parking lot to aircraft. The passenger should be able to park, or get out of a taxi, bus, or limousine, at a point near the ticket counter. Baggage is checked at this point. Then the passenger proceeds to the aircraft via a waiting room where rest rooms, telephones, concessions, and restaurant facilities should be available. At the loading position, there should be a hold room where the passenger may be processed for boarding the scheduled flight.

Deplaning passengers go directly from the aircraft to the baggage pickup area, then proceed to taxicab, bus, limousine, or parked automobile. Automobile rental counters should be near the baggage-pickup areas, and there should be telephones and rest rooms nearby.

Visitors should be provided with observation decks. The need for concession, restaurant, and office space will vary at each location. A greater variety of concession potential will obviously develop at the larger airports.

Airline facilities include ticket counters, ticket offices, baggage-processing areas (with baggage usually mechanically conveyed from the ticket counters), and operational space at the loading position. Inbound baggage should be available to the passenger at a convenient location, either placed by hand on a claim counter or mechanically conveyed by belt with spacers, diverters, or carousels for delivery to the arriving passenger.

At small airline terminals, the entire operation is at a single level. Larger terminals tend to have elevated roadways so that departing passengers enter the terminal at the second level and enter the aircraft at the same general elevation by means of a loading device. Deplaning passengers leave the aircraft at the second level and escalate down to the ground floor for baggage pickup and ground transportation. There are many variations of this scheme, but the pattern is the same.

The accommodation of FAA air-traffic-control quarters, as well as weather facilities, will vary from one location to another. There is a trend toward locating these government facilities in separate structures, away from the terminal but nearer to the general aviation activity. There is no fixed pattern. The need for such space in the terminal building varies from location to location.

18.16 Access Roads

In preparing a terminal-area layout, the engineer should recognize the importance of vehicular access. The area should be located so that full advantage is gained from freeways and other roads, planned or existing, that will expedite ground transportation to the airport.

Within the airport, the access-roadway system should provide a connection between the terminal area and the best routes to town. The system also should include roads for intercommunication

between the separate facilities. Separation of passenger and commercial traffic is desirable, as well as separation of patron, spectator, and employee traffic.

18.17 Hangars

The size of hangars depends on the dimensions and numbers of aircraft to be serviced. Airports for general aviation usually have one or more service hangars that hold several aircraft, for which repair and maintenance operations are conducted. These hangars are supplemented by nests of T hangars, which provide individual stalls for aircraft storage. At larger airports, the trend is toward cantilever hangars, capable of accommodating the largest aircraft.

Table 18.10 gives the gross weight, wing span, length, and height of typical aircraft. Jet transport aircraft are being produced in short-, medium-, and long-haul versions. The larger jets have become even larger, with fuselage increases in excess of 30 ft and with gross weights exceeding 350,000 lb. Supersonic aircraft exceed the length and weight of a stretched-out jet aircraft. One model has a length of about 300 ft, a wing span of about 120 ft, and a gross weight of nearly 500,000 lb. Hangars to serve such aircraft must have built-in flexibility.

18.18 Cargo and Service Buildings

At many airports, air cargo is handled through the terminal building. Where separate cargo facilities have been developed, they usually have been located adjacent to terminal areas. Size and type of cargo facilities vary, depending on local need. Most are long, low structures with truck docks on one side and aircraft parking on the other. The roadway level on the truck side should be depressed to provide a truck-high floor for easy loading and unloading.

These separate cargo buildings not only provide facilities to load cargo directly into aircraft on the adjacent apron but contain facilities for sorting out small freight shipments to be taken to the terminal area on small carts and placed aboard passenger aircraft.

Table 18.10 Physical Data for Selected Aircraft

Name and model	Gross weight, lb	Wing span	Length	Height
Single Engine, Prop.				
Beech Bonanza	3,125	33 ft 5 in	25 ft 2 in	7 ft 7 in
Cessna 210	2,900	36 ft 7 in	27 ft 9 in	8 ft 8 in
Piper Saratoga	2,900	36 ft 0 in	24 ft 11 in	7 ft 3 in
Multiengine, Prop.				
Aero Commander	8,000	49 ft 0 in	35 ft 1 in	14 ft 6 in
Beech Super King Air	12,500	54 ft 6 in	43 ft 10 in	15 ft 0 in
Cessna Conquest	9,925	49 ft 4 in	39 ft 0 in	13 ft 1 in
Piper Cheyenne	12,050	47 ft 8 in	43 ft 5 in	17 ft 0 in
Executive Jets				
Lockheed Jetstar	35,000	54 ft 5 in	60 ft 5 in	20 ft 5 in
Grumman Gulfstream II	51,340	68 ft 10 in	79 ft 11 in	24 ft 6 in
Learjet 25	13,300	35 ft 7 in	47 ft 7 in	12 ft 7 in
Rockwell Sabreliner	17,500	44 ft 5 in	43 ft 9 in	16 ft 0 in
Airline Transports				
Airbus A-300	330,700	147 ft 1 in	175 ft 6 in	55 ft 6 in
B-737-200	100,800	93 ft 0 in	100 ft 0 in	36 ft 9 in
B-727-200	173,000	108 ft 0 in	153 ft 2 in	34 ft 0 in
DC-9-30	109,000	93 ft 4 in	107 ft 0 in	27 ft 6 in
DC-8-63	358,000	148 ft 5 in	187 ft 5 in	43 ft 0 in
B-747	775,000	195 ft 8 in	229 ft 2 in	64 ft 8 in
B-757	225,000	124 ft 10 in	155 ft 4 in	45 ft 1 in
B-767	350,000	156 ft 1 in	180 ft 4 in	52 ft 7 in
L-1011	432,000	155 ft 4 in	178 ft 8 in	55 ft 10 in
DC-10-30	555,000	161 ft 4 in	181 ft 11 in	59 ft 7 in
MOC-MD-11	602,500	169 ft 10 in	201 ft 4 in	57 ft 10 in

At smaller airports, the cargo is handled at the terminal building and carried only by passenger planes.

At most airports with scheduled passenger service some form of aircraft rescue and fire-fighting facilities is required. They should be provided at a location having ready access to all parts of the airport. Other buildings that might be required are heating plant, utility buildings, maintenance buildings, equipment-storage buildings, electrical equipment, and transformer vaults.

18.19 Airport Lighting

Airport lighting provides illumination to keep the facilities available around the clock. Lighting, usually kept on from dusk to dawn, assists in location and identification of the airport, outlines the usable areas, and furnishes guidance to moving aircraft.

Basic lighting consists of beacons, lighted wind indicator, runway or trip lights, and such obstruction lights as are required. Figure 18.28 illustrates the basic lighting at a small airport. Airport-lighting equipment and systems are subject to considerable modification in concept and design. The latest FAA recommendations and practices should be followed.

Airport Beacon ▪ This is a double-end, rotating light situated on or near the airport and visible from considerable distances. Appropriate color coding of the two beacon lenses will identify the airport as an unlighted facility (both lenses clear), or equipped with runway lights, burning or readily available (clear-green).

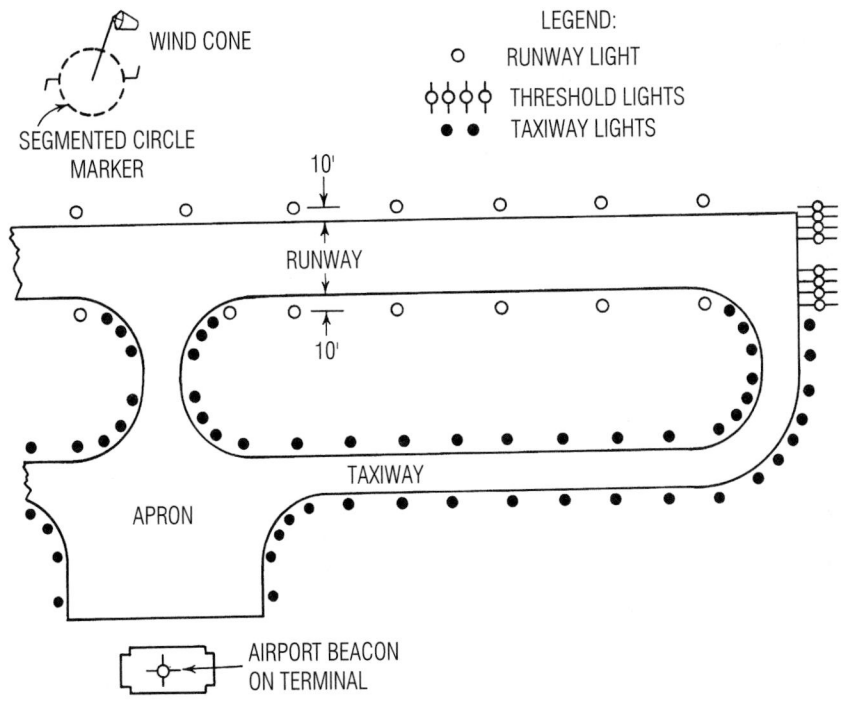

Fig. 18.28 Basic layout for airport lighting.

The beacon may be placed atop a structure or on a standard beacon tower. The beams of the beacon are set slightly above the horizontal and should clear all trees and obstructions in the vicinity.

Obstruction Lights ▪ These red lights mark objects that penetrate approach, horizontal, or conical surfaces (Fig. 18.1). Both steady-running and flashing obstruction lights are available for use according to requirements. The positions of lights will depend on the obstruction and its location with respect to the airport.

Wind Indicator ▪ Wind information is required at all times to permit aircraft to select the most favorable runway or landing strip for takeoff or landing. The simplest indicator is a wind cone, a free-swinging cloth cylinder which gives information as to wind direction and velocity. At larger airports, landing information is furnished by a wind tee. The cone and tee should be illuminated to provide information during hours of darkness.

18.19.1 Runway Lighting

These are low, elevated lights used to outline the edges of paved runways or to define unpaved runways. The smallest airports have lights mounted on driven stakes. At larger airports, the lights are mounted on heavy bases or small vaults of metal or concrete. The vault contains the isolating transformer for each fixture; otherwise, the transformer is buried alongside the runway light. The tops of bases and vaults are flush with the airport surface.

The lights are spaced 200 ft apart longitudinally and are usually 10 ft off the edge of the pavement (Fig. 18.28). They are fed from underground cables, either direct-burial or in ducts.

Medium-intensity runway lights are used on noninstrument runways and are adequate for visual operations. The intensity is controlled through a five-step regulator so that minimum intensity can be used in good weather. The lights have a Fresnel lens for optimum light distribution.

High-intensity runway lights are used on runways equipped for instrument landings or designated as instrument-landing runways by the FAA. These lights concentrate powerful beams down the longitudinal axis of the runway, in both directions. Intensity is controlled so that there is adequate guidance without undue glare.

Threshold Lights ▪ The effective end of each runway is indicated by runway lights (medium- or high-intensity) with green lenses. These lights are placed athwart the runway to mark its actual end, or outboarded beyond the edges of the runway for displaced thresholds. It is usual to place threshold lights at the actual runway ends (Fig. 18.28), except where clearance of obstructions in the approach dictates locating the threshold inward from the actual end.

In-Runway Lights ▪ Use of precision approach facilities to achieve lower weather minimums requires extensive electronic equipment in the aircraft, improved FAA navigation aids on the airport, and high-intensity runway lights, plus "in-runway" lighting. The last consists of center-line runway lights and touchdown-zone lighting. The lighting of the center line of taxiway turnoffs is desirable since it assists aircraft in clearing the runway during inclement weather.

Runway center-line lighting (Fig. 18.29) consists of fixtures installed at uniform intervals along the center line of a runway to give a continuous lighting reference from threshold to threshold. The lights are spaced at 50-ft intervals. Fixtures are installed in shallow holes drilled into the pavement. The lights are fed by cables installed in $\frac{1}{4}$in slots sawed 1 in deep. Isolating transformers are located at the sides of the runway.

Touchdown-zone lighting (Fig. 18.29) consists of 30 rows of transverse light bars at 100-ft intervals. Each row contains two bars. Set 30 ft on

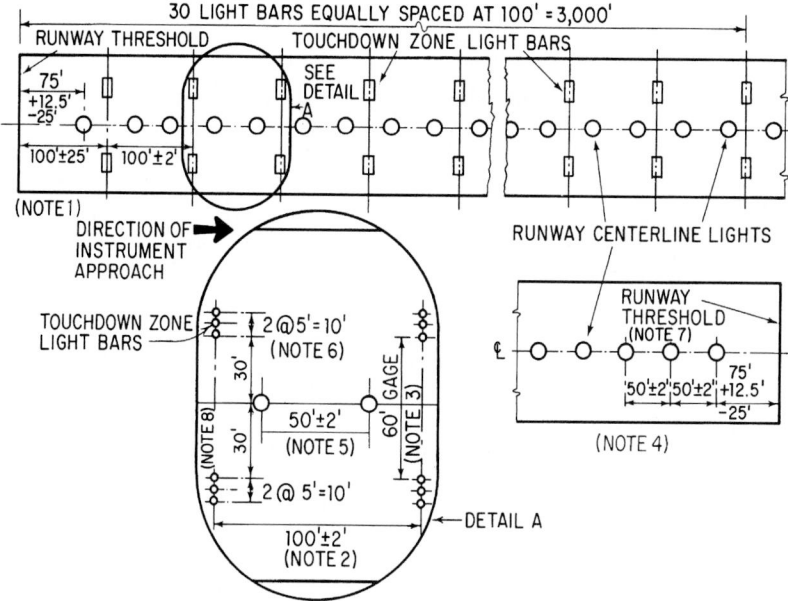

Fig. 18.29 Lighting layout for runway touchdown zone and center-line runway lights. NOTES: (1) In case of unusual joint location in concrete pavement, the first pair of light bars may be located 75 to 125 ft from the threshold. (2) Longitudinal tolerance should not exceed 2 ft. (3) Gage may be reduced to 55 ft to meet construction requirements. (4) Longitudinal installation tolerance for individual lights should not exceed 2 ft. (5) Center-line lights need not be aligned with transverse light bars. (6) Maximum uniform spacing of lights is 5 ft c to c. (7) Center-line lights may be located up to 2 ft from the runway center line to avoid joints. (8) Corresponding pairs of transverse light bars should lie along a line perpendicular to the runway center line. (*Federal Aviation Administration.*)

each side of the runway center line, each bar consists of three lights 5 ft apart, flush with the surface of the pavement, and aligned normal to the axis of the runway.

The fixtures are high-intensity lights installed in the pavement and fed through ducts or cemented into shallow holes drilled into the pavement and fed through cable installed in sawed joints. A variety of fixtures is available.

18.19.2 Taxiway Lighting

An airport with paved taxiways should have guidance lights if there is significant traffic at night. Taxiway lights are similar to medium-intensity runway edge lights, except that they are equipped with blue lenses. They are placed along the edges of taxiways to outline usable paved areas (Fig. 18.28). The longitudinal spacing varies with the taxiway configuration.

Taxiway Guidance Signs ▪ These are internally illuminated directional indicators placed low above the ground surface. They give abbreviated guidance to the ends of runways, terminal aprons, hangar areas, and other airport locations. Need for them depends on the volume of traffic and the complexity of airport layout and development.

Taxiway-turnoff lighting (Fig. 18.30) consists of lights installed, for relatively high-speed performance, along the center of a turnoff taxiway to indicate the exit path. The lights are spaced 50 ft

apart. The fixtures are similar to those used in the runway center line.

18.19.3 Airport-Lighting Control

All airport lights should be controlled from a single panel, readily accessible to an operator. At small airports, a regulator assembly with controls built into the same cabinet provides a simple solution for basic lighting. Automatic controls (photoelectric or astronomic time switches) may be used where it is not feasible to have an operator on duty, or for remote beacons, obstruction lights, or other equipment, where direct control lines would not be economically feasible.

At airports with more complex installations, relay control equipment is placed in a transformer vault and lights are remotely controlled from the airport traffic-control tower or other central source. The remote-control source should have an adequate control panel, usually mounted on the control-tower console, and should contain circuit-control and brightness-control switches.

18.20 Airport Electric Power Supply

Provision of electric power to an airport for general purposes, as well as for airport lighting, requires a determination of power requirements and study of power availability. Usually, a second source of power is desirable to ensure reliability of the

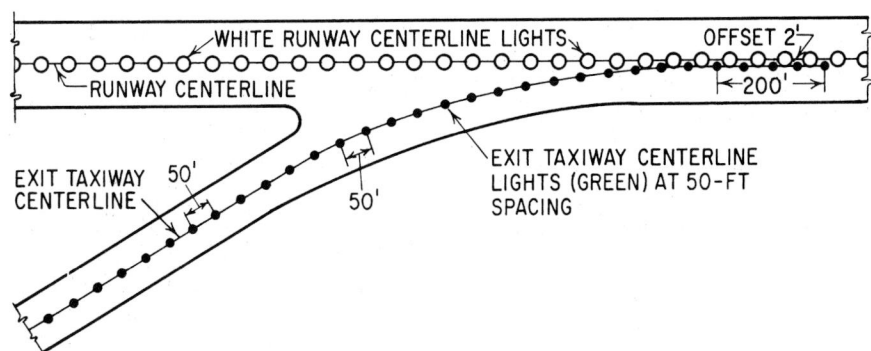

Fig. 18.30 Long-radius taxiway-turnoff lighting. A longitudinal tolerance may be necessary to avoid joints in rigid pavements. (*Federal Aviation Administration.*)

lighting system. The overall reliability of power from commercial sources will determine the possible need for standby service or equipment.

Electrical Ducts ▪ In the preparation of a master plan for an airport, provision of electrical ducts for all cable crossings under paved areas should be carefully studied. The various systems of lighting should be laid out in sufficient detail to permit cable runs to be determined. All lighting that can be contemplated as an ultimate requirement should be studied.

The FAA should be requested to furnish details of all installations that it might make so its cable requirements can be incorporated into the duct plan.

When runways, taxiways, or aprons are paved, care should be taken to ensure that adequate electrical ducts are provided to preclude costly jacking or cutting of pavements at some future date. A number of spare ducts might well be provided in all instances.

18.21 Airport Marking

In addition to airport lighting, marking of facilities assists in guidance day and night and enhances operations in periods of restricted visibility. Federal Aviation Administration national standards should be followed.

The basic marking at an airport consists of a segmented circle marker (Fig. 18.28) and a wind indicator. The segmented circle marker is placed just outside the usable landing area. It identifies an airport and provides a central location for such indicators as exist at that airport.

The marker is a broken circle 100 ft in diameter. At the center is a conventional wind cone. A tee, however, may be used as a landing direction indicator.

Radial extensions beyond the 100 ft circle show the orientation of landing strips or runways. Extensions of the radials to the left or right indicate the airport traffic pattern.

Obstructions should be day-marked for maximum visibility. Other marking includes the numbering and striping of runways for normal identification, striping of taxiways, marking of unusable areas, and special runway marking to facilitate operations during low-visibility weather.

18.22 Fuel Systems

Regardless of the volume of traffic at an airport, some system of supplying fuel to aircraft must be provided. The simplest system at a small airport is an underground tank and an elevated dispenser, not unlike a regular service station pump. Usually provided by a petroleum company, this system requires aircraft to taxi to a central location for service (Fig. 18.25a).

Generally, a single grade of fuel is available. Each additional grade and type of fuel requires a separate installation.

Airports with a medium volume of traffic normally use fuel-truck dispensers. These are serviced from local bulk stations if traffic is low and from airport storage if there is a sufficient volume of traffic to warrant. The busiest primary-system airports require fuel in such quantities that supply, storage, and distribution become special and complex problems.

Fuel Supply ▪ Depending on overall fuel requirements coupled with local conditions, fuel will be supplied to the airport by truck delivery from local sources; tank truck, rail, or barge deliveries from refineries or bulk-storage sources; direct pipeline delivery; or various combinations of these. The heavier the volume of traffic, the more varied the types of fuel required.

Even at some airports where large quantities of certain varieties of fuels come by pipeline, the demand for other varieties is so low that truck delivery is employed for them. It is necessary to make a forecast of the demand for various grades of fuel, to determine long-range sources of availability, and to study all possible methods of delivery as a prerequisite to the design of an airport fuel system.

Fuel Storage ▪ The bulk-storage system at the airport should provide for each type of fuel to be handled. Normal practice is also to maintain brand segregation. The capacity for each type should be adequate to accommodate fueling requirements for several days.

Delivery provisions should be flexible. There should be truck stands so that fuel can be unloaded from trucks and pumped into storage. Even where trucks are not the major source of supply to the airport, truck stands should be adequate to supply the entire fuel demand in an emergency, with pumping capacity sized accordingly. The same pumping capacity can be used for rail or barge supply.

Pipeline delivery will normally not require pumping capacity, inasmuch as fuel can be transferred under pressure to storage tanks. A waste tank should be provided for changing types of fuel without affecting type integrity.

The storage tanks should be interconnected to provide for interchange or transfer of fuel within the storage area. Tanks should be adequate to handle modern jet fuels. Usually, inert-gas explosion suppression is provided, or the tanks have floating roofs. The storage system should be capable of easy expansion or modification.

Fuel Transfer ▪ Fuel is pumped from storage tanks through filter separators to pits, hydrants, or truck-loading stands, either directly or through satellite storage areas. If the distance is great, the size and number of transfer pipes may be reduced by introducing one or more satellite storage areas.

It is usual to have a separate satellite area for each user or group of users. Pumps take fuel from satellite storage through filter separators and to pits, hydrants, or truck stands.

Fuel Delivery ▪ Trucks and pits are used for low-capacity delivery of fuel. High-capacity fuel delivery is accomplished by hydrants and hose carts.

Truck stands serve as loading points for fuel trucks, which deliver from the trucks directly into aircraft fuel tanks, through filter separators.

Pits contain booster pumps, filter separators, and coiled hose to deliver fuel directly into aircraft tanks similar to truck delivery.

Hydrants provide for quick connection to hose carts. These are powered vehicles equipped with filter separators and pressure regulators to deliver fuel at high rates, under pressure, through underwing loading.

18.23 Air-Traffic Control

Airports are developed through the initiative of local communities, but the control of air traffic is a function of the Federal government. It is usual for air-traffic-control facilities to be installed and operated wholly with Federal funds.

Some auxiliary facilities, such as high-intensity runway lights and in-runway lighting, are the responsibility of the local community owning the airport. Facilities that furnish guidance along the airways and assist in the transition from airway to airport are usually installed without local participation. The FAA has criteria based on volumes of traffic that are used to locate specific control facilities at an airport. Articles 18.23.1 to 18.23.3 provide general information concerning location and installation, but the FAA should be contacted for the latest revisions.

("Airport Design Requirements for Terminal Navigation Aids," Federal Aviation Administration.)

18.23.1 Instrument Landing System (ILS)

The ILS is an electronic facility that furnishes three-dimensional information in the final portion of an airport approach, to permit an aircraft to fly to the landing runway in inclement weather. The system consists of localized glide slope, outer marker, and middle marker.

Localizer equipment provides an electronic course down the projected center line of the runway for lateral guidance. The equipment is normally installed 1000 ft beyond the end of the runway opposite the approach direction. The area between the end of the runway and the localizer should be smooth, and within a circular area 500 ft from the localizer there should be no trees, buildings, roads, or fences.

Glide-slope reference is transmitted from equipment located 400 to 600 ft off the center line of the runway and 750 to 1250 ft in from the approach end. A smooth area is necessary for a considerable distance in front of the glide-slope unit to ensure the stability and accuracy of the electronic emissions.

Outer-marker equipment is located 4 to 7 mi from the airport on the projected center line of runway. The signals from the outer marker indicate distance from the runway end.

Middle-marker signals indicate a point about 3500 ft from the runway end.

("Airport Design Requirements for Terminal Navigation Aids," Federal Aviation Administration.)

18.23.2 Approach-Light Systems

This is a system of high-intensity lights that extend outward from the approach end along the projected center line of the runway. They provide visual reference to the instrument runway during the transition from instrument flight to visual flight.

The system consists of horizontal 12-ft bars of high-intensity lights spaced 100 ft apart longitudinally for a distance of 1400 to 3000 ft. Each bar contains, in addition, a condenser discharge light. These flash in sequence toward the runway.

An area 400×3200 ft is desirable for the installation of the approach-light system. The lights are placed on piers or towers as required to provide a uniform light line at a slope not exceeding 2% upward from the end of the runway or a slope of 1% downward.

Runway-End Identifier Lights ▪ This system consists of a pair of synchronized flashing lights. One is located on each side of the runway-landing threshold facing into the approach area. The lights are placed 40 ft outward from the runway edge lights. The flashing lights provide rapid and positive identification of the approach end of a particular runway.

Precision-Approach-Path Indicator ▪ This is a system of visual-approach indication, designed to provide visually the same information that a glide-slope unit provides electronically. Normally, four light units in one row are placed 1000 ft in from the runway threshold. The lights are placed 50 ft from the runway edges.

The light units have beams elevated so that a specific approach slope is indicated through the proper combination of red and white lights; for example, two red and two white. The approach slope may be set to clear a specific obstruction or to enhance noise-abatement procedures.

("Airport Design Requirements for Terminal Navigation Aids," Federal Aviation Administration.)

18.23.3 Other Airport Traffic Controls

Surveillance radar controls traffic within a considerable distance from the airport, about the same range as covered by the airport approach-service control. No unusual siting problems are involved.

Precision-approach radar system is used to monitor or control traffic approaching the instrument runway. It is located alongside the instrument runway 400 to 750 ft from the center line.

Terminal VOR omnirange is a terminal facility similar to the standard VOR (VHF omnirange) navigation device. When it is sited on an airport, there should be a clearance of 1200 ft in all directions to ensure true azimuth course indication.

Transmissometer is a device that furnishes visibility-measurement information for the runway touchdown area. The installation is located slightly more than 400 ft from the center line of the instrument landing system runway.

Airport traffic-control towers are provided by the FAA at those locations where new towers are required. The control tower should be located at a point from which all portions of the runways, taxiways, and aprons are visible. Requirements for each airport will vary; hence, they should be checked with the FAA.

Airport surface-detection equipment comprises a radar system that permits the observation of aircraft ground traffic on the airport. It is usually located on top of airport traffic-control towers.

("Airport Design Requirements for Terminal Navigation Aids," Federal Aviation Administration.)

18.24 Heliports

Helicopters in civil use vary in size, number of rotors, number of engines, and overall weight. Small helicopters usually employ a single rotor for lift and lateral control and a vertical tail rotor for pivotal or yaw control. Large civil helicopters have a single main rotor and vertical tail rotor or two main rotors located in tandem along the longitudinal axis of the helicopter. There are other potential configurations, including intermeshing main rotors placed normal to the main axis, and various models of vertical-lift devices and convertible aircraft that can take off vertically and, through variable aircraft geometry, fly horizontally at speeds greater than those possible for helicopters.

Helicopters rise vertically a few feet above the heliport surface when taking off. Then they accelerate upward and forward on a sloping path to climb-out speed and continue to en-route altitude.

Landing involves an approach on a sloping path to a hovering position a few feet above the heliport surface. Then, the craft descends vertically to a selected landing point. Sideward flight may be performed easily during the landing maneuver, so the helicopter will land in a precise position. Ability to operate vertically permits the helicopter to land and take off using areas only slightly larger than its own dimensions.

18.24.1 Heliport Classifications

Heliports, as described in FAA Advisory Circular 150/5390-2A, are classified by use as private, public—general aviation, public—transport, and hospital. Private-use heliports are developed for the exclusive use of the owner and person representing the owner. General aviation heliports are intended to accommodate individuals, corporations, and helicopter air-taxi operators. Scheduled passenger services may be available. Transport heliports are intended to accommodate air-carrier operators who provide service with large helicopters. Hospital heliports are limited to serving helicopters engaged in air ambulance or other hospital-related functions. Helistops are heliports with minimum support facilities; that is, no shelter, maintenance, or fueling.

Heliports are developed around a *design helicopter*, the largest helicopter expected to use the heliport during future years. For heliports located on airports, helicopters are classified as small, medium, or heavy for determining the distance between landing facilities.

Small helicopters (up to 4 passengers) generally weigh up to 6000 lb, are 30 to 40 ft long, 9 to 10 ft high, and have rotor diameters up to 35 ft. Medium helicopters have a takeoff weight between 6000 and 12,000 lb. Larger helicopters in general use weigh up to 20,000 lb, carry as many as 30 passengers, are 65 to 85 ft long, up to 17 ft high, and have rotor diameters up to 55 ft. Most small helicopters use a skid-type landing gear, but large helicopters use wheel landing gear with a three- or four-wheel configuration.

18.24.2 Final Approach and Takeoff Area

Although planning considerations are the same for the various classifications of heliports, requirements for physical characteristics differ slightly. Heliports have in common the final approach and takeoff area (FATO), which is an object-free area available for helicopter landings and takeoffs. The FATO may be at ground or water level or elevated, on a pier or rooftop. At least one clear path from the FATO aligned with the prevailing wind should permit approach and takeoff of a helicopter clear of all objects (Fig. 18.31.) A FATO should have a minimum dimension (length, width, or diameter) 1.5 times the overall length of the design helicopter. If a heliport is 1000 ft or more above mean sea level, consideration should be given to elongating the FATO in the direction of takeoff. The FATO should be smooth. Grades may range from 0.5% to a maximum of 5.0% to ensure good drainage, but they should not exceed 2% in any area where a helicopter is expected to land.

A safety area, free and clear of objects that could be struck by the main or tail rotor or that could catch the skids of an arriving or departing helicopter, should surround the FATO. The width of this area, measured outward from the FATO, should be at least one-third the rotor diameter but not less than 20 ft.

18.24.3 Touchdown and Liftoff Area

A FATO should have a touchdown and liftoff area (TLOF) with a paved or other hard surface, preferably centered in the FATO. The diameter of the TLOF should be at least that of the rotor of the design helicopter. (When the entire FATO is load-bearing, however, an identifiable TLOF may not be required.)

For ground-level heliports, the surface of the TLOF should be portland cement-concrete. An asphaltic concrete surface may also be used, but provision should be made for the possibility that ruts may form under wheels or skids due to hot climatic conditions or the repeated loads of landing and parking helicopters. (Ruts are suspected of being possible factors in some rollover incidents.) Pavements should be designed to support 1.5 times the maximum weight of the design helicopter. They should have a broomed finish to enhance safety of persons and helicopters on the TLOF.

An elevated TLOF also should be designed to support 1.5 times the maximum takeoff

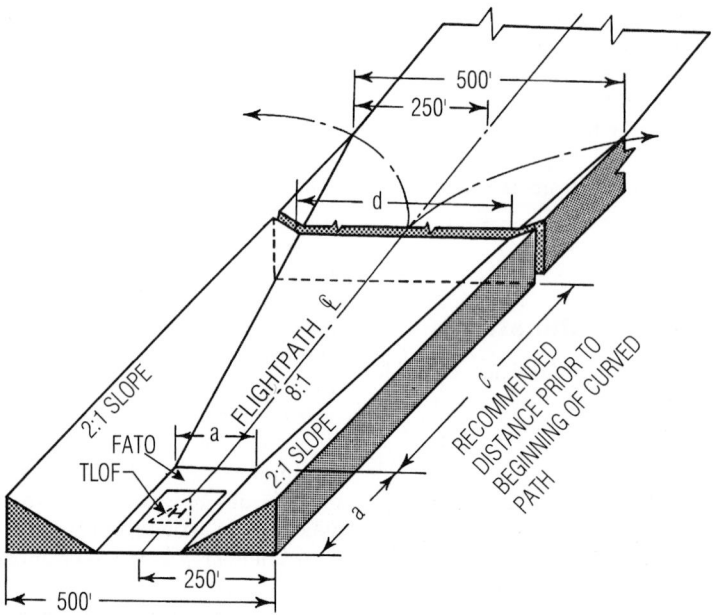

PERSPECTIVE VIEW OF APPROACH-DEPARTURE PATH

Fig. 18.31 Standard dimensions for heliports and approaches. FATO length and width *a* should be at least 1.5 times the overall length of the design helicopter. Straightway approach-departure length *c*, width *d* of the flight path at the wide end of the straightway, and the radius of the curved path should each be at least 300 ft.

weight of the design helicopter. The TLOF may be constructed of wood, metal, or concrete, but usually, a combination of steel and concrete is used. The gradient of an elevated TLOF should be about the same as that for a ground-based TLOF.

When a TLOF is elevated more than 30 in above its surroundings, a 5-ft-wide safety net or shelf should be provided in lieu of a railing. The safety net should have a load-carrying capacity of at least 25 lb/ft². Normally, it is installed with a slight slope upward and outward from the TLOF with the outer edge projecting above the TLOF 2 in or less.

Engineers should obtain information from local building officials on design loads for elevated heliports; fire-extinguishment requirements; and storage, handling, and dispensing of aircraft fuels. Information also should be obtained from the nearest FAA Airports District Office on requirements for heliport markers and markings, wind indicators, heliport lighting, and approach and navigational aids applicable to the type of heliport being designed. The FAA should also review

environmental impacts and compatibility of land uses in the vicinity of the proposed heliport.

18.24.4 Approach and Takeoff Paths

All public-use heliports should have more than one approach and takeoff path. One path should be aligned with the predominant wind during operations in visual meteorological conditions (VMC). Another path, if practicable, should be aligned with the prevailing winds for operations in instrument meteorological conditions (IMC). Visual approach and takeoff paths may curve to avoid objects or noise-sensitive areas or to utilize the airspace above public ways, such as freeways and rivers (Fig. 18.31).

A visual approach and takeoff surface should be centered on each approach and takeoff path. This surface should conform to the dimensions specified in the FAR Part 77 heliport approach surface.

Approach-departure paths should be laid out to offer the best lines of flight. It generally is necessary to have at least two flight paths, usually 180° apart, but the paths may be as little as 90° apart.

Curved paths are practicable but should be used with a minimum straightway approach-departure length of 300 ft (Fig. 18.31). The center-line radius of a curved path will vary, depending on local conditions and type of helicopter used. In general, however, the radius of the curved path should be at least 300 ft.

The approach-departure path has the same width as the contiguous edge of the landing and takeoff area and flares uniformly on each side of the center line to a width of 500 ft at the en route altitude. The slope of the path is 1 ft vertical for each 8 ft longitudinally (8:1). Objects that extend above this sloping plane are obstructions.

Transition areas are surfaces along the lateral boundaries of the landing and takeoff area and the approach-departure areas. The surfaces, or "side slopes," extend outward and upward from the edges of the heliport and approach-departure areas for a distance of 250 ft from the center line. The slope is 2:1 upward from the edge of the landing and takeoff area or from the edge of the sloping approach-departure plane.

Heliport proponents or owners should own or control property underlying the approach and takeoff surface outward to a distance where the surface is 35 ft above the heliport.

18.24.5 Helicopter Parking

Public heliports, not designed as helistops, should have an area designated for parking helicopters. The size required for this area, or apron, depends on the number of helicopters to be accommodated. The clear distance from any part of a helicopter on its intended path to another helicopter or any object should be at least one-third the rotor diameter, but not less than 10 ft. If a helicopter must turn more than 30° within a parking position, clearance of the tail rotor beyond one-third the rotor diameter or 10 ft. whichever is larger, may control location of parking positions. Parking pads should have a minimum dimension of 1.5 times the undercarriage length or width of the design helicopter.

Taxi routes or taxiways should connect the FATO to the area designated for helicopter parking. They should be designed to provide 20 ft of rotor tip clearance to objects and parked helicopters for

hover taxiing and 10 ft of clearance for ground taxiing. Paved surfaces of taxiways should have at least twice the undercarriage width of the design helicopter. If the surface is unpaved, it should be treated in some manner to prevent dirt and debris from being raised by the rotor wash of a taxiing helicopter.

18.24.6 Heliport Layout and Design

A heliport may be sited on the ground or on top of a building. For greatest utility to helicopters, the site should be as close as possible to the locale it serves. It should provide operational safety, have clear approaches, and be compatible with air traffic in the vicinity. It should fit in with area planning and not have an adverse impact on the community.

The small heliport may consist of only a designated area containing an unsurfaced landing and takeoff area (Fig. 18.32) or of an elaborate facility with a paved landing area, parking and service aprons, heliport terminal, and automobile parking (Fig. 18.33).

Standard grading and drainage practices should be employed. The rotor downwash of helicopter operations usually requires a stabilized landing and takeoff area at a minimum. A paved touch-down pad is desirable.

Ground locations for heliports usually permit less expensive construction than rooftop sites but are seldom available in congested areas. Rooftop locations usually have advantages of accessibility and clear approaches to counter the disadvantages of limited space, difficulty of locating emergency-landing areas, and the probable need to strengthen the structure. It is necessary to consider wind effects as well as local building codes, zoning, and fire regulations.

If the structure requires reinforcing, a load-distribution pad might be satisfactory. The pad need not be so large as the landing and takeoff area, but the full area should be a clear area. The pad can be as small as 20 × 20 ft for smaller helicopters, up to 50 × 50 ft for larger vehicles. The rooftop heliport should be of sufficient strength that it will not fail under unusual, high-impact landings. The landing surface should be designed for a concentrated load equal to 75% of the gross weight of the helicopter on any 1 ft^2 of the surface.

Wind conditions might require baffles to eliminate turbulence across the surface of the heliport.

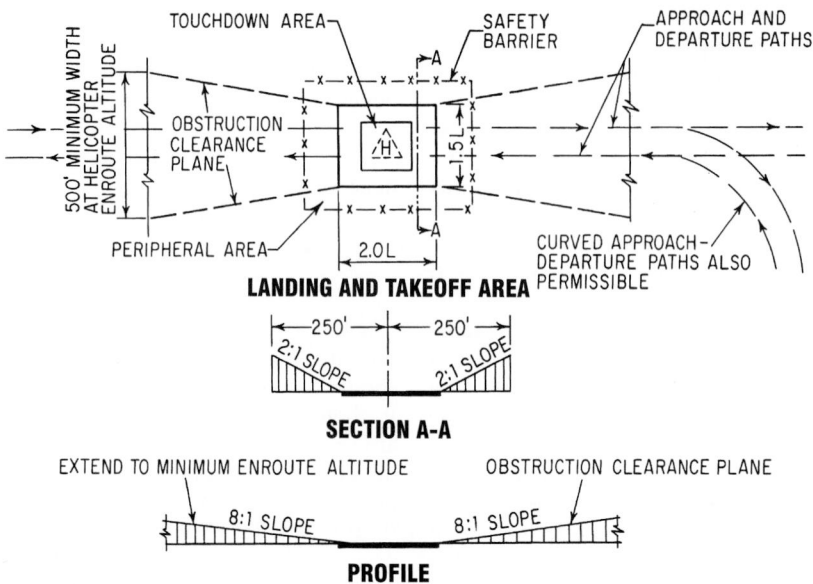

Fig. 18.32 Layout for a small heliport. L = overall length of design helicopter. ("*Heliport Design,*" *Federal Aviation Administration.*)

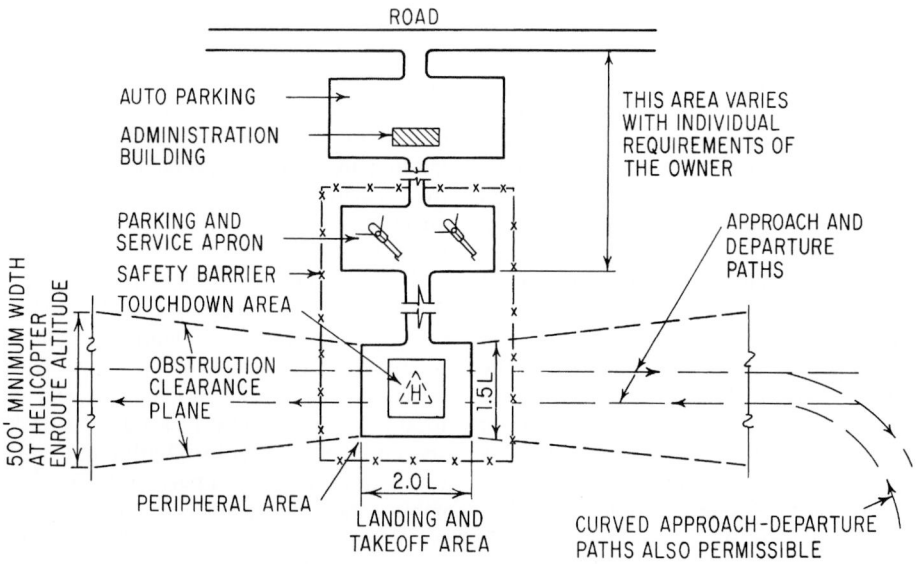

Fig. 18.33 Layout for a large heliport. L = overall length of design helicopter. (*Federal Aviation Administration.*)

Also, a safety device should be provided around elevated touchdown areas or landing pads. This should extend outward from the touchdown area.

18.24.7 Heliport Marking and Lighting

Standard heliport markers, placed near the center of the touchdown area, are shown in Fig. 18.34.

The touchdown area should be marked with a border at least 1 ft wide. The boundary of the landing and takeoff area should be made conspicuous by low markers spaced 25 ft apart. A wind indicator should be adjacent to the landing and takeoff area, located to provide true wind information.

Obstructions should be marked and lighted. Yellow boundary lights may be used to outline the

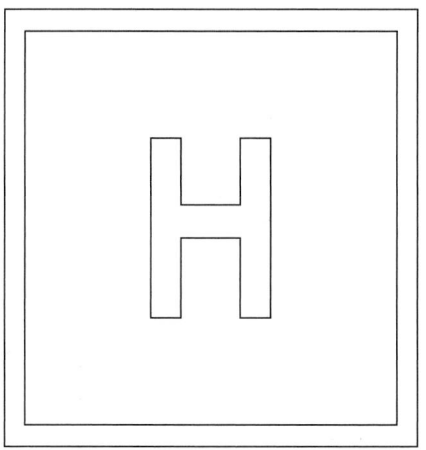

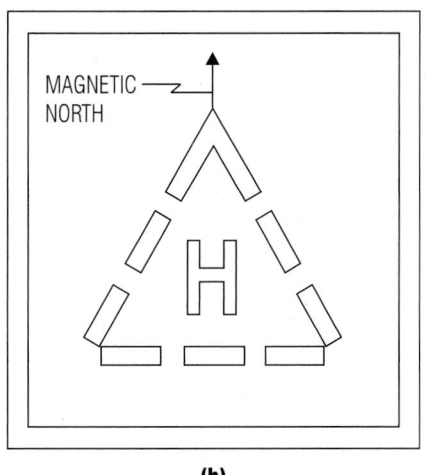

(a)   (b)

(BORDER AND SYMBOLS IN CONTRASTING COLORS TO ENHANCE CONSPICUOUSNESS)

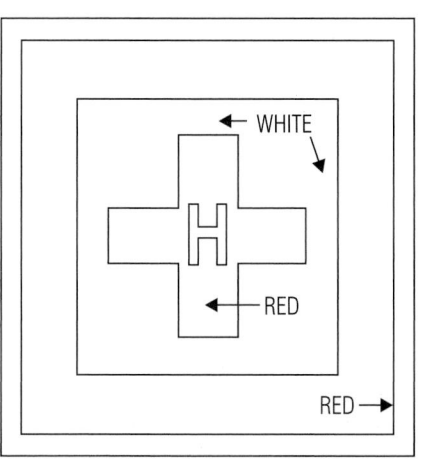

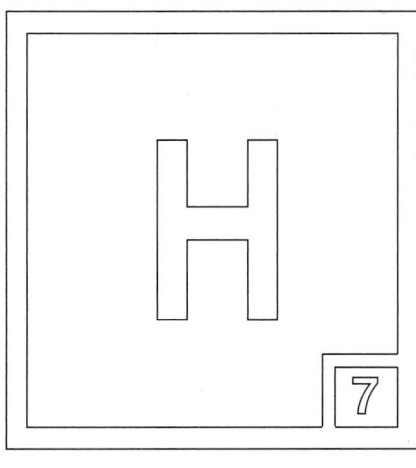

(c)   (d)

Fig. 18.34 Heliport markers: (*a*) Standard public heliport markers; (*b*) example of marker for private-use ports; (*c*) marker for hospital heliport; (*d*) weight-limiting marker (7000 lb indicated) for elevated heliports. ("*Heliport Design,*" *Federal Aviation Administration.*)

landing and takeoff area. Floodlighting will be effective. One method is to place low Fresnel-lens lights around the landing and takeoff area, with a sharp cutoff that will not bother the pilot.

("Heliport Design," Federal Aviation Administration.)

18.25 STOL Ports

There is a great potential for STOL (short take-off and landing) aircraft in short-haul transportation, serving stage distances of up to 500 mi. There is considerable advantage for city-center-to-city-center and intracity air-passenger carriers that can provide better service to passengers and relieve both air-space and ground congestion at large airports.

Criteria for STOL ports are tentative and subject to change as evaluation of proposed STOL aircraft and operational experience dictate. Significant future changes may be incorporated into revisions to the Federal Aviation Administration publication, "Planning and Design Criteria for Metropolitan STOL Ports." The STOL vehicle promises shorter runways, steeper approach paths, lesser real estate requirements, and the prospect of in-town airport locations, but more research and testing are needed.

Roger S. Boraas
Consulting Engineer
HNTB Corporation Denver,
Colorado

19

RAIL-TRANSPORTATION ENGINEERING*

Rail transportation is considered in this section as a system in which vehicles are supported and guided by rails or other guideways. Rail-transportation engineering deals with the need, planning, selection, design, and construction of such systems for movement of passengers and freight. It involves roadbed, track, bridges, trestles, culverts, yards, terminals, stations, office buildings, locomotive fueling facilities, environmental protection facilities, signals and communications, track-side protection devices, and railroad-car, locomotive, and transit-vehicle maintenance facilities. Engineers may be responsible for maintenance of way and structures. And they must be familiar with motive power, railway cars, and other equipment.

Rail transportation is the most effective way to handle increased transportation demands with relatively low power requirements, a low land requirement, little air pollution, and few accidents involving fatalities and injuries. As a result, as population and gross national product increase, rail transportation increases in importance. The U.S. Congress, in recognition of this, has passed legislation that, in particular, has added to the importance of rail transportation of freight and of passenger rail transit, for example, through deregulation under the Staggers Act in the early 1980s and the Intermodal Surface Transportation

Efficiency Act (ISTEA) about a decade later. In 1998, Congress passed the Transportation Equity Act for the 21st Century (TEA-21). A portion of this law authorizes Federal funding for grade crossing safety, transit and high speed rail systems and additional research on magnetic levitation technology. Additional legislation is being considered to fund development of viable high speed rail corridors.

19.1 Glossary

Following are terms commonly encountered in rail-transportation engineering:

Alignment. Horizontal location of a railroad as described by tangents and curves.

Apron, Car Ferry. Bridge structure supporting tracks and connecting the car deck of a car ferry to land. The apron is hinged at the shore end so that it is free to move vertically at the outboard end to accommodate varying elevations of the ferry.

Apron, Track. Railroad track along the waterfront edge of a pier or wharf for direct transfer of cargo between ship and car.

Ballast. Selected material, such as crushed stone, placed on the roadbed to hold the track in line and surface.

Batter Rail. The deformation of the surface of the head of a rail at the end.

Batter (Pile). Slope of inclined piles.

---

*Updated and revised from Sec. 19, Rail-Transportation Engineering, by D.L. McCammon in the fourth edition.

Branch Line. Secondary line or lines of a railway.

Branding. Identification markings hot-rolled in raised figures and letters on a rail web indicating weight of rail and section number, type of rail, kind of steel, name of manufacturer and mill, and year and month rolled.

Car, Light Rail (Trolley Car). A self-propelled vehicle operating on rails, generally in streets, and drawing electric power from overhead or underground conductors.

Car, Motor. A powered track car for transporting two to six people.

Car, Push. A four-wheeled railway work car designed to be pushed by hand or towed by a motorcar. It is used to transport materials too heavy to be carried on a motorcar.

Car, Track. Any car or machine operated on track, such as a motorcar, handcar, or trailer.

Car Retarder. Braking device, usually power-operated, built into a railway track to reduce the speed of cars. Brake shoes, when set in braking position, press against the sides of the lower portion of the car wheels.

Compromise Joint. Joint bars for connecting rails of different fishing height and section, or rails of the same section but with different joint drillings.

Cradle. Structure riding on an inclined track on a riverbank and having a horizontal deck with a track on it for transfer of railroad cars to and from boats at different water elevations.

Crib. Space between two successive ties.

Crossing, Grade. An at-grade crossing of a railroad and a highway, usually with protective devices such as warning signs, flashing lights, bells or gates.

Crossing (Track). Construction used where one track crosses another at grade; it consists of four connected frogs.

Crossing, Bolted Rail. A crossing in which all the running surfaces are of rolled rail. The parts are held together with bolts.

Crossing, Manganese-Steel Insert. A crossing in which a manganese-steel casting is inserted at each of the four intersections. Fitted into rolled rails, the casting forms the points and wings of the crossing frogs.

Crossing, Solid Manganese-Steel. A crossing in which the frogs are of the solid manganese-steel type.

Crossing, Movable-Point. A crossing of small angle in which each of the two center frogs consists essentially of a knuckle rail and two opposed movable center points with the necessary fixtures.

Crossing, Single-Rail. A crossing in which the connections between the end frogs and the center frogs consist of running rails only.

Crossing, Two-Rail. A crossing in which the connections between the end frogs and the center frogs consist of running rails and guardrails.

Crossing, Three-Rail. A crossing in which the connections between the end frogs and the center frogs consist of running rails, guardrails, and easer rails.

Crossing Plates. Plates interposed between a crossing and the ties or other timbers to protect the ties and to support the crossing better by distributing loads over larger areas.

Crossover. Two turnouts with the track between the frogs arranged to form a continuous passage between two nearby and generally parallel tracks (Fig. 19.15).

Crossover, Double. Two crossovers that intersect between the connected tracks; also two crossovers within a short distance that allow movements to connected tracks.

Curve, Compound. A continuous change in alignment effected with two or more contiguous, simple curves of different radii but with a common tangent at each junction (Fig. 19.5).

Curve, Degree of. (See Degree of Curve.)

Curve, Easement. A curve whose radius varies to provide gradual transition between a tangent and a simple curve or between two simple curves of different radii (Fig. 19.7).

Curve, Lead. Curve between switch and frog in a turnout (Fig. 19.15).

Curve, Reverse. Curve formed by two contiguous, simple curves with a common tangent but with centers of curvature on opposite sides of the tangent (Fig. 19.6).

Curve, Simple. A continuous change in alignment effected with an arc of constant radius and fixed center (Fig. 19.4).

Curve, Spiral. (See Curve, Easement.)

Curve, Vertical. An easement curve connecting intersecting grade (sloped) lines (Fig. 19.8).

Degree of Curve. Angle subtended at the center of a simple curve by a 100-ft chord.

Derail. A track structure for derailing rolling stock in an emergency.

Easer. (See Rail, Easer.)

Elevation of Curves (Superelevation). Height of outer rail above inner rail along a curve.

Fishing Space. Space between head and base of a rail occupied by a joint bar (Fig. 19.13).

Flangeway. Open way through a track structure that provides a passageway for wheel flanges (Fig. 19.16).

Flare. A tapered widening of the flangeway at the end of a guard line of a track structure. A flare may be at the end of a guardrail or at the end of a frog or crossing wing rail (Fig. 19.16).

Foot Guard. Filler for space between converging rails to prevent a foot from being accidentally wedged between the rails.

Frog. A track structure at the intersection of two running rails to provide support for wheels and passageways for their flanges, thus permitting wheels on either rail to cross the other (Fig. 19.16).

Frog Angle. The angle formed by the intersecting gage lines of a frog.

Frog, Bolted Rigid. A frog built of rolled rails with fillers between, held together with bolts.

Frog, Center. Either of the two frogs at the opposite ends of the short diagonal of a crossing.

Frog, End. Either of the two frogs at the opposite ends of the long diagonal of a crossing.

Frog, Flange. (See Frog, Self-Guarded.)

Frog, Heel of. The end of the frog farthest from the switch.

Frog, Movable Point. A frog with a movable point to eliminate the flangeway gap in locations with small frog angles.

Frog Number. Half the cotangent of half the frog angle.

Frog Point. That part of a frog lying between the extensions of the gage lines from their intersection toward the heel end (part farthest from the switch). The theoretical point is the intersection of the gage lines. The half-inch point is located at a distance from the theoretical point toward the heel equal, in inches, to half the frog number and at which the spread between the gage lines is $\frac{1}{2}$ in. Usually, measurements are made from the half-inch frog point.

Frog, Rail-Bound Manganese-Steel. A frog consisting essentially of a manganese-steel body casting fitted into and between rolled rails and held together with bolts (Fig. 19.16a).

Frog, Self-Guarded (Flange Frog). A frog with guides or flanges above its running surface to contact the tread rims of wheels to guide their flanges safely past the point of the frog (Fig. 19.16b).

Frog, Solid Manganese-Steel. A frog consisting essentially of a single manganese-steel casting (Fig. 19.16b).

Frog, Spring-Rail. A frog with a movable wing rail normally held against the point rail by springs. The rails thus form an unbroken running surface for wheels on one track, whereas the flanges of the wheels on the other track force the movable wing rail away from the point rail to provide a passageway. Viewed from the toe end toward the point, a right-hand frog has the movable wing rail on the right-hand side.

Frog, Toe of. The end of the frog closest to the switch.

Gage (Track). Distance between gage lines (Fig. 19.9). (Standard gage is 4 ft $8\frac{1}{2}$ in.)

Gage (Track Tool). A device by which the gage of a track is established or measured.

Gage Line. A line $\frac{5}{8}$ in below the top of the center line of head of running rail or corresponding location of tread portion of other track structures along that side nearer the center of track.

Grade Line. Line on profile representing tops of embankments and bottoms of cuts ready to receive ballast. This line is the intersection of the plane of the roadbed with a vertical plane through the center line.

Guard, Stock. A barrier between and along track rails to prevent passage of livestock on or along the track.

Guard Check Gage. Distance between guard and gage lines, measured perpendicular to gage lines across the track.

Guard Face Gage. Distance between guard lines, measured perpendicular to gage line across the track.

Guard Line. A line along that side of the flangeway nearer the center of track and at the same elevation as the gage line.

Guard Timber. A longitudinal timber placed outside the track rail to maintain tie spacing.

Guardrail. A rail or other structure parallel to the running rails of a track used to prevent wheels from being derailed, or to hold wheels in correct alignment to prevent their flanges from striking the points of turnouts, crossing frogs, or switches. Also, a guardrail is a rail or other structure laid parallel to the running rails of a track to keep derailed wheels adjacent to running rails.

Guardrail, Frog. A rail or other device to guide a wheel flange so that it is kept clear of the point of a frog.

Guardrail, Inner. A longitudinal member, usually a metal rail, secured on top of the ties inside the track rail to guide derailed wheels.

Guardrail, One-Piece. A guardrail consisting of a single component so designed that no auxiliary parts or fastenings other than spikes are required for its installation.

Hi-Rail Vehicle. A truck or other vehicle with special wheel assemblies that allow for travel on track in addition to highways.

Joint, Compromise. See Compromise Joint.

Joint, Insulated. A rail joint with insulating material to prevent the flow of electric current between abutting rail ends.

Joint, Rail. Splice uniting abutting ends of contiguous rails.

Joint Bar. A stiff steel member commonly used (in pairs) to join rail ends and to hold them firmly, evenly, and accurately in surface and gage-side alignment (Fig. 19.13).

Joint Gap. Distance between ends of contiguous rails in track, measured on the outside of the head $\frac{5}{8}$in below top of rail.

Lead. Distance between actual point of a switch and half-inch point of a frog. The actual lead is measured along the line of the parent track (Fig. 19.15). The curved lead is measured to the half-inch point of the frog but along the outside gage line of the turnout. The theoretical lead is the distance from the theoretical point of a uniform turnout curve to the theoretical point of the frog, measured along the line of the parent track.

Nosing. A transverse, horizontal motion of a locomotive that exerts a lateral force on the supporting structure.

Out of Face (Trackwork). Work, such as tie replacement, that proceeds completely and continuously over a given piece of track as distinguished from work at disconnected points.

Rail (Track). A rolled steel shape, commonly a T section, laid end to end, on crossties or other suitable supports, to form a track for railway rolling stock (Fig. 19.10).

Rail, Closure. Rail between the parts of any special trackwork layout, such as the rail between switch and frog in a turnout (sometimes called lead or connecting rail); also the rail connecting the frogs of a crossing or of adjacent crossings but not a part of the crossings (Fig. 19.15).

Rail, Compromise. Relatively short rail with two ends of different section to correspond with the ends of rails to be joined. It provides the transition between rails of different section.

Rail, Easer. A rail that provides a bearing for the portion of hollowed-out treads of worn wheels that overhangs a running rail. Sloped at the ends, an easer is laid with its head along the outside of and close to the head of the running rail.

Rail, Guard. (See Guardrail.)

Rail, Knuckle. A bent rail or equivalent structure forming an obtuse point at a movable-point crossing or slip switch. When set for traffic, the movable points of the crossing or switch rest against the obtuse point.

Rail, Reinforcing. A bent rail placed with its head outside of and close to the head of a knuckle rail to strengthen it and act as an easer rail; or a piece of rail similarly applied to a movable center point.

Rail, Running. Rail or surface on which the tread of a wheel bears.

Rail, Stock. Running rail against which the switch rail operates.

Rail, Switch (Switch Point or Switch-Point Rail). Tapered rail of a split switch (Fig. 19.17).

Rail, Welded. Two or more rails welded together to form a length less than 400 ft. When the length is 400 ft or more, the result is called a continuous welded rail.

Retarder, Car. See Car Retarder.

Retarder, Insert. A braking device without external power, built into a railway track to reduce the speed of cars with brake shoes against the sides of the lower portions of wheels. Sometimes, means are provided to open the retarder to nullify its braking effect.

Right-of-Way. Lands or rights used or held for railroad operation.

Shoulder. That portion of the ballast between the end of the tie and the toe of the ballast slope.

Siding. Track, auxiliary to the main track, used to permit trains to pass.

Spot Board. A sighting board placed above and across the track at a proposed elevation for the rails to indicate the new surface and insure its uniformity.

Stamping. Figures and letters indented, after hot sawing, in the center of the rail web, parallel with the direction of rolling, to indicate the serial heat number, ingot number as cast or rolled, and position of each rail relative to top of ingot.

Station, Loop. A form of through station in which the station track layout embraces a loop or part of a circle. Trains move in one direction only and turn relative to the station.

Station, Stub. Station with tracks connected at one end only.

Station, Through. Station with tracks connected at both ends.

Stock Pass. A culvert or bridge opening under a track primarily for passage of livestock.

Subballast. Material of superior character spread on the finished subgrade of a roadbed below the top-ballast to provide good drainage, prevent frost upheaval, and distribute the load over the roadbed (Figs. 19.1 and 19.9).

Subgrade. Finished surface of roadbed below ballast and track.

Surface, Running (Tread). Top part of structures on which the treads of wheels bear.

Switch. A track structure for diverting rolling stock from one track to another (Fig. 19.17).

Switch Rod. The rod connecting to the switch stand to enable movement of the switch points.

Switch, Slip. A combination of a crossing with left- and right-hand switches and curves between them within the limits of the crossing connecting the two intersecting tracks on both sides of the crossing without separate turnout frogs. A single slip switch combines a crossing with one right-hand and one left-hand switch; a double slip switch, with two right-hand and two left-hand switches.

Switch, Split. A switch consisting essentially of two movable-point rails with necessary fixtures (Fig. 19.17).

Switch, Spring. A switch with an operating mechanism incorporating a spring device to return the movable points automatically to their original or normal position. This action takes place after the points have been shifted by the flanges of trailing wheels passing along the track other than that for which the points are set for facing movements.

Switch Angle. Angle between the gage lines of a stock rail and the switch rail at its point.

Switch Detector Bar. Strip of metal, alongside the track rail, connected with the throwing mechanism of a switch to prevent moving of the switch under trains.

Switch Heel. End of a switch rail nearer the frog. Heel spread is distance at the heel between gage lines of stock and switch rails (standardized at $6\frac{1}{4}$in for straight switches).

Switch Point. (See also Rail, Switch.) Theoretically, the intersection of the gage line of the switch rail, extended, and the gage line of the stock rail. The actual point is that end of the switch rail farther from the frog; the point where the spread between the gage lines of stock and switch rails is sufficient for a practicable switch point (Fig. 19.17).

Switch Stand. Device for manual operation of switches or movable center points.

Switch Throw. Distance through which points of switch rails are moved sideways (standardized at $4\frac{3}{4}$ in). It is measured along the center line of the No. 1 switch rod or head rod.

Tangent. Straight rails or track; specifically, straight track contiguous with a curve.

Tie, Cross. The transverse member of the track structure to which rails are fastened to provide proper gage and to cushion and distribute traffic loads (Fig. 19.9). An adzed tie has plate-bearing areas on top made plane and smooth by machine. A bored tie has machine-made holes for spikes. A grooved tie has depressions machine-gouged across its top into which ribs on the bottom of a tie plate fit.

Tie, Heartwood. A tie with sapwood no wider than one-fourth the width of the tie top between 20 and 40 in from midlength.

Tie, Sapwood. A tie with sapwood wider than one-fourth the width of the tie top between 20 and 40 in from midlength.

Tie, Slabbed (Pole Tie, Round Tie). A tie sawed on ends, top, and bottom only.

Tie, Switch. A tie that functions as a crosstie but is longer and also supports a crossover or turnout.

Tie Plate. Plate interposed between a tie and rail or other track structure (Fig. 19.9).

Topballast. Material of superior character spread over a subballast to support the track structure, distribute the load, and provide good drainage (Fig. 19.1).

Track. Assembly of rails, ties, and fastenings over which cars, locomotives, and trains move.

Track, Body. Each of the parallel tracks of a yard on which cars are moved or stored.

Track, Connecting. Two turnouts with the track between the frogs arranged to form a continuous passage between one track and another intersecting or oblique track or another remote, parallel track.

Track, Crossover. (See Crossover.)

Track, Drill. A track connecting with a ladder track and over which locomotives and cars pass in switching.

Track, House. A track alongside or entering a freight house and used for cars receiving or delivering freight.

Track, Ladder. Track connecting the body tracks of a yard.

Track, Lead. An extended track connecting either end of a yard with the main track.

Track, Main. Track extending through yards and between stations and on which trains are operated by timetable or train order, or both, or the use of which is governed by block signals.

Track, Rider. A track in a hump yard on which a conveyance is operated for returning car riders to the summit of the hump.

Track, Running. A track reserved for movement through a yard.

Track, Side. A track auxiliary to the main track for use other than as a siding.

Track, Special. All rails, track structures, and fittings other than plain unguarded track that is neither curved nor fabricated before laying.

Track, Spur. A stub track diverging from another track.

Track, Stub. Track connected with another track only at one end.

Track, Team. Track on which cars are placed for transfer of freight between cars and highway vehicles.

Track, Transfer. A track so located with respect to other tracks and transferring facilities as to facilitate transfer of lading from one car to another.

Track, Wye. Triangular arrangement of tracks on which cars, locomotives, and trains may be turned.

Track Bolt. A buttonhead bolt with oval neck and threaded nut for fastening rails and joint bars.

Tread. Top surface of a railhead that contacts the wheels.

Turnout. Arrangement of a switch and frog for diverting rolling stock from one track to another (Fig. 19.15).

Turnout Number. Number corresponding to the frog number of the frog used in the turnout.

Turntable. A structure at the center of radial tracks that allows locomotives or cars to be turned and positioned for movement onto any of the tracks.

Wye. (See Track, Wye.)

Yard. System of tracks for such purposes as making up trains, storing cars, and sorting cars and over which movements not authorized by time-table or train order may be made, subject to prescribed signals and rules or special instructions.

Yard, Flat. A yard in which car movements are accomplished by locomotive, without material aid from gravity.

Yard, Gravity. Yard in which car classification is accomplished by locomotive, with material aid from gravity.

Yard, Hump. Yard in which car classification is accomplished by pushing the cars over a summit, beyond which they run by gravity.

Yard, Retarder. Hump yard equipped with retarders to control car speed during descent to classification tracks.

Yard, Sorting. Yard in which cars are classified in greater detail after they have passed through a classification yard.

American Railway Engineering and Maintenance of Way Association (AREMA)—8201 Corporate Drive, Suite 1125, Landover, Maryland 20785-2230 (301.459.3200) (www.arema.org).

19.2 Rail-Transportation Systems

There are three principal types of rail-transportation systems: intercity passenger and freight, commuter, and rapid transit. Outstanding attributes of each are safety, low energy requirements (a rolling resistance of 3 to 8 lb/ton for steel wheels on steel rails), ability to handle 1000 passengers or 10,000 tons of freight (or more) with one train, a minimum amount of land required for right-of-way, dependability of service under all weather conditions, and little atmospheric pollution. Other types of rail-transportation systems are personal rapid transit, which has the objective of taking passengers from one station on the line to any other station on the line with a minimum of waiting time for a car and no intervening stops, and monorail and magnetic-levitation fixed-guideway systems.

19.2.1 Freight Systems

Generally, freight railroads are private industries that own or lease their right-of-way, build and maintain their own track, structures, signal systems and communication systems and operate trains of owned or leased equipment. These railroads provide for the movement of freight between cities across the country and between bordering countries. Freight railroads move all types of goods. Some, such as bulk goods like coal and grain, are moved in unit trains, which are trains composed entirely of one type of car. Other goods are moved in mixed trains, which are comprised of many types of cars. Successful freight service depends on fair rates, consistent transit times and on-time delivery, freedom of lading damage, and ease of loading and unloading cargo. Good engineering and operation are required to provide profitable freight service.

19.2.2 Intercity and High Speed Passenger Systems

Standard intercity passenger service provides safe and reliable movement of people across the country at speeds up to 80 miles per hour. Typically, standard intercity passenger service uses tracks of freight rail companies and therefore shares the track with freight trains. Characteristic engineering requirements for a satisfactory passenger service include cars having trucks equipped with very long travel springs, snubbers, cross stabilizers, air conditioning, good lighting, attractive decor, comfortable and roomy seats, clean and adequate toilet facilities, convenient baggage storage, good dining car service reasonably priced, and vista-dome lounge cars (except for overnight service). Departure times, speed, on-time arrivals, and low fares are also important factors.

Currently, in the United States, most intercity passenger travel is by private automobile or airplane with only a small percentage travelling by rail. Until 1971, intercity rail passenger service was provided by freight railroads. Due to huge deficits incurred for this service, Congress established a quasi-public corporation—the National Railroad Passenger Corporation, usually called Amtrak—to operate a basic national passenger system. Railroads operating intercity passenger service were given the option to continue their own service or join Amtrak by contract. Railroads joining Amtrak, by payment of considerable fees, were relieved of all responsibility for provision of intercity rail passenger service. Amtrak began operations on

May 1, 1971 and operated about 60% of the intercity passenger trains that had existed before its creation. Eventually all railroads signed contracts with Amtrak. Amtrak continues to provide intercity rail passenger service on more than 22,000 route miles in 45 states and is subsidized by the federal government. Some state governments also subsidize Amtrak for specific trains that operate within their states.

High Speed Rail (HSR) service is generally accepted to be for speeds of 110 miles per hour and above. HSR typically has its own designated track and right-of-way separate from any freight service. Japan, Germany and France have had this service for many years and continue to build new routes. Recently, a French TGV traveled from Calais to Marseilles averaging 306 km/hr (190 mph). The United States is just starting to develop plans for HSR in corridors where such service will provide competitive timing to the airlines. Amtrak has started HSR service in the Northeast Corridor, serving major cities between Washington D.C. and Boston. High Speed Rail service uses trainsets composed of two power cars, one on each end, with baggage cars, multiple passenger cars and a food service car between the power cars. These trainsets are aerodynamically shaped to reduce airflow resistance.

19.2.3 Commuter Systems

These usually provide short-haul passenger service between a large city and its suburbs and operate as part of a larger rail system. Peak periods for transportation of workers occur during early morning and late afternoon. But some service must be provided throughout the day. Important requirements are reliability, minimum travel time, convenience, comfort, and economy. Trains typically travel on standard railroad track. They may incorporate self-powered cars or be moved by diesel-electric locomotives.

Automobile travel competes with commuter service, so it is important that engineers design a commuter system that will attract the maximum possible volume of travel. Attractions are trains at frequent intervals, protection from inclement weather, possible saving in travel time, potential improvement in air quality, and economy. Some studies indicate that rail transit in lieu of highways offers lower construction costs by decreasing requirements for right-of-way and travel lanes for automobiles. Use of double-deck, stainless-steel commuter cars, with air conditioning, good lighting, and comfortable seats; on-time performance and frequent scheduling; and push-pull operation have resulted in substantial increase in commuter travel, even with some increase in fares.

However, even with good commuter service, railroads have been unable to operate this service profitably. As a result, some states subsidize commuter operation where it is considered advantageous to do so. Some large urban areas have created transit districts that fund development and operation of commuter rail systems. In several localities existing underutilized or abandoned railroad freight lines have been purchased or refurbished to establish commuter service. The justification for such state and Federal aid has been saving of money and land that would otherwise be used for additional expressways, relief of automobile congestion in cities, reduction in the amount of parking space required in cities, fewer automobile accidents, and less noise and air pollution.

19.2.4 Rail Transit Systems

These are primarily intracity, although some provide service to nearby suburbs. Characteristic requirements are frequent and dependable service, quick loading and unloading, light weight for rapid acceleration and deceleration, low fares, and a degree of comfort consistent with the other requirements.

Rapid-transit vehicles are primarily propelled by some form of external electricity. During rush hours, passengers usually have to stand for a portion of the run. In congested areas, trackage traditionally has been located in subways or on elevated structures. In some congested areas, automobile-traffic lanes of a highway have been replaced by trackage. Also, in several central city areas and major transit hub areas, transit development has led to construction of pedestrian malls in conjunction with shopping areas.

With population growth, it becomes desirable to extend or add to the rapid-transit system in some cities, and in other cities that have no rapid-transit system, to study the desirability of providing a rapid-transit rail system or some other type of system to provide adequate transportation for the increased population. The advantages of rail rapid transit are much the same as those given for commuter lines. Although the likelihood of a rail rapid-transit system being self-supporting is not good, few cities are able to provide, on existing streets and highways, bus service that is self-supporting.

One principal difference between commuter systems and rapid transit is that rapid transit involves new construction in most cases. Therefore, studies should be made to determine location and station spacing that will be most compatible with feeder buses at stations and will also be most convenient for the maximum number of people.

Rapid transit is subsidized with Federal funds distributed by the Federal Transit Administration (FTA), Department of Transportation (DOT), and local or area transit authorities. The Intermodal Surface Transportation Efficiency Act (ISTEA) sponsors utilization of all modes of transportation, including rail transit. The FTA also sponsors research aimed at improvement of rapid-transit components and development of new transit concepts. European technology, especially that applicable to passenger vehicles, is used extensively for new at-grade systems.

19.2.5 Personal Rapid Transit (PRT)

Providing passengers with individualized service, PRT systems are also called Automated Guideway Transit (AGT) or People Movers. The PRT cars are relatively small. They are electrically operated. The best type of system allows a passenger to call a PRT car to a station by pushing a button or dialing. After boarding, the passenger can designate the station to which he or she wants to travel by pushing a button or dialing, and the car will proceed to that station without stopping at any intervening station. The objective is to minimize waiting and transit time. The operation is completely automatic. Interference with other cars operating on the same line is prevented by computer scheduling. Ticket selling and collecting are also automatic, using computer-controlled vending machines and turnstiles.

19.3 Cost-Benefit Analyses of Rail Transportation Systems

In the United States, new construction of freight systems consists mostly of line changes, grade revisions, and trackage to serve new industries, mines, quarries, and provide capacity for increased traffic. The justification for line changes is primarily reduction of curvature to permit higher speeds or shortening the line to reduce running time, to compete better for freight business. Reduction of curvature and shortening of the line also reduce maintenance costs, helping railroads to be more competitive with other transportation modes. The justification for grade reductions is to permit longer trains to be hauled with one crew or to eliminate the cost of helper engines. The benefits obtained by these measures may be determined from data given in Art. 19.19. The cost of the line changes or grade revisions should be estimated from the cost of right-of-way required and the cost of building the new line.

Benefits of the line or grade change should be estimated from reduced trip time, reduction in locomotives in inventory, decrease in rolling-equipment wear, fuel savings, reduction in track maintenance costs, decrease in track-component replacement costs, and elimination of fixed-plant facilities, including but not limited to sidings, stations, signal equipment, and road crossings. The benefits of new line construction for industries, mines, and quarries should be based on the added revenue the new lines may be expected to produce and balanced against the cost of construction, maintenance, and taxes for the added trackage.

Almost all new construction for intercity rail passenger service is contemplated to be for High Speed Rail. The cost benefit analysis is complex as dedicated right-of-way with no grade crossings is required. The location of terminals, length of the route and speed of operation determine total travel time from origin to destination. This, along with quality of on board service, must be comparable or better than equivalent airline travel.

For contemplated rail-transit, commuter, or personal rapid-transit systems, the cost-benefit analysis is more involved. There are also a number of nonquantifiable benefits that should influence the final decision. Cost and benefit comparisons should be made between alternative rail-transit systems as well as with other forms of mass transit and highway systems required to move an equivalent volume of persons efficiently and cost-effectively.

Quantifiable benefits are those that produce a net economic gain and are directly attributable to the rapid-transit system. These include land cost savings, increase in land values, savings in rider time, reduced auto operating and parking costs, less congestion for auto traffic, improvements in air quality, reduced adverse effects on the environment, decreased noise pollution, less pedestrian congestion in business districts, reduced need for a second or third car for some families, and cost savings for insurance and transportation.

The cost-benefit analysis should be made for a reasonable period of time into the future and include projected population growth, amortization of equipment at the interest rate that must be paid over a period of 25 to 30 years (some equipment has been used longer but should have been replaced because of obsolescence), and interest on the cost of rapid-transit fixed facilities (roadway, stations, and shops). It is assumed that maintenance and operating costs are included in the quantifiable transportation cost savings.

Nonquantifiable benefits include increased regional growth; development of community centers; attraction of new industrial development; added employment in the construction, maintenance, and operation of the system; adequate transportation for the young and aged; increased accessibility to educational, institutional, and recreational facilities; reduction in air pollution; and reduction in the total energy required.

Most transportation planning agencies use a 15-year horizon for project planning and transportation system modeling. Various scenarios are developed for analyses of alternative systems to determine the most cost-effective program for transportation improvement for a given locality. Rail-transportation alternatives, especially when the Intermodal Surface Transportation Efficiency Act is taken into account, are being selected in those localities where population growth and density justify the cost.

19.4 Route Selection

As stated in Art. 19.3, new construction for freight systems usually involves line changes, grade revisions, or trackage to new industries. The only consideration in route selection is to obtain the desired objective at lowest cost with minimum environmental damage. Since grading and bridge structures will probably be the only items that can be varied, use should be made of existing government topographical and geological maps to the extent they will suffice. If a considerable amount of trackage is involved, it will probably be desirable to have aerial

contour maps made (photogrammetry), first on a large scale to lay out one or more possible routes and then at small scale along each route to arrive at estimated grading quantities.

Maximum grade and rate of curvature must be established before a location can be chosen. Grade is expressed as the ratio of rise to distance in percent (a 1% grade rises 1 ft/100 ft). Rate of curvature is the central angle, degrees, subtended by a 100-ft chord. It is desirable that both grade and rate of curvature be kept to a minimum, but almost always a lower grade and rate of curvature mean increased construction cost and sometimes a longer time. Studies should be made of several routes having different grades and rates of curvature, taking into account the annual carrying charges on construction cost and the estimated costs for the anticipated train operation. From these studies, the grade and curvature may be selected to give minimum costs. A calculation of running time should be made and considered in making this decision.

Track gage must be decided on early. Standard gage for railway track in the United States (and many other countries) is 4 ft 8½ in, measured between the inner sides of the heads of the two rails of the track at a distance ⅝ in below the top of the rails. This gage should be used if equipment is to be interchanged with other railroads having standard gage. Locomotives, cars, and mechanized work equipment are commonly manufactured for this gage.

A roadway cross section must also be adopted. A minimum width of roadway crown of 24 ft is recommended for top of subgrade with subballast, ballast, and track placed on top. For sidings or multiple track, a minimum distance between track centers of 15 ft is recommended. On fills, the side slopes should be at least 1 on 1½ in earth, 1 on ½ in loose rock, and 1 on ¼ in solid rock (Fig. 19.1).

When aerial surveying techniques are used in conjunction with physically located control points along the chosen route, preliminary and final design can be accomplished with minimal field surveying. With computer programs developed for roadway design, engineers can prepare plans and determine earthwork quantities. Tying the control points to the local coordinate system also allows development of right-of-way information. Heavily wooded and bushy areas along the route, however, may cause some errors in ground elevations as well as hide some salient features critical to project success. Hence, a personal reconnaissance of the

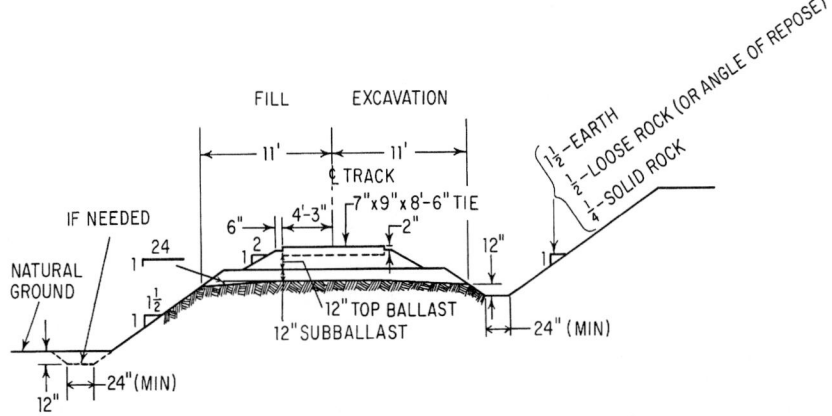

Fig. 19.1 Typical roadbed and ballast cross section for straight single track; top ballast, about 3700 yd^3/ mi; subballast, about 4900 yd^3/mi (includes 15% shrinkage).

site is advisable before a final alignment has been selected. Before construction starts, a final survey should be made to locate physically the control points and alignment and to stake the project, to provide adequate information for the construction contractor.

19.4.1 Intercity Systems

High Speed Rail systems being planned will require new dedicated right-of-way or substantial upgrading of existing rail routes. For new HSR systems, there are many considerations in route selection. These include a gentle alignment and profile, electric power sources, environmental effects of construction and avoidance of any at grade crossings.

19.4.2 Commuter and Rail-Transit Systems

Route selection for these shorter lines is determined by a number of factors. Since these systems are to serve people, a route that will be closest to the largest number of people connecting activity hubs is to be preferred. This may be done by taking into account the following:

Service to existing land use, which includes major employment areas; residential areas; institutions (hospitals, schools, churches, recreation, and other public facilities); and sports, zoo, parks, and other cultural and recreational areas.

Availability of right-of-way, an important factor in the cost. Alternative alignments that make use of existing right-of-way, vacant undeveloped land, and publicly owned land and streets will minimize acquisition costs and relocation of homes and businesses.

Current plans and proposals for public and private projects that are contemplated for the future.

Impacts of proposed transit on the environment, noise, neighborhoods passed through, safety, and opportunities to enhance neighborhood growth.

For rail-transit lines, it is possible to use steeper grades than for intercity passenger and freight lines, although here again minimum practical grades will afford operating economies. For horizontal curves, the degree of curvature should be kept to the minimum practicable. The maximum curvature permitted will depend on the desired running speed, the amount of superelevation provided, and characteristics of the rolling equipment. Consideration should also be given to the length of car to be operated in a subway because the sharper the curve, the greater the width required (due to overhang) and the higher the cost of the tunnel construction.

For one rail-transit system, the following track standards have been established:

Tangent: desired minimum length, 500 ft and absolute minimum, 75 ft; extension at stations, 100 ft beyond length of platform.

Curvature: desirable minimum radius for main-line track, 1000 ft; for yard tracks, 250 ft; minimum track radius for track in circular tunnels, 1000 ft.

Spiral or transition curves should be used between tangent and curvature of $1°$ or more, except in yard or slow-speed trackage; also should be used between compound curves.

Grades: maximum between stations, 3.5%; through stations and at terminal storage tracks, 0.3%; other storage tracks and yard areas, level; minimum length of constant-profile grade, 500 ft.

Vertical curves should be used between changes in grade; minimum curve length, 100 times the algebraic difference of the grades being connected, but not less than 200 ft. Where vertical and horizontal curves are combined and an unbalanced superelevation in excess of 1 in is present, the length should be doubled.

Reversed curves should not be used without incorporating the minimum tangent length or length required for the two runoffs of elevation, whichever is greater.

Superelevation should be the equilibrium for the speed permitted, with a maximum of 4 in. An unbalanced elevation of up to $1\frac{1}{2}$ in is permitted for speeds requiring more than 4-in elevation.

19.4.3 Right-of-Way

For intercity passenger and freight lines, the right-of-way should accommodate the number of tracks and the slope for the cuts, fills, and borrow pits. Unless a line is being located in a densely populated area or land cost is very high, a minimum right-of-way width of 50 ft each side of the track should be obtained. Allowance should also be made for any stations or yard facilities that may be required.

Location of receiving, classification, and departure yards is governed primarily by operating requirements. At one time, yards were provided between divisions having different ruling grades. With diesel power and the need to avoid delay due to switching, it is generally preferable to handle the same train from origin to destination, adding or taking off diesel units at intermediate points if conditions justify.

Rail-transit systems in subways, at grade, or on elevated structures are advantageous in densely populated areas. Construction costs, effects of construction on businesses and travel, as well as long-term maintenance costs and effects of the transit system on businesses and travel should be considered in determination of the best right-of-way for the system chosen for mass transit serving a specific area.

19.5 Track Location

Location of the tracks on prepared roadbed—including cut, fill, and sidehill cut and fill—or natural ground surface is the most economical and is to be preferred where practicable. However, in some instances, other locations are more desirable for reasons more important than the first cost. This applies particularly to rail-transit systems that are to be constructed.

In some instances, a city has provided space for trackage in the median strip of expressways when they were constructed, anticipating the construction of rapid-transit trackage at a later date when population growth would require it. In this case, the track location has already been decided. Otherwise, trackage for new rapid-transit systems should be constructed in open roadbed wherever practicable.

In residential areas, the trackage should be elevated or placed in open cuts to avoid street grade crossings. The choice between the two is largely a matter of which costs the least from an overall standpoint of first cost and maintenance. Open cuts will probably require reinforced concrete retaining walls on each side of the trackage, with a chain link fence and barbed-wire outriggers on top of each retaining wall to prevent children or others from falling into the cut. This could be avoided by use of a tunnel, but a tunnel is more costly to construct and maintain.

Elevated trackage is in most cases preferable to cuts. It has its disadvantages, principally in terms of aesthetics, and the effects of noise on nearby residents. However, in modern elevated-track construction, the track support is of either reinforced concrete or prestressed concrete, or a combination of the two, and of pleasing appearance. The elevated construction may have a ballast deck so that the trackage can be supported on ballast, which does much to reduce the noise, or the rails may be supported directly on the concrete floor, in which case special fastenings will be used with resilient pads between the rail and the deck to reduce the noise level. Figure 19.2, for example,

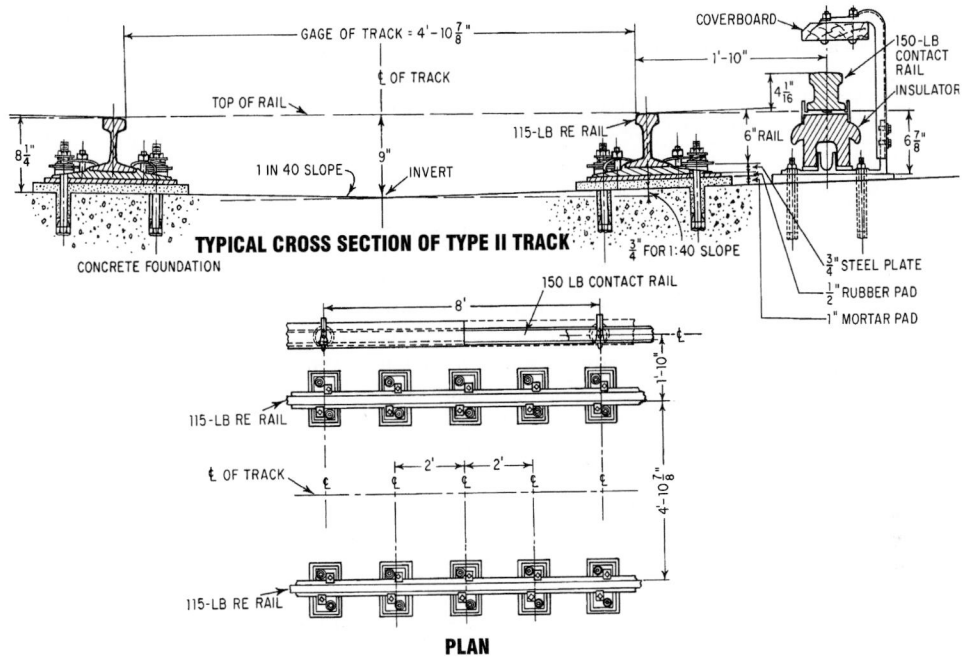

Fig. 19.2 Tangent track construction used in subways of the Toronto Transit Commission.

shows the rail fasteners used in tangent track construction in subways of the Toronto Transit Commission. A rubber pad is inserted under the steel rail plate for noise and vibration attenuation. Figure 19.3 shows the Landis direct-fixation rail

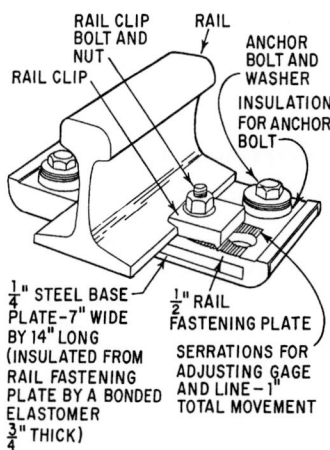

Fig. 19.3 Landis direct-fixation rail fastener.

fastener, developed for use on the San Francisco Bay Area Rapid Transit system and later used for other installations. This device for directly affixing rail to a rigid support structure incorporates a shear pad consisting of a $\frac{1}{2}$-in-thick steel rail fastening plate and a $\frac{1}{4}$-in-thick steel base plate bonded on opposite sides of a $\frac{3}{4}$-in-thick elastomeric pad. The base plate is bolted directly to the supporting structure. The elastomeric pad not only insulates the rail plate from the base plate but permits an elastic deflection of about $\frac{1}{4}$ in for attenuating noise and vibration.

A noise level of 70 to 75 dB(A) is comparable to noise frequently encountered in residential areas, and 70 to 80 dB(A) to usual noise in commercial and retail districts. In residential areas, tracks should not be closer to homes than 100 to 120 ft for elevated structures or ballast on grade or on fill. For speeds of 50 mi/h or above, a sound barrier should be placed between the track and any house within 120 ft. In commercial and retail areas, trackage can be as close to buildings as 30 ft if a sound barrier is provided. The sound barrier may be a vertical wall extending from the ground to 10 in above the bottom of the car side skirt with 8 to 10 in clearance, lined on the inside with acoustic material. (Most of the noise from the rapid-transit car comes from the trucks and wheel impacts.) Such a sound barrier will lower the noise level about 12 dB(A). Use of continuous welded rail with the running surface periodically ground smooth with a grinding car or a train of such cars is also helpful for reducing the noise level. It is also desirable to keep the car wheels ground smooth to reduce impact noise.

Placing a rapid-transit line in a subway in a main business district can eliminate the adverse effects of train noise and reduce street space requirements. Elevating the tracks also decreases the impact on street space but has the disadvantage of reducing the penetration of daylight to the street level. Construction and other costs of a subway or elevated line, however, are greater than those of a rapid-transit line at grade. Several rapid-transit systems are in successful operation with at-grade alignments with resilient rail supports and continuous welded rail. Designers of these systems took into account the effect on traffic. Use of resilient rail supports (Figs. 19.2 and 19.3), continuous welded rail periodically ground, and wheels kept ground smooth will minimize transmission of train vibrations through the ground to nearby buildings.

19.5.1 Intercity System Capacity

For intercity passenger and freight lines, the location of passing tracks and yard tracks should be taken into account in establishing the grade line. If the line is for a single track, the time it takes for train A to go from one passing track to the next and meet train B and then for train B to get to the first passing track determines the capacity of the railway in trains per day. Thus, passing tracks spaced close together afford larger line capacity than those spaced far apart.

The sidings should be long enough for the maximum length of train to be clear of the main line. If centralized traffic control is used, even longer passing tracks are desirable to permit passing without stopping the trains.

19.5.2 Commuter and Rail-Transit Systems Capacity

Commuter and rail-transit lines will be double-track in most cases. It is necessary to have cross-

overs suitably located to permit use of only one track at slack periods so that track repairs can be made, a disabled car or train bypassed, third rail or trolley repairs made, or for other reasons. However, the addition of crossovers cannot be expected to add much to traffic capacity of such a double-track line. Addition of a third and fourth track would be the most effective way to increase capacity, if that were needed.

In some areas with high demand at rush hour (morning and evening), commuter systems are operating successfully over existing freight railroad lines. These systems provide transit in the morning and evening, and freight traffic utilizes the lines when the commuter trains are not operating. One such system provides emergency taxi service during the day for patrons to travel back to the suburbs.

The capacity of a double-track system is normally about 40,000 passengers per track per hour. This is based on 10-car trains, with 300 passengers per car, operating at 5-min intervals. The determining factor is the time required for a train to come into a station, unload and load passengers, and depart from the station.

If capacities of this amount or more are contemplated, station design should be planned accordingly. Capacities in excess of 40,000 per track per hour are possible if the stations can be designed to handle passengers at the proposed rate. Normally, all the passengers for one train would not unload or load at a single station. The exceptions are stations that serve a baseball field, football stadium, or similar facility from which large numbers of passengers may be discharged in a short time, and some emergency that requires many passengers to leave a station swiftly.

19.6 Horizontal Curves for Railways

These include simple, compound, and reverse curves; superelevation required for such curves; and spiral curves as a means of introducing the superelevation on a gradual and uniform basis.

19.6.1 Simple Curves

A simple curve has a constant radius throughout. The **degree of curvature** generally is measured by the central angle subtended by a 100-ft-long chord. Radius R, ft, and degree of curve D are related by

$$R = \frac{50}{\sin(D/2)} \qquad (19.1)$$

For curves up to 7°, length measured along the curve is practically the same as that measured with 100-ft chords. Hence, the radius R of a curve is given approximately by

$$R = \frac{36,000}{2\pi D} = \frac{5730}{D} \qquad D < 7 \qquad (19.2)$$

For curves of more than 7°, the error in radius increases with the degree of curve.

In the location or staking of the center line of a simple curve, the tangents (to its ends) should be extended, if possible, to an intersection $P.I.$ and the intersection angle Δ measured (Fig. 19.4). The tangent distance T from the point of curve, $T.C.$, to $P.I.$ and from the end of curve, $C.T.$, to $P.I.$ may be determined from

$$T = R \tan \frac{\Delta}{2} \qquad (19.3)$$

Length of curve, ft, from $T.C.$ to $C.T.$ is given approximately by

$$L = \frac{100\Delta}{D} \qquad (19.4)$$

where Δ and D are in degrees.

Stakes should be driven and tacked to mark the $T.C.$ and $C.T.$ This can be done by setting a transit at $P.I.$ and sighting along each tangent. The transit then should be moved to the $T.C.$, sighted on $P.I.$,

and $\Delta/2$ turned for a check on $C.T.$ Next, stakes should be set every 50 ft for flat curves. The measurement should be made with 100-ft chords for curves over 7°. It is good practice to mark stations (100-ft intervals) around the curve and to set a stake at each station and at plus 50.

The transit deflections degrees (angle between tangent and line from $T.C.$ to point on the curve), for each stake equals

$$\alpha = \frac{LD}{200} \qquad (19.5)$$

where L = length of curve, ft

D = degree of curve (for α in minutes, multiply by 60)

Suppose, for example, a stake is to be set and tacked at $1108 + 50$ when the $T.C.$ comes at $1108 + 10.5$ and the degree of curve is 2° 30'. A length of $50 - 10.5 = 39.5$ ft should be taped from the $T.C.$ and a deflection angle of $39.5 \times 2.5 \times 60/200 = 30$ min turned with the transit, to set the stake at $1108 + 50$. For each succeeding stake, at 50-ft intervals, the increment of deflection is $50 \times 2.5 \times 60/200 = 37.5$ min.

19.6.2 Compound and Reverse Curves

A compound curve comprises two or more simple curves, each successive curve having a common tangent with the preceding curve (Fig. 19.5). The point of curve $T.C.$ and end of curve $C.T.$ are staked as for a simple curve, although calculation of the tangent distances is more involved. The transit should be moved to the beginning point of each simple curve to stake it. The degree and central angle for each simple curve of the compound curve have to be known or decided on in advance. Compound curves should be avoided, but they may be used where excessive excavation or fixed objects that must be cleared justify or require such a curve. (See also Spirals.)

A reverse curve (Fig. 19.6) is a combination of two simple curves with centers on opposite sides of a common tangent. Reverse curves are acceptable in slow-speed passing and yard tracks but should never be used in main line. A short tangent, at least 100 ft long, but preferably more, should be placed between curves of opposite direction in main line.

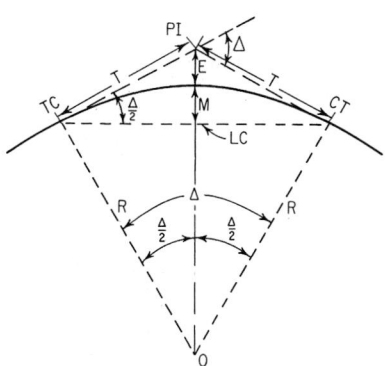

Fig. 19.4 Simple curve.

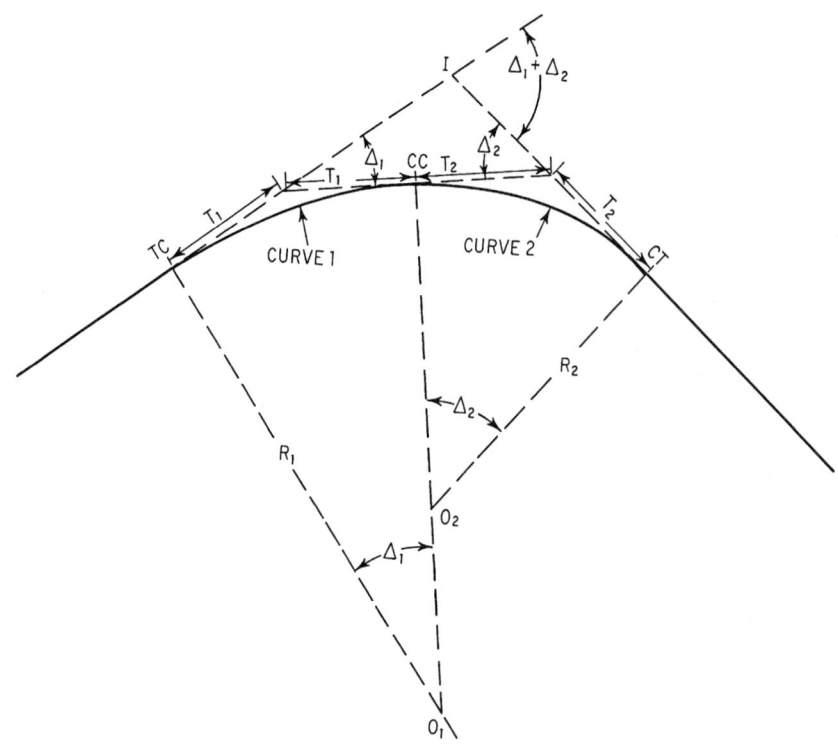

Fig. 19.5 Compound curve.

19.6.3 Superelevation of Curves

Elevation of the outer rail of a curve relative to the inner rail is desirable on main-line track. The amount of superelevation depends on degree of curvature and desired operating speed around the curve. However, the amount of superelevation is usually limited to 7 in to prevent undue tilting of the train if it stops on the curve. For sharp curves, it may be necessary to restrict train speed so that it will not exceed by too much the speed for which the curve is elevated.

The amount of superelevation to be provided on a curve, up to the 7-in maximum, is a matter of judgment, subject to change from service experience. Most freight railroads have their own criteria, which combine speed, curvature, amount of overbalance, and length of spiral to determine allowable superelevation. Passenger-train service on freight lines, however, affects superelevation requirements. Usually, and on single-track lines particularly, not all trains will go around a given curve at the same speed. If too little elevation is provided for the predominating traffic and speed, the outer rail will show excessive wear on the gage side from the wheel flanges. If too much elevation is provided, the inner rail will show excessive flow on the top of the railhead toward

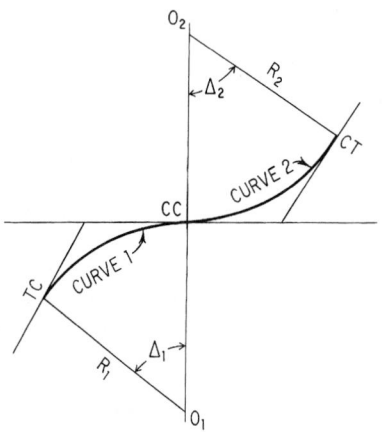

Fig. 19.6 Reverse curve.

the gage and field sides and sometimes surface corrugation.

Equilibrium speed is the speed at which outward centrifugal force from curvature is just balanced by the inward component of car weight resulting from elevation of the curve. For a given degree of curve and elevation

$$V = \sqrt{\frac{E}{0.000149gD}} \qquad (19.6)$$

where V = equilibrium speed, mi/h

E = superelevation of outer rail, in

D = degree of curve

g = gage of track, ft

Permissible speed somewhat in excess of equilibrium speed will not cause discomfort to passengers or other undesirable effects. This permissible speed may be obtained readily from Eq. (19.6) by adding 3 in to the actual elevation of curve. For example, for a 3° curve with 5-in superelevation and a track gage of 4.708 ft, the equilibrium speed is 49 mi/h. The permissible speed, however, is 62 mi/h (equilibrium speed for 8-in elevation). Thus, the permissible speed will have a deficiency in elevation of 3 in. This is acceptable for the type of equipment in general use in the United States. These requirements may change as high-speed passenger trains and "tilt trains" come into use. For passenger cars having antiroll devices, a somewhat higher deficiency is permissible (*Proceedings, American Railway Engineering Association*, vol. 56, p. 125). For some types of freight cars having a very high center of gravity (over 96 in above top of rail), a somewhat smaller deficiency may be desirable to guard against derailment.

19.6.4 Spirals

An easement curve or spiral should be placed between tangents and each end of a simple curve and between the simple curves of a compound curve. A spiral increases in curvature gradually, thus avoiding an abrupt change in the rate of lateral displacement of cars. It also provides a means of gradually elevating the high rail in proper relation to the degree of curvature.

Several forms of spiral may be used. The one generally used in the United States increases

degree of curvature with length.

$$d = ks \qquad (19.7)$$

where d = degree of curvature at any point

k = increase in degree of curvature per 100-ft station

s = length in 100-ft stations from beginning of spiral to any point

The central angle δ, degrees, from the beginning of spiral, $T.S.$ (Fig. 19.7), and the deflection a, degrees, from the tangent at $T.S.$ vary as the square of the length.

$$\delta = \frac{1}{2}ks^2 \qquad (19.8)$$

$$a = \frac{1}{6}ks^2 \qquad (19.9)$$

Also, the offset of the spiral, ft from either the tangent or the circular curve varies as the cube of the distance. Other key elements shown in Fig. 19.7 may be computed from

$$X_o = S(50 - 0.000508\Delta^2) \qquad (19.10)$$

$$T_s = X_o + (R + O)\tan\frac{I}{2} \qquad (19.11)$$

$$E_s = O + (R + O)\left(\sec\frac{I}{2} - 1\right) \qquad (19.12)$$

$$O = 0.1454\Delta S \qquad (19.13)$$

where S = total length of spiral in 100-ft stations

Δ = total central angle of spiral, degrees

R = radius of circular curve, ft

O = offset, ft, from tangent to circular curve extended at midlength of spiral

The deflection from the tangent to the end of the spiral, $S.C.$, with the transit set at $T.S.$, is one-third of Δ. When the transit is set at $S.C.$ and a backsight is taken on $T.S.$, a deflection of $\frac{2}{3}\Delta$ must be turned off to put the line of sight tangent to the circular curve. The deflections for the circular curve then should be turned from this tangent.

Stakes on the spiral should be set every 50 ft as for a simple curve. Deflections may be calculated to place the stakes on even stations and plus 50. Or if preferred to simplify calculation of deflections, the spiral may be divided into segments of equal length, say 10. Then, the deflection may be computed for the first segment, multiplied by 4

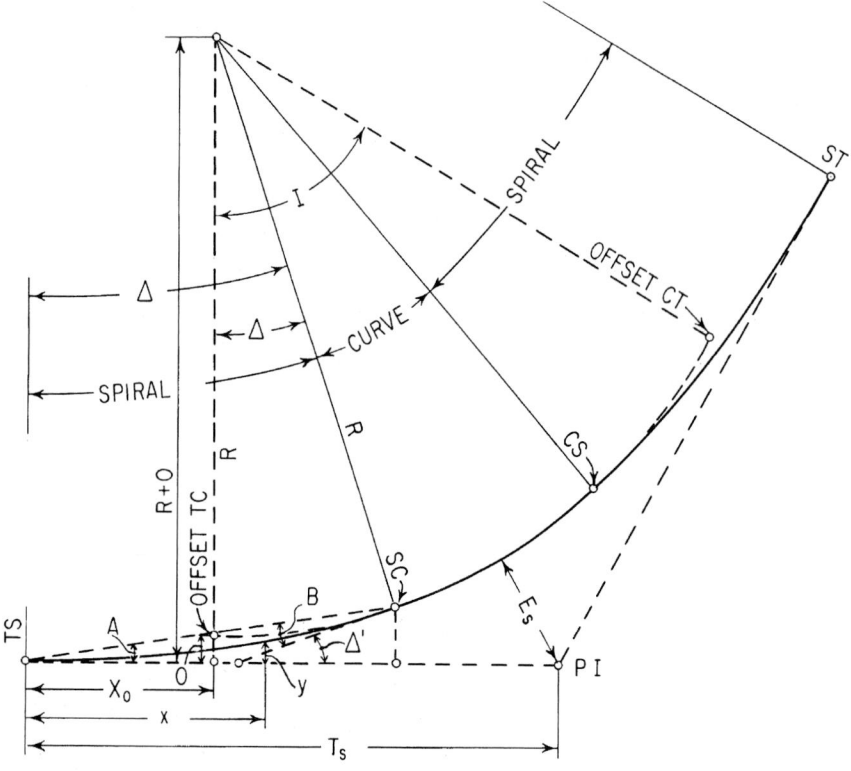

Fig. 19.7 Spiral provides transition between tangent and curved track.

(square of 2) to obtain the second deflection, by 9 (square of 3) for the third, by 16 for the fourth (square of 4), and so on.

To set the stakes for the spiral at a distance s with the transit at $S.C.$, subtract the deflection computed from Eq. (19.9) from the deflection computed for the same length of the circular curve extended. This deflection is then turned from the tangent at $S.C.$ to locate each stake.

The length of spiral should be such as to give passengers a time interval to adjust to the unbalanced centrifugal force without feeling a jerk on entering or leaving the curve. Also, the rate of change of elevation should be sufficiently gradual to prevent undue twisting of the car body. The desired minimum length of spiral, ft, is the greater of the lengths determined from

$$L = 1.63E_u V \qquad (19.14a)$$

$$L = 62E_a \qquad (19.14b)$$

where V = maximum train speed on curve, mi/h

E_u = unbalanced elevation (deficiency), in

E_a = elevation of outer rail, in

("Manual for Railway Engineering," American Railway Engineering and Maintenance-of-Way Association.)

19.7   Vertical Curves for Railways

At changes in grade on main line, a vertical curve of sufficient length should be provided to prevent excessive slack action in long freight trains or a sensation of discomfort to passengers at maximum speed. Experience has shown that the rate of change in grade per 100-ft station on vertical curves should not exceed 0.10% on summits or 0.05% in sags. Thus, if the grade changes from 0.20% descending to 0.20% ascending in a sag, the total change in grade is 0.40%, and a vertical curve $0.40 \times 100/0.05 = 800$ ft long should be provided. If a similar change in grade occurs on a summit, the length of vertical curve should be 400 ft.

Ordinarily, the length of vertical curve determined in this manner will not be an even number of stations. It is simpler and satisfactory to use a vertical curve of the next even number of stations; i.e., if the calculated length is 7.2 stations, use a vertical curve of 8 stations.

The form of the vertical curve is parabolic in a vertical plane. First, determine the elevations at the beginning and end of the vertical curve. Add these and divide by 2 to obtain the average. Determine the difference between this average and the elevation at the intersection of the two grades. One-half this difference is the offset from tangent, or correction, to be made at the middle of the vertical curve (Fig. 19.8). The correction at other points varies as the square of the ratio of the distance from the nearest end of the vertical curve to half the length of the curve. Table 19.1 illustrates the method of calculating a vertical curve on a summit.

(C. F. Allen, "Railroad Curves and Earthwork," and T. F. Hickerson, "Route Location and Design," McGraw-Hill Book Company, New York (books. mcgraw-hill.com).)

19.8   Track Construction

There are several different types of track construction used, depending on the type of rail-transportation service and the physical characteristics of the environment:

Monorail One line of a suitable type of rail and rail support, with the vehicles supported above or suspended below the monorail and electric-powered.

Table 19.1 Offsets from Tangent for Vertical Curve

$$\text{Length of curve} = \frac{+0.35-(-0.20)}{0.10} = 5.5 \text{ stations}$$

Use 6 stations, or 600-ft vertical curve.

Offset at $P.I. = \frac{1}{4}[939.65 - \frac{1}{4}(938.60 + 939.05)] = 0.41$ ft

Station	Grade elevation	Offset, ft*	Vertical-curve elevation
P.C. 1005 + 00	938.60	0.00	938.60
1006 + 00	938.95	0.05	938.90
1007 + 00	939.30	0.18	939.12
P.I. 1008 + 00	939.65	0.41	939.24
1009 + 00	939.45	0.18	939.27
1010 + 00	939.25	0.05	939.20
P.T. 1011 + 00	939.05	0.00	939.05

*Offset P.C. to P.I. (Fig. 19.8) varies as the square of the distance from P.C. Offset from P.T. to P.I. varies as the square of the distance from P.T.

A monorail supported on its under side may be used for elevated construction and in subway, but if used at ground level, it must have a grade separation at all highway and street crossings. A suspended monorail may be used for elevated construction. Enough clearance should be provided below the vehicle bottom for street and highway crossings. Its use in tunnel and subway construction would require that it be supported by the top of the opening, resulting in a high and costly opening. (The economics of a system in tunnel or subway construction depends a great deal on the area of the required opening.) Also, a monorail system has disadvantages in switching, weight support, construction cost, and ride quality.

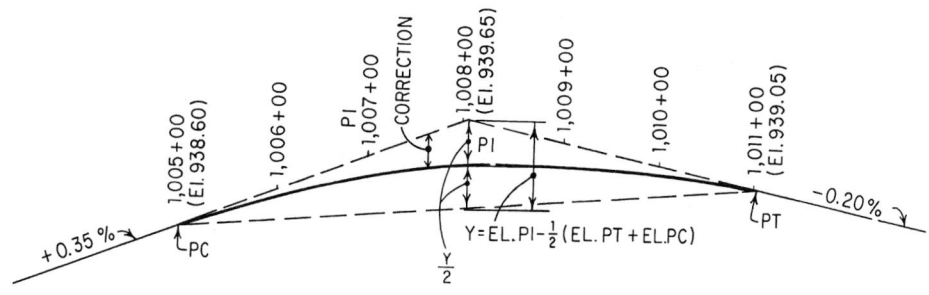

Fig. 19.8 Parabolic vertical curve connects two grades at a summit.

Dual Rail Two lines of parallel, steel running rails supported on tie plates, ties, and ballast (Fig. 19.9) for diesel-electric- or electric-powered vehicles.

Two parallel lines of steel or concrete beams (Fig. 19.19) to provide support and guidance for electric-powered rubber-tired vehicles.

Two parallel lines of suitably designed rails for the support of levitation-type vehicles, either air-cushion or magnetic (tracked air-cushion or Maglev vehicles), with linear-induction or turbojet motor power for traction.

The dual rail system with ties and ballast (Fig. 19.9) which is used for the bulk of the railway track miles in the United States, and its construction are covered in detail. Other systems either use some of the same components or are proprietary, in which case the details of construction can best be obtained from the owners.

The dual-rail systems with steel wheels on steel rails, with ties and ballast, or rubber-tired wheels on steel or concrete beams (Figs. 19.16 to 19.18) can be used for on-ground or elevated-track construction. However, in subway construction, the dual-rail steel-wheel-on-steel-rail system has the rails fastened to an insulated tie plate, which is bolted to the invert floor but separated by an insulating and cushioning pad and insulating thimbles and washers for the fastening bolts (Figs. 19.2 and 19.3). This fastening is more economical than the provision of additional tunnel or subway height to provide for tie and ballast depths, even if a depth of only 6 in of ballast is used under the tie. The dual-rail system with rubber tires on steel or concrete beams has some disadvantage in first cost because a larger-diameter tire must be used than for the steel wheel, thus resulting in increased height of tunnel or subway opening. This, however, is to a large extent offset by the narrower width of the opening required for the vehicles used in this system.

19.8.1 Roadbed

The roadbed, or subgrade, is a prepared ground on which to put the ballast, ties and rail. The subgrade

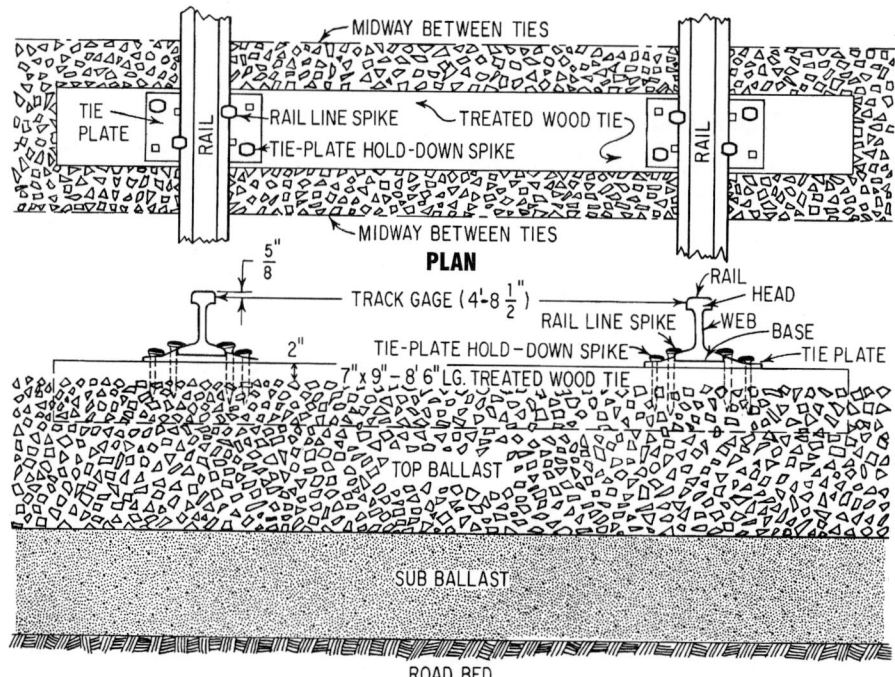

Fig. 19.9 Typical standard-gage, dual-rail track on tangent.

consists of compacted soil material that supports the track and train loading while transmitting and distributing the load with diminishing pressure to the natural ground below. The important dimensions to be considered in roadbed design are the top width of the roadbed, the height of fill (or depth of cut) and the side slopes of the cut or fill section. The top width of the subgrade must accommodate the track and ballast and may need to provide a walkway area outside of the ballast section. The top of the roadbed may be sloped downward away from the centerline to facilitate drainage from the ballast. The base of the roadbed must be wide enough to transmit the track and train loading within the allowable pressures of the natural ground material. A typical roadbed section used on freight railroads is shown in Figure 19.9.

Subballast is a finer graded material layer, between 6 and 12 inches thick, that is placed on the subgrade and acts as separation between the course rock ballast and the subgrade. The subballast can also provide protection to the subgrade against moisture infiltration from the track structure. It is most often used in mainline construction.

19.8.2 Drainage

Drainage of the roadbed and track structure is one of the most important aspects of good railroad construction and maintenance. Proper drainage helps to keep track alignment, profile and roadbed in good condition. Adequate side slopes, ditches and openings through the roadway should be provided. Soil that is wet or saturated has reduced bearing capacity. Undersized or blocked drainageways may cause flooding, washouts or settling of embankments. Poorly drained roadbed can cause accumulation of debris in the ballast, which leads to tie deterioration, pumping or low joints, and problems with track surface geometry. In cold weather, trapped moisture may lead to frost heaves, particularly at shallow culvert locations.

Intercept ditches and borrow ditches are primarily used to intercept surface water and carry it to nearby streams or other waterways. Typically, ditch gradients are determined by the track profile. However, a minimum gradient of 0.3% should be maintained to avoid sedimentation buildup. Conversely, gradients should not be so steep that erosion occurs due to high flow velocities. Where a high watertable is naturally occurring, subdrains made out of rock, with or without perforated pipe,

may be used to intercept the water and direct it away from the roadbed. Good drainage of groundwater in tunnels is an important factor in reducing track maintenance also.

Suitable drainage openings must be provided where the track construction crosses over waterways. The principal factors affecting the required size of a waterway opening are the area of the watershed, slope and characteristics of the ground within the watershed, and maximum intensity of rainfall that may be expected within a given period of time. In addition, culverts or drainage structures should be large enough to permit easy maintenance and cleaning.

Many methods are available to determine the maximum flow rates, velocities and backwater created for particular opening sizes and configurations. Each railroad has their own standard as to the minimum opening required and to the allowable effects of high flows on their right-of-way and facilities. For additional information on these methods and on waterway crossings in general, see the "Manual for Railway Engineering" produced by the American Railway Engineering and Maintenance of Way Association.

The new type of waterway opening provided may be galvanized steel or concrete pipe; concrete box or timber, concrete, or steel bridge. Culvert pipe and box culverts should have headwalls to prevent water erosion of fill. Abutment of bridges should have suitable wingwalls to contain fill and prevent erosion.

19.8.3 Tunnels

Tunnels are used to pass through mountains, under rivers or to bypass other topographic features that could hinder train operation. They are also used to allow trains to travel below ground in congested urban areas. The soil conditions will determine the type of construction necessary to keep the tunnel stable. Additional information on tunnels is found in Section 20.

19.8.4 Fencing

Right-of-way should be fenced if it is desired to keep off trespassers, livestock, or poultry. Posts should be not more than 16 ft 6 in apart. The fencing should be galvanized woven wire of No. 9 gage, or galvanized steel ribbon, smooth round, or barbed wire. The type and height of fencing are dictated by the local conditions and statutory requirements. (For details of fencing, see AREMA Manual.)

19.9 Rails and Rail Accessories

Rail serves three functions. It must resist contact pressure from the wheels, it must be capable of distributing the wheel load over several ties along the track and it must be able to do this repeatedly without breaking. To do these, rail must be hard and have sufficient stiffness, flexural strength and fatigue strength.

To provide flexural stiffness and strength, rail is shaped in section somewhat like an I beam. But the head is made narrower and deeper than the flange of an ordinary I beam to resist the contact pressure and wear from flanged wheels better. Table 19.2 and Fig. 19.10 show the principal dimensions and physical properties of sections that have been rolled in substantial rail tonnage or are being rolled today in the United States. The heavier sections are used for heavy traffic and high-speed lines.

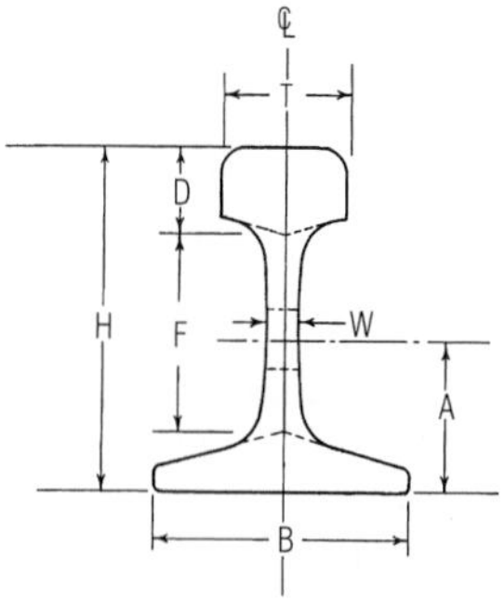

Fig. 19.10 Principal rail dimensions.

The standard length of rail in the United States and Canada is 39 ft.

The branding rolled in raised letters on one side of the rail web gives the weight of rail in pounds per yard, the section number, the mill, the year and

Table 19.2 Physical Properties of Rail Sections

Rail section	Weight: Lb per yd, Nominal	Weight: Lb per yd, Calculated	Weight: Net tons per mile	Dimensions, in (see Fig. 19.10): Height H	Base width B	Fishing F	Max head width T	Head depth D	Min web thickness W	Base to center of both holes, A	Cross sectional area, in^2	Moment of inertia, in^4	Section modulus: Head, in^3	Section modulus: Base, in^3
AREA (RE)	140	139.6	245.7	$7\frac{5}{16}$	6	$4\frac{1}{16}$	3	$2\frac{1}{16}$	$\frac{3}{4}$	3	13.9	95.9	24.3	28.6
AREA (RE)	136	135.8	239.0	$7\frac{5}{16}$	6	$4\frac{3}{16}$	$2\frac{15}{16}$	$1\frac{15}{16}$	$\frac{11}{16}$	$3\frac{3}{32}$	13.32	94.2	23.7	28.2
NYC	136	136.3	239.4	$7\frac{9}{32}$	$6\frac{1}{4}$	$4\frac{5}{32}$	$2\frac{15}{16}$	$1\frac{31}{32}$	$\frac{11}{16}$		13.36	93.9	23.9	28.1
AREA (RE)	133	133.4	234.8	$7\frac{1}{16}$	6	$3\frac{15}{16}$	3	$1\frac{15}{16}$	$\frac{11}{16}$	3	13.07	86.2	22.3	26.9
AREA (RE)	132	131.7	231.8	$7\frac{1}{8}$	6	$4\frac{3}{16}$	3	$1\frac{3}{4}$	$\frac{21}{32}$	$3\frac{3}{32}$	12.91	87.9	22.4	27.4
CB	122	122.5	215.6	$6\frac{25}{32}$	6	$3\frac{19}{32}$	$2\frac{15}{16}$	$1\frac{15}{16}$	$\frac{21}{32}$		12.01	74.0	20.6	23.3
AREA (RE)	119	118.7	208.9	$6\frac{13}{16}$	$5\frac{1}{12}$	$3\frac{13}{16}$	$2\frac{21}{32}$	$1\frac{7}{8}$	$\frac{5}{8}$		11.64	71.4	19.4	22.8
AREA (RE)	115	114.7	201.9	$6\frac{5}{8}$	$5\frac{1}{2}$	$3\frac{13}{16}$	$2\frac{23}{32}$	$1\frac{11}{16}$	$\frac{5}{8}$	$2\frac{7}{8}$	11.25	65.9	18.1	22.0
CF&I	106	106.6	187.6	$6\frac{13}{16}$	$5\frac{1}{2}$	$3\frac{3}{8}$	$2\frac{21}{32}$	$1\frac{3}{4}$	$\frac{19}{32}$		10.45	53.6	16.1	18.8
AREA (RE)	100	101.5	178.6	6	$5\frac{3}{8}$	$3\frac{9}{32}$	$2\frac{11}{16}$	$1\frac{21}{32}$	$\frac{9}{16}$	$2\frac{45}{64}$	9.95	49.0	15.1	17.8
ARA-A (RA-A)	100	100.4	175.6	6	$5\frac{1}{2}$	$3\frac{3}{8}$	$2\frac{3}{4}$	$1\frac{9}{16}$	$\frac{9}{16}$		9.84	48.9	15.0	17.8
ARA-B (RA-B)	100	100.5	176.9	$5\frac{41}{64}$	$5\frac{9}{64}$	$2\frac{55}{64}$	$2\frac{21}{32}$	$1\frac{45}{64}$	$\frac{9}{16}$		9.85	41.3	13.7	15.7
ASCE	100	100.4	175.6	$5\frac{3}{4}$	$5\frac{3}{4}$	$3\frac{5}{64}$	$2\frac{3}{4}$	$1\frac{45}{64}$	$\frac{9}{16}$		9.84	44.0	14.6	16.1
ARA-A (RA-A)	90	90.0	158.4	$5\frac{5}{8}$	$5\frac{1}{8}$	$3\frac{5}{32}$	$2\frac{9}{16}$	$1\frac{15}{32}$	$\frac{9}{16}$		8.82	38.7	12.6	15.2
ARA-B (RA-B)	90	90.5	159.3	$5\frac{17}{64}$	$4\frac{49}{64}$	$2\frac{5}{8}$	$2\frac{9}{16}$	$1\frac{39}{64}$	$\frac{9}{16}$	$2\frac{37}{64}$	8.87	32.3	11.5	13.2
ASCE	90	90.1	158.6	$5\frac{3}{8}$	$5\frac{3}{8}$	$2\frac{55}{64}$	$2\frac{5}{8}$	$1\frac{19}{32}$	$\frac{9}{169/16}$		8.83	34.4	12.2	13.5
ASCE	85	85.0	149.6	$5\frac{3}{16}$	$5\frac{3}{16}$	$2\frac{3}{4}$	$2\frac{9}{16}$	$1\frac{35}{64}$	$\frac{9}{169/16}$		8.33	30.1	11.1	12.2
ASCE	80	80.2	141.2	5	5	$2\frac{5}{8}$	$2\frac{1}{2}$	$1\frac{1}{211/2}$	$\frac{35}{64}$		7.86	26.4	10.1	11.1
ASCE	75	74.8	131.7	$4\frac{13}{16}$	$4\frac{13}{16}$	$2\frac{35}{64}$	$2\frac{15}{32}$	$1\frac{27}{64}$	$\frac{17}{32}$		7.33	22.9	9.1	9.9
ASCE	60	60.5	106.5	$4\frac{1}{4}$	$4\frac{1}{4}$	$2\frac{17}{64}$	$2\frac{3}{8}$	$1\frac{7}{32}$	$\frac{31}{64}$		5.93	14.6	6.6	7.1

month rolled, and the method of manufacture. A typical branding is as follows:

115	RE	CC	Manufacturer	1977	IIIII
(weight	(Type)	(If	(Mill	(Year	(Month
or section		controlled	brand)	rolled)	rolled)
number)		cooled)			

On the opposite side of the web, the rail is hot-stamped to show the heat number, rail letter (position in the ingot), and ingot number.

Currently, only rails with weights of 115 lbs per yard or higher are manufactured unless by special order. Rail specifications may be found in the "Manual for Railway Engineering," American Railway Engineering and Maintenance of Way Association. The chemical composition of typical rail is shown in Table 19.3. The minimum Brinell hardness specified for standard rails is 300. The Brinell hardness in high-strength rail (alloy or heat treated) normally ranges from 340 to 380. Rail is ultrasonically tested for internal defects and must meet specific dimensional tolerance requirements for sidesweep and upsweep over a 39 ft length.

Control cooling of rail (retarding the cooling rate under controlled conditions) is effective in preventing shatter cracks. These may lead to development of transverse fissures in service, so control cooling is included in rail specifications, except when rails are made from vacuum degassed steel.

On curves, many railroads use fully heat-treated rail, which has the top part of the head heat-treated, or alloy-steel rail, to withstand better the flange wear that occurs on the high rail of curves and the flow and corrugation that occur on the low rail.

19.9.1   Stress and Strain in Rails

Rail stresses and depressions for unusually heavy loads may be computed by considering a rail as a continuous beam on an elastic support (*American Railway Engineering Association Proceedings*, vol. 19, pp. 878-896). With the tie spacings in general use, the assumption that rail is continuously supported will not cause significant error. The modulus of elasticity of rail support u is the uniform load, lb/lin in of rail, required to depress the rail 1 in. It is further assumed that the pressure, lb/in, of the rail on its support at any point is

$$p = uy \tag{19.15}$$

where y = rail depression, in. Another significant term is the distance X_1, in, from point of application of a wheel load to the point where the bending moment caused by that load becomes zero and then reverses in direction.

$$X_1 = \frac{\pi}{4} \sqrt[4]{\frac{4EI}{u}} \tag{19.16}$$

where E = modulus of elasticity of rail steel (30,000 ksi)

I = moment of inertia of rail, in^4

For a single wheel load, the bending moment and rail depression along a rail may be determined in terms of M_o and Y_o from Fig. 19.11.

$$M_o = 0.318PX_1 \tag{19.17}$$

$$Y_o = -0.393 \frac{P}{uX_1} \tag{19.18}$$

where P = wheel load, lb

M_o = bending moment due to wheel load, in-lb

Y_o = rail depression under wheel load, in

Since there is always more than one wheel load, the master diagram may be used to determine the moment and depression at any point in the rail for all wheels by taking one wheel at a time and combining the effects algebraically. The maximum flexural stress in the rail base at this point may then be determined by dividing the total bending moment by the section modulus of the rail for the base. The tie load or reaction can be determined by calculating the average rail depression for the tie spacing and multiplying by the tie spacing and modulus u.

The value of u must be determined by actual measurement in track. This value ranges from 500

Table 19.3   Limitations on Chemical Content of Steel Rails, Percentage by Weight

Nominal weight of rail, lb/yd	115 or more
Carbon	0.72 to 0.82
Manganese	0.80 to 1.1
Silicon	0.10 to 0.60
Phosphorus, max.	0.035
Sulfur, max.	0.037

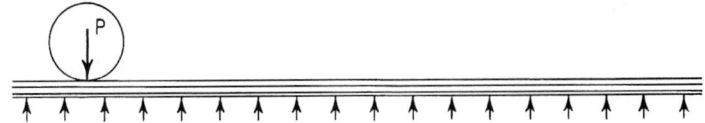

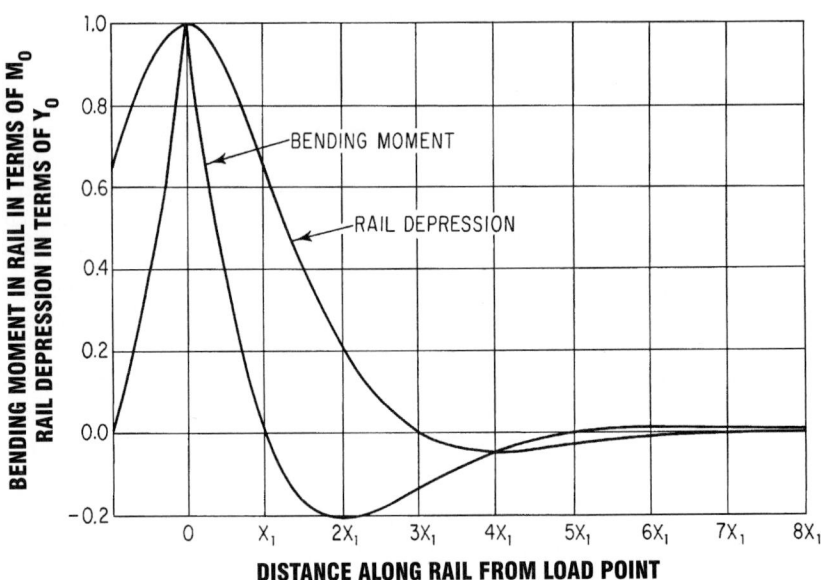

DISTANCE ALONG RAIL FROM LOAD POINT

Fig. 19.11 Diagram for calculating rail bending moment and depression under a single wheel load. (*ASCE-AREA Special Committee on Stresses in Railroad Track.*)

for track with little ballast and poorly compacted roadbed to 2000 or more on track with adequate ballast and well-compacted roadbed. The value of u is not critical in calculating rail stresses but is significant for rail depression.

There is no established impact effect or permissible working stress in rail because of variability of conditions on different railways. The following may be used as a guide: Multiply the stress for static loads by a percent impact factor of $33V/D$, where V is the speed, mi/h, and D the wheel diameter, in. Thus, with a 36-in-diameter wheel at 60 mi/h, the impact factor is 55%. A flexural stress at the extreme fiber of the base in jointed track of 35 ksi is permissible at speeds below 35 mi/h, or 30 ksi at higher speeds; in continuous welded rail, 25 ksi.

Figure 19.12 shows the bending stresses calculated by this method for a typical 100-ton freight car with four-wheel trucks. An approximate value of stress for other weights may be determined by multiplying the values shown by the ratio of the wheel weights on the rail.

19.9.2 Continuous Welded Rail

For new rail, most railways use continuous welded rail (CWR). It is usually placed in quarter-mile lengths, which are delivered to the job site in special trains. When in place in the track, the rails are welded end to end by a thermite welding process. In an alternative welding process, machines are used to butt weld in the field in the same manner as in the shop. Secondhand bolted rail is cropped to remove worn and battered ends and bolt holes and then is butt-welded before it is laid in track. However, any length of continuous rail that is 400 feet or longer is considered to be CWR.

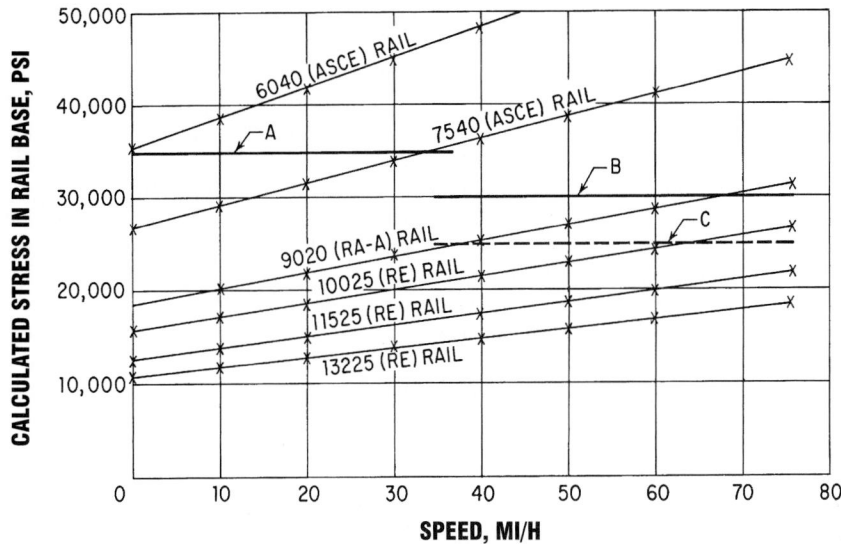

Fig. 19.12 Calculated rail stresses produced by a typical 100-ton-capacity hopper car (gross load of 263,000 lb). Recommended work stresses: (*a*) Jointed rail on branch line with speeds under 35 mi/h; (*b*) jointed rail on main line; (*c*) continuous welded rail on main line.

Expansion and contraction of continuous welded rail are controlled by the rail joints and the rail fastenings or anchors. The restraint stresses the rail. A tensile stress of 195 psi is produced in the rail by a 1 °F temperature drop. For example, if continuous welded rail is laid at 70 °F and the rail temperature drops to −30 °F, a tensile stress of 19,500 psi develops in the rail because it is restrained from shortening.

When the rail temperature increases above the laying temperature, compressive stresses develop in the rail due to the restrained ends of the rail string. The force developed by these stresses will cause the rail to move sideways. Unless this force is restrained, the rail and track will buckle. This movement is prevented by the use of rail anchors and sufficient ballast shoulders. The rail anchors provide a relatively uniform distribution of the lateral rail force to the ties and the ballast shoulder provides resistance to lateral tie movement. A minimum top of ballast shoulder width of 12 in is recommended. When the rail temperature decreases below the laying temperature, the rail tries to contract and induces tension at rail joints. The restraint at rail joints from track bolt shear strength and rail-joint friction may not be enough

to maintain the joint integrity and a pull-apart will occur. Additional rail anchors are used to restrain movement of the rail at joint locations (see Article 19.9.4).

An effort is made to lay continuous welded rail at about a mean temperature, which may require the rail to be heated or cooled. This is not always practical, so it may be desirable to adjust the rail length later if difficulty with track buckling or joint pull-aparts occurs.

19.9.3 Jointed Rail

Jointed rail is made up of short rail lengths (33 to 39 feet) joined together by joint bars with track bolts. This was the standard type of track before the development of continuous welded rail but is still widely used. Jointed rail in mainlines requires a higher level of maintenance than CWR. Wheel impacts at joints can cause rail end batter, loose or broken bolts and track surface geometry deterioration.

Rail-joint bars are used to join together abutting rails. As an alternative, rail is butt-welded into long lengths before it is laid in track. The welded strings are joined with rail-joint bars or thermite welds.

In jointed tracks, most railways use two 36-in joint bars with six bolts and spring washers per rail joint (Fig. 19.13). There are 271 rail joints per mile of jointed track.

In years past, joint bars were shaped somewhat like an angle in cross section and were called angle bars. Since about 1930, most joint bars have been shaped more like an I beam and are called joint bars or sometimes short-toe joint bars, to distinguish them from the long-toe angle bar. Headfree bars fit into the upper fillet between rail web and head. Take-up for contact-surface (fishing-surface) wear is provided in the base. Head contact bars have a slope on both the head and the base to match the fishing surfaces of the rail, and take-up for wear is provided at both head and base. Results with both types of bars have been equal in service tests. The "Manual for Railway and Maintenance of Way Engineering," American Railway Engineering Association, gives designs of joint bars for 115 RE, 119 RE, 132 RE, 133 RE, 136 RE, and 140 RE rail. Steel companies that roll joint bars can furnish design drawings of bars they are equipped to roll.

Most rail-joint bars are made of oil-quenched carbon steel, manufactured in accordance with specifications given in the AREMA Manual, or ASTM Standards. Carbon is specified at 0.35 to 0.60%; manganese, not over 1.20%; and phosphorus, not over 0.04%. Tensile strength of 100 ksi, yield point of 70 ksi, 12% elongation in 2 in, and 25% reduction of area are minimum requirements. A bend test is also required. Brinell hardness is not specified but usually varies from 225 to 275.

Rail-joint bars are punched with alternate oval and circular holes. Hence, the bars can be used on either side of the rail and always have an oval and circular hole match for the track bolt. Track bolts for 115 RE, 119 RE and 133 RE rail are 1-in diameter. Track bolts for 132 RE, 136 RE and 140 RE rail are 1-$\frac{1}{8}$ in diameter. Bar punching is spaced 6-6-7$\frac{1}{8}$-6-6 in/(AREMA Manual).

It is important that the bars be straight or cambered in the least harmful direction. For 36-in bars, a camber of $\frac{1}{16}$ in in either direction in the horizontal plane is acceptable. But in the vertical plane, the bar may not be low or more than $\frac{1}{16}$ in high at midlength.

19.9.3 Track Bolts

These are used for bolting a pair of joint bars in position. Most railways purchase heat-treated carbon-steel track bolts and carbon-steel nuts in accordance with specifications in the AREMA Manual or ASTM Standards. Track bolts have a forged button-type head with either an oval or elliptic neck to prevent turning in the joint bar. The threads are rolled. Most railways specify a Class 2 or finger-free fit. The bolt-and-nut design is in accordance with American National Standards Institute Standard B18.2. A minimum carbon of

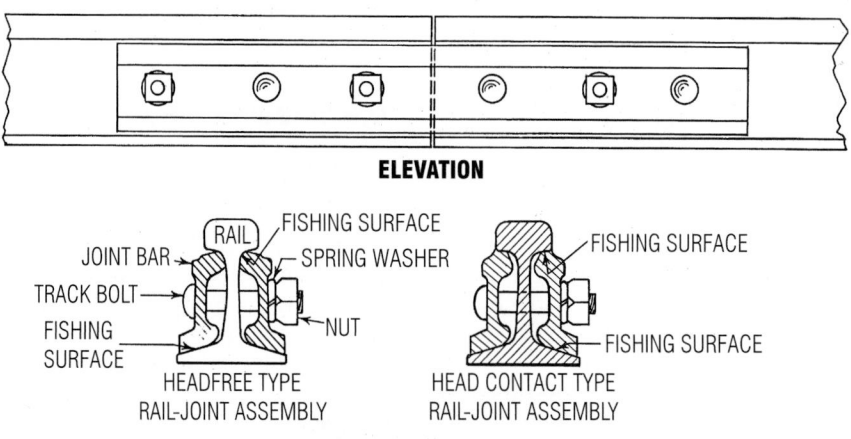

ELEVATION

CROSS SECTION

Fig. 19.13 Six-hole rail joint.

0.30%, maximum phosphorus of 0.04%, and maximum sulfur of 0.06% are specified. Tensile strength of 110 ksi, yield point of 80 ksi, 12% elongation in 2 in, and 25% reduction in area are minimum requirements. A bend test is specified, as is the minimum tension load that the bolt with nut fully engaged must withstand without stripping the nut or breaking the bolt. For the 1-in-nominal-diameter bolt, this is 66,560 lb; $1\frac{1}{16}$in, 76,360 lb; and $1\frac{1}{8}$-in, 83,900 lb. Initial bolt tension in track of 20,000 to 30,000 lb is recommended. Larger-diameter bolts have some value in resisting bending from the contraction force in the rail in cold weather.

Spring washers are used to maintain bolt tension and reduce the amount of bolt tightening required. Tests have shown that track bolts become loose because of fishing-surface wear, which permits the joint bars to move closer together, not because of vibration. Specifications for spring washers in the AREMA Manual require that, with a release of 0.03 in from an initial compression of 20,000 lb, spring washers will maintain a reactive force of at least 5000 lb. This amount of release is adequate for the fishing wear that occurs in a year's service, regardless of traffic, and a bolt tension of 5000 lb is sufficient to ensure proper functioning of the rail joint.

The high initial tension allows for proper seating of the joint bar and some subsequent relaxation of tension in service. Tension of in-service bolts should range from 15,000 to 25,000 lbs. Bolts must be checked for proper tension at regular intervals.

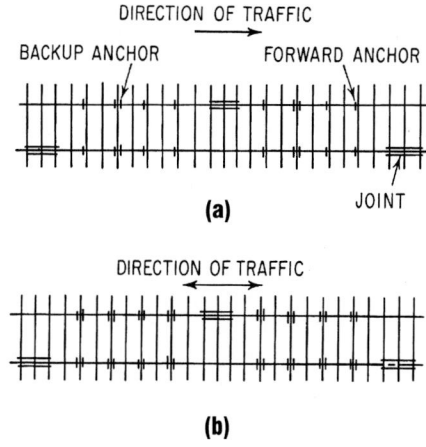

Fig. 19.14 Methods of anchoring jointed track. (*a*) For main track carrying traffic essentially in one direction. Under average conditions, with any type of ballast, use eight forward and two backup anchors per 39-ft rail length. (*b*) For main track carrying traffic in both directions. Under average conditions, with any type of ballast, use eight anchors per 39-ft rail length to resist movement in each direction, a total of 16.

changes and train movements. Where interruptions in CWR occur, such as at turnouts or crossings, every tie within 200 feet of the interruption, in both directions, should be box anchored.

19.9.4   Rail Anchors

A rail anchor is a device used to restrain lengthwise movement of rail. There are many different types in use. Most types engage the rail base by a spring clamping action and bear against the side of the tie or tie plate to restrain rail movement. The anchor should have sufficient holding power to move the tie in the ballast rather than permit the rail to slip through the anchor. Figure 19.14 shows a good method for anchoring jointed track with this type of anchor (AREMA Manual).

For continuous welded rail, every other tie should be have rail anchors applied on both rails on both sides of the tie. This is called box anchoring. Box anchoring will provide effective resistance to longitudinal stresses in the rail from temperature

19.9.5   Tie Plates

A rolled-steel plate is used between rail and tie to distribute the rail load, reduce tie abrasion and hold the rail to gage better. The trend has been toward larger tie plates and use of double shoulders, instead of just one shoulder to restrain the outer edge of the rail base. The plates, rolled to the desired cross section, are sheared to a width generally of $7\frac{3}{4}$ or 8 in. A cant of 1:40 is provided in the rail seat to incline the rail slightly inward. Tie plates having a length of 12 to 14 in are commonly used in the United States for rails having a base width of $5\frac{1}{2}$ in and a length of 13 to 16 in for rails with a base width of 6 in. A greater length of tie plate is provided on the field side of the rail than on the gage side (from $\frac{1}{2}$ to $2\frac{1}{2}$ in) to better resist the outward lateral forces on the rail on curves.

Generally, tie plates have four holes, $\frac{3}{4}$ in square, punched through the shoulders for spikes to hold the rail to line. The plates also have four holes, $\frac{11}{16}$ in square, punched near the corners for tie-plate fastening or hold-down spikes (Fig. 19.9). On tangent track, it is usual practice to use two line spikes in staggered holes for each tie plate. Sometimes, two hold-down spikes are used in oppositely staggered holes. On curves, two line and two hold-down spikes are used per plate. On curves of 6° and over with heavy traffic density, an additional line spike is used at the inner edge of the rail base. Some tie plates that are designed for use with elastic rail clips may have special configurations to allow for attachment of the clips to the plate.

Tie plates are made by various processes, generally by open-hearth or basic-oxygen. Carbon content varies from a minimum of 0.15% for low-carbon plates to a maximum of 0.85% for high-carbon plates. Low-carbon plates may be cold-worked; high-carbon plates must be hot-worked. Designs and specifications may be found in ASTM Standards and the AREMA Manual.

19.9.6 Rail Fastening

Rail fasteners consist of any device or system of components used to fasten the rail to the tie or other support. Many different types of rail fasteners are in use today. The most widely used is the steel cut spike. However, elastic or rigid steel clips and drive or screw spikes also are used. Rail fasteners provide restraint against vertical, lateral and rotational movement of the rail. They are also used to anchor tie plates to the ties. Some fasteners, such as elastic clips may provide some restraint against longitudinal movement of the rail as well.

Cut spikes are usually used to fasten rails to ties. They are formed with a wedge-shaped point to cut the tie fibers and prevent splitting. The head is rounded on top to facilitate driving; it is oval in shape and eccentric on the shank to provide a length of $\frac{11}{16}$ in to engage the top of the rail base. See also Art. 19.11.2.

Line spikes, for holding rails to gage, are commonly $\frac{5}{8}$ in square and 6 in long under the head. Hold-down spikes, for fastening tie plates to ties, are commonly $\frac{9}{16}$ in square and $5\frac{1}{2}$ in long under the head. A copper content of 0.20% is sometimes specified to give corrosion resistance.

For design and specifications, see ASTM Standards and the AREMA Manual.

Rail clips are designed to have contact with both the top surface of the rail base and the tie or rail support. Elastic clips are designed to deform measurably under load but return to their initial condition when unloaded. Rigid clips do not deform measurably under load. Clips may fit into special slots in the tie plate or be anchored to the support with bolts or screws.

Drive spikes, also known as screw spikes, are steel screws with square heads that may be used for tie plate hold downs, timber crossing hold downs and other timber applications. Drive spikes have material properties similar to cut spikes and come in diameters of $\frac{1}{2}$ to $\frac{3}{4}$ in.

19.10 Ties

Cross ties provide support for the rails and distribute rail loads to the ballast. Switch ties serve the same purpose as standard cross ties but are longer to support the widened sections of turnouts or railroad crossings. Ties are made of many types of materials, including wood, concrete, steel and composite materials. Most ties in railway track in the United States are of treated wood, mostly oak, gum, pine, or fir. For main-line track, the most common size is 7 × 9 in by 9 ft long. Smaller sizes are used for yard tracks, such as 6 × 8 in by 8 ft long. Ties are sawn rather than hewn and consist mostly of heartwood. This part of the tree is less desirable for lumber but more desirable for ties. Generally, ties containing the following will not be accepted by purchasers: decay; a hole more than 3 in deep and $\frac{1}{2}$ in in diameter when between 20 and 40 in from midlength, or more than one-fourth the width of the surface on which it appears when outside the sections of the tie between 20 and 40 in from its middle; a knot having an average diameter in excess of one-fourth the width of the surface on which it appears, except when outside the same 20- to 40-in zone; a shake larger than one-third the tie width; a split more than 5 in long; and a slant in grain in excess of 1 in 15 (AREMA manual).

When ties are received at the treating plant for seasoning, some railroads apply antisplitting devices, such as nail plates, to some or all ties. Ties should be seasoned prior to treatment with preservatives. Traditionally, air seasoning has been used wherein ties are stacked such that air flow is maintained around the ties to dry them to the proper moisture content. As this is a natural drying process, it is time consuming. Other seasoning methods include Boulton Drying, vapor drying and steam conditioning. Different types of wood respond differently to these methods and therefore care should be taken to select the best method.

Seasoning removes sufficient moisture from the wood to permit addition of a preservative. Before the seasoned ties are treated with preservatives, they should be adzed for the tie plates and bored for the track spikes.

Ties are treated with preservatives to prevent decay and extend the life of the tie. Coal Tar Creosote or mixtures with coal tar or heavy petroleum are the primary preservatives used. Treatment results are measured in pounds percubic foot of retention of the preservative. Depending on the wood type, retention of coal tar or petroleum-based preservatives should be between 6 and 8 pcf. Other preservatives, such as water-born salt or pentachlorophenol, may also be used. Tie treatments must meet specifications C-6 of the American Wood Preservers Association (AWPA).

Prestressed concrete ties are used in locations with heavy tonnage, high traffic volume or where steep grades or sharp curves are present. Prestressed concrete ties are manufactured with dimensions similar to wood ties. Inserts for rail fasteners are cast into the concrete. Tie plates are not used but insulating or cushioning pads are placed beneath the rail to absorb impact and prevent signal currents in the rail from entering the ties. Concrete strengths are generally 7000 psi and above.

Steel ties are often used in locations where vertical clearance is a factor. Due to the design of the steel ties, the distance from top of tie to bottom of ballast is less than for standard wood or concrete ties. Ties made of composite materials are being developed and tested to determine if they have the durability and flexibility to be used as replacement ties.

On most existing track, ties are renewed only as required or on a spot renewal basis. Prestressed-concrete ties should be placed out of face to give best results. Therefore, they are less economical for tie renewals in existing track.

19.11 Ballast

Ballast supports the ties, restricts movement of the ties and transmits rail loads to the subgrade or roadbed. Ballast also absorbs impact loads. Type and gradation of the material to be used for ballast and the cross section are important with respect to the cost of maintaining line and surface. This cost must be balanced against the original cost. In new track construction, best results can usually be obtained by placing a layer of subballast on top of the roadway and supporting the track structure, including the topballast, on this layer. The subballast should be small particles of a material that will not disintegrate. Its purpose is to provide drainage and keep the subgrade from penetrating up into the topballast while wet and under pressure. Stone or slag screenings, chat (residue after extracting ore from rock), and sand make acceptable subballast. Subballast should be placed in layers and thoroughly compacted.

The topballast may be of hard rock crushed to suitable size; crushed blast-furnace or properly processed open-hearth slag; or crushed gravel, if there is a sufficient quantity of angular material to prevent rolling. Individual railroads have different preferences for size of ballast. For complete specifications for ballast materials, see "Manual for Railway Engineering," American Railway Engineering and Maintenance of Way Association.

A recommended ballast section is shown in Fig. 19.11. A 12-in depth of top-ballast below the bottom of the ties and a 12-in depth of subballast will generally provide good track support for heavy loading and traffic (AREMA Manual). As the roadbed becomes further compacted by traffic, it will be necessary to add additional ballast to resurface the track from time to time. After several years of service, the depth of ballast under the ties will probably be considerably increased. If overhead clearances are reduced due to the additional ballast depth, undercutting may be done to remove excess ballast. Ballast may also become fouled from ballast particle degradation or other sources. Fouled ballast does not provide for drainage of water away from the ties and rail and should be cleaned or replaced.

19.12 Turnouts and Crossings

A turnout provides the means for trains to be directed from one track to another. A turnout is made up of a pair of switch points with accessories, a frog, a pair of guardrails, and a set of turnout ties (Fig. 19.15).

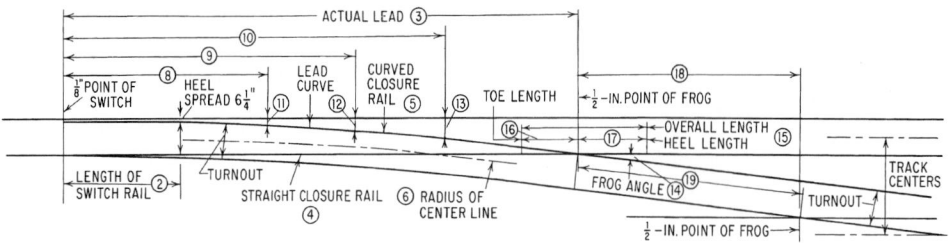

Fig. 19.15 Crossover consists of two turnouts and a crossover (connecting) track. Numbers indicate dimensions given in Table 19.4

19.12.1 Frogs

A **frog** is a special unit of trackwork that permits two rails to cross. It is designated by number and type.

The frog number is the ratio of the distance from the intersection of two gage lines to the spread, or distance between gage lines, at that distance. The number also is given by half the cotangent of half the frog angle. The frog number determines the frog angle, the degree of turnout curvature, and the lead, or distance from the point of switch to the point of frog. Table 19.4 gives these data for frog numbers from 5 to 20. Since speed is limited by curvature, the frogs with sharper turnout, the smaller-numbered ones, are used in yard tracks where speed is slow. The larger-numbered frogs are used in main-line locations to permit desired speed to the extent practical. Where extremely shallow frog angles are used, particularly in high speed operations, a movable point frog may be used. The movable point eliminates the long flangeway gap at the intersection of the flangeways caused by the shallow angle. In turnouts, the movable point control is tied to the switch control so that the frog point is always in the correct position for train movements through the switch.

Frogs are either rigid or spring-rail types. Rigid frogs are of bolted rail, rail-bound manganese-steel, or solid manganese-steel construction. In a bolted-rail rigid frog, components are made from regular rolled rail, planed or machined as required. The assembly is held together by bolts through the rail webs, with the components separated by filler blocks to form the flangeway. A rail-bound manganese-steel frog (Fig. 19.16) includes a cast insert of Hadfield manganese steel, which forms the point and wings, the locations most subject to impact, batter, and wear. (Hadfield manganese

steel is a high-manganese alloy, which when properly heat-treated increases in hardness with cold working, so it is especially well-suited to resist batter at frog corners.) The insert is supported by bent sections of rail, and the assembly is fastened together with bolts through the binding rails and the insert. A solid manganese-steel rigid frog (Fig. 19.16*b*) is made entirely of cast Hadfield manganese steel. It usually is self-guarded to save the cost of separate guardrails. The frog is joined to the two running rails at toe and heel by regular rail-joint bars and connecting bolts.

A spring-rail frog is made of machined rail sections. One side of the frog has a regular flangeway like a bolted-rail rigid frog. This is placed in the main running track. The other, or turnout, side, has a spring wing rail, which normally is held against the side of the frog point. Wheels passing through the turnout side force the spring rail out, against spring resistance, to provide a flangeway. The spring-rail frog provides a continuous running surface with a minimum of impact for the main running track.

Spring-type frogs are not recommended where there are many movements through the turnout side requiring frequent opening of the spring rail or on the outside of curves. Bolted-rail frogs cost the least, but they do not last as long and require more maintenance than the rail-bound or solid manganese. Self-guarded solid manganese frogs are used mostly for the smaller-numbered frogs in yard tracks where speeds are relatively slow.

19.12.2 Guardrails

A **guardrail** is fastened to each rail directly opposite the frog point, to contact the back of each passing wheel and prevent the flange of its mating

Table 19.4 Turnout and Crossover Data for Straight Split Switches*

1 Frog No.	2 Length of switch rail, Ft	2 In	3 Actual; lead, Ft	3 In	Closure distance: 4 Straight closure rail, Ft	4 In	Closure distance: 5 Curved closure rail, Ft	5 In	Lead curve: 6 Radius of center line, ft	Lead curve: 7 Degree of curve, Deg	7 Min	7 Sec	Gage line offsets: 8 Ft	8 In	9 Ft	9 In	10 Ft	10 In	11 In	12 In	13 Ft	13 In
5	11	0	42	6 1/2	28	0	28	4	177.80	32	39	56	18	0	25	0	32	0	11 13/16	20 5/8	2	8 7/8
6	11	0	47	6	32	9	33	0	258.57	22	17	58	19	2 1/4	27	4 1/2	35	6 3/4	12 3/8	21 5/8	2	10
7	16	6	62	1	40	10 1/2	41	1 1/4	365.59	15	43	16	26	2 1/4	35	10 1/2	45	6 3/4	11 3/8	19 9/16	2	6 7/8
8	16	6	68	0	46	5	46	7 1/2	487.28	11	46	44	27	7 1/4	38	8 1/2	49	9 3/4	11 7/8	20 9/16	2	8 5/16
9	16	6	72	3 1/2	49	5	49	7 1/4	615.12	9	19	30	28	10 1/4	41	2 1/2	53	6 3/4	12 5/16	21 3/8	2	9 7/16
10	16	6	78	9	55	10	56	0	779.39	7	21	24	29	11 3/4	43	5 1/2	56	11 1/4	12 1/4	21	2	8 5/8
11	22	0	91	10 1/4	62	10 1/4	63	0	927.27	6	10	56	37	8 1/2	53	5	69	1 1/2	12 1/4	21 3/8	2	9 3/4
12	22	0	96	8	66	10 1/2	67	0	1,104.63	5	11	20	38	8 1/2	55	5	72	1 1/2	12 7/16	21 5/8	2	9 7/8
14	22	0	107	0 3/4	76	5 1/4	76	6 3/4	1,581.20	3	37	28	41	1 1/4	60	2 1/2	79	3 3/4	12 7/8	22 5/16	2	10 1/2
15	30	0	126	4 1/2	86	11 1/2	87	0 3/4	1,720.77	3	19	48	51	9	73	6	95	3	12 1/8	21 1/4	2	9 3/4
16	30	0	131	4	91	11	92	0	2,007.12	2	51	18	53	0	76	0	99	0	12 7/16	21 13/16	2	10 5/16
18	30	0	140	11 1/2	99	11	100	0	2,578.79	2	13	20	55	0	80	0	105	0	12 3/4	22 1/8	2	10 7/16
20	30	0	151	11 1/2	110	11	111	0	3,289.29	1	44	32	57	9	85	6	113	3	13 1/16	22 11/16	2	11 3/16

1 Frog No.	Properties of frogs: 14 Frog angle, Deg	14 Min	14 Sec	15 Overall length, Ft	15 In	16 Toe length, Ft	16 In	17 Heel length, Ft	17 In	Crossover data: 18 Straight track, 13-ft track centers, Ft	18 In	19 Cross-over track, 13-ft track centers, Ft	19 In	For change of 12 in in track centers: Straight track, Ft	In	For change of 12 in in track centers: Cross-over track, Ft	In	Comfortable speed, mi/h
5	11	25	16	9	0	3	6 1/2	5	5 1/2	16	10 5/16	18	1 7/8	4	11 7/16	5	0 5/8	12
6	9	31	38	10	0	3	9	6	3	20	5 1/2	21	6 1/2	5	11 1/2	6	0 1/2	13
7	8	10	16	12	0	4	8 1/2	7	3 1/2	24	0 3/8	24	11 5/8	6	11 9/16	7	0 7/16	17
8	7	9	10	13	0	5	1	7	11	27	7 1/8	28	4 7/8	7	11 5/8	8	0 3/8	19
9	6	21	35	16	0	6	4 1/2	9	7 1/2	31	1 5/8	31	10 3/8	8	11 11/16	9	0 5/16	21
10	5	43	29	16	6	6	5	10	1	34	8 1/8	35	3 7/8	9	11 11/16	10	0 5/16	24
11	5	12	18	18	8 1/2	7	0	11	8 1/2	38	2 1/2	38	9 1/2	10	11 3/4	11	0 1/4	26
12	4	46	19	20	4	7	9 1/2	12	6 1/2	41	8 3/4	42	3 1/4	11	11 3/4	12	0 1/4	28
14	4	5	27	23	7	8	7 1/2	14	11 1/2	48	9 1/4	49	2 3/16	13	11 13/16	14	0 1/4	34
15	3	49	6	24	4 1/2	9	5	14	11 1/2	52	3 7/16	52	8 5/8	14	11 13/16	15	0 3/16	35
16	3	34	47	26	0	9	5	16	7	55	9 5/8	56	2 1/2	15	11 13/16	16	0 3/16	38
18	3	10	56	29	3	11	0 1/2	18	2 1/2	62	9 7/8	63	2 3/16	17	11 13/16	18	0 3/16	40
20	2	51	51	30	10 1/2	11	0 1/2	19	10	69	10	70	2	19	11 7/8	20	0 1/8	40

* Adapted from AREA Trackwork Plans. Comfortable speed added. Column numbers refer to dimensions in Fig. 19.15.
Calculated for turnouts from straight track for 4-ft 8½-in gage.
Turnouts and crossovers recommended: for main-line high-speed movements, No. 16 or No. 20; for mainline slow-speed movements, No. 12 or No. 10; for yards and sidings to meet general conditions, No. 8.

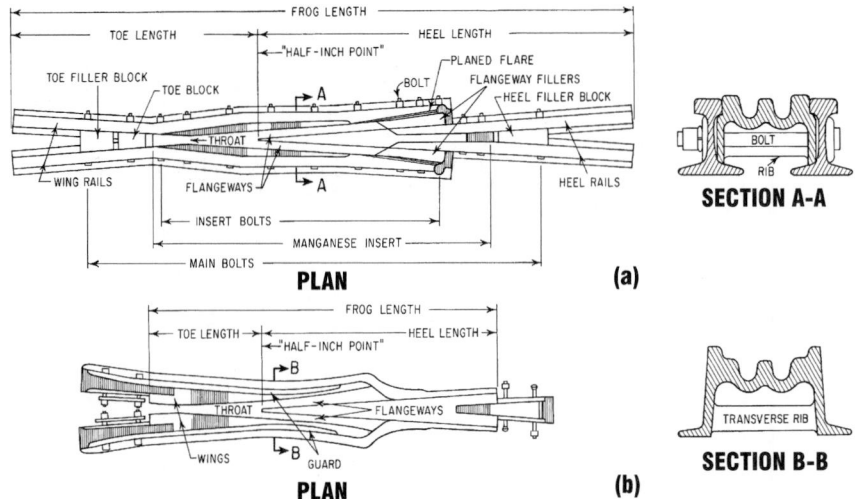

Fig. 19.16 Frogs used where rails intersect. (*a*) Rail-bound manganese-steel frog for main line. (*b*) Solid manganese-steel self-guarded frog for yard tracks.

wheel on the axle from going down the wrong side of the frog point. Guardrails are of rail or cast-manganese-steel construction. The ends are flared inwardly of the track to engage the back of the wheel flanges and guide the pair of wheels on each axle into proper lateral position in the track. It is important that the guardrails be long enough and properly positioned to ensure that the wheels are guarded past the frog point. It is also important that the guard check gage (distance between guard and gage lines) be maintained at not less than 4 ft 6⅝ in (for standard gage). Guardrails are not required with self-guarded frogs.

19.12.3 Switches

A **switch** consists of a pair of switch points, a set of switch slide plates with braces, main and connecting rods, and a manually or power-actuated switch stand (Fig. 19.17). The switch-point rails are planed from regular rolled rail and reinforced on each side of the web with steel straps riveted in place. Heel blocks are used to join each switch-point rail to the adjoining lead rail, and both are fastened to the other running rail.

The heel spread (distance between the two gage lines) is 6¼ in, so the switch angle is fixed by this distance and the length of the switch-point rail. A short switch point and large angle are satisfactory for slow-speed operation. For example, a 16-ft 6-in length of point is satisfactory for a No. 8 turnout. For a high-speed turnout, such as a No. 20, 30-ft points are used.

Usually, switch points are made straight. But for high speeds, the switch points are sometimes curved and 39 ft long for No. 18 and 20 turnouts. Comfortable operating speeds through turnouts are shown in Table 19.4.

Switch ties must be provided for turnouts. These are usually spaced on about 20-in centers. Two long ties must be provided at the switch point for the switch stand. Each tie thereafter is made long enough to extend from each outer rail base the same distance as on regular track. Whenever the switch tie becomes as long as twice the length of a regular tie, the switch ties are discontinued and regular ties used.

19.12.4 Crossings

A **crossing** of two tracks requires four crossing frogs, frog plates, and crossing ties. Crossing frogs are made of bolted rails with either regular control-cooled rail or heat-treated rail; of rail-bound manganese-steel castings; or of all manganese-steel castings. Each running rail has a guardrail with a 2-in-wide flangeway between. To ensure that such guardrails are effective in preventing the wheel flanges from entering the wrong side of the point, crossings should not be made with an angle of less

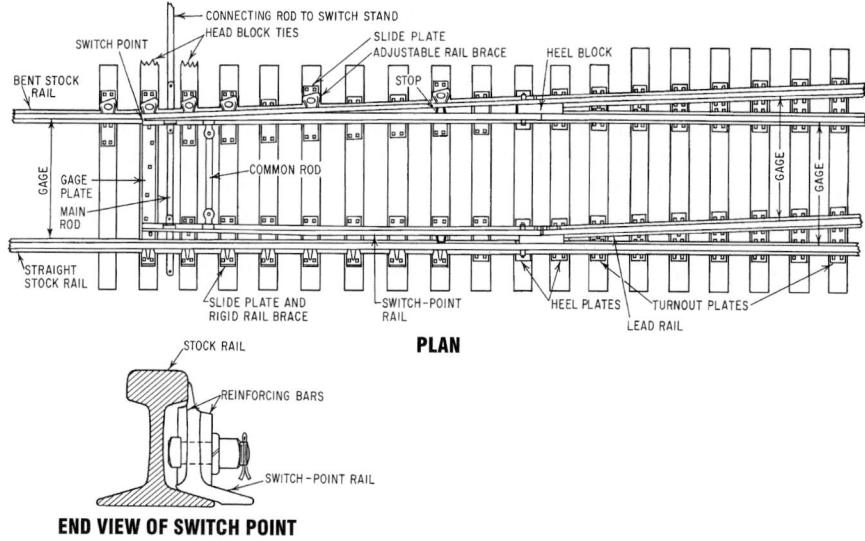

END VIEW OF SWITCH POINT

PLAN

Fig. 19.17 Left-hand, straight, split switch.

than 9° 36′ on tangents. (For curves, see American Railway Engineering and Maintenance of Way Association Trackwork Plan No. 820.)

It is desirable to locate crossings on tangent on both intersecting tracks, but when this is not practical, crossings can be made to fit any condition of curvature.

19.12.5 Trackwork Plans

Details and specification of material required for turnouts and crossings are given in the Trackwork Plans of the American Railway Engineering and Maintenance of Way Association (AREMA). Some major freight railroads, however, have their own standard plans and specifications for this material and these differ from those of the AREMA. When trackwork material is to be ordered for a railroad, first the standards to be used should be determined and then the railroad and plan number should be specified and specifications given. When crossings are to be ordered, the intersection angle, the curvature, if any, and the rail size should be specified.

19.13 Culverts, Trestles, and Bridges

Culverts provide waterway openings under tracks. Usually, culverts consist of galvanized corruga

ted pipe or arches, reinforced concrete pipe, or reinforced concrete rigid-frame boxes. They are cheaper to install and maintain than other types of openings.

Care must be exercised in placing the fill on the sides and over the larger-sized culverts because side pressure against the culvert is a large factor in its ability to support vertical pressure. Metal culverts of up to 180 ft^2 and reinforced concrete culverts of up to 300 ft^2 in opening area are in use.

Trestles often are built of treated-timber stringers supported on capped and braced treated-timber piles. Trestles have either an open deck or ballasted deck. Ballasted decks are more expensive in first cost but require less work to keep the track in line and surface and offer less of a fire hazard Treated-timber trestles are economical, have a life of 40 years or more, and require no painting.

Trestles are also constructed of steel or concrete piles, either reinforced or prestressed, with a concrete cap supporting steel or concrete stringers. Concrete trestles usually have ballasted decks. Steel trestles may have open or ballasted decks.

Bridges generally are built of steel, reinforced concrete, or prestressed concrete. Usually, the abutments and piers are of reinforced concrete. For steel bridges, rolled beams are generally used for spans up to 50 ft. Plate girders of bolted or welded construction may be used for spans up to about 150 ft, and trusses, either through or deck type, for longer spans. Open or ballasted decks are placed on steel bridges. Ballasted decks, however, are preferred for all types of bridges because of ease of track maintenance and reduction of impact.

"Manual for Railway Engineering," American Railway Engineering and Maintenance of Way Association, gives recommended designs and specifications for construction of all types of bridges, trestles, and culverts. These include recommendations for live load in terms of Cooper's *E* loading, allowance for impact effects, and permissible design stresses. (See also Sec. 17.)

19.14 Maintenance of Way

After a railway is completed, continual maintenance is required to keep it in condition for operation. Mechanized equipment is used for this to a major extent in the United States.

On-track equipment, such as Jordan spreaders and shovels, and off-track equipment, such as bulldozers, shovels, draglines, and special track-mounted ditching machines may be used for ditching. Side-dump cars may be used for transporting material from ditches in cuts for bank widening on fills. Chemical weed killers or track burners may be used to keep the ballast section clear of vegetation. Chemical weed and brush killers or power mowers or cutters may be used to control undesirable weeds and brush on the right-of-way. Ballast-cleaning equipment is available for cleaning ballast between ties, in the ballast shoulders, and under the ties.

19.14.1 Track Maintenance

Maintenance of track structure includes, tie renewals, periodic tightening of track bolts, raising the track to correct variations in surface and cross level, lining the track to correct deviations from alignment, and adding ballast and dressing the ballast section.

Use of continuous welded rail increases the likelihood of track buckling in very hot weather and rail pull-aparts in very cold weather. A study of a large number of track buckles that occurred over 3 years on a major railroad indicated the following:

The principal causes of track buckling are inadequate ballast, disturbing the tie bed during tie renewals or track surfacing, and improper rail laying or adjusting for temperature. Track buckles are more likely to occur during hot spells in the spring in the afternoon. They are also more likely to occur on curves than on tangent track, and the probability of occurrence increases as the degree of curvature increases.

Derailments caused by track buckles can be minimized by maintaining a full ballast section, using care not to disturb track in very hot weather, and inspecting track in the afternoons of the first days of early hot spells. In addition, appropriate measures should be taken by issuing slow orders and by track strengthening or stress relieving the rail when track buckles or impending track buckles are observed. (AAR Research and Test Department Report R-454, "An Investigation of Railroad Maintenance Practices to Prevent Track Buckling," Association of American Railroads, 50 F St., NW, Washington, DC 20001.

String lining is a convenient and satisfactory method of checking the alignment of curves. This may be done manually or with automated equipment found on some tamper/liner work equipment. The manual procedure requires that the outside rail be marked in 15.5 or 31 foot stations. Then the mid-ordinate of a 62 ft string line or chord is measured at each station. The mid-ordinate, measured in inches, indicates the degree of curvature at that station. Calculations can be made based on the mid-ordinates to determine the amount of track shift necessary to obtain a uniform curve. Automated equipment will document the measurements and translate the calculated shifts directly to the track liner. Automated equipment will create a uniformly smooth curve but the curve may not be at a specific degree of curvature.

19.14.2 Rail Life

This is usually expressed in million gross tons of traffic carried before rail must be replaced. Gross tons equal the total weight of the locomotives and cars and their loading, short tons.

Rail life is determined by a number of factors:

1. **Size of the Rail.** The heavier sections, such as the 140 RE, 136 RE, and 132 RE, will have a longer rail life than the lighter sections, such as the 100 RE, 115 RE, and 119 RE, under comparable operating conditions.

2. **Curvature.** In general, rails do not have as long a life on curves as on tangent track; the sharper the curve, the shorter is the rail life. Adequate lubrication of the outer rail on curves will greatly extend the rail life.

3. **Wheel Load.** The life of rail under 100-ton cars is appreciably less than under 70-ton cars but longer than under 125-ton cars. The contact pressures from the wheel loads of the heavier cars increase wear on the gage side of the outer rail and plastic flow on the top and sides of the head on the inner rail of curves faster than the increase in wheel loads. Also, the wheel loads from the heavier cars cause a disproportionate increase in the number of rail failures ("Comparison of Rail Behavior with 125-ton and 100-ton Cars," D. H. Stone, *AREA Proceedings*, vol. 81, p. 576.)

4. **Continuous Welded Rail (CWR).** This has made a decided improvement in the service life of rail on tangent track. With jointed rail, the principal factors determining rail life are fishing-surface wear and rail-end batter at rail joints. With CWR, there are no rail joints, only shop or field butt welds, except at insulated joints. (The relatively few insulated joints are now generally glued to prevent rail-end movement within the joint bars.) The rail life of CWR on tangent track and light curves (1° or under) is determined by the number of service and detected rail failures. These are mostly transverse, progressive fractures within the head, which increase in frequency with the cumulative number of wheel loadings. At some point, the frequency of occurrence of such failures makes it more economical to replace the rail than to cut out the failures or detected defects and weld in a length of repair rail.

5. **Metallurgy.** Most railways use alloy or heat-treated rail on very sharp curves. This decreases the rate of side wear on the outer rail and the plastic flow on the top and sides of the inner rail and thus materially extends the rail life.

6. **Need for Relay Rail.** This may be a factor in rail life. Either jointed rail or CWR may be removed although it would have further service life if there is a requirement for this rail for relay purposes on curves, branch lines, yard, or industry tracks.

7. **Corrugation.** This may be a factor in rail life, although grinding trains are usually used to correct this rail surface condition unless it gets too bad.

8. **Available Funds.** Availability of funds to purchase and lay new rails is a practical consideration in rail life. If funds are lacking, the rail life may be prolonged beyond the most economical life.

Track lubricators reduce flange wear of outer rails and curve resistance to train movement. The devices are fastened to the rails at curves to apply lubricant to the flange of each passing wheel. Generally, a track lubricator consists of a reservoir containing a suitable type of grease, an applicator, and a plunger activated by each passing wheel to pump a small quantity of grease into the applicator. The applicator is a steel member, several feet long, placed against the gage side of the rail and incorporating small holes through which the grease is pumped to contact wheel flanges. Several types of lubricators are available. Manufacturer's instructions should be followed with respect to location and type of lubricant.

Some lubricators are designed for application from hi-rail or track inspection vehicles. This method allows applications to be made at any location, not just fixed points, when the rail is too dry. With either method, care should be taken to minimize the chance of lubricant building up on the top of the rail where it may cause locomotive traction may be reduced.

19.14.3 Rail Defects

Rail-defect detection is an important factor in safe operation of railroads. Rail defects develop from service use and are classified as transverse fissure, compound fissure, detail fracture, engine-burn fracture, horizontal split head, vertical split head, crushed head, piped rail, split web, head and web separation, bolt-hole crack, broken base, rail weld failures or defects, and damaged rail. However, because of improvements in rail design, manufacture, and maintenance practices, the number of rail defects that develop is remarkably small. Most of the rail defects that do develop are in the head or web within the joint-bar area. Rail-defect-detection equipment is available with which such defects can generally be detected. It makes possible removal of defective rail before a service failure occurs. See also Art. 19.14.1.

Rail-defect-detector cars are available that travel over the track at testing speeds of 6 to 15 mi/h. By utilizing electrical magnetism or ultrasonic waves, equipment on board is able to locate internal defects in the railhead. Ultrasonic equipment is used to detect defects in rail webs, particularly at rail joints, inside road crossings and other paved areas, and in frogs. Most rail in main-line track in the United States is tested once a year or more often with detector cars.

19.14.4 Railroad Tie Renewals

Two methods are used for rail and tie replacements. The spot renewal method is used when only a few scattered ties or rail sections need to be replaced. Small gangs with limited equipment, such as boom trucks or backhoes, perform this work. Out-of-face renewals are used when large volumes of ties or continuous rail sections are replaced. This work is typically done with large gangs with many pieces of specialized equipment (see Article 19.14.6).

As rail on curves generally wears out before rail on tangents, often rail replacements just involve curve rail. Depending on the condition of the rail, the outside or high side rail may be replaced with new rail and then reused on the inside or low rail. The low rail may then be scrapped or reused in sidings or yards. Some tie renewals may be tied to track surfacing cycles as the combined cost is usually less than if the work is done separately, due to the cost of ballast and surfacing required for just tie renewals.

19.14.5 Structures Maintenance

At least annually, a detailed inspection should be made of all bridges, trestles, and culverts. Structures with deficiencies should be rated with respect to safe load-carrying capacity, while appropriate safety measures are in effect to allow movement of trains. This inspection should also establish maintenance requirements or replacement options. Special investigations, such as underwater inspections or detailed structural inspections, may be performed on an as-needed basis or as specified by the railroad's policy. Work not of an emergency nature, such as cleaning and painting, repairing concrete deterioration, and replace-

ment of any parts, should be scheduled. Periodic interim inspections noting the condition of bridges and trestles should be made at every opportunity in accordance with the railroad's policy. Structures susceptible to scour at the footings should be inspected more frequently. Patrolling and inspection of bridge and drainage structures may be required during storms and periods of high water level and after earthquakes. Recommended inspection practices are detailed in the "Manual of Railway Engineering," American Railway Engineering and Maintenance of Way Association.

Also, railroad buildings should be inspected to determine if maintenance and repair requirements are being met. Records of all inspections should be maintained, as required by the railroad. Computer systems for management of structures may be utilized to assist in decision making and planning.

19.14.6 Mechanized Work Equipment

When rails or ties are replaced out-of-face, mechanized equipment is available for handling almost every item or work. Mechanized, on-track equipment such as spike pullers, power wrenches, tie-adzers, tie removers, scarifiers, tie installers, rail cranes, spikers, track liners, tampers, ballast regulators are utilized in arranged sequences to remove and replace rail, remove and install ties and maintain track surface. There are also full production machines that can changeout both ties and rail in one continuous operation.

The rail-transportation engineer will find it helpful to have available a copy of the "Pocket List of Railroad Officials." This is published quarterly by Commonwealth Business Media, Inc., 400 Windsor Corporate Center, 50 Millsbury Road, Cranbury, NJ 08512. It contains an alphabetical listing of all railroad equipment supplies with the names and addresses of companies that manufacture or sell these products.

19.14.7 Track Safety Standards

Legislation was passed by the U.S. Congress in 1970 requiring the Federal Railroad Administration of the Department of Transportation to establish track safety standards and to see that the railroads complied with them. These standards established by the FRA prescribe minimum standards for the safe operation of trains with regard to drainage, vegetation, ballast, rail defects (including end mismatch and end batter), welds, joints, tie plates, spikes, shims, switches, frogs, track appliances, deviations and variability in track geometry (runoff of elevation, alignment, cross level, and surface), maximum elevation on curves, maximum unbalance on curves, and track inspection. The standards have been established for different classes of track, classification being determined by a range of permitted operating speed. These standards are revised as the FRA considers necessary or desirable and are issued as Title 49 Transportation, Code of Federal Regulations, Part 213 (49 CFR 213). The most recent revisions occurred in 1998. A volume including these and other railroad safety regulations is published annually. (Superintendent of Documents, Government Printing Office, Washington, DC 20402.)

19.15 Freight Terminals

On most railways, one or more yard facilities are required. Such a facility should have a receiving yard, classification yard, hold and repair tracks, engine servicing house, and departure yard ("Manual for Railway Engineering," American Railway Engineering and Maintenance of Way Association). Freight railroads should provide, in addition to the normal yard tracks, through tracks for trains that require minimal handling or do not require separating. The through tracks should be located to meet requirements for American Association of Railroad inspection, to accommodate change-train crews, and to allow trains to proceed with a minimum of delay. In some cases, such tracks have permitted a reduction in the number of yard tracks required for handling of trains.

A yard consists of a series of parallel tracks, called body tracks, on which cars are placed, and ladder track usually at each end. A turnout connects each body track to a ladder track. Thus, the ladder track is a means of placing cars on or removing them from each body track.

The **receiving yard** should be conveniently accessible from the main line, and its tracks should be long enough to hold the longest train without doubling (splitting) it into two tracks. The number of receiving tracks required depends on the spacing of train arrivals and time required for classification. A spacing of 18 ft should be provided between parallel ladder tracks, 15 ft between a ladder track and any parallel track, and not less than 14 ft between body tracks. Additional clearance may be required, depending on car-inspection needs and other requirements, such as car cleaning or repair operations. A gradient of not more than 0.15% is desirable to prevent the cars from rolling without setting the brakes.

The **classification yard** may be a flat yard if the number of trains and amount of switching are relatively small. A gravity or hump yard should be used otherwise.

A **hump yard** utilizes gravity to expedite switching of cars. The train of cars is pushed up an incline to a hump, at which point one or more cars are successively uncoupled while moving and allowed to roll down the incline from the hump into the classification yard. The height of the hump must be sufficient to impart enough velocity to overcome the rolling resistance of each car to the farthest point in the yard. Thus, if the distance from the hump to the farthest point is 3000 ft and the rolling resistance of the slowest-rolling car under adverse weather conditions is 10 lb/ton, equivalent to a 0.50% grade, then a minimum hump height of 15 ft would be required. Another requirement is that the decline from the hump be steep enough and long enough to separate the cars sufficiently to permit operation of switches and to clear the switches ahead of the following car. Usually, the hump height is from 16 to 20 ft. Two or three sets of retarders are provided for controlling the speed of the cars into the classification tracks. The retarders are set so that each car will roll the desired distance and couple to a standing car without undesirable impact (up to 4 mi/h).

Humping speed is about 1 mi/h. In a fully automated, or so-called push-button, yard, the operator pushes a button numbered to correspond to the track number into which a car is to go. When the car is uncoupled, it rolls down the hump and is weighed, if desired, on an electronic, uncoupled-in-motion, track scale. Also, the car's rolling resistance is measured by determining change in speed over a given length of track. This information goes to a control computer.

The approximate wheel load is measured by a track scale device. This reading goes to the computer, to limit the amount of retardation so that the wheels will not rise out of the retarder. The car speed is measured as it approaches each retarder, and this information goes to the computer. When the operator pushed the button for the track number, the computer was fed the total rolling resistance to the farthest point in that track. A wheel trip on each track corrects this value for the distance taken up by the number of cars that have already been placed in that track. From all these data, the control computer determines the speed the car should have as it leaves the last retarder to roll to the desired point and retards the car to that speed. Usually, a radar device is used to measure car speed. Retarders are pneumatically powered but electrically controlled. Switches are electrically set by the computer for the track number punched for the car.

The body tracks of a hump yard typically have a slight downgrade on the incoming end and a slight upgrade on the leaving end, giving them a bowl shaped profile. This profile assists in slowing and stopping the cars within the yard.

The **departure yard** should be long enough to accommodate the longest train and should be level, if possible. If the grade is adverse to the direction of starting, it should be at least 20% less than the ruling grade over which the train will operate. Track spacing should be the same as for the receiving yard.

Car-repair tracks should be provided to accommodate the number of cars to be repaired and the repair time. These tracks can be alternately spaced 18 ft and a width sufficient to accommodate mechanical equipment. A paved driveway of rail height should be provided between tracks. It is desirable to have a car-repair building, with the required number of tracks, through which cars can be moved by cables and "rabbits." This provides more efficient working conditions and mechanical equipment for repair work at minimum cost and delay. Some repair facilities employ a large truck-mounted vehicle as a car mover or a rubber-tired vehicle with rail wheels to move cars in and out of the shop.

Most railroads do major service and repair of locomotives at facilities that are centrally located on their system. These central engine houses or shops have platforms at cab floor height and provide access to the top of the locomotive. They also have below rail pits for access to the underside of locomotives. Engine servicing at most yards is limited to minor repairs, cab cleaning and resupplying of fuel, sand, water and lubricants. All facilities should provide means to collect and treat any spills of fuel or lubricants.

Other terminal facilities that may be required are:

Team tracks having an adjoining paved area for loading from trucks into the cars.

Mobile cranes for loading and unloading *piggyback* (truck trailers on flatcars) or containers on flatcars.

Stub tracks with a ramp at the ends. Narrow, car-floor-height platforms, with electric power outlets, should be placed between the tracks for loading and unloading piggybacks.

Elevating-type inclines for end loading of automobiles on auto-rack cars.

House tracks at the freight-station building for less-than-carload shipments.

Docks for loading cars or contents on boats.

Car dumpers that turn upside down and empty hopper-type cars.

Stockpiling facilities for coal and ore. Storage bins and elevators for grain.

Provision of overhead cranes, wheel grinders, wheel drop pits, paint shops, wash houses, sanitary-waste facilities for passengers, locomotives, and cabooses, industrial-waste facilities, and other accessories depends on the extent to which repairs will be made at the particular facility.

19.16 Passenger Terminals

Passenger terminal and station requirements vary with the type of passenger service provided.

19.16.1 Intercity Terminals

These comprise facilities for handling passengers and baggage and for servicing passenger cars and locomotives. The passenger facilities should include parking for automobiles, loading and unloading areas for taxis and buses, ticket windows, waiting rooms, baggage check room, concessions, food services, rest rooms, public telephones, and walkways, stairs or escalators and elevators to train concourses.

Tracks should be on 20-ft centers, with paved platforms between. The platforms should be covered with a suitable type of roof. A minimum width of passage for passengers of 6 ft should be allowed on platforms, stairways, and ramps. The incline of ramps should not exceed the maximum slope required for access by handicapped persons or as restricted by local building codes. Platforms at service points for trains passing through should have water hydrants, electrical outlets, steam connections, and brake shoes available. Handling of baggage, mail, and express requires separate platforms or wide platforms so that trucks may pass without interfering with passengers; ramps or conveyor belts; and adequate space for sorting and transferring to other trains or trucks.

Servicing of locomotives and cars requires a coach yard, preferably with a mechanical washer to wash all equipment as it enters the yard. Tracks in this yard should be level and on 20-ft centers. They should have concrete platforms between them. An inspection pit 36 in wide and 38 in below top of rail is desirable for some of the trackage. Preferably, this area should be covered over to facilitate work in bad weather. Jacking pads and wheel drop pits should be provided. Other facilities needed are water hydrants meeting U.S. Public Health Service requirements; hot water; low-pressure air connections for cleaning; high-pressure connections for air brakes; electrical service outlets, including 220-V alternating current for air conditioning equipment; steam supply lines; adequate lighting for night operation; a convenient supply of brake shoes and mounted car wheels; car pullers; commissary facilities for dining cars; service building providing offices, toilet, wash, locker, and lunch rooms; storehouse; repair shops; refuse disposal; fire protection; bottling plant for refilling gas cylinders; and fuel oil and sand supply for locomotives.

The extent to which all these facilities should be provided depends on the number of trains and passengers to be handled during peak periods, with an allowance for train delays. Detailed recommendations related to the number of passengers handled are in "Manual for Railway Engineering," American Railway Engineering and Maintenance of Way Association (AREMA), 8201 Corporate Drive, Suite 1125 Landover Maryland 20785 (www.arema.org).

19.16.2 Commuter Terminals

These should provide most of the facilities listed for passenger terminals, except for a baggage checkroom and food services which would probably not be needed. Vending machines may provide some food service.

Rapid-Transit Terminals ▪ Requirements for stations are given in Art. 19.6. Terminal facilities should be provided at the end of each line for car storage, and a shop should be available for emergency repairs. A shop for overall and scheduled maintenance of cars should be provided at the location most suitable from the standpoint of accessibility in operation, land availability, environmental factors, and so forth.

For storage tracks, the length required may be determined by calculating the length of the number of cars required for peak movements plus 10% for spare units to replace cars out of service for repairs. Storage tracks should be level, and grade for lead and other tracks should not exceed 0.3%. Curves should not be less than 200-ft radius but should be flatter if the car units are designed to require a longer radius.

One or more shops should be provided at the chosen shop location for repair of electrical, electronic, hydraulic, pneumatic, control, and undercar equipment (including drop pits); for mechanical repairs; for wheel grinding; and for painting and seat repair. An automatic car washer should be provided. It is desirable that all the above work areas, except the car washer, be located under cover; work should be scheduled on an assembly line basis; all work should be automated; and all workers should be provided with power tools to the extent such tools are available.

Storage and shop areas should be surrounded by a suitable fence to prevent trespassers from entering, for safety and to avoid pilferage. A guard or automatically operated gates should guard the entrance tracks and driveway to the storage and shop area. A 7-ft-high chain link fence with barbed-wire outriggers inside the right-of-way is well-suited for this purpose.

19.17 Station Location and Characteristics

Station locations have already been established in the United States for passenger trains and existing commuter and rapid-transit service. Many existing stations, however, are being remodeled or replaced to satisfy Federal laws requiring access for handicapped persons and for other safety measures. Commuter stations are located in suburbs or city areas a few miles apart, and where local or bus street transportation generates a large enough volume of passengers to justify a train stop. Usually, not all commuter trains stop at all the stations but a schedule will be established to provide reasonably frequent service, particularly in morning and evening rush hours, at stations with relatively low traffic volume.

19.17.1 Stations for Rail Transit

When a new system is being planned, several factors should be considered in deciding station locations:

Physical constraints: available space for the station, space for parking, space for bus and automobile circulation.

Accessibility: convenient location within network of freeways and arterial and feeder bus routes.

Service potential: number of persons, households, students, and jobs of various types located within 700 ft, 1500 ft, and 3000 ft of each station. Most people living or working within 1500 ft of a station will walk to or from it. In outlying, low-density areas, automobiles and feeder bus lines will expand the service area of a station.

Convenience to major institutions and centers: schools, hospitals, recreational areas (including sports facilities), and major industrial and commercial concentrations located within 700 ft of each station.

Development opportunities: joint development potential of vacant or deteriorated structures within 700 ft of each station.

Impact on neighborhood: localized traffic congestion, reinforcement of community centers and boundaries, and conformance with local development plans.

Projected ridership: number of riders coming to and from each station, projected for 15 to 25 years ahead, depending on transportation planning requirements. Passenger-seat-mile (1 passenger moving 1 mi) provides a useful measure for cost comparisons between the different modes of travel.

From the above, it will be possible to determine for each station the location (tentative) that will attract the maximum number of riders and give the best service. Stations should be placed closer together in areas expecting the greatest number of riders, not only to give better service but to avoid undue congestion within and outside the station.

19.17.2 Station Platforms

These should be as long as the longest train that will be operated. For passenger and commuter trains, the platforms or paving for loading and unloading are generally outside the track or two tracks. Also, most existing platforms are at top-of-rail level and 6 ft or more wide. However, in new construction for commuter service and rapid-transit service, platforms along or between tracks should meet the height required by the rolling stock selected and meet the requirements for access by handicapped persons. Platform height may range from rail level to as much as 42 in above the rails. In subways, platform width should in no case be less than 10 ft and should provide 8 ft^2 of occupancy space per person for maximum assembly crowds. In one subway system, a platform between tracks about 30 ft wide is provided. For at-grade systems, an existing sidewalk or ramp that meets building-code requirements may be utilized to provide access to the trains, depending on the type of car.

19.17.3 Provisions for Circulation in Stations

Important criteria, in addition to all appropriate safety measures, are traffic-handling capability, consistently available information, and orientation. Maps of the system showing all lines and station stops should be placed conspicuously, with the particular station clearly designated thereon. Stations should include a free area and a service area. Provision should be made for concessions, if any, in the free area and necessary facilities, such as electric outlets, water supply, and so on should be provided at a suitable location for the concessions. Automatic coin-operated dispensers should be located in the free area, but both these and concessions should be located so as not to interfere with circulation of passengers to and from the trains.

A minimum clearance of at least 8 ft should be provided throughout the station and platform to meet the requirements of the equipment used. Adequate space should be provided at ticket facilities to allow for a line of ticket purchasers without interference with the normal flow of passengers.

Passageways, stairs, ramps, and areas for handicapped persons should be located to provide balanced train loading and unloading—usually at each end of the platform, or at the platform mid-length, or both. An island-type platform between two tracks is preferable to a platform on each side of the two tracks.

To determine the space required for exit and entrance doors, which should open out or revolve, the maximum number of passengers that will pass in any 15-min peak period should be estimated. Enough space should then be allowed to clear the platform under normal conditions within the headway time between trains. For emergency evacuation of a train, provision should be made to clear the platform in 4 min. To do this, the designer may assume a crush capacity of 25 passengers per minute per foot width of passageways, 20 passengers per minute per foot width of stairways, and 100 passengers per minute for each 48-in escalator. Enough supplemental stairway width should be provided to permit evacuation if the escalator should become inoperable.

Telephones, toilets, storage lockers, service rooms, and toilets for station personnel should be provided at stations as warranted.

Wherever steps are utilized for access, a ramp, lift, or elevator must also be provided for access for persons with disabilities. Access should be illuminated when required for safety. Steps and ramps should be kept clear or sanded where exposed to the weather, and handholds meeting building-code requirements should be provided on both sides. All walking surfaces on stairs and in passageways should be kept dry and covered with a suitable nonskid material. Small sidewall depression "trenches" parallel to the side walls and suitably drained should be provided to accomplish this.

Escalators should be provided to carry passengers up whenever the stair height exceeds 12 ft and to carry them down when it exceeds 24 ft.

19.17.4 Environmental Considerations in Station Design

Station construction should comply with the applicable building codes. Suitable lighting, satisfac-

tory noise level, comfortable air conditioning, pleasant appearance from the standpoint of both decor and cleanliness, control of wind and odors, and clear circulation (by appropriate directional signs if needed) should be provided.

Lighting is of great importance for safety and the security of passengers. Table 19.5 is a guide for minimum illumination levels at different locations. Signs at street level, illuminated at night, should indicate clearly where the rapid-transit entrance, station, or stop is located. Steps up or down to the station should also be illuminated when required for safety.

Subway Ventilation ■ Objectives are to:

Provide a comfortable environment for patrons and staff.

Table 19.5 Recommended Minimum Illumination Levels in Passenger Stations*

Locations	Illumination, foot-candles
Platform, subway	20
Platform, under canopy, surface and aerial	15
Uncovered platform ends, surface	5
Mezzanine	20
Ticketing area, turnstile	30
Passageways	20
Stairs and escalators	25
Fare-collection kiosk	100
Concessions and vending machine areas	30
Elevator (interior)	20
Above-ground entry to subway (day)	30
(night)	10
Washrooms	30
Service and utility rooms	15
Electrical, mechanical, and train-control equipment rooms	20
Storage areas	5

* From "Guidelines for Design of Rapid Transit Facilities," American Public Transportation Association.

Provide, in the event of fire, control and removal of smoke and a supply of fresh air for evacuation of passengers and for fire-fighting personnel.

Provide for the removal of heat generated by normal train operation.

Provide control of condensate and haze and removal of objectionable or hazardous odors and gases.

The piston action of trains will provide a considerable amount of ventilation if suitable vent shafts are provided. If necessary, supplemental mechanical ventilation must be provided. Maximum piston-type ventilation can be obtained by having the tunnel or subway section as near the size of the train cross section as clearance requirements will permit and having a separate tunnel or subway for each track. Fan shafts must be located relative to the vent shafts and stations in such a way as to ensure that all sections of the subway and stations can be purged under emergency conditions. The ventilation rate should satisfy the purge rate. A minimum air velocity of 4 ft/s is recommended for determining the sizes of fans and appurtenances. Vent and other shaft openings on the surface should be located to draw in unpolluted air and protected by gratings or screens. Acoustic treatment of the shafts should be provided if needed. Fans for emergency ventilation should be connected to power feeders from two separate sources and should be operable through remote controls located at a control station.

("Subway Environmental Design Handbook," vol. 1, Federal Transit Administration, Washington, DC 20590.)

19.17.5 Security and Communications

Security can best be provided by closed-circuit television cameras suitably located at strategic locations in the station, passageways, and platforms. These instruments should be monitored at each station at which there is an agent and at a central control station. An alternative for trains would be to have the monitors for the cars where the train attendant can see them and provide the attendant with a means of communicating with the nearest agent and the central control office.

Telephone communication between each station and the central control office should be provided.

Portable radios or mobile telephones should be considered for use between security personnel and station agents and the central control office. This type of communication is not effective in subways or tunnels, so its provision to train attendants will depend on how much of the line is open or aerial structure. Or a special antenna line can be placed through the subway and tunnel sections to enable portable communication.

19.17.6 Fare Collection

This is generally accomplished on the trains in commuter service but at the stations in rapid transit. For this purpose, station personnel can be augmented by turnstiles, either coin-operated or using coded tickets or some other suitable method. In some commuter and rapid-transit service, coin (or currency or coin) vending machines are used for selling coded tickets, and computer-controlled turnstiles are used for collecting and monitoring them. Several transit systems allow customers to purchase tickets and do not use turnstiles or collection systems. The success of such methods depends on patrolling and customer honesty. Some systems establish a central zone with free travel but exact graduated fares with distance away from that zone. One system has experienced about 150 failures a month with 43 changemakers and well under the one-failure-per-machine-per-month guarantee for its computerized turnstiles.

19.17.7 Tracks between Subway Stations

Between subway stations, tracks usually are separated by the tunnel walls or a concrete wall. Walkways with a minimum width of 2 ft should be provided on one side of all line sections of tunnels and subways, and for high-speed train operation, a hand rail should be placed on the wall 3 ft above the walkway floor. Walkways should be placed on adjacent sides of the wall to permit cross connection between the walkways. Crossways should be placed not more than 1000 ft apart for workers and emergency evacuation of passengers.

("Guidelines for Design of Rapid Transit Facilities," American Public Transportation Association, 1666 K Street NW, Suite 1000 WA, DC, 20006, (www.apta.com).)

19.18 Vehicles for Rail Transportation

Except PRT cars, vehicles predominantly use steel wheels on steel rails because of the low rolling resistance and heavy weight that can be supported on a single wheel. A few rapid-transit systems utilize vehicles with rubber tires that run on concrete beams or "rails" and are self-guided. Disadvantages of such vehicles are the higher rolling resistance, greater operating cost, and lower weight-supporting capability. Another disadvantage of the rubber-tired system occurs in operation through turnouts. In one system (Fig. 19.18), the rubber tires are spaced far enough apart to permit a regular rail track structure with switch points and frog to be placed at the turnout location. The vehicle has two steel wheels and an axle at each end. As the vehicle approaches the turnout, the concrete rails are ramped downward so that the vehicle is supported on the steel wheels through the turnout, after which the concrete rails are ramped up to support the vehicle again. Vehicles must be operated at slow speed through the turnouts in this type of rubber-tired system.

PRT systems are designed for a specific purpose, and a type of vehicle is used that best serves that purpose. Mostly, these systems are used for transferring passengers at airports or in recreation centers. Rubber tires are preferred, partly because they provide traction for grades as steep as 10%. If the guideway is exposed to snow or ice, the running surface for the tires must be heated in cold weather. Magnetic levitation support is also being considered. Passenger capacity per vehicle varies from 4 to 20 seated. Mostly vehicles are operated as single or double units, but some operate in five- to eight-car trains.

19.18.1 Method of Traction

Trains for rail passenger and freight intercity systems are primarily moved by diesel-electric locomotives. Where traffic density warrants, electric locomotives are used with an overhead catenary or a third rail. Most commuter systems are powered by diesel-electric locomotives with push-pull controls in some of the cars so that the train does not have to be turned around at each terminal of the run. Several commuter systems are electrified, and each car has its own motor drive so that a separate locomotive is not required. All rail-transit systems are electrified, and each car has a driving motor for each axle to give sufficient adhesion for the rapid acceleration and deceleration required. Personal rapid-transit systems are also electrified. Some PRT systems, however, are propelled by linear induction motors set at intervals along the guideway. Propulsion is achieved with reaction plates on the bottom of the vehicles. Research sponsored by FTA on PRT systems includes the following guided minivehicle

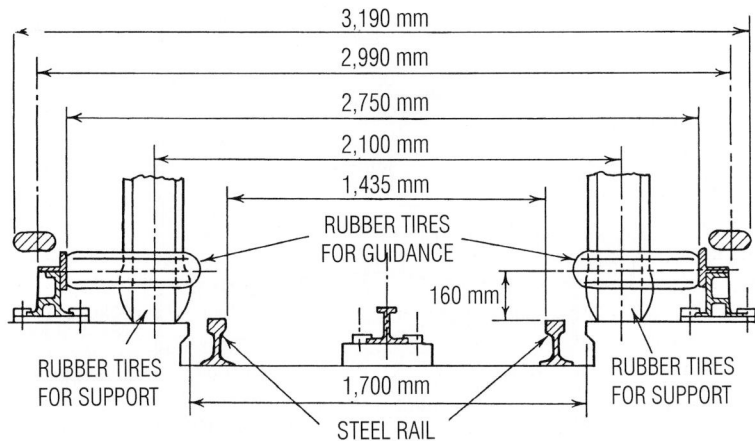

Fig. 19.18 Cross section of guideway and vehicle for Paris Metro rubber-tired rapid-transit vehicles. Steel rails and wheels are required for guidance through turnouts.

systems: suspended monorail, air levitation on a concrete guideway, rubber tires on an aluminum guideway, and rubber tires on a concrete guideway. No doubt other systems will be developed.

Since switching is such an important part of a personal rapid-transit system, the conventional dual-rail wheel system has an important advantage that will be difficult but not impossible to overcome. For example, the people-mover vehicle shown in Fig. 19.19 is supported on rubber-tired wheels and guided by another set of rubber-tired wheels bearing on a steel guide beam. For switching, one end of the guide beam can be moved back and forth to line up with the desired track. The personal-rapid-transit car in Fig. 19.20 is supported on four rubber-tired wheels and guided in the guideway by another set of four rubber-tired wheels. The guide wheels may be computer-controlled to make the vehicle follow either the left or right guiding surface. Thus, at a station, the car may be made to pass by directing the guide wheels to follow one guiding surface, or the car may be made to turn into the station track for a stop by directing the guide wheels to follow the other guiding surface. No moving parts are needed in the guideway to make a car bypass or stop at a station.

Supports for any type of system can be wheels (steel- or rubber-tired), air-cushion levitation (Fig. 19.21), or magnetic levitation (Fig. 19.22). Since either type of levitation is costly and complicated,

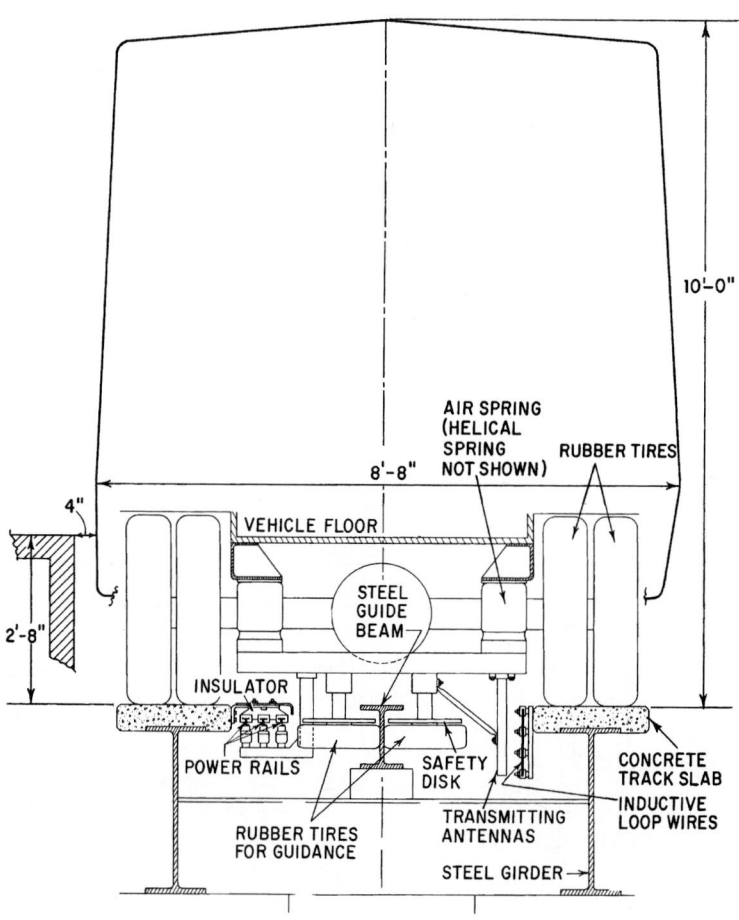

Fig. 19.19 Cross section of guideway and vehicle for Westinghouse people-mover system.

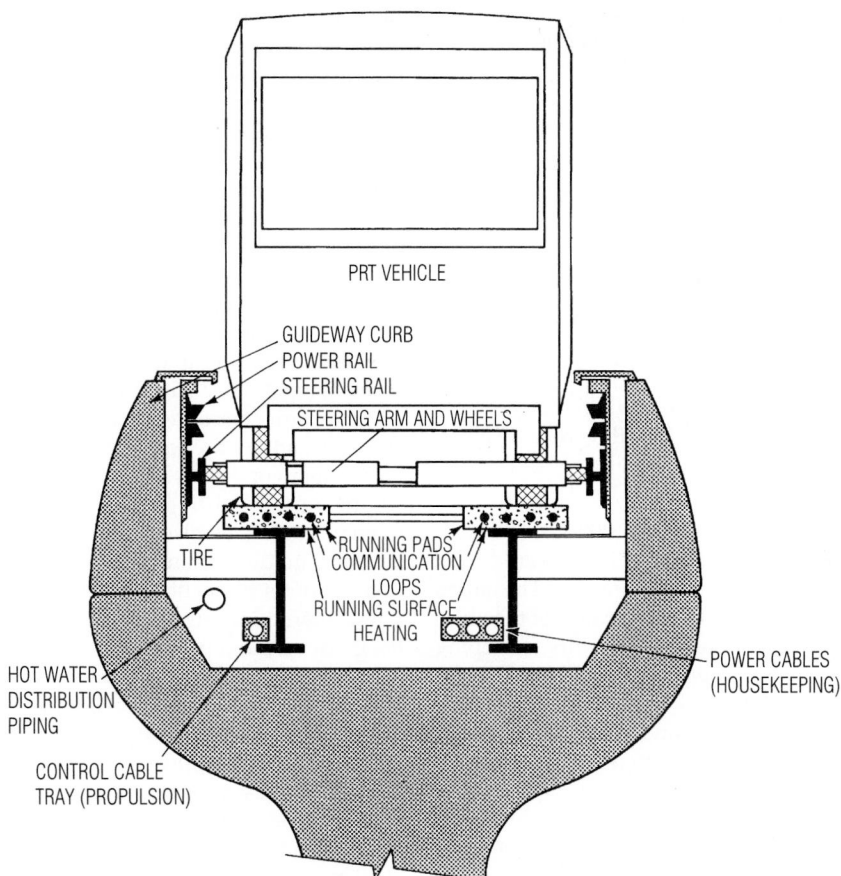

Fig. 19.20 Outline diagram of Boeing personal-rapid-transit system with single, elevated guideway at the University of West Virginia, Morgantown, W.V. Vehicles accommodate 8 seated and 13 standing. Speed ranges up to 30 mi/h. During peak hours, this system operates on a scheduled basis; otherwise, on a passenger-demand self-service basis. Tire running surfaces are heated to melt snow or ice.

there must be overriding advantages to justify this expense if such a system is to be used.

Power for a transportation system can be diesel-electric, electric, gas-turbine electric, gas-turbine hydraulic, jet propulsion, linear induction motor (Figs. 19.21 and 19.22), or pneumatic. The costs and characteristics of each must be taken into account in selection of the type of propulsion for any given transportation system. There has been much experience with the diesel-electric and electric motor drives; there has been some experience with the gas-turbine electric motor and gas-turbine hydraulic drive. This experience shows that it is difficult to compete with the

diesel-electric or the electric motor drive. So far, the efficiency of the turbo-electric or turbo-hydraulic drive has not been brought up to that of the other two.

For speeds over 100 mi/h, the electric motor drive has an advantage over the diesel-electric because the electric drive does not have to pull the weight of the electric generating plant; also, for short periods of time, it can draw a great deal of power from the catenary, whereas the diesel-electric has a fixed maximum power.

A speed of over 200 mi/h has been attained in trial runs with an electric-motor-powered vehicle supported on steel wheels on conventional track.

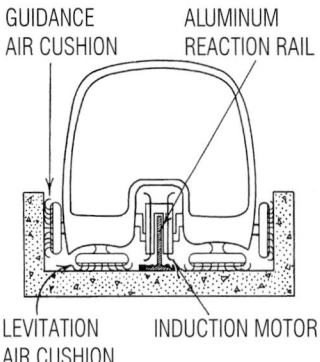

Fig. 19.21 Schematic illustrating the principle of tracked air-cushion vehicle with vertical reaction rail for linear induction motor.

However, to attain speeds of 200 to 300 mi/h regularly, vehicles may have to be powered by a linear induction motor or turbine jet. The latter, however, may be objectionable because of the noise level, and the former poses the difficulty of keeping the track reactor in accurate line and surface for such high speeds, as well as maintaining it free from windblown debris, sand, snow, and ice. At such speeds, the power required to overcome air drag is considerable.

19.18.2 Levitation Systems

Since 1965, much research, development, and testing have been conducted in the United States, Great

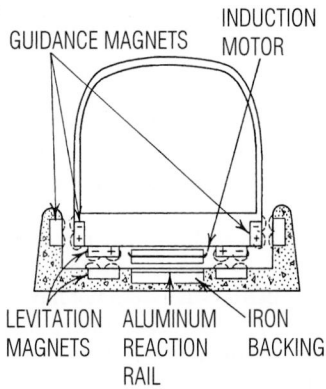

Fig. 19.22 Schematic illustrating the principle of the magnetic levitation vehicle with horizontal reaction rail for linear induction motor.

Britain, Germany, France, Japan, and Canada on levitated transportation systems. Two concepts have been studied: the tracked air-cushion vehicle (Fig. 19.21) and the magnetic levitation vehicle (Fig. 19.22). The air-cushion support system is not favored by many engineers working in the levitation field because of its high noise level, power requirements, weight of the air-cushion fans and motors, and lack of a suitable design for track switches.

One specialist states that a high-speed levitation system should offer these advantages: reduced travel time, comfort, safety, punctuality, competitive fares, minimum disruption of the environment, compatibility with other transit systems, and a minimum chance of failure. To these factors should be added the condition of being operable in all weather conditions.

The most promising type of magnetic levitation system requires no electric-current pick-up system between guideway and vehicle. Magnetic levitation supports the vehicle. For this purpose, three-phase alternating current is fed into coils located in the guideway and propels the vehicle. Levitation is controlled by changing the voltage of the magnets, and speed is controlled by changing the frequency of the three-phase winding. Speeds in the range of 250 to 300 mi/h appear practical of attainment, with an energy consumption per passenger-mile somewhat less than that of an airplane traveling 500 mi/h and somewhat more than an automobile at 60 mi/h.

Figure 19.23 shows a schematic of such a magnetic levitation system. The guideway may be elevated for practical reasons, although it could be placed underground. This system offers many advantages: no direct electrical or mechanical contact with the vehicle; no guiding, support, or propulsion friction; no moving parts to wear; high reliability; good passenger safety; low noise level; exceptional passenger ride comfort; and no atmospheric pollution. Indications are that its first cost as well as operating and maintenance costs can be competitive.

Of particular interest to civil engineers in this system are the construction of the guideway and supporting piers (or tunnels), the maintenance of the guideway alignment and surface, and obstruction warning devices.

Although levitation of the vehicle is generally associated with use in high-speed service, a PRT vehicle with magnetic suspension and linear propulsion that is used for slow speed and frequent

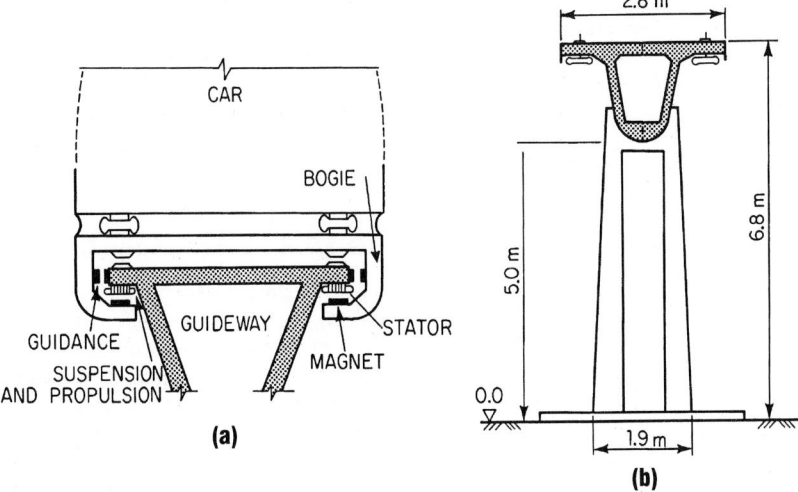

Fig. 19.23 Schematic of high-speed, magnetic levitated system developed for the Federal Republic of Germany by Transrapid EMS. (*a*) The vehicle levitation and propulsion system; (*b*) the guideway and support pier. Piers are spaced 25 m center to center.

stops has been developed by Automated Transportation Systems, Boeing Aerospace Company.

(R. D. Thornton, "Flying Low with Maglev," *IEEE Spectrum*, April 1973; Gerhart W. Heumann, "German High-Speed Railroads," *Machine Design*, Sept. 6, 1973; Klaus-Glatzel, G. Khordok, and D. Rogg, "The Development of the Magnetically Supported Transportation System in the Federal Republic of Germany," *Transactions on Vehicular Technology*, vol. VT-2, no. 1, February 1980, Institute of Electrical and Electronics Engineers, 345 E. 47th St., New York, NY 10017.)

19.18.3 Freight Cars

General types of freight cars include flat, box, stock, tank, hopper, covered hopper, gondola, refrigerator, and caboose. Some special types of freight cars include trailer on flat, double-stack container cars, auto rack, auto pack, container on flat, steel sheet, steel coil, and Hy-Cube. The objective of special cars is improved service; for example, auto-pack cars completely enclose the automobile and prevent the damage and pilferage that frequently occur with open auto-rack cars. Schnable cars are used to carry specially heavy and wide loads. The **auto train** combines auto-rack cars to transport the automobiles of passengers and conventional passenger cars of different types to transport

the passengers. The coupled length of freight cars ranges from 24 ft for ore hopper cars to 94 ft for Hy-Cube boxcars.

For freight cars to be freely interchanged in the United States, Canada, and Mexico, many components must have Association of American Railroads approval. These components include couplers, draft gear, center sill, air-brake system, wheels, axles, bearings, truck side frames, springs, snubbers, bolsters, and side bearings. The total rail load that is permitted is determined by the journal size. Table 19.6 lists maximum loads for several journal sizes for a car having four axles.

Width and height of freight cars must come within Plate B (Fig. 19.24) for unrestricted interchange and Plate C for interchange on most roads, as given in AAR Mechanical Division Specifica-

Table 19.6 Maximum Permissible Freight-Car Weight per Rail

Journal size, in	Weight, lb
5×9	142,000
$5\frac{1}{2} \times 10$	177,000
6×11	220,000
$6\frac{1}{2} \times 12$	263,000
7×12	315,000

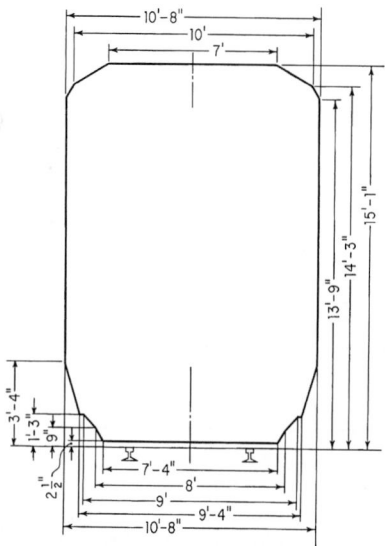

Fig. 19.24 Plate B clearance diagram for freight cars for unrestricted interchange service. Cars may be constructed to an extreme width of 10 ft 8 in and to the other limits of this diagram when truck centers do not exceed 41 ft 3 in. With truck centers of 41 ft 3 in, the swingouts of end of car should not exceed the swingout at center of car on a 13° curve. A car to these dimensions is defined as the base car. When truck centers exceed 41 ft 3 in, the car width should be reduced to compensate for the increased swingout at the center or ends of the car on a 13° curve, so that the extreme width of the car does not project beyond the center of the track more than the base car. The 2½-in clearance above top of rail is an absolute minimum. (*Mechanical Division, Association of American Railroads.*)

tions for Design, Fabrication, and Construction of Freight Cars. The dimensions of Plate B for width must be reduced for cars having truck centers in excess of 41 ft 3 in.

Most railroads have a clearance bureau for verifying load clearances and routes for cars carrying freight that exceed Plate B and Plate C limits. In addition, heavy loads that exceed weight restrictions are also cleared but restricted to designated routes and speeds, especially when travel over structures is necessary.

Freight cars of 45-ft coupled length can be operated around 45° curves coupled together.

Boxcars of 94-ft coupled length coupled to a short car can be operated around a curve of about 20°. Curve negotiability depends on clearance of the car corners and the free angling of the couplers in the draft gear pockets. ("Car and Locomotive Builders' Cyclopedia," Simmons-Boardman Publishing Corporation, Omaha, Neb.)

19.18.4 Passenger Cars

Types of passenger cars include baggage (baggage-dorm), coach, diner (cafe, dinette), lounge (club, parlor), sleeper, and combinations. Dimensions of passenger cars adopted as recommended practice by the Association of American Railroads (AAR) Mechanical Division are: coupled length, 85 ft; width, 10 ft; height, 13 ft 6 in (bi-level cars, 16 ft 6 in); and truck centers, 59 ft 6 in. Weight of these cars empty ranges from 100,000 to 160,000 lb. Seating capacity ranges from 44 to 89 in coaches and 23 to 48 in food service cars.

According to information provided by AMTRAK, a sleeping car has various combinations, mostly 10 roomettes and 6 bedrooms; a slumber-coach has 24 single roomettes and 8 double roomettes. (The "Car and Locomotive Builders' Cyclopedia," Simmons-Boardman Publishing Corporation, Omaha, Neb., contains photographs and floor plans of the latest passenger cars built.) Passenger cars must be constructed to meet AAR requirements for safety and interchange. Four-wheel trucks are generally used with 36-in-diameter wrought-steel wheels, roller bearings, helical or air-coil springs, snubbers or shock absorbers, cross stabilizers (lateral bumper), and load equalizers. Passenger-type air-brake equipment and air signal lines are provided. Electric air conditioning, heating, and lighting for the cars are powered through train lines from the head end.

Passenger cars are designed to negotiate a curve of 250-ft minimum radius when coupled together.

19.18.5 Commuter Cars

Several types of commuter cars are in use. One type is designed for push-pull operation by a separate locomotive. It is of semimonocoque design of aluminum with high-strength steel underframe. The vehicle is 85 ft long, 10 ft 6 in wide, and 12 ft 8 in high above top of rail. Truck centers are 59 ft 6 in, and wheel base is 8 ft 6 in. Trucks are inboard bearing, air suspension, with 32-in-diameter wheels, eight composition brake shoes, and electropneumatic braking. Weight is 74,000 lb. Seating capacity is 104. There are two 33-in-wide doors on each side near the car ends. Loading is from low platform level.

Another type is the bilevel or gallery-type push-pull car. A typical car of this type is 85 ft long, 10 ft wide, and 15 ft 10 in high. It seats 161 passengers in a trailer car, 155 in a cab car. The cab car weighs 128,500 lb; the trailer car, 123,400 lb. These cars are pulled or pushed in the train by a diesel-electric locomotive. A cab car with controls for the engineer is located at one end, the diesel-electric unit at the other end of the train. As many trailer cars as needed are coupled between them. These cars have four-wheel trucks 59 ft 6 in on centers. Double or triple doors at midlength of the cars expedite loading and unloading at low platform level.

A self-propelled rail diesel car is used to some extent in commuter service. A typical car is 85 ft long, 10 ft wide, and 14 ft 7 in high. It has two four-wheel trucks 59 ft 6 in on centers. The weight is 112,800 lb and seating capacity 89. A 550-hp diesel engine with electric drive powers the car. These cars are also used for mail, express, and passenger service on lines having light traffic. Loading is from low platform level.

The fourth type of commuter car is the electric multiple-unit (MU) coach. This type is used only for high traffic density. One design of MU car is 85 ft long, 10 ft wide, and 12 ft 6 in high. It has 59-ft 6-in truck centers; trucks have two axles spaced 8 ft 6 in on centers. The car weighs 105,600 lb and seats 122 passengers. Generally, several units are used in one train, but each has its own catenary trolley and is powered by four 156-hp motors. Loading is from floor level.

A fifth type is a double-deck MU car with cab controls at opposite ends of adjoining cars. This type is 85 ft long, 10 ft 5¾ in wide, and 15 ft 10 in above top of rail. Weight is 134,000 lb. Seating capacity is 156, equally divided between the double doors on each side near car midlength. It also has a single door on one side at the cab end. It operates from a 1500-V dc catenary system. The pantograph that collects the current for each car is located in a roof offset at the cab end, measuring 1 ft 10 in deep and 10 ft 4⁹⁄₁₆ in long. Loading is from floor level.

A sixth type of commuter car is constructed to operate off the third rail in electrified territory and from its own power supply in nonelectrified trackage. These cars are built as pairs with one power source. Each car is 85 ft long, weighs 140,000 lb, and seats 240 in a "married pair" of cars. Loading is from either floor or ground level. Power is supplied by two 550-hp gas turbine-electric generator units, mounted directly under the roof for easy maintenance. The two gas turbines drive alternators providing three-phase power at 420 Hz, 277 to 480 V. The rectified output is transmitted to a dc-dc chopper circuit that controls separately excited traction motors. The choppers (solid-state electronic switching devices) are advanced means of controlling dc traction-motor input power to provide smooth, efficient, jerkless acceleration for passenger comfort.

Electrically self-propelled commuter and rapid-transit cars may store energy developed by regenerative braking in storage batteries or in a high-speed flywheel for later use in train acceleration. This reserve energy supply could be used to operate the cars to the next station in the event of a power failure and, with storage batteries, to move cars in and out of yard and shop tracks and thus eliminate the need to electrify this trackage (resulting in less cost and greater safety).

All the types of commuter cars described previously have tinted glass windows, are air conditioned, and have comfortable seats, attractive decor, good lighting, racks for luggage or apparel, and toilets.

19.18.6 Rail-Transit Cars

Essential characteristics of rapid-transit cars are rapid acceleration and deceleration, quick entrance and exit, maximum seating capacity, and passenger comfort. These are provided, respectively, by high-horsepower motors, a combination of dynamic and air brakes, and lightweight construction; several doors per car; loading and unloading at floor level; seats and arrangement designed for best space utilization; and padded upholstered seats, air conditioning, good lighting, and attractive decor. Table 19.7 gives comparable car data for several rapid-transit systems. Cars can carry up to 350 passengers. Seats provided range from 56 to 83. Figure 19.25 illustrates the type of cars used on the Denver Light Rail System.

The Bay Area Rapid Transit (BART) cars are a good example of car design that offers excellent service, comfort, and safety. Type A cars have one

Table 19.7 Characteristics of Some Rapid Transit Cars

	Bay Area Rapid Transit District (BART)	New York City Transit Authority	Southeastern Pennsylvania Transit Authority (SEPTA)	Toronto Transit Commission	Washington Metropolitan Area Transit Authority (WMATA)	Metropolitan Atlanta Rapid Transit Authority (MARTA)
Capacity:						
Seats per car	72	72[c]	56	83	80	68[l]
Maximum passenger design	216	350	202	300	240	250[m]
Length over coupler faces	75 ft[a]	75 ft 0 in	55 ft 4 in	74 ft $9\frac{1}{8}$ in	75 ft 0 in	75 ft 0 in[n]
Height:						
Overall	10 ft 6 in	12 ft $1\frac{1}{2}$ in	12 ft 10 in	11 ft $11\frac{1}{2}$ in	10 ft $10\frac{1}{2}$ in	11 ft 10 in
Headroom	6 ft 9 in	6 ft $8\frac{3}{8}$ in[d]	7 ft 4 in	6 ft 11 in	6 ft 10 in	6 ft 10 in
Floor to top of rail	3 ft 3 in	3 ft $10\frac{3}{8}$ in	3 ft 10 in	3 ft $7\frac{1}{2}$ in	3 ft 4 in	3 ft 8 in
Width, maximum	10 ft 6 in	10 ft 0 in	9 ft 1 in	10 ft 4 in	10 ft $1\frac{3}{4}$ in	10 ft 6 in
Weight, total less passengers	56,500 lb[b]	87,000 lb[e]	48,760 lb	55,500 lb	72,000 lb	76,000 lb[o]
Trucks:						
Truck center distance	50 ft 0 in	54 ft 0 in	38 ft 0 in	54 ft 0 in	52 ft 0 in	52 ft 6 in
Wheel diameter	30 in	34 in	28 in	28 in	28 in	34 in
Track gage	5 ft 6 in	4 ft $8\frac{1}{2}$ in	5 ft $2\frac{1}{4}$ in	4 ft $10\frac{7}{8}$ in	4 ft $8\frac{1}{2}$ in	4 ft $8\frac{1}{2}$ in
Wheelbase	7 ft 0 in	6 ft 10 in	6 ft 8 in	6 ft 10 in	7 ft 3 in	7 ft 3 in
Minimum radius horizontal curve	500 ft	145 ft	140 ft	250 ft[f]	250 ft[h]	350 ft
Minimum radius vertical curve	1670 ft	2000 ft	3000 ft	2000 ft	[i]	1.5%/100 ft[p]
Number of motors	4	4	4	4	4	4
Horsepower per motor	150	115	100	116	160	160
Performance:						
Balancing speed, mi/h	80	80	55	55	75[j]	75[q]
Initial acceleration rate, mi/h·s	3.0	2.5	3.0	2.5[g]	3.0	3.0
Service braking rate, mi/h·s	3.0	3.0	2.75	2.8	3.0[k]	3.0[r]
Emergency braking rate, mi/h·s	3.3	3.2	3.0	3.0	3.2[k]	3.5[r]
Dynamic brake range	80–4	70–15	55–1	50–10	15 fade out	70–3
Doors:						
Number per side	2	4	3	4	3	3
Height	6 ft 4 in	6 ft 3 in	6 ft 3 in	6 ft $5\frac{1}{4}$ in	6 ft 4 in	6 ft 7 in
Width	4 ft 6 in	4 ft 2 in	4 ft 1 in	3 ft 9 in	4 ft 2 in	4 ft 2 in
Minimum number of cars per train	2	4	2	2	2	1
Maximum number of cars per train	10	8	10	6	8	8

[a]For A cars; B cars = 70 ft.
[b]For A cars; B cars = 55,000 lb.
[c]For A cars; B cars = 76
[d]Low ceiling; 7 ft $2\frac{3}{8}$ in for high ceiling.
[e]For A cars; B cars = 84,000 lb.
[f]Minimum desirable for main-line box structure and circular tunnels = 1000 ft.
[g]High rate; low rate = 1.9.
[h]For yard track; main line = 500 ft.
[i]Parabolic, min. length = $(G_1 - G_2)$ 100 ft, but not less than 200 ft.
[j]On 1% grade.
[k]Below 50 mi/h.
[l]For A and B cars; C cars = 62.
[m]For A and B cars; C cars = 235.
[n]For A and B cars; C cars = 75 ft 4 in.
[o]For A and B cars; C cars = 79,700 lb.
[p]Parabolic.
[q]Maximum overspeed.

Fig. 19.25 Typical light rail vehicle in service for the Regional Transportation District (RTD) in Denver, CO.

slanted end with a cab for a single attendant for train control (when needed), automatic train operation sensors, and a communications system. *A* cars are placed with the slanted end at the front and rear of the train, an arrangement that gives a pleasing, streamlined appearance. As many *B* cars as needed, up to 8, are placed between the two *A* cars. Vinyl-padded double seats are placed on each side of a middle aisle. The floors are carpeted; smoking is not permitted. The car interior is made of simple, durable, and fire-resistant construction and designed for ease of cleaning. No painting is required, and advertising signs are not used. Lighting fixtures use focusing lenses and provide 30 to 35 fc at reading height, 20 fc at floor level. At two locations in each car, a small push-to-talk intercom set permits passengers to report emergencies or seek information from the attendant. Either the attendant or the central office can make announcements to passengers from speakers in each car. A large enclosed passageway between cars with biparting doors and large panes of glass allows passengers to see seats in adjoining cars. This also facilitates observation of two cars by the attendant during night hours.

Each car has its own air conditioning system, which provides draft-free, uniform air distribution with fresh air infusion, 12-ton refrigeration, 30-W heating, and humidity control to below 60% relative humidity.

Automatic train control and cab signals are provided, but the attendant can override the train control in an emergency. Automatic couplers complete 24 electrical circuits throughout the train.

Wheels are designed for light weight and noise reduction. They have AAR wrought-steel, heat-treated rims and aluminum hubs. The car support and trucks include level-controlled air bellows, rubber "doughnuts" around the journal roller bearings, and hydraulic shock absorbers.

A dc chopper is used to control the 450-V direct current to each motor to give smooth starting and stopping. An automatic car identification system is used, with color-coded labels on each car. Scanners are located on yard leads to record miles run for maintenance purposes and to determine the location of each car.

Communication between trains and central control is by radio, using a line antenna through subway sections.

A more detailed description of the BART system is in "Modern Railroads Rapid Transit," February 1972; and "The Bay Area Rapid Transit Vehicle System," L. A. Irvin and J. R. Asmus, paper 680544, Society of Automotive Engineers.

19.18.7 Railway Service Cars

These are non-revenue and maintenance-of-way material and equipment cars and other special purpose cars. Maintenance-of-way cars include air dump cars for carrying rip rap and embankment material, ballast cars, tie cars, car sets with racks for carrying welded rail strings, clearance measuring cars, track geometry cars for measuring the quality of the track under loaded conditions, and flangers, spreaders and rotary snow plows. Other specialty equipment includes scale test cars, dynamometer cars for testing locomotives and wreck trains that contain tool cars, track material and cars and locomotive cranes of up to 250 ton capacity.

Instead of using wrecking cars, some major railroads employ contract services. The contractors use construction equipment, including track-mounted machinery, modified to handle cars and locomotives, to removed derailed equipment and restore trackage. The contractors generally move their equipment by truck and trailer and use additional trailers for cooking and sleeping facilities.

19.18.8 Locomotives

In the United States, almost all freight and passenger trains are moved and switching operations done with diesel-electric locomotives. Less than 1.5% of locomotives are electric; most of the remainder are diesel-electric. There are a few locomotives of other types in use, such as diesel-hydraulic and gas turbine-electric. Locomotive capacity is based on the rated horsepower, tractive effort, and rail adhesion characteristics of the locomotive. These determine the drawbar pull available for hauling cars.

Diesel-electric locomotives use on-board diesel engines to power electrical generators that feed current to electric motors that drive the axles. Early diesel-electric locomotives used Direct Current (DC) electrical systems. New diesels use either DC or Alternating Current (AC).

Diesel-electric can be operated in multiple units by one engine-man to afford the horsepower and

tractive effort required. It requires relatively few stops for fuel and water, and it has excellent starting characteristics because all the weight is on the driving wheels. In long unit trains, diesel-electric locomotives are used mid train and/or at the end of the train as helpers.

An electric locomotive has good efficiency. But since the electric power required is usually generated in a separate, immobile plant, there is some power loss in the line transmission, and the catenary system represents a considerable investment and maintenance expense. Electric locomotives, in general, are economical only on lines having fast and frequent train schedules. These locomotives have the advantages of being able to develop a high horsepower at high speed and requiring less maintenance than diesel-electric.

19.19 Propulsion Power Requirements for Trains

These vary with the type of service. For intercity passenger and freight trains, the power to pull the trains up grades and make the scheduled time is of paramount importance. For commuter and rapid-transit service, an important factor in the power requirement is the need to accelerate quickly. For personal rapid transit, the speed is relatively slow, but power must be adequate to accelerate quickly to the desired speed. For all types of service, power must be adequate to overcome grade, curve, and rolling resistance. Requirements for running time, frequency of service, and operating costs must all be considered.

19.19.1 Resistances to Train Movement

Grade resistance, offered by an ascending grade, equals 20 times the percent grade per ton of train. Thus, on a 1.5% grade, the grade resistance is 30 lb/ton; on a 1.0% grade, 20 lb/ton; and on a 0.5% grade, 10 lb/ton. On a descending grade, the same forces accelerate the train, which must be controlled by braking.

Curve resistance is the added resistance required to guide and slip the wheels in negotiating a curve. Curve resistance is generally considered equivalent to 0.04% grade per degree of curvature.

Thus, the curve resistance on a 4° curve would be $0.04 \times 20 \times 4 = 3.2$ lb/ton of train. It is customary on ruling grades to compensate for curvature by reducing the grade for the length of the curve. Thus, if the ruling grade on a line is 0.5%, compensated for curvature, no consideration need be given to curve resistance in calculating power requirements because it is already included in the grade. Use of rail lubricators reduces curve resistance by about one-half. Curve resistance tends to retard the train on descending grades.

Rolling or train resistance is the resistance to train movement on level tangent track. Train resistance is affected by speed, weight on the axle, and characteristics of the track. This last factor is usually neglected because it is relatively small. Starting resistance is less with roller bearings, but after a train starts, the train resistance is about the same for roller and solid bearings. For example, the starting resistance for a car with solid bearings might be as much as 20 lb/ton, but train resistance becomes 5 lb/ton as soon as the car is in motion. The same car on roller bearings would have the same starting resistance as when moving at slow speed, 5 lb/ton.

19.19.2 Formulas for Train Resistance

There are several formulas for calculating train resistance. The Davis formulas (W. J. Davis, Jr., "The Tractive Resistance of Electric Locomotives and Cars," *General Electric Review,* October 1926) are representative of results found by several investigators. According to the AREA "Manual for Railway Engineering," the Davis formulas have given satisfactory results for speeds between 5 and 40 mi/h. However, the increased dimensions and heavier loading of freight cars, the much higher operating speed of freight trains, and changes in types of cars since the formulas were developed have made it desirable to modify the constants in the Davis equation. Recent tests have shown improved results with the following modified Davis formula:

$$R = 0.6 + \frac{20}{W} + 0.01V + \frac{KV^2}{WN} \qquad (19.19)$$

where R = resistance, lb/ton

W = weight per axle, tons

N = number of axles per car

V = speed, mi/h

K = air resistance coefficient

= 0.07 for conventional freight-train equipment

= 0.16 for trailer on flatcar (piggyback)

= 0.0935 for containers on flatcars

The last term in this equation, KV^2/WN, represents the air drag due to train speed. At high speeds, this becomes a major factor in train resistance, and it is necessary to take into account the cross-sectional area of the car, aerodynamic properties of the car design, air density, and wind velocity and direction.

For a detailed treatment of this subject for high-speed passenger service, see J. L. Koffman, "Tractive Resistance of Multi-Unit and Locomotive-Hauled Passenger Trains," Rail Engineering International, April-May 1973. The author suggests the following formula as representative of modern passenger-train equipment on British and Continental railways:

$$R = 1.5W + (5.5 + n - 2)\left(\frac{V}{10}\right)^2 \qquad (19.20)$$

where R = total tractive resistance of conventional passenger train, kg

W = total weight of train, metric tons

n = number of coaches in train

V = train speed, km/h

and the effective locomotive and coach frontal area is taken to be 10 m². Equation (19.20) assumes an air drag coefficient of 0.6 based on experimental data on a locomotive at speeds up to 100 mi/h. With some car designs where little consideration was given to aerodynamic properties, the air-drag coefficient was found to be as much as 1.85 for an eight-car train. On the other hand, it was found to be as little as 0.97 for a 249.5-m-long, 10-coach Tokaido train, for which extensive model tests were made in a wind tunnel to obtain good aerodynamic performance. The Association of American Railroads continues to test new equipment and update values.

In design of vehicles to be operated at speeds over 100 mi/h, it is highly important that aerodynamic performance be considered because air drag causes most of the rolling resistance at these high speeds and increases as the square of the speed.

19.19.3 Calculating Running Time and Fuel Consumption

Running time and fuel consumption are useful data in comparing the relative desirability of various lines in new construction or in revisions of existing lines. Running time may be calculated by the velocity-profile method.

In this method, accelerating force, lb/ton, is computed by subtracting from the drawbar-pull characteristics of the locomotive the train resistance on level track. The computation is repeated for 5-mi/h increments from starting to maximum permitted operating speed. Since grade resistance is 20 lb/ton (see Art. 19.19.1), the accelerating force may be converted into an equivalent grade by dividing by 20. The actual profile of the line is plotted on a graph showing elevations versus distance. On the same graph, for each increment of speed, the equivalent grade is plotted between points for which the vertical difference between the actual and equivalent grades equals the velocity head. The velocity head, ft, for any speed is

$$VH = 0.035V^2 \qquad (19.21)$$

where V = speed, mi/h. This formula expresses the kinetic energy of a train due to its velocity and the rotating energy in its wheels as equivalent potential energy due to height. The same procedure is applied for braking to reduce speed or stop. The series of lines representing equivalent grades is the velocity profile. (Detailed instructions for train performance calculations are in "Manual for Railway Engineering," American Railway Engineering and Maintenance of Way Association.) After the velocity profile has been completed for the line, the running time is found by summing the time required to travel each increment of distance at the average speed for the increment. A computer may be used to facilitate and expedite the calculations required.

The time that a locomotive will be working at full capacity, part capacity, or drifting can be determined from the velocity profile. Multiplying each period of time by the corresponding rate at which fuel is used by a particular locomotive yields the fuel consumption. Another method that may be used to calculate fuel consumption is to first figure the total work done. This consists of the work done in overcoming rolling resistance, plus the resistance of gravity on ascending grades, plus the resistance due to curvature. From this sum should be subtracted the energy of gravity on descending grades, but the loss of energy (velocity head) due to application of brakes should be added to give total work.

The total work done, ft-lb, may be converted to gallons of diesel fuel by multiplying by 4 (efficiency of 25%) and dividing by 90 million (ft-lb of energy per gallon of diesel fuel).

A simpler method that will be sufficiently accurate for most purposes is as follows: Approximate a condensed profile of the line with a series of long grades. Calculate the speed at which the locomotive can handle the train over each grade. Obtain the time over each grade by dividing the distance by the speed and total these. Add an arbitrary 5 to 10 min for each stop and start, depending on the length of the train. This will give the approximate running time. The fuel consumption can be determined as in the velocity-profile method or from total work done.

19.19.4 Train Tonnage

The maximum tonnage that can be hauled over a line with a given locomotive is determined by the ruling gradient. However, the locomotive may not be able to handle this much tonnage at a high enough sustained speed to meet competitive traffic requirements or to avoid train-crew overtime. With diesel-electric locomotives, any number of units can be coupled together, but if the locomotives are placed at the head end of the train, trouble with broken couplers may be encountered if drawbar pull exceeds 200,000 lb.

If a train is made very long, for example, 200 cars, difficulty may be experienced from slack run-in, from excessive delays for replacement of broken couplers or setting out cars that have developed hot boxes, or from air-brake operation in very cold weather. A train of 100 cars is quite common in the United States. Occasionally, railroads operate trains with as many as 250 cars. Diesel-electric locomotives are sometimes added near midlength of a train and as pushers at the end on steep grades.

Generally, the more tonnage in a train, the lower the operating cost. Thus, train tonnage is a matter of economy, practicality of operation, and meeting competitive traffic requirements for speed and frequency of service.

Since train resistance varies with car weight and number of cars, a locomotive cannot handle as much tonnage in a train of empty cars as in one of loaded cars. Also, a locomotive cannot handle as much tonnage in cold weather as in warm weather. As a convenient means of compensating for these two factors, use may be made of the data in Table 19.8.

The adjusted tonnage rating may be considered as the sum of the weight of cars and contents, tons, and an adjustment, tons. For temperatures above 35 °F, this sum is called the adjusted tonnage A rating. The adjustment for computing the adjusted tonnage A rating is obtained by multiplying the adjustment factor given in Table 19.8 for a specific ruling grade by the number of cars in a train. For temperatures below 35 °F, a percentage of the A rating is used, as indicated in Table 19.8. The adjusted tonnage rating is independent of the number of cars in a train.

19.20 Train Control and Signal Systems

There are many methods for controlling the movement of trains on tracks, depending on the number of tracks and the characteristics of the traffic. The objective is to move the trains to conform to desired schedules between departure and destination points, with safety the paramount consideration.

Train orders and time schedules are used where only a few trains move over a line per day. Passenger train speeds are limited to 59 mi/h and freight train speeds to 49 mi/h by the Federal Railroad Administration (FRA) where this method of train control is used.

The **manual-block system** provides a safer operation. Operators stationed between blocks of track do not permit a train to enter the next block until notified by the operator at the other end of block that it is clear. Although safe, this method gives low track capacity, slow schedules, and high cost for block operators. With the manual-block system, speeds up to 79 mi/h are permitted by FRA order.

The **automatic-block signal system** provides for successive blocks of tracks to be separated electrically by insulated rail joints at both ends. Unless the rail is continuously welded, rail bonds are used at each bolted rail joint to ensure continuity of the electric circuit between rail ends. A three-position signal aspect is connected into the electric circuit for each block and for adjoining blocks. Many different types of signal aspects are used. A common signal aspect is a green light to show an approaching train that two blocks ahead are clear, a yellow light to show that the second block ahead is occupied, and a red light to show that the next block ahead is occupied. The block length should not be less than the service braking

Table 19.8 Data for Calculating Adjusted Tonnage Ratings

Ruling grade, %	Adjustment factor, tons per car	% of A rating		
		B, 20–35 °F	C, 0–20 °F	D, below 0 °F
0.1	29	84	70	57
0.2	20	89	78	66
0.3	15	91	82	72
0.4	12	93	85	76
0.5	10	94	87	79
0.7	8	95	90	84
1.0	5	97	93	87
1.5	4	98	95	91
2.0	3	98	96	93
2.5	2	99	97	94
3.0	2	99	97	95

distance required for the train speed. A block length of 1 mi is frequently used. Speeds up to 79 mi/h are permissible.

Automatic train control is provided by a wayside inductor located in advance of each block circuit over which the locomotive receiver passes. This receiver is mounted on the locomotive journal box to have $1\frac{1}{2}$-in clearance with the wayside inductor. An electric circuit is provided so that when a locomotive passes a restrictive signal, the engineman must acknowledge awareness by actuating a contactor; otherwise, the train brakes are automatically applied. With this system, train speed is not limited by FRA order to 79 mi/h.

Coded control has the advantage of using one pair of line wires to transmit the signals from blocks ahead instead of requiring many line wires for this purpose. An interruption of the dc voltage is used for different signal indications. For example, on track with one-direction movement, 180 interruptions per minute operates the "proceed" signal; 120, the "approach-medium"; 75, the "approach"; and no code, the "restrictive." Additional signals may be transmitted by combinations of reversed polarity.

Another advantage of coded control is that the code-following track relay must pick up with each pulse. Therefore, the train shunt need only be enough to reduce the track current at the relay below the pickup value, rather than below the dropout value. This permits higher track voltage to be used, and for given ballast resistance conditions, track circuits can be made twice as long.

Continuous cab signals are provided by using alternating current for the track circuits instead of direct current and placing inductive receivers in front of the leading wheels on the locomotive. Thus, the signal passing through the rails is transmitted by the receivers to give signals in the locomotive cab. Any change in signal aspect is immediately visible, whether or not the wayside signal is in sight of the engineman. With this system, wayside signals are not actually required. Coded control can be used with this system by interrupting the ac voltage in the same manner as for dc voltage. With continuous cab signals, train speed may exceed 79 mi/h.

Interlocking is usually provided at railroad grade crossings and at some turnouts. At crossings without interlocking, each train must first stop at the crossing and then proceed if the crossing is clear.

Mechanical interlocking operated by a towerman permits giving the right-of-way to one train, holding any on the track being crossed. Signal aspects are operated by levers and long pipe connectors, as are derails on each track.

Electric interlocking permits the operator to actuate signal aspects and derails electrically. The operator can also unlock and throw switches by electric control for crossovers or connecting tracks. Switches are thrown by electropneumatic or electric motor switch machines. Safety features are provided to prevent an operator from lining up signals, derails, and switches unless the track is clear for such movements.

For very complicated crossings involving many tracks and train movements, route interlocking is used. It is necessary for the operator to just push a button for the point where a train will enter the interlocking and another button for the point where the train is to leave. The best available route will then be automatically selected and lined up for the train. On simple crossings, train movements through an interlocking can be controlled automatically by an electric signal system.

Overlap and absolute permissive block signaling is required to avoid collision of trains moving in opposing directions on the same track. With automatic-block signals giving indications for only two blocks ahead, opposing trains could pass a clear signal simultaneously and find the next signal at stop but be unable to do so in time. This situation can be prevented by overlapping the advance blocks so that the stop aspect is displayed more than one block in advance of a train. With the absolute permissive block system, relays can be used to extend the blocks in advance but provide the normal block indications for following trains, thus expediting their movement. Where opposing trains are operated on the same track with absolute permissive block, the block control is extended far enough in advance to include a passing track or crossover so that the trains can pass.

Centralized traffic control (CTC) is officially designated by the FRA as the "traffic-control system." It is defined as "a block system under which train movements are authorized by block signals whose indications supersede the superiority of trains for both opposing and following movements on the same track."

With CTC, one operator directs the movement of all trains and usually of all switches and derails on the trackage under his or her control. For low-traffic-density lines, sometimes the switches are manually thrown by the trainman in accordance with a signal aspect at the switch. A panelboard shows the operator a diagrammatic layout of the trackage, with all turnouts identified and signal aspects at turnouts shown. Lights identify the location of all trains. Signals and switches at the ends of passing tracks are arranged as route-type interlocking. Automatic-block signaling controls movement between passing tracks. With only two line wires along the track and different frequencies for transmitting coded data, it is possible for one operator to control train movement over several hundred miles of trackage. The actual operation is performed on a small control board in front of the operator, who merely pushes buttons or turns small switches to send out the directing signal. Acknowledgment is indicated on the panelboard by a signal automatically sent back when the action has been completed.

Automatic train operation is the capability for complete scheduling and operation of trains, including starting and stopping, opening and closing doors, and so on, by computer command. Theoretically, a train attendant is not required. Actually, it is usually considered desirable if passenger traffic is involved to have an attendant on each train who can take over in an emergency and override the automatic operation with manual operation. For manual override, automatic-block signals or continuous cab signals are required. Also, presence of a train attendant may give the passengers a sense of security.

At the control center, one or more computers control the operation of each train according to schedule but have the capability to change the operation automatically as required by any delays that may occur. One or more "dispatchers" are provided to observe the control board showing the position of all trains and to take over manual operation in an emergency. Experience has shown that automatic train operation will get trains over a route in less time and with more comfort to passengers than can be obtained with manual operation.

The systems and procedures for train control to secure maximum performance have become so sophisticated that specialists in the field should be consulted for selection and design of a system for any given conditions.

Grade-crossing warning is an important signal function in train operation. Block-signal track circuits may be used to actuate flasher lights or crossing gates automatically to warn vehicles of approaching trains at highway grade crossings. Crossing gates are advantageous at crossings of two or more tracks because of the danger that a motorist will drive on the crossing after a train has cleared without waiting to see if a train may be approaching on another track.

Audiofrequency overlay circuits have been developed to actuate grade-crossing protection without the need for insulated rail joints.

Slide fences are frequently used at locations where falling rocks obstruct the track. These fences are drawn tight by spring tension at one end. The pressure of a rock at any point on the fence will cause end movement, which breaks an electric circuit, causes a control relay to become deenergized, and sets the block signal to a stop position.

Additional types of wayside detectors include those to find broken wheels, high-wide loads, dragging equipment and overheated bearings ("hot box"). High-wide load detectors use wires or radar type units to pick up any shifted or extra dimensional loads that may exceed the allowable clearance of a tunnel or through truss bridge. Dragging equipment detectors use boards placed both outside and inside the rails that rely on physical contact with anything that may be hanging below the normal equipment clearances. Hot box detectors use heat sensitive equipment to compare the heat of passing cars and locomotives' bearings to predetermined limits. Many of these detectors are now equipped with radio warning messages that are transmitted to the dispatcher and the train engineer.

19.21 Communication in Train Operation

Many types of communication are available to enhance the safety and performance of train operation and give passengers a feeling of security. These include simple communication by means of a train whistle or warning bell, wayside or cab signals, telephone, radio, microwave, and electronic direct circuit or inductance. Many of these have been discussed in Arts. 19.17, and 19.20. Specialists in the communications field should be consulted on selection and design of the most suitable communication system for a given railway operation.

In the late 1960s, some railroads tried an automatic car-identification system in which bar codes were placed on the sides of cars and read by optical scanners. Because of difficulties in keeping the codes clean, the scanners were not always successful in reading the codes, and the system was abandoned. The railroads also tried various systems in which personnel located in yard facilities used television cameras to verify car numbers as the cars entered or exited major yards. One current form of car identification consists of short range transponders to pick up signals from tags attached to the cars. In addition, satellite systems utilizing the global positioning system (GPS) are under development to monitor train movements.

Lars Christian F. Ingerslev, Arthur G. Bendelius
Parsons Brinckerhoff
New York, New York

20

TUNNEL ENGINEERING

Tunnel engineering makes possible many vital underwater and underground facilities. Unique design and construction techniques are involved because of the necessity of protecting the constructors and users of these facilities from alien environments. These facilities must be built to exclude the materials through which they pass, including water. Often, they have to withstand high pressures. And when used for transportation or human occupancy, tunnels must provide adequate lighting and a safe atmosphere, with means for removing pollutants.

Tunnels are constructed using many methods, depending upon the kind of soil and/or rock through which they will pass, their size, how deep they need to be, and the obstructions that may be encountered along the route. These methods include cut-and-cover construction, drill and blast, tunnel boring machine (TBM), immersion of prefabricated tunnels, and sequential excavation methods (SEM). More specialized methods, such as ground freezing and tunnel jacking, are used less frequently and often under very difficult conditions. Compressed air working has become uneconomical because of working hour restrictions, time for decompression that results from high working pressures (over 40 psi is not unusual), union labor agreements for work under compressed air, and high workmen's compensation and health benefit rates. Occasional entry under compressed air may still be required, such as to clear obstructions ahead of a tunnel boring machine, or to perform essential maintenance on parts of such a machine.

The design approach to underground and underwater structures differs from that of most other structures. Internal space, design life, and other requirements for the tunnel must first be defined. Geological and environmental data must then be collected. Critical design loading conditions must then be established, including acceptable conditions of the tunnel following extreme events (for example, how long before the tunnel is reusable). Appropriate construction methods are then evaluated to determine the most appropriate to meet the established criteria, conditions, and cost. The methods under consideration should include both temporary and permanent excavation support systems as well as the structures itself. Design standards and codes of practice apply primarily to above-ground structures, so that care should be used in their application to underground and underwater structures.

20.1 Glossary

Adit. A short, transverse tunnel between parallel tunnels or to the face of the slope in a sidehill tunnel.

Air Lock. A compartment in which air pressure can be varied between that of the compressed air used in shield tunneling and that of the outside air, to permit passage of workers or material.

Bench. Top of part of a tunnel section, with horizontal or nearly horizontal upper surface, temporarily left unexcavated.

Blowout. A sudden loss of a large amount of compressed air at the top of a tunnel shield.

Breast Boards. Timber planks to hold the face of tunnel excavation in loose soil.

Dry Packing. Filling a void with a stiff mortar, placed in small increments, each rammed into place.

Evasé Stack. An air-exhaust stack with a cross section increasing in the direction of air flow at a rate to regain pressure.

Face. The surface at the head of a tunnel excavation. A mixed face is a condition with more than one type of material, such as clay, sand, gravel, cobbles or rock.

Grommet. A ring of compressible material inserted under the head and nut of a bolt connecting tunnel liners to seal the bolt hole.

Heading. A small tunnel, or tunnels, excavated within a large tunnel cross section which will be enlarged to the full section.

Jumbo. A frame that rolls on tracks or rubber wheels and carries drills for excavation of rock tunnels.

Lagging. Timber planks or steel plates inserted above tunnel-supporting ribs to hold back rocks or soil.

Liner Plate. A steel segment to support the interior of a tunnel excavation.

Lining. A temporary or permanent structure made of concrete or other materials to secure and finish the tunnel interior or to support an excavation

Mucking. Removal of excavated or blasted material from face of tunnel.

Pilot Tunnel. A small tunnel excavated over part or the entire length to explore geological conditions and assist in final excavation.

Pioneer Bore. (See Pilot Tunnel.)

Poling Boards. Timber planks driven into soft soil, over timber supports, to hold back material during excavation.

Scaling. Removal of loose rocks from tunnel surface after blasting.

Shield. A steel cylinder of diameter equal to that of the tunnel, for excavation of tunnels in soft material to provide support at the face of the tunnel, to provide space for erecting supports, and to protect workers excavating and erecting supports.

Spiling. (See Poling Boards.)

20.2 Clearances for Tunnels

Clearance in a tunnel is the least distance between the inner surfaces of the tunnel necessary to provide space between the closest approach of vehicles or their cargo or pedestrian traffic and those surfaces. Minimum tunnel dimensions are determined by adding the minimum clearances established for a tunnel to the dimensions selected for the type of traffic to be accommodated in the tunnel and the space needed for other requirements, such as ventilation ducts and pipelines.

Clearances for Railroad Tunnels ■ Individual railroads have different standards to suit their equipment. But on tangent tracks, clearances for single- and double-track tunnels should not be less than those shown in Fig. 20.1. (Clearances shown are those in the "AREMA Manual" American Railway Engineering and Maintenance-of-Way Association, 8201 Corporate Drive, Suite 1125, Landover, MD 20785, (www.AREMA.org).

In rail tunnels, clearances for personnel are required on both sides where niches are not provided. These clearances should be at least 6 ft 8 in or 2 m high and 30 in wide each side of the vehicle clearance diagram, although a 24-in minimum is permitted on some lines. In highway tunnels, a 3 ft or 0.9 m clearance from face of curb is used where walkways are provided. In both road and rail tunnels, it is common practice to provide a walkway along the common wall between adjacent ducts to facilitate emergency evacuation between ducts and to prevent people from emerging directly into the path of oncoming traffic.

On curved tracks, the clearances should be increased to allow for overhang and tilting of an 85-ft-long car, 60 ft c to c of trucks, and a height of 15 ft 1 in above top of rail. (Distance from top of rails to top of ties should be taken as 8 in.)

The track should be superelevated at curves according to AREMA standards.

Clearances for pantograph, third-rail, or catenary construction should conform to diagrams published by the Electrical Section, Engineering Division of the Association of American Railroads.

The latest clearance standards of AREMA should be checked for new construction. Local legal requirements should govern if they exceed these standards.

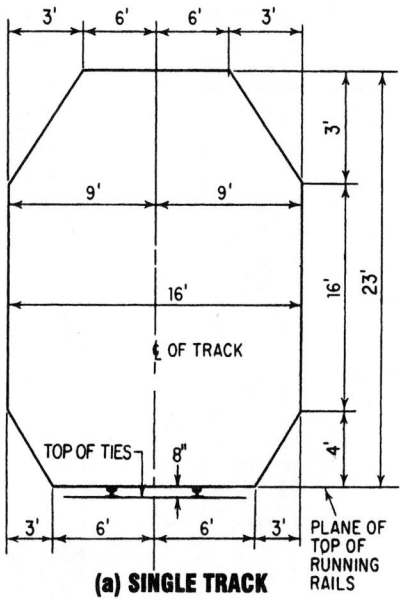

(a) SINGLE TRACK

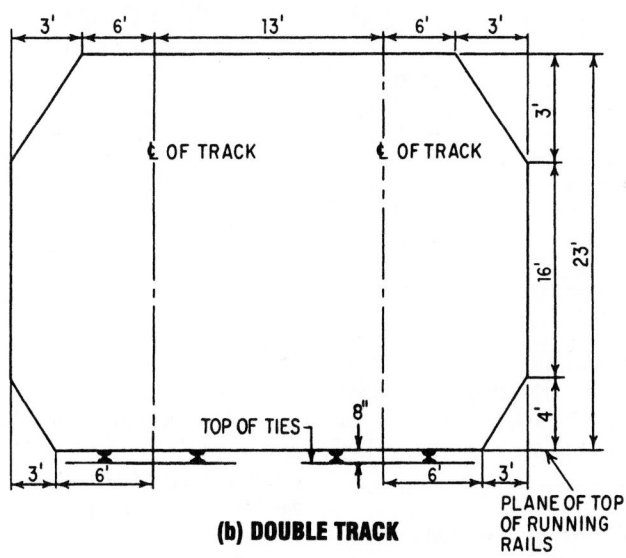

(b) DOUBLE TRACK

Fig. 20.1 Clearances specified by AREMA for railway tunnels on a tangent.

Circular tunnels should be fitted to the clearance diagrams, with such modifications as may be permissible.

Clearances for Rapid-Transit Tunnels ■
There are no general standards for clearances in rapid-transit tunnels. Requirements vary with size of rolling stock used in the system.

Figure 20.2 shows the normal-clearance diagram of the New York City BMT and IND Division −67 ft cars. Figure 20.3 gives the clearances established for the San Francisco Bay Area Rapid

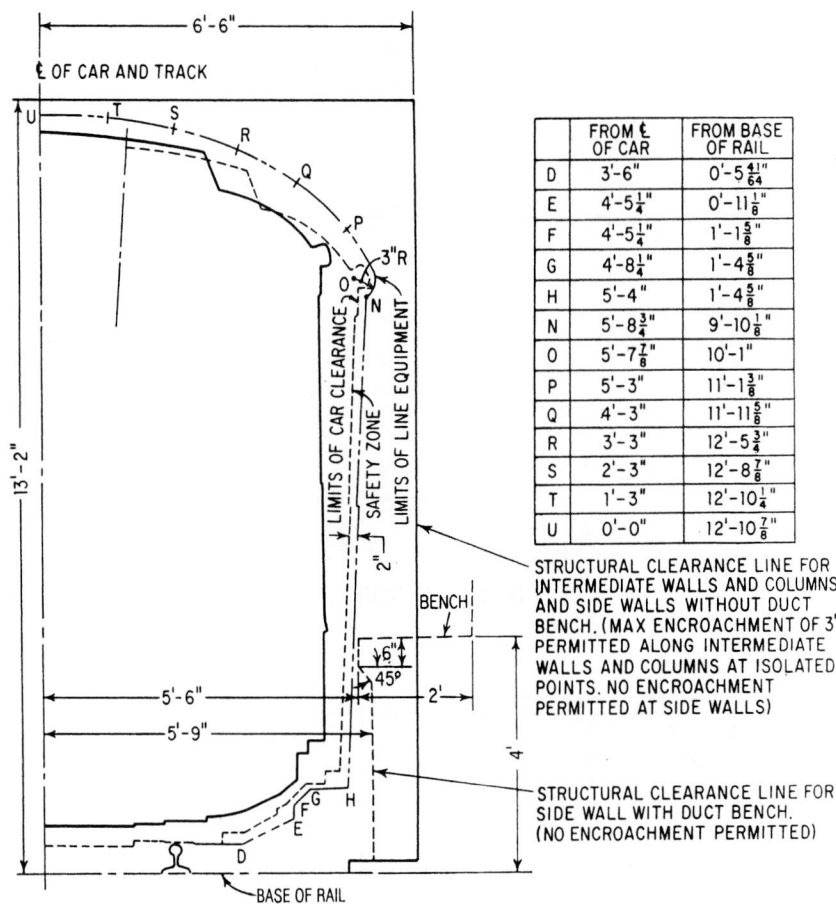

	FROM ℄ OF CAR	FROM BASE OF RAIL
D	3'-6"	0'-5$\frac{41}{64}$"
E	4'-5$\frac{1}{4}$"	0'-11$\frac{1}{8}$"
F	4'-5$\frac{1}{4}$"	1'-1$\frac{5}{8}$"
G	4'-8$\frac{1}{4}$"	1'-4$\frac{5}{8}$"
H	5'-4"	1'-4$\frac{5}{8}$"
N	5'-8$\frac{3}{4}$"	9'-10$\frac{1}{8}$"
O	5'-7$\frac{7}{8}$"	10'-1"
P	5'-3"	11'-1$\frac{3}{8}$"
Q	4'-3"	11'-11$\frac{5}{8}$"
R	3'-3"	12'-5$\frac{3}{4}$"
S	2'-3"	12'-8$\frac{7}{8}$"
T	1'-3"	12'-10$\frac{1}{4}$"
U	0'-0"	12'-10$\frac{7}{8}$"

STRUCTURAL CLEARANCE LINE FOR INTERMEDIATE WALLS AND COLUMNS AND SIDE WALLS WITHOUT DUCT BENCH. (MAX ENCROACHMENT OF 3" PERMITTED ALONG INTERMEDIATE WALLS AND COLUMNS AT ISOLATED POINTS. NO ENCROACHMENT PERMITTED AT SIDE WALLS)

STRUCTURAL CLEARANCE LINE FOR SIDE WALL WITH DUCT BENCH. (NO ENCROACHMENT PERMITTED)

Fig. 20.2 Clearance diagram for 67' car (BMT & IND Divisions). New York City Subway System.

Transit System, which has cars 10 ft wide and 75 ft long on a 5-ft 6-in gage track. The clearances allow not only for overhang of cars, tilting due to superelevation, and sway, but for a broken spring or defective car suspension.

Clearances for Highway Tunnels ■ The American Association of State Highway and Transportation Officials (AASHTO) has established standard horizontal and vertical clearances for various classes of highways. These have been modified and expanded for the Interstate Highway System under the jurisdiction of the Federal Highway Authority (FHWA) (Fig. 20.4).

For rural and most urban parts of the Interstate Highway System, a 16-ft vertical clearance is required.

Since construction costs of tunnels are high, clearance requirements are usually somewhat reduced. Although some older 2-lane tunnels have used roadway widths of 21 ft between curbs for unidirectional traffic and 23 ft for bi-directional traffic, usually with speed restrictions, these widths no longer meet current standards for 12 ft or 3.6 m lanes. Full width shoulders are rarely provided due to cost, but at least an additional 1 ft is provided adjacent to each curb. Wider shoulders or sight shelves may be required around horizontal curves to comply with sight distance requirements. A minimum distance between walls of 30 ft is a common requirement. Resurfacing within tunnels is rarely permitted without first removing the old surfacing, so no allowance for resurfacing is required for overhead clearance. It is usual in

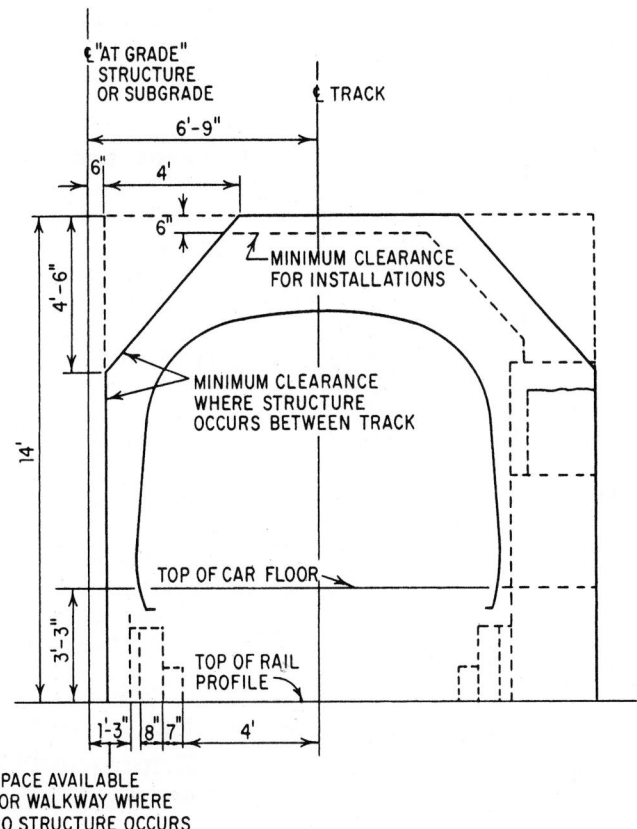

Fig. 20.3 Clearance diagram for San Francisco Bay Area Rapid Transit System.

tunnels to provide overhead lane signals to show which lanes are open to traffic in the direction of travel, so extra overhead allowance is required for these, and when appropriate also for lighting, overhead signs, jet fans for ventilation, and any other ceiling-mounted items. Minimum overhead traffic clearances depend upon which alternative routes are available for over-height vehicles and the classification of the highway, but accepted values usually lie between 14 ft and 5.1 m. Additional height may be required on vertical curves to allow for long trucks. Additional space may be required for ventilation, ventilation equipment, and ventilation ducts.

20.3 Alignment and Grades for Tunnels

Alignment of a tunnel, both horizontal and vertical, generally consists of straight lines connected by curves. Minimum grades are established to ensure adequate drainage. Maximum grades depend on

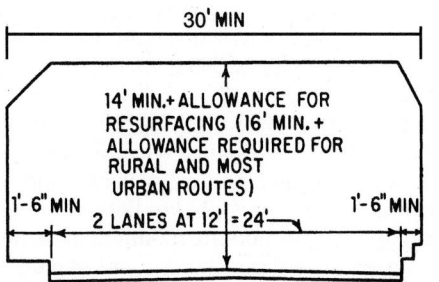

Fig. 20.4 Clearance diagram for interstate highway tunnels.

the purpose of the tunnel. Construction of a tunnel in the upgrade direction is preferred whenever possible, since this permits water to drain away from the face under construction.

Alignment and Grades for Railroad Tunnels

■ Straight alignments and grades as low as possible, yet providing good drainage, are desirable for train operation. But overall construction costs must be taken into account.

Grades in curved tunnels should be compensated for curvature, as is done for open lines. In general, maximum grades in tunnels should not exceed about 75% of the ruling grade of the line. This grade should be extended about 3000 ft below and 1000 ft above the tunnel.

Short (under 2500 ft), unventilated tunnels should have a constant grade throughout. Long, ventilated tunnels may require a high point near the center for better drainage during construction if work starts from two headings.

Radii of curves and superelevation of tracks are governed by maximum train speeds (Art. 19.9).

Alignment and Grades for Rapid-Transit Tunnels

■ Radii of curvature and limiting grades are governed by operating requirements. The New York City IND Subway has a 350-ft minimum radius, with transition curves for radii below 2300 ft. Maximum grades for this system are 3% between stations and 1.5% for turnouts and crossovers. The San Francisco BART system is designed for train speeds of 80 mi/h. Relation of speed to radius and superelevation of track for horizontal curves is determined by

$$E = \frac{4.65V^2}{R} - U \qquad (20.1)$$

where E = superelevation, in

R = radius, ft

V = train speed, mi/h

U = unbalanced superelevation, which should not exceed $2\frac{3}{4}$ in optimum or 4 in as an absolute maximum

For 80 mi/h design speed, the radius with an optimum superelevation would be 5000 ft. For a maximum permissible superelevation of $8\frac{1}{4}$ in, a minimum radius of 3600 ft would be required. The absolute minimum radius for yards and turnouts is

500 ft. Maximum line grade is 3.0% and 1.0% in stations. To ensure good drainage, grade should preferably be not less than 0.50%.

Alignment and Grades for Highway Tunnels

■ For tunnels under navigable water carrying heavy traffic, upgrades are generally limited to 3.5%; downgrades of 4% are acceptable. For lighter traffic volumes, grades up to 5% have been used for economy's sake. Between governing navigation clearances, grades are reduced to a minimum adequate for drainage, preferably not less than 0.25% longitudinally and a cross slope of 1.0%. For long rock tunnels with two-way traffic, a maximum grade of 3% is desirable to maintain reasonable truck speeds. Additional climbing lanes for slower traffic may be required when grades exceed 4%.

Radii of curvature should match tunnel design speeds. Short radii require superelevation and some widening of roadway to provide for overhang and sight distance.

20.4 Pavements and Equipment for Highway Tunnels

Roadway base is a reinforced concrete slab; on this is placed a renewable pavement. Well-designed bitumastic concrete has given good service and has good riding qualities.

Average daily traffic capacity of a two-lane two-directional tunnel is about 20,000 vehicles with a maximum of 1200 to 1500 vehicles per lane per hour. For single-direction traffic in both lanes, capacities are 10 to 15% higher.

Red-amber-green traffic lights are installed at about 1000-ft intervals, or at such spacing that the driver always sees at least one light. Telephones are placed in recesses about 500 ft apart for service and emergency calls.

Most tunnels, particularly those under water, are equipped with fire mains and hose outlets every 300 ft. Booster pumps in ventilation buildings raise supply pressure to 120 psi for use of foam. Fire extinguishers are mounted in recesses of hose outlets. Fire-alarm stations and phones are at the same locations. Emergency trucks with heavy hoists, fire hose, foam equipment, and emergency tools are kept in readiness at each portal.

20.5 Preliminary Investigations

Surveys should be made to establish all topographical features and locate all surface and subsurface structures that may be affected by the tunnel construction. For underwater tunnels, soundings should be made to plot the bed levels.

Knowledge of geological conditions is essential for all tunnel construction but is of primary importance for rock tunnels. Explorations by borings and seismic reflection for soft ground and underwater tunnels are readily made to the extent necessary. For rock tunnels, particularly long ones, however, possibilities for borings are often limited. A thorough investigation should be made by a geologist familiar with the area. This study should be based on a careful surface investigation and examination of all available records, including records of other construction in the vicinity, such as previous tunnels, mines, quarries, open cuts, shafts, and borings. The geologist should prepare a comprehensive report for the guidance of designers and contractors.

For soft ground and underwater tunnels, borings should be made at regular intervals. They should be spaced 500 to 1000 ft apart, depending on local conditions. Closer spacing should be used in areas of special construction, such as ventilation buildings, portals, and cut-and-cover sections. Spoon samples should be taken for soil classification, and undisturbed samples, where possible, for laboratory testing. Samples not needed in the laboratory, boring logs, and laboratory reports should be preserved for inspection by contractors. Density, shear and compressive strength, and plasticity of soils are of special interest.

All borings should be carried below tunnel invert. For pressure face tunnels, borings should be located outside the tunnel cross section.

For rock tunnels, as many borings as practicable should be made. Holes may be inclined, to cut as many layers as possible. Holes should be carried below the invert and may be staggered on either side of the center line, but preferably outside the tunnel cross section to prevent annoying water leaks. Where formations striking across the tunnel have steep dips, horizontal borings may give more information; borings 2000 ft in length are not uncommon. All cores should be carefully cataloged and preserved for future inspection. The ratio of core recovery to core length, called the rock quality designation (RQD), is an indicator of rock problems to be encountered.

Groundwater levels should be logged in all borings. Presence of any noxious, explosive, or other gases should be noted.

Where lowering of groundwater may be employed during construction of cut-and-cover or bored tunnels on land, the permeability of the ground should be tested by pumping tests in deep wells at selected locations. Rate of pumping and drawdown checked in observation wells at various distances should be recorded; as well as recovery of the water level after stopping the pumps.

Geophysical exploration to determine elevations of distinctive layers of soil or rock surfaces, density, and elastic constants of soil may be used for preliminary investigations. The findings should be verified by a complete boring program before final design and construction.

20.6 Tunnel Ventilation

Tunnels will be required to be ventilated to dilute or remove contaminants, control temperature, improve visibility and to control smoke and heated gases in the event of a fire in the tunnel.

20.6.1 Ventilation Requirements for Construction

Occupational Safety and Health Administration (OSHA) establishes standards, regulations, and procedures necessary to maintain safe, sanitary conditions for all workers on construction sites. Employers are required to initiate and maintain programs that will prevent accidents. Also, employers are advised to avail themselves of safety and health programs provided by OSHA and are required to instruct and train employees to recognize and avoid unsafe, unsanitary conditions, including prevention and spread of fires. OSHA requirements also cover underground construction. Following are some of the requirements applicable to ventilation.

Fresh air should be supplied to all underground work areas in sufficient quantities to prevent dangerous or harmful accumulation of dusts, fumes, mists, vapors, or gases. Unless natural ventilation meets this requirement, mechanical ventilation should be supplied. At least, 200 ft^3 of fresh air should be provided for each employee underground. The air flow should be at least 30 ft/min where blasting or rock drilling is conducted or where polluted air is likely to be present or developed. The direction of air flow should be reversible. After blasting, smoke and fumes should be immediately exhausted to outdoors before work is resumed in affected areas.

Underground operations are classified as gassy if air monitoring discloses for three consecutive days 10% or more of the lower explosive limit for methane or other flammable gases, measured about 12 in from work-area enclosure surfaces. Where such conditions occur, operations ether than those necessary for correcting the conditions should be discontinued. Ventilation systems should be made of fire-resistant materials. Controls for reversing air flow should be located above ground.

At normal atmospheric pressure underground, the air should contain at least 19.5% but not more than 22% oxygen. Test should be made frequently first for oxygen, then for carbon monoxide, nitrogen dioxide, hydrogen sulfide, and other pollutants. If hydrogen sulfide concentration reaches 20 ppm or 20% or more of the lower explosive limit for flammable gases is detected, precautions should be taken to protect or evacuate personnel.

Mobile diesel-powered equipment used underground in atmospheres other than gassy operations must either be approved by MSHA or the employer must demonstrate that it is fully equivalent to such MSHA-approved equipment. (30 CFR Part 32 MSHA).

For construction in compressed air, see Art. 20.16.

["Construction Industry: OSHA Standards for the Construction Industry (29 CFR 1926/1910)," Superintendent of Documents, Government Printing Office, Washington, DC 20402 (www.gpo.gov)].

A ventilation system dilutes and purges smoke and combustion and exhaust gases. Its capacity must be adequate to prevent irritating smoke or gas concentrations while a train passes through and to clear the air between train passages. Diesel content of nitrogen oxides may form corrosive acids in lungs when inhaled for long periods. The following systems are used by American railroads:

Injecting a stream of air at high velocity in the direction of train movement to keep smoke ahead of the train.

Injecting a high-speed high-volume air stream from the opposite end against the train motion to dilute smoke and clear the tunnel.

Addition of portal doors with the first injection system, to increase efficiency and prevent backflow in case of a stalled train. Doors are interlocked with signal systems (Moffat Tunnel).

Because of absence of smoke or exhaust gas when electric traction is used, ventilation by piston action of trains is adequate for tunnels for electric trains except under emergency conditions. Auxiliary exhaust fans should be installed to remove smoke in case of fire and to draw fresh air into the tunnel from the stations or portals. Fans may be installed in exhaust shafts between stations or in separate ventilation buildings in long underwater tunnels equipped with exhaust ducts. High-speed rapid-transit tunnels require air-relief shafts ahead of stations to prevent air blasts from entering the stations. In hot climates, heat dissipation in tunnels and stations requires special ventilation capacity and air conditioning. A computer program, the Subway Environment Simulation (SES), for system design has been developed. ("Subway Environmental Design Handbook," Urban Transportation Administration, Washington, DC 20590.)

20.6.2 Ventilation for Railroad and Rapid-Transit Tunnels

Short tunnels generally have no forced ventilation. Longer tunnels for diesel trains may need ventilation to purge smoke and exhaust gases. Tunnels for electric traction are adequately self-ventilated by piston action but may require emergency ventilation.

20.6.3 Emission Contaminants in Road Tunnels

Exhaust gases of gasoline internal combustion engines contain deadly carbon monoxide and irritating smoke and oil vapors. Diesel engines will also produce dangerous nitrogen oxides and aldehydes. The components of exhaust gases vary over a wide range.

The ventilation system also must be capable of controlling smoke and hot gases in case of fire (see Ventilation Systems for Road Tunnels following).

The Federal government or health authorities of states place restrictions on permissible carbon monoxide (CO) content. With new standards limiting contaminants in vehicle exhaust gases, however, it may eventually be possible to meet the CO limitations without extensive increase in ventilation. Engineers should check current rules at time of design.

Haze from vehicle exhaust gases, particularly from Diesel engined vehicles, does reduce visibility in the tunnel. In practice, when the CO level within the tunnel is maintained at levels as proposed in Table 20.1, adequate dilution of the irritating parts of exhaust gases and adequate visibility is assured.

New road tunnels built in the United States must comply with the time-weighted limits for concentration of CO established by the U.S. Environmental Protection Agency and the Federal Highway Administration. These limits are listed in Table 20.1.

Other countries may set other standards for carbon monoxide (CO) concentrations within their tunnels. The World Road Association (PIARC) publications provide documentation on the subject.

For tunnels in which traffic may incorporate a high percentage (10% or more) of diesel vehicles, the ventilation requirements for dilution of NO_x particles of nitrogen and particulates (smoke) become significant. The NO_x emitted by vehicles consists mainly of nitric oxide (NO), which oxidizes in the atmosphere to form nitrogen dioxide (NO_2). Based on exposure limits recommended by the American Conference of Governmental Industrial Hygienists and a typical 4-to-1 ratio for NO to NO_2, the maximum permissible concentration of NO_x is about 10 ppm.

Table 20.1 Limits on CO in Road Tunnels

Exposure time, min	Maximum CO concentration, ppm
0–15	120
16–30	65
31–41	45
46–60	35

Carbon Monoxide (CO) has been proven to be the umbrella pollutant in most road tunnels. That is, when the CO level in any given road tunnel is maintained at or below the levels shown in Table 20.1, all of the other vehicle pollutants will be within appropriate levels. The only exception to this is the case of particulate matter emitted by Diesel engined vehicles when the tunnel traffic stream contains on an average more than 15% Diesel engined vehicles.

The current method of determining the vehicle emissions to be considered for a road tunnel ventilation system design is to apply the United Stated Environmental Protection Agency's MOBIL series of computer programs. Mobil5B is the current version in use today.

20.6.4 Ventilation Systems for Road Tunnels

In straight tunnels up to about 1000 ft in length, natural air flow is usually sufficient, particularly with traffic in one direction. If a tunnel is exposed to heavy traffic congestion at times, installation of exhaust fans in a shaft or adit near the center for emergency ventilation is advisable if the length exceeds 500 ft.

Natural Ventilation ▪ Naturally ventilated tunnels rely primarily on atmospheric conditions to maintain airflow and a satisfactory environment in the tunnel. The piston effect of traffic provides additional airflow when the traffic is moving. Naturally ventilated tunnels over 1,000 feet (305 meters) long require emergency mechanical ventilation to extract smoke and hot gases generated during a fire as defined by NFPA 502 "Standard for Road Tunnels, Bridges, and Other Limited Access Highways". Tunnels with lengths between 800 and 1,000 feet (240 and 305 meters) will require the performance of an engineering analysis to determine the need for emergency ventilation. Because of the uncertainties of natural ventilation, especially the effect of adverse meteorological and operating conditions, reliance on natural ventilation, to maintain carbon monoxide (CO) levels, for tunnels over 800 ft (240 m) long should be thoroughly evaluated. If the natural ventilation is demonstrated to be inadequate, the installation of a mechanical system with fans should be considered for normal operations.

Smoke from a fire in a tunnel with only natural ventilation moves up the grade driven primarily by the buoyant effect of the hot smoke and gases. The steeper the grade the faster the smoke will move thus restricting the ability of motorists trapped between the incident and the portal at the higher elevation to evacuate the tunnel safely.

Mechanical Ventilation ▪ A tunnel that is sufficiently long, has heavy traffic flow, or experiences adverse atmospheric conditions requires mechanical ventilation with fans. Mechanical ventilation layouts in road tunnels are either of the longitudinal or transverse type.

Longitudinal Ventilation ▪ This type of ventilation introduces or removes air from the tunnel at a limited number of points, thus creating a longitudinal flow of air along the roadway. Ventilation is either by injection, or by jet fans.

Injection Longitudinal Ventilation is frequently used in rail tunnels and is also found in road tunnels. Air injected at one end of the tunnel mixes with air brought in by the piston effect of the incoming traffic. This type of ventilation is most effective where traffic is unidirectional. The air speed remains uniform throughout the tunnel, and the concentration of contaminants increases from zero at the entrance to a maximum at the exit. Injection longitudinal ventilation with the supply at a limited number of locations in the tunnel is economical because it requires the least number of fans, places the least operating burden on these fans, and requires no distribution air ducts.

Jet Fan Longitudinal Ventilation has been installed in a significant number of tunnels worldwide. Longitudinal ventilation is achieved with specially designed axial fans (jet fans) mounted at the tunnel ceiling. Such a system eliminates the spaces needed to house ventilation fans in a separate structure or ventilation building; however, it may require a tunnel of greater height or width to accommodate jet fans so that they are out of the tunnel's dynamic clearance envelope. This envelope, formed by the vertical and horizontal planes surrounding the roadway pavement in a tunnel, define the maximum limits of predicted vertical and lateral movement of vehicles traveling on the road at design speed. As the length of the tunnel increases, however, the disadvantages of longi-

tudinal systems, such as excessive air speed in the roadway and smoke being drawn the entire length of the roadway during an emergency, become apparent.

The longitudinal form of ventilation is the most effective method of smoke control in a road tunnel with unidirectional traffic as was determined in the Memorial Tunnel Fire Ventilation Test Program. A longitudinal ventilation system must generate sufficient longitudinal air velocity to prevent the backlayering of smoke. Backlayering is the movement of smoke and hot gases contrary to the direction of the ventilation airflow in the tunnel roadway. The air velocity necessary to prevent backlayering of smoke over the stalled motor vehicles is the minimum velocity needed for smoke control in a longitudinal ventilation system and is known as the critical velocity.

Transverse Ventilation ▪ Transverse ventilation includes systems that distribute supply air and collect exhaust air uniformly along the length of the tunnel. There are several such systems including the full transverse system which includes both supply and exhaust air uniformly distributed and collected. The semi- or partial transverse systems incorporate only one, either supply or exhaust air.

Semi transverse ventilation can be configured as either a supply system or an exhaust system. Semi transverse ventilation is normally used in tunnels up to about 7,000 feet (2,000 meters); beyond that length the tunnel air velocity speed near the portals may become excessive.

Supply semi transverse ventilation applied to a tunnel with bi-directional traffic produces a uniform level of contaminants throughout the tunnel because the air and the vehicle exhaust gases enter the roadway area at the same uniform rate. In a tunnel with unidirectional traffic, additional airflow is generated in the roadway by the movement of the vehicles, thus reducing the contaminant level in portions of the tunnel.

Because the tunnel airflow is fan-generated, this type of ventilation is not adversely affected by atmospheric conditions. The supply air travels the length of the tunnel in a tunnel duct fitted with supply outlets spaced at predetermined distances. If a fire occurs in the tunnel, the supply air initially dilutes the smoke, which was shown in the Memorial Tunnel Fire Ventilation Test Program to be an ineffective method for controlling smoke from larger fires. Supply semi transverse ventilation should be operated in a reversed mode for the emergency so that fresh air enters the tunnel through the portals to create a longitudinal flow of air equivalent to the critical velocity. It also provides a tenable environment for fire-fighting efforts and emergency egress.

Exhaust semi transverse ventilation installed in a unidirectional tunnel produces a maximum contaminant concentration at the exit portal. In a bi-directional tunnel, the maximum level of contaminants is located near the center of the tunnel.

In a fire emergency both the exhaust semi transverse ventilation system and the reversed semi transverse supply system create a longitudinal air velocity in the tunnel roadway thus extracting smoke and hot gases uniformly along the tunnel length.

Full Transverse Ventilation ▪ Full transverse ventilation has been used in extremely long tunnels and in tunnels with heavy traffic volume. Full transverse ventilation includes both a supply duct and an exhaust duct to achieve uniform distribution of supply air and uniform collection of vitiated air throughout the tunnel length. During a fire emergency the exhaust system in the incident zone should be operated at the highest available capacity while the supply system in the adjacent incident zone is operated. This mode of operation creates a longitudinal airflow (achieving the critical velocity) towards the incident zone and allows the smoke and heated gases to be extracted as close as possible to the fire and keep the upstream stopped traffic clear of smoke.

Other Ventilation Systems ▪ There are many variations and combinations of the systems described previously. Most of the hybrid systems are configured to solve a particular problem faced in the development and planning of the specific tunnel, such as excessive air contaminants exiting at the portal(s).

Ventilation System Enhancements ▪ A few enhancements are available for the systems described previously. The two major enhancements are single point extraction and oversized exhaust ports.

Single point extraction is an enhancement to a transverse system that adds large openings to the exhaust duct. These openings include devices that can be operated during a fire emergency to extract a large volume of smoke as close to the fire source as possible. Tests conducted as a part of the Memorial Tunnel Fire Ventilation Test Program concluded that this concept is extremely effective in reducing the temperature and smoke in the tunnel. The size of openings tested ranged from 100 to 300 ft^2 (9.3 to 28 m^2).

Oversized exhaust ports are simply an expansion of the standard exhaust port installed in the exhaust duct of a transverse or semi-transverse ventilation system. Two methods are used to create such a configuration. One is to install on each port expansion a damper with a fusible link; the other uses a material that when heated to a specific temperature melts and opens the airway. Several tests of such meltable material were conducted as part of the Memorial Tunnel Fire Ventilation Test Program but with limited success.

20.6.5 Elements of Road Tunnel Ventilation Systems

Major components of ventilation systems commonly used for road tunnels are described in the following.

Ventilation Buildings (Figs. 20.5 and 20.6) ▪ Fans, electrical transformers and switchgear, control board, and auxiliary equipment are housed in ventilation buildings. In short- and medium-length tunnels, one building at either portal is sufficient. Longer tunnels should have a building at each portal. A few of the longest have three or four buildings. For underwater tunnels, ventilation buildings may be at the water's edge, each building controlling a land and a river section of the tunnel.

Fresh air is taken in through large louver areas in the walls of the building. The louvers should be protected by bird screens. Louvers are usually aluminum and arranged for shedding water. Adequate drains should be provided in the fan room to remove rainwater, which may blow in through the louvers. Vitiated air is discharged through vertical stacks, which also should be covered by screens.

Tunnel Ducts are usually of constant area throughout their length. Concrete surfaces should be smooth for minimum friction. Obstructions,

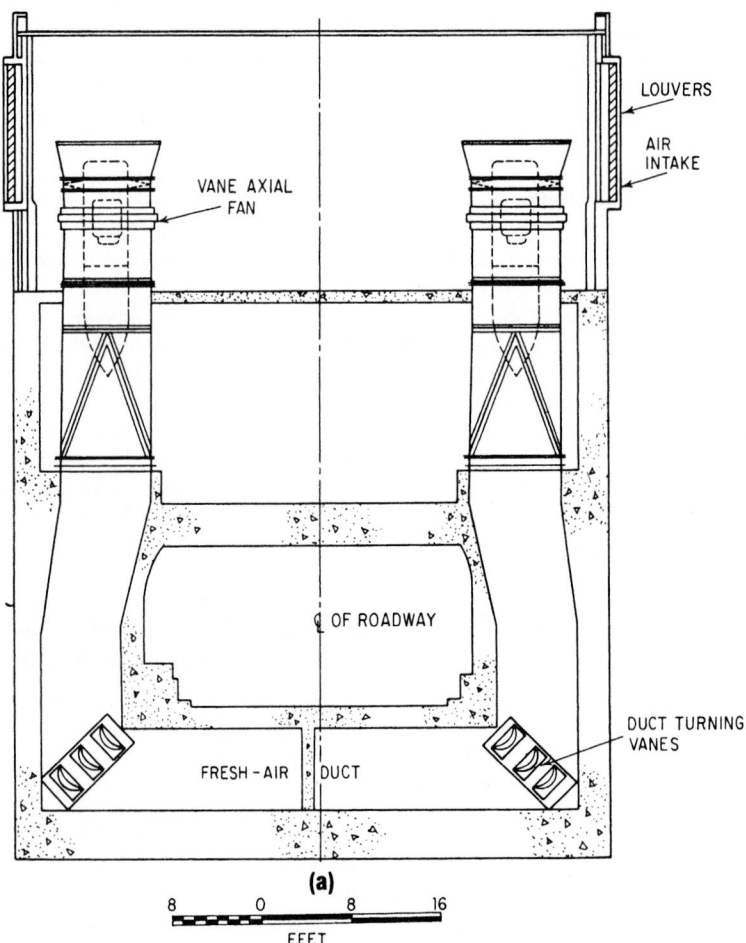

LOUVERS

AIR INTAKE

VANE AXIAL FAN

¢ OF ROADWAY

DUCT TURNING VANES

FRESH – AIR DUCT

(a)

8 0 8 16

FEET

Fig. 20.5 Sections through Hampton Roads Tunnel Ventilation Building. (*a*) Fresh-air supply system.

such as ceiling hangers, should be streamlined or at least rounded. Turns in ducts and shafts leading to the tunnel should be equipped with noncorrosive turning vanes for smooth air flow.

Flues spaced about 15 ft apart, extended from the ducts, supply fresh air slightly above roadway level. Ceiling ports are slanted at 45° in the direction of air flow in the ducts. All air openings should provide the means to adjust size to balance the air flow over the length of the tunnel.

Fans ▪ Two types of fans are available: centrifugal fans, used in all tunnels up to about

1938, and vane axial fans, a later development. Centrifugal fans have backward-curved blades and are nonoverloading. The efficiencies of well-designed fans of either type are about the same. For underwater tunnels, with vertical air shafts in the ventilation buildings, the vane axial fans require considerably less space and avoid the efficiency loss through the fan chambers usually associated with centrifugal fans. Blades for vane axial fans may have a fixed pitch or may be adjustable during operation. When reversed, the former type provides 80% of maximum capacity. The latter type may be adjusted from 0 to 100% of capacity for supply and exhaust, thus permitting

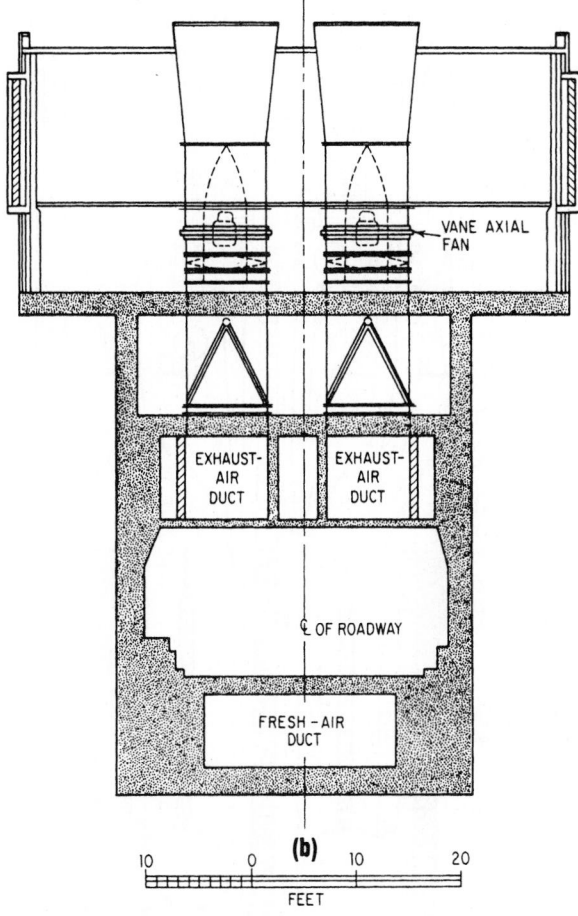

VANE AXIAL FAN

EXHAUST-AIR DUCT

EXHAUST-AIR DUCT

℄ OF ROADWAY

FRESH – AIR DUCT

(b)

10   0   10   20

FEET

Fig. 20.5 (*Continued*) (*b*) Exhaust-air system.

adjustments to meet variable demands for ventilation with fewer fans. The noise level of vane axial fans at maximum speed is somewhat higher than that for centrifugal fans because of greater tip speed. In sensitive surroundings, the noise from supply and exhaust fans can be dampened by sound baffles. Vane axial fans may have external drives or motors built into the hub of the impellers.

Centrifugal fans are operated by squirrel-cage motors through chain or multiple V-belt drives. The latter eliminate lubrication problems and wear on a multiplicity of parts (inherent in chain drives), give excellent service, and can be easily replaced.

Chains are enclosed in solid housings; belts are protected by wire guards.

For flexibility, the load is divided between several fans—at least two, sometimes as many as six—for each system. Four is a good number for demands exceeding about 600,000 ft^3/min.

To further adjust supply to variable demand, fan motors are equipped with two-speed windings. Three speeds with two motors have been used in earlier installations but are not necessary with an adequate number of fans. Spare fans may be provided as protection against breakdown, or total fan capacity may be increased by 10 or 15%. With good maintenance, fans are seldom out of

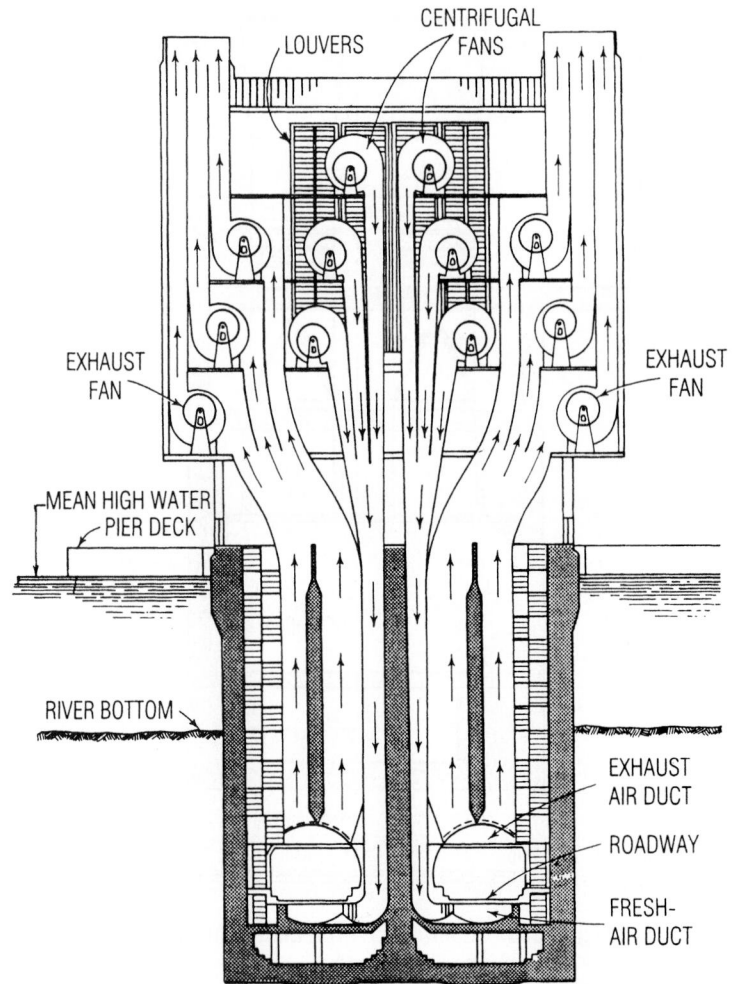

Fig. 20.6 Section through ventilation building of the Holland Tunnel.

commission, and the extra capacity of the system is sufficient to maintain acceptable conditions for limited periods with one unit out of service.

To protect the exhaust fans in case of a serious fire in the tunnel, automatic deluge sprinkler systems should be installed to cool the exhaust air.

Dampers ▪ All fans should be equipped with shutoff dampers to prevent short circuiting of air. Their operating motors should be interlocked with the control of the fan motors for automatic opening and closing. Trapdoor-type or multiblade dampers

are in use; the latter take less space and time to operate.

Fan Control ▪ In short, unattended tunnels, fans can be controlled automatically with carbon monoxide analyzers. Larger tunnels with heavy traffic may have operators stationed in the control room. They operate the fans to control conditions in the tunnel. At least two independent sources of electric power must be available, usually through feeders from different parts of the utility system. If these are not available, a diesel-engine emergency generator sufficient for minimum requirements should be installed.

Carbon Monoxide Analyzers ▪ These take continuous air samples from the tunnel and analyze them for CO content. The results are visually indicated and also recorded on paper tape, with time gradations. The recorders are mounted on the face of the control board, to guide the operator in selection of number of fans and speed necessary.

In a longitudinal or semitransverse supply system, air samples are taken from the tunnel proper at points of maximum concentration. In transverse systems, the samples may be taken from the exhaust ducts.

Haze Control ▪ To measure visibility in tunnels affected by haze from exhaust gases, instruments have been developed that give a reliable indication without excessive maintenance. Equipment manufactured for the Port Authority of New York and New Jersey uses the scattering of ultraviolet light by dust particles. Instruments protect the optics by recessing them in tubes through which filtered air is exhausted. Another type of instrument compares the intensities of two branches of a split light beam passing through the same optics, one going through a tube filled with clean air, the other through tunnel air.

Ventilation Power Requirements ▪ The power requirements and pressure losses are best evaluated using the prodedures contained in the Tunnel Engineering Handbook (J.O. Bickel and T.R. Kuesel, "Tunnel Engineering Handbook," Kluwer Academic Publishers, New York).

20.7 Tunnel Surveillance and Control

Emergency exhaust ventilations systems in short tunnels or tunnels with very light traffic may be activated by such instruments in the tunnel as carbon monoxide analyzers or fire-alarm or telephone boxes connected to the nearest fire and police departments. Emergency operation for other types of tunnels should be supervised by personnel in control centers.

Control of many newer tunnels is programmed for computer operation. The computers, however, may be bypassed for manual operation in an emergency.

To permit surveillance of tunnel traffic by personnel in the control room, monitors may be installed in that room to display views of the entire length of the roadways as transmitted by television cameras mounted in the tunnel. In a short tunnel, each camera covers a specific stretch of roadway and transmits to a specific monitor. For a long tunnel, to limit the number of monitors required to a convenient number, groups of cameras may be operated in sequence to transmit to their monitors. In an emergency, the sequence can be interrupted to permit a specific camera to focus on the region of concern.

Traffic Control ▪ Signal lights generally are mounted at the portals of a tunnel and at intervals in the interior such that at least one traffic light is plainly visible within a safe stopping distance. In a tunnel with two-way traffic, the signals facing traffic may incorporate red, amber, and green lights. Lights on the reverse side of those signals may be red and amber, to permit lane alternation in an emergency. In a tunnel with two-lane, one-way traffic, the signals facing traffic carry red, amber, and green lights, whereas lights on the reverse side of signals for the left lane may be amber and red, to permit two-way traffic in an emergency.

Traffic flow may be monitored by pairs of electric induction coils that are embedded in the pavement of each lane and that report the flow on indicators in the control room. If traffic velocity is too slow, for instance, less than 5 to 10 mi/h, the traffic lights are changed to amber, for caution. If traffic stops, the lights are changed to red. If necessary, for slow traffic, the traffic lights may be alternated between stop and go to space traffic flow into the tunnel.

Fire Control ▪ Automatic fire detectors may be installed in the ceiling throughout a tunnel. When a fire occurs, they indicate the location and send an alarm to an operator who alerts an emergency crew. If the operator verifies the alarm (which might have been activated by the heavy exhaust of a diesel engine rather than by a fire), an emergency program can be started: The emergency crew and vehicles are mobilized. For traffic moving toward a fire, signal lights are turned to red, while for traffic moving away from the fire, signals remain green, to permit evacuation. And the ventilation system for the affected part of the tunnel is converted to exhaust.

Hydrants generally are installed about 300 ft apart, in niches in the tunnel walls, to provide water for fire fighting. Water may be obtained from municipal water supplies, if available. Otherwise, the water mains may be connected to tanks providing about 10,000 gal of storage. The tanks may be located near each portal and supplied by pumps from local sources or from groundwater. Booster pumps may be installed to provide at least 125-psi pressure for application of water on fires. Fire alarms and fire extinguishers for control of minor fires may be installed next to the hydrants.

Communications ▪ Emergency telephones may be placed along the tunnel side walls for communication with an operator in the control room. An aerial in the tunnel will permit the operator to transmit messages to motorists through their car radios and allow them to receive other broadcasts while in the tunnel.

Power Supply ▪ Power should be supplied from two independent sources, for example, from two different utilities or independent substations of one utility. An alternative is a standby diesel generating plant capable of supplying power at least for ventilation and emergency lighting to keep the tunnel in operation. This equipment should be supplemented by storage batteries to supply instant power for the emergency lighting.

20.8 Tunnel Lighting

The Occupational Safety and Health Administration sets minimum requirements for illumination on construction sites:

5 ft-c—general construction-area lighting, warehouses, corridors, exitways, tunnels, and shafts

3 ft-c—concrete placement, excavation and waste areas, accessways, active storage areas, loading platforms, refueling, and field maintenance areas

10 ft-c—batch and screening plants, mechanical and electrical rooms, indoor work rooms, rigging lofts, indoor toilets, and tunnel and shaft headings during drilling, mucking, and scaling.

For other areas, follow illumination recommendations in "Practice for Industrial Lighting," IES RP7, the Illuminating Engineering Society of North America.

For emergency use, every employee underground should be equipped with a portable hand or cap lamp unless sufficient natural light or an emergency lighting system provides sufficient illumination along escape paths. Only portable lighting meeting OSHA requirements may be used within 50 ft of any heading during explosive handling. (See also Art. 20.16.)

["Construction Industry: OSHA Safety and Health Standards 29 CFR 1926/1910," Superintendent of Documents, Government Printing Office, Washington, DC 20402.]

Lighting for Tunnels in Service ▪ Since locomotives are equipped with strong headlights, railway tunnels are generally not lighted except for emergency evacuation. Subaqueous tunnels and other tunnels on electrified lines, particularly in cities, are equipped with a nominal amount of lights, especially in refuge niches.

Rapid-transit tunnels are lighted sufficiently to make obstructions on tracks visible and to facilitate maintenance work. The lights are installed and/or shielded to prevent glare in the motorman's eyes. Luminaires are installed in tunnels for emergency use.

For highway tunnels, the most troublesome lighting condition is the transition from bright light in the approach to the tunnel, the entrance (threshold zone) luminance, to the luminance in the interior. Guidelines for alleviating this condition have been issued by the American Association of State Highway and Transportation Officials (AASHTO) and the Illuminating Engineering Society of North America. Threshold zone illumination varies greatly with topography, orientation, sun exposure, and season and should be evaluated for the most critical condition. Daylight penetration through the portal into the threshold zone may assist the transition. In addition to the threshold zone, two or three transition zones gradually reduce the luminance to that of the interior. The length of each of these zones should be approximately one safe-stopping sight-distance (SSSD) at design speed. Reduction between zones should not exceed 3:1.

At night, a pavement luminance of $2-5$ cd/m^2 minimum is recommended for the entire length of the tunnel. The approach and exit roadways should have a luminance level of no less than one third the tunnel interior level for a distance of a SSSD.

There are four viable types of light sources used in tunnels, fluorescent, low-pressure sodium (LPS), high-pressure sodium (HPS), and metal halide (MH). The advantages and disadvantages of each are discussed in greater detail in ANSI/IESNA RP-22-96 8.1. These include restrike time in the event of momentary power interruption, linearity of source to reduce flicker, cost, color rendering, lamp size, lamp efficacy, control of light distribution, effects of air temperature, lumen depreciation with time, glare, the risk of lamp rupture, and keeping enclosures dust-tight and water tight. Florescent lamps frequently provide the lower illumination levels, combined with LPS at threshold and transition zones. Lower wattage LPS sources are also used in interior zones. HPS and MH lamps come in a wide selection of sizes, better lamp life, compact size and are easily optically controlled.

20.9 Tunnel Drainage

Most tunnels through hills and mountains have water problems. Surface water penetrates through fissures and percolates through permeable soils. Attempts to seal off the rock by grouting, with either cement or chemicals, usually are not completely successful since very high pressures may build up even if flows are low. Cast-in-place concrete linings may not be completely watertight. Water may find its way through shrinkage cracks in the linings into the interior of tunnels. There, it may freeze in cold weather and produce an unsightly appearance, objectionable in highway tunnels. Consequently, provision must be made to drain water from tunnels.

Fire fighting, washing of tunnel interiors, and flushing of pavements also introduce water that must be drained.

Although cut-and-cover tunnels can be water-proofed, this is difficult with bored tunnels. If the water problem is not serious, the most economical solution is to seal cracks in the lining that leak. With good concrete control, the number of these should be small. It is good practice to design tunnels assuming that they will leak and therefore provide appropriate drainage paths.

If water appears in considerable quantity during rock tunneling operations, tight steel lagging over the tunnel supports and grouting may prevent leakage. In serious cases, it may be necessary to dry-pack between the rock and the tunnel lagging to drain water. This is a slow, costly method requiring much manual labor. Dry pack behind the side walls can easily be placed and is effective in preventing the buildup of a hydrostatic head behind the lining. Longitudinal drain pipes should be installed behind the base of the side walls, with the laterals at regular intervals leading to the main drain (this is a large drain installed under the roadway, for roadway drainage). Water will flow through the dry packing and into the base drains.

In a rock tunnel, heavy flow of water coming through a drill hole indicates a water-bearing fault or seam. The flow may be stopped by drilling additional holes and injecting cement grout. Some holes should be slanted to reach beyond the periphery. If dense sand or rock flour in the fault prevents proper penetration of cement grout, chemical grouting may give satisfactory results. In special cases, it may be necessary to drill a pilot hole well ahead of the face to detect severe water conditions, especially substantial quantities under heavy pressure. This must be done for rock tunneling under deep bodies of water.

In highway tunnels, drainage inlets should be installed at regular intervals along the curbs, with cross connections to the main drain. The latter should be of generous size, in longer tunnels preferably large enough to provide crawl space to remove silt accumulations, particularly when grades are near horizontal. Traps at drainage inlets are undesirable, because of the danger in the event of a fuel spill.

Leakage in well-constructed underwater tunnels, either shield-driven or immersed, is usually minor. It can be controlled by calking joints in segmental liners or by injecting cracks where leaks appear. Main sources of water are washing of tunnel interior, fire fighting, drippings from vehicles, and rain collected in open approaches. Pumps are usually sized to handle the full flow from one fire hydrant.

Continuous open gutters recessed into the curbs have been used in many subaqueous tunnels. The gutters lead water to a low point, where it is collected in a sump. Drainage inlets, spaced about 50 ft apart along each curb and connected to longitudinal drain lines embedded in the concrete below the curbs, are desirable because they prevent propagation of fire by burning fuel in case of a serious accident. Drain lines should be at least 8 in in diameter. They should be equipped with cleanouts every 500 ft.

In straight, open approaches, transverse interceptors about 300 ft apart are most effective in preventing water from entering a tunnel. They are 18 in wide, extend from curb to curb, and are covered with gratings, with slots parallel to the center line of the roadway. An interceptor is placed in front of the tunnel portal and another about 10 ft inside.

In curved, superelevated approaches, drainage inlets should be installed at regular intervals along the low curb.

All drainage from open approaches should be collected inside the portals in sumps below the roadway. Each sump should be divided into a settling basin and a suction chamber. Easy access must be provided for cleaning out sediments. A minimum of three electrically driven, large-clearance drainage pumps should be installed, one as a standby. Alternating automatic controls rotate the pumps in service. High-water-level alarm circuits should be extended to the control room. Sump and pump capacity, with two pumps operating, should be designed for maximum, short-duration rainfall for the locality. An intensity of 4 in/h, based on a 15-min downpour at a rate of 8 in/h, is ample for most areas.

A smaller sump should be located at the low point of the tunnel. This sump also should be divided into a settling and a suction chamber. Two automatically controlled drainage pumps, with a capacity of 250 gal/min each, may be adequate. Their discharge should be carried to one of the portal sumps.

20.10 Water Tunnels

These may be diversion or intake tunnels for hydropower plants, or aqueducts bringing water to city and municipal distribution systems.

Diversion tunnels carry river water around dam sites during construction. They are designed to carry the maximum expected runoff during this period. They may also discharge excess water after the reservoir has been filled, or be converted to intake tunnels to a powerhouse located in the side of the valley below the dam. If they are not needed after completion of the project, the diversion tunnels are closed with concrete plugs. Extensive diversion tunnels have also been built to collect water from several watersheds for a central power plant.

Intake tunnels bring water from reservoirs to turbines or the heads of penstocks. The tunnels are mostly in rock and operate under a positive hydrostatic head. In pervious and fissured ground, they are lined with reinforced concrete or steel plate; in sound rock, a sprayed-concrete lining may be adequate to provide a smooth surface as long as there is sufficient overburden to exceed the internal pressure.

Many miles of aqueduct tunnels have been built for municipal or area water-distribution systems. These tunnels are, for the most part, in rock but may also contain stretches of soft-ground tunneling. They may be under large hydrostatic pressure, such as the New York City aqueduct, which crosses the Hudson River 600 ft below sea level.

Tunnels with small or no interior pressure generally have a horseshoe section; pressure tunnels are circular. Lining is concrete, 6 to 36 in thick, depending on size, pressure, and nature of rock. Welded steel tubes may be needed when pressures are particularly high. Grade tunnels may be lined with plain concrete, pressure tunnels with reinforced concrete. Diameters range from 7 ft for small aqueducts to 50 ft for Hoover Dam diversion tunnels. In very sound rock, sprayed-concrete lining has been used. Parts of the Colorado River aqueduct are lined with continuous steel shells against concrete backing, and the inside is protected by 2 in of reinforced sprayed concrete.

To expedite construction, long tunnels are subdivided into several headings by shafts or adits, about 2 to 5 mi apart.

20.11 Sewer and Drainage Tunnels

Large cities require miles of tunnels to carry off storm runoff and to conduct wastewater to treatment plants. These tunnels are built in a variety of soils. Some are constructed as box culverts by the cut-and-cover method, but most are tunneled with tunnel boring machines (TBMs). Size varies from about 7 to 15 ft. Drainage tunnels for storm water are usually less extensive since they can discharge into nearest open waters.

The cross section of sewer and drainage tunnels is usually horseshoe or circular, with concrete lining. Quality of concrete is of special importance to resist the detrimental effect of wastewater. Generally, they are grade tunnels, except for siphons under rivers, which are under pressure. A circular or egg-shaped section maintains velocity at low flow to prevent excessive settling of solids.

Alignment is dictated by location of treatment plants, soil conditions, and the street plan of the city. Continuous grades should be maintained except for siphons. A minimum grade should be maintained for gravity flow.

20.12 Cut-and-Cover Tunnels

Shallow-depth tunnels, such as rapid-transit lines under city streets, underpasses, land sections of underwater tunnels, and end sections of tunnels through hills, are built by cut-and-cover methods. A trench is excavated from the surface, within which a concrete tunnel is constructed. With bottom-up construction, the completed tunnel is covered up, and the surface reinstated. With top-down construction, the walls are constructed first, perhaps using bentonite slurry in narrow trenches. The roof is constructed next, backfilled and the surface reinstated. Excavation and construction of the floors below roof level then follow, using access from the ends or from glory holes. Both bottom-up and top-down construction almost always use cast-in place concrete. Depth of invert on subways and underpasses usually does not exceed 35 to 40 ft. For connections to subaqueous tunnels, cuts up to 100 ft have been used under special circumstances, and depths to 60 ft are not uncommon.

The Occupational Safety and Health Administration (OSHA) sets standards, regulations, and procedures for protection of personnel during excavation. OSHA requires that all surface encumbrances and underground utility installations, such as sewers, electrical and telephone conduits, and water pipes, be protected, supported, or removed as necessary to safeguard the workers. Also, structural ramps used for access or egress should be designed by a structural engineer and constructed in accordance with the design. For trench

Fig. 20.7 Taipei Metro Cut-and-Cover.

excavations that are 4 ft or more deep, a stairway, ladder, or ramp should be provided for egress so as to require no more than 25 ft of lateral travel for workers.

Among the measures that OSHA specifies for safeguarding personnel in excavations are the following: Precautions should be taken to prevent exposure of personnel to harmful levels of atmospheric contaminants (Art. 20.6). If natural lighting is inadequate for safe working conditions, illumination to meet OSHA requirements for excavation should be provided (Art. 20.8). Personnel should not be allowed to work in excavations in which water accumulates unless the workers are protected by safety harnesses and lifelines, water is being removed to control the water level within safe limits, and special supports or shields are used to protect against cave-ins.

To avoid exposure to falling objects, personnel should not be permitted below loads carried by

Fig. 20.8 63rd Street Tunnel Strutting and Tie-backs.

lifting or digging equipment. Unless excavations are entirely in stable rock or are less than 5 ft deep in stable soil, protection should be provided against cave-ins. Retaining devices may be required to prevent excavated or other materials or equipment from falling or rolling into the excavation.

Where space and depth of excavation permit and the ground is sufficiently firm, open slopes may be used along the sides of the excavation. For excavations up to 20 ft deep, OSHA limits the slope to a maximum of 1:1½ (34° with the horizontal), unless soil tests and analyses indicate steeper slopes will be stable. A registered professional engineer must design excavations deeper than 20 ft, and excavations must be monitored by a competent person as defined by OSHA. Personnel should be protected from loose rock or soil falling or rolling from the excavation face. For the purpose, loose material may be removed by scaling and protective barriers may be installed at intervals along the face. If the lower portion of the excavation has vertical sides, that region should be shielded or supported to a height least 18 in above the vertical sides.

Groundwater may be lowered, as needed, by tiers of wellpoints. This may lower the groundwater outside the excavation considerably and cause settlements. The lowering of the external groundwater can be reduced by the use of slurry walls, contiguous or overlapping bored piles, or steel sheet piling. Adjacent structures with a risk of settlement may require underpinning. Furthermore, where lowering of groundwater exposes wooden piles to air, deterioration may follow.

Where space permits, the sides of the trench may be sloped back to reduce the need to provide support to them. In confined or deep areas, support of excavation may be required. The excavation support may be temporary walls that are not part of the final structure, or they may form part of the final structure, especially when excavations are

Fig. 20.9 Tangent Piles.

deep. The ease and simplicity of constructing the final structure within temporary walls must be balanced against cost savings when they are incorporated into the final structure. Temporary excavation support may be steel sheet piles, soldier piles and lagging, and tangent or secant piles. For deeper excavations, concrete slurry walls may be constructed, usually 2 ft, 3 ft, or 4 ft thick, sometimes incorporating soldier piles or beams (SPTC walls), and may form part of the final structure.

Steel sheetpile walls, for depths to about 30 to 40 ft, supported by wales and cross bracing. The walls keep loss of ground to a minimum.

Soldier piles and lagging, made of steel H beams with wood or concrete lagging. These are used for greater depth. Lagging must be blocked tight against the earth to control loss of ground. Soldier piles may be combined with sheetpiles, instead of wood lagging, if tight bulkheads are required. Wales and cross bracing support the walls.

Concrete slurry walls built in bentonite-slurry trenches have been used to prevent loss of ground

and eliminate or reduce groundwater lowering. Sections of trenches about 20 ft long are excavated. The trenches are kept filled with bentonite slurry. Then, reinforcing cages are lowered into them, and concrete is placed to fill the trenches, displacing the slurry. Key sections are formed at the ends of the trenches. The walls serve as part of the final structure or as impervious bulkheads.

SPTC or California Wall, a combination of soldier piles and slurry wall. This was used on some stations of BART, and as part of the final structure for much of the Central Artery Project in Boston. Large wide-flange steel beams are inserted in slurry-filled bored holes, the space between the beams is excavated under slurry, and excavation and pipe holes are filled with concrete. Care must be used in excavation to have the concrete solidly keyed into the space between the flanges. The steel piles in the composite wall act as reinforcing and permit easy attachment of interior bracing.

The fundamental basis for the design of excavation support systems is consideration of how the soil being supported behaves, and perhaps also how the floor of the trench behaves, since substantial heave can occur under adverse conditions. Any movement of the support system can cause soil movement and hence settlement of adjacent structures. It is the amount of movement that can be tolerated at the adjacent structures that often dictates the type and stiffness of the excavation support system, with control of ground-water levels often being crucial. Structures may have to be designed for both short-term and long-term effects. One of the primary tools for this purpose is soil-structure interaction analysis. Frame structures supported by beam-on-an-elastic-foundation analysis are also commonly used. In most cases, a two-dimensional analysis is sufficient, although complex areas may require three-dimensional modeling, perhaps using finite element analysis.

Many subway and highway structures have been built using steel columns and beams with jack arches, but this is uncommon today. New structures are generally reinforced concrete box structures but when the excavation support system also forms part of the final structure, it may in practice be difficult to obtain full fixity between the walls and the slabs. Partial fixity may then be specified, and perhaps shear connections also provided. For high load on the roof or large spans, the composite action of a thin concrete slab on top of steel beams has been used. Haunches may help to reduce effective spans. Arched tunnel roofs are rare today.

Design loads include weight of overburden, self weight, live load surcharge, potential future construction, horizontal earth pressure, hydrostatic loads if below the water table, and seismic loads. Weight of submerged structures must be adequate to prevent floatation excluding all removable items from within the tunnel and above it while ignoring friction on the sides.

Tunnel Waterproofing ▪ Tunnels in dry soil need no waterproofing on base and walls, but roof slabs should have at least minimum waterproofing. Tunnels below groundwater level should be waterproofed all around.

Tunnels that are not waterproofed should be designed assuming that they will leak, and paths provided to remove any leakage water. Many tunnels have drainage channels adjacent to exterior walls and a second "false" wall a few inches from the first onto which the final tunnel finishes are attached. Joints are more likely to leak than other locations, so that special attention should be paid to waterproofing joints even if surface waterproofing is not applied.

Methods of surface waterproofing include membranes supplied in roll form with overlapping or welded joints, spray-applied membranes, blind-side waterproofing designed for the outside face of walls cast against existing ground or the support system, clay-based panels that swell on contact with water, and chemical additives to the concrete. Some methods of waterproofing require safety precautions during application, such as ventilation. Waterproofing membranes that adhere to the surface to which they are applied help to prevent the spread beneath the membrane of any water that could leak through a puncture. The repair of leaks appearing on the inside of the tunnel may then be as simple as injecting offending cracks locally, otherwise new leaks may appear as previous leaks are repaired.

For joints, a large number of extrusions are available that are designed to be buried in the concrete, half each side of the joint, some of which can be injected later if leaks still occur. Joints may also be waterproofed using hydrophilic materials, gaskets in compression within the joint or bolted to the surface each side of the joint, and surface applied injectable and reinjectable tubes.

Most waterproofing must be protected against mechanical damage (for example, during back-

filling or the placement of reinforcement), and against noxious gases (vehicle exhaust, for example) and fire within the tunnel. Protection methods have included a layer of concrete, plywood boards, and brick (against vertical faces). Heat resisting materials and cover plates with gaskets have been used to protect joints within tunnels.

To save on excavation width, waterproofing for walls may be applied to trench bulkheads and concrete placed against it.

20.13 Rock Tunneling

Standards, regulations, and procedures of the Occupational Safety and Health Administration should be adhered to in rock excavations, as in all construction operations. (See Arts. 20.6, 20.8, and 20.12.)

Tunneling in rock today is primarily by drill-and-blast or by using a TBM (tunnel boring machine). Drill-and-blast tunnels can be any shape, whereas most TBMs are only capable of drilling circular holes. A rule of thumb is that a tunnel requires at least one diameter of cover, although less may be possible. Where rock quality is particularly good, the tunnel may be unlined or may only need mesh, sprayed concrete and rock bolts or dowels. More fractured rock may require significant temporary ground support such as steel sets and lattice girders until a final lining is completed. Fracture zones may be particularly difficult to cross due to high flows of water under high pressure, and due to the quantities of loose material. For some purposes, lining may be required to promote flow or to prevent ingress of water.

(Kuesel, T. R., Tunnel Stabilization and Lining, in "Tunnel Engineering Handbook," Bickel, J. O., Kuesel, T. R., and King E. H., Editors, Chapman & Hall, 1996. U.S. Army Corps of Engineers Manual, 1997, Design of Tunnels and Shafts in Rock, EM 1110-2-2901.)

For rock excavation, the most important geological conditions to be anticipated are the presence of faults, usually involving areas of badly fractured rock; direction and degree of stratification; fissures and seams; presence of water, which may be cold or hot or contain corrosive or irritating ingredients; pockets of explosive or toxic gas; and rock strain. The petrography is of lesser importance unless the rock is highly abrasive, causing excessive wear of drills.

Too much information can never be provided for the engineer, to produce a realistic design, and for the contractors, to prepare sound bids. Even at best, unforeseen difficulties must be expected.

In addition to geological surveys and borings (Art. 20.5), engineers may use electric-resistivity measurements and gamma-ray absorption for information on depth and characteristics of rock formations. Information also may be obtained from the U.S. Geological Survey, which has extended its scope and geophysical studies beyond the mining field. Where geological conditions are particularly hard to evaluate or are especially severe, exploratory pilot tunnels, about 10 × 10 ft, may be driven part way from each end or for the entire length of a tunnel, prior to final design and advertising of construction. TBMs may be used for pilot tunnels

either used as a seperate service tunnel, or abandoned, or enlarged to form the final tunnel. In these pilot tunnels, internal rock stresses can be measured by pressure cells and strain gages inserted in transverse drill holes, and the nature of the rock, foliation, blockiness, and pressure of faults and water can be inspected.

Tunnel Boring Machines (TBM) ▪ The high initial cost of hard-rock TBMs tends to restrict their use to longer tunnels. Although ideally suited to circular tunnels due to the rotary motion of the cutting heads, variations of the excavated shape may be technically feasible, as have been done with a few soft-ground TBMs. Large grippers are jacked outwards, either to each side or top and bottom, and hold the main part of the machine and transfer the applied thrust to the adjacent firm rock. Disk

Table 20.2 Load H_p in Feet on Rock on Support in Tunnel*

Rock condition	H_p, ft	Remarks
1. Hard and intact	Zero	Light lining on rock bolts only if spalling or popping occurs
2. Hard, stratified, or schistose	0 to 0.5B	Light supports. Load may change erratically from point to point
3. Massive, moderately jointed	0 to 0.25B	
4. Moderately blocky and seamy[†]	0.35($B + H_t$) to 1.10($B + H_t$)	No side pressure
5. Very blocky and seamy	0.35($B + H_t$) to 1.10($B + H_t$)	Little or no side pressure
6. Completely crushed, chemically intact[†]	1.10($B + H_t$)	Considerable side pressure. Requires continuous support of lower ends of ribs or circular ribs
7. Squeezing rock, moderate depth	1.10($B + H_t$) to 2.10($B + H_t$)	Heavy side pressure; invert struts required. Circular ribs recommended.
8. Squeezing rock, great depth	2.10($B + H_t$) to 4.50($B + H_t$)	Same as for Type 7
9. Swelling rock	Up to 250 ft, regardless of value ($B + H_t$)	Circular ribs. In extreme cases, use yielding supports

* If depth of rock over tunnel is more than 1.5($B + H_t$), where B is width and H_t is height of tunnel. From R. V. Proctor and T. L. White, "Rock Tunnels and Steel Supports," Commercial Shearing & Stamping Co., Youngstown, Ohio.
  [†] If roof of tunnel is permanently above the water table, values for Types 4 and 6 can be reduced by 50%.

cutters mounted in the cutter-head face roll at high pressure on the exposed rock face and crush the rock, the tailings from which are mechanically removed. After an advance of up to six feet or so, the grippers must be retracted, moved forwards and jacked out again. Advance rates are very dependent on the hardness of the rock and its integrity, and the wear and tear that the rock may cause. In swelling rock, extra precautions must be taken to ensure that the machine does not get stuck. As the rock quality diminishes, the machine will need to resemble a soft ground TBM more and more. Rock TBMs tend to be launched from a mined chamber in which the whole machine can be assembled.

Headings ▪ In the past, when mucking was done by hand loading into mine cars, and drill equipment was cumbersome, excavation was advanced in drifts or headings. In weak rock or for very wide tunnels, this method is still used. A top heading may be advanced first. This permits installation of crown supports if needed. The rest is excavated by benching down from the top heading. These different levels make transportation of excavated material inconvenient. In wide tunnels, side headings may be advanced. In that case, legs of steel sets (supports for side walls and roof) are placed, where necessary. The side headings are followed by a top heading and erection of the arch supports. The remaining block can be attacked from the face or from the side drifts.

A bottom heading or pilot tunnel may be used instead. Enlargement proceeds at several places along the heading simultaneously. The pilot tunnel has to be large enough to allow in and out traffic and should be timbered to protect it.

In very long tunnels, a parallel heading, 40 ft or more from the tunnel axis, expedites excavation by providing access to several working faces through cross drifts. From this pilot tunnel, transverse headings are driven at several points to the main tunnel axis, from which tunnel excavation can proceed in both directions. The parallel heading carries all traffic to the different faces and serves as a drainage and ventilation tunnel. This method was used in the 12-mi Simplon tunnel, where the parallel drift was later enlarged to a full-sized single-track railroad clearance, and in the Moffat and New Cascade tunnels.

A center heading may also be used in large rock tunnels. From it, the section is enlarged to full size by radial drilling.

Full-Face Tunneling ▪ To save time and labor, full-face rock excavation is used wherever feasible, for efficient mechanization of the operation. Large track or rubber-wheel-mounted jumbo frames carry high-speed drills. As an alternative to drill and blast, roadheaders are sometimes used on weaker rock. They are smaller excavators equipped with ripper or point-attack teeth mounted on rotating balls attached to slewing and elevating arms. Being more maneuverable, roadheaders can excavate openings of almost any shape. Mucking (removal of excavation) is done by large, mechanized loaders. Muck is carried in diesel trucks, where permissible, or in trains of large mine cars pulled by battery-powered locomotives if laws prohibit use of internal combustion engines.

Excavation Limits ▪ Contract plans prescribe excavation profiles. An inner A line is the minimum theoretical section to be excavated; to this is added a tolerance, usually 6 in to the B line or payment line. Any overbreak beyond this is at the contractor's risk and has to be filled at the contractor's expense.

Blasting ▪ Drilling pattern and blasting charges are governed by the rock characteristics, fragmentation desired for mucking, and external conditions, such as proximity of sensitive structures. The procedure should be worked out by an experienced blasting expert and may have to be modified during construction. The center group of holes, fired first, are drilled convergent, so that a conical shape is blasted. Blasting proceeds toward the periphery with short-time delays. A 6- or 8-in-diameter center, or "burn" hole, without charge, acts as a relief opening, improving blasting effect. Rounds are usually about 10 ft deep but may be more or less, depending on the rock. Line drilling, a ring of straight holes, fairly closely spaced around the periphery, is used if as smooth a section as possible is desired.

Temporary Supports ▪ Practically all rock tunnels need some temporary supports. Timber may be used in pilot tunnels and small headings. For larger tunnel cross sections, steel sets are

Fig. 20.10 Drill and Blast (TARP).

more economical because of their strength and ease of installation. These are made of I beams cold-rolled into shape. For small tunnels with circular arches, the sets may be continuous frames. In larger tunnels or for flat arches, the sets consist of separate posts and arches (Fig. 20.11). Where roof supports only are necessary, the arches may be supported on plates resting on rock ledges. Steel sections are usually uniform for the entire tunnel, and spacing of sets is varied according to rock loads. Normal spacing is 4 ft. but spacing may be reduced to 2 ft or increased to as much as 6 ft.

The sets should be erected as soon as scaling of loose rock has been completed. Blocking should immediately be wedged between the steel and the rock surface at 3- to 5-ft intervals to prevent rock movement from starting. The steel frames should allow space at the crown, between the lower flange and the concrete surface, for a pipe for placing concrete.

Timber or steel lagging should be placed between the sets. The amount of lagging depends on rock conditions. Lagging may be practically solid, or there may be gaps of various widths between the sheets, as required by circumstances. Badly fragmented rock may require metal panning between sets if water is present. The pans are made of interlocking channels. The space between pans and rock should be dry-packed to allow water to run off into the drainage system.

The concentrated loads on the sets at blocking points produce bending moments in the frames. Table 20.2 presents formulas for loads on supports in rock tunnels (R. V. Proctor and T. L. White, "Rock Tunnels and Steel Supports," Commercial Shearing and Stamping Co., Youngstown, Ohio).

Through badly faulted rock or pressure areas, circular tunnel sections and ring supports are preferable, particularly in seismic areas (Fig. 20.12).

Rock Bolts ▪ In good rock, but also for some rock that may be classified as poor, rock bolts may be used to secure the excavation. They are usually 1 in in diameter and 8 ft long. They may be coupled, however. The bolts provide anchorage in sound rock, where they are held by wedges driven into split ends when the bolts are inserted or by expansion sleeves gripping the sides of the hole when the bolts are threaded in. The bolts are tested for pull-out and prestressed by nuts bearing against face plates on the rock surface. Untensioned deformed bars, bolts, or steel or glass-fiber tubes are used as rock reinforcement and fully grouted with cement or high-strength resin grout.

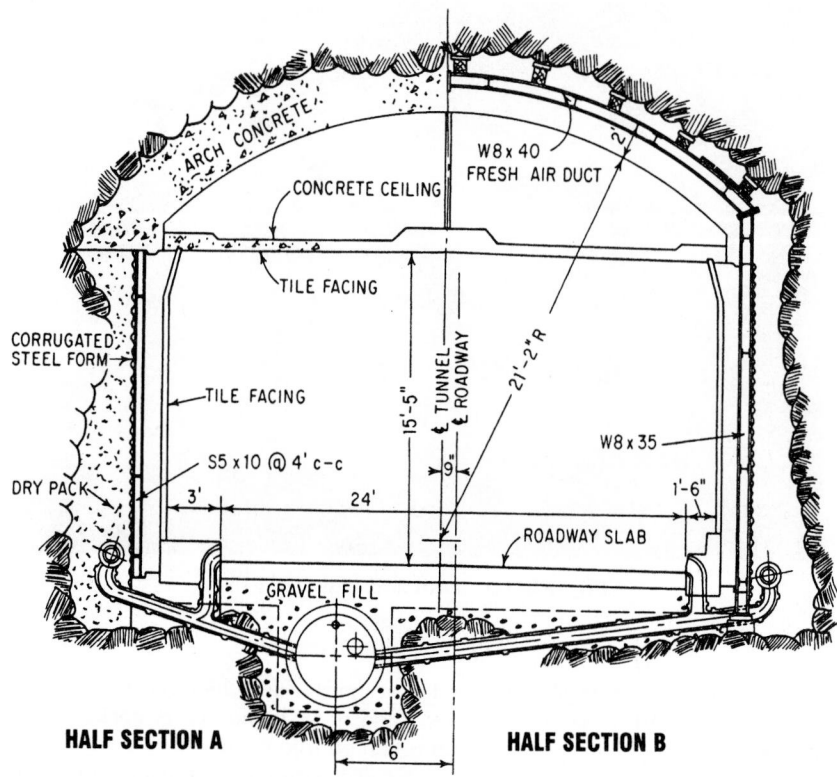

Fig. 20.11 Typical cross section through the Lehigh Tunnel on the Pennsylvania Turnpike Extension. Half section *B* shows the bracing, or sets.

They are stressed by deformation of the rock, which is monitored by extensometers and convergence measurement until equilibrium is reached. If necessary, additional untensioned or tensioned bolts are inserted. All rock bolts in permanent installations should be grouted as protection against corrosion.

Another type of bolt has a perforated sleeve, which is placed in a hole in the rock and filled with grout. As the bolt is pushed into the hole, the grout is squeezed through the perforations and against the rock. Bond between bolt, grout, and rock provides the holding force.

Shotcrete ■ Use of sprayed concrete (gunite or shotcrete) as preliminary tunnel support for rock tunnels was developed in Europe and has also been successful in North America. As soon as possible after blasting, while mucking is going on, a layer of concrete is sprayed on the roof. The concrete is made with a well-graded aggregate, up to $\frac{3}{4}$-in size, which is frequently dry-mixed with cement and an accelerating agent. The mixture is ejected through a nozzle under pressure by special pumps. Mixing water is added at the nozzle. Initial set takes place in about 30 to 120 s, final set in 12 min.

Also often used is a wet mix, for which aggregate, cement, and water are placed in the mixer and additive is injected as a liquid at the nozzle. Addition of about 5% by volume of microsilicate greatly improves adherence of shotcrete to the rock and reduces reinforcing steel requirements. Addition of $\frac{1}{2}$- to 1 $\frac{1}{2}$-in steel fiber to a mix in the amount of $\frac{1}{2}$ to 1% by weight considerably increases the ultimate strength and toughness of the shotcrete.

Thickness of the initial layer may vary from 2 to 4 in, depending on rock conditions. Additional

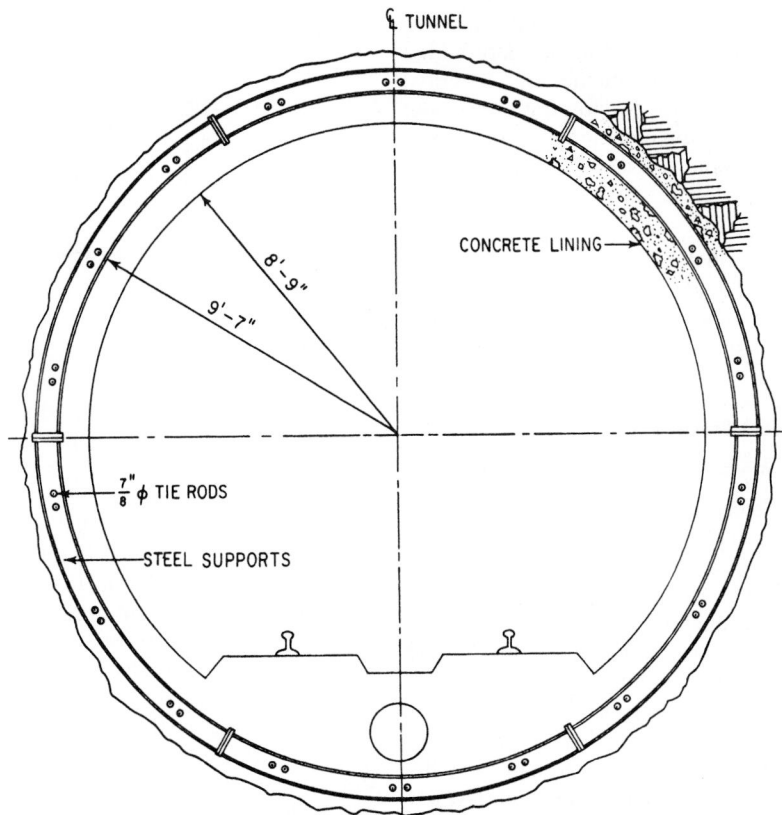

Fig. 20.12 Typical section through Berkeley Hill Rock Tunnel (heavily faulted rock) for the San Francisco Bay Area Rapid Transit.

layers may be sprayed on as needed. Total thickness may be as much as 8 in.

The nozzle may be held directly by an operator or attached to a boom manipulated by a worker stationed under the protective roof of the jumbo. Automatic application has been successful in a machine-bored tunnel (Heitersberg Tunnel in Switzerland). Robots, controlled by an operator on the jumbo, can be used to apply either dry or wet mix shotcrete.

Shotcrete is sprayed on the sidewalls after completion of mucking. Heavy water inflow must be intercepted and drained through inserts in the shotcrete. Well-trained operators and careful supervision and control are essential for good results. Properly executed, the method can be used successfully for fractured rock.

Strength of concrete in place reaches 200 to 250 psi in 2 h, 1400 to 1500 psi in 12 h. The ultimate compressive strength of 4000 to 5500 psi is about 15% less than that of the same concrete without accelerator.

Waterproofing ▪ Above the groundwater table, waterproofing is usually applied to ceilings in transportation tunnels to prevent dripping. Drainage paths may be provided along the base of the walls to handle any water that does appear. False walls with finishes are often used to hide walls where leakage is expected. For most tunnels below the groundwater table, a waterproofing membrane enveloping the tunnel is used between the initial ground support and the final lining. If the tunnel lining is undrained, the final lining will carry the full groundwater pressure and should be designed accordingly. Where drainage is provided outside the waterproofing membrane, the final lining may be designed for a reduced groundwater pressure. Waterproofing membranes may also need to resist deleterious gases or liquids expected to be present.

Leakage ▪ See Art. 20.13.

(J. O. Bickel and T. R. Kuesel, "Tunnel Engineering Handbook," Van Nostrand Reinhold Company, New York.)

20.14 Tunnels in Firm Materials

Occupational Safety and Health Administration requirements should be satisfied in underground excavations. (See Arts. 20.6, 20.8, and 20.12.)

Materials, other than rock, that may be encountered in tunneling are sands of various densities and grain sizes; sands mixed with silt or clay; clays, either pure or containing silt or sand, and varying from relatively plastic with high water content to firm and dry; and alluvial mixtures of sand and gravel or glacial till. To improve the properties of poorer ground, or to reduce water infiltration, ground improvement prior to mining may be undertaken. This may take the form of the injection of cementitious or chemical grout, or it might physically mix soil with these materials. Bentonite has also been used; environmental issues associated with the use of the selected material should be considered. If not subject to hydrostatic pressure of free water, materials may be excavated by mining. Temporary support is given by timber or steel framing in headings whose size and number depend on local conditions.

Mining of headings in all these materials requires the driving of poling boards, supported by cross timbers and posts to hold the roof. As excavation is advanced on a face as steep as the material will stand, these boards are driven further, with the rear supported by the frame, the front by the soil. A new support is set under the forward end of the poling boards and the process repeated. The sides of the heading are held by boards supported by the posts, as required. Figure 20.13 illustrates the basic procedure for this type of excavation.

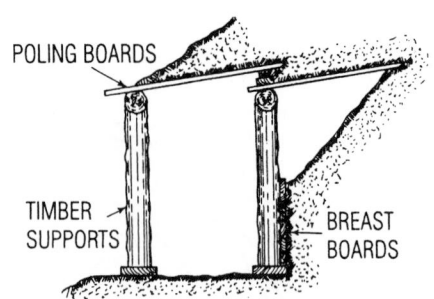

Fig. 20.13 Timber bents support poling boards in basic earth mining.

Steel supports are often used instead of timber, particularly for large headings. Steel lances, made of small wide-flange beams with wedge-shaped points, may be used instead of wood poling boards. The lances are long enough to be supported on two frames and driven by jacks or air hammers into the soft face for a distance equal to the support spacing.

In loose soil or running sand, the face is supported by breast boards. A shallow slot about 2 ft deep and one or two poling boards wide is excavated in the top of the face, and a short vertical breast board is placed immediately, to hold the face and support the forward end of the poling. After this slot has been excavated across the heading and all vertical breast boards set, a cap is installed, supported by short posts. The rest of the face may then be excavated downward and held by horizontal breast boards (see Fig. 20.14).

The size of the heading should be as large as soil characteristics allow, but not less than 5 ft wide by 7 ft high. Steel bents shaped to the tunnel arch are preferable to timber framing, if economical, considering both price and speed of operation. Poling may be timber or steel.

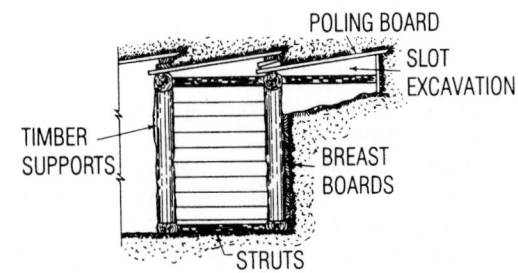

Fig. 20.14 Mining in running ground requires breast boards.

Steel liner plates are available in various shapes and sizes. They may be used to support the ground if a limited excavated area of the roof or arch will stand long enough for insertion of the liner plates, starting at the top of the arch and working down. The flange of each plate is bolted to the previously erected liner.

In small tunnels, ribbed or corrugated liner plates may give adequate support. In large tunnels or under heavier loads, the plates are backed up by steel ribs, against which they are blocked. Liner plates without flanges may also be used as lagging or poling. See also Art. 20.17.

To prevent settlement or unbalanced load, all voids behind the liner plates should be filled by injection of pea gravel or cement grout.

Small tunnels may consist of a single heading. For large tunnels, various combinations of headings are used. Some of these are known by the country of their origin, as American, Austrian, Belgian, English, German, or Italian methods, but are used in many variations. Originally, the methods required wood supports, but now steel supports are favored, where economical.

Sequential Excavation Method (SEM) ■
Also known as the New Austrian Tunneling Method (NATM), SEM was developed in Austria but is now used worldwide. It is a tunneling method adapted to the excavation of variable and non-circular cross-section reaches of tunnel, such as highway ramps and subway stations. This underground method of excavation divides the space (cross-section) to be excavated into segments, then mines the segments sequentially, one portion at a time. Excavation sequencing by the American method, Austrian system and Belgian method are outlined below.

The excavation can be carried out with common mining methods and equipment (often a backhoe), chosen according to the soil conditions; tunnel-boring machines are not used. Ground conditions are assessed at the face of the tunnel or from the side of a small tunnel, which helps to decide how to proceed in the best way and determines the choice of equipment and lining. It should be noted that the combination of ground treatment and SEM for the excavation of uniform cross-section tunnels would generally be more expensive than the use of pressurized face TBM construction under American underground construction labor and economic

conditions. Thus the application of SEM would be limited economically to variable geometry structures. However for shallow tunnels, such structures could probably be more economically constructed using cut-and-cover techniques.

SEM requires extremely dry conditions; dewatering is often necessary before the excavation can proceed. SEM involves careful sequencing of the excavation as well as installation of supports. Shotcrete (a kind of concrete sprayed from high-powered hoses) may be used to line the tunnel or support the face, and grouting (the injection of a cementing or chemical agent into the soil) may be used to increase the soil's strength and reduce its permeability. Because of the requirements of this method, the rate of excavation is slow. Use of this method in saturated, non-cohesive granular soils would require the use of groundwater control and ground improvement techniques. One real concern with the use of SEM in granular soils is sudden uncontrollable ground loss, often resulting in surface sinkholes. This can happen when the selected ground improvement method is unsuccessful because of localized variation in ground conditions.

One method often used to control groundwater is compressed air. However, the high air pressures often required might make the use of compressed air tunneling uneconomical in comparison to other possible methods. It is for similar reasons that shield tunneling using compressed air has been replaced by tunneling with pressurized face tunnel boring machines. Another commonly used method of controlling groundwater is dewatering. However, unrestricted dewatering can have a significant effect on adjacent foundations. An approach that has been used with variable success overseas has been to install groundwater cut-off walls (slurry walls, etc.) along both sides of the right-of-way and then dewatering inside the cut-off. When dewatering sands, running or fast-raveling ground, conditions may result so that some form of ground improvement, such as closely spaced groutable spiles or horizontal jet grouting above the crown of the excavation could be required. Two other ground improvement methods that could be used are jet grouting and chemical grouting. Each method would be used to create a block of stabilized ground through which the tunnels could be excavated.

American Method ■
As shown in Fig. 20.15*a*, excavation starts with (1) a top heading at the

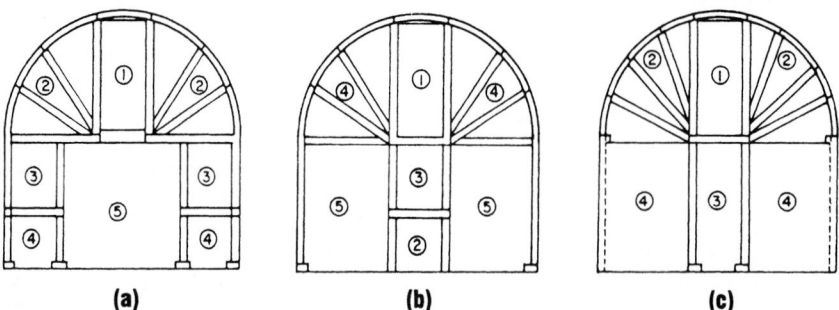

Fig. 20.15 Some excavation procedures for large tunnels: (*a*) American method. (*b*) Austrian method. (*c*) Belgian method.

tunnel crown, which is supported by poling, posts, and caps. Next, the excavation is widened between two bents and the top arch segments adjoining the crown are set, supported by extra posts or struts. (2) The excavation then is benched down along the sides, and another segment of ribs is set on each side. (3) and (4) These are doweled to the upper part and supported by temporary sills. This process is repeated to the invert sill. The bench finally is excavated to full section. (5) Ground between ribs is held by lagging, and voids are packed. This method is suitable in reasonably firm material.

Austrian System ▪ As shown in Fig. 20.15*b*, a full-height center heading is advanced. It either starts with a top heading and is cut down to the invert in short lengths or starts as separate bottom and top headings. (1) and (2) In the latter case, the core between the two is excavated for short distances, and the short posts replaced with long ones. (3) The arch section is widened in short lengths and is held by segmental arch ribs and longitudinal poling boards. (4) The arch ribs are supported by struts from the center-cut framing and sills at the spring line. The rest of the excavation is advanced to full face in short increments, and posts are set to support the sills. (5) This method is suitable for reasonably stable soil.

Belgian Method ▪ As shown in Fig. 20.15*c*, in firm ground, the upper half of the tunnel is excavated, starting with a center heading from the crown to the spring line. (1) This is widened to both sides, the ground being held by transverse polings.

These are supported by longitudinal timbers, in turn supported by struts extending fanlike from a sill in the center heading. (2) Next, a center cut is excavated to the invert (3), leaving benches to support the arch of the tunnel lining. Slots are cut into the benches at intervals to underpin the arches. The rest of the bench then is removed to complete the side walls (4), after which the invert is concreted. The excavation may be advanced a considerable distance before the tunnel lining need be inserted.

(J. O. Bickel and T. R. Kuesel, "Tunnel Engineering Handbook," Van Nostrand Reinhold Company, New York.)

20.15 Shield Tunnel in Free Air

This section describes shield tunneling where the face is essentially open and exposed at ambient air pressure, and Section 20.16 when exposed under compressed air. Both these methods are less common today than tunneling by the tunnel boring machines (TBM) described in Section 20.19, now widely used.

Shield tunneling is generally used in noncohesive, soft ground composed of loose sand, gravel, or silt and in all types of clay, or in mixtures of any of these. It is indispensable for tunneling in these materials below the water table.

Requirements of the Occupational Safety and Health Administration (OSHA) for underground construction should be complied with in operations with shields. OSHA requires the following, in particular: Lateral or other hazardous movement of a shield when subjected to a sudden lateral load should be restricted. Personnel accessing shielded areas should be protected against cave-ins. Personnel should not be permitted in shields when they are being installed, removed, or moved vertically. Excavations may extend up to 2 ft below a shield bottom, if the shield is designed to resist the forces from the full depth of the trench and if soil will not be lost from behind or below the bottom of the shield. (See also Arts. 20.6 and 20.8.)

The shield is a cylinder made of welded steel plate (Fig. 20.16). It has a diameter slightly larger than the outside of the tunnel lining. The plate is stiffened by two interior ring girders, the first one installed a short distance behind the cutting edge.

Depending on the diameter and loads, the girders are braced by horizontal and vertical steel struts. The cutting edge is beveled and reinforced by welded steel plates to a thickness of up to 3 or 4 in. For loose ground, the upper half of the shield is extended forward 12 to 18 in to form a protective hood.

The tail of the shield overlaps slightly the end of the finished lining and provides space for at least one liner ring, and for underwater tunnels is usually long enough to accommodate two rings. The inside of the tail clears the lining by about 1 in all around. For working in soft clay, the front of the shield may be closed by a steel bulkhead with door-

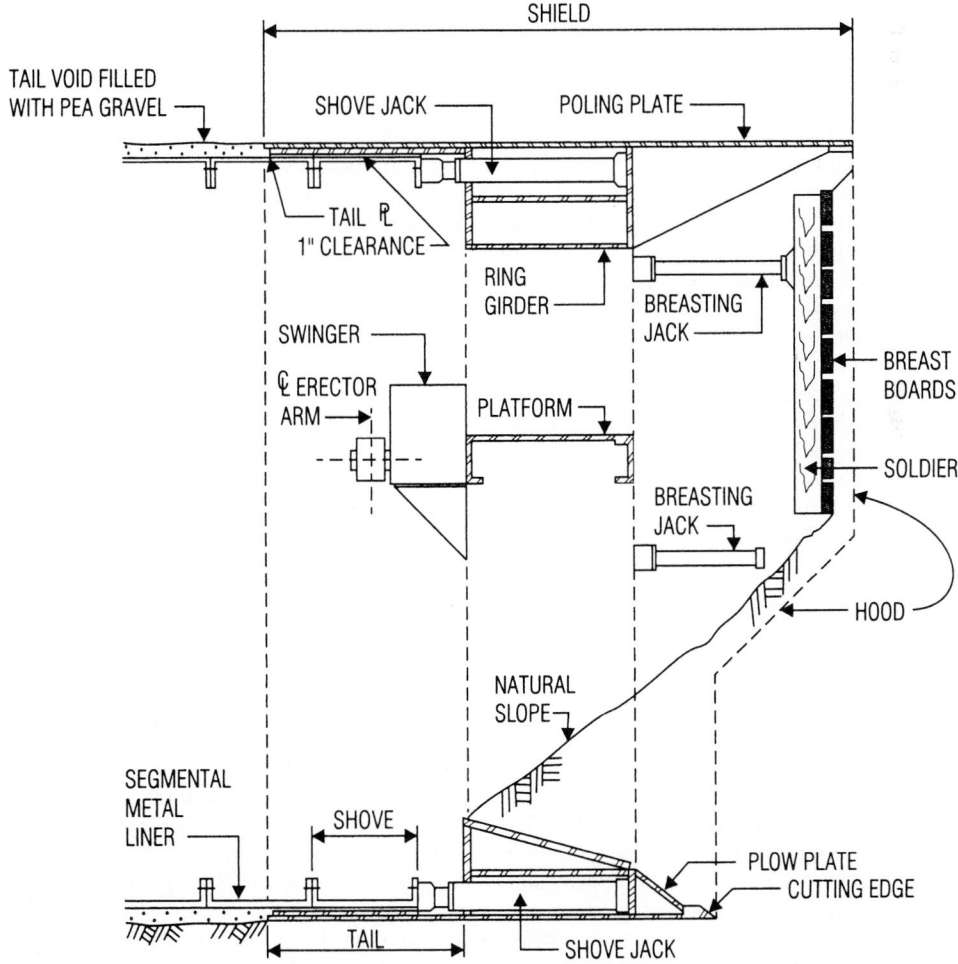

Fig. 20.16 Longitudinal section through a shield used for tunneling through soft ground in free air.

equipped openings through which material is excavated. Soft clay may be extruded through the openings while the shield advances.

Working platforms that can be advanced and retracted by hydraulic jacks are mounted on the shield bracing (Fig. 20.16). They give access to all parts of the face and, by keeping in contact with it, support it if necessary during shoving. Additional breasting jacks can be mounted in the bracing to hold breast boards against the face if it needs extensive support.

Shield Advancement ▪ Hydraulic jacks (Fig. 20.16) for advancing the shield are set on the webs of the ring girders close to the periphery of the shield. The shove jacks are evenly spaced around the perimeter and exert pressure against the forward ring girder, which is stiffened by brackets welded to the skin of the cutting edge. Jack plungers are equipped with shoes bearing against the tunnel lining. The stroke of the jacks is slightly more than the width of a liner ring.

A rotating erector arm is mounted inside the tail to pick up and place liner segments. Hydraulic pumps mounted behind the shield supply 5000- to 6000-psi pressure to the jacks, erector arm motor, and other hydraulically operated equipment. Control valves for these devices are mounted on a panel in the shield.

The method of operation, excavation, and speed of advance vary greatly according to the type of soil. In sand and gravel, the face usually has to be held by breast boards (Fig. 20.16), which are braced by telescoping struts, breasting jacks, or the working platforms. The breasting may have to be carried down to the invert of the face, which is excavated to the cutting edge of the hood. If compressed air is used, the breasting may be carried part way down to where the air pressure balances the hydrostatic head, the lower part of the face taking its natural slope. In firm materials, silty sand, or stiff clay, the full face may be excavated without breasting. Average progress for the 31-ft-diameter Queens Midtown Tunnel, New York City, in these materials was between 7 and 8 ft in 24 h.

Shields are not well-suited for rock tunneling, but rock or mixed faces, partly rock and partly soil, may be encountered in parts of soft-ground tunnels. If the rock is high enough, a bottom heading may be excavated ahead of the shield and a concrete cradle placed, with steel rails embedded, to exact line and grade to support the shield as it advances. A similar bottom heading may be used in a full rock face if the full cross section cannot be excavated. Then, the rock may be blasted around the periphery of the rest of the cutting edge to permit advancing the shield. Progress in mixed face in the Queens Midtown Tunnel averaged about 3 to 4 ft per 24-h day.

Best progress is made in plastic material through which the shield may be shoved blind, that is, without taking any soil into the inside, the volume being displaced by compressing or heaving of the surrounding material. To counteract the tendency of the tunnel to heave behind the shield, because of buoyancy, enough soil may be admitted through small openings in the face bulkhead and left in the invert to balance the forces until the interior lining is placed. This method is called shoving half-blind. In the first tube of the Lincoln Tunnel, New York City, about 20% of the material was taken in. If displacement or heaving of the soil may cause disturbance of adjacent structures, such as buildings or another tube nearby already in place, the openings should be adjusted to admit nearly all the displaced material. This was done in the second and third tubes of the Lincoln Tunnel, through openings aggregating 5 to 20% of the face area. Average daily progress was about 30 ft.

Shields are usually started from shafts sunk to the invert grade. These shafts may be specially constructed for this or may later be part of a ventilation building. An opening is provided in the shaft wall to fit the shield and is closed by a timber bulkhead during sinking operations. The shield is erected on a concrete cradle at the base of the shaft. The opposite shaft wall forms the abutment for the jacking forces. A few rings are erected behind the shield, which is advanced through the opening after removal of the bulkhead.

The shield is steered by varying the pressure of the shoving jacks around the periphery. On large tunnels, the total jacking force may be 3000 to 5000 tons. If the shield has a tendency to rise, more pressure is applied at the top than at the bottom. Similar corrections are made for other directions.

If the soil is relatively loose, it is excavated at the face by hand tools. In hard-packed silty sands or very stiff clay, air spades are used. Relatively soft clay may be cut by clay knives. The muck may be shoveled by hand on a short conveyor in the shield, but more commonly, a large hydraulic scraper or hoe is used to loosen the soil and scoop it onto the conveyor. From there, it is discharged on a loading conveyor mounted on a movable carriage behind the shield. The loading conveyor dumps it into mine cars, usually of about 4-yd^3 capacity in large tunnels. The muck trains are rolled back through the tunnels to an access shaft. The individual cars are hoisted up the access shaft and dumped into hoppers for discharging into trucks.

Tunnel Linings ▪ Except in very stiff or compact soils, segmental ring liners are used in shield tunnels. These used to be of cast iron but today steel or precast concrete is used. The segments are brought in by mine cars, unloaded by hoists mounted on the conveyor carriage, and deposited within reach of the erector arm. This is a telescoping, counterweighted arm pivoted on the center line of the tunnel for full rotation by a hydraulic motor (Fig. 20.17). A gripper at its outer end engages lugs or bars in the segments and places these, starting at the bottom. A short, tapered segment forms the key. See also Art. 20.17.

Packing ▪ Since the shield has a larger diameter than the lining, a void exists around the liner rings. This may permit a cave-in and cause settlement. The usual practice when segmental liners are used is to inject pea gravel into this void through grout holes in the liners immediately after the shield has been advanced (Fig. 20.16). Cement grout is later injected into the gravel to solidify it. In a section of the Victoria line of the London subway in deep, very stiff clay, an articulated cast-iron lining was installed and expanded against the clay behind the shield. The adjacent rings were pressed into contact by the jacking forces but were not bolted. Expansion of steel ribs with wood lagging has also been used to achieve tight fit against the soil.

Semicircular or semielliptical shields have been used as temporary supports for the roof or arch of excavations, mostly in dry or dewatered soils, for example, for tunnels at shallow depth where open-cut operations are prohibited by circumstances. They are advanced in a manner similar to that for circular shields.

(J. O. Bickel and T. R. Kuesel, "Tunnel Engineering Handbook," Van Nostrand Reinhold Company, New York.)

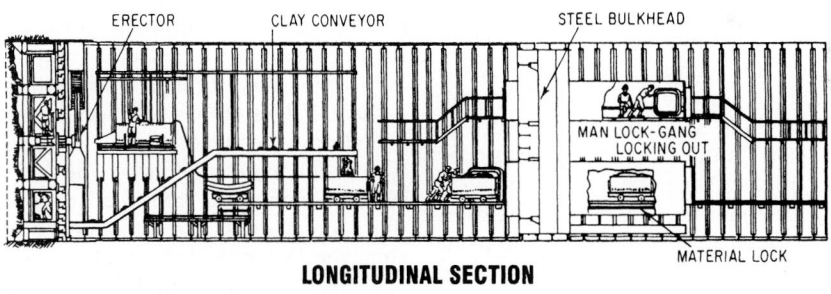

LONGITUDINAL SECTION

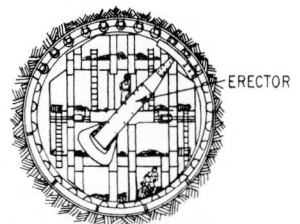

ELEVATION-CUTTING EDGE

Fig. 20.17 Section through a conventional shield (used in 1930 for the Detroit, Mich.—Windsor, Ont., Tunnel) for tunneling with compressed air.

20.16 Compressed-Air Tunneling

Although tunnel shields in free air are effective in naturally dry soil or ground that can be dewatered (Art. 20.15), compressed air is needed while tunneling below the water table, particularly in subaqueous tunnels. The air pressure counteracts the hydrostatic head. Also, the pressure reduces the water content of the soil at the face, making it more stable and safer to excavate.

Legal issues surrounding safety and health issues of compressed air tunneling have reduced use of this method. It still has to be used from time to time to remove obstructions in front of tunnel boring machines that they cannot handle, and to carry out maintenance and repairs on some of the machines.

Air Pressure ▪ Theoretically, the air pressure required to balance the hydrostatic head is 0.43 psi for fresh water and 0.44 psi for seawater per foot of depth. Actually, the pressure depends on the properties of the soil as well as on the method of excavation. In open material, such as pervious coarse sand and gravel, the full air pressure would be required, whereas in impervious soils, such as stiff clay, no pressure at all may be needed. A careful analysis of the soil at regular intervals along the alignment is needed to estimate the maximum air pressure and air quantities required. Closed shields for tunnels in the Hudson River silt operated with as little as 16-psi air pressure in depths up to 100 ft. In the sand and gravel under the East River in New York, the hydrostatic head was balanced for about one-quarter or one-third the diameter above the invert. To reduce loss of air at the top of the face, breast boards were plastered with clay.

Blowout Prevention ▪ With the air pressure balancing the head at the bottom of the face, there is an excess of pressure at the top. If the weight of cover over the tunnel is insufficient to hold the excess air pressure safely, a heavy clay blanket may be placed on the river bottom over the tunnel heading to prevent a blowout at the top of the face. If the air pressure equals the water pressure at the invert, the excess pressure at the top of a 30-ft-diameter face would be 13 psi in seawater, or 1870 lb/ft². For a 10-ft natural cover of 50-lb/ft³ material, the blanket would have to make up the deficiency of 1370 lb/ft². At 60 lb/ft³ submerged weight, 23 ft of clay would be required. Navigation requirements may make it necessary to remove the blanket after completion of the tunnel. Clay for this blanket should be relatively soft so that it will readily coalesce into an impervious layer.

Bulkheads ▪ When the shield is started from a shaft, an airtight deck is built above the tunnel, to hold the pressure until the shield is advanced some distance. An airtight bulkhead is then built into the tunnel a sufficient distance behind the shield to provide space for the loading conveyor and a few mine cars. To keep the volume filled with compressed air within reasonable limits and to comply with safety laws, new bulkheads are built as the tunnel advances and the old bulkheads are removed. Usually, regulations permit a maximum distance of 1000 ft between the face and the bulkhead, which may be constructed of steel or concrete.

Air Locks ▪ Worker and material locks are built into the airtight bulkhead. The locks are airtight cylinders at least 5 ft in diameter and have gasketed doors (Fig. 20.17). Worker locks should provide at least 30 ft³ of air space per occupant. Compressed air is admitted to the lock from the high-pressure side or from the compressed-air line and is exhausted through a connection to the free-air side. Valves of these connections are controlled from the inside in worker locks and generally from the outside in material locks. The door at the high-pressure side opens from the lock into the tunnel; the door at the free-air end opens into the lock chamber. Doors are held tight by the air pressure and cannot be opened until pressures on both sides are equalized. Pressure gages are provided in the locks as well as in the tunnel.

The material lock is at the level of the mine-car track. The lock should be large enough to accommodate several mine cars.

The worker lock is at a higher level and must have not less than 5 ft clear head space. This lock is equipped with benches for workers to sit down on. In large tunnels, two sets of locks may be used to speed up operations. If there is danger of rapid flooding, an extra worker lock may be placed as high as possible, and a hanging safety walk extended at this level from the lock to the shield.

A safety screen placed in the upper part of the tunnel near the heading will trap air above this safety walk in case of flooding and permit workers to escape. Some safety laws require installation of two worker locks.

A special decompression chamber capable of accommodating an entire shift of workers should be available when decompression time required is more than 75 min. A passageway should be provided to give workers in a man lock access to the special chamber.

Safety and Health ▪ For all compressed-air work, a well-equipped first-aid station and decompression chamber are required, staffed by a trained attendant at all times. A physician must be available at all times for emergency calls while work is in progress.

Most states and countries have laws regulating the working hours and locking rates for compressed-air work. Regulations of the U.S. Occupational Safety and Health Administration for work in compressed air as well as for construction in general and all underground operations should be observed. (See also Arts. 20.6, 20.8, and 20.15.) OSHA requires that a record be kept outside worker locks of the time in each shift that workers spend in compression and decompression. A copy should be given to the supervising physician.

During the first minute of compression in a lock, pressure may be increased up to 3 psig and should be held at that level, and again at 7 psig, long enough to determine if anyone in the lock is being adversely affected. After the first minute, pressure may be raised gradually at a rate up to 10 psi/min. If personnel experience discomfort, pressure should be reduced to atmospheric and distressed personnel should be evacuated from the lock. Except in emergencies, pressure in a lock should not exceed 50 psi. Temperature in a lock should be at least 70 °F but not more than 90 °F., whereas temperature in compressed-air working areas should not exceed 85 °F.

Unless air pressure in the working chamber is less than 12 psig, decompression in a worker lock should be automatic. Manual controls, however, should be provided inside and outside the lock to override the automatic mechanism in emergencies. The lock should have a window at least 4 in in diameter to permit observation of the occupants from the working chamber and free-air side of the lock.

OSHA requires also that at least 30 ft^3/min of ventilation air be supplied per worker in the working chamber. In addition, OSHA specifies that at least 10 ft-c be provided by electric lights on walkways, ladders, stairways, or working levels. Two independent supply sources should be used, including an emergency source that becomes operative if the regular source should fail. External parts of electrical equipment, including lighting fixtures, when installed within 8 ft of the floor, should be made of grounded metal or noncombustible, nonabsorptive, insulating material.

OSHA also requires that sanitary, comfortable dressing and drying rooms be provided for workers employed in compressed air. Facilities should include at least one shower for every 10 workers and at least one toilet for every 15 workers. Fire-fighting equipment should be available at all times in working chambers and worker locks.

OSHA requirements for total decompression time, which depends on the air pressure in the working chamber and the time of worker exposure to that pressure, are listed in Table 20.3. Decompression should take place in two or more stages, but not more than four. (Four stages are required for pressures of 40 psig or more.) In Stage 1, pressure may be reduced at a rate up to 5 psi/min from 10 to 16 psi, but not to less than 4 psig. In later stages, the rate of pressure reduction may not exceed 1 psi/min. Local rules, however, also should be checked. Limits of union contracts, though, are sometimes more stringent than legal requirements.

The amount of air required for compressed-air tunneling depends on so many variables that exact rules cannot be given. To determine the size of the compressor plant for a given job requires a great deal of judgment by the engineer, based on past experience. Low-pressure machines are installed for the tunnel air and high-pressure units for air tools. Adequate standby capacity must be provided by using a number of compressors. High-pressure air may be used as an emergency tunnel supply by interconnecting the compressors through reducing valves.

Shieldless Tunneling ▪ Some tunnels have been built in water-bearing ground by using compressed air in conjunction with liner plates,

Table 20.3 Total Decompression Time, Min, after Construction Work in Compressed Air*

Working-chamber, psig	Working period, h										
	½	1	1½	2	3	4	5	6	7	8	Over 8
9 to 12	3	3	3	3	3	3	3	3	3	3	3
14	6	6	6	6	6	6	6	6	16	16	33
16	7	7	7	7	7	7	17	33	48	48	62
18	7	7	7	8	11	17	48	63	63	73	87
20	7	7	8	15	15	43	63	73	83	103	113
22	9	9	16	24	38	68	93	103	113	128	133
24	11	12	23	27	52	92	117	122	127	137	151
26	13	14	29	34	69	104	126	141	142	142	163
28	15	23	31	41	98	127	143	153	153	165	183
30	17	28	38	62	105	143	165	168	178	188	204
32	19	35	43	85	126	163	178	193	203	213	226
34	21	39	58	98	151	178	195	218	223	233	248
36	24	44	63	113	170	198	223	233	243	253	273
38	28	49	73	128	178	203	223	238	253	263	278
40	31	49	84	143	183	213	233	248	258	278	288
42	37	56	102	144	189	215	245	260	263	268	293
44	43	64	118	154	199	234	254	264	269	269	293
46	44	74	139	171	214	244	269	274	289	299	318
48	51	89	144	189	229	269	299	309	319	319	
50	58	94	164	209	249	279	309	329			

* For normal conditions, as specified by the Occupational Safety and Health Administration in "Construction Industry: OSHA Safety and Health Standards (29CFR 1926/1910)," revised 1991.

without a shield. Considerable lengths of 7- to 12-ft-diameter interceptor sewers in New York were constructed this way. A few miles of Chicago subway were built in soft clay with steel ribs and liner plates under compressed air.

[J. O. Bickel and T. R. Kuesel, "Tunnel Engineering Handbook," Van Nostrand Reinhold Company, New York; "Construction Industry: OSHA Safety and Health Standards (29 CFR 1926/1910," Superintendent of Documents, Government Printing Office, Washington, DC 20402 (www.gpo.gov).]

20.17 Tunnel Linings

Unlined Tunnels ▪ Tunnels in very sound rock, not affected by exposure to air, humidity, or freezing, and where appearance is immaterial, are left unlined. This is the case with many railroad tunnels.

Unlined water tunnels in rock are susceptible to leakage either into or out of the tunnel, depending upon the relative pressures. There is therefore a risk that material could be washed out of weak zones and fissures, potentially leading to instability unless lined. However, Norwegian hydropower tunnels in good crystalline rock are often unlined for most of their length.

Shotcrete Lining ▪ Where rock is structurally sound but may deteriorate through contact with water or atmospheric conditions, it can be protected by coating with sprayed concrete, reinforced with wire fabric or fibers, or unreinforced (Art. 20.13). Such a lining may also be used in water tunnels in good rock to provide a smooth surface, reducing the friction factor and turbulence.

Cast-in-Place Concrete ▪ Most tunnels in rock, and all tunnels in softer ground, require a solid lining. Highway tunnels of any importance are always lined for appearance and better lighting conditions. Stone or brick masonry has been used to a great extent in the past, but currently concrete is preferred. The thickness of the permanent concrete lining is determined by the size of the tunnel, loading conditions, and the minimum required to embed the steel ribs of any primary lining.

The lining is placed in sections 20 to 30 ft long. Segmental steel forms are universally used and must be properly braced to support the weight of the fresh concrete. The walls are usually concreted first, up to the spring line. Next come the arch pours. It is important that the space between the forms and the rock or soil surface be completely filled. Grout pipes should be inserted in the arch concrete to permit filling any voids with sand-and-cement grout.

Concrete is placed through ports in the steel lining or pumped through a pipe introduced in the crown, a so-called slick line. Placement starts at the back of the pour, and the pipe is withdrawn slowly. A combination of both methods may be used. Concrete is either pumped or injected by slugs of compressed air. Admixtures are added to get an easily placed mix with low water content and to reduce concrete shrinkage. If there is leakage of water, it usually occurs at shrinkage cracks, which may be sealed with a plastic compound. Or the water may be carried off by copper drainage channels installed in chases cut in the concrete (Art. 20.9).

Footings for side walls in rock tunnels are cut into the rock below grade. They give adequate stability unless squeezing ground is encountered, in which case a concrete invert lining is placed. In soft ground, a concrete slab is placed, to serve as pavement in highway tunnels. If heavy side pressure exists, this slab may have to be made heavier to prevent buckling.

Unreinforced Concrete Lining ▪ A concrete lining is placed to protect the rock and provide a smooth interior surface. Where the concrete lining is exposed to compression stresses only, it may be unreinforced. Most shafts not subject to internal pressure are lined with unreinforced concrete. Shrinkage and temperature cracks are probable and may cause leakage. Where there is a risk of non-uniform loading, unreinforced liners are not used, such as in squeezing ground and through soil overburden.

(Recommendations in Respect of the Use of Plain Concrete in Tunnels, AFTES c/o SNCF, 17 Rue d'Amsterdam, F75008 Paris, France.)

Reinforced Concrete Lining ▪ In most cases, reinforcing steel will be required to withstand tension and bending stresses. Reinforcement is usually required at least on the inside face to resist temperature stresses and shrinkage, although reinforcement elsewhere may be needed to resist moments.

Linings for Shield Tunnels ▪ Linings for shield tunnels may be one-pass or two-pass. A one-pass lining system is when the final lining is also the initial lining, usually for tunnels in soil. With a two-pass lining system, an initial lining is installed behind the shield just sufficient to allow the shield to advance while a waterproofing membrane is installed and the final cast-in-place reinforced concrete lining is prepared. The advance rate is thus usually faster and costs fall. The initial lining may be segmental rings with minimal bolting for ease of erection (Fig. 20.18), or steel ribs with lagging. Precast concrete segments are now widely used and the use of cast iron and fabricated steel are rare due to their high cost. Although the initial lining may be designed as part of the final lining, any leakage through the seals would result in the full hydrostatic pressure acting on the inside final lining for which it should be designed.

Pipe in Tunnel ▪ Water and sewer tunnels up to 14 ft diameter are often provided with an internal pipe that forms the inner lining. After the pipe is secured against movement, the space between the initial ground support and the pipe is filled with cellular or mass concrete. Sewer pipes may require a further interior lining to protect against corrosive liquids and gases. Water tunnels with a high internal pressure exceeding the expected external pressures are usually provided with a steel lining if a reinforced concrete lining is insufficiently strong. Since the pipe may be dewatered, it must also be designed for the external pressure, which, if the pipe has leaked, may equal the internal pressure.

(U.S. Army Corps of Engineers Manual, 1997, Design of Tunnels and Shafts in Rock, EM 1110-2-2901.)

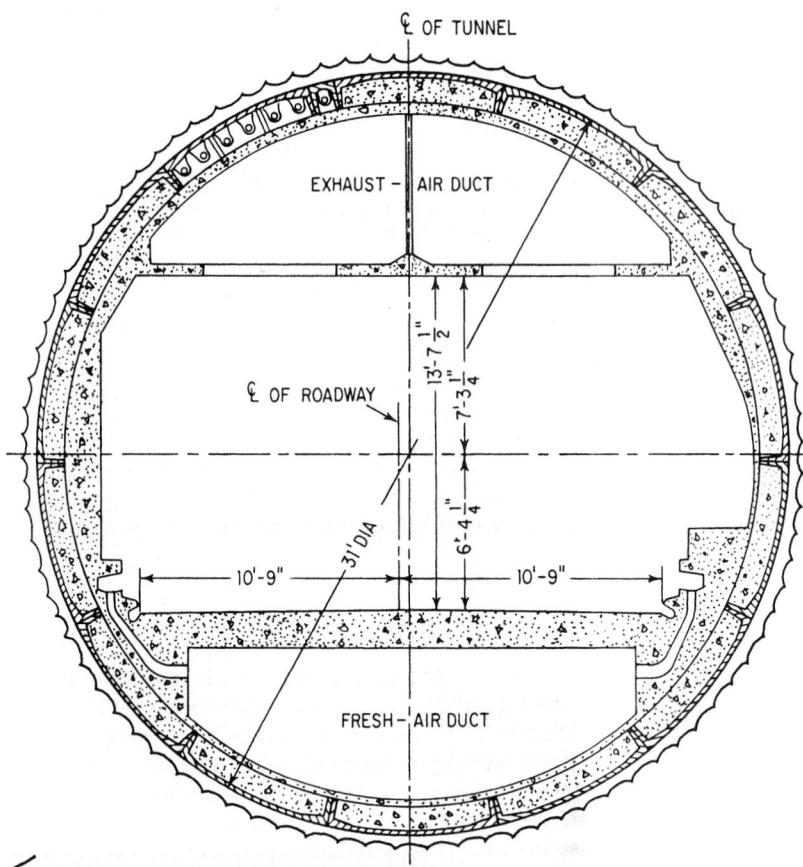

Fig. 20.18 Cross section shows typical segmental cast-iron lining for a tunnel (Lincoln Tunnel under the Hudson River).

In stiff soils, steel ribs, usually 4-in H beams, and wood lagging may be used as primary lining. The ribs are usually spaced 4 ft c to c and are erected in the tail of the shield. Precut and dressed wood lagging is placed solidly around the circumference between the flanges of the ribs. This lagging also transfers the jacking forces to the tunnel lining. Precast-concrete lagging has also been used successfully.

Segments are made as long as convenient handling permits, usually 6 to 7 ft. The width of the rings depends on the distance the face can be safely excavated ahead of the shield and weight to be handled. The wider the rings, the longer the tail of the shield and hence the more difficult the steering of the shield. Early tunnels had 18-in-wide rings. Recent tunnels have gone to 30 or 32 in.

Segments are made to close tolerances on all sides. They are connected by high-strength bolts. Longitudinal joints are offset in successive rings.

The flanges have recesses along their matching edges for calking. These grooves used to be filled with lead or impregnated asbestos calking strips, pounded in manually. Synthetic sealers, such as silicone rubber and polysulfides, can be injected into the grooves by calking guns. These compounds adhere to the metal sufficiently to form an effective seal under pressures usually encountered in underwater tunnels.

Each cast-iron segment is provided with a 2-in grout plug for injection of pea gravel and grout into the space between the lining and the soil. Bolt holes are sealed with grommets of impregnated fabric or plastic grommets, the latter being particularly effective. Bolts are tightened with hydraulic or pneumatic wrenches where possible otherwise with hand wrenches.

Welded steel segments, similar in shape to cast-iron segments, have been used for economic reasons in some subaqueous tunnels. They were welded in jigs to tolerances as close as practicable, but flanges were not machined and no calking grooves were provided. Difficulties were experienced in making them watertight with gaskets. An improved design includes calking grooves and fabrication tolerances similar to those for cast iron.

Precast Concrete ▪ Precast segments are essential to increasing the speed of machine tunneling. A compromise must be reached between the segment size and the number of segments to be installed, directly affecting the weight of the segments, the size of the equipment needed to handle the segments, and the number of operations to be carried out. The width of the segments is governed by the stroke of the jacks pushing the head of the shield, usually in the range of three to five feet. Tapered rings, narrower on one side, are used on bends. At least three segments per ring are required, with five to eight being more common. The closing segment in a ring is usually smaller and wedge shaped to facilitate insertion. Joints in adjacent rings are usually staggered so that all joints are discontinuous, helping to stiffen the rings.

Connection details to adjacent segments vary widely and can be flanged (Fig. 20.19). Straight bolts with nuts, washers and grommets are the most common, but the use of curved recessed bolts result in smaller pockets. Gaining popularity are straight bolts placed at an angle to minimize recesses; the bolts couple into sockets cast into the adjacent section. Dowels may also be used between adjacent rings. The bolts ensure that the rubber or neoprene seals between segments are compressed. The addition of a hydrophilic seal near the outside face may reduce leakage even further. Due to the very close tolerances needed to ensure seals remain watertight and that the diameter remains constant, a high degree of mechanization with steel forms is used. The segments must be installed within the shield tail and the space behind them (the tail void) grouted at a pressure at least equal to the external pressure, making lateral alignment modifications

very difficult. It is not uncommon for most bolts to be retrieved once the grout is set. Secondary linings are not essential.

Heavy, interlocking concrete blocks have been used successfully in relatively dry or impervious soil. They present difficulties when exposed to water pressure due to leakage.

Except where steel rings and lagging or concrete blocks are used as primary lining, no secondary concrete lining is used, unless required for appearance and interior finish of highway tunnels. In this case, a concrete lining of the minimum thickness practicable is placed. When the tunnel is to be faced with tile, provision should be made for attaching it. (To facilitate maintenance and improve lighting, walls and ceilings of highway tunnels are usually finished with ceramic tiles.) To provide good adherence of the scratch coat, scoring wires may be welded longitudinally on the steel forms for the lining to provide a rough concrete surface. Coating of smooth concrete surfaces with epoxy compound may result in satisfactory finishes at less cost.

See also Art. 20.18.

20.18 Design of Tunnel Linings

Article 20.17 discusses the types of linings usually used for tunnels. The following paragraphs describe design of a liner ring.

A liner ring is statically indeterminate. A one-pass lining is designed for transport and erection loads, loads during grouting, and ground loads including seismic. In lieu of computer analyses, which might be as simple as a two-dimensional analysis of a grid framework supported on springs, or as complex as finite element or finite difference three-dimensional analyses using soil-structure interaction for each step of the construction process, stresses in the liner ring may be computed after the ring is made statically determinate by a cut at the top and one end is fixed (Fig. 20.20).

For a circular ring of constant cross section symmetrically loaded the thrust at the crown C is

$$T_c = \frac{2}{\pi R} \int_0^\pi M \cos \phi \, d\phi \qquad (20.2)$$

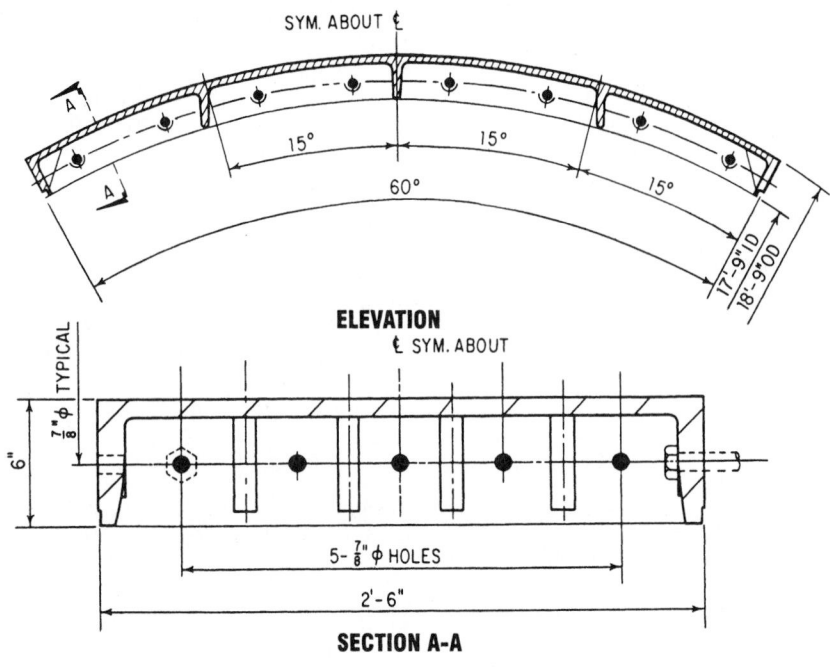

(a) CAST-IRON LINER SEGMENT

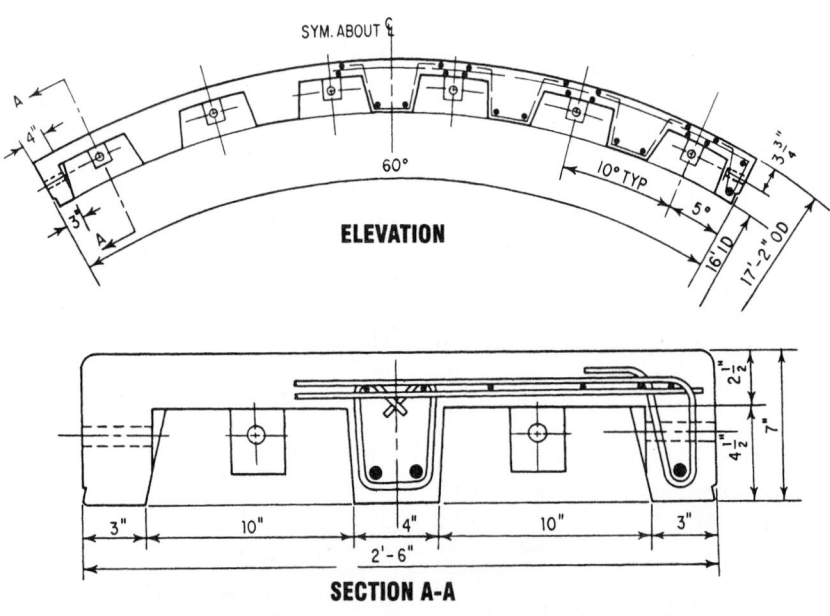

(b) PRECAST-CONCRETE LINER SEGMENT

Fig. 20.19 Typical liner segments used for rapid-transit tunnels.

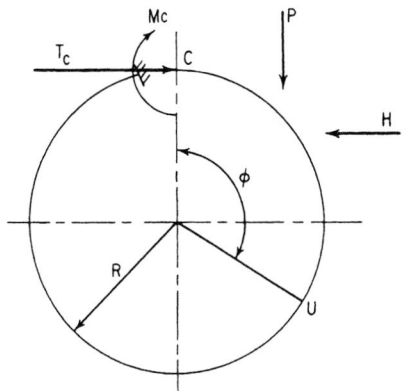

Fig. 20.20 Stresses in liner ring may be computed by assuming it cut at crown C.

The vertical shear at the crown is zero, and the moment is

$$M_c = -RT_c - \frac{1}{\pi}\int_0^\pi M\,d\phi \qquad (20.3)$$

where R = radius of ring

M = bending moment at any point U due to loads on CU

ϕ = angle between U and crown C

With the thrust and moment at the crown known, the stresses at any point on the ring can be computed, as for an arch (Art. 6.71).

(A set of equations is presented in Chapter 15B "Tunnel Structures, Structural Engineering Handbook," 2000 Update for ENGnetBASE, Edited by Wai-Fah Chen and Lian Duan, CRC Press, 2000 (www.crcpress.com).)

Loads on a lining include its own weight and internal loads, weight of soil above the tunnel (submerged soil for tunnels below water level), reaction due to vertical loads, uniform horizontal pressure due to soil and water above the crown, and triangular horizontal pressure due to soil and water below the crown.

Magnitude of loads on tunnel liners depends on types of soil, depth below surface, loads from adjacent foundations, and surface loads. These will require careful analysis, in which observations made on previous tunnels in similar materials will be most helpful.

In rock, the quality of the rock will affect the loads that are carried by the tunnel, and loads carried by any initial rock support may effect the

loads carried by the secondary lining. Compression of competent rock due to outward displacement of the tunnel lining in a pressure tunnel may also need to be considered. Often those linings must be designed to take the full internal pressure. Beyond 100 psi internal pressure, reinforced concrete liners may no longer be sufficient and steel liners may be needed. If a tunnel is watertight, the interior lining is usually designed to carry at least the full external water pressure, since leaks in any outer linings will eventually lead to full transfer of the hydraulic head. If the tunnel is drained, at least some of the hydrostatic head should be considered. Blasting may also disturb the rock locally, leading to loads different to those of a bored tunnel.

Following the derivation of moments, axial thrust and shears, the concrete cross section can be designed accordingly, and steel or fiber reinforcement placed accordingly as needed. Tension cracks in themselves do not necessarily result in failure, whereas through-cracks (often caused by shrinkage) can cause leakage and corrode exposed steel. It is usually undesirable for cracks to extend more than halfway through the section. Typical steel reinforcement for crack control may reach 0.28% or more of the section area. Restraints at the exterior face due to keying into an irregular rock surface may change the calculated behavior. Linings with irregular width are more likely to crack at the thinnest sections or at initial ground support embedments. Waterstops are used at construction joints to reduce leakage.

Because of flexibility, tunnel liner rings can offer only limited resistance to bending produced by unbalanced vertical and horizontal forces. The lining and soil will distort together until a state of equilibrium is obtained. If the deflection, in, exceeds more than $1.5D/10$, where D is the tunnel diameter, ft, the lining may have to be temporarily braced with tie rods when it leaves the shield until the final loading conditions and passive pressures have been developed. In certain soft materials, when shields were shoved blind (without material being excavated), initial horizontal pressures exceeded the vertical loads, so that the vertical diameter lengthened temporarily. Ultimately, the section reverted to approximately its initial circular configuration.

When a lining is in rock, determination of the loads imposed on the lining need to be done with care. Stable rock may distribute the stresses around the tunnel, and if impervious, may leave any tunnel lining virtually unloaded. Following excavation, rock that has not yet reached stability can still be moving, extreme examples of which are squeezing and swelling rock. Further displacements may be little affected by the presence of the lining in such cases, or may depend upon the relative stiffnesses of the two, so that the lining must be designed accordingly. Concrete cast against irregular rock may also be keyed into the rock and result in composite action. If the rock is anisotropic, material properties and movements may depend upon direction. Some external loads, such as groundwater pressure and some clays, are independent of displacements. Depending upon the porosity of the surrounding material, water pressure can eventually build up to the full hydrostatic load even in a rock tunnel.

Potable water supply tunnels may need to be made watertight when passing though areas where the inflow of groundwater is not acceptable. Where groundwater contains fine silt or chemicals that could clog drainage facilities on which the tunnel design is based, regular maintenance is required to keep drainage paths clear, or else new drainage paths must be provided or the tunnel designed as watertight. Sewage tunnels frequently generate hydrogen sulfide and so require extra protection against corrosion, such as using an internal PVC or HDPE membrane cast into the internal tunnel lining.

Precast Lining Segments ▪ Analysis and design must cover all aspects of fabrication, storage, transportation, installation, jacking loads, and expected loads in service. Allowance for creep and shrinkage during all stages is required. It is not uncommon for segments to be installed slightly askew, resulting in extra bolting forces and non-uniform loads. Curved bolts although easier to install, require extra reinforcement. Attention to reinforcement details at corners can help to reduce damage. Compressed gaskets tend to create tensile stresses that may require reinforcement. Durability of the tunnel lining is highly desirable, and may be enhanced by using low permeability crack-free concrete, the use of pozzolans to resist sulfate attack and microsilica to improve strength, protecting the bolts and reinforcement against corrosion such as by applying epoxy coating to the fabricated reinforcement, and by the use of external waterproofing, including quality grout and a bitumen coating to the exterior face.

[Kuesel, T. R., Tunnel Stabilization and Lining, in "Tunnel Engineering Handbook," Bickel, J. O., Kuesel, T. R., and King E. H., Editors, Chapman & Hall, 1996 (www.wkap.nl)] [U.S. Army Corps of Engineers Manual, 1997, Design of Tunnels and Shafts in Rock, EM 1110-2-2901 (www.USACE. ARMY.MIL/inet/USACE-docs/eng-manuals).] (The design, sizing and construction of precast concrete segments installed at the rear of a tunnel boring machine, 1997, translated into English 1999, French Tunneling Association AFTES c/o SNCF, 17 Rue d'Amsterdam, F75008 Paris, France.)

20.19 Machine Tunneling

To reduce costs and increase the speed of the ever-increasing amount of tunnel construction, a number of tunnel-boring machines (TBM) for rock and soft ground have been developed. Universal machines for mixed ground of rock and soft material (mixed face) are designed for each specific location and have opened up new possibilities for machine tunneling.

Hard Rock TBMs ▪ Rock-boring machines consist of a rotating head, either solid or with spokes, on which are mounted cutting tools suitable for the type of rock. The machines are mounted on large frames, which carry the driving machinery and auxiliaries, including a series of hydraulic jacks to exert heavy pressure against the face. Chisel cutters serve for soft rock, disk cutters break harder rock by wedge action, and toothed roller cutters with tungsten carbide inserts cut the hardest rocks. A critical factor in evaluating production is the amount of down time for maintenance and replacement of cutters and their cost. Most long tunnels in rock use hard-rock TBMs.

Soft Ground TBM ▪ Two main types of tunnel boring machine (TBM) are used in soft ground, a slurry TBM and an earth pressure balance (EPB) TBM. Both types operate with a sealed front compartment that is kept under sufficient pressure to stabilize the face and minimize ground movement. EPB TBMs have been limited to diameters under 33 ft due to the high torque needed to drive the rotating cutter head, although other forms of drive may overcome this limitation. Slurry TBMs have been built up to 50 ft diameter, and larger sizes are planned. Settlements at the surface in soft

Fig. 20.21 Preparing the Cairo Metro Shield for Assembly.

ground are directly related to the percentage loss of material outside the tunnel. Typical loss of material lies between 0.5% and 2.5%. Factors affecting the loss include the properties of the material traversed, the face pressure used, the design of the shield, and the rate of advance. Tunnels exist where the loss has been zero. Soft ground TBMs are generally launched from a relatively small shaft, with subsequent parts of the machine being added as progress is made.

The sealed front compartment of a slurry TBM is usually filled with a bentonite slurry held in equilibrium with the soil and groundwater pressures acting at the face. The equilibrium is often balanced by a compressed air reservoir and flow controls. The slurry also acts as a lubricant and holds loosened soil in suspension. The main disadvantage of a slurry TBM is that the slurry must be continuously circulated through a separator, often located on the surface, to remove the excavated material before returning the reconditioned slurry to the face. One main advantage is that the underground operators never come into contact with the excavated material. The slurry

TBM also has better face control, especially in mixed-face and bouldery ground. Many slurry TBMs also incorporate a boulder-crushing unit.

Soil excavated by the rotating cutter head of an EPM TBM falls behind the head and is removed by screw conveyor that discharges either onto a conveyor belt or directly into muck cars. The excavated material may be conditioned by the addition of water, clay or by biodegradable additives to assist in lubrication and to provide better pressure drop through the screw conveyor, thus preventing the direct exit of groundwater. The rate of advance must be closely coupled to the rate of advance to avoid excessive ground movement.

Flowing Ground ■ Tunneling machines have been developed for use in flowing ground, to confine the pressure to a small space between the face and a bulkhead behind the cutting wheel. They use a pressurized slurry composed of bentonite or of excavated material, to balance the pressure. The solids are settled out of the recirculated slurry, and their volume is accurately measured to determine the advance of the machine.

In very soft materials, the volume of material excavated must match the advance of the machine, or else sink holes may appear or mounds may be pushed up, in both cases causing unacceptably large ground movements.

20.20 Immersed Tunnels

Immersed tunnels should be considered for all water crossings. They are the shallowest form of tunnel, requiring only minimal protection against sinking ships and dropping anchors, with 5 ft of granular material and scour protection often being adequate. Because they are so shallow, they result in the shortest combined length of tunnel and approaches, and because they may serve their function better than other choices, the overall cost can be less. Approach gradients can also be flatter. Immersed tunnels can be constructed in ground conditions that would make bored tunneling difficult or expensive, such as the soft alluvial deposits characteristic of large river estuaries, but when rock has to be excavated under water, immersed tunneling may be less cost effective. Consequently, the ideal alignment for an immersed tunnel may not coincide with the ideal alignment for a bridge. Alignments do not need to be straight,

so immersed tunnels can be designed to suit design speeds, existing land uses, topography, and connections to existing road or rail systems.

(L.C.F. Ingerslev, "Water Crossings—the Options," *Tunneling and Underground Space Technology*, Vol. 13, No. 4, pp. 357–363, Elsevier Science Ltd., 1998.)

Immersed tunnels consist of very large precast concrete or concrete-filled steel tunnel elements. They are fabricated in convenient lengths on shipways, in drydocks, or in improvised floodable basins, sealed with bulkheads at each end, and then floated out. They may require outfitting at a pier close to their final destination before being towed to their final location, immersed, lowered into a prepared trench, and joined to previously placed tunnel elements. After any further foundation works have been completed as discussed below, immersed tunnels are backfilled and the bed reinstated (Fig. 20.22). Where intrusion into the water column is permitted, the final bed level may be higher than the original. The side slopes of the excavated trench depend upon the soil characteristics; often a slope of 1:1.5 is feasible under temporary conditions, although flatter grades and an allowance for possible sloughing may be required in softer materials.

Floating Tunnels ■ For particularly deep after-crossings at a number of locations, designs have been proposed for tunnels that remain totally exposed within the water column. These "floating" tunnels may be supported below the water surface on piers rising from the bed, unsupported if the distance between ends is short, supported from floating pontoons, or even held down by cables if positively buoyant. The text in Section 20.20 applies to floating tunnels as well as immersed tunnels, except for foundations and backfilling. Whereas immersed tunnels need only consider dynamic loads up to time that they are finally placed, floating tunnels have to be designed for dynamic loads throughout their life. They appear to be particularly attractive for deep narrow waterways, since the overall length of tunnel may be significantly shortened compared to other forms of tunnel. While most immersed tunnels are built for water depths of between 5 m and 20 m, concepts for 100 m depth have been prepared. Floating tunnels could avoid the need to be so

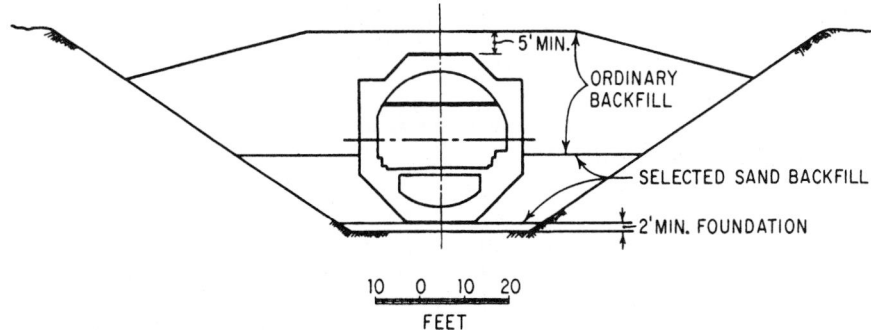

Fig. 20.22 Immersed tunnels are set in a trench, which is then backfilled.

deep, as long as risks from ship and submarine collision can be handled.

("International Tunneling Association Immersed and Floating Tunnels Working Group State-of-the-Art Report," Second Edition, Pergamon, April 1997.)

Internal Dimensions ▪ Highway design or client requirements should determine the required number of traffic lanes, tracks, or internal spaces. It is usual to avoid climbing lanes within immersed tunnels elements themselves, and nominal widths of emergency lanes or shoulders have almost always been used to minimize costs. If tunnels are particularly long, extra width may have to be provided at intervals to permit emergency stopping. Curbs, or more usually barriers, are provided to protect the walls from traffic impact. Barriers over 2 ft high may make the lane width seem narrower and slow motorists. Emergency access into an adjacent tunnel should be available, say at 300 ft intervals, which would require an emergency "walkway" at least 2 ft wide on top of the adjacent barrier. Such emergency "cross-passages" may need to be provided at intervals of say 100 meters. Extra space may be needed for tunnel and other utilities, construction and misalignment tolerances, lighting, lane signs, and highway signs, while keeping the clearance height as low as possible. Escape ducts, when provided, should be slightly over-pressurized relative to adjacent ducts to prevent entry of noxious fumes. With the minimum spaces determined, space allowances for any necessary ventilation system (such as for jet fans or additional ducts) can then be evaluated. The critical design case may be for moving or stalled

traffic, but a fuel fire usually governs. Permitted classes of vehicles may be restricted by legislation or owner requirements to limit it the potential size of the fire. Finally, additional air space may be needed for the tunnel elements to be able to just float when completed with bulkheads in place, and perhaps space for additional ballast, either inside or out, to stay submerged when completed and bulkheads removed. Floating tunnels that rely on buoyancy must have sufficient compartmentalized buoyancy to stay afloat in case of accidental damage to two adjacent compartments.

[L.C.F. Ingerslev et al, Chapter 15B Tunnel Structures, "Structural Engineering Handbook," 2000 Update for ENGnetBASE, Edited by Wai-Fah Chen and Lian Duan, CRC Press, 2000 (www.crcpress.com).]

Construction ▪ The technique of immersed tunneling is often less risky than bored tunneling, since tunnel element manufacture can be better controlled due to the construction of the elements in a controlled environment in the dry. As a result, immersed tunnels are nearly always much more watertight and therefore drier than bored tunnels.

Two main types of tunnel have emerged, known as steel and concrete. Steel tunnels use structural steel, usually in the form of stiffened plate, working compositely with the interior concrete, whereas concrete tunnels do not, relying on steel reinforcing bars or prestressing cables. The number of concrete tunnels is a almost twice that of steel tunnels. Steel tunnels can have a draft of as little as about 8 ft, whereas concrete tunnels have a draft of almost the full depth. Tunnel cross-sections may have flat sides or curved sides. Historically, concrete tunnels have been circular, or curved with a flat bottom, but the predominant shape has been rectangular (Fig. 20.23), which is particularly attractive for wide highways and combined road/rail tunnels. Steel tunnels have been circular, curved with a flat bottom, and rectangular (particularly in Japan), but the predominant shape in the past in the US has been a circular shell within an octagonal shape, with ventilation ducts above and below the roadway, either in single tube or binocular versions. This arrangement of ventilation ducts may change, since current techniques permit the use of longitudinal ventilation in much longer tunnels, often obviating the need for separate ventilation ducts. Steel tunnels can be categorized into three sub-types:

- Single shell, where the structural shell plate works compositely with the interior reinforced concrete and the shell plate requires corrosion protection (Fig. 20.25).

- Double shell, where the structural shell plate works compositely with interior reinforced concrete and is protected by external concrete placed within a non-structural form plate (Fig. 20.26). This shape has also been used in pairs for several tunnels, the most recent being Ted Williams in Boston, Massachusetts. Double shell tunnels are only found in the US.

- Sandwich, where structural steel plates, both internal and external, are connected by diaphragms and the internal space is filled with unreinforced self-compacting concrete.

Concrete Tunnels ▪ All but four of the concrete transportation tunnels built have been rectangular (Fig. 20.23). They are for road, rail, or for both road and rail. A number of other tunnels carrying pedestrians, utilities, sewage or water have also been built. Road and rail traffic are carried in separate ducts. Of all the road tunnels, only one carries four lanes in a single duct. All the others have three or less lanes per duct. In order to keep profiles as shallow as possible, any air ducts are usually located at the sides, rather than above or below the traffic duct. Because concrete tunnels are much heavier than steel tunnels when they are launched, they are usually constructed within dry docks or purpose-built casting basins (graving docks) capable of being flooded for removal of the elements. For many narrower tunnels, some form of catamaran lay barge has been used to support them during their immersion and placing, whereas some wider tunnels have had a pontoon placed on top near each end from which the tunnel was lowered. In most cases, watertightness has been assured by some form of exterior membrane, which itself may require protection. Ideally, the membrane should adhere to the concrete to limit the spread of any leakage through it. An outer steel membrane can yield and still remain watertight despite significant deformations.

An element is a length of tunnel that is floated and immersed as a single rigid unit. For a few Dutch tunnels and the Øresund tunnel (between Denmark and Sweden), the rigidity has been temporary and was later released, elements consisting of a number of discrete segments

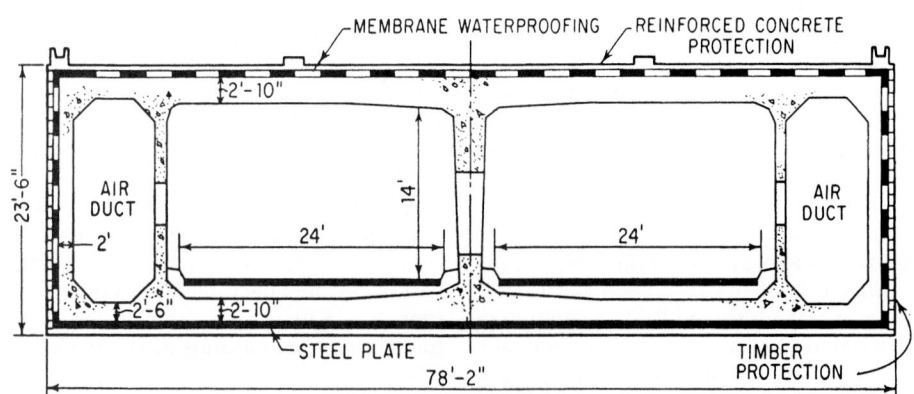

Fig. 20.23 Box-section concrete immersed tunnel (Deas Island Tunnel).

Fig. 20.24   Western Harbour Crossing, Hong kong.

stressed together longitudinally for ease of transportation and placing. After placing and release of the segments, each may act as a mini-element free to move at the segment joints. The ability to use discrete segments can depend upon subsurface conditions, acceptable displacements, and sufficient capacity to resist seismic effects.

Most tunnel elements are cast in bays, similar to segments, but continuous across the joints. Typically the floor slab is cast first. The walls and roof may be cast in either one or two operations. Special efforts must be made to reduce or preferably eliminate cracking in the concrete during fabrication. Testing and repair of leaks should be completed before submerging elements.

(L.C.F. Ingerslev, "Concrete Immersed Tunnels: The Design Process," Immersed Tunnel Techniques, The Institution of Civil Engineers, Telford, UK, 1989.)

Steel Single-Shell Tunnels ▪ Of the eight single-shell tunnels with some external curvature,

three are for rail in Tokyo, Japan, and three are for rail in the US including the unique two over two configuration of the 63rd Street Tunnel in Manhattan. The Baytown tunnel in Texas was circular with two lanes for highway, and the Cross Harbour Tunnel in Hong Kong is similar but binocular to give two lanes in each direction. The Detroit River tunnel (1910) and the Harlem River tunnel (1914) may not quite fit into this category, being the first two immersed transportation tunnels ever built, but do have similarities to single-shell tunnels and carry rail.

Figure 20.25 shows the cross-section of the San Francisco Bay Area Rapid Transit (BART) trans-bay tube with one track in each direction, separated by an exhaust air duct and a service passage. The 57 elements have a total length of about 19,000 ft. The steel shell is welded to make it continuous across the joints, as is the reinforced concrete lining, to provide security against major earthquake loading. At the landfall junctions with the ventilation

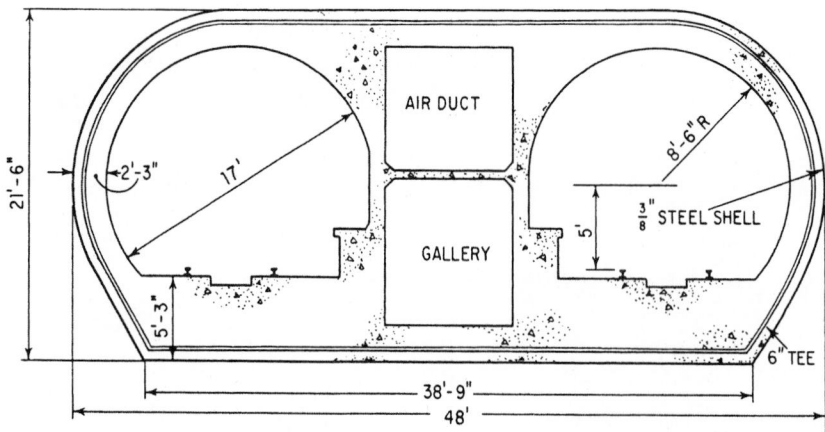

Fig. 20.25 Transbay Tube of the San Francisco Rapid Transit System (steel single shell).

buildings, special earthquake joints permit relative movement in all directions. The shell plate is protected against corrosion by a coating and a cathodic protection system.

All eight rectangular single-shell tunnels are located in Japan. They are very similar in layout to concrete tunnels, the difference being that the outer steel shell works compositely with the concrete, whereas even if an outer steel plate is present as a waterproofing membrane in a similar concrete tunnel, it is not considered in the strength design.

Steel Double-Shell Tunnels ▪ Fifteen of these tunnels have been built, of which five are binocular, the most recent being the Ted Williams Tunnel is Boston. Figure 20.26 shows a two-lane tunnel with a circular interior section. The steel shell plate was 31 ft diameter and about 300 ft long. Exterior diaphragms, approximately octagonal, are spaced about 15 ft apart, and longitudinal ribs of the bars and T sections stiffen the shell. Outside form plates were attached to the diaphragms and supported by angle struts extended from the shell stiffeners.

The tubes were erected on shipways. All welds were tested for watertightness with a compressed-air stream and soap solution. Before the ends were closed with welded watertight bulkheads, the reinforcing steel for the interior concrete lining was placed. The keel concrete in the space between the outside form plates and the bottom of the shell was cast before launching. The concrete lining and roadway slab were cast while the element was

floating at a fitting-out pier, by pumping concrete through hatches in the top of the shell into segmental steel forms. Pouring sequences were regulated to control increments of water pressure on the shell plate and longitudinal bending moments. The hatches were closed with welded plates, and the exterior concrete cap was cast and enough tremie concrete placed in the side pockets to reduce freeboard to about 1 ft. Most, if not all, of these tunnels have been immersed using a catamaran lay barge consisting of a barge each side of the tunnel element connected by two transverse beams from which the element is suspended (Fig. 20.28). Additional tremie concrete can then be added to give the required negative buoyancy for immersing, placing, and joining. More tremie concrete may be needed to achieve the final factor of safety against floating.

Figure 20.27 shows a combination of two such cylindrical sections for one of the two four-lane tubes of the Ft. McHenry Tunnel under Baltimore Harbor.

Steel Sandwich Tunnels ▪ Although postulated elsewhere many years earlier, steel sandwich tunnels have become a reality in Japan. Rectangular in shape, the principle behind this form of construction is that there is a steel skin inside and outside the tunnel, both acting compositely with the concrete between them. The plate is stiffened with flat and L-shaped ribs, and the interior is divided up into cells by diaphragms

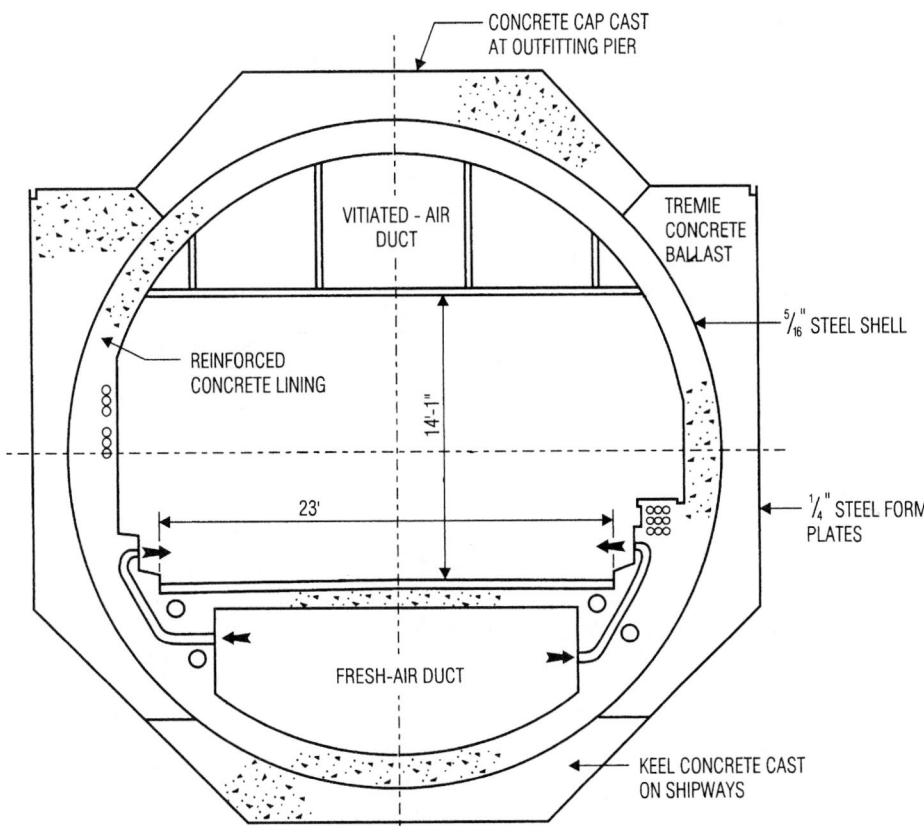

CONCRETE CAP CAST
AT OUTFITTING PIER

VITIATED - AIR
DUCT

TREMIE
CONCRETE
BALLAST

REINFORCED
CONCRETE LINING

$\frac{5}{16}"$ STEEL SHELL

14-1"

23'

$\frac{1}{4}"$ STEEL FORM
PLATES

FRESH-AIR DUCT

KEEL CONCRETE CAST
ON SHIPWAYS

Fig. 20.26 Cylindrical steel double-shell immersed tunnel (Hampton Roads Tunnel).

in two directions. Self-compacting non-shrink concrete is pumped into each cell though one hole while air is released through others. Osaka South Port and Kobe Port tunnels are being constructed by this method, the latter being the only steel tunnel so far to carry three lanes per duct.

Foundations ▪ There are two basic systems in use for supporting immersed tunnels on line and grade, a screeded foundation, and a pumped sand foundation. In addition, a few tunnels are founded on piles where soils are particularly soft or special conditions prevail. Such conditions can include earthquake where stone piles may help to dissipate excess pore water pressure and prevent soil liquefaction.

(LC.F. Ingerslev, "Immersed Tunnel Foundations," Comitato Organizzatore del Congresso,

"AITES-ITA 2001, World Tunnel Congress: Progress in Tunnelling after 2000," Proceedings pp 209–216, Milan, June 2001.)

With a screeded foundation (Fig. 20.22), the tunnel is founded on a leveled bed of sand or stone 2 ft to 3 ft thick, placed prior to the immersed tunnel. The leveling has been done by dragging either a heavy grid of steel beams or a steel box filled with the foundation material along the alignment, suspending them from a carriage on rails set parallel to the required grade. The material has also been placed in narrow passes transverse to the alignment using a pipe, the elevation of which was computer controlled.

For a pumped sand foundation, the tunnel is founded on a sand or mortar foundation of similar thickness, placed after the tunnel element is temporarily supported in place. The element can be set on two light pile bents that have been

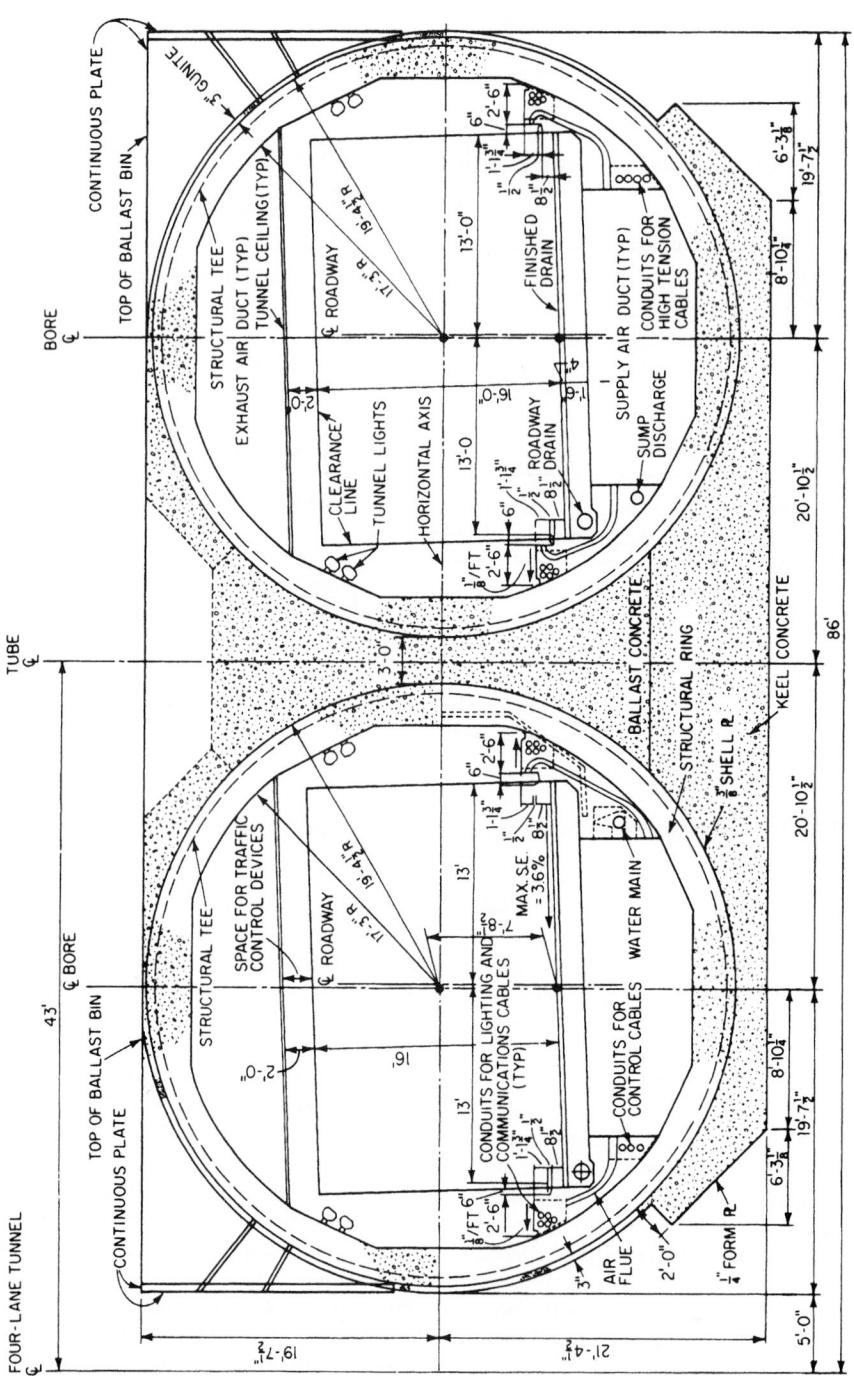

Fig. 20.27 Half of steel immersed tunnel combines two cylindrical sections (Ft. McHenry Tunnel).

constructed to the correct grade. The sand can be placed through movable pipes inserted and withdrawn from the sides beneath the structure (sand jetting), or through fixed pipes embedded with the structure (sand-flow) with connections either external on the roof or sides, or internal through valves.

One method of sand jetting, invented by the Danish firm of Christiani & Nielson, is particularly effective: A sand slurry is injected through a movable nozzle, and the surplus water is pumped off by another nozzle, the rolling motion depositing the sand in a compact layer. With sand flow, the rolling current deposits the sand around the discharge point in a firm circular layer (pancakes), often allowed to grow 20–26 ft radius. Discharge points are built into the underside of the element and are connected to pipes leading to convenient points at which the pumped sand supply may be connected.

As well as pile bents, elements have been temporarily supported by jacks, penetrating through the base of the section. The jacks bear on previously placed concrete blocks. By adjusting the jacks, the section is brought to exact grade. Then, the sand foundation course is flushed in.

Immersion, Placing and Joining ▪ Loosely termed the "sinking" operation, these three operations are performed with a high degree of control and therefore of accuracy. Immersion and lowering of each element is regulated by winches on special barges or pontoons, or by cranes, from which they are suspended. Alignment is controlled by instruments set on fixed points and sighting on targets mounted on temporary towers attached to the ends of the sections or by sonar to pre-installed targets to avoid using temporary towers. Steel-shell elements have historically been connected with short lengths of shell, which project beyond the end bulkheads. The gap between the ends was covered by hood plates extended from the lower and upper half of the shell extensions. Form plates were inserted into guides on the vertical edges of the bulkheads. The space around the joint was filled with tremie concrete as a preliminary seal. The inside of the joint was drained, and closure plates were welded to interior ribs of the shell extensions. Finally, the concrete lining was completed.

Making rigid immersion joints today using tremie concrete is unusual, with rubber-gasket joints being almost universally used. Flexible joints are generally sealed with a temporary immersion gasket or soft nosed gasket (Gina-type) in compression, attached to the end of one of the elements and mating with a flat steel face on the other. The use of a secondary independent flexible seal, capable of being replaced from within the tunnel, is common practice (often an omega-shaped seal). Each seal should be capable of resisting the external hydrostatic pressure and should allow for expected future movements. Jacks pull the tubes into contact to provide an initial seal. The joint is then drained, activating the full hydrostatic pressure on the opposite end of the tube. The pressure compresses the gaskets completely, providing a secure seal. Then, the bulkheads between the connected tubes can be opened and the joint completed from the inside.

Depending upon the construction sequence, the last element may need to be inserted in the remaining space, rather than appended to the end of the previous element. In order to achieve this, a small final gap will remain. This closure or final joint corresponds to a short length of tunnel that will need to be constructed in a special way. Methods used have include tremie concrete to seal a rigid joint, and for flexible joints:

- dewatering to complete the joint in the dry from the inside;

- terminal block where a short closure section is slid out from within one side until it meets the other and any remaining gap is closed with a rubber gasket in compression;

- wedge-shaped block dropped into the remaining gap until it is sealed against both sides.

Backfill ▪ Up to about half the height of the element, the trench is backfilled with well-graded self-compacting material to lock the elements securely into place. Ordinary backfill is placed to a depth of at least 5 ft over the top of the tunnel. If any part of the tunnel projects above the natural bottom, dikes should be built at least 50 ft away on both sides to a height of 5 ft above the tunnel. The space between the dikes should be filled with backfill, covered with a stone blanket to prevent scour where necessary.

Design ▪ Tunnel elements are designed as rigid structures to resist dead loads, live loads, exceptional loads and extreme loads. Dead load includes mean water level. Live load includes seabed erosion and siltation, and variations in water level, current, storm loads and earthquakes, each with a return period of 5 years or less. Exceptional loads include loss of support (subsidence) below the tunnel or to one side, and storms and extreme water levels with a probability of being exceeded once during the design life. Extreme loads include sunken ships, ship collision, water-filled tunnel, explosion (e.g. vehicular), fire, the design seismic event predicted for the location, and the resulting movement of soils. Conditions to be investigated should include normal, abnormal, extreme, and construction.

20.21 Shafts

In tunnel work, shafts are starting points for excavation in rock or firm material or shields. For long tunnels, such as aqueducts, several shafts are used to divide construction into shorter sections that can be worked simultaneously. For vehicular tunnels, especially subaqueous shield tunnels, the shafts are used as bases for ventilation buildings. In construction of shafts, regulations of the Occupational Safety and Health Administration should be observed (Arts. 20.6, 20.8, and 20.12 to 20.16).

Timber shafts are mined and braced in the same manner as tunnels in similar material. Usually, poling boards 5 to 6 ft long are driven into the ground and braced at regular intervals by rectangular timber frames. Then, the soil is excavated to the ends of the polings and a new frame installed at this level.

A relatively shallow shaft may be started oversize with sheeting 10 to 20 ft long driven vertically on the outside of the frame bracing. Intermediate frames are installed as the excavation proceeds. At the bottom of the tier of sheeting, the sides are stepped in to make room for the next tier of vertical sheeting.

In rock shafts, timbering is used to prevent loose rocks from falling off the walls. Its placement usually lags an appreciable distance behind the excavation.

Steel liner plates alone, or in combination with horizontal ribs, may be used in soft ground where excavation can be made in increments equal to the width of the liner plates. H beams driven vertically as soldier piles, with wood or steel lagging and horizontal bracing, may be used for rectangular

shafts. Enclosures of vertical steel sheetpiling, for round or rectangular shafts, are suitable for water-bearing ground.

Where ground conditions are poor and water-bearing, shafts may be constructed with a caisson (hollow box), with compressed air as needed to exclude water. Gravity pulls the caisson down as excavation proceeds. Since its weight is relatively small, the caisson may have to be temporarily ballasted or jetted for sinking. The depth to which a caisson may be sunk is limited by the high cost of compressed-air work, which results from the short working hours permitted under high pressure.

Open-bottom shafts with heavy walls, often circular or subdivided into compartments, may be built on the ground and sunk by excavating the ground underneath. In dry soil, the excavation may be done directly; if water is present, clamshell buckets and high-pressure jets may be used to loosen the soil and remove it. On reaching the proper depth, the bottom of the shaft is closed by tremie concrete.

As an alternative method for shaft construction, water-bearing ground may be frozen in a circular ring around the shaft location and the excavation made in the dry. Closed-end pipes are driven vertically into the ground around the periphery, and open-end smaller pipes inserted into them. A refrigerant, usually brine, is circulated at temperatures as low as $-30\,°F$ from the interconnected inner pipes into the larger ones and from them returned to the refrigeration plant. Several months may be required to freeze a deep ring solidly. The ventilation shaft of the Scheldt River Tunnel in Antwerp was built in this manner, as were a number of mine shafts in Germany and France.

(J. O. Bickel and T. R. Kuesel, "Tunnel Engineering Handbook," Van Nostrand Reinhold Company, New York.)

20.22 Seismic Analysis and Design

Earthquake loads, or more correctly seismic loads, are included among the loads on a structure that are required to be considered by most current design codes. Seismic effects can occur during the construction phase and should therefore also be considered during that period; an appropriate level of risk should be agreed with the owner. The seismic hazard at a tunnel site can be quantified by a project-specific seismic hazard assessment. Typically, a functional evaluation earthquake (FEE), likely to occur not more than once during the design life, is used first to design the structure for either limited or full performance following a seismic event, as agreed with the owner. Next as appropriate, either the safety evaluation earthquake (SEE) or the maximum credible earthquake (MCE), both concerned not only with life safety but also with the survivability of the structure under the most severe seismic event considered at the location, is checked to ensure compliance with minimum performance. If necessary, the strength of some parts of the structure may have to be enhanced to comply.

Structures buried in soil are generally constrained to follow the seismic deformations of the ground in which they are located. The stiffness of the tunnel is generally small relative to the soil, so that in all but soft soils, tunnel deformations will approximate to ground deformations, a conservative assumption. For a tunnel in rock, tunnel deformations will match those of the rock, but in softer soils, the tunnel will resist soil pressures. The response of the tunnel to the free-field soil displacements (as if the tunnel were absent) will depend upon both the stiffness of the tunnel and that of the soil. While the complex seismic analyses may be solved numerically using computers, some simplified procedures have been published. Simplified beam-on-elastic-foundation analysis can also be used to account for the soil-structure interaction effects of soil deformations, especially in soft soils. Horizontal shear S-waves, depending upon the angle of approach, cause transverse bending or axial waves and produce the largest strains that are usually governing. Compression P-waves should also be considered. At sites where there are deep deposits of soil, Rayleigh R-waves may govern the induced strains. Racking (ovaling)

Fig. 20.28 Lay Barge.

deformations in the plane of the cross-section can occur in tunnels, but not usually in vertical shafts, and is caused primarily by seismic waves propagating perpendicular to the tunnel longitudinal axis. Vertically propagating shear waves are generally considered the most critical type of waves for this mode of deformation. Axial and curvature deformations are induced by components of seismic waves that propagate along the longitudinal axis.

The effects of a seismic event on a tunnel as a whole can be integrated to give an effective acceleration at the tunnel location, expressed as a seismic coefficient times the acceleration due to gravity. Three seismic coefficients are usually obtained, for longitudinal, lateral and vertical effects. Internal elements, not in contact with the soil and with a natural frequency approaching that of the seismic waves, may need to be designed to substantially larger seismic coefficients. The vertical seismic coefficient can be reasonably assumed to be two-thirds of the design peak horizontal acceleration divided by the gravity.

Special precautions are needed for tunnels in soils that might liquefy or slip, especially so if crossing active faults. Liquefaction may cause tunnels to float up. Since it is virtually impossible to design against these conditions, the best policy is either to improve or replace the soil in question or to avoid it. Faults are best avoided, but if that

Fig. 20.29 TARP Calumet Shaft.

is not an option, then the tunnel must be designed to accommodate expected displacements, and perhaps the surrounding material may also need to permit relative movement of the tunnel. Locations of especially critical design in a tunnel are at changes of inertia and soil properties (where there will be contrasting responses), and at joints that might open up and cause flooding. Ductility in structures is particularly important for structures to survive and for life safety.

(Wang, J, "Seismic Design of Tunnels—A Simple State-of-the-Art Design Approach,"

Parsons Brinckerhoff Monograph No. 7, 1993.) (St. John, C. M., and Zahrah, T. F., Aseismic Design of Underground Structures, *Tunneling and Underground Space Technology*, Vol. 2, No. 2, 1987.) (Chapter 15B Tunnel Structures, "Structural Engineering Handbook," 2000 Update for ENGnetBASE, Edited by Wai-Fah Chen and Lian Duan, CRC Press, 2000 (www.crcpress.com).) (Earthquake Analysis, L. C. F. Ingerslev and O. Kiyomiya, International Tunneling Association Immersed and Floating Tunnels Working Group, "State-of-the Art Report," Second Edition, Pergamon, April 1997.)

21

M. Kent Loftin
President
Synint, Inc.
Hobe Sound, Florida

WATER RESOURCES ENGINEERING[*]

Water resources engineering is concerned with the protection, development, and efficient management of water resources for beneficial purposes. It involves planning, design, and construction of projects for supply of water for domestic, commercial, public, and industrial purposes, flood protection, hydroelectric power, control of rivers and water runoff, and conservation of water resources, including prevention of pollution.

Water resources engineering primarily deals with water sources, collection, flow control, transmission, storage, and distribution. For efficient management of these aspects, water resources engineers require a knowledge of fluid mechanics; hydraulics of pipes, culverts, and open channels; hydrology; water demand, quality requirements, and treatment; production of water from wells, lakes, rivers, and seas; transmission and distribution of water supplies; design of reservoirs and dams; and production of hydroelectric power. These subjects are addressed in the following articles.

21.1 Dimensions and Units

A list of symbols and their dimensions used in this section is given in Table 21.1. Table 21.2 lists conversion factors for commonly used quantities, including the basic equivalents between the English and metric systems. For additional conversions to the metric system (SI) of units, see the appendix.

Fluid Mechanics

Fluid mechanics describes the behavior of water under various static and dynamic conditions. This theory, in general, has been developed for an ideal liquid, a frictionless, inelastic liquid whose particles follow smooth flow paths. Since water only approaches an ideal liquid, empirical coefficients and formulas are used to describe more accurately the behavior of water. These empiricisms are intended to compensate for all neglected and unknown factors.

The relatively high degree of dependence on empiricism, however, does not minimize the importance of an understanding of the basic theory. Since major hydraulic problems are seldom identical to the experiments from which the empirical coefficients were derived, the application of fundamentals is frequently the only means available for analysis and design.

21.2 Properties of Fluids

Specific weight or **unit weight** w is defined as weight per unit volume. The specific weight of water varies from 62.42 lb/ft^3 at 32 °F to 62.22 lb/ft^3

[*] Revised and updated from Sec. 21, Water Engineering, by Samuel B. Nelson, in the third edition.

Table 21.1 Symbols, Dimensions, and Units Used in Water Engineering

Symbol	Terminology	Dimensions	Units
A	Area	L^2	ft^2
C	Chezy roughness coefficient	$L^{1/2}/T$	$ft^{1/2}/s$
C_1	Hazen-Williams roughness coefficient	$L^{0.37}/T$	$ft^{0.37}/s$
d	Depth	L	ft
d_c	Critical depth	L	ft
D	Diameter	L	ft
E	Modulus of elasticity	F/L^2	psi
F	Force	F	lb
g	Acceleration due to gravity	L/T^2	ft/s^2
H	Total head, head on weir	L	ft
h	Head or height	L	ft
h_f	Head loss due to friction	L	ft
L	Length	L	ft
M	Mass	FT^2/L	$lb \cdot s^2/ft$
n	Manning's roughness coefficient	$T/L^{1/3}$	$s/ft^{1/3}$
P	Perimeter, weir height	L	ft
P	Force due to pressure	F	lb
p	Pressure	F/L^2	psf
Q	Flow rate	L^3/T	ft^3/s
q	Unit flow rate	$L^3/T \cdot L$	$ft^3/(s \cdot ft)$
r	Radius	L	ft
R	Hydraulic radius	L	ft
T	Time	T	s
t	Time, thickness	T, L	s, ft
V	Velocity	L/T	ft/s
W	Weight	F	lb
w	Specific weight	F/L^3	lb/ft^3
y	Depth in open channel, distance from solid boundary	L	ft
Z	Height above datum	L	ft
ε	Size of roughness	L	ft
μ	Viscosity	FT/L^2	$lb \cdot s/ft$
ν	Kinematic viscosity	L^2/T	ft^2/s
ρ	Density	FT^2/L^4	$lb \cdot s^2/ft^4$
σ	Surface tension	F/L	lb/ft
τ	Shear stress	F/L^2	psi

Symbols for dimensionless quantities

Symbol	Quantity
C	Weir coefficient, coefficient of discharge
C_c	Coefficient of contraction
C_v	Coefficient of velocity
F	Froude number
f	Darcy-Weisbach friction factor
K	Head-loss coefficient
R	Reynolds number
S	Friction slope—slope of energy grade line
S_c	Critical slope
η	Efficiency
Sp. gr.	Specific gravity

Table 21.2 Conversion Table for Commonly Used Quantities

Area	Discharge
1 acr- e = 43,56- 0 ft²	1 ft³/s = 449 gal/min = 646,000 gal/day 1 ft³/s = 1.98 acre-ft/day = 724 acre-ft/year 1 ft³/s = 50 miner's inches in Idaho, Kansas, Nebraska, New Mexico, North Dakota, and South Dakota
1 m- i² = 640 a- cres	1 ft³/s = 40 miner's inches in Arizona, California, Montana, and Oregon
	1 MGD* = 3.07 acre-ft/day = 1120 acre-ft/year 1 MGD* = 1.55 ft³/s = 694 gal/min 1 million acre-ft/year = 1380 ft³/s

Volume

1 ft³ = 7.4805 gal
1 acre-ft = 325,850 gal
1 MG = 3.0689 acre-ft   Power
1 hp = 550 ft · lb/s
1 hp = 0.746 kW
1 hp = 6535 kWh/year Weight of water Pressure 1 ft³ weighs 62.4 lb 1 psi = 2.31 ft of water 1 gal weighs 8.34 lb = 51.7 mm of mercury 1 in of mercury = 1.13 ft of water 1 ft of water = 0.433 psi 1 atm† = 29.9 in of mercury = 14.7 psi Metric equivalents

1 ft = 0.3048 m	Area:
m²	Volume:
L	1 m³ = 264.17
gal	Weight: 1 lb = 0.4536 kg

Length:
1 acre = 4046.9
1 gal = 3.7854

* Prefix M indicates million; for example, MG = million gallons.
† atm indicates atmospheres.

at 80 °F but is commonly taken as 62.4 lb/ft³ for the majority of engineering calculations. The specific weight of sea water is about 64.0 lb/ft³.

Density ρ is defined as mass per unit volume and is significant in all flow problems where acceleration is important. It is obtained by dividing the specific weight w by the acceleration due to gravity g. The variation of g with latitude and altitude is small enough to warrant the assumption that its value is constant at 32.2 ft/s² in hydraulics computations.

The **specific gravity** of a substance is the ratio of its density at some temperature to that of pure water at 68.2 °F (20 °C).

Modulus of elasticity E of a fluid is defined as the change in pressure intensity divided by the corresponding change in volume per unit volume. Its value for water is about 300,000 psi, varying slightly with temperature. The modulus of elasticity of water is large enough to permit the assumption that it is incompressible for all hydraulics problems except those involving water hammer (Art. 21.13).

Surface tension and **capillarity** are a result of the molecular forces of liquid molecules. **Surface tension** σ is due to the cohesive forces between liquid molecules. It shows up as the apparent skin that forms when a free liquid surface is in contact with another fluid. It is expressed as the force in the liquid surface normal to a line of unit length drawn in the surface. Surface tension decreases with increasing temperature and is also dependent on the fluid with which the liquid surface is in contact. The surface tension of water at 70 °F in contact with air is 0.00498 lb/ft.

Capillarity is due to both the cohesive forces between liquid molecules and adhesive forces of liquid molecules. It shows up as the difference in liquid surface elevations between the inside and outside of a small tube that has one end submerged in the liquid. Since the adhesive forces of water molecules are greater than the cohesive forces between water molecules, water wets a surface and rises in a small tube, as shown in Fig. 21.1. Capillarity

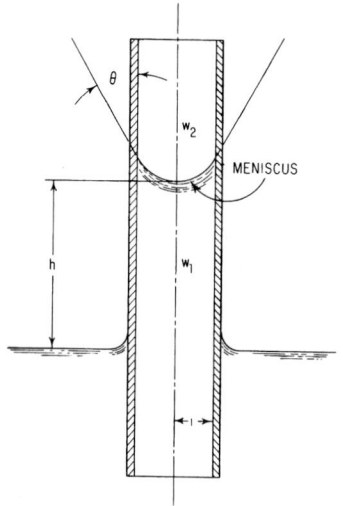

Fig. 21.1 Capillary action raises water in a small-diameter tube. Meniscus, or liquid surface, is concave upward.

is commonly expressed as the height of this rise. In equation form,

$$h = \frac{2\sigma \cos \theta}{(w_1 - w_2)r} \qquad (21.1)$$

where h = capillary rise, ft

σ = surface tension, lb/ft

w_1 and w_2 = specific weights of fluids below and above meniscus, respectively, lb/ft

θ = angle of contact

r = radius of capillary tube, ft

Capillarity, like surface tension, decreases with increasing temperature. Its temperature variation, however, is small and insignificant in most problems.

Surface tension and capillarity, although negligible in many water engineering problems, are significant in others, such as capillary rise and flow of liquids in narrow spaces, formation of spray from water jets, interpretation of the results obtained on small models, and freezing damage to concrete.

Atmospheric pressure is the pressure due to the weight of the air above the earth's surface. Its value at sea level is 2116 psf or 14.7 psi. The variation in atmospheric pressure with elevation from sea level to 10,000 ft is shown in Fig. 21.2. **Gage pressure**, psi,

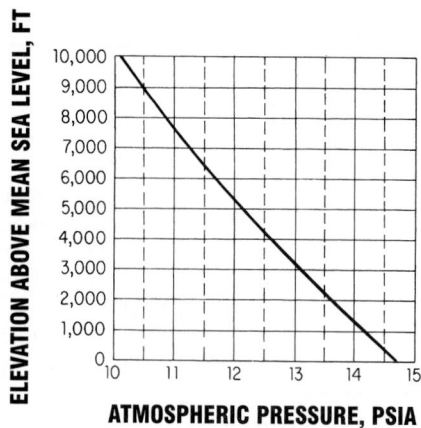

Fig. 21.2 Atmospheric pressure decreases with elevation above mean sea level. The curve is based on the ICAO standard atmosphere.

is pressure above or below atmospheric. **Absolute pressure**, psia, is the total pressure including atmospheric pressure. Thus, at sea level, a gage pressure of 10 psi is equivalent to 24.7 psia. Gage pressure is positive when pressure is greater than atmospheric and is negative when pressure is less than atmospheric.

Vapor pressure is the partial pressure caused by the formation of vapor at the free surface of a liquid. When the liquid is in a closed container, the partial pressure due to the molecules leaving the surface increases until the rates at which the molecules leave and reenter the liquid are equal. The vapor pressure at this equilibrium condition is called the saturation pressure. Vapor pressure increases with increasing temperature, as shown in Fig. 21.3.

Cavitation occurs in flowing liquids at pressures below the vapor pressure of the liquid. Cavitation is a major problem in the design of pumps and turbines since it causes mechanical vibrations, pitting, and loss of efficiency through gradual destruction of the impeller. The cavitation phenomenon may be described as follows:

Because of low pressures, portions of the liquid vaporize, with subsequent formation of vapor cavities. As these cavities are carried a short distance downstream, abrupt pressure increases force them to collapse, or implode. The implosion and ensuing inrush of liquid produce regions of very high pressure, which extend into the pores of

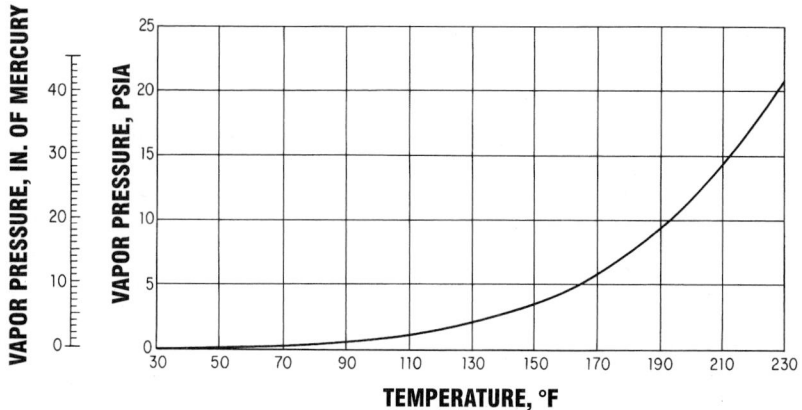

Fig. 21.3 Vapor pressure of water increases rapidly with temperature.

the metal. (Pressures as high as 350,000 psi have been measured in the collapse of vapor cavities by the Fluid Mechanics Laboratory at Stanford University.) Since these vapor cavities form and collapse at very high frequencies, weakening of the metal results as fatigue develops, and pitting appears.

Cavitation may be prevented by designing pumps and turbines so that the pressure in the liquid at all points is always above its vapor pressure.

Viscosity, μ of a fluid, also called the **coefficient of viscosity, absolute viscosity,** or **dynamic viscosity,** is a measure of its resistance to flow. It is expressed as the ratio of the tangential shearing stresses between flow layers to the rate of change of velocity with depth:

$$\mu = \frac{\tau}{dV/dy} \qquad (21.2)$$

where τ = shearing stress, lb/ft^2

V = velocity, ft/s

y = depth, ft

Viscosity decreases as temperature increases but may be assumed independent of changes in pressure for the majority of engineering problems. Water at 70 °F has a viscosity of 0.00002050 lb · s/ft^2.

Kinematic viscosity ν is defined as viscosity μ divided by density ρ. It is so named because its units, ft^2/s, are a combination of the kinematic units of length and time. Water at 70 °F has a kinematic viscosity of 0.00001059 ft^2/s.

In hydraulics, viscosity is most frequently encountered in the calculation of Reynolds number (Art. 21.8) to determine whether laminar, transitional, or completely turbulent flow exists.

21.3 Fluid Pressures

Pressure or **intensity of pressure** p is the force per unit area acting on any real or imaginary surface within a fluid. Fluid pressure acts normal to the surface at all points. At any depth, the pressure acts equally in all directions. This results from the inability of a fluid to transmit shear when at rest. Liquid and gas pressures differ in that the variation of pressure with depth is linear for a liquid and nonlinear for a gas.

Hydrostatic pressure is the pressure due to depth. It may be derived by considering a submerged rectangular prism of water of height Δh, ft, and cross-sectional area A, ft^2, as shown in Fig. 21.4. The boundaries of this prism are imaginary. Since the prism is at rest, the summation of all forces in both the vertical and horizontal directions must be zero. Let w equal the specific weight of the liquid, lb/ft^3. Then, the forces acting in the vertical direction are the weight of the prism $wA\Delta h$, the force due to pressure p_1, psf, on the top surface, and the force due to pressure p_2, psf, on the bottom surface. Summing these vertical forces and setting the total equal to zero yields

$$p_2 A - wA\Delta h - p_1 A = 0 \qquad (21.3a)$$

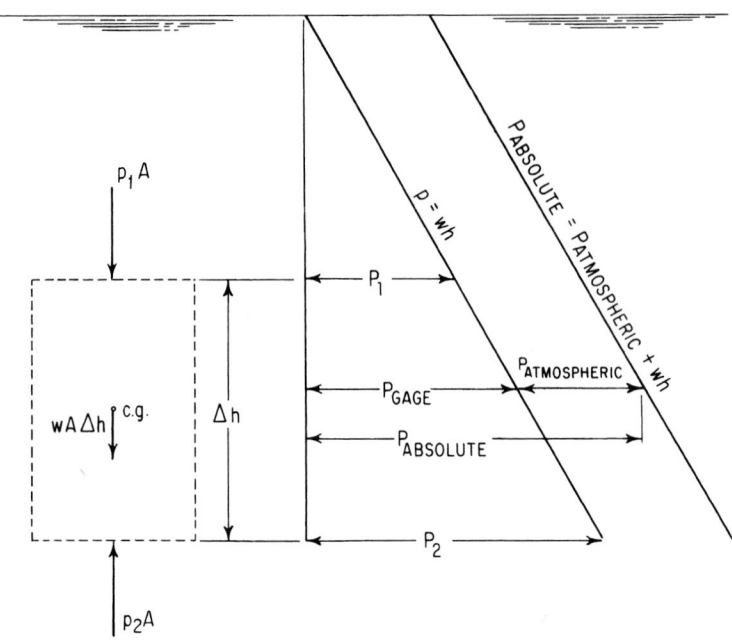

Fig. 21.4 Hydrostatic pressure varies linearly with depth.

Division of Eq. (21.3a) by A yields

$$p_2 = w\Delta h + p_1 \qquad (21.3b)$$

For the special case where the top of the prism coincides with the water surface, p_1 is atmospheric pressure. Since most hydraulics problems involve gage pressure, p_1 is zero (gage pressure is zero at atmospheric pressure). Taking Δh to be h, the depth below the water surface, ft, then p_2 is p, the pressure, psf, at depth h. Equation (21.3b) then becomes

$$p = wh \quad h = \frac{p}{w} \qquad (21.4)$$

Equation (21.4) gives the depth of water h of specific weight w required to produce a gage pressure p. By adding atmospheric pressure p_a to Eq. (21.4), absolute pressure p_{ab} is obtained as shown in Fig. 21.4. Thus,

$$p_{ab} = wh + p_a \qquad (21.5)$$

21.3.1 Pressures on Submerged Plane Surfaces

This is important in the design of weirs, dams, tanks, and other water control structures. For horizontal surfaces, the pressure-force determination is a simple matter since the pressure is constant. For determination of the pressure force on inclined or vertical surfaces, however, the summation concepts of integral calculus must be used.

Figure 21.5 represents any submerged plane surface of negligible thickness inclined at an angle θ with the horizontal. The resultant pressure force P, lb, acting on the surface is equal to $\int p \, dA$. Since $p = wh$ and $h = y \sin \theta$, where w is the specific weight of water, lb/ft^3,

$$P = w \int y \sin \theta \, dA \qquad (21.6)$$

Equation (21.6) can be simplified by setting $\int y \, dA = \bar{y}A$, where A is the area of the submerged surface, ft^2; and $\bar{y} \sin \theta = \bar{h}$, the depth of the centroid, ft. Therefore,

$$P = w\bar{h}A = p_{cg}A \qquad (21.7)$$

where p_{cg} is the pressure at the centroid, psf.

The point on the submerged surface at which the resultant pressure force acts is called the **center of pressure** (c.p.). It is below the center of gravity

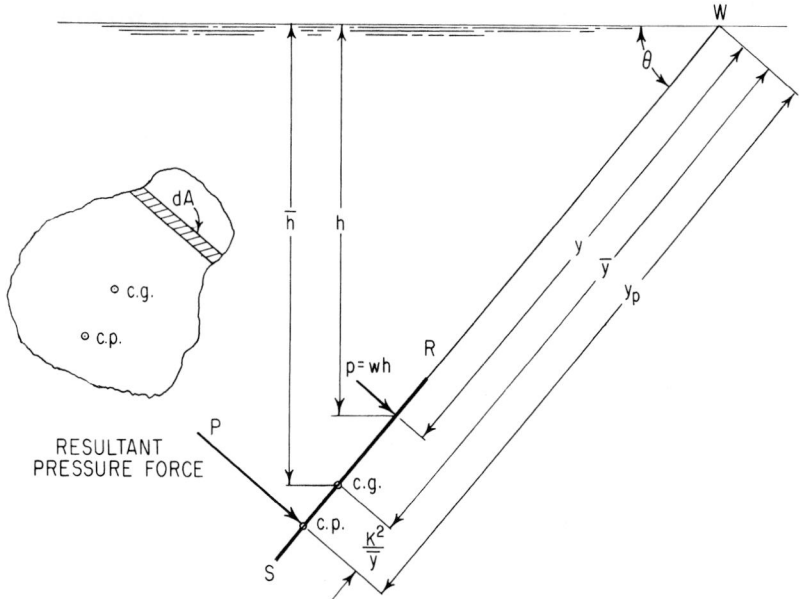

Fig. 21.5 Total pressure on submerged plane surface depends on pressure at the center of gravity (c.g.) but acts at a point (c.p.) that is below the c.g.

because the pressure intensity increases with depth. The location of the center of pressure, represented by the length y_p, is calculated by summing the moments of the incremental forces about an axis in the water surface through point W (Fig. 21.5). Thus, $Py_p = \int y\, dP$. Since $dP = wy \sin\theta\, dA$ and $P = w\int y \sin\theta\, dA$,

$$y_p = \frac{\int y^2\, dA}{\int y\, dA} \tag{21.8}$$

The quantity $\int y^2\, dA$ is the moment of inertia of the area about the axis through W. It also equals $AK^2 + A\bar{y}^2$, where K is the radius of gyration, ft, of the surface about its centroidal axis. The denominator of Eq. (21.8) equals $\bar{y}A$. Hence

$$y_p = \bar{y} + \frac{K^2}{\bar{y}} \tag{21.9}$$

and $K^2/\bar{y}$ is the distance between the centroid and center of pressure.

Values of K^2 for some common shapes are given in Fig. 21.6 (see also Fig. 6.29). For areas for which radius of gyration has not been determined, y_p may be calculated directly from Eq. (21.8).

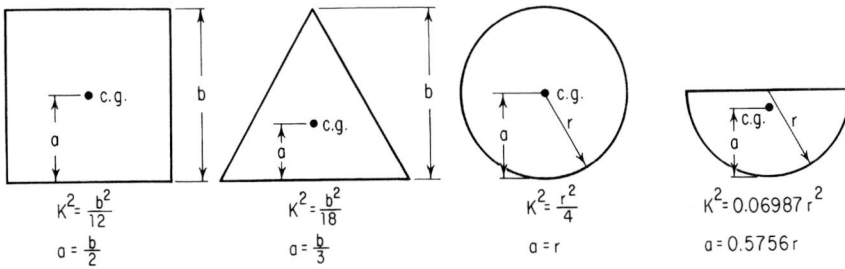

Fig. 21.6 Radius of gyration and location of centroid (c.g.) of common shapes.

The horizontal location of the center of pressure may be determined as follows: It lies on the vertical axis of symmetry for surfaces symmetrical about the vertical. It lies on the locus of the midpoints of horizontal lines located on the submerged surface, if that locus is a straight line. Otherwise, the horizontal location may be found by taking moments about an axis perpendicular to the one through W in Fig. 21.5 and lying in the plane of the submerged surface.

Example 21.1: Determine the magnitude and point of action of the resultant pressure force on a 5-ft-square sluice gate inclined at an angle θ of 53.2° to the horizontal (Fig. 21.7).

From Eq. (21.7), the total force $P = w\bar{h}A$, with

$$\bar{h} = [2.5 + \tfrac{1}{2}(5)]\sin 53.2° = 5.0 \times 0.8 = 4.0\,\text{ft}$$

$$A = 5 \times 5 = 25\,\text{ft}^2$$

Thus, $P = 62.4 \times 4 \times 25 = 6240\,\text{lb}$. From Eq. (21.9), its point of action is a distance $y_p = \bar{y} + K^2/\bar{y}$ from point G, and $\bar{y} = 2.5 + \tfrac{1}{2}(5.0) = 5.0\,\text{ft}$. Also, $K^2 = b^2/12 = 5^2/12 = 2.08$. Therefore, $y_p = 5.0 + 2.08/5 = 5.0 + 0.42 = 5.42\,\text{ft}$.

21.3.2 Pressure on Submerged Curved Surfaces

The resultant pressure force on submerged curved surfaces cannot be calculated from the equations developed for the pressure force on submerged plane surfaces because of the variation in direction of the pressure force. The resultant pressure force can be calculated, however, by determining its horizontal and vertical components and combining them vectorially.

A typical configuration of pressure on a submerged curved surface is shown in Fig. 21.8. Consider *ABC* a 1-ft-thick prism and analyze it as a free body by the principles of statics. Note:

1. The horizontal component P_H of the resultant pressure force has a magnitude equal to the pressure force on the vertical projection *AC* of the curved surface and acts at the centroid of pressure diagram *ACDE*.

2. The vertical component P_V of the resultant pressure force has a magnitude equal to the sum of the pressure force on the horizontal projection *AB* of the curved surface and the weight of the water vertically above *ABC*. The horizontal

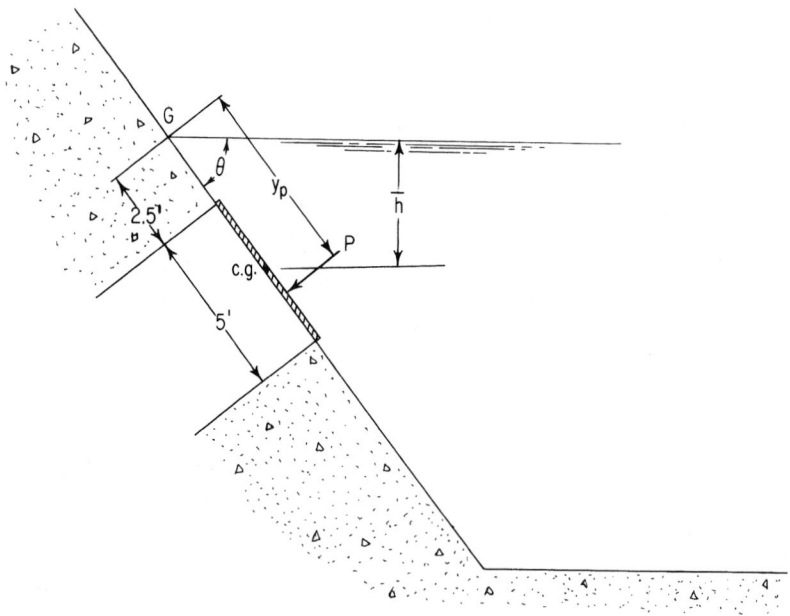

Fig. 21.7 Sluice gate (crosshatched) is subjected to hydrostatic pressure. (See Example 21.1.)

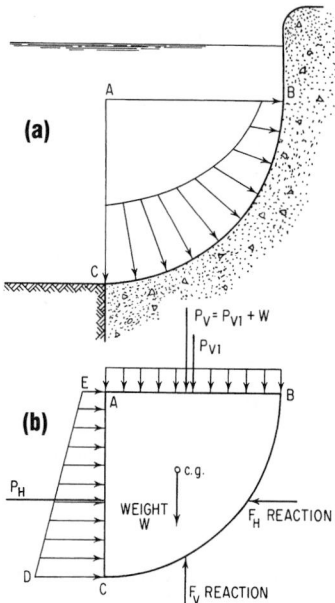

(a)

(b)

Fig. 21.8 Hydrostatic pressure on a submerged curved surface. (*a*) Pressure variation over the surface. (*b*) Free-body diagram.

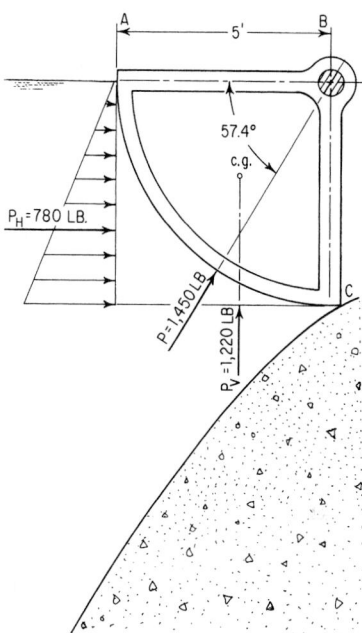

Fig. 21.9 Taintor gate has submerged curved surface under pressure. Vertical component of pressure acts upward. (See Example 21.2.)

location of the vertical component is calculated by taking moments of the two vertical forces about point C.

When water is below the curved surface, such as for a taintor gate (Fig. 21.9), the vertical component P_V of the resultant pressure force has a magnitude equal to the weight of the imaginary volume of water vertically above the surface. P_V acts upward through the center of gravity of this imaginary volume.

Example 21.2: Calculate the magnitude and direction of the resultant pressure on a 1-ft-wide strip of the semicircular taintor gate in Fig. 21.9.

The magnitude of the horizontal component P_H of the resultant pressure force equals the pressure force on the vertical projection of the taintor gate. From Eq. (21.7), $P_H = wh̄A = 62.4 \times 2.5 \times 5 = 780$ lb.

The magnitude of the vertical component of the resultant pressure force equals the weight of the imaginary volume of water in the prism ABC above the curved surface. The volume of this prism is $\pi R^2/4 = 3.14 \times 25/4 = 19.6$ ft^3, so the weight of the water is $19.6w = 19.6 \times 62.4 = 1220$ lb $= P_V$.

The magnitude of the resultant pressure force equals

$$P = \sqrt{P_H^2 + P_V^2} = \sqrt{780^2 + 1220^2} = 1450 \text{ lb}$$

The tangent of the angle the resultant pressure force makes with the horizontal $= P_V/P_H = 1220/780 = 1.564$. The corresponding angle is 57.4°.

The positions of the horizontal and vertical components of the resultant pressure force are not required to find the point of action of the resultant. Its angle with the horizontal is known, and for a constant-radius surface, the resultant must act perpendicular to the surface.

21.4 Submerged and Floating Bodies

The principles of buoyancy govern the behavior of submerged and floating bodies and are important in determining the stability and draft of cargo vessels.

The buoyant force acting on a submerged body equals the weight of the volume of liquid displaced.

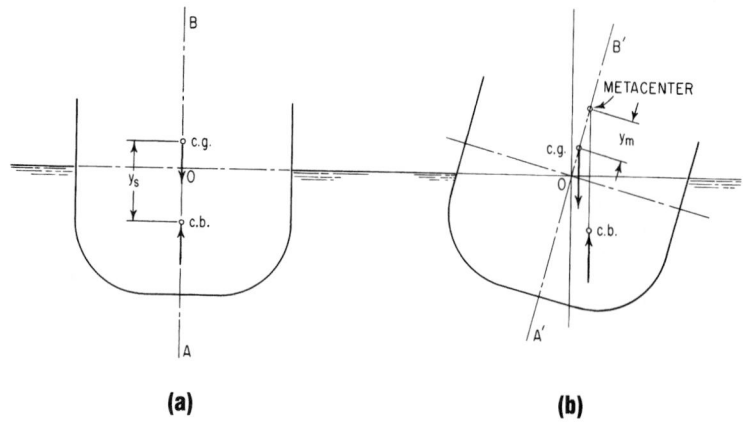

Fig. 21.10   Stability of a ship depends on the location of its metacenter relative to its center of gravity (c.g.).

A floating body displaces a volume of liquid equal to its weight.

The buoyant force acts vertically through the center of buoyancy c.b., which is located at the center of gravity of the volume of liquid displaced.

For a body to be in equilibrium, whether floating or submerged, the center of buoyancy and center of gravity must be on the same vertical line *AB* (Fig. 21.10a). The stability of a ship, its tendency not to overturn when it is in a nonequilibrium position, is indicated by the *metacenter*. It is the point at which a vertical line through the center of buoyancy intersects the rotated position of the line through the centers of gravity and buoyancy for the equilibrium condition *A′B′* (Fig. 21.10b). The ship is stable only if the metacenter is above the center of gravity since the resulting moment for this condition tends to right the ship.

The distance between the ship's metacenter and center of gravity is called the *metacentric height* and is designated by y_m in Fig. 21.10b. Given in feet by Eq. (21.10) y_m is a measure of degree of stability or instability of a ship since the magnitudes of moments produced in a roll are directly proportional to this distance.

$$y_m = \frac{I}{V} \pm y_s \qquad (21.10)$$

where I = moment of inertia of ship's cross section at waterline about longitudinal axis through 0, ft^4

V = volume of displaced liquid, ft^3

y_s = distance, ft, between centers of buoyancy and gravity when ship is in equilibrium

The negative sign should be used when the center of gravity is above the center of buoyancy.

21.5   Manometers

A manometer is a device for measuring pressure. It consists of a tube containing a column of one or two liquids that balances the unknown pressure. The basis for the calculation of this unknown pressure is provided by the height of the liquid column. All manometer problems may be solved with Eq. (21.4), $p = wh$. Manometers indicate h, the **pressure head,** or the difference in head.

The primary application of manometers is measurement of relatively low pressures, for which aneroid and Bourdon gages are not sufficiently accurate because of their inherent mechanical limitations. However, manometers may also be used in precise measurement of high pressures by arranging several U-tube manometers in series (Fig. 21.12c). Manometers are used for both static and flow applications, although the latter is most common.

Three basic types are used (shown in Fig. 21.11): piezometer, U-tube manometer, and differential manometer. Following is a brief discussion of the basic types.

The **piezometer** (Fig. 21.11a) consists of a tube with one end tapped flush with the wall of the

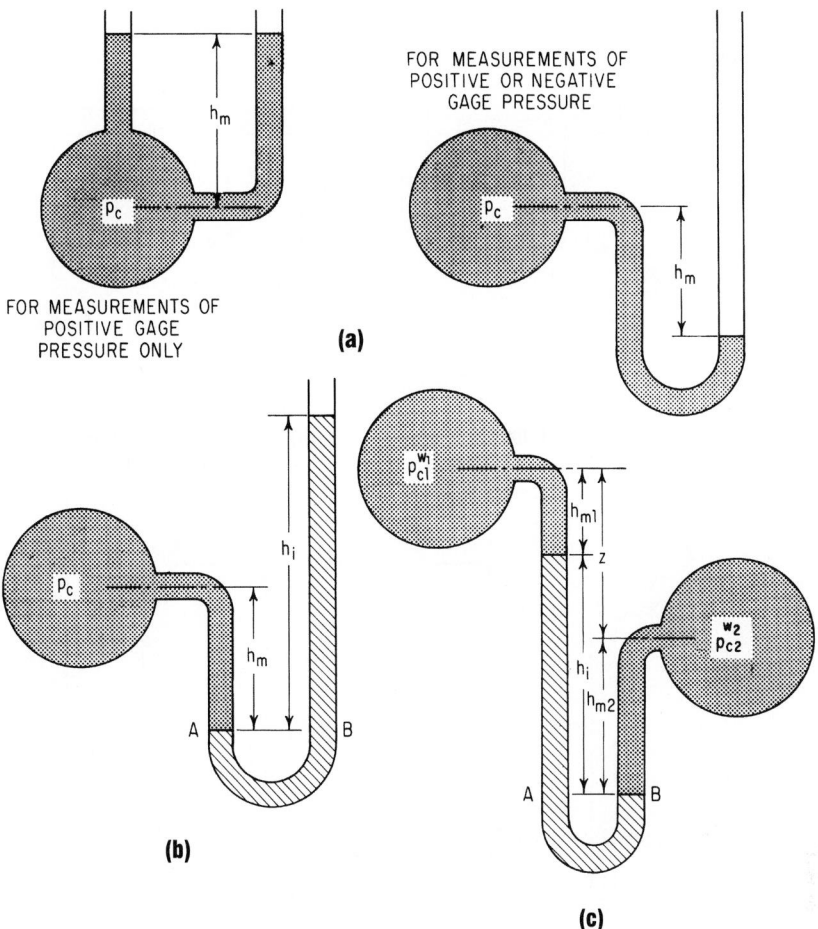

Fig. 21.11 Basic types of manometers. (*a*) Piezometers; (*b*) U-tube manometer; (*c*) differential manometer.

container in which the pressure is to be measured and the other end open to the atmosphere. The only liquid it contains is the one whose pressure is being measured (the **metered liquid**). The piezometer is a sensitive gage, but it is limited to the measurement of relatively small pressures, usually heads of 5 ft of water or less. Larger pressures would create an impractically high column of liquid.

Example 21.3: The gage pressure p_c in the pipe of Fig. 21.11*a* is 2.17 psi. The liquid is water with $w = 62.4$ lb/ft^3. What is h_m?

$$h_m = \frac{p_c}{w} = \frac{2.17 \times 144}{62.4} = 5.0 \, \text{ft}$$

For pressures greater than 5 ft of water, the **U-tube manometer** (Fig. 21.11*b*) is used. It is similar to the piezometer except that it contains an **indicating liquid** with a specific gravity usually much larger than that of the metered liquid. The only other criteria are that the indicating liquid should have a good meniscus and be immiscible with the metered liquid.

The U-tube manometer is used when pressures are either too high or too low for the piezometer. High pressures can be measured by arranging U-tube manometers in series (Fig. 21.12*c*). Very low pressures, including negative gage pressures, can be measured if the bottom of the U tube extends below the center line of the container of the

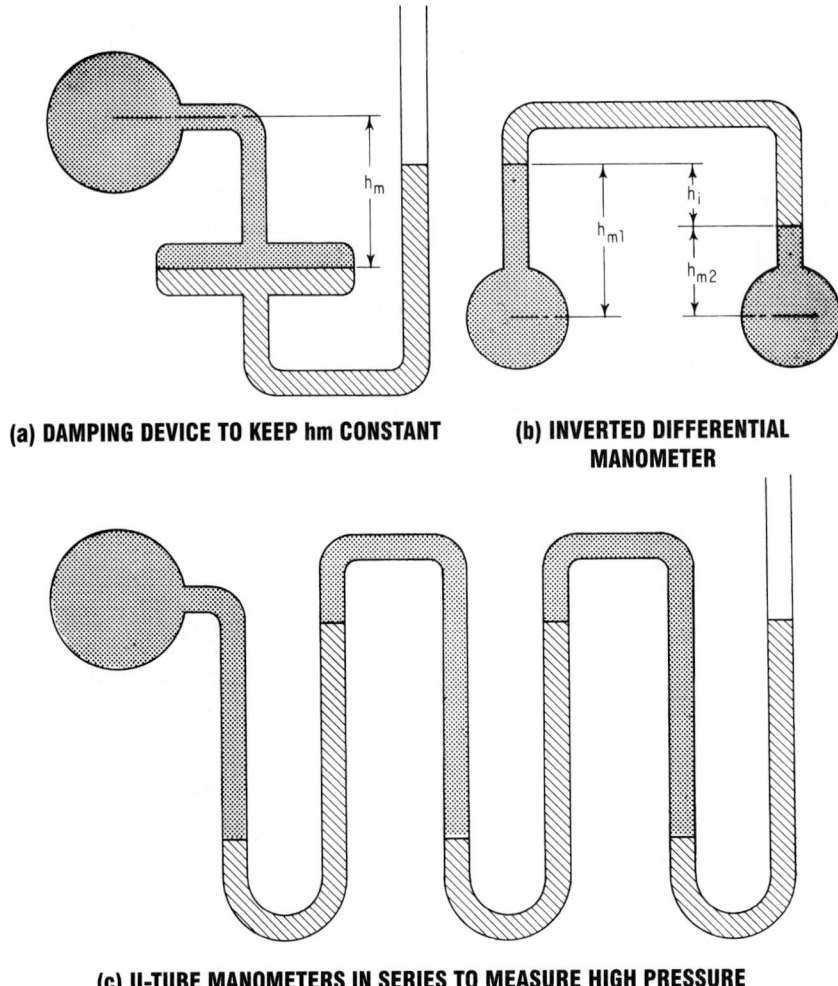

(a) DAMPING DEVICE TO KEEP hm CONSTANT

(b) INVERTED DIFFERENTIAL MANOMETER

(c) U-TUBE MANOMETERS IN SERIES TO MEASURE HIGH PRESSURE

Fig. 21.12 Manometer shapes: (*a*) Sump in manometer to damp flow disturbances. (*b*) Inverted U for measuring pressures on liquids with low specific gravity. (*c*) Series arrangement for measuring high pressures.

metered liquid. The most common use of the U-tube manometer is measurement of the pressures of flowing water. In this application, the usual indicating liquid is mercury.

A movable scale, as opposed to a fixed scale, facilitates reading the U-tube manometer. The scale is positioned between the two vertical legs and moved to adjust for the variation in distance h_m from the center line of the pressure vessel to the indicating liquid. This zero adjustment enables a direct reading of the heights h_i and h_m of the liquid

columns. The scale may be calibrated in any convenient units, such as ft of water or psi.

The **differential manometer** (Fig. 21.11*c*) is identical to the U-tube manometer but measures the difference in pressure between two points. (It does not indicate the pressure at either point.) The differential manometer may have either the standard U-tube configuration or an inverted U-tube configuration, depending on the comparative specific gravities of the indicating and metered liquids. The inverted U-tube configuration (Fig. 21.12*b*) is used when the indicating liquid has a lower specific gravity than the metered liquid.

Example 21.4: A differential manometer (Fig. 21.11c) is measuring the difference in pressure between two water pipes. The indicating liquid is mercury (specific gravity = 13.6), h_i is 2.25 ft, h_{m1} is 9 in, and z is 1.0 ft. What is the pressure differential between the two pipes?

$$h_{m2} = h_i + h_{m1} - z = 2.25 + 0.75 - 1.0 = 2.0 \, \text{ft}$$

The pressure at B, psf, is

$$p_B = p_{c2} + w_2 h_{m2} = p_{c2} + 62.4 \times 2.0 = p_{c2} + 125$$

The pressure at A, psf, is

$$p_A = p_{c1} + w_1 h_{m1} + w_i h_i$$
$$= p_{c1} + 62.4 \times 0.75 + 13.6 \times 62.4 \times 2.25$$
$$= p_{c1} + 1957$$

Since the pressure at A must equal that at B,

$$p_{c2} + 125 = p_{c1} + 1957$$

Hence, the pressure differential between the pipes is

$$p_{c2} - p_{c1} = 1832 \, \text{psf} = 12.7 \, \text{psi}$$

When small pressure differences in water are measured, if the specific gravity of the indicating liquid is between 1.0 and 2.0 and the points at which the pressure is being measured are at the same level, the actual pressure difference, when expressed in feet of water, is magnified by the differential manometer. For example, if the actual difference is 0.50 ft of water and the indicating liquid has a specific gravity of 1.40, the magnification will be 2.5; that is, the height of the liquid column h_i will be 1.25 ft of water. The closer the specific gravities of the metered and indicating liquids, the greater the magnification and sensitivity. This is true only up to a magnification of about 5. Above 5, the increased sensitivity may be deceptive because the meniscus between the two liquids becomes poorly defined and sluggish in movement.

Many factors affect the accuracy of manometers. Most of them, however, may be neglected in the majority of hydraulics applications since they are significant only in precise reading of manometers, such as might be required in laboratories. One factor, however, is significant: the existence of surges in the manometer caused by the pulsations and disturbances in the flow of water resulting from turbulence. These surges make reading of the manometer difficult. They may be reduced or eliminated by installing a large-diameter section, or sump, in the manometer, as shown in Fig. 21.12a. This sump will damp the pulsations and keep the distance from the center line of the conduit to the indicating liquid essentially at a constant value.

21.6 Fundamentals of Fluid Flow

For fluid energy, the law of conservation of energy is represented by the **Bernoulli equation**:

$$Z_1 + \frac{p_1}{w} + \frac{V_1^2}{2g} = Z_2 + \frac{p_2}{w} + \frac{V_2^2}{2g} \tag{21.11}$$

where Z_1 = elevation, ft, at any point 1 of flowing fluid above an arbitrary datum

Z_2 = elevation, ft, at downstream point in fluid above same datum

p_1 = pressure at 1, psf

p_2 = pressure at 2, psf

w = specific weight of fluid, lb/ft^3

V_1 = velocity of fluid at 1, ft/s

V_2 = velocity of fluid at 2, ft/s

g = acceleration due to gravity, 32.2 ft/s^2

The left side of the equation sums the total energy per unit weight of fluid at 1, and the right side, the total energy per unit weight at 2. Equation (21.11) applies only to an ideal fluid. Its practical use requires a term to account for the decrease in total head, ft. through friction. This term h_f, when added to the downstream side of Eq. (21.11), yields the form of the Bernoulli equation most frequently used:

$$Z_1 + \frac{p_1}{w} + \frac{V_1^2}{2g} = Z_2 + \frac{p_2}{w} + \frac{V_2^2}{2g} + h_f \tag{21.12}$$

The energy contained in an elemental volume of fluid thus is a function of its elevation, velocity, and pressure (Fig. 21.13). The energy due to elevation is the potential energy and equals $W Z_a$, where W is the weight, lb, of the fluid in the elemental volume and Z_a is its elevation, ft, above some arbitrary datum. The energy due to velocity is the kinetic energy. It equals $W V_a^2 / 2g$, where V_a is the velocity, ft/s. The pressure energy equals $W p_a / w$, where p_a is the

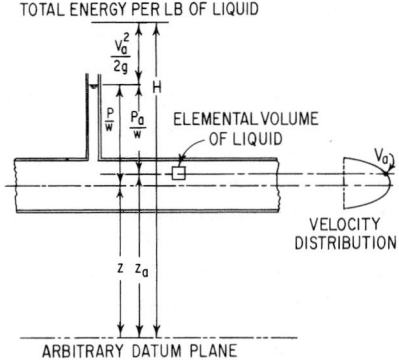

TOTAL ENERGY PER LB OF LIQUID

Fig. 21.13 Energy in a liquid depends on elevation, velocity, and pressure.

pressure lb/ft^2, and w is the specific weight of the fluid, lb/ft^3. The total energy, in the elemental volume of fluid is

$$E = WZ_a + \frac{Wp_a}{w} + \frac{WV_a^2}{2g} \qquad (21.13)$$

Dividing both sides of the equation by W yields the energy per unit weight of flowing fluid, or the **total head** ft:

$$H = Z_a + \frac{p_a}{w} + \frac{V_a^2}{2g} \qquad (21.14)$$

p_a/w is called **pressure head**; $V_a^2/2g$, **velocity head**.

As indicated in Fig. 21.13, $Z + p/w$ is constant for any point in a cross section and normal to the flow through a pipe or channel. Kinetic energy at the section, however, varies with velocity. Usually, $Z + p/w$ at the midpoint and the average velocity at a section are assumed when the Bernoulli equation is applied to flow across the section or when total head is to be determined. **Average velocity**, ft/s $= Q/A$, where Q is the quantity of flow, ft^3/s, across the area of the section A, ft^2.

Example 21.5: Determine the energy loss between points 1 and 2 in the 24-in-diameter pipe in Fig. 21.14. The pipe carries water flowing at $31.4 \, ft^3/s$.

Average velocity in the pipe $= Q/A = 31.4/3.14 = 10$ ft/s. Select point 1 far enough from the reservoir outlet that V_1 can be assumed to be 0. Since the datum plane passes through point 2, $Z_2 = 0$. Also, since the pipe has free discharge, $p_2 = 0$. Thus substitution in Eq. (21.12) yields

$$30 + 20 + 0 = 0 + 0 + \frac{10^2}{64.4} + h_f$$

where h_f is the friction loss, ft. Hence, $h_f = 50 - 1.55 = 48.45$ ft.

Note that in this example h_f includes minor losses due to the pipe entrance, gate valve, and any bends.

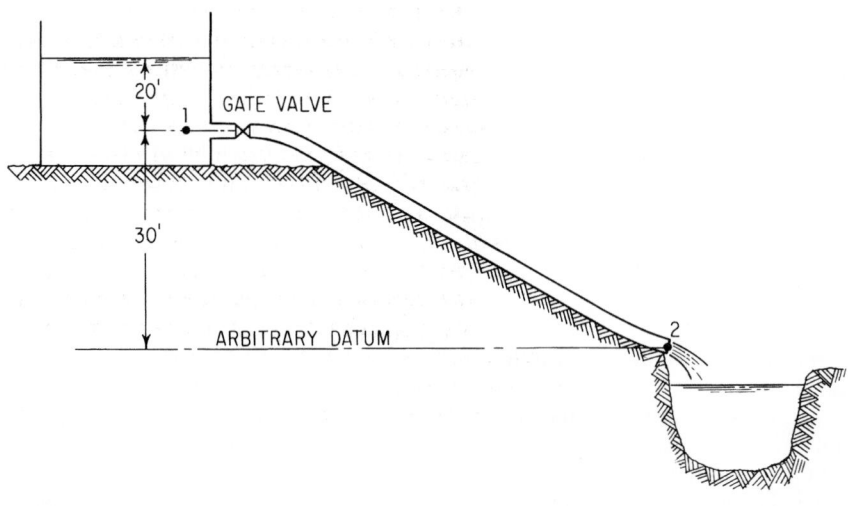

Fig. 21.14 Flow from an elevated reservoir—application of the Bernoulli equation. (See Example 21.5.)

The Bernoulli equation and the variation of pressure may be represented graphically, respectively, by energy and hydraulic grade lines (Fig. 21.15). The *energy grade line*, sometimes called the total head line, shows the decrease in total energy per unit weight H in the direction of flow. The slope of the energy grade line is called the **energy gradient** or friction slope. The **hydraulic grade line** lies a distance $V^2/2g$ below the energy grade line and shows the variation of velocity or pressure in the direction of flow. The slope of the hydraulic grade line is termed the **hydraulic gradient**. In open-channel flow, the hydraulic grade line coincides with the water surface, while in pressure flow, it represents the height to which water would rise in a piezometer (see also Example 21.7, Art. 21.9).

Momentum is a fundamental concept that must be considered in the design of essentially all waterworks facilities involving flow. A change in momentum, which may result from a change in either velocity, direction, or magnitude of flow, is equal to the impulse, the force F acting on the fluid times the period of time dt over which it acts. Dividing the total change in momentum by the time interval over which the change occurs gives the momentum equation, or impulse-momentum equation:

$$F_x = pQ\Delta V_x \qquad (21.15)$$

where F_x = summation of all forces in X direction per unit time causing change in momentum in X direction, lb

ρ = density of flowing fluid, $\text{lb} \cdot \text{s}^2/\text{ft}^4$ (specific weight divided by g)

Q = flow rate, ft^3/s

ΔV_x = change in velocity in X direction, ft/s

Similar equations may be written for the Y and Z directions. The impulse-momentum equation often

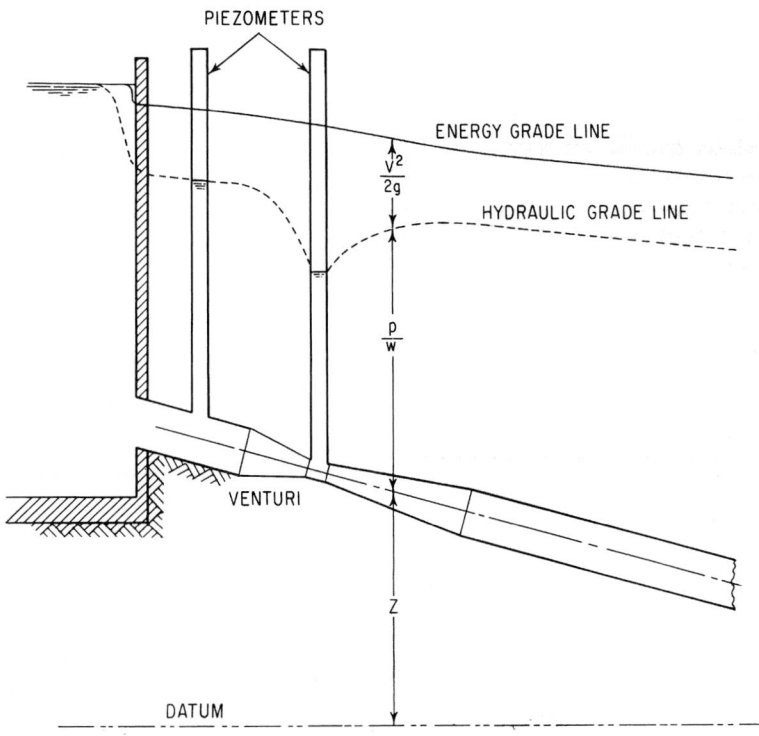

Fig. 21.15 Energy grade line and hydraulic grade line indicate variations in energy and pressure head, respectively, in a liquid as it flows along a pipe or channel.

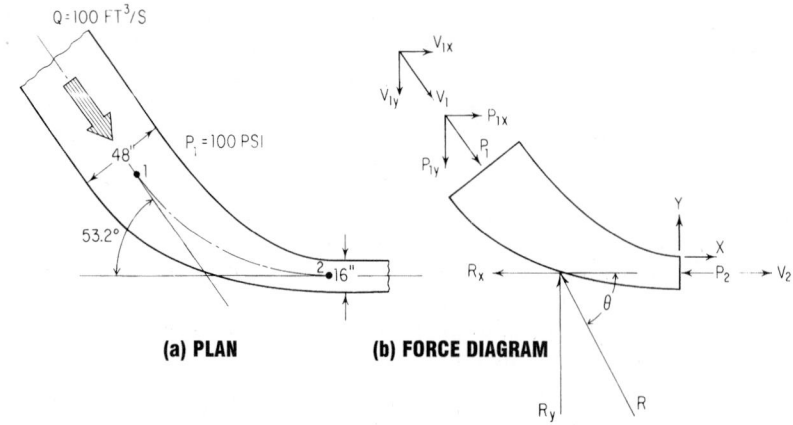

Fig. 21.16 Flow induces forces in a pipe at bends and at changes in size of section—application of momentum equation. (See Example 21.6.)

is used in conjunction with the Bernoulli equation [Eq. (21.11) or (21.12)] but may be used separately.

Example 21.6: Calculate the resultant force on the reducer elbow in Fig. 21.16. The pipe center line lies in a horizontal plane. The pipe reduces from 48 in in diameter to 16 in. The pressure at the upstream side of the reducer bend (point 1) is 100 psi, and the water flow is 100 ft³/s. (Neglect friction loss at the bend.)

Velocity at points 1 and 2 is found by dividing $Q = 100$ ft³/s by the respective areas: $V_1 = 100 \times 4/4^2\pi = 7.96$ ft/s and $V_2 = 100 \times 4/1.33^2\pi = 71.5$ ft/s.

With p_1 known, the Bernoulli equation for the flow in the elbow is:

$$0 + 100 \times \frac{144}{62.4} + \frac{7.96^2}{2 \times 32.2} = 0 + \frac{p_2}{62.4} + \frac{71.5^2}{2 \times 32.2}$$

Solution of the equation yields the pressure at 2:

$$p_2 = 9500 \text{ psf}$$

The total pressure force at 1 is $P_1 = p_1 A_1 = 181,000$ lb, and at 2, $P_2 = p_p A_2 = 13,200$ lb.

Let R be the force, lb, exerted by the pipe on the fluid (equal and opposite in direction to the force against the pipe, which is to be determined). Then, the force F changing the momentum of the fluid equals the vector sum $P_1 - P_2 + R$. To find F, apply Eq. (21.15) first in the X direction, then in the Y direction, and determine the resultant of the forces:

In the X direction, since $\Delta V_x = -(7.96 \sin 53.2° - 71.5) = 65.1$ and the density $\rho = 62.4/32.2 = 1.94$,

$$F_x = 181,000 \cos 53.2° - 13,200 + R_x$$
$$= 1.94 \times 100 \times 65.1$$
$$R_x = -82,600 \text{ lb}$$

In the Y direction, since $\Delta V_y = -(-7.96, \cos 53.2° - 0) = 4.78$,

$$F_y = -181,000 \sin 53.2° + R_y$$
$$= 1.94 \times 100 \times 4.78$$
$$R_y = 145,700 \text{ lb}$$

The resultant $R = \sqrt{R_x^2 + R_y^2} = 167,500$ lb. It acts at an angle θ with the horizontal such that $\tan \theta = 145,700/82,600$; so $\theta = 60.5°$. The force against the pipe acts in the opposite direction.

21.7 Water Resources Modeling

A model is a tool that can be used to determine the likely response of a system to a given set of stimuli without having to actually impose those stimuli on the system. In water resources engineering, models are used to determine the likely response of a system, such as a river, aquifer, or drainage basin, to a given set of stimuli, such as storm rainfall, droughts, alternative management schemes, or proposed works. Models are cost-effective and convenient for such investigations. See also Art. 1.7.

Models can typically be categorized as one of three major types:

Physical Models ▪ The system (prototype) is modeled with physical components that represent components of the system. Usually, scale factors are applied to set the model at only a fraction of the size and cost of the prototype. Physical models are expensive to build, operate, and maintain but are especially useful in analyzing complex phenomena that are not easy or presently possible to express mathematically.

Analog Models ▪ The system (prototype) is modeled with electronic circuits that represent components of the system. Some conveyance and resistance phenomena such as those found in transmission networks and groundwater analyses are easily modeled with analog techniques inasmuch as electric current flow and water flow behave similarly in certain instances. Analog models are an abstraction of the prototype. Popular before the advent of digital computers, analog models are now infrequently used in view of the efficiency and portability of mathematical models.

Mathematical Models ▪ The system (prototype) is modeled with sets of mathematical expressions that represent components of the system. Mathematical models are normally programmed in an appropriate computer language, and through execution of the computer program, simulations of prototype behavior are possible. Mathematical models are limited only by the model creator's ability to describe the prototype mathematically, the capability of the computing resources, or availability of data to support the modeling effort. They can be as simple or as complex as a given analysis requires and are among the most cost-effective means to perform certain analyses.

A fourth mode of modeling, hybrid modeling, employs both physical and mathematical models. It exploits the advantages of these types of models while avoiding their limitations. For instance, complex three-dimensional flow patterns, erosional scour, and sediment deposition occurring in the immediate vicinity of a bridge pier or water control structure can be best modeled with a physical model while the overall water surface, momentum, and velocity profile over the encompassing river reach can be best modeled by an appropriate mathematical model.

With hybrid models, one model often provides input to or verification of the other model. In the preceding example, the mathematical model would provide depth and velocity profile input to the physical model, and the physical model may be able to provide a more accurate estimate of local head loss at the pier or structure. In this way, the two models can be executed interactively until all common boundary conditions synchronize. The resulting hybrid model will consist of a mathematical model that properly accounts for overall hydraulic effects and local head loss at the pier or structure and a physical model that properly accounts for localized forces affecting the stability or performance of the pier or structure.

As computers become faster, numerical modeling in three dimensions, utilizing the full set of equations for hydrodynamic flow is making possible reliable solutions for very complex problems. Eventually, these sophisticated models may eliminate the need for physical modeling in most cases.

21.7.1 Similitude for Physical Models

A physical model is a system whose operation can be used to predict the characteristics of a similar system, or prototype, usually more complex or built to a much larger scale. A knowledge of the laws governing the phenomena under investigation is necessary if the model study is to yield accurate quantitative results.

Forces acting on the model should be proportional to forces on the prototype. The four forces usually considered in hydraulic models are inertia, gravity, viscosity, and surface tension. Because of the laws governing these forces and because the model and prototype are normally not the same size, it is usually not possible to have all four forces in the model in the same proportions as they are in the prototype. It is, however, a simple procedure to have two predominant forces in the same proportion. In most models, the fact that two of the four forces are not in the same proportion as they are in the prototype does not introduce serious error. The inertial force, which is always a predominant force, and one other force are made proportional.

Ratios of the forces of gravity, viscosity, and surface tension to the force of inertia are designated, respectively, Froude number, Reynolds number, and Weber number. Equating the Froude number of the model and the Froude number of the prototype ensures that the gravitational and inertial forces are in the same proportion. Similarly, equating the Reynolds numbers of the model and prototype ensures that the viscous and inertial forces will be in the same proportion. And equating the Weber numbers ensures proportionality of surface tension and inertial forces.

The **Froude number** is

$$\mathbf{F} = \frac{V}{\sqrt{Lg}} \qquad (21.16)$$

where $\mathbf{F}$ = Froude number (dimensionless)

V = velocity of fluid, ft/s

L = linear dimension (characteristic, such as depth or diameter), ft

g = acceleration due to gravity, 32.2 ft/s^2

For hydraulic structures, such as spillways and weirs, where there is a rapidly changing water-surface profile, the two predominant forces are inertia and gravity. Therefore, the Froude numbers of the model and prototype are equated:

$$\mathbf{F_m} = \mathbf{F_p} \qquad \frac{V_m}{\sqrt{L_m g}} = \frac{V_p}{\sqrt{L_p g}} \qquad (21.17a)$$

where subscript m applies to the model and p to the prototype. Squaring both sides of Eq. (21.17a) and grouping like terms yields

$$\frac{V_m^2}{V_p^2} = \frac{L_m}{L_p} \qquad (21.17b)$$

Let $V_r = V_m/V_p$ and $L_r = L_m/L_p$. Then

$$V_r^2 = L_r \qquad V_r = L_r^{1/2} \qquad (21.18)$$

The subscript r indicates ratio of quantity in model to that in prototype.

If the ratios of all the physical dimensions of a model to all the corresponding physical dimensions of the prototype are equal to the length ratio, the model is termed a **true model**. In a true model where the Froude number is the governing design criterion, the length ratio is the only variable.

Once the length ratio has been set, all the physical dimensions of the model are fixed. The discharge ratio is determined as follows:

$$Q_r = V_r A_r \qquad (21.19a)$$

Since $V_r = L_r^{1/2}$ and A_r = area ratio = L_r^2,

$$Q_r = V_r A_r = L_r^{5/2} \qquad (21.19b)$$

By this method all the necessary characteristics of a spillway or weir model can be determined.

The **Reynolds number** is

$$\mathbf{R} = \frac{VL}{\nu} \qquad (21.20)$$

$\mathbf{R}$ is dimensionless, and ν is the kinematic viscosity of fluid, ft^2/s. The Reynolds numbers of model and prototype are equated when the viscous and inertial forces are predominant. Viscous forces are usually predominant when flow occurs in a closed system, such as pipe flow where there is no free surface. The following relations are obtained by equating Reynolds numbers of the model and prototype:

$$\frac{V_m L_m}{\nu_m} = \frac{V_p L_p}{\nu_p} \qquad (21.21a)$$

$$V_r = \frac{\nu_r}{L_r} \qquad (21.21b)$$

The variable factors that fix the design of a true model when the Reynolds number governs are the length ratio and the viscosity ratio.

The **Weber number** is

$$\mathbf{W} = \frac{V^2 L \rho}{\sigma} \qquad (21.22)$$

where ρ = density of fluid, lb · s^2/ft^4 (specific weight divided by g)

σ = surface tension of fluid, psf

The Weber numbers of model and prototype are equated in certain types of wave studies, the formation of drops and air bubbles, entrainment of air in flowing water, and other phenomena where surface tension and inertial forces are predominant. The velocity ratio is determined as follows:

$$\frac{V_m^2 L_m \rho_m}{\sigma_m} = \frac{V_p^2 L_p \rho_p}{\sigma_p} \qquad (21.23a)$$

$$V_r^2 = \frac{\sigma_r}{\rho_r L_r} \qquad (21.23b)$$

The fluid properties and the length ratio fix the design of a model governed by the Weber number.

In some cases, such as a morning-glory spillway, inertial, viscous, and gravity forces all have an important effect on the flow. In these cases it is usually not possible to have both the Reynolds and Froude numbers of the model and prototype equal. The solution to this type of problem is mostly empirical and may consist of an attempt to evaluate the effects of viscosity and gravity separately.

For the flow of water in open channels and rivers where the friction slope is relatively flat, model designs are often based on the Manning equation. The relations between the model and prototype are determined as follows:

$$\frac{V_m}{V_p} = \frac{(1.486/n_m)R_m^{2/3}S_m^{1/2}}{(1.486/n_p)R_p^{2/3}S_p^{1/2}} \qquad (21.24)$$

where n = Manning roughness coefficient ($T/L^{1/3}$, T representing time)

R = hydraulic radius (L)

S = loss of head due to friction per unit length of conduit (dimensionless)

= slope of energy gradient

For true models, $S_r = 1$, $R_r = L_r$. Hence,

$$V_r = \frac{L_r^{2/3}}{n_r} \qquad (21.25)$$

In models of rivers and channels, it is necessary for the flow to be turbulent. The U.S. Waterways Experiment Station has determined that flow will be turbulent if

$$\frac{VR}{v} \geq 4000 \qquad (21.26)$$

where V = mean velocity, ft/s

R = hydraulic radius, ft

v = kinematic viscosity, ft^2/s

If the model is to be a true model, it may have to be uneconomically large for the flow to be turbulent. Another problem also encountered in true models is surface tension. In a true model of a wide river where the depth may be only a fraction of an inch, the surface tension will distort the flow to such an extent that the model may be useless. To overcome the effect of surface tension and to get turbulent flow, the depth scale is often made much larger

than the length scale. This type of model is called a **distorted model**.

The relations between a distorted model of a channel and a prototype are determined in the same manner as was Eq. (21.24). The only difference is that the slope ratio S_r equals the depth ratio d_r and the hydraulic-radius ratio is a function of the width ratio and depth ratio.

One type of model, called a movable-bed model, is used to study erosion and transportation of silt in riverbeds. Because the laws governing the transportation of material are not fully understood, movable-bed models are built largely on the basis of experience and give only qualitative results.

21.7.2 Types and Applications of Mathematical Models

Used in many applications of water resources engineering, mathematical models are, in particular, applied in hydrologic and hydraulic investigations of man-made and natural systems for both surface-water and groundwater purposes. The system (prototype) is modeled with sets of mathematical expressions that represent components of the system. These expressions, in turn, are linked together to represent the system as a whole.

Mathematical models are used for both analysis and design. They are normally programmed in an appropriate computer language, and through execution of the computer program, simulations of prototype behavior are possible. They may be single-purpose (for a specific site) or general purpose (applicable to a variety of sites).

Single-purpose models typically represent the specific temporal and spatial descriptions of the prototype directly in the computer code. For instance, the logical representation of prototypes, such as flow networks, catchment areas, and infiltration parameters, may be part of the source code and is said to be *hardwired* into the computer program. For such models, the software (the computer program code) and the application input codes (hydrologic and hydraulic parameters) are bound into one entity. This, however, usually has more disadvantages than advantages, especially when modifications of the model are required or when the model has to be applied by engineers who were not involved in the original program coding. The preferred approach in modeling is instead to develop general-purpose models by writing software that is essentially independent of application input code.

General-purpose models are used for specific analytical tasks. These may be as simple as determination of excess rainfall, given rainfall and rainfall-loss parameters, or as complex as long-period simulation of flow and pollutant transport in combined groundwater and surface-water systems.

Advances are continually being made in computer resources and use of models is becoming more widespread. As a result, the desirability of more uniformity of software packages and of object-oriented software has become apparent. In object-oriented software, every program component is generalized as much as feasible and the entire program is essentially a collection of modular software components. This approach, when fully implemented, will provide complete compatibility among all types of water resources software. Also, this approach will provide nearly complete compatibility of all databases, of all databases and software, and among water resources modelers in the government, academia, and private sectors. The result will be a reduction in duplication of the efforts of software developers and modelers and an increase in the efficiency of water-resources engineering investigations.

Typical applications of mathematical models include the following: stochastic processes; evaporation and irrigation; hydrodynamics; hydrologic forecasting; watershed hydrology; design of hydraulic structures; reservoir regulation; flood or drought impacts; flow routing; channel and river hydraulics; sediment or pollutant transport; quantity and quality of water supply; ecosystem impacts and restoration; impacts of dam breaks; wave or tidal analyses; landfill leachate analyses; and groundwater yield, seepage, or pollution.

Several different models varying in complexity or sophistication, or both, and in application type may be required in many types of investigations. As a general rule, if comparisons of different plans are required, the fewer the number of models employed in a given study, the greater the chance that meaningful results will be produced. The availability and quality of data for calibration and verification, the model output required for design or evaluation, and the general acceptance by the engineering community should be considered in selection of a model or group of models for any investigation.

Mathematical modeling is one of the fastest changing fields in engineering. Applications should be upgraded accordingly if their continued use is expected.

(D. R. Maidment, "Handbook of Hydrology," D. H. Hoggan, "Computer-Assisted Floodplain Hydrology and Hydraulics," N. S. Grigg, "Water Resources Planning," V. J. Zipparo and H. Hasen, "Davis' Handbook of Applied Hydraulics," McGraw-Hill, New York (books.mcgraw-hill.com).)

Pipe Flow

The term pipe flow as used in this section refers to flow in a circular closed conduit entirely filled with fluid. For closed conduits other than circular, reasonably good results are obtained in the turbulent range with standard pipe-flow formulas if the diameter is replaced by four times the hydraulic radius. But when there is severe deviation from a circular cross section, as in annular passages, this method gives flows significantly underestimated. (J. F. Walker, G. A. Whan, and R. R. Rothfus, "Fluid Friction in Noncircular Ducts," *Journal of the American Institute of Chemical Engineers*, vol. 3, 1957.)

21.8 Laminar Flow

In laminar flow, fluid particles move in parallel layers in one direction. The parabolic velocity distribution in laminar flow, shown in Fig. 21.17, creates a shearing stress $\tau = \mu \, dV/dy$, where dV/dy is the rate of change of velocity with depth and μ is the coefficient of viscosity (see Viscosity, Art. 21.2). As this shearing stress increases, the viscous forces become unable to damp out disturbances, and turbulent flow results. The region of change is

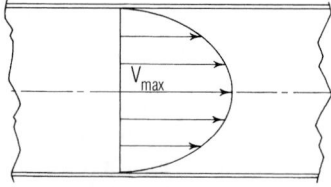

Fig. 21.17 Velocity distribution for lamellar flow in a circular pipe is parabolic. Maximum velocity is twice the average velocity.

dependent on the fluid's velocity, density, and viscosity and the size of the conduit.

A dimensionless parameter called the Reynolds number has been found to be a reliable criterion for the determination of laminar or turbulent flow. It is the ratio of inertial forces to viscous forces, and is given by

$$\mathbf{R} = \frac{VD\rho}{\mu} = \frac{VD}{\nu} \qquad (21.27)$$

where V = fluid velocity, ft/s

D = pipe diameter, ft

ρ = density of fluid, $\text{lb} \cdot \text{s}^2/\text{ft}^4$ (specific weight divided by g, 32.2 ft/s^2)

μ = viscosity of fluid $\text{lb} \cdot \text{s}/\text{ft}^2$

$\nu = \mu/\rho$ = kinematic viscosity, ft^2/s

For a Reynolds number less than 2000, flow is laminar in circular pipes. When the Reynolds number is greater than 2000, laminar flow is unstable; a disturbance will probably be magnified, causing the flow to become turbulent.

In laminar flow, the following equation for head loss due to friction can be developed by considering the forces acting on a cylinder of fluid in a pipe:

$$h_f = \frac{32\mu LV}{D^2 \rho g} = \frac{32\mu LV}{D^2 w} \qquad (21.28)$$

where h_f = head loss due to friction, ft

L = length of pipe section considered, ft

g = acceleration due to gravity, 32.2 ft/s^2

w = specific weight of fluid, lb/ft^3

Substitution of the Reynolds number yields

$$h_f = \frac{64}{\mathbf{R}} \frac{L}{D} \frac{V^2}{2g} \qquad (21.29)$$

For laminar flow, Eq. (21.29) is identical to the Darcy-Weisbach formula Eq. (21.30) since in laminar flow the friction $f = 64/\mathbf{R}$.

(E. F. Brater, "Handbook of Hydraulics," 6th ed., McGraw-Hill Book Company, New York (books. mcgraw-hill.com).)

21.9 Turbulent Flow

In turbulent flow, the inertial forces are so great that viscous forces cannot dampen out disturbances caused primarily by the surface roughness.

These disturbances create eddies, which have both a rotational and translational velocity. The translation of these eddies is a mixing action that affects an interchange of momentum across the cross section of the conduit. As a result, the velocity distribution is more uniform, as shown in Fig. 21.18, than for laminar flow (Fig. 21.17).

For a Reynolds number greater than 2000 but to the left of the dashed line in Fig. 21.19, there is a transition from laminar to turbulent flow. In this region, there is a laminar film at the boundaries that covers some of the smaller roughness projections. This explains why the friction loss in this region has both laminar and turbulent characteristics. As the Reynolds number increases, this laminar boundary layer decreases in thickness until, at completely turbulent flow, it no longer covers any of the roughness projections. To the right of the dashed line in Fig. 21.19, the flow is completely turbulent, and viscous forces do not affect the friction loss.

Because of the random nature of turbulent flow, it is not practical to treat it analytically. Therefore, formulas for head loss and flow in the turbulent regions have been developed through experimental and statistical means. Experimentation in turbulent flow has shown that:

The head loss varies directly as the length of the pipe.

The head loss varies almost as the square of the velocity.

The head loss varies almost inversely as the diameter.

The head loss depends on the surface roughness of the pipe wall.

The head loss depends on the fluid's density and viscosity.

The head loss is independent of the pressure.

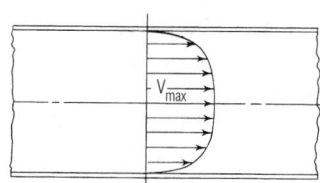

Fig. 21.18 Velocity distribution for turbulent flow in a circular pipe is more nearly uniform than that for lamellar flow.

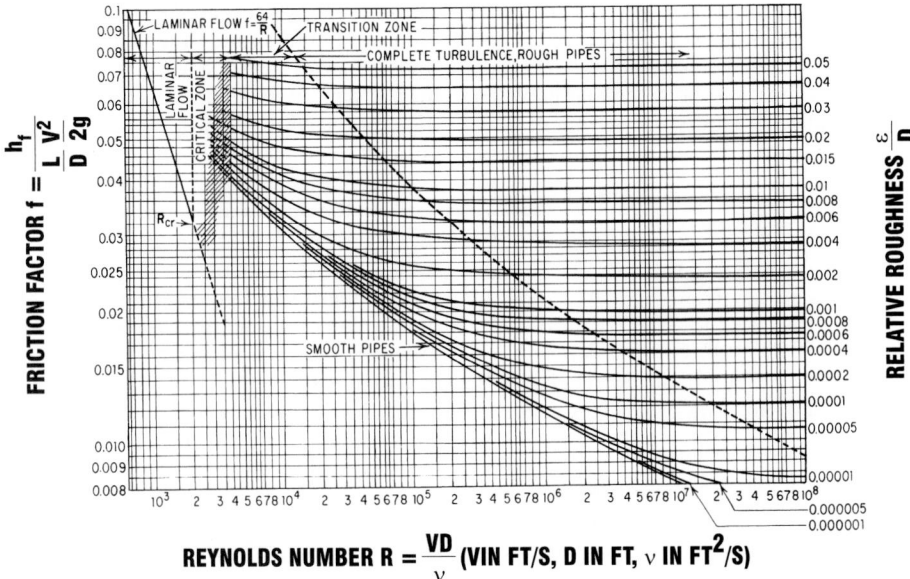

Fig. 21.19 Chart relates friction forces for flow in pipe to Reynolds numbers and condition of pipes.

21.9.1 Darcy-Weisbach Formula

One of the most widely used equations for pipe flow, the Darcy-Weisbach formula satisfies the above condition and is valid for laminar or turbulent flow in all fluids.

$$h_f = f \frac{L}{D} \frac{V^2}{2g} \qquad (21.30)$$

where h_f = head loss due to friction, ft

f = friction factor (see Fig. 21.19)

L = length of pipe, ft

D = diameter of pipe, ft

V = velocity of fluid, ft/s

g = acceleration due to gravity, 32.2 ft/s^2

It employs the Moody diagram (Fig. 21.19) for evaluating the friction factor f. (L. F. Moody, "Friction Factors for Pipe Flow," *Transactions of the American Society of Mechanical Engineers*, November 1944.)

Because Eq. (21.30) is dimensionally homogeneous, it can be used with any consistent set of units without changing the value of the friction factor.

Roughness values ε (ft) for use with the Moody diagram to determine the Darcy-Weisbach friction factor f are listed in Table 21.3.

The following formulas were derived for head loss in waterworks design and give good results for water-transmission and -distribution calculations. They contain a factor that depends on the surface roughness of the pipe material. The accuracy of these formulas is greatly affected by the selection of the roughness factor, which requires experience in its choice.

Table 21.3 Typical Values of Roughness for Use in the Moody Diagram (Fig. 21.19) to Determine f

	ε, ft
Steel pipe:	
Severe tuberculation and incrustation	0.03–0.008
General tuberculation	0.008–0.003
Heavy brush-coat asphalts, enamels, and tars	0.003–0.001
Light rust	0.001–0.0005
New smooth pipe, centrifugally applied enamels	0.0002–0.00003

(Table continued)

21.9.2 Chezy Formula

This equation holds for head loss in conduits and gives reasonably good results for high Reynolds numbers:

$$V = C\sqrt{RS} \qquad (21.31)$$

where V = velocity, ft/s

C = coefficient, dependent on surface roughness of conduit

S = slope of energy grade line or head loss due to friction, ft/ft of conduit

R = hydraulic radius, ft

Hydraulic radius of a conduit is the cross-sectional area of the fluid in it divided by the perimeter of the wetted section.

21.9.3 Manning's Formula

Through experimentation, Manning concluded that the C in the Chezy equation [Eq. (21.31)] should vary as $R^{1/6}$

$$C = \frac{1.486R^{1/6}}{n} \qquad (21.32)$$

where n = coefficient, dependent on surface roughness. (Although based on surface roughness, n in practice is sometimes treated as a lumped parameter for all head losses.) Substitution into Eq. (21.31) gives

$$V = \frac{1.486}{n} R^{2/3} S^{1/2} \qquad (21.33a)$$

Upon substitution of $D/4$, where D is the pipe diameter, for the hydraulic radius of the pipe, the

following equations are obtained for pipes flowing full:

$$V = \frac{0.590}{n} D^{2/3} S^{1/2} \qquad (21.33b)$$

$$Q = \frac{0.463}{n} D^{8/3} S^{1/2} \qquad (21.33c)$$

$$h_f = 4.66n^2 \frac{LQ^2}{D^{16/3}} \qquad (21.33d)$$

$$D = \left(\frac{2.159Qn}{S^{1/2}}\right)^{3/8} \qquad (21.33e)$$

where Q = flow, ft³/s.

Tables 21.4 and 21.11 (p. 21.47) give values of n for the foot-pound-second system. See also Table 22.3 for velocity and flow at various slopes.

21.9.4 Hazen-Williams Formula

This is one of the most widely used formulas for pipe-flow computations of water utilities, although it was developed for both open channels and pipe flow:

$$V = 1.318C_1 R^{0.63} S^{0.54} \qquad (21.34a)$$

For pipes flowing full:

$$V = 0.55C_1 D^{0.63} S^{0.54} \qquad (21.34b)$$

$$Q = 0.432C_1 D^{2.63} S^{0.54} \qquad (21.34c)$$

$$h_f = \frac{4.727}{D^{4.87}} L \left(\frac{Q}{C_1}\right)^{1.85} \qquad (21.34d)$$

$$D = \frac{1.376}{S^{0.205}} \left(\frac{Q}{C_1}\right)^{0.38} \qquad (21.34e)$$

Table 21.4   Values of n for Pipes, to Be Used with the Manning Formula

Material of Pipe	Variation		Use in designing	
	From	To	From	To
Clean cast iron	0.011	0.015	0.013	0.015
Dirty or tuberculated cast iron	0.015	0.035		
Riveted steel or spiral steel	0.013	0.017	0.015	0.017
Welded steel	0.010	0.013	0.012	0.013
Galvanized iron	0.012	0.017	0.015	0.017

(Table continued)

where V = velocity, ft/s

$\quad C_1$ = coefficient, dependent on surface roughness

$\quad R$ = hydraulic radius, ft

$\quad S$ = head loss due to friction, ft/ft of pipe

$\quad D$ = diameter of pipe, ft

$\quad L$ = length of pipe, ft

$\quad Q$ = discharge, ft^3/s

$\quad h_f$ = friction loss, ft

The C_1 terms in Table 21.5 are in the foot-pound-second system.

Determination of flow in branching pipes illustrates the use of friction-loss equations and the hydraulic-grade-line concept.

Example 21.7: Figure 21.20 shows a typical three-reservoir problem. The elevations of the hydraulic grade lines for the three pipes are equal at point D. The Hazen-Williams equation for friction loss [Eq. (21.34d)] can be written for each pipe

Table 21.5 Values of C_1 in Hazen and Williams Formula

Type of pipe		C_1
Cast iron:		
New	All sizes, 130	
5 years old	All sizes up to 24 in, 120	
	24 in and over, 115	
10 years old	12 in, 110	
	4 in, 105	
	30 in and over, 85	
40 years old	16 in, 80	
	4 in, 65	
Welded steel	Values the same as for cast-iron pipe, 5 years older	
Riveted steel	Values the same as for cast-iron pipe, 10 years older	
Wood stave	Average value, regardless of age, 120	
Concrete or concrete-lined	Large sizes, good workmanship, steel forms, 140	
	Large sizes, good workmanship, wood forms, 120	

(Table continued)

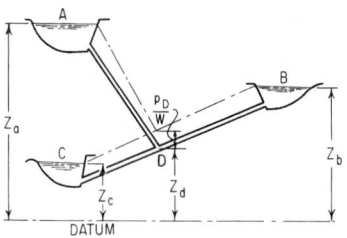

Fig. 21.20 Flow between reservoirs. (See Example 21.7)

meeting at D. With the continuity equation for quantity of flow, there are as many equations as there are unknowns:

$$Z_a = Z_d + \frac{pD}{w} + \frac{4.727L_A}{D_A^{4.87}}\left(\frac{Q_A}{C_A}\right)^{1.85} \qquad (21.35a)$$

$$Z_b = Z_d + \frac{P_D}{w} + \frac{4.727L_B}{D_B^{4.87}}\left(\frac{Q_B}{C_B}\right)^{1.85} \qquad (21.35b)$$

$$Z_c = Z_d + \frac{P_D}{w} + \frac{4.727L_C}{D_C^{4.87}}\left(\frac{Q_C}{C_C}\right)^{1.85} \qquad (21.35c)$$

$$Q_A + Q_B = Q_C \qquad (21.36)$$

where p_D = pressure at D

$\quad w$ = unit weight of liquid

With the elevations Z of the three reservoirs and the pipe intersection known, the easiest way to solve these equations is by trying different values of p_D/w in Eqs. (21.35) and substituting the values obtained for Q into Eq. (21.36) for a check. If the value of $Z_d + p_D/w$ becomes greater than Z_b, the sign of the friction-loss term is negative instead of positive. This would indicate water is flowing from reservoir A into reservoirs B and C. Flow in pipe network is easily determined with available computer programs, many of which are specialized to solve specific pipe design problems efficiently.

21.10 Minor Losses in Pipes

Energy losses occur in pipe contractions, bends, enlargements, and valves and other pipe fittings. These losses can usually be neglected if the length of the pipeline is greater than 1500 times the pipe's diameter. However, in short pipelines, because these losses may exceed the friction losses, minor losses must be considered.

21.10.1 Sudden Enlargements

The following equation for the head loss, ft, across a sudden enlargement of pipe diameter has been determined analytically and agrees well with experimental results:

$$h_L = \frac{(V_1 - V_2)^2}{2g} \qquad (21.37)$$

where V_1 = velocity before enlargement, ft/s

V_2 = velocity after enlargement, ft/s

$g = 32.2$ ft/s^2

It was derived by applying the Bernoulli equation and the momentum equation across an enlargement.

Another equation for the head loss caused by sudden enlargements was determined experimentally by Archer. This equation gives slightly better agreement with experimental results than Eq. (21.37):

$$h_L = \frac{1.1(V_1 - V_2)^{1.92}}{2g} \qquad (21.38)$$

A special application of Eq. (21.37) or (21.38) is the discharge from a pipe into a reservoir. The water in the reservoir has no velocity, so a full velocity head is lost.

21.10.2 Gradual Enlargements

The equation for the head loss due to a gradual conical enlargement of a pipe takes the following form:

$$h_L = \frac{K(V_1 - V_2)^2}{2g} \qquad (21.39)$$

where K = loss coefficient (see Fig. 21.21).

Since the experimental data available on gradual enlargements are limited and inconclusive, the values of K in Fig. 21.21 are approximate. (A. H. Gibson, "Hydraulics and Its Applications," Constable & Co., Ltd., London.)

21.10.3 Sudden Contraction

The following equation for the head loss across a sudden contraction of a pipe was determined by the same type of analytical studies as Eq. (21.37):

$$h_L = \left(\frac{1}{C_c} - 1\right)^2 \frac{V^2}{2g} \qquad (21.40)$$

where C_c = coefficient of contraction (see Table 21.6)

V = velocity in smaller-diameter pipe, ft/s

This equation gives best results when the head loss is greater than 1 ft. Table 21.6 gives C_c values

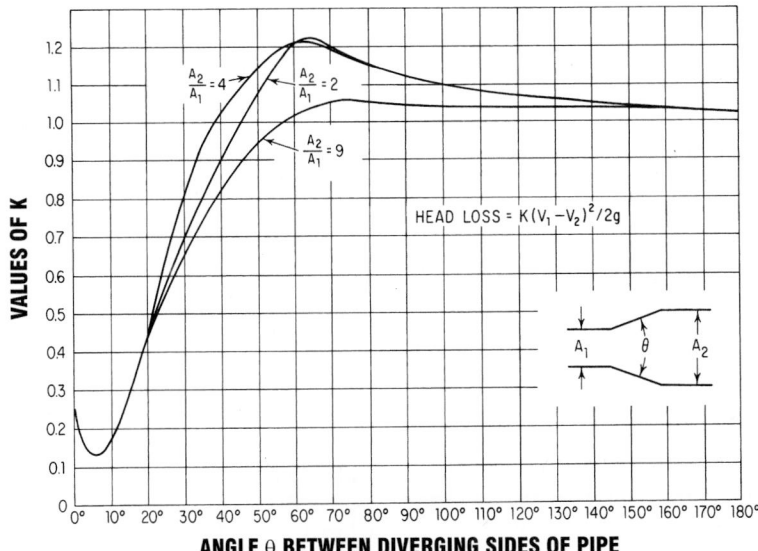

Fig. 21.21 Head-loss coefficients for a pipe with diverging sides depend on the angle of divergence of the sides.

Table 21.6 C_c for Contractions in Pipe Area from A_1 to A_2

A_2/A_1	0.1	0.2	0.3	0.4	0.5	0.6	0.7	0.8	0.9	1.0
C_c	0.62	0.63	0.64	0.66	0.68	0.71	0.76	0.81	0.89	1.0

for sudden contractions, determined by Julius Weisbach ("Die Experiments-Hydraulik").

Another formula for determining the loss of head caused by a sudden contraction, determined experimentally by Brightmore, is

$$h_L = \frac{0.7(V_1 - V_2)^2}{2g} \qquad (21.41)$$

This equation gives best results if the head loss is less than 1 ft.

A special case of sudden contraction is the entrance loss for pipes. Some typical values of the loss coefficient K in $h_L = KV^2/2g$, where V is the velocity in the pipe, are presented in Table 21.7.

21.10.4 Bends and Standard Fitting Losses

The head loss that occurs in pipe fittings, such as valves and elbows, and at bends is given by

$$h_L = \frac{KV^2}{2g} \qquad (21.42)$$

Table 21.8 gives some typical K values for these losses.

The values in Table 21.8 are only approximate. K values vary not only for different sizes of fitting but with different manufacturers. For these reasons, manufacturers' data are the best source for loss coefficients.

Experimental data available on bend losses cover a rather narrow range of laboratory experiments utilizing small-diameter pipes and do not give conclusive results. The data indicate the losses vary with surface roughness, Reynolds number, ratio of radius of bend r to pipe diameter D, and

Table 21.7 Coefficients for Entrance Losses

Pipe projecting into reservoir	$K = 0.80$

(Table continued)

angle of bend. The data are in agreement that the head loss, not including friction loss, decreases sharply as the r/D ratio increases from zero to around 4 or 5. When r/D increases above 4 or 5, there is disagreement. Some experiments indicate that the head loss, not including friction loss in the bend, increases significantly with an increasing r/D. Experiments on smooth pipes, indicate that this increase is very slight and that above an r/D of 4, the bend loss essentially remains constant. (H. Ito, "Pressure Losses in Smooth Pipe Bends," *Transactions of the American Society of Civil Engineers*, series D, vol. 82, no. 1, 1960.)

Because experiments have produced such widely varying data, bend-loss coefficients give only an approximation of losses to be expected. Figure 21.22 gives values of K for 90° bends for use with Eq. (21.42). (K. H. Beij, "Pressure Losses for Fluid Flow in 90° Pipe Bends," *Journal of Research, National Bureau of Standards*, vol. 21, July 1938.)

To obtain losses in bends other than 90°, the following formula may be used to adjust the K values given in Fig. 21.22:

$$K' = K\sqrt{\frac{\Delta}{90}} \qquad (21.43)$$

where Δ = deflection angle, deg

The K' value may be used in place of K in Eq. (21.42).

Table 21.8 Coefficients for Fitting Losses and Losses at Bends

Fitting	K
Globe valve, fully open	10.0
Angle valve, fully open	5.0
Swing check valve, fully open	2.5
Gate valve, fully open	0.2
Closed-return bend	2.2

(Table continued)

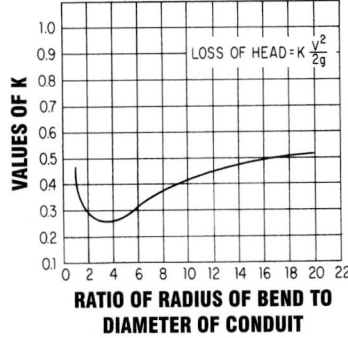

Fig. 21.22 Recommended values of head-loss coefficients K for 90° bends in closed conduits.

Minor losses are often given as the equivalent length of pipe that has the same energy loss for the same discharge. (V. J. Zipparo and H. Hasen, "Davis' Handbook of Applied Hydraulics," 4th ed., McGraw-Hill, Inc., New York (books.mcgraw-hill.com).)

21.11 Orifices

An orifice is an opening with a closed perimeter through which water flows. Orifices may have any shape, although they are usually round, square, or rectangular.

21.11.1 Orifice Discharge into Free Air

Discharge through a sharp-edged orifice may be calculated from

$$Q = Ca\sqrt{2gh} \qquad (21.44)$$

where Q = discharge, ft³/s

C = coefficient of discharge

a = area of orifice, ft²

g = acceleration due to gravity, ft/s²

h = head on horizontal center line of orifice, ft

Coefficients of discharge C are given in Table 21.9 for low velocity of approach. If this velocity is significant, its effect should be taken into account. Equation (21.44) is applicable for any head for which the coefficient of discharge is known. For low heads, measuring the head from the center line of the orifice is not theoretically correct; how-ever, this error is corrected by the C values.

Table 21.9 Smith's Coefficients of Discharge for Circular and Square Orifices with Full Contraction*

Dia. of circular orifices, ft				Head,	Side of square orifices, ft			
0.02	0.04	0.1	1.0	ft	0.02	0.04	0.1	1.0
	0.637	0.618		0.4		0.643	0.621	
0.655	0.630	0.613		0.6	0.660	0.636	0.617	
0.648	0.626	0 610	0.590	0.8	0.652	0.631	0.615	0.597
0.644	0.623	0.608	0.591	1	0.648	0.628	0.613	0.599
0.637	0.618	0.605	0.593	1.5	0.641	0.622	0.610	0.601
0.632	0.614	0.604	0.595	2	0.637	0.619	0.608	0.602
0.629	0.612	0.603	0.596	2.5	0.634	0.617	0.607	0.602
0.627	0.611	0.603	0.597	3	0.632	0.616	0.607	0.603
0.623	0.609	0.602	0.596	4	0.628	0.614	0.606	0.602
0.618	0.607	0.600	0.596	6	0.623	0.612	0.605	0.602
0.614	0.605	0.600	0.596	8	0.619	0.610	0.605	0.602
0.611	0.603	0.598	0.595	10	0.616	0.608	0.604	0.601
0.601	0.599	0.596	0.594	20	0.606	0.604	0.602	0.600
0.596	0.595	0.594	0.593	50	0.602	0.601	0.600	0.599
0.593	0.592	0.592	0.592	100	0.599	0.598	0.598	0.598

* Hamilton Smith, Jr., "Hydraulics," 1886.

The **coefficient of discharge** C is the product of the coefficient of velocity C_v and the coefficient of contraction C_c. The **coefficient of velocity** is the ratio obtained by dividing the actual velocity at the **vena contracta** (contraction of the jet discharged) by the theoretical velocity. The theoretical velocity may be calculated by writing Bernoulli's equation for points 1 and 2 in Fig. 21.23.

$$\frac{V_1^2}{2g} + \frac{p_1}{w} + Z_1 = \frac{V_2^2}{2g} + \frac{p_2}{w} + Z_2 \qquad (21.45)$$

With the reference plane through point 2, $Z_1 = h$, $V_1 = 0$, $p_1/w = p_2/w = 0$, and $Z_2 = 0$, and Eq. (21.45) becomes

$$V_2 = \sqrt{2gh} \qquad (21.46)$$

The actual velocity, determined experimentally, is less than the theoretical velocity because of the energy loss from point 1 to point 2. Typical values of C_v range from 0.94 to 0.99.

The **coefficient of contraction** C_c is the ratio of the smallest area of the jet, the vena contracta, to the area of the orifice. Contraction of a fluid jet will occur if the orifice is square-edged and so located that some of the fluid approaches the orifice at an angle to the direction of flow through the orifice. This fluid has a momentum component perpendicular to the axis of the jet which causes the jet to contract. Typical values of the coefficient of contraction range from 0.61 to 0.67.

If the water entering the orifice does not have this momentum, the contraction is completely suppressed. Figure 21.24a is an example of a partly suppressed contraction; no contraction occurs at the bottom of the jet. In Fig. 21.24b, the edges of the orifice have been rounded to reduce or eliminate the contraction. With a partly suppressed orifice, the increased area of jet caused by suppressing the contraction on one side is partly offset because more water at a higher velocity enters on the other sides. The result is a slightly greater coefficient of contraction.

21.11.2 Submerged Orifices

Flow through a submerged orifice may be computed by applying Bernoulli's equation to points 1 and 2 in Fig. 21.25.

$$V_2 = \sqrt{2g\left(h_1 - h_2 + \frac{V_1^2}{2g} - h_L\right)} \qquad (21.47)$$

where h_L = losses in head, ft, between 1 and 2.

Assuming $V_1 \approx 0$, setting $h_1 - h_2 = \Delta h$, and using a coefficient of discharge C to account for losses, Eq. (21.48) is obtained.

$$Q = Ca\sqrt{2g\Delta h} \qquad (21.48)$$

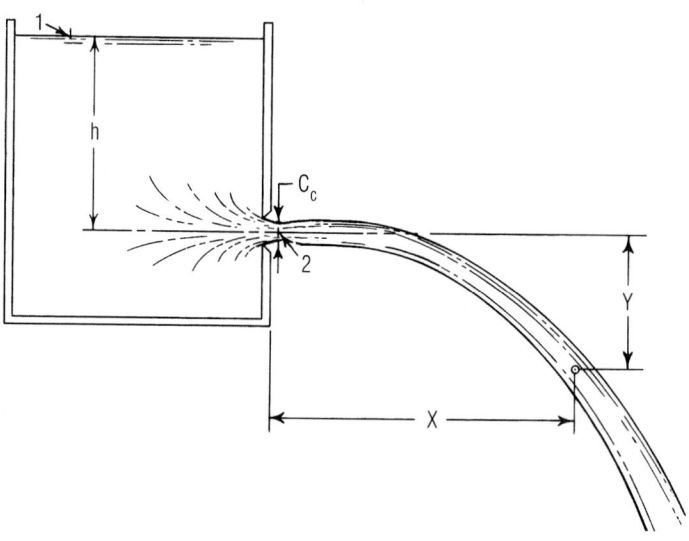

Fig. 21.23 Fluid jet takes a parabolic path.

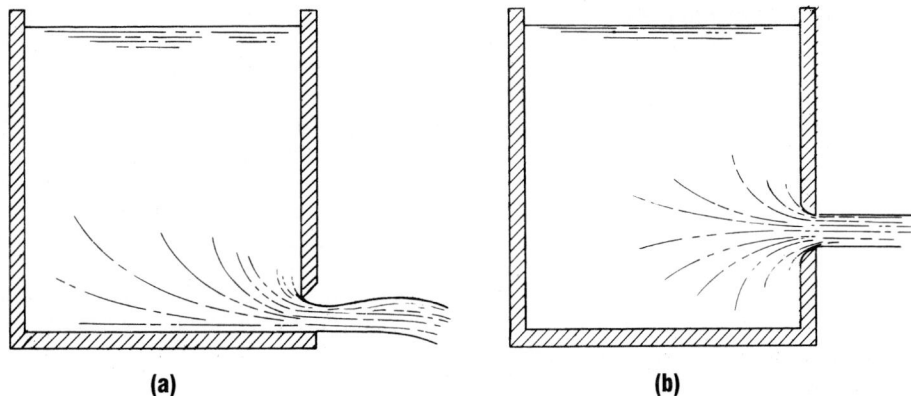

Fig. 21.24 Types of orifices: (*a*) Sharp-edged with partly suppressed contraction. (*b*) Round-edged with no contraction.

Values of C for submerged orifices do not differ greatly from those for nonsubmerged orifices. (For table of values of coefficients of discharge for submerged orifices, see E. F. Brater, "Handbook of Hydraulics," 6th ed., McGraw-Hill Book Company, New York (books.mcgraw-hill.com).)

21.11.3 Discharge under Falling Head

The flow from a reservoir or vessel when the inflow is less than the outflow represents a condition of falling head. The time required for a certain quantity of water to flow from a reservoir can be calculated by equating the volume of water that flows through the orifice or pipe in time dt to the volume decrease in the reservoir (Fig. 21.26):

$$Ca\sqrt{2gy}\,dt = A\,dy \qquad (21.49)$$

Solving for dt yields

$$dt = \frac{A\,dy}{Ca\sqrt{2gy}} \qquad (21.50)$$

where a = area of orifice, ft^2

A = area of reservoir, ft^2

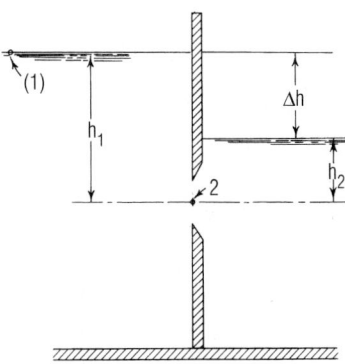

Fig. 21.25 Discharge through a submerged orifice.

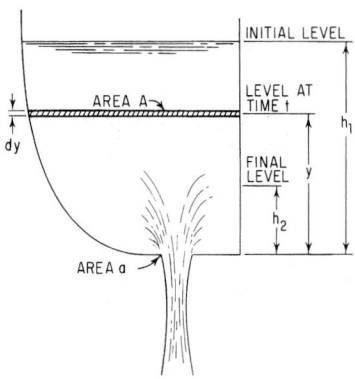

Fig. 21.26 Discharge from a reservoir with dropping water level.

y = head on orifice at time t, ft

C = coefficient of discharge

g = acceleration due to gravity, 32.2 ft/s²

Expressing the area as a function of y [$A = F(y)$] and summing from time zero, when $y = h_1$, to time t, when $y = h_2$, Eq. (21.50) becomes

$$t = \int_{h_2}^{h_1} \frac{F(y)\,dy}{Ca\sqrt{2gy}} \qquad (21.51)$$

If the area of the reservoir is constant as y varies, Eq. (21.51) upon integration becomes

$$t = \frac{2A}{Ca\sqrt{2g}}(\sqrt{h_1} - \sqrt{h_2}) \qquad (21.52)$$

where h_1 = head at the start, ft

h_2 = head at the end, ft

t = time interval for head to fall from h_1 to h_2, s

21.11.4 Fluid Jets

Where the effect of air resistance is small, a fluid discharged through an orifice into the air will follow the path of a projectile. The initial velocity of the jet is

$$V_0 = C_v\sqrt{2gh} \qquad (21.53)$$

where h = head on center line of orifice, ft

C_v = coefficient of velocity

The direction of the initial velocity depends on the orientation of the surface in which the orifice is located. For simplicity, the following equations were determined assuming the orifice is located in a vertical surface (Fig. 21.23). The velocity of the jet in the X direction (horizontal) remains constant.

$$V_x = V_0 = C_v\sqrt{2gh} \qquad (21.54)$$

The velocity in the Y direction is initially zero and thereafter a function of time and the acceleration of gravity:

$$V_y = gt \qquad (21.55)$$

The X coordinate at time t is

$$X = V_x t = tC_v\sqrt{2gh} \qquad (21.56)$$

The Y coordinate is

$$Y = V_{avg}t = \frac{gt^2}{2} \qquad (21.57)$$

where V_{avg} = average velocity over period of time t. The equation for the path of the jet [Eq. (21.58)], obtained by solving Eq. (21.57) for t and substituting in Eq. (21.56), is that for a parabola:

$$X^2 = C_v^2 4hY \qquad (21.58)$$

Equation (21.58) can be used to determine C_v experimentally. Rearranging Eq. (21.58) gives

$$C_v = \sqrt{\frac{X^2}{4hY}} \qquad (21.59)$$

The X and Y coordinates can be measured in a laboratory and C_v calculated from Eq. (21.59).

21.11.5 Orifice Discharge into Short Tubes

When water flows from a reservoir into a pipe or tube with a sharp leading edge, the same type of contraction occurs as for a sharp-edged orifice. In the tube or pipe, however, the water contracts and then expands to fill the tube. If the tube is discharging at atmospheric pressure, a partial vacuum is created at the contraction, as can be seen by applying the Bernoulli equation across points 1 and 2 in Fig. 21.27. This reduced pressure causes the flow through a short tube to be greater than that through a sharp-edged orifice of the same dimensions. If the head on the tube is greater than 50 ft and the tube is short, the water will shoot through the tube without filling it. When this happens, the tube acts as a sharp-edged orifice.

For a short tube flowing full, the coefficient of contraction $C_c = 1.00$ and the coefficient of velocity $C_v = 0.82$. Therefore, the coefficient of discharge $C = 0.82$. Solving for head loss as a proportion of final velocity head, a K value for Eq. (21.42) of 0.5 is obtained as follows: The theoretical velocity head with no loss is $V_T^2/2g$. Actual velocity head is $V_a^2/2g = (0.82V_T)^2/2g = 0.67V_T^2/2g$. The head loss $h_L = 1.00V_T^2/2g - 0.67V_T^2/2g = 0.33V_T^2/2g$. From $h_L = KV_a^2/2g$, where $V_a^2/2g$ is the actual velocity head,

$$K = \frac{2gh_L}{V_a^2} = \frac{(0.33V_T^2 \times 2g)}{(2g \times 0.67V_T^2)} = 0.5$$

For a reentrant tube projecting into a reservoir (Fig. 21.28), the coefficients of velocity and discharge equal 0.75, and the loss coefficient K equals 0.80.

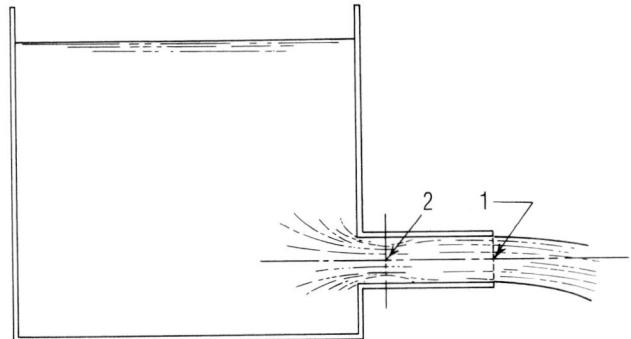

Fig. 21.27 Flow from a reservoir through a tube with a sharp-edged inlet.

21.11.6 Orifice Discharge into Diverging Conical Tubes

This type of tube can greatly increase the flow through an orifice by reducing the pressure at the orifice below atmospheric. Equation (21.60) for the pressure at the entrance to the tube is obtained by writing the Bernoulli equation for points 1 and 3 and points 1 and 2 in Fig. 21.29.

$$p_2 = wh\left[1 - \left(\frac{a_3}{a_2}\right)^2\right] \qquad (21.60)$$

where p_2 = gage pressure at tube entrance, psf

w = unit weight of water, lb/ft^3

h = head on center line of orifice, ft

a_2 = area of smallest part of jet (vena contracta, if one exists), ft^2

a_3 = area of discharge end of tube, ft^2

Discharge is also calculated by writing the Bernoulli equation for points 1 and 3 in Fig. 21.29.

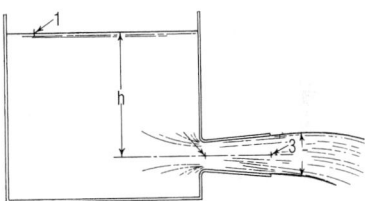

Fig. 21.29 Diverging conical tube increases flow from a reservoir through an orifice by reducing the pressure below atmospheric.

For this analysis to be valid, the tube must flow full, and the pressure in the throat of the tube must not fall to the vapor pressure of water. Experiments by Venturi show the most efficient angle θ to be around 5°.

21.12 Siphons

A siphon is a closed conduit that rises above the hydraulic grade line and in which the pressure at some point is below atmospheric (Fig. 21.30). The most common use of a siphon is the siphon spillway.

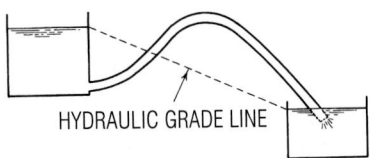

HYDRAULIC GRADE LINE

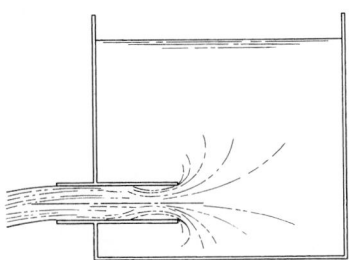

Fig. 21.28 Flow from a reservoir through a reentrant tube resembles that through a flush tube (Fig. 21.27) but the head loss is larger.

Fig. 21.30 Siphon between reservoirs rises above hydraulic grade line yet permits flow of water between them.

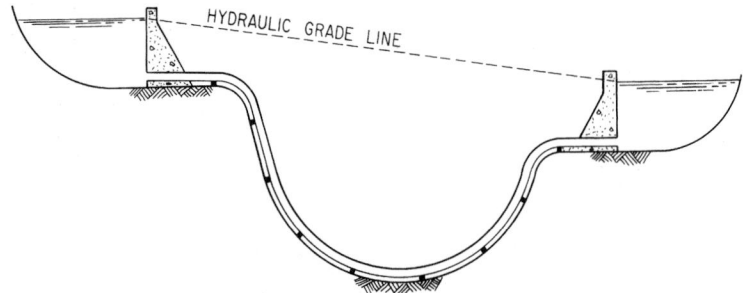

Fig. 21.31   Sag pipe permits flow between two reservoirs despite a dip and a rise.

Flow through a siphon can be calculated by writing the Bernoulli equation for the entrance and exit. But the pressure in the siphon must be checked to be sure it does not fall to the vapor pressure of water. This is accomplished by writing the Bernoulli equation across a point of known pressure and a point where the elevation head or the velocity head is a maximum in the conduit. If the pressure were to fall to the vapor pressure, vaporization would decrease or totally stop the flow.

The pipe shown in Fig. 21.31 is also commonly called a siphon or inverted siphon. This is a misnomer since the pressure at all points in the pipe is above atmospheric. The American Society of Civil Engineers recommends that the inverted siphon be called a **sag pipe** to avoid the false impression that it acts as a siphon.

21.13   Water Hammer

Water hammer is a change in pressure, either above or below the normal pressure, caused by a variation of the flow rate in a pipe. Every time the flow rate is changed, either increased or decreased, it causes water hammer. However, the stresses are not critical in small-diameter pipes with flows at low velocities.

The water flowing in a pipe has momentum equal to the mass of the water times its velocity. When a valve is closed, this momentum drops to zero. The change causes a pressure rise, which begins at the valve and is transmitted up the pipe. The pressure at the valve will rise until it is high enough to overcome the momentum of the water and bring the water to a stop. This pressure

buildup travels the full length of the pipe to the reservoir (Fig. 21.32).

At the instant the pressure wave reaches the reservoir, the water in the pipe is motionless, but at a pressure much higher than normal. The differential pressure between the pipe and the reservoir then causes the water in the pipe to rush back into the reservoir. As the water flows into the reservoir, the pressure in the pipe falls.

At the instant the pressure at the valve reaches normal, the water has attained considerable momentum up the pipe. As the water flows away from the closed valve, the pressure at the valve drops until differential pressure again brings the water to a stop. This pressure drop begins at the valve and continues up the pipe until it reaches the reservoir.

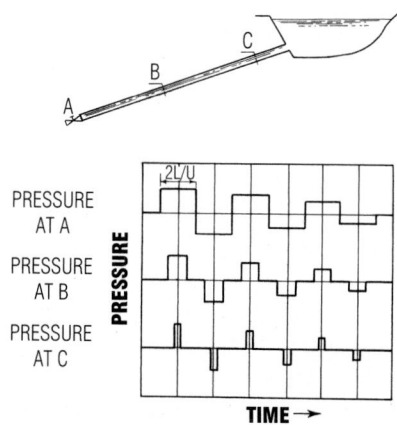

Fig. 21.32   Variation with time of pressure at three points in a penstock, for water hammer from instantaneous closure of a valve.

The pressure in the pipe is now below normal, so water from the reservoir rushes into the pipe. This cycle repeats over and over until friction damps these oscillations. Because of the high velocity of the pressure waves, each cycle may take only a fraction of a second.

The equation for the velocity of a wave in a pipe is

$$U = \sqrt{\frac{E}{\rho}}\sqrt{\frac{1}{1 + ED/E_p t}} \qquad (21.61)$$

where U = velocity of pressure wave along pipe, ft/s

E = modulus of elasticity of water, 43.2×10^6 psf

ρ = density of water, 1.94 lb · s/ft^4 (specific weight divided by acceleration due to gravity)

D = diameter of pipe, ft

E_p = modulus of elasticity of pipe material, psf

t = thickness of pipe wall, ft

21.13.1 Instantaneous Closure

The magnitude of the pressure change that results when flow is varied depends on the rate of change of flow and the length of the pipeline. Any gradual movement of a valve that is made in less time than it takes for a pressure wave to travel from the valve to the reservoir and be reflected back to the valve produces the same pressure change as an instantaneous movement. For instantaneous closure:

$$T < \frac{2L}{U} \qquad (21.62)$$

where L = length of pipe from reservoir to valve, ft

T = time required to change setting of valve, s

A plot of pressure vs. time for various points along a pipe is shown in Fig. 21.32 for the instantaneous closure of a valve. Equation (21.63a) for the pressure rise or fall caused by adjusting a valve was derived by equating the momentum of the water in the pipe to the force impulse required to bring the water to a stop.

$$\Delta p = -\rho U \Delta V \qquad (21.63a)$$

In terms of pressure head, Eq. (21.63a) becomes

$$\Delta h = -\frac{U \Delta V}{g} \qquad (21.63b)$$

where Δp = pressure change from normal due to instantaneous change of valve setting, psf

Δh = head change from normal due to instantaneous change of valve setting, ft

ΔV = change in the velocity of water caused by adjusting valve, ft/s

If the closing or opening of a valve is instantaneous, the pressure change can be calculated in one step from Eq. (21.63).

21.13.2 Gradual Closure

The following method of determining the pressure change due to gradual closure of a valve gives a quick, approximate solution. The pressure rise or head change is assumed to be in direct proportion to the closure time:

$$\Delta h_g = \frac{t_i \Delta h}{T} = \frac{2L \Delta V}{Tg} \qquad (21.64)$$

where Δh_g = head change due to gradual closure, ft

t_i = time for wave to travel from the valve to the reservoir and be reflected back to valve, s

T = actual closure time of valve, s

Δh = head rise due to instantaneous closure, ft

L = length of pipeline, ft

ΔV = change in velocity of water due to instantaneous closure, ft/s

g = acceleration due to gravity, 32.2 ft/s^2

Arithmetic integration is a more exact method for finding the pressure change due to gradual movement of a valve. The calculations can be readily programmed for a computer and are available in software packages. Integration is a direct means of studying every physical element of the process of water hammer. The valve is assumed to close in a series of small movements, each causing an individual pressure wave. The magnitude of these pressure waves is given by Eq. (21.63). The individual pressure waves are totaled to give the pressure at any desired point for a certain time.

The first step in this method is to choose the time interval for each incremental movement of the valve. (It is convenient to make the time interval some submultiple of L/U, such as L/aU, where a equals any integer, so that the pressure waves reflected at the reservoir will be superimposed upon the new waves being formed at the valve. The wave formed at the valve will be opposite in sign to the water reflected from the reservoir, so there will be a tendency for the waves to cancel out.) Assuming a valve is fully open and requires T seconds for closing, the number of incremental closing movements required is $T/\Delta t$, where Δt, the increment of time, equals L/aU.

Once the time interval has been determined, an estimate of the velocity change ΔV during each time interval must be made, to apply Eq. (21.63). A rough estimate for the velocity following the incremental change is $V_n = V_o(A_n/A_o)$, where V_n is the velocity following a certain incremental movement, V_o the original velocity, A_n the area of the valve opening after the corresponding incremental movement, and A_o the original area of the valve opening.

The change in head can now be calculated with Eq. (21.63). With the head known, the estimated velocity V_n can be checked by the following equation:

$$V_n = \frac{V_o A_n}{A_o} \sqrt{\frac{H_o + \Sigma \Delta h}{H_o}} \qquad (21.65)$$

where H_o = head at valve before any movement of valve, ft

$H_o + \Sigma \Delta h$ = total pressure at valve after particular movement; this includes pressure change caused by valve movement plus effect of waves reflected from reservoir, ft

A_n = area of valve opening after n incremental closings; this area can be determined from closure characteristics of valve or by assuming its characteristics, ft^2

If the velocity obtained from Eq. (21.65) differs greatly from the estimated velocity, then that obtained from Eq. (21.65) should be used to recalculate Δh.

(V. J. Zipparo and H. Hasen, "Davis' Handbook of Applied Hydraulics," 4th ed., McGraw-Hill, Inc., New York (books.mcgraw-hill.com).)

Example 21.8: The following problem illustrates the use of the preceding methods and compares the results: Steel penstock, length = 3000 ft, diameter = 10 ft, area = 78.5 ft^2, initial velocity = 10 ft/s, penstock thickness = 1 in, head at turbine with valve open = 1000 ft, and modulus of elasticity of steel = 43.2×10^8 psf.

(For penstocks as shown in Fig. 21.32, thickness and diameter normally vary with head. Thus, the velocity of the pressure waves is different in each section of the penstock. Separate calculations for the velocity of the pressure wave should be made for each thickness and diameter of penstock to obtain the time required for a wave to travel to the reservoir and back to the valve.)

Velocity of pressure wave, from Eq. (21.61), is

$$U = \sqrt{\frac{E}{\rho}} \sqrt{\frac{1}{1 + ED/E_p t}}$$

$$= \sqrt{\frac{43.2 \times 10^6}{1.94}}$$

$$\times \sqrt{\frac{1}{1 + 43.2 \times 10^6 \times 10 \times 12/(1 \times 43.2 \times 10^8)}}$$

$$= 3180 \text{ ft/s}$$

The time required for the wave to travel to the reservoir and be reflected back to the valve = $2L/U = 6000/3180 = 1.90$ s.

If closure time T of the valve is less than 1.90 s, the closure is instantaneous, and the pressure rise, from Eq. (21.63), is

$$\Delta h = \frac{U \Delta V}{g} = \frac{3180 \times 10}{32.2} = 990 \text{ ft}$$

Assuming $T = 4.75$ s, approximate equation (21.64) gives the following result:

$$\Delta h_g = \frac{t_i \Delta V}{T} = \frac{1.90 \times 990}{4.75} = 396 \text{ ft}$$

21.13.3 Surge Tanks

It is uneconomical to design long pipelines for pressures created by water hammer or to operate a valve slowly enough to reduce these pressures. Usually, to prevent water hammer, a surge tank is installed close to valves at the end of long conduits. A surge tank is a tank containing water and connected to the conduit. The water column, in effect, floats on the line.

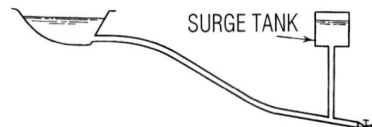

Fig. 21.33 Surge tank is placed near a valve on a penstock to prevent water hammer.

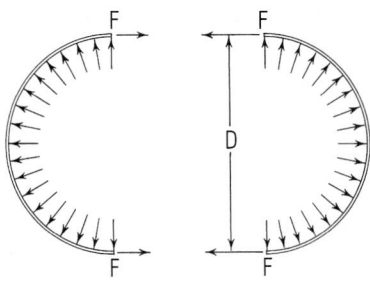

Fig. 21.34 Internal pipe pressure produces hoop tension.

When a valve is suddenly closed, the water in the line rushes into the surge tank. The water level in the tank rises until the increased pressure in the surge tank overcomes the momentum of the water. When a valve is suddenly opened, the surge tank supplies water to the line when the pressure drops. The section of pipe between the surge tank and the valve (Fig. 21.33) must still be designed for water hammer; but the closure time to reduce the pressures for this section will be only a fraction of the time required without the surge tank.

Although a surge tank is one of the most commonly used devices to prevent water hammer, it is by no means the only one. Various types of relief valves and air chambers are widely used on small-diameter lines, where the pressure of water hammer may be relieved by the release of a relatively small quantity of water.

Pipe Stresses

21.14 Pipe Stresses Perpendicular to the Longitudinal Axis

The stresses acting perpendicular to the longitudinal axis of a pipe are caused by either internal or external pressures on the pipe walls.

Internal pressure creates a stress commonly called hoop tension. It may be calculated by taking a free-body diagram of a 1-in-long strip of pipe cut by a vertical plane through the longitudinal axis (Fig. 21.34). The forces in the vertical direction cancel out. The sum of the forces in the horizontal direction is

$$pD = 2F \qquad (21.66)$$

where p = internal pressure, psi

D = outside diameter of pipe, in

F = force acting on each cut of edge of pipe, lb

Hence, the stress, psi, on the pipe material is

$$f = \frac{F}{A} = \frac{pD}{2t} \qquad (21.67)$$

where A = area of cut edge of pipe, ft^2

t = thickness of pipe wall, in

From the derivation of Eq. (21.67), it would appear that the diameter used for calculations should be the inside diameter. However, Eq. (21.67) is not theoretically exact and gives stresses slightly lower than those actually developed. For this reason the outside diameter often is used (see also Art. 6.10).

Equation (21.67) is exact for all practical purposes when D/t is equal to or greater than 50. If D/t is less than 10, this equation will usually be quite conservative and therefore will yield an uneconomical design. For steel pipes, Eq. (21.67) gives directly the thickness required to resist internal pressure.

For concrete pipes, this analysis is approximate, however, since concrete cannot resist large tensile stresses. The force F must be carried by steel reinforcing. The internal diameter is used in Eq. (21.67) for concrete pipe.

When a pipe has external pressure acting on it, the analysis is much more complex because the pipe material no longer acts in direct tension. The external pressure creates bending and compressive stresses that cause buckling.

(S. P. Timoshenko and J. M. Gere, "Theory of elastic Stability," 2nd ed., McGraw-Hill Book Company, New York (books.mcgraw-hill.com).)

21.15 Pipe Stresses Parallel to the Longitudinal Axis

If a pipe is supported on piers, it acts like a beam. The stresses created can be calculated from the bending moment and shear equations for a continuous circular hollow beam. This stress is usually not critical in high-head pipes. However, thin-walled pipes usually require stiffening to prevent buckling and excessive deflection from the concentrated loads.

21.16 Temperature Expansion of Pipe

If a pipe is subject to a wide range of temperatures, the pipe should be the stress due to temperature variation designed for or expansion joints should be provided. The stress, psi, due to a temperature change is

$$f = cE\Delta T \qquad (21.68)$$

where E = modulus of elasticity of pipe material, psi

ΔT = temperature change from installation temperature

c = coefficient of thermal expansion of pipe material

The movement that should be allowed for, if expansion joints are to be used, is

$$\Delta L = Lc\Delta T \qquad (21.69)$$

where ΔL = movement in length L of pipe

L = length between expansion joints

21.17 Forces Due to Pipe Bends

It is common practice to use thrust blocks in pipe bends to take the forces on the pipe caused by the momentum change and the unbalanced internal pressure of the water.

In all bends, there will be a slight loss of head due to turbulence and friction. This loss will cause a pressure change across the bend, but it is usually small enough to be neglected. When there is a change in the cross-sectional area of the pipe, there will be an additional pressure change that can be calculated with the Bernoulli equation (see Example 6, Art. 21.6). In this case, the pressure differential may be large and must be considered.

The force diagram in Fig. 21.35 is a convenient method for finding the resultant force on a bend. The forces can be resolved into X and Y components to find the magnitude and direction of the resultant force on the pipe. In Fig. 21.35:

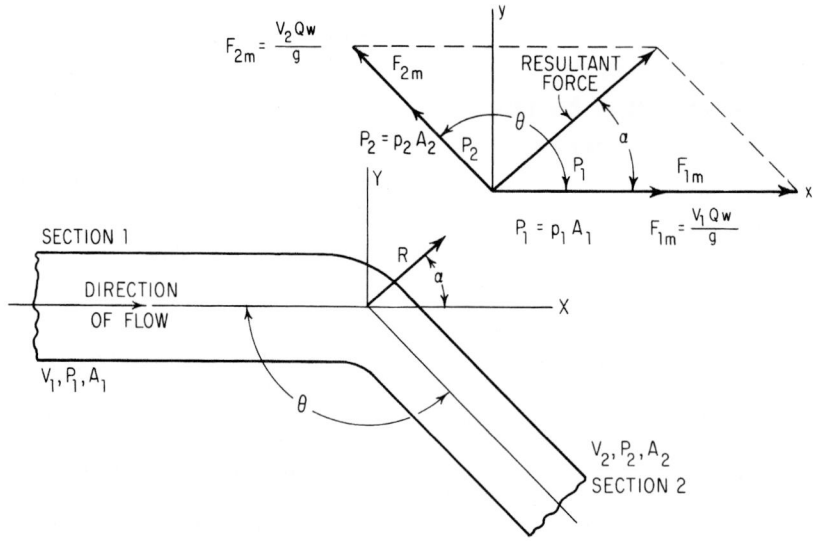

Fig. 21.35 Forces produced by flow at a pipe bend and change in diameter.

V_1 = velocity before change in size of pipe, ft/s

V_2 = velocity after change in size of pipe, ft/s

p_1 = pressure before bend or size change in pipe, psf

p_2 = pressure after bend or size change in pipe, psf

A_1 = area before size change in pipe, ft^2

A_2 = area after size change in pipe, ft^2

F_{2m} = force due to momentum of water in section $2 = V_2 Qw/g$

F_{1m} = force due to momentum of water in section $1 = V_1 Qw/g$

P_2 = pressure of water in section 2 times area of section $2 = p_2 A_2$

P_1 = pressure of water in section 1 times area of section $1 = p_1 A_1$

w = unit weight of liquid, lb/ft^3

Q = discharge, ft^3/s

If the pressure loss in the bend is neglected and there is no change in magnitude of velocity around the bend, Eqs. (21.70) and (21.71) give a quick solution.

$$R = 2A \left(w \frac{V^2}{g} + p \right) \cos \frac{\theta}{2} \qquad (21.70)$$

$$\alpha = \frac{\theta}{2} \qquad (21.71)$$

where R = resultant force on bend, lb

α = angle R makes with F_{1m}

p = pressure, psf

w = unit weight of water, 62.4 lb/ft^3

V = velocity of flow, ft/s

g = acceleration due to gravity, 32.2 ft/s^2

A = area of pipe, ft^2

θ = angle between pipes ($0° \leq \theta \leq 180°$)

Although thrust blocks are normally used to take the force on bends, in many cases the pipe material takes this force. The stress caused by this force is directly additive to other stresses along the longitudinal axis of the pipe. In small pipes, the force caused by bends can easily be carried by the pipe material; however, the joints must also be able to take these forces.

Culverts

A culvert is a closed conduit for the passage of surface drainage under a highway, a railroad, canal, or other embankment. The slope of a culvert and its inlet and outlet conditions are usually determined by the topography of the site. Because of the many combinations obtained by varying the entrance conditions, exit conditions, and slope, no single formula can be given that will apply to all culvert problems.

The basic method for determining discharge through a culvert requires application of the Bernoulli equation between a point just outside the entrance and a point somewhere downstream. An understanding of uniform and nonuniform flow is necessary to understand culvert flow fully. However, an exact theoretical analysis, involving detailed calculation of drawdown and backwater curves, is usually unwarranted because of the relatively low accuracy attainable in determining runoff. Neglecting drawdown and backwater curves does not seriously affect the accuracy but greatly simplifies the calculations.

21.18 Culverts on Critical Slopes or Steeper

In a culvert with a critical slope, the normal depth (Art. 21.22) is equal to the critical depth (Art. 21.23).

Entrance Submerged or Unsubmerged but Free Exit ▪ If a culvert is on critical slope or steeper, that is, the normal depth is equal to or less than the critical depth, the discharge will be entirely dependent on the entrance conditions (Fig. 21.36). Increasing the slope of the culvert past critical slope (the slope just sufficient to maintain flow at critical depth) will decrease the depth of flow downstream from the entrance. But the increased slope will not increase the amount of water entering the culvert because the entrance depth will remain at critical.

The discharge is given by the equation for flow through an orifice if the entrance is submerged, or by the equation for flow over a weir if the entrance is not submerged. Coefficients of discharge for weirs and orifices give good results, but they do not cover the entire range of entry conditions encountered in culvert problems. For this reason, computer software, charts, and nomographs have been

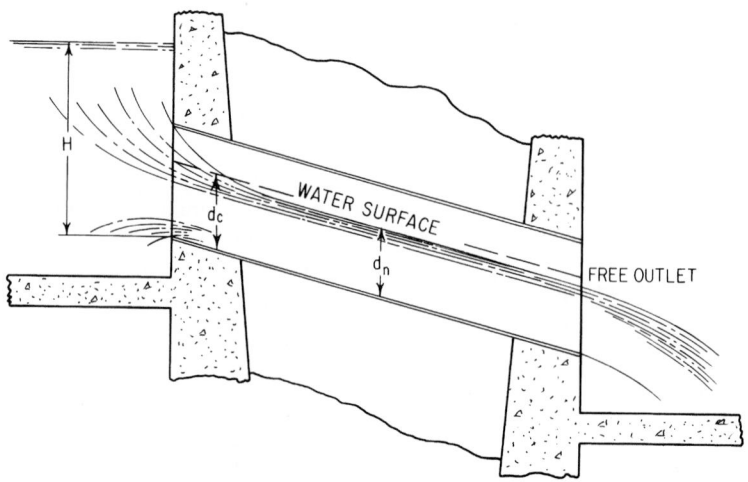

Fig. 21.36 Flow through a culvert with free discharge. Normal depth d_n is less than critical depth d_c; slope is greater than the critical slope. Discharge depends on the type of inlet and the head H.

developed and are used almost exclusively in design. ("Handbook of Concrete Culvert Pipe Hydraulics," EB058W, Portland Cement Association.)

Entrance Unsubmerged but Exit Submerged ■ In this case, the submergence of the exit will cause a hydraulic jump to occur in the culvert (Fig. 21.37). The jump will not affect the culvert discharge, and the control will still be at the inlet.

Entrance and Exit Submerged ■ When both the exit and entrance are submerged (Fig. 21.38), the culvert flows full, and the discharge is independent of the slope. This is normal pipe flow and is easily solved by using the Manning

or Darcy-Weisbach formula for friction loss [Eq. (21.33d) or (21.30)]. From the Bernoulli equation for the entrance and exit, and the Manning equation for friction loss, the following equation is obtained:

$$H = (1 + K_e)\frac{V^2}{2g} + \frac{V^2 n^2 L}{2.21 R^{4/3}} \quad (21.72)$$

Solution for the velocity of flow yields

$$V = \sqrt{\frac{H}{(1 + K_e/2g) + (n^2 L/2.21 R^{4/3})}} \quad (21.73)$$

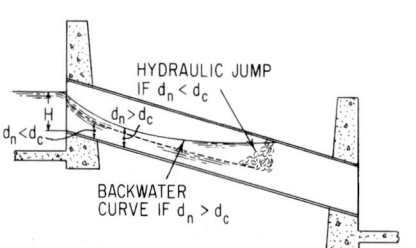

Fig. 21.37 Flow through a culvert with entrance unsubmerged but exit submerged. When slope is less than critical, open-channel flow takes place, and $d_n > d_c$. When slope exceeds critical, flow depends on inlet condition, and $d_n < d_c$.

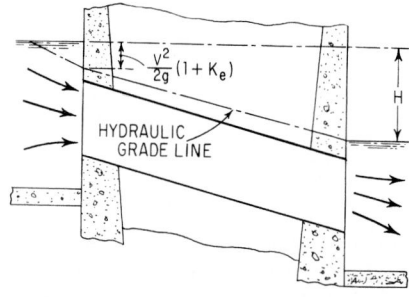

Fig. 21.38 With entrance and exit of a culvert submerged, normal pipe flow occurs. Discharge is independent of slope. The fluid flows under pressure. Discharge may be determined from Bernoulli and Manning equations.

where H = elevation difference between head-water and tailwater, ft

V = velocity in culvert, ft/s

g = acceleration due to gravity, 32.2 ft/s^2

K_e = entrance-loss coefficient (Art. 21.20)

n = Manning's roughness coefficient

L = length of culvert, ft

R = hydraulic radius of culvert, ft

Equation (21.72) can be solved directly since the velocity is the only unknown.

21.19 Culverts on Subcritical Slopes

Critical slope is the slope just sufficient to maintain flow at critical depth. When the slope is less than critical, the flow is considered subcritical (Art. 21.23).

Entrance Submerged or Unsubmerged but Free Exit ▪ For these conditions, depending on the head, the flow can be either pressure or open-channel.

The discharge, for the open-channel condition (Fig. 21.39), is obtained by writing the Bernoulli equation for a point just outside the entrance and a point a short distance downstream from the entrance. Thus,

$$H = K_e \frac{V^2}{2g} + \frac{V^2}{2g} + d_n \qquad (21.74)$$

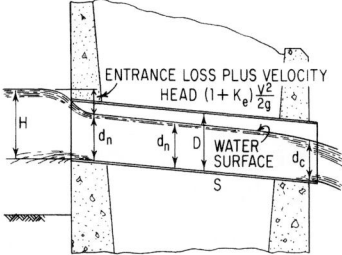

Fig. 21.39 Open-channel flow occurs in a culvert with free discharge and normal depth d_n greater than the critical depth d_c when the entrance is unsubmerged or slightly submerged. Discharge depends on head H, loss at entrance, and slope of culvert.

The velocity can be determined from the Manning equation:

$$V^2 = \frac{2.2SR^{4/3}}{n^2} \qquad (21.75)$$

Substituting this into Eq. (21.74) yields

$$H = (1 + K_e)\frac{2.2}{2gn^2}SR^{4/3} + d_n \qquad (21.76)$$

where H = head on entrance measured from bottom of culvert, ft

K_e = entrance-loss coefficient (Art. 21.20)

S = slope of energy grade line, which for culverts is assumed to equal slope of bottom of culvert

R = hydraulic radius of culvert, ft

d_n = normal depth of flow, ft

To solve Eq. (21.76), it is necessary to try different values of d_n and corresponding values of R until a value is found that satisfies the equation. If the head on a culvert is high, a value of d_n less than the culvert diameter will not satisfy Eq. (21.76). This means the flow is under pressure (Fig. 21.40), and discharge is given by Eq. (21.72).

When the depth of the water is slightly below the top of the culvert, there is a range of unstable flow fluctuating between pressure and open channel. If this condition exists, it is good practice to check the discharge for both pressure flow and open-channel flow. The condition that gives the lesser discharge should be assumed to exist.

Short Culvert with Free Exit ▪ When a culvert on a slope less than critical has a free exit, there will be a drawdown of the water surface at the exit and for some distance upstream. The magnitude of the drawdown depends on the friction slope of the culvert and the difference between the critical and normal depths. If the friction slope approaches critical, the difference between normal depth and critical depth is small (Fig. 21.39), and the drawdown will not extend for any significant distance upstream. When the friction slope is flat, there will be a large difference between normal and critical depth. The effect of the drawdown will extend a greater distance upstream and may reach the entrance of a short culvert (Fig. 21.41). This drawdown of the water level in the entrance of the culvert will increase the

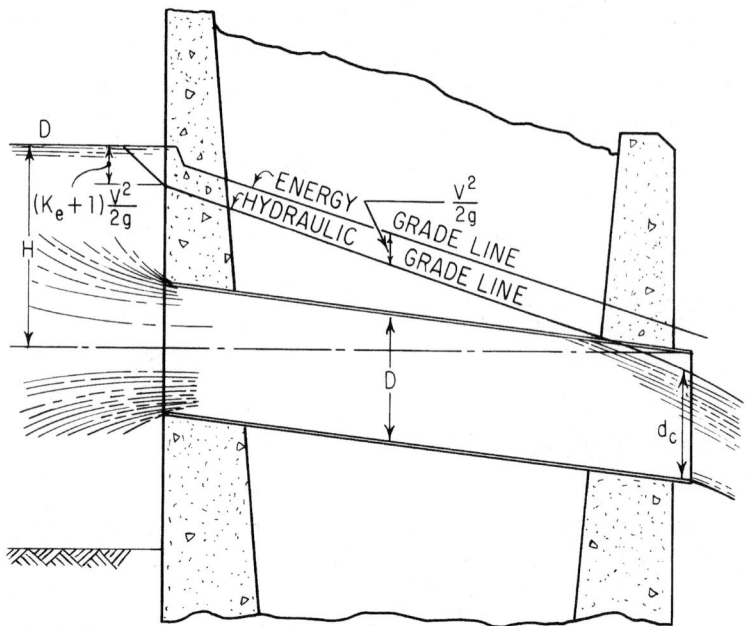

Fig. 21.40 Culvert with free discharge and normal depth d_n greater than critical depth d_c flows full when the entrance is deeply submerged. Discharge is given by equations for pipe flow.

discharge, causing it to be about the same as for a culvert on a slope steeper than critical (Art. 21.18). Most culverts, however, are on too steep a slope for the backwater to have any effect for an appreciable distance upstream.

Entrance Unsubmerged but Exit Submerged ▪ If the level of submergence of the exit is well below the bottom of the entrance (Fig. 21.37), the backwater from the submergence will not

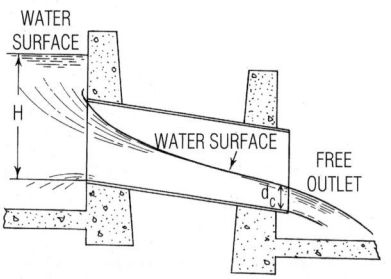

Fig. 21.41 Drawdown of water surface at a free exit of a short culvert with slope less than critical affects depth at entrance and controls discharge.

extend to the entrance. The discharge for this case will be given by Eq. (21.76).

If the level of submergence of the exit is close to the level of the entrance, it may be assumed that the backwater will cause the culvert to flow full and a pipe flow condition will result. The discharge for this case is given by Eqs. (21.72) and (21.73).

When the level of submergence falls between these two cases and the project does not warrant a trial approach with backwater curves, it is good practice to assume the condition that gives the lesser discharge.

21.20 Entrance Losses for Culverts

Flow in a culvert may be significantly affected by loss in head because of conditions at the entrance (Arts. 21.18 and 21.19). Table 21.10 lists coefficients of entrance loss K_e for some typical entrance conditions.

These values are for culverts flowing full. When the entrance is not submerged, the coefficients are usually somewhat lower. But because of the many

Table 21.10 Entrance Loss Coefficients for Culverts

Inlet condition	K_e
Sharp-edged projecting inlet	0.9
Flush inlet, square edge	0.5
Concrete pipe, groove or bell, projecting	0.15
Concrete pipe, groove or bell, flush	0.10
Well-rounded entrance	0.08

unknowns entering into determination of culvert flow, the values tabulated can be used for submerged or unsubmerged cases without much loss of accuracy.

Example 21.9: Given: Maximum head above the top of the culvert = 5 ft, slope = 0.01, length = 300 ft, discharge $Q = 40$ ft³/s, $n = 0.013$, and free exit. *Find:* size of culvert.

Procedure: First assume a trial culvert; then investigate the assumed section to find its discharge. Assume a 2 × 2 ft concrete box section. Calculate Q assuming entrance control, with Eq. (21.44) for discharge through an orifice. The coefficient of discharge C for a 2-ft-square orifice is about 0.6. Head h on center line of entrance = $5 + \frac{1}{2} \times 2 = 6$ ft. Entrance area $a = 2 \times 2 = 4$ ft².

$$Q = Ca\sqrt{2gh} = 0.6 \times 4\sqrt{64.4 \times 6} = 47.2 \text{ ft}^3/\text{s}$$

For entrance control, the flow must be supercritical and d_n must be less than 2 ft. First find d_n.

To calculate the hydraulic radius, assume the depth is slightly less than 2 ft since this will give the maximum possible value of the hydraulic radius for this culvert.

$$R = \frac{\text{area of flow}}{\text{wetted perimeter}} = \frac{2 \times 2}{6} = 0.67 \text{ ft}$$

Application of Eq. (21.33a) gives

$$V = \frac{1.486}{n}R^{2/3}S^{1/2} = \frac{1.486}{0.013} \times 0.67^{2/3} \times 0.01^{1/2}$$

$$= 8.76 \text{ ft/s}$$

$$d_n = \frac{Q}{V \times \text{width}} = \frac{47.2}{8.76 \times 2} = 2.69 \text{ ft}$$

Since d_n is greater than the culvert depth, the flow is under pressure, and the entrance will not control.

Since the culvert is under pressure, Eq. (21.72) applies. But

$$H = 5 + 0.01 \times 300 = 8 \text{ ft}$$

(see Fig. 21.40). The hydraulic radius for pipe flow is $R = 2^2/8 = \frac{1}{2}$. Substitution in Eq. (21.72) yields

$$8 = \frac{1.5V^2}{2g} + 0.0575V^2 = 0.0808V^2$$

$$V = \sqrt{8/0.0808} = 9.95 \text{ ft/s}$$

$$Q = Va = 9.95 \times 4 = 39.8 \text{ ft}^3/\text{s}$$

Since the discharge of tile assumed culvert section under the allowable head equals the maximum expected runoff, the assumed culvert would be satisfactory.

Open-Channel Flow

Free surface flow, or open-channel flow, includes all cases of flow in which the liquid surface is open to the atmosphere. Thus, flow in a pipe is open-channel flow if the pipe is only partly full.

21.21 Basic Elements of Open Channels

A **uniform channel** is one of constant cross section. It has **uniform flow** if the grade, or slope, of the water surface is the same as that of the channel. Hence, depth of flow is constant throughout. **Steady flow** in a channel occurs if the depth at any location remains constant with time.

The **discharge** Q at any section is defined as the volume of water passing that section per unit of time. It is expressed in cubic feet per second, ft³/s, and is given by

$$Q = VA \tag{21.77}$$

where V = average velocity, ft/s

A = cross-sectional area of flow, ft²

When the discharge is constant, the flow is said to be **continuous** and therefore

$$Q = V_1A_1 = V_2A_2 = \dots \tag{21.78}$$

where the subscripts designate different channel sections. Equation (21.78) is known as the continuity equation for continuous steady flow.

In a uniform channel, **varied flow** occurs if the longitudinal water-surface profile is not parallel with

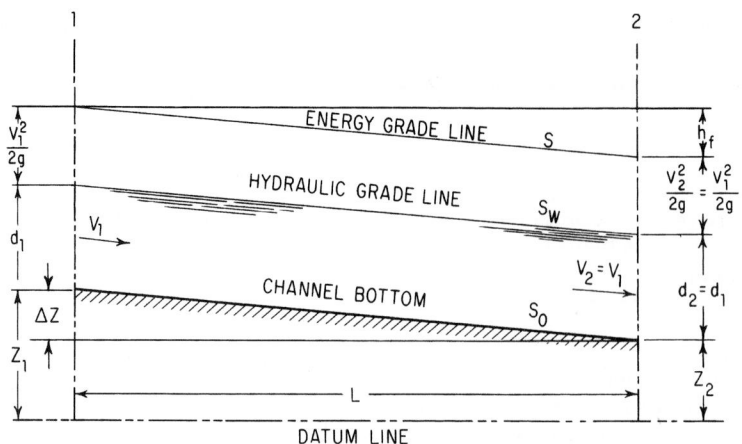

Fig. 21.42 Characteristics of uniform open-channel flow.

the channel bottom. Varied flow exists within the limits of backwater curves, within a hydraulic jump, and within a channel of changing slope or discharge.

Depth of flow d is taken as the vertical distance, ft, from the bottom of a channel to the water surface. The **wetted perimeter** is the length, ft, of a line bounding the cross-sectional area of flow, minus the free surface width. The **hydraulic radius** R equals the area of flow divided by its wetted perimeter. The **average velocity** of flow V is defined as the discharge divided by the area of flow,

$$V = \frac{Q}{A} \qquad (21.79)$$

The velocity head H_V, ft, is generally given by

$$H_V = \frac{V^2}{2g} \qquad (21.80)$$

where V = average velocity from Eq. (21.79), ft/s

g = acceleration due to gravity, 32.2 ft/s^2

Velocity heads of individual filaments of flow vary considerably above and below the velocity head based on the average velocity. Since these velocities are squared in head and energy computations, the average of the velocity heads will be greater than the average-velocity head. The **true velocity head** may be expressed as

$$H_{Va} = \alpha \frac{V^2}{2g} \qquad (21.81)$$

where α is an empirical coefficient that represents the degree of turbulence. Experimental data indi-

cate that α may vary from about 1.03 to 1.36 for prismatic channels. It is, however, normally taken as 1.00 for practical hydraulic work and is evaluated only for precise investigations of energy loss.

The total energy per pound of water relative to the bottom of the channel at a vertical section is called the **specific energy head** H_e. It is composed of the depth of flow at any point, plus the velocity head at the point. It is expressed in feet as

$$H_e = d + \frac{V^2}{2g} \qquad (21.82)$$

A longitudinal profile of the elevation of the specific energy head is called the **energy grade line**, or the **total-head line**. A longitudinal profile of the water surface is called the **hydraulic grade line**. The vertical distance between these profiles at any point equals the velocity head at that point.

Figure 21.42 shows a section of uniform open channel for which the slopes of the water surface S_w and the energy grade line S equal the slope of the channel bottom S_o.

Loss of head due to friction h_f in channel length L equals the drop in elevation of the channel ΔZ in the same distance.

21.22 Normal Depth of Flow

The depth of equilibrium flow that exists in the channel of Fig. 21.42 is called the normal depth d_n. This depth is unique for specific discharge and channel conditions. It may be computed by a trial-and-error process when the channel shape, slope, roughness, and discharge are known. A form of the Manning equation has been suggested for this calculation. (V. T. Chow. "Open-Channel Hydraulics," McGraw-Hill Book Company, New York.)

$$AR^{2/3} = \frac{Qn}{1.486S^{1/2}} \qquad (21.83)$$

where A = area of flow, ft^2

R = hydraulic radius, ft

Q = amount of flow or discharge, ft^3/s

n = Manning's roughness coefficient

S = slope of energy grade line or loss of head, ft, due to friction per lin ft of channel

$AR^{2/3}$ is referred to as a **section factor**. Depth d_n for uniform channels may be computed with computer software or for manual computations simplified by use of tables that relate d_n to the bottom width of

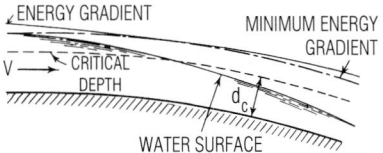

Fig. 21.43 Prismatic channel with gradually increasing bottom slope. Normal depth increases downstream as slope increases.

a rectangular or trapezoidal channel, or to the diameter of a circular channel. (See, for example, E. F. Brater, "Handbook of Hydraulics," 6th ed., McGraw-Hill Book Company, New York.)

In a prismatic channel of gradually increasing slope, normal depth decreases downstream, as shown in Fig. 21.43, and specific energy first decreases and then increases as shown in Fig. 21.44.

The specific energy is high initially where the channel is relatively flat because of the large normal depth (Fig. 21.43). As the depth decreases downstream, the specific energy also decreases. It

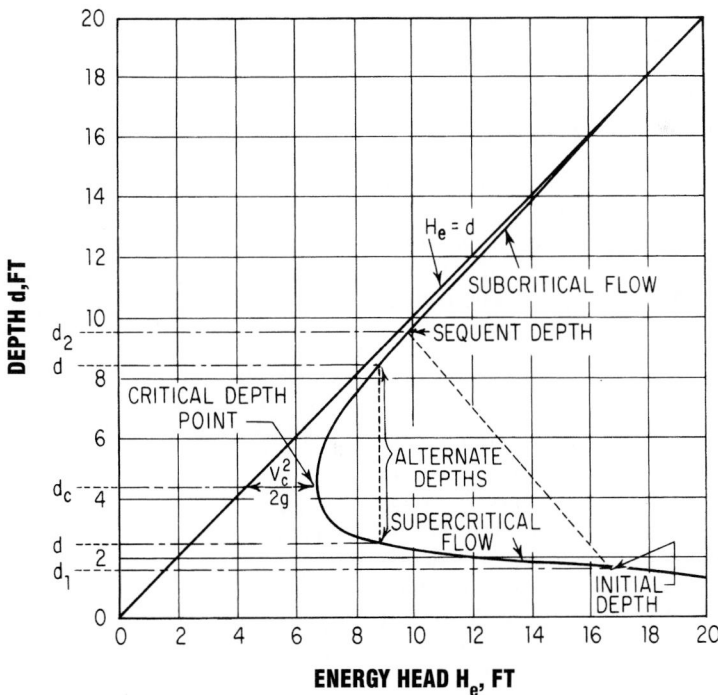

Fig. 21.44 Specific energy head H_e changes with depth for constant discharge in a rectangular channel of changing slope. H_e is a minimum for flow with critical depth.

reaches a minimum at the point where the flow satisfies the equation

$$\frac{A^3}{T} = \frac{Q^2}{g} \tag{21.84}$$

in which T is the top width of the channel, ft. For a rectangular channel, Eq. (21.84) reduces to

$$\frac{d}{2} = \frac{V^2}{2g} \tag{21.85}$$

where $V = Q/A$ = mean velocity of flow, ft^3/s

   d = depth of flow, ft

This indicates that the specific energy is a minimum where the normal depth equals twice the velocity head. As the depth continues to decrease in the downstream direction, the specific energy increases again because of the higher velocity head (Fig. 21.44).

21.23 Critical Depth of Open-Channel Flow

The depth of flow that satisfies Eq. (21.84) is called the **critical depth** d_c. For a given value of specific energy, the critical depth gives the greatest discharge, or conversely, for a given discharge, the specific energy is a minimum for the critical depth (Fig. 21.44).

In the section of mild slope upstream from the critical-depth point in Fig. 21.43, the depth is greater than critical. The flow there is called **subcritical flow**, indicating that the velocity is less than that at critical depth. In the section of steeper slope below the critical-depth point, the depth is below critical. The velocity there exceeds that at critical depth, and flow is **supercritical**.

Critical depth may be computed for a uniform channel once the discharge is known. Determination of this depth is independent of the channel slope and roughness since critical depth simply represents a depth for which the specific energy head is a minimum. Critical depth may be calculated by trial and error with Eq. (21.84), or it may be found directly from tables (E. F. Brater, "Handbook of Hydraulics," 6th ed., McGraw-Hill Book Company, New York). For rectangular channels, Eq. (21.84) may be reduced to

$$d_c = \sqrt[3]{\frac{Q^2}{b^2 g}} \tag{21.86}$$

where d_c = critical depth, ft

   Q = quantity of flow or discharge, ft^3/s

   b = width of channel, ft

Critical slope is the slope of the channel bed that will maintain flow at critical depth. Such slopes should be avoided in channel design because flow near critical depth tends to be unstable and exhibits turbulence and water-surface undulations.

Critical depth, once calculated, should be plotted for the full length of a uniform channel, regardless of slope, to determine whether the normal depth at any section is subcritical or supercritical. [As indicated by Eq. (21.85), if the velocity head is less than half the depth in a rectangular channel, flow is subcritical, but if velocity head exceeds half the depth, flow is supercritical.] If channel configuration is such that the normal depth must go from below to above critical, a hydraulic jump will occur, along with a high loss of energy. Critical depth will change if the channel cross section changes, so the possibility of a hydraulic jump in the vicinity of a transition should be investigated.

For every depth greater than critical depth, there is a corresponding depth less than critical that has an identical value of specific energy (Fig. 21.44). These depths of equal energy are called **alternate depths**. The fact that the energy is the same for alternate depths does not mean that the flow may switch from one alternate depth to the other and back again; flow will always seek to attain the normal depth in a uniform channel and will maintain that depth unless an obstruction is met.

It can be seen from Fig. 21.44 that any obstruction to flow that causes a reduction in total head causes subcritical flow to experience a drop in depth and supercritical flow to undergo an increase in depth.

If supercritical flow exists momentarily on a flat slope because of a sudden grade change in the channel (Fig. 21.52b, p. 21.56), depth increases suddenly from the depth below critical to a depth above critical in a hydraulic jump. The depth following the jump will not be the alternate depth, however. There has been a loss of energy in making the jump. The new depth is said to be sequent to the initial depth, indicating an irreversible occurrence. There is no similar phenomenon that allows a sudden change in depth from subcritical flow to supercritical flow with a corresponding gain in energy. Such a change occurs gradually, without turbulence, as indicated in Fig. 21.45.

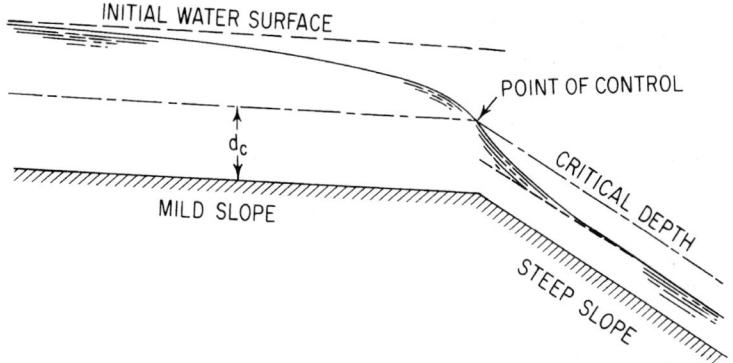

Fig. 21.45 Change in flow stage from subcritical to supercritical occurs gradually.

21.24 Manning's Equation for Open Channels

One of the more popular of the numerous equations developed for determination of flow in an open channel is Manning's variation of the **Chezy formula,**

$$V = C\sqrt{RS} \qquad (21.87)$$

where R = hydraulic radius, ft

V = mean velocity of flow, ft/s

S = slope of energy grade line or loss of head due to friction, ft/lin ft of channel

C = Chezy roughness coefficient

Manning proposed

$$C = \frac{1.486^{1/6}}{n} \qquad (21.88)$$

where n is the coefficient of roughness in the earlier Ganguillet-Kutter formula (see also Art. 21.25). When Manning's C is used in the Chezy formula, the Manning equation for flow velocity in an open channel results:

$$V = \frac{1.486}{n} R^{2/3} S^{1/2} \qquad (21.89)$$

Since the discharge $Q = VA$, Eq. (21.89) may be written

$$Q = \frac{1.486}{n} A R^{2/3} S^{1/2} \qquad (21.90)$$

where A = area of flow, ft^2

Q = quantity of flow, ft^3/s

Roughness Coefficient for Open Channels. Values of the roughness coefficient n for Manning's equation have been determined for a wide range of natural and artificial channel construction materials. Excerpts from a table of these coefficients taken from V. T. Chow, "Open-Channel Hydraulics," McGraw-Hill Book Company, New York (www.mcgraw-hill.com), are in Table 21.11. Dr. Chow compiled data for his table from work by R. E. Horton and from technical bulletins published by the U.S. Department of Agriculture.

Channel roughness does not remain constant with time or even depth of flow. An unlined channel excavated in earth may have one n value when first put in service and another when overgrown with weeds and brush. If an unlined channel is to have a reasonably constant n value over its useful lifetime, there must be a continuing maintenance program. Shallow flow in an unlined channel will result in an increase in the effective n value if the channel bottom is covered with large boulders or ridges of silt since these projections would then have a larger influence on the flow than for deep flow. A deeper-than-normal flow will also result in an increase in the effective n value if there is a dense growth of brush along the banks within the path of flow. When channel banks are overtopped during a flood, the effective n value increases as the flow spills into heavy growth bordering the channel. (Although based on surface roughness, n in practice is sometimes treated as a lumped parameter for all head losses.) The roughness of a lined channel experiences change with age because of both deterioration of the surface and accumulation of foreign matter; therefore, the average n values given in Table 21.11 are

Table 21.11 Values of the Roughness Coefficient n for Use in the Manning Equation

	Min	Avg	Max
A. Open-channel flow in closed conduits			
1. Corrugated-metal storm drain	0.021	0.024	0.030
2. Cement-mortar surface	0.011	0.013	0.015
3. Concrete (unfinished)			
a. Steel form	0.012	0.013	0.014
b. Smooth wood form	0.012	0.014	0.016
c. Rough wood form	0.015	0.017	0.020
B. Lined channels			
1. Metal			
a. Smooth steel (unpainted)	0.011	0.012	0.014
b. Corrugated	0.021	0.025	0.030
2. Wood			
a. Planed, untreated	0.010	0.012	0.014
3. Concrete			
a. Float finish	0.013	0.015	0.016
b. Gunite, good section	0.016	0.019	0.023
c. Gunite, wavy section	0.018	0.022	0.025
4. Masonry			
a. Cemented rubble	0.017	0.025	0.030
b. Dry rubble	0.023	0.032	0.035
5. Asphalt			
a. Smooth	0.013	0.013	
b. Rough	0.016	0.016	
C. Unlined channels			
1. Excavated earth, straight and uniform			
a. Clean, after weathering	0.018	0.022	0.025
b. With short grass, few weeds	0.022	0.027	0.033
c. Dense weeds, high as flow depth	0.050	0.080	0.120
d. Dense brush, high stage	0.080	0.100	0.140
2. Dredged earth			
a. No vegetation	0.025	0.028	0.033
b. Light brush on banks	0.035	0.050	0.060
3. Rock cuts			
a. Smooth and uniform	0.025	0.035	0.040
b. Jagged and irregular	0.035	0.040	0.050

recommended only for well-maintained channels. (See also Art. 21.9 and Table 21.4.)

21.25 Water-Surface Profiles for Gradually Varied Flow

Examples of various surface curves possible with gradually varied flow are shown in Fig. 21.46.

These surface profiles represent backwater curves that form under the conditions illustrated in examples (*a*) through (*r*).

These curves are divided into five groups, according to the slope of the channel in which they appear (Art. 21.23). Each group is labeled with a letter descriptive of the slope: M for mild (subcritical), S for steep (supercritical), C for critical, H for horizontal, and A for adverse. The two dashed lines in the left-hand figure for each

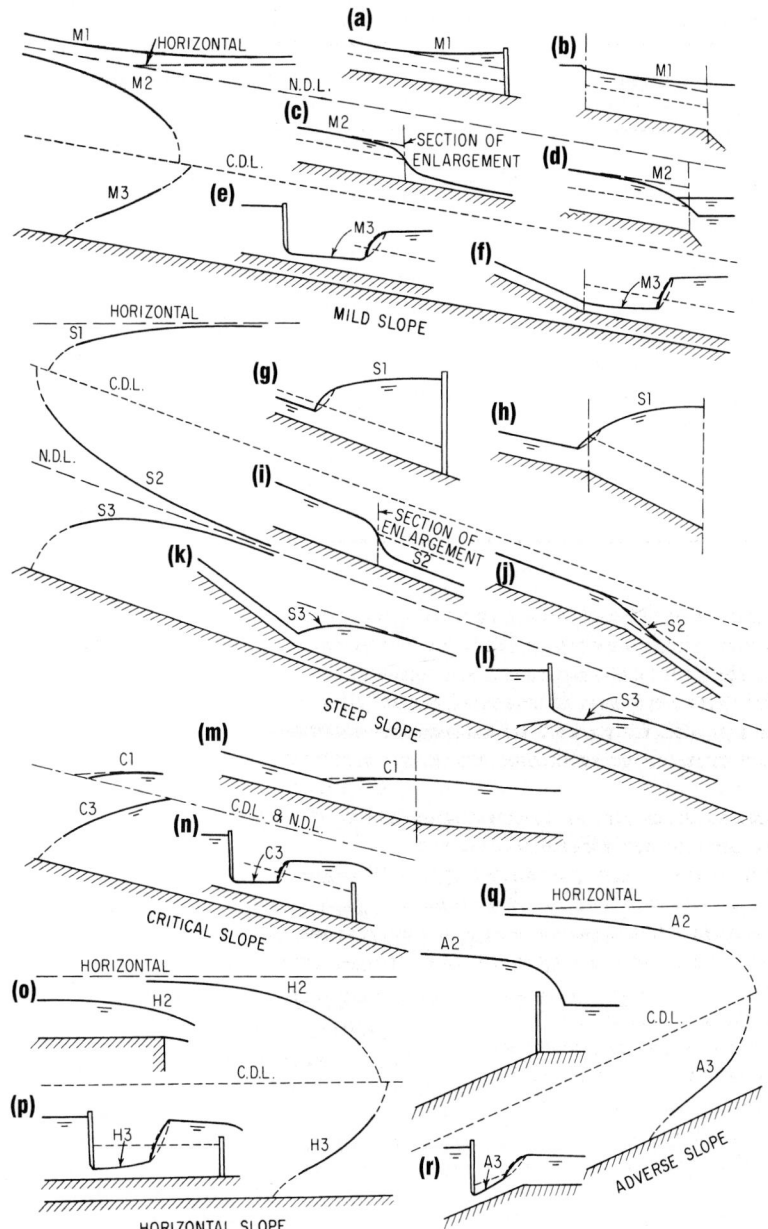

Fig. 21.46 Typical flow profiles for channels with various slopes. N.D.L. indicates normal-depth line; C.D.L., critical-depth line.

class are the **normal-depth line** N.D.L. and the **critical-depth line** C.D.L. The N.D.L. and C.D.L. are identical for a channel of critical slope, and the N.D.L. is replaced by a horizontal line, at an arbitrary elevation, for the channels of horizontal or adverse slope.

There are three types of surface-profile curves possible in channels of mild or steep slope, and two types for channels of critical, horizontal, and adverse slope.

The M1 curve is the familiar surface profile from which all backwater curves derive their name and is the most important from a practical point of view. It forms above the normal-depth line and occurs when water is backed up a stream by high water in the downstream channel, as shown in Fig. 21.46a and b.

The M2 curve forms between the normal- and critical-depth lines. It occurs under conditions shown in Fig. 21.46c and d, corresponding to an increase in channel width or slope.

The M3 curve forms between the channel bottom and critical-depth line. It terminates in a hydraulic jump, except where a drop-off in the channel occurs before a jump can form. Examples of the M3 curve are in Fig. 21.46e and f (a partly opened sluice gate and a decrease in channel slope, respectively).

The S1 curve begins at a hydraulic jump and extends downstream, becoming tangent to a horizontal line (Fig. 21.46g and h) under channel conditions corresponding to those for Fig. 21.46a and b.

The S2 curve, commonly called a drawdown curve, extends downstream from the critical depth and becomes tangent to the normal-depth line under conditions corresponding to those for Fig. 21.46i and j.

The S3 curve is of the transitional type. It forms between two normal depths of less than critical depth under conditions corresponding to those for Fig. 21.46k and l.

Examples in Fig. 21.46m through r show conditions for the formation of C, H, and A profiles.

The curves in Fig. 21.46 approach the normal-depth line asymptotically and terminate abruptly in a vertical line as they approach the critical depth. The curves that approach the bottom intersect it at a definite angle but are imaginary near the bottom since velocity would have to be infinite to satisfy Eq. (21.77) if the depth were zero. The curves are shown dotted near the critical-depth line as a reminder that this portion of the curve does not possess the same degree of accuracy as the rest of the curve because of neglect of vertical components of velocity in the calculations. These curves either start or end at what is called a point of control.

A **point of control** is a physical location in a prismatic channel at which the depth of steady flow may readily be determined. This depth is usually different from the normal depth for the channel because of a grade change, gate, weir, dam, free overfall, or other feature at that location that causes a backwater curve to form. Calculations for the length and shape of the surface profile of a backwater curve start at this known depth and location and proceed either up or downstream, depending on the type of flow. For subcritical flow conditions, the curve proceeds upstream from the point of control in a true backwater curve. The surface curve that occurs under supercritical flow conditions proceeds downstream from the point of control and might better be called a downwater curve.

The point of control is always at the downstream end of a backwater curve in subcritical flow and at the upstream end for supercritical flow. This is explained as follows: A backwater curve may be thought of as being the result of some disruption of uniform flow that causes a wave of disturbance in the channel. The wave travels at a speed, known as its **celerity**, which always equals the critical velocity for the channel. If a disturbance wave attempts to move upstream against supercritical flow (flow moving at a speed greater than critical), it will be swept downstream by the flow and have no effect on conditions upstream. A disturbance wave is held steady by critical flow and moves upstream in subcritical flow.

When a hydraulic jump occurs on a mild slope and is followed by a free overfall (Fig. 21.51), backwater curves form both before and after the jump. The point of control for the curve in the supercritical region above the jump will be located at the vena contracta that forms just below the sluice gate. The point of control for the backwater curve in the subcritical region below the jump is at the free overfall where critical depth occurs. Computations for these backwater curves are carried toward the jump from their respective points of control and are extended across the jump to help determine its exact location. But a backwater curve cannot be calculated through a hydraulic jump from either direction. The surface profiles involved terminate abruptly in a vertical line as they approach the critical depth, and a hydraulic jump always occurs across critical depth. See Art. 21.27.5.

(R. H. French, "Open-Channel Hydraulics," McGraw-Hill, Inc., New York (books.mcgraw-hill.com).)

21.26 Backwater-Curve Computations

The solution of a backwater curve involves computation of a gradually varied flow profile. Solutions available include the graphical-integration, direction-integration, and step methods. Explanations of both the graphical- and direct-integration methods are in V. T. Chow, "Open-Channel Hydraulics," McGraw-Hill Book Company, New York.

Two variations of the step method include the direct or uniform method and the standard method. They are simple and widely used and are available in many software packages.

For step-method computations, the channel is divided into short lengths, or reaches, with relatively small variation. In a series of steps starting from a point of control, each reach is solved in succession. Step methods have been developed for channels with uniform or varying cross sections.

Direct step method of backwater computation involves solving for an unknown length of channel between two known depths. The procedure is applicable only to uniform prismatic channels with gradually varying area of flow.

For the section of channel in Fig. 21.47, Bernoulli's equation for the reach between sections 1 and 2 is

$$S_o L + d_1 + \frac{V_1^2}{2g} = d_2 + \frac{V_2^2}{2g} + \bar{S}L \qquad (21.91)$$

where V_1 and V_2 = mean velocities of flow at sections 1 and 2, ft/s

d_1 and d_2 = depths of flow at sections 1 and 2, ft

g = acceleration due to gravity, 32.2 ft/s²

$\bar{S}$ = average head loss due to friction, ft/ft of channel

S_o = slope of channel bottom

L = length of channel between sections 1 and 2, ft

Note that $S_o L = \Delta z$, the change in elevation, ft, of the channel bottom between sections 1 and 2, and $\bar{S}L = h_f$, the head loss, ft, due to friction in the same reach. (For uniform, prismatic channels, h_i, the eddy loss, is negligible and can be ignored.) $\bar{S}$ equals the slope calculated for the average depth in

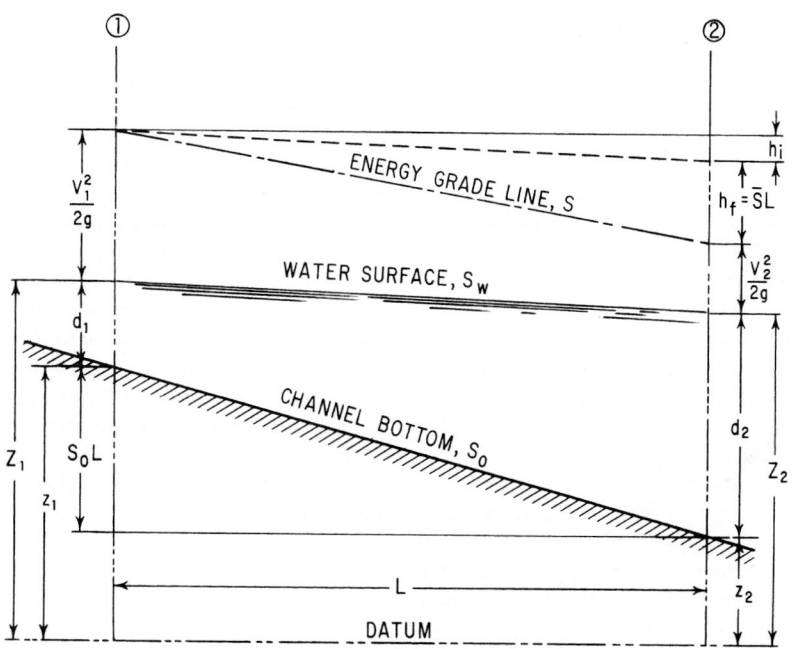

Fig. 21.47 Channel with constant discharge and gradually varying cross section.

the reach but may be approximated by the average of the values of friction slope S for the depths at sections 1 and 2.

Solving Eq. (21.91) for L gives

$$L = \frac{(d_2 + V_2^2/2g) - (d_1 + V_1^2/2g)}{S_o - \bar{S}}$$

$$= \frac{H_{e2} - H_{e1}}{S_o - \bar{S}} \qquad (21.92)$$

where H_{e1} and H_{e2} are the specific energy heads for sections 1 and 2, respectively, as given by Eq. (21.82). The friction slope S at any point may be computed by the Manning equation, rearranged as follows:

$$S = \frac{n^2 V^2}{2.21 R^{4/3}} \qquad (21.93)$$

where R = hydraulic radius, ft

n = roughness coefficient (Art. 21.24)

Note that the slope S used in the Manning equation is the slope of the energy grade line, not the channel bottom. Note also that the roughness coefficient n is squared in Eq. (21.93), and its value must therefore be chosen with special care to avoid an exaggerated error in the computed friction slope. The smaller the value of n, the longer the backwater curve profile, and vice versa. Therefore, the smallest n possible for the prevailing conditions should be selected for computation of a backwater curve if knowledge of the longest possible flow profile is required.

The first step in the direct step method involves choosing a series of depths for the end points of each reach. These depths will range from the depth at the point of control to the ending depth for the backwater curve. This ending depth is often the normal depth for the channel (Art. 21.22) but may be some intermediate depth, such as for a curve preceding a hydraulic jump. Depths should be chosen so that the velocity change across a reach does not exceed 20% of the velocity at the beginning of the reach. Also the change in depth between sections should never exceed 1 ft.

The specific energy head H_e should be computed for the chosen depth at each of the various sections and the change in specific energy between sections determined. Next, the friction slope S should be computed at each section from Eq. (21.93). The average of two sections gives the friction slope $\bar{S}$ between sections. Finally, the

difference between $\bar{S}$ and slope of channel bottom S_o should be computed and the length of reach determined from Eq. (21.92).

Standard step method allows computation of backwater curves in both nonprismatic natural channels and nonuniform artificial channels as well as in uniform channels. This method involves solving for the depth of flow at various locations along a channel with Bernoulli's energy equation and a known length of reach.

A surface profile is determined in the following manner: The channel is examined for changes in cross section, grade, or roughness, and the locations of these changes are given station numbers. Stations are also established between these locations such that the velocity change between any two consecutive stations is not greater than 20% of the velocity at the former station. Data concerning the hydraulic elements of the channel are collected at each station. Computation of the surface curve is then made in steps, starting from the point of control and progressing from station to station—in an upstream direction for subcritical flow and downstream for supercritical flow. The length of reach in each step is given by the stationing, and the depth of flow is determined by trial and error.

Nonprismatic channels do not have well-defined points of control to aid in determining the starting depth for a backwater curve. Therefore, the water-surface elevation at the beginning must be determined as follows:

The step computations are started at a point in the channel some distance upstream or downstream from the desired starting point, depending on whether flow is supercritical or subcritical, respectively. Then, computations progress toward the initial section. Since this step method is a converging process, this procedure produces the true depth for the initial section within a relatively few steps.

The energy balance used in the standard step method is shown graphically in Fig. 21.47, in which the position of the water surface at section 1 is Z_1 and at section 2, Z_2, referred to a horizontal datum. Writing Bernoulli's equation [Eq. (21.11)] for sections 1 and 2 Yields

$$Z_1 + \frac{V_1^2}{2g} = Z_2 + \frac{V_2^2}{2g} + h_f + h_i \qquad (21.94)$$

where V_1 and V_2 are the mean velocities, ft/s, at sections 1 and 2; the friction loss, ft, in the reach ($\bar{S}L$) is denoted by h_f; and the term h_i is added to account for eddy loss, ft.

Eddy loss, sometimes called **impact loss,** is a head loss caused by flow running contrary to the main current because of irregularities in the channel. No rational method is available for determination of eddy loss, and it is therefore often accounted for, in natural channels, by a slight increase in Manning's n. Eddy loss depends mainly on a change in velocity head. For lined channels, it has been expressed as a coefficient k to be applied as follows:

$$h_i = k\left(\frac{V_1^2}{2g} - \frac{V_2^2}{2g}\right) = k\left(\Delta \frac{V^2}{2g}\right) \qquad (21.95)$$

The coefficient k is 0.2 for diverging reaches, from 0 to 0.1 for converging reaches, and about 0.5 for abrupt expansions and contractions.

The total head at any section of the channel is

$$H = Z + \frac{V^2}{2g} \qquad (21.96)$$

where Z equals the elevation of the channel bottom above the given datum plus the depth of flow d at that section. Friction slope S is computed from Eq. (21.93). Then, $\bar{S}$, the average friction slope for the reach, is calculated as the mean of the slope for the section and the preceding section. Friction loss h_f is the product of $\bar{S}$ and the length of the reach L. Eddy loss h_i is found from Eq. (21.95). Next, total head H, ft, is obtained from Eq. (21.94), which, after substitution of H from Eq. (21.96), becomes

$$H_1 = H_2 + h_f + h_i \qquad (21.97)$$

where H_1 and H_2 equal the total head of sections 1 and 2, respectively. The value of total head computed from Eq. (21.97) must agree with the value of total head calculated previously for the section or the assumed water-surface elevation Z_1 is incorrect. Agreement is assumed if the two values of total head are within 0.1 ft in elevation. If the two values of total head do not agree, a new water-surface elevation must be assumed for Z_1 and the computations repeated until agreement is obtained. The value that finally leads to agreement gives the correct water-surface elevation.

Backwater curves for natural river or stream channels (irregularly shaped channels) are calculated in a manner similar to that described for regularly shaped channels. However, some account must be taken of the varying channel roughness and the differences in velocity and capacity in the main channel and the overbank or floodplain portions of the stream channel. The most expeditious way of determining the backwater curves is to plot the channel cross section to a scale convenient for measurement of lengths and areas; subdivide the cross section into main channels and floodplain areas; and determine the discharge, velocity, and friction slope for each subarea at selected water-surface elevations. Utilizing the above data, determine the total discharge (the sum of the subarea discharges), the mean velocity (the total discharge divided by the total area), and α (the energy coefficient or coriolis coefficient to be applied to the velocity head). Many of the available computer software packages that compute backwater profiles are applicable to irregular channels and flooded overbank areas.

The backwater curve is usually started by assuming normal depth at a point some distance downstream from the start of the reach under analysis. Several intermediate cross sections should be taken between the point where normal depth is assumed and the start of the reach for which a detailed water-surface profile is required. This allows the intermediate sections to "dampen out" any minor errors in the assumed starting water-surface elevation.

The accuracy or validity of the water-surface profile is contingent on an accurate evaluation of the channel roughness and judicious selection of cross-section location. A greater number of cross sections generally enhances the validity of the water-surface profile; however, because of the extensive calculations involved with each cross section, their number should be limited to as few as accuracy permits.

The effect of bridges, approach roadways, bridge piers, and culverts can be determined using procedures outlined in R. H. French, "Open-Channel Hydraulics," McGraw-Hill Book Company, New York, and J. N. Bradley, "Hydraulics of Bridge Waterways," Hydraulics Design Series no. 1, 2nd ed., U.S. Department of Transportation, Federal Highway Administration, Bureau of Public Roads, 1970.

21.27 Hydraulic Jump

This is an abrupt increase in depth of rapidly flowing water (Fig. 21.48). Flow at the jump

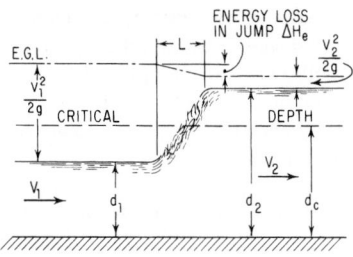

Fig. 21.48 Hydraulic jump.

changes from a supercritical to a subcritical stage with an accompanying loss of kinetic energy (Art. 21.23).

A hydraulic jump is the only means by which the depth of flow can change from less than critical to greater than critical in a uniform channel. A jump will occur either where supercritical flow exists in a channel of subcritical slope, as shown in Figs. 21.51 and 21.52b, or where a steep channel enters a reservoir. The first condition is met in a mild channel downstream from a sluice gate or ogee overflow spillway, or at an abrupt change in channel slope from steep to mild. The second condition occurs where flow in a steep channel is blocked by an overflow weir, a gate, or other obstruction.

A hydraulic jump can be either stationary or moving, depending on whether the flow is steady or unsteady, respectively.

21.27.1 Depth and Head Loss in a Hydraulic Jump

Depth at the jump is not discontinuous. The change in depth occurs over a finite distance, known as the length of jump. The upstream surface of the jump, known as the roller, is a turbulent mass of water, which is continually tumbling erratically against the rapidly flowing sheet below.

The depth before a jump is the **initial depth**, and the depth after a jump is the **sequent depth**. The specific energy for the sequent depth is less than that for the initial depth because of the energy dissipation within the jump. (Initial and sequent depths should not be confused with the depths of equal energy, or alternate depths.)

According to Newton's second law of motion, the rate of loss of momentum at the jump must equal the unbalanced pressure force acting on the moving water and tending to retard its motion. This unbalanced force equals the difference between the hydrostatic forces corresponding to the depths before and after the jump. For rectangular channels, this resultant pressure force is

$$F = \frac{d_2^2 w}{2} - \frac{d_1^2 w}{2} \qquad (21.98)$$

where d_1 = depth before jump, ft

d_2 = depth after jump, ft

w = unit weight of water, lb/ft^3

The rate of change of momentum at the jump per foot width of channel equals

$$F = \frac{MV_1 - MV_2}{t} = \frac{qw}{g}(V_1 - V_2) \qquad (21.99)$$

where M = mass of water, lb · s^2/ft

V_1 = velocity at depth d_1, ft/s

V_2 = velocity at depth d_2, ft/s

q = discharge per foot width of rectangular channel, ft^3/s

t = unit of time, s

g = acceleration due to gravity, 32.2 ft/s^2

Equating the values of F in Eqs. (21.98) and (21.99), and substituting $V_1 d_1$ for q and $V_1 d_1/d_2$ for V_2, the reduced equation for rectangular channels becomes

$$V_1^2 = \frac{g d_2}{2 d_1}(d_2 + d_1) \qquad (21.100)$$

Equation (21.100) may then be solved for the sequent depth:

$$d_2 = \frac{-d_1}{2} + \sqrt{\frac{2V_1^2 d_1}{g} + \frac{d_1^2}{4}} \qquad (21.101)$$

If $V_2 d_2/d_1$ is substituted for V_1, in Eq. (21.100),

$$d_1 = \frac{-d_2}{2} + \sqrt{\frac{2V_2^2 d_2}{g} + \frac{d_2^2}{4}} \qquad (21.102)$$

Equation (21.102) may be used in determining the position of the jump where V_2 and d_2 are known. Relationships may be derived similarly for channels of any cross section.

The head loss in a jump equals the difference in specific-energy head before and after the jump. This difference (Fig. 21.49) is given by

$$\Delta H_e = H_{e1} - H_{e2} = \frac{(d_2 - d_1)^3}{4 d_1 d_2} \qquad (21.103)$$

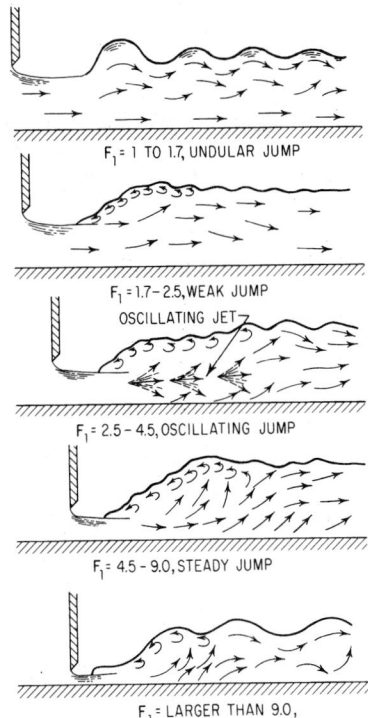

F₁ = 1 TO 1.7, UNDULAR JUMP

F₁ = 1.7 - 2.5, WEAK JUMP

OSCILLATING JET

F₁ = 2.5 - 4.5, OSCILLATING JUMP

F₁ = 4.5 - 9.0, STEADY JUMP

F₁ = LARGER THAN 9.0,
STRONG JUMP

Fig. 21.49 Type of hydraulic jump depends on Froude number.

where H_{e1} = specific-energy head of stream before jump, ft

H_{e2} = specific-energy head of stream after jump, ft

The specific energy for free-surface flow is given by Eq. (21.82).

The depths before and after a hydraulic jump may be related to the critical depth by the equation

$$d_1 d_2 \frac{d_1 + d_2}{2} = \frac{q^2}{g} = d_c^3 \qquad (21.104)$$

where q = discharge, ft³/s per ft of channel width

d_c = critical depth for the channel, ft

It may be seen from this equation that if $d_1 = d_c$, d_2 must also equal d_c.

21.27.2 Jump in Horizontal Rectangular Channels

The form of a hydraulic jump in a horizontal rectangular channel may be of several distinct types, depending on the Froude number of the incoming flow $\mathbf{F} = V/(gL)^{1/2}$ [Eq. (21.16)], where L is a characteristic length, ft; V is the mean velocity, ft/s; and g = acceleration due to gravity, ft/s². For open-channel flow, the characteristic length for the Froude number is made equal to the **hydraulic depth** d_h.

Hydraulic depth is defined as

$$d_h = \frac{A}{T} \qquad (21.105)$$

where A = area of flow, ft²

T = width of free surface, ft

For rectangular channels, hydraulic depth equals depth of flow.

Various forms of hydraulic jump, and their relation to the Froude number of the approaching flow $\mathbf{F}_1$, were classified by the U.S. Bureau of Reclamation and are presented in Fig. 21.49.

For $\mathbf{F}_1 = 1$, the flow is critical and there is no jump.

For $\mathbf{F}_1 = 1$ to 1.7, there are undulations on the surface. The jump is called an undular jump.

For $\mathbf{F}_1 = 1.7$ to 2.5, a series of small rollers develop on the surface of the jump, but the downstream water surface remains smooth. The velocity throughout is fairly uniform and the energy loss is low. This jump may be called a weak jump.

For $\mathbf{F}_1 = 2.5$ to 4.5, an oscillating jet is entering the jump. The jet moves from the channel bottom to the surface and back again with no set period. Each oscillation produces a large wave of irregular period, which, very commonly in canals, can travel for miles, doing extensive damage to earth banks and riprap surfaces. This jump may be called an oscillating jump.

For $\mathbf{F}_1 = 4.5$ to 9.0, the downstream extremity of the surface roller and the point at which the high-velocity jet tends to leave the flow occur at practically the same vertical section. The action and position of this jump are least sensitive to variation in tailwater depth. The jump is well-balanced, and the performance is at its best. The energy dissipation ranges from 45 to 70%. This jump may be called a steady jump.

For $F_1 = 9.0$ and larger, the high-velocity jet grabs intermittent slugs of water rolling down the front face of the jump, generating waves downstream and causing a rough surface. The jump action is rough but effective, and energy dissipation may reach 85%. This jump may be called a strong jump.

Note that the ranges of the Froude number given for the various types of jump are not clear-cut but overlap to a certain extent, depending on local conditions.

21.27.3 Hydraulic Jump as an Energy Dissipator

A hydraulic jump is a useful means for dissipating excess energy in supercritical flow (Art. 21.23). A jump may be used to prevent erosion below an overflow spillway, chute, or sluice gate by quickly reducing the velocity of the flow over a paved apron. A special section of channel built to contain a hydraulic jump is known as a **stilling basin**.

If a hydraulic jump is to function ideally as an energy dissipator, below a spillway, for example, the elevation of the water surface after the jump must coincide with the normal tailwater elevation for every discharge. If the tailwater is too low, the high-velocity flow will continue downstream for some distance before the jump can occur. If the tailwater is too high, the jump will be drowned out, and there will be a much smaller dissipation of total head. In either case, dangerous erosion is likely to occur for a considerable distance downstream.

The ideal condition is to have the *sequent-depth curve*, which gives discharge vs. depth after the jump, coincide exactly with the *tailwater-rating curve*. The tailwater-rating curve gives normal depths in the discharge channel for the range of flows to be expected. Changes in the spillway design that can be made to alter the tailwater-rating curve involve changing the crest length, changing the apron elevation, and sloping the apron.

Accessories, such as chute blocks and baffle blocks are usually installed in a stilling basin to control the jump. The main purpose of these accessories is to shorten the range within which the jump will take place, not only to force the jump to occur within the basin but to reduce the size and therefore the cost of the basin. Controls within a stilling basin have additional advantages in that they improve the dissipation function of the basin and stabilize the jump action.

21.27.4 Length of Hydraulic Jump

The length of a hydraulic jump L may be defined as the horizontal distance from the upstream edge of the roller to a point on the raised surface immediately downstream from cessation of the violent turbulence. This length (Fig. 21.48) defies accurate mathematical expression, partly because of the nonuniform velocity distribution within the jump. But it has been determined experimentally. The experimental results may be summarized conveniently by plotting the Froude number of the upstream flow F_1 against a dimensionless ratio of jump length to downstream depth L/d_2. The resulting curve (Fig. 21.50) has a flat portion in the range of steady jumps. The curve thus minimizes the effect of any errors made in calculation of the Froude number in the range where this information is most frequently needed. The curve, prepared by V. T. Chow from data gathered by the U.S. Bureau of Reclamation, was developed for jumps in rectangular channels, but it will give approximate results for jumps formed in trapezoidal channels.

For other than rectangular channels the depth d_1 used in the equation for Froude number is the hydraulic depth given by Eq. (21.105).

21.27.5 Location of a Hydraulic Jump

It is important to know where a hydraulic jump will form since the turbulent energy released in a jump can extensively scour an unlined channel or destroy paving in a thinly lined channel. Special reinforced sections of channel must be built to withstand the pounding and vibration of a jump and to provide extra freeboard for the added depth at the jump. These features are expensive to build; therefore, a great savings can be realized if their use is restricted to a limited area through a knowledge of the jump location.

The precision with which the location is predicted depends on the accuracy with which the friction losses and length of jump are estimated and on whether the discharge is as assumed. The method of prediction used for rectangular channels is illustrated for a sluice gate in Fig. 21.51.

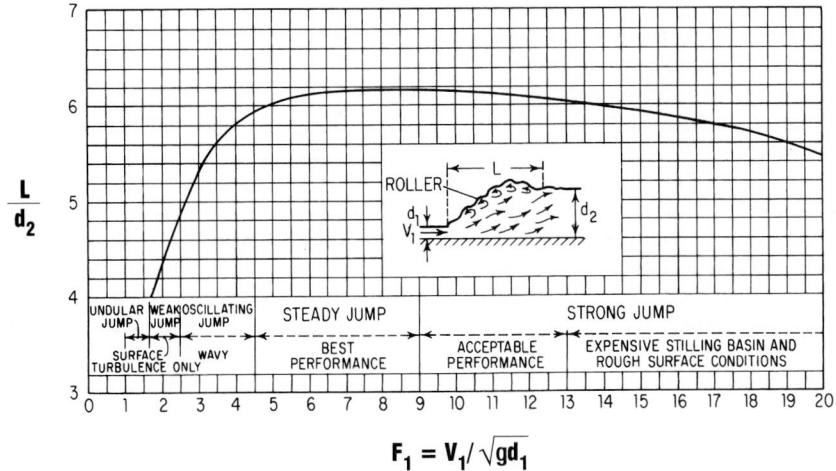

Fig. 21.50 Length of hydraulic jump in a horizontal channel depends on sequent depth d_2 and the Froude number of the approaching flow.

The water-surface profiles of the flow approaching and leaving the jump, curves AB and ED in Fig. 21.51, are type M3 and M2 backwater curves, respectively (Fig. 21.46e and c).

Backwater curve ED has as its point of control the critical depth d_c, which occurs near the channel drop-off. Critical depth does not exist exactly at the edge, as theory would indicate, but instead occurs a short distance upstream. The distance is small (from three to four times d_c) and can be ignored for most problems. The actual depth at the brink is 71.5% of critical depth, but it is normally assumed to be $0.7d_c$ for simplicity.

The point of control for backwater curve AB is taken as the depth at the vena contracta, which

forms just downstream from the sluice gate. The distance from the gate to the vena contracta L_e is nearly equal to the size of gate opening h. The amount of contraction varies with both the head on the gate and the gate opening. Depth at the contraction ranges from 50 to over 90% of h. The depth of flow at the vena contracta may be taken as $0.75h$ in the absence of better information.

Jump location is determined as follows: The backwater curves AB and ED are computed in their respective directions until they overlap, using the step methods of Art. 21.26. With values of d_2 obtained from Eq. (21.101), CB, the curve of depths sequent to curve AB, is plotted through the area where it crosses curve ED. A horizontal intercept

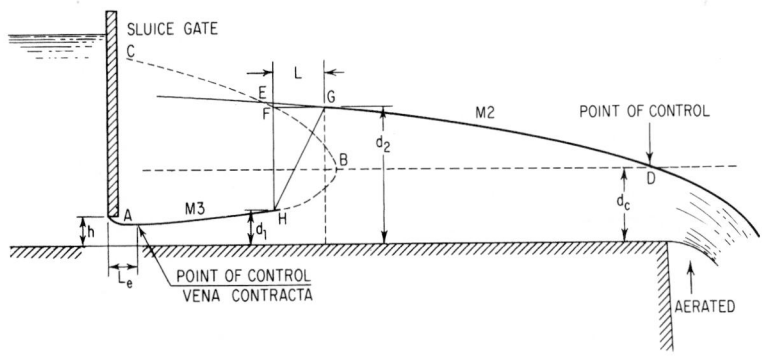

Fig. 21.51 Graphical method for locating hydraulic jump beyond a sluice gate.

FG, equal in length to L, the computed length of jump, is then fitted between the curves CB and ED. The jump may be expected to form between the points H and G since all requirements for the formation of a jump are satisfied at this location.

If the downstream depth is increased because of an obstruction, the jump moves upstream and may eventually be drowned out in front of the sluice gate. Conversely, if the downstream depth is lowered, the jump moves to a new location downstream.

When the slope of a channel has an abrupt change from steeper than critical (Art. 21.23) to mild, a jump forms that may be located either above or below the grade change. The position of the jump depends on whether the downstream depth d_2 is greater than, less than, or equal to the depth d'_1 sequent to the upstream depth d_1. Two possible positions are shown in Fig. 21.52.

It is assumed, for simplicity, that flow is uniform, except in the reach between the jump and the grade break. If the downstream depth d_2 is greater than the upstream sequent depth d'_1, computed from Eq. (21.101) with d_1 given, the jump occurs in the steep region, as shown in Fig. 21.52a. The surface curve EO is of the S1 type (Fig. 21.46) and is asymptotic to a horizontal line at O. Line CB' is a plot of the depth d'_1 sequent to the depth of approach line AB. The jump location is found by producing a horizontal intercept FG, equal to the computed length of the jump, between lines CB' and EO. A jump will form between H and G since all requirements are satisfied for this location. As depth d_2 is lowered, the jump moves downstream to a new position, as shown in Fig. 21.52b. If d_2 is less than d'_1, computed from Eq. (21.102), the jump will form in the mild channel and can be located as described for Fig. 21.51.

(R. H. French, "Open-Channel Hydraulics," McGraw-Hill, Inc., New York (books.mcgraw-hill.com).)

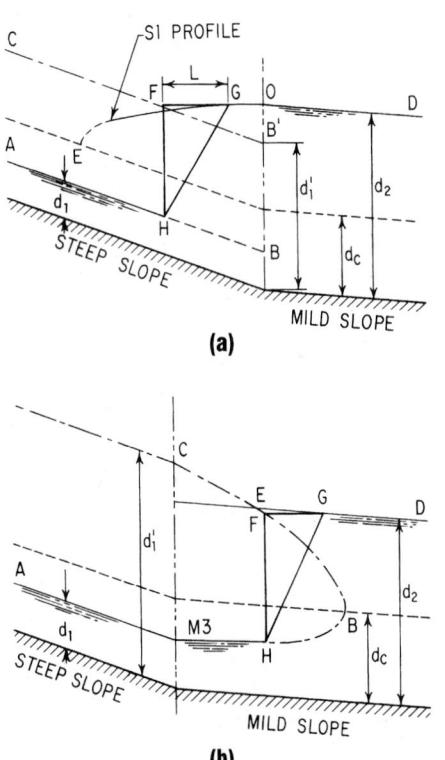

(a)

(b)

Fig. 21.52 Hydraulic jump may occur at a change in bottom slope, or (a) above it, or (b) below it.

21.28 Flow at Entrance to a Steep Channel

The discharge Q, ft^3/s, in a channel leaving a reservoir is a function of the total head H, ft, on the channel entrance, the entrance loss, ft and the slope of the channel. If the channel has a slope steeper than the critical slope (Art. 21.23), the flow passes through critical depth at the entrance, and discharge is at a maximum. If the channel entrance is rectangular in cross section, the critical depth $d_c = \frac{2}{3}H_e$ [according to Eqs. (21.82) and (21.85)], where H_e is the specific energy head, ft, in the reservoir and datum is the elevation of the lip of the channel (Fig. 21.53a).

From $Q = AV$, with the area of flow $A = bd_c = \frac{2}{3}bH_e$ and the velocity

$$V = \sqrt{2g(H_e - d_c)} = \sqrt{\frac{64.4H_e}{3}}$$

the discharge for rectangular channels, ignoring entrance loss, is

$$Q = 3.087bH_e^{3/2} \qquad (21.106)$$

where b is the channel width, ft.

If the entrance loss must be considered, or if the channel entrance is other than rectangular, the inlet

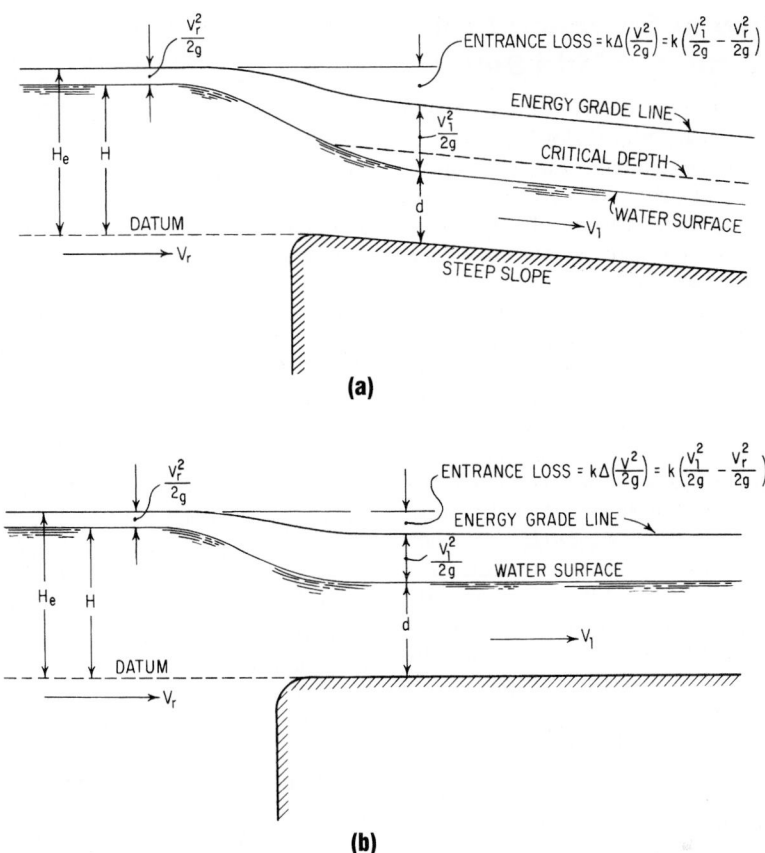

Fig. 21.53 Flow at entrance to (*a*) steep channel; (*b*) mild-slope channel.

depth must be solved for by trial and error since the discharge is unknown. The procedure for finding the correct discharge is as follows:

A trial discharge is chosen. Then, the critical depth for the given shape of channel entrance is determined (see those in E. F. Brater, "Handbook of Hydraulics," 6th ed., McGraw-Hill Book Company, New York.) Adding d_c to its associated velocity head gives the specific energy in the channel entrance, to which the resulting entrance loss is added. This sum then is compared with the specific energy of the reservoir water, which equals the depth of water above datum plus the velocity head of flow toward the channel. (This velocity head is normally so small that it may be taken as zero in most calculations.) If the specific energy computed for the depth of water in the reservoir equals the sum of specific energy and entrance loss determined for the channel entrance,

then the assumed discharge is correct; if not, a new discharge is assumed, and the computations continued until a balance is reached.

A first trial discharge may be found from $Q = A\sqrt{2g(H_e - d)}$, where $(H_e - d)$ gives actual head producing flow (Fig. 21.53). A reasonable value for the depth d would be $\frac{2}{3}H_e$ for steep channels and an even greater percentage of H_e for mild channels.

The entrance loss equals the product of an empirical constant k and the change in velocity head ΔH_v at the entrance. If the velocity in the reservoir is assumed to be zero, then the entrance loss is $k(V_1^2/2g)$, where V_1 is the velocity computed for the channel entrance. Safe design values for the coefficient vary from about 0.1 for a well-rounded entrance to slightly over 0.3 for one with squared ends.

21.29 Flow at Entrance to a Channel of Mild Slope

When water flows from a reservoir into a channel with slope less than the critical slope (Art. 21.23), the depth of flow at the channel entrance equals the normal depth for the channel (Art. 21.22). The entrance depth and discharge are dependent on each other. The discharge that results from a given head is that for which flow enters the channel without forming either a backwater or drawdown curve within the entrance. This requirement necessitates the formation of normal depth d since only at this equilibrium depth is there no tendency to change the discharge or to form backwater curves. (In Fig. 21.53b, d is normal depth.)

A solution for discharge at entrance to a channel of mild slope is found as follows: A trial discharge, ft^3/s, is estimated from $Q = A\sqrt{2g(H_e - d)}$, where $H_e - d$ is the actual head, ft, producing flow. H_e is the specific energy head, ft, of the reservoir water relative to datum at lip of channel; A is the cross-sectional area of flow, ft^2; and g is acceleration due to gravity, 32.2 ft/s^2. The normal depth of the channel is determined for this discharge from Eq. (21.83). The velocity head is computed for this depth-discharge combination, and an entrance-loss calculation is made (see Art. 21.33). The sum of the specific energy of flow in the channel entrance and the entrance loss must equal the specific energy of the water in the reservoir for an energy balance to exist between those points (Fig. 21.53b). If the trial discharge gives this balance of energy, then the discharge is correct; if not, a new discharge is chosen, and the calculations continued until a satisfactory balance is obtained.

21.30 Channel Section of Greatest Efficiency

If a channel of any shape is to reach its greatest hydraulic efficiency, it must have the shortest possible wetted perimeter for a given cross-sectional area. The resulting shape gives the greatest hydraulic radius and therefore the greatest capacity for that area. This can be seen from the Manning equation for discharge [Eq. (21.83)], in which Q is a direct function of hydraulic radius to the two-thirds power.

The most efficient of all possible open-channel cross sections is the semicircle. There are practical objections to the use of this shape because of the difficulty of construction, but it finds some use in metal flumes where sections can be preformed. The most efficient of all trapezoidal sections is the half hexagon, which is used extensively for large water-supply channels. The rectangular section with the greatest efficiency has a depth of flow equal to one-half the width. This shape is often used for box culverts and small drainage ditches.

21.31 Subcritical Flow around Bends in Channels

Because of the inability of liquids to resist shearing stress, the free surface of steady uniform flow is always normal to the resultant of the forces acting on the water. Water in a reservoir has a horizontal surface since the only force acting on it is the force of gravity.

Water reacts in accordance with Newton's first law of motion: It flows in a straight line unless deflected from its path by an outside force. When water is forced to flow in a curved path, its surface assumes a position normal to the resultant of the forces of gravity and radial acceleration. The force due to radial acceleration equals the force required to turn the water from a straight-line path, or mV^2/r_c for m, a unit mass of water, where V is its average velocity, ft/s, and r_c the radius of curvature, ft, of the center line of the channel.

The water surface makes an angle ϕ with the horizontal such that

$$\tan \phi = \frac{V^2}{r_c g} \qquad (21.107)$$

The theoretical difference y, ft, in water-surface level between the inside and outside banks of a curve (Fig. 21.54) is found by multiplying $\tan \phi$ by the top width of the channel T, ft. Thus,

$$y = \frac{V^2 T}{r_c g} \qquad (21.108)$$

where the radius of curvature r_c of the center of the channel is assumed to represent the average curvature of flow. This equation gives values of y smaller than those actually encountered because of the use of average values of velocity and radius, rather than empirically derived values more representative of actual conditions. The error will not be great, however, if the depth of flow is well above

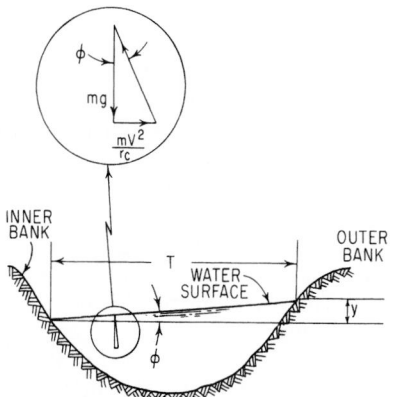

Fig. 21.54 Water-surface profile at a bend in a channel with subcritical flow.

critical (Art. 21.23). In this range, the true value of y would be only a few inches.

The difference in surface elevation found from Eq. (21.108), although it involves some drop in surface elevation on the inside of the curve, does not allow a savings of freeboard height on the inside bank. The water surface there is wavy and thus needs a freeboard height at least equal to that of a straight channel.

The top layer of flow in a channel has a higher velocity than flow near the bottom because of the retarding effect of friction along the floor of the channel. A greater force is required to deflect the high-velocity flow. Therefore, when a stream enters a curve, the higher-velocity flow moves to the outside of the bend. If the bend continues long enough, all the high-velocity water will move against the outer bank and may cause extensive scour unless special bank protection is provided.

Since the higher-velocity flow is pressed directly against the bank, an increase in friction loss results. This increased loss may be accounted for in calculations by assuming an increased value of the roughness coefficient n within the curve. Scobey suggests that the value of n be increased by 0.001 for each 20° of curvature in 100 ft of flume. His values have not been evaluated completely, however, and should be used with discretion. (F. C. Scobey, "The Flow of Water in Flumes," *U.S. Department of Agriculture, Technical Bulletin 393*.)

21.32 Supercritical Flow around Bends in Channels

When water, traveling at a velocity greater than critical (Art. 21.23), flows around a bend in a channel, a series of standing waves are produced. Two waves form at the start of the curve. One is a positive wave, of greater-than-average surface elevation, which starts at the outside wall and extends across the channel on the line *AME* (Fig. 21.55). The second is a negative wave, with a surface elevation of less-than-average height, which starts at the inside wall and extends across the channel on the line *BMD*. These waves cross at *M*, are reflected from opposite channel walls at *D* and *E*, recross as shown, and continue crossing and recrossing.

The two waves at the entrance form at an angle with the approach channel known as the wave angle β_o. This angle may be determined from the equation

$$\sin \beta_o = \frac{1}{\mathbf{F}_1} \qquad (21.109)$$

where $\mathbf{F}_1$ represents the Froude number of flow in the approach channel [Eq. (21.16)].

The distance from the beginning of the curve to the first wave peak on the outside bank is determined by the central angle θ_o. This angle may be found from

$$\tan \theta_o = \frac{T}{(r_c + T/2) \tan \beta_o} \qquad (21.110)$$

where T is the normal top width of channel and r_c is the radius of curvature of the center of channel. The depths along the banks at an angle $\theta < \theta_o$ are given by

$$d = \frac{V^2}{g} \sin^2 \left(\beta_o \pm \frac{\theta}{2} \right) \qquad (21.111)$$

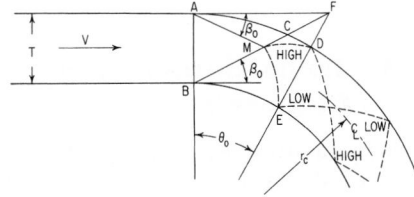

Fig. 21.55 Plan view of supercritical flow around a bend in an open channel.

where the positive sign gives depths along the outside wall and the negative sign, depths along the inside wall. The depth of maximum height for the first positive wave is obtained by substituting the value of θ_o found from Eq. (21.110) for θ in Eq. (21.111).

Standing waves in existing rectangular channels may be prevented by installing diagonal sills at the beginning and end of the curve. The sills introduce a counterdisturbance of the right magnitude, phase, and shape to neutralize the undesirable oscillations that normally form at the change of curvature. The details of sill design have been determined experimentally.

Good flow conditions may be ensured in new projects with supercritical flow in rectangular channels by providing transition curves or by banking the channel bottom. Circular transition curves aid in wave control by setting up counterdisturbances in the flow similar to those provided by diagonal sills. A transition curve should have a radius of curvature twice the radius of the central curve. It should curve in the same direction and have a central angle given, with sufficient accuracy, by

$$\tan \theta_t = \frac{T}{2r_c \tan \beta_o} \qquad (21.112)$$

Transition curves should be used at both the beginning and end of a curve to prevent disturbances downstream.

Banking the channel bottom is the most effective method of wave control. It permits equilibrium conditions to be set up without introduction of a counterdisturbance. The cross slope required for equilibrium is the same as the surface slope found for subcritical flow around a bend (Fig. 21.54). The angle ϕ the bottom makes with the horizontal is found from the equation

$$\tan \phi = \frac{V^2}{r_c g} \qquad (21.113)$$

21.33 Transitions in Open Channels

A transition is a structure placed between two open channels of different shape or cross-sectional area to produce a smooth, low-head-loss transfer of flow. The major problems associated with design of a transition lie in locating the invert and

determining the various cross-sectional areas so that the flow is in accord with the assumptions made in locating the invert. Many variables, such as flow-rate changes, wall roughness, and channel shape and slope, must be taken into account in design of a smooth-flow transition.

When proceeding downstream through a transition, the flow may remain subcritical or supercritical (Art. 21.23), change from subcritical to supercritical, or change from supercritical to subcritical. The latter flow possibility may produce a hydraulic jump.

Special care must be exercised in the design if the depth in either of the two channels connected is near the critical depth. In this range, a small change in energy head within the transition may cause the depth of flow to change to its alternate depth. A flow that switches to its subcritical alternate depth may overflow the channel. A flow that changes to its supercritical alternate depth may cause excessive channel scour. The relationship of flow depth to energy head can be shown on a plot such as Fig. 21.44, p. 21.43.

To place a transition properly between two open channels, it is necessary to determine the design flow and calculate normal and critical depths for each channel section. Maximum flow is usually selected as the design flow. Normal depth for each section is used for the design depth. After the design has been completed for maximum flow, hydraulic calculations should be made to check the suitability of the structure for lower flows.

The transition length that produces a smooth-flowing, low-head-loss structure is obtained for an angle of about 12.5° between the channel axis and the lines of intersection of the water surface with the channel sides, as shown in Fig. 21.56. The length of the transition L_t is then given by

$$L_t = \frac{\frac{1}{2}(T_2 - T_1)}{\tan 12.5°} \qquad (21.114)$$

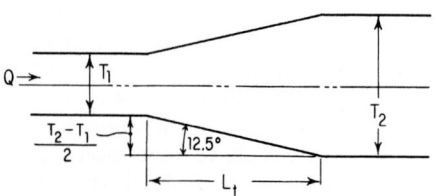

Fig. 21.56 Plan view of a transition between two open channels with different widths.

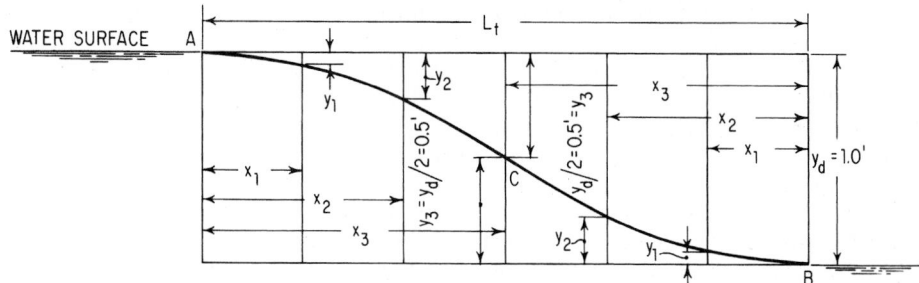

Fig. 21.57 Profile of reverse parabolic water-surface curve for well-designed transitions.

where T_2 and T_1 are the top widths of sections 2 and 1, respectively.

In design of an inlet-type transition structure, the water-surface level of the downstream channel must be set below the water-surface level of the upstream channel by at least the sum of the increase in velocity head, plus any transition and friction losses. The transition loss, ft, is given by $K(\Delta V^2/2g)$, where K, the loss factor, equals about 0.1 for an inlet-type structure; ΔV is the velocity change, ft/s; and $g = 32.2$ ft/s^2. The total drop in water surface y_d across the inlet-type transition is then $1.1[\Delta(V^2/2g)]$, if friction is ignored.

For outlet-type structures, the average velocity decreases, and part of the loss in velocity head is recovered as added depth. The rise of the water surface for an outlet structure equals the decrease in velocity head minus the outlet and friction losses. The outlet loss factor is normally 0.2 for well-designed transitions. If friction is ignored, the total rise in water surface y_r across the outlet structure is $0.8[\Delta(V^2/2g)]$.

Many well-designed transitions have a reverse parabolic water-surface curve tangent to the water surfaces in each channel (Fig. 21.57). After such a water-surface profile is chosen, depth and cross-sectional areas are selected at points along the transition to produce this smooth curve. Straight, angular walls usually will not produce a smooth parabolic water surface; therefore, a transition with a curved bottom or sides has to be designed.

The total transition length L_t is split into an even number of sections of equal length x. For Fig. 21.57, six equal lengths of 10 ft each are used, for an assumed drop in water surface y_d of 1 ft. It is assumed that the water surface will follow parabola AC for the length $L_t/2$ to produce a water-surface drop of $y_d/2$ and that the other half of the surface

drop takes place along the parabola CB. The water-surface profile can be determined from the general equation for a parabola, $y = ax^2$, where y is the vertical drop in the distance x, measured from A or B.

The surface drops at sections 1 and 2 are found as follows: At the midpoint of the transition, $y_3 = ax^2 = y_d/2 = 0.5 = a(30)^2$, from which $a = 0.000556$. Then $y_1 = ax_1^2 = 0.000556(10)^2 = 0.056$ ft and $y_2 = ax_2^2 = 0.000556(20)^2 = 0.222$ ft.

21.34 Weirs

A weir is a barrier in an open channel over which water flows. The edge or surface over which the water flows is called the **crest**. The overflowing sheet of water is the **nappe**.

If the nappe discharges into the air, the weir has *free discharge*. If the discharge is partly under water, the weir is **submerged** or drowned.

21.34.1 Types of Weirs

A weir with a sharp upstream corner or edge such that the water springs clear of the crest is a **sharp-crested weir** (Fig. 21.58). All other weirs are classed

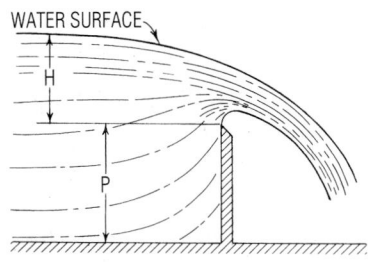

Fig. 21.58 Sharp-crested weir.

as **weirs not sharp-crested**. Sharp-crested weirs are classified according to the shape of the weir opening, such as rectangular weirs, triangular or V-notch weirs, trapezoidal weirs, and parabolic weirs. Weirs not sharp-crested are classified according to the shape of their cross section, such as broad-crested weirs, triangular weirs, and, as shown in Fig. 21.59, trapezoidal weirs.

The channel leading up to a weir is the **channel of approach**. The mean velocity in this channel is the **velocity of approach**. The depth of water producing the discharge is the **head**.

Sharp-crested weirs are useful only as a means of measuring flowing water. In contrast, weirs not sharp-crested are commonly incorporated into hydraulic structures as control or regulation devices, with measurement of flow as their secondary function.

21.34.2 Rectangular Sharp-Crested Weirs

Discharge over a rectangular sharp-crested weir is given by

$$Q = CLH^{3/2} \qquad (21.115)$$

where Q = discharge, ft^3/s

C = discharge coefficient

L = effective length of crest, ft

H = measured head = depth of flow above elevation of crest, ft

The head should be measured at least 2.5H upstream from the weir, to be beyond the drop in the water surface (surface contraction) near the weir.

Numerous equations have been developed for finding the discharge coefficient C. One such equation, which applies only when the nappe is fully

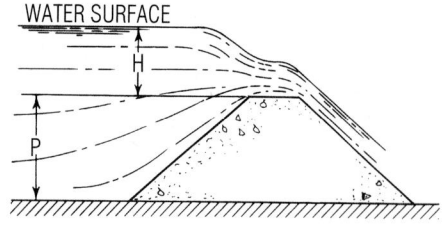

Fig. 21.59 Weir not sharp-crested.

ventilated, was developed by Rehbock and simplified by Chow:

$$C = 3.27 + 0.40\frac{H}{P} \qquad (21.116)$$

where P is the height of the weir above the channel bottom (Fig. 21.58) (V. T. Chow, "Open-Channel Hydraulics," McGraw-Hill Book Company, New York).

The height of weir P must be at least 2.5H for a complete crest contraction to form. If P is less than 2.5H, the crest contraction is reduced and said to be partly suppressed. Equation (21.116) corrects for the effects of friction, contraction of the nappe, unequal velocities in the channel of approach, and partial suppression of the crest contraction and includes a correction for the velocity of approach and the associated velocity head.

To be fully ventilated, a nappe must have its lower surface subjected to full atmospheric pressure. A partial vacuum below the nappe can result through removal of air by the overflowing jet if there is restricted ventilation at the sides of the weir. This lack of ventilation causes increased discharge and a fluctuation and shape change of the nappe. The resulting unsteady condition is very objectionable when the weir is used as a measuring device.

At very low heads, the nappe has a tendency to adhere to the downstream face of a rectangular weir even when means for ventilation are provided. A weir operating under such conditions could not be expected to have the same relationship between head and discharge as would a fully ventilated nappe.

A V-notch weir (Fig. 21.60) should be used for measurement of flow at very low heads if accuracy of measurement is required.

End contractions occur when the weir opening does not extend the full width of the approach channel. Water flowing near the walls must move toward the center of the channel to pass over the weir, thus causing a contraction of the flow. The nappe continues to contract as it passes over the crest. Hence, below the crest, the nappe has a minimum width less than the crest length.

The effective length L, ft, of a contracted-width weir is given by

$$L = L' - 0.1NH \qquad (21.117)$$

where L' = measured length of crest, ft

N = number of end contractions

H = measured head, ft

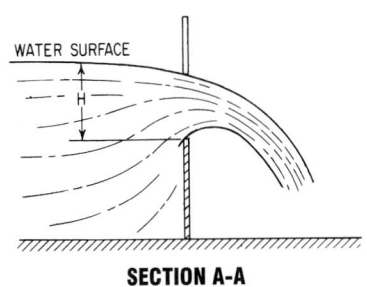

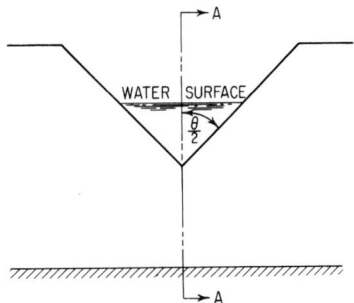

SECTION A-A

Fig. 21.60 V-notch weir.

If flow contraction occurs at both ends of a weir, there are two end contractions and $N = 2$. If the weir crest extends to one channel wall but not the other, there is one end contraction and $N = 1$. The effective crest length of a full-width weir is taken as its measured length. Such a weir is said to have its contractions suppressed.

21.34.3 Triangular or V-Notch Sharp-Crested Weirs

The triangular or V-notch weir (Fig. 21.60) has a distinct advantage over a rectangular sharp-crested weir (Art. 21.34.2) when low discharges are to be measured. Flow over a V-notch weir starts at a point, and both discharge and width of flow increase as a function of depth. This has the effect of spreading out the low-discharge end of the depth-discharge curve and therefore allows more accurate determination of discharge in this region.

Discharge is given by

$$Q = C_1 H^{5/2} \tan \frac{\theta}{2} \qquad (21.118)$$

where θ = notch angle

H = measured head, ft

C_1 = discharge coefficient

The head H is measured from the notch elevation to the water-surface elevation at a distance $2.5H$ upstream from the weir. Values of the discharge coefficient were derived experimentally by Lenz, who developed a procedure for including the effect of viscosity and surface tension as well as the effect of contraction and velocity of approach (A. T. Lenz, "Viscosity and Surface Tension Effects on V-Notch Weir Coefficients," *Transactions of the American*

Society of Civil Engineers, vol. 69, 1943). His values were summarized by Brater, who presented the data in the form of curves (Fig. 21.61) (E. F. Brater "Handbook of Hydraulics," 6th ed., McGraw-Hill Book Company, New York).

A V-notch weir tends to concentrate or focus the overflowing nappe, causing it to spring clear of the downstream face for even the smallest flows. This characteristic prevents a change in the head-discharge relationship at low flows and adds materially to the reliability of the weir.

21.34.4 Trapezoidal Sharp-Crested Weirs

The discharge from a trapezoidal weir (Fig. 21.62) is assumed the same as that from a rectangular weir and a triangular weir in combination.

$$Q = C_2 L H^{3/2} + C_3 Z H^{5/2} \qquad (21.119)$$

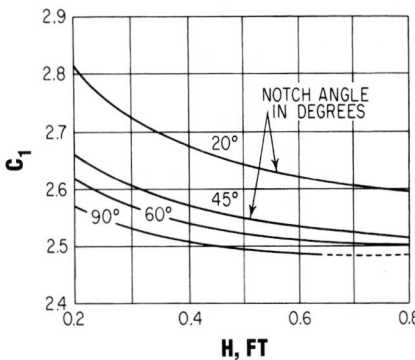

Fig. 21.61 Chart gives discharge coefficients for sharp-crested V-notch weirs. The coefficients depend on head and notch angle.

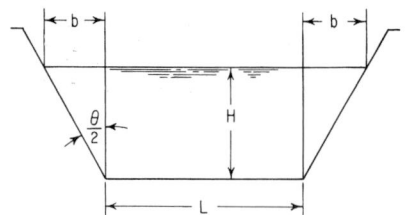

Fig. 21.62 Trapezoidal sharp-crested weir.

where Q = discharge, ft³/s

L = length of notch at bottom, ft

H = head, measured from notch bottom, ft

$Z = b/H$ [substituted for $\tan(\theta/2)$ in Eq. (21.118)]

b = half the difference between lengths of notch at top and bottom, ft

No data are available for determination of coefficients C_2 and C_3. They must be determined experimentally for each installation.

21.34.5 Submerged Sharp-Crested Weirs

The discharge over a submerged sharp-crested weir (Fig. 21.63) is affected not only by the head on the upstream side H_1 but by the head downstream H_2. Discharge also is influenced to some extent by the height P of the weir crest above the floor of the channel.

The discharge Q_s, ft³/s, for a submerged weir is related to the free or unsubmerged discharge Q, ft³/s, for that weir by a function of H_2/H_1. Villemonte expressed this relationship by the equation

$$\frac{Q_s}{Q} = \left[1 - \left(\frac{H_2}{H_1}\right)^n\right]^{0.385} \qquad (21.120)$$

where n is the exponent of H in the equation for free discharge for the shape of weir used. (The value of n

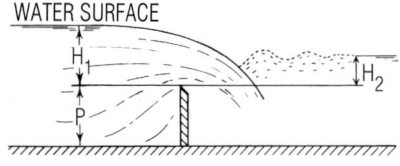

Fig. 21.63 Submerged sharp-crested weir.

is ³⁄₂ for a rectangular sharp-crested weir and ⁵⁄₂ for a triangular weir.) To use the Villemonte equation, first compute the rate of flow Q for the weir when not submerged, and then, using this rate and the required depths, solve for the submerged discharge Q_s. (J. R. Villemonte, "Submerged-Weir Discharge Studies," *Engineering News-Record*, Dec. 25, 1947, p. 866.)

Equation (21.120) may be used to compute the discharge for a submerged sharp-crested weir of any shape simply by changing the value of n. The maximum deviation from the Villemonte equation for all test results was found to be 5%. Where great accuracy is essential, it is recommended that the weir be tested in a laboratory under conditions comparable with those at its point of intended use.

21.34.6 Weirs Not Sharp-Crested

These are sturdy, heavily constructed devices, normally an integral part of hydraulic projects (Fig. 21.59). Typically, a weir not sharp-crested serves as the crest section for an overflow dam or the entrance section for a spillway or channel. Such a weir can be used for discharge measurement, but its purpose is normally one of control and regulation of water levels or discharge, or both.

The discharge over a weir not sharp-crested is given by

$$Q = CLH_t^{3/2} \qquad (21.121)$$

where Q = discharge, ft³/s

C = coefficient of discharge

L = effective length of crest, ft

H_t = total head on crest including velocity head of approach, ft

The head of water producing discharge over a weir is the total of measured head H and velocity head of approach H_v. The velocity head of approach is accounted for by the discharge coefficient for sharp-crested weirs but must be considered separately for weirs not sharp-crested. Thus, for such weirs, Eq. (21.115) is rewritten in the form

$$Q = CL\left(H + \frac{V^2}{2g}\right)^{3/2} \qquad (21.122)$$

where H = measured head, ft

V = velocity of approach, ft/s

$V^2/2g = H_v$, velocity head of approach, ft, neglecting degree of turbulence given by Eq. (21.81)

g = acceleration due to gravity, 32.2 ft/s^2

Since velocity and discharge are dependent on each other in this equation and both are unknown, discharge must be found by a series of approximations, which may be done as follows: First, compute a trial discharge from the measured head, neglecting the velocity head. Then, using this discharge, compute the velocity of approach, velocity head, and finally total head. From this total head, compute the first corrected discharge. This corrected discharge will be sufficiently accurate if the velocity of approach is small. But the process should be repeated, starting with the corrected discharge, where approach velocities are high.

The discharge coefficient C must be determined empirically for weirs not sharp-crested. If a weir of untested shape is to be constructed, it must be calibrated in place or a model study made to determine its head-discharge relationship. The problem of establishing a fixed relation between head and discharge is complicated by the fact that the nappe may assume a variety of shapes in passing over the weir. For each change of nappe shape, there is a corresponding change in the relation between head and discharge. The effect is most critical for low heads. A nappe undergoes several changes in succession as the head varies, and the successive shapes that appear with an increasing stage may differ from those pertaining to similar stages with decreasing head. Therefore, care must be exercised when using these weirs for flow measurement to ensure that the conditions are similar to those at the time of calibration.

Large weirs not sharp-crested often have piers on their crest to support control gates or a roadway. These piers reduce the effective length of crest by more than the sum of their individual widths because of the formation of flow contractions at each pier. The effective crest length for a weir not sharp-crested is given by

$$L = L' - 2(NK_p + K_a)H_t \qquad (21.123)$$

where L = effective crest length, ft

L' = net crest lengths, ft = measured length minus width of all piers

N = number of piers

K_p = pier-contraction coefficient

K_a = abutment-contraction coefficient

H_t = total head on crest including velocity head of approach, ft

(U.S. Department of the Interior, "Design of Small Dams," Government Printing Office, Washington, DC 20402.)

The pier-contraction coefficient K_p is affected by the shape and location of the pier nose, thickness of pier, head in relation to design heads, and approach velocity. For conditions of design head H, the average pier-contraction coefficients are as shown in Table 21.12.

The abutment-contraction coefficient K_a is affected by the shape of the abutment, the angle between the upstream approach wall and the axis of flow, the head in relation to the design head, and the approach velocity. For conditions of design head H_d, average coefficients may be assumed as shown in Table 21.13.

21.34.7 Submergence of Weirs Not Sharp-Crested

Spillways and other weirs not sharp-crested are submerged when their tailwater level is high enough to affect their discharge. Because of the surface disturbance produced in the vicinity of the crest, such a spillway or weir is unsatisfactory for accurate flow measurement.

Approximate values of discharge may be found by applying the following rules proposed by E. F. Brater: (1) If the depth of submergence is not greater than 0.2 of the head, ignore the submergence and treat the weir as though it had free discharge. (2) For narrow weirs having a sharp upstream leading edge, use a submerged-weir formula for sharp-crested weirs. (3) Broad-crested weirs are not affected by submergence up to approximately 0.66 of the head. (4) For weirs with narrow rounded crests, increase discharge

Table 21.12 Pier-Contraction Coefficients

Condition	K_p
Square-nosed piers with corners rounded on a radius equal to about 0.1 of the pier thickness	0.02
Rounded-nosed piers	0.01
Pointed-nosed piers	0

Table 21.13 Abutment-Contraction Coefficients

Condition	K_a
Square abutment with headwall at 90° to direction of flow	0.20
Rounded abutments with headwall at 90° to direction of flow when $0.5H_d \geq r^* \geq 0.15H_d$	0.10
Rounded abutments where $r^* > 0.5H_d$ and headwall is placed not more than 45° to direction of flow	0

$^* r$ = radius of abutment rounding.

obtained by a formula for submerged sharp-crested weirs by 10% or more. Of the above rules, 1, 2, and 3 probably apply quite accurately, while 4 is simply a rough approximation.

21.34.8 The Ogee-Crested Weir

The ogee-crested weir was developed in an attempt to produce a weir that would not have the undesirable nappe variation normally associated with weirs not sharp-crested. A shape was needed that would force the nappe to assume a single path for any discharge, thus making the weir consistent for flow measurement. The ogee-crested weir (Fig. 21.64) has such a shape. Its crest profile conforms closely to the profile of the lower surface of a ventilated nappe flowing over a rectangular sharp-crested weir.

The shape of this nappe, and therefore of an ogee crest, depends on the head producing the discharge. Consequently, an ogee crest is designed for a single total head, called the design head H_d. When an ogee weir is discharging at the design

head, the flow glides over the crest with no interference from the boundary surface and attains near-maximum discharge efficiency.

For flow at heads lower than the design head, the nappe is supported by the crest and pressure develops on the crest that is above atmospheric but less than hydrostatic. This crest pressure reduces the discharge below that for ideal flow. (Ideal flow is flow over a fully ventilated sharp-crested weir under the same head H.)

When the weir is discharging at heads greater than the design head, the pressure on the crest is less than atmospheric, and the discharge increases over that for ideal flow. The pressure may become so low that separation in flow will occur. According to Chow, however, the design head may be safely exceeded by at least 50% before harmful cavitation develops (V. T. Chow, "Open-Channel Hydraulics," McGraw-Hill Book Company, New York (books.mcgraw-hill.com)).

The measured head H on an ogee-crested weir is taken as the distance from the highest point of the crest to the level of the water surface at a distance $2.5H$ upstream. This depth coincides with the depth measured between the upstream water level and the bottom of the nappe, at the point of maximum contraction, for a sharp-crested weir. This relationship is shown in Fig. 21.65.

Discharge coefficients for ogee-crested weirs are therefore determined from sharp-crested-weir coefficients after an adjustment for this difference in head. These coefficients are a function of the approach velocity, which varies with the ratio of height of weir P to actual total head H_t, where discharge is given by Eq. (21.122). Figure 21.66 for

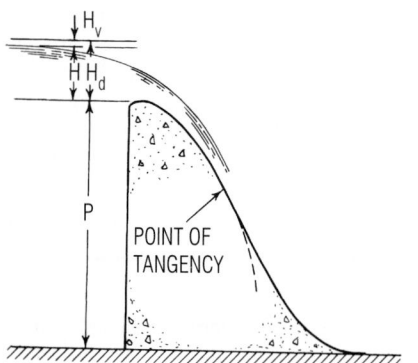

Fig. 21.64 Ogee-crested weir with vertical upstream face.

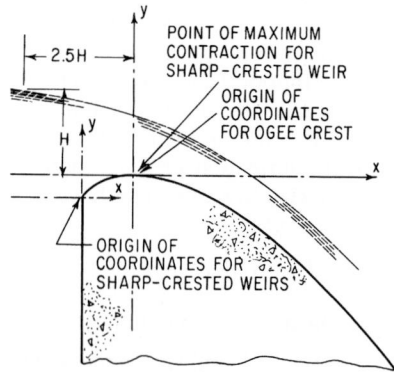

Fig. 21.65 Location of origin of coordinates for sharp-crested and ogee-crested weirs.

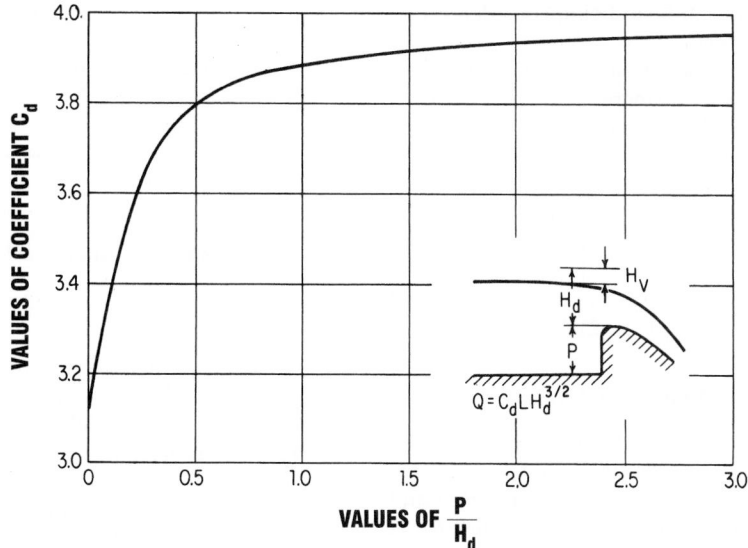

Fig. 21.66 Chart gives discharge coefficients at design head H_d for vertical-faced ogee-crested weirs. (*From "Design of Small Dams," U.S. Bureau of Reclamation.*)

an ogee weir with a vertical upstream face gives coefficient C_d for discharge at design head H_d. (U.S. Department of the Interior, "Design of Small Dams," Government Printing Office, Washington, DC 20402. This manual and V. T. Chow, "Open-Channel Hydraulics," McGraw-Hill Book Company, New York (books.mcgraw-hill.com), present methods for determining the shape of an ogee crest profile.) When the weir is discharging at other than the design head, the flow differs from ideal, and the discharge coefficient changes from the discharge coefficient given in Fig. 21.66.

Figure 21.67 gives values of the discharge coefficient C as a function of the ratio H_t/H_d,

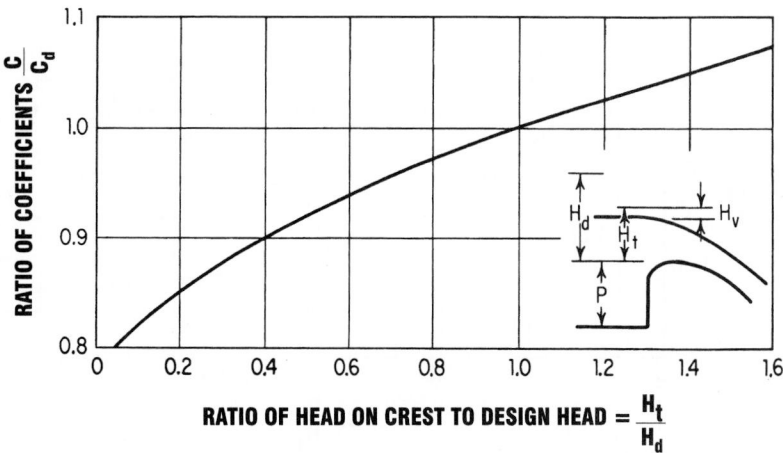

Fig. 21.67 Chart gives discharge coefficients for vertical-faced ogee-crested weirs at heads H_t other than design head H_d. (*From "Design of Small Dams," U.S. Bureau of Reclamation.*)

where H_t is the actual head being considered and H_d is the design head.

If an ogee weir has a sloping upstream face, there is a tendency for an increase in discharge over that for a weir with a vertical face. Figure 21.68 shows the ratio of the coefficient for an ogee weir with a sloping face to the coefficient for a weir with a vertical upstream face. The coefficient of discharge for an ogee weir with a sloping upstream face, if flow is at other than the design head, is determined from Fig. 21.66 and is then corrected for head and slope with Figs. 21.67 and 21.68.

21.34.9 Broad-Crested Weir

This is a weir with a horizontal or nearly horizontal crest. The crest must be sufficiently long in the direction of flow that the nappe is supported and hydrostatic pressure developed on the crest for at least a short distance. A broad-crested weir is nearly rectangular in cross section. Unless otherwise noted, it will be assumed to have vertical faces, a plane horizontal crest, and sharp right-angled edges.

Figure 21.69 shows a broad-crested weir that, because of its sharp upstream edge, has contraction of the nappe. This causes a zone of reduced pressure at the leading edge. When the head H on a broad-crested weir reaches one to two times its breadth b, the nappe springs free, and the weir acts as a sharp-crested weir.

Discharge over a broad-crested weir is given by Eq. (21.115) since the velocity of approach was

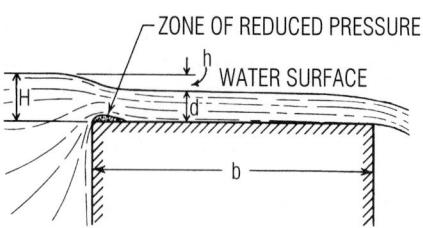

Fig. 21.69 Broad-crested weir.

ignored in experiments performed to determine the coefficient of discharge. These coefficients probably apply more accurately, therefore, where the velocity of approach is not high. Values of the discharge coefficient, compiled by King, appear in Table 21.14. (E. F. Brater, "Handbook of Hydraulics, 6th ed., McGraw-Hill Book Company, New York (books.mcgraw-hill.com).)

21.34.10 Weirs of Irregular Section

This group includes those weirs whose cross section deviates from typical broad-crested or ogee-crested weirs. Weirs of irregular section, fairly common in waterworks projects, are used as spillways and control structures. Experimental data are available on the more common shapes. (See, for example, E. F. Brater, "Handbook of Hydraulics," 6th ed., McGraw-Hill Book Company, New York (books.mcgraw-hill.com).)

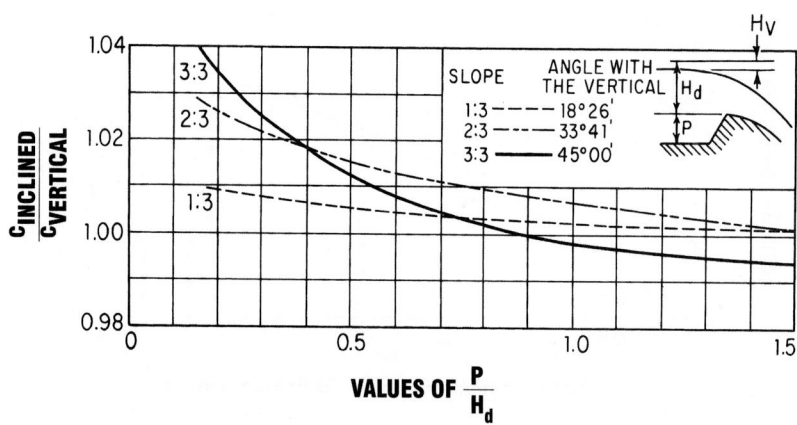

Fig. 21.68 Chart gives design coefficients at design head H_d for ogee-crested weirs with sloping upstream face. (*From "Design of Small Dams," U.S. Bureau of Reclamation.*)

Table 21.14 Values of C in $Q = CLH^{3/2}$ for Broad-Crested Weirs

Mea-sured Head H, ft	Breadth of crest of weir, ft										
	0.50	0.75	1.00	1.50	2.00	2.50	3.00	4.00	5.00	10.00	15.00
0.2	2.80	2.75	2.69	2.62	2.54	2.48	2.44	2.38	2.34	2.49	2.68
0.4	2.92	2.80	2.72	2.64	2.61	2.60	2.58	2.54	2.50	2.56	2.70
0.6	3.08	2.89	2.75	2.64	2.61	2.60	2.68	2.69	2.70	2.70	2.70
0.8	3.30	3.04	2.85	2.68	2.60	2.60	2.67	2.68	2.68	2.69	2.64
1.0	3.32	3.14	2.98	2.75	2.66	2.64	2.65	2.67	2.68	2.68	2.63
1.2	3.32	3.20	3.08	2.86	2.70	2.65	2.64	2.67	2.66	2.69	2.64
1.4	3.32	3.26	3.20	2.92	2.77	2.68	2.64	2.65	2.65	2.67	2.64
1.6	3.32	3.29	3.28	3.07	2.89	2.75	2.68	2.66	2.65	2.64	2.63
1.8	3.32	3.32	3.31	3.07	2.88	2.74	2.68	2.66	2.65	2.64	2.63
2.0	3.32	3.31	3.30	3.03	2.85	2.76	2.72	2.68	2.65	2.64	2.63
2.5	3.32	3.32	3.31	3.28	3.07	2.89	2.81	2.72	2.67	2.64	2.63
3.0	3.32	3.32	3.32	3.32	3.20	3.05	2.92	2.73	2.66	2.64	2.63
3.5	3.32	3.32	3.32	3.32	3.32	3.19	2.97	2.76	2.68	2.64	2.63
4.0	3.32	3.32	3.32	3.32	3.32	3.32	3.07	2.79	2.70	2.64	2.63

(Table continued)

21.35 Sediment Transfer and Deposition in Open Channels

Sediment from open channels has many undesirable effects: Reservoirs have a reduced useful life because of loss of storage through the accumulation of silt. Sediment causes a hazard in navigable channels and harbors and an increase in frequency of flooding due to aggravation of rivers and flood channels. Silting of arable land by flooding rivers destroys fertility when the silt originates from bank or gully erosion rather than from surface, or soil, erosion. The cost of operating irrigation systems is increased by the need for frequent dredging. Water-supply facilities have increased costs because of the necessity of providing desilting works and because of the wear on mechanical equipment, such as gates, valves, and turbines.

21.35.1 Sediment Deposition in Reservoirs

Deposition of silt results when the transporting forces of a river are dissipated as the river enters a body of still water, such as a reservoir. Heavier silt sizes, those forming the bed load, are deposited in a delta as the river enters calm water. The smaller silt sizes, those carried in suspension, travel farther into the reservoir before deposition.

This incoming water, with its load of suspended silt, has a specific gravity greater than that of the clear water in the reservoir and may form a **density current**, rather than mixing immediately with the clear water. A density current, once formed, quickly moves to the bottom and flows in a dense cloud down the slopes of the reservoir until it is blocked by a dam. The dense flow then spreads out in this deeper area, where the stilling effect of the basin eventually causes deposition of the sediment. Deposits of fine sediment form about one-third of the volume of silt deposits in a reservoir. Much of all of this fine sediment is transported to its final location by density currents. The visible delta formed by the coarse sediments frequently distracts attention from the unseen bottom deposits of fine sediment, which are often of equal consequence.

Most reservoirs trap from 70 to almost 100% of the incoming sediment, depending on whether the reservoir is used for flood control or storage.

Flood-control reservoirs are normally emptied shortly after a storm, so the suspended materials are carried out with the water before settling can occur. This procedure reduces new deposits by almost 30% after each storm. Storage reservoirs used for water supply or power generation purposes, on the other hand, normally retain any inflow long enough for settlement of all suspended matter to occur. Their discharges are regulated to allow generation of power or to produce a uniform flow downstream with no thought to the venting of silt-laden storm flows.

The greater part of the annual suspended silt load in a stream may be carried in a relatively short time. The stream runs comparatively clear during the remainder of the year.

Venting of much of the annual suspended silt load is feasible through the use of density currents. These currents are stable, once formed, and often extend to the reservoir outlet. If density currents are observed and their time of arrival at the outlet determined, appropriate gates can be opened and much of the fine sediment entering a storage reservoir can be vented before it has time to form permanent deposits. This venting operation can extend the life of a reservoir by many years.

Numerous phenomena can destroy a reservoir, such as loss of storage capacity by landslide and loss of the dam by earthquake, landslide, over-topping, or failure of materials. The most common manner of destruction, however, is through loss of storage by deposition of silt. Redemption of reservoir capacity lost through silting is almost always economically unfeasible because of the wide distribution of deposits in a reservoir and the large quantity present. The most practicable means of avoiding a loss in reservoir capacity are to prevent formation of permanent deposits by taking advantage of density currents and to control rate of sediment production from eroding areas. When neither can be done, sufficient storage space must be provided in the design of the reservoir to compensate for depletion by silting during a reasonable economic lifetime.

Sediment production and its transportation to reservoirs or navigable waters cannot be prevented at costs proportionate to the resulting benefits. However, nature may be economically improved at times through a program of erosion control, reducing sediment production to less than that normally found under virgin conditions.

The deposits of silt that form in a storage reservoir are categorized into two distinct types: **Delta deposits**, formed from the **bed load**, are coarse-grained, with an in-place weight of about 80 lb/ft^3. Deposits produced from the **suspended load** are fine-grained, with an average weight of about 30 lb/ft^3. They constitute about one-sixth of the total weight of sediment delivered but account for about one-third of the volume of all deposits in a storage reservoir because of their low density. If sediment deposits are periodically above water, because of fluctuations in the reservoir water level, their density increases and the volume ratios given above for continued submergence no longer apply.

21.35.2 Prediction of Sediment-Delivery Rate

Two methods of approach are available for predicting the rate of sediment accumulation in a reservoir; both involve predicting the rate of sediment delivery.

One approach depends on historical records of the silting rate for existing reservoirs and is purely empirical. By this method, the silting records of a reservoir may be used to predict either the silting rate for that reservoir or the probable pattern of silt accumulation for a proposed reservoir in a similar area. This method allows transposition of data from one watershed to another because the measured annual sediment accumulation of a reservoir is expressed as a **rate of sediment delivery** per unit area of its watershed. Of course, the rate is not uniform during the year, or from year to year, because of variations in rainfall, but it should average the computed annual amount over the life of the project. The annual silt accumulation in a reservoir is determined by surveying exposed deltas and taking depth soundings. The resulting volume is adjusted to account for any silt loss through sluice gates or over the spillway and is then expressed as silt delivery per square mile of drainage area. This silt-delivery figure is further adjusted for rainfall and runoff conditions, to give a figure that could reasonably be expected during a year of average rainfall. If this adjusted figure is to be transposed to a neighboring drainage basin, adjustments should be made to account for both soil cover and rainfall differences between the basins. (For a discussion of the factors upon which this adjustment is based, see Art. 21.39.)

Silt-delivery measurements or estimates do not give total silt production for an area because part of the silt produced in a basin is deposited on floodplains and in channels before it reaches a reservoir. The difference between the amount of silt produced and that delivered increases as the size of the drainage area increases because of the increased chance that the silt will be deposited before it reaches the reservoir. Therefore, if a silt-delivery measurement or estimate is to be transposed to a basin of different size, an adjustment should be made to account for this discrepancy as well. The information for this adjustment can come only from field reconnaissance of the two areas to determine differences that might account for a variation in deposition of silt along the water courses.

The second general method of calculating sediment-delivery rate involves determining the rate of sediment transport as a function of stream discharge and density of suspended silt. The total sediment inflow for the year is then computed from these relationships and the recorded stream-discharge data.

The total quantity of sediment carried by a river is assumed transported either as suspended load or as bed load (Art. 21.35.1). The division is based on particle size but depends on velocity of flow as well. There is no sharp line of demarcation between the two classes. According to Witzig, about 80% of the volume of all sediment is produced by streambank erosion; the remaining 20% is produced by land-surface erosion. Constant erosion of the streambanks keeps the streambed well supplied with the coarse silt that travels as bed load. The fine silt that travels in suspension is produced in small amounts by streambank erosion. But for the most part, this silt comes from land-surface erosion, which generally occurs only during a storm.

The total quantity of sediment in suspension is not necessarily related directly to discharge at all times. The quantity is affected by seasonal variations in the supply and source of fine sediment and by distribution of rainfall and runoff from the watershed. Therefore, measurement of the sediment load for a given discharge does not necessarily indicate the amount that may be carried by an equal discharge at another time.

The bed load consists of the silt particles too large to be held in suspension. This size range includes particles of coarse sand, gravel, and boulders. The bed-load particles are moved by rolling along the bed of the stream. Some of the finer bed-load particles are moved in a series of steps or jumps representing a transition between transportation as bed load and suspended load.

The quantity of bed load is considered a constant function of the discharge because the sediment supply for the bed-load forces is always available in all but lined channels. An accepted formula for the quantity of sediment transported as bed load is the Schoklitsch formula:

$$G_b = \frac{86.7}{D_g^{1/2}} S^{3/2}(Q_i - bq_o) \qquad (21.124)$$

where G_b = total bed load, lb/s

D_g = effective grain diameter, in

S = slope of energy gradient

Q_i = total instantaneous discharge, ft^3/s

b = width of river, ft

q_o = critical discharge, ft^3/s per ft of river width

   = $(0.00532/S^{4/3})D_g$

An approximate solution for bed load by the Schoklitsch formula can be made by determining or assuming mean values of slope, discharge, and a single grain size representative of the bed-load sediment. A mean grain size of 0.04 in in diameter (about 1 mm) is reasonable for a river with a slope of about 1.0 ft/mi.

The size of grains moving on the bed of a river depends on velocity of flow, which varies with both slope and discharge. Therefore, the mean grain size changes as the flow increases during a storm or as the river changes slope along its course. It is obvious that considerable error could result from the use of Eq. (21.124) if it is necessary to guess at a mean grain diameter in the absence of carefully collected field data. Frequently, however, if insufficient data or lack of money prevent more thorough investigations, this shortcut can give results of sufficient accuracy.

Numerous formulas have been developed to represent the condition of flow involved in transportation of suspended sediment. These formulas express the degree of turbulent energy involved in suspension of the sediment and the mode of transfer of this energy to the silt and other fluid particles. The formulas require a number of empirical constants but are based on a sound physical and rational foundation. They require information as to the sediment composition by grain size, the actual quantity of silt in suspension at a given depth, and the stream velocity. (See H. A. Einstein, "The Bed-Load Function for Sediment Transportation in Open-Channel Flows," U.S. Department of Agriculture.)

An approximate determination of suspended load may be made without using these complicated formulas. The weight of suspended sediment transported by a river in an average year normally equals about 20% of the weight transported as bed load. The total weight of material annually moved by a river is therefore equal to 120% of the weight of material transported as bed load during the year as computed from Eq. (21.124).

(W. H. Graf, "Hydraulics of Sediment Transport," McGraw-Hill Book Company, New York (books.mcgraw-hill.com).)

21.36 Erosion Control

The various methods used in erosion control are collectively called *upstream engineering*. They consist of soil conservation measures such as reforestation, check-dam construction, planting of burned-over areas, contour plowing, and regulation of crop and grazing practices. Also included are measures for proper treatment of high embankments and cuts and stabilization of streambanks by planting or by revetment construction.

One phase of reforestation that may be applied near a reservoir is planting of vegetation screens. Such screens, planted on the flats adjacent to the normal stream channel at the head of a reservoir, reduce the velocity of silt-laden storm inflows that inundate these areas. This stilling action causes extensive deposition to occur before the silt reaches the main cavity of the reservoir. Use of vegetation screens, debris barriers, or desilting basins above a reservoir should be planned with future development in mind. For instance, if the dam is raised at a later date, the accumulated silt in this area would detract from the added storage that might otherwise have been obtained.

Hydrology

Hydrology is the study of the waters of the earth, their occurrence, circulation, and distribution, their chemical and physical properties, and their reaction with their environment, including their relation to living things. A major concern is the circulation, on or near the land surface, of water and its constituents throughout the hydrologic cycle. In this cycle, water evaporation from oceans, rivers, lakes, and other sources is carried over the earth and precipitated as rain or snow. The precipitation forms runoff on the land, infiltrates into the soil, recharges groundwater, discharges into streams, and then flows into the oceans and lakes, from which evaporation restarts the cycle. Thus hydrology deals with precipitation, evaporation, infiltration, groundwater flow, runoff, and stream flow.

21.37 Precipitation

The primary concern with precipitation in water resources engineering is forecasting it. The means for doing so are based on either current or past data, or a combination of the two.

Current data, in the form of synoptic weather charts, are published daily by the U.S. Weather Bureau. These charts summarize the various meteorological factors, such as wind, temperature, and pressure, through whose interaction precipitation is produced.

Past data are primarily in the form of rainfall records for a standard period, such as an hour, day, or year. They are the major source of data for determination of the recurrence interval for storms of a definite magnitude and the magnitude of storms in a definite recurrence interval.

Rainfall records are obtained from rain gages, which are of two types. The first type is a recording or automatic gage. It continually records, by ink pen and revolving drum, or digital microchip technology, the variation in rainfall intensity as well as the total rainfall volume. The second type is a nonrecording gage; it measures only the total rain volume that fell during the period between observations. The standard observation time for nonrecording gages for the U.S. Weather Bureau is 24 h.

Corrections must be made to rain-gage records to account for the mean precipitation over the entire drainage basin, for hourly rainfall rates when only daily volumes are given, and for errors arising out of the location of the gage. Most methods used in runoff determinations are based on the assumption that rainfall is uniform over the entire drainage basin. This necessitates development of a correction factor to balance out the rainfall variation caused by various topographical features in the watershed. Rain gages tend to give rainfall volumes that are too small. This error is caused by the movement of wind around the gage, and it increases as wind velocity increases. This "windage" error is much more pronounced when the rain gage is near the top or bottom of a cliff or near other big obstructions. Care must be exercised in placement of rain gages to ensure accuracy.

The probable maximum precipitation is the greatest rainfall intensity or volume that could ever be expected to occur in a specific drainage basin. This rainfall magnitude is frequently used as the design storm for major hydraulic structures to serve the basin when the rainfall records are short and extrapolation to the desired design-storm frequency could be grossly inaccurate. The magnitude of probable maximum precipitation is based on simultaneous occurrence of the maximum values of the meteorological factors that combine to form precipitation. The two most important factors are wind and air-mass moisture content. An idea of the magnitude of the probable maximum precipitation can also be obtained by transposing the greatest rainfall that has occurred in a meteorologically homogeneous region. For methods for determining the probable maximum precipitation, see D. R. Maidment, "Handbook of Hydrology," McGraw-Hill, Inc., New York.

Not all rain reaches the ground. A portion may evaporate as it falls, while another portion may be caught on leaves, branches, and other vegetation surfaces. This phenomenon, called **interception**, is a loss from a runoff standpoint since the rain evaporates and never reaches the ground. Interception may be significant for small-intensity storms occurring with little or no wind over an area with heavy vegetation growth.

21.38 Evaporation and Transpiration

These are processes by which moisture is returned to the atmosphere. In evaporation, water changes from liquid to gaseous form. In transpiration, plants give off water vapor during synthesis of plant tissue.

Evapotranspiration, commonly termed **consumptive use**, refers to the total evaporation from all sources such as free water, ground, and plant-leaf surfaces. On an annual basis, the consumptive use may vary from 15 in/year for barren land to 35 in/year for heavily forested areas and 40 in/year in tropical and subtropical regions. Evapotranspiration is important because, on a long-term basis, precipitation minus evapotranspiration equals runoff.

Evaporation may occur from free-water, plant, or ground surfaces. Of the three, free-water surface evaporation is usually the most important. It must be considered in the design of a reservoir, especially if the reservoir is shallow, has a relatively large surface area, and is located in a semiarid or arid region. Evaporation is a direct function of the wind and temperature and an inverse function of atmospheric pressure and amount of soluble solids in the water.

The rate of evaporation is dependent on the vapor-pressure gradient between the water surface and the air above it. This relation is known as Dalton's law. The **Meyer equation** [Eq. (21.125)], developed from Dalton's law, is one of many evaporation formulas and is popular for making evaporation-rate calculations.

$$E = C(e_w - e_a)\psi \qquad (21.125)$$
$$\psi = 1 + 0.1w \qquad (21.126)$$

where E = evaporation rate, in 30-day month

C = empirical coefficient, equal to 15 for small, shallow pools and 11 for large, deep reservoirs

e_w = saturation vapor pressure, in of mercury, corresponding to monthly mean air temperature observed at nearby stations for small bodies of shallow water or corresponding to water temperature instead of air temperature for large bodies of deep water

e_a = actual vapor pressure, in of mercury, in air based on monthly mean air temperature and relative humidity at nearby stations for small bodies of shallow water or based on information obtained about 30 ft above the water surface for large bodies of deep water

w = monthly mean wind velocity, mi/h at about 30 ft above ground

ψ = wind factor

As an example of the evaporation that may occur from a large reservoir, the mean annual evaporation from Lake Mead is 6 ft.

Evaporation from free-water surfaces is usually measured with an evaporation pan. This pan is a standard size and is located on the ground near the body of water whose evaporation is to be determined. The depth of water in this pan is checked periodically and corrections made for factors other than evaporation that may have raised or lowered the water surface. A pan coefficient is then applied to the measured pan evaporation to get the reservoir evaporation.

The **standard evaporation pan** of the National Weather Service, called a Class A Level Pan, is in widespread use. It is 4 ft in diameter and 10 in deep. It is positioned 6 in above the ground. Its pan coefficient is commonly taken as 0.70, although it may vary between 0.60 and 0.80, depending on the geographical region. Annual evaporation from the pan ranges from 25 in in Maine and Washington to 120 in along the Texas-Mexico and California-Arizona borders.

Evaporation rates from reservoirs may be reduced by spreading thin molecular films on the water surface. Hexadeconal, or cetyl alcohol, is one such film that has been effective on small reservoirs where there is little wind. On large reservoirs, wind tends to push the film to the shore. Since hexadeconal is removed by wind, birds, insects, aquatic life, and biologic attrition, it must be applied periodically for maximum effectiveness. Hexadeconal appears to have no adverse effects on either humans or wildlife.

Evaporation from ground surfaces is usually of minor importance, except in arid, tropical, and subtropical regions having high water tables and where it pertains to the determination of initial soil-moisture conditions in a runoff analysis.

(D. R. Maidment, "Handbook of Hydrology," McGraw-Hill, Inc., New York (books.mcgraw-hill.com).)

21.39 Runoff

This is the residual precipitation remaining after interception and evapotranspiration losses have been deducted. It appears in surface channels, natural or manmade, whose flow is perennial or intermittent. Classified by the path taken to a channel, runoff may be surface, subsurface, or groundwater flow.

Surface flow moves across the land as *overland flow* until it reaches a channel, where it continues as *channel* or *stream flow*. After joining stream flow, it combines with the other runoff components in the channel to form *total runoff*.

Subsurface flow, also known as *interflow, subsurface runoff, subsurface storm flow*, and *storm seepage*, infiltrates only the upper soil layers without joining the main groundwater body. Moving laterally, it may continue underground until it reaches a channel or returns to the surface and continues as overland flow. The time for subsurface flow to reach a channel depends on the geology of the area. Commonly, it is assumed that subsurface flow reaches a channel during or shortly after a storm. Subsurface flow may be the major portion of total runoff for moderate or light rains in arid regions since surface flow under those conditions is reduced by unusually high evaporation and infiltration.

Groundwater flow, or **groundwater runoff**, is that flow supplied by deep percolation. It is the flow of the main groundwater body and requires long periods, perhaps several years, to reach a channel. Groundwater flow is responsible for the dry-weather flow of streams and remains practically constant during a storm. Groundwater flow is primarily the concern of water-supply engineers. Surface and subsurface flow are of interest to flood-control engineers.

In practice, **direct runoff** and **base flow** are the only two divisions of runoff used. The basis for this classification is travel time rather than path. Direct runoff leaves the basin during or shortly after a storm, whereas base flow from the storm may not leave the basin for months or even years.

Runoff is supplied by precipitation. The portion of precipitation that contributes entirely to direct runoff is called *effective precipitation*, or *effective rain* if the precipitation is rain. That portion of the precipitation which contributes entirely to surface runoff is called *excess precipitation*, or *excess rain*. Thus, effective rain includes subsurface flow, whereas excess rain is only surface flow.

The two major characteristics that affect runoff are climatic and drainage-basin factors. The number of factors is an indication of the complexity of accurately determining runoff:

1. **Climatic characteristics**

   a. Precipitation—form (rain, hail, snow, frost, dew), intensity, duration, time distribution, seasonal distribution, areal distribution, recurrence interval, antecedent precipitation, soil moisture, direction of storm movement

b. Temperature—variation, snow storage, frozen ground during storms, extremes during precipitation

c. Wind—velocity, direction, duration

d. Humidity

e. Atmospheric pressure

f. Solar radiation

2. **Drainage-basin characteristics**

a. Topographic—size, shape, slope, elevation, drainage net, general location, land use and cover, lakes and other bodies of water, artificial drainage, orientation, channels (size, shape of cross section, slope, roughness, length)

b. Geologic—soil type, permeability, groundwater formations, stratification

21.40 Sources of Hydrologic Data

The importance of exhausting all possible sources of hydrologic data, both published and unpublished, as the first step in design of a hydraulic project cannot be overemphasized. The majority of hydrologic data is collected and published by government agencies, those of the Federal government being the largest and most important. The principal source of precipitation data is the U.S. Weather Bureau. Its extensive system of gages supplies complete precipitation data as well as all other types of hydrologic data. These data are compiled and presented in monthly and yearly summaries in the Bureau's "Climatological Data." In addition to the monthly and yearly summaries, special-interest items, such as rainfall intensity for various durations and recurrence intervals, are published in Weather Bureau technical papers.

Other sources are *Water Bulletins* of the International Boundary Commission, the U.S. Agricultural Research Service, and various state and local agencies.

The principal source of runoff data is the *Water Supply Papers* of the U.S. Geological Survey. These papers contain records of daily flow, mean flow, yearly flow volume, extremes of flow, and statistical data pertaining to the entire record. Also included in the *Papers* are lists of reports covering

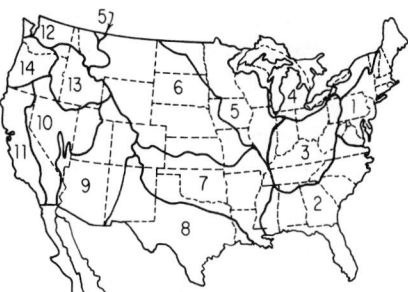

Fig. 21.70 Drainage subdivisions of the United States for stream-flow records published in "Water Supply Papers," U.S. Geological Survey.

unusually large floods and records of discharge collected by agencies other than the U.S. Geological Survey. The *Water Supply Papers* are published yearly in 14 parts; each part is for an area whose boundaries coincide with natural-drainage features, as shown in Fig. 21.70.

Other agencies that collect and publish stream-flow and flood records are the Corps of Engineers, TVA, International Boundary Commission, and Weather Bureau. The Corps of Engineers publishes data on floods in which loss of life and extensive property damage occurred. Less obvious sources of stream-flow data are water-right decrees by district courts, county records of water-right filings and State Engineer permits, and annual reports of various interstate-compact commissions.

21.41 Methods for Runoff Determinations

The method selected to determine runoff depends on its applicability to the area of concern, the quantity and type of data available, the detail required in the final answer, and the accuracy desired. Applicability depends on the characteristics of the particular area and the assumptions from which the method was developed. Quantity and type of data available refer to the length, detail, and completeness of the hydrologic records, which may be either precipitation or stream flow. An example of the variation of detail in the final result may be found in the determination of flood runoff. Several methods yield only peak discharge; others give the complete hydrograph. Accuracy is limited by the cost of performing analyses and assumptions made in the development of a method.

The methods that follow are a convenient means for solving typical runoff problems encountered in water resources engineering. One method pertains to minor hydraulic structures, the second to major hydraulic structures. A minor structure is one of low cost and of relatively minor importance and presents small downstream damage potential. Typical examples are small highway and railroad culverts and low-capacity storm drains. Major hydraulic structures are characterized by their high cost, great importance, and large downstream damage potential. Typical examples of major hydraulic structures are large reservoirs, deep culverts under vital highways and railways, and high-capacity storm drains and flood-control channels.

21.41.1 Method for Determining Runoff for Minor Hydraulic Structures

The most common means for determining runoff for minor hydraulic structures is the **rational formula**

$$Q = CIA \qquad (21.127)$$

where Q = peak discharge, ft^3/s

C = runoff coefficient = percentage of rain that appears as direct runoff

I = rainfall intensity, in/h

A = drainage area, acres

The assumptions inherent in the rational formula are:

1. The maximum rate of runoff for a particular rainfall intensity occurs if the duration of rainfall is equal to or greater than the time of concentration. The time of concentration is commonly defined as the time required for water to flow from the most distant point of a drainage basin to the point of flow measurement.

2. The maximum rate of runoff from a specific rainfall intensity whose duration is equal to or greater than the time of concentration is directly proportional to the rainfall intensity.

3. The frequency of occurrence of the peak discharge is the same as that of the rainfall intensity from which it was calculated.

4. The peak discharge per unit area decreases as the drainage area increases, and the intensity of rainfall decreases as its duration increases.

5. The coefficient of runoff remains constant for all storms on a given watershed.

Since these assumptions apply reasonably well for urbanized areas with simple drainage facilities of fixed dimensions and hydraulic characteristics, the rational formula has gained widespread use in the design of drainage systems for these areas. Its simplicity and ease of application have resulted in its being used for more complex urban systems and rural areas where the assumptions are not so applicable.

The rational formula is criticized for expressing runoff as a fraction of rainfall rather than as rainfall minus losses and for combining all the complex factors that affect runoff into a single coefficient. Although these and similar criticisms are valid, use of a more complicated formula is not justified because the time and money spent to obtain the necessary data would not be warranted for minor hydraulic structures.

Numerous refinements have been developed for the runoff coefficient. As an example, the Los Angeles County Flood Control District gives runoff coefficients as a function of the soil and area type and of the rainfall intensity for the time of concentration. Other similar refinements are possible if the resources are available. Careful selection of the runoff coefficient C will give values of peak runoff consistent with project significance. The values of C in Table 21.15 for urban areas are commonly recommended design values (V. T. Chow, "Hydrologic Determination of Waterway Areas for the Design of Drainage Structures in Small Drainage Basins," *University of Illinois Engineering Experimental Station Bulletin 426*, 1962).

After selection of the design-storm frequency of occurrence, for example, a 50- or 100-year-frequency storm, the rainfall intensity I may be determined from any of a number of formulas or from a statistical analysis of rainfall data if enough are available. Chow lists 24 rainfall-intensity formulas of the form

$$I = \frac{KF^{n_1}}{(t+b)^n} \qquad (21.128)$$

Table 21.15 Common Runoff Coefficients

Type of Drainage Area	Runoff Coefficient C
Business:	
Downtown areas	0.70–0.95
Neighborhood areas	0.50–0.70
Residential:	
Single-family areas	0.30–0.50
Multiunits, detached	0.40–0.60
Multiunits, attached	0.60–0.75
Suburban	0.25–0.40
Apartment dwelling areas	0.50–0.70
Industrial:	
Light areas	0.50–0.80
Heavy areas	0.60–0.90
Parks, cemeteries	0.10–0.25
Playgrounds	0.20–0.35
Railroad-yard areas	0.20–0.40
Unimproved areas	0.10–0.30
Streets:	
Asphaltic	0.70–0.95
Concrete	0.80–0.95
Brick	0.70–0.85
Drives and walks	0.75–0.85
Roofs	0.75–0.95
Lawns:	
Sandy soil, flat, 2%	0.05–0.10
Sandy soil, avg, 2–7%	0.10–0.15
Sandy soil, steep, 7%	0.15–0.20

(Table continued)

where I = rainfall intensity, in/h

$K, b, n,$ and n_1 = respectively, coefficient, factor, and exponents depending on conditions that affect rainfall intensity

F = frequency of occurrence of rainfall, years

t = duration of storm, min

= time of concentration

Perhaps the most useful of these formulas is the **Steel formula**:

$$I = \frac{K}{t + b} \qquad (21.129)$$

where K and b are dependent on the storm frequency and region of the United States (Fig. 21.71 and Table 21.16).

Equation (21.129) gives the average maximum precipitation rates for durations up to 2 h.

The time of concentration T_c at any point in a drainage system is the sum of the overland flow time; the flow time in streets, gutters, or ditches; and the flow time in conduits. Overland flow time may be determined from any number of formulas developed for the purpose. (See D. R. Maidment, "Handbook of Hydrology," McGraw-Hill, Inc., New York.) The flow time in gutters, streets, ditches, and conduits can be determined from a calculation of the average velocity using the Manning equation [Eq. (21.89)]. The time of concentration is usually expressed in minutes.

After determining the time of concentration, calculate the corresponding rainfall intensity from either Eq. (21.128) or Eq. (21.129), or any equivalent method. Then select the runoff coefficient from Table 21.15 and determine the peak discharge from Eq. (21.127).

Since the rational formula assumes a constant uniform rainfall for the time of concentration over the entire area, the area A must be selected so that this assumption applies with reasonable accuracy. Adhering to this assumption may necessitate subdividing the drainage area.

21.41.2 Method for Determining Runoff for Major Hydraulic Structures

The unit-hydrograph method, pioneered in 1932 by LeRoy K. Sherman, is a convenient, widely accepted procedure for determining runoff for major

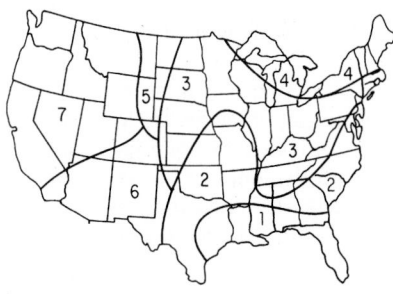

Fig. 21.71 Regions of the United States for use with the Steel formula.

Table 21.16 Coefficients for Steel Formula

Fre-quency, years	Coeffi-cients	Region						
		1	2	3	4	5	6	7
2	K	206	140	106	70	70	68	32
	b	30	21	17	13	16	14	11
4	K	247	190	131	97	81	75	48
	b	29	25	19	16	13	12	12
10	K	300	230	170	111	111	122	60
	b	36	29	23	16	17	23	13
25	K	327	260	230	170	130	155	67
	b	33	32	30	27	17	26	10
50	K	315	350	250	187	187	160	65
	b	28	38	27	24	25	21	8
100	K	367	375	290	220	240	210	77
	b	33	36	31	28	29	26	10

hydraulic structures. (Leroy K. Sherman, "Streamflow from Rainfall by Unit-Graph Method," *Engineering News-Record*, vol. 108, pp. 501–505, January-June 1932.) It permits calculation of the complete runoff hydrograph from any rainfall after the unit hydrograph has been established for the particular area of concern.

The **unit hydrograph** is defined as a runoff hydrograph resulting from a **unit storm**. A unit storm has practically constant rainfall intensity for its duration, termed a *unit period*, and a runoff volume of 1 in (water with a depth of 1 in over a unit area, usually 1 acre). Thus, a unit storm may have a 2-in/h effective intensity lasting $\frac{1}{2}$ h or a 0.2-in/h effective intensity lasting 5 h. The significant part of the definition is not the volume but the constancy of intensity. Adjustments can be made within unit-hydrograph theory for situations where the runoff volume is different from 1 in, but corrections for highly variable rainfall rates cannot be made.

The unit hydrograph is similar in concept to determining a set of factors for a specific drainage basin. The set consists of one factor for each variable that affects runoff. The unit hydrograph is much quicker, easier, and more accurate than any such set of factors. The method is summarized by the formula

Effective rain × unit hydrograph = runoff

(21.130)

The unit hydrograph thus is the link between rainfall and runoff. It may be thought of as an integral of the many complex factors that affect runoff. The unit hydrograph can be derived from rainfall and stream-flow data for a particular storm or from stream-flow data alone.

Assumptions made in the development of the unit-hydrograph theory are:

1. Rainfall intensity is constant for its duration or a specified period of time. This requires that a storm of short duration, termed a unit storm, be used for the derivation of the unit hydrograph.

2. The effective rainfall is uniformly distributed over the drainage basin. This specifies that the drainage area be small enough for the rainfall to be essentially constant over the entire area. If the watershed is very large, subdivision may be required; the unit-hydrograph theory is then applied to each subarea.

3. The base of the hydrograph of direct runoff is constant for any effective rainfall of unit duration. This needs no clarification except that the base of a hydrograph, that is, the time of storm runoff, is largely arbitrary since it depends on the method of base-flow separation.

4. The ordinates of the direct runoff hydrographs of a common base time are directly proportional to the total amount of direct runoff represented by each hydrograph. Illustrated in Fig. 21.72,

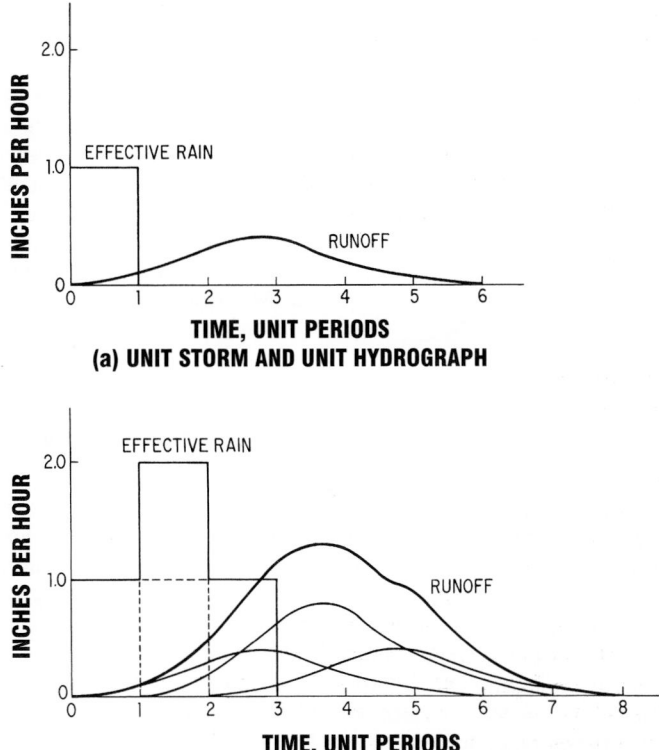

TIME, UNIT PERIODS
(a) UNIT STORM AND UNIT HYDROGRAPH

TIME, UNIT PERIODS
(b) COMPOSITE STORM AND COMPOSITE HYDROGRAPH

Fig. 21.72 Unit hydrograph (*a*) prepared for a unit storm is used to develop a composite hydrograph (*b*) for any storm.

this is basically the principle of superposition or proportionality. It enables calculation of the runoff for a storm of any intensity or duration from a unit storm, which is of fixed intensity and duration. A given storm may be resolved into a number of unit storms. Then, the runoff may be calculated by superimposing that number of unit hydrographs.

5. The hydrograph of direct runoff for a given period of rainfall reflects all the combined physical characteristics of the basin (commonly referred to as the *principle of time invariance*). This assumption implies that the characteristics of the drainage basin have not changed since the unit hydrograph was derived. Because this applies with varying degrees of accuracy to watersheds, the characteristics of the drainage basin must be fixed or specified. Daily and weekly variations in initial soil moisture are

probably the greatest source of error in this method since they are largely unknown. Man-made alterations and stream-flow conditions can be accounted for much more easily.

For ease of manipulation, the unit hydrograph is frequently expressed in histogram form as a **distribution graph** (Fig. 21.73), which illustrates the percentages of total runoff that occur during successive unit periods. The ordinate for each unit period is the mean value of runoff for that period.

Since the unit hydrograph is derived for a unit storm of specific duration, it may be used only for storms divided into unit periods of that length. Usually, because of storm variations, the unit period must be different from that for which the unit hydrograph was derived. This requires the recalculation of the unit hydrograph for the new unit period. This is accomplished by offsetting two *S* hydrographs by a time equal to the duration

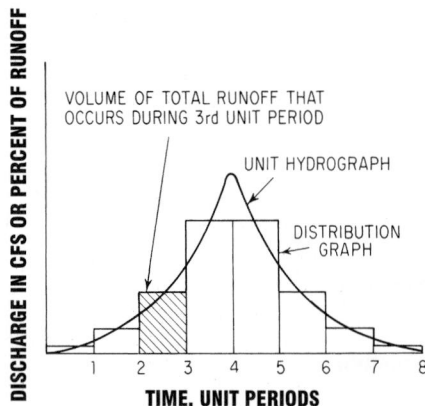

Fig. 21.73 Distribution graph represents a unit hydrograph as a histogram.

of the desired unit period (Fig. 21.74). An **S hydrograph** is a representation of the cumulative percentages of runoff that occur during a storm which has a continuous constant rainfall. It is calculated by cumulatively plotting the distribution percentages that make up the distribution graph. The distribution percentages for the new unit hydrograph are determined by taking the difference between mean ordinates for the two offset S hydrographs and dividing by the new unit period.

Transposition of a unit hydrograph from one basin to another similar basin may be made by correlating their respective shape and slope factors. This method was developed by Franklin F. Snyder (*Transactions of the American Geophysical Union*, vol. 19, pt. I, pp. 447–454). Also, since S hydrographs are a characteristic of a drainage basin, those from

various basins may be compared to obtain an idea of the variations that might exist when transposing data from one basin to another.

In the application of the unit-hydrograph method, a loss rate must be established to determine effective rain. This loss, during heavy storms, is usually considered to be entirely infiltration. The infiltration capacity of a soil may be determined experimentally by lysimeter or infiltrometer (R. K. Linsley et al., "Hydrology for Engineers," 3rd ed., McGraw-Hill, Inc., New York (books.mcgraw-hill.com).)

21.42 Groundwater

Groundwater is subsurface water in porous strata within a zone of saturation. It supplies about 20% of the United States water demand. Where groundwater is to be used as a water-supply source, the extent of the groundwater basin and the rate at which continuing extractions may be made should be determined.

Aquifers are groundwater formations capable of furnishing an economical water supply. Those formations from which extractions cannot be made economically are called **aquicludes**.

Permeability indicates the ease with which water moves through a soil and determines whether a groundwater formation is an aquifer or aquiclude.

The rate of movement of groundwater is given by **Darcy's law:**

$$Q = KIA \qquad (21.131)$$

where Q = flow rate, gal/day

K = hydraulic conductivity, ft/day or m/day

I = hydraulic gradient, ft/ft or m/m

A = cross-sectional area, perpendicular to direction of flow, ft^2 or m^2

Hydraulic conductivity is a measure of the ability of a soil to transmit water. It is a nonlinear function of volumetric soil water content and varies with soil texture. Many methods are available for determining hydraulic conductivity. (See D. R. Maidment, "Handbook of Hydrology," McGraw-Hill, Inc., New York (books.mcgraw-hill.com).)

Transmissibility is another index for the rate of groundwater movement and equals the product of

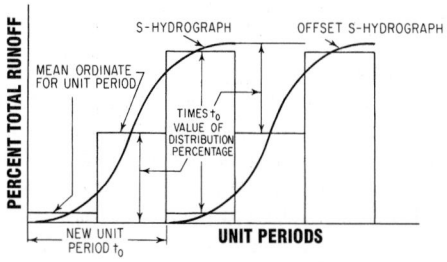

Fig. 21.74 Distribution percentages are determined from an offset *S* hydrograph.

hydraulic conductivity and the thickness of the aquifer. Transmissibility indicates for the aquifer as a whole what hydraulic conductivity indicates for the soil.

An aquifer whose water surface is subjected to atmospheric pressure and may rise and fall with changes in volume is a *free* or *unconfined aquifer.* An aquifer that contains water under hydrostatic pressure, because of impermeable layers above and below it, is a *confined* or *artesian aquifer.* If a well is drilled into an artesian aquifer, the water in this well will rise to a height corresponding to the hydrostatic pressure within the aquifer. Frequently, this hydrostatic pressure is sufficient to cause the water to jet beyond the ground surface into the atmosphere. An artesian aquifer is analogous to a large-capacity conduit with full flow in that extractions from it cause a decrease in pressure, rather than a change in volume. This is in contrast to a free aquifer, where extractions cause a decrease in the elevation of the groundwater table.

Groundwater Management ■ With increasing use being made of groundwater resources, effective groundwater management is an absolute necessity. Adequate management should include not only quantity but quality. Quantity management consists of effective control over extractions and replenishment. Quality management consists of effective control over groundwater pollution resulting from waste disposal, recycling, poor-quality replenishment waters, or other causes.

Several steps or investigations are necessary for developing an effective management program. First is a comprehensive geologic investigation of the groundwater basin to determine the characteristics of the aquifers. Second is a qualitative and quantitative hydrologic study of both surface water and groundwaters to determine historical surpluses and deficiencies, safe yield, and overdraft. (**Safe yield** is the magnitude of the annual extractions from an aquifer that can continue indefinitely without bringing some undesirable result. Deteriorating water quality, need for excessive pumping lifts, or infringement on the water rights of others are examples of undesirable results that could define safe yield. Regardless of how it is defined, safe yield applies only to a specific set of conditions based largely on judgment as to what is desirable. Extractions in excess of the safe yield are termed **overdrafts**.) In conjunction with the hydrologic study, present and future water demands should be determined. A detailed water-quality study should be made not only of the groundwater within the basin but also of all surface waters, wastewaters, and other waters that replenish the groundwater basin. Undesirable water-quality and -quantity conditions should be identified.

Following the preceding preliminary work, alternative management plans should be formulated. These management plans should consider variations in the quantity of extractions; groundwater levels; quality, quantity, and location of artificial replenishment; source, quantity, and quality of water supply; and methods of wastewater disposal. All alternative plans must recognize all legal and jurisdictional constraints.

The final step is the operational-economic evaluation of the alternatives and the selection of a recommended groundwater management plan. Operations and economic studies are normally conducted by superimposing present and future conditions in each alternative plan on historical hydrologic conditions that occurred during a base period. (A **base period** is a period of time, usually a number of years, specifically chosen for detailed hydrologic analysis because conditions of water supply and climate during the period are equivalent to a mean of long-term conditions and adequate data for such hydrologic analysis are available.)

Economic evaluation of alternative plans should consider cost of water-supply facilities, cost of replenishment water, cost of wastewater-disposal facilities, cost of pumping groundwater at the various operational levels considered, and indirect water-quality use costs, among others. (*Indirect water-quality use costs* are those indirect costs incurred by water distributors and consumers as a result of using water of different qualities. These costs include increased soap costs, water softening costs, and costs associated with the more rapid deterioration of plumbing and waterworks equipment—all of which increase as the hardness and salinity of the water increase.)

Operational studies should determine the most efficient manner of joint operation of surface and groundwater systems (conjunctive use).

Use of computers and the development of a mathematical model for the groundwater basin are almost essential because of the number of repetitive calculations involved.

Upon completion of the operational and economic studies, the most favorable management scheme should be selected as the recommended plan. This selection should be based not only on economic and operational considerations but on social, institutional, legal, and environmental factors. The plan should be capable of being readily implemented, flexible enough to accommodate different growth rates, financially feasible, and generally acceptable to the water and wastewater agencies operating in the basin.

An operating agency should be designated or formed to implement the recommended plan. The agency should have adequate powers to control or cooperate in the control of surface-water supplies groundwater recharge sites, surface-water delivery facilities, amount and location of groundwater extractions, and wastewater treatment and disposal facilities. The operating agency should develop a comprehensive monitoring network and a data collection and evaluation program to determine the effectiveness of the management plan and to implement any changes in the plan deemed necessary. This monitoring network may consist of selected wells where groundwater levels and chemical characteristics are measured and certain surface-water sampling locations where both quantitative and qualitative factors are measured. The program should also include quantitative evaluation of extractions, water used, wastewater disposed, and natural and artificial replenishment. Integration of the above data with the computer model of the groundwater basin is an efficient method of evaluating the groundwater management scheme.

("Ground Water Management," Manual and Report on Engineering Practice, no. 40, American Society of Civil Engineers, 1987; J. Bear, "Hydraulics of Ground Water," N. S. Grigg, "Water Resources Planning," A. I. Kashef, "Groundwater Engineering," R. K. Linsley et al., "Hydrology for Engineers," 3rd ed., McGraw-Hill Book Company, New York (books.mcgraw-hill.com).)

Water Supply

A waterworks system is created or expanded to supply a sufficient volume of water at adequate pressure from the supply source to consumers for domestic, irrigation, industrial, fire-fighting, and sanitary purposes. A primary concern of the engineer is estimation of the quantity of potable water to be consumed by the community since the engineer must design adequately sized components of the water-supply system. Water-supply facilities consist of collection, storage, transmission, pumping, distribution, and treatment works.

To assure continuous service to the consumer for fire-fighting and sanitary purposes in the event of an earthquake, fire, flood, or other unforeseen emergency, careful consideration must be given to the selection of standby equipment and alternative supplies of water. Maximum protection must be given to power sources and pumps that must be available to operate continuously during emergency conditions. A dependable supply with sufficient pressure for fighting fires considerably increases capital expenditures for system construction. The smaller the system, the larger the percentage of the total cost chargeable to dependable fire flow.

21.43 Water Consumption

The size of a proposed water-supply project is usually based on an average annual per capita consumption rate. Therefore, forecasts of population for the design period are of the greatest importance and must be made with care to ensure that components for the project are of adequate size. Estimation of future population, however, is a very difficult task.

Several mathematical methods are available for use in predicting populations of cities. Some methods commonly used are arithmetical increase, percentage increase, decreasing percentage increase, graphical comparison with other cities, and the ratio method of comparing a community with a state or country of which the community is a part. Great care and judgment must be exercised in population prediction since many factors, such as industrial development, land speculation, geographical boundaries, and age of the city, may drastically alter mathematical estimates.

The total water supply of a city is usually distributed among the following four major classes of consumers: domestic, industrial, commercial, and public.

Domestic use consists of water furnished to houses, apartments, motels, and hotels for drinking, bathing, washing, sanitary, culinary, and lawn-sprinkling purposes. Domestic use accounts for between 30 and 60% (50 to 60 gal per capita per day) of total water consumption in an average city.

Commercial water is used in stores and office buildings for sanitary, janitorial, and air conditioning purposes. Commercial use of water amounts to about 10 to 30% of total consumption.

Industrial uses of water are diverse but consist mainly of heat exchange, cooling, and cleaning. No direct relationship exists between the amount of industrial water used and the population of the community, but 20 to 50% of the total quantity of water used per capita per day is normally charged to industrial usage. Usually the larger-sized cities have a high degree of industrialization and show a correspondingly greater percentage of total consumption as industrial water.

Public use of water for parks, public buildings, and streets contributes to the total amount of water consumed per capita. Fire demands are usually included in this class of water use. The total quantity of water used for fire fighting may not be large, but because of the high rate at which it is required, it may control the design of the facilities. About 5 to 10% of all water used is for public uses.

Waste and miscellaneous usage of water include that lost because of leakage in mains, meter malfunctions, reservoir evaporation, and unauthorized uses. About 10 to 15% of total consumption may be charged to waste and miscellaneous uses.

Water Demand Rate ■ Many factors, such as the climate, size of the city, standard of living, degree of industrialization, type of service (metered or unmetered), lawn sprinkling, air conditioning, cost, pressure, and quality of the water, influence the demand rate for water.

Presence of industries usually increases the total per capita use of water but decreases the demand fluctuation. A good estimate of the potential industrial water demand can be made by relating demand to the percent of land zoned for industrial use.

Small cities frequently have a low per capita demand for water, especially if portions of the city are unsewered. Fluctuations in demand are greater in small cities, mainly because of the lack of large industries. High standards of living increase water demand and fluctuations in rate of use.

Warm and dry climates have a higher rate of water consumption because of sprinkling and air conditioning. Cold weather sometimes increases consumption because water is allowed to run to prevent pipes from freezing.

Demand for water is related to water-service meters, cost, quality, and pressure. Metering water reduces the quantity of water consumed by 10 to 25% because of the usual increase in total cost of consumers if they continue to use water at the unmetered rate. High water pressures increase demand because of greater losses at leaking mains, valves, and faucets. Normally, if the cost of water increases, the demand for it decreases. Demand for water usually increases with an improvement in quality.

Demand rates vary with time of day, month, and year. Table 21.17 is a comparison between water-demand rates for the city of Los Angeles and a national average calculated from data in a *U.S. Public Health Service Report*. The national demand-rate data, as presented in Table 21.17, are the average of a range of values, including some very high and very low rates due to variations in climatic conditions, degree of industrialization, and time of day. Examples of divergent average daily demand rates for various United States cities are: 230 gal per capita per day for Chicago, 210 gal per capita per day for Denver, 150 gal per capita per day for Baltimore, and 135 gal per capita per day for Kansas City, Mo.

The "California Water Atlas," 1979, State of California Office of Planning and Research, presents average monthly demand rates for water use

Table 21.17 Water-Demand Rates

	National avg		Los Angeles, Calif.	
	Gal per capita per day	% of avg annual rate	Gal per capita per day	% of avg annual rate
Avg day	160	100	175	100
Max day	265	165	280	160
Max h	400	250	420	240

Table 21.18 Required Fire Flow, Hydrant Spacing, and Fire Reserve Storage*

Population	Fire Flow		Duration, h	Reserve storage, MG†	Avg area served per hydrant in high-value districts, ft*	
	gal/min	MGD†			Direct streams	Engine streams
1,000	1,000	1.4	4	0.3	100,000	120,000
2,000	1,500	2.2	6	0.6	90,000	
4,000	2,000	2.9	8	1.0	85,000	110,000
10,000	3,000	4.3	10	1.8	70,000	100,000
17,000	4,000	5.8	10	2.4	55,000	90,000
28,000	5,000	7.2	10	3.0	40,000	85,000
40,000	6,000	8.6	10	3.6	40,000	80,000
80,000	8,000	11.5	10	4.8	40,000	60,000
125,000	10,000	14.4	10	6.0	40,000	48,000
200,000	12,000	17.3	10	7.2	40,000	40,000

in four coastal and four inland cities for the period 1966 to 1970. In the atlas, the effect of warm, dry climatic conditions is indicated for each location by the ratio of average monthly use to the annual average gallon per capita per day. The maximum monthly water-demand rates ranged from 119 to 141% of the annual rate for the coastal cities and from 144 to 187% of the annual rate for the dry, inland, valley cities. Moreover, the annual demand rates for the inland areas averaged 78% higher than those for the coastal cities. The difference is due primarily to the great amount of lawn sprinkling in Los Angeles. Past water-demand records of both the city being considered and other cities of similar size, industrialization, climate, and so on should be considered and incorporated in demand-rate projections for water systems.

The total quantity of water used for fighting fires is normally quite small, but the demand rate is high. The fire demand as established by the American Insurance Association is

$$G = 1020\sqrt{P}(1 - 0.01\sqrt{P}) \qquad (21.132)$$

where G = fire-demand rate, gal/min

P = population, thousands

The required fire flows computed from this formula are listed in Table 21.18. When calculating the total flow to be used in design, fire flow should be added to the average consumption for the maximum day.

21.44 Water-Supply Sources

The major sources of a water supply are surface water and groundwater. In the past, surface sources have included only the commonly occurring natural fresh waters, such as lakes, rivers, and streams, but with rapid population expansion and increased per capita water use associated with a higher standard of living, consideration must be given to desalination and waste-water reclamation as well.

In selection of a source of supply, the various factors to be considered are adequacy and reliability, quality, cost, legality, and politics. The criteria are not listed in any special order since they are, to a large extent, interdependent. Cost, however, is probably the most important because almost any source could be used if consumers are willing to pay a high enough price. In some local areas, as increasing demands exceed the capacity of existing sources, the increasing cost of each new supply focuses attention on reclamation of local supplies of wastewater and desalination.

Adequacy of supply requires that the source be large enough to meet the entire water demand. Total dependence on a single source, however, is frequently undesirable, and in some cases, diversification is essential for reliability. The source must also be capable of meeting demands during power outages and natural or created disasters. The most desirable supplies from a reliability standpoint, in order, are (1) an inexhaustible supply, whether from surface or groundwater, which flows by gravity through the distribution system; (2) a gravity source supplemented by storage reservoirs; (3) an inexhaustible source that requires pumping; and (4) sources that require both storage and pumping. As demand increases and supplies become overtaxed, conservation practices in everyday use become a valuable management tool.

Quality of the source determines both acceptability and cost; it varies considerably between regions. Preliminary estimates of quality can be made by examining the source, geology, and culture of the area.

Legality of supply is determined by doctrines and principles of water rights, such as appropriation, riparian, and ownership rights. Appropriation right gives the first right priority over later rights: "first in time means first in right." Riparian right permits owners of land adjacent to a stream or lake to take water from that stream or lake for use on their land. Ownership right gives a landowner possession of everything below and above the land. Legality is especially important for groundwater supplies or where there is transfer of water from one watershed to another.

A political problem with water supply exists because political boundaries seldom conform to natural-drainage boundaries. This problem is especially acute in extensive water-importation plans, but it even exists in varying forms for wastewater reclamation and desalination projects.

Desalination processes are of two fundamental types: those that extract salt from the water, such as electrodialysis and ion exchange, and those that extract water from the salt, such as distillation, freezing, and reverse osmosis. The energy cost of the former processes is dependent on the salt concentration. Hence, they are used mainly for brackish water. The energy costs for the water-extraction processes are essentially independent of salinity. These processes are used for seawater conversion. Very large dual-purpose nuclear power and desalination plants, which take advantage of the economies realized by enormous facilities, have been proposed, but such plants are feasible only for those large urban areas located on coasts. Transmission and pumping costs make inland use uneconomical. Although desalination may have advantages as a local source, it is not at present a panacea that will irrigate the deserts.

Acceptance of wastewater reclamation as a water source for direct domestic use is hindered by public opinion and uncertainty regarding viruses.

Much effort has been expended to solve these problems. But until such time as they are solved, wastewater reclamation will have only limited use for water supply. In the meantime, reclaimed water is being used for irrigation in agricultural and landscaping applications.

(D. W. Prasifka, "Current Trends in Water-Supply Planning," Van Nostrand Reinhold, New York.)

21.45 Quality Standards for Water

The Safe Drinking Water Act of 1974 mandated that nationwide standards be established to help ensure that the public receives safe water throughout the United States. National Interim Primary Drinking Water Standards were adopted in 1975, based largely on the 1962 U.S. Public Health Service Standards (Publication no. 956), which were used for control of water quality for interstate carriers. These earlier standards had been widely adopted voluntarily by both public and private utilities and received the immediate endorsement of the American Water Works Association as a minimum standard for all public water supplies in the United States. Similar standards were developed by the World Health Organization as standards for drinking-water quality at international ports ("International Standards for Drinking Water," World Health Organization, Geneva, Switzerland). Heightened concern over our changing environment and its health effect on water supplies was a major cause of the change from voluntary to mandatory water-quality standards. The law was amended in 1986 and 1996. The law applies to every public water system in the United States except private wells serving fewer than 25 individuals.

The Safe Drinking Water Act defines **contaminant** as any physical, chemical, biological, or radiological substance or matter in water. **Maximum contaminant level** (MCL) indicates the maximum permissible level of a contaminant in water that is delivered to any user of a public water system. The act clearly delineates between health-related quality contaminants and aesthetic-related contaminants by classifying the former as *primary* and the latter as *secondary* contaminants.

Primary Standards ▪ Table 21.19 lists maximum contaminant levels required by the National Primary Drinking Water Standards.

Table 21.19 National Primary Drinking Water Standards (Safe Drinking Water Act, 1974 and as amended in 1986 and 1996 and specified in EPA 816-F-02-013, July 2002. See USEPA's website: www.epa.gov/safewater/)

Contaminant	MCLG[1] (mg/L)[2]	MCL or TT[1] (mg/L)[2]	Potential health effects from exposure above the MCL	Common sources of contaminant in drinking water
MICROORGANISMS				
Cryptosporidium	zero	TT[3]	Gastrointestinal illness (e.g., diarrhea, vomiting, cramps)	Human and fecal animal waste
Giardia lamblia	zero	TT[3]	Gastrointestinal illness (e.g., diarrhea, vomiting, cramps)	Human and animal fecal waste
Heterotrophic plate count (HPC)	n/a	TT[3]	HPC has no health effects; it is an analytic method used to measure the variety of bacteria that are common in water. The lower the concentration of bacteria in drinking water, the better maintained the water system is	HPC measures a range of bacteria that are naturally present in the environment
Legionella	zero	TT[3]	Legionnaire's Disease, a type of pneumonia	Found naturally in water; multiplies in heating systems
Total coliforms (including fecal coliform and *E. Coli*)	zero	5.0%[4]	Not a health threat in itself; it is used to indicate whether other potentially harmful bacteria may be present[5]	Coliforms are naturally present in the environment; as well as feces; fecal coliforms and *E. coli* only come from human and animal fecal waste
Turbidity	n/a	TT[3]	Turbidity is a measure of the cloudiness of water. It is used to indicate water quality and filtration effectiveness (e.g., whether disease-causing organisms are present). Higher turbidity levels are often associated with higher levels of disease-causing microorganisms such as viruses, parasites and some bacteria. These organisms can cause symptoms such as nausea, cramps, diarrhea, and associated headaches.	Soil runoff
Viruses (enteric)	zero	TT[3]	Gastrointestinal illness (e.g., diarrhea, vomiting, cramps)	Human and animal fecal waste
DISINFECTION BYPRODUCTS				
Bromate	zero	0.010	Increased risk of cancer	Byproduct of drinking water disinfection

(Table continued)

Table 21.19 (*Continued*)

Contaminant	MCLG[1] (mg/L)[2]	MCL or TT[1] (mg/L)[2]	Potential health effects from exposure above the MCL	Common sources of contaminant in drinking water
Chlorite	0.8	1.0	Anemia; infants & young children: nervous system effects	Byproduct of drinking water disinfection
Haloacetic acids (HAA5)	n/a[6]	0.060	Increased risk of cancer	Byproduct of drinking water disinfection
Total Trihalomethanes (TTHMs)	none[7] n/a[6]	0.10 0.080	Liver, kidney or central nervous system problems; increased risk of cancer	Byproduct of drinking water disinfection
DISINFECTANTS	MRDL[1] (Mg/L)[2]	MRDL[1] (Mg/L)[2]		
Chloramines (as Cl_2)	MRDLG = 4	MRDL = 4.0[1]	Eye/nose irritation; stomach discomfort, anemia	Water additive used to control microbes
Chlorine (as Cl_2)	MRDLG = 4	MRDL = 4.0[1]	Eye/nose irritation; stomach discomfort	Water additive used to control microbes
Chlorine dioxide (as ClO_2)	MRDLG = 0.8[1]	MRDL = 0.8[1]	Anemia; infants & young children: nervous system effects	Water additive used to control microbes
INORGANIC CHEMICALS				
Antimony	0.006	0.006	Increase in blood cholesterol; decrease in blood sugar	Discharge from petroleum refineries; fire retardants; ceramics; electronics; solder
Arsenic	0[7]	0.010 as of 1/23/06	Skin damage or problems with circulatory systems, and may have increased risk of getting cancer	Erosion of natural deposits; runoff from orchards, runoff from glass & electronics production wastes
Asbestos (fibers > 10 micrometers)	7 million fibers per liter (MFL)	7 MFL	Increased risk of developing benign intestinal polyps	Decay of asbestos cement in water mains; erosion of natural deposits
Barium	2	2	Increase in blood pressure	Discharge of drilling wastes; discharge from metal refineries; erosion of natural deposits

(*Table continued*)

Table 21.19 (*Continued*)

Contaminant	MCLG[1] (mg/L)[2]	MCL or TT[1] (mg/L)[2]	Potential health effects from exposure above the MCL	Common sources of contaminant in drinking water
Beryllium	0.004	0.004	Intestinal lesions	Discharge from metal refineries and coal-burning factories; discharge from electrical, aerospace, and defense industries
Cadmium	0.005	0.005	Kidney damage	Corrosion of galvanized pipes; erosion of natural deposits; discharge from metal refineries; runoff from waste batteries and paints
Chromium (total)	0.1	0.1	Allergic dermatitis	Discharge from steel and pulp mills; erosion of natural deposits
Copper	1.3	TT[8]; Action Level = 1.3	Short term exposure: Gastrointestinal distress Long term exposure: Liver of kidney damage People with Wilson's Disease should consult their personal doctor if the amount of copper in their water exceeds the action level	Corrosion of household plumbing systems; erosion of natural deposits
Cyanide (as free cyanide)	0.2	0.2	Nerve damage or thyroid problems	Discharge from steel/metal factories; discharge from plastic and fertilizer factories
Fluoride	4.0	4.0	Bone disease (pain and tenderness of the bones); Children may get mottled teeth	Water additive which promotes strong teeth; erosion of natural deposits; discharge from fertilizer and aluminum factories

(*Table continued*)

Table 21.19 (*Continued*)

Contaminant	MCLG[1] (mg/L)[2]	MCL or TT[1] (mg/L)[2]	Potential health effects from exposure above the MCL	Common sources of contaminant in drinking water
Lead	zero	TT[8]; Action Level = 0.015	Infants and children: Delays in physical or metal development; children could show slight deficits in attention span and learning abilities Adults: Kidney problems; high blood pressure	Corrosion of household plumbing systems; erosion of natural deposits
Mercury (inorganic)	0.002	0.002	Kidney damage	Erosion of natural deposits; discharge from refineries and factories; runoff from landfills and croplands
Nitrate (measured as Nitrogen)	10	10	Infants below the age of six months who drink water containing nitrate in excess of the MCL could become seriously ill and, if untreated, may die. Symptoms include shortness of breath and blue-baby syndrome.	Runoff from fertilizer use; leaching from septic tanks, sewage; erosion of natural deposits
Nitrate (measured as Nitrogen)	1	1	Infants below the age of six months who drink water containing nitrate in excess of the MCL could become seriously ill and, if untreated, may die. Symptoms include shortness of breath and blue-baby syndrome.	Runoff from fertilizer use; leaching from septic tanks, sewage; erosion of natural deposits
Selenium	0.05	0.05	Hair or fingernail loss; numbness in fingers or toes; circulatory problems	Discharge from petroleum refineries; erosion of natural deposits; discharge from mines
Thallium	0.0005	0.002	Hair loss; changes in blood; kidney, intestine, or liver problems	Leaching from ore-processing sites; discharge from electronics, glass, and drug factories
ORGANIC CHEMICALS				
Acrylamide	zero	TT[9]	Nervous system or blood problems; increased risk of cancer	Added to water during sewage/wastewater treatment
Alachlor	zero	0.002	Eye, liver, kidney or spleen problems; anemia; increased risk of cancer	Runoff from herbicide used on row crops

(*Table continued*)

Table 21.19 (*Continued*)

Contaminant	MCLG[1] (mg/L)[2]	MCL or TT[1] (mg/L)[2]	Potential health effects from exposure above the MCL	Common sources of contaminant in drinking water
Atrazine	0.003	0.003	Cardiovascular system or reproductive problems	Runoff from herbicide used on row crops
Benzene	zero	0.005	Anemia; decrease in blood platelets; increased risk of cancer	Discharge from factories; leaching from gas storage tanks and landfills
Benzo(a)pyrene (PAHs)	zero	0.0002	Reproductive difficulties; increased risk of cancer	Leaching from linings of water storage tanks and distribution lines
Carbofuran	0.04	0.04	Problems with blood, nervous system, or reproductive system	Leaching of soil fumigant used on rice and alfalfa
Carbon tetrachloride	zero	0.005	Liver problems; increased risk of cancer	Discharge from chemical plants and other industrial activities
Chlordane	zero	0.002	Liver or nervous system problems; increased risk of cancer	Residue of banned termiticide
Chlorobenzene	0.1	0.1	Liver or kidney problems	Discharge from chemical and agricultural chemical factories
2,4-D	0.07	0.07	Kidney, liver, or adrenal gland problems	Runoff from herbicide used on row crops
Dalapon	0.2	0.2	Minor kidney changes	Runoff from herbicide used on rights of way
1,2-Dibromo-3-chloropropane (DBCP)	zero	0.0002	Reproductive difficulties; increased risk of cancer	Runoff/leaching from soil fumigant used on soybeans, cotton, pineapples, and orchards
o-Dichlorobenzene	0.6	0.6	Liver, kidney, or circulatory system problems	Discharge from industrial chemical factories
p-Dichlorobenzene	0.075	0.075	Anemia; liver, kidney or spleen damage; changes in blood	Discharge from industrial chemical factories
1,2-Dichloroethane	zero	0.005	Increased risk of cancer	Discharge from industrial chemical factories
1,1-Dichloroethylene	0.007	0.007	Liver problems	Discharge from industrial chemical factories

(*Table continued*)

Table 21.19 *(Continued)*

Contaminant	MCLG[1] (mg/L)[2]	MCL or TT[1] (mg/L)[2]	Potential health effects from exposure above the MCL	Common sources of contaminant in drinking water
cis-1,2-Dichloro-ethylene	0.07	0.07	Liver problems	Discharge from industrial chemical factories
trans-1,2-Dichloro-ethylene	0.1	0.1	Liver problems	Discharge from industrial chemical factories
Dichloromethane	zero	0.005	Liver problems; increased risk of cancer	Discharge from drug and chemical factories
1,2-Dichloropropane	zero	0.005	Increased risk of cancer	Discharge from industrial chemical factories
Di(2-ethylhexyl) adipate	0.4	0.4	Weight loss, liver problems, or possible reproductive difficulties	Discharge from chemical factories
Di(2-ethylhexyl) phthalate	zero	0.006	Reproductive difficulties; liver problems; increased risk of cancer	Discharge from rubber and chemical factories
Dinoseb	0.007	0.007	Reproductive difficulties	Runoff from herbicide used on soybeans and vegetables
Dioxin (2,3,7,8-TCDD)	zero	0.00000003	Reproductive difficulties; increased risk of cancer	Emissions from waste incineration and other combustion; discharge from chemical factories
Diquat	0.02	0.02	Cataracts	Runoff from herbicide use
Endothall	0.1	0.1	Stomach and intestinal problems	Runoff from herbicide use
Endrin	0.002	0.002	Liver problems	Residue of banned insecticide
Epichlorohydrin	zero	TT[9]	Increased cancer risk, and over a long period of time, stomach problems	Discharge from industrial chemical factories; an impurity of some water treatment chemicals
Ethylbenzene	0.7	0.7	Liver or kidneys problems	Discharge from petroleum refineries
Ethylene dibromide	zero	0.00005	Problems with liver, stomach, reproductive system, or kidneys; increased risk of cancer	Discharge from petroleum refineries
Glyphosate	0.7	0.7	Kidney problems; reproductive difficulties	Runoff from herbicide use
Heptachlor	zero	0.0004	Liver damage; increased risk of cancer	Residue of banned termiticide
Heptachlor epoxide	zero	0.0002	Liver damage; increased risk of cancer	Breakdown of heptachlor

(Table continued)

Table 21.19 (*Continued*)

Contaminant	MCLG[1] (mg/L)[2]	MCL or TT[1] (mg/L)[2]	Potential health effects from exposure above the MCL	Common sources of contaminant in drinking water
Hexachlorobenzene	zero	0.001	Liver or kidney problems; reproductive difficulties; increased risk of cancer	Discharge from metal refineries and agricultural chemical factories
Hexachlorocyclopent adiene	0.05	0.05	Kidney or stomach problems	Discharge from chemical factories
Lindane	0.0002	0.0002	Liver or kidney problems	Runoff/leaching from insecticide used on cattle, lumber, gardens
Methoxychlor	0.04	0.04	Reproductive difficulties	Runoff/leaching from insecticide used on fruits, vegetables, alfalfa, livestock
Oxamyl (Vydate)	0.2	0.2	Slight nervous system effects	Runoff/leaching from insecticide used on apples, potatoes, and tomatoes
Polychlorinated biphenyls (PCBs)	zero	0.0005	Skin changes; thymus gland problems; immune deficiencies; reproductive or nervous system difficulties; increased risk of cancer	Runoff from landfills; discharge of waste chemicals
Pentachlorophenol	zero	0.001	Liver of kidney problems, increased cancer risk	Discharge from wood preserving factories
Picloram	0.5	0.5	Liver problems	Herbicide runoff
Simazine	0.004	0.004	Problems with blood	Herbicide runoff
Styrene	0.1	0.1	Liver, kidney, or circulatory system problems	Discharge from rubber and plastic factories; leaching from landfills
Tetrachloroethylene	zero	0.005	Liver problems; increased risk of cancer	Discharge from factories and dry cleaners
Toluene	1	1	Nervous system, kidney, or liver problems	Discharge from petroleum factories
Toxaphene	zero	0.003	Kidney, liver, or thyroid problems; increased risk of cancer	Runoff/leaching from insecticide used on cotton and cattle
2,4,5-TP (Silvex)	0.05	0.05	Liver problems	Residue of banned herbicide
1,2,4-Trichlorobenzene	0.07	0.07	Changes in adrenal glands	Discharge from textile finishing factories

(*Table continued*)

Table 21.19 *(Continued)*

Contaminant	MCLG[1] (mg/L)[2]	MCL or TT[1] (mg/L)[2]	Potential health effects from exposure above the MCL	Common sources of contaminant in drinking water
1,1,1-Trichloroethane	0.20	0.2	Liver, nervous system, or circulatory problems	Discharge from metal degreasing sites and other factories
1,1,2-Trichloroethane	0.003	0.005	Liver, nervous system, or circulatory problems	Discharge from industrial chemical factories
Trichloroethylene	zero	0.005	Liver problems; increased risk or cancer	Discharge from metal degreasing sites and other factories
Vinyl chloride	zero	0.002	Increased risk of cancer	Leaching from PVC pipes; discharge from plastic factories
Xylenes (total)	10	10	Nervous system damage	Discharge from petroleum factories; discharge from chemical factories

RADIONUCLIDES

Alpha particles	none[7]	15 picocuries per Liter (pCi/L)	Increased risk of cancer	Erosion of natural deposits of certain minerals that are radioactive and may emit a form of radiation known as alpha radiation
Beta particles and photon emitters	none[7]	4 millirems per year (mrem/yr)	Increased risk of cancer	Decay of natural and man-made deposits of certain minerals that are radioactive and may emit forms of radiation known as photons and beta radiation
Radium 226 and Radium 228 (combined)	none[7]	5 pCi/L	Increased risk of cancer	Erosion of natural deposits
Uranium	zero	30 μg/L as of 12/08/03	Increased risk of cancer, kidney toxicity	Erosion of natural deposits

NOTES

1—Definitions
- Maximum Contaminant Level Goal (MCLG)—The level of a contaminant in drinking water below which there is no known or expected risk to health. MCLGs allow for a margin of safety and are non-enforceable public health goals.
- Maximum Contaminant Level (MCL)—The highest level of a contaminant that is allowed in drinking water. MCLs are set as close to MCLGs as feasible using the best available treatment technology and taking cost into consideration. MCLs are enforceable standards.
- Maximum Residual Disinfectant Level Goal (MRDLG)—The level of a drinking water disinfectant below which there is no known or expected risk to health. MRDLGs do not reflect the benefits of the use of disinfectants to control microbial contaminants.

- Maximum Residual Disinfectant Level (MRDL)—The highest level of a disinfectant allowed in drinking water. There is convincing evidence that addition of a disinfectant is necessary for control of microbial contaminants.
- Treatment Technique (TT)—A required process intended to reduce the level of a contaminant in drinking water.

2—Units are in milligrams per liter (mg/L) unless otherwise noted. Milligrams per liter are equivalent to parts per million (ppm).

3—EPA's surface water treatment rules require systems using surface water or ground water under the direct influence of surface water to (1) disinfect their water, and (2) filter their water or meet criteria for avoiding filtrations so that the following contaminants are controlled at the following levels:

- *Cryptosporidium* (as of 1/1/02 for systems serving >10,000 and 1/14/05 for systems serving <10,000) 99% removal.
- *Giardia lamblia*: 99.9% removal/inactivation
- Viruses: 99.9% removal/inactivation
- *Legionella*: No limit, but EPA believes that if *Giardia* and viruses are removed/inactivated, *Legionella* will also be controlled.
- Turbidity: At no time can turbidity (cloudiness of water) go above 5 nephelolometric turbidity units (NTU); systems that filter must ensure that the turbidity go no higher than 1 NTU (0.5 NTU for conventional or direct filtration) in at least 95% of the daily samples in any month. As of January 1, 2002, turbidity may never exceed 1 NTU, and must not exceed 0.3 NTU in 95% of daily samples in any month.
- HPC: No more than 500 bacterial colonies per milliliter
- Long Term 1 Enhanced Surface Water Treatment (Effective Date: January 14, 2005); Surface water systems or (GWUDI) systems serving fewer than 10,000 people must comply with the applicable Long Term 1 Enhanced Surface Water Treatment Rule provisions (e.g. turbidity standards, individual filter monitoring, Cryptosporidium removal requirements, updated watershed control requirements for unfiltered systems).
- Filter Backwash Recycling; The Filter Backwash Recycling Rule requires systems that recycle to return specific recycle flows through all processes of the system's existing conventional or direct filtration system or at an alternate location approved by the state.

4—No more than 5.0% samples total coliform-positive in a month. (For water systems that collect fewer than 40 routine samples per month, no more than one sample can be total coliform-positive per month.) Every sample that has total coliform must be analyzed for either fecal coliforms or *E. coli* if two consecutive TC-positive samples, and one is also positive for *E. coli* fecal coilforms, system has an acute MCL violation.

5—Fecal coliform and *E. coli* are bacteria whose presence indicates that the water may be contaminated with human or animal wastes. Disease-causing microbes (pathogens) in these wastes can cause diarrhea, cramps, nausea, headaches, or other symptoms. These pathogens may pose a special health risk for infants, young children, and people with severely compromised immune systems.

6—Although there is no collective MCLG for this contaminant group, there are individual MCLGs for some of the individual contaminants:

- Haloacetic acids: dichloroacetic acid (zero); trichloroacetic acid (0.3 mg/L)
- Trihalomethanes: bromodichloromethane (zero); bromoform (zero); dibromochloromethane (0.06 mg/L)

7—MCLGs were not established before the 1986 Amendments to the Safe Drinking Water Act. The standard for this contaminant was set prior to 1986. Therefore, there is no MCLG for this contaminant.

8—Lead and copper are regulated by a Treatment Technique that requires systems to control the corrosiveness of their water. If more than 10% of tap water samples exceed the action level, water systems must take additional steps. For copper, the action level is 1.3 mg/L, and for lead is 0.015 mg/L.

9—Each water system must certify, in writing, to the state that when it uses acrylamide and/or epichlorohydrin to treat water, the combination (or product) of dose and monomer level does not exceed the levels specified, as follows: Acrylamide = 0.05% dosed at 1 mg/L (or equivalent); Epichlorohydrin = 0.01% dosed at 20 mg/L (or equivalent).

Enforcement responsibility rests with the U.S. Environmental Protection Agency or with those states electing to take primary responsibility for ensuring compliance with the regulations. The EPA updates standards periodically.

Secondary Standards ▪ The aesthetic contaminants are covered by the secondary drinking water regulations. The limits are called secondary maximum contaminant levels (SMCL) and are listed in Table 21.20. These levels represent reasonable goals for drinking-water quality but are not Federally enforceable. The states may use these SMCL as guidelines and establish higher or lower levels that may be appropriate, dependent on local conditions, such as unavailability of alternate source waters or other compelling factors, if public health and welfare are not adversely affected.

Source Protection ▪ The U.S. Public Health Service Drinking Water Standards recognized the need for protecting the source of water supplies, as indicated by the following extract:

The water supply should be obtained from the most desirable source feasible, and effort should be made to prevent or control pollution of the source. If the source is not adequately protected against pollution by natural means, the supply shall be adequately protected by treatment.

Sanitary surveys shall be made of the water-supply system, from the source of supply to the connection of the customer's service piping, to locate and correct any health hazards that might exist. The frequency of these surveys shall depend upon the historical need.

Adequate capacity shall be provided to meet peak demands without development of low pressures and the possibility of backflow of polluted water from customer piping.

Case histories and monitoring programs have been reported indicating that active source protection can enhance water quality with minimal extra expense. Preventing contamination of drinking water supplies is part of EPA's mission. (See R. B. Pojasek, "Drinking-Water Quality Enhancement through Source Protection," Ann Arbor Science Publishers, Inc., Ann Arbor, Mich.)

Water Treatment

Water is treated to remove disease-producing bacteria, unpleasant tastes and odors, particulate and colored matter, and hardness and to lower the levels of any contaminants when necessary to meet water-quality standards. Some of the more common methods of treatment are plain sedimentation and storage, coagulation-sedimentation, slow and rapid sand filtration, disinfection, and softening (see also Art. 21.51).

Long-term storage of water reduces the amount of disease-producing bacteria and particulate matter. But economic conditions usually compel water purveyors to use more efficient methods of treatment, such as those mentioned above.

Table 21.20 Secondary Drinking-Water Standards

Contaminant	Secondary Standard
Aluminum	0.05 to 0.2 mg/L
Chloride	250 mg/L
Color	15 (color units)
Copper	1.0 mg/L
Corrosivity	noncorrosive
Fluoride	2.0 mg/L
Foaming Agents	0.5 mg/L
Iron	0.3 mg/L
Manganese	0.05 mg/L
Odor	3 threshold odor number
pH	6.5–8.5
Silver	0.10 mg/L

(Table continued)

21.46 Sedimentation Processes

Sedimentation or clarification is a process of removing particulate matter from water through gravity settlement in a basin by reducing the flow-through velocity. Factors that affect the settling rate of particulate matter suspended in water are size, shape, and specific gravity of the suspended particles; temperature and viscosity of the water; and size and shape of the settling basin.

The settling velocity v_s of spherically shaped particles in a viscous liquid can be found by use of Stokes' law if the Reynolds number $\mathbf{R} = v\rho d/\mu$, calculated with $v = v_s$, is equal to or less than 1.

$$v_s = \frac{g(\rho_1 - \rho)d^2}{18\mu} \qquad (21.133)$$

where v_s = settling velocity of particle, m/s

g = acceleration due to gravity, m/s^2

μ = absolute viscosity of the fluid, Pa · s

ρ_1 = density of particle, g/mm^3

ρ = density of fluid, g/mm^3

d = particle diameter, mm

If $\mathbf{R} \geq 2000$, Newton's law applies:

$$v_s = \frac{4g(\rho_1 - \rho)d}{3\rho C_D} \tag{21.134}$$

where C_D is the drag coefficient. Figure 21.75 shows a plot of C_D values vs. Reynolds numbers, to be used in Eq. (21.134).

In the region where $1.0 < \mathbf{R} < 2000$, there is a transition from Stokes' law to Newton's. The settling velocity in this region is somewhere between the values given by Newton's law and those given by Stokes' law; however, no exact expression has been developed to give the velocity.

Figure 21.76 shows the relationship of settling velocity to diameter of spherical particles with specific gravity S between 1.001 and 5.0.

21.46.1 Plain Sedimentation

The ideal settling basin (Fig. 21.77) is a sedimentation tank in which flow is horizontal, velocity is constant, and concentration of particles of each size is the same at all points of the vertical cross section at the inlet end. The basin has a volumetric capacity C, depth h_o, and width B. The surface loading rate or overflow velocity v_o, equal to the settling velocity of the smallest particle to be completely removed, can be determined by dividing the flow rate Q by the settling surface area A. For this ideal basin, the overflow velocity therefore is $v_o = Q/A = Q/BL_o$, where $Q = Bh_oV$ and L_o is the length of settling zone, V the flow-through velocity. (Usually, v_o is expressed in gallons per day per square foot of surface area.) The detention time $t = h_o/v_o = L_o/V$ also equals the volumetric capacity C divided by the rate of flow Q.

Particles with a settling velocity $v_s \geq v_o$, and those that enter the settling zone between f and j (at left in Fig. 21.77) with a settling velocity v_s larger than $(v_1 = h_1V/L_o)$ but less than v_o, are removed in this basin. The particles with a settling velocity $v_s \leq v_1$ that enter the settling zone between f and e are not removed in this basin.

The efficiency of a sedimentation basin is the ratio of the flow-through period to the detention time. The flow-through period is the time required for a dye, salt, or other indicator to pass through the basin. Settling-basin efficiencies are reduced by many factors such as cross currents, short circuiting, and eddy currents. A well-designed tank should have an efficiency of 30 to 50%.

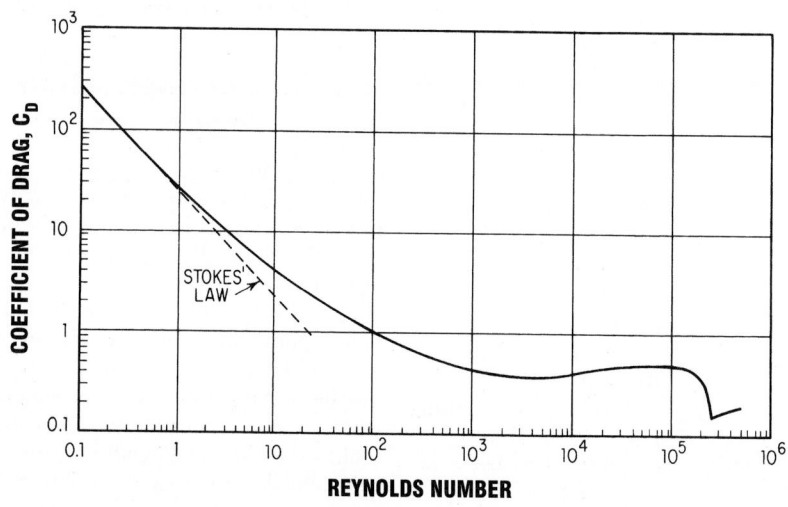

Fig. 21.75 Newton drag coefficients for spheres in fluids. (*Observed curves, after Camp, Transactions of the American Society of Civil Engineers, vol. 103, p. 897, 1946.*)

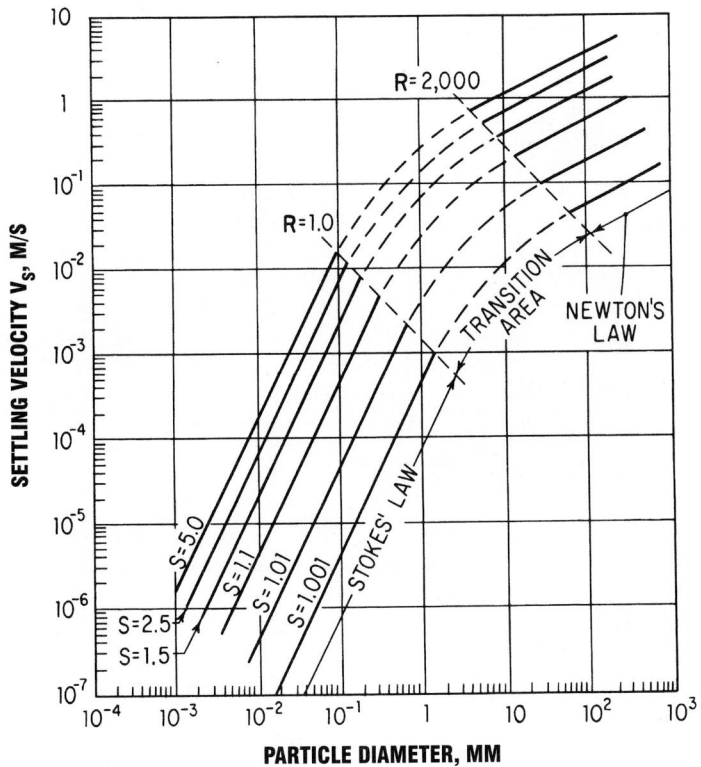

Fig. 21.76 Chart gives settling velocities of spherical particles with specific gravities S, at $10\,°C$.

Some design criteria for sedimentation tanks are:

Period of detention—2 to 8 h

Length-to-width ratio of flow-through channel—3:1 to 5:1

Depth of basin—10 to 25 ft (15 ft average)

Width of flow-through channel—not over 40 ft (30 ft most common)

Diameter of circular tank—35 to 200 ft (most common, 100 ft)

Flow-through velocity—not to exceed 1.5 ft/min (most common velocity, 1.0 ft/min)

Surface loading or overflow velocity, gal per day per ft² of surface area—between 500 and 2000 for most settling basins

Sedimentation tanks may be built in any of a variety of shapes, for example, rectangular

(Fig. 21.78a) or circular (Fig. 21.78b). Multistory tanks, such as the two-story basin with a single tray in Fig. 21.78c, occupy less site area than the single-story basin. The tubular settler (Fig. 21.78d) with parallel flow upward provides very high surface areas.

(American Water Works Association and American Society of Civil Engineers, "Water Treatment Plant Design," McGraw-Hill, Inc., New York (books. mcgraw-hill.com); G. M. Fair, J. C. Geyer, and D. A. Okun, "Water and Wastewater Engineering," John Wiley & Sons, Inc., New York (www.wiley. com).)

21.46.2 Coagulation-Sedimentation

To increase the settling rate and remove finely divided particles in suspension, coagulants are added to the water. Without coagulants, finely divided particles do not settle out because of their high ratio of surface area to mass and the presence

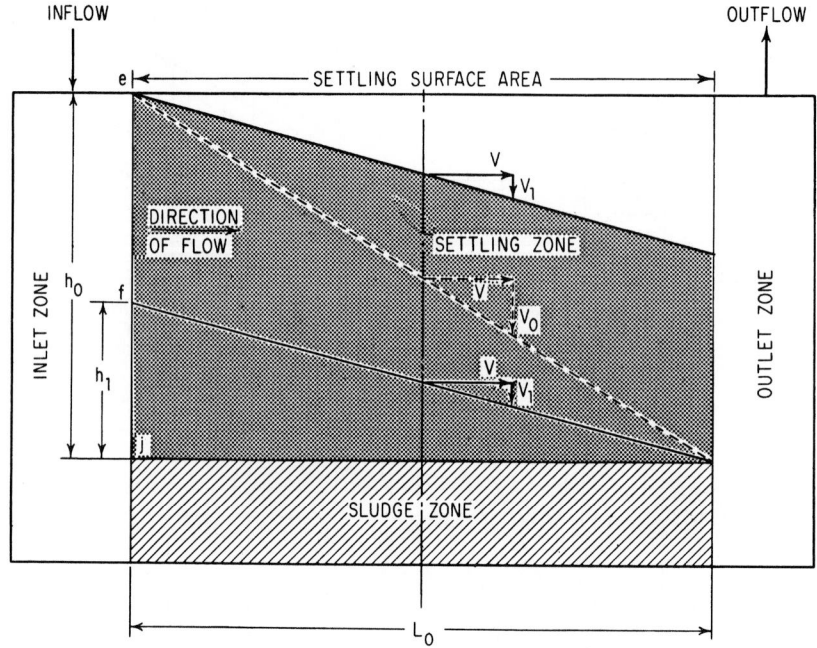

Fig. 21.77 Longitudinal section through an ideal settling basin.

of negative charges on them. The velocity at which drag and gravitational forces are equal is very low, and the negative charges on the particles produce electrostatic forces of repulsion that tend to keep the particles separated and prevent agglomeration. When coagulating chemicals are mixed with water, however, they introduce highly charged positive nuclei that attract and neutralize the negatively charged suspended matter.

Iron and aluminum compounds are commonly used as coagulants because of their high positive ionic charge. The alkalinity of the water being treated must be high enough for an insoluble hydroxide or hydrate of these metals to form. These insoluble flocs of iron and aluminum, which combine with themselves and other suspended particles, precipitate out when a floc of sufficient size is formed.

The more common coagulants are aluminum sulfate, commonly known as alum $[Al_2(SO_4)_3 \cdot 18H_2O]$; ferrous sulfate $(FeSO_4 \cdot 7H_2O)$; ferric chloride $(FeCl_3)$; and chlorinated copperas (a mixture of ferric chloride and ferrous sulfate). The type and amount of coagulant necessary to clarify a specified water depend on the qualities of water to be treated,

such as pH, temperature, turbidity, color, and hardness. Jar tests are usually made in a laboratory to determine the optimum amount of coagulant.

Some organic polymers are alternatives to the metallic coagulants. Polymers are long-chain, high-molecular-weight, organic polyelectrolytes. They are available in three types: cationic, or positively charged; anionic, or negatively charged; and non-ionic, or neutral in charge. Cationic polymers are generally the most suitable for use as primary coagulants. Anionic polymers, however, are often used as flocculant aids in conjunction with an iron or aluminum salt to cause the formation of larger floc particles. Thereby, lesser amounts of metallic salt are needed to effect good coagulation.

Because of differences in the characteristics of the suspended matter found in natural waters, not all waters can be treated with equal success with the same polymer or the same dosages. Jar tests should be run with several dosages of the various polymers available to aid in selecting the material best suited for each water supply, considering both cost and performance.

There are several reasons for considering the use of polymers: increased settling rate and improved

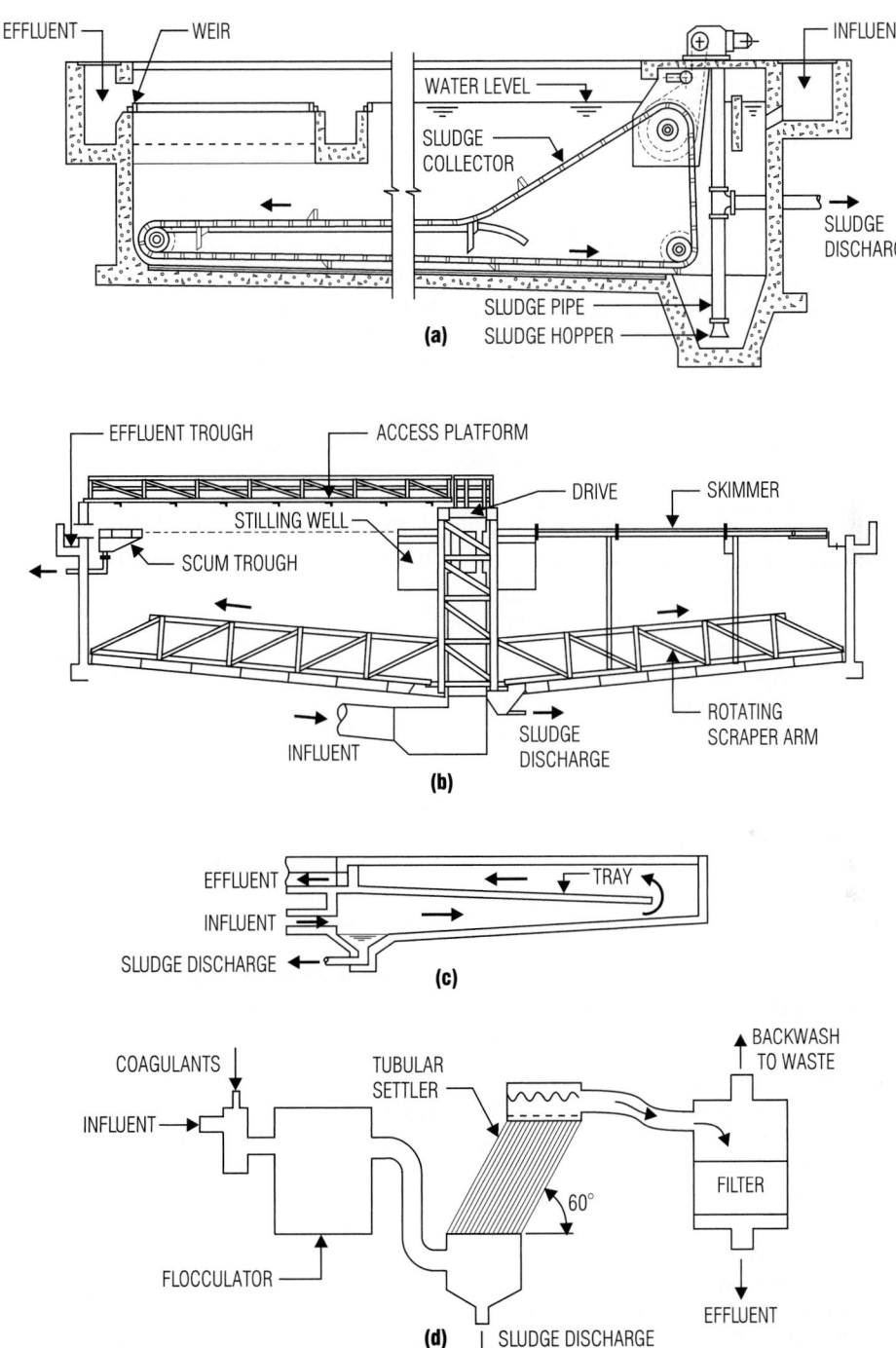

Fig. 21.78 Types of sedimentation tanks: (*a*) Rectangular settling basin. (*b*) Circular clarifier. (*c*) Two-story sedimentation basin. (*d*) Tubular settler.

filtrability of the floc, production of a smaller volume of sludge, and easier dewatering. Also, polymers have a minor effect on pH; consequently, the need for final pH adjustment in the finished water may be reduced.

Process Steps ▪ The complete clarification process is usually divided into three stages: (1) rapid chemical mixing; (2) flocculation or slow stirring, to get the small floc to agglomerate; and (3) coagulation-sedimentation in low-flow-velocity settling basins. Rapid chemical mixing may be accomplished with many devices, such as mechanical stirrers, centrifugal pumps, and air jets. The time necessary for mixing ranges from a few seconds to 20 min. Flocculation or slow stirring increases floc size and speeds up settling. The speed of the agitators must be great enough, however, to cause contact between the small floc but not so great that the larger floc is broken up. Flocculator detention time should be in the 20- to 60-min range. The coagulation-sedimentation process takes place in a **clarifier** basin nearly identical to a plain sedimentation basin. The detention period for a clarifier should be between 2 and 8 h.

(G. L. Culp and R. L. Culp, "New Concepts in Water Purification," Van Nostrand Reinhold Company, New York; American Water Works Association, "Water Quality and Treatment," 4th ed., T. J. McGhee, "Water Supply and Sewerage," R. A. Corbitt, "Standard Handbook of Environmental Engineering," McGraw-Hill, Inc., New York (books.mcgraw-hill.com).)

21.47 Filtration Processes

Passing water through a layer of sand removes much of the finely divided particulate matter and some of the larger bacteria. The filtering process has many components, such as physical straining, chemical and biological reactions, settling, and neutralization of electrostatic charges.

Direct Filtration ▪ It is possible by use of direct filtration to eliminate the sedimentation step, in some instances, for treatment of raw waters that are low in turbidity, color, coliform organisms, plankton, and suspended solids, such as paper fiber. Direct filtration is a water-treatment process in which raw water is not settled prior to the filtration step. It usually includes addition of a coagulant to destabilize the colloidal particles and a polymer as a flocculant aid. The process requires rapid mixing, agitation in a well-designed flocculator for 10 to 30 min, addition of a polymer as a filter aid, and dual- or mixed-media filtration.

Pilot plant tests are essential for selecting the best combination of coagulant and flocculant aid to obtain a strong floc and to provide criteria for design of the filtration units.

The principal advantages of direct filtration are its lower capital and operation costs. Elimination of settling basins can result in capital cost savings of 20 to 30%, and operational cost may be cut 10 to 30% by reduced chemical doses. Direct filtration merits investigation before construction of new facilities if the turbidity of the source water averages less than 25 TU.

Slow Sand Filters ▪ These consist of an underdrained, watertight container containing a 2- to 4-ft layer of sand supported by a 6- to 12-in layer of gravel. The effective size of the sand should be in the 0.25- to 0.35-mm range. (The **effective size** is the size of a sieve, in millimeters, that will pass 10%, by weight, of the sand. The **uniformity coefficient** is the ratio of the size of a sieve that will pass 60% of the sample to the effective size.) The uniformity coefficient of the sand should be less than 3. The sand is normally submerged under 4 or 5 ft of water. The water passes through the filter at a rate of 3 to 6 MG per acre per day, depending on the turbidity. The slow filter is not as versatile or as efficient as rapid sand filters.

Rapid Sand Filtration ▪ This is normally preceded by chemical treatment, such as flocculation-coagulation and disinfection, so the water can be passed through the sand at a higher rate. Usually, the effluent from a rapid filter needs further disinfection or chlorination because the bacteria are not completely removed in this process. A diagram of a typical gravity-type rapid sand filter is shown in Fig. 21.79.

The normal order of flow through the varying components of the filter is from the clarifiers (settling tanks) to the top of the sand layer, through the sand and gravel layers, through the underdrain laterals to the main drain, and then through the controller to the clear well for storage. Wash (cleaning) water flow takes place in a reverse direction after the filter effluent line has been closed. The wash-water flow is through the main

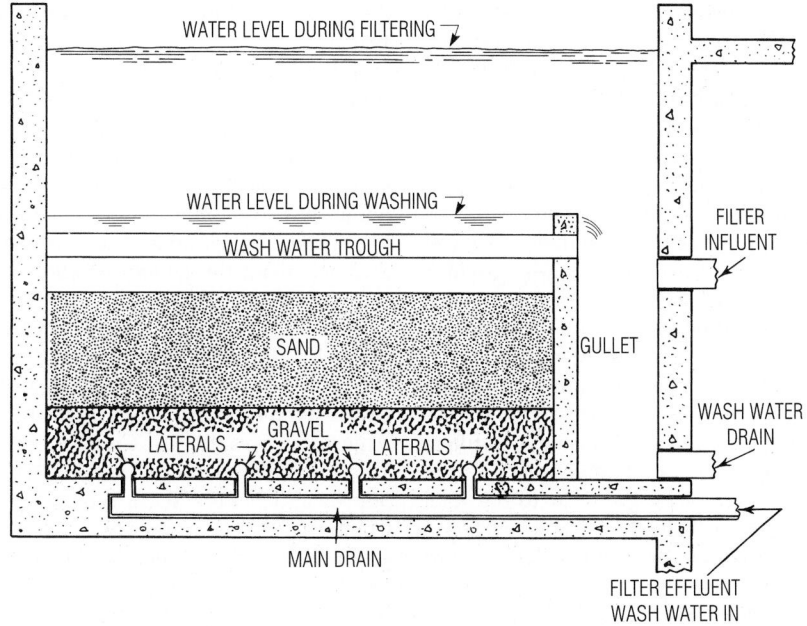

Fig. 21.79 Gravity-type rapid sand filter.

drain to the laterals, from the laterals upward through the gravel and sand to the wash-water troughs. The troughs carry the water to the gullet, which is drained to waste.

Some common design factors for rapid sand filters are:

Effective grain size—0.35 to 0.55 mm

Uniformity coefficient—1.20 to 1.75

Thickness of sand layer—24 to 30 in depending on grain size

Thickness of gravel layer—15 to 24 in

Gravel size—from $\frac{1}{8}$ to $1\frac{1}{2}$ in

Filtration rate—2 to 4 gal/min · ft² (125 to 250 MG per acre per day)

Total depth of each basin—8 to 10 ft

Maximum head loss allowed before washing sand—8 to 10 ft

Sand expansion during washing—25 to 50%

Wash-water rate—15 to 20 gal/min · ft²

Distance from top edge of wash-water trough to top of unexpanded sand—24 to 30 in

Length of filter runs between washings—12 to 72 h

Spacing between wash-water troughs—4 to 6 ft

Ratio of length to width of each basin—1.25 to 1.35

Rapid sand filters are operated until the particulate matter and unsettled floc cover the openings between the sand grains, creating a high head loss across the filter. This high head loss slows down the flow rate and may force some of the particulate matter through the sand and gravel layers. Filters are usually backwashed when the particulate-matter concentration increases in the filter effluent or when the head loss reaches 8 to 10 ft. Backwashing a filter consists of forcing filtered water through the filter from the drains upward to the wash-water troughs. The light-weight sediment is washed from the sand grains by the moving water and sometimes by other agitating devices, such as rakes, water sprays, and air jets. Filters must be washed thoroughly or difficulties with mud balls, bed cracking, or sand incrustation will be encountered.

Immediately after washing, filters pass water at a high rate, which produces an undertreated effluent. Either manual or automatic rate control must be used to prevent such an occurrence. Many treatment plants control the rate of filtration by using venturi tube devices, which throttle the filter effluent line when there is high-velocity flow. As clogging begins to occur in the filter, the velocity of flow in the effluent line decreases, and the rate controller then opens to increase the velocity.

A negative head is produced on the filter when the head loss across the filter is greater than the depth of water on the sand. Negative heads can produce a condition known as air binding, which is caused by removal of dissolved gases from the water and formation between sand grains of bubbles that decrease filter capacity.

The underdrains of a filter are commonly made of perforated pipe or porous plates. The underdrains should be arranged so that each area filters and distributes its proportionate share of water. The ratio of total area of perforations to the total filter-bed area is normally in the 0.002:1 to 0.005:1 range. The diameter of the perforations varies between $\frac{1}{4}$ and $\frac{3}{4}$ in.

Wash-water troughs should be evenly spaced, and water should not have to travel more than 3 ft horizontally to get to a wash-water gutter. The depth of water flow in a horizontal gutter may be calculated from

$$Q = 1.72by^{3/2} \qquad (21.135)$$

where Q = total flow received by trough, gal/min

b = width of trough, in

y = water depth at upstream end of trough, in

The total gutter depth can be found by adding 2 or 3 in of freeboard to the calculated depth y.

Other Processes ▪ Anthracite coal may be used in place of sand in gravity-type filters. The effective grain size is greater than that of sand, thus permitting higher filtration rates and longer filter runs.

Dual-media, mixed-media, or deep coarse-media filters, however, may be more advantageous. They operate at the higher filtration rates of 4 to 8 gal/min · ft^2.

A **pressure filter** is composed of a gravity-filter medium enclosed in a watertight vessel. The filtering medium may be sand, diatomaceous earth, or anthracite coal. Pressure filters are primarily supplemental and are used for specialized industrial uses and for clarifying swimming pool water.

Filter galleries are made up of horizontal, perforated, or open-joint pipes, placed in shallow sand or gravel aquifers. Galleries typically are fed by diversion or pumping from streams into spreading basins with gravel or sand bottoms. Some, however, may be located in aquifers with high groundwater table. The filtered water may be pumped from the gallery or allowed to flow out one end by gravity.

(G. L. Culp and R. L. Culp, "New Concepts in Water Purification," Van Nostrand Reinhold Company, New York; American Water Works Association, "Water Quality and Treatment," 4th ed., and American Society of Civil Engineers, "Water Treatment Plant Design," and T. J. McGhee, "Water Supply and Sewerage," 6th ed., McGraw-Hill Book Company, New York (books.mcgraw-hill.com); G. M. Fair, J. C. Geyer, and D. A. Okun, "Water and Wastewater Engineering," John Wiley & Sons, Inc., New York (www.wiley.com).)

21.48 Water Softening

Presence of the bicarbonates, carbonates, sulfates, and chlorides of calcium and magnesium in water causes hardness. Three major classifications of hardness are: (1) carbonate (temporary) hardness caused by bicarbonates, (2) noncarbonate (permanent) hardness, and (3) total hardness. Municipal treatment plants generally use either the lime-soda (precipitation) process or the base-exchange (zeolite) process to reduce the hardness of the water to below 100 mg/L (about 100 ppm) of $CaCO_3$ equivalence.

In the lime-soda process, lime (CaO), hydrated lime [$Ca(OH)_2$], and soda ash (Na_2CO_3) are added to the water in sufficient quantities to reduce the hardness to an acceptable level. The amounts of lime and soda ash required for softening to a residual hardness can be determined by use of chemical-equivalent weights, taking into account that commercial grades of lime and hydrated lime are about 90 and 68% CaO, respectively. Residual hardness of 50 to 100 mg/L as $CaCO_3$ remains in the treated water because of the very slight solubility of both $CaCO_3$ and $Mg(OH)_2$. Hardness of water is normally expressed in grains per gallon (gpg) or mg/L of $CaCO_3$, where 1 gpg = 17.1 mg/L.

Chemical equations for the common lime-soda softening processes are

$$CO_2 + CaO \rightarrow CaCO_3 \downarrow \qquad (21.136)$$

$$Ca(HCO_3)_2 + CaO \rightarrow 2CaCO_3 \downarrow + H_2O \quad (21.137)$$

$$MgSO_4 + CaO + H_2O$$
$$\rightarrow Mg(OH)_2 \downarrow CaSO_4 \text{ (soluble)} \quad (21.138)$$

$$CaSO_4 \text{ (soluble)} + Na_2CO_3 \rightarrow CaCO_3 \downarrow + NaSO_4$$
$$(21.139)$$

Since the carbonate and magnesium hydroxide particles settle out in sedimentation basins, facilities must be provided for particle removal and disposal. Deposition of $CaCO_3$ and $Mg(OH)_2$ on sand grains, in clear wells, and in distribution pipes can be prevented by recarbonation with CO_2 before sand-filter treatment.

Hardness in water can be reduced to zero by passing the water through a base-exchange or zeolite material. These materials remove cations, such as calcium and magnesium, from water and replace them with soluble sodium and hydrogen cations. Calcium can be removed from water as shown by the following reaction:

$$Ca^{2+} + Na_2R \leftrightharpoons CaR + 2Na^+ \qquad (21.140)$$

where Ca^{2+} is the calcium hardness ion removed, Na^+ is the sodium ion replacing the Ca^{2+} in water, and R is the zeolite material. The reaction can be reversed (from right to left) by increasing the Na^+ concentration to a high value, as generally is done in regeneration of the softening unit.

Sodium chloride (table salt) is commonly used to regenerate the unit. Regeneration requires between 0.3 and 0.5 lb of salt per 1000 grains of hardness removed.

Some hardness-removal capacities per cubic foot of base-exchange material are: natural zeolite—2500 to 3000 grains, synthetic zeolite—5000 to 30,000 grains (1 lb = 7000 grains).

(American Water Works Association, "Water Quality and Treatment," 4th ed., and American Society of Civil Engineers, "Water Treatment Plant Design," McGraw-Hill Book Company, New York (books.mcgraw-hill.com).)

21.49 Disinfection with Chlorine

Chlorine in either the liquid, gas, or hypochlorite form is frequently used for destroying bacteria in water supplies. Other disinfectants are iodine, bromine, ozone, chlorine dioxide, ultraviolet light, and lime.

The reaction of chlorine with water is

$$Cl_2 + H_2O \leftrightharpoons H^+ + Cl^- + HOCl \qquad (21.141)$$

The hypochlorous acid (HOCl) reacts with the organic matter in bacteria to form a chlorinated complex that destroys living cells. The amount of chlorine (*chlorine dose*) added to the water depends on the amount of impurities to be removed and the desired residual of chlorine in the water. Chlorine residuals of 0.1 or 0.2 mg/L are normally maintained in water-treatment-plant effluent streams as a factor of safety for the water as it travels to the consumer.

The concern over trihalomethane formation following chlorination of waters containing appreciable amounts of natural organic materials (Art. 21.62) has led to use of alternate disinfectants. The prime candidates are ozone and chlorine dioxide. The benefits of ozone should be investigated for new or modified treatment plants, particularly if there are color or taste and odor problems in the raw water.

(American Water Works Association and American Society of Civil Engineers, "Water Treatment Plant Design," and T. J. McGhee, "Water Supply and Sewerage," McGraw-Hill, Inc., New York (books.mcgraw-hill.com).)

21.50 Carbonate Stability

Water may either corrode or place a protective carbonate film on the interior surfaces of pipes. Which it does depends on the nature and amount of chemicals dissolved in the water.

An approximation of the stability of a water supply can be obtained by adding an excess of calcium carbonate powder to one-half of a water sample. Stir or shake each half sample at 5-min intervals for about 1 h. Filter both solutions; then, either take the pH or determine the methyl orange alkalinity of each sample. If the untreated water has a higher alkalinity or pH than the $CaCO_3$-treated water, the water is saturated with carbonate and may deposit protective films in pipes. If the untreated water has a lower pH or alkalinity value than the treated water, the water is unsaturated with carbonate and may be corrosive. If the pH or alkalinity is the same in both samples, the water is in equilibrium in regard to carbonates.

The greater the difference in either alkalinity or pH between the two samples, the greater the amount of either unsaturation or saturation with respect to carbonates. If the untreated water has a much higher pH or alkalinity than the treated water, the water is highly saturated with carbonates. It can cause a problem with heavy carbonate deposits in pipes and appurtenances of the purveyor and consumer.

(G. M. Fair, J. C. Geyer, and D. A. Okun, "Water and Wastewater Engineering," John Wiley & Sons, Inc., New York (www.wiley.com).)

21.51 Miscellaneous Treatments

Many different methods of treatment are used to remove such undesirable elements as color, taste, odor, excessive fluorides, detergents, iron, manganese, and substances exceeding the water-quality maximum contaminant levels (Art. 21.45).

Activated carbon is commonly used for taste and odor removal. The carbon can be applied as a powder to the water and later removed by a sand filter, or the water can be passed through a bed of carbon to remove natural and synthetic organic chemicals.

Treatment techniques for removal of inorganic contaminants include conventional coagulation, lime softening, cation exchange, anion exchange, activated carbon, reverse osmosis, and electro-dialysis. Concerns over the potential for lead poisoning from lead in drinking water passing through lead pipes installed long ago but still in use or from leaded solder used for pipe joints have encouraged abandonment of such practices. Where the presence of lead is detected in a water supply, despite its low solubility, its concentration can be nearly completely removed with lime softening or alum and ferric sulfate coagulation.

(American Water Works Association and American Society of Civil Engineers, "Water Treatment Plant Design," McGraw-Hill, Inc., New York (books.mcgraw-hill.com).)

Water Collection, Storage, and Distribution

21.52 Reservoirs

The basic purpose of impounding reservoirs is to hold runoff during periods of high runoff and release it during periods of low runoff. The specific functions of reservoirs are hydroelectric, flood control, irrigation, water supply, and recreation (see also Art. 21.52.1). Many large reservoirs are multipurpose, as a consequence of which the specific functions may dictate conflicting design and operating criteria. Also, equitable cost allocation is more difficult.

Sizing of a reservoir for a project where the demand for water is much greater than the mean stream flow is an economic balance between benefits and costs. A preliminary study of available reservoir sites should be made to obtain the relative costs for various size reservoirs. The dependable flow that can be obtained from various-size reservoirs can be determined from the mass diagram for stream flow. An economic comparison should then be made of the benefits of various flows and the costs of various reservoirs. The reservoir size that will give the maximum benefit should be selected.

When the demand rate is known, as is the case for many water-supply projects, the required size of the reservoir can be determined directly from a mass diagram of stream flow.

The **mass diagram** (Fig. 21.80) is a graphical plot of total stream-flow volume against time. The slope of the curve is the rate of flow.

Selection of the critical period of years for a mass curve depends on the function of the reservoir. For a water-supply or hydroelectric development, minimum flows will be critical, whereas for flood-control reservoirs, maximum flows will govern.

Reservoir capacity for a certain demand can be obtained by drawing a line with a slope equal to the demand tangent to the mass curve at the beginning of a selected dry period, as shown by lines *AB* and *AC* in Fig. 21.80. The ordinates *d* and *e* represent the storage required to maintain demands *AB* and *AC*.

Once a reservoir site has been selected, **area-volume curves** (Fig. 21.81) are drawn to give the characteristics of the site. The plot of volume vs. water elevation is determined by planimetering the area of selected contours within the reservoir site and multiplying by the contour interval. Aerial mapping has made it possible to obtain accurate

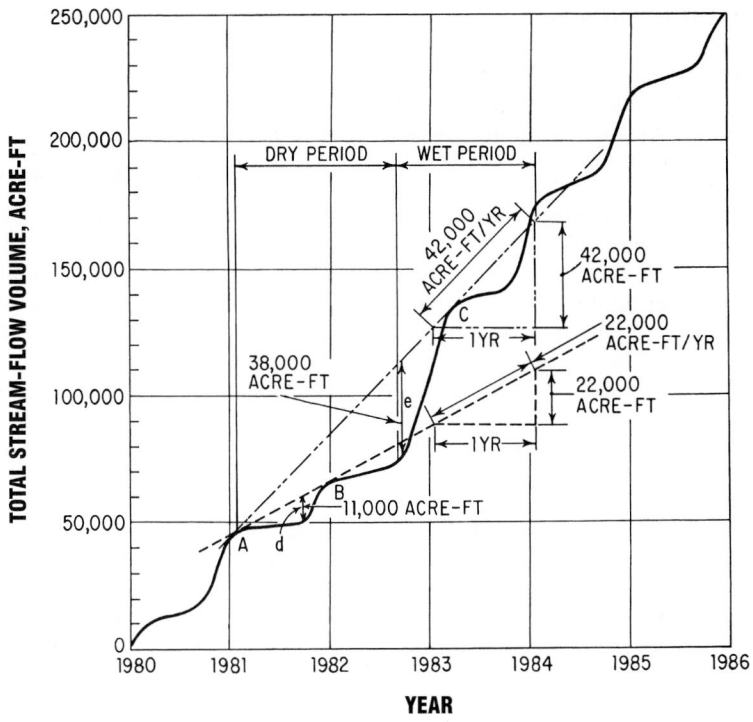

Fig. 21.80 Mass diagram of stream flow.

contour maps at only a fraction of the costs of older methods.

Another important consideration in the design of reservoirs is deposition of sediment (see Arts. 21.35 and 21.52.2).

In selection of a site for a water-supply reservoir, give special attention to water quality. If possible, the watershed should be relatively uninhabited to reduce the amount of treatment required. (Water from practically all sources should be disinfected in

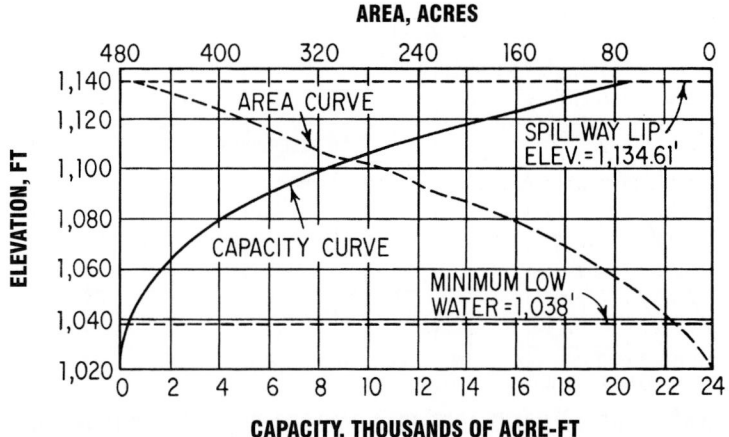

Fig. 21.81 Area-capacity curves for a reservoir.

the distribution system to ensure against pollution and contamination.) Shallow reservoirs usually give more problems with color, odor, and turbidity than deep reservoirs, particularly in warm climates or during warm seasons of the year. Runoff heavily laden with silt and debris should be diverted from the reservoir or treated before it is mixed with the water supply. Alum is mixed into reservoirs to reduce turbidity, and copper sulfate is used to kill vegetation.

In deep reservoirs, during the summer months the upper part of the reservoir will be warmed, while below a certain level the temperature may be many degrees cooler. The zone where the abrupt temperature change takes place, which may be only a few feet thick, is called the **thermocline**. The waters above and below the thermocline circulate, but there is no circulation across this zone. The water in the lower level becomes low in dissolved oxygen and develops bad tastes and odors. When the temperature drops in the fall, the water at the upper level becomes heavier than the water at the lower level and the two levels become intermixed, causing bad tastes and odors in the entire reservoir. To oxidize organic matter and prevent poor water quality in lower levels of reservoirs during summer months, chlorine or compressed air should be released at various points on the bottom of the reservoirs.

21.52.1 Distribution Reservoirs

The two main functions of distribution reservoirs are to equalize supply and demand over periods of varying consumption and to supply water during equipment failure or for fire demand. Major sources of supply for some cities, such as New York, San Francisco, and Los Angeles, are large distances from the city. Because of the large cost of aqueducts, it is usually economical to size them for the mean annual flow and provide terminal storage for daily and seasonal fluctuations of demand. Terminal storage is also necessary because of the possibility of a failure along an aqueduct.

It is usually economical to have equalizing reservoirs at various points in the distribution system so that main supply lines, pumping plants, and treatment plants can be sized for maximum daily instead of maximum hourly demand. During

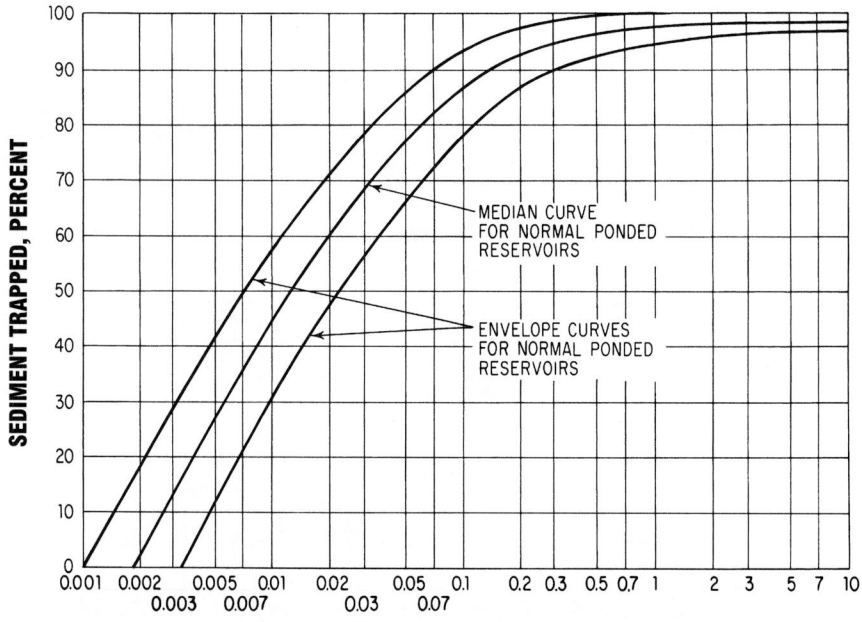

CAPACITY-INFLOW RATIO, ACRE-FEET CAPACITY PER ACRE-FOOT ANNUAL INFLOW

Fig. 21.82 Chart indicates percentage of incoming sediment trapped in reservoirs.

hours of maximum demand, water flows from these reservoirs to the consumers. When the demand drops off, the flow refills the reservoir. A mass diagram (Fig. 21.80) can be used to determine the required capacity of the reservoir.

Equalizing reservoirs are usually built at the opposite end of the system from the source of supply, so that during peak flows the maximum distance from the supply to the consumer is cut in half. It is necessary for an equalizing reservoir to have an elevation high enough to provide adequate pressure throughout the system served. For the correct hydraulic grade, it is necessary to build the reservoir above the area it serves. If the topography will not allow a surface reservoir, a standpipe or an elevated tank must be constructed. Standard elevated tanks are available in capacities up to 2 MG.

21.52.2 Reservoir Trap Efficiency

The methods of Art. 21.35.2 for determining the quantities of sediment delivered to a reservoir require knowledge of the trap efficiency of the reservoir before the percentage of the incoming silt that will remain to reduce storage can be determined. Studies of trap efficiency were made by G. M. Brune, who developed a curve to express the relationship between trap efficiency and what he called the *capacity-inflow ratio* for a reservoir

(Fig. 21.82) (G. M. Brune, "Trap Efficiency of Reservoirs," *Transactions of the American Geophysical Union*, vol. 34, no. 3, June 1953).

The higher the capacity-inflow ratio, acre-feet of storage per acre-foot of annual inflow, the greater the percentage of sediment trapped in a reservoir. For any given storage reservoir, the trap efficiency decreases with time since the capacity-inflow ratio decreases as sediment builds up. The rate of silting of a storage reservoir decreases when the capacity is reduced to an amount such that some spillage of silt-laden water occurs with each major storm. This rate decrease occurs because an increasing percentage of the annual suspended silt load is vented before sedimentation can occur.

21.53 Wells

A gravity well is a vertical hole penetrating an aquifer that has a free-water surface at atmospheric pressure (Fig. 21.83). A pressure or artesian well passes through an impervious stratum into a confined aquifer containing water at a pressure greater than atmospheric (Fig. 21.84). A flowing artesian well is an artesian well extending into a confined aquifer that is under sufficient pressure to cause water to flow above the casing head. A gallery or horizontal well is a horizontal or nearly

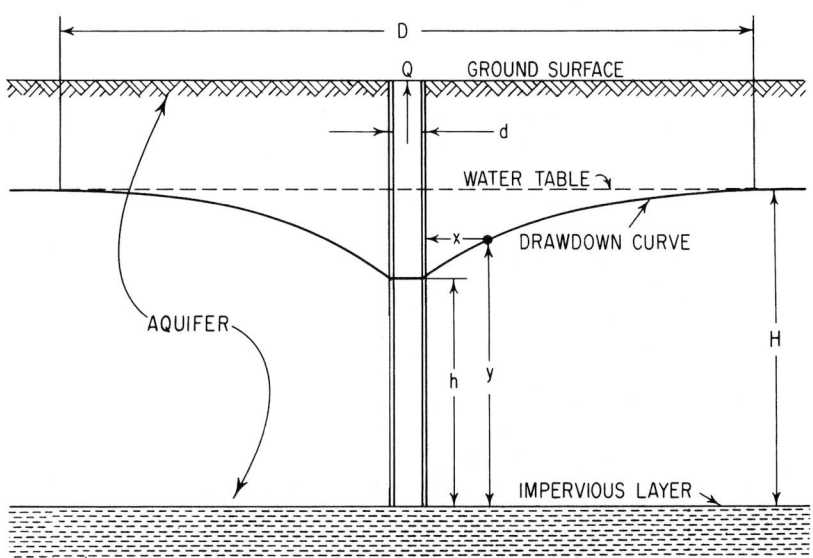

Fig. 21.83 Gravity well in a free aquifer.

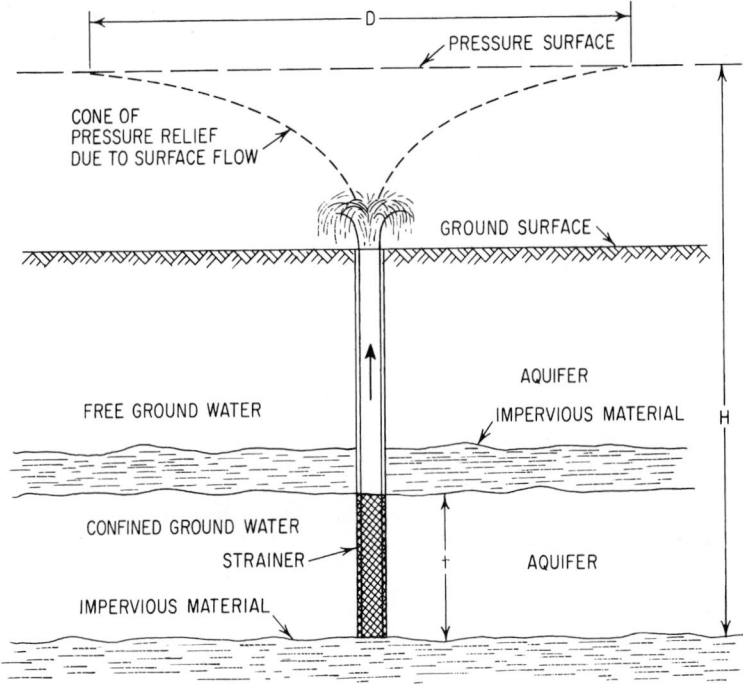

Fig. 21.84 Artesian well in a pressure aquifer.

horizontal tunnel, ditch, or pipe placed normal to groundwater flow in an aquifer.

21.53.1 Drawdown

When water is pumped from a well, the water level around the well draws down and forms a *cone of depression* (Fig. 21.83). The line of intersection between the cone of depression and the original water surface is called the *circle of influence.*

Interference between two or more wells is caused by the overlapping of circles of influence. Drawdown for each interfering well is increased and the rate of water flow is decreased for each well in proportion to the degree of interference. Interference between two or more closely spaced wells may increase to the extent that the system of wells produces one large cone of depression.

Since nearly all soils are heterogeneous, pumping tests should be made in the field to determine the value of the hydraulic conductivity *K*. A permeability analysis of a soil sample that is not representative of the soil throughout the aquifer would produce an unreliable value for *K*.

21.53.2 Flow From Wells

The steady flow rate Q can be found for a gravity well by using the **Dupuit formula**:

$$Q = \frac{1.36K(H^2 - h^2)}{\log(D/d)} \qquad (21.142)$$

where Q = flow, gal/day

K = hydraulic conductivity, ft/day under 1:1 hydraulic gradient

H = total depth of water from bottom of well to free-water surface before pumping, ft

h = H minus drawdown, ft

D = diameter of circle of influence, ft

d = diameter of well, ft

The steady flow, gal/day, from an artesian well is given by

$$Q = \frac{2.73Kt(H - h)}{\log(D/d)} \qquad (21.143)$$

where t is the thickness of confined aquifer, ft (Fig. 21.84).

A long time elapses between the beginning of pumping and establishment of a steady-flow condition (a circle of influence with constant diameter). Hence, correct values for drawdown and the circle of influence can be obtained only after long periods of continuous pumping.

A nonequilibrium formula developed by Theis and a modified nonequilibrium formula produced by Jacob are used in analyzing well flow conditions where equilibrium has not been established. Both methods utilize a storage coefficient S and the coefficient of transmissibility T to eliminate complications due to the time lag before reaching steady flow. (C. V. Theis, "The Significance of the Cone of Depression in Groundwater Bodies," *Economic Geology*, vol. 33, p. 889, December 1938; C. E. Jacob, "Drawdown Test to Determine Effective Radius of Artesian Well," *Proceedings of the American Society of Civil Engineers*, vol. 72, no. 5, p. 629, 1940.) Computer software packages are available for analysis of groundwater flow with finite-element models.

21.53.3 Excavation of Wells

Wells may be classed by the method by which they are constructed and their depth. Shallow wells (less than 100 ft deep) are usually dug, bored, or driven. Deep wells (depth greater than 100 ft) are usually drilled by either the standard cable-tool, waterjet, hollow-core, or hydraulic rotary methods.

21.53.4 Well Equipment

Essential well equipment consists of casing, screen, eductor or riser pipe, pump (Art. 21.57), and motor. The casing keeps the wall material and polluted water from entering the well and prevents the leakage of good water from the well.

The screen is placed below the casing to contain the walls of the aquifer, to allow water to pass from the aquifer into the well, and to stop movement of the larger sand particles into the well. The pump, motor, and eductor pipe are utilized to move the water from the aquifer to the collecting lines at the ground surface.

(G. M. Fair, J. C. Geyer, and D. A. Okun, "Water and Wastewater Engineering," John Wiley & Sons, Inc., New York (www.wiley.com); T. J. McGhee,

"Water Supply and Sewerage," 6th ed., McGraw-Hill, Inc., New York (books.mcgraw-hill.com).)

21.54 Water Distribution Piping

A water-distribution system should reliably provide potable water in sufficient quantity and at adequate pressure for domestic and fire-protection purposes. To provide adequate domestic service, the pressure in the main at house service connections usually should not be below 45 psi. But if oversized plumbing is provided, 35 psi is adequate. In steep hillside areas, the system is usually divided into several different pressure zones, interconnected with pumps and pressure regulators. Since each additional zone causes increased expenses and decreased reliability, it is desirable to keep their number to a minimum. The American Water Works Association has recommended 60 to 75 psi as a desirable range for pressures; however, in areas of steep topography where local elevation differences may be over 1000 ft, such a narrow range is not practical.

House plumbing is designed to withstand a maximum pressure of between 100 and 125 psi. When the pressure in distribution lines is above 125 psi, it is necessary to install pressure regulators at each house to prevent damage to appliances, such as water heaters and dishwashers.

21.54.1 Water for Fire Fighting

Pressure requirements for fire fighting depend on the technique and equipment used. Four methods of supplying fire protection are:

1. Use of mobile pumpers which take water from a hydrant. This method is used in most large communities that have full-time, well-trained fire departments. The required pressure in the immediate area of the fire is 20 psi.

2. Maintenance of adequate pressure at all times in the distribution system to allow direct connection of fire hoses to hydrants. This technique is commonly used in small communities that do not have a full-time fire department and mobile pumpers. The pressure in the distribution system in the vicinity of a fire should be between 50 and 75 psi.

3. Use of stationary fire pumps located at various points in the distribution system, to boost the pressure during a fire and allow direct connection of hoses to hydrants. This method is not so reliable or so widely used as the first two.

4. Use of a separate high-pressure distribution system for fire protection only. There are only rare instances in high-value districts of large cities where this method is used because the cost of a dual distribution system is usually prohibitive.

21.54.2 Hydraulic Analysis of Distribution Piping

Distribution systems are usually laid out on a gridiron system with cross connections at various intervals. Dead-end pipes should be avoided because they cause water-quality problems.

Economic velocities are usually around 3 to 4 ft/s, although during fires they can be much higher. Two- and four-inch-diameter pipe can be used for short lengths in residential areas; however, the American Insurance Association (AIA) requires 6-in pipe for fire service in residential areas. Also, maximum length between cross connections is limited to 600 ft. In high value districts, the AIA requires an 8-in pipe, with cross connection at all intersecting streets. The AIA standards also require that gate valves be located so that no single case of pipe breakage, outside main arteries, requires shutting off from service an artery or more than 500 ft of pipe in high valued districts, or more than 800 ft in any area. All small distribution lines branching from main arteries should be equipped with valves. ("Standard Schedule for Grading Cities and Towns of the United States with Reference to Their Fire Defenses and Physical Conditions," American Insurance Association.)

Adequate service requires a knowledge of the hydraulic grade at many points in a distribution system for various flows. Several methods, based on the following rules, have been developed for analysis of complex networks:

1. The head loss in a conduit varies as a power of the flow rate.

2. The algebraic sum of all flows into and out of any pipe junction equals zero.

3. The algebraic sum of all head losses between any two points is the same by any route, and the algebraic sum of all head losses around a loop equals zero.

A convenient device for simplifying complex networks of various size pipes is the *equivalent pipe*. For a series of different size pipes or several parallel pipes, one pipe of any desired diameter and one specific length or any desired length and one specific diameter can be substituted; this will give the same head loss as the original for all flow rates if there are no take-outs or inputs between the two end points. In complex networks, the equivalent pipe is used mainly to simplify calculation.

Example 21.10: Determine the equivalent 8-in-diameter pipe that will have the same loss of head as the sections of pipe from *A* to *D* in Fig. 21.85*a*.

First, transform pipes *CD*, *AB*, and *BD* into equivalent lengths of 8-in pipe; then, transform the resulting sections into a single 8-in pipe with the same head loss. The head loss may be calculated from Eqs. (21.34*d*).

Assume any convenient flow through *CD*, say 500 gal/min. Equation (21.34*d*) indicates that loss of

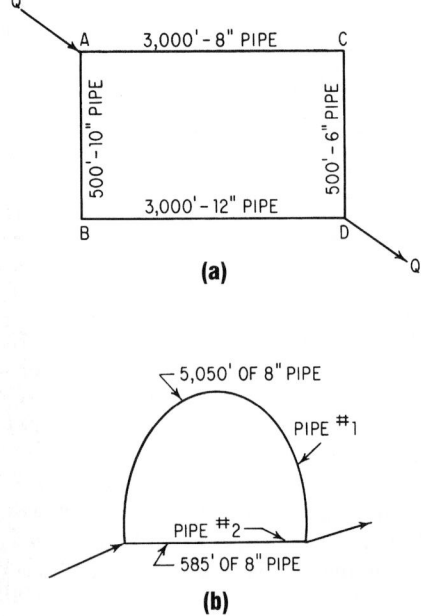

(a)

(b)

Fig. 21.85 Distribution loop (*a*) may be replaced by equivalent loop (*b*).

head in 1000 ft of 6-in pipe is 32 ft and in 1000 ft of 8-in pipe, 7.8 ft. Then, the equivalent length of 8-in pipe for CD is $500 \times 32/7.8 = 2050$ ft. Similarly, the equivalent pipe for AB should be 165 ft long, and for BD, 420 ft long. The network of 8-in pipe is shown in Fig. 21.85b. It consists of pipe 1, $3000 + 2050 = 5050$ ft long, connected in parallel to pipe 2, $165 + 420 = 585$ ft long.

To reduce the parallel pipes to an equivalent 8-in pipe, assume a flow of 1000 gal/min through pipe 2. For this flow, the head loss in an 8-in pipe per 1000 ft is 29 ft. Hence the head loss in pipe 2 would be $29 \times 585/1000 = 17$ ft. Since the pipes are connected in parallel, the head loss in pipe 1 also must be 17 ft, or 3.37 ft/1000 ft. The flow that will produce this head loss in an 8-in pipe is 310 gal/min [Eq. (21.34c)]. The equivalent pipe, therefore, must carry $1000 + 310 = 1310$ gal/min with a head loss of 17 ft. For a flow of 1310 gal/min, an 8-in pipe would have a head loss of 48 ft in 1000 ft, according to Eq. 21.34d. For a loss of 17 ft, an 8-in pipe would have to be $1000 \times 17/48 = 350$ ft long. So the pipes between A and D in Fig. 21.85a are equivalent to a single 8-in pipe 350 ft long.

Pipe Network Equations ▪ For hydraulic analysis of a water distribution system, it is convenient to represent the network by a mathematical model. Generally, it is useful to include in the model only the major elements needed for a mathematical description of the basic network. (For models that are to be used for such conditions as low pressures in a small service region, however, it may be necessary to include all the distribution mains in the system.) The three analysis rules on p. 21.110 can then be used to develop a system of simultaneous equations that can be solved for flow and pressure in the network.

Typically, either the Darcy-Weisbach or the Hazen-Williams formula is used to relate the characteristics of each pipe in the system. Consequently, the equations for each pipe are nonlinear. As a result, a direct solution generally is not available. In practice, the equations are solved by an iteration process, in which the values of some variables are assumed to make the equations linear and then the equations are solved for the other variables. The initial assumptions are corrected and used to develop new linear equations, which are solved to obtain more accurate values of the variables.

One example of this technique is the **Hardy Cross method**, a controlled trial-and-error method, which was widely used before the advent of computers. Flows are first assumed; then consecutive adjustments are computed to correct these assumed values. In most cases, sufficient accuracy can be obtained with three adjustments; however, there are rare cases where the computed adjustments do not approach zero. In these cases, an approximate method must be used.

Assumed flows in a loop are adjusted in accordance with the following equation:

$$\Delta Q = \frac{\Sigma KQ^n}{\Sigma nKQ^{n-1}} \qquad (21.144)$$

where $KQ^n = h_f =$ loss of head due to friction. When the Hazen-Williams equation, used in Example 21.10, is put in the form $h_f = KQ^n$ then $K = 4.727L/D^{4.87}C_1^{1.85}$ and $n = 1.85$. The expression ΣnKQ^{n-1} equals $\Sigma(nKQ^n/Q)$. In the Hazen-Williams formula $n = 1.85$ for all pipes and can therefore be taken outside the summation sign. Hence, the adjustment equation becomes

$$\Delta Q = \frac{\Sigma h_f}{n\Sigma(h_f/Q)} \qquad (21.145)$$

It is important that a consistent set of signs be used. The sign convention chosen for the following example makes clockwise flows and the losses from these flows positive; counterclockwise flows and their losses are negative.

Approaches generally used for formulating the equations for analysis of a water distribution network include the following:

Flow method, in which pipe flows are the unknowns.

Node method, in which pressure heads at the pipe end points are the unknowns.

Loop method, in which the energy in each independent loop is expressed in terms of the flows in each pipe in the loop. In turn, the actual flow in each pipe is expressed in terms of an assumed flow and a flow correction factor for each loop.

Computer software packages are available for analysis of networks by such methods. They can perform not only steady-state analyses of pressures and flows in pipe networks but also time-dependent analyses of pressure and flow under changing system demands and of flow patterns and basic water quality.

(V. J. Zipparo and H. Hasen, "Davis' Handbook of Applied Hydraulics," McGraw-Hill, Inc., New York (books.mcgraw-hill.com); AWWA, "Distribution Network Analysis for Water Utilities," Manual of Water Supply Practices M32, American Water Works Association, Denver, Colo. (www.awwa.org); T. M. Walski, "Analysis of Water Distribution Systems," Van Nostrand Reinhold, New York.)

21.54.3 Cover over Buried Pipes

The cover required over distribution pipes depends on the climate, size of main, and traffic. In northern areas, frost penetration, which may be as deep as 7 ft. is usually the governing factor. In frost-free areas, a minimum of 24 in is required by the AIA. If large mains are placed under heavy traffic, the stress produced by wheel loads should be investigated.

21.54.4 Maintenance of Water Pipes

Maintenance of distribution systems involves keeping records, cleaning and lining pipe, finding and repairing leaks, inspecting hydrants and valves, and many other functions necessary to eliminate problems in operation. Valves should be inspected annually and fire hydrants semiannually. Records of all inspections and repairs should be kept.

Unlined distribution pipes, after years of usage, lose much of their capacity because of corrosion and incrustations. Cleaning and lining with cement mortar restores the original capacity. Dead-end pipes should be flushed periodically to reduce the accumulation of rust and organic matter.

21.54.5 Economic Sizing of Distribution Piping

When designing any major project, the designer should choose the most economical of numerous alternatives. Most of these alternatives can be separated and studied individually. An example of two alternatives for a distribution system is one serving peak hourly demands totally by pumps and one doing it by pumps and equalizing reservoirs. The total costs of each plan should be compared by an annual or present-worth cost analysis.

A method of determining minimum cost that can readily be adapted to many conditions is setting the first derivative of the total cost, taken with respect to the variable in question, equal to zero. In the sizing of pipes in a distribution system supplied by pumps, the total costs of the pipes, pumping plant, and energy may be expressed as an equation. To find the most economical diameter of pipe, the first derivative of the total cost, taken with respect to the pipe diameter, should be set equal to zero. The following equation for the most economical pipe diameter was derived in that manner:

$$D = 0.215\left(\frac{fbQ_a^3 S}{aiH_a}\right)^{1/7} \quad (21.146)$$

where D = pipe diameter, ft

f = Darcy-Weisbach friction factor

b = value of power, dollars/hp per year

Q_a = average discharge, ft^3/s

S = allowable unit stress in pipe, psi

a = in-place cost of pipe, dollars/pound

i = yearly fixed charges for pipeline (expressed as a fraction of total capital cost)

H_a = average head on pipe, ft

21.54.6 Pipe Materials

Cast iron, steel, concrete, and plastics, such as polyvinyl chloride, polyethylene, polybutylene, and glass-fiber-reinforced thermosetting resins, are the most common materials used in distribution pipes. Wood pipelines are still in existence, but wood is rarely used in new installations. Copper, lead, zinc, brass, bronze, and plastic are materials used in small pipes, valves, pumps, and other appurtenances. Common pipe-joint materials are: cement, sand, rubber, plastic, and sulfur compounds.

Cast iron is the most common material for city water mains. Standard sizes range from 2 to 24 in in diameter. Cast iron is resistant to corrosion and usually has good hydraulic characteristics. If it is cement-lined, the Hazen-Williams C value may be as high as 145. In unlined pipes, however, iron tubercles may form and seriously affect flow capacity. Tuberculation can be prevented by lining with cement or tar materials. The relatively high cost of cast-iron pipe is only a slight disadvantage, largely offset by the long average life of trouble-free service. Bell-and-spigot and flange are the most common joints in cast-iron pipe.

Steel is commonly used for large pipelines and trunk mains but rarely for distribution mains. Steel pipes with either longitudinal or spiral joints are formed at steel mills from flat sheets. The tranverse joints between pipe sections are usually made by welding, riveting, bell-and-spigot with rubber gasket, sealed flanges, or Dresser-type couplings. Since steel is stronger than iron, thinner and lighter pipes can be used for the same pressures. Some disadvantages of thin steel pipe are inability to carry high external loads, possibility of collapse due to negative gage pressures, and high maintenance costs due to higher corrosion rates and thinner pipe walls. Steel pipes are usually corrosion-protected on both the outside and inside with coal tar or cement mortar. Under favorable conditions, the life of steel pipe is between 50 and 75 years.

Concrete pipe may be precast in sections and assembled on the job or cast in place. A machine that produces a monolithic, jointless concrete pipe without formwork has been developed for gravity-flow and low-pressure applications. Most of the precast-concrete pipe is reinforced or prestressed with steel. Concrete pipe may be made watertight by insertion of a thin steel cylinder in the pipe walls. High-strength wire is frequently wound around the thin steel cylinder for reinforcement. Concrete is placed inside and outside the steel cylinder to prevent corrosion and strengthen the pipe. Some advantages of concrete pipe are low maintenance cost, resistance to corrosion under normal conditions, low transportation costs for materials if water and aggregate are available locally, and ability to withstand external loads. Some disadvantages to be considered are leaching of free lime from the concrete, the tendency to leak under pressure due to the cracking and permeability of concrete, and corrosion in strong acids or alkalies.

21.55 Corrosion in Water Distribution Systems

Many millions of dollars are expended every year to replace pipes, valves, hydrants, tanks, and meters destroyed by corrosion. Some causes of corrosion are the contact of two dissimilar metals with water or soil, stray electric currents, impurities and strains in metals, contact between acids and metals, bacteria in water, or soil-producing compounds that react with metals.

Electrochemical corrosion of a metal occurs when an electrolyte and two electrodes, an anode and a cathode, are present. (Water may serve as an electrolyte.) At the anode, the metal in contact with the electrolyte changes into a positively charged particle, which goes into solution or forms an oxide film. (The ease with which a metal changes to a metallic ion when it is in contact with water depends on its oxidation potential or solution pressure. Metals can be arranged in an electromotive series of decreasing oxidation potentials. Metals high in the electromotive series corrode more readily than metals located in a lower position.) For an iron pipe exposed to water, for example, the anode reaction is Fe (metal) → $Fe^{2+} + 2e$, where e is an electron. At the cathode, the metal having the excess electrons gives them up to a charged particle, such as hydrogen in solution: $2H^+ + 2e \rightarrow H_2$ (gas). If the hydrogen gas produced at the cathode is removed from the cathode by reaction with oxygen to produce water molecules or by water movement (depolarization), the corrosion process continues (Fig. 21.86). Indications of corrosion in an inaccessible iron or steel pipeline are discharges of rusty-colored water (due to the loosening of rust and scale) and metallic-tasting water.

A marked decrease in capacity and pressure in a pipe section usually indicates tuberculation inside the line. Tuberculation is caused by the deposition and growth of insoluble iron compounds inside a pipe. Iron-consuming bacteria in water can produce ferrous oxide directly if the iron concentration is about 2 ppm. A continuous supply of soluble iron in the presence of iron-consuming bacteria or dissolved oxygen and basic substances in the water increases the size of the tubercles. Tubercles may become so large and decrease the capacity in the pipe to such an extent that it has to be cleaned or replaced.

Several factors influence the type and quantity of metallic corrosion:

Presence of protective films. Some metals form oxide films that act as protective layers for the metal. Aluminum, zinc, and chromium are examples of this type of metal.

Strains, cracks, and undissolved impurities in a metal act as sites for corrosion.

Agitation or movement of water increases the corrosion rate of a metal because the oxygen supply

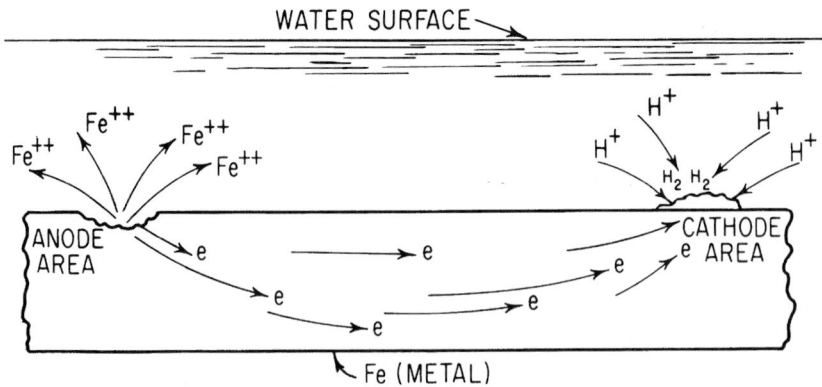

Fig. 21.86 Electrochemical corrosion of iron in low-pH water.

rate to the cathode and the removal rate of metal ions from the anode are increased. The presence of ionic compounds in the water speeds up corrosion because the ions act as conductors of electricity, and the more ions, the faster electrons can move through the water.

Alternate wetting and drying tends to break up the rust or oxide film, thus facilitating penetration of the film by oxygen and water and lead to increased corrosion.

High hydrogen-ion concentrations increase corrosion rates because of the greater accessibility of the hydrogen ions to the cathode.

Corrosion may be prevented or retarded by proper selection of materials, use of protective coatings, and treatment of the water. When selecting materials, the engineer should take into account the characteristics of the water and soil conditions encountered. Protective coatings for metals may be metallic or nonmetallic and applied on both the inside and outside surfaces of the pipe. Common nonmetallic coatings are cement and asphalt. Zinc is an example of metallic coating materials used. Steel pipe dipped in zinc (galvanized) or copper tubing is commonly used for small service lines.

Also, to prevent corrosion, water may be treated with bases, such as soda ash, caustic soda, and lime, to reduce hydrogen-ion concentration and to induce precipitation of thin films of carbonates, hydroxides, oxides, and so on on the walls of the pipes. These thin films reduce the ability of water to corrode otherwise unprotected metal surfaces. Corrosion, however, normally precedes deposition of scale because iron must be in solution to react with the basic substances and dissolved oxygen in the water to form scale.

Electrochemical corrosion of external surfaces of pipelines and water tanks can be retarded by application of a direct current to the metal to be protected and to another metal that acts as a sacrificial anode (Fig. 21.87). The potential applied to or produced by the two metal surfaces must be large enough to make the protected metal act as a cathode. The sacrificial anode corrodes and must be replaced periodically. Zinc, magnesium, graphite, and aluminum alloys are commonly used for anode materials.

(American Water Works Association, "Water Quality and Treatment," 4th ed., McGraw-Hill, Inc., New York.)

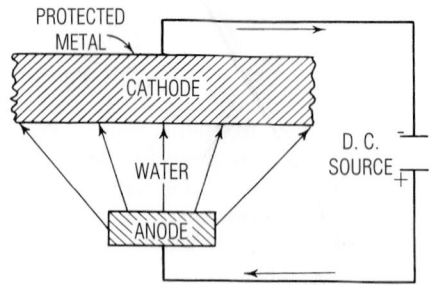

Fig. 21.87 Cathodic protection of a metal.

21.56 Centrifugal Pumps

The purpose of any pump is to transform mechanical or electrical energy into pressure energy. The centrifugal pump, the most common waterworks pump, accomplishes that in two steps. The first transforms the mechanical or electrical energy into kinetic energy with a spinning element, or impeller. The kinetic energy is then converted to pressure energy by diffuser vanes or a gradually diverging discharge tube, called a volute (Fig. 21.88).

Water enters at the center, or eye, of the impeller and is forced outward toward the casing by centrifugal force. The discharge head of a centrifugal pump is a function of the impeller diameter and speed of rotation.

Design factors requiring consideration in the selection of a centrifugal pump are net positive suction head required, efficiency, horsepower, and the head-discharge relationship.

Net positive suction head (NPSH) is the energy in the liquid at the center line of the pump. To have practical meaning, it must be referred to as either the required or available NPSH. Required NPSH is a characteristic of the pump and is given by the manufacturer. Available NPSH is a characteristic of

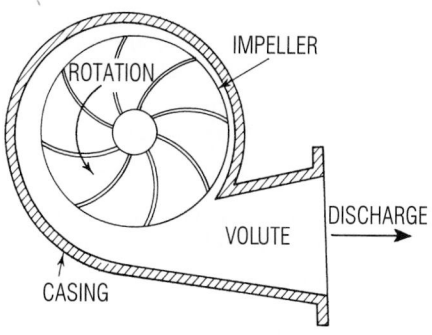

Fig. 21.88 Volute-type centrifugal pump.

the system and is determined by the engineer. It is the pressure in the liquid over and above its vapor pressure at the suction flange of the pump and is given, in feet, by

$$\text{Available NPSH} = 144\frac{p_a - p_v}{w} - h_f + z \quad (21.147)$$

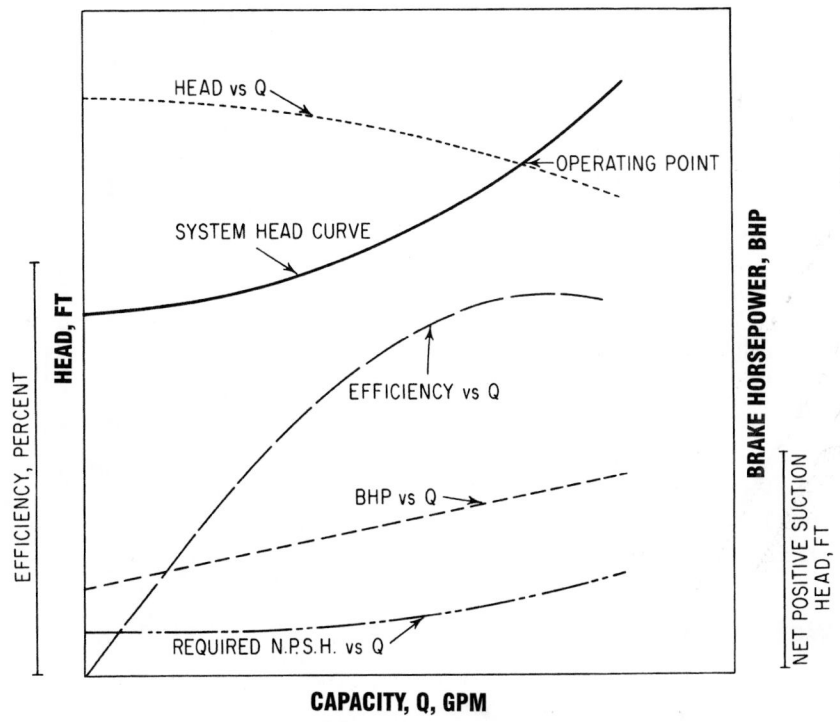

Fig. 21.89 Curves used in selection of a centrifugal pump.

where p_a = pressure, psia, on free-water surface or at center line of closed conduit

p_v = vapor pressure, psia, of water at its pumping temperature

h_f = friction loss in suction line, ft of water

z = elevation difference, ft, between pump center line and water surface

w = unit weight of liquid, lb/ft³

If the suction water surface is below the pump center line, z is negative. To prevent cavitation, it is necessary to have the available NPSH always greater than the required NPSH. For that reason, it is customary to analyze a required NPSH vs. discharge curve with the brake horsepower, head, and efficiency curves when selecting a pump.

The operating point of a centrifugal pump is determined by the intersection of the pump's head-capacity curve and the *system head curve*, as shown in Fig. 21.89. (Also included in Fig. 21.89 are the other curves used in pump selection.) A system head curve is a plot of the head losses in the system vs. pump discharge. This curve shows the head differential that must be supplied by the pump. In a typical water-system analysis, there may be three or four pertinent system head curves corresponding to various consumption rates. The intersection of these curves with the head vs. Q curve define a range of operation rather than a single point.

Selection of a centrifugal pump is largely a matter of matching one of the many pumps available to the system characteristics. An important consideration is that the point of maximum efficiency should be at or near the operating point. Centrifugal pumps are available in almost any capacity desired, with lifts of up to 700 ft per stage. Efficiencies may be as high as 93% for large pumps.

See also Art. 21.57 and check valves in Art. 21.58.

(I. J. Karassik et al., "Pump Handbook," 2nd ed., McGraw-Hill Book Company, New York (books. mcgraw-hill.com).)

21.57 Well Pumps

These are classified as centrifugal, propeller, jet, helical, rotary, reciprocating, and air lift. Although centrifugal pumps (Art. 21.56) are the most common for both shallow-well and deep-well

pumps, circumstances may dictate one of the other types.

Centrifugal pumps are used in wells over 6 in in diameter. They have capacities up to 4000 or 5000 gal/min and heads up to 1200 ft, depending on the number of stages. Efficiencies may be as high as 90% for the larger capacities; however, below 200 gal/min, the maximum efficiency is 75 to 80%.

Propeller pumps are an axial-flow type. They are used in high-capacity low-head applications.

Jet pumps (Fig. 21.90) operate by discharging water through a nozzle and diverging conical tube, which are located at the well bottom. The diverging conical tube creates lift by converting the high-

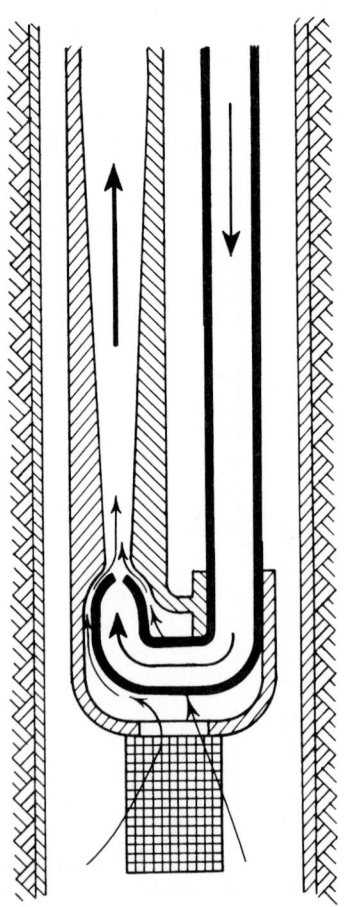

Fig. 21.90 Section through a jet pump (simplified).

velocity head to pressure head. The suction connection is made between the nozzle and entrance to the diverging tube. Jet pumps have low efficiencies. They are used in small-capacity low-lift applications, especially where the water contains sand or other impurities.

Helical pumps are a positive-displacement type with a metal helical rotor rotating inside a rubber helical stator. The screw action of the rotor forces water through the pump and up the discharge pipe. Helical pumps are small-capacity high-lift pumps. They may be used in wells over 4 in in inside diameter.

Rotary pumps are also of the displacement type. They have a fixed chamber in which gears, vanes, cams, or pistons rotate with very close tolerances. These pumps have relatively constant partial-load efficiencies. Full-load efficiencies range from 50 to 85%. Because of the close tolerances, they can be used only for sediment-free water.

Reciprocating pumps, either hand- or motor-driven, utilize piston action to move water. Their present-day use is primarily for small-capacity low-lift private applications.

Air-lift pumps generate lift by using air bubbles to decrease the specific weight of the column of water in the discharge pipe below that of the surrounding water in the well and create a pressure differential that forces the water out of the well. Air-lift pumps are the simplest and most foolproof of well pumps since they have no submerged moving parts. They can be used in any well but have the disadvantage of efficiencies below 50%.

Specific speed N_s is a widely used criterion for pump selection. It is the impeller speed corresponding to a discharge of 1 gal/min at 1 ft of head for the most efficient design.

$$N_s = nQ^{1/2}h^{-3/4} \qquad (21.148)$$

where n = impeller speed, r/min

Q = discharge, gal/min

H = head, ft

The favorable design range of N_s for radial-flow (centrifugal) pumps is from 1500 to 4100. For N_s between 4100 and 7500, mixed-flow pumps having both radial and axial characteristics should be used, and for N_s above 7500, axial-flow (propeller) pumps should be used.

Shallow-well pumps have their motors and impellers at ground level, so that the entire lift is

suction. Since excessive suction lifts cause cavitation, the lift is limited by atmospheric pressure and the velocity head at the impeller, which is a function of specific speed. At sea level, the maximum practical lift for a shallow-well pump is about 25 ft.

Deep-well pumps have their impellers close enough to the water surface to eliminate cavitation. The motor may be at ground level with a long shaft connecting it to the impellers, or it may be at the bottom of the well, below and directly adjacent to the impellers. The former type is called a *deep-well turbine pump* and the latter a *submersible pump*. Deep-well turbine pumps can be used only for straight wells. The pump shaft is supported at intervals of about 10 ft by rubber or metal bearings, which are water- or oil-lubricated, respectively. If sand is carried out with the water, an enclosed-shaft or submersible pump must be used to prevent bearing damage. Submersible pumps may be used in crooked wells. Other advantages include ease of increasing the well depth or lift and silent operation. One disadvantage is that the motors are difficult to reach for repairs.

(I. J. Karassik et al., "Pump Handbook," 2nd ed., McGraw-Hill Book Company, New York (books. mcgraw-hill.com).)

21.58 Valves

Water facilities use many different types of valves. These are generally classified according to the function they perform. The two major water-valve classifications are isolating and controlling.

Isolating valves are used for separating or shutting off sections of pipe, pumps, and control devices from the rest of the system for inspection and repair purposes. The major types of isolating valves are gate, plug, sluice gate, and butterfly.

A control valve is normally used for continuously controlling pressures and flow rates. Check, needle, globe, air-relief, pressure-regulating, pressure-relief, and altitude valves are usually considered as control valves.

Gate valves are the isolating valves used most often in distribution systems, primarily because of their low cost, availability, and low head loss when fully open. They have limited value as control or throttling devices because of seat wear and the downstream deflection and chatter of the gate disk. Also, the open area and rate of flow through the valve are not proportional to the percentage opening of the valve when partly open. Corrosion, solids deposition, tubercle formation, large pressure differences, and thermal expansion produce difficulties in opening normally closed gate valves or in closing normally open gate valves. Periodic inspection and operation of valves that are infrequently operated will prevent many operational difficulties. Some of the larger gate valves have gear-reduction drives to permit manual operation. Very large valves are operated by hydraulic and electric power.

A **plug valve** may be used for both control and isolation purposes. It consists of a cylindrically shaped plug (with a rectangular slot or circular orifice) placed in a close-fitting cylindrical seat perpendicular to the direction of flow. Cone and spherical valves are special types of plug valves. Plug, cone, and spherical valves can all be fully closed or opened by a 90° rotation of the plug. The valves may or may not be lubricated (large iron valves usually are). Hydraulic or electric power is commonly utilized for operating the larger valves. Small plug valves are commonly used for isolation purposes on domestic and commercial service connections and are known as service, curb, or corporation cocks. Usually, because the meter is not directly adjacent to the distribution pipe, three valves must be used, one at the service connection, one just upstream of the meter, and one between the meter and the customer's service line. Plug and cone valves are also used for throttling and remote-control shutoff. Low head loss, in-service lubrication features, and easy, fast operation, even in the presence of unequal pressures across the valve, are the major advantages of plug-type valves. But these valves cost more than gate, globe, and butterfly valves.

Butterfly valves can be used for throttling and isolation purposes. The butterfly-valve mechanism consists of a relatively thin circular disk pivoted on a horizontal shaft. Hand or motor power, applied through a gear-reduction device, rotates the disk. Simplicity of construction and quick, easy operation are reasons why these valves are replacing sluice gates and gate valves in many locations. Los Angeles has replaced many sluice gates in reservoir towers with butterfly valves having seats of corrosion-resistant metal, rubber, or Neoprene. A disadvantage of butterfly valves is the higher cost relative to sluice gates or gate valves.

Sluice gates are mainly used on the sides of reservoir control towers and in open-channel structures where pressure on one side of the gate helps to seat it and prevent water leakage. Difficulties with leakage and corrosion of gate frames and stems are the main disadvantages of sluice gates. Low cost and ease of operation in open-channel flow conditions are the major advantages.

A **needle valve** is made of a streamlined plug or needle that fits into a small orifice with a carefully machined seat. Needle valves are used for accurate control of water flow because a large movement of the needle is necessary before any measurable change of flow rate takes place. Needle valves are not normally used for isolating purposes because of the high head losses produced by water flow through the small orifices. Large-sized needle valves are used for flow regulation under high heads, such as for free discharge from reservoirs. Interior-differential, tube, and hollow-jet are three of the most common types of large needle valves.

Globe valves are commonly used in smaller sizes for domestic purposes. The valve mechanism consists of a screw-operated disk that is forced down on a circular seat. Because of high head losses, globe valves are rarely used for isolation purposes, but they are commonly used for pressure regulation in water-distribution systems. Many automatic control valves, such as pressure regulators and altitude, check, and relief valves, have globe-valve bodies with various types of control mechanisms.

Pressure-regulating valves are used to reduce pressures automatically. An air-relief and inlet valve serves the dual purpose of allowing air to either escape or enter a pipeline. Air that accumulates at high points in a pipe impedes water flow and should be allowed to escape through an air-relief valve placed at this location. Furthermore, draining water from low elevations in a pipeline may cause negative pressures at higher elevations and collapse a pipe. Air should be allowed to enter through air-relief and inlet valves at the high points to prevent this.

Pressure-relief valves are used to release excess pressure in an enclosure. Often, these excess pressures are caused by sudden closure of a valve.

Altitude valves are used to control the water level of elevated reservoirs. A pressure-activated control closes the altitude valve when the tank is full and opens the valve to allow water to flow from the tank when pressure below the valve decreases.

Check valves are used in pipelines to allow for one-directional flow only. Check valves placed in centrifugal-pump suction lines are called foot valves. These valves hold water in the suction line and pump case so that the pump will not need manual priming when started. The most common check valve is the swing type.

21.59 Fire Hydrants

A fire hydrant normally consists of a cast-iron barrel and a gate or compression-type shutoff valve, which connects the barrel to the main. Two or more hose outlets are normally located in the barrel above the ground surface. Usually, an additional gate valve is required between the hydrant and the main to allow for shutoff and repair of the hydrant.

The number of $2\frac{1}{2}$-in-diameter hose outlets on a hydrant determines its class. For example, a hydrant with two hose outlets is called a two-way hydrant.

Fire-hydrant construction standards have been established by the American Water Works Association and the American Insurance Association. These standards relate the diameter of the barrel to the size of the main-valve opening. A barrel diameter of at least 4 in is required for a two-way hydrant, 5 in for a three-way hydrant, and 6 in for a four-way hydrant. A minimum of two hose outlets is required on a fire hydrant. Where pumper service is necessary for adequate water pressure, a large pumper outlet must be furnished. This may take the place of one of the smaller $2\frac{1}{2}$in hose outlets. The minimum allowable diameter for the pipe connection between the main and the hydrant is 6 in.

Fire hydrants usually are either dry or wet barrel, depending on the location of the main valve in the hydrant. The main valve in the dry-barrel type should be located below the frost line. When the valve is in a closed position, a drain should be open to prevent freezing of water in the barrel. The wet-barrel, or California type, hydrants have the main valve located near the hose outlets. Many fire hydrants have a safety joint above the ground surface to permit removal of the upper part of the barrel with a minimum loss of water.

Hose connections $3\frac{1}{16}$ in in diameter with $7\frac{1}{2}$ threads per inch have been selected by the American Insurance Association as standard to allow for interchange of fire-fighting equipment between cities.

Friction losses should not exceed $2\frac{1}{2}$psi in a hydrant and 5 psi between the main and outlet when flow is 600 gal/min.

21.60 Metering Devices

Metering devices are classified as either velocity or displacement types. Velocity types measure the velocity of flow either directly by current-measuring devices or indirectly by venturi-principle devices and are usually calibrated to indicate the flow rate directly. The velocity-type metering devices are applied to measurement of flows in streams, rivers, and large pipes, such as trunk lines of distribution systems. Displacement-type metering devices indicate flow rate directly, by recording and integrating the rate at which their measuring chambers are filled and emptied. Weighing meters are also displacement-type metering devices, but they are used primarily in laboratories. Displacement types are used for the smaller flows in distribution systems, such as meters for individual customer connections.

Criteria for selection of a type of water meter include accuracy and range of measurement, amount of head loss through the meter, durability, simplicity and ease of repairs, and cost.

Velocity-Type Metering Devices ▪ Venturi meters, or modifications thereof, are the most common velocity-type devices. These meters produce a regular and predictable fall in the hydraulic grade line that is related to flow rate. Three devices that operate on this principle are the venturi, nozzle, and orifice plate meters shown in Fig. 21.91.

Straightening vanes are installed upstream from these and other velocity-type meters if the pipe is of insufficient length to eliminate helical flow components caused by bends or other fittings.

The standard venturi meter (Fig. 21.91a) was developed to provide a device with minimum head loss. Since most of the loss is associated with the diffuser section, its angle is the major factor in determining the head loss.

Flow through a venturi meter is given by

$$Q = cKd_2^2\sqrt{h_1 - h_2} \qquad (21.149)$$

$$K = \frac{4}{\pi}\sqrt{\frac{2g}{1 - (d_2/d_1)^2}} \qquad (21.150)$$

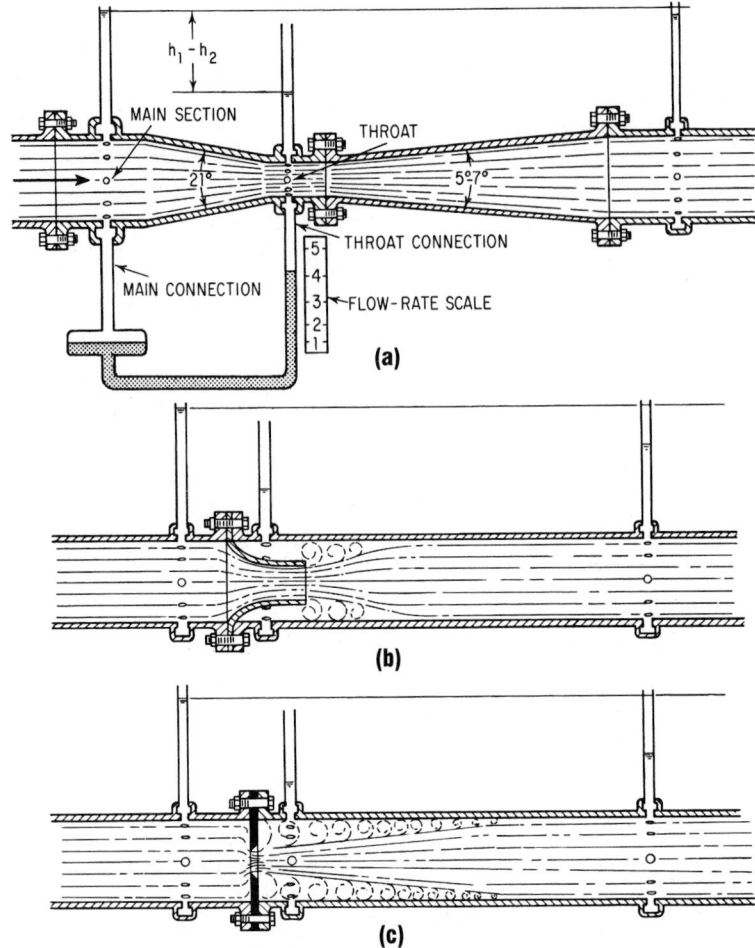

Fig. 21.91 Venturi-type metering devices: (*a*) Standard venturi meter. (*b*) Nozzle meter. (*c*) Orifice-plate meter.

where Q = flow rate, ft^3/s

c = empirical discharge coefficient dependent on throat velocity and diameter

d_1 = diameter of main section, ft

d_2 = diameter of throat, ft

h_1 = pressure in main section, ft of water

h_2 = pressure in throat section, ft of water

(For values of c and K for various throat diameters and velocities, see E. F. Brater, "Handbook of Hydraulics," 6th ed., McGraw-Hill Book Company, New York (books.mcgraw-hill.com).)

As in venturi meters, flows through nozzle and orifice-plate meters are calculated from the pressure difference across the meters. Nozzle and orifice-plate meters are used where conservation of head is not the prime concern or where head dissipation is desired.

Current meters consist of either a propeller or a series of cups or vanes mounted on a shaft free to rotate under the action of the flowing water. The propeller type has its axis of rotation horizontal and will not give accurate measurement unless the current velocity is parallel to the axis of rotation. The cup-type meter, called a Price meter, has a vertical axis of rotation and measures currents whose velocity is in any direction in a horizontal plane. However, vertical velocity components, which do not affect propeller meters, cause the Price meter to indicate greater-than-actual velocities. A clicking noise, made by the making and breaking of an electrical contact and picked up by a set of earphones, indicates the speed of rotation of the meter. The clicking noise occurs either once each revolution or once each five revolutions. Current meters are used almost exclusively for stream flow, although the propeller type is occasionally substituted for a venturi meter in pipe flow.

Displacement-Type Meters ■ These may be piston, rotary, or nutating-disk types. The nutating disk is used, almost to the exclusion of the two other types, for metering domestic-service connections. Its widespread use stems from its simplicity of construction and long-term accuracy. The nutating-disk meter derives its name from the disk's nodding motion, which is similar to that of a top before it stops. The disk is kept in motion by successive volumes of water which enter above and below it. A hard rubber that softens at high temperature is usually used for the disk, so a backflow-prevention device is required between a nutating-disk meter and a water heater. Error of nutating-disk meters is about 1.5% within the normal test-flow limits.

Compound meters contain separate measuring devices for both low and high flows. They are usually a nutating-disk meter and a propeller-type current meter, respectively. An automatic pressure-sensing device directs the flow through the appropriate meter.

21.61 Water Rates

The interests of the public and individual customers of water-supply systems can best be served by self-sustained, utility-type enterprises. Rates charged to finance these systems should be based on sound engineering and economic principles and designed to avoid discrimination between classes of customers. Gross revenue should cover operating and maintenance expenses, fixed charges on capital investment, and development of the system. Billings for water should be based on metered use

and such fixed charges as are required. Rate structures are typically based on demand, load factors, fire use, peak rates of use, seasonal use, and similar items. The system of accounting should conform to the legally established system of accounting prescribed for the utility, if any, or to some other recognized system.

Rates most commonly used today are flat rate, step rate, and block rate.

Flat rate is a monthly or quarterly charge that does not vary with the amount of water used. This type of charge tends to encourage waste. Although it has been commonly employed in small communities where water is not metered, flat rate is falling into disuse.

With *step rate*, customers are charged at one rate per 1000 gal for all water used. The rate a customer pays decreases as the total quantity used increases. The major objection to this method is that a customer who uses a quantity slightly less than the point of rate change will pay more than the customer who uses a little more.

The *block rate schedule* consists of one price per 1000 gal for the first volume or block of water used per billing period and lesser rates for additional blocks. This type of pricing tends to discourage waste but does not restrict usage unnecessarily. Both the step and block rates can have a monthly service charge.

When fixing a system of rates, the supplier should consider the following factors: (1) cost of collection facilities, treatment chemicals, pumping energy, and, where applicable, buying water from a wholesale supplier; (2) cost of distribution and treatment facilities; and (3) cost, including metering and billing, of serving an individual customer. Cost component 1, called the commodity component, is directly dependent on total usage and therefore should be distributed equally to all water sold. Cost component 2, called the demand component, depends on the peak usage of a customer. If a customer's usage is zero during peak hour, it will not appreciably affect the cost or design of distribution facilities. Since peak-hour demands usually govern the design of a distribution system, this is a good criterion for allocating distribution costs. It is generally recognized that residential areas, where the majority of small users are, have very high ratios of peak demand to total usage and should therefore pay a major share of the demand component. Both the step and block rates attempt to allocate this cost to the small user by charging a higher rate for the first water sold to a customer and charging decreasing rates with increased usage. For most distribution systems, a large share of the demand component also should be allocated to fire service. The portion attributed to fire service is usually paid by taxes. Cost component 3, called the customer component, is usually distributed to the customer by a monthly service charge that depends only on the size of service. This charge is usually small.

Hydroelectric Power and Dams

Hydroelectric plants, which generate electric power from water dropping a sufficient vertical distance to drive large hydraulic turbines, supply an appreciable portion of the electric power consumed in the U.S. Hydroelectric generation is an attractive power source because it is a renewable resource and a nonconsumptive use of water. A typical hydroelectric plant consists of a dam to divert or store water from a river or stream; canals, tunnels, penstocks, and a forebay to convey water to turbines; draft tube, tunnel, or tailrace to return water downstream to the river or stream; turbines and governors; generators and exciters; equipment such as protective devices and regulators; a building to house the machinery and equipment; and transformers, switching equipment, and power transmission lines to deliver the power produced to a load center for distribution to consumers.

21.62 Hydroelectric-Power Generation

Hydroelectric power is electrical power obtained from conversion of potential and kinetic energy of water. The potential energy of a volume of water is the product of its weight and the vertical distance it can fall:

$$PE = WZ \qquad (21.151)$$

where PE = potential energy

W = total weight of the water

Z = vertical distance water can fall

Because the kinetic energy of the supply source is very small or zero in most *hydropower* (hydroelectric power) developments, the kinetic-energy term does not appear in power formulas.

Power is the rate at which energy is produced or utilized.

$$1 \text{ horsepower (hp)} = 550 \text{ ft} \cdot \text{lb/s}$$
$$1 \text{ kilowatt (kW)} = 738 \text{ ft} \cdot \text{lb/s}$$
$$1 \text{ hp} = 0.746 \text{ kW}$$
$$1 \text{ kW} = 1.341 \text{ hp}$$

Power obtained from water flow may be computed from

$$\text{hp} = \frac{\eta Qwh}{550} = \frac{\eta Qh}{8.8} \qquad (21.152a)$$

$$\text{kW} = \frac{\eta Qwh}{738} = \frac{\eta Qh}{11.8} \qquad (21.152b)$$

where kW = kilowatts

hp = horsepower

Q = flow rate, ft^3/s

w = unit weight of water = 62.4 lb/ft^3

h = effective head = total elevation difference minus line losses due to friction and turbulence, ft

η = efficiency of turbine and generator

Hydroplants can be classified on a basis of reservoir capacity and use as run-of-river hydro without storage, base-load plants, run-of-river plants with storage, and peak-load plants.

Run-of-River Hydro without Storage ■ This type of plant has no storage facilities. Power generation is totally dependent on the flow of the river. A development of this type is usually built for some other purpose, such as navigation, power production being only incidental.

The economics of a run-of-river hydroplant depend on the minimum flow of the river. If the minimum flow is very low, it will be necessary to invest money in steam-generation facilities to provide supplemental power during low-flow periods. Therefore, the value of the plant will be only the fuel saved that would otherwise be required for steam generation.

Base-Load Hydro Plants ■ This type is also a run-of-river hydroplant without storage, but it is located on a river that provides a minimum flow capable of serving the power demand without supplementary steam-generating facilities. The reliable plant capacity is set below the expected minimum flow in the river. This type of run-of-river hydroplant utilizes only a small proportion of the flow of a river. It must pass not only high seasonal flows but also the water it cannot utilize during hours of low power demand.

Cost of a base-load plant can be compared with the cost of the steam capacity that would be necessary to serve power demands if hydrogeneration were not developed.

Run-of-River Plants with Storage ▪ A small amount of storage can greatly increase the reliable capacity of a hydroplant. The water not required for generation during hours of low power demand can be stored and used for generation during periods of peak demand.

Storage can be provided for a daily, weekly, or seasonal cycle. On a daily cycle, the required reservoir capacity is less than the river's daily flow volume. On a weekly cycle, the flow during the periods of low power demand on weekends is also stored to give additional capacity for peak periods

during the week. On a seasonal cycle, the high flood flows are stored to be used during periods of low flow. The seasonal operation requires many times the storage necessary for weekly or daily operation and therefore may be uneconomical unless the reservoir is multipurpose. Then, part of its cost can be underwritten by flood-control or irrigation projects.

Peak-Load Plants ▪ The power demand on an electrical system fluctuates from a daily high to a nightly low. Depending on the size of the utility and type of customers served, peak demands may be several times the magnitude of the low demands encountered at night. These fluctuations in demand necessitate generation facilities whose full capacity is used only a few hours a day, during periods of peak power demand (Fig. 21.92).

Capacity factor is the percentage of the time the full capacity of a plant is used or the ratio of the average power the plant produces to the plant's capacity. It can be computed on a daily, weekly, or yearly basis.

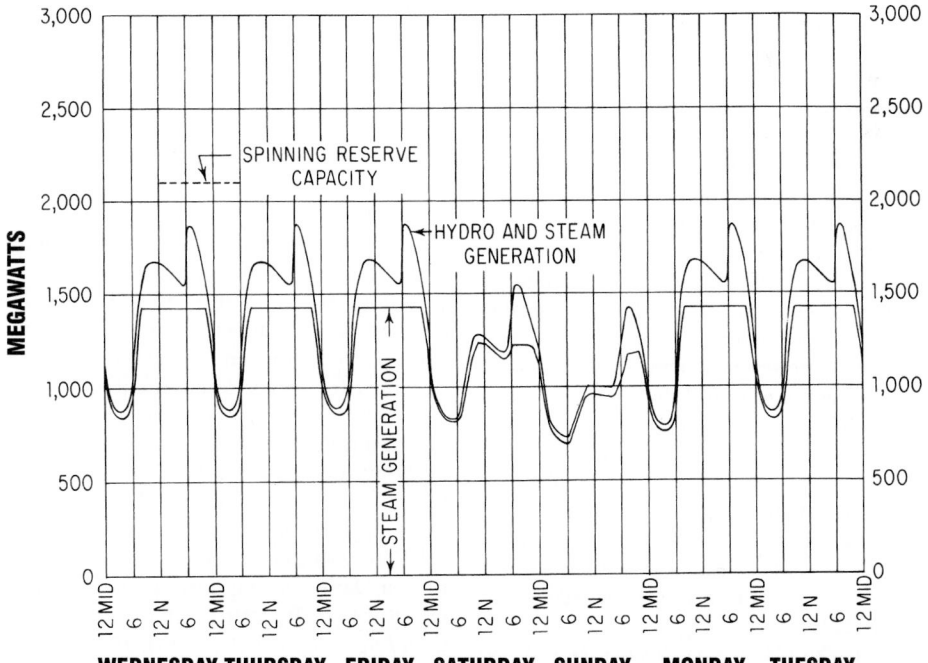

Fig. 21.92 Daily load curves for generating plants. (*Department of Water and Power, Los Angeles, Calif.*)

Hydroplants that are used mainly to supply power for the periods of peak demand are generally called peak-load plants. The main classes of peak-load plants are pumped-storage plants and run-of-river plants with storage.

If sufficient generating capacity and reservoir storage are planned for a run-of-river hydroplant, only a relatively small supply of water is needed to produce a high generation capacity for a few hours duration. This enables a large utility to use steam generation at a high capacity factor where it is most efficient and to supply peak demands from hydroplants.

Pumped Storage ■ This is a means of storing large quantities of energy, generated during periods when excess generating capacity is available, to be used at some future time. Water is pumped from a low reservoir to a higher one by energy from steam or base-load hydro when power demand is low. When needed, the water generates power by flowing through a turbine back into the low reservoir. Because of friction loss in the penstock and losses due to the imperfect efficiencies of pumps and turbines, only two-thirds of the energy required to pump the water is recovered.

The balance of energy between pumping and generating can be on a daily or weekly basis. But because the weekly cycle requires several times more reservoir storage than the daily cycle, it usually is not as economical.

When pumped storage is operated at a high capacity factor to transfer large quantities of electric energy from off-peak to peak, the energy loss may make it uneconomical. This undesirable energy-loss feature of pumped storage is overcome when it is used as reserve capacity.

Electrical systems require what is called *spinning reserve*, which is capacity above that necessary to serve the expected maximum load, ready instantly to generate power in case of failure of generating equipment or an unanticipated high power demand. Many utilities keep a spinning reserve capacity equal to the size of their largest single generating unit, or 15% of their maximum demand (Fig. 21.92).

(V. J. Zipparo and H. Hasen, "Davis' Handbook of Applied Hydraulics," 4th ed., McGraw-Hill Book Company, New York (books.mcgraw-hill.com).)

21.63 Dams

Dams are usually classified on the basis of the type of construction material or the method used to resist water pressure. The main classifications are gravity, arch, buttress, earth, and rock-fill.

Gravity dams are concrete or masonry dams that resist the forces acting on them entirely by their weight. Figure 21.93 shows the forces that act on a typical gravity dam. The largest force is usually the hydrostatic force of the water F_1. Its distribution is triangular, varying from zero at the top to full hydrostatic at the bottom. Force F_2 represents silt pressure, which results from deposition of silt at the base of the dam. This silt pressure can be calculated by Rankine's theory for earth pressure using the submerged weight of the silt.

Force F_3 represents ice pressure against the face of the dam. In cold climates, ice, which forms on the reservoir surface, expands when the temperature rises and exerts a force on the top of a dam. In the past, ice pressures as high as 50,000 psf have been used for the design of dams in the north; however, today it is realized these values are much too high. A method of calculating these forces, presented by Edwin Rose, gives values ranging from 2000 to 10,000 psf, depending on the rate of temperature rise and restraining conditions at the edges of the reservoir. (E. Rose, "Thrust Exerted by Expanding Ice," *Proceedings of the American Society of Civil Engineers*, May 1946.)

Practically all regions in the United States are subject to earthquakes of varying intensity. Earthquakes cause vertical and horizontal accelerations of the earth, which create forces on any object resting on it. The magnitude of these forces equals the mass of the object times the acceleration from the earthquake. These accelerations occur in every direction, so the effect of the forces must be analyzed for all directions. Most dams in seismically active regions in the United States have been designed for an acceleration equal to 0.1 g, where g is the acceleration due to gravity. The effect of accelerations on the dam is represented in Fig. 21.93 by forces F_4 and F_5. Force F_6 represents the inertial force of the water on the face of the dam. A close approximation of the force, given by Eq. (21.153), was developed by von Karman. ("Pressure on Dams During Earthquakes," discussion by von Karman, *Transactions of the American Society of Civil Engineers*, vol. 98, p. 434, 1933.)

$$F_6 = 0.555awh^2 \qquad (21.153)$$

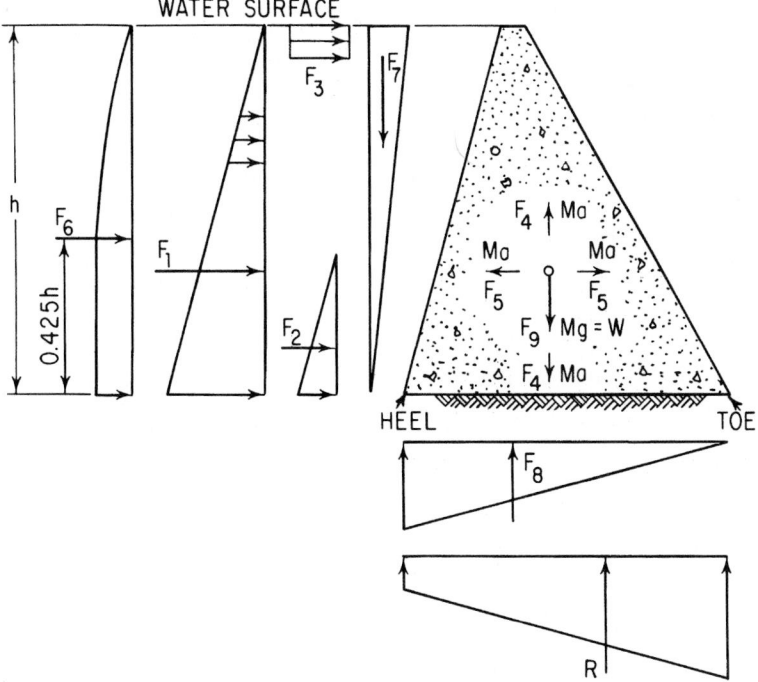

Fig. 21.93 Forces acting on a concrete gravity dam.

where w = unit weight of water, lb/ft^3

a = acceleration due to earthquake, ft/s^2

h = depth of water behind dam, ft

The force F_6 acts at a point $0.425h$ above the base.

Force F_7 is due to the weight of water on an inclined face. Gravity dams usually have an inclined upstream face to facilitate construction.

Force F_8 represents an uplift force that acts on the undersurface of any section taken through the dam or under the base of the dam. This uplift is caused by the seepage of water through pores or imperfections in the foundations or through imperfectly bonded construction joints in the masonry. In the past, engineers assumed that, because of bearing contact, this pressure acted only on some percentage of the total area. Recent belief, however, is that uplift acts on 100% of the area of the base.

A process used to reduce uplift pressures calls for grouting along the heel and use of drains behind the grout. When the base is not drained, the uplift pressure is assumed to vary linearly from between full and one-half hydrostatic pressure at the heel to the full tailwater pressure at the toe.

Force F_9 represents the weight of the dam. It acts at the centroid of the cross-sectional area of the dam.

Summation of the vertical forces and of moments about any point yields the foundation pressure. The foundation pressure at the heel of the dam should be compressive. Hence, the resultant of all forces acting on the dam should fall within the middle third of the base of the dam.

The basic modes of failure possible for a gravity dam are by sliding along a horizontal plane, overturning by rotating about the toe, or failure of the foundation material. The first two modes depend mainly on the cross-sectional shape of the dam, whereas the third depends on both the cross-sectional shape and the foundation material.

Gravity dams can be built on earth foundations, but their height in these cases has been limited to around 65 ft. The main reason gravity dams are used is that they can pass large flood flows over their crest without damage. Their first cost and maintenance cost are usually greater than those of earth or rock-fill dams of comparable height and crest length.

Arch dams are concrete dams that carry the force of the water through arch action. Stresses in an arch dam may be determined with computers by the finite-element method or by an approximate method in which the water force is divided between elements: a series of horizontal arches that span between the abutments and a series of vertical cantilevers fixed at the foundation. The distribution of load between the arches and cantilevers is determined by the trial-load method. First, a division of the load is assumed and the deflections in the arches and cantilevers are computed. The deflection of an arch at any point must equal the deflection of the cantilever at the same point. If the deflections are not equal, a new division of the load is assumed and the deflections recalculated. This process is continued until equal deflections are obtained.

The external forces an arch dam must resist are basically the same as those on a gravity dam; however, their relative importance is much different. On arch dams, uplift is not so important, but ice loads and temperature stress are much more critical. Arch dams require much less concrete than gravity dams and usually have a much lower first cost. They are not suited to most sites, however, since they must be located in a relatively narrow canyon supported by good rock abutments.

Buttress dams consist of a watertight membrane supported by a series of buttresses at right angles to the axis of the dam. Although there are many types of buttress dams, those widely used are the flat-slab and the multiple-arch. These differ in that the water-supporting membrane for the flat-slab type is a continuous concrete slab spanning the buttresses. In the multiple-arch, the membrane is a series of concrete arches. The multiple-arch requires less reinforcing steel and can span longer distances between buttresses, but its formwork is more expensive.

The upstream face of a buttress dam is usually inclined at about 45°. The weight of the water on the face is necessary to increase the dam's resistance to sliding and overturning.

The forces acting on a buttress dam are exactly the same as those that act on a gravity dam. However, the vertical load of the water is much greater on a buttress dam, and uplift forces are smaller. The modes of failure are also the same, but the structural design is much more critical.

Although buttress dams usually require less than half the volume of concrete required by gravity dams, they are not necessarily less expensive because of the large amount of formwork and reinforcing steel required. With the rapidly increasing cost of labor over the past several decades, the buttress dam has lost much of its earlier popularity.

Earth dams are designed to utilize materials available at the dam site. They can be constructed of almost any material with very primitive construction equipment. Successful earth dams have been built of gravel, sand, silt, rock flour, and clay. If a large quantity of pervious material, such as sand and gravel, is available and clayey materials must be imported, the dam would have a small impervious clay core, the material available locally making up the bulk of the dam. Concrete has been used for an impervious core, but it does not provide the flexibility of clay materials. If pervious material is not available, the dam can be constructed of clayey materials with underdrains of imported sand or gravel under the downstream toe to collect seepage and relieve pore pressures.

Slopes of an earth dam are rarely greater than 2 horizontal to 1 vertical and are usually about 3 to 1. The governing criterion is usually the stability of the slopes against slide-out failure. Stability under the action of seismic forces is especially critical. For soils in which pore pressure changes develop as a result of shear strain induced by an earthquake, determination of appropriate values for yield acceleration is very difficult. For some types of soil, no well-defined yield acceleration exists; displacements occur over a wide range of accelerations.

Another factor that sometimes determines the steepness of the slopes is the amount of seepage that can be tolerated. If the dam is on a pervious foundation, it may be necessary to increase the base width to reduce seepage. The seepage may also be reduced by placing an impervious blanket on the upstream side of the dam to increase the seepage path or by using a cutoff wall in the foundation, such as sheetpiling or a clay-filled trench.

Earth dams can be built to almost any height and on foundations not strong enough for concrete dams. Improvements in earth-moving equipment have resulted in a decreased cost for earth dams, and rising labor costs have increased the cost for concrete dams.

Rock-fill dams usually consist of a dumped rock fill, a rubble cushion of laid-up stone on the upstream face, bonding into the dumped rock, and an upstream impervious facing, bearing on the rubble cushion, with a cutoff wall extending into the foundation. The dumped rock fill may consist of rocks varying in size from small fragments to boulders weighing as much as 25 tons. The fill is usually compacted by dropping the rock, sometimes from as high as 175 ft. onto the fill. Sluicing of the fill with high-pressure hoses is also used to wash fines from between contact points of the rock and reduce settlement. The rubble cushion consists of rocks individually placed to reduce the voids and provide support for the impervious facing. The facing is usually concrete, or wood over concrete, although steel has been used occasionally. The cutoff wall is usually concrete.

Rock-fill dams are generally designed empirically. Low rock-fill dams may have an upstream face as steep as $\frac{1}{2}$ horizontal on 1 vertical. The downstream face is usually 1.3 on 1, the natural angle of repose of rock. For dams over 200 ft high, both the upstream and downstream faces are usually on a slope of 1.3 on 1.

The major problem encountered in rock-fill dams is large settlements that occur after construction when the reservoir is first filled. Vertical settlements and horizontal displacements in excess of 5% of the height of the dam have occurred; therefore, the impervious facing must be very flexible or damage will occur during settlement. One solution to this problem has been to put a temporary facing on the dam and to replace it with a permanent facing after settlement has taken place. Temporary facings are usually of wood.

Rock-fill dams are used extensively in remote locations where cement is expensive and the materials for an earth dam are not available. Their cost compares favorably with that of concrete dams. Leakage should be expected, but rock-fill dams are very stable and have been overtopped without suffering major damage.

(V. J. Zipparo and H. Hasen, "Davis' Handbook of Applied Hydraulics," 4th ed., McGraw-Hill Book Company, New York (books.mcgraw-hill.com); "Design of Small Dams" and "Embankment Dams," U. S. Bureau of Relamation; "Earth and Rockfill Dams: General Design and Construction Considerations," EM 1110-2-2300, U. S. Army Corps of Engineers (www.usace.army.mil/inet/usace-docs/eng-manual/em.htm).)

21.64 Hydraulic Turbines

In the past, hydraulic power-generating machines meant a large number of different types of equipment. Today, however, the turbine is the only type of importance in hydraulic power generation. Its function is transformation of the kinetic and potential energy of water into useful work.

Turbines are classified as impulse turbines and reaction turbines.

Impulse turbines utilize the energy of water by first transforming it into kinetic energy, by free discharge of the water through a nozzle. The nozzle is directed at buckets positioned along the perimeter of a water wheel. The force of the water striking these buckets causes the wheel to rotate, providing power.

The only type of water wheel used today in impulse turbines was developed in 1880 by Pelton—the **Pelton wheel** (Fig. 21.94). The wheel is covered by a housing to prevent splashing and to guide the discharge after the water strikes the wheel.

In most impulse turbines, the water wheel rotates on a horizontal shaft and is acted on by the discharge from one or two nozzles. But vertical shafts may be used with as many as six nozzles, to obtain a high efficiency for very low loads. In such installations, efficiencies of 92% for full load and slightly below 90% for loads as low as one-quarter of full load have been obtained.

Impulse turbines are commonly used for heads greater than 1000 ft. (An impulse turbine at the Reisseck Power Plant in Austria operates under a net effective head of 5800 ft.) There is no lower limit of head for impulse turbines. They have been used for heads as low as 50 ft; however, the reaction turbine is usually better suited to low heads at large flows.

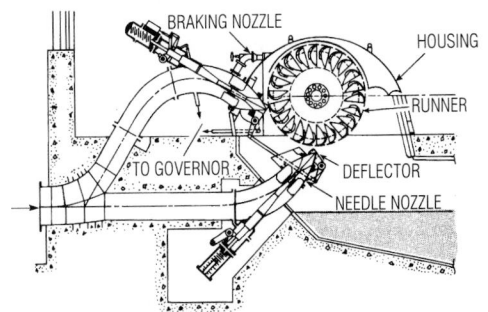

Fig. 21.94 Impulse (Pelton) type of hydraulic turbine.

Reaction Turbines ▪ Types of reaction turbines include the Francis (Fig. 21.95a), the propeller-type (Fig. 21.95b) and the axial flow (Fig. 21.95c). In these, the flow from the headwater to the tailwater is in a closed conduit system.

The **Francis turbine** usually consists of four essential parts: scroll case, wicket gates, runner, and draft tube.

The *scroll case* transfers the water from the penstock (supply pipe) to the wicket gates and

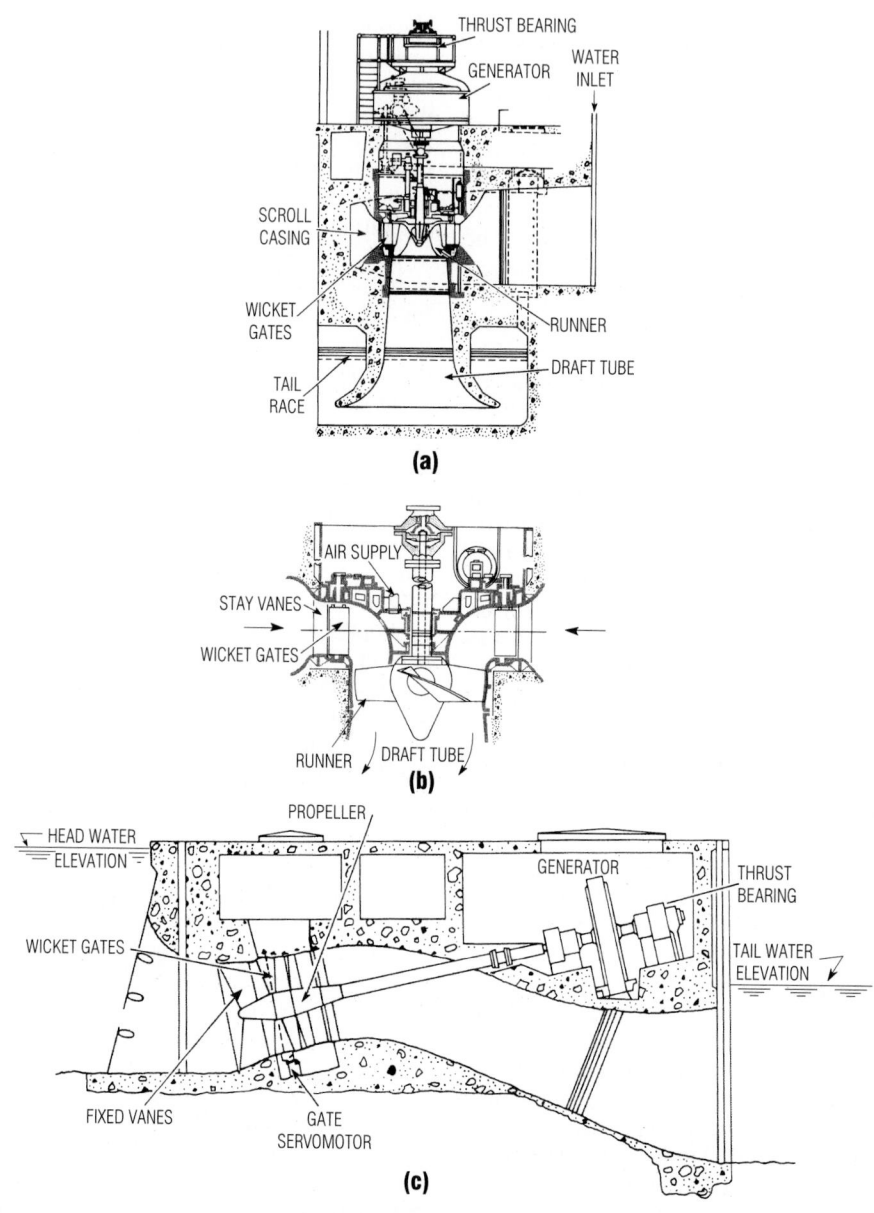

Fig. 21.95 Reaction types of hydraulic turbines: (*a*) Francis; (*b*) Kaplan; (*c*) axial flow.

runner. It distributes the water so that all points on the perimeter of the runner receive the same quantity of water.

The *wicket gates*, located just outside the perimeter of the runner, control the amount of water that enters the turbine. When the power demand on the turbine changes, a governor actuates a mechanism that opens or closes the gates.

The *runner* is the part of the turbine that transforms the pressure and kinetic energy of the water into useful work. As the water flows through the turbine, it changes direction. This creates a force on the runner, causing it to rotate and turn the generator.

The *draft tube* is a conical tube with diverging sides. It decelerates the flow discharged from the runner, so that the remaining kinetic energy may be regained by conversion into suction head.

Francis turbines have a maximum efficiency of about 94% when operated at or close to full load. However, if the load drops below 50%, their efficiency decreases rapidly. Francis turbines are commonly used for heads between 100 and 1000 ft. At heads above 1000 ft. problems are encountered in controlling cavitation and in building a scroll case to take the high pressures. At heads below 100 ft. the propeller-type turbine is usually more efficient.

Propeller Turbines ▪ There are two types of propeller turbines: the movable-blade type, such as the **Kaplan turbine**, and the fixed-blade type. The only difference between the two is that the pitch of the propeller blades is adjustable in a Kaplan turbine.

The propeller turbine (Fig. 21.95) has the same basic parts as the Francis turbine: scroll case, wicket gates, runner, and draft tube. The basic difference between the Francis turbine and the propeller turbine is in the shape of the runner. The runner of a propeller-type turbine operates in the same manner as a fan or a ship's propeller: The water moving past the blades creates a force that causes the runner to rotate.

Propeller-type turbines are used for heads ranging from a few feet to about 100 ft. The Kaplan turbine has an efficiency of about 94% for full load and drops only to 92% for 40% load. The fixed-blade-type turbine also has an efficiency of about 94% for full load; however, its efficiency drops off rapidly below full load.

Axial-Flow Turbines ▪ These provide enhanced performance for operation under low-head and large capacity.

(V. J. Zipparo and H. Hasen, "Davis' Handbook of Applied Hydraulics," 4th ed., McGraw-Hill Book Company, New York (books.mcgraw-hill.com).)

21.65 Methods for Control of Flows from Reservoirs

Any reservoir with an appreciable drainage area must have a spillway to discharge flood flows without damage to the dam and to keep the reservoir water surface below some predetermined level.

21.65.1 Spillways

An **overflow spillway** allows water to pass over the crest of a section of the dam. This type of spillway is widely used for concrete dams because, if designed correctly, the dam will not be damaged by the water. To use an overflow spillway for earth or rock-fill dams, it is necessary to make the spillway a concrete gravity section. This may not be possible for high earth dams because the foundation may not be able to support a high concrete gravity section.

The discharge over an overflow spillway is given by the equation for discharge over a weir (Art. 21.34). Since the discharge varies as the head to the $\frac{3}{2}$ power, overflow spillways keep the water level within close limits even when there is a large variation in flows.

It is desirable for an overflow spillway to have the form of the underside of the nappe of a sharp-crested weir. This type of spillway, called an **ogee spillway**, should be designed—as should all spillways—so that separation of the water from the face of the spillway will not occur. Thus, the danger of cavitation will be eliminated.

In a **chute spillway**, water flows over a crest into a steeply sloping, lined, open channel. The flow is made supercritical to keep the size and length of the chute to a minimum. Gradual vertical curves should be used in the chute to avoid separation of the flow from the channel bottom.

Chute spillways are commonly used for earth and rock-fill dams where the topography allows a chute to carry the water away from the toe to eliminate the danger of undermining. The discharge over the crest is given by the equations for discharge over a weir or the entrance to an open channel.

In a **side-channel spillway**, the flow passes over a crest into a channel parallel to this crest. The crest is usually a concrete gravity section, although it can be concrete laid on the natural embankment. Side-channel spillways are often used in narrow canyons where it is not possible to obtain sufficient crest length for overflow or chute spillways. The flow in the channel parallel to the crest is determined by applying the momentum principle in the direction of flow and assuming the energy of the water flowing over the crest is completely dissipated (U.S. Bureau of Reclamation, "Design of Small Dams," Government Printing Office, Washington, DC 20402).

In a **shaft spillway**, sometimes called a **morning-glory spillway**, the water flows over a circular weir into a vertical shaft. The shaft terminates in a horizontal conduit that carries the water past the dam. The weir can be sharp-crested, flared, or ogee in cross section. (This type of spillway should not be constructed over or through earth dams.) If the topography is not suitable for a chute or side-channel spillway, a shaft spillway may be the best alternative.

There are two conditions of discharge for a shaft spillway, both depending on the head on the weir. When the head is relatively low, the discharge is governed by the flow over the weir, which is directly proportional to the $\frac{3}{2}$ power of the head on the weir. As the head increases, at some point the discharge will no longer be controlled by the amount of water that can flow over the weir but by the amount of water that can flow through the conduit. The discharge for this condition is directly proportional to the $\frac{1}{2}$ power of the elevation difference between the reservoir water level and the level of discharge of the spillway conduit. Once this second condition is reached, a large increase in head will cause only a small increase in flow. Since analytical analysis of discharge does not give good results on this type of spillway, model tests are usually employed.

A **siphon spillway** (Fig. 21.96) is a closed conduit for discharging water over or through a dam. The entrance to a siphon spillway is usually

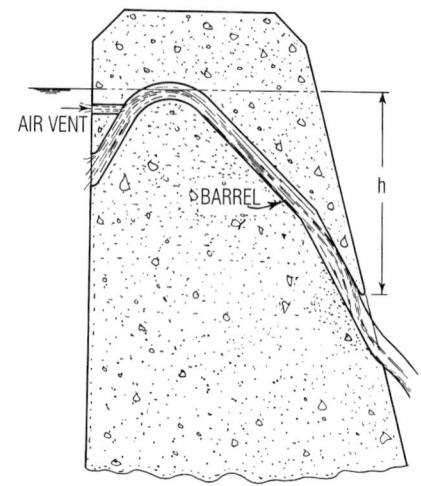

Fig. 21.96   Siphon spillway.

submerged below the normal water level so that it will not clog with debris or ice. The discharge end of the siphon is usually sealed by deflecting the flow across the barrel or by submerging it so that air cannot enter.

The air vent shown in Fig. 21.96 determines the reservoir level at which the siphon flow begins. When the reservoir water level rises above the vent, the siphon's intake is sealed. Water flowing over the crest of the siphon removes the air in the siphon and full flow begins. Because the flow depends on the siphoning action, siphon spillways hold the water level of a reservoir within close limits. But they are not good for handling large variations in flows because their discharge is directly proportional to the square root of the head. They are relatively expensive because of the cost of forming the barrel.

21.65.2   Intake Structures

The various functions an intake structure may serve include permitting withdrawal of water from various levels of a reservoir, controlling flow, excluding debris and ice from a conduit, and providing support for the conduit. The type of intake structure required depends on the functions and characteristics of the reservoir. The simplest type of intake is a block of concrete supporting the end of a conduit equipped with a bar screen to exclude foreign matter. In contrast, the intake towers at Hoover Dam, which serve 30-ft-diameter penstocks, are 395-ft-high concrete towers, with two 32-ft-diameter cylinder gates under a maximum head of over 300 ft.

Intake towers are commonly used where there is a large fluctuation in the water level of a reservoir or where it is necessary to control the quality of water used for a domestic supply. They are usually made of concrete and have ports at various levels to permit selection of water from different elevations. The ports are usually provided with gates or valves and some type of trash rack.

The main hydraulic consideration in the design of an intake is to keep losses to a minimum. To do this, the velocities through the trash racks should be kept less than 0.5 ft/s, and the standard rules for reducing hydraulic losses should be observed.

21.65.3   Crest Gates

These include a number of different types of permanent and temporary devices that operate on the crest of spillways to increase the storage of a reservoir temporarily while control of spillway flows is retained. During periods of low flow when the full spillway capacity is not required, the additional head and storage gained with crest gates may be very valuable.

Flashboards and **stop logs** are the most common types of crest gates used for small installa-

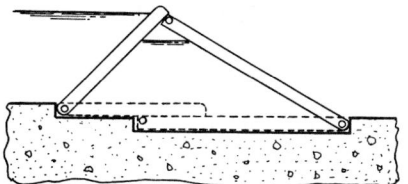

Fig. 21.98   Bear-trap gate.

tions under low head. Flashboards are usually wood planks that span between vertical pipes that cantilever above the spillway crest. When the reservoir water surface reaches some predetermined level, the pipes fail, allowing the full capacity of the spillway to be utilized. Stop logs are wood planks that span between slotted vertical piers which cantilever above the spillway crest.

On large stop-log installations, the hydrostatic force creates large frictional forces between the sliding element and the vertical guide, making removal difficult. These frictional forces make it necessary to use a type of gate that depends on rolling rather than sliding friction and operates freely under hydrostatic pressure.

Taintor gates and **sliding gates** mounted on low-friction roller bearings are the most widely used types of crest gates on major installations. In a taintor gate (Fig. 21.97), the friction is concentrated in the trunnion and does not affect the operation. Since flow passes under taintor and slide gates, there is a tendency for ice and trash to pile up against them, causing damage and hampering operation.

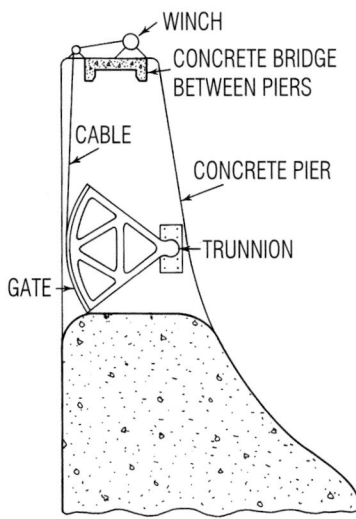

Fig. 21.97   Taintor gate.

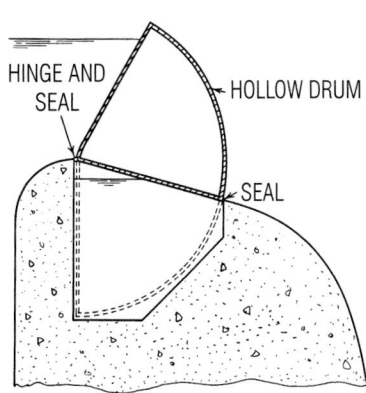

Fig. 21.99   Drum gate.

Bear-trap and **drum gates** allow the flow to pass over the top. The bear-trap gate consists of two leaves hinged, as shown in Fig. 21.98. To raise a bear-trap gate, water is admitted to the space under the leaves to force the leaves up. The drum gate (Fig. 21.99) consists of a segment of a cylinder that is lowered into a recess in the crest when not in use. Because of the large recess required in the dam, drum gates are not suited to small dams.

(V. J. Zipparo and H. Hasen, "Davis' Handbook of Applied Hydraulics," 4th ed., and H. E. Babbitt, J. J. Doland, and J. L. Cleasby, "Water Supply Engineering," McGraw-Hill Book Company, New York (books.mcgraw-hill.com).)

22

G. Raymond Schulte
Consulting Engineer,
Johnson, Mirmiran and
Thompson, Inc. Baltimore, MD

Thomas E. Wilson
Consulting Engineer,
Barrington, IL

ENVIRONMENTAL ENGINEERING

Environmental engineers are concerned with works developed to protect and promote public health, improve the environment, and prevent degradation of land, water, and air. Their practice includes surveys, reports, designs, reviews, management, operation, and investigations of such works. They also engage in research in engineering sciences and such related sciences as chemistry, physics, and microbiology to advance the objectives of protecting public health and controlling environment.

Environmental engineering deals with treatment and distribution of water supply; collection, treatment, and disposal of wastewater; control of pollution in surface and underground waters; collection, treatment, and disposal of solid and hazardous wastes; housing and institutional sanitation; rodent and insect control; control of atmospheric pollution; limitations on exposure to radiation; limitations on noises; and other environmental factors affecting the health, comfort, safety, and well-being of people. This section, while covering primarily the aspects related to handling of liquid wastes, also deals briefly with other environment-related tasks, such as solid-waste handling and air pollution. (See also environmental discussions in Sec. 14 and subsequent sections.)

22.1 Prevention of Environmental Pollution

Because of public concern over accelerating deterioration of the natural environment, Congress established the Environmental Protection Agency (EPA) and passed legislation to control disposal of solid wastes and discharges to water and air. The following legislation is of particular significance to environmental engineers.

National Environmental Policy Act ▪ All agencies of the Federal government and state and municipal agencies executing programs supported by Federal funds are required to carefully consider the environmental consequences of major actions, including proposed construction projects, and proposed legislation. The objectives are:

1. Fulfill the responsibilities of each generation as trustee of the environment for the succeeding generation.

2. Assure for all Americans safe, healthful, productive, and esthetically and culturally pleasing surroundings.

3. Attain the widest range of beneficial uses of the environment without degradation, risk to health or safety, or other undesirable and unintended consequences.

4. Preserve important historic, cultural, and natural aspects of our national heritage, and maintain, wherever possible, an environment that supports diversity and variety of individual choice.

5. Achieve a balance between population and resource use that will permit high standards of living and a wide sharing of life's amenities.

6. Enhance the quality of renewable resources and approach the maximum attainable recycling of depletable resources.

Clean Water Act (Federal Water Pollution Control Act) ▪ The objective is to restore and maintain the chemical, physical, and biological integrity of the nation's waters. The act directs EPA to establish technology-based limitations and standards for industrial discharges. The states set water-quality standards for their waters. Control is achieved principally by issuance of permits by EPA or delegated states under the National Pollutant Discharge Elimination System (NPDES).

Safe Drinking Water Act ▪ EPA is required to establish regulations for public drinking water supplies. Primary regulations set maximum allowable levels for contaminants in drinking water and establish criteria for water treatment. Secondary regulations deal with taste, odor, and appearance of drinking water. Other regulations protect groundwater through controls over injection wells under the Underground Injection Control Program. EPA delegates primary responsibility for enforcement to the states and supports state programs with grants.

Resource Conservation and Recovery Act ▪ The objectives are to improve management of solid wastes, protect the environment and human health, and conserve valuable material and energy resources. The act also provides for state programs regulating hazardous wastes from generation to disposal, including disposal of industrial sludges containing toxic materials. The states regulate disposal of solid wastes on land in accordance with Federal criteria.

Marine Protection, Research and Sanctuaries Act ▪ EPA is required to protect the oceans from indiscriminate dumping of wastes and to designate safe sites for dumping. An objective is an ultimate halt in ocean dumping of wastes. The Corps of Engineers issues, subject to EPA approval, permits for dredging, filling of wetlands, or dumping of dredged material.

Superfund (Comprehensive Environmental Response, Compensation and Liability Act) ▪ The Federal government is authorized to remove and safely dispose of pollutants in hazardous waste sites, underground water supplies and other facilities. The act establishes a Hazardous Waste Response Fund to pay for cleanup and damage claims. EPA designates substances that may present substantial hazards to public health or welfare or to the environment. The National Response Center should be notified of releases of hazardous substances.

Clean Air Act ▪ The objective is to protect public health and welfare from the harmful effects of air pollution. EPA promulgates National Ambient Air Quality Standards. To meet these standards, the states prepare State Implementation Plans and plans for enhancement of visibility and prevention of significant deterioration of air quality in areas where the standards have been attained. EPA also develops New Source Performance Standards, to reduce pollutant emissions, and National Emission Standards for Hazardous Air Pollutants, applicable to pollutants that will cause an increase in mortality or incapacitating illness. In addition, EPA sets limits on emissions from moving sources of air pollution.

(R. A. Corbitt, "Standard Handbook of Environmental Engineering," McGraw-Hill Publishing Company (www.books.mcgraw-hill.com).)

22.2 Major Sources of Water Pollution

There are two major sources of water pollution: point sources and nonpoint sources. The former consists of sources that discharge pollutants from a well-defined place, such as outfall pipes of sewage-treatment plants and factories. Nonpoint sources, in contrast, cannot be located with such precision. They include runoff from city streets, construction sites, farms, or mines. Therefore, prevention of water pollution requires a mixture of controls on discharges from both point and nonpoint sources.

Domestic wastewater and industrial discharges are major point sources. The Clean Water Act and the Marine Protection, Research and Sanctuaries Act (Art. 22.1) aim at elimination of discharge of pollutants in navigable waters and the ocean.

Wastewater is the liquid effluent of a community. This spent water is a combination of the liquid and water-carried wastes from residences, commercial buildings, industrial plants, and institutions, plus groundwater, surface water, or storm water.

Wastewater may be grouped into four classes:

Class 1 ▪ Effluents that are nontoxic and not directly polluting but liable to disturb the physical nature of the receiving water; they can be improved by physical means. They include such effluents as cooling water from power plants.

Class 2 ▪ Effluents that are nontoxic but polluting because they have an organic content with high oxygen demand. They can be treated for removal of objectionable characteristics by biological methods. The main constituent of this class of effluent usually is domestic sewage. But the class also includes storm water and wastes from dairy product plants and other food factories.

Class 3 ▪ Effluents that contain poisonous materials and therefore are often toxic. They can be treated by chemical methods. When they occur, such effluents generally are included in industrial wastes, for example, those from metal finishing.

Class 4 ▪ Effluents that are polluting because of organic content with high oxygen demand and, in addition, are toxic. Their treatment requires a combination of chemical, physical, and biological processes. When such effluents occur, they generally are included in industrial wastes, for example, those from tanning.

Domestic wastewater is collected from dwelling units, commercial buildings, and institutions of the community. It may include process wastes of industry, groundwater infiltration, surface-water inflow, and miscellaneous waste liquids. It is primarily spent water from building water supply, to which have been added the sanitary waste materials of bathroom, kitchen, and laundry. (See Art. 22.14.)

Storm water is precipitation collected from property and streets and carrying with it the washings from surfaces.

Industrial wastes are primarily the specific liquid waste products collected from industrial processing but may contain small quantities of domestic sewage. Such wastes vary with the process and contain some quantity of the material being processed or chemicals used for processing purposes. Industrial cooling water when mixed with process water is also called industrial waste.

Industrial wastes, as distinguished from domestic wastes, are related directly to processing operations and usually are the liquid fraction of processing that has no further use in recovery of a product. These wastes may contain substances that, when discharged into surface water or groundwater, cause some biological, chemical, or physical change in the water.

Organic substances exert a biochemical oxygen demand (BOD) of relatively high proportion compared with domestic waste. It is not unusual in food processing to have wastes with a BOD of 1,000 to 5,000 mg/L or in the processing of edible oils to have 10,000 to 25,000 mg/L BOD.

The wastes may cause discoloration of a receiving stream, as in the release of dyes, or increase the temperature of the water, as in the case of a cooling tower or process-cooling water discharges.

Chemicals in the waste may be toxic to aquatic life, animals, or human populations using the water, or may in some way affect water quality by imparting taste or odor. Phenols introduced into water in the parts-per-billion range can produce such marked taste that the water becomes unusable for many purposes. Nitrogen and phosphorus, referred to as nutrients, may stimulate aquatic growth, and large concentrations of algae and other microorganisms in the receiving stream may be increased. Some algae and other microorganisms are detrimental to water quality since they too produce taste, odor, color, and turbidity in the water.

Industrial wastes that contain large quantities of solids may produce objectionable and dangerous levels of sludge on the bottom of a stream or along the banks. These add to the chemical, biological, and physical degradation of the stream. Discharges containing grease or oil may render bathing beaches useless, interfere with nesting water fowl, and present extra problems of removal in water-treatment processes.

Wastes containing acids or alkalies may attack pier structures and water craft and produce serious toxic effects on aquatic life.

Some wastes, such as those containing copper, interfere with the normal processes of wastewater treatment and may, if mixed with municipal waste, render the whole treatment process inoperative. Pretreatment of industrial wastes is often required to protect the sewers and treatment plant maintained by a municipal agency. Toxic pollutants are controlled by EPA General Pretreatment

Regulations, which contain limits on specific substances discharged by various industries. Most municipalities have industrial pretreatment ordinances in effect. Treatment of industrial wastes to the degree required to protect a receiving body of water is a requirement in all states; it may range from neutralization and other simple primary treatments to complete treatment or, in some instances, even an advanced stage of treatment to remove trace chemicals (see also Art. 22.31).

Combined wastes are the mixed discharge of domestic waste and storm water in a single pipeline. Industrial waste may or may not be found in a combined waste.

(W. W. Eckenfelder, Jr., "Industrial Water Pollution Control," and R. A. Corbitt, "Standard Handbook of Environmental Engineering," McGraw-Hill Publishing Company, New York (www.books.mcgraw-hill.com); N. L. Nemerow, "Liquid Wastes of Industry: Theories, Practices and Treatment," and R. L. Culp et al., "Advanced Wastewater Treatment," Van Nostrand Reinhold Company, New York.)

22.3 Types of Sewers

A sewer is a conduit through which wastewater, storm water, or other wastes flow. Sewerage is a system of sewers. The system may comprise sanitary sewers, storm sewers, or a combination of both. Usually, it includes all the sewers between the ends of building-drainage systems and sewage-treatment plants or other points of disposal.

Sanitary sewers carry mostly domestic wastewater. They may also receive some industrial wastes. But they are not designed for storm water or groundwater.

Storm sewers are designed specifically to convey storm water, street wash, and other surface water to disposal points.

Combined sewers are sewers that carry both domestic wastewater and storm water. They were constructed, generally in larger communities, in the early half of the 20th century, to save the higher construction cost of separate sanitary and storm sewers. Combined sewers include diversion and overflow structures, which, when the combined sewer reaches its carrying capacity, discharge sanitary sewerage diluted with storm water, to a watercourse. Combined sewers are no longer permitted to be built in the U.S. The EPA has issued a national policy statement (40 CFR Part 122) entitled "Combined Sewer Overflow (CSO) Control Policy." This policy establishes a consistent national approach for controlling discharges from CSOs to the Nation's waters through the National Pollutant Discharge Elimination System (NPDES) permit program. Contained in the Policy are provisions for developing appropriate, site-specific NPDES permit requirements for all combined sewer systems (CSS) that overflow as a result of wet weather events.

Building sewers, or **house connections**, are pipes carrying wastewater from the plumbing systems of buildings to a sewer or disposal plant. In urban areas, the flow goes to a **common sewer**, which serves abutting property. This conduit may be a **lateral**, one that receives wastewater only from house sewers. A **submain**, or **branch sewer**, takes the flow from two or more laterals. A **main**, or **trunk sewer**, handles the flow from two or more submains or a submain plus laterals. An **outfall** sewer extends from the end of a collection system or to a treatment plant disposal point.

An **intercepting sewer** receives dry-weather flow and specific, limited quantities of storm water from several combined sewers. A **storm-overflow sewer** carries storm-flow excess from a main or intercepting sewer to an independent outlet.

A **relief sewer** is one built to relieve an existing sewer with inadequate capacity.

Usually, domestic-wastewater or storm-water flow does not completely fill the conduit. But all sewers may be filled at some time and must be capable of withstanding some hydraulic pressure. Some types are always under pressure. **Force mains** flow full under pressure from a pump. **Inverted siphons**, conduits that dip below the hydraulic grade line, also flow full and under pressure.

22.4 Estimating Wastewater Flow

Before a sewer is designed, the community or area to be served should be studied for the purpose of estimating the type and quantity of flow to be handled. Design should be based on the flow estimated at some future time, 25 to 50 years ahead, or at completion of the development.

The quantity and flow patterns of domestic wastewater are affected principally by population and population increase; population density and density change; water use, water demand, and water consumption; industrial requirements; commercial requirements; expansion of service geographically; groundwater geology of the area; and topography of the area.

The quantity of domestic wastewater, however, generally is less than water consumption since some portion of water used for firefighting, lawn irrigation, street washing, industrial processing, and leakage does not reach the sewer. Some of these losses, however, may be offset by addition of water from private wells, groundwater infiltration, and illegal connections from roof drains and sump pumps. If the community to be served by the sewerage system already exists, the wastewater flow may be estimated from the gallons per capita per day (gcd) of water being consumed. For a planned community, the estimate may be based on the gcd of water being consumed by an existing similar community. Table 22.1 lists reported flows for several large United States cities. Although flow may range from 70 to 130% of water consumption, designers often assume the average flow equal to the average water consumption or, for estimating purposes, 100 to 110 gcd. The peak flow often is several times larger than the average.

The rate of flow of domestic wastewater varies with water use. But short-term fluctuations tend to dampen out inasmuch as there is a time lag from the time of water use to the time the flow reaches the sanitary-sewer mains. Hourly, daily, and seasonal fluctuations, though, affect design of sewers, pumping stations, and treatment plants.

Daily and seasonal variations depend largely on community characteristics. In a residential district, greatest use of water is in the early morning. A pronounced peak usually occurs about 9 A.M. in the laterals. In commercial and industrial districts, where water is used all day, a peak may occur during the day, but it is less pronounced. At the outfall, the peak flow probably will occur about noon. Wherever possible, measure flow in existing sewers and at treatment plants to determine actual variations in flow.

For residences housing families with both spouses working, weekend flows may be higher than weekday flows. Industrial operations of a seasonal nature influence the seasonal average. The seasonal and annual averages often are about equal in May and June. The seasonal average may rise to about 125% of the annual average in late summer and drop to about 90% at winter's end.

Peak flows may exceed 300% of average in laterals and at the treatment plant. Several state health departments require that laterals and submains be designed for a minimum of 400 gcd, including normal infiltration (see below), and main, trunk, and outfall sewers, for a minimum of 250 gcd, including normal infiltration, and any known substantial amounts of industrial waste.

Inflow into Sewers ▪ Water may inflow into a sewer system and service connections from such sources as roof, cellar, yard, area, and foundation drains, cooling-water discharges, drains from springs and swamps, manhole covers, cross connections from storm and combined sewers, catch basins, storm water, surface runoff, and street washes or drainage. Inflow does not include infiltration into sewers.

Infiltration into Sewers ▪ Water may infiltrate sewers through poor joints, cracked pipes, and walls of manholes. Sewers in wet ground with a high water table or close to streambeds will have more infiltration than sewers in other locations. Since infiltration increases the sewage load, it is undesirable. The sewer design should specify joints that will allow little or no infiltration, and the joints should be carefully made in the field.

Some specifications limit infiltration to 500 gal per day per inch diameter per mile. Often, enforcement agency specifications and requirements call for leakage tests. Some states limit the net leakage to 500 gal/day per inch diameter per mile for any section of the system.

22.5 Sewer Design

Before a sanitary-sewer system can be designed, the quantities of wastewater to be handled and the rates of flow must be estimated. This requires a comprehensive study of the community or area to be served (Art. 22.4). Then, a preliminary layout of the sewerage can be made. Also, pipe sizes, slopes, and depths below grade can be tentatively selected. Preliminary drawings should include a plan of the proposed system and show, in elevation and plan, location of roads, streets, water courses, buildings, basements, underground utilities, and geology. In addition, construction costs should be estimated.

After the preliminary design has been accepted, a survey should locate, in plan and elevation, all

Table 22.1 Municipal Discharges*

City	State	Population served	Design flow mgd	Design, gcd
Bismarck	ND	37,000	4.95	133
Boise	ID	75,000	10	133
Bozeman	MT	21,000	5.2	247
Chicago	IL			
West S.W.		2,900,000	1200	413
Calumet		604,000	310	513
North Side		1,243,330	410	330
Cleveland	OH			
East		819,101	123	150
South		635,000	96	151
Des Moines	IA	201,200	35	174
Detroit	MI	2,400,000	1290	538
Houston	TX			
North Side and 69th		460,000	55	120
Sims Bayou		359,463	48	134
Southwest		173,433	30	173
Indianapolis	IN			
S. Belmont Rd		539,108	120	223
Southport Rd		205,516	57	277
Jacksonville	FL	164,000	17.5	107
Kansas City	MO	418,000	85	203
Los Angeles	CA			
Hyperion		3,000,000	420	140
Terminal Island		115,000	14	122
Minneapolis	MN	434,000	218	502
New York	NY			
Wards Island		1,270,000	180	142
Hunts Point		770,000	150	195
Bowery Bay		725,000	120	166
Tallman's Island		460,000	60	130
Newtown Creek		2,100,000	310	148
Oakwood Beach		105,000	15	143
Oklahoma City (South)	OK	218,900	30	137
Philadelphia	PA			
Northeast		1,240,000	175	141
Southwest		925,000	136	147
Portland	OR	377,800	100	265
Reno, Sparks	NV	110,000	20	182*
Salt Lake City	UT	181,650	45	248
San Francisco	CA			
North Point		353,840	150	424
Richmond, Sunset		220,030	30	136
South East		177,450	37	208
Schenectady	NY	77,985	15.1	194
Seattle (West Point)	WA	494,000	125	253
St. Louis	MO			

(Table continued)

Table 22.1 Continued

City	State	Population served	Design flow mgd	Design, gcd
Le May		849,783	240	282
Bistle Point		988,357	251	254
Washington	DC	1,780,000	240	135
Wichita	KS	275,000	45	164

*From Computer Run 1974: National Water Quality Inventory, app. C, vol. II, Office of Water Planning and Standards, EPA 440/9-74-001.

existing structures and underground utilities that may affect the design. Preferably, borings should be taken to determine soil characteristics along the alignment and at sites for structures in the system. Physical characteristics of the area, including contours, should be shown on a topographic map. Scale may be 1 in to 200 ft, unless the number of details requires a larger scale. Contours at 5- or 10-ft intervals usually are satisfactory. Elevations of streets should be noted at intersections and abrupt changes in grade.

Sufficient depth of soil cover is necessary to prevent damage from traffic loads. Also, the sewers must be below the frost level. Municipal and state regulations on cover should always be reviewed before a design for a specific location is undertaken.

The location of the sewers should be shown in elevation on profiles. Horizontal scale may be 1 in in 40 ft or 1 in in 100 ft, depending on the amount of detail. Vertical scale generally is 10 times the horizontal.

The final design should include a general map of the whole area showing location of all sewers and underground utilities and the drainage areas; detailed plans and profiles of sewers showing ground levels, sizes of pipe and slopes, and location of appurtenances; detailed plans of all appurtenances and structures; a complete report with necessary charts and tables to make clear the exact nature of the project; complete specifications; and a confidential estimate of costs for the owner or agency responsible for the project.

Extensive plans require tabulation of data beginning at the upper end of the system and proceeding downstream from manhole to manhole. The addition to flow from connecting sewers should be included.

Approval of a supervising government agency, such as a county, parish, city, or state agency,

usually must be obtained for the plans. Sewer designers should be familiar with requirements for sewers in the locale in which work is to be done.

Design Flows ▪ Unless force mains are required because sewage must be pumped, or inverted siphons are necessary because of a drop in terrain or encounters with obstacles, sewers usually are sized for open-channel flow. Maximum flow occurs when a conduit is not completely full. For example, for a circular pipe, maximum discharge takes place at about 0.9 of the total depth of the section. Sewers, however, should be designed to withstand some hydraulic pressure.

For storm sewers, common practice is to permit pipe to carry design flow at full depth. Sanitary sewers should be designed to carry peak design flow with a depth from half full for the smallest sewers to full for the larger sewers. For example, sewers under 15 in in diameter are usually designed to flow half full during peak flow periods, whereas sewers from 15 to 60 in in diameter may be designed to flow three-quarters full and sewers larger than 60 in, to flow full. Laterals may be designed for ultimate flow of the area to be served. Submains may be designed for 10 to 40 years ahead. Trunk sewers may be planned for long periods, with provision made in design for parallel or separate routings of trunks of smaller size to be constructed as the need arises. Appurtenances may have a different life since replacement of mechanical equipment will be necessary. Usually, they are designed for 20 to 25 years ahead, and a timetable of additions during that period is then scheduled in an overall improvement plan.

In general, flow may be assumed uniform in straight sewers. Velocity changes, however, will occur at obstacles and changes in sewer cross section and should be considered in making hydraulic computations.

Velocity Formulas ▪ Velocity of flow, ft/s, in straight sewers without obstructions may be estimated with satisfactory accuracy from the Manning formula

$$V = \frac{C}{n} R^{2/3} S^{1/2} \qquad (22.1)$$

where n = coefficient dependent on roughness of conduit surface

R = hydraulic radius, ft = area, ft^2, of fluid divided by wetted perimeter, ft

S = energy loss, ft/ft of conduit length; approximately the slope of the conduit invert for uniform flow

C = 1.486 (conversion factor to account for change from metric units used in development of the formula)

(See also Art. 21.9.) A common value for n is 0.013, suitable for well-laid brickwork, smooth concrete pipe, and vitrified clay pipe with liner plates. For vitrified clay pipes without liner plates, and plastic and resin-lined pipes, 0.011 may be used for n for design purposes. For corrugated-metal pipe, n may range from 0.011 for a spun asphalt lining to 0.02 for the plain pipe or pipe with a paved invert. Smaller values of n than the preceding may be used for new smooth pipe, but the roughness, and value of n, is likely to increase with age. The quantity of flow, ft^3/s, is given by

$$Q = AV \qquad (22.2)$$

where A = cross-sectional area of flow, ft^2.

Minimum Velocity ▪ Velocity should be at least 2 ft/s, and preferably 2.5 ft/s, in sanitary sewers to prevent settlement of solids. Slopes and cross sections of sewers should be chosen to achieve this or a larger velocity for design flows. Where sewers are sized for lower velocities than recommended minimums, provision for flushing and removal of obstructions should be made in the design.

Slopes ▪ Pipe slopes generally should exceed the minimum desirable for maintaining minimum velocity for design flow since actual flows, especially before a development reaches its ultimate size, may be much smaller than design flow. Actual velocity then may be less than the self-cleaning velocity. For example, suppose a circular pipe is sized and sloped to handle design flow when flowing full at 3 ft/s. This velocity will also be maintained when the pipe is flowing half full to full. But if the depth of flow drops to one-third the diameter, the velocity will decrease to about 2.4 ft/s; and at a depth 0.2 of the diameter, velocity declines to about 1.8 ft/s.

Table 22.2 gives the hydraulic characteristics of circular pipe. It enables the quantity and velocity of flow to be computed for a circular pipe flowing partly full, when the respective values for the pipe flowing full are known. The quantity, ft^3/s, for flow full may be estimated from

$$Q = \frac{0.463}{n} d^{8/3} S^{1/2} \qquad (22.3)$$

and the velocity for flow full from

$$V = \frac{0.59}{n} d^{2/3} S^{1/2} \qquad (22.4)$$

where d = inside diameter of pipe, ft.

Table 22.3 lists the quantities and slopes given by these formulas for various velocities and diameters. Information such as that in Tables 22.1 to 22.3 may be stored in computer memories for design use. Programs for application of the data are available commercially.

Minimum Pipe Size ▪ In many cities, 8 in is the minimum diameter of sewer permitted, and in large cities and metropolitan areas 10 in may be the minimum. In any case, pipe smaller than 6 in in diameter should not be used because of the possibility of stoppages.

Maximum Velocities ▪ High velocities in sewers also should be avoided because the solids carried in the flow may erode the conduit. A usual upper limit for sanitary sewers is 10 ft/s. For velocities in that range, though, lining at least the lower portion of the sewers with abrasion-resistant material, such as vitrified-clay blocks, is advisable.

Energy Losses ▪ The assumption of uniform, open-channel flow in sewer design implies that the hydraulic grade line, or water surface, will parallel the sewer invert. This may quite often be true. But where conditions exist that change the slope of the

Table 22.2 Hydraulic Characteristics of a Circular Pipe

Depth of flow / Inside diameter	Partial area / Total area	Quantity, ft³/s, partly full / Quantity, ft³/s, flowing full	Velocity partly full / Velocity flowing full
0	0	0	0
0.05	0.019	0.005	0.25
0.10	0.052	0.021	0.40
0.15	0.094	0.049	0.52
0.20	0.143	0.088	0.62
0.25	0.196	0.137	0.70
0.30	0.252	0.195	0.77
0.35	0.312	0.262	0.84
0.40	0.374	0.336	0.90
0.45	0.437	0.416	0.95
0.50	0.500	0.500	1.00
0.60	0.627	0.671	1.07
0.70	0.748	0.837	1.12
0.80	0.858	0.977	1.14
0.90	0.950	1.067	1.12
0.95	0.982	1.075	1.09
1.00	1.000	1.000	1.00

water surface, the carrying capacity of the sewer will change, regardless of the constancy of the invert slope. This should be taken into account in hydraulic computations for flow near intersections of large sewers, any structure combining the flow from two or more sources, interchange of velocity and pressure head, and submerged outlets at outfalls.

In curved sewer lines, allowance must be made for larger energy losses than in straight sewers. The energy losses may be determined by application of formulas found in references on hydraulics.

To account for the energy loss due to change in direction of sewers at manholes, the invert in the manhole may be dropped about 0.04 ft. If the sewer increases in size at the manhole, the design-flow-depth points of the pipes may be set at the same elevation. The invert drop also may offset head losses due to size changes. Thus, it reduces the danger of the flow backing up and building up pressure. If the sewer size decreases at the manhole, pipe invert elevations may be kept the same.

Sewer Shapes ▪ In selection of a sewer shape, designers sometimes favor one that permits higher velocities at both small and large flows. For example, an egg shape, with the small end down, offers a rapidly decreasing cross-sectional area for decreasing flows. Since, for a given quantity of flow, velocity is inversely proportional to area, velocity in an egg shape does not fall off so rapidly with decreasing flow as in other shapes. But cost of constructing such curved sections may be higher than that for simpler shapes. Often, a compromise shape is chosen, one that has favorable hydraulic characteristics and relatively low cost.

For this reason, circular sewers generally are used, especially for prefabricated conduit. This shape provides the maximum cross-sectional area for the volume of material in the wall and has fair hydraulic properties (Table 22.2). But because of the roundness, there is added cost in bedding circular pipe compared with shapes with a flat bottom.

Figure 22.1 shows some typical shapes that have been used for large reinforced concrete sewers. The inverts usually are curved or incorporate a *cunette*, or small channel, to concentrate small flows to obtain desirable velocities.

Sewer Materials ▪ Sewers should be made of materials resistant to corrosion and abrasion and with sufficient strength to resist hydraulic pressure,

Table 22.3 Quantities, Velocities, and Slopes for Circular Sewers, Flowing Full*

Dia, in		Velocity, ft/s						
		2.0	3.0	4.0	5.0	6.0	7.0	8.0
8	$Q^\ddagger$	0.70	1.1	1.4	1.8	2.1	2.4	2.8
	$S^\ddagger$	3.3	7.5	13.3	20.8	30.0	40.7	53.2
10	Q	1.1	1.6	2.2	2.7	3.3	3.8	4.4
	S	2.5	5.6	9.9	15.5	22.3	30.3	39.6
12	Q	1.6	2.4	3.1	3.9	4.7	5.5	6.3
	S	1.9	4.4	7.8	12.1	17.5	23.8	31.0
15	Q	2.5	3.7	4.9	6.1	7.4	8.6	9.8
	S	1.4	3.2	5.8	9.0	13.0	17.8	23.0
18	Q	3.5	5.3	7.1	8.8	10.6	12.4	14.2
	S	1.1	2.5	4.5	7.1	10.1	13.8	18.1
21	Q	4.8	7.2	9.6	12.0	14.4	17.8	19.2
	S	0.92	2.1	3.7	5.8	8.3	11.3	14.7
24	Q	6.3	9.4	12.6	15.7	18.8	22.0	25.2
	S	0.77	1.7	3.1	4.8	7.0	9.5	12.4
27	Q	8.0	11.9	15.9	19.9	23.9	27.9	31.9
	S	0.66	1.5	2.6	4.1	5.9	8.1	10.5
30	Q	9.8	14.7	19.6	24.5	29.4	34.4	39.3
	S	0.57	1.3	2.3	3.6	5.2	7.0	9.2
33	Q	11.9	17.8	23.8	29.7	35.7	41.7	47.6
	S	0.50	1.1	2.0	3.1	4.5	6.2	8.1
36	Q	14.1	21.2	28.3	35.4	32.4	49.5	56.6
	S	0.45	1.1	1.8	2.8	4.0	5.5	7.2
42	Q	19.2	28.9	38.4	48.1	57.7	67.3	76.9
	S	0.36	0.82	1.5	2.3	3.3	4.5	5.8
48	Q	25.2	37.7	50.3	62.8	75.4	88.0	101
	S	0.30	0.68	1.2	1.9	2.7	3.7	4.9
54	Q	31.8	47.7	63.6	79.5	95.4	111	127
	S	0.26	0.59	1.0	1.6	2.4	3.2	4.2
60	Q	39.2	58.8	78.5	98.1	118	137	157
	S	0.23	0.51	0.90	1.4	2.0	2.8	3.6
66	Q	47.6	71.3	95.1	119	143	166	190
	S	0.20	0.45	0.80	1.2	1.8	2.4	3.2
72	Q	56.5	84.7	113	141	170	198	226
	S	0.17	0.40	0.71	1.1	1.6	2.2	2.8
78	Q	66.4	99.5	133	166	199	232	266
	S	0.16	0.36	0.64	0.99	1.4	2.0	2.5
84	Q	77.0	115	154	192	231	270	308
	S	0.14	0.33	0.58	0.91	1.3	1.8	2.3
90	Q	88.4	133	177	221	265	309	353
	S	0.13	0.30	0.53	0.83	1.2	1.6	2.1
96	Q	101	151	201	252	302	352	402
	S	0.12	0.27	0.48	0.76	1.1	1.5	1.9
108	Q	127	191	254	318	381	444	508
	S	0.10	0.23	0.41	0.64	0.93	1.3	1.7
120	Q	157	236	314	392	471	549	628
	S	0.09	0.20	0.36	0.56	0.81	1.1	1.5

*From Manning formula [Eqs. (22.3) and (22.4)] for $n = 0.013$. For other values of n, multiply slopes given in the table by $n/0.013$; multiply quantities and velocities by $0.013/n$. Velocities less than 2 ft/s are not recommended.

[†]Q = quantity of flow ft^3/s.

[‡]S = slope, ft/1000 ft.

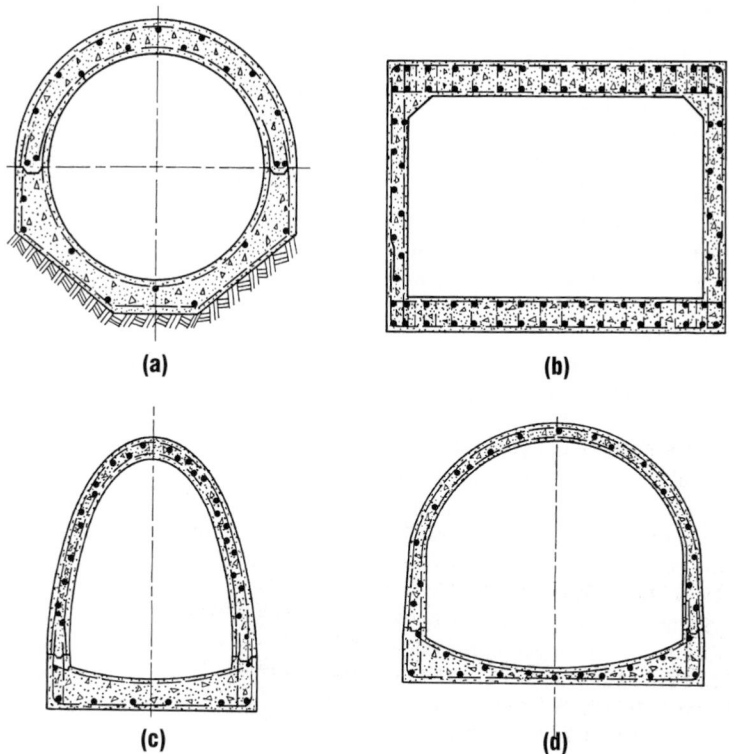

(a)

(b)

(c)

(d)

Fig. 22.1 Some shapes used for large reinforced concrete sewers: (*a*) Circular; (*b*) rectangular; (*c*) semi-elliptical; (*d*) horseshoe.

handling, and earth and traffic loads with economy. Materials meeting these requirements include vitrified clay, reinforced concrete, cast iron, galvanized iron, brick, asbestos-cement, coated steel, bituminized fiber, and plastics formulated for the purpose. Sewer pipe is covered by Federal standards and specifications of the American Public Works Association.

Loads on Sewers ▪ Sewers must be designed with adequate strength to withstand superimposed loads without crushing, collapsing, or through cracks. Usually, the loads are produced by earth pressure or loads transmitted through earth and may be assumed to be uniformly distributed.

Vertical earth loads on sewers may be estimated as indicated in Art. 7.28. Stresses in large sewers may be computed by elastic theory, and the sewers

can be sized to resist these stresses. Standard culverts and sewer pipe generally may be selected with the aid of allowable-load tables prepared by the manufacturers.

(G. M. Fair, J. C. Geyer, and D. A. Okun, "Elements of Water Supply and Wastewater Disposal," John Wiley & Sons, Inc., New York (www.wiley.com); Metcalf & Eddy, Inc., "Wastewater Engineering," 3rd ed., T. J. McGhee, "Water Supply and Sewerage," 6th ed., and H. W. King and E. F. Brater, "Handbook of Hydraulics," McGraw-Hill Publishing Company, New York (www.books.mcgraw-hill.com); "Design and Construction of Sanitary and Storm Sewers," and "Gravity Sanitary Sewer Design and Construction," Manuals and Reports on Engineering Practice, No. 37 and 60, respectively, American Society of Civil Engineers (www.asce.org).)

22.6   Storm-Water Inlets

An inlet is an opening in a gutter or curb for passing storm-water runoff to a drain. In urban areas, inlets usually are positioned at street intersections to remove storm water before it reaches pedestrian crossings and so that water is never required to cross over the street crown to reach an inlet. If the distance between intersections is more than 500 ft long, an inlet may be placed near the midpoint. Along rural highways, inlets generally are installed at low points. Spacings generally range from 300 ft for flat terrain and expressways to 600 ft. Often, however, the capacity of an inlet is increased by permitting some of the water to flow past to an inlet at a lower level. A common practice is to provide three inlets in each sag vertical curve, one at the low point and one at each side of it where gutter elevation is about 0.2 ft higher. Several inlets also are necessary to reduce pondage where the drainage area would be too large for a single inlet in a valley.

Flow through an inlet is directed by a concrete or masonry enclosure to a pipe at the bottom (Fig. 22.2). The size of the enclosure generally is determined by the inlet length, which in turn is determined by the quantity of runoff to be drained, depth of water in the gutter at the inlet, and slope of the gutter. Runoff quantity can be estimated by use of the rational formula [see Sec. 21.41.1].

An inlet may be a curb opening, a gutter grating, or a combination of the two. Capacity of the curb-opening type when diverting 100% of gutter flow may be computed from

$$Q = 0.7L(a + y)^{3/2} \qquad (22.5)$$

where Q = quantity of runoff, ft^3/s

L = length of opening, ft

a = depression in curb inlet, ft

y = depth of flow at inlet, ft

In practice, the gutter may be depressed up to 5 in below the normal gutter line along the length of the inlet. Slope of the gutter commonly is 1 in 12 in. The depression may extend up to 3 ft from the curb. Depth of flow in the gutter may be estimated from the Manning formula.

Grate inlets should be placed with bars parallel to the flow. Length of opening should be at least 18 in to allow the flow to fall clear of the downstream end of the slot. For depths of flow

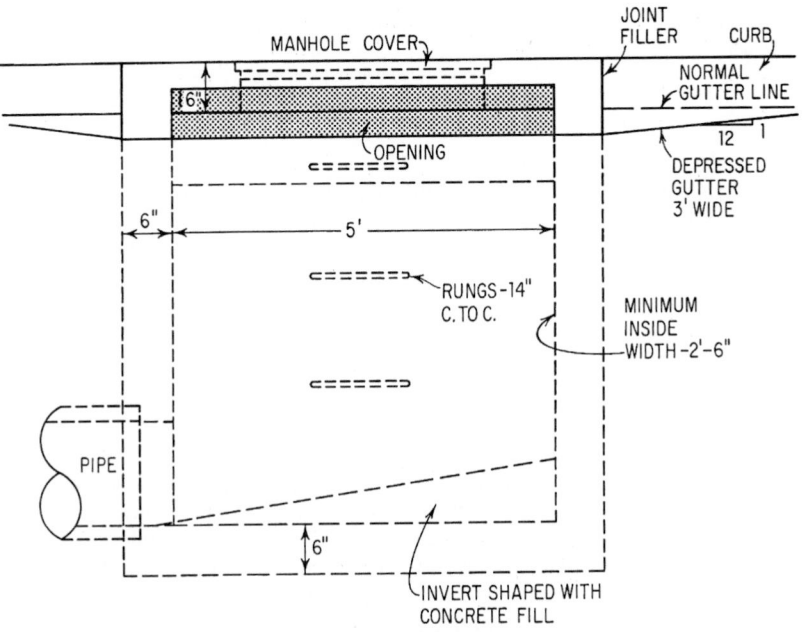

Fig. 22.2   Storm-water inlet with opening in a curb.

up to 0.4 ft capacity of inlet may be calculated from the weir formula

$$Q = 3Py^{3/2} \qquad (22.6)$$

where P = perimeter, ft, of grate opening over which water may flow, ignoring the bars. For depths of flow greater than 1.4 ft, capacity may be computed from the orifice formula

$$Q = 0.6A\sqrt{2gy} \qquad (22.7)$$

where A = total area of clear opening, ft^2
g = acceleration due to gravity, 32 ft/s^2

At depths between 0.4 and 1.4 ft, neither formula may be applicable because of turbulence. A rough estimate may be made by using the smaller of the values of Q obtained from Eqs. (22.6) and (22.7).

Combination inlets are desirable, especially at low points, because the curb opening provides relief from flooding if the gate becomes clogged. If the gutter grate is efficient, the combination inlet will have a capacity only slightly greater than a similar inlet with grate alone. Hence, only the grate capacity should be depended on in designing a combination inlet.

Catch basins (Fig. 22.3) are inlets with enclosures that permit debris to settle out before the water enters the drain. With good drain grades and careful construction, however, catch basins are unnecessary because flow will be adequate to prevent debris from clogging the drain. Also, since water trapped in catch basins may permit mosquitoes to hatch and may be a source of bad odors, simple inlets are preferable. Furthermore, catch basins are more expensive to maintain because they must be cleaned frequently.

("Design and Construction of Sanitary and Storm Sewers," Manual 37, American Society of Civil Engineers (www.asce.org); Metcalf & Eddy, Inc., "Wastewater Engineering," 3rd ed., McGraw-

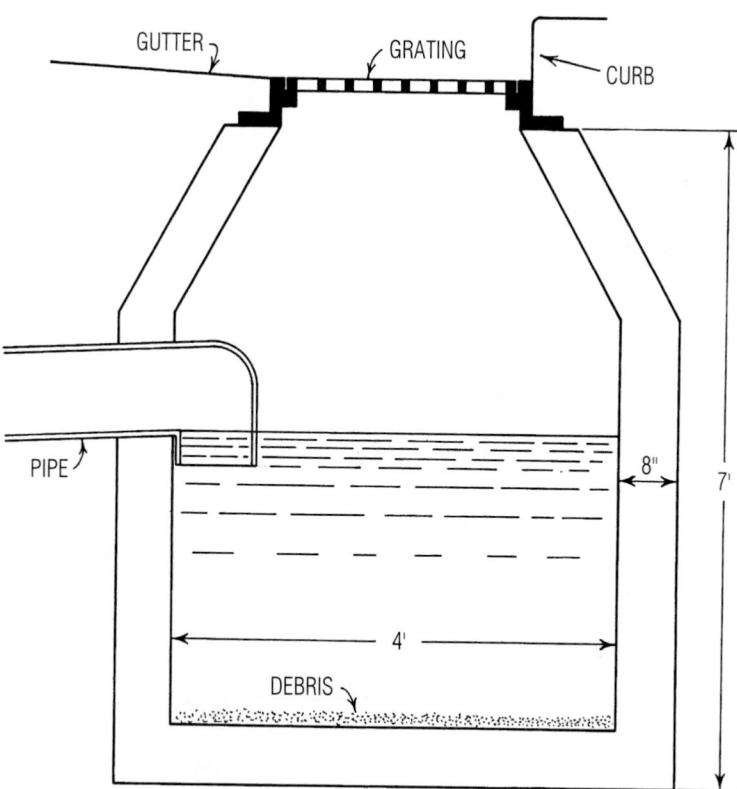

Fig. 22.3 Catch basin with grating inlet in a gutter.

Hill Publishing Company, New York (books. mcgraw-hill.com); G. M. Fair et al., "Water and Wastewater Engineering," John Wiley & Sons, Inc., New York (www.wiley.com).)

22.7 Manholes

A manhole is a concrete or masonry enclosure for providing access to a sewer. The lower portion usually is cylindrical, with an inside diameter of at least 4 ft to allow adequate space for workers. The upper portion generally tapers to the opening to the street. About 2 ft in diameter, the opening is capped with a heavy cast-iron cover seated on a cast-iron frame. Figure 22.4a shows a typical manhole for sewers up to about 60 in in diameter, and Fig. 22.4b shows one type used for larger sewers.

Sewers are interrupted at manholes to permit inspection and cleaning. The flow passes through the manholes in channels at the bottom. Stainless steel or plastic-coated rungs on the manhole walls enable workers to climb down to the sewers.

For sewers up to about 60 in in diameter, manholes are spaced 300 to 600 ft apart. They also are placed where sewers intersect or where there is a significant change in direction, grade, or pipe size. Since workers can walk through larger sewers, manholes for these may be spaced farther apart.

Drop manholes are used where one sewer joins another several feet below. The lower sewer enters the manhole at the bottom in the usual manner. The upper sewer, however, turns down sharply just outside the manhole and enters it at the bottom, where a channel feeds the flow to the main channel. To permit cleaning of the upper sewer from the manhole, the upper sewer also extends to the manhole at constant slope past the sharp drop through which the wastewater flows. Although the upper sewer could be brought down to the lower one more gradually, use of the drop manhole permits a more reasonable slope and thus saves

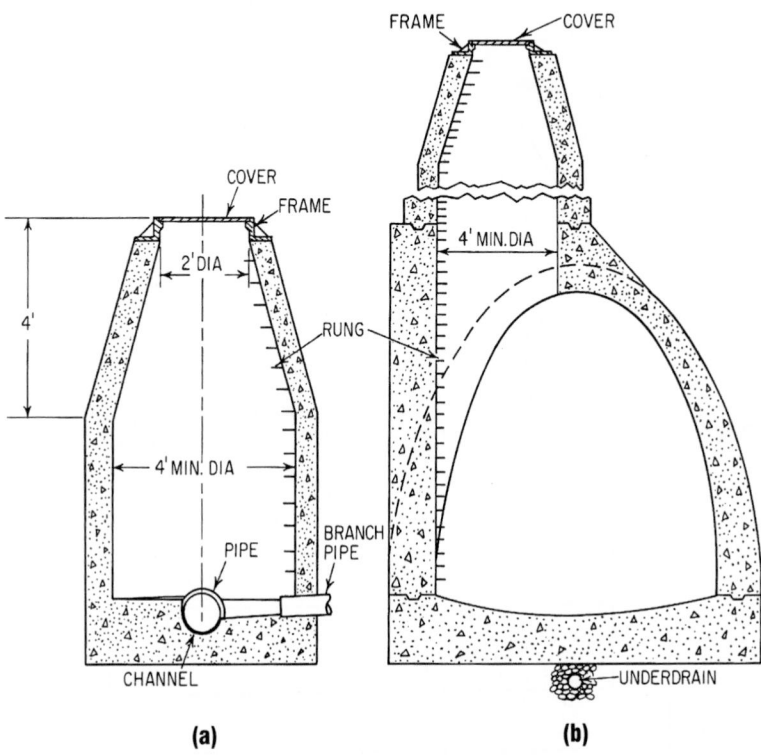

Fig. 22.4 Concrete manholes: (a) For sewers under 60 in in diameter; (b) for larger sewers.

considerable excavation. If the drop is less than 2 ft, however, a steeper slope for the upper sewer is usually more economical.

Where a large quantity of sewage must be dropped a long distance, a wellhole may be used. The fall may be broken by staggered horizontal plates in the shaft or by a well or sump at the bottom from which the sewage overflows to a lower level sewer. In a flight sewer, concrete steps break the fall.

Most street and highway departments and departments of public works have standard plans for manholes. Use of deep manholes, however, is decreasing, because of objections of maintenance workers. Deep manholes must include intermediate landings or other means for safe ingress and egress, in conformance with OSHA safety codes.

("Gravity Sanitary Sewer Design and Construction," Manual and Report on Engineering Practice No. 60, American Society of Civil Engineers; R. A. Corbitt, "Standard Handbook of Environmental Engineering," McGraw-Hill Publishing Company, New York (books.mcgraw-hill.com).)

22.8 Sewer Outfalls

Type of outfall depends on quantity of sewage to be discharged, degree of treatment of the sewage, and characteristics of the disposal source. The outlet should be located to avoid pollution of water supplies and creation of a nuisance. Submerged outlets away from shore are preferable to discharge along a shore or bank, which may create an unsightly appearance and odors. Currents should be strong enough to prevent buildup of sludge near the outlet. It should be protected against scour by its location or suitable construction. A flap valve or automatically closing gate is desirable at the outlet to prevent entrance of water into the sewer during highwater stages.

Outfalls in tidal waters require special investigations to ensure suitable dispersion of the wastewater and to avoid floating wastes at the water surface. These outfalls often are constructed with a multiple discharge at the end, thus spreading the effluent over a large area and through a large volume of water. Depth of water over the outfall must be sufficient to accomplish dispersion before currents can transport the concentrated effluent streams shoreward, over shellfish beds, or into shallow water.

The outfall may be laid on the bottom. For protection against waves and scour, the pipe may be set in a trench or between two rows of piles and securely anchored.

An outlet discharging treated wastewater into a small stream should be protected, by a concrete head wall and a concrete apron on the bank, against undercutting by the flow of the stream or wastewater. A similarly protected outlet may be used at a river bank for storm-water discharge.

(G. M. Fair et al., "Water and Wastewater Engineering," John Wiley & Sons, Inc., New York (www.wiley.com).)

22.9 Inverted Siphons (Sag Pipes)

These are sewers that dip below the hydraulic grade line. They are used to avoid such obstructions as waterways, open-cut railways, subways, and extensive utility piping and structures. After passing under an obstruction, the pipe is brought to grade to permit open-channel flow in the continuation, to keep down the amount of cut and thus the cost of installing the sewer. The portion of the sewer below the hydraulic grade line flows full under pressure. Hence, it must have tight joints and be made of a suitable material and it must be designed for the maximum expected pressure.

To prevent solids from being deposited and obstructing an inverted siphon, it should be sized and sloped to keep flow velocities as much above 3 ft/s as feasible. Although experience has been good with a single pipe, 12 to 24 in in diameter, carrying flows with such velocities, a pipe big enough to handle the maximum flow at an adequate velocity may carry small flows at undesirably low speeds. In that case, two or more parallel pipes may be used instead of a single pipe.

An inlet chamber is constructed at the upstream end of the inverted siphon and an outlet chamber at the downstream end (Fig. 22.5). These chambers may be concrete enclosures, which may be entered through manholes extending to grade. The inlet chamber for a multiple-pipe inverted siphon usually incorporates flow-regulating devices to control the flow to each pipe. As a safety measure, the inlet chamber may also incorporate a bypass, or overflow, pipe to relieve the inlet should the inverted siphon be overloaded or obstructed. In the

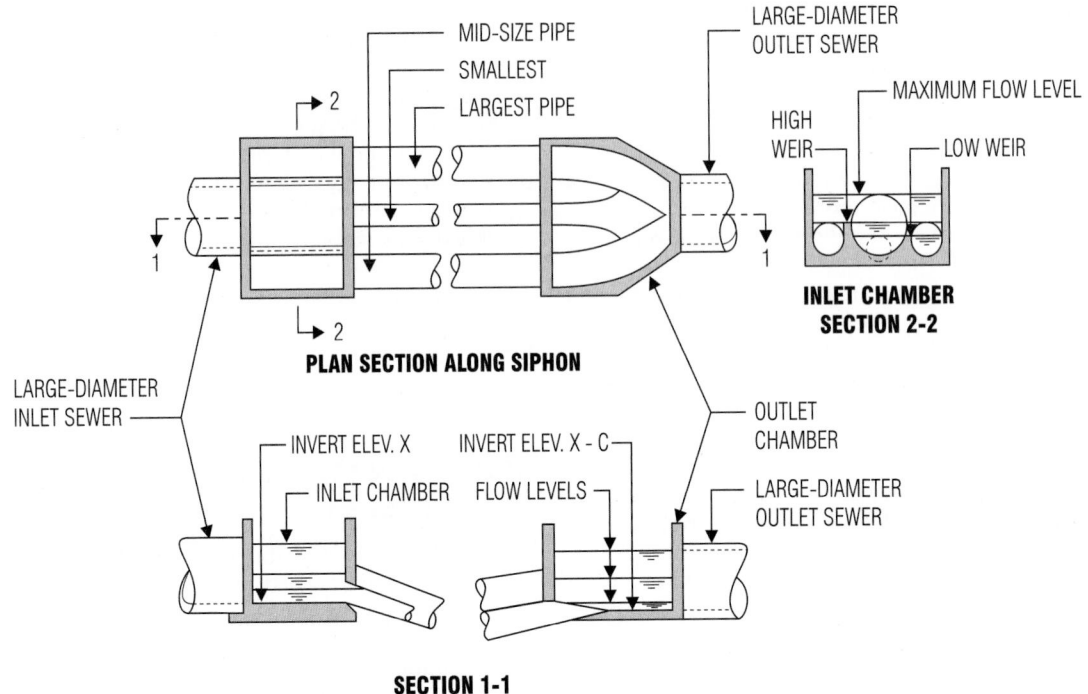

Fig. 22.5 Three-pipe inverted siphon.

outlet chamber, the inverts of the pipes merge into a single channel, which becomes the invert of the continuing sewer. Provision should be made in the chambers for cleaning and repairing for pipes and for draining them for these purposes. The designer should always investigate the hydraulic heads required in the inlet chamber to avoid surcharge on the upstream pipes.

A three-pipe system may be used for a large sewer. As indicated in Fig. 22.5, the smallest pipe may be assigned the minimum dry-weather flow; a larger pipe, the excess up to a specified percentage of the maximum flow; and the largest pipe, the remainder of the flow. Built-in weirs may be used to regulate the flow to each pipe.

For large sewers, where the venting of the air entrapped upstream of the siphon would be undesirable because of odor, another pipeline may be required to transfer the air to the downstream siphon manhole. The pipeline transferring the entrapped air should be about one-fourth the diameter of the siphon pipeline and may span or undercross the obstacle that necessitated the siphon.

(T. J. McGhee, "Water Supply and Sewerage," 6th ed., McGraw-Hill Publishing Company, New York (books.mcgraw-hill.com).)

22.10 Flow-Regulating Devices in Sewers

Sewerage systems often require some means for controlling flow, such as weirs, spillway siphons, and gates and valves. The devices may be used to divert flow from one conduit to another or to distribute flow among several pipes.

Side Weirs ▪ A simple device for such an application is a side weir, an overflow weir along the side wall of the sewer (Fig. 22.6a). Diversion,

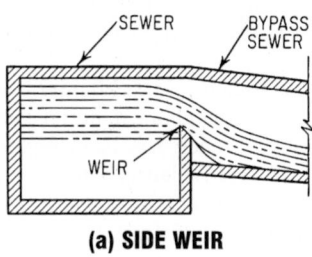

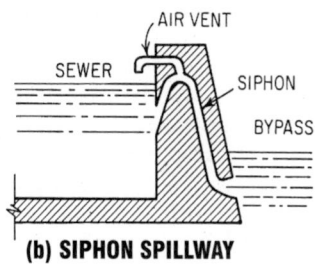

(a) SIDE WEIR

(b) SIPHON SPILLWAY

Fig. 22.6 Flow-regulating devices for sewers.

ft^3/s, may be estimated from the Engels formula:

$$Q = 3.32l^{0.83}h^{1.67} \qquad (22.8)$$

where l = length of weir, ft
 h = depth of flow over weir at downstream end, ft

Siphon Spillways ▪ Although simple to construct, side weirs may not control flow as closely as desired. Siphon spillways (Fig. 22.6*b*) are more effective, especially for large flows. The outlet may be placed considerably below the inlet (differences in elevation up to 33.9 ft at sea level under standard atmospheric conditions may be used). Siphons operate under higher heads than weirs and permit much larger flows. Control is better because siphons can be constructed to start or stop discharge at any desired depth of flow in the combined sewer.

Area, ft², of the siphon throat can be determined from

$$A = \frac{Q}{c\sqrt{2gh}} \qquad (22.9)$$

where Q = discharge, ft^3/s

 c = coefficient of discharge, which varies from 0.6 to 0.8

 g = acceleration due to gravity = 32.2 ft/s^2

 h = head, ft

For proper operation, the air vent should have an area of about $A/24$. The siphon inlet should be shaped to minimize entrance losses. The outlet should be completely submerged or sealed by the discharge.

Gates and Valves ▪ Diversion of flow also may be accomplished with float-actuated gates and valves. For example, low flows may be permitted to reach an outfall through an opening controlled by a gate. When water reaches a predetermined level in a float chamber, a float valve closes the gate to divert the water to a bypass. The designer must provide access to diversion chambers for cleaning since debris carried into the sewer will fill in the channel, clog the openings, and otherwise defeat the purpose of the flow-regulating device.

22.11 Sewer-Construction Methods

Sewers usually are placed in trenches, but occasionally sewers may be constructed or installed in tunnels or laid at grade and covered with embankment.

In trench construction, the sewer line is located with respect to an offset line, laid out with survey instruments sufficiently far away to avoid disturbance. The trench then is marked or staked out on the ground and excavated. Sewers are generally laid with the use of laser instruments to control grade.

Trench excavation may be done by hand or with powered equipment as described in Sec. 13. In rock, explosives should be avoided or used with great caution to avoid collapsing the trench or damaging nearby structures or utilities.

Experience generally will indicate whether the depth and type of soil require that the sides of the trench be laid back sheeted and braced, or trenching boxes be employed. OSHA safety regulations regarding trenching must be followed. If there is any doubt, sheeting should be used so as not to endanger workers. The sheeting methods described in Art. 7.24 are applicable to trench construction. Unless this is forbidden by specifications to prevent possible failures, the sheeting may be salvaged as backfilling proceeds.

Water may be drained, except in quicksand, by leading it to sumps and pumping it out. Wellpoints may be necessary to prevent quicksand from forming in a sandy trench bottom or to dry out the bottom.

The support for the sewer should be shaped to the conduit bottom, whether the support be the subgrade, a granular fill, or a concrete cradle. In rock, excavation should be carried to a depth of one-fourth the conduit diameter below the bottom of the conduit, but not less than 4 in below. The space between the trench bottom and the conduit should be refilled with $\frac{3}{4}$-in gravel or lean concrete $(1:4\frac{1}{2}:9$ mix), so that at least $120°$ of the pipe will be supported on it.

Pipelaying usually proceeds upgrade. Pipe is laid with bell ends upstream, to receive the spigots of subsequent sections. Invert elevation usually is required to be within 12-in of that specified.

Joints between lengths of pipe usually are calked with a plastic or rubber-compound gasket and a filling of plastic, bitumen, or portland cement mortar (1:1 mix). Resilient joints are preferable to rigid types, which differential settlement may crack.

Feeder sewers come with Y or T stub branches for house sewer connections. If these connections are not made when the feeder is installed, a disk stopper is mortared into the bell of the stubs. Field notes should record the location of each branch so that it can be found when a connection has to be made in the future. The location is usually referenced by stationing to a nearby manhole.

Backfilling should start as soon as possible. Earth should be placed and tamped evenly around the pipe to avoid disturbance of newly made joints and creation of high or unbalanced side pressures on the pipe. Material should be placed in layers not exceeding 6 in in thickness and tamped lightly until the fill covers the top of the pipe. The initial trench backfill around flexible pipe is critical and should be evaluated for each installation. Proper bedding of flexible pipe is required to avoid excessive vertical deflection and subsequent collapse.

The upper portion of the backfill should be heavily tamped, to reduce future settlement, if the surface over the trench is to be paved. Backfill must be carefully placed throughout, and materials that may permit excessive settlement should not be used.

For trenches in fields, the backfill need not be tamped. After all the material previously excavated from the trench has been replaced, the resulting mound may be left to settle naturally.

Large sewers in trenches generally are constructed of reinforced concrete, cast in long, reusable forms. Often, the invert is concreted first. Then, the forms for the upper portion are supported on the hardened invert concrete.

For sewers in tunnels, the methods described in Sec. 20 may be used.

("Gravity Sanitary Sewer Design and Construction," Manual and Report on Engineering Practice No. 60, American Society of Civil Engineers, New York (www.asce.org).)

22.12 Sewer Force Mains

Sewage conveyed under pressure travels in sewer force mains. As with gravity sewers, the velocity in sewer force mains must be sufficient to prevent deposition of grit and other solids. The minimum velocity in a sewer force main should be 2.0 ft/s, with 2.5 ft/s recommended, where possible. Full pipe flow velocity can be determined from

$$V = \frac{Q}{A} \qquad \text{(Form of equation 22.3)} \qquad (22.10)$$

Calculation of pressure in a sewer force main is necessary for proper selection of pipe class and also for pump selection. When calculating pipe pressure, it is necessary first to determine where the pipeline will be running under full pipe flow and where the pipeline will run partially full. If a sewer force main has a continuous positive (uphill) slope from a pumping station to a discharge point, such as a sewer manhole, then the pipe will run full to the discharge point. If a sewer force main has intermediate high and low points, with air release/vacuum relief valves at the high points, then the pipeline could experience a combination of both full and partially full flow. Development of a pipeline hydraulic grade line (HGL) is very useful in analyzing this condition. To visualize HGL, consider if an open top vertical riser pipe were attached to a sewer force main, wastewater would rise to a point in that riser pipe to the sewer force main HGL.

HGL can be calculated as follows. Start at the downstream end of the pipe. The HGL at this point is the elevation of the free water surface. For points upstream, at the design flow Q, compare the pipe slope S_P to the energy loss slope S, calculated from

$$S = \left[\frac{Qn}{0.59Ad^{2/3}}\right]^2 \tag{22.11}$$

(Form of equations 22.3 and 22.5)

Where S_P remains equal to or steeper than S, the pipe is running partially full, and the HGL is the water surface in the pipe. This is shown grapically in Fig. 22.7.

22.13 Friction Loss in Full Pipe Flow

Friction loss in full pipe flow consists of losses from pipe friction, H_P as well as losses from pipe fittings, valves, and similar losses, generally referred to as minor losses, H_M. Pipe friction losses for full pipe flow can be calculated from the Hazen-Williams formula (adapted for circular pipe):

$$S = \left[\frac{V}{1.318C(D/4)^{0.63}}\right]^{1.8519} \tag{22.12}$$

where S = energy loss, ft/ft of pipe length

V = velocity of flow, ft/s (see equation 22.3)

C = coefficient of friction

D = pipe inside diameter, ft

Coefficient of friction, C, varies based on the type of pipe material and its age. Typical values for C are:

130   for new ductile iron pipe
100   for old ductile iron pipe
150   for new PVC pipe
140   for old PVC pipe.

Pipe friction loss, H_P, ft, can be determined for a given length of pipe L, ft, by:

$$H_P = SL \tag{22.13}$$

Minor losses, H_M, ft, are generally a function of the velocity head (see equation 21.14). The following equation is used to calculate minor losses:

$$H_M = K\frac{V^2}{2g} \tag{22.14}$$

where K = minor loss coefficient

V = pipe velocity, ft/s

g = acceleration due to gravity = 32.2 ft/s^2

Minor loss coefficient, K, values are available in various hydraulics texts, including publications by

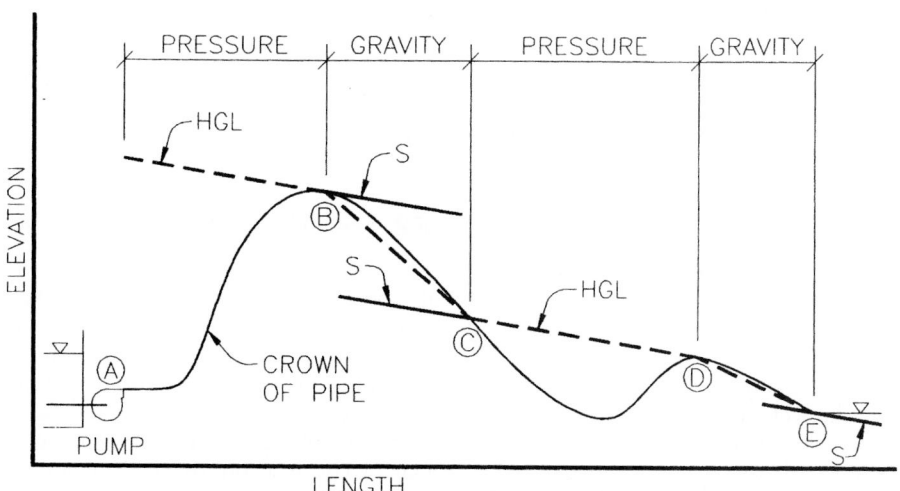

Fig. 22.7   Force main profile with HGL.

the Crane Company, the Hydraulics Institute, and others.

22.14 Pumping Stations for Wastewater

Lift stations are used where it is necessary to pump sewage to a higher level. The installation may be underground or above grade, housed in a building. (For a discussion of sewage pumps, see Art. 22.5.)

Most installations have at least two pumps. One is available as a standby, ready to take over if the first should fail. Main pumping stations should have at least three pumps; with the largest pump out of service, the other two should be able to handle the design flow. Several pumps with different capacities permit flexibility of operation. The smallest pump should be able to handle minimum flow. The others can be brought on-stream in succession as flow increases.

Main pump stations typically have two or more pumps equipped with devices to vary the pump speed, and thus vary the pumped flow rate, to match the rate of flow into the pump station. The most common device for this purpose is the variable frequency drive (VFD), which varies the speed of an alternating current electric pump motor.

A submersible pump, consisting of a pump close coupled to a submersible motor, can be inserted directly into a manhole. Submersible sewage pumps are typically installed with guide rails that allow the pump to be inserted and removed without entering the manhole or wet well. The rails guide the pump discharge onto a base elbow, which is secured to the manhole floor and is connected to the discharge force main pipe.

At a small pumping station, sewage may flow into a manhole or a tank. A pump may be installed in a "dry" compartment alongside the manhole; or a vertical pump, on the roof of the tank (Fig. 22.8). At a large pumping station, sewage flows into a wet well. The pumps may be installed above or in an adjacent dry well.

Often, the pumps operate automatically when the liquid in the wet well reaches a selected level. (See, for example, Fig. 22.7.) The motors may be started and stopped by switches operated by a float rod, which rises and falls with the liquid level; by plastic encapsulated float switches, located within the liquid; by pressure switches located on a tube, with compressed air bubbling through the outlet at the bottom of the wet well; or by an ultrasonic liquid level sensor, located above the liquid. Usually, two sources of electric power are provided to ensure continuity of operation. If no attendants are present in an automatic station, provision should be made for an alarm to be sounded and recorded at a remote location when a pump fails or the liquid level rises above a selected elevation.

Seepage in a dry well should be directed to a sump. It can be drained by one of the sewage pumps or by a special pump. This pump may also have a suction line to the wet well to drain it for cleaning and repair.

The wet well usually is small, to preclude septic action in the sewage; however, the well must be designed to handle maximum flow without flooding. The well should be vented to the outside, to prevent accumulation of odors. It may be divided into two interconnected compartments, which may be isolated for cleaning and repair by closure of a gate.

Pumps, though nonclogging, should be protected against debris in the sewage by a screen. For that purpose, a basket screen may be placed at the entering sewer or a bar screen ahead of the wet well. If the screen clogs frequently, a comminutor may be installed to grind the clogging materials prior to pumping.

(T. J. McGhee, "Water Supply and Sewerage," 6th ed., and Metcalf & Eddy, Inc., "Wastewater Engineering: Collection and Pumping of Wastewater," McGraw-Hill Publishing Company, New York (www.books.mcgraw-hill.com); "Design of Wastewater and Stormwater Pumping Stations," MOP No. FD-4, Water Environmental Federation, Arlington, Va (www.wef.org).)

22.15 Wastewater Pumps

Although wastewater generally flows by gravity through conduit and treatment plants, pumping sometimes is required. Pumping may be the most economical means of conveying wastewater past a hill, or the only way to get wastewater from a cellar to a sewer at a higher level. Where desirable invert slopes would place a sewer far underground, making construction costs high, a more economical method is to raise the wastewater in a pumping plant and then let it flow by gravity. Similarly, pumping may be necessary to give sufficient head

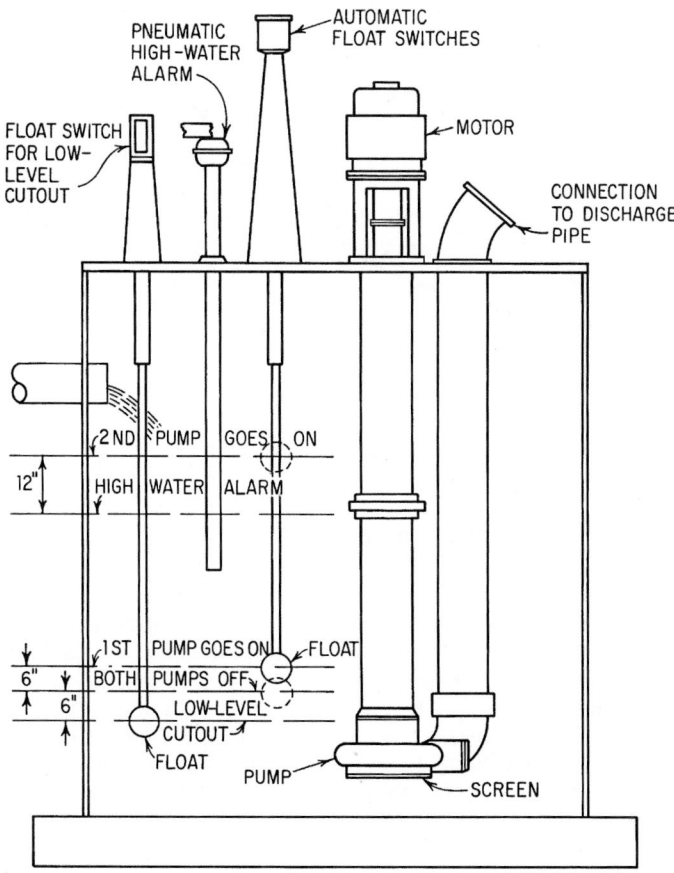

Fig. 22.8 Small automatic wastewater pumping station.

for wastewater to flow by gravity through a treatment plant.

"Nonclogging" centrifugal pumps are generally used. They are capable of passing solids with a maximum size of about 80% of the inside diameter of the pump suction and discharge pipes. These single-suction volute pumps may be bladeless or have two vanes. In some cases, grit chambers may be desirable ahead of pumps, to prevent accelerated wear in the pumps, and bar screens, perhaps mechanically cleaned, may be justified.

The pumps generally are driven by electric motors. Types preferred have a high efficiency over a wide range of operating conditions, but dependability is the most important characteristic. Also,

slow-speed pumps are desirable for long life and less noise.

The shaft of the pump may be horizontal or vertical. Vertical pumps permit installation of motors above the pump pit, where they are less likely to be damaged by floods.

Wastewater ejectors operated by compressed air are an alternative to nonclogging centrifugal pumps. In buildings where compressed air is available, such ejectors may be used as sump pumps.

In a commonly used type of wastewater ejector, the wastewater flows into a storage chamber until it is full. During this stage, air is exhausted from the chamber as the liquid level rises. A float rod closes

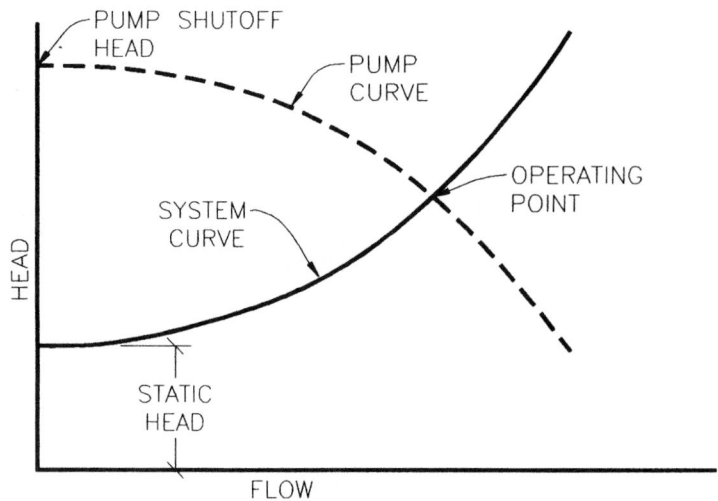

Fig. 22.9 Pump-system curve.

the air exhaust and opens a compressed-air inlet. The compressed air forces the wastewater up the discharge pipe. When the storage chamber is emptied, the float valve shuts the compressed-air valve and opens the air exhaust. Check valves in inlet and discharge pipes prevent back flow.

(T. G. Hicks, "Pump Application Engineering," and I. J. Karassik et al., "Pump Handbook," 2nd ed., McGraw-Hill Publishing Company, New York (www.books.mcgraw-hill.com).) See also Art. 22.

22.16 Work and Pump Efficiency

The head against which a pump works is defined as pumping head, which is calculated as follows:

$$P_H = H_P + H_M + S \qquad (22.15)$$

where P_H = pumping head, ft
H_P, ft (see equation 22.13)
H_M, ft (see equation 22.14)
S = static head, ft

Static head, S, is the elevation of the water surface at the end of the force main or the elevation of the crown (inside top) of the force main where gravity flow commences (whichever governs), minus the water surface elevation at the pump suction. Please refer to Section 22.12 for a discussion of this topic.

In Fig. 22.7 S is the elevation of the crown of the pipe at location B, minus the elevation of the water surface at the pump wet well, at location A. In Fig. 22.7, if the HGL were higher than the crown of pipe throughout the length of the force main, then the elevation of the water surface at the end of the force main would be used in determination of the value of S. This condition would occur if the crown of the force main pipe at the end, location E, were higher than the elevation at location B. The water surface elevation at the end of the force main would be the crown of the pipe, if the force main terminated at a gravity sewer manhole, or would be the pool water surface elevation if the end of the force main were submerged.

22.17 Pump and Pump Motor Selection

To select a pump and pump motor it is necessary to develop a system curve. A system curve is a plot of flow, Q, along the abscissa, versus pumping head, P_H, along the ordinate (see equation 22.15). The system curve is developed by calculating P_H at various flow rates, both below and above the target design pumping flow. For pump selection, pipe friction coefficients, C, are generally selected for old pipe, when developing the system curve. This procedure accounts for additional friction loss as the force main pipe ages. A pump curve is also a plot of head versus flow, which shows the performance of a particular pump. The pump manufacturers furnish pump curves. If the pump curve is superimposed over the system curve, the intersection of the two curves is the operating point of the pump. Figure 22.9 depicts a pump-system curve. Note that at zero flow condition on the system curve, the pump head is equal to the static head. The zero flow condition of the pump curve is known as the pump shutoff head.

A manufacturer furnished pump curve typically shows a series of curves, for various size impellers available for the pump. The curve also shows the pump efficiency, E_P, at various flows. Select a pump/impeller size that produces a design point with a head at or above the design flow, at a pump efficiency at or near the optimum for the pump.

When two identical centrifugal pumps are operating simultaneously, the combined flow rate is typically less than twice the flow rate produced by a single operating pump. To determine the flow rate of multiple pumps in simultaneous operation, superimpose on the system curve a composite pump curve, developed by multiplying the flow rate times the number of pumps in service, for various values of head. The operating point is the intersection of the system curve and the composite pump curve.

The power required to operate a centrifugal pump, P, is calculated as follows:

$$P = \frac{QP_H}{3960E_P} \qquad (22.16)$$

where P = power, HP

Q = flow rate, gpm

P_H = pump head, ft

E_P = pump efficiency, dimensionless

In sizing a pump motor, it is important to calculate the pump power requirement, P, at a condition where the highest power draw will occur. For a centrifugal pump, the power draw generally increases with the flow rate. Referring to Fig. 22.9, on the system curve, the maximum flow, Q, will occur at a condition of minimum head, caused by minimum friction loss, and minimum static head. Generally, the maximum pump power condition is with new pipe, at a maximum pump station wet well water surface elevation. Develop a system

curve with these conditions to identify the operating point, for the selected pump, for maximum pump power draw. Calculate P for the Q and P_H at this operating point, and select a pump motor whose rated power output is equal to or larger than the maximum pump power requirement.

The efficiency of a pumping system, E_S, also known as wire to water efficiency, can be calculated as follows:

$$E_S = E_P E_M \qquad (22.17)$$

where E_S = system efficiency, dimensionless

E_P = pump efficiency, dimensionless

E_M = motor efficiency, dimensionless

Motor efficiencies are available from the motor manufacturers.

22.18 Characteristics of Domestic Wastewater

Usually, wastewater contains less than 0.1% of solid matter. Much of the flow looks like bath or laundry effluent, with garbage, paper, matches, rags, pieces of wood, and feces floating on top. Within a few hours at temperatures above 40°F, wastewater becomes stale. Later, it may become septic, often with the odors of hydrogen sulfide, mercaptans, and other sulfur compounds predominating. The more putrescible compounds there are in wastewater, the greater is its concentration or strength. In general, strength will vary with the amount of organic matter, water consumption per capita, and amount of industrial wastes.

Solids ▪ Total solids in wastewater comprise suspended and dissolved solids. About one-third of the total solids usually are in suspension. Suspended solids are those that can be filtered out. Usually more than half of these solids are organic material.

Suspended solids include settleable solids and colloids. Settleable solids precipitate out in sedimentation tanks in the usual detention periods. Colloids, mostly organic material, are smaller than 0.0001 mm in diameter and can remain in suspension indefinitely. They can pass through filter paper but are retained on a filtering membrane. Elimination of suspended solids from wastewater is desirable because they contain insoluble organic and inorganic pollutants and harbor bacteria and viruses.

Dissolved solids are the residue from evaporation after removal of suspended solids. Excessive dissolved solids can have adverse effects on living things, taste, irrigation, water softness, and water reuse after treatment.

Solids also may be classified as volatile or fixed. The loss of weight when dried solids are burned is attributed to the volatile solids, which are considered to be organic material. The residue comprises fixed solids such as salts, which are assumed to be inorganic.

Organic Content ▪ The organic content of wastewater may be classified as nitrogenous and nonnitrogenous. Principal nitrogenous compounds include proteins, urea, amines, and amino acids. Principal nonnitrogenous compounds comprise soaps, fats, and carbohydrates.

Analyses of Wastewater ▪ Tests are made on wastewater to determine its strength, potential harmful effects in its disposal, and progress made in treating it. The most commonly made tests measure:

Suspended solids

Biochemical oxygen demand (BOD)

Amounts of ammonia, which decrease as wastewater is treated (NH_4^+-N)

Total Kjeldahl Nitrogen (TKN) includes NH_4-N and organic N

Nitrites and nitrates, which increase as wastewater is treated ($NO_2^- + NO_3^-$)-N)

Dissolved oxygen (O.O)

Ether-soluble matter, or fats and greases, which can form a heavy scum (FOG)

pH value, which decreases, indicating greater acidity, as wastewater becomes stale

Chemical oxygen demand, which approximates the total oxidizable carbonaceous content (COD)

Phosphorus, which can stimulate undesirable algae growths in lakes and streams (P)

Heavy metals, such as mercury, silver, and lead, which are toxic

Total organic carbon (TOC), which may be determined in large laboratories or in industrial plants; small laboratories are not equipped with suitable apparatus to run the required test

Chlorine demand, the amount of chlorine added to wastewater to produce a residual after a certain time, usually 15 min

Chlorine Residual, the amount of chlorine in wastewater usually measured at the point of discharge to receiving stream

Bacteria and other microorganisms

Coliform tests are usually required. Fecal coliform tests may be required when effluent is discharged into bathing, drinking waters, or tidal waters.

Bacteria ▪ These may be aerobic, requiring air for survival; anaerobic, thriving without air; or facultative, carrying on with or without air. (An important category of facultative organisms are anoxic organisms which can extract oxygen from nitrate and nitrite ions and are important for the removal of nitrogen from wastewater. Some may be pathogenic, causers of intestinal diseases. If they are present, the effluent may have to be chlorinated or otherwise treated to eliminate such bacteria, depending on the method of disposal.) Bacteria are useful in stabilizing wastewater, breaking it down into substances that do not decompose further.

Anaerobic bacteria are used in sludge digestion, the stabilization of organic material removed from wastewater by sedimentation. Anaerobic stabilization, is more sensitive to environmental conditions, and can produce more disagreeable odors. Because the process is lengthy, it usually is not carried to complete stability but to a stage where further decomposition proceeds slowly. Stabilization is part of a cycle in which the products of decomposition become food for plants, then in turn food for people and animals, and finally are reconverted into wastes (Fig. 22.10a).

Aerobic bacteria serve in self-purification of streams, and suspended growth and attached growth biological processes method of treatment. In streams, oxygen may become available from several sources: absorption of air at the water surface; release by algae, which absorb carbon dioxide and release oxygen; and production by decomposition of such compounds as nitrates. In some attached growth biological processes, oxygen is supplied by allowing wastewater to pass over or through media while air circulates through the voids, in other attached growth processes, the

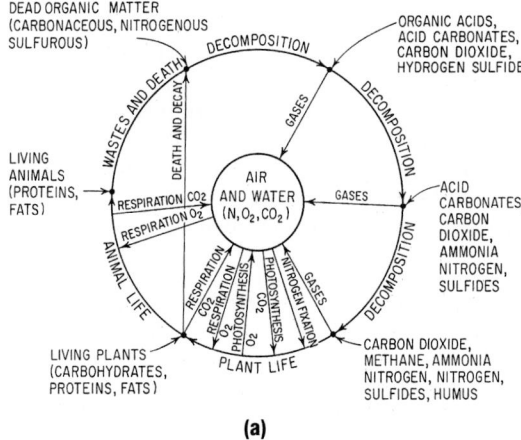

(a)

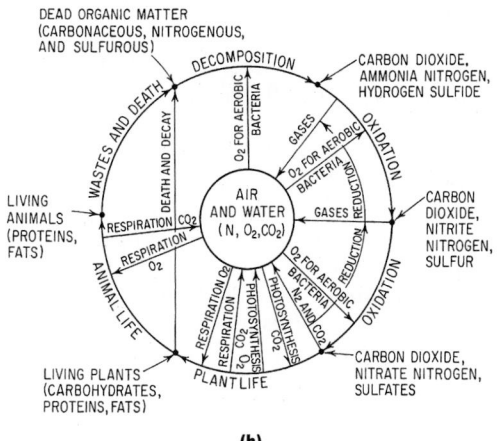

(b)

Fig. 22.10 Carbon nitrogen, and sulfur cycles in (*a*) anaerobic decomposition; (*b*) aerobic decomposition. (*From E. W. Steel and T. J. McGhee, "Water Supply and Sewerage," McGraw-Hill Book Company, New York.*)

media is flooded and air is introduced by an aeration system. In the activated-sludge process, oxygen is furnished by passing air through a mixture of wastewater and previously activated sludge and by strongly agitating the mixture to dissolve air into the liquid. In aerobic stabilization also, decomposition occurs in steps and is part of a cycle (Fig. 22.10*b*). If the supply of oxygen is inadequate, however, anaerobic action will occur and disagreeable odors may be produced.

BOD, CBOD, and COD ▪ The amount of oxygen used during decomposition of organic material is the **biochemical oxygen demand (BOD)**. It is a measure of the amount of biodegradable organic material present. If the BOD of wastewater discharged into a stream or lake exceeds the oxygen content of that water, the oxygen will be used up and the stream or lake will become septic at the discharge area. Fish and aquatic plants cannot survive in such conditions.

BOD is determined by diluting a wastewater sample with water with known dissolved-oxygen content and storing the mixture for 5 days at 20°C. The oxygen content at the end of the period is measured, and the difference is reported as the BOD.

At the end of a period of t days at 20°C

$$BOD = O(1 - 100^{-K_1 t}) \qquad (22.18)$$

where O = oxygen demand when $t = 0$ or at start of any oxidation period

K_1 = deoxygenation coefficient, usually about 0.1 for wastewater, but may range from less than 0.05 to more than 0.2. For temperatures other than 20°C, multiply K_1 for 20°C by 1.047^{T-20}. Regression analyses of laboratory data should be used to determine K_1 for the wastewater being tested.

T = temperature, °C.

To obtain the initial oxygen demand at temperatures other than 20°C, multiply O for 20°C by $0.02T + 0.6$.

The load on a receiving body of water may be estimated from the size of contributing population. For example, the 5-day BOD, lb/day per capita, may be assumed as 0.2 for domestic wastewater, 0.3 for combined wastewater and storm water, and 0.5 for when the combination contains large amounts of industrial wastewater.

Sometimes, the wastewater concentration is expressed as **population equivalent**, the number of persons required to create the total oxygen demand of the wastewater per day. For example, suppose domestic wastewater has a BOD of 5000 lb/day. The population equivalent then may be taken as $5000/0.2 = 25,000$ persons.

As an example of the use of BOD, consider a residential community of 100,000 persons producing a wastewater flow of 25 mgd to be disposed of in a river with no BOD and a dissolved-oxygen content of 10 ppm. Permissible oxygen content downstream is 6.5 ppm. What should the flow in the river be?

The total oxygen demand may be assumed to be $100,000 \times 0.2 = 20,000$ lb/day. Since a gallon of water weighs 8.33 lb, the total demand for 25 mgd of wastewater is equivalent to

$$\frac{20,000}{25 \times 8.33} = 96 \text{ ppm or mg/L}$$

The required river flow Q, mgd, must supply this oxygen. Hence,

$$8.33(10 - 6.5)Q = 20,000 \quad \text{and} \quad Q = 686 \text{ mgd}$$

CBOD is a carbonaceous biochemical oxygen demand. Its similar to BOD, except that the chemicals are added to the test to prevent oxidation of the nitrogenous compounds. This is used to distinguish the oxygen demand associated with the carbon oxidation from that used for oxygenation of the TKN. It is most commonly used as a measure in well treated effluents.

Some of the organic material, such as pesticides, in wastewater may not be biologically degradable. They are not measured by BOD. Some of these materials may have adverse long-term effects on living things and can create undesirable taste, odors, and colors in a receiving body of water. **Chemical oxygen demand** (COD) is a measure of the quantities of such materials present in the water. COD, however, as measured in a COD test, also includes the demand of biologically degradable materials because more compounds can be oxidized chemically than biologically. Hence, the COD is larger than the BOD. Treatments are available for removing COD and BOD from wastewater.

Since concentration and composition of wastewater vary considerably throughout a day, care must be taken to obtain a representative sample for each type of test. Sampling and analyses should be made as directed in Standard Methods for the Examination of Water and Wastewater, American Public Health Association, 1015 18th St., NW, Washington, DC 20036 (www.apha.org); American Water Works Association, 6666 Quincy Ave., W., Denver, CO 80235 (www.awwa.org); Water Environmental Federation, Arlington, Va (www.wef.org).

(R. A. Corbitt, "Standard Handbook of Environmental Engineering," Metcalf & Eddy, Inc., "Wastewater Engineering," T. McGhee, "Water Supply and Sewerage," 3rd ed., McGraw-Hill Publishing Company, New York (www.books.mcgraw-hill.com); G. M. Fair, J. C. Geyer, and D. A. Okun, "Elements of Water Supply and Wastewater Disposal," John Wiley & Sons, Inc., New York (www.wiley.com); L. D. Benefield and C. W. Randall, "Biological Process Design for Wastewater Treatment," Prentice-Hall, Inc., Englewood Cliffs, N.J.)

22.19 Wastewater Treatment and Disposal

Because of the objectionable characteristics of raw wastewater (Art. 22.18), it must be treated before disposal, and disposal requires consideration of many factors, especially health hazards; odors, appearance, and other nuisance conditions; and economics. Selection of the type and degree of treatment depends on the nature of the raw wastewater, effluent quality after treatment, initial cost of the treatment plant, costs of operation and maintenance, process reliability, capability for disposal of sludge produced, potential for air pollution from pollutants removed, treatment chemicals required, energy consumed in the process, space requirements for the treatment plant, and potential hazards within the plant and in the surrounding area if the plant should malfunction or during transport of materials to and from the plant.

Wastewater has a capacity for self-purification or it can be treated by several methods including: oxidation ponds, or lagoons, and attached growth and suspended growth biological processes.

Oxidation ponds are artificial lagoons of wastewater utilizing natural forces for purification. Properly designed and maintained, they provide satisfactory treatment even for raw wastewater. Effluents may be equal to those from wastewater treatment plants providing secondary treatment.

Oxidation ponds are suitable for use where large areas of land are available at low cost. Successful operation, however, usually requires relatively high temperatures and sunshine. Ponds nevertheless are in use in northern states. When the water surface freezes, effluents are poor, but the ice prevents odors.

Wastewater treatment in oxidation ponds depends on aerobic decomposition of organic matter (Art. 22.14). Bacterial decomposition of this matter releases carbon dioxide. Algae develop, consume carbon dioxide, ammonia, and other waste products, and under proper climatic conditions release oxygen during daylight. Oxygen also is dissolved from the atmosphere at the lagoon surface. Hence, a large ratio of surface area to volume of liquid is desirable. Aeration, however, may be used to increase the supply of oxygen, which decreases substantially at night and in cold weather where algae are depended on heavily for oxygen. Aerated ponds are not so susceptible to climatic conditions as ordinary lagoons.

Pond depths normally range from 2.5 to 4 ft. With greater depths, septic conditions may develop at the bottom. Shallower ponds permit vegetation to emerge. Unless controlled, this encourages mosquito breeding and obstructs movement of the water, which is desirable for solution of atmospheric oxygen. When sludge deposits on the pond bottom become sufficiently deep, they are removed by dredging.

Size of pond required may be conservatively estimated for southern areas at 0.003 acre per capita for raw sewage and 0.002 acre per capita for wastewater with primary treatment. Another basis for design is a strength-surface-loading relationship in which 50-lb BOD per acre per day is considered satisfactory. For large installations, 220 BOD per acre per day is maximum for shallow aerobic ponds. About double these areas are required in northern regions.

For large installations, two or more ponds may be operated in series or in parallel. With series operation, aerobic conditions in the first pond can be improved by returning some of the effluent from the second pond.

An acceptable location for the inlet to a pond is its center. Effluent can be discharged at a convenient point along a bank.

Oxidation ponds may fail to meet requirements of secondary wastewater treatment for removal of suspended solids because of algae carried in the pond effluent. Methods for removing the algae include filtration through sand beds at low rates, filtration through a rock bed, or a combination of settling and chemical treatment of the effluent. The degree of treatment thus achieved may exceed the requirements of secondary treatment for both BOD and suspended solids.

(R. A. Corbitt, "Standard Handbook of Environmental Engineering," T. McGhee, "Water Supply and Sewerage," 6th ed., McGraw-Hill Publishing Company, New York (www.books.mcgraw-hill.com); "Wastewater Treatment Plant Design," Manual 36, American Society of Civil Engineers (www.asce.org).)

Self-Purification ▪ Wastewater, with or without extensive treatment, has been disposed of by dilution in a natural body of water. Partial or complete treatment then takes place in the water. Sometimes self-purification occurs; more often, if the wastewater has not had adequate treatment, the body of water becomes polluted. It may be unsafe for water supply and swimming, may contaminate or kill fish and shellfish, and may produce odors and have an unpleasant appearance. Therefore, treatment consistent with the self-purification characteristics of the body of water is desirable and is usually required by law. Secondary treatment now is required in most states. Requirements for tertiary treatment may be imposed to protect stream-water quality.

In polluted water, decomposition of organic matter utilizes oxygen from the water. If there is an adequate supply of oxygen, the BOD may be satisfied while enough dissolved oxygen remains to support fish life. If not, anaerobic decomposition will occur (Art. 22.14); the water becomes septic and malodorous and unable to support fish life.

Unpolluted water usually is saturated with oxygen. Table 22.4 shows the amount of oxygen that fresh water can hold in solution at various temperatures. The saturation quantity also depends on the concentration of dissolved substances. Salt water, for example, holds about 80% as much oxygen as fresh water.

Irrigation is of importance because it permits reclamation of the water content, to replenish the groundwater. Surface, flood, or subsurface irrigation may be used: *Surface irrigation* discharges wastewater on the ground. Part evaporates and part percolates into the ground, but a sizable amount remains on the surface and must be collected in surface drainage channels. For domestic wastewater, the method is not efficient. A modification, *spray irrigation*, however, has been used successfully for some industrial wastes. *Flood irrigation* also discharges the wastewater on the ground, but the wastewater seeps down and is

Table 22.4 Solubility of Oxygen in Fresh Water at Sea Level

Temperature		Dissolved oxygen, ppm or mg per liter	Temperature		Dissolved oxygen, ppm or mg per liter
°C	°F		°C	°F	
1	33.8	14.23	16	60.8	9.95
2	35.6	13.84	17	62.6	9.74
3	37.4	13.48	18	64.4	9.54
4	39.2	13.13	19	66.2	9.35
5	41.0	12.80	20	68.0	9.17
6	42.8	12.48	21	69.8	8.99
7	44.6	12.17	22	71.6	8.83
8	46.4	11.87	23	73.4	8.68
9	48.2	11.59	24	75.2	8.53
10	50.0	11.33	25	77.0	8.38
11	51.8	11.08	26	78.8	8.22
12	53.6	10.83	27	80.6	8.07
13	55.4	10.60	28	82.4	7.92
14	57.2	10.37	29	84.2	7.77
15	59.0	10.15	30	86.0	7.63

usually collected in underdrains. The soil acts as a filter and partly purifies the waste. But unless the wastewater is treated before irrigation, odors and insects may be produced, the soil may become clogged by grease or soap, and surface and groundwater may become contaminated. Surface irrigation sometimes is used for watering and fertilizing crops. This application, however, may create potential health hazards unless treatment has stabilized and disinfected the effluent. Another form of irrigation, *subsurface irrigation*, often is used with cesspools (Art. 22.24) and septic tanks (Art. 22.23).

Oxygen deficit D is the difference between saturation content and actual content, ppm or mg/L. As oxygen is removed from the water, the loss is offset by absorption of atmospheric oxygen at the surface. The rate at which this reaeration occurs depends on deficit D, the amount of turbulence, and the ratio of volume of water to the surface area. At any time t, days,

$$D = \frac{K_1 O}{K_2 - K_1}(10^{-K_1 t} - 10^{-K_2 t}) + 10^{-K_2 t}D_o \quad (22.19)$$

where $K_1 =$ coefficient of deoxygenation [see Eq. (22.18)]

$K_2 =$ reaeration coefficient, which ranges from 0.05 to 0.5 at 20°C, depending on depth, velocity, and turbulence of the water. For temperatures other than 20°C, multiply K_2 by 1.047^{T-20}

$T =$ temperature, °C

$O =$ oxygen demand at $t = 0$, ppm or mg/L

$D_o =$ oxygen deficit at point of pollution, or $t = 0$, ppm or mg/L

(H. W. Streeter, "The Role of Atmospheric Reaeration of Sewage-Polluted Streams," *Transactions, American Society of Civil Engineers*, vol. 89, p. 1355, 1926.)

In Fig. 22.11, the deoxygenation curve indicates the amount of dissolved oxygen remaining at any time as wastewater with initial demand O stabilizes, if the supply of oxygen is not replenished. The reaeration curve shows the amount of new oxygen dissolved during the same period. The oxygen sag curve represents at any given time the dissolved oxygen present, the sum of the remaining oxygen after deoxygenation and the oxygen from reaeration. The oxygen deficit D, as given by Eq. (22.19), is the ordinate of the oxygen sag curve measured from the horizontal line representing oxygen content at saturation.

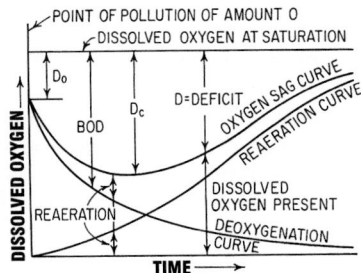

Fig. 22.11 Curves show the variation in oxygen content of a stream below a point of pollution.

The lowest or critical point of the sag curve indicates the occurrence of minimum dissolved oxygen, or maximum deficit. The time at which this occurs may be calculated from

$$t_c = \frac{1}{K_1(f-1)} \log f \left[1 - \frac{D_o}{O}(f-1) \right] \quad (22.20)$$

where $f = K_2/K_1$ = self-purification coefficient.
When $f = 1$,

$$t_c = \frac{0.434}{K_1} \left(1 - \frac{D_o}{O} \right) \quad (22.21)$$

The critical deficit is given by

$$D_c = \frac{O}{f} 10^{-K_1 t_c} \quad (22.22)$$

The pollution load O that a stream may absorb depends on the value of D_c, coefficients f and K_1, and the initial deficit D_o. The allowable value of D_c usually is established by law. The initial deficit is determined by existing pollution. The coefficients may be estimated from tests on the wastewater and the receiving body of water, or values may be assigned based on experience. Seasonal variations in temperature and water level or stream flow affect the amount of oxygen the water can hold and the amount of water available for dilution. Hence, the most critical conditions usually occur during summer, when rainfall is low and temperatures are high.

Self-purification is slower in lakes than in streams because of the low rate of dispersion of wastewater. With turbulence usually not present, mixing of water and wastewater in lakes depends mostly on currents and wind. Outfalls should be designed to take advantage of conditions encoura-

ging dispersion, to prevent sludge buildup at the discharge.

In estuaries, tides complicate dispersion. They carry pollutants back and forth many times. Salinity, density, and currents may change with time. These factors may also affect dispersion in ocean waters. Special care is necessary in outfall design to promote mixing and to take advantage of currents.

Quality Standards ▪ Present legal standards for water quality for recreation and water supply are not uniform. A typical standard may limit coliforms to an average of 10/mL; 5-day BOD to an average of 3 and a maximum of 6.5 mg/L; and phenols to a maximum of 0.001 mg/L. Dissolved oxygen may be required to be at least 5 and average 6.5 mg/L, while pH must be between 6.5 and 8.5. (See also Art. 21.45.)

Stream Capacity ▪ A rough approximation of the capacity of a stream to absorb a pollutional load may be based on the **dilution factor**, the ratio of amounts of diluting water to wastewater. The significance of this factor is questionable. Use of Eq. (22.21) is preferred.

Types of Treatment ▪ Any of several degrees of treatment of wastewater may be used to satisfy disposal requirements.

Wastewater treatment is any process to which wastewater is subjected to remove or alter its objectionable constituents and thus render it less offensive or dangerous. Treatment may be classified as preliminary, primary, secondary, or tertiary or advanced, depending on the degree of processing.

Preliminary treatment or pretreatment may be the conditioning of industrial waste before discharge to remove or neutralize substances injurious to sewers and treatment processes, or it may be unit operations that prepare the wastewater for major treatment. Enhanced primary treatment is defined as the use of chemicals to enhance precipitation of suspended solids. Chemicals frequently used in enhanced primary treatment include: ferric chloride, alum, and polymer.

Primary treatment is the first and sometimes the only treatment of wastewater. This process removes floating solids and suspended solids, both fine and coarse. If a plant provides only primary treatment, the effluent is considered only partly treated.

Secondary treatment applies biological methods to the effluent from primary treatment. Organic matter still present is stabilized by aerobic processes.

Tertiary or complete treatment removes a high percentage of suspended, colloidal, and organic matter. The wastewater also may be disinfected.

Advanced waste treatment is any physical, chemical, or biological process that accomplishes a degree of treatment higher than secondary. This commonly includes biological nitrification and denitrification and both biological and chemical phosphous removal.

Efficiency of treatment depends on quality of plant design and operation and on type and strength of wastewater. Details of treatment methods are given in the following articles.

(R. A. Corbitt, "Standard Handbook of Environmental Engineering," Metcalf & Eddy, Inc., "Wastewater Engineering," 3rd ed., T. J. McGhee, "Water Supply and Sewerage," 6th ed., McGraw-Hill Publishing Company, New York (www.books.mcgraw-hill.com); R. L. Culp, G. M. Wesner, and G. L. Culp, "Advanced Wastewater Treatment," Van Nostrand Reinhold Company, New York; H. W. Parker, "Wastewater Systems Engineering," Prentice-Hall, Inc., Englewood Cliffs, N.J. (www.prenhall.com); "Design of Municipal Wastewater Treatment Plants," Manual 36, "Sanitary Landfill," Manual 39, and "Glossary—Water and Wastewater Control Engineering," American Society of Civil Engineers (www.asce.org); P. E. Moffa et al., "Control and Treatment of Combined Sewer Overflows," Van Nostrand Reinhold, New York.)

22.20 Wastewater Pretreatment

The purpose of pretreatment of wastewater is to remove coarse materials that may interfere with treatment, or do not respond to treatment, or may damage or clog pumps, pipes, valves, and nozzles. Various types of screening devices are used for this purpose. Generally, they are the first units in a treatment plant.

Racks are fixed screens composed of parallel bars, set vertically or sloped in the direction of flow, to catch debris. Coarse racks have spaces between the bars of 2 in or more. They usually are used for large plants to protect sewage pumps. Medium racks, used more frequently, have bar spacings of 12 to $1\frac{1}{2}$ in. They may be fixed or movable. Movable racks are three-sided cages. Wastewater enters through the open side and leaves through the bars. One cage is periodically hoisted to the surface for manual cleaning, while wastewater passes through a second cage. Fixed-bar racks may be manually or mechanically cleaned. The bars may be curved to the horizontal at the top to facilitate cleaning.

While a minimum velocity of about 2 ft/s is desirable in the approach channel to prevent sediment from clogging it, velocity through a rack should be lower, perhaps 0.5 to 1 ft/s, so that objects should not be forced through. This requires enlargement of the conduit in the vicinity of the rack. To allow for head loss through a rack, the conduit bottom may be lowered below the rack 3 to 6 in.

Fine screens, with uniform-size openings or slots $\frac{1}{8}$ in wide or less, are useful for removal of bulky and fibrous materials and are becoming quite common. Generally, fine screens are movable and mechanically cleaned. Various types are used: rotating disk or drum, band, plate; or vibratory screens.

Screenings may be disposed of by burial, incineration, or digestion. Digestion of sludge proceeds normally when fine screenings are added in sludge-digestion tanks. In some treatment plants, screenings are passed through a grinder and returned to the flow, to settle out subsequently in a sedimentation tank. Screening and cutting are combined in such devices as comminutors, barminutors, and griductors. Their high-speed rotating edges cut through the sewage flow and chop and shred the solids, which then pass on to a sedimentation tank. Shearing-type units should be located after a grit chamber to prevent excessive wear of cutting edges.

Skimming tanks are rarely used for domestic wastewater treatment but can be advantageous ahead of sedimentation tanks. Skimmers remove oil and grease, which tend to form scum, clog fine screens, obstruct filters, and reduce the efficiency of activated sludge. Compressed air, applied through porous plates in the bottom of the tank, coagulates the grease and oil and causes them to rise to the surface. About 0.1 ft^3 of air is required per gallon. Detention period ranges from 5 to 15 min. About 2 mg/L of chlorine increases the efficiency of grease removal. After the effluent reaches the sedimentation tank, the coagulated material is removed with the scum or settled solids.

(T. J. McGhee, "Water Supply and Sewerage," 6th ed., Metcalf & Eddy, Inc., "Wastewater Engineering," 3rd ed., McGraw-Hill Publishing Company, New York (www.books.mcgraw-hill.com); G. Fair et al., "Water and Wastewater Engineering," John Wiley & Sons, Inc., New York (www.wiley.com).)

22.21 Sedimentation

At most wastewater-treatment plants, sedimentation is the means of primary treatment. In activated-sludge plants, secondary sedimentation is required after oxidation. It also is used after oxidation of wastewater for some attached growth processes such as RBCS and biotowers. Secondary sedimentation tanks are discussed under suspended grate processes.

The major objective of sedimentation is removal of settleable solids. But often, some floating materials also are removed by clarifiers, skimming devices usually are built into sedimentation tanks. These processes occur while wastewater moves slowly through a settling basin.

Efficiency of a sedimentation tank depends on particle size, specific gravity, and settling velocity and on several other factors: concentration of suspended matter, temperature, surface area of the liquid, retention period, depth and shape of basin, baffling, total length of flow, wind, and biological effects. Density currents and short circuiting may negate theoretical detention computations. Improper baffling may reduce the effective surface area of the liquid and create dead or nonflow areas within the tank.

Settling velocity of a particle is a function of the specific gravity and diameter of the particle, and specific gravity and viscosity of fluid. Settling rates of particles larger than $200\,\mu m$ are determined empirically. Sizes less than $200\,\mu m$ settle in accordance with Stokes' law for drag of small settling spheres in a viscous fluid [Eq. (21.133)]. (See also Art. 21.46.)

Theoretically, if the forward motion of the water is less than the vertical settling rate of all the particles, they will settle some distance below the surface in a given time interval while in the tank. After that period, if the surface layer of water were removed, it would contain no solids.

Surface settling rate, or overflow rate, gal/ft^2 of surface area per day, is a measure of the rate of flow through the basin when the rate of flow, ft^3/s, equals the surface area, ft^2, times the settling velocity, ft/s, of the smallest particle to be removed. Hence, selection of an overflow or surface settling rate establishes a relationship between flow and area.

Detention period is the theoretical time water is detained in a basin. The average detention period is V/Q, where Q is the flow, mgd or ft^3/s, and V, the basin volume. Since most of the settleable solids will settle out in 1 to 2 h, long detention periods are not advantageous. In fact, they are undesirable because the wastewater may become septic.

The flowing-through period is the time required for wastewater to pass through the basin. This time may be estimated by adding sodium chloride (or dye) to the influent and testing the effluent for increase of chloride or checking for dye. The flowing-through period should be at least 30% of the theoretical detention period. Dye may be used to follow the flow pattern.

Grit chambers (Fig. 22.12) are settling basins used to remove coarse inorganic solids. They are

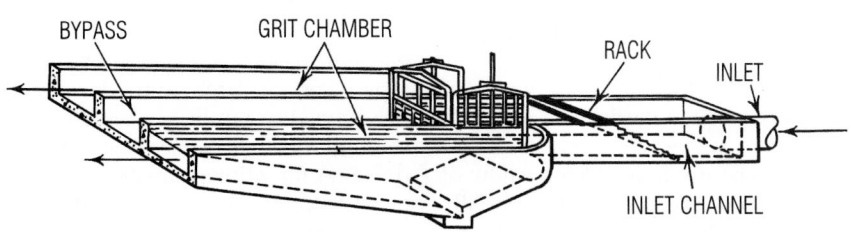

Fig. 22.12 Grit chamber.

provided to (1) protect moving mechanical equipment from abrasion and accompanying abnormal wear; (2) reduce formation of heavy deposits in pipelines, channels, and conduits; and (3) reduce the frequency of digester cleaning caused by excessive accumulations of grit. The removal of grit is essential ahead of centifuges, heat exchangers, and high-pressure diaphragm pumps.

There are three general types of grit chambers: horizontal-flow, of either a rectangular or a square configuration; aerated; or vortex type. In the horizontal-flow type, the flow passes through the chamber in a horizontal direction and the straight-line velocity of flow is controlled by the dimensions of the unit, an influent distribution gate, and a weir at the effluent end. The aerated type consists of a spiral-flow aeration tank where the spiral velocity is induced and controlled by the tank dimensions and quantity of air supplied to the unit. The vortex type consists of a cylindrical tank in which the flow enters tangentially creating a vortex flow pattern; centrifugal and gravitational forces cause the grit to separate. Design of grit chambers is commonly based on the removal of grit particles having a specific gravity of 2.65 and a wastewater temperature of 15.5°C (60°F). However, analysis of grit-removal data indicates the specific gravity ranges from 1.3 to 2.7.

Design of a grit chamber should ensure settlement of all particles over 0.2 mm in size but should not remove organic solids. Flow should be fast enough to secure this result but without scouring solids already deposited. Scour will occur if the horizontal velocity, ft/s, of the wastewater exceeds

$$v = 2.2\sqrt{\frac{gd}{f}(s-1)} \qquad (22.23)$$

where f = roughness coefficient (Darcy formula for flow in pipes) for chamber (see Fig. 21.19)

g = acceleration due to gravity, 32.2 ft/s^2

d = particle diameter, ft

s = specific gravity of particle

Usually, grit chambers are designed for a flow of about 1 ft/s. Flow may be controlled by specially shaped gates or weirs to keep velocity constant. The material settling out may be removed manually or mechanically. Also, devices may be added to mechanically cleaned units to wash most of the organic material out of the grit.

A **plain sedimentation tank** is a settling basin where sedimentation is not aided by coagulants and the settled solids, or sludge, are not retained for digestion. Generally, sludge and scum are removed mechanically. Any of several methods may be used to remove light, suspended material.

Flocculent suspensions have little or no settling velocity. Although they may occur in raw wastewater, more frequently they are encountered when effluents from activated-sludge units undergo secondary settling. The suspensions may be removed by passing inflowing wastewater upward through a blanket of the flocculent material (vertical-flow sedimentation tank). The objective is to produce a mechanical sweeping action in which small particles attach to larger particles, which then have sufficient weight to settle. Another removal method employs an inner chamber equipped with baffles that rotate and stir the liquid, to aid formation of larger, heavier floc. The same results also may be achieved by agitation with air. Some of the settled sludge is raised by air lift and mixed with the floc, to form a conglomerate with better settling characteristics. Variations utilizing the preceding principles have been introduced by several manufacturers, for example, the up-flow tube clarifier.

Design of a sedimentation tank should be based on the settling velocity of the smallest particle to be removed. Depth should be no larger than necessary for preventing scour and to accommodate cleaning mechanisms. Almost all treatment plants use mechanically cleaned sedimentation tanks of standardized circular or rectangular design (see Fig. 22.13). The selection of the type of sedimentation unit for a given application is governed by the size of the installation, by rules and regulations of local control authorities, by local site conditions, and by the experience and judgment of the engineer. Two or more tanks should be provided so that the process may remain in operation while one tank is out of service for maintenance and repair work. At large plants, the number of tanks is determined largely by size limitations. The surface-settling-rate requirement generally is 600 gal/(ft$^2 \cdot$ day) for primary treatment when used alone or when activated sludge is handled by them and 800 to 1200 ahead of secondary treatment. The detention period normally is 2 h. These three design parameters must be adjusted since each is

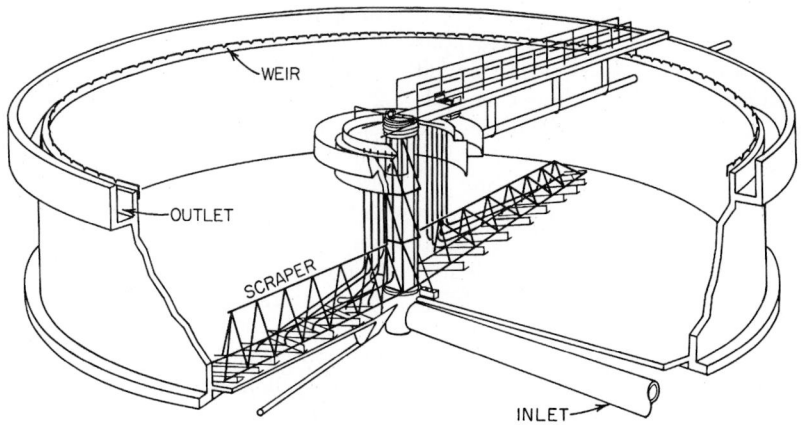

Fig. 22.13 Circular sedimentation tank.

dependent on the other for a given design flow (average daily flow for a plant). Modern design calls for at least 12′ foot side wall depth for most applications and 10–16 foot depth have been used historically.

Rectangular tanks are built in units with common walls. Width per unit ranges up to 25 ft. Minimum length should be at least 10 ft. The length-width ratio should not exceed 5 : 1. Circular tanks can be 10–200 feet in diameter and are most commonly 40–150 feet in small tanks. Final sizes may be determined by dimensions of available sludge-removal equipment.

In circular tanks (Fig. 22.13) radial blades scrape the bottom to move the sludge to a central sludge hopper. In rectangular tanks, the hopper is located near the inlet end since the heaviest sludge accumulation occurs in that region. Blades moving along the bottom against the flow of sewage push the sludge to the hopper. In some tanks, the same blades may be lifted to the surface and, traveling with the sewage flow, move scum to the outlet end. There, the scum may be trapped by a baffle until taken out by a scum-removal device.

Many mechanical aids for use with sedimentation tanks are available commercially. Manufacturers' literature should be carefully studied and specifications written to ensure procurement of equipment exactly meeting design requirements.

Actual flowing-through time is influenced by inlet and outlet construction. For circular tanks, inlets are submerged, at the center (Fig. 22.11). Wastewater rises inside a baffle extending down-

ward, to still the currents. The outlet device nearly always is a circumferential weir adjusted to level after installation. The weir may be sharp-edged and level or provided with V notches about 1 ft or less apart. The notches permit more constant flow since they are less affected by local differences in weir elevation and surface tension. For rectangular tanks, inlets also may be submerged, but at one end. More often, the wastewater is brought to a trough that has a weir extending the width of the tank. The flow then moves forward with less short circuiting. At the outlet, to provide enough weir length, a launder is used. This consists of a series of fingerlike shallow conduits set at water level and receiving flow from both sides. Each finger is connected to a common discharge trough. Normal weir loading should not exceed 10,000 gal/ft of weir per day in small plants, or 15,000 in units handling more than 1 mgd.

Chemical precipitation sometimes is used to improve the effluent from sedimentation. The process is similar to that for water clarification. The high cost of chemicals and the intermediate grade of treatment obtained with chemicals have kept the process from general use. Chemical precipitation has, however, been found useful in specialized treatment. Phosphorus removal, preparation of sludges for filtration or dewatering, and removal of trace metals are examples of such treatment.

Alum, ferric chloride, ferric sulfate, lime, sodium aluminate, ferrous chloride, ferrous sulfate, and polyelectrolytes are chemicals used to expedite precipitation. The coagulation resulting is, actually, the result of a complex group of reactions involving the hydrolysis products of the added chemicals. Effectiveness of the various chemicals depends on the conditions under which they are used and the types of wastes.

There has to be an optimum pH and an optimum dosage for efficient wastewater coagulation. Consequently, dosages are often determined by trial (jar tests). Measurement of zeta potential (an electrical potential related to particle stability and hence useful in controlling coagulation) and of phosphate content is also desirable.

Design requirements include rapid mixing, mixer-blade peripheral speeds less than 5 ft/s, control of slurry concentration, minimum sludge blanket levels, and controlled horizontal movement of clearer water by launder or weir spacing and by weir overflow rate control. (See also Art. 21.46.2.)

(T. J. McGhee, "Water Supply and Sewerage," 6th ed., and Metcalf & Eddy, Inc., "Wastewater Engineering," 3rd ed., McGraw-Hill Publishing Company, New York (www.books.mcgraw-hill.com); G. Fair, et al., "Water and Wastewater Engineering," E.J. and E.T. Martin, "Technologies for Small Water and Wastewater Systems," John Wiley & Sons, Inc., New York (www.wiley.com); "Wastewater Treatment Plant Design," Manual 36, American Society of Civil Engineers (www.asce.org).)

22.22 Attached Growth Biological Processes

Secondary treatment of wastewater frequently employs oxidation to decompose and stabilize the putrescible matter remaining after primary treatments. Attached growth biological processes are one type of these secondary treatments. Others include the attached growth processes, oxidation ponds, and irrigation. The latter two have been discussed in Article 22.19. These oxidation methods bring organic matter in wastewater into immediate contact with microorganisms under aerobic conditions. In attached growth systems the microorganisms coat the filtering media. As the wastewater flows through, adsorption occurs, and most of the organic materials are removed by contact with the coating. The organisms decompose organic nitrogen compounds and destroy carbohydrates. Efficiency of the method, as measured by reduction of BOD, is high.

Intermittent sand filters are sand beds, usually $2\frac{1}{2}$ to 3 ft deep, with underdrains for collecting and carrying off the effluent. They are used in small wastewater treatment plants. Settled wastewater, the effluent from a sedimentation tank, is applied to the sand surface in intermittent doses. A rest period between doses allows time for air to assist in oxidation of the organic matter. Application rates generally range from 20,000 gallons per acre per day (gad) to 125,000 gad when the filters serve as a secondary treatment. Rates may go as high as 0.5 million gallons per acre per day (mgad) for tertiary treatments.

Sand for an intermittent filter should have a uniformity coefficient of 5 or less; 3.5 is preferred. (**Uniformity coefficient** is the ratio of the sieve size that will pass 60% of the material to the effective size of the sand. **Effective size** is the size, mm, of the sieve that passes 10%, by weight, of the sand.) The effective size of the sand should be between 0.2 and 0.5 mm. A bed of gravel 6 to 12 in thick usually underlies the sand.

A mat of solids forms on the filter surface and must be removed periodically. Generally, the mat can be scraped off when dry, but occasionally the top 6 in or so of the filter material must be replaced.

In winter, there is danger that the sand surface will freeze. To keep the filter in operation, the bed may be ridged on 3-ft centers to support the ice while the wastewater flows underneath it.

Trickling filters are beds of coarse aggregate over which settled wastewater is dropped or sprayed and through which the wastewater trickles to underdrains (Fig. 22.14). Filter media include gravel, crushed rock, ceramic chapes, slag, redwood slats, or plastics. Stone and crushed rock that do not fragment, flour, or soften on exposure to wastewater are widely used. Generally, rock sizes are kept between 2- and 4-in nominal diameter. Underdrains collect and carry off the effluent. Filters may be ventilated through the underdrain system or by other means, to supply air to the aerobic organisms that grow on the media surfaces.

Since suspended solids can clog filters, sedimentation of the wastewater is desirable before it is fed to the filters. When, however, a waste, such as milk waste, contains a concentration of dissolved solids, it may be applied directly to a filter. In that case, preaeration is desirable, so that the waste contains some dissolved oxygen.

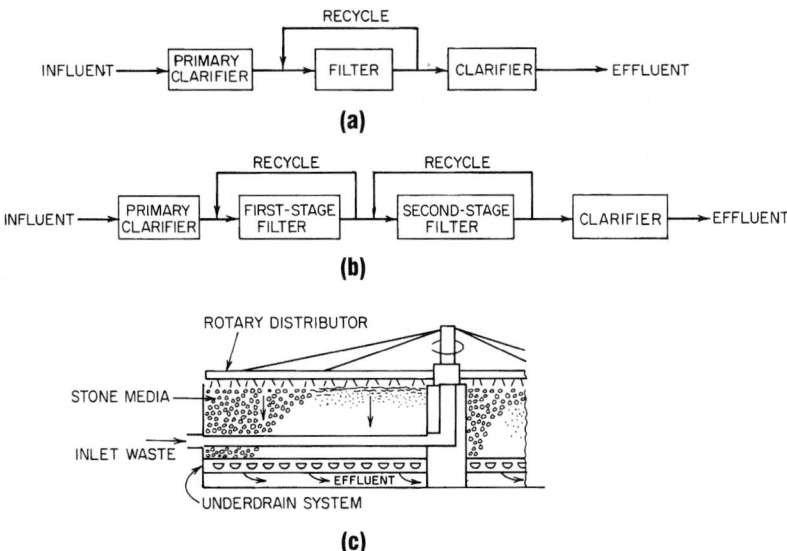

Fig. 22.14 Biotowers (trickling filters) supply bacteria for consumption of organic matter in wastewater. (*a*) Schematic of process with single-stage filtration. (*b*) Schematic of process with two-stage filtration. (*c*) Cross section of a trickling filter with a rotary distribution of wastewater.

In time, oxidized matter breaks away from the filter media and is flushed from the filter with the effluent. Hence, the effluent is passed through a secondary settling basin, or **clarifier**. Design of these basins is similar to that of primary sedimentation tanks. Efficiency, or percent reduction of BOD, of a trickling filter generally is measured for both the filter and final sedimentation.

Trickling filters are classified as standard or low-rate, high-rate, and controlled.

Standard filters were introduced in the United States early in the twentieth century. They consisted of an underdrained bed of stones, 6 to 8 ft deep. Settled wastewater was distributed over the surface through fixed nozzles. Later, the fixed nozzles were superseded by a rotary distributor. This type of distributor has two or four radial arms supported on a center pedestal (Fig. 22.14*c*). Jets of wastewater from nozzles on the arms cause rotation. Thus, the filter surface is sprayed as the arms revolve. Dosing, as a result, is intermittent, though the interval between doses is short, often not more than 15 s. A distributor may be kept rotating continuously by feeding the nozzles from a weir box or a dosing tank, with siphons or pumps.

To accommodate rotary distribution, standard filters are built round in plan.

These low-rate filters are dosed at a rate of 1 to 4 mgad, substantially lower than that for high-rate filters. Loading also may be expressed in terms of 5-day BOD, lb/acre-ft·day. Some state health departments limit the load on a standard filter to 400 to 600 lb/acre-ft·day. The approximate load w to be applied to a filter, lb/day·acre-ft of filter volume, when the BOD of the wastewater is known and a limit is specified for the BOD of the effluent, may be computed from

$$w = 13,840\left(\frac{B}{A-B}\right)^2 \qquad (22.24)$$

where A = 5-day BOD of the influent, mg/L
B = specified maximum BOD of effluent, mg/L

High-rate filters receive a load three or more times greater than that usually applied to standard filters. Usual rate is about 20 mgad, but rates from 9 to 44 mgad have been used. Some state health departments limit the load to 2000 to 5000 lb of BOD per acre-ft per day.

Such high rates are feasible because the effluent is recirculated through the filter (Fig. 22.14a). Recirculation reduces the load on the filter, seeds the media continuously with organisms, allows continuous dosage, offsets fluctuations in wastewater flow, and reduces odors by freshening the influent. Several recirculation alternatives may be used. For example, part of the filter effluent may be returned directly to the filter. (Proponents of this method of recirculation claim direct return intensifies biological oxidation.) Or part of the effluent of the filter or the final clarifier may be combined with the influent to the primary sedimentation tank. Sometimes, dual recirculation is used: The filter effluent is returned to the primary sedimentation tank, while part of the final clarifier effluent is sent back through the filter. In some cases, the sludge from the final clarifier is recirculated through the primary clarifier.

Two-stage filtration (Fig. 22.14b) may be used when a better effluent is desired than can be obtained from a single filter. For this purpose, two filters are connected in series. Various recirculation methods may be used in this case also.

The recirculation ratio, or ratio of returned effluent to sewage influent, ranges from 1:1 to about 5:1. At each passage, the amount of BOD removed decreases because response to treatment decreases. If the ratio of the decrease per passage to the BOD is given by k, then the number of effective passages of sewage through a filter may be computed from

$$F = \frac{1 + R}{(1 + kR)^2} \qquad (22.25)$$

where R = recirculation ratio. Under normal conditions, k may have a value of about 0.1.

The approximate load, lb of BOD per day per acre-ft of filter volume, to be applied to a single-stage high-rate filter or the first filter of a two-stage system, when the BOD of the sewage is known and a limit is specified for the BOD of the effluent, may be computed from

$$w = 13,840F \left(\frac{B}{A - B}\right)^2 \qquad (22.26)$$

where A = 5-day BOD of influent, mg/L

B = specified maximum BOD of effluent, mg/L

The approximate load for a second-stage filter may be estimated from

$$w = 13,840 \left(\frac{B_1}{A_1}\right)^2 \left(\frac{B_2}{A_2 - B_2}\right)^2 F \qquad (22.27)$$

where A_1 = 5-day BOD of influent of first-stage filter, mg/L

B_1 = specified maximum BOD of effluent of first-stage filter, mg/L

A_2 = 5-day BOD of influent of second-stage filter, mg/L

B_2 = specified maximum BOD of effluent of final clarifier, mg/L

F = number of effective passages through second-stage filter

Equations (22.24) to (22.27) are based on formulas recommended by a committee of the National Research Council ("Sewage Treatment at Military Institutions," *Sewage Works Journal*, vol. 18, no. 5, p. 794, September 1946).

Wastewater is sprayed over high-rate filters by rotary distributors or by a motor-driven disk that rains wastewater continuously and uniformly over the surface. Hence, the filters are built circular.

Biodisks or rotating biological contactors. Wastewater secondary-treatment quality can be achieved with a method that has characteristics of both trickling filters and contact stabilization. A series of closely spaced, 10- to 12-ft-diameter, plastic disks (biodisks) is mounted vertically on a horizontal shaft and rotated slowly with about half their surface area continuously immersed in a reservoir of wastewater (Fig. 22.15).

The disks provide a surface for buildup of attached microbial growth, subject this growth alternately to submersion in the wastewater and the air, and aerate both the wastewater and biological growth suspended in the wastewater. As the disk surfaces emerge from the reservoir during their rotation, they expose to the air a film of wastewater that adheres to them. On returning to the reservoir, this film adds oxygen to that already present in the reservoir. Also, microorganisms from the wastewater adhere to the rotating surfaces and grow in number until the disks are covered with a thin layer of biological slime. As these microorganisms pass through the reservoir, they absorb and break down other organic substances. Excessive growth of microorganisms is sheared from the

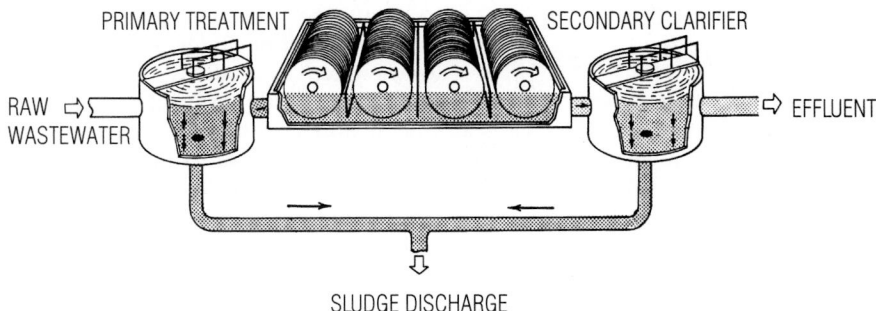

Fig. 22.15 Schematic of biodisk, or biological contactor, process.

disks as they move through the reservoir and is carried out with treated wastewater to a settling basin for removal. Loading rates range from 2 to $4 \, gal/day \cdot ft^2$ of effective media area. With several sets of disks in series, it is possible to achieve higher degrees of treatment, including biological conversion of ammonia to nitrates.

(H. N. Parker, "Wastewater Systems Engineering," Prentice-Hall, Inc., Englewood Cliffs, N.J.; R. A. Corbitt, "Standard Handbook of Environmental Engineering," McGraw-Hill Publishing Company, New York (www.books.mcgraw-hill.com).)

Submerged attached growth processes. These are also sometimes referred to as BAFs (Biological Aerated Filters) or BAFFs (Biological Aerated Flooded Filters). Aerobic submerged fixed-film processes consist of three phases: a packing, biofilm, and liquid. The BOD and/or NH_4-N removed from the liquid flowing past the biofilm is oxidized. Oxygen is supplied by diffused aeration into the packing or by being predissolved into the influent wastewater. Aerobic fixed-film processes include downflow packed-bed reactors, upflow packed-bed reactors, and upflow fluidized-bed reactors. The type and size of packing is a major factor that affects the performance and operating characteristics of submerged attached growth processes. Designs differ by their packing configuration and inlet and outlet flow distribution and collection. No clarification is used with aerobic submerged attached growth processes, and excess solids from biomass growth and influent suspended solids are trapped in the system and must be periodically removed. Most designs require a

backwashing system much like that used in a water filtration plant to flush out accumulated solids, usually on a daily basis.

The major advantages of submerged attached growth processes are their relatively small space requirements, the ability to effectively treat dilute wastewaters, no sludge settling issues as in activated-sludge process, and aesthetics. Also for many processes solids filtration occurs to produce a high-quality effluent. Such fixed-film systems have equivalent hydraulic retention times of less than 1 to 1.5 h, based on their empty tank volumes. Their disadvantages include a more complex system in terms of instrumentation and controls, limitations of economies of scale for application to larger facilities, and generally a higher capital cost than activated-sludge treatment.

A wide variety of submerged attached growth processes have been used. The purpose of this section is to describe the more common processes used and their design loadings and performance capability. Virtually all are proprietary processes including the downflow Biocarbone® process, the upflow Colox®, Biofor® and Biostyr® process, and the aerobic upflow fluidized-bed reactor. See also *Filtration* in Art. 22.30.

(Metcalf & Eddy, Inc., "Wastewater Engineering," 3rd ed., McGraw-Hill Publishing Company, New York (www.books.mcgraw-hill.com); "AWWA Standard for Filtering Material," American Water Works Association (www.awwa.org); "Wastewater Treatment Plant Design," Manual 36, American Society of Civil Engineers (www.asce.org); G. M. Fair et al., "Water and Wastewater Engineering," John Wiley & Sons, Inc., New York (www.wiley.com).)

22.23 Suspended Growth Biological Processes

The most common suspended growth biological process is the activated-sludge process which is a form of biological treatment in which a mixture of wastewater and a sludge of microorganisms is agitated and aerated and from which the solids are subsequently removed and returned to the aeration process as required.

Passing air bubbles through wastewater coagulates colloids and grease, satisfies some of the BOD, and reduces ammonia nitrogen a little. Aeration also may prevent wastewater from becoming septic denitrifying (floating solids because of anonix conditions) in a following sedimentation tank. But if wastewater is mixed with previously aerated sludge and then aerated, as is done in activated-sludge methods, the effectiveness of aeration is considerably improved. Reduction of BOD and suspended solids in the conventional activated-sludge process, including primary sedimentation and final sedimentation, may range from 80 to 95% typically producing an effluent with well less than 25 mg/L BOD and SS in it and well designed and operated systems can achieve less than 15 mg/L.

In a conventional activated-sludge plant (Fig. 22.16), incoming wastewater first passes through a primary sedimentation tank. Activated sludge is added to the effluent from the tank, usually in the ratio of 1 part of sludge to 3 or 4 parts of settled wastewater, by volume, depending upon mass concentration and (SVI) settling characteristics, and the mixture goes through an aeration tank. In that tank, atmospheric air is mixed with the liquid by mechanical agitation, or compressed air is diffused in the fluid by various devices: filter plates, filter tubes, ejectors, and jets. Modern practice almost exclusively uses membrane or ceramic fine pored disks. In either method, the wastewater thus is brought into intimate contact with microorganisms contained in the sludge. In the first 15 to 45 min, the sludge adsorbs suspended and colloidal solids. As the organic matter is adsorbed, biological oxidation occurs. The organisms in the sludge decompose organic nitrogen compounds and destroy carbohydrates. The process proceeds rapidly at first, then falls off gradually for 2 to 5 h. After that, it continues at a nearly uniform rate for several hours. Generally, the aeration period ranges from 6 to 8 or more hours.

Most plants are designed to meet either USEPA secondary treatment standards (effluent BOD_5 and $SS \leq 30$ mg/L) or more stringent advanced wastewater treatment parameters (design of the latter are discussed more in article 22.30). Modern practice uses SRT (Solids Retention Time—sometimes called MCRT (Mean Cell Retention Time)) to size

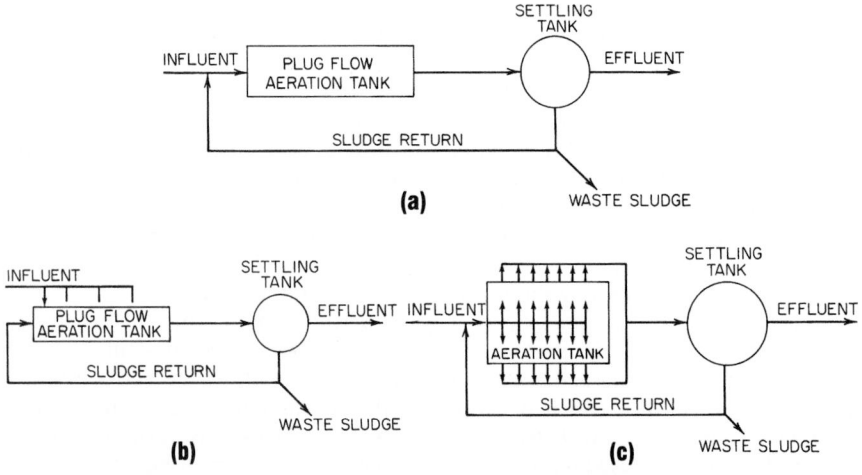

(a)

(b)

(c)

Fig. 22.16 Schematics of activated-sludge processes: (*a*) Conventional; (*b*) step aeration; (*c*) complete mix. (*From "Environmental Pollution Control Alternatives: Municipal Wastewater," Environmental Protection Agency, Cincinnati, Ohio.*)

aeration tanks:

$$SRT = 8.34 \times MLSS \times V_a/W \qquad (22.28)$$

or

$$V_a = SRT \times W/(8.34 \times MLSS) \qquad (22.29)$$

where MLSS = Mixed Liquor Concentration, mg/L typically 2000 mg/L for secondary systems and 3000–5000 mg/L for extended aeration and nitrification (advanced) systems

V_a = Volume of aeration basin, MG

W = Solids wasted (including effluent suspended solids), lbs/d; typically about 0.9 times lbs BOD_5 applied for secondary systems (about 0.7 for nitrifying systems and about 0.6 times lbs BOD_5 applied for extended air systems)

The design SRT chosen is typically at least 3 days in cold waters and at least 2 days in warm waters, for secondary treatment. For nitrification, a formula that is typically used to choose design SRT is:

$$SRT = 5.56 \times \exp(0.116 \times (15 - T)) \qquad (22.30)$$

where T is MLSS temperature, in °C.

Extended Aeration systems are generally simply designed with a 24 hour aeration time.

Most States do not accept designs with aeration tanks smaller than 35–60 lbs/d BOD_5/1000 cf, except for **oxygen activated sludge** plant designs, which may be higher.

The aeration-tank effluent goes to a secondary sedimentation tank, where the fluid is detained, usually from 2 to 3 h, to settle out the sludge. The effluent from this tank is completely treated and, after chlorination, may be safely discharged.

A flow equal to about 25 to 35% of the influent flow (more for extended air or BNR) of the settled sludge from the final sedimentation tank is returned for recirculation with incoming wastewater. Typical design allows for 50–100% return rates for conventional systems and at least 100% for BNR and extended air systems. Sludge should not be detained in the tank. Frequent removal (at intervals of less than 1 h) or continuous removal is necessary to avoid denitrification.

Overflow rates for final sedimentation normally range from about 200–400 gal/(ft² · day) for small plants to 400–800 for plants of over 2-mgd capacity. Most states allow a maximum of 1000–1200 gpd/s.f. at peak flow. Weir loadings preferably should not exceed 10,000 gal/(lin ft · day). When tank

volume required exceeds 2500 ft³, multiple sedimentation tanks are desirable, though this number is often safely exceeded in many large plants. A more common measure is diameter of circular clarifiers. These are commonly in 40–140′ range, though some are at least in the 40–140′ range. Some are as large as 220 feet and appear to work well.

Modern secondary clarifier design recognizes solids loading as the most important factor. The allowable design solids loading is a function of MLSS settlability, and typically ranges from less than 10 lbs/d/sf for poor settling sludges to as high as 50 lbs/d/sf for excellent settling sludges. One method of calculating design SOR is to use the equation:

Design SOR (gpd/sf)

$$= (179.5 \times V_{max}/CSF) \times \exp(-4 \times SVI$$

$$\times MLSS \times 10^{-6}) \qquad (22.31)$$

where V_{max} = maximum settling velocity, fph (typically 12–30 fph—roughly equal to MLSS tem-perature in degrees C) 23 is a commonly used value

CSF = Clarifier safety factor, usually equal to at least 1.1 times ratio of maximum to design flow to clarifier and typically is about 2.0

SVI = Sludge Volume Index, usually assumed to be 150 in plants without selectors and 100–120 for plants with selectors

MLSS = MLSS concentration fed to clarifier, mg/L, typically 2000 mg/L for secondary treatment systems and 3000–5000 mg/L for extended aeration and BNR plants

Ref: *Activated Sludge Process Design and Control,* second edition, Technomic Publishing, 1998 (www.sagepub.co.uk).

Multiple aeration tanks are required when total tank volume exceeds 5000 ft³, but again this number is often exceeded in large plants. Aeration tanks, in which compressed air is used, generally are long and narrow. To conserve space, the channel may be turned 180° several times, with a common wall between the flow in opposing directions. Historically, an air main is generally run along the top of the wall to feed diffusers (Fig. 22.17a and b) or porous plates (Fig. 22.17c) along its length. The air sets up a spiral motion in the liquid as it flows through the tanks. Modern practice typically uses complete floor coverage for diffusers.

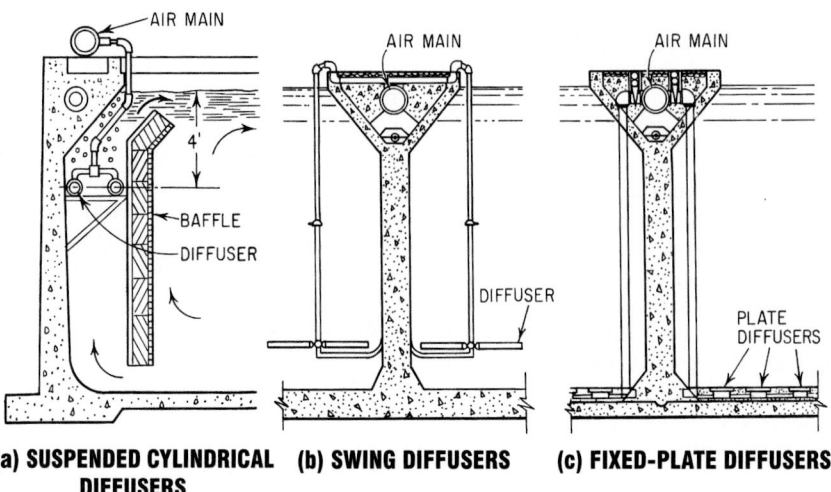

(a) SUSPENDED CYLINDRICAL   **(b) SWING DIFFUSERS**   **(c) FIXED-PLATE DIFFUSERS**
DIFFUSERS

Fig. 22.17 Air main atop aeration-tank walls supplies air to diffusers in adjoining channels in which the mixture of activated sludge and sedimentation-tank effluent flows.

Width of channel ranges from 15 to 30 ft. Depth is about 15 ft.

Dissolved oxygen should be maintained at 2 ppm (mg/L) or more. Air requirements normally range from 0.2 to 1.5 ft³/gal of wastewater treated. Most state authorities require a minimum of 1000 ft³ of air per lb of applied BOD per day, though usually less is allowed with more efficient fine-pore diffusers.

Mechanical aeration may be done in square, rectangular, or circular tanks, depending on the mechanism employed for agitation. In some plants, the fluid may be drawn up vertical tubes and discharged in thin sheets at the top, or the liquid may pass down draft tubes while air is bubbled through it. In both methods, agitation at the surface produced by the movement of the liquid increases aeration.

Several modifications of the activated-sludge method, seeking to improve performance or cut costs, are in use. These include modified, activated, tapered aeration, step aeration oxidation ditches, and oxygen activated sludge and complete-mix aeration, and the Kraus, biosorption, and bioactivation processes.

Modified aeration decreases the aeration period to 3 h or less and holds return sludge to a low proportion. Results are intermediate between primary sedimentation and full secondary treatment.

Activated aeration places aeration tanks in parallel. The activated sludge from one final sedimentation tank or group of such tanks is added to the influent of the aeration tanks. Other sludge is concentrated and removed. With much less air, results are better than with modified aeration.

Tapered aeration differs from conventional in that air diffusers are not uniformly spaced. Instead, more diffusers are placed near the inlet end of the aeration tanks than near the outlet. The theory is that oxygen demand is greater near the inlet, and so the efficiency of the treatment should be improved if more air is supplied there. However, results depend on degree of longitudinal mixing, rate of sludge return, and characteristics of recirculated matter, for example, air content of sludge or mixed liquor.

Step Aeration (Step Feed) (Fig. 22.15*b*) adds settled wastewater at several points (typically 3 or 4) along an aeration tank. Each portion of such an aeration tank is usually referred to as a pass. This leads to a more even distribution of the aeration along the length of the tank. For a given aeration volume and SRT, it also creates a lower solids loading on the secondary clarifier and, in some cases, allows designing either smaller aeration tanks, smaller clarifiers or both. It also is an operational tool allowing operator to reduce loading to secondary clarifiers during high flows or bulking conditions by changing step feed points. For example it is common practice at some plants to shift feed to mostly, or only, the last pass during storm events. Established originally in large NYC plants, it is now gaining favor in smaller plants, some as small as 5 mgd. It also has found a niche in advanced waste treatment, being the bases for step denitrification (see article 22.) (Refs: "Activated Sludge Process Design and Control", 2nd ed., Technomic Publishing, 1998; "Design of Municipal Wastewater Treatment Plants", *MOP 8*, 4th ed., WEF, 1998 and "Wastewater Engineering", 4th ed., McGraw-Hill, 2003 (www.books.mcgraw-hill.com).

Extended Aeration is similar to conventional activated sludge except that detention times are longer (typically 24 hrs vs. 6–8 hrs) and no primary clarification is typically used. Final clarifiers are usually larger (200–400 gpd/sf vs. 600–800 gpd/sf) also. It is used primarily in smaller (<5 mgd) plants and provides nitrification and aerobic sludge digestion in a single aeration tank. (Refs: "Activated Sludge Process Design and Control", 2nd ed., Technomic Publishing, 1998 (www.sagepub.co.uk); "Design of Municipal Wastewater Treatment Plants", *MOP 8*, 4th ed., WEF, 1998 and "Wastewater Engineering", 4th ed., McGraw-Hill, 2003 (www.books.mcgraw-hill.com)).

Oxidation Ditches and SBRs (Sequencing Batch Reactors) have become the most common form of activated sludge in smaller (<3 mgd) plants, although there have been successful designs for plants at least as large as 50 mgd. Both are typically designed as extended aeration plants. **Oxidation Ditches** typically are proprietary designs, typically in some sort of continuous loop and aerated with various types of horizontal and vertical aerators. Examples are Carrousel™ and Orbal™® processes. SBRs are batch reactors with clarification and aeration both in a single tank. These are also generally proprietary; examples are ICEAS™ and the CASS™ processes. More details about these can be found in "Wastewater Engineering", 4th ed., McGraw-Hill, 2003 (www.books.mcgraw-hill.com).

Modified Aeration, Activated Aeration, Completely Mixed Activated Sludge (CMAS) and the Krause Process were once popular versions of activated sludge but are no longer commonly used

for domestic wastewater treatment. Similarly, **Contact Stabilization** also is not commonly used anymore except as an operational mode in **Step Aeration** plants.

Oxygen Activated Sludge ▪ Excellent results have been obtained by substituting oxygen for air in the activated-sludge process. For efficient use of the oxygen, the aeration tanks are covered. The oxygen may then be recirculated through several stages, entering the first stage of the process and flowing through the oxygenation basin with the wastewater being treated. Pressure under the tank cover is close to atmospheric and enough to maintain control and prevent backmixing of successive stages. Within each stage, mixing may be achieved with surface aerators or a submerged turbine. Pure oxygen permits use of smaller tanks, and oxygenation time may be $1\frac{1}{2}$ to 2 h instead of the conventional 6 to 8 h. The activated sludge produced often settles more easily and is easier to dewater than that from conventional processes.

Activated-sludge plants should be closely controlled for optimum performance. This requires frequent checking of the sludge content of the mixed liquor. Solids usually are limited to 1500 to 2500 ppm (mg/L) in diffused-air plants and about 1000 ppm when mechanical agitation is used. Settling characteristics of the sludge are indicated by the Mohlman index more commonly referred to as Sludge Volume Index (SVI):

$$SVI = \frac{V_{set}}{MLSS} \qquad (22.32)$$

where V_{set} = volume of sludge settled in 30 min, cc/L

MLSS = mixed liquor conc., gm/L

A good settling sludge has an index below 100. Generally sludges with SVIs over 150 are considered "bulky" and difficult to settle. These sludges typically have an excess of undesirable filamentous organisms. An alternative measure is the sludge density index (SDI), 100 divided by the Mohlman index. Operating control may be maintained by holding a constant mixed-liquor suspended-solids (MLSS) or volatile-suspended-solids (MLVSS) concentration, by holding a constant ratio of food to microorganisms (F:M), or by holding a constant SRT or MCRT.

Sludge age is a term used mostly on the East Coast, and is similar to SRT, but is defined as the ratio of dry weight of sludge in the aeration tank, lb, to the suspended-solids load, lb/day, of the incoming wastewater. In a well-operated activated-sludge plant, sludge age is 3 to 5 days.

(Metcalf & Eddy, Inc., "Wastewater Engineering," 3rd ed., and L. Rich, "Low Maintenance Wastewater Treatment Systems," McGraw-Hill Publishing Company, New York (www.books.mcgraw-hill.com); "Wastewater Treatment Plant Design," Manual 36, American Society of Civil Engineers (www.asce.org); G. M. Fair et al., "Water and Wastewater Engineering," John Wiley & Sons, Inc., New York (www.wiley.com).)

See also Arts. 22.19 and 22.27.

(R. A. Corbett, "Standard Handbook of Environmental Engineering," McGraw-Hill Publishing Company, New York (www.books.mcgraw-hill.com).)

22.24 Sludge Treatment and Disposal

Sludge comprises the solids and accompanying liquids removed from wastewater in screening and treating it. Solids are removed as screenings, grit, primary sludge, secondary sludge, and scum. Often sludge treatment is necessary to make possible safe, economical disposal of these wastes. The treatment to be selected depends on quantity and characteristics of the sludge, nature and cost of disposal, and cost of treatment.

Screenings are putrescible and offensive. They may be disposed of by burning, burial, grinding and return to wastewater, or grinding and transfer to a sludge digester. The quantity of screenings is variable and dependent on wastewater characteristics. Coarse screenings may range from 0.3 to 5 ft^3/million gal. Fine screenings may range from 5 to 35 ft^3/million gal.

Sand and other gritty materials also may be present in widely varying amounts. Normally, the volume will be between 1 and 10 ft^3/million gal.

Sludge varies in quantity and characteristics with the characteristics of the wastewater and plant operations. Usually, more than 90% is water containing suspended solids with a specific gravity of about 1.2. Roughly, there may be about 0.20 lb of these solids per capita daily in sanitary wastewater; 0.22 lb if a moderate amount of industrial wastes is present; 0.25 lb in effluents of combined sewers if considerable industrial wastes are present; and 0.32

to 0.36 lb if the wastewater contains ground garbage also.

Primary sludge, derived from sedimentation tanks or the influent of digestion chambers of Imhoff tanks, is putrescible and odorous. It is composed of gray, viscous identifiable solids and has a moisture content of 95% or more. Primary treatment of 1 million gal of wastewater may produce about 2500 gal of this sludge.

Trickling filter sludge is black or dark brown, granular or flocculent, and partly decomposed. It is not highly odorous when fresh. Moisture content may be about 93%. Passage of 1 million gal of wastewater through a trickling filter may produce about 500 gal of this sludge.

Activated sludge is dark to golden brown, granular or flocculent, and partly decomposed. It has an earthy odor when fresh. Moisture content may be about 98%. Influent to an activated-sludge plant may yield about 13,500 gal of waste sludge per million gal.

Chemical-precipitation sludge may have a solids content more than double that of sludge from primary sedimentation. Normally, chemical precipitation from 1 million gal of wastewater will yield about 5000 gal of sludge with moisture content of 95%.

Digested sludge, from septic, Imhoff, or separate digestion tanks, is very dark in color and has a homogeneous texture. When wet, it has a tarry odor. Roughly, treatment of 1 million gal of wastewater will produce 800 gal of digested sludge with a moisture content of about 90%.

The sludges removed in wastewater treatment may contain as much as 97% water. The objective of sludge treatment is to separate the solids from the water and return that water to a wastewater-treatment plant for processing. Sludge treatment may require:

1. **Conditioning**. Sludge is treated with chemicals or heat so that the water may be readily separated.

2. **Thickening**. Removal of as much water as possible by gravity or flotation.

3. **Stabilization**. Processes known as sludge digestion are employed to stabilize (make less odorous and less putrescible) the organic solids in the sludge so that they can be handled or used as soil conditioners without creating a nuisance or health hazard.

4. **Dewatering**. Further removal of water by drying the sludge with heat or suction.

5. **Reduction**. The solids are converted into a stable form by incineration or wet oxidation processes.

Sludge Conditioning ▪ This may employ any of several methods to facilitate separation of the water from the solids in sludge. Historically many approaches have been used, but now the only common one is adding polymers to the sludges to improve their separation and thickening.

Sludge Thickening ▪ This may be accomplished in one of several ways: gravity thickening, flotation thickening, belt thickening, or centrifuge thickening. Simple and inexpensive, gravity thickening is essentially a sedimentation process, employing a tank similar in appearance and action to a circular clarifier used in primary and secondary sedimentation (Fig. 22.18a). Best results are obtained with sludges from primary wastewater treatment. In flotation thickening (Fig. 22.18b), air is injected into the sludge under pressures of 40 to 80 psi. Containing large amounts

of dissolved air, the sludge flows into an open tank. There, under atmospheric pressure, the dissolved air comes out of solution as minute air bubbles. These attach themselves to solids in the sludge and float them to the surface, where a skimming mechanism removes them. This method is effective on activated sludge, which is difficult to thicken by gravity, as are belt thickening, drum thickening or centrifuge thickening which use proprietary equipment to thicken. Current practice generally favors use of belt drum thickening in smaller plants and centrifuge thickening in larger plants.

Sludge Digestion ▪ This employs the biological decomposition of organic matter, which makes up about 70% of total solids, by weight, in sludge to achieve gasification, liquefaction, and mineralization of the solids. It can be applied to treatment-process sludges other than chemical sludges and those containing substances toxic to sludge organisms, such as cyanides and chromium. Advantages of sludge digestion include production of a stable, inoffensive sludge (if the process is continued long enough); 35 to 45% reduction of suspended solids; 40 to 60% reduction

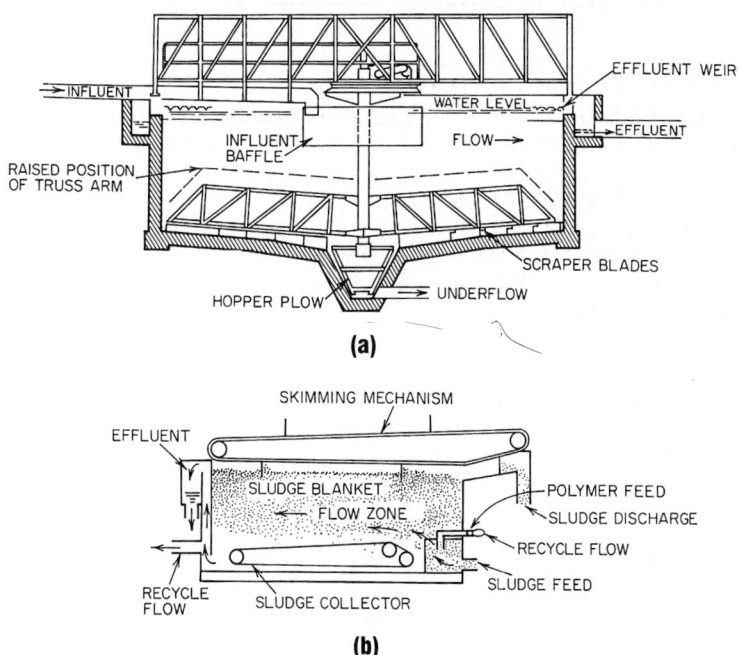

Fig. 22.18 Cross sections of sludge-thickening equipment: (*a*) Gravity thickener; (*b*) flotation thickener.

in dry weight of volatile matter; reduction in moisture content; and production of a sludge from which water may be more easily removed. The digested sludge may be used as a soil conditioner and weak fertilizer under certain conditions. Furthermore, gases produced during digestion may be used as fuel. (If the sludge is to be dewatered and incinerated, digestion is not usually employed.) Digestion may be anaerobic, performed in closed tanks devoid of oxygen, or aerobic, with air injected into the sludge.

Anaerobic Sludge Digestion ▪ Sludges are transferred to separate digestion tanks, unless Imhoff-type tanks or septic tanks are used. While sludge decomposes in a digester, fresh sludge is added periodically. Anaerobic bacteria attack the carbohydrates first, forming organic acids. After this initial acid fermentation, acid digestion occurs. Organisms living in the acid environment attack the organic acids and nitrogenous matter. Then, a period of digestion, stabilization, and gasification takes place, in which the anaerobic bacteria feed on proteins and amino acids, such as acetic or butyric acid. Volatile acids are reduced, and the pH rises. In the final stage, methane fermentation occurs, with methane as the principal gaseous product. Speed of digestion is indicated by the rate of gas formation. Periodic removal of liquefied matter, excess liquor

(or supernatant liquor), and digested solids makes room for fresh sludge.

Supernatant liquor, the liquid fraction in a digester, is high in solids and biochemical oxygen demand. It has an offensive odor. Withdrawn from a digester in small quantities at a level where the liquor contains relatively few solids, it is disposed of by insertion in the influent to a primary sedimentation tank.

Sludge-gas production under good operating conditions is about 12 ft^3/lb of volatiles destroyed. The gas is 60 to 70% methane, 20 to 30% carbon dioxide, plus minor amounts of other gases, including hydrogen sulfide. Fuel value of sludge gas usually ranges from 600 to 700 Btu/ft^3. The gas may be used at the treatment plant to operate auxiliary engines and provide heat for sludge-heating systems. Excess gas is burned.

In conventional designs, stages of the anaerobic process proceed simultaneously in the tank. Mixing of well-digested sludge with fresh sludge provides balance. If the pH holds between 7.2 and 7.4, conditions for digestion will be most favorable. Once achieved, balance may usually be maintained if addition of fresh solids is held to less than 4%, by weight, of the solids in the tank.

Speed of anaerobic digestion depends on temperature (Fig. 22.19). In conventional sludge digestion, as illustrated in Fig. 22.20a (mesophilic range), 100°F is the optimum temperature. Between 110 and 140°F (thermophilic range), thermophilic,

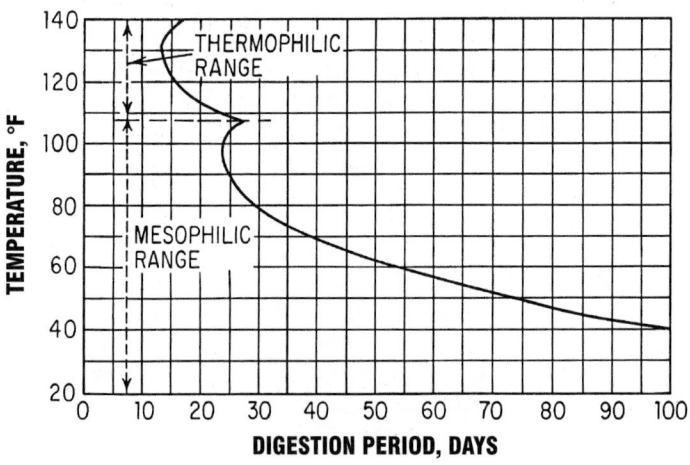

Fig. 22.19 Digestion period decreases with increasing temperature, reaching a minimum in the mesophilic range at about 100°F and in the thermophilic range at about 130°F.

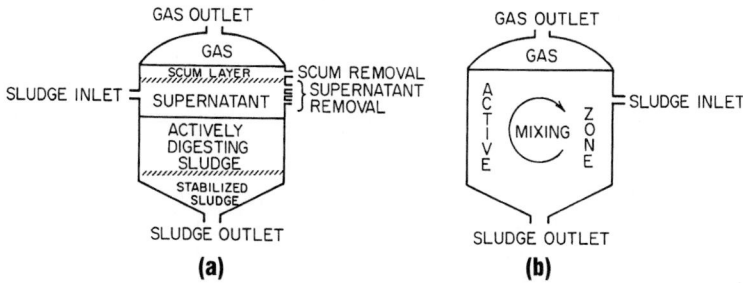

Fig. 22.20 Sludge digestion. (*a*) Standard-rate digestion—unheated, detection time 30 to 60 days, loading 0.03 to 0.10 lb of volatile suspended solids per ft³·day, intermittent feeding and withdrawal, and stratification. (*b*) High-rate digestion—heated to between 85 and 95°F, detection 15 days or less, loading 0.10 to 0.50 lb of volatile suspended solids per ft³·day, continuous or intermittent feeding or withdrawal, and homogeneous.

or heat-loving, bacteria become active and speed digestion even more, with an optimum temperature of 130°F. Tanks are heated to hasten digestion (Fig. 22.20*b*).

Most states have established schedules of capacity requirements for digestion tanks, depending on type of sludge and whether or not the tanks are heated. Typical requirements set a capacity, ft³ per capita, for heated tanks of 2 to 3 for primary sludges, 3 to 4 for mixtures of primary and standard-filter sludges, and 4 to 6 for activated sludge or mixtures of primary and high-rate filter sludges. Capacities of unheated tanks should be twice as great for each type of sludge.

Sludge-digestion tanks are either cylinders, or more recently "egg shaped". They generally provide a means of manipulating the sludge. The system also may include preheater and heater equipment, recirculation pumps with sludge suction at several levels, supernatant-liquor drawoff at several levels, gas dome or collector, stirring mechanism, sludge rakes, and drawoff. The tank cover may be floating or fixed. With a fixed cover, when fresh sludge is added to a tank kept full, an equal volume of supernatant liquor must be removed. Addition of sludge creates currents, as a result of which the liquor being removed may carry off some of the sludge. A floating cover allows the liquor to be withdrawn before or after the fresh sludge enters the tank.

In simple conventional multistage anaerobic digestion, two or more digesters are placed in series. The sludge drawoff of each is fed to a subsequent one, and digested sludge is removed from the last (Fig. 22.21). The system provides flexibility in manipulating and mixing sludges and in controlling supernatant liquor. Also, it may be possible to use a smaller tank than required for single-stage operation or, for a given-size tank, to retain solids longer. In two-stage digestion, good results may be obtained if less than 20%, by volume, of material transferred from the first to the second tank is the best-digested sludge and more than 80% is supernatant liquor with the lowest solid content.

Advanced multistage anaerobic systems are becoming more common. These generally require less tankage, are more stable, destroy more volatile solids and produce more gas; in some cases they can improve the dewatering properties of the product sludge (called **Biosolids** as per current convention for all treated sludges). When a thermophilic stage is included, these can be designed to produce an USEPA Class A Biosolid. The most common of these are:

• **TPAD** (Temperature Phased Anaerobic Digestion)—usually consist of a 5 day thermophilic digester followed by a 10 day mesophilic digester. This is a patented process and a license fee must be paid.
• **AG** (Acid/Gas Anaerobic Digestion or Two Phased Digestion)—commonly consists of 1.5–2 day acid phase reactor (where almost no gas is produced and the pH is biochemically lowered to <6) followed by a 12+ day gas phase digester

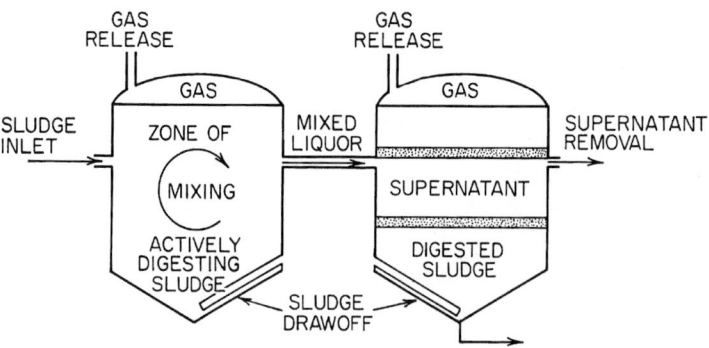

Fig. 22.21 Simple conventional two-stage anaerobic digestion.

(where 0.95% of gas is produced). Typically the first stage is mesophilic 9 (though a proprietary version uses a thermophilic first stage) and the second stage can be mesophilic or thermophilic.

Several new proprietary pretreatment systems (before anaerobic digestion) are just coming to the market and include thermal hydrolysis (CAMBI®) and various ultrasonic pre-treatments. These generally claim similar benefits to the advanced systems above, including production of a Class A Biosolids for CAMBI® and even smaller tankage.

Aerobic Digestion ■ Organic sludges are aerated in an open tank similar to an activated-sludge aeration tank. The process provides about the same reduction in solids as the anaerobic process but is more stable in operation and recycles fewer pollutants to the wastewater-treatment plant. Aerobic digestion, however, has higher power costs and does not produce fuel gases.

Stabilization ■ Primary and secondary waste-water sludges may be stabilized for reuse as sludge conditioners by composting. In this process, the sludge is mixed with a bulking material, such as wood chips or refuse. Placed in piles or windrows about 7 ft high, the mixture undergoes biological action that stabilizes the sludge and heats it sufficiently to kill most disease-causing organisms in it. The composting takes about 3 weeks, after which the mixture usually is cured for about another month before its reuse. Other competing methods are also common, including several proprietary processes.

Sludge Dewatering ■ Before disposal, digested biosolids from relatively small treatment plants may be concentrated in drying beds. Area needed for this purpose is about 2 to 3 ft² per capita (about three-fourths as much if the beds are covered). Beds consist of up to 12 in of coarse sand over 12 to 18 in of gravel. The natural earth bottom is sloped to underdrains, usually spaced about 30 ft apart. A bed may be from 20 to 30 ft wide and up to 125 ft long. It may be bounded or separated from an adjacent bed by a concrete wall extending about 15 in above the sand surface.

The bed is dosed with sludge to a depth of 9 to 12 in and allowed to drain and dry. A well-digested, granular sludge drains easily and reduces to a depth of 3 to 4 in when dry (60 to 70% moisture content). Sludge removed from the bed has little or no odor. It may be used as a weak fertilizer or may be landfilled.

For relatively large treatment plants, mechanical dewatering systems are advantageous because they are more compact and more controllable. Such systems include belt filter presses, centrifuging, and pressure filtration. Vacuum filtration, once very popular, is seldom used anymore.

Belt Filter Press ■ Belt-filter presses are continuous-feed dewatering devices that use the principles of chemical conditioning, gravity drainage, and mechanically applied pressure to dewater sludge (Fig. 22.22). The belt-filter press was introduced in the United States in the early 1970s and has become one of the predominant sludge-dewatering devices. It has proved to be

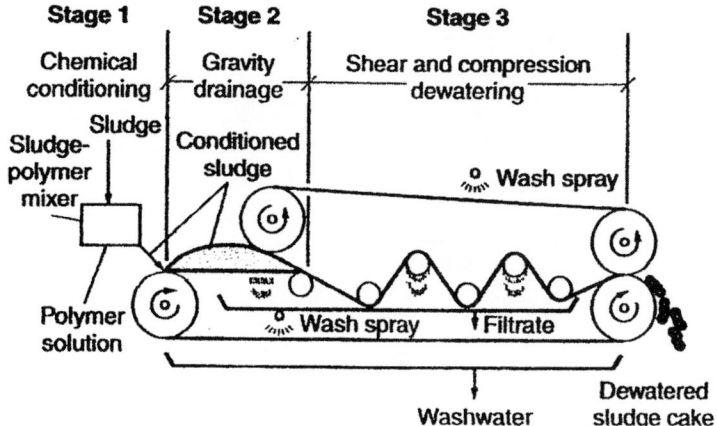

Fig. 22.22 Belt-press dewatering: three basic stages of belt-press dewatering.

effective for almost all types of municipal wastewater sludge and biosolids.

In most types of belt-filter presses, conditional sludge is first introduced on a gravity drainage section where it is allowed to thicken. In this section, a majority of the free water is removed from the sludge by gravity. On some units, this section is provided with vacuum assistance, which enhances drainage and may help to reduce odors. Following gravity drainage, pressure is applied in a low-pressure section, where the sludge is squeezed between opposing porous cloth belts. On some units, the low-pressure section is followed by a high-pressure section where the sludge is subjected to shearing forces as the belts pass through a series of rollers. The squeezing and shearing forces thus induce release of additional quantities of water from the sludge. The final dewatered sludge cake is removed from the belts by scraper blades.

Centrifuge dewatering of sludge is accomplished in a horizontal drum rotated at 1600 to 2000 rpm. Sludge is pumped into the centrifuge and injected with polymers for sludge conditioning. As the drum turns, the solids are spun to the outside of the drum and removed by a conveyer (Fig. 22.23). Costs and results are similar to those obtained with vacuum filtration.

Pressure filtration is accomplished by pumping sludge at pressures up to 225 psi through filters attached to a series of plates. The plates are supported in a frame between a fixed and a moving end. When sludge is forced into the chambers between plates, the liquid passes through the filters while the solids are retained. When the filter chambers fill up with solids, the sludge feed is stopped. The filter cake is dislodged by shifting the moving end so that the plates can be moved. Pressure filtration provides the driest cake obtained by mechanical dewatering methods, produces a clear filtrate, and often reduces chemical conditioning costs.

Sludge Soil Conditioning ▪ EPAs biosolids regulations encourage the land application of biosolids. Because sludge from municipal wastewater treatment contains some essential plant nutrients, it can be used as a fertilizer or soil conditioner. For that purpose, however, it is desirable that the sludge be first stabilized. It is often also dewatered.

Some cities apply liquid sludge to croplands. This eliminates dewatering costs but requires transporting of large amounts of sludge, and for health reasons, the sludge cannot be used for root crops or crops eaten raw. In the Chicago area, crops fertilized with liquid sludge include corn, soybeans, and winter wheat.

In some cases, sludge is dried in high-heat flash driers to reduce the volume substantially. Flash driers operate by mixing a portion of dried sludge with incoming wet sludge cake and introducing a high-velocity high-temperature gas stream. The

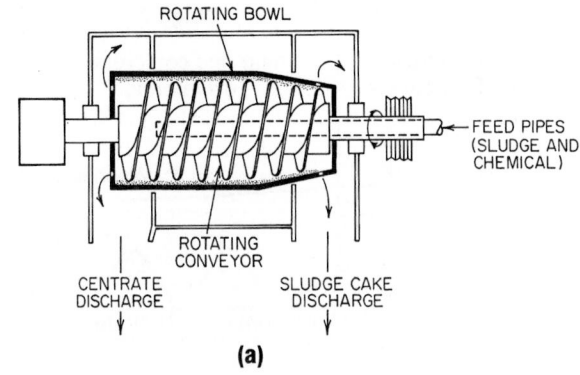

ROTATING BOWL

FEED PIPES
(SLUDGE AND
CHEMICAL)

ROTATING
CONVEYOR

CENTRATE
DISCHARGE

SLUDGE CAKE
DISCHARGE

(a)

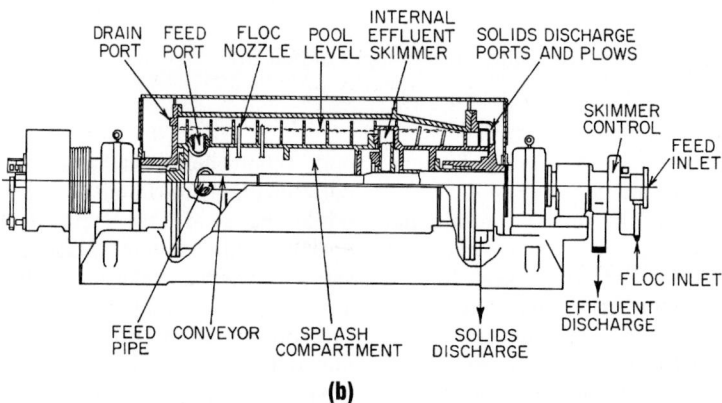

DRAIN
PORT / FEED
PORT / FLOC
NOZZLE / POOL
LEVEL / INTERNAL
EFFLUENT
SKIMMER / SOLIDS DISCHARGE
PORTS AND PLOWS

SKIMMER
CONTROL

FEED
INLET

FLOC INLET

EFFLUENT
DISCHARGE

FEED
PIPE / CONVEYOR / SPLASH
COMPARTMENT / SOLIDS
DISCHARGE

(b)

Fig. 22.23 Centrifuge equipment for dewatering sludge: (*a*) Continuous countercurrent, solid-bowl, screw-conveyor-discharge centrifuge. (*b*) Concurrent-flow, solid-bowl, conveyor-discharge centrifuge.

dried material is separated from the gas in a cyclone separator and moved to storage. If a refuse incinerator is located at the wastewater-treatment site, it can provide heat for sludge drying.

Sludge Reduction ▪ If sludge is not to be used as a soil conditioner and if a landfill disposal site is not available, the sludge may be reduced to a more innocuous and more easily handled form by incineration, chemical oxidation, or wet oxidation.

During incineration, the moisture in the sludge is completely evaporated and the organic solids are burned to a sterile ash. Digested as well as undigested sludge, however, may be disposed of by incineration. Auxiliary heat is needed because the moisture content of the filter cake is high. Gas,

including digester gas, oil, or coal, may be used as fuel.

In the past, incinerators used to burn sludge were multiple-hearth. Fed initially to the top hearth, sludge is pushed down to the next hearth by agitator arms as it dries. The heat drives off water and volatile gases, which are ignited by the high temperature. To avoid excessive odors, the temperature should be maintained at 1500°F or more. Ash residue, if it meets state standards, may be used for fill or cover on sanitary landfill. Flue gases are passed through a scrubber to limit air pollution.

If digester gas is not available for fuel, cost of sludge incineration may be high. As an alternative, filter cake may be mixed with solid wastes and burned in a municipal incinerator, if it adjoins the treatment plant.

A fluidized-bed incinerator is an alternative (see Art. 22.33.)

(Metcalf & Eddy, Inc., "Wastewater Engineering," 3rd ed., L. Rich, "Low Maintenance, Mechanically Simple Wastewater Treatment Systems," C. R. Brunner, "Handbook of Incineration Systems," and R. A. Corbitt, "Standard Handbook of Environmental Engineering," McGraw-Hill Publishing Company, New York (www.books.mcgraw-hill.com); G. M. Fair et al., "Water and Wastewater Engineering," John Wiley & Sons, New York (www.wiley.com); "Wastewater Treatment Plant Design," Manual 36, American Society of Civil Engineers (www.asce.org); W. F. Ettlich et al., "Operations Manual—Sludge Handling and Conditioning," Environmental Protection Agency, Cincinnati, Ohio (www.epa.gov); M. J. Satriana, "Large-Scale Composting," Noyes Data Corp., Park Ridge, N.J.; "Sludge Thickening," MOP 15 no. FD-1, Water Environmental Federation, Arlington, Va (www.wef.org).)

22.25 Imhoff Tanks

Developed by Karl Imhoff in Germany for the Emscher sewage district, this type of tank was widely used in the United States from 1907 until about 1970 for primary treatment of wastewater. The tank permits both sedimentation and sludge digestion to take place. Sludge comprises the settled solids in wastewater, and sludge digestion is the anaerobic decomposition of organic matter in sludge (Art. 22.21).

Efficiency of Imhoff tanks is about the same as for plain sedimentation tanks. Imhoff effluents are suitable for treatment in trickling filters. Sludge digestion, however, may proceed much more slowly in an Imhoff tank than in a separate digester. In an Imhoff tank, sludge digestion takes place without heat. Since rate of digestion decreases with drop in temperature (Fig. 22.19), lack of temperature control is a disadvantage, especially in regions where winters are cold.

Imhoff sludge has a tarlike odor and a black, granular appearance. It is dense. When withdrawn from a tank, it may have a moisture content of 90 to 95%. It dries easily, and when dry, it is comparatively odorless. It is an excellent humus but not a fertilizer.

Imhoff tanks are compartmented (Fig. 22.24). Sedimentation occurs in an upper, or flowing-through, chamber. Sludge settles into a lower chamber for digestion. To facilitate transfer of the settling solids, the flowing-through chamber has a smooth, sloping bottom (about 60° with the horizontal) with a slot at the lowest level. After particles pass through the slot, they are trapped in the lower chamber. Their path is obstructed either by overlapping walls at the slot, as shown in the cross section in Fig. 22.24, or by a triangular beam with an apex just below the slot.

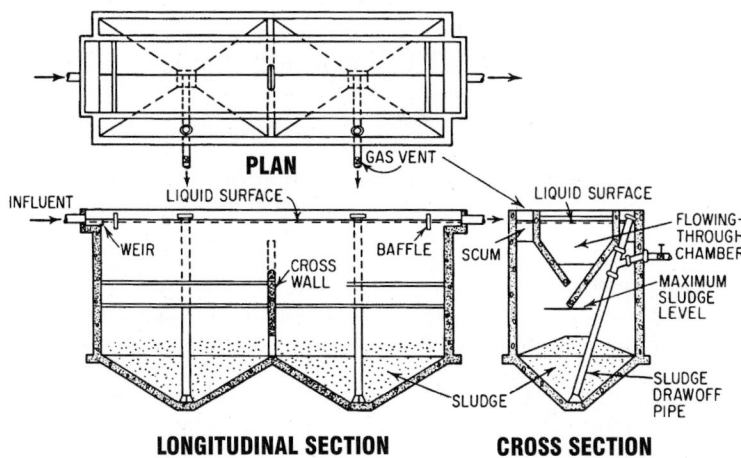

Fig. 22.24 Imhoff tank permits sedimentation of wastewater in upper compartments and sludge digestion in lower compartments.

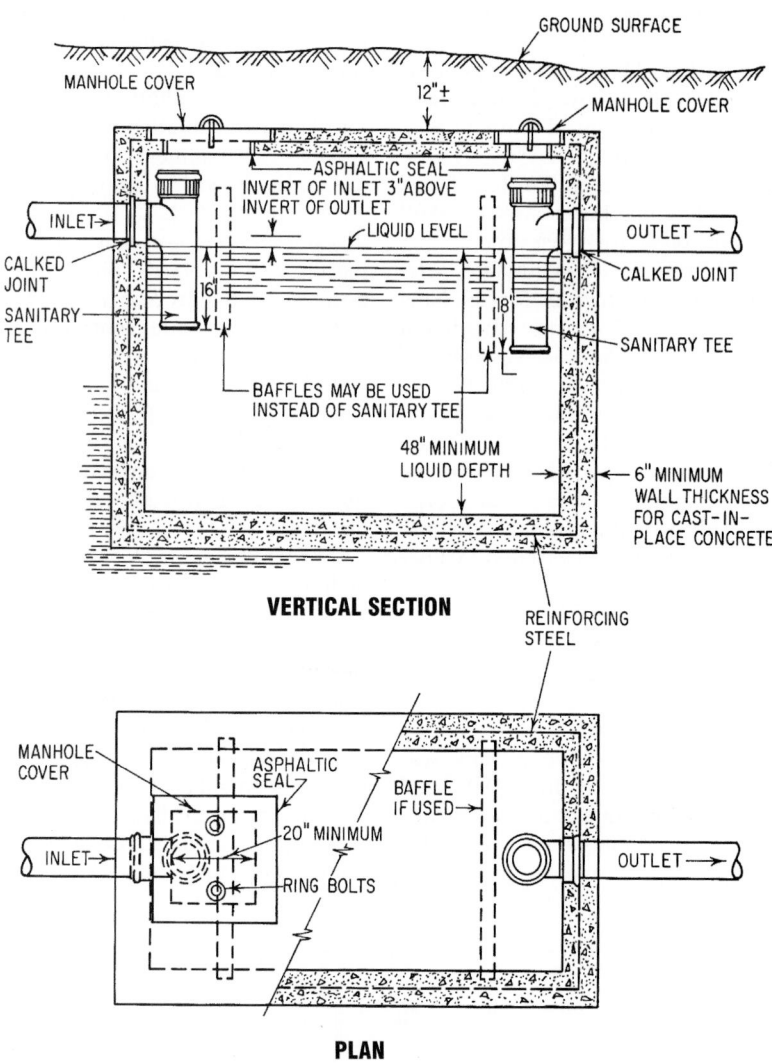

Fig. 22.25 Septic tank permits sedimentation and sludge digestion in the same compartment.

As digestion proceeds in the lower chamber, scum is formed by rising sludge in which gas is trapped. The scum is directed to a scum chamber and gas vent alongside the upper chamber. As gases escape, sludge sinks back from the scum chamber to the lower chamber. (The gas vents occasionally may give off offensive odors.) The scum chamber should have a surface area 25 to 30% of the horizontal surface of the digestion chamber. Vents should be at least 24 in wide. And top

freeboard should be at least 2 ft to contain the scum. If foaming occurs at a gas vent, it can be knocked down with a water jet from a hose.

In the digestion chamber, sludge settles to the sloped bottom. After sufficient time has elapsed for anaerobic decomposition, the sludge is removed through drawoff pipes. Since the height of a tank usually is 30 to 40 ft, the sludge can be expelled under the hydraulic pressure of the liquid in the tank. Ordinarily, sludge withdrawals are made twice a year. With such a schedule, the digestion chamber may be designed for a capacity of 3 to 5 ft^3 per capita of connected wastewater load. If, however, sludge removal is less frequent, or if industrial wastes with large quantities of solids are present in the wastewater, the capacity should be greater. Some chambers have been constructed with capacities up to 6.5 ft^3 per capita.

Large tanks are provided with means for reversing flow in the upper chamber. Since sedimentation generally is largest near an inlet, flow reversal permits a more even distribution of settled solids over the digestion chamber.

Detention period in the upper chamber usually is about 2½ h. Surface settling rate generally is 600 gal/(ft^2 · day). The weir overflow rate normally does not exceed 10,000 gal/lin ft of weir per day. Velocity of flow is held below 1 ft/s.

Length-width ratios of Imhoff tanks range from 3 : 1 to 5 : 1. Depth to slot is about equal to the width.

Multiple units are preferable to a single large tank. Sometimes it also is expedient to set two flowing-through chambers above one digestion chamber.

22.26 Septic Tanks

Like Imhoff tanks (Art. 22.25), septic tanks permit both sedimentation and sludge digestion. But unlike Imhoff tanks, septic tanks do not provide separate compartments for these processes. While undergoing anaerobic decompositions, the settled sludge is in immediate contact with wastewater flowing through the tank.

Septic tanks have limited use in municipal treatment. Their effluents are odorous, high in biochemical oxygen demand, and dangerous because of possible content of pathogenic organisms. Septic tanks, however, are widely used for treatment of wastewater from individual residences. Such tanks also are used by isolated schools and institutions and for treatment of sanitary wastewater at small industrial plants.

State and local health departments regulate the size of a residential septic tank, typically based on the number of bedrooms or bathrooms in the house. Minimum size for residential septic tanks is usually 1,000 gallons. Many regulatory agencies require a second compartment of approximately 300 gallons capacity, separated from the first compartment by a vertical partition. The partition

has a horizontal slot, about 6 in high, to permit passage of effluent from the first compartment.

A septic tank may be constructed of coated metal or reinforced concrete and should be watertight. It should have a minimum liquid depth of 4 ft. Length of a rectangular tank may be about twice the width. Cast-in-place concrete tanks should be at least 6 in thick, unless completely reinforced. The top slab, at least, should be reinforced to support 150 psf. The tank top should be between 12 and 24 in below finished grade. An opening at least 16 in in diameter should be provided for a manhole. The underside of the tank top should be at least 1 in above the tops of partitions and baffles. The invert of the inlet pipe should be at least 1 in, preferably 3 in, above the invert of the outlet. When the length of a tank exceeds 9 ft, two compartments should be used. Figure 22.25 shows a typical tank.

Residential septic tanks usually are buried in the ground and forgotten until the system gives trouble because of clogging or overflow. Actually, sludge should be pumped from the septic tank on a yearly basis.

Commercial scavenger companies are available for sludge removal in most areas. Using a tank truck equipped with pumps, they remove the contents of a septic tank and cart them to a sewer manhole or a treatment plant for disposal. In rural areas, the sludge may be buried in an isolated site.

Municipal and institutional septic tanks are designed to hold 12- to 24-h flow, plus stored sludge. For camps for 40 or more persons, septic tanks should have a liquid capacity of at least 25 gal per person served. For day schools, the capacity may be two-thirds as large.

For residential units, the main vent for the house plumbing normally provides adequate ventilation. For large septic tanks, however, separate vents for the tanks are desirable.

Septic-tank effluent may be disposed of in a leaching cesspool (Art. 22.27) or a tile field. The latter consists of lines of open-jointed tile or perforated pipe laid in trenches 18 to 30 in deep. The lines receive the effluent from a distribution box, which distributes the liquid equally. From the box, the lines spread out, so that they are at least 6 ft apart. Lines should be of equal length, but none should be over 60 ft long.

Laid on a slight slope, not more than $\frac{1}{16}$ in/ft, the tile or pipe is firmly set in a bed of crushed stone or washed gravel. The aggregate should extend 12 in below and 2 in above the conduit. The effluent, discharging from the openings, disperses over the entire trench bottom and seeps into the ground. The size of the tile field should be determined from the results of soil-percolation tests (Table 22.5).

At least two soil-percolation tests should be made in the area of the tile field. To perform a test, dig a hole 8 in in diameter or 12 in square. It should extend 6 in below the trench bottom or about 30 in below the final ground surface. Place 2 in of coarse sand or fine gravel in the bottom of the hole. Presoak the hole by filling it with water several hours before the test and again at the time of test and allowing the water to seep away. Remove any soil that falls into the hole. Pour clean water to a depth of 6 in in the hole. Record the time, minutes, required for the water to drop 1 in. Repeat until the time for the water to drop from the 6- to 5-in levels is about the same for two successive tests. Use the results of the last test as the stabilized rate. Alternative percolation-test methods have been developed for use where peculiar soil conditions exist.

Lots with less than 10 ft of soil above a rock formation usually are not suitable for construction of both wastewater systems and well-water supplies because of contamination hazards. Tile fields should not be constructed under driveways. The fields should be more than 100 ft away from any source of water supply, 20 ft from house foundation walls, and 10 ft from property lines. Trench bottoms should be at least 2 ft above groundwater, 5 ft above rock. Roof, footing, and basement drains should not be connected to septic tanks or they will be overloaded with water not requiring treatment. Water from roof gutters and other storm water should be routed away from the tile field. This water would saturate the soil and interfere with proper operation of the field.

Where soil is impervious or nearly so, an underdrained tile field may be used. This, in reality, is a buried sand filter placed below the tile drainage system. The drainage tile is laid in trenches filled with gravel or other porous media. Underdrains at the bottom collect and convey the effluent to a central collection point. There, the waste may be either drained out by gravity, chlorinated and discharged to a body of water, or pumped to a discharge point.

("Sewage Disposal Systems for the Home," Part III, *Bulletin* 1, Department of Health, State of New York, Albany, N.Y.; "Studies on Household Sewage Disposal System," Parts 1 to 3, Robert A. Taft Sanitary Engineering Center, U.S. Public Health Service; L. Rich, "Low Maintenance Mechanically Simple Wastewater Treatment Systems," McGraw-Hill Publishing Company, New York (www.books.mcgraw-hill.com); W. J. Jewell and R. Swan, "Water Pollution Control in Low-Density Areas," University of Vermont.)

22.27 Cesspools and Seepage Pits

A cesspool is a lined and covered hole in the ground into which wastewater is discharged. It is

Table 22.5 Suggested Sizes of Tile Fields for Septic-Tank Effluent

Soil-percolation rate, min*	Wastewater application, gal per ft² per day	Trench width, in	Trench length, lin ft, for wastewater loads, gal/day, of			
			300	450	600	1000
0–5	2.4	24	63	94	125	209
6–7	2.0	24	73	110	146	244
8–10	1.7	36	59	88	118	196
11–15	1.3	36	77	116	154	256
16–20	1.0	36	95	143	191	317
21–30	0.8	36	125	188	250	417
31–45†	0.6	36	167	250	334	555
46–60†	0.4	36	250	375	500	834

*Time for 1-in drop in water level in soaked hole.

†If the percolation rate exceeds 60 min, the system is not suitable for a tile field. A rate over 30 min indicates borderline suitability for soil absorption; special care should be used in design and construction.

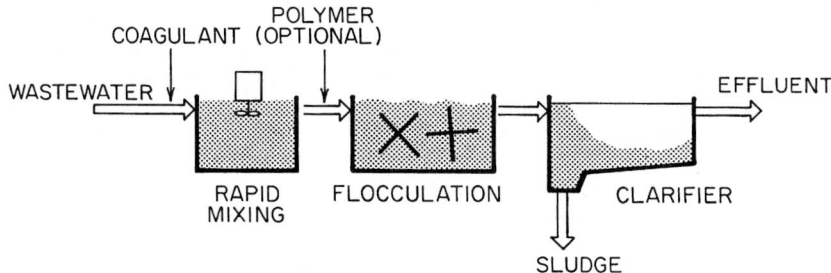

Fig. 22.26 Schematic of coagulation-sedimentation process.

used only when a sewerage system is not available. It may be watertight or leaching. A watertight cesspool retains wastewater until it is removed, by pumps or buckets. This type of cesspool is used only where no drainage into surrounding soil or rock is permitted. A leaching cesspool allows wastewater to seep into the surrounding ground.

Seepage pits of similar construction may be used to supplement tile fields (Art. 22.26) or instead of such fields where conditions are favorable. The pits also may be used in series with cesspools or septic tanks, to drain overflow liquid into the surrounding soil. Results are similar to those obtained with septic tanks (Art. 22.26).

Use of a leaching cesspool for direct disposal should be restricted to a small family in a remote location where there is absorptive soil and no danger of groundwater pollution. Leaching cesspools and seepage pits should never be attempted in clay soils.

The bottom of a seepage pit should be at least 2 ft above groundwater and 5 ft above rock. Lots with less than 10 ft of soil above a rock formation generally are not suitable for construction of both seepage pits and well-water supplies because of contamination hazards. Pits should be located more than 100 ft from a source of water supply, 20 ft from buildings, and 10 ft from property lines. Clear distance between two pits should be at least two times the diameter of the larger pit.

Size of seepage pit should be determined on the basis of 75 gal per person per day or 150 gal per bedroom per day. When bedrooms are used as a criterion, allowance should be made for future conversion of some rooms into bedrooms. The pit lining should be open-jointed or perforated to permit liquid to leak out. Wall area should be large enough to allow the soil to absorb the liquid without the pit overflowing. The required wall area, or effective absorption area, should

be determined from soil-percolation tests (Table 22.6).

Percolation tests for seepage pits are the same as for tile fields (Art. 22.26). The tests should be made, however, at half the depth and at the full estimated depth of the seepage pit. A larger excavation may be made for the upper portion of the hole, to facilitate execution of the test.

When the required absorption area has been obtained from Table 22.6, the outside diameter and effective depth of pit may be obtained from Table 22.7. The lining generally is made of concrete block or precast-concrete sections. Thickness should be at least 8 in. With rectangular block, the bottom should not be more than 10 ft below grade; with interlocking block, not more than 15 ft. For deeper pits, the lining should be structurally designed to resist saturated-earth pressures. The top should have a watertight manhole and concrete cover.

Coarse gravel should be placed in the bottom of the pit to a depth of 6 in. Backfill around the lining in the absorption area should be clean crushed stone or gravel, $1\frac{1}{2}$ to 2 in in diameter, to a thickness of at least 6 in. A 2 in-thick layer of straw should be placed on top of the gravel before soil is backfilled.

When a seepage pit is used at the end of a tile field, the pit wall should be at least 6 ft from the end of the trench. The pipe connecting the end of the line with the pit should have tight joints.

("Sewage Disposal for the Home," Part III, *Bulletin* 1, Department of Health, State of New York, Albany, N.Y.)

22.28 Disinfection

The last step in secondary treatment of wastewater is disinfection of the effluent to kill pathogenic

Table 22.6 Suggested Absorption Areas for Seepage Pits

Soil-percolation rate, min*	Wastewater application, gal per ft² per day	Required absorption area, ft², for wastewater loads, gal/day, of			
		300	450	600	1000
0–5	3.2	94	141	188	313
6–10	2.3	130	196	261	435
11–15	1.8	167	250	334	555
16–20	1.5	200	300	400	666
21–30	1.1	273	409	545	911
31–45†	0.8	375	562	750	1250
46–60†	0.5	600	900	1200	2000

*Time for 1-in drop in water level in soaked hole.

†If the percolation rate exceeds 60 min, the system is not suitable for a seepage pit. A rate over 30 min indicates borderline suitability for soil absorption; special care should be used in design and construction.

(disease-causing) bacteria and viruses. For the purpose, chlorine or ultraviolet radiation is generally used.

Chlorination ▪ The major purpose of chlorinating treated wastewater is to destroy pathogenic organisms. **Chlorine demand** of domestic or industrial wastewater is the difference between the amount of chlorine added and the residual after a short time. This interval usually is taken as 15 min since this is the time required to kill nearly all the objectionable bacteria. Sufficient chlorine should be added to treatment effluent to satisfy the demand and provide a residual of 2 ppm (mg/L). The contact period should be at least 15 min at peak hourly flow or maximum pumping rate and 30 min at average daily flow.

The following dosages, ppm or mg/L, may be required for disinfection of treated wastewater: primary sedimentation effluent, 20 or more; trickling-filter-plant effluent, 15; activated-sludge-plant effluent, 8; and sand-filter effluent, 6. Such disinfection is desirable and often mandatory where discharge of the effluent may pollute water supplies, shellfish beds, or beaches.

The 5-day BOD of wastewater is reduced about 2 ppm for each ppm of chlorine added. A BOD reduction of 15 to 35% may be expected with residuals of 0.2 to 0.5 ppm after 10 min.

Table 22.7 Seepage-Pit Dimensions* for Required Absorption Area, Ft²

Depth, ft	Outside diameter, ft							
	5	6	7	8	9	10	11	12
3	47	57	66	75	85	94	104	113
4	63	75	88	101	113	126	138	151
5	79	94	110	126	141	157	173	188
6	94	113	132	151	169	189	207	226
7	110	132	154	176	197	220	242	263
8	126	151	176	201	225	252	276	302
9	141	170	198	226	254	283	310	339
10	157	189	220	251	282	314	346	377
11	173	207	242	276	310	346	380	415
12	188	226	263	302	339	377	415	453

*Outside diameter and effective depth. Bottom area excluded from computations.

Chlorinators usually are used to feed chlorine to the treatment effluent. Chlorine gas normally is dissolved in water, and the solution is pumped into the effluent in measured amounts, proportional to the flow. In small plants and some large plants, hypochlorinators may be used. These may feed sodium hypochlorite (laundry bleach) or calcium hypochlorite.

Chlorination should be done in a baffled contact tank, unless there will be sufficiently long contact time in a conduit or outfall before the chlorinated effluent is discharged. The accuracy of the chemical feeders should be checked daily by determining the weight of chlorine or hypochlorites used. In addition, the efficacy of dosages applied should be checked frequently by bacteriological tests.

Chlorine also may be useful in preventing odors at wastewater treatment plants. For this purpose, it may be added on line or to primary influent. Chlorination before primary sedimentation is not detrimental to sludge digestion.

Other uses of chlorine include neutralization of hydrogen sulfide, or prevention of its formation, where it may corrode concrete sewerage or structures; increasing the efficiency of air in grease removal in skimming tanks; control of ponding and filter-fly larvae on trickling filters; conditioning of sludge before dewatering; and treatment of industrial wastes.

Some states place rigid restrictions on discharge of effluents containing chlorine that may form trihalomethane, a potential cancer-causing agent, in receiving waters used for drinking. A tentative maximum contaminant level of 100 mg/L has been proposed. Check with state authorities for limitations on free available chlorine in discharges.

Dechlorination, now commonly required, conventionally accomplished by injection of alum, sodium bisulfite, sodium sulfite, or sulfur dioxide.

Ultraviolet Disinfection ▪ Another alternative to chlorine is use of ultraviolet light to kill bacteria and viruses. The wastewater is passed over horizontal glass cylinders, inside of which are ultraviolet light sources. A circular windshield wiper keeps the tube surfaces clean.

(American Water Works Association, Inc., "Water Quality and Treatment," 4th ed., McGraw-Hill Publishing Company, New York (www. books.mcgraw-hill.com); G. C. White, "Handbook of Chlorination," 2nd ed., Van Nostrand Reinhold Company, New York; R. A. Corbitt, "Standard Handbook of Environmental Engineering," McGraw-Hill Publishing Company, New York.)

22.29 Advanced Wastewater Treatment

Wastewater secondary treatment and disinfection generally produce an acceptable effluent for disposal on land or a large body of water in that more than 85% of the BOD and suspended solids and nearly all pathogens are removed from the wastewater. This treatment, however, usually removes only small percentages of some pollutants, such as phosphorus, nitrogen, and heavy metals. Where these pollutants in an effluent are of major concern, advanced, or tertiary, wastewater treatment should be applied. The following processes are capable of improving the effluent from secondary treatment to the degree that it is adequate for many reuse purposes.

Coagulation-Sedimentation ▪ When used as a tertiary treatment, coagulation-sedimentation improves overall treatment of wastewater by providing a means for removal of the excessive quantities of solids that may escape occasionally from the biological processes. Coagulation-sedimentation also may remove high percentages of phosphorus, heavy metals, bacteria, and viruses.

In this treatment, coagulants, such as lime, alum (aluminum sulfate), or ferric chloride, are injected into the wastewater. They speed settlement of the solids in the wastewater because they cause the solids to clump together. This action is accelerated by addition of a polymer as a settling aid and by flocculating, or slowly stirring, the wastewater. After flocculation, the wastewater flows to a sedimentation tank, or clarifier, where the solids settle to the bottom, from where they are removed (Fig. 22.27).

Filtration ▪ In tertiary treatment, filtration is used to remove suspended solids from a secondary effluent or from the effluent from a coagulation-sedimentation process. Filtration may be performed in an open concrete structure by gravity flow or in steel vessels by pressure. Plain filtration (Art. 22.22) can reduce the suspended solids and the particulate fraction of the BOD_5 in activated-

(a) Modified Ludzack-Ettinger (MLE)

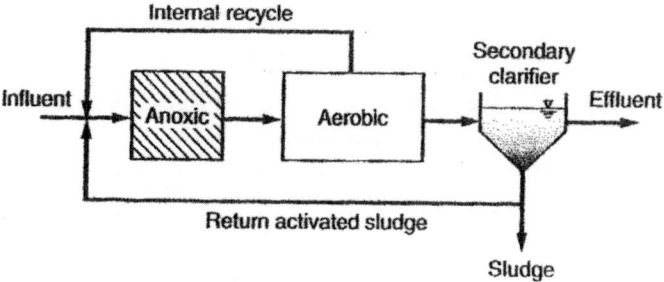

(b) Step-denite

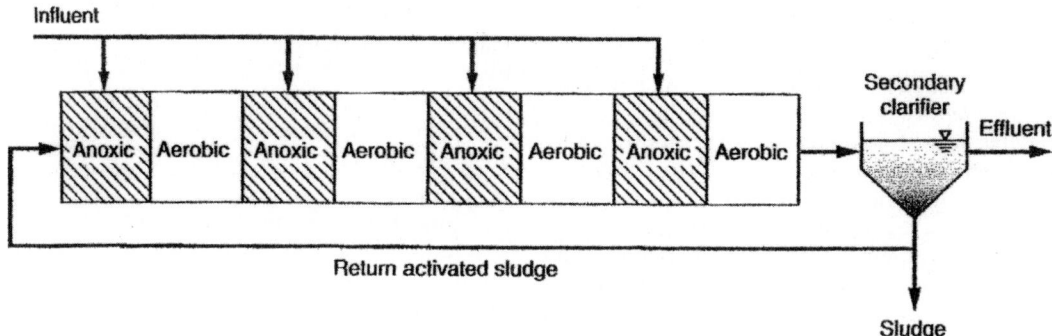

Fig. 22.27 Schematic of suspended growth processes for nitrogen removal. (*a*) Modified Ludzack-Ettinger (MLE). (*b*) Step-denite.

sludge effluent from 50 to 75%. Effective filtration of the effluent from tertiary coagulation-sedimentation can reduce phosphorus to 0.1 mg/L or less and eliminate suspended solids.

The filters may be multimedia, composed of a mixture of different materials, such as coal, sand, and garnet or deep bed coarse (1–1½ in) monomedia filters typically with 5 to 6 feet of media or shallow fine monomedia filters. The filters are coarse in the upper layers and become uniformly finer with depth. The wastewater is passed downward during normal operation, but flow is reversed to clean the filters. A newer type of effluent filter introduced in the 1980s is a continuous backwash filter (CBF). This filter is an upflow filter and uses 1.0–1.6 mm sand in nominal 6′ feet deep beds. As the name implies, the media is continuously backwashed by removing small amounts of the filter, cleaning and redepositing it in the filter bed. This kind of filter is quite common in smaller plants, but has been used in plants as large as 15 mgd.

Carbon Adsorption ▪ Activated carbon has the capacity for removing from wastewater refractory organics, organic substances resistant to biological breakdown, which are responsible for the color of secondary effluent. These substances adhere to the surfaces of the porous carbon particles and can be removed by heating the carbon in a furnace with very low levels of oxygen. The activated carbon can then be reused. Carbon adsorption has had very limited success in domestic wastewater treatment, but where very high degrees of treatment are desired, after secondary treatment, coagulation-sedimentation, and filtration. This combination of processes can produce a colorless, odorless effluent, free of bacteria and viruses, with a BOD of less than 1 mg/L and a COD of less than 10 mg/L, suitable for many reuse purposes. In any case, the wastewater to be treated is passed through beds of granular carbon particles, about 0.8 mm in diameter, arranged like a gravity filter or in columns 20 to 25 ft deep. Time for contact between carbon and wastewater may range from 20 to 40 min.

Nitrogen Reduction Treatments ▪ Nitrogen contained in wastewater is converted into ammonia during conventional biological secondary treatment. Ammonia, although not toxic to humans, is toxic to fish and is objectionable also because it consumes dissolved oxygen, corrodes copper fittings, and increases the amount of chlorine needed for disinfection. The amount of ammonia retained in wastewater can be reduced by biological or physical-chemical methods. The latter include ammonia stripping, selective ion exchange, and breakpoint chlorination and are rarely used any more. Both carbon adsorption and nitrogen reduction should be tried out on the wastewater to be treated in a pilot plant before the prototype is built.

Biological nitrification-denitrification first biologically converts the ammonia nitrogen into nitrates (nitrification). This is accomplished by injection into the wastewater of sufficient oxygen (about 4.5 lb per pound of ammonia nitrogen in the wastewater). The next step is denitrification, biological conversion of the nitrates to gaseous nitrogen, which escapes to the atmosphere. Denitrification can be performed in an anaerobic activated-sludge process (suspended growth system) or a fixed-film system.

While there are many different ways to biologically nitrify the most common ways are to use a version of the **MLE** (Modified Ludzack–Ettinger) process or a version of the **step-denite** (or **step feed BNR**) process. The **MLE** process as shown on Fig. 22.27a, consists of a nitrifying activated sludge tank preceded by a small (typically 2–3 hr) anoxic tank. Nitrified Mixed liquor is continuously re-circulated (at a flow of 2–4 times influent flow) brings the nitrate to the head of the anoxic tank where it mixes with the raw wastewater organic carbon needed for denitrification. The step denite process, as shown on Fig. 22.27b, is essentially a step feed plant operated in the nitrifying mode with the first part of each pass (usually about a third of the pass) operating as anoxic denitrification tanks. In this process the raw sewage organic carbon is brought to the nitrate, the energy costs are lower and the tankage generally is smaller. **SBRs, oxidation ditches** may also be designed and operated in a nitrification–denitrification mode. All of these are generally capable of reducing the total N (TN) to about 8–10 mg/L, without chemical addition, and, under some circumstances, a little lower. These all return about half the alkalinity lost in nitrification back to the process (about 7.2 lb of alkalinity (as $CaCO_3$) is lost per lb of N oxidized, and in low alkalinity waters must otherwise be replaced). In general most **attached growth systems** can either nitrify or denitrify, but not in the same tank—a two stage system is typically needed for these.

In order to reach lower levels of TN, usually an external carbon source, commonly methanol (about 3–4 lb/lb of oxidized N denitrified), must be used. The most common way of doing this is to add the methanol ahead of deep bed monomedia filters (see above). These filters thus serve the dual purposes of effluent filtration and denitrification. They are usually sized at a loading of 2 ± 0.5 gpm/sf. Oxidized N level of <0.5 mg/L are easily achieved this way and in most circumstances an affluent TN of 3.0 (depending on amount of undegradable, soluble organic N in wastewater) can be maintained. Other types of filters have also been used, but less commonly, for this purpose.

Bio-P Removal (BPR) ▪ Chemical removal of P has been covered in the coagulation section (above). **BPR** is done one of several originally proprietary processes (AO (Phoredox), Bardenpho (5-stage), UCT, VIP etc.—see Fig. 22.28). Each uses a small (usually ≤1 hr) anaerobic tank ahead of anoxic and/or aeration tank to select for microorganisms that biologically can release P and then pick it (plus more) up biologically in the following aeration tanks. These are generally considered capable of reducing P to below 1.0 mg/L, though some claim as low as 0.5 mg/L after filtration. To get lower levels (to as low as <0.1 mg/L) these processes must be combined with chemical P removal.

(a) Modified bardenpho (5-stage)

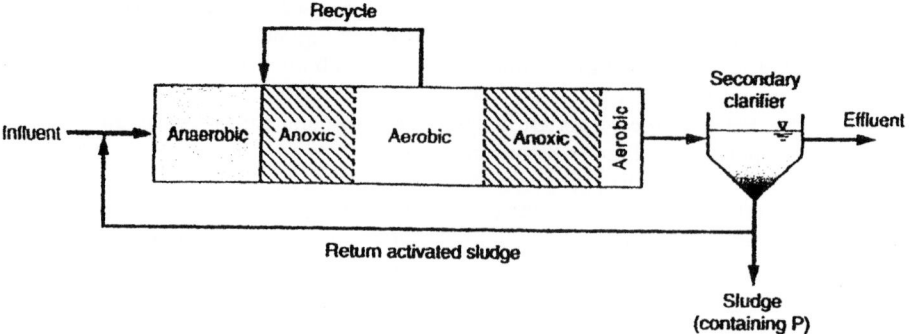

(b) UCT (standard and modified)

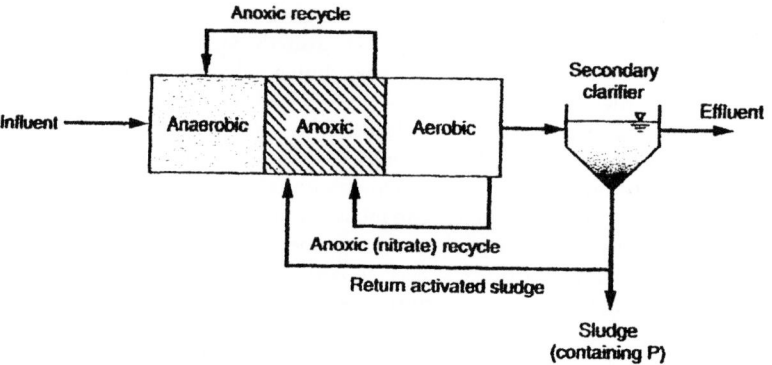

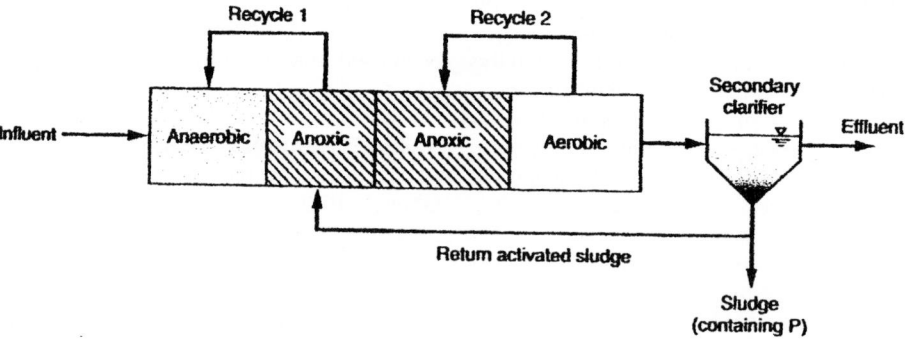

Fig. 22.28 Schematic of suspended growth process for phosphorus removal. (*a*) Modified Bardenpho (5-stage). (*b*) UCT (standard and modified). (*c*) VIP. (*d*) Johannesburg process. (*e*) Phoredox (A/O). (*f*) A^2/O.

(c) VIP

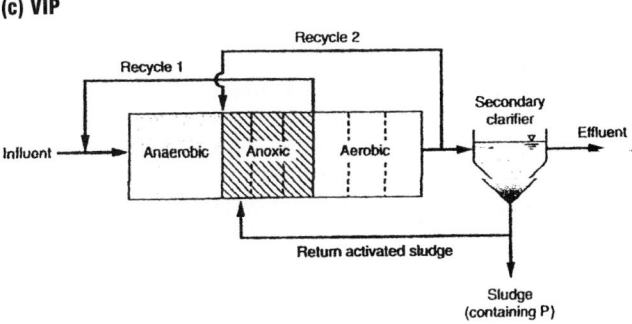

(d) Johannesburg process

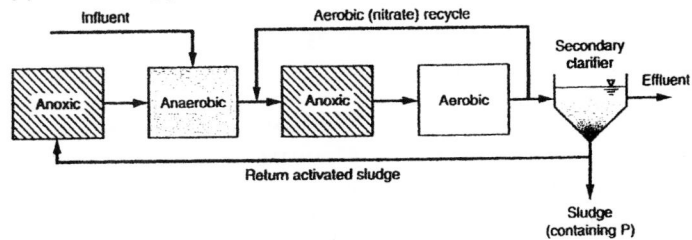

(e) Phoredox (A/O)

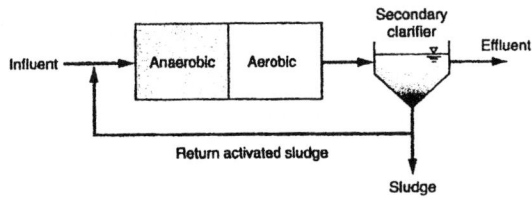

(f) A²/O

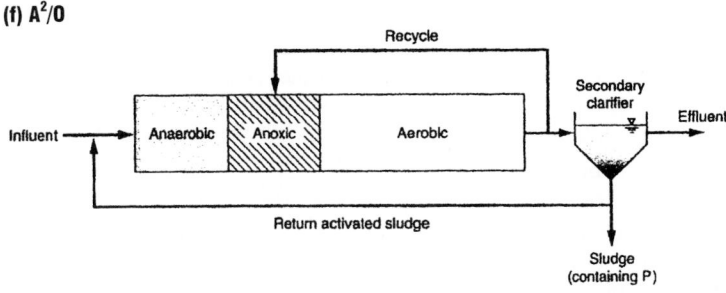

Fig. 22.28 (*continued*)

22.30 Industrial Waste Treatment

The treatment of industrial wastes (see Art. 22.2) is highly specialized. Selection of treatment processes must be engineered to the peculiar characteristics of a process waste. It is desirable, whenever possible, to reduce the volume of wastewater requiring treatment or to separate wastes requiring intensive treatment from those requiring little or no treatment. Cooling water, for example, can be segregated from high-strength wastes, thereby reducing the size of the treatment plant.

Process wastes have a wide range of flow from hour to hour, depending on the operation. Hence, it may be necessary to provide equalization or holding tanks to produce a more uniform flow to be treated over a 24-h period. This is more efficient than treatment units designed to handle maximum flows produced during an 8-h shift. It is also possible with equalization tanks to mix wastes of different characteristics, such as acids and alkalies, and obtain a neutralized waste. Peak solids production and BOD may also be reduced or regulated.

Industrial wastes may be placed in general classifications, such as food processing, textile and apparel manufacture, chemical manufacture, and basic materials manufacture, including pulp and paper, iron and steel, metal plating, oil processing, glass, plastic, and rubber production and processing. Table 22.8 indicates some characteristics of waste typical of the several classifications. When preparing for treatment of any specific waste, an engineer should see that the waste is sampled over a sufficient time period to include major variations introduced by process operation.

Treatment of process wastes may require a series of methods selected to accomplish certain degrees of treatment that would ultimately produce an effluent acceptable for discharge to a receiving stream. These methods may include, in order:

Pretreatment to reduce temperature, neutralize the wastes, and remove fibers and other coarse solids by screening

Primary treatment to remove settleable solids

Secondary treatment by biological processes applied to biodegradable wastes

Secondary treatment with chemicals for chemical conversion, precipitation and removal of solids, and oxidation or reduction of substances contained in the waste

Preconditioning or secondary treatment by anaerobic digestion to produce a biochemical conversion of substances

Ion exchange, dialysis, reverse osmosis, or evaporation to remove inorganic solids or recover chemicals

Chlorination for oxidation or disinfection purposes

Various forms of irrigation, lagooning, or algal oxidation ponds

It is frequently necessary to select theoretically best combinations of treatment for a process waste and to follow up the selection with pilot-plant operations to establish the parameters of design for the full-scale treatment plant. Employment of advanced waste-treatment methods may be necessary for specified purposes, such as removal of trace metals, control of phosphorus- and nitrogen-bearing compounds, and reduction of excessive amounts of suspended solids. Several methods of treatment are described in Arts. 22.19 to 22.23, 22.25, 22.26, and 22.27.

Discharge Permits ▪ In accordance with the Clean Water Act (Art. 22.1), anyone discharging wastewater to the waters of the United States is required to obtain a permit for that purpose from the Environmental Protection Agency (EPA) or a designated state, under the National Pollutant Discharge Elimination System (NPDES). Permits usually are written for a specific term (up to 5 years) and contain effluent limitations and monitoring requirements for each discharge point. One objective is to have industry apply the best available technology economically achievable for controlling toxic pollutants and the best conventional pollutant-control technology for conventional pollutants. (Industries discharging to municipal sewer systems, however, are not required to obtain NPDES permits. Control of pollutants from these sources is achieved through EPA General Pretreatment Regulations, which set specific industry-by-industry standards with specific limits on effluents.)

EPA has established a list of hazardous wastes from specific industries, such as electroplating wastes or air-pollution-control scrubber sludges from coke ovens and blast furnaces. If a waste is on this general list, the producer must treat it as a hazardous waste. EPA has also compiled a list of toxic chemicals that are often contained in industrial wastes. Wastes containing any of these chemicals must be treated as a hazardous waste. But even if a waste is not on either list, it should be considered hazardous if it is radioactive, ignitable, corrosive, reactive, or toxic.

Table 22.8 Types and Characteristics of Industrial Wastes*

Type of waste	Unit	Volume, gal per unit	BOD, lb per unit	Suspended solids, lb per unit	Population equivalent per unit
Canning					
Corn products	Ton	12,000	19.5	30.0	186
Beans	Case no. 2 cans	35	200.0	60.0	0.35
Peaches	Ton	2,610	29.2	13.0	280
Tomatoes	Ton	227	8.4	2.9	82
Milk products					
General dairy	1,000 lb raw milk	340	570	540	10
Fermentation					
Brewing	1 barrel beer	204	1.2	0.6	12
Laundry	100 lb dry wash	400±	1,250±	500±	20–25
Roofing					
Paperboard	Ton	36,075	18.2	144.0	125
General slaughterhouse	1 animal	360	7.7	3.2	74
Paper mill					
Paperboard	Ton pulp	14,000	121		84
Textile					
Cotton sizing	1,000 lb goods processed	60.0			2
Basic dyeing	1,000 lb goods processed	18,000			90
Rayon viscose	1,000 lb product	140	110	9.6	800
Wool dyeing and scouring	1,000 lb product	240,000	125		1,500
Vegetable oils					
Acidulating waste	1 ton oil	385	1	0.5	10

*From E. B. Besselievre and M. Schwartz, "The Treatment of Industrial Wastes," 2d ed., McGraw-Hill Book Company, New York.

Hazardous Wastes ■ Options for disposal of hazardous wastes include recovery for reuse, incineration, landfills with the option of fixation before landfilling, land treatment, mine storage, and deep-well injection.

Radioactive wastes are subject to severe restrictions when the receiving body of water may be used for human consumption, recreational bathing, fish propagation for food, or plant irrigation. Federal and state regulations should be reviewed whenever radioactive wastes are to be disposed. Permissible concentrations of radioactive material in water are usually specified in microcuries per milliliter of water. Procedures that have been used for the treatment of radioactive wastes include concentration and storage, and dilution and disposal. Burial after required decay may follow the first and discharge to sewers or streams may follow the latter. Low-activity material may be diluted, while high-activity material, requiring long storage periods, may be safely enclosed in containers and buried or stored in isolated caves or other underground facilities.

Concentration of radioactive wastes before storage may be accomplished by coprecipitation. The radioactive sludge concentrate is then removed, packaged, and buried.

Evaporation is widely used for concentration of low-activity wastes. Condensate may be released to a sewer. The sludge is transferred to polyethylene-lined drums for burial. Cation exchange with synthetic resins may be used on small liquid volumes having low solids concentration and low radioactivity levels.

Land treatment is suitable for wastes that can be biodegraded, chemically altered, immobilized, or deactivated by interaction with soil. Studies should be made to determine acceptable waste-loading rates and monitoring requirements. Provisions should be made to prevent discharge from the site of untreated water or for treatment of such water. Unless contaminated soil is to be removed and transported to a treatment or disposal facility, applications of wastes containing metals should be controlled so that no toxic hazard will result. After the last load of waste has been placed, the site should be stabilized with vegetation or capped as required for a landfill to control infiltration and erosion.

For deep-well disposal, liquid wastes are injected through shafts into deep subsurface geologic formations, where the wastes will be contained. The technique has long been used by the oil industry for disposal of brine. Care must be taken that groundwater which might be required for use above ground will not be polluted.

For landfills, see Art. 22.31; for incineration, see Art. 22.32.

(W. W. Eckenfelder, Jr., "Industrial Water Pollution Control," H. M. Freeman, "Standard Handbook of Industrial Waste Treatment and Disposal," and S. C. Reed and E. J. Middlebrooks, "Natural Systems for Waste Management and Treatment," McGraw-Hill Publishing Company, New York (www.books.mcgraw-hill.com); J. R. Conner, "Chemical Fixation and Solidification of Hazardous Wastes," R. L. Culp et al., "Advanced Wastewater Treatment," J. Devinney et al., "Subsurface Migration of Hazardous Wastes," E. J. Martin and J. H. Johnson, "Hazardous Waste Management Engineering," and N. Nemerow and A. Dasgupta, "Industrial and Hazardous Waste Treatment," Van Nostrand Reinhold Company, New York.)

22.31 Sanitary Landfills

Refuse collected from households, commercial establishments, and industrial plants must be disposed of at minimum cost and without creating health hazards or nuisances. One solution is a sanitary landfill, which requires daily compaction of refuse and daily placement of an earth cover 6 to 12 in thick. The cover is increased to 2 ft when filling has been completed.

The method is suitable where low-cost land is available within convenient hauling distance of the contributing population and good soil is available for the earth cover. Other factors to consider in selection of a landfill site are possible adverse effects on quality of surface water, groundwater, and air and potential for subsurface migration of leachates. Also, a site should not be located within 200 ft of a fault nor within a 100-year floodplain.

Refuse comprises all solid wastes except body wastes. It may consist of garbage, ashes, rubbish, street cleanings, dead animals, abandoned automobiles and solid market and industrial wastes. Garbage consists of putrescible wastes resulting from processing, handling, preparation, cooking, and consumption of food. Rubbish consists of solid wastes other than ashes, body wastes, and garbage from domestic, commercial, and institutional sources.

About 14 acre-ft, including cover, per 10,000 population per year of operation will be required for sanitary landfill. Sufficient land should be available to ensure area for a preplanned period of 5 to 10 years. The area needed can be derived from an estimate of volume required computed from

$$V = \frac{R}{D}\left(1 - \frac{P}{100}\right) + C_v \qquad (22.33)$$

where V = volume, yd^3 per capita per year, of sanitary landfill

R = weight of refuse, lb per capita per year, to be handled at landfill

D = average density of refuse, lb/yd^3

P = percentage reduction of refuse volume from compaction

C_v = volume, yd^3, of cover material required (6- to 12-in-thick intermediate layers, temporary sides, front slope, and top, and at least 24 in on all finished surfaces)

C_v varies from 17% of the refuse volume for deep fills to 33% for shallow fills. It may be assumed at 25% in estimates. For this value, the required volume of landfill may be estimated from

$$V = 1.25 \frac{R}{D}\left(1 - \frac{P}{100}\right) \qquad (22.34)$$

Drainage of the site before, during, and after filling should be planned in advance. Provision should be made for windbreaks to keep dust, paper, and other light objects from being blown away from dumping areas and becoming a nuisance. Also, the final disposition of the site should be planned in advance.

Parks, recreational areas, and outdoor storage are suitable end uses for landfills. Choice of end use should be influenced by the uncertain settlement characteristics of such fills and the objectionable odors that may be released where excavations are made. A covered fill may be odorless, but excavation may be hazardous and expensive because of the presence of obnoxious toxic and flammable gases produced by decomposing refuse.

Soil used for cover should not have a high proportion of sand or clay or operation of trucks will be hindered. Clay also is difficult to handle, and when dry, it cracks, providing openings for rodents, insects, and air. A sand-clay-loam mixture with about 50% sand has been found satisfactory.

Landfills must meet a number of restrictive requirements imposed by state regulatory agencies. An impermeable liner is typically placed at the bottom of the landfill to prevent leachate from entering and contaminating groundwater. A drainage system consisting of perforated pipe in a gravel drainage layer is placed at the bottom of the landfill, above the liner, for leachate collection. Leachate is either treated on site or transported to a wastewater treatment plant for treatment. The specific requirements in each state should be ascertained by the engineer. Usually, the engineer is required to submit a plan and report on the specific areas to be filled, schedule of filling, site preparation, sources and types of materials to be used as cover, and subbase. The plan should also include details on application of cover material; composition of waste; final grades; handling of surface water and fill drainage, including the method of collection and treatment of leachate to prevent groundwater or surface-water pollution; erosion control; nuisance control; air-pollution prevention measures; method of record keeping; and, in general, any data required to ensure that environmental impact (Art. 22.34) will not be adverse or unacceptable to the enforcement agency.

Landfills should be provided with means for controlling leachate and runoff. Leachate is contained by placing an impermeable liner under and around the site. Choice of liner depends on the nature of the wastes to be discharged. Liners often are concrete, synthetic fabrics, or impermeable clay.

(R. A. Corbitt, "Standard Handbook of Environmental Engineering," McGraw-Hill Publishing Company, New York (books.mcgraw-hill. com); American Public Works Association Committee on Refuse Disposal, "Municipal Refuse Disposal," APWA, 2345 Grand Boulevard, Suite 500, Kansas City, MO 64108-2641 (www.APWA. net); D. G. Wilson, "Handbook of Solid Waste Management," and A. Bagchi, "Design, Construction and Monitoring of Sanitary Landfills," Van Nostrand Reinhold Company, New York.)

22.32 Incineration of Refuse and Hazardous Wastes

Where land is costly or unavailable for sanitary landfill, municipalities may resort to incineration for refuse disposal. Refuse comprises all solid wastes except body wastes. The material is not homogeneous, and its characteristics vary considerably. Fuel value may range from 600 to 6500 Btu per pound of refuse, as fired. Moisture content influences this value significantly.

Controlled high-temperature (1600°F or more) incineration is an effective alternative to traditional methods of disposal of hazardous wastes. Such incineration is capable of converting many hazardous wastes into innocuous gases and ash and often recovering some of the energy produced by combustion. The process must be controlled to prevent emission to the atmosphere of hazardous combustion products or products of incomplete combustion. EPA regulations require incinerator operators to obtain a permit to burn the specific wastes to be treated. Basic standards call for 99.99% destruction and removal efficiency for each principal hazardous component of the waste; 99% removal of HCl from the exhaust, when the wastes contain more than 0.5% of organically bound chlorine; and emission not exceeding 180 mg/m³ of exhaust gas.

Refuse ▪ In incineration of refuse, volatiles are driven off by destructive distillation. They ignite from the heat of a combustion chamber (Fig. 22.29). Gases produced pass through a series of oxidation changes in which time-temperature relationship is important. They must be heated above 1400°F to destroy odors. Combustion products ultimately discharge from a stack at 800°F or less, usually after passing through an expansion chamber, fly-ash collector, and wet scrubbers. Normally, only submicron- and the smaller micron-size particles should escape with the flue gases. Dust emissions may be in the range of 2 to 3 lb per ton of refuse charged for a well-operated unit equipped with scrubbers.

Air needed ranges from 5 to 8 lb per pound of refuse burned. For nonhomogeneous wastes, up to 200% of theoretical requirements may be needed for combustion. The air provides oxygen for combustion of organic matter, helps dry wet refuse, and mixes with organic gases. But the air cools the gases if too much dry material is being burned. Air should be passed over the refuse and through it from under the grates.

Incinerators generally are rated in accordance with the estimated weight of refuse they are capable of burning in 24 h. Loading rates range up to slightly over 100 lb of refuse per hour per square foot of grate area for incinerators with mechanical stoking. Small incinerators for apartment buildings and institutions are loaded at much lower rates. Standards of the Incinerator Institute of America suggest loading rates for domestic refuse, lb/(h·ft^2), of 20 in 100-lb/h burning units to 30 in 1000-lb/h burning units.

Several types of incinerators are available from manufacturers. Kiln shape may be round or rectangular. The kiln may be stationary or rotate about a horizontal axis. The hearth may be horizontal and fixed, with grates; traveling, with grates; multiple; step movement; or barrel-type rotary (Fig. 22.29). Some types have drying hearths. Feed may be continuous, stoker, gravity, or batch.

For rational design of incinerators, the engineer should know or estimate such characteristics of the refuse to be burned as weight, water content, percentage of combustible and inert material, and Btu content. Available heat from the refuse must be balanced against heat losses due to radiation, excess air, flue gas, and ash. Heat balance can be calculated from several estimates based on averages. Manufacturers of each type of incinerator recommend sizes for various conditions. Furnace volume may be approximated by allowing 20,000 Btu/ft^3, and grate area by allowing 300,000 Btu/ft^2. Secondary combustion chambers permit combustion to continue to completion. Volumes of such chambers range from 10 to 25 ft^3/ton of rated capacity. Expansion chambers and other air-cleaning devices remove fly ash and other particles carried out of the furnace by gases. Expansion chambers are desirable where a stack serves more than one furnace. Gas velocities in secondary chambers should not exceed 10 ft/s.

Stacks should be designed for gas velocities of about 25 ft/s with maximum air. As a rough approximation, 0.3 ft^2 of stack area may be required per ton of rated capacity. Stack heights usually range from 100 to 180 ft. Height is desirable for

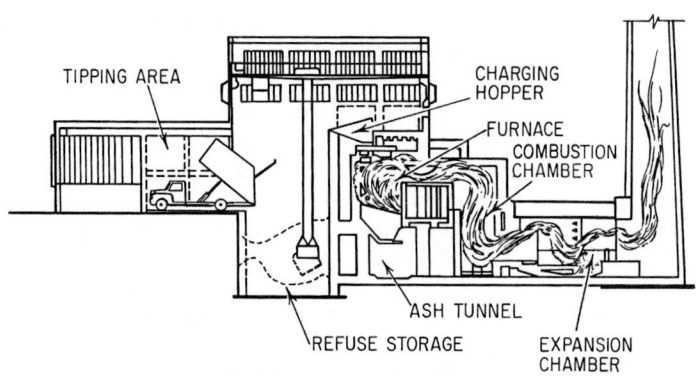

Fig. 22.29 Schematic of refuse incinerator.

creating natural draft and for dispersion of gases in the atmosphere.

Hazardous Wastes ▪ For incineration of hazardous wastes, liquid-injection incinerators are usually used. Vertically aligned units are generally used for wastes high in organic salts and yielding large quantities of ash. Horizontal incinerators are preferable for wastes producing small quantities of ash.

A fluidized-bed incinerator is one alternative (Fig. 22.30). It is a vertical steel cylinder with a grid supporting hot sand through which combustion air flows at a velocity high enough to keep the sand in suspension. The sand is heated by an air preheat system, plus fuel-fired combustion. Dewatered sludge, injected into the sand, is burned at temperatures between 1400 and 1500°F if it does not contain hazardous wastes and over 1600°F if it does. Ash is carried off with the exhaust gases and is captured in air-pollution-control devices. This equipment makes more efficient use of fuel than the multiple-hearth furnace (Art. 22.24), which, however, is simpler to operate and maintain.

Rotary kiln incinerators are another alternative. They can be used to burn solid and containerized wastes, slurries, and liquids. Other alternatives include starved-air/pyrolysis incinerators, incineration in high-temperature industrial facilities, such as cement kilns and industrial boilers; incineration at sea on special ships or off-shore platforms; and mobile incineration employing special heavy-duty truck trailers.

Because facilities that incinerate wastes and produce energy for other applications are exempt from Resource Conservation and Recovery Act

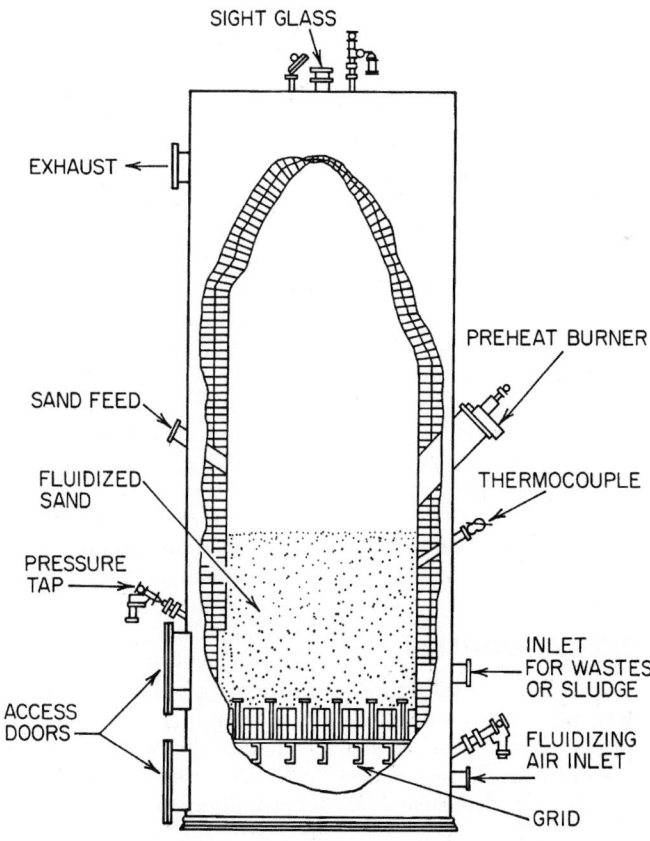

Fig. 22.30 Schematic of fluidized-bed incinerator.

emission-control regulations, there is an incentive to cofire wastes in industrial facilities. Studies indicate that organic wastes can be used to replace on the average up to 15% of cement-kiln fuel.

(E. T. Oppelt, "Thermal Destruction Options for Controlling Hazardous Wastes," *Civil Engineering*, September 1981.)

Incinerators should be designed to ensure that the wastes, auxiliary fuel if needed, hot combustion gases, and combustion air come into intimate contact and that the wastes stay long enough in the combustion chamber to be destroyed. For this purpose, high-efficiency burners should be used, liquid-waste feed should be atomized for insertion in the combustion chamber, and excess combustion air should be supplied and controlled to maintain turbulence in that chamber.

Emission-control devices should be used to limit emission of hazardous exhausts. Afterburners may be employed to provide additional combustion volume at high temperatures to burn incompletely combusted exhaust products. Scrubbers (Art. 22.33) are advantageous for removal of particulates, acid gases, and residual organics from the exhaust.

("Engineering Handbook on Hazardous Waste Incineration," Environmental Protection Agency, National Service Center for Environmental Publications, P.O. Box 42419, Cincinnati, OH 45242 (www.epa.gov/epahome/publications.htm); C. R. Brunner, "Handbook of Incineration Systems," and G. Tchobanoglous, "Solid Wastes: Engineering Principles and Management Issues," McGraw-Hill Publishing Company, New York (www.books.mc-graw-hill.com); D. G. Wilson, "Handbook of Solid Waste Management," and E. J. Martin and J. H. Johnson, Jr., "Hazardous Waste Management Engineering," Van Nostrand Reinhold Company, New York; H. B. Palmer and J. M. Beer, "Combustion Technology," Academic Press, New York (www.academicpress.com).)

22.33 Air-Pollution Control

Air pollution exists when one or more substances, such as dust, fumes, gas, mist, odor, smoke, or vapor, are present for a sufficient time in the atmosphere in quantities and with characteristics injurious to life or property, or detrimental to comfortable enjoyment of life and property.

These pollutants derive from numerous sources. They may be roughly classified as natural, industrial, transportation, agricultural, commercial and domestic heat and power, municipal activities, and fallout.

Natural sources include water droplets or spray evaporation residues, windstorm dusts, meteoric dusts, surface detritus, and pollen from weeds.

Industrial sources include process waste discharges, ventilation products from local exhaust systems, and heat, power, and waste disposal by combustion processes.

Transportation sources include discharges from motor vehicles, rail-mounted vehicles, airplanes, and vessels.

Agricultural sources include applications of insecticides and pesticides and burning of vegetation.

Commercial and domestic heat and power sources include gas-, oil-, and coal-fired furnaces used to produce heat for dwellings, commercial establishments, and utilities.

Municipal activity sources include refuse disposal, liquid-waste disposal, road and street paving, and fuel-fired combustion operations. Fallout comprises radioactive pollutants suspended in the air after a nuclear explosion.

Since pollutants are contributed by many sources, air pollution is always present but in varying degrees. In effect, pollution from natural sources is a base line with which total pollution can be compared. The major correctable sources of pollution are associated with community activity, rather than rural activity, because community air generally is more polluted.

Environment is made less desirable by pollutants. Hence, there is ample reason to conserve air as a resource, in many ways parallel to the need for conservation of water.

Air-pollution control requires knowledge of what constitutes an ideal atmosphere. This leads to establishment of criteria for clean air and standards setting limits on the permissible degree of pollution. Control also requires means for precise measurement of pollutants and practical methods for treating polluting sources to prevent undesirable emissions.

In addition to its adverse effects on health, air pollution also is objectionable because of its contribution to reduced visibility. In many parts of the world, burning of soft coal yields particles that combine with fog to produce smog, a mixture that at times reduces visibility to zero. Smog is created when microscopic water droplets condense about nucleating substances in the air to form aerosols. These are liquid or solid, submicron-size particles dispersed in a gaseous medium. In an atmosphere with an aerosol concentration of about 1 mg/m^3, visibility may be limited to 1600 ft. There would be about 16,000 particles per milliliter restricting visibility by scattering light.

Coal is only one source of nucleating particles that are responsible for smog. Chemical conversion of reaction products in the air also produces nucleating substances that grow large enough to cause light scattering. Converted sulfur dioxide too becomes a nucleating substance as it oxidizes and hydrolizes to form sulfuric acid mist.

The most desirable means of controlling air pollution is to prevent contaminants from getting into the atmosphere. Complete elimination of air pollution, however, is not always practicable. But there are many means for reducing it. Sulfur dioxide release, for example, can be decreased by use of a fuel with low sulfur content. An industrial process with a gaseous effluent can be changed to eliminate the gaseous waste. Aerosols and particles can be removed from a gas stream by air-cleaning equipment.

Air-Quality Standards ▪ In accordance with the Clean Air Act (Art. 22.1), the Environmental Protection Agency (EPA) develops National Ambient Air Quality Standards. These list the maximum amount of an air pollutant that can be present with an adequate margin of safety in protection of public health and welfare and that will not cause significant deterioration of air quality in areas where ambient standards have been attained. Check with EPA for the latest criteria because they are subject to change when appropriate for public protection.

EPA also develops National Emission Standards for Hazardous Air Pollutants to limit emissions that cause or contribute to air pollution. The standards apply to both new and existing sources. For example, for restricting emissions of inorganic arsenic from smelters, EPA requires high-efficiency particulate controls operated at optimum temperature for arsenic condensation for process gas streams, effective capture systems, and high-efficiency particulate controls for several sources of fugitive emissions.

In addition, EPA develops New Source Performance Standards based on the best systems that have been demonstrated to reduce emissions continually, taking into account costs and energy requirements. The standards apply to new sources and existing sources that have been modified after establishment of EPA criteria. For example, EPA has issued limitations for the following:

SO_2 and NO_2 emissions from industrial boilers

NO_2 emissions from diesel engines

Hydrocarbon emissions from dry cleaning equipment

Emission of volatile organics from numerous processes and storage units

Evaporative emissions from metal cleaning and degreasing operations

Particulate emissions from numerous processes, including battery manufacturing, processing of minerals prior to metal reduction, phosphate rock processing, coke ovens, manufacture of asphalt roofing and gypsum, and combustion of wood, municipal solid wastes, refuse-derived fuels, and bagasse—alone or combined with fossil fuels.

Air-Cleaning Devices ▪ Sizes of substances to be eliminated (Table 22.9) are a major factor in selection of air-cleaning devices. Coarse solids can be removed by screens. Particles down to 10 μm in diameter can be settled out in settling chambers with expanding cross section for velocity reduction to under 10 ft/s. Particles between 10 and 200 μm can be removed in cyclone separators, with an efficiency of 50 to 90%. In this equipment, the gas to be cleaned is injected tangentially into a cylindrical chamber. The gas spirals downward, then upward through the vortex at high velocity, and exits at the top. Before the gas leaves, however, particles are centrifuged out, hit the side walls, and drop to the conical bottom of the chamber.

Particles 10 μm in diameter or smaller may be removed with filters made of cloth, metal, or glass fiber. But air or gas velocities leaving such filters are low. For dry fiber filters, efficiency may be only about 50%. The efficiency of such filters, however, may be increased by application of a viscous coating, such as an oil with low volatility. Filters made of cloth usually are tubular bags, which trap particles as air or gas passes through. Many bags may be enclosed in a large chamber. When loaded with dust, they are shaken, and the dust falls into a

Table 22.9 Approximate Sizes of Particles in Aerosols, Dusts, and Fumes

Type of Particle	Size Range, Microns
Tobacco smoke	0.01–0.2
Rosin smoke	0.01–1.1
Carbon black	0.01–0.3
Zinc oxide fumes	0.01–0.4
Magnesium oxide smoke	0.01–0.5
Metallurgical fumes	0.01–1.3
Viruses	0.01–0.05
Oil smoke	0.03–1.0
Pigments	0.09–8
Ammonium chloride fumes	0.1–1.4
Alkali fumes	0.1–1.6
Metallurgical dust	0.1–200
Sulfate mist	0.5–3
Spray-dried milk	1–10
Bacteria	1–12
Pulverized-coal fly ash	1–60
Fog	1–50
Sulfuric acid concentrator	1.1–11
Cement dust	5–200
Sulfide ore for flotation	8–300
Foundry dusts	8–1000+
Stoker fly ash	10–90
Pulverized coal	10–500
Ground limestone	30–900
Mist	50–600

hopper. Bag filters remove 99% of particles larger than 10 μm. Filters packed with activated charcoal are used to absorb gases.

Wet collectors or scrubbers remove particles 1 to 5 μm in size. These devices also may remove water-soluble gases. In a scrubber, the gas to be cleaned may pass through a countercurrent water flow. The water may be sprayed or atomized. The scrubber may have deflectors to improve mixing of the gas and water. Chemicals may be added to the liquid to improve absorption.

Wet collectors often are used to clean air from kilns, roasters, and driers. They also are used for processes producing fine dust, films, vapors, and mists in food, chemical, foundry, metalworking, and ceramic industries. Scrubbers may be classified as dynamic precipitators, centrifugal collectors, orifice collectors, collectors with high-pressure nozzles, and packed towers.

In dynamic precipitators, dynamic or centrifugal forces, aided by water, clean the air. In centrifugal collectors, centrifugal forces throw particles in the air against wetted collector surfaces. After striking the surfaces, the particles fall to the bottom of the device and are removed. Orifice collectors deliver large quantities of water to a collecting zone where dust is removed from the air by centrifugal force, impingement, or collision. In collectors with high-pressure nozzles, air at 20,000 ft/min or more and water under 250 psi or more jet through venturi tubes. The water breaks into a fine mist, increasing the probability of contact with tiny particles. The turbulence disperses the water, causes quick impact with dust in the air, and removes the particles. In packed towers, dust particles are removed when air flows upward through the packing, which usually is in the form of ceramic saddles, while water flows downward.

Ionizable aerosols and particles down to 0.1 μm in size can be removed by electrostatic precipitators with an efficiency of 80 to 99%. These devices ionize particles in a gas passing by high-voltage electrodes. Oppositely charged plates trap the particles. To rid the plates of the particles, the current to the plates is interrupted or the plates are rapped.

Dispersal of Pollutants ▪ When pollutants cannot be completely eliminated at the source, air pollution may be reduced by keeping the concentration of the pollutants low by dispersing them. Whether atmospheric dilution is a suitable solution depends on the meteorology of a region, local topography, and building configurations. Basic meteorological conditions of the atmosphere that must be considered include wind speed, direction, and gustiness, and vertical temperature distribution. Under some conditions, humidity also is important. In general, diffusion theories predict that ground concentration of a gas or fine-particle effluent with very low subsidence velocity is inversely proportional to the mean wind speed. Vertical temperature distribution determines the distance from a stack of given height at which maximum ground concentration occurs. Raising the temperature of gas leaving a stack is equivalent to increasing stack height.

Gas does not normally come to the ground under inversion conditions. It may accumulate aloft when the atmosphere is calm or nearly so and be brought down to the surface as the sun heats the ground in early morning.

Turbulence caused by buildings and topography usually is so complex that theoretical computation of the effect is impractical. In some cases, however, model studies in wind tunnels have been used successfully to make predictions based on measurements of gas concentration and visible patterns of smoke.

Air Sampling and Monitoring ▪ The degree of air pollution at any time and place is determined by taking air samples and analyzing them. Air-sampling methods may be classified as those sampling particles, inorganic metals and salts, inorganic gases, organic substances, and mixed miscellaneous substances.

Many automatic, recording, air-monitoring instruments are available. They can be operated with few attendants and little manipulation. It is generally necessary to calibrate automatic instruments against a standard wet chemical or physical measurement method. Subsequent field calibration before, during, and after use may also be essential to maintain reliable test results. Although there are many variations, particle-sampling devices generally use gravity or suction-type collection and pass the sample through thermal or electrostatic precipitators, impingers and impactors, cyclones, absorption and adsorption media, scrubbing apparatus, or filters of various materials, such as paper, glass, plastic, or cloth.

Several types of units with air pumps drawing air through paper tapes mounted on a spool are available. The tape is moved automatically so that successive samples are taken for timed intervals on fresh paper.

In addition to standard wet chemical methods of measuring gases, there are many automatic or semiautomatic instruments designed to measure a spectrum (mass spectrometer) of one or more specific gases. These employ many different analytical principles, such as electrical conductivity; potentiometry; coulometry; flame ionization, thermal conductivity; heat of combustion; colorimetry; infrared, ultraviolet, and visible light photometry; gas chromatography; atomic absorption and electron capture.

Stack sampling requires special techniques and usually a train of sampling devices to measure particles and gases.

High-volume samplers are used at many sampling network stations in the United States. Electron microscopes may be used to examine aerosols and submicron particles. Photoelectric meters are used to control alarm systems connected to stacks. Combination instruments may be used for general sampling and location of emission sources. Such devices measure wind direction and velocity and direct air samples into multiple sample units, each representing a wind-direction sector.

(R. A. Corbitt, "Standard Handbook of Environmental Engineering," M. L. Davis and D. A. Cornwell, "Introduction to Environmental Engineering and Technology," and H. S. Peavey and D. R. Rowe, "Environmental Engineering," McGraw-Hill Publishing Company, New York (www.books.mcgraw-hill.com); P. O. Warner, "Analysis of Air Pollutants," and W. L. Faith and A. A. Atkinson, "Air Pollution," John Wiley & Sons, Inc., New York (www.wiley.com); R. O. Gilbert, "Statistical Methods for Environmental Pollution Monitoring," Van Nostrand Reinhold, New York.)

22.34 Environmental Impact Statements

In accordance with the National Environmental Policy Act (Art. 22.1), Federal agencies, departments, and establishments are required to prepare environmental impact statements in connection with proposals for legislative and other major Federal activities significantly affecting the quality of human environment.

It is essential that a draft statement be prepared, as early as possible, by the project engineer or other appropriate authorized person for review and comment. The actions may include all or any of the following:

1. Agency recommendations on their own proposals for legislation

2. Agency reports on legislation initiated elsewhere but concerning subject matter for which the agency has primary responsibility

3. Projects and continuing activities that may be

    a. Undertaken directly by an agency

b. Supported in whole or in part through Federal contracts, grants, subsidies, loans, or other forms of funding assistance

c. Part of a Federal lease, permit, license, certificate, or other entitlement for use

4. Decisions of policy, regulation, and procedure making

Although it is possible that there can be exceptions, the following actions are generally considered major or environmentally significant:

1. Actions whose impact is significant or highly controversial on environmental grounds

2. Actions that are precedents for much larger actions that may have considerable environmental impact

3. Actions that are decisions in principle about major future courses of action

4. Actions that are major because of the involvement of several Federal agencies, even though a particular agency's individual action is not major

5. Actions whose impact includes environmentally beneficial as well as environmentally detrimental effects

Contents of Environmental Impact Statements ▪ Environmental impact statements must assess in detail the potential environmental impact of a proposed action. The purpose of the statement is to disclose the environmental consequences of a proposed action. That disclosure is designed to alert the agency decision maker (local, state, or Federal, or any combination of these), the public, and, perhaps on major works, Congress and the President to environmental risks involved.

Environmental impact statements should present:

1. A detailed description of the proposed action, including information and technical data adequate to permit a careful assessment of environmental impact

2. Discussion of the probable impact on the environment, including any impact on ecological systems and any direct or indirect consequences that may result from the action

3. Any adverse environmental effects that cannot be avoided

4. Alternatives to the proposed action that might avoid some or all of the adverse environmental effects, including analysis of costs and environmental impacts of these alternatives

5. An assessment of the cumulative, long-term effects of the proposed action, including its relationship to short-term use of the environment versus the environment's long-term productivity

6. Any irreversible or irretrievable commitment of resources that might result from the action or that would curtail beneficial use of the environment

When the final statement is prepared, it must also include any discussions, objections, or comments presented by Federal, state, and local agencies, private organizations, and individuals that addressed the subject during review of the draft statement.

Impact Statement Review ▪ In general, any Federal, state, or local agency that has jurisdiction by law or specific expertise with respect to any environmental impact involved must be consulted for comments. Agencies to be consulted include those having responsibilities for the following (state or local agencies may have additional agency review requirements):

Water quality

Air quality

Weather modification

Environmental aspects of electric energy generation and transmission

Toxic materials, pesticides, and herbicides

Transportation and handling of hazardous materials

Wetlands, estuaries, waterfowl refuges, beaches

Historic and archeological sites

Flood plains and watersheds

Mineral land reclamation

Parks, forests, outdoor recreational areas and wildlife

Soil and plant life, sedimentation, erosion, and hydrologic conditions

Noise control and abatement

Food additives, food sanitation, and chemical contamination of food products

Microbiological contamination

Radiation and radiological health

Sanitation and waste systems

Transportation and air and water quality

Environmental effects with special impact on low-income neighborhoods

Rodent control

Urban planning, congestion in urban areas, housing and building displacement

River and canal regulation and stream channelization

In areas of environmental engineering activity, the principal government agency having responsibilities for reviewing impact statements is the Environmental Protection Agency. As a matter of fact, any Federal agency having a jurisdiction that centers around air and water pollution, drinking water supplies, solid waste, pesticides, radiation, and noise may be involved. Hence, engineers should ascertain specifically any agencies in addition to EPA that may have review responsibilities.

Engineers should also determine to what extent the state agencies dealing with the above areas have jurisdiction. In addition, engineers should check with the appropriate regional and municipal planning agencies. (See also Sec. 14.)

How to Prepare an Impact Report ▪ There are several alternate formats of report that would contain all the pertinent information required under the Federal guidelines. One method that has had technical acceptance is the base matrix, in which a series of actions that are part of a proposed project are related to the characteristics and conditions of the environment that are affected. Under each of the actions proposed, a ranking from 1 to 10 is placed to indicate impact magnitude, 10 being the highest order. Correspondingly, under a diagonal in the box, a ranking from 1 to 10 can be inserted concerning the importance of a specific impact as related to an environmental condition. Any suitable form of text that will discuss the significance of these two interrelated indices should be acceptable. A sample matrix illustrating these points is shown in Fig. 22.31.

One of the more complete diagrams for an information matrix was prepared by the United States Geological Survey in 1971. It appears as a separate attachment in Geological Survey Circular No. 645. The basis for the preparation of this matrix is indicated in Table 22.10.

In dealing with any particular project, the engineer can select from the matrix in Table 22.10 on either margin those conditions and actions applicable to the project. It is then possible for the engineer to prepare an environmental impact document so that reviewing agencies can provide comments in an orderly fashion. Within the format, it is important to present both present conditions and current trends, the alternate action proposed, and the impact either favorable or unfavorable that will result with and without the proposed action. If unavoidable harm may result from the proposed action, the procedures for reducing the harmful effect, together with the ultimate benefits resulting even though some harm may be done, should be presented in full detail with objective substantiation of all statements.

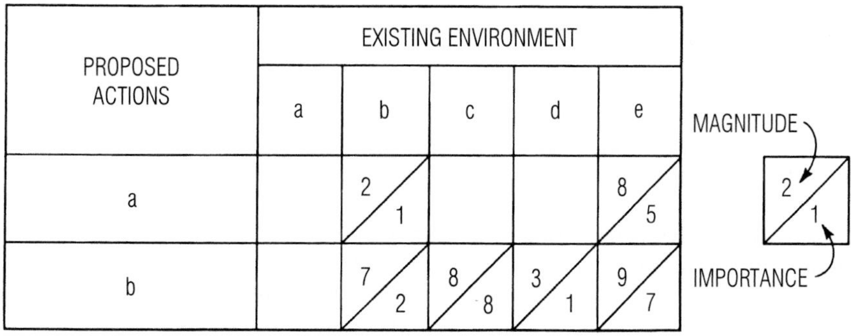

Fig. 22.31 Matrix used to demonstrate the environmental impact of proposed actions.

Table 22.10 Items for Inclusion in Environment Impact Matrix

Top Margin: Proposed Actions That may Cause Environmental Impact	Left Margin: Existing Characteristics and Conditions of the Environment
A. Modification of Regime	**A.** Physical and Chemical Characteristics
a. Exotic flora	**1.** Earth
b. Biological controls	a. Mineral resources
c. Modification of habitat	b. Construction material
d. Alteration of ground cover	c. Soils
e. Alteration of groundwater hydrology	d. Land form
f. Alteration of drainage	e. Force field and background radiation
g. River control and flow modification	f. Unique physical features
h. Canalization	**2.** Water
i. Irrigation	a. Surface
j. Weather modification	b. Ocean
k. Burning	c. Underground
l. Surface or paving	d. Quality
m. Noise and vibration	e. Temperature
B. Land Transformation and Construction	f. Recharge
a. Urbanization	g. Snow, ice, and permafrost
b. Industrial sites and buildings	**3.** Atmosphere
c. Airports	a. Quality (gases, particulates)
d. Highways and bridges	b. Climate (micro, macro)
e. Roads and trails	c. Temperature
f. Railroads	**4.** Processes
g. Cables and lifts	a. Floods
h. Transmission lines, pipelines, and corridors	b. Erosion
i. Barriers including fencing	c. Deposition (sedimentation, precipitation)
j. Channel dredging and straightening	d. Solution
k. Channel revetments	e. Sorption (ion exchange, complexing)
l. Canals	f. Compaction and settling
m. Dams and impoundments	g. Stability (slides, slumps)
n. Piers, seawalls, marinas, and sea terminals	h. Stress-strain (earthquake)
o. Offshore structures	i. Air movements
p. Blasting and drilling	**B.** Biological Conditions
q. Cut and fill	**1.** Flora
r. Tunnels and underground structures	a. Trees
C. Resource Extraction	b. Shrubs
a. Blasting and drilling	c. Grass
b. Surface excavation	d. Crops
c. Subsurface excavation and retorting	e. Microflora
d. Well drilling and fluid removal	f. Aquatic plants
e. Dredging	g. Endangered species
f. Clear cutting and other lumbering	h. Barriers
g. Commercial fishing and hunting	i. Corridors
D. Processing	**2.** Fauna
a. Farming	a. Birds
D. Processing (*Continued*)	b. Land animals including reptiles
b. Ranching and grazing	c. Fish and shellfish
	2. Fauna (*Continued*)
	d. Benthic organisms

Table 22.10 Continued

Top Margin: Proposed Actions That may Cause Environmental Impact	Left Margin: Existing Characteristics and Conditions of the Environment
c. Feed lots	e. Insects
d. Dairying	f. Microfauna
e. Energy generation	g. Endangered species
f. Mineral processing	h. Barriers
g. Metallurigcal industry	i. Corridors
h. Chemical industry	C. Cultural Factors
i. Textile industry	1. Land Use
j. Automobile and aircraft	a. Wilderness and open spaces
k. Oil refining	b. Wetlands
l. Food	c. Forestry
m. Lumbering	d. Grazing
n. Pulp and paper	e. Agriculture
o. Product storage	f. Residential
E. Land Alteration	g. Commercial
a. Erosion and control and terracing	h. Industrial
b. Mine sealing and waste control	i. Mining and quarrying
c. Strip mining rehabilitation	2. Recreation
d. Landscaping	a. Hunting
e. Harbor dredging	b. Fishing
f. Marsh fill and drainage	c. Boating
F. Resource Renewal	d. Swimming
a. Reforestation	e. Camping and hiking
b. Wildlife stocking and management	f. Picnicking
c. Groundwater recharge	g. Resorts
d. Fertilization application	3. Aesthetics and Human Interest
e. Waste recycling	a. Scenic views and vistas
G. Changes in Traffic	b. Wilderness qualities
a. Railway	c. Open space qualities
b. Automobile	d. Landscape design
c. Trucking	e. Unique physical features
d. Shipping	f. Parks and reserves
e. Aircraft	g. Monuments
f. River and canal traffic	h. Rare and unique species or ecosystems
g. Pleasure boating	i. Historical or archaeological sites and objects
h. Trails	j. Presence of misfits
i. Cables and lifts	4. Cultural Status
j. Communication	a. Cultural patterns (lifestyle)
k. Pipeline	b. Health and safety
H. Waste Emplacement and Treatment	c. Employment
a. Ocean dumping	d. Population density
b. Landfill	5. Constructed Facilities and Activities
c. Emplacement of tailing, spoil, and overburden	a. Structures
	b. Transportation network (movement, access)
d. Underground storage	c. Utility networks
e. Junk disposal	d. Waste disposal
f. Oil well flooding	e. Barriers

(Table continued)

Table 22.10 Continued

Top Margin: Proposed Actions That may Cause Environmental Impact	Left Margin: Existing Characteristics and Conditions of the Environment
H. Waste Emplacement and Treatment (*Continued*)	**5.** Constructed Facilities and Activities (*Continued*)
g. Deep well emplacement	f. Corridors
h. Cooling water discharge	**D.** Ecological Relationships Such As:
i. Municipal waste discharge including spray irrigation	a. Salinization of water resources
j. Liquid effluent discharge	b. Eutrophication
k. Stabilization and oxidation ponds	c. Disease-insect vectors
l. Septic tanks, commercial and domestic	d. Food chains
m. Stack and exhaust emission	e. Salinization of surficial material
n. Spent lubricants	f. Brush encroachment
I. Chemical Treatment	g. Other
a. Fertilization	**E.** Others
b. Chemical deicing, of highways, etc.	
c. Chemical stabilization of soil	
d. Weed control	
e. Insect control (pesticides)	
J. Accidents	
a. Explosions	
b. Spills and leaks	
c. Operational failure	
K. Others	

It is very important that engineers present the multiphasic effects of the proposed project on air, water, and land characteristics, the biota, and on constructed structures, if any. It is also important to relate, in the discussion of impact, environmental interests related to recreation, education, science, history, and culture as well as to overall community well-being. Health and safety considerations, both within the project and in any exterior community relationship, must be discussed. Specific guidance on format and content can be found in the Council on Environmental Quality guidelines, and more specifically, in each sponsoring federal agency guidelines, such as Department of Transportation, Federal Aviation Administration or the Army Corps of Engineers.

As required by the Environmental Protection Agency, the report must assess:

1. The probable impact of the action

2. The adverse environmental effect should the project be implemented

3. The alternatives

4. The relationship between the local short-term effect on environment, and maintenance of or increased benefit to the environment over the long term

5. The commitments of resources that might be considered irreversible if the proposed action should take place

(E. T. Chanlett, "Environmental Protection," 2nd ed., R. A. Corbitt, "Standard Handbook of Environmental Engineering," and J. F. Rau and D. C. Woolen, "Environmental Impact Analysis Handbook," McGraw-Hill Publishing Company, New York (www.books.mcgraw-hill.com); J. E. Heer, Jr., and D. J. Hagerty, "Environmental Assessments and Statements," and D. C. Rona, "Environmental Permits," Van Nostrand Reinhold, New York; S. J. Rosen, "Manual for Environmental Impact Evaluation," Prentice-Hall, Inc., Englewood Cliffs, N.J. (www.prenhall.com).)

23

Scott L. Douglass
Professor
Department of Civil Engineering
University of South Alabama
Mobile, Alabama

Robert A. Nathan
Moffat & Nichol Engineers
Tampa, Florida

Jeffrey D. Malyszek
Moffat & Nichol Engineers
Tampa, Florida

COASTAL AND PORT ENGINEERING

C oastal and port engineering encompasses planning, design, and construction of projects to satisfy society's needs and concerns in the coastal environment, such as harbor and marina development, shore protection, beach nourishment, and other constructed systems in the coastal wave and tide environment.

Over time, the scope of this field of engineering has broadened from only navigation improvement and property protection to include recreational beaches and environmental considerations. It takes into account the environmental conditions unique to the coastal area, including wind, waves, tides, and sand movement. Thus, coastal engineering makes extensive use of the sciences of oceanography and coastal geomorphology as well as of geotechnical, environmental, structural, and hydraulic engineering principles.

23.1 Risk Level in Coastal Projects

Because of the nature of littoral drift, or longshore sand transport along the coasts, erosion caused by coastal engineering projects along adjacent shorelines, sometimes several miles away, has been a recurring problem. Tools for prediction and evaluation of such shoreline dynamics are continually improving but are still limited, in part because of nature's unpredictability. Hence, postconstruction monitoring of the response of nearby beaches is often a required component of coastal engineering projects.

The design level of risk in many coastal engineering projects may be higher than in other civil engineering disciplines because the price of more effective design is often not warranted.

The design environment is very challenging. It varies with time, since design conditions are often affected by storms that contain much more energy and induce very different loadings from those normally experienced. Also, because the physical processes are so complex, often too complex for theoretical description, the practice of coastal engineering is still much of an art. Consequently, practitioners should have a broad base of practical experience and should exercise sound judgment.

The practice of coastal engineering has changed rapidly in the last several decades owing to increases in natural pressures, such as that created by sea-level rise, and societal pressures, such as those from growing populations along the coast with greater environmental awareness. The changes are recorded in the proceedings of specialty conferences, such as those of the American Society of Civil Engineering (ASCE), including Coastal Engineering Practice; Dredging, Ports, Coastal Sediments, Coastal Zone, International Coastal Engineering Conference, and the Florida Shore and

Beach Preservation Association's Beach preservation Technology Conference series.

Coastal Hydraulics and Sediments

Waves often apply the primary hydraulic forces of interest in coastal engineering. Tides and other water-level fluctuations control the location of wave attack on the shoreline. Waves and tides generate currents in the coastal zone. Breaking waves provide the forces that drive sand transport along the coast and can cause beach changes, including erosion due to coastal engineering projects.

23.2 Characteristics of Waves

Water waves are caused by a disturbance of the water surface. The original disturbance may be caused by wind, boats or ships, earthquakes, or the gravitational attraction of the moon and sun. Most of the waves are initially formed by wind. Waves formed by moving ships or boats are **wakes**. Waves formed by earthquake disturbances are **tsunamis**. Waves formed by the gravitational attraction of the moon and sun are **tides**.

After waves are formed, they can propagate across the surface of the sea for thousands of miles.

The properties of propagating waves have been the subject of various wave theories for over a century. The most useful wave theory for engineers is the linear, or small-amplitude, theory.

23.2.1 Linear Wave Theory

Essentially, linear wave theory treats only a train of waves of the same length and period in a constant depth of water. As in optics, this is called a monochromatic wave train. Linear wave theory relates the length, period, and depth of waves as indicated by Eq. (23.1).

$$L = \frac{gT^2}{2\pi} \tan h \frac{2\pi d}{L} \qquad (23.1)$$

where L = wavelength, ft, the horizontal distance between crests

d = vertical distance, ft, between mean or still water level and the bottom

g = acceleration due to gravity, 32.2 ft/s

T = wave periods, the time required for propagation of a wave crest over the wavelength (Fig. 23.1)

Wave height H, the fourth value needed to completely define a monochromatic wave train, is an independent value in linear wave theory, but not for higher-order wave theories (Art. 23.2.2).

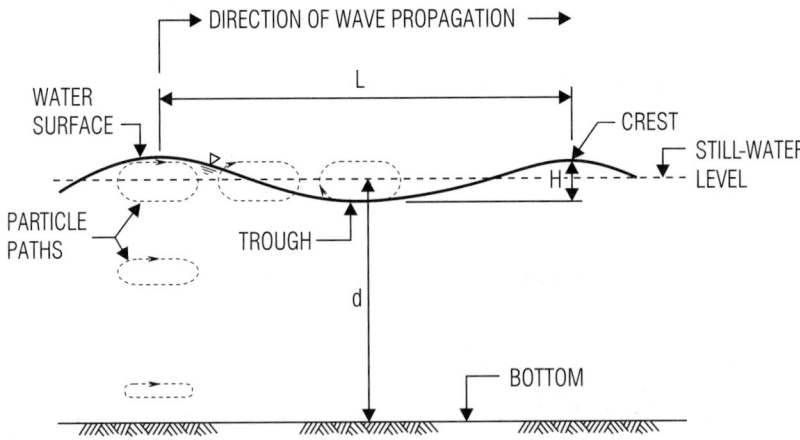

Fig. 23.1 Wave in shallow water. Water particles follow an elliptical path. L indicates length of wave, crest to crest; H wave height, d depth from still-water level to the bottom. The wave period T is the time for a wave to move the distance L.

Equation (23.1), implicit in terms of L, requires an iterative solution except for deep or shallow water. When the relative depth d/L is greater than $\frac{1}{2}$, the wave is in deep water and Eq. (23.1) becomes

$$L = \frac{gT^2}{2\pi} \qquad (23.2)$$

For shallow water, $d/L < \frac{1}{25}$, eq. (23.1) reduces to

$$L = T\sqrt{gd} \qquad (23.3)$$

Individual water particles follow a closed orbit. They return to the same location with each passing wave. The orbits are circular in deep water and elliptical in shallow water. Linear wave theory equations for the water-particle trajectories, the fluctuating water-particle velocities and accelerations, and pressures under wave trains are given in R. G. Dean and R. A. Dalrymple, "Water Wave Mechanics for Scientists and Engineers," Prentice-Hall, Englewood Cliffs, N. J. (www.prenhall.com); R. M. Sorenson, "Basic Wave Mechanics: For Coastal and Ocean Engineers," John Wiley & Sons, Inc., New York (www.wiley.com).)

23.2.2 Higher-Order Wave Theories

The linear wave theory provides adequate approximations of the kinematics and dynamics of wave motion for many engineering applications. Some areas of concern to civil engineers where the linear theory is not adequate, however, are very large waves and shallow water. Higher-order wave theories, such as Stokes' second order and cnoidal wave theories, address these important situations. Numerical wave theories, however, have the broadest range of applicability. Useful tables from stream-function wave theory, a higher-order, numerical theory, are given in R. G. Dean, "Evaluation and Development of Water Wave Theories For Engineering Applications," Special Report No. 1, U.S. Army Coastal Engineering Research Center, Ft. Belvoir, Va.

Determination of the water surface elevations for large waves or waves in shallow water requires use of a higher-order wave theory. A typical waveform is shown in Fig. 23.2. The crest of the wave is more peaked and the trough of the wave is flatter than for the sinusoidal water surface profile in linear wave theory. For a horizontal bottom, the height of the wave crest above the still-water level is a maximum of about $0.8d$.

("Shore Protection Manual," 4th ed., U.S. Army Coastal Engineering Research Center, Government Printing Office, Washington, D.C. (www.gpo.gov); "Coastal Engineering Manual," (www.usace.mil/inet/usace-docs/eng-manuals/em-htm).)

23.2.3 Wave Transformations

As waves move toward the coast into varying water depths, the wave period remains constant (until breaking). The wavelength and height, however, change because of shoaling, refraction, diffraction, reflection, and wave breaking.

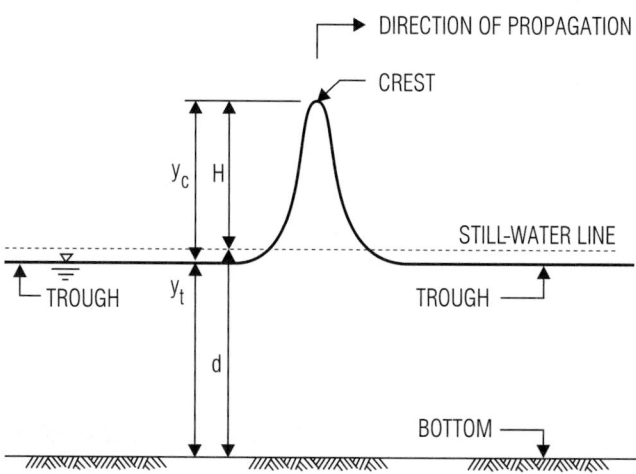

Fig. 23.2 Water surface for a large wave in shallow water.

Shoaling ▪ As a wave moves into shallower water the wavelength decreases, as indicated by Eq. (23.1), and the wave height increases. The increase in wave height is given by the shoaling coefficient K_s.

$$K_s = \frac{H}{H'_o} \qquad (23.4)$$

where H = wave height in a specific depth of water

H'_o = deep-water unrefracted wave height

K_s varies as a function of relative depth d/L as shown in Table 23.1. For an incident wave train of period T, Table 23.1 can be used to estimate the wave height and wavelength in any depth with Eq. (23.2) for L_o.

Refraction ▪ This is a term, borrowed from optics, for the bending of waves as they slow down. As waves approach a beach at an angle, a portion of the wave is in shallower water and moving more slowly than the rest. Viewed from above, the wave crest appears to bend.

Refraction changes the height of waves as well as the direction of propagation. Refraction can cause wave energy to be focused on headlands and defocused from embayments.

There are two general types of refraction models. Wave-ray models trace the path of wave rays, lines perpendicular to the wave crests. The other type of computer refraction model computes solutions to differential equations for the wave-height field. The physics simulated varies slightly from model to model.

Table 23.1 Shoaling Coefficient and Wavelength Changes as Waves Move into Shallower Water

d/L_o	d/L	K_s
0.005	0.028	1.70
0.010	0.040	1.43
0.020	0.058	1.23
0.030	0.071	1.13
0.040	0.083	1.06
0.050	0.094	1.02
0.10	0.14	
0.20	0.22	
0.30	0.31	
0.50	0.50	1.0

Diffraction ▪ Another term borrowed from optics, this is the spread of energy along a wave crest. An engineering example of wave diffraction is the spreading of energy around the tip of a breakwater into the lee of the breakwater. The wave crest wraps around the tip of a breakwater and appears to be propagating away from that point. Diffraction also occurs in open water where refraction occurs. It can reduce the focusing and bending due to refraction.

Reflection ▪ Waves are reflected from obstructions in their path. Reflection of wave energy is greatest at vertical walls, 90% to 100%, and least for beaches and rubble structures. Undesirable wave-energy conditions in vertical-walled marinas can often be reduced by placing rubble at the water line.

Breaking ▪ This happens constantly along a beach, but the mechanics are not well modeled by theory. Thus, much of our knowledge of breaking is empirical. In shallow water, waves break when they reach a limiting depth for the individual wave. This depth-limited breaking is very useful in coastal structure design and surf-zone dynamics models.

For an individual wave, the limiting depth is about equal to the water depth and lies in the range given by Eq. (23.5.).

$$0.8 < \left(\frac{H}{d}\right)_{\text{max}} < 1.2 \qquad (23.5)$$

where $(H/d)_{\text{max}}$ = maximum ratio of wave height to depth below mean water level for a breaking wave. The variation in $(H/d)_b$ (the subscript b means breaking) is due to beach slope and wave steepness H/L.

Equation (23.5) is often useful in selecting the design wave height for coastal structures in shallow water. Given an estimate of the design water depth at the structure location, the maximum wave height H_{max} that can exist in that depth of water is about equal to the depth. Any larger waves would have already broken farther offshore and been reduced to H_{max}.

23.2.4 Irregular Waves

The smooth water surfaces of monochromatic wave theories are not realistic representations of the real surf zone. Particularly under an active wind, the water surface will be much more irregular. Two different sets of tools have been developed by oceanographers to describe realistic sea surfaces. One is a statistical representation and one is a spectral representation.

Statistics of Wave Height ▪ The individual waves in a typical sea differ in height. The heights follow a theoretical Rayleigh distribution in deep water. In shallow water, the larger individual waves break sooner, and thus the upper tail of the distribution is lost.

A commonly used, single wave-height parameter is the significant wave height $H_{1/3}$. This is the average of the highest one-third of the waves. Other wave heights used in design can be related to $H_{1/3}$ via the Rayleigh distribution as indicated in Table 23.2.

23.2.5 Wave Spectra

Spectral techniques are available that describe the amount of energy at the different frequencies or wave periods in an irregular sea. They provide more information about the irregular wave train and are used in some of the more advanced coastal-structure design methods. A wave-height parameter that is related to the total energy in a sea is H_{m_o}. (H_{m_o} is often called significant wave height also.)

Significant wave height H_s is a term that has a long history of use in coastal engineering and oceanography. As indicated above and in Art. 23.2.4, two fundamentally different definitions for significant wave height are used in coastal engineering. One is statistically based and the other is energy- or spectral-based. Since they are different, the notations, $H_{1/3}$ and H_{m_o} are recommended to avoid confusion in use of H_s:

$$H_{1/3} = \text{statistical significant wave height}$$
$$H_{m_o} = \text{spectral significant wave height}$$

In deep water, H_{m_o} is approximately equal to $H_{1/3}$. In shallow water, and in particular in the surf zone, the two parameters diverge. (There is little that is truly significant about either parameter. Few of the waves in an actual wave train will have the significant height. It is basically a statistical artifact.)

Transformations of actual wave seas such as shoaling, refraction, diffraction, and breaking are not completely understood and not well modeled. Although the monochromatic wave transformations are well modeled, as described in the preceding, in actuality the individual waves and wave trains interact with each other and change the wave field. (These wave-wave interactions are the subject of significant research efforts.) Thus, the more realistic conditions, that is, irregular seas, are the least understood. However, models that account for the transformation of wave spectra across arbitrary bottom contours are available.

23.2.6 Wave Generation by Wind

Waves under the influence of the winds that generated them are called **sea**. Waves that have propagated beyond the initial winds that generated them are called **swell**.

Fetch is the distance that a wind blows across the water. For enclosed bays, this is the distance across the water body in the direction of the wind. **Duration** is the time that a wind at a specific speed blows across the water. The waves at any spot may be fetch-limited or duration-limited. When a wind

Table 23.2 Wave Heights Used in Design

Symbol	Description	Multiple of $H_{1/3}$
$H_{1/3}$	Average height of highest one-third of waves	1.0
H_{av}	Average wave height	0.6
H_{10}	Average height of highest 10% of waves	1.3
$H_{1\%}$	Wave height exceeded 1% of the time	1.6
H_{sin}	Height of simple sine waves with same energy as the actual irregular height wave train	0.8

starts to blow, wave heights are limited by the short time that the wind has blown; in other words, they are duration-limited. Seas not duration-limited are *fully arisen.* If the waves are limited by the fetch, they are fetch-limited.

For enclosed bay and lake locations, simple parametric models can provide useful wave information. Table 23.3 gives wave height and wave period estimates for deep water for different fetch distances and different wind speeds. The values are based on the assumption that the wind blows for a sufficient time to generate fully arisen conditions. In shallow water, the wave heights will be less.

On the open ocean, waves are almost never fetch-limited. They are free to continue to move after the wind ceases or changes. Swell wave energy can propagate across entire oceans. The waves striking the beach at any moment in time may include swell from several different locations plus a local wind sea. Thus, for an open-ocean situation, numerical models that grid the entire ocean are required to keep track of wave-energy propagation and local generation.

Wave-generation models can forecast waves for marine construction operations. They can also hindcast, that is, estimate waves based on measured or estimated winds at times in the past, for wave climatology studies, probabilistic design, or historic performance analysis. The U.S. Army Corps of Engineers "Wave Information Study (WIS)" has hindcast 40 years of data, 1956–1995, to generate probabilistic wave statistics for hundreds of locations along the coasts of the United States. The wave statistics are available in tabular form, and the actual time sequence of wave conditions is available in digital form.

(J. B. Herbich, "Handbook of Coastal and Ocean Engineering," Gulf Publishing Company, Houston, Tex (www.gulfpub.com).)

23.2.7 Ship and Boat Wakes

Ship wakes are sometimes the largest waves that occur at a location and thus become the design wave. Vessel wakes from large ships can be up to 6 ft high and have wave periods less than 3 s. Ship wakes can be estimated with methods presented in J. R. Weggel and R. M. Sorensen, "Ship Wave Prediction for Port and Channel Design," Proceedings, Port Conference, 1986, ASCE. Approaches for estimating the wakes due to recreational boats are presented in ASCE Manual 50, "Planning and Design Guidelines for Small-Craft Harbors," and R. R. Bottin et al., "Maryland Guide Book for Marina Owners and Operators on Alternatives Available for the Protection of Small Craft against Vessel Generated Waves," U.S. Army Corps of Engineers Coastal Engineering Research Center, Washington, D.C.

Table 23.3 Spectral Significant Heights and Periods for Wind-Generated Deep-Water Waves*

Wind speed, knots	Fetch length, statute miles				
	0.5	1	2	10	50
20					
H_{m_o}, ft	0.6	0.8	1.1	2.2	4.1
T_p, s	1.3	1.6	2.0	3.2	4.7
40					
H_{m_o}, ft	1.3	1.8	2.5	5.4	11
T_p, s	1.7	2.2	2.7	4.5	7
60					
H_{m_o}, ft	2.2	3.1	4.2	9.1	18
T_p, s	2.1	2.6	3.2	5.4	8

*Based on method presented in S. L. Douglass et al., "Wave Forecasting for Construction in Mobile Bay," Proceedings, Coastal Engineering Practice, 1992, pp. 713–727, American Society of Civil Engineers. H_{m_o} = spectral significant wave height and T_p = wave period.

23.3 Design Coastal Water Levels

The design water level depends on the type of project. For design of some protective coastal structures, for example, a water level based on a recurrence interval such as a 10-year or 100-year return period often is selected. The Federal Emergency Management Agency (FEMA) "Flood Insurance Rate Maps (FIRM)" are based on such a concept. They provide a first estimate of high-water levels along the U.S. coastlines. Since the design of some coastal structures can be extremely sensitive to the design water level, more in-depth analysis may be justified. For engineering projects concerned with normal water levels, for example, where dock elevations and beachfill elevations are determined by the water level, an estimate of the normal water level and the normal range around that mean is needed. All coastal engineering projects should be designed to take into account the full range of potential water levels.

The water level at any time in a specific location is influenced by the tides, mean sea-level elevation, storm surge, including wind influence, and other local influences, such as fresh-water inflow in estuaries.

Tides ▪ The tide is the periodic rise and fall of ocean waters produced by the attraction of the moon and sun. Generally, the average interval between successive high tides is 12 h 25 min, half the time between successive passages of the moon across a given meridian. The moon exerts a greater influence on the tides than the sun. Tides, however, are often affected by meteorological conditions, including propagation of storm tides from the sea into coastal waters.

The highest tides, which occur at intervals of half a lunar month, are called **spring tides**. They occur at or near the time when the moon is new or full, i.e., when the sun, moon, and earth fall in line, and the tide-generating forces of the moon and sun are additive. When the lines connecting the earth with the sun and the moon form a right angle, i.e., when the moon is in its quarters, then the actions of the moon and sun are subtractive, and the lowest tides of the month, the **neap tides**, occur.

Tidal waves are retarded by frictional forces as the earth revolves daily around its axis, and the tide tends to follow the direction of the moon. Thus, the highest tide for each location is not coincident with conjunction and opposition but occurs at some constant time after new and full moon. This interval, known as the *age* of the tide, may amount to as much as $2\frac{1}{2}$ days.

Large differences in tidal range occur at different locations along the ocean coast. They arise because of secondary tidal waves set up by the primary tidal wave or mass of water moving around the earth. These movements are also influenced by the depth of shoaling water and configuration of the coast. The highest tides in the world occur in the Bay of Fundy, where a rise of 100 ft has been recorded. Inland and landlocked seas, such as the Mediterranean and the Baltic, have less than 1 ft of tide, and the Great Lakes are not noticeably influenced.

Tides that occur twice each lunar day are called **semidiurnal tides**. Since the lunar day, or time it takes the moon to make a complete revolution around the earth, is about 50 min longer than the solar day, the corresponding high tide on successive days is about 50 min later. In some places, such as Pensacola, Florida, only one high tide a day occurs. These tides are called **diurnal tides**. If one of the two daily high tides is incomplete, i.e., if it does not reach the height of the previous tide, as at San Francisco, then the tides are referred to as **mixed diurnal tides**. Table 23.4 gives the spring and mean tidal ranges for some major ports.

There are other exceptional tidal phenomena. For instance, at Southampton, England, there are four daily high waters, occurring in pairs, separated by a short interval. At Portsmouth, there are two sets of three tidal peaks per day. **Tidal bores**, a regular occurrence at certain locations are high-crested waves caused by the rush of flood tide up a river, as in the Amazon, or by the meeting of tides, as in the Bay of Fundy.

The rise of the tide is referred to some established datum of the charts, which varies in different parts of the world. In the United States, it is mean lower low water (MLLW).

Mean high water is the average of the high water over a 19-year period, and **mean low water** is the average of the low water over a 19-year period. **Higher high water** is the higher of the two high waters of any diurnal tidal day, and **lower low water** is the lower of the two low waters of any diurnal tidal day. **Mean higher high water** is the average height of the higher high water over a 19-year period, and **mean lower low water** is the

Table 23.4 Mean and Spring Tidal Ranges for Some of the World's Major Ports*

	Mean range, ft	Spring range, ft
Anchorage, Alaska	26.7	29.6[†]
Antwerp, Belgium	15.7	17.8
Auckland, New Zealand	8.0	9.2
Baltimore, Md	1.1	1.3
Bilboa, Spain	9.0	11.8
Bombay, India	8.7	11.8
Boston, Mass.	9.5	11.0
Buenos Aires, Argentina	2.2	2.4
Burntcoat Head, Nova Scotia (Bay of Fundy)	41.6	47.5
Canal Zone, Atlantic side	0.7	1.1[†]
Canal Zone, Pacific side	12.6	16.4
Capetown, Union of South Africa	3.8	5.2
Cherbourg, France	13.0	18.0
Dakar, Africa	3.3	4.4
Dover, England	14.5	18.6
Galveston, Tex	1.0	1.4[†]
Genoa, Italy	0.6	0.8
Gibraltar, Spain	2.3	3.1
Hamburg, Germany	7.6	8.1
Havana, Cuba	1.0	1.2
Hong Kong, China	3.1	5.3[†]
Honolulu, Hawaii	1.2	1.9[†]
Juneau, Alaska	14.0	16.6[†]
La Guaira, Venezuela		1.0[†]
Lisbon, Portugal	8.4	10.8
Liverpool, England	21.2	27.1
Manila, Philippines		3.3[†]
Marseilles, France	0.4	0.6
Melbourne, Australia	1.7	1.9
Murmansk, U.S.S.R.	7.9	9.9
New York, N.Y.	4.4	5.3
Osaka, Japan	2.5	3.3
Oslo, Norway	1.0	1.1
Quebec, Canada	13.7	15.5
Rangoon, Burma	13.4	17.0
Reikjavik, Iceland	9.2	12.5
Rio de Janeiro, Brazil	2.5	3.5
Rotterdam, Netherlands	5.0	5.4
San Diego, Calif.	4.2	5.8[†]
San Francisco, Calif.	4.0	5.7[†]
San Juan, Puerto Rico	1.1	1.3
Seattle, Wash.	7.6	11.3[†]
Shanghai, China	6.7	8.9
Singapore, Malaya	5.6	7.4

Table 23.4 (Continued)

	Mean range, ft	Spring range, ft
Southampton, England	10.0	13.6
Sydney, Australia	3.6	4.5
Valparaiso, Chile	3.0	3.9
Vladivostok, U.S.S.R.	0.6	0.7
Yokohama, Japan	3.5	4.7
Zanzibar, Africa	8.8	12.4

* "Tide Tables," National Ocean Service.
[†] Diurnal range.

average height of the lower low waters over a 19-year period (tidal epoch). **Highest high water** and **lowest low water** are the highest and lowest, respectively, of the spring tides of record. **Mean range** is the height of mean high water above mean low water. The mean of this height is generally referred to as **mean sea level (MSL)**. **Diurnal range** is the difference in height between the mean higher high water and the mean lower low water.

The National Ocean Service annually publishes tide tables that give the time and elevation of the high and low tides at thousands of locations around the world and that can be used to forecast water levels at all times. The tide tables forecast the repeating, astronomical portions of the tide for specific locations but do not directly account for the day-to-day effects of changes in local winds, pressures, and other factors. Along most coasts, the tide table forecasts are within 1 ft of the actual water level 90% of the time.

Relative sea-level rise is gradually changing all of the epoch-based datum at any coastal site. Although, the datum that is used for design and construction throughout an upland area is not particularly important, the relation between construction and actual water levels in the coastal zone can be extremely important. The level of the oceans of the world has been gradually increasing for thousands of years. The important change is the relative sea-level change, the combined effect of water level and land-mass elevation changes due to subsidence (typical of the U.S. Atlantic and Gulf coasts) or rebound or emergence (Pacific coast of the U.S.). Measured, long-term tide data for major U.S. ports show that the relative sea-level rise differs from location to location. For example, at Galveston, Tex., there has been about 1 ft of relative sea-level rise during the last 50 years. At Anchorage, Alaska, there has been about 2 ft of relative sea-level fall during the last 50 years.

The impact of long-term sea-level rise has rarely been taken into account in design, except when it has already impacted the epoch-based tidal datum, such as MLLW. The National Geodetic Vertical Datum (NGVD) was established at the mean sea level (MSL) of 1929. Since sea-level rise has continued since then, the NGVD is now below the current day MSL along much of the U.S. Atlantic and Gulf coasts. At many locations, it is between the MSL and the MLLW. For accurate location of the NGVD relative to the MSL or MLLW, analysis with data from a local tide gage is required. For some harbor and coastal design, a staff gage is installed for recording water levels for a sustained period of time to confirm the relation between the local surveyor's elevation datum, the assumed tidal datum, and the actual water surface elevation.

Storm Surge ▪ This can be defined broadly to include all the effects involved in a storm, including wind stress across the continental shelf and within an estuary or body of water, barometric pressure, and wave-induced setup. The combined influence of these effects can change the water level by 5 to 20 ft depending on the intensity of the storm and coastal location. Engineers can use return-period analysis curves to estimate the likelihood of any particular elevation. The Federal Emergency Management Agency and the various Corps of Engineer Districts have developed such curves based on historic high-water-mark elevations and numerical models of the hydrodynamics of the continental shelf.

23.4 Coastal Sediment Characteristics

Most beach sediments are sand. The day-to-day dynamics of the surf zone usually ensure that most fines, silts, and clays will be washed away to more quiescent locations offshore. Some beaches have layers of cobbles, rounded gravel, or shingles, flattened gravel.

The size and composition of beach sands varies around the world and even along adjacent shorelines. Essentially, the beach at any particular site consists of whatever loose material is available.

Quartz is the most common mineral in beach sands. Other constituents in sands include feldspars and heavy minerals. Some beaches have significant portions of seashell fragments and some beaches are dominated by coral carbonate material.

Beach sands are usually described in terms of grain-size distribution. The median diameter d_{50} is a common measure of the central size of the distribution. The range of the distribution of sand sizes around this median is usually discussed in terms of sorting.

The color of the sand depends primarily on the composition of the grains. The black sand beaches of Hawaii are derived from volcanic lava. The white sands of the panhandle of Florida are quartz that has developed a white color owing to miniature surface abrasions and bleaching.

23.5 Nearshore Currents and Sand Transport

As wave energy enters the surf zone, some of the energy is transformed to nearshore currents and expended in sand movement. The nearshore current field is dominated by the incident wave energy and the local wind field. The largest currents are the oscillatory currents associated with the waves. However, several forms of mean currents (longshore currents, rip currents associated with nearshore circulation cells, and downwelling or upwelling associated with winds) can be important to sand transport.

Longshore current is the mean current along the shore between the breaker line and the beach that is driven by an oblique angle of wave approach. The waves provide the power for the mean longshore current and also provide the wave-by-wave agitation to suspend sand in the current. The resulting movement of sand is **littoral drift** or **longshore sand transport**. This process is referred to as a *river of sand* moving along the coast. Although the river-of-sand concept is an effective, simple explanation of much of the influence of engineering on adjacent beaches, the actual sand transport paths are more complex. This is particularly so near inlets with large ebb-tidal shoals that influence the incident wave climate.

Even on an open coast with straight and parallel offshore bottom contours, the longshore-sand-transport direction changes constantly in response to changes in the incident wave height, period, and direction. The so-called **CERC equation**, or **energy flux method**, provides a rough estimate of the instantaneous longshore sand transport rate. The instantaneous rate can vary from nil up to several million cubic yards per year in either direction during storms on some coasts. The variations in the instantaneous rate are so significant that 90% of the sand transport occurs during only 10% of the time—the storms. Even when averaged over a year, the net sand transport rate can vary significantly from year to year and can even change directions. The daily variability in the longshore-sand-transport rate follows a joint, log-normal distribution. This can be used to simulate rates for the design of sand bypassing systems.

Knowledge of the long-term transfer of sand across the continental shelf to or from beaches is limited. For short time frames, the cross-shore transport of sand during storms can be modeled to estimate the beach and dune erosion and the shoreline recession that will occur during some design storm.

Harbor and Marina Engineering

A **harbor** is a bay, cove, inlet, or recess of the sea or a lake, or the mouth of a river in which ships can enter and be sheltered from wind and waves. A **port** is a harbor with facilities for the docking of ships, cargo handling and storage, and transfer of passengers between land and waterborne transportation. A **marina** is a shallow draft harbor for small, predominately recreational craft. Small-craft harbors accommodate commercial operations or waterborne transportation operations, or both, as well as recreational boats of various sizes. Harbor and marina engineering is concerned with the design of navigable waterways in harbors, protective structures, docks, and the facilities for servicing boats or ships.

23.6 Types of Ports and Harbors

Harbors may be classified as natural, seminatural, or artificial, and as harbors of refuge, military harbors, or commercial harbors. Commercial harbors may be either municipal or privately owned.

A **natural harbor** is an inlet or water area protected from storms and waves by the natural configuration of the land. Its entrance is so formed and located as to facilitate navigation while ensuring comparative quiet within the harbor. Natural harbors are located in bays, tidal estuaries, and river mouths. Well-known natural harbors are New York, San Francisco, and Rio de Janeiro.

A **seminatural harbor** may be an inlet or a river sheltered on two sides by headlands requiring artificial protection only at the entrance. Next to a purely natural harbor, it forms the most desirable harbor site, other things being equal. Plymouth and Cherbourg take advantage of their natural location to become well-protected harbors by the addition of detached breakwaters at the entrances.

An **artificial harbor** is one protected from the effect of waves by breakwaters or one created by dredging. Buffalo, New York; Matarani, Peru; Hamburg, Germany; and Le Havre, France, are examples of artificial harbors.

A **harbor of refuge** may be used solely as a haven for ships in a storm, or it may be part of a commercial harbor. Sometimes an outer harbor serves as an anchorage, while a basin within the inner breakwater constitutes a commercial harbor. The essential features are good anchorage and safe and easy access from the sea during any condition of weather and state of tide. Well-known harbors of refuge are the one at Sandy Bay, near Cape Ann, Massachusetts, and that at the mouth of Delaware Bay. A fine example of a combined harbor of refuge and commercial harbor exists at Dover, England.

A **military harbor** or naval base accommodates naval vessels and serves as a supply depot. Guantanamo, Cuba; Hampton Roads, Virginia; and Pearl Harbor, Hawaii, are some well-known naval bases.

A **commercial harbor** is one in which docks are provided with the necessary facilities for loading and discharging cargo. Drydocks are sometimes provided for ship repairs. Many commercial harbors are privately owned and operated by companies representing the steel, aluminum, copper, oil, coal, timber, fertilizer, sugar, fruit, chemical, and other industries. Municipal- or government-controlled harbors, often operated by port authorities, exist in many countries and are usually part of extensive port works, such as the harbors in New York, Los Angeles, and London.

A **port** is a harbor where marine terminal facilities are provided. These consist of piers or wharves at which ships berth while loading or unloading passengers and cargo, transit sheds and other storage areas where ships may discharge incoming cargo, and warehouses where goods may be stored for longer periods while awaiting distribution or sailing. The terminal must be served by railroad, highway, or inland-waterway connections. In this respect the area of influence of the port reaches out for a considerable distance beyond the harbor.

A **port of entry** is a designated location where foreign goods and foreign citizens may be cleared through a custom house.

23.7 Harbor Layout

The number and size of ships using a harbor determine its size to a large extent, but existing site conditions are also an important influence. Generally, unless the harbor is a natural one, its size will be kept as small as feasible for safe and reasonably comfortable operations to take place. Use of tugs to assist maneuvering of ships in docking may also influence the size of the harbor.

23.7.1 Turning Basins

A turning basin is a water area inside a harbor or an enlargement of a channel to permit the turning of a ship. When space is available, the area should have a radius of at least twice the length of the ship to permit either free turning or turning with the aid of tugs, if wind and water conditions require. When space is limited, the ship may be turned by warping around the end of a pier or turning dolphin, either with or without the use of its lines. In those cases, the turning basin will be much smaller and of a more triangular or rectangular shape. The minimum diameter should be at least 20% greater than the length of the largest ship to be turned.

The usual minimum harbor area is the space required for docks plus a turning basin in front of them. In some layouts, where a ship is turned by warping it around the end of the pier or turning dolphin, the harbor may be even smaller. For instance, a minimum harbor with a single pier and turning basin and a long approach channel from the open sea (Fig. 23.3) can accommodate two 500-ft ships. This artificial harbor may be formed by dredging a channel through shallow water, protected by offshore reefs and islands, and enlarging the inshore end to provide the minimum area of harbor that will meet the shipping requirements specified for the project. In leaving its berth, a ship must warp itself around the end of the pier so as not to have to back out through the long approach channel.

Another, less restricted type of harbor is nearly square, protected by two breakwater arms, with one opening. The harbor has several docks and a turning basin with an area sufficient to inscribe a

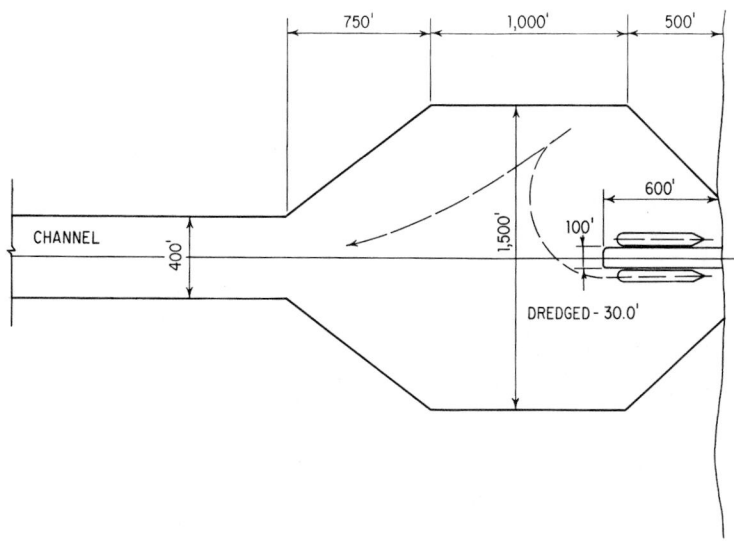

Fig. 23.3 Typical layout for a small artificial harbor.

turning circle with a radius equal to at least twice the length of the largest ship. This is the smallest radius a ship can comfortably turn on, under continuous headway, without the help of a tug. Figure 23.4 shows such a harbor.

23.7.2 Arrangements of Breakwaters

Breakwaters are required for protection of artificial and seminatural harbors. Their location and extent depend on the direction of maximum waves, configuration of the shoreline, and minimum size of harbor required for the anticipated traffic in the port. They may consist of two "arms" out from the shore, plus a single breakwater, more or less parallel to the shore, thereby providing two openings to the harbor; or the harbor may be protected with a single arm out from shore. Or the

harbor may be protected by two arms converging near their outshore ends and overlapping to form a protected entrance to the harbor.

Selection of the most suitable arrangement of breakwaters depends principally on the direction of the maximum waves. The effectiveness of the chosen arrangement in quieting the harbor may be checked by model tests. For comfortable berthing, the wave height should not exceed 2 ft and winds should not exceed 10 to 15 mi/h. But wave heights up to 4 ft have been allowed where bulk cargo is being handled and where the wind direction is such as to hold a docked ship off the dock. In general, winds and current are more bothersome in docking a vessel when it is light than are relatively small harbor waves and may necessitate use of a tug.

Rarely will a location be found where the waves are from one direction only. Generally, it is better

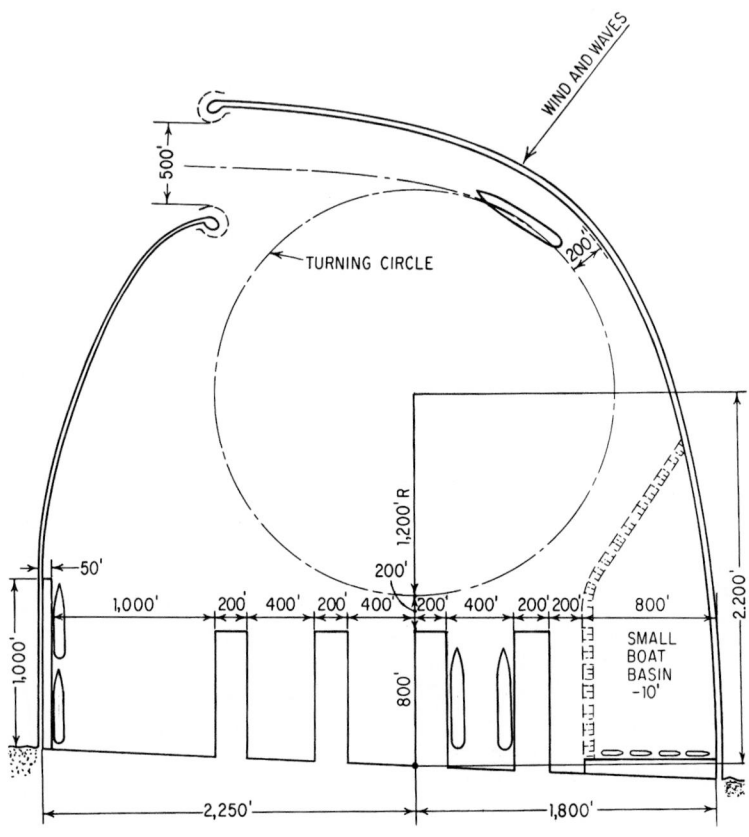

Fig. 23.4 Medium-sized artificial harbor with full-sized turning basin.

in a harbor having two openings for ships to enter from the direction of the minimum wind and waves and to leave toward the direction of the maximum wind and waves. On leaving the harbor, the ships usually have open water in which to maneuver, whereas on entering the harbor, they are immediately in a restricted area and must approach the docks at reduced speed and at a certain inclination to its face.

A single breakwater arm may be used where the waves predominate from one direction only. It also may serve where the configuration of the shoreline reduces the fetch in the opposite direction to such an extent that the wave-generating area is not sufficient to permit formation of bothersome waves within the harbor.

23.7.3 Use of Moorings and Anchorages

An **offshore mooring** is provided usually where it is not feasible or economical to construct a dock or provide a protected harbor. Such an anchorage consists of a number of anchorage units, each consisting of one or more anchors, chains, sinkers, and buoys to which the ship will attach its mooring lines. These anchorages are supplemented in most cases by the ship's bow anchors. Bulk cargo is usually transported to or from the ship by pipeline or trestle conveyor; other cargo may be transferred by lighter.

An **anchorage area** is a place where ships may be held for quarantine inspection, to await docking space, sometimes while removing ballast in preparation for taking on cargo, or to await favorable weather conditions. Special anchorages are sometimes provided for ships carrying explosives or dangerous cargo and are usually so designated on harbor maps by name and depth of water.

23.7.4 Harbor Entrances

To reduce wave height within a harbor, entrances should be no wider than necessary for safe navigation and for preventing dangerous currents when the tide is coming in and going out. The entrance width should be in proportion to the size of the harbor and the ships using it. In general, the following widths will be satisfactory: small harbors, 300 ft; medium harbors, 400 to 500 ft; and large harbors, 500 to 1000 ft. When the entrance is between breakwaters with sloping faces, the width is measured at the required harbor or channel depth below low water. Thus, the entrances will be appreciably wider than the recommended widths at low-water level. In such cases, it is advisable to mark the full harbor depth of the entrance with buoys, placing one or more on each side of the entrance channel.

The entrance should be on the lee side of the harbor, where possible. If the entrance must be located at the windward end of the harbor, breakwaters should overlap so vessels may pass through the restricted entrance and be free to turn with the wind before being hit broadside by the waves. Also, the interior of the harbor will be protected from the waves.

When the entrance to a harbor is unobstructed, storm waves from the sea pass through the opening into the harbor. Unless they are reflected by a vertical surface, they will gradually decrease in height as they progress away from the entrance and as the harbor widens relative to the entrance width. Model tests will give an indication of wave conditions and are essential for studying various arrangements of breakwaters for important harbors.

In tidal harbors where there are strong currents, the entrance width should be sufficient to prevent the velocity of the current through the opening at ebb tide from exceeding 4 ft/s; otherwise, it may affect navigation of ships and create scour at the base of adjacent breakwaters.

If waves pass through an entrance and strike a vertical face on the opposite side of the harbor, they reflect. The result is an increase in wave height within the harbor. This condition can be corrected by building wave-absorbing beaches, flat slopes of rock or granular material, in front of the vertical surface. However, when the vertical surface is a wharf or bulkhead used for berthing of ships, it is impossible to use a beach. Where conditions will permit use of a hollow structure, the outside vertical wall may be perforated or slotted and the energy of the waves absorbed in the chamber in back of the wall. Other means may be resorted to, such as installation of short wave-deflecting walls or wave traps along the approach channel to the dock.

23.7.5 Channel Depth

The harbor and approach channel for ideal operating conditions should be of sufficient depth to permit navigation at lowest low water when ships are fully loaded. This depth must include an allowance for the surge of the ship, which is about one-half the wave height, the out-of-trim or squat when in motion, and from 2- to 4-ft clearance under the keel, the larger figure being used when the bottom is of hard material such as rock. When there is a very soft mud bottom, a keel may at times touch bottom because of surge and squat without damaging the ship, but it would be disastrous to have its fully loaded weight bump a hard rock bottom. Therefore, greater allowances must be made in computing the depth when the bottom is hard. Also, the harbor and approach channel or approach sea lanes must be carefully swept to make sure there are no obstructions, such as reefs or rocky pinnacles, boulders, or sunken ships, above the required depth for safe navigation.

In some harbors, some ships arrive or depart on the rising tide. But the advent of supertankers with deadweight tonnage over 300,000 tons and a draft exceeding 80 ft creates a need for deeper harbors. Otherwise, the cargo has to be transferred to smaller tankers or piped ashore through submarine lines from offshore anchorages.

Tides significantly influence harbor depth. Table 23.4 (p. 23.8) gives the tidal ranges in feet at principal ports throughout the world. Note that the tidal range along the coasts of the United States seldom exceeds 10 ft, and therefore the harbors are dredged to provide the required depth for navigation at lowest low water. The condition is entirely different in the British Isles and on the western coast of Europe, where the port of Liverpool, England, has a spring tidal range of 27 ft; London, England, 20 ft; Calais, France, 20 ft; and others have even greater variations. In most cases, this fluctuation in sea level has resulted in the use of wet docks in all stages of the tide. These dock systems require entrance locks with massive gates, heavy swing or bascule bridges and the machinery for working them, pumping equipment, and other accessories. Since all this results in great cost, as well as continuing operational and maintenance expense, the question arises as to the limiting range for the natural tidal working of ports without recourse to enclosed docks. Generally, about 10 to 15 ft is considered the dividing point.

23.7.6 Channel Width

Width of a channel may be measured between the toes of the bordering side slopes or at the design depth.

Minimum nominal width required for a channel depends on many factors. The following are the most important:

1. Maximum beam of traversing ships
2. Length and maneuverability of the longest vessels
3. Accuracy and reliability of navigational aids
4. Speed, volume, and nature of traffic
5. Nature, intensity, and variation of currents along the channel
6. Ability and experience of mates and pilots
7. Channel depth and curvature
8. Whether ships are to pass each other

Because of insufficient knowledge of the influence of these factors when determining a minimum design width, standard channel widths have not been established. The Permanent International Association of Navigation Congress, however, recommended in 1978 that the nominal width of a one-way channel under ideal conditions should be at least five times the beam of the largest ship expected to use the channel. On a curve, this width should be increased by $L^2/8R$, to account for the effect on the width of the ship's path because of the length L of the vessel and the radius R of the curve. The width on a curve should be additionally increased to take into account maneuvering difficulties. If ships are to pass in a one-way channel, the nominal width of the channel should be at least 12 times the largest beam (ship width). For a two-way channel, the width should be adequate to provide a distance between passing ships of at least two times the beam of the larger vessel. The preceding recommended minimum widths should be increased, if necessary, to allow for the effects of crosswinds and crosscurrents.

23.7.7 Channel Alignment

Basically, a channel should provide a path for a ship through the harbor entrance directly to a berth without requiring difficult maneuvering or subjecting the ship to strong crosscurrents or high waves. If the channel must be curved, the radius of the curves should be at least five times the length of the longest ship expected to use the channel.

Straight tangents should be used between successive curves. (S curves should be avoided.)

The tangents should have a length of at least 10 times that of the longest vessel.

In a sea strait, the channel preferably should coincide with the deepest troughs of the strait and have as few curves as possible. Sight distance should be at least 0.5 mi.

23.7.8 Port Structures

A marine terminal is that part of a port or harbor that provides docking, cargo-handling, and storage facilities. When only passengers embark and disembark along with their baggage and miscellaneous small cargo, generally from ships devoted mainly to the carrying of passengers, it is called a *passenger terminal*. When the traffic is mainly cargo carried by freighters, although many of these ships may carry also a few passengers, the terminal is commonly referred to as *freight* or *cargo terminal*. In many cases, it is known as a *bulk cargo terminal*, where such products as petroleum, cement, and grain are stored and handled.

Docking facilities may consist of a single pier or as many as 1000 piers. The number of berths depends on the anticipated number of ships that will use the port and the time it will take to discharge and take on cargo or passengers. This will vary for different kinds of cargo, but usually a vessel will not be in port more than 48 h. Many bulk cargo ships are loaded in 24 h or less.

Wharves and piers should be located in the most sheltered part of the harbor or along the lee side of the breakwaters. Where possible, piers should be so oriented as to have ships alongside headed as nearly into the wind and waves as possible. This is particularly important if the harbor is not well-protected.

Onshore marine-terminal facilities may consist of one or more of the following, depending on the size of the port and the service it renders:

Transit sheds are located immediately in back of the apron on a pier or wharf. Their function is to store for a short period of time cargo awaiting loading or distribution after being unloaded from ships.

Warehouses may replace transit sheds at some marine terminals. But when used to supplement sheds, warehouses are usually located inland and not on the pier structure.

Bulk storage may be in open piles over conveyor tunnels, which may be covered with sheds when protection from the elements is required; in bins and silos or elevators (for grain storage); or in storage tanks (for liquids). These should be located as near the waterfront as possible, and sometimes directly alongside the wharf or pier, to enable direct loading into the hold of the ship.

A **terminal building** houses port-administration personnel and custom officials if a separate custom house is not provided. The terminal building should be located in a prominent and convenient location with respect to the docks.

Guardhouses are located at strategic points in the port area, such as the entrance gates of highways and railways, entrances to piers or terminal areas, bonded storage, and so on.

Stevedores' warehouses house cargo-handling gear, wash and locker rooms, and other facilities for stevedores.

Miscellaneous buildings and structures include a fire house and fire-fighting equipment, power plant, garages, repair shops, drydocks, marine railways, fishing piers, or yacht basins.

(P. Bruun, "Port Engineering," 4th ed., Gulf Publishing Company, Houston, Tex (www.gulfpub.com).)

23.8 Hydrographic and Topographic Surveys

After preliminary layouts of a port have been completed and before the final design is started, it is necessary in most instances to obtain additional site information.

A **hydrographic survey**, if not already available, should be made to determine the elevations of the bottom of the body of water and should extend over an area somewhat larger than the proposed channel and harbor. In addition, the survey should locate the shoreline at low and high water and all structures or obstructions in the water and along the shore, such as sunken ships, reefs, or large rocks.

Determination of the relief of the bottom of the body of water is made by soundings or by the use of a fathometer designed for hydrographic surveys. The latter method is being used by the National Ocean Survey and has superseded lead-line soundings to a large extent. A fathometer or depth-recording instrument is usually mounted in a motorboat, which is kept on course on established range lines or its location is followed by use of an electronic positioning system as the recording chart registers a natural profile of the bottom. The fathometer, when operated by experienced personnel and properly adjusted and calibrated daily, is superior to lead-line soundings in both accuracy and speed with which a survey can be made.

The depths of soundings are referred to water level at the time made and later corrected to the datum water level with tide gages or tide tables. Therefore, it is important to keep a record of the time and day the soundings are made and to have a concurrent tide- or water-level gage operating in the immediate vicinity.

Soundings should be made at about 25-ft intervals along lines from 50 to 100 ft apart, depending on the irregularity of the bottom. Closer spacing may be needed where greater detail is required to determine sharp changes in the profile of the bottom or to outline obstructions.

Soundings are plotted, usually relative to low-water datum, on a drawing (hydrographic map), which should show the datum, high- and low-water lines, contour lines of equal depth interpolated from the soundings, and principal land and water features. Contour depths may be in either feet, meters, or fathoms, although the last is not used generally for making harbor and marine-terminal studies and layouts. Since the sea bottom is usually less precipitous and the slopes more gentle and uniform than those on land, the scale of the hydrographic map may be somewhat smaller than would normally be used for plotting land topography. Unless the harbor area is very large, a scale of 1 in = 200 ft or 1:2000 in a proportional scale is satisfactory. It is desirable to have all the hydrography on one sheet because this gives a better overall picture of the harbor. In general, the scale should be large enough so that not more than 10 contour lines, in 2-ft intervals, occur within 1 in.

If dredging of a harbor or channel is required, the material is usually measured in place to determine the quantity for payment. To determine this quantity, soundings on fixed sections are taken before and after dredging, and the changes in cross sections are determined by computation or planimeter. It is usually specified that payment will be made for material removed to a maximum of 2 ft below the required dredged bottom, but all material must be removed to at least the minimum depth specified.

A **topographic survey** of the marine-terminal area should be made, to obtain ground contours at 2- to 5-ft intervals. The larger figure is used where terrain is rough and in areas where there is to be little or no construction of importance. In building areas, elevations on 25-ft centers in two directions, with additional elevations at abrupt changes in ground, provide satisfactory information. Where there is dense ground cover, the cross-profile method is most suitable. The profiles may be made with level and tape or stadia, on about 100-ft centers, by clearing paths to permit an unobstructed line of sight. The ground between the 100-ft profiles should be examined, as far as possible, and any prominent irregularities in ground level estimated and noted, so that contours, which are interpolated from elevations along the profiles, can be estimated for the areas in between.

Topographic maps, besides showing the contours of the ground, should locate all borings and test pits, buildings, utilities, and any prominent landmarks. Contours generally are referred to high-water datum. The map scale should be such that the contour lines are not spaced closer than 30 to the inch. Where considerable detail is involved the scale should be 1 in = 100 ft or 1:1000 or less, but for small-scale maps 1 in = 1000 ft or 1:10,000 or more may be used.

(H. Agerschou et al., "Planning and Design of Ports and Marine Terminals," John Wiley & Sons, Inc., New York (www.wiley.com); J.B. Gerbich, "Handbook of Coastal and Ocean Engineering," Gulf Publishing Company, Houston, Tex. (www.gulfpub.com); A. D. F. Quinn, "Design and Construction of Ports and Marine Structures," McGraw-Hill Publishing Company, New York (books.mcgraw-hill.com).)

23.9 Ship Characteristics

The length, beam (breadth), and draft of ships that will use a port have a direct bearing on the design of the approach channel, harbor, and marine-terminal facilities. The last is affected also by the type of vessel, its wind area, and its capacity or tonnage. The **draft** of a ship, expressed in relation to its displacement of water as either loaded or light draft, is the depth of the keel of the ship below water level for the particular condition of loading.

Displacement tonnage is the actual weight of the vessel, or the weight of water displaced when afloat, and may be either "loaded" or "light." **Displacement loaded** is the weight, in metric tons, of the ship and its contents when fully loaded with cargo to the Plimsoll mark, or load line, painted on the hull of the ship (1 ton = 2205 lb). The **Plimsoll mark**, used on British ships, and the **load line**, commonly used on American vessels, designate the depth under the maritime laws to which a ship may be loaded in different bodies of water during various seasons of the year. **Displacement light** is the weight, in metric tons, of the ship without cargo, fuel, and stores.

Deadweight tonnage is the carrying capacity of a ship in metric tons and the difference between displacement light and displacement loaded to the Plimsoll mark or load line. It is the weight of cargo, fuel, and stores a ship carries when loaded to the load line, as distinguished from loaded to space capacity. This tonnage varies with latitude and season. It also depends on salinity of the water because of the effect of temperature and salinity on the specific gravity and buoyancy of the water in which the vessel is operating. Unless otherwise indicated, deadweight tonnage is the mean of tropical, summer, and winter deadweight. Deadweight tonnage is indicated by weight, and gross tonnage by volume measurement; both indicate carrying capacity.

Ships are registered with gross or net tonnage expressed in units of 100 ft^3. **Gross tonnage** is the entire internal cubic capacity of a ship, and **net tonnage** is the gross tonnage less the space provided for the crew, machinery, engine room, and fuel.

Cargo or **freight tonnage**, a commercial expression, is the basis of the freight charge. This tonnage may be measured by either weight or volume.

An ordinary seagoing vessel that can carry a nominal deadweight of 8000 tons of cargo, fuel, and stores will have a displacement of about 11,500 tons, a gross of about 5200 tons, and a net of about 3200 tons.

Ballast is the weight added in the hold or ballast compartments of a ship to increase its draft after it has discharged its cargo and to improve its stability. It usually consists of water and is expressed in long tons. In an oceangoing tanker, saltwater ballast replaces a certain amount of petroleum when the ship is unloaded, whereas a dry-cargo or passenger vessel has separate compartments for ballast.

23.10 Types of Ship Mooring Structures

Facilities for mooring of ships are major elements of ports and harbors. Such facilities include docks, wharves, bulkheads, piers, dolphins, fixed mooring berths, and ground tackle in fixed positions for attachment of a ship's mooring lines. Appurtenances for these facilities include fenders for absorption of ship impact during mooring or departure, trestles, catwalks, bitts, bollards, cleats, chocks, hooks, and capstans.

A **dock**, in general, is a marine structure for mooring or tying up of vessels, loading and unloading cargo, or embarking and disembarking passengers. Often, piers, wharves, bulkheads, and, in Europe, jetties, quays, or quay walls, are called docks. In Europe also, where there are large variations in tide level, a dock is commonly considered an artificial basin for vessels and is called a **wet dock**. When the basin is pumped out, it is termed a **dry dock**.

A **wharf** or **quay** is a dock that parallels the shore. It is generally contiguous with the shore but may not necessarily be so. On the other hand, a **bulkhead** or **quay wall**, although similar to a wharf and often referred to as such, is backed up by ground; the name is derived from the very nature of holding or supporting ground in back of it.

In many locations where industrial plants are to be built adjacent to water transportation, the ground is low and marshy; it is therefore necessary to fill it in. The fill is often obtained by dredging the adjacent waterway, creating a navigable channel or harbor along the property. To retain the made ground, which will now be at a much higher elevation along the waterway, a bulkhead is usually installed. This, or a part of its length, may be used as a wharf for docking vessels if mooring appurtenances, paving, and facilities for handling and storing cargo are added. It is then called a **bulkhead wharf**.

A **pier** or **jetty** is a dock that projects into the water. Sometimes it is referred to as a **mole**. When built in combination with a breakwater, it is termed a **breakwater pier**. In contrast with a wharf, which can be used for docking on one side only, ships may use a pier on both sides. But there are instances where only one side is used, owing to either the physical conditions of the site or the lack of need for additional berthing space.

A pier may be more or less parallel to the shore and connected to it by a mole or trestle, generally at right angles to the pier. In this case, the pier is commonly referred to as a **T-head pier** or **L-shaped pier**, depending on whether the approach is at the center or at the end.

Dolphins are marine structures for mooring vessels. They are commonly used in combination with piers and wharves to shorten the length of these structures. Dolphins are a principal part of the fixed-mooring-berth type of installation used extensively in bulk-cargo loading and unloading installations. Also, they are used for tying up ships and for transferring cargo between ships moored along both sides of the dolphins. There are two types of dolphins: breasting and mooring.

Breasting dolphins, usually the larger of the two types, are designed to take the impact of a ship when docking and to hold the ship against a broadside wind. Therefore, they are provided with fenders to absorb the impact of the ship and to protect the dolphin and ship from damage. Breasting dolphins usually have bollards or mooring posts to take the ship's lines, particularly springing lines for moving a ship along the dock or holding it against the current. These lines are not very effective in a direction normal to the dock, particularly when a ship is light.

To hold a ship against a broadside wind blowing in a direction away from the dock, additional dolphins must be provided off the bow and stern, some distance in back of the face of the dock. These are called *mooring dolphins.* They are not designed for the impact of the ship since they are away from the face of the dock, where they will not be hit. If two mooring dolphins are to be used, they should be located about 45° off the bow and stern and permit mooring lines not less than 200 ft or more than 400 ft long. The largest ships may require two additional dolphins, off the bow and stern. These dolphins are usually located so that the mooring lines will be normal to the dock, which makes them most effective for holding the ship against an offshore wind. Mooring dolphins are provided with bollards or mooring posts and with capstans when heavy lines are to be handled. The maximum pull usually should not exceed 50 tons on a single line, or 100 tons on a single bollard if two lines are used.

A **fixed mooring berth** is a marine structure consisting of dolphins for tying up a ship and a platform for supporting the cargo-handling equipment. The platform is usually set back 5 to 10 ft from the face of the dolphins so that the ship will not come in contact with it. Therefore, the platform does not have to be designed to take the impact of the ship when docking.

Offshore moorings for ships consist of ground tackle placed in fixed positions for attaching a ship's lines. Each unit of ground tackle consists of one or more anchors with chain, sinker, and buoy to which the ship's line is attached. These mooring units are usually located so as to take the bow and stern lines and, if the ship is large, one or more breasting lines. For some moorings, where the wind is in one direction, the ship may use its own bow anchor and the fixed tackle off the bow may be omitted.

(P. Bruun, "Port Engineering," Gulf Publishing Company, Houston, Tex. (www.gulfpub.com).)

23.11 Dock and Appurtenance Design for Ship Mooring

Wharves, piers, bulkheads, and fixed mooring berths fall generally into two broad classifications: docks of open construction with their decks supported by piles or cylinders; and docks of closed or solid construction, such as sheetpile cells, bulkheads, cribs, caissons, and gravity (quay) walls.

The following codes, references, and standards are recommended as a basis for analysis and design of a wharf structure, fill containment structure, mooring and fender devices, and associated equipment.

Primary Codes for Design

● American Concrete Institute (ACI) "Building Code Requirements for Reinforced Concrete"

● American Institute of Steel Construction (AISC) "Manual of Steel Construction"

● Uniform Building Code (UBC)

● American Society of Civil Engineers (ASCE) "Minimum Design Loads for Buildings and Other Structures"

Information and Reference

● American Association of State Highway and Transportation Officials (AASHTO) "Standard Specification for Highway Bridges" (Reference Document)

- Military Handbook "Piers and Wharves" (MIL-HDBK-1025/1) (Reference Document)

- Naval Facilities Engineering Command (NAV-FAC) Design Manual "Fixed Moorings" (DM 26.4) (Reference Document)

- Naval Facilities Engineering Command (NAV-FAC) Design Manual "Seawalls, Bulkheads, and Quaywalls" (DM 25.04) (Reference Document)

- U.S. Army Corps of Engineers (USACE) "Shore Protection Manual"

- U.S. Army Corps of Engineers (USACE) "Coastal Engineering Manual"

Two structure classifications are considered for wharf design:

- Open-Type Marginal Wharf

- Solid-Type Marginal Wharf

These options and their methods of construction are discussed below.

23.11.1 Open-Type Marginal Wharf

A marginal wharf is a wharf structure that is connected to the upland shore area along its full length. Such an arrangement is adaptable to the land transfer of containers, luggage, and passengers for cruise and/or cargo transport. The alternatives that lend themselves to this arrangement are the open-type marginal wharf with reveted slope, and the solid-type marginal wharf.

For an open-type marginal wharf with reveted slope, the wharf geometry is dictated by the relatively deep-water conditions at the face of the berth.

The following five pile supported structure configurations for use as open-type marginal wharf structures with reveted slope.

- Precast Box Beam

- Cast-in-Place Flat Plate

- Composite Panel Flat Plate

- Cast-in-Place Slab and Bent

- Ballasted Flat Plate

A summary of advantages and disadvantages of the open type marginal wharf alternatives are provided in Table 23.5.

Precast Box Beam Wharf ▪ The first marginal open wharf alternative presented consists of a composite concrete deck surface supported by precast box beams which in turn are supported by concrete bent caps and prestressed concrete piles. A schematic drawing of the precast box beam alternative can be found in Fig. 23.5.

The main working surface is a concrete composite deck composed of precast panels and a cast-in-place concrete topping slab. The deck would be supported by standard AASHTO precast prestressed concrete box beams which would be spaced to optimize structure costs.

The advantages of this alternative include minimization of the total number of piles supporting the wharf structure and maximization of the use of precast construction, which will reduce the field labor and construction time required. Precast deck panels would serve as a bottom form for the cast-in-place concrete deck thus reducing costly formwork placement and removal. The slab design is such that the cast-in-place deck slab will be composite with the precast bottom form.

The disadvantages of this alternative are related to the use of precast elements. The size of the precast box beams will require specialized material handling equipment to place the box beams. The use of precast elements results in the need for tighter construction tolerances. Unlike with cast-in-place construction where adjustments are easily made to accommodate misalignments or elevation variations, precast construction requires very accurate construction and constant monitoring of precast element placement. Finally, precast construction inherently has two problems—special detailing required for continuity of connections between individual precast elements, and high cost of specialty elements. Continuity is required because vertical and lateral loads must have a continuous, structural pathway between precast elements to allow transfer of these loads to the piles and subsequently to the soils.

Cast-in-Place Flat Plate Wharf ▪ The second marginal, open wharf type is a pile supported cast-in-place concrete flat plate slab. The cast-in-place concrete deck provides two-way slab action to resist vertical and lateral loading. This type of construction is relatively new in the heavy waterfront construction industry but has been gaining in popularity due to the initial construction cost savings that can be achieved.

Table 23.5 Open Type Marginal Wharf Alternatives

Precast Box Beam	Cast-in-Place Flat Plate	Composite Panel	Cast-in-Place Slab and Bent	Ballasted Cast-in-Place Flat Plate
		ADVANTAGES		
Minimizes number of piles required	Affords economy of materials	Maximizes use of precast elements	Construction tolerances are not as critical	Affords economy of materials
Maximizes use of precast elements	Provides highest degree of structural continuity	Minimizes field labor	Conventional type of construction	Provides highest degree of structural continuity
Minimizes field labor	Minimizes formwork costs	Precast elements are relatively small	Easy to form recessed utility pits, trenches, and other items	Minimizes formwork costs
Reduced construction time	Less specialized equipment required for construction	Requires no special equipment for construction	No special equipment required for construction	Less specialized equipment required for construction
Reduces need for formwork		Reduced construction time		Flexibility for adding rail and utilities
		DISADVANTAGES		
Requires special materials handling equipment	Requires highly qualified labor	Requires tighter construction tolerances	Highly dependent on qualified labor	Requires highly qualified labor
Requires tighter construction tolerances	Labor intensive construction	Requires special detailing for continuity of connections	Labor intensive construction	Labor intensive construction
Requires deeper pile embedment	Increased construction time	Requires a large member of piles	Increased construction time	Increased construction time
Requires special detailing for continuity of connections	Requires special detailing for ductility	Difficult to expand for rail	Requires special concrete casting sequence	Requires special detailing for ductility
Difficult to expand for rail	Requires special concrete casting sequence		Formwork relatively costly	Requires special concrete casting sequence
	Difficult to expand for rail		Difficult to expand for rail	

Note: All open-type wharves require excavation and complete demolition of existing structures.

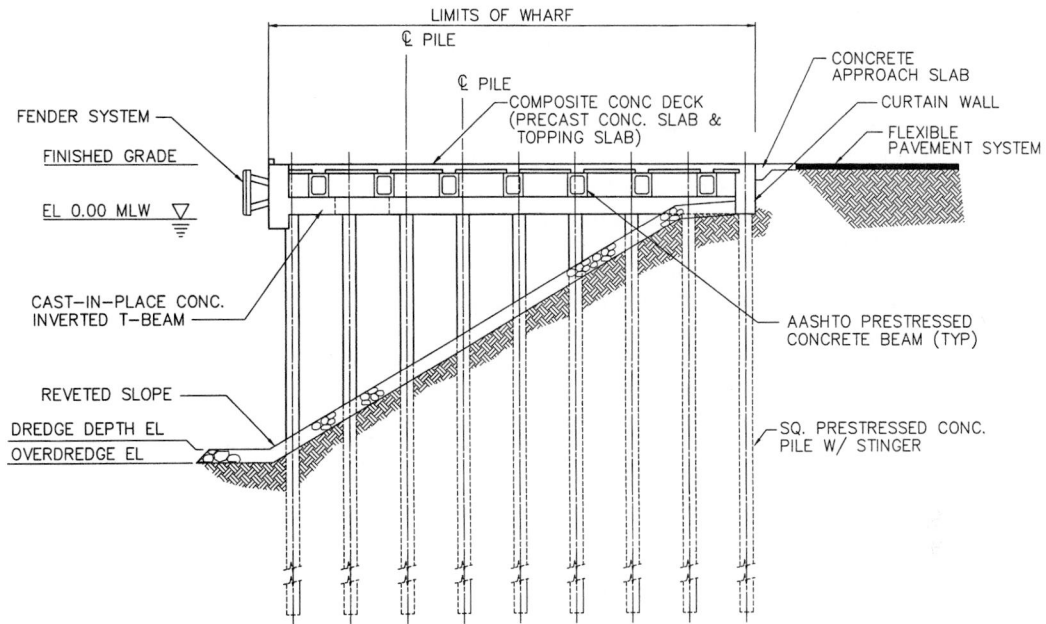

Fig. 23.5 Precast box beam open wharf.

The flat plate design consists of a cast-in-place concrete deck, which provides two-way slab action to resist vertical and lateral loading. A drop-panel fascia on the waterside face of the wharf accommodates the dock fender system and the mooring devices, as well as providing a transition area that would accommodate any required utility or service pits.

The flat plate slab would be supported by precast prestressed concrete piles. Piles are spaced to match slab and pile capacities. The advantages of the flat plate alternatives include minimization of forming costs for cast-in-place concrete construction and economy of materials. The driven piles act as supports for the bottom form, which is level. Because the structure is principally cast-in-place construction, less specialized equipment is required during construction. Most general contractors can perform the necessary forming and concrete placement required. Finally, since the structure consists only of cast-in-place concrete and precast piles, there are fewer structural components to integrate together, and thus, this design provides a high degree of structural continuity.

The disadvantages associated with this design are related to the extensive use of cast-in-place concrete. The techniques employed for construction are labor intensive. This includes the placement of formwork and the need to strip the formwork and move it to the next slab unit being placed. To minimize the time on the site and to account for shrinkage effects, a special concrete casting sequence must be used. A large pool of relatively skilled workmen will be required. In addition, this type of construction requires more time in the field than a precast alternative as the contractor must wait for adequate strength development of the concrete slab before form stripping can begin. Also, with this type of wharf future expansion for rail is very difficult and costly.

Note that the pile spacing for this type of structure is controlled primarily by span limitations of the cast-in-place concrete deck slab under vertical loading. Therefore, pile spacing is selected independent of lateral resistance requirements of the substructure. Lateral resistance is incorporated into the substructure by selecting a pile size capable of resisting the lateral design loads.

Composite Panel Wharf ▪ The third marginal open wharf design consists of a composite concrete deck surface supported by either a cast-in-place or precast pile cap which in turn is supported by a precast prestressed concrete pile. A schematic drawing of the composite panel alternative can be found in Fig. 23.6.

The main working surface of this design is a concrete composite deck composed of precast concrete panels and cast-in-place concrete topping slab. Precast reinforced or precast prestressed concrete deck panels would be placed upon masonite bearing pads, which in turn would be supported by precast or reinforced cast-in-place concrete pile caps. Each pile cap would in turn be supported by a square precast prestressed concrete pile. A cast-in-place concrete drop-panel fascia on the waterside face of the wharf accommodates the dock fender system and the mooring devices, as well as providing a transition area that would accommodate any required utility or service pits.

One of the main advantages of composite panel construction is the minimization of labor intensive concrete formwork over water, resulting in reduced field labor costs and generally a reduction in the construction time. Additionally, precast items theoretically afford a greater degree of quality control. Modular type construction is also inherently less time consuming than more labor intensive methods such as the cast-in-place concrete alternatives. Most of the precast elements for this alternative are of the size and configuration that would facilitate casting on site.

The disadvantages of the composite panel alternative include the relative discontinuity of adjacent structural members. This results in the need for tighter construction tolerances and special connection details. This alternative requires a large volume of material and special connection details to make up for the relatively flexible nature of this type of structure and to effectively transfer lateral loads from the deck elevation into the supporting pile system. Also, with this type of wharf future expansion for rail is very difficult and costly.

Cast-in-Place Slab with Bent Wharf ■ The fourth, marginal open wharf design considered is a cast-in-place slab and bent system. This is a more conventional means of construction that utilizes pile supported cast-in-place concrete bent caps and deck slab. A schematic drawing of the cast-in-place slab with bent design can be found in Fig. 23.7.

This alternative consists of a cast-in-place concrete deck that provides one-way slab action to resist vertical and lateral loading. The slab is cast integrally with a series of pile supported concrete bent caps. A drop-panel fascia on the waterside face of the wharf accommodates the dock fender

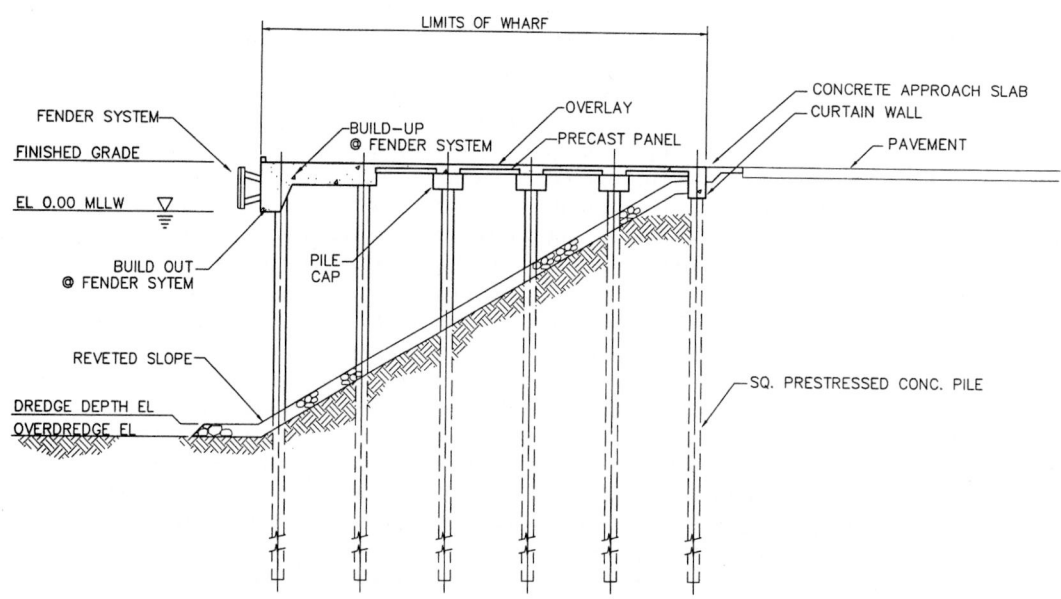

Fig. 23.6 Composite panel open wharf.

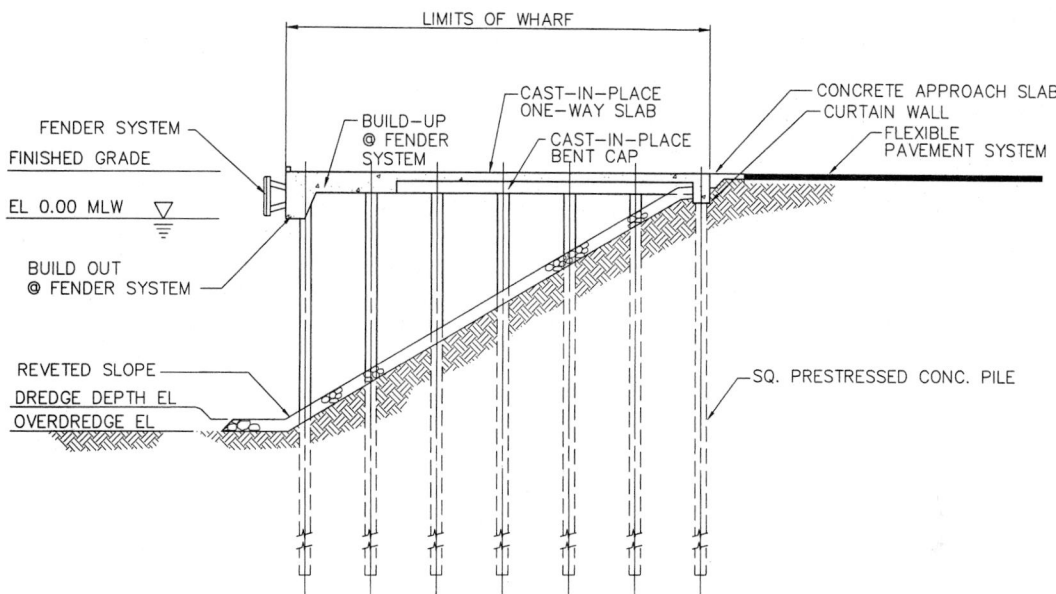

Fig. 23.7 Cast-in-place slab with bent.

system and the mooring devices, as well as providing a transition area that would accommodate any required utility or service pits. The transverse beams would be supported by prestressed concrete piles.

The advantage of this cast-in-place slab and bent configuration is that construction tolerances are generally not as critical as that of precast or modular systems. While cast-in-place construction is usually more time consuming and labor intensive than precast methods, it may be feasible alternative where there is an available supply of low cost labor. In addition, this type of construction can be completed by most general contractors and requires no special construction equipment or handling techniques.

The disadvantages of this type of construction principally involve the use of cast-in-place construction. This type of work is extremely labor intensive, so a dependable supply of low cost labor must be available. In addition, this type of construction requires more time in the field than a precast alternative as the contractor must wait for adequate strength development of the concrete slab before form stripping can begin. Finally, to minimize the time on the site and to account for shrinkage effects, a special concrete casting sequence must be used.

Ballasted, Cast-in-Place Flat Plate Wharf ▪ The fifth, marginal open wharf design considered is a ballasted open marginal wharf, which can be constructed with or without a toe wall. This structure is identical to the cast-in-place flat plate wharf described, but the deck is constructed at a lower elevation, then covered with a granular fill material, which can then be paved with asphalt. A schematic drawing of the ballasted flat plate alternative can be found in Fig. 23.8.

This consists of an 18-inch thick, cast-in-place concrete deck, which provides two-way slab action to resist vertical and lateral loading. An upset fascia system on the waterside face of the wharf retains the fill and accommodates the dock fender system and the mooring devices. Utility trenches can be constructed in the fill material between the deck and the asphalt.

The main advantages of the ballasted construction are the flexibility provided by asphalt paving and the lower deck elevation saves a row of piles. The advantages of the flat plate deck construction include minimization of forming costs for cast-in-place concrete construction and economy of materials. Because the structure is principally cast-in-place construction, less specialized equipment is required during construction. Most general contractors can perform the necessary finishing and

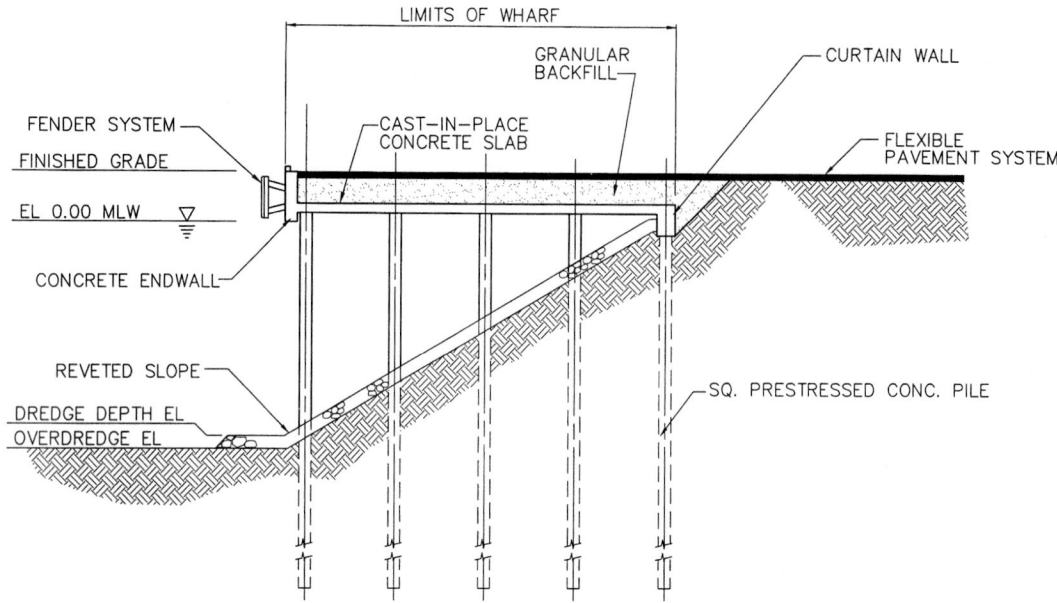

Fig. 23.8 Ballasted cast-in-place slab open wharf.

concrete placement required. Finally, since the structure consists only of cast-in-place concrete and precast piles, there are fewer structural components to integrate together, and thus, this alternative provides a high degree of structural continuity.

The disadvantages associated with this alternative are related to the extensive use of cast-in-place concrete. To minimize the time on the site and to account for shrinkage effects, a special concrete casting sequence must be used. A large pool of relatively skilled workmen will be required.

Note that the pile spacing for this alternative is controlled primarily by span limitations of the cast-in-place concrete deck slab under vertical loading. Lateral resistance is incorporated into the substructure by selecting a pile size capable of resisting the lateral design loads.

23.11.2 Solid-Type Marginal Wharf

The term solid-type marginal wharf refers to a wharf that has a continuous vertical face at or near the pierhead line. Such structures would include cellular cofferdams, either steel or concrete sheet piles walls, or some other type of structure that can contain backfill while providing a safety factor against sliding and overturning.

The six solid-type wharf alternatives considered here as follows:

● Cellular Cofferdam

● Open Cell Cofferdam

● Steel King Pile Bulkhead

● Steel Sheet Pile Bulkhead with Relieving Platform

● Precast Concrete Caisson

● Precast Concrete Gravity Wall

Each of these alternatives is best located so that the vertical load bearing elements of the structures are used. With slid-type marginal wharf structures, the width of the wharf is not constrained by geometry as with the open, pile-supported wharves where the minimum width is controlled by the reveted slope and cut-off wall. Solid-type wharves are only limited by the width of the cap.

A summary of advantages and disadvantages of the solid type marginal wharf alternatives are provided in Table 23.6. A description of each alternative follows thereafter.

Cellular Cofferdam ▪ A cellular cofferdam alternative can be considered a solid-type marginal

Table 23.6 Solid Type Marginal Wharf Alternatives

Cellular Cofferdam	Open Cell Cofferdam	Steel King Pile Bulkhead	Steel Pile with Relieving Platform	Precast Concrete Caisson	Precast Concrete Gravity Quaywall
ADVANTAGES					
Facilitates backfilling and vibro-compaction	Facilitates backfilling and vibro-compaction	Combines vertical and horizontal load bearing elements	Combines vertical and horizontal load bearing elements	Precast Elements Afford Better Quality Control	Precast Elements Afford Better Quality Control
Reduces width of wharf	Reduces width of wharf	Reduces initial dredging	Reduces initial dredging	Minimizes Field Labor	Combines Vertical and Horizontal Load Bearing Elements
Stable in seismic event	Stable in seismic event	Minimizes structure requirements— fewer driven piles	Minimizes structure requirements— fewer driven piles	Combines Vertical and Horizontal Load Bearing Elements	Reduces Width of Wharf
Reduced settlements	Reduced settlements			Reduced Settlements	Reduced Settlements
Constructed from land	Constructed from land				
DISADVANTAGES					
Sheet piling requires corrosion protection	Sheet piling requires corrosion protection	Highly dependent on quality of backfill material	Highly dependent on quality of backfill material	Special equipment required	Special equipment required
Requires complicated concrete closure pours for fascia	Requires complicated concrete closure pours for fascia	Requires geosynthetics for materials segregation	Requires geosynthetics for materials segregation	Requires use of dry-type casting yard	Requires use of dry-type casting yard
Increased time for construction	Dependent on quality of soils	Increased settlement potential		Complicated construction sequence	Requires geosynthetics for materials segregation
Highly dependent on quality of soils	Requires partial demolition of existing structure			Requires demolition prior to construction	Requires removal of existing structure prior to construction
Requires partial demolition of existing structure				Construction from barge	Requires construction from barge
				Large excavation and backfill quantities	Requires large quantities from excavation and backfill

Note: All solid-type wharves require less excavation and demolition than open-type wharves.

wharf containment structure. Although normally associated with providing a dry work area in waterfront construction, the inherent mass and stability of a cofferdam structure lends itself to fill containment for construction staging. Cofferdams may also be integrated with landside and water-side structures to provide a working platform for port operations.

Cellular cofferdams are generally constructed using straight web, steel sheet piling with con-trolled fill materials placed within the cells. Several geometric configurations of cellular cofferdams are available, including circular, diaphragm, and parallel wall configurations with open anchor cells. Circular cells are generally preferred for offshore construction and their circular shape favors single unit stability, east of forming templates for con-struction, and economy of material.

Cellular cofferdams depend on the interaction of the steel sheet piling and the material used to fill the cell to provide stability. Quality of the fill material has a direct effect on the size and, therefore, the expense of the sheet pile cofferdam. An ideal fill material would be a dense, granular, free-draining material with a high angle of internal friction and a low percentage of silt and clay materials. Fill materials that do not meet all of these requirements may be improved somewhat by vibratory compaction after placement within the cell. Unsuitable soils can be removed and replaced with suitable soils.

A schematic drawing of a cellular cofferdam alternative, intended for use as a solid-type mar-ginal wharf, and combined with dock fendering devices is provided as Fig. 23.9.

A cellular cofferdam structure is generally most feasible when the underlying materials are dense gravel, rock, or other similar dense and hard materials. Founding the coffer structure in this type of material allows the structure to develop

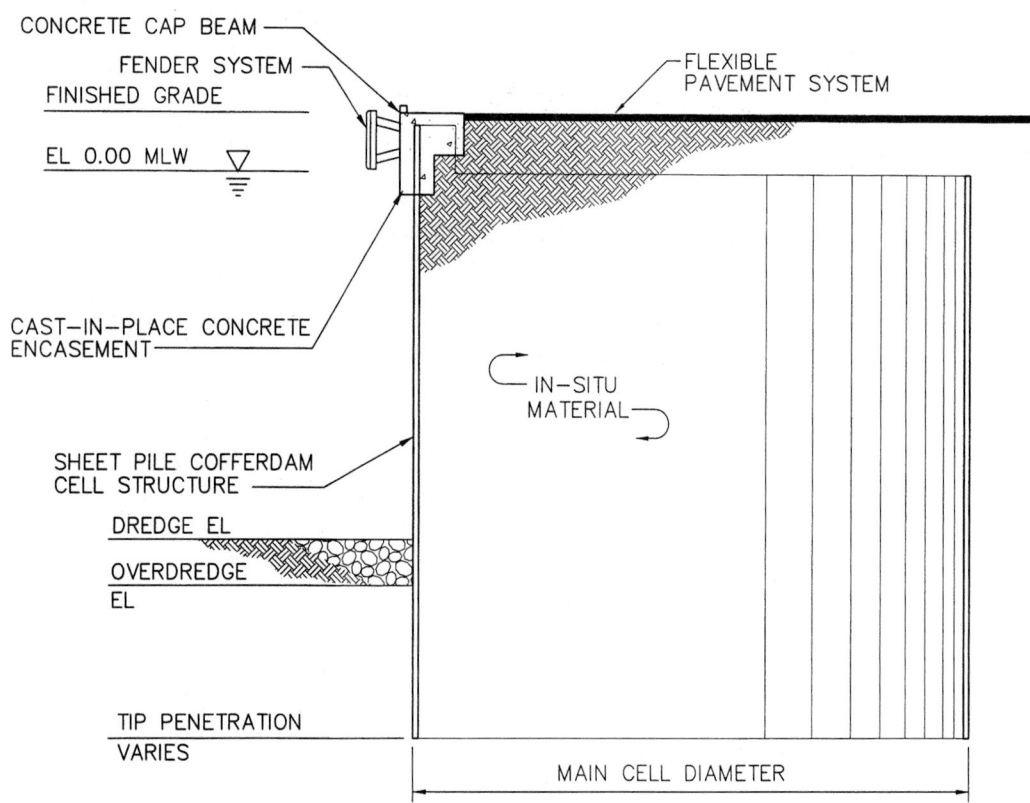

Fig. 23.9 Cellular cofferdam.

sufficient strength to resist the overturning and sliding forces that act on such a gravity structure. The drawback is that pre-drilling is required when shallow rock layers are encountered.

Transfer of the anticipated horizontal loads from mooring and berthing would be accomplished through special details which connect the facing structure to the cofferdam. These details could consist of a continuous reinforced concrete closure pour and additional horizontal restraining features such as longitudinal and transverse reinforced concrete shear keys.

Open Cell Cofferdam ■ An open-cell cofferdam is a variation of the cellular cofferdam and presented as a solid type marginal wharf. For a sketch of the configuration of an open-cell cofferdam, see Fig. 23.10.

The tail walls anchor the system and extend landward from the curved segments of sheet pile wall. The driving tolerances and template requirements are not as tight as those for the cellular cofferdam structures. Embedment depths on this type of structure are shallower than for the other structures discussed, reducing material costs. The tail walls can be manipulated to avoid obstructions illustrating the exceptional tolerances inherent in the system.

Open cell cofferdams depend on the interaction of the steel sheet piling and the material used to fill the cell to provide stability. Quality of the fill material has a direct effect on the size and, therefore, the expense of the sheet pile cofferdam. An ideal fill material would be a dense, granular, free draining material with a high angle of internal friction and a low percentage of silt and clay materials. Fill materials that do not meet all of these requirements may be improved somewhat by vibratory compaction after placement within the cell. Unsuitable soils can be removed and replaced with suitable soils.

In evaluating an open cell cofferdam as a solid-type marginal wharf structure, consideration should be given to construction sequencing, and vertical and horizontal load carrying capability of the structure.

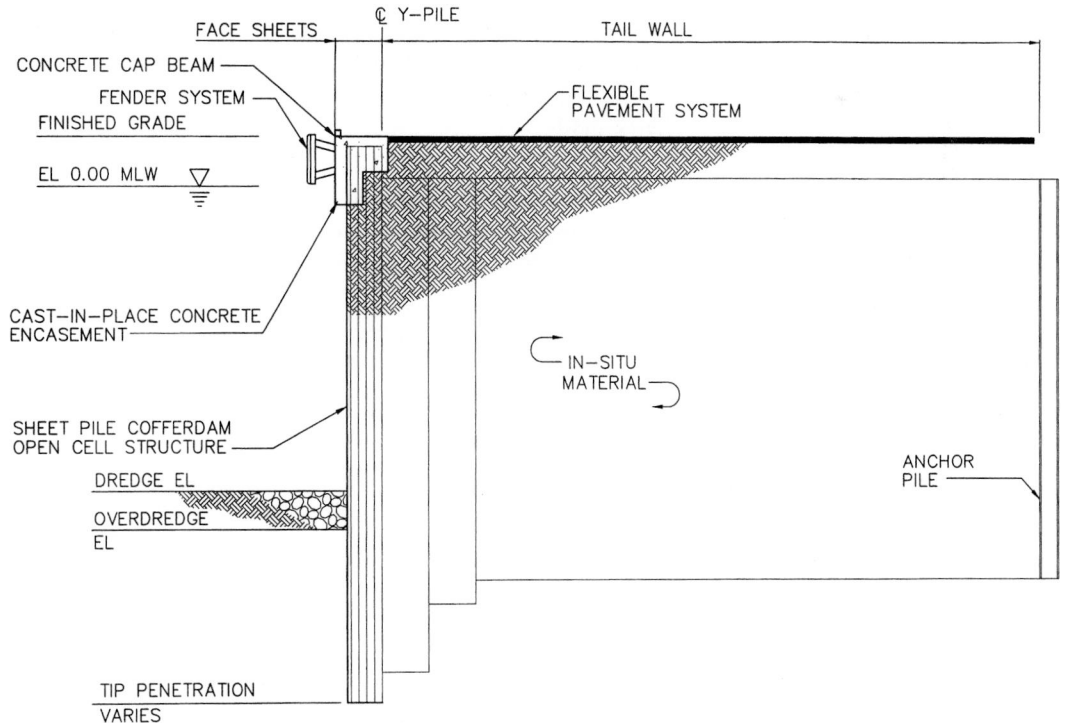

Fig. 23.10   Open cell cofferdam.

An open cell cofferdam structure is generally most feasible when the underlying materials are dense gravel, rock, or other similar dense and hard materials. Founding the coffer structure on this type of material allows the structure to develop sufficient strength to resist the overturning and sliding forces, which act on such a gravity structure. The drawback is that pre-drilling is required when shallow rock is encountered, which drives up the cost of construction.

Transfer of the anticipated horizontal loads from mooring and berthing would be accomplished through special details that connect the facing structure to the cofferdam. These details could consist of a continuous, reinforced concrete closure pour and additional horizontal restraining features such as longitudinal and transverse reinforced concrete shear keys.

Steel King Pile Bulkhead ▪ A steel king pile bulkhead is another form of a solid-type marginal wharf. For this type of structure, the bulkhead face is located as close as practical to the pierhead line. For a sketch of a steel king pile bulkhead, see Fig. 23.11.

The sheet pile bulkhead can consist of a high-modulus, H-piles interlocked with "Z" sheeting.

This will develop a significant vertical capacity, accomplished by driving the H-piles into any underlying hardrock layer. The "Z" sheeting can be driven to considerably less depth than the vertical load bearing elements (H-piles). This form of sheet piling is widely used, especially in Europe, with a great deal of success. Its main drawback is the need to provide an effective corrosion protection system for the steel. With proper corrosion protection measures, the steel sheet piling will provide a service life that meets or exceeds the service design life requirements of the port facility. Depending on the depth of intermediate sheet piles, this configuration can be susceptible to scour. In order to provide lateral capacity to resist the loads generated from retained fills and moored ships, the use of a conventional tieback system of tie rods and deadmen is typically incorporated.

Sheet Pile Bulkhead with Relieving Platform ▪ A sheet pile bulkhead with relieving platform is another solid-type marginal wharf containment structure. In order to reduce the size of the bulkhead structure required, a relieving platform would be incorporated into the system. The relieving platform transfers the vertical live loads

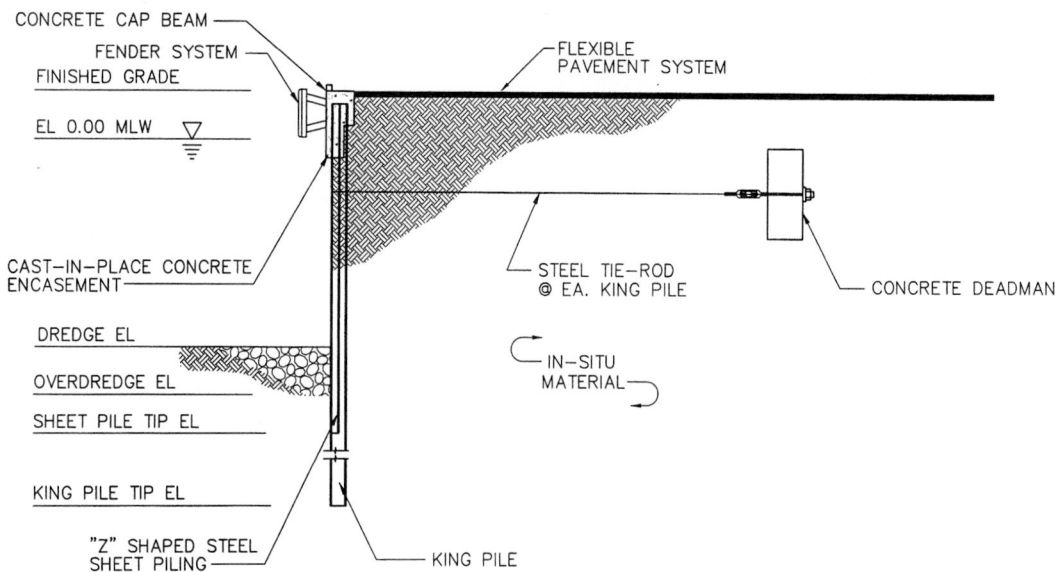

Fig. 23.11 King pile bulkhead with tie-rod and deadman.

from the deck directly into the underlying soils. Thus, the bulkhead is not required to resist the lateral surcharge, which would be generated from the live loads. To minimize the material requirements, the bulkhead has been developed to support some of the vertical loads from the wharf deck to supplement to relieving platform load bearing elements. A schematic of this is shown in Fig. 23.12. This design uses "Z" configuration steel sheet piles.

The relieving platform portion of this design is a cast-in-place concrete structure supported on prestressed concrete piles. The bottom of the relieving platform deck would be constructed at or near the mean high water elevation. At the outboard face of the relieving platform a gravity-type retaining wall would serve to contain the pavement supporting fill material.

In order to provide lateral capacity to resist the loads generated from retained fills and moored ships, it would be possible to incorporate a series of battered piles into the relieving platform structure. Another alternative for lateral restraint involves the use of a conventional tieback system of tie rods and deadmen.

Precast Concrete Caisson ▪ The most common of the available configurations of precast concrete caissons is that of a multicell box-type design. Structures of this type are usually constructed in drydocks, cofferdams, or other similar dry-type basins. Upon completion, the caissons are then floated into position, where they are flooded and lowered onto a prepared leveling course. The construction sequence is completed by the filling of the cells with a suitable ballast material. The inherent mass and stability of a large caisson structure lends itself to fill containment during construction staging. Caissons may also be integrated with landside and waterside structures to provide a working platform for port operations. For a sketch of a proposed configuration of a precast concrete caisson, see Fig. 23.13.

One of the most important aspects in the utilization of a caisson structure is that of preparation of a proper foundation course. It is also important to design into the structure tolerances that allow for differential settlement between adjacent caisson units. Joint design also merits careful consideration so that the structure is soil-tight.

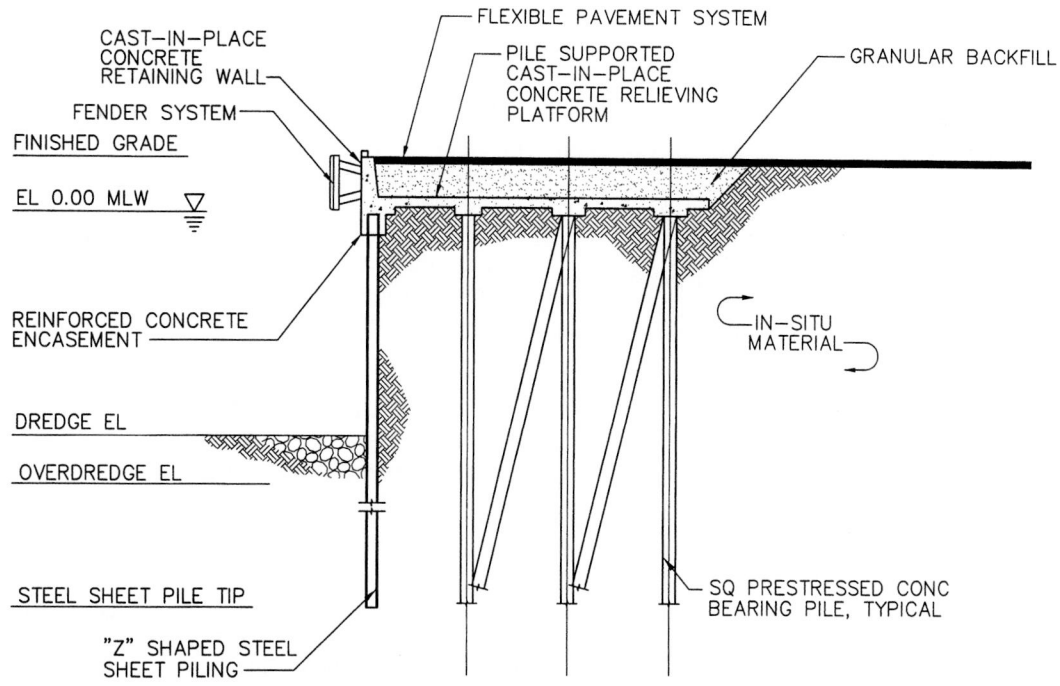

Fig. 23.12 Sheet pile bulkhead with relieving platform.

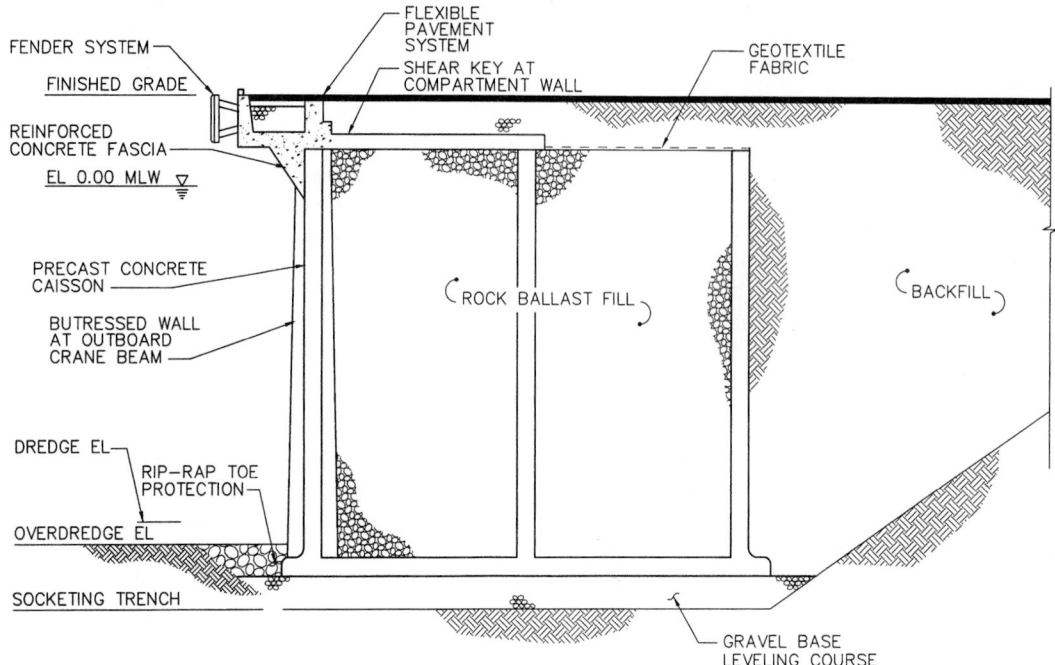

Fig. 23.13 Precast concrete caisson.

The anticipated sequence of construction for a precast concrete caisson solid-type wharf structure is as follows:

- Perform the initial dredging of existing river silt material in the entire area of the caisson structure. Dredge a keyway within the limits of the caisson.
- Construct the compartmentalized precast concrete caisson. The length of caisson will be dependent on the configuration of the proposed casting facility.
- Provide a gravel base leveling course in the socketing trench and graded to suit the dimensions of caisson selected.
- Transfer the caisson unit(s) from the casting yard to the site using tugs. Maneuver the precast concrete caisson into place, and begin lowering the unit by filling with water.
- Align and adjust the structure while it is buoyant. When alignment is achieved, begin placement of the rock ballast fill. Adjust the leveling and placing process by controlled placement of ballast.

- Provide a base protection material of graded rip-rap. At this point, the terminal side fill material may be placed.

In order to accomplish the transfer of the anticipated horizontal loads from mooring and berthing, special details would need to be developed to connect the facing structure to the caisson. These details could consist of a continuous reinforced concrete closure pour and additional horizontal restraining features such as reinforced concrete shear keys.

Precast Concrete Gravity Quaywall ▪ A precast concrete gravity quaywall design is presented as a solid-type marginal wharf structure. Components of these structures are usually cast in work yards and then transported to the site on barges, where they are lowered onto a prepared leveling course. The inherent mass and stability of a large gravity wall structure lends itself to fill containment during construction staging. In order to achieve a properly interlocked structure, the separate concrete elements are usually match-cast and marked before transport to the field for placement. For a sketch of a configuration of a precast concrete gravity quaywall, see Fig. 23.14.

One of the most important aspects in the utilization of a precast concrete gravity structure is that of providing a properly prepared foundation course. It is also important to design into the structure tolerances that allow for differential settlement between adjacent structure units. Joint design and soil key interlock design also merit careful consideration so that the structure is stable and soil-tight.

The anticipated sequence of construction for a precast concrete gravity containment structure would be as follows:

- Perform the initial dredging of existing materials in the entire area of the wall structure. Dredge a keyway within the limits of the gravity wall.
- Construct the precast concrete wall sections. The size of the blocks will be dependent on the capabilities of materials handling equipment, cost for transport, and other similar factors.

- Provide a gravel base leveling course in a socketed trench graded to suit the dimensions of the precast concrete gravity wall.
- Transport the concrete blocks to the site, and lower the blocks onto the prepared bed. Backfill the area behind the wall with select materials so that the resultant active earth pressures on the wall are minimized.
- Provide a base protection material of graded rip-rap.

Transfer of the anticipated horizontal loads from the wharf structure to the gravity quaywall would be required. This load transfer would be accomplished through the use of a continuous concrete cap block that links the wharf to the wall. Additional horizontal restraining features may also include transverse reinforced concrete shear keys. An additional advantage of this type of system is that the wall structure may serve as a vertical load bearing element for a waterside crane beam, thereby reducing some of the structure costs associated with the crane beam support.

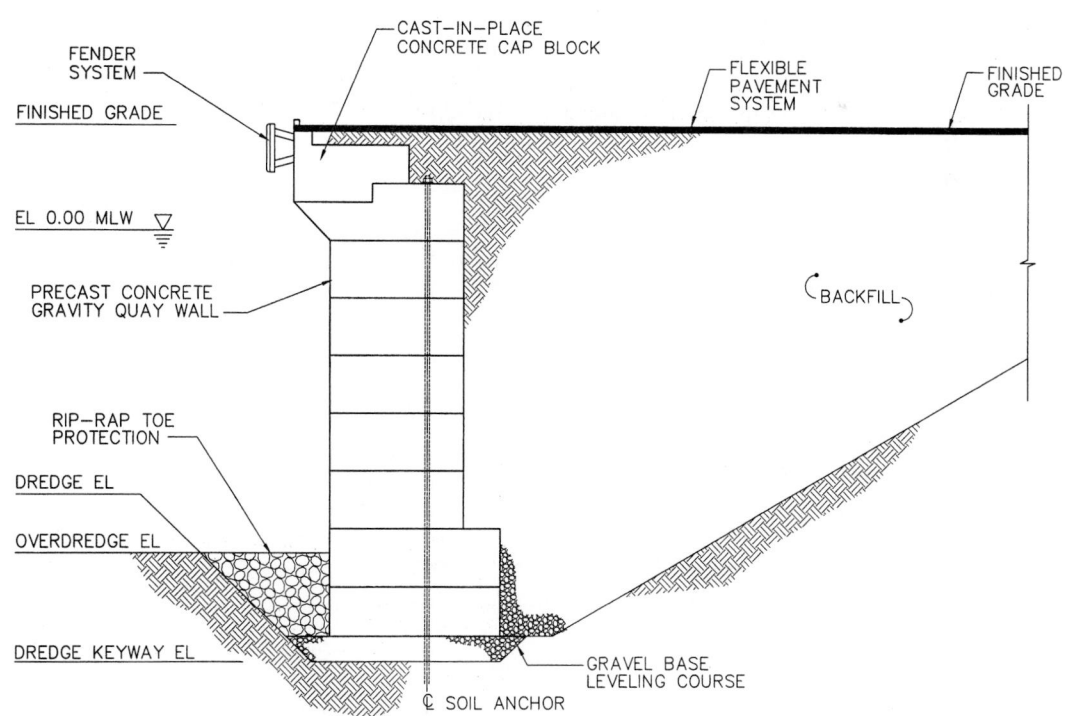

Fig. 23.14 Precast concrete gravity quaywall.

23.11.3 Loads on Docks

When the type of dock and its general construction features have been determined, it is necessary to establish the lateral and vertical loads for which the dock is to be designed. These consist of the following:

Wind Forces ▪ Mooring lines, which pull the ship into or along the dock or hold it against the force of the wind or current, exert lateral forces on a dock. The maximum wind force equals the exposed area, ft^2, of the broadside of the ship in a light condition, multiplied by the wind pressure, psf, to which a shape factor 1.3 is applied. This is a combined factor that takes into consideration a reduction due to height and an increase for suction on the leeward side of the ship. The wind force varies with the location and local building codes.

Current Forces ▪ The force of the current, psf, equals $wv^2/2g$, where w is weight, lb/ft^3, of water, v is the velocity of the current, ft/s, and g is 32.2 ft/s^2. For salt water this results in a pressure, psf, equal to v^2. The velocity of current usually varies between 1 and 4 ft/s, which results in pressures of 1 to 16 psf, respectively. Current pressure is applied to the area of a ship below the water line when the ship is fully loaded. Since ships are generally berthed parallel to the current, this force is seldom a controlling factor in design of the structure. However, currents are important in fender system design.

Impact ▪ Docking impact is caused by a ship striking the dock when berthing. The assumption is usually made that the maximum impact to be considered is that produced by a ship fully loaded (displacement tonnage) striking the dock at an angle of 10° with the face of the dock, with a velocity normal to the dock of 0.25 to 0.5 ft/s (Fig. 23.15). A few installations have been designed for as much as 1.0 ft/s, but this may be excessive; it corresponds to a velocity of approach of about $3\frac{1}{2}$ knots at an angle of 10° to the face of the dock, and such an impact could damage a ship.

Fender systems are designed to absorb the docking energy of impact. The resulting force to be resisted by a dock depends on the type and construction of the fender and the deflection of the dock if it is designed as a flexible structure.

Earthquake Forces ▪ These have to be considered if a dock is in an area where seismic disturbances may occur. The horizontal force, applied at the dock's center of gravity, may vary between 0.025 and 0.15 of the acceleration of gravity g times the mass. The force also can be expressed as 0.025 to 0.15 of the weight, respectively. The weight to be used is the total dead load plus one-half the live load. Unless the dock is of massive or gravity-type construction, seismic effect on the design will usually be small since the allowable stress, when combined with dead- and live-load stresses, may be increased by $33\frac{1}{3}\%$.

Gravity Loads ▪ These consist of the dead weight of the structure, or dead load, and the live load, which usually consists of a uniform load and

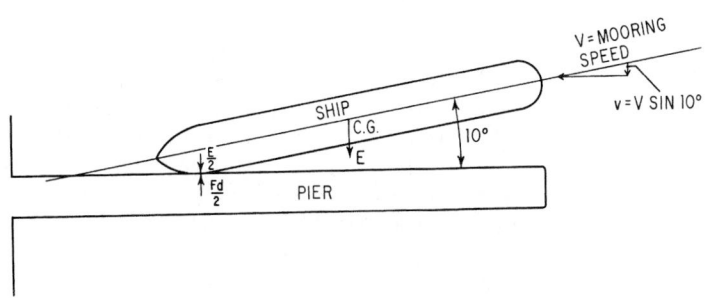

Fig. 23.15 Ship is assumed to strike pier at 10° angle for design against impact.

wheel loads from trucks, railroad cars or locomotives, cargo-handling cranes, and equipment. The uniform live load may vary from 250 to 1000 psf on the deck area. The smaller figure is used for oil docks and similar structures that handle bulk materials by conveyor or pipeline and where general cargo is of secondary importance. General-cargo piers usually are designed for heavier live loads, ranging from 600 to 800 psf. Piers handling heavy metals, such as copper ingots, may be designed for 1000 psf or more. The uniform live load controls design of the piles and pile caps, whereas the concentrated wheel loads, including impact, usually control the design of the deck slab and beams. A reduction of $33\frac{1}{3}\%$ is sometimes made in the uniform live load in figuring the pile loads and designing the pile caps or girders, based on the assumption that the entire deck area of adjoining bays will not be fully loaded at one time.

23.11.4 Dock Fenders

Fenders act as the interface between the vessel and structure. They serve to protect both the structure and the vessel from damage while the vessel is moored. There are several categories of commercially available fenders including buckling fenders with low friction panels, foam filled and pneumatic floating fenders, cylindrical side loaded fenders, and molded arch type fenders with low friction panels. Low friction panels are coated with an ultra high molecular weight polyethylene (UHMW-PE).

Buckling fenders include cell fenders and Π-shaped fenders. A small length to height ratio is preferred for Π-shaped fenders. The panels distribute the reaction force over a large area thereby exerting a low hull pressure while absorbing a high quantity of energy. These fender systems are often used for cruise ships and container vessels. They are moderately priced.

Foam filled and pneumatic floating fenders are most widely used for ship-to-ship operations and in areas where tidal variations are extreme. They are easily moved, absorb a high quantity of energy, and exert very low hull pressures. Due to their buoyancy, they move vertically with the vessel when the water level changes and allow berthing of several types of surface vessels. The fender requires a backing system to distribute the load.

Cylindrical side loaded fenders are used predominately for berthing small ships and pleasure craft. They are usually constructed of rubber and attached to the structure with chains. Cylindrical fenders absorb less energy than other fender systems and are often not used alone. They exert a moderately high hull pressure on the vessel and do not allow the ship to slide as well as low friction (UHMW-PE) panels. They are moderately priced.

Molded arch type fenders with low friction panels are similar to buckling rubber element fenders, but they are smaller and do not absorb as much energy due to their increased length to height ratio. The rubber fender is attached to the berthing structure and steel panels topped with UHMW-PE provide the contact surface for the vessels.

References for the design and selection of fender systems are the Military Handbook "Piers and Wharves" MIL-HDBK-1025/1 and PIANC 1984 Report of the International Commission for Improving the Design of Fender Systems.

23.11.5 Dolphins

Dolphins are designed principally for horizontal loads of impact, wind, and current forces from a ship when docking or moored. These forces are determined in the same manner as for design of docks.

Dolphins may be of the flexible or rigid type. Wood-pile clusters are examples of the former type. These are driven in clusters of 3, 7, 19, etc., piles, which are wrapped with galvanized cable (Fig. 23.16). The center pile of each cluster is usually permitted to extend about 3 ft above the other piles to provide a means of attaching a ship's mooring lines. A modification of this type of dolphin arranges the piles symmetrically and on a slight batter. They are bolted to wood cross members located just above low-water level, with wood framing at the top. Large steel cylinders and groups of steel-pipe piles have also been used to provide flexible dolphins.

In general, dolphins of the flexible type have been used for mooring small vessels, not exceeding 5000 DWT (deadweight tonnage), as an outer defense for protection of docks, or for breasting off somewhat larger vessels from loading platforms and structures not designed to take the impact of ships. Bottom soil conditions must be suitable for a flexible-type installation. If the soil is too soft, the dolphins or pile clusters will not rebound to their original positions after being struck by a vessel,

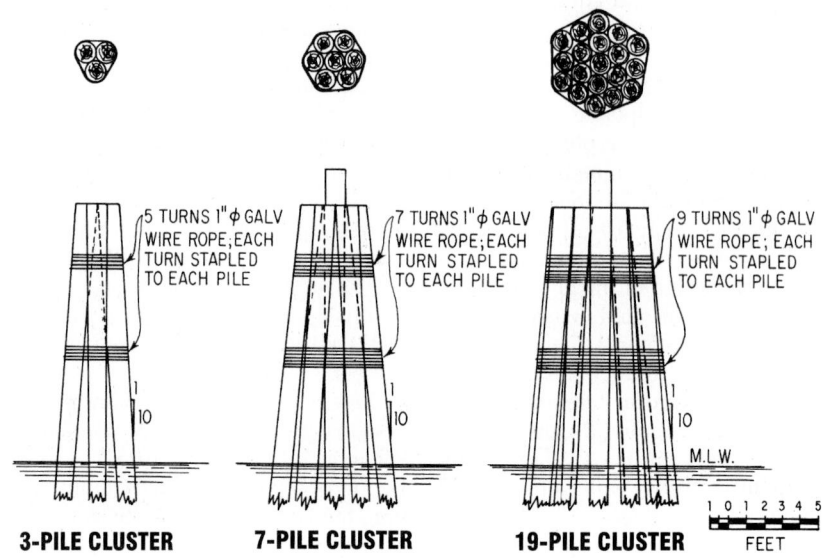

Fig. 23.16 Typical wood-pile dolphins in clusters of 3, 7, and 19 piles.

and their energy-absorbing ability, which depends on deflection, will be gradually dissipated.

For larger cargo ships and tankers of 9000 to 17,000 DWT, a wood-platform type of rigid dolphin, utilizing wood batter piles, may be used for mooring and breasting. Since the wood platform is relatively lightweight, its lateral stability depends to a large extent on the pullout value of the wood piles. In general, a lateral force of about 40 to 50 tons is about the most that a dolphin of this type can resist without becoming too large and unwieldy.

If bottom soil conditions are suitable, sheetpile cells make excellent dolphins. They can be designed to withstand the forces from the largest ships, if provided with adequate fenders. Cells, because of their circular shape, are well suited for **turning dolphins**, for warping or turning a ship around at the end of a dock. Cellular dolphins are usually capped with a heavy concrete slab, to which the mooring post or bollard is anchored. When large ships are to be handled, a powered capstan should be provided to draw in the heavy wire-rope mooring lines.

For big ships, dolphins may be designed with heavy concrete platform slabs supported by vertical and batter piles, usually of steel or precast concrete. This type of dolphin with low-reaction-force, high-energy-absorption rubber fenders can take the docking and mooring forces from the largest supertankers. For this purpose, a large number of batter piles are required. The uplift from these piles, in turn, makes it necessary to have a considerable amount of deadweight since the vertical piles will in general resist only a relatively small portion of the uplift. This deadweight is supplied by the concrete slab, which may be 5 to 6 ft thick. A sufficient number of vertical piles must be provided to support this deadweight. In addition, the vertical piles must not be spaced too far apart; otherwise, it will be difficult and expensive to provide forms for the concrete. When the depth of slab exceeds 4 ft 6 in, it is usually economical to cast the slab in two lifts, with horizontal shear keys at the construction joint. This greatly reduces the cost of the forms.

23.12 Marina Layout and Design

The layout of marinas varies according to local markets, customs, and site geometry. Marinas provide berthing space in calm water, access to utilities for boats, parking, fueling and sometimes repair service, boat ramps, stores, launch hoists, and storage sheds or areas. Marinas should be designed to meet the needs of the physically impaired as required by the Americans with Disabilities Act.

Selection of slip length is based on economics and the local boating community's typical lengths. Fairway aisles widths are 1.5 to 2 times the slip length (Fig. 23.17), depending on whether or not over-length boats are allowed in the slips. The width of single slips is about 4 ft wider than the boat beam or about half the slip length for slips less than 30 ft long. Piers can be either fixed or floating with ramps down from the surrounding grade level. Fixed-elevation piers are typically used in areas with a tide range of less than 3 ft. The width of piers gives at least 5 or 6 ft of clearance (inside of boat and utility boxes). Finger-pier widths are less. Wider piers are used for long piers that have more traffic.

Channel-depth requirements are often dictated by local physical conditions and boat fleet requirements. Dredging of marinas to provide adequate depth is a planning challenge because of the permit requirements related to the dredging and disposal of dredged material. (See J. B. Herbich, "Handbook of Coastal and Ocean Engineering," Gulf Publishing Company, Houston, Tex (www.gulfpub.com).)

23.12.1 Design Considerations for Megayachts

Recreational vessels typically 80 feet in length or greater are referred to as megayachts. Requirements for berthing facilities for these larger vessels differ from the requirements for berthing the smaller recreational boats in the 25 to 50 foot length.

Dock systems can be either fixed or floating, but require higher freeboard than normal to allow access from the vessel. The dock width should also be wider since most megayachts have their own gangway or steps that extend onto the dock. The fendering system along the dock to protect the expensive finishes on the vessels is more substantial than a typical marina dock. The wind loads exerted on the dock from a megayacht are also higher.

Utility requirements for megayachts are substantial. Electrical requirements may range from single phase, 50-amp to three phase, 400-amp. The

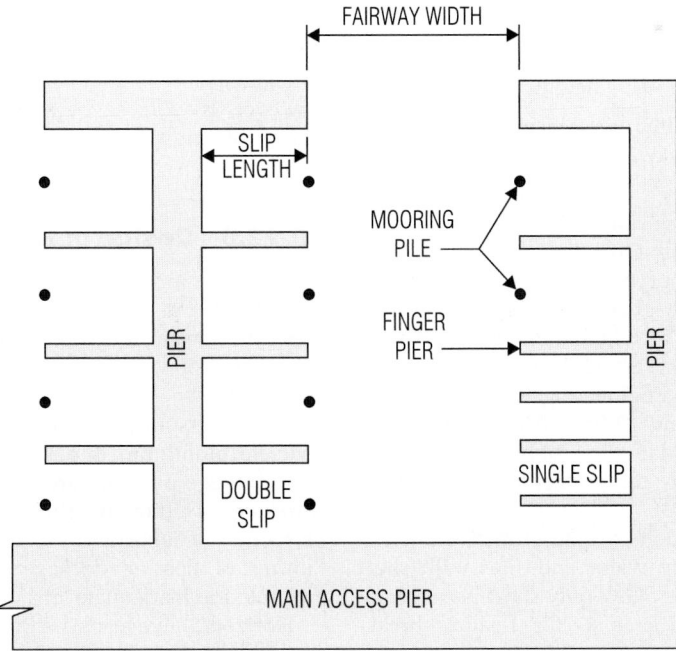

Fig. 23.17 Layout of marina with single and double slips.

larger water holding tanks on the yachts may require 1- to 3-inch water mains to fill the tanks within a reasonable time.

23.12.2 Breakwaters for Marinas

Some form of breakwater may be required to reduce wave heights to an acceptable level, usually less than 1 ft. (Although they may be necessary, breakwaters usually do not directly produce income for the marina.) Breakwater layout should provide adequate protection as well as allow safe movement into and out of the marina. The width of the entrance channel is often dictated by the local boat fleet. Enough room should be provided for safe passage in two-way traffic. This leads to use of typical entrance widths of 75 to 125 ft of navigable depth. Alternative breakwater types include rubble-mound, caissons, floating breakwaters, vertical walls, or wave fences. (See R. R. Bottin et al., "Maryland Guidebook for Marina Owners and Operators on Alternatives Available for the Protection of Small Craft against Vessel Generated Waves," U.S. Army Engineers Coastal Engineering Research Center, Washington, D.C.)

Wave protection of a breakwater is expressed in terms of the wave transmission coefficient k_t:

$$k_t = \frac{H_t}{H_i} \qquad (23.6)$$

where H_i = incident wave height

H_t = transmitted wave height in the lee of the breakwater

Transmission through breakwaters is caused by water moving through the interstices of the rocks and by splash over the top. The wave energy that rounds the edges of breakwaters is diffraction energy. The wave heights in the lee of a structure will be the result of a combination of transmission and diffraction. The "Shore Protection Manual," U.S. Army Coastal Engineering Research Center, Government Printing Office, Washington, D.C., provides design guidance for rubble-mound structures used primarily for protection against large waves.

Floating breakwaters are an attractive alternative for use in deep water and sites with short fetches for wind that allow only short-period seas. Transmission coefficients k_t for floating breakwaters depend on the ratio of the width of the float to the incident wavelength; the ratio of the depth of penetration below water level to the water depth; and the stiffness of the mooring system. For relatively deep water, k_t ranges from 0.2 to 0.4 when the ratio of float width to wavelength is greater than 0.5. For example, for a float width of 15 ft, this ratio gives a wavelength $L = 30$ ft. This L corresponds to a wave period T of about 2.4 s [Eq. (23.2)]. For a windspeed of 40 knots, this corresponds to a fetch distance of about 1 to 2 miles (Table 23.3). Thus, for fetches of about 1.5 miles and design wind conditions of 40 knots, the incident wave height of 2.2 ft would be reduced to less than 1 ft by a 15-ft-wide floating breakwater. Transmission coefficients can be much higher for narrower widths or longer waves. (See J. W. Gaythwaite, "Design of Marine Facilities for the Berthing, Mooring, and Repair of Vessels," Van Nostrand Reinhold, New York, and "Planning and Design Guidelines for Small-Craft Harbors," Manual 50, American Society of Civil Engineers.

Transmission at complete, impermeable, vertical walls is zero. However, diffraction and wave reflection allow energy to penetrate breakwater gaps into lee areas.

Wave fencelike structures are built to allow some water flow through them. These structures also are most effective with short-period wind-generated seas from limited fetches. For significant reduction of wave energy, the gap spacing between fence boards should be such that the total gap area is less than 10% of the total area. These small gaps, however, may close because of biofouling and defeat the original purpose.

23.12.3 Design of Docks for Marinas

Dock design includes design of the dock structure and the related, desired utilities, such as electrical, water (domestic and fire protection), telephone, cable television, and sanitary sewer hookups or pumpouts.

Fixed piers may be constructed of timber, steel, concrete, aluminum, or plastic. Manufactured floating dockage systems are available. Flotation is provided by either an airtight compartment, foam, or wood. Reliable anchorage is essential; many failures of floating dockage systems have been attributed to inadequate anchorage.

Horizontal live loads are the sum of current and wave loads and wind loads due to the exposed cross section of the dock and boats.

Wood decking on piers is often 2 × 6-in pressure-treated lumber with galvanized screws. Bolting of cleats to the structural frame of the deck is recommended (ASCE Manual 50 (www.asce.org).)

23.12.4 Boat Ramps

Boat ramps are an inclined paved surface that extends into the water that allows trailerable boats to be launched and retrieved. The top elevation of the ramp should be a least 2 feet above the highest expected water level the ramp is designed to operate and at least 4 feet below the lowest expected water level that the ramp is designed to operate. The number of lanes is dependent upon the use and demand. The lane width should be 12 feet minimum; 15 feet is preferred. Studies have indicated that a single lane can handle up to 50 launches and 50 retrievals per day under average conditions. The grade of the ramp should be uniform and between 12.5% and 15%. A smooth transition with a vertical curve should be made between the top of the ramp and the approach to the ramp.

Ramps are typically constructed of reinforced concrete or precast reinforced concrete slabs placed over an aggregate base. It is important to provide a non-skid surface to allow for maximum traction for vehicles launching and retrieving boats. Concrete ramps should be finished with a V-groove surface with the grooves placed at an angle of 60 degrees from the axis of the ramp.

Boarding docks, fixed or floating, are provided for access to the boats after launching and before retrieving.

(B. O. Tobiasson and R. C. Kollmeyer, "Marinas and Small-Craft Harbors," Van Nostrand Reinhold, New York; "Planning and Design Guide for Small-Craft Harbors," ASCE Manual 50 (www.asce.org).)

23.12.5 Ice in Marinas

Ice is a significant marina engineering concern, including horizontal and vertical ice loadings on structures and boats. The most common adverse effect is the vertical jacking of piles as ice grips them and then rises with the tide. The severity of the condition depends on site location and seasonal use of the marina. (See C. A. Wortley, "Ice Engineering Manual for Design of Small-Craft Harbors and Structures," University of Wisconsin Sea Grant Institute; ASCE Manual 50 (www.asce.org).)

23.12.6 Marina Flushing Analysis

Marina facilities may pose a threat to the health of aquatic systems if they are poorly planned or managed. The water in a marina must meet applicable water quality standards. Thus, water quality considerations are a part of the planning and permit application process for marina construction and modification. Pollutants at marinas include boat grease, oil, and gas, head discharges, bilge discharges, and land runoff. Runoff can introduce to the marine water typical nonpoint-source runoff pollutants, such as automobile oil and grease from parking lots and fertilizers, pesticides, and herbicides from adjacent grass areas. Proper runoff management practices can be critical to protection of marina water quality.

A simple tidal-prism analysis tool for estimating the flushing of a proposed marina is described in "Coastal Marinas Assessment Handbook," Environmental Protection Agency (EPA). The fundamental concept is the same as that for estuary or bay flushing: A pollutant concentration in a marina will decrease with each tidal dilution but will never completely flush. A typical analysis goal is to estimate the time needed to reduce the concentration of a pollutant to 10% of its initial concentration. The tidal-prism analysis can be extended to a dissolved oxygen or nutrient balance to address the question of whether or not a significant reduction will occur in dissolved oxygen or nutrient level over a tidal cycle.

The EPA methodology assumes that tidal exchange is the dominant flushing mechanism. It can be used with fresh-water inflow. It does not, however, specifically account for other water exchanges, such as wind- and current-generated circulation. A more limiting assumption is that the water is well mixed inside the marina.

There are several commercially available numerical modeling software systems capable of marina flushing analysis. Examples of such programs are the RMA2/RMA4 programs in the Surface-Water Modeling Systems (SMS) created by Brigham Young University or the Water Quality or Spill Analysis modules in MIKE21 developed by DHI. Typical input parameter for numerical models include bathymetry and hydrodynamic data, boundary conditions and extents, pollutant concentrations and type, and simulation length. There are also many input variables available for each model to customize simulations to specific criteria.

23.13 Beach Nourishment

Also called beachfill, this refers to mechanically or hydraulically placed sand on a beach. The arguments for beach nourishment are threefold:

1. Enhanced recreation and increased property value. A sandy beach is a desirable recreation area. This is a very important consideration for developments along many coastlines and is probably the primary reason for most beachfills. The **strand**, the sand (and maybe also the dunes) between the water and hard structures such as houses, has economic and aesthetic value, although it may not be easily documented.

2. Increased property protection. Beach nourishment can provide a volume of sand that protects properties behind the beach from the sea. This protection may be in the form of long, high sand dunes that are designed to erode somewhat during storms without breaching or low, wide beaches, which reduce wave heights, or both.

3. Beneficial disposal of dredged material. Beachfills can serve as disposal sites for sediment dredged from nearby coastal construction projects. Sometimes, the purpose of the placement of dredged material on adjacent beaches is to continue the natural littoral drift or *river of sand* by keeping dredged inlet sands in the littoral system. Sometimes, the beachfill is for convenience in obtaining construction permits or to avoid more expensive disposal options.

With respect to the first two contentions, beach nourishment is justified if the annualized economic benefits exceed the annualized cost of the beachfill. Note that beach nourishment should be perceived as an ongoing or maintenance requirement. **Periodic renourishment** is a requirement of many nourishment projects.

Many very large beachfills (several million cubic yards each) have been placed on open ocean beaches as well as in more sheltered locations, such as bay shorelines. The purposes of some of these beachfills have been wetlands habitat protection and creation as well as recreation and property protection.

Beachfill Behavior ▪ One of the fundamental design criteria for beachfills is estimated project life. There is not at present a consensus on what this value should be.

One of the best ways to estimate the behavior of a planned beachfill is to evaluate the behavior of the last beachfill at that site. This, however, is not always possible, but it is justification for implementation of a monitoring system. If a beach is now being nourished, it probably will need nourishing in the near future. If a beachfill was previously placed at that site, the result with more beachfill, however, will not be the same, even if the beachfill design is exactly the same as before. Fill response depends on time variations in waves and longshore transport. Alterations in the volume, length, width, configuration, or size of the sand will also change the backfill behavior. However, qualitative and quantitative estimates based on previous fill behavior are valuable if the designs are similar enough.

The behavior of beachfills follows some general patterns. These can be viewed as the constructed beach moving into dynamic equilibrium with the wave and water-level climate. The volume of sand per unit length of beach is the sum of the following:

1. Volume of sand required to create an equilibrated beach of the desired width (profile equilibration)

2. Volume expected to move out of the project limits in the longshore direction (planform equilibration) during the design life

This sum should be multiplied by an overfill factor for winnowing due to source-native sand-size difference (size equilibration).

Cross-shore or profile equilibration is the most rapid adjustment of a beachfill. Beaches change shape constantly in response to changes in incident waves and water levels. Natural beach slopes are mild, from 1:10 to 1:100. Often, the constructed beach has a much steeper beach face than that which occurs naturally (Fig. 23.18).

Sand is pulled offshore from the dry beach onto the portions of the beach that are below the water surface. Since the dry beach area rapidly decreases, this change is often incorrectly perceived as a loss of the beachfill. Although the public may be unimpressed by the unseen, underwater portions of a beachfill, they are no less important than the

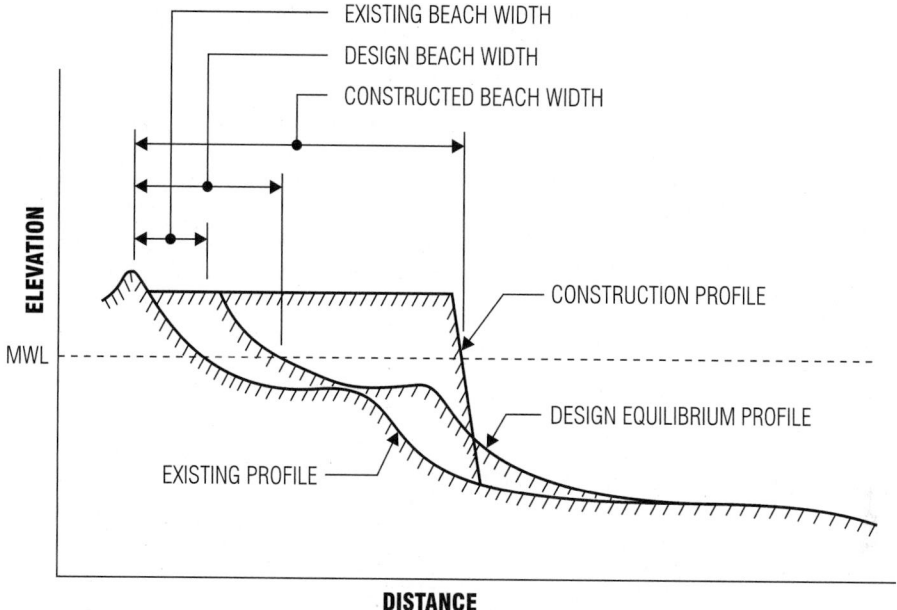

Fig. 23.18 Vertical cross section through a beach showing the existing beach profile, beachfill construction profile, and the effects of the cross-shore equilibration process.

underground foundation of a building. The full volume of sand required to widen the dry beach some distance must account for the underwater portion of the profile. Given that the dry beach is often a primary reason for benefits, however, there does not seem to be much reason for attempting to construct a more natural, milder beach slope. This would be difficult anyway. Public education, including construction-site signs and public meetings, are often used to inform the public that the initially constructed beach width is much greater than the intended design beach width.

A common technique for estimating the required volume of sand to build a desired beach width is based on the assumption that the post-construction beach slopes will match the native beach slopes but will be offset seaward by the design beach width (Fig. 23.18). There is often some *ideal* or *healthy* **composite beach profile** that represents the features, primarily slopes, of the beaches along a coast. This is essentially an estimated profile based on site-specific data on the equilibrium beach profile. The data include the actual wave climate and grain-size distribution. Often, a number of profiles estimated for several

locations along the beach and based on a common point (HWL, MWL, toe of dunes) can be overlain to develop the composite profile. Care should be taken to account for overstarved profiles that have been previously armored and are not representative of healthy profiles.

When the source material has a different size or size distribution from that of the native sand, a theoretical equilibrium approach based on an equilibrium-beach-slope concept can be used. The time required to reach cross-shore equilibrium of a beachfill can be roughly estimated with some form of a cross-shore transport model.

Longshore transport or **planform equilibration** moves sand from the constructed project limits in the alongshore direction. Natural longshore sand transport processes will move sand from the constructed location to adjacent beaches or inlets. Although the sand is not lost from the littoral system, it is lost from the project. There are three general ways to estimate the rate of planform equilibration:

1. Use of some historical or *background* erosion rate. Historical profile and shoreline change

data can be used to estimate volumetric erosion rates. Vital to this estimate is a clear understanding of the local coastal processes, particularly the cause of erosion at the site. This approach assumes that the beachfill sand will leave the project area at the same rate as sand has been leaving the existing beach. However, constructed beaches may change faster than the existing beach has been, because the constructed beaches are not in equilibrium with adjacent beaches and shoals and the wave climate.

2. Use of a *diffusion-type* model for planform equilibration. The spread of sand away from a location may be computed from the same differential equation as that for the spread of heat in a bar, the classic heat or diffusion equation. One theoretical result of this approach is that if the length of a beachfill is doubled, the beachfill longevity will be quadrupled. This type of approach is most appropriate for long, straight coastlines away from the influence of inlets. It suffers from the constraint that it is based on some single, representative-wave-height parameter and does not directly account for the time variability of longshore sand transport.

3. Use of a time-dependent shoreline change model could be applied to estimate the fate of the fill under assumed incident-wave conditions.

Winnowing of fines or **sand-size equilibration** is the wave-induced sorting and winnowing of any fines that may be in the fill material. Many dredging projects include large portions of fine material that will not remain in the active beach environment. The type of dredging used influences this process. In general, hydraulic dredging removes much of the fines during the washing involved in the construction process itself. This can create a suspended sediment plume from the pipe outfall area. The alternative, mechanical placement, probably reduces the construction plume and leaves more room for fill deflation due to winnowing. Alternative sources of beachfill sand are nearby inlets, offshore shoals, upland quarries, back-bay dredging, and sand hauled from a distant source. The amount of fine-sand winnowing can be estimated with the overfill ratio method described in the "Shore Protection Manual," 4th ed., U.S. Army Coastal Engineering Research Center, Washington, D.C.

The duration of each portion of the equilibration process can range from months to years, depending on the size of the beachfill. Larger beachfills in milder wave climates take longer to respond fully.

Because of the public nature of most beachfills and the confusion concerning the *success* or *failures* of beachfills, postconstruction monitoring should be a prearranged, funded component of the project. A well-designed monitoring program will allow a more complete and rational evaluation of beachfill behavior. Also, monitoring results will be valuable for design of future beachfills at the site.

On some coastlines, structures can be built to extend the life of beachfill projects. Offshore segmented breakwaters with a beach nourishment or marsh construction project in the lee are one alternative for constraining beach erosion that can provide environmental habitat benefits. Rosati and Weggel provide design guidance for offshore segmented breakwaters. Bender outlines a headland breakwater concept that may be used to construct pocket beaches with beachfill.

[T. Bender, "An Overview of Segmented Offshore/Headland Breakwater Projects Constructed by the Buffalo District," in "Coastal Engineering Practice '92," Proceedings of a Specialty Conference on the Planning, Design, Construction, and Performance of Coastal Engineering Projects, American Society of Civil Engineers (ASCE). J. D. Rosati, "Functional Design of Breakwaters for Shore Protection: Empirical Methods," Technical Report CERC-90-15, U.S. Army Engineer Waterways Experiment Station, Vicksburg, Miss.; D. K. Stauble and N. C. Kraus, "Beach Nourishment Engineering and Management Considerations," ASCE. L. S. Tait, Proceedings of the Annual National Conference on Beach Preservation Technology, The Florida Shore & Beach Preservation Association, Tallahassee, Fla. *Shore Protection Manual*, 4th ed., U.S. Army Coastal Engineering Research Center, Government Printing Office, Washington, D.C.; J. R. Weggel, "Coastal Groins and Nearshore Breakwaters," Engineer Manual EM 1110-2-1617, U.S. Army Corps of Engineers, Washington, DC 20314-1000.]

23.14 Monitoring Programs for Coastal Engineering Projects

Postconstruction monitoring of coastal engineering projects should be standard coastal engineering practice. In addition to engineering benefits from such practice, there are public relations and scientific reasons for monitoring.

A formal monitoring program consists of response measurements, such as surveys, and forcing-function monitoring, such as wave climatology. If well designed, the program facilitates a verifiable, fact-based evaluation of project functioning.

Monitoring is especially important for projects that have a potential for adverse impact on nearby beaches, waterways, coastal structures, or dredging. Beachfill projects also should be monitored to obtain information on how fast the sand is leaving project boundaries.

The primary purpose of monitoring programs is to obtain data for future management decisions relating to the project site. Though present ability to model quantitatively the complex natural processes of the coastal zone are limited, project-specific monitoring is a proven method for development of cost-effective engineering solutions. Monitoring provides the data for an improved understanding of how a beach responds to engineering and also why it responds that way. A monitoring program is often funded with project construction.

Coastal Structures

23.15 Effects of Coastal Structures on Beaches

One of the recurring problems with engineered structure on beaches is evaluation of their impact on adjacent beaches. Because the littoral system is interconnected, what is done on one stretch of beach can have significant impacts on nearby beaches. The impact is related to the coastal processes of the area, in particular the longshore sand-transport rate. Longshore sand transport along beaches can extend for many miles. Inlet shoals are part of the same littoral system as the adjacent beaches if sand can move from the beaches to the shoals and vice versa.

Coastal structures by themselves can only function to redistribute the sand that is in the littoral systems. They do not create sand.

Structures such as groins that are perpendicular to the shore cause a buildup of sand on the updrift side by reducing the sand supply to the downdrift side. A field of groins will essentially realign the shoreline toward the dominant wave climate. Information on groin fields is limited but there is some indication that they are probably beneficial in stabilizing some stretches of coast. They probably provide wave protection and slow the local longshore transport rate because of the local realignment in shoreline angle. Groins, however, reduce the sand movement to the downdrift side. Also, they can induce nearshore circulation cells with rip currents capable of removing sand from the beach face into the sand-bar system. The net result is probably a beach system with a more stable shoreline. The beach neither comes nor goes as much as it would without the groins.

Inlet jetties can have impacts on adjacent beaches as do groins because of trapping of sand on the updrift side. They also can realign and change the volume of sand that is stored in the tidal shoal system.

Mechanical bypassing of sand is used to maintain the littoral drift system at some engineered inlets. The bypassing can consist of either periodic dredging with disposal on downdrift beaches or construction of a fixed bypassing plant. Design of a bypassing scheme is based on estimates of both the rate and variability of longshore sand transport at the site.

Dredging of navigation channels, even without inlet jetties, can contribute to beach erosion through two mechanisms:

1. The sand dredged from the ebb-tidal shoal is often disposed offshore in deeper water and thus removed from the littoral system.

2. The dredging can realign the main ebb-tidal channel, which results in realignment of the ebb-tidal shoals and thus affects the sheltering of the adjacent beaches provided by those shoals.

Seawalls, bulkheads, and revetments protect the land behind the wall, not the beach in front of the wall. There is little clear evidence that seawalls actually cause erosion, except for the impact of the reduction in sediment available for the littoral system if the shoreline erodes to the seawall. However, a seawall constructed on a beach that is receding for other reasons, the usual site for a seawall, will contribute to loss of beach. ("Shore Protection Manual," 4th ed., U.S. Army Coastal Engineering Research Center, Government Printing Office, Washington, D.C., "Coastal Engineering

Manual," (www.USACE.mil/net/asce-docs/eng-manuals/em-htm).)

23.16 Revetment and Seawall Design

A revetment, or seawall, is built to protect property from erosion. Many revetments are constructed as a rubble mound, because the technique has proved to be effective in the harsh and dynamic wave environment. One particularly attractive feature of rubble-mound structures is that, when designed correctly, they continue to function even after damage due to storms larger than the design storm.

There are four typical failure mechanisms for revetments: inadequate armoring, flanking, toe scour, and splash. All are related to inadequate protection of the underlying soil on the downstream side of the structure. Armor, or front-face, design includes selection of the armor unit size and layer thickness plus any underlayers required. For rubble-mound structures (Fig. 23.19), the armor-unit rock size can be calculated from Hudson's equation:

$$W = \frac{w_r H^3}{K_D (S_r - 1)^3 \cot \theta} \qquad (23.7)$$

where W = median weight of armor units

H = design wave height

S_r = specific weight of armor unit material

w_r = unit weight of armor unit

θ = slope of structure face

K_D = empirical coefficient that includes shape, roughness, and interlocking ability of the armor units

Hudson's equation is based on a 5% damage level observed in laboratory tests with monochromatic waves. There is no factor of safety inherent in Hudson's equation.

For rough, angular, randomly placed quarrystone placed in at least two layers, K_d varies from 1.5 to 4, depending on the structure slope, whether the stones are exposed to breaking or nonbreaking waves, and whether the stone is at the head of a pointed structure or in the main body of a long revetment. For breaking waves on the main body of a revetment of 1:3 slope, $K_D = 2.0$ when the gradation of the individual units is within the fairly narrow range $0.75W < W < 1.25W$. For small design wave heights ($H < 5$ ft), a wider range of stone size, riprap with a weight range of $0.25W_{50} < W < 4W_{50}$, can be used with a stability coefficient of 2.2 to 2.5. W_{50} is the median stone weight. The stability coefficient can be higher for concrete armor units, such as tetrapods, dolosse, or quadrapods.

The design wave height H used in Eq. (23.7) should be the average of the heights of the highest 10% H_{10} or 5% H_5 of the waves. The wave height is often depth-limited.

Slopes of 1:2 (vertical to horizontal) or 1:3, for heavy wave action, are recommended. Although

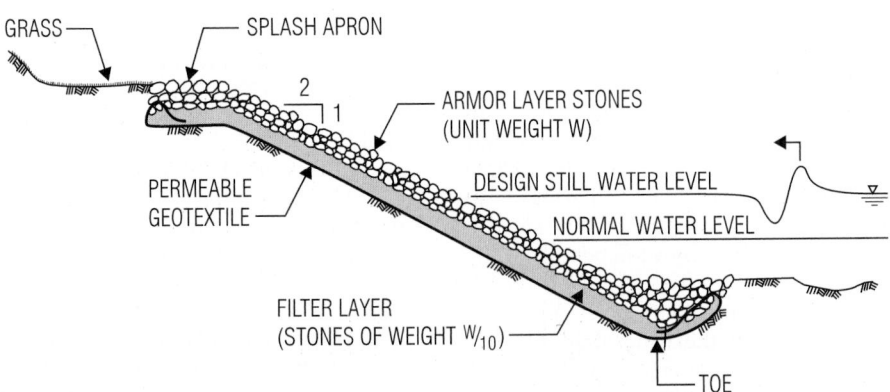

Fig. 23.19 Rubble-mound revetment with armor layer stones, underlain by a stone filter layer and permeable geosynthetic membrane.

slopes as high as 1:1.5 have been tested with Hudson's equation and much steeper slopes can be built by careful crane operators, more failures have occurred on these steeper slopes. Alternatives for steeper slopes include vertical sheetpile bulkheads or gabions. **Gabions**, galvanized wire baskets with rocks, are used in very mild ($H < 1$ ft) wave environments.

Underlayers should be sized so that the stones do not get pulled through the gaps in the overlaying layer. To meet this requirement, underlayer stones should have a minimum median weight of 10% of the overlying median weight. Permeable geotextiles may be installed between the revetment layers and the underlying soil. The toe design shown in Fig. 23.19 allows armor units to roll down into any scour hole that may form.

An alternative to the type of design in Fig. 23.19 with its layers and large armor units is a **dynamic revetment** with a larger volume of stones of a smaller size and wider gradation. The objective of the dynamic revetment is to allow the revetment to move in response to storm waves, much as a cobble beach responds.

The splash of storm waves over a revetment can cause failure by removing underlying soils. The level of wave runup above the design still-water level can exceed the equivalent of one to two full wave heights, depending on the structure slope and roughness and the incident wave period. Run-up elevations of irregular waves vary following a form of a Rayleigh distribution, much as wave heights do. Revetments are commonly designed for a certain level of runup and are provided with a splash apron of stones or grass, or both, for protection against the waves that overtop the structures. Estimating the amount of water washing over a seawall or bulkhead is difficult because of limited test data and the extreme sensitivity of overtopping to still-water level.

Flanking due to shoreline recession at the ends of a revetment can be avoided by either tying it into another adjacent seawall or revetment or constructing return walls perpendicular to the shore. Return walls should be long enough to protect against long-term shoreline recession, any storm recession, and excess storm erosion due to the presence of the seawall. Since a structurally sound seawall protects the underlying sediment from erosion during a storm, excess erosion may occur adjacent to walls during storms. This excess erosion is roughly 20% of the length of the wall.

Bulkhead Design ▪ Design of vertical bulkheads is controlled primarily by geotechnical considerations. Two coastal engineering considerations are the potential for scour at the wall base (due to wave action) and wave-induced pumping action at improperly joined seams. Scour at the toe of a vertical wall can approach 1 to 1.5 times the wave height. Inasmuch as waves are often depth limited and, on a horizontal bottom, the maximum wave height is about equal to the water depth, the scour elevation below the mudline is about 1 to 1.5 times the depth of the water above the original mudline. Scour can be accounted for by designing the wall for a lowered mudline.

Another alternative is to design a rubble-mound toe. The size of the armor units at the base of the rubble mound can be determined from Eq. (23.7) for toes that will be exposed to breaking waves. For submerged toes, the design median weight of the rocks should be

$$W = \frac{w_r H^3}{(S_r - 1)^3 N_s^3} \qquad (23.8)$$

where N_s, the stability number, varies from $1.8 < N_s < 5$. H is depth limited or H_{10} or H_5. Other terms are defined as given for Eq. (23.7).

(Y. Goda, "Random Seas and Design of Maritime Structures," University of Tokyo Press, Tokyo. J. B. Herbich, "Handbook of Coastal and Ocean Engineering," Gulf Publishing Company, Houston, Tex (www.gulfpub.com). "Seawall, Bulkheads, and Revetment Design," Engineer Manual EM-1110-2-1614, U.S. Army Corps of Engineers, Washington, DC 20314-1000 (www.USACE.army.mil/inet/USA-CE-docs/eng-manuals/em-htm).)

23.17 Use of Physical and Numerical Models in Design

Physical and numerical models are used in coastal engineering for a variety of reasons. The traditional physical model is used to account for turbulent processes that limit the ability of equations to predict phenomena. With small-scale physical models, engineers have the ability to address problems with site-specific and design-specific geometry, but the models have some scaling constraints that limit their usefulness. Numerical models, or systems of equations, have been developed to address a wide variety of coastal engineering related problems. Use of any model should consist of two distinct phases, calibration and verification, before the application.

(M. V. Cialone, "The Coastal Modeling System (CMS): A Coastal Processes Software Package," *Journal of Coastal Research*, vol. 10, no. 3, pp. 576–587; S. A. Hughes, "Physical Models and Laboratory Techniques in Coastal Engineering," World Scientific Publishing Co., River Edge, Nev.)

APPENDIX*

Frederick S. Merritt
Consulting Engineer

FACTORS FOR CONVERSION TO THE METRIC SYSTEM (SI) OF UNITS

ongress committed the United States to conversion to the metric system of units when it passed the Metric Conversion Act of 1975. This act states that it shall be the policy of the United States to change to the metric system in a coordinated manner and that the purpose of this coordination shall be to reduce the total cost of the conversion. Although conversion has already taken place in some industries and in some engineering disciplines, it is taking place in short steps at long intervals in civil engineering and construction. Consequently, conventional units are used throughout this handbook. The metric system is explained and factors for conversion to it are presented in this appendix, to guide and assist those who have need to apply metric units in design or construction.

The system of units being adopted in the United States is known as the International System of units, or SI, an abbreviation of the French *Le Système International d'Unités*. This system, intended as a basis for worldwide standardization of measurement units, was developed and is being maintained by the General Conference on Weights and Measures (CGPM).

For engineering, the SI has the advantages over conventional units of being completely decimal and of distinguishing between units of mass and units of force. With conventional units, there sometimes is confusion between use of the two types of units; for example, lb or ton may represent either mass or force.

SI units are classified as base, supplementary, or derived units. There are seven base units (Table A.1), which are dimensionally independent, and two supplementary units (Table A.2), which may be regarded as either base or derived units.

Derived units are formed by combining base units, supplementary units, and other derived units in accordance with algebraic relations linking the corresponding quantities. Symbols for derived units represent the mathematical relationships between the component units. For example, the SI unit for velocity, metre per second, is represented by m/s or $m \times s^{-1}$; for acceleration, metres per second per second, by m/s^2; and for bending moment, newton-metres, by $N \times m$. Figure A.1 indicates how units may be combined to form derived units.

As indicated in Fig. A.1, some of the derived units have been given special names; for example, the unit of energy, $N \times m$, is called joule and the unit of pressure or stress, N/m^2, is called pascal. Table A.3 lists and defines derived SI units that have special names and symbols approved by CGPM. Some derived units used in civil engineering are in Table A.4; others are listed with the conversion factors in Table A.6.

Prefixes

Except for the unit of mass, the kilogram (kg), names and symbols of multiples of SI units by powers of 10, positive or negative, are formed by adding a prefix to base, supplementary, and derived units. Table A.5 lists prefixes approved by CGPM. For historical reasons, kilogram has been retained as a base unit. Nevertheless, for units of mass, prefixes are attached to gram, 10^{-3} kg. Thus, from Table A.5, 1 Mg $= 10^3$ kg $= 10^6$ g.

*With revisions and additions by Jonathan T. Ricketts, Consulting Engineer, Palm Beach Gardens, FL

Table A.1 SI Base Units

Quantity	Unit	Symbol
Length	metre	m
Mass	kilogram	kg
Time	second	s
Electric current	ampere	A
Thermodynamic temperature	kelvin	K
Amount of substance	mole	mol
Luminous intensity	candela	cd

Table A.2 Supplementary SI Units

Quantity	Unit	Symbol
Plane angle	radian	rad
Solid angle	steradian	sr

The prefixes should be used to indicate orders of magnitude without including insignificant digits in whole numbers or leading zeros in decimals. Preferably, a prefix should be chosen so that the numerical value associated with a unit lies between 0.1 and 1000. Preferably also, prefixes representing powers of 1000 should be used. Thus, for civil engineering, units of length should be millimetres, mm; metres, m; and kilometres, km. Units of mass should be milligrams, mg; gram, g; kilogram, kg; and megagram, Mg.

When values of a quantity are listed in a table or when such values are being compared, it is desirable that the same multiple of a unit be used throughout.

To form a multiple of a compound unit, such as of velocity, m/s, only one prefix should be used, and, except when kilogram occurs in the denominator, the prefix should be attached to a unit in the numerator. Examples are kg/m and MJ/kg; do not use g/mm or kJ/g, respectively. Also, do not form a compound prefix by juxtaposing two or more prefixes; for example, instead of Mkm, use Gm. If values outside the range of approved prefixes are required, use a base unit multiplied by a power of 10.

To indicate that a unit with its prefix is to be raised to a power indicated by a specific exponent, the exponent should be applied after the unit; for example, the unit of volume, $mm^3 = (10^{-3}\,m)^3 = 10^{-9}\,m^3$.

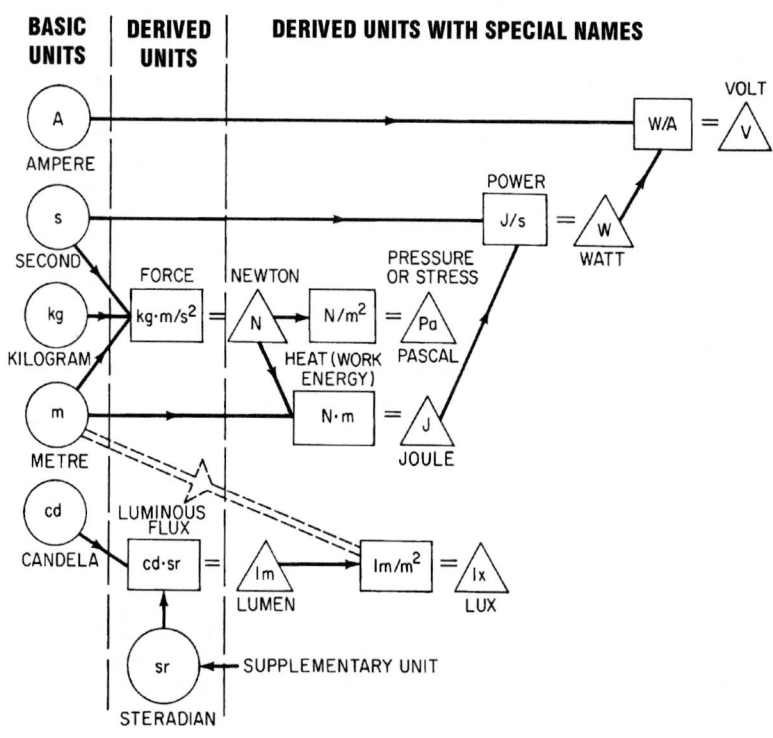

Fig. A.1 How SI units of measurement may be combined to form derived units.

Table A.3 Derived SI Units with Special Names

Quantity	Unit	Symbol	Formula	Definition
Celsius temperature	degree Celsius	°C	K − 273.15	The *degree Celsius* is equal to the kelvin and is used in place of the kelvin for expressing Celsius temperature (symbol t) defined by the equation $t = T - T_o$ where T is the thermodynamic temperature and $T_o = 273.15$ K by definition.
Electric capacitance	farad	F	C/V	The *farad* is the capacitance of a capacitor between the plates of which there appears a difference of potential of one volt when it is charged by a quantity of electricity equal to one coulomb.
Electric conductance	siemens	S	A/V	The *siemens* is the electric conductance of a conductor in which a current of one ampere is produced by an electric potential difference of one volt.
Electric inductance	henry	H	Wb/A	The *henry* is the inductance of a closed circuit in which an electromotive force of one volt is produced when the electric current in the circuit varies uniformly at a rate of one ampere per second.
Electric potential difference, electromotive force	volt	V	W/A	The *volt* (unit of electric potential difference and electromotive force) is the difference of electric potential between two points of a conductor carrying a constant current of one ampere, when the power dissipated between these points is equal to one watt.
Electric resistance	ohm	Ω	V/A	The *ohm* is the electric resistance between two points of a conductor when a constant difference of potential of one volt, applied between these two points, produces in this conductor a current of one ampere, this conductor not being the source of any electromotive force.
Energy	joule	J	N · m	The *joule* is the work done when the point of application of a force of one newton is displaced a distance of one metre in the direction of the force.
Force	newton	N	kg · m/s^2	The *newton* is that force which, when applied to a body having a mass of one kilogram, gives it an acceleration of one metre per second squared.
Frequency	hertz	Hz	1/s	The *hertz* is the frequency of a periodic phenomenon of which the period is one second.

(Table continued)

Table A.3 (*Continued*)

Quantity	Unit	Symbol	Formula	Definition
Illuminance	lux	lx	lm/m^2	The *lux* is the illuminance produced by a luminous flux of one lumen uniformly distributed over a surface of one square metre.
Luminous flux	lumen	lm	$cd \cdot sr$	The *lumen* is the luminous flux emitted in a solid angle of one steradian by a point source having a uniform intensity of one candela.
Magnetic flux	weber	Wb	$V \cdot s$	The *weber* is the magnetic flux which, linking a circuit of one turn, produces in it an electromotive force of one volt as it is reduced to zero at a uniform rate in one second.
Magnetic flux density	tesla	T	Wb/m^2	The *tesla* is the magnetic flux density given by a magnetic flux of one weber per square metre.
Power	watt	W	J/s	The *watt* is the power which gives rise to the production of energy at the rate of one joule per second.
Pressure or stress	pascal	Pa	N/m^2	The *pascal* is the pressure or stress of one newton per square metre.
Quantity of electricity	coulomb	C	$A \cdot s$	The *coulomb* is the quantity of electricity transported in one second by a current of one ampere.

Units in Use with SI

Where it is customary to use units from different systems of measurement with SI units, it is permissible to continue the practice, but such uses should be minimized. For example, for time, the SI unit is the second, but it is customary to use minutes (min), hours (h), days (d), and so on. Thus, velocities of vehicles may continue to be given as kilometres per hour, km/h. Similarly, for angles, the SI unit for plane angle is the radian, but it is permissible to use degrees and decimals of a degree. As another example, for volume, the cubic metre is the SI unit, but litre (L), mL, or mL may be used for measurements of liquids and gases. Also, for mass, while Mg is the appropriate SI unit for large quantities, short ton, long ton, or metric ton may be used for commercial applications.

For temperature, the SI unit is the kelvin, K, whereas degree Celsius, °C (formerly centigrade) is widely used. A temperature interval of 1°C is the same as 1 K, and °C = K − 273.15, by definition.

SI Units Preferred for Construction

Preferred units for measurement of length for relatively small structures, such as buildings and bridges, are millimetres and metres. Depending on the size of the structure, a drawing may conveniently note: "All dimensions shown are in millimetres" or "All dimensions shown are in metres." By convention, the unit for numbers with three digits after the decimal point, for example, 26.375 or 0.425 or 0.063, is metres, and the unit for whole numbers; for example, 2638 or 425 or 63, is millimetres. Hence, it may not be necessary to show unit symbols. For large-size construction, such as highways, metres and kilometres may be used for length measurements and millimetres for width and thickness.

For area measurements, square metres, m², are preferred, but mm² are acceptable for small areas. $1 \, m^2 = 10^6 \, mm^2$. For very large areas, square

Table A.4 Some Common Derived Units of SI

Quantity	Unit	Symbol
Acceleration	metre per second squared	m/s^2
Angular acceleration	radian per second squared	rad/s^2
Angular velocity	radian per second	rad/s
Area	square metre	m^2
Density, mass	kilogram per cubic metre	kg/m^3
Energy density	joule per cubic metre	J/m^3
Entropy	joule per kelvin	J/K
Heat capacity	joule per kelvin	J/K
Heat flux density	watt per square metre	W/m^2
Irradiance	watt per square metre	W/m^2
Luminance	candela per square metre	cd/m^2
Magnetic field strength	ampere per metre	A/m
Moment of force	newton-metre	$N \cdot m$
Power density	watt per square metre	W/m^2
Radiant intensity	watt per steradian	W/sr
Specific energy	joule per kilogram	J/kg
Specific entropy	joule per kilogram kelvin	$J/(kg \cdot K)$
Specific heat capacity	joule per kilogram kelvin	$J/(kg \cdot K)$
Specific volume	cubic metre per kilogram	m^3/kg
Surface tension	newton per metre	N/m
Thermal conductivity	watt per metre kelvin	$W/(m \cdot K)$
Velocity	metre per second	m/s
Viscosity, dynamic	pascal second	$Pa \cdot s$
Viscosity, kinematic	square metre per second	m^2/s
Volume	cubic metre	m^3

Table A.5 SI Prefixes

Multiplication factor	Prefix	Symbol
$1,000,000,000,000,000,000,000 = 10^{21}$	zetta	Z
$1,000,000,000,000,000,000 = 10^{18}$	exa	E
$1,000,000,000,000,000 = 10^{15}$	peta	P
$1,000,000,000,000 = 10^{12}$	tera	T
$1,000,000,000 = 10^{9}$	giga	G
$1,000,000 = 10^{6}$	mega	M
$1,000 = 10^{3}$	kilo	k
$100 = 10^{2}$	hecto*	h
$10 = 10^{1}$	deka*	da
$0.1 = 10^{-1}$	deci*	d
$0.01 = 10^{-2}$	centi*	c
$0.001 = 10^{-3}$	milli	m
$0.000\ 001 = 10^{-6}$	micro	μ
$0.000\ 000\ 001 = 10^{-9}$	nano	n
$0.000\ 000\ 000\ 001 = 10^{-12}$	pico	p
$0.000\ 000\ 000\ 000\ 001 = 10^{-15}$	femto	f
$0.000\ 000\ 000\ 000\ 000\ 001 = 10^{-18}$	atto	a
$0.000\ 000\ 000\ 000\ 000\ 000\ 001 = 10^{-21}$	zepto	z

* To be avoided where practical.

kilometres, km^2, or hectares, ha, may be used. $1\ ha = 10^4\ m^2 = 10^{-2}\ km^2$.

For volume measurements, the preferred unit is the cubic metre, m^3. The volume of liquids, however, may be measured in litres, L, or millilitres, mL. $1\ L = 10^{-3}\ m^3$. For flow rates, cubic metres per second, m^3/s, cubic metres per hour, m^3/h, and litres per second, L/s, are preferred.

For concentrated gravity loads, the force units newton, N, or kilonewton, kN, should be used. For uniformly distributed wind and gravity loads, kN/m^2 is preferred. (Materials weighed on spring scales register the effect of the force of gravity, but for commercial reasons, the scales may be calibrated in kilograms, the units of mass. In such cases, the readings should be multiplied by g, the acceleration of a mass due to gravity, to obtain the load in newtons.) For dynamic calculations, the force in newtons equals the product of the mass, kg, by the acceleration a, m/s^2, of the mass. The recommended value of g for design purposes in the U.S. is $9.8\ m/s^2$.

The standard international value for g is 9.806650 m/s^2, whereas it actually ranges between 9.77 and 9.83 m/s^2 over the surface of the earth.

For both pressure and stress, the SI unit is the pascal, Pa (1 Pa = 1 N/m^2). Because section properties of structural shapes are given in millimetres, it is convenient to give stress in newtons per square millimetre (1 N/mm^2 = 1 MPa). For energy, work, and quantity of heat, the SI unit is the joule, J (1 J = 1 N $\cdot$ m = 1 W $\cdot$ s). The kilowatthour, kWh (more accurately, kW $\cdot$ h) is acceptable for electrical measurements. The watt, W, is the SI unit for power.

Dimensional Coordination

The basic concept of dimensional coordination is selection of the dimensions of the components of a building and installed equipment so that sizes may be standardized and the items fitted into place with a minimum of cutting in the field. One way to achieve this is to make building components and equipment to fit exactly into a basic cubic module or multiples of the module, except for the necessary allowances for joints and manufacturing tolerances. For this purpose, a basic module of 4 in is widely used in the United States. Larger modules often used include 8 in, 12 in, 16 in, 2 ft, 4 ft, and 8 ft.

For modular coordination in the SI, Technical Committee 59 of the International Standards Organization has established 100 mm (3.937 in) as the basic module. In practice, where modules of a different size would be more convenient, preferred dimensions have been established by agreements between manufacturers of building products and building designers. For example, in Great Britain, the following set of preferences have been adopted:

1st preference	300 mm (about 12 in)
2nd preference	100 mm (about 4 in)
3rd preference	50 mm (about 2 in)
4th preference	25 mm (about 1 in)

Accordingly, for a dimension exceeding 100 mm, the first preference is a multimodule of 300 mm; second choice is the basic module of 100 mm.

The preferred multimodules for horizontal dimensioning are 300, 600 (about 2 ft), and 1200 (about 4 ft) mm, although other multiples of 300 are acceptable. Preferred modules for vertical dimensioning are 300 and 600 mm, but increments of 100 mm are acceptable up to 3000 mm. The submodules, 25 and 50 mm, are used only for thin sections.

Some commonly used dimensions, such as the 22 in used for unit of exit width, cannot be readily converted into an SI module. For example, 22 in = 558.8 mm. The nearest larger multimodule is 600 mm (23$\frac{5}{8}$), and the nearest smaller multimodule is 500 mm (19$\frac{11}{16}$). The use of either multimodule would affect the sizes of doors, windows, stairs, and so forth. For conversion of SI to occur, designers and product manufacturers will have to agree on preferred dimensions.

Conversion Factors

Table A.6 lists factors with seven-digit accuracy for converting conventional units of measurement to SI units. To retain accuracy in a conversion, multiply the specified quantity by the conversion factor exactly as given in Table A.6, then round the product to the appropriate number of significant digits that will neither sacrifice nor exaggerate the accuracy of the result. For that purpose, a product or quotient should contain no more significant digits than the number with the smallest number of significant digits in the multiplication or division.

In Table A.6, the conversion factors are given as a number between 1 and 10 followed by E (for exponent), a plus or minus, and two digits that indicate the power of 10 by which the number should be multiplied. For example, to convert lbf/in^2 (psi) to pascals (Pa), Table A.6 specifies multiplication by 6.894 757 $\times$ 10^3. For conversion to kPa, the conversion factor is 6.894 757 $\times$ 10^3 $\times$ 10^{-3} = 6.894 757.

["Standard for Metric Practice," E 380, and "Practice for Use of Metric (SI) Units in Building Design and Construction," E 621, ASTM, 1916 Race St., Philadelphia, PA 19103; "NBS Guidelines for Use of the Metric System," NBS LC 1056, November 1977, and "The International System of Units (SI)," NBS Specification Publication 330, 1977, Superintendent of Documents, Government Printing Office, Washington, DC 20402; "Metric Guide for Federal Construction," No. 5071-2, "Metric Design Guide," No. 5090-1, and "M2: Metric Design Guide," No. 5110-1, National Institute of Building Sciences, Washington, DC 20005.]

Table A.6 Factors for Conversion to SI Units of Measurement

To convert from	To	Multiply by
acre foot	cubic metre, m^3	1.233489E+03
acre	square metre, m^2	4.046873E+03
angstrom	metre, m	1.000000*E−10
atmosphere (standard)	pascal, Pa	1.013250*E+05
atmosphere (technical = 1 kgf/cm^2)	pascal, Pa	9.806650*E+04
bar	pascal, Pa	1.000000*E+05
barrel (for petroleum, 42 gal)	cubic metre, m^3	1.589873E−01
board-foot	cubic metre, m^3	2.359737E−03
British thermal unit (mean)	joule, J	1.05587E+03
Btu (International Table)·in/ (h)(ft^2)($^\circ$F) (k, thermal conductivity)	watt per metre kelvin, W/(m · K)	1.442279E−01
Btu (International Table)/h	watt, W	2.930711E−01
Btu (International Table)/ (h)(ft^2)($^\circ$F) (C, thermal conductance)	watt per square metre kelvin, W/(m^2 · K)	5.678263E+00
Btu (International Table)/lb	joule per kilogram, J/kg	2.326000*E+03
Btu (International Table)/ (lb)($^\circ$F) (c, heat capacity)	joule per kilogram kelvin, J/(kg · K)	4.186800*E+03
Btu (International Table)/ft^3	joule per cubic metre, J/m^3	3.725895E+04
bushel (U.S.)	cubic metre, m^3	3.523907E−02
calorie (mean)	joule, J	4.19002E+00
cd/in^2	candela per square metre, cd/m^2	1.550003E+03
centimetre of mercury (0°C)	pascal, Pa	1.33322E+03
centimetre of water (4°C)	pascal, Pa	9.80638E+01
chain	metre, m	2.011684E+01
circular mil	square metre, m^2	5.067075E−10
day	second, s	8.640000*E+04
day (sidereal)	second, s	8.616409E+04
degree (angle)	radian, rad	1.745329E−02
degree Celsius	kelvin, K	$T_K = t_C + 273.15$
degree Fahrenheit	degree Celsius	$t_C = (t_F − 32)/1.8$
degree Fahrenheit	kelvin, K	$T_K = (t_F + 459.67)/1.8$
degree Rankine	kelvin, K	$T_K = T_R/1.8$
($^\circ$F)(h)(ft^2)/Btu (International Table) (R, thermal resistance)	kelvin square metre per watt, K · m^2/W	1.761102E−01
($^\circ$F)(h)(ft^2)/(Btu (International Table)·in) (thermal resistivity)	kelvin metre per watt, K · m/W	6.933471E+00
dyne	newton, N	1.000000*E−05
fathom	metre, m	1.828804E+00
fluid ounce (U.S.)	cubic metre, m^3	2.957353E−05
foot	metre, m	3.048000*E−01
foot(U.S. survey)	metre, m	3.048006E−01

(Table continued)

Table A.6 (*Continued*)

To convert from	To	Multiply by
foot of water (39.2°F) (pressure)	pascal, Pa	2.98898E+03
ft^2	square metre, m^2	9.290304*E−02
ft^2/h (thermal diffusivity)	square metre per second, m^2/s	2.580640*E−05
ft^2/s	square metre per second, m^2/s	9.290304*E−02
ft^3 (volume or section modulus)	cubic metre, m^3	2.831685E−02
ft^3/min	cubic metre per second, m^3/s	4.719474E−04
ft^3/s	cubic metre per second, m^3/s	2.831685E−02
ft^4 (area moment of inertia)	metre to the fourth power, m^4	8.630975E−03
ft/min	metre per second, m/s	5.080000*E−03
ft/s	metre per second, m/s	3.048000*E−01
ft/s^2	metre per second squared, m/s^2	3.048000*E−01
footcandle	lux, lx	1.076391E+01
footlambert	candela per square metre, cd/m^2	3.426259E+00
ft · lbf	joule, J	1.355818E+00
ft · lbf/min	watt, W	2.259697E−02
ft · lbf/s	watt, W	1.355818E+00
ft-poundal	joule, J	4.214011E−02
free fall, standard g	metre per second squared, m/s^2	9.806650*E+00
gallon (Canadian liquid)	cubic metre, m^3	4.546090E−03
gallon (U.K. liquid)	cubic metre, m^3	4.546092E−03
gallon (U.S. dry)	cubic metre, m^3	4.404884E−03
gallon (U.S. liquid)	cubic metre, m^3	3.785412E−03
gallon (U.S. liquid) per day	cubic metre per second, m^3/s	4.381264E−08
gallon (U.S. liquid) per minute	cubic metre per second, m^3/s	6.309020E−05
grad	degree (angular)	9.000000*E−01
grad	radian, rad	1.570796E−02
grain	kilogram, kg	6.479891*E−05
gram	kilogram, kg	1.000000*E−03
hectare	square metre, m^2	1.000000*E+04
horsepower (550 ft · lbf/s)	watt, W	7.456999E+02
horsepower (boiler)	watt, W	9.80950E+03
horsepower (electric)	watt, W	7.460000*E+02
horsepower (water)	watt, W	7.46043E+02
horsepower (U.K.)	watt, W	7.4570E+02
hour	second, s	3.600000*E+03
hour (sidereal)	second, s	3.590170E+03
inch	metre, m	2.540000*E−02
inch of mercury (32°F) (pressure)	pascal, Pa	3.38638E+03
inch of mercury (60°F) (pressure)	pascal, Pa	3.37685E+03
inch of water (60°F) (pressure)	pascal, Pa	2.4884E+02
in^2	square metre, m^2	6.451600*E−04

(*Table continued*)

Table A.6   (*Continued*)

To convert from	To	Multiply by
in^3 (volume or section modulus)	cubic metre, m^3	1.638706E−05
in^4 (area moment of inertia)	metre to the fourth power, m^4	4.162314E−07
in/s	metre per second, m/s	2.540000*E−02
kelvin	degree Celsius	$t_C = T_K - 273.15$
kilogram-force (kgf)	newton, N	9.806650*E+00
kgf · m	newton metre, N·m	9.806650*E+00
kgf · s^2/m (mass)	kilogram, kg	9.806650*E+00
kgf/cm^2	pascal, Pa	9.806650*E+04
kgf/m^2	pascal, Pa	9.806650*E+00
kgf/mm^2	pascal, Pa	9.806650*E+06
km/h	metre per second, m/s	2.777778E−01
kWh	joule, J	3.600000*E+06
kip(1000 lbf)	newton, N	4.448222E+03
kip/in^2 (ksi)	pascal, Pa	6.894757E+06
knot (international)	metre per second, m/s	5.144444E−01
lambert	candela per square metre, cd/m	3.183099E+03
liter	cubic metre, m^3	1.000000*E−03
maxwell	weber, Wb	1.000000*E−08
mho	siemens, S	1.000000*E+00***
microinch	metre, m	2.540000*E−08
micron	metre, m	1.000000*E−06
mil	metre, m	2.540000*E−05
mile (international)	metre, m	1.609344*E+03
mile (U.S. statute)	metre, m	1.609347E+03
mile (international nautical)	metre, m	1.852000*E+03
mile (U.S. nautical)	metre, m	1.852000*E+03
mi^2 (international)	square metre, m^2	2.589988E+06
mi^2 (U.S. statute)	square metre, m^2	2.589998E+06
mi/h (international)	metre per second, m/s	4.470400*E−01
mi/h (international)	kilometre per hour, km/h	1.609344*E+00
mi/min (international)	metre per second, m/s	2.682240*E+01
mi/s (international)	metre per second, m/s	1.609344*E+03
millibar	pascal, Pa	1.000000*E+02
mho	siemens, S	1.000000*E+00
microinch	metre, m	2.540000*E−08
micron	metre, m	1.000000*E−06
mil	metre, m	2.540000*E−05
mile (international)	metre, m	1.609344*E+03
mile (U.S. statute)	metre, m	1.609347E+03
mile (international nautical)	metre, m	1.852000*E+03
mile (U.S. nautical)	metre, m	1.852000*E+03
mi^2 (international)	square metre, m^2	2.589988E+06
mi^2 (U.S. statute)	square metre, m^2	2.589998E+06
mi/h (international)	metre per second, m/s	4.470400*E−01
mi/h (international)	kilometre per hour, km/h	1.609344*E+00

(*Table continued*)

Table A.6 (*Continued*)

To convert from	To	Multiply by
millibar	pascal, Pa	1.000000*E+02
millimetre of mercury (0°C)	pascal, Pa	1.33322E+02
minute (angle)	radian, rad	2.908882E−04
minute	second, s	6.000000*E+01
minute (sidereal)	second, s	5.983617E+01
ounce (avoirdupois)	kilogram, kg	2.834952E−02
ounce (troy or apothecary)	kilogram, kg	3.110348E−02
ounce (U.K. fluid)	cubic metre, m^3	2.841307E−05
ounce (U.S. fluid)	cubic metre, m^3	2.957353E−05
ounce-force	newton, N	2.780139E−01
ozf · in	newton metre, N · m	7.061552E−03
oz (avoirdupois)/ft^2	kilogram per square metre, kg/m^2	3.051517E−01
oz (avoirdupois)/yd^2	kilogram per square metre, kg/m^2	3.390575E−02
perm (0°C)	kilogram per pascal second metre, $kg/(Pa \cdot s \cdot m)$	5.72135E−11
perm (23°C)	kilogram per pascal second metre, $kg/(Pa \cdot s \cdot m)$	5.74525E−11
perm·in (0°C)	kilogram per pascal second metre, $kg/(Pa \cdot s \cdot m)$	1.45322E−12
perm·in (23°C)	kilogram per pascal second metre, $kg/(Pa \cdot s \cdot m)$	1.45929E−12
pint (U.S. dry)	cubic metre, m^3	5.506105E−04
pint (U.S. liquid)	cubic metre, m^3	4.731765E−04
poise (absolute viscosity)	pascal second, Pa · s	1.000000*E−01
pound (lb avoirdupois)	kilogram, kg	4.535924E−01
pound (troy or apothecary)	kilogram, kg	3.732417E−01
lb · in^2 (moment of inertia)	kilogram square metre, $kg \cdot m^2$	2.926397E−04
lb/ft·s	pascal second, Pa · s	1.488164E+00
lb/ft^2	kilogram per square metre, kg/m^2	4.882428E+00
lb/ft^3	kilogram per cubic metre, kg/m^3	1.601846E+01
lb/gal (U.K. liquid)	kilogram per cubic metre, kg/m^3	9.977633E+01
lb/gal (U.S. liquid)	kilogram per cubic metre, kg/m^3	1.198264E+02
lb/h	kilogram per second, kg/s	1.259979E−04
lb/in^3	kilogram per cubic metre, kg/m^3	2.767990E+04
lb/min	kilogram per second, kg/s	7.559873E−03
lb/s	kilogram per second, kg/s	4.535924E−01
lb/yd^3	kilogram per cubic metre, kg/m^3	5.932764E−01
poundal	newton, N	1.382550E−01
pound · force (lbf)	newton, N	4.448222E+00
lbf · ft	newton-metre, N·m	1.355818E+00
lbf/ft	newton per metre, N/m	1.459390E+01
lbf/ft^2	pascal, Pa	4.788026E+01
lbf/in	newton per metre, N/m	1.751268E+02
lbf/in^2 (psi)	pascal, Pa	6.894757E+03
quart (U.S. dry)	cubic metre, m^3	1.101221E−03
quart (U.S. liquid)	cubic metre, m^3	9.463529E−04

(*Table continued*)

Table A.6 (*Continued*)

To convert from	To	Multiply by
rod	metre, m	5.029210E+00
second (angle)	radian, rad	4.848137E−06
second (sidereal)	second, s	9.972696E−01
square (100 ft^2)	square metre, m^3	9.290304*E+00
ton (assay)	kilogram, kg	2.916667E−02
ton (long, 2240 lb)	kilogram, kg	1.016047E+03
ton (metric)	kilogram, kg	1.000000*E+03
ton (refrigeration)	watt, W	3.516800E+03
ton (register)	cubic metre, m^3	2.831685E+00
ton (short 2000 lb)	kilogram, kg	9.071847E+02
ton (long)/yd^3	kilogram per cubic metre, kg/m^3	1.328939E+03
ton (short)/yd^3	kilogram per cubic metre, kg/m^3	1.186553E+03
ton-force (2000 lbf)	newton, N	8.896444E+03
tonne	kilogram, kg	1.000000*E+03
Wh	joule, J	3.600000*E+03
yard	metre, m	9.144000*E−01
yd^2	square metre, m^2	8.361274E−01
yd^3	cubic metre, m^3	7.645549E−01
year (365 days)	second, s	3.153600*E+07
year (sidereal)	second, s	3.155815E+07

* Exact value. From "Standard for Metric Practice," E380, ASTM.

INDEX

I.4 Index

I.12 Index